LDL	low-density lipoprotein
Leu (L)	leucine
Lys (K)	lysine
Man	mannose
MHC	major histocompatability complex
Met (M)	methionine
NAD^+	nicotinamide-adenine dinucleotide (oxidized form)
NADH	nicotinamide-adenine dinucleotide (reduced form)
$NADP^+$	nicotinamide-adenine dinucleotide phosphate (oxidized form)
NADPH	nicotinamide-adenine dinucleotide phosphate (reduced form)
NDP	nucleoside-5′-diphosphate
NAM	*N*-acetylmuramic acid
NMR	nuclear magnetic resonance
NTP	nucleoside-5′-triphosphate
Phe (F)	phenylalanine
P_i	inorganic orthophosphate
PEP	phosphoenolpyruvate
PFK	phosphofructokinase
PG	prostaglandin
2PG	2-phosphoglycerate
3PG	3-phosphoglycerate
PIP_2	phosphatidylinositol-4,5-bisphosphate
PK	pyruvate kinase
PLP	pyridoxal-5-phosphate
PP_i	inorganic pyrophosphate
Pro (P)	proline
PRPP	phosphoribosylpyrophosphate
PS	photosystem
Q	ubiquinone or plastoquinone
QH_2	ubiquinol or plastoquinol
RER	rough endoplasmic reticulum
RF	release factor or replicative form
RFLP	restriction-fragment length polymorphism
RNA	ribonucleic acid
hnRNA	heterogeneous nuclear RNA
mRNA	messenger RNA
rRNA	ribosomal RNA
snRNA	small nuclear RNA
tRNA	transfer RNA
snRNP	small ribonucleoprotein
RNase	ribonuclease
Ru1,5P	ribulose-1,5-bisphosphate
Ru5P	ribulose-5-phosphate
R5P	ribose-5′-phosphate
RSV	Rous sarcoma virus
s	Svedberg constant
SAM	*S*-adenosylmethionine
SDS	sodium dodecyl sulfate
Ser (S)	serine
S7P	sedoheptulose-7-phosphate
SRP	signal recognition particle
T	thymine
THF	tetrahydrofolate
Thr (T)	threonine
TLC	thin-layer chromatography
TMV	tobacco mosaic virus
TPP	thiamine pyrophosphate
Trp (W)	tryptophan
TTP	thymidine-5′-triphosphate
Tyr (Y)	tyrosine
U	uracil
UDP	uridine-5′-diphosphate
UDPG	UDP-glucose
UMP	uridine-5′-monophosphate
UQ	ubiquinone
Val (V)	valine
VLDL	very-low-density lipoprotein
XMP	xanthosine-5′-monophosphate
Xu5P	xylulose-5′-phosphate

BIOCHEMISTRY

THIRD EDITION

BIOCHEMISTRY

THIRD EDITION

GEOFFREY ZUBAY
COLUMBIA UNIVERSITY

Book Team

Editor *Kevin Kane*
Developmental Editor *Megan Johnson*
Production Editor *Sherry Padden*
Designer *Mark Elliot Christianson*
Art Editor *Miriam J. Hoffman/Janice M. Roerig*
Photo Editor *Lori Gockel*
Permissions Editor *Vicki Krug*
Visuals/Design Developmental Consultant *Marilyn A. Phelps*

Wm. C. Brown Publishers
A Division of Wm. C. Brown Communications, Inc.

Vice President and General Manager *Beverly Kolz*
National Sales Manager *Vincent R. Di Blasi*
Assistant Vice President, Editor-in-Chief *Edward G. Jaffe*
Director of Marketing *John W. Calhoun*
Marketing Manager *Carol J. Mills*
Advertising Manager *Amy Schmitz*
Director of Production *Colleen A. Yonda*
Manager of Visuals and Design *Faye M. Schilling*
Design Manager *Jac Tilton*
Art Manager *Janice Roerig*
Publishing Services Manager *Karen J. Slaght*
Permissions/Records Manager *Connie Allendorf*

Wm. C. Brown Communications, Inc.

Chairman Emeritus *Wm. C. Brown*
Chairman and Chief Executive Officer *Mark C. Falb*
President and Chief Operating Officer *G. Franklin Lewis*
Corporate Vice President, President of WCB Manufacturing *Roger Meyer*

Biochemistry: © Orion/West Light
Artistic rendition of the double helix
Volume One: Energy, Cells, and Catalysis: Photograph by A. M. Lesk
Model for hemoglobin
Volume Two: Catabolism and Biosynthesis: Photograph by A. M. Lesk
Model for phosphofructokinase
Volume Three: Genetics and Physiology: Photograph by A. M. Lesk
Model for λ cro protein binding to DNA

Copyedited by *Emily Arulpragasam*

Freelance permissions editor *Karen Dorman*

The credits section for this book begins on page C-1 and is considered an extension of the copyright page.

Library of Congress Catalog Card Number: 91–76759

ISBN 0–697–14267–1 **Biochemistry** (Casebound)
0–697–14878–5 **Volume One: Energy, Cells, and Catalysis**
0–697–14879–3 **Volume Two: Catabolism and Biosynthesis**
0–697–14880–7 **Volume Three: Genetics and Physiology**
0–697–14877–7 **Biochemistry** (Boxed Set)

Printed in the United States of America by Wm. C. Brown Communications, Inc., 2460 Kerper Boulevard, Dubuque, IA 52001

10 9 8 7 6 5 4 3 2 1

Publisher's Note

Biochemistry, third edition, is available as a full-length casebound text, boxed set, or as three paperbound separates

Binding Option	Description	ISBN
Biochemistry (Casebound)	The full-length text, Chapters 1–36	0-697-14267-1
Biochemistry, Volume One (Paperbound)	Chapters 1–11	0-697-14878-5
Biochemistry, Volume Two (Paperbound)	Chapters 12–24	0-697-14879-3
Biochemistry, Volume Three (Paperbound)	Chapters 25–36	0-697-14880-7
Biochemistry, Boxed Set (Paperbound)	The full-length text in an attractive boxed set of all three paperback "splits".	0-697-14877-7

This book is dedicated to Ed Jaffe, the late Editor-in-Chief of the William C. Brown Publishing Company.

Ed passed away suddenly just before this text went to press. I am deeply saddened by his passing and that he will not see the text he backed so strongly. He was a man of wisdom and integrity who maintained an excellent rapport with the WCB staff and with its authors.

Brief Contents

Volume One

Volume Two

PART 4

Catabolism and the Generation of Chemical Energy 305

PART 5

Biosynthesis of the Building Blocks 473

Volume Three

PART 6

Storage and Utilization of Genetic Information 689

PART 7

Physiological Biochemistry 929

Contents

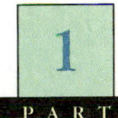

PART

Structure and Function of Major Components of the Cell 45

Chapter 11

Vitamins, Coenzymes, and Metal Cofactors 278

Perry A. Frey

PART

Catabolism and the Generation of Chemical Energy 305

Chapter 12

Metabolic Strategies 307

Daniel E. Atkinson and Geoffrey Zubay*

Chapter 13

Glycolysis, Gluconeogenesis, and the Pentose Phosphate Pathway 321

Daniel E. Atkinson and Geoffrey Zubay*

Chapter 14

The Tricarboxylic Acid Cycle 354

Daniel E. Atkinson and Geoffrey Zubay*

Chapter 15

Electron Transport and Oxidative Phosphorylation 379

William W. Parson

*Revised by Geoffrey Zubay from material in the preceding edition by Daniel E. Atkinson.

Chapter 16

Photosynthesis and Other Processes Involving Light 414

William W. Parson

Chapter 17

Metabolism of Fatty Acids 444

Dennis E. Vance

PART

Biosynthesis of the Building Blocks 473

Chapter 18

Biosynthesis of Amino Acids 475

H. Edwin Umbarger and Geoffrey Zubay

Chapter 19

The Metabolic Fate of Amino Acids 513

H. Edwin Umbarger and Geoffrey Zubay

Chapter 20

Nucleotides 547

Raymond L. Blakley

Chapter 21

Chapter 22

Chapter 23

Chapter 24

Integration of Metabolism and Hormone Action 659

Richard Palmiter and Geoffrey Zubay

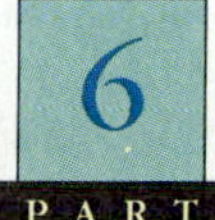

PART

Storage and Utilization of Genetic Information 689

Chapter 25

Structures of Nucleic Acids and Nucleoproteins 691

Geoffrey Zubay and Julius Marmur

Chapter 26

DNA Replication, Repair, and Recombination 721

Geoffrey Zubay and Julius Marmur

Chapter 27

DNA Manipulation and Its Applications 757

Geoffrey Zubay

Chapter 28

RNA Synthesis and Processing 791

Richard R. Burgess, Ann Baker Burgess, and Geoffrey Zubay

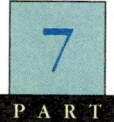

Chapter 34

Chapter 35

Chapter 36

List of Supplementary Material

List of Contributors

Raymond L. Blakley
Chapter 20
Division of Biochemical and Clinical Pharmacology
St. Jude Children's Research Hospital
Memphis, TN 38101

James W. Bodley
Chapter 29
Biochemistry Department
University of Minnesota
Minneapolis, MN 55455

Ann Baker Burgess
Chapter 28
McArdle Laboratory for Cancer Research
University of Wisconsin
Madison, WI 53706

Richard R. Burgess
Chapter 28
McArdle Laboratory for Cancer Research
University of Wisconsin
Madison, WI 53706

Perry A. Frey
Chapter 11
Institute for Enzyme Research
University of Wisconsin
Madison, WI 53706

Irving Geis
Chapter 5
4700 Broadway
Apt. 4B
New York, NY 10040

Lloyd L. Ingram (Retired)
Chapter 2
Department of Biochemistry and Biophysics
University of California at Davis
Davis, CA 95616

Gary R. Jacobson
Chapters 7, 32, 35
Department of Biology
Boston University
Boston, MA 02215

Julius Marmur
Chapters 25, 26, 31
Department of Biochemistry
Albert Einstein College of Medicine
Bronx, NY 10461

Richard Palmiter
Chapter 24
Department of Biochemistry
University of Washington
Seattle, WA 98195

William W. Parson
Chapters 2, 8, 9, 10, 15, 16, 36
Department of Biochemistry
University of Washington
Seattle, WA 98195

Milton H. Saier, Jr.
Chapters 7, 32, 35
Department of Biology
University of California at San Diego
Lajolla, CA 92093

Pamela Stanley
Chapters 6, 21
Department of Cell Biology
Albert Einstein College of Medicine
Bronx, NY 10461

H. Edwin Umbarger
Chapters 18, 19
Department of Biological Sciences
Purdue University
West Lafayette, IN 47907

Dennis E. Vance
Chapters 7, 17, 22, 23
Lipid and Lipoprotein Group
Department of Biochemistry
University of Alberta
Edmonton, Alberta
Canada, T6G 2C2

Geoffrey Zubay
Chapters 1, 2, 3, 4, 5, 6, 12, 13, 14, 18, 19, 21, 24, 25, 26, 27, 28, 30, 31, 33, 34
Department of Biological Sciences
Columbia University
New York, NY 10027

Guided Tour through the Biochemistry Learning System

CHAPTER OUTLINES

Each chapter begins with an outline. These will allow students to tell at a glance how the chapter is organized and what major topics have been included in the chapter.

Thermodynamics in Biochemistry

he primary usefulness of thermodynamics to biochemists lies in predicting whether particular chemical reactions could occur spontaneously. A simple illustration is to predict what compounds could possibly serve as energy sources for an organism. You are aware from everyday experience that oxidation of organic molecules by molecular oxygen releases energy. For example, wood or coal burns with a large output of heat. Similarly, organisms can obtain energy by oxidizing carbohydrates, fats, or proteins. Some organisms oxidize hydrocarbons, some oxidize reduced forms of sulfur, and others oxidize iron. But no organisms live by oxidizing molecular nitrogen, and the explanation lies in thermodynamics. The reaction cannot occur spontaneously. This example illustrates the importance of thermodynamics in controlling all life. Because organisms live by extracting chemical energy from their surroundings, thermodynamics is not an esoteric subject. It is a matter of life or death.

We say that thermodynamics determines whether a process "could" occur, because thermodynamics tells us only whether the process is possible, not whether it actually will occur in a finite period of time. The rate at which a thermodynamically possible reaction occurs depends on the detailed mechanism of the process. For a biochemical process to occur rapidly, appropriate enzymes must be available. The distinction between spontaneity and speed is more critical for biochemists than it is for chemists, because if a chemical reaction does not proceed rapidly a chemist can change the pressure or temperature or increase the concentration of the reactants. A living organism is under more rigid constraints: It must function at a fixed temperature and pressure and within a limited range of concentrations of reactants.

In this chapter we will elaborate on the thermodynamic quantities that we introduced in chapter 1. We will then expand the discussion to show how the concept of free energy is used in predicting biochemical pathways, and we will explore the central role of ATP in providing energy for biochemical reactions.

Thermodynamic Quantities

The properties of a substance can be classified as either intensive or extensive. Intensive properties, which include density, pressure, temperature, and concentration, do not depend on the amount of the material. Extensive properties, such as volume and weight, do depend on the amount. Most of the thermodynamic properties we will be discussing are extensive properties. These include energy (*E*), enthalpy (*H*), entropy (*S*), and free energy (*G*).

Energy, enthalpy, entropy, and free energy are all properties of the state of a substance. This means that they do not depend on how the substance was made or how it reached a particular state. In a chemical reaction, it is the difference between the initial and final states that is important; the pathway that is taken to get from the initial state to the final state has no bearing on whether the overall reaction releases or consumes energy (fig. 2.1).

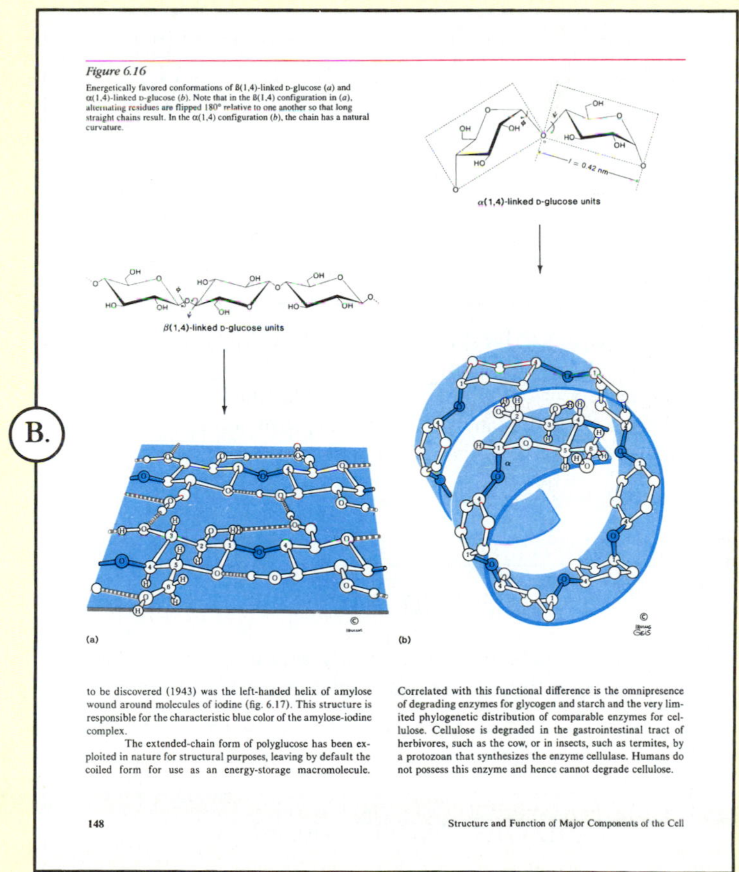

Figure 6.16

Energetically favored conformations of ß(1,4)-linked D-glucose (*a*) and α(1,4)-linked D-glucose (*b*). Note that in the ß(1,4) configuration in (*a*), alternating residues are flipped 180° relative to one another so that long straight chains result. In the α(1,4) configuration (*b*), the chain has a natural curvature.

to be discovered (1943) was the left-handed helix of amylose wound around molecules of iodine (fig. 6.17). This structure is responsible for the characteristic blue color of the amylose-iodine complex.

The extended-chain form of polyglucose has been exploited in nature for structural purposes, leaving by default the coiled form for use as an energy-storage macromolecule. Correlated with this functional difference is the omnipresence of degrading enzymes for glycogen and starch and the very limited phylogenetic distribution of comparable enzymes for cellulose. Cellulose is degraded in the gastrointestinal tract of herbivores, such as the cow, or in insects, such as termites, by a protozoan that synthesizes the enzyme cellulase. Humans do not possess this enzyme and hence cannot degrade cellulose.

148 Structure and Function of Major Components of the Cell

DRAMATIC VISUALS PROGRAM

Colorful and informative photographs, illustrations, and tables enhance the learning program.

C.

Measurement of Ultraviolet Absorption in Solution

he general quantitative relationship that governs all absorption processes is called the Beer-Lambert law:

$$I = I_0 10^{-\epsilon cd}$$

where I_0 is the intensity of the incident radiation, I is the intensity of the radiation transmitted through a cell of thickness d (in centimeters) that contains a solution of concentration c (expressed either in moles per liter or in grams per 100 ml), and ϵ is the extinction coefficient, a characteristic of the substance being investigated (see fig. 3.7).

Light absorption is measured by a spectrophotometer as shown in the illustration. The spectrophotometer usually is capable of directly recording the absorbance A, which is related to I and I_0 by the equation

$$A = \log_{10} (I_0/I)$$

Hence $A = \epsilon cd$, and A is a direct measure of concentration. We can see from figure 3.7 that the ϵ values are largest for tryptophan and smallest for phenylalanine.

Since protein absorption maxima in the near-ultraviolet (240–300 nm) are determined by the content of the aromatic amino acids and their respective values, most proteins have absorption maxima in the 280-nm region. By contrast, absorption in the far-ultraviolet (around 190 nm) is shown by all polypeptides regardless of their aromatic amino acid content. The reason is that absorption in this region is due primarily to the peptide linkage.

Figure 1

Schematic diagram of a spectrophotometer for measuring light absorption. Laboratory instruments for making measurements are much more complex than this, but they all contain the same basic components: a light source, a monochromator, a sample, and a detector. λ is the wavelength of the light, I_0 and I are the incident light intensity and the transmitted light intensity, respectively, and d is the thickness of the absorbing solution.

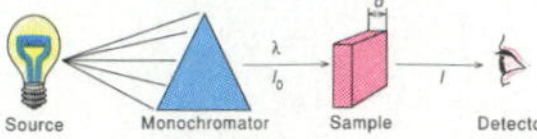

Another convention for referring to configurations is called the R, S convention. As the R, S convention is not as popular for amino acids or sugars as it is for other types of biomolecules, such as lipids, we will not discuss this notation until chapter 11 (see box 11A).

Peptides and Polypeptides

D.

Amino acids can link together by a covalent peptide bond between the α-carboxyl end of one amino acid and the α-amino end of another. Formally, this bond is formed by the loss of a water molecule, as shown in figure 3.9. The peptide bond has partial double-bond character owing to resonance effects; as a result, the C—N peptide linkage and all of the atoms directly connected to C and N lie in a planar configuration called the amide plane. In the following chapter we will see that this amide plane, by limiting the number of orientations available to the polypeptide chain, plays a major role in determining the three-dimensional structures of proteins.

Any number of amino acids can be joined by successive peptide linkages, forming a polypeptide chain. The polypeptide chain, like the dipeptide, has a directional sense. One end, called the N-terminal or amino-terminal end, has a free α-amino group, whereas the other end, the C-terminal or carboxyl-terminal end, has a free α-carboxyl group. The sequence of main-chain atoms from the N-terminal end to the C-terminal end is C_α—C—N—C_α, etc., and in the opposite direction it is C_α—N—C—C_α, etc. Short polypeptide chains, up to a length of about 20 amino acids, are called peptides or oligopeptides if they are fragments of whole polypeptide chains. A small protein molecule may contain a polypeptide chain of only 50 amino acids; a large protein may contain chains of 3,000 amino acids or more. One of the larger single polypeptide chains is that of the muscle protein myosin, which consists of approximately 1,750 amino acid residues. Figure 3.10 shows a section of a polypeptide chain as a linear array with α carbons and planar amides alternating as repeating units of the main chain. Different side chains are attached to each α carbon.

In addition to the covalent peptide bonds formed between adjacent amino acids within a polypeptide chain, covalent disulfide bonds can be formed within the same polypeptide chain or between different polypeptide chains (fig. 3.11). Such disulfide linkages have an important stabilizing influence on the structures formed by many proteins (see chapter 4).

54 Structure and Function of Major Components of the Cell

SUPPLEMENTARY BOX MATERIAL

Throughout the text, supplemental information on experimental procedures, mechanisms, and methods of evaluation are presented in boxes

UNDERLINED KEY CONCEPTS

Important terms and concepts are highlighted by underlining.

END-OF-CHAPTER SUMMARIES

Each summary offers a concise review of the material covered in the chapter. Students can use the summary to review the chapter or to preview the important topics.

SELECTED READINGS

Each chapter concludes with carefully selected references that contain further information on the topics covered in that chapter.

END-OF-CHAPTER PROBLEMS

These problems will challenge students' mastery of the chapter's basic concepts. Odd-numbered problems are answered briefly in the back of the text.

E.

Summary

In this chapter we have dealt with some of the fundamental properties of amino acids and polypeptide chains. The following points are especially important.

1. Nineteen of the twenty amino acids commonly found in proteins have a carboxyl group and an amino group attached to an α-carbon atom; they differ in the side chain attached to the same α carbon.
2. All amino acids have acidic and basic properties. The ratio of base to acid form at any given pH can be calculated from the pK with the help of the Henderson-Hasselbach equation.
3. All amino acids except glycine are asymmetric and therefore can exist in at least two different stereoisomeric forms.
4. Peptides are formed from amino acids by the reaction of the α-amino group from one amino acid with the α-carboxyl group of another amino acid.
5. Polypeptide formation involves a repetition of the process involved in peptide synthesis.
6. The amino acid composition of proteins can be discovered by first breaking down the protein into its component amino acids and then separating the amino acids in the mixture for quantitative estimation.
7. The amino acid sequences of proteins can be discovered by breaking down the protein into polypeptide chains and then partially degrading the polypeptide chains. For each polypeptide chain fragment, the sequence is determined by stepwise removal of amino acids from the amino terminal end of the polypeptide chain. Two different methods of forming polypeptide chain fragments are used so as to produce a map of overlapping fragments, from which the sequence of undegraded polypeptide chains in the proteins can be deduced.
8. Polypeptide chains with a predetermined amino acid sequence can be synthesized by chemical methods involving carboxyl-group activation.

Selected Readings

Barrett, G. C. (ed.). *Chemistry and Biochemistry of Amino Acids.* New York: Chapman and Hall, 1985. A recent and authoritative volume on this classical subject.

Gray, W. R., End group analysis using dansyl chloride. *Methods in Enzymology* 25:121–138, 1972. This volume of *Methods in Enzymology* contains several chapters on end-group analysis.

Hunkapiller, M. W., J. E. Strickler, and K. J. Wilson, Contemporary methodology for protein structure determination. *Science* 226:304–311, 1984.

Kent, S. B. H., Chemical synthesis of peptides and proteins. *Ann. Rev. Biochem.* 57:957–989, 1988. Comprehensive and up-to-date.

Merrifield, B., Solid phase synthesis. *Science* 232: 341–347, 1986.

Sanger, R., Sequences, sequences and sequences. *Ann. Rev. Biochem.* 57:1–28, 1988.

G.

Problems

1. (a) A 10-mM solution of a weak monocarboxylic acid has a pH of 3.00. Calculate K_a and pK_a for this carboxylic acid.
   (b) You add 0.06 g NaOH (M_r = 40) to 1,000 ml of the acid solution in part (a). Calculate the final pH, assuming no volume change.
2. Given the pK_a values in the text, predict how the titration curves for glutamic acid and glutamine would differ.
3. You have 50 ml of 10-mM fully protonated histidine. How many millimoles of base must be added to bring the histidine solution to a pH that is equivalent to the pI?
4. Calculate the isoelectric point for histidine, aspartic acid, and arginine. Calculate the fractional charge for each ionizable group on aspartate at pH equal to pI. Do the results verify the isoelectric point of aspartic acid?
5. Which of the naturally occurring amino acid side chains are charged at pH 2? pH 7? pH 12? (Consider only those amino acids whose side chains have >10% charge at the pH indicated.)
6. Amino acids are sometimes used as buffers. Indicate the appropriate pH value(s) of a buffer containing aspartic acid, histidine, and serine.
7. Ten ml of a 10-mM solution of lysine was adjusted to pH 11.20. Draw the structures of the principal ionized forms present in solution. Use the pK_a values shown in table 3.3 and calculate the concentration of each principal form.
8. For the tripeptide shown below, the numbers in parentheses are the pK_a values of the ionizable groups.

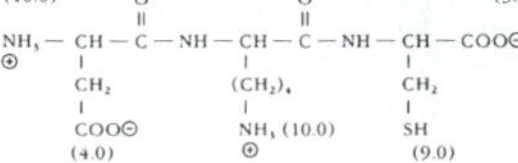

   (a) Estimate the net charge at pH 1 and pH 14.
   (b) Estimate the isoelectric pH.
9. Polyhistidine is insoluble in water at pH 7.8 but is soluble at pH 5.5. Explain the observation. Would you expect the polymer to be soluble at pH 10?

The Building Blocks of Proteins: Amino Acids, Peptides, and Polypeptides 67

Preface

What makes biochemistry exciting? What makes biochemistry unique? What makes biochemistry so important? The answer to all three of these questions is virtually the same. Each and every reaction in biochemistry serves a function; that function is the maintenance and propagation of the living system. To understand biochemistry is to appreciate the place that each reaction occupies in the system.

Students of the previous generation discovered the principles that govern biochemistry, the basic reactions, and an immense body of facts consistent with these principles. Future biochemists to emerge from this generation of students will uncover the mysteries of developmental biology and physiology. They will find cures for cancers, cystic fibrosis, sickle cell anemia and many other diseases. My goal in putting this text together has been to convey my enthusiasm for this subject to this generation of students, to acquaint them with what is known and to prepare them for the discoveries of the future.

While the main principles of biochemistry are understood, the number of new facts that keep adding to the subject is immense. The field is becoming more and more subdivided as the volume of knowledge increases. No single person could directly cope with all the new information that is continuously appearing in the scientific literature. For the purpose of writing an authoritative up-to-date textbook on biochemistry, it would be ideal to involve a team of experts with hands-on experience. The contributing authors of this text are all research specialists who have made their mark in a specific area of biochemistry; they are teachers as well. Each contributing author could write an entire book on the chapter he/she has contributed to in this text. Recognizing that the primary goal of this text is to meet the needs of students facing the subject for the first time, each contributing author has had to pare down what he/she knows emphasizing principles, but not ignoring important facts. Each chapter has been carefully crafted with that goal in mind. My responsibility as coordinating author of this text has been to mold the contributions into a smooth reading, well integrated product.

This is the third edition of *Biochemistry.* Whereas I have had the same goal throughout this project, it is only with this edition, that the long-hoped-for product has finally emerged from our team approach. Reviewers tell us that this edition is not only consistently authoritative, it now reads like a cohesive, consistent single author textbook.

Each chapter emphasizes principles. Since the overriding principle is that every reaction serves a function, there is a strong emphasis on analysis of biochemical pathways and how different pathways are interrelated. In this pursuit balanced emphasis is given to chemistry and mechanisms, bioenergetics, metabolic regulation, and methods of biochemical analysis.

We have organized this text along conventional lines for teaching purposes. In Part 1 an overview is followed by an explanation of the importance of thermodynamics in biochemistry. In Part 2 we consider the structure and function of the major components of the cell. In Part 3 we discuss enzymes and coenzymes and the mechanisms of catalysis. Part 4 deals with energy producing catabolic processes, and Part 5 deals with energy consuming synthetic processes. In Part 6 we focus on nucleic acids and the way in which the biochemical information encoded in the chromosome is translated into the amino acid sequences found in proteins. Finally, in Part 7 we approach some of the applications of biochemistry to physiology.

Organizational Changes

The overall organization of the previous edition has been changed significantly:

- The chapter on methods for characterization and purification of proteins (formerly chapter 3) has been eliminated. Much of the material from this chapter has been disseminated into other chapters, appearing at the points where it is most relevant. The same goes for the discussion of methods in general, and there is a special appendix indicating where specific methods are discussed.

- The chapter on nucleic acid structure (formerly chapter 7) has been moved to a later part of the text (chapter 25). This has been done because nucleic acids are not discussed in any depth until Part 6. The general significance of nucleic acids is adequately discussed in chapter 1 so this should create no problem for the rare student who has not encountered this subject already.
- The chapter on origin of life has been eliminated. A few points about this subject are now made in chapters 1 and 28. This is a delightful subject and very close to my major research interests. However, it simply is not mainstream biochemistry and space would not allow us to keep this chapter.
- After many years of agonizing appraisal I have decided to place thermodynamics as the second chapter. Bioenergetics is so central to everything in biochemistry that I thought it should be discussed as soon as possible. At the same time an esoteric chapter on this subject could easily turn off many students. To avoid this we have considerably simplified this chapter to make it as palatable as possible without losing any of the significance of a more esoteric chapter.
- A new chapter on functional diversity of proteins (chapter 5) has been added. This gives us an opportunity to dwell on the very important subject of the relationship between protein structure and function. In addition we have used this chapter as a location to discuss the subject of protein purification.
- The chapters on lipids and membrane structures have been combined into one chapter. Students are more attentive to a description of lipids if they can see the important role they play in membrane structure.
- The chapters on lipid catabolism and synthesis have been combined in this edition (chapter 17). This makes it easier to discuss regulation of lipid metabolism.
- At the end of the treatment of intermediary metabolism (chapter 24) we have added a chapter on the integration of metabolism. In this chapter we have included most of the coverage of hormone action that was presented much later in the second edition. It is more relevant here as hormones are the key to the integration of metabolism in multicellular organisms.
- Two short chapters have been added to Part 7; chapter 33, on immunobiology, and chapter 34 on carcinogenesis. These are two hot topics that are of great interest to students. The chapters are kept short and simple for those who would choose to use them in a two term course. It would be best to consider them after the core chapters of Part 6 (i.e., chapters 25, 26, 28 and 29).

Content Features of This Edition

The coverage of biochemistry in this text is determined by the consensus feedback that we have received as well as the limitations of the present state of our knowledge. Coverage of the biochemistry of bacteria and mammals is emphasized over plant biochemistry.

Part 1 is entitled "An Overview of Biochemistry and Energy Considerations." The first chapter "Overview of Biochemistry," is very similar to the introductory chapter in the second edition. It presents the basic principles in chemistry and biology that relate to biochemistry. The second chapter, "Thermodynamics in Biochemistry," has been considerably rewritten and simplified over the corresponding chapter in the second edition, which had appeared as chapter 12. Chapter 2 elaborates on the thermodynamic quantities introduced in chapter 1, to show how the concept of free energy is used to predict the possibility that given reactions can occur. It also describes the central role of ATP in providing energy for biochemical reactions.

Part 2, "Structure and Function of Major Components of the Cell," contains five chapters describing the structure and function of proteins, carbohydrates, and lipids. The contents of Part 2 are similar to the contents of Part 1 in the previous edition, except that the discussion of nucleic acids has been moved to Part 6, and a new chapter 5, "Functional Diversity of Proteins," has been introduced.

Chapters 3, 4, and 5 present the basic structural and chemical properties of proteins. Chapter 3, "The Building Blocks of Proteins: Amino Acids, Peptides, and Polypeptides," is very similar to the comparable chapter in the second edition. Chapter 4, "The Three-Dimensional Structure of Proteins," describes the secondary, tertiary, and quaternary structures of proteins, the rules for which govern the folding of polypeptide chains and the ways in which these structures are determined. In this edition there is more coverage of the relationship between symmetry and quaternary structures than in the second edition. Chapter 5, "Functional Diversity of Proteins," focuses on the question of how protein structure relates to function and describes techniques used to isolate proteins.

Chapter 6, "Carbohydrates, Glycoproteins, and Cell Walls," describes the structural properties of sugars, oligosaccharides, and polysaccharides, along with the various roles that these compounds play as energy-storage molecules, components of glycoproteins, and in cell walls. This chapter's segment on glycoproteins has been considerably updated. Finally, in chapter 7, "Lipids and Membranes," two chapters from the second edition have been fused into one cohesive chapter. In it, the structural properties of lipids are discussed with reference to the roles lipids play as energy-storage molecules and in membrane structures.

Part 3, "Catalysis," is divided into four chapters, the first three of which have been completely rewritten, while the last has been simplified considerably, with discussions of lipid-soluble vitamins and the role of metals as cofactors added.

Chapter 8, "Enzyme Kinetics," features a description of kinetic methods for analyzing enzyme-catalyzed reactions. Chapter 9, "Mechanisms of Enzyme Catalysis," focuses on the way in which the structure of the enzyme active site is related to enzyme activity under different conditions. In this chapter a general discussion of factors determining enzyme specificity is followed by several detailed examples. Chapter 10, "Regulation of Enzyme Activities," describes enzymes regulated by structural modifications or the binding of specific factors at allosteric sites, and chapter 11, "Vitamins, Coenzymes, and Metal Cofactors," covers the function of small molecules acting in conjunction with enzymes as cocatalysts.

Part 4 "Catabolism and the Generation of Chemical Energy," is divided into six chapters. Since both Parts 4 and 5 are concerned with intermediary metabolism, it's not easy, or even appropriate, to separate the subject into artificially tidy discussions of catabolism and anabolism. Suffice it to say that the emphasis in Part 4 is on catabolism and the generation of chemical energy, while in Part 5 the emphasis is on biosynthesis. Chapter 12, "Metabolic Strategies," which initiates Part 4, relates strongly to both Parts 4 and 5, as does chapter 24, "Integration of Metabolism and Hormone Action," which concludes Part 5.

Chapter 12, "Metabolic Strategies," deals with the general ways in which metabolism is organized and the ways in which it is regulated. Chapters 13, 14, and 15 are devoted to carbohydrate metabolism in relationship to the generation of ATP. Chapter 13, "Glycolysis, Gluconeogenesis, and the Pentose Phosphate Pathway," describes anabolic as well as catabolic processes—namely, the synthesis of simple sugars and simple sugar polymers, as well as their breakdown. Both synthesis and breakdown are covered in the same chapter in order to facilitate meaningful discussion of the regulation of these two processes, which are intimately related.

Chapter 14, "The Tricarboxylic Acid Cycle," describes the further catabolism of sugars possible under aerobic conditions. Chapter 15, "Electron Transport and Oxidative Phosphorylation," complements chapter 14 because it shows how electrons released by the aerobic oxidation of sugars are channeled down the electron-transport system, ultimately to synthesize ATP. Chapter 16, "Photosynthesis and Other Processes Involving Light," appears immediately after the chapter on electron transport and the production of ATP because both chapters describe energy-generating metabolism. The complex light-powered processes are explained so that a minimal background in physics and chemistry is needed.

Part 4 concludes with Chapter 17, "Metabolism of Fatty Acids," which represents a fusion of two chapters from the second edition. The reasons for discussing both synthesis and breakdown of lipids in the same chapter are exactly the same as the reasons for doing this with sugars (as in chapter 13).

In many schools a one-semester treatment of biochemistry would end with chapter 17. Even though the first half of the text is tightly structured, instructors may still wish to seek additional material or skip around. Two recommendations involve chapter 25 and chapter 32. Chapter 25, "Structures of Nucleic Acids and Nucleoproteins," could be treated as a chapter in Part 2. If this is done, instructors should make sure students read the opening pages of chapter 20, which deal with the structure of nucleotides. Likewise, some instructors may wish to assign readings from chapter 32, which deals with mechanisms of membrane transport, after they have assigned chapter 7 on lipids and membranes.

Part 5, "Biosynthesis of the Building Blocks," contains seven chapters, mainly describing the biosynthesis and utilization of amino acids, nucleotides, complex carbohydrates, and lipids. Although there is a great deal of useful information in these chapters, most of Part 5, except for chapter 24, can be skipped if there is a strong desire to get to Part 6 as quickly as possible.

Chapter 18, "Biosynthesis of Amino Acids," describes the biosynthesis of amino acids. Because twenty amino acids found in proteins are made in bacteria and plants (whereas only a few amino acids are synthesized *de novo* in mammals), and because these pathways are better understood in bacteria, most of this chapter focuses on biosynthesis in bacteria. Nonprotein amino acids are discussed briefly. Finally, the problem of nitrogen fixation in biological systems is examined.

Chapter 19 is entitled "The Metabolic Fate of Amino Acids." It discusses the way in which amino acids, when in excess or when present as the only available carbon or energy source, are reduced to carbohydrates, which can serve as a source of energy or carbon skeletons for the synthesis of other molecules. A few products derived from amino acid metabolism are also discussed in detail, including porphyrins, biological active amines, glutathione, and peptide antibiotics.

Chapter 20, "Nucleotides," deals with the structures, synthesis, and breakdown of nucleotides, with considerable attention devoted to the role of nucleotide derivatives as inhibitors of nucleic acid synthesis and the important role they play in the medical field. The biosynthesis of nucleotide-containing coenzymes is also discussed in this chapter.

Chapter 21, "Biosynthesis of Complex Carbohydrates," describes the synthesis of complex carbohydrates that serve structural roles and the synthesis of oligosaccharides that exist in combination with proteins. A great deal of new information on this subject appears in this chapter, as does a detailed discussion of bacterial cell-wall synthesis.

Chapter 22, "Biosynthesis of Membrane Lipids and Related Substances," describes the metabolism of complex lipids. Phospholipids and sphingolipids are viewed in terms of their important structural roles in membranes, while eicosanoids are described in terms of functioning local hormones.

Chapter 23, "Metabolism of Cholesterol," describes the biosynthesis of cholesterol and some of its derivatives—bile acids and steroids. The chapter devotes considerable attention to the transport of these compounds, along with the problems associated with defects in cholesterol metabolism.

Chapter 24, "Integration of Metabolism and Hormone Action," concludes the two parts on intermediary metabolism. This chapter emphasizes the way in which intermediary metabolism is regulated in multicellular organisms.

Part 6, "Storage and Utilization of Genetic Information," features seven chapters describing the metabolism of nucleic acids and proteins. The entire section has been updated to keep pace with this rapidly changing area. Where useful, key genetic phenomena have also been explained.

This part begins with chapter 25, "Structures of Nucleic Acids and Nucleoproteins," which describes structures and functions of DNA and deoxyribonucleoproteins. It has intentionally been placed after structures of nucleotides have been discussed (in chapter 20) and before the structures of RNA and ribonucleoproteins are discussed (in chapter 28).

In chapter 26, "DNA Replication, Repair, and Recombination," the emphasis is on the metabolism of replication. The treatment is equally balanced between replication in prokaryotes and eukaryotes including animal viruses.

Chapter 27, "DNA Manipulation and Its Applications," begins with a discussion of the procedures for determining DNA sequence, proceeds with a discussion of the techniques for manipulating DNA sequences, and concludes by describing two important applications of DNA manipulation—the mapping of the globin gene family and the mapping of a gene responsible for the dreaded disease cystic fibrosis.

Chapter 28, "RNA Synthesis and Processing," which describes RNA synthesis and processing equally, presents a considerable amount of new information on transcription factors in animals cells and a variety of ways in which RNA is processed.

Chapter 29, "Protein Synthesis, Targeting, and Turnover," has been significantly expanded over the corresponding chapter in the second edition, particularly in the area of targeting, where many new discoveries have been made recently.

Chapter 30, "Regulation of Gene Expression in Prokaryotes," includes a discussion of several well-characterized gene systems found in *E coli*. It also features a treatment of the sequence of regulatory events involving bacteriophage λ infection and lysogeny.

Chapter 31, "Regulation of Gene Expression in Eukaryotes," begins by discussing several different gene systems in the intensively studied unicellular eukaryote *Saccharomyces cerevisiae*. Then it shifts to multicellular eukaryotes, focusing on the complex assemblages of protein factors involved in regulating various genes. Both this chapter and the previous one cover in considerable detail the structures of proteins that regulate gene expression and the ways in which they interact with DNA. The chapter concludes with a special section on the ways in which gene expression is regulated during early development.

Part 7, "Physiological Biochemistry," contains five chapters: "Mechanisms of Membrane Transport" (chapter 32), "Immunobiology" (chapter 33), "Carcinogenesis and Oncogenes" (chapter 34), "Neurotransmission" (chapter 35), and "Vision" (chapter 36). Except for the first one, all of these chapters are intentionally short, so that it will be convenient to assign them at different points in the course, depending on the taste of the instructor. The titles of the chapters are self-explanatory.

Learning Aids

We have included a number of in-text learning aids to help both the professor and the student navigate their way through the text.

The goal of this edition is to stress, not merely the facts, but especially the principles of biochemistry. In as many cases as is practical, explanations are given for how certain facts have been revealed by scientific investigations. Biochemists are sleuths always devising new methods to discover Nature's secrets. We felt it best to integrate discussion of the methods of biochemistry directly in the text at points where the methods are actually employed in biochemical investigations. To aid in location of a particular method description, we have placed a section entitled *A Student's Guide to Methods of Biochemical Analysis* in the appendices.

Great care has been taken to make explanations clear and to inject the maximum significance into topical headings. *Key terms* are always explained the first time they are introduced and *Key concepts* are underlined . Nevertheless, it is easy to forget the meaning of a particular term on occasion. To overcome this frustration we have included a *glossary* just before the index and a list of *common biochemical abbreviations,* we have placed a list just inside the back book cover.

As an additional learning aid, a *brief introduction* to each of the book's seven parts is included. Furthermore, each chapter begins with an *outline of contents* and an *introduction.*

All chapters have *summaries* at the end, and the longer chapters have periodic summaries within them. For those who would like to know more about any specific topic, *extensive references* are included at the end of each chapter.

The *exercise problems* at the end of the chapters are intended to reinforce what has been learned by the reading. *Brief answers* to selected problems are given at the end of the book, and instructors can obtain a *Solutions Manual* that provides complete solutions to all of the problems in the text (see the following section on ancillary materials).

Ancillary Material

In choosing the order of our presentation, we have been strongly influenced by extensive feedback from instructors of biochemistry. As always, however, some instructors would have preferred a different ordering of the subject matter. For their benefit we have suggested alternative possibilities for using the text in an accompanying instructor's manual. This manual also includes additional problems and solutions not offered in the text.

As an aid to course presentations, a set of 100 overhead color transparencies and 200 black-and-white transparency masters is available upon adoption of the text. Also available upon request to professors is a solutions manual, including worked-out solutions to all of the end-of-chapter problems. This manual is available for sale to students, but only at the instructor's request.

Acknowledgments

This preface is full of "I"s. That's because I, the coordinating author, wrote it. By contrast, the text itself is the brainchild of some twenty biochemists, of which I was only one. Nevertheless, because the manuscripts submitted by the contributing authors were modified by me, I must take the blame for any errors. If you should find errors, please notify me so that I can make corrections in the second printing. I would be most grateful for your input.

I would like to make a special note that Chapters 12, 13, and 14 were written by me based on Dr. Daniel Atkinson's material from the second edition of *Biochemistry*. So again, any errors should be considered my responsibility.

My fellow authors and I consulted many experts during the writing of each specific chapter, largely on an informal, "over the phone" basis. Although the list of their names is too long to include here, we are indebted to all of them. In addition, we had the benefit of formal reviews of chapters and whole sections of the text. I am deeply grateful to each of the reviewers, whose aid we will certainly wish to solicit for future editions. The reviewers included:

Teh-hui Kao *Penn State University*
Todd P. Silverstein *Willamette University*
Paul Melius *Auburn University*
Galen Mell *University of Montana*
Edward J. Miller *University of Alabama at Birmingham*
Rodney Cate *Midwestern State University*
Ricki Lewis *SUNY Albany*
Ronald A. MacQuarrie *University of Missouri-Kansas City*
Gerald W. Hart *Johns Hopkins University School of Medicine*
Larry Kirk *California State University at Chico*
Ralph C. Jacobson *California State University-San Luis Obispo*
Peter D. Jones *University of Tennessee*
C. Reynold Verret *Tulane University*
Arthur L. Haas, Ph.D *Medical College of Wisconsin*
Hans Gunderson *Northern Arizona University*
Ezio A. Moscatelli *University of Missouri at Columbia*
Gail Dinter-Gotleib *Drexel University*
David Gross *University of Massachusetts*
Edye E. Groseclose *Southeastern University of Health Sciences*
Maria O. Longas *Purdue University Calumet*
Bruce Howard Weber *California State University, Fullerton*
Donald R. Halenz *Pacific Union College*
Anthony T. Tu *Colorado State University*
Gary D. Small *University of South Dakota School of Medicine*
Celia Marshak *San Diego State University*
D. J. Davis *University of Arkansas*
Karl A. Wilson *SUNY at Binghamton*
Harry R. Matthews *University of California, Davis School of Medicine*
Richard E. Ebel *Virginia Polytechnic Institute and State University*
Malcolm Potts *Virginia Polytechnic Institute and State University*
Robert Lindquist *San Francisco State University*
Jeff Velten *New Mexico State University*

I am also very grateful to many workers at the Wm. C. Brown Publishing Company. First of all to Kevin Kane, who shared my vision and made a tremendous investment in our text. I am overwhelmed by the tremendous help received from Meg Johnson, who gave me advice on many aspects of the text, who obtained a multitude of top-notch reviews that were most useful in preparing the final manuscript, and who stuck by me during the hectic period of production. At times production problems became so great that it wasn't easy to keep going. I was very lucky to have a production coordinator, Sherry Padden, from whom I could get valuable advice on a day-by-day basis. She functioned well beyond the normal responsibilities of her job to help me and the other production workers to pull this project together and bring it to a successful conclusion.

Last but not least I once again had the opportunity to benefit from the superb copy editing of Emily Arulpragasam, who smoothed out bumps in the prose and helped me spot errors before they could be set in type. Her understanding of the subject was a major asset in making changes in the text that were always scientifically accurate.

An Overview of Biochemistry and Energy Considerations

1

PART

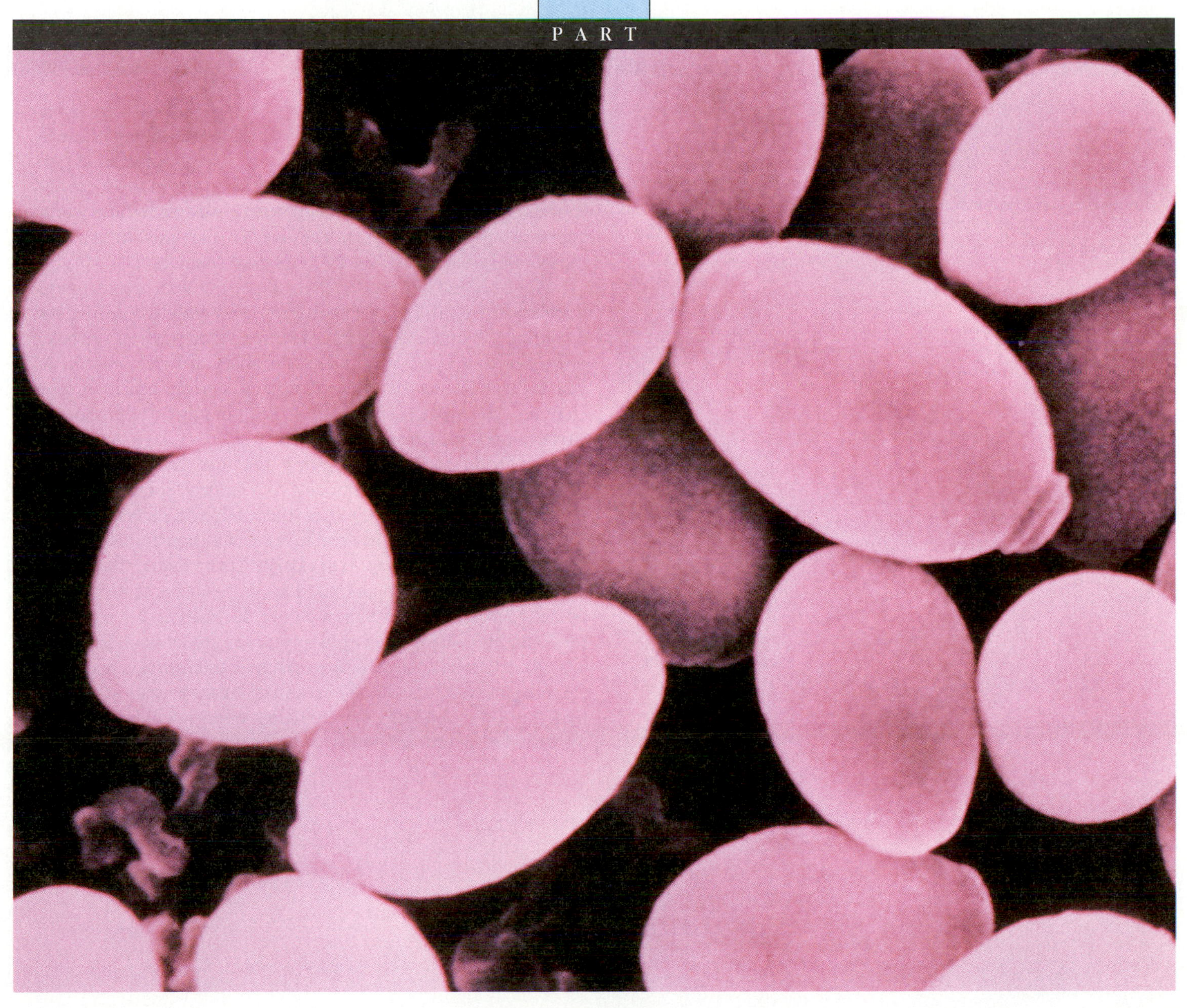

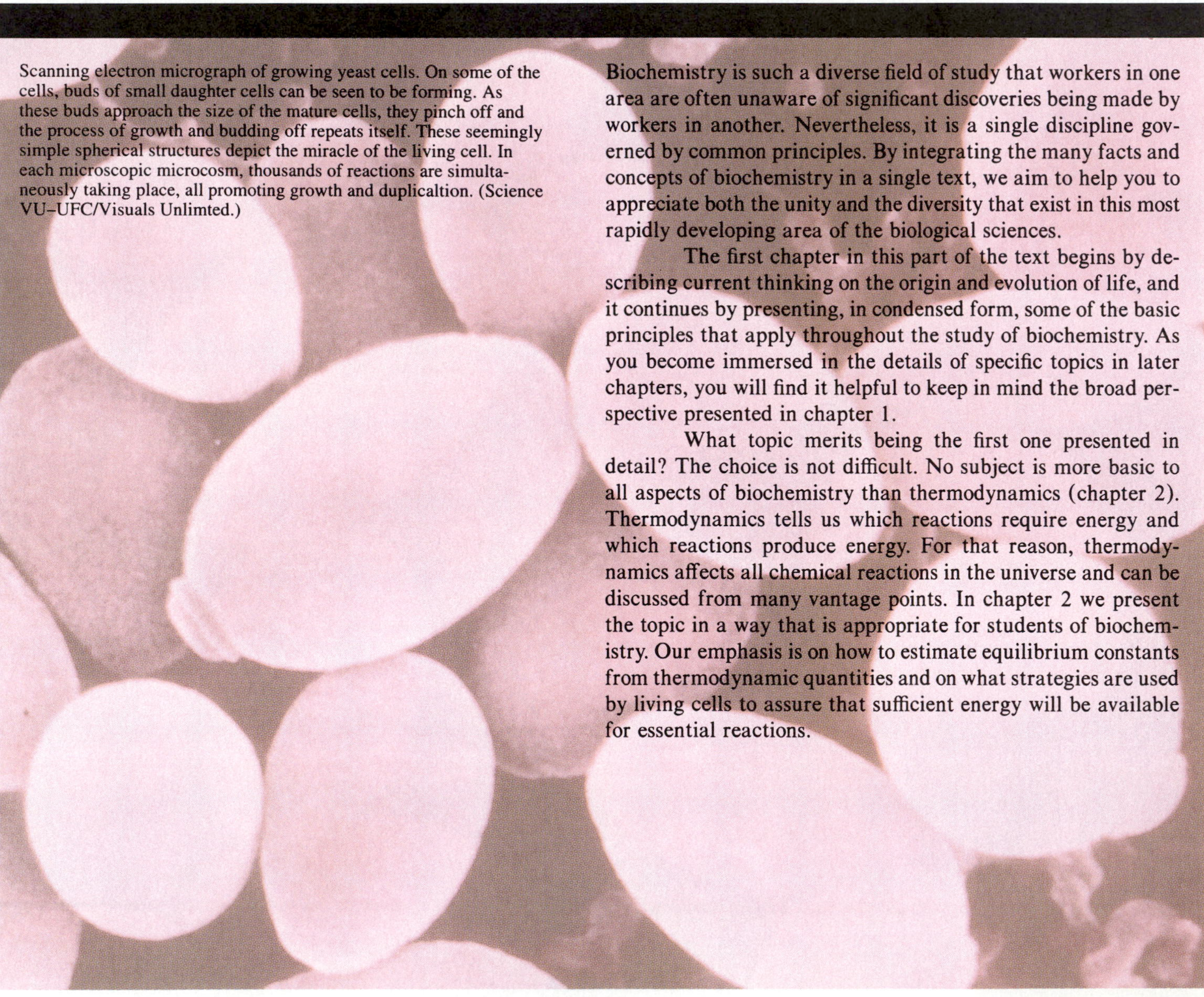

Scanning electron micrograph of growing yeast cells. On some of the cells, buds of small daughter cells can be seen to be forming. As these buds approach the size of the mature cells, they pinch off and the process of growth and budding off repeats itself. These seemingly simple spherical structures depict the miracle of the living cell. In each microscopic microcosm, thousands of reactions are simultaneously taking place, all promoting growth and duplicaltion. (Science VU–UFC/Visuals Unlimted.)

Biochemistry is such a diverse field of study that workers in one area are often unaware of significant discoveries being made by workers in another. Nevertheless, it is a single discipline governed by common principles. By integrating the many facts and concepts of biochemistry in a single text, we aim to help you to appreciate both the unity and the diversity that exist in this most rapidly developing area of the biological sciences.

The first chapter in this part of the text begins by describing current thinking on the origin and evolution of life, and it continues by presenting, in condensed form, some of the basic principles that apply throughout the study of biochemistry. As you become immersed in the details of specific topics in later chapters, you will find it helpful to keep in mind the broad perspective presented in chapter 1.

What topic merits being the first one presented in detail? The choice is not difficult. No subject is more basic to all aspects of biochemistry than thermodynamics (chapter 2). Thermodynamics tells us which reactions require energy and which reactions produce energy. For that reason, thermodynamics affects all chemical reactions in the universe and can be discussed from many vantage points. In chapter 2 we present the topic in a way that is appropriate for students of biochemistry. Our emphasis is on how to estimate equilibrium constants from thermodynamic quantities and on what strategies are used by living cells to assure that sufficient energy will be available for essential reactions.

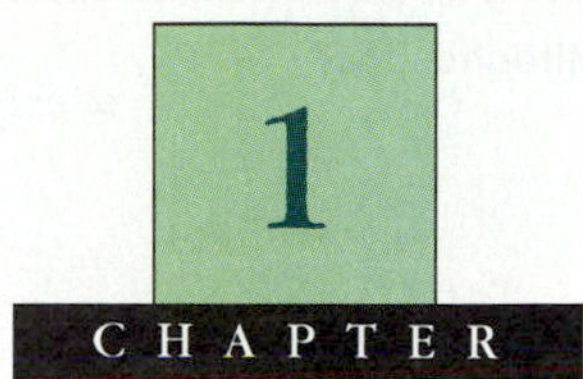

Overview of Biochemistry

Sometime between three and four billion years ago, a chemical change occurred on the young earth that was to have a profound effect on the planet's future. In the ancient seas, one of the several types of macromolecules that spontaneously formed from free-floating precursors acquired a striking new talent—the ability to replicate itself, functioning as a template to assemble component parts into a faithful copy of the original. This self-replicating molecule had another important characteristic: It was composed of a sequence of building blocks, so that its replication was at the same time an act of duplication and one of transmission of information. A likely candidate for this first molecule leading to life was ribonucleic acid (RNA), which is the only known molecule today that can indeed replicate itself without assistance from another molecule.

Yet a single type of molecule, no matter how complex its properties, was not in itself sufficient to qualify as the first example of the thing we call life. Over the millenia the RNA, or whatever the primordial molecule was, associated with other informational polymers—most likely its chemical cousin deoxyribonucleic acid (DNA) and various proteins. An aggregation between nucleic acid and protein may represent a bridge of sorts between the nonliving and the living—a structure that persists today in viruses, which consist of just a nucleic acid wrapped in a protein coat.

At first, the molecules that led to life obtained their building blocks—nutrients—from their surroundings. As more proteins, particularly enzymes, joined the nucleic acid–protein assemblages, these early protocells became capable of capturing energy, perhaps from the covalent bonds of carbohydrates, and of using that energy to convert one nutrient molecule into another. Gradually, metabolic pathways arose. When a third type of macromolecule, the lipids, joined the activity, the raw material for building membranes was introduced to the recipe for life. And finally—perhaps through random, trial-and-error combinations—from proteins, lipids, carbohydrates, and the founding nucleic acids there arose the first cells, the first units of life.

So successful was the simplest cell—a membrane housing nucleic acids, proteins, carbohydrates, and lipids, with some arranged as additional membranes—that it persists today in the form of bacteria and other prokaryotes. However, the development of life did not stop there. The original theme was elaborated on, first to generate the more complex eukaryotic cells, distinguished by their "sacs-within-sacs" division of labor, and then, much later, to produce multicelled organisms.

From chemistry arose life, and life today remains based on chemistry, studied in the subdiscipline of biochemistry. Our cells today must obtain nutrients, must extract energy from their

Figure 1.1

Cells are the fundamental units in all living systems, and they vary tremendously in size and shape. All cells are functionally separated from their environment by the plasma membrane that encloses the cytoplasm. Generalized representations of the internal structures of animal and plant cells (eukaryotic cells). Plant cells have two structures not found in animal cells: a cellulose cell wall, exterior to the plasma membrane, and chloroplasts. The many different types of bacteria (prokaryotes) are all smaller than most plant and animal cells. Bacteria, like plant cells, have an exterior cell wall, but it differs greatly in chemical composition and structure from the cell wall in plants. Like all other cells, bacteria have a plasma membrane that functionally separates them from their environment. Some bacteria also have a second membrane, the outer membrane, exterior to the cell wall.

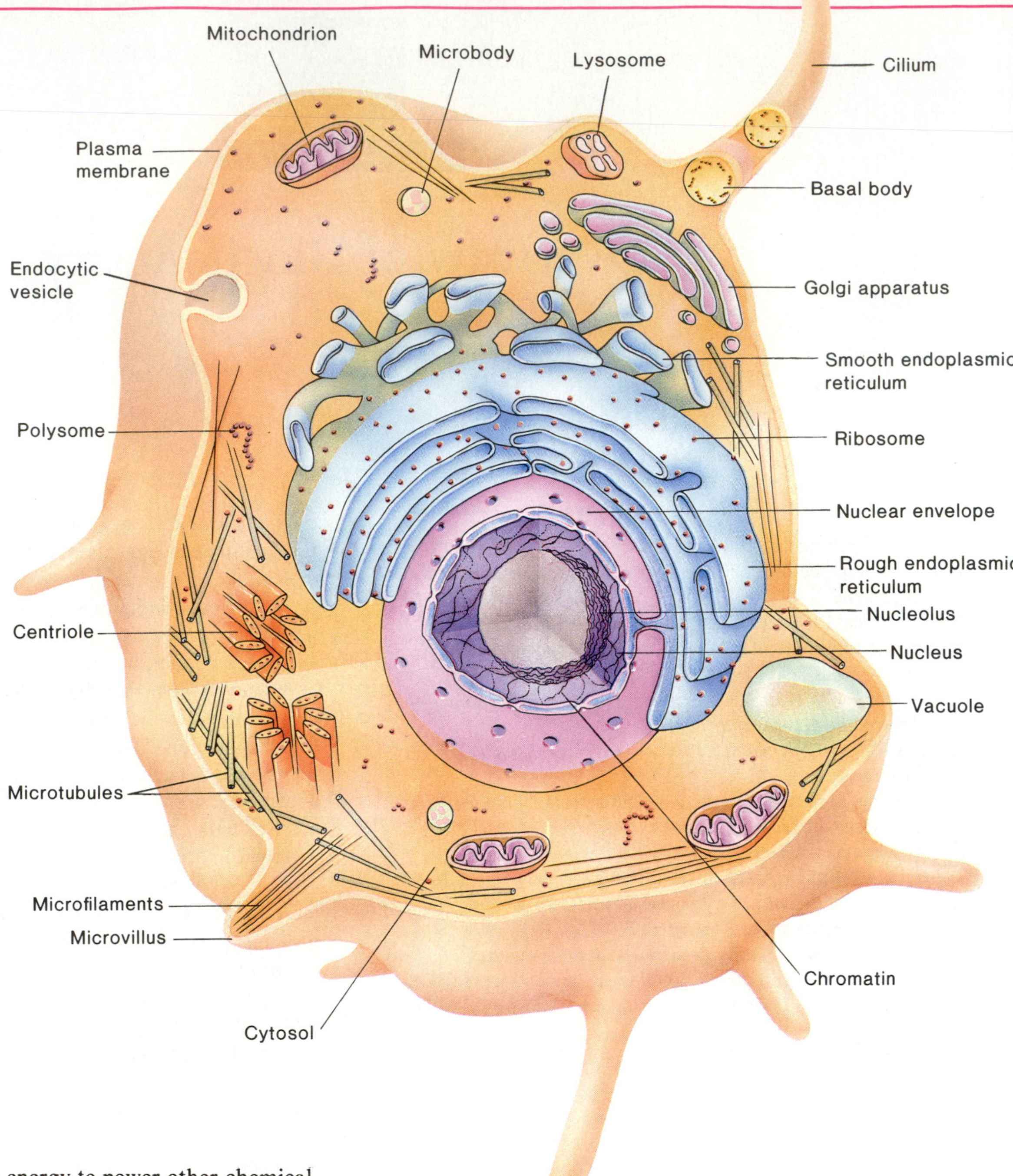

chemical bonds, must use that energy to power other chemical reactions, and, like that very first protocell, must replicate themselves, passing along, in the language of biochemistry, the information that distinguishes the living from the nonliving.

In the introduction to this part of the text, we said that certain common principles underlie all of biochemistry. Let's look briefly now at some of these principles.

The Cell Is the Fundamental Unit of Life

Microscopic examination of any organism will reveal that it is composed of membrane-enclosed structures called cells. The enclosing membrane is called the cell membrane or the plasma membrane. Cells vary enormously in size and shape. Figure 1.1 shows prototypical animal and plant cells, along with some common shapes and sizes of bacteria. Bacteria are single-celled organisms, but sometimes they are connected into long chains. In multicellular organisms the cells associate to form specialized tissues.

The plasma membrane is a delicate, semipermeable, sheetlike covering for the entire cell. By forming an enclosure it prevents gross loss of the intracellular contents; its semipermeable character permits the selective absorption of nutrients and the selective removal of metabolic waste products. In many plant and bacterial (but not animal) cells, a cell wall encompasses the plasma membrane. The cell wall is a more porous

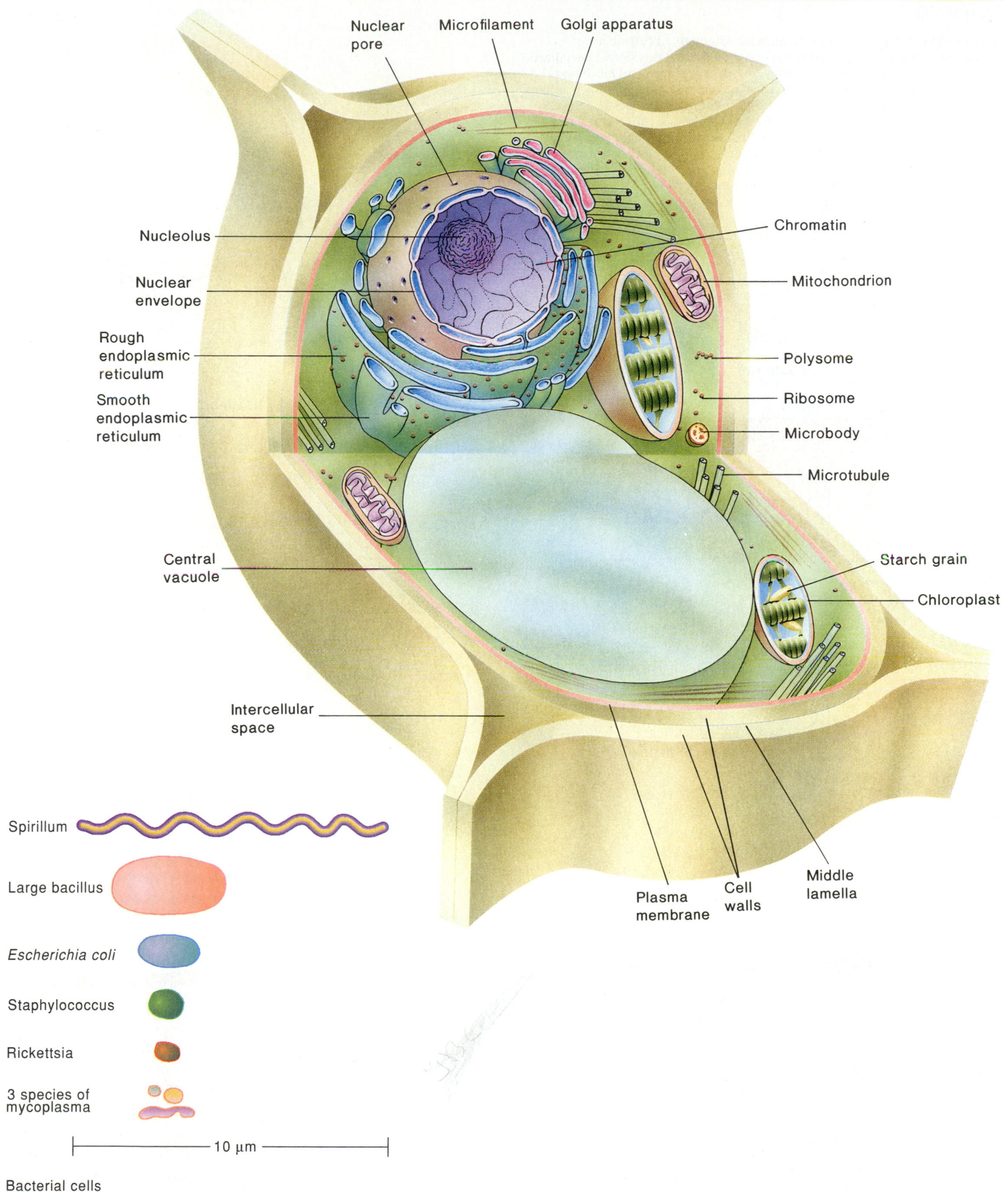
Nuclear pore
Microfilament
Golgi apparatus
Nucleolus
Chromatin
Nuclear envelope
Mitochondrion
Rough endoplasmic reticulum
Polysome
Smooth endoplasmic reticulum
Ribosome
Microbody
Microtubule
Central vacuole
Starch grain
Chloroplast
Intercellular space
Plasma membrane
Cell walls
Middle lamella
Spirillum
Large bacillus
Escherichia coli
Staphylococcus
Rickettsia
3 species of mycoplasma
10 μm
Bacterial cells

Figure 1.2

Specialized cell types found in the human. Although all cells in a multicellular organism have common constituents and functions, specialized cell types have unique chemical compositions, structures, and biochemical reactions that establish and maintain their specialized functions. Such cells arise during embryonic development by the complex processes of cell proliferation and cell differentiation. Except for the sex cells, all cell types contain the same genetic information, which is faithfully replicated and partitioned to daughter cells. Cell differentiation is the process whereby some of this genetic information is activated in some cells, resulting in the synthesis of certain proteins and not other proteins. Thus specialized cells come to have different complements of enzymes and metabolic capacities.

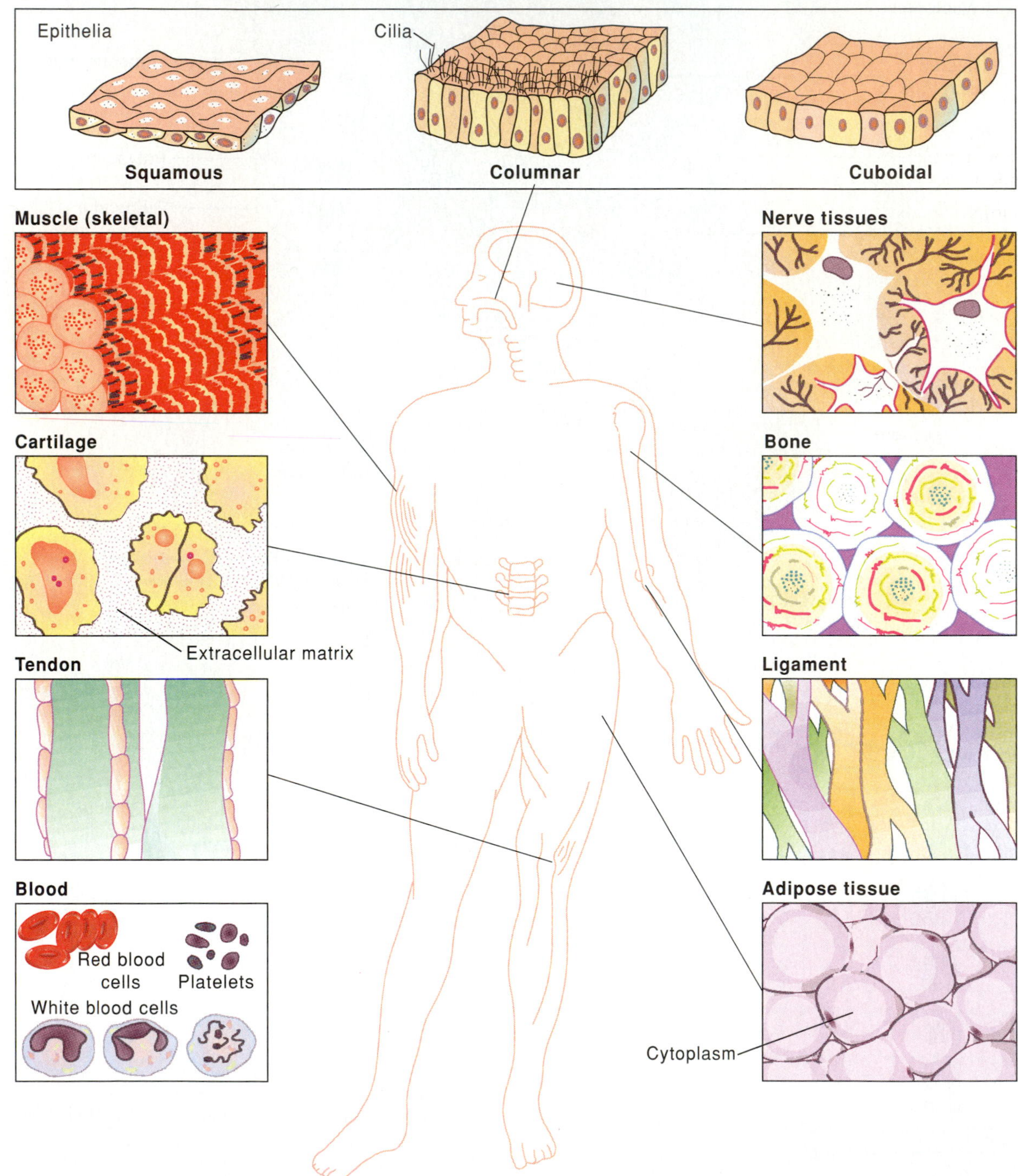

structure than the plasma membrane, but it is mechanically stronger because it is constructed of a covalently cross-linked, three-dimensional network. The cell wall maintains a cell's three-dimensional form when it is under stress.

The contents enclosed by the plasma membrane constitute the cytoplasm. The purely liquid portion of the cytoplasm is called the cytosol. Within the cytoplasm are a number of macromolecules and larger structures, many of which can be seen by high-power light microscopy or by electron microscopy. Some of the structures are membranous and are called organelles. Organelles commonly found in plant and animal cells include the nucleus, the mitochondria, the endoplasmic reticulum, the Golgi apparatus, the lysosomes, and the peroxisomes (see fig. 1.1). Chloroplasts are an important class of organelles found in many plant cells but never in animal cells. Each type of organelle is a specialized biochemical factory in which certain biochemical products are synthesized. In addition to organelles, animal and plant cells contain a collection of filamentous structures termed the cytoskeleton, which is important in maintaining the three-dimensional integrity of the cell.

As we will see, the evolutionary tree is bisected into a lower, or prokaryotic, domain and an upper, or eukaryotic, domain. The terms prokaryote and eukaryote refer to the most basic division between cell types. The fundamental difference is that in eukaryotes the cell contains a nucleus, whereas in prokaryotes it does not. The cells of prokaryotes usually lack most of the other membrane-bounded organelles as well. Plants, fungi, and animals are eukaryotes, and bacteria are prokaryotes. The biochemical functions associated with organelles are frequently present in bacteria, but they are carried out in a different way.

Cells are organized in a variety of ways in different living forms. Prokaryotes of a given type produce cells that are very similar in appearance. A bacterial cell replicates by binary fission, a process in which two identical daughter cells arise from an identical parent cell. Simple eukaryotes can also exist as single nonassociating cells. Eukaryotes of increasing complexity can contain many cells with specialized structures and functions. For example, humans contain about 10^{14} cells of more than a hundred different types. Specialized cells make up the skin, connective tissue, nervous tissue, muscle, blood, sensory functions, and reproductive organs (fig. 1.2). In such a complex organism, the capacity of different cells for replication is limited. When a skin cell or a muscle cell precursor replicates, it makes more cells of the same type. The only cells capable of reproducing an entire organism are the germ cells, that is, the sperm and the egg.

Cells Are Composed of Small Molecules, Macromolecules, and Organelles

Of the many different types of molecules in the various organelles and the cytosol that constitute the living cell, water is by far the most abundant, constituting about 70% by weight of most living matter (table 1.1). As a result, most other components are essentially in an aqueous environment.

Table 1.1
The Approximate Chemical Composition of a Bacterial Cell

	Percent of Total Cell Weight	Number of Types of Each Molecule
Water	70	1
Inorganic ions	1	20
Sugars and precursors	3	200
Amino acids and precursors	0.4	100
Nucleotides and precursors	0.4	200
Lipids and precursors	2	50
Other small molecules	0.2	~200
Macromolecules (proteins, nucleic acids, and polysaccharides)	22	~5,000

Except for water, most of the molecules found in the cell are macromolecules, which can be classified into four different categories: lipids, carbohydrates, proteins, and nucleic acids. Each type of macromolecule possesses distinct chemical properties that suit it for the functions it serves in the cell.

Lipids are primarily hydrocarbon structures (fig. 1.3). They tend to be poorly soluble in water, and are therefore particularly well suited to serve as a major component of the various membrane structures found in cells. Lipids also serve as a compact means of storing chemical energy to drive the metabolism of the cell.

Carbohydrates, like lipids, contain a carbon backbone, but they also contain many polar hydroxyl (—OH) groups and are therefore very soluble in water. Large carbohydrate molecules called polysaccharides consist of many small, ringlike sugar molecules, the sugar monomers, attached to one another by glycosidic bonds in a linear or branched array to form the sugar polymer (fig. 1.4). In the cell, such polysaccharides often form storage granules that may be readily broken down into their component sugars. With further chemical breakdown these sugars release chemical energy and may also provide the carbon skeletons for the synthesis of a variety of other molecules. Important structural functions are also served by polysaccharides. Linear polysaccharides form a major component of plant cell walls, and bacterial cell walls are composed of linear polysaccharides that are cross-linked by polypeptide chains.

Proteins are the most complex macromolecules found in the cell. They are composed of linear polymers called polypeptides, which contain amino acids connected by peptide bonds

Figure 1.3

The structures of common lipids. (*a*) The structures of saturated and unsaturated fatty acids, represented here by stearic acid and oleic acid. (*b*) Three fatty acids covalently linked to glycerol by ester bonds form a triacylglycerol. (*c*) The general structure for a phospholipid consists of two fatty acids esterified to glycerol, which is linked through phosphate to a polar head group. The polar head group may be any one of several different compounds—for example, choline, serine, or ethanolamine.

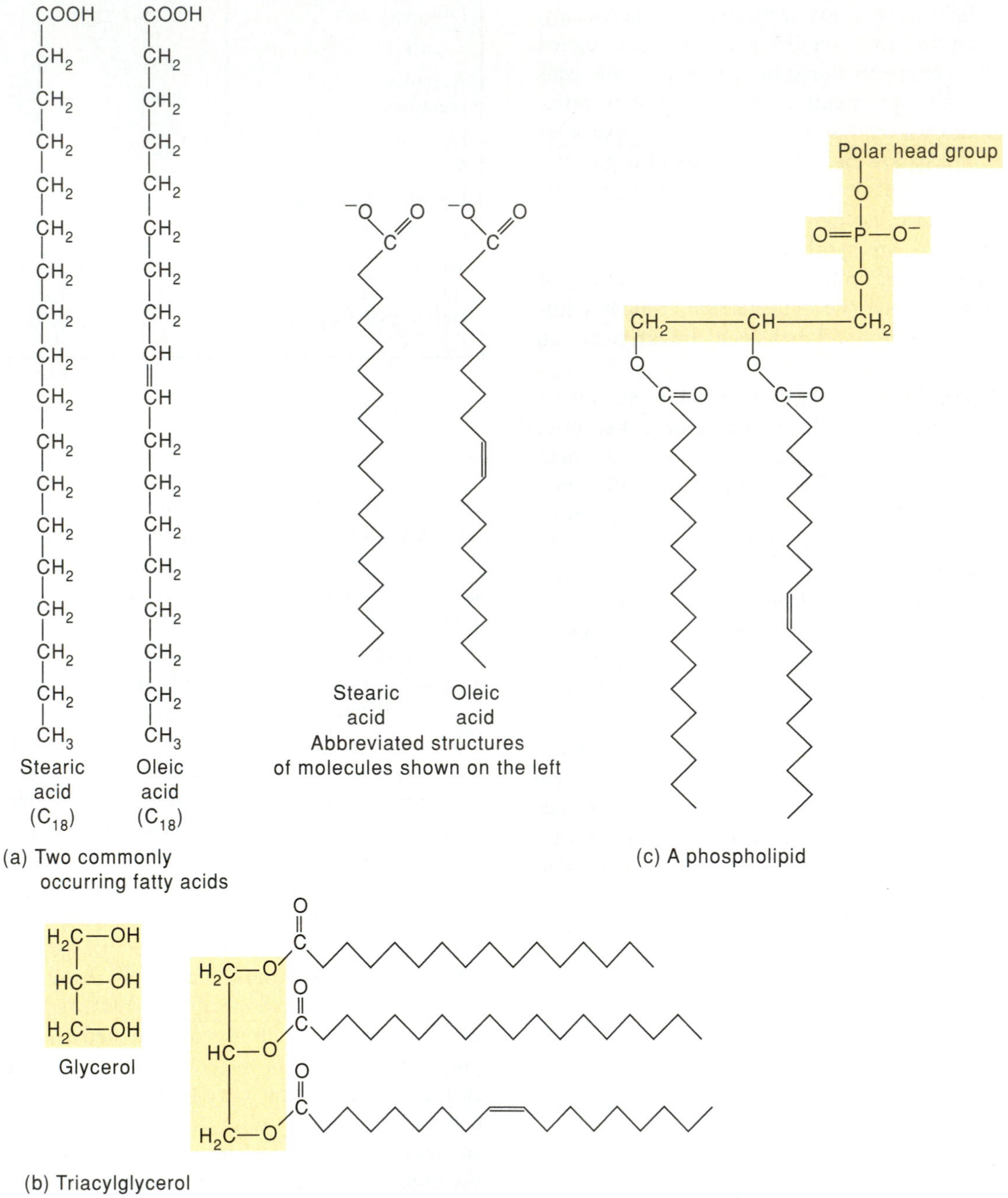

(fig. 1.5). Each amino acid contains a central carbon atom attached to four substituents: (1) a carboxyl group, (2) an amino group, (3) a hydrogen atom, and (4) an R group. The R group gives each amino acid its unique characteristics. There are twenty different amino acids in proteins. Some R groups are charged, some are neutral but still polar, and some are apolar.

The linear polypeptide chains of a protein fold in a highly specific way that is determined by the sequence of amino acids in the chains. Many proteins are composed of two or more polypeptides. Certain proteins function in structural roles. Some structural proteins interact with lipids in membrane structures. Others aggregate to form part of the cytoskeleton that gives the

Figure 1.4

Monomers and polymers of carbohydrates. (*a*) The most common carbohydrates are the simple six-carbon (hexose) and five-carbon (pentose) sugars. In aqueous solution, these sugar monomers circularize to form ring structures. (*b*) Polysaccharides are usually composed of hexose monosaccharides covalently linked together by glycosidic bonds to form long straight-chain or branched-chain structures.

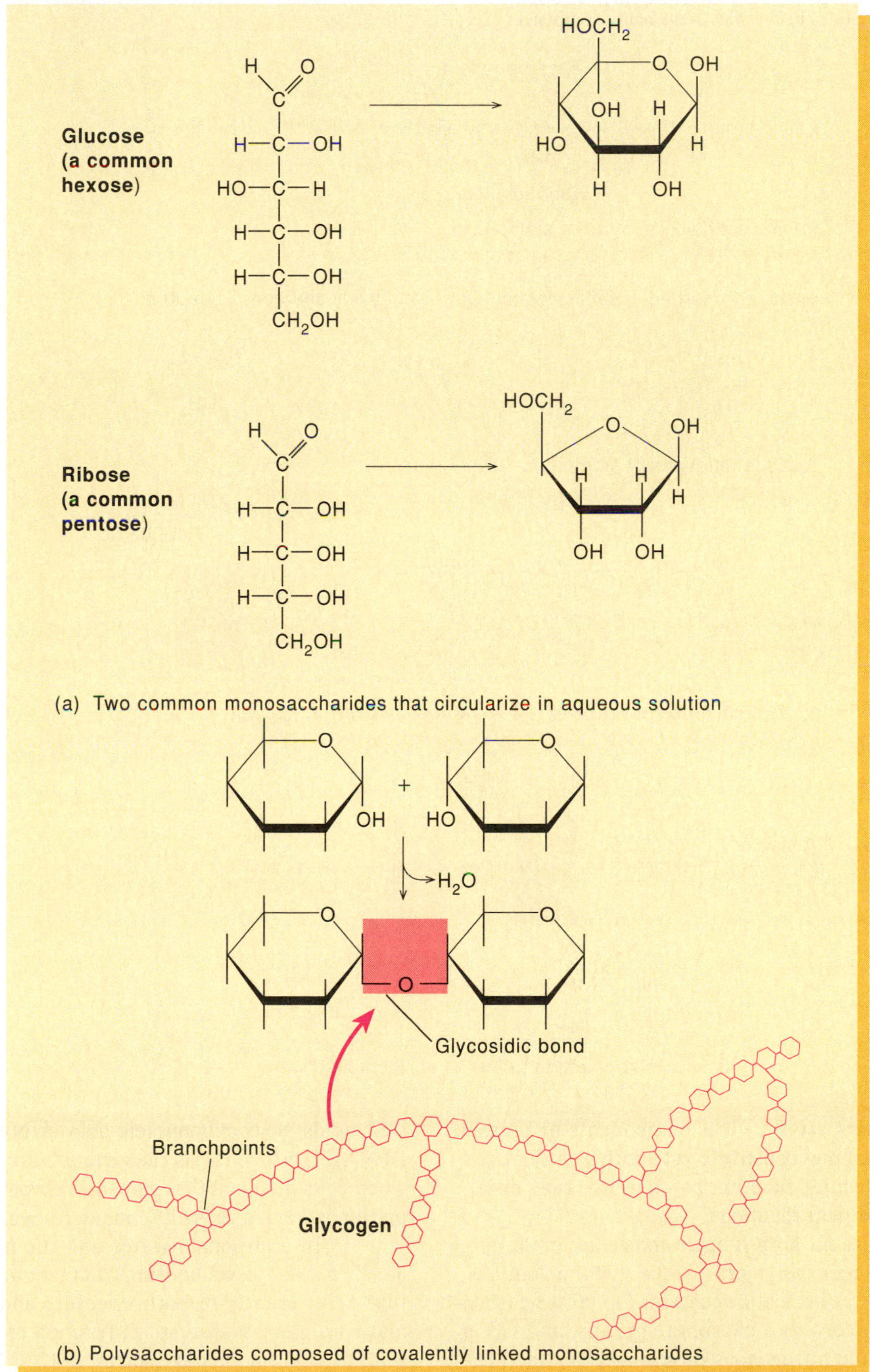

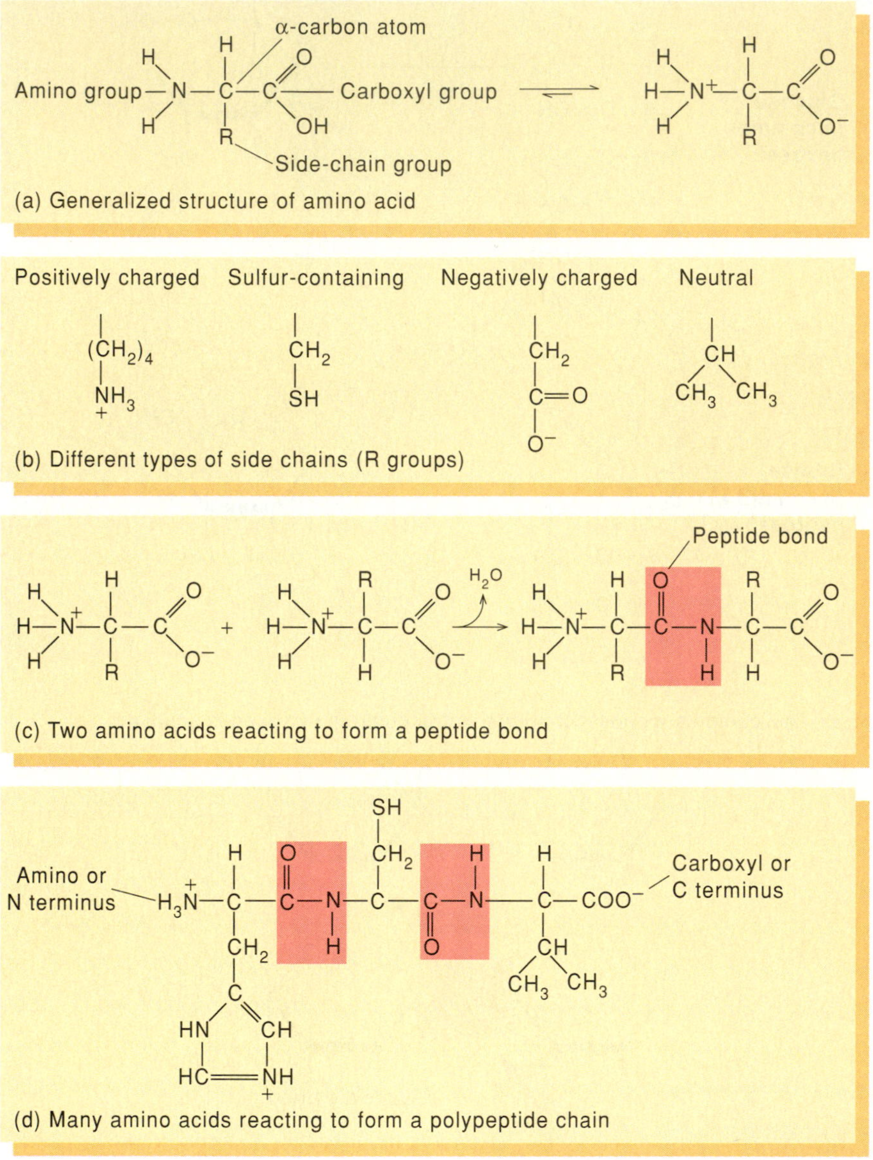

Figure 1.5

Amino acids and the structure of the polypeptide chain. Polypeptides are composed of L-amino acids covalently linked together in a sequential manner to form linear chains. (*a*) The generalized structure of the amino acid. The zwitterion form, in which the amino group and the carboxyl group are ionized, is strongly favored. (*b*) Structures of some of the R groups found for different amino acids. (*c*) Two amino acids become covalently linked by a peptide bond, and water is lost. (*d*) Repeated peptide bond formation generates a polypeptide chain, which is the major component of all proteins.

cell its shape. Still others are the chief components of muscle or connective tissue. Enzymes constitute yet another major class of proteins. These molecules function as catalysts that accelerate and direct biochemical reactions.

Nucleic acids are the largest macromolecules in the cell. They are very long linear polymers, called polynucleotides, composed of nucleotides. A nucleotide contains (1) a five-carbon sugar molecule, (2) one or more phosphate groups, and (3) a nitrogenous base. It is the nitrogenous base that gives each nucleotide a distinct character (fig. 1.6). There are five different types of nitrogenous bases found in the two main types of nucleic acids, deoxyribonucleic acid (DNA) and ribonucleic acid (RNA). DNA contains the genetic information that is inherited when cells divide and organisms reproduce. This genetic information is used in the cell to make ribonucleic acids and proteins.

In addition to water and the macromolecules and organelles, the cytosol contains a large variety of small molecules that differ greatly in both structure and function. These never make up more than a small fraction of the total cell mass despite their great variety (see table 1.1). One class of small molecules consists of the monomer precursors of the different types of macromolecules. These monomers are derived by chemical

Figure 1.6

The structural components of nucleic acids. Nucleic acids are long linear polymers of nucleotides, called polynucleotides. (*a*) The nucleotide consists of a five-carbon sugar (ribose in RNA or deoxyribose in DNA) covalently linked at the 5′ carbon to a phosphate, and at the 1′ carbon to a nitrogenous base. (*b*) Nucleotides are distinguished by the types of bases they contain. These are either of the two-ring purine type or of the one-ring pyrimidine type. (*c*) When two nucleotides become linked they form a dinucleotide, which contains one phosphodiester bond. Repetition of phosphodiester bond formation leads to a polynucleotide.

Base

Phosphate

Sugar

(a) Generalized structure of a nucleotide

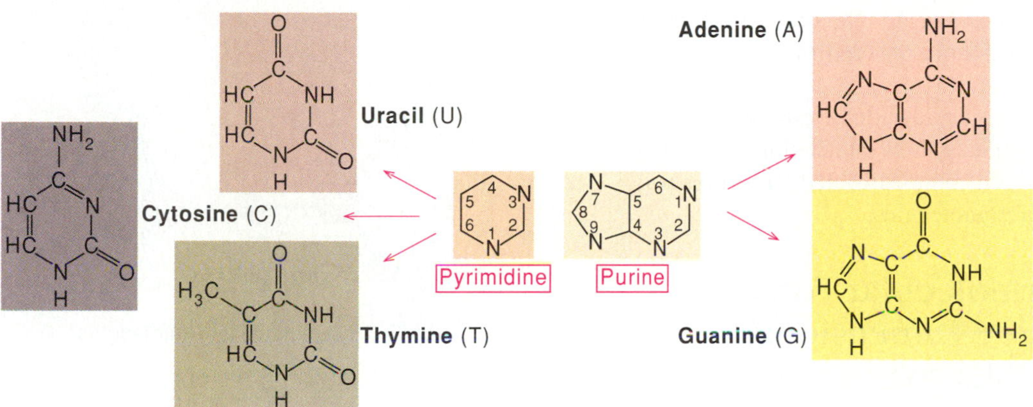

(b) Different bases found in nucleotides

Base

Base

Base

Phosphodiester linkage

Base

(c) Two nucleotides reacting to form a dinucleotide

modification from the nutrients absorbed through the cell membrane. Rarely are the nutrients themselves the actual monomers used by the cell. As a rule each nutrient must undergo a series of enzymatically catalyzed alterations before it is suitable for incorporation into one of the biopolymers. The intermediate molecules between nutrients and monomers are also present in small concentrations in the cytosol. Another varied class of molecules found in the cytosol includes molecules formed as side products in important synthetic reactions and as breakdown products of the macromolecules. Finally, the cytosol contains small bioorganic molecules known as coenzymes, which act in concert with the enzymes in a highly specific manner to catalyze a wide variety of reactions.

All Biochemical Processes Obey the Laws of Thermodynamics

Reactions can occur only if there is an overall decrease in free energy (ΔG). You may recall from earlier courses that ΔG is a composite of two very different terms that dictate the spontaneity with which a transition between two states can occur:

$$\Delta G = \Delta H - T\,\Delta S$$

In this equation T is the temperature, ΔH is the enthalpy, which is approximately equal to the difference in bond energies, and ΔS is the entropy, which is a measure of the disorder in the system. For a given transition, a negative ΔH (favorable) indicates that the system is going from a weaker set to a stronger set of bonded interactions. A positive ΔS (favorable) indicates that the system is going from a more ordered to a less ordered state.

Thermodynamics plays a role in determining which reactions will proceed spontaneously, how biochemical reaction systems are designed, and which complex folded structures will be adopted by biological macromolecules.

Noncovalent Intermolecular Forces Determine the Three-Dimensional Folding Pattern of Macromolecules

The complex folding of biomolecules rarely entails making or breaking covalent linkages, but noncovalent linkages (to be described shortly) are important. The thermodynamic parameters ΔH and ΔS for the transition from unfolded to folded states have rarely been measured for macromolecules. However, in the few cases studied, it appears that the overall entropy of folding is slightly negative (unfavorable) and the overall enthalpy is also slightly negative (favorable). Thus, thermodynamically, folding is opposed by entropy but favored by the enthalpy change and occurs because the latter factor outweighs the former.

Interaction with Water Is a Primary Factor in Determining the Type of Structures that Form

Water, as we have seen, is the major component of living systems, and it interacts with many biomolecules. Some molecules are water-loving or hydrophilic, others are water-abhorring or hydrophobic, and still others are amphipathic or in between. What properties of a molecule make it hydrophilic or hydrophobic? First, consider the molecular properties of water and how water interacts with itself.

An individual water molecule has a significant dipole that is due to the greater electronegativity of the oxygen atom over the hydrogen atoms (fig. 1.7). This dipole leads to strong interactions between water molecules, in the form of hydrogen bonds. A hydrogen bond is a noncovalent interaction between polar molecules, one of which is an unshielded proton. In solid water or ice the polar forces hold the individual molecules together in a regular three-dimensional lattice (fig. 1.8). Most of the hydrogen bonds present in ice are also present in liquid water. Hence water is a highly hydrogen-bonded structure, not too different from ice, but with a somewhat less regular structure in which the individual molecules have greater mobility.

The dipolar properties of water molecules affect the interaction between water and other molecules that dissolve in water. For example, a favorable interaction accounts for the high solubility of sodium chloride in water (fig. 1.9). The kinds of

Figure 1.7

The structure of water and the interaction of water with other water molecules.

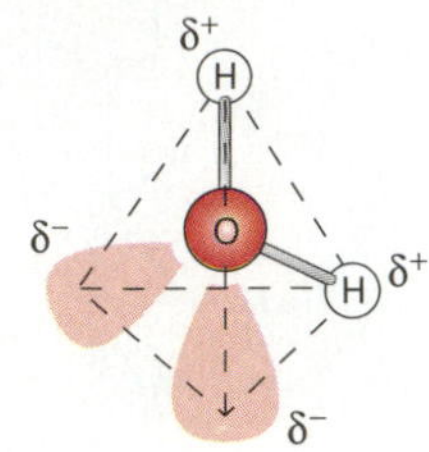

(a) Single water molecule

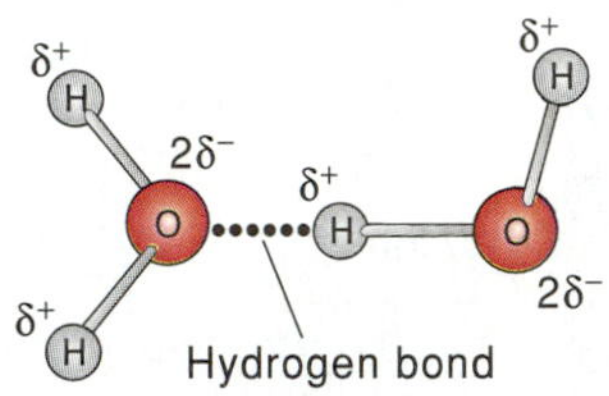

(b) Two interacting water molecules

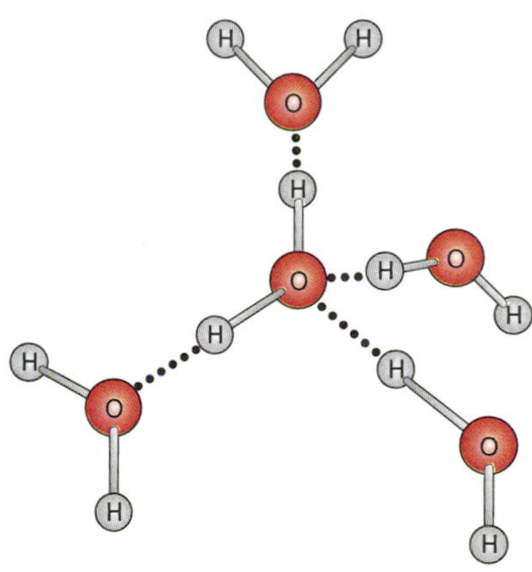

(c) Cluster of interacting water molecules

Figure 1.8

The arrangement of molecules in an ice crystal. Water molecules are orientated so that one proton along each oxygen–oxygen axis is closer to one or the other of the two oxygen atoms.

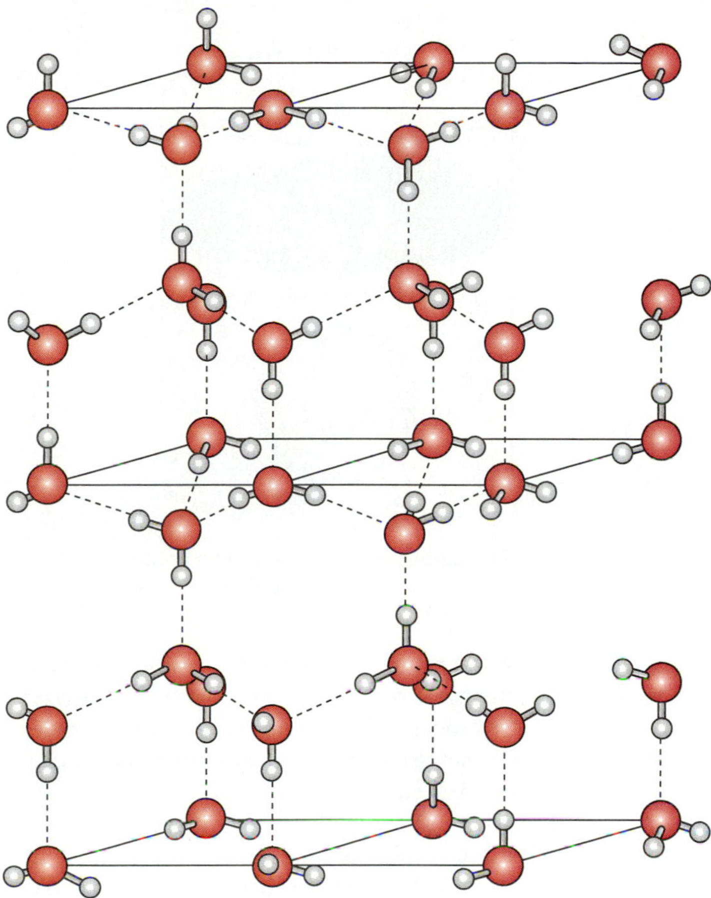

Figure 1.9

The water molecule is composed of two hydrogen atoms covalently bonded to an oxygen atom with tetrahedral (sp^3) electron orbital hybridization. As a result, two lobes of the oxygen sp^3 orbital contain pairs of unshared electrons, giving rise to a dipole in the molecule as a whole. The presence of an electric dipole in the water molecule allows it to solvate charged ions because the water dipoles can orient to form energetically **favorable** electrostatic interactions with charged ions.

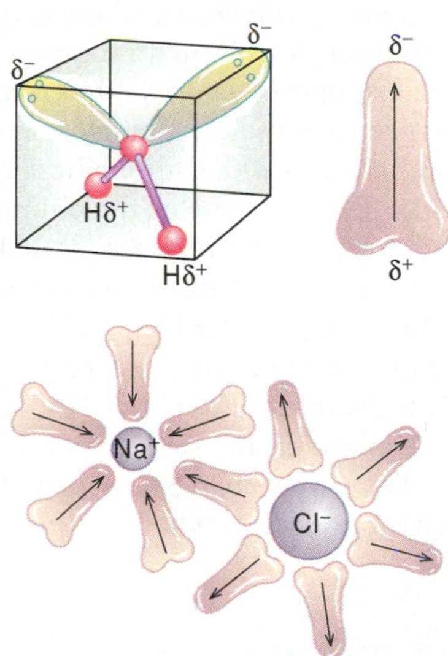

Figure 1.10

Clathrate structures are ordered cages of water molecules around hydrocarbon chains. A portion of the cage structure of $(nC_4H_9)_3 S^+F^-$, 23 H_2O is shown. The trialkyl sulfur ion nests within the hydrogen-bonded framework of water molecules. In the intact framework, each oxygen is tetrahedrally coordinated to four others. One such oxygen atom and its associated hydrogens are shown by the arrow.

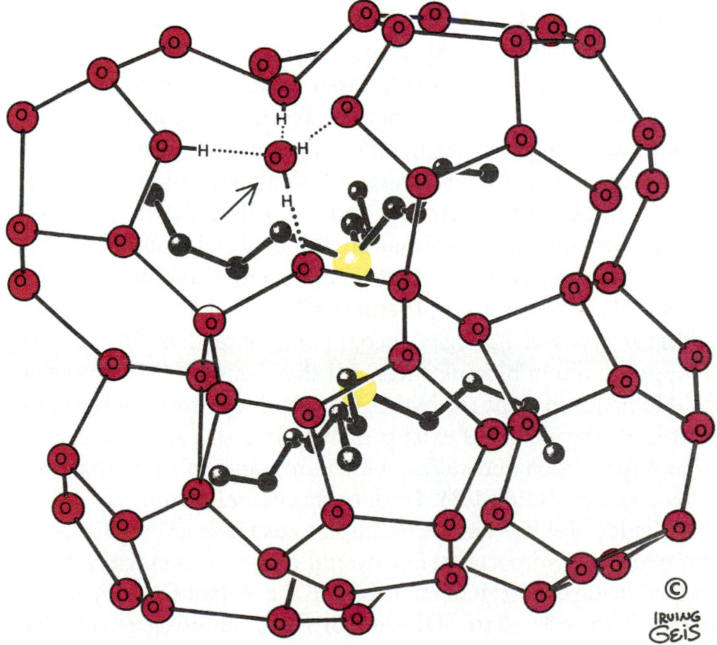

ion–dipole interactions that take place between water and simple ions such as Na^+ and Cl^- are also important in the interactions between the charged or polar groups on biomolecules and water. Thus biomolecules that contain charged residues, hydrogen-bond-forming substituents, or other kinds of polar groups are hydrophilic. In the form of small molecules such groups tend to be very soluble in water. When attached to biopolymers they determine which parts of the molecule will be oriented on the exposed surface, where they will make contact with water.

Apolar groups such as neutral hydrocarbon side chains do not contain significant dipoles or the capacity for forming hydrogen bonds. Consequently, they have nothing to gain by interacting with water, as evidenced by their poor solubility in water. When such hydrophobic molecules are present in water, the water forms a rigid clathrate (cagelike) structure around them (fig. 1.10). Apolar groups in biopolymers tend to bury themselves within the structure of the biopolymer, where they will be in the proximity of other apolar groups and will avoid contacts with water.

Some structures illustrating these principles for macromolecular interaction are shown in figures 1.11 through 1.13. Phospholipids (see fig. 1.3*c*), which have a hydrophilic polar group on one end and long hydrophobic side chains attached to it, form multimolecular aggregates in an aqueous environment (fig. 1.11). These phospholipid aggregates form monomolecular layers at the air–water interface, or micelles or bilayer vesicles within the water. In all of these structures, the polar head groups of the lipid are in contact with water, whereas the apolar side chains are excluded from the solvent structure.

As another example of polarity effects on macromolecular structure, consider polypeptide chains, which usually contain a mixture of amino acids with hydrophilic and hydrophobic side chains. Enzymes fold into complex three-dimensional globular structures with hydrophobic residues located on the inside of the structure and hydrophilic residues located on the surface, where they can interact with water (fig. 1.12).

DNA forms a complementary structure of two helically oriented polynucleotide chains (fig. 1.13). The polar sugar and phosphate groups are on the surface, where they can interact with water; the nitrogenous bases from the two chains form intermolecular hydrogen bonds in the core of the structure.

Figure 1.11

Structures formed by phospholipids in aqueous solution. Phospholipids may form a monomolecular layer at the air–water interface, or they may form spherical aggregations surrounded by water. A vesicle consists of a double molecular layer of phospholipids surrounding an internal compartment of water.

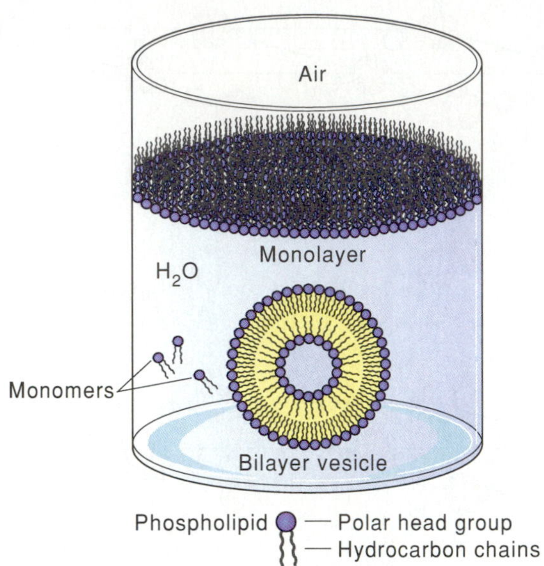

Biochemical Reactions Are a Subset of Ordinary Chemical Reactions

Even though the total number of biochemical reactions is very large, it is still much smaller than the potential number of reactions that occur in ordinary chemical systems. This simplification results partly from the fact that only a limited number of elements account for the vast majority of substances found in living cells. The elements of major importance, in order of decreasing numerical abundance, are hydrogen (H), carbon (C), oxygen (O), nitrogen (N), phosphorus (P), and sulfur (S). Certain metal ions are also important; these include Na^+, K^+, Mg^{2+}, Ca^{2+}, Zn^{2+}, and Fe^{2+} or Fe^{3+}. Other metals and elements that are needed in very small amounts are iodine, cobalt, molybdenum, selenium, vanadium, nickel, chromium, tin, fluorine, silicon, and arsenic. In some cases we don't know the biological roles of these "trace elements" but only that they are needed by some organisms for normal growth or development.

The types of covalent linkages most commonly found in biomolecules are also quite limited (table 1.2). Only sixteen different types of linkages account for more than 95% of the linkages found in biomolecules. All the elements form single or double bonds, except for hydrogen, which can form only single bonds; all the elements exist primarily in a single valence state except for carbon and sulfur, which are found in more than two valence states (table 1.3). Despite this overall simplicity, many other valence states can be found in unusual cases, and some of these are very important. For example, the biochemistry of nitrogen involves consideration of all the valence states of nitrogen from +5 to 0 to −3. A major source of nitrogen available to biosystems is gaseous nitrogen found in the atmosphere (valence state 0). Biochemical reactions convert gaseous nitrogen into other forms of nitrogen. These reactions are very complex and occur only in a select group of microorganisms that possess the necessary enzyme systems.

Biochemical reactions involving the different classes of substances use a limited number of functional groups, which are illustrated in figure 1.14. Most of the reactive groups in biomolecules contain one or more of these functional groups or closely related ones. Many cellular reactions involving these functional groups are closely related to reactions that take place outside the cell under different conditions and are studied in organic chemistry.

For example, alcohols, which contain a hydroxyl functional group, can undergo dehydration reactions with either carboxylic or phosphoric acid to form esters.

$$\underset{\text{Alcohol}}{R{-}OH} + \underset{\text{Carboxylic acid}}{HO{-}\overset{O}{\overset{\|}{C}}{-}R'} \rightleftharpoons \underset{\text{Ester}}{R{-}O{-}\overset{O}{\overset{\|}{C}}{-}R'} + \underset{\text{Water}}{H_2O}$$

$$\underset{\text{Alcohol}}{R{-}OH} + \underset{\text{Phosphoric acid}}{HO{-}\overset{O}{\overset{\|}{\underset{O^-}{\underset{|}{P}}}}{-}O^-} \rightleftharpoons \underset{\text{Phosphoric acid ester}}{R{-}O{-}\overset{O}{\overset{\|}{\underset{O^-}{\underset{|}{P}}}}{-}O^-} + \underset{\text{Water}}{H_2O}$$

Figure 1.12

A graphic representation of a three-dimensional model of the protein cytochrome *c*. Amino acids with nonpolar, hydrophobic side chains (color) are found in the interior of the molecule, where they interact with one another. Polar, hydrophilic amino acid side chains (gray) are on the exterior of the molecule, where they interact with the polar aqueous solvent.

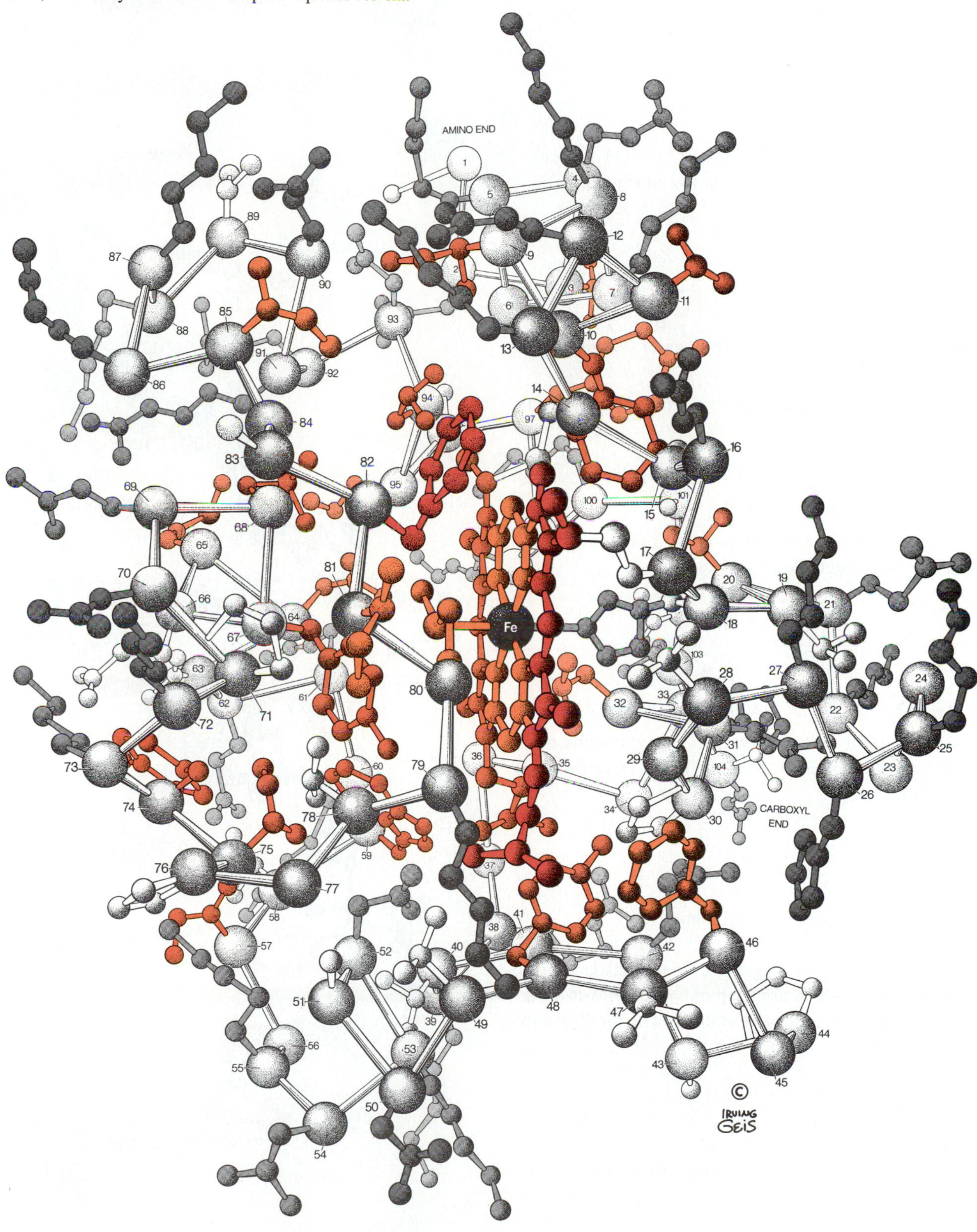

Figure 1.13

The right-handed helical structure of DNA. DNA normally exists as a two-chain structure held together by hydrogen bonds (···) formed between the bases in the two chains. Along the chain the planar surfaces of these bases interact and, together with the hydrogen bonds, contribute to the stability of the two-chain structure. The negatively charged phosphate groups are on the outside of the structure, where they interact with water, ions, or charged molecules.

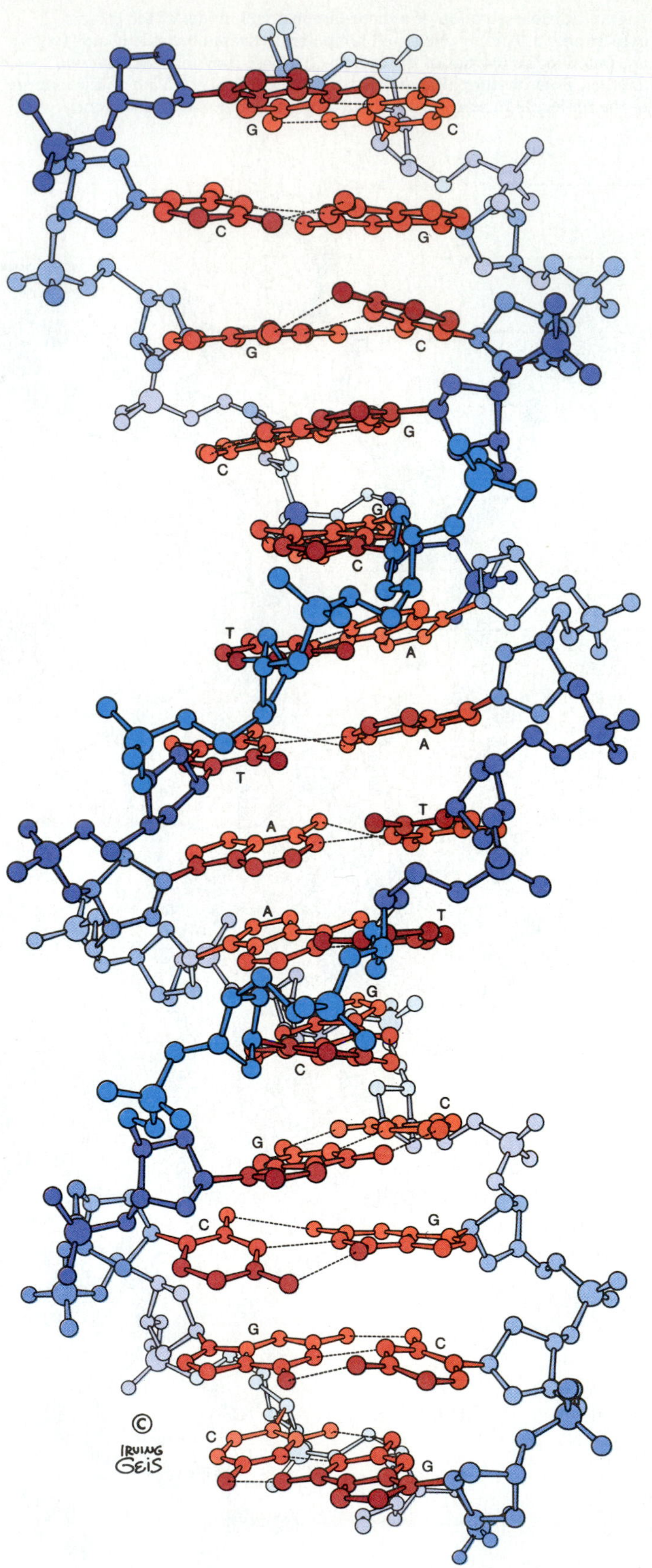

Thiols, containing sulfhydryl groups (—SH), can substitute for alcohols in some reactions, leading to the formation of thiol esters.

$$\underset{\textbf{Thiol}}{R{-}SH} + \underset{\textbf{Carboxylic acid}}{HO{-}\overset{O}{\overset{\|}{C}}{-}R'} \rightleftharpoons \underset{\textbf{Thiol ester}}{R{-}S{-}\overset{O}{\overset{\|}{C}}{-}R'} + \underset{\textbf{Water}}{H_2O}$$

Two alcohols can react with one another to form an ether.

$$\underset{\textbf{Alcohol}_1}{R{-}OH} + \underset{\textbf{Alcohol}_2}{HO{-}R'} \longrightarrow \underset{\textbf{Ether}}{R{-}O{-}R'} + H_2O$$

Alcohols can also undergo a dehydrogenation reaction to form a carbonyl derivative (aldehyde or ketone).

$$\underset{\textbf{Alcohol}}{R{-}CH(R){-}OH} \underset{+2H}{\overset{-2H}{\rightleftharpoons}} \underset{\textbf{Aldehyde or ketone}}{R_2C{=}O}$$

Amines undergo reactions with carboxylic acids comparable to the formation of esters from alcohols. The product is known as an amide.

$$\underset{\textbf{Primary amine}}{R{-}\overset{+}{N}H_3} + \underset{\textbf{Carboxylic acid}}{{}^{-}O{-}\overset{O}{\overset{\|}{C}}{-}R'} \rightleftharpoons \underset{\textbf{Amide}}{R{-}\overset{H}{\overset{|}{N}}{-}\overset{O}{\overset{\|}{C}}{-}R'} + H_2O$$

Amines can also undergo dehydrogenation reactions leading to the formation of imines, which are frequently unstable in water and hydrolyze to ketones or, in cases where one of the R groups is an H, to aldehydes:

$$\underset{\textbf{Amine}}{R{-}CH(R){-}NH_2} \xrightarrow{-2H} \underset{\textbf{Imine}}{R_2C{=}NH} \xrightarrow{+H_2O} \underset{\textbf{Ketone}}{R_2C{=}O} + \underset{\textbf{Ammonia}}{NH_3}$$

Table 1.2
Types of Covalent Linkages Most Commonly Found in Biomolecules

	H	C	O	N	P	S
H						
C	—C—H	—C—C— C=C				
O	—O—H	—C—O— C=O				
N	N—H	—C—N= C=N—	—			
P	—	—	P—O P=O	—		
S	—S—H	—C—S—	S—O— S=O	—	—	—S—S—

Table 1.3
Most Common Valences Displayed by Atoms in Covalent Linkages

Element	Valence
H	+1
C	−4 to +4
O	−2
P	+5
N	−3
S	+6, −2, −1

Aldehydes and ketones both may be reduced to alcohols by hydrogenation (see the alcohol dehydrogenation reaction). Aldehydes may react with either water or alcohol to form aldehyde hydrates, or hemiacetals, respectively. Reaction of an aldehyde with two molecules of alcohol leads to acetal formation.

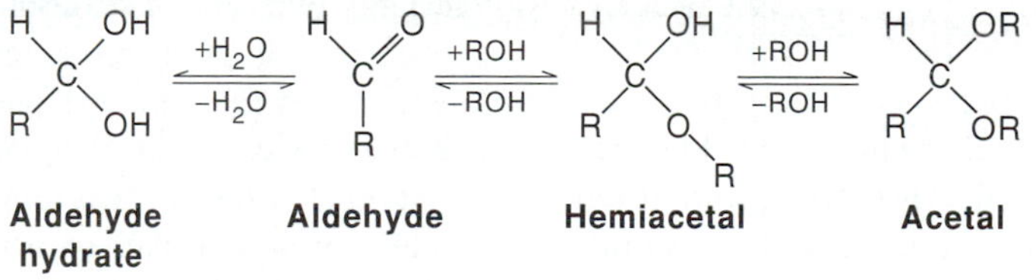

$$\underset{\text{Aldehyde hydrate}}{\mathrm{RCH(OH)_2}} \underset{+H_2O}{\overset{-H_2O}{\rightleftharpoons}} \underset{\text{Aldehyde}}{\mathrm{RCHO}} \underset{-ROH}{\overset{+ROH}{\rightleftharpoons}} \underset{\text{Hemiacetal}}{\mathrm{RCH(OH)OR}} \underset{-ROH}{\overset{+ROH}{\rightleftharpoons}} \underset{\text{Acetal}}{\mathrm{RCH(OR)_2}}$$

Figure 1.14

Different functional groups found in biomolecules. This figure includes the major functional groups. Other functional groups are found in minor amounts.

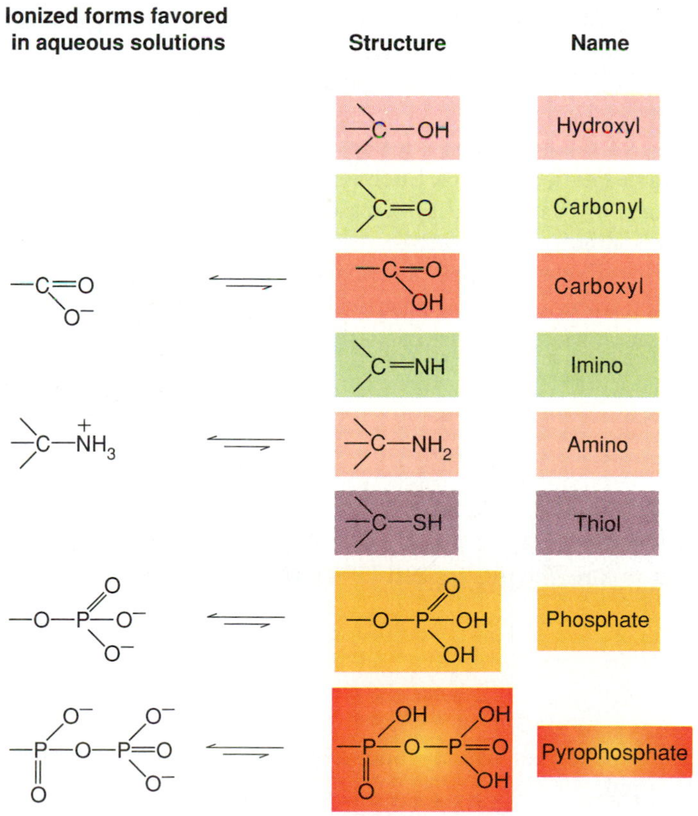

Ionized forms favored in aqueous solutions		Structure	Name
		$\geq$C—OH	Hydroxyl
		>C=O	Carbonyl
—C(=O)O⁻	⇌	—C(=O)OH	Carboxyl
		>C=NH	Imino
$\geq$C—NH_3^+	⇌	$\geq$C—NH_2	Amino
		$\geq$C—SH	Thiol
—O—P(=O)(O⁻)—O⁻	⇌	—O—P(=O)(OH)—OH	Phosphate
—P(=O)(O⁻)—O—P(=O)(O⁻)—O⁻	⇌	—P(=O)(OH)—O—P(=O)(OH)—OH	Pyrophosphate

Dehydrogenation of an aldehyde hydrate leads to carboxylic acid formation.

$$\text{R-CH(OH)}_2 \xrightarrow{-2H} \text{R-C(=O)OH}$$

Aldehyde hydrate **Carboxylic acid**

Aldehydes and ketones may also isomerize to the enol form as long as the adjacent carbon atom contains at least one H atom. In the reaction a hydrogen migrates and the double bond shifts.

$$\text{R-CHR-C(R)=O} \rightleftharpoons \text{R}_2\text{C=C(R)-OH}$$

Keto form **Enol form**

Pyrophosphates may hydrolyze to inorganic phosphoric acid (phosphate) and an organophosphoric acid.

$$\text{R-O-P(=O)(O}^-\text{)-O-P(=O)(O}^-\text{)-O}^- \underset{-H_2O}{\overset{+H_2O}{\rightleftharpoons}} \text{R-O-P(=O)(O}^-\text{)-O}^- + \text{HO-P(=O)(O}^-\text{)-O}^- + H^+$$

Organopyrophosphate **Organophosphoric acid** **Phosphoric acid**

Hydrolysis reactions of this sort yield considerable energy, which can be utilized in biosynthesis.

All of the functional groups that we have described are electrostatically neutral in organic solvents. However, in water many of these functional groups either lose or gain protons to become charged species. Such ionization reactions are very important in biochemical systems because they frequently influence solubility and reactivity.

Carboxylic and phosphoric acids lose one or more protons in water to become negatively charged. The ionized forms are stabilized by resonance as shown:

$$\text{R-C(=O)OH} \underset{+H^+}{\overset{-H^+}{\rightleftharpoons}} \left[\text{R-C(=O)O}^- \leftrightarrow \text{R-C(O}^-\text{)=O}\right]$$

$$\text{(HO)}_2\text{P(=O)OH} \underset{+H^+}{\overset{-H^+}{\rightleftharpoons}} \left[\text{}^-\text{O-P(=O)(HO)(OH)} \leftrightarrow \text{O=P(O}^-\text{)(HO)(OH)}\right] \underset{+H^+}{\overset{-H^+}{\rightleftharpoons}} \text{}^-\text{O-P(=O)(}^-\text{O)(OH)} \underset{+H^+}{\overset{-H^+}{\rightleftharpoons}} \left[\text{}^-\text{O-P(=O)(}^-\text{O)(O}^-\text{)} \leftrightarrow \text{}^-\text{O-P(O}^-\text{)(=O)(O}^-\text{)}\right]$$

Amines usually add a proton to become positively charged.

$$\text{R}_3\text{C-NH}_2 \underset{-H^+}{\overset{+H^+}{\rightleftharpoons}} \text{R}_3\text{C-}\overset{+}{\text{N}}\text{H}_3$$

Near neutrality (10^{-7} M H^+), where most biochemical systems function, the carboxyl group exists mainly in the negatively charged form, phosphoric acid exists mainly in the diionized form, and amino groups exist mainly in the positively charged form. This fact has interesting consequences for amino acids, since they contain one amino group and one carboxyl group. The amino acids are usually neutral overall, even though they contain two charged groups, one resulting from the deprotonation of the carboxyl group and the other resulting from the protonation of the amino group. Amino acids existing as dipolar ions are called zwitterions.

$$\text{H}_2\text{N-CH(R)-C(=O)OH} \rightleftharpoons \text{H}_3\overset{+}{\text{N}}\text{-CH(R)-C(=O)O}^-$$

Uncharged **Zwitterion**

These are some of the more important reactions involving covalent bond breakage or formation in biochemistry. By now two things should be apparent about biochemical reactions: (1) as stated at the outset, the number of reactions in biochemistry is much more limited than in ordinary chemistry; (2) as far as the reactants and products are concerned, biochemical reactions can be understood in the same terms as ordinary chemical reactions.

Conditions for Biochemical Reactions Differ from Those for Ordinary Chemical Reactions

Although biochemical reactions resemble ordinary chemical reactions, they differ in some important ways. Chemical reactions are frequently carried out in nonaqueous solvents, using elevated temperatures and pressures, acids or bases, or other harsh reagents, conditions that would destroy the functional organization of a living cell. Biochemical reactions usually take place under very mild conditions in aqueous solution. However, many chemical reactions do not proceed at reasonable rates under such conditions. Biochemical reactions proceed at substantial rates because of the very special nature of the enzyme catalysts that accelerate them.

Enzymes are structurally complex, highly specific catalysts; each enzyme usually catalyzes only one type of reaction. The enzyme surface binds the interacting molecules, or substrates, so that they are favorably disposed to react with one another (fig. 1.15). The specificity of enzyme catalysis also has a selective effect, so that only one of several potential reactions will take place. For example, a simple amino acid can be utilized in the synthesis of any of the four major classes of macromolecules or can be simply secreted as waste product (fig. 1.16). The fate of the amino acid is determined as much by the presence of specific enzymes as by its reactive functional groups.

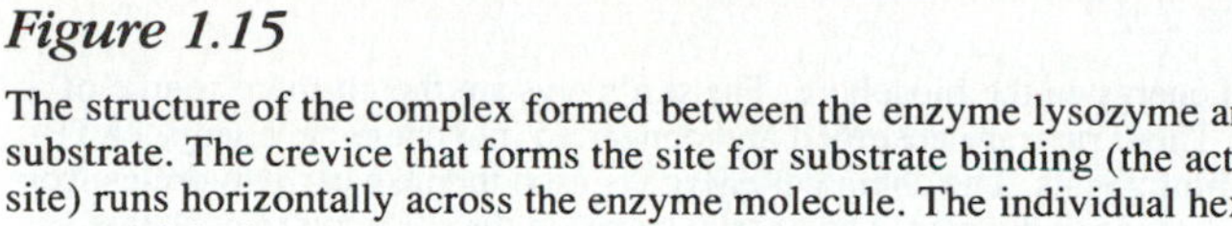

Figure 1.15

The structure of the complex formed between the enzyme lysozyme and its substrate. The crevice that forms the site for substrate binding (the active site) runs horizontally across the enzyme molecule. The individual hexose sugars of the hexasaccharide substrate are shown in a darker color and labeled A–F.

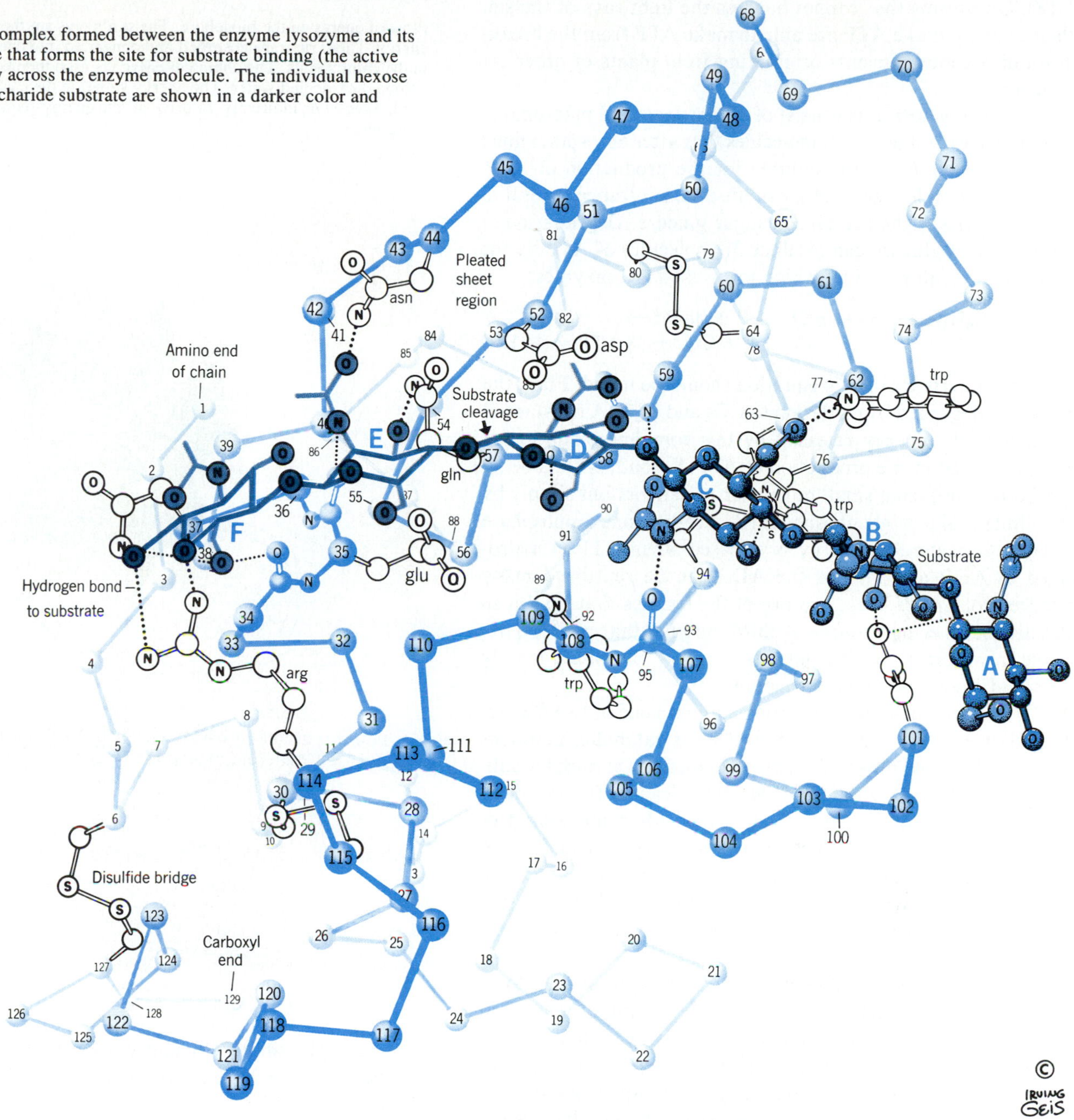

Figure 1.16

The different fates of an amino acid. Depending on which enzymes are present and active and on the needs of the organism, an amino acid can be metabolized in different ways. Each of these conversions involves one or more steps, and usually each step requires a specific enzyme.

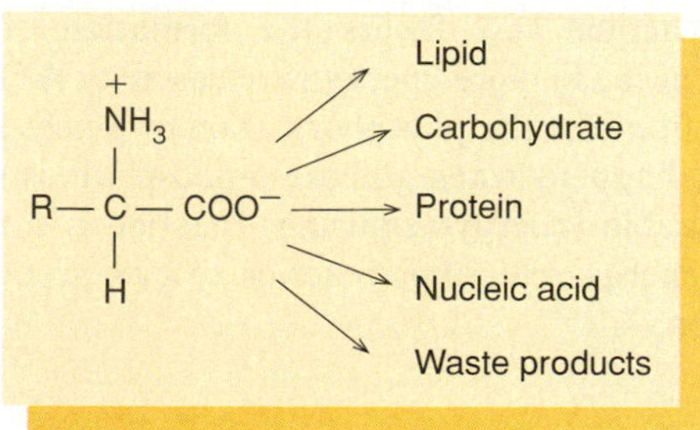

Many Biochemical Reactions Require Energy

An appreciable amount of energy is needed to build a cell. Even maintaining a cell in a steady nongrowing state requires energy input. Chemical energy is needed to drive many biochemical reactions, to do mechanical work, and for transport of substances across the plasma membrane. The ultimate source of energy that drives a cell's reactions is sunlight (fig. 1.17). Light energy is converted into chemical energy in the chloroplasts of plant cells or in the photosynthetic grana of certain microorganisms. The main form of chemical energy produced in the chloroplast is a nucleotide containing three phosphoric acid groups attached in sequence, adenosine triphosphate or ATP (fig. 1.18). Organisms that cannot harness the light rays of the sun themselves to make ATP are able to make ATP from the breakdown of organic nutrients originating from plants or other organisms.

Many nutrients consist of partly degraded macromolecules or various other small molecules that after absorption must be converted into a form suitable for the production of ATP. One of the simplest and yet most effective substances useful in ATP synthesis is the six-carbon sugar glucose. Degradation of a molecule of glucose can produce 38 molecules of ATP by the following overall reaction, which involves many enzymes:

$$38\ H^+ + C_6H_{12}O_6 + 6\ O_2 + 38\ ADP + 38\ P_i^* \rightarrow 6\ CO_2 + 38\ ATP + 44\ H_2O$$

Two characteristics of this equation should be noted. First, the glucose is degraded by oxidation to CO_2 and H_2O. A substantial fraction of the energy released by this complete oxidation of glucose is used in the production of ATP. Second, ATP is being synthesized not from small molecular precursors but simply by the addition of a single phosphate (P_i) to adenosine diphosphate (ADP). Considerable energy is released when ATP is hydrolyzed to ADP and P_i, and the ADP can be reutilized many hundreds of times. These are two of the reasons that ATP is so effective as an energy source. A third reason is that ATP is quite stable in water; it does not lose its terminal phosphate readily except in enzyme-catalyzed reactions.

Most biochemical reactions fall into one of two classes: degradative or synthetic. Degradative, or catabolic, reactions result in the breakdown of organic compounds to simpler substances. Synthetic, or anabolic, reactions lead to the assembly of biomolecules from simpler molecules. Most anabolic processes require energy to drive them. This energy is usually supplied by coupling the energy-requiring biosynthetic reactions to energy-releasing catabolic reactions. Most frequently the energy-releasing reaction involves ATP hydrolysis, either directly or indirectly.

As an example of coupling the hydrolysis of ATP to an energy-requiring reaction, let us examine the use of glucose in an anabolic reaction, the synthesis of a polysaccharide containing glucose as the monomer. The first step in glucose utilization involves its conversion to glucose-6-phosphate. The reaction of glucose with P_i to form glucose-6-phosphate is energetically unfavorable. Another way of saying this is that the equilibrium for the reaction favors the reactants, not the products.

$$\text{Glucose} + P_i \leftrightharpoons \text{glucose-6-phosphate}$$

Thus, even in the presence of a catalyst for this reaction, very little glucose-6-phosphate would be formed. In order to make

*P_i is the inorganic phosphate ion, which has the structural formula

$$^*P_i = HO{-}\overset{\overset{\displaystyle O}{\|}}{\underset{\underset{\displaystyle O^-}{|}}{P}}{-}O^-$$

Figure 1.17

Flow of energy in the biosphere. The sun's rays are the ultimate source of energy. These rays are absorbed and converted into chemical energy (ATP) in the chloroplasts. The chemical energy is used to make carbohydrates from carbon dioxide and water. The energy stored in the carbohydrates is then used, directly or indirectly, to drive all the energy-requiring processes in the biosphere.

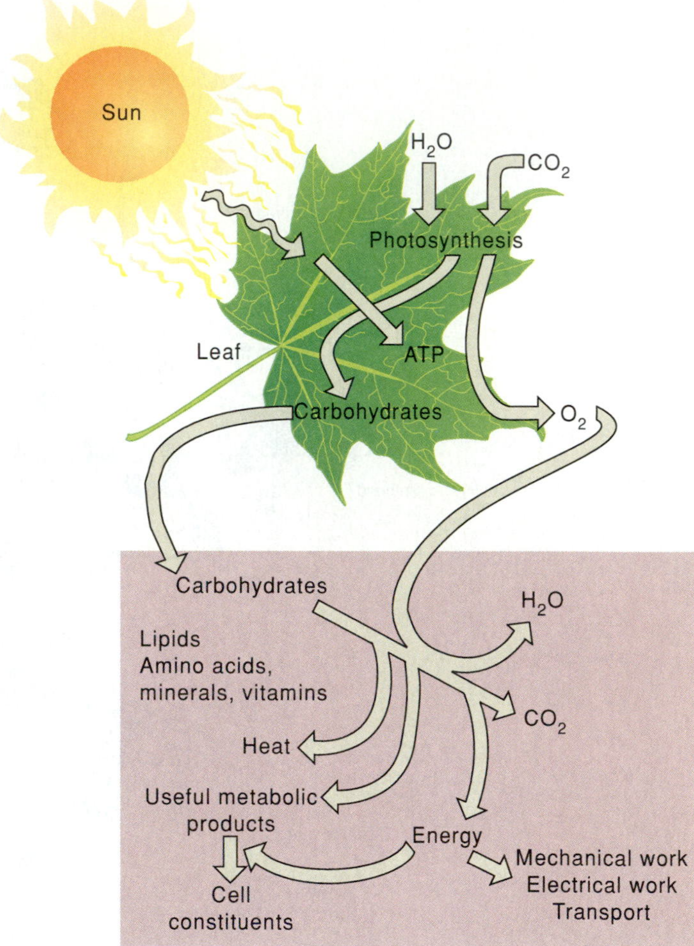

glucose-6-phosphate efficiently, its formation from glucose is coupled to the energetically favorable reaction of ATP breakdown:

$$ATP \rightleftharpoons ADP + P_i$$

When these two reactions are biochemically coupled, the net reaction is

$$\text{Glucose} + ATP \rightleftharpoons \text{glucose-6-phosphate} + ADP$$

The net reaction now favors the formation of glucose-6-phosphate, because more energy is released by ATP hydrolysis than is required for the phosphorylation of glucose. In the cell this is what happens when glucose-6-phosphate is synthesized. The unfavorable (energy-requiring) reaction is coupled to the favorable (energy-releasing) reaction to give an overall favorable reaction.

Figure 1.18

The structures of ATP and ADP and their interconversion. The two compounds differ by a single phosphate group.

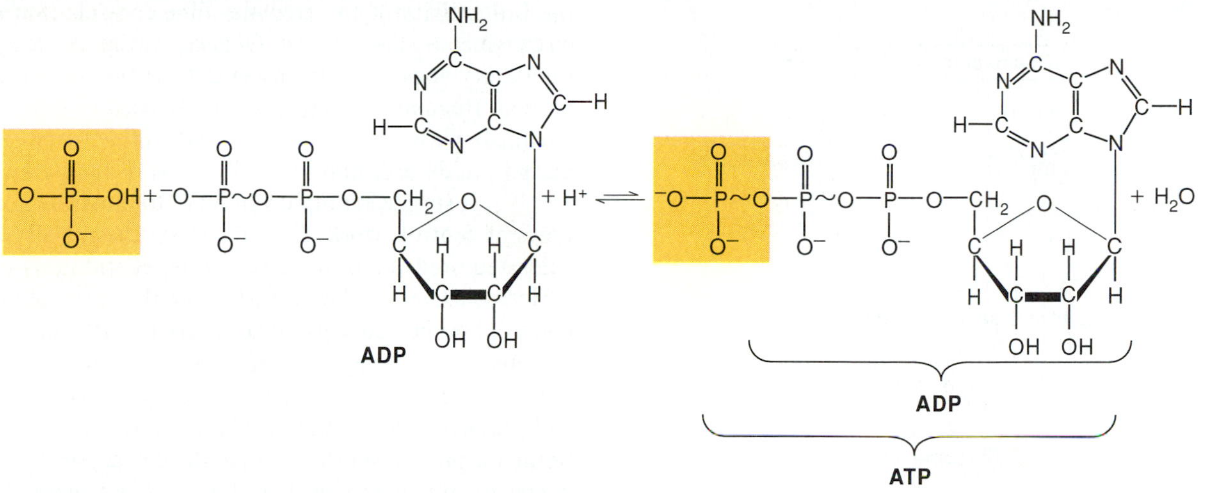

Biochemical Reactions Are Localized in the Cell

Biochemical reactions are organized so that different reactions occur in different parts of the cell. This organization is most apparent in eukaryotes, where membrane-bounded structures are visible proof for the localization of different biochemical processes. For example, the synthesis of DNA and RNA takes place in the nucleus of a eukaryotic cell. The RNA is subsequently transported across the nuclear membrane to the cytoplasm, where it takes part in protein synthesis. Proteins made in the cytoplasm are used in all parts of the cell. A limited amount of protein synthesis also occurs in chloroplasts and mitochondria. Proteins made in these organelles are used exclusively in organelle-related functions. Most ATP synthesis occurs in chloroplasts and mitochondria. A host of reactions that transport nutrients and metabolites occur in the plasma membrane and the membranes of various organelles. The localization of functionally related reactions in different parts of the cell concentrates reactants and products at sites where they can be most efficiently utilized.

Biochemical Reactions Are Organized into Pathways

Most biochemical reactions are integrated into multistep pathways utilizing several enzymes. For example, the breakdown of glucose into CO_2 and H_2O controls a series of reactions that begins in the cytosol and continues to completion in the mitochondrion. A complex series of reactions like this is referred to as a biochemical pathway (fig. 1.19). Synthetic reactions, such as the biosynthesis of amino acids in the bacterium *Escherichia coli,* are similarly organized into pathways (fig. 1.20). Frequently pathways have branchpoints. For example, the synthesis of the amino acids threonine and lysine starts with oxaloacetate. After three steps, a branchpoint is reached with the formation of the organic compound aspartic-β-semialdehyde. One branch of this pathway leads to the synthesis of the amino acid lysine, and another branch leads to the synthesis of the amino acids methionine, threonine, and isoleucine.

Figure 1.19

Summary diagram of the breakdown of glucose to carbon dioxide and water in a eukaryotic cell. As depicted here, the process starts with the absorption of glucose at the plasma membrane and its conversion into glucose-6-phosphate. In the cytosol, this six-carbon compound is then broken down by a sequence of enzyme-catalyzed reactions into two molecules of the three-carbon compound pyruvate. After absorption by the mitochondrion, pyruvate is broken down to carbon dioxide and water by a sequence of reactions that requires molecular oxygen.

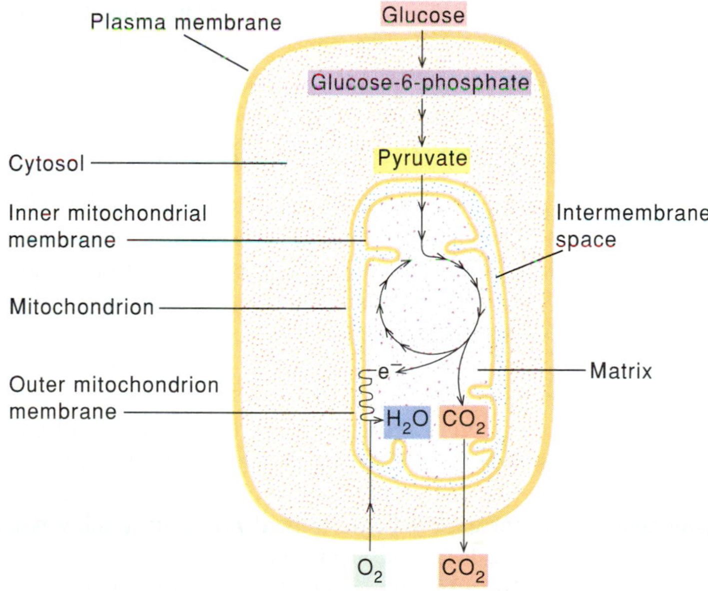

Figure 1.20

Synthesis of various amino acids from oxaloacetate. Each arrow represents a discrete biochemical step requiring a unique enzyme. Thus aspartic acid is produced in one step from oxaloacetate, whereas isoleucine is produced in five steps from threonine.

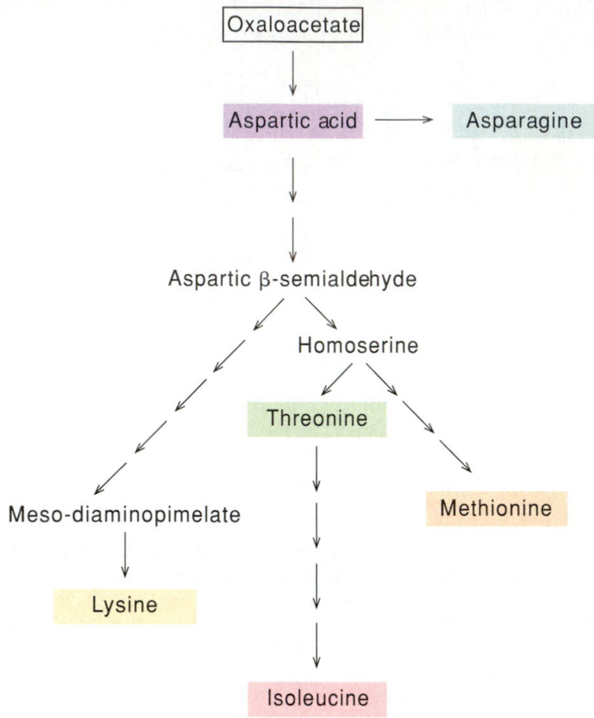

To understand the role of each biochemical reaction we must identify its position in a pathway and also consider how that pathway interacts with others.

Biochemical Reactions Are Regulated

Hundreds of biochemical reactions take place even in the cells of relatively simple microorganisms. Living systems have evolved a sophisticated hierarchy of controls that permits them to maintain a stable intracellular environment. These controls insure that substances required for maintenance and growth are produced in adequate amounts but without huge excesses. Biochemical controls have developed in such a way that the cell can make adjustments in response to a changing external environment. Adjustments are needed because the temperature, ionic strength, acid concentration, and concentration of nutrients present in the external environment vary over much wider limits than could be tolerated inside the cell.

The rate of intracellular reactions is a function of the availability of substrates and enzymes. Enzyme activity is controlled at two different levels. First and foremost, the rate of a catalyzed reaction is regulated by the amount of the catalyzing enzyme that is present in the cell. Control of enzyme amounts is usually accomplished by regulating the rate of enzyme synthesis; in some cases the rate of enzyme degradation is also regulated. We can think of controls that regulate the total amount of enzyme present as coarse controls. They define the limits of possible enzyme activity as being anywhere from 0 to 100% of the full activity of the enzyme. Fine controls that act directly on enzymes are also present. Only certain special enzymes, called regulatory enzymes, are susceptible to this second type of regulation. Regulatory enzymes usually occupy key points in biochemical pathways, and their state of activity frequently is decisive in determining the utilization of the pathway.

A simple example will serve to illustrate how these two types of controls work. *E. coli* can synthesize all of the amino acids required for protein synthesis. Histidine is one of these amino acids; its synthesis starts from the sugar phosphate compound phosphoribosylpyrophosphate (PRPP) and requires ten enzymes. Each enzyme catalyzes one reaction in a ten-step pathway. The synthesis of all ten enzymes is regulated by the end product of the pathway, histidine. If there is sufficient histidine for protein synthesis, then the cell ceases to make the enzymes for this pathway. The shutdown is triggered by controls that sense the level of histidine in the cell and as a result turn off synthesis of the messenger RNA required to make the enzymes. The histidine pathway also provides an example of a regulatory enzyme. The activity of the first enzyme in the pathway is directly inhibited by histidine. Thus, when there is sufficient histidine present in the cell, the activity of the first enzyme in the pathway is inhibited, and no more material or energy is funneled into the synthesis of unneeded histidine. Finally, if there is abundant histidine available from the external environment, the extracellular histidine is absorbed into the cell and the synthesis of the enzymes of the histidine pathway is brought to a halt. Bacterial cells grown in a histidine-rich growth medium for several generations contain only trace amounts of these enzymes.

The underlying principle in regulation is to maintain a favorable intracellular environment in the most economical manner. The cell makes products in the amounts that are needed. Each pathway is regulated in a somewhat different way, assuring that biochemical energy and substrates are efficiently utilized.

Organisms Are Biochemically Dependent on One Another

As we stated at the beginning of this chapter, 3–4 billion years ago the first self-replicating molecules appeared on earth. These entities had to have the capacity for extracting nutrients from the chemical compounds that existed in prebiotic times. We have some general notions about what types of substances were present at that time. One of the most important substances that was not present at that time in significant amounts was molecular oxygen, O_2. Currently this form of oxygen is required by all forms of life visible to the naked eye.

The O_2 that is used by most organisms is ultimately converted by them into CO_2. Oxygen is utilized at a rapid rate and it would soon disappear if it were not for special classes of photosynthetic organisms that are constantly producing more O_2 by the oxidation of water.

The oxygen story is an example of the dependence of one class of organisms on another for certain chemicals. A similar situation exists with the elements carbon and nitrogen, which must be converted from gaseous forms, CO_2 and N_2, to organic forms utilizable by most organisms. Reduced carbon compounds are constantly being lost by oxidation to gaseous CO_2. The supply of organic carbon compounds required by all forms of life is replenished by photosynthetic organisms; these include most plants and certain microorganisms. Similarly, nitrogen in organic molecules is constantly being lost to the atmosphere in the form of gaseous nitrogen. The reactions required for the conversion of nitrogen to a reduced form more usable to the majority of organisms occurs in only a limited number of microorganisms; yet without these nitrogen-fixing organisms life as we know it would soon vanish.

As we ascend the evolutionary tree, we find increasingly complex multicellular forms. Such organisms generally require more complex nutrients, which must ultimately be supplied to them by simpler living forms. Bacteria like *E. coli* can make all of their own amino acids from a reduced form of nitrogen, such as NH_3, and a reduced form of carbon, such as glucose. Humans, on the other hand, must receive most of their amino acids as nutrients. Humans and other complex organisms have gained new biochemical capacities, which permit them to synthesize the components associated with highly specialized differentiated tissues. At the same time, they have lost many of the biochemical systems required to survive on simpler nutrients.

Many biochemical reactions of great importance take place in only a limited number of organisms. This fact increases the complexity of the study of biochemistry. We must learn many reactions; we must also be aware of the biochemical potentials of different organisms. This is the only way we can understand the biochemical interdependency of organisms.

Information for the Synthesis of Proteins Is Carried by the DNA

DNA contains the genetic information transmitted to each daughter cell when cells divide. The DNA usually exists in the form of nucleoprotein (DNA-protein) complexes called chromosomes. A prokaryotic cell contains a single chromosome. Prior to cell division this chromosome duplicates and segregates so that an identical complement of DNA goes to each of two newly formed daughter cells.

Eukaryotic cells are more complex than prokaryotic cells and usually contain more DNA, which is partitioned between several chromosomes. In both prokaryotes and eukaryotes, almost all cells of the same organism contain the same number of chromosomes. In eukaryotes most of the chromosomes are localized in the nucleus. Thus the DNA is isolated from the main body of the cytoplasm—a unique feature of eukaryotes and the primary distinction between prokaryotes and eukaryotes. Some organelles, notably the mitochondria and the chloroplasts, also contain a single circular chromosome.

Eukaryotic chromosomes are detectable by light microscopy at the stage just prior to cell duplication. At this stage, called mitosis, chromosomes appear as elongated refractile structures that can be seen to segregate in equal numbers and types to each of the daughter cells before cell division (fig. 1.21). Each chromosome carries specific hereditary (genetic) information necessary for the synthesis of specific compounds essential for cell maintenance, growth, and replication. Each chromosome contains a single very long DNA molecule composed of 10^6 or more nucleotides in a specific sequence. The sequence of nucleotides in the chromosomal DNA determines the sequence of amino acids in the protein polypeptide chains of the organism. The relationship between base sequences and resultant amino acids is known as the genetic code. Each grouping of three bases, called a triplet, represents a specific amino acid and is called a codon. The genetic code ensures that the organism's characteristics will be reflected by the sequence of nucleotides in its DNA. When chromosomes replicate, the DNA replicates precisely, so that the same nucleotide sequence is passed along to each of the daughter cells resulting from mitosis and cell division.

The DNA does not transfer its genetic information directly to protein. Rather, this information passes through an intermediary, the messenger RNA (mRNA). The mRNA is made on a DNA template in the nucleus of a eukaryotic cell and then passes into the cytoplasm, where it serves in turn as a template for the synthesis of the polypeptide chain. The overall process of information transfer from DNA to mRNA (transcription) and from mRNA to protein (translation) is depicted in figure 1.22.

Biochemical Systems Have Been Evolving for Almost Four Billion Years

Biochemical systems are conservative and opportunistic. That is because they tend to evolve one step at a time, using a readily accessible route that leads to an advantage. To appreciate biochemical systems today it is useful to have some understanding of how they came to be.

The earth was formed by a process of accretion about 4.6 billion years ago. Initially it was a molten mass lacking the gravitational pull to retain its gases at the prevalent elevated temperatures. And yet, within a mere 700 million years of the planet's birth, as calculated from the isotopic record of sediments, cellular life almost certainly existed. What raw materials were available to bring about this miraculous turn of events?

Figure 1.21

Mitosis and cell division in eukaryotes. After DNA duplication has occurred, mitosis is the process by which quantitatively and qualitatively identical DNA is delivered to daughter cells formed by cell division. Mitosis is traditionally divided into a series of stages characterized by the appearance and movement of the DNA-bearing structures, the chromosomes.

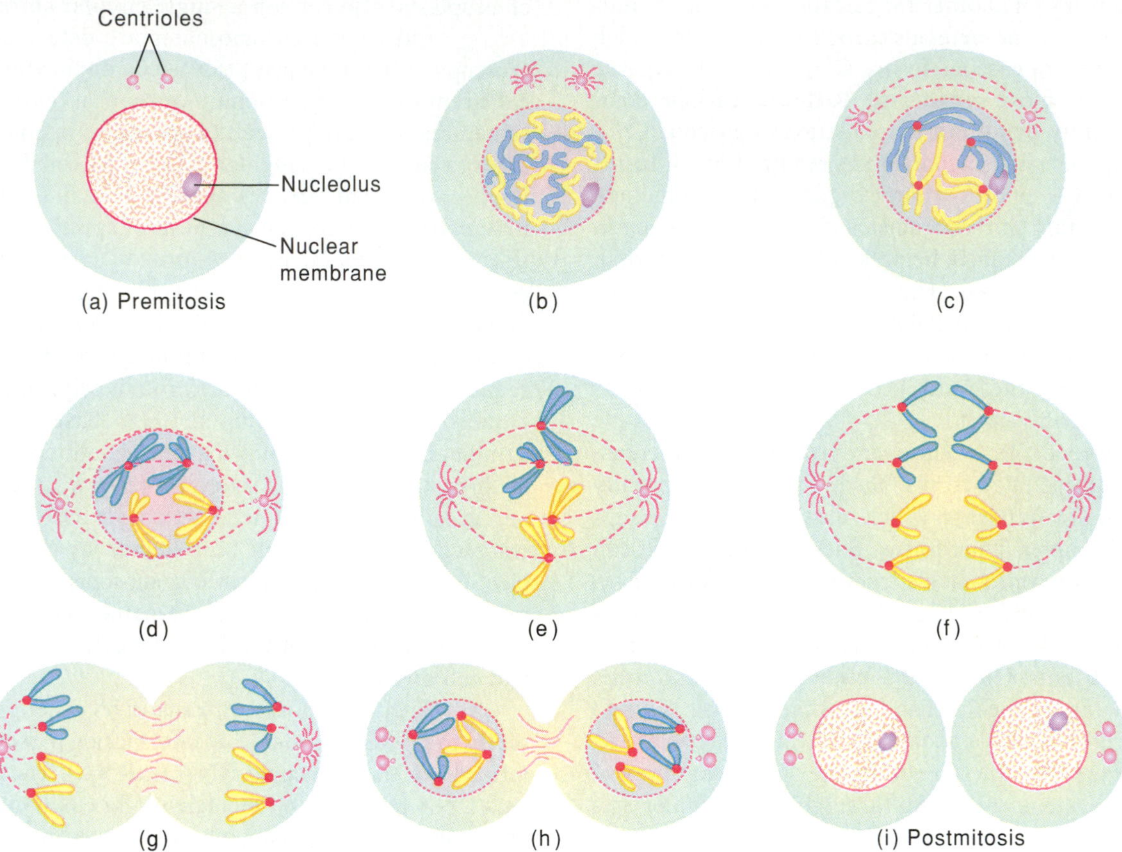

What were the sources of energy used to drive the necessary reactions? Where did the important reactions take place? Was it in the atmosphere, in the oceans, on dry land, or all three?

Table 1.4 shows a distribution of the major elements found in the earth's crust, the ocean water, and the human body. The composition of the human body, which is reasonably representative of living organisms, differs appreciably from that of the earth's crust. The four most abundant elements in the human body are hydrogen, carbon, nitrogen, and oxygen. Of these, only oxygen belongs to the highest-abundance class of elements making up the earth's crust. Nevertheless, the elements needed to make living things were present in sufficient quantities in the earth's crust and its primitive oceans and the atmosphere.

The original water and air associated with the newly formed planet were lost because of the high temperatures. As the earth cooled, water and various gases on the surface and in the atmosphere were produced by an outgassing process. Today the earth is only about 0.5% by weight water, but because of water's low density, most of it is present on the earth's surface, where it has a major impact on the environment. Water cycles through the gaseous form in the atmosphere and the liquid form in the oceans and bodies of fresh water.

Geologic evidence indicates that appreciable amounts of the total water mass have always been present as liquid water. Thus the temperature of the planet has for the most part been between 0 and 100° C, a range that is conducive to the formation of biomolecules and the origin and propagation of life. The earth has also been kind to living things in other ways. The buffering action of various clays and minerals is believed to have maintained the pH level of the oceans between 8.0 and 8.5, which is close to the pH inside living cells.

Unlike the temperature and pH of the surface water, the composition of the atmosphere has changed drastically since the origin of life. In fact, the processes taking place in living things are primarily responsible for these changes. Today's atmosphere, which is about one millionth of the mass of the earth

Figure 1.22

Transfer of information from DNA to protein. The nucleotide sequence in DNA specifies the sequence of amino acids in a polypeptide. DNA usually exists as a two-chain structure. The information contained in the nucleotide sequence of only one of the DNA chains is used to specify the nucleotide sequence of the messenger RNA molecule (mRNA). This sequence information is used in polypeptide synthesis. A three-nucleotide sequence in the messenger RNA molecule codes for a specific amino acid in the polypeptide chain.

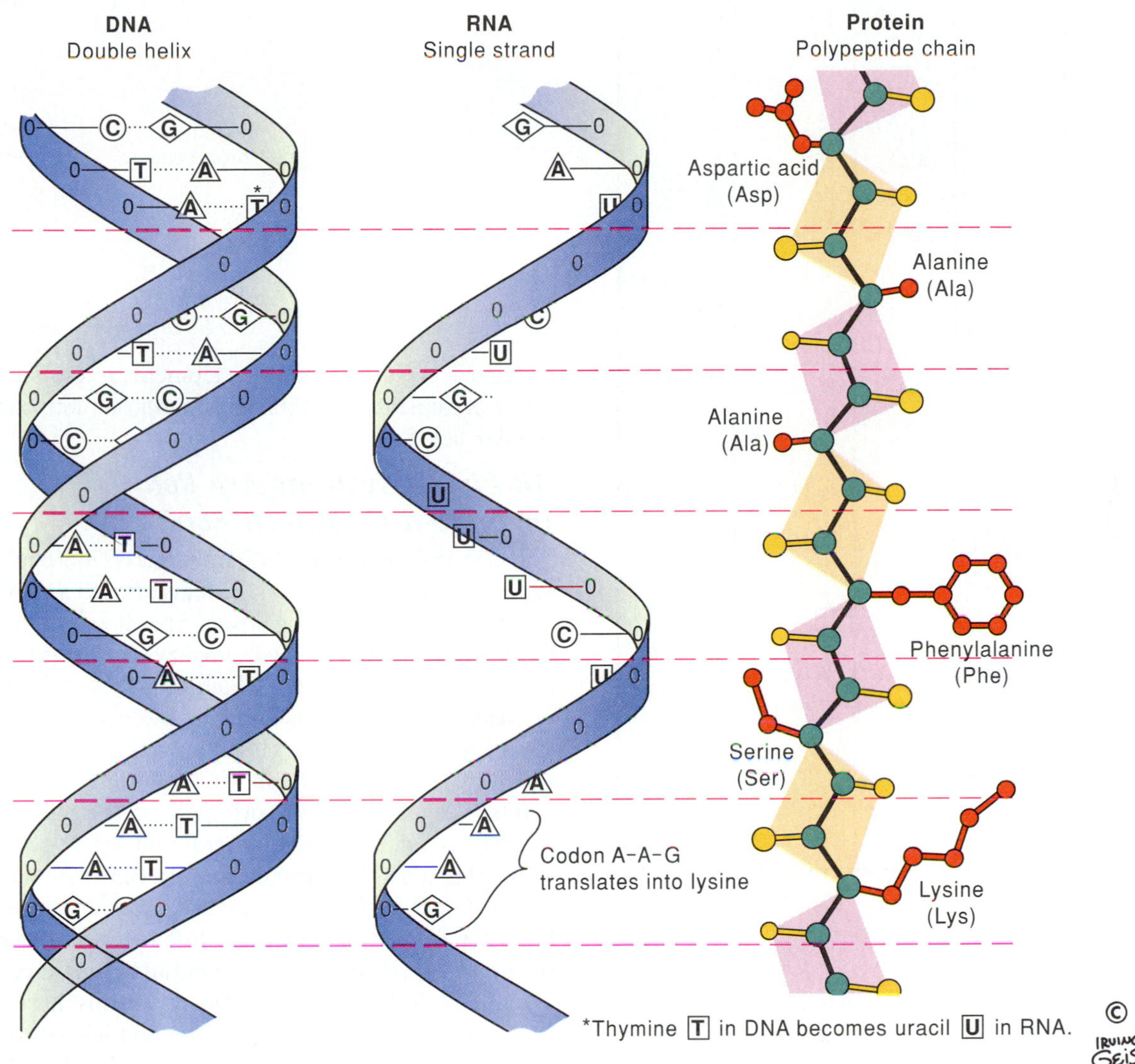

itself, is mainly composed of nitrogen (78%) and oxygen (21%). Most of the remaining atmosphere is argon (0.9%), water (variable up to 4%), and carbon dioxide (0.034%). The gases that made up the primitive atmosphere were quite different; especially conspicuous was the absence of gaseous oxygen. Most of the oxygen in the present atmosphere is due to the oxidation of water by photosynthetic organisms. The main forms of carbon and nitrogen in the primitive atmosphere were probably CO_2 and N_2 as they are today. In addition, and probably of great significance to the origin of life, small amounts of the more reduced forms of carbon and H_2 gas were present. Thus the primitive earth probably had a weakly reducing atmosphere as contrasted with today's highly oxidizing atmosphere. This situation was most favorable to the origin of life, as organic compounds that enter the biomass tend to be in a reduced state and they are readily oxidized in the presence of gaseous oxygen. It is generally believed that the first organics formed in this primitive atmosphere and then rained down to a more concentrated form to make larger bioorganic molecules.

The origin of life probably occurred in three phases (fig. 1.23): (1) The earliest phase was a period of chemical evolution during which the compounds needed for the nucleation of life must have been formed. These compounds include the most important class of biological macromolecules, the nucleic acids. In

Table 1.4
Distribution of the 24 Elements Used in Biological Systems[a]

Element	Atomic No.	Earth's Crust	Ocean	Human Body
Hydrogen (H)	1	2,882	66,200	60,562
Carbon (C)	6	56	1.4	10,680
Nitrogen (N)	7	7	<1	2,440
Oxygen (O)	8	60,425	33,100	25,670
Fluorine (F)	9	77	<1	<1
Sodium (Na)	11	2,554	290	75
Magnesium (Mg)	12	1,784	34	11
Silicon (Si)	14	20,475	<1	<1
Phosphorus (P)	15	79	<1	130
Sulfur (S)	16	33	17	130
Chlorine (Cl)	17	11	340	33
Potassium (K)	19	1,374	6	37
Calcium (Ca)	20	1,878	6	230
Vanadium (V)	23	4	<1	<1
Chromium (Cr)	24	8	<1	<1
Manganese (Mn)	25	37	<1	<1
Iron (Fe)	26	1,858	<1	<1
Cobalt (Co)	27	1	<1	<1
Nickel (Ni)	28	3	<1	<1
Copper (Cu)	29	1	<1	<1
Zinc (Zu)	30	2	<1	<1
Selenium (Se)	34	<1	<1	<1
Molybdenum (Mo)	42	<1	<1	<1
Iodine (I)	53	<1	<1	<1

[a]Amounts are given in atoms per 100,000.

Figure 1.23

The origin of life probably occurred in three overlapping phases: In Phase I, chemical evolution, involved the noninstructed synthesis of biological macromolecules. In Phase II, biological macromolecules self-organized into systems that could reproduce. In Phase III, organisms evolved from simple genetic systems to complex multicellular organisms. The arrow pointing from left to right emphasizes the unidirectional nature of the overall process.

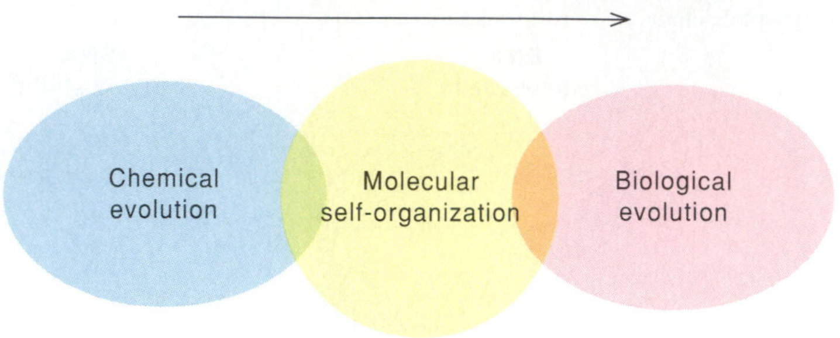

this phase of evolution, the synthesis of nucleic acids was "noninstructed." (2) As soon as some nucleic acids were present, physical forces between them must have led to an "instructed" synthesis, in which the already formed molecules served as templates for the synthesis of new polymers. It seems likely that "feedback loops" selected out certain nucleic acids for preferential synthesis. At some point during this period of instructed synthesis more nucleic acids and possibly protein macromolecules were formed. The products of this phase of molecular self-organization must sooner or later have begun to resemble the complex organized units that we observe in the self-reproducing biosynthetic cycles of living cells. (3) In the final phase of the origin of life we find the beginnings of the divergent process of biological evolution, the development of the simplest single-celled organisms, and their differentiation into complex multicellular beings.

All Living Systems Are Related through a Common Evolution

The classical view of evolution, based on morphological differences among organisms, can be diagrammed as a branching tree in which all existing organisms are shown at the tips of the branches (fig. 1.24). An evolutionary tree starts from the simple ancestral prokaryotic cell, which branches off in three main directions into the archaebacteria, the eubacteria, and the eukaryotes. Each of these kingdoms has continued to branch in elaborate ways; we have shown only some of the main branch points in figure 1.24. Prokaryotes for the most part have remained as relatively undifferentiated single-celled organisms containing a single chromosome. By contrast, eukaryotes have changed dramatically. Although the organisms on many branches of the eukaryotic part of the evolutionary tree have remained as relatively undifferentiated single-cell forms, significant numbers of eukaryotic organisms have evolved into multicellular forms in which the individual cells of the total organism have differentiated to serve different functions. This process has given rise to plants, animals, and fungi.

A somewhat different view of biological evolution has arisen from a comparison of nucleic acid sequences in different organisms. The best-known sequences for such studies have come from the 16S ribosomal RNAs in different organisms. This comparative study has provided us with the evolutionary pattern shown in figure 1.25, where distances are proportional to sequence differences. A most striking characteristic of this unrooted tree is the distinctness of the three primary kingdoms as evidenced by the large sequence distances that separate each of

Figure 1.24

Classical evolutionary tree. All living forms have a common origin, believed to be the ancestral prokaryote. Through a process of evolution some of these prokaryotes changed into other organisms with different characteristics. The evolutionary tree indicates the main pathways of evolution.

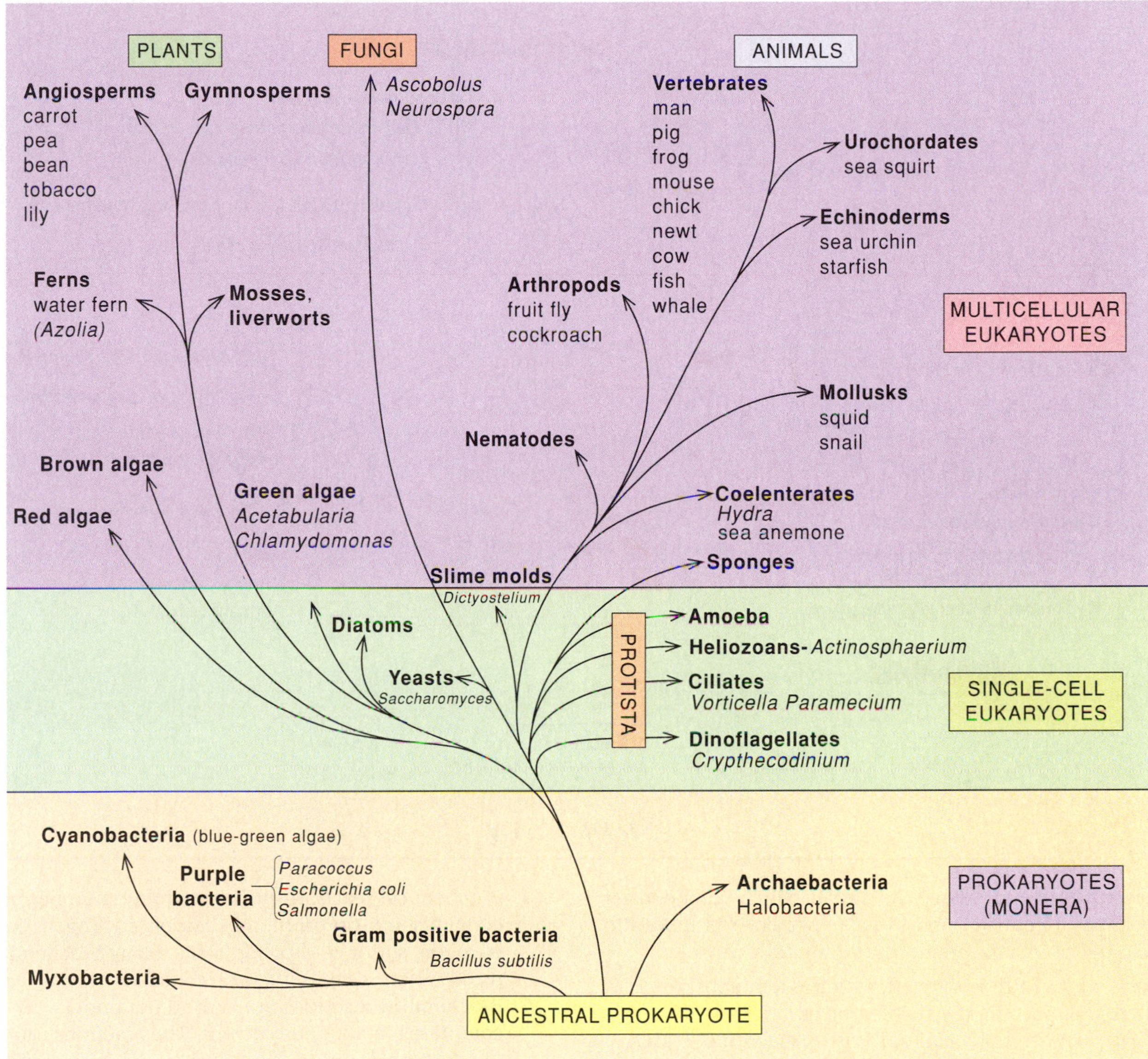

the kingdoms from the others. Although we do not know the position of the root of the tree, it seems likely that the archaebacteria are closer to the common ancestor of all three kingdoms than are the eubacteria or the prokaryotes.

The organisms most familiar to us, the multicellular plants and animals, occupy a shallow domain within the eukaryotic line of descent. True, the developmental programs of the multicellular forms have generated an incredible diversity in form and function. Nevertheless, both bacteria and unicellular eukaryotes span far greater evolutionary histories. This fact is reflected in the greater biochemical diversity that we find in the unicellular microorganisms.

Figure 1.25

An evolutionary tree can be constructed by comparing the complete sequences of 21 different 16S and 16S-like ribosomal RNAs (rRNAs). The scale bar represents the number of accumulated nucleotide differences (mutations) per sequence position in the rRNAs of the various organisms. (Source: N. R. Pace, G. J. Olsen, and C. R. Woese, in *Cell* 45:325, 1986. Copyright © 1986 Cell Press, Cambridge, Mass.)

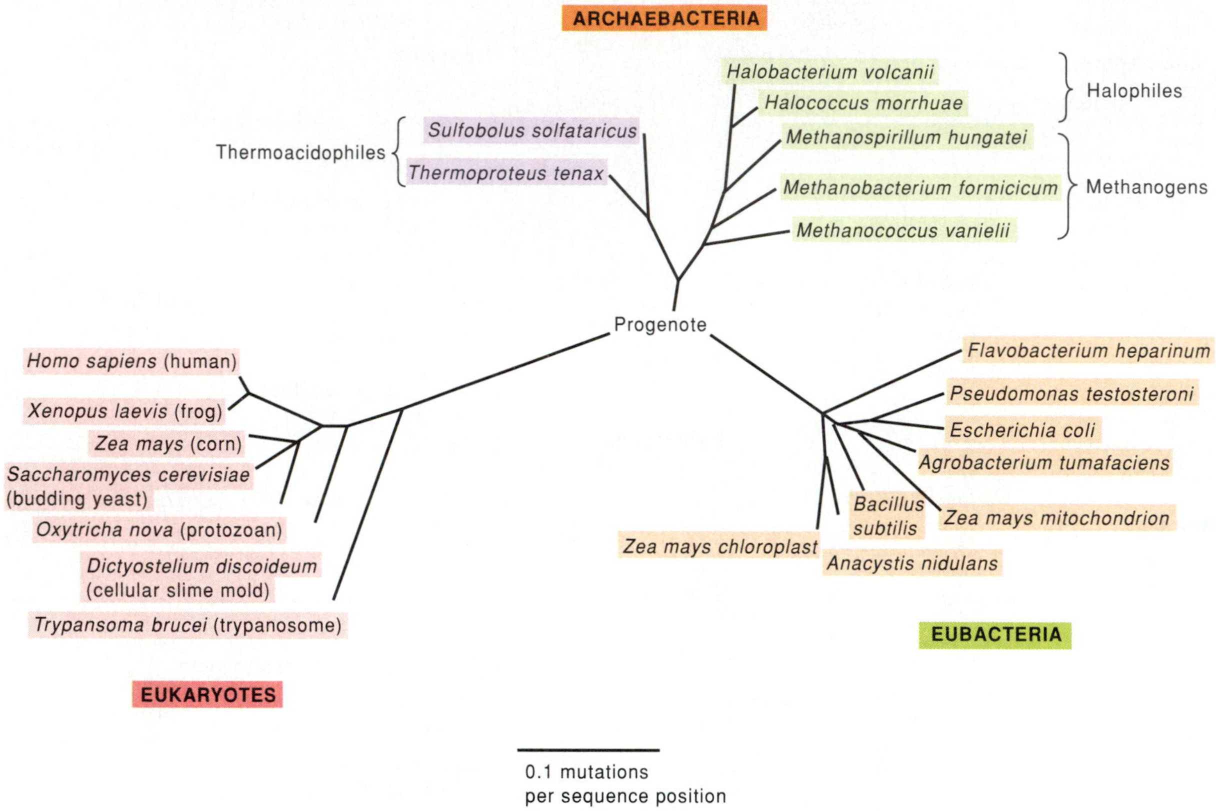

ARCHAEBACTERIA
Halobacterium volcanii
Halococcus morrhuae
Halophiles
Thermoacidophiles
Sulfobolus solfataricus
Thermoproteus tenax
Methanospirillum hungatei
Methanobacterium formicicum
Methanococcus vanielii
Methanogens
Progenote
Homo sapiens (human)
Xenopus laevis (frog)
Zea mays (corn)
Saccharomyces cerevisiae (budding yeast)
Oxytricha nova (protozoan)
Dictyostelium discoideum (cellular slime mold)
Trypansoma brucei (trypanosome)
EUKARYOTES
Flavobacterium heparinum
Pseudomonas testosteroni
Escherichia coli
Agrobacterium tumafaciens
Bacillus subtilis
Zea mays mitochondrion
Zea mays chloroplast
Anacystis nidulans
EUBACTERIA
0.1 mutations per sequence position

Summary

In this chapter we have discussed the ways in which biochemistry parallels ordinary chemistry and those in which it is quite different. The chief points to remember are the following.

1. The basic unit of life is the cell, which is a membrane-enclosed, microscopically visible object.
2. Cells are composed of small molecules, macromolecules, and organelles. The most prominent small molecule is water, which constitutes 70% by weight of the cell. Other small molecules are present only in quite small amounts; they are precursors or breakdown products of macromolecules or coenzymes. There are four types of macromolecules: lipids, carbohydrates, proteins, and nucleic acids.
3. Noncovalent intermolecular forces largely determine the folded structure adopted by a macromolecule, particularly the relative affinity of different groupings on the macromolecule for water. In general, hydrophobic groupings are buried within the folded macromolecular structure, while hydrophilic groupings are located on the surface, where they can interact with water.
4. Biochemical reactions utilize a limited number of elements, most prominently carbon, hydrogen, oxygen, nitrogen, sulfur, and phosphorus. Many biochemical reactions are simple organic reactions.
5. Biochemical reactions are carried out under very mild conditions in aqueous solvent. The reactions can proceed under these conditions because of the highly efficient nature of protein enzyme catalysts.
6. Biochemical reactions frequently require energy. The most common source of chemical energy used is adenosine triphosphate (ATP). The splitting of a phosphate from the ATP molecule can provide the energy needed to make an otherwise unfavorable reaction go in the desired direction.
7. Biochemical reactions of different types are localized to different parts in the cell.
8. Biochemical reactions are frequently organized into multistep pathways.
9. Biochemical reactions are regulated according to need by controlling the amount and activity of enzymes in the system.

10. Most organisms depend on other organisms for their survival. Frequently this is because a given organism cannot make all of the compounds needed for its growth and survival.
11. The specific properties of any protein are due to the specific sequence of amino acids in its polypeptide chains. This sequence is determined by the genetic information carried by the sequence of DNA nucleotides. DNA transfers the information to messenger RNA, which serves as the template for protein synthesis.
12. Shortly after the earth was formed and had cooled to a reasonable temperature, chemical processes produced compounds that would be used in the development of living cells. Nucleic acids are the most important compounds for living cells, and it is believed that they played the central role in the origin of life.
13. Evolutionary trees based on morphology or biochemical differences indicate that all living systems are related through a common evolution.

Selected Readings

Becker, W. M., *The World of the Cell.* Menlo Park, Calif.: Benjamin/ Cummings, 1986. A very readable cell biology book that could be referred to while taking biochemistry.

de Duve, C., *Blueprint For A Cell.* Burlington, N.C.: California Biological Supply Co., 1991. A short book on the origin of life that contains an excellent reference list.

Dickerson, R. E., Chemical evolution and the origin of life. *Sci. Am* 239(3):70–86, 1978.

Doolittle, R. F., The genealogy of some recently evolved vertebrate proteins. *Trends Biochem. Sci.* 10:233–237, 1985.

Kimura, M., The neutral theory of molecular evolution. *Sci. Am.* 241(5):98–126, 1979.

Schopf, J. W., The evolution of the earliest cells. *Sci. Am.* 229(3):10–138, 1978. An authoritative account from a foremost geologist.

Stillinger, F. H., Water revisited. *Science* 209:451–457, 1980. A reminder of the central importance of water to the origin of life.

Wilson, A. C., The molecular basis of evolution. *Sci. Am.* 253(4):164–173, 1985.

2

CHAPTER

Thermodynamics in Biochemistry

The primary usefulness of thermodynamics to biochemists lies in predicting whether particular chemical reactions could occur spontaneously. A simple illustration is to predict what compounds could possibly serve as energy sources for an organism. You are aware from everyday experience that oxidation of organic molecules by molecular oxygen releases energy. For example, wood or coal burns with a large output of heat. Similarly, organisms can obtain energy by oxidizing carbohydrates, fats, or proteins. Some organisms oxidize hydrocarbons, some oxidize reduced forms of sulfur, and others oxidize iron. But no organisms live by oxidizing molecular nitrogen, and the explanation lies in thermodynamics. The reaction cannot occur spontaneously. This example illustrates the importance of thermodynamics in controlling all life. Because organisms live by extracting chemical energy from their surroundings, thermodynamics is not an esoteric subject. It is a matter of life or death.

We say that thermodynamics determines whether a process "could" occur, because thermodynamics tells us only whether the process is possible, not whether it actually will occur in a finite period of time. The rate at which a thermodynamically possible reaction occurs depends on the detailed mechanism of the process. For a biochemical process to occur rapidly, appropriate enzymes must be available. The distinction between spontaneity and speed is more critical for biochemists than it is for chemists, because if a chemical reaction does not proceed rapidly a chemist can change the pressure or temperature or increase the concentration of the reactants. A living organism is under more rigid constraints: It must function at a fixed temperature and pressure and within a limited range of concentrations of reactants.

In this chapter we will elaborate on the thermodynamic quantities that we introduced in chapter 1. We will then expand the discussion to show how the concept of free energy is used in predicting biochemical pathways, and we will explore the central role of ATP in providing energy for biochemical reactions.

Thermodynamic Quantities

The properties of a substance can be classified as either intensive or extensive. Intensive properties, which include density, pressure, temperature, and concentration, do not depend on the amount of the material. Extensive properties, such as volume and weight, do depend on the amount. Most of the thermodynamic properties we will be discussing are extensive properties. These include energy (E), enthalpy (H), entropy (S), and free energy (G).

Energy, enthalpy, entropy, and free energy are all properties of the state of a substance. This means that they do not depend on how the substance was made or how it reached a particular state. In a chemical reaction, it is the difference between the initial and final states that is important; the pathway that is taken to get from the initial state to the final state has no bearing on whether the overall reaction releases or consumes energy (fig. 2.1).

Figure 2.1

The change in energy of a system depends only on the initial and final states, not on the path by which the system gets from one state to the other. This diagram illustrates the conversion of a phosphate ester of glucose (glucose-6-phosphate) to free glucose and inorganic phosphate ion (P_i) by two different pathways. Although the two routes proceed through intermediate compounds that differ in energy (A, B, C, and D), the overall energy change (ΔE) is the same. However, the work done and the amount of heat absorbed or released generally will not be the same for the two paths.

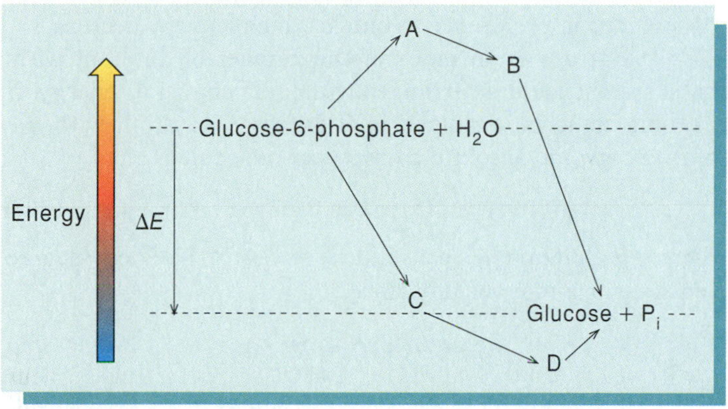

Figure 2.2

A system and its surroundings. Heat flow into the system is designated as a positive quantity (q), and work that the system does on the surroundings is designated as a positive quantity (w). The first law of thermodynamics relates q and w to changes in the energy of the system. Any change in the energy of the system (ΔE_{sys}) is balanced by an opposite change in the energy of the surroundings (ΔE_{sur}), so that the overall energy change (ΔE_{tot}) is zero.

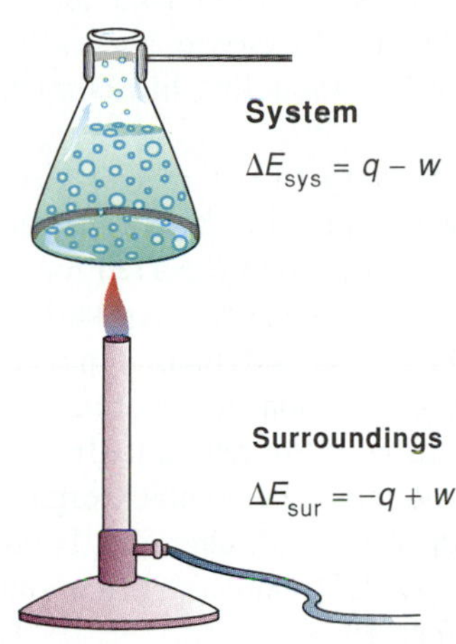

The First Law of Thermodynamics: Any Change in the Energy of a System Requires an Equal and Opposite Change in the Surroundings

Of the thermodynamic properties we have just listed, energy is probably the most familiar. Energy is the capacity to do work. The energy of a molecule includes the internal nuclear energies and the molecular electronic, translational, rotational, and vibrational energies. Electronic energies, which reflect the interactions among the electrons and nuclei, usually are much larger than the translational, rotational, and vibrational energies.

In biochemistry, we are concerned not so much with the absolute energies of molecules as with changes in energy that occur in the course of reactions. It is easier to evaluate a change in energy (ΔE) than to calculate the absolute energies of the reactants or products, because many of the terms that contribute to the total energy do not change much during a chemical reaction. Electronic energies usually dominate ΔE, as they do the total energies. Good estimates of the energy change resulting from a chemical reaction can usually be obtained by calculating the difference between the bond energies of the reactants and products (see table 4.2).

The first law of thermodynamics says that the total amount of energy in the universe is constant. Energy can undergo transformations from one form to another; for example, the chemical energy of a molecule can be transformed into thermal, electrical, or mechanical energy. But any change in the total energy of one part of the universe is matched by an equal and opposite change in another part, so that the overall energy of the universe remains constant.

To apply the first law to a chemical reaction, we must take into account all the energy changes that occur. This means including any changes in the surroundings, as well as in the system of interest (fig. 2.2). The system might be a reaction occurring in a test tube or a living cell; the surroundings are all the rest of the universe. In practical terms, however, we usually need to consider only the immediate surroundings, because only these are likely to be influenced by what happens in the system. According to the first law, the overall energy remains constant even though energy in some form may flow from the system to its surroundings or from the surroundings to the system.

If the amount of matter in a system is constant, there are only two means by which the system can gain or lose energy: transfer of heat or performance of work. The energy of a system increases if the system absorbs heat from its surroundings or if the surroundings do work on the system; it decreases if the system gives off heat to the surroundings or does work. A common way of stating the first law of thermodynamics is to equate the change in energy of the system (ΔE) to the difference between the heat absorbed by the system (q) and the work done by the system (w):

$$\Delta E = q - w \tag{1}$$

Both q and w depend on the path of the reaction, but their difference, ΔE, is independent of the path and therefore defines a state function.

The relative energy of a compound can be measured in a bomb calorimeter, a device in which the compound is thoroughly combusted and the resulting heat is measured. Energies (and enthalpies, discussed next) are usually given in units of kilocalories per mole or kilojoules per mole. One kilojoule (kJ) is the amount of energy needed to apply 1 newton of force over 1 km; 1 kilocalorie (kcal) is the heat needed to raise the temperature of 1 kg of water from 14.5 to 15.5° C. One kcal is equivalent to 4.184 kJ. Physicists generally prefer to use kilojoules; most chemists, including biochemists, prefer to use kilocalories.

It is customary to record energies of molecules with reference to the energies of their constituent elements. The standard energy of formation reported for a molecule, ΔE°_f, is equal to the energy change associated with the formation of the molecule from the elements in their standard states. The energy change in any other reaction can be obtained by subtracting the energies of formation of the reactants from the energies of formation of the products. The complex organic molecules found in cells have relatively weak chemical bonds, and thus higher energies, than H_2O, CO_2, and the other small molecules from which they are formed.

Enthalpy is a function of state that is closely related to energy but is usually more pertinent for describing the thermodynamics of chemical or biochemical reactions. The change in enthalpy (ΔH) is related to the change in energy (ΔE) by the expression

$$\Delta H = \Delta E + \Delta(PV) \qquad (2)$$

where $\Delta(PV)$ is the change in the product of the pressure (P) and volume (V) of the system. ΔH is the amount of heat that is absorbed from the surroundings if a reaction occurs at constant pressure and no work is done other than the work of expansion or contraction of the system. (The work done when a system expands by ΔV against a constant pressure P is $P\,\Delta V$. This type of work is generally not very useful in biochemical systems.) In most biochemical reactions, there is little change in either pressure or volume, so the difference between ΔH and ΔE is relatively small.

In most chemical reactions that occur spontaneously, the enthalpy of the system decreases. If no work is done, the system gives off heat to the surroundings. But changes in enthalpy or energy do not provide a reliable way of determining whether a reaction can proceed spontaneously. For example, although LiCl and $(NH_4)_2SO_4$ both dissolve readily in water, the former process releases heat, whereas the latter absorbs heat. A mixture of solid $(NH_4)_2SO_4$ and water proceeds spontaneously to a state of higher enthalpy. This reaction is driven by an increase in the entropy of the system.

The Second Law of Thermodynamics: In Any Spontaneous Process the Total Entropy of the System and the Surroundings Increases

The second law of thermodynamics is that the universe inevitably proceeds from states that are more ordered to states that are more disordered. This phenomenon is measured by a thermodynamic function called entropy, which is denoted by the symbol S. A reaction in which entropy increases (ΔS is positive) will proceed in preference to one in which entropy decreases.

Entropy is an index of the number of different ways that a system could be arranged without changing its energy. If a system could be arranged in Ω different ways, all with the same energy, the absolute entropy per molecule is

$$S = k \ln \Omega \qquad (3)$$

where k is Boltzmann's constant ($k = 3.4 \times 10^{-24}$ cal/degree Kelvin). For a mole of substance,

$$S = Nk \ln \Omega = R \ln \Omega \qquad (4)$$

Here N is the number of molecules in a mole (6×10^{23}) and R is the gas constant ($R \approx 2$ cal/(degree K · mole). Quantitative values for entropies are usually given in entropy units (1 eu = 1 cal/degree K).

The underlying idea here is that the more ways a particular state could be obtained, the greater is the probability of finding a system in that state. A system that is highly disordered could be obtained in many different arrangements that are all energetically equivalent. Thus a state in which molecules are free to move about and rotate into many different orientations or conformations will be favored over a state in which motion is more restricted. The second law makes the remarkably general assertion that the total entropy change in any reaction that occurs spontaneously must be greater than zero. But note that this statement specifies the total entropy, which means that we must consider the entropy change in the surroundings as well as that in the system. The entropy of the system can decrease if the entropy of the surroundings increases by a greater amount.

The absolute entropies of small molecules can be calculated by statistical mechanical methods. Table 2.1 shows the results of such calculations for liquid propane. The largest contributions to the entropy come from the translational and rotational freedom of the molecule, and much smaller contributions from vibrations; electronic terms are insignificant. Although exact calculations of this type become intractable for large biological molecules, the relative sizes of the contributions from different types of motions are similar to those in small molecules. Thus entropy is associated primarily with translation and

Table 2.1
Contributions to the Entropy of Liquid Propane at 231 K

	kcal/(degree K · mole)
Translational entropy	36.04
Rotational entropy	23.38
Vibrational entropy	1.05
Electronic entropy	0.00
Total	60.47

rotation. This is very different from enthalpy, in which electronic terms are dominant and translational and rotational energies are comparatively small.

Statistical mechanical calculations show that the translational entropy of a molecule depends on $\frac{3}{2}R \ln M_r$ (plus some smaller terms), where R is the gas constant and M_r is the molecular weight. Suppose that a molecule, Y, undergoes a dimerization reaction so that its molecular weight doubles:

$$2\,Y \rightarrow Y_2$$

Intuitively, we expect dimerization to decrease the entropy because the two monomeric units can no longer move independently. We can calculate the effect quantitatively as a function of the molecular weight as follows. If M_r is the molecular weight of the monomer, then the change in translational entropy in going from the monomeric state to the dimer is approximately

$$\begin{aligned} \Delta S &\approx \frac{3}{2}R \ln 2M_r - 2\left(\frac{3}{2}R \ln M_r\right) \\ &= \frac{3}{2}R \ln 2 - \frac{3}{2}R \ln M_r \\ &= -\frac{3}{2}R \ln (M_r/2) \end{aligned} \tag{5}$$

The decrease of the translational entropy resulting from dimerization is a logarithmic function of the molecular weight.

Structural features that make molecules more rigid reduce rotational and vibrational contributions to entropy. Thus the formation of a double bond or ring decreases the entropy even when the molecular weight is unchanged. The formation of comparatively rigid macromolecular structures from flexible polypeptide or polynucleotide chains also requires an entropy decrease, although this can be offset by increases in the entropy of the surrounding water molecules (see chapter 4).

The entropy of a compound depends strongly on the physical state of the material. A gas has more translational and rotational freedom than a liquid, and a liquid has more freedom than a solid. As a result, entropy increases when a solid melts or a liquid vaporizes.

It can be shown that the increase in the entropy of a system that undergoes an isothermal, reversible process is

$$\Delta S = \frac{\Delta H}{T} \tag{6}$$

where T is the absolute temperature in degrees kelvin. An isothermal process is one that occurs at constant temperature. A reversible process is one that proceeds infinitely slowly through a series of intermediate states in which the system is always at equilibrium. For any real process occurring at a finite rate, the system is not strictly at equilibrium, and ΔS is larger than the value given by equation (6).

From equation (6), the entropy increase on vaporization or melting can be determined simply from the heat of vaporization divided by the boiling point, or the heat of fusion divided by the melting point. The entropy increase on vaporization of water is 26 eu/mole and that on melting of ice is 5.3 eu/mole. These values are consistent with our intuition that the increase in translational and rotational freedom is much greater in going from a liquid to a gas than in going from a solid to a liquid.

The entropy of a solution is increased by mixing of solvents, and it is decreased by interactions among the solvent molecules or interactions of solutes with the solvent. The mixing of two miscible liquids is a thermodynamically favorable process because it increases the number of positions that are available to the molecules. The entropy change on going from the unmixed liquids to the mixed state can be calculated from the expression

$$\begin{aligned} \Delta S &= n_a R \ln \frac{1}{X_a} + n_b R \ln \frac{1}{X_b} \\ &= -n_a R \ln X_a - n_b R \ln X_b \end{aligned} \tag{7}$$

where n_a and n_b are the number of moles of A and B that are mixed, and X_a and X_b are the corresponding mole fractions in the final solution. Because X_a and X_b are always less than 1, ln X_a and ln X_b will be negative. This means that the dilution of each component resulting from the mixing makes a positive contribution to the entropy. Equation (7) applies to the mixing of ideal solutions, in which there are no interactions among the molecules. Any intermolecular interactions will decrease the entropy by restricting the system's translational and rotational freedom.

Solvation, the interaction of a solute with the solvent, makes an important negative contribution to the entropy of a solution. Solvation can take the form of hydrogen bonding to donor or acceptor groups on the solute, or of a looser clustering of solvent molecules oriented around the solute (fig. 2.3). In general, the entropy of solvation by water becomes more negative with an increase in the charge or polarity of the solute. Small ions are solvated more strongly than large ions with the same charge, and anions are solvated more strongly than cations.

Figure 2.3

The entropy decrease resulting from solvation. When a salt is dissolved in water, the entropy of the dissociated cations and anions increases because of the increased possibilities for translation and rotation. But at the same time the movement of water molecules becomes restricted in the vicinity of the ions. The net effect is frequently a decrease in the entropy of the solution. Such a decrease in entropy can occur if the solution releases heat to the surroundings, because this will increase the entropy of the surroundings.

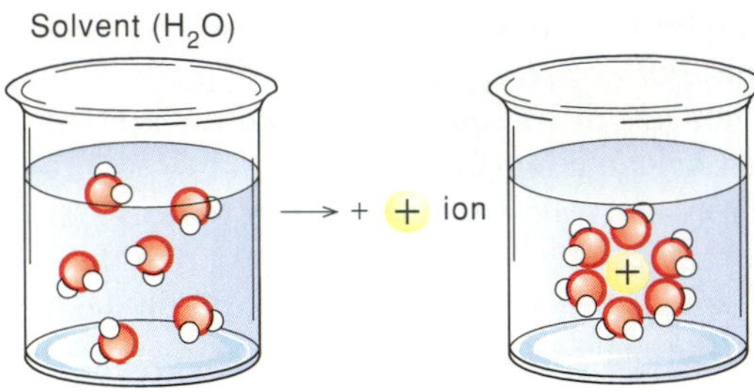

It is noteworthy that although enthalpy depends mainly on electronic interactions, whereas entropy depends mainly on translation and rotation, solvation affects both enthalpy and entropy. Enthalpies and entropies of solvation usually tend to oppose each other. For charged species, the more negative (favorable) the enthalpy of solvation, the more negative (unfavorable) the entropy of solvation.

From our earlier discussion, you might expect that the dissociation of a proton from a carboxylic acid, which increases the number of independent particles, would lead to an increase in entropy. However, this effect is more than counterbalanced by solvation effects. The charged anion and proton both "freeze out" many of the surrounding molecules of water (fig. 2.4). Thus the ionization of a weak acid decreases the number of mobile molecules, and so leads to a decrease in entropy. The entropy of ionization of a typical carboxylic acid in water is about -22 eu/mole. The entropy of dissociation of a proton from a quaternary ammonium group ($R—NH_3^+ \rightarrow R—NH_2 + H^+$) is usually smaller, because in this case the dissociation does not alter the number of charged species in the solution.

A different type of solvation effect occurs when an apolar molecule is added to water. The result is a decrease in entropy, but not because of favorable interactions between the molecule and the solvent. The water orients on the surface of the apolar molecule to form a relatively rigid cage held together by hydrogen bonds (see fig. 1.10). This effect plays important roles in governing the folding of proteins and determining the structure of biological membranes (see chapter 4).

Substantial changes in entropy can occur when a small molecule binds to a protein or other macromolecule. Of particular interest is binding of a substrate or inhibitor to an enzyme. It is instructive to compare the entropy changes here with those that accompany the binding of gas molecules to the surface of a solid catalyst. When gas molecules adsorb on a solid surface, there is a large decrease in entropy (fig. 2.5). The translational entropy of the gas disappears, and the thermodynamics of the adsorbed molecules becomes more like that of a solid. This negative change in entropy is an obstacle to the industrial use of solid catalysts for reactions of gases. To make the reaction proceed, the unfavorable entropy change must be overcome by increasing the pressure or tailoring the catalyst so that the enthalpy of interaction with the gas is strongly negative. In contrast, the binding of a substrate to an enzyme frequently has a positive ΔS. The reason is that water molecules are displaced when the substrate binds (fig. 2.5). Binding of an ester substrate to pepsin,

Figure 2.4

An ionization reaction often decreases the entropy of a solution, instead of increasing it as one might at first expect, because clustering of water molecules around the ions can result in a net decrease in the number of free particles.

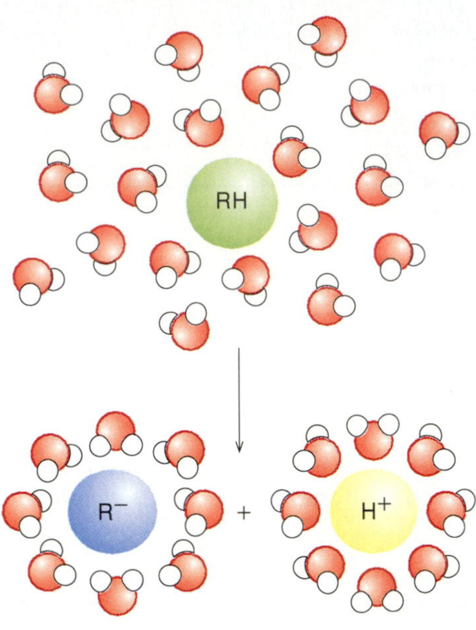

Figure 2.5

The entropy of binding of gas molecules to a solid catalyst is negative because of the restricted movements of the adsorbed molecules. By contrast, the entropy change on binding of a substrate to an enzyme is frequently positive. This effect arises because the restricted movement of the substrate is more than compensated for by the release of bound water from the enzyme and the substrate.

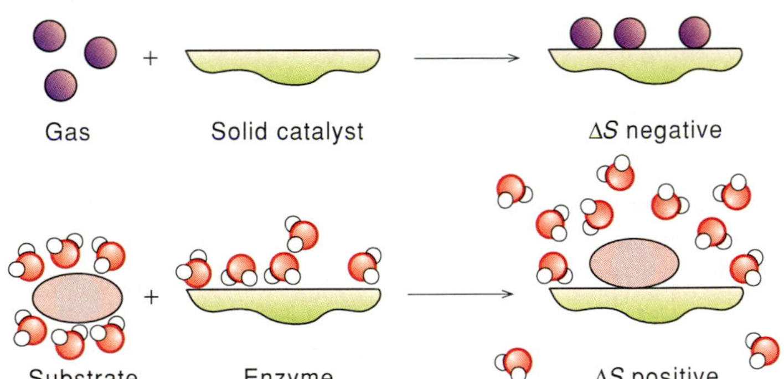

Table 2.2

Standard Enthalpy and Entropy Changes on Forming Complexes between Cadmium Ion and Methylamine or Ethylenediamine (en)

Reaction[a]	$\Delta H°$ (kcal/mole)	$\Delta S°$ (eu/mole)	$T\,\Delta S°$ (kcal/mole)	$\Delta G°$ (kcal/mole)
$Cd^{2+} + 4\ CH_3NH_2 \rightarrow Cd(CH_3NH_2)_4^{2+}$	−13.7	−16.0	−4.77	−8.94
$Cd^{2+} + 2\ en \rightarrow Cd(en)_2^{2+}$	−13.5	−3.3	+0.98	−14.50

[a]en = ethylenediamine; temperature = 25° C.

Source: Data from Spike and Parry, "Thermodynamics of chelation I. The statistical factor in chelate ring formation" in *Journal American Chemical Society* 75:2726, 1953.

for example, produces an entropy increase of 20.6 eu/mole, and binding of urea to urease produces an entropy increase of 13.3 eu/mole. Because of the favorable entropy change, the binding of the substrate may occur spontaneously even if ΔH is unfavorable.

Another important entropy effect when an enzyme and substrate combine could be called the chelation effect. This effect is best understood by discussing a relatively simple case of metal chelation. Cadmium ion tends to be quadrivalent, so if there is one amino group in a ligand molecule, as in methylamine, the cadmium can bind to four molecules. If there are two amino groups, as in ethylenediamine, the cadmium will combine with two ligand molecules, as shown in table 2.2. Notice that the entropy change is much more unfavorable for the combination with four methylamines than for that with two ethylenediamines. Water molecules are released from the cadmium ion when the ligands bind, but the entropy increase from this release is about the same whether methylamine or ethylenediamine is added. The less favorable entropy change resulting from association with methylamine is due to the larger number of molecules that must attach to the cadmium in this case.

In general, the chelation effect means that a molecule with *n* points of attachment to another molecule will bind more strongly than *n* molecules with one point of attachment, even though the enthalpy change upon binding at each point is the same. The chelation effect is important in substrate binding to enzymes, where there typically are multiple points of attachment. Several weak interactions can produce an overall tight binding because of the additive contributions of the small, favorable enthalpy changes and the lack of a proportional decrease in entropy. This effect is even greater in the binding of two proteins or the binding of a protein to a nucleic acid, because there usually are many points of interaction between these large molecules.

The final column in table 2.2 indicates the changes in free energy accompanying the reactions of cadmium ion with the two amine compounds. Free energy is a function of both enthalpy and entropy; it provides the most useful criterion as to whether a reaction could proceed spontaneously, as explained in the next section.

Free Energy Provides the Most Useful Criterion for Spontaneity

We have seen that a system tends toward the lowest enthalpy and the highest entropy. The tendency for the enthalpy of a system to decrease can be explained simply by the second law. If no work is done, an enthalpy decrease means a transfer of heat from the system to the surroundings, and if the surroundings are at a lower temperature such a flow of heat will be driven by an increase in the entropy of the surroundings. But enthalpy changes do not afford a reliable rule for determining whether a reaction can proceed spontaneously, because the enthalpy of a system can increase if the entropy of the system also increases. The second law does provide a reliable rule, but its application is often difficult because it requires that we consider the entropy changes in the surroundings as well as the system.

A more convenient function for predicting the direction of a reaction was discovered by Josiah Gibbs. He was the first to appreciate that in reactions occurring at equilibrium and constant temperature, the change in entropy of the system is numerically equal to the change in enthalpy divided by the absolute temperature. This relationship is the one already presented in equation (6). The equation can be transposed to

$$\Delta H - T\,\Delta S = 0 \quad \textbf{(8)}$$

In search of a criterion for spontaneity, Gibbs proposed a new function called the free energy, defined by the equation

$$\Delta G = \Delta H - T\,\Delta S \quad \textbf{(9)}$$

Here ΔH and ΔS are the changes in enthalpy and entropy in the system alone, not including the surroundings. For a reaction occurring at equilibrium, such as the melting of ice at 0° C, the change in free energy is zero. For the same reaction occurring at a higher temperature, say 10° C, the term $T\,\Delta S$ is larger, making ΔG negative. Ice at 10° C melts spontaneously. In the reverse reaction, conversion of water to ice at 10° C, there would be a positive change in free energy. This process does not occur spontaneously. Gibbs proposed that a reaction can occur spontaneously if, and only if, ΔG is negative. If ΔG is zero, the system is in equilibrium and no net reaction will occur in either direction.

Free energy, like energy, enthalpy, and entropy, is a state function and an extensive property of a system. If the free energy change is favorable (negative), and a good pathway exists, a reaction will occur. If no pathway exists for the conversion, a catalyst may be added that provides an acceptable pathway. However, if the free energy change is unfavorable (positive), no catalyst can ever make the reaction proceed.

Applications of the Free Energy Function

The free energy function dominates most discussions of thermodynamics in biochemistry. Not only does the sign of ΔG determine the direction in which a reaction will proceed, but the magnitude of ΔG indicates just how far the reaction must proceed before the system comes to equilibrium. This is because the standard free energy change, $\Delta G°$, has a simple relationship to the equilibrium constant. We will elaborate on these points in the following sections. Despite its usefulness, however, many people find the free energy function difficult to grasp intuitively. The reason is that ΔG is a composite of enthalpic and entropic terms, which often make opposite contributions.

Values of Free Energy Are Known for Many Compounds

The standard free energy of formation of a compound, $\Delta G°_f$, is the difference between the free energy of the compound in its standard state and the total free energies of the elements of which the compound is composed, again when the elements are in their standard states. The standard states usually are chosen to be the states in which the elements or molecules are stable at 25° C and 1 atmosphere pressure. For oxygen and nitrogen, these are the gases O_2 and N_2; for solid elements such as carbon, they are the pure solids. For most solutes, the standard states are taken to be 1 M solutions. However, in biochemistry the standard state for hydrogen ion in solution is usually defined as a 10^{-7} M solution because this is close to the concentration in most systems of interest to biochemists.

Standard free energies of formation are known for thousands of compounds. They usually are given in units of kcal/mole or kJ/mole. The values for a few compounds of biological interest are collected in table 2.3. By subtracting the sum of the free energies of formation of the reactants from the sum of the free energies of formation of the products, it is possible to calculate the standard free energy change in any reaction for which all the free energies of formation are known.

From the values listed in table 2.3, we can calculate the standard free energy change for the reaction

$$\text{Oxaloacetate}^{2-} + H^+ (10^{-7}\text{ M}) \rightarrow CO_2(g) + \text{pyruvate}^-$$

as

$$\Delta G° = -113.44 - 94.45 - (-9.87 - 190.62)$$
$$= -7.4 \text{ kcal/mole}$$

Table 2.3
Standard Free Energies of Formation of Some Compounds of Biological Interest

Substance	$\Delta G°_f$ (kcal/mole)	$\Delta G°_f$ (kJ/mole)
Lactate ions	−123.76	−516
Pyruvate ions	−113.44	−474
Succinate dianions	−164.97	−690
Glycerol (1 M)	−116.76	−488
Water	−56.69	−280
Acetate anions	−88.99	−369
Oxaloacetate dianions	−190.62	−797
Hydrogen ions (10^{-7} M)	−9.87[a]	−41
Carbon dioxide (gas)	−94.45	−394
Bicarbonate ions	−140.49	−587

[a]This is the value for hydrogen ions at a concentration of 10^{-7} M. The free energy of formation at unit activity (1 M) is 0.

The free energy change, when all the reactants and products are in their standard states (1M oxaloacetate dianion and pyruvate anion, 10^{-7}M hydrogen ion, and 1 atm CO_2), is −7.4 kcal/mole. The negative value of $\Delta G°$ means that the reaction would proceed spontaneously under these conditions. However, some of the concentrations are not very realistic. At pH 7, carbon dioxide would be present partly in the form of the bicarbonate anion, rather than as gaseous CO_2. To take this into account, we can add the standard free energy change for the reaction of CO_2 with water to give the bicarbonate anion plus a proton:

$$CO_2(g) + H_2O \rightarrow HCO_3^- + H^+$$

This calculation yields a correction of $-140.49 - 9.87 - (-56.69 - 94.45) = 0.8$ kcal/mole. The free energy change for the reaction of oxaloacetate to form pyruvate and 1M bicarbonate ions instead of CO_2 is $-7.4 + 0.8 = -6.6$ kcal/mole.

The preceding calculation illustrates the point that the standard free energy change for a reaction can be found by adding or subtracting the free energies of any other reactions that combine to give the desired reaction. Another example is the calculation of the standard free energy of hydrolysis of ATP at pH 7. This calculation can be done by combining the free energy change for the hydrolysis of glucose-6-phosphate with the free energy change for forming glucose-6-phosphate from glucose and ATP, as shown in table 2.4. We will return to these reactions in a later section.

Table 2.4
Calculating the Standard Free Energy of ATP Hydrolysis ($\Delta G^{\circ\prime}$) by Adding the Free Energies of Two Other Reactions

Reaction[a]	$\Delta G^{\circ\prime}$ (kcal/mole)
Glucose + $ATP^{4-} \rightleftharpoons$ glucose-6-phosphate^{2-} + ADP^{3-} + H^+	−5.4
Glucose-6-phosphate^{2-} + $H_2O \rightleftharpoons$ glucose + HPO_4^{2-}	−3.0
$ATP^{4-} + H_2O \rightleftharpoons ADP^{3-} + HPO_4^{2-} + H^+$	−8.4

[a]The values of $\Delta G^{\circ\prime}$ are for reactions at pH 7 in the absence of Mg^{2+}. In the presence of 10 mM Mg^{2+}, $\Delta G^{\circ\prime}$ for ATP hydrolysis is about −7.5 kcal/mole.

The Standard Free Energy Change in a Reaction Is Related Logarithmically to the Equilibrium Constant

In biochemistry we are most concerned with reactions occurring in aqueous solution. Suppose we have a chemical reaction with the stoichiometry

$$aA + bB \rightleftharpoons cC + dD$$

where *a, b, c,* and *d* refer to the moles of A, B, C, and D, respectively. The free energy change in the reaction is

$$\Delta G = G_{\text{final state}} - G_{\text{initial state}} \tag{10}$$

If the reaction occurs at constant temperature and pressure, equation (10) can be expressed as the difference between the standard free energies of the products and reactants, ΔG°, plus a correction for the concentrations:

$$\Delta G = \Delta G^\circ + RT \ln \frac{[C]^c[D]^d}{[A]^a[B]^b} \tag{11}$$

The last term in equation (11) is the correction for concentration and as such is an entropic contribution to ΔG. It is derived by using equation (7) to find the entropy changes associated with diluting the reactants and products from their standard states (1 M) to the actual concentrations in the solution. (In a rigorous treatment, we should use activities instead of concentrations in this formula, but for simplicity we will ignore the difference, keeping in mind that it can be substantial in some cases.) If the concentrations of the reactants exceed those of the products, so that the ratio $[C]^c[D]^d/[A]^a[B]^b$ is less than 1, the logarithm will be negative, making ΔG more negative than ΔG° and favoring the reaction in the forward direction. A concentration ratio greater than 1 will favor the reverse reaction.

When the reaction comes to equilibrium,

$$\frac{[C]^c[D]^d}{[A]^a[B]^b} = K_{eq} \tag{12}$$

where K_{eq} is the equilibrium constant for the reaction. We also know that at equilibrium $\Delta G = 0$. Therefore, from equation (11),

$$\Delta G^\circ = -RT \ln K_{eq} \tag{13}$$

Table 2.5
Relationship between ΔG° and K_{eq} (at 25° C)

ΔG° (kcal/mole)[a]	K_{eq}
−6.82	10^5
−5.46	10^4
−4.09	10^3
−2.73	10^2
−1.36	10
0	1
1.36	10^{-1}
2.73	10^{-2}
4.09	10^{-3}
5.46	10^{-4}
6.82	10^{-5}

[a]ΔG° values at 25° C are calculated from the equation

$$\begin{aligned}\Delta G^\circ &= -RT \ln K_{eq}\\ &= -1.98 \times 298 \times 2.3 \log K_{eq}\\ &= -1364 \log K_{eq}\end{aligned}$$

Thus the standard free energy change for a reaction can be used to obtain the equilibrium constant. Conversely, if we know K_{eq}, we can find ΔG°. Because of the logarithmic relationship, K_{eq} has a very steep dependence on ΔG° (table 2.5). A reaction that proceeds to 99% completion is, for most practical purposes, a quantitative reaction. It requires an equilibrium constant of 100, but a standard free energy change of only −2.7 kcal/mole, which is little more than half the standard free energy change for the formation of a hydrogen bond.

Equations (11) and (13) are two of the most important thermodynamic relationships for biochemists to remember. If the concentrations of reactants and products are at their equilibrium values, there is no change in free energy for the reactions going in either direction. Living cells, however, maintain some compounds at concentrations far from the equilibrium values, so that their reactions are associated with large changes in free energy. We will amplify on this point in chapter 12.

We have mentioned that biochemists usually define the standard state of protons as 10^{-7} M and report values of free energy and equilibrium constants for solutions at pH 7. These values are designated by a prime and written as $\Delta G^{\circ\prime}$, $\Delta G'$ and K'_{eq}. *Unprimed symbols are used to designate values based on a standard state of 1* M for protons (pH 0). For a reaction that releases one proton, the relationship between K'_{eq} and K_{eq} is $K'_{eq} = 10^7 K_{eq}$. In evaluating the standard free energies $\Delta G^{\circ\prime}$ and ΔG°, it is critical to use the equilibrium constants K'_{eq} and K_{eq}, respectively, because these can be very different quantities.

Free Energy Is the Maximum Energy Available for Useful Work

The free energy change gives a quantitative measure of the maximum amount of useful work that could be obtained from a reaction that occurs at constant temperature and pressure. By "useful" work, we mean work other than the unavoidable work of expansion or contraction against the fixed pressure of the surroundings. If ΔG is zero, the system is at equilibrium, which means that we could not obtain any useful work from the process. If ΔG is less than zero, the process could yield useful work as the system proceeds spontaneously toward equilibrium. If ΔG is greater than zero, the process is headed away from equilibrium, and we would have to perform work on the system in order to drive it in this direction. The farther the reactants are from equilibrium, the larger is the value of $-\Delta G$, and the larger is the amount of work that we might obtain from the reaction. However, $-\Delta G$ gives only the maximum amount of useful work. Remember that work and heat, unlike ΔG, ΔH, and ΔS, are not functions of state. The amount of work that is actually obtained depends on the path that the process takes, and it can be zero even if $-\Delta G$ is large.

Biological Systems Perform Various Kinds of Work

To sustain and propagate life requires that cells do various types of work. This work takes three major forms, related to three broad categories of cellular activities:

1. Mechanical work: changes in location or orientation. Mechanical work is done whenever an organism, cell, or subcellular structure moves against the force of gravity or friction. As examples, consider the contracting muscles that propel a runner up a hill, the swimming of a flagellated protozoan in a pond, the migration of chromosomes toward the opposite poles of the mitotic spindle, and the movement of a ribosome along a strand of messenger RNA.
2. Concentration and electrical work: movements of molecules and ions across membranes. Concentration work, the movement of a molecule or ion across a membrane against a prevailing concentration gradient, establishes the localized concentrations of specific materials on which most essential life processes depend. Concentration work is sometimes referred to as osmotic work. Examples include the uptake of amino acids from the blood by muscle cells, pumping of sodium ions out of a marine microorganism, and movement of nitrate from the soil into the cells of a plant root. Electrical work is required to move a charged species across a membrane against an electrical potential gradient. Although the most dramatic example of this is the generation of large potential differences in the electric organ of the electric eel, electrical work is done by almost all types of cells. It underlies the mechanisms of excitation of nerve and muscle cells and the conduction of impulses along axons.
3. Synthetic work: changes in chemical bonds. Synthetic work is necessary for the formation of the complex organic molecules of which cells are composed. As we have seen, these are in general molecules of higher enthalpy and lower entropy than the simple molecules that are available to organisms from their environment, so that free energy must be expended in their synthesis. Synthetic work is most obvious during periods of growth of an organism, but it also occurs in nongrowing, mature organisms, which must continuously repair and replace existing structures. The continuous expenditure of energy to elaborate and maintain ordered structures that were created out of less-ordered raw materials is one of the most characteristic properties of living cells.

Favorable Reactions Can Drive Unfavorable Reactions

In table 2.4, we made use of the principle that the free energies of all the components of a solution are additive. In general, if the free energy changes associated with two reactions $A \rightleftharpoons B$ and $C \rightleftharpoons D$ are ΔG_{AB} and ΔG_{CD}, the free energy change accompanying the combined process $A + C \rightleftharpoons B + D$ is simply $\Delta G_{AB} + \Delta G_{CD}$. It is this principle that allows living organisms to synthesize complex molecules with high enthalpies and low entropies. Thermodynamically unfavorable reactions can be driven by coupling them to favorable processes.

From equation (13) you can see that whereas free energy changes combine additively, equilibrium constants combine multiplicatively. If the equilibrium constants for the reactions $A \rightleftharpoons B$ and $C \rightleftharpoons D$ are K_{AB} and K_{CD}, the equilibrium constant for $A + C \rightleftharpoons B + D$ is $K_{AB}K_{CD}$.

There are numerous ways to achieve a coupling of favorable and unfavorable reactions. As an example, let's return to the formation of glucose-6-phosphate and water from glucose and inorganic phosphate ion (P_i):

$$\text{Glucose} + P_i \rightleftharpoons \text{glucose-6-phosphate} \quad \textbf{(14)}$$

This reaction has an unfavorable $\Delta G^{\circ\prime}$ of $+3.0$ kcal/mole at 298 K (K_{eq} is 0.0062); the reaction will not occur spontaneously. On the other hand, the reaction

$$\text{ATP} + H_2O \rightleftharpoons \text{ADP} + P_i + H^+ \quad \textbf{(15)}$$

Figure 2.6

The formation of glucose-6-phosphate (G-6-P) has a positive $\Delta G^{\circ\prime}$; the hydrolysis of ATP and ADP has a negative $\Delta G^{\circ\prime}$. If the two reactions are combined, the overall $\Delta G^{\circ\prime}$ is negative.

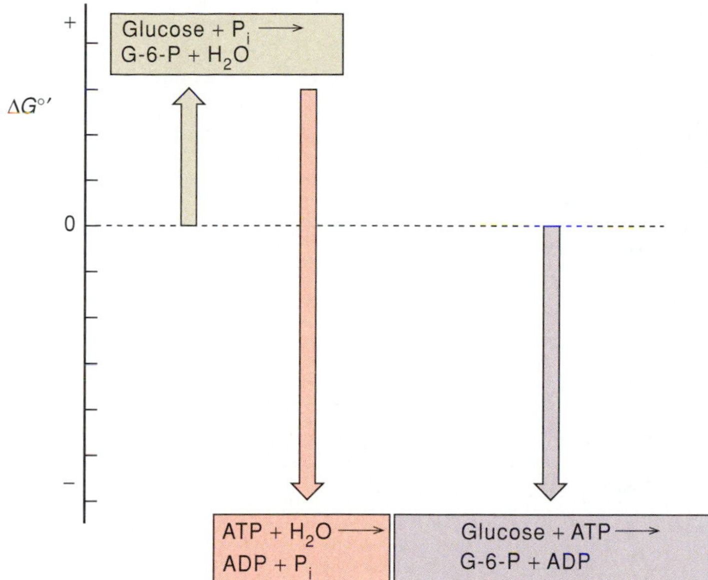

has a highly favorable $\Delta G^{\circ\prime}$ of about -8.4 kcal/mole ($K_{eq} \approx 1.35 \times 10^6$). If reactions (14) and (15) are combined to give the reaction

$$\text{Glucose} + \text{ATP} \rightleftarrows \text{glucose-6-phosphate} + \text{ADP} \quad \textbf{(16)}$$

the overall $\Delta G^{\circ\prime}$ is $3.0 - 8.4$, or -5.4 kcal/mole, and the overall equilibrium constant is 8.7×10^3. The combined reaction is thermodynamically favorable (fig. 2.6). A cell could use this reaction to synthesize glucose-6-phosphate, provided that it has a source of ATP.

To take advantage of such a thermodynamic combination of favorable and unfavorable processes, a cell must have a catalytic mechanism for actually linking the two reactions. The breakdown of ATP to ADP and P_i, reaction (15), would be fruitless if it occurred independently of reaction (13). In many cells, the combined reaction (16) is catalyzed by an enzyme that facilitates the transfer of phosphate from ATP directly to glucose (hexokinase). This is a common motif in biosynthetic processes. But the coupling mechanism does not have to be so direct.

Another common mechanism for coupling an unfavorable reaction to a favorable one is simply to arrange for one of the reactions to precede or follow the other. If the free energy changes for the reactions A $\rightleftarrows$ B and B $\rightleftarrows$ C are ΔG_{AB} and ΔG_{BC}, the free energy change for A $\rightleftarrows$ C is $\Delta G_{AB} + \Delta G_{BC}$. As an example, consider the following sequence of reactions:

$$\text{Acetyl-CoA} + \text{oxaloacetate} \xrightarrow{1} \text{citryl-CoA} \xrightarrow{2} \text{citrate} + \text{coenzyme A} \quad \textbf{(17)}$$

The first step has a $\Delta G^{\circ\prime}$ of -0.05 kcal/mole, which is close to zero; it will not occur to any great extent unless the concentrations of acetyl-CoA and oxaloacetate are greater than the concentration of citryl-CoA. The second step, however, has a highly favorable $\Delta G^{\circ\prime}$ of -8.4 kcal/mole. When the two steps are combined, $\Delta G^{\circ\prime}$ for the overall reaction is about -8.3 kcal/mole and the equilibrium constant lies far in the forward direction. These two reactions are catalyzed by the enzyme citrate synthase, by a mechanism that insures that they always occur together.

In the case of reactions that occur sequentially, with one step pulling or pushing the other, it is not necessary for the two steps to be catalyzed by the same enzyme, although this can help to speed up the overall sequence. For example, in many organisms the reactions shown in (17) are preceded by a reaction in which malate is converted to oxaloacetate:

$$\text{Malate}^{2-} + \text{NAD}^+ \rightarrow \text{oxaloacetate}^{2-} + \text{NADH} + \text{H}^+ \quad \textbf{(18)}$$

This step is catalyzed by a separate enzyme, malate dehydrogenase. It has a very unfavorable $\Delta G^{\circ\prime}$ of about $+7$ kcal/mole. When this reaction is followed by reaction (17), the combined $\Delta G^{\circ\prime}$ is about -1.3 kcal/mole, so the equilibrium constant for the overall sequence is favorable. You can view this as simply an illustration of the principle of mass action: (17) pulls (18) along by removing one of the products.

This last point deserves additional emphasis. It is important to keep in mind that what determines whether or not a process will occur spontaneously is ΔG, not ΔG°. As we saw in equation (11), the actual free energy change depends on the concentrations of the reactants and products. The hydrolysis of ATP (reaction 15), for example, has a $\Delta G^{\circ\prime}$ of about -8.4 kcal/mole, but the actual ΔG in the cytosol of living cells is typically more negative than this by between 5 and 6 kcal/mole because the concentration ratio, $[\text{ADP}][\text{P}_i]/[\text{ATP}]$, is much less than 1.

ATP as the Main Carrier of Free Energy in Biochemical Systems

Virtually all living organisms use ATP for transferring free energy between energy-producing and energy-consuming systems. Processes that proceed with large negative changes in free energy, such as the oxidative degradation of carbohydrates or fatty acids, are used to drive the formation of ATP, and the hydrolysis of ATP is used to drive biosynthetic reactions and other processes that require increases in free energy. In the human body, about 2.3 kg of ATP is formed and consumed every day in the course of these reactions.

The Hydrolysis of ATP Yields a Large Amount of Free Energy

Adenosine triphosphate (ATP) can be hydrolyzed in two different ways, as shown in figure 2.7. Hydrolysis of the linkage between the β and γ phosphate groups yields ADP and P_i. Hydrolysis between the α and β phosphates gives adenosine monophosphate (AMP) and pyrophosphate ion ($HP_2O_7^{3-}$). The standard free energy change ($\Delta G^{\circ\prime}$) is about -8.4 kcal/mole in either case. The formation of AMP and pyrophosphate, however, can be pulled forward by hydrolysis of the pyrophosphate

Figure 2.7

Alternative routes of ATP hydrolysis. The charged species shown are the main ones present at physiological pH and ionic strength. The phosphate groups of ATP are referred to as α, ß, and γ as indicated. Under physiological conditions, ATP and ADP also bind Mg^{2+} (not shown).

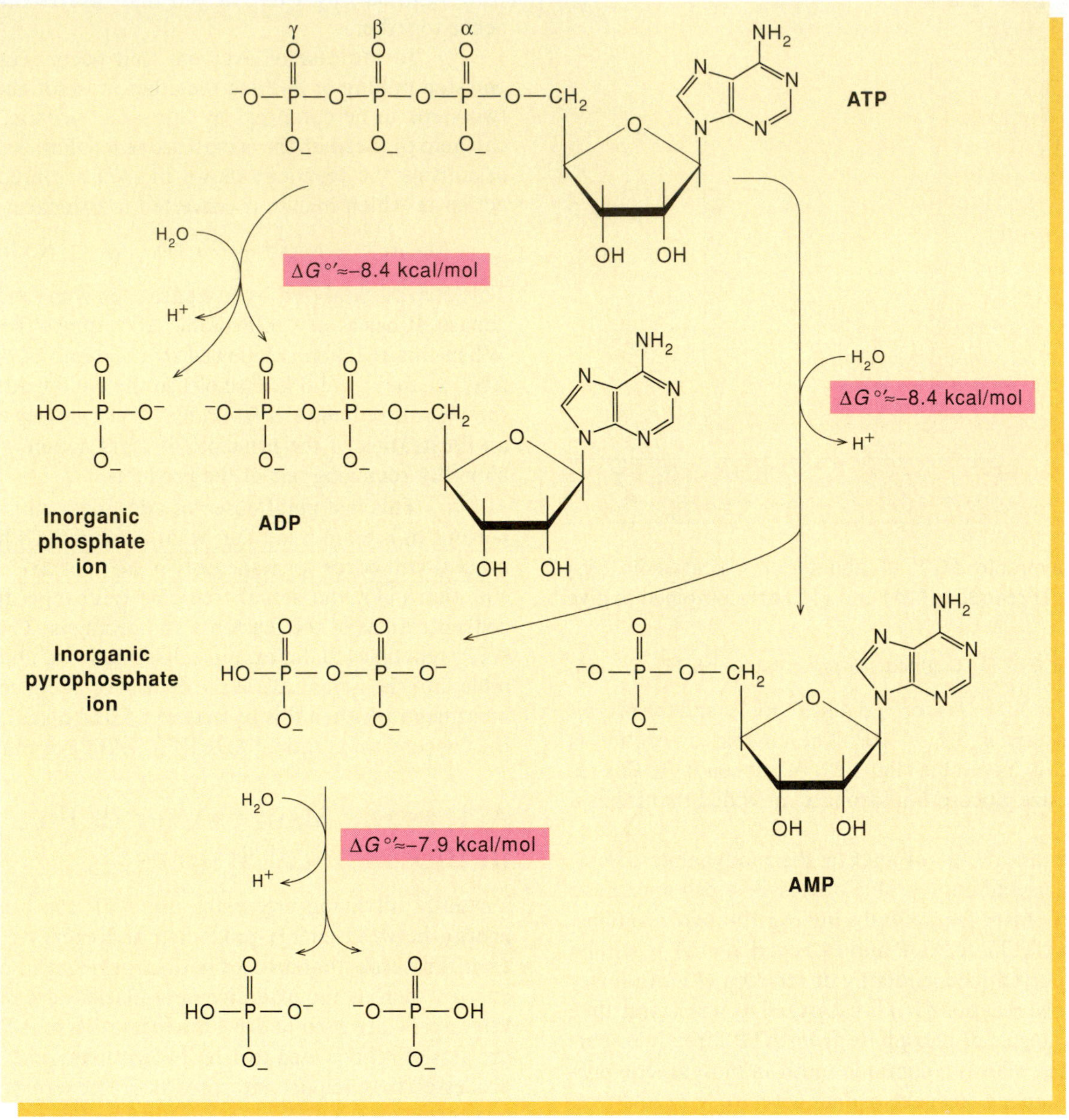

to give two equivalents of P_i. This secondary reaction has a $\Delta G°'$ of about -7.9 kcal/mole, and is catalyzed by pyrophosphatase enzymes present in most types of cells. The overall $\Delta G°'$ of about -16.3 kcal/mole for breakdown of ATP to AMP and 2 P_i makes this process effectively irreversible ($K_{eq} \approx 1 \times 10^{12}$). Cells commonly use this series of reactions in biosynthetic processes in which a reversal would be intolerable, such as in the synthesis of nucleic acids. The simpler hydrolysis to produce ADP and P_i allows some reversibility but has the advantage that less free energy is needed to resynthesize ATP from ADP than from AMP.

The standard free energies for the hydrolysis of ATP, ADP, or pyrophosphate depend on the pH, on the ionic strength, and also on the concentration of Mg^{2+}, which binds to both the reactants and the products and is required as a cosubstrate by most enzymes that use ATP. At pH 7, the principal ionic forms of ATP, ADP, and P_i have net charges of -4, -3, and -2, respectively (see fig. 2.7). Because the hydrolysis of ATP^{4-} to give $ADP^{3-} + HPO_3^{2-}$ is accompanied by release of a proton, raising the pH makes the hydrolysis more favorable. Increasing the Mg^{2+} concentration makes the hydrolysis less favorable.

$\Delta G°'$ decreases in magnitude from -8.4 kcal/mole in the absence of Mg^{2+} to about -7.7 kcal/mole in the presence of 1 mM Mg^{2+}, and to about -7.5 kcal/mole at 10 mM. The value -7.5 kcal/mole is probably close to the $\Delta G°'$ under typical physiological conditions, and will be used elsewhere in this text. But remember that the low ratio of [ADP][P_i] to [ATP] can make ΔG for hydrolysis of ATP in living cells substantially more negative than the $\Delta G°'$. In resting cells, the enzymatic reactions that synthesize ATP usually are more than adequate to keep up with the processes that consume it.

Why is $\Delta G°'$ for the hydrolysis of ATP so negative? The answer involves several different factors. At pH 7 in the presence of 10 mM Mg^{2+}, $\Delta H°'$ and $-T\,\Delta S°'$ for ATP hydrolysis are both negative and the two terms make similar contributions to the overall value of $\Delta G°'$. The favorable entropy change results partly from the fact that the proton released in the reaction is diluted into a solution with a very low proton concentration. In addition, solvent water molecules probably are less highly ordered around ADP and P_i than they are around ATP. The negative $\Delta H°'$ results partly from the fact that the negatively charged oxygen atoms in ATP tend to repel each other. The phosphoric anhydride bond in ATP also is weakened by competition between the phosphorus atoms, which both tend to pull electrons away from the bridging oxygen. Another consideration is that the major products of ATP hydrolysis at pH 7 (ADP^{3-} and $HOPO_3^{2-}$) have a larger total number of resonance forms than the reactant ATP^{4-} does (fig. 2.8). The increased resonance lowers the energy of ADP^{3-} and $HOPO_3^{2-}$ relative to ATP^{4-}. The magnitudes of all of these effects vary with the pH, because protonation of the oxygen atoms in ATP, ADP, or P_i relieves some of the repulsive electrostatic interactions, decreases the contributions that some of the resonance forms make to the structure, and decreases the ordering of nearby water molecules. At very low pH, when ATP, ADP, and P_i are all fully protonated, $\Delta G°$ probably becomes slightly positive, favoring ATP formation rather than hydrolysis.

In addition to ATP and other nucleoside triphosphates, cells use a variety of other organic phosphate compounds in energy metabolism. These include acetyl phosphate, glycerate-1,3-bisphosphate, phosphoenolpyruvate, phosphocreatine, and phosphoarginine. Figure 2.9 shows the structures of these compounds and some simpler phosphate esters. The compounds are ranked in the figure in order of their standard free energies of hydrolysis, with the materials having the most negative values of $\Delta G°'$ at the top. Given equal concentrations of reactants and products, any compound in the figure could be synthesized, in principle, at the expense of any of the compounds above it. Thus ADP could be phosphorylated to ATP by phosphocreatine, glycerate-1,3-bisphosphate, or phosphoenolpyruvate, and ATP could be used to convert glucose to glucose-6-phosphate or glycerol to glycerol phosphate. The reverse processes will occur only if the ratio of reactants to products is sufficiently high, but this is not necessarily an unusual circumstance. In cells of some tissues, the reaction

$$\text{Creatine} + \text{ATP} \rightleftharpoons \text{creatine phosphate} + \text{ADP} + \text{H}^+ \quad \textbf{(19)}$$

occurs readily in either direction in response to changes in the concentrations of the reactants and products. The same is true of the reaction

$$\text{Glycerate-3-phosphate} + \text{ATP} \rightleftharpoons \text{glycerate-1,3-bisphosphate} + \text{ADP} + \text{H}^+ \quad \textbf{(20)}$$

Figure 2.8

The phosphate groups of ATP, ADP, and P_i can be written in a variety of resonance forms. This figure shows the major resonance forms of the ß phosphate group in ATP and of the same group in ADP. The hydrolysis of ATP to ADP results in an increase in the number of resonance forms available to this group. This increase contributes to the negative $\Delta H°'$ of hydrolysis of ATP. At physiological pH there are also favorable contributions to $\Delta G°'$ from the release of a proton and from the disordering of water around the polyphosphate chain.

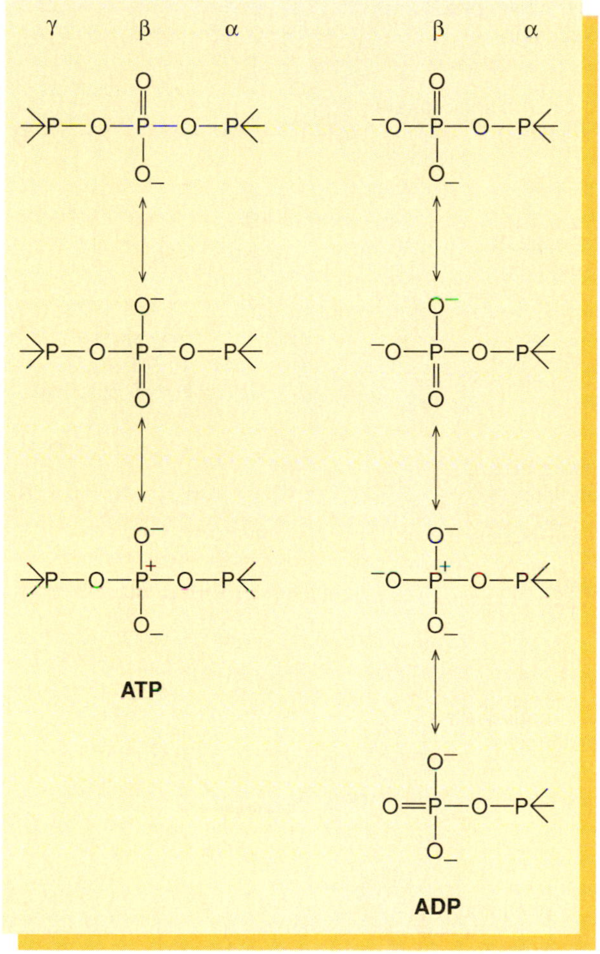

Figure 2.9

Standard free energies of hydrolysis for some common phosphorylated compounds. The $\Delta G^{\circ\prime}$ value refers to hydrolysis of the phosphate group indicated by the symbol Ⓟ. Note that ATP occupies an intermediate position among these compounds. Given equal concentrations of the reactants and products, ADP can accept a phosphate group from any of the compounds above it, and ATP can donate a phosphate to the unphosphorylated forms of the compounds below it.

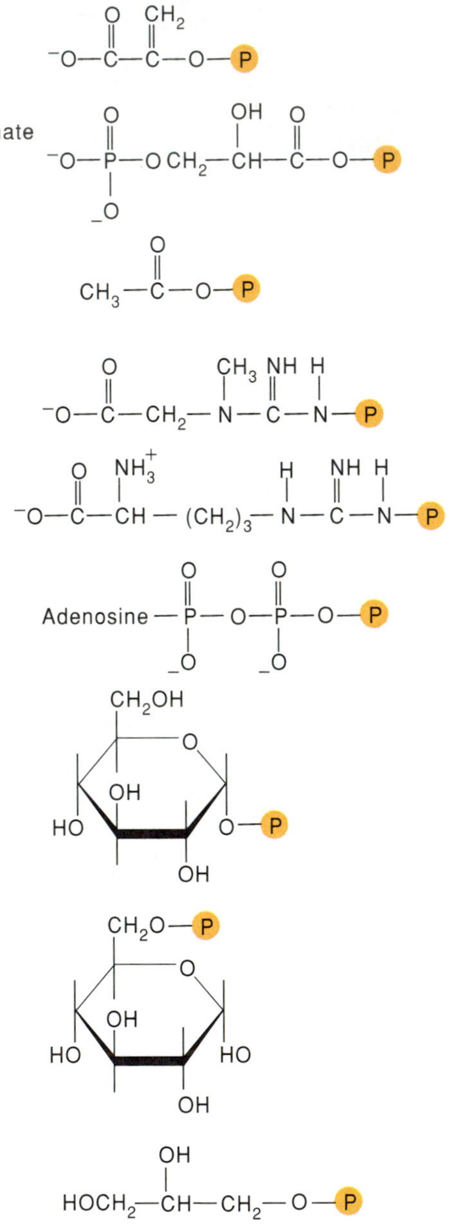

Phosphoenolpyruvate (–14.8 kcal/mol)

Glycerate-1,3-bisphosphate (–11.8 kcal/mol)

Acetyl phosphate (–11.3 kcal/mol)

Phosphocreatine (–10.3 kcal/mol)

Phosphoarginine (–9.1 kcal/mol)

ATP (–7.5 kcal/mol)

Glucose-1-phosphate (–5.0 kcal/mol)

Glucose-6-phosphate (–3.3 kcal/mol)

Glycerol-3-phosphate (–2.2 kcal/mol)

Most of the phosphorylated materials shown in figure 2.9 participate in only a few biochemical reactions, and many of them are formed only in certain types of cells or organisms. ATP, in contrast, is formed by all living things, and it participates in literally hundreds of different enzymatic reactions. What is it about ATP that has made this molecule such a universal currency of free energy in biology? Several considerations are relevant here. First, $\Delta G^{\circ\prime}$ for hydrolysis of ATP to ADP is large enough so that this reaction releases a substantial amount of free energy, enough to drive many of the reactions that are important for biosynthetic pathways. At the same time, the $\Delta G^{\circ\prime}$ is small enough so that ATP itself can be synthesized readily at the expense of available nutrients. We touched on this point above in discussing the relative merits of hydrolysis at the α-β or β-γ positions in ATP.

The second consideration returns us to the distinction between spontaneity and speed. Although the hydrolysis of ATP has a large negative $\Delta G^{\circ\prime}$, it also is critical that ATP is a relatively stable compound in aqueous solution. It does not hydrolyze rapidly under physiological conditions of pH and temperature. The hydrolysis, though far downhill thermodynamically, is slowed by a substantial activation barrier. But the barrier must be of such a nature that it can easily be overcome enzymatically. This feature allows the free energy of hydrolysis to be channeled quickly and selectively into reactions where it is needed, but to be conserved when energy is not in demand.

Third, the products of the hydrolysis of ATP provide opportunities for coupling to a wide variety of chemical reactions. We have seen how the phosphate group is incorporated into glucose-6-phosphate, and later chapters will provide many illustrations of similar enzymatic processes. We also will see reactions in which the pyrophosphate group or the adenylyl group (AMP) is incorporated into the products. Such a broad array of reactions could not be driven by a molecule that decomposed to release an inert material such as N_2.

Finally, the adenine and ribosyl groups of ATP, ADP, and AMP provide additional structural features that allow these molecules to bind to enzymes, and thus to participate in regulating enzymatic activities. This may be part of the reason that no known organisms base their energy-transfer reactions entirely on inorganic pyrophosphate or other polyphosphate compounds without a nucleoside moiety.

Summary

In this chapter we have discussed some principles of thermodynamics as they relate to biochemical reactions. The following points are of greatest importance.

1. Thermodynamics is useful in biochemistry for predicting whether a given reaction could occur and, if so, how much work a cell could obtain from the process.
2. The thermodynamic quantities energy, enthalpy, entropy, and free energy are properties of the state of a system. Changes in these quantities depend only on the difference between the initial and final states, not on the mechanism whereby the system goes from one state to the other.
3. Energy is the capacity to do work.
4. The first law of thermodynamics says that energy cannot be created or destroyed in a chemical reaction. If the energy of a system increases, the surroundings must lose an equivalent amount of energy, either by the transfer of heat or by the performance of work.
5. The energy of a molecule includes translational, rotational, and vibrational energy, as well as electronic and nuclear energy. Electronic terms usually account for most of the change in energy, ΔE, in a chemical reaction.
6. The change in enthalpy, ΔH, is given by the expression $\Delta H = \Delta E + \Delta(PV)$. For most biochemical reactions, ΔE and ΔH are nearly equal. The organic molecules found in cells generally have much higher enthalpies than the simpler molecules from which they are built.
7. In most reactions that proceed spontaneously, the enthalpy of the system decreases. If no work is done, the system gives off heat to the surroundings. But in some spontaneous reactions, heat is absorbed in the absence of work and the enthalpy of the system increases. Such reactions invariably show an increase in the entropy of the system.
8. Entropy is a measure of the order in a system: Systems that are highly ordered have low entropies. The entropy of a molecule depends mainly on translational and rotational freedom. Biological macromolecules generally have much lower entropies than their building blocks.
9. The second law of thermodynamics states that there must be an overall increase in the entropy of the system and its surroundings in any process that occurs spontaneously. An isolated system proceeds spontaneously to states of increasingly greater entropy (greater disorder).
10. The change in the free energy of a system is defined as $\Delta G = \Delta H - T\Delta S$, where T is the absolute temperature. A reaction at constant pressure and temperature can occur spontaneously if, and only if, ΔG is negative. The maximal amount of useful work that can be obtained from a reaction is equal to $-\Delta G$.
11. For a reaction in solution, ΔG depends on the standard free energy change ($\Delta G°$) and on the concentrations of the reactants and products. The standard free energy change is related to the equilibrium constant by the expression $\Delta G° = -RT \ln K_{eq}$. Increasing the concentration of the reactants relative to the concentration of the products makes ΔG more negative.
12. Reactions that are thermodynamically unfavorable can be coupled to favorable reactions. The coupling of the reactions may be direct or sequential, as in a biochemical pathway.
13. ATP is the main coupling agent for free energy in living cells. The free energy provided by the hydrolysis of ATP is used to drive many reactions that would not occur spontaneously by themselves.
14. Several features make ATP particularly well suited for its role. First, hydrolysis of ATP to ADP and P_i or to AMP and PP_i releases a considerable amount of free energy. Second, ATP does not hydrolyze rapidly by itself, but it can be hydrolyzed readily in enzymatically catalyzed reactions. This difference allows the free energy of hydrolysis to be channeled into reactions where it is needed, but to be conserved when energy is not in demand. Third, the products of the hydrolysis of ATP provide opportunities for coupling to a wide variety of chemical reactions. Finally, the adenine and ribosyl groups of ATP, ADP, and AMP provide additional structural features that allow these molecules to bind to a large number of enzymes, and thus to participate in regulating enzymatic activities.

Selected Readings

Alberty, R. A., and F. Daniels, *Physical Chemistry,* 5th ed. New York: Wiley, 1975.

Cantor, C. R., and P. R. Schimmel, *Biophysical Chemistry.* San Francisco: Freeman, 1980.

Ingraham, L. L., and A. B. Pardee, Free Energy and Entropy in Metabolism. In *Metabolic Pathways,* Vol. 1 (D. M. Greenberg, ed.). New York: Academic Press, 1967.

Tinoco, I., Jr., K. Sauer, and J. C. Wang, *Physical Chemistry, Principles and Applications in Biological Sciences,* 2d ed. Englewood Cliffs, N.J.: Prentice-Hall, 1985.

Van Holde, K. E., *Physical Biochemistry,* 2d ed. Englewood Cliffs, N.J.: Prentice-Hall, 1985.

Problems

1. How do conditions in the cell limit the manipulation of reaction conditions?
2. What is meant by a state function? Why is enthalpy a state function?
3. What is the basic difference between intensive and extensive thermodynamic parameters?
4. Why can we equate energy and enthalpy for most biochemical reactions?
5. A reaction mechanism cannot be defined by free energy considerations. Explain.
6. Transfer of a hydrophobic molecule (e.g., a hydrophobic amino acid side chain) from an aqueous to a nonaqueous environment is entropically favorable. Explain.
7. As we will see in chapter 14, oxaloacetate is formed by oxidation of malate. The reaction

   $$\text{L-malate} + \text{NAD}^+ \rightarrow \text{Oxaloacetate} + \text{NADH} + \text{H}^+$$

   has a $\Delta G°'$ of $+7.0$ kcal/mole. Suggest reasons that the reaction proceeds in the direction of oxaloacetate production in the cell.
8. Cite three factors that make ATP ideally suited to transfer energy in biochemical systems.
9. You wish to measure the $\Delta G°'$ for the hydrolysis of ATP,

   $$\text{ATP} \rightarrow \text{ADP} + \text{P}_i$$

   but the equilibrium for the hydrolysis lies so far toward products that analysis of the ATP concentration at equilibrium is neither practical nor accurate. However, you have the following data that will allow calculation of the value indirectly.

   Creatine phosphate + ADP →
   ATP + creatine $K'_{eq} = 59.5$ **(P1)**

   Creatine + P_i → creatine phosphate
   $\Delta G°' = +10.5$ kcal/mole **(P2)**

   Assume that $2.3RT = 1.36$ kcal/mole.
   (a) Calculate the value of $\Delta G°'$ for reaction (P1).
   (b) Calculate the $\Delta G°'$ for hydrolysis of ATP.
10. As we will see in a later chapter, mitochondria establish a proton gradient across the inner mitochondrial membrane. Translocation of protons from the matrix (or inner surface of the inner membrane) to the outer surface of the inner membrane establishes the gradient. Translocation of the protons is an endergonic process. Explain. (Information in chapter 15 is necessary to answer this problem.)
11. What is the minimum number of moles of protons that must be translocated across a membrane to drive phosphorylation of ADP → ATP if $\Delta\psi$ is -150 mV and ΔpH is 0.5? Assume that phosphorylation of ADP requires 11 kcal/mole.
12. The hydrolysis of lactose (D-galactosyl-β (1,4) D-glucose) to D-galactose and D-glucose occurs with a $\Delta G°'$ of -4.0 kcal/mole.
   (a) Calculate K'_{eq} for the hydrolytic reaction.
   (b) What are the $\Delta G°'$ and K'_{eq} for the synthesis of lactose from D-galactose and D-glucose?
   (c) Lactose is synthesized in the cell from UDP-galactose plus D-glucose and is catalyzed by lactose synthase. Given that $\Delta G°'$ of hydrolysis of UDP-galactose is -7.3 kcal/mole, calculate for $\Delta G°'$ and K'_{eq} for the reaction

   $$\text{UDP-galactose} + \text{D-glucose} \rightarrow \text{Lactose} + \text{UDP}$$

13. Dihydrolipoamide dehydrogenase catalyzes the reaction

   $$\text{Dihydrolipoamide} + \text{NAD}^+ \rightleftharpoons \text{lipoamide} + \text{NADH} + \text{H}^+$$

   $\Delta G°'$ is $+1.38$ kcal/mole. Calculate the steady-state ratio of lipoamide/dihydrolipoamide if the ratio of NADH/NAD^+ is (a) 1:10 and (b) 10:1.
14. For each of the following reactions, calculate $\Delta G°'$ and indicate whether the reaction is thermodynamically favorable as written.
   (a) Glycerate-1,3-bisphosphate + creatine → phosphocreatine + 3-phosphoglycerate
   (b) Glucose-6-phosphate → glucose-1-phosphate
   (c) Phosphoenolpyruvate + ADP → pyruvate + ATP
   (d) Glycerol phosphate + ADP → glycerol + ATP
15. Although ATP is an important phosphate donor, in biosynthetic reactions the AMP portion of the molecule is often transferred to an acceptor with the release of pyrophosphate. Such a transfer occurs as an intermediate step in the reaction.

   $\text{R—COO}^- + \text{ATP} + \text{CoASH} \rightarrow \text{R—CO—SCoA} + \text{AMP}$
   + pyrophosphate **(P3)**

   Inorganic pyrophosphatases hydrolyze the pyrophosphate, yielding two molecules of inorganic phosphate. Assume that the hydrolysis of R—CO—SCoA to R—COO^- + CoASH proceeds with $\Delta G°'$ of -10 kcal/mole and that hydrolysis of a pyrophosphate anhydride bond yields -7.5 kcal/mole. Calculate the K'_{eq} for reaction (P3) in the presence and in the absence of inorganic pyrophosphatase. What role do pyrophosphatases play in biosynthetic reactions dependent on adenylate transfer?

Structure and Function of Major Components of the Cell

2

PART

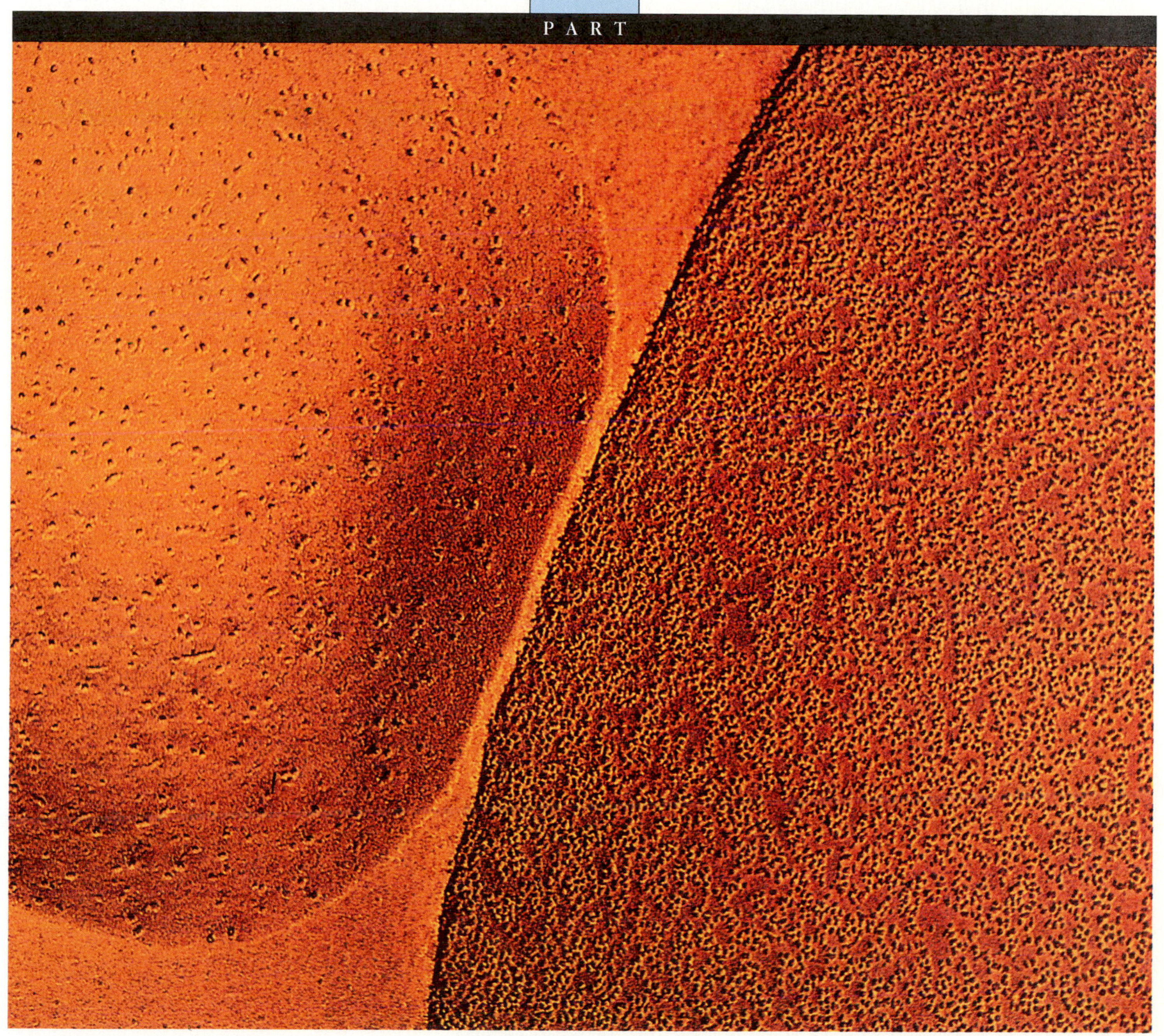

Scanning electron micrograph of human erythrocyte plasma membrane prepared by freeze fracture (120,000×). The micrograph has been color-enhanced to match the normal yellow tinge of blood plasma. When embedded in ice, membranes usually fracture between the monolayer leaflets of the lipid bilayer. The outer fracture face shows a large number of protrusions (presumably proteins); there is a relative lack of such particles on the inner fracture face. (© Harold H. Edwards/Visuals Unlimited.)

The major components of the living cell, other than water, are organic polymers: nucleic acids, proteins, carbohydrates, and lipids. Part 2 describes the structures and functions of three of these molecular species and their building blocks. Nucleic acid structure and function is treated later, with other closely related topics (see chapter 25). However, some instructors may prefer that you turn to these aspects of nucleic acids after the chapter on carbohydrates, treating that material as part of this section.

Proteins are composed of one or more polypeptides, and each polypeptide is composed of many amino acids regularly linked by peptide bonds into long chains. The polypeptide chains aggregate into long fibrous molecules that provide structural support, or into compact globular molecules important in metabolism (chapters 3–5).

Carbohydrates form both linear and branched-chain structures that are sometimes complexed with proteins (chapter 6). In their pure form, carbohydrate polymers with straight- and branched-chain structures store readily usable chemical energy. Linear and cross-linked polysaccharides also form cell walls in bacteria and plants. Short straight- or branched-chain oligosaccharides, linked to asparagine or serine side chains in proteins, usually enhance the recognition properties of proteins. This feature facilitates transport of proteins to an appropriate location in the cell or extracellular milieu and increases their specificity once they get there.

Lipids (chapter 7) are even more varied in their roles than carbohydrates. Some lipids are stored in fat deposits, where they can be called upon to satisfy the energy requirements of the cell when more accessible sources of biochemical energy are exhausted. Other lipids are used in membrane construction. Finally, the members of a select class of complex lipids function as hormones.

3

CHAPTER

The Building Blocks of Proteins: Amino Acids, Peptides, and Polypeptides

In the middle of the nineteenth century, the Dutch chemist Geradus Mulder extracted a substance common to animal tissues and the juices of plants, which he believed to be "without doubt the most important of all substances of the organic kingdom, and without it life on our planet would probably not exist." At the suggestion of the famous Swedish chemist Berzelius, Mulder named this substance protein (from the Greek *proteios,* meaning "of first importance"), and assigned to it a specific chemical formula ($C_{40}H_{62}N_{10}O_{12}$). Although he was wrong about the chemistry of proteins, he was right about their being indispensable to living organisms. The term "protein" endures.

Proteins are the most abundant of cellular components. They include enzymes, antibodies, hormones, transport molecules, and even components for the cytoskeleton of the cell itself. Proteins are also informational macromolecules, the ultimate heirs of the genetic information encoded in the sequence of nucleotide bases within the chromosomes. Structurally and functionally, they are the most diverse and dynamic of molecules and play key roles in nearly every biological process. Proteins are complex macromolecules with exquisite specificity; each is a specialized player in the orchestrated activity of the cell. Together they tear down and build up molecules, extract energy, repel invaders, act as delivery systems, and even synthesize the genetic apparatus itself.

In the first three chapters of part 2 we will discuss the basic structural and chemical properties of proteins. In this chapter we will concentrate on the structural and chemical properties of amino acids, peptides, and polypeptides—the building blocks of proteins. From our presentation you will learn the following:

1. Certain acidic and basic properties are common to all amino acids found in proteins except for the amino acid proline.
2. Side chains give amino acids their individuality. These side chains serve a variety of structural and functional roles.
3. The alpha carboxyl group of one amino acid can react with the alpha amino group of another amino acid to form a dipeptide.
4. Many amino acids, reacting in a similar way, can become linked to form a linear polypeptide chain.
5. The amino acid sequence in a polypeptide can be determined by a process of partial breakdown into manageable fragments, followed by stepwise analysis proceeding from one end of the chain to the other.
6. Polypeptide chains with a prespecified sequence can be synthesized by well-established chemical methods.

Amino Acids

Every protein molecule can be viewed as a polymer of amino acids. There are twenty common amino acids. Figure 3.1*a* shows the structure of a single amino acid. At the center is a tetrahedral carbon atom called the alpha (α) carbon (C_α). It is covalently bonded on one side to an amino group (NH_2) and on the other side to a carboxyl group (COOH). A third bond is always hydrogen, and the fourth bond is to a variable side chain (R). In neutral solution (pH 7), the carboxyl group loses a proton and the amino group gains one. Thus an amino acid in solution, while neutral overall, is a double charged species called a zwitterion (fig. 3.1*b*).

Figure 3.1

Amino acid anatomy. (*a*) Uncharged amino acid. (*b*) Doubly charged zwitterion.

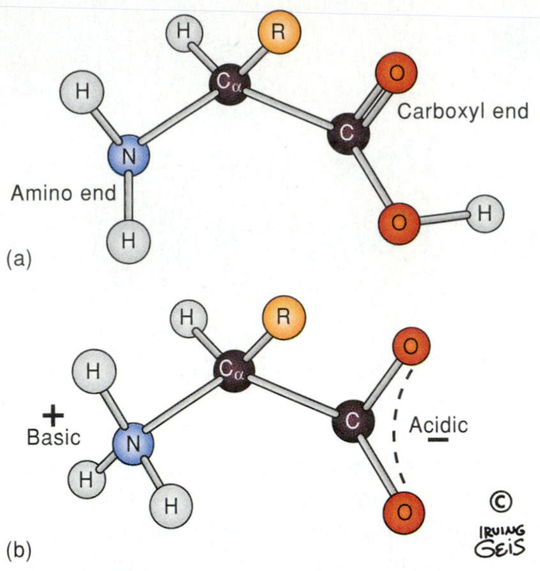

The structures of the twenty amino acids commonly found in proteins are listed in table 3.1. All of these amino acids except proline have an α ammonium ion ($—N^+H_3$) attached to the α carbon. In proline one of the N—H linkages is replaced by an N—C linkage forming part of a cyclic structure. Various ways of classifying amino acids according to their R groups have been proposed. In table 3.1 we have divided the amino acids into three categories. The first category contains eight amino acids with relatively apolar R groups; the second category contains seven amino acids with uncharged polar R groups; and the third category contains five amino acids with R groups that normally exist in the charged state.

Amino acids are often abbreviated by three-letter symbols; when this proves to be too cumbersome (as in certain kinds of charts and figures), one-letter symbols are used. Both designations are given in table 3.1, together with the molecular weight (M_r) of each amino acid.

In addition to the twenty commonly occurring α-amino acids, a variety of other amino acids are found in minor amounts in proteins and in nonprotein compounds. The unusual amino acids found in proteins result from modification of the common amino acids. In a few cases these amino acids are incorporated directly into the polypeptide chains during synthesis (see box 29A for selenocysteine). Most frequently the amino acid is modified after incorporation (see box 18B for the modification of proline to hydroxyproline). The unusual amino acids found in nonprotein compounds are extremely varied in type and are formed by a number of different metabolic pathways (see chapter 18).

Amino Acids Have Both Acid and Base Properties

The charge properties of amino acids are very important in determining the reactivity of certain amino acid side chains and in the properties they confer on proteins. The charge properties of amino acids in aqueous solution may best be considered under the general treatment of acid–base ionization theory. We will find this treatment useful at other points in the text as well.

Recall that water can be considered a weak acid (or a weak base) because it dissociates into a proton and a hydroxide ion, according to the equilibrium

$$H_2O \rightleftarrows H^+ + OH^- \qquad (1)$$

The equilibrium expression for this reaction is

$$K_{eq} = \frac{[H^+][OH^-]}{[H_2O]} \qquad (2)$$

Because water dissociates to such a small extent, the concentration of undissociated water is high and does not vary significantly for chemical reactions in aqueous solution. Therefore, the denominator in this equation is effectively constant, with a value of 55.5. The constant K_w for the dissociation of water is redefined by the expression

$$K_w = [H^+][OH^-] = 10^{-14}\,(\text{mole/liter})^2 \qquad (3)$$

at 25° C.

In pure water we expect equal amounts of H^+ ("hydrogen ion") and OH^- ("hydroxide ion"). From equation (3) we can calculate the concentration of H^+ or OH^- in pure water to be 10^{-7} M. Therefore, a solution with a H^+ concentration of 10^{-7} M is defined as neutral. A H^+ concentration greater than 10^{-7} M indicates an acidic solution; a H^+ concentration less than 10^{-7} M indicates a basic solution. Rather than deal with exponentials, it is convenient to express the H^+ concentration on a pH scale, the term pH being defined by the equation

$$\text{pH} = \log(1/[H^+]) = -\log[H^+] \qquad (4)$$

According to this definition a neutral solution has a pH of 7. Other values of pH and corresponding H^+ and OH^- concentrations are given in table 3.2.

The most common equilibria that biochemists encounter are those of acids and bases. The dissociation of an acid may be written as

$$HA \rightleftarrows H^+ + A^- \qquad (5)$$

The equilibrium constant for this reaction is called the acid dissociation constant, K_a, written as

$$K_a = \frac{[H^+][A^-]}{[HA]} \qquad (6)$$

Table 3.1

Structure of the Twenty Amino Acids Found in Proteins

Group I. Amino Acids with Apolar R Groups	R Groups	Group II. Amino Acids with Uncharged Polar R Groups	R Groups	Group III. Amino Acids with Charged R Groups	R Groups
Alanine Ala A M_r 89[a]	$CH_3-\overset{H}{\underset{NH_3^+}{C}}-COO^-$	Glycine Gly G M_r 75	$H-\overset{H}{\underset{NH_3^+}{C}}-COO^-$	Aspartic acid Asp D M_r 133	$^-O(O=)C-CH_2-\overset{H}{\underset{NH_3^+}{C}}-COO^-$
Valine Val V M_r 117	$(CH_3)_2CH-\overset{H}{\underset{NH_3^+}{C}}-COO^-$	Serine Ser S M_r 105	$HO-CH_2-\overset{H}{\underset{NH_3^+}{C}}-COO^-$	Glutamic acid Glu E M_r 147	$^-O(O=)C-CH_2-CH_2-\overset{H}{\underset{NH_3^+}{C}}-COO^-$
Leucine Leu L M_r 131	$(CH_3)_2CH-CH_2-\overset{H}{\underset{NH_3^+}{C}}-COO^-$	Threonine Thr T M_r 119	$CH_3-\underset{OH}{CH}-\overset{H}{\underset{NH_3^+}{C}}-COO^-$	Lysine Lys K M_r 146	$H_3\overset{+}{N}-(CH_2)_4-\overset{H}{\underset{NH_3^+}{C}}-COO^-$
Isoleucine Ile I M_r 131	$CH_3-CH_2-\underset{CH_3}{CH}-\overset{H}{\underset{NH_3^+}{C}}-COO^-$	Cysteine Cys C M_r 121	$HS-CH_2-\overset{H}{\underset{NH_3^+}{C}}-COO^-$	Arginine Arg R M_r 174	$H_2N-\underset{NH_2^+}{\overset{\Vert}{C}}-NH-(CH_2)_3-\overset{H}{\underset{NH_3^+}{C}}-COO^-$
Proline Pro P M_r 115	Ring: H_2C, $C H_2$, H_2C, $\overset{+}{N}H_2$, $C(H)-COO^-$	Tyrosine Tyr Y M_r 181	$HO-C_6H_4-CH_2-\overset{H}{\underset{NH_3^+}{C}}-COO^-$	Histidine (at pH 6.0) His H M_r 155	$HC=C-CH_2-\overset{H}{\underset{NH_3^+}{C}}-COO^-$ (ring: HN, NH, $\overset{+}{C}$H)
Phenylalanine Phe F M_r 165	$C_6H_5-CH_2-\overset{H}{\underset{NH_3^+}{C}}-COO^-$	Asparagine Asn N M_r 132	$NH_2(O=)C-CH_2-\overset{H}{\underset{NH_3^+}{C}}-COO^-$		
Tryptophan Trp W M_r 204	Indole ring (C, CH, N–H) $-CH_2-\overset{H}{\underset{NH_3^+}{C}}-COO^-$	Glutamine Gln Q M_r 146	$NH_2(O=)C-CH_2-CH_2-\overset{H}{\underset{NH_3^+}{C}}-COO^-$		
Methionine Met M M_r 149	$CH_3-S-CH_2-CH_2-\overset{H}{\underset{NH_3^+}{C}}-COO^-$				

[a]Molecular weights in this text are expressed in units of grams per mole.

Table 3.2
The pH Scale

pH	$[H^+]$	$[OH^-]$
0	10^0	10^{-14}
1	10^{-1}	10^{-13}
2	10^{-2}	10^{-12}
3	10^{-3}	10^{-11}
4	10^{-4}	10^{-10}
5	10^{-5}	10^{-9}
6	10^{-6}	10^{-8}
7	10^{-7}	10^{-7}
8	10^{-8}	10^{-6}
9	10^{-9}	10^{-5}
10	10^{-10}	10^{-4}
11	10^{-11}	10^{-3}
12	10^{-12}	10^{-2}
13	10^{-13}	10^{-1}
14	10^{-14}	10^0

Strong acids in aqueous solution dissociate completely into anions and protons. The concentration of hydrogen ion $[H^+]$ is therefore equal to the total concentration C_{HA} of the acid HA that is added to the solution. Thus the pH of the solution of a strong acid is simply $-\log C_{HA}$.

The pH of the solution of a weak acid is a function of both the C_{HA} and the acid dissociation constant. The dissociation constant of a weak acid may be written in terms of the species present in the equation for the acid dissociation constant.

First solving equation (6) for $[H^+]$ gives

$$[H^+] = \frac{K_a[HA]}{[A^-]} \quad (7)$$

Taking the logarithm of both sides and changing signs gives us

$$-\log[H^+] = -\log K_a + \log\frac{[A^-]}{[HA]} \quad (8)$$

Substituting pH for $-\log[H^+]$ and pK_a for $-\log K_a$ in equation (8), we obtain the Henderson-Hasselbach equation:

$$\text{pH} = pK_a + \log\left[\frac{A^-}{HA}\right] = pK_a + \log\left[\frac{\text{base}}{\text{acid}}\right] \quad (9)$$

The Henderson-Hasselbach equation is useful for calculating the molar ratio of base (proton acceptor) to acid (proton donor) for a given pH and p*K* or for calculating the p*K* given the ratio of base (proton acceptor) to acid (proton donor). It can be seen that when the concentration of anion or base is equal to the concentration of undissociated acid (i.e., when the acid is half neutralized), the pH of the solution is equal to the p*K* of the acid.

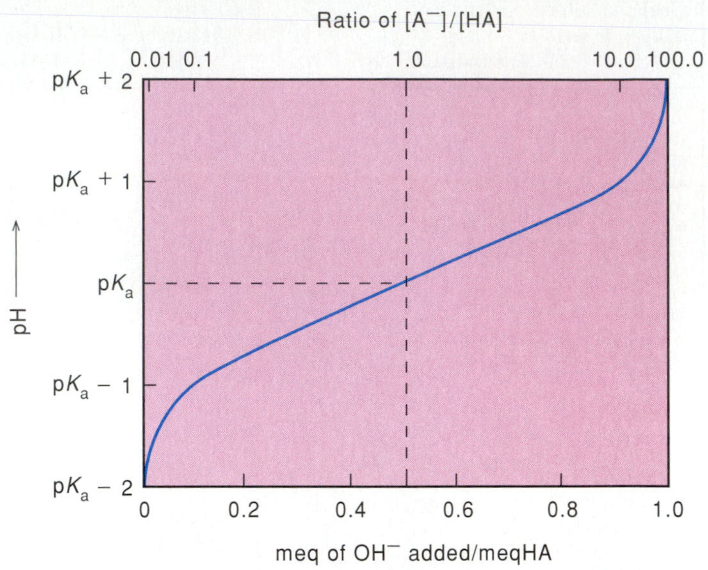

Figure 3.2

The dependence of pH on the equivalents of base added to a typical weak acid. Note that at the pK_a, $[A^-] = [HA]$.

The values of p*K* for a particular molecule are determined by titration. A typical pH dependence curve for the titration of a weak acid by a strong base is shown in figure 3.2. The concentration of the anion equals the concentration of the acid when the acid is exactly half neutralized. Note that at this point on the curve, the pH is least sensitive to the quantity of added base (or acid). Under these conditions, the solution is said to be buffered. Biochemical reactions are typically highly dependent on the pH of the solution. Therefore, it is frequently advantageous to study reactions in buffered solutions. The ideal buffer is one that has a p*K* numerically equivalent to the working pH.

A simple amino acid with a nonionizable R group gives a complex titration curve with two inflection points. For an example, see the titration of alanine, shown in figure 3.3. At very low pH, alanine carries a single positive charge on the α-amino group. The first inflection point occurs at a pH of 2.3. This is the p*K* for titration of the carboxyl group, pK_1 ($\text{—COOH} \rightarrow \text{—COO}^-$). At a pH of 6.0, alanine has an equal amount of positive and negative charge. This value is referred to as the isoelectric point (pI) or the isoelectric pH. As the titration continues, a second inflection point is reached at a pH of 9.7. The p*K* at this point, pK_2, represents the equilibrium for the titration of the proton from the amino group ($\text{—N}^+\text{H}_3 \rightarrow \text{NH}_2$).

Amino acids with an ionizable R group show even more complex titration curves, indicative of three ionizable groups (fig. 3.4). The p*K* for the ionizable side chain, pK_R, is usually readily distinguishable from the p*K* values for the ionizable α-carboxyl and α-amino groups, pK_1 and pK_2, respectively, as the latter have numerical values close to the comparable p*K* values of alanine (see fig. 3.3 and table 3.3). Note that the only ionizable R group with a pK_R in the vicinity of 7, where most biological systems

Figure 3.3

Titration curve of alanine. The predominant ionic species at each cardinal point in the titration is indicated.

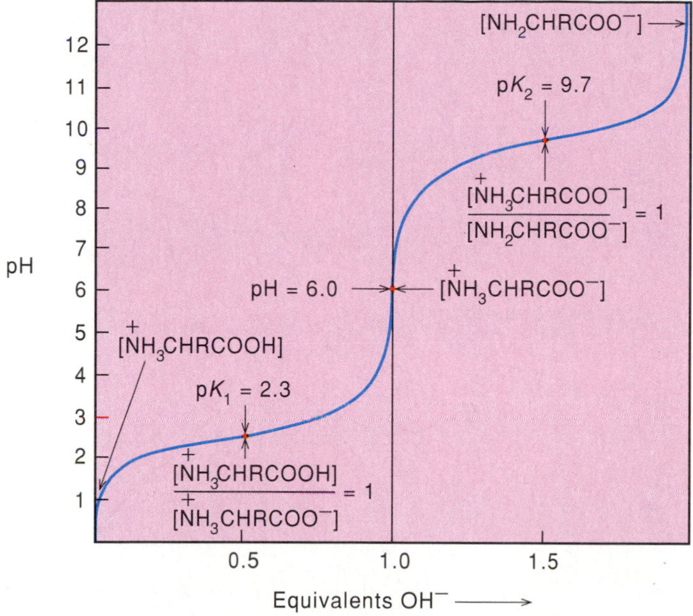

Figure 3.4

Titration curves of glutamic acid, lysine, and histidine. In each case, the pK of the R group is designated pK_R.

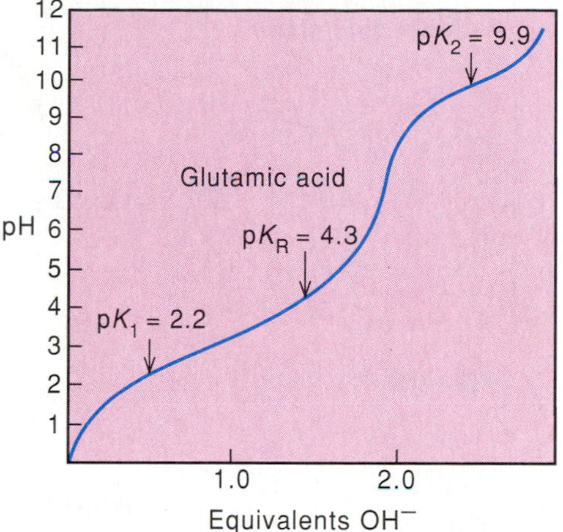

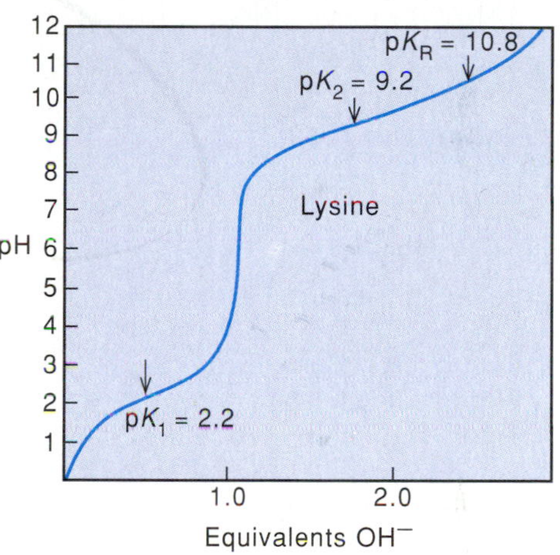

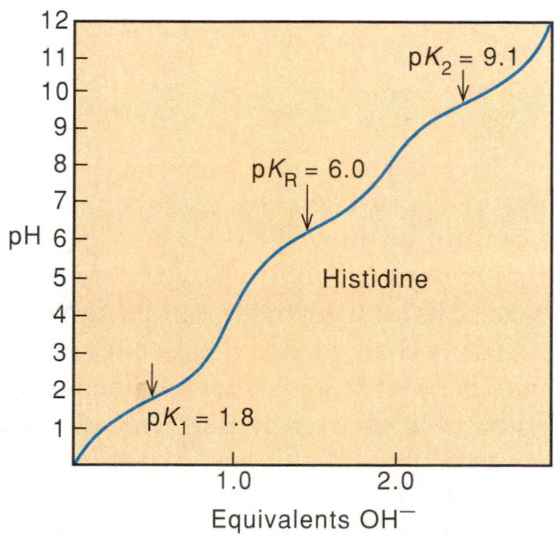

Table 3.3

Values of pK for the Ionizable Groups of the Twenty Amino Acids Commonly Found in Proteins

Amino Acid	pK_1 (α —COOH)	pK_2 (α —NH_3^+)	pK_R (R Group)
Alanine	2.35	9.87	—
Arginine	1.82	8.99	12.48
Asparagine	2.1	8.84	—
Aspartic acid	1.99	9.90	3.90
Cysteine	1.92	10.78	8.33
Glutamic acid	2.10	9.47	4.07
Glutamine	2.17	9.13	—
Glycine	2.35	9.78	—
Histidine	1.80	9.33	6.04
Isoleucine	2.32	9.76	—
Leucine	2.33	9.74	—
Lysine	2.16	9.18	10.79
Methionine	2.13	9.28	—
Phenylalanine	2.16	9.18	—
Proline	1.95	10.65	—
Serine	2.19	9.21	~13
Threonine	2.09	9.10	~13
Tryptophan	2.43	9.44	—
Tyrosine	2.20	9.11	10.13
Valine	2.29	9.74	—

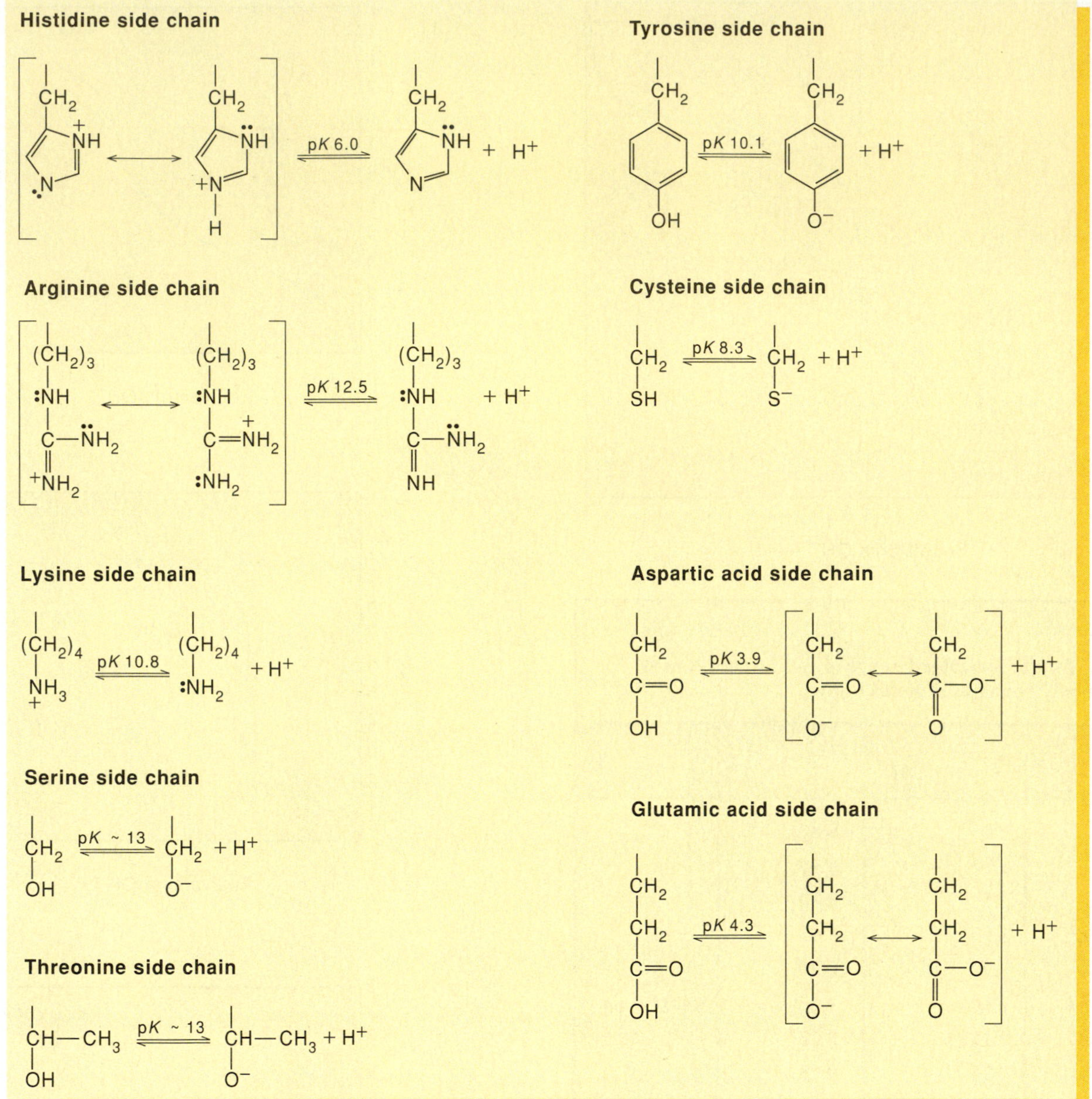

Figure 3.5

Equilibrium between charged and uncharged forms of amino acid side chains.

function, is that for histidine. This means that although other ionizable groups are usually fully charged under biological conditions, the side chain of histidine can be fully charged, uncharged, or partially charged, depending on the precise situation. This variability has major implications for the way the histidine side chain functions in enzyme catalysis. The side chain can serve as either a proton donor or a proton acceptor (see discussion in chapter 9).

An additional point should be noted from table 3.3. Whereas the amino acid side chains (R groups) that are normally charged at physiological pH are restricted to five amino acids (aspartic acid, glutamic acid, lysine, arginine, and sometimes histidine), a number of potentially ionizable R groups are part of other amino acids. These include cysteine, serine, threonine, and tyrosine. The ionization reactions for all of the potentially ionizable side chains are indicated in figure 3.5.

The acidic and basic groups within a protein can be titrated just like free amino acids to determine their number and their pK_a values. A titration curve for β-lactoglobulin is shown in figure 3.6. This protein contains 94 potentially ionizable groups. The protein is positively charged at low pH and negatively charged at high pH. At intermediate pH values a

Figure 3.6

Titration curve of ß-lactoglobulin. At very low values of pH (<2) all ionizable groups are protonated. At a pH of about 7.2 (indicated by horizontal bar) 51 groups (mostly the glutamic and aspartic amino acids and some of the histidines) have lost their protons. At pH 12 most of the remaining ionizable groups (mostly lysine and arginine amino acids and some histidines) have lost their protons as well. (Source: R. H. Haschenmeyer and A. E. V. Haschenmeyer, *A Guide to Study by Physical and Chemical Methods*. Copyright ©1973, John Wiley & Sons, Inc., New York, N.Y.)

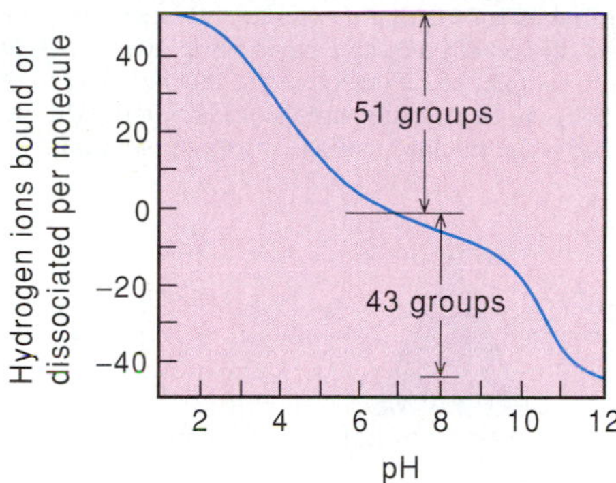

Figure 3.7

Ultraviolet absorption spectra of tryptophan (Trp), tyrosine (Tyr), and phenylalanine (Phe) at pH 6. The molar absorptivity is reflected in the extinction coefficient, with the concentration of the absorbing species expressed in moles per liter. (Source: D. B. Wetlaufer, *Advances in Protein Chemistry*. 17:303-390, 1962. Copyright ©1962 Academic Press Inc., San Diego, Calif.)

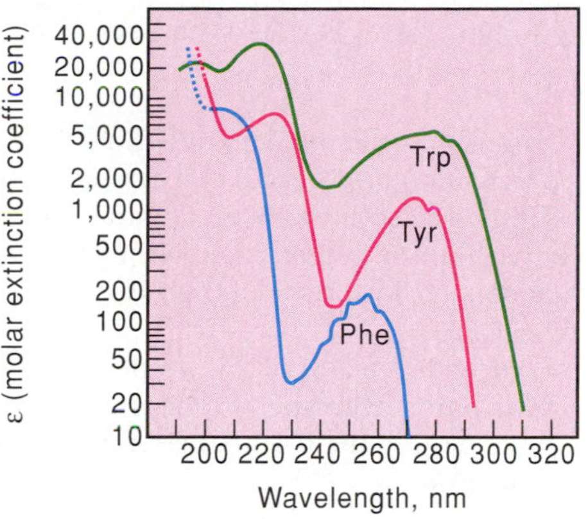

point is found where the sum of the positive side-chain charges exactly equals the sum of the negative charges, so that the net charge on the protein is zero. This value, as we have noted, is the isoelectric point (pI) of the protein; for β-lactoglobulin the pI is about 5.2. The isoelectric point is not an invariant quantity. The binding of charged species present in the solution could raise or lower the pI, depending on their charge.

Aromatic Amino Acids Absorb Light in the Near-Ultraviolet

The aromatic amino acids phenylalanine, tyrosine, and tryptophan all possess absorption maxima in the near-ultraviolet (fig. 3.7). These absorption bands arise from the interaction of radiation with electrons in the aromatic rings. The near-ultraviolet absorption properties of proteins are determined solely by their content of these three aromatic amino acids. In solution, UV absorption can be quantified with the help of a conventional spectrophotometer and used as a measure of the concentration of proteins (see box 3A).

All Amino Acids Except Glycine Show Asymmetry

One of the most striking and significant properties of amino acids is their chirality or handedness. The word "chiral" is related to the Greek word meaning hand. Just as the right hand is related to the left hand by a mirror image, so, in general, a naturally occurring amino acid is related to a stereoisomer by its mirror image. This observation is true of nineteen out of the twenty amino acids; the one exception is glycine.

The chirality of amino acids stems from the chiral or asymmetric center, the α-carbon atom. The α-carbon atom is a chiral center if it is connected to four different substituents. Thus

Figure 3.8

The covalent structure of alanine, showing the three-dimensional structure of the Ⓛ and Ⓓ stereoisomeric forms.

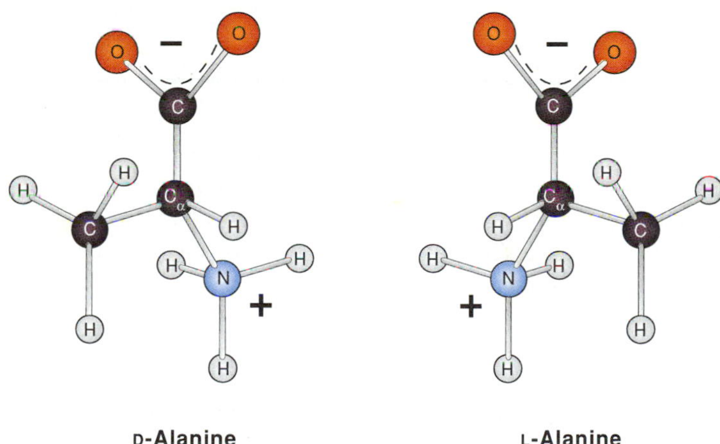

glycine has no chiral center. Two of the amino acids, isoleucine and threonine, possess additional chiral centers because each has one additional asymmetric carbon. You should be able to locate these carbons by simple inspection.

Two structures that constitute a stereoisomeric pair are referred to as enantiomers. The two enantiomers for alanine are illustrated in figure 3.8. These two isomers are called L-alanine and D-alanine, according to the way in which the substituents are arranged about the asymmetric carbon atom. The naming by L and D (for "dextrorotatory" and "levorotatory"; see chapter 6) refers to a convention established by Emil Fischer many years ago. According to this convention all amino acids found in proteins are of the L form. Some D-amino acids are found in bacterial cell walls and certain antibiotics.

3A BOX

Measurement of Ultraviolet Absorption in Solution

The general quantitative relationship that governs all absorption processes is called the Beer-Lambert law:

$$I = I_0\, 10^{-\epsilon cd}$$

where I_0 is the intensity of the incident radiation, I is the intensity of the radiation transmitted through a cell of thickness d (in centimeters) that contains a solution of concentration c (expressed either in moles per liter or in grams per 100 ml), and ϵ is the extinction coefficient, a characteristic of the substance being investigated (see fig. 3.7).

Light absorption is measured by a spectrophotometer as shown in the illustration. The spectrophotometer usually is capable of directly recording the absorbance A, which is related to I and I_0 by the equation

$$A = \log_{10} (I_0/I)$$

Hence $A = \epsilon cd$, and A is a direct measure of concentration. We can see from figure 3.7 that the ϵ values are largest for tryptophan and smallest for phenylalanine.

Since protein absorption maxima in the near-ultraviolet (240–300 nm) are determined by the content of the aromatic amino acids and their respective values, most proteins have absorption maxima in the 280-nm region. By contrast, absorption in the far-ultraviolet (around 190 nm) is shown by all polypeptides regardless of their aromatic amino acid content. The reason is that absorption in this region is due primarily to the peptide linkage.

Figure 1

Schematic diagram of a spectrophotometer for measuring light absorption. Laboratory instruments for making measurements are much more complex than this, but they all contain the same basic components: a light source, a monochromator, a sample, and a detector. λ is the wavelength of the light, I_0 and I are the incident light intensity and the transmitted light intensity, respectively, and d is the thickness of the absorbing solution.

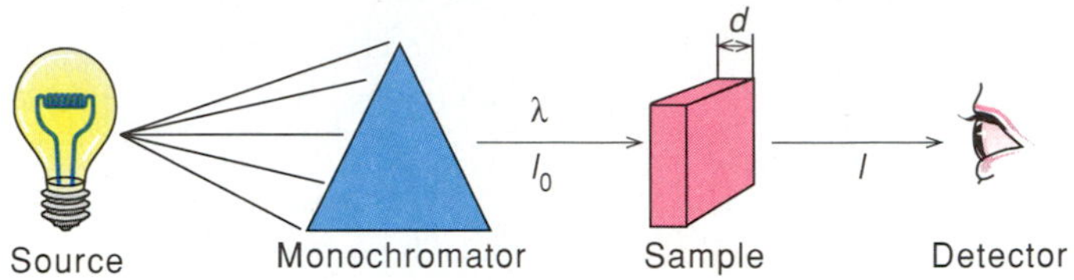

Another convention for referring to configurations is called the R, S convention. As the R, S convention is not as popular for amino acids or sugars as it is for other types of biomolecules, such as lipids, we will not discuss this notation until chapter 11 (see box 11A).

Peptides and Polypeptides

Amino acids can link together by a covalent peptide bond between the α-carboxyl end of one amino acid and the α-amino end of another. Formally, this bond is formed by the loss of a water molecule, as shown in figure 3.9. The peptide bond has partial double-bond character owing to resonance effects; as a result, the C—N peptide linkage and all of the atoms directly connected to C and N lie in a planar configuration called the amide plane. In the following chapter we will see that this amide plane, by limiting the number of orientations available to the polypeptide chain, plays a major role in determining the three-dimensional structures of proteins.

Any number of amino acids can be joined by successive peptide linkages, forming a polypeptide chain. The polypeptide chain, like the dipeptide, has a directional sense. One end, called the N-terminal or amino-terminal end, has a free α-amino group, whereas the other end, the C-terminal or carboxyl-terminal end, has a free α-carboxyl group. The sequence of main-chain atoms from the N-terminal end to the C-terminal end is C_α—C—N—C_α, etc., and in the opposite direction it is C_α—N—C—C_α, etc. Short polypeptide chains, up to a length of about 20 amino acids, are called peptides or oligopeptides if they are fragments of whole polypeptide chains. A small protein molecule may contain a polypeptide chain of only 50 amino acids; a large protein may contain chains of 3,000 amino acids or more. One of the larger single polypeptide chains is that of the muscle protein myosin, which consists of approximately 1,750 amino acid residues. Figure 3.10 shows a section of a polypeptide chain as a linear array with α carbons and planar amides alternating as repeating units of the main chain. Different side chains are attached to each α carbon.

In addition to the covalent peptide bonds formed between adjacent amino acids within a polypeptide chain, covalent disulfide bonds can be formed within the same polypeptide chain or between different polypeptide chains (fig. 3.11). Such disulfide linkages have an important stabilizing influence on the structures formed by many proteins (see chapter 4).

Figure 3.9

Formation of a dipeptide from two amino acids. (*a*) Two amino acids. (*b*) A peptide bond (CO—NH) links amino acids by joining the α-carboxyl group of one with the α-amino group of another. A water molecule is lost in the reaction. It is conventional to draw dipeptides and polypeptides so that their free amino terminal is to the left and their free carboxyl terminal is to the right. The amide plane refers to six atoms that lie in the same plane.

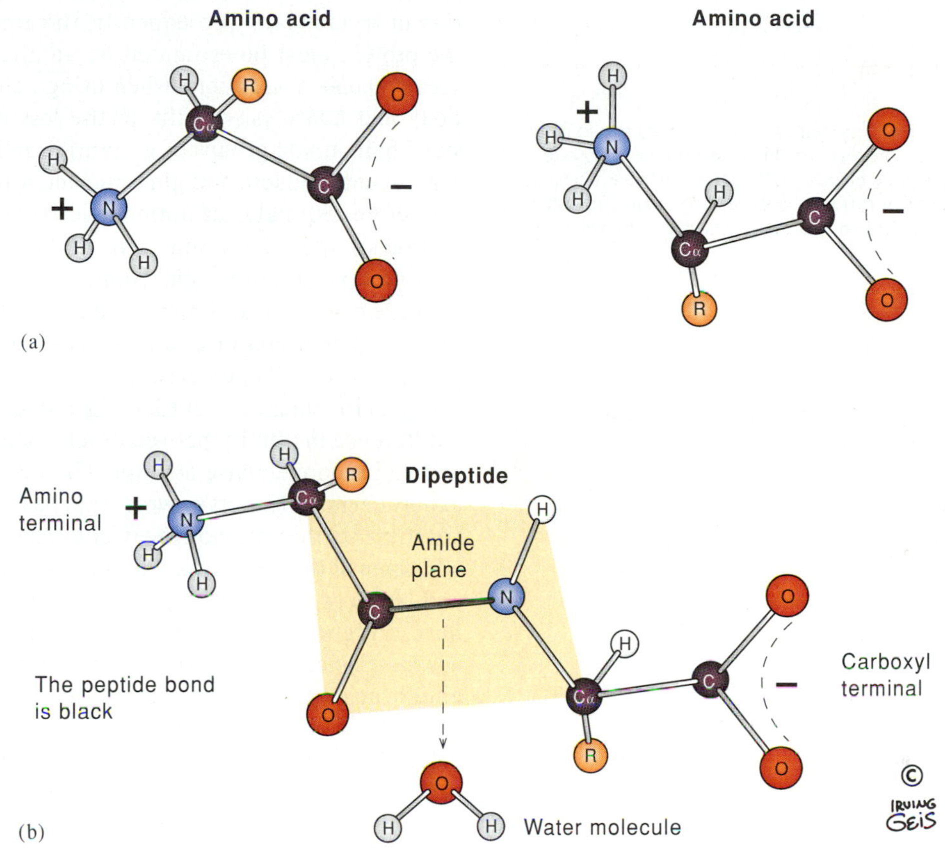

Figure 3.10

A polypeptide chain, with the backbone shown in color and the amino acid side chains in outline.

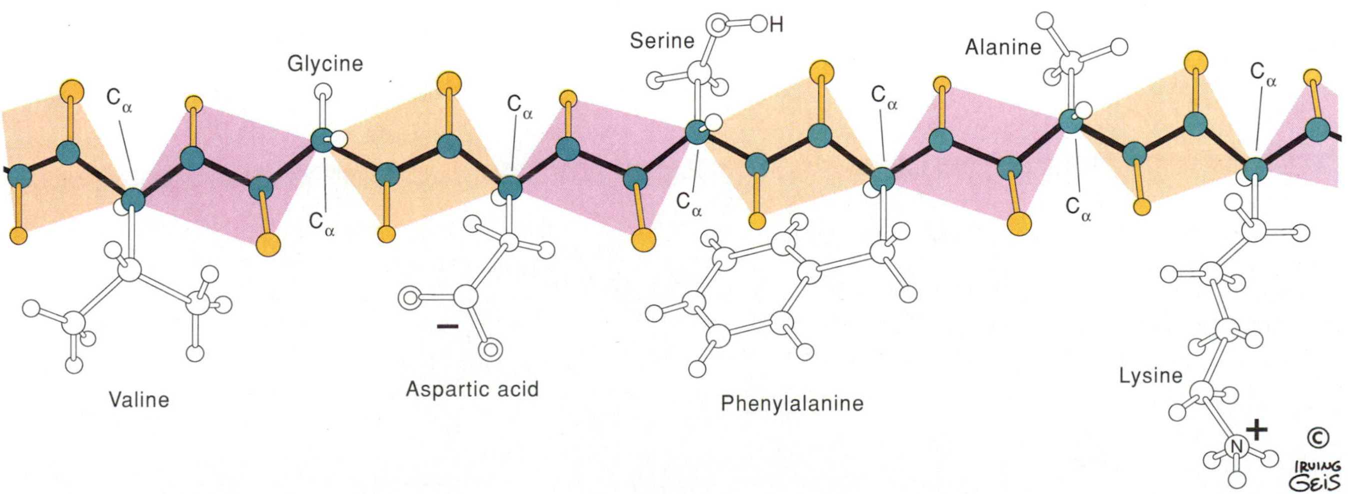

Determination of Amino Acid Composition of Proteins

Each protein is uniquely characterized by its amino acid composition and sequence. A protein's amino acid composition is defined simply as the number of each type of amino acid composing the polypeptide chain. In order to discover a protein's amino acid composition it is necessary to (1) break down the polypeptide chain into its constituent amino acids, (2) separate the resulting free amino acids according to type, and (3) measure the quantities of each amino acid.

Figure 3.11

Disulfide bonds can form between two cysteines. The cysteines can exist in the cytosol as free amino acids (as shown), in which case they give rise to cystine, or they can be on polypeptide chains. In the latter instance, they can be on the same polypeptide chains or different polypeptide chains. In either case the formation of covalent disulfide bonds stabilizes structural relationships.

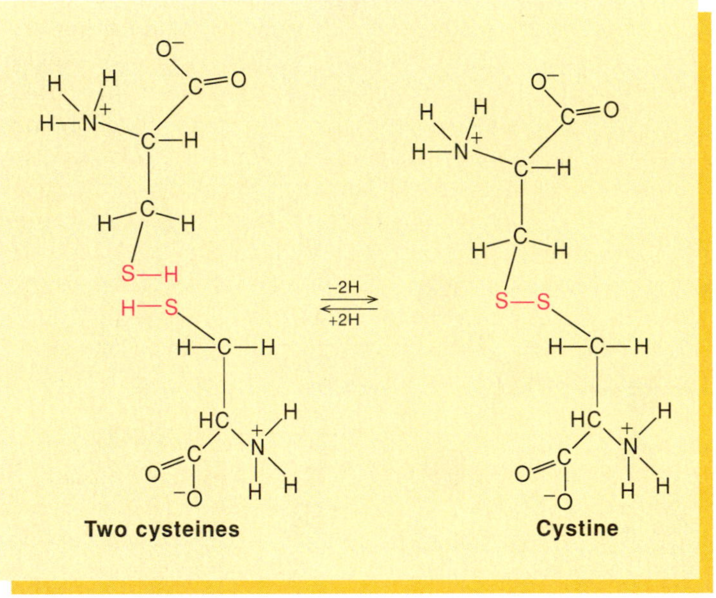

Cleavage of the peptide bonds is usually achieved by boiling the protein in 6-*N* HCl; this treatment causes hydrolysis of the peptide bonds and the consequent release of free amino acids (fig. 3.12). Although acid hydrolysis is the most frequently used means of breaking a protein into its constituent amino acids, it results in the partial destruction of the indole ring of tryptophan. Consequently, the amount of tryptophan in the protein must be estimated by an alternative method (e.g., spectroscopic absorption) when using acid hydrolysis. In addition, acid hydrolysis results in the loss of ammonia from the side-chain amide groups of glutamine and asparagine, with the consequent production of glutamic and aspartic acids (fig. 3.13). Therefore, estimates of amino acid composition based on acid hydrolysis show glutamine and glutamic acid combined and measured as glutamic acid. Similarly, asparagine and aspartic acid are combined and measured as aspartic acid.

Separation of amino acids for quantitative analytical purposes is usually achieved by ion-exchange chromatography. The general efficacy of chromatographic techniques is based on a difference in affinity between each compound to be separated and an immobile phase or resin. The resin consists of some relatively chemically inert polymer that has weakly basic side-chain constituents that are positively charged at pH 7. If we were to add some of this resin to a solution containing free aspartic acid and lysine at pH 7, the negatively charged aspartic acid would have a higher affinity for the resin than would the positively charged lysine. If we were then to pump a solution of these two amino acids through a column containing such a positively charged resin, the progress of the aspartic acid through the column would be retarded relative to the lysine, owing to the greater affinity of the aspartic acid for the resin (fig. 3.14).

Ion-exchange resins have been developed that have differential binding affinities for all the naturally occurring amino acids. Such resins are effective in separating a solution of amino

Figure 3.12

Acid hydrolysis of a protein or polypeptide to yield amino acids.

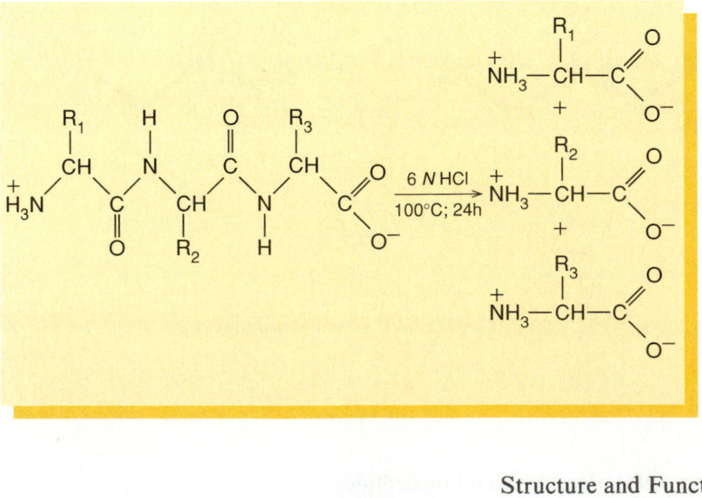

Figure 3.13

Acid hydrolysis of protein converts glutamine to glutamic acid. A similar reaction occurs for asparagine, which possesses an identical side-chain functional group.

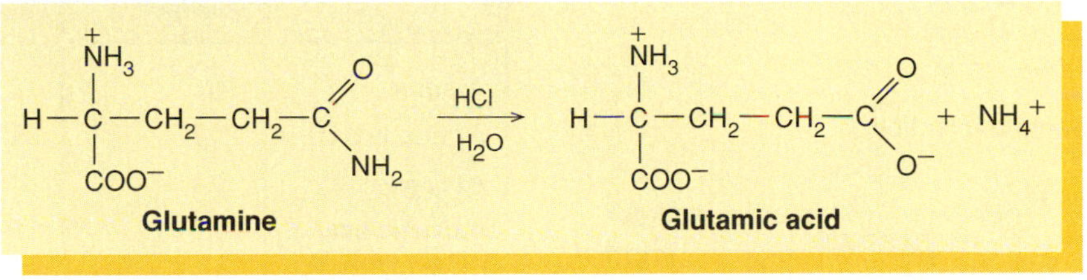

$$\underset{\textbf{Glutamine}}{H-\overset{\overset{+}{NH_3}}{\underset{COO^-}{C}}-CH_2-CH_2-C(=O)NH_2} \xrightarrow[H_2O]{HCl} \underset{\textbf{Glutamic acid}}{H-\overset{\overset{+}{NH_3}}{\underset{COO^-}{C}}-CH_2-CH_2-C(=O)O^-} + NH_4^+$$

Figure 3.14

Migration of aspartic acid ⊖ and lysine ⊕ through a column with a higher affinity for aspartic acid. Views show the column at successively increasing time intervals after starting the elution.

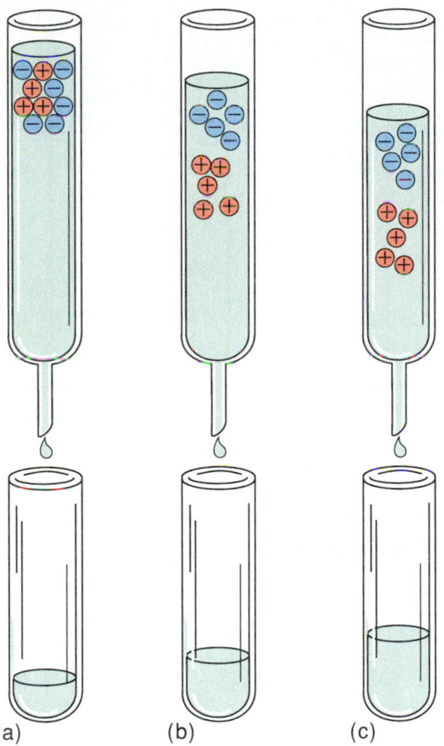

acids into its components. We must emphasize that the details of the forces responsible for the differential binding of amino acids to an ion-exchange resin are quite complicated and depend additionally on side-chain polarity, on subtle differences in the pK values of α-amino and α-carboxyl groups, on solvation effects, and on other factors. To enhance the separation properties of the column, such separation techniques frequently exploit changes in the pH of the solution buffer (eluting buffer) used to remove the compounds of interest. For example, a column might initially be run with the eluting buffer at a pH that results in some amino acids being so strongly bound to the resin that they are essentially immobile. However, after the separation and elution of the less strongly bound amino acids, the pH of the eluting buffer can be appropriately shifted to lessen the charge difference between the resin and the strongly bound amino acids. These amino acids can then be eluted and separated according to the newly established pattern of resin-binding affinities.

Quantitative determination of the separated amino acids is achieved by their reaction with ninhydrin to produce a colored reaction product. This product is measured spectrophotometrically. As shown in figure 3.15, the ninhydrin reaction abstracts an amino group from each amino acid, so that the amount of colored product formed is proportional to the amount of amino acid initially present.

Quantitative measurements of amino acid composition are usually carried out on an amino acid analyzer, a device that automates the previously described operations. As illustrated in

Figure 3.15

Reaction of ninhydrin with an amino acid yields a colored complex. The ninhydrin reaction permits qualitative location of amino acids in chromatography and quantitative assay of separated amino acids.

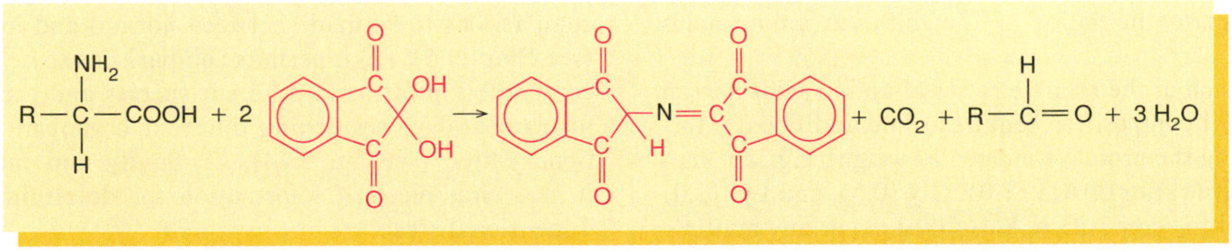

$$R-\overset{NH_2}{\underset{H}{C}}-COOH + 2\,(\text{ninhydrin}) \longrightarrow (\text{indanedione})\!-\!\overset{}{\underset{H}{C}}\!-N=C(\text{indanedione}) + CO_2 + R-\overset{H}{C}=O + 3\,H_2O$$

Figure 3.16

Schematic diagram of an amino acid analyzer. The amino acids are passed through an ion-exchange column and thereby separated. Eluted fractions are mixed and reacted with ninhydrin. The intensity of the resulting colored product is measured in a spectrophotometer and the results are displayed on a recording chart.

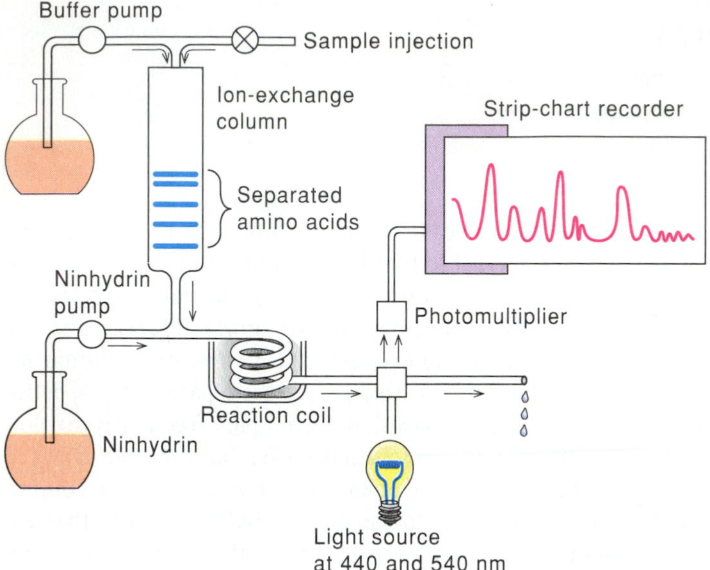

figure 3.16, the amino acid analyzer consists of an ion-exchange column through which the appropriate eluting buffer is pumped after the amino acids are introduced at the top of the column. As the separated amino acids emerge, they are mixed with ninhydrin solution and passed through a heated coil of tubing to allow the formation of the colored ninhydrin reaction product. The separated ninhydrin reaction products then pass through a cell that measures their optical absorbance at 540 and 440 nm and plots the results on a strip-chart recorder. The absorbance is measured at two wavelengths because proline, which is substituted at its amino group, forms a different ninhydrin reaction product, with an absorption maximum that is correspondingly different from that of the remaining amino acids.

Usually the amino acid analyzer is first standardized by running through it a sample containing known quantities of amino acids, in order to account for any differences in their ninhydrin reaction properties. In this way it is possible to directly relate the amount of amino acid present to the amount of colored product formed, as measured by the area under the "peak" produced on the strip-chart recorder (see fig. 3.16). Similarly, the amino acid hydrolysate of a protein of unknown composition can be run through the analyzer, and the relative peak areas can be used to estimate the ratios of the different amino acids present.

Conversion of the relative ratios of amino acids into an estimate of actual composition requires some additional information concerning the protein's molecular weight; e.g., an analysis giving relative ratios of Ala (1.0), Gly (0.5), and Lys (2.0) could correspond to composition Ala_2-Gly-Lys_4 or any multiple thereof. The required information is usually available, and in any case, an estimation of composition based on a minimum molecular weight of the protein is always possible. Results for three proteins are shown in table 3.4.

Table 3.4

Amino Acid Content of Proteins (in percent)

Constituent	Insulin (Bovine)	Ribonuclease (Bovine)	Cytochrome (Equine)
Alanine	4.6	7.7	3.5
Amide NH_3	1.7	2.1	1.1
Arginine	3.1	4.9	2.7
Aspartic acid	6.7	15.0	7.6
Cysteine	0	0	1.7
Cystine	12.2	7.0	0
Glutamic acid	17.9	12.4	13.0
Glycine	5.2	1.6	5.6
Histidine	5.4	4.2	3.4
Isoleucine	2.3	2.7	5.4
Leucine	13.5	2.0	5.6
Lysine	2.6	10.5	19.7
Methionine	0	4.0	2.1
Phenylalanine	8.6	3.5	4.5
Proline	2.1	3.9	3.3
Serine	5.3	11.4	0
Threonine	2.0	8.9	8.4
Tryptophan	0	0	1.5
Tyrosine	12.6	7.6	4.9
Valine	9.7	7.5	2.4

Determination of Amino Acid Sequence of Proteins

The most important properties of a protein are determined by the sequence of amino acids in the polypeptide chain. This sequence is called the primary structure of the protein. We know the sequences for thousands of peptides and proteins, largely through the use of methods developed in Fred Sanger's laboratory and first used to determine the sequence of the peptide hormone insulin in 1953. Knowledge of the amino acid sequence is extremely useful in a number of ways: (1) it permits comparisons to be made between normal and mutant proteins (see chapter 5); (2) it permits comparisons to be made between comparable proteins in different species and thereby has been instrumental in positioning different organisms on the evolutionary tree (see fig. 1.24); (3) finally and most important, it is a vital piece of information for determining the three-dimensional structure of the protein.

Figure 3.17

Steps involved in the sequence determination of the B chain of insulin. Amino acids are represented here by their single-letter codes (see table 3.1).

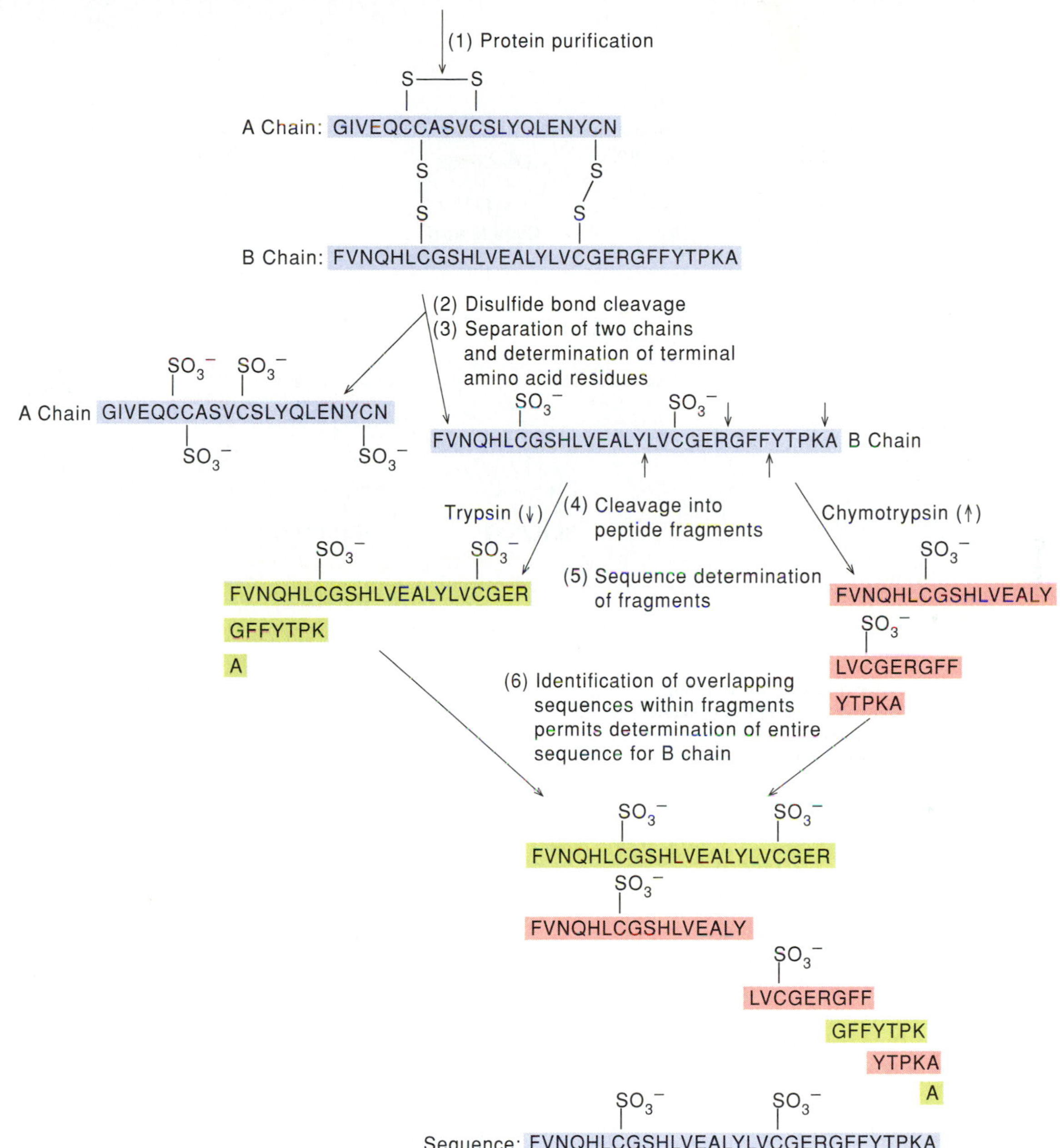

Determining the order of amino acids involves the sequential removal and identification of successive amino acid residues from one or the other free terminal of the polypeptide chain. However, in practice it is extremely difficult to get the required specific cleavage reaction of the desired products to proceed with 100% yield. This obstacle becomes significant when sequencing long polypeptides, because the fraction of the total material of minimum polypeptide chain length becomes constantly smaller as the successive removal of terminal residues continues. Conversely, the amino acid released from the polypeptide chain becomes increasingly contaminated with amino acids released from previously unreacted chains.

Because of this fundamental chemical limitation, the polypeptide chain must be broken down into sequences short enough for the chemistry to produce reliable results. The short sequences are then reassembled to obtain the overall sequence. The steps actually involved in protein sequencing (fig. 3.17) are (1) purification of the protein; (2) cleavage of all disulfide

Figure 3.18

Disulfide cleavage reactions. Prior to sequence analysis inter- and intrachain disulfide linkages are irreversibly cleaved by one of the two procedures shown.

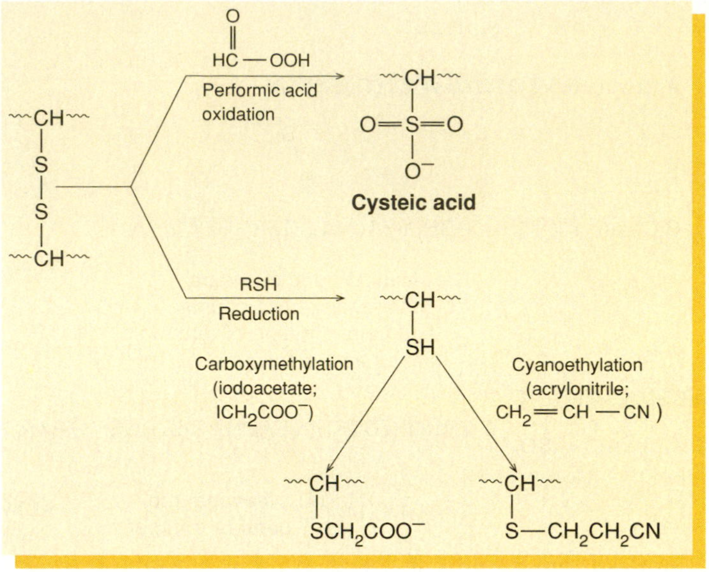

Figure 3.19

Polypeptide chain end-group analysis. (*a*) Amino-terminal group identification. A more sensitive method, the dansyl chloride method, is described in box 3B. (*b*) Carboxyl-terminal group indentification. Identification of this amino acid is considerably more difficult.

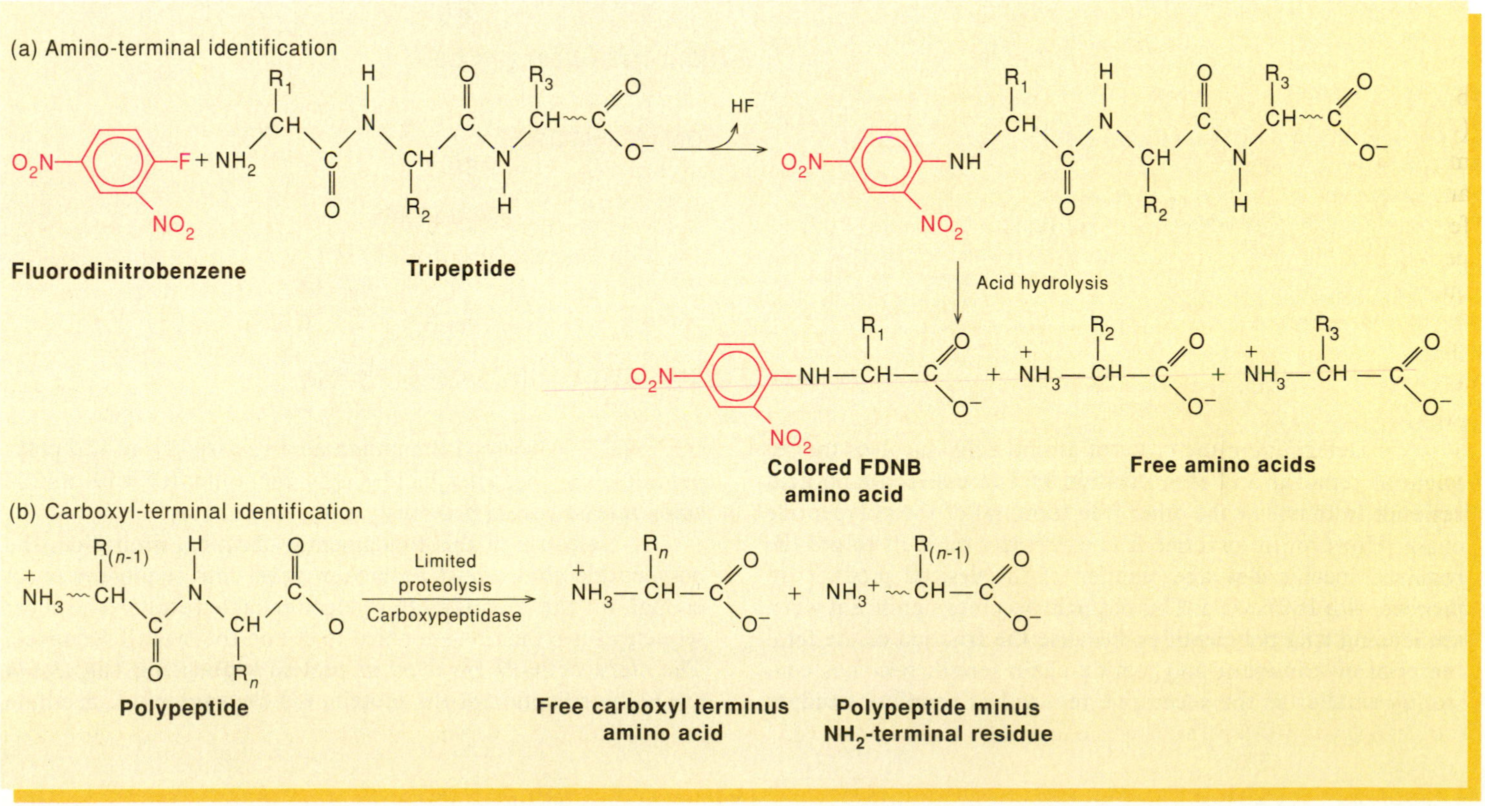

Figure 3.20

Site of action of some endopeptidases used for polypeptide chain cleavage prior to sequence analysis. Of the four different enzymes used, trypsin is used most frequently because of its high specificity.

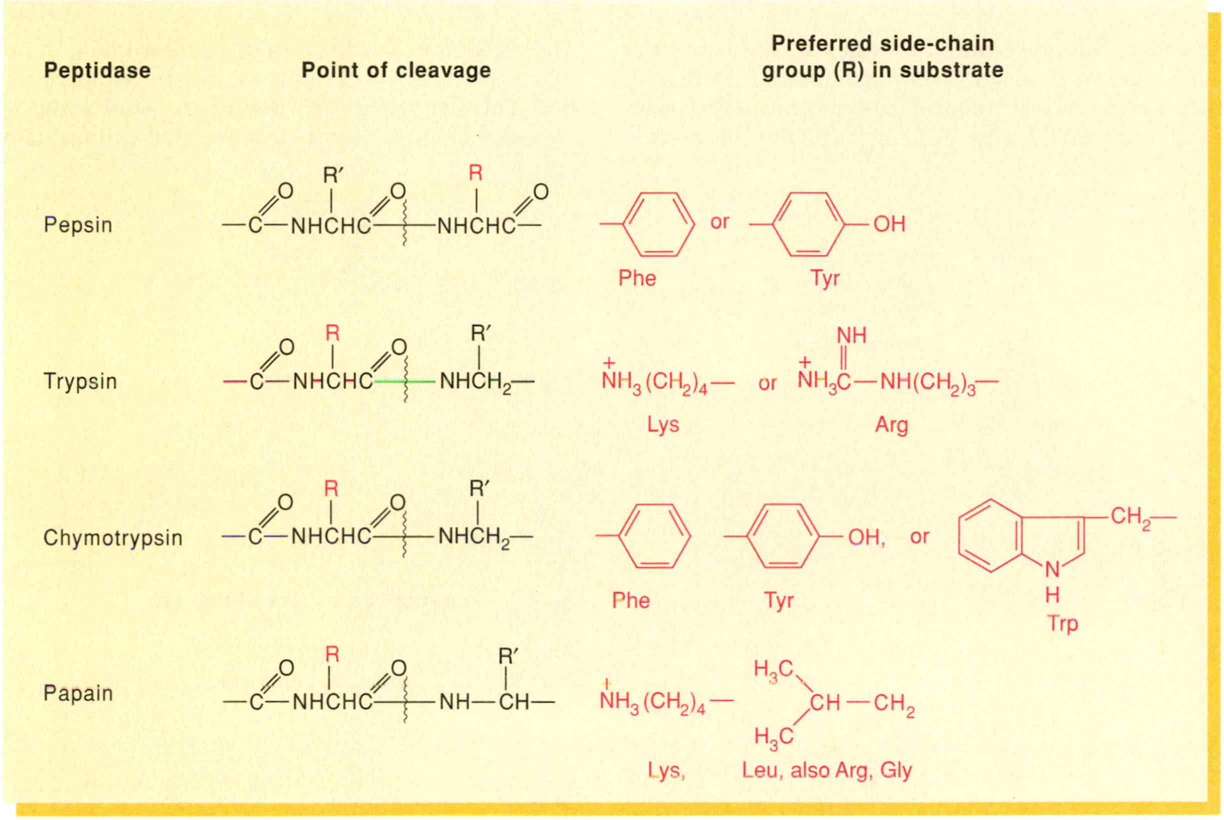

Peptidase	Point of cleavage	Preferred side-chain group (R) in substrate
Pepsin	—C(=O)—NHCH(R′)C(=O)—⸾—NHCH(R)C(=O)—	Phe or Tyr ($—C_6H_4—OH$)
Trypsin	—C(=O)—NHCH(R)C(=O)—⸾—NHCH(R′)—	$\overset{+}{N}H_3(CH_2)_4—$ Lys or $\overset{+}{N}H_3C(=NH)—NH(CH_2)_3—$ Arg
Chymotrypsin	—C(=O)—NHCH(R)C(=O)—⸾—NHCH(R′)—	Phe, Tyr ($—C_6H_4—OH$), or Trp (indolyl$—CH_2—$)
Papain	—C(=O)—NHCH(R)C(=O)—⸾—NH—CH(R′)—	$\overset{+}{N}H_3(CH_2)_4—$ Lys, $(H_3C)_2CH—CH_2$ Leu, also Arg, Gly

bonds; (3) determination of the terminal amino acid residues; (4) specific cleavage of the polypeptide chain into small fragments in at least two different ways; (5) independent separation and sequence determination of peptides produced by the different cleavage methods; and (6) reassembly of the individual peptides with appropriate overlaps to determine the overall sequence.

The first step, protein purification, will be discussed in chapter 5. Once the protein is pure, sequence analysis can begin, with cleavage of the disulfide bonds. Cleavage is achieved by oxidizing the disulfide linkages with performic acid (fig. 3.18). Sometimes this step results in the production of two or more polypeptide chains, in which case the individual chains must be separated.

The third step is to determine the polypeptide-chain end groups. If the polypeptide chains are pure, then only one N-terminal and one C-terminal group should be detected. The amino-terminal amino acid can be identified by reaction with fluorodinitrobenzene (FDNB) (fig. 3.19). Subsequent acid hydrolysis releases a colored DNP-labeled amino-terminal amino acid, which can be identified by its characteristic migration rate on thin-layer chromatography or paper electrophoresis. A more sensitive method of end-group determination involves the use of dansyl chloride (see box 3B).

Chemical methods for carboxyl end-group determination are considerably less satisfactory. Treatment of the peptide with anhydrous hydrazine at 100° C results in conversion of all the amino acid residues to amino acid hydrazides except for the carboxyl-terminal residue, which remains as the free amino acid and can be isolated and identified chromatographically. Alternatively, the polypeptide can be subjected to limited breakdown (proteolysis) with the enzyme carboxypeptidase. This results in release of the carboxyl-terminal amino acid as the major free amino acid reaction product. The amino acid type can then be identified chromatographically.

Step 4 involves breaking down the polypeptide chain into shorter, well-defined fragments for subsequent sequence analysis. Fragmentation can be achieved by the use of endopeptidases, which are enzymes that catalyze polypeptide-chain cleavage at specific sites in the protein. Figure 3.20 shows the specificity of four endopeptidases commonly used for this purpose. Another specific chemical method for polypeptide-chain

3B
BOX

The Dansyl Chloride Method for N-Terminal Amino Acid Determination

The dansyl chloride method provides an alternative to the Sanger method for N-terminal amino acid determination. Because it is considerably more sensitive than the Sanger method, it has become the method of choice. The reaction is diagrammed in the illustration. A polypeptide is treated with dansyl chloride to give an *N*-dansyl peptide derivative. This derivative is hydrolyzed to yield a highly fluorescent *N*-dansyl-amino acid, which is detected chromatographically.

Polypeptide

Dansyl chloride

***N*-dansyl derivative of peptide**

Hydrolysis

$$H_3\overset{\oplus}{N}-CH(R_2)-CO_2H + H_3\overset{\oplus}{N}-CH(R_3)-CO_2H - \cdots +$$

***N*-dansyl-amino acid**
[highly fluorescent]

cleavage involves reaction with cyanogen bromide. This reaction cleaves specifically at the methionine residues, with the accompanying conversion of free carboxyl-terminal methionine to homoserine lactone (fig. 3.21). Although this methionine reaction product differs from the twenty naturally occurring amino acids, it is nevertheless readily identified by subsequent conversion to homoserine.

Peptides resulting from cleavage of the intact protein are generally separated by ion-exchange chromatographic methods. The isolated peptides may then be analyzed (step 5) to determine both their amino acid composition and their sequence. Sequence determination involves the stepwise removal and identification of successive amino acids from the polypeptide amino terminal by means of the Edman degradation (fig. 3.22). This process is carried out by reacting the free amino-terminal group with phenylisothiocyanate to form a peptidyl

Figure 3.21

The cleavage of polypeptide chains at methionine residues by cyanogen bromide. The cleavage reaction is accompanied by the conversion of the newly formed free carboxyl-terminal methionine to homoserine lactone.

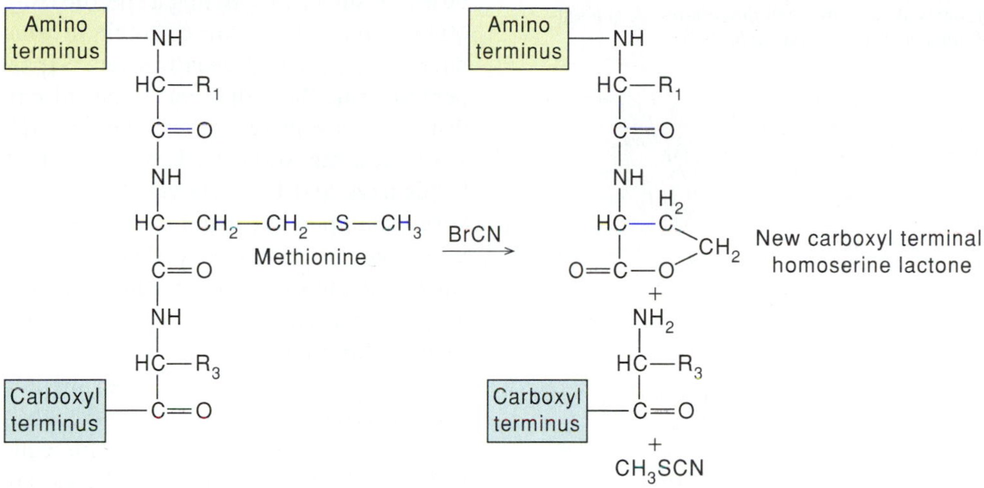

Figure 3.22

The Edman degradation method for polypeptide sequence determination. The sequence is determined one amino acid at a time, starting from the amino-terminal end of the polypeptide. First the polypeptide is reacted with phenylisothiocyanate to form a polypeptidyl phenylthiocarbamoyl derivative. Gentle hydrolysis releases the amino-terminal amino acid as a phenylthiohydantoin (PTH), which can be separated and detected spectrophotometrically. The remaining intact polypeptide, shortened by one amino acid, is then ready for further cycles of this procedure. A more sensitive reagent, dimethylaminoazobenzene isothiocyanate, can be used in place of phenylisothiocyanate. The chemistry is the same.

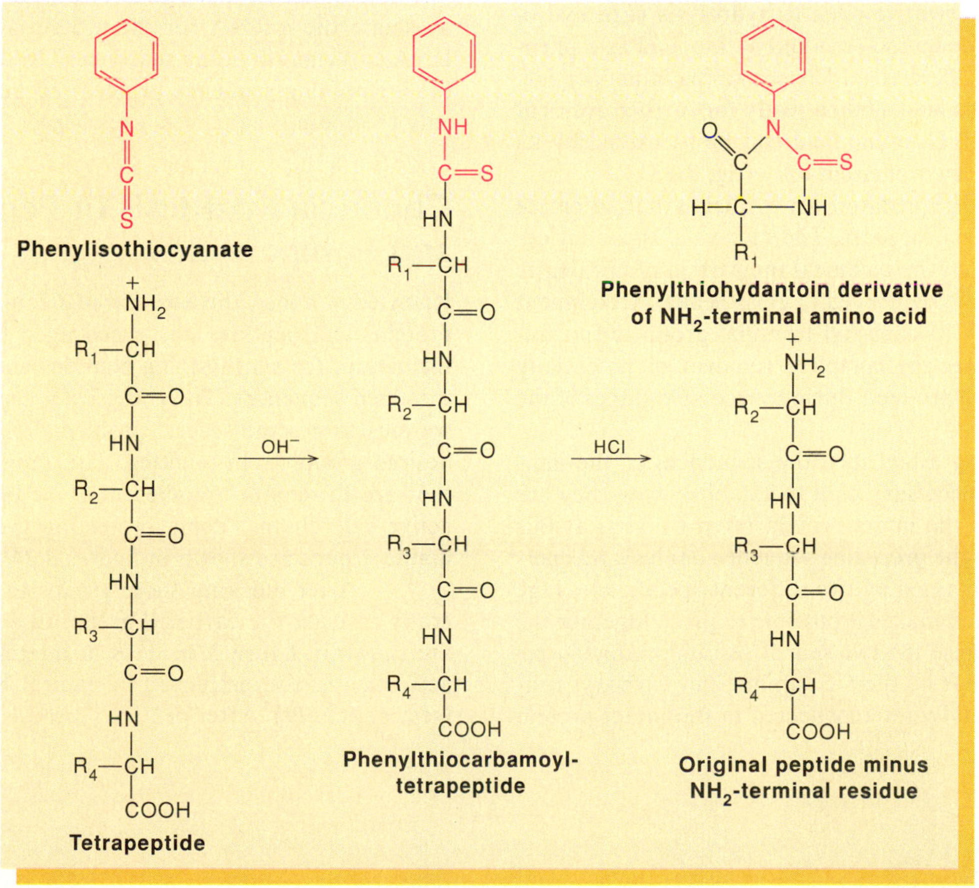

Figure 3.23

Thin-layer chromatography of amino acid–phenylthiohydantoin derivatives on silica gel plates. (*a*) Separation is done in a 98:2 mixture of chloroform and ethanol. (*b*) This is followed by further separation using an 88:2:10 mixture of chloroform, ethanol, and methanol. More sophisticated procedures, using column chromatography, give superior resolution and improved sensitivity. Automated sequencers always use such procedures. A general description of the use of columns is given in chapter 5.

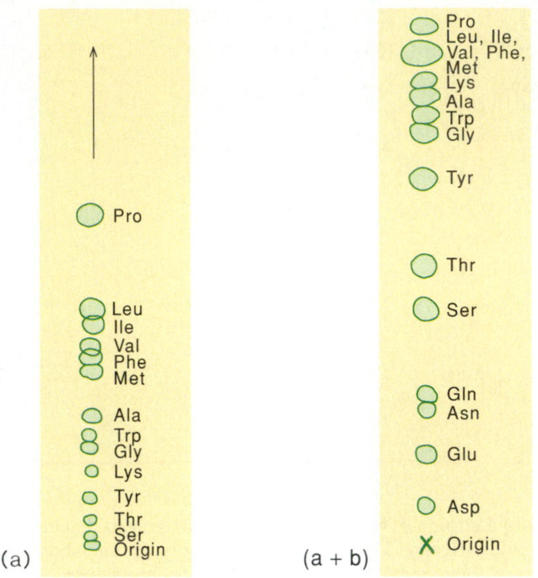

phenylthiocarbamyl derivative. Gentle hydrolysis with hydrochloric acid releases the amino-terminal amino acid as a phenylthiohydantoin (PTH) derivative. The remaining intact peptide, shortened by one amino acid, is then ready for further cycles of this procedure. The PTH-amino acid can be identified by its properties on thin-layer chromatography (fig. 3.23).

Devices called sequenators are available that automate the Edman degradation procedure. The success of these devices depends in large part on the technical innovation of covalently linking the peptide to be sequenced to glass beads. Attachment of the peptide through its carboxyl-terminal group to this immobile phase facilitates the complete removal of potentially contaminating reaction products during successive stages of the degradation.

Finally, having established the sequences of the individual peptides, it is necessary only to establish how they are connected together in the intact protein (step 6). It is at this stage that we see why the preceding sequence analysis was performed on peptides obtained by two different specific cleavage methods. This approach makes it possible to piece together the overall sequence, because the two sets of results produce overlapping sequences. That is, the free amino and carboxyl residues of peptides originally interconnected in the intact protein and liberated by one specific cleavage method will recur in the internal sequences of the peptides liberated by a second specific method.

Once the protein's primary sequence has been determined, the location of disulfide bonds in the intact protein can be established by repeating a specific enzymatic cleavage on another sample of the same protein in which the disulfide bonds have not previously been cleaved. Separation of the resulting peptides will show the appearance of one new peptide and the disappearance of two other peptides, when compared with the enzymatic digestion product of the material whose disulfide bonds have first been chemically cleaved. In fact, these difference techniques are generally useful in the detection of sites of mutations in protein molecules of previously known sequence, since a single substitution will generally affect the chromatographic properties of only a single peptide released during proteolytic digestion.

Great progress has been made in recent years in devising procedures for sequencing the DNA that encodes for proteins (see chapter 27). Knowing the sequence of coding triplets in DNA allows us to read off the amino acid sequence of the corresponding protein. Nevertheless, such studies have produced the remarkable observation that some eukaryotic DNA sequences coding for proteins are not continuous, but instead contain untranslated intervening DNA sequences. Although these results have profound implications for protein evolution, they obviously confound the general applicability of DNA-sequencing methods for the purposes of protein primary-structure determination. In cases like this, the usual solution has been to isolate the mRNA for the protein and use this to make a DNA carrying the same sequence. This procedure circumvents the intervening-sequence problem because the mRNA carries only the coding sequences (see chapter 27).

Chemical Synthesis of Peptides and Polypeptides

Knowledge about the structure-function interrelationships in proteins and peptides has encouraged biochemists to develop techniques for synthesizing peptides and proteins with predetermined sequences. To synthesize a peptide in the laboratory, we must overcome several problems related to preventing undesired groups from reacting. The amino and carboxyl groups that are to remain unlinked must be blocked; so must all reactive side chains. Some protecting groups for carboxyl and amino groups are shown in figures 3.24 and 3.25, respectively.

After blocking those groups to be protected, we generally activate the carboxyl group; two methods for doing this are shown in figure 3.26. It is of interest that carboxyl-group activation is also employed in natural biosynthesis in the cell (see chapter 29). After peptide synthesis, the protecting groups

Figure 3.24

Carboxyl protecting groups used in peptide synthesis. The symbol ▲ in the amino acid structure on the left, stands for one of the protecting groups (middle) leading to the named compound indicated on the right. The protecting group prevents the carboxyl group from participating in subsequent reactions involved in peptide synthesis.

Amino acid	Carboxyl-terminal protecting group ▲	Esters formed
$RCH(NH_2)C(=O)-$▲	CH_3CH_2O-	Ethyl
	$O_2N-C_6H_4-CH_2-O-$	Nitrobenzyl
	$(CH_3)_2(H_3C)C-O-$	*t*-Butyl

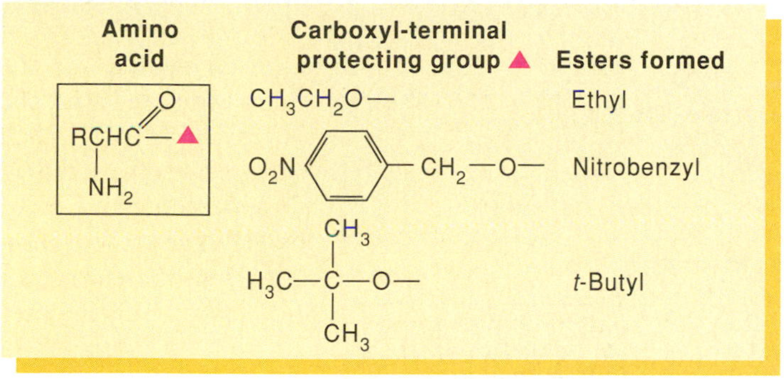

Figure 3.25

Amino protecting groups used in peptide synthesis.

Amino acid	Amino-terminal protecting group ■	Name
$RCHCOOH$ with $NH-$■	$C_6H_5-CH_2OC(=O)-$	Carbobenzoxy
	$(CH_3)_3COC(=O)-$	*t*-Butyloxycarbonyl
	$CH_3-C_6H_4-S(=O)_2-$	Tosyl
	$F_3CC(=O)-$	Trifluoroacetyl

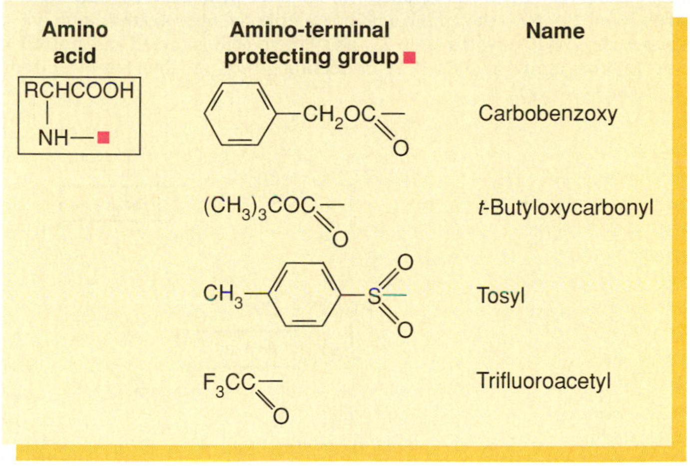

Figure 3.26

Different ways of activating the carboxyl group for peptide synthesis. Activated amino acids will react spontaneously with most α-amino acids as illustrated in figure 3.27.

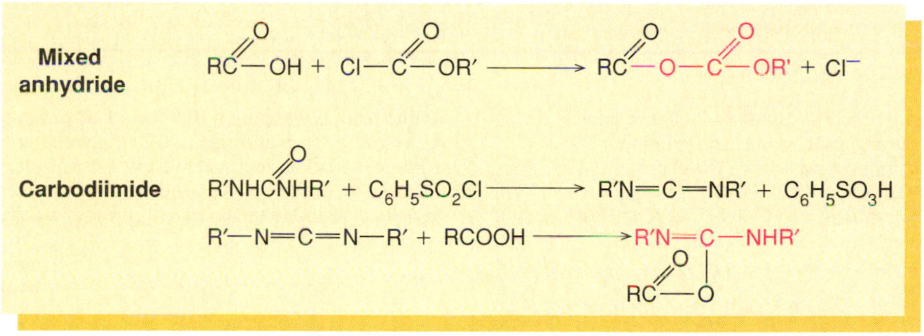

Mixed anhydride

$$RC(=O)-OH + Cl-C(=O)-OR' \longrightarrow RC(=O)-O-C(=O)-OR' + Cl^-$$

Carbodiimide

$$R'NHC(=O)NHR' + C_6H_5SO_2Cl \longrightarrow R'N{=}C{=}NR' + C_6H_5SO_3H$$

$$R'-N{=}C{=}N-R' + RCOOH \longrightarrow R'N{=}C(-O-C(=O)R)-NHR'$$

must be removed by a mild method. The overall process—comprising protection, activation, coupling, and unblocking—is shown in figure 3.27.

An important variation of the usual methods of peptide synthesis involves attaching a protected (*t*-butyloxycarbonyl group) amino acid to a solid polystyrene resin; removal of the amino protecting group; condensation with a second protected amino acid; and so on. In the last step, the finished peptide is cleaved from the resin. This method (outlined in figure 3.28) has the advantage that cumbersome purification between steps, often resulting in serious losses, is replaced by mere washing of the insoluble resin. Since each reaction is essentially quantitative, very long peptides, and even proteins, can be synthesized by this method. Indeed, Li synthesized a 39-amino-acid protein hormone, adrenocorticotropic hormone, by this method, and Merrifield synthesized bovine pancreatic ribonuclease, which contains 129 amino acids in a single polypeptide chain. A number of variants of ribonuclease that contain one or more changes in amino acid sequence also have been made by this method. The importance of the Merrifield process was underscored by the awarding of a Nobel Prize to Merrifield in 1984.

Figure 3.27

Schematic diagram illustrating the chemical method for peptide synthesis. First the amino acids to be linked are selected. The carboxyl group and the amino group that are to be excluded from peptide synthesis are protected (steps 1 and 1′). Next the amino acid containing the unprotected carboxyl group is carboxyl-activated (step 2). This amino acid is mixed and reacted with the other amino acid (step 3). Protecting groups are then removed from the product (step 4).

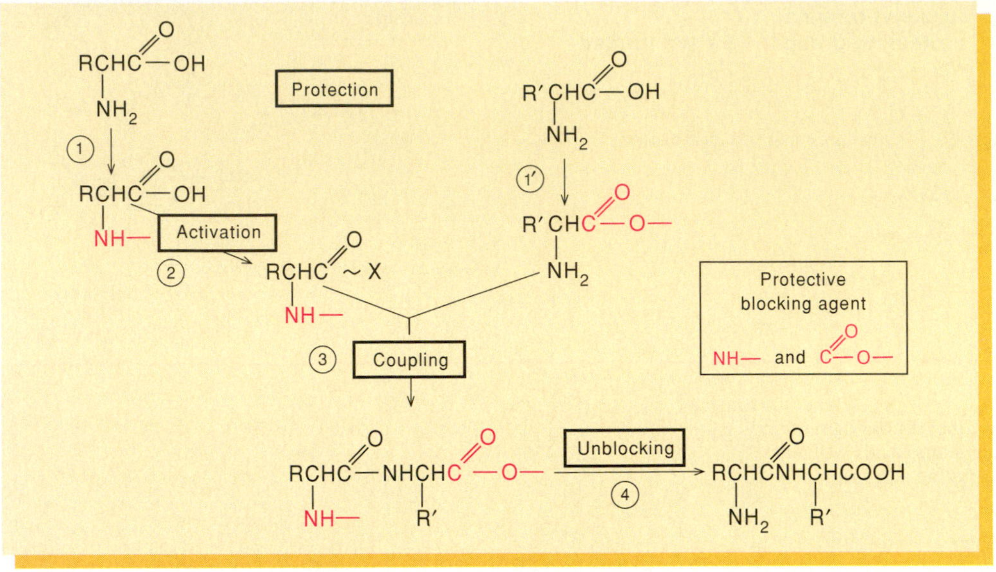

Figure 3.28

Merrifield procedure for solid-state dipeptide synthesis. (1) Polymer is activated. (2) Amino acid containing BOC protecting group is carboxyl-linked to polymer. This amino acid will be the carboxyl-terminal amino acid in the final peptide. (3) The BOC protecting group is removed from the polymer-linked amino acid. (4) A second amino acid, containing a BOC on its α-amino group and a dicyclohexylcarbodiimide (DCC) activated group, is reacted with the column-bound amino acid to form a dipeptide. (5) The dipeptide is released from the polymer and the BOC protecting group by adding hydrogen bromide (HBr) in trifluoroacetic acid.

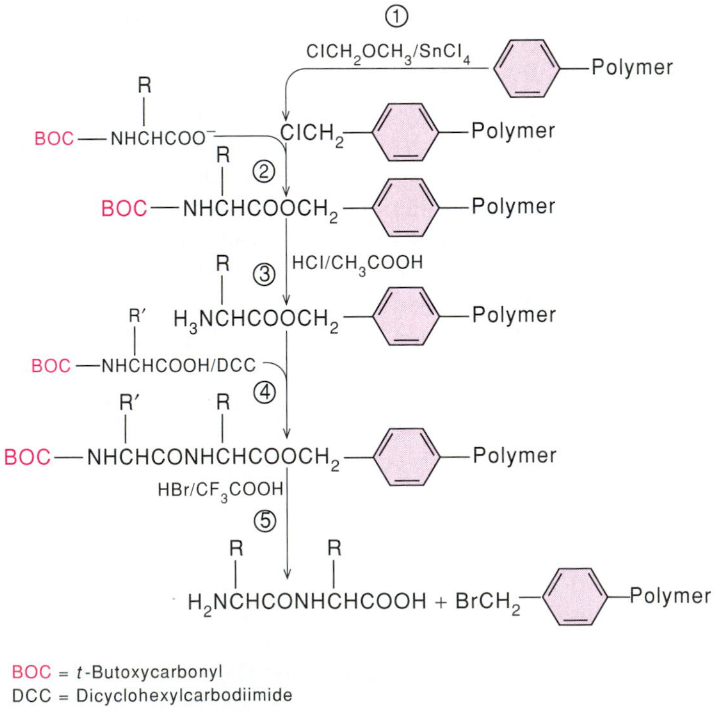

Summary

In this chapter we have dealt with some of the fundamental properties of amino acids and polypeptide chains. The following points are especially important.

1. Nineteen of the twenty amino acids commonly found in proteins have a carboxyl group and an amino group attached to an α-carbon atom; they differ in the side chain attached to the same α carbon.
2. All amino acids have acidic and basic properties. The ratio of base to acid form at any given pH can be calculated from the pK with the help of the Henderson-Hasselbach equation.
3. All amino acids except glycine are asymmetric and therefore can exist in at least two different stereoisomeric forms.
4. Peptides are formed from amino acids by the reaction of the α-amino group from one amino acid with the α-carboxyl group of another amino acid.
5. Polypeptide formation involves a repetition of the process involved in peptide synthesis.
6. The amino acid composition of proteins can be discovered by first breaking down the protein into its component amino acids and then separating the amino acids in the mixture for quantitative estimation.
7. The amino acid sequences of proteins can be discovered by breaking down the protein into polypeptide chains and then partially degrading the polypeptide chains. For each polypeptide chain fragment, the sequence is determined by stepwise removal of amino acids from the amino-terminal end of the polypeptide chain. Two different methods of forming polypeptide chain fragments are used so as to produce a map of overlapping fragments, from which the sequence of undegraded polypeptide chains in the proteins can be deduced.
8. Polypeptide chains with a predetermined amino acid sequence can be synthesized by chemical methods involving carboxyl-group activation.

Selected Readings

Barrett, G. C. (ed.). *Chemistry and Biochemistry of Amino Acids.* New York: Chapman and Hall, 1985. A recent and authoritative volume on this classical subject.

Gray, W. R., End group analysis using dansyl chloride. *Methods in Enzymology* 25:121–138, 1972. This volume of *Methods in Enzymology* contains several chapters on end-group analysis.

Hunkapiller, M. W., J. E. Strickler, and K. J. Wilson, Contemporary methodology for protein structure determination. *Science* 226:304–311, 1984.

Kent, S. B. H., Chemical synthesis of peptides and proteins. *Ann. Rev. Biochem.* 57:957–989, 1988. Comprehensive and up-to-date.

Merrifield, B., Solid phase synthesis. *Science* 232: 341–347, 1986.

Sanger, R., Sequences, sequences and sequences. *Ann. Rev. Biochem.* 57:1–28, 1988.

Problems

1. (a) A 10-mM solution of a weak monocarboxylic acid has a pH of 3.00. Calculate K_a and pK_a for this carboxylic acid.
   (b) You add 0.06 g NaOH (M_r = 40) to 1,000 ml of the acid solution in part (a). Calculate the final pH, assuming no volume change.
2. Given the pK_a values in the text, predict how the titration curves for glutamic acid and glutamine would differ.
3. You have 50 ml of 10-mM fully protonated histidine. How many millimoles of base must be added to bring the histidine solution to a pH that is equivalent to the pI?
4. Calculate the isoelectric point for histidine, aspartic acid, and arginine. Calculate the fractional charge for each ionizable group on aspartate at pH equal to pI. Do the results verify the isoelectric point of aspartic acid?
5. Which of the naturally occurring amino acid side chains are charged at pH 2? pH 7? pH 12? (Consider only those amino acids whose side chains have >10% charge at the pH indicated.)
6. Amino acids are sometimes used as buffers. Indicate the appropriate pH value(s) of a buffer containing aspartic acid, histidine, and serine.
7. Ten ml of a 10-mM solution of lysine was adjusted to pH 11.20. Draw the structures of the principal ionized forms present in solution. Use the pK_a values shown in table 3.3 and calculate the concentration of each principal form.
8. For the tripeptide shown below, the numbers in parentheses are the pK_a values of the ionizable groups.

```
(10.0)        O                O                 (1.8)
              ||               ||
NH3 — CH — C — NH — CH — C — NH — CH — COO⊖
⊕      |                 |                 |
      CH2             (CH2)4              CH2
       |                 |                 |
      COO⊖            NH3 (10.0)          SH
      (4.0)           ⊕                  (9.0)
```

   (a) Estimate the net charge at pH 1 and pH 14.
   (b) Estimate the isoelectric pH.
9. Polyhistidine is insoluble in water at pH 7.8 but is soluble at pH 5.5. Explain the observation. Would you expect the polymer to be soluble at pH 10?

10. Protamines are basic proteins whose sulfate salts are often used as a precipitating agent added to crude cellular extracts. What types of biomolecules would you expect to be precipitated by protamine sulfate at pH 7? What is the physical basis for the precipitation?

11. A mixture of alanine, glutamic acid, and arginine was chromatographed on a weakly basic ion-exchange column (positively charged) at pH 6.1. Predict the order of elution of the amino acids from the ion-exchange column. Are the amino acids separated from each other? Explain.
Suppose you have a weakly acidic ion-exchange column (negatively charged), also at pH 6.1. Predict the order of elution of the amino acids from this column. Propose a strategy to separate the amino acids using one or both columns. Explain your rationale. (Assume only ionic interactions between the amino acids and the ion exchange resin.)

12. For the following peptide sequences, determine the products resulting from the following treatments:
(a) Trypsin digestion. (b) Treatment of the peptide with succinic anhydride, then trypsin. (c) Reaction with ethyleneimine followed by trypsin. (d) Chymotrypsin.
(e) Cyanogen bromide (CNBr).

Ala-Glu-Lys-Phe-Val-Cys-Tyr-Met-Gly-Phe

13. You have a peptide that is a potent inhibitor of nerve conduction and you wish to obtain its primary sequence. Amino acid analysis reveals the composition to be Ala (5); Lys; Phe. Reaction of the intact peptide with FDNB releases free DNP-alanine on acid hydrolysis. (ϵ-DNP-lysine but not α-DNP-lysine is also found.) Trypsin digestion gives a tripeptide (composition Lys, Ala_2) and a tetrapeptide (composition Ala_3, Phe). Chymotryptic digestion of the intact peptide releases a hexapeptide and free alanine. Derive the peptide sequence.

14. Performic acid oxidation followed by acid hydrolysis of a decapeptide yielded the following amino acid composition: Ala(1); Asp(2); cysteic acid(2); Gly(2); methionine sulfone(1); Phe(1); Val(1). In addition, one mole of ammonium ion was detected per mole of peptide hydrolyzed. The following results were obtained.
(i) Carboxypeptidase A released Asn, then Ala.
(ii) Two cycles of sequential Edman degradation released the phenylthiohydantoins of Asp and Val, in that order.
(iii) Chymotrypsin released two peptides whose compositions were determined after acid hydrolysis: CHT A (Ala, Asp, Cys, Gly, Met); and CHT B (Asp, Cys, Gly, Phe, Val).
(iv) Treatment of the original peptide with 2-bromoethylamine and then with trypsin released three peptides whose compositions were determined after acid hydrolysis: T-1 (Ala, Asp);
T-2 (Asp, modified Cys, Gly, Val); and T-3 (modified Cys, Gly, Met, Phe).
Deduce the primary sequence of the peptide. If there is a portion of the sequence whose assignment is ambiguous or if there is a disulfide bond, so indicate. If there is an ambiguity, what other single cleavage might you use to assign an unambiguous sequence?

15. You have isolated from a rare fungus an octapeptide that prevents baldness and you wish to determine its amino acid sequence. The amino acid composition is Lys_2, Asp, Tyr, Phe, Gly, Ser, Ala. Reaction of the intact peptide with FDNB yields DNP-alanine plus 2 moles ϵ-DNP-lysine upon acid hydrolysis. Cleavage with trypsin yields peptides whose compositions are: (Lys, Ala, Ser) and (Gly, Phe, Lys) plus a dipeptide. Reaction with chymotrypsin releases free aspartic acid, a tetrapeptide with the composition (Lys, Ser, Phe, Ala), and a tripeptide whose composition following acid hydrolysis is (Gly, Lys, Tyr). What is the sequence?

4

CHAPTER

The Three-Dimensional Structure of Proteins

The enormous structural diversity of proteins begins with the amino acid sequences of polypeptide chains. Each protein consists of one or more unique polypeptide chains, and each of these polypeptide chains is folded into a three-dimensional structure. The final folded arrangement of the polypeptide chain in the protein is referred to as its conformation. Most proteins exist in unique conformations exquisitely suited to their function. It is the availability of a wide variety of conformations that permits proteins as a group to perform a broader range of functions than any other class of biomolecules.

In this chapter we will deal primarily with the structural properties of proteins, and in the following chapter we will consider the functional diversity of proteins. Traditionally proteins have been divided into two groups: fibrous and globular. Fibrous proteins aggregate to form highly elongated structures having the shape of fibers or sheets. Each protein unit going into these aggregated structures is built from a repeating structural motif, and for that reason the molecules have a basically simple structure that is relatively easy to analyze. By contrast, globular proteins are more complex, containing one or more polypeptide chains folded back on themselves many times to give an approximately spherical shape. We will consider the relatively simple fibrous proteins first.

Fibrous Protein Structures

Two names stand out above all others in the history of the discovery of fibrous protein structure. These are the names Linus Pauling and Robert Corey. Before they ventured into the world of proteins, Pauling and Corey spent a great deal of time studying the simple crystals formed by amino acids and low-molecular-weight peptides. From their crystallographic investigations, Pauling and Corey formulated two rules that describe the ways in which amino acids and peptides interact with one another to form noncovalently bonded crystalline structures. These rules remain central to our understanding of how amino acids interact with one another in protein polypeptide chains.

The first rule was that the peptide C—N linkage and the four atoms to which the C and the N atoms are immediately linked always form a planar structure, as though the C—N linkage were a double bond rather than a single bond as normally written. Pauling reasoned that the C—N linkage is a res-

Figure 4.1

Resonance and the planar structure of the peptide bond. (*a*) Two major hybrids contribute to the structure of the peptide bond. In structure 1 the C—N bond is a single bond with no overlap between the nitrogen lone electron pair and the carbonyl carbon. The carboxyl carbon is sp^2- hybridized and is therefore planar, while the nitrogen is sp^3-hybridized and pyramidal. By contrast, in structure 2 there is a double bond between the amide nitrogen and the carbonyl carbon; also, the nitrogen atom bears a charge of +1 and the carbonyl oxygen bears a charge of −1. Both the carboxyl carbon and the amide nitrogen are sp^2-hybridized, both are planar, and all six atoms lie in the same plane. (*b*) The structure of the peptide bond is a compromise between the two resonating hybrids, structures 1 and 2. (*c*) Dimensions of the peptide bond and surrounding linkages. The C—N bond length of 1.325 Å is significantly less than the length of a single C—N bond, 1.47 Å.

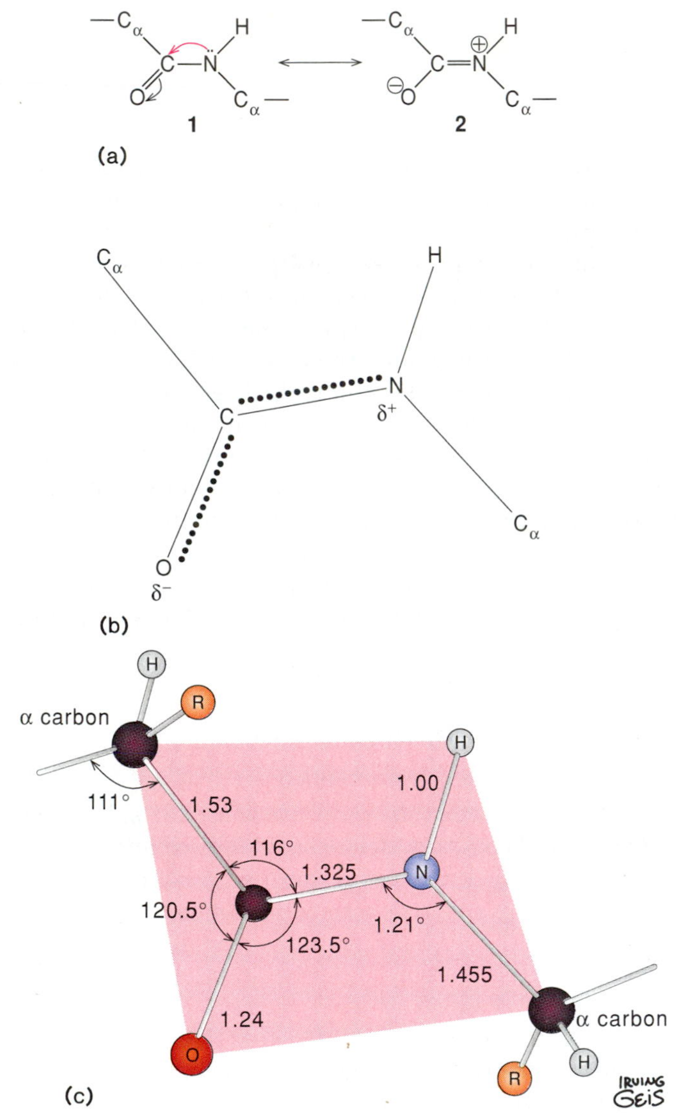

(a)

(b)

α carbon 111° 1.53 116° 1.325 1.00 120.5° 123.5° 1.21° 1.455 1.24 α carbon

(c)

onating structure with partial double-bond character (fig. 4.1), so that it locks the peptide grouping into a planar conformation. This characteristic is extremely important because it greatly reduces the flexibility in the polypeptide chain. The only flexibility remaining in the polypeptide backbone results from rotation about the α carbon that joins two peptide planar groups (fig. 4.2).

Figure 4.2

Basic dimensions of a dipeptide. The conformational degrees of freedom of a polypeptide chain are restricted to rotations about the single-bond connections between the adjacent planar transpeptide groups to C_α, i.e., the C_α—C_2 and C_α—N_1 single bonds. The corresponding rotations are represented by ψ and ϕ, respectively, which have values of 180° for the fully extended configuration shown. (Reprinted with permission from R. E. Dickerson and I. Geis, *The Structure and Action of Proteins*, Benjamin/Cummings, Menlo Park, Calif., 1969.)

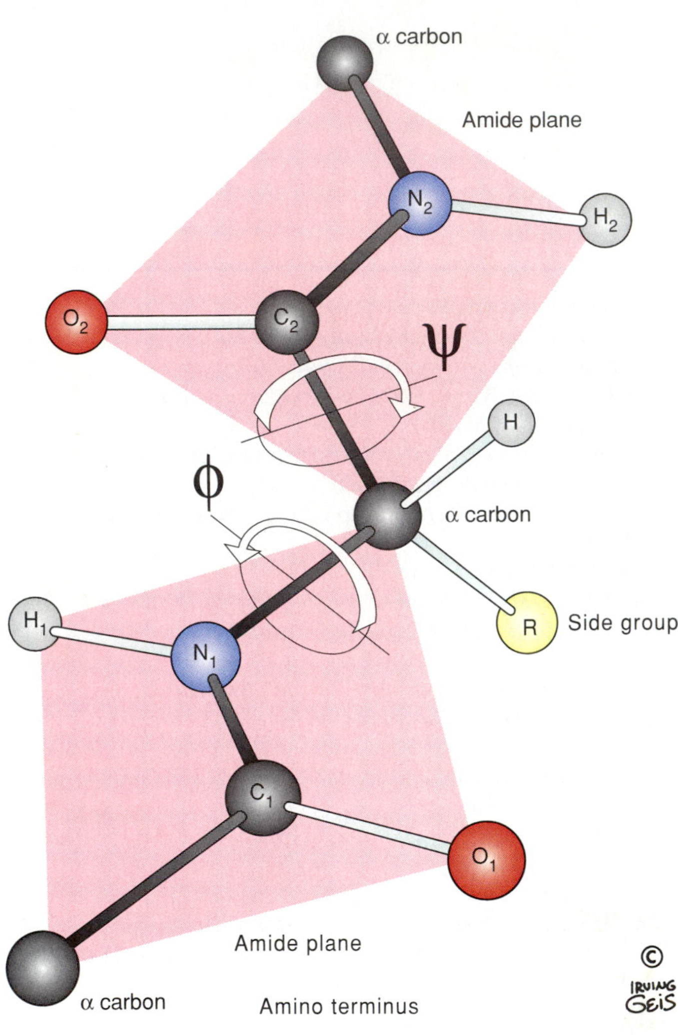

The second rule that Pauling and Corey formulated was that peptide carbonyl and NH groups always form the maximum number of hydrogen bonds. Recall that hydrogen bonds are formed by a partially unshielded proton from one molecule and a nitrogen or oxygen atom possessing a lone pair of electrons that originates from another molecule (see fig. 1.7). The attraction between the two groups is strongest along the lone-pair orbital axis of the hydrogen bond acceptor group. As a rule the angle between the N or O acceptor and the N—H or

Figure 4.3

Major hydrogen-bond donor and acceptor groups found in proteins. Note that the angle between the O acceptor and the N—H donor is 180° in the hydrogen-bond complex.

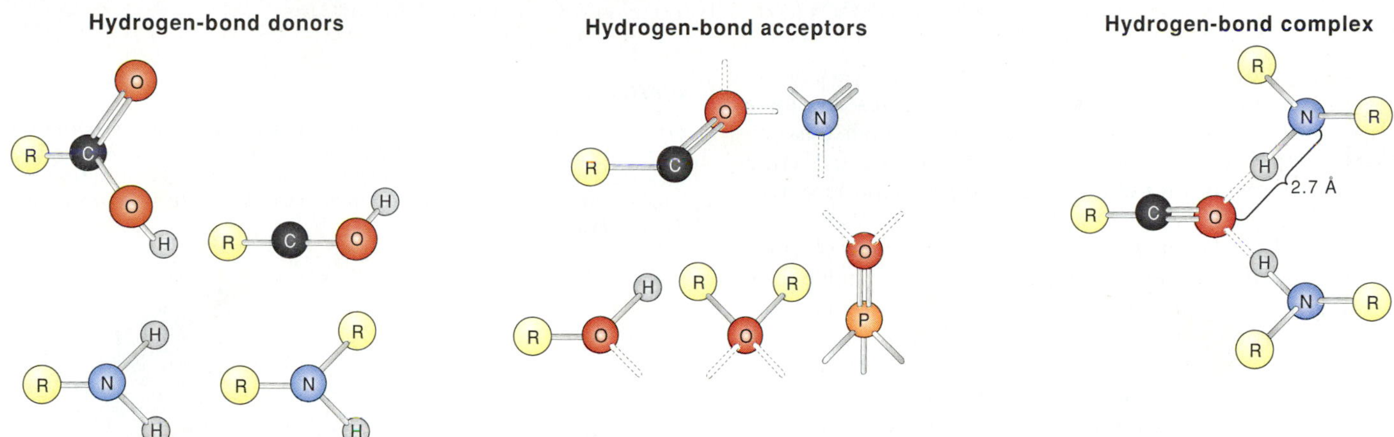

O—H donor is close to 180° (fig. 4.3). Thus a hydrogen bond brings two interacting groups close together and orients them in a certain way. This second rule of Pauling and Corey further limits the number of conformations available to polypeptide chains.

With these rules in their arsenal, Pauling and Corey examined the x-ray diffraction patterns of a number of fibrous proteins. Despite the vast number of fibrous proteins that exist in nature, the majority of them give diffraction patterns that fall into one of three types: the α pattern (box 4A), the β pattern, and the collagen pattern. Fiber diffraction gives information about the repeating units of a protein structure only, but since fibrous proteins are arranged as simple repetitious units, this was just the information that was needed. The first type of diffraction pattern observed for a subgroup of the keratins was consistent with a helical arrangement of the polypeptide chains and an advance per turn along the helix axis of 5.4 Å. (The advance per turn is called the pitch of the helix.) If the fibers are tilted by about 30° with respect to the x-ray beam, they show a prominent diffraction spot indicative of an advance per amino acid along the helix axis of 1.5 Å.

The α-Keratins Contain Helically Arranged Polypeptide Chains Held Together by Intramolecular Hydrogen Bonds

From diffraction data, their two rules, and a set of precisely constructed space-filling molecular models, Pauling and Corey went to work testing the structural possibilities. Their models were designed so that covalently linked atoms were accurately spaced according to well-known dimensions for such groups (table 4.1). Moreover, the individual atoms were made of a size so that nonbonded atoms could get no closer to one another than their van der Waals radii would normally allow (see table 4.1). Recall from chemistry that the average separation between two nonbonded atoms is known as their van der Waals separation. There is a very high repulsion if nonbonded atoms get closer than this, and there is a rapid fall-off in favorable interaction if the distance is larger. With their carefully constructed "toys" and a boyish enthusiasm, the two learned scientists tried to arrange the polypeptide chains so as to maximize the number of peptide hydrogen bonds in a way that was consistent with the x-ray diffraction data. By trial and error they came to the conclusion that the most acceptable structure was a polypeptide chain arranged in a right-handed helical coil. This structure, known as the alpha (α) helix, has a rigid, regularly repeating backbone structure with an advance per amino acid residue along the helix axis of 1.5 Å, corresponding to the prominent spot in the diffraction pattern. The diffraction spots resulting from the helix itself could be explained by a spacing between adjacent turns of the polypeptide backbone along the helix axis of 5.4 Å.

In figure 4.4 we see three different ways of representing the structure of the α helix. In (*a*) we see a ball-and-stick model in which the planar groups are highlighted and the hydrogen bonds are represented by dashed lines. In (*b*) and (*c*) we see a space-filling model and a wire model, respectively, representing the same structure. The latter two models are shown in both side views and top view. The space-filling model, similar to the one used by Pauling and Corey, reveals that the α helix is a very tightly packed structure with no unfilled cavities, whether one looks at a profile or down the helix. The wire model (*c*) illustrates the helical structure best. But for purposes of discussion the ball-and-stick model (*a*) is the most suitable. Careful inspection shows that the polypeptide backbone follows the path of a right-handed helical spring in which each residue's carbonyl group forms a hydrogen bond with the amide NH group of the residue four amino acids further along the polypeptide chain. All residues in the α helix have nearly identical conformations, so they lead to a regular structure in which each 360° of helical turn incorporates approximately 3.6 amino acid residues and rises 5.4 Å along the helix axis direction. This arrangement gives the observed advance per amino acid residue along the helix axis (5.4/3.6 = 1.5 Å).

4A
BOX

X-Ray Diffraction: Applications to Fibrous Proteins*

X-ray diffraction played a major role in discovery of the structure of fibrous proteins. In most cases the fibers under study are oriented in two dimensions by stretching. In this box we illustrate how the technique is used to study the α form of the synthetic polypeptide poly-L-alanine.

A stretched fiber containing many poly-L-alanine molecules is suspended vertically and exposed to a collimated monochromatic beam of CuK_α x-rays, as shown in figure 1(*a*) of the illustration. Only a small percentage of the x-ray beam is diffracted, while most of the beam travels through the specimen with no change in direction. A photographic film is held in back of the specimen. A hole in the center of the film allows the incident undiffracted beam to pass through.

Coherent diffraction occurs only in certain directions specified by Bragg's law: $2d\sin\theta = n\lambda$, where d is the distance between identical repeating structural elements, θ is the angle between the incident beam and the regularly spaced diffracting planes, λ is the wavelength of x-rays used, and n is the order of diffraction, which may equal any integer but is usually strongest for $n = 1$. For small θ, $\sin\theta \approx \theta$ and $d \approx 1/\theta$, so that a spot far out on the photographic film is indicative of a repeating element of small dimension.

Figure 1(*b*) shows the diffraction pattern that is obtained when the fiber axis is normal to the beam. Note the strong off-vertical reflection at 5.4 Å (arrow). A different diffraction pattern (*c*) is obtained when the fiber axis is inclined to the beam at 31°. Note the strong reflection at 1.5 Å in the upper part of the diagram (arrow).

*Source: C. H. Bamford et al., *Synthetic Polypeptides.* Copyright © 1956, Academic Press, Orlando, Fla.

Figure 1

(*a*) Experimental arrangement for obtaining x-ray diffraction pattern shown in (*b*). (*b*) Diffraction pattern of the α form of a cluster of poly-L-alanine molecules oriented vertically. (*c*) Diffraction pattern of the same fiber bundle with the fiber axis inclined to the beam at 31°. (*b* and *c* Brown & Trotter, 1956.)

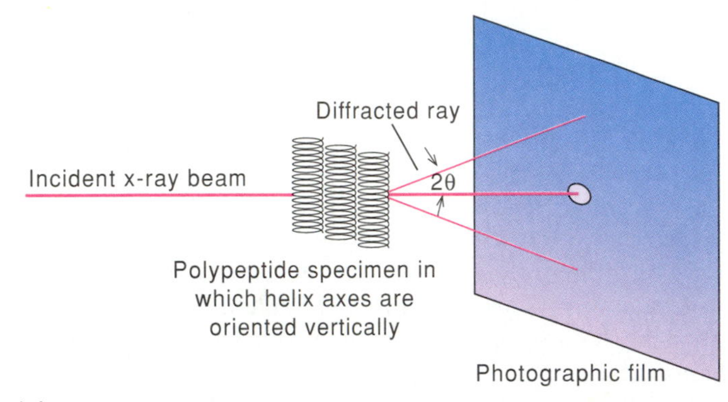

(a)

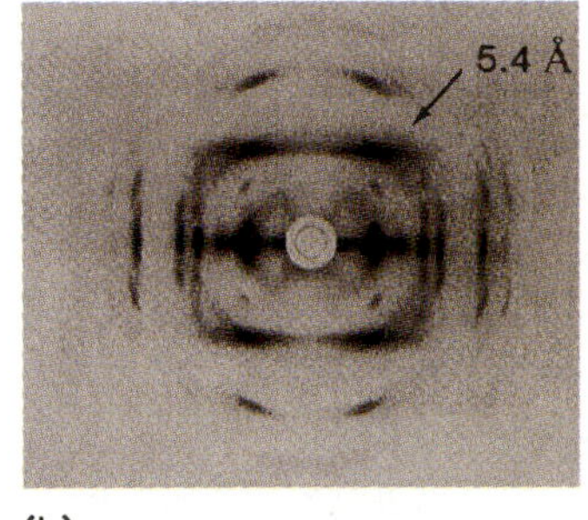

(b)

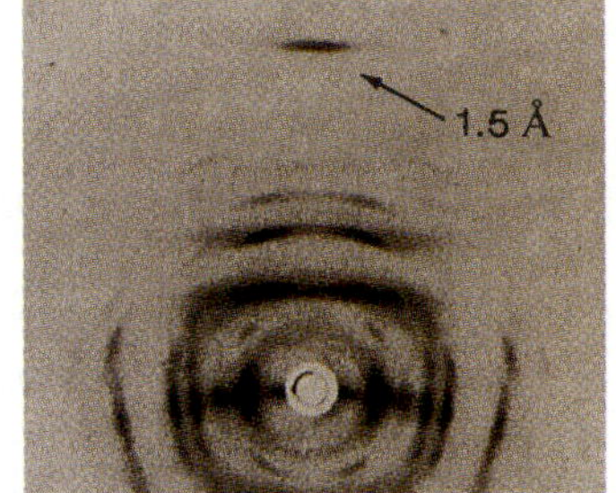

(c)

Table 4.1
Radii for Covalently Bonded and Nonbonded Atoms

Covalent Bond Radii (in Å)				Van der Waals Radii (in Å)			
Element	*Single Bond*	*Double Bond*	*Triple Bond*	*Element*			
Hydrogen	0.30			Hydrogen	1.2		
Carbon	0.77	0.67	0.60	Carbon	2.0		
Nitrogen	0.70			Nitrogen	1.5		
Oxygen	0.66			Oxygen	1.4		
Phosphorus	1.10			Phosphorus	1.9		
Sulfur	1.04			Sulfur	1.8		

Figure 4.4

Three ways of projecting the α-helix. (*a*) This simple ball-and-stick model highlights the planar peptides. The interpeptide hydrogen bonds are shown by dashed lines and the amino acid side chains are indicated by R groups. Approximately two turnings of the helix are shown. There are about 3.6 residues per turn. In (*b*) and (*c*) we see identical projections of a side view using space-filling and wire models, respectively. The space-filling models use van der Waals radii for the atoms.

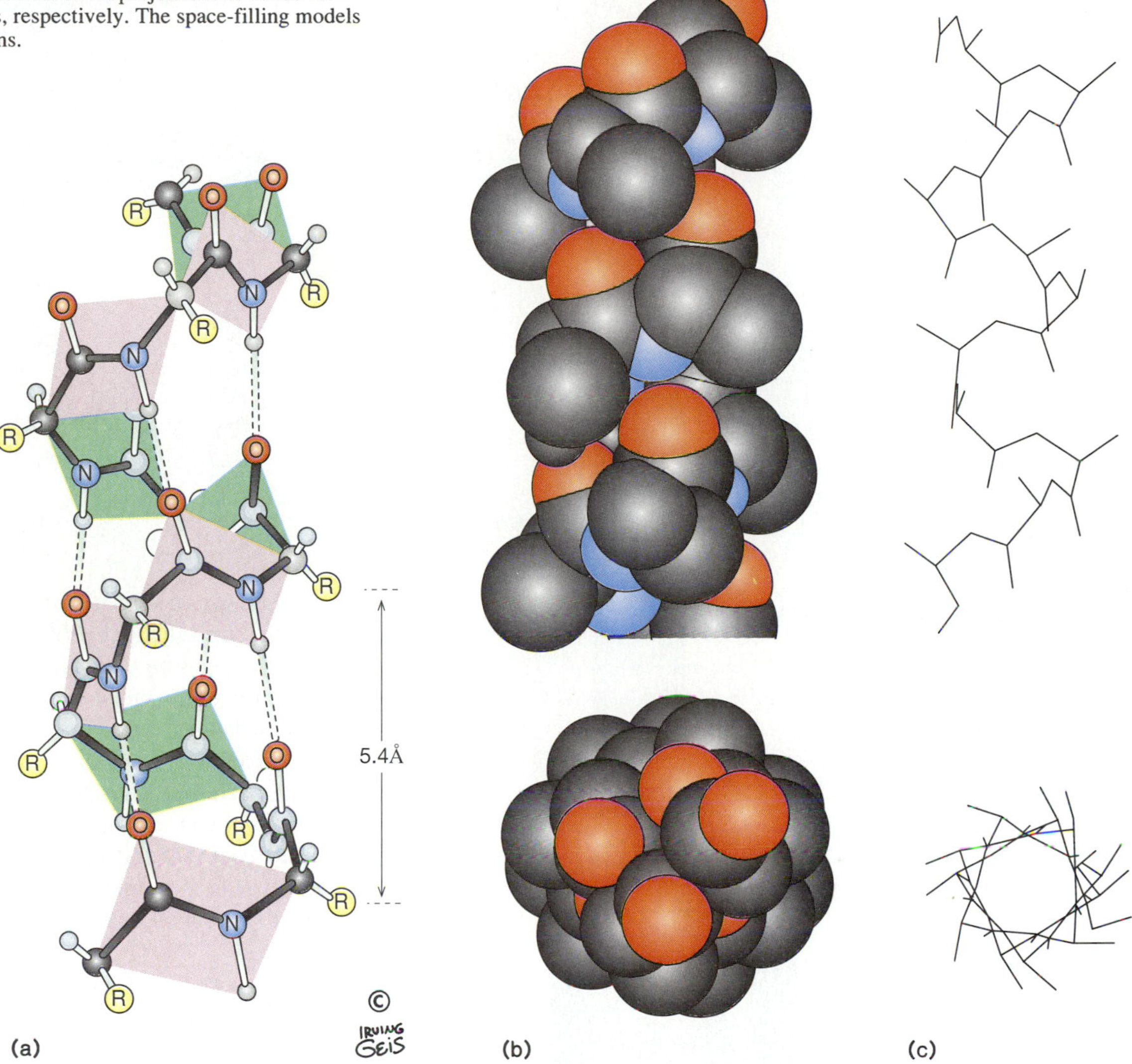

Although alternative helical arrangements having different hydrogen-bonding patterns and different geometries are conformationally possible, the α helix is by far the most commonly observed helical arrangement found for polypeptide chains in proteins. The special stability of the helix is probably related not only to the formation of stable hydrogen bonds between all the carbonyl and NH groups, but also to the tight packing achieved in folding the chain to form the structure.

Fibrous proteins in which the α helix is a major structural component are found in hair, scales, horns, hooves, wool, beaks, nails, and claws; these proteins are referred to as α-keratins. When individual α helices aggregate in side-by-side fashion to produce such proteins, they usually form long cables in which the individual helices are spirally twisted so that the resulting cable has an overall left-handed twist (fig. 4.5). The formation of such a cable appears to result from optimization of packing among the amino acid side-chain residues between helices. In figure 4.6 we see how the side-chain residues of an α helix are arranged in a spiral fashion so that residues falling on the same side of a helix generally do not lie along a line parallel to the helix axis. The packing together of helices is consequently optimized when the helices interact at an angle of about 18°. Obviously, if the α helices involved in such a packing interaction were straight, they would soon become separated. However, with the left-twisted cable structure characteristic of the keratins, their packing interaction can be preserved. The coiled-coil character of such fibers consequently represents a trade-off between some local deformations that coil the α helix and the optimization of extended side-chain packing interactions in the cable as a whole.

Figure 4.5

The assembly of hair α-keratin from one α helix to a protofibril, to a microfibril, and finally, to a single hair.

Figure 4.6

Coiling of α helices in α-keratins. Residues on the same side of an α helix form rows that are tilted relative to the helix axis. Packing helices together in fibers is optimized when the individual helices wrap around each other so that rows of residues pack together along the fiber axis. Helices in coiled coil (*c*) are oriented in parallel.

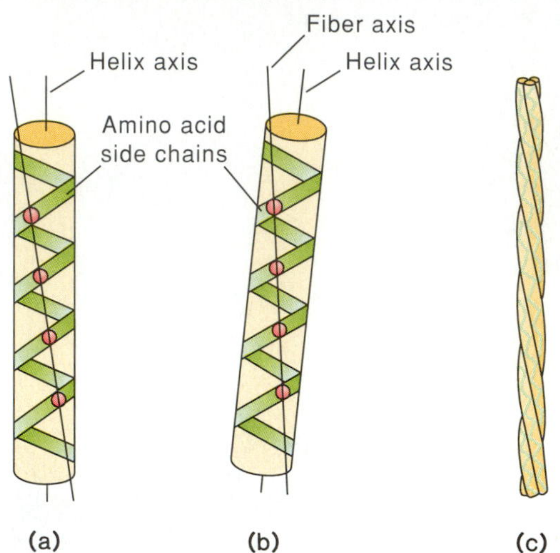

The springiness of hair and wool fibers results from the tendency of the α-helical cables to untwist when stretched and spring back when the external force is removed. In many forms of keratin, the individual α helices or fibers are covalently linked by disulfide bonds formed between cysteine residues of adjacent polypeptide chains. In addition to giving added strength to the fibers, the pattern of these covalent interactions serves to influence and fix the extent of curliness in the hair fiber as a whole. Chemical reactions and mechanical processes involving reductive cleavage, reorganization, and reoxidation of these interhelix disulfide bonds form the basis of the "permanent wave."

The β-Keratins Form Sheetlike Structures with Extended Polypeptide Chains

Pauling and Corey noticed that certain fibrous proteins give radically different diffraction patterns and behave macroscopically as sheets rather than fibers. The diffraction patterns yielded by fibrous proteins from different sources revealed a repeat along the direction of the extended polypeptide chain at intervals of either 13.0 or 14.0 Å. The similarity of intervals suggested two closely related structures. Pauling and Corey interpreted these repeating patterns as resulting from extended polypeptide chains lying side-by-side in either a parallel or an antiparallel fashion (fig. 4.7). Although both structures exist in nature, the antiparallel sheet is more common in fibrous proteins. In such proteins the structure is known as the antiparallel β-pleated sheet (fig. 4.8). Regular hydrogen bonds form between the peptide backbone amide NH and carbonyl oxygen groups of adjacent chains. The β sheet can be easily extended into a multistranded structure simply by adding successive chains in the appropriate direction to the sheet.

Figure 4.7

Two forms of the ß-sheet structure: (*a*) the antiparallel and (*b*) the parallel ß sheet. The advance per two amino acid residues is indicated for each structure.

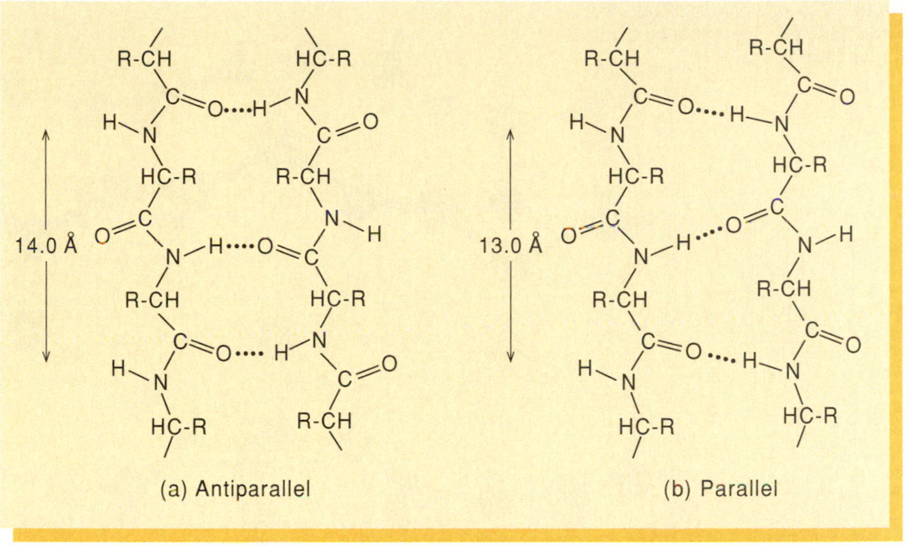

(a) Antiparallel

(b) Parallel

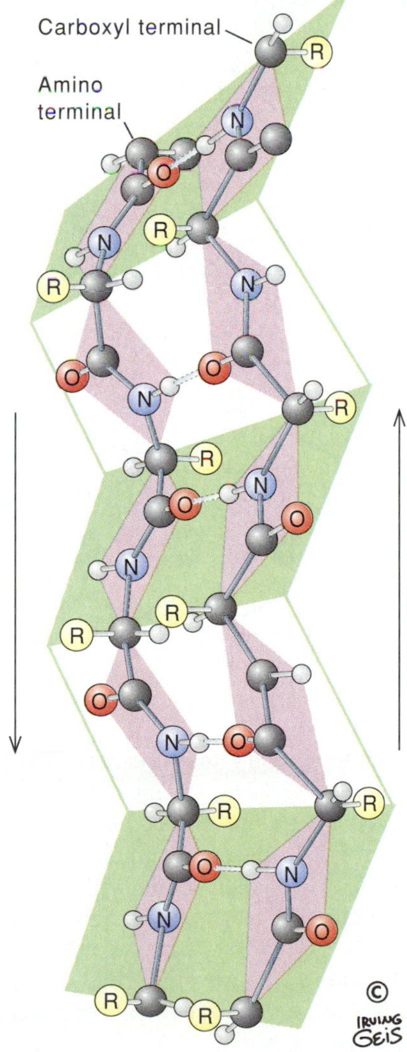

Figure 4.8

The antiparallel ß sheet. This structure is composed of two or more polypeptide chains in the fully extended form, with hydrogen bonds formed between the chains. Hydrogen bonds are shown as dashed lines.

Parallel and antiparallel β-pleated sheets are both composed of polypeptide chains that have conformations pointing alternate R groups to opposite sides of the sheet, but have their peptide planes nearly in the sheet plane to allow good interchain hydrogen bonding. Nevertheless, the chain conformation that produces the best interchain hydrogen bonding in parallel sheets is slightly less extended than that for the antiparallel arrangement. As a result, the parallel sheet has both a shorter advance per amino acid residue, 6.5 Å (versus 7.0 Å for the antiparallel structure), and a more pronounced pleat.

The best-known β-sheet structure in nature is silk, a variety of fibrous proteins produced by certain insects. Silks are composed of stacked antiparallel β-pleated sheets (fig. 4.9). Sequence analysis of silk proteins shows them to be largely composed of glycine, serine, and alanine, where every alternate residue is glycine. Since the side-chain groups of a flat antiparallel sheet point alternately upward and downward from the plane of the sheet, all the glycine residues are arranged on one surface of each sheet and all the substituted amino acids on the other. Two or more such sheets can consequently be intimately packed together to form an arrangement in which two adjacent glycine-substituted or alanine-substituted sheet surfaces interlock with each other (see fig. 4.9*b*). Owing to both the extended conformations of the polypeptide chains in the β sheets and the interlocking of the side chains between sheets, silk is a mechanically rigid material that resists stretching.

Collagen Forms a Unique Triple-Stranded Structure with Interstrand Hydrogen Bonds

Collagen is a particularly rigid and inextensible protein that serves as a major constituent of tendons and many connective tissues. In the electron microscope it can be seen that collagen fibrils have a distinctive banded pattern with a periodicity of

Figure 4.9

The three-dimensional architecture of silk (*a*). The side chains of one sheet nestle quite efficiently between those of neighboring sheets (*b*).

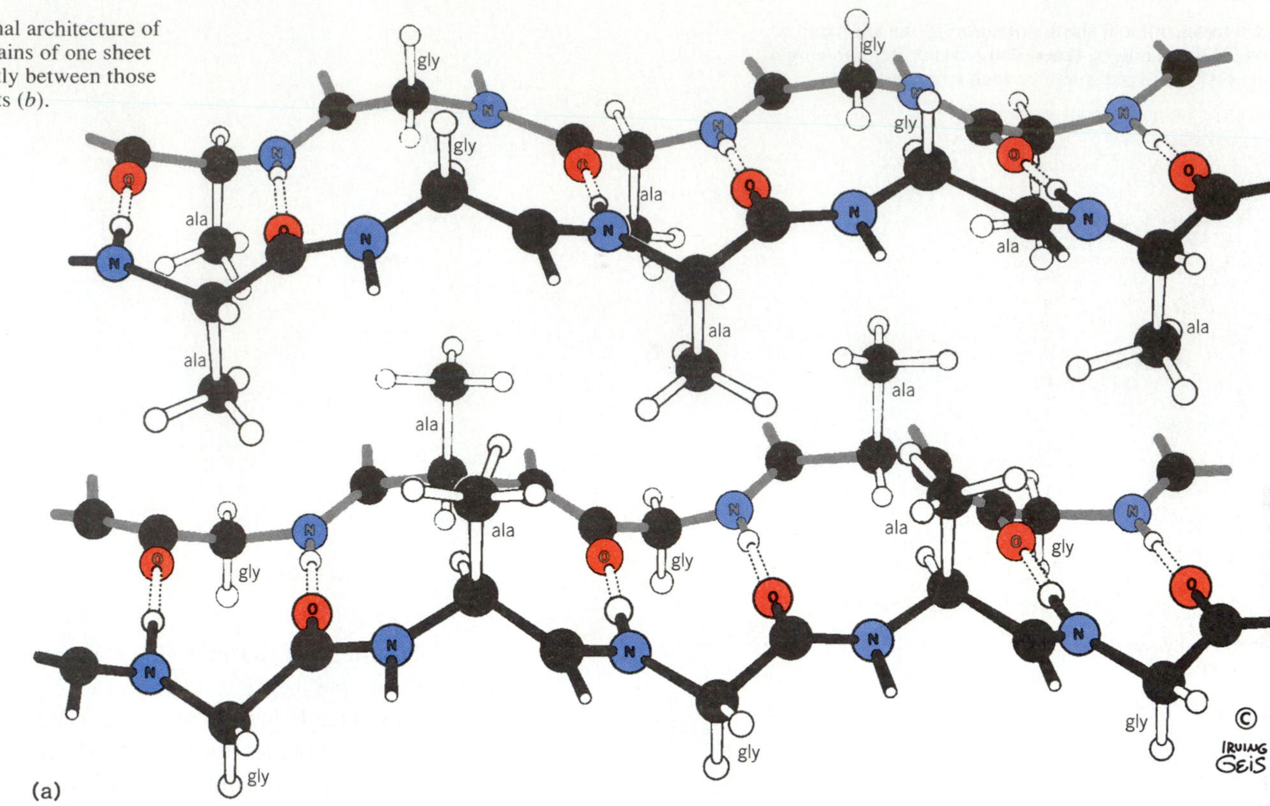

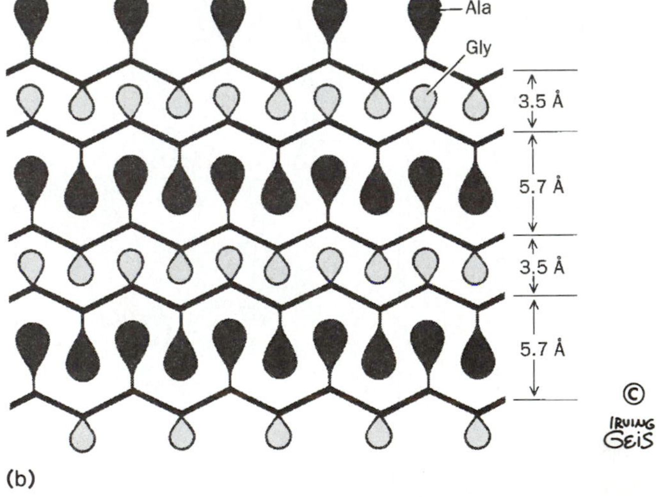

680 Å (fig. 4.10). These fibrils are of varying thickness, depending on the source and the mode of preparation. Individual fibrils are composed of collagen molecules 3,000 Å long that aggregate in a staggered side-by-side fashion (fig. 4.11).

Analysis of collagen indicated a most unusual amino acid composition in which glycine, proline, and hydroxyproline are the dominant amino acids. Further characterization of the polypeptide chains showed that these amino acids are arranged in a repetitious tripeptide sequence, Gly-X-Y, where X is frequently a proline and Y is frequently a hydroxyproline. This unusual amino acid sequence and the unique diffraction pattern of collagen suggested that collagen is a totally different type of fibrous protein. Pauling attempted to determine its structure by his molecular model approach but failed. At a meeting in Cambridge in 1958 where the correct structure was presented, Pauling suggested, more or less in jest, that the structure he had derived might be more stable than the one found in nature. The correct structure was in fact found by Ramachandran, whose interpretation was supported by Crick and Rich.

The repeating proline residue in collagen excluded the possibility that the polypeptide chains could adopt either an α-helical or a β-sheet conformation. Instead, individual collagen polypeptide chains assume a left-handed helical conformation and aggregate into three-stranded cables with a right-handed twist (fig. 4.12). When viewed down the polypeptide chain axis (fig. 4.13*b*), the successive side-chain groups can be seen to point toward the corners of an equilateral triangle. The glycine at every third residue is required because there is no room for any other amino acid inside the triple helix, where the glycine R groups are located. The three collagen chains do not form hydrogen bonds among residues of the same chain. Instead, the collagen chains within each three-stranded cable form interchain hydrogen bonds. This arrangement produces a highly interlocked fibrous structure that is admirably suited to its biological role of providing rigid connections between muscles and bones as well as structural reinforcement for skin and connective tissues.

Figure 4.10

An electron micrograph of collagen fibrils from skin. (Courtesy of Jerome Gross, Massachusetts General Hospital.)

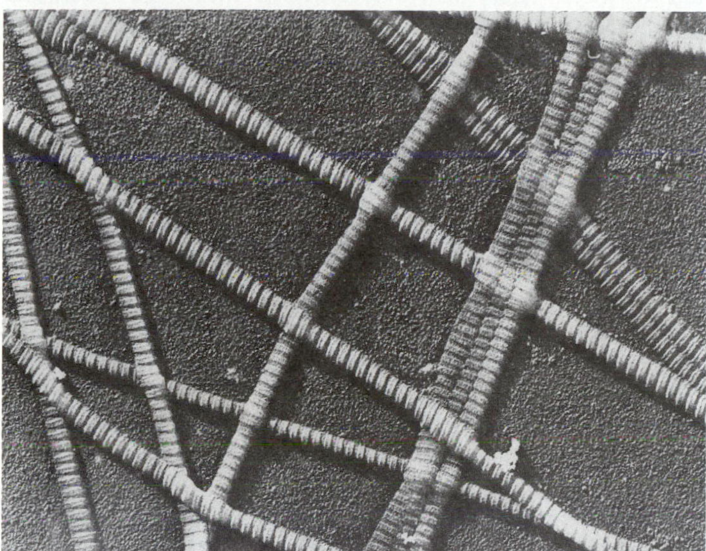

Figure 4.11

The banded appearance of collagen fibrils in the electron microscope arises from the schematically represented staggered arrangement of collagen molecules (*above*) that results in a periodically indented surface. *D*, the distance between cross striations, is ~680 Å so that the length of a 3,000-Å-long collagen molecule is 4.4*D*. (©Michael C. Webb/Visuals Unlimited.)

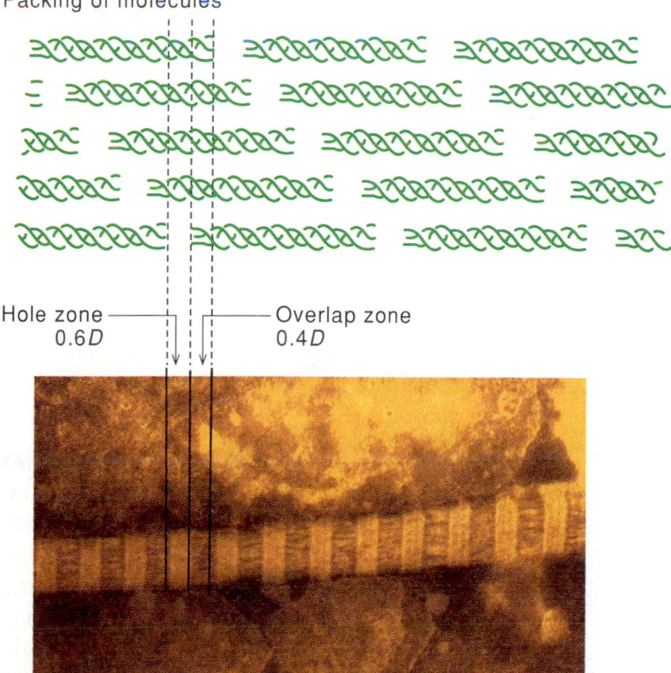

Figure 4.12

The triple helix of collagen.

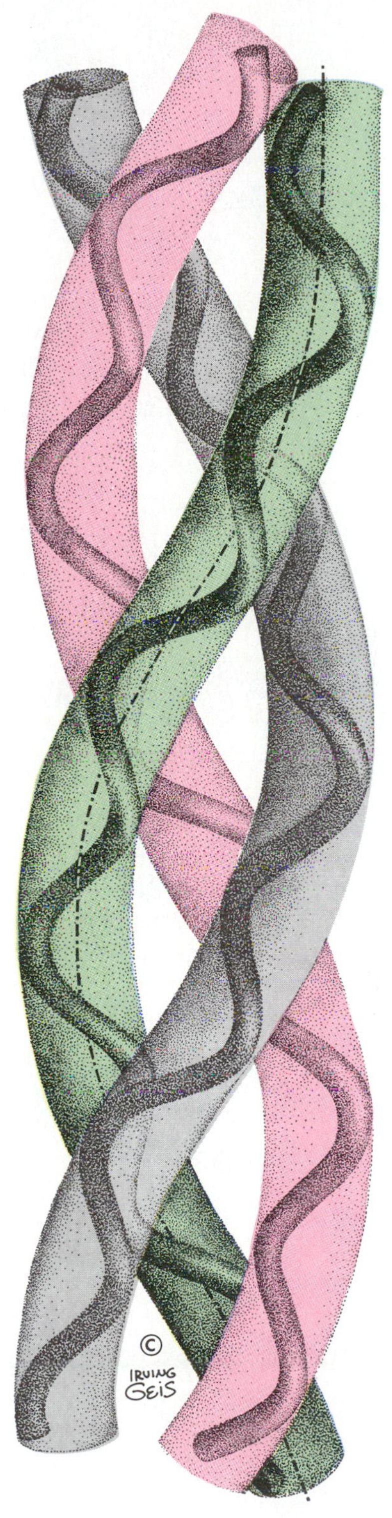

Figure 4.13

The basic coiled-coil structure of collagen. Three left-handed single-chain helices wrap around one another with a right-handed twist. (*a*) Ball-and-stick single-collagen chain. (*b*) View from top of helix axis. Note that glycines are all on the inside. In this structure the C=O and N—H groups of glycine protrude approximately perpendicularly to the helix axis so as to form interchain hydrogen bonds.

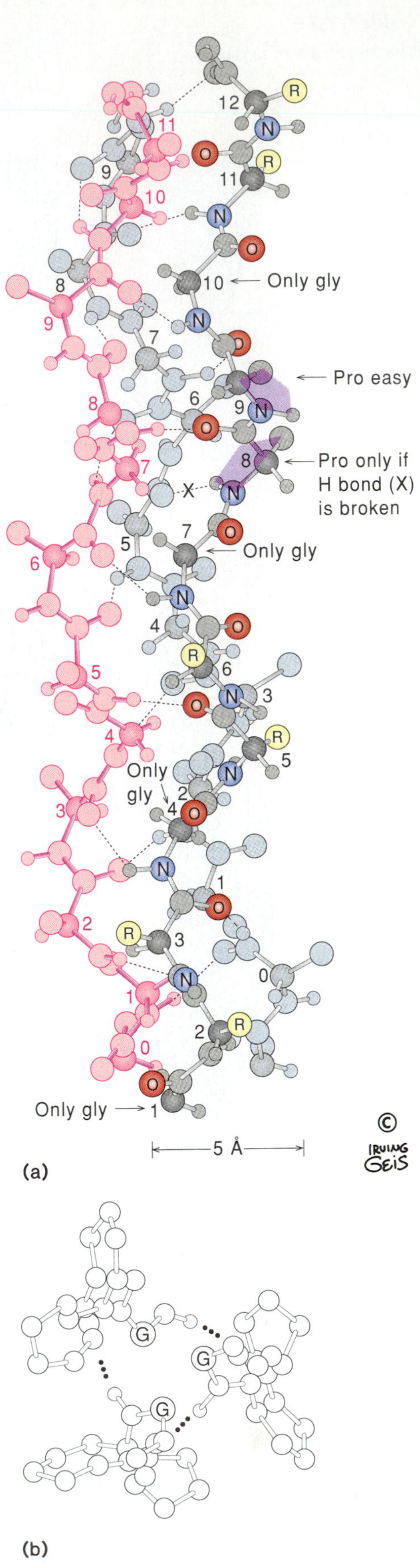

Although there exist in living organisms additional types of fibrous proteins, as well as polysaccharide-based structural motifs, we have focused here on those three arrangements whose structural properties are currently the best understood and the most widely distributed. Two of these, the α-keratins and the silks, incorporate polypeptide secondary structures that also commonly occur in globular proteins. Collagen, in contrast, is a protein that evolution has developed to play a more specialized role.

The Use of Ramachandran Plots to Predict Sterically Permissible Structures

In figure 4.2 we saw a ball-and-stick model of a short section of a polypeptide chain. As we noted, many geometric features of this structure are fixed as a result of bonded interactions between adjacent atoms. The bond lengths and bond angles are nearly constant for all proteins. Additionally, the backbone peptide bond has substantial double-bond character so that all the atoms of the peptide bond, together with the connected α-carbon atoms (conventionally labeled C_α), lie in a common plane with the carbonyl oxygen and amide hydrogen in the *trans* configuration. Consequently, the only adjustable geometric features of the polypeptide-chain backbone involve rotations about the single covalent bonds that connect each residue's C_α to the adjacent planar peptide groups.

Rotations about the C_α—N bond are labeled with the Greek letter ϕ (phi), and rotations about the C_α-carbonyl carbon are labeled ψ (psi) (fig. 4.14). All possible conformations of a

Figure 4.14

The conformation corresponding to $\phi = 0°$, $\psi = 0°$. This conformation is disallowed by the steric overlap between the H and O atoms of adjacent peptide planes. Rotation of both ϕ and ψ by 180° gives the fully extended conformation seen in figure 4.2. Curved arrows for ϕ and ψ indicate positive variations in angle.

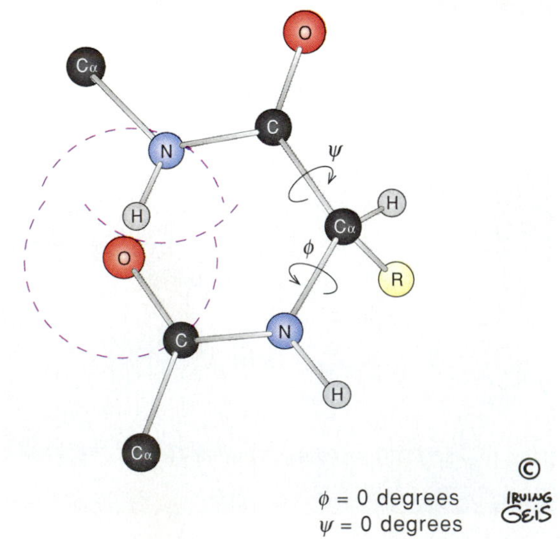

polypeptide chain can be described in terms of their ϕ,ψ conformational angles, a description that automatically takes account of the fixed geometric features of the polypeptide backbone. Thus any polypeptide conformation can be represented as a point on a plot of ϕ versus ψ, where ϕ and ψ have values that range from $-180°$ to $+180°$. By convention, the formation corresponding to $\phi = 0, \psi = 0$ is one in which both peptide planes that are connected to a common C_α atom lie in the same plane, as shown in figure 4.14. Positive variations in ϕ correspond to clockwise rotations of the preceding peptide about the C_α—N_1 bond when viewed from C_α toward N_1 (see fig. 4.2). Positive variations in ψ correspond to clockwise rotations of the succeeding peptide about the C_α—C_2 bond when viewed from C_α toward C_2 (again, see fig. 4.2).

Experiments with models that approximate the polypeptide atoms as hard spheres, with appropriate van der Waals radii, quickly reveal that many ϕ,ψ angular combinations are impossible because of steric collisions between atoms along the backbone or between backbone atoms and the side-chain R groups. For example, it is clear that the $\phi = 0, \psi = 0$ conformation shown in figure 4.14 is impossible. The reason is that this conformation results in noncovalently bonded interatomic contacts that are considerably less than the sum of the van der Waals radii of the atoms involved. In fact, of all the possible ϕ,ψ combinations, only a relatively restricted number of conformations are sterically allowed. The Ramachandran plot (fig. 4.15) shows explicitly how the accessible regions of ϕ,ψ space

Figure 4.15

Ramachandran plot, showing which atomic collisions (using a hard-sphere approximation) produce the restrictions of the main-chain angles ϕ and ψ. The cross-hatched regions are allowed for all residues, and each boundary of a prohibited region is labeled with the atoms that collide in that conformation. Additional shaded regions are for glycine residues only. The numbering scheme for amide atoms used in the derivation diagram is given in figure 4.2. Each boundary of a prohibited region is labeled with the atoms that collide in that conformation. For an explanation of the various labeled structures, see the text.

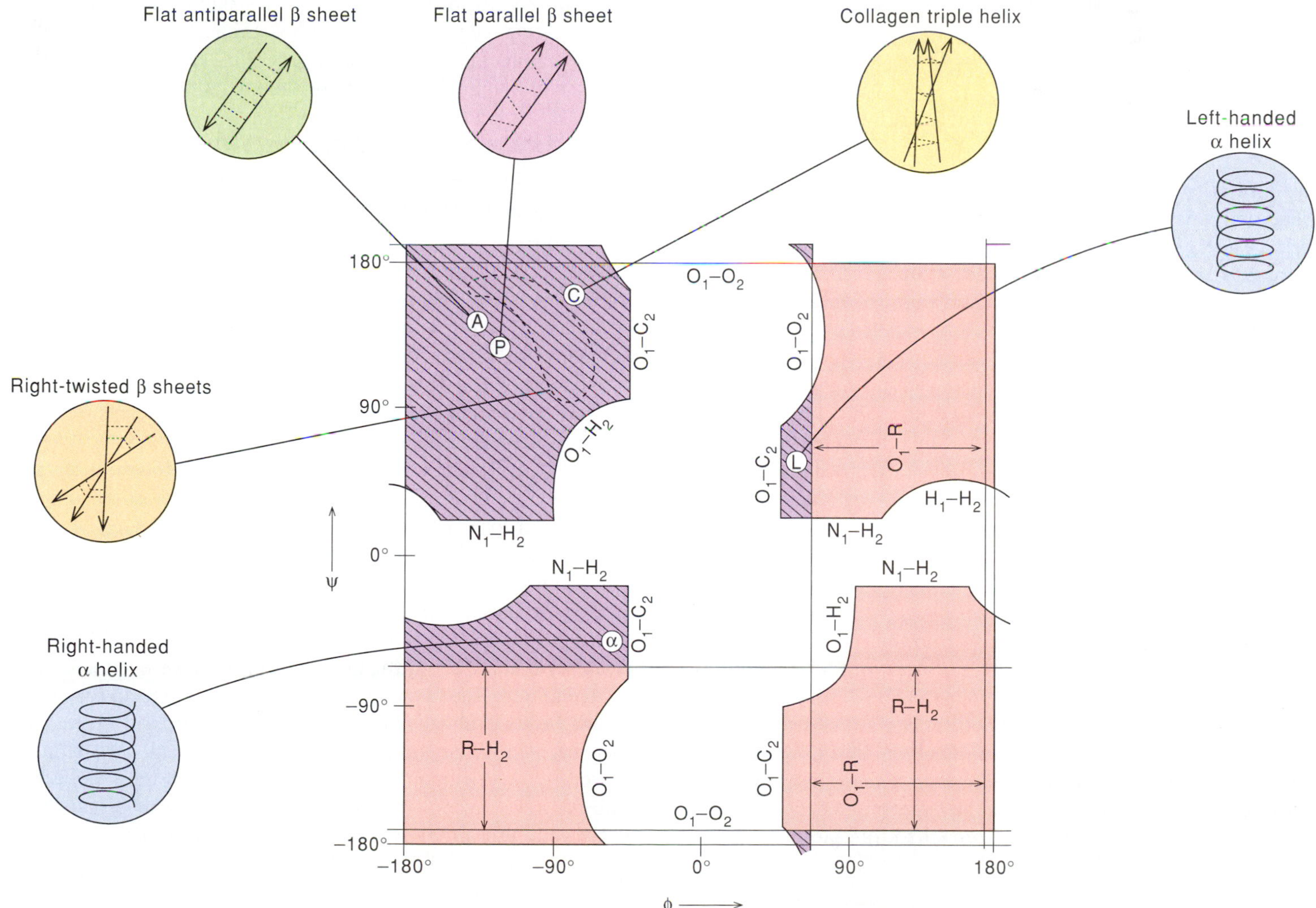

Figure 4.16

Ramachandran plot of main-chain angles ϕ and ψ, experimentally determined for approximately 1,000 nonglycine residues in eight proteins whose structures have been refined at high resolution (chosen to be representative of all categories of tertiary structure).

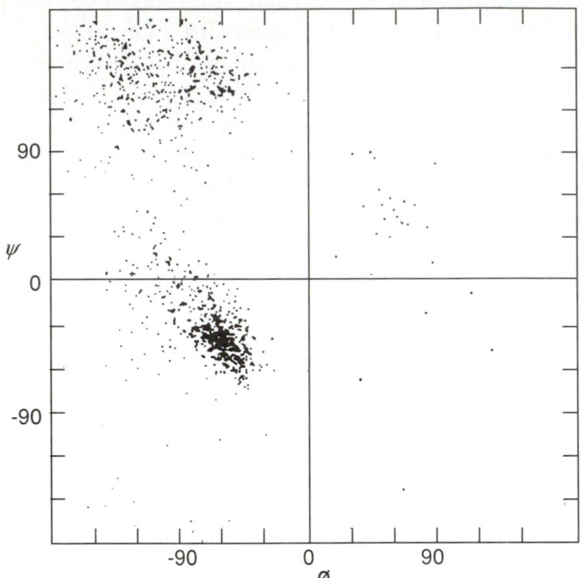

are limited by steric interactions among the polypeptide backbone and side-chain groups, assuming that the atomic groups behave as rigid spheres having appropriate van der Waals radii. In reality, the atoms in molecules do not behave as rigid spheres, so real proteins span a slightly greater range of values than suggested by this plot.

Figure 4.16 shows the distribution of some observed conformational values for proteins whose three-dimensional structures are known from crystallography. The great majority of these lie within the bounds defined by allowable steric interactions. The exceptional residues are usually glycines. Glycine frequently can assume conformations that are sterically hindered in other amino acids because its R group, a hydrogen atom, is considerably smaller than the CH_2 or CH_3 groups connected to C_α in all other amino acids.

In summary, owing to the basic geometric properties of the polypeptide chain, the sterically allowed conformations are greatly restricted by the occurrence of unfavorable steric interactions between various atomic groups. As a result, fibrous proteins with regularly repeating structures can be defined by single values for the coordinates ϕ and ψ. Proteins that have less regular structures would require more than a single set of coordinates but nevertheless they would be expected to obey the same set of limits established by the Ramachandran plot.

Analysis of Globular Proteins

It is not surprising that repeating structures with long-range order were the first protein structures to be understood. The demands on the available technology were minimal. Much more sophisticated technology was required to interpret the diffraction patterns of most proteins that have less long-range repetition. A host of crystallographers struggled over these structures, but the outstanding pioneers were Jonathan Kendrew and Max Perutz. They led their research teams to discovery of the structures of myoglobin and hemoglobin, the first globular-protein structures to be deciphered.

Today, enormous advances in protein chemistry and computer technology have systematized the necessary research and greatly reduced the amount of work and time required to investigate a protein structure. Accurate structure determinations have now been made on over 500 different proteins. It appears that detailed three-dimensional structures of proteins also may be deduced either by a combination of electron diffraction and low-dose imaging in the electron microscope or by high-resolution two-dimensional nuclear magnetic resonance (NMR) spectroscopy. Nevertheless, essentially all the information summarized here comes from the results of x-ray crystallography. From this wealth of data, patterns of structure are becoming apparent that suggest, among other things, that the overall folding arrangements of proteins may some day be predictable from the amino acid sequences of the polypeptide chain.

Three-Dimensional Protein Structure Can Be Analyzed by X-Ray Diffraction of Protein Crystals

The quality of the information potentially available through x-ray diffraction depends on the degree or extent of order of the protein molecules in a given sample. A sample under investigation usually consists of a hydrated purified protein. If the individual protein molecules are packed in a random order relative to one another, the distances between atoms or groups of atoms can be derived, but we cannot tell how these spacings are ordered in three dimensions.

If the protein is fibrous, it is often possible to learn its two-dimensional order. For example, when a fibrous sample consisting of long α helices is stretched, the helix axes become oriented in the direction of stretching. The resulting fibrous bundle gives a characteristic x-ray diffraction pattern indicating an ordered molecular arrangement along the helix axis. As we have seen, the pitch of the α helix (5.4 Å) and the advance per residue along the helix axis (1.5 Å) were detected by this technique. But the orientation of specific atoms in the helix cannot be determined from diffraction patterns of stretched fibers because of the lack of three-dimensional order. To deduce the correct three-dimensional structure for the α helix (and the β sheet), it was necessary to work with molecular models. (Currently, computer programs are available for such purposes.) By

Figure 4.17

Schematic diagram of the procedures followed for image reconstruction in light microscopy (top) and x-ray crystallography (bottom).

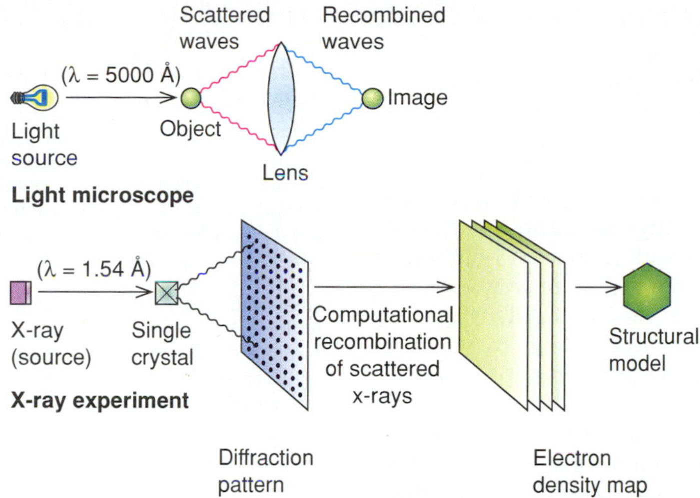

trial and error, model structures were built that were consistent with the steric limitations of the polypeptide chain that had the helix pitch and the advance per residue indicated by the diffraction pattern of stretched fibers.

The most information about a protein's structure is obtained from ordered three-dimensional protein crystals; this is the main interest of x-ray crystallographers. The goal in x-ray crystallography is to obtain a three-dimensional image of a protein molecule in its native state at a sufficient level of detail to locate its individual constituent atoms. The way this is done can most easily be appreciated by considering the more familiar problem of how we obtain a magnified image of an object in a conventional light microscope. In a visible-light microscope, light from a point source is projected on the object we wish to examine. When the light waves hit the object, they are scattered so that each small part of the object essentially serves as a new source of light waves. The important point is that the light waves scattered from the object contain information about its structure. The scattered waves are collected and recombined by a lens to produce a magnified image of the object (fig. 4.17).

Given this picture, we might ask what prevents us from simply putting a protein molecule in place of our object and viewing its magnified image. The basic problem here is one of resolution. The resolution, or extent of detail, that can be recovered from any imaging system depends on the wavelength of light incident upon the object. Specifically, the best resolution obtainable equals $\lambda/2$, or one-half the wavelength of the incident light. Since λ lies in the range of 4,000–7,000 Å for visible light, a visible-light microscope clearly does not have the resolving power to distinguish the atomic structural detail of molecules. What we need is a form of incident radiation with a wavelength comparable to interatomic distances. X-rays emitted from excited metal atoms, with wavelengths in the range of one to a few angstroms, would be most suitable.

However, simply replacing a visible-light source with an x-ray source does not solve all the problems. For example, to get a three-dimensional view of a protein, some provision must be made for looking at it from all possible angles, an obvious impossibility when dealing with a single molecule. Furthermore, when x-rays interact with proteins, very few of the rays are scattered. Most x-rays pass through the protein, but a relatively large number of them interact destructively with the protein, so that a single molecule would be destroyed before scattering enough x-rays to form a useful image. Both these problems are overcome by replacing a single protein molecule with an ordered three-dimensional array of many molecules, which scatters x-rays essentially as if it were one molecule. This ordered array of protein molecules forms a single crystal, so the general technique is called protein x-ray crystallography.

The problems do not end here, because although the protein crystal readily scatters incident x-rays, there are no lens materials available that can recombine the scattered x-rays to produce an image. Instead, the best that can be done is to directly collect the scattered x-rays in the form of a diffraction pattern. Although recording the diffraction pattern results in loss of some important information, experimental techniques have been developed for recovering the lost information. Eventually the scattered waves can be mathematically recombined in a computational analog of a lens. By collecting the diffraction pattern of the crystal in many orientations, it is possible to construct a three-dimensional image of the protein molecule (see box 4B).

4B

BOX

Interpretation of Diffraction Patterns from Protein Crystals

Crystals suitable for protein x-ray studies may be grown by a variety of techniques, which generally depend on solvent perturbation methods for rendering proteins insoluble in a structurally intact state. The trick is to induce the molecules to associate with each other in a specific fashion to produce a three-dimensionally ordered array. A typical protein crystal useful for diffraction work is about 0.5 mm on a side and contains about 10^{12} protein molecules (an array 10^4 molecules long along each crystal edge). Note especially that, because protein crystals are from 20 to 70% solvent by volume, crystalline protein is in an environment that is not substantially different from free solution.

The x-ray radiation usually employed for protein crystallographic studies is derived from the bombardment of a copper target with high-voltage (50 kV) electrons, producing characteristic copper x-rays with $\lambda = 1.54$ Å. Figure 1 shows, in schematic fashion, the x-ray diffraction pattern from a protein crystal. Several features about this pattern bear explanation. First, as you can see, the diffraction pattern consists of a regular lattice of spots of different intensities. The spots are due to destructive interference of waves scattered from the repeating unit of the crystal. For the crystal whose diffraction pattern is shown, the repeating unit (or crystal unit cell) contains four symmetrically arranged protein molecules. Corresponding symmetrical features appear in the spot-intensity pattern. Further, the lattice spacing of the diffraction spots is inversely proportional to the actual dimensions of the crystal's repeating unit or unit cell. Consequently, both the crystal's unit-cell dimensions and general molecule packing arrangement can be derived from inspection of the crystal's diffraction pattern.

Information concerning the detailed structural features of the protein is contained in the intensities of the diffraction spots. All the atoms in the protein structure make individual contributions to the intensity of each diffraction spot. Therefore, to deduce the three-dimensional structure, all the spots must be measured, either by scanning the x-ray films with a densitometer or by measuring the diffraction spots individually with a scintillation counter.

Initial studies of a protein's tertiary structure are generally carried out at low resolution, that is, using intensity data near the origin (center) of the diffraction pattern. Diffraction data near the origin reflect large-scale structural features of the molecule, while those nearer the edge correspond to progressively more detailed features. Figure 2 provides examples of electron-density maps calculated at different resolutions to show how various levels of structural detail appear at different degrees of resolution.

A powerful aspect of protein crystallography is that once the native structure is known, various cofactors or enzyme substrate

Figure 1

Schematic view of an x-ray diffraction pattern. The spacing of the spots is reciprocally related to the dimensions of the repeating unit cell of the crystal. The symmetry of the spots (e.g., the mirror planes in the sample shown) and the pattern of missing spots (alternating spots along the mirror axes) give information on how molecules are arranged in the unit cell. Information concerning the structure of the molecule is contained in the intensities of the spots. Spots closest to the center of the film arise from large-scale or low-resolution structural features of the molecule, while those farther out correspond to progressively more detailed features. Circles show 5-Å and 3-Å regions of resolution. Mirror axes are labeled *m*. Spacing of vertically oriented spots, b^*, and horizontally oriented spots, a^*, are reciprocally related to b and a, the dimensions of the unit cell.

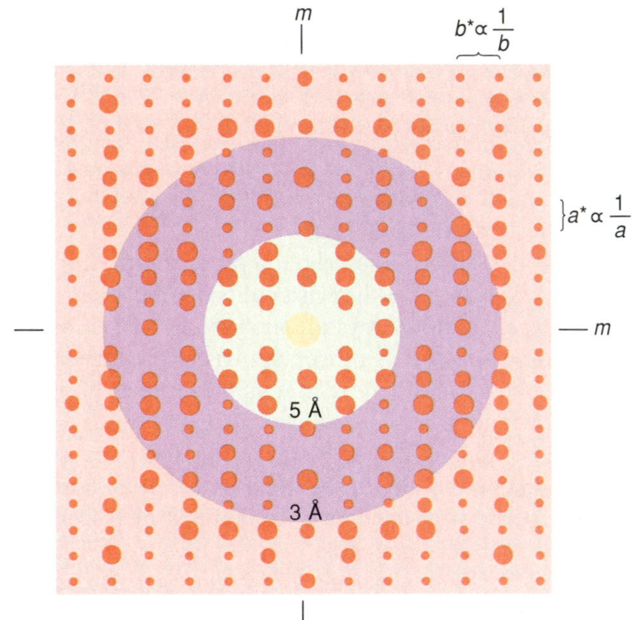

The Hierarchy of Globular Protein Structure

In our analysis of fibrous protein structures, we said little about the importance of interaction with water in determining the final folded structures of the proteins. This is because most of the side chains in fibrous proteins are routinely exposed to water except when they interact with each other to form multimolecular aggregates. Then the difference in their affinity for other, like side chains and for water becomes a major issue. In the case of globular proteins, the interaction of amino acid side chains is a major issue almost from the start, since globular proteins have a large number of their amino acid side chains buried in the interior of their folded structures. Hence in our analysis of

Figure 2

Views of crystallographic electron-density maps, showing how the structural detail revealed depends on the resolution of the data used to compute the maps. The actual molecular structure is inserted in its true position in the electron-density maps.

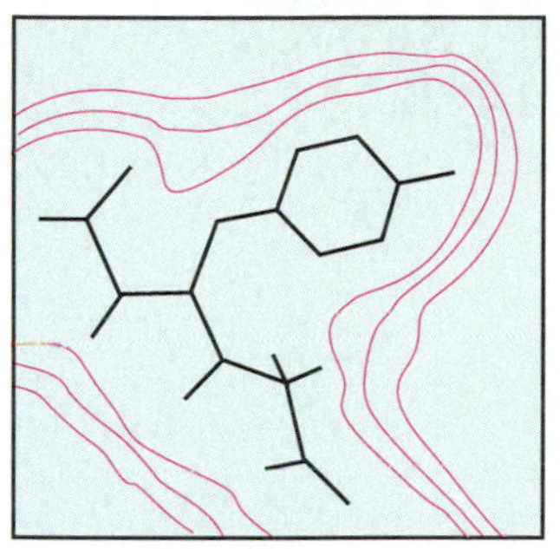

5-Å resolution

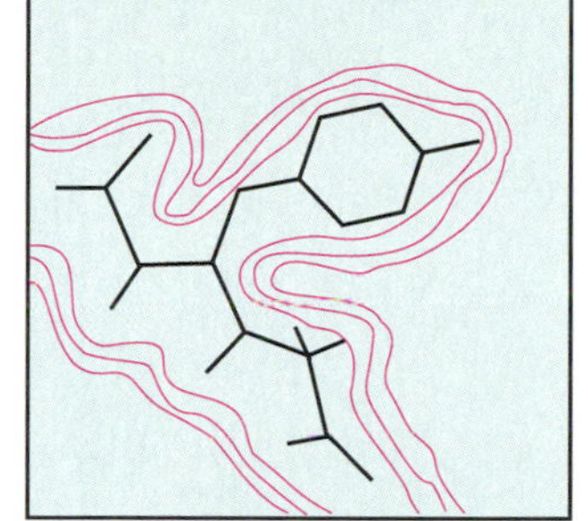

3-Å resolution

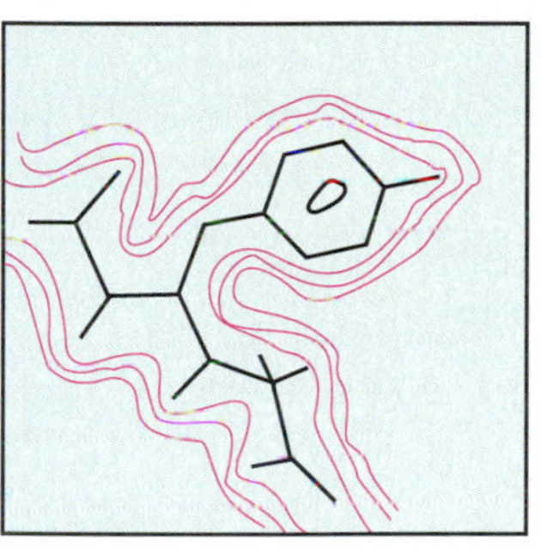

2-Å resolution

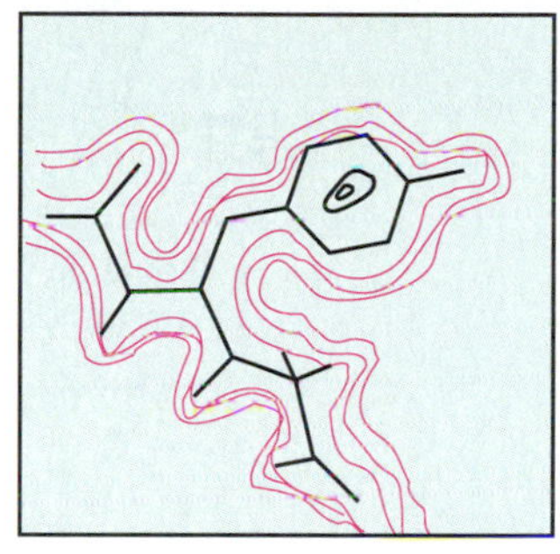

1.5-Å resolution

analogs can be bound to the molecule in the crystal. By simply measuring the diffraction intensities, we can compute a new map that allows direct and explicit examination of the structural interactions between the native protein and its substrate or cofactor molecules. Detailed analysis of these interactions has provided much of the foundation for our current understanding of many protein catalytic and functional properties.

the higher orders of structure of globular proteins we must be aware, not only of those structural considerations that were important in the analysis of fibrous proteins, but also of those considerations, first raised in chapter 1, that relate to the interaction of amino acid side chains with water.

We may think of the structure of globular proteins at four levels (fig. 4.18). Steric interactions restrict accessible conformations and reflect features of the protein's amino acid sequence, or primary structure. The requirements for hydrogen-bond preservation in the folded structure result in the cooperative formation of regular structural regions in proteins. This situation arises principally because of the regularly repeating geometry of the hydrogen-bonding groups of the polypeptide backbone, which leads to the formation of regular hydrogen-bonded secondary structures. Association between elements of secondary structure in turn results in the formation of structural domains, whose properties are determined both by chiral properties of the polypeptide chain and packing requirements that effectively minimize the molecule's hydrophobic surface area. Further association of domains results in the formation of the protein's tertiary structure, or overall spatial arrangement of the polypeptide chain in three dimensions. Finally, fully folded protein subunits can pack together to form quaternary structures, which may either serve a structural role or provide a structural basis for modification of the protein's functional properties.

Because globular proteins have nonrepeating structures, it is essential to have a means for displaying the entire three-dimensional structure with sufficient detail and yet not too much detail, so that the overall structural design can be appreciated (see box 4C).

Primary Structure Determines Tertiary Structure

Throughout the discussion of protein structures we have assumed that structures form because they represent the most stable way of arranging the polypeptide chains. The first direct support for this notion for a globular protein came from the studies of Cris Anfinsen. What Anfinsen did was to completely unfold pancreatic ribonuclease, an enzyme containing 124 amino acid residues with four disulfide bridges, and then to find conditions under which it could be refolded into its original native structure. First, the enzyme was denatured in a solution containing the hydrogen-bond-breaking reagent urea and β-mercaptoethanol, a thiol reagent that reduces disulfides to sulfhydryls, thus cleaving the covalent cross-links (fig. 4.19). These conditions have been used since as a general means of denaturing proteins, by completely disrupting the conformation without giving rise to coagulation or precipitation. The reduced, denatured ribonuclease was enzymatically inactive because its native structure had been destroyed. Renaturation was carried out by removing the urea, a step that caused the protein to refold. Finally the mercaptoethanol was removed to permit the air oxidation of the reduced disulfides back to disulfide cross-links. The result of this series of manipulations was an almost complete recovery of the enzymatic activity of the original ribonuclease molecule.

Thus it appears that the information for folding to the native conformation is embodied in its amino acid sequence, for of the many disulfide-paired ribonuclease isomers that are possible, only one was formed in major yield. Further studies have indicated that a similar result is obtained with many other proteins.

Figure 4.18

Hierarchies of protein structures.

(a) Primary structure (amino acid sequence in the protein chain)

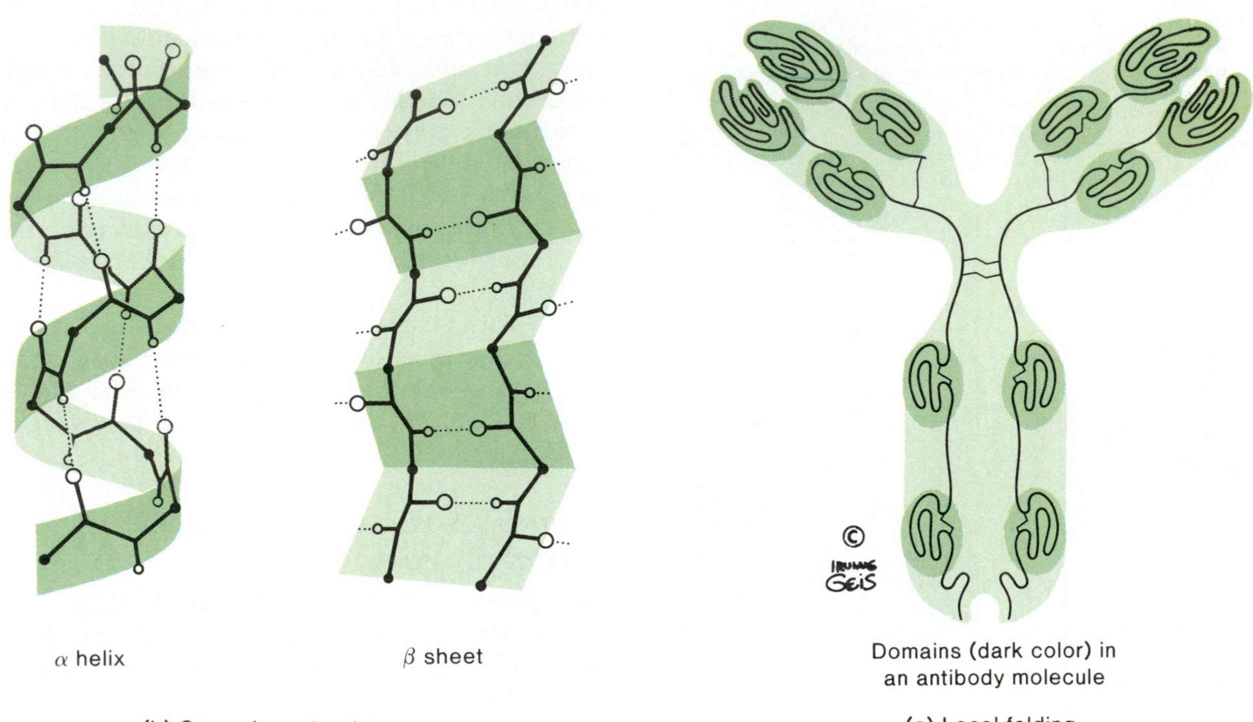

α helix

β sheet

Domains (dark color) in an antibody molecule

(b) Secondary structure

(c) Local folding

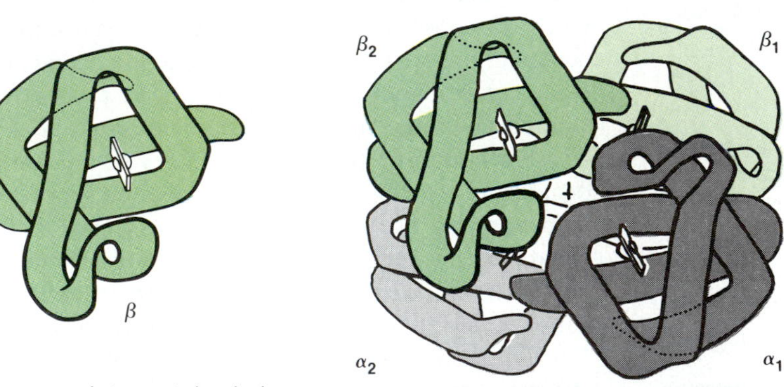

One complete protein chain (β chain of hemoglobin)

(d) Tertiary structure

The four separate chains of hemoglobin assembled into an oligomeric protein

(e) Quaternary structure

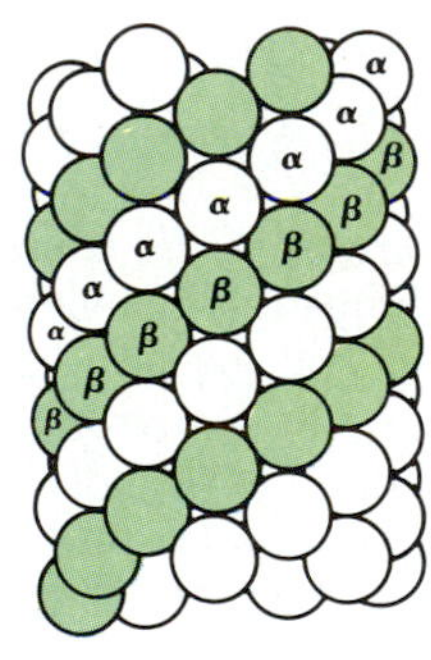

σ (white) and β (color) tubulin molecules in a microtubule

(f) Quaternary structure

Figure 4.19

Schematic representation of an experiment to demonstrate that the information for folding into a biologically active conformation is contained in the protein's amino acid sequence.

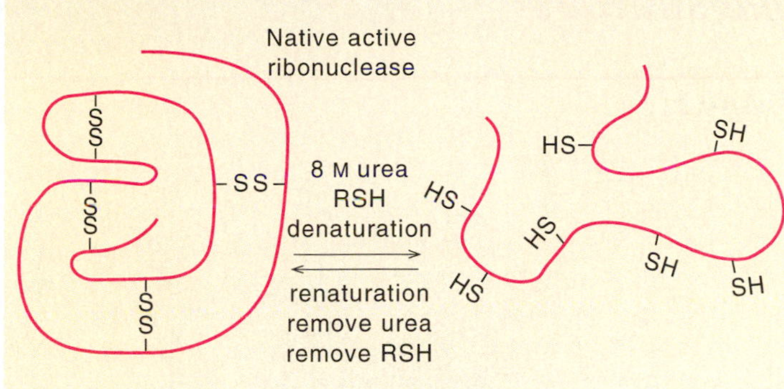

Despite the elegance and simplicity of the Anfinsen experiment, current indications are that there are groups of proteins that accelerate the folding process and that possibly return damaged or incorrectly folded proteins to their native state. These polypeptide-chain-binding proteins (PCB proteins) fall into two families: proteins that are related to the heat shock protein that has a subunit molecular weight of 70,000 (designated hsp70), and the GroEL family of proteins. The GroEL proteins are found in bacteria, mitochondria, and chloroplasts. The hsp70 proteins appear to be ubiquitous, occurring in bacteria, mitochondria, and the eukaryotic cytosol. They are especially abundant in the endoplasmic reticulum.

Hydrophobic Forces Play A Major Role in Determining Folding Patterns

The studies of Anfinsen not only show that folding can be a spontaneous process predetermined by the primary amino acid sequence, they suggest that folded structures are thermodynamically more stable. Thus the folding of globular proteins should be understandable in thermodynamic terms. Although x-ray diffraction cannot be used for studying proteins in solution, there are other optical methods available for the purpose (see box 4D).

Hydrogen bonding and van der Waals forces are of great importance in determining the secondary structures formed by fibrous proteins. To understand the complex folded structures found in globular proteins, additional types of interactions between amino acid side chains and water must also be considered. The so-called hydrophobic forces that lead to the interaction of hydrophobic groups in proteins are the hardest type of noncovalent interactions to appreciate. Whereas H bonding and van der Waals forces are due primarily to enthalpic factors, hydrophobic forces relate primarily to entropic factors. Furthermore, the entropic factors mainly concern the solvent, not the solute.

As a first step to understanding the nature of these forces it is useful to consider the interaction between water and a small hydrophobic molecule like hexane. It is tempting to ascribe the water insolubility of hexane to van der Waals attractive forces between these small hydrophobic molecules. However, thermodynamic measurements indicate that this explanation is wrong. The hydrophobic molecule hexane has a small favorable enthalpy for solution in water; however, it is highly insoluble in water because of an unfavorable entropic factor. What is this mysterious factor? Owing to the weak enthalpic interactions between hexane and water, the water withdraws slightly in the region of the apolar hydrophobic molecule and forms a relatively rigid hydrogen-bonded network with itself (for example, see the clathrate structure illustrated in figure 1.10). The network effectively restricts the number of possible orientations of water molecules directly opposed to the dissolved hexane molecules. This ordering of water constitutes an energetically unfavorable entropic effect.

The same type of entropic effect plays a major role in directing the folding of globular proteins. About half of the amino acid side chains in proteins are hydrophobic (e.g., alanine, valine, isoleucine, leucine, and phenylalanine). For these side chains, entropic effects strongly favor internal locations, where they are free from contacts with water (e.g., see figure 1.12, which shows the location of polar and apolar side chains for cytochrome *c*).

The native folded state of a globular protein reflects a delicate balance between opposing energetic contributions of large magnitude. Whereas entropic factors favor the folding of hydrophobic side chains into the interior regions of globular proteins, enthalpic factors favor interaction of hydrophilic side chains on the surface of the protein, where they interact with water. In the limited number of cases for which data are available, it appears that the overall entropy of folding is slightly negative (unfavorable) and the overall enthalpy is also slightly negative (favorable). On balance, folding is opposed by the entropy but favored by the enthalpy change and occurs because the latter factor outweighs the former.

Secondary Valence Forces Are the Glue that Holds Polypeptide Chains Together

When forces involved in determining protein conformation are being considered, we can ignore covalent bond energies, despite their large magnitude (table 4.2). This is because covalent bonds (except for disulfide bonds) are not made or broken when polypeptide chains fold into their native three-dimensional conformations. The bonds that are affected (made or broken) on folding are, by and large, of the noncovalent type, involving secondary valence forces. Typically the intermolecular bond energies between noncovalently linked atoms range in value from 0.1 to 6 kcal/mol. The intermolecular forces between noncovalently linked atoms may be grouped into four categories: hydrophobic forces, electrostatic forces, van der Waals forces, and hydrogen

Visualizing Molecular Structures

The primary data of protein crystallography yield a three-dimensional electron-density map, which must be interpreted in terms of a complex three-dimensional model of all atom positions in the protein. Such modeling is now generally done by computer graphics.

The difficulty with complete models is that even for relatively small proteins, their complexity is almost impossible to comprehend. Therefore, it is more common to show only selected parts of a molecule (such as only the polypeptide backbone) and to highlight particular features of interest.

Here, we show four different presentations for ribonuclease, in which each model is seen from the same perspective. Each method of presentation has its advantages. The space-filling model (figure 1) is excellent for displaying the volume occupied by molecular constituents and the shape of the outer surface. Figure 2 shows a stereo pair of a space-filling model from exactly the same perspective as figure 1. When the pair is viewed with stereo glasses, the three-dimensional illusion is striking. Figure 3 shows the polypeptide chain, with N and O atoms labeled and with dotted lines representing hydrogen bonds. In addition, all α-carbon positions are numbered. Figure 4 shows an abstraction of the polypeptide chain in which the β strands are characterized as flat arrows and the α helices as spiral ribbons. This simplified style has proved useful in classifying and comparing proteins according to their secondary- and tertiary-structure folding patterns.

Figure 1

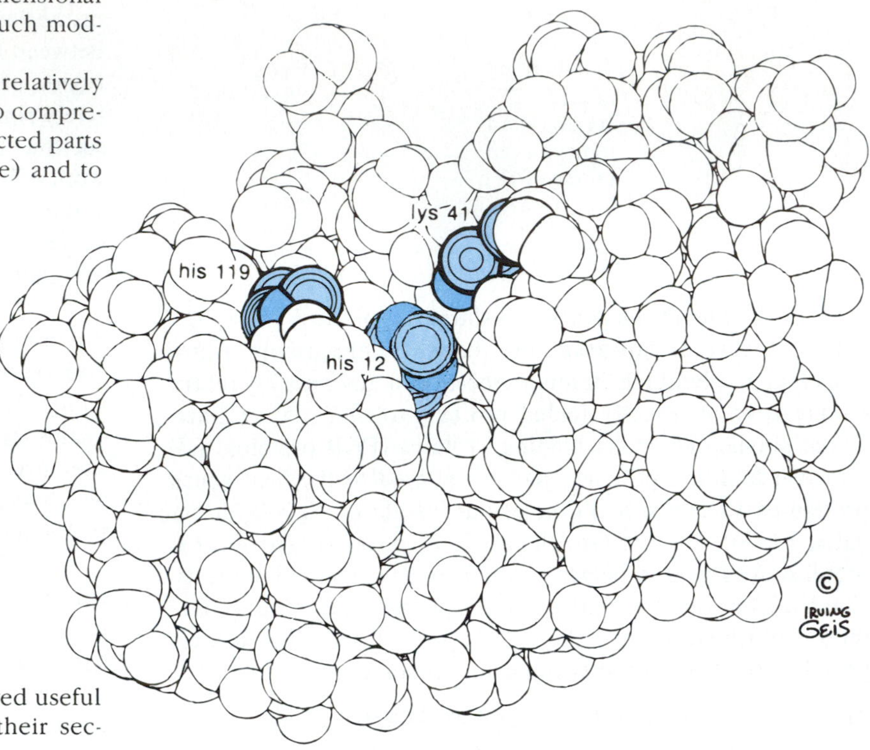

Figure 2

The 3-D illusion can be seen without stereo glasses by the following method. With your eyes about 10 inches from the page, stare at the drawings below as if you were looking straight ahead at a far-away object. A double image will form and the central pair should drift together and fuse. Then the illusion becomes apparent. Adjust the page, if necessary, so that a horizontal line is perfectly parallel with the eyes. As an aid to seeing one image with each eye, use two cardboard tubes from toilet-paper rolls. Close the right eye and focus the left eye on the left image. Do the same for the other eye. Now, with both eyes together, the 3-D illusion should appear.

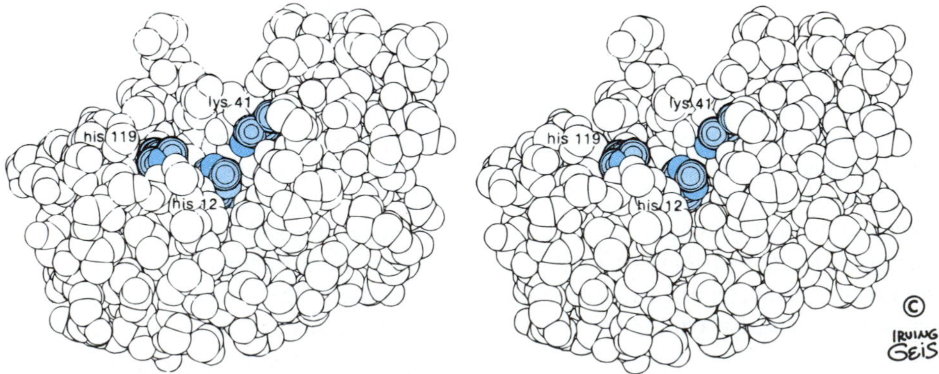

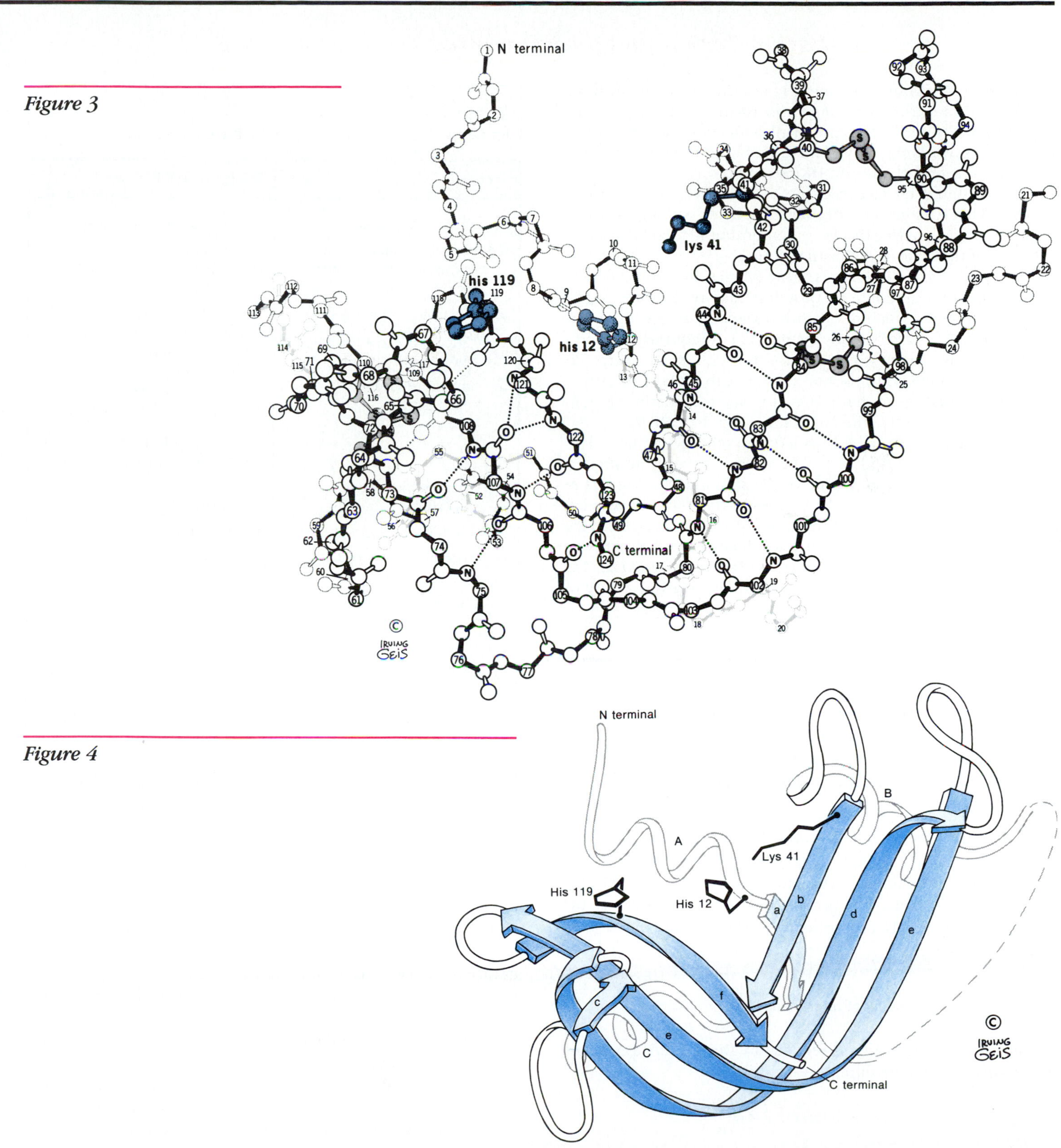

Figure 3

Figure 4

4D
BOX

Radiation Techniques for Examining Protein Structure

The Anfinsen experiment raises the question, Must we always use x-ray diffraction to determine a protein's structure? In the case of ribonuclease the structure is known from x-ray diffraction, but at the time of Anfinsen's investigation it was not. He relied heavily on the fact that the reconstituted enzyme had the same activity as the native enzyme. This remains an acceptable approach as long as one is working with an enzyme that has a demonstrable catalytic activity, but what of the many proteins that do not?

Because of such limitations, x-ray diffraction is still the most powerful tool for determining protein structure. However, apart from the enormous amount of work it requires to use, x-ray diffraction suffers from two disadvantages: (1) it requires that a protein be available in the crystalline form, which is not always the case; and (2) there is no absolute assurance that a protein's structure in the crystalline form is the same as its structure in solution, which may be more like the environment in the cell.

Fortunately, there are at least a dozen other techniques, less fussy in their demands, that may be used to investigate protein structure either in the solid state or the solution state (see table 4D). The information obtained by these procedures is very extensive. As you can see from the table, it covers virtually every aspect of protein structure, with considerable overlap between what is yielded by different techniques. Excellent references explaining the use of other radiation techniques are given at the end of this chapter.

Table 4D
Radiation Techniques for Examining Protein Structure

Technique	Information Obtained
(a) X-ray diffraction	Detailed atomic structure
(b) Nuclear magnetic resonance spectrometry (NMR)	Structure of specific sites; ionization state of individual residues
(c) Electron paramagnetic resonance	Structure of specific sites; this includes structures of small molecules whether they be carbohydrates, lipids, small proteins or complexes between segments of DNA and proteins
(d) Spectrometry (EPR)	
(e) Optical absorption spectroscopy	Measurement of concentrations or rate of reactions
(f) Infrared absorption spectroscopy	Type and extent of secondary structure
(g) Light scattering	Molecular weight and size
(h) Ultraviolet absorption spectroscopy	Concentration and conformation
(i) Fluorescence	Proximity between specific sites
(j) Raman scattering	Structure of specific sites
(k) Optical rotatory dispersion (ORD)	Type and extent of secondary structure
(l) Circular dichroism (CD)	Type and extent of secondary structure
(m) Fluorescence polarization	Molecular weight, shape, flexibility and orientation of secondary structure units

bonds (H bonds). Hydrophobic forces have already been discussed in detail. We considered some aspects of the other types of secondary valence forces in the preceding section and in chapter 1; we will consider additional properties of these forces here.

Electrostatic Forces Electrostatic forces are of three types: charge–charge interactions, charge–dipole interactions, and dipole–dipole interactions. The energy of interaction between two charges Q_1 and Q_2 is proportional to the product of the charges and inversely proportional to the distance R between them (table 4.3):

$$\text{energy of interaction} \propto \frac{Q_1Q_2}{R}$$

In solution this interaction is reduced by the dielectric constant of the surrounding medium:

$$\text{energy of interaction} \propto \frac{Q_1Q_2}{\epsilon R}$$

If the two charges in question are buried within a protein, their interaction energy can be substantially increased because the dielectric constant in the regions inaccessible to water is much lower than the dielectric constant of water. The long-range nature of charge–charge interactions has led to the speculation that such forces can be important in accelerating interaction between proteins, between proteins and nucleic acids, and between proteins and small molecules such as coenzymes and substrates (see chapter 9).

Favorable charge–charge interactions between oppositely charged amino acids are less significant in determining protein folding than are the ion–dipole interactions between the

Table 4.2
Bond Energies between Some Atoms of Biological Interest

Energy Values for Single Bonds (kcal/mole)					
C—C	82	C—H	99	S—H	81
O—O	34	N—H	94	C—N	70
S—S	51	O—H	110	C—O	84
Energy Values for Multiple Bonds (kcal/mole)					
C═C	147	C═N	147	C═S	108
O═O	96	C═O	164	N≡N	226

Table 4.3
Dependence of Energy of Interaction on the Distance of Separation of the Interacting Species

Range of Interaction	Type of Interaction
$1/R$	Charge–charge
$1/R^2$	Charge–dipole
$1/R^3$	Dipole–dipole
$1/R^6$	Van der Waals (dipole–induced dipole) attractive forces
$1/R^{12}$	Van der Waals repulsive forces

charged groups of amino acid side chains and water. With very few exceptions, side chains containing charged groups as well as polar side chains are located on the protein surface at the protein–water interface.

Van der Waals Forces We referred to van der Waals forces earlier in the chapter, in connection with the models constructed by Pauling and Corey. Here we will take a more careful look.

Van der Waals interactions are of two types, one attractive and one repulsive. Attractive van der Waals forces involve interactions among induced instantaneous dipole moments that arise from fluctuations in the electron charge densities of neighboring nonbonded atoms. Such interactions amount to 0.1 to 0.2 kcal/mole; despite their small size, the large number of such interactions that occur when molecules come close together makes the interactions quite significant. As we will see, van der Waals forces favor close packing in folded protein structures.

Repulsive van der Waals interactions occur when noncovalently bonded atoms or molecules come very close together. An electron–electron repulsion arises when the charge clouds between two molecules begin to overlap. If two molecules are held together exclusively by van der Waals forces, their average separation will be governed by a balance between the van der Waals attractive and repulsive forces. This distance is known as the van der Waals separation. Some van der Waals radii for biologically important atoms are given in table 4.1. The van der Waals separation between two nonbonded atoms is given by the sum of their respective van der Waals radii.

Hydrogen Bonds We have discussed many characteristics of H bonds already. Nevertheless, there are still some points to be made. Evidence for a significant H bond comes from the observation of a decreased distance between the donor and acceptor groups making the H bond. Thus, from the van der Waals radii given in table 4.1, we can calculate that the distances between nonbonded H and O atoms and between nonbonded H and N atoms are 2.6 and 2.7 Å, respectively. When an H bond is present, the distances are usually reduced by about 0.8 Å. Some of the more important H-bond donors and acceptors are shown in figure 4.2.

Polypeptides carry a number of H-bond donor and acceptor groups, both in their backbone structure and in their side chains. Water also contains a hydroxyl donor group and an oxygen acceptor group for making H bonds (see fig. 1.8). Formation of the maximum number of H bonds between a polypeptide chain and water would require the complete unfolding of the polypeptide chain. However, it is not obvious that such an unfolding would result in a net energy gain. The reason is that water is a highly H-bonded structure, and for every H bond formed between water and protein, an H bond within the water structure itself must be broken. The strategy followed by most proteins is to maximize the number of intramolecular H bonds between the backbone peptide groups, but to keep most of the potential H-bond-forming side chains of the protein near the protein–water interface, where they can interact directly with water. It seems likely that such side-chain–water interactions will involve both H bonds, on the one hand, and charge–dipole interactions and dipole–dipole interactions on the other.

β Bends Are Useful for Building Compact Globular Proteins

Thus far we have described the geometry of protein secondary structures that resemble long rods or flat sheets. Obviously, to fold a polypeptide chain such as RNase to a compact globular form, there must be some way to change the direction of the polypeptide chain. Such folding might, for example, be required to connect adjacent ends of the polypeptide chains in an antiparallel β sheet. A commonly observed and particularly efficient way to do this is by formation of a tight loop in which a residue's carbonyl group forms a hydrogen bond with the amide NH group of the residue three positions farther along the polypeptide chain. The resulting so-called β bend (fig. 4.20) reverses the direction of the polypeptide chain.

Figure 4.20

The two major types of tight turn of ß bends (I and II). In type II, R_3 is generally glycine.

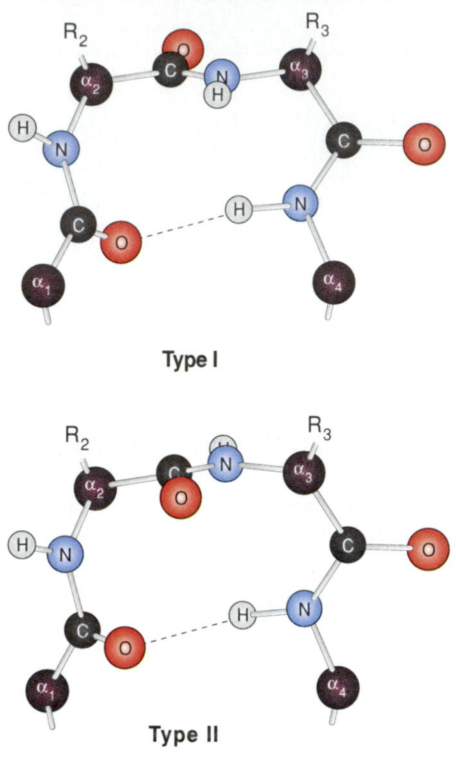

Figure 4.21

Lysozyme. In this and succeeding figures the polypeptide backbone is represented as a ribbon to allow the polypeptide-chain course to be followed easily.

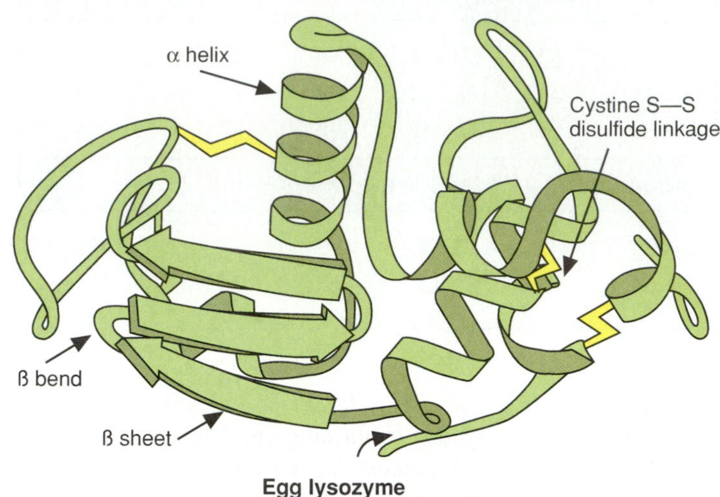

Several conformational variations of the β bend have been observed that are a function of the amino acid sequence in the bend. In particular, it has been observed that the amino acids glycine and proline occur frequently in β bends. Because of its small size, glycine is conformationally more flexible than other amino acids. It can therefore serve as a flexible hinge between regions of polypeptide chains whose steric interactions would otherwise keep them in more extended conformations. Proline, in contrast, is more conformationally restricted than other amino acids, since its cyclically bonded structure fixes its conformational degrees of freedom. In a sense, then, part of the geometry that results in bend formation is performed in proline-containing sequences. It appears probable that either situation might promote the formation of a bend during initial stages of protein folding and so cause structures such as antiparallel β sheets to assemble cooperatively, in a manner resembling the closure of a zipper.

Chirality and Packing Influence the Tertiary Structure of Globular Proteins

There is a seemingly endless array of different folding patterns that can be found for globular proteins. Despite this complexity there are rules that govern the folding process. We have already considered some basic rules that relate directly to thermodynamics. It will be helpful now to consider some additional rules relating to thermodynamic or kinetic aspects of the folding process.

Globular proteins, as their name implies, differ from fibrous proteins in that they generally have a more or less spherical shape. Nevertheless, three-dimensional structural studies of globular proteins show that they incorporate many of the secondary structural features that typify the fibrous proteins. Figure 4.21, for example, illustrates the first enzyme whose three-dimensional structure was determined, the 120-residue protein lysozyme. This protein has local regions of ordered α-helical and antiparallel β-sheet secondary structures, but it has, in addition, several extended regions incorporating β bends and extended polypeptide chains with a less regular conformation.

Among the approximately 500 proteins whose structures have been determined, there is a rich variety of alternative structural arrangements. Each different polypeptide sequence is associated with the formation of a unique tertiary structure. Careful comparisons have shown, however, that many proteins share some fundamental structural similarities. Furthermore, the similarities recur among proteins that show little similarity in sequence or function. This fact suggests that the recurring features have common physical origins. They appear, in fact, to arise from two different sorts of physical effects.

The first of these is the chiral effect, meaning the tendency for extended structural arrangements in proteins to be "handed," because their constituent polypeptide chains are composed of chiral L-amino acids. Chiral effects manifest themselves both in the manner in which regions of secondary structure are interconnected in globular proteins and in the geometric properties of globular protein sheets. The second effect of im-

Figure 4.22

The natural tendency for the polypeptide chain to twist in the right-hand direction produces structures with an overall right-handed connectivity. The structure represents a single fully extended polypeptide chain.

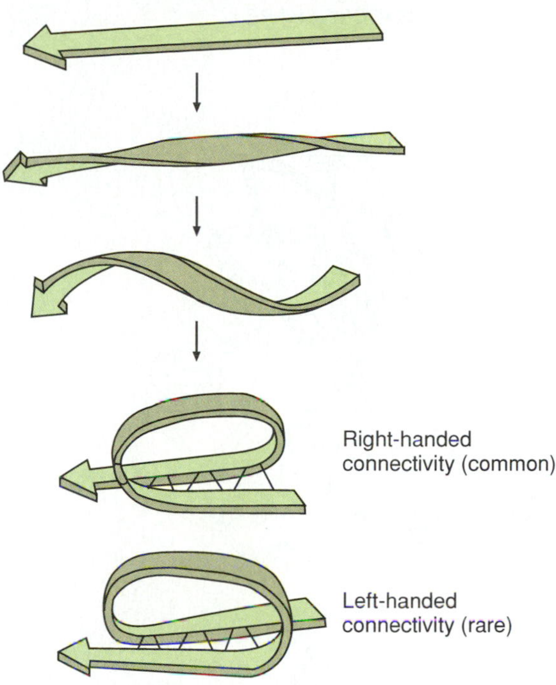

Figure 4.23

Three ways of making connections between ß strands. (*a*) A hairpin same-end connection is commonly found for ß strands in the antiparallel orientation. (*b*) A right-handed crossover connection is commonly found for ß strands in the parallel orientation. (*c*) A left-handed crossover connection is rarely found.

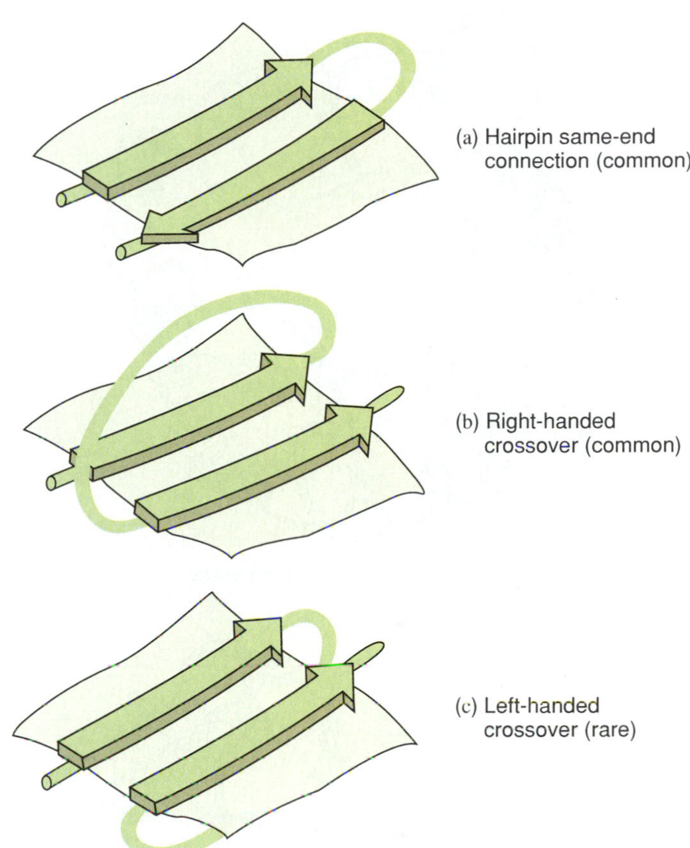

portance in tertiary structural organization relates to how secondary structural regions, such as α helices and β sheets, can most efficiently pack together so as to minimize the protein's solvent-accessible surface area.

How Chiral Properties Influence Conformation In the preceding discussion we noted that many potential polypeptide conformations are ruled out by unfavorable steric interactions. However, the frequent occurrence of structures such as α helices suggests that certain of these arrangements not only are allowed, but are particularly stable. As in the case of the α helix, the relative stability of a particular conformation is governed by the details of the interaction forces among the atoms comprising the polypeptide chain. Given the fact that proteins are composed primarily of chiral L-amino acids, it is not surprising that the most stable conformations of extended polypeptide chains are not straight. Instead, detailed conformational energy calculations have shown that extended polypeptide chains prefer to be slightly twisted in a right-handed sense when viewed down the polypeptide chain axis. Since the residues of straight, extended polypeptide chains alternate in position by 180° (e.g., see fig. 4.7), the cumulative effect of this tendency toward right twisting is to produce extended structures that are coiled in a right-handed direction (fig. 4.22).

The effects of this tendency for extended-chain structures to form right-handed coiled or twisted structures are revealed in two related but different structural features common to virtually all known proteins. The first is the kind of connection that occurs between the ends of parallel polypeptide strands that form β sheets in globular proteins. The connection from one β strand to the next can occur at the same end of the sheet in a simple hairpin turnaround only in antiparallel β sheets. In parallel β sheets a "crossover" connection to the other end of the sheet is required; theoretically, this can be either right-handed or left-handed (fig. 4.23). All crossover connections observed in protein structures are right-handed, irrespective of whether the strands they join are adjacent. This invariant pattern is most likely a result of the energetically favored nature of the right-handed crossover.

The second feature that reveals the influence of right-handed twisting is the geometry of globular protein parallel β sheets. The β sheets of globular proteins are always twisted in a right-handed sense when viewed along the polypeptide chain direction. The twisting behavior of β sheets is an important feature in protein structural architecture, since twisted β sheets frequently constitute the backbone of protein structures. Figure

Figure 4.24

A comparison of parallel ß-sheet structures forming the backbone structures in different enzymes (or parts of enzymes): (*a*) ß-barrel arrangement, (*b*) saddle shape.

ß-BARREL SHAPE

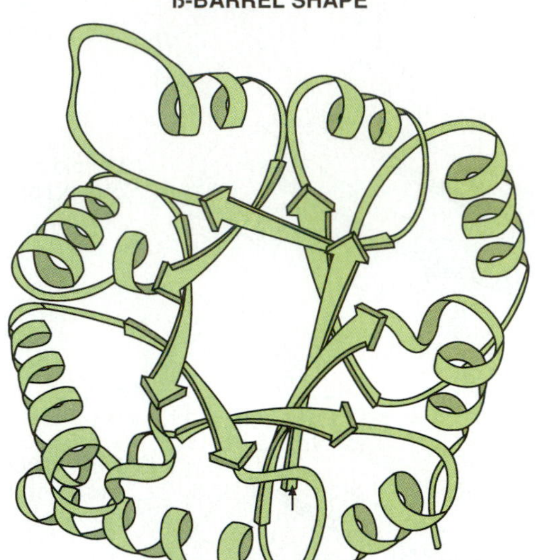

Triose phosphate isomerase

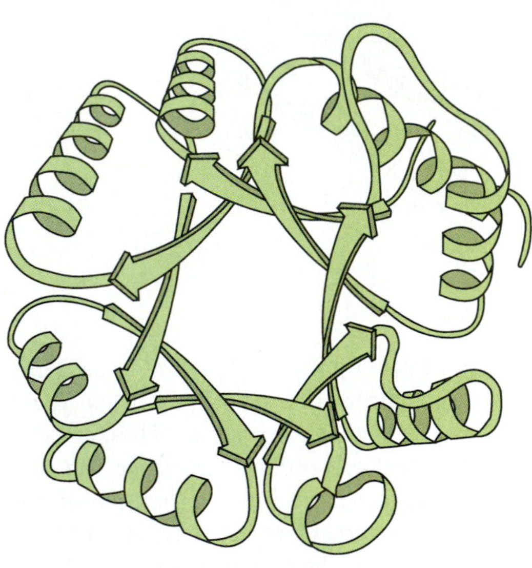

Pyruvate kinase domain 1

(a)

SADDLE SHAPE

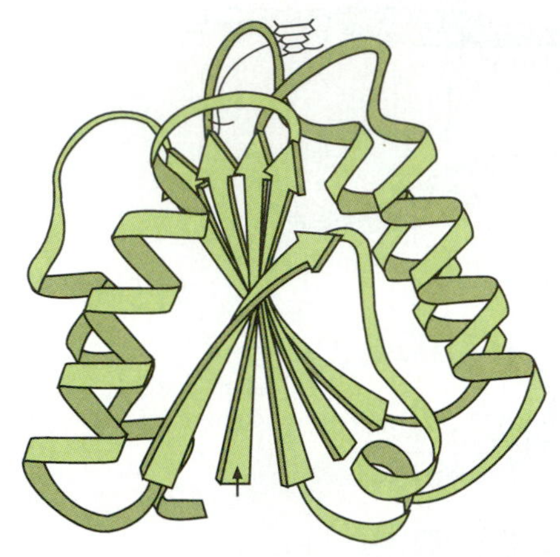

Flavodoxin

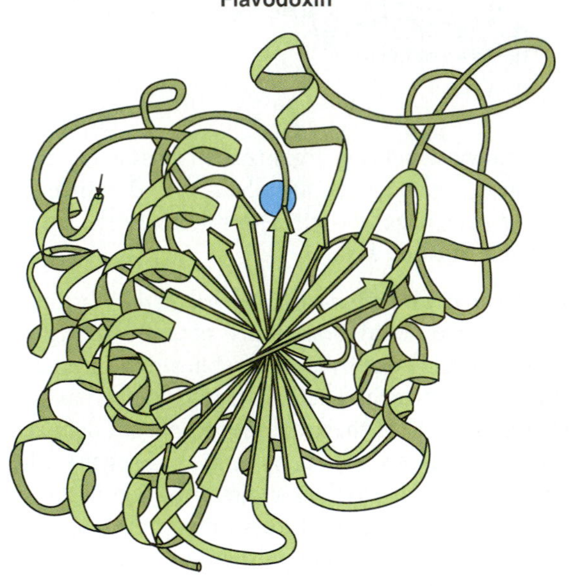

Carboxypeptidase

(b)

4.24 shows the polypeptide-chain folding of four proteins that incorporate twisted parallel or mixed β sheets. These include the exoprotease carboxypeptidase A, the electron-transport protein flavodoxin, and the glycolytic enzyme triose phosphate isomerase. Although these proteins all have right-twisted β sheets, it is clear that their overall geometries differ. That is, the β sheets in carboxypeptidase and flavodoxin are smoothly twisted to form saddle-shaped surfaces, while the β sheets in triose phosphate isomerase take the form of a cylinder or β barrel.

Within each overall type of parallel β sheet organization, the detailed hydrogen-bond pattern can be understood in terms of the forces acting within and between the polypeptide chains (fig. 4.25). In the case of the roughly rectangular sheets in carboxypeptidase and flavodoxin, the observed geometry reflects a competition between the tendency of the individual chains to twist and the tendency of the interchain hydrogen bonds to remain firm. Basically, the interchain hydrogen bonds tend to stretch when the sheet is twisted and so resist introduc-

Figure 4.25

Origin of ß-barrel and ß-sheet conformations. The observed geometry in ß-sheet structures represents a competition between the tendency of the individual chains to twist in a right-handed way and the tendency of the interchain hydrogen bonds to be preserved. Rectangularly arranged sheets give rise to the saddle shape. Staggered sheets give rise to the ß-barrel shape.

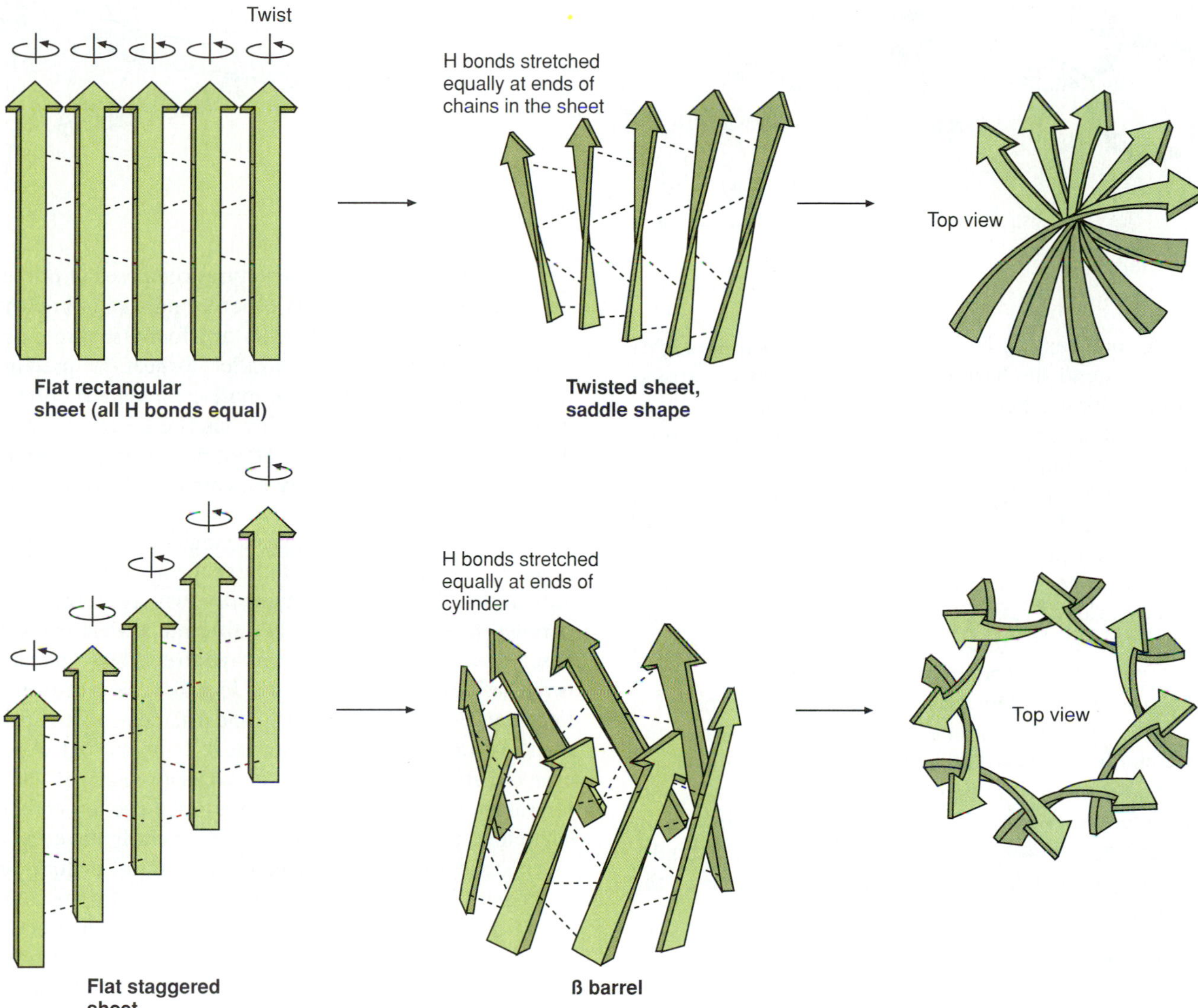

tion of twist into the sheet. The observed saddle-shaped geometry reflects the uniform distribution of these conflicting forces throughout the sheet, as shown in figure 4.25.

The β sheet that forms the barrel in triose phosphate isomerase has an hourglass-shaped surface with cylindrical curvature. Twisted β strands with a staggered hydrogen-bond pattern (see fig. 4.24*a*) automatically produce a cylindrical curvature. Conversely, twisted strands on a cylindrical surface necessitate a staggered hydrogen-bond pattern. Again, a compromise occurs between twisting and hydrogen bonding, leading to approximately straight chains with somewhat stretched hydrogen bonds at the top and bottom, which produce the hourglass shape. The differences in the geometries of rectangular and staggered plane sheets therefore result from differences in how adjacent sheet strands are hydrogen-bonded together. In either case, the operative forces are similar, and the final result reflects a compromise between chain twisting and preservation of good interchain hydrogen bonds.

Chiral preferences affect the connectivity as well as the sheet geometry in parallel β proteins. The right-handed crossover in parallel β barrels cannot go down the center, which is only large enough to accommodate the hydrophobic side chains.

Figure 4.26

Highly simplified sketches of (*a*) a singly wound parallel ß barrel, (*b*) a doubly wound ß sheet. Thin arrows next to the diagrams show the direction in which the chain is progressing from strand to strand in the sheet. The α and ß labels in the figure refer to regions of α helix and ß sheet.

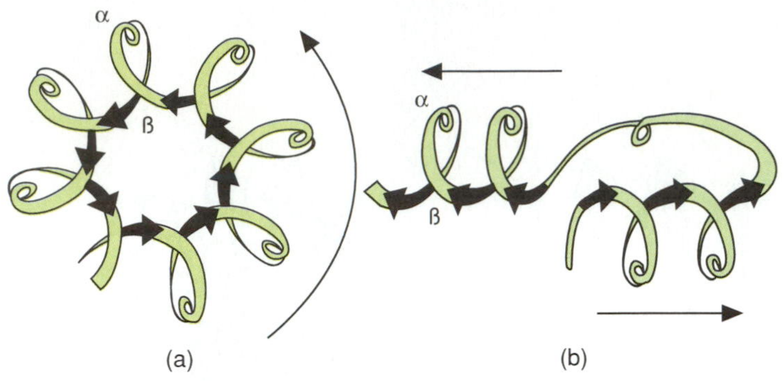

Figure 4.27

A ßαß loop. This arrangement forms the basis of many of the more extended structural arrangements, such as those shown in figure 4.24.

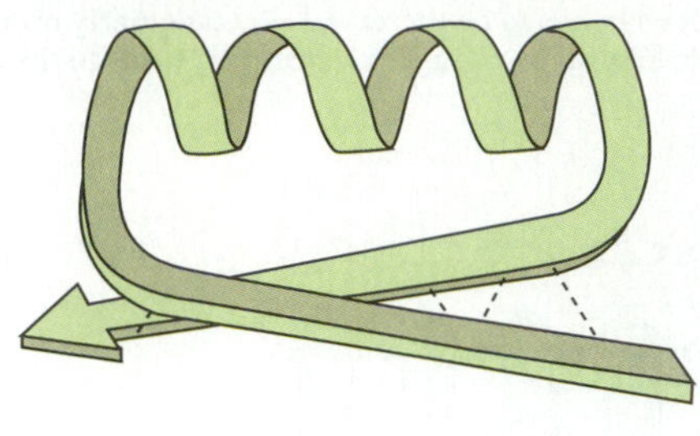

As a rule, the polypeptide backbone winds in a simple right-handed spiral around the barrel, moving over by one β strand at a time and packing helices or loops around the outside. Thus, although these structures tend to be large, their organization (fig. 4.26*a*) is very simple.

The saddle-shaped parallel β sheets, such as those in carboxypeptidase or flavodoxin (see fig. 4.24*b*), have a layer of helices and loops on each side. In order to accomplish this with right-handed crossover connections, the polypeptide chain must sometimes move along the sheet in one direction and sometimes in the other direction. The most common organizational pattern found in known protein structures starts in the middle of the sheet and winds toward one edge, with right-handed crossovers packing a layer of helices on one side of the sheet. Then the polypeptide chain returns to the middle of the sheet and winds out to the opposite edge, packing helices against the other side of the sheet (see fig. 4.26*b*). This pattern is often known as a nucleotide-binding domain, since most of these proteins bind a mononucleotide or dinucleotide cofactor in the middle of the C-terminal end of the β sheet.

Limits on Tertiary Structure Imposed by Efficiency of Packing between Secondary Structures An additional important factor in protein structural organization is the efficient packing together of secondary structural elements to form larger units. Figure 4.27 shows one of the most commonly observed arrangements, called a $\beta\alpha\beta$ loop. This is a special case of the right-handed crossover connection described in the previous section. A $\beta\alpha\beta$ loop is composed of a pair of adjacent hydrogen-bonded parallel β strands that are right-connected, with a stretch of α helix that packs tightly on the surface of the sheet. Generally, the β-sheet strands in a local $\beta\alpha\beta$ loop are part of a much larger β sheet. Structures such as triose phosphate isomerase (see fig. 4.24*a*) can be viewed as a series of overlapping $\beta\alpha\beta$ structural units. The overlapping produces a final structural arrangement in which an inner barrel, composed of β-sheet strands, packs tightly within an outer barrel composed of α helices.

Although tertiary structures composed of $\beta\alpha\beta$ loops are perhaps the most commonly observed pattern in known protein structures, other arrangements are found in proteins having either predominantly antiparallel β-sheet or predominantly α-helical conformations. The most common structural organization in antiparallel β proteins has two layers of β sheets (fig. 4.28). Other antiparallel β proteins have a single twisted β sheet covered on only one side by a layer of helices and loops (fig. 4.29).

No protein is stable as a single-layer structure, since it requires at least two layers to bury the hydrophobic core. Thus antiparallel β proteins are typically two-layer structures; antiparallel sheets are apparently quite stable when one side is exposed to solvent. Parallel sheets require at least one additional layer besides the sheet in order to make the crossover connections between the parallel β strands. In contrast to antiparallel β sheets, they apparently cannot tolerate solvent exposure on even one side and are always found as a structural "backbone" in protein interiors, with other layers of structure on both sides. Therefore, proteins with $\beta\alpha\beta$ loops generally have either three layers, as in the nucleotide-binding domains, or four layers, as in the parallel β barrels (see fig. 4.24). Usually the outer layers are formed of α helices, which must pack against one another and also against the surface formed by the β-sheet side chains.

The structural geometry of proteins that have only an α-helical secondary structure is also largely determined by requirements for efficient packing between the helices. We have already encountered one particularly stable interhelical packing arrangement in the discussion of fibrous proteins (see fig. 4.6). In the α-keratins, adjacent helices pack together with an interaction angle between helices of about 18°. This extended-helix interaction pattern forms the basis for a protein structural motif frequently seen in globular proteins, the 4-α-helical bundle. In this arrangement, four α helices, sequentially connected to their nearest neighbors, pack together to form an array with a roughly square cross section. Since each helix interacts with its neighbors at an angle of about 18°, the overall bundle has a left-handed twist (fig. 4.30). This commonly observed folding domain clearly represents a minimum accessible surface area arrange-

Figure 4.28

Examples of proteins containing ß-sheet domains.

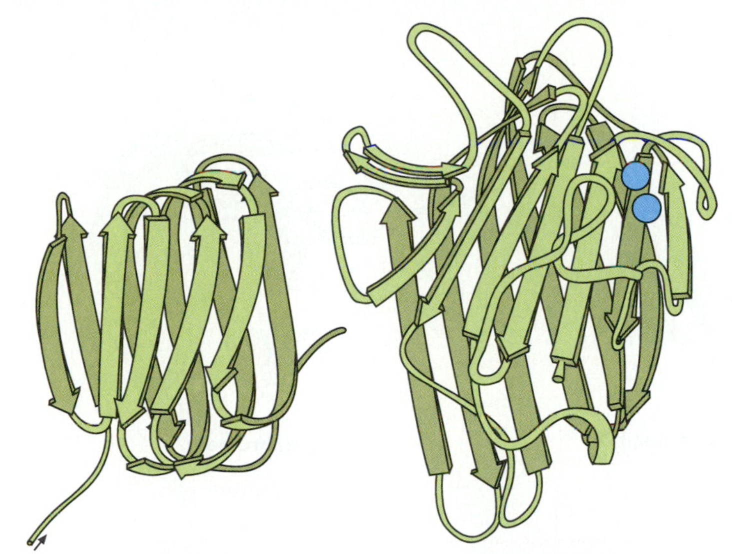

Figure 4.29

Examples of antiparallel ß proteins that are covered on only one side by larger helices and loops.

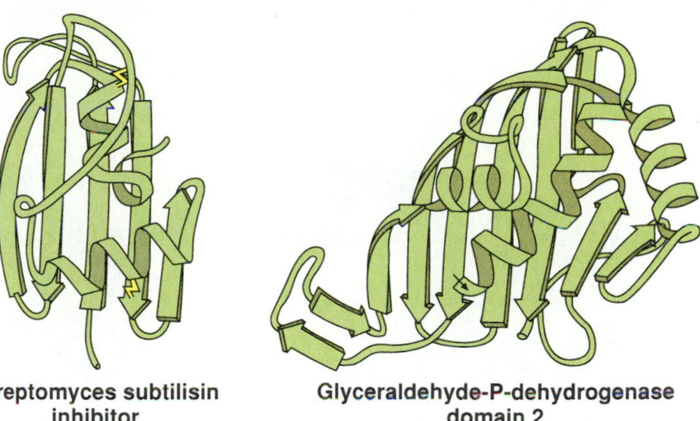

Figure 4.30

Examples of some proteins that share a common structural motif of four α helices.

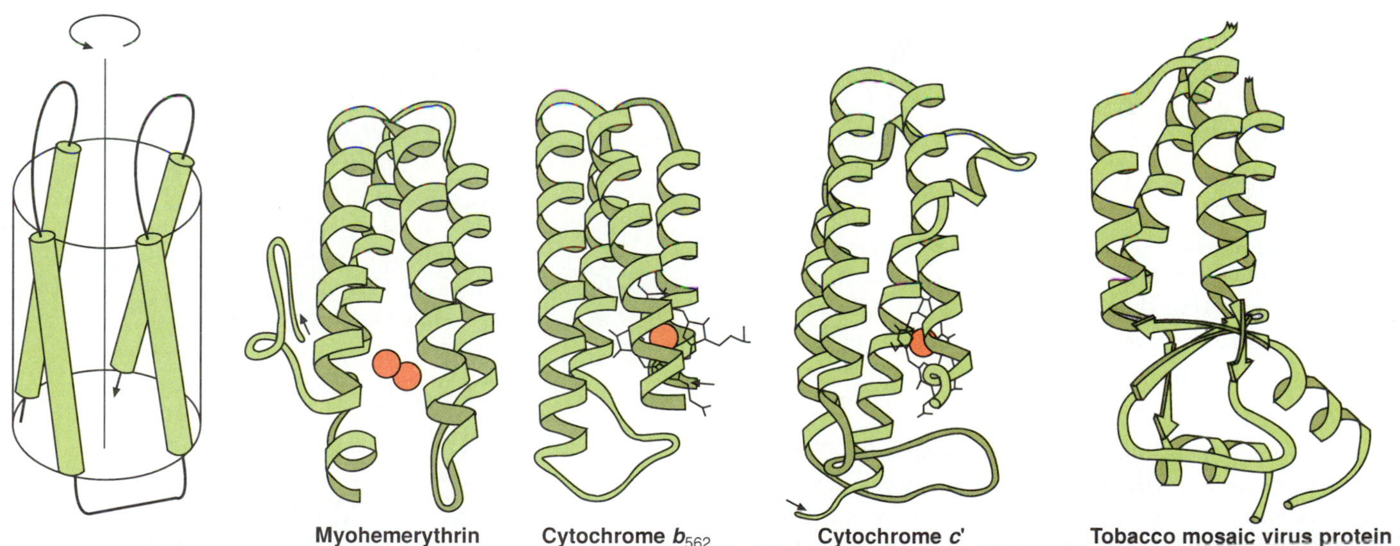

ment for four sequentially connected α helices of approximately equal length. Many α-helical proteins that lack these features have more complex and irregular geometries. However, even in these cases it appears that the relative orientations of adjacently packed helices reflect geometric restrictions that accompany close packing between helices.

In addition to the packing of elements of protein secondary structure, which is a dominant feature in most proteins, there are some cases, especially among the smallest structures, where the geometry and packing of disulfide bonds or nonpeptidyl groups are a dominant factor. Figure 4.31 shows examples of this sort, in which the secondary structures are short and irregular and cannot assume their native structures if the disulfides are broken (*a, b*), or if the nonpeptidyl groups are missing (*c*).

In summary, examination of a large number of protein tertiary structures has shown that they incorporate several different sorts of recurring structural arrangements. In general,

Figure 4.31

Examples of some small proteins or domains in which disulfide bonds. (*a, b*) or a porphyrin group (*c*) are a dominant factor holding the structure together. Porphyrins are depicted in figure 5.11 and, in greater detail, in figure 11.20.

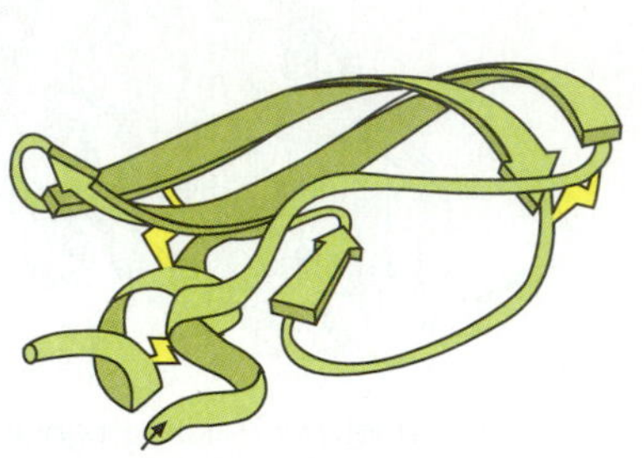

Pancreatin trypsin inhibitor
(a)

Wheat germ agglutinin domain 2
(b)

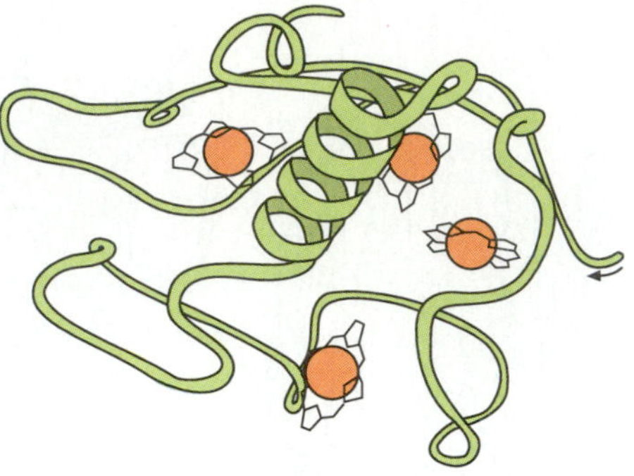

Cytochrome c_3
(c)

Figure 4.32

Papain, a protein in which the domains are very different from one another.

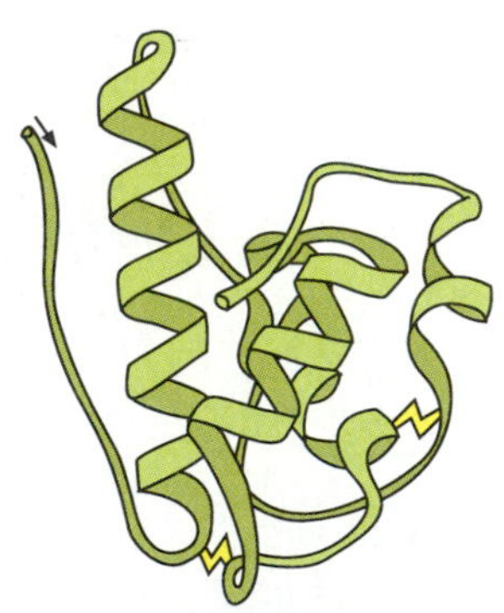

Papain domain 1

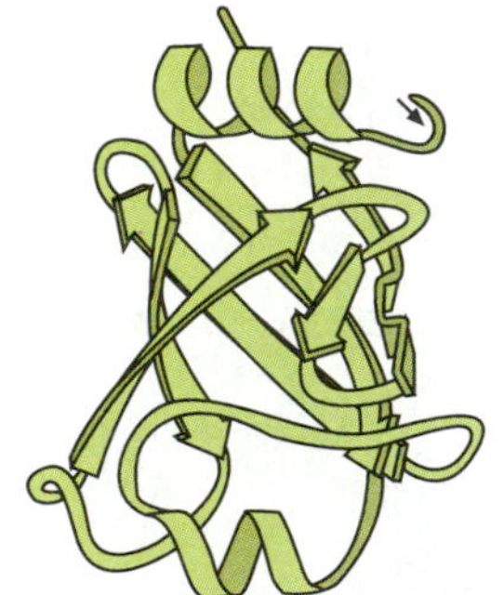

Papain domain 2

these arrangements owe their origins to physical effects, some of which predispose the most stable conformations of the polypeptide chain, and some of which govern the formation of intimately packed tertiary structures.

Domains as Functional Units of Tertiary Structure The patterns of tertiary structure described in this and previous sections frequently constitute the entire protein. However, within a single folded chain or subunit, contiguous portions of the polypeptide chain often fold into compact local units called domains, each of which might consist, for example, of a four-helix cluster or a barrel or an antiparallel β sheet. Sometimes the domains within a protein are very different from one another, as within the protease papain (fig. 4.32), but often they resemble each other very closely, as in rhodanase (fig. 4.33).

The separateness of two domains within a subunit varies all the way from independent globular domains joined only by a flexible length of polypeptide chain, to domains with tight and extensive contact and a smooth globular surface for the outside of the entire subunit, as in the proteolytic enzyme elastase (fig. 4.34). An intermediate level of domain separateness, characterized by a definite neck or cleft between the domains, is found in phosphoglycerate kinase (fig. 4.35).

Figure 4.33

Rhodanese domains 1 and 2 as an example of a protein with two domains that resemble each other extremely closely. Rhodanese is a liver enzyme that detoxifies cyanide by catalyzing the formation of thiocyanate from thiosulfate and cyanide.

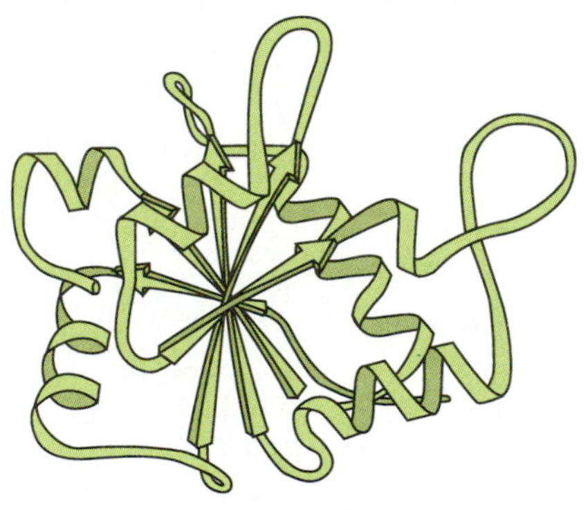

Rhodanese domain 1

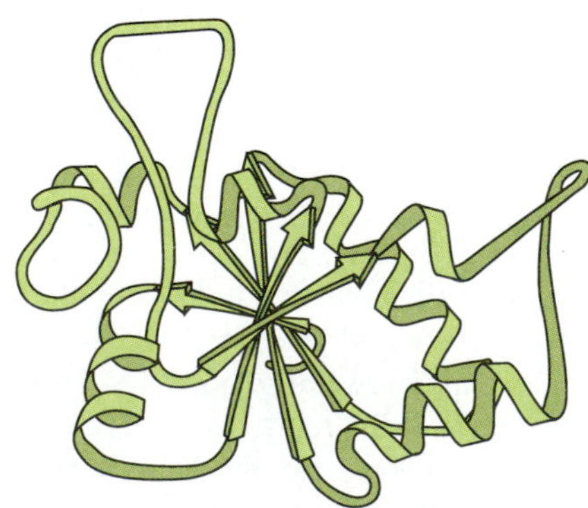

Rhodanese domain 2

Figure 4.34

Schematic backbone drawing of the elastase molecule, showing the similar ß-barrel structures of the two domains.

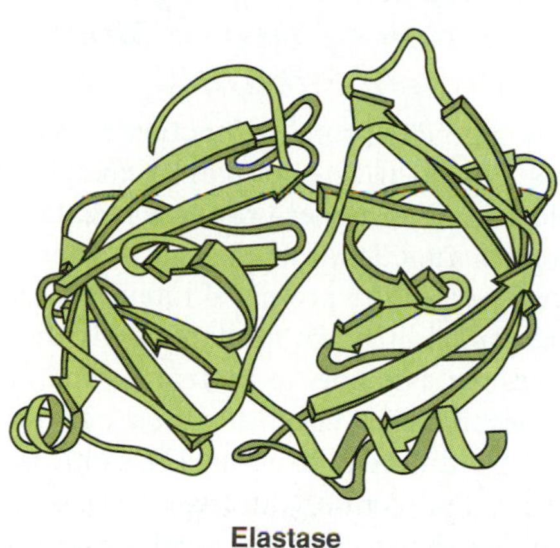

Figure 4.35

The dumbbell domain organization of phosphoglycerate kinase, with a relatively narrow neck between two well-separated domains.

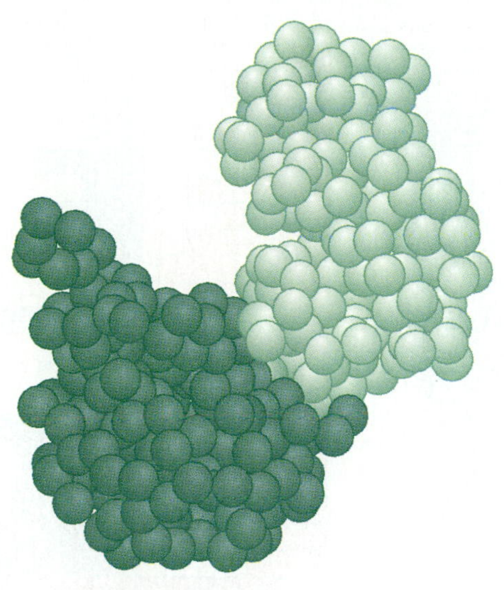

Domains as well as subunits can serve as modular bricks to aid in efficient assembly of the native conformation. Undoubtedly the existence of separate domains is important in simplifying the protein-folding process into separable, smaller steps, especially for very large proteins. There is no strict upper limit on folding size. Indeed, known domains vary in size all the way from about 40 residues to over 400.

Another important function of domains is to allow for movement. Completely flexible hinges would be impossible between subunits because they would simply fall apart. However, flexible hinges can exist between covalently linked domains. Limited flexibility between domains is often crucial to substrate binding, allosteric control (discussed in chapter 10), or assembly of large structures. In hexokinase, the two domains within the individual subunits hinge toward each other upon binding of the substrate glucose, enclosing it almost completely (fig. 4.36). In this manner glucose can be bound in an environment that excludes water as a competing substrate (see chapter 13 for further details on the hexokinase reaction).

Figure 4.36

Schematic representation of the change in conformation of the hexokinase enzyme on binding substrate. E and E′ are the inactive and active conformations of the enzyme, respectively. G is the sugar substrate. Regions of protein or substrate surface excluded from contact with solvent are indicated by a crinkled line. Figure 9.3 presents a more detailed view of the hexokinase molecule. (Source: W. S. Bennett and T. A. Steitz, "Glucose-induced conformational change in yeast hexokinase," *Proceedings. National Academy of Sciences USA*, 75:4848, 1978. Copyright ©1978 National Academy of Sciences, Washington, D.C.)

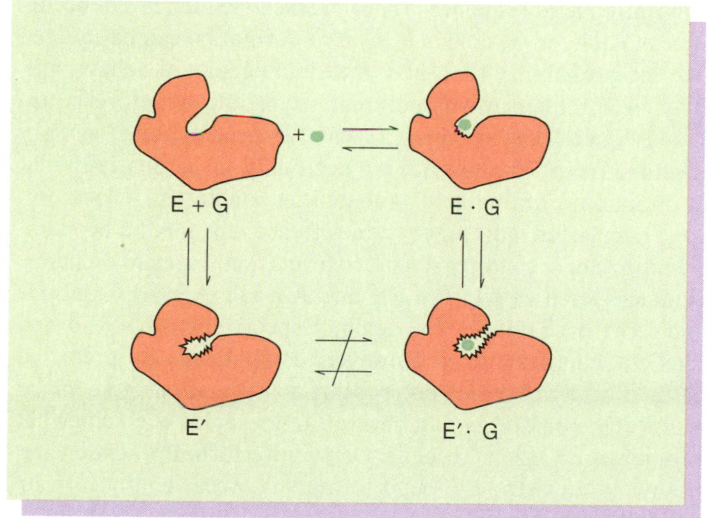

The Prediction of Secondary and Tertiary Structures

We began our discussion of globular protein tertiary structure by pointing out that the tertiary structure is determined by the primary structure, probably because in many cases the native tertiary structure is the most stable structure that can be formed from a given primary structure. If this is so, then it might be possible to predict a protein's structure from its primary sequence alone. Right now, however, x-ray diffraction and nuclear magnetic resonance (NMR) are yielding information on tertiary structures at such a rate that efforts in that direction may not be necessary—especially since most proteins are made of a limited number of domains, which tend to reappear in many proteins. Thus in the future we may be able to predict the structures of many proteins by combining the limited amount of structure information accumulated from x-ray diffraction or NMR studies with the vast data on amino acid sequence of proteins whose tertiary structure is unknown.

Figure 4.37

Relative probabilities that any given amino acid will occur in the α-helical, ß-sheet, or ß-hairpin-bend secondary structural conformations.

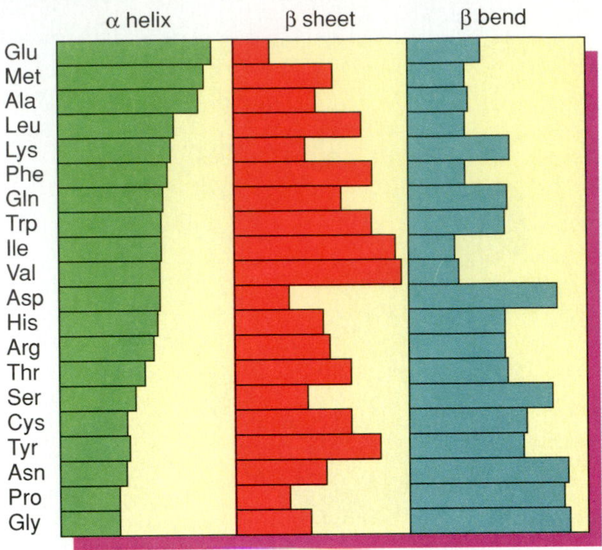

At this juncture it seems clear that various amino acids have tendencies to form different secondary structures. As shown in figure 4.37, glutamic acid, methionine, and alanine appear to be the strongest α-helix formers and valine, isoleucine, and tyrosine the most probable β-sheet formers, while proline, glycine, asparagine, aspartic acid, and serine occur most frequently in β-bend conformations. This type of information is of value in the prediction of secondary structural regions of proteins from their amino acid sequences. The observed frequencies of occurrence of each amino acid in a given conformation can be equated with the probability that the same amino acid will behave similarly in a sequence whose actual secondary structure is unknown. Therefore, in theory at least, to predict the secondary structure from the sequence we need only to sequentially plot the individual amino acid probabilities, or better, a local average over a few adjacent residues to account for the cooperative nature of secondary-structure formation. In such a scheme, sequences such as Gly-Pro-Ser and Ala-His-Ala-Glu-Ala give high joint probabilities for being, respectively, in β-bend and α-helical conformations. However, comparisons of predicted versus directly observed polypeptide conformations give mixed results. One reason is that several amino acids are somewhat ambiguous in their secondary-structure-forming tendencies; another is that strong β-bend formers do occasionally turn up in the middle of α helices.

Nevertheless, it is clear that the structural arrangement of a protein in its folded state must depend ultimately on its sequence. The information given in figure 4.37—that certain amino acids have general tendencies toward forming particular sorts of structures—is a basic point of departure. The attainment of a particular folding arrangement depends on details of both the short- and long-range interactions that uniquely characterize and stabilize each protein's structure.

Quaternary Structure Depends on the Interaction of Two or More Proteins or Protein Subunits

Although many globular proteins function as monomers, biological systems abound with examples of complex protein assemblies (table 4.4). This higher-order organization of globular subunits to form a functional aggregate is referred to as the quaternary structure of the protein. Protein quaternary structures can be classified into two fundamentally different types. The first involves the assembly of proteins (sometimes referred to as subunits because they constitute a part of the final structure) that are very different structures. Examples range from dimeric molecules that contain different molecular subunits to complex assemblies such as ribosomes, which contain twenty or more nonidentical protein subunits in addition to one or more RNA components. The organization of these sorts of quaternary structures depends on the specific nature of each interaction made between the different molecular subunits and their neighbors. Each intermolecular interaction generally occurs only once within a given aggregate arrangement, so that the overall complex structure has a highly irregular geometry. A widely used approach for determining the state of aggregation of proteins in solution is by sedimentation and diffusion analysis (see box 4E).

A second, commonly observed pattern of quaternary structure is typified by molecular aggregates composed of multiple copies of one or more different kinds of subunits. Owing to the recurrence of specific structural interactions between the subunits, such aggregates typically form regular geometric arrangements. Given that proteins are fundamentally asymmetrical objects (because they incorporate chiral L-amino acids), it is clear that the simplest pattern of quaternary structure involves formation of a linear aggregate. As illustrated in figure 4.38*a, b,* the formation of such an aggregate results from the repetition of one sort of specific structural interaction between adjacent subunits of the assembly. Structures of this type can be extended simply by the addition of successive subunits.

Somewhat more frequently observed than linear arrangements are helical arrangements of identical molecular subunits. As is the case for the amino acid residues in the α helix, in helical quaternary structures the individual subunits display different, local interactions with their nearest neighbors (see fig. 4.38*c, d*). However, the pattern of nearest-neighbor interactions is repeated for each subunit.

Helical molecular aggregates are frequently associated with self-assembling molecular structures. Some outstanding examples of such aggregates are the rodlike and filamentous viruses, in which the helical aggregation of the protein-coat subunits forms a cylindrical container for the virus's nucleic acid (fig. 4.39). Since the entire coat is assembled from multiple

Table 4.4
Molecular Weight and Subunit Composition of Selected Proteins

Protein	Molecular Weight	Number of Subunits	Function
Glucagon	3,300	1	Hormone
Insulin	11,466	2	Hormone
Cytochrome *c*	13,000	1	Electron transport
Ribonuclease A (pancreas)	13,700	1	Enzyme
Lysozyme (egg white)	13,900	1	Enzyme
Myoglobin	16,900	1	Oxygen storage
Chymotrypsin	21,600	1	Enzyme
Carbonic anhydrase	30,000	1	Enzyme
Rhodanese	33,000	1	Enzyme
Peroxidase (horseradish)	40,000	1	Enzyme
Hemoglobin	64,500	4	Oxygen transport
Concanavalin A	102,000	4	Unknown
Hexokinase (yeast)	102,000	2	Enzyme
Lactate dehydrogenase	140,000	4	Enzyme
Bacteriochlorophyll protein	150,000	3	Enzyme
Ceruloplasmin	151,000	8	Copper transport
Glycogen phosphorylase	194,000	2	Enzyme
Pyruvate dehydrogenase (*E. coli*)	260,000	4	Enzyme
Aspartate carbamoyltransferase	310,000	12	Enzyme
Phosphofructokinase (muscle)	340,000	4	Enzyme
Ferritin	440,000	24	Iron storage
Glutamine synthase (*E. coli*)	600,000	12	Enzyme
Satellite tobacco necrosis virus	1,300,000	60	Virus coat
Tobacco mosaic virus	40,000,000	2,130	Virus coat

Figure 4.38

Linear and helical quaternary aggregates of protein molecules. (*a*) A linear arrangement of hypothetical protein subunits (illustrated as simplified right shoes). The interactions in such a linear arrangement are all identical, and the structure lends itself to the formation of an indefinitely long linear structure whose subunits are related by translation in one dimension. (*b*) A helical arrangement in which equivalently interacting subunits are related by unit translations along the helix axis followed by a 180-degree rotation about the helix axis. (*c*) A helical arrangement in which subunits are related by unit translation plus a rotation of 360 degrees/*n* to give an *n*-fold helix. (*d*) An *n*-fold multiple-start helix, an arrangement in which subunits form different equivalent interactions with their nearest neighbors. Red arrows indicate two types of identical contact points.

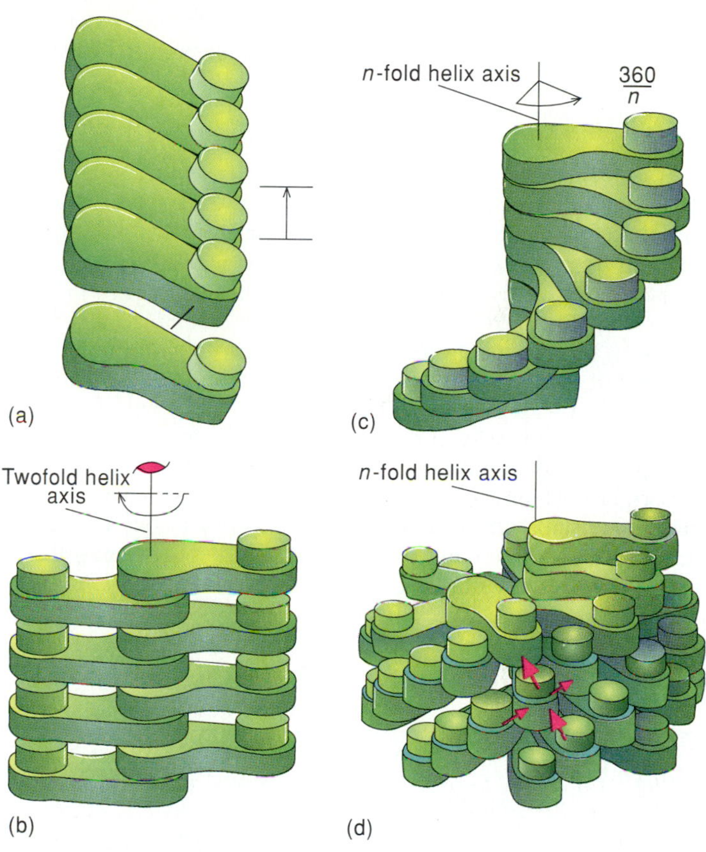

copies of the same protein, this arrangement represents a very efficient utilization of the information content of the virus nucleic acid.

Helical quaternary structures, then, are characterized by a repeating interaction that results in structures whose subunits are related both by some rise along and twist around a central axis. They are therefore similar to linear arrangements in that they are, at least potentially, indefinitely extendable. In fact, in helical viruses the length of the coat-protein structure is determined, not by the protein, but rather by the fixed length of the virus nucleic acid.

4E
BOX

The Use of Sedimentation and Diffusion to Calculate Molecular Weights in Solution

In a centrifugal force field, protein molecules slowly migrate toward the bottom of a centrifuge tube at a rate that is proportional to their molecular weight (figure 1). The rate of sedimentation may be recorded by optical methods that do not interfere with the operation of the centrifuge. From this rate, we can obtain the sedimentation constant *s*. This constant equals the rate at which a molecule sediments, divided by the gravitational field (angular acceleration in a spinning rotor), as defined by the equation

$$s = \frac{dx/dt}{\omega^2 x} \qquad \text{(B1)}$$

where dx/dt is the rate at which the particle travels at distance x from the center of rotation, ω is the angular velocity of the rotor in radians per second (hence $\omega^2 x$ is the angular acceleration), and t is the time of centrifugation in seconds. The sedimentation constant is usually given in Svedberg units (S); one S = 10^{-13} *s*.

From the sedimentation constant we can obtain the molecular weight, provided we have certain other information. This additional information includes the frictional coefficient (f) of the protein and the density of the protein. The coefficient f is bigger for larger proteins and, for proteins of the same molecular weight, it is larger for elongated, rodlike molecules. The density of the protein is important because of the buoyancy factor, $1 - \bar{v}_p\rho_s$, which takes into account the density difference between solvent (ρ_s) and the volume of water displaced per gram of protein ($\bar{v}_p$). The equation that relates *s, M,* and *f* is

$$s = \frac{M(1 - \bar{v}_p\rho_s)}{Nf} \qquad \text{(B2)}$$

Figure 1

Apparatus for analytical ultracentrifugation. (*a*) The centrifuge rotor and method of making optical measurements. (*b*) The optical recordings as a function of centrifugation time. As the light-absorbing molecule sediments, the solution becomes transparent.

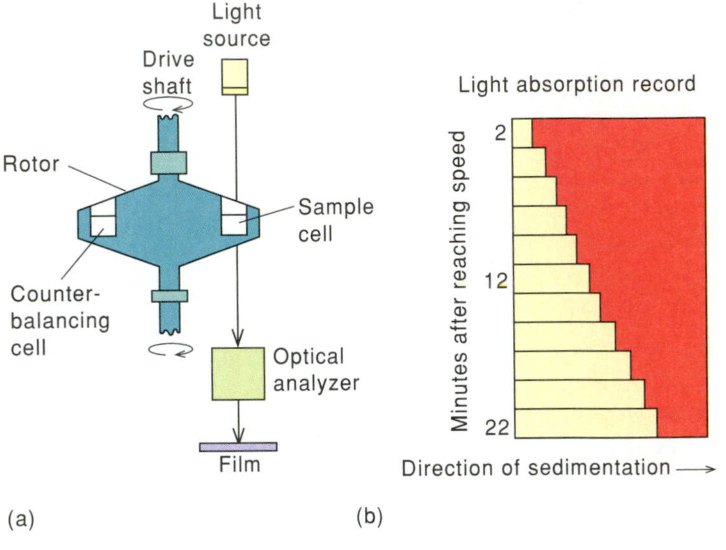

Table 4E
Physical Constants of Some Proteins

Protein	Molecular Weight	Diffusion Constant ($D \times 10^7$)	Sedimentation Constant (*s*)	pI[a] (Isoelectric)
Cytochrome *c* (bovine heart)	13,370	11.4	1.17	10.6
Myoglobin (horse heart)	16,900	11.3	2.04	7.0
Chymotrypsinogen (bovine pancreas)	23,240	9.5	2.54	9.5
β-Lactoglobulin (goat milk)	37,100	7.5	2.9	5.2
Serum albumin (human)	68,500	6.1	4.6	4.9
Hemoglobin (human)	64,500	6.9	4.5	6.9
Catalase (horse liver)	247,500	4.1	11.3	5.6
Urease (jack bean)	482,700	3.46	18.6	5.1
Fibrinogen (human)	339,700	1.98	7.6	5.5
Myosin (cod)	524,800	1.10	6.4	—
Tobacco mosaic virus	40,590,000	0.46	198	—

[a]pI = −log *I* = the isoelectric point, that is, the pH at which the protein carries no net charge (see box 5B).

where N is Avogadro's number. Thus in order to estimate the molecular weight from the sedimentation constant we must have a means of determining $\bar{v}_p$ and f.

For most proteins $\bar{v}_p$ is about 0.75 cc/g, so its value does not present much of a problem. The frictional coefficient, however, is a sensitive function of the shape, varying over a wide range, and we must usually know its value if we need a serious estimate. The value of f is usually found by working with the diffusion constant D, which is related to the frictional constant by

$$D = \frac{RT}{Nf} \text{ or } f = \frac{RT}{ND} \tag{B3}$$

where R is the gas constant and T is the absolute temperature. Substituting this expression for f in equation (B2) and transposing leads to

$$M_r = \frac{RTS}{D(1 - \bar{v}_p\rho_s)} \tag{B4}$$

The diffusion constant that we need if we want to use equation (B4) to calculate the molecular weight is usually obtained with the help of Fick's first law of diffusion:

$$\frac{dn}{dt} = -DA\left(\frac{dc}{dx}\right)_t \tag{B5}$$

This equation states that the amount dn of a substance crossing a given area A in time dt is proportional to the concentration gradient dc/dx across that area. The diffusion constant D, which is a function of both molecular weight and shape, is the proportionality constant. It can be measured by observing the spread of an initially sharp boundary between the protein solution and a solvent as the protein diffuses into the solvent layer. Once we know the value of the diffusion constant, we can combine the information with the sedimentation data and calculate the molecular weight of the protein.

Representative sedimentation and diffusion data, together with the calculated molecular weights, are presented in table 4E. The sedimentation constants are usually larger the greater the molecular weight. Likewise, the diffusion constants are usually inversely proportional to the molecular weights. Exceptions arise because of proteins with unusual shapes. For example, the globular protein urease and the rodlike protein myosin have similar molecular weights. Yet their sedimentation and diffusion constants both differ by about a factor of 3. Rapid diffusion can be a highly desirable property for an enzyme that must pass rapidly from one point to another. Such a protein would benefit from a globular shape, which is quite common for enzymes. By contrast, a rodlike protein could be advantageous to create a cytoskeletal boundary within the cytoplasm. Myosin is used for just such purposes in many cell types.

Figure 4.39

Tobacco mosaic virus structure. (*a*) Diagram of TMV structure, an example of a helical virus. The nucleocapsid (protein shell) is composed of a helical assembly of 2,130 identical protein subunits (protomers) with the RNA of the virus spiraling on the inside. (*b*) An electron micrograph of the negatively stained helical capsid (400,000×). In negative staining the virus is immersed in a pool of a heavy metal salt that is much more electron-dense than the virus. The result is that the darker portions of the figure that surround the virus appear more dense than those parts of the figure where the less dense nucleoprotein is located. (©Dennis Kunkel/Phototake.)

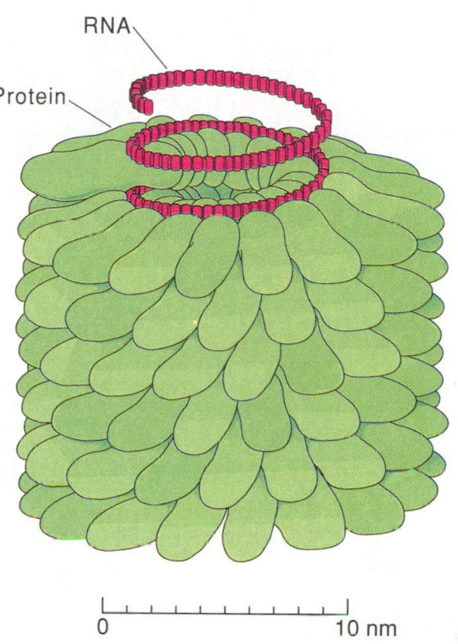

(a)

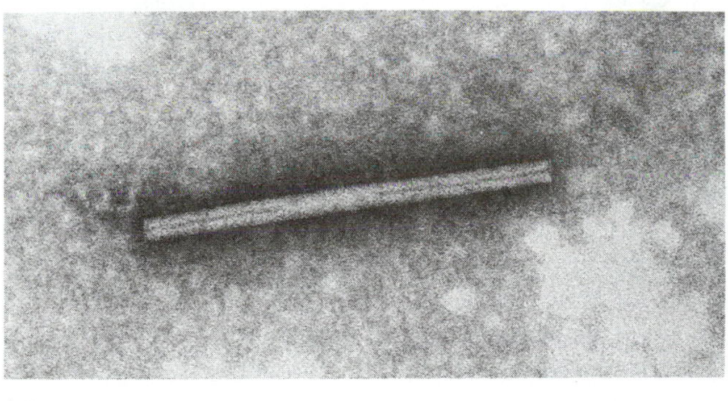

(b)

We can also imagine patterns of repeating molecular interactions that involve only twists between the subunits and so result in the formation of quaternary structures that are essentially like flat rings (fig. 4.40*a–e*). Such cyclically repeating interactions typically give rise to symmetrical molecular dimers and trimers. Larger aggregates, in contrast, most frequently do not form flat-ring structures, since flat rings would not provide enough total contact surface to stabilize an open, extended arrangement. Instead, they form arrangements that resemble geometric polyhedra. The formation of polyhedral aggregates

Figure 4.40

Quaternary structure with rotational and polyhedral symmetry. (*a*) Arrangements of two molecules related by twofold rotational symmetry to form symmetrical dimers. (*b*) Symmetrical trimers. In these arrangements, the intersubunit contacts are all identical. (*c*) The most common arrangement for tetrameric molecules (as in hemoglobin), where each subunit makes three different interactions with its neighbors: a "side-by-side" interaction, a "toe-to-toe" interaction, and a "heel-to-heel" interaction. (*d*) A common arrangement for hexameric molecules. (*e*) Octameric molecules. (*f*) A cubic quaternary structure with 24 subunits as found in some iron-storage proteins. (*g*) An icosahedral quaternary structure with 60 subunits, 3 to each triangular face. This pattern is frequently seen in viruses.

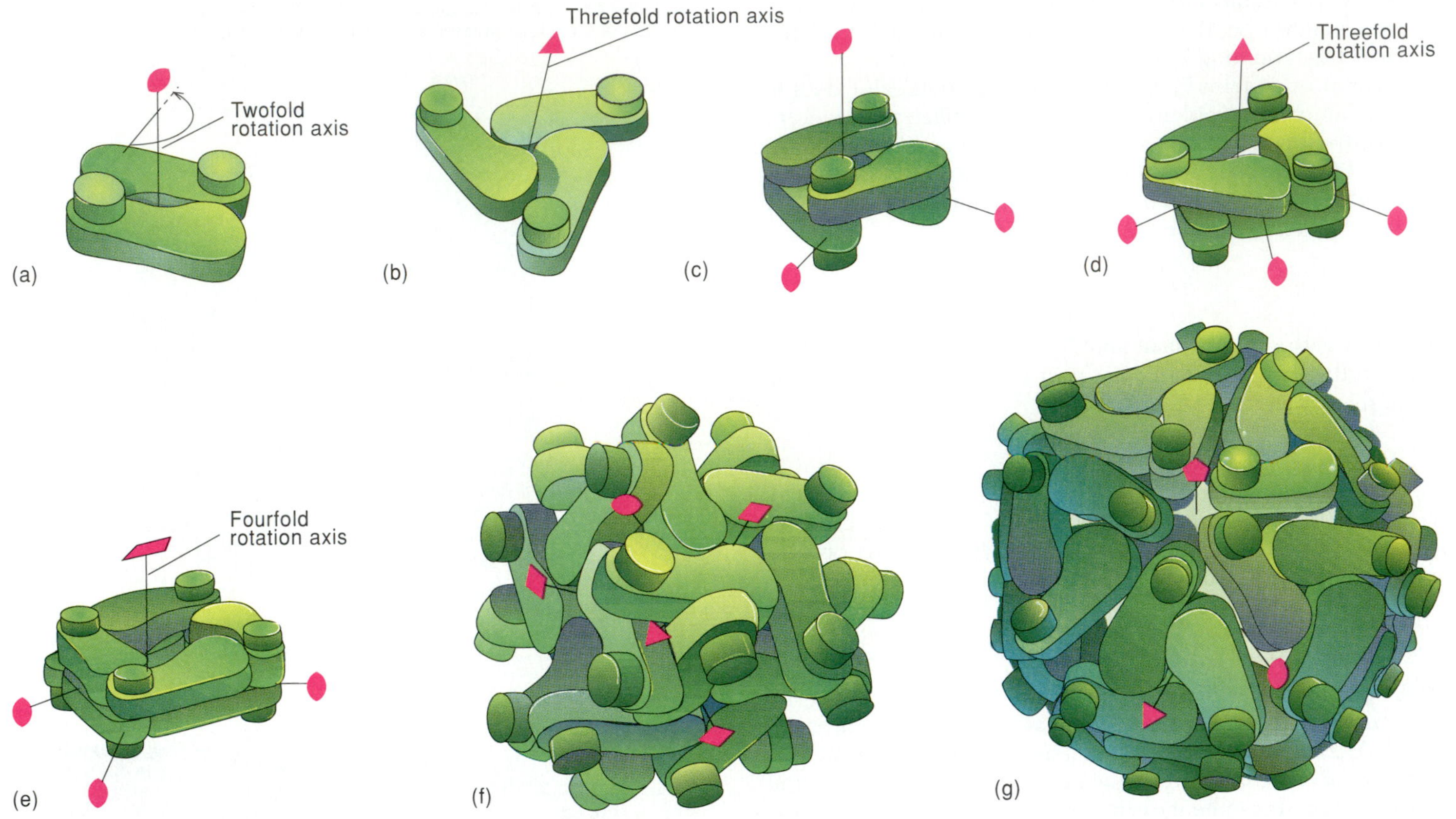

reflects the fact that the molecular subunits can have more than one type of intermolecular interaction. Figure 4.40*f, g* illustrates some of the types of structures observed. Such arrangements can be composed of identical subunits or of different types of subunits. One property that distinguishes the polyhedral and ring quaternary structures from linear and helical types is that they incorporate fixed numbers of subunit copies.

The structures of helical and polyhedral viruses demonstrate that quaternary structures play a central role in the self-assembly of very large biological structures from individual molecular subunits. Structural stabilization occurs when all the subunits interact in geometrically similar ways, i.e., essentially like the atoms in a salt crystal. Surprisingly, however, x-ray crystallography reveals that many quaternary interactions are not symmetrical or equivalent, even when they pertain to chemically identical subunits. The simplest sort of nonequivalence occurs at some dimer contacts, where individual side chains close to the twofold axis (which in such cases is only approximate) are forced to take up different positions in order to avoid overlapping. The departures from exact symmetry are usually local, in which case they probably have no functional consequences, but sometimes the nonequivalence extends to other parts of the subunit (e.g., in insulin and in malate dehydrogenase), where it can produce such effects as different binding constants for ligands. It is even easier, of course, for contacts between nonidentical subunits to be asymmetrical. For instance, in hemoglobin the contact between the two β chains is wider than that between the two α chains and produces the binding site for several important effector molecules (see chapter 5).

An even more extreme case of asymmetrical association occurs in the dimer of yeast hexokinase, where in place of the pure 180° rotation of a twofold axis, the subunits are related by a rotation of 156° plus a translation of 13.8 Å. Although this is basically a helical contact relationship, it cannot be extended past the dimer because if a third subunit were to bind by the same rule it would collide with the first one, as illustrated in figure 4.41. In hexokinase the asymmetrical association creates a significant conformational difference between two initially equivalent subunits so that the two active sites in the dimer have quite different functional properties.

Figure 4.41

A schematic drawing of a heterologous dimer interaction (lavender and green structures) in which infinite polymerization is sterically prevented. Addition of further subunits (pink and beige structures) to the free binding sites on the heterologous dimer is prevented by overlap of proteins. This arrangement of subunits is observed in the hexokinase dimer. (Adapted from a drawing obtained from T. A. Steitz.)

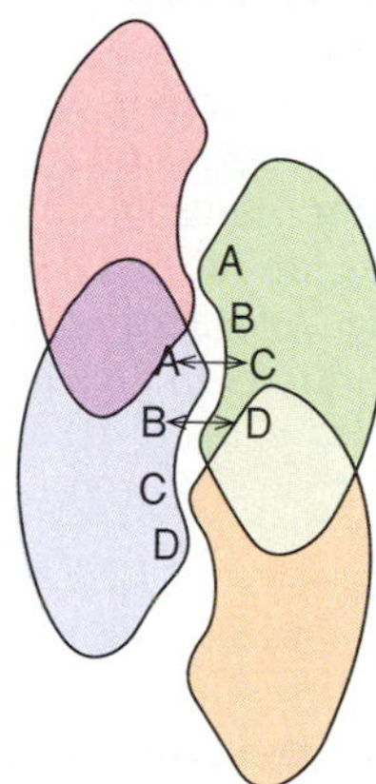

Figure 4.42

The structure of the capsid (protein shell) for an icosahedral virus such as tomato bushy stunt virus. Pentons (P) are located at the twelve vertices of the icosahedron. Hexons (H), of which there are 20, form the edges and faces of the icosahedron. Each penton is composed of 5 protein subunits and each hexon is composed of six protein subunits. In all, the structure contains 180 protein subunits.

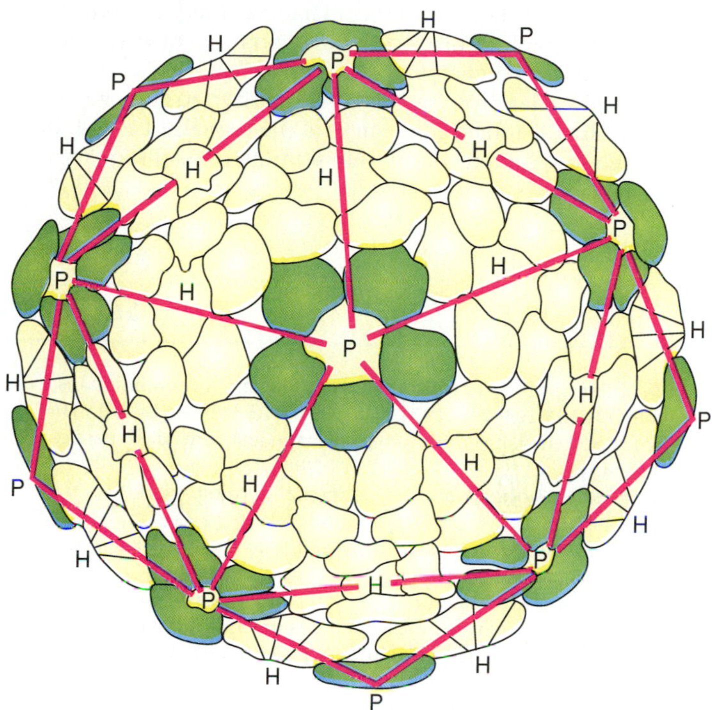

Yet another type of nonequivalent association occurs in icosahedral viruses, with only 60 symmetry-equivalent positions (see fig. 4.40*g*) but with more than 60 subunits. One way of reconciling this apparent contradiction is shown by the 180 subunits of tomato bushy stunt virus, which are placed so that five subunits are in contact around each fivefold axis, while six other subunits have a distinct but similar contact around each threefold axis (fig. 4.42). The versatility that permits such nonequivalent associations thus allows assembly of larger and more complex structures, and it may be even more common in biological structures that are too large to have been examined crystallographically.

The formation of a subunit aggregate can have extremely important functional results. In particular, as we will show later, the contact interactions provide a means of communication between the individual subunits (e.g., see chapters 5, 10, and 32). As a result, the interaction of a ligand or substrate molecule with one subunit of an aggregate can influence the course of subsequent events in other subunits of the aggregate. Such interactions, which form the basis for cooperativity in biochemical systems, are of great importance because they provide many of the control mechanisms for regulating biochemical processes.

Summary

In this chapter we have introduced the subject of the three-dimensional structures of proteins. This will be an important subject to keep in mind throughout the text. Our discussion focused on the following points.

1. Most proteins may be divided into two groups, fibrous and globular. Fibrous proteins usually serve structural roles. Globular proteins function as enzymes and in many other capacities.
2. The three most prominent groups of fibrous proteins are the α-keratins, the β-keratins, and collagen.
3. The α-keratins are composed of right-handed helical polypeptide chains in which all the peptide NH and carbonyl groups form intramolecular hydrogen bonds. When these helical coils interact they form left-handed coiled coils.
4. The β-keratins consist of extended polypeptide chains in which adjacent polypeptides are oriented in either a parallel or an antiparallel fashion. Sheets formed from such extended polypeptide chains may be stacked on top of one another.
5. Collagen fibrils are composed of extended polypeptide chains that are coiled in a left-handed manner. Three of these chains interact by hydrogen bonding and coil together into a right-handed cable. Collagen fibrils are composed of a staggered array of many such cables interacting in a side-by-side manner.
6. The structures of fibrous proteins are determined by the amino acid sequence, by the principle of forming the maximum number of hydrogen bonds, and by the steric limitations of the polypeptide chain, in which the peptide grouping is in a planar conformation.

7. X-ray diffraction provides data from which we can deduce the dimensions of the polypeptide chains in proteins. The use of x-ray techniques is, however, limited to molecules that can be oriented to achieve two- or three-dimensional order.
8. Fibrous proteins may achieve two-dimensional order, but they never achieve three-dimensional order. Therefore, the diffraction pattern of fibrous proteins gives information about the regularly repeating elements along the long axis of the fibers but tells us very little about the orientation of amino acid side chains.
9. Many globular proteins can be crystallized to achieve three-dimensional order. Study of the crystals of a globular protein can lead to a complete solution of its three-dimensional structure.
10. The forces that hold globular proteins together are the same as those that hold fibrous proteins together, but there is less emphasis on regularity and more emphasis on burying the hydrophobic regions in the interior of the protein.
11. The secondary structures found in the keratins recur in smaller patches in globular proteins. Such regions of secondary structure are folded into a seemingly endless array of tertiary structures.
12. Tertiary structures can be understood in terms of a limited number of domains. Chirality and packing considerations are major factors in determining the geometry of domains.
13. Quaternary structures are formed between nonidentical subunits to give irregular macromolecular complexes or between identical subunits to give geometrically regular structures.

Selected Readings

Anfinsen, B. C., Principles that govern the folding of protein chains. *Science* 181:223–230, 1973. Nobel Prize recounting by the man who showed that proteins fold spontaneously into their native structures.

Baron, M., D. G. Norman, and I. D. Campbell, Protein modules. *TIBS* 16:13–17, 1991.

Branden, Carl, and John Tooze, *Introduction to Protein Structure.* New York and London: Garland Publishing, 1991.

Cantor, C. R., and P. R. Schimmel, *Biophysical Chemistry,* vols. 1, 2, and 3. New York: Freeman, 1980. Includes several chapters (2, 5, 13, 17, 20, and 21) on the principles of protein folding and conformation.

Chothia, C., Principles that determine the structures of proteins. *Ann. Rev. Biochem.* 53:537–572, 1984.

Chothia, C., and A. Leak, Helix movements in proteins. *Trends Biochem. Sci.* 10:116–118, 1985.

Chothia, C., and A. V. Finkelstein, The classification and origins of protein folding patterns. *Ann. Rev. Biochem.* 59:1007–39, 1990.

Cohen, C., and D. A. D. Parry, α-Helical coiled coils—a widespread motif in protein. *Trends Biochem. Sci.* 11:245–248, 1986.

Creighton, T. E., *Proteins, Structures and Molecular Principles.* New York: Freeman, 1984. Very readable and reasonably comprehensive.

Dorit, R. L., L. Schoenbach, and W. Gilbert, How big is the universe of exons? *Science* 250:1377–1381, 1990. Predicts that there are between 1,000 and 7,000 different kinds of domains in all proteins found in nature.

Farber, G. K., and G. A. Petsko, The evolution of α/β barrel enzymes. *TIBS* 15:228–234, 1990.

Fasman, G. D., Protein conformation prediction. *TIBS* 14:295–299, 1989.

Fersht, A. R., The hydrogen bond in molecular recognition. *TIBS* 12:301–304, 1987.

Hogle, J. M., M. Chow, and D. J. Filman, The structure of polio virus. *Sci. Am.* 256(3):42–49, 1987.

Karplus, M., and J. A. McCannon, The dynamics of proteins. *Sci. Am.* 254(4):42–51, 1986. A reminder that proteins are not rigid inflexible structures.

Pauling, L., *The Nature of the Chemical Bond,* 3d ed. Ithaca: Cornell University Press, 1960. A classic on molecular structure.

Pauling, L., and R. B. Corey, Configurations of polypeptide chains with favored orientations around single bonds: two new pleated sheets. *Proc. Natl. Acad. Sci. USA* 37:729–740, 1953. Classic paper.

Pauling, L., R. B. Corey, and H. R Branson, The structure of proteins: two hydrogen-bonded helical configurations of the polypeptide chain. *Proc. Natl. Acad. Sci. USA* 27:205–211, 1951. Another classic paper.

Richardson, J. S., and D. C. Richardson, The *de novo* design of protein structures. *TIBS* 14:304–309, 1989.

Rose, C. D., A. R. Geselowizt, G. J. Lesser, R. H. Lee, and M. H. Zehfus, Hydrophobicity of amino acid residues in globular proteins. *Science* 229:834–838, 1985.

Rossman, M. G., and P. Argos, Protein folding. *Ann. Rev. Biochem.* 50:497–532, 1981.

Rossman, M. G., and J. E. Johnson, Icosahedral RNA virus structure. *Ann. Rev. Biochem.* 58:533–573, 1989.

Sali, A., J. P. Overington, M. S. Johnson, and T. L. Bundell, From comparisons of protein sequences and structures to protein modelling and design. *TIBS* 15:235–240, 1990.

Tonegawa, S., The molecules of the immune system. *Sci. Am.* 253(4):122–130, 1985.

Valegard, K., L. Liljas, K. Fridborg, and T. Unge, The three-dimensional structure of the bacterial virus MS2. *Nature* 345:36–41, 1990.

Wuthrich, K., Protein structure determination in solution by nuclear magnetic resonance spectroscopy. *Science* 243:45–50, 1989. The most effective technique for determining protein fine structure in cases where x-ray diffraction cannot be used.

Wright, P. E., What can two-dimensional NMR tell us about proteins? *TIBS* 14:255–259, 1989.

Yang, J. T., Protein secondary structure and circular dichroism: a practical guide. *Chemtracts, Biochemistry and Molecular Biology* 1:484–490, 1990.

Problems

1. The principal force driving the folding of some proteins is the movement of hydrophobic amino acid side chains out of an aqueous environment. Explain.
2. Outline the hierarchy of protein structural organization.
3. What is the role of loops or short segments of "random" structure in a protein whose structure is primarily α-helix?
4. What are some consequences of changing a hydrophilic residue to a hydrophobic residue on the surface of a globular protein? What are the consequences of changing an interior hydrophobic to a hydrophilic residue in the protein?
5. Some proteins are anchored to membranes by insertion of a segment of the N-terminal into the hydrophobic interior of the membrane. Predict (guess) the probable structure of the sequence (Met-Ala-(Leu-Phe-Ala)$_3$-(Leu-Met-Phe)$_3$-Pro-Asn-Gly-Met-Leu-Phe). Why would this sequence be likely to insert into a membrane?
6. Suppose that every other Leu residue in the peptide shown in problem 5 were changed to Asp. Would that necessarily alter the secondary structure? Explain whether insertion into the membrane would be altered.
7. Amino acid side chains coordinate to the metal cofactor in metalloproteins. Examples of these coordination ligands include Asp, Glu, His, and Cys. In most of the proteins studied, the side chains directly surrounding the ligand amino acid are highly conserved among homologous proteins isolated from different organisms, while nonconservative alterations in amino acid sequence are found at sites distant from the metal binding site. How do these observations fit the argument that biological structure dictates function?
8. Proteins that span biological membranes to provide ion-conducting channels or pores frequently have multiple α-helical segments aligned parallel to each other. Proteins that span the membrane with a single α-helical segment do not allow conduction of ions. Explain why ion conduction through a single α-helical segment does not occur.
9. An investigator purified a protein (protein X) from *E. coli.* She injected protein X into rabbits to generate antibodies that recognize and bind to protein X. Using an electrophoretic technique, she separated the proteins from a crude cell extract of *E. coli* and used the antibody to locate protein X on the gel. To her surprise, the antibody bound not only with protein X but also with a second, unrelated protein (protein Y). When proteins X and Y were sequenced, she found that the sequence of residues 67–78 in protein X and that of residues 120–131 in protein Y were identical. Help the investigator rationalize the data, recognizing that antigenic determinants (epitopes) of proteins are clusters of amino acids.
10. "Left- and right-handed α-helices of polyglycine are equally stable." Defend or refute the statement. (Consider glycine's chirality or lack thereof.)
11. Molecular weight analysis of a protein yields the following information:

Solvent	M_r
Dilute buffer	200,000
6 M Guanidinium chloride (GuHCl)	100,000
6 M GuHCl + 100 mM 2-mercaptoethanol	75,000 and 25,000

(Guanidinium chloride is a chaotropic (denaturing) reagent and 2-mercaptoethanol can reduce disulfide bonds.) What can you deduce about the protein's quaternary structure?

12. Using the Ramachandran diagram in the text, explain why polypeptides assume only a limited number of regular structures.
13. It might be argued that in protein structure, as in everyday life, it is a "right-handed world." Use examples of protein structure discussed in the chapter to support this contention.
14. Recombinant DNA technology allows the amino acid sequences in proteins to be altered. However, not all of these "genetically engineered" proteins yield stable, catalytically active proteins. Why?
15. Write all the quaternary forms possible for a hexameric protein composed of A and B type subunits. (Homohexamers are allowed.) What forces likely bind the subunits to each other?

5

CHAPTER

Functional Diversity of Proteins

Now that we have described the chief types of protein structure, let's turn to the question of how these structures relate to the function for which they were designed. We will begin by introducing the proteins that occupy the different parts of the cell or extracellular environment. (The treatment will be brief at this point, since we will be discussing many of these proteins later in the text.) We will then examine two protein systems in some detail, to provide a perspective on how structure relates to function. Next, we will consider protein design from the evolutionary viewpoint. As we will see, small refinements arise from point mutations that lead to single amino acid changes, and grosser changes arise by a reshuffling of domains. Finally, recognizing that individual proteins cannot be characterized unless they are first isolated, we will conclude this chapter with a discussion of protein purification techniques.

Spatial Localization and Functional Diversity

The cell is a highly organized factory in which the constituent parts are assembled in different locations and specialized machinery exists for specific purposes. Thus single-cell organisms are compartmentalized so that specific reactions occur in unique locations. In multicellular organisms the localization of reactions is even greater. The workers in the biochemical factory of the organism are the proteins.

Proteins Are Directed to the Regions Where They Are Utilized

Our first consideration is how proteins get to their final destination, that is, the locations where they function. All proteins are made in the cytoplasm, but their final location depends on a variety of signals. We will give a brief overview of this subject here, reserving a consideration of the mechanisms for chapters 21 and 29.

All proteins are made on ribosomes. Except for a small number of ribosomes located inside the organelles themselves, the vast majority of proteins are made on ribosomes in the cytosol. Some of the ribosomes are freely floating in the cytosol and some are attached to the endoplasmic reticulum. The ribosomes that remain free account for the proteins that are targeted to locations in the cytosol, the nucleus, the peroxisomes, the mitochondria, and the chloroplasts (fig. 5.1). Ribosomes that are bound to the endoplasmic reticulum make proteins that are deposited in the lumen of the endoplasmic reticulum. From there

Figure 5.1

The different routes traveled by proteins during and after synthesis. In a typical eukaryotic cell, proteins are synthesized on free polysomes or in the endoplasmic reticulum on membrane-bound polysomes. Built into the proteins are amino acid sequences that determine in which of these two locations they will be synthesized. Arrows indicate location to which proteins are transported after synthesis. Some proteins synthesized on free polysomes remain in the cytosol; others become incorporated into mitochondria, chloroplasts (not shown), peroxisomes, or the nucleus. Some proteins synthesized on membrane-bound polysomes remain in the endoplasmic reticulum; others are transported to the Golgi. Some proteins transported to the Golgi remain there; others are transported to lysosomes, secretory vesicles, or the plasma membrane. Arrows indicate the directions of protein transport.

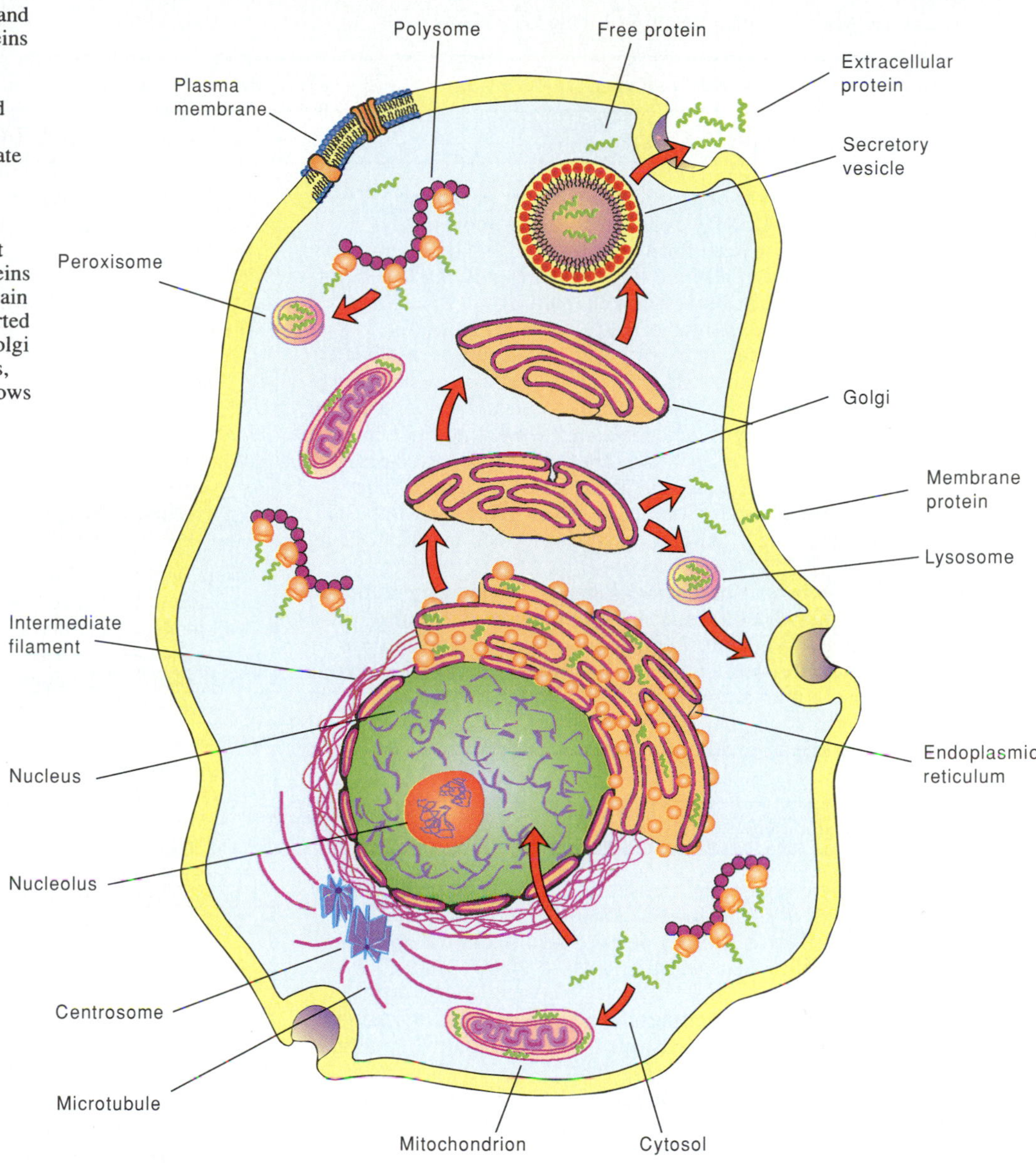

the newly synthesized proteins may be transferred to the Golgi apparatus while undergoing modifications of various sorts. At some point parts of the Golgi pinch off and the modified proteins that do not remain in the Golgi are transferred to specific locations such as the lysosomes, the plasma membrane, and the secretory granules. Those proteins targeted to the secretory granules are eventually exported.

Classification of Proteins According to Location Emphasizes Functionality

Because of the great structural and functional diversity of proteins, it is difficult to capture the important features or the whole range of them within any one classification scheme. For our present purposes we will classify proteins according to the locations they occupy when they are fully functional. This is a useful classification scheme because it emphasizes functional interrelatedness—proteins that go together work together. The structures and functions of many proteins found in different locations, both inside and outside the cell, are listed in table 5.1, which also notes points in the text where the various proteins are discussed in greater detail.

Protein Structure Is Suited to Protein Function

We have seen that highly elongated fibrous proteins are well suited for compartmentalization, for giving stable form to organellar and cellular structures, and for processes involving movement of the organism. Because of their generally low mobility, fibrous proteins are rarely associated with enzyme activity or used for transport purposes. For those functions globular proteins are more suitable. In this section we will consider two classical examples of protein assemblages that are ideally designed for the roles they play in the cell: hemoglobin and the skeletal muscle system.

Table 5.1
Some of the Main Proteins Found in Living Organisms

Protein/*Characteristics and Functions*
I. Main proteins of the cytoskeleton
Actin Bihelical filaments of aggregated globular monomers (monomer M_r = 42,000). Form cross-linked networks. In combination with myosin form actinomyosin, which functions in muscular contraction. Muscle is discussed in this chapter. Actin in the cytoskeleton is discussed in chapter 7. *Tubulin* Hollow tube composed of 13 protofilaments; each protofilament contains an extensive linear aggregate of globular tubulin dimers (monomer M_r = 50,000). Tubulin exists mainly as single filaments emanating from the centrosome (see fig. 5.1) to locations throughout the cytoplasm. Involved in maintenance of cell shape. The mitotic apparatus that directs chromosomes to opposite poles of a dividing cell is composed of tubulin. The cilium is a special case of tubulin in which many microtubules interact to produce a complex apparatus for cell motility (see fig. 4.18). *Intermediate filaments* Interrupted α-helical proteins interacting in a side-by-side twisted manner to form ropelike structures (monomer M_r = 40,000–75,000). Intermediate filaments are more abundant in cells subject to mechanical stress; they occupy locations near membranes where they appear to exert a protective function. *Spectrin* Cytoskeletal protein particularly abundant in erythrocytes (see chapter 7).
II. Human plasma proteins
Albumin Osmotic regulation; transports acids and other substances (monomer M_r = 66,000). Most abundant serum protein. *α-Globulins* A mixture of many proteins involved in transport and possibly other functions (monomer M_r = 20,000–400,000). *β-Globulins* Transferin Binds and transports iron (monomer M_r = 76,500). β_2-Microglobulin Associated with the histocompatibility antigen (see chapter 33). *α-Globulins* Antibodies (see chapter 33) (monomer M_r = 150,000). *Fibrinogen* Circulating soluble protein, which, after proteolysis by thrombin, forms fibrin polymers of the blood clot (monomer M_r = 340,000) (see chapter 9). *Complement* A mixture of about 11 proteins that work together to complement the immune system (see chapter 33) (monomer M_r = 80,000–200,000).
III. Some proteins of the extracellular matrix
Glycosaminoglycans Occupy large amounts of space forming hydrated gels (see chapters 6 and 21). *Proteoglycans* Long glycosaminoglycans covalently linked to a core protein (see chapters 6 and 21). *Collagen* (about 12 major types) The major proteins in the extracellular matrix (see chapters 4 and 29). *Types I–III* Assemble into fibrils organized to meet the needs of the tissue. *Type IV* Assembles into a laminar network. *Elastin* Cross-linked random coil protein that gives elasticity to tissues. *Fibronectin* A glycoprotein that helps to mediate cell-matrix adhesion (see chapter 7). *Integrins (several)* Integral membrane protein that helps to bind cells to the extracellular matrix. Each protein usually consists of two different subunits.
IV. Digestive enzymes of the gastrointestinal tract
Amylase Degrades starch to disaccharides. *Pepsin* Degrades proteins to large peptides. *Amylase* As above. *Peptidases* Split large peptides to small peptides. *Trypsin* Degrades proteins to large peptides (see chapter 9). *Chymotrypsin* Degrades proteins to large peptides (see chapter 9). *Lipase* Degrades lipids into fatty acids and glycerol (see chapters 17, 22). *Ribonuclease* Degrades RNA to oligonucleotides (see chapter 9). *Peptidases* Degrade peptides to amino acids. *Disaccharidases* Degrade disaccharides to monosaccharides.
V. Proteins of the cytosol
Many (between 300 and 1,000) Synthesis of most small molecules required by the cell. Synthesis of proteins, carbohydrates, and lipids (see chapters 11, 22, 23, 24).
VI. Proteins of the nucleus
Histones (5) Proteins that complex with DNA to make chromosomes (see chapter 26). There are five major histones. *Nucleic acid polymerizing enzymes (5 to 10)* For DNA and RNA synthesis (see chapters 26 and 28). There are between 5 and 10 nucleic acid polymerases in different cells.
VII. Proteins of the mitochondria and the chloroplasts
Many (100 to 300) Proteins involved in energy production from metabolites or light (see chapters 14, 15, 16).
VIII. Proteins of the endoplasmic reticulum and the Golgi
Many (50 to 200) Enzymes involved in protein modification and in oligosaccharide and lipid synthesis (see chapters 17, 21, 22, 23).
IX. Proteins of the lysosomes and the peroxisomes
Many (30 to 100) Enzymes involved in a wide variety of degradation processes for removing undesired compounds. (See chapter 17 for role of peroxisomes in fatty acid degradation. See chapter 14 for role of peroxisomes or glyoxyosomes in utilization of C-2 carbon source.)
X. Proteins of the plasma membrane
Many (100 to 500) Proteins involved in transport across membranes and for transmission of important metabolic signals across the plasma membrane (see chapters 7, 24, 31).

Hemoglobin—An Allosteric Oxygen-Binding Protein

Hemoglobin is the best-known transport protein. Its chief function is to pick up oxygen in the lungs, where it is plentiful, and deliver it to tissues throughout the body. Figure 5.2 shows the central feature of hemoglobin (and myoglobin as well). This is a water-free pocket for the heme, with its central iron atom located where oxygen is bound. (Note: Heme is a complex of Fe^{2+} and protoporphyrin IX; its structure is presented in figure 5.11 and in atomic detail in chapters 11 and 15.) The hydrophobic character of the heme binding cavity is dictated by the apolar side chains that line it. This environment is particularly suited to binding the hydrophobic porphyrin ring and creates an environment where iron (Fe^{2+}) can bind oxygen reversibly without itself being oxidized to Fe^{3+}.

Hemoglobin consists of two α subunits, each with 141 amino acids, and two β subunits, each with 146 amino acids. Each subunit is capable of binding a single molecule of oxygen. In muscle cells, a reserve oxygen store is provided by the myoglobin molecule, which is similar in structure to hemoglobin except that it exists as a monomer. While the components of and hemoglobin are remarkably similar, their physiological responses are very different. On a weight basis, each molecule binds about the same amount of oxygen at high oxygen tensions (pressures). At low oxygen tensions, however, hemoglobin gives up its oxygen much more readily. These differences are reflected in the oxygen-binding curves of the purified proteins in aqueous solution (fig. 5.3).

The oxygen-binding curve for myoglobin (Mb) is hyperbolic in shape, as would be expected for simple one-to-one association of myoglobin and oxygen:

$$\mathrm{Mb} + \mathrm{O_2} \rightleftarrows \mathrm{MbO_2}$$

$$K_f = \frac{[\mathrm{MbO_2}]}{[\mathrm{Mb}][\mathrm{O_2}]} = \text{equilibrium formation constant} \qquad (7)$$

If y is the fraction of myoglobin molecules saturated, and if we express the oxygen concentration in terms of the partial pressure of oxygen $[O_2]$,* then

$$K_f = \frac{y}{[1-y][\mathrm{O_2}]} \text{ and } y = \frac{K_f[\mathrm{O_2}]}{1 + K_f[\mathrm{O_2}]} \qquad (8)$$

This is the equation of a hyperbola, as shown in figure 5.3.

Hemoglobin (Hb) behaves differently. Its sigmoidal binding curve can be fitted by an association-constant expression with a greater-than-first-power dependence on the oxygen concentration:

$$K_f = \frac{[\mathrm{HbO_2}]}{[\mathrm{Hb}][\mathrm{O_2}]^n} \text{ and } y = \frac{K_f \mathrm{O_2^n}}{1 + K_f \mathrm{O_2^n}} \qquad (9)$$

Under physiological conditions the value of n is around 2.8, indicating that the binding of oxygen molecules to the four hemes in hemoglobin is not independent and that binding to any one heme is affected by the state of the other three. (A fuller discussion of the equations for multiple binding and the procedure for determining n from experimental data are presented in box 5A.) The first oxygen attaches itself with the lowest affinity, and successive oxygens are bound with a higher affinity. The exact value of n for hemoglobin is a function of the extent of oxygen binding as well as the presence of other factors discussed below.

*Partial pressure is usually indicated by a lower case p to the left. For simplicity the p has been omitted.

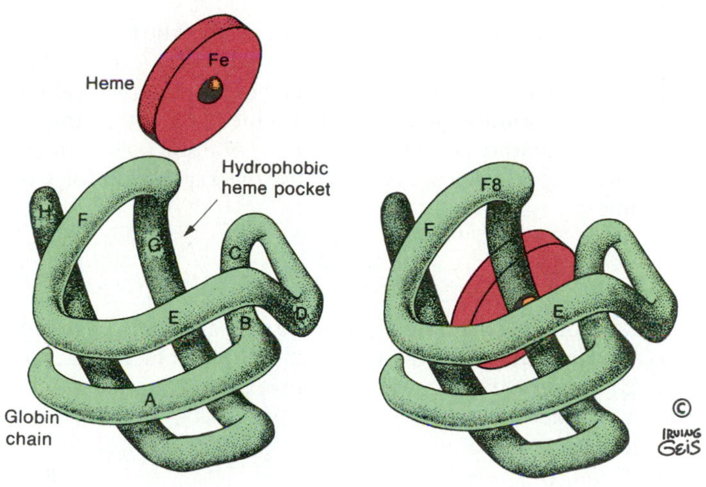

Figure 5.2

The heme pocket. The helices of hemoglobin (and myoglobin) form a hydrophobic pocket for the heme and provide an environment where the iron atom can be oxygenated when it reversibly binds oxygen. The chemical structure of heme is shown in figure 5.11 and is described in atomic detail in chapters 11 and 15.

Figure 5.3

Equilibrium curves measure the affinity for oxygen of hemoglobin and of the simpler myoglobin molecule. Myoglobin, a protein of muscle, has just one polypeptide chain and resembles a single subunit of hemoglobin. The vertical axis gives the amount of oxygen bound to one of these proteins, expressed as a percentage of the total amount that can be bound. The horizontal axis measures the partial pressure of oxygen in a mixture of gases with which the solution is allowed to reach equilibrium. For myoglobin, the equilibrium curve is hyperbolic. Myoglobin absorbs oxygen readily, but becomes saturated at a low pressure. The hemoglobin curve is sigmoidal. Initially hemoglobin is reluctant to take up oxygen, but its affinity increases with oxygen uptake. At arterial oxygen pressure, both molecules are nearly saturated, but at venous pressure, myoglobin would give up only about 10% of its oxygen, whereas hemoglobin releases roughly half. At any partial pressure, myoglobin has a higher affinity than hemoglobin, which allows oxygen to be transferred from blood to muscle.

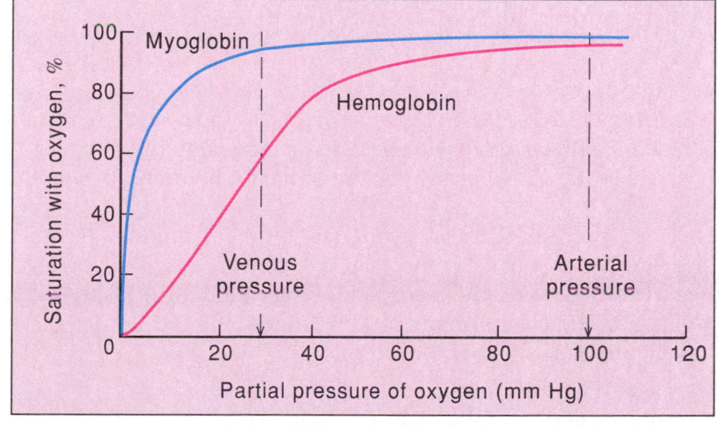

5A

BOX

Theory of Multiple Binding and the Hill Plot

Binding studies require a measurement of the free ligand, B, in the presence of the macromolecule, A. From this we may determine the average binding number, called y. The value of y is equal to the ratio of the total number of B molecules bound to the total number of binding sites. If there is only one binding site per A molecule, the expression for y is rather simple:

$$y = \frac{[AB]}{[A] + [AB]} \quad \text{(B1)}$$

The value of [A] + [AB] is known from the total amount of A added to the solution. The value of [AB] is equal to the total amount of B added, minus the experimentally observed concentration of free B after A is added:

$$y = \frac{\text{total B} - \text{free B}}{\text{total A}} \quad \text{(B2)}$$

Let us first consider the simple binding of one ligand B to a protein molecule A with a formation constant K_f:

$$A + B \overset{K_f}{\rightleftharpoons} AB; \; K_f = \frac{[AB]}{[A][B]} \quad \text{(B3)}$$

From Equations (B2) and (B3) we can express y in terms of K_f, the formation constant, or K_d, the dissociation constant:

$$y = \frac{K_f[B]}{1 + K_f[B]} \quad \text{(B4)}$$

or, since $K_f = 1/K_d$,

$$y = \frac{[B]}{[B] + K_d} \quad \text{(B5)}$$

Taking the reciprocal of both sides, we obtain

$$\frac{1}{y} = 1 + K_d\left(\frac{1}{[B]}\right) \quad \text{(B6)}$$

By plotting the experimentally determined values of $1/y$ against $1/[B]$, we obtain a straight line with a slope equal to the dissociation constant K_d. The intercept on the ordinate should be 1 (figure 1).

When there is more than one site on A for binding B, the equations become more complex and in general the plots are not linear. If we consider the average binding number for n sites on a molecule, the total average binding number is the sum of the binding numbers for each of these sites.

$$y = \sum_{i=1}^{n} \frac{K_{fi}[B]}{1 + K_{fi}[B]} \quad \text{(B7)}$$

Figure 1

A binding plot to determine the dissociation constant K_d for the simple situation where there is one ligand binding site. y is the average binding number and [B] is the concentration of ligand.

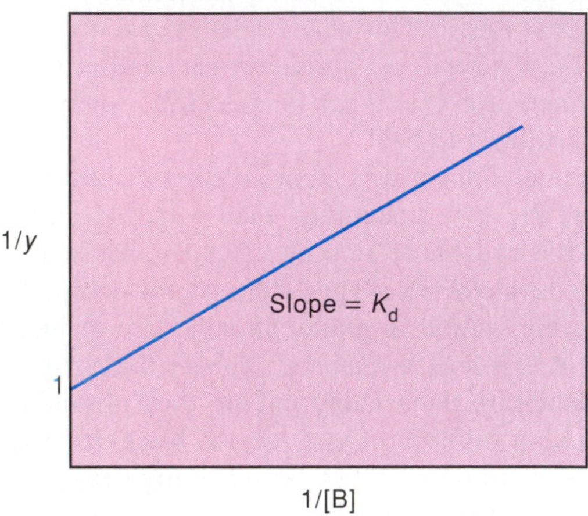

There are two situations in which the data for multiple binding may be treated rather simply. First, when all binding sites bind B with the same energy, all terms are equal and the solution to the sum is simply n times each term.

$$y = \frac{nK_f[B]}{1 + K_f[B]} \quad \text{(B8)}$$

Again, since $K_f = 1/K_d$,

$$y = \frac{n[B]}{[B] + K_d} \quad \text{(B9)}$$

and taking the reciprocal of both sides,

$$\frac{1}{y} = \frac{1}{n} + \frac{1}{n}K_d\left(\frac{1}{[B]}\right) \quad \text{(B10)}$$

If we plot $1/y$ against $1/[B]$, we find that the slope is equal to K_d/n and the intercept (when $1/[B]$ approaches 0) is $1/n$. Alternatively, we may plot the data as y versus $y/[B]$ (figure 2), in which case the slope is $-K_d$ and the intercept at $y/[B] = 0$ is n, since

$$y = n - \frac{yK_d}{[B]} \quad \text{(B11)}$$

Figure 2

A binding plot to determine the dissociation constant K_d and the number of binding sites n for the situation where there are n ligand binding sites per macromolecule with identical binding affinities.

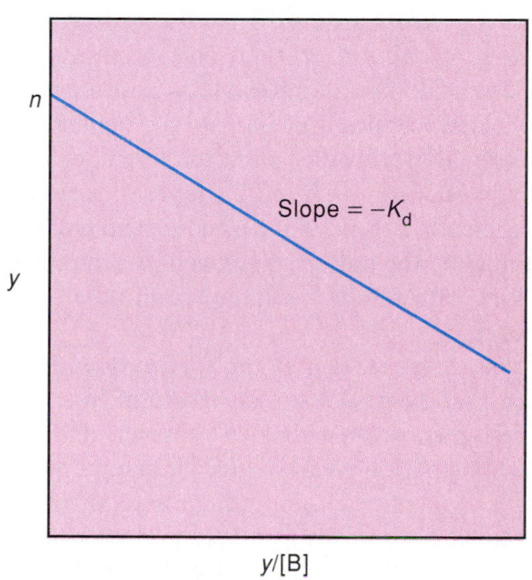

The second situation arises when the sites interact so strongly that only the fully saturated product AB_n is formed. The data may then be treated in the following way. The formation of only one major product means that the binding of the ligand molecule to the macromolecule greatly enhances the further binding of additional ligands such that at equilibrium, only three species exist in significant concentrations: A, B, and AB_n:

$$A + nB \rightleftharpoons AB_n$$

$$y = \frac{n[AB_n]}{[AB_n] + A} \tag{B12}$$

$$y = \frac{nK_f[B]^n}{1 + K_f[B]^n} \tag{B13}$$

$$\frac{1}{y} = \frac{1}{n} + \frac{1}{nK_f[B]^n} \tag{B14}$$

Figure 5.4

The structure of glycerate-2, 3-bisphosphate, an allosteric effector for hemoglobin oxygen release.

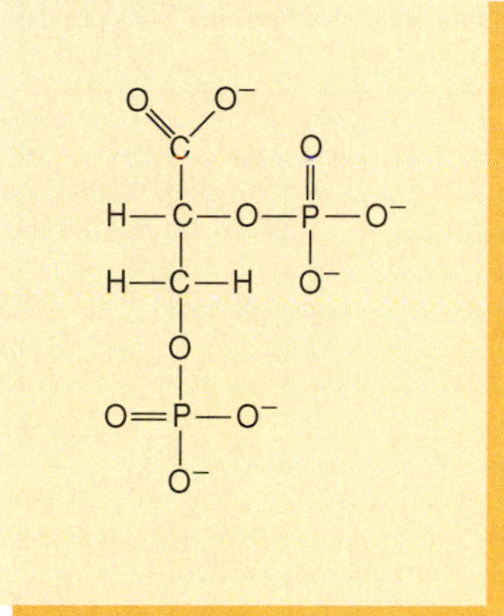

In general, a value of $n > 1$ indicates cooperative binding (or positive cooperativity) between small-molecule ligands, a value of $n < 1$ indicates anticooperative binding (or negative cooperativity), and a value of $n = 1$ indicates no cooperativity.

The cooperative binding of oxygen by hemoglobin is ideally suited to the conditions involved in oxygen transport. Thus in the lung, where the oxygen tension is relatively high, hemoglobin can become nearly saturated with oxygen, while in the tissues, where the oxygen tension is relatively low, hemoglobin can release about half its oxygen (see fig. 5.3). If myoglobin were used as the oxygen transporter, less than 10% of the oxygen would be released under similar conditions. The positive cooperativity associated with oxygen binding to hemoglobin is a special case of allostery in which the binding of "substrate" to one site stimulates the binding of "substrate" to another site on the same multisubunit protein. Although this is a special case of allostery, it is quite commonly observed in regulatory proteins (see chapter 10).

The Binding of Certain Factors to Hemoglobin Influences Oxygen Binding in a Negative Way

The combination of hemoglobin with oxygen depends not only on oxygen tension but also on pH, CO_2, and glycerate-2,3-bisphosphate (GBP or BPG). GBP (fig. 5.4) binds preferentially to the deoxygenated form of hemoglobin with a dissociation constant of about 10^{-5} M^{-1}. Its dissociation constant with HbO_2 is only about 10^{-3} M^{-1}. Since the concentrations of GBP and hemoglobin are both about 5 mM in the erythrocyte, we would expect most of the deoxy form to be complexed with GBP and most of the oxyhemoglobin to be free of GBP. The net effect of the GBP is to shift the oxygen-binding curve to higher oxygen tensions (fig. 5.5). This shift is not sufficient to lower the binding of oxygen at the high oxygen tensions in the capillaries of the

Figure 5.5

Oxygen binding curve for hemoglobin as a function of the partial pressure of oxygen. Two curves are shown, one in the absence and one in the presence of glycerate-2, 3-bisphosphate (GBP). GBP decreases the affinity between oxygen and hemoglobin, as shown by the displacement of the binding curve to high oxygen concentrations in its presence.

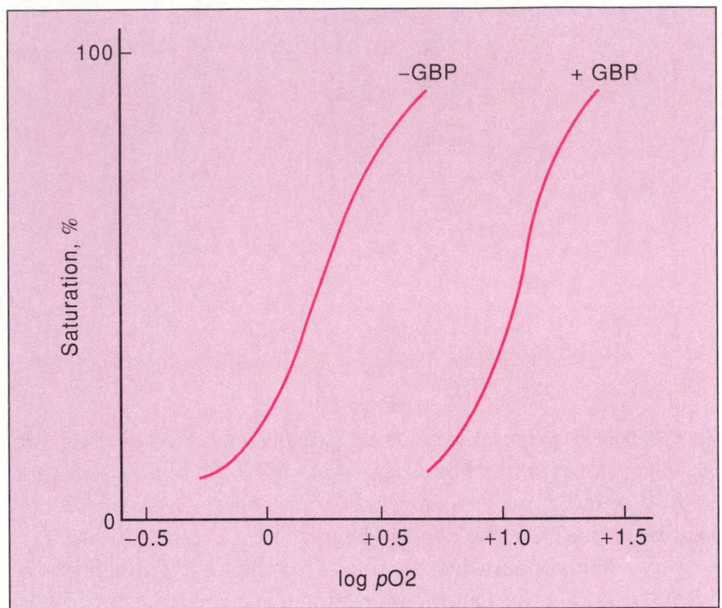

lungs, but it is sufficient to cause a substantially greater release of oxygen at the lower oxygen tensions that exist where it is needed, that is, in the rest of the body tissues.

The negative influences of H^+ and CO_2 on oxygen binding are causally related and together are known as the Bohr effect. The CO_2 that diffuses across the capillary wall from the tissue is largely in solution as CO_2 molecules, since hydration to form carbonic acid (H_2CO_3) is a slow reaction with a half-time of about 10 s. The substantial amount of CO_2 generated by the various decarboxylation reactions of intermediary metabolism diffuses from cells through interstitial fluid into the blood plasma, with only a small fraction becoming hydrated as carbonic acid. In the erythrocytes, however, hydration of CO_2 occurs very rapidly, catalyzed by carbonic anhydrase, which is concentrated in the erythrocytes. At the pH of blood (about 7.4) the carbonic acid largely dissociates into H^+ and HCO_3^-:

$$CO_2 + H_2O \overset{\text{Carbonic anhydrase}}{\rightleftharpoons} H_2CO_3 \rightleftharpoons H^+ + HCO_3^-$$

The production of protons through these two consecutive reactions explains why an increase in concentrations of CO_2 causes a lowering of pH (an increase in the H^+ concentration). The net result is a lowering of the affinity of hemoglobin for oxygen, since protons, like GBP, bind preferentially to deoxy hemoglobin. For example, at pH 7.6 and 40 mm Hg of oxygen tension, hemoglobin retains more than 80% of its oxygen; at pH 6.8 it retains only 45%. The negative effect of CO_2 on oxygen binding is mainly due, then, to the tendency of CO_2 to lower the pH (i.e., raise the H^+ concentration), and the negative effects of the protons on oxygen binding are qualitatively similar to the negative effects of GBP.

The Bohr effect is also closely related to the major role that hemoglobin plays in disposing of the CO_2 produced in tissues, and in controlling the blood pH. While oxygen is being delivered to the tissues in the venous blood, the CO_2 is being absorbed from the tissues (fig. 5.6). This process would stop very quickly if it were not for the erythrocytes and the hemoglobin. As we have just seen, the CO_2 that diffuses into the erythrocytes is rapidly converted into carbonic acid, which in turn dissociates into H^+ and HCO_3^-. The protons produced by this dissociation would lower the pH and reverse this dissociation if it were not for the buffering action of the hemoglobin. Loss of oxygen increases the acid dissociation constant of the hemoglobin so that it picks up the excess protons. This change serves two purposes; it helps to keep the pH constant and it enables the system to absorb more CO_2. The additional CO_2 is ultimately disposed of when the blood reaches the lung and the hemoglobin again becomes oxygenated. Another point to remember is that the majority of the HCO_3^- produced in the erythrocyte diffuses into the venous blood. The pH of the blood system is controlled within narrow limits by the buffering action of the bicarbonate and the hemoglobin, with minor assistance from other proteins in the bloodstream.

One must marvel at the way various factors work in concert so that hemoglobin can be useful in so many roles—oxygen deliverer, carbon dioxide remover, and pH stabilizer. From the explanation we have given it should be clear why the hemoglobin is confined to a cellular structure in the plasma rather than being present as a free plasma protein. Carbonic anhydrase and GBP are essential for efficient hemoglobin function, and their presence in the bloodstream at adequate concentrations would probably be unattainable or intolerable to the blood system. Furthermore, the protons released as a result of the carbonic anhydrase reaction are rapidly picked up by the deoxygenated hemoglobin before they have a chance to cause a potentially harmful lowering of the pH of the plasma.

X-Ray Diffraction Studies Reveal Two Conformations for Hemoglobin

X-ray diffraction studies on fully oxygenated hemoglobin and deoxygenated hemoglobin have shown that the molecule is capable of existing in two states, with significant differences in tertiary and quaternary structures (fig. 5.7). Further studies on partially oxygenated hemoglobin may indicate additional intermediate structures between these two extremes. Until these can be characterized in structural terms, the two-state model serves as a useful conceptual framework for explaining the allosteric mechanism of the hemoglobin system.

The hemoglobin tetramer is composed of two identical halves (dimers), with the $\alpha_1\beta_1$ subunits in one dimer and the $\alpha_2\beta_2$ subunits in the other. The subunits within the dimers are tightly held together; the dimers themselves are capable of motion with respect to one another (fig. 5.8). The interface between the movable dimers contains a network of salt bridges and hydrogen bonds when hemoglobin is in the deoxy conformation (fig. 5.9). The quaternary transformation that takes place on binding of oxygen causes the breakage of these bonds.

Figure 5.6

Transport of oxygen (O_2) and carbon dioxide (CO_2) in the circulatory system. In most tissues O_2 is released and CO_2 is withdrawn by the red blood cells; in the lungs these processes are reversed.

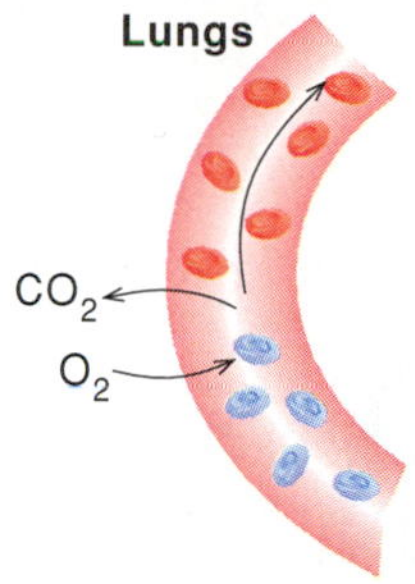

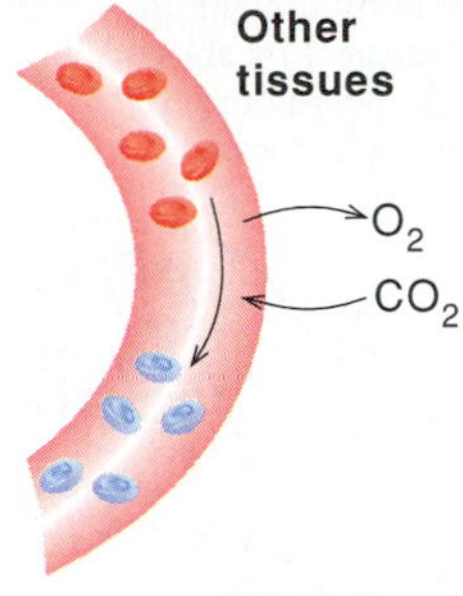

Lungs

$$HCO_3^- + H^+ \xrightleftharpoons{\text{Carbonic anhydrase}} H_2CO_3 \rightleftharpoons CO_2 + H_2O$$

$$O_2 + HHb^+ \rightleftharpoons H^+ + HbO_2$$

$$HHb^+ + O_2 + HCO_3^- \rightleftharpoons HbO_2 + CO_2 + H_2O$$

Other tissues

$$H^+ + HbO_2 \rightleftharpoons HHb^+ + O_2$$

$$CO_2 + H_2O \xrightleftharpoons{\text{Carbonic anhydrase}} H_2CO_3 \rightleftharpoons H^+ + HCO_3^-$$

$$HbO_2 + CO_2 + H_2O \rightleftharpoons HHb^+ + O_2 + HCO_3^-$$

Figure 5.7

Three-dimensional structure of oxy- and deoxyhemoglobin as determined by x-ray crystallography. This is a view down the dyad (two fold) axis, with the ß chains on top. In the oxy-deoxy transformation (quaternary motion) $\alpha_1\beta_1$ and $\alpha_2\beta_2$ dimers move as units relative to each other. This allows glycerate-2, 3-bisphosphate to bind to the larger central cavity in the deoxy conformation. A close-up of the binding site is shown in figure 5.10.

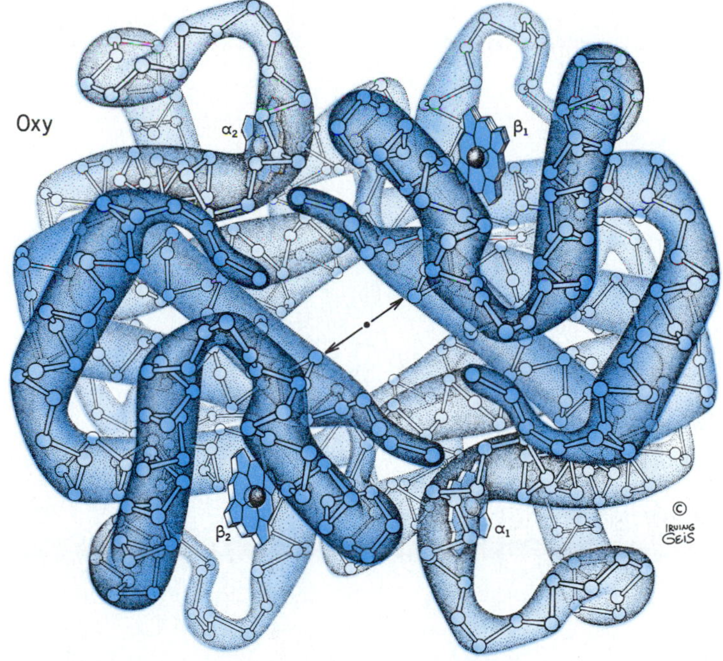

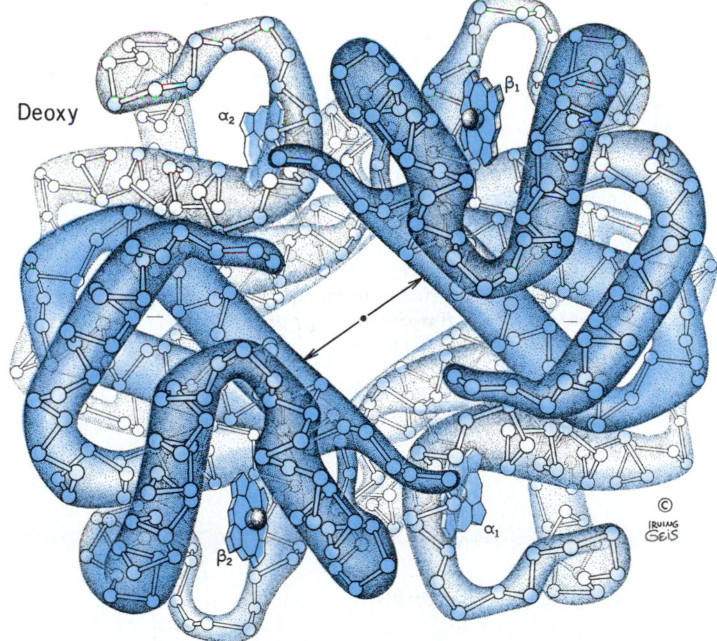

The effects of H^+, CO_2, and glycerate-2,3-bisphosphate on oxygen binding can be understood in terms of their stabilizing effect on the deoxy conformation. The decreased oxygen binding as the pH is lowered from 7.6 to 6.8 suggests the involvement of histidine side chains, because these are the only side-chain groups in proteins that have a p*K* in this pH range. Certain histidines in the charged form make salt linkages that contribute to the stability of the deoxy form (see fig. 5.9). As the pH is lowered, these histidines tend to become charged, which increases the stability of the deoxy form. Such a change should inhibit a structural transition to the oxy form and thereby lower the affinity of the protein for oxygen. Similarly, glycerate-2,3-bisphosphate binds most strongly to the deoxy form (fig. 5.10) and thereby discourages the transition to

Figure 5.8

The deoxy-to-oxy shift upon binding oxygen in one hemoglobin molecule. The projection shown in this figure is approximately perpendicular to the one shown in figure 5.7. The $\alpha_1\beta_1$ dimer moves as a unit relative to the $\alpha_2\beta_2$ dimer. The interface between the two dimers is crucial to the cooperativity effect in hemoglobin. The interface is not visible in this figure (see figure 5.9).

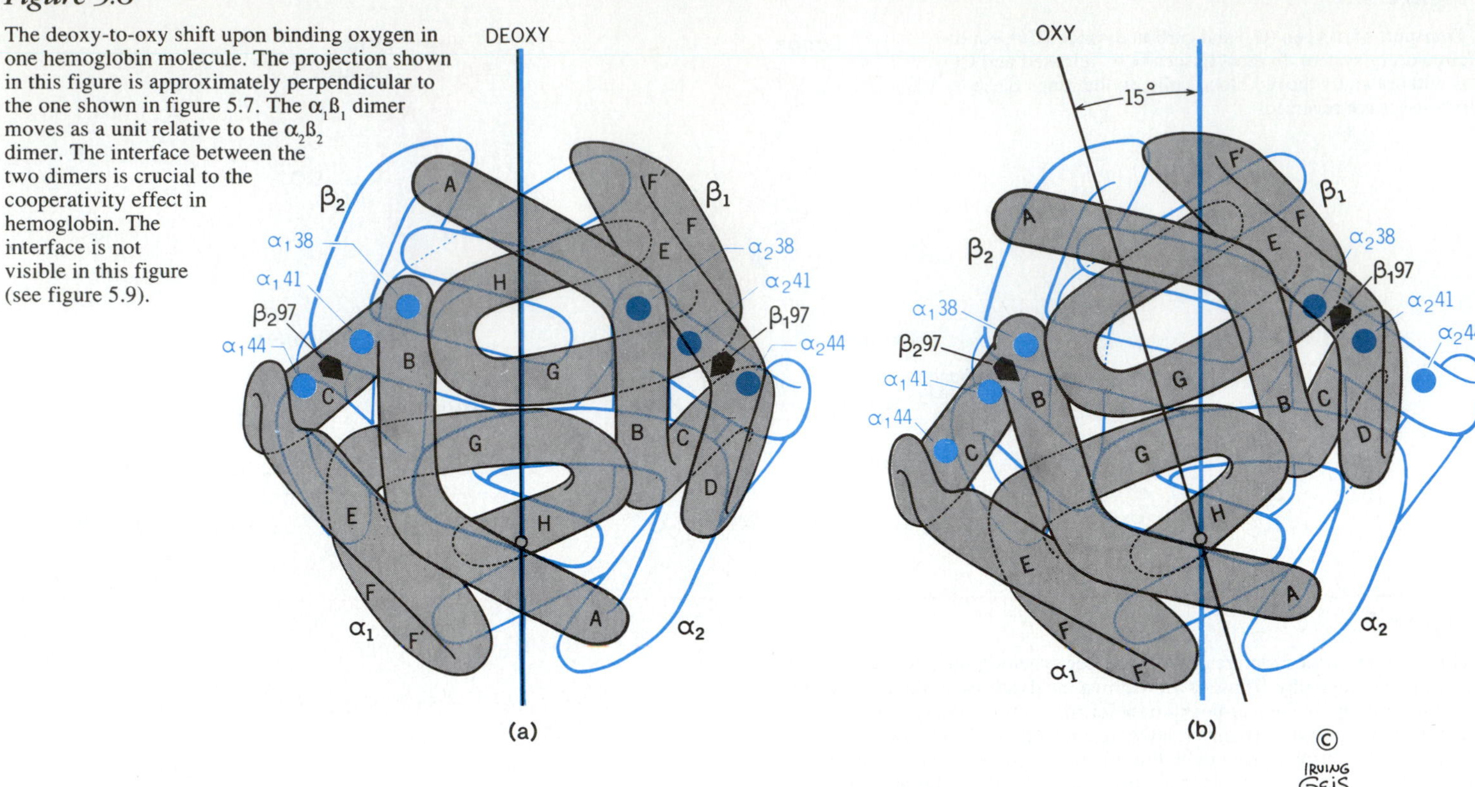

Figure 5.9

(*a*) The $\alpha_1\beta_2$ (and $\alpha_2\beta_1$) interface is shown schematically at the lower left and in detail (*b*). This is the regulatory zone of the hemoglobin molecule, which contains crucial hydrogen bonds and salt bridges. (*b*) All the hydrogen bonds and salt bridges shown here (dotted lines) exist only in the deoxy state, with the exception of $\alpha_1 41–\beta_2 40$ and $\alpha_1 94–\beta_2 102$, which exist only in the oxy state. Only the α carbons are shown in the backbone structure of the hemoglobin, except in the region of the interface.

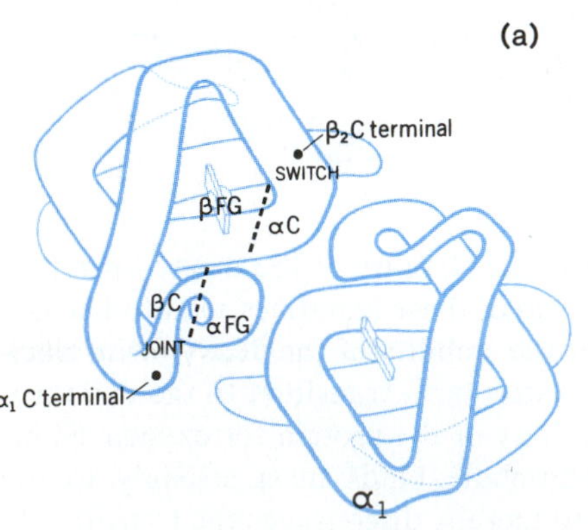

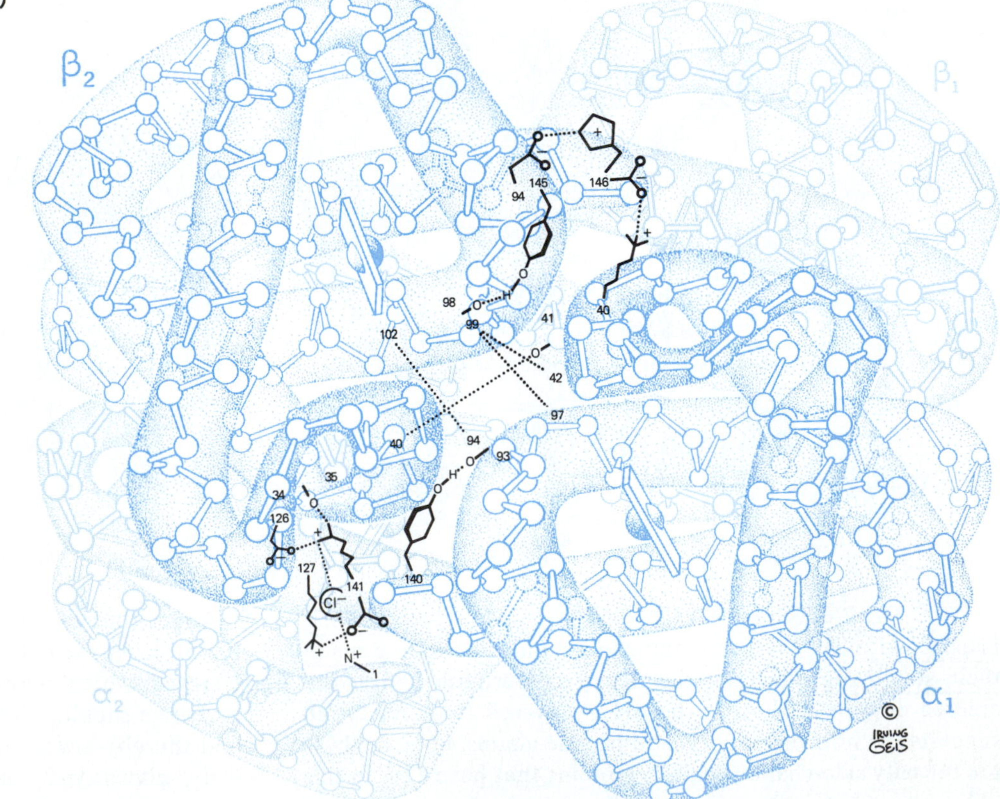

Figure 5.10

The binding of glycerate-2, 3-bisphosphate in the central cavity of deoxyhemoglobin between ß chains. The surrounding positively charged residues are the amino terminal, His 2, Lys 82, and His 143.

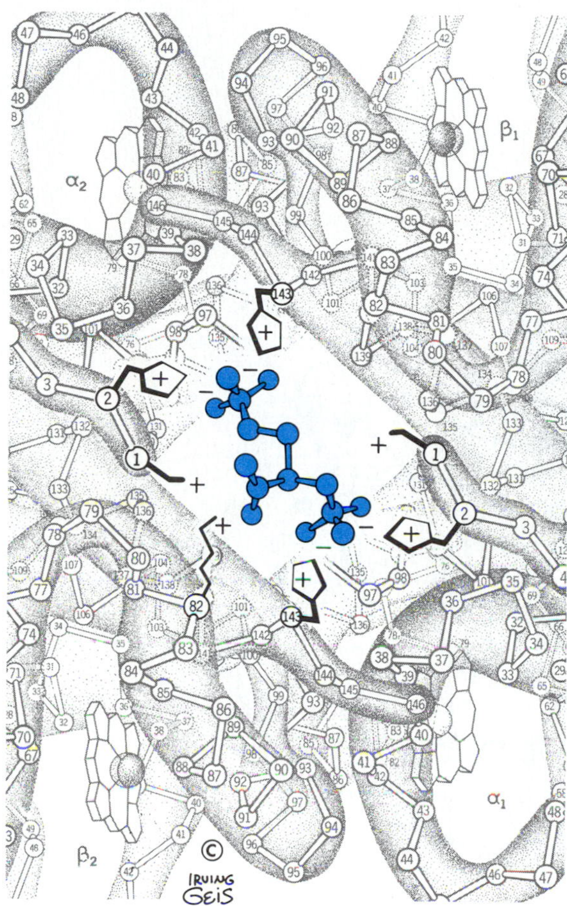

the oxy form, which lowers the affinity for oxygen. Carbon dioxide binds as bicarbonate to the α-amino groups in hemoglobin; this binding also favors the deoxy conformation. Binding of CO_2 is freely reversible, being favored by the high CO_2 tensions in the tissues. As a result hemoglobin becomes a carrier of CO_2 from tissues to the lungs, where it is discharged.

Changes in Conformation Are Initiated by Oxygen Binding

The oxygen binding at the heme group itself initiates the changes in tertiary and quaternary structure that are responsible for the cooperative effect seen on oxygen binding. The heme group contains an Fe^{2+} ion located near the center of a porphyrin ring. The Fe^{2+} makes four single bonds to the nitrogens in the heme ring, and a fifth bond to a histidine side chain of the F helix, F8 histidine (fig. 5.11). When oxygen is present it binds at the sixth coordination position of the iron on the other side of the heme. Movement of the iron upon oxygen binding (or release) pulls the F8 histidine and the F helix to which it is covalently attached (fig. 5.12). The tertiary-structure change in the F helix

Figure 5.11

A close-up view of the iron-porphyrin complex with the F helix in deoxyhemoglobin. Note that the iron atom is displaced slightly above the plane of the porphyrin.

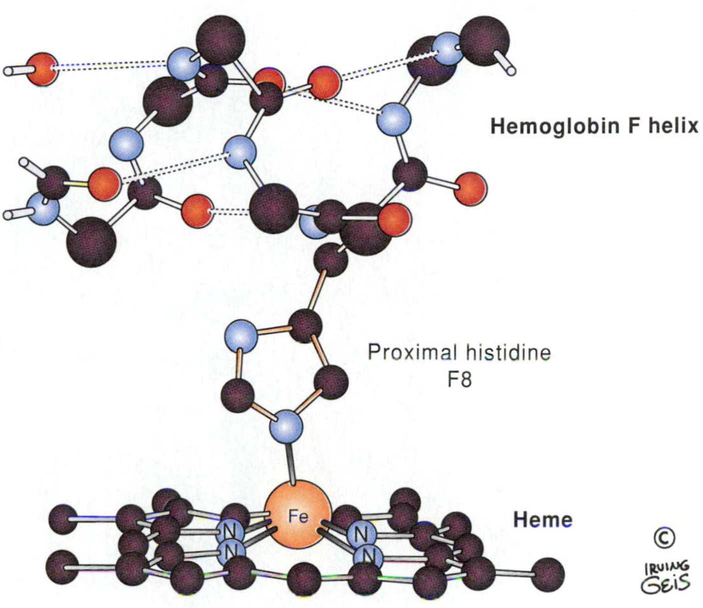

Figure 5.12

Downward movement of the iron atom and the complexed polypeptide chain on binding oxygen. The structure is shown before (black) and after (blue) binding oxygen. Movement of His F8 is transmitted to FG5 valine, straining and breaking the hydrogen bond to the penultimate tyrosine. Only the α chain is shown here.

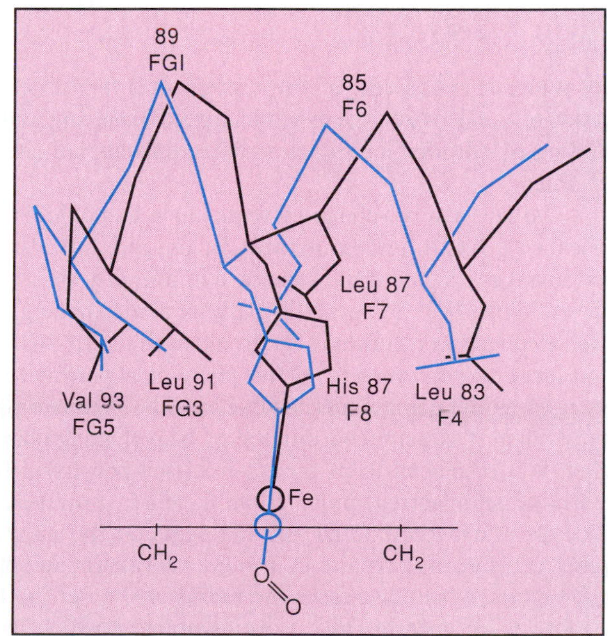

Figure 5.13

Invariant residues in the α and ß chains of mammalian hemoglobin. The colored dots, indicating the positions of invariant residues, line the heme pockets as well as the crucial $\alpha_1\beta_2$ interface. The invariant residues have been found in about 60 species. There are 43 invariant positions in the hemoglobin molecule.

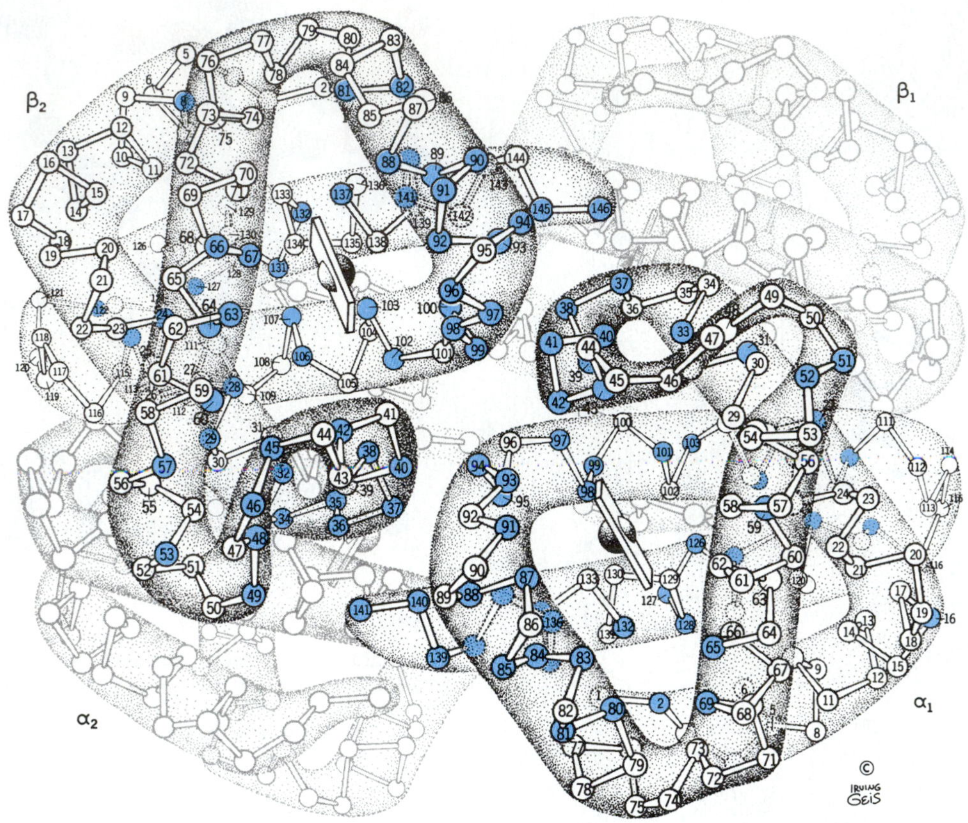

induces a strain in the rest of the protein that facilitates the conversion of the deoxy to the oxy structure. This change favors the binding of additional oxygen at other unoccupied sites in the tetramer.

The Fe^{2+} ion is well suited to its job in hemoglobin, not only because it has a natural affinity for oxygen, but also because it changes its electronic structure in a highly significant way in so doing. Fe^{2+} is normally paramagnetic, having four unpaired electrons in its outer *d* electronic orbitals. In this state it is too large to sit precisely in the plane of a porphyrin, as studies with model compounds have shown. Fe^{2+} is also paramagnetic when it is pentacoordinated in deoxyhemoglobin; as expected, it is displaced from the plane of the porphyrin by a few tenths of an angstrom unit. When O_2 binds, however, the Fe^{2+} becomes hexacoordinated and diamagnetic (no unpaired electrons). This change results in a major reorganization of its outer *d* orbitals, which decreases the radius of the Fe^{2+} so that it can move to an energetically more favorable position in the center of the porphyrin (see fig. 5.12).

The structural arguments advanced here to explain oxygen binding by hemoglobin are supported by amino acid sequences of α and β chains for a large number of hemoglobins from different species. Data from 60 species of α chains and 66 species of β chains reveal 43 invariant positions in the hemoglobin molecule. These invariants are plotted on a map of the hemoglobin structure in figure 5.13, where the invariant positions are shown by colored dots. In a sense, the dots provide a diagram of the working machinery of hemoglobin, for the invariant positions line the heme pockets where oxygen is bound and the crucial $\alpha_1\ \beta_2$ interface, which changes its orientation when oxygen binds. Electrostatic forces and hydrogen bonds stitch the interface together in the deoxy conformation when the molecule gives up its oxygen to the tissues. If there are changes, resulting from mutation at any of these positions, then we would expect trouble to develop. This is just what happens, as can be seen in figure 5.14, which shows the positions of pathological mutations in hemoglobin. Where there are changes in the heme pockets or in the $\alpha_1\ \beta_2$ interface, hemoglobin abnormalities occur; many of these are associated with serious diseases.

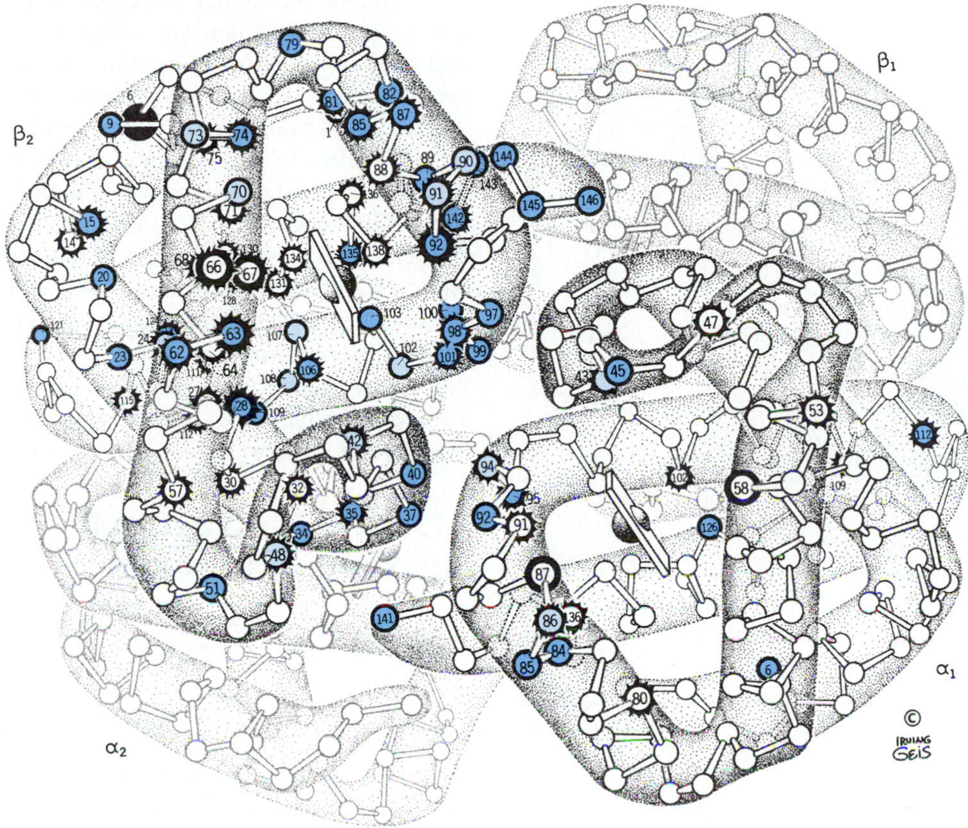

Figure 5.14

Positions of mutations in hemoglobin that produce a pathological condition. Comparison with figure 5.13 shows that these mutations, in general, show the same pattern as the distribution of the invariant positions. Dark circles indicate positions of abnormal residues, solid black dot indicates the valine β_6 mutation in sickle-cell anemia, heavy circles indicate M (Met) hemoglobin, and jagged perimeter indicates unstable hemoglobin. Dark color indicates increased oxygen affinity; light color indicates decreased oxygen affinity.

As a rule, invariant amino acids are the only critical loci for hemoglobin function. One striking exception occurs at position 6 of the β chain. A hydrophobic valine residue is substituted for glutamic acid (a charged side chain) with disastrous results. The specific consequence of the β_6 alteration is to cause hemoglobin tetramers to aggregate when they are in the deoxy state. This is because the β_6 valine fits into a hydrophobic pocket in an adjoining hemoglobin molecule. The aggregates form long fibers that stiffen the normally flexible red blood cell. The resulting distortion of the red cells leads to capillary occlusion, which prevents proper delivery of oxygen to the tissues. This pathological condition is known as sickle-cell anemia.

Two Models Have Been Proposed for the Way Hemoglobins and Other Allosteric Proteins Work

We have spent a good deal of time describing the function of hemoglobin because it is the best-understood regulatory protein and provides us with a model system for understanding in general terms how other allosteric proteins work. The first indication that hemoglobin was an allosteric protein came from the sigmoidal shape of its oxygen-binding curve (see fig. 5.3). Allosteric proteins are usually composed of two or more subunits. Different ligands may bind to quite different sites, or to quite similar sites as in the case of hemoglobin.

Two quite different models were proposed about twenty-five years ago to explain the unusual nature of the hemoglobin oxygen-binding curve. These models could also be used as a starting point for discussing other allosteric proteins (fig. 5.15). The first model, introduced by Monod, Wyman, and Changeux in 1965, is called the symmetry model. In this model hemoglobin can exist in only two conformations, one with all four of the subunits within a given tetramer in the low-affinity form and one with all four subunits in the high-affinity form (see fig. 5.15*a*). Also, in this model, the hemoglobin molecule is always symmetrical, i.e., all the subunits are either in one state or the other, and all of the binding sites have identical affinities. The binding of oxygen to one of the subunits favors the transition to the high-affinity form. The greater the number of oxygens binding to the tetramer, the more likely it is that the transition from the low-affinity form to the high-affinity form will occur.

Figure 5.15

Alternate models for hemoglobin allostery. (*a*) In the symmetry model hemoglobin can exist in only two states. (*b*) In the sequential model hemoglobin can exist in a number of different states. Only the subunit binding oxygen must be in the high-affinity form.

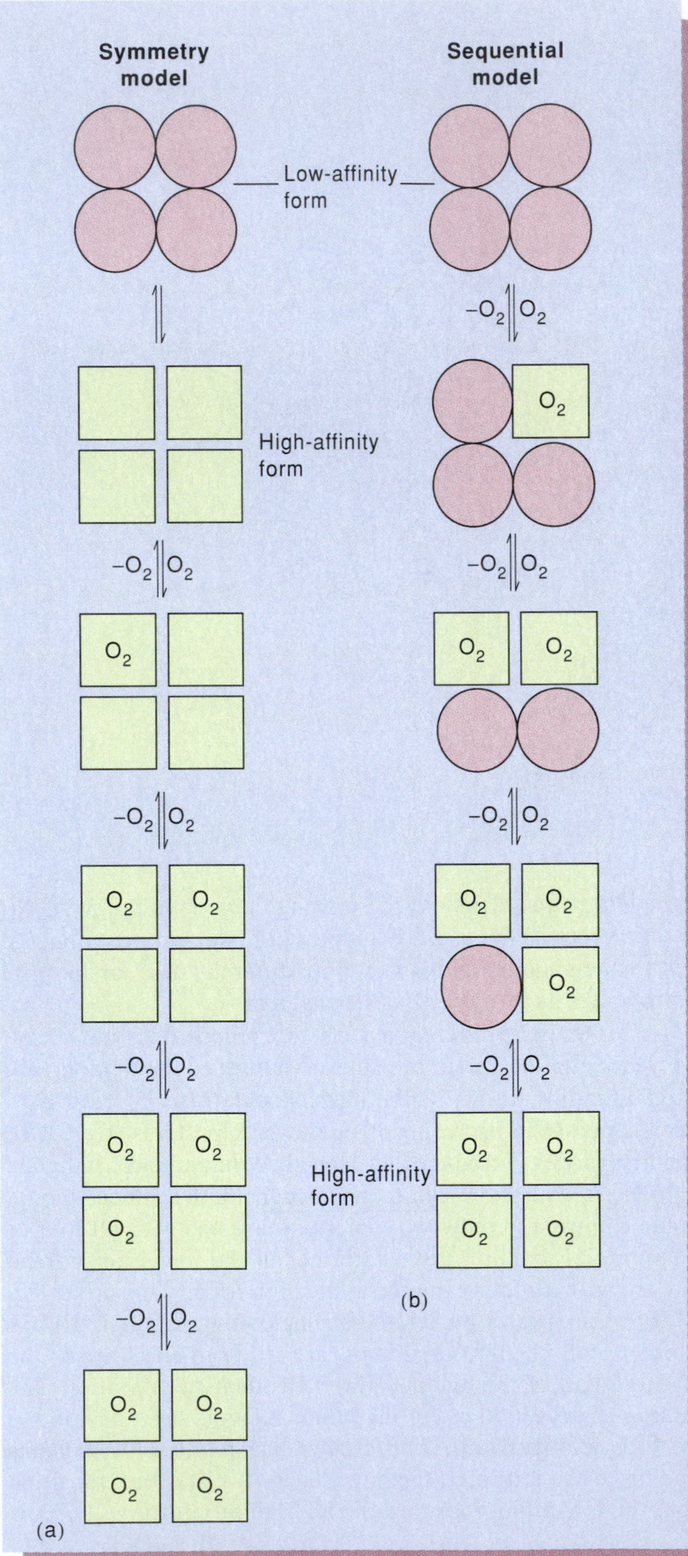

The second model, proposed by Koshland, Nemethy, and Filmer in 1966, is referred to as the sequential model (see fig. 5.15*b*). In this model the binding of an oxygen molecule to a given subunit causes that subunit to change its conformation to the high-affinity form. Because of its molecular contacts with its neighbors, the change increases the probability that another subunit in the same molecule will switch to the high-affinity form and bind a second oxygen more readily. The binding of the second oxygen has the same type of enhancing effect on the remaining unoccupied oxygen-binding sites.

Either of these models (or something in between) could account for the sigmoidal oxygen-binding curve of hemoglobin, and either is consistent with the fact that deoxygenated and fully oxygenated hemoglobin have different conformations. The only way to rigorously discriminate between these two models is to obtain structural information on partially oxygenated hemoglobin. This information is still lacking, so no final judgment can yet be made. Even when the situation is fully resolved for hemoglobin, there is no assurance that other allosteric proteins work in the same way.

Muscle—An Aggregate of Proteins Involved in Contraction

Vertebrate skeletal muscle represents a remarkable example of a supermolecular aggregate capable of undergoing a reversible reorganization. Voluntary muscle tissue is arranged into fibers that are surrounded by an electrically excitable membrane called the sarcolemma (fig. 5.16). Each fiber is composed of many myofibrils, which when viewed in the light microscope present a striated and banded appearance. As shown in figure 5.17, a myofibril exhibits a longitudinally repeating structure called the sarcomere. This 23,000-Å-long repeating unit is characterized by the appearance of several distinct bands, the less optically dense band being referred to as the I band and the more dense one as the A band. Furthermore, a dense line appears in the center of the I band, called the Z line; and a dense narrow band somewhat similar in appearance also occurs in the center of the A band, called the M line. Adjacent to the M line are regions of the A band that appear less dense than the remainder; these are referred to as the H zone.

Transverse sections of the sarcomere reveal that these patterns result from the interdigitation of two sets of filaments (fig. 5.17). For example, when a sarcomere is sectioned in the I band, a somewhat disordered arrangement of thin filaments (about 70 Å in diameter) is seen. In contrast, when sectioned in the H zone, a hexagonal array of thick filaments (about 150 Å in diameter) is apparent. The substantive observation is that a transverse section in the dense region of the A band shows a regularly packed array of interdigitating thick and thin filaments. This observation led Hugh Huxley to propose that the process of muscle contraction involves sliding the thick and thin filaments past each other (fig. 5.18).

Figure 5.16

The hierarchy of muscle organization. A voluntary muscle such as the bicep is a composite of many muscle fibers, connected to tendons at both ends. Each muscle fiber is composed of several myofibrils that are surrounded by an electrically excitable muscle fiber membrane (sarcolemma). Myofibrils exhibit longitudinally repeating structures called sarcomeres. The fine structure of the sarcomere is described in figure 5.17.

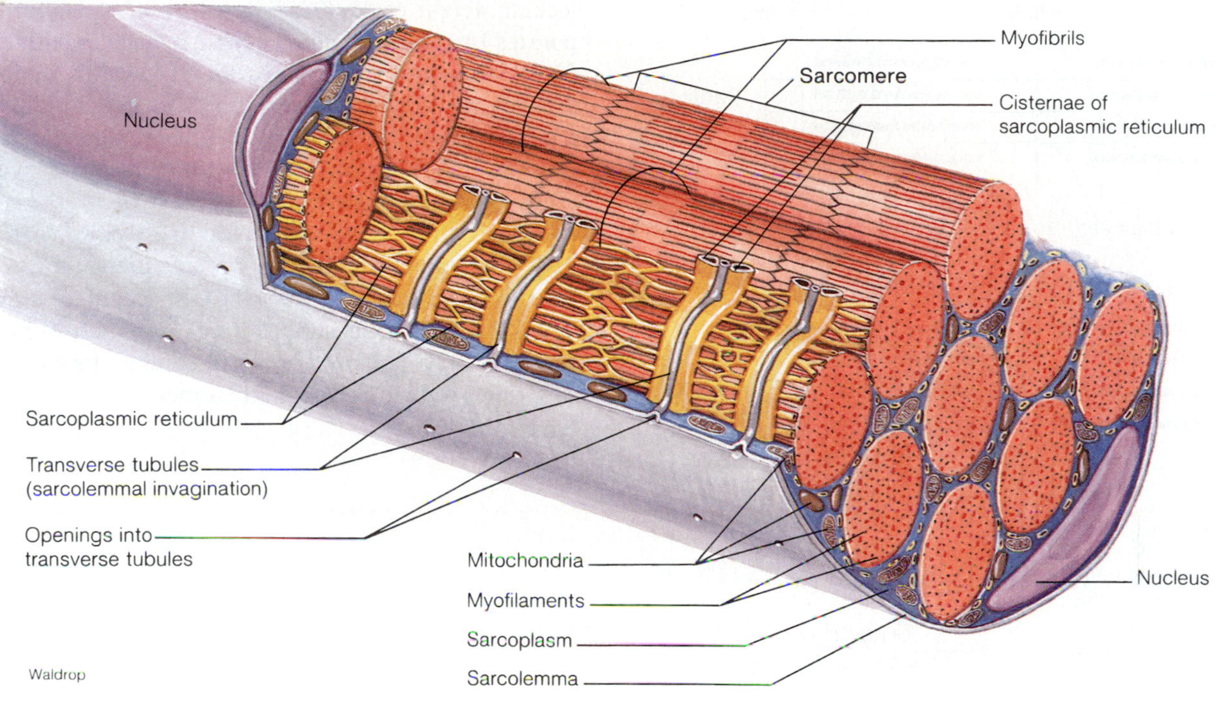

Figure 5.17

Electron micrograph of a striated muscle sarcomere showing the appearance of filamentous structures when cross-sectioned at the locations illustrated below. (Electron micrograph courtesy of Dr. Hugh Huxley, Brandeis University.)

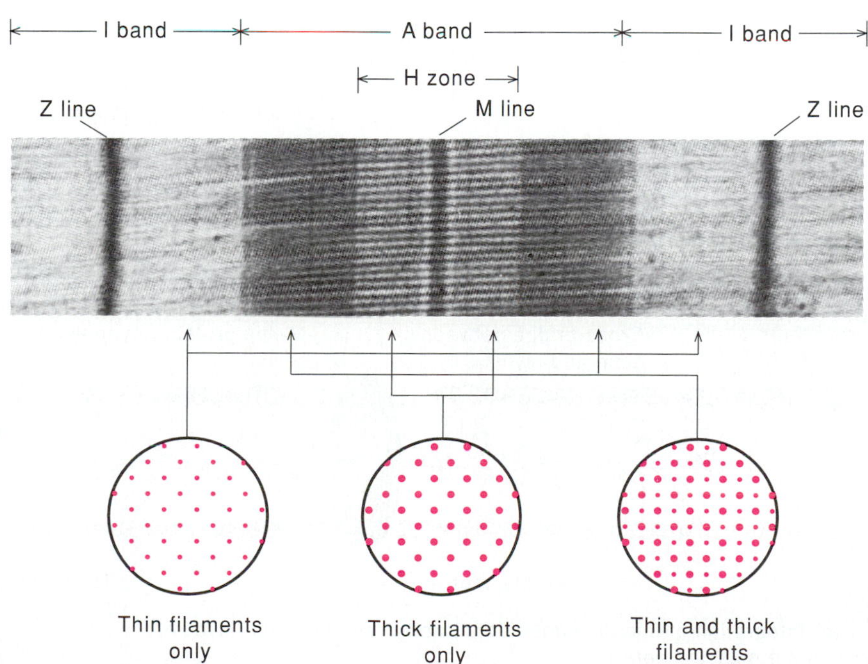

Figure 5.18

The sliding-filament model of muscle contraction. During contraction, the thick and thin filaments slide past each other so that the overall length of the sarcomere becomes shorter.

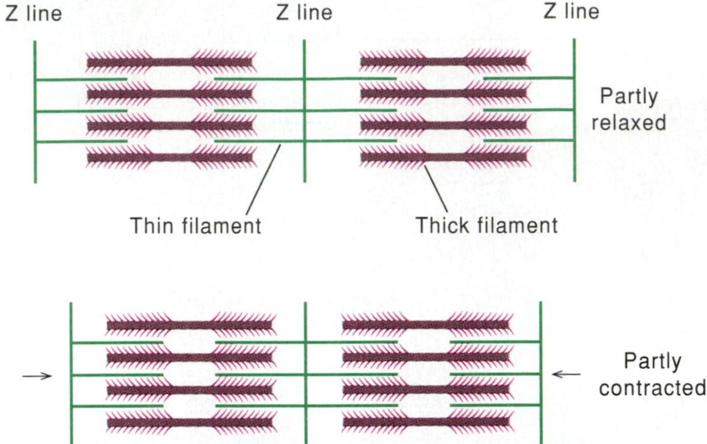

Subsequent analyses have shown that the thin filaments are composed of three proteins (fig. 5.19 and table 5.2). The main filamentous structure consists of an aggregate of globular actin molecules which takes on the form of a right-handed double helix. The individual actin molecules have a molecular weight of 42,000. Every turn of the actin helix incorporates 14 actin molecules and two molecules of the 360-Å-long filamentous protein tropomyosin (TM) that fit into the two grooves created by the actin double helix. The TM molecule is a dimer of two identical α-helical chains that wind around each other in a coiled coil. Each TM dimer spans seven actin monomers, and a succession of TM dimers extends the full length of the thin filament. Two molecules of troponin (TN) bind to the actin filament at each helical repeat. Troponin is a complex of three nonidentical subunits: TN-C, a calcium binding subunit; TN-T, a TM binding subunit; and TN-A, an "inhibitory" subunit. The TN-TM proteins form a regulatory complex whose properties we will discuss in a moment.

Figure 5.19

A molecular view of muscle structure. Formation of a thick filament involves the lateral aggregation of many myosin molecules to form the central bipolar structure shown in (*c*).

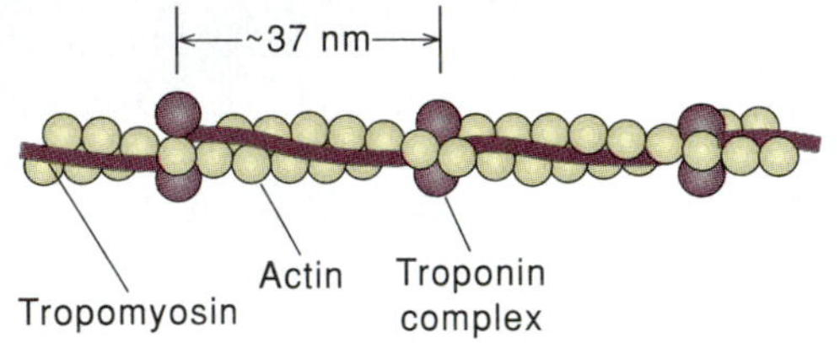

(a) Segment of actin-tropomyosin-troponin

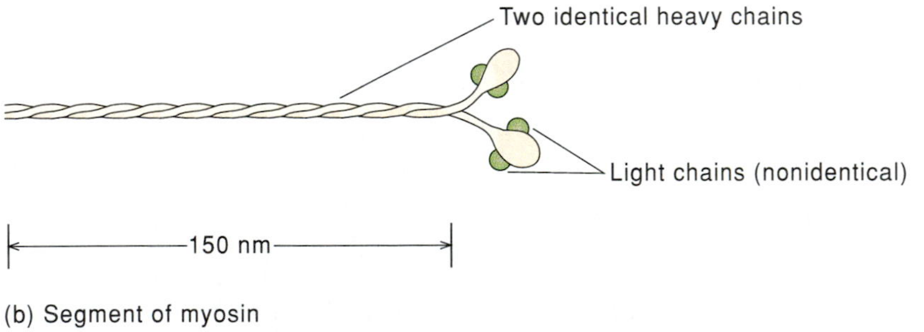

(b) Segment of myosin

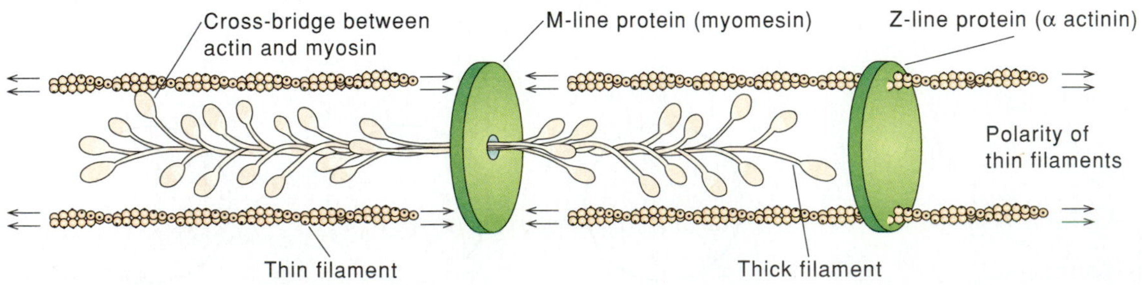

(c) Integration of thin filaments (actin) and thick filaments (myosin) in a muscle fiber.

Thick filaments are composed of myosin, a large molecule containing two identical heavy chains ($M_r = 223$) and two light chains ($M_r = 22$ and 18). The structural organization of myosin is illustrated in figure 5.19. The molecule has two identical globular head regions that incorporate the light chains and a significant fraction of the heavy chains. The tails of the heavy chains form very long α helices that wrap around each other to form left-handed coiled coils. The individual myosin molecules can be cleaved into fragments by partial degradation with various proteases. By separation of such fragments, it has been demonstrated that the binding sites of myosin for actin and the ATPase activity of myosin are located in the globular head regions. The α-helical coiled coils form the backbone of the thick filament, while the remainder forms an arm that can provide a flexible extension or hinge for the globular head away from the body of the thick filament. The thick filament contains many myosin molecules oriented in a staggered bipolar fashion.

Granted that this is a marvelous piece of molecular architecture, but how does it actually work? The answer lies in the observation that actin cyclically binds the globular myosin head group to form cross-bridges in a reaction that depends on the myosin-catalyzed hydrolysis of adenosine triphosphate (ATP). The cyclic binding of actin to myosin is driven by the energy-releasing hydrolysis of ATP, catalyzed by the myosin head group in a manner that causes rearrangement of the actin-myosin cross-bridges. When muscle is completely relaxed, there is a minimum number of cross-bridges and the muscle is fully stretched. However, when the muscle is activated and under tension, it contracts and more cross-bridges are formed as the region of overlap between actin and myosin increases. At each stage of the contraction process it is essential to break the existing bridges with the help of ATP hydrolysis before new ones can be formed. It is important to realize that although ATP encourages more bridges to be formed, the ATP is required to break the bridges, not to form them. The breaking of bridges is required so that new and more numerous bridges can be formed. Thus cross-bridge formation is energetically favored.

A likely scenario for the contraction process is shown in figure 5.20. The ATP that binds to myosin causes the bridges to break or weaken. This bound ATP is rapidly hydrolyzed to ADP and P_i, but the hydrolysis products are not immediately released by the myosin. The P_i is released first, but its release requires effective contact with the actin. This is the regulated step in muscular contraction. Once the P_i has been released, a strong bridge forms between the myosin and the actin. This is followed by a structural change in the myosin that leads to the translocation of the myosin relative to the actin filament and finally to ADP dissociation. The translocation step is referred to as the power stroke in muscular contraction because it is at this point that the energy ultimately donated by the ATP is expended in the form of a complex structural change that is still not fully understood. Further contraction requires fresh ATP, followed by bridge dissociation and ATP hydrolysis. If conditions are right the P_i dissociates and the bridges reform, but this time they reform at points further along the actin. Cyclic repetition of this process results in a net increase in the number of actin-myosin bridges and further contraction of the sarcomere.

As we indicated earlier, the process of contraction is regulated or triggered by the TN-TM system. Since voluntary muscles are under the conscious control of the animal, one would expect a signal from the central nervous system to initiate the process of contraction. A nerve impulse communicated to the muscle causes a depolarization of the sarcolemma membrane that surrounds the muscle fibers. This in turn causes a release of Ca^{2+} from the endoplasmic reticulum in the cytoplasm of the muscle cell (fig. 5.21). The Ca^{2+} ions form a complex with the TN-C component of the troponin molecule (see table 5.2). This induces changes within the TN complex which overcomes the inhibitory effect of the TN-I subunit. Then, through TN-T, a signal is sent to TM that triggers the contraction event. The precise nature of this signal from troponin to tropomyosin is unclear; it appears to involve a movement of the tropomyosin on the actin surface that leads to an allosteric transition of the actin that encourages more favorable contact between the complementary binding sites on actin and myosin. As a result a strong bridge is formed and the P_i is released from the myosin. The remaining steps in muscular contraction have already been described. It is noteworthy that for some time after death, the muscle enters a state of rigor in which the muscle is fully contracted and the maximum number of bridges is formed. This is probably due to excessive neuronal firing and a considerable discharge of calcium from the sarcoplasmic reticulum. Normally the cytosolic Ca^{2+} concentration is restored to resting levels within 30 ms of receiving a signal and the myofibrils relax.

Table 5.2
Principal Proteins of Vertebrate Skeletal Muscle

Protein	M_r	Subunits	Function
Myosin	510,000	2 × 223,000 (heavy chains) 22,000–18,000 (light chains)	Major component of thick filaments
Actin	42,000	One type	Major component of thin filaments
Tropomyosin	64,000	2 × 32,000	Rodlike protein that binds along the length of actin filaments
Troponin	78,000	30,000 (TN-T) 30,000 (TN-I) 18,000 (TN-C)	Complex of three protein subunits involved in the regulation of muscle contraction

Figure 5.20

Steps in the contraction process. Since contraction is a cyclical process, the choice of a starting point is somewhat arbitrary Five frames are shown; the first two and the last two frames are identical to make the cyclical nature of the process clear. In the first frame (*a*), the myosin head groups contain the hydrolysis products of a single ATP molecule, ADP and P_i. A structural transition in the actin leads to contact between the actin and the myosin and the release of P_i. The release of the P_i is the rate-limiting step in muscular contraction. In the second frame (*b*), strong bridges form between actin and myosin. This is followed by a structural alteration in the myosin molecules and an effective translocation of the thick filament relative to the thin filament in (*c*). During this process the ADP is released. After the translocation step, the bridge structure is broken by the binding of ATP, which is rapidly hydrolyzed to ADP and P_i. Each thick filament has about 500 myosin heads and each head cycles about five times per second in the course of a rapid contraction.

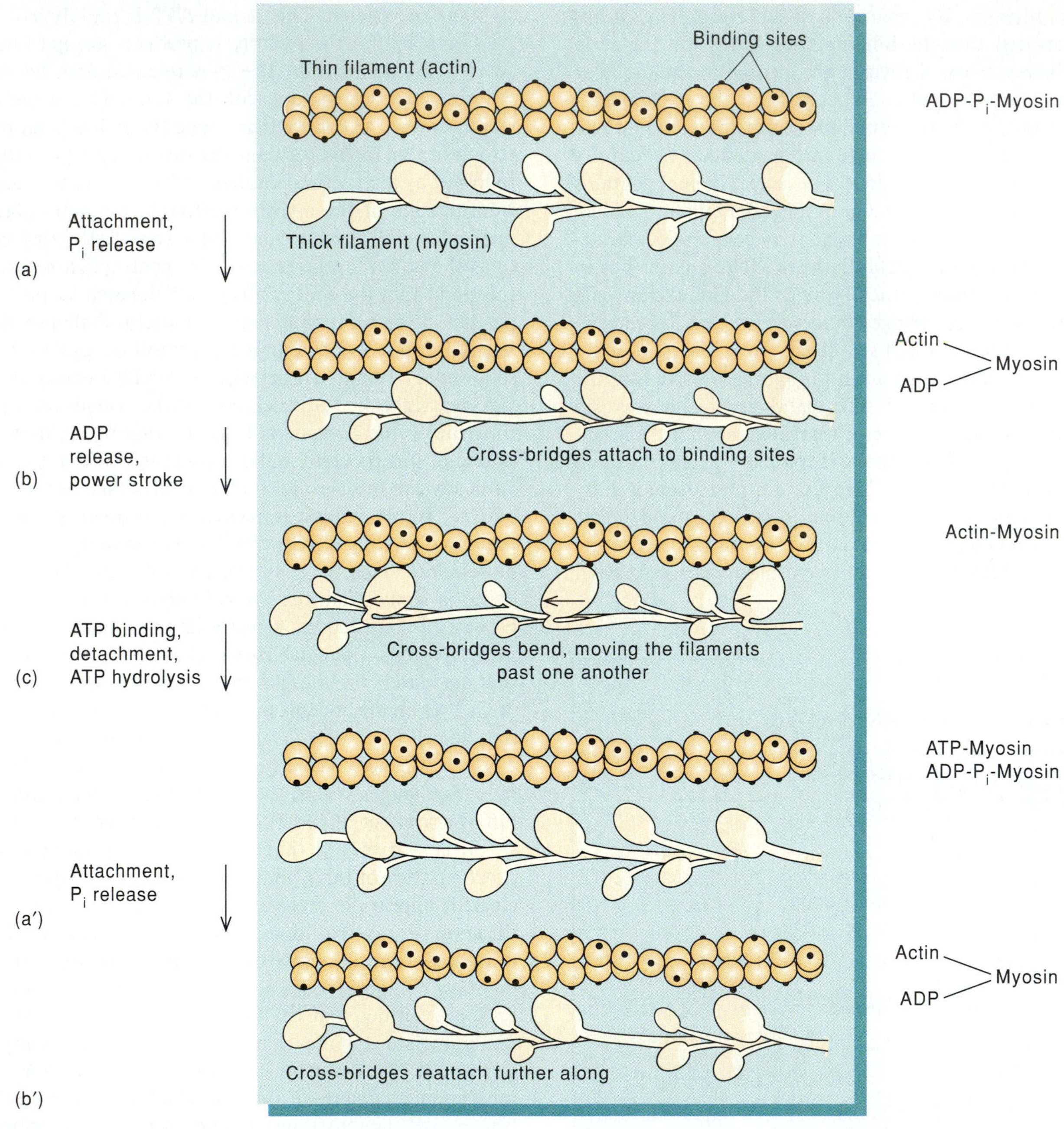

Protein Diversification as a Result of Evolutionary Pressures

We can gain further insight into protein function by asking the question, How do selective evolutionary pressures lead to the diversification of protein structure and function? The answer to this question is multifaceted.

Proteins that serve similar or identical biological functions in different living organisms are typically very similar in both their amino acid sequences and their tertiary structures. Such related families of molecules are generally assumed to reflect processes of divergent evolution. That is, they are thought to have evolved through gradual point-by-point modification of one ancestral molecule.

Figure 5.21

The effect of calcium on muscle contraction. Binding of calcium to the TN-TM-actin complex produces a shift in the location of TM, which produces an allosteric transition in actin. The allosteric transition in actin facilitates the release of P_i from myosin, which strengthens the interaction between actin and myosin.

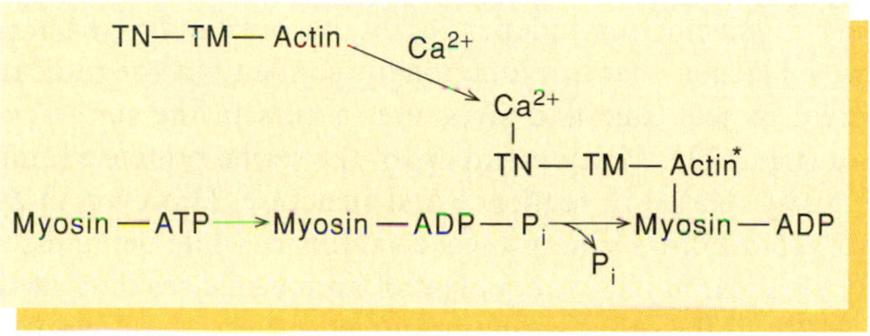

TN—TM—Actin + Ca^{2+} → Ca^{2+}–TN—TM—Actin*

Myosin—ATP → Myosin—ADP—P_i → Myosin—ADP + P_i

One of the most extensively studied protein families is the cytochrome *c* family. Cytochrome *c* proteins function as electron carriers in the mitochondrial electron-transport chains of all multicellular organisms (see chapter 15). Sequence comparisons of mitochondrial cytochrome *c* from organisms as diverse as humans and green plants reveal an extraordinary degree of sequence conservation. This fact suggests that the functional role of the molecule was highly refined by selective evolutionary pressures prior to the emergence of the first multicellular organisms. Some positions in the sequence are quite variable, whereas others are essentially invariant. Structural and chemical modification studies have shown that some of the invariant amino acid residues are associated with functionally important heme interactions, while others are important in governing the interactions of cytochrome *c* with its physiological oxidase and reductase.

If the sequence and structure of mitochondrial cytochrome *c* were indeed highly refined prior to the emergence of multicellular organisms, then its evolutionary precursors should still exist in prokaryotic organisms. In fact, in virtually all prokaryotic organisms that use oxidative or photosynthetic electron-transport chains to synthesize the high-energy intermediate adenosine triphosphate (ATP), we find molecules that are strikingly similar to mitochondrial cytochrome *c*. However, as might be expected, cytochrome *c* proteins from prokaryotes exhibit much more sequence diversity than the proteins typically found in higher organisms. In particular, the prokaryotic cytochrome *c* proteins often contain multiple amino acid insertions or deletions relative to mitochondrial cytochrome *c*. Nevertheless, from tertiary-structure determination of several prokaryotic proteins, we find that these molecules are all variations on a basic structural theme (fig. 5.22). Further, the prokaryotic molecules all show a strong conservation of those amino

Figure 5.22

Examples of structural diversification in prokaryotic cytochromes *c*. Three cytochromes are shown; cytochrome c_{550} from the denitrifying bacterium *P. denitrifican;* cytochrome *c* from mitochondria; cytochrome c_3 from the photosynthetic bacterium *P. rubrum*. The prokaryotic cytochromes contain more residues in their polypeptide chains (shaded) than does cytochrome *c*. Despite these variations there is a strong conservation of those amino acid residues that interact in functionally important ways with the protein's heme group.

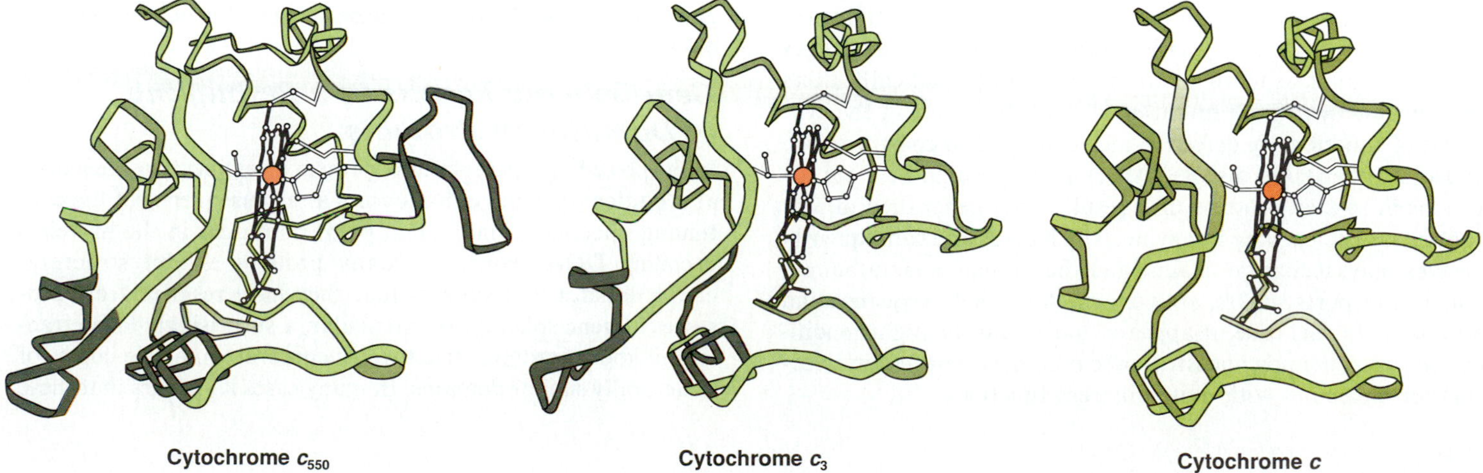

Figure 5.23

Schematic illustration of two serine proteases, elastase and subtilisin. Their molecules differ totally in sequence and tertiary structure but have catalytic sites that are nearly identical. The configuration of the active site for elastase is described in chapter 9.

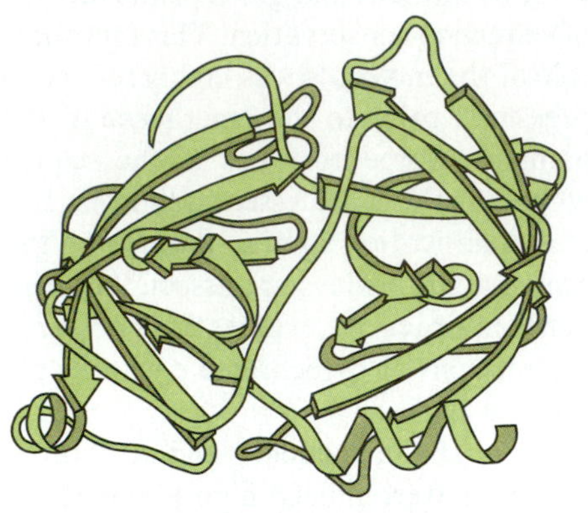

Elastase

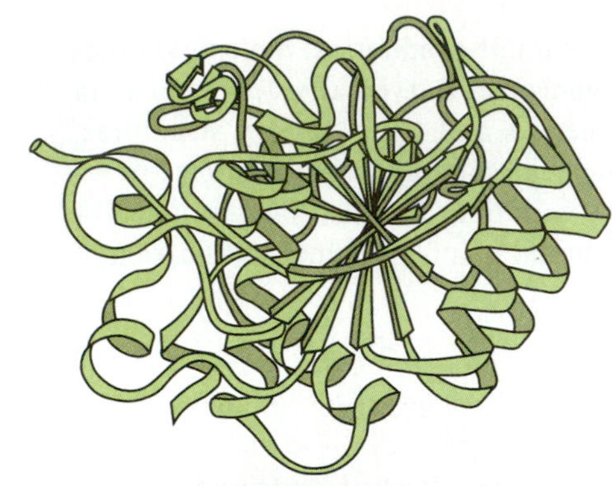

Subtilisin

acid residues that interact in functionally important ways with the protein's heme group. The slight variations of the basic structural theme of cytochrome *c* that are observed appear to optimize the molecule's function in different organisms.

In addition to the cytochrome *c* proteins, several other families of proteins have been found to share similarities in amino acid sequence and tertiary structure. Again, the observed differences in sequence and structure among individual members reflect evolutionary pressures that modified a basic structural arrangement in order to diversify the functional properties of the molecules. Examples of such structurally and functionally related families include the oxygen-binding globins that we discussed earlier, the serine protease enzyme families (see chapter 9), and the dehydrogenases. Generally, related members within a given enzyme family catalyze chemically similar reactions but exhibit varying specificities for structurally different substrate molecules. For example, while all serine proteases catalyze the hydrolytic cleavage of peptide bonds, different members of this molecule family cleave polypeptides at different locations, depending on the nature of the amino acid side chains adjacent to the cleavage site (see chapter 9).

Although divergent evolutionary processes usually produce gradual changes in a given protein function, some changes are less gradual, as when a mutation occurs that radically alters protein function. Such mutations frequently result in the synthesis of functionally defective molecules and so constitute one cause of inheritable disease. Alternatively, amino acid substitutions in related proteins may result in the generation of new functions. The enzyme lysozyme, which binds and subsequently cleaves polysaccharide chains, and the protein α-lactalbumin, which transports sugars, are very similar in both sequence and structure. In this case, it appears that relatively slight modifications of a common ancestral precursor have resulted in selection for molecules with quite different functions.

Not all functionally related families of proteins arose by divergent evolution from a common ancestor. In some cases, proteins that have extensive functional or structural similarities appear to have arisen independently. An outstanding example of two molecules that are functionally similar, but are radically different in sequence and structure, occurs in the serine proteases (fig. 5.23). Many members of the serine protease family are closely related in sequence and structure. However, in *Bacillus subtilis* the serine protease subtilisin, while being essentially identical in its arrangement of amino acid residues at the active site to the other serine proteases, otherwise differs from them completely in sequence and tertiary structure. This situation presumably reflects convergent evolution on a particular active site arrangement required for the protein's catalytic function.

More frequently, proteins that differ completely in sequence and function have quite similar tertiary structures. In these cases, the observed structural similarities most probably reflect selection of a particularly stable structural arrangement. Examples of such structurally related molecules include those with similarly twisted β sheets (see fig. 4.24) and proteins organized as a bundle of four closely packed α helices (see fig. 4.30).

Gene Splicing Results in a Reshuffling of Domains in Proteins

In the preceding descriptions of protein diversification, we looked at evolutionary change that occurs as a consequence of the continuing selection of individual point mutations in the protein's encoding DNA. However, many proteins exhibit structural characteristics that suggest that they have resulted from processes of gene splicing. In particular, a surprisingly large fraction of known protein structures incorporate multiple copies of structurally similar domains. In many cases it appears that these

Figure 5.24

Pyruvate kinase domains 1, 2, and 3 as an example of a protein whose domains show no structural resemblance whatsoever.

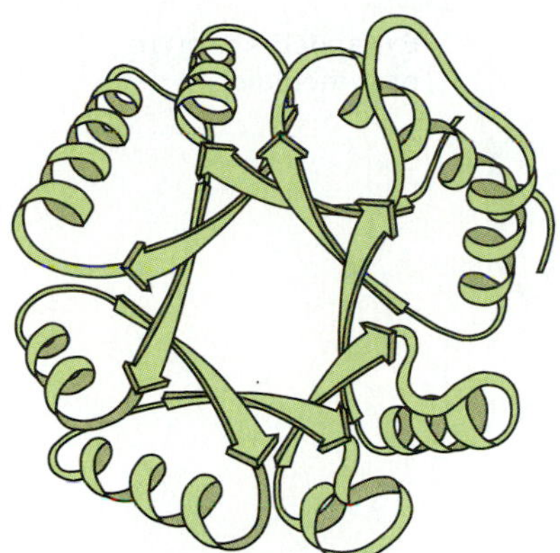

Pyruvate kinase domain 1

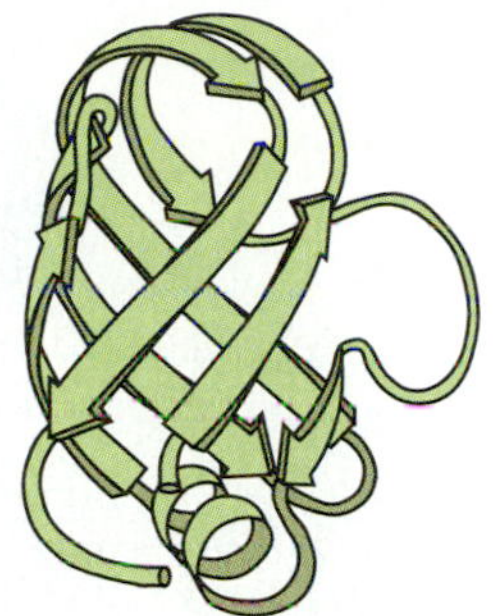

Pyruvate kinase domain 2

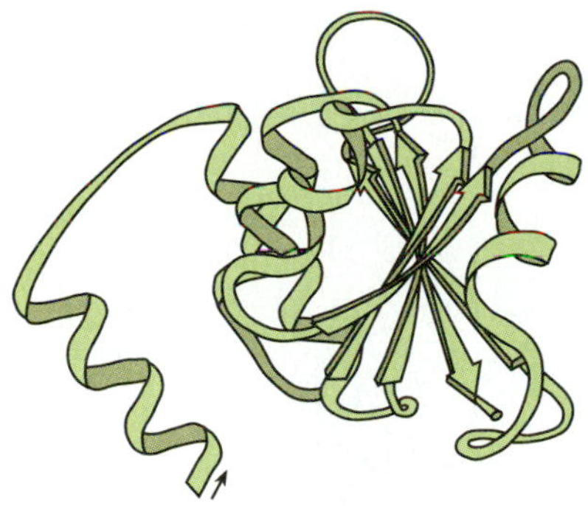

Pyruvate kinase domain 3

molecules have arisen by the splicing together of duplicate or multiple copies of a gene coding for a given structural domain, followed by the essentially independent fixation of mutations throughout the spliced genome. The eventual result is a protein composed of sequentially different but structurally similar repeating domains.

Additional evidence for the role of gene splicing in protein evolution comes from the observation that some large proteins are composed of several different structural domains, each of which may structurally resemble parts or the entirety of other known proteins. A good example is the glycolytic enzyme pyruvate kinase (fig. 5.24). This large protein is organized as three structural domains, two of which show convincing structural similarities to, respectively, the β barrel of triose phosphate isomerase (domain 1; see fig. 4.24) and a twisted β-sheet domain common to many dehydrogenases (domain 3; see fig. 4.24). The third domain of pyruvate kinase (domain 2) also has convincing similarity to a common structural type, the antiparallel β barrel.

Evolutionary Diversification Is Directly Involved in Antibody Formation

In most cases the fixation of new mutations is a relatively infrequent event, resulting in the gradual evolution of proteins such as cytochrome *c*. By contrast, one of the important biological defense mechanisms of higher organisms, the immune response, depends on the rapid generation of structurally novel molecules that can recognize and bind foreign substances that may be harmful to the organism. The molecules responsible for the initial recognition and binding of foreign substances are the immunoglobulins. These molecules are composed of two pairs of polypeptide chains of different length that are interconnected by covalent cysteine disulfide linkages (fig. 5.25). Sequence studies of various immunoglobulins have shown that both the heavy and light polypeptide chains contain repeating homologous sequences that are about 110 residues in length. Structural studies of the immunoglobulin molecule show that the sequentially homologous regions fold individually into similar structural domains, arranged as a bilayer of antiparallel sheets. The molecule in its entirety is formed of twelve similar structural domains, of which eight are formed by the two heavy chains and four by the two light chains.

Immunoglobulins that are specific for binding to different foreign substances vary greatly in the sequences found in the amino-terminal domains of both the heavy and light chains. It is these variable regions that form the binding sites between the immunoglobulin molecule and the foreign substances that trigger the immune response. The remarkable property of this system is that it can rapidly diversify the sequence of variable regions by mutation, gene splicing, and RNA splicing. The net result is that the organism can produce an enormous variety of antibodies from a quite limited amount of informational DNA originating in the germ-line tissue.

Protein Purification Procedures

To characterize the proteins that we have discussed so far, we must usually first isolate the protein under study from the complex mixture of proteins found in the organism. Once we have decided to purify a particular protein, we must weigh several factors. For example, how much material is needed? What level of purity is required? The starting material should be readily available and should contain the desired protein in relative abundance. If the protein is part of a larger structure, such as the nucleus, the mitochondria, or the ribosome, then it is advisable to isolate the large structure first from a crude cell extract.

Figure 5.25

The pattern of disulfide cross-links in immunoglobulin G. The subscripts L and H refer to light and heavy chains, respectively; C and V refer to regions of the sequence that are relatively constant or quite variable, respectively, in different IgG species. Each block of each sequence folds into an independent tertiary-structure domain, so that the final structure resembles a protein with 12 subunits, even though it is a single covalent entity.

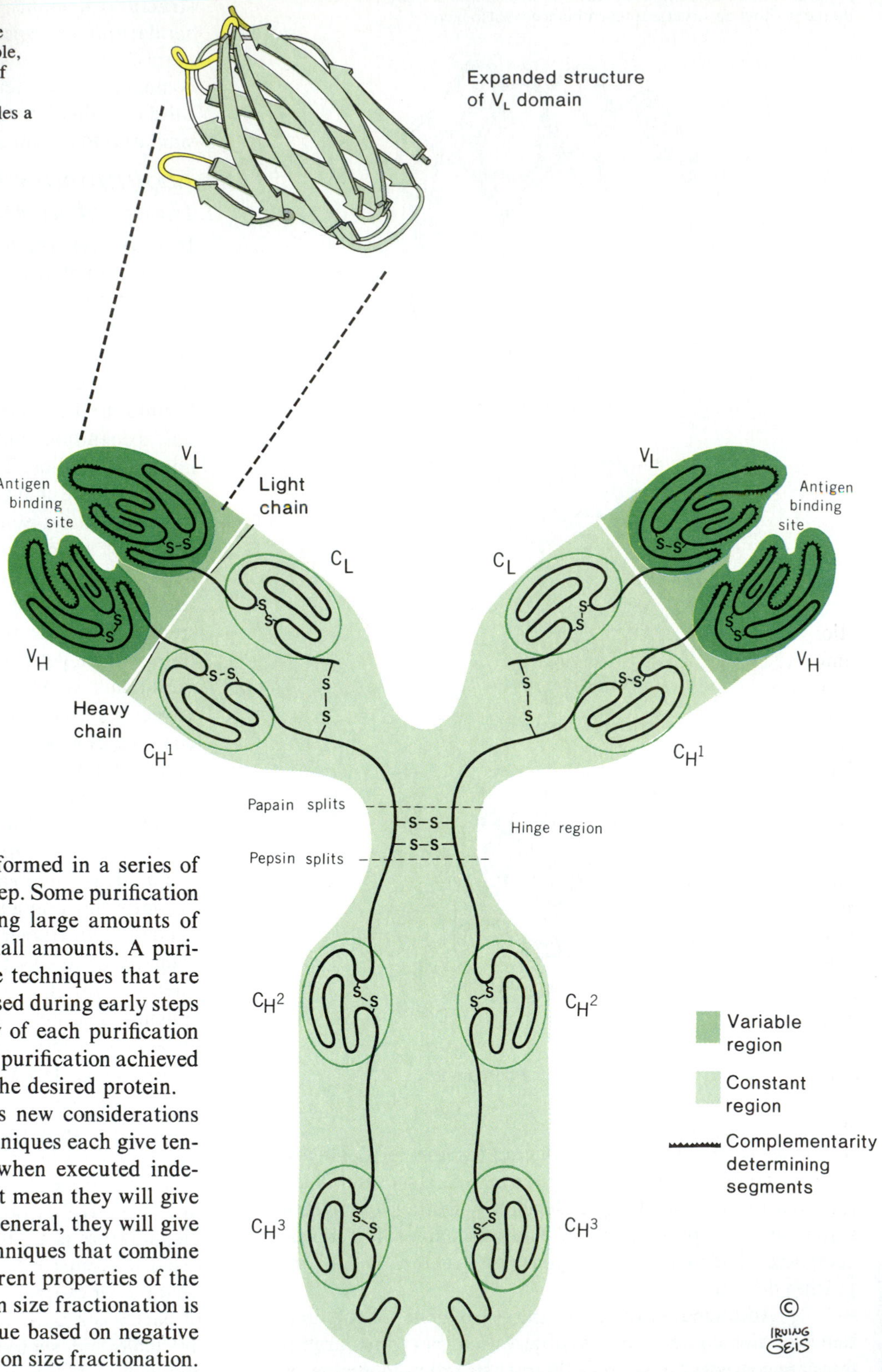

Purification must usually be performed in a series of steps, using different techniques at each step. Some purification techniques are more useful when handling large amounts of material, whereas others work best on small amounts. A purification procedure is arranged so that the techniques that are best for working with large amounts are used during early steps in the overall purification. The suitability of each purification step is evaluated in terms of the amount of purification achieved by that step and the percent recovery of the desired protein.

Combining techniques introduces new considerations and new problems. If two purification techniques each give tenfold enrichment for the desired protein when executed independently on a crude extract, this does not mean they will give 100-fold enrichment when combined. In general, they will give somewhat less. As a rule, purification techniques that combine most effectively usually are based on different properties of the protein. For example, a technique based on size fractionation is more effectively combined with a technique based on negative charge than with another technique based on size fractionation.

Table 5.3
Outline of Purification of UMP Synthase from Ehrlich Ascites Carcinoma

Fraction	Volume (ml)	Protein (mg)	OMPDase[a]			OPRTase[a]			Ratio of OMPDase to OPRTase
			Units[b]	Sp. Act.[c]	Percent Recovery	Units[b]	Sp. Act.[c]	Percent Recovery	
1. Streptomycin fraction	1040	11,700	40.4	0.0034		20.5	0.0018		2.0
2. Dialyzed $(NH_4)_2SO_4$ fraction	144	311	24.3	0.0078	60	8.7	0.0028	42	2.8
3. Affinity column eluate (concentrated)	0.475	0.51	4.0	7.8[d]	10	0.35	0.69	3.3	11.4

[a]OMPDase = OMP decarboxylase; OPRTase = orotate PRTase.
[b]Units refer to total amount of enzyme activity.
[c]Specific activity refers to the units of enzyme activity divided by the total protein.
[d]This value represents a 2,300-fold enrichment from fraction 1.

Throughout the purification we must have a convenient means of assaying for the desired protein, so we can know the extent to which it is being enriched relative to the other proteins in the starting material. In addition, a major concern in protein purification is stability. Once the protein is removed from its normal habitat, it becomes susceptible to a variety of denaturation and degradation reactions. Specific inhibitors are sometimes added to minimize attack by proteases on the desired protein. During purification it is usual to carry out all operations at 5° C or below. This temperature control minimizes protease degradation problems and decreases the chances of denaturation.

In their natural habitat, proteins are usually surrounded by other proteins and organic factors. When these are removed or diluted as during purification, the protein becomes surrounded by water on all sides. Proteins react differently to a pure aqueous environment; many are destabilized and rapidly denatured. A common remedial measure is to add 5 to 20% glycerol to the purification buffer. The organic surface of the glycerol is believed to simulate the environment of the protein in the intact cell. Two other ingredients that are most frequently added to purification buffers are mercaptoethanol and ethylenediamine tetraacetate (EDTA). The mercaptoethanol inhibits the oxidation of protein —SH groups, and the EDTA chelates divalent cations. The latter, even in trace amounts, can lead to aggregation problems or activate degradative enzymes.

The following two examples of purification show how various techniques can be effectively combined to produce purified proteins with a minimum of effort and loss of activity.

Purification of An Enzyme with Two Catalytic Activities

The last two steps in the biosynthesis of the mononucleotide uridine 5′-monophosphate (UMP) are catalyzed by (1) orotate phosphoribosyltransferase (OPRTase) and (2) orotate 5′-monophosphate (OMP) decarboxylase.

Mary Ellen Jones and her colleagues set out to purify the enzyme or enzymes involved in these two reactions. Their main goal was to determine whether the two reactions are carried out by one protein or more than one. Their findings indicated that the two reactions were both catalyzed by the same enzyme, consisting of a single polypeptide chain. To demonstrate this fact, it was necessary to monitor both enzyme activities at each step in the purification and show that both activities copurified. For this purpose, Jones used specific enzyme assays for both enzyme activities. All fractions were assayed for both enzymatic activities at each stage of the purification.

The main data associated with the purification are summarized in table 5.3. This table indicates the total protein obtained in each step, the number of enzyme units* for each enzyme, and the ratio of enzyme units to total protein, called the specific activity. In the absence of enzyme inactivation, the specific activity should be directly proportional to the enrichment. The percent recovery refers to the amount of enzyme activity in the indicated fraction, as compared with the amount present in fraction 1. This number is usually less than 100%. The apparent losses may reflect actual losses of enzyme during purification, or they may reflect inactivation (usually due to unknown causes) of the enzyme during purification.

The nine steps involved in the purification of UMP synthase from starting tissue are summarized in figure 5.26. All steps were carried out at 0–5° C. About 200 g of Ehrlich ascites cells, a mammalian tumor rich in the desired enzymes, was suspended in buffer and processed in a tissue homogenizer, which mechanically breaks down the tissue and the cell membranes (step 1). Then EDTA and an —SH reagent were added to this total cell lysate. Solid streptomycin sulfate was also added with stirring (step 2). Streptomycin sulfate aggregates nucleic acids so that they may be more easily removed by centrifugation. The

*Enzyme units are proportional to the amount of enzyme activity. The relationship between enzyme units and absolute amount of enzyme need not concern us here.

Figure 5.26

Outline of purification scheme for UMP synthase from Ehrlich ascites tumor cells of mice.

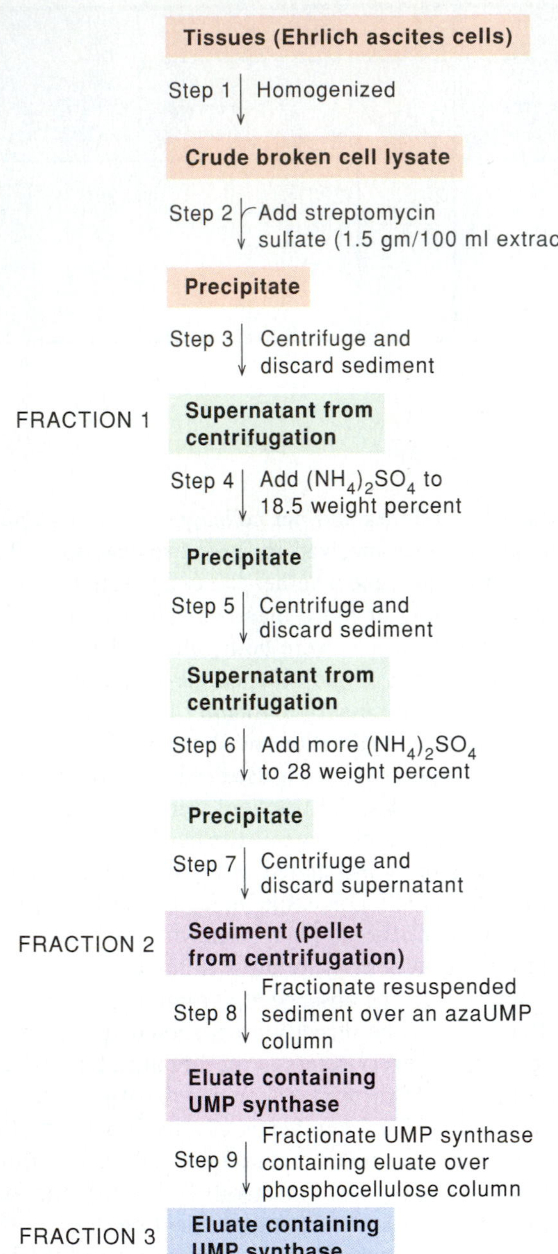

resulting slurry was subjected to high-speed centrifugation, and the resulting supernatant (see table 5.3, fraction 1) was carefully decanted for further processing (step 3).

Preliminary experiments had shown that the desired enzymes were in the 18.5–28% $(NH_4)_2SO_4$ fraction. (The reasoning behind use of $(NH_4)_2SO_4$ in purification is described in box 5B.) This knowledge served as the basis for the next three steps. First 239 g of solid $(NH_4)_2SO_4$ was added to 1040 ml of supernatant (step 4). The resulting precipitate was removed by centrifugation (step 5). Then an additional 120 g of $(NH_4)_2SO_4$ was added to the supernatant (step 6). The resulting slurry was centrifuged. This time the supernatant was discarded after centrifugation, leaving the sediment containing the enzyme activity for further processing (step 7). The sediment was resuspended in a dilute buffer for column chromatography (see table 5.3, fraction 2). Inspection of table 5.3 indicates only about a twofold increase in specific activity between fractions 1 and 2.

The main purification was achieved by two column steps, carried out in series. (Applications of column chromatography to protein purification are described in box 5C.) The first column was an affinity column containing an analog of UMP, 6-azauridine 5′-monophosphate (azaUMP), covalently attached to an agarose column support system. In dilute buffer, greater than 99% of the protein in fraction 2 is retained on this column. After thorough rinsing of the column with dilute buffer, 5×10^{-5} M azaUMP was added to the buffer. This addition resulted in the elution of the UMP synthase (step 8). Then the column eluant carrying the two enzyme activities associated with UMP synthase was resuspended in pure buffer and passed over a phosphocellulose column (step 9). Phosphocellulose was chosen because the negatively charged phosphate groups result in the retention of proteins by electrostatic attraction alone, but they also resemble the phosphate groups in the naturally occurring enzyme substrate, OMP. Thus the phosphocellulose column may be thought of as an ion-exchange column and an affinity column combined. In ordinary ion-exchange chromatography, the protein, after column loading, is eluted by increasing the ionic strength with a simple inorganic salt. In this example, Jones used a more specific method to elute the enzyme, which involved adding 10^{-5} M azaUMP and 2×10^{-5} M OMP to the original loading buffer. The addition does not substantially increase the ionic strength of the buffer. Therefore the only phosphocellulose-bound proteins that are likely to be eluted by this treatment are those with an especially high affinity for either of these nucleotides (azaUMP or OMP). This fact should greatly favor selective elution of those enzymes that carry specific sites for binding these nucleotides. The two column steps together resulted in an enzyme preparation (see table 5.3, fraction 3) that was approximately 2,300-fold purified over the starting material, as measured by the increase in specific activity.

The final product (see table 5.3, fraction 3) was examined by SDS gel electrophoresis and found to contain a single band with an estimated molecular weight of 51,000. (Gel electrophoresis is discussed in box 5D.) The denaturing conditions of an SDS gel would be expected to dissociate a multisubunit protein. Hence the SDS gel result indicated that the enzyme contains one type of polypeptide chain, but it does not tell us whether the enzyme contains one or more of these chains. Sedimentation analysis on a sucrose density gradient in a nondenaturing buffer indicated a single band with an estimated molecular weight of about 50,000. These two results taken together demonstrate that the enzyme in its native state contains a single polypeptide chain. The purity of the enzyme was also confirmed by isoelectric focusing and two-dimensional electrophoresis, with isoelectric focusing in the first direction followed by SDS gel electrophoresis in the second direction.

5B

BOX

Methods of Protein Purification: Differential Precipitation with $(NH_4)_2SO_4$

Because proteins differ in their solubility, it is possible to separate them by differential precipitation. In this box we examine some of the factors that bear on the choice of solvents used in this procedure.

The solubility of a protein reflects a delicate balance between different energetic interactions, both internally within the protein and between the protein and the surrounding solvent. Consequently, the choice of solvent can affect both the solubility and the structure of a protein.

Proteins typically have charged amino acid side chains on their surfaces that undergo energetically favorable polar interactions with the surrounding water. The total charge on the protein is the sum of the side-chain charges. However, the actual charge on the weakly acidic and basic side-chain groups also depends on the solution pH. In fact, the acidic and basic groups within the protein can be titrated just like free amino acids (see fig. 3.6) to determine their number and their p*K* values.

Proteins tend to show a minimum solubility at their isoelectric pH. This fact is apparent for β-lactoglobulin in figure 1. The decrease in solubility at the isoelectric pH reflects the fact that the individual protein molecules, which would all have similar charges at pH values away from their isoelectric points, cease to repel each other. Instead, they coalesce into insoluble aggregates.

Proteins also show a variation in solubility that depends on the concentration of salts in the solution. These frequently complex effects may involve specific interactions between charged side chains and solution ions or, particularly at high salt concentrations, may reflect more comprehensive changes in the solvent properties. For the effect of salt concentration on the solubility of β-lactoglobulin, again see figure 1. Most globulins are sparingly soluble in pure water. The increase in solubility that occurs upon adding salts such as sodium chloride is often referred to as salting in.

The effects of four different salts on the solubility of hemoglobin at pH 7 can be seen in figure 2. All four salts produce the salting-in effect with this protein; two of them, sodium sulfate and ammonium sulfate, also produce a greatly decreased solubility of the protein at high salt concentrations. This result is called salting out and occurs with salts that effectively compete with the protein for available water molecules. At high salt concentrations, the protein molecules tend to associate with each other because protein–protein interactions become energetically more favorable than protein–solvent interactions.

Each protein has a characteristic salting-out point, and we can exploit this fact to make protein separations in crude extracts. For this purpose $(NH_4)_2SO_4$ is the most commonly used salt because it is very soluble and is generally effective at lower concentrations than many other salts.

Figure 1

Solubility of ß-lactoglobulin as a function of pH and ionic strength. The isoelectric pH (pI) for this protein is about 5.2. This corresponds to the point of minimum solubility.

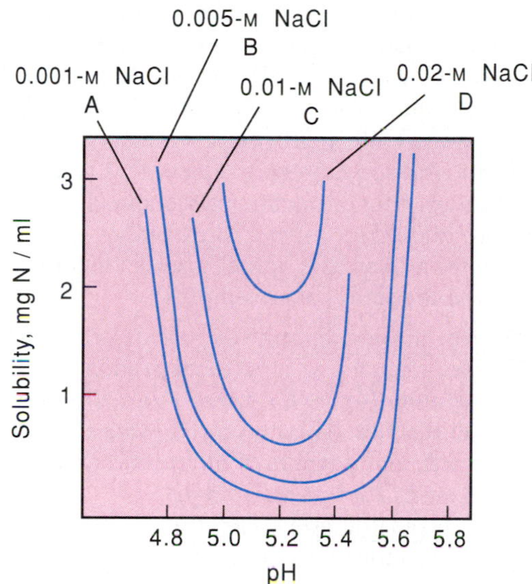

Figure 2

Solubility of horse carbon monoxide hemoglobin in different salt solutions. The addition of a moderate amount of salt (salting in) is required to solubilize this protein. At high concentrations, certain salts compete more favorably for solvent, decreasing the solubility of the protein and thus leading to its precipitation (salting out). (Source: E. J. Cohn and J. T. Edsall, *Proteins, Amino Acids, and Peptides as Ions and Dipolar Ions*. Copyright © 1942, Reinhold, New York, N.Y.)

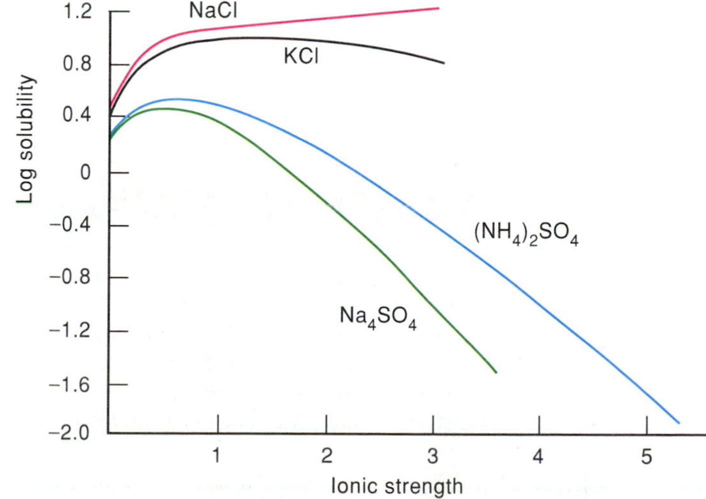

Methods of Protein Purification: Column Chromatography

Column procedures are the most effective and most varied of purification methods. Common to all column procedures is the use of a glass cylinder with an opening at the top and bottom. The cylinder is filled with a column of hydrated material, and the protein sample is applied to the top of the column. Then further buffer is passed through the column. In most column procedures, proteins bind differentially to the column material, and a change in the elution buffer causes them to be eluted differentially according to their degree of affinity for the column material. The eluant exiting from the bottom of the column is collected in equal-size fractions with the help of an automatic fraction collector (figure 1). Each fraction is analyzed, and fractions containing an appreciable amount of the desired protein are pooled for further purification.

Many variations on this basic procedure are in common use, as may be seen from the following examples.

1. In gel-exclusion chromatography a cross-linked dextran without any special attached functional groups is used for the column substrate (figure 2). Large molecules flow more rapidly through this type of column than small ones. The dextrans have different degrees of cross-linking, making them effective over different size ranges.
2. Ion-exchange chromatography makes use of the fact that proteins differ enormously in their affinity for positively or negatively charged columns. However, the cross-linked resins used in amino acid fractionation (see chapter 3) are rarely used for protein separations because proteins are too large to penetrate the resin beads. Instead, finely divided celluloses containing either positively or negatively charged groups are most commonly used (table 5C). The affinity of a protein for the column material is proportional to the salt concentration required to release the protein from the material. Typically, a column is loaded with protein solution at a low ionic strength so that most of the proteins bind to the column. After loading, elution is initiated by gradually increasing the salt concentration of the elution buffer. Proteins are eluted in the order of increasing affinity.
3. Finely divided celluloses may also be used in the column in conjunction with attached hydrophobic groups such as octyl alcohol. In this case, it is proteins with exposed hydrophobic centers that bind to the column with varying affinities. These proteins may be eluted in order of decreasing affinity for the column by increasing the level of free octyl alcohol in the eluting buffer.
4. Affinity chromatography makes use of chemical groups that have a special binding affinity for the proteins that are being sought. For example, many enzymes bind preferentially to cofactors such as adenosine triphosphate or pyridine nucleotides. Often, the separation of such enzymes from other proteins can be achieved on a column that has one of these cofactors attached. As the mixture passes through the column, proteins that have specific binding affinities for the cofactor bind to the column while other proteins pass through. The cofactor-binding proteins can then be eluted with a solution containing the same soluble cofactor.
5. High-performance liquid chromatography (HPLC) is not so much a new type of chromatography as a new way of applying old chromatographic techniques, using extremely high pressures. The same principles are involved, but the column materials usually consist of more finely divided particles made of physically stronger materials, which can withstand pressures of 5,000–10,000 psi without changing their structure. The column apparatus itself must also be designed to withstand high pressures.

Figure 1

Collecting fractions during column chromatography. Column material and elution procedure are chosen to effect optimal separation of the desired protein.

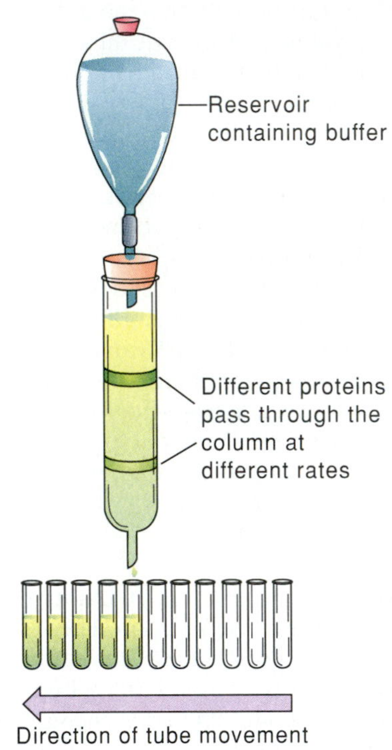

The conclusion drawn from these results was that both of the enzyme activities associated with UMP synthase—OMPDase and OPRTase—are contained in a single protein. The basis for the conclusion was that both enzyme activities are always present in the same fractions throughout the multistep purification. However, inspection of table 5.3 indicates a possible objection to this interpretation. In the columns showing percent recovery, it can be seen that substantial amounts of enzyme activity are lost for both enzymes during purification, but that considerably more activity is lost for the OPRTase. These losses could be due to actual loss of enzymes during purification or to some sort of inactivation of the enzyme sites. The preferential loss of OPRTase activity is emphasized by the last column in table 5.3, which gives the ratio of the two enzyme activities in the different fractions. Considered alone, these data could indicate that a separate catalytic unit of orotate PRTase is lost during purification. Jones thinks this is unlikely for two reasons: (1) the activity appears in no fractions other than with OMP decarboxylase during purification, and (2) the orotate PRTase activity is notably unstable. It appears, then, that both enzyme activities exist at distinct sites on a single protein. The greater loss in activity of one enzyme activity over the other can be attributed to a greater sensitivity of one reaction site over the other on the enzyme surface.

Purification of a Membrane-Bound Protein

The second purification procedure we will look at illustrates an unusual approach to the purification of a membrane-bound protein. The lactose carrier protein of *E. coli* is normally tightly bound to the plasma membrane. This protein is involved in the active transport of the dissaccharide lactose across the cytoplasmic membrane. When lactose carrier protein is present, the intracellular concentration of lactose can achieve levels 1,000-fold higher than those found in the external medium. Ron Kaback devised a simple yet elegant procedure for the purification of this protein.

Purification of the membrane-bound lactose carrier protein is a very different problem from the purification of the soluble OMP synthase. Both the approach to purification and the assays for the protein during purification are quite novel. The assay involves reconstituting a transport system with membranes that are free of lactose carrier protein, then adding the partially purified carrier protein and radioactively labeled lactose. The activity in this assay system is proportional to the transport of radioactive lactose across the membrane in the cell-free reconstituted system.

The results of the purification steps are tabulated in table 5.4, and the purification procedure is outlined in figure 5.27. In this procedure advantage was taken of the fact that the carrier protein in its native state is firmly bound to the cytoplasmic membrane. Thus the first step consisted of isolating these membranes from the rest of the cell constituents. Starting from the membrane fraction only 35-fold purification was required to achieve pure carrier protein. This rapid result was possible

Figure 2

Polydextran column showing separation of small and large molecules. The column material is immersed in solvent, which penetrates the gel particles. A separation is initiated by layering a small sample containing different-size proteins on the top of the column. This sample is pushed through the column by opening the stopcock at the bottom and adding further solvent at the top to keep up with the flow. As shown, the small protein molecules can penetrate the gel particles but the big ones cannot. Therefore the big proteins move through the column much more rapidly, and a separation of the two proteins results.

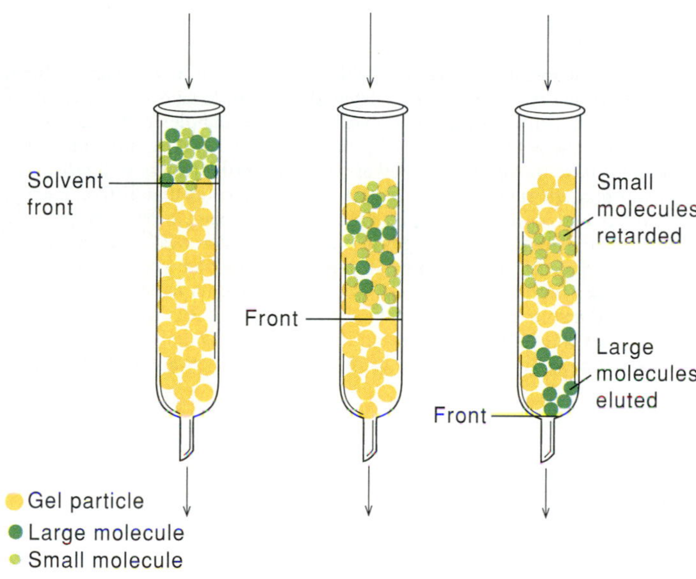

Table 5C
Some Column Materials for Ion-Exchange Chromatography of Proteins

Matrix	Functional Groups on Column
Phosphocellulose (PC)	$-PO_3^-$
Carboxymethyl cellulose (CMC)	$-CH_3-COO^-$
Diethylaminoethyl cellulose (DEAE)	$-(CH_2)_2-N(CH_2-CH_3)_2$

5D BOX

Methods of Protein Purification: Gel Electrophoresis

Gel electrophoresis is the best way to analyze mixtures and assess purity. Gel electrophoresis separates proteins according to their size and their charge. It is almost always performed in aqueous solution supported by a gel system. The gel is a loosely cross-linked network that functions to stabilize the protein boundaries between the protein and the solvent, both during and after electrophoresis, so that they may be stained or otherwise manipulated.

The most popular method of electrophoretic separation is called SDS gel electrophoresis. This method not only gives an index of protein purity but yields an estimate of the protein subunit molecular weights. The mixture of proteins to be characterized is first completely denatured by adding sodium dodecyl sulfate (a detergent) and mercaptoethanol and by briefly heating the mixture. Denaturation is caused by the association of the apolar tails of the SDS molecules with protein hydrophobic groups. Any cystine disulfide bonds are cleaved by a disulfide interchange reaction with mercaptoethanol.

The resulting unfolded polypeptide chains have relatively large numbers of SDS molecules bound to them. The success of the technique for molecular-weight estimation depends on two facts: (1) Each bound SDS molecule contributes one negative charge to the denatured protein complex, so that the charge of the protein in its native state is effectively masked by the more numerous charged groups of the associated detergent molecules. (2) The total number of detergent molecules bound is proportional to the polypeptide-chain length or, equivalently, the protein's molecular weight. As a result, the SDS-denatured protein molecules acquire net negative charges that are approximately proportional to their molecular weights (figure 1).

Another electrophoretic method frequently used for characterizing proteins is based on differences in their isoelectric points; this method is called isoelectric focusing. The apparatus usually consists of a narrow tube containing a gel and a mixture of ampholytes, which are small molecules with positive and negative charges. The ampholytes have a wide range of isoelectric points, and are allowed to distribute in the column under the influence of an electric field. This step creates a pH gradient from one end of the gel to the other, as each particular ampholyte comes to rest at a position coincident with its isoelectric point. At this stage, a solution of proteins is introduced into the gel. The proteins migrate in the electric field until each reaches a point at which the pH resulting from the ampholyte gradient exactly equals its own isoelectric point. Isoelectric focusing provides a way of both accurately determining a protein's isoelectric point and effecting separations among proteins whose isoelectric points may differ by as little as a few hundredths of a pH unit.

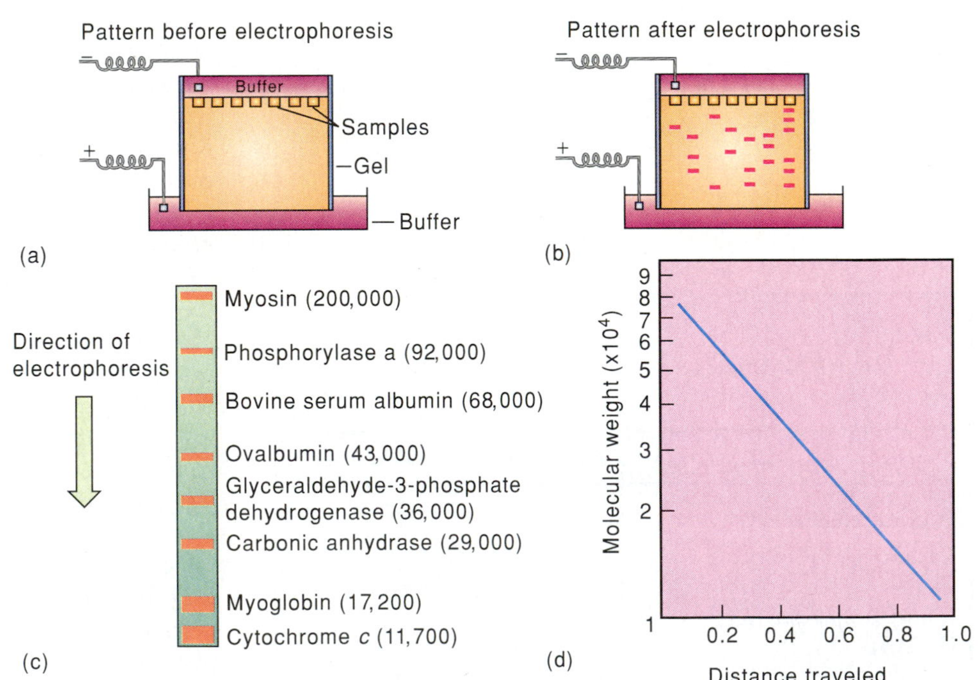

Figure 1

Gel electrophoresis for analyzing and sizing proteins. (*a*) Apparatus for slab-gel electrophoresis. Samples are layered in the little slots cut in the top of the gel slab. Buffer is carefully layered over the samples and a voltage is applied to the gel for a period of usually 1 to 4 h. (*b*) After this time the proteins have moved into the gel at a distance proportional to their electrophoretic mobility. The pattern shown indicates that different samples were layered in each slot. (*c*) Results obtained when a mixture of proteins was layered at the top of the gel in phosphate buffer, pH 7.2, containing 0.2% SDS. After electrophoresis the gel was removed from the apparatus and stained with Coomassie blue. The protein and its molecular weight are indicated next to each of the stained bands. (*d*) The logarithm of the molecular weight against the mobility (distance traveled) shows an approximately linear relationship. (Data of K. Weber and M. Osborn.)

Table 5.4
Purification of the Lactose Carrier Protein

Fraction	Protein (mg)	Percent Recovery (Total Protein)	Percent Recovery (Carrier Protein)	Purification Factor
1. Membrane fraction	12.5	100	100	1.0
2. Urea-extracted membrane	5.6	45	76	1.7
3. Urea/cholate-extracted membrane	2.6	21	61	2.9
4. Octylglucoside extract	0.4	3.2	38	12
5. DEAE column peak	0.056	0.4	14	35

Figure 5.27

Outline of purification procedure for lactose carrier protein from *E. coli*.

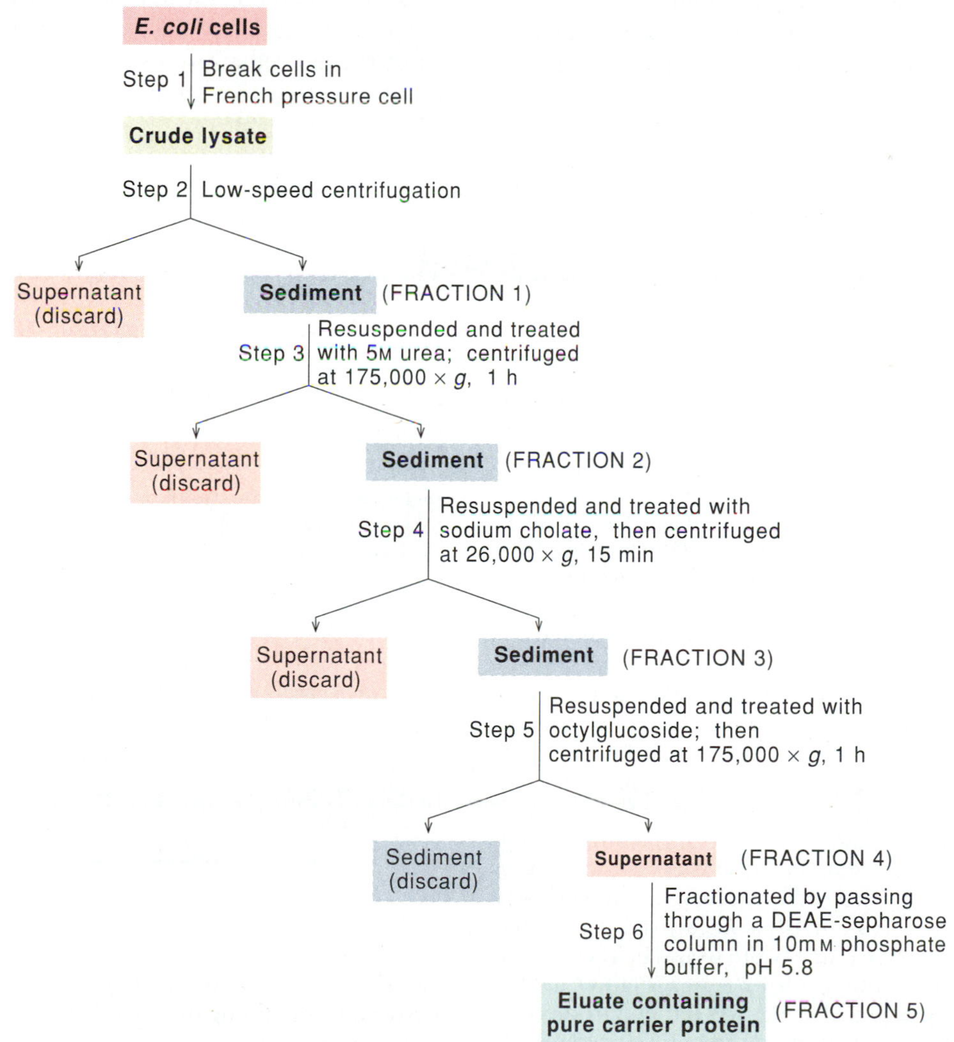

because a special strain of *E. coli,* containing about 100 times the normal carrier protein, was used as starting material for the purification. The high initial content was engineered by putting the carrier protein gene on a multicopy plasmid, which was then inserted into the cell—a procedure described in chapter 27.

Bacterial cells are much tougher than mammalian cells, requiring a more stringent procedure for cell disruption. In this case, the cells were placed in a so-called French pressure cell and bled through an orifice from very high pressures (~ 10,000 psi) to atmospheric pressure (step 1). Under these conditions the cells literally explode, fragmenting their membranes and releasing the cytoplasmic contents. The membranes were pelleted by a brief centrifugation, leaving a supernatant containing DNA, ribosomes, and cytoplasmic protein, which was removed by decantation (step 2). (See box 5E for differential centrifugation techniques.) The pelleted membrane fraction was next resuspended and extracted with 5 M urea and then reextracted with 6% sodium cholate. The pellet obtained by high-speed centrifugation from these two extractions still contained most (61%) of the carrier protein in the more rapidly sedimenting membrane fraction (fraction 3), although 79% of the total membrane-bound protein was released by these treatments (steps 3 and 4).

At this point, the carrier protein was released from a suspension of the membranes by addition of the hydrophobic reagent octylglucoside in the presence of *E. coli* phospholipid (step 5). It is believed that the octyl part of octylglucoside competes effectively with the membrane for binding to hydrophobic centers on the carrier protein. The *E. coli* phospholipid facilitates dissociation of the carrier protein from the membrane fraction. The solubilized octylglucoside-containing extract was fourfold enriched in carrier protein after a high-speed centrifugation to remove residual membrane and membrane-bound proteins.

Finally, the octylglucoside-containing extract was passed over a positively charged diethylaminoethyl (DEAE) sepharose column (sepharose is a form of cross-linked polydextran) in a buffer containing 10 mM potassium phosphate, 20 mM lactose, and 0.25 mg of washed *E. coli* lipid per milliliter. The carrier protein passed through this column as a symmetrical peak of protein (step 6). Most of the remaining protein in the extract adsorbed to the positively charged column. The fractions containing the bulk of the carrier protein activity were judged to be pure by SDS gel electrophoretic analysis. The purified protein contained a single polypeptide chain with an estimated molecular weight of 33,000.

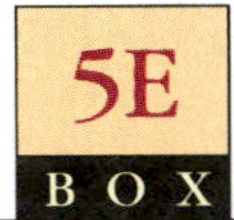

5E BOX

Methods of Protein Purification: Differential Centrifugation

A typical crude broken cell preparation contains broken cell membranes, cellular organelles, and a large number of soluble proteins, all dispersed in an aqueous buffered solution. The membranes and the organelles can usually be separated from one another and from the soluble proteins by differential centrifugation. Differential centrifugation divides a sample into two fractions: the pelleted fraction, or sediment, and the supernatant fraction, that is, the fraction that does not sediment. The two fractions may then be separated by decantation.

Differential centrifugation involves the use of different speeds and different times of centrifugation (table 5E). For example, if the protein of interest were in the mitochondrial fraction, the crude lysate should be centrifuged first at 4000 × *g* for 10 min to remove cell membranes, nuclei, and (in the case of plant material) chloroplasts. The supernatant from this step contains, among other elements, the mitochondria, and would be decanted and recentrifuged at 15,000 × *g* for 20 min to obtain a sediment primarily containing mitochondria. If ribosomes instead of mitochondria were the goal, then the crude lysate would be centrifuged at 30,000 × *g* for 30 min and the resulting supernatant would be decanted and centrifuged at 100,000 × *g* for 180 min. If the soluble protein fraction were the goal, then the entire lysate would be centrifuged at 100,000 × *g* for 180 min and the resulting supernatant, containing the soluble protein, would be decanted for further processing.

Table 5E
Sedimentation Conditions for Different Cellular Fractions

Fraction Sedimented	Centrifugal Force (× *g*)	Time (min)
Cells (eukaryotic)	1,000	5
Chloroplasts; cell membranes; nuclei	4,000	10
Mitochondria; bacteria cells	15,000	20
Lysosomes; bacterial membranes	30,000	30
Ribosomes	100,000	180

The two purifications described here are as different as the two proteins involved. No two purifications are exactly alike, but the principles of purification, as stated at the outset, are quite similar. Fortunately, an almost endless variety of purification techniques exists. This variety is both helpful and challenging, as a great deal of knowledge and creativity are required to exploit it. In addition to professional expertise, the two most important things required to make a purification possible are an unambiguous assay for the protein in question and a means of stabilizing the protein during purification.

Summary

In this chapter we have considered the relationship between structural and functional properties of proteins and the means of purification so that individual proteins may be studied in isolation. The following points are the most important.

1. The polypeptide chains of proteins are synthesized on ribosomes. Ribosomes that remain free in the cytosol are associated with the synthesis of proteins that are targeted to locations in the cytosol, the nucleus, the peroxisomes, the mitochondria and the chloroplasts. Other ribosomes, which are attached to the endoplasmic reticulum, are engaged in the synthesis of proteins that are targeted to the endoplasmic reticulum, the Golgi, the lysosomes, the plasma membrane, and the secretory granules.
2. Proteins that cooperate in a particular biochemical process are usually found in the same location within the cell or extracellular milieu.
3. Each protein is carefully designed to carry out a specific function, as revealed by our discussion of hemoglobin and muscle.
4. Hemoglobin is a tetramer made of two almost identical subunits. The function of hemoglobin is threefold: to transport O_2 from the lungs to the tissues where it is consumed, to transport CO_2 from the tissues where it is produced to the lungs, where it is expelled, and to maintain the blood pH over a narrow range. The structure of hemoglobin is designed so that it will pick up the maximum amount of oxygen at high oxygen tensions in the lung tissue and will deliver the maximum amount of oxygen in the oxygen-consuming tissues.
5. Muscle is an aggregate of several different proteins involved in contraction. The main protein components of muscle are organized as overlapping filaments of two types: thin filaments, composed mainly of actin molecules, and thick filaments, composed of myosin molecules.
6. The process of muscular contraction entails a sliding of the two types of filaments past one another. In a fully contracted component of muscle tissue the actin and myosin filaments show a maximum overlap with one another. The contraction process involves the breakage and reformation of bridges between the actin and the myosin molecules in a reaction that requires the expenditure of ATP.
7. Protein diversification is the product of a long evolutionary process. It can be studied by comparing the structures of different proteins in the light of their functions. It can also be studied by comparing structurally related proteins with similar functions in different organisms, or by comparing structurally related domains between proteins that have different functions.
8. Antibodies provide the unique opportunity for studying diversification within a single organism, for antibodies with unique specificities are the result of an ongoing evolutionary process that takes place within the organism.
9. Whereas proteins must be studied *in vivo* in their normal milieu, to characterize them in great detail they must also be isolated in pure form. Protein purification is a complex art with a great variety of purification methods usually being applied in sequence for the purification of any protein.
10. Frequently the first step in protein purification is differential sedimentation of broken cell parts. In this way soluble proteins may be separated from organelle-sequestered proteins.
11. Column procedures are useful in fractionating proteins with different affinity properties and sizes. By the use of different column materials in conjunction with specific eluting solutions, highly purified protein preparations can be obtained.
12. Gel electrophoresis, which separates proteins according to their size and charge, can be used in purification or as a means of assaying the purity during a purification procedure.
13. The molecular weights of soluble proteins can be estimated by gel electrophoresis or can be rigorously determined by sedimentation and diffusion techniques.

Selected Readings

Akers, G. K., M. C. Doyle, D. Myers and M. A. Daugherty, Molecular code for cooperativity in hemoglobin. *Science* 255:54–63, 1992. A new model for hemoglobin allostery.

Allen, R. D., The microtubule as an intracellular engine. *Sci. Am.* 238(2):42–49, 1987.

Baldwin, J., Structure and cooperativity of haemoglobin. *Trends Biochem. Sci.* 5:224–228, 1980.

Berg, H. C., How bacteria swim. *Sci. Am.* 233(2):36–44, 1975.

Cantor, C. R., and P. Schimmel, *Biophysical Chemistry.* New York: Freeman, 1980. Especially see volume 2, entitled *Techniques for Study of Biophysical Structure and Function.*

Caplan, A. I., Cartilage. *Sci. Am.* 251(4):84–94, 1984.

Dickerson, R. E., and I. Geis, Hemoglobin. Menlo Park, Calif.: Benjamin/Cummings, 1983. A magnificent presentation of every facet of hemoglobin biochemistry and genetics.

Doolittle, R., Proteins. *Sci. Am.* 253(4):88–96, 1985. Overview emphasizing evolutionary considerations.

Eisenberg, D., and D. Crothers, *Physical Chemistry and Its Applications to the Life Sciences.* Menlo Park, Calif.: Benjamin/Cummings, 1979.

Eyre, D. R., M. A. Pdaz, and P. M. Gallop, Cross-linking in collagen and elastin. *Ann. Rev. Biochem.* 53:717–748, 1984.

Gething, M. J., and J. Sambrook, Protein folding in the cell. *Nature* 355:33–45, 1992.

Huxley, H. E., Sliding filaments and molecular motile systems. *J. Biological Chemistry* 265:8347–8352, 1990.

Hynes, R. O., Fibronectins. *Sci. Am.* 254(6):42–51, 1986.

Ingram, V. M., Gene mutation in human haemoglobin: the chemical difference between normal and sickle-cell haemoglobin. *Nature* 180:326–328, 1957. A classic paper.

Karplus, M., and J. A. McCammon, The dynamics of proteins. *Sci. Am.* 254(4):42–51, 1986. A reminder that proteins are dynamic structures.

Lawn, R. M., and G. A. Vehar, The molecular genetics of hemoglobin. *Sci. Am.* 254(3):48–65, 1986.

Martin, G. R., R. Timpl, P. K. Muller, and K. Kuhn, The genetically distinct collagens. *Trends Biochem. Sci.* 10:285–287, 1985. A brief, authoritative account of the most abundant protein found in vertebrates.

Methods in Enzymology. New York: Academic Press. A continuing series of over 250 volumes that discuss most methods at the professional level but are still understandable for students.

Pauling, L., H. A. Itano, S. J. Singer, and I. C. Wells, Sickle-cell anemia: a molecular disease. *Science* 110:543–548, 1949. A classic paper.

Pollard, T. D., and J. A. Cooper, Actin and actin-binding proteins. *Ann. Rev. Biochem.* 55:987–1036, 1986. Discusses structure and function.

Salemme, R., Structure and function of cytochromes *c. Ann. Rev. Biochem.* 46:299–329, 1977.

Scopes, R., *Protein Purification: Principles and Practice,* 2d ed. New York: Springer-Verlag, 1987. A recent general treatment of this subject.

Steinert, P. M., and D. R. Roop, Molecular and cellular biology of intermediate filaments. *Ann. Rev. Biochem.* 57:593–626, 1988.

Tonegawa, S., The molecules of the immune system. *Sci. Am.* 253(4):122–130, 1985.

Wilson, A. C., The molecular basis of evolution. *Sci. Am.* 253(4): 164–170, 1985.

Problems

Step	**Volume (ml)**	**Total Protein (mg)**	**Total Units**	**Specific Activity (U/mg)**	**Yield (%)**	**Purification (*n*-fold)**
Cell extract	2,800	70,000	2,700			
($(NH_4)_2SO_4$) fractionation	3,000	25,400	2,300			
Heat treatment	3,000	16,500	1,980			
DEAE chromatography	80	390	1,680			
CM-cellulose chromatography	50	47	1,350			
Bio-Gel A chromatography	7	35	1,120			

1. A method for the purification of 6-phosphogluconate dehydrogenase from *E. coli* is summarized in the table. For each step, calculate the specific activity, percentage yield, and degree of purification (*n*-fold). Indicate which step results in the greatest purification. Assume that the protein is pure after gel exclusion (Bio-Gel A) chromatography. What percentage of the initial crude cell extract protein was 6-phosphogluconate dehydrogenase?
2. Although used effectively in the 6-phosphogluconate dehydrogenase isolation procedure, heat treatment cannot be used in the isolation of all enzymes. Explain.
3. Assume that the isoelectric point (pI) of the 6-phosphogluconate dehydrogenase is 6. Explain why the buffer used in the DEAE cellulose chromatography must have a pH greater than 6 but less than 9 in order for the enzyme to bind to the DEAE.
4. Will the 6-phosphogluconate dehydrogenase bind to the CM-cellulose in the same buffer pH range used with the DEAE-cellulose? Explain. In what pH range might you expect the dehydrogenase to bind to CM-cellulose? Explain.
5. Examine the isolation procedure shown in problem 1 and explain why gel exclusion chromatography is used as the final step rather than as the step following the heat treatment.
6. A student isolated an enzyme from an anaerobe and subjected a sample of the protein to SDS-polyacrylamide gel electrophoresis. A single band was observed upon staining the gel for protein. His advisor was excited about the result, but suggested that the protein be subjected to electrophoresis under nondenaturing (native) conditions. Electrophoresis under nondenaturing conditions revealed two bands after the gel was stained for protein. Assuming the sample had not been mishandled, offer an explanation for the observations.
7. A salt-precipitated fraction of ribonuclease contained two contaminating protein bands in addition to the ribonuclease. Further studies showed that one contaminant had a molecular weight of about 13,000 (similar to ribonuclease) but an isoelectric point 4 pH units more acidic than the pI of ribonuclease. The second contaminant had an isoelectric point similar to ribonuclease but had a molecular weight of 75,000. Suggest an efficient protocol for the separation of the ribonuclease from the contaminating proteins.

8. You have a mixture of proteins with the following properties:
Protein 1: M_r 12,000, pI = 10
Protein 2: M_r 62,000, pI = 4
Protein 3: M_r 28,000, pI = 8
Protein 4: M_r 9,000, pI = 5

Predict the order of emergence of these proteins when a mixture of the four is chromatographed in the following systems.
   (a) DEAE-cellulose at pH 7, with a linear salt gradient elution.
   (b) CM-cellulose at pH 7, with a linear salt gradient elution.
   (c) A gel exclusion column with a fractionation range of 1,000–30,000 M_r, at pH 7.
9. The absorbance at 280 nm of an 0.5 mg ml^{-1} solution of protein A was 0.75. In addition, a shoulder on the absorbance spectrum at 288 nm was observed. An absorbance of 0.2 at 280 nm was measured for protein B in the same buffer and at the same concentration as protein A, but no spectral feature at 288 nm was observed. Amino acid analysis revealed that each protein contained approximately the same molar quantities of Tyr and Phe. Suggest a reason for the differences in absorbance at 280 nm and 288 nm.
10. The absorbance of a protein solution measured at 278 nm was 0.846. The protein content of that solution, calculated from quantitative amino acid analysis, was 460 μg/ml. Calculate the extinction coefficient of the protein in units of ml mg^{-1} cm^{-1}. Assume the molecular weight of the protein to be 42,000. Calculate the molar extinction coefficient.
11. You have isolated a manganese-containing pyrophosphatase and wish to determine its molecular weight. A Sephadex G-100 gel exclusion column was calibrated using proteins whose molecular weight had been established, with the following results.

Protein	Molecular Weight	Elution Volume (ml)
Serum albumin	69,000	84
Ovalbumin	43,000	92
Carbonic anhydrase	29,000	100
Chymotrypsinogen	23,000	104
Myoglobin	17,000	110
Blue dextran	2,000,000	60

The pyrophosphatase eluted from the calibrated column in 94 ml. Determine the molecular weight of the enzyme.

12. Individual proteins of known subunit molecular weight and the pyrophosphatase whose *native* molecular weight had been determined (problem 11) were denatured in buffer containing both SDS and 2-mercaptoethanol. A sample of each protein was electrophoretically separated by SDS-polyacrylamide gel electrophoresis. The marker dye migrated 12 cm. Staining the gel for protein revealed the following:

Protein	Molecular Weight (subunit)	Distance Migrated (cm)
Serum albumin	69,000	1.6
Catalase	60,000	2.2
Ovalbumin	43,000	3.5
Carbonic anhydrase	29,000	5.2
Myoglobin	17,000	7.4

The pyrophosphatase migrated 6.7 cm. However, when the 2-mercaptoethanol was omitted from the sample buffer, the pyrophosphatase migrated 3.8 cm. Determine the subunit molecular weight and comment on the quaternary structure of the enzyme.
13. A mutant form of alkaline phosphatase was focused to its isoelectric point and was found to differ from the native (nonmutant) alkaline phosphatase by an amount corresponding to one additional positive charge relative to the native enzyme. Assuming that the mutant enzyme's isoelectric point reflects the substitution of a single amino acid residue, what are the possibilities?
14. You wish to purify an ATP-binding enzyme from a crude extract that contains several contaminating proteins. In order to purify the enzyme rapidly and to the highest purity, you must consider some sophisticated strategies, among them affinity chromatography. Explain how affinity chromatography may be applied to this separation and explain the physical basis of the separation.
15. You have isolated a protein complex that sediments in the ultracentrifuge similarly to a hemoglobin marker. If the ultracentrifugation is repeated under identical conditions except that 2-M NaC1 is added to the dilute buffer, the protein sediments similarly to a myoglobin marker. What conclusion can you reach about the properties of the protein complex?
16. Predict the effect on O_2 transport of a mutant form of hemoglobin with a markedly decreased affinity for glycerate-2,3-bisphosphate.
17. The concentration of glycerate-2,3-bisphosphate (GBP) is approximately 5 mM in the erythrocyte. Calculate the concentration of hemoglobin in the erythrocyte and compare it with that of GBP. Assume that the hemoglobin content of blood is 14 gm/dl and that the erythrocytes occupy 45% of whole blood volume. For the calculations, assume that hemoglobin (M_r = 64,500) occupies 100% of the erythrocyte volume.

6

CHAPTER

Carbohydrates, Glycoproteins, and Cell Walls

Carbohydrates are made from carbon skeletons richly laced with hydroxyl groups. The hydrophilic hydroxyl groups give carbohydrates the potential for strong interaction with water; they also give them functional groups to which various substituents can add. Finally, the hydroxyl groups provide the possibility of strong intra- or interchain interaction via hydrogen bonds. The simplest carbohydrates contain only carbon, hydrogen, and oxygen. Derivatives of carbohydrates contain nitrogen, phosphorus, and even sulfur compounds. Carbohydrates can interact with other macromolecules to form glycoproteins or glycolipids, and they are also an important component of nucleic acids.

Small carbohydrates such as glucose, fructose, and pyruvate occupy key roles in energy metabolism and supply carbon skeletons for the synthesis of other compounds. Polymeric carbohydrates are important as short-term energy-storage compounds and also as major structural compounds in plant and bacterial cell walls and in the extracellular matrix. Branched-chain polymeric carbohydrates covalently linked to proteins are used to give proteins certain surface characteristics that may be exploited to signal protein targeting and in a variety of cell–cell recognition processes.

Carbohydrates play so many roles that it is almost easier to discuss the things they can't do rather than the things they can do. There are two areas where they seem to be lacking in potential, at least by themselves. They cannot store genetic information and they cannot function as enzymes (so far as is known). In this chapter we will describe the structural and functional properties of carbohydrates.

Monosaccharides and Related Compounds

The simplest carbohydrates, sometimes referred to as monosaccharides or sugars, are either polyhydroxyaldehydes (aldoses) or polyhydroxyketones (ketoses). They can be derived from polyalcohols (polyols) by oxidation of one carbinol group to a carbonyl group. For example, the simple three-carbon triol glycerol can be converted either to the aldotriose, glyceraldehyde, or to the ketotriose, dihydroxyacetone, by loss of two hydrogens (fig. 6.1).

Since the middle carbon of glyceraldehyde is connected to four different substituents, it is a chiral center leading to two possible forms of glyceraldehyde. D-Glyceraldehyde is illustrated in figure 6.1 in the Fischer projection formula, in which the —OH group attached to the central carbon atom points to the right. If the central carbon were in the plane of the paper with tetrahedrally arranged substituents, the H and

Figure 6.1

Loss of two hydrogens by glycerol leads to the formation of glyceraldehyde or dihydroxyacetone, depending on whether the two hydrogens are lost from the end or middle position, respectively.

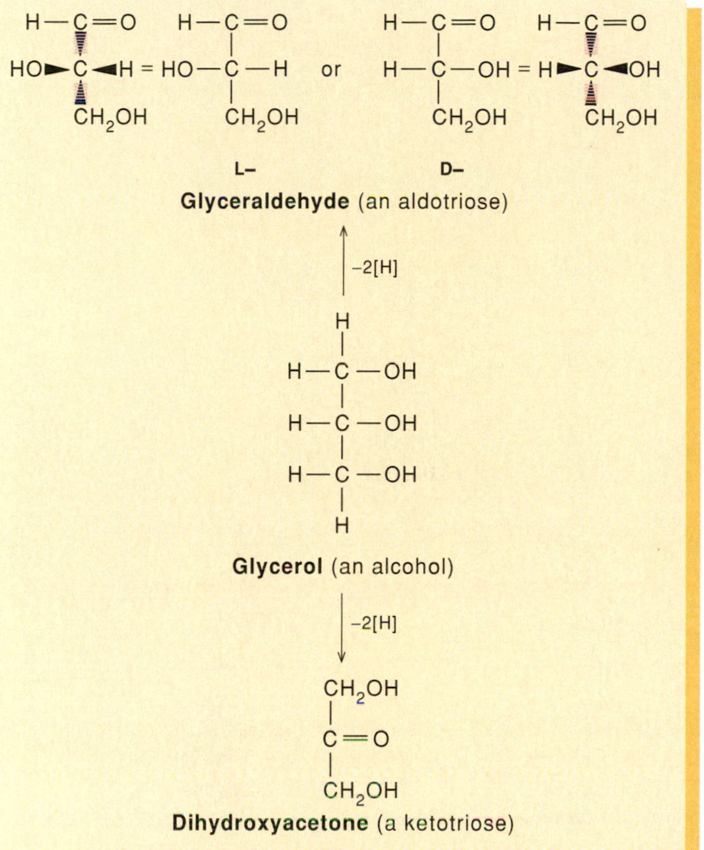

L– D–

Glyceraldehyde (an aldotriose)

–2[H]

Glycerol (an alcohol)

–2[H]

Dihydroxyacetone (a ketotriose)

OH connected to it would project above the plane of the paper and the other two substituents would project below the plane of the paper. For larger sugars, Fischer projections are written with the most highly oxidized carbon, C-1, at the top.

A molecule such as glyceraldehyde, having one center of asymmetry (chiral center), is optically active, and the two forms of the molecule can be described as the dextrorotatory and levorotatory forms, according to the way in which they rotate plane-polarized light. The symbol *d* or (+) refers to dextrorotatory rotation, and the symbol *l* or (−) refers to levorotatory rotation. Rotation can be measured by making a solution of an optically active compound and measuring the rotation of plane-polarized light that passes through it (see box 6A). A mixture containing equal amounts of the two forms is optically inactive and is referred to as a racemic mixture.

The stereochemistry of sugars is based on configurational properties. The actual sign of rotation may still be indicated by the italic letters *d* and *l*, but the absolute configuration of the four different substituents around the asymmetric carbon atom is designated by the prefix symbols D and L. For the common sugars, the prefixes D and L refer to that center of asymmetry most remote from the aldehyde or ketone end of the molecule. By convention, all optically active centers are related to the asymmetric carbon of glyceraldehyde. Isomers stereochemically related to D-glyceraldehyde are designated D, and those related to L-glyceraldehyde are designated L. We may visualize the four-, five-, and six-carbon sugars as arising from the trioses through the stepwise condensation of formaldehyde to either glyceraldehyde or dihydroxyacetone. The biosynthesis of the sugars occurs by other means.

Families of Monosaccharides Are Structurally Related

A phosphorylated derivative of D-glyceraldehyde is an intermediate in the degradation of carbohydrates. This aldose is reversibly converted to its ketose isomer, dihydroxyacetone (see fig. 6.1), by an enzyme. The tetroses, pentoses, and hexoses related to D-glyceraldehyde are shown in figure 6.2. The ketoses (e.g., fructose) are similarly related to dihydroxyacetone (fig. 6.3).

Monosaccharides Cyclize to Form Hemiacetals

Aldehydes can add hydroxyl compounds to the carbonyl group. If a molecule of water is added, the product is an aldehyde hydrate, as shown in figure 6.4. If a molecule of alcohol is added, the product is a hemiacetal; the addition of a second alcohol results in an acetal. Sugars form intramolecular hemiacetals in cases where the resulting compound has a five- or six-membered ring.

Hemiacetal formation first became apparent from optical studies of D-glucose. The optical rotation of a freshly dissolved sample of D-glucose changes with time because there are two different hemiacetals that are convertible in solution (fig. 6.5). They are referred to as α-D-glucose, where $[\alpha] = 112°$, and β-D-glucose, where $[\alpha] = 19°$. A freshly prepared solution of either of these compounds will eventually approach an intermediate value that depends on the equilibrium between the two forms.

The convention for numbering hexoses is shown in the central structure of figure 6.5. The α designation for the D series indicates that the aldehyde or C-1 hydroxyl group is on the same side of the structure as the ring oxygen in the Fischer projection, and the β designation indicates that it is on the opposite side. For the L series the reverse is the case. When the sugar is dissolved in water, the hemiacetal is in equilibrium with the straight-chain hydrated form. The straight-chain form can produce either hemiacetal, α or β. Conversion of one isomer to another in solution is referred to as mutarotation. The different forms are referred to as anomers, and the anomeric carbon is the carbon that contains the reactive carbonyl. In the case of glucose this is the C-1 carbon. Equilibrium is reached without added catalyst in a few hours at room temperature. The open-chain form usually represents only a small fraction of the total (see figure 6.5 for actual percentages).

Figure 6.2

Configurational relationships of the D-aldoses. The most important sugars are starred. Note that in the two D series the configuration about the chiral center farthest from the carbonyl is the same.

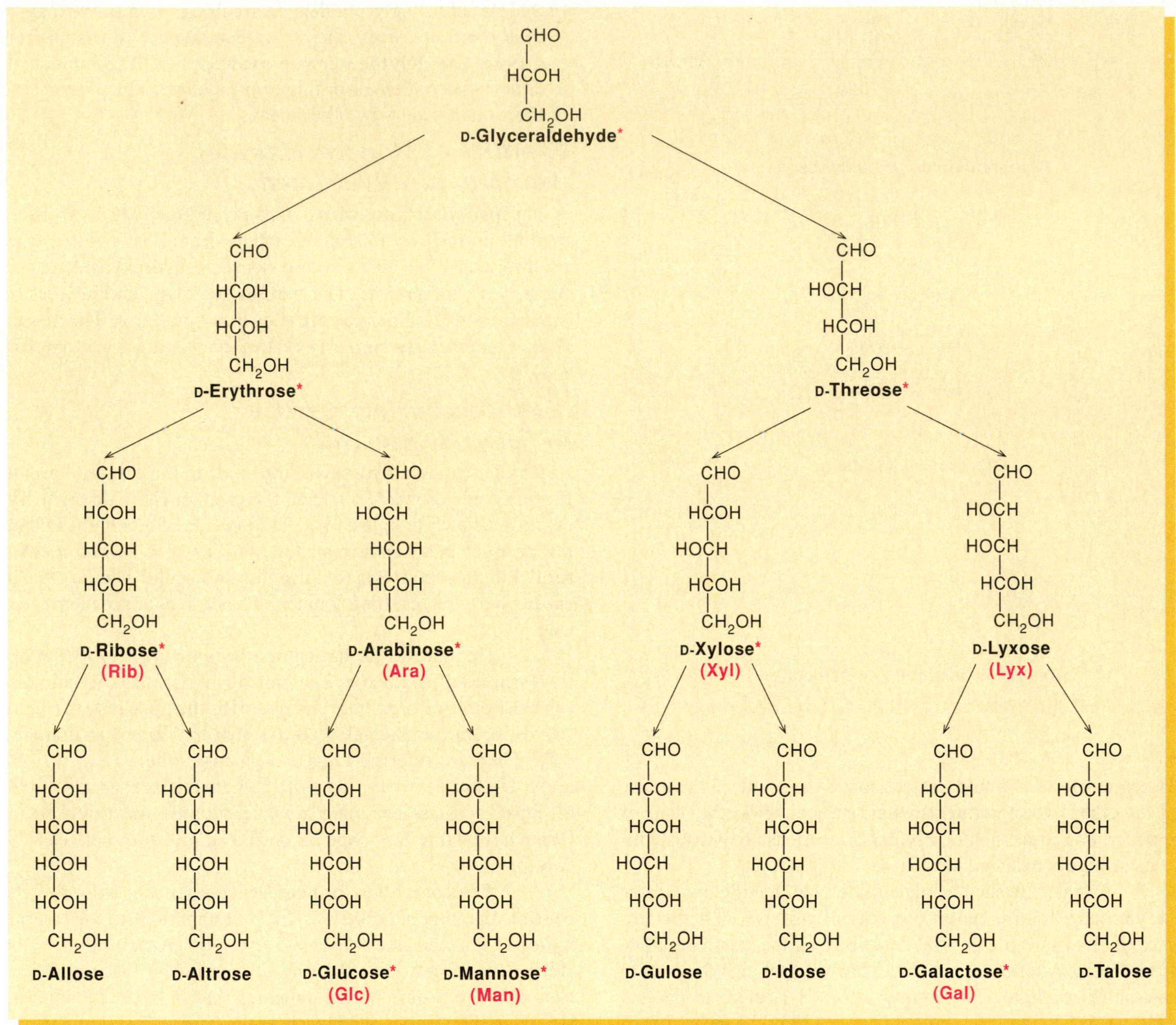

Figure 6.3

Configurational relationships of the D-ketoses. The most important sugars are starred.

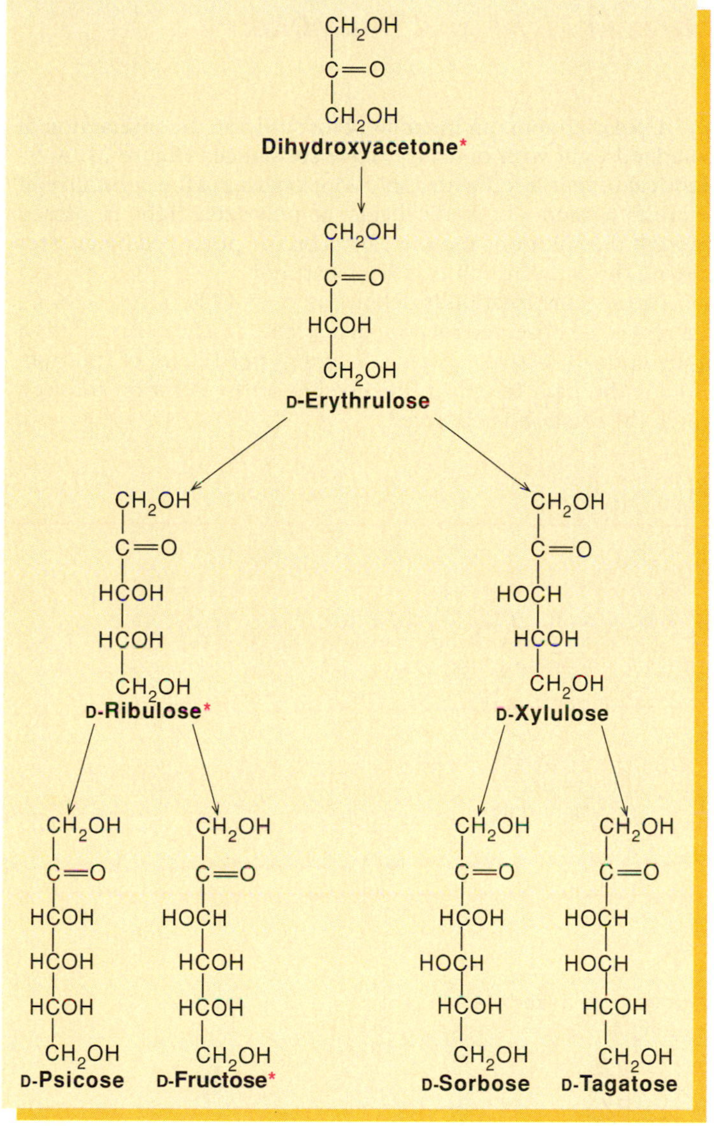

Figure 6.5

Different forms of glucose that result from dissolving glucose in water. At 25°C in water, glucose reaches an equilibrium containing about 0.02% free aldehyde, 38% α-pyranose form ($[\alpha]_D$ = 120), 62% ß-pyranose form ($[\alpha_D]$ = 19), and less than 0.5% of the furanose forms. The anomeric carbon is shown in color.

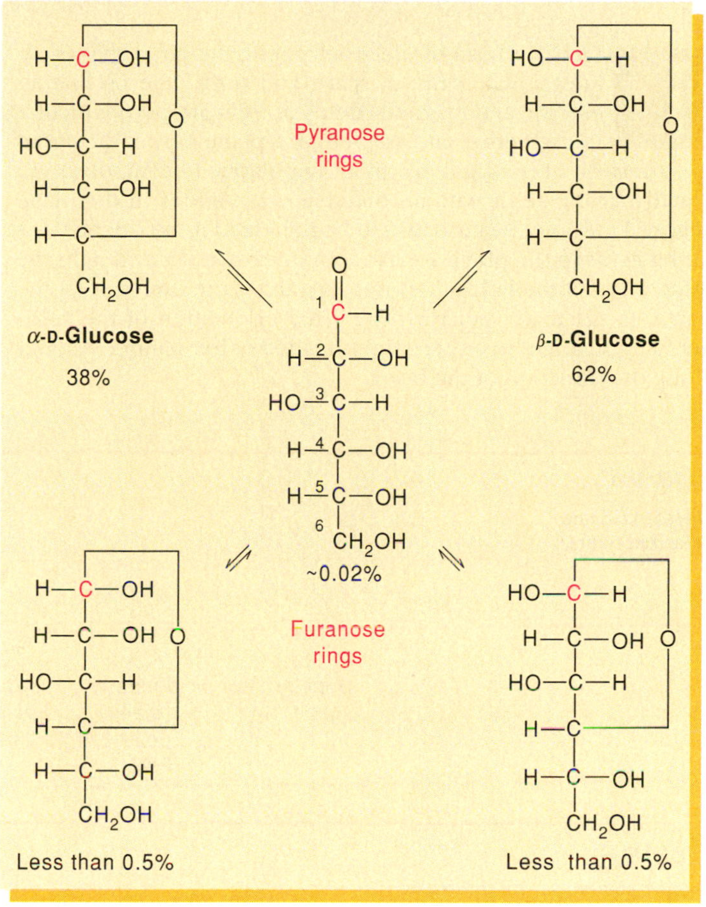

Figure 6.4

Aldehydes can add H_2O to form hydrates or can add alcohols to form hemiacetals and acetals.

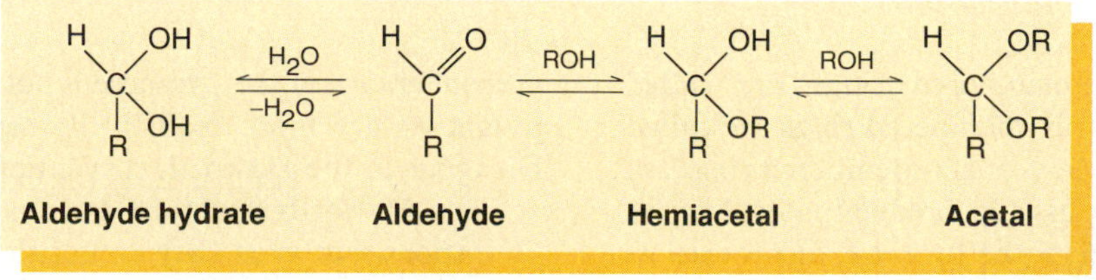

6A
BOX

Methods for Structural Analysis: Polarized Light and Polarimetry

Light is a form of electromagnetic radiation that oscillates sinusoidally in space and time. The oscillating electric and magnetic fields of light are perpendicular to each other and are both in a plane perpendicular to the direction of the light ray. In an unpolarized beam, there are equally strong fields with all different orientations in the plane (figure 1). A light beam is said to be polarized if the orientations of the fields in the plane are fixed. In linearly polarized light the orientation of the fields does not vary along the direction of the beam. In circularly polarized light the orientation of the fields gradually changes in either a right-handed or left-handed manner along the direction of the beam.

A polarimeter is an instrument for studying the interaction of polarized light with optically active substances (figure 2). A cylindrical tube is filled with a solution containing the substance of interest. A monochromatic beam of polarized light is passed through the solution, and the effects on the polarized beam after passing through the solution are measured.

The specific rotation is defined as $[\alpha] = (100 \times A)/(c \times l)$, where A is the observed rotation in degrees, c is the concentration of the optically active substance in grams per 100 ml of solution, and l is the path length in decimeters of the solution through which the rotation is observed.

Figure 1

Unpolarized and polarized light.

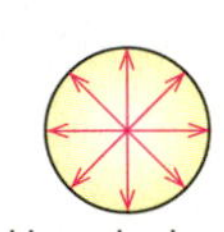

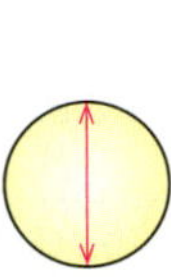

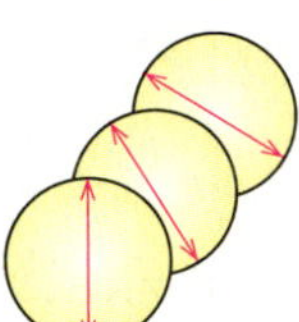

Figure 2

Simple polarimeter for measuring rotation of linearly polarized light.

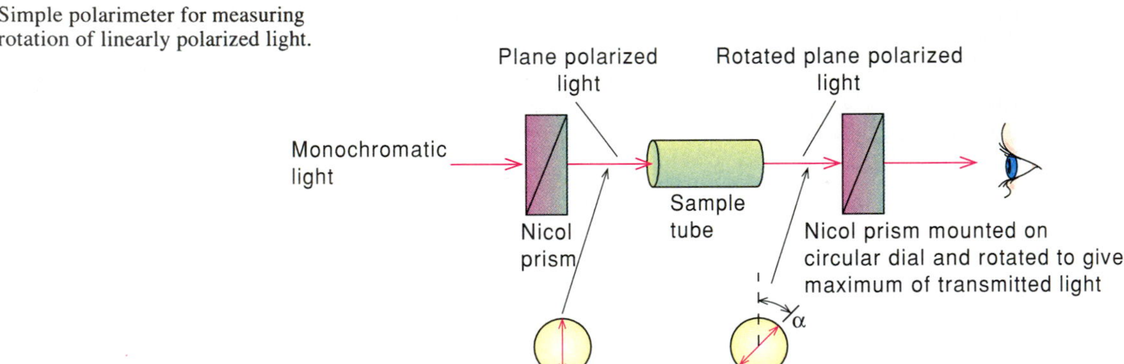

Hemiacetals with five-membered rings are called furanoses, and hemiacetals with six-membered rings are called pyranoses. In cases where either five- or six-membered rings are possible, the six-membered ring usually predominates. For example, for glucose less than 0.5% of the furanose forms exist at equilibrium (see fig. 6.5, bottom). The reason for the general preponderance of the pyranose is not known. Both furanoses and pyranoses are more realistically represented by pentagons or hexagons in the Haworth convention, as shown in figure 6.6.

Haworth structures are unambiguous in depicting configurations, but even they do not show correctly the spatial relationship of groups attached to rings. The normal valence angle

Figure 6.6

Comparison of the Fischer and Haworth projections for α- and ß-D-glucose. The Haworth projection is a step closer to reality.

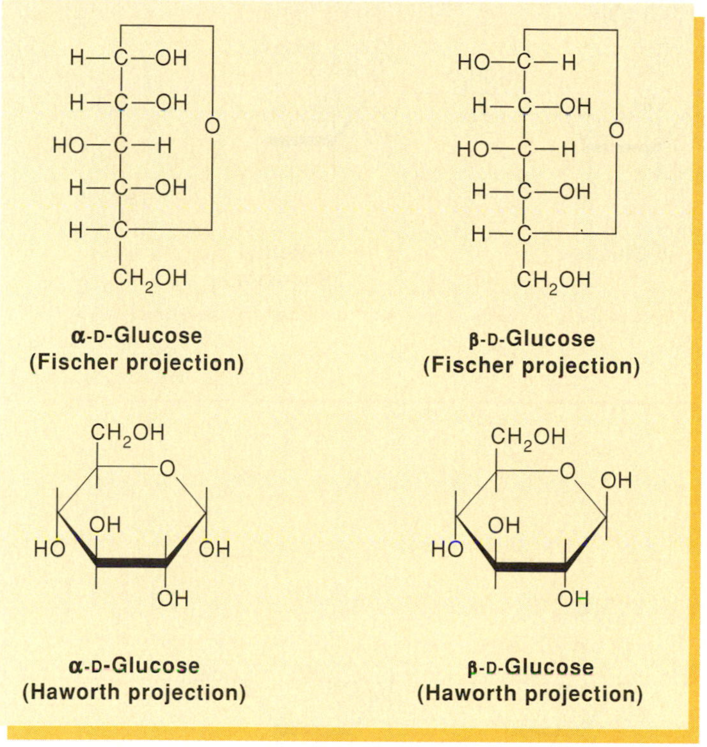

α-D-Glucose (Fischer projection)

ß-D-Glucose (Fischer projection)

α-D-Glucose (Haworth projection)

ß-D-Glucose (Haworth projection)

of saturated carbon (109°) prevents a stable planar arrangement for cyclohexane or the related pyranose molecule. The two most likely conformations* are the so-called chair and boat forms (fig. 6.7). Usually the chair form is considerably more stable than the boat form. The twelve substituent atomic groups of the ring carbons fall into two classes, those that are approximately perpendicular to the plane of the ring, i.e., axial, and those that are parallel to the plane of the ring, i.e., equatorial. As a rule, a substituent is at a lower energy state in the equatorial position because there is less chance of steric hindrance with other substituents. This fact becomes more important with larger substituents. The two anomers for the favored chair form of D-glucose are shown in figure 6.8. In sugar chemistry, formulating the conformations of aldohexoses is important in interpreting reactivity of hydroxyl groups and other sizable substituents. The most stable conformation for a particular aldohexose is the chair form, which places the maximum number of substituents larger than hydrogen in the equatorial position.

*When discussing sugars we make frequent use of the terms "configuration" and "conformation." These terms have different meanings. Two conformations of the same molecule are interconvertible without breakage of any chemical bonds; two configurations of the same molecule are not. For example, the chair and boat forms of α-D-glucose represent two different conformations of the same molecule. But α-D-glucose and β-D-glucose represent two different configurations of D-glucose.

Figure 6.7

Chair and boat forms for a generalized pyranose ring structure. Structures of this type are more realistic than the Haworth structure, as the carbon–carbon bond angle is correct. The chair form is usually favored over the boat form. The substituents are labeled "a" (axial) and "e" (equatorial). The axis of symmetry is labeled for both forms. Axial bonds are parallel to this axis of symmetry. Equatorial bonds are parallel to the nonadjacent sides of the rings. A large substituent generally prefers to be in an equatorial location.

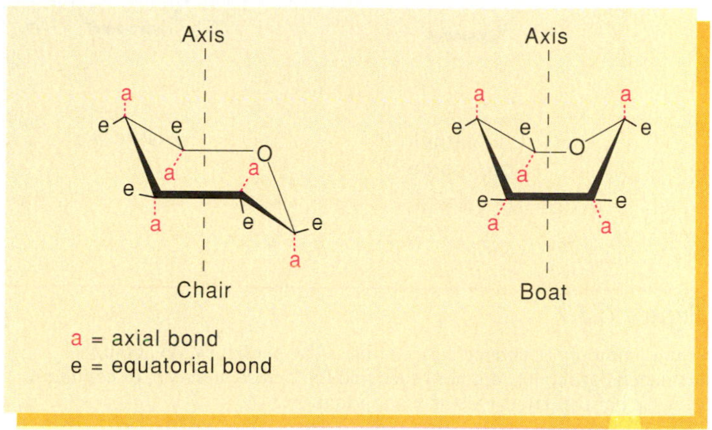

Figure 6.8

Chair configurations for the two anomers of D-glucose. Note that the largest substituent, —CH_2OH, is in an equatorial location in both structures. The differences between the two anomers are shown in color.

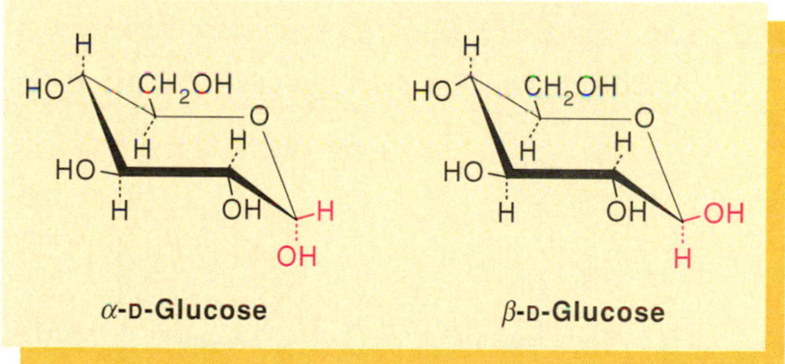

The furanose ring is nonplanar also and can exist in more than one conformation. The conformations for D-ribose (β-D-ribofuranose) and D-2-deoxyribose (β-D-2-deoxyribofuranose), the two pentoses found in all nucleic acids, will be discussed in chapter 25.

Monosaccharides Are Linked by Glycosidic Bonds

Warming glucose in methanol and acid produces a mixture of two new substances, α- and β-methylglucoside; the comparable derivatives of galactose are referred to as galactosides, and so on. Generally the bond between a sugar and an alcohol is referred to as a glycosidic bond, and the compound is known by

Figure 6.9

Formation of methyl glucosides. Glucosides (or glycosides) are quite stable in alkali but they hydrolyze readily in dilute acid.

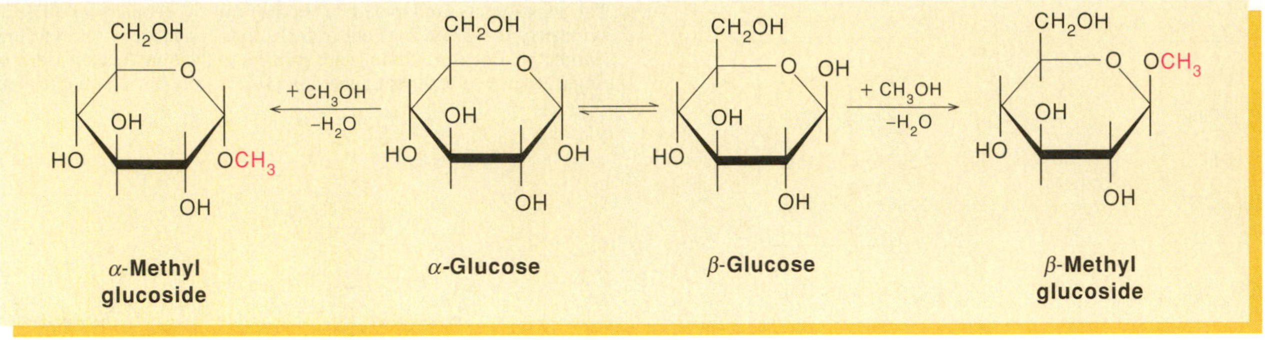

α-Methyl glucoside ⇌ α-Glucose ⇌ β-Glucose ⇌ β-Methyl glucoside ($+CH_3OH$, $-H_2O$)

Figure 6.10

Four commonly occurring disaccharides. The configuration about the hemiacetal group has not been specified for lactose, maltose, or cellobiose because both anomers exist in equilibrium.

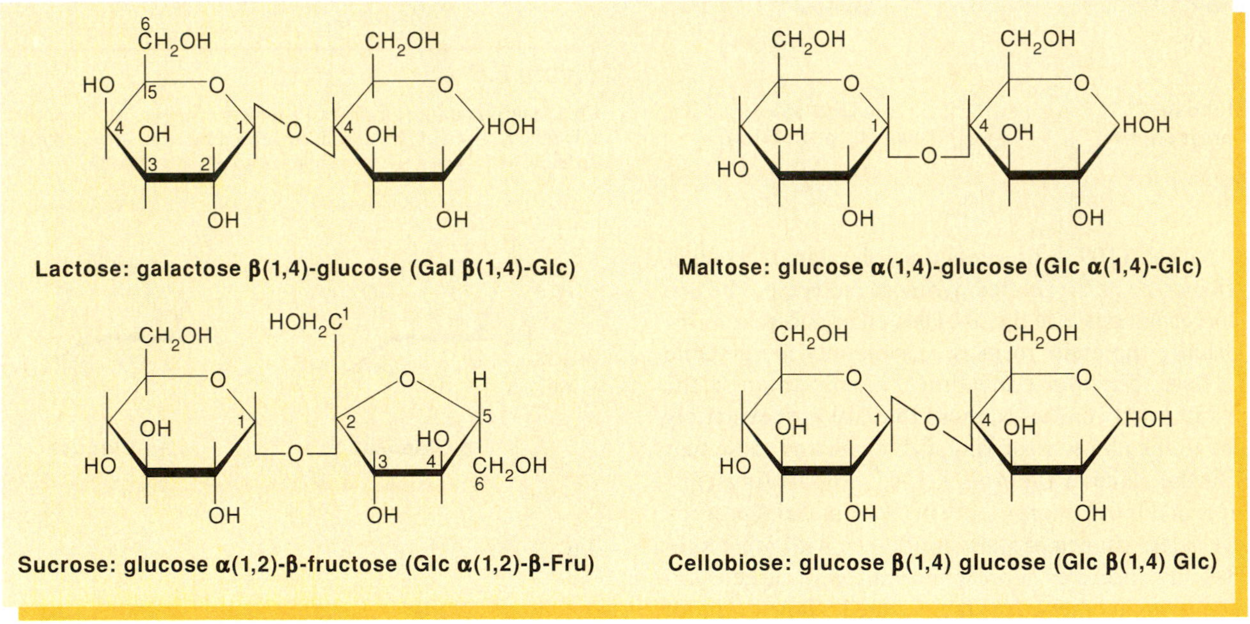

Lactose: galactose β(1,4)-glucose (Gal β(1,4)-Glc)

Maltose: glucose α(1,4)-glucose (Glc α(1,4)-Glc)

Sucrose: glucose α(1,2)-β-fructose (Glc α(1,2)-β-Fru)

Cellobiose: glucose β(1,4) glucose (Glc β(1,4) Glc)

the generic name glycoside. The formation of a glycoside from a sugar and methanol by acid catalysis is identical to the formation of an acetal from an aldehyde and an alcohol (fig. 6.9). While the two forms of glucose in solution are in equilibrium through mutarotation, the corresponding glycosides are locked into one configuration. This is understandable, because mutarotation requires that, in the intermediate, the anomeric carbon adopt a carbonyl structure, which is not possible in a glycoside. A glycoside can be formed with aliphatic alcohols, phenols, and hydroxy carboxylic acids, as well as with another sugar.

Disaccharides and Polysaccharides

The most important glycosides are those formed with other sugars. Monosaccharides are glycosidically linked to form disaccharides (fig. 6.10). For instance, the disaccharide maltose contains a glycosidic bond between the C-1 of one glucose molecule and the C-4 of another glucose molecule. The compound is said to have an α(1,4) glycosidic linkage because the anomeric C-1 carbon of one sugar is connected to the C-4 of another sugar and the configuration about the anomeric carbon is α. Maltose possesses one potentially free aldehyde group and is

Figure 6.11

Some of the sugar building blocks found in polysaccharides.

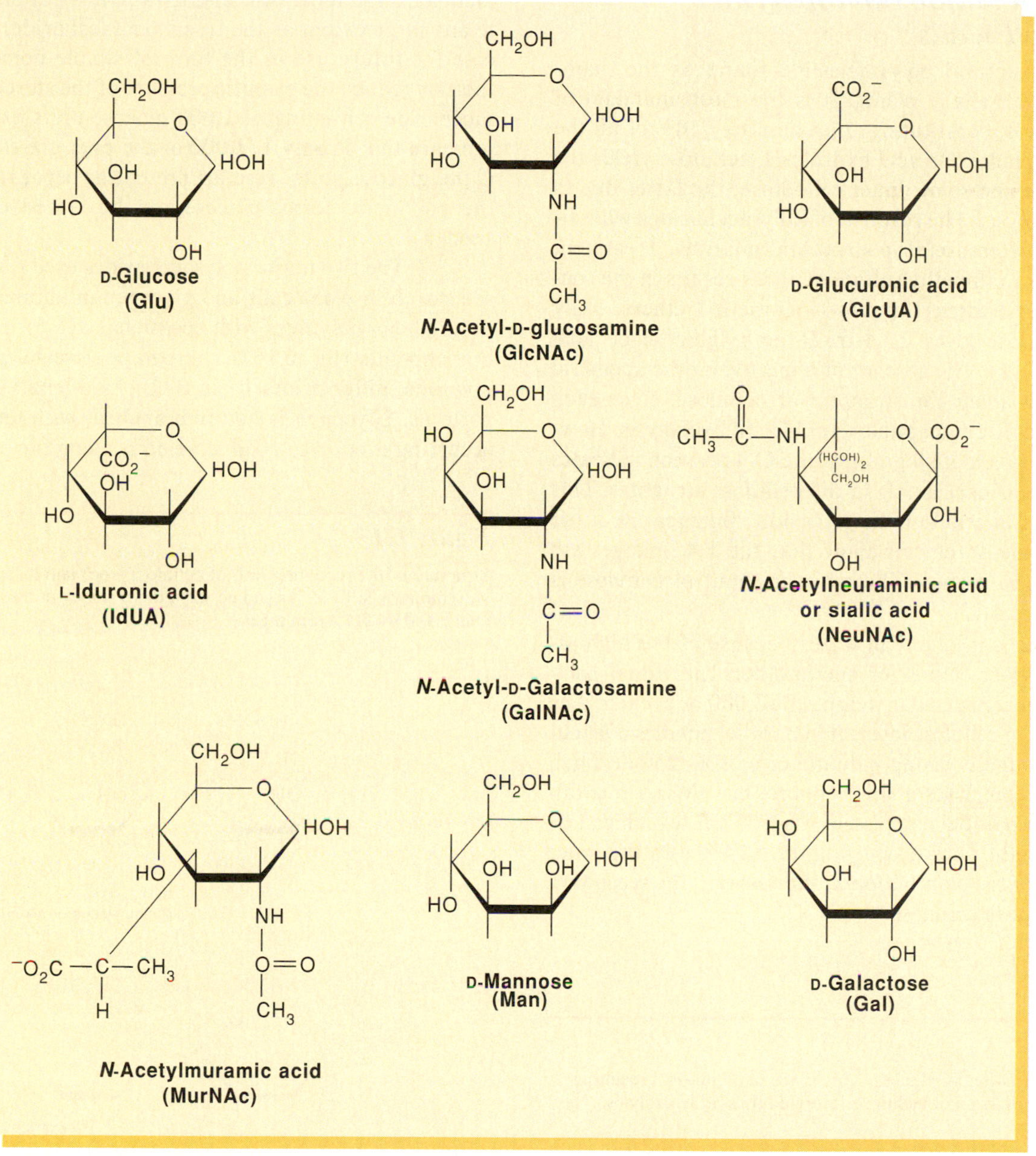

D-Glucose (Glu)

N-Acetyl-D-glucosamine (GlcNAc)

D-Glucuronic acid (GlcUA)

L-Iduronic acid (IdUA)

N-Acetyl-D-Galactosamine (GalNAc)

N-Acetylneuraminic acid or sialic acid (NeuNAc)

N-Acetylmuramic acid (MurNAc)

D-Mannose (Man)

D-Galactose (Gal)

therefore referred to as a reducing sugar. The configuration about the hemiacetal hydroxyl group has not been specified in figure 6.10 because it can undergo mutarotation. Maltose is most familiar as a degradation product of starch.

The disaccharide cellobiose is identical with maltose except for having a $\beta(1,4)$ glycosidic linkage. Cellobiose is a degradation product of cellulose. Lactose is a disaccharide found exclusively in the milk of mammals. Lactose contains a $\beta(1,4)$ glycosidic linkage between galactose and glucose. Sucrose is found in abundance in sugar beets and sugar cane. On acid hydrolysis it yields equivalent amounts of D-glucose and D-fructose. Sucrose contains an $\alpha(1,2)\beta$ glycosidic linkage.

Most carbohydrates in nature exist as high-molecular-weight polymers called polysaccharides. Polysaccharides are composed of simple or derived sugars connected by glycosidic bonds. The most common building block used in polysaccharides is D-glucose. Other sugars that are used include D-mannose, D- and L-galactose, D-xylose, L-arabinose, D-glucuronic acid, D-galacturonic acid, D-mannuronic acid, D-glucosamine, D-galactosamine, and neuraminic acid. Some of these sugar building blocks are illustrated in figure 6.11; others are illustrated in figure 6.2. Polymers composed of a single type of building block are called homopolymers; those composed of more than one type are called heteropolymers.

Polysaccharides function in two quite distinct roles; some serve as a means for storage of chemical energy and others serve a structural function.

Cellulose Is a Major Homopolymer Found in Cell Walls

Cellulose is a structural polysaccharide found as the major component of cell walls in plants. It is the most abundant of organic compounds, constituting approximately 50% of all the carbon found in plants. On acid hydrolysis, cellulose yields the monomer glucose and some dimer cellobiose, the latter due to incomplete hydrolysis. The reaction of polysaccharides with dimethylsulfate is often useful in structural analysis. Treatment of a saccharide with dimethylsulfate in alkali results in the conversion of all free hydroxyl groups to *O*-methyl ethers. Fully methylated cellulose gives 2,3,6-tri-*O*-methylglucose on acid hydrolysis (fig. 6.12). The absence of a methyl on the anomeric C-1 says nothing about the structure of cellulose, since such methyl derivatives are susceptible to mild acid hydrolysis. However, the absence of a methyl group at the C-4 position indicates that this position is inaccessible in the cellulose structure. This fact suggests that in cellulose the glycosidic linkages are of the 1,4 type. Other measurements show that the 1,4 linkages are about the anomeric carbon. The repeating unit of cellulose is indicated in figure 6.13.

Cellulose is insoluble in water because of the high affinity of the polymer chains for one another. Individual polymeric chains have a molecular weight of 50,000 or greater. The molecular chains of cellulose interact in parallel bundles of about 2,000 chains, a bundle having a diameter of 100–250 Å. Each bundle of 2,000 comprises a single microfibril. Many microfibrils arranged in parallel comprise a macrofibril, which can be seen under the light microscope. Figure 6.14 shows the inner secondary walls of the plant *Valonia;* the fibrils in the secondary wall are almost pure cellulose.

Figure 6.12

Structure of 2,3,6-tri-*O*-methylglucose. This is the main product resulting from exhaustive methylation of cellulose followed by acid hydrolysis.

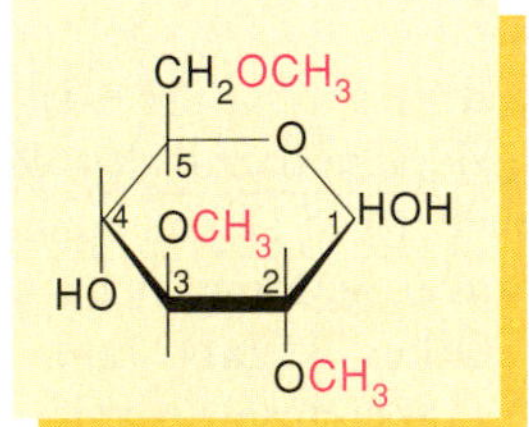

2,3,6-tri-*O*-methylglucose

Starch and Glycogen Are Major Energy-Storage Polysaccharides

Although glucose is the most important sugar involved in energy metabolism in most cells and tissues, it is not present in the cell to any large extent as the free monosaccharide. Cells store glucose for future use in the form of simple homopolymers, and thereby reduce the osmotic pressure of the stored sugar. A polysaccharide consisting of 1,000 glucose units exerts an osmotic pressure that is only 1/1,000 of the pressure that would result if the glucose units were all present as separate molecules. In the polymeric form, glucose can be stored compactly until needed.

The two major polysaccharides used for energy storage are starch in plant cells and glycogen in animal cells. Both are $\alpha(1,4)$ homopolymers with occasional $\alpha(1,6)$ linkages to make branchpoints (fig. 6.15). The two polysaccharides, starch and glycogen, differ primarily in their chain lengths and branching patterns. Glycogen is highly branched, with an $\alpha(1,6)$ linkage occurring every 8 to 10 glucose units along the backbone, giving

Figure 6.13

Structure of the repeating unit of cellulose (top) and its analysis by reaction with dimethylsulfate. As can be seen, the residues in the native structure are connected by ß(1,4) linkages.

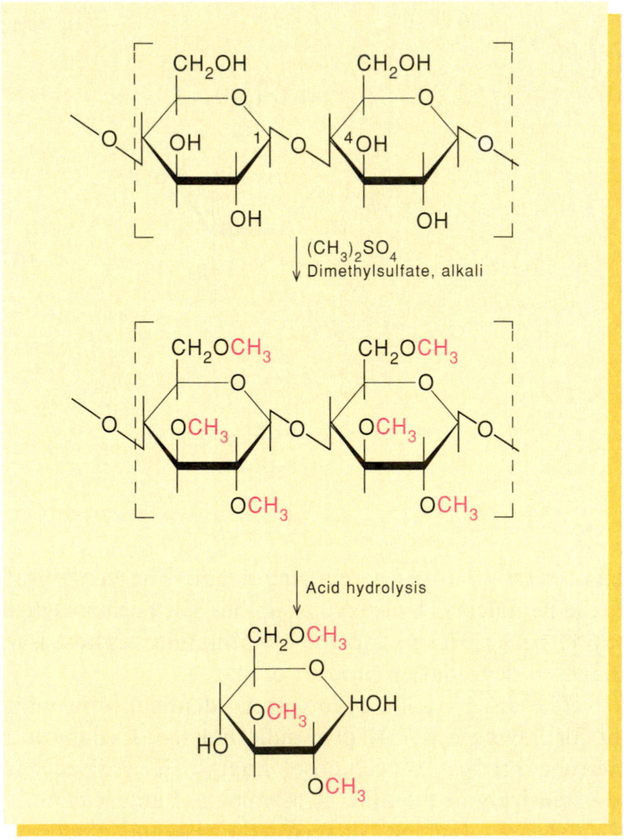

rise in each case to short side chains of about 8 to 12 glucose units each. Starch occurs both as unbranched amylose and as branched amylopectin. Like glycogen, amylopectin has $\alpha(1,6)$ branches, but these occur less frequently along the molecule (once every 12 to 25 glucose residues) and give rise to longer side chains (lengths of 20 to 25 glucose units are common). Starch deposits are usually about 10–30% amylose and 70–90% amylopectin.

Figure 6.14

Fibril arrangements in the cell wall of *Valonia* (12,000×). (Electron micrograph from A. Frey-Wyssling and K. Mühlethaler, *Ultrastructural Plant Cytology,* 1965, p. 298. Reprinted with permission from Elsevier Science Publishers.)

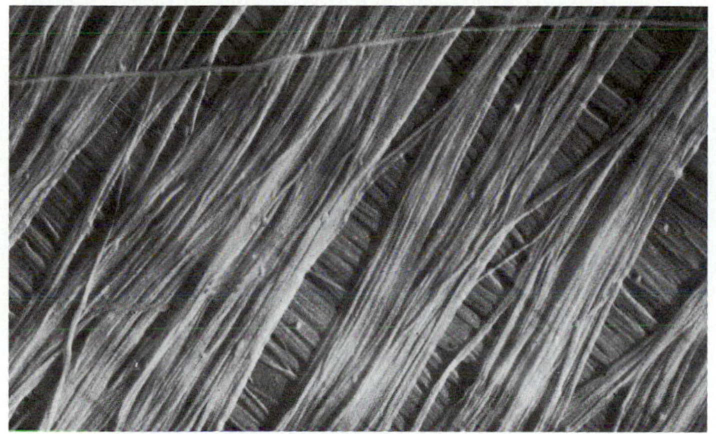

The Configurations of Glycogen and Cellulose Dictate Their Roles

It is a remarkable fact that the main energy-storage polysaccharides and the main structural polysaccharides found in nature both have a primary structure of (1,4)-linked polyglucose. Why should two such closely related compounds be used in totally different roles? A closer look at the stereochemistry of the α and β glycosidic linkage for polyglucose indicates why this is so.

Recall that D-glucose exists in the chair form of a pyranose ring (see fig. 6.8). The ring has a rigid character to it. We can think of it as a structural building block in a polysaccharide chain, just as we think of the rigid planar peptide grouping as a structural building block in a polypeptide chain. It is also possible to specify two torsional angles ϕ and ψ for rotation about the glycosidic C—O linkage (fig. 6.16). These angles are used extensively to discuss polypeptide configuration in proteins (see chapter 4). They have limited value in discussing polysaccharide structures, because much less information is available about such structures. Nevertheless, it is clear that only the $\beta(1,4)$-linked polyglucose has the capacity to form straight chains (see fig. 6.16*a*). A straight chain can be created by flipping each glucose unit by 180° relative to the previous one. This process should result in an almost fully extended polysaccharide chain, which is known to be characteristic of the structure of cellulose. Evidently this conformation is energetically favored. By contrast, $\alpha(1,4)$-linked units in a polyglucose cause a natural turning of the chain (see fig. 6.16*b*). Consistent with this fact is the observation that amylose adopts a coiled helical configuration. Indeed, one of the first helical structures

Figure 6.15

Structure of the storage polysaccharides glycogen and starch. The main chain is $\alpha(1,4)$-linked. Side chains are connected to the main chain by $\alpha(1,6)$ linkages.

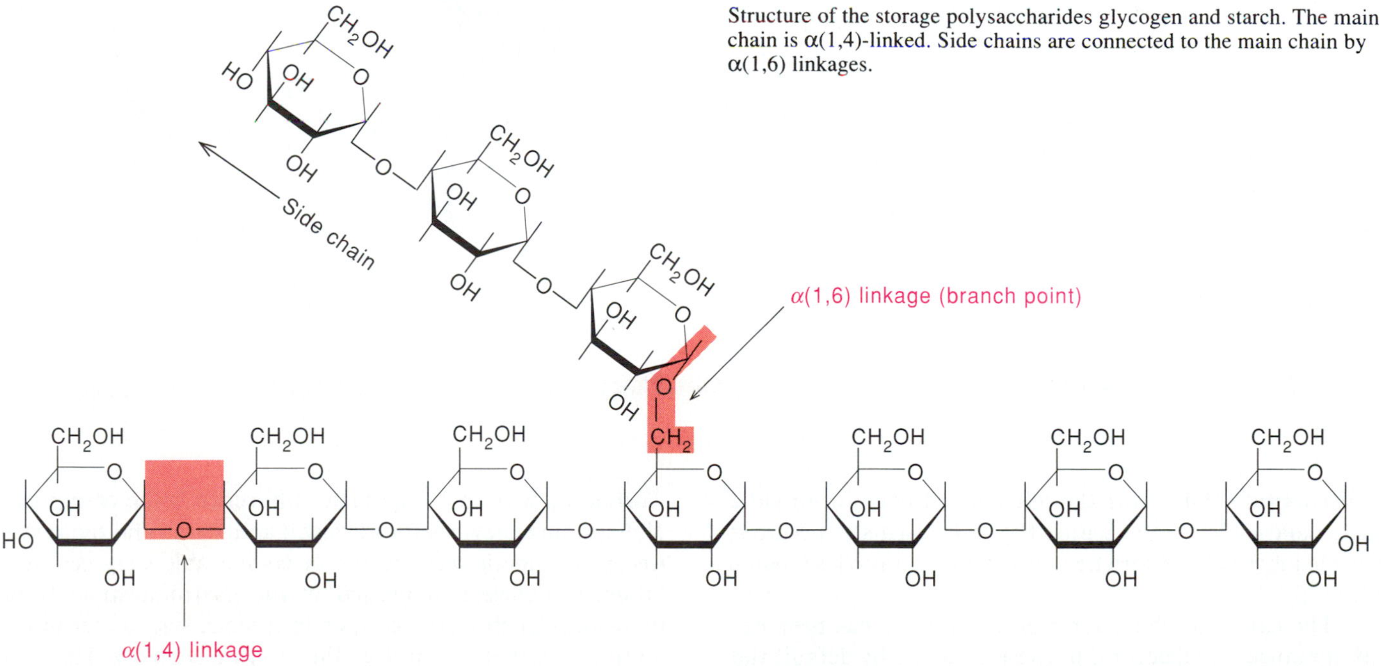

Figure 6.16

Energetically favored conformations of β(1,4)-linked D-glucose (*a*) and α(1,4)-linked D-glucose (*b*). Note that in the β(1,4) configuration in (*a*), alternating residues are flipped 180° relative to one another so that long straight chains result. In the α(1,4) configuration (*b*), the chain has a natural curvature.

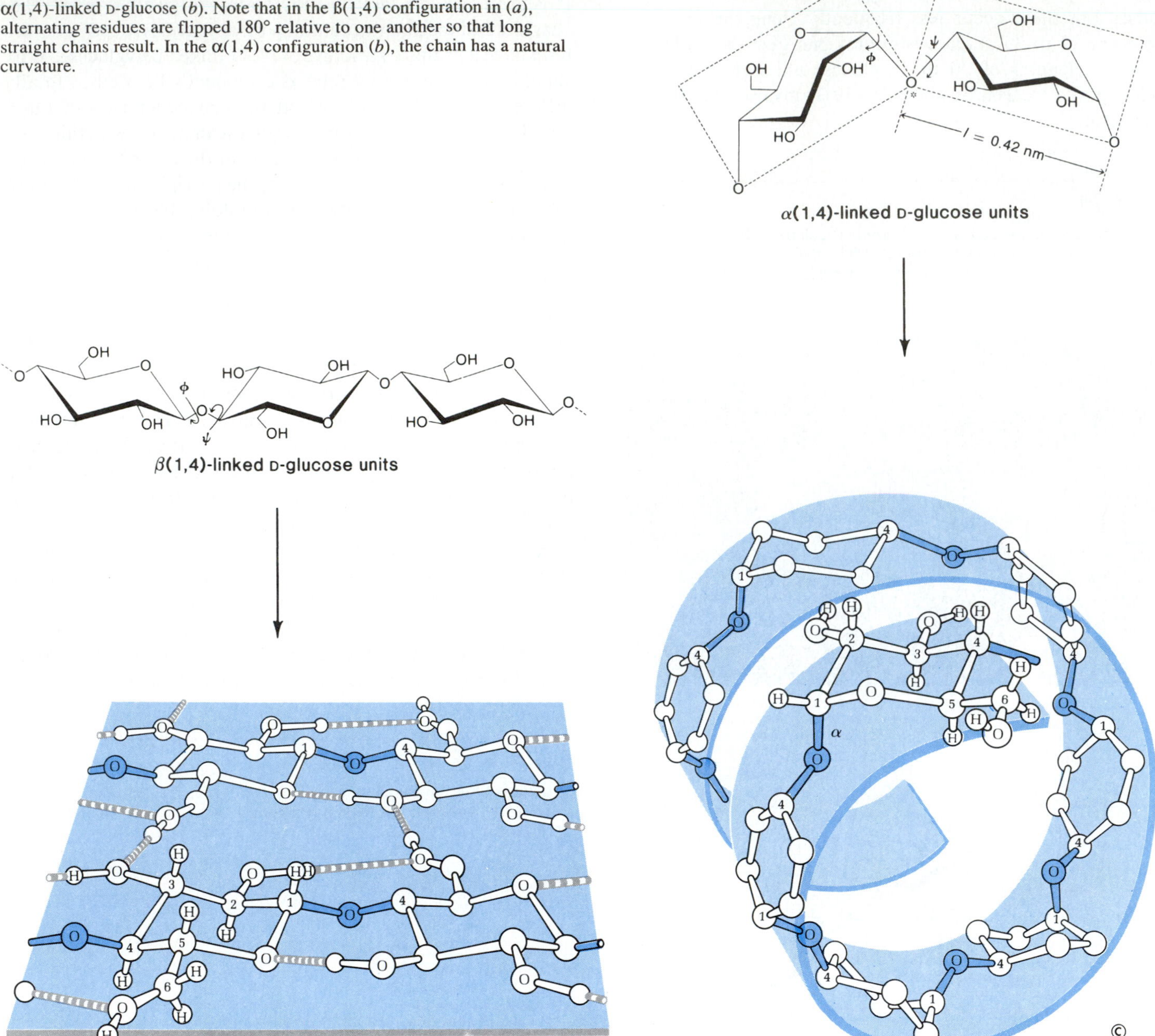

to be discovered (1943) was the left-handed helix of amylose wound around molecules of iodine (fig. 6.17). This structure is responsible for the characteristic blue color of the amylose-iodine complex.

The extended-chain form of polyglucose has been exploited in nature for structural purposes, leaving by default the coiled form for use as an energy-storage macromolecule. Correlated with this functional difference is the omnipresence of degrading enzymes for glycogen and starch and the very limited phylogenetic distribution of comparable enzymes for cellulose. Cellulose is degraded in the gastrointestinal tract of herbivores, such as the cow, or in insects, such as termites, by a protozoan that synthesizes the enzyme cellulase. Humans do not possess this enzyme and hence cannot degrade cellulose.

Figure 6.17

Structure of the helical complex of amylose with iodine (I_2). The amylose forms a left-handed helix with six glucosyl residues per turn and a pitch of 0.8 nm. The iodine molecules (I_2) fit inside the helix parallel to its long axis.

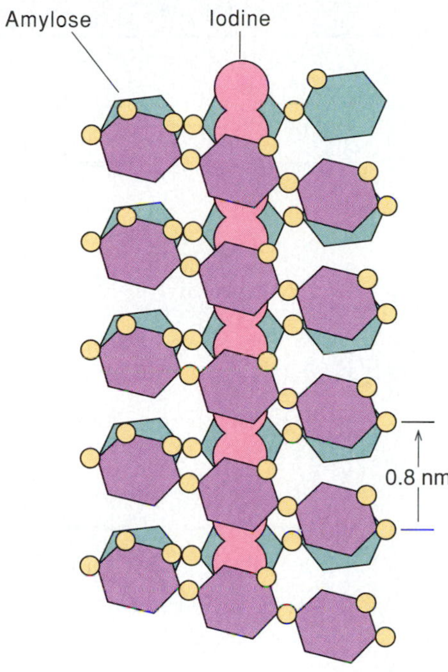

Chitin Contains a Different Building Block

Many polysaccharides consist of sugars other than glucose or different combinations of sugars. In some cases the new sugar is merely a derivative of glucose. Chitin is an example of a structural polysaccharide that uses a modified derivative of glucose. Chitin is found in the shells of crustaceans and insects and in the cell walls of fungi; it is a linear $\beta(1,4)$ polymer of *N*-acetyl-D-glucosamine (fig. 6.18). A major rigid component of bacterial cell walls, the peptidoglycan, which we will discuss later, could be regarded as a substituted chitin.

Heteropolysaccharides Contain More than One Building Block

A large variety of carbohydrates contain either modified sugars, like the *N*-acetylglucosamine found in chitin, or two or more different sugars in straight-chain or branched-chain linkages.

Structurally, the simplest and best known of the heteropolysaccharides are the glycosaminoglycans. These are long, unbranched polysaccharide chains composed of repeating disaccharide subunits in which one of the two sugars is either *N*-acetylglucosamine or *N*-acetylgalactosamine (table 6.1). Glycosaminoglycans are highly negatively charged because of the presence of carboxyl or sulfate groups on many of the sugar residues. The high negative charge causes the polymeric chains to adopt a stretched or extended conformation. Their extended structure gives a high viscosity to the surrounding region, even in a dilute solution of the polysaccharide. Glycosaminoglycans are usually found in extracellular space in multicellular organisms, where they produce a viscous extracellular matrix that resists compression. Such an environment can be beneficial to the organism in various ways; it can provide a passageway for cell migration, supply lubrication between joints, or help maintain certain structural shapes such as the ball of the eye.

Figure 6.18

Structure of the repeating unit of chitin: a ß(1,4) homopolymer of *N*-acetyl-D-glucosamine.

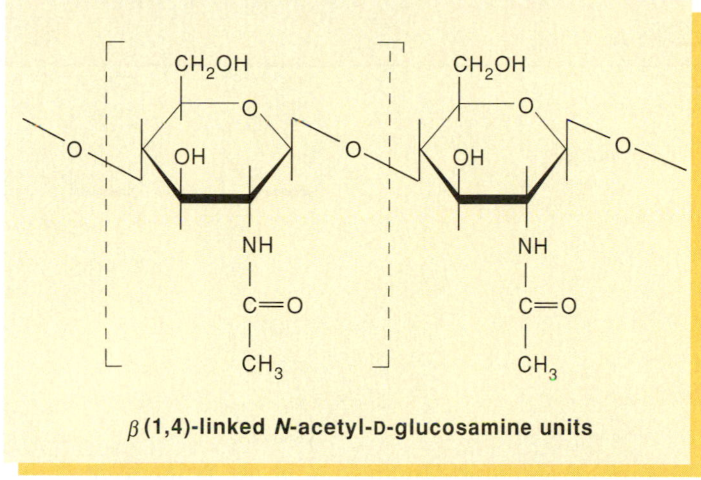

Figure 6.19

Structure of the repeating unit of hyaluronic acid: GlcUA ß(1,3)-GlcNAc ß(1,4).

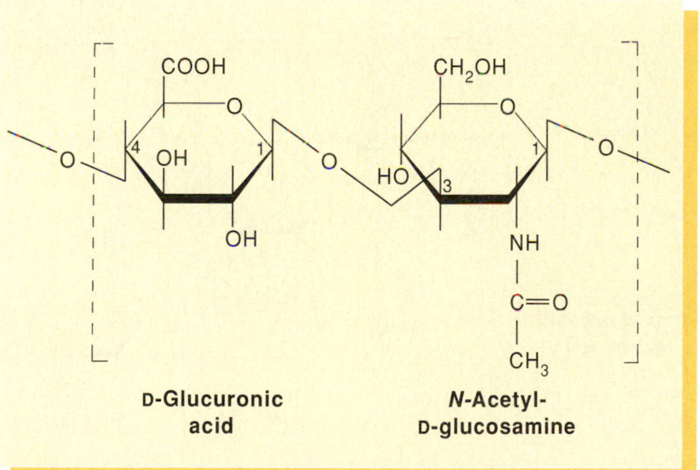

Hyaluronic acid is a copolymer of D-glucuronic acid and *N*-acetyl-D-glucosamine (fig. 6.19). It is much larger than other glycosaminoglycans, reaching molecular weights in excess of 10^6.

Table 6.1
Structure of Glycosaminoglycans

Polysaccharide	Monosaccharide Units[a] A	Monosaccharide Units[a] B	Substituents	Repeating Unit
Hyaluronate	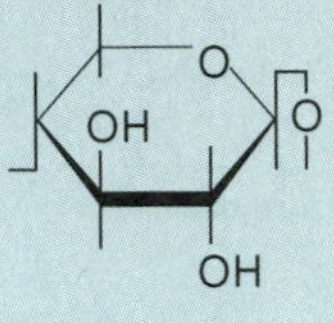 β-D-GlcUA	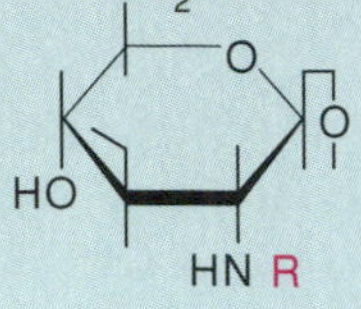 β-D-GlcN	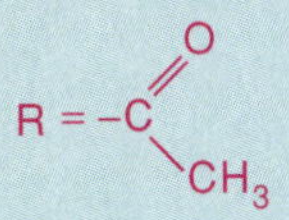 R = –C(=O)–CH_3	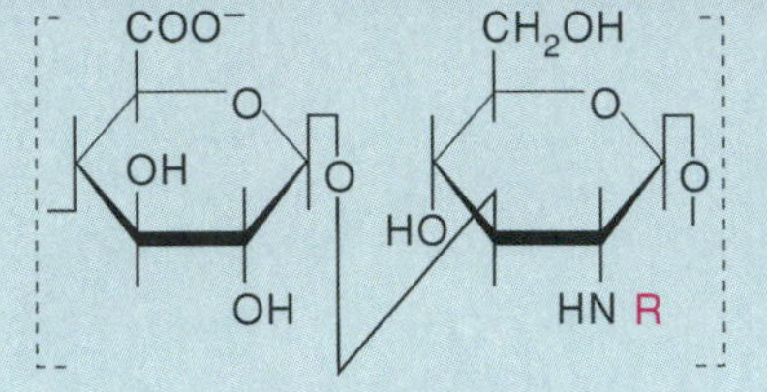
Chondroitin sulfates Dermatan sulfate	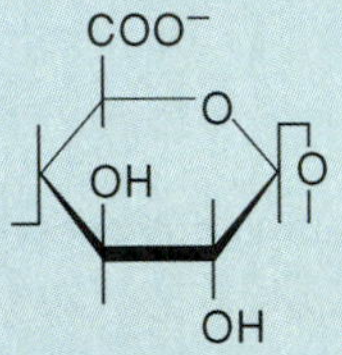 β-D-GlcUA 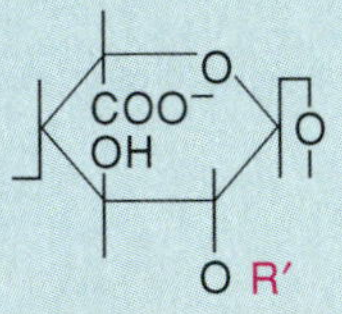 α-L-IdUA	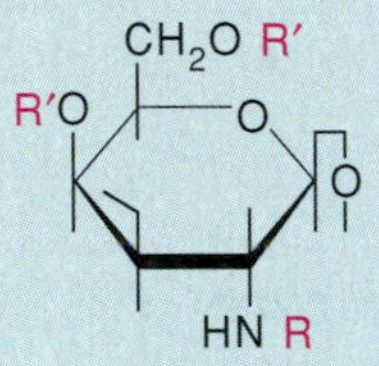 β-D-GalN	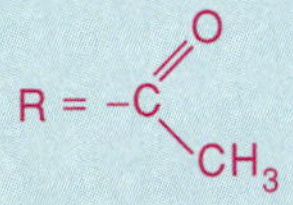 R = –C(=O)–CH_3 R′ = –H or $-SO_3^-$	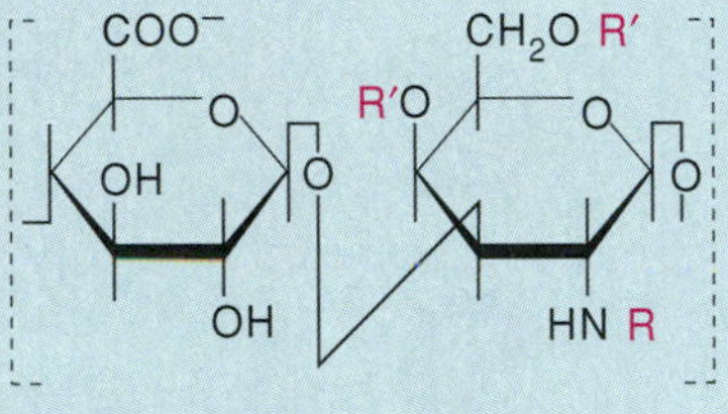
Heparan sulfate and heparin	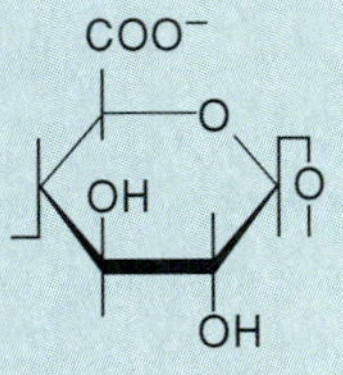 β-D-GlcUA 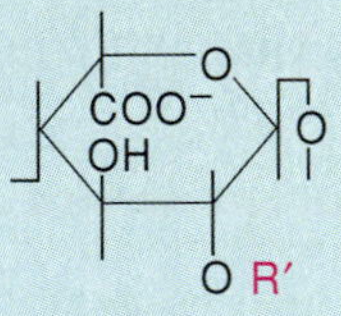 α-L-IdUA	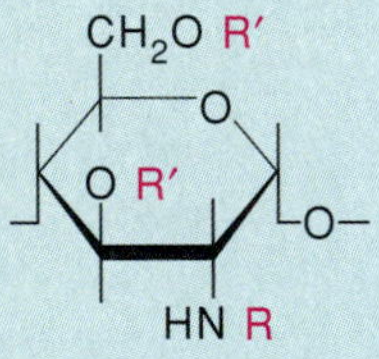 α-D-GlcN	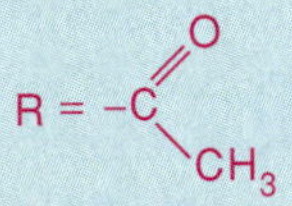 R = –C(=O)–CH_3 R′ = –H or $-SO_3^-$	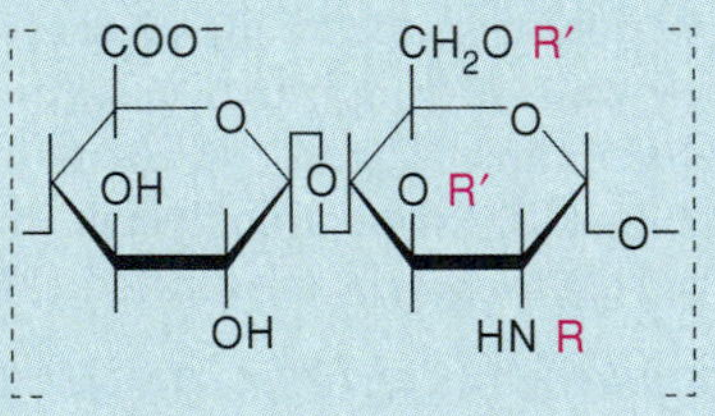
Keratan sulfate	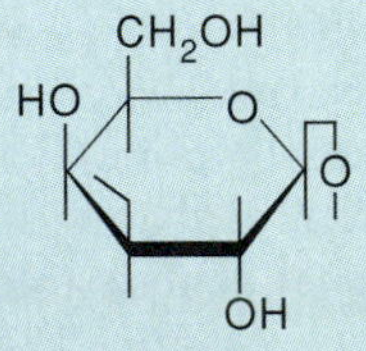 β-D-Gal	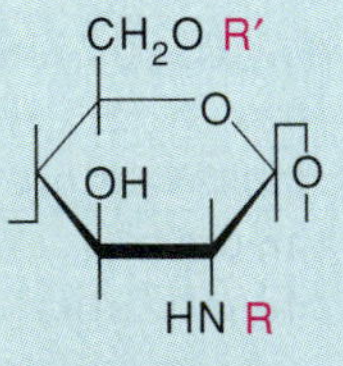 β-D-GlcN	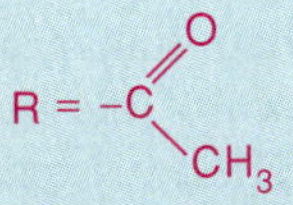 R = –C(=O)–CH_3 R′ = –H or $-SO_3^-$	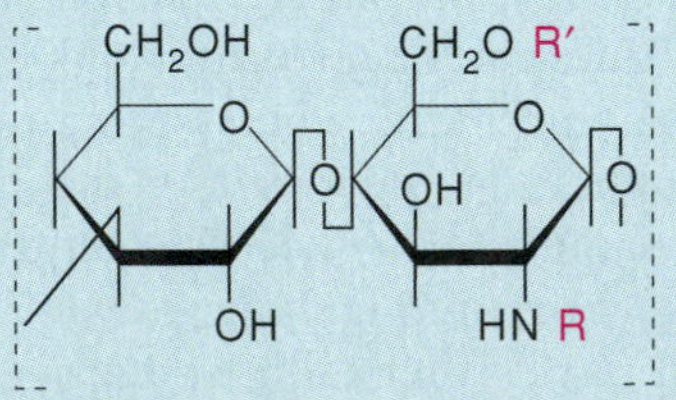

[a]The polysaccharides are depicted as linear polymers of alternating A and B monosaccharide units.
Abbreviations: GlcUA, glucuronic acid; IdUA, iduronic acid; GalN, galactosamine; GlcN, glucosamine; Gal, galactose.

Figure 6.20

(*a*) Structure of a typical proteoglycan. (*b*) The attachment site between a serine in the core protein and the glycosaminoglycan.

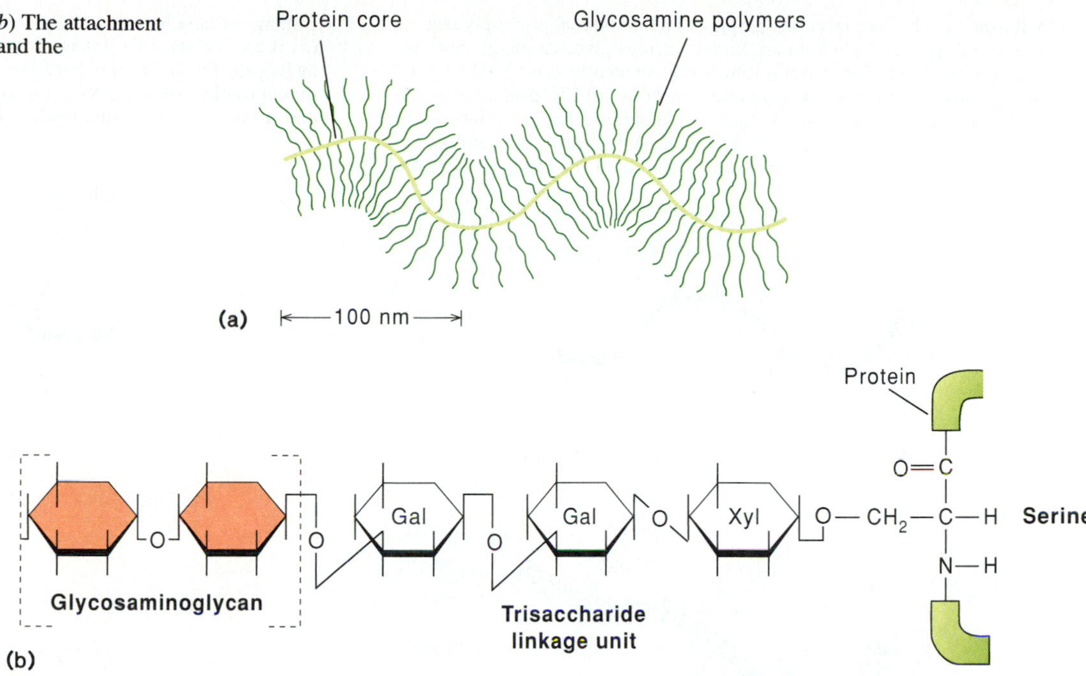

Proteoglycans

Most glycosaminoglycans are linked to a core protein as lateral extensions, forming a proteoglycan with a highly extended, brushlike structure (fig. 6.20). The linkage between the glycosaminoglycans and the core protein within the proteoglycan is mediated by a specific trisaccharide unit that is linked on one side to the repeating disaccharide unit of the glycosaminoglycan and on the other side to a serine hydroxyl group of the core protein (see fig. 6.20). In the extracellular matrix, hyaluronic acid and the other proteoglycans form aggregates with each other as well as with other macromolecular components, such as serum glycoproteins, growth factors, collagen, elastin, or the outer plasma membrane of cells. Hyaluronic acid, when complexed with other proteoglycans through their core proteins, produces very large complexes (fig. 6.21).

Thus far we have seen that glycosaminoglycans and their proteoglycans form highly extended aggregates that impart a rigid, gel-like structure to the extracellular matrix. It seems likely that many more specific reactions occur involving these compounds. However, it should be emphasized that relatively little is known about the organization or reactions of these molecules in the extracellular matrix. One exception is the case of the glycosaminoglycan heparin. We know that this compound functions as a highly specific anticoagulant. First it forms a complex with the plasma protein antithrombin III. This complex in turn inhibits the serine proteases of the blood clotting system.

Glycoproteins

Proteins are frequently adorned by straight-chain or branched oligosaccharides, in which case they are called glycoproteins. This type of modification can serve a variety of functions. It can stabilize the protein, facilitate its correct folding, be part of a lipid anchor for attaching the protein to a membrane, and provide the protein with surface characteristics that facilitate its recognition.

The carbohydrate moiety in a glycoprotein is attached to the polypeptide chain by a covalent connection to certain amino acid side chains. Two types of glycosidic linkages are commonly found: the O-glycosidic linkage involves attachment of the carbohydrate to the hydroxyl group of serine, threonine, or hydroxylysine; the N-glycosidic linkage involves attachment to the amide group of asparagine (fig. 6.22). Conformational constraints appear to be a major factor in determining the addition, although other factors also play a role. Thus, in the case of O-linked sugars, clusters of serines or threonines may be a preferred substrate. For N-glycosidic linkage, only asparagine residues of the sequence Asn-X-Ser(Thr) will accept carbohydrates. In this sequence X may be any amino acid except proline.

Glycoproteins are found in all cellular compartments and are also secreted from the cell. Those found in the cytoplasm have the simplest type of modification, consisting of a single *N*-acetylglucosamine that is O-glycosidically linked to Ser(Thr). Collagen, a secreted glycoprotein of the extracellular matrix, also has simple carbohydrates—the disaccharide Glcβ(1,2)Gal linked to hydroxylysine.

Figure 6.21

A hyaluronic acid–proteoglycan complex. The individual proteoglycans contain a core protein by which are linked various glycosaminoglycans as shown in figure 6.20. The core protein is noncovalently associated with a single hyaluronic acid molecule via two link proteins. Electron micrograph below shows a large aggregate (*A*) and a small aggregate (*B*) whose sizes are determined mainly by the length of the hyaluronate central filament of the individual apprepates. (From Buckwalter, J. A. and Rosenberg, L. Coli. Rel. Res. (1983) 3, 489–504.)

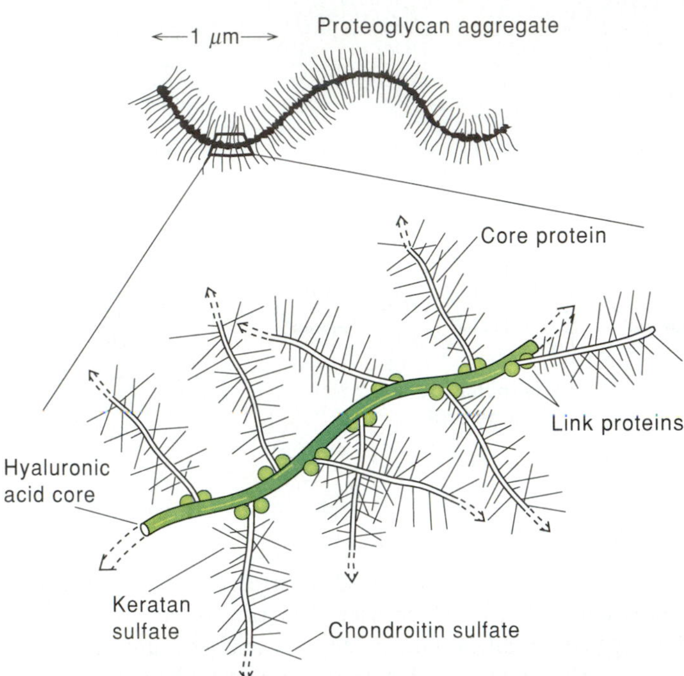

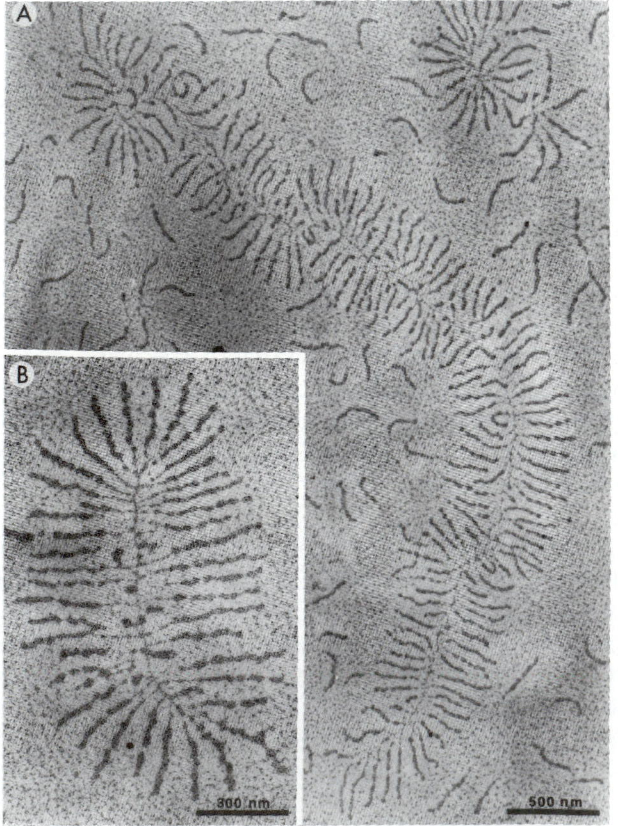

Figure 6.22

Linkages found between oligosaccharides and proteins in glycoproteins: (*a*) is an O-glycosidic linkage found in many glycoproteins and mucins; an analogous linkage occurs between *N*-acetylglucosamine and serine in glycoproteins; (*b*) is an N-glycosidic linkage found in many glycoproteins; and (*c*) is an O-glycosidic linkage found only in collagen.

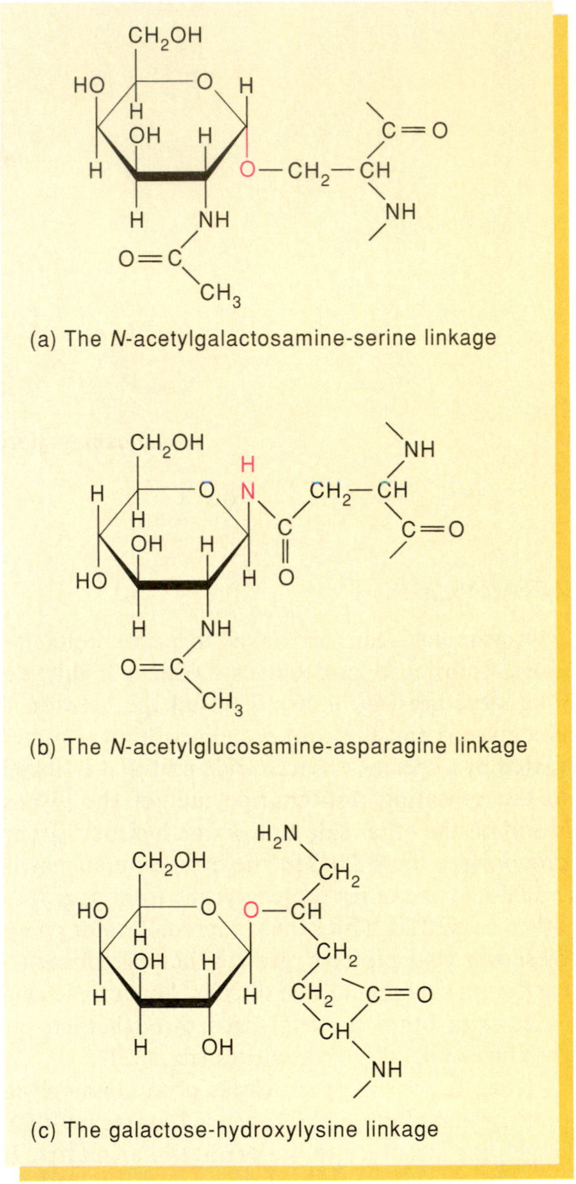

(a) The *N*-acetylgalactosamine-serine linkage

(b) The *N*-acetylglucosamine-asparagine linkage

(c) The galactose-hydroxylysine linkage

Most secreted and membrane-bound glycoproteins have more complicated oligosaccharides containing between 4 and 30 sugar residues. A typical O-linked carbohydrate of a membrane glycoprotein is shown in figure 6.23. This structure can be further extended and branched by the addition of galactose, *N*-acetylgalactosamine, *N*-acetylglucosamine, fucose, or sialic acid in different linkages. Larger O-linked structures with branching occur on proteins with high Ser(Thr) content, forming mucins. These molecules are found in body fluids and carry the blood group determinants (see chapter 21).

Figure 6.23

Representative carbohydrate structures found on mammalian glycoproteins. (*a*) An O-glycosidically-linked carbohydrate from a red blood cell membrane glycoprotein; (*b*) an oligomannosyl N-linked carbohydrate; (*c*) a hybrid N-linked carbohydrate; and (*d*) a complex, lactosamine-containing carbohydrate. The latter three structures are found on many membrane and secreted glycoproteins of mammalian cells. Structure (*d*) may be much more complex, with the addition on the core mannose residues of extra branches of various lengths terminating with different sugars. It should be noted that all the N-linked carbohydrates (structures *b*, *c*, and *d*) have a common core of five sugars. (NeuNAc = sialic acid; Gal = galactose; GlcNAc = *N*-acetylglucosamine; Man = mannose; Fuc = fucose; Asn = asparagine; Ser = serine.)

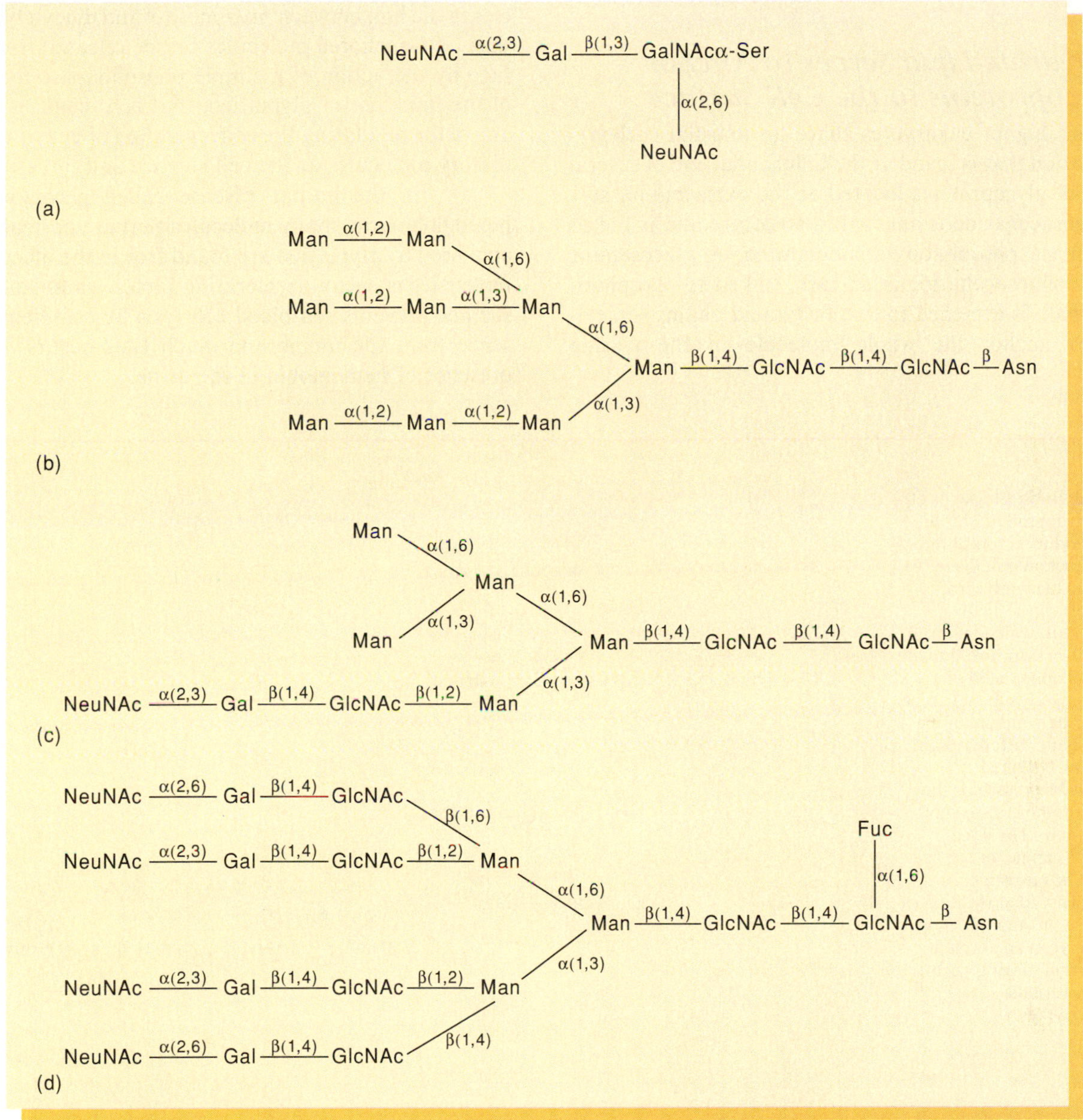

In contrast to the usual O-linked oligosaccharides of mammalian glycoproteins, N-linked carbohydrates always contain mannose and *N*-acetylglucosamine, and may also contain galactose, fucose, and sialic acid in combinations that vary as a result of differences in branching and linkage relationships. Three general types of structure form the basis of the variations in N-linked carbohydrates. These are termed oligomannosyl, hybrid, and lactosamine-containing (or complex) structures (see fig. 6.23). They all have a common core of five sugars attached to Asn because they are all synthesized by a common pathway. In fact, the wide variety of N-linked carbohydrates found in glycoproteins reflects intermediates of the biosynthetic pathway. The particular structures associated with a completed glycoprotein are a result of the conformation of that protein during biosynthesis, the availability of specific glycosyltransferase enzymes in the host cell, and the speed with which the glycoprotein travels through the secretory pathway along which glycosylation enzymes are located (see chapter 21).

Yeast glycoproteins carry only the oligomannosyl type of N-linked carbohydrates with mature structures that may contain hundreds of mannose residues. Thus far such structures have been found only in yeast. Yeast do not synthesize hybrid or lactosamine-containing structures, although the initial steps of N-linked carbohydrate biosynthesis appear to be the same in all eukaryotes (see chapter 21).

A Carbohydrate-Lipid Serves to Anchor Some Glycoproteins to the Cell Surface

Both yeast and higher eukaryotes share an unusual carbohydrate modification that is found at the C-terminal end of several diverse types of glycoproteins located at the extracellular surface. The oligosaccharide is unusual in structure and is linked to the protein via phosphatidylethanolamine. A glucosamine residue of the oligosaccharide is, in turn, linked to phosphatidylinositol, which is attached to two fatty acid chains (diacylglycerol) that anchor the whole molecule in the plasma membrane. This complex modification is termed a glycosylphosphatidylinositol (GPI) anchor, and proteins that carry such an anchor are said to be glypiated. The structure of a GPI anchor from a mammalian cell surface glycoprotein is shown in figure 6.24. The presence of this anchor in a glycoprotein may be ascertained by treatment with nitrous acid, which cleaves the glucosamine-*myo*-inositol linkage, or with phospholipases that cleave the link between *myo*-inositol and diacylglycerol. The fact that GPI-anchored molecules can be released from the cell surface by the action of a simple phospholipase may explain one of the functions of glypiation; that is, it could provide a mechanism for regulating the concentration of enzymes or other regulatory molecules at the cell surface and in tissue fluids.

In the human disease called paroxysmal noctornal hemoglobinuria, many molecules that are normally membrane-anchored by glypiation are found free in the blood. One of these molecules is decay-accelerating factor whose action at the cell surface prevents red blood cell lysis by complement. In its absence from the membrane much lysis occurs, leading to the presence of hemoglobin in the urine.

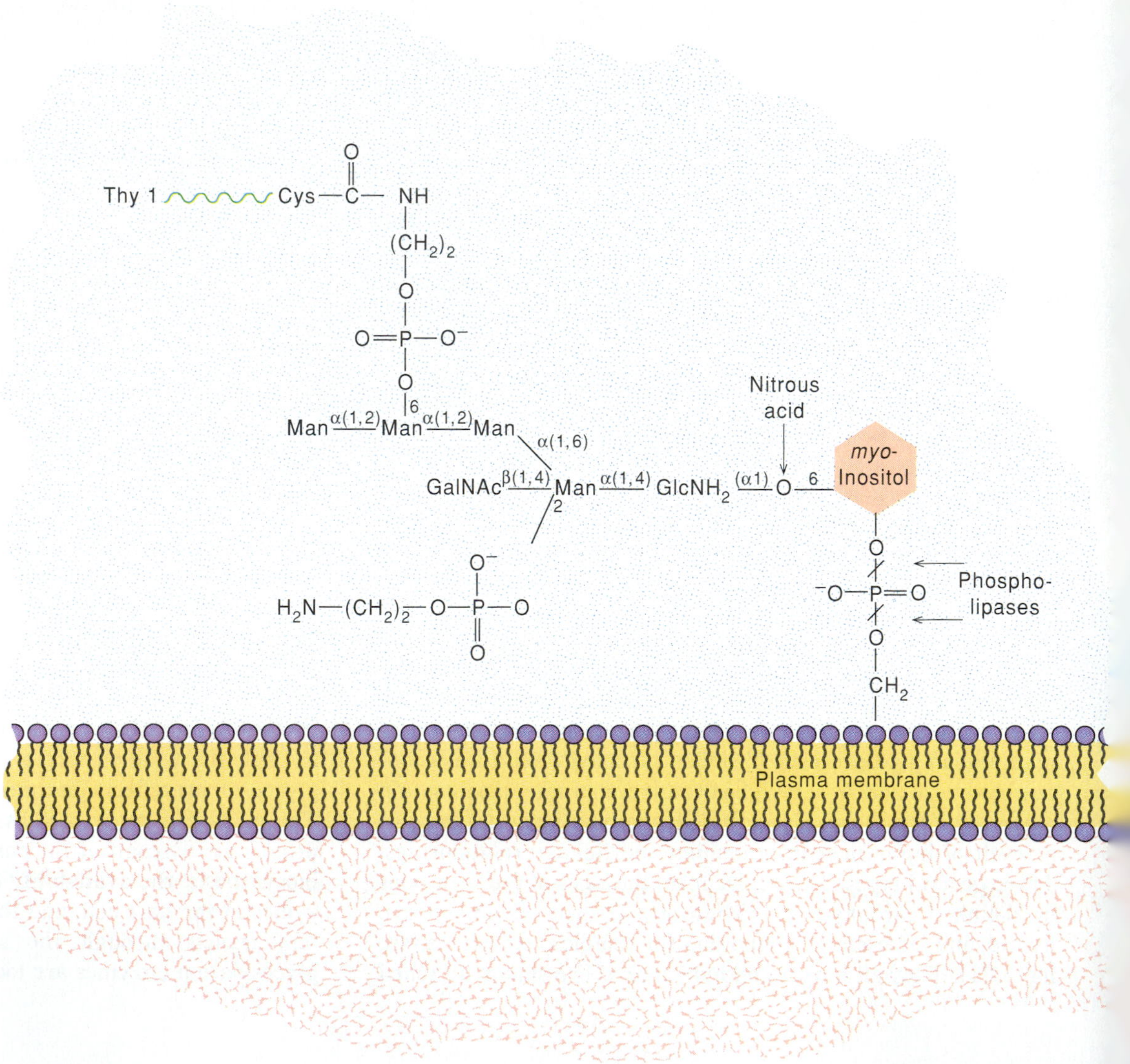

Figure 6.24

Glycophosphatidylinositol anchor of the membrane glycoprotein Thy 1. Many enzymes and receptors at the cell surface are anchored in the membrane via the diacylglycerol (DAG) portion of phosphatidylinositol, which is linked to a glycan that is in turn linked to the carboxyl-terminal amino acid of a protein via phosphatidylethanolamine. The oligosaccharide structure of the glycan region is quite different from that of O-linked or N-linked oligosaccharides. Most novel is the presence of glucosamine instead of *N*-acetylglucosamine. The bond between the glucosamine and *myo*-inositol can be cleaved by nitrous acid allowing identification of proteins that carry this modification. In addition, the glycan and protein can be released from diacylglycerol by cleavage with various phospholipase enzymes.

The GPI anchor serves to locate glycoproteins on the outer leaflet of the plasma membrane, where they are significantly more mobile than membrane proteins that span the bilayer. These glycoproteins are thus more accessible to other extracellular molecules and are more readily released by a phospholipase.

Carbohydrate Modification Is Important in Targeting Certain Enzymes to the Lysosomes

Soluble lysosomal hydrolases are targeted to lysosomes by a specific carbohydrate recognition marker that they acquire in the Golgi complex. Oligomannosyl carbohydrates on soluble lysosomal enzymes carry one or two phosphate residues at the 6-position of mannose (Man-6-P). These phosphorylated mannose residues are recognized by a glycoprotein called the Man-6-P receptor, which binds and transports lysosomal enzymes via several cellular compartments. In an acidic, prelysosomal compartment the binding between the Man-6-P receptor and the lysosomal enzyme is disrupted. The lysosomal enzyme continues in a vesicle destined to fuse to and thereby deliver its contents to the lysosome. Fibroblasts from patients with a lysosomal storage disease called I-cell disease cannot add the carbohydrate recognition marker and consequently their lysosomal hydrolases are largely secreted instead of being targeted to the lysosome. As a result, many molecules that are normally degraded by lysosomal hydrolases accumulate in the lysosomes. Morphologists have termed these dense lysosomes inclusion bodies, hence the name I-cell disease.

Thus far no other cases of protein targeting by carbohydrate modification are known, although their existence is a distinct possibility.

Carbohydrates of the Plasma Membrane

Carbohydrates appear prominently on the outside leaflet of the plasma membrane in the form of N- and O-linked glycoproteins, glycolipids, proteoglycans, and GPI-anchored proteins (fig. 6.25). In addition to being accessible for recognition by carbohydrate-binding proteins, cell-surface carbohydrates also appear to be important for cell shape and cell–cell interactions. Many infectious agents, such as bacteria, viruses, or parasites, recognize and bind to host cells via specific carbohydrate structures. For example, influenza virus binds to cells via specific types of sialic acids. Mutant mammalian cells with truncated carbohydrates are more rounded and aggregate with each other. Similar features are typical of red blood cells from patients with a rare blood disorder called HEMPAS (congenital dyserythropoetic anemia type II). In these patients, carbohydrates of red blood cell-surface glycoproteins are truncated, with the result that the red cells clump together. The removal of red cells from the circulation by this means is the cause of the anemia.

Figure 6.25

Diagram of a eukaryotic cell plasma membrane. Oligosaccharides are found on integral, transmembrane glycoproteins, glycolipids, glycophosphatidylinositol (GPI)-anchored glycoproteins, and glycoconjugates adsorbed at the cell surface. All carbohydrates on these molecules face the outside of the cell except O-linked *N*-acetylglucosamine, which may be found on the cytoplasmic portion of glycoproteins. Since many carbohydrates terminate in sialic acid, the external surface of the plasma membrane is negatively charged. The oligosaccharides of membrane glycoconjugates form a sugar coat, "glycocalyx," for the mammalian cell, which can be readily seen in the electron microscope by staining with ruthenium red or other sugar-binding dyes. Proteoglycans are found in the extracellular matrix and provide the support substance for cells in all tissues. Proteoglycan chains have also been found on certain membrane glycoproteins. Many membrane glycoproteins are receptors that bind other glycoproteins (e.g., growth factors), absorbing them to the cell surface.

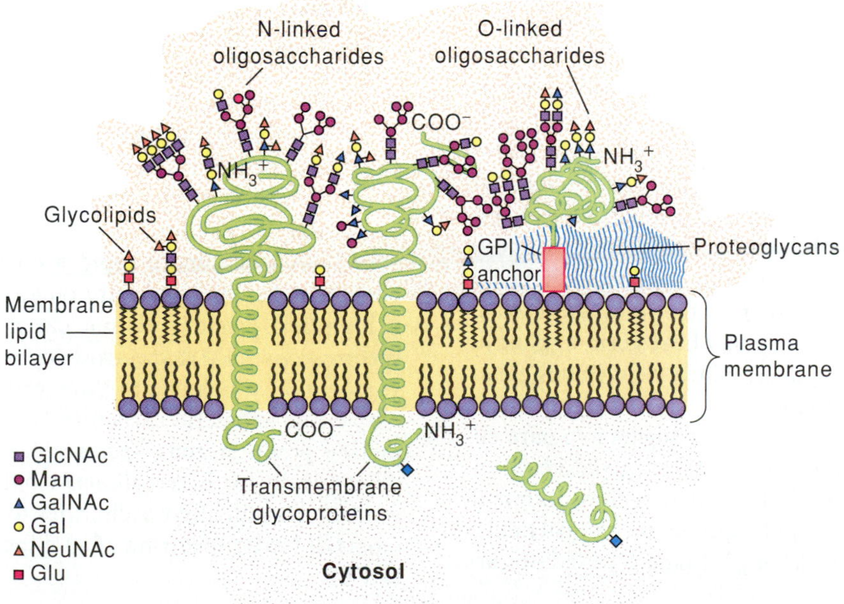

Figure 6.26

Separation of N-linked carbohydrates by lectin-affinity chromatography. Radiolabeled carbohydrates are released from a glycoprotein by an endoglycosidase (*N*-glycanase) that cleaves between the Asn of the protein and the first GlcNAc residue of the carbohydrate. They can then be separated into branched or biantennary (two branches) lactosamine-containing species or oligomannosyl species by affinity chromatography on concanavalin A-sepharose. Branched carbohydrates do not bind to the column (nor do O-linked oligosaccharides), biantennary carbohydrates bind and are eluted by 10-mM α-methylglucoside (α-MG), while oligomannosyl carbohydrates bind more strongly to the column and are eluted by increasing concentrations of α-methylmannoside (α-MM; 10 mM and 100 mM). Although several types of oligosaccharides are completely separated by this method, each species represented by a peak may include a mixture of related structures. For example, hybrid structures that may contain GlcNAc, Gal, and sialic acid attached to the $Man_5GlcNAc_2$ as shown are also eluted with α-MM(10).

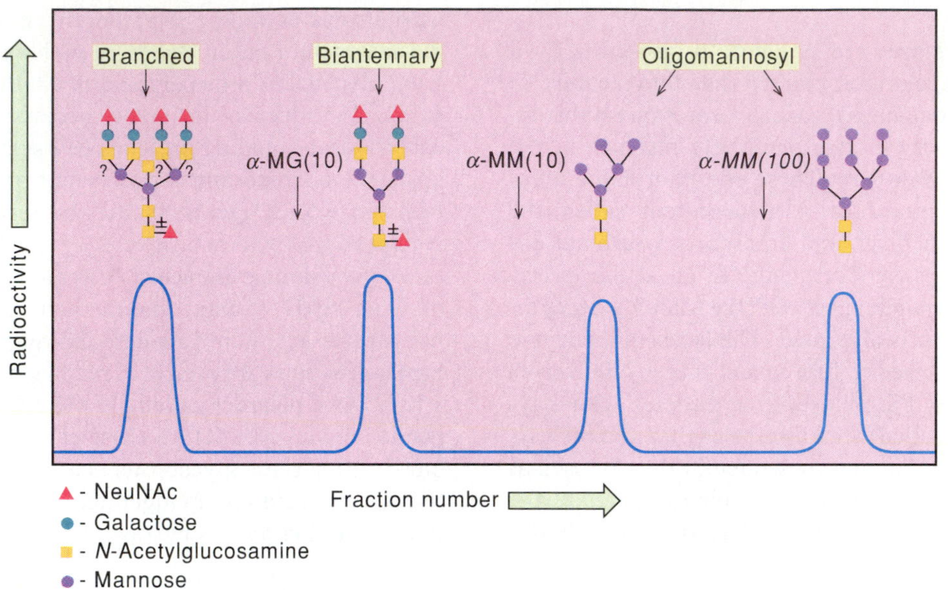

Structural Analysis of Carbohydrates

In order to correlate the structure of carbohydrates with their function, we must determine the precise arrangements of sugars in an oligosaccharide. For glycoproteins, often we must also determine the types of oligosaccharide structures at each glycosylation site. To do this, we first separate protease-generated fragments of the protein each containing one glycosylation site, then release the oligosaccharide from the peptide. Oligosaccharide moieties can be obtained free of protein by chemical or enzymic release from the protein backbone or by exhaustive proteolysis to give glycopeptides. Oligosaccharide mixtures can be separated on the basis of size or charge by standard chromatographic procedures and on the basis of structure by lectin-affinity chromatography (fig. 6.26). Lectins are proteins (or glycoproteins) that bind specific carbohydrate structures. For example, concanavalin A from the jack bean binds oligomannosyl N-linked carbohydrates but does not bind O-linked moieties or branched N-linked structures. Other lectins that are used in lectin-affinity chromatography include wheat germ agglutinin (binds sialic acid and *N*-acetylglucosamine), ricin (binds galactose), and lotus lectin (binds fucose).

We can determine the sequence of sugars in a pure oligosaccharide by sequential digestion with exoglycosidase enzymes (fig. 6.27). These are hydrolases that remove terminal, nonreducing sugars. They are highly specific for the particular sugar and its anomeric linkage (α or β). Some glycosidases are also specific for the type of linkage, cleaving a $\beta(1,2)$, for example, but not an $\alpha(1,4)$ linkage. Thus by sequential exoglycosidase digestion in combination with methods for detecting sugar removal (conventional or lectin-affinity chromatography), it is possible to determine the general structure of an oligosaccharide. Endoglycosidases are also most useful for carbohydrate structural analysis because they are able to recognize certain structural elements. Different endoglycosidases (D, F, or H) can be used to distinguish between complex, oligomannosyl, or hybrid structures as shown in figure 6.27*b*. The composition of an oligosaccharide can be determined after acid hydrolysis by gas chromatography or by separation of sugar oxyanions formed at high pH. The types of linkages between sugars can be determined by methylation analysis as described in figure 6.12. With combined information a complete structure can usually be deduced.

A much more rapid method of structural analysis is possible if ≥ 100 μg of pure oligosaccharide is available. This is high field proton (^{1}H) nuclear magnetic resonance (NMR) spectroscopy. NMR spectroscopy is nondestructive and gives a spectrum in which one to three protons of each sugar—for example, the anomeric carbon (C-1) proton or the C-2 proton or methyl group protons—are resolved. The resolution improves dramatically with two-dimensional (2D) NMR, allowing all protons in small oligosaccharides to be resolved. An example of a partial NMR spectrum of an oligosaccharide is interpreted in box 6B.

Figure 6.27

Glycosidase enzymes used for structural analysis of oligosaccharides. (*a*) Exoglycosidase enzymes remove terminal, nonreducing sugars with specificity for the sugar and whether it is in α or ß linkage. Used sequentially and in conjunction with various separation techniques, exoglycosidases reveal the sequence of sugars in an oligosaccharide. Endoglycosidases may be used to remove the intact oligosaccharide from the protein backbone. (*b*) Different endoglycosidases vary in their specificity for cleaving N-linked structures. The enzyme that cleaves the GlcNAc-Asn bond is an amidase or peptide-*N*-glycosidase called PGNase F. It requires the glycosylated asparagine to be substituted on both sides (i.e., to be within a peptide) for its action and it leaves aspartic acid in the peptide after removal of the oligosaccharide. The enzyme that cleaves between the GlcNAc residue is an *N*-glycosidase called endoglycosidase F. It leaves one GlcNAc residue linked to the asparagine which need not be substituted for the carbohydrate to be a substrate.

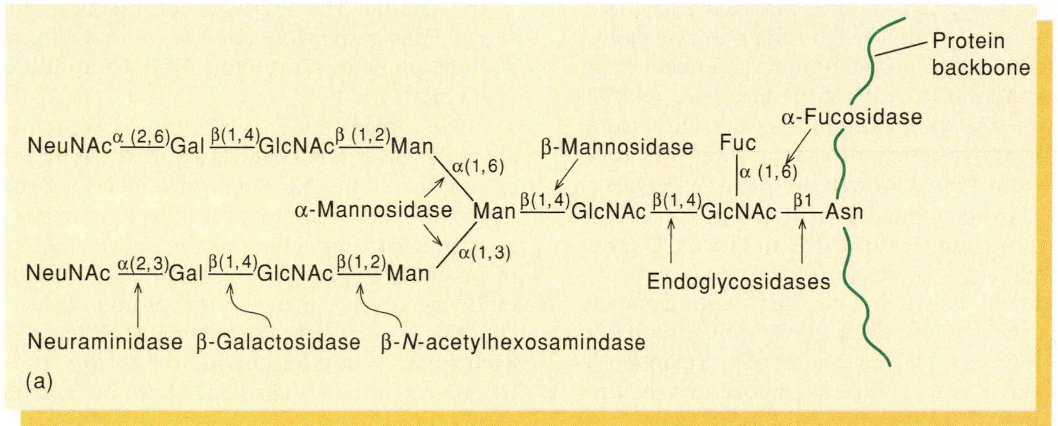

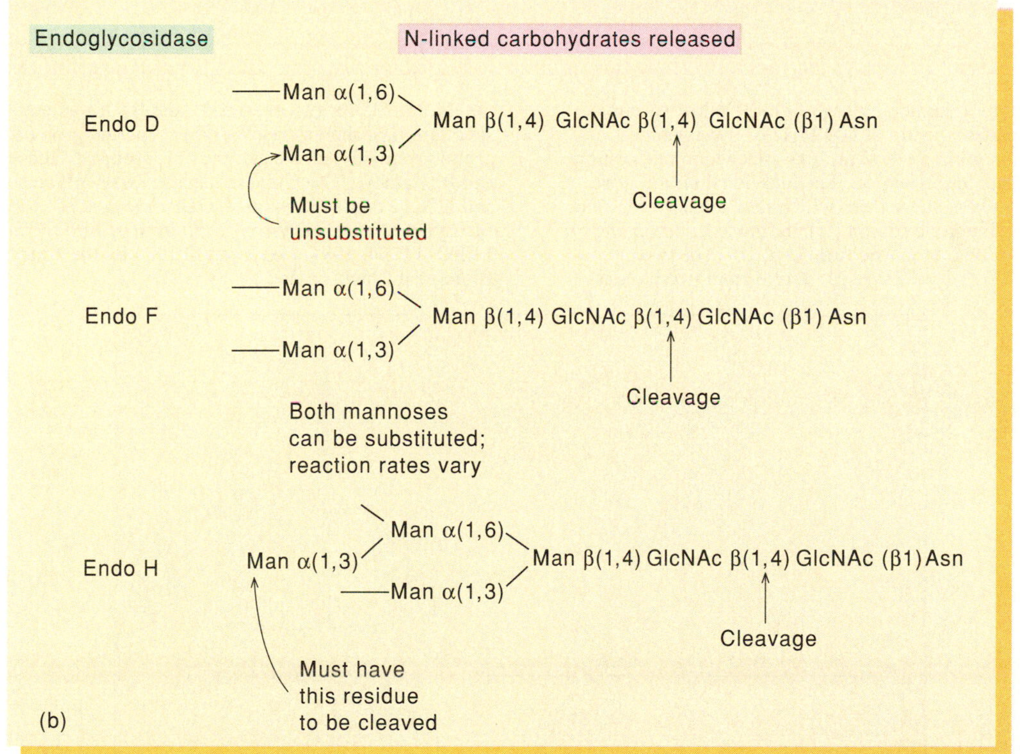

6B
BOX

Methods for Structural Analysis: Nuclear Magnetic Resonance (NMR) Spectroscopy

Nuclear magnetic resonance (NMR) spectroscopy exploits the fact that when a spinning, paramagnetic, charged particle (e.g., a proton) is placed in a magnetic field, it aligns mainly with the field and precesses about the field with a frequency (the Lamar frequency) dependent on the particle properties and the strength of the magnetic field. To obtain an NMR spectrum, a sample of protons is placed in a strong magnetic field (generated by the magnet of the NMR spectrometer) and is irradiated with a range of radiofrequency energies at 90° to the main field. This treatment causes all the protons in the sample to absorb energy at their characteristic frequency, flipping their magnetic orientations 90° with respect to their original state. After the applied puise field is switched off, the protons gradually relax to precess about the main field. Receiver coils in a probe surrounding the sample detect the frequencies of precessing protons as a set of oscillating electric currents, induced by the precessing magnetic vectors, which constitute the NMR signal. The magnitude of the induced voltage decays exponentially, giving rise to a free induction decay (FID). An FID is actually a mixture of sine waves arising from each of the chemical classes of protons in the sample. The FID is called the time domain of the NMR signal. Fourier transformation decodes the frequencies in the FID so they can be displayed in a plot of amplitude (amount) versus frequency.

The Lamar frequency of a proton is precisely dictated by its chemical environment and is expressed as a chemical shift in parts per million (ppm) Hz. For oligosaccharides, therefore, the chemical shift of an anomeric proton of a particular sugar (e.g., galactose) will vary depending on the structure of the oligosaccharide in which the sugar exists (see figure 1). Powerful NMR spectrometers can resolve many of the protons in a complex oligosaccharide. By identifying the chemical shifts of specific protons for each sugar in oligosaccharides of known structure (determined by other methods), data banks have been acquired which allow complete structures of unknown compounds to be deduced.

Figure 1

Oligosaccharide structural determination by high field ^{1}H-NMR spectroscopy. These traces are the partial spectra at 500 MHz of two related oligosaccharides from human milk. Several protons attached to one or more carbons of each sugar resonate in this region. The individual proton peaks are numbered to correspond to the sugar from which they are derived. Two peaks are obtained for the monomeric proton (H1) of glucose, depending on whether it derives from the α or ß anomeric form (1α, 1ß). The two resonances seen for fucose (5) and galactose (2) derive from two protons—the H1 and H5 for fucose, the H1 and H4 for galactose. Other regions of the spectrum (not shown) resolve other reported protons (e.g., protons of the *N*-acetyl group of GlcNAc and the CH_3 group of fucose). (Source: C. Campbell and P. Stanley, "The Chinese hamster ovary glycosylation mutants LEC11 and LEC12 express two novel GDP-fructose: *N*-acetylglucosaminide 3-α-L-fucosyltransferase enzymes" in *Journal of Biological Chemistry*, 259(18): 11208–11284, 1984. Copyright © 1984 by the American Society of Biological Chemists, Inc.)

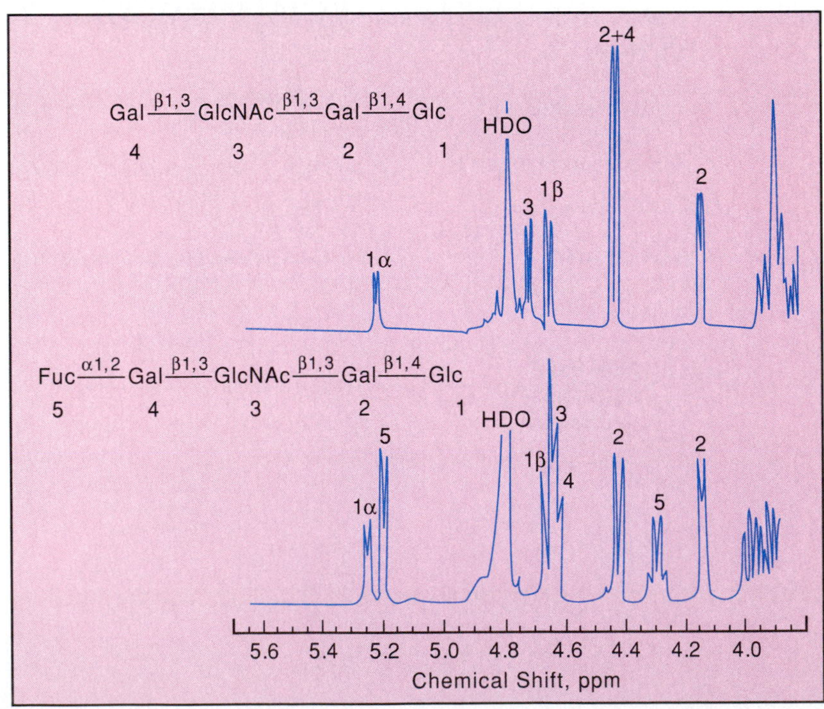

Figure 6.28

Diagram of a Gram negative cell envelope. The trimers of matrix protein of the outer membrane are associated with lipoprotein and with lipopolysaccharide (of variable polysaccharide length), and lipoprotein is covalently bound to peptidoglycan. Diagram also illustrates some general properties of membranes. Phospholipid molecules are illustrated with a circle for the polar groups and a line for each fatty acid acyl moiety. (Courtesy M. Inouye.)

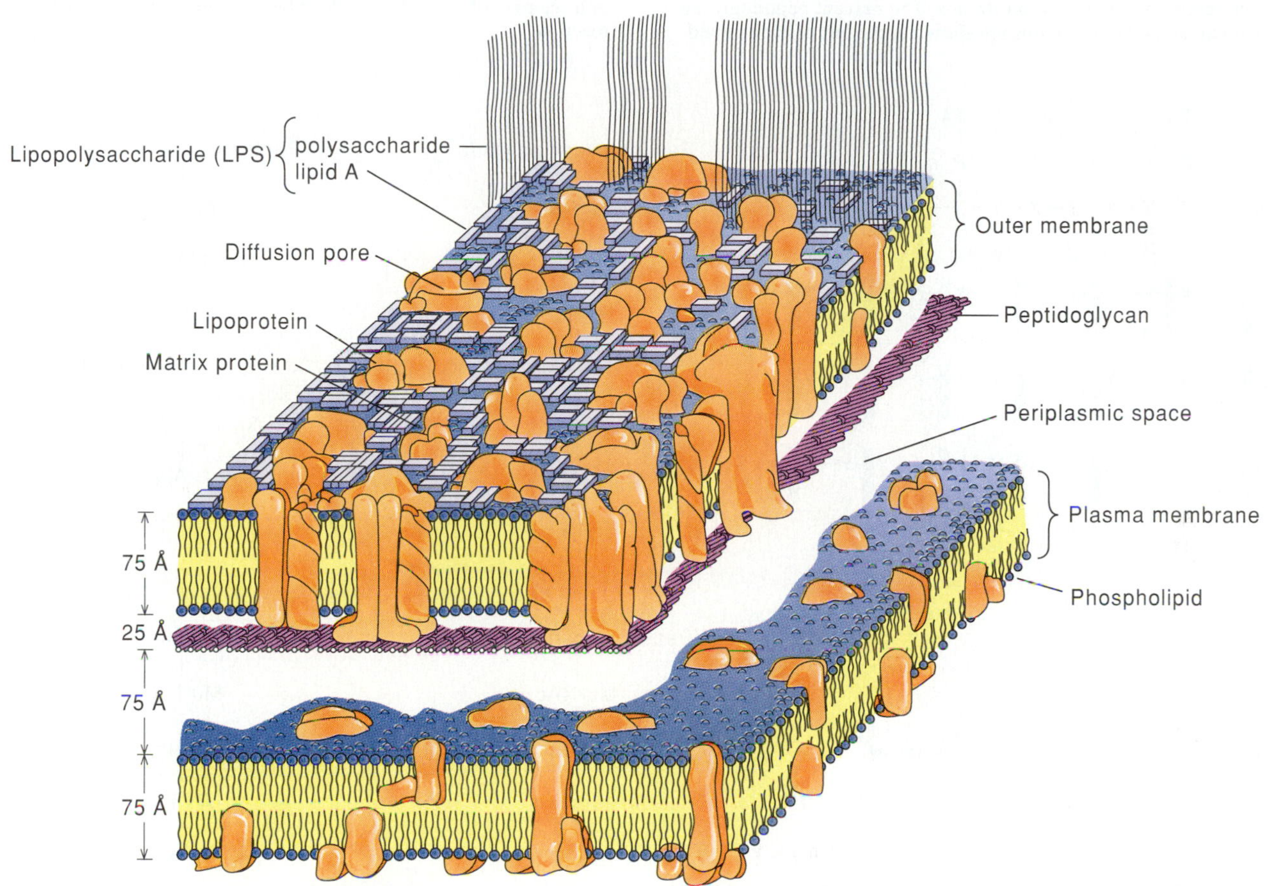

Peptidoglycans

As we noted in chapter 1, a unique feature of bacteria is the cell wall that surrounds the plasma membrane and provides the mechanical strength that enables bacteria to resist shear and osmotic shock. The cell wall is composed of a network of linear heteropolysaccharides cross-linked by peptides. A structure of this sort is called a peptidoglycan. Some bacterial cells (Gram negative cells) also possess an outer membrane composed of lipids, proteins, and polysaccharides. The main structural features of a Gram negative bacterial cell envelope, which is a composite of the two membranes and the cell wall, are illustrated in figure 6.28.

Here we will discuss the structural aspects of the cell wall, leaving a description of its biosynthesis to chapter 21. The peptidoglycan that constitutes the cell wall is a polymeric structure consisting of a heteropolysaccharide composed of amino sugars in one dimension, cross-linked through branched polypeptides in the other (fig. 6.29). The amino sugars alternate in the polymer, forming the glycan strands (see fig. 6.29). The carboxyl group of the lactic acid moiety of the acetylmuramic acid is substituted by a tetrapeptide, which in the Gram positive bacterium *Staphylococcus aureus* has the sequence L-alanyl-D-γ-glutamyl-L-lysyl-D-alanine. In this tetrapeptide the glutamyl residue is attached through its γ-COOH rather than its α-COOH. All the muramic acids are substituted in this way to form peptidoglycan strands. Variations of the same basic structure occur in all bacterial species. The peptidoglycan strands are further linked to each other by means of an interpeptide bridge. In *S. aureus,* this bridge is a pentaglycine chain that extends from the terminal carboxyl group of the D-alanine residue of one tetrapeptide to the ϵ-NH_2 group of the third amino acid, L-lysine, in another tetrapeptide (see fig. 6.29). The third dimension is probably built up by bridges extending in different planes. This gigantic macromolecule has the mechanical stability required for the cell wall.

Figure 6.29

Structure of the peptidoglycan of the cell wall of *Staphylococcus aureus*. (*a*) In this representation, X (*N*-acetylglucosamine) and Y (*N*-acetylmuramic acid) are the two sugars in the peptidoglycan. Light green circles represent the four amino acids of the tetrapeptide L-alanyl-D-glutamyl-L-lysyl-D-alanine. Dark green circles are pentaglycine bridges that interconnect peptidoglycan strands. The nascent peptidoglycan units bearing open pentaglycine chains are shown at the left of each strand. TA—P is the teichoic acid antigen of the organism, which is attached to the polysaccharide through a phosphodiester linkage. Teichoic acids are discussed in chapter 7 (see fig. 7.22). (*b*) The structure of X (*N*-acetylglucosamine) and Y (*N*-acetylmuramic acid), are connected by ß(1,4) linkages that alternate in the glycan strand. (*c*) The structure of a segment of the peptidoglycan before and after the final cross-linking reaction.

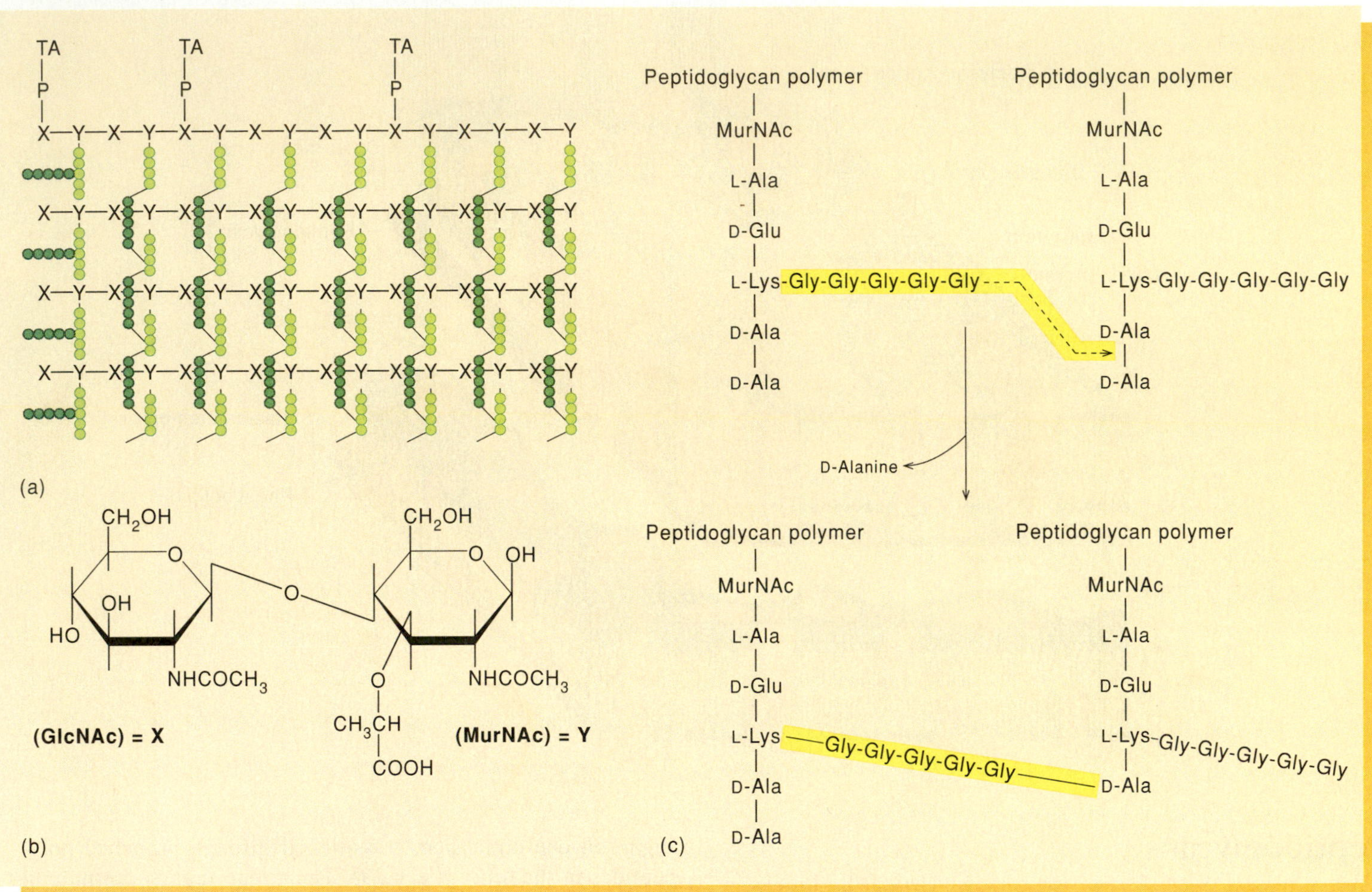

Summary

Carbohydrates are the most abundant organic substances. They are an important source of carbon compounds for all biomolecules and they also serve important energy and structural needs. In this chapter we have focused on the following points.

1. Carbohydrates may be divided into monosaccharides, oligosaccharides, and polysaccharides.
2. The monosaccharides are either polyhydroxyaldehydes or polyhydroxyketones. All monosaccharides are optically active because of the presence of one or more asymmetrical carbon atoms (chiral centers).
3. Straight-chain sugars in aqueous solution tend to form ring structures known as intramolecular hemiacetals, especially when the resultant is a five-membered (furanose) ring or a six-membered (pyranose) ring. Depending on which way the ring forms about the anomeric carbon, the structure is called an α or β hemiacetal.
4. For a given hemiacetal several conformations are possible. Usually the chair form is favored over the boat form because of the lower steric repulsion produced by side chains in the chair form. Polymer synthesis locks in a particular hemiacetal configuration.

5. Monosaccharides are linked by glycosidic bonds to form oligosaccharides and polysaccharides.
6. Polysaccharides function in two quite distinct roles: Some serve as a means for storage of chemical energy and others serve a structural function.
7. Polymers that use one type of building block (monomer) are called homopolymers, and those that use more than one are called heteropolymers.
8. Cellulose is the best known and most abundant structural polysaccharide. It is a homopolymer of glucose with $\beta(1,4)$ linkages between adjacent monomeric residues.
9. Starch and glycogen are energy-storage polysaccharides. They also are homopolymers of glucose but with $\alpha(1,4)$ linkages between adjacent residues. In addition, they contain branches with $\alpha(1,6)$ linkages.
10. Many other polysaccharides use different sugars as well as different combinations of sugars or modified sugars. Most of these function in a structural capacity.
11. Many heteropolysaccharides are linked to peptides (peptidoglycans) or proteins (proteoglycans or glycoproteins). Glycoproteins are mostly protein, with short, highly branched carbohydrate chains. By contrast, in proteoglycans the protein is the minor component.
12. Carbohydrates are also found linked to lipids in certain membrane structures.
13. Recent advances make it possible to determine structure/function relationships of oligosaccharides from a particular glycosylation site in a glycoconjugate.
14. The main component in bacterial cell walls is a peptidoglycan. The peptidoglycan is a polymeric structure made of a glycosaminoglycan extending in one dimension, cross-linked through oligopeptides extending in the other dimension. This cross-linked network completely surrounds the bacterial cell and provides it with the mechanical strength needed to resist osmotic shock and other mechanical stresses.

Selected Readings

Albersheim, P., and A. G. Darvill, Oligosaccharins. *Sci. Am.* 253(3):58–64, 1985.

Ferguson, M. A. J., and A. F. Williams, Cell surface anchoring of proteins via glycosylphosphatidylinositol structures. *Ann. Rev. Biochem.* 57:285–320, 1988.

Fransson, L.-A., Structure and function of cell-associated proteoglycans. *Trends Biochem. Sci.* 12:406–411, 1987. A recent account on a fast-moving subject.

Ginsberg, V., and P. Robbins, eds., *Biology of Carbohydrates.* New York: Wiley, 1984.

Hassell, J. R., J. H. Kimina, and L. Cantly, Proteoglycan core protein families. *Ann. Rev. Biochem.* 55:539–568, 1986.

Homans, S. W., M. A. J. Ferguson, R. A. Dwek, T. W. Rademacher, R. Anand, and A. F. Williams, Complete structure of the glycosylphosphatidylinositol membrane anchor of rat brain Thy-1 glycoprotein. *Nature* 333:269–272, 1988.

Lennarz, W. J., ed., *The Biochemistry of Glycoproteins and Proteoglycans.* New York, London: Plenum, 1980.

Lis, H. and N. Sharon, Lectins as molecules and as tools. *Ann. Rev. Biochem.* 55:35–67, 1986.

Maley, F., R. B. Trimble, A. L. Tarentino, and T. H. Plummer, Jr., Characterization of glycoproteins and their associated oligosaccharides through the use of endoglycosidases. *Anal. Biochem.* 180:195–204, 1989.

McNeil, M., A. G. Darvill, S. C. Fry, and P. Albersheim, Structure and function of the primary cell walls of plants. *Ann. Rev. Biochem.* 53:625–664, 1984.

Quiocho, F. A., Carbohydrate-binding proteins: tertiary structure and protein–sugar interactions. *Ann. Rev. Biochem.* 55:287–316, 1986.

Rademacher, T. W., R. B. Parekh, and R. A. Dwek, Glycobiology. *Ann. Rev. Biochem.* 57:785–838, 1988.

Ruoslahti, E., Structure and biology of proteoglycans. *Ann. Rev. Cell. Biol.* 4:229–255, 1988.

Sweeley, C. C., and H. Nuñez, Structural analysis of glycoconjugates by mass spectrometry and nuclear magnetic resonance spectroscopy. *Ann. Rev. Biochem.* 54:765–801, 1985.

Vliegenthart, J. F. G., L. Dorland, and H. van Halbeek, High-resolution, ^{1}H-nuclear magnetic resonance spectroscopy as a tool in the structural analysis of carbohydrates related to glycoproteins. *Adv. Carbohydr. Chem. Biochem.* 41:209–374, 1983.

von Figura, K., and A. Hasilik, Lysosomal enzymes and their receptors. *Ann. Rev. Biochem.* 55:167–193, 1986.

Problems

1. Compare the Haworth projections of D-glucose, D-mannose, and D-galactose. Indicate the differences between the structures.
2. Indicate the chiral carbons in D-glucose and determine the number of possible stereoisomers.
3. Is it possible that carbohydrates could have produced the diversity required to catalyze the myriad cellular reactions now relegated primarily to proteins?
4. In solution, D-glucose has a specific rotation of $[\alpha]^{20}{}_D = +52.7°$. The specific rotation of pure β-D-glucose is $+18.7°$ and that of pure α-D-glucose is $+112.2°$. Calculate the fraction of the α and β anomers in solution.
5. If either pure crystalline α- or β-D-glucose is dissolved in water, the final solution will contain the same fraction of α and β anomers as determined in problem 4. Explain chemically how this mutarotation occurs.
6. You are given two containers of polysaccharide by a colleague who has labeled one container as cellulose and one as glycogen. In his haste he may have mislabeled them and has asked you to verify which is which by methylation analysis. Indicate what product(s) you would expect from exhaustive methylation and mild acid hydrolysis of each polysaccharide and how the products could be used to differentiate the samples.

7. (a) Glycogen, starch, and cellulose are polymers of glucose. Suggest reasons, based on structure, that the physical form of each is appropriate to its role in nature. Why are the polymer forms of starch and glycogen much more desirable than an equivalent amount of free glucose in the cell?
   (b) Suggest how a cell might selectively synthesize starch but not cellulose.
8. Consider the packing of lipid triglycerides in adipocytes and of glycogen granules in the liver. Comment on the feasibility of using only glycogen, rather than lipid, as sole energy reserve. (See chapter 7, "Some fatty acids are stored as an energy reserve in triglycerides.")
9. Given the trisaccharide D-mannose-$\beta(1,3)$D-glucose-$\alpha(1,6)$D-galactose, draw the structure of the trisaccharide using Haworth projections. Name and draw the structures of the products of exhaustive methylation with dimethyl sulfate methylation and mild acid hydrolysis of the trisaccharide.
10. Chemical degradation of glycosaminoglycans causes reduced viscosity of the synovial fluid and subsequent damage to joints. Explain.
11. A tetrasaccharide has the following composition: D-Man(2), D-Gal(1), D-Glu(1). The tetrasaccharide gave a positive reducing sugar test in which the glucose residue was oxidized. Exhaustive methylation and mild acid hydrolysis released 2,3,4,6-tetra-*O*-methylmannose, 2,3-di-*O*-methylgalactose, and 2,3,6-tri-*O*-methylglucose. Treatment of the tetrasaccharide with an α-mannosidase released mannose and a trisaccharide that yielded the following methylation products: 2,3,4,6-tetra-*O*-methylmannose, 2,3,6-tri-*O*-methylgalactose, and 2,3,6-tri-*O*-methylglucose. Deduce the sequence and specificity of anomeric linkages and indicate any ambiguity.
12. Which enzymes can be used to ascertain the presence of sialic acid, galactose, *N*-acetylglucosamine, and mannose in a complex carbohydrate?
13. How might a cell synthesizing an N-linked glycoprotein specifically glycosylate only two of ten asparagine residues in the protein?
14. What structural features of oligosaccharides complicate the determination of their sequence as compared with the sequence determination of proteins?

7

CHAPTER

Lipids and Membranes

Lipids are biological molecules that are soluble in organic solvents. They have four major biological functions: (1) in all cells, the major structural elements of membranes are composed of lipids; (2) certain lipids, the triacylglycerols, serve as efficient reserves for the storage of energy; (3) many of the vitamins and hormones found in animals are lipids or derivatives of lipids; and (4) the bile acids help to solubilize the other lipid classes during digestion.

In this chapter, we will focus on the structures and functions of the major membrane lipids. From the structures of these compounds, we will see how they form the structural scaffolding of biological membranes and how proteins embedded in, or associated with, these membranes are arranged so as to carry out specific and essential functions in all cells. Lipid biosynthesis, membrane biogenesis, and the structures and roles of other classes of lipids are considered in part V (chapters 22 and 23), while one major function of biological membranes, transmembrane transport, is discussed in more detail in part VII (chapter 32). Nothing in those chapters depends on the intervening material, so if you prefer you can proceed to them directly after reading this chapter.

Fatty Acids

Compounds with the structural formula $CH_3(CH_2)_nCOOH$ that contain no carbon–carbon double bonds are known as saturated fatty acids. The two most abundant saturated fatty acids are palmitic and stearic acids (table 7.1). Some other saturated fatty acids present in smaller quantities in mammalian tissues are also shown in table 7.1. The sphingolipids, which we will consider later, contain longer-chain fatty acids ($n = 20$–24), as well as palmitic and stearic acids. Some tissues also contain short-chain fatty acids, such as decanoic acid ($n = 10$), found in milk.

Fatty acids with double bonds in the aliphatic chain are called unsaturated fatty acids. Monounsaturated fatty acids have one double bond, while polyunsaturated fatty acids contain more than one double bond. The double bonds in naturally occurring fatty acids are *cis*. The double bonds in polyunsaturated fatty acids are always separated by one methylene group.

Mammalian tissues contain all of the unsaturated fatty acids listed in table 7.1 with the exception of vaccenic acid, which is present in *E. coli* and other bacteria. However, *E. coli* and most other bacteria do not contain polyunsaturated fatty acids. While oleic acid is the most common unsaturated fatty acid in mammals, two other unsaturated fatty acids, linoleic and linolenic acids, are not synthesized by mammals and are therefore

Table 7.1
Fatty Acids

Common Name	Systematic Name	Structure	Abbreviation[a]
Saturated Fatty Acids			
Myristic acid	*n*-Tetradecanoic acid	$CH_3(CH_2)_{12}COOH$	14:0
Palmitic acid	*n*-Hexadecanoic acid	$CH_3(CH_2)_{12}CH_2CH_2COOH$	16:0
Stearic acid	*n*-Octadecanoic acid	$CH_3(CH_2)_{12}CH_2CH_2CH_2CH_2COOH$	18:0
Arachidic acid	*n*-Eicosanoic acid	$CH_3(CH_2)_{12}CH_2CH_2CH_2CH_2CH_2CH_2COOH$	20:0
Behenic acid	*n*-Docosanoic acid	$CH_3(CH_2)_{12}CH_2CH_2CH_2CH_2CH_2CH_2CH_2CH_2COOH$	22:0
Lignoceric acid	*n*-Tetracosanoic acid	$CH_3(CH_2)_{12}CH_2CH_2CH_2CH_2CH_2CH_2CH_2CH_2CH_2CH_2COOH$	24:0
Cerotic acid	*n*-Hexacosanoic acid	$CH_3(CH_2)_{12}CH_2CH_2CH_2CH_2CH_2CH_2CH_2CH_2CH_2CH_2CH_2CH_2COOH$	26:0
Unsaturated Fatty Acids			
Palmitoleic acid	*cis*-9-Hexadecenoic acid	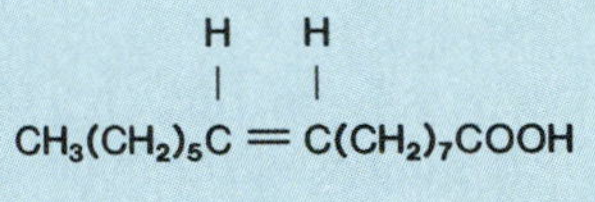 $CH_3(CH_2)_5\overset{H}{C}=\overset{H}{C}(CH_2)_7COOH$	$16{:}1^{\Delta 9}$
Oleic acid	*cis*-9-Octadecenoic acid	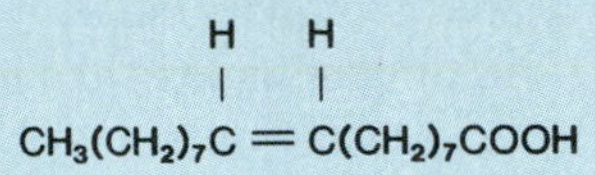 $CH_3(CH_2)_7\overset{H}{C}=\overset{H}{C}(CH_2)_7COOH$	$18{:}1^{\Delta 9}$
Vaccenic acid	*cis*-11-Octadecenoic acid	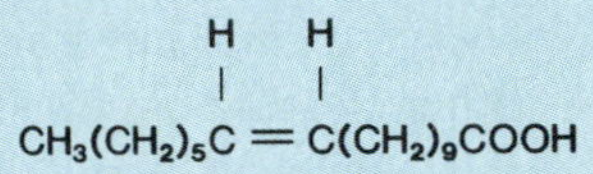 $CH_3(CH_2)_5\overset{H}{C}=\overset{H}{C}(CH_2)_9COOH$	$18{:}1^{\Delta 11}$
Linoleic acid	*cis,cis*-9,12-Octadecadienoic acid	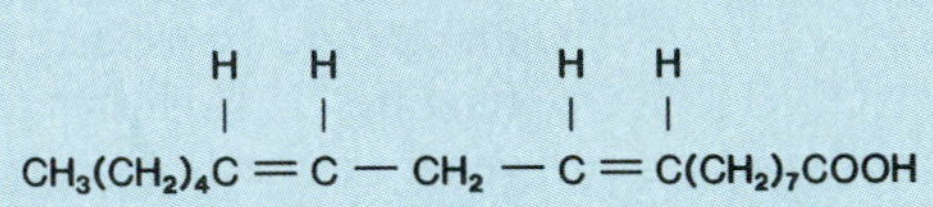 $CH_3(CH_2)_4\overset{H}{C}=\overset{H}{C}-CH_2-\overset{H}{C}=\overset{H}{C}(CH_2)_7COOH$	$18{:}2^{\Delta 9,12}$
α-Linolenic acid	All-*cis*-9,12,15-Octadecatrienoic acid	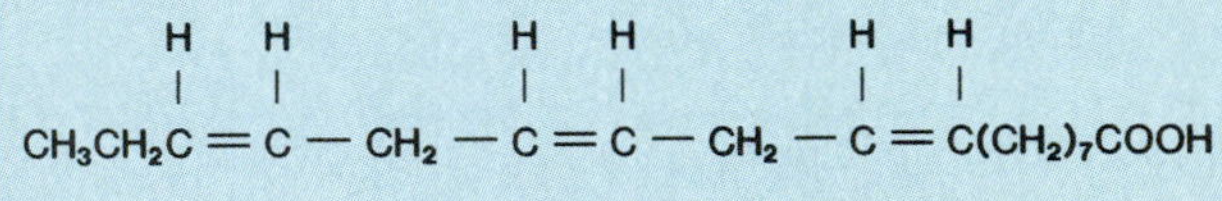 $CH_3CH_2\overset{H}{C}=\overset{H}{C}-CH_2-\overset{H}{C}=\overset{H}{C}-CH_2-\overset{H}{C}=\overset{H}{C}(CH_2)_7COOH$	$18{:}3^{\Delta 9,12,15}$
Arachidonic acid	All-*cis*-5,8,11,14-Eicosatetraenoic acid	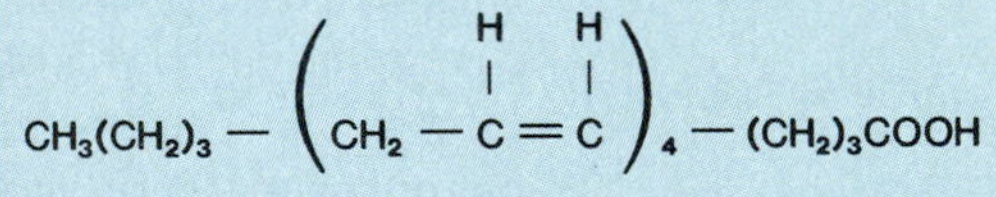 $CH_3(CH_2)_3-\left(CH_2-\overset{H}{C}=\overset{H}{C}\right)_4-(CH_2)_3COOH$	$20{:}4^{\Delta 5,8,11,14}$
	All-*cis*-4,7,10,13,16,19-Docosahexaenoic acid	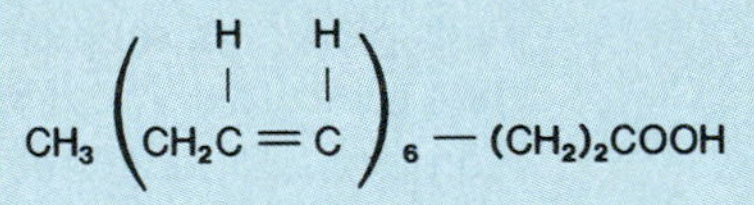 $CH_3\left(CH_2\overset{H}{C}=\overset{H}{C}\right)_6-(CH_2)_2COOH$	$22{:}6^{\Delta 4,7,10,13,16,19}$
Some Unusual Fatty Acids			
	2,4,6,8-Tetramethyl decanoic acid	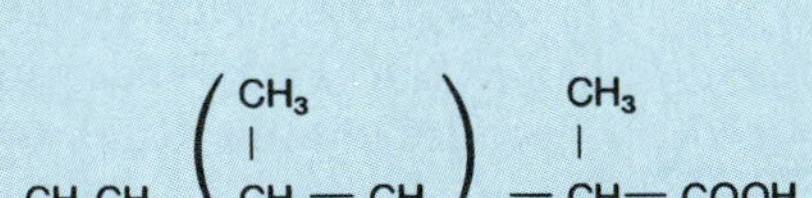 $CH_3CH_2\left(\overset{CH_3}{CH}-CH_2\right)_3-\overset{CH_3}{CH}-COOH$	
Lactobacillic acid		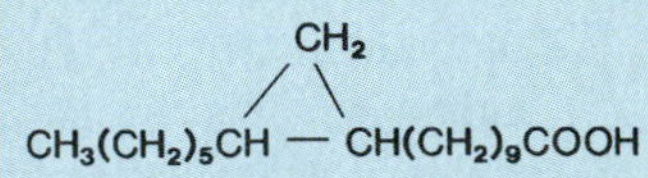 $CH_3(CH_2)_5CH-CH(CH_2)_9COOH$ (with CH_2 bridging the two CH groups)	
An α-mycolic acid		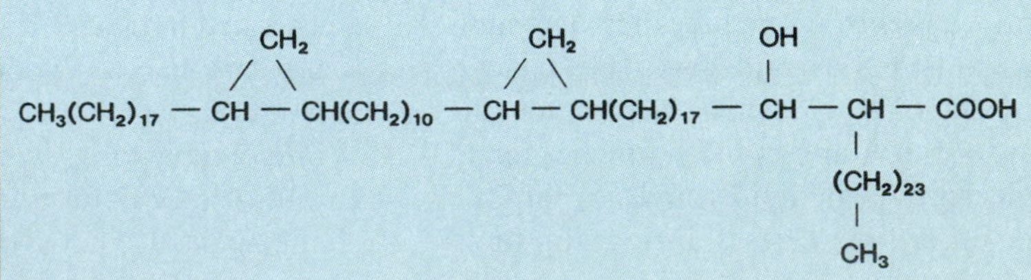 $CH_3(CH_2)_{17}-CH-CH(CH_2)_{10}-CH-CH(CH_2)_{17}-\overset{OH}{CH}-\underset{(CH_2)_{23}CH_3}{CH}-COOH$ (each CH–CH pair bridged by CH_2)	

[a]In these abbreviations the number to the left of the colon is the number of carbon atoms, and the number to the right is the number of double bonds. Δ9 signifies that there are 8 carbons between carboxyl group and double bond.

Figure 7.1

Space-filling and conformational models of (*a*) stearic and (*b*) linolenic acids. Each of these fatty acids has 18 carbon atoms, but the three double bonds in linolenic acid create a more rigid, curved molecule that interferes with tight packing in membrane structures.

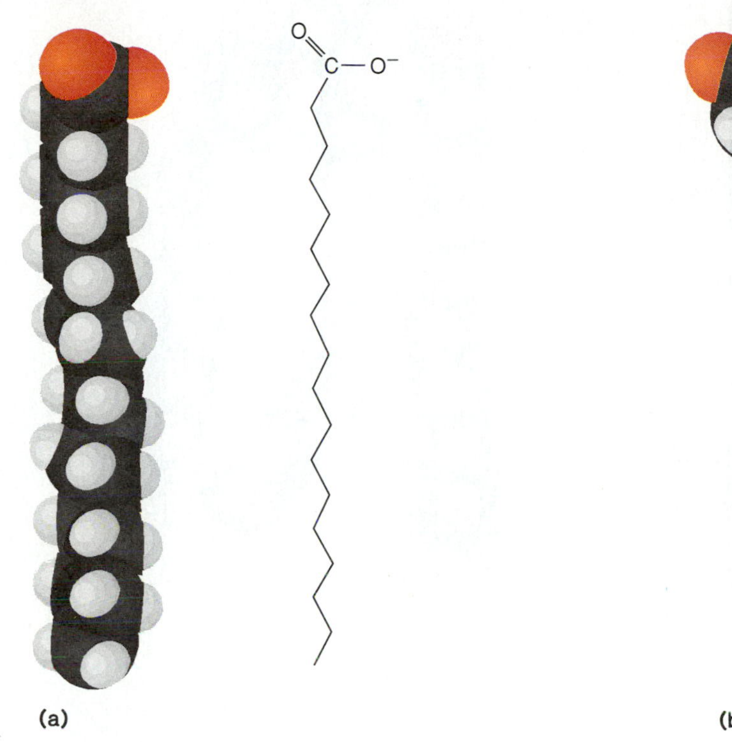

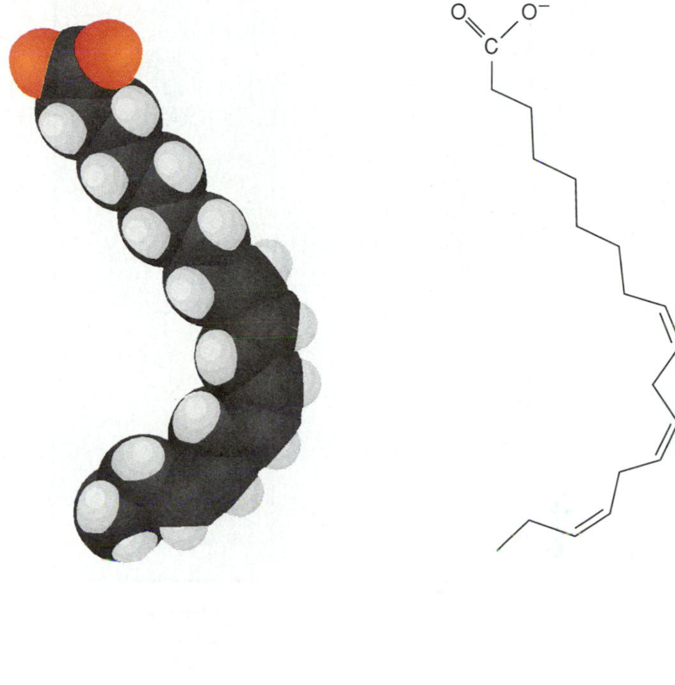

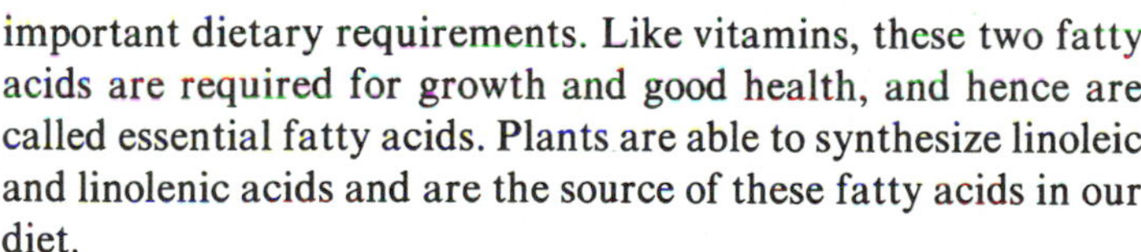

important dietary requirements. Like vitamins, these two fatty acids are required for growth and good health, and hence are called essential fatty acids. Plants are able to synthesize linoleic and linolenic acids and are the source of these fatty acids in our diet.

In addition to the commonly occurring fatty acids, many structural variations have evolved. There are over 100 other fatty acids found in various organisms, often associated with specialized functions. For instance, branched-chain fatty acids are found in many different tissues. The uropygial gland of the duck produces such a fatty acid (2,4,6,8-tetramethyldecanoic acid). The duck uses the fatty acids secreted by this gland to preen its feathers and thereby ensure that water continues to "run off its back." Other examples are fatty acids with a cyclopropane ring in the alkyl chain, found in many bacteria. The bacterium that causes tuberculosis, *Mycobacterium tuberculosis,* produces a family of complex fatty acids known as mycolic acids, which contain cyclopropane rings. One class of these is the α-mycolic acids (an example is given in table 7.1), and many structurally related α-mycolic acids are found in the mycobacteria and other related organisms (nocardiae and corynebacteria). These compounds appear to have a structural function in the outer part of the bacterial cell wall. There is much evidence to suggest that a major drug used in the treatment of tuberculosis, Isoniazid, functions by inhibiting an early reaction of α-mycolic acid biosynthesis.

Fatty acids are usually found as components of complex lipids, and only rarely as unesterified (free) fatty acids. Nevertheless, the pK_a for dissociation of the acid proton is around 4.7. Therefore, at pH 7.0, the fatty acid exists primarily in the dissociated form ($RCOO^-$):

$$CH_3(CH_2)_nCOOH \rightleftharpoons CH_3(CH_2)_nCOO^- + H^+$$

Because it exists as an anion at neutral pH, a fatty acid is not easily extracted from an aqueous medium by organic solvents such as hexane. However, if the pH is lowered by the addition of HCl or another strong acid, the fatty acid becomes protonated and is easily extracted by organic solvents.

Another property of fatty acids that you should note is the variation in their physical form at room temperature. If n equals 8 or less, the fatty acid is a liquid, whereas if n equals 10 or more, the fatty acid is a solid. If a fatty acid has a double bond, it has a lower melting point than the saturated fatty acid with the same number of carbons. *Cis*-unsaturated fatty acids are more condensed in length than the corresponding saturated fatty acids, and also contain one or more inflexible kinks (fig. 7.1). These properties explain the lowered melting temperatures of unsaturated fatty acids as well as the fact that they pack less tightly within membranes than saturated fatty acids, a point to which we will return later.

Table 7.2
Neutral Glycerides[a]

Common Name	Systematic Name	Structure
Triglyceride	1,2,3-Triacyl-*sn*-glycerol	$\begin{array}{l} \quad\quad\ \ CH_2OC(=O)R \\ R'—C(=O)OCH \\ \quad\quad\ \ CH_2OC(=O)R' \end{array}$
Diglyceride	1,2-Diacyl-*sn*-glycerol	$\begin{array}{l} \quad\quad\ \ CH_2OC(=O)R \\ R'—C(=O)OCH \\ \quad\quad\ \ CH_2OH \end{array}$
Monoglyceride	1-Monoacyl-*sn*-glycerol	$\begin{array}{l} CH_2OC(=O)R \\ HOCH \\ CH_2OH \end{array}$

[a]Because the substituents esterified to the first and third carbons of these glycerol derivatives are usually different, the second carbon atom is asymmetric. In naming and numbering these compounds, a special convention has been adopted: The prefix *sn*- (for *s*tereospecifically *n*umbered) immediately precedes "glycerol" and differentiates the naming of the compound from other approaches, such as the R S system described in chapter 11. The glycerol derivative is drawn in a Fischer projection with the secondary hydroxyl to the left of the central carbon, and the carbons are numbered 1, 2, and 3 from the top to the bottom. The prefix *rac*- (for *rac*emo) precedes the name if the compound is an equal mixture of antipodes. If the configuration is unknown or not specified, *x*- precedes the name.

Fatty acids are commonly analyzed by gas chromatography of the methyl esters. These esters are formed by esterification of the fatty acids with methanol (R represents hydrogen (H) or any group to which the fatty acid is esterified):

$$CH_3(CH_2)_nC(=O)—OR + CH_3OH \xrightarrow[\text{HCl or BF}_3]{} CH_3(CH_2)_nC(=O)OCH_3 + ROH$$

Some Fatty Acids Are Stored as an Energy Reserve in Triacylglycerols

Fatty acids serve two major roles: they are major components of the triacylglycerols (table 7.2 and fig. 7.2) and of most of the complex lipids present in membranes. Triacylglycerols are the major uncharged glycerol derivatives found in animals and they are stored as an energy reserve. Monoacylglycerols and diacylglycerols are metabolites of triacylglycerols and of phospholipids (see chapters 17 and 22), and are normally present in cells in very small quantities.

Although triacylglycerols are found in the liver and intestine, they are primarily found in adipose tissue (fat), which functions as a storage depot for this lipid. The specialized cell in adipose tissue is called the adipocyte. Its cytoplasm is full of lipid vacuoles that are almost exclusively triacylglycerols (fig. 7.3) and that serve as an energy reserve for mammals. At times when the diet or glycogen reserves are insufficient to supply the body's need for energy, the fuel stored as fatty acyl components of the triacylglycerols is mobilized and transported to other tissues in the body. A second important function of adipose tissue is insulation of the body from cold. This function is most obvious in such cold-water mammals as the arctic (Beluga) whales, which have vast stores of fat (blubber).

Figure 7.2

Space-filling and conformational models of a triacylglycerol. Note the uncharged nature of this molecule, which serves as a means of storing fatty acids for future energy needs.

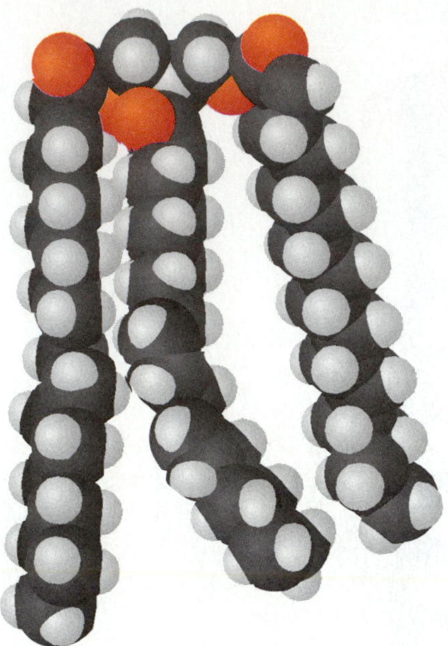

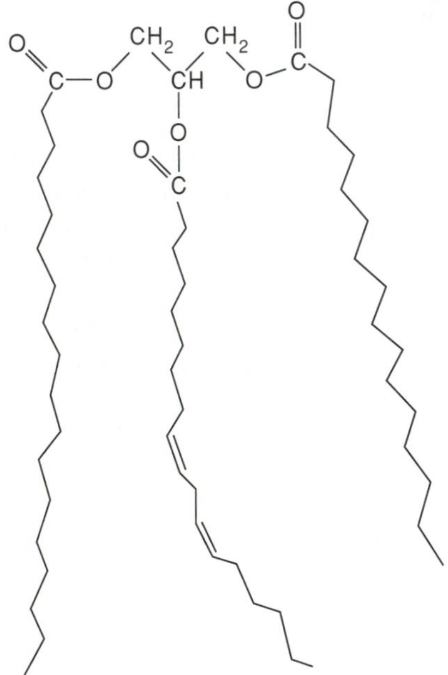

Triacylglycerols are structurally related to lipids that are found in membranes. They differ in one major respect; they are neutral, whereas most lipids found in membranes are charged at one end. This lack of charge has a profound effect on the role they can play; it suits them for compact storage in vesicles but makes them totally unsuitable for structural components in membranes.

Figure 7.3

Scanning electron micrograph of white adipocytes from rat adipose tissue (600X). (Courtesy of Dr. A. Angel and Dr. M. J. Hollenberg of the University of Toronto.)

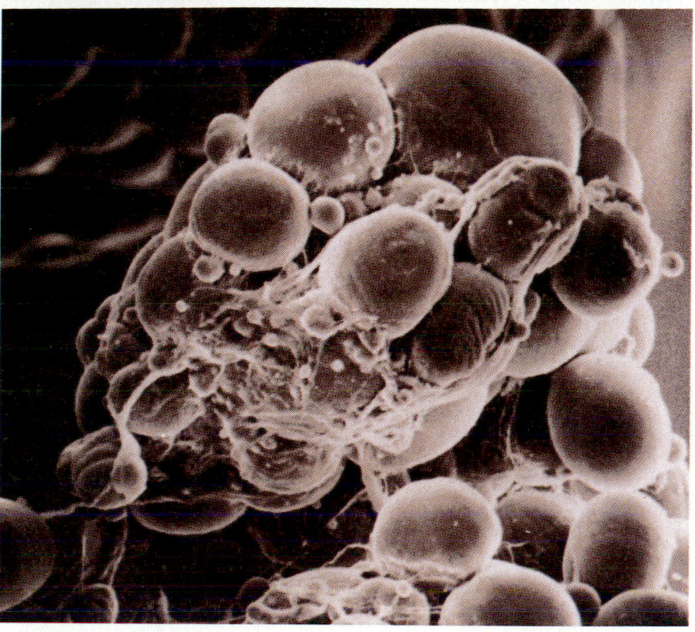

Membrane Lipids

The major lipids found in biological membranes include the phospholipids (phosphoglycerides and sphingomyelin), the glycosphingolipids, and cholesterol. Phosphoglycerides, quantitatively the most important structure group, contain, in addition to phosphate, a glycerol backbone and esterified fatty acids and alcohols. While the phosphoglycerides are the predominant lipids in biological membranes, cholesterol and the sphingolipids are important components of some cellular membranes, especially in eukaryotic cells. The special importance of phospholipids to living organisms is underscored by the nearly complete lack of genetic defects in the metabolism of these lipids in humans. Presumably, any such defects are lethal at early stages of development and therefore are never observed. Isolation and analysis of phospholipids are described in box 7A.

7A BOX

Isolation and Analysis of Phospholipids

Lipids are extracted from cells or tissues with organic solvents (for example, $CHCl_3$). Phospholipids are resolved from uncharged lipids (neutral lipids) by adsorption column chromatography (usually silicic acid) or by adsorption thin-layer chromatography. The various classes of phospholipids are usually separated on a small scale by thin-layer chromatography, as shown in the illustration. Large-scale preparations usually involve column chromatography or high-pressure liquid chromatography.

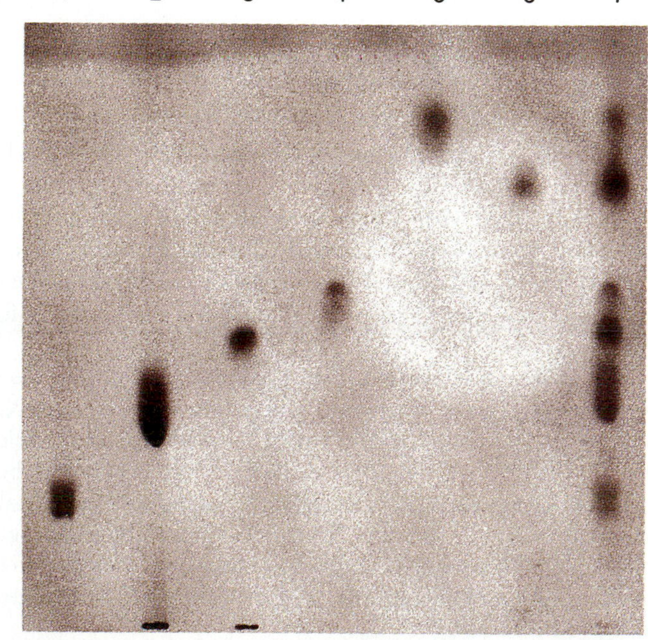

Figure 1

Thin-layer chromatography of the major phospholipids. Lanes 1 to 6 contain 0.1 mg of each lipid. Lane 7 is a mixture of the six phospholipids. PE = phosphatidylethanolamine; PG = phosphatidylglycerol; PS = phosphatidylserine; PI = phosphatidylinositol; PC = phosphatidylcholine; S = sphingomyelin. Each compound was spotted on a silica gel G60 thin-layer plate that had been activated at 100°C 1 h before the analysis. The plate was developed in a solvent that contained $CHCl_3$:CH_3OH:CH_3COOH:H_2O (50:25:8:4 by volume). The compounds were visualized after spraying the plate with dilute sulfuric acid and heating in the oven.

Figure 7.4

The structure of a phosphatidic acid, a phosphoglyceride. The cluster of polar and charged oxygens gives phosphatidic acid its amphipathic properties. The fatty acids attached to carbons 1 and 2 are saturated and unsaturated, respectively.

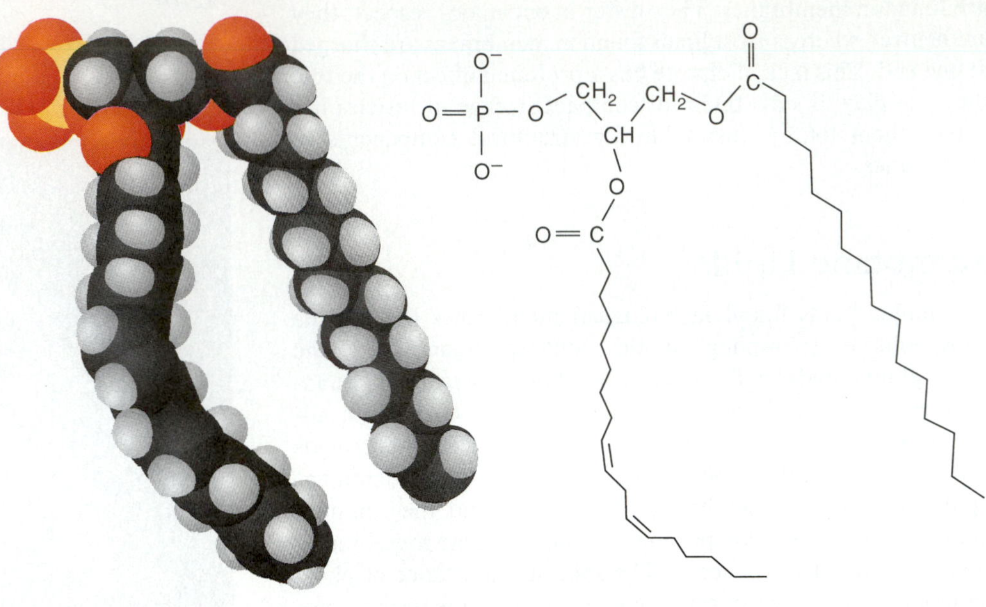

Figure 7.5

Phospholipids with alkyl or alkenyl ether substituents. The structures with alkenyl ether substituents are also called plasmalogens.

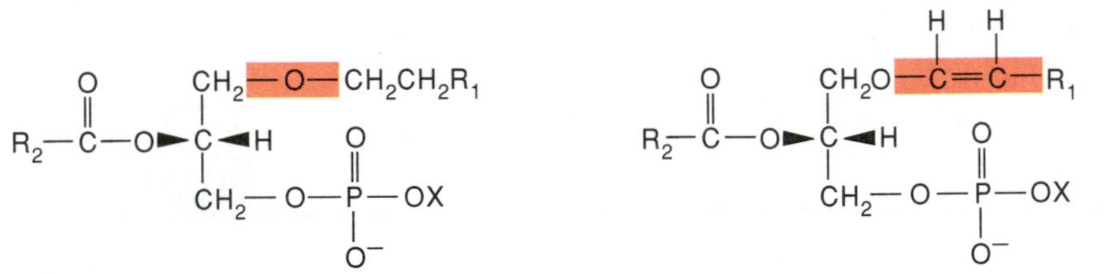

Phospholipid with an alkyl ether

Phospholipid with an alkenyl ether

Phosphoglycerides Have a Glycerol-3-Phosphate Backbone

All phosphoglycerides have a glycerol-3-phosphate backbone, as shown in figure 7.4. The hydroxyls on carbons 1 and 2 are usually acylated with fatty acids, and in most phospholipids the fatty acid substituent at carbon 1 is saturated, while the one at carbon 2 is unsaturated. In some instances, the substituent on carbon 1 is an alkyl ether or an alkenyl ether as shown in figure 7.5.

Phosphoglycerides are classified according to the substituent (X) on the phosphate group (see table 7.3). If X is a hydrogen, the compound is called 3-*sn*-phosphatidic acid. If X—OH is choline, the lipid is called 3-*sn*-phosphatidylcholine (lecithin); this is the most abundant phospholipid in animal tissues (fig. 7.6). In addition to its role in membrane structure, phosphatidylcholine is an important structural component of the plasma lipoproteins and bile. The other major phosphoglycerides are listed in table 7.3.

An important subclass of phosphoglycerides is the lysophospholipids, in which one of the acyl substituents (usually from position 2) is missing, as shown in figure 7.7. If the acyl substituent at carbon 1 were removed, the acyl group from position 2 would spontaneously migrate to position 1. We can differentiate between deacylation at position 1 or 2 by analysis of the fatty acids derived from the lysophospholipid. If the fatty acid on the molecule is saturated, it is likely that the fatty acid from position 2 of the phospholipid has been cleaved. However, if the fatty acid on the lysophospholipid is mostly unsaturated, the fatty acid from position 1 has probably been cleaved and migration of the fatty acid from position 2 has occurred. The lysophospholipids are named simply by adding the prefix *lyso-* to the name of the original phospholipid (e.g., lysophosphatidylcholine). The lysophospholipids account for only 1–2% of the total phospholipids in animal cells.

Table 7.3
Major Classes of Phosphoglycerides

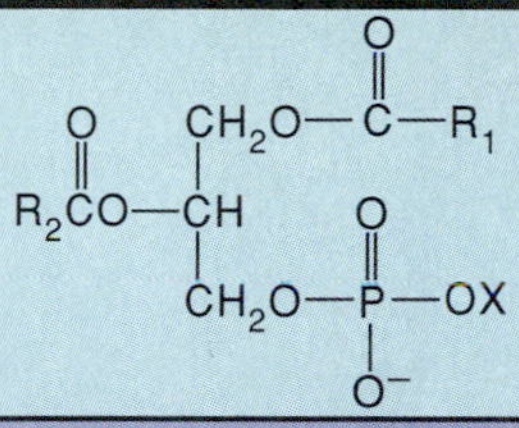

X Substituent		
Name of X—OH	**Formula of X**	**Name of Phospholipid**
Water	$-H$	Phosphatidic acid
Choline	$-CH_2CH_2\overset{+}{N}(CH_3)_3$	Phosphatidylcholine (lecithin)
Ethanolamine	$-CH_2CH_2\overset{+}{N}H_3$	Phosphatidylethanolamine
Serine	$-CH_2-CH(\overset{+}{N}H_3)COO^-$	Phosphatidylserine
Glycerol	$-CH_2CH(OH)CH_2OH$	Phosphatidylglycerol
Phosphatidylglycerol	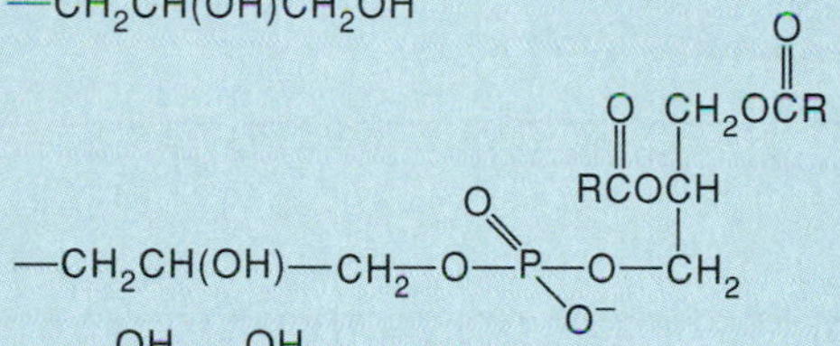 $-CH_2CH(OH)-CH_2-O-P(O)(O^-)-O-CH_2CH(OCOR)CH_2OCOR$	Diphosphatidylglycerol (cardiolipin)
myo-Inositol	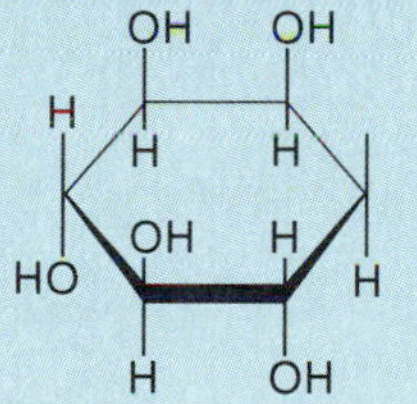	Phosphatidylinositol

Sphingolipids Contain a Long-Chain, Hydroxylated Secondary Amine

In addition to the phosphoglycerides, the sphingolipids are commonly found in eukaryotic cell membranes. Sphingomyelin contains phosphate and therefore is also a phospholipid, while the glycosphingolipids generally lack phosphate but contain sugar or oligosaccharide residues. The common structural feature of sphingolipids is a long-chain, hydroxylated secondary amine. There are three major long-chain bases (table 7.4) that contain 18 carbons and a number of other bases that differ in chain length, number of double bonds, or branching of the alkyl chain. Sphingosine (4-sphingenine) is quantitatively the most important long-chain base (usually 90% or more) in animal cells, whereas phytosphingosine (4-hydroxysphinganine) is characteristically found in plant tissues. Most bacteria, including *E. coli,* do not contain sphingolipids, whereas yeast cells do.

Figure 7.6

Structure of a phosphatidylcholine (lecithin).

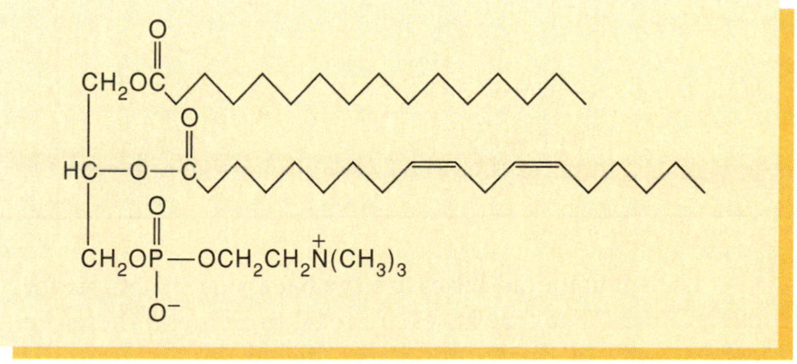

Figure 7.7

The basic structure for lysophospholipids. These phospholipids usually lack a substituent at position 2.

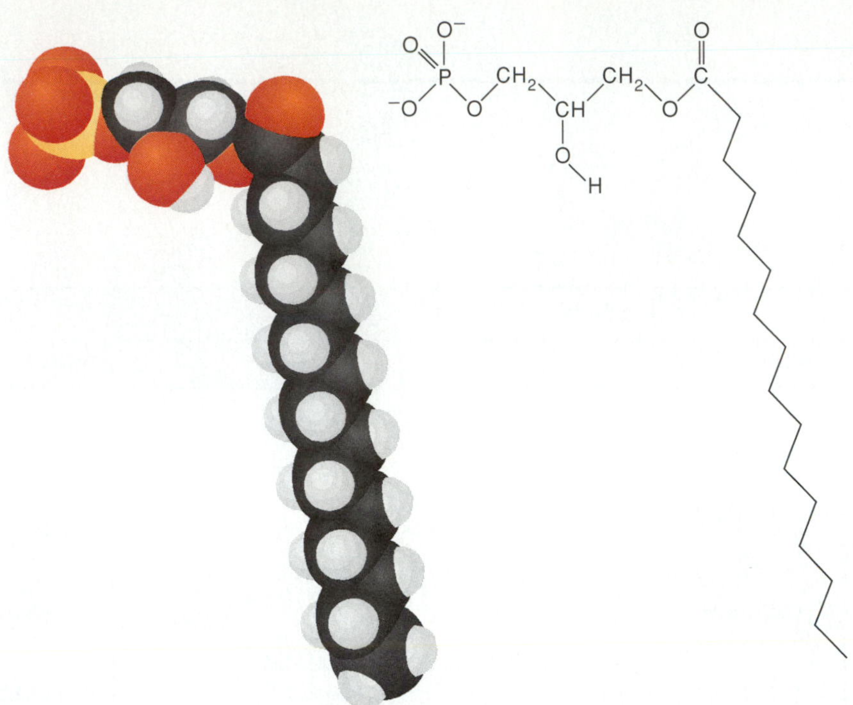

Table 7.4
Three Important Sphingolipid Bases

Structure	Systematic Name	Common Name
$CH_3(CH_2)_{12}CH{=}CH{-}CH(OH){-}CH(\overset{+}{N}H_3){-}CH_2OH$	4-Sphingenine	Sphingosine
$CH_3(CH_2)_{12}CH_2CH_2{-}CH(OH){-}CH(\overset{+}{N}H_3){-}CH_2OH$	Sphinganine	Dihydrosphingosine
$CH_3(CH_2)_{12}CH_2CH(OH){-}CH(OH){-}CH(\overset{+}{N}H_3){-}CH_2OH$	4-Hydroxysphinganine	Phytosphingosine

The sphingolipid bases are the backbone structures for all sphingolipids. The free bases are toxic to cells and therefore are present only in trace quantities. The sphingolipid base is acylated on the amine with a fatty acid to give ceramide (fig. 7.8), which is common to all the sphingolipids. The fatty acid substituents are mainly C_{16}, C_{18}, C_{22}, or C_{24}, saturated or monounsaturated. In many cases, the acyl group also may contain an α-hydroxyl residue. Ceramide is further modified on the primary hydroxyl group to give the final sphingolipid structures; modification with phosphocholine gives sphingomyelin (fig. 7.9), while modification with carbohydrate gives the class

Figure 7.8

Structure of ceramide with sphingosine as the long-chain base. Sphingosine (shown in color) is very toxic to cells and is usually found only in trace amounts.

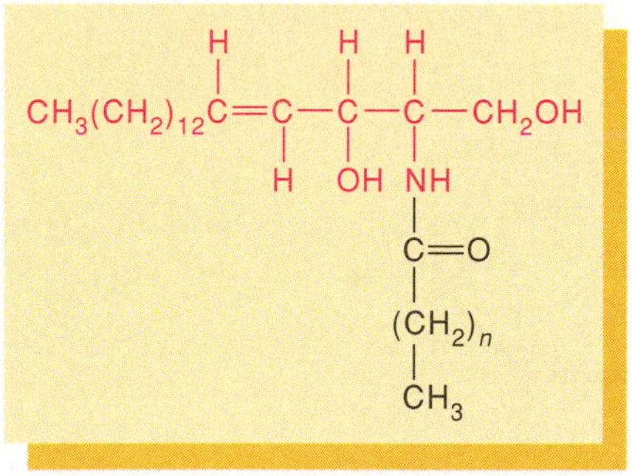

Figure 7.10

Structure of α-*N*-acetylneuraminic acid (sialic acid). Gangliosides are a subclass of glycosphingolipids that always contain one or more molecules of *N*-acetylneuraminic acid. This compound is also commonly found as a terminal residue of glycoproteins (see chapter 6).

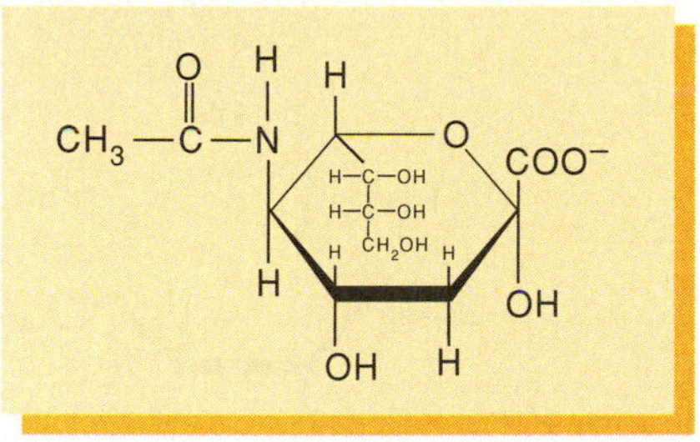

Figure 7.9

Structure of a sphingomyelin.

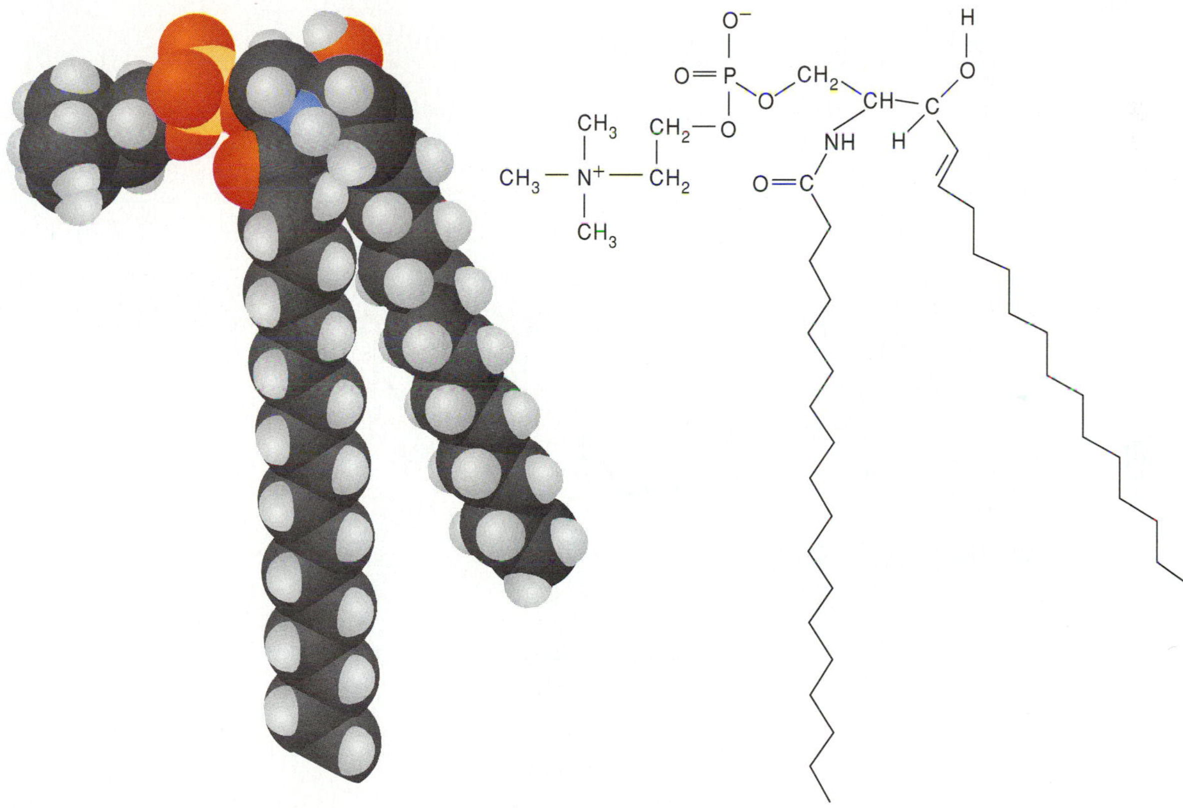

called glycosphingolipids. The carbohydrates most often associated with the glycosphingolipids are glucose, galactose, *N*-acetylglucosamine, and *N*-acetylgalactosamine. There is a subdivision of the glycosphingolipids called gangliosides, and these also contain one or more molecules of *N*-acetylneuraminic acid (sialic acid) (fig. 7.10) in addition to other carbohydrates. As the name implies, the gangliosides were first isolated from nerve tissue; subsequently, however, they were found in most other animal tissues. The structures of two important glycosphingolipids, globoside and GM_2, are shown in figures 7.11 and 7.12. Over 50 separate classes of glycosphingolipids have been identified on the basis of differences in the structure of the oligosaccharide.

Figure 7.11

Structure of globoside, a glycosphingolipid.

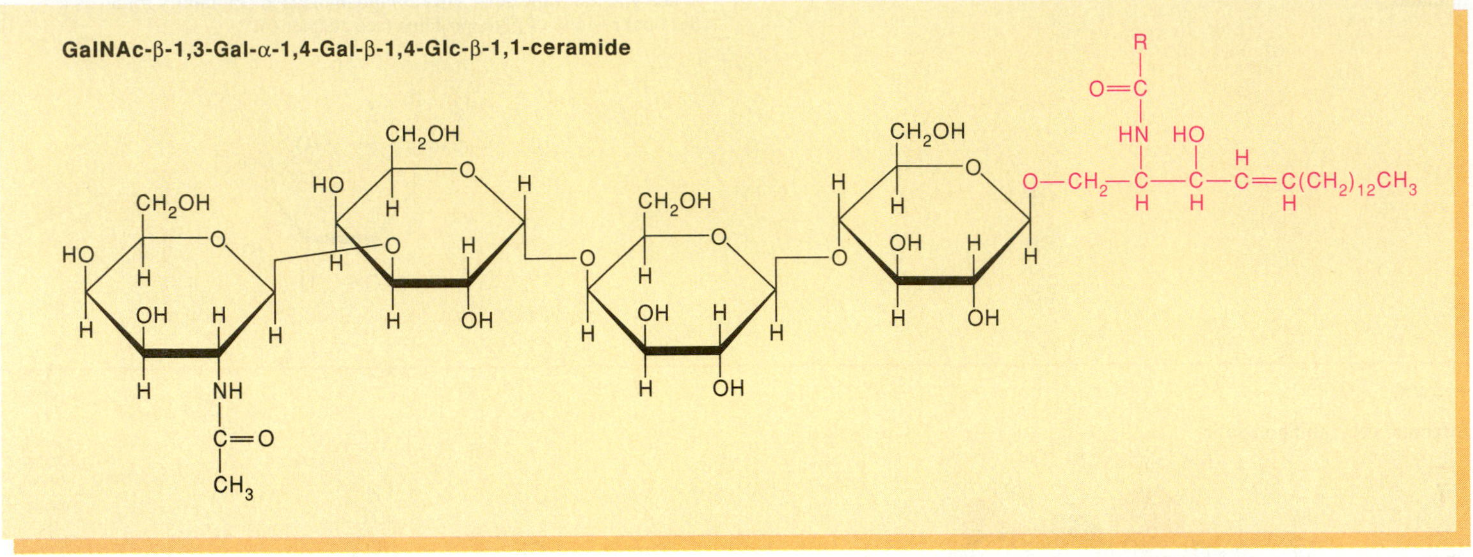

GalNAc-β-1,3-Gal-α-1,4-Gal-β-1,4-Glc-β-1,1-ceramide

Figure 7.12

Structure of Tay-Sachs ganglioside (GM_2).

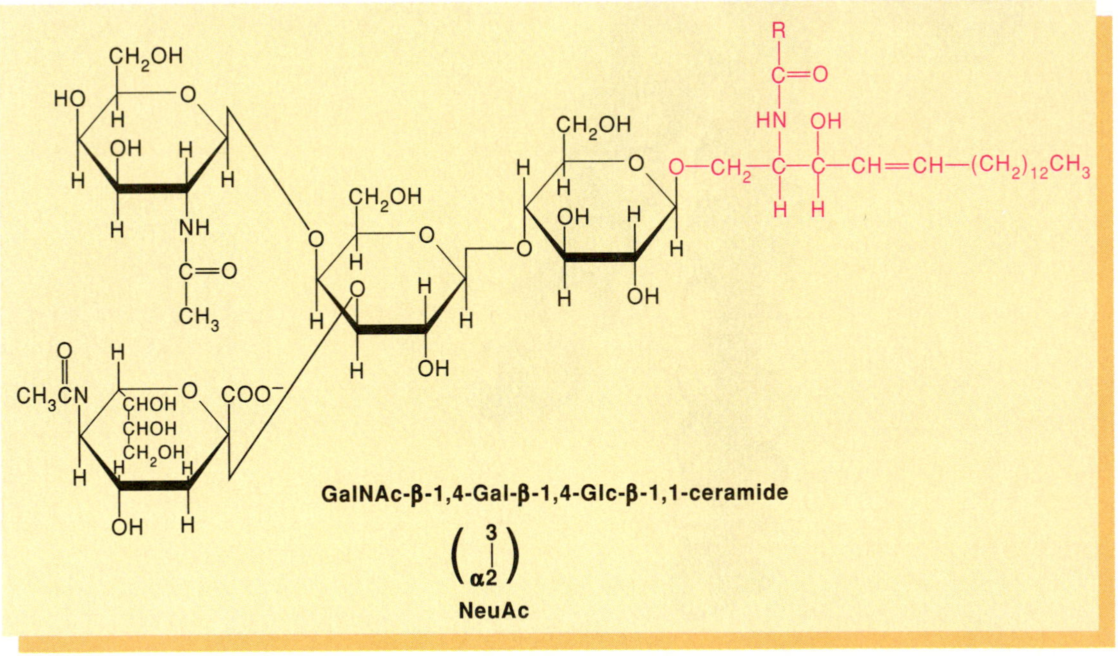

GalNAc-β-1,4-Gal-β-1,4-Glc-β-1,1-ceramide
(3 | α2) NeuAc

Sphingolipids are important components of the myelin sheath, a multilayered membranous structure that protects and insulates nerve fibers (see chapter 35). The lipids in human myelin contain 5% sphingomyelin (the original source of this lipid, as the name implies) and 15% galactosylceramide (galactocerebroside) (fig. 7.13). In addition, a sulfate derivative, 3′-sulfate-galactosylceramide, makes up 5% of the myelin lipid. The sphingolipids also are found in blood plasma as components of lipoproteins, primarily low-density lipoproteins, which are discussed in chapter 23.

Cholesterol Is a Steroid Found in Eukaryotic Membranes

Cholesterol is the most prominent member of the steroid family of lipids. Probably best known for its association with cardiovascular disease, it is an important structural component in some eukaryotic membranes, but is generally absent in most bacterial membranes.

Figure 7.13

Structure of galactosylceramide. This lipid comprises 15% of human myelin lipid.

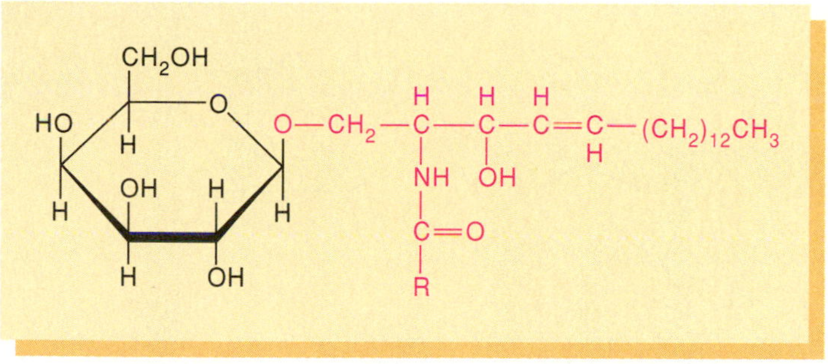

Figure 7.14

Structures of phenanthrene and perhydrocyclopentanophenanthrene.

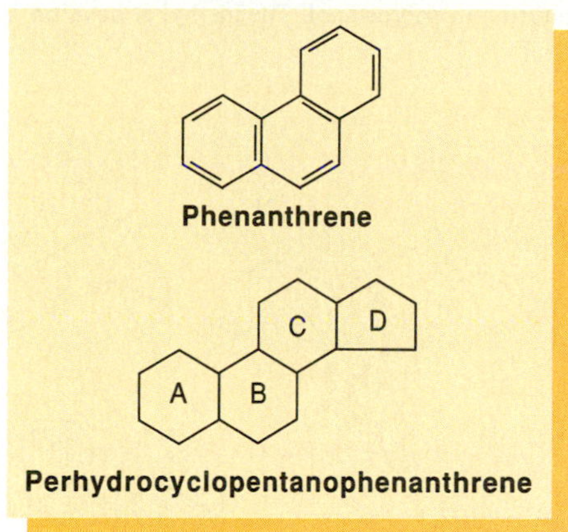

Figure 7.15

Structure of cholesterol, in three different views. The conventional projection is shown at the top center. The more realistic space-filling and conformational models are shown at the lower left and the lower right, respectively.

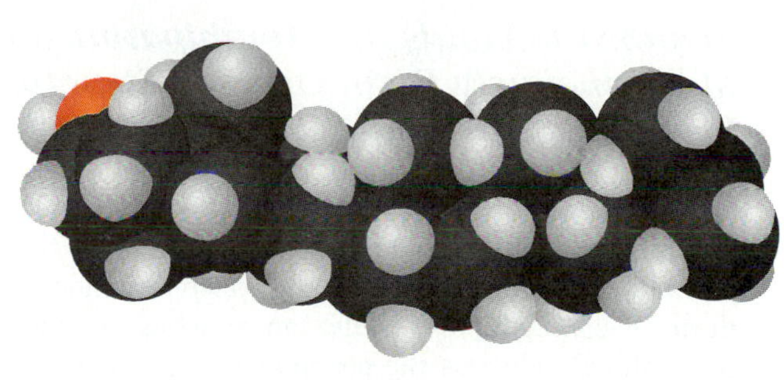

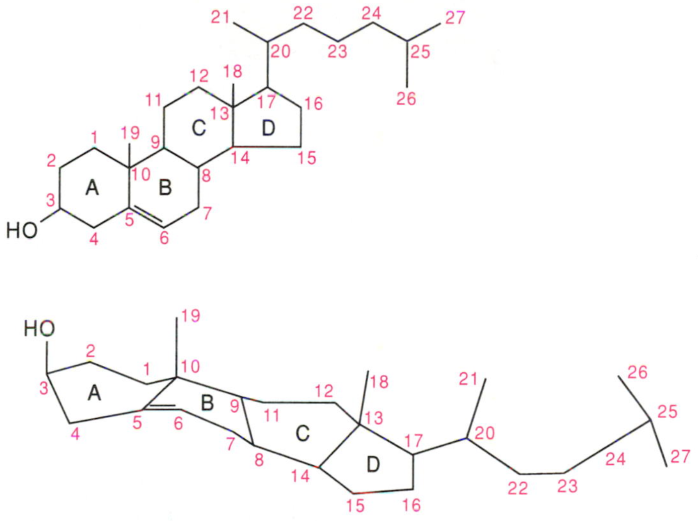

Cholesterol, like other steroids, is a derivative of the tetracyclic hydrocarbon perhydrocyclopentanophenanthrene (fig. 7.14). The four rings are identified by the first four letters of the alphabet, and the carbons are numbered in the sequence shown in figure 7.15. In addition to the basic ring structure, cholesterol contains a hydroxyl group at C-3, an aliphatic chain at C-17, methyl groups at C-10 and C-13, and a Δ^5 double bond.

The cyclohexane rings of steroids can adopt either the chair or boat conformation. The chair conformation is more stable and is the preferred conformation of steroids. The conformations of cholestanol and coprostanol, the two saturated derivatives of cholesterol, are shown in figure 7.16. The A and B rings can be joined in a *trans* configuration, as in cholestanol, or in a *cis* configuration, as in coprostanol. As you can see, the spatial orientation of the A and B rings of these two stereoisomers is very different.

Although it is important that we recognize the three-dimensional structure of steroids, such structural representations are too cumbersome for most uses in biochemistry. Hence a configurational convention for steroids has been adopted in which structural formulas are more easily drawn and recognized. The substituents of the steroid rings are related to the CH_3 group at position 10, which by definition projects above the plane of the rings. This methyl, which is said to be a β substituent, is indicated in the structural formulas by a solid line (—). Similarly, other groups that are above the plane of the rings are referred to as β. Those substituents below the plane of the rings are called α and are indicated in structural formulas by a dashed line (---). Examples of α and β substituents are shown in figure 7.16.

Figure 7.16

Conformational and conventional structures of cholestanol and coprostanol. The A and B rings are joined in a *trans* configuration in cholestanol and in a *cis* configuration in coprostanol. The methyl at position 10 is located above the plane of the rings and is said to be in the ß orientation. Other substituents are labeled ß or α, depending on whether they are above or below the planes of the rings.

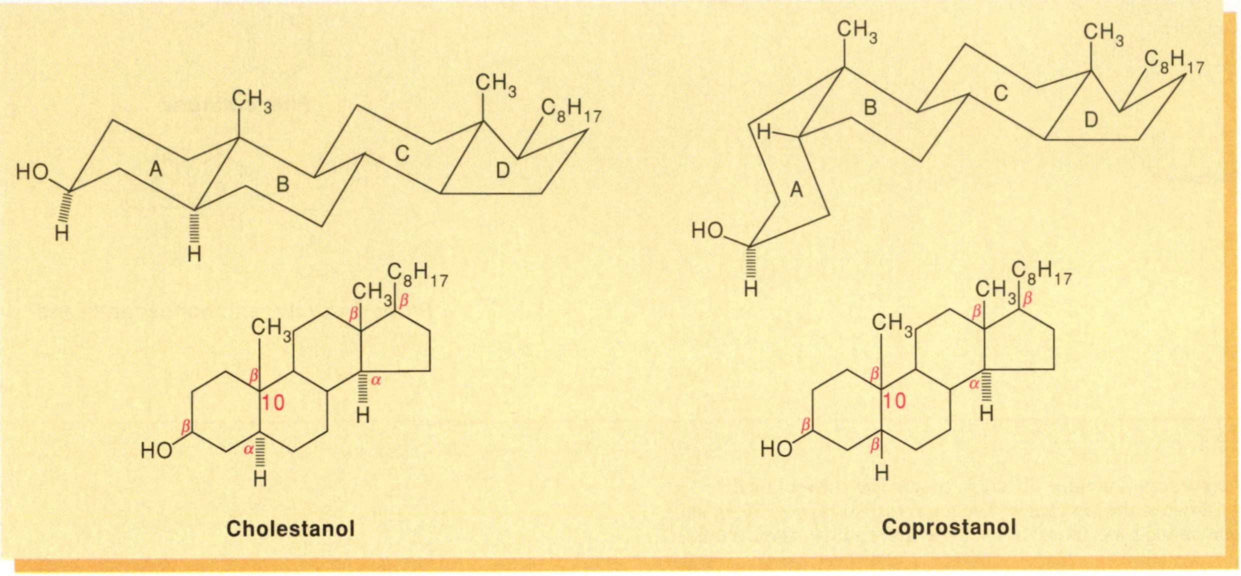

Cholestanol

Coprostanol

Figure 7.17

Structures formed by phospholipids in a cross section of aqueous solution. Each molecule is depicted schematically as a polar head group (●) attached to one or two fatty acyl hydrocarbon chains. The monolayer at the air–water interface is the first to form. When this interface has become saturated, further phospholipid forms bilayer vesicles, or in the case of lysophospholipids (one fatty acyl chain), micelles.

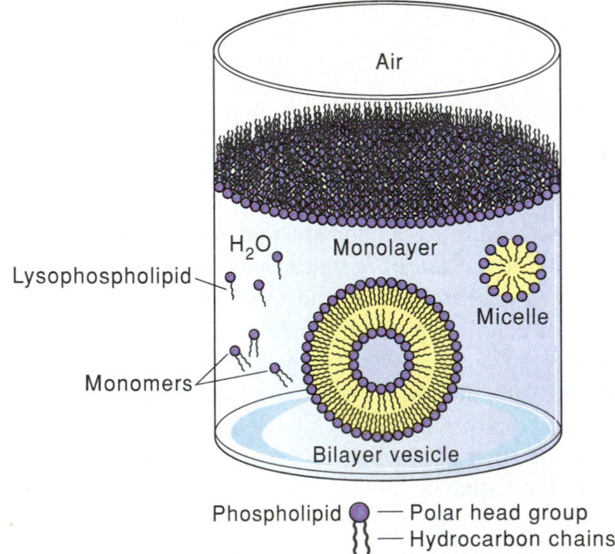

Membrane Lipids Are Amphipathic and Spontaneously Form Ordered Structures

Membrane lipids as a group are amphipathic ("having dual sympathy") molecules because they have both polar and nonpolar portions. The polar head groups of the phosphoglycerides and sphingolipids (e.g., the phosphate group plus the X-substituent of phosphoglycerides or the phosphocholine or carbohydrate of sphingolipids) prefer an aqueous environment (are hydrophilic), whereas the nonpolar (hydrophobic) acyl substituents ("tails") are excluded from aqueous environments. Even cholesterol has hydrophobic and hydrophilic "sides," although it differs altogether in chemical structure from the fatty-acid–containing lipids. It is this amphipathic property that causes most phospholipids to arrange spontaneously into ordered structures when suspended in an aqueous environment.

When phospholipid is added to water, very few lipid molecules exist freely in solution as monomers because of the large hydrophobic surface of the molecule. Instead, a "film" of phospholipid first forms on the water–air interface. Physical studies have shown that this film is a monolayer of phospholipid arranged such that the polar head groups are in contact with water, while the hydrocarbon tails extend up into the air phase (fig. 7.17). When more phospholipid is added to the solution, saturating the air–water interface, other assemblages of phospholipids are formed including bilayers, the structure preferred by most naturally occurring phospholipids, and micelles, which are usually formed in substantial quantity only by the lysophospholipids (see fig. 7.17). Both of these structures maximize hydrophobic interactions between the fatty acyl chains, effectively excluding water from their vicinity, and allow the polar head groups to interact with water molecules. Monolayers, bilayers, and micelles are the favored forms of the various phospholipids in aqueous solution because their formation results in an increase in entropy (positive ΔS), which is due to the fact that water molecules need not order themselves around the hydrophobic hydrocarbon tails of the phospholipid monomer (see fig. 1.10).

As shown in figure 7.17, phospholipid bilayers in aqueous solution are actually spherical "bubbles" or vesicles with water inside and out. This structure is favored over a planar bilayer because exposed hydrocarbon tails, which would occur around the periphery of a planar sheet of phospholipid, are not present. In vesicular structures, no hydrophobic groups need to be exposed to water molecules. Most naturally occurring phospholipids prefer to form vesicular bilayers instead of micelles in water solution because more efficient packing of the molecules can take place in the bilayer vesicle. Lysophospholipids, as well as free fatty acids and detergents (which we will discuss later), form micelles more readily than bilayers because of their geometry, which includes a smaller hydrophobic surface area relative to the diacyl-phospholipids.

The ability of phospholipids and glycolipids to spontaneously form these ordered structures is the basis for their role as the major components and structural determinants of biological membranes. Because lipid bilayers, with their hydrophobic interiors, are relatively impermeable to most hydrophilic molecules, some specific functions of membranes, including the transmembrane transport of hydrophilic molecules, are usually carried out instead by the other major membrane components, proteins. In the remainder of this chapter, we will consider the contributions of lipids and proteins to the structures and functions of biological membranes.

The Diversity of Biological Membranes

Typically, a biological membrane contains lipid, protein, and carbohydrate in ratios varying with the source of the membrane (table 7.5). Nearly always, the carbohydrate is covalently associated with protein (glycoproteins) or with lipid (glycolipids and lipopolysaccharides). Thus the membrane can be thought of as a lipid-protein matrix in which specific functions are carried out by proteins, while the permeability barrier and the structural integrity of the membrane are provided by lipids.

The one membrane structure common to all cells is the plasma membrane. This membrane encapsulates the cytoplasm and creates internal compartments in which essential functions are carried out. In addition to its role as a physical barrier that maintains the integrity of the cell, the plasma membrane provides functions necessary for the survival of a cell, including exclusion of harmful substances, acquisition of nutrients and energy sources, disposal of unusable and toxic materials, reproduction, locomotion, and interaction with components in the environment. All these functions require coordination both for short-range processes, such as sensation, and for long-range processes, such as growth and differentiation. If we look at a typical plasma membrane in the electron microscope, we see a trilaminar (three-layered) structure: two dark lines, representing the polar surfaces of the lipid bilayer, separated by a light region corresponding to the hydrophobic interior of the bilayer (fig. 7.18).

Table 7.5
Chemical Compositions of Some Cell Membranes

Membrane	Protein (%)	Lipid (%)	Carbohydrate (%)
Myelin	18	79	3
Human erythrocyte plasma membrane	49	43	8
Amoeba plasma membrane	54	42	4
Mycoplasma cell membrane	58	37	1.5
Halobacterium purple membrane	75	25	0

Source: G. Guidotti, "Membrane Proteins" in *Annual Review of Biochemistry,* 41:731, 1972. Copyright © 1972 Annual Reviews Inc., Palo Alto, Calif.

Figure 7.18

Cross section of microvilli of cat intestinal epithelial cells, showing the trilaminar (three-legged) structure of the cytoplasmic membranes (165,000X). (Courtesy of Dr. S. Ito.)

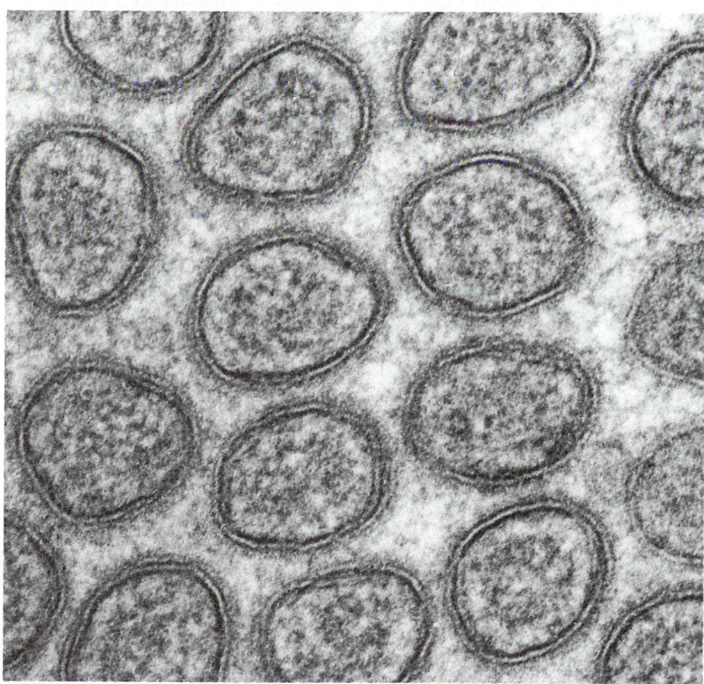

Eukaryotic cells contain numerous organelles of widely differing structure, each of which is specialized in its function—digestion (lysosomes), respiration (mitochondria), photosynthesis (chloroplasts), secretion (endoplasmic reticulum and Golgi apparatus), or nucleic acid biosynthesis (nucleus). Each organelle is surrounded by its own specialized membrane system, which has evolved to participate in its respective function (fig. 7.19). In contrast, prokaryotic cells (bacteria) typically have all functions integrated into the plasma membrane and lack specialized intracellular organelles. These differences in cell structure do not necessarily indicate different biochemical mechanisms, but merely the presence or absence of compartments specifically designed to fulfill separate functions. In the generally larger eukaryotic cells, each process is performed in a spatially isolated domain, whereas in the smaller prokaryotic cell the processes can operate for the most part within a single compartment.

From research done over the last few decades, a number of fundamental principles have emerged that appear to apply to most membrane systems that have been studied. For this reason, we will be able to examine aspects of membrane structure and function in systems as seemingly divergent as bacteria and mammalian mitochondria. In chapters 15 and 32 we will find that in both cases, the biosynthesis of ATP is a membrane-associated process that occurs by a similar mechanism. To take another example, certain bacterial membranes contain proteins that behave in artificial membrane systems much like nerve-cell membrane channels, which are partially responsible for propagation of the nerve impulse in vertebrates (chapter 35). Thus many mechanisms responsible for complex membrane phenomena are undoubtedly used repeatedly throughout the living kingdom.

Different Membrane Structures Can Be Separated According to Their Density

As with the structure of any biological entity, in order to study the structure of biological membranes we must first isolate them in a more or less intact form from the cell. In eukaryotic cells, this problem is complicated by the existence of several different membrane systems in addition to the plasma membrane, each surrounding a specific organelle. To separate membrane fractions, we must first disrupt the plasma membrane under conditions that leave subcellular organelles intact. One common procedure involves mild homogenization in a medium in which the osmolarity is below the normal physiologic value (hypotonic medium). Another method, nitrogen cavitation, involves forcing nitrogen gas into the cells under pressure and then rapidly releasing the pressure to "explode" the cell membrane.

Organelles can be isolated from disrupted cells by differential centrifugation, which separates them on the basis of their size (fig. 7.20). Ruptured plasma membrane fragments can be purified from the same mixture by equilibrium-density-gradient (isopycnic) centrifugation because of their low density (high lipid content) relative to intact organelles (table 7.6, column 4). This technique relies on centrifuging the sample into a preformed gradient of a solute, such as sucrose. When equilibrium is reached, each type of membrane or organelle is found in the region of the gradient corresponding to its own density (see fig. 7.20). Gradients of synthetic sucrose polymers (Ficoll) or colloidal silica particles (Percoll) also are used in these separations because of their inertness, ability to form stable gradients, and impermeability to biological membranes.

The properties of isolated rat liver organelles are summarized in table 7.6. The entries in column 2 provide some idea of the relative proportions of these organelles in the mammalian liver. For example, mitochondria represent 25% of the total cell protein, while lysosomes comprise about 2%. Interestingly, the plasma membrane, which completely surrounds the cell, also represents only 2% of the total protein. In addition to the soluble protein that is membrane-free, there is considerable soluble protein within the organelles, leaving somewhat less than 50% of the total cell protein in the membrane-associated form.

Figure 7.19

Electron micrograph of a cell from the rat pancreas, showing several different intracellular organelles. (PM = plasma membrane; NE = nuclear envelope; Nu = nucleolus; M = mitochondrion; ER = endoplasmic reticulum; Go = Golgi apparatus; arrows show pore complexes in the nuclear envelope; 24,000X.) (From S. L. Wolfe, *Biology of the Cell,* 2d ed., © Copyright 1981, Wadsworth Publishing Co.)

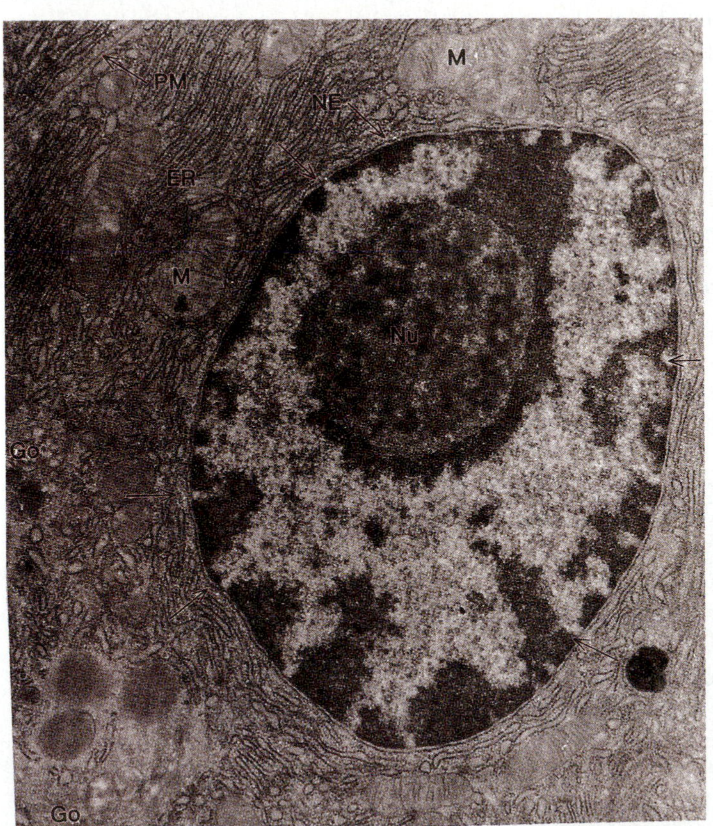

In column 3 of table 7.6, the relative sizes of the organelles are indicated. Note that the sedimentation behavior of an organelle in a sucrose density gradient (column 4) does not correlate with its size, but rather with its density, which is determined by its chemical composition (see fig. 7.20). Nucleic acid ($\rho \sim 1.7$) is more dense than protein ($\rho = 1.25$), and protein is more dense than lipid ($\rho \approx 0.9$–1.1). These facts account for the relatively high density of nuclei and the low density of the Golgi apparatus, which has a high lipid content.

Since each organelle has a specific function, it must also possess a unique complement of enzymes. This prediction is verified by the subcellular localization of numerous enzymes (see table 7.6, column 5), and these specific associations have greatly facilitated the assay and isolation of organelles from eukaryotic cells.

Once the subcellular organelles have been separated, their membranes can be isolated. For those organelles enclosed by a single membrane, treatment in hypotonic buffer (osmotic shock) followed by centrifugal separation of the membrane fragments from the intraorganellar soluble proteins allows us to study membrane composition. Nuclei and mitochondria, however, possess two membranes (inner and outer), and these must be separated before their individual chemical and physical properties can be studied. In these cases, selective solubilization of the outer membrane can be obtained by treatment with appropriate detergents (see later discussion), allowing purification of intact inner membranes. Procedures such as these have made it possible to analyze in detail the lipid and protein contents of organellar membranes as summarized in table 7.7. They have also provided experimental systems in which to study the structures and functions of each different membrane system.

Figure 7.20

Comparison of differential centrifugation (left), which separates on the basis of size, and isopycnic centrifugation (right), which separates on the basis of density. ρ is the density in grams per milliliter.

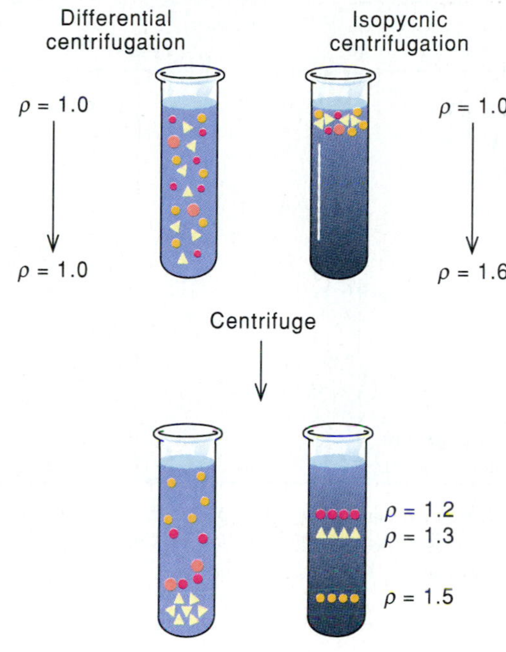

Table 7.6
Properties of Rat Liver Organelles

Organelle	Percent of Cell Protein	Diameter (μm)	Equilibrium Density in Sucrose (g/ml)	Organelle-Specific Enzyme Marker
Liver cell	100	20	1.20	—
Nuclei	15	5–10	1.32	DNA polymerase
Golgi apparatus	2	2	1.10	Glycosyl transferases
Mitochondria	25	1	1.20	Monoamine oxidase (outer membrane); cytochrome *c* (inner membrane)
Lysosomes	2	0.5	1.20	Acid phosphatase
Endoplasmic reticular vesicles	20	0.1	1.15	Cytochrome b_5 reductase and cytochrome b_5; glucose-6-phosphatase
Cytoplasmic membrane	2	—	1.15	Na^+-K^+ ATPase; viral receptors
Soluble protein	30	<0.01	—	—

Source: From M. H. Saier, Jr., and C. D. Stiles, *Molecular Dynamics in Biological Membranes,* Heidelberg Science Library, Vol. 22, Copyright © 1975, Springer Verlag, New York, N.Y. Reprinted with permission.

Table 7.7
Protein and Lipid Content of Organellar Membranes

Membrane	Approximate Protein/Lipid Ratio (wt/wt)	Approximate Cholesterol/Other Lipids (Molar Ratio)
Golgi apparatus	0.7	0.08
Liver plasma membrane	1.0	0.40
Endoplasmic reticulum	1.0	0.06
Mitochondrial outer membrane	1.0	0.05
Mitochondrial inner membrane	3.0	0.03
Nuclear membrane	3.0	0.11
Lysosomal membrane	3.0	0.16

Source: M. H. Saier, Jr., and C. D. Stiles, *Molecular Dynamics in Biological Membranes,* Heidelberg Science Library, Vol. 22, Copyright © 1975, Springer Verlag, New York, N.Y.

Figure 7.21

Electron micrographs of sections through the surface layers of (*a*) a Gram positive and (*b*) a Gram negative bacterium. (cm = cytoplasmic membrane; om = outer membrane; pg = peptidoglycan; ta = teichoic acid.) Note the thick cell wall in (*a*), compared with the distinct inner and outer trilaminar membranes separated by a thin peptidoglycan layer in (*b*). (Courtesy of J. Stolz; 150,000×.)

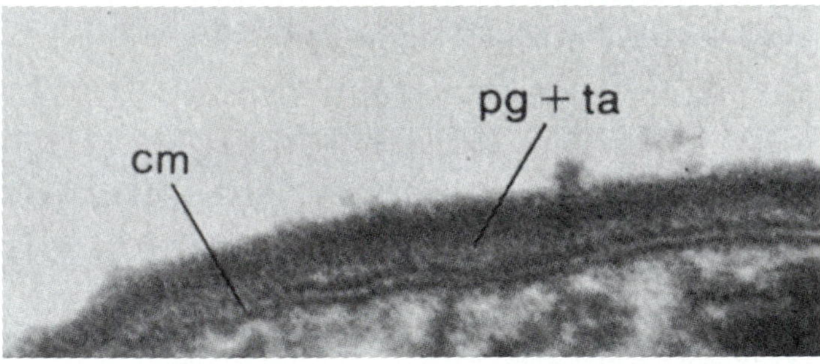

(a)

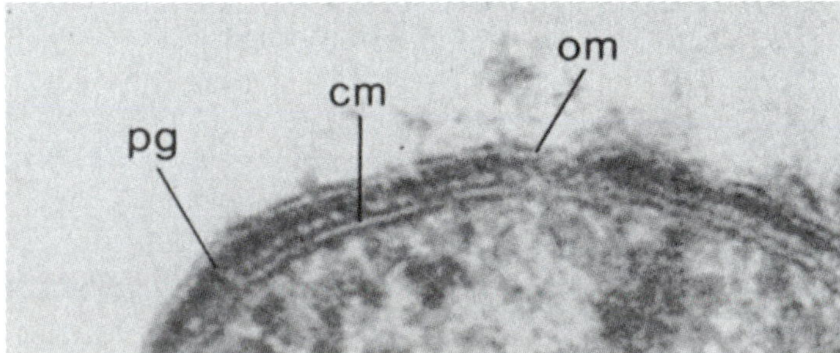

(b)

Bacteria Are Surrounded by One or Two Membranes and a Cell Wall

In contrast to animal cells, most prokaryotic cells are surrounded by a cell wall, which allows bacteria to live in a hypotonic environment without bursting and confers upon these cells their characteristic shape (rod, sphere, or spiral). In 1884, Christian Gram discovered that bacteria could be divided into those that retained a crystal violet-iodine dye complex after washing with alcohol (Gram positive) and those that did not (Gram negative). Even today, the Gram stain reaction is a useful tool in classifying bacteria, and this difference in staining has been found to correlate with a fundamental difference in cell-wall structure between Gram positive and Gram negative cells (fig. 7.21). Gram positive cells are surrounded by a cytoplasmic membrane and a thick cell wall consisting of a sugar–amino acid heteropolymer, or peptidoglycan (see chapter 6), and polyol phosphate polymers called teichoic acids (fig. 7.22*a*). Gram negative bacteria have a much thinner cell wall, consisting mainly of peptidoglycan and associated proteins; this cell wall is surrounded by a second, outer membrane composed of lipid, protein, and lipopolysaccharide (see fig. 7.22*b;* also see fig. 6.29). The biosynthesis of peptidoglycan and lipopolysaccharide is discussed in chapter 21. In Gram negative bacteria, the space between the inner and outer membranes, called periplasmic space, also contains proteins that have a variety of functions.

The two cell layers of Gram negative bacteria can be separated by treatment of the cells with lysozyme (which hydrolyzes peptidoglycan) and EDTA (which destabilizes the outer membrane) in isoosmotic sucrose solutions (fig. 7.23). Periplasmic proteins that are released by this first step can be separated by sedimenting the resulting spheroplasts, which have lost any nonspherical shape characteristic of the original cell because their cell wall has been digested. Then the spheroplasts can be treated with high-frequency sound (sonication) to rupture both the inner and the outer membranes, which quickly reseal into smaller spherical, closed vesicles (see fig. 7.23). Because of their higher carbohydrate content, outer membrane vesicles have a higher density than ones derived from the inner membrane and thus can be separated from them in a sucrose density gradient.

By these techniques, workers have discovered that electron-transport chains, ATP-synthesizing enzymes, many transport proteins, and other enzymes are located on the inner membrane of Gram negative bacteria, while the outer membrane harbors receptors for bacteriophage and bacteriocins, certain other transport proteins, and various phospholipases. The periplasmic space contains hydrolytic enzymes as well as nutrient-binding proteins involved in transmembrane transport and chemotaxis (chapter 32).

The organization of proteins in Gram positive bacteria is usually much simpler because these cells are surrounded by a single membrane. Thus soluble and cytoplasmic membrane proteins carry out functions similar to those of Gram negative bacteria, while macromolecular hydrolases are exported into the extracellular medium, where they function to scavenge nutrients from the environment.

Figure 7.22

Structures of some bacterial cell envelope constituents. (*a*) Some teichoic acids of Gram positive bacteria: (i) *Lactobacillus casei* (R = D-alanine); (ii) *Actinomyces antibioticus* (R = D-alanine); (iii) *Staphylococcus lactis* (R = D-alanine); (iv) *Bacillus subtilis* (R = glucose). Compounds (i)–(iii) are composed of repeating glycerol units, while (iv) is a ribitol teichoic acid to which D-alanine may be attached at either position 3 or 4 of the pentitol. (Adapted from R. Y. Stanier, E. A. Adelberg, and J. L. Ingraham, *The Microbial World,* 4th ed., Prentice-Hall, Englewood Cliffs, N.J., 1976. Used with permission.) (*b*) Schematic illustration of the structure of lipopolysaccharide in the outer membrane of *Salmonella typhimurium.* (EtN = ethanolamine; KDO = 2-keto-3-deoxyoctonic acid; Hep = L-glycero-D-mannoheptose; Abe = abequose; Man = mannose; Rha = rhamnose.) (Source: H. Nikaido, "Biosynthesis and assembly of lipopolysaccharide" in *Bacterial Membranes and Walls,* edited by L. Leive. Copyright © 1973 Dekker, New York, N.Y.)

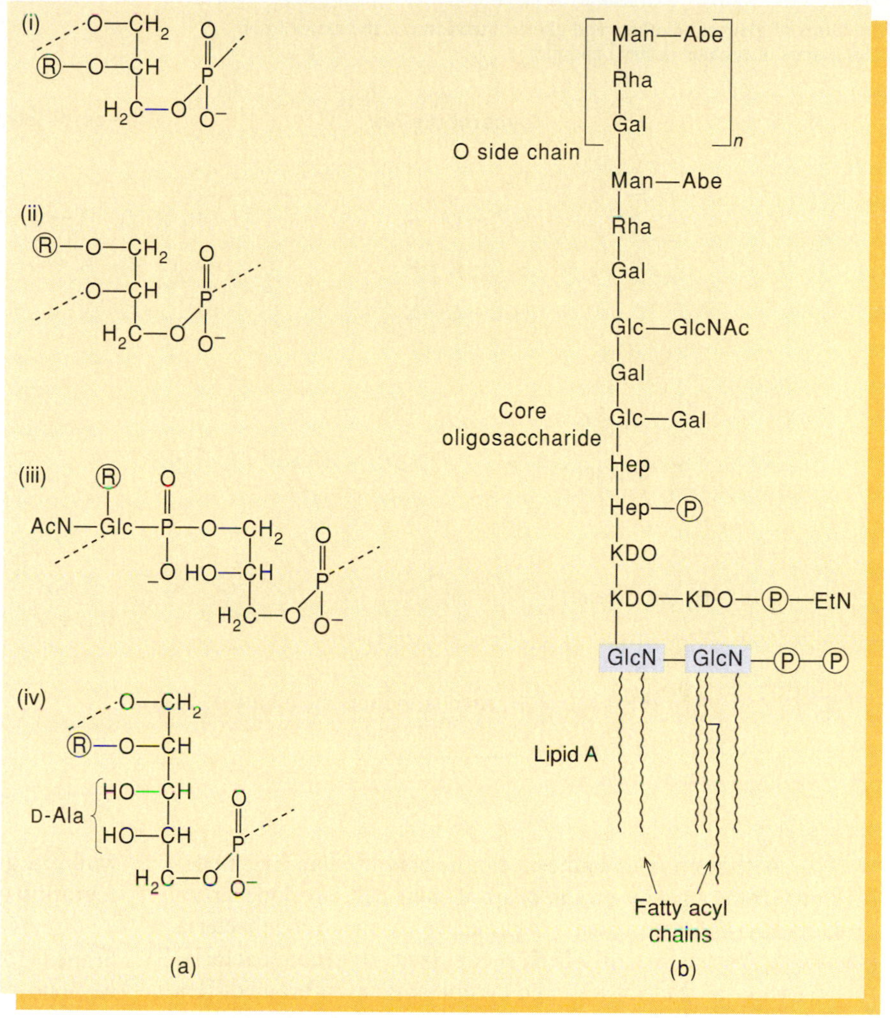

Figure 7.23

Separation of the periplasmic proteins and inner and outer membranes of a Gram negative bacterium. The periplasmic proteins are those proteins that are concentrated in the space between the inner and outer membranes. By treating intact cells with lysozyme and EDTA in an isotonic solution, the periplasmic proteins are released, leaving the spheroplasts. The proteins are removed by sedimentation; then the spheroplasts are suspended in a buffered saline solution and sonicated. This treatment ruptures the outer and inner membranes into smaller spherical closed vesicles. These may be separated by centrifugation on a sucrose density gradient. The outer membrane vesicles sediment to a lower point in the centrifuge tube because of their higher density.

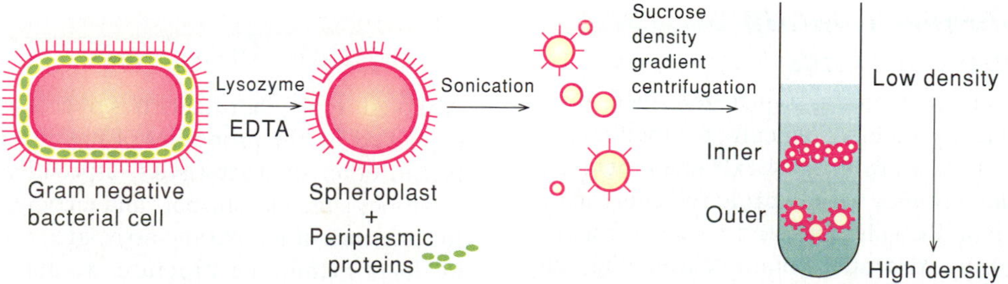

Figure 7.24

Structures of glycerol diethers and glycerol tetraethers, the major lipid components of archaebacterial membranes.

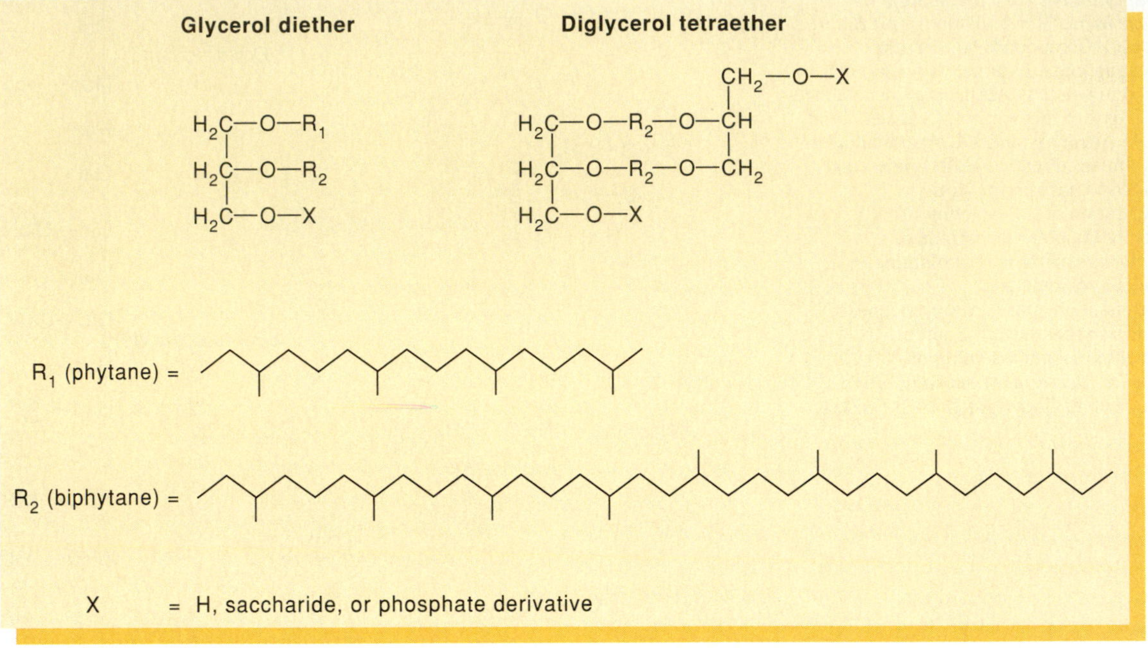

Glycerol diether

$$\begin{array}{l} H_2C{-}O{-}R_1 \\ \quad | \\ H_2C{-}O{-}R_2 \\ \quad | \\ H_2C{-}O{-}X \end{array}$$

Diglycerol tetraether

$$\begin{array}{l} \qquad\qquad\qquad\quad CH_2{-}O{-}X \\ \qquad\qquad\qquad\quad | \\ H_2C{-}O{-}R_2{-}O{-}CH \\ \quad | \qquad\qquad\qquad | \\ H_2C{-}O{-}R_2{-}O{-}CH_2 \\ \quad | \\ H_2C{-}O{-}X \end{array}$$

R_1 (phytane) =

R_2 (biphytane) =

X = H, saccharide, or phosphate derivative

Although most bacteria can be classified as Gram positive or Gram negative on the basis of their cell envelope structure, another interesting class of prokaryotes, the archaebacteria (see fig. 1.25) has recently been recognized. In archaebacteria, the structure of the cell wall and cytoplasmic membrane differs considerably from that in most prokaryotes. The cell wall may consist of a peptidoglycanlike structure called pseudomurein, but more often it is composed of a complex polysaccharide or of glycoproteins or lipoproteins. The single membrane surrounding the archaebacterial cell is composed of unusual lipids that are mainly glycerol diethers or glycerol tetraethers. In these compounds, the first two carbon atoms of glycerol are attached to long-chain polyisoprenoid alcohols called phytane or biphytane (also see chapter 22) via ether linkages, while the X-substituent on the third carbon may be hydrogen, sugar, or a phosphate derivative (fig. 7.24).

Biological Membranes Contain a Complex Mixture of Lipids

Once isolated by any of the procedures just described, membrane fractions can be analyzed biochemically for their content of lipids and proteins. Lipids are usually extracted from proteins by treating the membranes with organic solvents such as chloroform and methanol. Phospholipids are resolved from uncharged (neutral) lipids by adsorption column chromatography (usually the adsorbant is silicic acid) or by adsorption thin-layer chromatography. The various classes of phospholipids are usually separated on a small scale by thin-layer chromatography, while large-scale preparations involve either column chromatography or high-pressure liquid chromatography (HPLC).

Table 7.8 compares the lipid compositions of membranes from a number of biological sources. Several generalizations can be drawn from the data in this table. First, phosphatidylcholine is the chief phospholipid found in membranes of animal cells, while phosphatidylethanolamine predominates in bacteria. Second, in addition to cholesterol, both sphingomyelin and glycolipids (except for lipopolysaccharides) are usually absent from prokaryotic membranes. Despite their widely varying lipid compositions, most of the membranes represented in the table have a characteristic trilaminar appearance in the electron microscope (see fig. 7.18). This similarity reflects the fact that the major structural components of biological membranes, the phospholipids, all form bilayer structures regardless of their exact composition.

Membrane Proteins

Although some characteristics of biological membranes can be explained by the properties of membrane lipids in aqueous solution, other characteristics, especially the ability to perform functions such as transport and enzymatic activities, depend on the presence of membrane-associated proteins. Therefore, in any model of membrane structure, we must also consider the properties of proteins found in biological membranes.

Table 7.8
Lipid Compositions of Membrane Preparations

Source	Lipid Composition (lipid percent)[a]									
	Cholesterol	**PC**	**SM**	**PE**	**PI**	**PS**	**PG**	**DPG**	**PA**	**Glycolipids**
Rat liver										
Cytoplasmic membrane	30.0	18	14.0	11	4.0	9.0	—	—	1.0	—
Endoplasmic reticulum (rough)	6.0	55	3.0	16	8.0	3.0	—	—	—	—
Endoplasmic reticulum (smooth)	10.0	55	12.0	21	6.7	—	—	1.9	—	—
Mitochondria (inner)	3.0	45	2.5	25	6.0	1.0	2.0	18.0	0.7	—
Mitochondria (outer)	5.0	50	5.0	23	13.0	2.0	2.5	3.5	1.3	—
Nuclear membrane	10.0	55	3.0	20	7.0	3.0	—	—	1.0	—
Golgi	7.5	40	10.0	15	6.0	3.5	—	—	—	—
Lysosomes	14.0	25	24.0	13	7.0	—	—	5.0	—	—
Rat brain										
Myelin	22.0	11	6.0	14	—	7.0	—	—	—	21
Synaptosome	20.0	24	3.5	20	2.0	8.0	—	—	1.0	—
Rat erythrocyte	24.0	31	8.5	15	2.2	7.0	—	—	0.1	3
Rat rod cell (outer segment)	3.0	41	—	37	2.0	13.0	—	—	—	—
E. coli cytoplasmic membrane	0	0	—	80	—	—	15.0	5.0	—	—
Bacillus megaterium cytoplasmic membrane	0	0	—	69	—	—	30.0	1.0	—	Trace

[a]PC = phosphatidylcholine; SM = sphingomyelin; PE = phosphatidylethanolamine; PI = phosphatidylinositol; PS = phosphatidylserine; PG = phosphatidylglycerol; DPG = diphosphatidylglycerol (cardiolipin); PA = phosphatidic acid.

Source: M. K. Jain and R. C. Wagner, ***Introduction to Biological Membranes.*** © 1980 John Wiley & Sons, Inc., New York; and from J. E. Rothman and E. P. Kennedy, "Asymmetrical distribution of phospholipids in the membrane of ***Bacillus megaterium***" in ***Journal of Molecular Biology.*** 110:603, 1977. Copyright © 1977, Academic Press Ltd., London England.

Membrane Proteins Are Integral or Peripheral

Membrane proteins are classified as peripheral or integral. Peripheral proteins are probably bound to the membrane as a result of specific interactions with exposed, hydrophilic portions of integral membrane proteins. As a consequence they can be dissociated from isolated membranes by agents that disrupt ionic or hydrogen bonds, such as high salt, EDTA (which chelates divalent cations), or urea. In contrast, integral membrane proteins appear to be deeply embedded in the membrane. They can be released from the membrane only by disrupting the hydrophobic interactions of membrane lipids with organic solvents or detergents. Significant hydrophobic interactions with membrane lipids and proteins probably are responsible for the interaction properties of integral membrane proteins. Figure 7.25 illustrates the differences between integral and peripheral membrane proteins.

To purify integral membrane proteins for study, the proteins first must be dissociated from the membrane matrix, a step that is usually accomplished with detergents. Detergents are amphipathic molecules that disrupt membranes by intercalation into the membrane matrix and solubilization of the component lipids and proteins. Examples of detergents that are

Figure 7.25

Schematic illustration of integral and peripheral membrane proteins. Peripheral membrane proteins may be removed by mild reagents without disruption of the membrane structure. Integral membrane proteins can be released only if the membranes are disrupted.

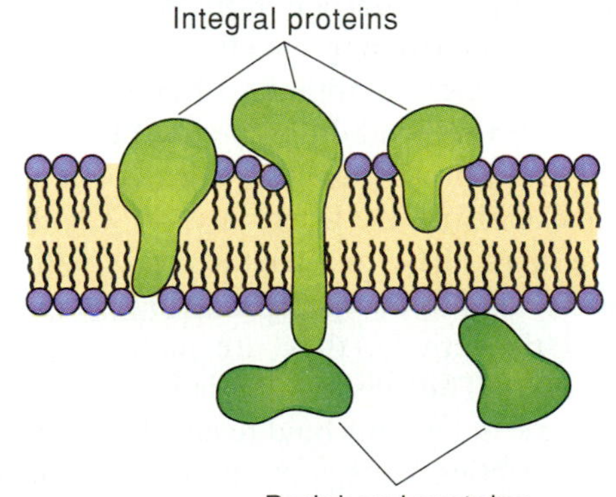

Figure 7.26

Structure of sodium deoxycholate, a steroid that is a naturally occurring ionic detergent.

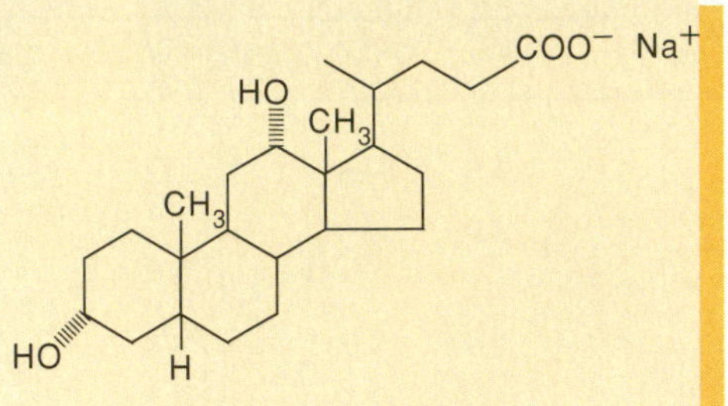

Sodium deoxycholate

Figure 7.27

Two synthetically derived nonionic detergents. These detergents are useful for dissolving membranes.

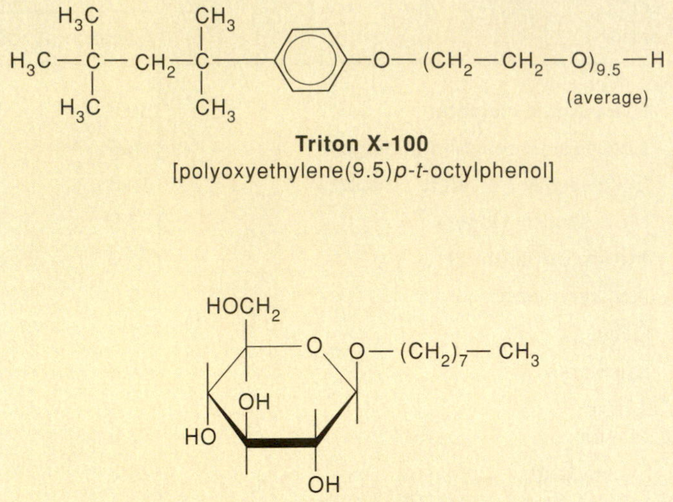

Triton X-100
[polyoxyethylene(9.5)*p*-*t*-octylphenol]

Octylglucoside
(octyl-β-D-glucopyranoside)

natural products include lysolecithin and sodium deoxycholate, a steroid bile salt (fig. 7.26). Both are ionic detergents. Synthetic detergents include Triton X-100 and octylglucoside (fig. 7.27), two nonionic detergents, and the ionic compounds cetyl trimethylammonium bromide and sodium dodecyl sulfate (SDS) (fig. 7.28), which is also an extremely effective protein denaturant.

Many detergents, including the nonionic ones and lysolecithin, dissolve membranes by forming detergent-lipid and detergent-lipid-protein mixed micelles (fig. 7.29). These detergents prefer to form micelles rather than bilayers because of their particular geometries. Thus an excess of these compounds, when added to a membrane suspension, will tend to shift the equilibrium of all amphipathic molecules in the mixture from bilayer to mixed micelle (see fig. 7.29).

Each detergent has a characteristic critical micellar concentration (CMC) above which it exists in aqueous solution almost entirely in a micellar form, and a characteristic hydrophilic-lipophilic balance (HLB), which is defined as the ratio of the molecular weight of the hydrophilic portion of the molecule to that of the hydrophobic portion. Both these properties appear to be important in determining the effectiveness of a particular detergent in dissolving membranes, although different membrane systems often respond differently to the same detergent.

Some of the properties of detergents commonly used in membrane biochemistry are listed in table 7.9. Because ionic detergents are more likely to alter the conformation of hydrophilic portions of membrane proteins, which are often responsible for catalytic activity, they are more likely to destroy biological function. This observation is in agreement with the fact that only ionic detergents bind to and denature soluble proteins. Little interaction is usually observed between most hydrophilic proteins and nonionic detergents.

Once dissociated from the membrane with detergent, individual membrane proteins can be isolated by a variety of separation techniques, provided an assay is available for the protein of interest. Ion-exchange chromatography (chapter 5) can be useful in separating solubilized membrane proteins if the solvent is a nonionic detergent, such as Triton X-100. Gel filtration and other procedures that separate proteins on the basis of size and shape (chapter 5) are not particularly useful because inclusion of integral membrane protein in detergent-lipid micelles, which can have molecular weights of 10^5 or greater, tends to mask individual size differences. Only with detergents possessing a high CMC, such as deoxycholate and octylglucoside, are separations based on size generally possible. These difficulties have led to the search for techniques that might be especially suited to membrane protein isolation. One such technique is hydrophobic interaction chromatography, which separates proteins on the basis of their relative hydrophobicities. This technique uses insoluble supports, such as agarose or polyacrylamide, to which hydrophobic alkyl or aryl groups have been covalently attached. Proteins bound to the resin by means of hydrophobic interactions can be eluted sequentially, according to the strength of this interaction, by gradual changes in hydrophobicity, ionic strength, pH, or temperature of the eluting buffer (fig. 7.30).

The molecular weights of membrane proteins and the complexity of a specific membrane can be estimated by the technique of polyacrylamide gel electrophoresis, after the membrane has been completely solubilized in the ionic detergent SDS (chapter 5). An example of this technique for examining proteins of the human erythrocyte membrane is shown in figure 7.31. In table 7.10 you will find a comparison of some physical and chemical properties of a number of integral membrane proteins that have been purified to apparent homogeneity. Some of

Figure 7.28

Two additional synthetically derived ionic detergents that can dissolve membranes. Sodium dodecyl sulfate is also an extremely effective protein denaturant and is commonly used in protein gel electrophoresis (see chapter 5).

$$H_3C-(CH_2)_{15}-N^{+}(CH_3)_2-CH_3Br^{-}$$

Cetyl trimethylammonium bromide

$$H_3C-(CH_2)_{11}-O-S(=O)_2-O^{-}\ Na^{+}$$

Sodium dodecyl sulfate (SDS)

Figure 7.29

Detergent solubilization of biological membranes yields detergent-lipid-protein mixed micelles. The detergent disrupts the membrane and makes a complex with the hydrophobic portion of the integral membrane protein and the membrane lipid.

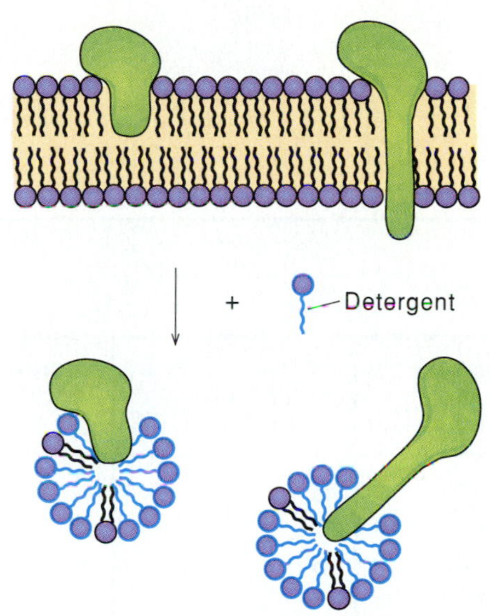

Figure 7.30

Hydrophobic interaction chromatography. Proteins bound by hydrophobic interactions to insoluble alkyl agarose beads can be differentially removed by changes in ionic strength, hydrophobicity, pH, or temperature. General chromatographic procedures are discussed in chapter 5.

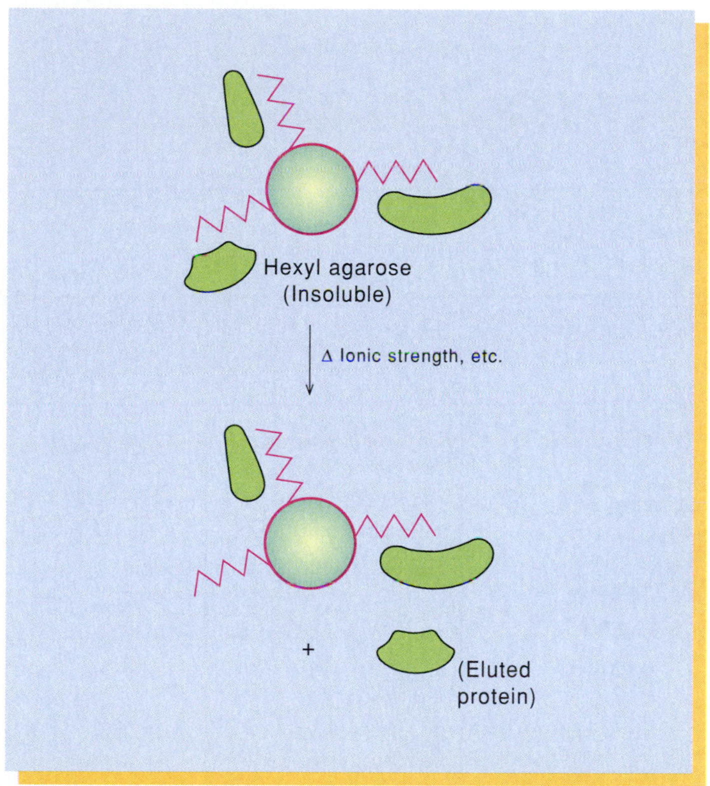

Table 7.9
Properties of Some Detergents

Detergent	Monomer M_r	CMC (mM)[a]	Micellar M_r (average)[a]
Nonionic			
Triton X-100	625	0.24	90,000
Octyl-β-D-glucoside	292	25	—
Ionic			
Deoxycholate (anion)	392	4	800
Lysolecithin (egg; mixture)	500–600	0.02–0.2	95,000
SDS	288	8	17,000

[a]Both CMC and micellar size are dependent on a number of factors, including ionic strength and pH. These numbers should therefore be used only for general comparisons, since each was determined under slightly different conditions.

these proteins, such as bacteriorhodopsin from *H. halobium,* and lactose permease from *E. coli,* are highly enriched in hydrophobic amino acids. This property probably reflects the proportion of the polypeptide chain embedded in the hydrophobic portion of the membrane (see the following discussion). One property, however, that seems to be common to many isolated membrane proteins is that to maintain their biologically active structures they usually require a hydrophobic environment. Therefore, techniques that remove detergent and residual phospholipid from these proteins often lead to inactivation of any enzymatic or other biological activity they possess. This inactivation is often accompanied by aggregation or precipitation.

Figure 7.31

Proteins of the human erythrocyte membrane resolved by polyacrylamide gel electrophoresis in the presence of sodium dodecyl sulfate. Protein bands were stained with the dye Coomassie brilliant blue. The bands so stained are numbered 1 to 7. (After G. Fairbanks, T. L. Steck, and D. F. H. Wallach, Electrophoretic analysis of the major polypeptides of the human erythrocyte membrane, *Biochemistry* 10:2606, 1971.) PAS-1 to PAS-4 are sialoglycoproteins, which stain heavily with a reagent that detects carbohydrate. On this electrophoretogram, the anion channel (band 3) and glycophorin A (PAS-1) are not completely resolved. (From V. T. Marchesi, H. Furthmayr, and M. Tomita, "The red cell membrane"," Reproduced with permission from the Annual Review of Biochemistry, 45:667 © 1976 by Annual Reviews, Inc."

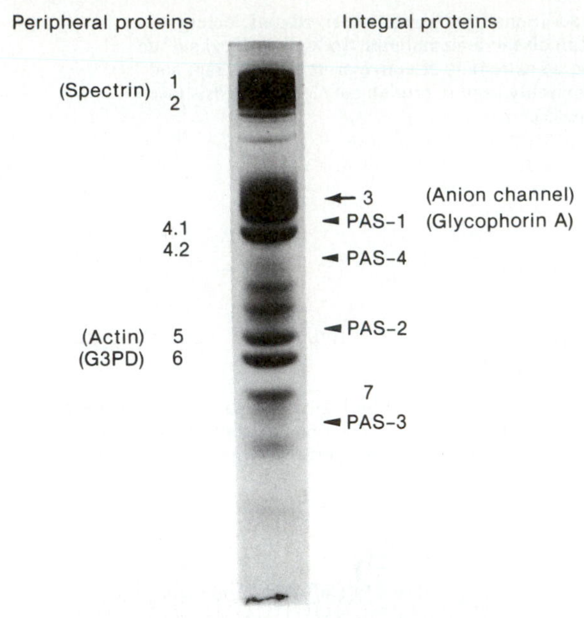

Table 7.10
Properties of Some Purified Membrane Proteins

Protein	Source	Monomer M_r	Subunit Structure	Percent Hydrophobic Amino Acids[a]	Covalent Carbohydrate
Cytochrome b_5	Liver endoplasmic reticulum	16,000	Dimer (0.4% deoxycholate)	40	−
Cytochrome b_5 reductase	Liver endoplasmic reticulum	43,000	Monomer (0.4% deoxycholate)	48	−
Anion-transport (band 3) protein	Human erythrocytes	95,000	Dimer	48	+
Glycophorin	Human erythrocytes	31,000	Dimer	38	++
Bacteriorhodopsin	*Halobacterium halobium*	27,000	Trimer	57	−
Lactose permease	*E. coli* plasma membrane	46,000[b]	Monomer or Dimer	59[b]	−
Mannitol permease	*E. coli* plasma membrane	68,000[c]	Dimer	46[c]	−
Porin	*E. coli* outer membrane	36,000	Trimer	34	−

[a]Mole percent of Pro, Ala, ½ Cys, Val, Met, Ile, Leu, Phe, and Trp.

[b]Deduced from the DNA sequence of the *lacY* gene. (D. E. Büchel, B. Gronenborn, and B. Müller-Hill, "Sequence of the lactose permease gene." *Nature* 283:541, 1980.)

[c]Deduced from the DNA sequence of the *mtlA* gene (C. A. Lee and M. H. Saier, Jr., "Mannitol-specific enzyme II of the bacterial phosphotransferase system III. The nucleotide sequence of the permease gene." *Journal of Biological Chemistry,* 258:10761, 1983.

Figure 7.32

Topography of glycophorin in the mammalian erythrocyte membrane. Carbohydrate residues (red hexagons) are all in the N-terminal domain on the outside of the cell and are attached mainly to the hydroxyl groups of serine and threonine residues of the protein. (Source: J. T. Segrest and L. D. Kohn, "Protein-lipid interactions of the membrane penetrating MN-glycoprotein from the human erythrocyte" in *Protides of the Biological Fluids,* 21st colloquium edited by H. Peeters. Copyright © 1973, Pergamon Press, New York, N.Y.)

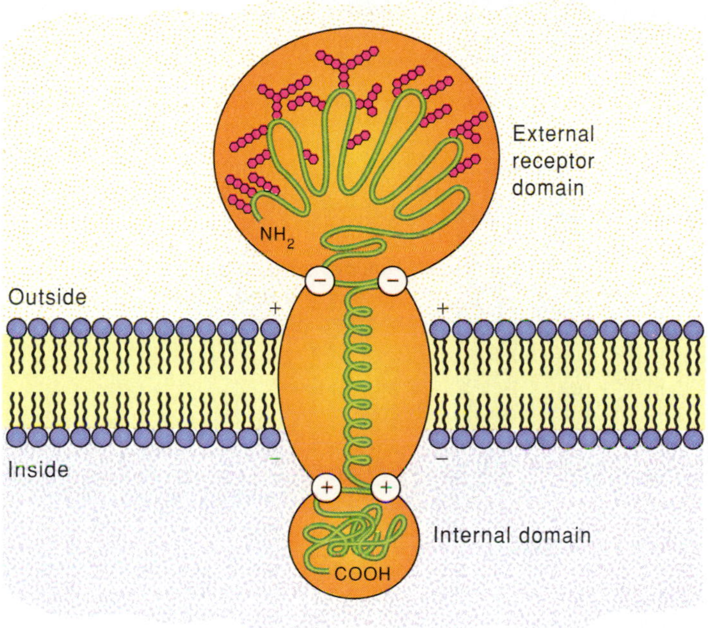

Figure 7.33

Possible disposition of the anion-channel (band 3) protein in the human erythrocyte membrane. Each identical subunit consists of 929-amino-acid residues, and the N-terminal methionine has been shown to be acetylated (NAc). Glycosyl residues are linked almost exclusively on an extracellular asparagine residue. The figure also shows the positions of cleavage by chymotrypsin and trypsin, giving rise to the fragments discussed in the text. (Source: D. Jay and L. Cantley, "Structural aspects of the red cell anion exchange protein" in *Annual Review of Biochemistry*. 55:511, 1986. Copyright © 1986 Annual Reviews Inc., Palo Alto, Calif.)

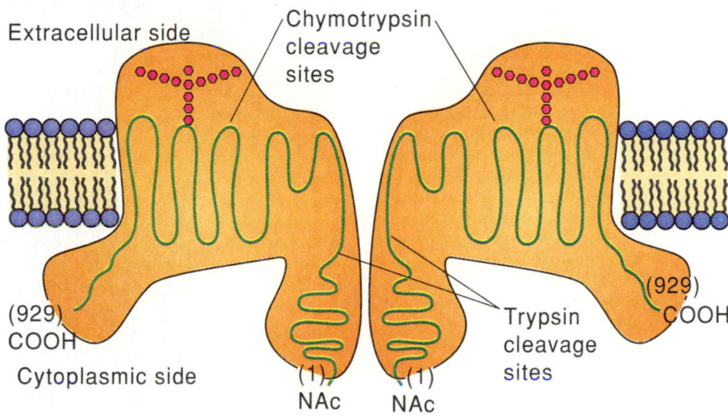

Integral Membrane Proteins Interact with Lipids through Their Hydrophobic Side Chains

In order to understand how proteins contribute to the structure of biological membranes, we need to examine their three-dimensional structures with respect to the membrane system in which they are found. A number of studies have focused on proteins that are major components of particular biological membrane systems. In this section we will describe some of the better-characterized integral membrane proteins.

The protein glycophorin has been isolated from the red blood cell membrane. It has a molecular weight of about 31,000 and contains a very large percentage of carbohydrate (60%). This carbohydrate bears the ABO- and MN-blood group antigenic specificities of the cell and is also the receptor for influenza virus. The probable structure of glycophorin is shown in figure 7.32. It is a 131-amino-acid polypeptide chain on which short carbohydrate chains are covalently attached to an asparagine residue (N-linked) or to serine or threonine residues (O-linked) that are present in the N-terminal region of the protein. This portion of the protein is highly polar, because of the presence of sugar and hydrophilic amino acid residues. The carboxyl end of the molecule is also rich in polar amino acids, but the central region of the protein contains a hydrophobic stretch of 19-amino-acid residues. Labeling experiments have shown that the carbohydrate-containing N terminal is localized on the external surface of the red blood cell and that this region comprises over 80% of the mass of glycophorin. Because the carboxyl terminal of the protein has been shown to be exposed to the cytoplasm, each molecule of glycophorin must span the membrane.

Other studies have shown that the hydrophobic portion of glycophorin has a high affinity for phospholipids and cholesterol, the two principal lipid components of the red blood cell membrane. Moreover, this portion of the molecule forms a very stable α helix in a hydrophobic environment. The length of the helix is about 40 Å, slightly less than the known width of biological membranes. These observations provide strong experimental evidence for the structural model in figure 7.32. Though much is known about the structure of glycophorin, its function is still unclear.

A second well-characterized integral membrane protein is the anion-channel, or band 3, protein of the human erythrocyte membrane. This protein plays an important role in CO_2 transport by enabling HCO_3^- to be exchanged for Cl^-. Labeling studies using intact cells and unsealed membrane ghosts (membranes with their normal contents removed) have shown that this 929-amino-acid glycoprotein also spans the membrane with the carbohydrate residues on the outside of the cell (fig. 7.33). In contrast to glycophorin, however, nearly half of the mass of the polypeptide, including the N terminal, is exposed to the cytoplasm of the cell. The C-terminal third of the molecule, including most of the amino acids to which sugar residues are attached, is at least partially available to labels from the outside of the cell. In addition, a 17,000-M_r segment, delineated by a

Figure 7.34

A model for the structure of bacteriorhodopsin. (*a*) Appearance of the membrane-exposed and membrane-removed helical sides; the hydrophobic residues are concentrated on the membrane-exposed side. (*b*) The way in which the helical units are believed to be clustered.

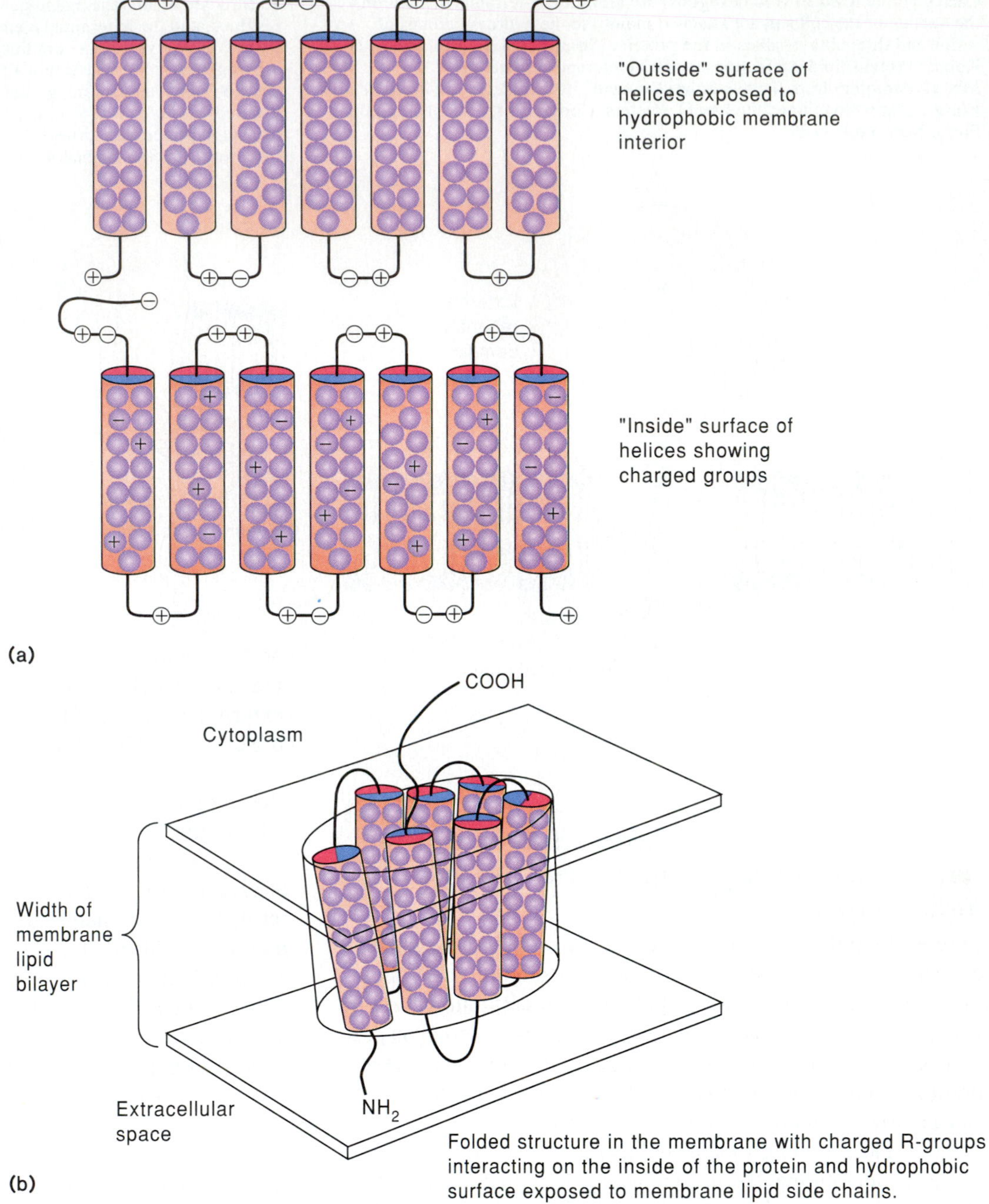

chymotrypsin-sensitive site on the outside of the cell and a tryptic cleavage site on the inside, has been shown to be tightly embedded in the membrane. Both of these latter domains, as compared with the completely water-soluble cytoplasmic domain, are rich in hydrophobic amino acids. The functional structure of the anion-channel protein appears to be a dimer of identical subunits. Recent studies suggest that within the hydrophobic portions of the molecule, the polypeptide chain of the anion channel traverses the membrane at least eight times, and possibly as many as twelve (see fig. 7.33).

In a few instances, it has been possible to isolate a biological membrane that contains only a single kind of protein. The so-called purple membrane from the halophilic archaebacterium *Halobacterium halobium* consists of lipid and one protein, bacteriorhodopsin. This protein functions as a proton pump in response to light and it is important to the organism for ATP synthesis under anaerobic conditions (chapter 32).

Because bacteriorhodopsin forms a highly ordered hexagonal lattice within the plane of the purple membrane, it has been possible to deduce its overall structure from electron micrographs and diffraction patterns. Initial studies showed that the 248-residue polypeptide chain comprising the molecule is organized as a bundle of seven α helices whose long axes are roughly perpendicular to the membrane surfaces (fig. 7.34).

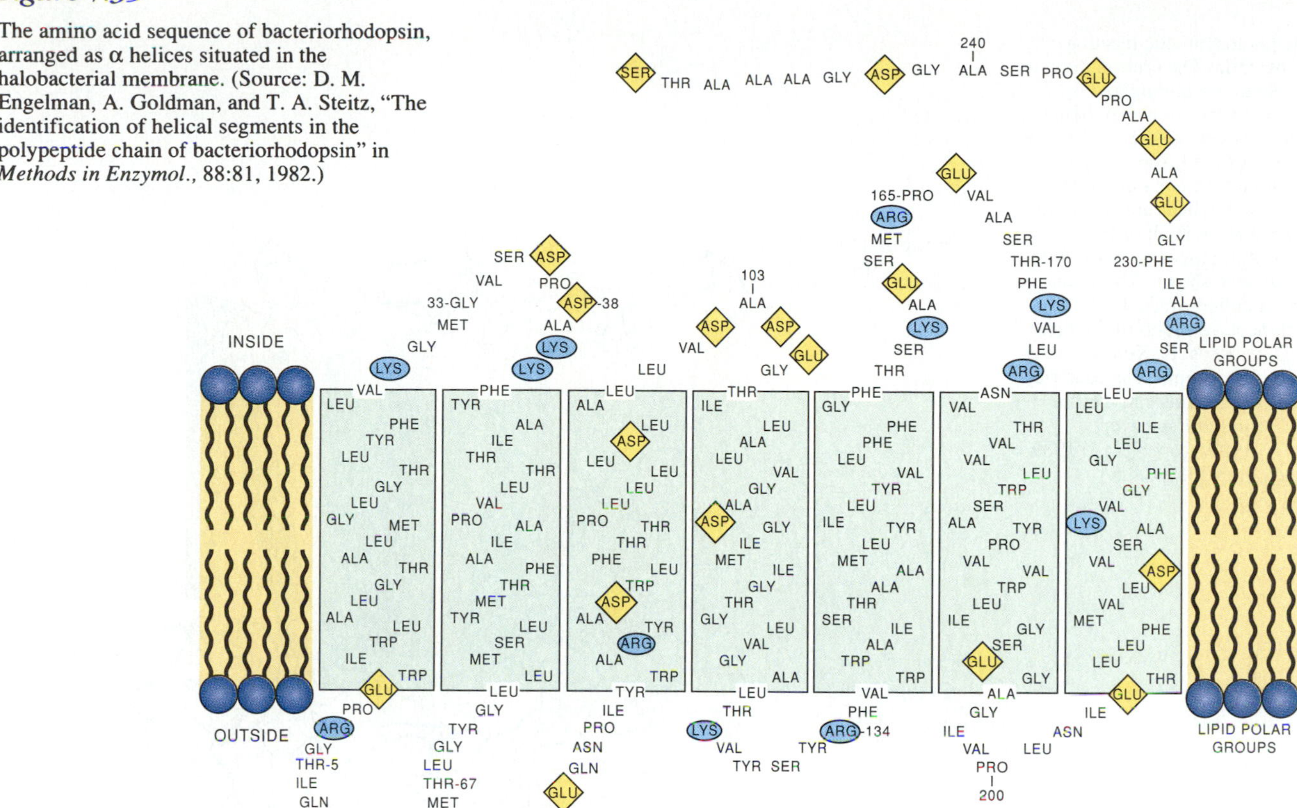

Figure 7.35

The amino acid sequence of bacteriorhodopsin, arranged as α helices situated in the halobacterial membrane. (Source: D. M. Engelman, A. Goldman, and T. A. Steitz, "The identification of helical segments in the polypeptide chain of bacteriorhodopsin" in *Methods in Enzymol.*, 88:81, 1982.)

From estimates of the number of residues required to make seven α helices long enough to each span the membrane (about 40 Å), it appears that virtually all of the polypeptide chain is required to form the helices, so that the interconnections between them must be quite short.

In more recent studies, the structure of bacteriorhodopsin has been determined to near-atomic resolution. These studies have led to several important suggestions concerning the structure of membrane proteins. First is the observation that the regions of structure within the membrane appear to be primarily α-helical. As outlined in chapter 4, α helices are secondary structures whose backbones are completely hydrogen-bonded, except at their ends. Consequently, they might be readily inserted into membranes, since there are no unsatisfied hydrogen-bonded groups whose stabilization would require the formation of hydrogen-bonded interactions with water. Second, the observed pattern of helices shows that only hydrophobic side chains are exposed to the hydrophobic interior of the membrane (figs. 7.34 and 7.35). The charged and polar amino acid side chains that do occur within the helices probably interact with each other or with water and are oriented toward the interior of the bundle of helices (see fig. 7.34). Put simply, the membrane protein bacteriorhodopsin is inside-out relative to globular proteins that exist in a polar water environment.

The first structure of a membrane protein at atomic resolution was determined by H. Michel, J. Diesenhofer, R. Huber, and co-workers. They were able to crystallize a protein comprising the photosynthetic reaction center (RC) from the purple sulfur bacterium *Rhodopseudomonas viridis* (also see chapter 16), and to determine its structure to a resolution of 2.3 Å using x-ray crystallography. This landmark achievement revealed for the first time many of the principles of membrane protein folding that previously had only been inferred for proteins such as bacteriorhodopsin.

The membrane-spanning RC complex consists of three protein subunits called L, M, and H, and associated with these are various cofactors for photosynthesis, including four molecules of bacteriochlorophyll, two of bacteriopheophytin, two quinone molecules, a nonheme iron, and a cytochrome molecule. The structure (fig. 7.36) clearly shows an array of roughly parallel α helices, five each from subunits L and M and one from subunit H, that are undoubtedly the eleven membrane-spanning regions of this protein. The amino acid residues comprising the helices are almost entirely hydrophobic, while the more hydrophilic regions of the protein form domains that are exposed on either side of the membrane-spanning region as shown in figure 7.36. We will consider the implications of this structure for the mechanism of photosynthesis in chapter 16.

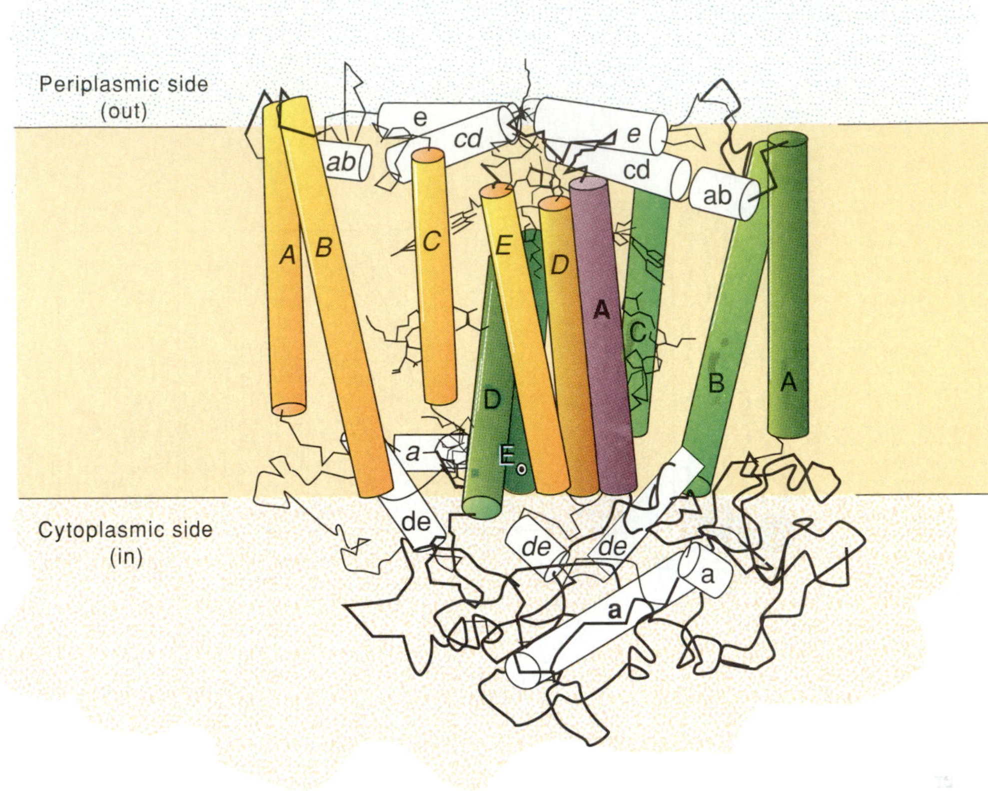

Figure 7.36

The structure of the photosynthetic reaction center from purple bacteria. The presumptive transmembrane α helices are highlighted in color: *yellow (A–E)* = the five transmembrane helices of the M subunit; green (A–E) = the five transmembrane helices of the L subunit; purple (**A**) = the one transmembrane helix of the H subunit. Other cylindrical helices are also found on either hydrophilic surface in all of the subunits. Some of the cofactors associated with the reaction center are also shown. The solid lines connecting the α helices trace the backbone of the polypeptide chains of the three subunits. (Source: D. G. Rees, H. Komiya, T. O. Yeates, J. P. Allen and G. Feher, "The bacterial photosynthetic reaction center as a model for membrane proteins" *Annual Review of Biochemistry.* 58:607, 1989. Copyright © 1989 Annual Reviews Inc., Palo Alto, Calif.)

From the structures of integral membrane proteins that have been determined or inferred so far, a number of generalizations now appear valid:

1. A significant proportion of the polypeptide must interact with membrane lipids through hydrophobic interactions.
2. α helices are common intramembrane structures of polypeptides.
3. The main carbohydrate portions of plasma membrane glycoproteins are always found on the outside of the cell.
4. Hydrophilic regions of membrane proteins are exposed at the membrane surface, although the sizes and locations of these domains (inside or outside) are highly variable among different membrane proteins.

Given these ground rules, many workers have tried to predict the intramembrane structures of membrane proteins from their primary amino acid sequences alone, and they have had some success. For example, hydrophobic, membrane-spanning α helices can be predicted by looking for hydrophobic stretches of amino acids in the primary sequence that are long enough to span the membrane (about twenty residues). Computer algorithms have been developed to identify such regions in membrane proteins; these algorithms yield hydropathy plots, an example of which is shown in figure 7.37 for bacteriorhodopsin. In this case, the hydropathy plot reasonably accurately predicts the seven membrane-spanning helices inferred from the electron-diffraction structure.

Most Integral Membrane Proteins Can Move within the Membrane Structure

The currently accepted model of biological membrane structure, called the fluid mosaic model, was proposed by S. J. Singer and G. L. Nicolson in 1972 (fig. 7.38). In the fluid mosaic model, the essential structural repeating unit is the phospholipid molecule, in a bilayer arrangement with a thickness of about 50 Å. Integral membrane proteins are "dissolved" in the bilayer in a seemingly random fashion. Some proteins (such as certain mitochondrial cytochromes) are localized at one or the other of the two surfaces of the lipid bilayer; other proteins (such as glycophorin and the anion channel) span the membrane but with considerable mass extending into the aqueous surroundings; and still others (such as bacteriorhodopsin) are largely embedded in the hydrophobic matrix. A most important aspect of the fluid mosaic model is its dynamic character, with most components capable of relatively rapid lateral diffusion and of rotational motion about an axis perpendicular to the plane of the bilayer.

Figure 7.37

Example of a hydropathy plot for bacteriorhodopsin. The water-to-oil transfer free energy (ΔG) for successive 19-amino-acid segments is plotted as a function of the first amino acid in the sequence of a given segment. Regions below the line for $\Delta G = 0$ thus correspond to favorable partitioning of a region of the polypeptide into the lipid bilayer. Points A through G are the proposed start points of the transmembrane α helices of bacteriorhodopsin. This analysis was, in part, used to formulate the model shown in figure 7.35. (Source: D. M. Engelman, A. Goldman, and T. A. Steitz, "The identification of helical segments in the polypeptide chain of bacteriorhodopsin" in *Methods Enzymology* 88:81, 1982.)

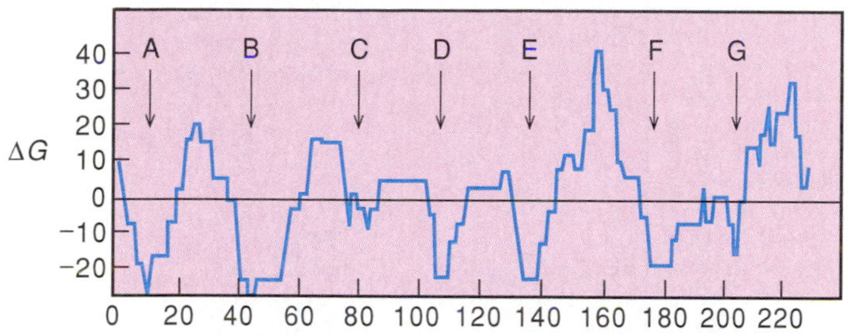

Figure 7.38

The fluid mosaic membrane as envisioned by Singer and Nicolson. The proteins within the bilayer have lateral mobility about an axis perpendicular to the plane of the membrane. Some rotational movement about an axis perpendicular to the plane of the membrane is also possible.

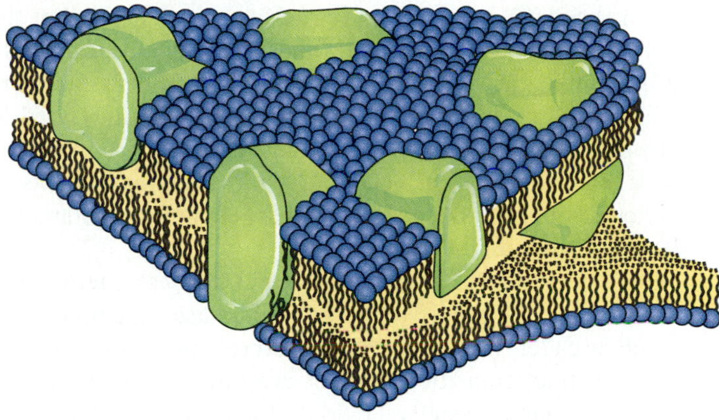

Rotation of lipids and proteins through the plane of the bilayer, however, is proposed to be a rare event. Figure 7.38 therefore depicts a hypothetical biological membrane at one point in space and time.

The abilities of membrane proteins and lipids to diffuse laterally, parallel to the plane of the bilayer, have been amply demonstrated by a variety of techniques. One of the more straightforward of these is the fluorescence photobleach recovery technique. In this method, fluorescently labeled membrane components are bleached (their fluorescence is destroyed) in a small area of the membrane by a short, intense flash of light. By monitoring this region as a function of time, using a fluorescence microscope, one can observe a reappearance of fluorescence (recovery), which is due to unbleached molecules diffusing into the treated area. If a specific membrane component is so labeled, its diffusion coefficient in the plane of the membrane can be calculated from the time course of fluorescence recovery. This technique has revealed lateral movements of both proteins and lipids in a number of biological membranes. For example, rhodopsin, a light-receptor protein of vertebrate retinal membranes (chapter 36), has proved to be highly mobile within the lipid bilayer. Other proteins vary enormously in their intramembrane diffusion coefficients (see the following discussion).

In an elegant experiment also designed to test protein mobility in biological membranes, L. Frye and M. Edidin fused a mouse cell and a human cell with the aid of a virus, called Sendai virus, that facilitates such fusion. Membrane proteins of the mouse cell were labeled with specific antibodies that fluoresced green, while similar labeling of the human cells was with red fluorescent antibody. If membrane proteins failed to diffuse, then the heterokaryon resulting from the fusion of these two cells should remain "half red" and "half green" even after long periods of incubation at physiological temperatures (37° C). In this experiment, however, significant intermixing of red and green labels was observed in less than 30 min, and complete mosaics of human mouse proteins were seen in all heterokaryons within 1 h. These observations provided direct evidence for the lateral mobility of integral membrane proteins and indirect evidence for the movement of the lipid components of the bilayer.

The specific orientations maintained by integral membrane proteins with respect to the bilayer suggest that rotation of these molecules through the plane of the bilayer (flip-flop or transverse motion) does not occur. Since most integral membrane proteins have at least some hydrophilic surface area exposed at one or both membrane faces, a transverse rotation would be an energetically unfavorable event, requiring transient interaction of these polar groups with the hydrophobic interior of the bilayer. Indeed, physical and chemical measurements have failed to detect such motions of proteins in biological membranes. Comparable measurements for phospholipids in model membranes have revealed that flip-flop of these molecules can occur, but that half-times for this process can be as long as days at physiological temperatures. These rates are many orders of magnitude smaller than those of lateral diffusion and again can be explained by the high energies required for these processes. As we will see, the asymmetric topography of biological membranes is essential to many membrane functions. This topography is maintained by the amphipathic nature of the membrane's lipid and protein constituents.

Temperature, Composition, and Extramembrane Interactions Affect Dynamic Properties

Diffusion rates in lipid bilayers are a function both of temperature and of the composition of the membrane being examined. Bilayers consisting of a single type of phospholipid typically show an abrupt change in physical properties over a characteristic and narrow temperature range (T_m). These temperature-dependent phase transitions are due to an organizational change in the fatty acyl side chains and can be detected by a variety of physical techniques. For example, differential scanning calorimetry measures heat absorption as a function of temperature and shows that over the temperature range of the phase transition, a relatively large amount of heat is absorbed per degree of temperature change compared with temperature ranges well above or below the transition (fig. 7.39). It is thought that the membranes pass from a state in which the fatty acyl chains are highly ordered (gel phase) to one in which they are far more mobile (liquid crystalline phase). This process is illustrated in figure 7.39. It is accompanied by increased rotational motion about the carbon–carbon bonds of the hydrocarbon chains of the phospholipids, allowing them to assume more random, disordered conformations.

In contrast to pure phospholipid bilayers, membranes isolated from cells usually undergo such phase transitions over a much broader (~10° C) temperature range. In some instances, distinct transitions cannot even be distinguished, because of the heterogeneity of lipids found in most biological membranes and because integral membrane proteins may decrease the mobility of lipids in their immediate vicinity. In general, lipids bearing short or unsaturated fatty acyl chains undergo phase transitions at lower temperatures than those containing long-chain saturated fatty acids. This is because short hydrocarbon chains have a smaller surface area with which to undergo hydrophobic interactions that stabilize the gel state and because *cis* unsaturation introduces "kinks" in the fatty acyl chain (see fig. 7.1) that also lead to more disorder in the bilayer. These effects of fatty acid composition on phase transition temperatures of model bilayers are apparent from the data presented in table 7.11, and in part explain the fact that broad phase transitions are a general characteristic of cellular membranes.

From the data in table 7.11, it can further be seen that the midtransition temperature of a pure phospholipid suspension also depends on the nature of the polar head group. Thus the dipalmitoyl esters of phosphatidic acid and phosphatidylethanolamine "melt" at temperatures about 20° C higher than the same derivatives of phosphatidylcholine and phosphatidylglycerol. Divalent cations, such as Ca^{2+} and Mg^{2+}, also affect membrane fluidity, presumably because of ionic interactions with neighboring phosphoryl head groups, tending to "tie" phospholipid molecules together and decrease their mobility. This is undoubtedly one of the reasons that divalent cations are well-known stabilizers of biological membranes, and their removal often facilitates lysis of cells as well as dissociation of peripheral membrane proteins.

Finally, cholesterol is a well-known modulator of membrane structure in animal cells and in one type of bacteria, the mycoplasmas. In these cells cholesterol intercalates among the fatty acyl chains, with its polar hydroxyl group interacting with the polar head groups of membrane lipids. At low concentrations of cholesterol in phospholipid bilayers, separate domains, or patches, of cholesterol plus phospholipid and pure phospholipid appear to exist, with a resultant broadening of the phase transition profile compared with phospholipid alone. Its effects

Figure 7.39

(*Top*) Differential scanning calorimetry of various phospholipids dispersed in water. Heat absorption is indicated by a trough in the plot relating differential heat flow to temperature. The lowest point in the trough is the phase transition temperature (T_m): (*a*) dipalmitoyl phosphatidylethanolamine; (*b*) dimyristoyl lecithin; (*c*) dipalmitoyl lecithin; (*d*) egg lecithin (plus ethylene glycol to prevent freezing).
(*Top*) (Reproduced, with permission, from the *Annual Review of Biophysics and Bioengineering*.,Vol. 5, © 1976 by Annual Reviews, Inc.)
(*Bottom*) Molecular interpretation of the heat-absorbing reaction during the phase transition.

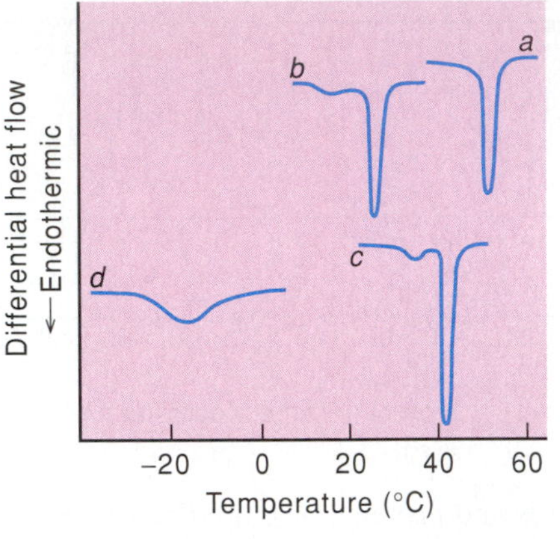

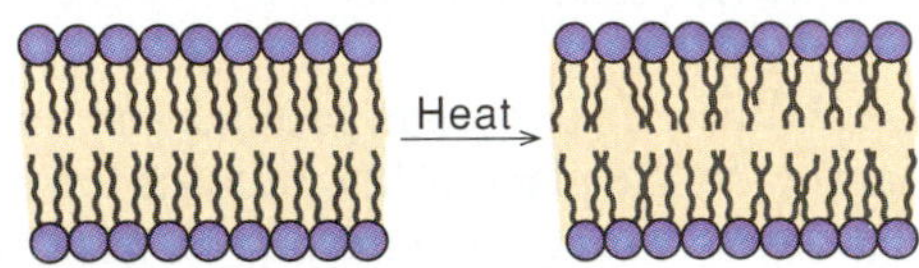

on various physical parameters of membranes depends on its proportion to other lipid components, as well as on the membrane system examined. However, a general property of membranes containing a high concentration of cholesterol appears to be an inhibition of processes dependent on a fluid environment. For example, the permeability of vesicles made from purified egg phosphatidylcholine to both water and glucose decreases when greater than 20 mol % cholesterol is incorporated into the vesicles. Intercalated cholesterol probably restricts the freedom of motion of the phospholipid hydrocarbon side chains, thereby decreasing the mobilities of membrane constituents.

Most cells maintain a lipid composition that allows for a relatively rapid lateral diffusion of many membrane components at normal growth temperatures. Membrane-associated processes, such as the vectorial reactions catalyzed by some transmembrane transport systems (chapter 32) and endo- and exocytosis, rely on a semifluid environment for their operation. Consequently, organisms have evolved intricate mechanisms to maintain this environment under a variety of conditions. One of the most remarkable of these is the ability of plant, animal, and bacterial cells to increase the proportion of membrane unsaturated fatty acids in response to a decrease in temperature. This ensures proper functioning of the membrane at the lower temperature. In bacteria, this modulation appears to be the result

Table 7.11
Midtransition Temperatures for Aqueous Suspensions of Phospholipids

Phospholipid[a]	T_m (°C)
Di-14:0 PC	24
Di-16:0 PC	41
Di-18:0 PC	58
Di-22:0 PC	75
Di-18:1 PC	~22
1-18:0, 2-18:1 PC	3
Di-14:0 PE	51
Di-16:0 PE	63
Di-14:0 PG	23
Di-16:0 PG	41
Di-16:0 PA	67

[a]Phospholipid abbreviations are as in table 7.8; additionally, Di-14:0, for example, refers to dimyristoyl (14 carbons, 0 double bonds).

Source: M. K. Jain and R. C. Wagner, *Introduction to Biological Membranes.* Copyright © 1980 John Wiley & Sons, Inc., New York, N.Y.

of temperature effects on the activities and/or the induction of synthesis of enzymes involved in the biosynthesis of phospholipids containing unsaturated fatty acyl chains. In plants and yeast, the increased solubility of O_2 at low temperatures apparently increases the proportion of unsaturated fatty acids because O_2 is a substrate of the desaturase enzyme that leads to their biosynthesis. Other factors that affect unsaturated fatty acid biosynthesis, and thus membrane fluidity, in various systems include light (in plants), nutrition, developmental stage, and aging.

Some Membrane Proteins Are Immobilized by External Forces

Despite the large body of evidence in favor of the lateral mobility of many membrane constituents at physiological temperatures, it is an oversimplification to view a biological membrane only as a random "sea" of lipids with proteins floating about aimlessly in them. Nearly all eukaryotic cells contain within their cytoplasm a cytoskeleton, made up in part of microtubules and microfilaments, which consist primarily of the proteins tubulin and actin (see table 5.1), respectively. Microfilaments, which are structurally similar to actin filaments of muscle cells, have been shown to form bundles just beneath the plasma membrane of many cells (fig. 7.40). These bundles are believed to have an important role in such processes as locomotion and phagocytosis, which involve local or general changes in the shape of the cell surface and thus in the plasma membrane. In many cases, direct or indirect association of microfilaments with the plasma membrane has been demonstrated. For example, the filamentous protein fibronectin, a peripheral, cell-surface glycoprotein in many animal cells, is believed to have a role in cell–cell and cell–substratum adhesion. A transmembrane association of fibronectin with cytoskeletal microfilaments has been deduced from immunofluorescent microscopy (fig. 7.41). Lateral diffusion of fibronectin has been shown to be at least 5,000 times slower than that of freely diffusible membrane proteins and

Figure 7.40

Membrane-associated cytoskeletal components of cultured mouse cells. (PM = plasma membrane, MF = microfilaments, MT = microtubules; 54,000×.) (From G. L. Nicolson, Transmembrane control of the receptors on normal and tumor cells. I. Cytoplasmic influence over cell surface components, *Biochem. Biophys. Acta* 457:57, 1976. Reprinted with permission from Elsevier Science Publishers.)

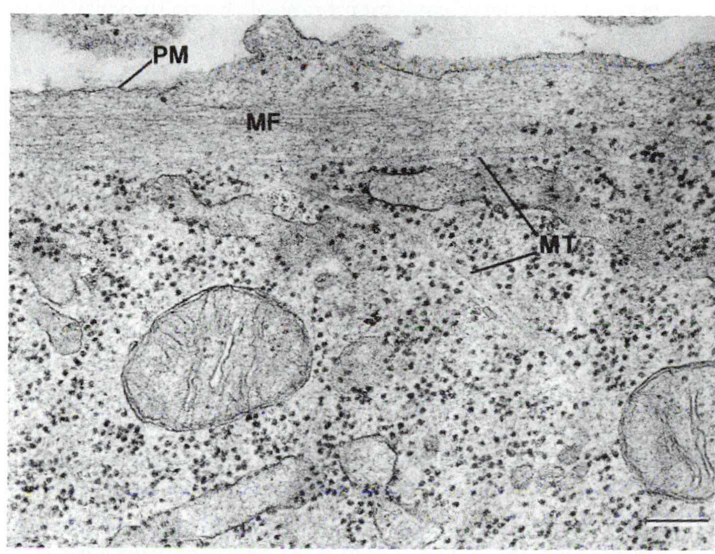

Figure 7.41

Colinearity of actin and fibronectin fibrils in cultured hamster fibroblast cells as seen by immunofluorescence in the light microscope. (*a*) The fluorescence is due to actin antibodies bound to intracellular microfilaments: These antibodies were produced in rabbits. Their location of binding is made visible by staining with fluorescent goat antirabbit antibodies. (*b*) The fluorescence is that of the fluorescent antifibronectin antibodies bound to extracellular fibronectin fibrils. The correspondence between the arrangement of actin and fibronectin filaments strongly suggests a transmembrane association between the two proteins. (From R. O. Hynes and A. T. Destree, "Relationships between fibronectin (LETS protein) and actin," *Cell* 15:875, 1978 © Cell Press.)

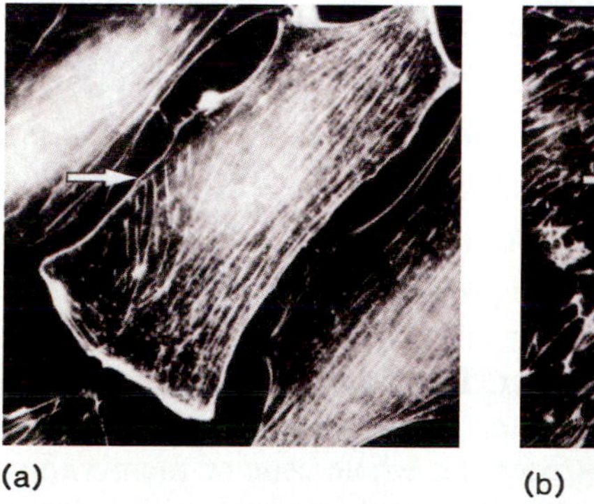

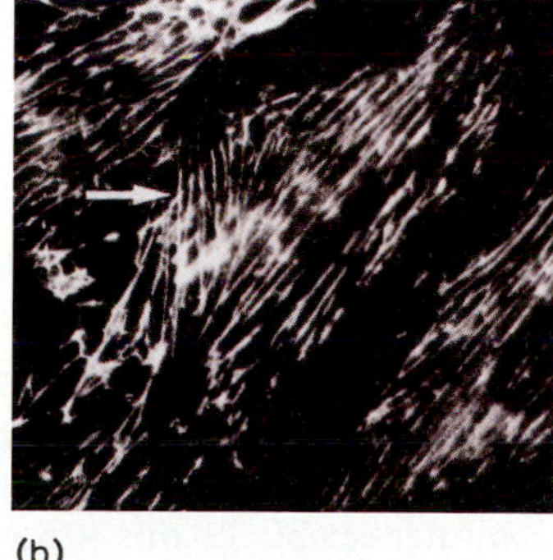

lipids. A number of transmembrane and membrane-associated proteins have been implicated in the attachment of extracellular fibronectin to intracellular actin filaments, including a transmembrane fibronectin receptor and the intracellular proteins talin, vinculin, and α-actinin. One possible arrangement of these proteins is depicted in figure 7.42.

A second well-characterized membrane cytoskeleton is that just beneath the plasma membrane of the erythrocyte. This scaffoldlike structure is believed to have a role in maintaining the biconcave shape of the red blood cell and in protecting it from the large shear forces it encounters in the peripheral circulation, while still conferring a degree of flexibility on the cell surface. The major structural protein of the red-cell membrane skeleton is called spectrin. It consists of α (260 kDa) and β (225 kDa) polypeptide-chain subunits. Each subunit folds into a rather extended structure consisting of 106-amino-acid repeating units, each of which in turn contains three α helices. An $\alpha\beta$ dimer of spectrin is formed when an α and a β subunit twist about one another in parallel; in the erythrocyte membrane, these dimers are joined end-to-end to produce an $\alpha_2\beta_2$ tetramer with a length of about 200 nm (fig. 7.43*a*). Just beneath the erythrocyte membrane, spectrin tetramers are indirectly attached to the membrane through the protein ankyrin, which cross-links spectrin with the anion-channel (band 3) protein, and through protein 4.1 (see fig. 7.31), which cross-links spectrin with glycophorin. In addition, each molecule of protein 4.1 also assists in the binding of several tetramers of spectrin to actin filaments from the cytoskeleton of the red cell. The result is a mesh-work of spectrin tetramers linked to both the cytoplasmic membrane and the cytoskeleton as depicted in figure 7.43*b*. As a consequence, the plasma membrane is structurally stabilized, and the mobilities of the anion channel and glycophorin are likely to be greatly reduced compared with those of other membrane proteins.

From the foregoing discussion, we can conclude that temperature, ionic environments, and composition all can affect the general physical state of a biological membrane, while local mobilities of membrane components can be influenced by protein–protein, lipid–protein, and lipid–lipid interactions.

Both Proteins and Lipids Are Asymmetrically Arranged in Biological Membranes

Physical chemical probes have provided ample evidence for the unidirectional, asymmetric orientation of membrane proteins with respect to the lipid bilayer. This asymmetry also has been more directly observed by the technique of freeze-fracture electron microscopy. In this procedure, whole cells or membranes are rapidly frozen, and the specimen is then struck with a sharp knife called a microtome. Very often, the fracture plane actually passes between the outer and inner monolayers (leaflets) of the membrane lipid bilayer because the relatively weak hydrophobic interactions between the fatty acyl chains of the two leaflets offer a "path of least resistance" (fig. 7.44). The inner surfaces of the two leaflets can then be viewed in the electron microscope after heavy-metal shadowing of the fractured specimen. Many biological membranes, when split in half by freeze-fracture, can be seen to have quite different morphologies of the inner and outer leaflet surfaces, indicating preferential adherence of some membrane proteins to one of the two monolayers (see fig. 7.44). This feature presumably reflects the asymmetric and unidirectional disposition of the polypeptides across the lipid bilayer. In a related method, freeze-etching electron microscopy, ice is sublimed away from the sample after freeze-fracture, to allow a view of the outer surfaces of the two leaflets. Both procedures are especially useful in examining membrane morphology, because they avoid the potentially destructive fixing, embedding, and staining steps of more conventional sample preparation procedures.

Proteins are not the only membrane constituents to show asymmetric orientations. Many lipids also have been shown to be unequally distributed between the inner and outer leaflets of many biological membranes. In 1972, M. S. Bretscher first showed that most of the phosphatidylethanolamine and all of the phosphatidylserine of human erythrocyte membranes are inaccessible to chemical modification from outside the cell. Both phosphoglycerides, however, could be modified in erythrocyte ghosts in which both the inner and outer leaflets were exposed

Figure 7.42

Schematic illustration of a so-called focal contact, showing how extracellular fibronectin is believed to be indirectly attached to the intracellular cytoskeleton through a transmembrane fibronectin receptor and several other peripheral membrane proteins. (Source: B. Alberts et al., *Molecular Biology of the Cell,* 2d ed. Copyright © 1989, Garland Publishing, New York, N.Y.)

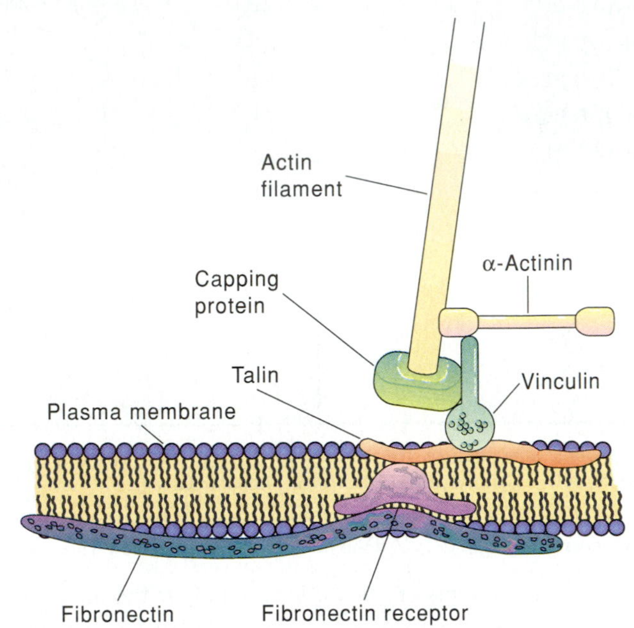

Figure 7.43

The structure of spectrin and the location of spectrin in the cytoskeleton. (*a*) An αß dimer of spectrin. Both α and ß subunits are extended structures consisting of end-to-end domains of 106 amino-acyl residues folded into three α helices (left); the subunits twist about one another loosely as shown. (*b*) The erythrocyte membrane skeleton. Spectrin tetramers ($\alpha_2\beta_2$), shown in yellow, are linked to the cytoplasmic domain of the anion channel (blue) by the protein ankyrin (red), and to glycophorin and actin filaments by protein 4.1. This structure lends structural stability to the red cell membrane while maintaining sufficient flexibility to allow erythrocytes to withstand substantial shear forces in the peripheral circulation.

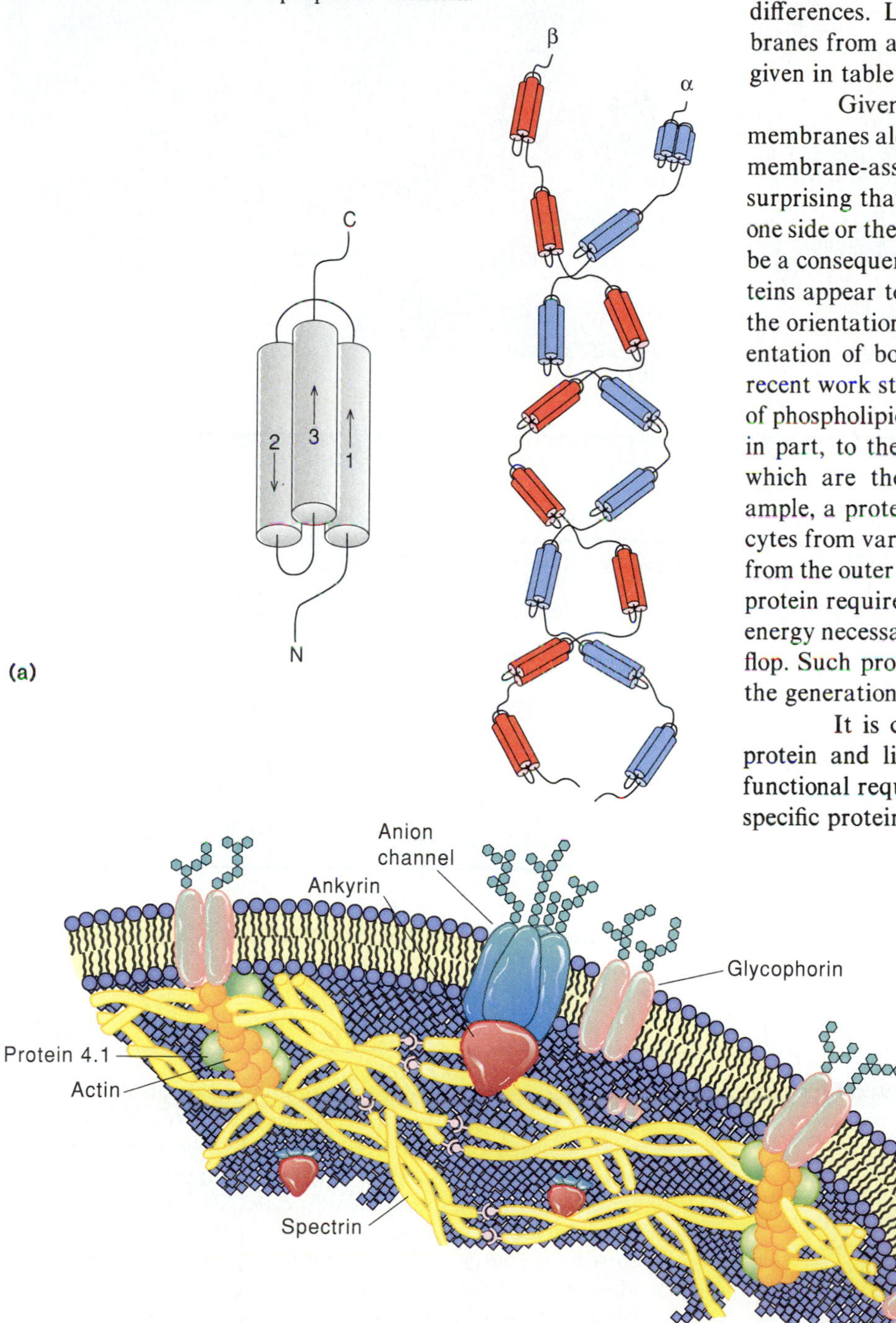

to the chemical reagent. These results suggested that lipid asymmetry also could exist in a biological membrane. Subsequent experiments using enzymes that degrade phospholipids (e.g., phospholipases and sphingomyelinase) showed that phosphatidylcholine and sphingomyelin were preferentially found in the outer leaflet of human red blood cell membranes (table 7.12). Thus, although total membrane lipid is equally distributed between outer and inner monolayers in the erythrocyte membrane, the lipid composition of each leaflet shows striking differences. Lipid asymmetry has now been found in membranes from a variety of biological sources; some examples are given in table 7.12.

Given the asymmetric arrangements of proteins in membranes along with the nearly unidirectional nature of many membrane-associated processes (chapter 32), it is perhaps not surprising that certain lipids might be found predominantly on one side or the other of the bilayers. This could at least partially be a consequence of the fact that some integral membrane proteins appear to associate preferentially with certain lipids, and the orientation of the protein could therefore influence the orientation of bound lipid with respect to the bilayer. However, recent work strongly suggests that the asymmetric partitioning of phospholipids into the two bilayer leaflets is also due, at least in part, to the activities of specific phospholipid transporters, which are themselves integral membrane proteins. For example, a protein has been identified in membranes of erythrocytes from various sources that translocates phosphatidylserine from the outer to the inner membrane leaflet. This 31,000-dalton protein requires ATP for its activity, presumably to provide the energy necessary to catalyze this energetically unfavorable flip-flop. Such proteins may therefore be generally responsible for the generation of lipid asymmetry in biological membranes.

It is clear from energetic arguments how membrane protein and lipid asymmetry can be maintained, and, from functional requirements, why this asymmetry exists. Although specific proteins may be responsible for lipid asymmetry in biological membranes, a much more complex question is, How do integral membrane proteins adopt their specific asymmetric orientations in membranes? Although this question has not yet been fully answered, it is clearly related to the broader question of how biological membranes are assembled in the first place (membrane biogenesis). We will consider this topic in Part 5 (chapter 22).

Figure 7.44

Freeze-fracture electron microscopy. (*a*) When struck with a sharp knife, membranes embedded in ice usually fracture between the monolayer leaflets of the lipid bilayer. (*b*) Freeze-fracture electron micrograph of the plasma membrane of *Streptococcus faecalis*, showing a large number of protrusions (presumably proteins) on the outer fracture face and the relative lack of such particles on the inner fracture face (inset). (From H. C. Tsien and M. L. Higgins, "Effect of temperature on the distribution of membrane particles in *Streptococcus faecalis* as seen by the freeze-fracture technique," *J. Bacteriol.* 118:725, 1974. Reprinted with permission from American Society for Microbiology.)

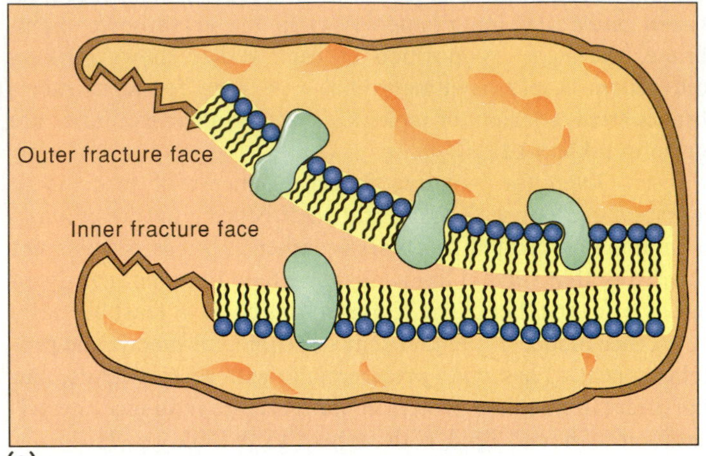

(a)

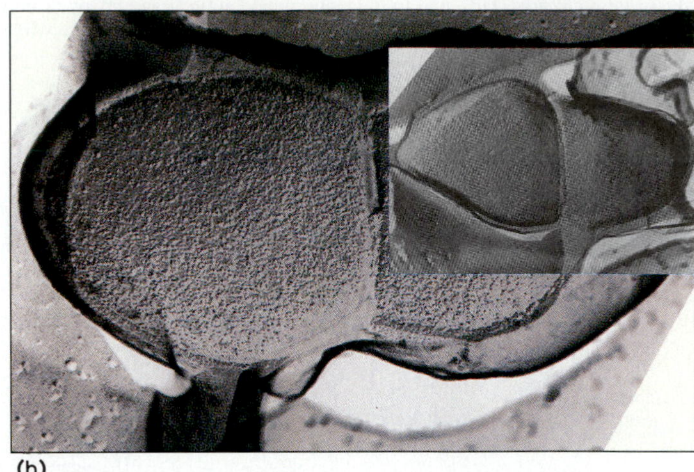

(b)

Table 7.12
Lipid Asymmetry in Biological Membranes

Membrane	Preferential Outside	Preferential Inside	Equal	Membrane	Preferential Outside	Preferential Inside	Equal
Various erythrocytes	PC, SM, glycolipids, cholesterol	PE, PS	—	*E. coli* outer membrane	—	PE	—
				Bacillus megaterium	—	PE	—
Rabbit sarcoplasmic reticulum	PE	PS	PC lyso PC	*Micrococcus lysodeikticus*	PG	PI	DPG
Mouse LM cell plasma membrane	SM	PE	PC	PC/PE artificial vesicles	PC	PE	—

Note: Abbreviations are as in table 7.8.

Source: J. A. F. Op den Kamp, "Lipid asymmetry in membranes" in *Annual Review of Biochemistry,* 48:47, 1979. Copyright © 1979 Annual Reviews, Inc. Palo Alto, Calif.

Summary

In this chapter we have focused on the structures and functions of lipids and membranes. The following points are central to this subject.

1. Lipids are biological molecules that are soluble in organic solvents. They comprise a wide diversity of structures and they serve four major biological functions: (a) some are major structural elements of membranes; (b) others are efficient reserves for the storage of energy; (c) many vitamins and hormones are lipids or lipid derivatives; and (d) the bile acids help to solubilize the other lipid classes during digestion.
2. Fatty acids are the major structural components of many membrane lipids and are the biological molecules in which energy can be stored most efficiently. They exhibit a large number of structural variations, primarily in chain length and in the number and location of double bonds. Triacylglycerols are the molecules in which fatty acids are stored for energy and are found mostly in adipose tissue.
3. Major lipids found in biological membranes include the phospholipids (phosphoglycerides and sphingomyelin), the glycosphingolipids, and cholesterol.
4. The phospholipids and glycosphingolipids contain a polar head group and a nonpolar tail portion composed of long-chain acyl or alkyl groups. The amphipathic nature of these

lipids allows them to form lipid bilayers spontaneously in an aqueous medium. Hence they play a major structural role in all biological membranes.

5. Cholesterol, a derivative of the tetracyclic hydrocarbon perhydrocyclopentanophenanthrene, is the most prominent member of the steroid family of lipids. Cholesterol is an important structural component of some eukaryotic membranes, but is generally absent from bacterial membranes.
6. Biological membranes are composed of lipids, proteins, and carbohydrates in various combinations.
7. Membrane lipids are complex mixtures that reflect the variety both of fatty acids and of polar head groups found in these molecules. The predominant lipid in eukaryotic membranes is phosphatidylcholine, while in bacteria it is phosphatidylethanolamine.
8. Membrane proteins are either peripheral or integral. Peripheral proteins may be dissociated from the membrane in a water-soluble form by relatively mild procedures that disrupt hydrogen or ionic bonds. Integral proteins can be brought into solution only by disrupting the bilayer structure, for example with detergents.
9. Physical and chemical analyses of purified integral membrane proteins show that they vary widely in structure, but that most of them require a hydrophobic environment for maintenance of their biologically active structures.
10. Some integral proteins, such as bacteriorhodopsin, have most of their mass buried within the lipid bilayer. Others have a large proportion of the polypeptide extending into the aqueous environment on either or both sides of the membrane. Examples of this second type are the anion-channel protein, glycophorin, and the photosynthetic reaction center of purple bacteria.
11. Many integral membrane proteins have been shown to span the lipid bilayer. The membrane-spanning regions of these proteins are often hydrophobic α helices.
12. If carbohydrates are covalently bound to a plasma membrane protein, they are always found on the external surface of the cell.
13. In 1972, Singer and Nicolson proposed the fluid mosaic model for membrane structure. In this model, the essential structural unit is a relatively impermeable lipid bilayer in which protein molecules are embedded. Both protein and lipid molecules are capable of rapid lateral diffusion. They are also capable, at physiological temperatures, of rotational motion about an axis perpendicular to the plane of the bilayer, but spontaneous rotation through the plane of the bilayer (transverse or flip-flop motion) is very rare. Thus in general the membrane maintains an asymmetric structure with respect to both protein orientation and the distributions of specific lipids between inner and outer leaflets. Biophysical techniques such as x-ray diffraction, and fluorescence spectroscopy, have confirmed these essential features of the fluid mosaic model.
14. Many factors have been shown to affect the general mobility of membrane components, including temperature, the fatty acid compositions of phospholipids and glycolipids, and the presence or absence of cholesterol. Local mobilities within a given membrane system are a function of protein–protein, lipid–protein, and lipid–lipid interactions.
15. In response to environmental changes, many cells can regulate the composition of their membranes to maintain the overall semifluid environment necessary for many membrane-associated functions.
16. The asymmetric orientation of integral membrane proteins with respect to the bilayer is necessary for their proper functioning. Similarly, the unequal partitioning of some membrane lipids between the two leaflets of the bilayer may also be necessary for the functions of some membrane systems. This partitioning may be influenced by both protein asymmetry and the activities of specific lipid transporters.

Selected Readings

Bennett, V., The membrane skeleton of human erythrocytes and its implications for more complex cells. *Ann. Rev. Biochem.* 54:273, 1985. A review of the organization of the red blood cell surface, including a discussion of the possible roles of the cytoskeleton.

Deuel, H. J., *The Lipids: Biochemistry,* vol. 3. New York: Interscience, 1957. A comprehensive and classical treatise on the biochemistry of lipids until the mid-1950s.

Diesenhofer, J., O. Epp, N. Miki, R. Huber, and H. Michel, X-ray structure analysis of a membrane protein complex. Electron density map at 3-Å resolution and a model of the chromophores of the photosynthetic reaction center from *Rhodopseudomonas viridis. J. Mol. Biol.* 180:385, 1984. First x-ray structure of an integral membrane protein.

Fleischer, S., and L. Packer (eds.), *Methods in Enzymology,* vols. 31 and 32. New York: Academic Press, 1974. Volume 31 of this invaluable series focuses on subcellular fractionation and membrane isolation techniques, including isopycnic and differential centrifugation. Volume 32 deals with the composition and characterization of various membranes and membrane components, and includes articles on model membrane systems.

Fleischer, S., and B. Fleischer (eds.), *Methods in Enzymology,* vol. 172. New York: Academic Press, 1989. The more recent volume updates volumes 31 and 32, and deals with methods of membrane separation, analysis and physical characterization.

Hawthorne, J. N., and G. B. Ansell (eds.), *New Comprehensive Biochemistry,* vol. 4: *Phospholipids.* Amsterdam: Elsevier, 1982. Detailed collection of review articles on the structures, functions, and biosynthesis of the phospholipids.

Helenius, A., and K. Simons, Solubilization of membranes by detergents. *Biochim. Biophys. Acta* 415:29, 1975. Lengthy review of the properties of detergents and their use for membrane solubilization.

Henderson, R., J. M. Baldwin, T. A. Ceska, F. Zemlin, E. Beckmann, and K. H. Downing, Model for the structure of bacteriorhodopsin based on high resolution electron cryo-microscopy. *J. Mol. Biol.,* 213:899, 1990. The structure of bacteriorhodopsin is determined to near-atomic resolution in this article.

Jain, M. K., *Introduction to Biological Membranes,* 2d ed. New York: Wiley, 1988. Lucid introduction to the organization and functions of biological membranes.

Jay, D., and L. Cantley, Structural aspects of the red cell anion exchange protein. *Ann. Rev. Biochem.* 55:511, 1986. Review of techniques that have been used to determine the intramembrane structure of the band 3 (anion-channel) protein of erythrocytes.

Jennings, M. L., Topography of membrane proteins. *Ann. Rev. Biochem.* 58:999, 1989. Review of membrane protein structure and methods of structure prediction and determination.

Razin, S., and S. Rottem (eds.), *Current Topics in Membranes and Transport,* vol. 17: *Membrane Lipids of Prokaryotes.* New York: Academic Press, 1982. A comprehensive summary of lipid structure and function in bacteria. Other recent volumes in this continuing series contain up-to-date review articles on topics of current interest in membrane structure and function.

Rees, D. C., H. Komiya, T. O. Yeates, J. P. Allen, and G. Feher, The bacterial photosynthetic reaction center as a model for membrane proteins. *Ann. Rev. Biochem.* 58:607, 1989. Discusses theoretical and practical implications of the RC structure on the possible structures of other membrane proteins.

Singer, S. J., The molecular organization of membranes. *Ann. Rev. Biochem.* 43:805, 1974. Excellent review of the earlier literature on membrane structure through 1973.

Vance, D. E., and J. E. Vance (eds.), *Biochemistry of Lipids, Lipoproteins and Membranes.* Amsterdam: Elsevier, 1991. Compilation of articles on lipid and membrane structure, metabolism, and function.

Problems

1. "Individuals whose diet is devoid of plant tissue may develop essential fatty acid deficiency." Defend or refute this statement.
2. Compare the relative efficiency of extraction of free fatty acid, fatty acid methylester, and triacylglycerol into organic solvent. Which component(s) have a pH dependency of extraction? Why?
3. Both triacylglycerols and phospholipids have fatty acid ester components, but only one group can be considered amphipathic. Indicate which is amphipathic and explain why.
4. The term phosphatidylcholine (PC) defines a class of phospholipids. What portion of the phosphatidylcholine molecule is common to all members of the class? What portion of the molecule is variable among the PCs?
5. Use Haworth formulas to represent the structure of the glycosphingolipid referred to as Trihexosylceramide (D-Gal-α-1,4-D-Gal-β-1,4-D-Glc-β-1,1-ceramide). What type of linkage bonds the carbohydrate to the ceramide?
6. If phosphatidylcholine were dispersed in an aqueous buffer, what types of molecular associations would yield the most stable suspension?
7. Triton X-100 (M_r = 625, CMC = 0.24 mM) and sodium deoxycholate (M_r = 414, CMC = 4 mM) are each used to solubilize proteins from membranes before application of further isolation techniques.
   (a) Which of these detergents would be more easily removed by dialysis from an 0.1% (wt/vol) aqueous solution?
   (b) In which of these detergents would ion-exchange chromatography be more likely successful for protein purification?
   (c) In which of the detergents would gel exclusion chromatography be likely to approximate the true molecular weight of an integral membrane protein in 0.1% aqueous solution of the detergent? Why?
8. Quinones that are structurally similar to vitamins K_1 and K_2 (see fig. 11.24) are associated with the membranes of mitochondria and chloroplasts. How does the structure represented by vitamin K_2 explain the molecule's association with a hydrophobic membrane?
9. The relative orientation of polar and nonpolar amino acid side chains in integral membrane proteins is "inside out" relative to that of the amino acid side chains of water-soluble globular proteins. Explain.
10. What physical properties are conferred on biological membranes by phospholipids? How could the charge characteristics of the phospholipids affect binding of peripheral proteins to the membrane? What role might divalent metal ions play in the interaction of peripheral membrane proteins with phospholipids?
11. Predict the effects of the following on the phase transition temperature (T_m) and/or phospholipid mobility in pure dipalmitoylphosphatidylcholine vesicles (T_m = 41° C).
    (a) Raising the temperature from 30° to 50° C.
    (b) Introducing dipalmitoylphosphatidylcholine into the vesicle.
    (c) Introducing dimyristoylphosphatidylcholine into the vesicles.
    (d) Incorporating integral membrane proteins into the vesicles.
12. Differentiate between peripheral and integral membrane proteins with respect to location, orientation, and interactions that bind the protein to the membrane. What are some strategies used to differentiate peripheral and integral proteins by means of detergents or chelating agents?
13. Frequently, integral membrane proteins are glycosylated with complex carbohydrate arrays. Explain how glycosylation further enhances the asymmetric orientation of integral proteins.
14. In examining a cell extract from a culture of Gram negative bacteria, you identified a protein suspected to be periplasmic. Upon performing the osmotic shock experiment described in the text, you find a small amount (app. 10%) of the total enzymatic activity in the extracellular fraction. What control or other experiments might you do to convince yourself that the protein is periplasmic? If you had only data from the experiment described, what other possibilities could explain your results?
15. You are characterizing a protein in a membrane fraction that was dissolved in octylglucoside. You have estimated the molecular weight to be approximately 60,000. However, upon exhaustive treatment to remove most of the detergent, the protein elutes from a 100,000-M_r cutoff gel exclusion column in the void (excluded) volume. What can you conclude from these data?
16. Integral transmembrane proteins often contain helical segments of the appropriate length to span the membrane. These helices are composed of hydrophobic amino acid residues. In transmembranous proteins with multiple segments that span the membrane, you may find some hydrophilic residue side chains. Why are hydrophilic side chains not favored in single-span membrane proteins? How may the hydrophilic side chains be accommodated in multiple-span proteins?

Catalysis

PART

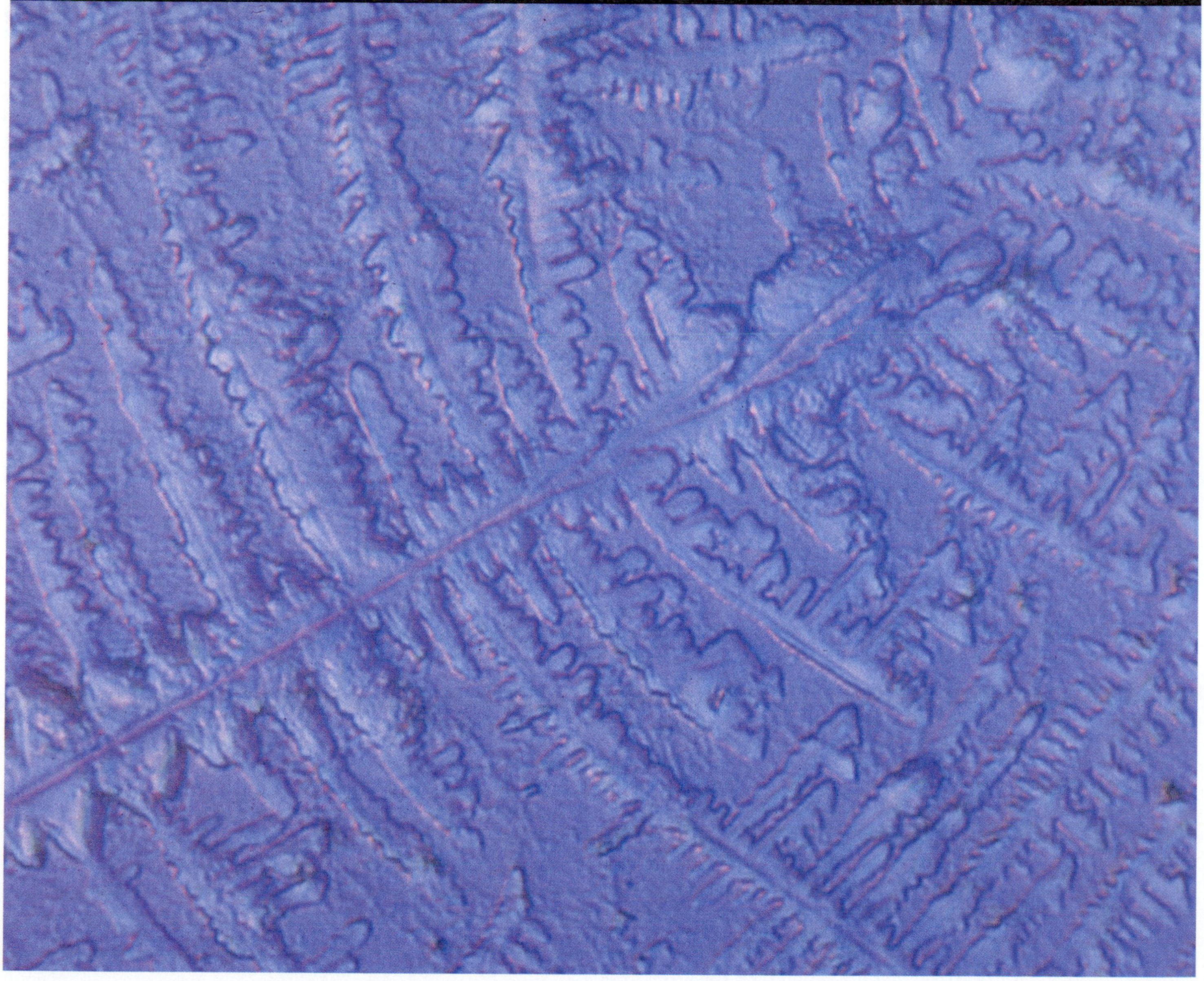

Light micrograph (160X) of human insulin crystals. The insulin was synthesized in bacteria using plasmids prepared by standard recombinant DNA technology methods. (© Visuals Unlimited.)

Most biochemical reactions are catalyzed by protein enzymes that function under very mild conditions so as not to disrupt the delicate fabric of the cell. The surface of the enzyme bears specific sites for binding the substrates and reactive groups that actually execute the reactions (chapters 8 and 9). Inhibitors also bind to the enzyme surface and either influence the binding of substrate or the activity of the enzyme. Enzyme kinetics is studied for two reasons: (1) it is a practical concern to determine the activity of the enzyme under different conditions; (2) frequently the analysis of enzyme kinetics gives information about the mechanism of enzyme action.

Most enzymes spontaneously process substrates when present, as long as inhibitory factors do not prevent this from happening. A few enzymes, known as regulatory enzymes, do not react spontaneously with their substrate(s) unless other metabolic signals give them the go-ahead. These metabolic signals usually consist of small molecules that bind to the enzyme and influence the structure of the active site (chapter 10). Regulatory enzymes function at strategic locations in metabolic pathways, usually at the first step in a linear pathway and at the first steps after a branchpoint in a forked pathway.

Frequently enzymes act in concert with small-molecule coenyzmes or cofactors, which supplement the reactive groups on the enzyme (chapter 11). Coenzymes or cofactors are distinguished from substrates by the fact that they function as catalysts. They are distinguishable from inhibitors or activators in that they participate directly in the reaction.

8

CHAPTER

Enzyme Kinetics

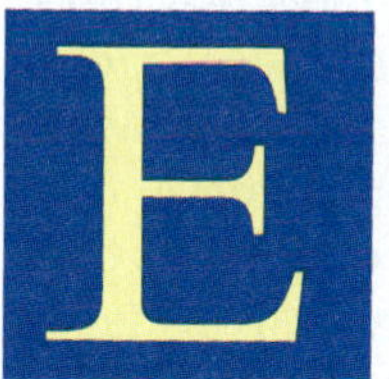

Enzymes are biochemical catalysts. By a "catalyst" we mean a substance that accelerates a chemical reaction without itself undergoing any net change. The catalytic activities of some enzymes are extraordinary. For example, carbonic anhydrase, an enzyme found in red blood cells, catalyzes the reaction

$$CO_2 + H_2O \rightarrow H_2CO_3 \quad (1)$$

In the presence of the enzyme, this reaction occurs about 10^7 times more rapidly than it does in the absence of the enzyme. One molecule of carbonic anhydrase can hydrate about 10^6 molecules of CO_2 a second. Catalase, an enzyme that occurs in many different tissues, is even faster. It catalyzes the reaction

$$2\ H_2O_2 \rightarrow 2\ H_2O + O_2 \quad (2)$$

Each molecule of catalase can decompose more than 10^7 molecules of H_2O_2 a second! Catalase and carbonic anhydrase are among the fastest enzymes known. At the other extreme, lysozyme, which catalyzes the hydrolysis of glycosidic bonds in bacterial cell walls, succeeds in splitting a bond only about once every two seconds.

In this and the next two chapters, we will explore how enzymes work and how cells are able to regulate their activities. We will focus first on the kinetics of enzymatic reactions. By examining how the rates of enzymatic reactions depend on pH and temperature and on the concentrations of the reactants, products and inhibitors, we can learn a great deal about how enzymes operate.

The Discovery of Enzymes

The existence of catalysts in biological materials was recognized as early as 1835 by Jons Berzelius, who discovered several elements, introduced the way of writing chemical symbols, and coined the term "catalysis." Berzelius found that potatoes contained something that catalyzed the breakdown of starch, and he suggested that all natural products are formed under the influence of such catalysts. But the chemical nature of biological catalysts was unknown, and it remained a mystery for many years. In the period from 1850 to 1860, Louis Pasteur demonstrated that fermentation, the anaerobic breakdown of sugar to CO_2 and ethanol, occurred in the presence of living cells, and did not occur in a flask that was capped after any cells that it contained had been killed by heat. Then, in 1897, Eduard Buchner discovered quite by accident that fermentation was catalyzed by a clear juice that he had prepared by grinding yeast with sand and filtering out the unbroken cells. Looking for a way to preserve the juice, Buchner had added sugar. His reasoning was that cooks used sugar to preserve jam. It probably was a disappointment to him that the sugar was broken down rapidly and the mixture frothed with CO_2. In spite of his disappointment, Buchner's discovery made it possible to explore metabolic processes such as fermentation in a greatly simplified system, without having to deal with the complexities of cell growth and multiplication, and without the barriers imposed by cell walls or membranes. Arthur Harden and William Young soon showed that yeast extracts contained two different types of molecules, both of which were necessary in order for fermentation to occur. Some were small, dialyzable, heat-stable molecules such as inorganic phosphate; others were much larger, nondialyzable molecules, the enzymes, which were destroyed easily by heat.

Although early investigators surmised that enzymes might be proteins, the point remained in dispute until 1927, when James Sumner succeeded in purifying and crystallizing the enzyme urease from beans. In the 1930s, John Northrup isolated and characterized a series of digestive enzymes, generalizing Sumner's conclusion that enzymes are proteins. Since then, thousands of different enzymes have been purified, and the structures of many of them have been solved to atomic resolution; almost all of these molecules have proved to be proteins. Surprisingly, however, recent work has shown that some RNA molecules have enzymatic activity.

One of the most striking features of enzymes is their specificity. Each enzyme catalyzes only one type of reaction, showing a high selectivity for both reactants and products. The proteolytic enzyme trypsin, for example, catalyzes the cleavage of polypeptide chains next to a lysine or arginine residue (chapter 1). Chymotrypsin catalyzes the same type of reaction, but is specific for bonds following a phenylalanine, tyrosine, or tryptophan. Enzymes also are very specific with regard to the stereochemistry of a reaction. For example, fumarase, which adds water to the double bond of fumaric acid, always generates the L isomer of malic acid.

Enzyme Terminology

The reactants that enter into an enzymatic reaction are termed substrates. Many enzymes require additional small molecules called cofactors for their activity. Cofactors can be simple inorganic ions such as Mg^{2+}, or complex organic molecules known as coenzymes. In many cases the cofactor binds tightly to a specific site on the enzyme. An enzyme lacking an essential cofactor is called an apoenzyme, and the intact enzyme with the bound cofactor is called a holoenzyme.

Enzymes often are known by common names obtained by adding the suffix *-ase* to the name of the substrate or to the reaction that they catalyze. Thus glucose oxidase is an enzyme that catalyzes the oxidation of glucose; glucose-6-phosphatase catalyzes the hydrolysis of phosphate from glucose-6-phosphate; and urease catalyzes the hydrolysis of urea. Common names also are used for some groups of enzymes. For example, an enzyme that transfers a phosphate group from ATP to another molecule is usually called a "kinase," instead of the more formal "phosphotransferase."

Table 8.1
Classification of Two Representative Enzymes

Number	Systematic Name	Common Name	Reaction
1	Oxidoreductases (enzymes catalyzing oxidation-reduction reactions)		
1.1	Acting on CH—OH group of donors		
1.1.1	With NAD^+ or $NADP^+$ as acceptor		
1.1.1.1	Alcohol:NAD oxidoreductase	Alcohol dehydrogenase	Alcohol + $NAD^+ \leftrightharpoons$ aldehyde or ketone + NADH + H^+
1.1.3	With O_2 as acceptor		
1.1.3.4	β-D-glucose:oxygen oxidoreductase	Glucose oxidase	β-D-glucose + $O_2 \leftrightharpoons$ D-glucono-*d*-lactone + H_2O_2

A systematic scheme for classifying enzymes has been adopted by the International Union of Biochemistry. In this scheme, each enzyme is designated by four numbers separated by periods: the main class, the subclass, the sub-subclass, and the serial number of the enzyme in its sub-subclass. There are six main classes: (1) oxidoreductases, (2) transferases, (3) hydrolases, (4) lyases, (5) isomerases, and (6) ligases or synthases. Oxidoreductases catalyze oxidation-reduction reactions. Transferases catalyze the transfer of a functional group from one molecule to another. Hydrolases catalyze bond cleavage by the introduction of water. Lyases catalyze the removal of a group to form a double bond, or the addition of a group to a double bond. Isomerases catalyze intramolecular rearrangements. And ligases catalyze reactions that join two molecules. Table 8.1 illustrates the divisions of these main classes into subdivisions with two examples.

Basic Aspects of Chemical Kinetics

Before we dig into enzyme kinetics, we need to discuss some of the basic principles that apply to the kinetics of both enzymatic and nonenzymatic reactions. Let's first consider a simple nonenzymatic reaction in which a single reactant (A) is transformed into a product (P):

$$A \rightarrow P \tag{3}$$

To measure the rate, or velocity, of the reaction (v), we could plot the concentration of A as a function of time (fig. 8.1). The rate at any particular time t is

$$v = -\frac{d[A]}{dt} \quad (4)$$

where $d[A]/dt$ is the slope of the plot at that time. The minus sign is needed because v is defined by convention to be a positive number, whereas $d[A]/dt$ is always negative (A is disappearing with time). However, we might equally well choose to measure the increase in the concentration of the product as a function of time, in which case we could express the rate as $v = d[P]/dt$.

For a simple reaction of this type, the rate at any given time usually is found to be proportional to the remaining concentration of the reactant:

$$v = k[A] \quad (5)$$

The proportionality constant, k, is the rate constant. The rate constant is independent of the concentration of the reactant, but it can depend on other parameters such as temperature or pH, and as we will see it may be altered by a catalyst. It has dimensions of reciprocal seconds (s^{-1}). Combining equations (4) and (5), we have

$$\frac{d[A]}{dt} = -k[A] \quad (6)$$

This equation has the mathematical solution

$$[A] = [A]_0 e^{-kt} \quad \text{or} \quad \ln[A] = \ln[A]_0 - kt \quad (7)$$

where $[A]_0$ is the initial concentration of A and e is the base of the natural logarithm. Equation (7) says that [A] will decay exponentially from $[A]_0$ toward zero (see fig. 8.1). When $t = 1/k$, the reaction will have gone about 63% of the way to completion.

A reaction of this type is said to follow first-order kinetics, because the rate is proportional to the concentration of a single species raised to the first power (equation 5). An example is the decay of a radioactive isotope such as ^{14}C. The rate of decay at any time (the number of radioactive disintegrations per second) is simply proportional to the amount of ^{14}C that is present. The rate constant for this extremely slow nuclear reaction is $8 \times 10^{-12}\ s^{-1}$. Another example is the initial electron-transfer reaction that occurs when photosynthetic organisms are excited with light. In this case there actually are two reactants, the electron donor and the acceptor, but they are held close together on a protein so that they react as a unit; the excited complex simply decays spontaneously to a more stable state. This occurs with a rate constant of $3 \times 10^{11}\ s^{-1}$, which makes it one of the fastest reactions known. We will discuss this reaction in more detail in chapter 16.

A slightly more complicated situation arises if the reaction is reversible:

$$A \underset{k_r}{\overset{k_f}{\rightleftharpoons}} P \quad (8)$$

Figure 8.1

The kinetics of a first-order reaction in which a single reactant (A) is converted irreversibly to a product (P). The concentrations of A and P are plotted as functions of time. The rate (v) at any given time can be obtained from the slope of either curve: $v = -\frac{d[A]}{dt} = \frac{d[P]}{dt}$. The time-dependence of [A] is described by an exponential function, $[A] = [A]_0 e^{-kt}$, where $[A]_0$ is the initial concentration and k is the rate constant. When t is equal to $1/k$, [A] has decreased from $[A]_0$ to $[A]_0 \cdot e^{-1}$ (about $0.37[A]_0$).

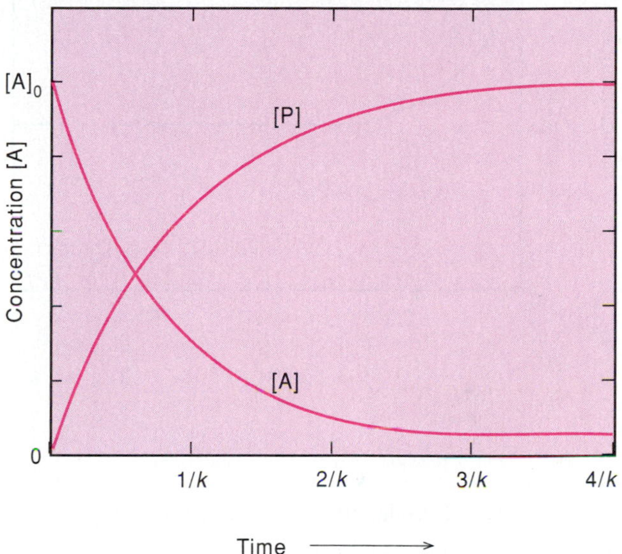

Here k_f is the rate constant for the forward reaction and k_r is that for the reverse. In a reversible reaction, the rate equation becomes

$$\frac{d[A]}{dt} = -k_f[A] + k_r[P] \quad (9)$$

The term $-k_f[A]$ is the same as in equation (6) and the term $k_r[P]$ describes the formation of A from P. In this case, [A] and [P] will proceed from their initial values ($[A]_0$ and $[P]_0$) to their final, equilibrium values, which generally will not be zero. At equilibrium, $d[A]/dt$ must go to zero, which requires that

$$k_f[A]_{eq} = k_r[P]_{eq} \quad (10)$$

where $[A]_{eq}$ and $[P]_{eq}$ are the equilibrium concentrations. Rearranging this expression gives

$$\frac{[P]_{eq}}{[A]_{eq}} = \frac{k_f}{k_r} = K_{eq} \quad (11)$$

where K_{eq} is the equilibrium constant. The solution to equation (9) is that [A] and [P] approach $[A]_{eq}$ and $[P]_{eq}$ exponentially with an overall rate constant that is the sum of k_f and k_r (fig. 8.2).

Now consider a reaction involving two reactants, A and B:

$$A + B \rightarrow P \quad (12)$$

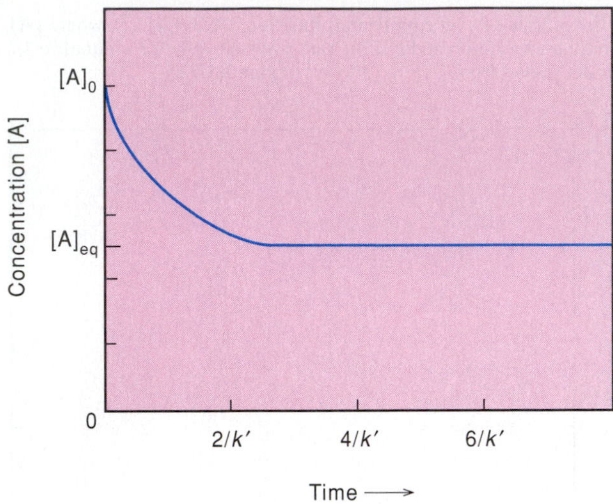

Figure 8.2

The kinetics of a reversible first-order reaction. The solution to the rate equation for such a reaction is $([A] - [A]_{eq}) = ([A]_0 - [A]_{eq})\, e^{-k't}$, where $k' = k_f + k_r$ and $[A]_{eq} = [A]_0/(1 + k_f/k_r)$. The curve drawn here is obtained when $k_f = k_r$.

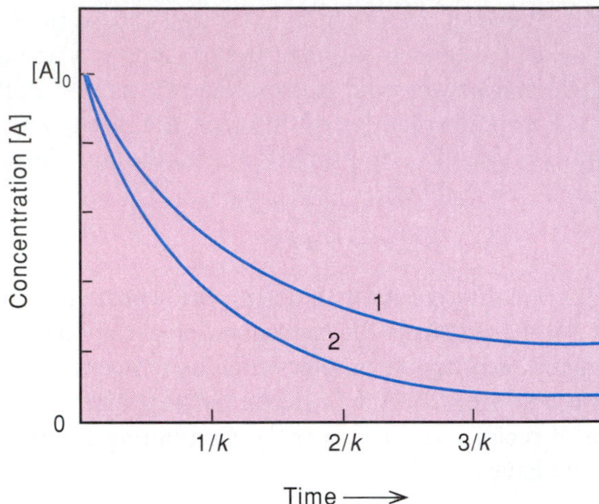

Figure 8.3

Curve 1: Kinetics of a second-order reaction between two reactants starting at the same concentration ($[A]_0$). The curve is a plot of the expression $[A] = 1/(kt + 1/[A]_0)$, which is the solution to the rate equation $d[A]/dt = -k[A]^2$. Curve 2: Kinetics of a second-order reaction when one reactant (B) is present in 100-fold excess. The rate is given by $v = k_{app}[A]$ with $k_{app} = k[B]$. The value of k was decreased by a factor of 100 relative to that used for curve 1. Curve 2 has essentially the same form as the curve for [A] in figure 8.1.

The rate of such a bimolecular reaction usually depends on the concentrations of both reactants:

$$\frac{d[A]}{dt} = -k[A][B] \qquad (13)$$

The reaction is said to follow second-order kinetics, because its rate is proportional to a product of two concentrations. The kinetics are first order in either [A] or [B] alone, but second order overall. The rate constant for a second-order reaction has dimensions of $\text{M}^{-1}\,\text{s}^{-1}$. If the concentrations of A and B are the same at the start of the reaction, they will remain the same as they both decrease toward zero. The rate then would be proportional to $[A]^2$ (see curve 1 in figure 8.3). On the other hand, if B is present in a large excess, then its concentration will not change very much while [A] decreases. In this case, the rate will be approximately proportional to [A] (curve 2 in figure 8.3).

Molecular Interpretations of Rate Constants: A Critical Amount of Energy Is Needed for the Reactants to Reach the Transition State

Part of the rationale behind equation (13) is simply that reactants A and B have to collide in order to react. Collisions occur as a result of random diffusion of the reactants in the solution. The number of collisions that occur per second is proportional to the product of the two concentrations, and the second-order rate constant k includes the proportionality factor for this relationship. But not every collision will result in a reaction. The rate constant also must include a factor that gives the fraction of the collisions that are effective. If every collision does result in a reaction, so that this second factor is 1, the rate constant for the reaction of two small molecules in aqueous solution will typically be about $10^{11}\ \text{M}^{-1}\,\text{s}^{-1}$. Because of their large masses, proteins diffuse relatively slowly, so the frequency at which a protein and a small molecule will collide is lower than the collision frequency for two small molecules. The maximum possible value of the rate constant for a reaction between a protein and a small molecule thus is typically on the order of 10^8 to 10^9 $\text{M}^{-1}\,\text{s}^{-1}$.

What determines the fraction of the collisions that result in a reaction? A partial answer is that the colliding species must have a certain critical energy in order to surmount a barrier that separates the reactants from the products. This principle is illustrated schematically in figure 8.4. The surface in figure 8.4*a* represents the energy of a system in which a proton can be bound to either of two molecules. Suppose that the proton is initially on molecule A, and we are interested in how rapidly it moves to molecule B. As the proton moves from one place to the other, its electrostatic interactions with molecule A become less favorable, while its interactions with B improve. The energy of the system goes through a maximum when the proton is at an intermediate position. At this point, the system is said to be in the transition state. The probability that a collision will lead

Figure 8.4

Schematic energy diagrams for a reaction in which a proton moves from one molecule to another. In the perspective drawing (*a*), coordinates in the plane at the bottom represent the location of the proton. The proton moves in three-dimensional space, but only its positions in two of these dimensions can be indicated in the drawing. The energy of the system for any particular set of coordinates is represented by the distance of the cuplike surface above the plane. The two minima in the energy surface indicate the positions of the proton when it is bound optimally to one molecule or the other. The best route along the surface from one of these minima to the other goes through a pass or saddle point. Drawing (*b*) shows a plot of the energy as a function of distance along the optimal route over this pass. The activation energy of the reaction (ΔE_a) is the difference between the energies at the pass and at the starting point.

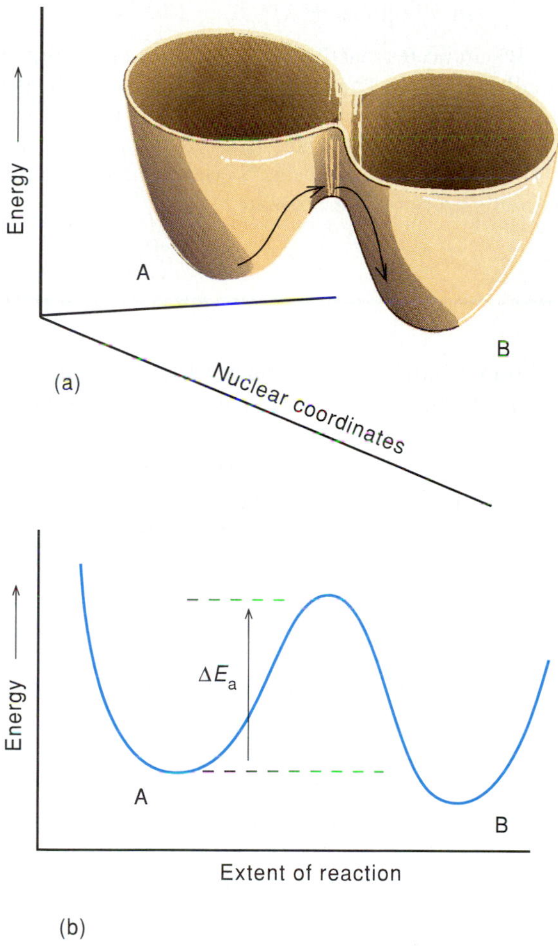

to a reaction depends, in part, on the probability that the molecules collide with enough energy to reach this state. The difference between the lowest possible energy of the reactants and the energy of the transition state is termed the activation energy.

The probability that the collision complex of the reactants will have enough energy to reach the transition state can be calculated from an expression that was first applied to chemical kinetics by Svante Arrhenius. Assume that in order to reach the transition state the energy of the collision complex must exceed the minimum possible energy by an amount ΔE_a. The fraction of the complexes whose energy meets this requirement

Figure 8.5

The temperature dependence of a reaction rate, shown in the form of an Arrhenius plot. The natural logarithm of the rate constant is plotted as a function of $1/T$, where T is the absolute temperature. (Temperature decreases from left to right in the graph.)

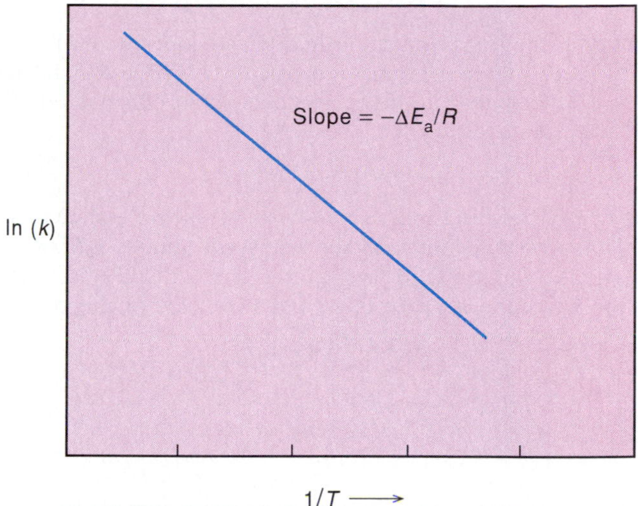

is proportional to $e^{-\Delta E_a/RT}$, where R is the gas constant (2 cal mole^{-1} deg^{-1}) and T is the absolute temperature. We therefore can write the rate constant as

$$k = k_c e^{-\Delta E_a/RT} \quad \text{or} \quad \ln k = \ln k_c - \Delta E_a/RT \qquad (14)$$

where k_c is a proportionality constant. Arrhenius pointed out that equation (14) provides a simple explanation for why many reactions speed up dramatically with increasing temperature. A plot of ln k versus $1/T$, which is termed an Arrhenius plot, frequently is found to be linear and to have a negative slope, as shown in figure 8.5. According to equation (14), the slope of such a plot is $-\Delta E_a/R$.

In figure 8.4, note that the nuclear geometry of the molecules in the transition state is intermediate between the optimal geometries of the reactants and the products. To reach the transition state from either the reactant side or the product side requires a nuclear distortion of the molecules, and this distortion must be in the right direction to connect the reactants with the products. In fact, we can view the transition state as a state in which the geometries of the reactants and the products have become the same.

The requirement for a certain nuclear geometry in the transition state is an additional factor that is superimposed on the requirement for the activation energy, ΔE_a. These two requirements can be combined by replacing ΔE_a in equation (14) by an activation free energy, $\Delta G^{\ddagger}$:

$$k = k_0 e^{-\Delta G^{\ddagger}/RT} \qquad (15)$$

In this expression, k_0 is an intrinsic rate constant for the conversion of the transition state into the products. The factor $e^{-\Delta G^{\ddagger}/RT}$ can be interpreted as an effective equilibrium constant

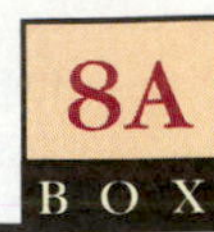

Activation Free Energies, Enthalpies, and Entropies

In the transition-state theory, the concentration of the reactant in the transition state (A^*) is related to the total concentration of the reactant by an effective equilibrium constant, $K^‡$:

$$[A^*] = K^‡[A] \qquad (B1)$$

If the rate of the reaction is given by $k[A]$, and we equate this to $k_0[A^*]$, the observed rate constant k must be equal to $k_0K^‡$. $K^‡$ can be related to the standard free energy change ($\Delta G^‡$) associated with the formation of A^* by using equation (13) of chapter 2:

$$\Delta G^‡ = -RT \ln K^‡$$

or

$$K^‡ = e^{-\Delta G^‡/RT} \qquad (B2)$$

Combining equation (B2) with equation (16) of this chapter gives:

$$k = k_0 e^{-\Delta G^‡/RT} = k_0 e^{-(\Delta H^‡ - T\Delta S^‡)/RT} \qquad (B3)$$

$$= k_0[e^{-\Delta H^‡/RT}][e^{\Delta S^‡/R}] \qquad (B4)$$

or

$$\ln k = [\ln k_0 + \Delta S^‡/R] - \Delta H^‡/RT \qquad (B5)$$

Equation (B5) indicates that the slope of an Arrhenius plot ($-\Delta E_a/R$) is actually $-\Delta H^‡/R$. The activation enthalpy ($\Delta H^‡$) usually is similar to the difference in energy between the reactants and the transition state, but it also takes into account any changes in volume. The activation entropy ($\Delta S^‡$) contributes to the intercept of the Arrhenius plot on the ordinate.

---

for the formation of the transition state from the reactants (see box 8A). From the definition of free energy (equation 9 of chapter 2), we also can write

$$\Delta G^‡ = \Delta H^‡ - T\Delta S^‡ \qquad \mathbf{(16)}$$

where $\Delta H^‡$ and $\Delta S^‡$ are the changes in enthalpy and entropy that are needed in order to reach the transition state. Thus $\Delta G^‡$ includes the change in entropy associated with arranging the nuclei of the molecules in the particular way that is required in the transition state. If reaching this state requires going to a more restricted nuclear geometry, as is generally the case, $\Delta S^‡$ will be negative and therefore will decrease the rate constant.

Similar considerations apply to a unimolecular chemical reaction of the type expressed by equation (3). Although there are no collisions between separate reactants in that case, the requirements that the reactant have sufficient energy and undergo appropriate nuclear distortions are just the same. The magnitude of the rate constant depends on how frequently these conditions are met in the course of the random thermal fluctuations of the molecule and the solvent.

Although many reactions can be sped up by increasing the temperature, this is not a very useful option for living organisms. Many organisms have little or no control over the ambient temperature. In addition, most species can survive only within a rather narrow range of temperatures, partly because proteins and other macromolecules become denatured at higher temperatures. Finally, changing the temperature is not a very selective way to control reaction rates, since the great majority of chemical reactions have positive activation energies; all of these reactions will speed up when the temperature is raised.

From equation (15), it appears that there are two other possible ways to increase the rate constant for a reaction, in addition to changing the temperature: increase k_0 or decrease $\Delta G^‡$. However, changing k_0 is not a very realistic option because there usually is no way that it can be done to any significant extent. $\Delta G^‡$ is another matter. Because rate constants depend *exponentially* on $-\Delta G^‡/RT$, relatively small changes in $\Delta G^‡$ can change reaction rates by many orders of magnitude. At physiological temperature, it takes a decrease of only 1.36 kcal/mole to speed up a reaction by a factor of 10, and a decrease by 8.16 kcal/mole will increase the rate by a factor of 10^6. The formation of a single H bond can release anywhere from 4 to 10 kcal/mole. In addition, because $\Delta G^‡$ depends critically on the detailed structures of the reactants and products and the detailed nature of the reaction, it clearly affords an opportunity to control reaction rates with a great deal of specificity. An enzyme, then, must work by decreasing the activation free energy for the specific reaction that it catalyzes (fig. 8.6). This could occur if the structure of the active site is, in some way, complementary to the structure of the transition state. If, for example, the formation of the transition state requires the accumulation of negative electrical charge on a particular atom of the substrate, the free energy of the transition state could be lowered if a positive charge was placed nearby on the enzyme.

From figure 8.6, note that an enzyme does not alter the free energies of the substrates or the products of the reaction. Thus an enzyme cannot change the overall equilibrium constant of a reaction. An enzyme, or any catalyst, for that matter, affects only the speed with which a reaction approaches equilibrium.

Figure 8.6

An enzyme speeds up a reaction by decreasing $\Delta G^{\ddagger}$. The enzyme does not change the free energy of the substrate (S) or product (P); it lowers the free energy of the transition state. The two vertical arrows indicate the activation free energies ($\Delta G^{\ddagger}$) of the catalyzed and uncatalyzed reactions. This figure neglects the enzyme-substrate and enzyme-product complexes that are intermediates in the reaction.

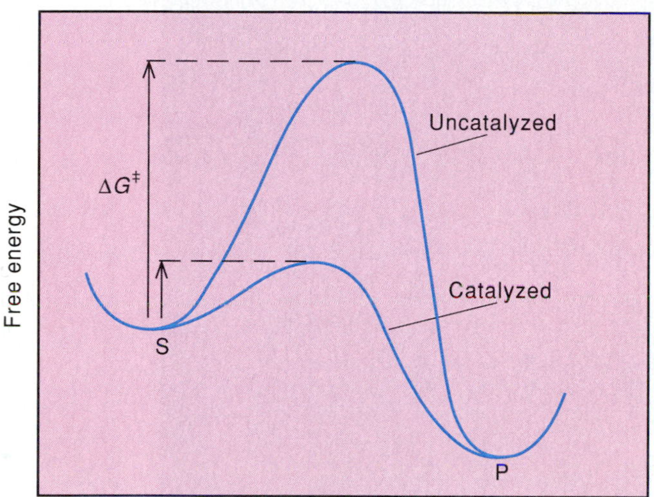

Figure 8.7

The rates of many enzyme-catalyzed reactions have a hyperbolic dependence on the substrate concentration. V_{max} is the maximum rate, and K_m is the substrate concentration at which the rate is half-maximal.

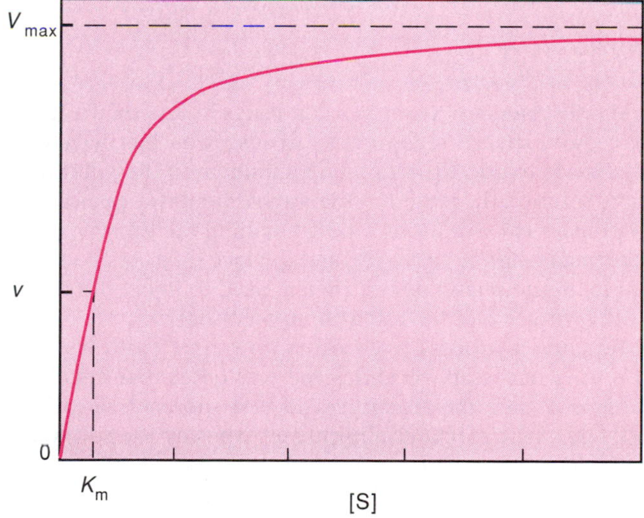

Kinetics of Enzyme-Catalyzed Reactions

The rate equations that we discussed in the preceding section imply that the velocity of an uncatalyzed reaction would increase indefinitely with an increase in the concentration of the reactants. With enzyme-catalyzed reactions, something very different is observed. The rate usually increases linearly with the substrate concentration at low concentrations, but then levels off at high concentrations (fig. 8.7). This behavior is described as a hyperbolic dependence on concentration, because the curve has the form of a hyperbola. The explanation is straightforward. In order for an enzyme to affect $\Delta G^{\ddagger}$, the substrate must bind to a special site on the protein, the active site (fig. 8.8). At very low concentrations of substrate, the active sites of most of the enzyme molecules in the solution will be unoccupied. Under these conditions, increasing the substrate concentration will bring more enzyme molecules into play, and the reaction will speed up. At high concentrations, on the other hand, most of the enzyme molecules will have their active sites occupied, and the observed rate will depend only on the rate at which the bound reactants are converted into products. Further increases in the substrate concentration then will have little effect.

Techniques for measuring rates of enzyme-catalyzed reactions are described in box 8B.

The Michaelis-Menten Equation Provides a Means to Analyze Enzyme-Catalyzed Reactions in Terms of Rate Constants

We can analyze the kinetics of an enzymatic reaction in more detail as follows. Consider a reaction involving a single substrate (S), one product (P), and an enzyme (E). Suppose that the substrate binds to the enzyme with rate constant k_1 to give an enzyme-substrate complex (ES). Suppose further that ES is converted to P with rate constant k_2, but that it also can dissociate to release S with rate constant k_{-1}:

$$\mathrm{E} + \mathrm{S} \underset{k_{-1}}{\overset{k_1}{\rightleftharpoons}} \mathrm{ES} \xrightarrow{k_2} \mathrm{E} + \mathrm{P} \quad \textbf{(17)}$$

We can ignore any reversibility of the step in which ES is converted to E and P, provided that we measure the rate quickly enough after we mix E and S so that the concentration of P is still very small.

Let's express the velocity of the reaction in terms of the rate of formation of P:

$$v = \frac{d[\mathrm{P}]}{dt} = k_2[\mathrm{ES}] \quad \textbf{(18)}$$

The problem is to relate [ES], the concentration of the enzyme-substrate complex, to the concentration of the substrate, [S]. We can do this by writing a rate equation that describes the formation and breakdown of ES:

$$\frac{d[\mathrm{ES}]}{dt} = k_1[\mathrm{E}][\mathrm{S}] - k_{-1}[\mathrm{ES}] - k_2[\mathrm{ES}] \quad \textbf{(19)}$$

Figure 8.8

Thermolysin, an enzyme that hydrolyzes peptide bonds, with a structural analog of a substrate bound at the active site. The enzyme is shown in orange; the substrate analog (phosphoramidon), in yellow. In the active site, the substrate interacts with a Zn^{2+} ion (white) and amino acid residues that participate in the catalytic mechanism. The side chains of two key residues, Glu 143 and His 231, are shown in turquoise. (Based on the crystal structure described by D. E. Tronrud, A. F. Monzingo, and B. W. Matthews.) The mechanism of action of thermolysin is discussed in chapter 9.

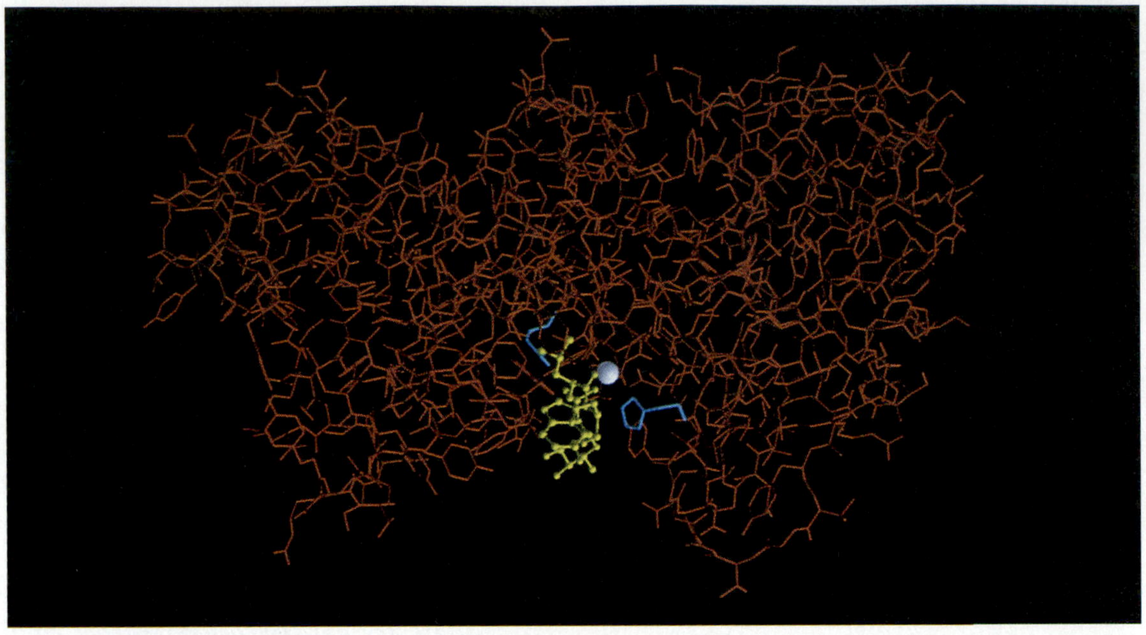

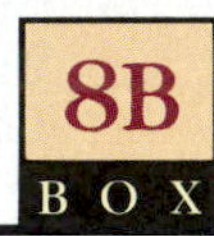

8B BOX

Techniques for Measuring Enzymatic Reaction Rates

In a typical measurement of enzyme kinetics, we prepare a solution containing all components of the reaction mixture except one. The missing component could be either the enzyme or a substrate. We then add this material, allow the reaction to proceed for a fixed period of time (usually about a minute), and measure the change in the concentration of a reactant or product.

Spectroscopic analytical techniques lend themselves well to kinetic studies, because the measurements can be made essentially instantaneously. In many cases, it is not necessary to stop the reaction whose concentration is to be determined. For example, NADH, has an optical absorption band at 340 nm and a fluorescence emission band at 450 nm. Neither of these properties is shared by the oxidized form of the molecule, NAD^+. The kinetics of any reaction in which NADH is oxidized to NAD^+ can be measured from the change of absorbance at 340 nm or fluorescence at 450 nm.

Measurements of reactions that occur in less than a few seconds require special techniques, because it ordinarily takes several seconds to add the limiting component to the solution and mix the solution thoroughly. One way to circumvent this problem is to prepare two solutions, one containing the enzyme and the other containing the substrate, and to place them in two separate syringes. A pneumatic device then is used to inject the contents of both syringes rapidly into a common chamber.

To study a process that occurs on a time scale faster than 0.01 s, it is necessary to find some way other than mechanical mixing to initiate the reaction. The best approach usually is to create one of the reactants abruptly by exposing the solution to a brief flash of light.

In most studies of enzyme-catalyzed reactions the concentration of the enzyme is very low compared with the concentration of the substrate. Under these conditions, the conversion of some or even all of the enzyme into ES will cause only a relatively minute decrease in [S]. When the enzyme is first added to the solution of the substrate, there will be a brief period while [ES] increases and [E] decreases, but the system will soon reach a steady state in which [ES] is relatively constant (fig. 8.9). In

Figure 8.9

Concentrations of free enzyme (E), substrate (S), enzyme-substrate complex (ES), and product (P) over the time course of a reaction. The shaded portion of the top graph is shown in expanded form in the bottom graph. After a brief initial period (usually less than a few seconds) the concentration of ES remains approximately constant for an extended period. The steady-state approximation is applicable during this second period. Most measurements of enzyme kinetics are made in the steady state. The measurement still must be made quickly enough so that the substrate concentration is present at approximately its initial concentration, [P] is close to zero, and the rate of the reaction ($-d[S]/dt$ or $d[P]/dt$) is more or less constant. At later times, it is necessary to consider the reverse reaction, P → S.

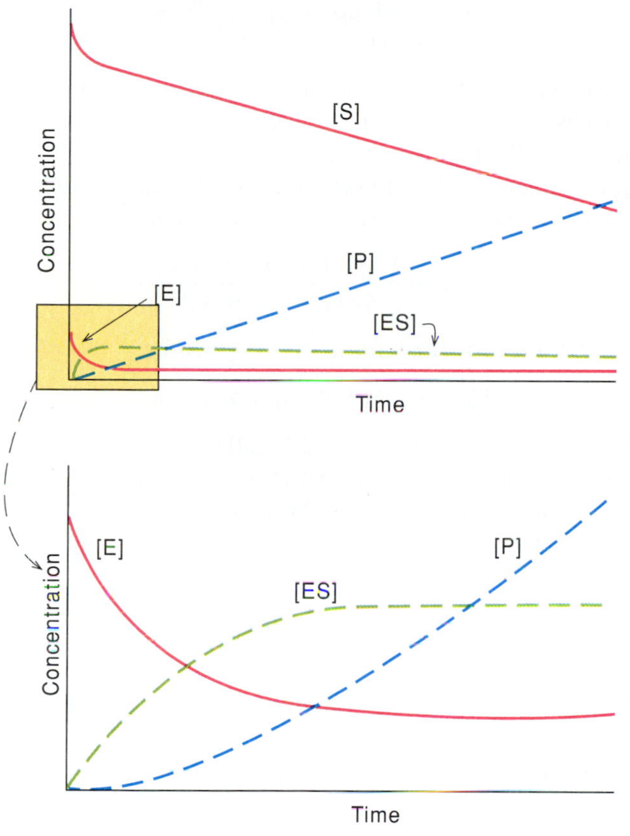

the steady state, the rates of formation and breakdown of ES have become essentially equal, and $d[ES]/dt \approx 0$. From equation (19), we then have

$$k_1[E][S] = k_{-1}[ES] + k_2[ES] \quad (20)$$

or

$$[ES] = [E][S]k_1/(k_{-1} + k_2) \quad (21)$$

In equations (19)–(21), note that [E] refers to the concentration of free enzyme, not the total concentration. But if we represent the total concentration of enzyme by $[E_T]$, we can write

$$[E] = [E_T] - [ES] \quad (22)$$

Combining equations (21) and (22) gives

$$[ES] = \frac{\{[E_T] - [ES]\}[S]k_1}{k_{-1} + k_2} \quad (23)$$

This expression can be simplified by defining a single constant, K_m, made up from the three rate constants:

$$K_m = \frac{k_{-1} + k_2}{k_1} \quad (24)$$

With this substitution, equation (23) becomes

$$[ES] = \frac{\{[E_T] - [ES]\}[S]}{K_m} \quad (25)$$

which can be rearranged to

$$[ES] = \frac{[E_T][S]}{[S] + K_m} \quad (26)$$

Combining equations (18) and (26) gives

$$v = \frac{k_2[E_T][S]}{[S] + K_m} \quad (27)$$

The product $k_2[E_T]$ in the numerator of equation (27) is the *maximum* rate of the reaction at the particular concentration of the enzyme designated by $[E_T]$. Suppose that [S] is infinitely large ($>> K_m$). The denominator of equation (27) then becomes approximately equal to [S], and v approaches $k_2[E_T]$. If we define another constant $V_{max} = k_2[E_T]$, equation (27) can be written as

$$v = \frac{V_{max}[S]}{[S] + K_m} \quad (28)$$

Equation (28) accounts for the hyperbolic relationship between v and [S]. It is known as the Michaelis-Menten equation because it was first derived, in a slightly simpler form, by Leonor Michaelis and Maude Menten in 1913. The constant K_m is called the Michaelis constant. To see the meaning of K_m, suppose that $[S] = K_m$. The denominator in equation (28) then is equal to 2[S], and $v = V_{max}/2$. Thus, K_m is the substrate concentration at which the velocity is half-maximal (see fig. 8.7). Note that K_m has the same dimensions as a concentration (M), because k_{-1} and k_2, the two rate constants in the numerator of equation (24), are first-order rate constants and have units of s^{-1}, whereas k_1 is a second-order rate constant with units of M^{-1} s^{-1}.

If [S] is much smaller than K_m, the denominator of equation (28) becomes approximately equal to K_m, and v is approximately equal to $V_{max}[S]/K_m$. So the Michaelis-Menten equation also accounts for the linear dependence of v on [S] at very low substrate concentrations (see fig. 8.7).

A common way of plotting data that fit the Michaelis-Menten equation is to plot $1/v$ versus $1/[S]$. Such a plot is known as a Lineweaver-Burk plot or a double-reciprocal plot. Taking the reciprocals of both sides of equation (28) gives

$$\frac{1}{v} = \frac{1}{V_{max}} + \frac{K_m}{V_{max}}\frac{1}{[S]} \quad (29)$$

This expression indicates that a plot of $1/v$ versus $1/[S]$ will fit a straight line with a slope of K_m/V_{max} (fig. 8.10). The intercept of the line on the ordinate occurs at $1/v = 1/V_{max}$, and the intercept on the abscissa occurs at $1/[S] = -1/K_m$. V_{max} and K_m thus can be determined simply from the graph. However, most experimenters now use a computer program to fit data directly to equation (28) because this approach provides a better analysis of the experimental uncertainties.

The Michaelis Constant Is a Function of Three or More Rate Constants

We have seen that K_m is the substrate concentration at which an enzyme-catalyzed reaction occurs at half its maximal rate. K_m thus provides an indication of the substrate concentration range at which the enzyme is most effective at increasing the rate of the reaction. K_m values for a number of different enzyme-substrate pairs are given in table 8.2. The values typically range between 10^{-6} and 10^{-1} M. With enzymes that act on several different substrates, K_m can vary substantially from substrate to substrate. With chymotrypsin, for example, the K_m for glycyltyrosinamide is about 50 times that for *N*-benzoyltyrosinamide (see table 8.2).

Does K_m give us any information on how tightly a particular substrate binds to the active site of an enzyme? Not necessarily. In equation (24), we defined K_m as $(k_{-1} + k_2)/k_1$. Note that the numerator in this expression includes both k_{-1}, the rate constant for dissociation of the enzyme-substrate complex, and k_2, the rate constant for the conversion of ES into products. If k_{-1} is much greater than k_2, K_m will be approximately equal to k_{-1}/k_1. The ratio k_{-1}/k_1 is the dissociation constant of ES, which is the reciprocal of the equilibrium constant for the formation of ES. (See equation 11.) Thus, in some cases, K_m is approximately equal to the dissociation constant of the enzyme-substrate complex. Under these circumstances, a smaller K_m means a smaller dissociation constant, which implies tighter binding to the enzyme. But whether or not this is a valid approximation depends on the enzyme and the substrate. In general, we can say only that the dissociation constant must be smaller than K_m.

The relationship between K_m and the dissociation constant of the enzyme-substrate complex becomes even more tenuous if the reaction mechanism is more complex than that shown in equation (17). Suppose, for example, that there are two different, interconvertible complexes on the enzyme, one involving the substrate and the other involving the products:

$$E + S \underset{k_{-1}}{\overset{k_1}{\rightleftharpoons}} ES \underset{k_{-2}}{\overset{k_2}{\rightleftharpoons}} EP \overset{k_3}{\rightarrow} E + P \tag{30}$$

It turns out that the rate of such a reaction still follows the Michaelis-Menten equation (equation 28), and still gives a linear Lineweaver-Burk plot (equation 29), but the observed K_m and V_{max} are made up of more complicated algebraic combinations of the individual rate constants. The equation for K_m is

$$K_m = \frac{k_{-1}k_3 + k_{-1}k_{-2} + k_2k_3}{k_1(k_2 + k_{-2} + k_3)} \tag{31}$$

Similarly, V_{max} for this mechanism is given by

$$V_{max} = k_2k_3[E_T]/(k_2 + k_{-2} + k_3) \tag{32}$$

If k_3 is much greater than k_2 and k_{-2}, these expressions collapse to the expressions that were derived for the simpler mechanism. The rate-determining step of the reaction then would be the conversion of ES to EP.

Figure 8.10

The Michaelis-Menten equation accounts for the hyperbolic dependence of velocity on substrate concentration. A plot of the reciprocal of the rate ($1/v$) as a function of the reciprocal of the substrate concentration ($1/[S]$) fits a straight line. Extrapolating the line to its intercept on the ordinate (infinite substrate concentration) gives $1/V_{max}$. Extrapolating to the intercept on the abscissa gives $-1/K_m$.

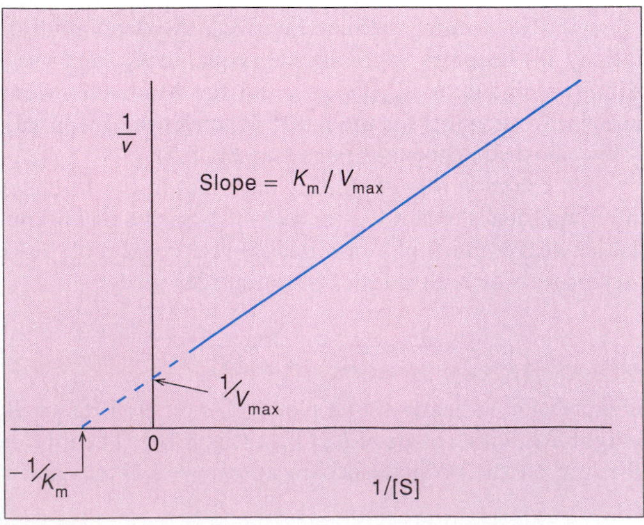

Table 8.2
The Michaelis Constants for Some Enzymes

Enzyme and Substrate	K_m (M)
Catalase	
H_2O_2	1.1
Hexokinase	
Glucose	1.5×10^{-4}
Fructose	1.5×10^{-3}
Chymotrypsin	
N-Benzoyltyrosinamide	2.5×10^{-3}
N-Formyltyrosinamide	1.2×10^{-2}
N-Acetyltyrosinamide	3.2×10^{-2}
Glycyltyrosinamide	1.2×10^{-1}
Aspartate aminotransferase	
Aspartate	9.0×10^{-4}
α-Ketoglutarate	1.0×10^{-4}
Fumarase	
Fumarate	5.0×10^{-6}
Malate	2.5×10^{-5}

If additional intermediate complexes are added to the mechanism, the Michaelis-Menten equation continues to hold, but the relationships of K_m and V_{max} to the microscopic rate constants become more and more complex. One lesson here is that kinetic measurements can never prove that a particular reaction mechanism is correct, because many different mechanisms could result in the same observed kinetics. Kinetic measurements can, however, often be used to rule out a possible mechanism, and thus to distinguish between several alternative mechanisms.

The Specificity Constant Is Usually the Best Index of Enzyme Effectiveness

The turnover number of an enzyme, k_{cat}, is the maximum number of moles of substrate that are converted to product each second, per mole of enzyme (or per mole of active sites if the enzyme has more than one active site). Because the maximum rate is obtained at high substrate concentrations, when all the active sites are occupied with substrate, k_{cat} is a measure of how rapidly an enzyme can operate once the active site is filled. It is given simply by $k_{cat} = V_{max}/[ET]$. As was the case with K_m, the relationship of k_{cat} to microscopic rate constants such as k_2 and k_3 depends on the details of the reaction mechanism. Some representative turnover numbers are given in table 8.3.

Table 8.3
Values of k_{cat} for Some Enzymes

Enzyme	k_{cat} (s^{-1})
Catalase	40,000,000
Carbonic anhydrase	1,000,000
Acetylcholinesterase	14,000
Penicillinase	2,000
Lactate dehydrogenase	1,000
Chymotrypsin	100
DNA polymerase I	15
Lysozyme	0.5

Under physiological conditions, enzymes usually do not operate at saturating substrate concentrations. More typically, the ratio of the substrate concentration to K_m is in the range of 0.01 to 1.0. We have seen that the rate of an enzyme-catalyzed reaction at a low substrate concentration is given by $v = V_{max}[S]/K_m$. Under these conditions, the number of moles of substrate converted to product per second per mole of enzyme is $(V_{max}[S]/K_m)/[E_T]$ which is the same as $(k_{cat}/K_m)[S]$. The ratio k_{cat}/K_m is therefore a measure of how rapidly an enzyme can work at low [S]. This ratio is referred to as the specificity constant. Values of k_{cat}/K_m for some particularly active enzymes are given in table 8.4.

The specificity constant, k_{cat}/K_m, is useful for comparing the relative abilities of different compounds to serve as a substrate for the same enzyme. If the concentrations of two substrates are the same, and are small relative to the values of K_m, the ratio of the rates with the two substrates is equal to the ratio of the specificity constants.

Another use of the specificity constant is to compare the rate of an enzyme-catalyzed reaction with the rate at which the random diffusion of the enzyme and substrate brings the two molecules into collision. We mentioned earlier that if every collision between a protein and a small molecule resulted in a reaction, the maximum possible value of the second-order rate constant would typically be on the order of 10^8 to 10^9 M^{-1} s^{-1}. Some of the values of k_{cat}/K_m given in table 8.4 are in this range. These enzymes have achieved an astonishing state of perfection. The reactions they catalyze proceed at nearly the maximum possible speed, given a fixed, low concentration of substrate and given the restriction that the enzyme and substrate have to find each other by diffusion. The only ways to go much faster would

Table 8.4
Enzymes for Which k_{cat}/K_m Is Close to the Diffusion-controlled Association Rate

Enzyme	Substrate	k_{cat} (s^{-1})	K_m (M)	k_{cat}/K_m (M^{-1} s^{-1})
Acetylcholinesterase	Acetylcholine	1.4×10^4	9×10^{-5}	1.6×10^8
Carbonic	CO_2	1×10^6	0.012	8.3×10^7
anhydrase	HCO_3^-	4×10^5	0.026	1.5×10^7
Catalase	H_2O_2	4×10^7	1.1	4×10^7
Crotonase	Crotonyl-CoA	5.7×10^3	2×10^{-5}	2.8×10^8
Fumarase	Fumarate	800	5×10^{-6}	1.6×10^8
	Malate	900	2.5×10^{-5}	3.6×10^7
Triosephosphate isomerase	Glyceraldehyde 3-phosphate	4.3×10^3	4.7×10^{-4}	2.4×10^8
β-Lactamase	Benzylpenicillin	2.0×10^3	2×10^{-5}	1×10^8

Source: From *Enzyme Structure and Mechanism,* 3rd Edition by Alan Fersht. Copyright © 1985 by W. H. Freeman & Company. Reprinted with permission.

be to have the substrate generated right on the enzyme or in its immediate vicinity, so that little diffusional motion is necessary, or to decrease the size of the enzyme, so that it can diffuse more rapidly. The first of these possibilities arises when two or more enzymes are combined in a multienzyme complex. The product of one enzymatic reaction then can be released close to the active site of the next enzyme. It appears to be more difficult to decrease the sizes of enzymes because a certain amount of tertiary structure is necessary in order to create the proper geometry of the active site. In addition, the large sizes of many enzymes are dictated by the need for secondary binding sites for other molecules that act to regulate enzymatic activity.

Kinetics of Enzymatic Reactions Involving Two Substrates

Enzymes that catalyze reactions with two or more substrates work in a variety of ways. In some cases, the intermolecular reaction occurs when all of the substrates are bound in a common enzyme-substrate complex; in others, the substrates bind and react one at a time. A frequent application of kinetic measurements is to distinguish between such alternatives.

Consider a reaction in which two substrates, S_1 and S_2, are converted to products P_1 and P_2. One possible way for the reaction to occur is for S_1 to bind to the enzyme first, forming the binary complex ES_1. This could be followed by the binding of S_2 to give the ternary complex ES_1S_2, which then undergoes conversion to the products:

$$E \xleftrightarrow{S_1} ES_1 \xleftrightarrow{S_2} ES_1S_2 \xrightarrow{P_1 + P_2} E \quad (33)$$

This alternative is referred to as an ordered mechanism. Another possibility is that the two substrates can bind to the enzyme in either order:

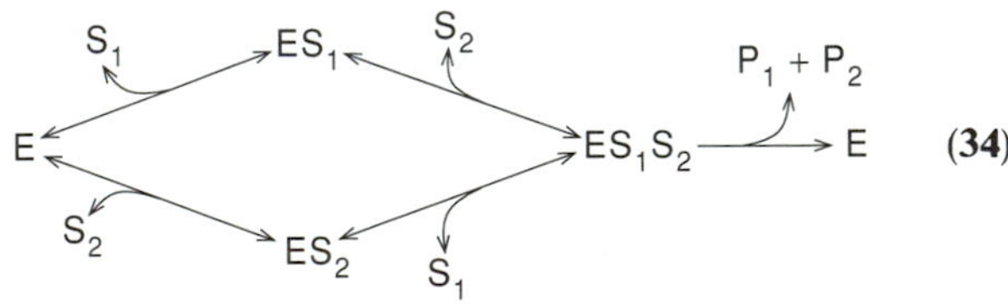

(34)

This alternative is a random mechanism. The ordered mechanism can be viewed as the limiting case of a random mechanism in which the upper path is much more favorable than the lower. Hexokinase (chapter 13) and citrate synthase (chapter 14) exhibit random, or nearly random, mechanisms. Lactate dehydrogenase (chapter 10) and aspartate carbamoyltransferase (chapter 10) use an ordered pathway, or at least display a marked preference for one route over the other.

In either the random or the ordered mechanism, substrates S_1 and S_2 both have to bind to the enzyme before either of the products is released. A different type of mechanism would be for one substrate, say S_1, to bind to the enzyme and be converted to P_1, leaving the enzyme in an altered form, E′. S_2 then could bind to E′ and be converted to P_2, returning the enzyme to its original form:

$$E \xleftrightarrow{S_1} ES_1 \xleftrightarrow{P_1} E' \xleftrightarrow{S_2} E'S_2 \xrightarrow{P_2} E \quad (35)$$

This is called a Ping-Pong mechanism to emphasize the bouncing of the enzyme between two states, E and E′. Ping-Pong pathways are commonly observed with enzymes that contain tightly bound coenzymes, such as pyridoxal phosphate or flavin groups (see chapter 11 for the structures of these coenzymes). The interconversion of the enzyme between the two forms usually involves a modification of the coenzyme. In the case of pyridoxal enzymes, the coenzyme switches between an aldehyde (pyridoxal phosphate) and an amine (pyridoxamine phosphate). An example is glutamate transaminase (chapters 11 and 19). In some enzymes with bound flavin coenzymes, the flavin alternates between oxidized and reduced states.

Kinetic equations for these different mechanisms can be worked out in the same way as those for reactions involving only one substrate. Techniques for doing so are described in the references given at the end of the chapter; here we will simply give some of the results. For the Ping-Pong mechanism (equation 35), the double-reciprocal form of the final kinetic expression is

$$\frac{1}{v} = \frac{1}{V_{max}}\left(1 + \frac{K_{m2}}{[S_2]}\right) + \frac{K_{m1}}{V_{max}}\frac{1}{[S_1]} \quad (36)$$

where K_{m1} is the Michaelis constant for S_1, and K_{m2} is that for S_2. This expression is similar to that for a single-substrate reaction (equation 29), except that the first term on the right is multiplied by the factor $1 + (K_{m2}/[S_2])$. If we measure the rate of the reaction as a function of $[S_1]$, keeping $[S_2]$ constant, a plot of $1/v$ versus $1/[S_1]$ will be linear, but the intercept on the ordinate (the apparent V_{max}) will depend on $[S_2]$. Increasing $[S_2]$ will increase the apparent V_{max} (fig. 8.11). From a series of such plots, measured at different values of $[S_2]$, we could find the true V_{max}, in addition to K_{m1} and K_{m2}. As with a reaction with one substrate but involving several steps (equations 30 to 32), the two values of K_m and V_{max} for a two-substrate reaction are made up of combinations of the microscopic rate constants.

The ordered and random mechanisms (equations 33 and 34) both give a kinetic expression of the form

$$\frac{1}{v} = \frac{1}{V_{max}}\left(1 + \frac{K_{m2}}{[S_2]} + \frac{K_{m1}}{[S_1]} + \frac{K_{m2}}{[S_2]}\frac{K_{d1}}{[S_1]}\right) \quad (37)$$

where K_{d1} is the dissociation constant for ES_1. A plot of $1/v$ versus $1/[S_1]$ at constant $[S_2]$ is still linear, but now both the slope and the intercept depend on $[S_2]$ (fig. 8.12). Again, all of the macroscopic constants can be obtained from a series of such plots measured at different S_2 concentrations. However, because of the symmetry of equation (37), additional measurements have to be made in order to determine whether the mechanism is ordered, and if so whether S_1 or S_2 binds to the enzyme first. In some cases, the substrate that binds first can be shown to bind to the enzyme to give a stable enzyme-substrate complex in the absence of the other substrate. Lactate dehydrogenase, for example, will bind NAD^+ in the absence of lactate, but will not bind lactate in the absence of NAD^+.

All of the kinetic models that we have discussed lead to hyperbolic dependences of the rate on the concentrations of the substrates, and to linear double-reciprocal plots. There are, however, many enzymes that behave in more complex ways. In

Figure 8.11

Double-reciprocal plots ($1/v$ versus $1/[S_1]$) for the Ping-Pong mechanism. Measurements made at different values of $[S_2]$ give a set of parallel straight lines. K_{m1}, K_{m2} and V_{max} can be obtained by replotting the intercepts as a function of $1/[S_2]$.

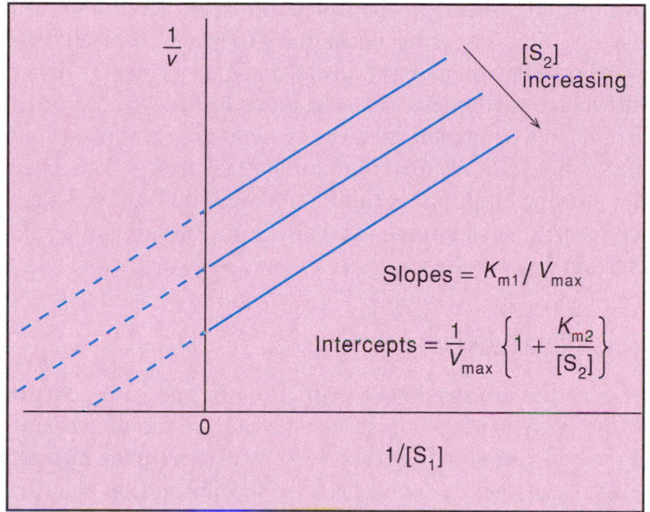

Figure 8.12

Double-reciprocal plots for an ordered or random mechanism. Measurements made at different fixed values of $[S_2]$ give a set of lines that intersect to the left of the ordinate. The two values of K_m, as well as V_{max} and K_{d1}, can be obtained by replotting the slopes and intercepts of these lines as functions of $1/[S_2]$. Ordered and random mechanisms can be distinguished by making such measurements for the reverse reaction ($P_1 + P_2 \rightarrow S_1 + S_2$) in addition to the forward reaction.

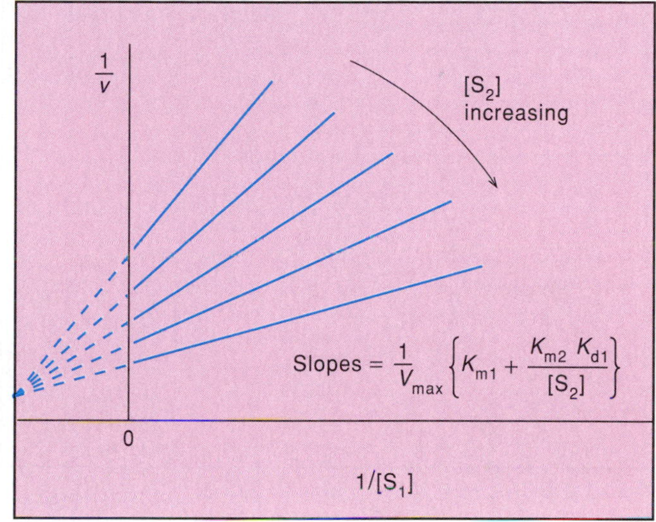

enzymes that have multiple subunits, the binding of substrate to one subunit may increase or decrease the activity of another subunit, just as the binding of O_2 to one of the subunits of hemoglobin increases the affinity of the other subunits for O_2 (see chapter 5). Such effects are seen in an important group of enzymes called allosteric enzymes, which are discussed in chapter 10.

Effects of Temperature, pH, and Isotopic Substitutions

With most enzymes, the turnover number increases with temperature until a temperature is reached where the enzyme is no longer stable (fig. 8.13). Above this point, there is a precipitous drop in activity that usually is irreversible. At lower temperatures, where the enzyme is stable, the apparent activation energy of the reaction (ΔE_a or $\Delta H^‡$) can be obtained from an Arrhenius plot of $\ln k_{cat}$ versus $1/T$. The activation energy varies considerably from case to case, but with many enzymes a 10° rise in temperature increases k_{cat} by about a factor of 2, which translates into an apparent activation energy on the order of 12 kcal/mole. If k_{cat} is equal simply to k_2, as it is in the simple mechanism described by equation (17), $\Delta H^‡$ can be related to the formation of the transition state between ES and the products. More generally, as we saw in equation (32), k_{cat} is made up of a combination of microscopic rate constants, and it is difficult to relate the activation energy to any particular step. If the mechanism involves a sequence of several steps, the kinetics are usually most sensitive to the slowest, or rate-determining, step.

Figure 8.13

This graph shows how the activity of a typical enzyme depends on temperature. The turnover number (k_{cat}) increases with temperature until a point is reached where the enzyme is no longer stable. The apparent activation energy of the rate-determining step could be obtained by replotting the data for low temperatures in an Arrhenius plot. The temperature at which k_{cat} is greatest should not be interpreted as the "optimum temperature" for the enzyme, because the position of the maximum depends partly on how quickly the experimenter is able to assay the enzyme's activity. (Denaturation of the enzyme occurs continuously during the measurement and increases in rate with increasing temperature.)

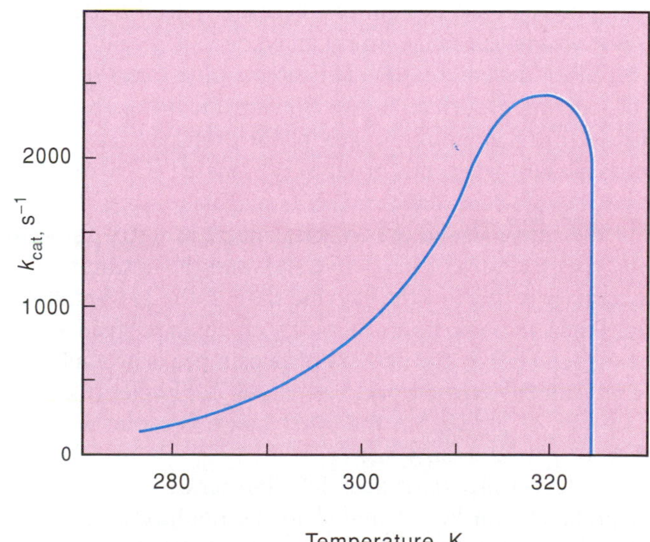

Figure 8.14

Enzyme activity (k_{cat}/K_m) as a function of pH for three different enzymes. The optimum pH usually is a characteristic of the enzyme and not the particular substrate. Often the pH sensitivity is an indication of an ionizable group at the active site, but it can also reflect changes in the tertiary structure of the enzyme.

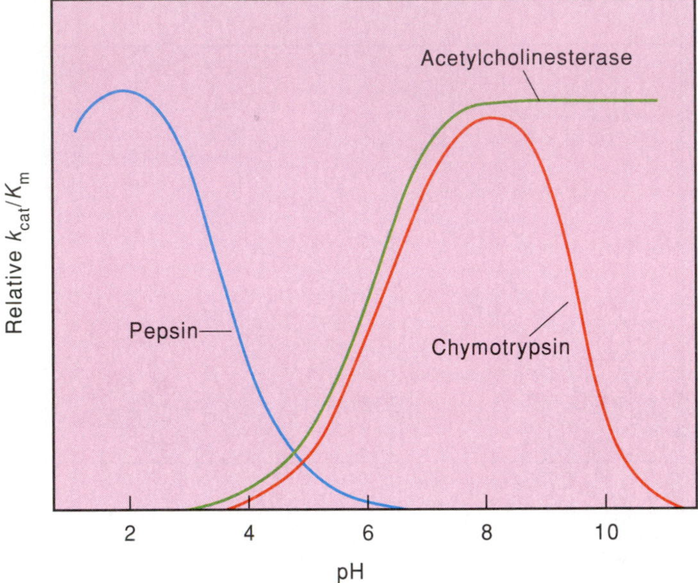

One way to explore the nature of the rate-determining step in an enzymatic reaction is to investigate the effect of replacing a particular H atom of the substrate by deuterium (^{2}H, or D). If a bond to this atom is broken in the critical step, replacing the H atom by the heavier D usually will result in a decrease in k_{cat} by a factor of between 3 and 10. Similar experiments can be done with isotopes of O or N, although the effects usually are smaller than those obtained with D.

Enzymes, like other proteins, are stable over only a limited range of pH. Outside this range, changes in the charges on ionizable amino acid residues result in modifications of the tertiary structure of the protein and eventually lead to denaturation. But within the range where an enzyme is stable, both k_{cat} and K_m often depend on pH. The effects of pH can reflect the pK_a of ionizing groups on either the enzyme or a substrate. A substrate that has an amine group, for example, may bind to the enzyme best when this group is protonated. In many cases, however, the pH dependence reflects ionizable residues that form part of the active site on the enzyme, or are essential for maintaining the structure of the active site, and the optimum pH is a characteristic more of the enzyme than of the particular substrate. Thus the maximum activity of chymotrypsin always occurs around pH 8, the activity of pepsin peaks around pH 2, and acetylcholinesterase works best at pH 7 or higher (fig. 8.14). The activity of papain, on the other hand, is essentially independent of pH between 4 and 8.

The sensitivity of acetylcholinesterase to pH (see fig. 8.14) probably can be attributed to the imidazole group of a critical histidyl residue, which must be unprotonated in order for the enzyme to operate. The histidine probably removes a proton from the bound substrate in the rate-determining step. Chymotrypsin's bell-shaped activity curve reflects the pK_a values of *two* critical groups, one of which must be protonated and the other unprotonated. As we will see in chapter 9, an unprotonated histidyl residue is essential in the catalytic step, and the free amino group of the N-terminal isoleucine apparently must be protonated. Although the isoleucine itself is not part of the active site, deprotonating its amino group causes a change in the tertiary structure of the protein, which evidently disrupts the structure of the site. The decrease in k_{cat}/K_m for chymotrypsin at low pH is due to a decrease in k_{cat}, whereas the decrease at high pH results from an increase in K_m. The activity of the enzyme fumerase exhibits a similar bell-shaped dependence on pH, but in this case both limbs of the curve reflect changes in k_{cat}.

Enzyme Inhibition

Most enzymes are sensitive to inhibition by specific agents that interfere with the binding of a substrate at the active site or with the conversion of the enzyme-substrate complex into products. Study of these effects can provide information about how an enzyme operates. In many cases, an inhibitor is found to resemble the substrate structurally, and to bind reversibly at the same site on the enzyme. This effect is called competitive inhibition because the inhibitor and the substrate compete for binding (fig. 8.15*a*); the inhibitor is prevented from binding if the site is already occupied by the substrate. This type of inhibition is exemplified by the effect of malonic acid on the enzyme succinate dehydrogenase. Malonate, whose structure resembles that of the substrate, succinate, competes with it for binding at the active site (fig. 8.16).

Inhibitors of a different sort can bind at separate sites where they do not compete directly with the substrate. Instead, they act by interfering with the reaction of the enzyme-substrate complex. An inhibitor that binds to an enzyme whether or not the active site is occupied by the substrate is termed a noncompetitive inhibitor (see fig. 8.15*b*). A third possibility is that the inhibitor binds only after formation of the enzyme-substrate complex (see fig. 8.15*c*). This effect is called uncompetitive inhibition. Uncompetitive inhibition is most common in reactions involving more than one substrate.

Kinetic measurements are useful for distinguishing between different types of inhibition as well as providing quantitative information on the effectiveness of various inhibitors. This information is essential for an understanding of how cells regulate their enzymatic activities. Comparisons of the effects of a series of inhibitors also can help in mapping the structure of an enzyme's active site, and such studies are a key step in the rational design of therapeutic drugs. Finally, competitive inhibitors are useful in x-ray crystallographic studies for pinpointing the active site in a crystal structure, and thus revealing how the surrounding amino acid residues interact with a bound molecule. Crystallographic studies usually cannot be carried out with the substrate itself because the enzyme-substrate complex is converted too rapidly into products.

Figure 8.15

Types of enzyme inhibition. (*a*) A competitive inhibitor competes with the substrate for binding at the same site on the enzyme. (*b*) A noncompetitive inhibitor binds to a different site, but blocks the conversion of the substrate to products. (*c*) An uncompetitive inhibitor binds only to the enzyme-substrate complex. (E = enzyme; S = substrate.)

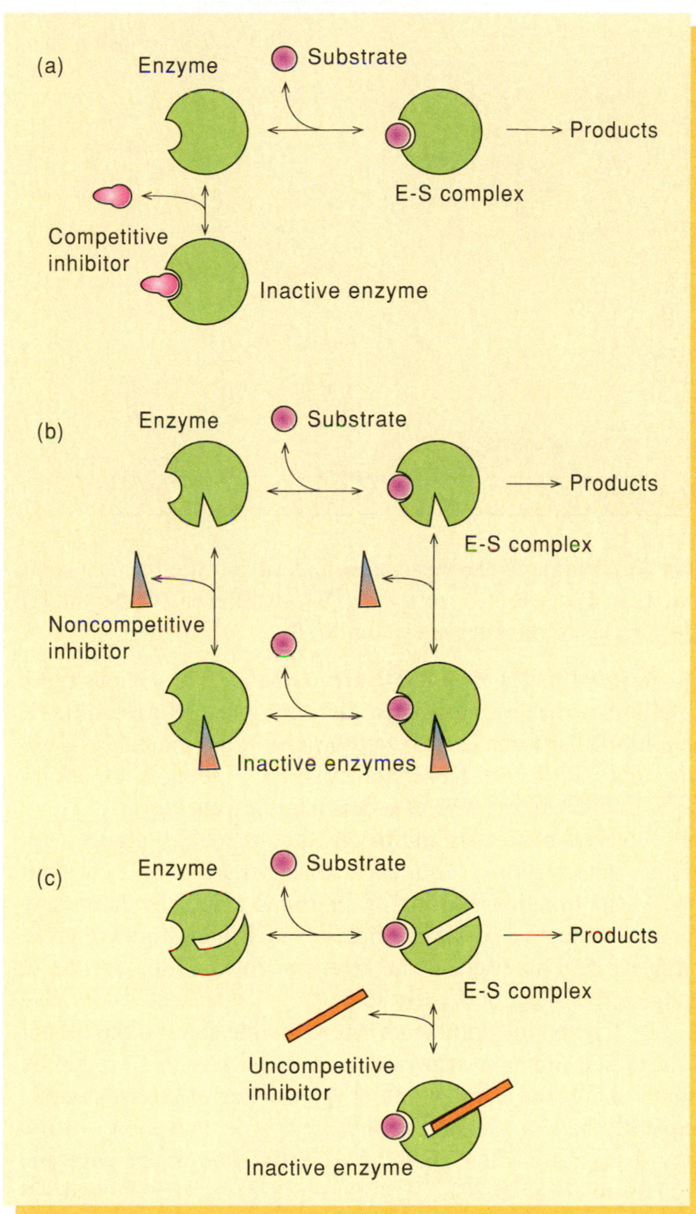

Figure 8.16

Malonate is a competitive inhibitor of succinate dehydrogenase. It is structurally similar to the substrate, succinate.

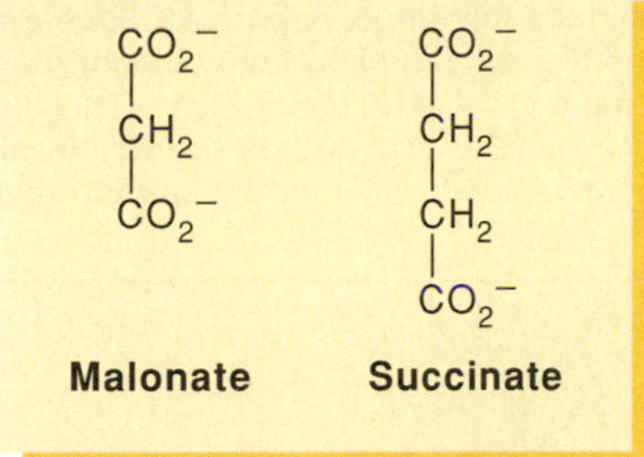

Competitive Inhibition An expression describing enzyme kinetics in the presence of a competitive inhibitor can be derived straightforwardly. Consider the simple, one-substrate reaction that we treated earlier:

$$E + S \underset{k_{-1}}{\overset{k_1}{\rightleftharpoons}} ES \overset{k_2}{\rightarrow} E + P \qquad (17')$$

We now have the additional feature that the enzyme can react reversibly with the inhibitor (I) to give an inactive complex (EI):

$$E + I \underset{k_{-1}}{\overset{k_1}{\rightleftharpoons}} EI \qquad (38)$$

The derivation proceeds just as in equations (18)–(29) above, except that in place of equation (22) we have

$$[E] = [E_T] - [ES] - [EI] \qquad (39)$$

The inhibitor simply decreases the amount of free enzyme that is available to react with S. The concentration of the enzyme-inhibitor complex depends on the concentration of the free inhibitor and on a dissociation constant, K_I:

$$\frac{[E][I]}{[EI]} = \frac{k_{-1}}{k_1} = K_I \qquad (40)$$

As a result, in place of equations (28) and (29) we end up with

$$v = \frac{V_{max}[S]}{[S] + K_m(1 + [I]/K_I)} \qquad (41)$$

and

$$\frac{1}{v} = \frac{1}{V_{max}} + \frac{K_m}{V_{max}}\frac{1}{[S]}\left(1 + \frac{[I]}{K_I}\right) \qquad (42)$$

According to equation (42), a plot of $1/v$ versus $1/[S]$ will be linear and will pass through the same intercept on the ordinate as the plot obtained in the absence of the inhibitor ($1/V_{max}$). This is equivalent to saying that the effect of the inhibitor disappears at high substrate concentration, which is just what we might have expected. The slope of the double-reciprocal plot, however, depends on the product $(K_m/V_{max})(1 + [I]/K_I)$, instead of simply K_m/V_{max} (fig. 8.17). By measuring the slope as a function of [I] we can determine K_I.

Competitive inhibitors can be designed to take advantage of the fact that an enzyme stabilizes the transition state of a reaction more than it does the initial enzyme-substrate complex. An inhibitor that is structurally similar to the transition state often will bind to the enzyme particularly tightly. Such an inhibitor is termed a transition-state analog. For example, many enzymes that hydrolyze phosphate diesters are very sensitive to inhibition by derivatives of vanadium in which the vanadium atom is surrounded by five oxygens. These inhibitors appear to mimic a transition state in which phosphorus has a similar pentacovalent geometry.

Figure 8.17

Competitive inhibition. A series of double-reciprocal plots ($1/v$ versus $1/[S]$) measured at different concentrations of the inhibitor (I) all intersect at the same point ($1/V_{max}$) on the ordinate. The slopes of the plots, and the intercepts on the abscissa, are simple, linear functions of $[I]/K_I$, where K_I is the dissociation constant of the inhibitor-enzyme complex.

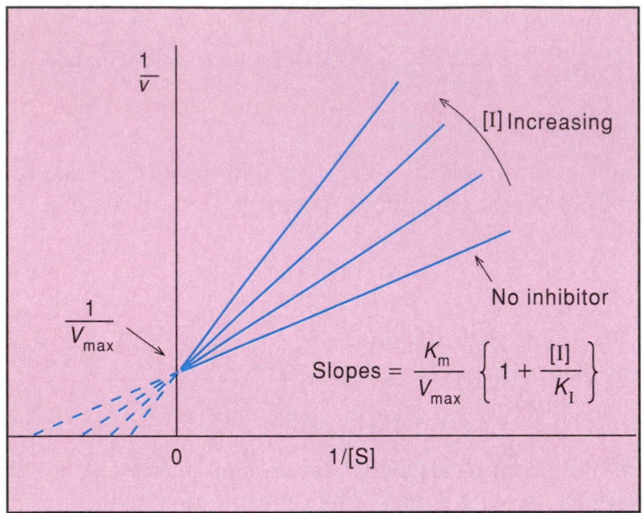

Figure 8.18

Noncompetitive inhibition. The double-reciprocal plots pass through different points on the ordinate, but intersect at the same point ($-1/K_m$) on the abscissa. The slopes and the intercepts on the ordinate are linear functions of $[I]/K_I$.

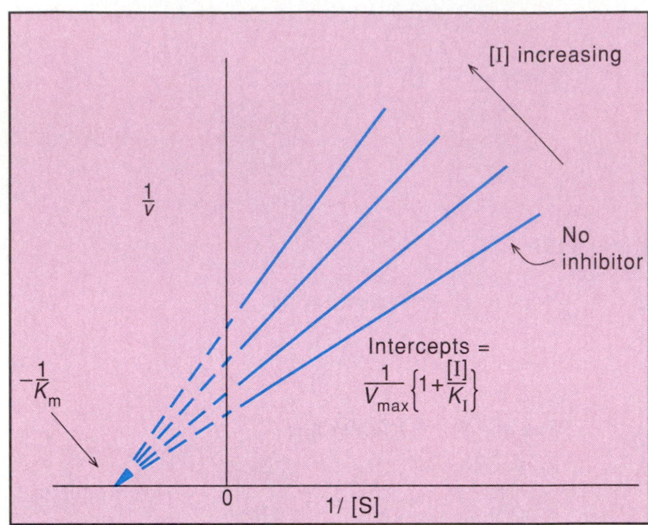

Noncompetitive Inhibition Noncompetitive inhibition can be treated in the same manner except that the inhibitor now can react with the enzyme even if S is already bound, so that in addition to equation (38) we have:

$$\mathrm{ES} + \mathrm{I} \underset{k_{-\mathrm{IS}}}{\overset{k_{\mathrm{I\,S}}}{\rightleftharpoons}} \mathrm{ESI} \qquad \textbf{(43)}$$

In the simplest situation, the binding of S has no effect at all on the binding of I, so that the dissociation constant of ESI is the same as that of ES (K_i). The final double-reciprocal expression for this case is

$$\frac{1}{v} = \left(\frac{1}{V_{max}} + \frac{K_m}{V_{max}}\frac{1}{[S]}\right)\left(1 + \frac{[I]}{K_I}\right) \qquad \textbf{(44)}$$

This expression indicates that a noncompetitive inhibitor decreases the maximum velocity, but does not affect K_m. The inhibitor removes a certain fraction of the enzyme from operation, no matter what the concentration of the substrate is. Plots of $1/v$ versus $1/[S]$ in the presence of different concentrations of the inhibitor intersect at the same point on the abscissa ($-1/K_m$), but pass through the ordinate at different points (fig. 8.18). Again, K_I can be found by measuring the intercepts on the ordinate as a function of [I]. If the dissociation constant for S from ESI differs from that of ES, the double-reciprocal plots will intersect above or below the abscissa.

Uncompetitive Inhibition An uncompetitive inhibitor leads to a double-reciprocal kinetic expression of the form

$$\frac{1}{v} = \frac{1}{V_{max}}\left(1 + \frac{[I]}{K_I}\right) + \frac{K_m}{V_{max}}\frac{1}{[S]} \qquad \textbf{(45)}$$

Here K_I pertains to the dissociation of I from the ternary complex, ESI. Plots of $1/v$ versus $1/[S]$ at different values of [I] give a series of parallel lines (fig. 8.19).

Irreversible Inhibitors and Affinity Labels The various types of inhibition that we have been discussing are all reversible. If the inhibited enzyme is dialyzed to remove the inhibitor, its activity increases again. Reversibility of the binding of the inhibitor is implicit in our use of a dissociation constant, K_I. There are, however, numerous inhibitors that react essentially irreversibly with enzymes, usually by the formation of a covalent bond to the functional group of an amino acid side chain or to a bound coenzyme. Some examples of such inhibitors are given in table 8.5. The effect of an irreversible inhibitor can be to change either V_{max} or K_m, or both.

Irreversible inhibitors often provide clues to the nature of the active site on an enzyme. Enzymes that are inhibited by organic mercurial compounds or by iodoacetate, for example, frequently have a cysteine in the active site, and the cysteinyl sulfhydryl group often plays an essential role in the catalytic mechanism (fig. 8.20). An example is glyceraldehyde-3-phosphate dehydrogenase, in which the catalytic mechanism begins with a reaction of the cysteine with the aldehyde substrate to form a thiohemiacetal. The mechanism of action of this enzyme will be discussed in more detail in chapter 13. Diisopropylfluorophosphate reacts irreversibly with a critical serine residue in many proteolytic enzymes, including trypsin and chymotrypsin (see fig. 8.20). The reaction of the serine group destroys the catalytic activity. The mechanism of action of trypsin and chymotrypsin is discussed in the following chapter.

In the case of glyceraldehyde-3-phosphate dehydrogenase, further studies have confirmed that the active site does

Figure 8.19

Uncompetitive inhibition. The double-reciprocal plots are parallel.

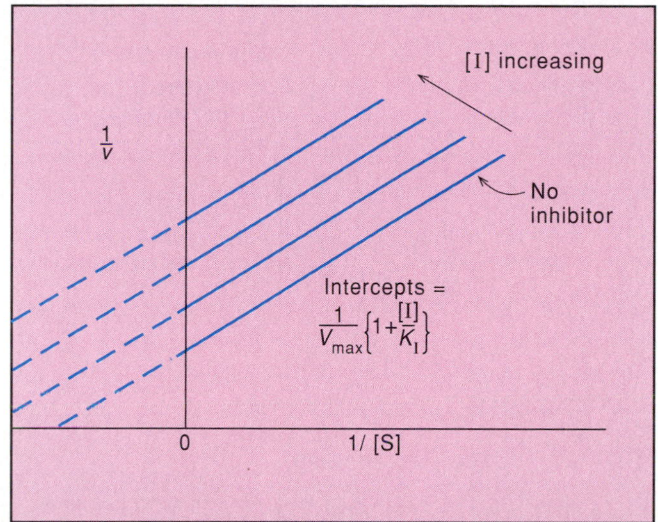

Figure 8.20

Iodoacetamide is an irreversible inhibitor of many enzymes that contain a cysteine residue in the active site. Diisopropylfluorophosphate is an irreversible inhibitor of trypsin, chymotrypsin, and several related enzymes. It reacts with a serine residue at the active site.

$$\underset{\textbf{Iodoacetamide}}{\mathrm{ICH_2C(=O)NH_2}} + \mathrm{Enz{-}SH} \xrightarrow{\ \ \mathrm{HI}\ \ } \underset{\textbf{Alkylated enzyme}}{\mathrm{Enz{-}S{-}CH_2C(=O)NH_2}}$$

$$\underset{\textbf{Diisopropylfluorophosphate}}{\mathrm{F{-}P(=O)[O{-}CH(CH_3)_2]_2}} + \mathrm{Enz{-}OH} \xrightarrow{\ \ \mathrm{HF}\ \ } \underset{\textbf{Diisopropyl-phosphate derivative of enzyme}}{\mathrm{Enz{-}O{-}P(=O)[O{-}CH(CH_3)_2]_2}}$$

Table 8.5

Some Inhibitors of Enzymes that Form Covalent Linkages with Functional Groups on the Enzyme

Inhibitor	Enzyme Group that Combines with Inhibitor
Cyanide	Fe, Cu, Zn, other transition metals
p-Mercuribenzoate	Sulfhydryl
Diisopropylfluorophosphate	Serine hydroxyl
Iodoacetate	Sulfhydryl, imidazole, carboxyl, thioether

contain an essential cysteine residue, and the same is true of the essential serine in trypsin and chymotrypsin. But in exploring a new enzyme it is important to keep in mind that the chemical modification of an amino acid side chain generally causes some perturbation of the secondary or tertiary structure of a protein. A reaction involving an amino acid residue well outside the active site thus could have a long-range disruptive effect that alters the structure of the active site sufficiently to inhibit the enzyme. This possibility is less of a concern with a reversible, competitive inhibitor because in that case the competition with the substrate supports the conclusion that the inhibitor binds directly in the active site.

An irreversible inhibitor often can be designed for the active site of a particular enzyme by incorporating a reactive group in a molecule that resembles a substrate. For example, 3-bromoacetol phosphate is a structural analog of dihydroxyacetone phosphate, which is a substrate for triosephosphate isomerase (fig. 8.21*a*). The inhibitor binds to the active site of the enzyme, and then reacts irreversibly with the carboxyl group of a nearby glutamic acid residue. Binding to the active site greatly increases the selectivity of the inhibitor for reaction with this particular residue, in preference to glutamyl residues elsewhere in the protein. Labeling the inhibitor with a radioisotope such as ^{14}C or ^{3}H facilitates the identification of the derivatized amino acid residue after the protein has been split into smaller peptides, making it possible to locate the reactive residue in the amino acid sequence. Such a reagent is called an affinity label. Photoaffinity labels are a particularly useful group of reagents of this type, in which the covalent attachment to the protein can be triggered by light after the reagent has bound to the enzyme (see fig. 8.21*b*). Another related technique is to use a reagent that is not intrinsically reactive, but becomes reactive after it has been modified chemically by the enzyme itself (see fig. 8.21*c*). Such a reagent is termed a mechanism-based inhibitor or suicide substrate, to emphasize that the enzyme brings about its own inhibition.

Metal-ion Chelators Enzymes that require metal ions as cofactors often are inhibited by chelators that bind to the metal. Examples of such metalloenzymes are lactate dehydrogenase from muscle and aldolase from yeast, both of which contain Zn^{2+}. Chelators inhibit aldolase by removing the required metal. The inhibition is not reversed simply by dialysis, but it can be reversed by adding Zn^{2+} to the depleted enzyme. In the case of lactate dehydrogenase the enzyme holds the Zn^{2+} more tightly, and the metal-chelator complex remains attached to the inhibited enzyme.

Figure 8.21

Affinity labels and suicide inhibitors. (*a*) 3-bromoacetol phosphate is a structural analog of dihydroxyacetone phosphate. It binds to the active site of triosephosphate isomerase, and then reacts to form a covalent bond with the carboxyl group of a nearby glutamyl residue. Bromoketone groups have been incorporated into many molecules to make similar affinity labels for other enzymes. (*b*) A photoaffinity label can be made by attaching a diazoacetyl group to a molecule (R) that resembles the substrate for a particular enzyme. After the reagent binds to the active site, it is exposed to light. This causes it to break down, forming a carbene derivative. The carbene reacts rapidly with any of several amino acid residues to form a covalent bond to the enzyme. (*c*) Vinylglycine can be used as a mechanism-based inhibitor for some enzymes that catalyze modifications of amino acids. It is not intrinsically a reactive compound, but is converted into a reactive allyl-imine in the course of the reaction catalyzed by the enzyme.

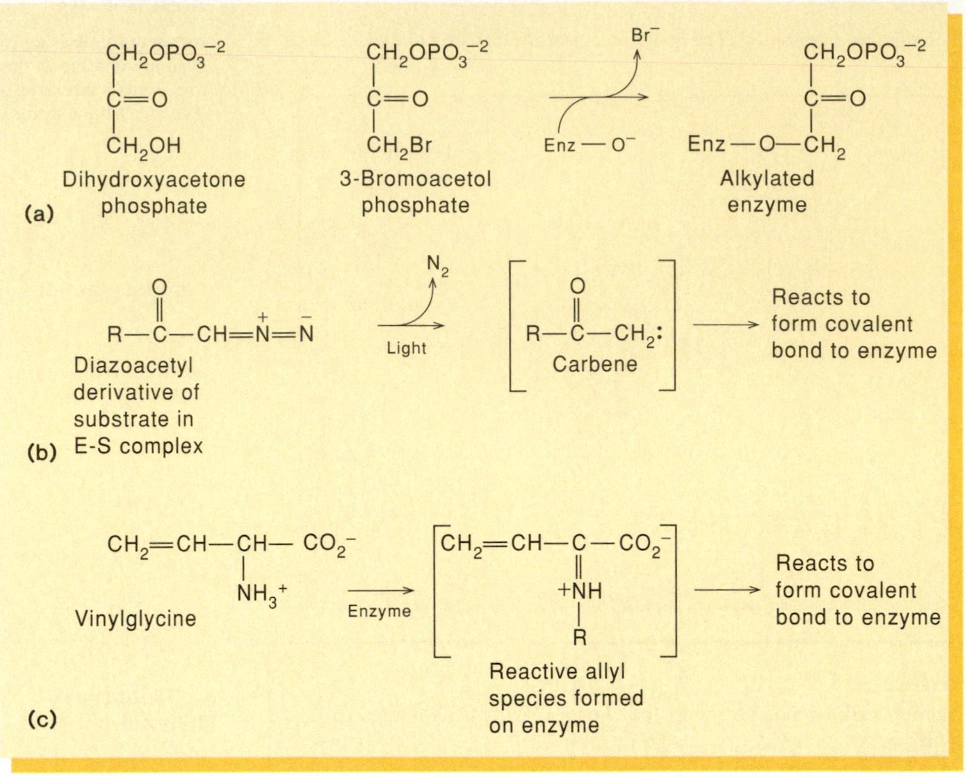

(a)
$CH_2OPO_3^{-2}$—C=O—CH_2OH Dihydroxyacetone phosphate
$CH_2OPO_3^{-2}$—C=O—CH_2Br 3-Bromoacetol phosphate
Enz—O^- → Br^-
$CH_2OPO_3^{-2}$—C=O—CH_2—O—Enz Alkylated enzyme

(b)
R—C(=O)—CH=$\overset{+}{N}$=$\overset{-}{N}$ Diazoacetyl derivative of substrate in E-S complex
Light → N_2
[R—C(=O)—CH_2: Carbene] → Reacts to form covalent bond to enzyme

(c)
CH_2=CH—CH(NH_3^+)—CO_2^- Vinylglycine
Enzyme →
[CH_2=CH—C(=+NH—R)—CO_2^-] Reactive allyl species formed on enzyme → Reacts to form covalent bond to enzyme

Summary

Enzymes are biological catalysts. Most enzymes are proteins and are highly specific in the reactions they catalyze. Kinetic analysis is one of the most broadly used tools for characterizing enzymatic reactions. In this chapter we have focused on the following points.

1. The rates of both enzymatic and nonenzymatic reactions are functions of the frequency of collisions between the reacting species and the fraction of the collisions that produce products. The former is determined by the temperature and the concentrations of the reactants; the latter depends on the temperature and the activation free energy, $\Delta G^{\ddagger}$. $\Delta G^{\ddagger}$ can be interpreted as the free energy that is needed to convert the reactants to a transition state. A catalyst increases the reaction rate by lowering $\Delta G^{\ddagger}$.
2. Enzymes have localized catalytic sites. The substrate (S) binds at the active site to form an enzyme-substrate complex (ES). Subsequent steps transform ES into an enzyme-product complex, and dissociation of the product regenerates the free enzyme. The overall speed of the reaction is proportional to the concentration of ES. Shortly after the enzyme and substrate are mixed, the concentration of ES becomes approximately constant and remains so for a significant period of time known as the steady state.
3. Enzymatic reaction rates usually are studied by leaving one essential ingredient out of the reaction and then adding this component and measuring the formation of product or the disappearance of a reactant with time. Special techniques are necessary to measure very fast reactions. It is common to measure the rate as a function of the concentrations of substrates and products, and to examine its dependence on pH and temperature.
4. The rate (v) of an enzymatic reaction in the steady state usually has a hyperbolic dependence on the concentration of the substrate. It is proportional to [S] at low concentrations but approaches a maximum (V_{max}) at high concentrations, when the enzyme is fully charged with substrate. The Michaelis constant, K_m, is the substrate concentration at which the rate is half-maximal. K_m and V_{max} often can be obtained from a plot of $1/v$ versus 1/[S]. If the dissociation of ES occurs rapidly relative to the conversion of the complex into products, K_m is approximately equal to the dissociation constant for the complex. Under other conditions, K_m is a function of all the rate constants involved in the formation of ES and its conversion to products. In general, K_m is larger than the dissociation constant.
5. The turnover number, k_{cat}, is the maximum number of moles of substrate converted to product per unit time per mole of enzyme, and is equal to V_{max} divided by the enzyme concentration. The specificity constant, k_{cat}/K_m, is a measure of how rapidly an enzyme can work at low substrate concentrations. This is usually the best index of the effectiveness of an enzyme with different substrates.
6. Enzymes that catalyze reactions of two or more substrates work in a variety of ways that can be distinguished by kinetic analysis. Some enzymes bind the substrates in a fixed order; others bind their substrates in random order. Sometimes the binding of one substrate gives a partial reaction before the second substrate binds.

7. Many enzymes are sensitive to inhibition by specific agents that interfere with the binding of substrate at the active site or with conversion of the enzyme-substrate complex into products. Study of these effects can provide information about how an enzyme operates.
8. Reversible inhibitors are classified as competitive, noncompetitive, or uncompetitive. A competitive inhibitor competes with substrate for binding to the enzyme. Consequently, a sufficiently high concentration of substrate can eliminate the effect of a competitive inhibitor. A noncompetitive inhibitor binds to the enzyme at a separate site and interferes with the reaction regardless of whether or not the active site is occupied by substrate. An uncompetitive inhibitor binds to the enzyme-substrate complex, but not to the free enzyme. The different forms of reversible inhibition are distinguishable by measuring the rate as a function of the concentrations of the substrate and the inhibitor.
9. Irreversible inhibitors often provide information on the active site by forming covalent complexes that can be characterized.

Selected Readings

Advances in Enzymology. New York: Academic Press. An ongoing, annually published volume containing monographs on selected topics.

Boyer, P. D. (ed.), *The Enzymes.* New York: Academic Press. A continuing series with more than 16 volumes of monographs on selected enzymes. See particularly the chapter entitled "Steady State Kinetics" by W. W. Cleland in vol. 2.

Fersht, A., *Enzyme Structure and Mechanism,* 2d ed. New York: Freeman, 1985.

Frost, A. A., and R. G. Pearson, *Kinetics and Mechanism* (2d ed.). New York: Wiley, 1961. An excellent introduction to general chemical kinetics.

Purich, D. L., *Contemporary Enzyme Kinetics and Mechanism.* New York: Academic Press, 1983. A good source of detailed information on how to analyze kinetic data, and on effects of temperature, pH, and inhibitors. The chapters are selected from several volumes of *Methods in Enzymology* (New York: Academic Press), an ongoing series of monographs.

Segal, I. H., *Enzyme Kinetics.* New York: Wiley, 1975.

Problems

1. Explain what is meant by the order of a reaction, using the reaction below as an example. What is the reaction order for each reactant? (Consider the forward and reverse reaction.) What is it for the overall reaction?

$$A + B \rightleftharpoons 2C$$

2. In a first order reaction a substrate is converted to product so that 87% of the substrate is converted in 7 min. Calculate the first-order rate constant. In what time would 50% of the substrate be converted to product?
3. K_m is frequently equated with K_S, the [ES] dissociation constant. However, there is usually a disparity between those values. Why? Under what conditions are K_m and K_S equivalent?
4. Differentiate between the enzyme-substrate complex and the transition-state intermediate in an enzymatic reaction.
5. An enzyme was assayed with a substrate concentraton of twice the K_m value. The progress curve of the enzyme (product produced per minute) is shown here. Give two possible reasons why the progress curve becomes nonlinear.

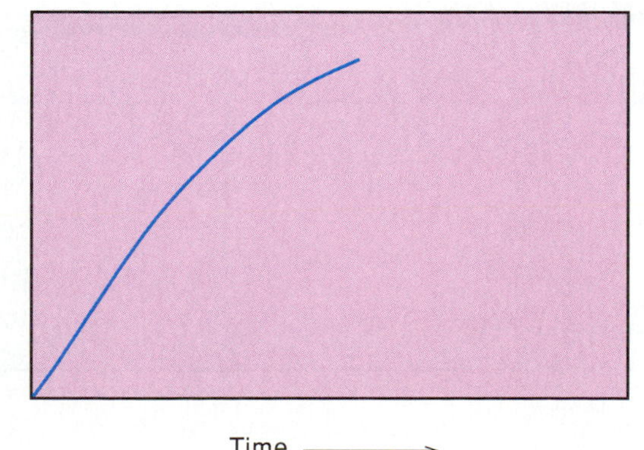

6. What is the steady-state approximation and under what conditions is it valid?
7. Assume that an enzyme-catalyzed reaction follows Michaelis-Menten kinetics with a K_m of 1 μM. The initial velocity is 0.1 μM/min at 10 mM substrate. Calculate the initial velocity at 1 mM, 10 μM, and 1 μM substrate. If the substrate concentration were increased to 20 mM, would the initial velocity double? Why or why not?
8. If the K_m for an enzyme is 1.0×10^{-5} M and the K_I of a competitive inhibitor of the enzyme is 1.0×10^{-6} M, what concentration of inhibitor is necessary to lower the reaction rate by a factor of 10 when the substrate concentration is 1.0×10^{-3} M? 1.0×10^{-5} M? 1.0×10^{-6} M?
9. Assume that an enzyme-catalyzed reaction follows the scheme shown:

$$E + S \underset{k_2}{\overset{k_1}{\rightleftharpoons}} ES \underset{k_4}{\overset{k_3}{\rightleftharpoons}} E + P$$

where $k_1 = 10^9$ M^{-1} s^{-1}, $k_2 = 10^5$ s^{-1}, $k_3 = 10^2$ s^{-1}, $k_4 = 10^7$ M^{-1} s^{-1}, and $[E_T]$ is 0.1 nM. Determine the value of each of the following.

(a) K_m
(b) V_{max}
(c) Turnover number
(d) Initial velocity when $[S]_0$ is 20 μM.

10. A colleague has measured the enzymatic activity as a function of reaction temperature and obtained the data shown in this graph. He insists on labeling point A as the "temperature optimum" for the enzyme. Try, tactfully, to point out the fallacy of that interpretation.

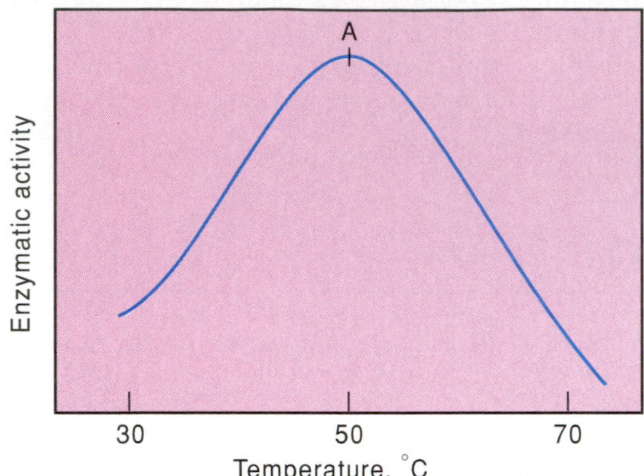

11. You have isolated an enzyme that catalyzes a bimolecular reaction:

$$A + B \rightleftharpoons P + Q$$

The initial velocity data yielded intersecting double-reciprocal plots with [A] varied at fixed [B] or [B] varied at fixed [A]. Which kinetic pattern—sequential (ordered or random), or Ping-Pong—might you rule out?

12. You have isolated a tetrameric NAD^+-dependent dehydrogenase. You incubate this enzyme with iodoacetamide in the absence or presence of NADH (at ten times the K_m concentration) and you periodically remove aliquots of the enzyme for activity measurements and amino acid composition analysis. The results of the analyses are shown.

(a) What can you conclude about the reactivities of the cysteinyl and histidyl residues of the protein?

(b) Which residue could you implicate in the catalytic active site? On what do you base the choice? Are the data conclusive concerning the assignment of a residue to the active site? Why or why not?

(c) After 1 h you dilute the enzyme incubated with iodoacetamide but no NADH. Would you expect the enzyme activity to be restored? Explain.

	(No NADH Present)			(NADH Present)		
Time (min)	Activity (U/mg)	His (Residues/mole)	Cys (Residues/mole)	Activity (U/mg)	His (Residues/mole)	Cys (Residues/mole)
0	1,000	20	12	1,000	20	12
15	560	18.2	11.4	975	20	11.4
30	320	17.3	10.8	950	20	10.8
45	180	16.7	10.4	925	19.8	10.4
60	100	16.4	10.0	900	19.6	10.0

13. The following initial velocity data were obtained for an enzyme.

[S] (nM)	Velocity ($M\ s^{-1}$) $\times 10^7$
0.10	0.96
0.125	1.12
0.167	1.35
0.250	1.66
0.50	2.22
1.0	2.63

Each assay at the indicated substrate concentration was initiated by adding enzyme to a final concentration of 0.01 nM. Derive K_m, V_{max}, k_{cat}, and the specificity constant.

14. You have measured the initial velocity of an enzyme in the absence of inhibitor and with inhibitor A or inhibitor B. In each case, the inhibitor is present at 10 μM. The data are shown in the table.

[S] (mM)	Velocity ($M\ s^{-1}$) $\times 10^7$ Uninhibited	Velocity ($M\ s^{-1}$) $\times 10^7$ Inhibitor A	Velocity ($M\ s^{-1}$) $\times 10^7$ Inhibitor B
0.333	1.65	1.05	0.794
0.40	1.86	1.21	0.893
0.50	2.13	1.43	1.02
0.666	2.49	1.74	1.19
1.0	2.99	2.22	1.43
2.0	3.72	3.08	1.79

(a) Determine K_m and V_{max} of the enzyme.

(b) Determine the type of inhibition imposed by inhibitor A and calculate $K_{I(S)}$.

(c) Determine the type of inhibition imposed by inhibitor B and calculate $K_{I(S)}$.

15. Irreversible inactivation of an enzyme by a compound may be confused with noncompetitive inhibition. Why? How could you distinguish between a reversible noncompetitive inhibitor and an irreversible inactivator? Enzyme supply is not limiting.

9

CHAPTER

Mechanisms of Enzyme Catalysis

In chapter 8 we saw that enzymes can increase the rates of reactions by many orders of magnitude. We noted that enzymes work under mild conditions of temperature, pH, and pressure, and that they are highly specific in the types of reactions they catalyze and in the particular substrates they accept. In this chapter we will explore the mechanisms of several enzyme-catalyzed reactions in greater detail. Our goal is to relate the activity of each of these enzymes to the structure of the active site, where the functional groups of amino acid side chains, the polypeptide backbone, or bound cofactors must interact with the substrates in such a way as to favor the formation of the transition state. We will be exploring enzyme catalytic mechanisms in many subsequent chapters as well, but usually in less detail than here.

Five Themes that Recur in Discussing Enzymatic Reactions

Several broad themes recur frequently in discussing enzymatic reaction mechanisms. Among the most important of these are (1) the proximity effect, (2) electrostatic effects, (3) general-acid and general-base catalysis, (4) nucleophilic or electrophilic catalysis by enzymatic functional groups, and (5) structural flexibility. For all known enzymes at least one of these themes is relevant, and in most cases more than one. We will start by discussing the five themes in general terms, and then see how they apply to some representative enzymes.

The Proximity Effect: Enzymes Bring Reacting Species Close Together

The idea of the proximity effect is that an enzyme can accelerate a reaction between two species simply by holding the two reactants close together in an appropriate orientation. It has long been known that intramolecular reactions between groups that are tied together in a single molecule are usually much faster than the corresponding intermolecular reactions between two independent molecules. The cyclization of succinic acid to form succinyl anhydride (equation 1), for example, is much more rapid than the formation of acetic anhydride from two molecules of acetic acid (equation 2):

$$HO_2C{-}CH_2{-}CH_2{-}CO_2H \longrightarrow \text{(cyclic)}\ CH_2{-}C({=}O){-}O{-}C({=}O){-}CH_2 + H_2O \qquad (1)$$

$$CH_3CO_2H + CH_3CO_2H \longrightarrow CH_3C({=}O){-}O{-}C({=}O)CH_3 + H_2O \qquad (2)$$

It is not possible to compare the rate constants for these two reactions directly, because they are expressed in different units. The intramolecular reaction (equation 1) is kinetically first order, while the intermolecular reaction (equation 2) is second order. But suppose that one of the two reactants in the intermolecular reaction is present in great excess over the other reactant, so that the process is effectively first order in the concentration of the second, limiting reactant (see fig. 8.3). Then we can ask what the molarity of the more abundant species would have to be, in order to make the effective first-order rate constant of the intermolecular reaction the same as the measured first-order rate constant for the intramolecular reaction. For the reactions of succinic and acetic acids (equations 1 and 2), the answer is 3×10^5 M. This is far above any concentration that could be obtained, even if the intermolecular reaction were carried out in pure acetic acid. (The concentration of CH_3COOH in glacial acetic acid is only 17.5 M.) In the related reaction

$$\begin{array}{c} CH_3C{-}CO_2H \\ \| \\ CH_3C{-}CO_2H \end{array} \longrightarrow \begin{array}{c} CH_3C{-}C(=O) \\ \| \qquad\quad \backslash \\ \qquad\qquad\quad O \\ \| \qquad\quad / \\ CH_3C{-}C(=O) \end{array} + H_2O \qquad (3)$$

the result is even more dramatic. For the corresponding intermolecular reaction to match this intramolecular process the concentration of the fixed reactant would have to be 2×10^{12} M!

These examples from organic chemistry show that tying two reactants together in a single molecule can have an enormous effect on the rate of a reaction. This effect is, for the most part, due simply to differences between the entropy changes that accompany the inter- and intramolecular reactions. The formation of the product involves a much larger loss of translational and rotational entropy in the intermolecular molecular reaction than it does in the corresponding intramolecular reaction. A negative change in entropy increases both the overall free energy change in the reaction ($\Delta G = \Delta H - T\,\Delta S$), and the activation free energy ($\Delta G^{+} = \Delta H^{+} - T\,\Delta S^{+}$) for the formation of the transition state (see equation 9 in chapter 2 and equation 16 in chapter 8). In the intramolecular reaction much of this entropy decrease has already occurred during the preparation of the reactant.

Enzymes that catalyze intermolecular reactions take advantage of the proximity effect by binding the reactants close together in the active site, so that the reactive groups are oriented appropriately for the reaction. Once the substrates are fixed in this way, the subsequent reaction behaves kinetically like an intramolecular process. The entropy decrease associated with the formation of the transition state has been moved to an earlier step, the binding of the substrates to form the enzyme-substrate complex. This step often is driven by an enthalpy decrease associated with electrostatic interactions between polar or charged groups of the substrates and the enzyme. There are, however, exceptions to this generalization, particularly in reactions involving hydrophobic substrates. As we discussed in chapter 2, the removal of a hydrophobic molecule from aqueous solution is favored by an entropy increase, and the binding of hydrophobic substrates to enzymes can be driven in this way.

General-Base and General-Acid Catalysis Provide Ways of Avoiding the Need for Extremely High or Low pH

Chemical bonds are formed by electrons, and the rearrangement or breakage of bonds requires the migration of electrons. In broad terms, reactive chemical groups can be said to function either as electrophiles or as nucleophiles. Electrophiles are electron-deficient substances that react with electron-rich substances; nucleophiles are electron-rich substances that react with electron-deficient substances. The task of a catalyst often is to make a potentially reactive group more reactive by increasing its intrinsic electrophilic or nucleophilic character. In many cases the simplest way to do this is to add or remove a proton. As an example, consider the hydrolysis of an ester (fig. 9.1). Because the electronegativity of the oxygen atom in the C=O group is greater than that of the carbon, the oxygen has a fractional negative charge, δ^-, and the carbon has a fractional positive charge, δ^+. Hydrolysis of an ester in neutral aqueous solution can occur if the oxygen atom of H_2O, acting as a nucleophile, attacks the positively charged carbon. The initial product is an intermediate in which the carbon atom has four substituents in a tetrahedral arrangement. The reaction is completed by the rapid breakdown of the tetrahedral intermediate to release the alcohol.

Water is intrinsically a comparatively weak nucleophile, and its reaction with esters in the absence of a catalyst is very slow. The hydrolysis of esters occurs much more rapidly at high pH, when the negatively charged hydroxide ion replaces water as the reactive nucleophile (see fig. 9.1*a*). But the nucleophilic character of water itself also can be increased by interaction with a basic group other than OH^- (see fig. 9.1*b*). The base offers a pair of electrons to one of the protons of the water, and thus increases the electron density on the oxygen.

The term general base is used to describe any substance that is capable of binding a proton in aqueous solution. Enzymes use a variety of functional groups to fill this role. There are two factors that make free hydroxide ions unsuitable for enzymatic catalysis, and that dictate the choice of a general base. First, the low concentration of OH^- limits its availability at physiological pH. In contrast, proteins contain numerous functional groups that can serve as general bases at moderate pH, or even under mildly acidic conditions. The only requirement is that the base start out mainly in its unprotonated form, which will be the case as long as the ambient pH is above the pK_a of the conjugate acid. This condition can easily be met by selecting a basic group from among the ionizable or polar amino acid side chains, from an amino-terminal —NH_2 group or a carboxyl-terminal carboxylate ion, or from the oxygen or nitrogen atom

Figure 9.1

Several ways that the hydrolysis of an ester can occur. A colored, curved arrow represents the movement of an electron pair from an electron donor to an acceptor. (*a*) Catalysis by free hydroxide ion. (*b*) General-base catalysis. (*c*) General-acid catalysis.

(a) Hydroxide ion catalysis

(b) General-base catalysis

(c) General-acid catalysis

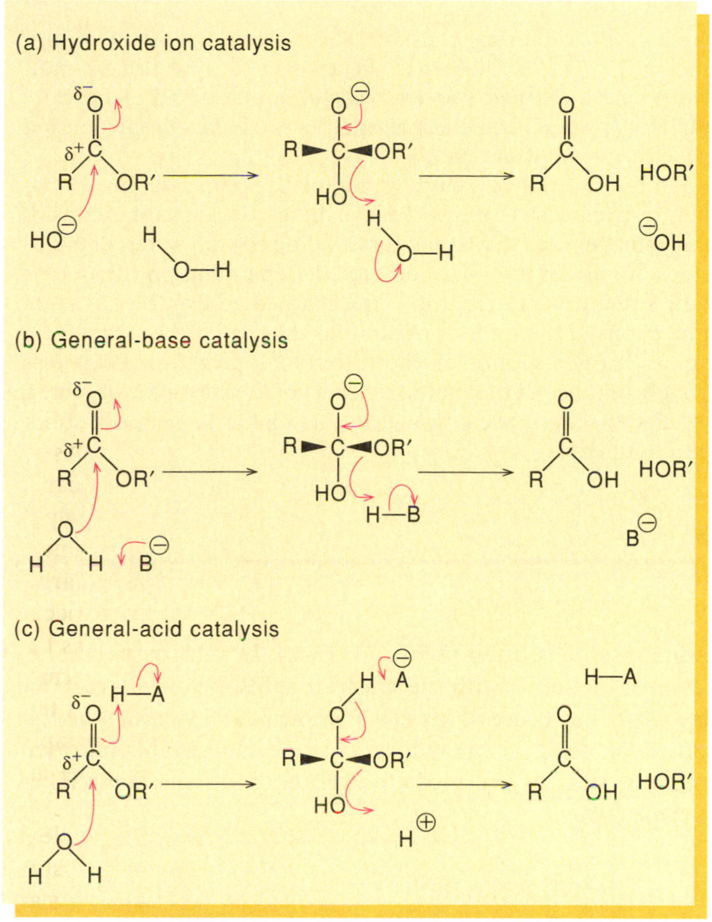

of a peptide bond (see table 3.3). The pK_a of any of these groups can vary over a considerable range, depending on the local environment in the enzyme. The second advantage of using a general base instead of OH^- is that a basic group that is provided by the protein can be positioned precisely with respect to the substrate in the active site, allowing the proximity effect to come into play. Free hydroxide ions tend to be much more mobile. In exceptional cases in which a hydroxide ion acts as a nucleophile in an enzymatic reaction, it usually is tightly bound to a metal ion.

The hydrolysis of an ester also can be catalyzed by an acid (see fig. 9.1*c*). The acid donates a proton to the oxygen of the ester's C=O group, increasing the positive charge on the carbon and increasing the susceptibility of the ester to attack by a nucleophile. Again, the term general acid is used to refer to any substance that is capable of releasing a proton, and enzymes almost always use such proton donors in preference to free protons or hydronium ions, presumably because a general acid can operate at moderate pH and is easy to fix in position. In this case, the requirement is that the pH be below the pK_a.

An important point to note in figure 9.1 is that the same general acid or base that catalyzes the formation of the tetrahedral intermediate also can participate in the decomposition of the intermediate. When a general acid (HA) donates a proton to the ester oxygen, it becomes the conjugate base (A^-), which can retrieve the proton as the intermediate breaks down. When a general base (B^-) removes a proton from water, it becomes the conjugate acid (BH), which can provide a proton to the alcohol. Note also that general-acid and general-base catalysis are not mutually exclusive: They could both occur in a concerted manner in the same step of a reaction.

Electrostatic Interactions Can Promote the Formation of the Transition State

The frequent use of general acids and general bases in enzymatic reaction mechanisms illustrates the underlying principle that enzymes act by stabilizing the distribution of electrical charge in transition states. In the enzymatic hydrolysis of an ester, the key transition state probably is structurally similar to the tetrahedral intermediates shown in figure 9.1. To form such an intermediate, electrons must move from the attacking nucleophile, through the carbon atom of the C=O group, to the oxygen of the C=O. There is thus a net movement of negative charge from the nucleophile to the substrate. In the absence of a general acid or base, a charge approaching +1 would appear on the nucleophile and a charge approaching −1 would appear on the C=O oxygen. A general base can stabilize this new distribution of charge by offering electrons to the nucleophile, so that some of the positive charge moves to the base. By providing a proton to the C=O oxygen, a general acid can delocalize the negative charge at this end of the system. But there are other ways that an enzyme could achieve a similar stabilization. Suppose that the active site included a positively charged amino acid side chain, such as that of lysine or arginine, located near the oxygen atom of the C=O group. A fixed positive charge in this region would favor the formation of the tetrahedral intermediate, even if there were no transfer of a proton from the charged species to the oxygen. A fixed negative charge in the region of the nucleophile would have a similar effect. The interactions of such fixed charges are termed electrostatic effects. The magnitude of electrostatic effects in proteins is discussed in box 9A.

Electrostatic interactions can be significant even between groups whose net formal charge is zero. This is because charge distributions within molecular groups are not uniform, but rather varies from atom to atom. We alluded to this point earlier in discussing the partial charges on the oxygen and carbon atoms of an ester (see fig. 9.1). Similar considerations apply to other functional groups, including even methylene and methyl groups: The electron distributions around the nuclei leave each atom with a small net positive or negative charge, even though the overall sum of these charges is zero. In an alcoholic $-CH_2OH$ group, for example, the oxygen atom has a negative charge of approximately −0.4 atomic charge units, and the hydrogen has a charge of about +0.4.

The Magnitude of Electrostatic Effects in Proteins

The energy of the electrostatic interaction between two charges Q_1 and Q_2 separated by a distance r Å is (in kcal/mole)

$$V = 332\frac{Q_1Q_2}{r} \quad \text{(B1)}$$

From this expression, it is clear that electrostatic effects can be appreciable even at relatively large distances; the interaction energy of a set of opposite charges 10 Å apart would be −33.2 kcal/mole. However, equation (B1) refers to charges in a vacuum. In a polar solvent such as water, electrostatic interactions are weaker because they are screened by the dielectric effect of the solvent. To take this into account, equation (B1) is often replaced by an expression of the form

$$V = 332\frac{Q_1Q_2}{\epsilon r} \quad \text{(B2)}$$

where ϵ is the dielectric constant. The dielectric constant of pure water at 25°C is 78. Dielectric effects arise because solvent molecules near a charged species become oriented and electrically polarized, so that each charged species is effectively surrounded by a cloud of opposite charges.

In the interior of a protein, molecular reorientation is relatively restricted, so that the effective dielectric constant is considerably smaller than that of the surrounding solvent water. Effective values of ϵ are in the range of 2 to 10, depending on the details of the structure in the region of the charged groups. Electrostatic interactions in the interior of proteins thus can be comparatively strong. Charged groups on the surface of a protein interact less strongly because of the dielectric effect of the surrounding water; the effective dielectric constant in this region is probably about 40 in most cases.

As a reacting substrate is transformed into a transition state, the changing charges on its atoms interact with the charges on all of the other atoms in the surrounding protein, and also with the charges on any nearby water molecules. The energy difference between the initial state and the transition state thus depends critically on the details of the protein structure. We will see illustrations of this principle in the serine proteases and the other enzymes that we discuss later in the chapter. Modern computational techniques, when taken with the wealth of structural information that has become available from x-ray crystallography and other biophysical studies, have made it possible to calculate the contributions that various components of an enzyme's active site make to the activation free energy ΔG^{+}, and to predict quantitatively how ΔG^{+} might be altered by modifications of the protein. These predictions can be tested experimentally by modifying the gene that encodes the protein, a technique termed "site-directed mutagenesis" (see chapter 27). This combination of biophysical, computational, and molecular biological techniques has opened exciting new frontiers for exploring the detailed mechanisms of enzymic catalysis.

Enzymatic Functional Groups Provide Nucleophilic and Electrophilic Catalysts

Another strategy for catalyzing the hydrolysis of an ester or an amide is to replace water by a stronger nucleophilic group that is part of the enzyme's active site. The $HOCH_2$— group of a serine residue is often used in this way. In such cases, the reaction of the serine with the substrate splits the overall reaction into a two-step process. Instead of immediately yielding the free carboxylic acid, the breakdown of the initial tetrahedral intermediate yields an intermediate ester that is covalently attached to the enzyme:

$$\begin{aligned} &R—C(=O)—NHR' + HOCH_2—\text{Enzyme} \longrightarrow \\ &R—C(=O)—OCH_2—\text{Enzyme} + R'NH_2 \end{aligned} \quad \textbf{(4)}$$

The acyl-enzyme ester intermediate must be hydrolyzed by a second reaction, in which water becomes the nucleophile:

$$\begin{aligned} &R—C(=O)—OCH_2—\text{Enzyme} + H_2O \longrightarrow \\ &R—C(=O)—OH + HOCH_2—\text{Enzyme} \end{aligned} \quad \textbf{(5)}$$

The proteolytic enzymes trypsin, chymotrypsin, and elastase, discussed in a later section, all work in this way. The two-step pathway requires that the intermediate be more susceptible to nucleophilic attack by water than the original ester or amide. This is likely if the original substrate is an amide, because amides are generally less reactive than esters.

Nucleophilic groups on enzymes participate in a variety of other types of reactions in addition to hydrolytic reactions. An example is acetoacetic acid decarboxylase, which catalyzes the reaction

$$CH_3 - \overset{\overset{\displaystyle O}{\parallel}}{C} - CH_2 - CO_2H \rightarrow CH_3 - \overset{\overset{\displaystyle O}{\parallel}}{C} - CH_3 + CO_2 \quad (6)$$

The reaction proceeds by the formation of a Schiff base intermediate, in which the substrate is covalently attached to the ϵ-amino group of a lysine residue at the enzyme's active site:

$$CH_3 - \overset{\overset{\displaystyle O}{\parallel}}{C} - CH_2 - CO_2H + \text{Enzyme} - NH_2 \rightarrow$$

$$CH_3 - \overset{\overset{\displaystyle N - \text{Enzyme}}{\parallel}}{C} - CH_2CO_2H + H_2O \quad (7)$$

This intermediate is formed by a nucleophilic attack of the amino group on the carbonyl carbon, followed by the splitting out of water. Protonation of the nitrogen atom of the Schiff base introduces a positive charge that pulls electrons from the nearby carbon–carbon bond, causing decarboxylation (fig. 9.2). This is an extreme example of an electrostatic effect: The enzyme introduces a charged group, not just nearby in the active site, but into the substrate itself! Aldolase and transaldolase, two enzymes that catalyze steps in the breakdown of carbohydrates, use lysine residues in a similar manner.

A basic feature of both of the mechanisms outlined in equations (4)–(7), and of other instances of nucleophilic catalysis by enzymes, is the formation of an intermediate state in which the substrate is covalently attached to a nucleophilic group on the enzyme. In addition to the $-CH_2OH$ group of serine and the ϵ-amino group of lysine, the $-CH_2SH$ of cysteine is often used as a nucleophile. The carboxylate of aspartate or glutamate participates in reactions involving the hydrolysis of ATP, and the imidazole group of histidine can play a similar role. Some enzymes take advantage of bound coenzymes such as thiamine, biotin, pyridoxamine, or tetrahydrofolate to obtain additional nucleophilic reagents (see chapter 11).

There also are numerous enzymes that use bound metal ions to form complexes with substrates. In these enzymes, the metal ion generally serves as an *electrophilic,* rather than a nucleophilic, functional group. Carbonic anhydrase, for example, contains a Zn^{2+} ion that binds one of the substrates, hydroxide ion, as a ligand. The bound OH^- reacts with the other substrate, CO_2. In alcohol dehydrogenase, and in the proteolytic enzymes thermolysin and carboxypeptidase A, a Zn^{2+} ion in the active site forms a complex with the carbonyl oxygen atom of the aldehyde or peptide substrate. The withdrawal of electrons by the Zn^{2+} increases the partial positive charge on the carbonyl carbon atom, and thus promotes the reaction of the carbon with a nucleophile. We will discuss such enzymes in more detail in a later section.

Figure 9.2

In acetoacetic acid decarboxylase, the positive charge of a protonated Schiff base intermediate pulls electrons from a nearby carbon–carbon bond, thereby releasing CO_2.

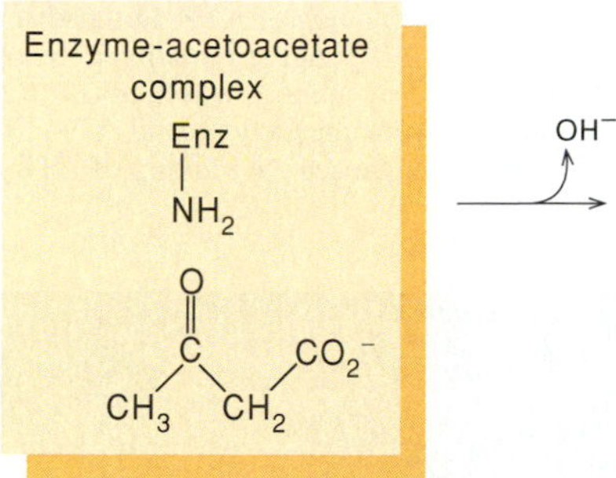

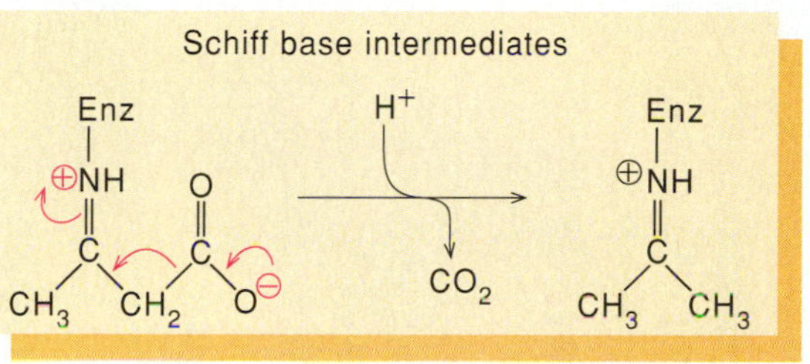

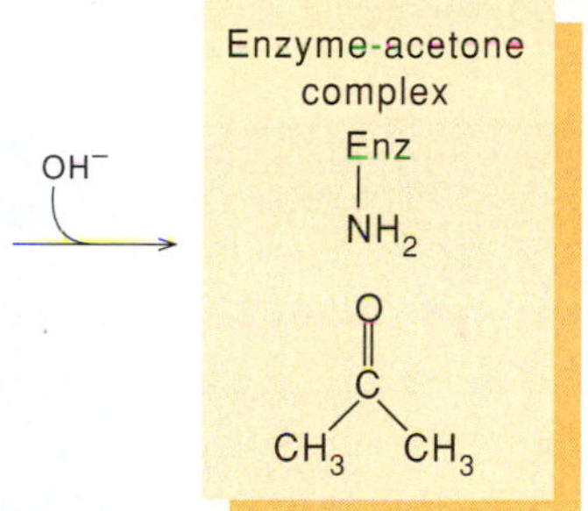

Structural Flexibility Can Increase the Specificity of Enzymes

Although precise positioning of the reactants is a fundamental aspect of enzyme catalysis, some enzymes undergo major structural rearrangements when they bind substrates or inhibitors. An example is hexokinase, which catalyzes the transfer of a phosphate group from ATP to glucose:

$$\text{ATP} + \text{glucose} \rightarrow \text{ADP} + \text{glucose-6-phosphate} \quad (8)$$

When hexokinase binds glucose, it undergoes a structural reorganization that brings together the elements of the active site (fig. 9.3). The enzyme literally closes like a set of jaws around the substrate! Such a structural change is often referred to as an induced fit.

Figure 9.3

Models of the crystallographic structure of hexokinase in the "open" (*a*) and "closed" (*b*) conformations. The enzyme (shown in blue) adopts the open conformation in the absence of substrates, but switches to the closed conformation when it binds glucose (red). Hexokinase also has been crystallized with a bound analog of ATP. In the absence of glucose, the enzyme with the bound ATP analog remains in the open conformation. The structural change caused by glucose would result in the formation of additional contacts between the enzyme and ATP. This could explain why the binding of glucose enhances the binding of ATP. (Courtesy of Dr. Thomas A. Steitz.)

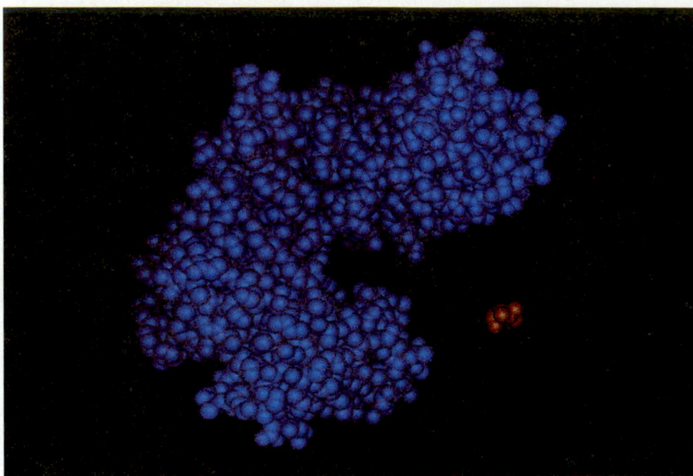

(a)

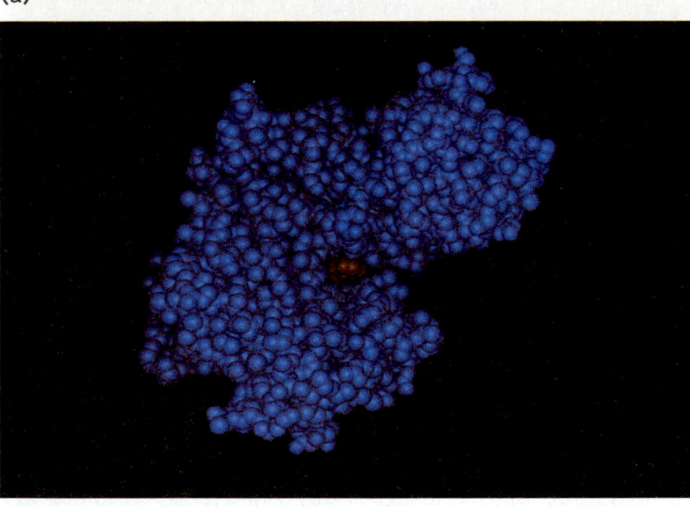

(b)

Carboxypeptidase A, which we will discuss in more detail later, is another enzyme that undergoes a major structural change when it binds its substrate. In this case, the rearrangement of the protein effectively pulls the hydrophobic part of the substrate out of the aqueous solution by surrounding it with nonpolar portions of the protein. Enfolding a substrate in this way can be beneficial in several ways. First, it can serve to maximize the favorable entropy change associated with removing a hydrophobic molecule from water. Second, it should allow the enzyme to control and intensify the electrostatic effects that promote the formation of the transition state. The substrate is forced to respond to the directed electrostatic fields from the enzyme's functional groups, instead of the disordered fields from the solvent.

Structural changes also can help to explain the high specificity of some enzymatic reactions. In hexokinase, for example, the structural change induced by glucose promotes the binding of the other substrate, ATP (see fig. 9.3). ATP does not bind to the enzyme properly unless glucose is already present in the catalytic site. If ATP were to bind in the absence of glucose, the enzyme might have a tendency to catalyze the transfer of phosphate from ATP to water, resulting in a wasteful loss of ATP:

$$\text{ATP} + \text{H}_2\text{O} \rightarrow \text{ADP} + \text{P}_i \quad \textbf{(9)}$$

Hexokinase does not catalyze this side-reaction; it waits for glucose to bind first.

As another example, the enzyme serine hydroxymethylase (see chapter 18) catalyzes the removal of formaldehyde from serine, forming glycine. In a second step, the enzyme transfers the formaldehyde to a bound coenzyme, tetrahydrofolic acid (THF). If the removal of formaldehyde from serine proceeded even in the absence of THF, the formaldehyde might be set free in the cytosol, where it could enter into other, undesirable reactions. (Free formaldehyde is highly reactive and is toxic to cells.) To prevent this potential disaster, serine hydroxymethylase does not catalyze the first reaction until after THF is bound. The binding of the coenzyme causes the enzyme to fold so that the first step can occur, even though the THF is not yet involved directly in the chemistry.

The structural changes that occur in hexokinase, carboxypeptidase A, and serine hydroxymethylase bring home the point that enzyme crystal structures give static snapshots of molecules that, in many cases, actually are highly flexible. In solution, the structure of an enzyme undergoes fluctuations that vary widely in amplitude and frequency from place to place in the protein. Vibrations and rotations involving only a few atoms occur on time scales of 10^{-13} to 10^{-11} s. Somewhat larger motions, such as the flipping of the aromatic ring of a tyrosine or tryptophan, typically occur on scales of 10^{-9} to 10^{-8} s. Major reorganizations may take 10^{-6} to 10^{-3} s. All of these types of motions can be important in catalysis.

Detailed Mechanisms of Enzyme Catalysis

In the foregoing sections, we have discussed five themes that are related to enzyme reaction mechanisms. We will now examine several representative enzymes in finer detail. We will focus on enzymes for which crystal structures have been obtained, because the most decisive advances in our understanding of enzyme reaction mechanisms have come by inspecting such structures. Crystals of many enzymes have been shown to be enzymatically active, and it appears that in most cases the three-dimensional structures of crystalline enzymes are close to the structures of the proteins in solution. It is important to keep in mind, however, that crystal structures provide pictures of enzymes in relatively stable states. To fill in the intermediates and transition states between these resting states requires a variety of other techniques, including studies of related nonenzymatic reaction mechanisms.

There are over 1,500 known enzymes, each with its own unique structure, specificity, and catalytic mechanism. However, the situation is less complicated than this number might suggest, because many enzymes can be grouped in families that share certain basic features. In some cases the enzymes that make up a family appear to have diverged from a common evolutionary ancestor. Family members that arose in this way are apt to retain similar secondary and tertiary structures, and they typically have the same amino acid residue at between 20 and 50% of the corresponding positions in their primary sequences. In other cases enzymes with diverse ancestral origins appear to have converged on structural features that are well suited for catalyzing particular types of reactions. Such enzymes resemble each other in their active sites, but may have little in common elsewhere in their structures.

Serine Proteases Are a Diverse Group of Enzymes that Use a Serine Residue for Nucleophilic Catalysis

The serine proteases are a large family of proteolytic enzymes that use the reaction mechanism for nucleophilic catalysis outlined in equations (4) and (5), with a serine residue as the reactive nucleophile. The best known members of the family are three closely related digestive enzymes, trypsin, chymotrypsin, and elastase. These enzymes are synthesized in the mammalian pancreas as inactive precursors termed zymogens. They are secreted into the small intestine, where they are activated by proteolytic cleavage in a manner that will be described in chapter 10. Many of the enzymes that participate in blood coagulation also are serine proteases; these enzymes circulate in the blood as inactive zymogens and are activated by proteolytic cleavage when blood vessels are damaged (see chapter 10). The serine protease family also includes many enzymes from bacteria and other nonmammalian organisms.

In the digestive system, trypsin, chymotrypsin, and elastase work as a team. They are all endopeptidases, which means that they cleave protein chains at internal peptide bonds, but each preferentially hydrolyzes bonds adjacent to a particular type of amino acid residue (fig. 9.4). Trypsin cuts just past the carbonyl groups of basic residues (lysine or arginine); chymotrypsin cuts next to aromatic residues (phenylalanine, tyrosine, or tryptophan); elastase is less discriminating, but prefers small, hydrophobic residues such as alanine.

About half of the amino acid residues of trypsin are identical to the corresponding residues in chymotrypsin, and about a quarter of the residues are conserved in all three of the pancreatic endopeptidases (fig. 9.5). The structural similarities of trypsin, chymotrypsin, and elastase are even more evident in the crystal structures. As is shown in figure 9.6, the folding of the polypeptide chain is essentially the same in all three enzymes, with the only substantial variations occurring in the external loops. These enzymes are classical illustrations of diverging evolution from a common ancestor. The structures of some of the bacterial serine proteases also are homologous to those of the mammalian enzymes. On the other hand, subtilisin,

Figure 9.4

Trypsin, chymotrypsin, and elastase, three members of the serine protease family, catalyze the hydrolysis of proteins at internal peptide bonds adjacent to different types of amino acids. Trypsin prefers lysine or arginine residues; chymotrypsin, aromatic side chains; and elastase, small, nonpolar residues. Carboxypeptidases A and B, which are not serine proteases, cut the peptide bond at the carboxyl-terminal end of the chain. Carboxypeptidase A preferentially removes aromatic residues; carboxypeptidase B, basic residues.

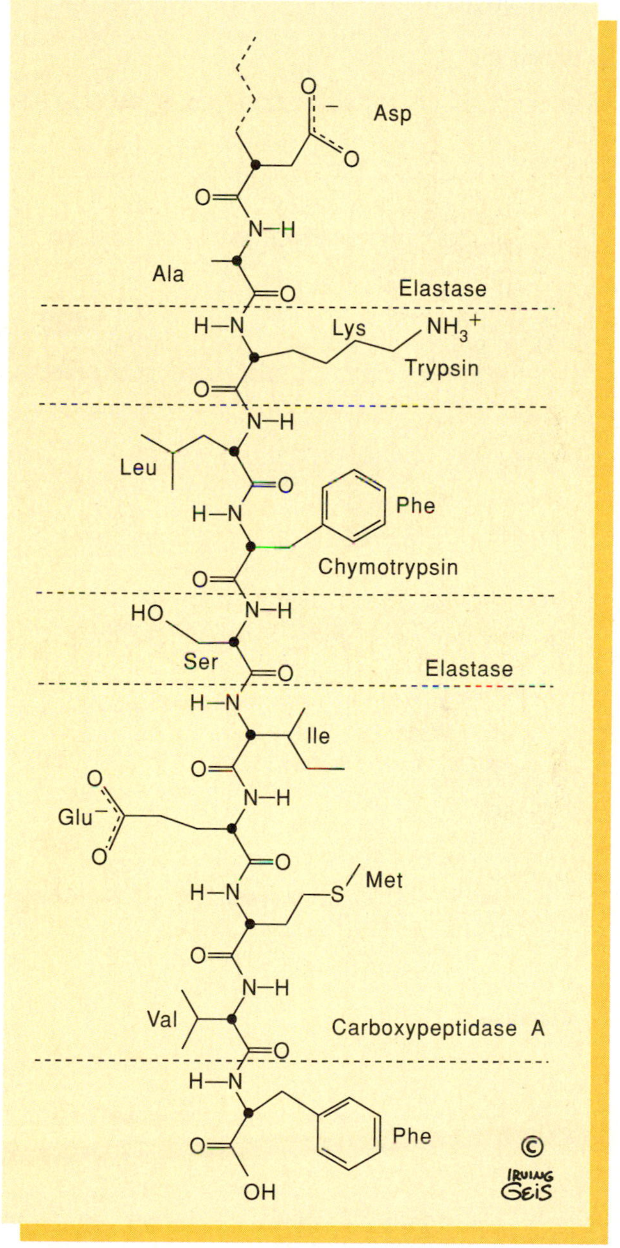

Figure 9.5

Schematic diagrams of the amino acid sequences of chymotrypsin, trypsin, and elastase. Each circle represents one amino acid. Amino acid residues that are identical in all three proteins are in solid color. The three proteins are of different lengths, but have been aligned to maximize the correspondence of the amino acid sequences. All of the sequences are numbered according to the sequence in chymotrypsin. Long connections between nonadjacent residues represent disulfide bonds. Locations of the catalytically important histidine, aspartate and serine residues are marked. The links that are cleaved to transform the inactive zymogens to the active enzymes are indicated by parenthesis marks. After chymotrypsinogen is cut between residues 15 and 16 by trypsin, and is thus transformed into an active protease, it proceeds to digest itself at the additional sites that are indicated; these secondary cuts have only minor effects on the enzyme's catalytic activity.

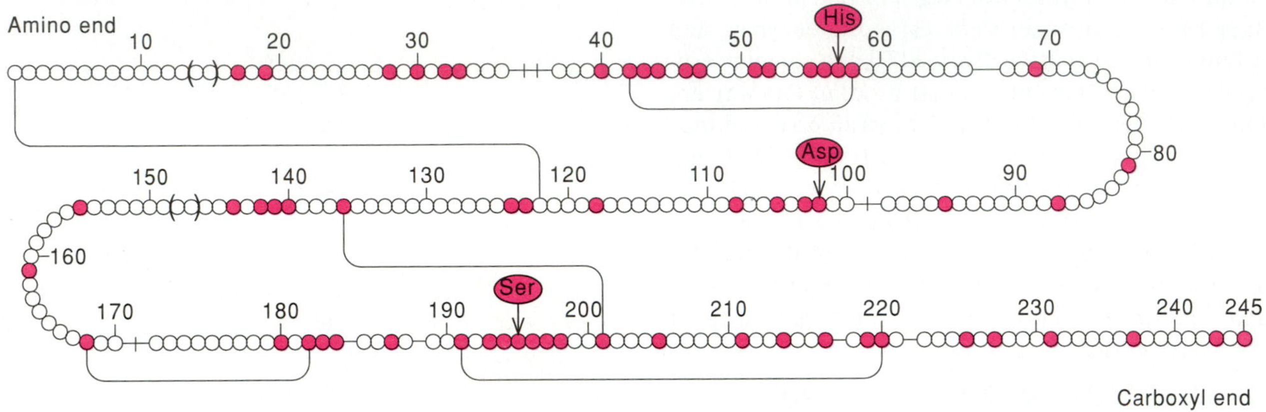

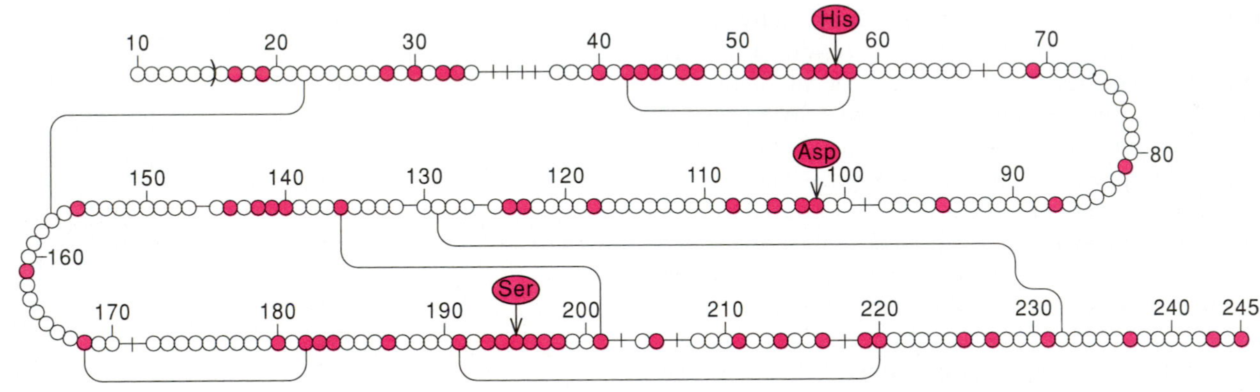

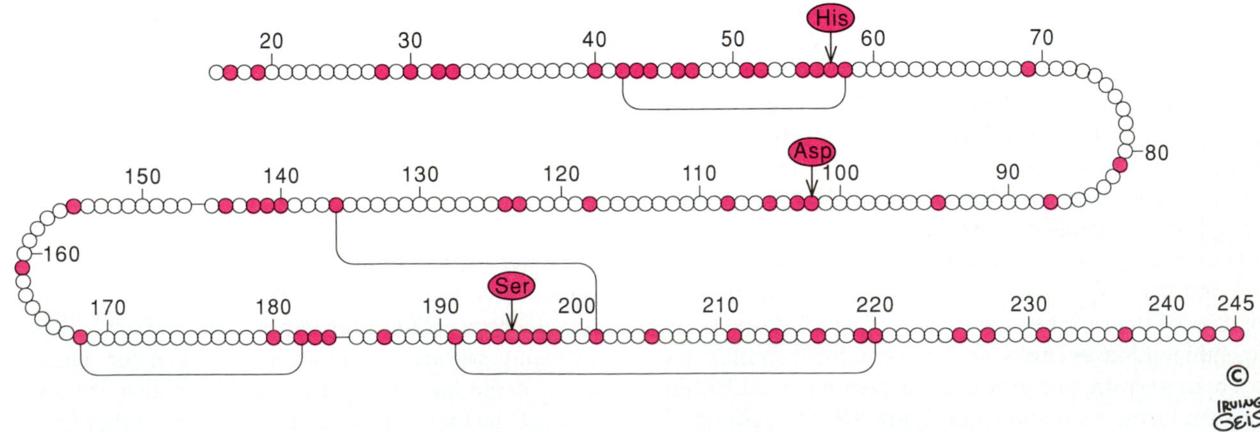

Figure 9.6

Crystal structures of (*a*) trypsin, (*b*) chymotrypsin, and (*c*) elastase. The yellow ribbons superimposed on the structures show the similar folding patterns of the α-carbon chains in the three structures. Three residues that play important roles in the catalytic mechanism (Asp 101, His 57, and Ser 195) are shown in red; other residues are in blue. The structure for trypsin includes a bound inhibitor, benzamidine, in yellow. The positively charged nitrogen atom of benzamidine fits into the bottom of the pocket that determines the specificity of trypsin for positively charged amino acid residues (lysine and arginine). (Based on the crystal structures described by W. Bode, P. Schwager, and J. Walter for trypsin, H. Tsukada and D. Blow for chymotrypsin, and L. C. Sieker and D. L. Hughes for elastase.)

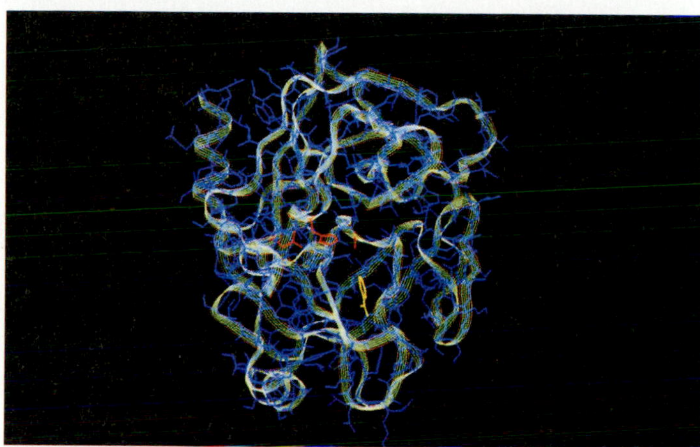

(a)

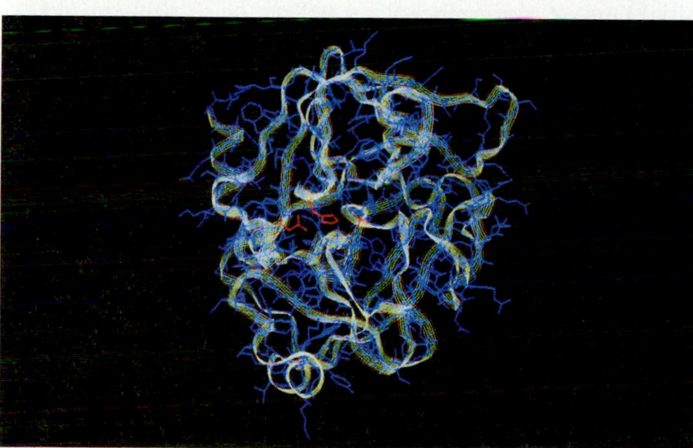

(b)

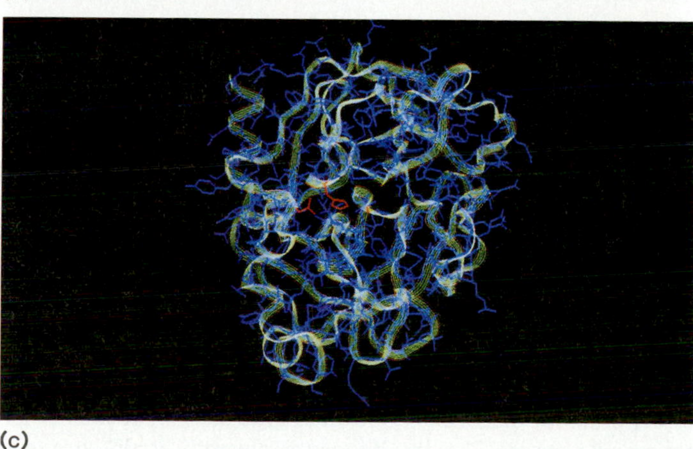

(c)

Figure 9.7

(*a*) A space-filling model of the crystal structure of trypsin, seen from the same perspective as in figure 9.6*a*. Hydrogen atoms are not shown. The coloring of the amino acid residues and the inhibitor (benzamidine) is as in figure 9.6. (*b*) A close-up view of the active site of trypsin with bound benzamidine. The amide NH hydrogens of Ser 195 and Gly 193, the OH hydrogen of Ser 195, and the imidazole NH hydrogen of His 57 are shown in green; other atoms are colored as in figure 9.6. In the oxyanion intermediate, the negatively charged oxygen atom of the substrate interacts with the amide hydrogens of Ser195 and Gly193. Benzamidine does not have an analogous oxygen atom.

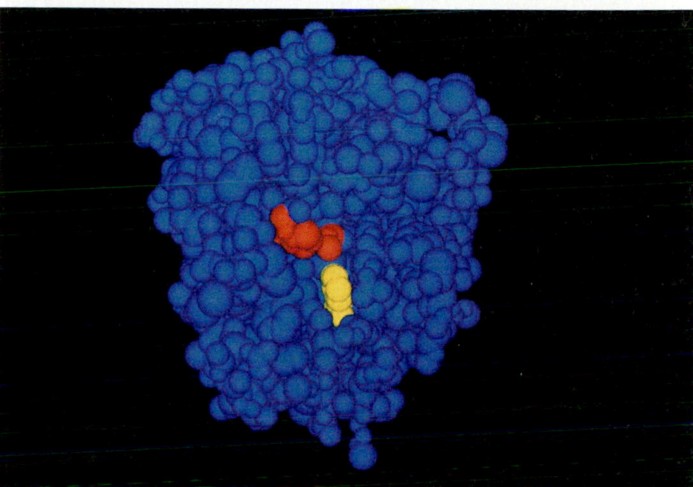

(a)

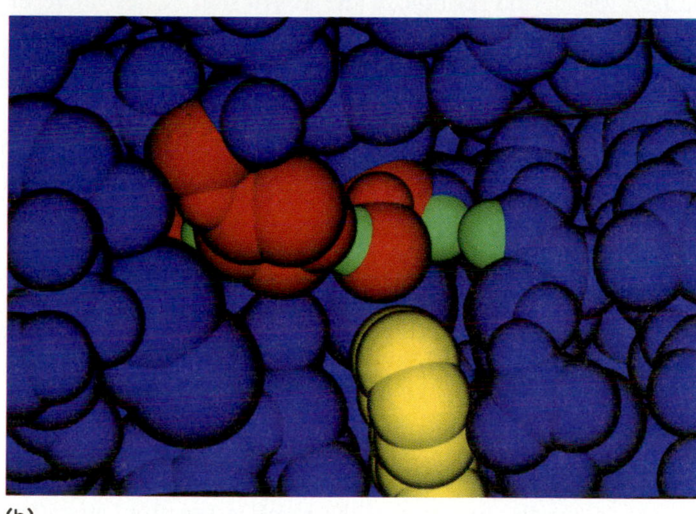

(b)

a serine protease obtained from *Bacillus subtilis,* has an amino acid sequence that seems totally unrelated to the mammalian sequences. Its three-dimensional structure also is very different from those of the mammalian enzymes (see fig. 5.23). Subtilisin therefore is likely to have joined the serine protease family by convergent evolution. Remarkably, there is a small set of critical amino acid residues that come together in the folded structure to form the essential elements of the active site in all of these proteases. In the chymotrypsin numbering system, these are His 57, Asp 102 and Ser 195. The locations of these residues in the three-dimensional structures of trypsin, chymotrypsin, and elastase can be seen in figures 9.6 and 9.7.

That Ser 195 played an important role in the catalytic mechanism was known from early studies on the enzyme inhibitor diisopropylfluorophosphate (see chapter 8). This inhibitor reacts irreversibly with chymotrypsin or trypsin to form an inactive derivative in which the diisopropylphosphate group is covalently attached to the serine residue (see fig. 8.20). The derivative, a phosphate ester of the serine, is similar in structure to the acyl-enzyme

$$(\mathrm{R}-\overset{\overset{\displaystyle \mathrm{O}}{\|}}{\mathrm{C}}-\mathrm{OCH_2}-\text{Enzyme})$$

intermediate in equations (4) and (5). A variety of other inhibitors have been found to react in a parallel manner with this particular serine residue. As a rule, these inhibitors do not react with other serines in the enzyme, or with serines in enzymes that are not part of the serine protease family. Exceptions to the rule are a number of enzymes that catalyze the hydrolysis of esters; these enzymes have a similar, reactive serine residue, and their catalytic mechanism appears to be very similar to that of the serine proteases. The reactivity of Ser 195 thus is not a property of serine residues in general, but depends on the special surroundings of this residue in the protein. We will see shortly that it is the juxtaposition of Ser 195 with His 57 and Asp 102 that makes this serine especially reactive.

Although His 57 is far removed from Ser 195 in the primary sequence, studies with affinity labels showed that it must be near the serine residue in the active site. A derivative of phenylalanine containing a reactive chloromethyl ketone group was found to inhibit chymotrypsin irreversibly by reacting with the histidine. The inhibition can be prevented by the presence of other aromatic molecules that bind competitively at the active site.

The initial evidence for the formation of an acyl-enzyme ester intermediate came from studies of the kinetics with which chymotrypsin hydrolyzed various analogs of its normal polypeptide substrates. The enzyme turned out to hydrolyze esters, as well as peptides and simpler amides. Of particular interest was the reaction with the ester *p*-nitrophenyl acetate. This substrate is well suited for kinetic studies because one of the products of its hydrolysis, *p*-nitrophenol, has a characteristic yellow color in aqueous solution, whereas *p*-nitrophenyl acetate itself is colorless. The change in the absorption spectrum makes it easy to follow the progress of the reaction spectrophotometrically. When rapid mixing techniques were used to add the substrate to the enzyme, it was found that an initial burst of *p*-nitrophenol was released within the first few seconds, before the reaction settled down to a constant rate (fig. 9.8). The amount of *p*-nitrophenol that appeared in the burst was approximately equal to the amount of enzyme present in the solution. These observations suggested that the overall enzymatic reaction occurs in two distinct steps, as shown in figure 9.9. In the first step, *p*-nitrophenol is released and the acetyl group is transferred to the enzyme, forming an acyl-enzyme intermediate. In the second step, the intermediate is hydrolyzed, and acetate is released. Diisopropylfluorophosphate prevents the

Figure 9.8

p-Nitrophenol formation as a function of time during the hydrolysis of *p*-nitrophenyl acetate by chymotrypsin. A rapid initial burst of *p*-nitrophenol is followed by a slower, steady-state reaction. The amount of *p*-nitrophenol released in the burst is approximately equal to the amount of enzyme present.

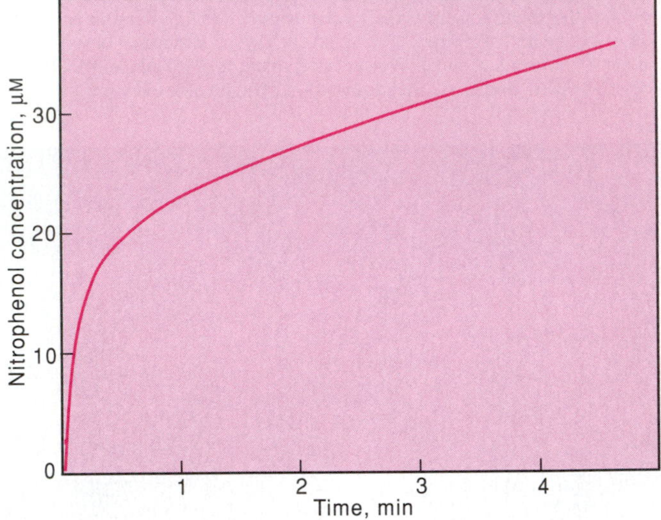

initial burst of *p*-nitrophenol, as well as the subsequent steady-state reaction, a fact suggesting that the enzymatic group that forms the ester intermediate is Ser 195.

When the crystal structures of trypsin, chymotrypsin, and elastase with bound substrate analogs were solved, the substrate analogs were indeed found to be located close to Ser 195 (see figs. 9.6 and 9.7). Histidine 57, the second of the three residues mentioned above, is located nearby, in an orientation suggesting that the OH group of Ser 195 forms a hydrogen bond to the imidazole side chain of the histidine. Aspartic acid 102 sits on the opposite edge of the imidazole ring, where its negatively charged carboxylate group could interact with the proton on the other nitrogen of the ring. The side chains of the aspartate and histidine residues thus appear to be oriented so as to facilitate removal of the proton from the serine's OH group:

$$-\overset{\overset{\displaystyle \mathrm{O}}{\|}}{\mathrm{C}}-\mathrm{O}^- \cdots \mathrm{HN}\overset{\mathrm{C=C}}{\frown}\mathrm{N}\cdots\mathrm{HOCH_2}- \tag{10}$$

Because withdrawing the proton would increase the nucleophilic character of the oxygen, this arrangement explains why Ser 195 is exceptionally reactive. The same arrangement of aspartate, histidine, and serine residues has been found in all of the serine proteases that have been examined. It is often referred to as a "charge relay system." The dependence of the enzyme kinetics on pH agrees with this picture. Enzymatic activity appears to depend on the presence of a basic group with a pK_a of about 6.8, which is in the range consistent with a histidine side chain. The k_{cat} decreases abruptly if this group is

Figure 9.9

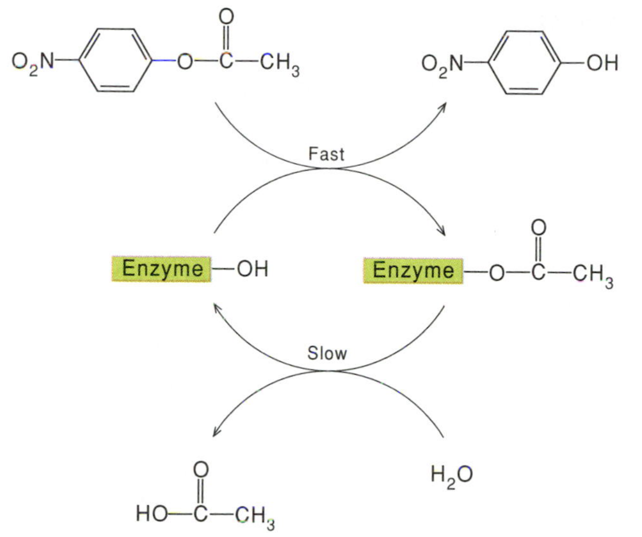

Steps in the hydrolysis of *p*-nitrophenyl acetate by chymotrypsin. In the hydrolysis of this and most other esters, the breakdown of the acyl-enzyme intermediate is the rate-determining step. In the hydrolysis of peptides and amides, the rate-determining step usually is the formation of the acyl-enzyme intermediate. This makes the transient formation of the intermediate more difficult to study, because the intermediate breaks down as rapidly as it forms.

protonated (fig. 9.10). This is because protonating His 57 prevents the histidine from forming a hydrogen bond to Ser 195, and thus greatly decreases the nucleophilic reactivity of the serine.

The crystal structures also provided a simple explanation for the different substrate specificities of trypsin, chymotrypsin, and elastase. In both trypsin and chymotrypsin, the side chain of the substrate fits snugly into a pocket (see figs. 9.6*a*, 9.6*b*, and 9.7). At the far end of the pocket in trypsin is the carboxylate group of an aspartic acid residue. The negative charge of the carboxylate would favor the binding of the positively charged side chain of lysine or arginine. In chymotrypsin the aspartic acid residue is replaced by serine, creating a less polar environment that suits the side chain of tyrosine, phenylalanine, or tryptophan. In elastase (see fig. 9.6*c*), the binding pocket is obstructed by the bulky side chains of a valine and a threonine residue, so that it can accommodate only small substrates such as alanine.

Another important feature of the substrate binding site in serine proteases is that the carbonyl oxygen atom of the scissile peptide bond is hydrogen-bonded to one or more NH groups. In trypsin, chymotrypsin, and elastase, these are the amide NH groups of Ser 195 and of Gly 193. The interaction with the two protons would favor an increase in the negative charge on the oxygen, facilitating the formation of a tetrahedral intermediate state, as we discussed earlier in connection with general-acid catalysis and electrostatic effects. Figure 9.7*b* shows a space-filling model of this part of the active site in trypsin.

Figure 9.10

The turnover number (k_{cat}) and the Michaelis constant (K_m) as a function of pH for the hydrolysis of *N*-acetyl-L-tryptophanamide by chymotrypsin at 25°C. The decrease in k_{cat} as the pH is lowered between 8 and 6 probably reflects the protonation of His 57. The increase in K_m above pH 9 probably reflects the deprotonation of Ile 16, which results in the rotation of Gly 193 out of the substrate-binding site (see fig. 10.1).

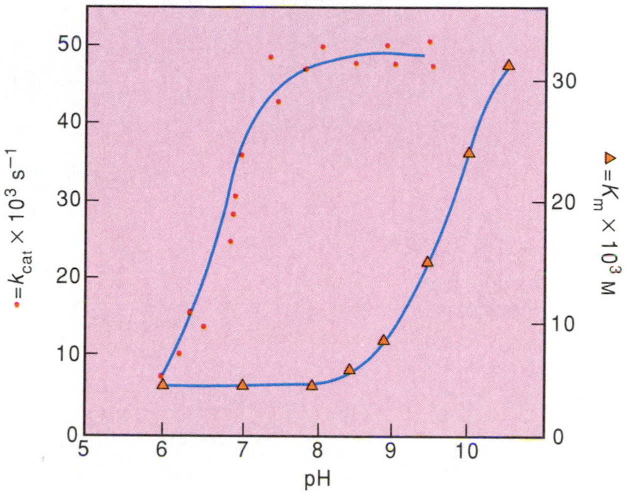

The overall reaction mechanism of chymotrypsin is sketched in figure 9.11. Part (*a*) shows the enzyme-substrate complex, with an aromatic side chain of the substrate seated in the binding pocket and the carbonyl oxygen atom hydrogen-bonded to the amide NH hydrogens of Gly 193 and Ser 195. (The crystal structure suggests that several additional hydrogen bonds contribute to fixing the substrate in an optimal orientation on the enzyme; these have been left out of the figure for clarity.) Aspartate 102 and His 57 are aligned as described above, with Ser 195 forming a hydrogen bond with the histidine. Part (*b*) shows a tetrahedral intermediate state. Serine 195 has released its proton to His 57 and launched a nucleophilic attack on the carbonyl carbon of the substrate. The histidine residue thus acts as a general-base catalyst in the formation of the intermediate. The movement of negative charge to the carbonyl oxygen creates what is often called an "oxyanion," which is stabilized largely by electrostatic interactions with the amide protons of Ser 195 and Gly 193. The tetrahedral oxyanion intermediate is not stable enough to be isolated, and the evidence for its existence is inferential. In part (*c*), this fleeting intermediate has decomposed to form the more stable acyl-enzyme intermediate, in which the serine is linked as an ester to the carboxylic part of the substrate and the amine product has been released. Histidine 57 probably facilitates this decomposition by acting as a general acid.

To complete the reaction, the breakdown of the acyl-enzyme probably occurs as shown in parts (*d*), (*e*), and (*f*) of figure 9.11. The steps here are essentially a reversal of the steps through parts (*a*), (*b*), and (*c*), except that water replaces the amino part of the substrate. The reaction probably proceeds by way of a tetrahedral intermediate (*e*) that is similar to one shown in (*b*) except for this substitution.

Figure 9.11

The probable mechanism of action of chymotrypsin. The six panels show the initial enzyme-substrate complex (*a*), the first tetrahedral (oxyanion) intermediate (*b*), the acyl-enzyme (ester) intermediate with the amine product departing (*c*), the same acyl-enzyme intermediate with water entering (*d*), the second tetrahedral (oxyanion) intermediate (*e*), and the final enzyme-product complex (*f*). In the transition states between these intermediates, there probably is a more even distribution of negative charge between the different oxygen atoms attached to the substrate's central carbon atom.

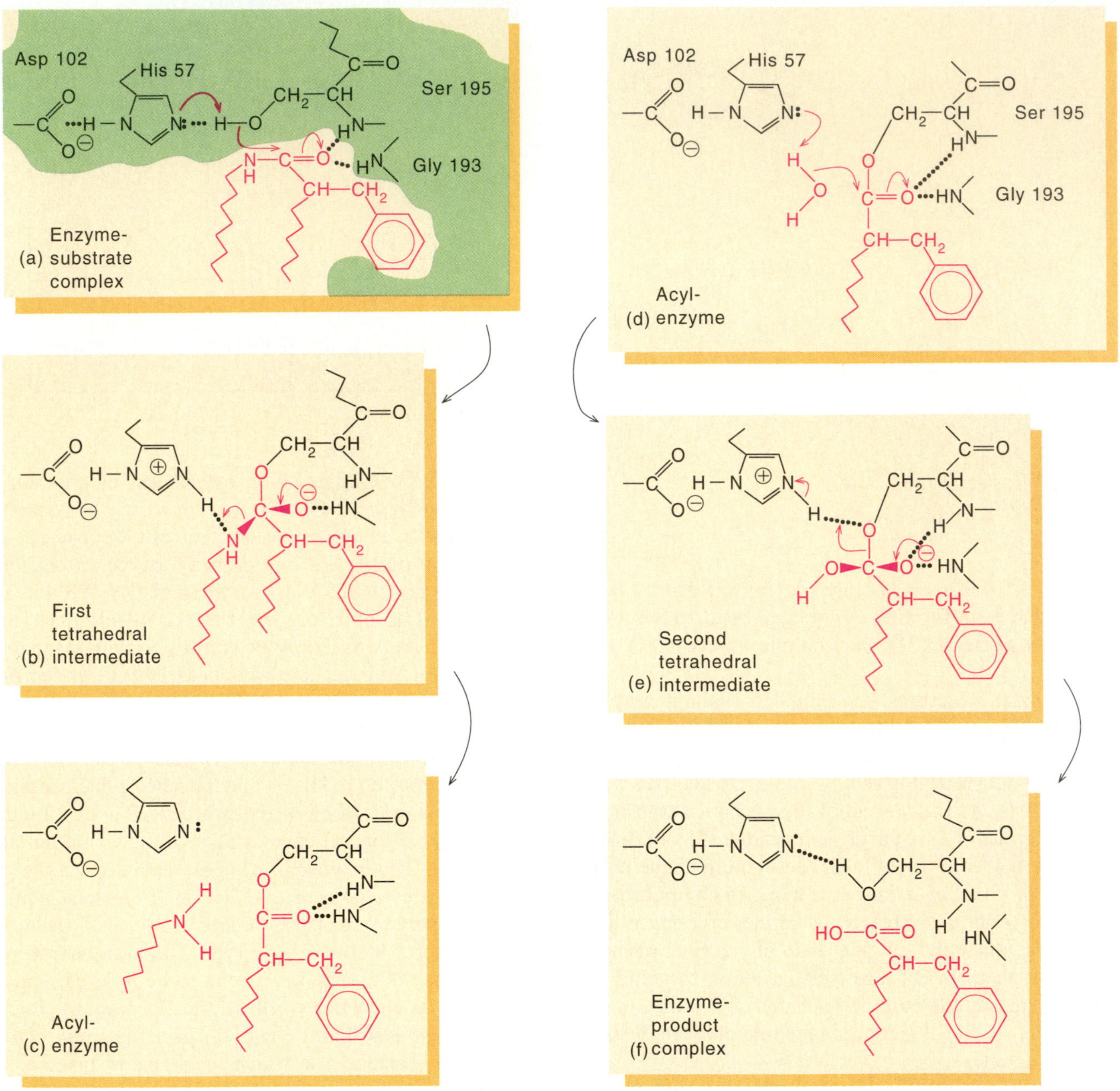

In figure 9.11, note that Asp 102 does not actually remove a proton from His 57 in the formation of either of the tetrahedral intermediates. Recent calculations indicate that the effect of the aspartate's negatively charged carboxyl group probably is best viewed simply as an electrostatic effect that favors the movement of positive charge from Ser 195 to His 57. The stabilization of the oxyanion intermediate by the amide protons of Gly 193 and Ser 195 can be described in a similar manner, as a favorable electrostatic interaction of the negatively charged oxygen with nearby atoms that tend to have positive charges, rather than as the actual transfer of a proton to the oxygen. The carboxyl group of Asp 102 also would help to align the imidazole ring of His 57 in the proper orientation for removing the proton from Ser 195.

One way to test a scheme such as that shown in figure 9.11 is to investigate the effects of modifying the amino acid residues that play important roles. We have already mentioned the inhibitory effect of the reaction of Ser 195 with diisopropylfluorophosphate. To examine the importance of His 57, this residue also was modified by methylation, which would disrupt the interaction with Asp 102. This treatment decreased the activity of chymotrypsin by a factor of more than 10^3. The importance of Asp 102 was tested recently by site-directed mutagenesis of trypsin and subtilisin. Replacing the aspartate by asparagine reduced k_{cat} by a factor of about 10^4.

Zinc Provides an Electrophilic Center in Some Proteases

In the preceding discussion, we described a group of closely related proteases secreted by the pancreas—trypsin, chymotrypsin, and elastase. The pancreas also secretes two proteases that hydrolyze oligopeptides one residue at a time from the C-terminal end, carboxypeptidases A and B. Carboxypeptidase A prefers aromatic residues (see fig. 9.4) and carboxypeptidase B has a preference for basic residues, providing a complementarity similar to that of chymotrypsin and trypsin. The two carboxypeptidases also resemble chymotrypsin and trypsin in being secreted as zymogens that are processed to form the active enzymes in the intestine. The carboxypeptidases are considerably more effective catalysts than the serine proteases. Values of k_{cat} for carboxypeptidase A with its best peptide substrates are on the order of 100 times greater than the k_{cat} values for trypsin with *its* best peptide substrates.

Carboxypeptidases A and B are members of a family of proteolytic enzymes that contain bound zinc. The zinc proteases also include enzymes that digest collagen (collagenases), and a number of bacterial enzymes. The most thoroughly studied of the bacterial zinc proteases is thermolysin, an enzyme obtained from *Bacillus thermoproteolyticus*. Thermolysin hydrolyzes internal peptide bonds adjacent to hydrophobic residues; collagenases cut internal bonds preferably adjacent to glycine residues. As in the case of serine proteases, the zinc protease family appears to have resulted from a combination of divergent and convergent evolution. Carboxypeptidases A and B have very similar structures indicative of a common ancestry; thermolysin has a very different structure overall, but resembles the carboxypeptidases in the amino acid residues that bind the zinc and in a few critical residues that probably participate in the enzymatic mechanism.

In the crystal structure of thermolysin, the Zn^{2+} ion is bound to the imidazole rings of two histidine residues and the carboxylate of a glutamic acid (fig. 9.12*a*). The fourth ligand of the Zn^{2+} is a molecule of water. Another glutamic acid residue (Glu 143) and a third histidine (His 231) are located nearby (see fig. 8.8). Thermolysin also has been crystallized with several different bound inhibitors, one of which, phosphoramidon, contains an amide bond between phosphoric acid and an amino acid (see figs. 8.8 and 9.12*b*). Because the oxygen and nitrogen substituents of the phosphorus atom are arranged in a tetrahedron, phosphoramidon seems likely to mimic the tetrahedral intermediate that would be formed in the course of the hydrolysis of a peptide (see box 9B). In the crystal, one of the oxygen atoms of the phosphate is linked directly to the Zn^{2+} ion, displacing the molecule of water. Glutamic acid 143 appears to form a hydrogen bond to the hydroxyl oxygen, and one of the nitrogen atoms of His 231 forms a hydrogen bond with the inhibitor's NH group (see fig. 9.12*b*).

Figure 9.12

Functional groups in the active site of thermolysin. (*a*) The Zn^{2+} ion is bound to two histidines and a glutamic acid, and has a molecule of water as its fourth ligand. (*b*) Phosphoramidon, an inhibitor that probably mimics the tetrahedral intermediate formed during hydrolysis of a peptide, binds to the Zn^{2+} by displacing the molecule of water. Residue Glu 143 forms a hydrogen bond with the hydroxyl group of the bound inhibitor, and His 231 is located close to the NH group.

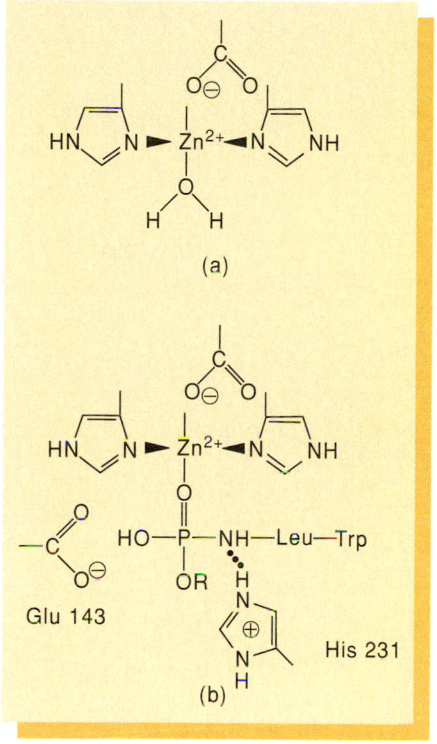

The structure of the enzyme-phosphoramidon complex suggests that the enzymatic hydrolysis of a peptide by thermolysin proceeds as shown in figure 9.13. In this scheme, binding of the peptide carbonyl oxygen atom to the Zn^{2+} withdraws electrons from the carbonyl group, increasing its susceptibility to nucleophilic attack by water. Glutamic acid 143 acts as a general base to remove a proton from the water. Histidine 231 then could act as a general acid, providing a proton to the amine group as the tetrahedral intermediate decomposes.

Several variations on this mechanism have been proposed. One suggestion is that the water molecule that attacks the peptide carbonyl is the one that starts out as a ligand of the Zn^{2+}, and that the Zn^{2+} goes through a stage in which it has five ligands—both the peptide oxygen and the water, in addition

Figure 9.13

The reaction mechanism of thermolysin probably involves polarization of the peptide carbonyl group by interaction with the Zn^{2+} ion and general-base catalysis by Glu 143 to form a tetrahedral intermediate. His 231 probably acts as a general acid to promote the decomposition of the intermediate.

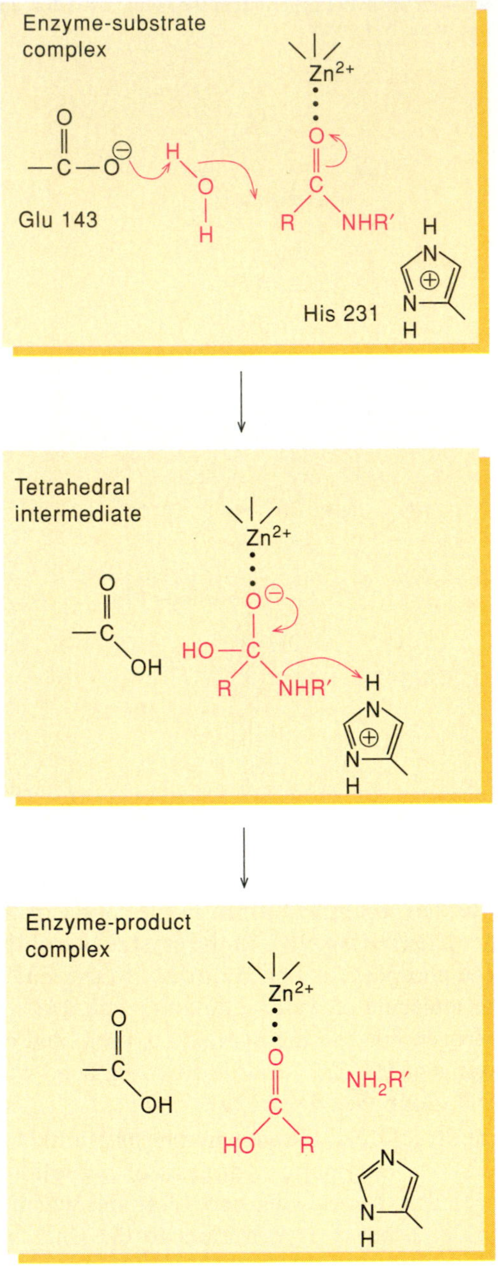

to the original two histidines and glutamic acid. It also has been suggested that the general acid that facilitates breakdown of the tetrahedral intermediate is Glu 143, rather than His 231. At present, the experimental data do not distinguish clearly between these alternatives. However, the central feature of all the proposed mechanisms is the electrostatic effect of the Zn^{2+} on the peptide C=O group.

The Zn^{2+} atom of carboxypeptidase A or B resembles that of thermolysin in having two histidines (His 69 and His 196), a glutamic acid (Glu 72), and a molecule of water as its ligands. Again a second glutamic acid residue (Glu 270) is found nearby, in a very similar position with respect to the metal ion. Carboxypeptidase A also has been crystallized with a bound inhibitor, glycyl-L-tyrosine. This dipeptide is hydrolyzed only very slowly, possibly because it interacts with the enzyme in a somewhat more complicated manner than the oligopeptides that are the normal substrates. The free N-terminal amino group of the dipeptide is linked indirectly to the carboxyl group of Glu 270 by way of a hydrogen-bonded molecule of water, as shown in figure 9.14. This could prevent the glutamic acid from participating as a general base in the hydrolytic reaction. In a normal oligopeptide substrate, the N-terminal amino group would be too far away to interact with the glutamic acid in this way.

As in thermolysin, the carbonyl oxygen of the peptide binds directly to the Zn^{2+} atom of carboxypeptidase A, displacing the water (see fig. 9.14). The active site also includes an arginine residue (Arg 145), which interacts closely with the negatively charged, C-terminal carboxyl group of glycyl-L-tyrosine and presumably would do the same with a better substrate. This explains the specificity of the carboxypeptidases for pruning off carboxyl-terminal amino acid residues. Carboxypeptidase A differs from thermolysin in that there is no histidine imidazole close enough to the bound dipeptide to serve as a general acid in the hydrolysis. The phenolic OH group of Tyr 248 *is* located nearby (see fig. 9.14), but we will see shortly that the role of the tyrosine residue appears to be rather different.

A remarkable finding in the carboxypeptidase A crystal structures, which was not seen with the serine proteases, is that the binding of the substrate analog (glycyl-L-tyrosine) causes the protein to undergo a major structural rearrangement (fig. 9.15). Arginine 145 and Glu 270 both move by about 2 Å to come into position near the substrate, and Tyr 248 swings down by about 12 Å so that its phenolic group is close to the NH group of the scissile bond. At least four molecules of H_2O are expelled in the process, so that the substrate finds itself surrounded largely by hydrophobic groups of the enzyme instead of by water. The binding of the substrate to the enzyme thus is favored by the entropy increase associated with removing the substrate's aromatic side chain from water, in addition to the electrostatic interactions between the terminal carboxyl group and Arg 145.

The most plausible mechanism for the operation of the carboxypeptidases is essentially the same as the scheme we just discussed for thermolysin: The Zn^{2+} ion pulls electrons from the oxygen atom of the peptide bond, increasing the positive charge on the carbonyl carbon, and Glu 270 acts as a general base, increasing the nucleophilic character of a molecule of water by removing a proton. Tyrosine 248 then might serve as a general acid to facilitate the decomposition of the tetrahedral intermediate state. Several experimental observations support this scheme. First, the pH dependence of the kinetics indicates that

9B BOX

Transition-State Analogs

Important clues to an enzyme mechanism often can be obtained by studying the inhibitory effects of compounds that resemble possible intermediates or transition states. Since the enzyme must stabilize the transition state of the reaction, a compound that mimics this state might be expected to bind to the enzyme particularly tightly. Such transition-state analogs frequently do prove to be strong competitive inhibitors. Thus the inhibitory effect of phosphoramidon on thermolysin supports the view that the enzymatic reaction proceeds by way of a state or intermediate in which a tetrahedral carbon atom is surrounded by a nitrogen and several oxygen atoms. A closely related example is the compound

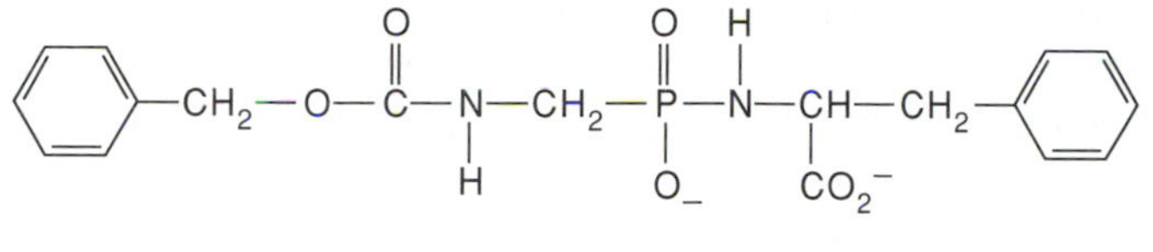

$$C_6H_5{-}CH_2{-}O{-}C(=O){-}NH{-}CH_2{-}P(=O)(O^-){-}NH{-}CH(CO_2^-){-}CH_2{-}C_6H_5$$

which is a potent competitive inhibitor of carboxypeptidase A. Here again, the tetrahedral phosphorus atom resembles the tetrahedral carbon in the intermediate that probably forms in the course of the enzymatic reaction. Similarly, we will see in this chapter that uridine vanadate, which contains a vanadium atom surrounded by five oxygen atoms, is a competitive inhibitor of ribonuclease. This behavior is consistent with the formation of an intermediate in which a phosphorus atom of the substrate takes on a pentavalent geometry (see fig. 9.22).

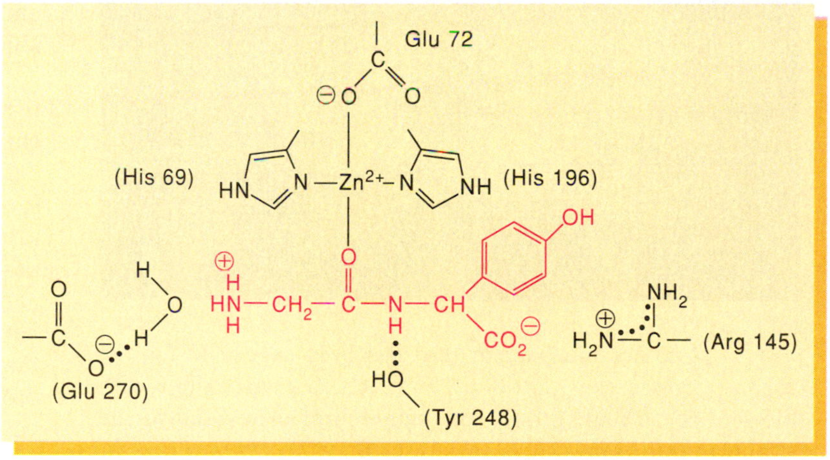

Figure 9.14

The functional groups surrounding the Zn^{2+} ion in carboxypeptidase A are similar to those in thermolysin. Glycyl-L-tyrosine, shown here in color, is a competitive inhibitor of carboxypeptidase A. The tyrosine side chain of glycyl-L-tyrosine fits into a hydrophobic pocket on the enzyme. The peptide oxygen binds directly to the Zn^{2+} ion, the carboxyl group forms a salt bridge with Arg 145, and the amide NH forms a hydrogen bond with Tyr 248. The salt bridge between the amino group and Glu 270 probably explains why glycyl-L-tyrosine is not a good substrate, because it would prevent the glutamic acid residue from acting as a general base.

both basic and acidic functional groups are involved in the reaction. The rate of the reaction decreases if a group with an apparent pK_a of about 6.2 is protonated, or if a group with an apparent pK_a of about 9.0 is deprotonated. The former group probably can be identified with Glu 270, and the latter could be Tyr 248. Second, chemical modifications of either Glu 270 or Tyr 248 destroy or modify the enzymatic activity. Acetylation of the tyrosine, for example, makes the enzyme essentially inactive. But when Tyr 248 was changed to phenylalanine by site-directed mutagenesis, it was found to decrease k_{cat} by only a factor of about 2, indicating that the tyrosine phenolic group is not essential for enzymatic activity. The substitution of phenylalanine for Tyr 248 did increase the K_m, suggesting that the tyrosine contributes primarily to the binding of the substrate.

Figure 9.15

Crystal structures of carboxypeptidase A with no substrate bound (*a*) and with bound glycyl-L-tyrosine (*b*). The glycyl-L-tyrosine or, in (*a*), the molecule of water that binds to the zinc in place of the substrate is shown in yellow. The side chains of Glu 270 and Tyr 248 are in orange, and the histidine ligands of the zinc are in red. Note how the large change in the position of the tyrosine, together with smaller changes in other residues, folds the enzyme around the substrate. The figures are based on the crystal structures described by W.N. Lipscomb.

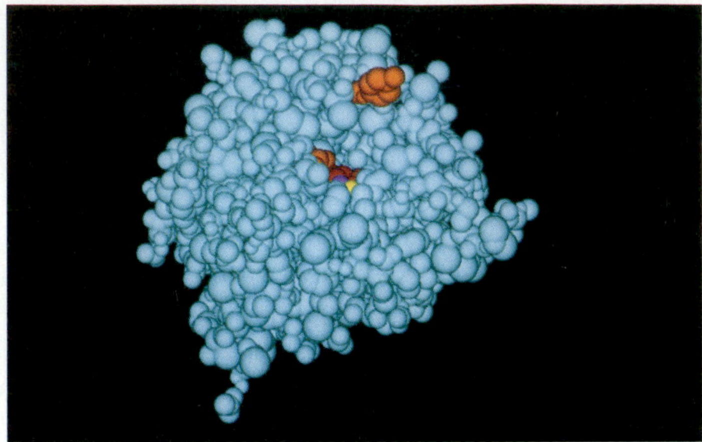

(a)

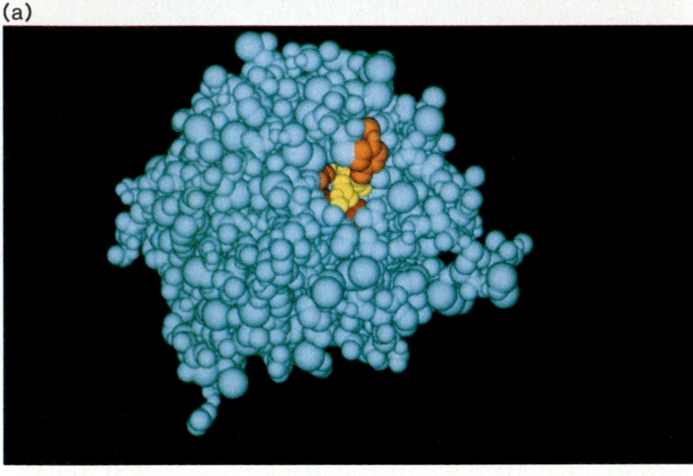

(b)

Another mechanism that has been suggested for the carboxypeptidases is that Glu 270 reacts as a nucleophile with the peptide carbonyl group. This would form an acyl-anhydride attached to the enzyme as an intermediate:

$$\mathrm{R{-}\overset{\displaystyle O}{\overset{\|}{C}}{-}NHR' + HO{-}\overset{\displaystyle O}{\overset{\|}{C}}{-}Enzyme \rightarrow}$$
$$\mathrm{R{-}\overset{\displaystyle O}{\overset{\|}{C}}{-}O{-}\overset{\displaystyle O}{\overset{\|}{C}}{-}Enzyme + R'NH_2} \qquad \textbf{(11)}$$

$$\mathrm{R{-}\overset{\displaystyle O}{\overset{\|}{C}}{-}O{-}\overset{\displaystyle O}{\overset{\|}{C}}{-}Enzyme + H_2O \rightarrow}$$
$$\mathrm{R{-}\overset{\displaystyle O}{\overset{\|}{C}}{-}OH + HO{-}\overset{\displaystyle O}{\overset{\|}{C}}{-}Enzyme} \qquad \textbf{(12)}$$

There is, however, little evidence in favor of this scheme. For example, reagents that might be expected to react with an anhydride intermediate do not alter the course of the enzymatic reaction. Also, if the second step of the mechanism (equation 12) is reversible, the enzyme should catalyze an exchange of ^{18}O back and forth between H_2O and the carboxylate group of the product (see box 9C); attempts to detect such an exchange have given negative results.

Ribonuclease A: An Example of Concerted Acid–Base Catalysis

Ribonucleases are a widely distributed family of enzymes that hydrolyze RNA by cutting the P — O ester bond attached to a ribose 5′ carbon. The reaction occurs in two steps, with a 2′,3′-phosphate cyclic diester intermediate (fig. 9.16). The intermediate can be identified relatively easily, because its breakdown is much slower than its formation. Ribonucleases do not hydrolyze DNA, which lacks the 2′-hydroxyl group needed for the formation of the cyclic intermediate. The best-studied of the ribonucleases, the pancreatic enzyme ribonuclease A, is specific for a pyrimidine base (uracil or cytosine) on the 3′ side of the phosphate bond that is cleaved.

When the amino acid sequence of bovine ribonuclease A was determined in 1960 by Sanford Moore and William Stein, it was the first enzyme and only the second protein to be sequenced. Ribonuclease A also was one of the first enzymes whose three-dimensional structure was elucidated by x-ray diffraction, and it was the first to be synthesized completely from its amino acids. The synthetic protein proved to be enzymatically indistinguishable from the native enzyme.

Ribonuclease A (RNase A) is a relatively small protein, with a molecular weight of 13,680 and a single polypeptide chain of 124 amino acid residues. An early discovery that turned out to be extremely useful was that the protein could be cleaved specifically between residues 20 and 21 by the bacterial serine protease subtilisin. The resulting two polypeptides could be separated and purified. They were found to be enzymatically inactive individually, but to regain the complete activity of the native enzyme when they were recombined. This work showed that there are strong, noncovalent interactions that can hold protein chains together even when one of the peptide links is cut. It also made it possible to modify specific amino acid residues of the two polypeptide chains independently, and to explore how each residue contributed to the reassembly of the protein and the recovery of enzymatic activity.

RNase A is completely inhibited if either of two different histidine residues (His 12 or His 119) is modified by carboxymethylation with iodoacetate, a fact suggesting that both of these histidines play important roles in the active site (fig. 9.17). In support of this conclusion, the reaction of iodoacetate with His 12 or His 119 is inhibited by cytidine-3′-phosphate or other small molecules that bind at the active site. Lysine 41 has been implicated similarly in the active site by the observation that the reaction of fluorodinitrobenzene with the ϵ-amino group of this residue destroys enzymatic activity (see fig. 9.17). A

Isotopic Exchange Reactions

One way to detect an intermediate state in an enzymatic reaction is to look for a transfer of an isotopic label between the solvent and a reactant or product. Suppose that the reaction scheme of equations (11) and (12) is correct for carboxypeptidase A, and consider a reversal of these reactions. If the enzyme is incubated with both of the products ($R'NH_2$ and R — COOH), a small amount of

$$\mathrm{R-\overset{\overset{\displaystyle O}{\|}}{C}-NHR'}$$

will be formed, depending on the equilibrium constant for the overall reaction. If the enzyme is incubated with R — COOH alone,

$$\mathrm{R-\overset{\overset{\displaystyle O}{\|}}{C}-NHR'}$$

cannot be formed but the reverse reaction should still proceed as far as the anhydride intermediate,

$$\mathrm{R-\overset{\overset{\displaystyle O}{\|}}{C}-O-\overset{\overset{\displaystyle O}{\|}}{C}-Enzyme}$$

If the incubation is continued, the intermediate will be formed and broken down again many times. The formation of the anhydride requires the splitting out of H_2O, so whenever the intermediate breaks down, the elements of H_2O must be reincorporated into R — COOH and the carboxylate group on the enzyme. Thus if the enzyme is incubated with R — COOH in water that is labeled with ^{18}O, the isotope will be incorporated into the carboxylate group. This prediction can be tested by isolating the R — COOH and examining it in a mass spectrometer. For carboxypeptidase A, no such isotopic exchange was detected. This result favors the view that Glu 270 in carboxypeptidase does not react with the substrate to form an anhydride intermediate, and more likely functions as a general base.

Figure 9.16

A portion of an RNA chain, indicating points of cleavage by pancreatic ribonuclease. "Pyr" refers to a pyrimidine; "Base" can be either a purine or a pyrimidine. The 2′, 3′ and 5′ carbon atoms are labeled. The enzymatic reaction proceeds in two steps, with a cyclic 2′, 3′-phosphate diester as an intermediate.

Figure 9.17

When RNase A is treated with iodoacetate ($ICH_2CO_2^-$), the two major products obtained are carboxymethylated derivatives of His 12 and His 119. Both of these enzymes are severely inhibited. This result suggests that both His 12 and His 119 are important in the active site. The enzyme also is completely inhibited by the reaction of Lys 41 with fluorodinitrobenzene.

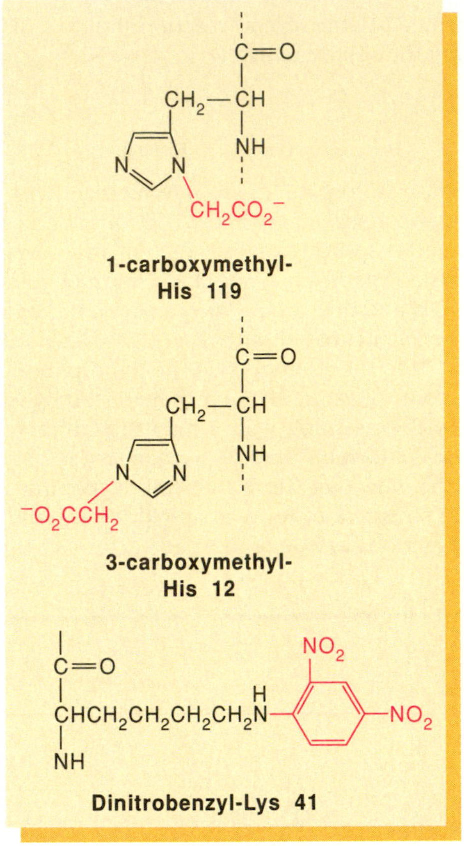

1-carboxymethyl-His 119

3-carboxymethyl-His 12

Dinitrobenzyl-Lys 41

second lysine (Lys 7) also appears to be important for substrate binding, although the enzyme retains some activity when this residue is modified.

The enzymatic activity of RNase A shows a bell-shaped dependence on pH, with an optimum near pH 7 (fig. 9.18). The operation of the enzyme appears to require that a dissociable group with a pK_a of about 6.3 be in the protonated form, and that a group with a pK_a of about 8 be unprotonated. These groups have been identified as His 12 and His 119 by measuring the nuclear magnetic resonance (NMR) spectrum of the enzyme as a function of pH. The spectral peaks due to histidine residues undergo characteristic shifts as the imidazole ring is protonated. Analysis of the NMR spectra, though not straightforward, was aided by the fact that bovine RNase A contains only four histidine residues, one of which (His 12) can be separated from the others by cutting the protein with subtilisin. Histidines 12 and 119 have pK_a values of 5.8 and 6.2 in the free enzyme, but both pK_a values shift to a more alkaline range when the enzyme binds its substrate.

Figure 9.18

The dependence of k_{cat} of RNase A on pH. The bell-shaped curve suggests that one histidine residue must be in the protonated state and another must be unprotonated. Similar pH dependences are found for the hydrolysis of either RNA or pyrimidine nucleoside-2′, 3′-phosphate cyclic diesters, a fact indicating that the two steps of the overall reaction both require concerted general-acid and general-base catalysis.

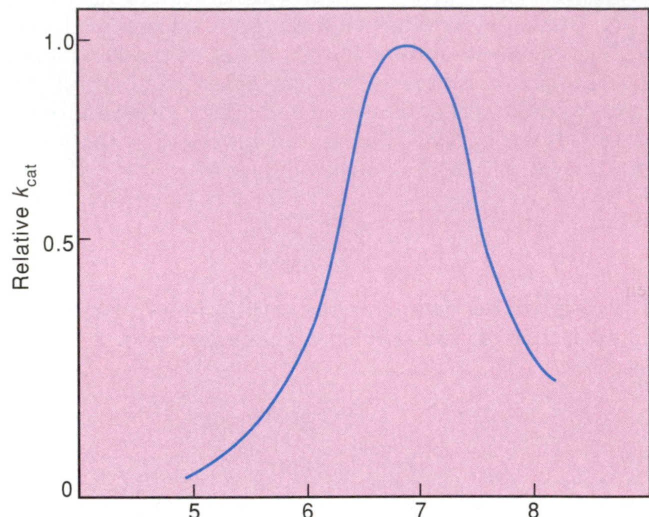

The pH dependence of the kinetics, taken with the information that histidines 12 and 119 are essential for enzymatic activity, suggests that the two histidines participate in the reaction by concerted general-base and general-acid catalysis. This supposition is supported by the crystal structure of the enzyme. Figure 9.19 shows a model of the crystal structure of RNase S (the active enzyme obtained by cutting ribonuclease with subtilisin and recombining the two polypeptides) with a bound substrate analog, "UpCH_2A." The analog is a competitive inhibitor of the enzyme. It resembles the dinucleotide UpA but cannot be hydrolyzed because it has a methylene carbon in place of the oxygen between the P and the 5′-CH_2 of the ribose of adenosine (fig. 9.20). Similar crystal structures have been obtained with a variety of other substrate analogs. In all of the structures, His 12 and His 119 are found on opposite sides of the substrate's phosphate group, with His 12 positioned close to the 2′ OH group. Lysine residues 7, 14, and 66 also are located nearby, where their positively charged amino groups would have favorable electrostatic interactions with the negatively charged phosphate. Near His 119 is the carboxyl group of an acidic residue, Asp 121. The electrostatic effect of Asp 121 thus would favor the transfer of a proton from a molecule of water to His 119, much as Asp 102 in chymotrypsin induces a proton to move from Ser 195 to His 57.

The pyrimidine ring on the 3′ side of the substrate fits snugly into a groove on RNase A. Valine 43 is located on one side of the pyrimidine ring, and Phe 120 on the other. The specificity of the enzyme for pyrimidine nucleotides appears to result largely from hydrogen bonding to Thr 45. The threonine side chain can form a pair of complementary hydrogen bonds with either uracil or cytosine (fig. 9.21). Crystallographic studies have

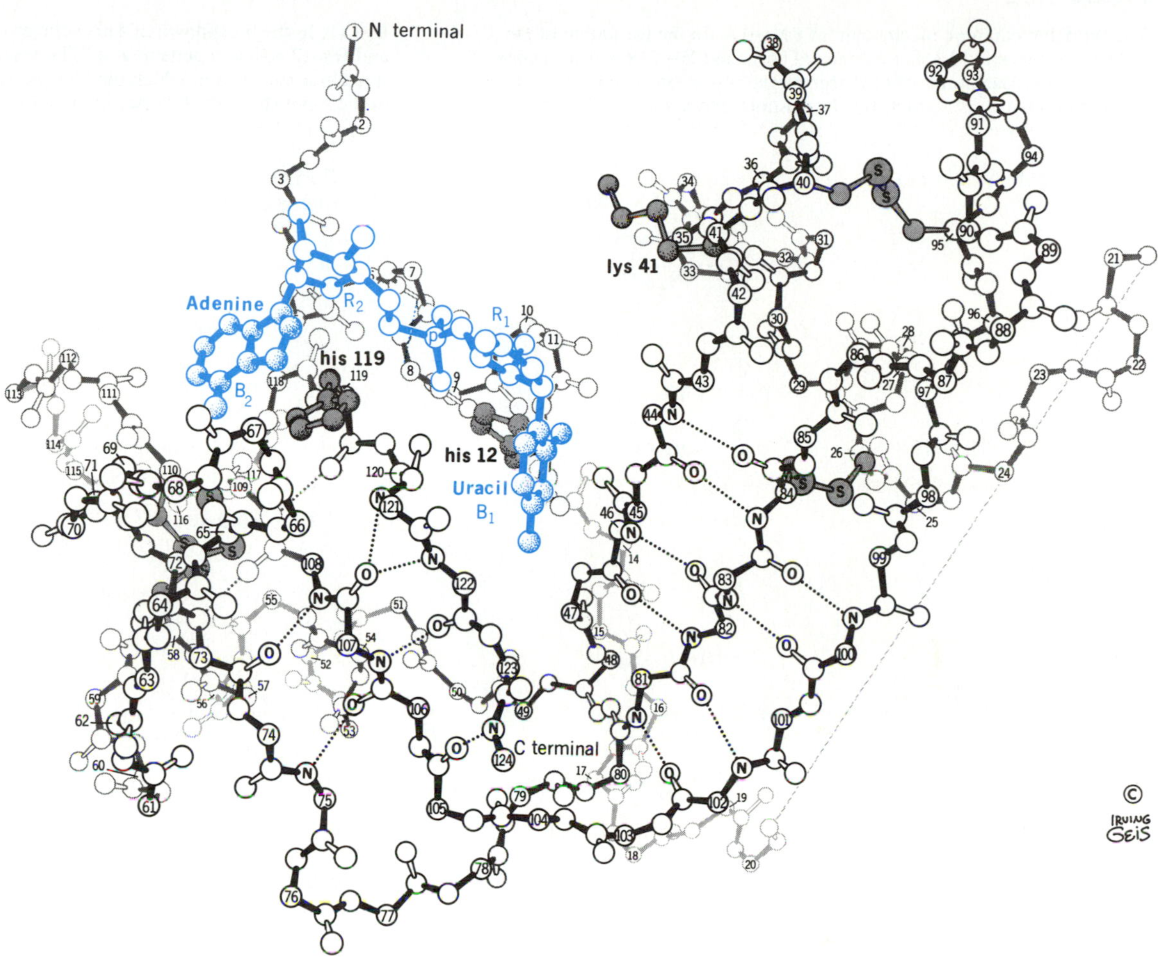

Figure 9.19

A model of the binding of the substrate analog $UpCH_2A$ (blue) to RNase S. (The structure of $UpCH_2A$ is shown in figure 9.20.) RNase S is an active enzyme, although it differs from the native enzyme because of a peptide bond cleavage between residues 20 and 21 (light dashed line). The two histidines and the lysine that are crucial to the active site (His 12, His 119, and Lys 41) are indicated in gray. The dotted lines indicate some of the hydrogen bonds that maintain the protein structure.

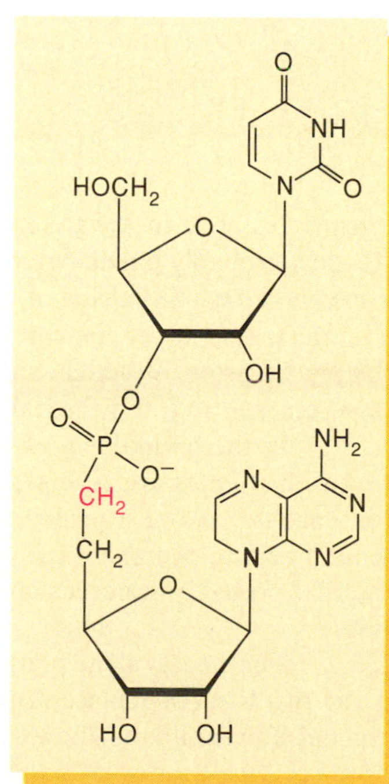

Figure 9.20

Structure of $UpCH_2A$, a competitive inhibitor of RNase A. $UpCH_2A$ differs from the dinucleotide substrate UpA in having a methylene group instead of oxygen between the phosphorus and C-5′ of the adenosine.

Figure 9.21

The side chain and amide NH group of Thr 45 in RNase A can form a pair of complementary hydrogen bonds with either uracil or cytosine. These probably account for the specificity of RNase A for pyrimidine nucleotides.

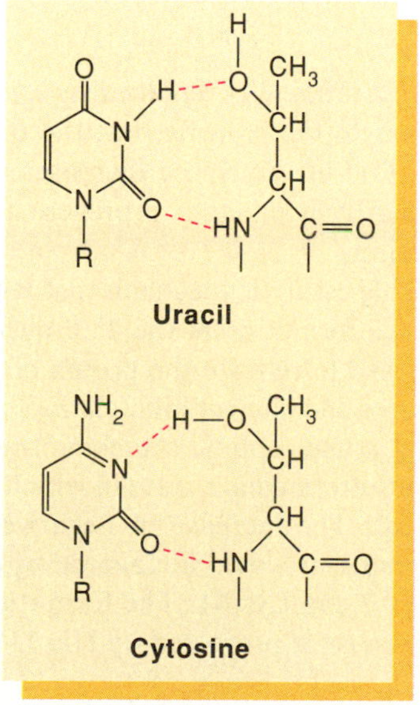

Uracil

Cytosine

Figure 9.22

The probable catalytic mechanism of RNase A. In the formation of the 2′, 3′-cyclic ester, His 12 acts as a general base, and His 119 acts as a general acid. The increase in negative charge on the phosphate's oxygen atoms in the transition state is favored by electrostatic interactions with Lys 7 and Lys 41. In the breakdown of the cyclic ester, His 119 acts as a general base and His 12 acts as a general acid. The reaction proceeds through two short-lived intermediates in which the phosphorus atom is pentavalent, in addition to the more stable cyclic ester intermediate. Py = pyrimidine.

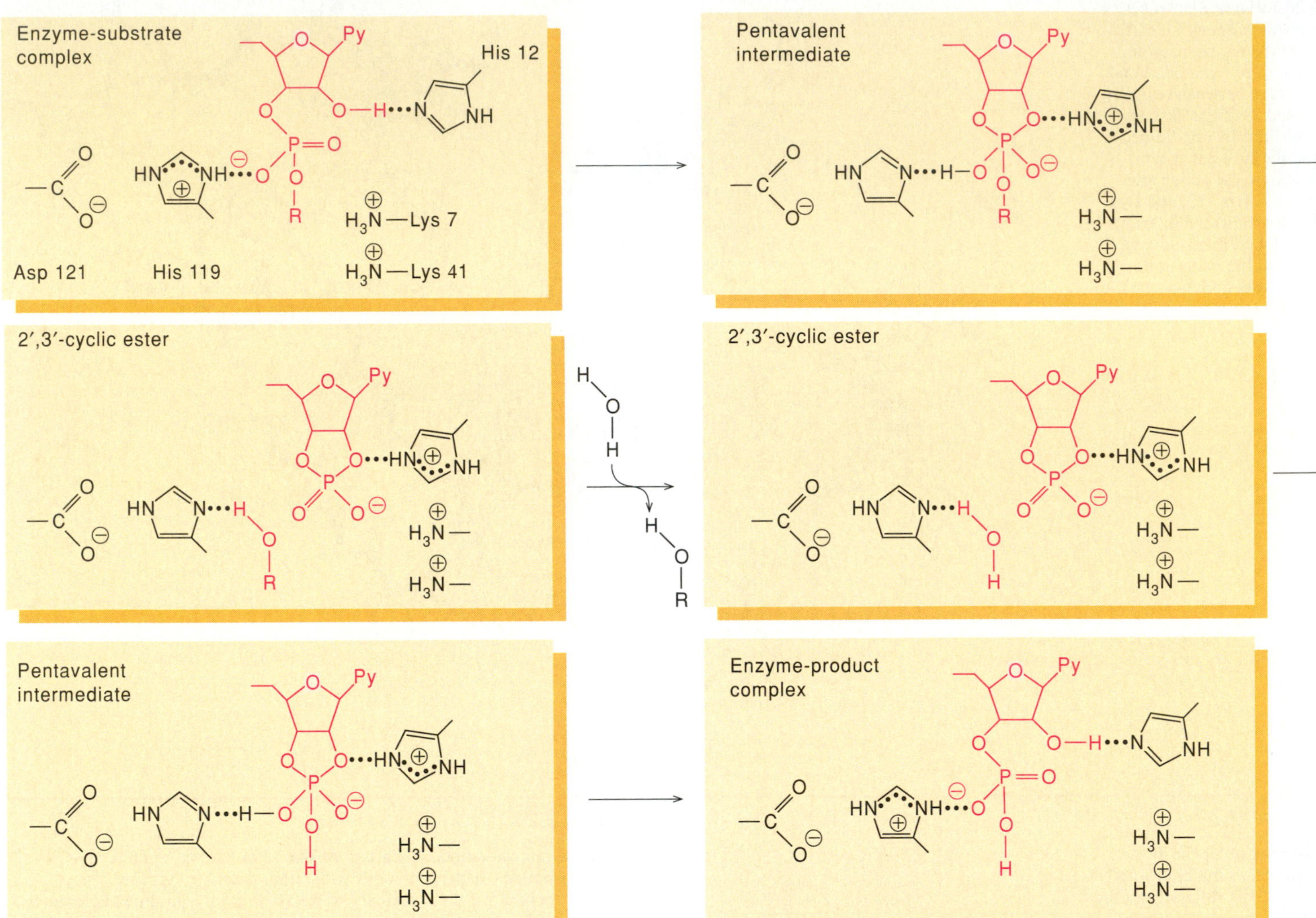

shown that if the pyrimidine is replaced by a purine, the compound can still bind to the enzyme, but the distance between His 12 and the 2′ OH of the ribose increases by about 1.5 Å. This increase evidently is enough to prevent the catalytic reaction from occurring.

The probable catalytic mechanism of RNase A is shown in figure 9.22. In the formation of the 2′,3′-cyclic ester, His 12 acts as a general base to remove the proton of the 2′ hydroxyl group. This increases the nucleophilic character of the oxygen atom, precipitating a nucleophilic attack on the phosphate and creating a transient intermediate state in which the phosphorus atom is pentavalent. The increase in negative charge on the phosphate's oxygen atoms would be favored by electrostatic interactions with Lys 7 and Lys 41. The formation of the pentavalent intermediate also is promoted by His 119, which acts as a general acid to protonate one of the phosphate oxygens. His 119 then probably participates in the transfer of a proton to the 5′ oxygen atom. Expulsion of the protonated 5′ OH group converts the transient pentavalent intermediate into the more stable 2′,3′-cyclic ester, in which the substituents of the phosphorus have returned to a tetrahedral geometry.

In the breakdown of the 2′,3′-cyclic ester, the roles of the two histidines are reversed. Histidine 119 now acts as general base to remove a proton from H_2O, and His 12 acts as a general acid to protonate the substrate's 2′ oxygen. Again, the reaction probably proceeds by way of a pentavalent intermediate.

The pentavalent phosphoryl intermediates that figure in the reactions of RNase A probably have the geometry of a trigonal bipyramid, as shown in the upper part of figure 9.23.

Figure 9.23

Two routes leading to phosphoryl group transfer. Both routes pass through a pentavalent intermediate, in which three of the oxygen atoms lie in a plane with the phosphorus atom. In the "in-line" mechanism, the entering and leaving groups ($^-$OR′ and RO$^-$) are on opposite sides of this plane. In the "adjacent" mechanism the entering and leaving groups are in neighboring positions. The in-line mechanism results in stereochemical inversion; the adjacent mechanism gives retention. Three of the oxygen atoms have been labeled with numbers to indicate their stereochemical relationships in the figure. Experimentally, the stereochemical outcome of the RNase reaction can be tested by using a substrate that is labeled stereospecifically with a combination of ^{17}O and ^{18}O. Substrates in which one of the oxygens is replaced by sulfur also can be used. (The adjacent mechanism requires one additional step not shown here: before it can depart, the leaving group has to move to an axial orientation perpendicular to the plane. This rearrangement is called *pseudorotation.* Several of the substituents move in a concerted manner in the pseudorotation step, so that their stereochemical relationships are maintained.)

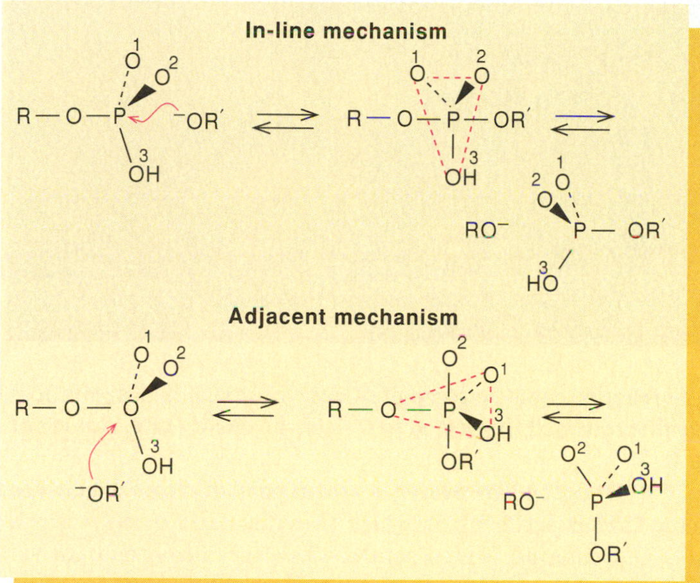

Three of the phosphorus atom's substituents lie in a plane. The entering and leaving oxygen atoms are on either side of the plane, forming a straight line with the phosphorus. One observation that supports the formation of such pentavalent intermediates is that RNase A is inhibited strongly by uridine vanadate, in which the vanadium atom is surrounded by five oxygens in a similar geometry. Figure 9.24 shows the structure of the complex of the enzyme with the inhibitor. The disposition of histidines 12 and 119 on either side of the vanadate group in the crystal structure is consistent with the linear arrangements of the entering and leaving oxygen atoms on either side of the phosphorus atom in figures 9.22 and 9.23. Uridine vanadate can be viewed as a transition-state analog (box 9B).

The "in-line" reaction mechanism also can be distinguished experimentally from the alternative "adjacent" mechanism (see fig. 9.23) by studying substrates in which one of the oxygen atoms is replaced by sulfur and another is labeled stereospecifically with ^{18}O. An in-line mechanism results in stereochemical inversion of the configuration around the phosphorus, whereas an adjacent arrangement of the entering and departing groups results in retention of the configuration (see fig. 9.23). The reactions catalyzed by RNase A and a number of other ribonucleases have been shown to proceed with inversion. Many of the other enzymes that transfer phosphoryl groups from one substrate to another exhibit retention.

Figure 9.24

The crystal structure of the complex between RNase A and uridine vanadate. Uridine vanadate, which contains a vanadium atom surrounded by five oxygens, probably mimics the pentavalent transition state formed by a phosphoryl substrate. It binds to RNase A about 10^4 times more tightly than the corresponding uridine 2′, 3′-cyclic phosphate ester. The V atom of uridine vanadate is in purple, and the other atoms of the inhibitor in yellow. A smoothed representation of the α-carbon chain of the protein is shown in blue, with the side chains of histidines 12 and 119 indicated in red. (Based on the crystal structure described by A. Wlodawer.)

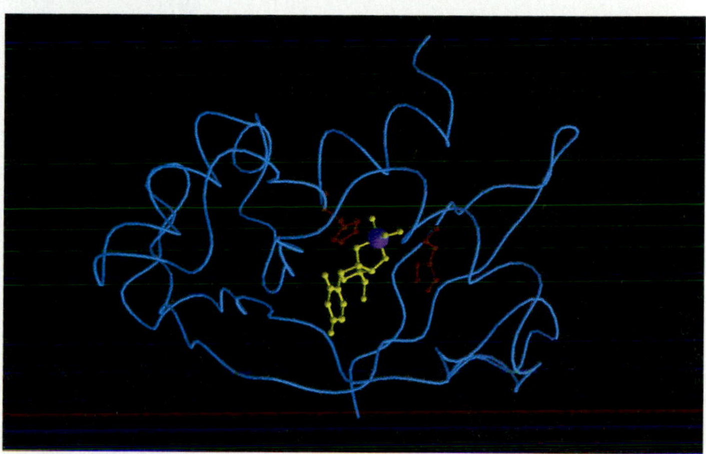

Triosephosphate Isomerase Has Approached Evolutionary Perfection

Triosephosphate isomerase catalyzes the interconversion of dihydroxyacetone phosphate and glyceraldehyde-3-phosphate:

$$\begin{array}{ccc} CH_2OH & & H-C=O \\ | & & | \\ C=O & \rightleftarrows & H-C-OH \\ | & & | \\ CH_2OPO_3^{2-} & & CH_2OPO_3^{2-} \\ \text{Dihydroxyacetone phosphate} & & \text{Glyceraldehyde-3-phosphate} \end{array} \quad (13)$$

The reaction requires moving a proton from carbon 1 of dihydroxyacetone phosphate to carbon 2, moving another proton from the oxygen atom at C-1 to the oxygen at C-2, and moving a pair of electrons in the same direction. The interconversion of the two triosephosphates is an essential step in the catabolism of carbohydrates. Triosephosphate isomerase is among the enzymes that have approached evolutionary perfection in the sense that the specificity constant k_{cat}/K_m is greater than 10^8 M^{-1} s^{-1}, which is near the limit set by the rates of diffusion of the substrate and the protein (see table 8.4). The enzyme achieves this high value for the specificity constant by having both a relatively high k_{cat} and a relatively low K_m.

Figure 9.25

Crystal structure of triosephosphate isomerase from chicken muscle. This figure shows one of the two identical subunits of the enzyme. The side chains of Glu 165, His 95, and Lys 13 are in purple, red, and yellow, respectively. (Based on the crystal structure described by D. W. Banner, A. C. Bloomer, G. A. Petsko, D. C. Phillips, and I. A. Wilson.)

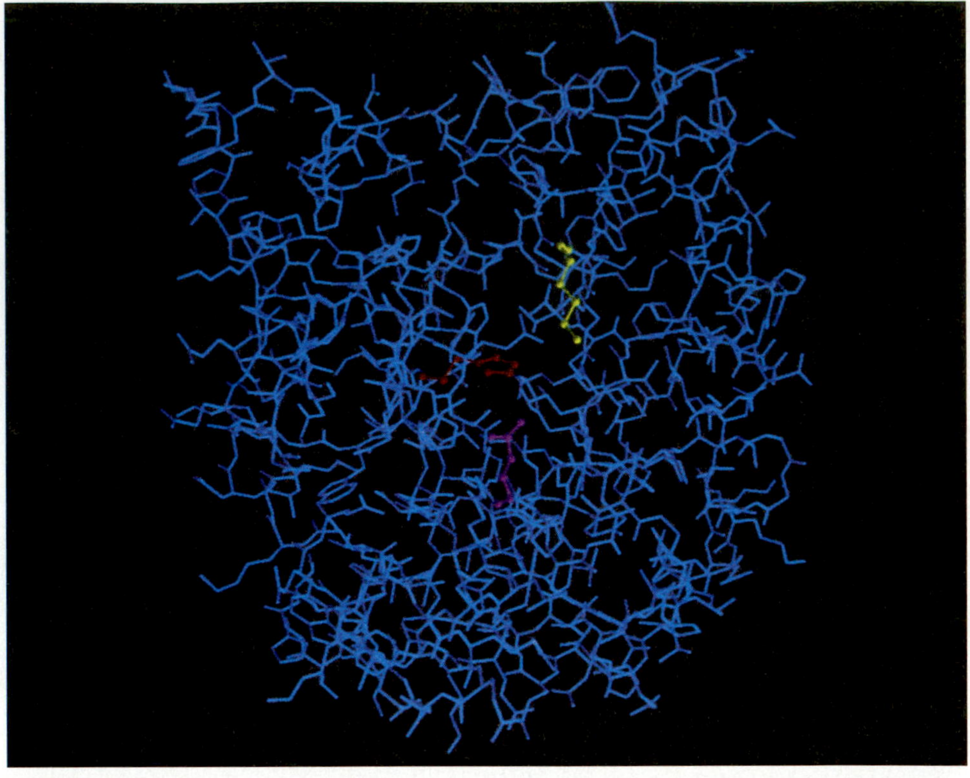

Crystal structures have been obtained for triosephosphate isomerase purified both from yeast and from chicken muscle. Although the proteins from the two organisms differ at about half of the positions in their amino acid sequences, their three-dimensional structures are nearly identical, and the same sets of amino acids have been implicated in their catalytic mechanisms. The overall structure is a β barrel, with eight parallel β sheets linked by eight α helices (fig. 4.24 and fig. 9.25). In both enzymes, a glutamic acid residue (Glu 165) reacts with affinity labels that probably bind at the active site, such as chloroacetone phosphate. (Chloroacetone phosphate is the same as dihydroxyacetone phosphate except that a chlorine atom replaces the hydroxyl group on C-1.) Crystal structures of the complexes of the enzyme with bound dihydroxyacetone phosphate or a competitive inhibitor show that the carboxyl group of Glu 165 is located close to C-1 of the substrate. When Glu 165 was changed to aspartate by site-directed mutagenesis, the rate of the isomerization reaction catalyzed by the enzyme was decreased approximately 1,000-fold. This is a remarkable effect, considering that the substitution of Asp for Glu probably changes the position of the carboxyl group by less than 1 Å.

The rate of the isomerase reaction decreases if a basic group is protonated by bringing the pH below 6.5. It seems clear that this is the carboxyl group of Glu 165, because the reaction of Glu 165 with affinity labels drops out in parallel with the enzymatic activity. This suggests that, in the enzymatic reaction, Glu 165 acts as a general base to remove the proton from C-1 of the dihydroxyacetone phosphate, converting the substrate into an ene-diolate intermediate as shown in figure 9.26. The reaction could be completed if the protonated glutamic acid residue returned the proton to C-2 of the ene-diolate, instead of to C-1.

Independent evidence in support of this scheme has come from experiments in which the enzymatic reaction is run in D_2O. Deuterium is incorporated stereospecifically onto carbon 2 of the product:

$$\begin{array}{c} CH_2OH \\ | \\ C=O \\ | \\ CH_2OPO_3^{-2} \end{array} \xrightarrow{D_2O} \begin{array}{c} H-C=O \\ | \\ D-C-OH \\ | \\ CH_2OPO_3^{-2} \end{array} \qquad \textbf{(14)}$$

In the absence of the enzyme, the carbon-bound hydrogen atom on glyceraldehyde-3-phosphate does not exchange with deuterium atoms of the solvent at any significant rate. (We are not concerned here with an exchange of the proton that is bound to the alcohol oxygen in the reactant or product. This proton does exchange rapidly with the solvent in either the presence or the absence of an enzyme, but the deuteron that is incorporated here can be removed immediately by putting the product back in H_2O.) The incorporation of the deuterium isotope can be explained as follows: After a proton is transferred from C-1 of the substrate to the carboxyl group of Glu 165, it has an opportunity to escape into the solution and to be replaced by a deuteron. A proton attached to a carboxylic acid oxygen usually exchanges very rapidly with the solvent. The deuteron on Glu 165 then can be transferred to C-2 of the product.

Figure 9.26

The probable reaction mechanism of triosephosphate isomerase. The γ-carboxylate group of Glu 165 acts as a general base to remove a proton from C-1 of the substrate, dihydroxyacetone phosphate (DHAP). This generates a planar ene-diolate intermediate that has two tautomeric forms. After an interconversion of ene-diolate tautomers, the protonated glutamic acid residue returns a proton to C-2. The electrostatic effects of His 95 and Lys 12 facilitate the formation of the negatively charged intermediate. A possible variation on the mechanism would be for His 95 or Lys 12 to act as a general acid. If this happened, the intermediate would be a neutral ene-diol. However, the pH-dependence of the kinetics, and the hydrogen-bonding pattern of His 95 in the crystal structure are more consistent with the scheme shown here.

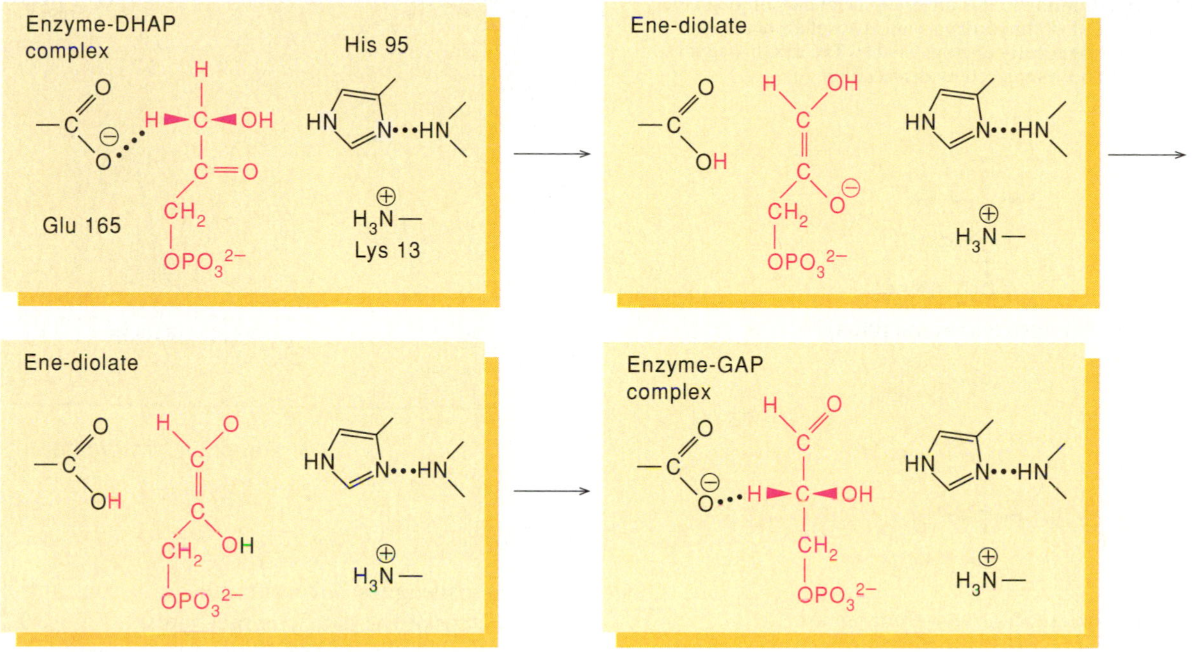

Enzyme-DHAP complex | His 95 | Glu 165 | Lys 13 | Ene-diolate | Ene-diolate | Enzyme-GAP complex

Additional studies of the kinetics of such exchange processes indicate that the proton probably moves from Glu 165 directly back to the ene-diolate in one step, rather than hopping first to another acid–base group and proceeding from there to the ene-diolate. This is consistent with the crystal structure, which shows the carboxylate group of Glu 165 in a good position to interact with a proton on either C-1 or C-2 of the substrate.

We have not yet considered the itinerary of the proton that must move from one of the oxygen atoms to the other. In the crystal structure there are two other nearby acid–base groups that could participate in the travels of this proton; these are His 95 and Lys 13 (see fig. 9.25). (The latter number is for the chicken enzyme; in yeast, the homologous lysine is residue 12.) His 95 appears to play an important role in the enzymatic reaction, because replacing this residue with glutamine by site-directed mutagenesis decreases k_{cat} by a factor of about 400. Considering the position of His 95 in the crystal structure, a possible scheme for the overall reaction would be for this residue to act together with Glu 165 to provide concerted acid–base catalysis in both steps of the reaction. In the formation of the intermediate state, Glu 165 would remove a proton from C-1 of dihydroxyacetone phosphate, and His 95 would add a proton to the carbonyl oxygen. This means that the intermediate would be a neutral ene-diol, rather than the ene-diolate ion shown in figure 9.26. In the breakdown of the ene-diol, Glu 165 would return a proton to C-2, and His 95 would extract a proton from the oxygen atom on C-1. Histidine 95 is in a better position than Lys 13 to do this because the ϵ-amino group of the lysine, though close to the oxygen atom on C-2, seems too far away from the C-1 oxygen.

There are several difficulties with the idea that His 95 works in this way. First, if His 95 has to donate a proton to the substrate, the rate of the reaction should decrease when the histidine is deprotonated by raising the pH. In fact, k_{cat} is essentially independent of pH from 6.5 up to almost 10. A pK_a in the region of 10 would seem more consistent with a lysine residue than with a histidine. In addition, as is indicated in figure 9.26, His 95 is hydrogen-bonded to an amide NH group, which means that it is unlikely to be protonated in the pH range where the enzyme is active.

These observations suggest that the roles of His 95 and Lys 12 may not be to act as general acids, but rather to stabilize the negative charge on an ene-diolate intermediate by electrostatic effects. Several additional observations are consistent with this view. The carbonyl bond of dihydroxyacetone phosphate appears to be strongly polarized when the substrate binds to the enzyme, because its infrared absorption band shifts to a markedly longer wavelength and its reactivity with sodium borohydride is greatly increased. The most likely mechanism for triosephosphate isomerase thus appears to involve a combination of the electrostatic effects of His 95 and Lys 12 with the general-base catalysis by Glu 165, as shown in figure 9.26.

Figure 9.27

The substrates of triosephosphate isomerase can be labeled specifically with deuterium. A kinetic isotope effect of about 3 is obtained with [1(R)-^{2}H]-dihydroxyacetone phosphate (*a*). This indicates that the formation of the ene-diolate intermediate is the rate-limiting step in the conversion of dihydroxyacetone phosphate to glyceraldehyde-3-phosphate. Little or no isotope effect is obtained when the reaction is run in the opposite direction, starting with [2-^{2}H]-glyceraldehyde-3-phosphate (*b*). The breakdown of the ene-diolate must be rate-limiting in this direction. Note that although the two hydrogen atoms on C-1 of dihydroxyacetone phosphate are chemically identical, they are stereochemically distinguishable. The enzyme always removes the one that has been replaced by deuterium in (*a*).

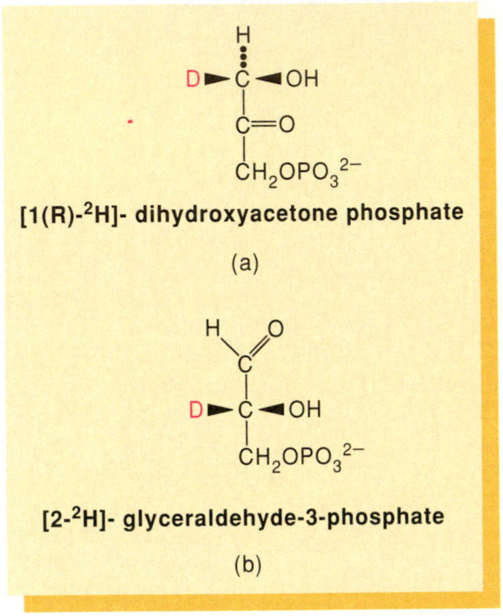

[1(R)-^{2}H]- dihydroxyacetone phosphate

(a)

[2-^{2}H]- glyceraldehyde-3-phosphate

(b)

Assuming that the scheme shown in figure 9.26 is correct, how could we determine whether the formation or the breakdown of the ene-diolate intermediate is the rate-limiting step in the overall reaction? The best approach is to study the kinetics using substrates that are specifically labeled with deuterium. With [1(R)-^{2}H]-dihydroxyacetone phosphate (fig. 9.27*a*), k_{cat} is about three times smaller than the k_{cat} obtained with the ^{1}H analog. An isotope effect of this magnitude is consistent with the view that the bond holding the H or deuterium atom to C-1 must be broken in the rate-limiting step. This means that the formation of the ene-diolate is rate-limiting. Similar measurements with [2-^{2}H]-glyceraldehyde-3-phosphate (see fig. 9.27*b*) show that there is little or no isotope effect on k_{cat} for the reverse direction, from glyceraldehyde-3-phosphate to dihydroxyacetone phosphate. In this direction, the formation of the ene-diolate intermediate must be faster than its breakdown.

By measuring the kinetics of the isotopic exchange reactions catalyzed by the enzyme, along with the overall kinetics of the isomerization reaction in both directions, Jeremy Knowles and his colleagues have obtained rate constants for the individual steps of the catalytic mechanism. Figure 9.28 shows the complete pathway in the form of a free energy profile. In this diagram, the free energy changes accompanying the binding of the reactant or product to the enzyme are calculated on the assumption that dihydroxyacetone phosphate and glyceraldehyde-3-phosphate are both present at a concentration of 40μM, which is approximately their concentration in muscle cells. At this low concentration of substrate, when the overall rate of the reaction is described by the specificity constant k_{cat}/K_m, the rate-limiting step appears to be the formation of the enzyme-substrate complex. This step occurs with a rate constant of about 10^8 M^{-1} s^{-1} and evidently is set by the rate of diffusion of the enzyme and substrate. It can be slowed down by increasing the viscosity of the solution. Because the activation free energies for the additional steps in the interconversion of the enzyme-substrate and enzyme-product complexes are low enough so that these steps do not limit the overall rate of the reaction, there is no evolutionary pressure to increase the rate constants for these steps any further. Furthermore, evolutionary changes that led to tighter binding of the substrate could actually be harmful, because lowering the free energy of the enzyme-substrate complex could result in an increase in the activation free energy for one or more of the steps in the conversion of this complex into the product. Given the fixed, low concentration of its substrate, triosephosphate isomerase appears to have reached perfection!

Figure 9.28

The calculated free energy profile of the reaction catalyzed by triosephosphate isomerase. (Enz = enzyme; DHAP = dihydroxyacetone phosphate; GAP = glyceraldehyde-3-phosphate.) The free energy changes associated with binding of DHAP and GAP to the enzyme are calculated on the assumption that DHAP and GAP are present at concentrations of 40 μM.

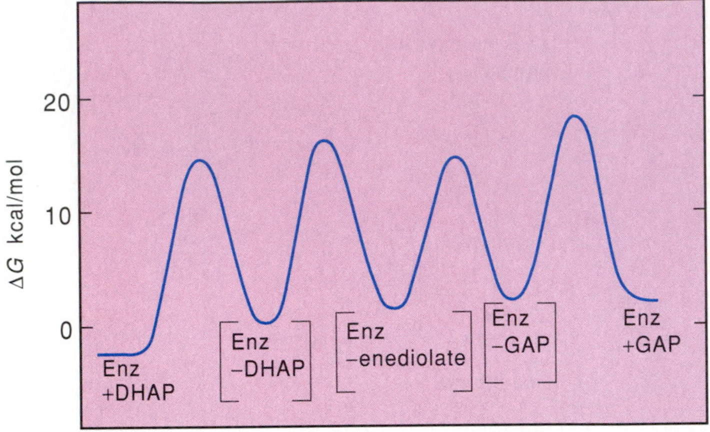

Figure 9.29

Cell-wall polysaccharide, the substrate of lysozyme. (*a*) Conventional drawing of the hexose rings. (*b*) Drawing showing the conformations of the rings. In (*b*), alternating hexose units are flipped over by 180° relative to their neighbors; this is the preferred conformation for polysaccharides with ß(1,4) linkages (see chapter 6). The bond that is cleaved by lysozyme is indicated.

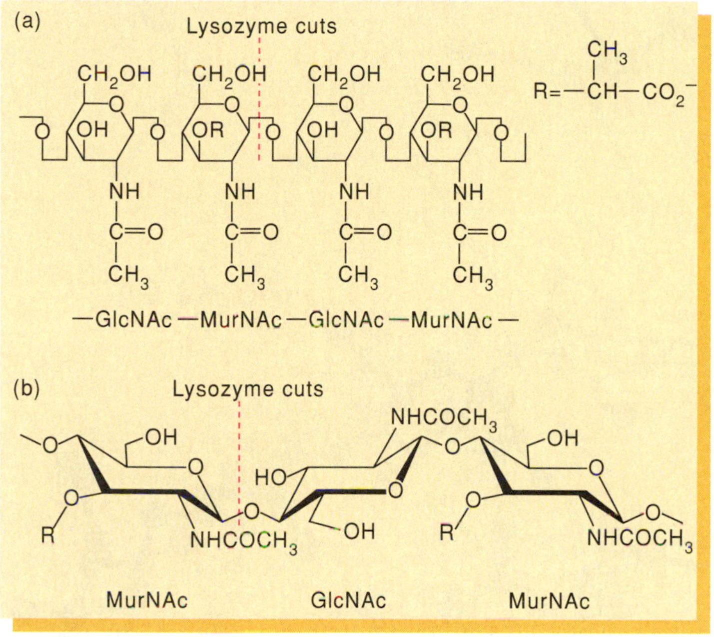

Lysozyme Hydrolyzes Complex Polysaccharides Containing Five or More Residues

Lysozyme catalyzes the hydrolysis of polysaccharide chains that form structural elements of bacterial cell walls. It was discovered first in the whites of chicken eggs, and subsequently was found to occur widely in biological tissues and secretions such as tears. Related glycosidases have been obtained from a variety of organisms including fungi, bacteria, and the bacteriophage T4. The enzyme from hen eggwhite, a small, readily purified protein with a molecular weight of 14,500, was the first enzyme to have its crystal structure solved by x-ray diffraction. Although essentially nothing was known at the time about how lysozyme worked, the structure quickly suggested a likely mechanism.

Lysozyme's principal substrate, the bacterial cell-wall polysaccharide, is a polymer of alternating *N*-acetylglucosamine (GlcNAc) and *N*-acetylmuramic acid (MurNAc) residues connected by $\beta(1,4)$ glycosidic linkages (fig. 9.29). In the cell walls, the MurNAc residues are cross-linked by short polypeptides to form a two-dimensional network termed a peptidoglycan (see chapter 6). Lysozyme cuts the polysaccharide chain at C-1 of a MurNAc residue. It also will hydrolyze some shorter oligosaccharides such as $(GlcNAc\text{-}MurNAc)_3$ or $(GlcNAc)_6$, but it does not accept $(MurNAc)_6$ or other homopolymers of MurNAc. Thus the active site appears to be specific for a GlcNAc residue next to the bond that is cleaved. Very short oligomers of GlcNAc, such as $(GlcNAc)_3$, bind to the enzyme, but are poor substrates (table 9.1).

Table 9.1

Rates of Reaction and Cleavage Patterns Shown by Different Substrates of Lysozyme

Compound	Turnover Number k_{cat} (s^{-1})	Cleavage Pattern
$(GlcNAc)_3$	8.3×10^{-6}	$X_1 \downarrow X_2 \downarrow X_3$
$(GlcNAc)_4$	6.6×10^{-5}	$X_1 \downarrow X_2 \downarrow X_3 \downarrow X_4$
$(GlcNAc)_5$	0.033	$X_1 \downarrow X_2 \downarrow X_3 \downarrow X_4 \downarrow X_5$
$(GlcNAc)_6$	0.25	$X_1 - X_2 - X_3 - X_4 \downarrow X_5 - X_6$
$(GlcNAc\text{-}MurNAc)_3$	0.50	$X_1 - X_2 - X_3 - X_4 \downarrow X_5 - X_6$

The crystal structure of eggwhite lysozyme was solved both with and without bound $(GlcNAc)_3$. Although longer oligosaccharides were hydrolyzed too rapidly to afford crystal structures, it was possible to build a model for the enzyme with bound $(GlcNAc)_6$ by starting with the structure of the complex with $(GlcNAc)_3$. Figure 9.30 shows a drawing of the model. The oligosaccharide sits in a shallow crevice that contains recognizable binding sites for six hexose units. These are labeled A through F in the figure. All of the sites appear to be tailored to bind the acetamide

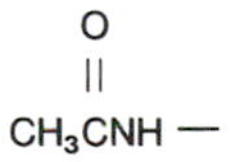

$$CH_3\overset{\overset{O}{\|}}{C}NH-$$

side chains, but sites A, C, and E would not be spacious enough to accommodate the lactyl

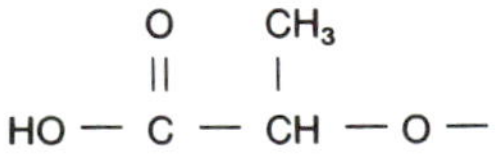

$$HO-\overset{\overset{O}{\|}}{C}-\overset{\overset{CH_3}{|}}{CH}-O-$$

group of MurNAc. The restrictions are particularly severe in site C because of the bulky side chain of Ile 98. This suggests that the normal substrate of repeating (GlcNAc-MurNAc) units would bind with GlcNAc units occupying sites A, C, and E and with MurNAc units in sites B, D, and F, as shown schematically in figure 9.31. When taken with the observation that lysozyme hydrolyzes $(GlcNAc)_6$ but not $(GlcNAc)_3$, and the fact that $(GlcNAc\text{-}MurNAc)_3$ is hydrolyzed at C-1 of a MurNAc residue (see fig. 9.29 and table 9.1), the model indicates that the glycosidic linkage that is cleaved must fall between sites D and E.

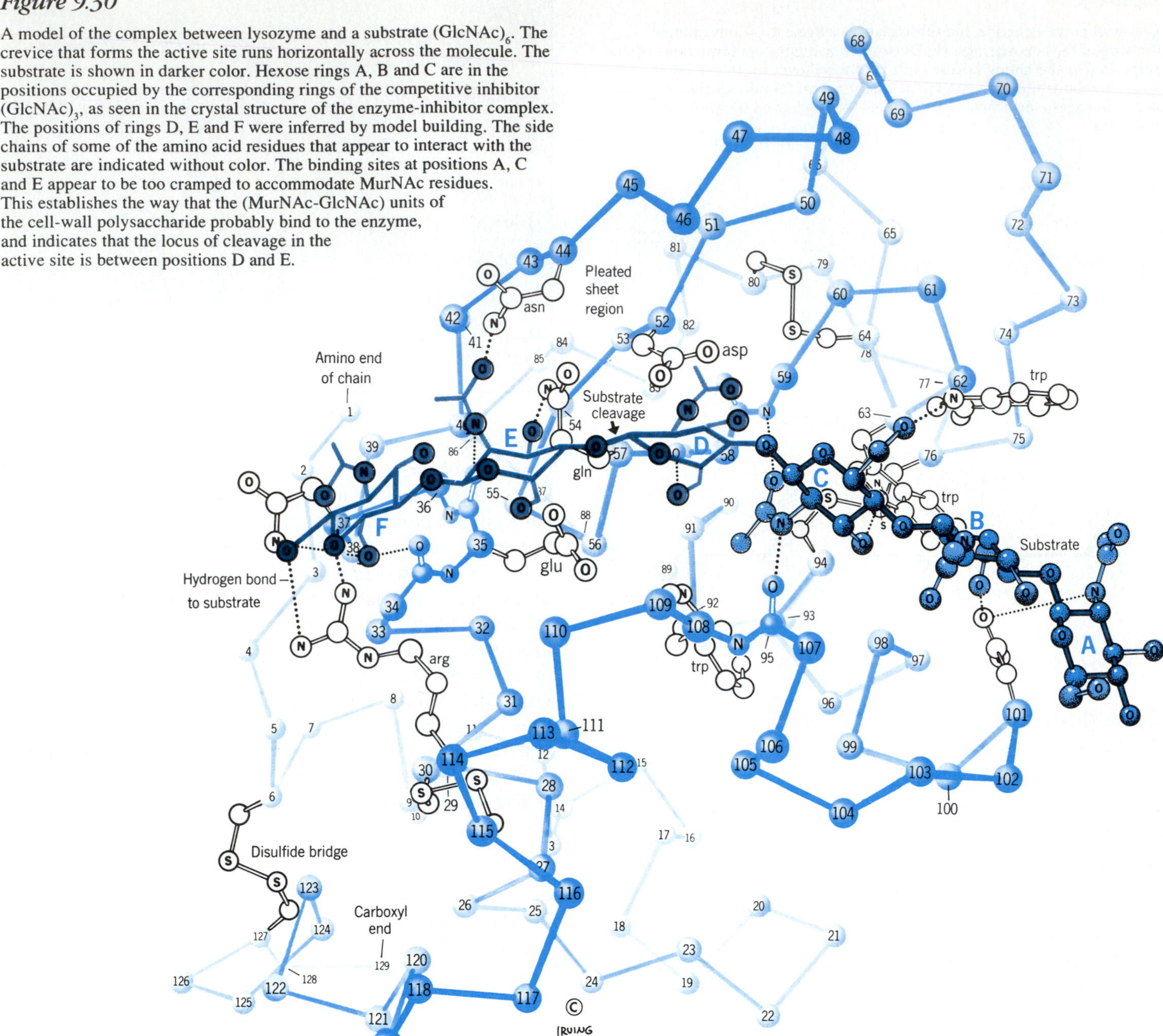

Figure 9.30

A model of the complex between lysozyme and a substrate $(GlcNAc)_6$. The crevice that forms the active site runs horizontally across the molecule. The substrate is shown in darker color. Hexose rings A, B and C are in the positions occupied by the corresponding rings of the competitive inhibitor $(GlcNAc)_3$, as seen in the crystal structure of the enzyme-inhibitor complex. The positions of rings D, E and F were inferred by model building. The side chains of some of the amino acid residues that appear to interact with the substrate are indicated without color. The binding sites at positions A, C and E appear to be too cramped to accommodate MurNAc residues. This establishes the way that the (MurNAc-GlcNAc) units of the cell-wall polysaccharide probably bind to the enzyme, and indicates that the locus of cleavage in the active site is between positions D and E.

There was another reason for focusing attention on the region of the enzyme between binding sites D and E. The amino acid side chains here included two potentially reactive groups, the carboxyl groups of Asp 52 and Glu 35. When the model for the $(GlcNAc)_6$ substrate was positioned so that sites A–E were all occupied, the glycosidic bond that would be cleaved was located in between the two carboxyl groups (see fig. 9.30). Homologous acidic amino acid residues subsequently were found in the active site of bacteriophage T4 lysozyme, and mutations that perturb the activity of the bacteriophage enzyme were found to cluster in this region. In eggwhite lysozyme, chemical conversion of the carboxyl group of Asp 52 to $—CH_2OH$ destroys enzymatic activity. Studies with affinity labels also have implicated Asp or Glu residues in other bacterial and fungal glucosidases.

Lysozyme works best under mildly acidic conditions; its pH optimum is about 5. The pH dependence of the kinetics indicates that the reaction depends on an acidic group with a pK_a of about 6 and on a basic group with a pK_a of about 4.5. The latter group was identified as Asp 52 by studying how specific chemical modification of this residue affects the pH titration curve of the protein. The group with a pK_a of about 6 is probably Glu 35.

Figure 9.31

A schematic diagram showing the specificity of the hen eggwhite lysozyme for its cell-wall polysaccharide substrate. Six subsites (A–F) on the enzyme bind the hexose units. Alternate sites (A, C, and E) interact with the acetamide side chains (a). These sites are unable to accommodate MurNAc residues with their lactyl side chains (Lac). The glycosidic linkage that is cleaved is between sites D and E.

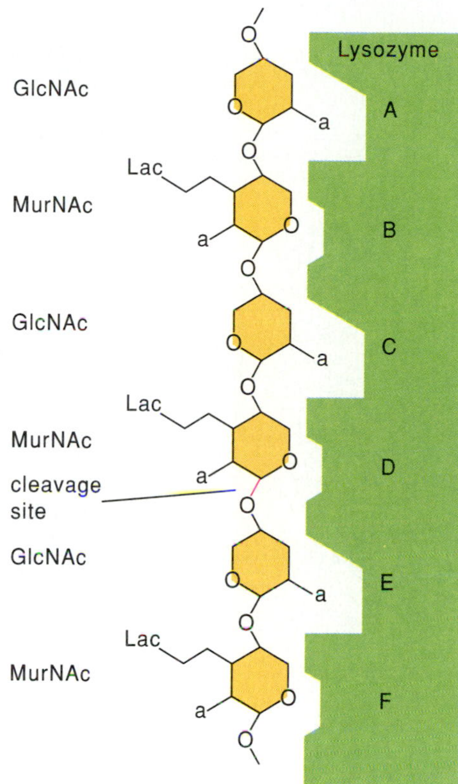

Figure 9.32

A direct nucleophilic (S_N2) attack by a hydroxyl ion on the ß-glycosidic bond in the cell-wall polysaccharide would leave the MurNAc product with an α-hydroxyl group. The observed product has a ß-hydroxyl group. This suggests that the reaction involves an intermediate step in which a derivative of the MurNAc residue remains associated with the enzyme while the alcohol product is replaced by water, as shown in figure 9.33.

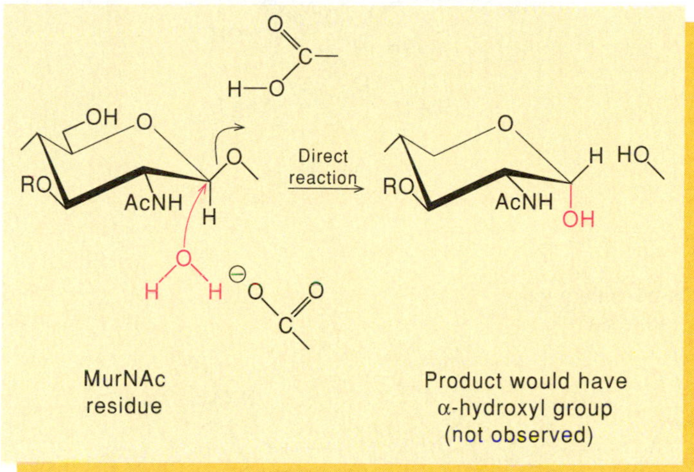

Given the information that the enzymatic reaction requires the carboxyl group of Glu 35 to be protonated and that of Asp 52 to be unprotonated, the first reaction mechanism that comes to mind might be a direct nucleophilic attack by H_2O, aided by concerted general-acid and general-base catalysis (fig. 9.32). In this scheme, Asp 52 could remove a proton from the water molecule, and Glu 35 could provide a proton to the departing alcoholic group. But this mechanism is at odds with the stereochemistry of the reaction. If H_2O or HO^- attacked C-1 of the MurNAc residue from one side of the tetrahedral carbon atom, and the alcoholic group of GlcNAc departed from the other side, the result would be an inversion of configuration around the C-1 carbon. The β-glycosidic bond would be replaced by a hydroxyl group in the α configuration. Contrary to this expectation, the product is found to have the hydroxyl group in the β configuration. This indicates that the reaction probably proceeds in two distinct steps. The first step evidently removes the alcohol from the β side of the MurNAc residue, but leaves an intermediate derivative of the MurNAc residue associated with the enzyme. Water or HO^- then must enter the active site on the same side of the MurNAc residue, replacing the alcohol.

Figure 9.33 shows a mechanism that meets these requirements. In this scheme, the intermediate is a carboxonium ion, in which a positive charge is distributed between C-1 of the MurNAc and the attached oxygen atom. The formation of this species would involve general-acid catalysis by Glu 35, supported by a favorable electrostatic interaction between the carboxonium ion and the negatively charged carboxylate of Asp 52. In the breakdown of the carboxonium intermediate, Glu 35 could act as a general base to remove a proton from water.

Another plausible mechanism would be for Asp 52 to launch a direct nucleophilic attack on the glycosidic bond. This could generate an enzyme-bound, carboxylic ester of the MurNAc as an intermediate. As yet, the experimental information is not sufficient to distinguish decisively between this mechanism and the one shown in figure 9.33, although measurements of kinetic isotope effects favor the latter.

The original formulation of the carboxonium-intermediate mechanism (see fig. 9.33) included an additional feature that was suggested by the crystallographic structure. In the model of $(GlcNAc)_6$ bound to the enzyme, it seemed difficult to squeeze even a GlcNAc residue into site D without some distortion of the hexose ring. Experimental measurements of the binding energies for a series of substrates also indicated that, whereas the binding of a GlcNAc residue to any of sites A, B, C, E, or F was energetically favorable, the binding to site D was very weak, or even energetically unfavorable. These observations suggested that the tight binding of hexose units to sites A, B, C, E, and F forced the hexose unit at site D to sit uncomfortably in a distorted geometry. The geometric distortion appeared to push the hexose ring from its normal boat configuration into a chair configuration. In the chair configuration, carbons 1 and 2 and the internal oxygen atom of the pyranose ring all lie

Figure 9.33

Proposed mechanism for the reaction catalyzed by lysozyme. In the carboxonium ion intermediate, carbons 1 and 2 and the internal oxygen of the pyranose ring all lie in a plane. The formation of this intermediate is favored by electrostatic interactions with Asp 52, along with general-acid catalysis by Glu 35.

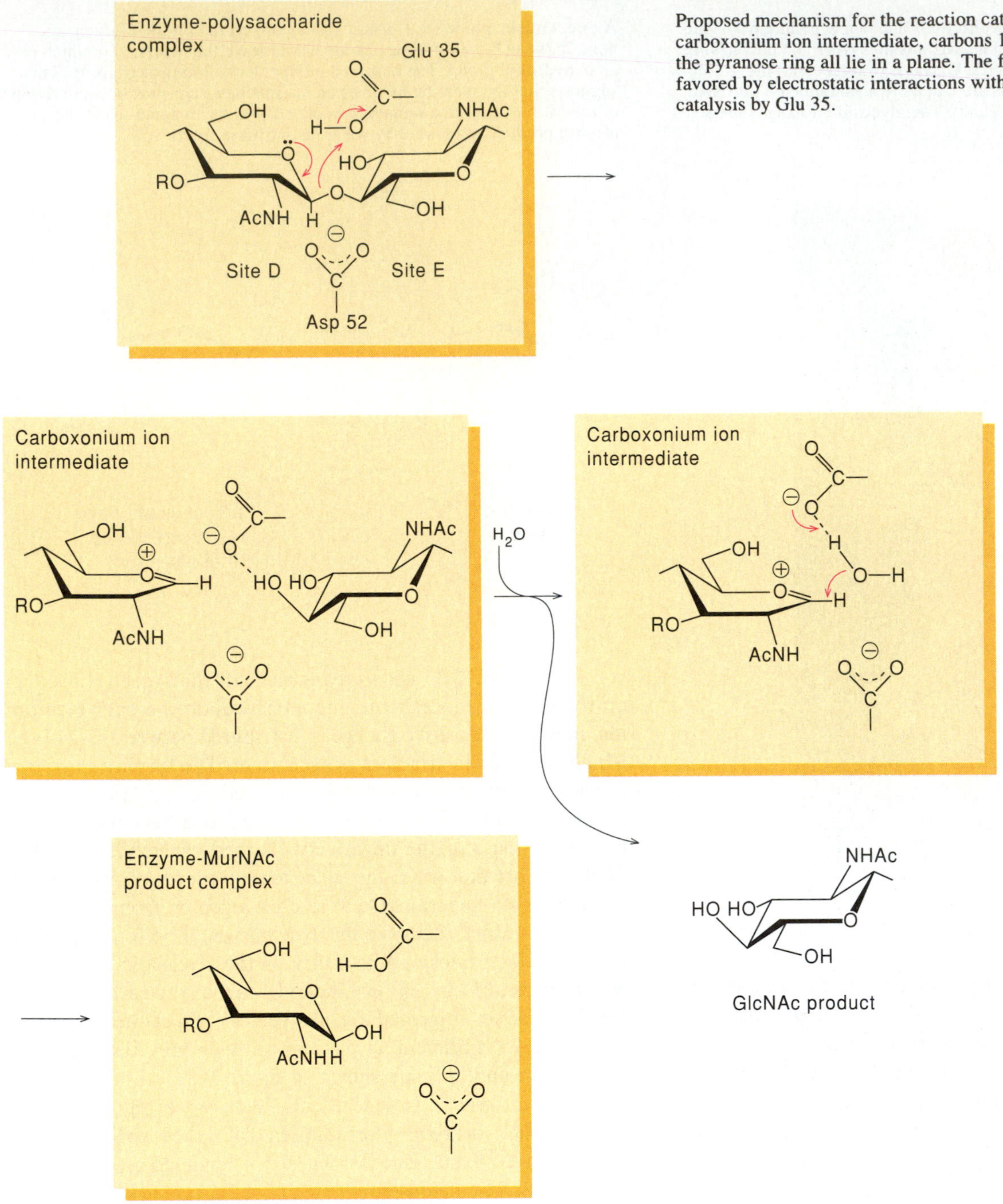

in a plane. Because these atoms would have a similar, planar configuration in the carboxonium intermediate (see fig. 9.33), it was suggested that geometric "strain" in the bound substrate contributed significantly to pushing the substrate in the direction of the transition state. The enzyme thus would decrease the activation free energy for the reaction (ΔG^{+}) partly by raising the energy of the bound substrate.

Although the notion that an enzyme can exert a geometric strain on a bound substrate is an appealing one, it is not supported by computer modeling that takes into account the flexibility of the enzyme. Calculations of the energies of the enzyme-substrate complex in various conformations indicate that the hexose unit that binds to side D on lysozyme probably remains in the boat conformation and is not significantly strained.

Figure 9.34

Structures of NAD^+ and NADH.

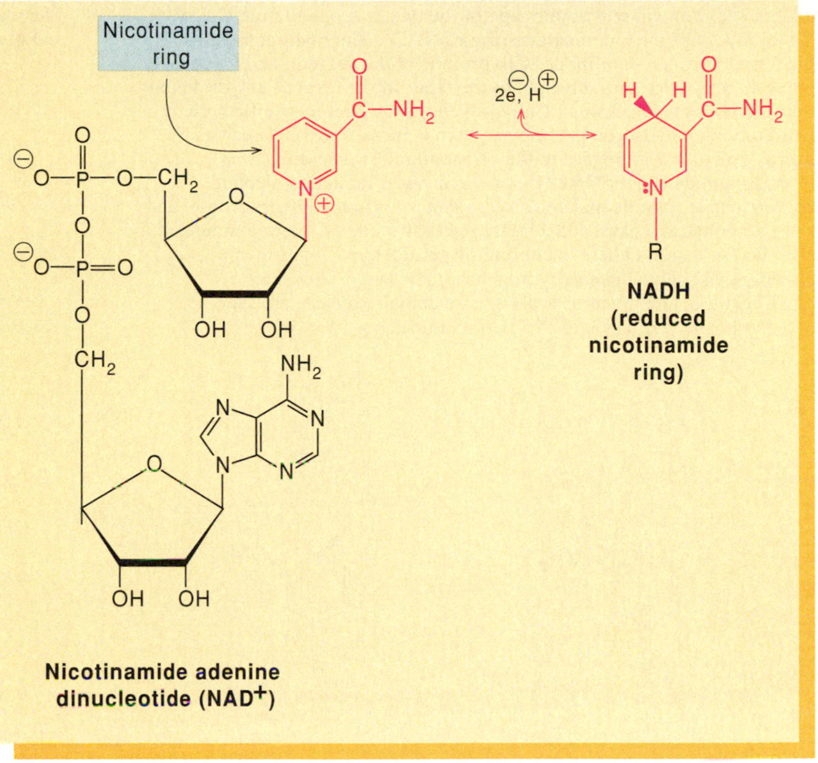

The experimental finding that the binding of a hexose unit to this site is energetically unfavorable can be attributed to the displacement of two water molecules that are bound to the carboxylate of Asp 52 in the free enzyme. Thus the largest contribution to lowering ΔG^+ probably comes from the electrostatic effect of Asp 52, and not from geometric strain in the substrate.

Alcohol Dehydrogenase and Lactate Dehydrogenase: Enzymes that Catalyze the Transfer of Electrons

Numerous enzymes catalyze the transfer of electrons from one substrate to another. In many of these oxidation-reduction reactions, one or more protons also are removed from the substrate that becomes oxidized and are added to the substrate that is reduced. The oxidation of an alcohol to an aldehyde, for example, requires the removal of two protons along with two electrons:

$$RCH_2OH \rightleftharpoons R-\overset{\displaystyle H}{\overset{|}{C}}=O + 2\,e^- + 2\,H^+ \qquad (15)$$

The most common electron acceptor in the biological oxidation of alcohols is nicotinamide adenine dinucleotide (NAD^+), a coenzyme whose structure is shown in figure 9.34. When NAD^+ is reduced to nicotinamide adenine dinucleotide hydrogen (NADH), it picks up one proton and two electrons:

$$NAD^+ + 2\,e^- + H^+ \rightleftharpoons NADH \qquad (16)$$

The oxidation of an alcohol by NAD^+ thus results in the transfer of two electrons and the net release of one proton:

$$RCH_2OH + NAD^+ \rightleftharpoons R-\overset{\displaystyle H}{\overset{|}{C}}=O + NADH + H^+ \qquad (17)$$

Because the reaction written in equation (17) is formally equivalent to the removal of two hydrogen atoms from the alcohol, enzymes that catalyze such oxidation-reduction reactions generally are called dehydrogenases. The term "oxidase," which also might be a reasonable name for an oxidation-reduction enzyme, is reserved for reactions in which the oxidant is molecular oxygen (O_2).

The atomic or subatomic particles that a dehydrogenase actually moves from one substrate to another are usually not neutral hydrogen atoms (H•), as the name "dehydrogenase" suggests, nor are they necessarily free electrons and protons, as equations (15) and (16) suggest. In the reduction of NAD^+ a proton is released to the solution, but the species that is transferred from the reducing substrate to the nicotinamide ring of the coenzyme is probably a hydride ion (H^-), which is a proton with two electrons. This conclusion is based largely on model studies of nonenzymatic reactions of molecules resembling the coenzymes. In addition, it has been shown that in the enzymatic reactions, the hydride ion is transferred stereospecifically to a particular side of the nicotinamide ring, without equilibrating with protons of the solvent (see figs. 9.35 and 11.7). The strategy of all NAD^+-dependent dehydrogenases appears to be to orient

Figure 9.35

Alcohol dehydrogenase transfers one of the two hydrogens from C-1 of ethanol to C-4 of the nicotinamide ring of NAD^+. The hydrogen that is transferred does not equilibrate with protons of the solvent, and it probably moves directly from one substrate to the other in the form of a hydride ion (a proton with two electrons). Dehydrogenases are stereospecific with regard to which hydrogen is removed from a primary alcohol such as ethanol, and also with regard to the orientation of the hydrogen on the nicotinamide ring of NADH. Some enzymes, including alcohol dehydrogenase, soluble malate dehydrogenase, lactate dehydrogenase, and isocitrate dehydrogenase, add a hydrogen to the side of the ring indicated here (the "A" side). Others, including glyceraldehyde-3-phosphate dehydrogenase, glutamate dehydrogenase, glucose-6-phosphate dehydrogenase, and glycerol-3-phosphate dehydrogenase, are specific for the opposite ("B") side of the nicotinamide ring.

Figure 9.36

The reactions catalyzed by malate dehydrogenase, lactate dehydrogenase, and glyceraldehyde-3-phosphate dehydrogenase.

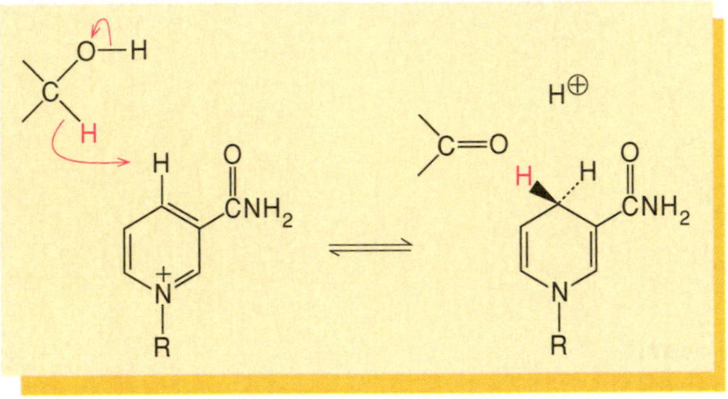

the coenzyme and substrate on the enzyme surface so that C-4 of the nicotinamide ring is close to the reactive hydrogen of the substrate, and to provide a functional group that can facilitate a redistribution of electrical charge in the substrate's carbon–oxygen bond.

The crystal structures of several dehydrogenases have been determined in the presence and absence of bound coenzymes and competitive inhibitors. The best-known structures are those of alcohol dehydrogenase, which catalyzes the interconversion of ethanol and acetaldehyde (equation 17 with R $= CH_3$), glyceraldehyde-3-phosphate dehydrogenase, lactate dehydrogenase, and soluble malate dehydrogenase. The last three enzymes catalyze the oxidation of glyceraldehyde-3-phosphate to glyceric acid 1,3-bisphosphate, lactic acid to pyruvic acid, and malic acid to oxaloacetic acid (fig. 9.36). All four enzymes play important roles in carbohydrate metabolism. (The term "soluble" attached to malate dehydrogenase is used to distinguish a cytosolic enzyme from the enzyme that catalyzes the same reaction in mitochondria.)

Lactate dehydrogenase and the soluble malate dehydrogenase probably arose from a common ancestor, because their overall structures are very similar. There is more variability in the structures of alcohol dehydrogenase and glyceraldehyde-3-phosphate dehydrogenase. In all four enzymes, however, the region of the protein that binds NAD^+ contains similar structural elements. The nucleotide-binding domain is made up of six strands of parallel β sheet connected by α-helical stretches running antiparallel to the β strands (fig. 9.37). NAD^+ binds to the enzyme in an extended conformation near the ends of the β strands, although the details of its contacts with enzyme vary among the different dehydrogenases. The negatively charged pyrophosphate group of the coenzyme usually is sandwiched between a positively charged residue such as arginine and the end of an α helix (see fig. 9.37). The helix is oriented so that its internal hydrogen bonds create a net excess of positive charge close to the pyrophosphate.

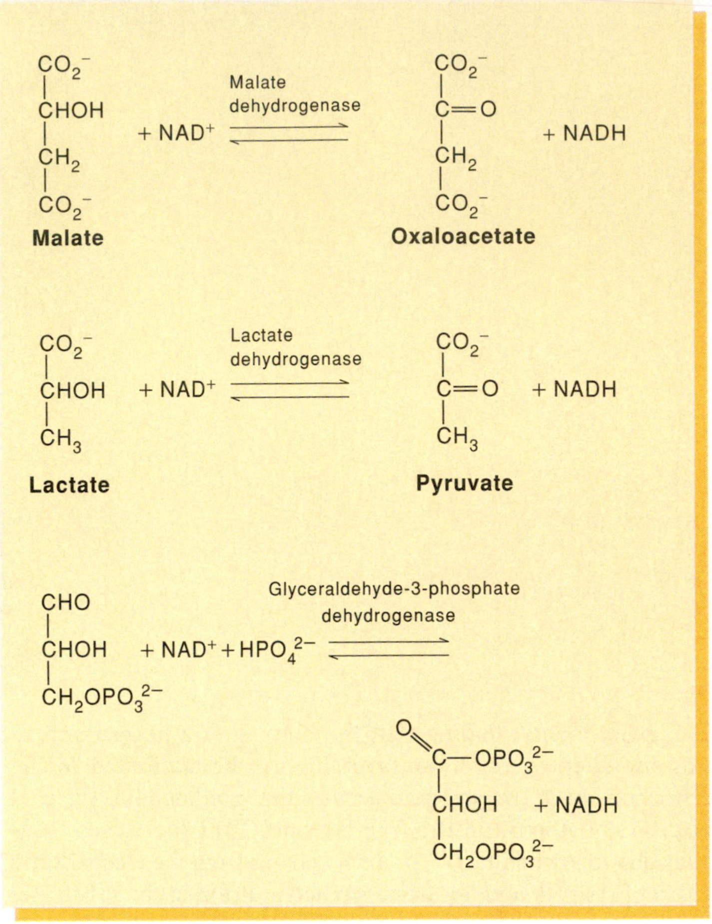

As we discussed in chapter 8, there are several possible pathways for an enzymatic reaction involving two substrates: (1) a random pathway, in which either substrate can bind to the enzyme first; (2) an ordered pathway, in which one substrate must bind before the other; and (3) a reactive-intermediate pathway, in which the first substrate binds and reacts with the enzyme to form one of the products, leaving the enzyme in a modified form that then reacts with the second substrate. Kinetic studies have shown that the reaction mechanism of lactate dehydrogenase follows an ordered sequence. In the oxidation of lactate, NAD^+ binds to the enzyme first, followed by lactate. The transfer of the hydride ion then occurs rapidly in

Figure 9.37

Crystal structures of the nucleotide-binding regions of (*a*) lactate dehydrogenase and (*b*) alcohol dehydrogenase. Bound NAD^+ is shown in yellow in (*b*), and a covalently linked adduct of NAD^+ and the substrate lactate ("S-lac-NAD") in (*a*). The arginine residue that interacts with the pyrophosphate of the NAD^+ is in red; other amino acid residues are in brown. The yellow ribbons show the folding of the α-carbon chain. The figures are based on crystal structures described by U. M. Grau and M. S. Rossmann for lactate dehydrogenase, and H. Eklund, J.-P. Sanama, L. Wallen, C.-I. Branden, A. Akeson, and T. A. Jones for alcohol dehydrogenase.

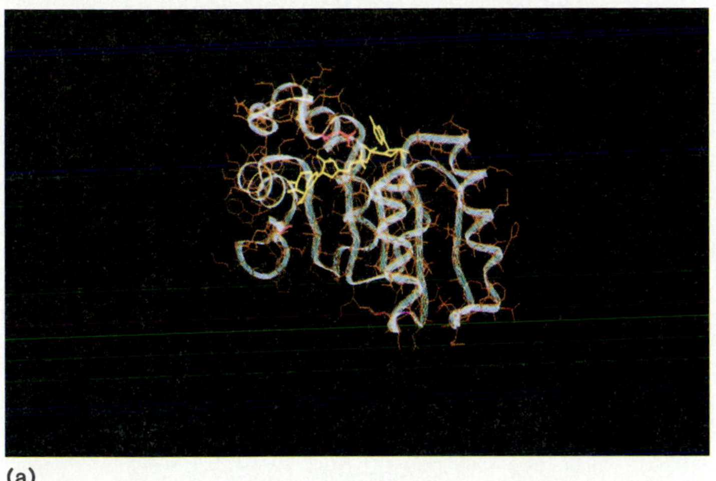

(a)

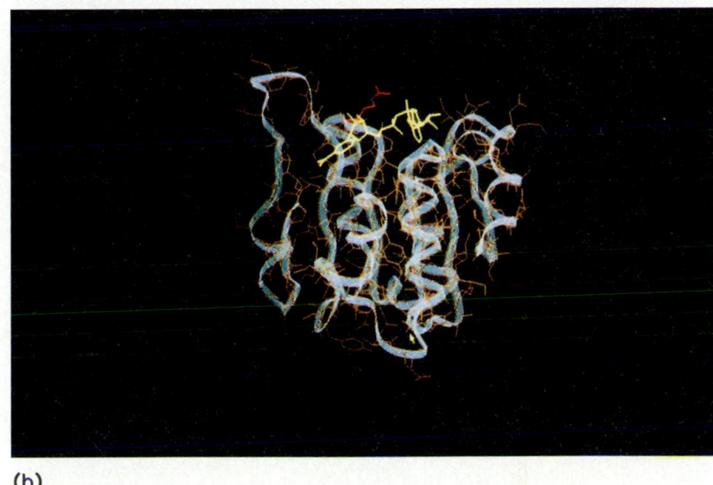

(b)

Figure 9.38

The catalytic mechanism of lactate dehydrogenase. His 195 probably acts as a general base in the conversion of lactate to pyruvate, and as a general acid in the conversion of pyruvate to lactate. Arg 171 probably serves to orient the substrate.

either direction, giving an equilibrium mixture of the two ternary complexes, enzyme-NAD^+-lactate and enzyme-NADH-pyruvate. Finally, pyruvate dissociates from the enzyme, followed by NADH. The rate-limiting step in the overall steady-state reaction is the dissociation of NADH. The obligatory sequences of the binding and dissociation steps can be explained by structural changes that result from the binding of NAD^+ or NADH. The binding of either form of the nucleotide causes a major structural rearrangement that brings together components of the binding site for lactate or pyruvate. The structural changes include a particularly large movement of the loop of the protein between two of the β strands. Arginine 101, the residue that interacts with the pyrophosphate of NAD^+ or NADH, is in the middle of this loop (see fig. 9.37*a*). In the free enzyme the loop extends out into the solvent; in the ternary complexes it is pulled in around the coenzyme and substrate.

The substrate-binding site of lactate dehydrogenase contains an arginine residue (Arg 171) that serves to orient lactate or pyruvate by forming an ion pair with the carboxylate group, and a histidine (His 195) that forms a hydrogen bond to the alcohol or keto group. In the oxidation of lactate to pyruvate, His 195 probably acts as a general base to remove a proton from the substrate's hydroxyl group, as shown in figure 9.38.

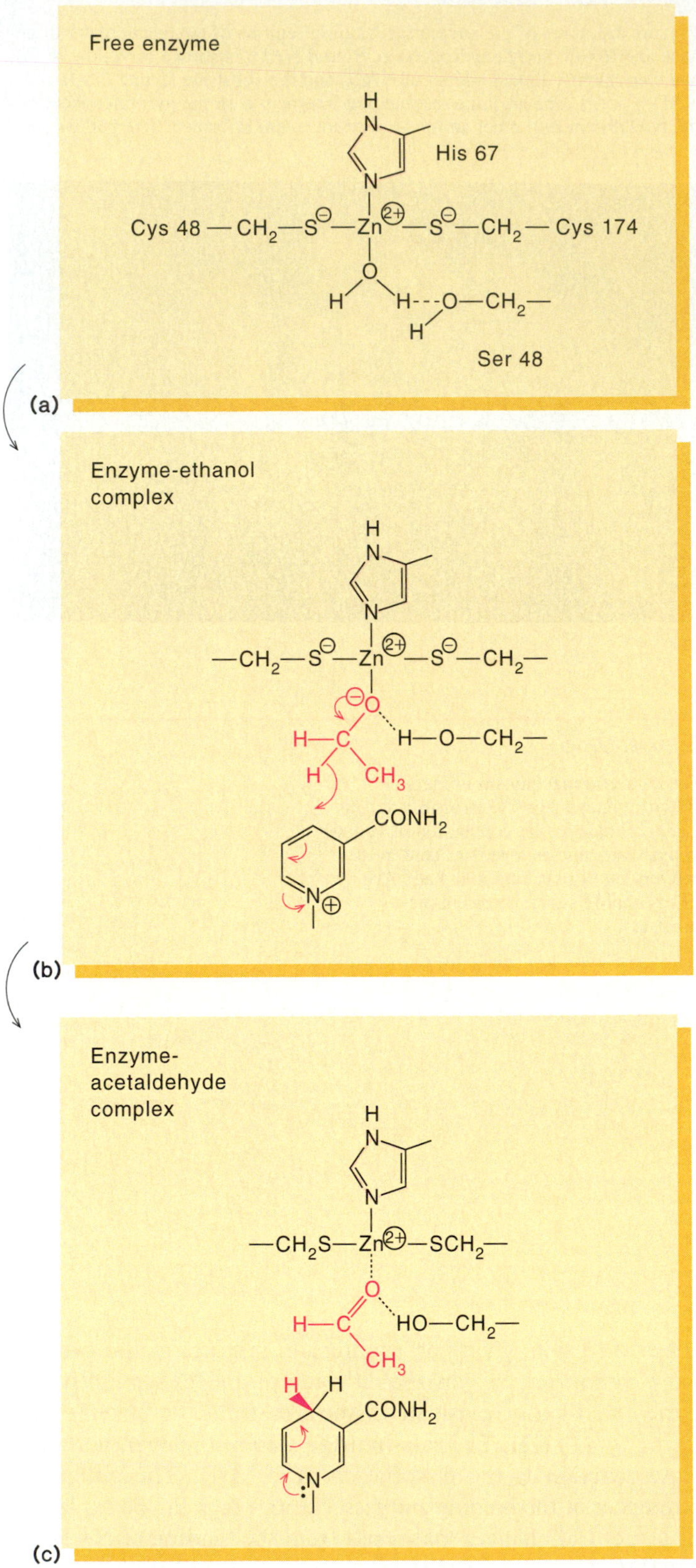

Figure 9.39

Functional groups at the active site of alcohol dehydrogenase from horse liver. (*a*) In the absence of substrate, the ligands of the Zn^{2+} ion are two cysteines, a histidine, and a molecule of water that is hydrogen-bonded to Ser 48. (*b*) The substrate ethanol probably binds to the Zn^{2+} as the alcoholate anion, displacing the molecule of water and forming a hydrogen bond to Ser 48.

This would increase the electron density on the adjacent carbon atom, easing the release of the hydride ion. In the reverse reaction, the imidazolium group of the histidine would act as a general acid to protonate the carbonyl oxygen of pyruvate. This would increase the positive charge on the carbon atom, making the enzyme-bound substrate more receptive to a hydride ion.

Alcohol dehydrogenase also follows an ordered kinetic pathway, in which the nucleotide binds to the enzyme first and dissociates last. Again, the binding of the nucleotide causes substantial structural changes that evidently enhance the enzyme's ability to bind ethanol or acetaldehyde. However, alcohol dehydrogenase differs from lactate dehydrogenase in having a Zn^{2+} ion at its active site, and its enzymatic mechanism appears to be based on electrostatic effects of the metal ion rather than on general-acid or -base catalysis.

Alcohol dehydrogenases have been crystallized from a variety of sources, including horse liver and yeast. The crystal structures show that the Zn^{2+} is attached to a histidine residue and the thiol groups of two cysteines (fig. 9.39). In the absence of the substrate, the fourth ligand of the Zn^{2+} is a molecule of water that forms a hydrogen bond to a nearby serine residue. When the substrate adds to the enzyme, the alcoholic or carbonyl oxygen atom appears to coordinate directly to the Zn^{2+}, displacing the molecule of water. Studies of the pH dependence of binding of a variety of alcohols indicate that the alcohol probably binds in the form of the alcoholate anion, as shown in figure 9.39. The negative charge on the oxygen atom makes the alcoholate a better donor of a hydride ion for the reduction of NAD^+, compared with the undissociated alcohol. For the reaction in the opposite direction, the electrostatic effect of the Zn^{2+} ion would facilitate the reduction of acetaldehyde.

Summary

In this chapter we have looked at mechanisms of enzyme catalysis, first from a general standpoint, and then in some detail. The following points are the most important.

1. Like ordinary chemical catalysts, enzymes associate directly with the reacting species and interact with them in a manner that lowers the free energy of the transition state. Enzyme-substrate interactions usually are highly specific, and can depend on critically positioned amino acid side chains, the atoms of the polypeptide backbone, or bound cofactors such as metal ions.
2. All known enzymatic reaction mechanisms depend on one or more of the following themes: proximity effects (enzymes hold the reactants close together in an appropriate orientation); electrostatic effects (charged, polar, or polarizable groups of the enzyme are positioned to favor the redistribution of electrical charges that occur as the substrate evolves into the transition state); general-acid or general-base catalysis (acidic or basic groups of the enzyme donate or remove protons, and often do first one and then the other); nucleophilic or electrophilic catalysis (nucleophilic or electrophilic functional groups of the enzyme or a cofactor react with complementary groups of the substrate to form covalently-linked intermediates); and structural flexibility (changes in the protein structure can increase the specificity of enzymatic reactions by insuring that substrates bind or react in an obligatory order and by sequestering bound substrates in pockets that are protected from the solvent).
3. Trypsin, chymotrypsin, and elastase are very similar in structure, but have substrate-binding pockets that are tailored for different types of amino acid side chains. The active site of each enzyme contains three critical residues: serine, histidine, and aspartate. The side chains of these residues are oriented so that the serine hydroxyl group becomes a strong nucleophilic reagent for attacking the substrate's peptide carbonyl group adjacent to the preferred amino acid. This reaction displaces the amine member of the peptide bond, and generates an acyl-ester intermediate, in which the carboxyl member is linked covalently to the serine group of the enzyme. Formation and hydrolysis of the acyl-ester intermediate probably are promoted both by electrostatic interactions that stabilize tetrahedral transition states and by general-acid and general-base catalysis.
4. Thermolysin and carboxypeptidase A contain a bound Zn^{2+} ion, which interacts with the carbonyl oxygen of the peptide bond that is to be cleaved. The electrostatic effect of the Zn^{2+} probably facilitates formation of a tetrahedral transition state in the direct nucleophilic attack of H_2O or OH^- at the carbonyl carbon. Acidic and basic amino acid side chains again may serve as general-acid and general-base catalysts in the reaction. Carboxypeptidase A undergoes a substantial change in structure when it binds its substrate.
5. Ribonuclease A hydrolyzes RNA adjacent to pyrimidine bases. The reaction proceeds through a 2′,3′-phosphate cyclic diester intermediate. The formation and breakdown of the cyclic diester appear to be promoted by concerted general-base and general-acid catalysis by two critical histidine residues, and by electrostatic interactions with two lysines. These reactions proceed through pentavalent phosphoryl intermediates. The geometry of the oxygens surrounding the phosphorus atom in these intermediates resembles the geometry of vanadate compounds that act as inhibitors of the enzyme.
6. Triosephosphate isomerase interconverts dihydroxyacetone phosphate and glyceraldehyde-3-phosphate. A glutamic acid residue probably acts as a general base to remove a proton from the substrate, forming an ene-diolate intermediate. Nearby histidine and lysine side chains would stabilize this intermediate electrostatically. Triosephosphate isomerase appears to have reached evolutionary perfection, in the sense that it catalyzes its reaction at the maximum possible rate, given the concentration of the substrate in the cell. The rate-limiting step is the collision of the enzyme and substrate by diffusion through the solution.
7. Lysozyme hydrolyzes complex polysaccharides containing five or more hexose residues. Its substrate-binding site includes a crevice that can accommodate six such residues. The enzyme probably first releases the truncated polysaccharide from one side of the bond that is cleaved, leaving the polysaccharide on the other side bound noncovalently to the enzyme in the form of a positively charged carboxonium intermediate. The formation of this intermediate is promoted by general-acid catalysis by a glutamic acid residue, and by the electrostatic effect of a nearby aspartate.
8. Lactate dehydrogenase and alcohol dehydrogenase catalyze the reversible transfer of a hydride ion (H^-) to NAD^+ from, respectively, lactate and ethanol. Both enzymes bind their substrates in an obligatory order. NAD^+ binds first, causing a structural change that sets up the binding site for the alcohol. The alcohol substrate then binds with the hydrogen that will be transferred close to C-4 of the nicotinamide ring. In lactate dehydrogenase, a histidine residue probably acts as a general base to remove a proton from the —OH group. This would facilitate the release of the negatively charged hydride ion. In alcohol dehydrogenase, a Zn^{2+} ion attached to the enzyme probably encourages the ethanol to bind in the form of the alcoholate anion, which again is a good donor of a hydride ion.

Selected Readings

Albery, W. J., and J. R. Knowles, Free-energy profile of the reaction catalyzed by triosephosphate isomerase. *Biochem.* 15:5588, 5627, 1976.

Blackburn, P., and S. Moore, Pancreatic ribonuclease. *The Enzymes* 15:317, 1982.

Blow, D., Structure and mechanism of chymotrypsin. *Acc. Chem. Res.* 9:145, 1976.

Branden, C. -I., H. Jornvall, H. Eklund, and B. Furugren, Alcohol dehydrogenases. *The Enzymes* 11:104, 1975.

Breslow, R., How do imidazole groups catalyze the cleavage of RNA in enzyme models and enzymes? Evidence from "negative catalysis." *Acc. Chem. Res.* 24:317, 1991.

Eklund, H., B. V. Plapp, J. -P. Samama, and C. -I. Branden, Binding of substrate in a ternary complex of horse liver alcohol dehydrogenase. *J. Biol. Chem.* 257:14349, 1982.

Fersht, A., *Enzyme Structure and Mechanism,* 2d ed. New York: Freeman, 1985.

Fersht, A. R., D. M. Blow, and J. Fastrez, Leaving group specificity in chymotrypsin-catalyzed hydrolysis of peptides: A stereochemical interpretation. *Biochem.* 12:2035, 1973.

Findlay, D., D. G. Herries, A. P. Mathias, B. R. Rabin, and C. A. Ross, The active site and mechanism of pancreatic ribonuclease. *Nature* 190:781, 1961.

Holbrook, J. J., A. Liljas, S. J. Steindel, and M. G. Rossmann, Lactate dehydrogenase. *The Enzymes* 11:191, 1975.

Holmes, M. A., D. E. Tronrud, and B. W. Matthews, Structural analysis of the inhibition of thermolysin by an active-site-directed irreversible inhibitor. *Biochem.* 22:236, 1983.

Imoto, T., L. N. Johnson, A. C. T. North, D. C. Phillips, and J. A. Rupley, Vertebrate lysozymes. *The Enzymes* 7:665, 1972.

Kelly, J. A., A. R. Sielecki, B. D. Sykes, M. N. G. James, and D. C. Phillips, X-ray crystallography of the binding of the bacterial cell wall trisaccharide NAM-NAG-NAM to lysozyme. *Nature* 282:875, 1979.

Kraut, J., How do enzymes work? *Science* 242:533, 1988.

Lolis, E., T. Alber, R. C. Davenport, D. Rose, F. C. Hartman, and G. A. Petsko, Structure of yeast triosephosphate isomerase at 1.9-Å resolution. *Biochem.* 29:6609, 1990.

Markley, J. L., Correlation proton magnetic resonance studies at 250 MHz of bovine pancreatic ribonuclease. I. Reinvestigation of histidine peak assignments. *Biochem.* 14:3546, 1975.

Page, M. I. (ed.), *Enzyme Mechanisms.* London: Royal Society of Chemistry, 1987. See particularly the chapters by M. I. Page (Theories of Enzyme Catalysis), A. L. Fink (Acyl Group Transfer—The Serine Proteinases), D. S. Auld (Acyl Group Transfer—Metalloproteinases), P. M. Cullis (Acyl Group Transfer—Phosphoryl Transfer), M. L. Sinnott (Glycosyl Group Transfer), and J. P. Richard (Isomerization Mechanisms through Hydrogen and Carbon Transfer).

Reeke, G. N., J. A. Hartsuck, M. L. Ludwig, F. A. Quiocho, T. A. Steitz, and W. N. Lipscomb, The structure of carboxypeptidase A: Some results at 2.0-Å resolution, and the complex with glycyl tyrosine at 2.8-Å resolution. *Proc. Natl. Acad. Sci. USA* 58:2220, 1967.

Rose, I. A., Mechanism of the aldose-ketose isomerase reactions. *Adv. Enzymol.* 43:491, 1975.

Shoham, M., and T. Steitz, Crystallographic studies and model building of ATP at the active site of hexokinase. *J. Mol. Biol.* 140:1, 1980.

Warshel, A., G. Naray-Szabo, F. Sussman, and J. -K. Hwang, How do serine proteases really work? *Biochem.* 28:3629, 1989.

Problems

1. (a) In what ways are the mechanistic features of chymotrypsin, trypsin, and elastase similar?
   (b) If the mechanisms of these enzymes are similar, what features of the enzyme active site dictate substrate specificity?
2. (a) If you monitor the lactate dehydrogenase reaction by the formation of NADH (increase in absorbance at 340 nm), should increasing pH make it easier to measure the dehydrogenation of lactate to pyruvate?
   (b) Would a chemical trapping agent for pyruvate serve the same purpose at lower pH values? Why?
   (c) Dehydrogenase activity can be measured by reoxidizing the NADH and reducing a tetrazolium dye. The reduced dye is intensely colored. What are the advantages of measuring the LDH reaction by means of the tetrazolium dye system?
3. For many enzymes, V_{max} is dependent on pH. At what pH would you expect V_{max} of RNase to be optimal? Why?
4. If a lysine were substituted for the aspartate in the trypsin side chain binding crevice, would you expect the enzyme to be functional? If it were functional, what effect would you predict the substitution to have on substrate specificity?
5. Carboxypeptidase A preferentially cleaves C-terminal aromatic residues from proteins. When the aromatic substrate side chain is bound, water is expelled from the active site. How does the release of water stabilize binding of substrate in the active site?
6. You have isolated a metalloenzyme that preferentially cleaves basic amino acids from the carboxyl terminal of proteins. Would you expect the enzyme to retain an arginine in the active site as does carboxypeptidase A? Why or why not? What other residues would you predict to be in the substrate binding site for the new enzyme? How would these residues dictate cleavage specificity?
7. RNase can be completely denatured by boiling or by treatment with chaotropic agents (e.g., urea), yet can refold to its fully active form upon cooling or removal of the denaturant. By contrast, when enzymes of the trypsin family and carboxypeptidase A are denatured, they do not regain full activity upon renaturation. What aspects of trypsin and carboxypeptidase A structure preclude their renaturation to the fully active form?

8. (a) The amino acids in the active site of the protease papain are shown. Predict a feasible reaction mechanism for papain.

His 159
Asn 175
NH_2
Cys 25
Papain active site
O·····H—N N·····H—S
R_1 O
N—C ← Substrate
H R_2

(b) *N*-ethylmaleimide (NEM) reacts rapidly with cysteine thiolate anion via a Michaelis addition. What is the product of the reaction between NEM and cysteine? Would you expect the rate of R–SH reaction with *N*-ethylmaleimide to be more rapid at pH 5 or pH 7.5? Why?

(c) Would you expect cysteine 25 (see part a) to be more reactive with *N*-ethylmaleimide than any of the other cysteine residues in the protein? Explain.

9. Why do structural analogs of the transition-state intermediate of an enzyme inhibit the enzyme competitively and with low K_I values?

10. Transition-state analogs of a specific chemical reaction have been used to elicit antibodies with catalytic activity. These catalytic antibodies have great promise as experimental tools as well as having commercial value. Why is it reasonable to assume that the binding site for the transition-state analog on the antibody would mimic the enzyme active site? What difficulties might be encountered if a catalytic antibody were sought for a reaction requiring a cofactor (coenzyme)?

11. Superoxide dismutases catalyze the reaction

$$O_2^- + O_2^- + 2H^+ \rightarrow O_2 + H_2O_2$$

The catalytic mechanism of the superoxide dismutases involves active site transition metals (Cu, Fe, Mn) that undergo valence changes in the catalytic cycle. Write equations that represent the catalytic cycle of each of these transition metals. (Remember, in each case you must finish with the catalyst in the same state as when you began.) (See Fridovich, I. 1986. *Adv. Enzymol.* 58:61–97.)

12. The superoxide dismutase isolated from most sources has an isoelectric point around 5.

(a) What problem does the acidic pI of superoxide dismutase present to the catalytic disproportionation of superoxide anion at pH 7? (The pK_a of $HO_2\cdot$ is 4.8.)

(b) How might strategically placed basic residues assist in catalysis?

(c) Would the acidic pI of the enzyme or presence of basic residues at the active site impede release of product from the active site? Why?

13. Using site-directed mutagenesis techniques, you have isolated a series of recombinant enzymes in which specific lysine residues were replaced with aspartate residues. The enzymatic assay revealed the following.

Enzyme Form	Activity (U/mg)
Native enzyme	1,000
Recombinant Lys 21 → Asp 21	970
Recombinant Lys 86 → Asp 86	100
Recombinant Lys 101 → Asp 101	970

(a) What might you infer about the role(s) of Lys 21, 86, and 101 in the catalytic mechanism of the native enzyme?

(b) Speculate on the location of Lys 21 and Lys 101. Would you expect these residues to be conserved in an evolutionary sense?

(c) Would you expect Lys 86 to be evolutionarily conserved? Why or why not?

14. You have isolated a microorganism capable of metabolizing the deoxysugar shown below. You find that the compound is phosphorylated by a specific kinase on the C-1 OH group and then the bond between C-3–C-4 is cleaved, yielding dihydroxyacetone phosphate and acetaldehyde. You also find that the cleavage is catalyzed by a zinc-containing enzyme.

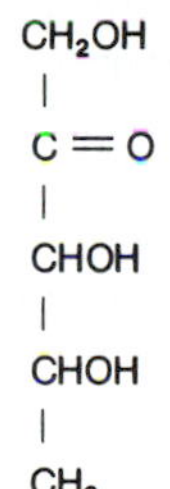

(a) The mechanism of cleavage of the sugar includes generation of a carbanion on C-3. Explain how zinc might stabilize the carbanion. (Consider the role of zinc in the carboxypeptidase A mechanism.)

(b) Would you expect the nonphosphorylated deoxysugar to be a substrate? Why or why not?

15. During catalysis, covalent chemical bonds are broken and formed. In that sense, almost all enzymes perform covalent catalysis. However, the terms "covalent" and "noncovalent" have particular meanings in an enzyme mechanism. Define the difference between those terms as they apply to catalysis.

10

CHAPTER

Regulation of Enzyme Activities

Perhaps the characteristic that most distinguishes living things from inanimate objects is their ability to control their own metabolic activities. Living cells achieve this control by regulating both the concentrations and the activities of specific enzymes. Because the great majority of enzymes are proteins, enzyme concentrations depend on a balance between the rates of protein synthesis and degradation. The synthesis of specific proteins is regulated at several steps, including the synthesis of mRNA and the translation of the RNA message into protein. Rates of protein degradation are controlled by varying the concentrations or activities of other, degradative enzymes, and by modifying particular proteins in a way that tags them for destruction. These controls on protein synthesis and degradation will be discussed in chapters 28, 29, and 30.

Because protein synthesis and turnover are relatively slow processes, adjustments of their rates are used mainly to respond to long-term changes of conditions. The levels of several enzymes that participate in fatty acid biosynthesis, for example, increase markedly within a few days after mammals are switched to a diet that is rich in carbohydrates but low in fats. These enzymes also increase for periods of weeks or months during lactation. But living organisms often need to respond quickly to more sudden changes in conditions, and they do so by regulating enzyme *activities*. This will be the focus of the present chapter.

In principle, the activities of many enzymes could be altered by changes in pH. Cells do take advantage of this possibility in a few cases. Lysozyme, for example, is most active in the pH region of 5 that is characteristic of some extracellular secretions, and is much less active in the intracellular pH region near 7. The activity of lysozyme thus remains low until the enzyme is secreted. But this is not a very practical solution to the problem of regulating the activity of an enzyme that must remain in the cell, because in most cells the intracellular pH must be held within rather narrow limits. There are two strategies that are much more widely applicable. The first is to modify the covalent structure of the enzyme in such a way as to alter either K_m or k_{cat}. The second strategy is to use an inhibitor or activator molecule, an *effector,* that binds reversibly to the enzyme and, again, alters either the K_m or k_{cat}. Such an effector may bind either at the active site itself, or at some more distant site on the enzyme. In the latter case, it is termed an allosteric effector, from the Greek *allos* ("other") and *stereos* ("solid," or "space"), and enzymes that are regulated by such effectors are called allosteric enzymes. Although allosteric effectors usually are small molecules such as ATP, some proteins are inhibited or activated when they bind to another protein.

In the following sections, we will first consider some of the general features of covalent and allosteric control mechanisms, and then turn to a more detailed discussion of how several important enzymes are regulated.

Partial Proteolysis: An Irreversible Covalent Modification

In chapter 9, we mentioned that the pancreas secretes trypsin, chymotrypsin, elastase, and the carboxypeptidases as inactive zymogens, which are activated extracellularly when other proteases cleave them at specific peptide bonds. Trypsin is activated when the intestinal enzyme enterokinase cuts off an N-terminal hexapeptide. Trypsin in turn activates chymotrypsin by cutting it at the N-terminal end between Arg 15 and Ile 16. Delaying the activation in this way prevents the digestive enzymes from destroying the pancreatic cells in which they are synthesized.

The crystal structures of chymotrypsin and its zymogen precursor chymotrypsinogen have provided an explanation for the increase in catalytic activity that results from trimming off the N-terminal end of the zymogen. The sluggishness of chymotrypsinogen can be attributed primarily to a much higher K_m for peptide substrates in the zymogen than in the active enzyme. Although the charge relay system of Asp 102, His 57, and Ser 195 has a similar structure in chymotrypsinogen and chymotrypsin, the substrate-binding pocket is not properly formed in the zymogen (fig. 10.1*a*). The NH group of Gly 193 is not in position to form a hydrogen bond with the carbonyl oxygen of the substrate. The importance of this bond for the activity of the enzyme was discussed in chapter 9 (see figs. 9.7 and 9.11). The major constraint preventing the completion of the binding pocket in the zymogen appears to be a hydrogen bond between the negatively charged carboxylate group of the neighboring residue, Asp 194, and His 40. Rotation of Gly 193 into the correct orientation occurs when the zymogen is cleaved between Arg 15 and Ile 16 and the new N-terminal —NH_3^+ group of Ile 16 forms a salt bridge with Asp 194 (see fig. 10.1*b*). A similar salt bridge between the N-terminal group and Asp 194 occurs in trypsin.

The enzymes that participate in blood clotting also are activated by partial proteolysis, and again this serves to keep them in check until they are needed. The blood-coagulation system involves a cascade of at least seven serine proteases, each of which activates the subsequent enzyme in the series (fig. 10.2). Because each molecule of enzyme that is activated can, in turn, activate many molecules of the next enzyme, initiation of the process by factors that are exposed in damaged tissue leads explosively to the conversion of prothrombin to thrombin, the final serine protease in the series. Thrombin then cuts another protein, fibrin, into peptides that stick together to form a clot.

Figure 10.1

A schematic drawing of the structural changes that occur when chymotrypsinogen is converted to chymotrypsin. In chymotrypsinogen, the carboxylate group of Asp 194 forms a salt bridge to His 40; in chymotrypsin, the bridge goes to the new N-terminal—NH_3^+ group of Ile 16. This change evidently allows Gly 193 to swing around so that its amide NH comes closer to the NH of Ser 195. The two amide groups form essential hydrogen bonds to the substrate in the enzyme-substrate complex (see figs. 9.7 and 9.11).

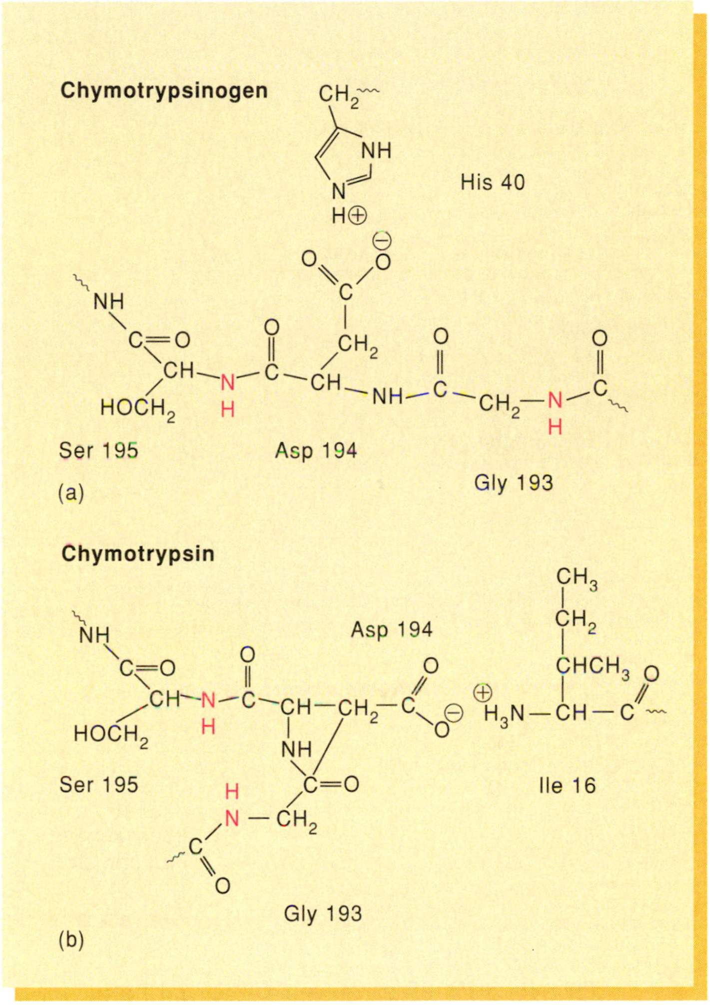

Table 10.1 lists some other enzymes that are activated by partial proteolysis. In general, the peptide bond that is cleaved is located in a loop connecting two different domains of the protein, and cutting this bond probably relieves a constraint that interferes with the formation of the active site, much as it does in chymotrypsin or trypsin.

Figure 10.2

The blood-coagulation cascade. Each of the curved red arrows represents a proteolytic reaction, in which a protein is cleaved at one or more specific sites. With the exception of fibrinogen, the substrate in each reaction is an inactive zymogen; except for fibrin, each product is an active protease that proceeds to cleave another member in the series. Many of the steps also depend on interactions of the proteins with Ca^{2+} ions and phospholipids. The cascade starts when factor XII and prekallikrein come into contact with materials that are released or exposed in injured tissue. (The exact nature of these materials is still not fully clear.) When thrombin cleaves fibrinogen at several points, the trimmed protein (fibrin) polymerizes to form a clot.

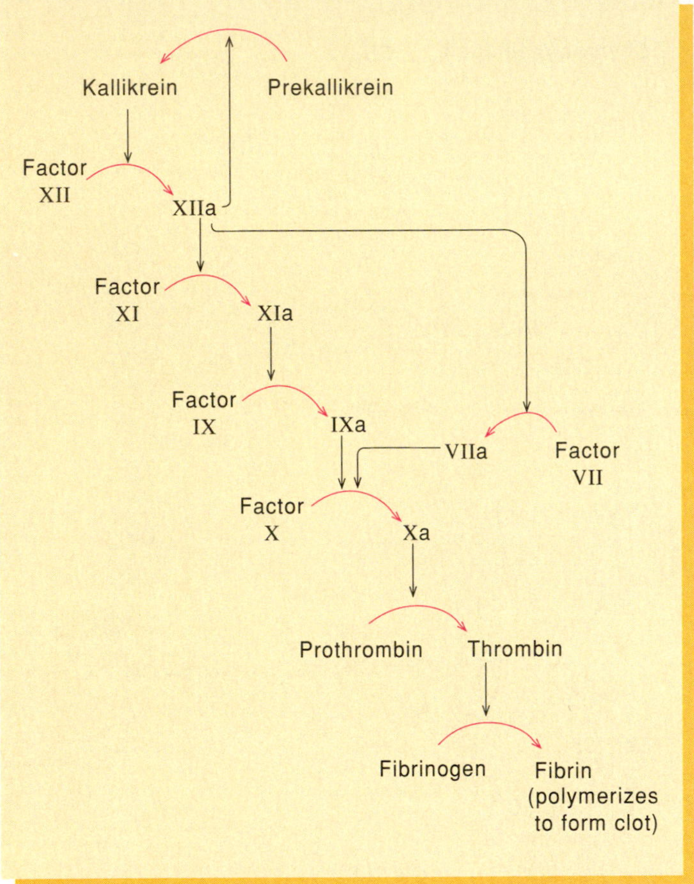

The activation of an enzyme by partial proteolysis is an irreversible process. Once the enzyme is cut, it remains active until it is degraded or inhibited by some other means. This is fine for the digestive enzymes, but it clearly raises a problem in blood coagulation: There must be a mechanism to prevent the clot from spreading away from the site of the injury and taking over the entire blood supply. Several mechanisms work to this end. The activated enzymes are diluted by the flow of the blood, degraded in the liver, and inhibited by other blood proteins that bind tightly to the active enzymes and occlude their active sites. Blood coagulation thus depends on a delicate balance between a rapid, irreversible mechanism of enzyme activation, and rapid, irreversible mechanisms for disposing of the active enzymes. We will return later to the role that specific inhibitory proteins play in maintaining this balance.

Table 10.1
Some Enzymes that Are Regulated by Partial Proteolysis

Digestive Enzymes
Trypsin, chymotrypsin, carboxypeptidase A and B, elastase, pepsin, phospholipase
Blood-Coagulation Enzymes
Factors VII, IX, X, XI, and XIII, kallikrein, thrombin
Enzymes Involved in Dissolving Blood Clots
Plasminogen, plasminogen activator
Enzymes Involved in Programmed Development
Chitin synthetase, cocoonase, collagenase

Phosphorylation, Adenylylation, and Disulfide Reduction: Reversible Covalent Modifications

A type of covalent modification that is used more frequently than partial proteolysis is phosphorylation of the side chains of serine, threonine, or tyrosine residues. Phosphorylation differs from partial proteolysis in being reversible. The introduction and removal of the phosphate group are catalyzed by separate enzymes (phosphorylation by a protein kinase, and dephosphorylation by a phosphatase), which are themselves generally under metabolic regulation (fig. 10.3).

In eukaryotic organisms, phosphorylation is used to control the activities of literally hundreds of enzymes, a few of which are listed in table 10.2. These enzymes generally are phosphorylated, or dephosphorylated, in response to *extracellular* signals such as hormones or growth factors. The adrenal hormone epinephrine, for example, is transmitted through the blood to muscle and adipose tissue when there is an immediate need for muscular exertion. On reaching the target tissue, epinephrine initiates a chain of events that lead to the activation of a protein kinase. The kinase (cAMP-dependent protein kinase) then catalyzes the phosphorylation of one or more specific enzymes, depending on the tissue. Some enzymes are activated when they are phosphorylated, and inactivated when the phosphate is removed; others are inactivated by phosphorylation. In adipose tissue, phosphorylation switches on triacylglycerol lipase, an enzyme that breaks down esters of fatty acids (triacylglycerols). In muscle, phosphorylation activates glycogen phosphorylase, an enzyme that breaks down glycogen, and it stops the synthesis of glycogen by switching off glycogen synthase.

Phosphorylation also can increase or decrease an enzyme's sensitivity to particular allosteric effectors. Phosphorylation of glycogen synthase greatly increases its sensitivity to inhibition by glucose-6-phosphate or P_i. Phosphorylation of glycogen phosphorylase makes this enzyme much less dependent

Figure 10.3

Phosphorylations of serine side chains in enzymes are catalyzed by kinases, and dephosphorylation by phosphatases. Threonine and tyrosine side chains undergo similar reactions, but these are less common than the phosphorylation of serine.

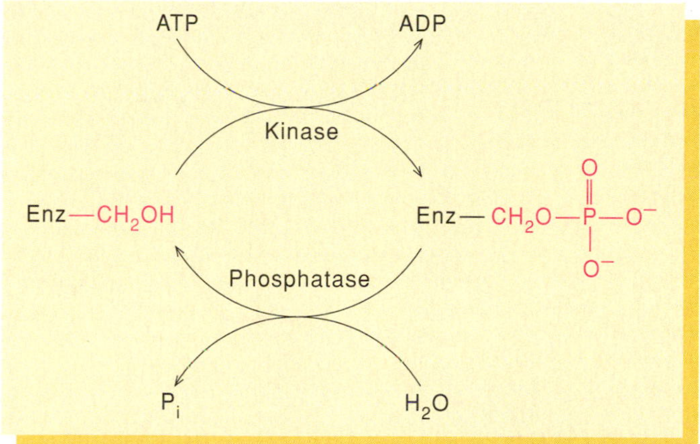

Table 10.2 Some Enzymes that Are Regulated by Phosphorylation
Enzymes of Carbohydrate Metabolism
Glycogen phosphorylase
Phosphorylase kinase
Glycogen synthase
Phosphofructokinase-2
Pyruvate kinase
Pyruvate dehydrogenase
Enzymes of Lipid Metabolism
Hydroxymethylglutaryl-CoA reductase
Acetyl-CoA carboxylase
Triacylglycerol lipase
Enzymes of Amino Acid Metabolism
Branched-chain ketoacid dehydrogenase
Phenylalanine hydroxylase
Tyrosine hydroxylase

on the allosteric activator AMP. Thus a covalent modification triggered by an extracellular signal can, in some cases, override the influence of intracellular allosteric regulators; the response is relatively insensitive to the exact concentrations of the local agents. In other cases, variations in the concentrations of intracellular allosteric effectors can cause the response to the covalent modification to vary, depending on the metabolic state of affairs in the cell.

As yet, glycogen phosphorylase is the only enzyme for which crystal structures of both the phosphorylated and the nonphosphorylated forms are known. Because glycogen phosphorylase also is regulated by allosteric effectors, we will postpone a discussion of this complex enzyme until after we have considered some simpler examples of allosteric regulation. The manner in which hormones lead to an increase in the activity of cAMP-dependent protein kinase by increasing the concentration of its allosteric effector 3′,5′-cyclicAMP (cAMP) is discussed in chapters 13 and 24.

The covalent addition of an adenylyl (adenosine monophosphate) group to a tyrosine residue is another form of reversible, covalent modification (fig. 10.4). In *E. coli,* adenylylation is used to regulate glutamine synthase, an enzyme that plays a major role in nitrogen metabolism (see chapter 18). The tyrosine residue that accepts the adenylyl group on glutamine synthase is located close to the active site. The addition of the bulky and negatively charged adenylyl group inhibits the enzyme, perhaps simply by occluding the active site.

Other groups that can be attached covalently to enzymes include fatty acids, isoprenoid alcohols such as farnesol, and carbohydrates. Although such modifications appear to be common, our understanding of how cells use them to regulate enzymatic activities is still fragmentary.

A reversible covalent modification used extensively in plants is the reduction of cystine disulfide bridges to sulfhydryls. Many of the enzymes that participate in photosynthetic carbohydrate synthesis are activated when a disulfide bond is reduced to produce two cysteines (table 10.3). Some of the enzymes involved in carbohydrate breakdown are inactivated by the same mechanism. The reductant for this reaction is a small protein called thioredoxin, which contains two nearby cysteine residues (fig. 10.5). In plants, reduced thioredoxin becomes available as a result of electron-transfer reactions that are driven by sunlight, and it serves as a signal to switch carbohydrate metabolism from carbohydrate breakdown to synthesis. In one of the enzymes that are regulated (phosphoribulokinase), the cysteines that are freed when the disulfide bridge is reduced are spaced 39 amino acid residues apart, and one of them probably forms part of the catalytic active site. In two other cases (NADP-malate dehydrogenase, and fructose-1,6-bisphosphatase), the cysteines are spaced only four or five residues apart, but both appear to be located at some distance from the catalytic site. Exactly how the structural changes that result from the reduction are transmitted to the active sites in these enzymes is not yet known.

Covalent modifications of specific enzymes allow a cell to regulate its metabolic activities more rapidly, and in much more intricate ways, than would be possible by changing the absolute concentrations of the same enzymes. They still do not provide a mechanism for a truly instantaneous response to a change in conditions, however, because each such modification requires the action of another enzyme, which must itself be subject to regulation. There also is a lag in responding to the removal or inactivation of the enzyme that causes the modification, because reversing the modification requires still another enzyme.

Figure 10.4

The transfer of the adenylyl group from ATP to a tyrosine residue is used to regulate some enzymes. The other product of the adenylylation reaction (*top*) is inorganic pyrophosphate. Hydrolysis of the adenylyl-tyrosine ester bond releases AMP (*bottom*).

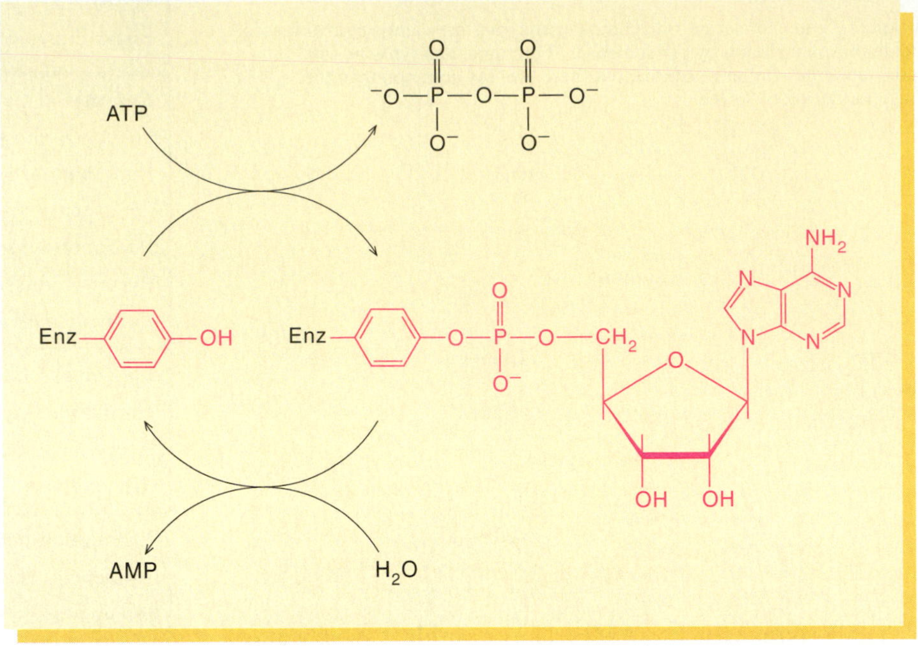

Table 10.3 Some Enzymes that Are Regulated by Disulfide-Reduction in Plants
Activated by Reduction
Fructose-1,6-bisphosphatase
Sedoheptulose-1,7-bisphosphatase
Glyceraldehyde-3-phosphate dehydrogenase
NADP-malate dehydrogenase
Phosphoribulokinase
Thylakoid ATP-synthase
Inhibited by Reduction
Phosphofructokinase

Figure 10.5

Reduction of the disulfide bond of cystine is used to activate enzymes of photosynthetic carbohydrate biosynthesis in plants. The reductant is a small protein called thioredoxin. Thioredoxin also serves as a reductant for the biosynthesis of deoxynucleotides in animals and microorganisms as well as in plants.

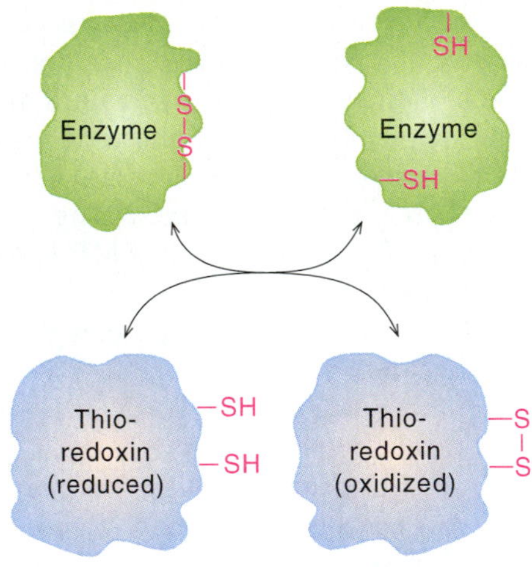

Allosteric Regulation

Whereas eukaryotic organisms generally use phosphorylation to handle responses to extracellular signals, both prokaryotic and eukaryotic organisms commonly use allosteric regulation in responding to changes in conditions within a cell. A typical circumstance that might demand such a response would be a surplus or deficit of ATP, a particular amino acid, or some other metabolic intermediate that the cell is equipped to synthesize or consume. Allosteric regulation enables a cell to adjust a specific enzymatic activity almost instantaneously in response to a change in the concentration of a particular metabolic intermediate because, unlike covalent modification, it does not require the action of an intermediate enzyme. If the intermediate acts as an allosteric effector, the activity of the enzyme can increase (or decrease) as soon as the concentration of the effector rises, and can decrease (or increase) again as soon as the concentration falls.

Regulation of enzymes by allosteric effectors is considerably more common than regulation by compounds that bind directly at the active site. There probably are two main reasons for this. First, whereas an agent that binds at the active site will usually act as an inhibitor, a compound that binds at an allosteric site can act as either an inhibitor or an activator, depending on the structure of the enzyme. Second, a substance that binds to an allosteric site does not need to have any structural relationship to the substrate. Consider, for example, the metabolic pathway of histidine biosynthesis in plants and bacteria. The pathway requires nine enzymes that work one after another. If histidine is already present in abundant supply, it is advantageous for a cell to cut off the entire pathway at the first step to avoid wasting energy or accumulating the products of the intermediate steps. The first step in the pathway is the reaction between phosphoribosyl pyrophosphate and ATP, neither of which even vaguely resembles histidine, and yet the enzyme that catalyzes this step is strongly inhibited by histidine. Binding of histidine to an allosteric site on the enzyme causes a structural change that is transmitted to the active site.

The control of the pathway leading to histidine illustrates a recurrent theme in enzyme regulation: Enzymes that are regulated often occupy key positions in metabolic pathways. Typical control points are the first step of a pathway, or the first step of a branch leading to an alternate product. Many key enzymes are regulated by ATP, ADP, AMP, or P_i. The relative concentrations of these materials provide a cell with an index as to whether energy is abundant or in short supply. Because ATP, ADP, AMP, and P_i often are chemically unrelated to the substrate of the enzyme that must be regulated, they generally bind to an allosteric site rather than to the active site.

The Kinetics of Allosteric Enzymes Typically Exhibit a Sigmoidal Dependence on Substrate Concentration

In chapter 5 we saw that the binding of glycerate-2,3-bisphosphate to hemoglobin causes a decrease in the affinity of the protein for O_2, and that the binding of O_2 to any one of the four subunits increases the affinity of the other subunits for O_2. These effects can be related to cooperative changes in the tertiary and quaternary structure of the protein. The binding of O_2 alters the interactions between the different subunits in such a way that the entire protein tends to flip into a state with increased O_2 affinity; binding of glycerate-2,3-bisphosphate favors a transition in the opposite direction. When Jacques Monod, Jeffreys Wyman, and Jean-Pierre Changeux first advanced the idea of allosteric enzymes in 1963, they suggested that these enzymes might often contain multiple subunits, and that the changes in catalytic activity resulting from the binding of allosteric effectors might reflect alterations in quaternary structure. This suggestion has turned out to be remarkably accurate. Although there is no reason why an enzyme consisting of a single subunit cannot be sensitive to allosteric effectors, most of the enzymes that are regulated in this way do have multiple subunits, and in many cases the changes in enzymatic activity can be related to interactions among the subunits.

One indication that allosteric effectors often involve cooperative interactions within multisubunit proteins is that many allosteric enzymes do not obey the classical Michaelis-Menten kinetic equation. A plot of the rate of reaction as a function of substrate concentration is not hyperbolic, as described by the Michaelis-Menten equation, but rather sigmoidal, resembling the curve for the cooperative binding of O_2 to hemoglobin (see fig. 5.3). Further, allosteric effectors often cause the kinetics to change from one of these forms to the other, much as glycerate-2,3-bisphosphate, bicarbonate, and protons affect the degree of cooperativity in the binding of O_2 to hemoglobin. Figure 10.6 shows an illustration of these effects for phosphofructokinase, an enzyme that catalyzes the formation of fructose-1,6-bisphosphate by transferring a phosphate group from ATP to fructose-6-phosphate:

Figure 10.6

Kinetics of the reaction catalyzed by phosphofructokinase. In the presence of 1.5 mM ATP, the rate has a sigmoidal dependence on the concentration of the substrate fructose-6-phosphate. Although ATP also is a substrate for the reaction, the sigmoidal kinetics seen under these conditions are associated with the binding of ATP to an inhibitory allosteric site. The kinetics become hyperbolic if a low concentration of AMP is added.

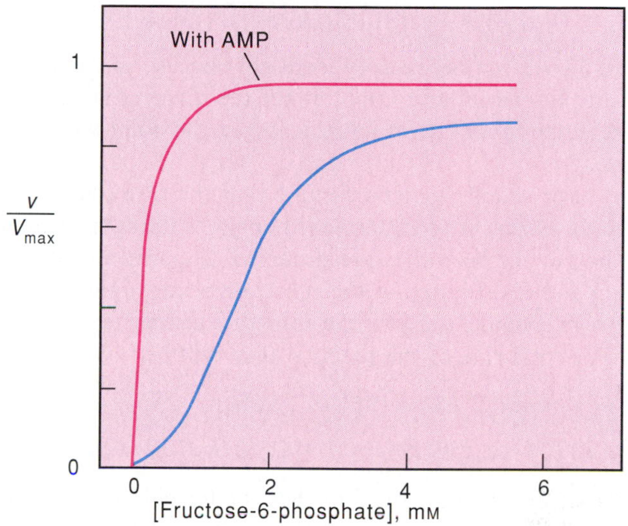

$$\text{Fructose-6-phosphate} + \text{ATP} \xrightarrow{\text{phosphofructokinase}} \text{fructose-1,6-bisphosphate} + \text{ADP} \quad (1)$$

In the presence of 1.5-mM ATP, the kinetics have a sigmoidal dependence on the concentration of the substrate fructose-6-phosphate; at very low concentrations of ATP, or in the presence of the allosteric effector AMP, the kinetics become hyperbolic.

Phosphofructokinase has four identical subunits. To explore how sigmoidal kinetics can arise in such an enzyme, consider an enzyme that has just two such subunits, each with its own catalytic active site. The binding of a substrate to the enzyme can be schematized as follows:

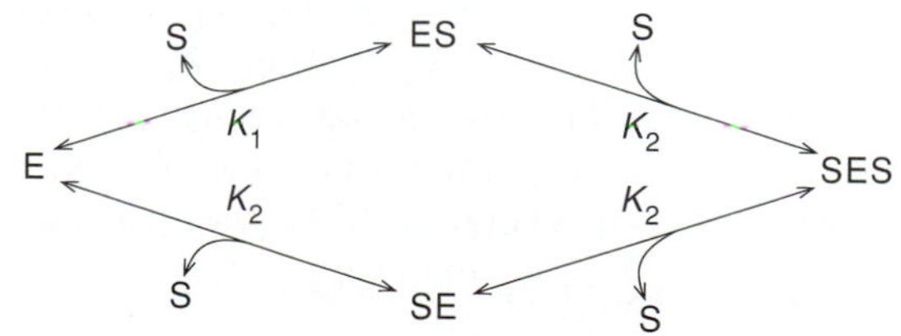

Here ES and SE represent the binary complexes of the enzyme (E) with substrate (S) on the two different subunits, and SES represents the ternary complex in which both binding sites are occupied. Using the dissociation constants K_1 and K_2, we can relate the concentrations of any of these complexes to the concentrations of the free enzyme and substrate:

$$[\mathrm{ES}] = [\mathrm{SE}] = [\mathrm{E}][\mathrm{S}]/K_1 \tag{2}$$

$$[\mathrm{SES}] = [\mathrm{ES}][\mathrm{E}]/K_2 = [\mathrm{SE}][\mathrm{S}]/K_2 = [[\mathrm{E}][\mathrm{S}]/K_1][[\mathrm{S}]/K_2] \tag{3}$$

K_1 and K_2 are not necessarily identical because the binding of substrate to one subunit could affect the dissociation constant for the other subunit. In fact, this is just the point we want to explore.

For simplicity, let's assume that the rate constant k_{cat} for the formation of products at either site is the same, whether only that site or both sites are occupied. Suppose also that the binding and dissociation of the substrate occur rapidly relative to the conversion of the enzyme-substrate complexes into products. The total rate of the reaction then will be

$$\begin{aligned} v &= k_{cat}\{[\mathrm{ES}] + [\mathrm{SE}] + 2[\mathrm{SES}]\} \\ &= 2k_{cat}\{[\mathrm{E}][\mathrm{S}]/K_1 + [[\mathrm{E}][\mathrm{S}]/K_1][[\mathrm{S}]/K_2]\} \end{aligned} \tag{4}$$

The factor of 2 reflects the fact that the formation of products occurs independently (and with the same rate constant) at the two sites. Because the total enzyme concentration $[\mathrm{E_T}]$ is [E] + [ES] + [SE] + [SES], the maximum rate of the reaction is

$$\begin{aligned} V_{max} = 2k_{cat}[\mathrm{E_T}] = 2k_{cat}\{&[\mathrm{E}] + [\mathrm{E}][\mathrm{S}]/K_1 \\ &+ [\mathrm{E}][\mathrm{S}]/K_1 + [[\mathrm{E}][\mathrm{S}]/K_1][[\mathrm{S}]/K_2]\} \end{aligned} \tag{5}$$

Combining equations (4) and (5) gives

$$\frac{v}{V_{max}} = \frac{\dfrac{[\mathrm{S}]}{K_1}\left[1 + \dfrac{[\mathrm{S}]}{K_2}\right]}{\left[1 + \dfrac{[\mathrm{S}]}{K_1}\right] + \dfrac{[\mathrm{S}]}{K_1}\left[1 + \dfrac{[\mathrm{S}]}{K_2}\right]} \tag{6}$$

If the binding of S to one subunit does *not* affect binding to the other, then $K_2 = K_1$, and equation (6) reduces to

$$\frac{v}{V_{max}} = \frac{[\mathrm{S}]}{K_d + [\mathrm{S}]} \tag{7}$$

where $K_d = K_1 = K_2$. This is simply the Michaelis-Menten equation for the limiting case in which substrate release is much faster than the conversion of the enzyme-substrate complex into products. (See equations 24 and 28 in chapter 8.) A plot of v versus [S] according to equation (7) is hyperbolic. Such a plot is shown again as the solid curve in figure 10.7. The dashed curve in figure 10.7 is a similar plot of equation (6) for the case $K_2 = K_1/25$. To facilitate comparison with the solid curve, K_1 is taken to be $5K_d$ and K_2 to be $K_d/5$, where K_d has the same value for the two curves. This means that the overall free energy change for the formation of the ternary complex SES from E + 2S is the same. (The overall equilibrium constant for this process is $(1/K_1)(1/K_2)$, or $1/K_d^2$. Remember that the standard free energy change is $-2.3RT$ times the $\log_{10}$ of the equilibrium constant, where T is the temperature and $R = 2$ cal/mole/degree.) Note that the dashed curve is decidedly sigmoidal in shape. It starts out rising more slowly than the solid curve, rises steeply in the region of $[\mathrm{S}] \approx K_d$, where $v/V_{max} \approx 0.5$, and then continues rising more slowly.

Figure 10.7

Kinetics of an enzyme with two identical subunits. The curves were calculated with equation (6). (Substrate dissociation is assumed to be rapid relative to the catalytic step.) The abscissa is the ratio of the substrate concentration [S] to K_d, where K_d is the geometric mean of K_1 and K_2 ($K_d = \sqrt{K_1K_2}$). K_d is taken to be the same for all three curves. For the solid curve, the dissociation constants K_1 and K_2 for the first and second molecule of substrate were assumed to be identical ($K_1 = K_2 = K_d$); equation (6) then reduces to equation (7). For the dashed curve, $K_2 = K_1/25$ ($K_1 = 5K_d$ and $K_2 = K_d/5$). For the dotted curve, $K_2 = 10^{-4}K_1$ ($K_1 = 100K_d$ and $K_2 = K_d/100$); equation (6) then reduces to equation (8). The three curves intersect at the point where $[\mathrm{S}] = K_d$ and $v = V_{max}/2$.

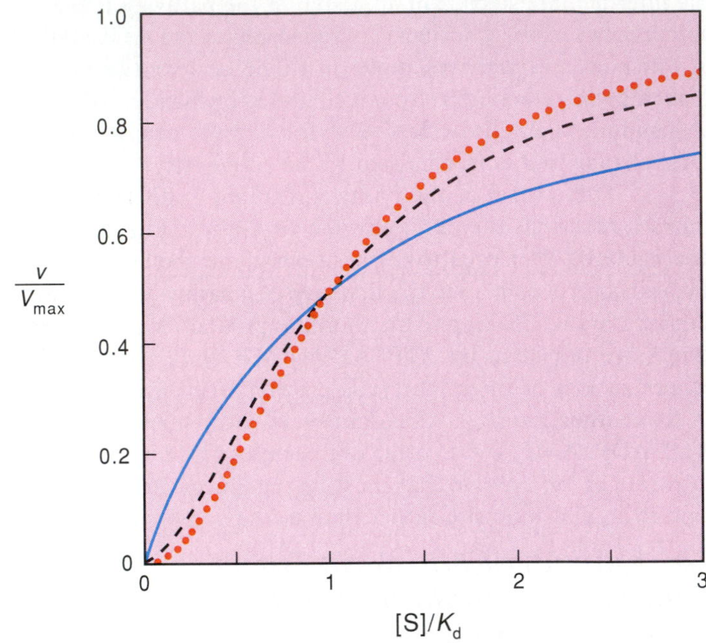

The ratio of 25 between the dissociation constants K_1 and K_2 used in this illustration means that the standard free energy change for binding the second molecule of substrate to the enzyme is only 1.9 kcal/mole more favorable than that for binding the first molecule. This is the order of magnitude of the free energy change associated with forming a single hydrogen bond in the interior of a protein. Evidently, a sigmoidal kinetic curve similar to the dashed curve in figure 10.7 might be obtained if binding the substrate caused a relatively minor change in the conformation of the protein. In chapter 9, we noted that the binding of a substrate or coenzyme to hexokinase, carboxypeptidase A, or lactate dehydrogenase results in pronounced conformational changes that can be seen in the crystal structures. But in order to yield sigmoidal kinetics, it is essential that events occurring at two different binding sites for the substrate be coupled: Binding at one site must cause a decrease in the

Figure 10.8

Symmetry model for allosteric transitions of a dimeric enzyme. The model assumes that the enzyme can exist in either of two different conformations (T and R), which have different dissociation constants for the substrate (K_T and K_R). Structural transitions of the two subunits are assumed to be tightly coupled, so that both subunits must be in the same state. L is the equilibrium constant (T)/(R) in the absence of substrate. If the substrate binds much more tightly to R than to T ($K_R << K_T$), the binding of a molecule of substrate to either subunit will pull the equilibrium between T and R in the direction of R. Because both subunits must go to the R state, the binding of the second molecule of substrate will be promoted.

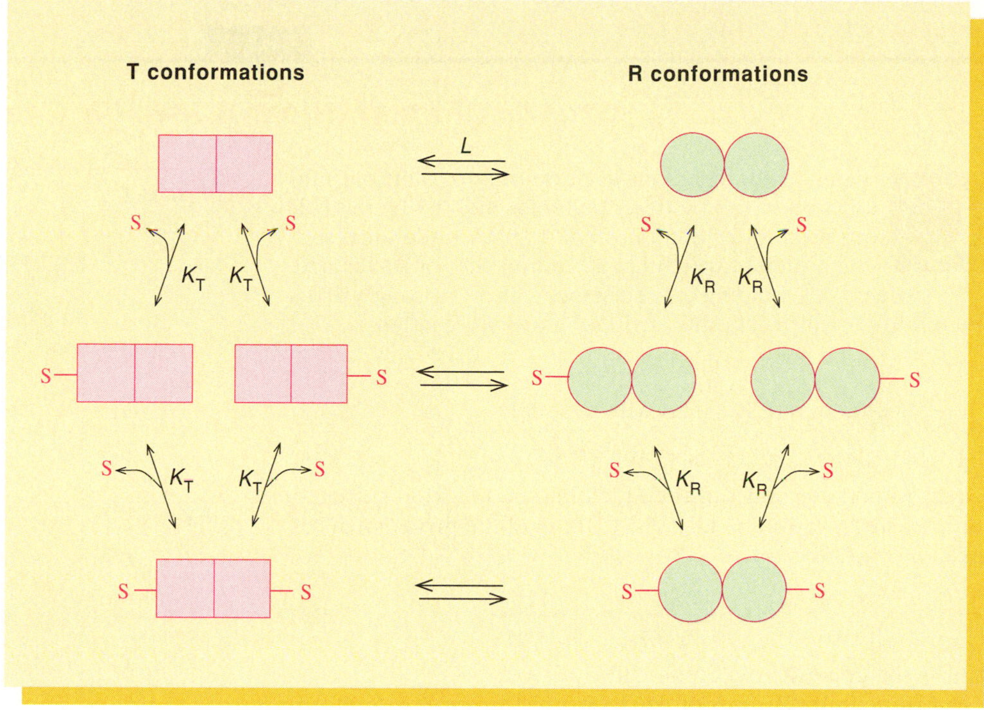

dissociation constant at the other site. If there is no such coupling, the kinetics will follow equation (7) and will be hyperbolic even if the enzyme does have multiple subunits. This situation is observed with many enzymes, including several that we discussed in chapter 9. Thus, although lactate dehydrogenase exists as a tetramer of four identical subunits, and triosephosphate isomerase exists as a dimer, the kinetics in both cases are hyperbolic.

The dotted curve in figure 10.7 shows a plot of equation (6) with $K_2 = 10^{-4} K_1$, which is equivalent to a difference of 5.5 kcal/mole between the $\Delta G°$ values for binding the first and second molecules of substrate. When the ratio of the dissociation constants is this large, the kinetics can be described equally well by the simpler expression

$$\frac{v}{V_{max}} = \frac{[S]^2}{K_1K_2 + [S]^2} \tag{8}$$

This is the limiting form of equation (6) for the situation $K_2 << [S] << K_1$. It is essentially the same as the equation that we used in chapter 5 to describe the cooperative binding of O_2 to hemoglobin, except that the exponent for [S] was larger than 2 in that case because the binding of O_2 to hemoglobin involves cooperative interactions among four subunits. The general equation of this form, with an arbitrary exponent n, is called the Hill equation, and the exponent is referred to as the Hill coefficient (see box 10A).

Although we have considered an enzyme with just two subunits, it is not uncommon for enzymes that are regulated allosterically to have four or even more subunits. The active form of acetyl-CoA carboxylase, for example, consists of linear strings of ten or more identical subunits. In general, the larger the number of subunits that interact in a manner such that the binding of substrate to one subunit promotes binding to others, the more steeply the enzyme's kinetics will depend on [S] in the region where $v/V_{max} \approx 0.5$. This point is developed in more detail in box 10A. We will return to sigmoidal enzyme kinetics in chapter 12, where we consider their implications for the regulation of metabolic pathways. At the moment, our concern is to explore what types of structural changes can couple the binding of substrates on different subunits of a protein and how this cooperativity can be modified by allosteric effectors.

The Symmetry Model Provides a Useful Framework for Relating Conformational Transitions to Allosteric Activation or Inhibition

Several theoretical models have been developed for relating changes in dissociation constants to changes in the tertiary and quaternary structures of oligomeric proteins. Two such models were discussed in chapter 5 in connection with the oxygenation of hemoglobin (see fig. 5.15), and one of them is shown again in a slightly different form in figure 10.8. The basic idea is that the protein's subunits can exist in either of two distinct conformations, R and T. The substrate is assumed to bind more tightly to the R form than to the T form, which means that binding of the substrate favors the transition from the T conformation to R. The conformational transitions of the individual subunits are assumed to be tightly linked, so that if one subunit flips from T

10A
BOX

Graphical Evaluation of the Hill Coefficient

Cooperative interactions in proteins with multiple subunits often can be described conveniently by the Hill equation. Consider a protein, E, with n identical subunits, each of which has a binding site for a substrate, S. Suppose that the binding of the first molecule of S strongly favors the binding to all n subunits, giving the overall reaction

$$\mathrm{E} + n\,\mathrm{S} \rightleftarrows \mathrm{ES}_n \tag{B1}$$

The concentration of ES_n then will be simply

$$[\mathrm{ES}_n] = [\mathrm{E}][\mathrm{S}]^n/K_\mathrm{h} \tag{B2}$$

where K_h is the product of the individual dissociation constants for all n steps leading to ES_n. The fraction of the protein that has taken up the substrate is

$$y \approx \frac{[\mathrm{ES}_n]}{[\mathrm{E}] + [\mathrm{ES}_n]} = \frac{[\mathrm{S}]^n}{K_\mathrm{h} + [\mathrm{S}]^n} \tag{B3}$$

This is the general form of the Hill equation.

Figure 1 shows plots of equation (B3) for $n = 1$, 2, and 4. The curve for $n = 1$ has the same hyperbolic shape as the solid curve in figure 10.7, and the curve for $n = 2$ has essentially the same sigmoidal shape as the dotted curve in that figure. Increasing the value of n to 4 makes the curve rise still more steeply in the region where $[\mathrm{S}] \approx K_\mathrm{h}$.

The Hill equation can be rearranged to

$$\frac{y}{1-y} \approx \frac{[\mathrm{S}]^n}{K_\mathrm{h}} \tag{B4}$$

By taking the logarithm of both sides of this expression we find

$$\log\frac{y}{1-y} \approx n \log[\mathrm{S}] - \log K_\mathrm{h} \tag{B5}$$

A plot of $\log[y/(1-y)]$ versus $\log[\mathrm{S}]$ thus should give a straight line with a slope of n. A set of such plots for $n = 1$, 2, and 4 is shown in figure 2. For an enzymatic reaction in which binding and dissociation of the substrate are rapid relative to the catalytic step, the quantity y is equivalent to v/V_{max}.

Equation (B3), like equation (8), is an approximation and not an exact expression; the denominator neglects the concentrations of the intermediate species that have some, but not all, of the binding sites occupied. However, plots like that shown in illustration 2 often are used to find an effective value of the Hill coefficient as a phenomenological measure of the cooperativity in the binding. The binding of O_2 to hemoglobin, for example, can be described reasonably well by taking n to be approximately 2.8. This value is obtained by plotting $\log[y/(1-y)]$ versus $\log[\mathrm{S}]$ and taking the slope at the point where $y = 0.5$. In the case of phosphofructokinase, which has four subunits, the dependence of the rate on the concentration of fructose-6-phosphate at 1 mM ATP is described well by the Hill equation with $n \approx 3.8$.

Figure 1

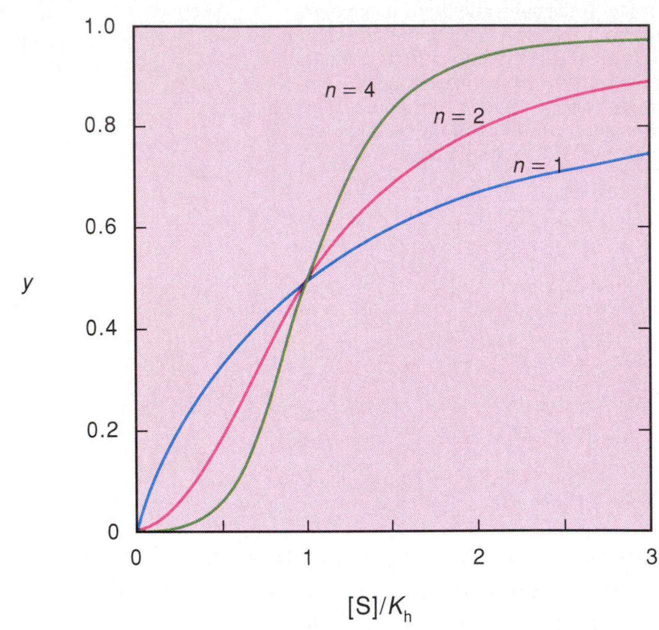

Figure 2

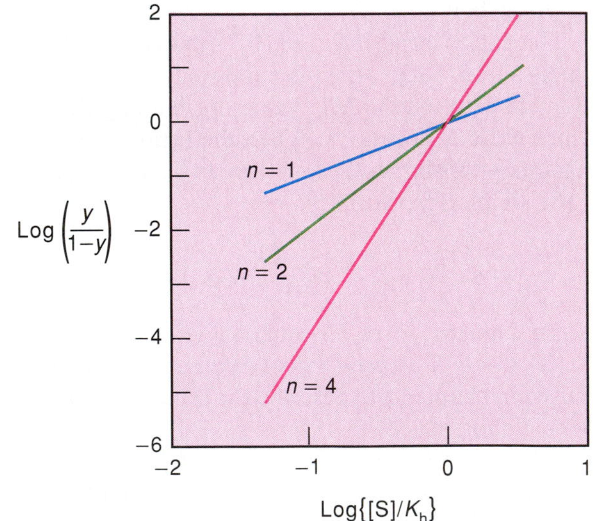

to R the others must do the same. The binding of the first molecule of substrate thus promotes the binding of the second. Because the concerted transition of all of the subunits from T to R or back preserves the overall symmetry of the protein, this model is called the symmetry model. The symmetry model can be elaborated to include allosteric activators by assuming that these compounds also bind preferentially to the R state and thus stabilize the conformation that is more effective at binding the substrate. Allosteric inhibitors would act by stabilizing the T state.

Given the symmetry model, the fraction of the binding sites that will be occupied at any given substrate concentration can be described mathematically with an expression that includes the substrate dissociation constants K_R and K_T for the two conformations and the equilibrium constant between the T and R conformations in the absence of substrate, $L = (T)/(R)$. Thus the symmetry model attempts to explain the difference between the two dissociation constants K_1 and K_2 in equation (6) by introducing a third independent parameter.

Considering that equation (6) would fit the experimental data for a dimeric enzyme satisfactorily with only two parameters, what do we gain by using a more complicated equation? The usefulness of the symmetry model is that it can easily be generalized to enzymes that have a larger number of subunits. Although the general expression for the extent of binding is more complex than equation (6), it still contains only four independent parameters: K_R, K_T, L, and the total number of subunits (n). But this simplicity comes with a cost: We have to assume that the entire oligomeric protein is restricted to just two different conformational states. Would it not be more realistic to assume that the protein also can adopt a variety of intermediate states? The answer depends in part on the particular protein, and in part on our goals in using any type of model. Models that allow additional conformational states have been developed, but they of course require even more independent parameters. The sequential model shown in figure 5.15*b* is one of these more elaborate models. Although the sequential model is more general than the symmetry model, and thus is probably more realistic, the available experimental data in most cases do not justify a distinction between the two.

The symmetry model is useful even if it does oversimplify the situation, because it provides a conceptual framework for discussing the relationships between conformational transitions and the effects of allosteric activators and inhibitors. In the following sections we will consider three oligomeric enzymes that are under metabolic control, and we will see that substrates and allosteric effectors do tend to stabilize each of these enzymes in one or the other of two distinctly different conformations.

Phosphofructokinase: Allosteric Control of Glycolysis Is Consistent with the Symmetry Model

Phosphofructokinase catalyzes the transfer of a phosphate group from ATP to the —OH group on carbon 1 of fructose-6-phosphate (equation 1). This is the major site of regulation of glycolysis, the metabolic pathway by which glucose breaks down to pyruvate (see chapter 13). As we saw in figure 10.6, the kinetics of phosphofructokinase are strongly cooperative with respect to the substrate fructose-6-phosphate. The kinetics are noncooperative with respect to the other substrate, ATP, at low concentrations, but at concentrations in the range of 0.5 to 1 mM or higher, ATP acts as an inhibitor. The inhibitory effect results from binding to an allosteric site that is distinct from the substrate-binding site for ATP. Phosphofructokinase also is inhibited by phosphoenolpyruvate and by citrate, a key intermediate in two other metabolic pathways that embark from pyruvate. On the other hand, it is *stimulated* by ADP, AMP, GDP, cAMP, fructose-2,6-bisphosphate, and a variety of other compounds, depending to some extent on the organism from which the enzyme is purified. Most if not all of the activators bind to the same allosteric site as ATP, and may work simply by preventing the inhibitory effect of ATP. The effects of ATP, ADP, and AMP are such that phosphofructokinase is restrained when the cell's needs for energy have been satisfied, as reflected in a high ratio of [ATP] to [ADP] and [AMP], and is unleashed when additional energy is needed. The effects of cAMP and fructose-2,6-bisphosphate relate to the hormonal control of carbohydrate metabolism in higher organisms, and will be discussed further in chapters 13 and 24. Our focus here will be on how fructose-6-phosphate and some of the major allosteric effectors alter the structure of the enzyme.

Phosphofructokinase was one of the first enzymes to which Monod and his colleagues applied the symmetry model of allosteric transitions. It contains four identical subunits, each of which has both an active site and an allosteric site. The very high cooperativity of the kinetics suggests that the enzyme can adopt two different conformations (T and R) that have similar affinities for ATP but differ markedly in their affinity for fructose-6-phosphate. The binding for fructose-6-phosphate is calculated to be about 2,000 times tighter in the R conformation than in T. When fructose-6-phosphate binds to any one of the subunits, it appears to cause all four subunits to flip from the T conformation to R, just as the symmetry model specifies. The allosteric effectors ADP, GDP, and phosphoenolpyruvate do not alter the maximum rate of the reaction, but change the dependence of the rate on the fructose-6-phosphate concentration in a manner suggesting that they change the equilibrium constant (L) between the T and R conformations.

Philip Evans and his co-workers have determined the crystal structures of phosphofructokinase from two species of bacteria, *E. coli* and *Bacillus stearothermophilus*. By crystallizing the enzyme in the presence and absence of the substrate and several allosteric effectors, they obtained detailed views of both the T and R conformations. This work led to an explanation of why phosphofructokinase appears to be constrained largely to all-or-nothing transitions between these two states, rather than adopting a series of intermediate conformations.

Figure 10.9 shows the crystal structure of one of the four identical subunits of phosphofructokinase from *B. stearothermophilus*. In the complete enzyme, the subunits are disposed symmetrically about three mutually perpendicular axes.

Figure 10.9

Structure of phosphofructokinase from *Bacillus stearothermophilus*. The figure shows the α-carbon chain of one of the four identical subunits of the enzyme. It is based on the crystal structure described by P. R. Evans and P. J. Hudson. The enzyme was crystallized in the R conformation in the presence of the substrate fructose-6-phosphate (yellow) and the allosteric activator ADP (blue). A molecule of ADP at the catalytic site is shown in red. Mg^{2+} ions bound to the ADP molecules are in white.

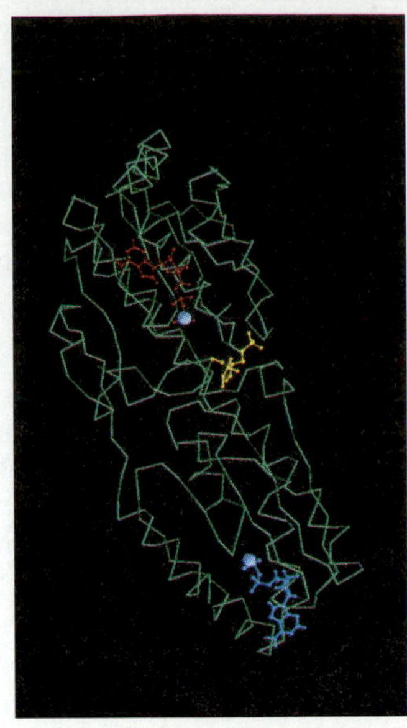

Each of these axes is a twofold symmetry axis, which means that rotating the entire structure by one-half of a full circle around the axis results in an identical structure. This effect is shown diagrammatically in figure 10.10. Because ADP is a product of the enzymatic reaction as well as an allosteric activator, it binds at both the catalytic and allosteric sites. The locations of both sites can be seen in figure 10.9 and are indicated in figure 10.10. In each subunit, the catalytic site for fructose-6-phosphate is at the interface of the subunit with one of its neighbors, and the allosteric site is at the interface with a different neighbor.

In the transition between the T and R conformations, the four subunits rotate by about 7° with respect to each other (see fig. 10.10). This rotation is associated with coupled rearrangements of the structures at the interfaces between adjacent subunits. Figure 10.11 shows how these rearrangements affect the binding site for fructose-6-phosphate. The most significant structural change in this region is an inversion of the orientation of the side chains of Glu 161 and Arg 162. In the R structure (fig. 10.11*b*), Arg 162 forms a hydrogen bond to the phosphate

Figure 10.10

Outlines of phosphofructokinase in the T (solid lines) and R (dashed lines) conformations. The enzyme contains four identical subunits (A, B, C, and D). The locations of the catalytic and allosteric sites are indicated in the two subunits closest to the viewer (A and D). The binding sites for fructose-6-phosphate (F6P) are at the interface of these subunits; the allosteric sites are at the interfaces of A with B, and of D with C. Two of the three perpendicular symmetry axes are labeled *p* and *q*. A 180° rotation about axis *q* interchanges the positions of subunits A and C, and also interchanges B and D. A similar rotation about *p* interchanges A with D, and B with C. (Source: T. Schirmer and P. R. Evans, "Structural basis of the allosteric behaviour of phosphofructokinase" in *Nature*, 343:140, 1990. Copyright © 1990 Macmillan Magazines Ltd., London, England.)

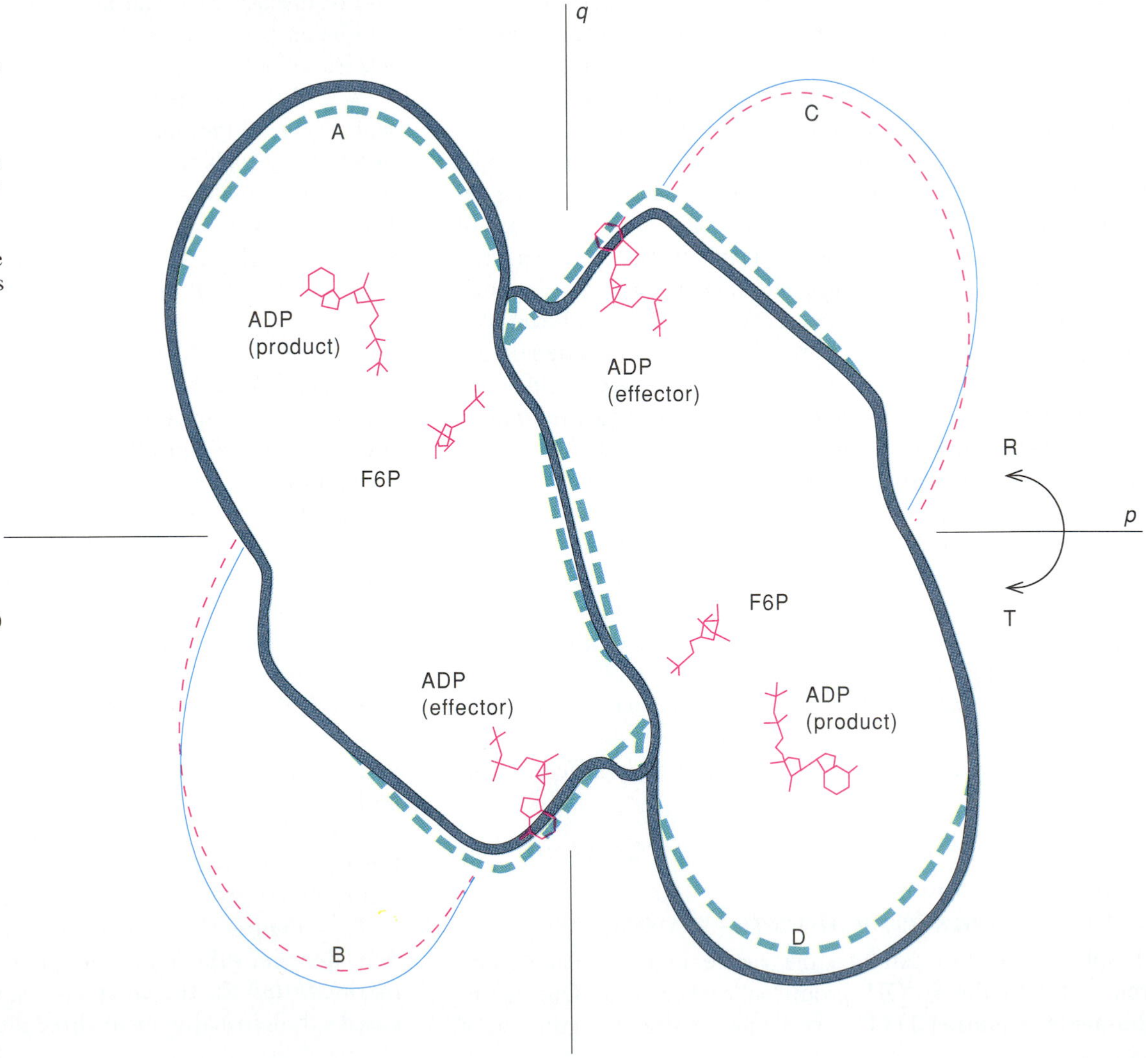

Figure 10.11

Interface between subunits A and D of phosphofructokinase near the catalytic site in (*a*) the T and (*b*) the R structures. Crystals of the enzyme in the R state were obtained in the presence of fructose-6-phosphate and ADP (see fig. 10.9); crystals in the T state were obtained in the presence of a nonphysiological allosteric inhibitor, 2-phosphoglycolate. The wavy line represents part of the boundary between subunits A and D. The heavy line indicates the polypeptide backbone. The side chains of Glu 161 and Arg 162 are shown in color. Note the inversion of the positions of these side chains in the two structures. (Source: T. Schirmer and P. R. Evans, "Structural basis of the allosteric behaviour of phosphofructokinase" in *Nature* 343:140, 1990. Copyright © 1990 Macmillan Magazines Ltd., London, England.)

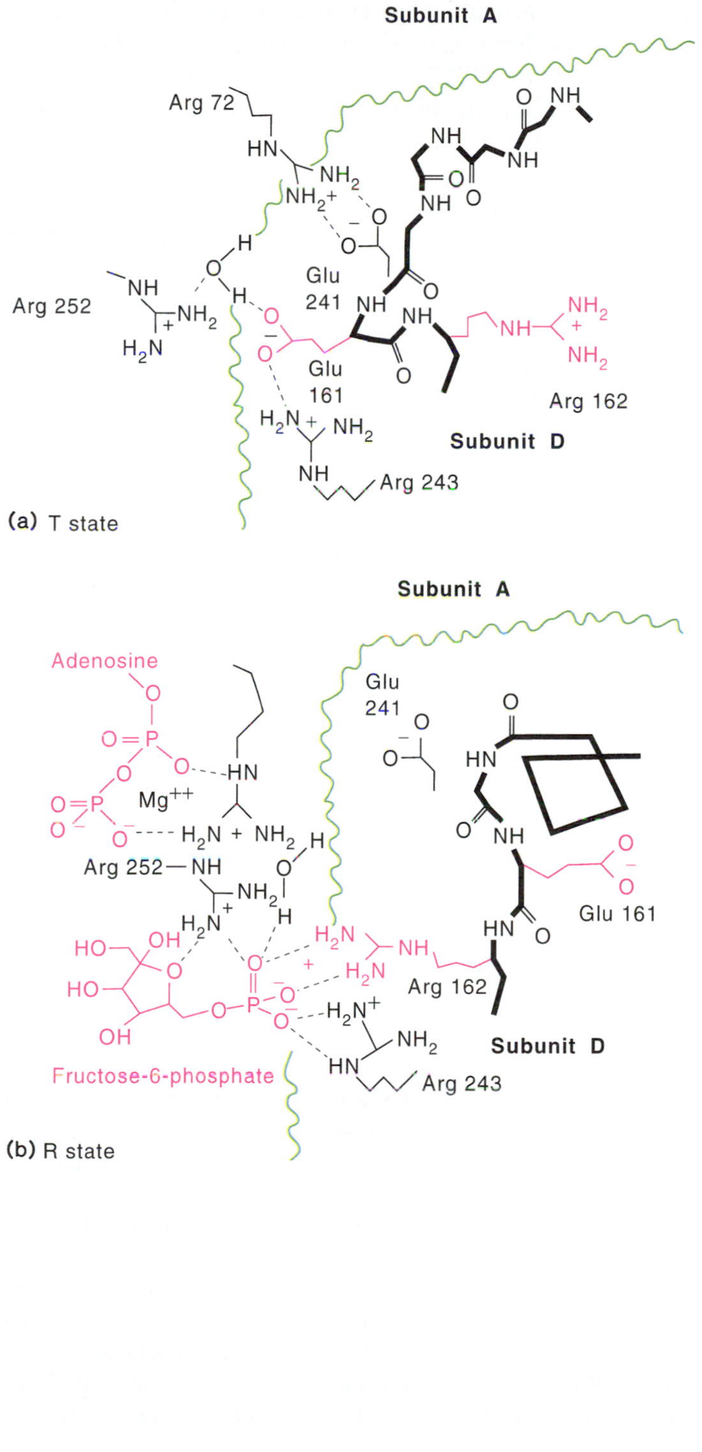

group of fructose-6-phosphate, while Glu 161 points in the opposite direction. In the T structure (fig. 10.11*a*), Arg 162 points away from the binding site, while Glu 161 inserts a negative charge into the site, where it forms a hydrogen bond with Arg 243. The change in the orientations of the negatively charged Glu 161 and the positively charged Arg 162 probably accounts for most of the difference between the dissociation constants for fructose-6-phosphate in the two states. Note that although the molecule of fructose-6-phosphate that binds in the R state is located on subunit A in figure 10.11*b*, Glu 161 and Arg 162 are residues of subunit D. Arginines 252 and 243 also contribute to the binding site from opposite sides of the boundary. Structural changes that occur on one of the subunits thus are intricately linked to changes on the other. The substrate binding site for ATP, on the other hand, is made up of residues from only one subunit (subunit A in figure 10.11). This may explain why the binding of fructose-6-phosphate to the enzyme is strongly cooperative, whereas the binding of ATP as a substrate is not cooperative.

In the T structure, Glu 161 and Arg 162 are located at the end of a stretch of polypeptide that winds up into a helical turn in the transition to the R structure (see fig. 10.11). This coiling is linked to a significant structural change that occurs in an adjacent region of the interface between the A and D subunits. The interface here includes a pair of antiparallel β strands, each of which is hydrogen-bonded to a parallel strand in its own subunit, as shown in figure 10.12. In the T structure (fig. 10.12*a*), the two antiparallel β strands from the different subunits are hydrogen-bonded together directly across the interface. In the R structure (fig. 10.12*b*), the strands have moved apart, and the region between them is filled by a row of six hydrogen-bonded water molecules.

The insertion of water at the interface between subunits A and D is an essential component of the rotation of the subunits with respect to each other, and it would appear to be an all-or-none effect. Intermediate conformations in which only some of the water molecules are present would have a less extensive network of hydrogen bonds, and thus probably would be less stable than either the R or the T conformation. The same might be said of the winding of the helical turn between residues 155 and 161; intermediates in which the helical turn is partially unwound probably would be destabilized by steric crowding, or would force the structure to expand in a way that would leave empty spaces in other regions. These considerations, taken with the close coupling between the individual subunits, seem to explain why all of the subunits undergo concerted transitions from the R to the T state or back, without giving appreciable concentrations of intermediate states.

Although the crystal structures explain why the binding of fructose-6-phosphate is strongly cooperative, they leave unclear why different allosteric effectors tend to stabilize the protein in different conformational states. The allosteric binding site, like the site for fructose-6-phosphate, lies at the interface of different subunits, but the binding of either an activator or

Figure 10.12

Hydrogen bonds of the peptide backbone and the side chain of threonine 245 at the interface between subunits A and D of phosphofructokinase, in (*a*) the T and (*b*) the R structures. Note the additional molecules of water between the two subunits in the R structure. The portions of the polypeptide chains shown here are almost contiguous with those shown in figure 10.11. (Source: T. Schirmer and P. R. Evans, "Structural basis of the allosteric behaviour of phosphofructokinase" in *Nature* 343:140, 1990. Copyright © 1990 Macmillan Magazines Ltd., London, England.)

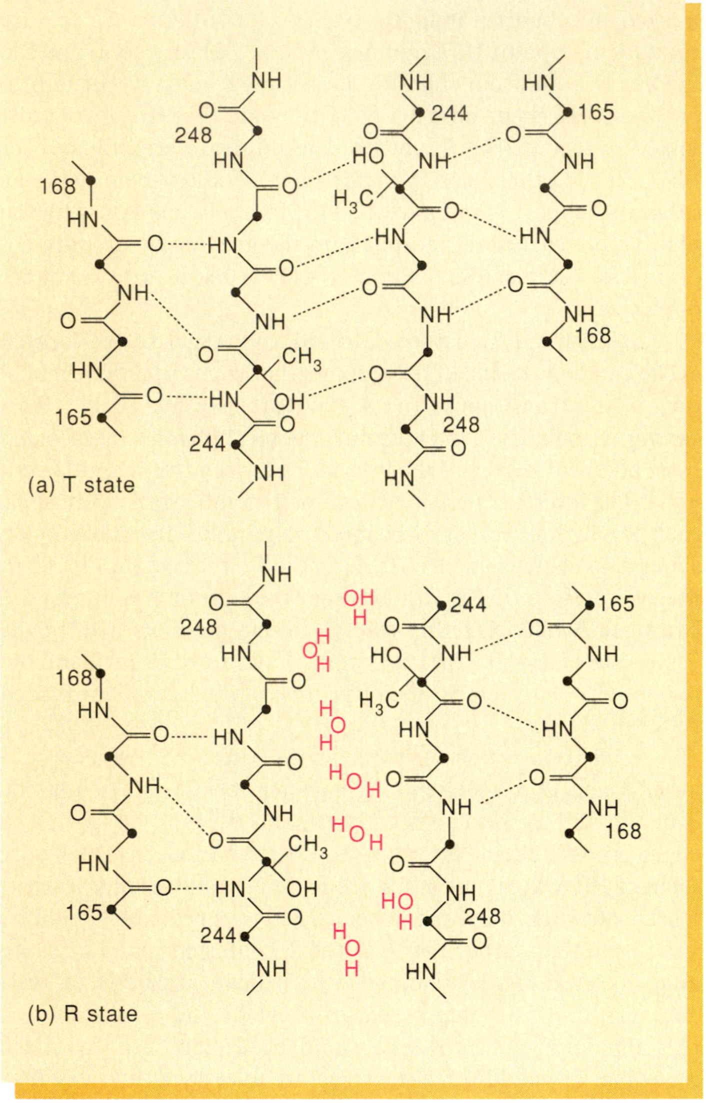

(a) T state

(b) R state

an inhibitor appears to cause only minor changes of the structure in this region of the protein. The side chain of a glutamic acid residue does rotate by about 140° to take up different positions, depending on whether the allosteric effector is an activator or an inhibitor. Whether this rearrangement is sufficient to account for a large shift in the equilibrium between the R and T conformations is unclear. However, the crystal structures that have been elucidated so far include only a single example of an allosteric inhibitor, and this was not ATP but rather a nonphysiological inhibitor, 2-phosphoglycolate. Further studies with ATP or other inhibitors should provide a broader basis for identifying structural changes that are essential to the allosteric inhibition.

Aspartate Carbamoyltransferase: Allosteric Control of Pyrimidine Biosynthesis

Aspartate carbamoyltransferase, or aspartate transcarbamylase, as it is frequently called, catalyzes the transfer of a carbamoyl group

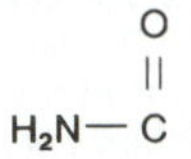

$$H_2N-\overset{\overset{\displaystyle O}{\|}}{C}-$$

from carbamoyl phosphate to the amino group of aspartic acid to form carbamoyl aspartate (fig. 10.13). This step commits aspartate to the biosynthetic pathway for pyrimidines. Aspartate carbamoyltransferase is regulated by feedback inhibition by cytidine triphosphate (CTP) and uridine triphosphate (UTP), the end products of the pathway, and it is stimulated by ATP. The opposing effects of CTP, UTP, and ATP serve to keep the biosynthesis of pyrimidines in balance with that of purines. This balance is important because cells need purines and pyrimidines in approximately equal amounts for the synthesis of nucleic acids.

The kinetics and physical properties of aspartate carbamoyltransferase from *E. coli* have been studied in considerable detail by Howard Schachman and his colleagues. The rate of the reaction has a sigmoidal dependence on the concentration of aspartate, as shown in figure 10.14. CTP shifts the kinetic curve to the right. The enzyme thus is inhibited strongly by CTP at low concentrations of aspartate but not at high concentrations. ATP reverses the effect of CTP, or in the absence of CTP eliminates the positive cooperativity altogether, making the kinetics hyperbolic instead of sigmoidal. UTP (not shown in the figure) acts similarly to CTP.

Direct evidence for the allosteric nature of these effects came from the finding that the regulatory behavior disappeared when the enzyme was treated with an organic mercurial compound such as *p*-hydroxymercuribenzoate (see fig. 10.14). After this treatment, the binding of aspartate no longer showed positive cooperativity, and ATP and CTP had no effect on its activity. Exposure to mercurials was found to cause the enzyme to dissociate into two types of fragments, one of which retained the enzymatic activity but was no longer affected by CTP or ATP. The other fragment bound CTP and ATP, but had no enzymatic activity. In the native enzyme, each molecule is a complex of six catalytic (*c*) subunits, and six regulatory (*r*) subunits (fig. 10.15). The fragments that result from treatment with mercurials consist of trimers of the *c* subunit and dimers of *r*. When the intact c_6r_6 complex was reconstituted from the c_3 and r_2 fragments, the enzyme regained its sigmoidal kinetics and its sensitivity to CTP and ATP. These observations provided a conclusive demonstration that the regulatory agents bind at an allosteric site on the enzyme, and not at the active site. The two sites are on totally different subunits!

The binding of substrate analogs causes changes in several physical and chemical properties of aspartate carbamoyltransferase. An analog that has been particularly useful for studying these effects is *N*-phosphonacetyl-L-aspartate (PALA). PALA is structurally similar to a covalently linked adduct of

Figure 10.13

The reaction catalyzed by aspartate carbamoyltransferase, and the feedback inhibition of this enzyme in *E. coli* by the end product of the pathway, CTP. The series of small arrows represents additional reaction steps in the pathway from carbamoyl aspartate to CTP. These steps are discussed in chapter 20. The upward arrow with the negative sign indicates the feedback inhibition.

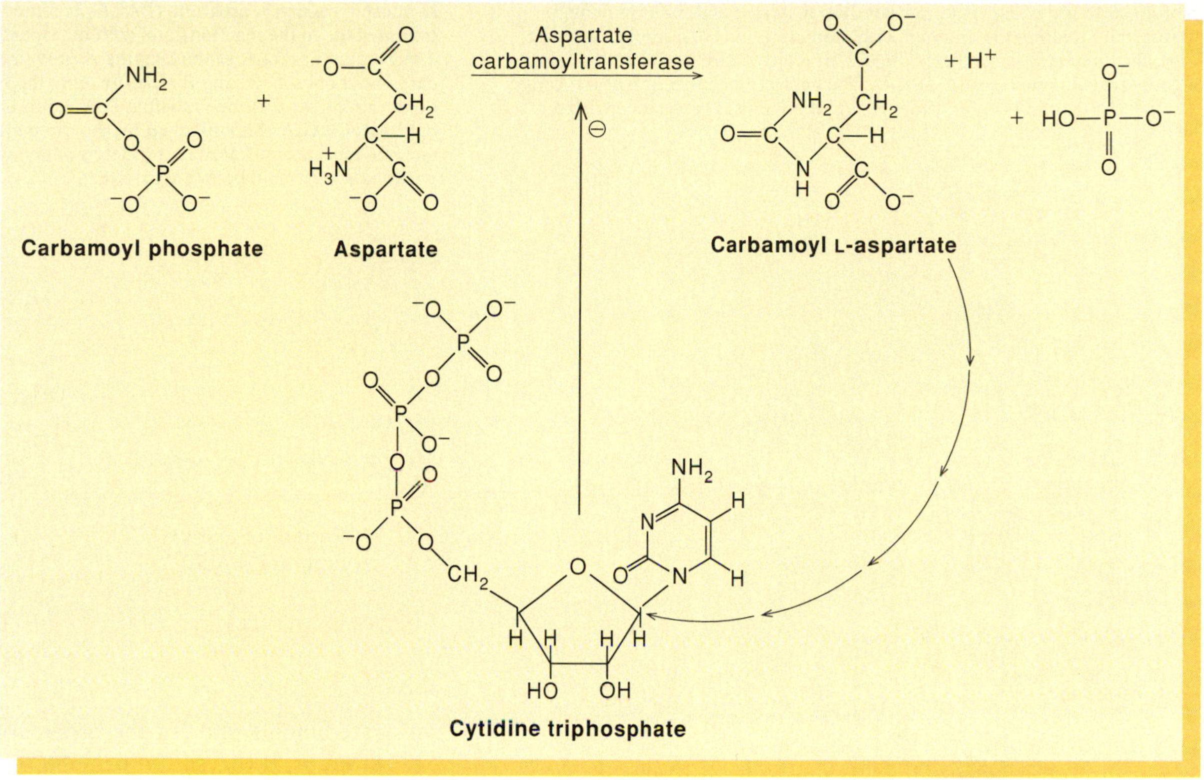

Figure 10.14

Effects of CTP, ATP, and mercurials on the rate of the reaction catalyzed by aspartate carbamoyltransferase. In the absence of CTP and ATP, the sigmoidal kinetics show positive cooperativity with respect to aspartate. CTP augments the positive cooperativity; ATP reverses the effect of CTP. An organic mercurial, or ATP in the absence of CTP, eliminates the cooperativity, converting the curve from sigmoidal to hyperbolic.

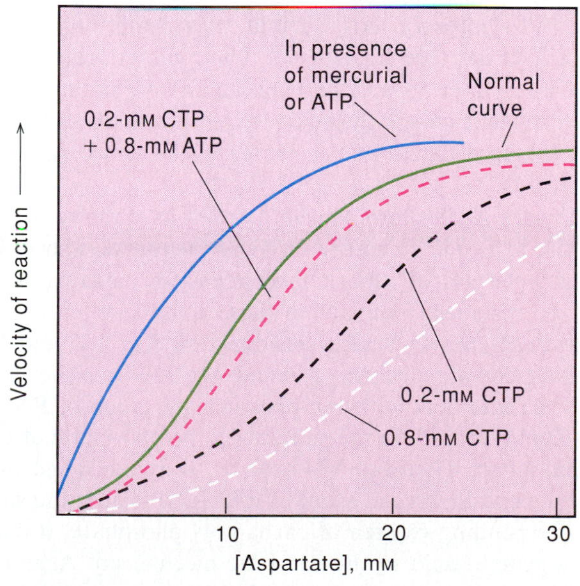

carbamoyl phosphate and aspartate, and thus resembles an intermediate that is likely to be formed in the course of the enzymatic reaction (fig. 10.16). It binds to the enzyme in a highly cooperative manner, with a Hill coefficient of 2.0, and its binding is promoted by ATP and opposed by CTP. (ATP decreases the Hill coefficient to 1.4, and CTP raises it to 2.3.) The binding of PALA causes a decrease in the sedimentation and diffusion coefficients of the enzyme, indicating that the protein expands or changes shape. It also causes a decrease in the chemical reactivity of a cysteine residue in each *c* subunit, and an increase in the reactivity of several cysteines in each *r* subunit. CTP opposes these effects.

The changes in sedimentation coefficient and chemical reactivity caused by PALA can be interpreted by the model that the enzyme exists in two distinct conformations (T and R), and that the binding of PALA to only one or two of the *c* subunits in the c_6r_6 complex causes the entire enzyme to flip from the T to the R state. The dissociation constant for PALA is higher in the T state. The equilibrium constant $L = (T)/(R)$ has been calculated to be 250 in the absence of substrates and allosteric effectors, 70 in the presence of ATP alone, and 1,250 in the presence of CTP alone. Thus ATP shifts the equilibrium toward the conformational state that favors the binding of the substrate, and CTP shifts the equilibrium in the direction of weaker binding.

Figure 10.15

Subunit structure of aspartate carbamoyltransferase and the fragments produced by treating the enzyme with mercurials. In the complete enzyme (top), the three sets of regulator dimers are sandwiched between two trimers of catalytic subunits (see fig. 10.17). The approximate location of the active site in each *c* subunit of the trimer facing the viewer is indicated with a *c*.

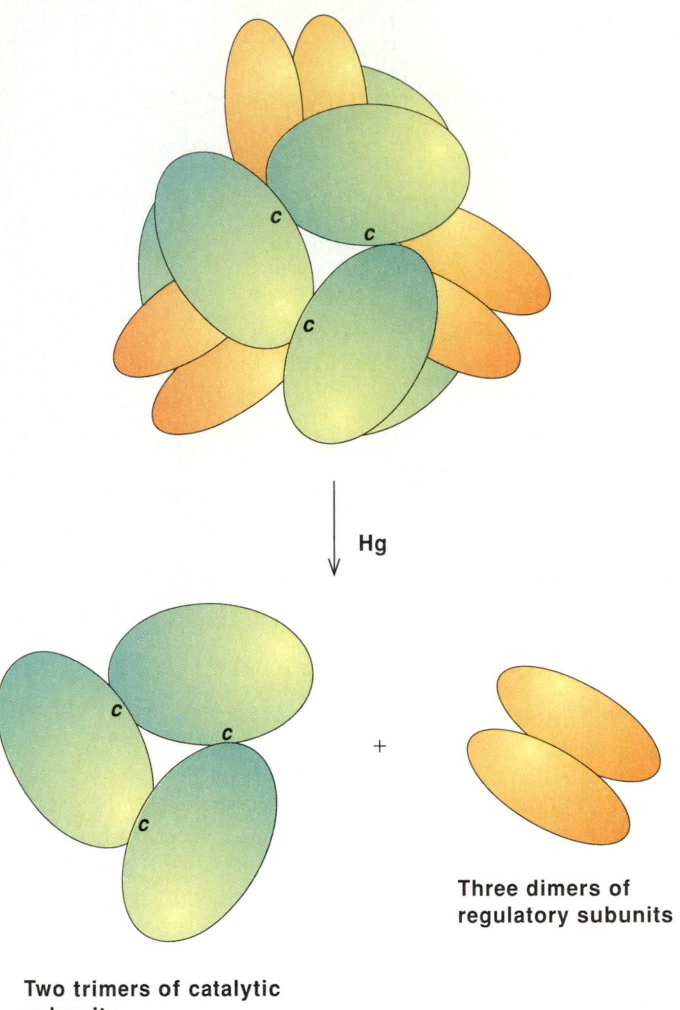

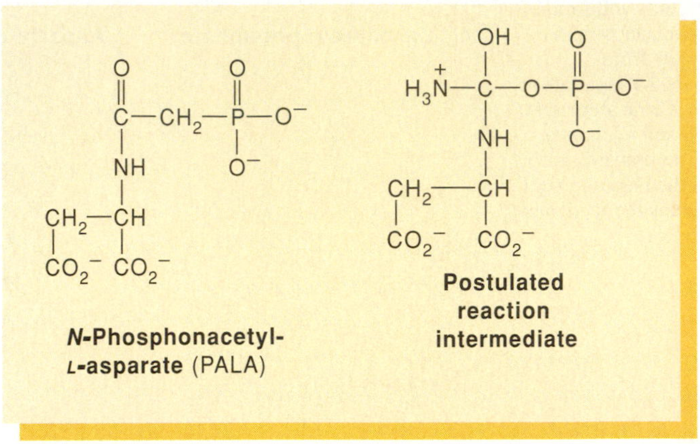

The crystallization of aspartate carbamoyltransferase with and without bound PALA or CTP, and the elucidation of the crystal structures by William Lipscomb and his colleagues, led to detailed pictures of the structural changes that accompany the transition of the enzyme between the R and T states. Figures 10.17 and 10.18 show the crystal structures from two different perspectives. In both the R and the T conformations, the six *c* subunits are arranged in two equilateral trimers, one of which is inverted and stacked on top of the other (see figs. 10.15 and 10.17). The three *c* units within each trimer are related to each other by a threefold axis of symmetry. (Rotating the structure by one-third of a circle about this axis results in an identical structure.) The substrate-binding site on each *c* subunit is located in a pocket between two domains of the polypeptide. At the end of one of these domains the *c* subunit interacts with another *c* subunit in the same trimer, close to *its*

Figure 10.16

N-phosphonacetyl-L-aspartate (PALA) is structurally similar to a likely intermediate in the reaction catalyzed by aspartate carbamoyltransferase. The binding of PALA to the enzyme is prevented competitively by carbamoyl phosphate, and PALA prevents the binding of aspartate. These observations support the view that PALA binds at the catalytic site. The binding of PALA is not blocked by aspartate alone, probably because the enzyme mechanism follows an ordered pathway in which carbamoyl phosphate must bind before aspartate.

***N*-Phosphonacetyl-L-asparate** (PALA)

Postulated reaction intermediate

substrate-binding site. In the other domain, it interacts more extensively with a *c* subunit in the other c_3 trimer. The six *r* subunits are arranged in three sets of dimers that form another equilateral triangle about the same threefold axis of symmetry. Like the *c* subunits, each *r* subunit is folded into two domains: a peripheral domain where it interacts with its companion *r* subunit in the dimer, and a smaller domain that interacts with two adjacent *c* subunits. In the latter region, the *r* subunit binds an atom of Zn^{2+} that evidently plays a purely structural role in the enzyme. The binding site for the allosteric effectors CTP and ATP is located in the peripheral domain of the *r* subunit, at a considerable distance from the active sites.

In the transition from the T to the R conformation, the two *c* trimers rotate slightly with respect to each other about the threefold symmetry axis, so that they come into a more eclipsed alignment (see fig. 10.17). The *r* dimers rotate with respect to each other about a perpendicular axis, and appear to act as a lever that moves the two *c* trimers apart by about 12 Å and opens up a cavity at the center of the entire structure (see fig. 10.18).

Figure 10.19 shows some of the details of the substrate-binding site in the R structure. As was mentioned above, the binding site consists of a pocket between two domains of a *c* subunit, and residues from each of these domains interact with the substrate. Some of the key residues are Arg 105 and His 134 from one domain, and Arg 167 and Arg 229 from the other. Arginine 105 interacts with the phosphonate group of PALA, and presumably would do the same with the phosphate of carbamoyl phosphate. Histidine 134 appears to provide a hydrogen bond to the peptide oxygen atom of PALA. If it does the same to the corresponding oxygen of carbamoyl phosphate, it could serve as a general acid in the catalytic mechanism. Arginines 167 and 229 interact with the two carboxylate groups of PALA, and presumably would interact similarly with aspartate.

Figure 10.17

Structures of aspartate carbamoyltransferase in the T conformation (*a*), and the R conformation (*b*), viewed along the threefold symmetry axis. The enzyme contains two c_3 clusters and three r_2 clusters. The α-carbon chains of one of the c_3 groups are shown in aqua; those of the other c_3 group are in blue. One of the *r* subunits in each of the r_2 groups is shown in orange; the other, in red. The enzyme was crystallized in the T form in the absence of substrate or allosteric effectors; the R structure was obtained with bound PALA. The PALA molecules are seen in yellow in (*b*). Zinc ions bound to the *r* subunits are shown in white. (Based on the crystal structures described by W. N. Lipscomb and his colleagues.)

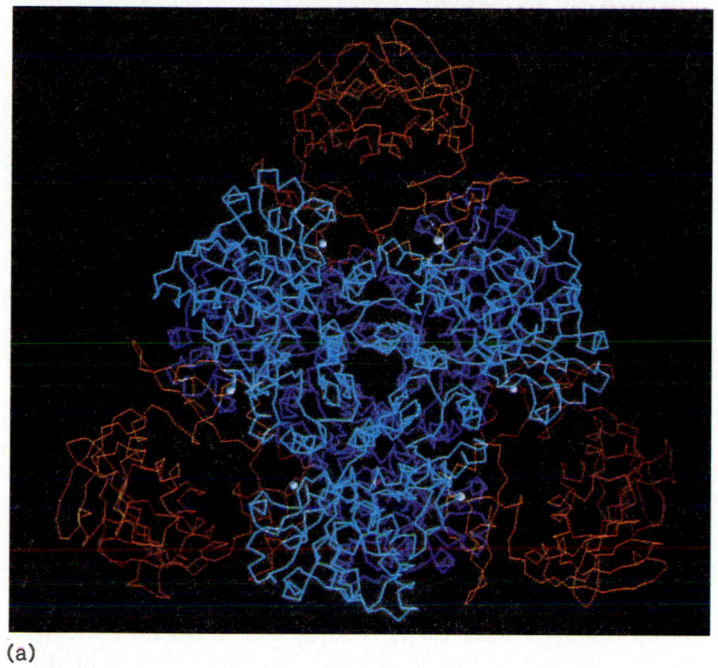
(a)

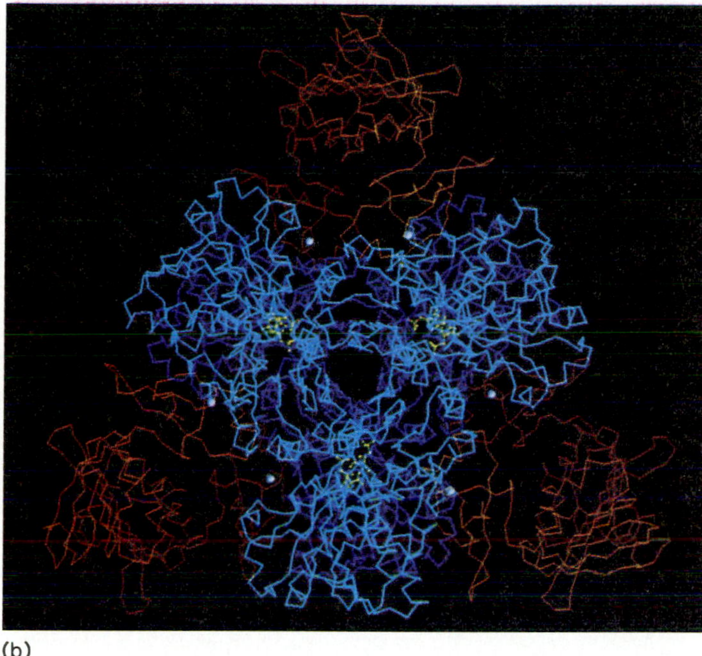
(b)

Figure 10.18

Structures of aspartate carbamoyltransferase in the T conformation (*a*), and the R conformation (*b*), viewed along an axis perpendicular to the threefold symmetry axis. The structures and the color coding are the same as in figure 10.17. Note the expansion of the cavity between the upper and lower c_3 groups in the R structure.

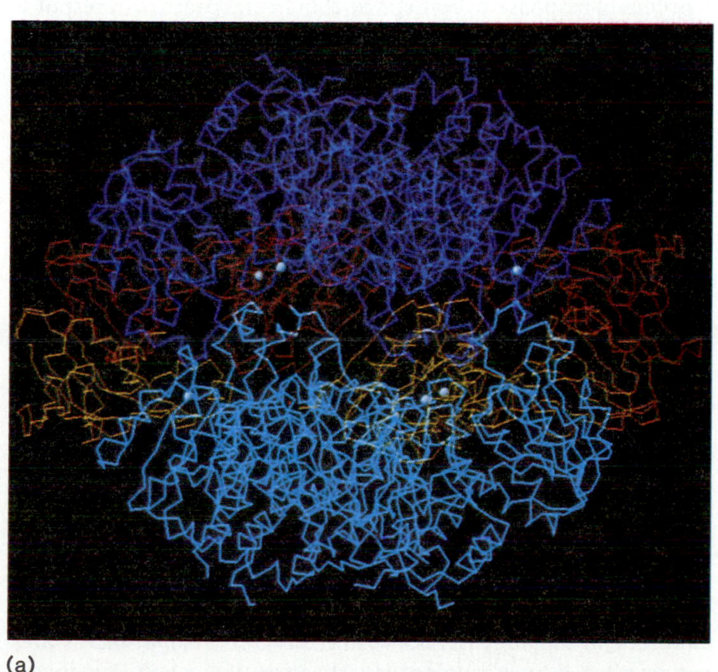
(a)

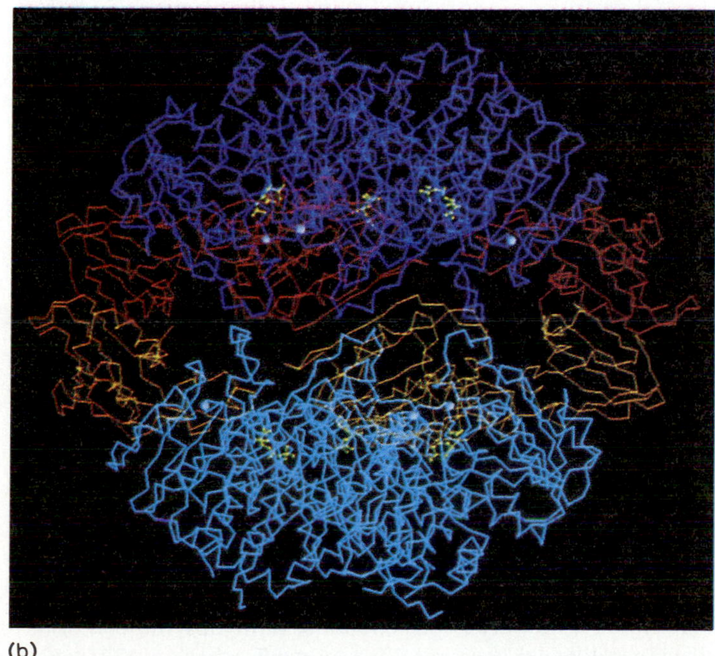
(b)

Figure 10.19

The binding of PALA to the R conformation of aspartate carbamoyltransferase. Arg 105 and His 134 are provided by one domain of a *c* subunit, and Arg 167 and Arg 229 by the other domain. Ser 80 and Lys 84 are part of a loop of protein from a different *c* subunit. The bound PALA is shown in color.

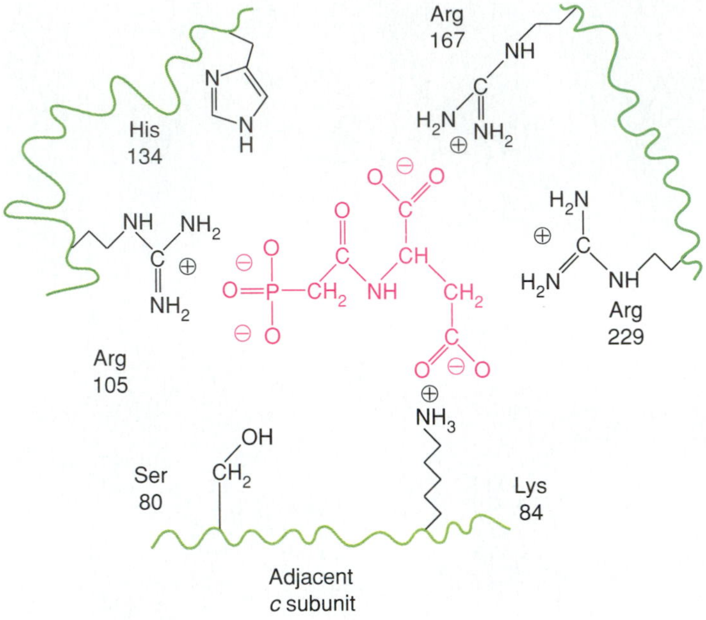

In addition to these residues, the active site also includes two residues from a different *c* subunit, Ser 80 and Lys 84 (see fig. 10.19). Serine 80 appears to be hydrogen-bonded to the phosphonate group of PALA, and Lys 84 interacts with one of the carboxylate groups. As with phosphofructokinase, the location of the active site at the interface between two subunits provides a clue to how the binding of substrate to one subunit can affect the binding at another.

The structural transition to the T state disrupts the active site in two major ways. First, the domain of the *c* subunit that includes Arg 105 and His 134, which must interact with carbamoyl phosphate, is pulled away from the domain that must interact with aspartate, because some of the residues in both domains are tied up in alternative sets of hydrogen bonds. Arginine 105 is hydrogen-bonded to Glu 50 in the same domain, instead of to the substrate, and His 134 interacts with a residue in the other *c* trimer. In addition, the loop of the *c* subunit that contains Ser 80 and Lys 84 is pulled out of the active site by a set of hydrogen bonds involving still another *c* subunit. A baroque network of interrelationships thus links the catalytic sites of all the different *c* subunits in the complex.

The transition between the R and T states also involves large changes in the conformation of the *r* subunits. These changes include both the peripheral domain where CTP or ATP binds and the domain that interfaces with the *c* subunits (see figs. 10.17 and 10.18). However, it still is not clear how the binding of CTP to the peripheral domain tips the conformational equilibrium constant L in favor of T, whereas ATP, which binds to the same site as CTP, favors the formation of R. The crystal structures of the enzyme with bound ATP or CTP are very similar, both at the catalytic site and at the allosteric site.

Glycogen Phosphorylase: Control of Glycogen Breakdown both by Allosteric Effectors and by Phosphorylation

Glycogen phosphorylase catalyzes the removal of a terminal glucose residue from glycogen. The glycosidic bond is cleaved by a reaction with inorganic phosphate ion (a "phosphorolysis") instead of simply by hydrolysis, so that the product is glucose-1-phosphate instead of free glucose:

$$\text{Glycogen with } n \text{ glucose units} + P_i \longrightarrow \text{glycogen with } n - 1 \text{ glucose units} + \text{glucose-1-phosphate} \quad (9)$$

This is the first step in the metabolic breakdown of glycogen to pyruvate.

In the early 1940s, Carl and Gerty Cori discovered that phosphorylase exists in two forms, *a* and *b*, which differ greatly in their catalytic activities. As shown in figure 10.20, phosphorylase *b* has virtually no activity in the absence of AMP. It is activated by AMP with an apparent dissociation constant of about 40 μM, but the activation is inhibited competitively by ATP. At the concentrations of AMP and ATP that prevail in resting muscle tissue, phosphorylase *b* is essentially inactive. Phosphorylase *a*, on the other hand, has about 80% of its maximal activity in the absence of AMP, and becomes fully active at very low concentrations of AMP. It also is relatively insensitive to inhibition by ATP (see fig. 10.20).

The Coris found that the interconversion of phosphorylase *a* and *b* is catalyzed by another enzyme, and subsequent work by Earl Sutherland showed that this process is under hormonal control. In muscle, the conversion of phosphorylase *b* to *a* occurs in response to epinephrine; in liver, it occurs in response to glucagon, a hormone elaborated by the pancreas. However, the structural basis for the difference between the two forms of the enzyme remained unknown until the late 1950s, when Edwin Krebs and Edmond Fischer showed that phosphorylase *a* and *b* differ by a single covalent modification: Phosphorylase *a* has a phosphate on Ser 14. Krebs and Fischer also showed that the kinase that catalyzes the addition of the phosphate group to the serine residue (phosphorylase kinase) is itself regulated by a phosphorylation catalyzed by another kinase (the cAMP-dependent protein kinase).

The main effect of AMP on either phosphorylase *b* or phosphorylase *a* is to decrease the K_m for P_i. The K_m for glucose-1-phosphate also is increased for the reaction in the reverse direction. These changes can be interpreted much as we have interpreted the actions of allosteric effectors on phosphofructokinase and aspartate carbamoyltransferase, on the model that the enzyme can exist in either of two different conformational states (R and T) with different affinities for the substrate. However, phosphorylase presents the additional complexity that the equilibrium constant (L) between the two conformational states can be altered by a covalent modification of the enzyme. In the

Figure 10.20

The rate of the reaction catalyzed by glycogen phosphorylase, as a function of the concentration of its main allosteric activator, AMP. The curves shown in color were obtained in the presence of ATP. Phosphorylase *b* (lower two curves) is almost completely inactive in the absence of AMP. Its activity is half-maximal at an AMP concentration of about 40 μM. ATP greatly increases the concentration of AMP required for activity. Phosphorylase *a* (upper two curves) has about 80% of its maximal activity in the absence of AMP, and reaches full activity at very low AMP concentrations; it also is relatively insensitive to inhibition by ATP. (Source: N. B. Madsen in *The Enzymes,* 3d ed., vol. XVII, Edited by P. D. Boyer and E. G. Krebs. Copyright © 1986 Academic Press, New York, N.Y.)

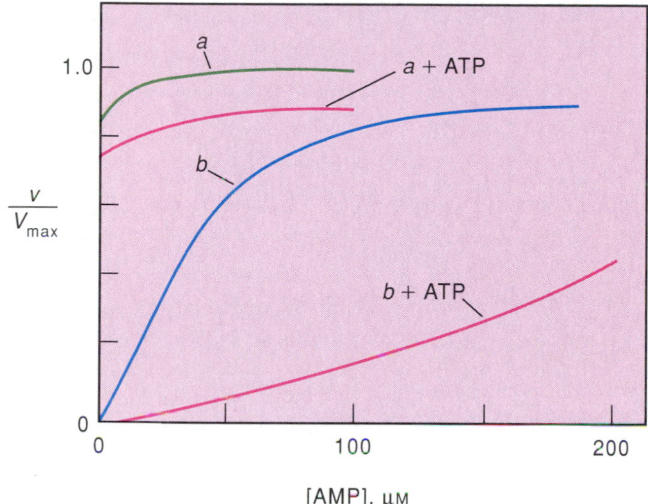

absence of substrates, (T)/(R) is estimated to be greater than 3,000 in phosphorylase *b,* but to decrease to about 10 in phosphorylase *a.* The phosphorylation thus appears to reduce the free energy difference between the two conformations by at least 3.5 kcal/mole.

In addition to activation by AMP and inhibition by ATP, both forms of phosphorylase are sensitive to inhibition by glucose or glucose-6-phosphate. Glucose inhibits by binding at the catalytic site; glucose-6-phosphate binds predominantly at the same allosteric site as AMP and ATP. There also is a separate inhibitory allosteric site that binds adenine, adenosine, or (much more weakly) AMP.

Louise Johnson and her co-workers have determined structures for both the T and the R forms of muscle phosphorylase *b,* and also of the R form of phosphorylase *a.* In parallel with this work, Robert Fletterick and his co-workers determined the crystal structure for the T form of muscle phosphorylase *a.* The crystal structures provide an incisive look at the structural changes that accompany the transition from the T to the R form and the conversion of nonphosphorylated form of the enzyme to the phosphorylated.

In keeping with the complexity of its allosteric and covalent regulation, phosphorylase is a large, complex enzyme. It consists of a dimer of two identical subunits, each with a molecular weight of about 97,400 (fig. 10.21). The catalytic site is buried near the center of each subunit, at the end of a tunnel about 15 Å in length. The tunnel opens to a concave surface

Figure 10.21

(*a*) Ribbon diagram of the crystal structure of phosphorylase *a* in the R state. The view is along the twofold rotational symmetry axis of the dimer, with the allosteric sites and the phosphoserine (Ser-P) on each subunit facing forward. Regions where the positions of the C_α carbons differ by more than 1 Å between the R and T states are shown in orange for subunit 1 (*bottom*) and in pink for subunit 2 (*top*). More fixed regions of the polypeptide chain are in green for subunit 1, and in blue for subunit 2. The N-terminal residues (10–23) and the C-terminal residues (837–842) are in white (residues 1–9 are disordered, and cannot be seen in the crystal structure). The bound pyridoxal-phosphate (PLP) indicates the location of the catalytic site. The allosteric effector site is occupied by AMP. The first two helices at the N-terminal end (labeled α_1 and α_2 in subunit 1) are connected by a loop (Cap) that forms one of the interfaces between the two subunits. (*b*) Ribbon diagram of phosphorylase *b* in the T state. The orientation of the dimer is the same as in (*a*). Regions where the C_α positions differ by more than 1Å between the R and T states are represented in red for subunit 1 (*bottom*) and in yellow for subunit 2 (*top*); more fixed regions are in cyan and purple. The N-terminal residues and the C-terminal residues are in white. PLP is at the catalytic site and AMP at the allosteric effector site, as in (*a*). Maltopentaose is bound at the glycogen-storage site. There also is a molecule of glucose-1-phosphate bound at the catalytic site, and (not clearly visible) a second molecule of AMP at the nucleoside inhibitor site. (From D. Barford, S.-H. Hu, and L. N. Johnson, "Structural mechanism for glycogen phosphorylase control by phosphorylation and AMP," *J. Mol. Biol.* 218:233, 1991. © 1991 Academic Press LTD., London England.)

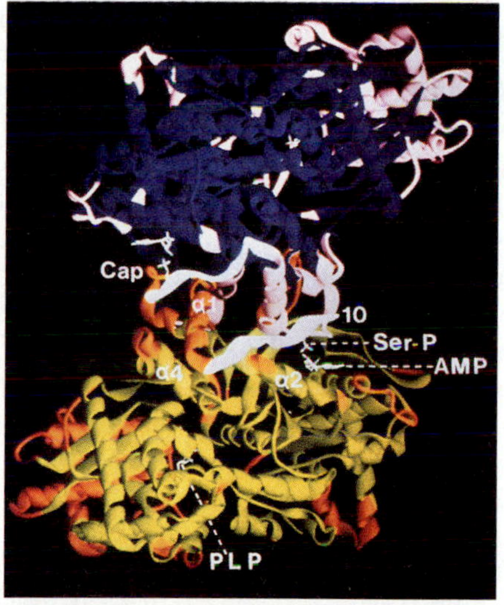

(a)

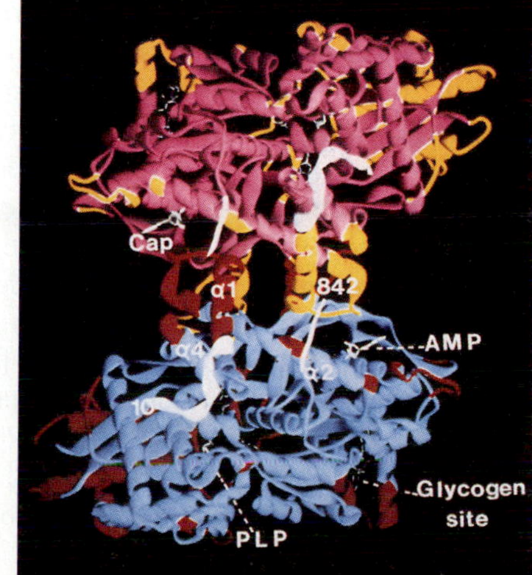

(b)

Figure 10.22

Pyridoxal phosphate, a prosthetic group in glycogen phosphorylase, is covalently attached to a lysine side chain of the enzyme. The phosphate group of the pyridoxal phosphate probably acts as a general acid to transfer a proton to inorganic phosphate in the enzymatic mechanism. (Other parts of the pyridoxal molecule can be modified chemically with little effect on the enzymatic activity, but modifying the phosphate group destroys the activity.) Other biochemical reactions of pyridoxal phosphate are discussed in chapter 11.

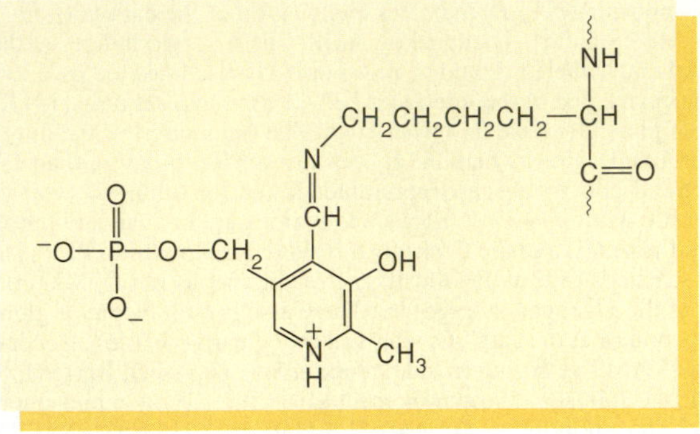

Figure 10.23

(*a*) In the T state of phosphorylase *b*, a loop of the polypeptide chain between residues 282 and 286 obstructs the active site. Asp 283, near the middle of this loop, inserts its negatively charged side chain into the binding site for phosphate. The locations of the aspartate and the pyridoxal phosphate prosthetic group (PLP) are indicated in red. (*b*) In the R state, the subunits rotate with respect to each other. The loop from residues 282 to 286 is disordered, and Asp 283 leaves the phosphate-binding site. (Source: D. Barford and L. N. Johnson, "The allosteric transition of glycogen phosphorylase" in *Nature*, 340:609, 1989. Copyright © 1989 Macmillan Magazines Ltd., London, England.)

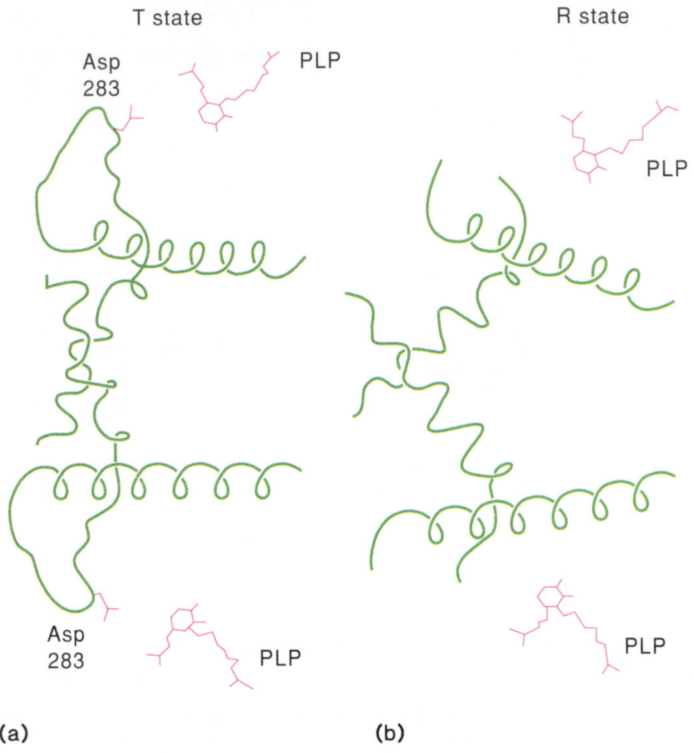

whose curvature is similar to the curvature of a glycogen particle. A binding site for glycogen can be recognized by the location of a small oligosaccharide (maltoheptaose or maltopentaose) in the crystal structure. Because the glycogen-attachment site is about 30 Å from the catalytic site, the enzyme evidently clings to one branch of the glycogen particle while it chews on another branch. Near the catalytic site there is a covalently bound molecule of pyridoxal phosphate (fig. 10.22), which appears to participate in the catalytic mechanism by acting as a general acid.

The binding site for the allosteric effectors AMP, ATP, and glucose-6-phosphate is about 30 Å from the catalytic site, at one of the interfaces between the two subunits (see fig. 10.21). Serine 14, the locus of the covalent modification that converts phosphorylase *b* to *a*, is in the same region of the molecule, about 15 Å from the allosteric site.

When phosphorylase *b* undergoes the transition from the T to the R form, structural changes occur in the N-terminal region of each subunit (see fig. 10.21). There also are significant changes in a loop between residues 282 to 286, which connects two α-helical chains (fig. 10.23). In the T form, this loop is well ordered, and is located so that it obstructs the substrate-binding site for phosphate. In the R form, the loop is pulled out of the phosphate site, and is disordered. Some of the details of the structural changes in the active site are shown in figure 10.24. In the R form, the substrate phosphate ion is bound to Arg 569, the amide NH of Gly 135, Lys 574 (not shown in the figure), and the pyridoxal phosphate. In the T form, the side chain of Asp 283 sits in this region, and Arg 569 is pulled away by hydrogen bonding to the amide NH of Pro 281. The replacement of the positively charged arginine side chain by the negatively charged aspartate explains why the affinity for phosphate is much lower in the T form than in the R form.

The transformation from phosphorylase *b* to phosphorylase *a* includes structural changes that tighten the interactions between the two subunits of the enzyme. Figures 10.21 and 10.25 show some of these changes. In phosphorylase *a*, the phosphate group attached to each Ser 14 is hydrogen-bonded to Arg 69 of its own subunit, but also to Arg 43 of the other subunit. There also are hydrogen bonds between Arg 10 and Leu 115 of different subunits, and between Glu 72 and Asp 42. All of these intersubunit hydrogen bonds are missing in phosphorylase *b*. Instead, there is a bond between Arg 43 and leucine in the same subunit, and one intersubunit hydrogen bond between His 36 and Asp 838. The N-terminal portion of the chain is ejected from this region of the structure in phosphorylase *b*, and is replaced by residues from the C-terminal end, including Asp 838. At the nearby allosteric effector site, the pulling together of the two subunits in phosphorylase *a* enhances the binding of AMP, but disfavors the binding of the inhibitor glucose-6-phosphate.

Figure 10.24

Residues in the region of the substrate-binding site for phosphate in phosphorylase *b* in the T state (*a*) and the R state (*b*). The side chain of Asp 283 leaves the binding site in the R structure, and the side chain of Arg 569 becomes available to interact with the phosphate. (Source: D. Barford and L. N. Johnson, "The allosteric transition of glycogen phosphorylase" in *Nature,* 340:609, 1989. Copyright © 1989 Macmillan Magazines Ltd., London, England.)

(a) T state

(b) R state

Figure 10.25

Structural changes that accompany the conversion of phosphorylase *b* to phosphorylase *a*. This figure shows the interface between the two subunits in the region of Ser 14, the residue that gains a phosphate group in phosphorylase *a*. Portions of the polypeptide backbone and the side chains of some residues from one of the subunits are drawn in color; the other subunit is drawn in black. A larger number of hydrogen bonds link the two subunits in phosphorylase *a* (*top*) than in phosphorylase *b* (*bottom*). (Source: S. R. Sprang et al., "Structural changes in glycogen phosphorylase induced by phosphorylation" in *Nature,* 336:215, 1988. Copyright © 1989 Macmillan Magazines Ltd., London, England.)

Phosphorylase *a*

Phosphorylase *b*

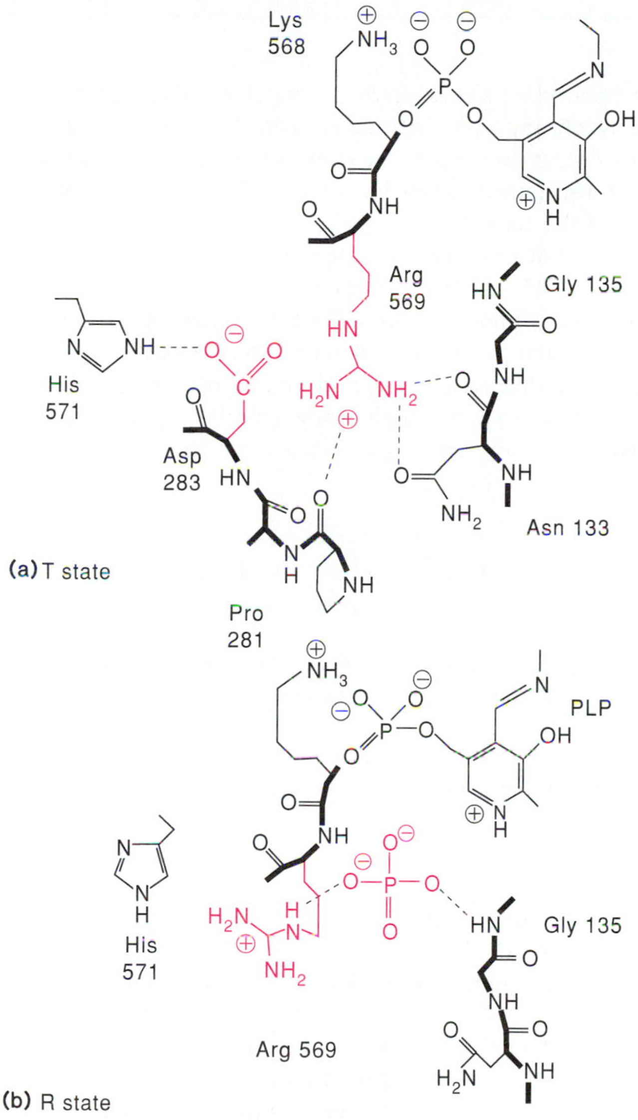

Calmodulin Mediates Regulation of Some Enzymes by Ca^{2+}

Phosphorylase kinase, the enzyme that converts phosphorylase *b* to phosphorylase *a,* also is controlled by both covalent modification and allosteric effectors. When muscle cells are stimulated by epinephrine, or liver cells by glucagon, the cAMP-dependent protein kinase catalyzes the phosphorylation of phosphorylase kinase at multiple serine residues. (The modification of more than one serine actually is typical of enzymes

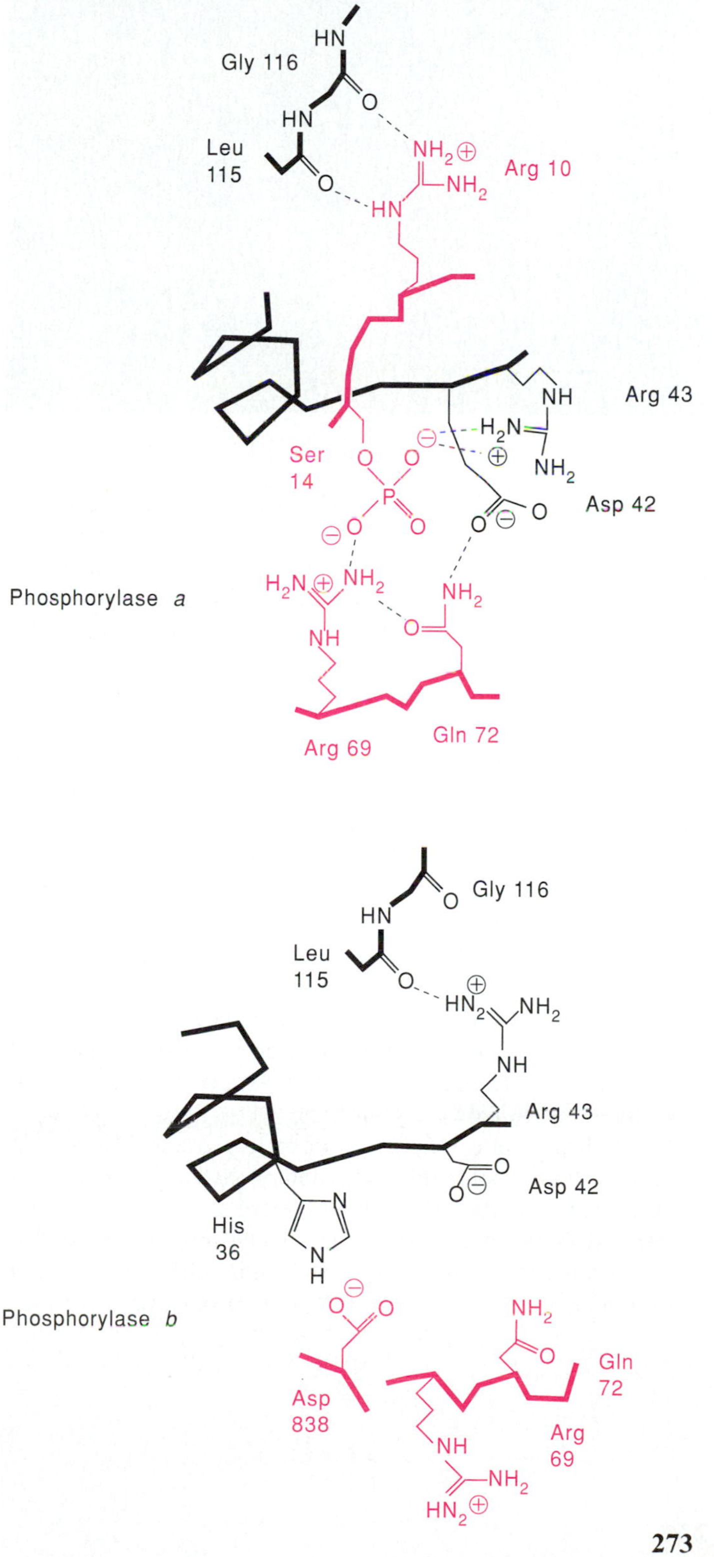

Figure 10.26

The structure of calmodulin, a small protein that binds Ca^{2+} and regulates the activities of many enzymes in response to changes in the intracellular Ca^{2+} concentration. The bound Ca^{2+} ions are shown in yellow. (Based on the crystal structure described by Y. S. Babu, C. E. Bugg, and W. J. Cook.)

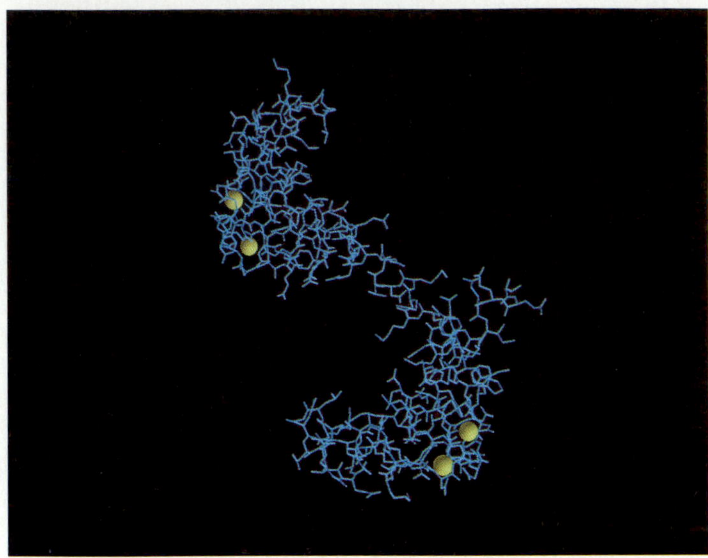

that are controlled by phosphorylation; phosphorylase is unusual in that only a single serine residue is modified.) Muscle phosphorylase kinase contains four nonidentical subunits (α, β, γ, and δ), which form an $(\alpha\beta\gamma\delta)_4$ tetramer with a total molecular weight of about 1.2×10^6. The catalytic sites are on the γ subunits, whereas the serines that become phosphorylated are on the α and β subunits. The δ subunit has turned out to be of particular interest because it is a polypeptide called calmodulin that also is associated with many other enzymes. Calmodulin appears to modify the activities of these enzymes in response to changes in the intracellular concentration of Ca^{2+} ions.

Calmodulin occurs in virtually all eukaryotic cells. As shown in figure 10.26, it is a relatively small protein ($M_r = 17{,}000$) that has two globular domains connected by an extended α helix. Each of the globular regions contains two subdomains that form binding sites for Ca^{2+} ions. Ca^{2+} binds to calmodulin with a dissociation constant of about 10^{-6} M, causing the globular domains to undergo conformational changes that evidently can be transmitted to the enzyme with which the calmodulin is associated.

The role of calmodulin in muscle phosphorylase kinase appears to be to enable the activation of the enzyme whenever the Ca^{2+} concentration rises above a level of about 1 μM. Ca^{2+} is released from intracellular stores when muscles are stimulated neurally, and it plays a critical role in initiating muscular contraction. The binding of Ca^{2+} to the calmodulin associated with phosphorylase kinase serves to switch on glycogen breakdown in synchrony with contraction, with the ultimate effect of providing energy to support the contraction. Calmodulin-Ca^{2+} complexes also activate an enzyme in the cell membrane to pump Ca^{2+} out of the cell.

Table 10.4
Some Enzymes that Are Regulated by Ca^{2+}-Calmodulin

Muscle phosphorylase kinase
Myosin light-chain kinase
Ca^{2+}-ATPase (transmembrane Ca^{2+} pump)
Ca^{2+}/calmodulin-dependent protein kinase

Some of the other enzymes that are regulated by calmodulin and Ca^{2+} are listed in table 10.4. The regulatory systems that use calmodulin resemble the systems that use the cAMP-dependent protein kinase in that they generally respond to signals that come from outside the cell. However, calmodulin also is essential in yeast, which are unicellular organisms. The roles that it plays in these organisms are not fully understood.

Inhibitory Proteins that Block the Actions of Proteolytic Enzymes

In discussing the activation of digestive enzymes and blood-coagulation enzymes by partial proteolysis, we noted the importance of keeping these enzymes in an inactive state unless they are needed. Tissues, blood, and other biological fluids of both vertebrate and invertebrate organisms contain a large array of specific inhibitory proteins that participate in this task. Similar inhibitory proteins have been purified from bacteria, especially from various species of *Streptomyces*. They also are abundant in plants, where they probably play a role in combatting insect pests.

Protease inhibitors typically bind very tightly to the active site of a particular protease. Bovine pancreatic trypsin inhibitor, for example, binds to trypsin with a dissociation constant of about 10^{-13} M. Crystal structures have been obtained for several such enzyme-inhibitor complexes, including the complexes of trypsin with soybean trypsin inhibitor and bovine pancreatic trypsin inhibitor, and the complex of subtilisin with *Streptomyces* subtilisin inhibitor. Figure 10.27 shows the crystal structure of the complex of bovine trypsin with the bovine pancreatic trypsin inhibitor. Inspection of the crystal structure in the region of the active site reveals that the inhibitor presents itself to the enzyme as a substrate. The carbonyl carbon atom of Lys 15 of the inhibitor is close to the OH oxygen of trypsin's reactive Ser 195, and the carbonyl oxygen atom of the lysine points toward the NH hydrogen atoms of Ser 195 and Gly 193. The carbonyl carbon thus seems poised to react with the serine oxygen, but for reasons that are not entirely clear, the reaction occurs only at a very low rate. The inhibitor behaves like a substrate that reacts with an extremely low K_m, but also with a very low k_{cat}. This also applies to the other serine-protease inhibitors that have been studied. A possible explanation for the low value of k_{cat} is that the enzyme-inhibitor complex is too rigid to complete the structural distortion that is needed for hydrolysis of the peptide bond.

Specific inhibitory proteins also have been found for enzymes other than proteases, including α-amylases, deoxyribonucleases, protein kinases, and phospholipases. However, these inhibitors are greatly outnumbered by the protein inhibitors that act on proteases, and perhaps this is not surprising. Because proteins are the natural substrates of proteases, it may take only minor modifications to turn them into inhibitors.

Figure 10.27

The complex of trypsin with bovine pancreatic trypsin inhibitor. The enzyme is in purple; the inhibitor, in red. (Based on the crystal structure described by R. Huber and J. Deisenhofer.)

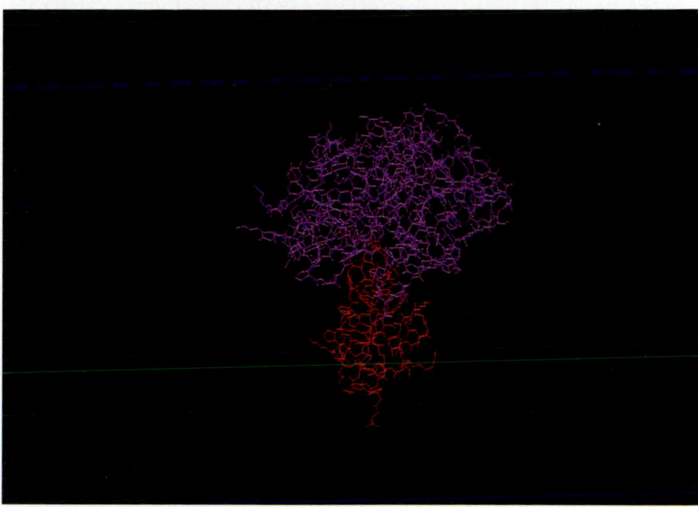

Summary

In this chapter we have been concerned with the regulation of enzyme activities. The chief points in our discussion are the following.

1. Cells regulate their metabolic activities by modulating the rates of synthesis and degradation of enzymes and by adjusting the activities of specific enzymes. Enzyme activities vary in response to changes in pH, temperature, and the concentrations of substrates or products, but also can be controlled by covalent modifications of the protein structure or by interactions with activators or inhibitors.
2. Partial proteolysis, an irreversible process, is used to activate proteases and other digestive enzymes after their secretion, and to switch on cascades of enzymes that cause blood coagulation or dissolve blood clots. Common types of reversible covalent modification include phosphorylation of serine or tyrosine residues, adenylylation, and disulfide reduction.
3. Allosteric effectors are inhibitors or activators that bind to enzymes at sites distinct from the active sites. Allosteric regulation allows cells to adjust enzyme activities rapidly and reversibly in response to changes in the concentrations of substances that are structurally unrelated to the enzyme substrates or products.
4. The initial steps in a biosynthetic pathway commonly are inhibited by the end products of the pathway, and numerous enzymes are regulated by ATP, ADP, or AMP.
5. The activities of allosteric enzymes typically show a sigmoidal dependence on substrate concentration, rather than the hyperbolic dependence predicted by the Michaelis-Menten equation. These enzymes usually have multiple subunits, and the sigmoidal kinetics can be ascribed to cooperative interactions of the subunits. Binding of the substrate to one subunit changes the dissociation constant for substrate on another subunit. The extent of the cooperativity can be described phenomenologically by the Hill equation or by equations based on the symmetry model or the more general sequential model.
6. The symmetry model postulates that the enzyme can exist in either of two conformations, T and R. It is assumed that the substrate binds more tightly to the R conformation than to T, that the binding of the substrate or an allosteric effector can change the equilibrium between these conformations, and that cooperative interactions among the subunits make all of the subunits switch from one conformation to the other in a concerted manner. Although the symmetry model generally oversimplifies the situation, it provides a useful conceptual framework that is consistent with the behavior of many allosteric enzymes.
7. Phosphofructokinase, the key regulatory enzyme of glycolysis, has four identical subunits. It exhibits sigmoidal kinetics with respect to fructose-6-phosphate and is inhibited by ATP and stimulated by ADP and numerous other metabolites. The crystal structures show that the four subunits rotate with respect to each other in the transition between the R and T conformations. Accompanying this rotation, there is a rearrangement of the binding site for fructose-6-phosphate, which is located at an interface between different subunits. At another interface, antiparallel β strands of two subunits are hydrogen-bonded together in the T structure but are separated by a row of water molecules in the R structure. This structural feature explains the cooperative nature of the conformational transition in all of the subunits: Intermediate structures probably would be much less stable than either the R or the T structure.

8. Aspartate carbamoyltransferase, the first enzyme in the biosynthesis of pyrimidines from aspartate, is inhibited allosterically by CTP, an end product of the pathway, and is stimulated by ATP. It has six identical catalytic subunits and six regulatory subunits. The different types of subunits can be separated by treating the enzyme with mercurials. Binding of a substrate analog (PALA) to one of the catalytic subunits causes the entire c_6r_6 complex to flip from T to R; binding of CTP to a regulatory subunit favors the T conformation. Again, the crystal structures show that the substrate-binding sites are located at interfaces between subunits, and that the T $\rightarrow$ R transition results in a rotation of the subunits and brings together components of the substrate-binding site.
9. Glycogen phosphorylase breaks down glycogen to glucose-1-phosphate. It exists in two forms that differ by a covalent modification. Phosphorylase *a* is phosphorylated on a serine residue and is the more active form under typical cellular conditions. Phosphorylase *b*, which lacks the phosphate, is less active under cellular conditions because of strong allosteric inhibition by ATP, although it is stimulated allosterically by AMP. The enzyme consists of two large, identical subunits. The binding site for allosteric effectors is at the interface; the substrate-binding site is about 30 Å away. In the T $\rightarrow$ R transition there is a repositioning of arginine and aspartate residues at the binding site for the substrate phosphate. Conversion of phosphorylase *b* to phosphorylase *a* tightens the interactions between the subunits and favors the transition to the R form.
10. Some proteins are activated or inhibited by interactions with other proteins. Calmodulin, a Ca^{2+}-binding protein, activates numerous proteins in response to changes in intracellular Ca^{2+} concentrations. Proteases generally are held in check by specific inhibitor proteins.

Selected Readings

Babu, Y. S., J. S. Sack, T. J. Greenough, C. E. Bugg, A. R. Means, and W. J. Cook, Three-dimensional structure of calmodulin. *Nature* 315:37, 1985.

Barford, D., S. -H. Hu, and L. N. Johnson, Structural mechanism for glycogen phosphorylase control by phosphorylation and AMP. *J. Mol. Biol.* 218:233, 1991.

Barford, D., and L. N. Johnson, The allosteric transition of glycogen phosphorylase. *Nature* 340:609, 1989.

Cséke, C., and B. B. Buchanan, Regulation of the formation and utilization of photosynthate in leaves. *Biochem. Biophys. Acta.* 853:43, 1986.

Furie, B., and B. C. Furie, The molecular basis of blood coagulation. *Cell* 53:505, 1988.

Laskowski, M., Jr., and I. Kato, Protein inhibitors of proteinases. *Ann. Rev. Biochem.* 49:593, 1980.

Lipscomb, W. N., Structure and function of allosteric enzymes. *Chemtracts-Biochem. Mol. Biol.* 2:1, 1991.

Perutz, M. F., Mechanisms of cooperativity and allosteric regulation in proteins. *Quart. Revs. Biophys.* 22:139–51, 1989.

Schachman, H. R., Can a simple model account for the allosteric transition of aspartate transcarbamoylase? *J. Biol. Chem.* 263:18583, 1988.

Schirmer, T., and P. R. Evans, Structural basis of the allosteric behaviour of phosphofructokinase. *Nature* 343:140, 1990.

Sprang, S. R., K. R. Acharya, E. J. Goldsmith, D. I. Stuart, K. Varvill, R. J. Fletterick, N. B. Madsen, and L. N. Johnson, Structural changes in glycogen phosphorylase induced by phosphorylation. *Nature* 336:215, 1988.

Stevens, R. C., J. E. Gouaux, and W. N. Lipscomb, Structural consequences of effector binding to the T state of aspartate carbamoyltransferase: Crystal structures of the unligated and ATP- and CTP-complexed enzymes at 2.6 Å resolution. *Biochem.* 29:7691, 1990.

Problems

1. Show by writing the reactions how the combination of a protein kinase and the corresponding phosphoprotein phosphatase, if unregulated, theoretically forms a futile cycle.
2. The cAMP-dependent protein kinases phosphorylate specific Ser (Thr) residues on target proteins. Given the availability of serine and threonine residues on the surface of globular proteins, how might a protein kinase select the "correct" residues to phosphorylate?
3. Assume that the flow diagram shown represents an amino acid biosynthetic pathway where G, J, and H are amino acids and A is a common precursor. Products G, H, and J are required by the cell. Enzymes catalyzing the steps are numbered. Suggest a plausible scheme for the regulation of specific enzymes by their products.

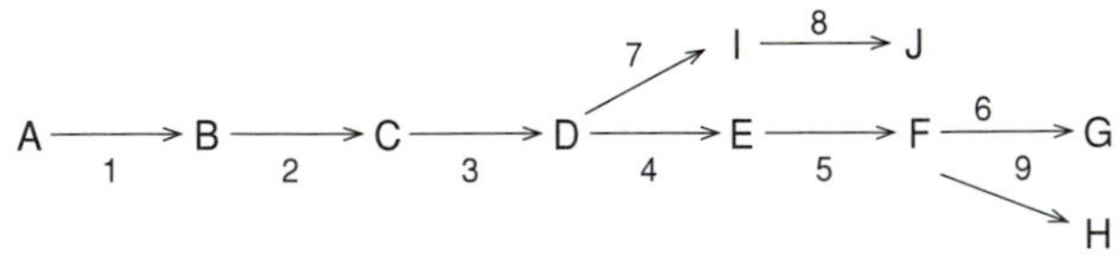

4. Aspartate carbamoyltransferase is an allosteric enzyme in which the active sites and the allosteric effector binding sites are on different subunits. Explain how it might be possible for an allosteric enzyme to have both kinds of sites on the same subunit.

5. ATP is both a substrate and an inhibitor of the enzyme phosphofructokinase (PFK). Although the substrate fructose-6-phosphate binds cooperatively to the active site, ATP does not bind cooperatively. Explain how ATP may be both a substrate and an inhibitor of PFK.
6. Calculate the substrate concentration [S] in terms of K_m for a hyperbolically responding enzyme when the velocity is 10% V_{max} or 90% V_{max}. What is the ratio of substrate concentrations that affect the ninefold velocity change ($S_{0.9}/S_{0.1}$)? How would the ($S_{0.9}/S_{0.1}$) ratio differ for an allosterically responding enzyme?
7. The substrate concentration yielding half-maximal velocity is equal to K_m for hyperbolically responding enzymes. Is this relationship true for allosterically or sigmoidally responding enzymes? How do positive or negative allosteric effectors change the substrate concentration required for half-maximal velocity?
8. Examine the relationship of aspartate carbamoyltransferase (ACTase) activity to aspartate concentration shown in the figure. Estimate the ($S_{0.9}/S_{0.1}$) ratio for the reaction under the following conditions: (a) normal curve, (b) plus 0.2-mM CTP, (c) plus 0.8-mM ATP. Do these ratios differ significantly? Explain.

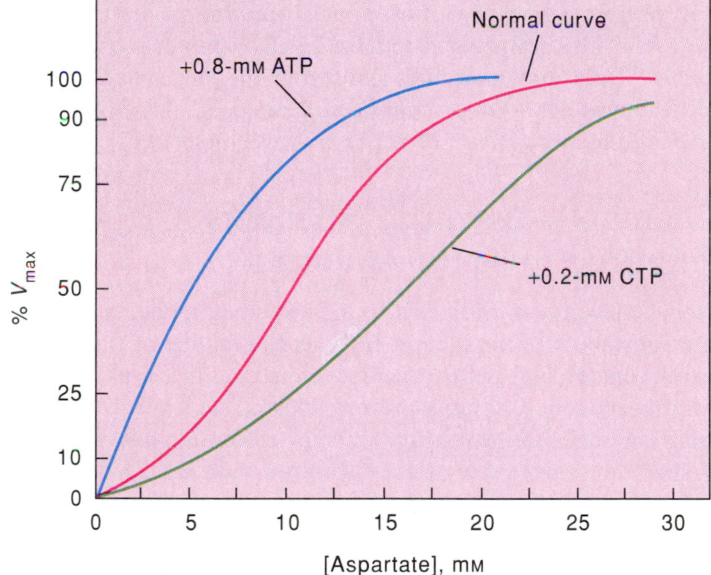

9. If you separated hemoglobin into dimers of ($\alpha + \beta$) subunits, would you expect the dimers to bind more or less O_2 at low O_2 tensions? Explain. What effect would 2,3-bisphosphoglycerate have on oxygen binding? (Review O_2 binding to hemoglobin in chapter 5.)
10. Treatment of ACTase with mercurials causes loss of allosteric regulation by ATP and CTP and eliminates positive cooperativity with aspartate. Can you suggest a strategy to "lock" an allosterically regulated enzyme into the active form without dissociating the subunits?
11. NAD^+ binding to the dimeric liver alcohol dehydrogenase (LADH) causes rearrangement of the active site residues to foster binding of the substrate, yet LADH shows no cooperativity in binding either NAD^+ or substrate. What element of the conformational change is lacking in the LADH that is present in allosteric enzymes?
12. Light-dependent activation of key enzymes in photosynthetic CO_2 fixation involves activation by thioredoxin-mediated reduction of critical disulfides on the enzymes. Write a reaction linking the reductant ferredoxin (a single-electron donor) to the reduction of protein disulfides using thioredoxin as an intermediate.
13. Assuming that the thioredoxin-dependent activation of proteins resulting from disulfide reduction is reversible, what possibilities exist for the "deactivation" of the enzymes? What are the metabolic and energetic ramifications to the cell in utilizing irreversibly activated or inactivated enzymes?
14. In some instances, protein kinases are activated or inhibited by low-molecular-weight modifiers. Explain how a metabolite may more effectively regulate an enzyme by modifying a protein kinase rather than directly inhibiting the target enzyme.
15. Cite some advantages in having calmodulin tightly associated with the Ca^{2+}-regulated enzyme rather than in an uncomplexed form in the cell.

11

CHAPTER

Vitamins, Coenzymes, and Metal Cofactors

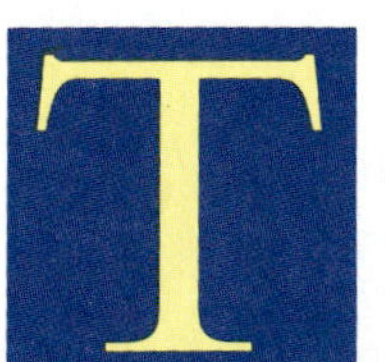

The types of chemical reactions that can be catalyzed by proteins alone are limited by the chemical properties of the functional groups found in the side chains of nine amino acids: the imidazole ring of histidine; the carboxyl groups of glutamate and aspartate; the hydroxy groups of serine, threonine, and tyrosine; the amino group of lysine; the guanidinium group of arginine; and the sulfhydryl group of cysteine. These groups can act as general acids and bases in catalyzing proton transfers and as nucleophilic catalysts in group transfer reactions.

Many metabolic reactions involve chemical changes that could not be brought about by the structures of the amino acid side chain functional groups in enzymes acting by themselves. In catalyzing these reactions, enzymes act in cooperation with other smaller organic molecules or metallic cations, which possess special chemical reactivities or structural properties that are useful for catalyzing reactions. In this chapter we will introduce these small molecules and survey the range of reactions that they catalyze. Our emphasis will be on the principles underlying the mechanisms of action of these molecules.

Vitamins—Essential Nutrients Required in Small Amounts

People knew it was important to include small amounts of certain substances in the diet as early as the middle of the eighteenth century, long before the biochemistry of these substances was understood. Vitamins are organic molecules essential in small quantities for healthy nutrition in rats or humans. The list of such molecules grew as they were purified from foodstuffs and shown to cure various disorders in animals maintained on deficient diets. The name "vitamine" was given in 1911 to the first vitamin to be isolated, thiamine. When it became clear that a number of essential organic micronutrients were not amines, the *-e* was dropped.

Vitamins are divided into water-soluble and lipid-soluble groups. In addition to vitamins there are vitaminlike nutrients that are required in small amounts by the organism and frequently function in similar capacities in the organism. These compounds are not classified as vitamins because rats and humans have a limited capacity to synthesize them, provided that the diet contains the essential precursors. Table 11.1 lists both vitamins and vitaminlike nutrients.

Most Coenzymes Are Modified Forms of Vitamins

In this chapter we will be concerned, not primarily with vitamins *per se,* but with coenzymes. Most coenzymes are modified forms of vitamins. The modifications take place in the organism

Table 11.1
Vitamins and Vitaminlike Nutrients

Vitamin	Function
Water-Soluble Vitamins	
Thiamine (B_1)	Precursor of the coenzyme thiamine pyrophosphate. Deficiency can cause beriberi.
Riboflavin (B_2)	Precursor of the coenzymes flavin mononucleotide and flavin adenine dinucleotide. Deficiency leads to growth retardation.
Pyridoxine (B_6)	Precursor of the coenzyme pyridoxal phosphate. Deficiency causes dermatitis in rats.
Nicotinic acid (niacin)	Precursor of the coenzymes nicotinamide adenine dinucleotide and nicotinamide adenine dinucleotide phosphate. Deficiency leads to pellagra.
Pantothenic acid	Precursor of coenzyme A (CoA). Deficiency leads to dermatitis in chickens.
Biotin	Precursor of the coenzyme biocytin. Deficiency leads to dermatitis in humans.
Folic acid	Precursor of the coenzyme tetrahydrofolic acid. Deficiency causes anemias.
Vitamin B_{12}	Precursor of the coenzyme deoxyadenosyl cobalamin. Deficiency leads to pernicious anemia.
Vitamin C	Cosubstrate in the hydroxylation of proline in collagen. Deficiency leads to scurvy.
Lipid-Soluble Vitamins	
Vitamin A	Vision, growth, and reproduction (see chapter 36).
Vitamin D	Regulation of calcium and phosphate metabolism (see chapter 24).
Vitamin E	Antisterility factor in rats.
Vitamin K	Important for blood coagulation.
Vitaminlike Nutrients	
Inositol	Mediator of hormone action (see chapters 22 and 24).
Choline	Important for integrity of cell membranes and lipid transport (see chapter 22).
Carnitine	Essential for transfer of fatty acids to mitochondria (see chapter 17).
α-Lipoic acid	Coenzyme in the oxidative decarboxylation of keto acids (this chapter).
p-Aminobenzoate (PABA)	Component of folic acid (this chapter).
Coenzyme Q (ubiquinones)	Important for electron transport in mitochondria (see chapter 15).

after ingestion of the vitamins. Coenzymes act in concert with enzymes to catalyze biochemical reactions. Tightly bound coenzymes are sometimes referred to as prosthetic groups. A coenzyme usually functions as a major component of the active site on the enzyme, which means that understanding the mechanism of coenzyme action usually requires a complete understanding of the catalytic process.

Water-Soluble Vitamins and Their Coenzymes

As we have noted, some vitamins are soluble in water, while others are soluble in lipids. In the following sections we will survey the range of biochemical reactions in which water-soluble coenzymes participate.

Thiamine Pyrophosphate (TPP) Is Involved in C—C and C—X Bond Cleavage

The structure of thiamine pyrophosphate is given in figure 11.1. The vitamin, thiamine or vitamin B_1, lacks the pyrophosphoryl group. Thiamine pyrophosphate (TPP) is the essential coenzyme involved in the actions of enzymes that catalyze cleavages

Figure 11.1

(*a*) Structure of thiamine pyrophosphate and (*b, c*) the bonds it cleaves or forms. Reactive part of the coenzyme and the bonds subject to cleavage in (*b*) and (*c*) are indicated in color.

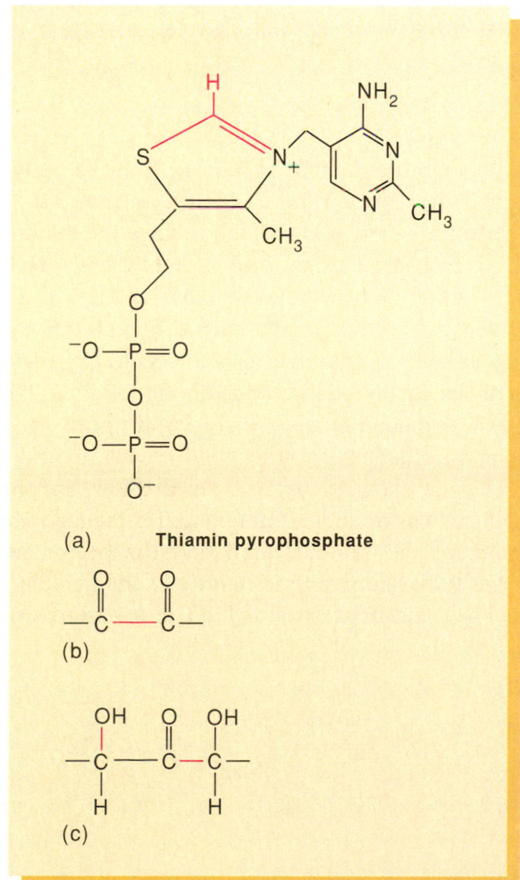

Figure 11.2

Mechanism of thiamine pyrophosphate action. Intermediate (*a*) is represented as a resonance-stabilized species. It arises from the decarboxylation of the pyruvate-thiamine pyrophosphate addition compound shown at left of (*a*) and in equation (2). It can react as a carbanion with acetaldehyde, pyruvate, or H^+ to form (*b*), (*c*), or (*d*), depending on the specificity of the enzyme. It can also be oxidized to acetyl-thiamine pyrophosphate (*e*) by other enzymes, such as pyruvate oxidase. The intermediates (*b*) through (*e*) are further transformed to the products shown by the actions of specific enzymes.

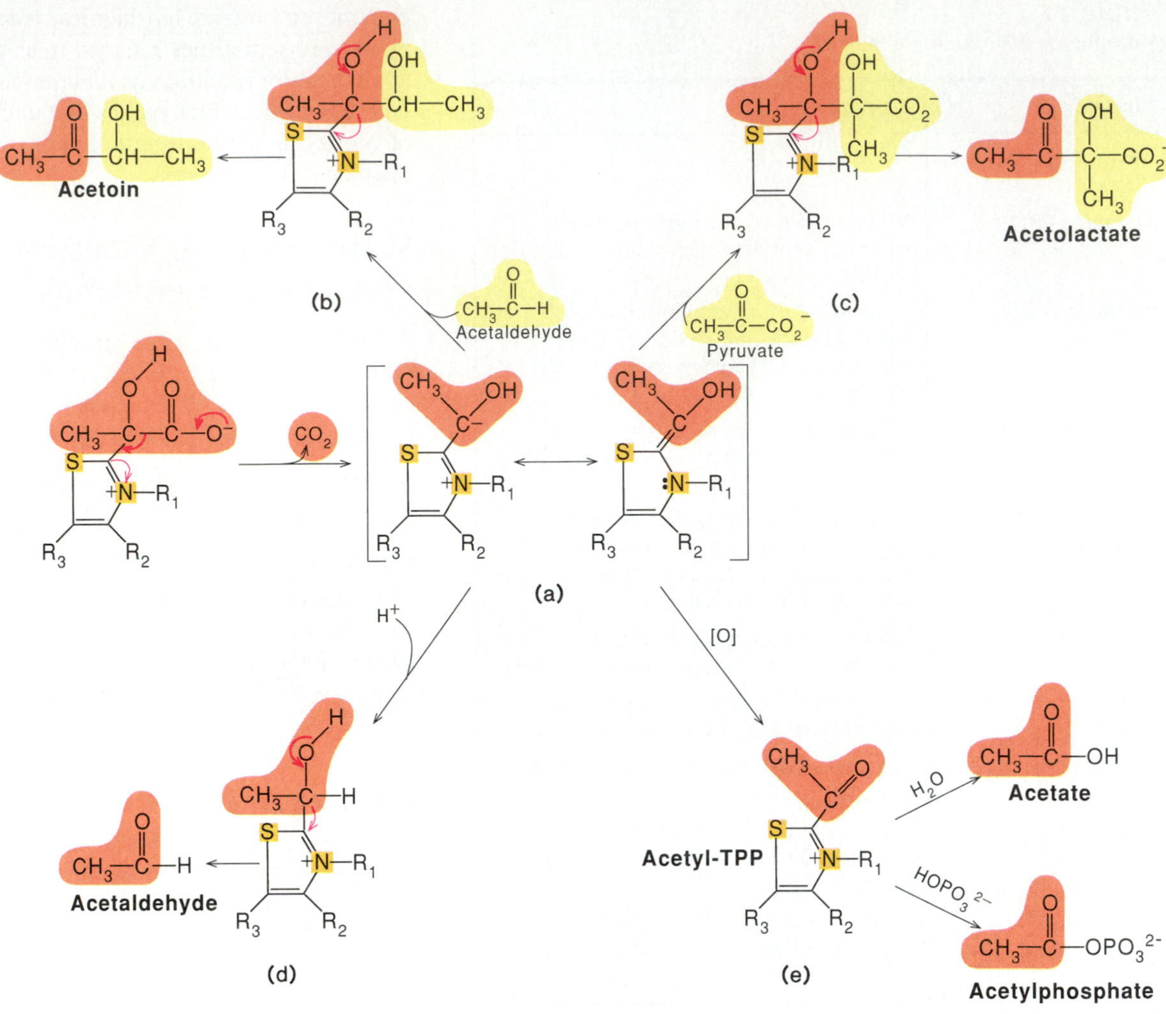

of the bonds indicated in color in figure 11.1. The bond scission in figure 11.1*b* is representative of those in many α-keto acid decarboxylations, nearly all of which require the action of TPP. The phosphoketolase reaction involves both cleavages shown in figure 11.1*c*, while the transketolase reaction (see fig. 13.22) involves the cleavage of the carbon–carbon bond but not the elimination of —OH. Acetolactate and acetoin arise by the formation of the carbon–carbon bond in figure 11.1*c* (the structures of these two compounds are shown in figure 11.2).

The mechanism for the bond cleavages indicated in figure 11.1*b* was clarified by Ronald Breslow. In one of the earliest applications of nuclear magnetic resonance to biochemical mechanisms, he demonstrated that the proton bonded to C—2 in the thiazolium ring is readily exchangeable with the protons of H_2O and deuterons of D_2O in a base-catalyzed reaction:

$$\text{(thiazolium, C2–H; } R_1, R_2, R_3) + OD^- \rightleftharpoons \text{(thiazolium C2 carbanion)} + HOD \rightleftharpoons \text{(thiazolium, C2–D)} + OH^- \quad (1)$$

The active intermediate shown in equation (1) undergoes nucleophilic addition to the bond of polar carbonyl groups in substrates to produce intermediates such as

$$CH_3-\overset{O}{\overset{\|}{C}}-CO_2^- + \text{(thiazolium C2 carbanion)} \underset{-H^+}{\overset{+H^+}{\rightleftharpoons}} CH_3-\overset{OH}{\overset{|}{C}}(-CO_2^-)-\text{(thiazolium C2)} \quad (2)$$

Figure 11.3

Structures of vitamin B_6 derivatives and the bonds cleaved or formed by the action of pyridoxal phosphate (*a*). The reactive part of the coenzyme is shown in color in (*a*). The bonds shown in color in (*d*) are the types of bonds in substrates that are subject to cleavage.

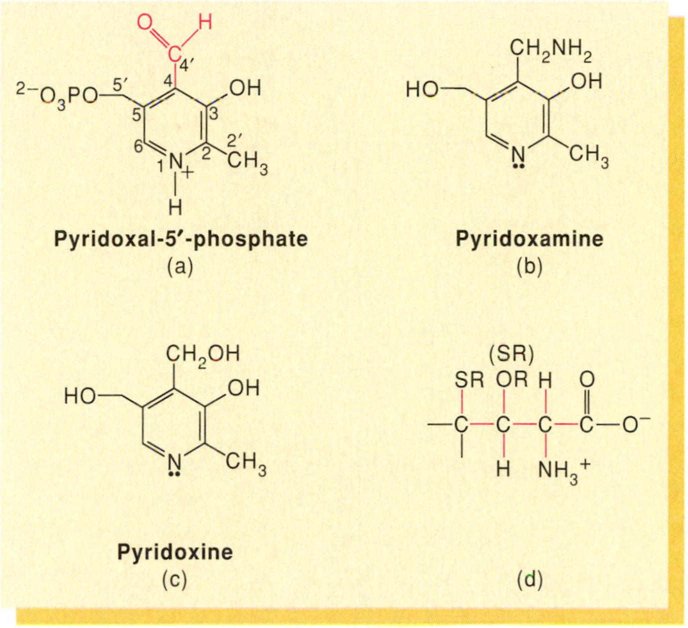

Intermediates of this type have the necessary chemical reactivity for cleaving the bonds indicated in figure 11.1*b* and *c*. The decarboxylated product of the pyruvate adduct shown in equation (2) is resonance-stabilized by the thiazolium ring (fig. 11.2*a*). This intermediate may be protonated to α-hydroxyethyl thiamine pyrophosphate (fig. 11.2*d*); alternatively, it may react with other electrophiles, such as the carbonyl groups of acetaldehyde or pyruvate, to form the species in figure 11.2*b* and *c;* or it may be oxidized to acetyl-thiamine pyrophosphate (fig. 11.2*e*). The fate of the intermediate depends on the reaction specificity of the enzyme with which the coenzyme is associated.

Pyridoxal-5′-Phosphate Is Required for a Variety of Reactions with α-Amino Acids

Pyridoxal-5′-phosphate is the coenzyme form of vitamin B_6, and has the structure shown in figure 11.3. The name vitamin B_6 is applied to any of a group of related compounds lacking the phosphoryl group, including pyridoxal, pyridoxamine, and pyridoxine.

Pyridoxal-5′-phosphate participates in many reactions with α-amino acids, including transaminations, α-decarboxylations, racemizations, α,β eliminations, β,γ eliminations, aldolizations, and the β decarboxylation of aspartic acid. The following equations illustrate several reactions in which pyridoxal-5′-phosphate acts as a coenzyme.

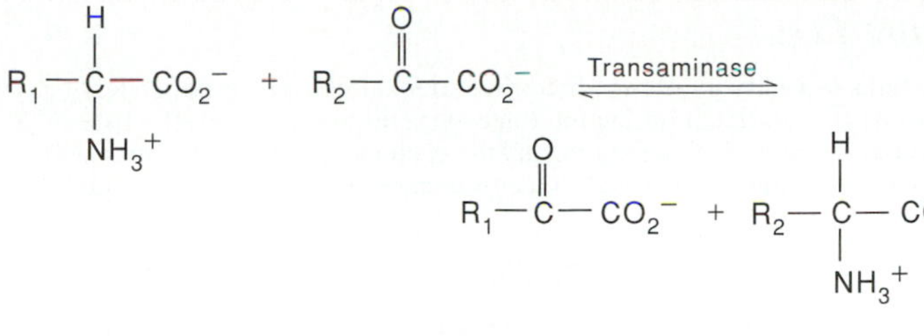

$$R_1-CH(NH_3^+)-CO_2^- + R_2-C(=O)-CO_2^- \xrightleftharpoons{\text{Transaminase}} R_1-C(=O)-CO_2^- + R_2-CH(NH_3^+)-CO_2^- \quad (3)$$

$$R-CH(NH_3^+)-CO_2^- + H^+ \xrightleftharpoons{\text{Decarboxylase}} CO_2 + R-CH_2-NH_3^+ \quad (4)$$

$$R-\overset{H}{C}(NH_3^+)-CO_2^- \xrightleftharpoons{\text{Racemase}} R-\overset{NH_3^+}{C}(H)-CO_2^- \quad (5)$$

$$R-CH(OH)-CH(NH_3^+)-CO_2^- \xrightleftharpoons{\text{Dehydratase}} R-CH_2-C(=O)-CO_2^- + NH_4^+ \quad (6)$$

$$R-CH(OH)-CH(NH_3^+)-CO_2^- \xrightleftharpoons{\text{Aldolase}} R-CHO + CH_2(NH_3^+)-CO_2^- \quad (7)$$

$$H_2O + S(CH_2-CH(NH_3^+)-CO_2^-)-CH_2-CH_2-CH(NH_3^+)-CO_2^- \xrightleftharpoons{\text{Cystathionase}} HS-CH_2-CH(NH_3^+)-CO_2^- + CH_3CH_2-C(=O)-CO_2^- + NH_4^+ \quad (8)$$

$$^-O_2C-CH_2-CH(NH_3^+)-CO_2^- + H^+ \xrightleftharpoons{\text{Aspartate-}\beta\text{-decarboxylase}} CO_2 + CH_3-CH(NH_3^+)-CO_2^- \quad (9)$$

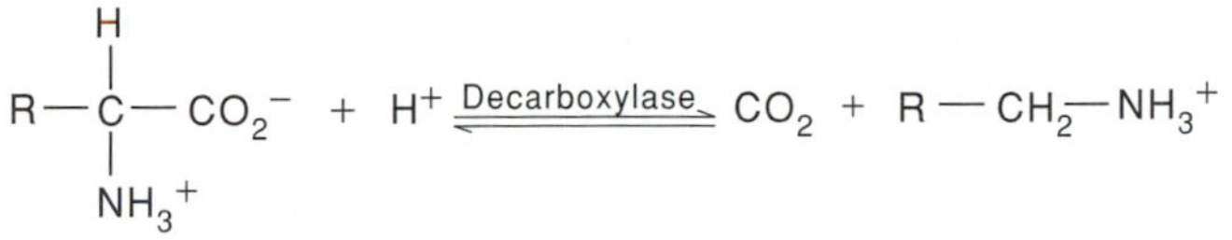

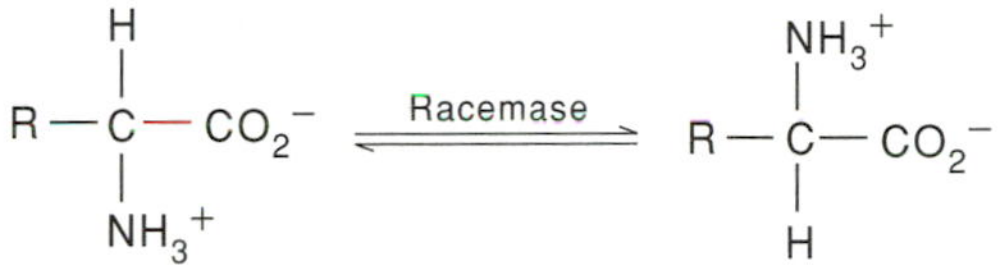

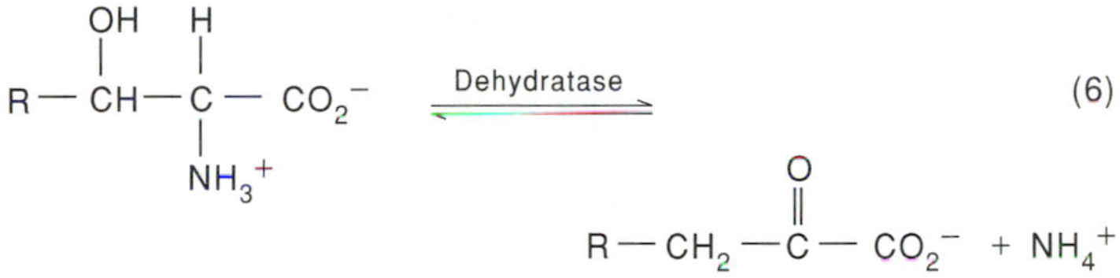

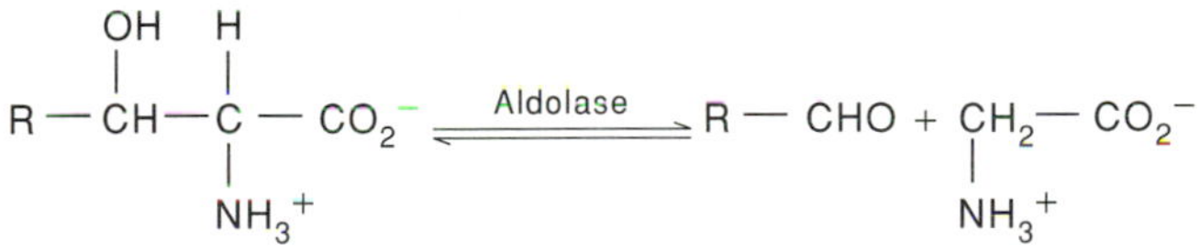

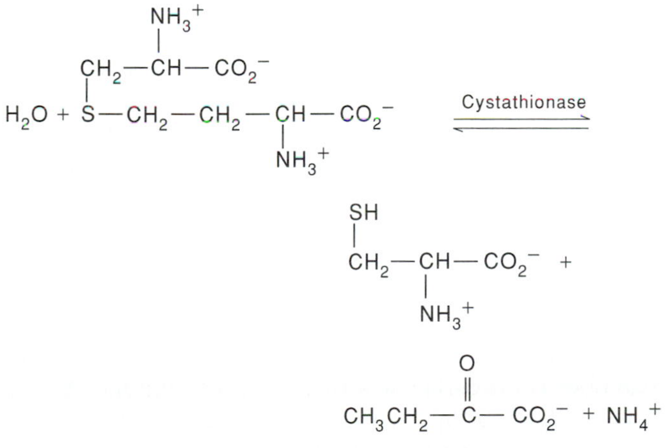

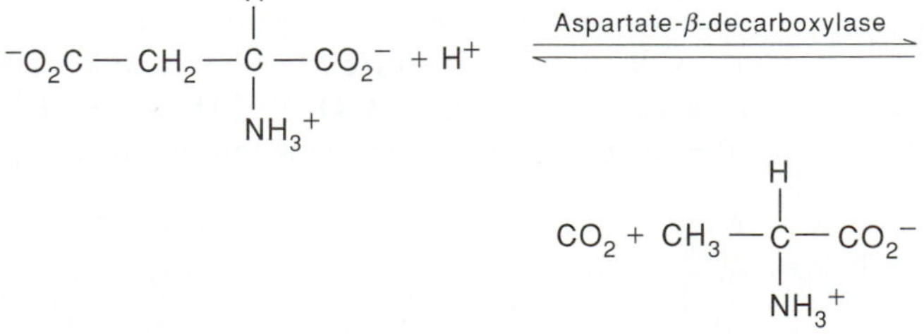

Figure 11.4

Structures of catalytic intermediates in pyridoxal-phosphate–dependent reactions. The initial aldimine intermediate resulting from Schiff's base formation between the coenzyme and the α-amino group of an amino acid (*a*). This aldimine is converted to the resonance-stabilized intermediate (*b*) by loss of a proton at the alpha carbon. Further enzyme-catalyzed proton transfers to intermediates (*c*) and (*d*) may occur, depending upon the specificity of a given enzyme. The enzymes use their general acids and bases to catalyze these proton transfers.

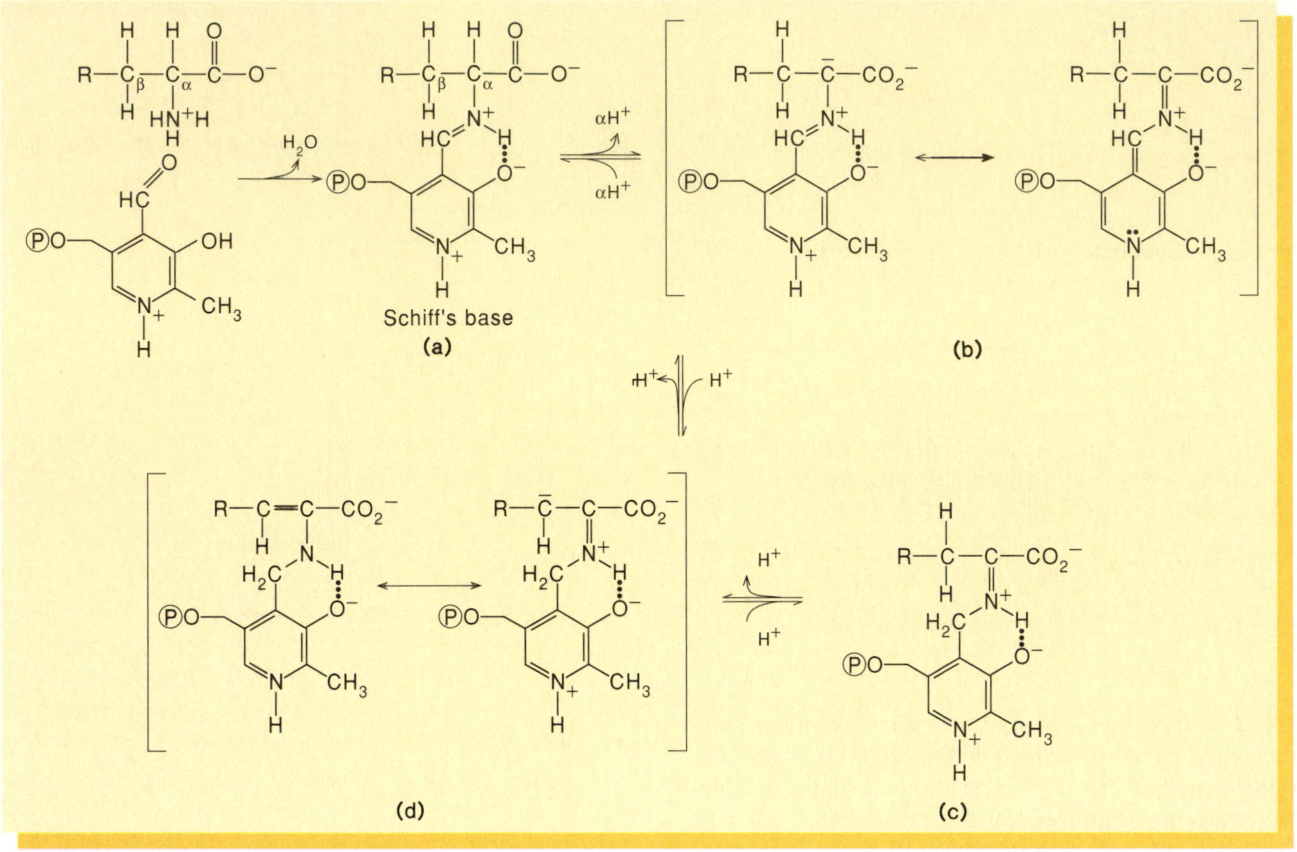

These equations involve bond cleavages of the type shown in color in figure 11.3*d*. Pyridoxal-5′-phosphate promotes these heterolytic bond cleavages by stabilizing the resulting electron pairs at the α- or β-carbon atoms of α-amino acids. To do this, the aldehyde group of the coenzyme first reacts with the α-amino group of an amino acid to produce an aldimine (fig. 11.4*a*) or Schiff's base, which is internally stabilized by H bonding. Loss of the α hydrogen as H^+ produces a resonance-stabilized species (fig. 11.4*b*) in which the electron pair is delocalized into the pyridinium system. This active intermediate may undergo further reactions at the carbon to form products determined by the reaction specificity of the enzyme. If, for example, the enzyme is a racemase, the species resulting from the loss of the proton from the α carbon may accept a proton from the opposite side to produce, ultimately, the enantiomer of the amino acid.

When the substrate is substituted at the β carbon with a potential leaving group, such as —OH, —SH, $—OPO_3^{3-}$ (see fig. 11.3*d*), the corresponding α-carbanion intermediate (see fig. 11.4*b*) can eliminate the group. This is an essential step in α,β eliminations, such as the dehydratase reaction (equation 6; also see fig. 19.11). Upon hydrolysis, the elimination intermediate produces pyridoxal-5′-phosphate and the substrate-derived enamine, which spontaneously hydrolyzes to ammonia and an α-keto acid.

The full series of intermediates in a transamination is shown in figure 11.5*a*. After protonation at the aldimine carbon of pyridoxal-5′-phosphate (step 3), hydrolysis (step 4) forms an α-keto acid and pyridoxamine-5′-phosphate. The reverse of this sequence with a second α-keto acid (steps 5 through 8) completes the transamination reaction.

An intermediate analogous to that in figure 11.4*b*, but generated from glycine and so lacking the β and γ carbons, can react as a carbanion with an aldehyde to produce a β-hydroxy-α-amino acid. These reactions are catalyzed by aldolases such as threonine aldolase or serine hydroxymethyl transferase (see fig. 19.11).

β-Decarboxylases (fig. 11.5*b*) generate intermediates analogous to that in figure 11.4*b* by catalyzing the elimination of CO_2 instead of H^+ from the intermediate in figure 11.4*a* (step 4, fig. 11.5*b*). Protonation of the α-carbanionic intermediates by protons from H_2O, followed by hydrolysis of the resulting imines, produces the amines corresponding to the replacement of the carboxylate group in the substrate by a proton (steps 5 through 8, fig. 11.5*b*).

The stability of the resonance hybrid (see fig. 11.4*b*) accounts for the catalytic action of pyridoxal-5′-phosphate in the reactions shown in equations (3) through (7).

Returning to the intermediate in figure 11.4*c*, we see that elimination of a β proton produces a β-carbanion (fig. 11.4*d*) that is stabilized by resonance with the neighboring protonated imine. This carbanion can eliminate a good leaving group from the carbon, exemplified by the γ-cystathionase-catalyzed elimination of cysteine from cystathionine in equation (8). The β decarboxylation of aspartate (equation 9) proceeds by elimination of a β-carbanionic intermediate like that in figure 11.4*d* from the ketimine, analogous to the intermediate produced by loss of the α proton from the aldimine of aspartate with pyridoxal-5′-phosphate.

The fundamental biochemical function of pyridoxal-5′-phosphate is the formation of aldimines with α-amino acids that stabilize the development of carbanionic character at the α and β carbons of α-amino acids in intermediates such as those in figures 11.4*b* and *c*. Enzymes acting alone cannot stabilize these carbanions and so cannot, by themselves, catalyze reactions requiring their formation as intermediates.

Nicotinamide Coenzymes Are Used in Reactions Involving Hydride Transfers

Nicotinamide adenine dinucleotide (NAD^+) is one of the two coenzymatic forms of nicotinamide (fig. 11.6). The other is nicotinamide adenine dinucleotide phosphate ($NADP^+$), which differs from NAD^+ by the presence of a phosphate group at C-2′ of the adenosyl moiety.

The nicotinamide coenzymes are biological carriers of reducing equivalents, i.e., electrons. The most common function of NAD^+ is to accept two electrons and a proton (H^- equivalent) from a substrate undergoing metabolic oxidation to produce NADH, the reduced form of the coenzyme. This then diffuses or is transported to the terminal-electron transfer sites of the cell and reoxidized by terminal-electron acceptors, O_2 in aerobic organisms, with the concomitant formation of ATP (chapter 15). Equations (10), (11), and (12) are typical reactions in which NAD^+ acts as such an acceptor.

$$NAD^+ + CH_3CH_2OH \xrightleftharpoons{\text{Alcohol dehydrogenase}} CH_3{-}\overset{\displaystyle O}{\overset{\|}{C}}H + NADH + H^+ \qquad (10)$$

$$^-O_2C(CH_2)_2\overset{\displaystyle NH_3^+}{\overset{|}{C}}HCO_2^- + NAD^+ + H_2O \xrightleftharpoons{\text{Glutamate dehydrogenase}} {}^-O_2C(CH_2)_2\overset{\displaystyle O}{\overset{\|}{C}}CO_2^- + NADH + NH_4^+ + H^+ \qquad (11)$$

$$HPO_4^{2-} + {}^{2-}O_3POCH_2\overset{\displaystyle OH}{\overset{|}{C}}H{-}\overset{\displaystyle O}{\overset{\|}{C}}H + NAD^+ \xrightleftharpoons{\text{Glyceraldehyde-3P dehydrogenase}} {}^{2-}O_3POCH_2\overset{\displaystyle OH}{\overset{|}{C}}H\overset{\displaystyle O}{\overset{\|}{C}}OPO_3^{2-} + NADH + H^+ \qquad (12)$$

The chemical mechanisms by which NAD^+ is reduced to NADH in equations (10) through (12) are probably similar, as represented in generalized forms in equation (13).

$$R{-}^+N\text{(nicotinamide ring, 4-H; }{-}C(=O)NH_2) \;\; H{-}\overset{|}{\underset{|}{C}}{-}O{-}H \;\; \ddot{B}{-}E \rightleftharpoons R{-}\ddot{N}\text{(dihydronicotinamide ring, 4-}H_2\text{; }{-}C(=O)NH_2) \;\; \rangle C{=}O \;\; H{-}^+B{-}E \qquad (13)$$

According to this formulation, the immediate oxidation product in equation (11), where $-NH_2$ replaces $-OH$ in equation (13), would be the imine of α-ketoglutarate, which would quickly undergo hydrolysis to α-ketoglutarate and ammonia in aqueous solution. The oxidation of an aldehyde group catalyzed by glyceraldehyde-3-phosphate dehydrogenase [equation (12)] also can be understood on the basis of this formulation once it is realized that there is an essential —SH group at the active site that is transiently acylated during the course of the reaction. The —SH group reacts with the aldehyde group of glyceraldehyde-3-phosphate according to equation (14), forming a thiohemiacetal which becomes oxidized. The resulting acyl-enzyme then reacts with phosphate to produce glycerate-1,3-bisphosphate.

$$E{-}SH + \overset{\displaystyle O}{\overset{\|}{C}}H(CHOH)(CH_2OPO_3^{2-}) \rightleftharpoons E{-}S{-}\overset{\displaystyle OH}{\overset{|}{C}}{-}H(CHOH)(CH_2OPO_3^{2-}) \xrightleftharpoons{NAD^+ \;\to\; NADH + H^+} E{-}S{-}\overset{\displaystyle O}{\overset{\|}{C}}(CHOH)(CH_2OPO_3^{2-}) \xrightarrow{HPO_4^{2-}} E{-}SH + \overset{\displaystyle O}{\overset{\|}{C}}{-}OPO_3^{2-}(CHOH)(CH_2OPO_3^{2-}) \qquad (14)$$

Figure 11.5

Mechanisms of action of pyridoxal phosphate: (*a*) in glutamate-oxaloacetate transaminase, and (*b*) in aspartate ß-decarboxylase.

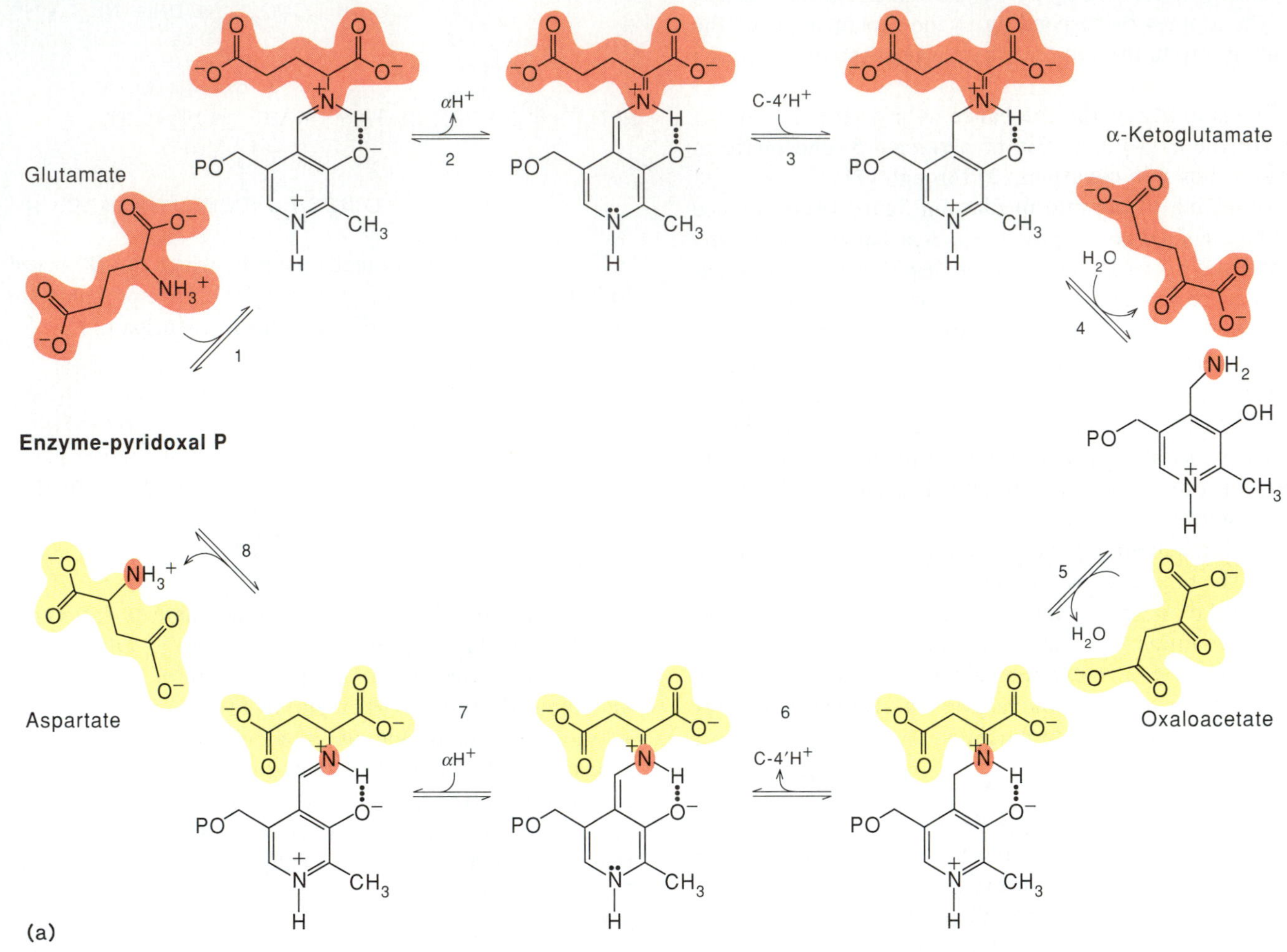

Figure 11.6

Structures of nicotinamide and nicotinamide coenzymes. The reactive sites of the coenzymes are shown in color.

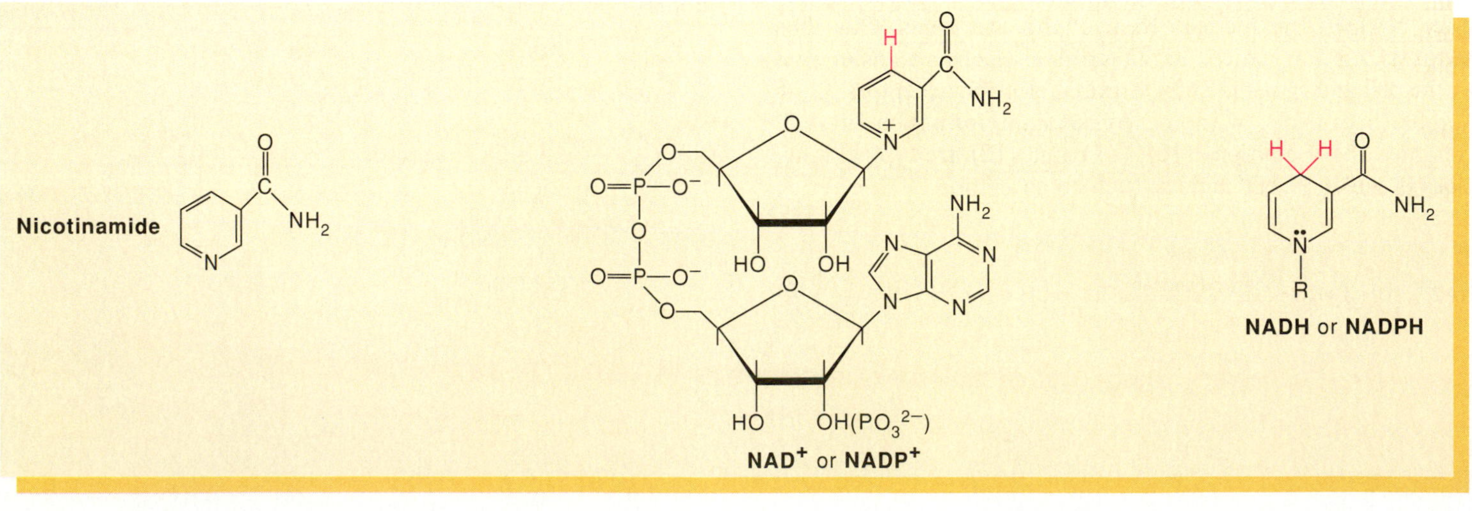

Nicotinamide

NAD⁺ or **NADP⁺**

NADH or **NADPH**

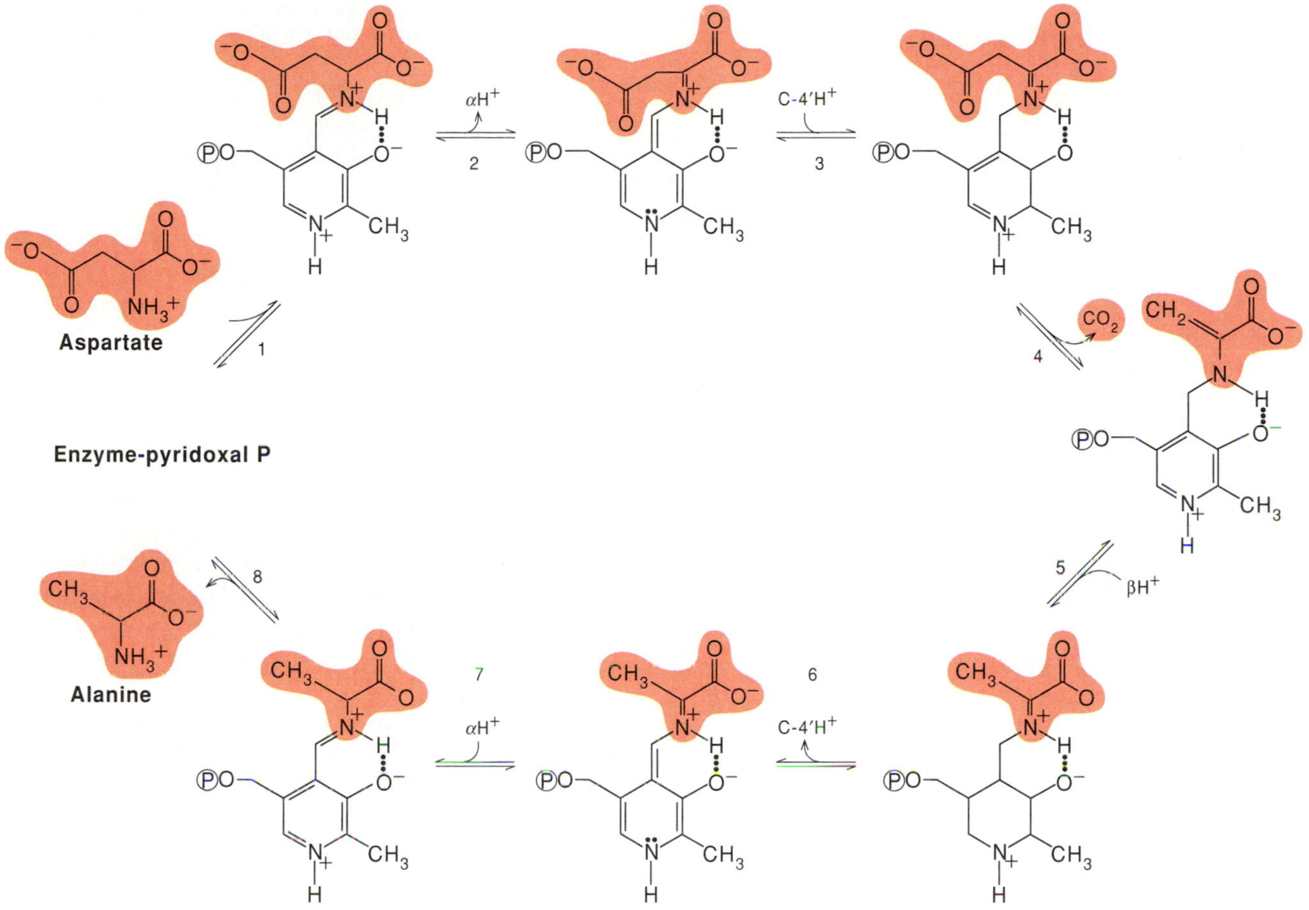

(b)

Equation (13) implies that the hydrogen atom and two electrons are transferred in a concerted process, i.e., as a hydride equivalent, with the quaternary nitrogen in the pyridinium ring serving as an electron sink. The hydrogen atom is certainly transferred directly; this reaction is discussed in the following section.

In a series of classic experiments, Westheimer, Vennesland, and co-workers demonstrated that hydrogen transfer between NAD^+ and substrates such as those in equations (10) through (12) are direct hydrogen transfers and occur with stereospecificity. The experiments with alcohol dehydrogenase were the first to establish these points, and they were the earliest to define the remarkable stereospecificity of enzymatic action at prochiral centers involving chemically equivalent hydrogen atoms. (For a discussion of prochiral centers, see box 11A.) Once the stereospecificity of hydrogen transfer to NAD^+ and the absolute configurations of the molecules were established, it became possible to formulate these processes as set forth in equations (15) through (17) for alcohol dehydrogenase.

$$CH_3{-}CD_2{-}OH + R{-}^{+}N\text{(nicotinamide, 4-H, 3-CONH}_2) \longrightarrow CH_3{-}\overset{O}{\overset{\|}{C}}{-}D + R{-}N\text{:(4-D, 4-H, 3-CONH}_2) + H^+ \quad (15)$$

$$H^+ + CH_3{-}\overset{O}{\overset{\|}{C}}{-}H + R{-}N\text{:(4-D, 4-H, 3-CONH}_2) \longrightarrow CH_3{-}C(OH)(D)(H) + R{-}^{+}N\text{(4-H, 3-CONH}_2) \quad (16)$$

$$H^+ + CH_3{-}\overset{O}{\overset{\|}{C}}{-}D + R{-}N\text{:(4-H, 4-H, 3-CONH}_2) \longrightarrow CH_3{-}C(OH)(H)(D) + R{-}^{+}N\text{(4-H, 3-CONH}_2) \quad (17)$$

The tracing of deuterium through these transformations established quite clearly that enzymes are stereospecific in abstracting chemically equivalent hydrogens from prochiral centers and in transferring hydrogens specifically to one face of planar molecules, even in molecules as small as acetaldehyde. This specificity is thought to be a natural consequence of the fact that enzymes are asymmetrical molecules that form highly stereoselective complexes with their substrates, even some that have planes of symmetry or are themselves planar molecules. The situation is illustrated schematically in figure 11.7, which shows how specific binding interactions can lead to stereospecific hydrogen transfer between acetaldehyde and NADH.

In NADH, the two hydrogens bonded to nicotinamide C-4 are chemically equivalent, so that from a purely chemical standpoint either could be transferred to an aldehyde or ketone. It is their topographic inequivalence that leads to stereospecificity in enzymatic reactions: however, the enzymes do not all exhibit the same stereospecificity for catalyzing hydrogen transfer from this center or in forming this center from NAD^+.

Figure 11.7

Stereospecificity of hydrogen transfer in nicotinamide coenzymes. This example shows that a highly stereoselective complex with substrates can result in a stereospecific reaction even when the substrate has a plane of symmetry. The arrows represent hypothetical enzyme-binding interactions. Because of the way in which the coenzyme and the substrate are bound and oriented toward one another, one hydrogen is predisposed for transfer, the other is not. The subscripted stereochemical symbols R and S are explained in box 11A.

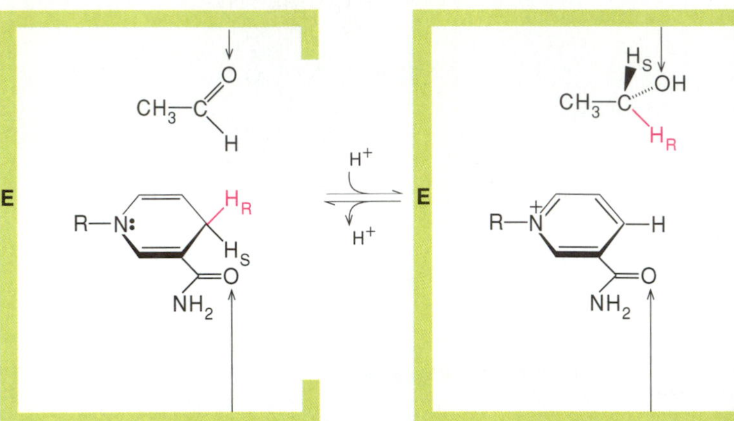

Those enzymes such as alcohol dehydrogenase which catalyze transfer of the pro-A hydrogen in NADH (see fig. 11.7) are known as *R*-side-specific enzymes, and those transferring the other hydrogen, the pro-S hydrogen, are known as *S*-side-specific enzymes.

In addition to acting as a cellular electron carrier, NAD^+ also acts as a true coenzyme with certain enzymes. Enzymes are sometimes confronted with the problem of catalyzing such reactions as epimerizations, aldolizations, and eliminations on substrates lacking the intrinsic chemical reactivities required for these reactions to occur at significant rates. Sometimes such reactivities can be introduced into the substrate by oxidizing an appropriate alcohol group to a carbonyl group, and the enzyme is then found to contain NAD^+ as a tightly bound coenzyme. NAD^+ functions coenzymatically by transiently oxidizing the key alcohol group to the carbonyl level, producing an oxidatively activated intermediate whose further transformation is catalyzed by the enzyme. In the last step, the carbonyl group is reduced back to the hydroxyl group by the transiently formed NADH. A reaction of this type is illustrated in figure 11.8 for the enzyme UDP-galactose-4-epimerase, which contains tightly bound NAD^+.

Flavins Are Used in Reactions Where One or Two Electron Transfers Are Involved

Flavin adenine dinucleotide (FAD) (fig. 11.9) and flavin mononucleotide (FMN) are the coenzymatically active forms of vitamin B_2, riboflavin. Riboflavin is the N^{10}-ribityl isoalloxazine portion of FAD, which is enzymatically converted into its coenzymatic forms first by phosphorylation of the ribityl C-5′ hydroxy group to FMN and then by adenylylation to FAD. FMN and FAD appear to be functionally equivalent coenzymes, and the one that is involved with a given enzyme appears to be a matter of enzymatic binding specificity.

Figure 11.8

Mechanism of NAD⁺ action in UDPgalactose-4-epimerase. There is no net oxidation or reduction. Only the intermediate is oxidized.

UDPgalactose-4-epimerase

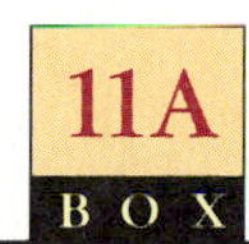

The R, S *System for Naming Stereoisomer Configurations*

"Chiral" and "prochiral" are derived from the Greek word χειρ, meaning "hand," and as we have seen, they refer to "handedness" in stereoisomeric or potentially stereoisomeric centers. In earlier chapters we encountered the use of the prefixes L and D for distinguishing between stereoisomers with left-handed and right-handed twist, respectively. Another set of stereochemical symbols frequently used for designating two stereoisomers is the pair R and S. These symbols are assigned as follows.

For a tetrahedral carbon (or other tetrahedral atom), the four different substituents are assigned relative priorities by applying rules that generally accord higher priority to groups having the larger summation of atomic weights. (You will find these rules set out in the article by G. Popják that is listed in the Selected Readings at the end of this chapter.) Once the priorities are assigned, the atom is viewed from the side opposite the lowest priority group, and the symbol R, for "rectus," is assigned if the remaining groups appear in clockwise order from highest to lowest priority. The symbol S, for "sinister," is assigned if they appear in counterclockwise order.

For atoms whose substituent groups are $a < b < c < d$ in order of increasing priority, the following structures are those of the R and S isomers:

Atoms in which two of the groups are identical are said to be prochiral, since the elevation of one of the identical groups to a higher priority would lead to a chiral center. Carbon-1 in ethanol and nicotinamide C-4 in NADH are prochiral centers, since they each have two hydrogens and two other substituents. In figure 11.7, the two hydrogens are subscripted R or S for the configurational designations that would result from according higher priority to one or the other. One means of granting such priority is by the use of heavy isotopes of hydrogen. Thus deuterium has a higher priority than protium, so that the configuration of deuterioethanol in equation (16) is R, while in equation (17) it is S.

The catalytically functional portion of the coenzymes is the isoalloxazine ring, specifically N-5 and C-4a (see fig. 11.9*b*), which are thought to be the immediate locus of catalytic function, although the entire chromophoric system extending over N-5, C-4a, C-10, N-1, and C-2 should be regarded as an indivisible catalytic entity, as are the nicotinamide, pyridinium, and thioazolium rings of NAD⁺, pyridoxal phosphate, and thiamine pyrophosphate.

Flavin-containing enzymes are known as flavoproteins and, when purified, normally contain their full complements of FAD or FMN. The bright yellow color of flavoproteins is due to the isoalloxazine chromophore in its oxidized form. In a few flavoproteins, the coenzyme is known to be covalently bonded to the protein by means of a sulfhydryl or imidazole group at the C-8 methyl group and in at least one case at C-6. In most flavoproteins, the coenzymes are tightly but noncovalently bound, and many can be resolved into apoenzymes that can be reconstituted to holoenzymes by readdition of FAD or FMN.

Figure 11.9

Structures of the vitamin riboflavin (*a*) and the derived flavin coenzymes (*b*). Like NAD^+ and $NADP^+$, the coenzyme pair FMN and FAD are functionally equivalent coenzymes, and the coenzyme involved with a given enzyme appears to be a matter of enzymatic binding specificity. The catalytically functional portion of the coenzymes is shown in color.

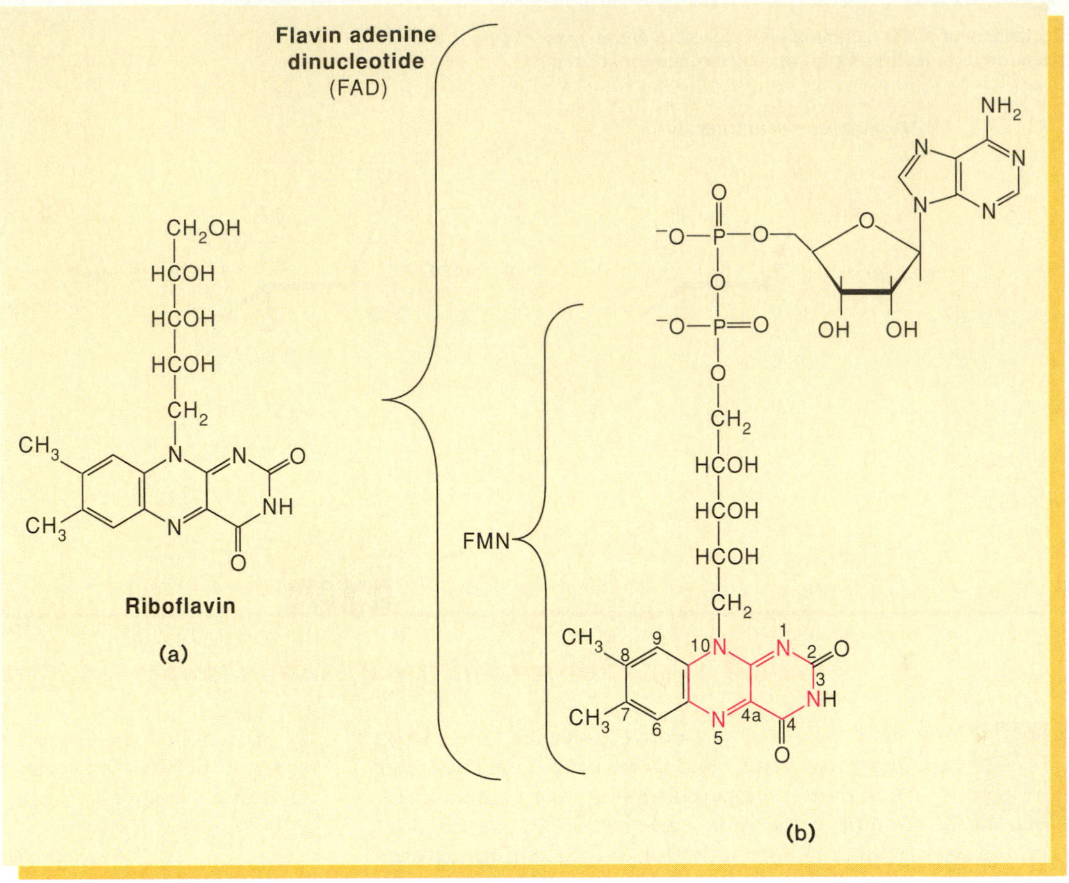

Flavin coenzymes exist in three spectrally distinguishable oxidation states that account in part for their catalytic functions; the yellow oxidized form, the red or blue one-electron-reduced form, and the colorless two-electron-reduced form. Their structures are depicted in figure 11.10. These and other less well-defined forms often have been detected spectrally as intermediates in flavoprotein catalysis.

Flavins are very versatile redox coenzymes. Flavoproteins are dehydrogenases, oxidases, and oxygenases that catalyze a variety of reactions on an equal variety of substrate types. Since these classes of enzymes do not consist exclusively of flavoproteins, it is difficult to define catalytic specificity for flavins. Biological electron acceptors and donors in flavin-mediated reactions can be two-electron acceptors, such as NAD^+ or $NADP^+$, or a variety of one-electron acceptor systems, such as cytochromes (Fe^{2+}/Fe^{3+}) and quinones, and molecular oxygen is an electron acceptor for flavoprotein oxidases as well as the source of oxygen for oxygenases. The only obviously common aspect of flavin-dependent reactions is that all are redox reactions.

Typical reactions catalyzed by flavoproteins are listed in table 11.2, which groups flavoproteins into those that do not utilize molecular oxygen as a substrate and those that do. You can best appreciate the significance of this difference when you realize that $FADH_2$, a likely intermediate in many flavoprotein reactions, spontaneously reacts with O_2 to produce H_2O_2.In the case of the dehydrogenases, therefore, either $FADH_2$ is not an intermediate or it is somehow prevented from reacting with O_2. Among the dehydrogenases are several that utilize the two-electron acceptor substrates NAD^+ or $NADP^+$, and it is reasonable to suppose that the two-electron reduction of NAD^+ by an intermediate $E \cdot FADH_2$ might be involved. Also listed in table 11.2 are other dehydrogenases for which the electron acceptors from $E \cdot FADH_2$ are not given. These enzymes are membrane-bound and transfer electrons directly to membrane-bound acceptors, mainly one-electron acceptors such as quinones and cytochromes (Fe^{2+}/Fe^{3+}). The stability of the flavin semiquinone, FAD · and FMN · in figure 11.10, gives flavins the capability to interact with one-electron acceptors in electron-transport systems.

The other classes of flavoproteins in table 11.2 interact with molecular oxygen either as the electron-acceptor substrates in redox reactions catalyzed by oxidases or as the substrate sources of oxygen atoms for oxygenases. Molecular oxygen also serves as an electron acceptor and source of oxygen for metalloflavoproteins and dioxygenases, which are not listed in table

Figure 11.10

Oxidation states of flavin coenzymes. The flavin coenzymes exist in three spectrally distinguishable oxidation states that account in part for their catalytic functions. They are the yellow oxidized form, the red or blue one-electron-reduced form, and the colorless two-electron-reduced form.

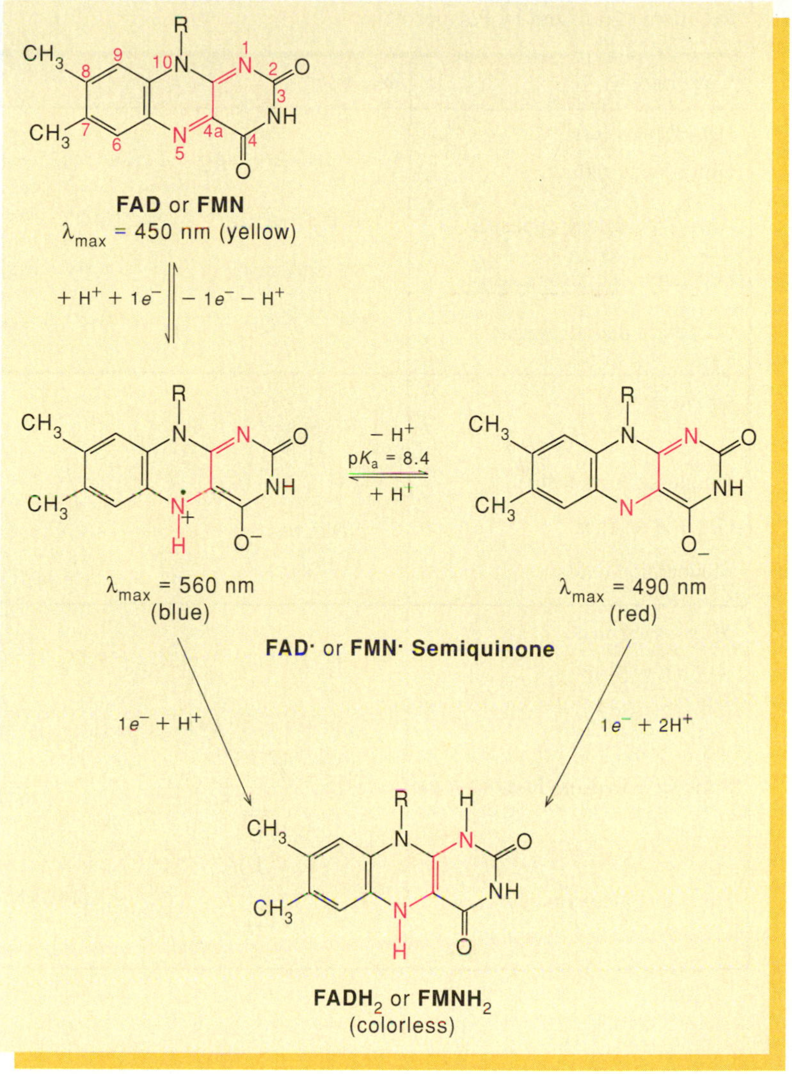

11.2. These enzymes catalyze more complex reactions, involving catalytic redox components such as metal ions and metal-sulfur clusters in addition to flavin coenzymes.

The mechanisms of action of flavin coenzymes are currently under active investigation. A recurrent theme appears to be the probable involvement of $FADH_2$ or $FMNH_2$ as transient intermediates in a variety of flavoprotein reactions. Figure 11.11 illustrates a reasonable catalytic pathway for the first enzyme listed in table 11.2; this reaction shows the likely involvement of E · $FADH_2$ in each case. The mechanisms by which E · FAD is reduced to E · $FADH_2$ by NADPH in the forward direction and by glutathione in the reverse direction are undoubtedly different. The mechanism shown is one recently proposed.

The biochemical importance of flavin coenzymes appears to be their versatility in mediating a variety of redox processes, including electron transfer and the activation of molecular oxygen for oxygenation reactions. The detailed mechanisms of oxygen activation are not well understood. An especially important manifestation of their redox versatility is their ability to serve as the switch point from the two-electron processes, which predominate in cytosolic carbon metabolism, to the one-electron transfer processes, which predominate in membrane-associated terminal electron-transfer pathways. In mammalian cells, for example, the end products of the aerobic metabolism of glucose are CO_2 and NADH (see chapter 14). The terminal electron-transfer pathway is a membrane-bound system of cytochromes, nonheme iron proteins, and copper-heme proteins—all one-electron acceptors that transfer electrons ultimately to O_2 to produce H_2O and NAD^+ with the concomitant production of ATP from ADP and P_i. The interaction of NADH with this pathway is mediated by NADH dehydrogenase, a flavoprotein that couples the two-electron oxidation of NADH with the one-electron reductive processes of the membrane.

Table 11.2
Reactions Catalyzed by Flavoproteins

Flavoprotein	Reaction
Dehydrogenases	
Glutathione reductase	$H^+ + GSSG + NADPH \rightleftharpoons 2\,GSH + NADP^+$
Acyl-CoA dehydrogenases	$RCH_2CH_2COSCoA + NAD^+ \rightleftharpoons RCH{=}CHCOSCoA + NADH + H^+$
Succinate dehydrogenase	$^-O_2CCH_2CH_2CO_2^- + E \cdot FAD \rightleftharpoons {}^-O_2CCH{=}CHCO_2^- + E \cdot FADH_2$
D-Lactate dehydrogenase	$CH_3{-}CHOH{-}CO_2^- + E \cdot FAD \rightleftharpoons CH_3{-}CO{-}CO_2^- + E \cdot FADH_2$
Oxidases	
Amino acid oxidases	$R{-}CH(NH_3^+){-}CO_2^- + O_2 + H_2O \longrightarrow R{-}CO{-}CO_2^- + H_2O_2 + NH_4^+$
Glucose oxidase	D-Glucose $+ O_2 \longrightarrow$ D-gluconolactone $+ H_2O_2$
Monoamine oxidase	$R{-}CH_2\overset{+}{N}H_3 + O_2 + H_2O \longrightarrow R{-}CHO + H_2O_2 + \overset{+}{N}H_4$
Monooxygenases	
Lactate oxidase	$CH_3{-}CHOH{-}CO_2^- + O_2 \longrightarrow CH_3{-}CO_2^- + CO_2 + H_2O$
Salicylate hydroxylase	$2H^+ +$ 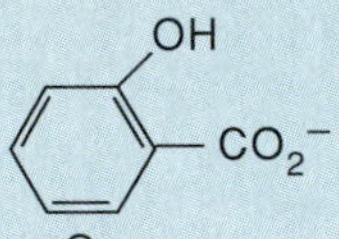 salicylate (CO_2^-, OH on benzene ring) $+ O_2 + NADH \longrightarrow$ 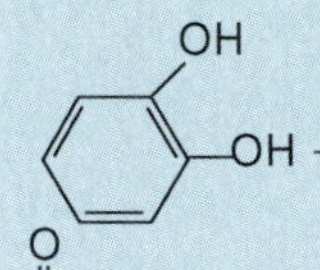 catechol (OH, OH on benzene ring) $+ CO_2 + NAD^+ + H_2O$
Ketone monooxygenase	$H^+ +$ 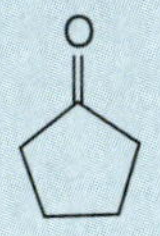 cyclopentanone $+ O_2 + NADPH \longrightarrow$ 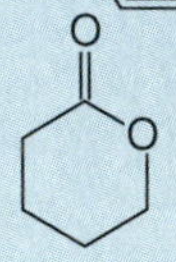 δ-valerolactone $+ NADP^+ + H_2O$

Figure 11.11

Mechanism of the flavin-dependent glutathione reductase reaction. The first steps, not shown, involve the reduction of FAD to $FADH_2$ by NADPH and the binding of glutathione (glutathione is a sulfhydryl compound, see figure 19.32). The mechanism by which oxidized glutathione is reduced by the E•$FADH_2$ is shown.

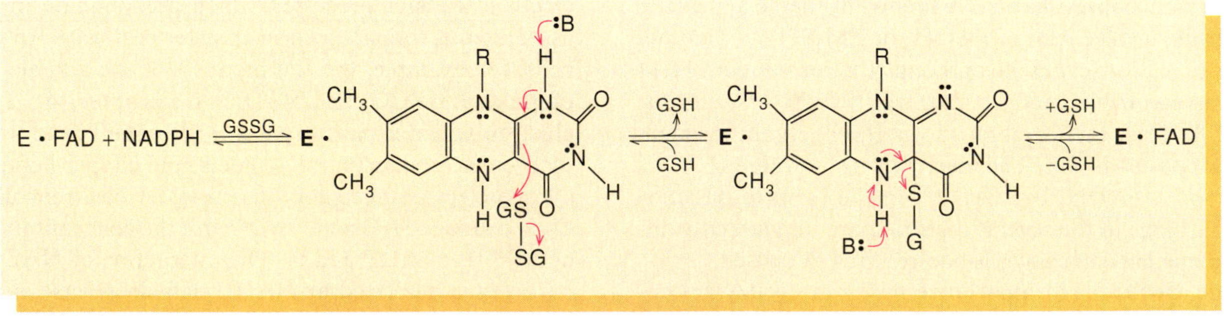

Figure 11.12

Structures of the vitamin pantothenic acid and coenzyme A. The terminal —SH is the reactive group in coenzyme A (CoASH).

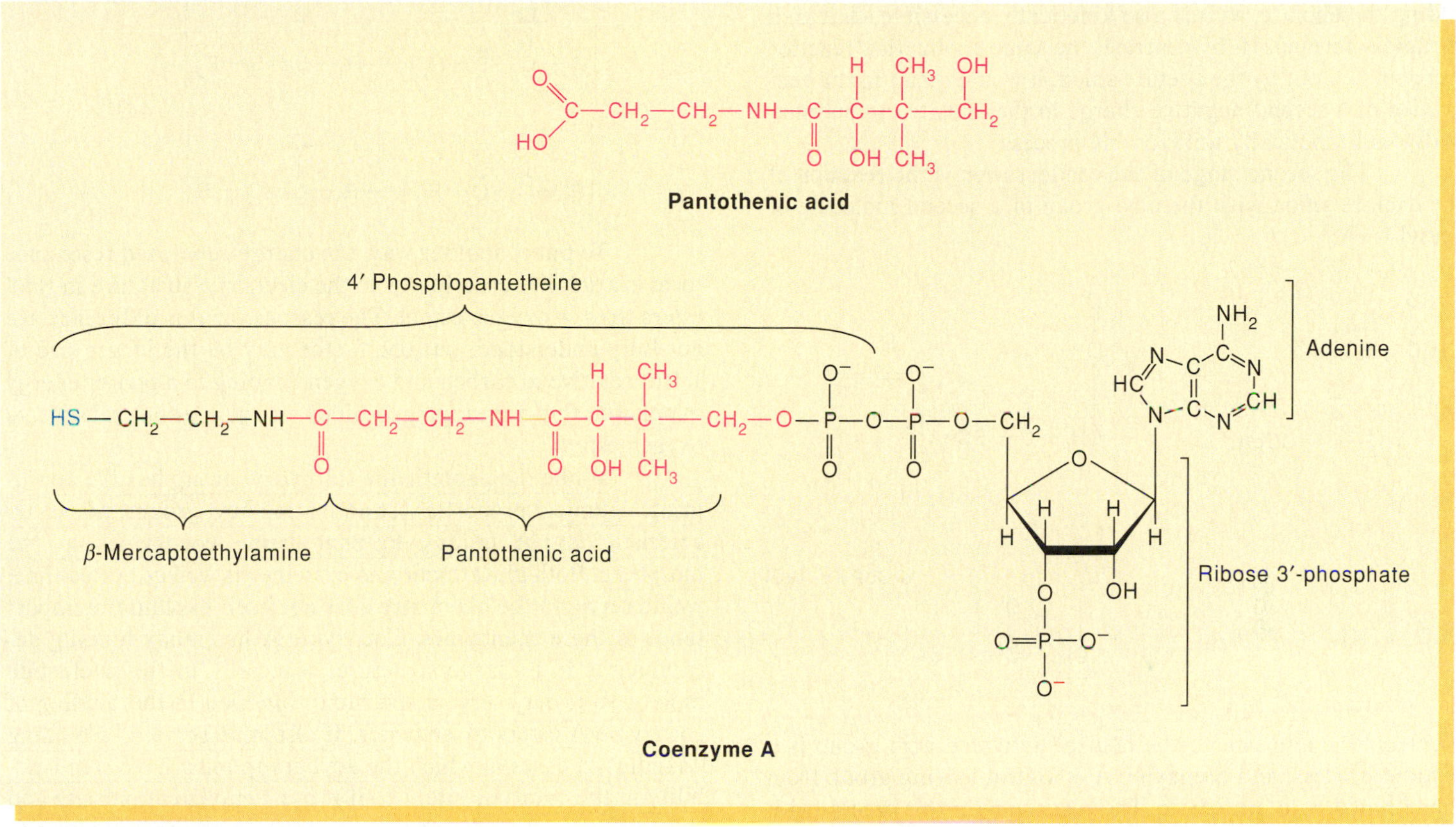

$$HOOC-CH_2-CH_2-NH-\overset{O}{\overset{\|}{C}}-CH(OH)-C(CH_3)_2-CH_2OH$$

Pantothenic acid

4′ Phosphopantetheine

$$HS-CH_2-CH_2-NH-\overset{O}{\overset{\|}{C}}-CH_2-CH_2-NH-\overset{O}{\overset{\|}{C}}-CH(OH)-C(CH_3)_2-CH_2-O-\overset{O^-}{\underset{O}{P}}-O-\overset{O^-}{\underset{O}{P}}-O-CH_2\text{-ribose 3'-phosphate-adenine}$$

β-Mercaptoethylamine — Pantothenic acid — Adenine — Ribose 3′-phosphate

Coenzyme A

Reactions Requiring Acyl Activation Frequently Use Phosphopantetheine Coenzymes

4′-Phosphopantetheine coenzymes are the biochemically active forms of the vitamin pantothenic acid. In figure 11.12, 4′-phosphopantetheine is shown as covalently linked to an adenylyl group in coenzyme A; or it can also be linked to a protein such as a serine hydroxyl group in acyl carrier protein (ACP). It is also found bonded to proteins that catalyze the activation and polymerization of amino acids to polypeptide antibiotics. Coenzyme A was discovered, purified, and structurally characterized by Fritz Lipmann and colleagues in work for which Lipmann was awarded the Nobel Prize in 1953.

The sulfhydryl group of the β-mercaptoethylamine (or cysteamine) moiety of phosphopantetheine coenzymes is the functional group that is directly involved in the enzymatic reactions for which they serve as coenzymes. From the standpoint of the chemical mechanism of catalysis, it is the essential functional group, although it is now recognized that phosphopantetheine coenzymes have other functions as well. Many reactions in metabolism involve acyl-group transfer or enolization of carboxylic acids that exist as unactivated carboxylate anions at physiological pH. The predominant means by which these acids are activated for acyl transfer and enolization is esterification with the sulfhydryl group of pantetheine coenzymes.

The mechanistic importance of activation is exemplified by the condensation of two molecules of acetyl-coenzyme A to acetoacetyl-coenzyme A catalyzed by β-ketothiolase:

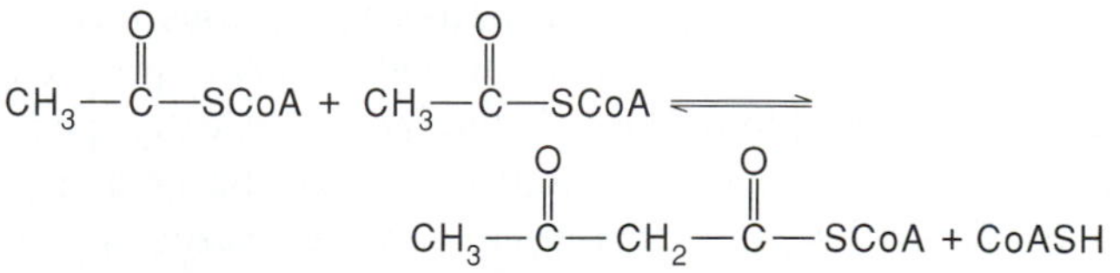

$$CH_3-\overset{O}{\overset{\|}{C}}-SCoA + CH_3-\overset{O}{\overset{\|}{C}}-SCoA \rightleftharpoons CH_3-\overset{O}{\overset{\|}{C}}-CH_2-\overset{O}{\overset{\|}{C}}-SCoA + CoASH \quad (18)$$

The two important steps of the reaction depend on both acetyl groups being activated, one for enolization and the other for acyl-group transfer. In the first step, one of the molecules must be enolized by the intervention of a base to remove an α proton, forming an enolate:

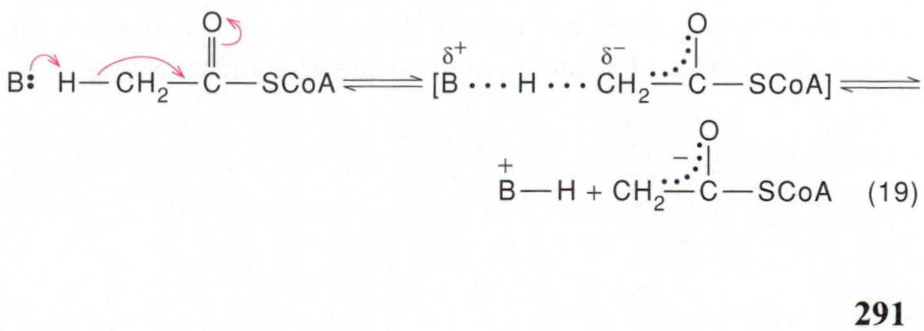

$$B\!:\ H-CH_2-\overset{O}{\overset{\|}{C}}-SCoA \rightleftharpoons [\overset{\delta^+}{B\cdots H}\cdots \overset{\delta^-}{CH_2}\cdots\overset{O}{\overset{\vdots}{C}}-SCoA] \rightleftharpoons \overset{+}{B}-H + CH_2\cdots\overset{O^-}{\overset{\vdots}{C}}-SCoA \quad (19)$$

The enolate is stabilized by delocalization of its negative charge between the α carbon and the acyl oxygen atom, making it thermodynamically accessible as an intermediate. Moreover, this developing charge is also stabilized in the transition state preceding the enolate, so it is also kinetically accessible; that is, it is rapidly formed. If, by contrast, the same enolization reaction were carried out by the acetate anion, it would result in the generation of a second negative charge in the enolate, an energetically and kinetically unfavorable process.

The second stage of the condensation is the reaction of the enolate anion with the acyl group of a second molecule of acetyl-CoA:

$$CH_3C(=O)-SCoA + {}^{-}CH_2-C(=O)-SCoA \rightleftharpoons \left[CH_3C(O^-)(SCoA)-CH_2-C(=O)-SCoA\right] \overset{H^+}{\rightleftharpoons} CH_3C(=O)-CH_2-C(=O)-SCoA + CoASH \quad (20)$$

Nucleophilic addition to the neutral activated acyl group is a favored process, and coenzyme A is a good leaving group from the tetrahedral intermediate. The occurrence of this process with the acetate anion, i.e., acetate reacting with an enolate anion, again provides a sharp contrast with the process of equation (20), for it would entail the nucleophilic addition of an anion to an anionic center, generating a dianionic transition state, an unfavorable process from both thermodynamic and kinetic standpoints. Moreover, the resulting intermediate would not have a very good leaving group other than the enolate anion itself, so the transition-state energy for acetoacetate formation would be high. Finally, the K_{eq} for the condensation of 2 moles of acetate to 1 mole of acetoacetate is not favorable in aqueous media, whereas the condensation of 2 moles of acetyl-CoA to produce acetoacetyl-CoA and coenzyme A is thermodynamically spontaneous. The maintenance of metabolic carboxylic acids involved in enolization and acyl-group transfer reactions as coenzyme A esters provides the ideal lift over the kinetic and thermodynamic barriers to these reactions.

The foregoing discussion, in emphasizing the purely electrostatic energy barriers, does not address the question of whether there is an activation advantage in thiol esters relative to oxygen esters. Why thiol esters in preference to oxygen esters? Thiol esters are more readily enolized than oxygen esters. They are more "ketonelike" because of their electronic structures, in which the degree of resonance-electron delocalization from the sulfur atom to the acyl group resulting from overlapping of the occupied *p* orbitals of sulfur with the acyl-π bond is less than that of oxygen esters.

$$\left[R_1-C(=O)-\ddot{S}-R_2 \longleftrightarrow R_1-C(O^-)=\overset{+}{S}-R_2\right]$$
$$\left[R_1-C(=O)-\ddot{O}-R_2 \longleftrightarrow R_1-C(O^-)=\overset{+}{O}-R\right] \quad (21)$$

To put it another way, the charge-separated resonance form is a smaller contributor to the electronic structure in thiol esters than in oxygen esters. The reasons for this difference are not fully understood, but one factor may be the larger size of sulfur relative to carbon and oxygen, leading to a poorer energy match for the overlapping orbitals in thiol esters relative to oxygen esters.

While the pantetheine sulfhydryl group has the appropriate chemical properties for activating acyl groups, this characteristic is not unique to pantetheine coenzymes in the biosphere. Both glutathione and cysteine, as well as cysteamine, would serve, so the chemistry does not itself explain the importance of these coenzymes. Coenzyme A has many binding determinants in its large structure, especially in the nucleotide moiety, so it may serve a specificity function in the binding of coenzyme A esters by enzymes. It also may serve as a binding "handle" in cases in which the acyl group must have some mobility in the catalytic site, i.e., if it must enolize at one site and then diffuse a short distance to undergo an addition reaction to a ketonic group of a second substrate.

One system in which pantetheine almost certainly performs such a carrier role is the fatty acid synthase from *E. coli*, in which 4′-phosphopantetheine is a component of the acyl carrier protein (see chapter 17).

α-Lipoic Acid Is the Coenzyme of Choice for Reactions Requiring Acyl Group Transfers Linked to Oxidation-Reduction

α-Lipoic acid is the internal disulfide of 6,8-dithiooctanoic acid, whose structural formula is given in figure 11.13. It is the coupler of electron and group transfers catalyzed by α-keto acid dehydrogenase multienzyme complexes. The pyruvate and α-ketoglutarate dehydrogenase complexes are centrally involved in the metabolism of carbohydrates by the glycolytic pathway (chapter 13) and the tricarboxylic acid cycle (chapter 14). They catalyze two of the three decarboxylation steps in the complete oxidation of glucose, and they produce NADH and activated acyl compounds from the oxidation of the resulting ketoacids:

$$R-C(=O)-CO_2^- + NAD^+ + CoASH \longrightarrow CO_2 + NADH + R-C(=O)-SCoA \quad (22)$$

Figure 11.13

Structure of the α-lipoyl enzyme, showing the reactive disulfide in color. The lipoic acid is commonly covalently bonded through amide linkage with the ε amino group of a lysine residue as shown.

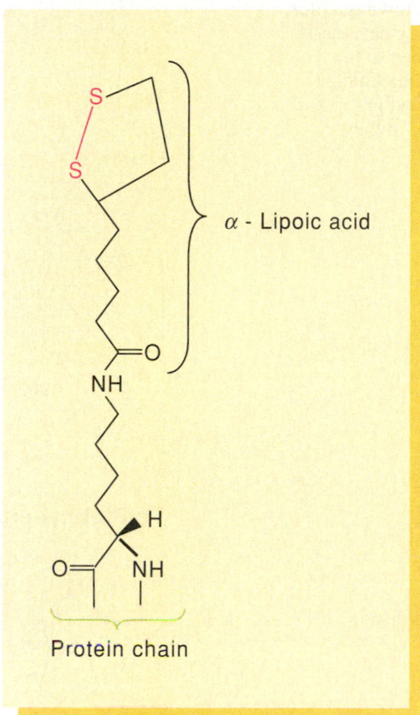

Figure 11.14

Interactions of α-lipoyl groups in the pyruvate dehydrogenase complex. The cubic structure represents the 24 subunits of dihydrolipoyl transacetylase, which constitutes the core of the complex. Two of the 48 lipoyl groups in the core are shown interacting with one of the 24 pyruvate decarboxylases (E_1•TPP) and one of the 12 dihydrolipoyl dehydrogenase (E_2•FAD) subunits. Note the interaction of the lipoyl groups in relaying electrons over the long distance between TPP and FAD. The lipoic acid must interact at active sites over distances too long to be covered by a single fully extended coenzyme. This problem is solved by use of a shuttle system involving two α-lipoyl groups for each transfer.

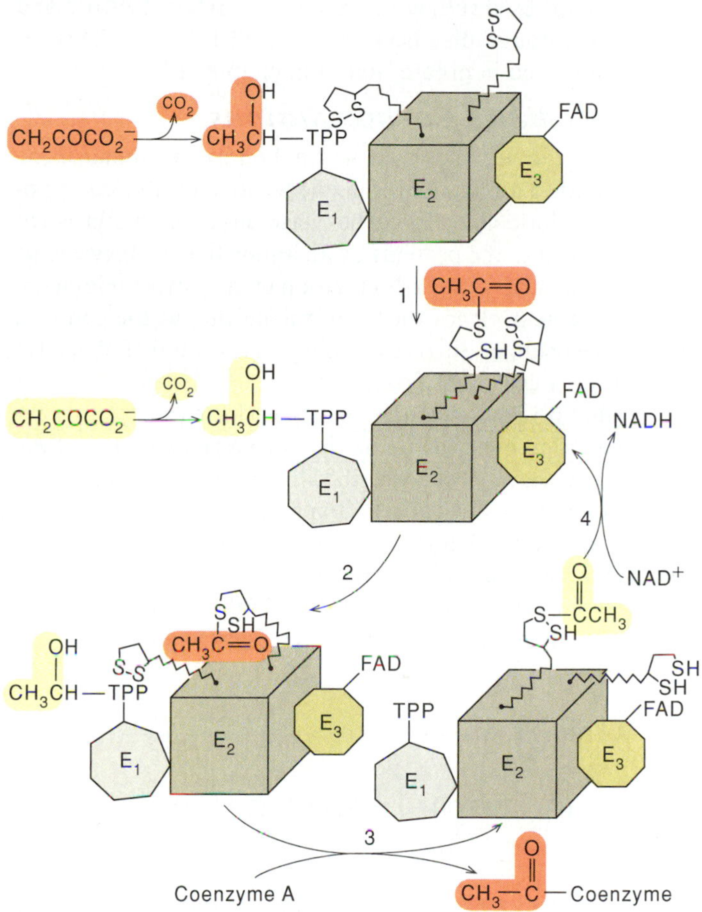

The chemical aspect of the coenzymatic action of α-lipoic acid is to mediate the transfer of electrons and activated acyl groups resulting from the decarboxylation and oxidation of α-keto acids within the complexes. In this process, lipoic acid is itself transiently reduced to dihydrolipoic acid, and this reduced form is the acceptor of the activated acyl groups. Its dual role of electron and acyl-group acceptor enables lipoic acid to couple the two processes.

The interactions of α-lipoic acid in the *E. coli* pyruvate dehydrogenase complex exemplify its coenzymatic functions. The complex consists of three proteins: a pyruvate decarboxylase, which is thiamine-pyrophosphate-dependent and designated $E_1 \cdot$ TPP; dihydrolipoyl transacetylase, designated E_2-lipoyl-S_2, which contains α-lipoic acid covalently bonded through amide linkage with the ε amino group of a lysine residue (see fig. 11.14); and dihydrolipoyl dehydrogenase, a flavoprotein designated $E_3 \cdot$ FAD. A single particle of the complex consists of at least 24 chains of each of the first two enzymes and 12 of the flavoprotein.

Lipoic acid must interact at active sites on all three enzymes. α-Lipoic acid bonded to E_2 is, as shown in figure 11.13, bonded through an amide linkage to a lysyl-ε-NH_2 group that places the reactive disulfide at the end of a flexible chain of atoms with rotational freedom about as many as ten single bonds. When fully extended, this chain is 1.4 nm (14 Å) long, giving α-lipoic acid the potential capacity to sweep out a space having a spherical diameter of 2.8 nm. This distance turns out to be inadequate to account fully for the transport of electrons in the complex because the average distance between TPP on E_1 and FAD on E_3 has been estimated at between 4.5 and 6.0 nm by fluorescence energy-transfer measurements. The problem of long-distance interactions is overcome in the complex by the fact that each E_2 subunit contains two lipoyl groups that interact with each other and with α-lipoyl groups on other subunits. This interaction facilitates the transport of electrons and acetyl groups through a network of α-lipoyl groups encompassing the entire core of E_2. Moreover, the lipoyl-bearing domains of E_2 are conformationally mobile; this mobility is thought to extend the "reach" of the lipoyl groups. The relay process, illustrated schematically in figure 11.14, shows how two or more *S*-acetyldihydrolipoyl groups can interact to transport electrons over the large distances separating the sites for TPP on pyruvate dehydrogenase and FAD on dihydrolipoyl dehydrogenase.

The coenzymatic capabilities of α-lipoyl groups result from a fusion of its chemical and physical properties, the ability to act simultaneously as both electron and acyl-group acceptor, the ability to span long distances to interact with sites separated by up to 2.8 nm, and the ability to act cooperatively with other α-lipoyl groups by disulfide interchange to relay electrons and acyl groups through distances that exceed its reach. This reaction is discussed in greater detail in chapter 14.

Biotin Mediates Carboxylations

The biotin structure shown in figure 11.15 is an imidazolone ring *cis*-fused to a tetrahydrothiophene ring substituted at position 2 by valeric acid. In carboxylase enzymes, biotin is covalently bonded to the proteins by an amide linkage between its carboxyl group and a lysyl-ϵ-NH_2 group in the polypeptide chain. This arrangement places the imidazolone ring at the end of a long flexible chain of atoms extending a maximum of about 1.4 nm from the α carbon of lysine.

Biotin is the essential coenzyme for carboxylation reactions involving bicarbonate as the carboxylating agent. Several reactions have been described in which ATP-dependent carboxylation occurs at carbon atoms activated for enolization by ketonic or activated acyl groups. One reaction is known where a nitrogen atom of urea is carboxylated.

A general formulation of the ATP-dependent carboxylation of an α carbon by ^{18}O-enriched bicarbonate is

$$RCH_2-\overset{\displaystyle O}{\overset{\|}{C}}-SCoA + ATP + HC^{18}O_3^- \xrightarrow{\text{Biotinyl carboxylase}} H^+ + R-\underset{\displaystyle C^{18}O_2^-}{\underset{|}{CH}}-\overset{\displaystyle O}{\overset{\|}{C}}-SCoA + ADP + H^{18}OPO_3^{2-} \quad (23)$$

The appearance of ^{18}O in inorganic phosphate verifies that the function of ATP in the reaction is essentially the "dehydration" of bicarbonate.

The ATP-dependent carboxylation of biotin by bicarbonate is believed to control the transient formation of carbonic-phosphoric anhydride, or "carboxyphosphate," as an active carboxylation intermediate:

$$HO-\overset{\displaystyle O}{\overset{\|}{C}}-O^- \xrightarrow[\text{ADP}]{\text{ATP}} {}^-O-\underset{\displaystyle OH}{\underset{|}{\overset{\displaystyle O}{\overset{\|}{P}}}}-O-\overset{\displaystyle O^-}{\overset{\|}{C}}-O^- \xrightarrow[\text{P}_i]{\text{Biotinyl-E}} N^1\text{-carboxybiotinyl-E} \quad (24)$$

The coenzymatic function of biotin appears to be to mediate the carboxylation of substrates by accepting the ATP-activated carboxyl group and transferring it to the carboxyl acceptor substrate. There is good reason to believe that the enzymatic sites of ATP-dependent carboxylation of biotin are physically separated from the sites at which N^1-carboxybiotin transfers the carboxyl group to acceptor substrates, i.e., the transcarboxylase sites. In fact, in the case of the acetyl-CoA carboxylase from *E. coli* (see chapter 17), these two sites reside on two different subunits, while the biotinyl group is bonded to a third, a small subunit designated biotin carboxyl carrier protein. Transcarboxylase is also a multisubunit protein, one subunit being a small biotinyl protein. The carboxylation of acetyl-CoA by carboxybiotin is discussed in chapter 17.

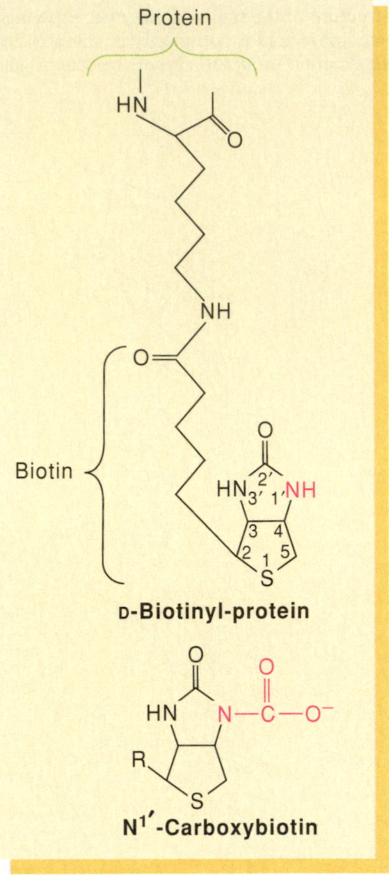

Figure 11.15
Structures of biotinyl enzyme and N^1-carboxybiotin. The reactive portions of the coenzyme and the active intermediate are shown in color. In carboxylase enzymes, biotin is covalently bonded to the proteins by an amide linkage between its carboxyl group and a lysyl-ε-NH_2 group in the polypeptide chain.

Biotin appears to have just the right chemical and structural properties to mediate carboxylation. It readily accepts activated carboxyl groups at N^1 and maintains them in an acceptably stable yet reactive form for transfer to acceptor substrates. Since biotin is bonded to a lysyl group, the N^1-carboxyl group is at the end of a 1.6-nm chain with bond rotational freedom about nine single bonds, giving it the capability to transport activated carboxyl groups through space from the carboxyl activation sites to the carboxylation sites.

Folate Coenzymes Are Used in Reactions for One-Carbon Transfers

Tetrahydrofolate and its derivatives N^5,N^{10}-methylenetetrahydrofolate, N^5,N^{10}-methenyltetrahydrofolate, N^{10}-formyltetrahydrofolate, and N^5-methyltetrahydrofolate are the biologically active forms of folic acid, a four-electron-oxidized form of

Figure 11.16

Structures and enzymatic interconversions of folate coenzymes. The reactive centers of the coenzymes are shown in red. The most active forms of the coenzyme contain oligo- or polyglutamyl groups.

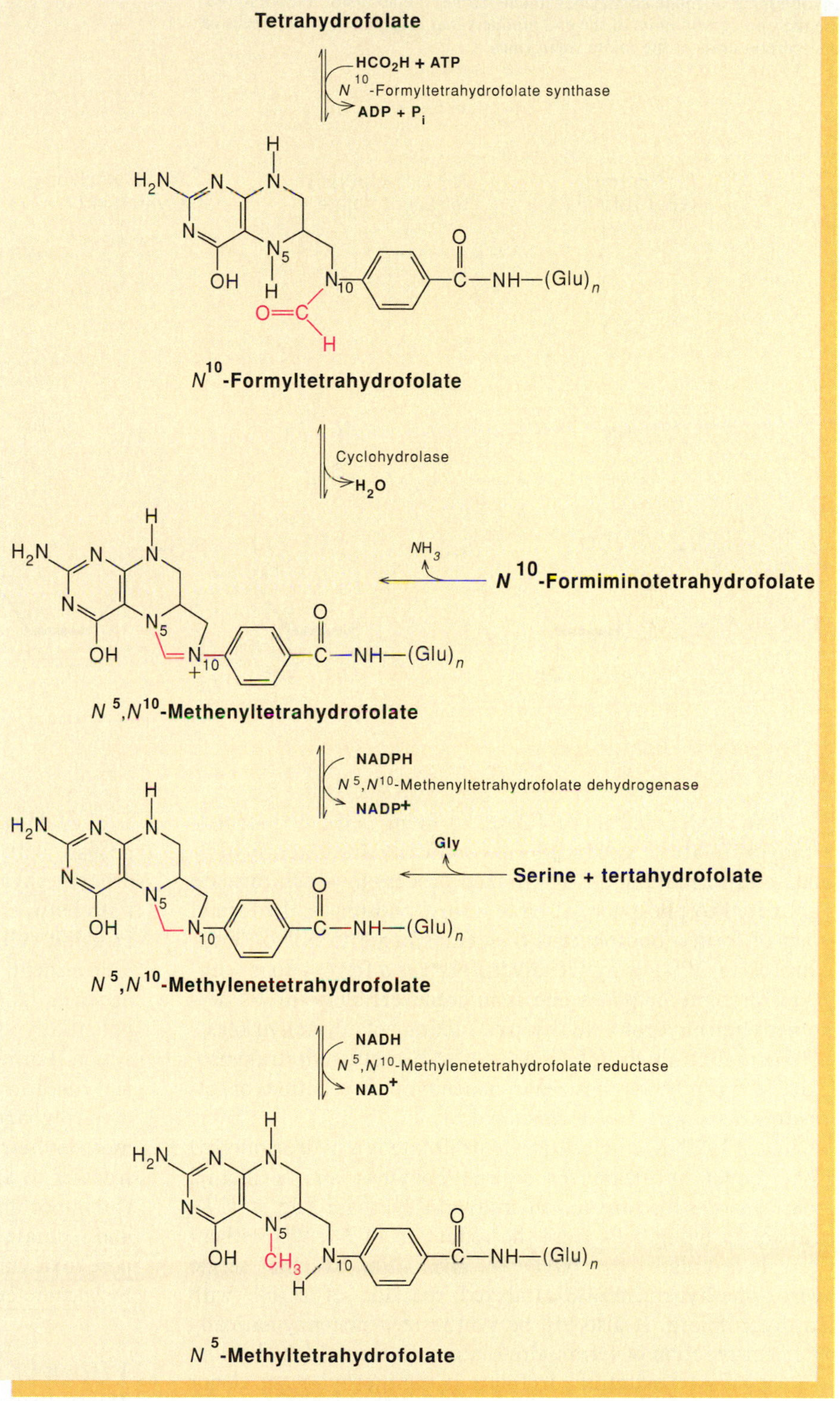

Tetrahydrofolate

HCO_2H + ATP
N^{10}-Formyltetrahydrofolate synthase
ADP + P_i

N^{10}-Formyltetrahydrofolate

Cyclohydrolase
H_2O

NH_3 ← **N^{10}-Formiminotetrahydrofolate**

N^5,N^{10}-Methenyltetrahydrofolate

NADPH
N^5,N^{10}-Methenyltetrahydrofolate dehydrogenase
$NADP^+$

Gly ← **Serine + tertahydrofolate**

N^5,N^{10}-Methylenetetrahydrofolate

NADH
N^5,N^{10}-Methylenetetrahydrofolate reductase
NAD^+

N^5-Methyltetrahydrofolate

tetrahydrofolate. The structural formulas are given in figure 11.16, which also shows how they arise from tetrahydrofolate. The structures are shown glutamylated on the carboxyl group of the *p*-amino benzoyl group; the most active forms contain oligo- or polyglutamyl groups, linked through the γ-carboxyl groups.

The tetrahydrofolates do not function as tightly enzyme-bound coenzymes. Rather they function as cosubstrates for a variety of enzymes associated with one-carbon metabolism. N^{10}-Formyltetrahydrofolate is a formyl-group donor substrate for several transformylases. The methenyl derivative is an intermediate between N^{10}-formyltetrahydrofolate and

Figure 11.17

Involvement of folate coenzymes in one-carbon metabolism. Shown in red are the one-carbon units of the end products that originate with the reactive one-carbon units of the folate coenzymes.

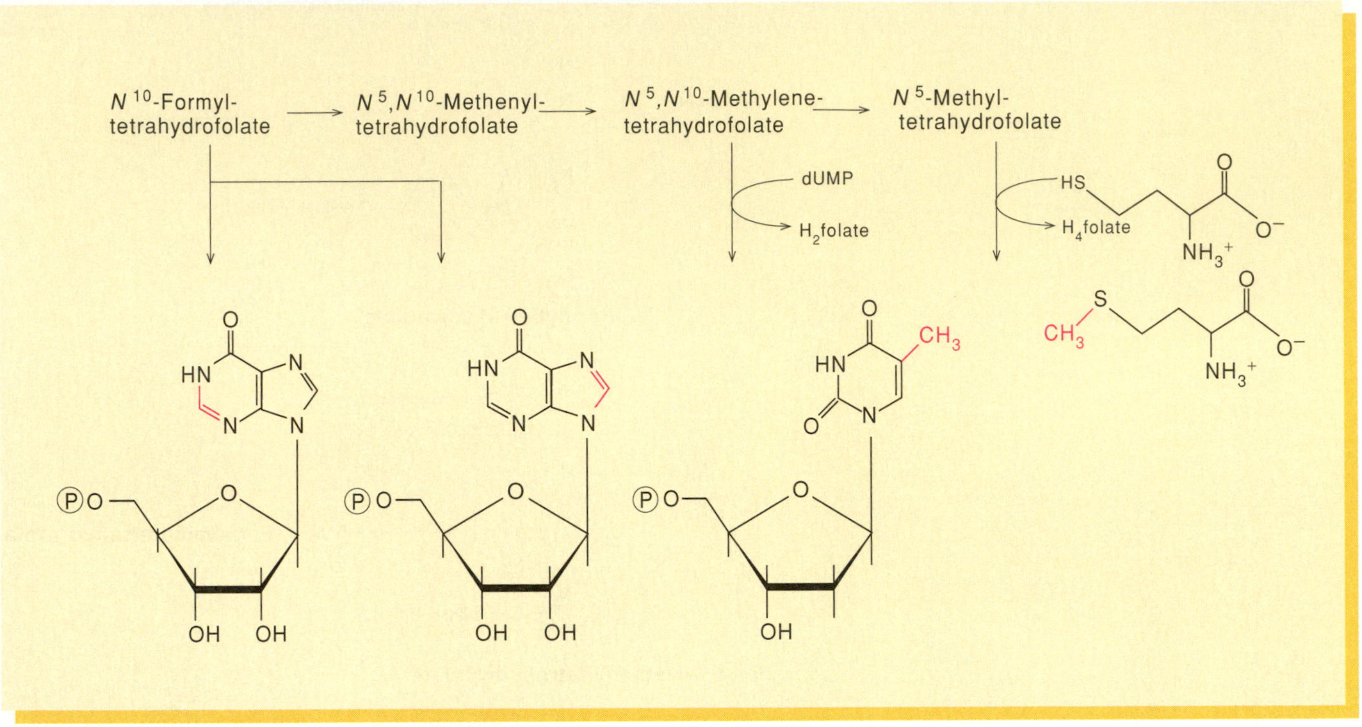

N^5,N^{10}-methylenetetrahydrofolate. In living cells, N^{10}-formyltetrahydrofolate is produced enzymatically from tetrahydrofolate and formate in an ATP-linked process in which formate is activated by phosphorylation to formyl phosphate: the formyl group of formyl phosphate is then transferred to N^{10} of tetrahydrofolate. N^{10}-Formyltetrahydrofolate is a formyl donor substrate for some enzymes and is interconvertible with N^5,N^{10}-methenyltetrahydrofolate by the action of cyclohydrolase. N^{10}-Formyltetrahydrofolate and N^5,N^{10}-methenyltetrahydrofolate also can be synthesized in nonenzymatic reactions of tetrahydrofolate with free formic acid.

N^5,N^{10}-Methylenetetrahydrofolate is a hydroxymethyl group donor substrate for several enzymes and a methyl group donor substrate for thymidylate synthase (fig. 11.17). It arises in living cells from the reduction of N^5,N^{10}-methenyltetrahydrofolate by NADPH and also by the serine hydroxymethyltransferase-catalyzed reaction of serine with tetrahydrofolate. It also can be synthesized nonenzymatically by direct reaction of tetrahydrofolate with formaldehyde.

N^5-Methyltetrahydrofolate is the methyl-group donor substrate for methionine synthase, which catalyzes the transfer of the five-methyl group to the sulfhydryl group of homocysteine. This and selected reactions of the other folate derivatives are outlined in figure 11.17, which emphasizes the important role tetrahydrofolate plays in nucleic acid biosynthesis by serving as the immediate source of one-carbon units in purine and pyrimidine biosynthesis.

Formaldehyde is a toxic substance that reacts spontaneously with amino groups of proteins and nucleic acids, hydroxymethylating them and forming methylene-bridge crosslinks between them. Free formaldehyde therefore wreaks havoc in living cells and could not serve as a useful hydroxymethylating agent. In the form of N^5,N^{10}-methylenetetrahydrofolate, however, its chemical reactivity is attenuated but retained in a potentially available form where needed for specific enzymatic action. Formate, however, is quite unreactive under physiological conditions and must be activated to serve as an efficient formylating agent. As N^{10}-formyltetrahydrofolate and N^5,N^{10}-methenyltetrahydrofolate it is in a reactive state suitable for transfer to appropriate substrates. The fundamental biochemical importance of tetrahydrofolate is to maintain formaldehyde and formate in chemically poised states, not so reactive as to pose toxic threats to the cell but available for essential processes by specific enzymatic action.

Vitamin B_{12} Coenzymes Are Associated with Rearrangements on Adjacent Carbon Atoms

The principal coenzymatic form of vitamin B_{12} is 5′-deoxyadenosylcobalamin, whose structural formula is given in figure 11.18. The structure includes a cobalt–carbon bond between the 5′ carbon of the 5′-deoxyadenosyl moiety and the cobalt (III) ion of cobalamin. [Note: The metal oxidation state may be denoted either as Co^{3+} or as Co(III). The former stresses

Figure 11.18

Structure of 5′-deoxyadenosylcobalamin coenzyme (vitamin B_{12}). The reactive groups are shown in color.

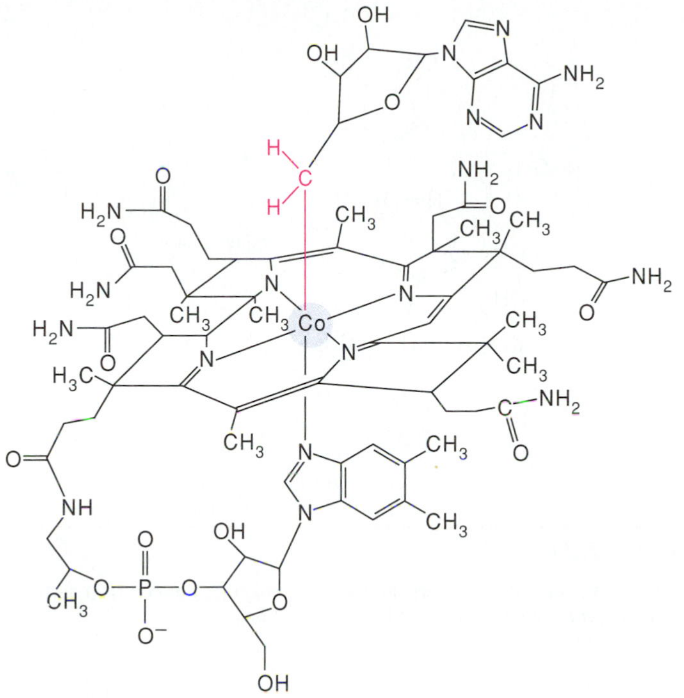

the free-ion character, while the latter stresses the bound character.] Vitamin B_{12} itself is cyanocobalamin, in which the cyano group is bonded to cobalt in place of the 5′-deoxyadenosyl moiety. Other forms of the vitamin have water (aquocobalamin) or the hydroxyl group (hydroxycobalamin) bonded to cobalt.

The vitamin was discovered in liver as the anti–pernicious anemia factor in 1926, but discovery of its complete structure had to await its purification, chemical characterization, and crystallization, which required more than twenty years. Even then the determination of such a complex structure proved to be an elusive goal by conventional approaches of that day and had to await the elegant x-ray crystallographic study of Lenhert and Hodgkin in 1961, for which Dorothy Hodgkin was awarded the Nobel Prize in 1964.

Most 5′-deoxyadenosylcobalamin-dependent enzymatic reactions are rearrangements that follow the pattern of equation (25), in which a hydrogen atom and another group (designated X) bonded to an adjacent carbon atom exchange positions, with the group X migrating from C_α to C_β:

$$a-\underset{X}{\overset{b}{C_\alpha}}-\underset{H}{\overset{c}{C_\beta}}-d \rightleftharpoons a-\underset{H}{\overset{b}{C_\alpha}}-\underset{X}{\overset{c}{C_\beta}}-d \qquad (25)$$

Three specific examples of rearrangement reactions are given in equations (26) through (28). It is interesting and significant that the migrating groups —OH, —COSCoA, and —CH$(NH_3^+)CO_2^-$ have little in common and that the hydrogen atoms migrating in the opposite direction are often chemically unreactive.

$$^-O_2C-CH_2CH_2-\underset{NH_3^+}{CH}-CO_2^- \xrightleftharpoons{\text{Glutamate mutase}} {}^-O_2C-\overset{CH_3}{CH}-\underset{NH_3^+}{CH}-CO_2^- \qquad (26)$$

$$^-O_2C-CH_2CH_2-\overset{O}{\overset{\|}{C}}-SCoA \xrightleftharpoons{\text{Methylmalonyl-CoA mutase}} {}^-O_2C-\overset{CH_3}{CH}-\overset{O}{\overset{\|}{C}}-SCoA \qquad (27)$$

$$CH_3\underset{OH}{CH}CH_2OH \xrightarrow{\text{Dioldehydrase}} CH_3CH_2\underset{OH}{CH}-OH \xrightarrow{\text{Dioldehydrase}} CH_3CH_2CHO + H_2O \qquad (28)$$

Indeed, hydrogen migrations in all the B_{12} coenzyme-dependent rearrangements proceed without exchange with the protons of water; i.e., isotopic hydrogen in substrates is conserved in the products. This fact plus spectroscopic evidence implicating Co(II) and organic radicals as catalytic intermediates in the reaction have led to the proposal of the mechanism illustrated in figure 11.19.

The reaction begins by homolytic cleavage of the Co—C bond (fig. 11.19), generating Co(II) and 5′-deoxyadenosyl free radical (step 1). The radical abstracts a hydrogen atom from the substrate, the migrating hydrogen in equation (25), generating 5′-deoxyadenosine and a substrate-derived free radical as intermediates (step 2). The substrate-radical undergoes rearrangement to a product-derived free radical, which abstracts a hydrogen atom to form the final product and regenerate the coenzyme (steps 3–6). Much evidence supports the involvement of the intermediates shown in figure 11.19; however, quantitative aspects of available data suggest the possible existence of additional species and a more complex hydrogen transfer process.

The most fundamental property of 5′-deoxyadenosylcobalamin leading to its unique action as a coenzyme is the weakness of the Co—C bond. This bond has a low dissociation energy, less than 30 kcal/mole, strong enough to be essentially stable in free solution but weak enough to be broken as a result of strain induced by multiple binding interactions between the enzymic binding sites and the adenosyl and cobalamin portions of the coenzyme. The radicals resulting from cleavage of this bond and abstraction of hydrogen from substrates undergo the rearrangements characteristic of B_{12}-dependent reactions.

Figure 11.19

Hypothetical partial mechanism of vitamin B_{12}-dependent rearrangements. The designations Co(III) and Co(II) refer to species that are spectrally and magnetically similar to Co^{3+} and Co^{2+}, respectively. Co(III) is diamagnetic and red, while Co(II) is paramagnetic (unpaired electron) and yellow. The metal does not undergo a change in electrostatic charge when the cobalt–carbon bond breaks homolytically (i.e., without charge separation), since one electron remains with the metal and the other with 5′-deoxyadenosine. One or more unknown intermediates, symbolized by the brackets, may be involved in the rearrangement.

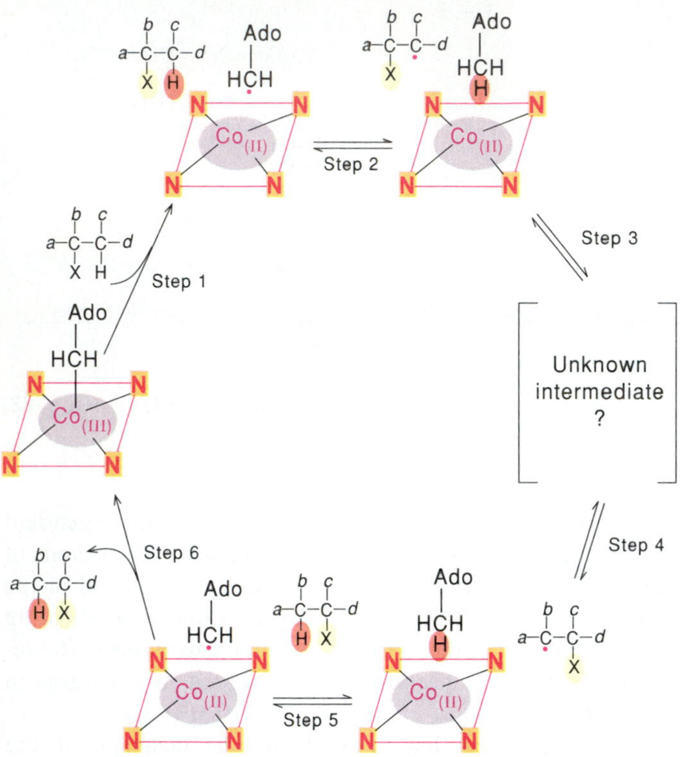

Figure 11.20

Structure of protoporphyrin IX. This coenzyme acts in conjunction with a number of different enzymes involved in oxidation and reduction reactions.

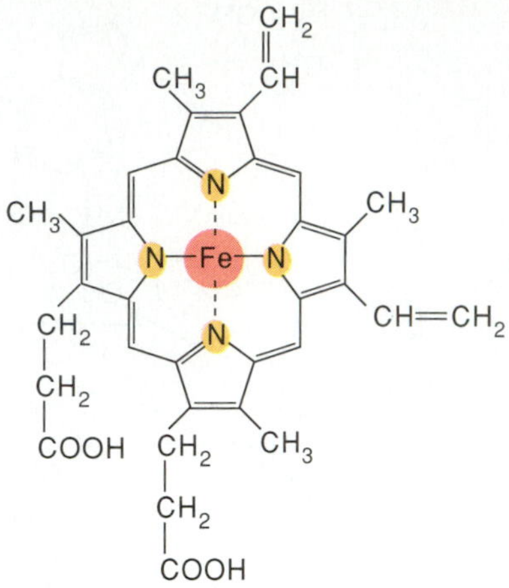

Figure 11.21

Structures of iron-sulfur clusters. Many redox enzymes contain iron-sulfur clusters that mediate one-electron transfer reactions.

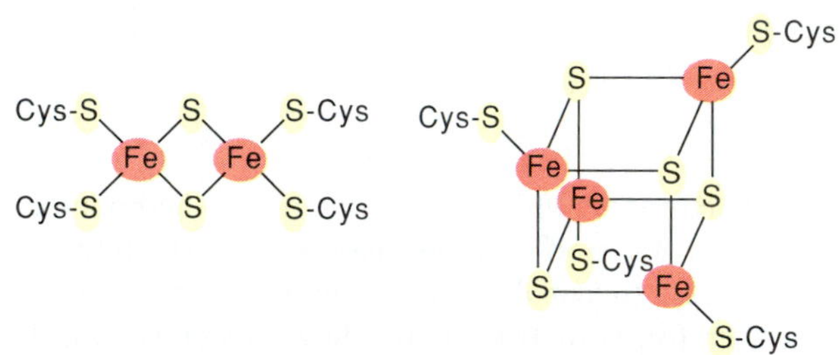

Iron-Containing Coenzymes Are Frequently Involved in Redox Reactions

Iron as a cofactor in catalysis is receiving increasing attention. The metal exists in two oxidation states; Fe^{2+} and Fe^{3+}. Iron complexes are nearly all octahedral, and practically all are paramagnetic (as a result of unpaired electrons in the 3*d* orbital). The most common form of iron in biological systems is heme. Heme groups (Fe^{2+}) and hematin (Fe^{3+}) most frequently involve a complex with protoporphyrin IX (fig. 11.20). They are the coenzymes (prosthetic groups) for a number of redox enzymes, including catalase, which catalyzes dismutation of hydrogen peroxide (equation (29)), and peroxidases, which catalyze the reduction of alkyl hydroperoxides by such reducing agents as phenols, hydroquinones, and dihydroascorbate (represented as AH_2 in equation (30)).

$$2\ H_2O_2 \rightleftharpoons 2\ H_2O + O_2 \qquad (29)$$

$$R{-}O{-}O{-}H + AH_2 \longrightarrow A + R{-}O{-}H + H_2O \qquad (30)$$

Heme proteins exhibit characteristic visible absorption spectra as a result of protoporphyrin IX; their spectra differ depending on the identities of the lower axial ligand donated by the protein and the oxidation state of the iron as well as the identities of the upper axial ligands donated by the substrates. Spectral data show clearly that the heme coenzymes participate directly in catalysis; however, the mechanisms of action of hemes are not as well understood as those of other coenzymes.

Many redox enzymes contain iron-sulfur clusters that mediate one-electron transfer reactions. The clusters consist of two or four irons and an equal number of inorganic sulfide ions clustered together with the iron, which is also liganded to cysteinyl-sulfhydryl groups of the protein (fig. 11.21). The enzyme nitrogenase, which catalyzes the reduction of N_2 to 2 NH_3 contains such clusters in which some of the iron has been replaced by molybdenum (see chapter 18). Electron-transferring proteins involved in one-electron transfer processes often contain iron-sulfur clusters. These proteins include the mitochondrial membrane enzymes NADH dehydrogenase and succinate dehydrogenase (chapter 14), which are flavoproteins, and the small-molecular-weight proteins ferredoxin, rubredoxin, adrenodoxin, and putidaredoxin (chapters 15, 16, and 23).

Heme coenzymes, iron-sulfur clusters, flavin coenzymes, and nicotinamide coenzymes cooperate in multienzyme systems to catalyze the chemically remarkable hydroxylations of hydrocarbons such as steroids (chapter 23). In these hydroxylation systems, the heme proteins constitute a family of proteins known as cytochrome P450, named for the wavelength corresponding to the most intense absorption band of the carbon monoxide-liganded heme, an inhibited form. The reactions catalyzed by these systems are represented in generalized form by equation (31), which also shows the fate of the two oxygens from $^{18}O_2$.

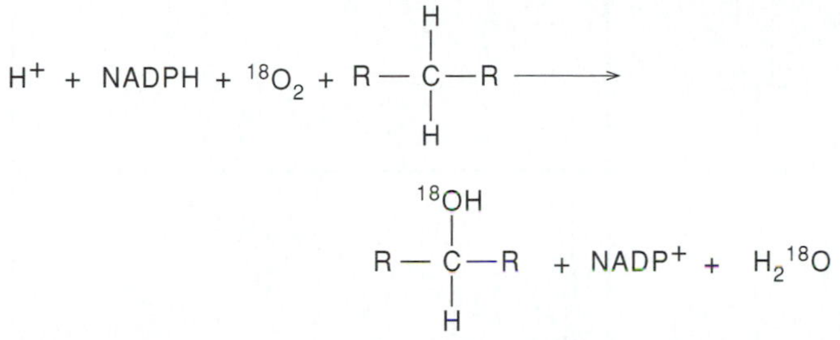

$$H^+ + \text{NADPH} + {}^{18}O_2 + R\text{—}\underset{H}{\overset{H}{C}}\text{—}R \longrightarrow R\text{—}\underset{H}{\overset{^{18}OH}{C}}\text{—}R + \text{NADP}^+ + H_2{}^{18}O \qquad (31)$$

One oxygen of O_2 is incorporated into the hydroxyl group of the product, whereas the other is incorporated into water. The enzymes usually include a cytochrome P450, an iron-sulfur cluster-containing protein such as adrenodoxin or putidaredoxin, and a flavoprotein reductase.

In the mechanism of oxygenation by these enzymes, the flavoproteins and iron-sulfur proteins supply reducing equivalents in one-electron units from NADPH to cytochrome P450, and reduced cytochrome P450 reacts directly with O_2. These reactions generate a Fe(III)-peroxide complex, shown in figure 11.22, which undergoes a further dehydration to an oxygenating species of cytochrome P450. The oxygenating species is thought to be an oxo-complex of Fe(IV), in which the porphyrin ring has been oxidized to a radical-cation by loss of an electron. The positive charge and unpaired electron in figure 11.22 are stabilized by delocalization through the conjugated π-electron system of the porphyrin ring (see fig. 11.20).

The most widely accepted mechanism for cytochrome P450-catalyzed oxygenation of substrates is illustrated for a hydrocarbon substrate in figure 11.23. The oxygenating species, the oxo-Fe(IV) radical-cation, abstracts a hydrogen atom from the hydrocarbon to form a hydrocarbon radical and a hydroxy-Fe complex. Hydrogen abstraction proceeds by the pairing of one electron from the oxo-Fe(IV) species and one electron from the carbon–hydrogen bond of the hydrocarbon. In the second step of oxygenation the hydrocarbon radical abstracts the hydroxy group from the hydroxy-Fe(III) complex by a pairing of the unpaired radical electron with one electron of the hydroxyl group, leaving cytochrome P450 in the +3 oxidation state, where it began in figure 11.22. The mechanism shown in figure 11.22 is known as the "rebound" mechanism of oxygenation. The second step of the mechanism is the rebound step, and the rate constant for this step in a microsomal cytochrome P450 has been estimated to be greater than 10^{10} s^{-1}. The magnitude of this rate constant explains why the radical and hydroxy-Fe intermediates have not been detected by presently available spectroscopic methods.

Figure 11.22

Steps in the formation of the oxygenating species of cytochrome P450. The oxygenating species of cytochrome P450 is generated by the transfer of two electrons in discrete steps from NADPH via the flavoprotein reductase and iron-sulfur clusters to the iron porphyrin complex, together with the reaction with oxygen. The Fe(III)-peroxide then undergoes a further expulsion of water to form the oxygenating species shown in brackets. The oxygenating species is not directly observed, since it reacts quickly, but it is thought to contain Fe(IV) and a delocalized radical-cation in the porphyrin ring.

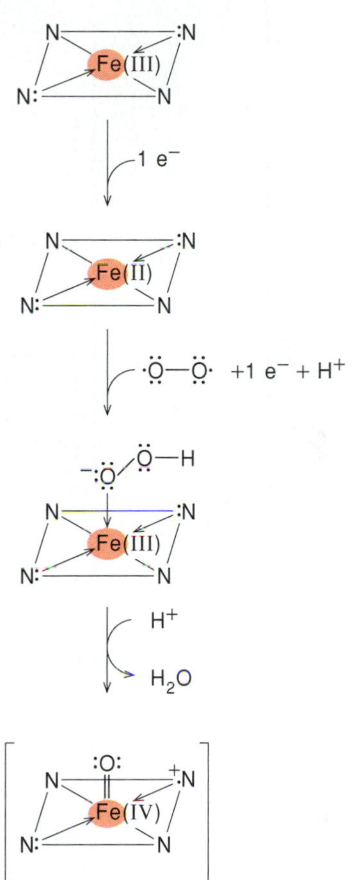

Figure 11.23

The rebound mechanism for oxygenation of hydrocarbons by cytochrome P450. The oxygenating species of cytochrome P450 in figure 11.22 is a highly reactive paramagnetic species that can abstract an unactivated and unreactive hydrogen from a hydrocarbon in the first step to form a hydrocarbon radical. In the rebound step the hydrocarbon radical abstracts a hydroxyl radical from the hydroxy-porphyrin intermediate.

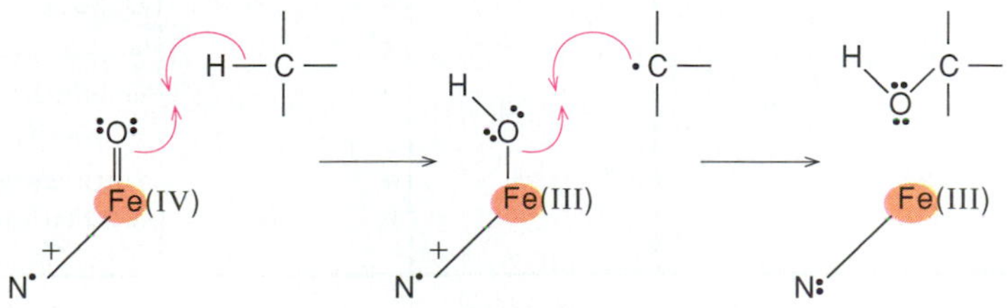

Table 11.3
Complexing Properties of Transition Metals of Biologic Importance

Metal and Valence State	Electron Configuration	Coordination Number	Stereo Chemistry	Examples
Manganese (Mn)				
Mn(II)	$3d^5$	4	Tetrahedral 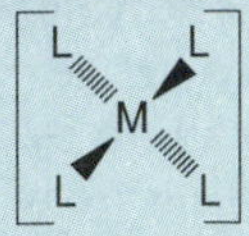	Pyruvate decarboxylase (see chapters 11, 14, and 15)
		4	Square planar 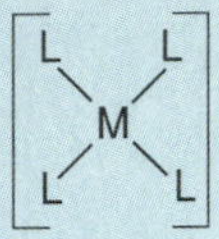	
Mn(III)	$3d^4$	6	Octahedral	
Iron (Fe)	$3d^6$	4	Tetrahedral	
Fe(II)		5	Trigonal bipyramidal 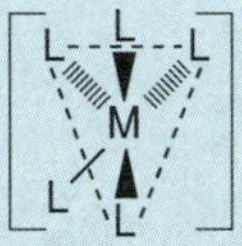	
		6	Octahedral	Hemoglobin (chapter 5)
Fe(III)	$3d^5$	4	Tetrahedral	Iron sulfur coenzymes and cytochromes (see chapters 11 and 15)
		6	Octahedral 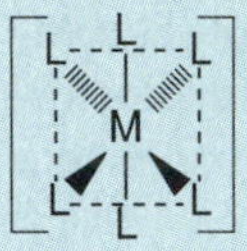	
Cobalt (Co)				
Co(I)	$3d^8$	4 5	Tetrahedral Trigonal bipyramidal	
Co(II)	$3d^7$	4 5 6	Tetrahedral; square planar Trigonal bipyramidal Octahedral	
Co(III)	$3d^6$	6	Octahedral	Vitamin B_{12} (this chapter)
Copper (Cu)				
Cu(I)	$3d^{10}$	2 4	Linear [L—M—L] Tetrahedral	
Cu(II)	$3d^9$	4 6	Tetrahedral; square pyramidal	Cytochrome (see chapter 15)
Zinc(Zn)				
Zn(II)	$3d^{10}$	4 5 6	Tetrahedral Trigonal bipyramidal Octahedral	Carboxypeptidase (chapter 9) Nucleic acid polymerase (chapter 26) Regulatory proteins (chapters 30 and 31)
Molybdenum (Mo)				
Mo(III)	$4d^3$	6 8	Octahedral Dodecahedral	
Mo(V)	$4d^1$	5 6 8	Trigonal bipyramidal Octahedral Dodecahedral 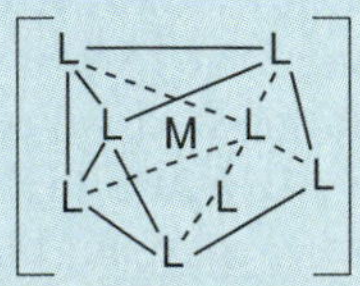	Nitrogenase (chapter 18)
Mo(VI)	$4d^0$	4 6	Tetrahedral Octahedral	

Metal Cofactors

In addition to cobalt and iron (discussed previously), other metals frequently function as cofactors in enzyme-catalyzed reactions. Like coenzymes, their usefulness is due to the fact that they offer something not available in amino acid side chains. The most important features of metals in this regard are their high concentrations of positive charge, their directed valences for interacting with two or more ligands, and their ability to exist in two or more valence states.

The alkali metal and alkaline earth metal ions are spherically symmetrical with respect to charge, so they do not usually show directed valences. There are notable exceptions, such as the case of Mg^{2+} in chlorophyll (see chapter 16), which adopts an approximately square planar distribution of orbitals. Alkali metals, Na^{+} and K^{+}, with their single positive charges and lack of *d*-electronic orbitals for sharing, almost never make tight complexes with proteins. On rare occasions the alkaline earth divalent cations, Mg^{2+} and Ca^{2+}, can make strong complexes. The alkali metal and alkali earth cations are asymmetrically distributed in the organism, with K^{+} and Mg^{2+} being concentrated in the cytosol and Na^{+} and Ca^{2+} being concentrated in the organelles and the blood.

Most of the remaining metals found in biological systems are from the first transition series, in which the 3*d* orbitals are only partially filled. These metals all prefer structures with multiple coordination numbers, and, except for Zn^{2+}, which has a filled 3*d* orbital, they can exist in more than one oxidation state, a feature making them potentially useful in oxidation-reduction reactions. The stability of zinc's electronic state combined with its preference for a coordination number of 4 to 6 makes zinc singularly useful in enzyme reactions, where it acts as a Lewis acid, and in structural situations where it chelates nitrogen (usually in histidine side chains) and/or sulfur-containing amino acid side chains (usually cysteines) to rigidify the structural domain of a protein.

The involvement of transition metals in biochemical reactions is discussed in later chapters as indicated in table 11.3, which also indicates the coordination numbers and stereochemistries of the most significant oxidation states of these metals.

Lipid-Soluble Vitamins

We have seen that most water-soluble vitamins are converted by single or multiple steps to coenzymes. Our understanding of the lipid-soluble vitamins (see table 11.1), and of how they are utilized by the organism, is much less extensive. In this section we will briefly discuss the structure and functions of some lipid-soluble vitamins.

Vitamin D_3 (cholecalciferol) can be made in the skin from 7-dehydrocholesterol in the presence of ultraviolet light (see fig. 24.13). Vitamin D_3 is formed by the cleavage of ring B of 7-dehydrocholesterol. Vitamin D_3 made in skin or absorbed from the small intestine is transported to the liver and hydroxylated at C-25 by a microsomal mixed-function oxidase. 25-Hydroxyvitamin D_3 appears to be biologically inactive until it is hydroxylated at C-1 by a mixed-function oxidase in kidney mitochondria. The 1,25-dihydroxyvitamin D_3 is delivered to target tissues for the regulation of calcium and phosphate metabolism. The structure and mode of action of 1,25-dihydroxyvitamin D_3 is analogous to that of the steroid hormones (see chapter 24).

Figure 11.24

Structures of vitamins K_1 and K_2. K_1 is found in plants, K_2 in animals and bacteria.

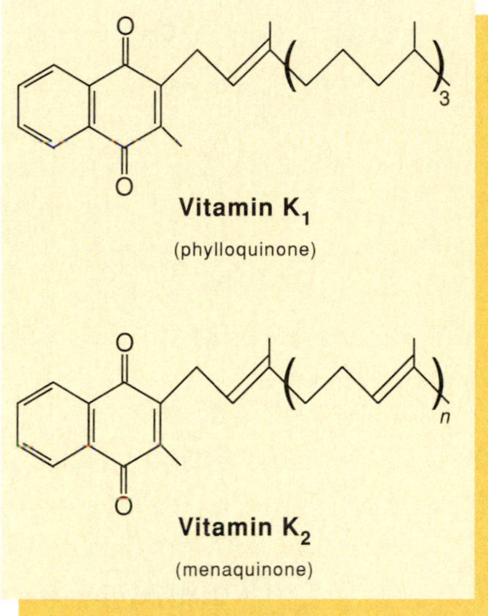

Vitamin K was discovered by Henrik Dam in Denmark in the 1920s as a fat-soluble factor important in blood coagulation (K is for "koagulation"). The structures of vitamins K_1 and K_2 (fig. 11.24) were elucidated by Edward Doisy. Vitamin K_1 is found in plants, vitamin K_2 in animals and bacteria. How this vitamin functions in blood coagulation eluded scientists until 1974, when vitamin K was shown to be needed for the formation of γ-carboxyglutamic acid (fig. 11.25) in certain proteins. γ-Carboxyglutamic acid specifically binds calcium, which is important for blood coagulation. Such modified glutamic acid residues appear to be important in many other processes involving calcium transport and calcium-regulated metabolic sequences.

Vitamin E (α-tocopherol) (fig. 11.26) was recognized in 1926 as an organic-soluble compound that prevented sterility in rats. The function of this vitamin still has not been clearly established. A favorite theory is that it is an antioxidant that prevents peroxidation of polyunsaturated fatty acids. Tocopherol certainly prevents peroxidation *in vitro,* and it can be replaced by other antioxidants. However, other antioxidants will not relieve all the symptoms of vitamin E deficiency.

Vitamin A (*trans*-retinol) is called an isoprenoid alcohol because it consists, in part, of units of a single five-carbon compound called isoprene:

Figure 11.25

Vitamin-K-dependent carboxylation of a glutamic acid residue in a protein. This reaction is essential to blood clotting.

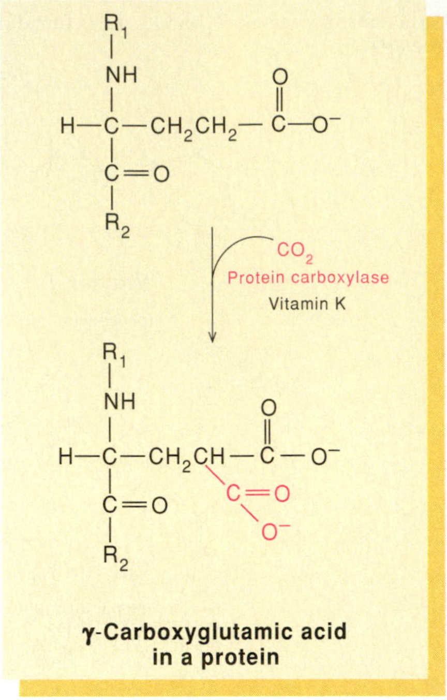

γ-Carboxyglutamic acid in a protein

Figure 11.26

Structure of vitamin E (α-tocopherol).

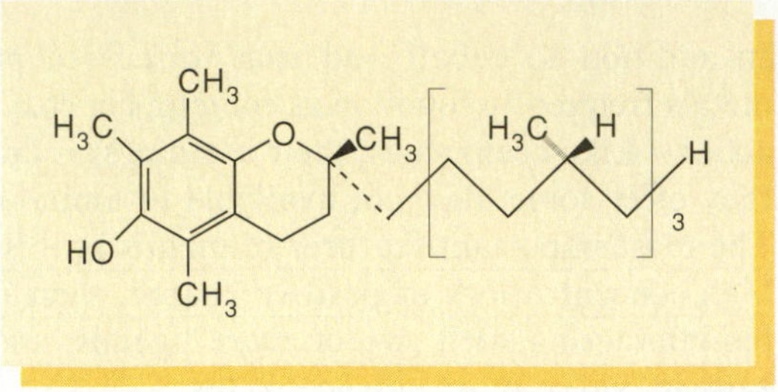

Isoprene is also a precursor of steroids and terpenes. This relationship will become clear when we examine the biosynthesis of these compounds (see chapter 23). Vitamin A is either biosynthesized from β-carotene (see fig. 36.4) or absorbed in the diet. Vitamin A is stored in the liver predominantly as an ester of palmitic acid. For many decades, it has been known to be important for vision and for animal growth and reproduction. The form of vitamin A active in the visual process is 11-*cis*-retinal, which combines with the protein opsin to form rhodopsin. Rhodopsin is the primary light-gathering pigment in the vertebrate retina (see chapter 36).

Summary

Coenzymes are molecules that act in cooperation with enzymes to catalyze biochemical processes that require functions that enzymes are otherwise chemically or physically ill-equipped to carry out. Our discussion in this chapter has focused on the following points.

1. Although a few coenzymes, such as hemes, lipoic acid, and iron-sulfur clusters, are biosynthesized in the body, most coenzymes are derivatives of the water-soluble vitamins.
2. Each coenzyme plays a unique chemical or physical role in the enzymatic processes of living cells, and the chemical structure and reactivity of each are suited to its biochemical function.
3. Thiamine pyrophosphate promotes the decarboxylation of α-keto acids and the cleavage of α-hydroxyl ketones by reacting with the ketone groups of substrates to produce intermediates in which electron pairs resulting from carbon–carbon bond cleavages can be stabilized by the thiazolium ring of the coenzyme.
4. Pyridoxal-5′-phosphate promotes decarboxylations, racemizations, transaminations, aldol cleavages, and α,β and β,γ eliminations in amino acid substrates in which electron pairs resulting from cleavages of covalent bonds are stabilized by the pyridine ring of the coenzyme.
5. Nicotinamide coenzymes act as intracellular electron carriers in transporting reducing equivalents from metabolic intermediates to the terminal electron-transport systems in cellular and mitochondrial membranes. Nicotinamide coenzymes also act as cocatalysts with enzymes by transiently oxidizing substrates to form reactive intermediates.
6. Flavin coenzymes are oxidation-reduction coenzymes th act as cocatalysts with enzymes in a large variety of biochemical redox reactions, most of which involve O_2.
7. Phosphopantetheine coenzymes facilitate the enolizatio and acyl-group transfer reactions of acyl groups to whicl they are bonded by thioester linkages.
8. Lipoic acid in α-keto acid dehydrogenase complexes mediates electron transfer and acyl-group transfer amon the active sites in the complex by undergoing transient reduction and acylation.
9. Biotin mediates the carboxylation of substrates by undergoing transient carboxylation to N^1-carboxybiotin.
10. Phosphopantetheine, lipoic acid, and biotin, by virtue o their flexible chainlike structures, facilitate the physical translocation of chemically reactive species among catal sites or subunits.
11. Tetrahydrofolates are cosubstrates for a variety of enzym associated with one-carbon metabolism. Tetrahydrofolat maintain formaldehyde and formate in chemically poise states available for essential processes by specific enzym action.
12. Heme coenzymes play essential roles in a variety of enzymatic reactions involving reductions of peroxides. Iron-sulfur clusters, composed of Fe and S in equal numbers together with cysteinyl side chains of proteins, mediate electron-transfer processes in a variety of enzymatic reactions, including the reduction of N_2 to 2 NH_3. Nicotinamide, flavin, and heme coenzymes act cooperatively with iron-sulfur clusters in multienzyme systems to catalyze hydroxylations of hydrocarbons.

13. 5′-Deoxyadenosylcobalamin (vitamin B_{12} coenzyme) is involved as the essential cofactor in intramolecular rearrangements in which an unreactive hydrogen exchanges positions with a group bonded to an adjacent carbon. These are radical rearrangements in which the coenzyme initiates the formation of substrate radicals by virtue of the weakness of the cobalt–carbon bond in the coenzyme. Homolytic scission of this bond generates a 5′-deoxyadenosyl-5′-radical that initiates radical formation by abstraction of a hydrogen atom.

14. Metals often perform the role of enzyme cofactors. Examples include iron in porphyrins and iron-sulfur compounds. The most important features of metals as cofactors are their high concentrations of positive charge, their directed valences for interacting with two or more ligands, and their ability to exist in two or more valence states.

15. A number of vitamins are lipid-soluble, including vitamins D, K, E, and A. In general, less is known about the mechanisms of action of lipid-soluble vitamins and their derivatives.

Selected Readings

Bruice, T. C., and S. J. Benkovic, *Bioorganic Chemistry,* vols. 1 and 2. Menlo Park, Calif.: Benjamin, 1966. A detailed discussion of the mechanisms of bioorganic reactions, including those involving coenzymes.

DiMarco, A. A., T. A. Bobik, and R. S. Wolfe, Unusual coenzymes of methanogenesis. *Ann. Rev. Biochem.* 59:355 (1990). A review of the recently characterized coenzymes required for the biosynthesis of methane in methanogenic bacteria.

Dolphin, D. (ed.), *B_{12},* vols. 1 and 2. New York: Wiley-Interscience, 1982. Chemistry and mechanism of action of Vitamin B_{12}.

Dolphin, D., R. Poulson, and O. Avamovic (eds.), *Pyridoxal Phosphate, Part A & Part B.* New York: Wiley-Interscience, 1987. Chemistry and mechanism of action of pyridoxal phosphate.

Dolphin, D., R. Poulson, and O. Avamovic (eds.), *Pyridine Nucleotide Coenzymes, Part A & Part B.* New York: Wiley-Interscience, 1987. Chemical, biochemical, and medical aspects of pyridine nucleotide coenzymes.

Frausto de Silva, J. J. K., and R. J. P. Williams, *The Biological Chemistry of the Elements.* Oxford: Clarendon Press, 1991.

Frey, P. A., The importance of organic radicals in enzymatic cleavage of unactivated C—H bonds. *Chem. Rev.* 90:1343, 1990. A brief review of coenzymes required to cleave unreactive C—H bonds.

Jencks, W. P., *Catalysis in Chemistry and Enzymology.* New York: McGraw-Hill, 1969. A detailed analysis of mechanisms of enzymatic and nonenzymatic reactions, including those involving coenzymes.

Knowles, J. R., The mechanism of biotin-dependent enzymes. *Ann. Rev. Biochem.* 58:195, 1989. Review of the chemical mechanism of biotin-dependent carboxylation reactions.

Ortiz de Montellano, P. R. (ed.), *Cytochrome P-450: Structure, Mechanism and Biochemistry.* New York: Plenum, 1986. A treatise on the structure and function of cytochrome P450 monooxygenases.

Phipps, D. A., *Metals and Metabolism.* Oxford Chemistry Series. Oxford: Clarendon Press, 1976. Examines the importance of metal ions in metabolic processes.

Popják, G., Stereospecificity of enzymic reactions. In P. D. Boyer (ed.), *The Enzymes,* vol. 2. New York: Academic Press, 1970. P. 115.

Walsh, C. T., *Enzymatic Reaction Mechanisms.* San Francisco: Freeman, 1977. Provides an up-to-date discussion of the mechanisms of enzymatic reactions. An indepth treatment of coenzymes.

Problems

1. What structural features of biotin and lipoic acid allow these cofactors to be covalently bound to a specific protein in a multienzyme complex yet participate in reactions at active sites on other enzymes of the complex?

2. The following reactions are catalyzed by pyridoxal-5′-phosphate-dependent enzymes. Write a reaction mechanism for each, showing how pyridoxal-5′-phosphate is involved in catalysis.

(a) $CH_3—C(=O)—COO^- + R—CH(NH_3^+)—COO^- \rightarrow CH_3—CH(NH_3^+)—COO^- + R—C(=O)—COO^-$

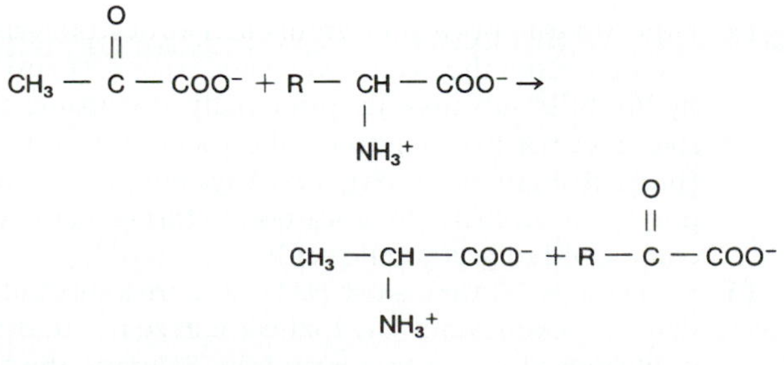

(b) $H_2N—(CH_2)_4—CH(NH_2)—CO_2^- \rightarrow CO_2 + H_2N—(CH_2)_5—NH_2$

(c) $^-O_2C—CH_2—CH(NH_2)—CO_2^- \rightarrow CO_2 + CH_3—CH(NH_2)—CO_2^-$

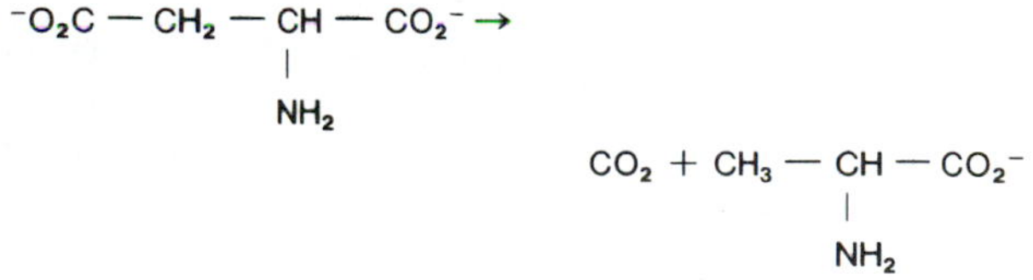

3. $NADP^+$ differs from NAD^+ only by phosphorylation of the C-2′ OH group on the adenosyl moiety. The redox potentials differ only by about 5 mV. Why do you suppose it is necessary for the cell to employ two such similar redox cofactors?

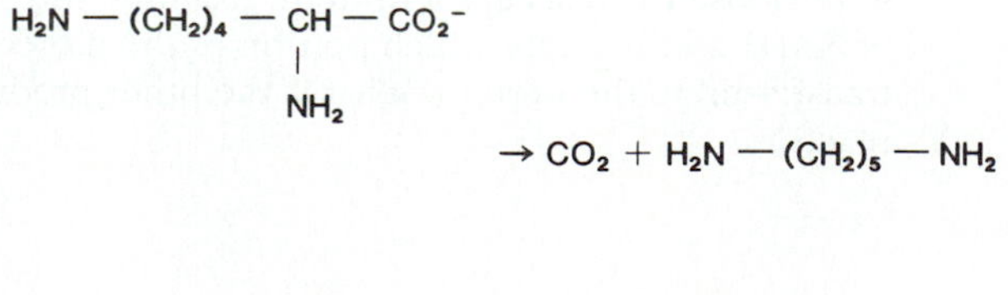

4. Thiamine-pyrophosphate-dependent enzymes catalyze the reactions shown below. Write a chemical mechanism that shows the catalytic role of the coenzyme.

(a) $CH_3—C(=O)—CO_2H \rightarrow CO_2 + CH_3—C(=O)—H$

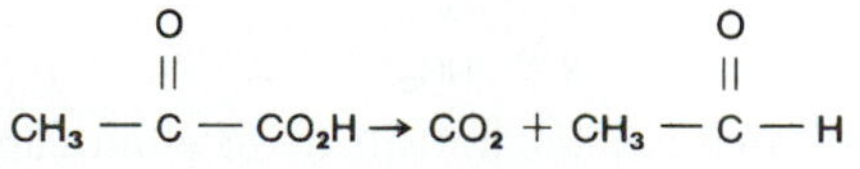

(b) ⓅOCH$_2$—(CHOH)$_2$—CH(OH)—C(=O)—CH$_2$OH + HOPO$_3^{2-}$ → ⓅO—CH$_2$—(CHOH)$_2$—CHO + CH$_3$—C(=O)—OPO$_3^{2-}$ + H$_2$O

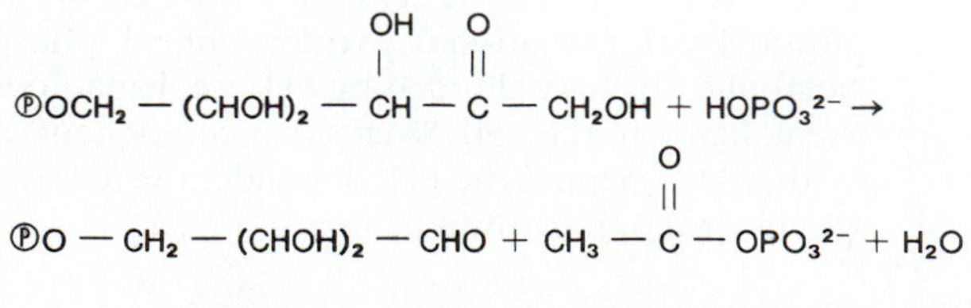

5. Malate synthase catalyzes the condensation of acetyl-CoA with glyoxalate to form L-malate.
   (a) Write a chemical reaction illustrating the reaction catalyzed by malate synthase.
   (b) Explain the chemical basis for using acetyl-CoA rather than acetate in the condensation reaction.
   (c) The product released from the enzyme is L-malate rather than malyl-CoA. Explain the role of thioester hydrolysis in the condensation reaction.
6. Write the mechanism showing how NAD(P)$^+$ is involved in the following reactions.
   (a) Malate dehydrogenase:

$$\mathrm{HOOC{-}CHOH{-}CH_2{-}COOH} + \mathrm{NAD^+} \rightarrow \mathrm{HOOC{-}C(=O){-}CH_2{-}COOH} + \mathrm{NADH} + \mathrm{H^+}$$

   (b) Malate dehydrogenase (decarboxylating; malic enzyme)

$$\mathrm{HOOC{-}CHOH{-}CH_2{-}COOH} + \mathrm{NADP^+} \rightarrow \mathrm{HOOC{-}C(=O){-}CH_3} + \mathrm{NADPH} + \mathrm{CO_2} + \mathrm{H^+}$$

7. What chemical features allow flavins (FAD, FMN) to mediate electron transfer from NAD(P)H to cytochromes or iron-sulfur proteins?
8. Write the mechanisms that show the involvement of biotin in the following reactions.

   (a) $$\mathrm{CH_3{-}C(=O){-}SCoA} + \mathrm{HCO_3^-} + \mathrm{ATP} \rightarrow \mathrm{HO_2C{-}CH_2{-}C(=O){-}SCoA} + \mathrm{ADP} + \mathrm{HOPO_3^{2-}}$$

   (b) $$\mathrm{CH_3{-}CH_2{-}C(=O){-}SCoA} + \mathrm{HO_2C{-}CH_2{-}C(=O){-}CO_2H} \rightleftarrows \mathrm{HO_2C{-}CH(CH_3){-}C(=O){-}SCoA} + \mathrm{CH_3{-}C(=O){-}CO_2H}$$

9. (a) What metabolic advantage is gained by having flavin cofactors covalently or tightly bound to the enzyme?
   (b) Would covalently bound NAD$^+$ (NADP$^+$) be a metabolic advantage or disadvantage?
10. Given the amino acid of the general structure

$$\mathrm{CH_3{-}CH(X){-}CH(NH_3^+){-}COO^-}$$

we could use pyridoxal-5′-phosphate to eliminate X, decarboxylate the amino acid, or oxidize the α carbon to a carbonyl with formation of pyridoxamine-5′-phosphate. The metabolic diversity afforded by PLP, unchanneled, could wreak havoc in the cell. What other components are required to channel the PLP-dependent reaction along specific reaction pathways?

11. Amino acid oxidases catalyze the flavin-dependent reaction

$$\mathrm{R{-}CH(NH_3^+){-}COO^-} + \mathrm{O_2} + 2\,\mathrm{H^+} \rightarrow \mathrm{R{-}C(=O){-}COO^-} + \mathrm{H_2O_2} + \mathrm{NH_4^+}$$

What advantages are gained by the cell in using the PLP-dependent transamination reaction rather than the FAD-dependent deamination reaction to convert α-amino acids to the corresponding α-keto acids?
12. How is flavin adenine dinucleotide involved in the following reaction? Show possible mechanisms.

$$\alpha\text{-D-glucose} + \mathrm{O_2} \rightarrow \text{D-gluconolactone} + \mathrm{H_2O_2}$$

13. For each of the following enzymatic reactions, identify the coenzyme involved.

   (a) $$\mathrm{R{-}CH(NH_3^+){-}COOH} + \mathrm{H_2O} + \mathrm{O_2} \rightarrow \mathrm{R{-}C(=O){-}COOH} + \mathrm{NH_4^+} + \mathrm{H_2O_2}$$

   (b) $$\mathrm{HO{-}CH_2{-}CH(NH_3^+){-}CO_2H} \rightarrow \mathrm{CH_3{-}C(=O){-}CO_2H} + \mathrm{NH_4^+}$$

   (c) $$\mathrm{CH_3{-}CH_2{-}C(=O){-}SCoA} + \mathrm{HCO_3^-} + \mathrm{ATP} \rightarrow \mathrm{HO_2C{-}CH(CH_3){-}C(=O){-}SCoA} + \mathrm{ADP} + \mathrm{P_i}$$

   (d) $$\mathrm{CH_2(O\text{Ⓟ}){-}CH(OH){-}CH(OH){-}CH(OH){-}C(=O){-}CH_2(OH)} + \mathrm{CH_2(O\text{Ⓟ}){-}CH(OH){-}CHO} \rightleftarrows \mathrm{CH_2(O\text{Ⓟ}){-}CH(OH){-}CH(OH){-}CHO} + \mathrm{CH_2(O\text{Ⓟ}){-}CH(OH){-}CH(OH){-}C(=O){-}CH_2(OH)}$$

14. Lipid-soluble polycyclic aromatic hydrocarbons are in part excreted from the body conjugated to one of several hydrophilic substituents, principally glucuronic acid. What role does the liver microsomal cytochrome P450 (polysubstrate monooxygenase) system play in converting a polycyclic aromatic hydrocarbon to forms that could be conjugated to glucuronic acid?
15. Some bacterial toxins use NAD$^+$ as a true substrate rather than as a coenzyme. The toxins catalyze the transfer of ADP-ribose to an acceptor protein. Examine the structure of NAD$^+$ and indicate which portion of the molecule is transferred to the protein. What is the other product of the reaction?

Catabolism and the Generation of Chemical Energy

4

PART

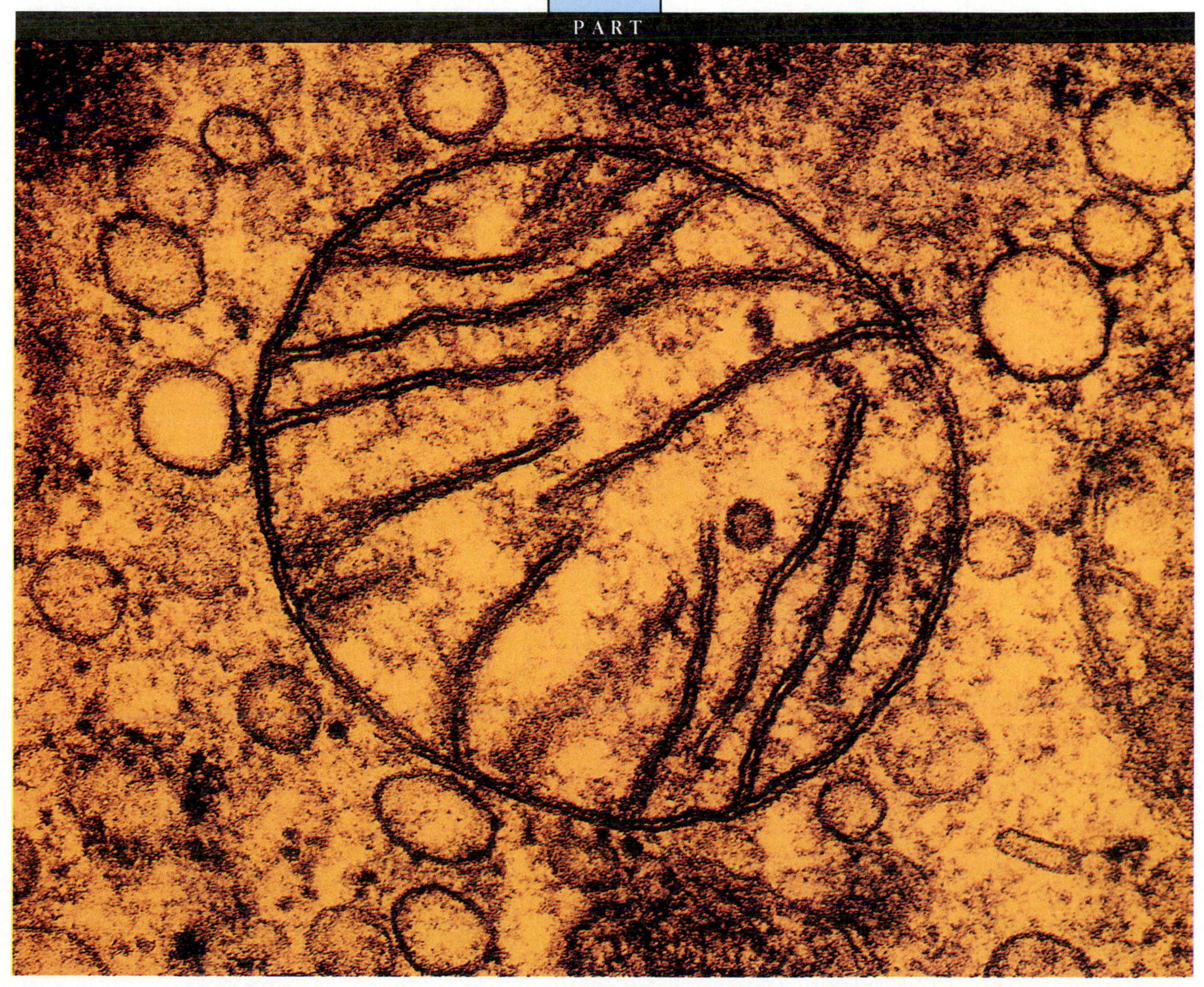

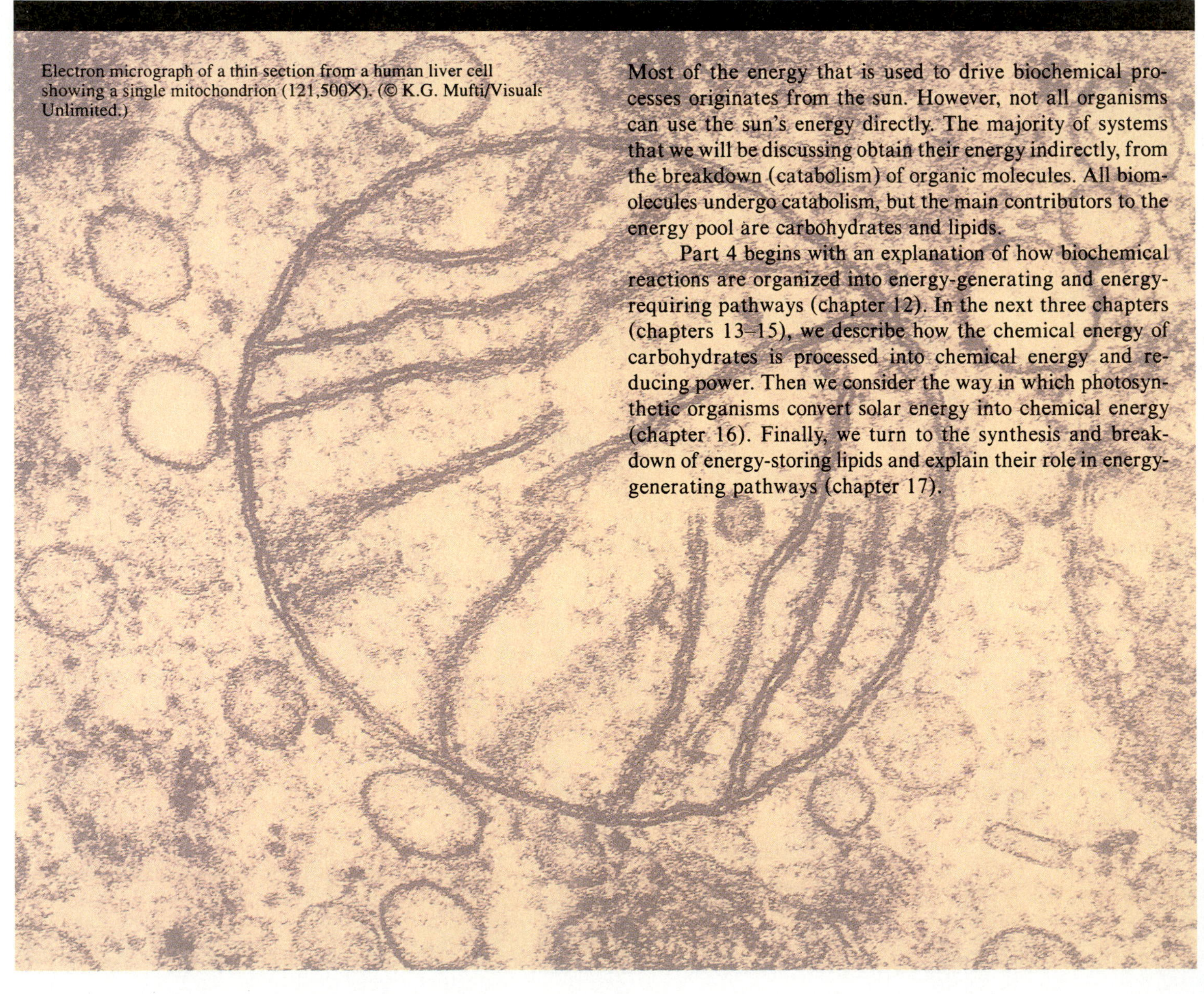

Electron micrograph of a thin section from a human liver cell showing a single mitochondrion (121,500×). (© K.G. Mufti/Visuals Unlimited.)

Most of the energy that is used to drive biochemical processes originates from the sun. However, not all organisms can use the sun's energy directly. The majority of systems that we will be discussing obtain their energy indirectly, from the breakdown (catabolism) of organic molecules. All biomolecules undergo catabolism, but the main contributors to the energy pool are carbohydrates and lipids.

Part 4 begins with an explanation of how biochemical reactions are organized into energy-generating and energy-requiring pathways (chapter 12). In the next three chapters (chapters 13–15), we describe how the chemical energy of carbohydrates is processed into chemical energy and reducing power. Then we consider the way in which photosynthetic organisms convert solar energy into chemical energy (chapter 16). Finally, we turn to the synthesis and breakdown of energy-storing lipids and explain their role in energy-generating pathways (chapter 17).

12

CHAPTER

Metabolic Strategies

Living systems have evolved to the point where we are inclined to view them as totally different from the nonliving chemical systems from which they were derived. The precellular intermediates to living cells, which we discussed briefly in chapter 1, are gone without a trace, and we have little hope of reconstructing the early steps in prebiotic evolution or the first living things. In spite of these enormous gaps in our knowledge, it is useful to consider how living materials differ from nonliving materials, as we find them at present on the earth. Several unique properties are displayed by the chemistry of living systems:

1. Typically, several hundred to several thousand reactions proceed simultaneously in the confines of a living cell.
2. High rates of reaction are achieved under mild conditions of temperature, pH, and concentrations of starting materials and intermediates.
3. Reactions show a high degree of chemical specificity.
4. Reactions are functionally related.

Of these extraordinary properties of biochemical reactions, the last is of greatest value in studying and understanding how biological systems work. Biochemical activities have a purpose, the maintenance and propagation of the system. When new biochemical reactions are discovered, we immediately try to see what role they play in the overall process. Just as frequently, we start with the hypothesis that in light of a known function, a certain reaction must exist, and this conviction gives the impetus that leads to discovery of that reaction. Throughout this text we will be studying biochemical reactions from the standpoint of the functions they serve.

General Principles of Intermediary Metabolism

In parts IV and V of this text we will be considering the synthesis and degradation of small molecules in the living cell. These processes are collectively referred to as intermediary metabolism because they operate on the small-molecule intermediates of biological macromolecules. We will start, in this chapter, by discussing some general principles governing intermediary metabolism. In later sections we will see how metabolic pathways are regulated and what procedures are useful in pathway analysis.

Intermediary Metabolism Maintains a Steady Supply of Biochemical Intermediates and Energy

Intermediary metabolism, the synthesis and degradation of small molecules, serves two functions: it supplies the energy needed for the synthesis of macromolecules or other energy-requiring processes, and it furnishes these processes with the necessary starting materials—amino acids for protein synthesis, fatty acids for lipid synthesis, nucleoside triphosphates for nucleic acid synthesis, and sugars for polysaccharide synthesis.

The rates at which different biological processes require energy and materials vary widely. To compensate for these fluctuating needs, the rates of the reaction sequences in intermediary metabolism must be adjustable over broad ranges so that energy and intermediates will always be available to meet a given need. Quite miraculously, these rates are adjusted so that the concentrations of key metabolites are remarkably stable. This constancy of key intermediates is achieved by maintaining a rate of synthesis for each intermediate that is equivalent to its rate of utilization. The term steady state is used to describe this nonequilibrium situation.

Organisms Differ in Sources of Energy, Reducing Power, and Starting Materials for Biosynthesis

To maintain the steady state and to permit growth and reproduction as well, all living cells require energy (obtained from ATP) and starting materials for biosynthesis. They also require reducing power (obtained from NADPH), since most biosynthesis involves converting compounds to a more reduced state. Although the needs of different organisms for these three components are quite similar, the way in which organisms satisfy their needs can be quite different. Indeed, the most fundamental metabolic distinction between organisms relates to the ways in which they satisfy their basic metabolic needs (table 12.1).

Autotrophs exploit the inorganic environment without depending on compounds produced by other organisms. Thus all of their carbon compounds are synthesized from the only widely available inorganic compounds of carbon—carbonates or carbon dioxide. Different kinds of chemoautotrophs can obtain reducing power by oxidation of a wide variety of inorganic materials such as hydrogen or reduced compounds of sulfur or nitrogen (each species is usually quite specific as to the electron donors that it can utilize). The same reduced compounds supply electrons for regeneration of ATP by electron-transfer phosphorylation. Photoautotrophs obtain ATP by electron-transfer phosphorylation during the cycling of a photochemically excited electron back to ground-state chlorophyll. They are also able to use these excited electrons for the reduction of $NADP^+$, while obtaining electrons from water to replace them in the chlorophyll system. Photoautotrophs have thereby achieved independence of all sources of energy except for the sun and of all sources of electrons except for water. For this reason, photoautotrophic fixation of carbon dioxide is the predominant base of the food chain in the biosphere.

Common heterotrophs depend on preformed organic compounds for all three primary needs. Although some carbon dioxide is fixed in heterotrophic metabolism, notably in the synthesis of amino acids, the heterotrophic cell thrives at the expense of compounds formed by other cells, and is not capable of the net conversion (fixation) of carbon dioxide into organic compounds. Some bacteria, referred to as photoheterotrophs, are able to regenerate ATP photochemically, but they cannot supply electrons to $NADP^+$ as a result of photochemical reactions. Such organisms are like other heterotrophs in that they are dependent on preformed organic compounds. But because of their photochemical apparatus, the photoheterotrophs are able to use available organic food more efficiently than common heterotrophs.

Although all heterotrophs are dependent on preformed organic compounds, they differ markedly in the numbers and types of compounds they require. Some species can make all

Table 12.1
Metabolic Classification of Organisms

Type of Organism	Source of ATP	Source of NADPH[a]	Source of Carbon	Examples
Chemoautotroph	Oxidation of inorganic compounds	Oxidation of inorganic compounds	CO_2	Hydrogen, sulfur, iron, and denitrifying bacteria
Photoautotroph	Sunlight	H_2O	CO_2	Higher plants, blue-green algae, photosynthetic bacteria
Photoheterotroph	Sunlight	Oxidation of organic compounds	Organic compounds	Nonsulfur purple bacteria
Heterotroph	Oxidation of organic compounds	Oxidation of organic compounds	Organic compounds	All higher animals, most microorganisms, nonphotosynthetic plant cells

[a]NADPH is the main source of reducing power.

required compounds when supplied with a single carbon source; others have lost some or many biosynthetic capabilities. Mammals must obtain about half of their amino acids from external sources and are also unable to make several metabolic cofactors, as evidenced by their well-known nutritional requirements for vitamins. Some bacteria and some parasites have even more extensive nutritional requirements.

Reactions Are Organized into Sequences or Pathways

To appreciate how biochemical reactions are organized to serve common functions, we must examine the organization of biochemical reactions in some detail. Most of the enzyme-catalyzed reactions in living cells are organized into sequences or pathways. It is the sequence that serves a function, not the individual reaction. For example, the conversion of the organic compound chorismate to tryptophan occurs in five discrete steps, each requiring a specific enzymatic activity (fig. 12.1). The intermediates between chorismate and tryptophan serve no function except as precursors of tryptophan. The amino acid tryptophan has many functions, including its use as a building block in protein synthesis.

The role of breakdown of the six-carbon sugar glucose to the three-carbon compound pyruvate is to supply ATP to the cell. But, as in the case of the tryptophan pathway, the function of each reaction in glucose catabolism can be appreciated only by considering the overall sequence. This pathway involves nine enzymatically catalyzed steps (fig. 12.2). In the first three steps ATP is actually consumed. It is only in the last four steps that the starting amount of ATP is recovered and further ATP is synthesized. Hence the function of this sequence is not fully served until each substrate has passed through the entire sequence.

Figure 12.1

The biosynthesis of tryptophan. Tryptophan is synthesized in five steps from chorismate. Each step requires a specific enzyme activity. The four intermediates between chorismate and tryptophan serve no function other than as precursors of tryptophan. The detailed steps of each reaction in this sequence are described in chapter 18.

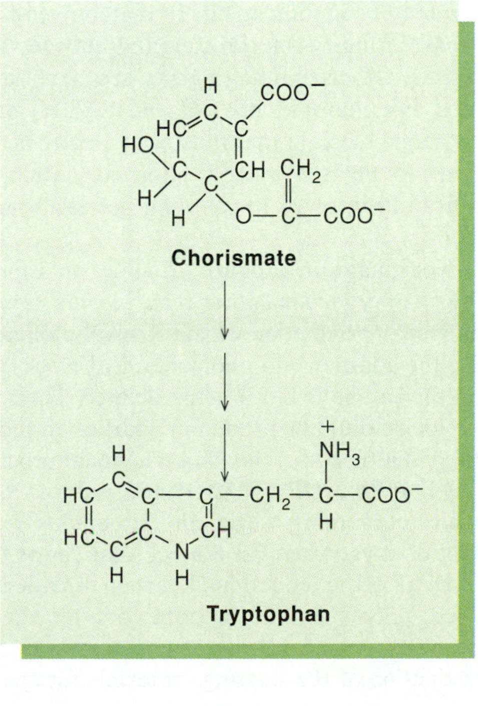

Chorismate

Tryptophan

Figure 12.2

The breakdown of glucose to pyruvate. The conversion of glucose to pyruvate requires nine steps and one enzyme for each step. In steps 1 and 3, ATP is consumed. In steps 6 and 9, ATP is produced. The net production of ATP is 2 moles for each mole of glucose consumed. Unlike the intermediates in the tryptophan pathway, many of the intermediates between glucose and pyruvate serve as useful substances for other purposes. A full description of all the steps in this sequence is given in chapter 13.

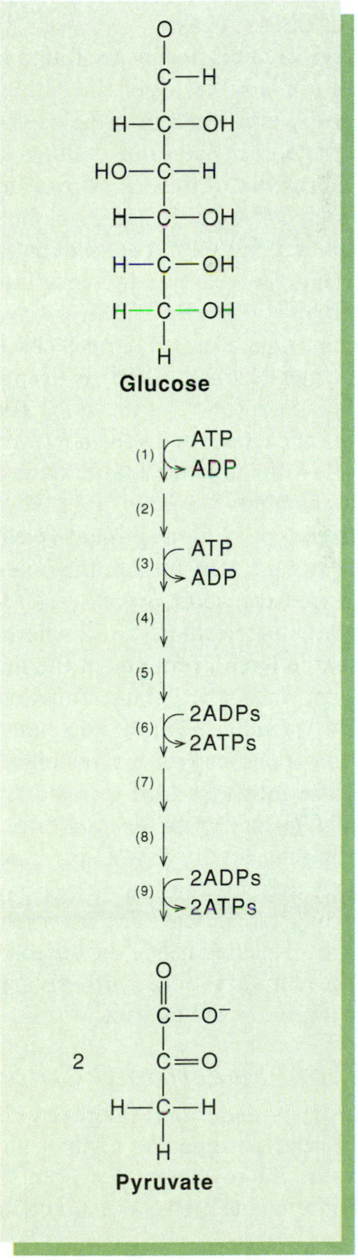

Glucose

(1) ATP → ADP

(2)

(3) ATP → ADP

(4)

(5)

(6) 2ADPs → 2ATPs

(7)

(8)

(9) 2ADPs → 2ATPs

2 Pyruvate

In glucose breakdown and tryptophan biosynthesis we have examples of the simplest types of pathways, consisting of a linear series of reactions. Some pathways contain branches, so that common intermediates can follow more than one sequence after the branchpoint. Pathways may also be organized so that reactions can go in either the forward or the reverse directions in certain cases.

Sequentially Related Enzymes Are Frequently Clustered

A fundamental principle of biochemical organization is that enzymes that catalyze different steps in a sequence are usually clustered together in the cell. As a result, an intermediate resulting from the first reaction in a pathway can be directly shunted to the second enzyme in the pathway, and so on. As you might expect, such an arrangement accelerates synthesis of end product and minimizes the loss of intermediates. It also facilitates regulation of the pathway.

Three degrees of clustering are found (fig. 12.3). In the simplest situation (fig. 12.3*a*), all of the catalytic activities for a particular pathway are found in proteins that exist as independent soluble proteins in the same cellular compartment. In such cases the intermediates must get from one enzyme to the next in the sequence by free diffusion through the cytosol or by transfer after contact between two sequentially related enzymes, one carrying the reactive intermediate and the other ready to receive it.

The second type of arrangement observed for sequentially related enzymes (see fig. 12.3*b*) is exemplified by the *E. coli* fatty acid synthase (see chapter 17). This synthase is a complex of most of the enzymes involved in fatty acid synthesis. The intermediates in this case are bound to the enzyme complex until synthesis is complete.

In the third type of arrangement (see fig. 12.3*c*), functionally related enzymes are membrane-bound. An example occurs in the electron-transport processes associated with oxidative phosphorylation (see chapter 15), where many of the enzymes are membrane-bound proteins in the mitochondrion.

Finally, enzymes with related functions are often segregated in specific organelles. This arrangement enhances metabolic efficiency in various ways. First, the higher concentrations of enzymes and intermediates lead to faster rates of reaction. Second, packaging in an organelle facilitates regulation of a pathway by limiting the supply of starting materials or of a key intermediate. Fatty acid metabolism in eukaryotes superbly illustrates this type of arrangement. The enzymes involved in fatty acid catabolism are all located inside the mitochondrion, whereas the enzymes involved in fatty acid synthesis are all located outside the mitochondrion, in the cytosol.

Pathways Show Functional Coupling

Metabolic chemistry is sometimes taught by reference to large charts that have been designed to display all the major biochemical pathways. Such charts, however, include a bewildering array of interconnections; the functional relationships between different pathways tend to be obscured by the complexity of presentation. The overall operational aspects of metabolism are easier to grasp when they are presented in simpler block diagrams that omit details and focus on functional relationships. Such a diagram for a typical heterotrophic aerobic cell is shown in figure 12.4, where the metabolism of the cell is symbolized by two functional blocks:

1. Catabolism, or degradative metabolism. Foods are oxidized to carbon dioxide. Most of the electrons liberated in this oxidation are transferred to oxygen, with concomitant production of ATP (electron-transport phosphorylation). Other electrons are used in the regeneration of NADPH, the most frequently used reducing agent for biosynthesis. The major pathways in this block are the glycolytic sequence and the tricarboxylic acid (TCA) cycle. These same sequences, with the pentose phosphate pathway, also supply carbon skeletons for the cell's biosynthetic processes.
2. Biosynthesis, or anabolism. This block includes much greater chemical complexity than the first. The starting materials produced by glycolysis and the TCA cycle are converted into hundreds of cell components, with NADPH serving as a reducing agent when necessary, and with ATP as the universal coupling agent or energy-transducing compound. This conversion is rapidly followed by the synthesis of macromolecules and growth.

The starting materials consumed in biosynthetic sequences must be continuously replaced by catabolic processes. The energy sources and reducing sources, ATP and NADPH, are used in very different ways. When they contribute to biosynthesis, these coupling agents are converted to $NADP^+$ and ADP or AMP. They must be regenerated, reduced, or rephosphorylated at the expense of substrate oxidation in the catabolic block of reactions. As figure 12.4 indicates, ATP and NADPH are the fundamental coupling agents in metabolism—the compounds that have primary roles in coupling between the major functional blocks. Other coupling agents are essential but less broad in scope. For example, NADH and $FADH_2$ participate within the catabolic block in the transfer of electrons from substrate to oxygen in electron-transfer phosphorylation, but they are not significantly involved in coupling between major functional blocks.

The block diagram of figure 12.4 does not represent the total metabolism of organisms other than aerobic heterotrophs. The photochemical production of ATP in photoheterotrophs would require the addition of a photochemical block, with ATP production in the catabolic block being deleted. For a photoautotroph, two blocks would be needed in addition to those shown in figure 12.4: (1) a block for the photochemical processes that regenerate ATP from ADP or AMP and reduce $NADP^+$ to NADPH, while consuming water and producing a stoichiometric amount of oxygen and (2) a block that shows the use of ATP and NADPH in the reduction of carbon dioxide to various products. These photosynthetic products provide the input to the catabolic block, which in this case serves only one function—the production of the starting materials for synthesis.

Figure 12.3

The organization of functionally related enzymes. Functionally related enzymes are organized in three ways: (*a*) as unlinked proteins soluble in the aqueous milieu of the same cellular compartment; (*b*) as components in a multiprotein complex; or (*c*) as components on a membrane.

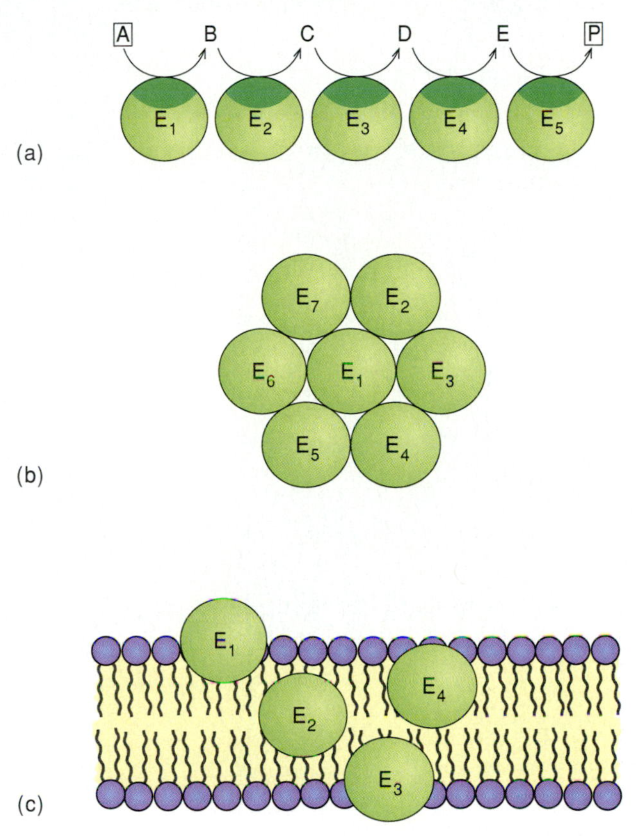

Figure 12.4

A schematic block diagram of the metabolism of a typical aerobic heterotroph. The block labeled catabolism represents pathways by which nutrients are converted to small-molecule starting materials. Catabolism also supplies the energy (ATP) and reducing power (NADPH) needed for activities that occur in the second block; these compounds shuttle between the two boxes. The block labeled "Biosynthesis" represents the synthesis of low- to medium-molecular-weight components of the cell as well as the synthesis of proteins, nucleic acids, lipids, and carbohydrates and the assembly of membranes, organelles, and the other structures of the cell.

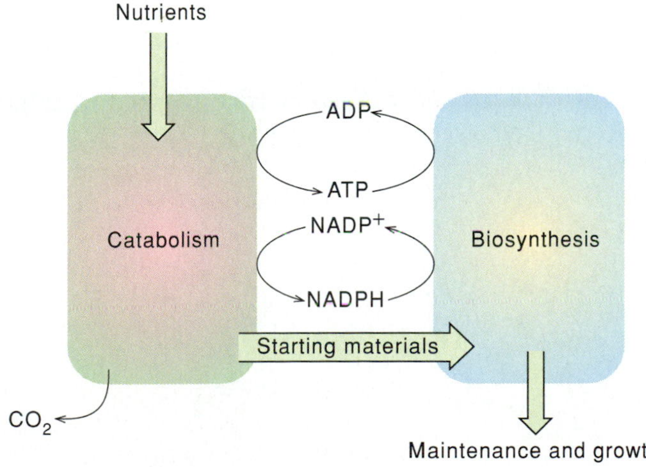

A more detailed way to look at metabolism is depicted in figure 12.5. Here, metabolism is divided into three nested boxes. The innermost box shows the central metabolic pathways and the interconversions of various low-molecular-weight carbohydrates. These substances serve as the starting materials for all other metabolism. Some of the compounds formed from these intermediates are shown in the middle box. These compounds serve as building blocks for the final products indicated in the outermost box. Each of the next twelve chapters will deal with an expanded segment of figure 12.5, and we will repeatedly refer to this figure in the introductory passages of those chapters.

The ATP-ADP System Assures Conversions in Both Directions

Figure 12.4 emphasizes the coupling of catabolic sequences and biosynthetic (anabolic) sequences to the ATP-ADP cycle. In fact, the involvement of the ATP-ADP system is much greater than indicated. Most metabolic sequences within the two major blocks are linked with the ATP-ADP system. To appreciate the way in which this is done we must look at individual sequences in greater detail.

In analyzing metabolism we must distinguish between a sequence and a conversion. A sequence is a group of reactions, arranged sequentially, that carry out a biological conversion. In other words, a conversion is the metabolism of starting material A to end product Z or of Z to A; a sequence is the specific set of reactions by which the conversion is carried out. As a rule, a conversion in one direction is catabolic and produces energy in the form of ATP, whereas a conversion in the reverse direction is anabolic and requires the expenditure of energy. Any conversion could be carried out by many different sequences of reactions, which might be coupled to nearly any number of ATP-to-ADP or ADP-to-ATP interconversions, depending on the specificities of the enzymes that catalyze the component reactions. The values of the overall free energy change and of the overall equilibrium constant are determined by the stoichiometry of this coupling—that is, by the number of moles of ATP converted to ADP (or of ADP converted to ATP) per mole of Z that is produced or consumed.

As we have seen (chapter 2), any conversion can be made thermodynamically favorable by coupling it to a sufficient number of ATP-to-ADP conversions. That fact is exploited in the design of metabolic sequences. Metabolic sequences commonly occur in pairs. The two sequences of a pair connect the same compounds, which in the general case we are calling A and Z. As a rule, the reactions of the two oppositely directed sequences are catalyzed by different enzymes, the formation of most or all of the intermediates is catalyzed by different enzymes, and most or all of the intermediates are different. However, the most significant difference between the sequences is that, because of the specificities of their component enzymes, they are coupled to different numbers of ATP-to-ADP conversions. Remember that coupling an ATP-to-ADP conversion into a metabolic sequence changes the equilibrium ratio of products by a very large factor. In each pair of unidirectional sequences, one sequence contains a number of ATP-to-ADP conversions such that, under any conditions that could arise in a living cell,

Figure 12.5

The overall metabolism of an aerobic heterotroph, represented as three concentric boxes. The central metabolic pathways are represented in the inner box. It is here that carbon compounds are degraded to produce energy, reducing power, and starting materials for biosynthesis. The glycolytic pathway, which results in the breakdown of glucose to pyruvate, produces a limited amount of energy and reducing power. The main energy production takes place in the tricarboxylic acid cycle, which starts with the conversion of pyruvate into acetyl-CoA and the subsequent condensation of acetyl-CoA with oxaloacetate. Under aerobic conditions the TCA cycle functions only in the clockwise direction. By contrast, the glycolytic pathway can be reversed, in which case the process is called gluconeogenesis. Just as glycolysis yields energy, so does gluconeogenesis require energy. Some key intermediates in the glycolytic pathway and the TCA cycle are shown. Essentially every intermediate in the central metabolic pathways can be converted into any intermediate that is in short supply. Glucose-6-phosphate can also be converted to a pentose that is used for carbon fixation in the photosynthetic process. Most of the compounds represented in the central metabolic pathways serve as starting materials for other compounds, some of which are shown in the middle concentric box. Thus glucose-6-phosphate is the precursor of glucose-1-phosphate, acetyl-CoA is the precursor of both mevalonate and fatty acids, α-ketoglutarate is the precursor of glutamate, succinyl-CoA is the precursor of aminolevulinate, and oxaloacetate is the precursor of aspartate. Nucleotides are built from C-5 sugars and aspartate and glutamate. In the outer box are shown several of the end products of biosynthesis: nucleic acids, heme, protein, lipids, cholesterol, steroids, bile acids, glycogen, and starch. This diagram is greatly simplified so that it is possible to get an overview of what is happening in metabolism without getting submerged in detail. For example, protein is represented as being derived from glutamate. This is the main function of glutamate, but nineteen other amino acids also are required for protein synthesis. All of these amino acids are formed from reactions that begin with the carbohydrate compounds shown in the central metabolic pathways. The acetyl-CoA used in mevalonate synthesis is formed indirectly from citrate (see fig. 14.16). Reactions connected by two half-arrows probably use the same enzyme in both directions. Reactions (or conversions) connected by two slightly curved whole arrows use different enzymes. Conversions connected by one whole arrow go in only one direction.

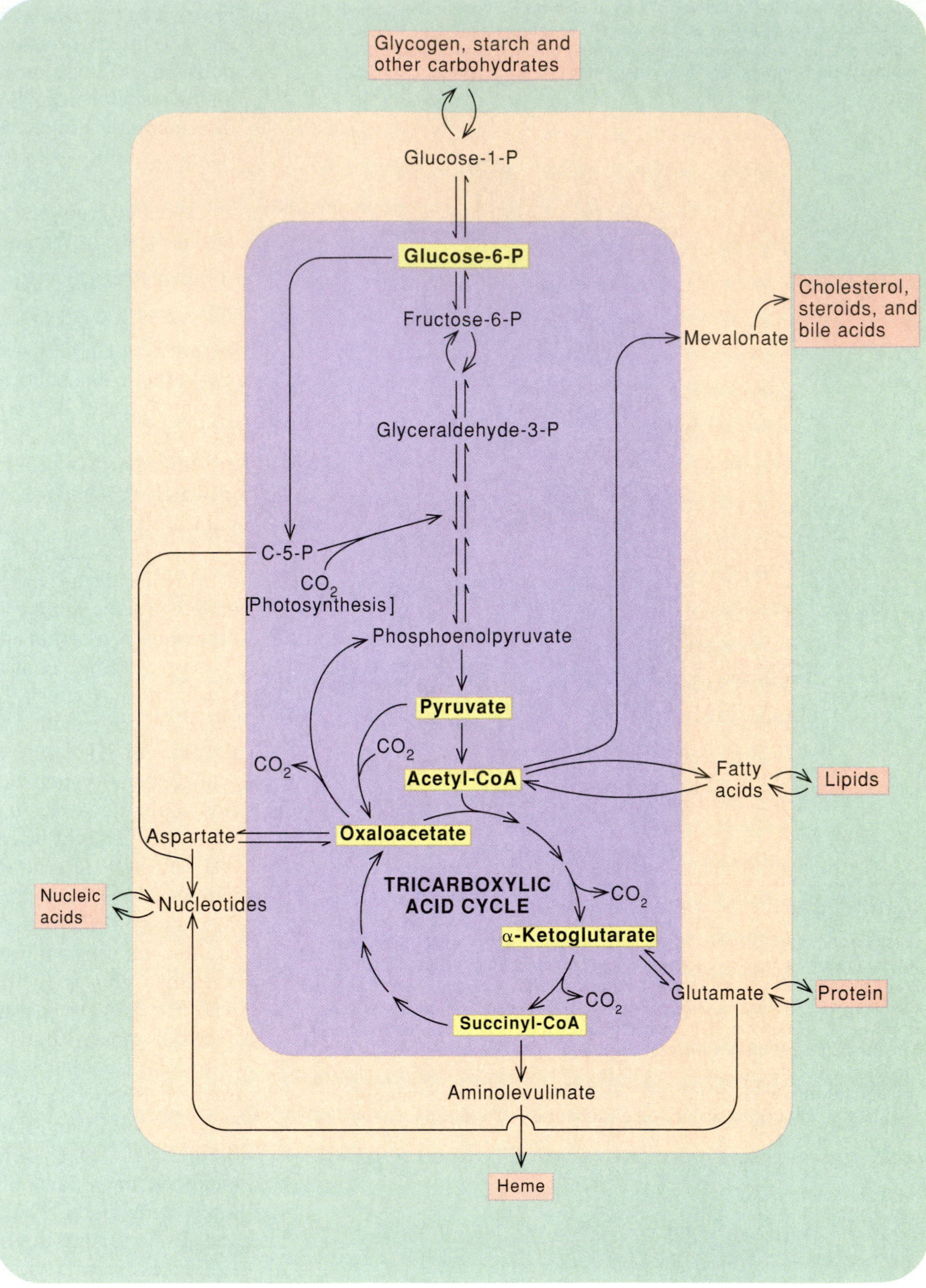

Figure 12.6

Schematic illustration of the organization of metabolic sequences into oppositely directed pairs. The two sequences result in opposite conversions. The values for the overall equilibrium constant are a function of the conversion and the number of ATP-to-ADP conversions to which each sequence is coupled.

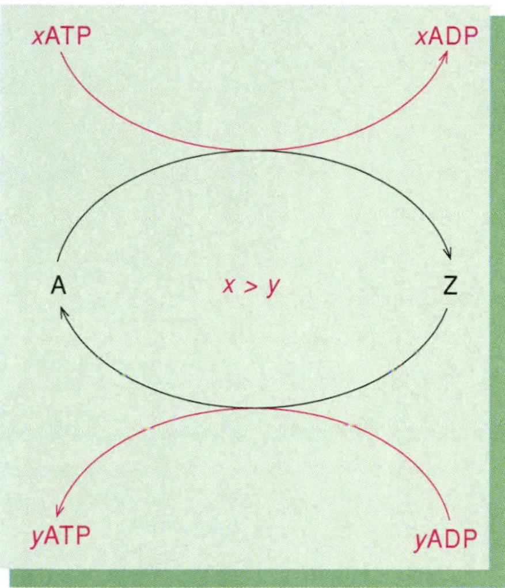

Figure 12.7

The interconversion of fructose-6-phosphate and fructose-1,6-bisphosphate. This reaction exemplifies the type of interconversion shown in figure 12.6, where $x = 1$, $y = 0$, A = fructose-6-phosphate, and Z = fructose-1,6-bisphosphate.

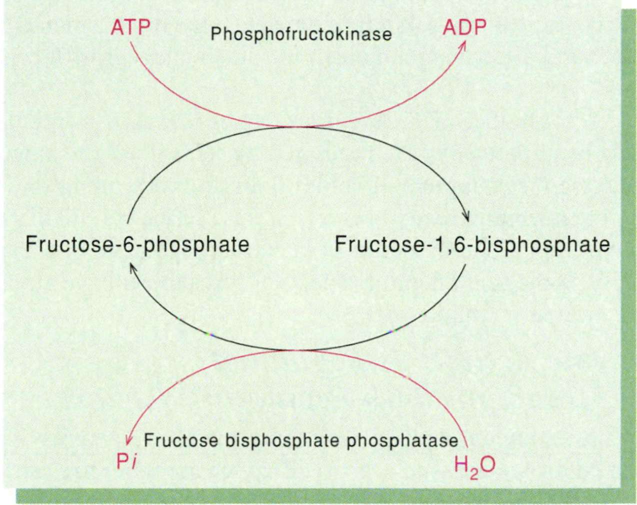

the equilibrium constant favors the conversion of A to Z. For the other sequence, the ATP-ADP stoichiometry is such that the equilibrium constant favors the conversion of Z to A. The general situation is illustrated schematically in figure 12.6. It is the difference between x and y that allows the conversions in both directions to be thermodynamically favorable at all times.

Conversions Are Kinetically Regulated

Pairs of oppositely directed sequences look like cycles (see fig. 12.6). Indeed, if both sequences were simultaneously active to a significant degree, they would form a true cycle. But since the end products of one sequence are the starting materials for the other, such a cycle would have no metabolic consequences except that ATP would be wasted. This is because more ATP is used in one sequence than is regenerated in the other. Because their cyclic operation would have no value, such oppositely directed pairs of sequences are often called futile cycles or pseudocycles. Pseudocycles almost never function as true cycles, thanks to kinetic constraints. These kinetic constraints result in the regulation of one or more enzymes for each sequence of a pair so that conversions can occur in only one direction at any given time.

The interconversion of fructose-6-phosphate and fructose-1,6-bisphosphate provides an especially instructive example of the importance of ATP coupling and the regulation of an interconversion. Each "sequence" of this pseudocycle consists of only one reaction, and the "sequences" differ by one ATP-to-ADP conversion. The reactions are shown by two equations:

$$\text{Fructose-6-phosphate} + \text{ATP} \rightarrow \text{fructose-1,6-bisphosphate} + \text{ADP}$$

$$\text{Fructose-1,6-bisphosphate} + H_2O \rightarrow \text{fructose-6-phosphate} + P_i$$

Figure 12.7 shows these reactions in the form of a pseudocycle. The equilibrium constant for the hydrolysis of fructose-1,6-bisphosphate to fructose-6-phosphate is about 10^4. Thus this reaction will always be favorable at any feasible physiological ratio of the two compounds. The coupling of an ATP-to-ADP interconversion changes the ratio of product to substrate under physiological conditions by a factor of approximately 10^8. Thus the ratio of fructose-6-phosphate to the bisphosphate could be about 10^4 if the phosphatase reaction were at equilibrium under physiological conditions, while the ratio of fructose-1,6-bisphosphate to fructose-6-phosphate could be about 10^4 if the phosphofructokinase reaction were at equilibrium. At any ratio of concentrations that can reasonably be expected in a metabolizing cell, both reactions are far from equilibrium and in the direction that is highly favorable for reaction.

Clearly, if both reactions in this conversion were allowed to function at the same time, the result would be nothing except the loss of ATP. This metabolic disaster is prevented from happening by the design of the enzymes that carry out the two conversions. Both enzymes are regulatory enzymes whose activities are in turn regulated by small molecules that serve as signals of the metabolic state of the cell. The system is designed so that only one of the enzymes is active at any given time. Which enzyme is active depends on the metabolic needs of the cell. The precise mechanisms for the regulation of these enzymes are taken up in chapters 13 and 15. Some general aspects of the regulation of pathways by control of enzyme activity are taken up in the following section.

Regulation of Pathways

The two most important criteria governing the regulation of biochemical pathways are economy and flexibility. Organisms must regulate their metabolic activities economically to avoid significant deficiencies or excesses of metabolic products. When there is a significant change in the environment, such as a shift in the concentration or kinds of nutrients, the organism must be flexible and alter its metabolism to suit the new growth conditions.

A number of different strategies are used to regulate metabolism; pathways are regulated by controlling the amount of enzyme (see chapters 30 and 31) and by controlling the activity of enzymes already present. In this section we will discuss pathway regulation by means of the direct regulation of enzyme activity. Some fundamental aspects of this subject have already been covered in chapter 10.

Enzyme Activity Is Regulated by Interaction with Regulatory Factors

The most common way of regulating metabolic activity is by direct control of enzyme activity. Enzyme activities are usually regulated by noncovalent interaction with small-molecule regulatory factors (see chapter 10) or by a reversible covalent modification such as that brought about by phosphorylation or adenylylation (e.g., see chapters 10, 11, and 24) of an amino acid side chain. The effect of the regulatory factor in specific instances may be to increase or decrease the activity of the enzyme.

Regulatory Enzymes Occupy Key Positions in Pathways

Enzymes that are susceptible to direct regulation occupy key positions in metabolic pathways. They are usually called regulatory enzymes to emphasize the fact that they have specific sites for binding regulatory factors, which are distinct from their substrate-binding sites. In a multienzyme pathway the first enzyme in the pathway is usually regulated, while the others are not (fig. 12.8*a*). In the case of CTP synthesis, discussed in chapter 10, we saw that the enzyme aspartate carbamoyltransferase is negatively regulated by CTP. That is, excess CTP binds to the regulatory sites on this multiprotein enzyme, thereby inhibiting its activity. Aspartate carbamoyltransferase catalyzes the first step in a multistep pathway leading to CTP. Since CTP is the end product of the pathway, this type of regulation is referred to as end-product inhibition. End-product inhibition is very common for anabolic pathways.

The mechanism is clearly flexible and economic. If there is sufficient end product, there is no point in processing substrates down the pathway. To do so would be a waste of both energy and materials. Furthermore, it is most effective to block the first enzyme in the pathway; this by itself will rapidly reduce activity of the entire pathway. It would be redundant to block any other enzymes in the pathway, since no intermediates will be available to these enzymes if the first step is blocked. Blocking only a middle enzyme in the pathway, instead of the first enzyme, would also be quite wasteful. The result would be accumulation of an intermediate that would serve no useful function. Indeed, such intermediates in excess frequently have harmful effects.

In a branched-chain pathway, end-product inhibition results in inhibition of the first enzyme after the branchpoint (fig. 12.8*b*). This arrangement leads to control of either pathway after the branchpoint. If the supply of the branchpoint substrate (D in figure 12.8*b*) is limiting, the inhibition of one pathway after the branchpoint could increase the metabolic flow of the other pathway. This effect is often highly significant. *Apropos* of this point, anabolic sequences frequently arise as branchpoints from catabolic pathways (fig. 12.9). In such instances the control factors affect whether a catabolic intermediate will be further degraded or will serve as a substrate for biosynthesis.

Figure 12.8

Two patterns for end-product inhibition. In end-product inhibition the first reaction in the pathway (*a*) or the first reactions after a branchpoint (*b*) are inhibited by the specific end products.

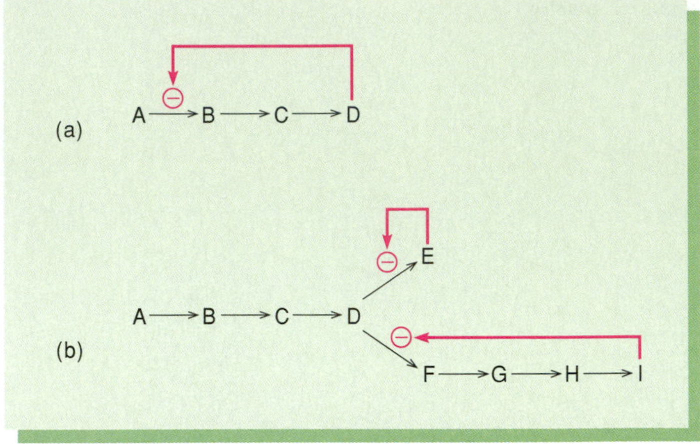

Figure 12.9

Branchpoints in metabolic pathways occur at locations where an intermediary, S, can follow more than one route. In this case S can proceed along a catabolic sequence or an anabolic sequence. The first reactions after the branchpoints are catalyzed by enzymes B and A, respectively. Once the first step after the branchpoint has been taken, the metabolic intermediate is irreversibly committed to follow that pathway.

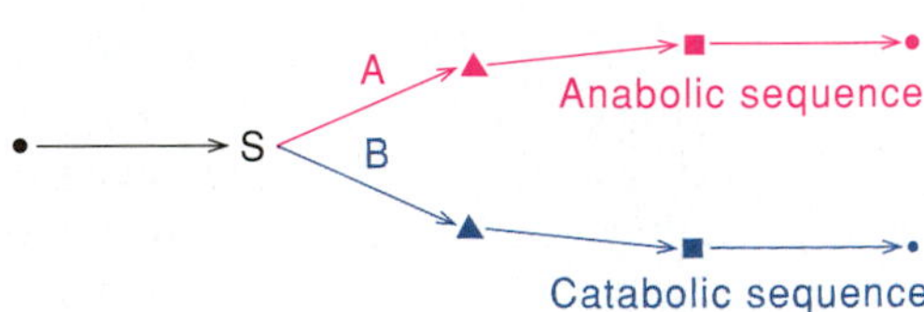

Figure 12.10

Reaction velocity as a function of substrate concentration for a first-order enzymic reaction (left) and for a cooperative enzyme with fourth-order kinetics (right). Substrate concentration is expressed as $[S]/[S_{0.5}]$ and velocity as a fraction of maximum velocity. At a $[S]/[S_{0.5}]$ value of 1 the substrate concentration is equal to $[S_{0.5}]$ and the reaction velocity is half of the maximum velocity.

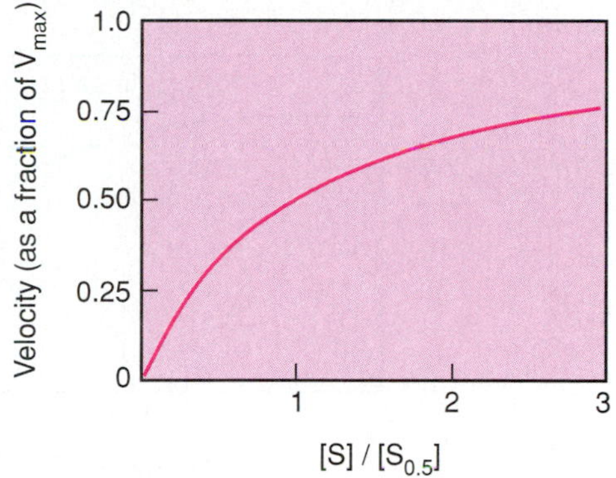

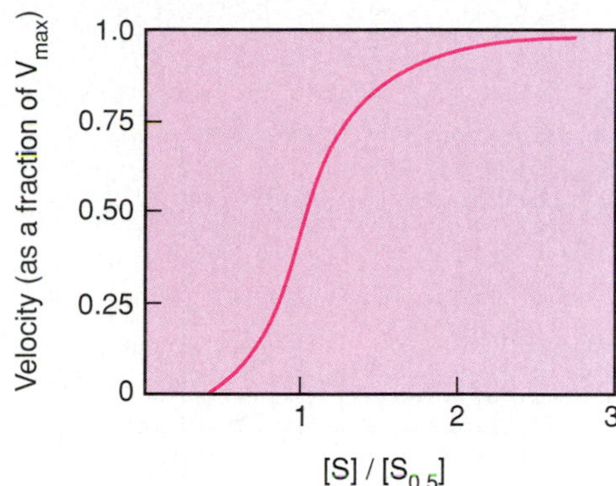

Regulatory Enzymes Often Show Cooperative Behavior

It is clear from what we said in the previous subsection that control of the rates of metabolic sequences is most effective when it is exerted at branchpoints. As you might expect, the enzymes that catalyze reactions at branchpoints have evolved features to make the partitioning very sensitively responsive to metabolic signals. A striking feature of a typical enzyme at a branchpoint (the first enzyme in a biosynthetic sequence, for example) is that the reaction that it catalyzes is of high kinetic order with respect to its substrate (see chapter 10). Whereas other enzymes along the pathway typically exhibit normal Michaelis kinetic responses to the concentrations of their substrates (first-order or hyperbolic kinetics), branchpoint enzymes usually catalyze reactions of higher order (cooperative or sigmoidal kinetics). A fourth-order response is quite common. The two types of behavior are shown in figure 12.10. Any enzyme-catalyzed reaction must level off (to a kinetic order of zero) at high concentration because of saturation of the catalytic sites, but for the "normal Michaelis" enzyme the kinetic order at very low concentrations of substrate is one, whereas for the cooperative enzyme it has a higher value, frequently four.

A major advantage of this higher order is that it makes the system very sensitive to changes in substrate concentration. If the first reaction in the sequence responds to the fourth power of the concentration of its substrate (the starting material for the sequence), the flow of material into the sequence will be much more sensitive to small changes in the concentration of the starting material than if the reaction were first order with respect to substrate, and intermediates in catabolic sequences will not be diverted to biosynthetic use unless they are in good supply.

Perhaps an even more important advantage of high kinetic order is that the regulatory responses of the enzyme are greatly strengthened. We have discussed kinetic regulation of enzymes at metabolic branchpoints by negative feedback, in which the end product of the sequence binds to a regulatory site and decreases the affinity of the catalytic site for the substrate (the branchpoint metabolite). When the enzyme is constructed so as to have cooperative interactions between sites, the sensitivity of negative feedback control is greatly sharpened.

Both Anabolic and Catabolic Pathways Are Regulated by the Energy Charge

It is not always an advantage for the concentration of an amino acid or other end product to be the only factor that determines the rate at which that end product is synthesized. If a cell is starved for energy or catabolic intermediates, it may not be able to afford to synthesize amino acids, even if their concentrations are quite low. Continued biosynthesis under such circumstances might deplete the already scarce supply of energy to the point where essential functions, such as maintenance of concentration gradients across membranes, would be impaired. It would clearly be advantageous for the rate of biosynthesis to be regulated by the general energy status of the cell as well as the need for the specific end product.

Since it is the adenine nucleotide system ATP-ADP (and occasionally ATP-AMP) that couples energy into biosynthetic sequences, we might expect that this system would also supply the necessary signal indicating the energy status of the cell, to be sensed by the kinetic control mechanisms of biosynthesis. It is helpful to have a term that quantitatively expresses the energy status of the cell. The term used is the energy charge, which is defined as the effective mole fraction of ATP in the ATP/ADP/AMP pool:

$$\text{Energy charge} = (\text{ATP} + 0.5\ \text{ADP})/(\text{ATP} + \text{ADP} + \text{AMP})$$

In this equation the 0.5 in the numerator takes into account the fact that ADP is about half as effective as ATP at carrying chemical energy. Thus two ADPs can be converted into one ATP and one AMP by a reaction with an equilibrium constant of about 1.

$$\text{ATP} + \text{AMP} \rightleftharpoons 2\ \text{ADP},\ K_{eq} \simeq 1$$

Figure 12.11

Relative concentrations of ATP, ADP, and AMP as a function of the adenylate energy charge. The adenylate kinase reaction was assumed to be at equilibrium, and a value of 1.2 was used for its effective equilibrium constant in the direction shown in the equation in the text.

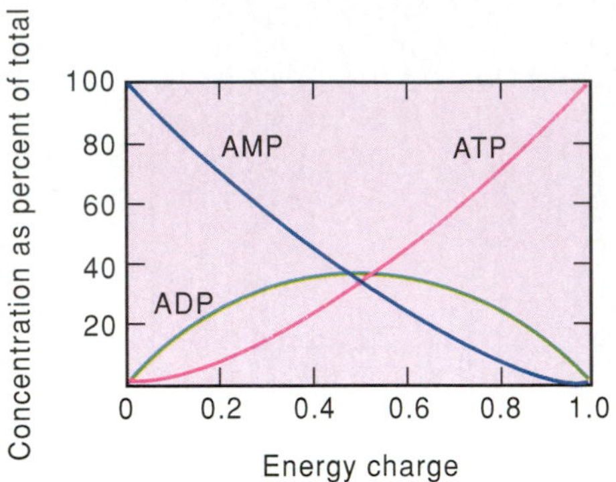

Figure 12.12

Variation in reaction rates as a function of the energy charge. As the energy charge increases, the rate of catabolic reactions decreases. Meanwhile, the rate of anabolic reactions increases. The combined effect is to stabilize the energy charge at a value around 0.9.

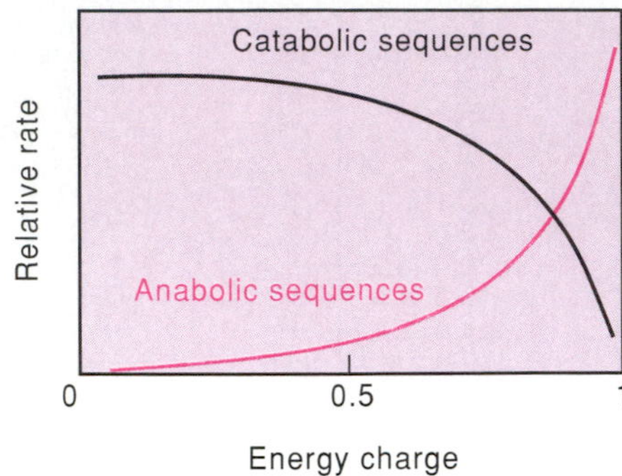

The values for the energy charge could conceivably vary from 0 to 1, as illustrated in figure 12.11. In fact, however, the values for real cells are usually held within very narrow limits. The result is a general stabilizing effect (homeostatic effect) on the cellular metabolism. In much the same way, voltage regulation is necessary for the effective operation of many electronic devices. Now let us see how the limits are maintained.

Several reactions in anabolic and catabolic pathways have been found to respond to variation in the value of the energy charge. As might be expected, the enzymes in catabolic pathways respond in a direction opposite to that of enzymes in anabolic pathways. The two responses are compared in figure 12.12. It is clear that if catabolic sequences, which lead to the generation of ATP, respond as shown by the upper curve, and biosynthetic sequences, which use ATP, respond as shown by the lower curve, then the charge will be strongly stabilized at a value of about 0.9, where the two curves intersect. A tendency for the charge to fall will be resisted by the resulting general increase in the rate of catabolic sequences and the general decrease in the rate of biosynthetic sequences. A tendency for the charge to rise will be resisted by the opposite effects.

Regulation of Pathways Involves the Interplay of Kinetic and Thermodynamic Factors

The general properties of paired unidirectional sequences and their advantages to the organism are in one sense very simple. But on further consideration, we see that they depend on a rather complex interplay between kinetic and thermodynamic factors and effects.

In any kind of negative feedback system, cause-and-effect relationships are circular. We see the circularity of negative feedback at a simple level in the response of a biosynthetic sequence to the concentration of its end product. The concentration of the end product is a major factor controlling the rate of synthesis, and the rate of synthesis is a major factor controlling the concentration of the end product.

For pairs of oppositely directed metabolic sequences, greater sophistication is needed than for simple feedback control of a synthetic pathway. In our discussion of pseudocycles thus far, we have assumed that the ATP/ADP ratio is held at a value very far from equilibrium. But of course, in the absence of highly effective controls, the ATP/ADP ratio would not remain far from equilibrium; it would rapidly approach its equilibrium value. What are the controls that keep this from happening? The answer is the enzymes that catalyze first steps in pseudocycles—the same controls that we have already discussed.

The rates of reactions that control sequences in which there is net conversion of ATP to ADP (such as A to Z in our generalized pseudocycle) are high when the energy charge (or the ATP/ADP ratio) is high, and they decrease sharply with a decrease in those parameters. Rates of regulatory enzymes in sequences in which ATP is regenerated (see fig. 12.12) are high when the energy charge or ATP/ADP ratio is low, and decrease sharply as the charge increases. The curves intersect at an energy charge value of about 0.9 (which corresponds to an ATP/ADP ratio of 5). These kinetic effects play a major role in stabilizing the energy charge and the ATP-ADP system *in vivo*. So kinetic control of rates of conversion depends on the value of the ATP/ADP ratio being far from equilibrium (that is, on thermodynamic factors); but at the same time the ATP/ADP ratio itself depends on reciprocal regulation of oppositely directed sequences (that is, on kinetic factors).

As we have seen, the characteristic and essential feature of paired, oppositely directed sequences is that they differ in their ATP stoichiometries. But that difference is meaningful only because the ATP/ADP ratio is far from equilibrium in the cell. If ATP, ADP, and P_i were at equilibrium, it would not matter how many ATP-to-ADP conversions were coupled to any metabolic sequence. For any system at equilibrium ΔG is zero, and coupling to another reaction or sequence would have no effect on the free energy change or the position of equilibrium

Figure 12.13

The tryptophan biosynthetic pathway in *E. coli*. There are five enzymatically catalyzed reactions involved in tryptophan biosynthesis and five different polypeptides associated with these reactions. Polypeptides E and D normally make a tetrameric complex, which catalyzes the first two reactions in the pathway. The next two reactions are catalyzed by a single polypeptide, C. The final reaction is catalyzed by a tetrameric complex composed of the B and A polypeptides in equal numbers.

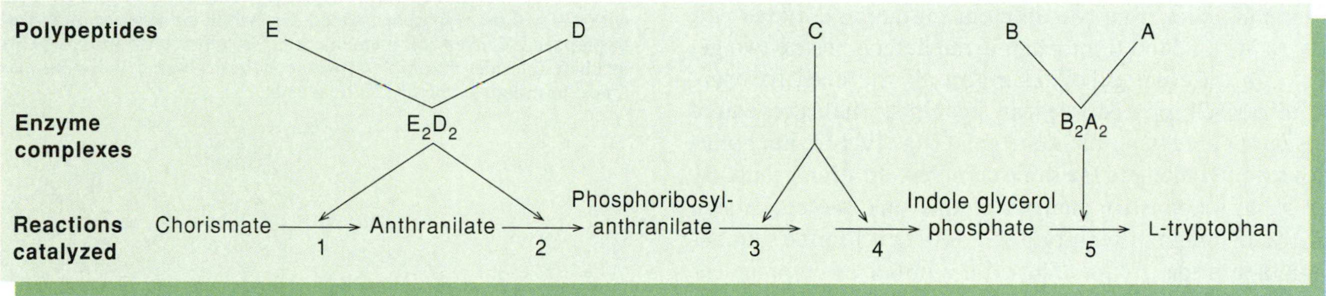

of the reaction or sequence. So the differential ATP stoichiometry would mean nothing if it were not for the kinetic controls that hold the ATP/ADP ratio at a steady-state value far from equilibrium. But kinetic regulation could not control directions of conversions and thus regulate the balance between ATP utilization and regeneration if it were not for the thermodynamic difference between the sequences of each pair, which depends on the ATP/ADP ratio. Neither the thermodynamic nor the kinetic features can be said to be more fundamental. Metabolic correlation and control, and hence life, depend on an intricate interplay between thermodynamic and kinetic factors.

Strategies for Pathway Analysis

In this section we will look at some approaches that are used in analyzing biochemical pathways.

Different Procedures Are Needed to Analyze Simple and Complex Pathways

Precursor–product relationships vary in complexity. Some pathways, such as the conversion of phenylalanine to tyrosine, involve only one enzyme-catalyzed reaction, one precursor, and one product. Other pathways, such as the utilization of CO_2 in autotrophic organisms, illustrate the most complex type of relationship between precursor (CO_2) and products (carbon-containing compounds of the cell), a multistep pathway with many branchpoints. A description of all of the reactions involved in CO_2 utilization would include most of the synthetic reactions of the organism. Analysis of this type of pathway is accordingly more complex.

Analysis of Single-Step Pathways To investigate a single-step pathway, we usually begin with a crude cell-free extract from a source believed to be abundant in the enzyme that catalyzes the conversion. Two of the most popular sources of cells for many biochemical studies are rat liver and *E. coli*. To investigate a particular reaction, we add the precursor (substrate) to the extract and determine the amount of product formed as a function of time. Once we have developed an assay for disappearance of precursor and appearance of product, we can process the crude extract into fractions and test them to see which are active in the conversion. Through further fractionation and assays we should ultimately be able to purify the enzyme of interest. Some procedures followed in enzyme purification were discussed in chapter 5, and many procedures used to determine the mechanisms of action of the purified enzymes were considered in chapters 9 and 10.

The process of assaying for a particular reaction during purification may be complicated in cases where there are requirements for cosubstrates, coenzymes, or cofactors. Usually we discover these additional requirements by the trial-and-error procedure of adding test substances to the reaction mixture and observing whether they accelerate the reaction or lead to the formation of more product(s). If more than one step and several enzymes are involved, then we must also consider intermediates between precursor and product, and the whole process of analysis increases in complexity.

Analysis of Multistep Pathways Most pathways consist of a series of enzyme-catalyzed steps, as exemplified by the five-step pathway leading from the precursor chorismate to the product tryptophan (fig. 12.13). Since tryptophan synthesis does not occur in vertebrates, this reaction cannot be studied in rat liver extracts, but it can be studied in extracts made from *E. coli* bacteria. Bacteria and many other microorganisms have the additional advantage of being good subjects for supplementary genetic methods.

In general, genetic methods are most useful at the beginning of a study, when the goal is to determine the number of genes and related enzyme-catalyzed reactions of a pathway. Genetic methods are also useful to test any conclusions drawn from *in vitro* studies by making parallel measurements *in vivo*. In the case of the tryptophan biosynthetic pathway, we could use genetic methods to determine the number of gene-encoded proteins required to synthesize tryptophan. One common procedure, known as complementation analysis, is performed as follows. First, we isolate a large number of bacterial mutants that cannot grow unless tryptophan is added to the growth medium. Such mutants usually have defects in the genes required for *de novo* tryptophan biosynthesis. To ensure a collection with mutants in each of the genes required for tryptophan biosynthesis, we must isolate a hundred or more tryptophan-requiring mutants. Then we perform pairwise matings between mutants to create partial diploid cells that carry a complete set of tryptophan genes from two different mutants. A mated pair does not require added tryptophan if the defects in the two genomes are in different genes. Therefore, if the need for tryptophan is not eliminated, we can conclude that the paired genomes carry defects in the same gene (fig. 12.14). Such mutants are said to belong to the same complementation group. By completing an exhaustive analysis of this sort, we can assign most of the tryptophan-requiring mutants to a limited number of complementation groups. The total number of complementation groups should correspond to the number of genes required to make a functional tryptophan pathway. Since most genes encode a single polypeptide chain, the number of genes should reflect the number of polypeptide chains involved in the biosynthesis of tryptophan.

The next phase in analysis of a multistep pathway involves biochemical procedures. We isolate as many intermediates as possible, determine their structures and the order of reactions, and then isolate and characterize the enzymes that catalyze the different reaction steps. Mutants that have been isolated for the complementation analysis are also valuable aids for many of the biochemical experiments, because each mutation introduces a specific block at some point in the pathway. A bacterium with one nonfunctioning enzyme in a pathway will accumulate the intermediate just before the defect and make little or no intermediates (or final product) for the remaining steps in the pathway (fig. 12.15). Specific intermediates are much easier to isolate from mutants than from normal cells because of their abundance and stability. Ideally, we would identify all intermediates isolated from different mutant cells and analyze their structures. The information resulting from such observations should give a picture of the overall pathway.

From this point on, there are numerous ways of proceeding with the analysis. For example, we could make crude extracts from different mutant cells, then use the extract from one mutant to complement the extract from another mutant in tryptophan synthesis. This analysis could lead to an assay for a particular enzyme carried by one mutant that is missing in the other mutant. The goal at this juncture is to purify each and every member enzyme of the pathway so that we can scrutinize individually the properties of the enzymes and the reactions they catalyze. All of the tryptophan enzymes have been isolated from *E. coli*. Partial studies on other systems indicate a remarkable similarity for the operation of this pathway in different microorganisms and plants.

Radio-Labeled Compounds Facilitate Pathway Analysis

Before the advent of radioactively labeled substrates, detection of intermediates and products was often quite difficult. Radiolabeled compounds greatly increase the sensitivity of detection of products and especially of intermediates in a pathway. Moreover, they also make it easier to determine the order of reactions in a pathway. Thus we would expect a radio-labeled precursor, if administered to a reaction mixture for a short time, to preferentially label the early intermediates, whereas we would expect the same compound, if administered for a long time, to accumulate in the final product of the pathway. This is just what happens with an extract prepared from a wild-type cell.

Figure 12.14

Complementation occurs only between genomes that are defective in different genes. On the left are shown two different types of gene pairs that might arise in a mating to test for complementation. The genes *E, D, C, B,* and *A* are intended to represent the five genes of the tryptophan pathway, which are clustered in *E. coli*. A minus sign indicates a defective gene. On the right are shown the results of plating ten cells on a minimal medium with and without tryptophan added. Growth in the absence of added tryptophan indicates complementation. Growth is observed after overnight incubation. Under favorable growth conditions each cell divides many times, forming a clone of identical cells.

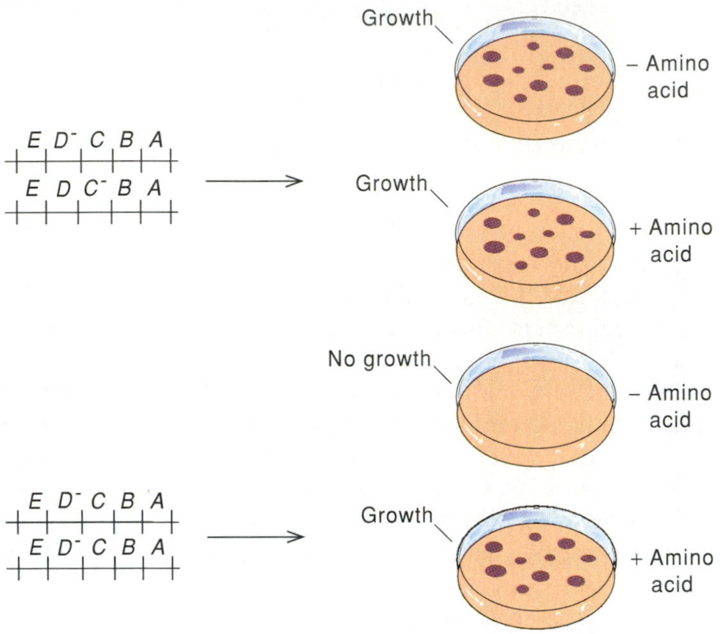

Specific Inhibitors Serve the Same Role as Genetic Blocks in the *in Vitro* Analysis of a Pathway

Genetic methods were rarely used for pathway analysis until the 1940s. Nevertheless, many important pathways were elucidated earlier. For example, the complex multistep pathway known as the Krebs cycle was discovered in the 1930s. In his analysis, Hans Krebs used cell-free extracts from pigeon flight muscle, which are especially rich in the enzymes of the cycle (see chapter 14). Although relevant mutants were not available for imposing specific blocks on the passage of intermediates in this pathway, Krebs discovered a strong inhibitor of one of the reactions that played a crucial role in determining the general

Figure 12.15

Comparison of the relative amounts of pathway intermediates and final products in normal and mutant organisms containing a defective enzyme in the pathway. The precursor, intermediates, and final product of the pathway are labeled A, B-D, and E, respectively. The enzymes 1, 2, 3, and 4 are indicated by the numbers over the reaction arrows. A cross through the reaction arrow indicates a defective enzyme. In the wild-type organism with no defective enzymes (*top*), intermediates are present at low concentrations compared with final product. In the mutant organism with a defective enzyme in the pathway (*bottom*), the intermediate just before the defective enzyme accumulates to an abnormally high concentration. Intermediates after the block, as well as product synthesized through the enzymes in the pathway, are practically nonexistent. If the final product of the pathway is required for viability, it would have to be supplied directly from external sources.

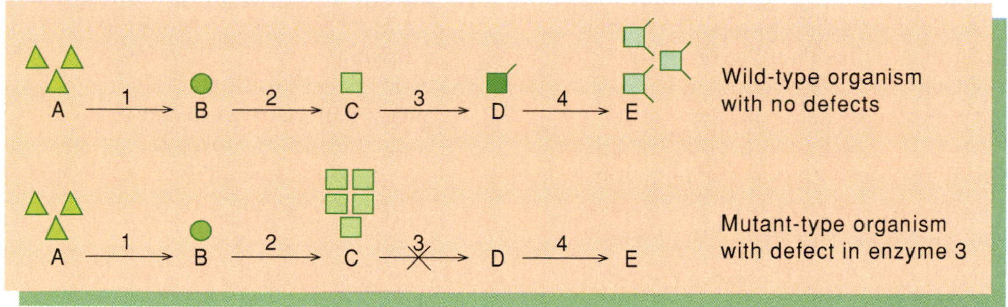

nature of the pathway. Frequently, even today, inhibitors are highly useful for pathway analysis, especially in cases where mutants are not available. In some studies inhibitors have an advantage over the use of mutant extracts for *in vitro* analysis because they can be added at various times after starting the reaction or can be omitted altogether in parallel reactions.

Pathways Are Usually Studied Both In Vitro *and* In Vivo

Although biochemists spend most of their time studying reactions *in vitro,* the ultimate proof of the significance of a reaction or a series of reactions is confirmation that it is used *in vivo.* For the purpose of making parallel measurements, isotopes and genetic blocks are invaluable. Isotopes permit parallel intermediates and products to be detected *in vitro* and *in vivo.* Genetic mutants permit parallel observations to be made of the effect of a specific enzyme deficiency on the specified pathway. If *in vitro* and *in vivo* analyses lead to different conclusions, usually it is because the significance of the *in vitro* analysis has not been correctly assessed, and further work is needed. A classic example of this situation occurred in early investigations of *E. coli* DNA polymerase (see chapter 26). In that case, workers had isolated an enzyme that could catalyze the replication of DNA *in vitro,* but subsequent mutant studies showed that this activity was not required for replication of DNA *in vivo.*

The brief description of pathway analysis that we have provided gives more emphasis to the genetic approach than most biochemists have given it in the past. A new approach that is sweeping the fields of genetics and biochemistry is use of the techniques of DNA recombinant technology. Such methods are revolutionizing pathway analysis, because they have made it possible to isolate virtually any gene from any cell type, even those for which genetic methods have not been suitable in the past. DNA recombinant technology methods are discussed in chapter 27.

Summary

In this chapter we have introduced the study of catabolism by presenting some principles of intermediary metabolism and by indicating how metabolic pathways are regulated. The following points are the highlights of our discussion.

1. All cells need energy and starting materials for synthesis. Ultimately these are supplied by autotrophic organisms, especially green plants; in plants the starting materials are made from CO_2 and the supply of chemical energy and reducing power is dependent on the absorption of light energy. In a heterotrophic organism, the role of catabolism is to supply those basic needs from conversions (usually oxidation) of foodstuff.
2. Metabolic chemistry is characterized by functionality. Each reaction is important because of its participation in a sequence of reactions, and each sequence interacts functionally with other sequences.
3. Sequences may be broadly classified into two main types: biosynthetic or anabolic, and degradative or catabolic. Anabolic sequences are usually energy-requiring and catabolic sequences are usually energy-producing.
4. Metabolic regulatory mechanisms have evolved so as to stabilize concentrations of key metabolites over a broad range of conditions.
5. Energy is coupled from energy-producing catabolic sequences to energy-requiring activities of a cell by the ATP-ADP system. In a similar manner, reducing power is coupled by the NADPH-$NADP^+$ system.
6. The stoichiometry of coupling to ATP-to-ADP conversions determines the overall equilibrium constant of a sequence and therefore the direction of conversion that will be thermodynamically favorable. Any conversion can be made favorable by coupling to an appropriate number of ATP-to-ADP conversions.

7. Metabolic sequences occur in oppositely directed pairs that are controlled by regulatory enzymes.
8. Regulatory enzymes respond to signals in such a way that rates of biosynthesis are controlled by the need for product.
9. Metabolism is regulated primarily by adjustment of the ratios by which intermediates are partitioned at metabolic branchpoints.
10. Different procedures are used for analysis of simple and complex pathways. Analysis of single-step pathways often begins with the isolation and characterization of the enzyme involved. During enzyme isolation each purification step is monitored by a specific assay that measures the conversion of substrate to product. Multistep pathway analysis ideally begins with complementation analysis, a genetic technique that entails the isolation of mutants with genetic blocks in each step of the pathway. Once the numbers of enzymes and intermediates have been established, each enzyme can be isolated with the help of a specific assay.
11. Radio-labeled compounds are most useful for pathway analysis in two respects: first, they permit detection of pathway intermediates with great sensitivity, and second, they can be used as tracers for determining the order of intermediates in a pathway.
12. Inhibitors that block specific steps in a pathway play the same role as genetic blocks in the *in vitro* analysis of a pathway.
13. A complete understanding of a pathway requires parallel investigations *in vitro* and *in vivo*. *In vitro* analyses permit a detailed study of isolated components. *In vivo* observations sustantiate the biologic significance of the *in vitro* observations.

Selected Readings

Atkinson, D. E., *Cellular Energy Metabolism and Its Regulation.* New York: Academic Press, 1977. A general discussion, covering some topics in this and the two following chapters in somewhat greater depth than the treatment in this book.

Cohen, P., *Control of Enzyme Activity,* 2d ed. London and New York: Chapman and Hall, 1983. Brief discussion of some types of regulation of activity of metabolic enzymes, emphasizing regulation by covalent modification of the enzymes.

Herman, R. H., R. M. Cohn, and P. D. McNamara, *Principles of Metabolic Control in Mammalian Systems.* New York and London: Plenum Press, 1980. Discusses various aspects of metabolic control in mammals, mainly at the intracellular level. Many references.

Hochachka, P. W., and G. N. Somera, *Biochemical Adaptation.* Princeton: Princeton U. Press, 1984. An excellent and extensive discussion of how biochemical processes, including many discussed in this book, are adapted by various types of organisms in fitting themselves for survival under specific and often difficult conditions.

Problems

1. Explain why thermodynamics and kinetics are each important in life processes.
2. Explain why each of the following statements is false in terms of efficient metabolic regulation.
   (a) Most enzymes operate *in vivo* near V_{max}.
   (b) End-product inhibition usually occurs at the last or next-to-last enzyme in a metabolic pathway.
   (c) Catabolic pathways tend to diverge from a single metabolite.
   (d) The enzymes regulated in a metabolic pathway usually exhibit simple Michaelis-Menten kinetics.
   (e) Energy charge is unimportant in regulation of anabolic sequences but is of primary importance in the regulation of catabolic sequences.
   (f) Enzymes that catalyze a sequence of reactions are rarely grouped in multienzyme complexes.
   (g) Compartmentalization of metabolic pathways is seldom a regulatory strategy the cell uses.
3. Catabolic and anabolic pathways often differ in the specific reductant used (NADH or NADPH) and in the use of an energy source. How is each of these factors a metabolic advantage?
4. How is it possible that both the glycolytic degradation of glucose to lactate and the reverse process, formation of glucose from lactate (gluconeogenesis), are energetically favorable?
5. What is the metabolic advantage of having the "committed step" of a pathway under strict regulation?
6. Theoretically, the reactions shown below constitute a futile cycle. Explain.

$$\text{Glucose} + \text{ATP} \xrightleftharpoons{\text{Glucokinase}} \text{G-6-P} + \text{ADP}$$

$$\text{G-6-P} + \text{HOH} \xrightleftharpoons{\text{glucose-6-phosphatase}} \text{glucose} + \text{phosphate}$$

In the liver cell, the enzymes are spatially separated, glucokinase in the cytosol and glucose-6-phosphatase in the endoplasmic reticulum. Does this separation influence the futile cycle?

13

CHAPTER

Glycolysis, Gluconeogenesis, and the Pentose Phosphate Pathway

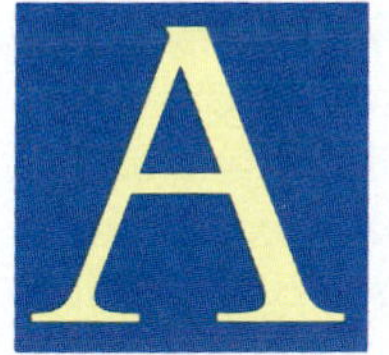
As we have seen, carbohydrates play the central role in energy metabolism in most cells, and polymers of glucose are important energy-storage compounds that can be used for ATP production when needed. In this and the next three chapters our central concern will be with the relationship between carbohydrate metabolism and energy production.

This chapter is divided into four parts, corresponding to the conversions outlined in figure 13.1. In the first part we will examine the breakdown of starches and simple sugars by the glycolytic pathway. This process yields energy in the form of ATP and ends with three carbon compounds that can be further degraded or used as carbon skeletons in biosynthesis. The main steps in glycolysis are identical for a wide variety of sugars. It is only the beginning steps and the terminating steps that differ appreciably in different organisms. In the second part of this chapter we will discuss the reverse process. Many of the enzymes that are used in glycolysis are also used in gluconeogenesis. Only certain key steps, those that determine whether synthesis or breakdown will be favored, employ different enzymes. With a description of the basic reactions of degradation

Figure 13.1

Outline of the main reactions involved in carbohydrate metabolism that will be considered in this chapter. Parts of this figure shown in black are taken from the master diagram of figure 12.5. Additional reactions, not shown in figure 12.5, are drawn in red. These include the formation of neutral glucose, the end products of glycolysis formed under anaerobic conditions, and interconversions of sugars containing from three to seven carbon atoms. Reactions connected by antiparallel half arrows use the same enzyme in either direction with little change in free energy. (C-3-P = ribulose-3-phosphate; C-4-P = ribulose-4-phosphate, etc.)

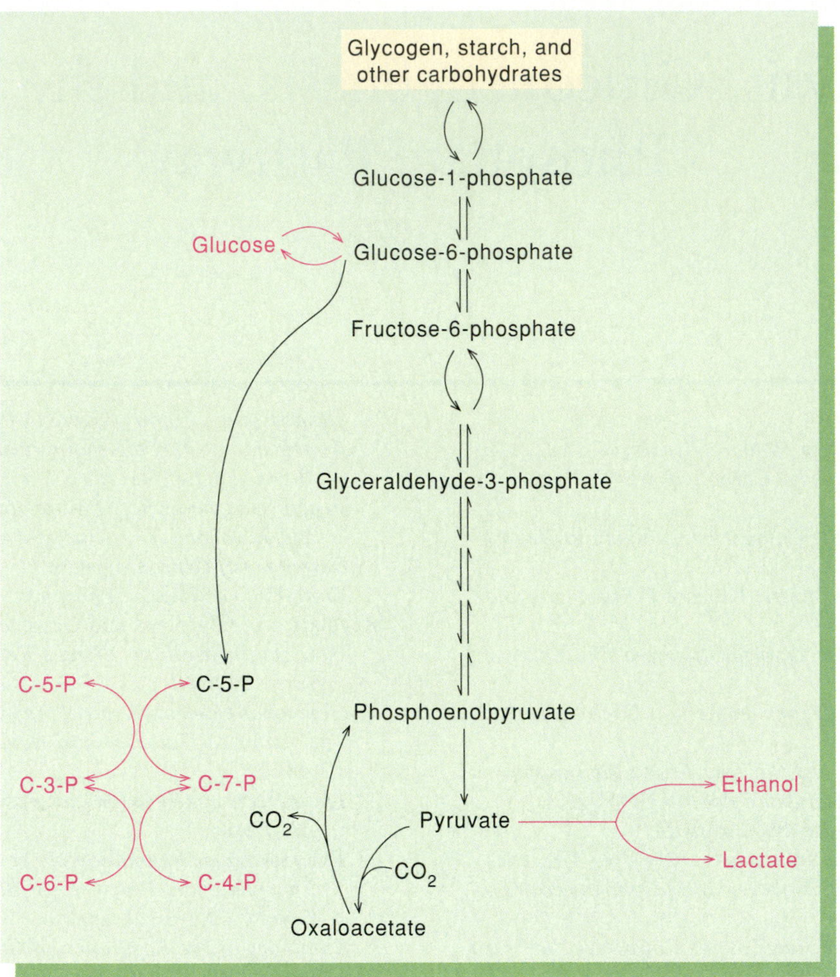

and synthesis in hand, we will be in a position to consider how these processes are regulated, and that is the topic covered in the third part of this chapter. Finally, we will finish the chapter with a section on the formation of pentoses and the interconversions of pentoses and other sugars containing from three to seven carbon atoms.

Glycolysis

Whether aerobic or anaerobic, the process of glucose catabolism involves a reaction sequence that is the single most ubiquitous pathway in all energy metabolism. Glycolysis occurs in almost every living cell. It is generally regarded as a primitive process. Thus it occurs in the cytosol rather than being compartmentalized into a specific organelle within the eukaryotic cell. Furthermore, it is thought to have arisen early in biological history, before the advent of eukaryotic organelles and before oxygen was a prominent component in the atmosphere.

The glycolytic pathway was the first major metabolic sequence to be elucidated. Most of the decisive work was done in the 1930s by the German biochemists G. Embden, O. Meyerhof, and O. Warburg, two of whom gave the sequence its alternative name, the Embden-Meyerhof pathway. The essence of the process is suggested by the name, since glycolysis comes from the Greek roots *glykos,* meaning "sweet," and *lysis,* meaning "loosing." Literally, then, glycolysis is the loosing or splitting of something sweet, which is, of course, the starting sugar. From figure 13.2 it is clear that the actual splitting occurs at the aldolase step. It is at this point that a six-carbon sugar is cleaved to yield two three-carbon compounds, one of which, glyceraldehyde-3-phosphate, is the only oxidizable molecule in the whole pathway. Subsequent to the cleavage, two ATP-generating steps occur. These represent the energy payoff of the process, since they are the only ATP-yielding reactions of the pathway under anaerobic conditions.

Figure 13.2

The glycolytic pathway from glucose to pyruvate, indicating two anaerobic options (ethanol or lactate) and one aerobic option (TCA cycle). The red arrow indicates how NADH formed in glycolysis could be reoxidized to NAD^+ when pyruvate is converted to end product of ethanol or lactate under anaerobic conditions.

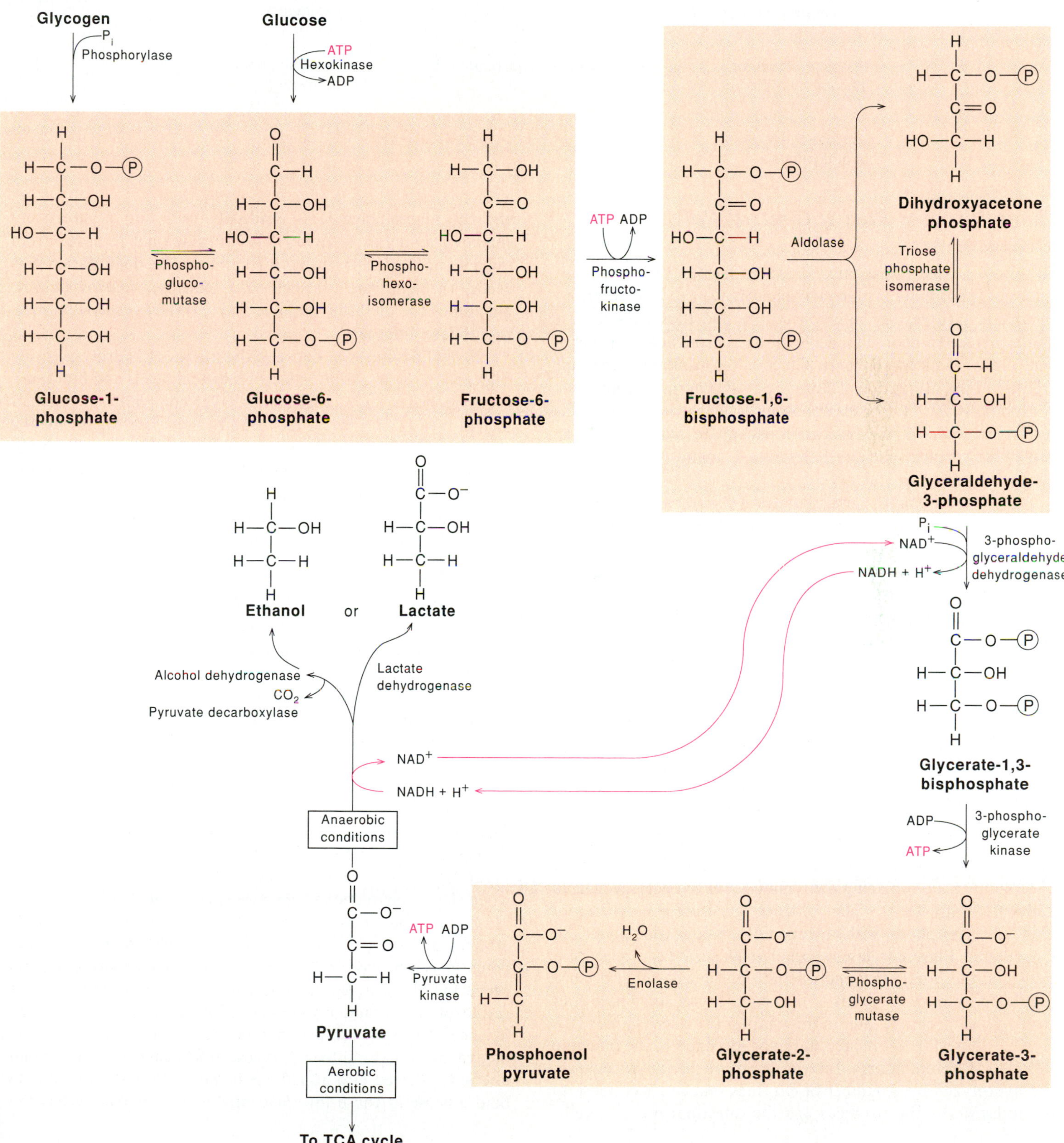

Figure 13.3

The hexose monophosphate pool. The equilibrium percentages of the three hexose monophosphates are indicated. Horizontal flow arrows indicate major routes of replenishment and utilization of the three hexose monophosphates.

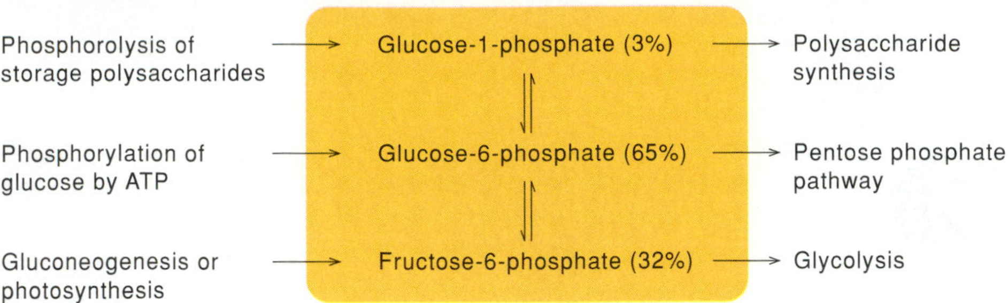

Three Hexoses Play a Central Role in Glycolysis

The three hexose phosphates shown in figures 13.2 and 13.3, glucose-1-phosphate, glucose-6-phosphate, and fructose-6-phosphate, are readily interconvertible and thus constitute a single metabolic pool. The pool can be replenished by generation of any of its components. Glucose-1-phosphate is the first product in the utilization of storage polysaccharides; glucose-6-phosphate is the first hexose phosphate formed when free glucose is metabolized; and fructose-6-phosphate is the first hexose phosphate formed when carbohydrate is made *de novo* (from noncarbohydrate precursors).

Production of carbohydrates *de novo* is of two types. Gluconeogenesis, the synthesis of sugars from such precursors as the three-carbon compound lactate, is a process that can be carried out by many kinds of cells, heterotrophic as well as autotrophic, including the cells of several organs of our bodies. Only autotrophic organisms are capable of producing sugars *de novo* from CO_2 (see chapter 16). In both cases fructose-6-phosphate is the first member of the hexose monophosphate pool to be formed. We will now see how these members of the hexose monophosphate pool are formed and interconverted.

Phosphorylase Converts Storage Carbohydrates to Hexose Phosphates

The first step in the cell's use of starch or glycogen is the removal of a terminal glucose residue, a reaction catalyzed by the enzyme phosphorylase. Overall, the reaction involves same-side displacement at C-1 of the terminal residue, with an incoming phosphate group replacing the remainder of the polysaccharide molecule (fig. 13.4). The product generated is therefore glucose-1-phosphate, with retention of configuration about the C-1 residue. Phosphorylase is part of an elaborate control system by which the production and utilization of polysaccharides are regulated to meet the needs of the cell or organism. Those controls will be discussed later in this chapter (also see chapter 10).

Digestion of dietary starch or glycogen in the intestine follows a different course. Interior bonds of the polysaccharide are hydrolyzed by a number of enzymes that are secreted into the intestine by the pancreas and the intestinal mucosal cells, and the macromolecule is gradually split into progressively smaller fragments. Individual residues are not removed from the polysaccharide or its major fragments; the degradation proceeds through the disaccharide maltose, glucosyl-α-1,4-glucoside. The final step in the intestine is the hydrolysis of maltose to free glucose.

The difference between the modes of hydrolysis inside cells and in the intestine reflects biological needs and functions. Phosphorylated sugars, like other polar compounds, cross biological membranes much less readily than free sugars. In the cell, phosphorylated hexoses are desirable because they will not be lost by diffusion out of the cell. In contrast, the very reason for hydrolysis of polysaccharides in the intestine is to make their component residues available for absorption into the body, so unphosphorylated glucose is preferable.

Hexokinase and Glucokinase Convert Free Sugars to Hexose Phosphates

Free glucose—for example, that obtained by mammals from the diet through intestinal hydrolysis of lactose, sucrose, glycogen, or starch—is brought into the hexose phosphate pool through the action of *hexokinase* (see fig. 9.3). This enzyme catalyzes phosphorylation at the oxygen attached to C-6 of glucose (fig. 13.5). As usual in metabolism, the source of the phosphate group is ATP:

$$\text{Glucose} + \text{ATP} \rightarrow \text{glucose-6-phosphate} + \text{ADP} + \text{H}^+ \qquad (1)$$

The standard free energy change is about -5 kcal/mole, and the equilibrium constant is about 4,000. Hexokinase is thus able to catalyze the conversion of glucose to glucose-6-phosphate even when the concentration of glucose is very low.

Of course, as we saw in chapter 2, equilibrium considerations indicate only the potential for a reaction to occur; the potential can be converted to reality only by appropriate kinetic properties. That is, the favorable thermodynamic properties of the hexokinase reaction would be of limited use to the cell if the Michaelis constant of the enzyme were large, since the thermodynamic potential to use glucose at low concentrations could not be translated into actual use if the enzyme were unable to bind glucose at those low concentrations. Hexokinase has the

Figure 13.4

The reaction catalyzed by phosphorylase. An oxygen of a phosphate ion attacks C-1 of the terminal glucosyl unit of starch or glycogen, displacing the macromolecule and generating glucose-1-phosphate. The reaction proceeds with retention of configuration, a fact suggesting that it may involve an oxonium ion intermediate as in the case of lysozyme hydrolysis of cell wall polysaccharide (see chapter 9). Phosphorylase is a complex regulatory enzyme whose allosteric properties were discussed in chapter 10.

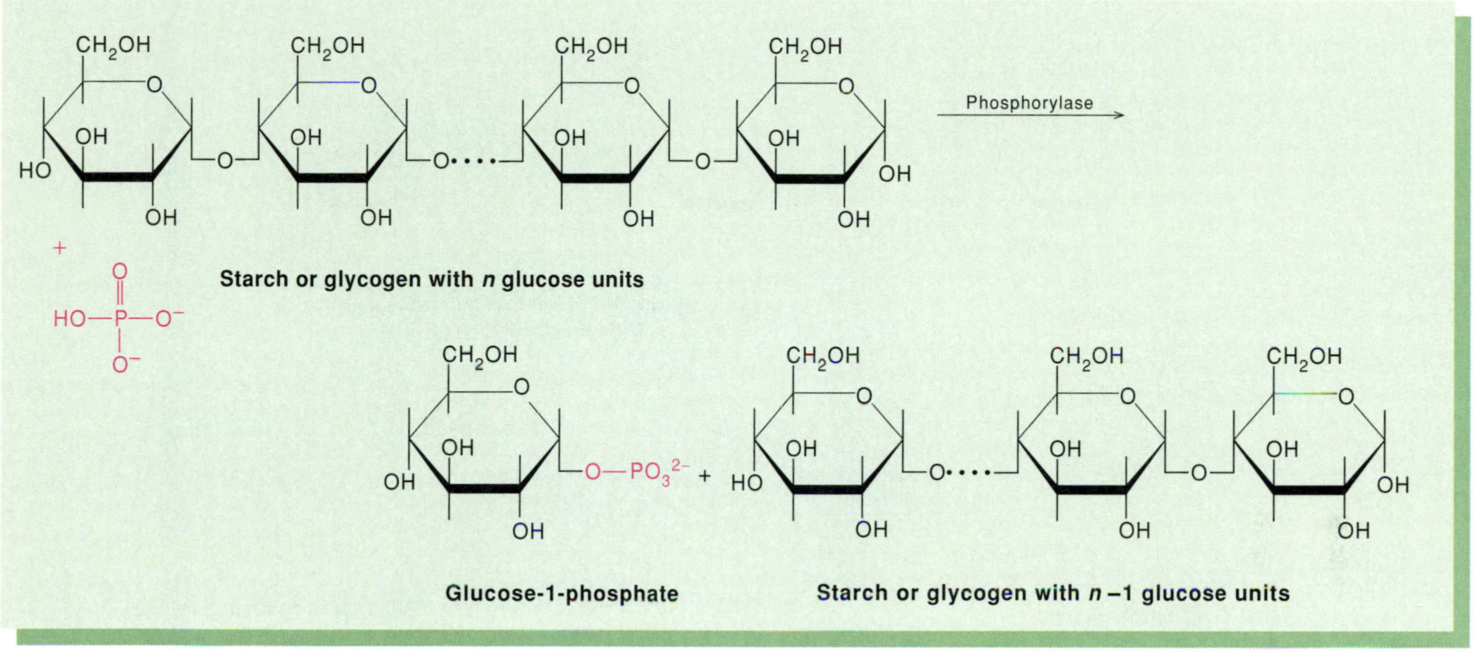

Figure 13.5

The reaction catalyzed by hexokinase. Attack on the terminal phosphorus atom of ATP is probably facilitated by proton removal by a negatively charged group in the catalytic site of the enzyme (:B in the figure). It is also facilitated by the fact that the terminal phosphate of ATP is an excellent leaving group.

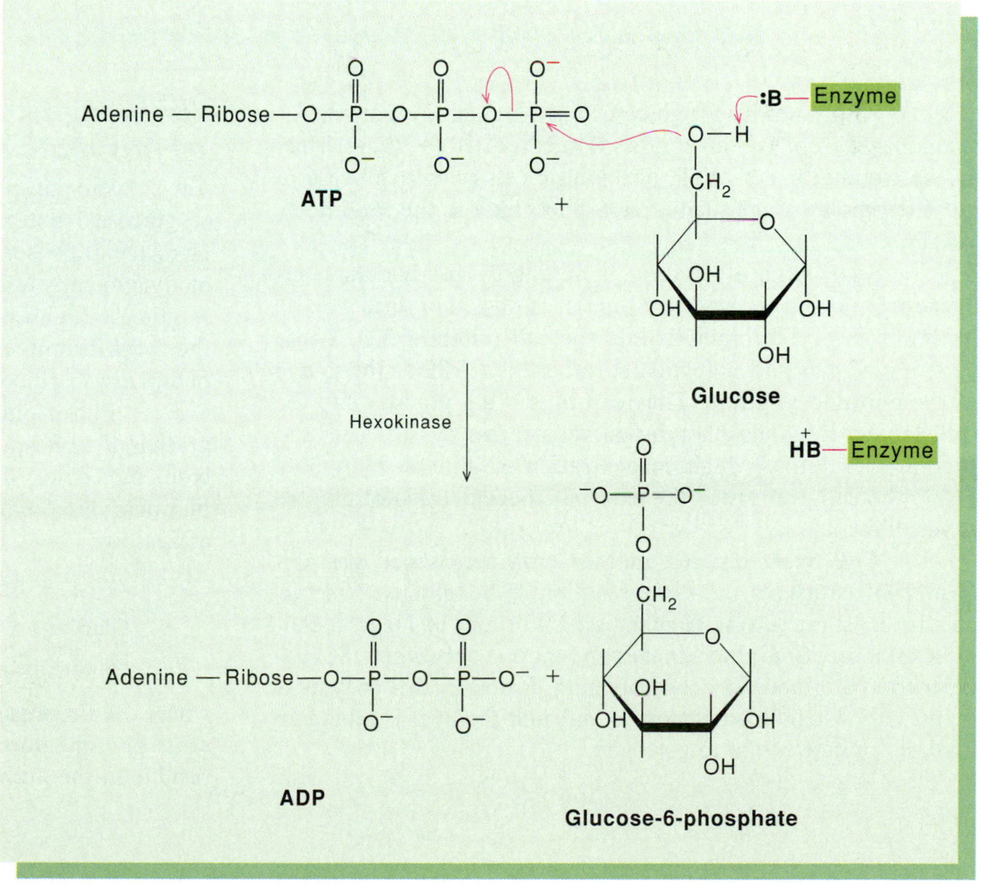

Figure 13.6

The reaction catalyzed by phosphoglucomutase. The enzyme apparently can bind glucose phosphates in two ways, allowing it to transfer phosphoryl groups to and from either the oxygen atom at C-1 or the oxygen at C-6. The direction of reaction will be driven by mass action, depending on the relative concentrations of glucose-1-phosphate and glucose-6-phosphate.

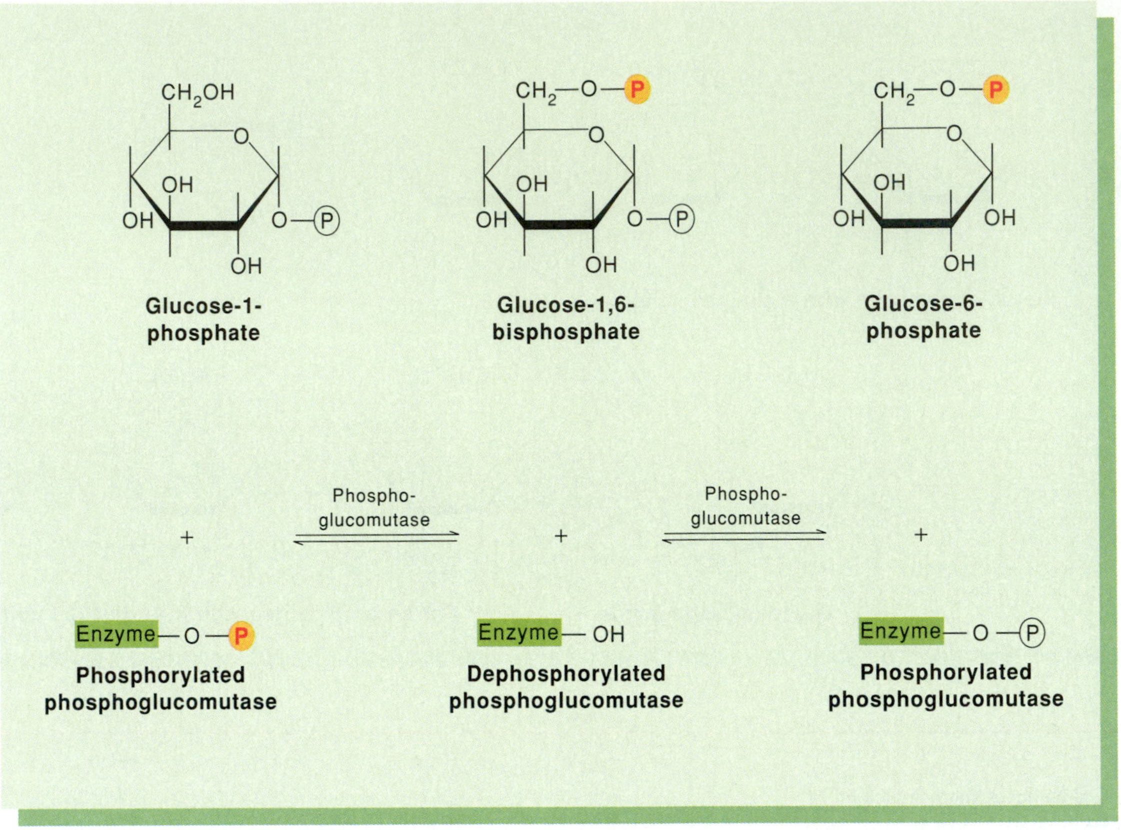

necessary high affinity for glucose. The Michaelis constants for hexokinases from various sources range from 10 to 20 μM. Thus by the expenditure of ATP, hexokinase can convert glucose in the micromolar range to glucose-6-phosphate in the millimolar range.

Some kinds of bacteria, including *E. coli,* contain complex enzymic and transport systems that phosphorylate glucose in the process of bringing it into the cell (chapter 32). These systems use phosphoenolpyruvate rather than ATP as the source of the phosphoryl group. They can thus bring glucose into the cell, protect it by phosphorylation against loss by diffusion, and produce a relatively high concentration of glucose phosphate inside the cell even when the external concentration of glucose is very low.

The liver contains another enzyme, named glucokinase, that catalyzes the same reaction as hexokinase. Its Michaelis constant (about 10 mM) is 1,000 times as large as that of hexokinase; thus glucokinase can function only when the concentration of glucose is relatively high. Probably this enzyme is active only when blood glucose is high and the liver is taking up glucose for conversion to glycogen.

Phosphoglucomutase Interconverts Glucose-1-Phosphate and Glucose-6-Phosphate

The enzyme phosphoglucomutase catalyzes the interconversion of glucose-1-phosphate and glucose-6-phosphate (fig. 13.6). Hemiacetal phosphates, such as glucose-1-phosphate, are thermodynamically less stable than ordinary phosphate esters, such as glucose-6-phosphate. The difference is small, however, and the equilibrium constant for the conversion of glucose-1-phosphate to glucose-6-phosphate is only about 19.

Phosphoglucomutase does not catalyze the direct migration of a phosphoryl group between the two ends of the glucose molecule. The transfer is indirect, by way of a phosphorylated enzyme and glucose-1,6-bisphosphate. The reactions are

$$\text{G-1-P} + \text{Enz-P} \rightleftharpoons \text{glucose-1,6-bisphosphate} + \text{Enz} \quad (2)$$

$$\text{Glucose-1,6-bisphosphate} + \text{Enz} \rightleftharpoons \text{G-6-P} + \text{Enz-P} \quad (3)$$

$$\text{Sum: Glucose-1-phosphate} \rightleftharpoons \text{glucose-6-phosphate} \quad (4)$$

where Enz represents phosphoglucomutase and Enz-P represents phosphoglucomutase phosphorylated at a specific serine residue in the active site. The enzyme cycles between the free

Figure 13.7

Mechanism of the interconversion of glucose-6-phosphate and fructose-6-phosphate. Loss of a proton from the oxygen attached to C-2 of the intermediate enediol leads to fructose-6-phosphate. A and B represent catalytic groups on the enzyme. It is not always known what specific groups are involved in a catalysis. In this case the HA group originates from a glutamate on the enzyme.

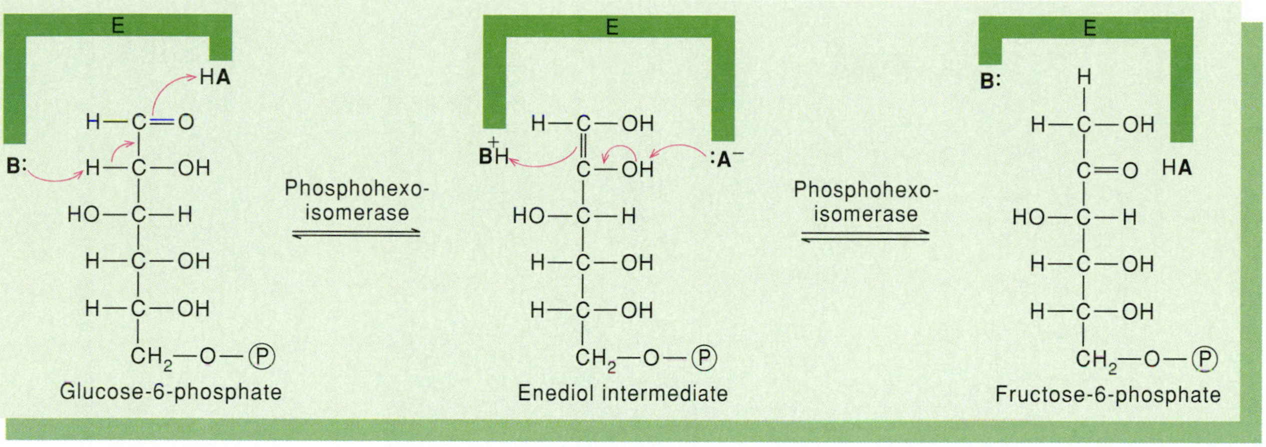

and the phosphorylated states at each conversion of glucose-1-phosphate to glucose-6-phosphate (or of glucose-6-phosphate to glucose-1-phosphate).

Phosphohexoisomerase Interconverts Glucose-6-Phosphate and Fructose-6-Phosphate

Glucose-6-phosphate and fructose-6-phosphate interconvert readily in weakly alkaline solution. Presumably this is because the intermediate enediol is stabilized by ionization (fig. 13.7). Glucose-6-phosphate and fructose-6-phosphate are interconverted specifically by the action of the enzyme phosphohexoisomerase. The reaction presumably resembles the nonenzymic conversion in going by way of the enediol intermediate.

The free energies of formation of aldoses and ketoses are very similar, so at equilibrium the concentration ratio of glucose-6-phosphate to fructose-6-phosphate is about 2. From this value and the equilibrium constant of about 19 for the phosphoglucomutase reaction, we can calculate that at equilibrium the hexose monophosphate pool consists of approximately 3% glucose-1-phosphate, 65% glucose-6-phosphate, and 32% fructose-6-phosphate. This pool, as noted earlier (see fig. 13.3), can be replenished by phosphorolysis of storage polysaccharide (forming glucose-1-phosphate), by phosphorylation of glucose (forming glucose-6-phosphate), or by gluconeogenesis or photosynthesis (forming fructose-6-phosphate). The pool is depleted in two ways. One is by catabolic sequences such as glycolysis, which uses fructose-6-phosphate as the starting point, and the pentose phosphate pathway, which uses glucose-6-phosphate. The other is by storage as polysaccharide or by use in synthesis of disaccharides or complex carbohydrates and other macromolecules that contain hexose derivatives. Glucose-1-phosphate is the starting point for these storage and biosynthetic uses. Thus each component of the hexose monophosphate pool serves both as an entry point and as an exit point. In vertebrate liver, glucose-6-phosphate can also be hydrolyzed to form free glucose for export to maintain the normal blood glucose level. These sequences that replenish or deplete the hexose-phosphate pool will be discussed in this chapter. We turn our attention now to reactions leading from the hexoses to pyruvate.

Formation of Fructose-1,6-Bisphosphate Is a Commitment to Glycolysis

The cleavage of hexose phosphate to yield triose phosphate is the basis for the name glycolysis ("sugar splitting"), by which the whole metabolic sequence between glucose and pyruvate, lactate, or ethanol is commonly known. Fructose-6-phosphate is converted to triose phosphates in two steps. A second phosphoryl group is added to fructose-6-phosphate before the actual cleavage occurs.

Fructose-6-phosphate is converted to fructose-1,6-bisphosphate by transfer of a phosphoryl group from ATP in the reaction catalyzed by phosphofructokinase (fig. 13.8). Since net regeneration of ATP is a major function of the catabolism of carbohydrates, it may seem strange that the early steps of glucose breakdown consume two molecules of ATP for each molecule of glucose. (Only one molecule of ATP is used for each glucosyl residue when the starting material is storage glycogen or starch.) As we have noted, phosphorylation of substrates helps to protect against loss by diffusion out of the cell, but the hexose phosphates are already protected. Why should the second phosphoryl group be added? One answer is again related to prevention of loss of intermediates by diffusion. If fructose-6-phosphate were the substrate for aldolase, which cleaves the six-carbon

Figure 13.8

The reaction catalyzed by phosphofructokinase. The mechanism of this reaction is very similar to the hexokinase reaction shown in figure 13.5. :B is a proton acceptor at the active site.

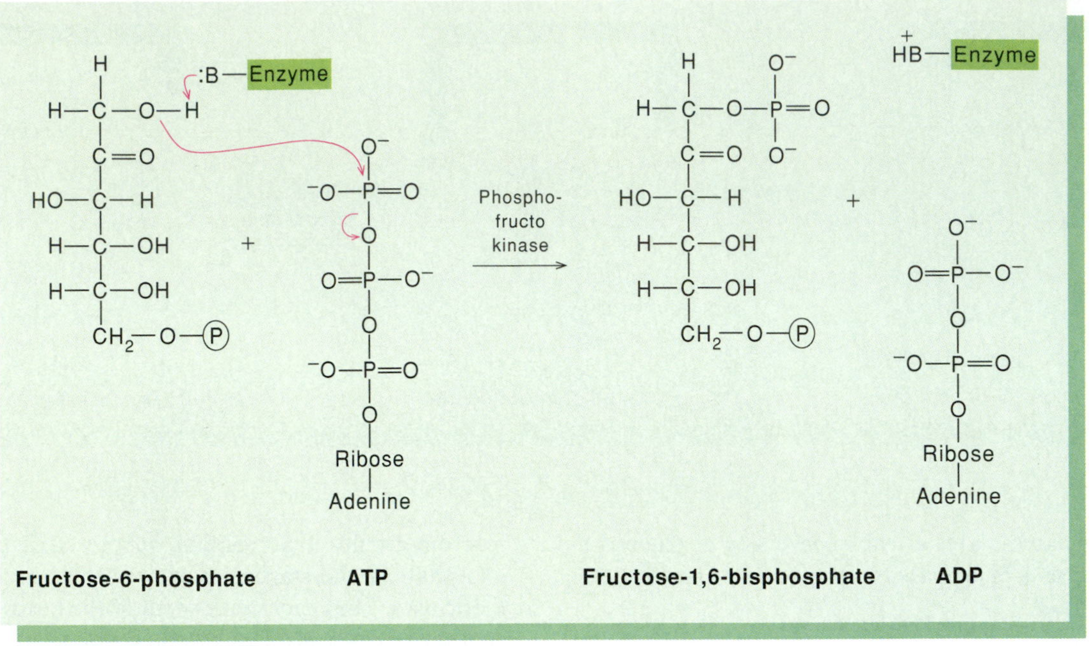

sugar to two trioses, only one of the products would be phosphorylated. The unphosphorylated product might need to be phosphorylated immediately for protection against loss by diffusion, and phosphorylation before cleavage is even more effective.

But there is another advantage to phosphorylation before cleavage that is important. The rate at which hexose phosphates are used in glycolysis must be regulated so that the needs of the cell or organism will be met without wasteful use of more than the necessary amount of substrate. The reaction by which material is removed from the hexose phosphate pool is the point at which such control must be exerted for maximal effectiveness. Metabolic regulation is possible only at steps for which the physiological ratio is far from the equilibrium ratio, so that kinetic control mechanisms can cause increases and decreases in the rates of reactions without thermodynamic constraints.

Under physiological conditions the concentration ratio of fructose bisphosphate to fructose-6-phosphate varies considerably, depending on metabolic conditions, but is probably between 5 and 0.2. At equilibrium of the phosphofructokinase reaction the ratio would be about 10^4. When the concentration ratio is so far from equilibrium, there are no thermodynamic limitations on the reaction. By leading to a very large value of the equilibrium constant, the use of ATP thus allows the phosphofructokinase reaction to be controlled by kinetic factors that affect the behavior of the enzyme. We will see that both of the phosphoryl groups of fructose-1,6-bisphosphate are recovered as ATP in a later step of glycolysis. Thus the decrease in free energy of a later step is in effect fed back by way of ATP to build up the concentrations of hexose phosphates (the hexokinase reaction) and to make control of the early steps in the sequence possible (the phosphofructokinase reaction).

Like other kinases, phosphofructokinase catalyzes a simple attack by a specific oxygen atom of the substrate (in this case, the O attached to C-1) on the terminal phosphorus atom of ATP, with displacement of ADP (see fig. 13.8), and like other kinases it requires Mg^{2+} ion, which may participate in a chelated transition state, not shown in the figure. However, as we will discuss later in this chapter, phosphofructokinase is far more than a simple kinase. It plays a major role in the regulation of the rate of glycolysis, and responds in a complex way to a number of metabolic signals. It may be because of this regulatory complexity that phosphofructokinase is quite large, in contrast to the relatively small size of many kinases. The enzyme from muscle, for example, is a tetramer with a total molecular weight of about 360,000.

Fructose-1,6-Bisphosphate and the Two Triose Phosphates Constitute the Second Metabolic Pool in Glycolysis

The metabolic pool that consists of fructose-1,6-bisphosphate and the two triose phosphates, glyceraldehyde-3-phosphate and dihydroxyacetone phosphate (DHAP), is somewhat different from the other two pools of intermediates in glycolysis because of the nature of the chemical relationships between these compounds. In the other pools the relative concentrations of the component compounds at equilibrium are independent of ab-

13A

BOX

Calculation of the Equilibrium Constant, K_{eq}, When the Numbers of Reactants and Products Are Not Equal

In the expression for the equilibrium constant each concentration term actually represents a ratio of the concentration to the concentration in the standard state. Thus the thermodynamic equilibrium constant is a dimensionless quantity even when the numbers of reactants and products are not equal. Since it is customary to express concentrations in units of molarity and the standard state is usually 1 molar, the equilibrium constant is usually expressed in terms of the actual concentrations. It is also customary to indicate a unit such as M^1 or M^{-1} when appropriate in the numerical expression for the equilibrium constant, since this value depends on the unit used to express concentration even though it is a dimensionless quantity. The unit appears only when the numbers of products and reactants differ. Thus there is no need to use the unit M^0 when stating the numerical value for the equilibrium constant.

Another potential complication in evaluating the equilibrium constant is how to treat water when it appears in the equilibrium expression. The problem seldom arises, since water is assumed to be in its standard state (as far as concentration is concerned) at all times. Thus it need not appear in the equilibrium expression. But clearly, we can imagine cases where water would not be in its standard state—for example, reactions occurring in the relatively anhydrous environment of a lipoprotein membrane. However, we will not treat this type of complication here.

Figure 13.9

Cleavage of fructose-1,6-bisphosphate, an aldolase-catalyzed reaction. The aldolase reaction entails a reversal of the familiar aldol condensation. The first step involves abstraction of the hydrogen of the C-4 hydroxyl group, followed by elimination of an enolate anion.

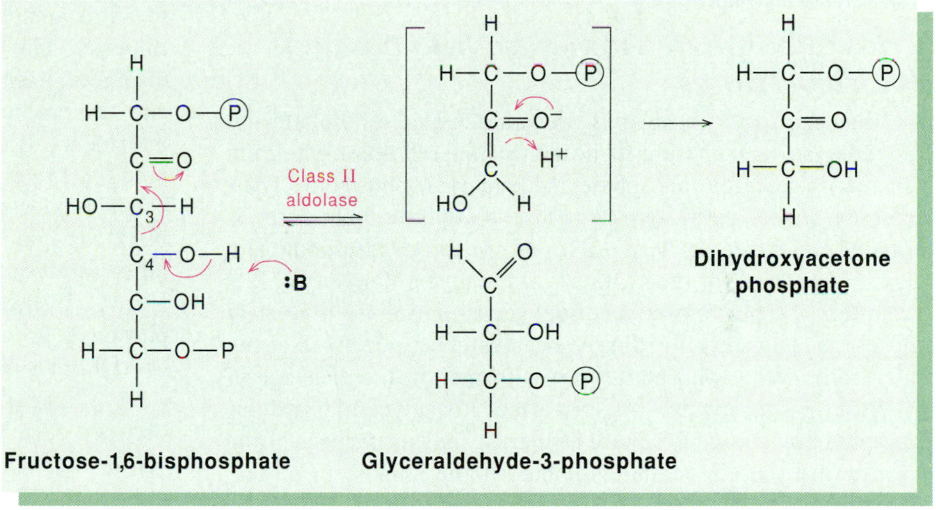

solute concentrations. Because of the cleavage of one substrate into two products, the relative concentrations of fructose-1,6-bisphosphate and the triose phosphates are functions of the actual concentrations. For such reactions, the relative concentrations of the split products must increase with dilution. (For the reaction A $\rightleftharpoons$ B + C, the equilibrium constant is equal to [B][C]/[A]. If the concentration of A decreases, for example, by a factor of 4, equilibrium is restored when the concentrations of B and C decrease by a factor of only 2; see box 13A.) But except for that difference, interconversions of the compounds in this pool are similar to those in the other pools in that reactions within the pool are close to equilibrium and can go in either direction, depending on which components are added to the pool and which are removed.

Aldolase Cleaves Fructose-1,6-Bisphosphate

Fructose-1,6-bisphosphate is cleaved to two molecules of triose phosphate by aldolase in a reversal of an aldol condensation (fig. 13.9). For some aldolases, the cleavage reaction is preceded by a condensation reaction between the carbonyl of the substrate and the amino group of a specific lysine residue at the catalytic site of the enzyme. Other aldolases (as shown in figure 13.9) merely bind the substrate noncovalently. Most aldolases are highly specific for the "upper" end of the substrate molecule, requiring phosphorylation at C-1, a carbonyl at C-2, and a specific one of the four possible steric configurations at C-3 and C-4. The nature of the remainder of the molecule is relatively unimportant.

Whatever the substrate, the reaction is always the same cleavage between C-3 and C-4 (see fig. 13.9), and carbons 1, 2, and 3 of the substrate are always converted to dihydroxyacetone phosphate (DHAP). The identity of the other product depends on the nature of the substrate. The specificity is similar for the reverse reaction, so ketose phosphates containing five to eight carbon atoms can be produced by condensations between dihydroxyacetone phosphate and appropriate aldoses. For example when fructose-1,6-bisphosphate is the substrate, the products of aldolase-catalyzed cleavage are the two isomeric triose phosphates dihydroxyacetone phosphate and glyceraldehyde-3-phosphate (see fig. 13.9).

The equilibrium constant of the aldolase reaction is about 10^{-4} when the concentrations are expressed in the usual units of molarity. From that value, it appears that this cleavage reaction might pose thermodynamic difficulties. The problem is more apparent than real, however, because the relative concentrations of the products are increased with dilution. The steady-state concentration of the reactant fructose-1,6-bisphosphate in a cell may be around 1 mM. With an equilibrium constant of 10^{-4}, this gives 0.32 mM for the concentration of each of the triose phosphate products. At lower concentrations, the ratio of hexose to triose would be even smaller. Thus cleavage of fructose bisphosphate presents no problem as far as thermodynamics is concerned.

Triose Phosphate Isomerase Interconverts the Two Trioses

The isomeric triose phosphates, glyceraldehyde-3-phosphate and dihydroxyacetone phosphate, bear the same relationship to each other as do glucose-6-phosphate and fructose-6-phosphate. Their interconversion, catalyzed by triose phosphate isomerase, is equally facile (see fig. 13.2). Dihydroxyacetone phosphate is a starting material for the synthesis of the glycerol moiety of fats (chapter 17), but only glyceraldehyde-3-phosphate is used in glycolysis. Thus under ordinary circumstances nearly all of the dihydroxyacetone phosphate that is formed in the cleavage of fructose bisphosphate is converted to glyceraldehyde-3-phosphate by triose phosphate isomerase. All six carbons of the hexoses are thereby made available for the later steps of carbohydrate catabolism.

At equilibrium, the concentration ratio of dihydroxyacetone phosphate to glyceraldehyde-3-phosphate is about 22. That ratio, together with the K_{eq} of about 10^{-4} for the aldolase reaction, allows us to calculate the composition of this pool at equilibrium. As noted above, the composition varies with absolute concentrations, but over the likely physiological range of concentrations the pool contains enough of each component to allow free conversion in either direction. If the concentration of fructose-1,6-bisphosphate were 0.5 mM, the pool would contain 31% fructose-1,6-bisphosphate, 66% dihydroxyacetone phosphate, and 3% glyceraldehyde-3-phosphate at equilibrium. At 5 mM fructose-1,6-bisphosphate, the values would be 58%, 40%, and 2%, respectively.

The mechanism for the triose phosphate isomerase reaction is described in chapter 9.

The Conversion of Triose Phosphates to Phosphoglycerates Occurs in Two Steps

The oxidation of glyceraldehyde-3-phosphate to glycerate-3-phosphate is the first energy-yielding (ATP-producing) reaction in the glycolytic pathway. The mechanism by which an inorganic phosphate ion is taken up and transferred to ADP during the conversion of glyceraldehyde-3-phosphate to glycerate-3-phosphate is indirect and rather elaborate, requiring two separate enzyme-catalyzed reactions.

The first reaction is catalyzed by the enzyme 3-phosphoglyceraldehyde dehydrogenase, sometimes called triose phosphate dehydrogenase.

$$\text{Glyceraldehyde-3-phosphate} + NAD^+ + P_i \rightarrow \text{glycerate-1,3-bisphosphate} + NADH + H^+ \quad \mathbf{(5)}$$

The reaction is initiated by a condensation of the —SH group of a specific cysteine residue at the catalytic site of the enzyme with the aldehyde carbonyl, forming a sulfhydryl adduct or thiohemiacetal (fig. 13.10, step 1). A pair of electrons, along with a proton (thus in effect a hydride ion) are then donated to a NAD^+ molecule that is tightly bound nearby, converting the tetrahedral hemiacetal into a thioester (step 2). Thioesters provide considerably more free energy on hydrolysis than ordinary oxygen esters. That facilitates the next step (step 3), an attack by an oxygen of a phosphate ion on the carbonyl carbon of the thioester. The sulfur atom, and thus the enzyme molecule, is displaced from covalent linkage to the reaction intermediate, and the product, glycerate-1,3-bisphosphate, is released. The hydride ion is passed on to an NAD^+ molecule in solution that binds to the enzyme and dissociates after reduction. Thus an NADH molecule is generated in solution and the tightly bound NAD^+ is again in the oxidized form, ready to participate in another catalytic cycle.

In the second step leading to glycerate-3-phosphate, a phosphate group is transferred from glycerate-1,3-bisphosphate to ADP. This reaction is catalyzed by 3-phosphoglycerate kinase.

The glycerate-1,3-bisphosphate is a mixed anhydride between a carboxylic acid and phosphoric acid. Like related compounds such as acyl halides, which are often used as acylating agents in organic syntheses, acyl phosphates are good acylating or phosphorylating agents. The standard free energy of hydrolysis of glycerate-1,3-bisphosphate to glycerate-3-phosphate is about −12 kcal/mole. The transfer of a phosphate group to ADP, forming ATP, thus goes with a significant decrease in standard free energy and a favorable equilibrium constant of about 2,000.

$$H^+ + \text{glycerate-1,3-bisphosphate} + ADP \rightarrow \text{glycerate-3-phosphate} + ATP \quad \mathbf{(6)}$$

In the two steps catalyzed by 3-phosphoglyceraldehyde dehydrogenase and 3-phosphoglycerate kinase, the oxidation of glyceraldehyde-3-phosphate to glycerate-3-phosphate is coupled to the regeneration of ATP. The equilibrium constant for the overall conversion is about 160.

Figure 13.10

The reaction catalyzed by glyceraldehyde-3-phosphate dehydrogenase (3-phosphoglyceraldehyde dehydrogenase). This interesting and complex reaction consists of several steps. The enzyme first catalyzes a reaction of the substrate with a sulfhydryl group of a cysteine residue of the enzyme itself. The substrate is then oxidized from the aldehyde level of oxidation to the carboxylic acid level while still attached covalently to the enzyme. Displacement of the enzyme by inorganic phosphate ion liberates the product, glycerate-1,3-bisphosphate. The bound NADH of the enzyme, which became reduced when the substrate was oxidized, then transfers a pair of electrons to an unbound NAD^+, and the enzyme is ready for another catalytic cycle.

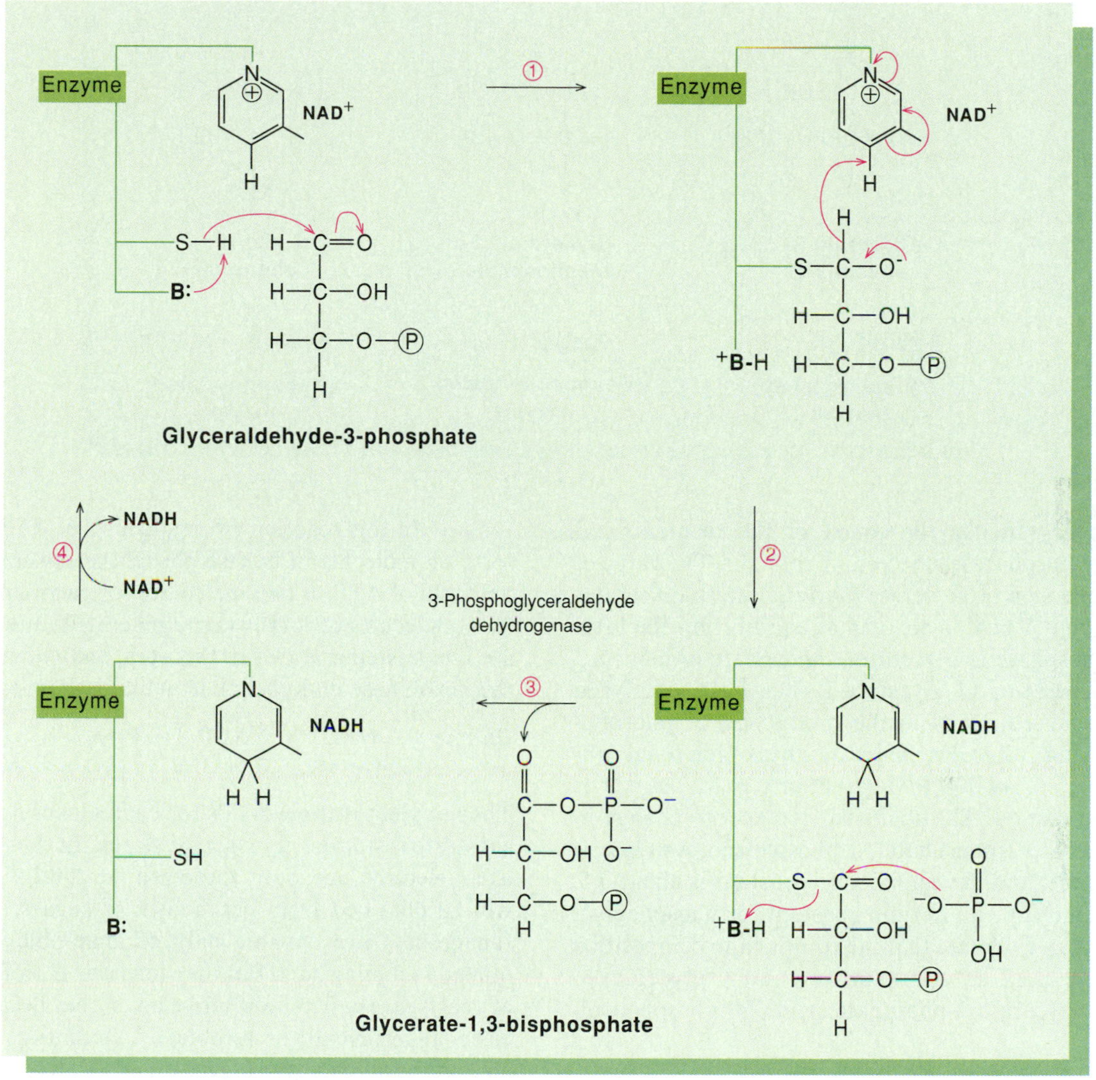

Since two molecules of glyceraldehyde-3-phosphate are produced from each molecule of hexose, the oxidation of the aldehyde to glycerate-3-phosphate leads to the production of two molecules of ATP per molecule of glucose or other hexose consumed. At this point the energy (ATP) gained is equal to the energy invested if the starting material was free hexose. The two phosphate groups that were introduced in the hexokinase and phosphofructokinase reactions are still contained in glycerate-3-phosphate. If storage glycogen or starch is the starting material, only one phosphate group is supplied by ATP, and one of the two molecules of ATP regenerated in the phosphoglycerate kinase step represents a net gain.

The Three-Carbon Phosphorylated Acids Constitute the Third Metabolic Pool

Glycerate-3-phosphate, glycerate-2-phosphate, and phosphoenolpyruvate (PEP) make up another equilibrium group of metabolites that, like the hexose phosphates or the triose phosphates, may be considered to be in effect a single metabolic pool (see fig. 13.2).

The first interconversion is the apparent migration of the phosphoryl group from C-3 to C-2 of the phosphoglycerate molecule (fig. 13.11). The actual reaction is not intramolecular, however; the reaction path is similar to that of the reaction catalyzed by phosphoglucomutase (see fig. 13.6). The enzyme is

Figure 13.11

Interconversion of glycerate-3-phosphate and glycerate-2-phosphate, catalyzed by phosphoglyceromutase. The reaction closely resembles that catalyzed by phosphoglucomutase (see fig. 13.6) except that the phosphate binds to a histidine side chain instead of a serine side chain. The enzyme can transfer a covalently bound phosphoryl group either to the oxygen on C-2 of glycerate-3-phosphate or to the oxygen on C-3 of glycerate-2-phosphate. The resulting glycerate-2,3-bisphosphate, in turn, can donate either of its phosphoryl groups to the enzyme. These catalytic capabilities provide for interconversion of the two monophosphoglycerates in either direction.

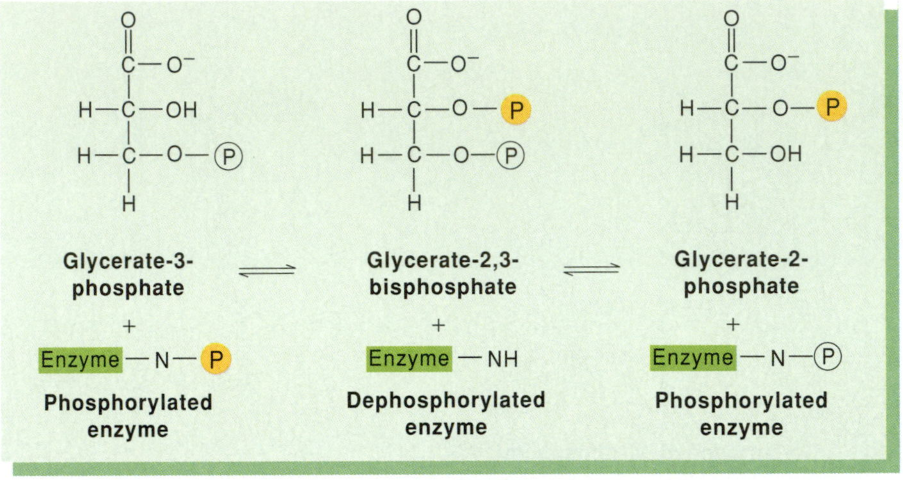

reversibly phosphorylated in the course of the reaction, and glycerate-2,3-bisphosphate is a necessary cofactor. The value of $\Delta G°'$ for the conversion of glycerate-3-phosphate to glycerate-2-phosphate is about 1 kcal/mole, and at equilibrium the ratio of glycerate-3-phosphate to glycerate-2-phosphate is about 6.

The dehydration of glycerate-2-phosphate, catalyzed by the enzyme enolase, leads to the production of phosphoenolpyruvate (see fig. 13.2). Enolase is a dimer of identical subunits (subunit M_r = 44,000). Magnesium ion (or Mn^{2+}) is required for the reaction. The standard free energy change in going from glycerate-2-phosphate to phosphoenolpyruvate is about 0.4 kcal/mole, and the equilibrium constant is about 0.5. From this value and the equilibrium constant for phosphoglycerate mutase, we can calculate that the equilibrium composition of this metabolic pool is approximately 80% glycerate-3-phosphate, 13% glycerate-2-phosphate, and 7% phosphoenolpyruvate.

Conversion of Phosphoenolpyruvate to Pyruvate Generates ATP

The enzyme pyruvate kinase was named for the reverse of the reaction it normally catalyzes in glycolysis, the phosphorylation of pyruvate. Because of the thermodynamics of the system, that reaction could never occur in a living cell. The standard free energy of hydrolysis of phosphoenolpyruvate is the most negative among the phosphorylated intermediates of the central pathways, about −14 kcal/mole. Thus even when the phosphoryl group is transferred to ADP instead of water, the standard free energy change is still about −7.5 kcal/mole (the difference between the standard free energies of hydrolysis of phosphoenolpyruvate and ATP), and the equilibrium constant is of the order of 10^6.

In this reaction two molecules of ATP are regenerated for each molecule of hexose phosphate consumed, bringing the net yield of ATP to two molecules for each molecule of glucose (two molecules of ATP were regenerated in the phosphoglycerate kinase step and two in this step, and two were consumed in the hexokinase and phosphofructokinase steps).

$$\text{Glucose} + 2\ \text{NAD}^+ + 2\ \text{ADP} + 2\ \text{P}_i \rightarrow 2\ \text{pyruvate} + 2\ \text{NADH} + 2\ \text{H}^+ + 2\ \text{ATP} + 2\ \text{H}_2\text{O} \quad (7)$$

The net yield is three ATPs for each glucosyl residue when the substrate is storage glycogen or starch. In the absence of oxygen as an electron acceptor, these are the total yields of ATP that will be obtained from catabolism of sugars. Many organisms (anaerobes) are capable only of anaerobic metabolism, and others, including such familiar forms as *E. coli* and many kinds of yeasts, are called facultative anaerobes because they can live either anaerobically or aerobically. Organisms of either type are able, in the absence of oxygen, to satisfy all their metabolic needs for growth and reproduction from the conversion of sugars to pyruvate (or from related anaerobic sequences).

When oxygen is not available as an ultimate or terminal electron acceptor, pyruvate is, in terms of energetics, the end product of the glycolytic sequence. No more energy is available from its further metabolism under these conditions. But pyruvate still serves an essential function: that of contributing to oxidation-reduction balance (fig. 13.12). In the reaction catalyzed by 3-phosphoglyceraldehyde dehydrogenase, a molecule of NAD^+ was reduced for each molecule of phosphoglyceraldehyde that was oxidized. The NADH that was produced must be oxidized to regenerate NAD^+ if glycolysis is to continue.

Figure 13.12

Regeneration of NAD^+ by reduction of pyruvate to lactate. Since NAD^+ is a necessary participant in the oxidation of glyceraldehyde-3-phosphate to glycerate-1,3-bisphosphate, glycolysis is possible only if there is a way by which NADH can be reoxidized.

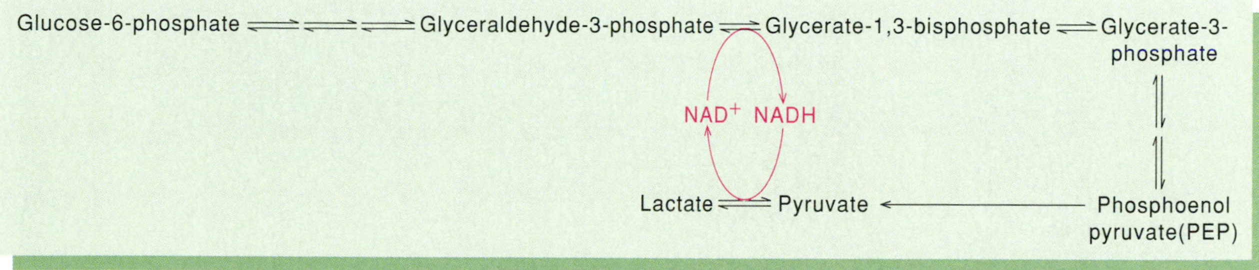

Since equimolar amounts of NADH and pyruvate are produced in glycolysis, the simplest way to oxidize the NADH would be by transfer of electrons to pyruvate, forming lactate. That is the solution that is employed by many kinds of organisms. Several types of anaerobic bacteria—for example, some species of the genus *Lactobacillus* that usually live in decaying plant material—can satisfy all their needs for energy and starting materials for biosynthesis from the conversion of glucose to lactate. The reduction of pyruvate is catalyzed by the enzyme lactate dehydrogenase.

$$\text{Pyruvate} + \text{NADH} + \text{H}^+ \rightarrow \text{lactate} + \text{NAD}^+ \quad \textbf{(8)}$$

The mechanism of this reaction was discussed in chapter 9 (see fig. 9.38). The equilibrium constant of the reaction is about 2×10^4. This large value reflects the ease with which aldehydes and ketones are reduced to alcohols.

The conversion of a mole of glucose to two moles of lactate involves no net oxidation or reduction. The standard free energy change for this conversion alone is about -47 kcal/mole of glucose; as it actually occurs in metabolism, coupled with the conversion of two moles of ADP to ATP per mole of glucose, the standard free energy change is about $-47 + (2 \times 7.5) = -32$ kcal/mole, which corresponds to an equilibrium constant for the sequence of nearly 10^{23}. A simple calculation shows that, assuming the ratio [ATP]/[ADP] to have its normal physiological value of about 5 and the concentration of P_i to be about 10 mM, one molecule of glucose would be in equilibrium with nearly one gram of lactic acid. Such high values of equilibrium constants are common in metabolism.

This same conversion is used also in mammalian muscles when the rate at which oxygen can be supplied becomes a limiting factor in the regeneration of ATP by oxidative phosphorylation (chapters 14 and 15). Although only two moles of ATP are produced per mole of glucose used, the rate of glucose utilization can be very rapid, so that intense muscular activity can be supported for a short time by production of lactic acid. Alternatively, the pyruvate may be transaminated to alanine (see chapter 19).

Yeasts and some other kinds of organisms produce ethanol and CO_2, rather than lactic acid, from the anaerobic catabolism of sugars:

$$\text{Pyruvate} + \text{NADH} + 2\ \text{H}^+ \rightarrow \text{ethanol} + \text{CO}_2 + \text{NAD}^+ \quad \textbf{(9)}$$

In these species pyruvate is decarboxylated to acetaldehyde through the action of pyruvate decarboxylase, which requires thiamin pyrophosphate as a cofactor (fig. 13.13). α-Keto acids, such as pyruvic acid, do not lose CO_2 readily because of the high energy of the carbanion of the keto carbon that would be a potential intermediate. As in all of the cases where thiamin pyrophosphate is used, its function is to add to the carbonyl carbon, leading to a reaction intermediate in which a negative charge on that carbon is stabilized (see fig. 13.13).

In organisms that engage in ethanol fermentation, NADH from the phosphoglyceraldehyde dehydrogenase step is reoxidized to NAD^+ by reaction with acetaldehyde, catalyzed by ethanol dehydrogenase:

$$\text{CH}_3\text{CHO} + \text{NADH} + \text{H}^+ \rightleftharpoons \text{CH}_3\text{CH}_2\text{OH} + \text{NAD}^+ \quad \textbf{(10)}$$

The overall standard free energy change for the conversion of glucose to ethanol and CO_2 is about -40 kcal/mole, so that when coupled to the regeneration of two moles of ATP this fermentation, too, has a very large equilibrium constant for the overall sequence.

Unlike true facultative anaerobes such as *E. coli,* which can live anaerobically for an indefinite period, yeasts can live only for a few generations in the total absence of oxygen because they require molecular oxygen for synthesis of membrane components. Rather than a true alternative life style, for yeasts alcohol fermentation appears to be part of a very effective competitive strategy. When fruits ripen and fall from the tree or bush, yeasts, which are widespread in nature, are likely to be among the first invaders. In the anaerobic interior of the fruit, yeasts rapidly convert sugars to ethanol, which they excrete. As ethanol accumulates, the growth of most other microorganisms is discouraged, but yeasts can continue to grow until the concentration of ethanol reaches about 12%. When this level is

Figure 13.13

The conversion of pyruvate to acetaldehyde and carbon dioxide, catalyzed by pyruvate decarboxylase with thiamin pyrophosphate as bound cofactor. Decarboxylation is facilitated by stabilization of the product of the decarboxylation step, the anion I, by the strong electron-withdrawing power of the thiazolium ring. Alternatively, intermediate I can be said to be stabilized by the existence of the resonance form II without charge separation.

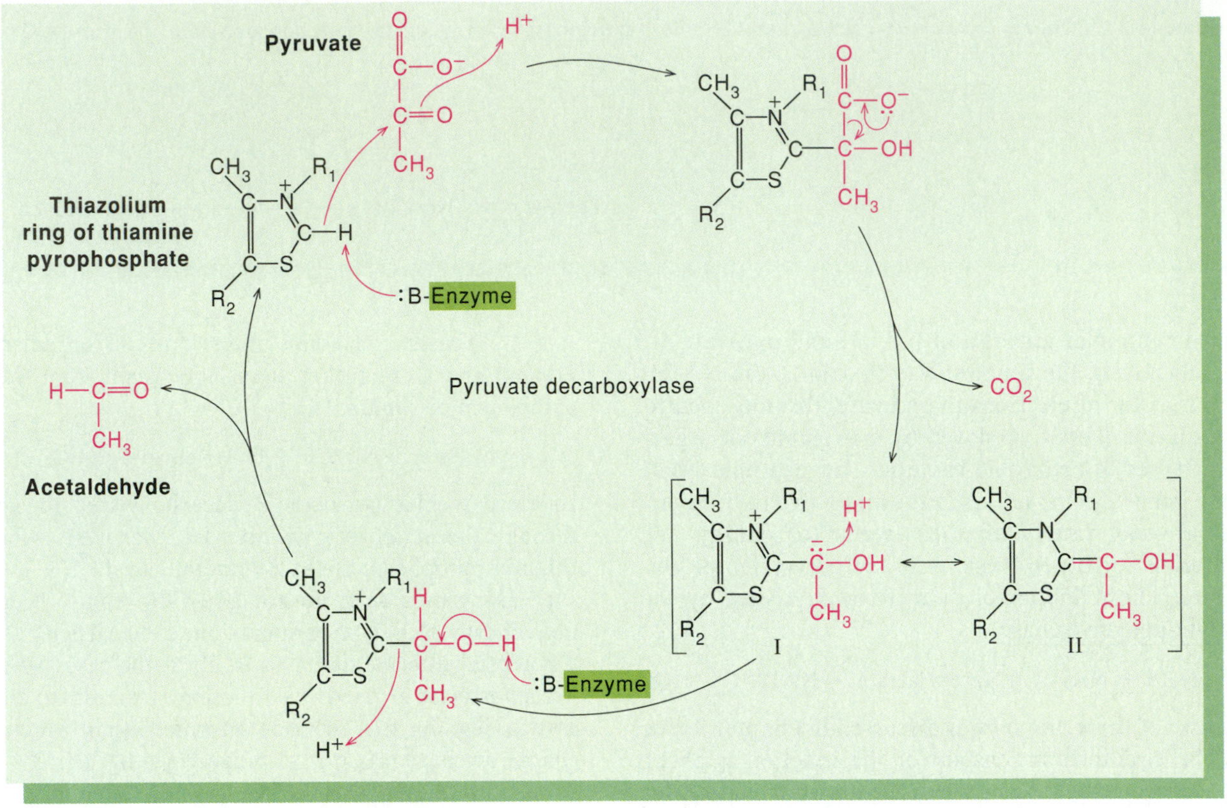

Pyruvate

Thiazolium ring of thiamine pyrophosphate

:B-Enzyme

Pyruvate decarboxylase

CO_2

Acetaldehyde

I

II

:B-Enzyme

reached, and when the softened fruit breaks open, there is a thorough-going change in the basic energy metabolism of the yeast cells. They begin to take up the ethanol that they had previously excreted and oxidize it to CO_2, with a much larger production of ATP than in the anaerobic stage (about 18 moles of ATP per mole of ethanol, as compared with two moles of ATP per mole of glucose in the anaerobic stage). By rapidly converting the available sugars into a compound that cannot be metabolized for energy by most other organisms and in fact is toxic to them, yeasts gain an advantage over competing microorganisms, and also gain energy in the process. They then can exploit the waste product of the first stage of growth as the main nutrient for the later aerobic stage. This example illustrates that the organization of even the most central metabolic pathways can, like other properties of an organism, be shaped by competition.

Summary of Glycolysis

Glycolysis consists of a chain of ten reactions that starts from glucose and ends with the three-carbon compound pyruvate (table 13.1). All of the intermediates in this pathway are phosphorylated and in the process of degradation two ATP molecules are made from two ADP molecules. This is the energy payoff of the pathway. In effect, the energy of the carbon–carbon bond is expended to create the immediately useful form of chemical energy, ATP. This pathway does not require oxygen and it is found in nearly all cells, whether they are aerobic or anaerobic. Frequently glucose is not the starting material for glycolysis and so differences in organisms, or in the same organism in different metabolic states, show up at the starting point of the pathway. The trick is to convert polysaccharides or sugars, whatever they may be, into glucose-6-phosphate, the first intermediate in the pathway. There is one oxidation step in the pathway, which results in conversion of the coenzyme NAD^+ into NADH. Under strictly anaerobic conditions this conversion could create a problem if there was no simple way to regenerate the NAD^+ so that the pathway could continue to operate. The usual solution is to have a reduction step at the end of the pathway, which involves the conversion of pyruvate into either ethanol or lactate and the concomitant reoxidation of the NADH to NAD^+.

Energetically, the glycolytic pathway resembles a series of lakes connected by short rivers. This pattern is reflected in the ways that functional metabolic relationships have evolved. All of the branchpoint compounds that serve as starting materials for biosynthesis are found in the "lakes," or metabolic

Table 13.1
Reactions and Enzymes for Steps in the Glycolytic Pathway

Step	Reaction	Enzyme
1	Glucose + ATP $\rightarrow$ glucose-6-phosphate + ADP + H^+	Hexokinase
2	Glucose-6-phosphate $\rightleftharpoons$ fructose-6-phosphate	Phosphohexose isomerase
3	Fructose-6-phosphate + ATP $\rightarrow$ fructose-1,6-bisphosphate + ADP + H^+	Phosphofructokinase
4	Fructose-1,6-bisphosphate $\rightleftharpoons$ dihydroxyacetone phosphate + glyceraldehyde-3-phosphate	Aldolase
5	Dihydroxyacetone phosphate $\rightleftharpoons$ glyceraldehyde-3-phosphate	Triose phosphate isomerase
6	Glyceraldehyde-3-phosphate + P_i + NAD^+ $\rightleftharpoons$ glycerate-1,3-bisphosphate + NADH + H^+	Phosphoglyceraldehyde dehydrogenase
7	H^+ + glycerate-1,3-bisphosphate + ADP $\rightleftharpoons$ glycerate-3-phosphate + ATP	3-Phosphoglycerate kinase
8	Glycerate-3-phosphate $\rightleftharpoons$ glycerate-2-phosphate	Phosphoglyceromutase
9	Glycerate-2-phosphate $\rightleftharpoons$ phosphoenolpyruvate + H_2O + H^+	Enolase
10	Phosphoenolpyruvate + ADP + H^+ $\rightarrow$ pyruvate + ATP	Pyruvate kinase

Net reaction: $C_6H_{12}O_6 + 2\ NAD^+ + 2\ ADP + 2\ P_i \rightarrow 2\ C_3H_4O_3 + 2\ NADH + 2\ H^+ + 2\ ATP + 2\ H_2O$

pools. Reactions in which ATP and ADP are involved are in the interconnecting reactions, or "rivers." That clearly is where they are to be expected; an ATP-linked reaction within a metabolic pool would make no more sense than a hydroelectric power plant in the middle of a lake. Regulatory enzymes are also found in the segments of the pathways that connect the pools. Again it is clear that that is where they must be to be effective. Kinetic controls cannot be applied to reactions that are at or close to equilibrium, as is the case for reactions within the pools.

We will defer further discussion of the kinetics and regulation of glycolysis until after we have considered the reverse of glycolysis, that is, the synthesis of glucose from pyruvate.

Gluconeogenesis

Gluconeogenesis refers to the production of sugars from nonsugar precursors such as lactate or amino acids.

Glycolysis and gluconeogenesis make up a pseudocycle, or pair of oppositely directed reaction sequences, of the type that were discussed in the preceding chapter (fig. 13.14). In vertebrates and many other kinds of heterotrophs, the fluxes (rates of flow of intermediates) through these sequences are much greater than those through any other pair of oppositely directed sequences. Perhaps for that reason, the relationship between glycolysis and gluconeogenesis is different from that of any other pseudocycle.

In other pseudocycles, the two sequences are often entirely separate. In this case, however, the intermediates in both pathways are the same (except for two added intermediates, oxaloacetate and UDP-glucose, in the conversion of three-carbon compounds to glycogen). Most of the enzymes—those that catalyze interconversions within the metabolic pools—function both in glycolysis and in gluconeogenesis. Only at three points, all outside the pools, are reactions of glycolysis and gluconeogenesis distinct and catalyzed by different enzymes (see fig. 13.14).

Gluconeogenesis Consumes ATP

It is the organization of glycolysis and gluconeogenesis as a series of metabolic pools that allows most of the same enzymes to function in both of the oppositely directed sequences. The intermediates in a metabolic pool are at near-equilibrium concentrations and can proceed in either direction as a result of small changes in concentrations. The reactions that connect the metabolic pools are those that involve large changes in standard free energy, are coupled to the ATP-ADP system, and are the sites for regulatory interactions that determine the net direction of flux and the rates of conversion. Thus the properties of the energy-dependent and kinetically regulated segments between the pools will determine which components of the pools will be consumed and which will be replenished under a particular set of conditions.

Flux within a pool is determined by simple thermodynamic and mass-action considerations. Thus when glycerate-3-phosphate is being produced in the phosphoglycerate kinase reaction and phosphoenolpyruvate is being removed by action of pyruvate kinase, the flux in the three-carbon carboxylate pool will be in the glycolytic direction, from glycerate-3-phosphate toward phosphoenolpyruvate. When the pattern of activation of the regulatory enzymes is different, so that phosphoenolpyruvate is being produced and glycerate-3-phosphate is being consumed, the flux in the same pool, catalyzed by the same enzymes, will flow in the gluconeogenic direction, from phosphoenolpyruvate to glycerate-3-phosphate. Similarly for the other pools, the direction of net flux through the pool will depend on which

Figure 13.14

Relationships in glycolysis and gluconeogenesis. Points at which ATP is produced or consumed are indicated. Compounds in the same metabolic pool are indicated by a colored box. There are three small pseudocycles (I, II, III) in the paired sequences between glycogen and pyruvate, or between glycogen and glucose. Only enzymes that are unique to either glycolysis or gluconeogenesis are indicated.

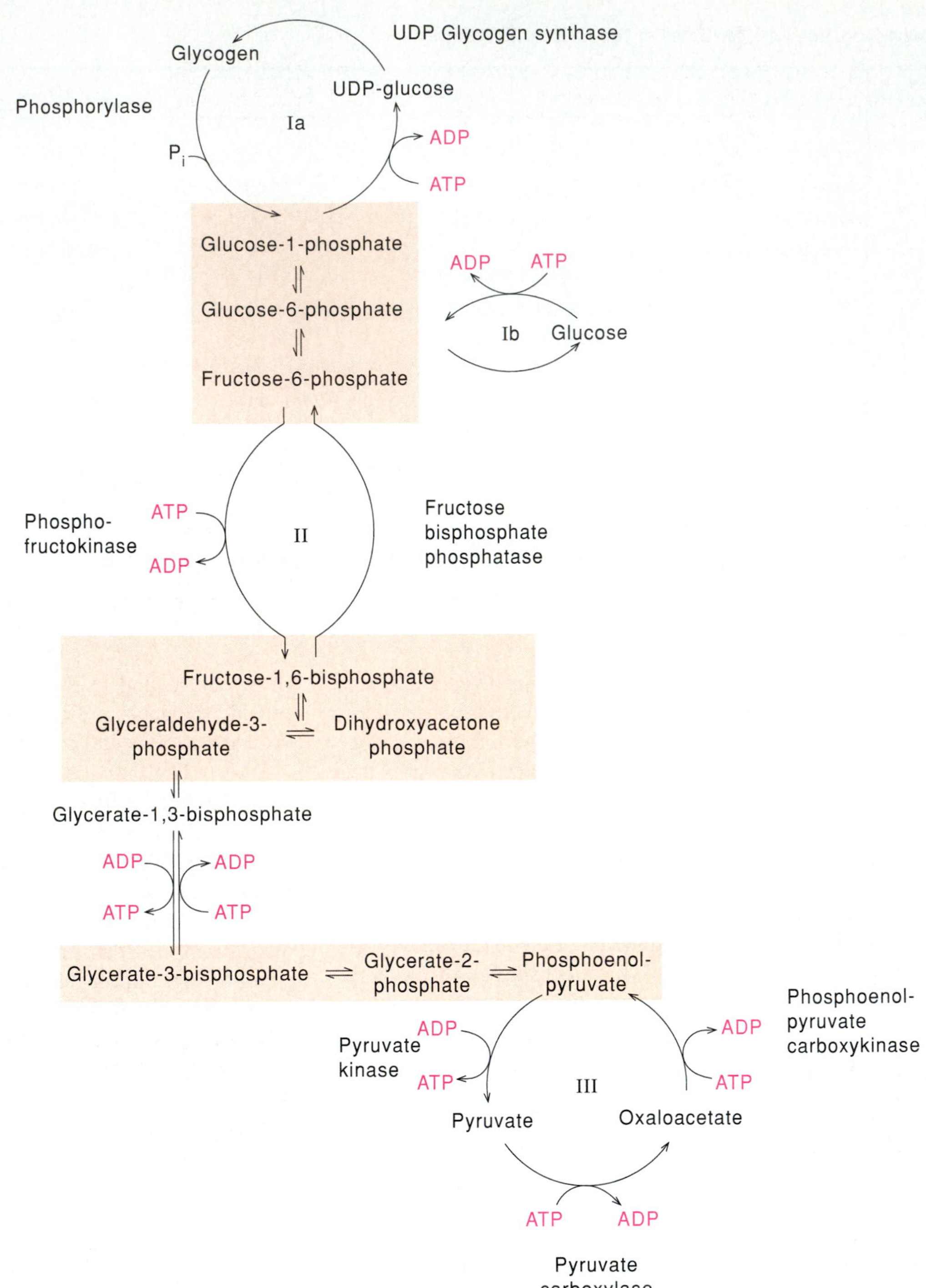

components are flowing into the pool and which are being removed; those rates, of course, depend in turn on the regulatory properties of the enzymes that catalyze the reactions that connect the pools.

In the direction of glycolysis there is a sizable drop in the standard free energy of formation of the component compounds in going from one metabolic pool to the one below it (see figs. 13.14 and 13.2). In two cases (glyceraldehyde-3-phosphate to glycerate-3-phosphate and phosphoenolpyruvate to pyruvate), one molecule of ADP is phosphorylated to ATP for each molecule of intermediate that flows from one pool to the next. In our lake-and-river analogy, there is a hydroelectric generating plant on the river, and useful energy is obtained from the flow between lakes. But unlike a hydroelectric plant, the ATP-generating machinery is quantized. Even if there is a large enough free energy drop to make 1.5 molecules of ATP, such

Figure 13.15

The pyruvate-phosphoenolpyruvate pseudocycle. One molecule of ATP is regenerated in the conversion of phosphoenolpyruvate to pyruvate, catalyzed by pyruvate kinase, which occurs in glycolysis. Conversion of pyruvate to phosphoenolpyruvate, which is necessary in gluconeogenesis, must be coupled to two ATP-to-ADP conversions if it is to be thermodynamically feasible. The first ATP is used indirectly, in the production of the active carboxyl transfer agent, biotin-COO⁻ (chapter 11). Transfer of a carboxyl group to pyruvate produces oxaloacetate. Decarboxylation of oxaloacetate aids in the attack by the carbonyl oxygen on the terminal phosphorus atom of ATP or GTP, producing phosphoenolpyruvate and ADP or GDP. This figure expands on the reactions shown in the lower part of figure 13.14 (III).

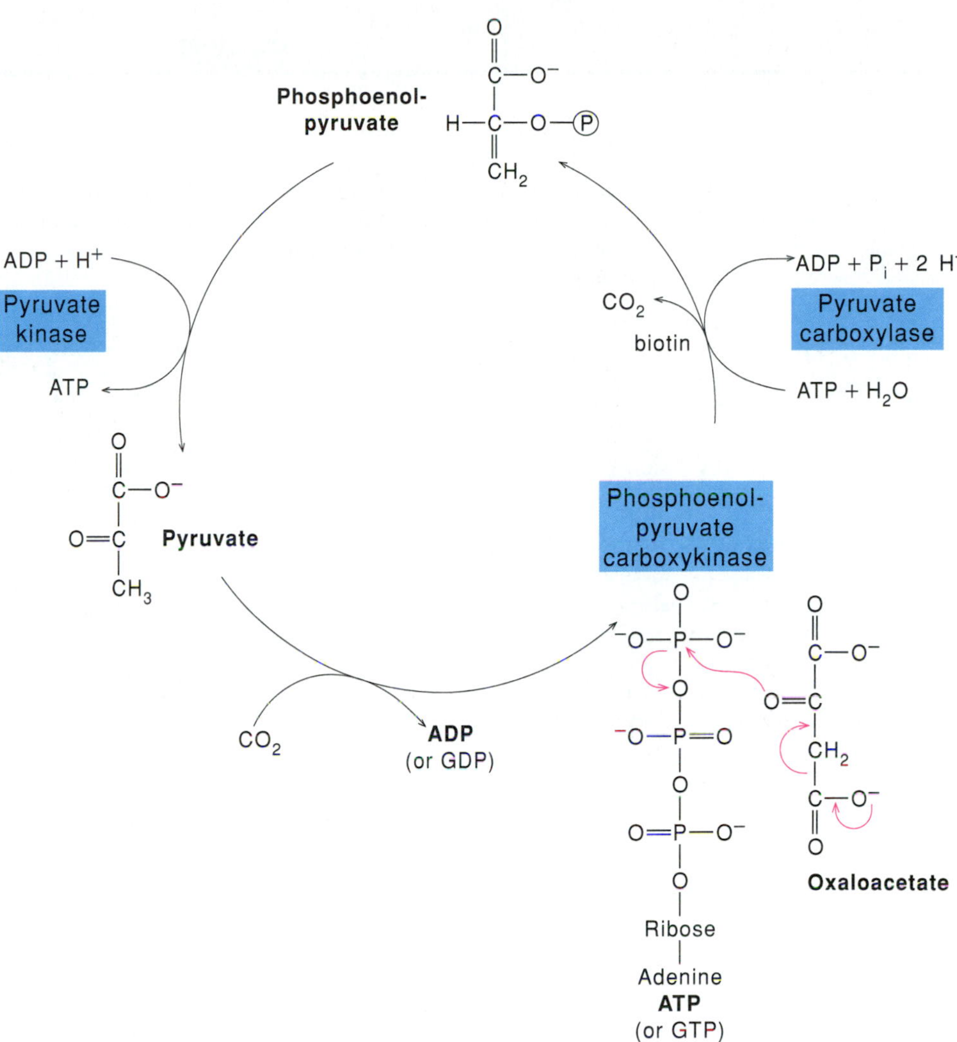

fractional stoichiometry is not possible, and only one molecule can actually be made. In such a case, the equilibrium constant for the reaction is therefore large, and the reaction is effectively irreversible. We have seen that the equilibrium constant for the pyruvate kinase reaction, for example, is about 10^6, making it impossible for that reaction ever to go in the reverse direction to any significant amount in a living cell. The conversion could be reversed only if it were coupled to two ATP-to-ADP conversions. This is a specialized restatement of the principle that oppositely directed sequences have different ATP stoichiometries (see chapter 12).

Conversion of Pyruvate to Phosphoenolpyruvate Requires ATP and GTP

In gluconeogenesis it is necessary to form phosphoenolpyruvate from pyruvate that is produced by oxidation of lactate or by catabolism of alanine or other amino acids. Here the principle that any conversion can be made thermodynamically favorable by coupling in ATP-to-ADP conversions is applicable. Reversal of the pyruvate kinase reaction, which would use one ATP for each phosphoenolpyruvate produced, is not thermodynamically favorable, so it is necessary that the conversion be coupled to at least two ATP-to-ADP conversions. The means by which that is done is shown in figure 13.15.

The first step in the conversion, the carboxylation of pyruvate to form oxaloacetate, is catalyzed by the enzyme pyruvate carboxylase. As in many other enzymic carboxylation reactions, the immediate CO_2 donor is a carboxylated derivative of the cofactor biotin (see chapter 11). ATP is used in the carboxylation of biotin; the stoichiometry for the production of oxaloacetate is

$$\text{Pyruvate}^- + \text{ATP}^{4-} + \text{HOCO}_2^- \rightarrow \text{oxaloacetate}^{2-} + \text{ADP}^{3-} + \text{P}_i^{2-} + \text{H}^+ \quad \textbf{(11)}$$

The detailed mechanism of this reaction is considered in box 13B. In addition to its role as an intermediate in gluconeogenesis, oxaloacetate is also an intermediate in the tricarboxylic acid (TCA) cycle; we will defer further discussion of this fascinating enzyme until the next chapter, where we can discuss its key role more fully.

13B

BOX

Pyruvate Carboxylase, a Biotin-Dependent Carboxylation

Pyruvate carboxylase is one of those carboxylation enzymes that utilizes bicarbonate as the source of CO_2 for carboxylating a substrate. Carbon 3 of pyruvate is the acceptor of the carboxyl group:

$$CH_3{-}\overset{O}{\overset{\|}{C}}{-}CO_2^- + HOCO_2^- + ATP \xrightarrow{Mg^{2+}} {}^-O_2C{-}CH_2{-}\overset{O}{\overset{\|}{C}}{-}CO_2^- + ADP + P_i$$

This reaction depends on the dehydration of bicarbonate ($HOCO_2^-$) to CO_2 and on cleavage of a carbon–hydrogen bond in pyruvate to form a carbanion. Carbon dioxide is an electrophilic molecule that readily reacts with nucleophiles, whereas bicarbonate is unreactive toward nucleophiles. The relevant nucleophile in the carboxylation of pyruvate is the C-3 carbanion, which is readily formed by enolization of pyruvate to the enolate anion. Carboxylation can then occur by the following process:

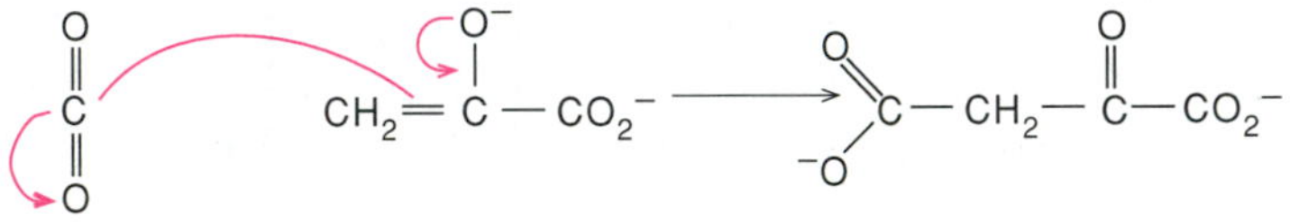

$$O{=}C{=}O \quad CH_2{=}\overset{O^-}{\overset{|}{C}}{-}CO_2^- \longrightarrow {}^-O(O{=})C{-}CH_2{-}\overset{O}{\overset{\|}{C}}{-}CO_2^-$$

Pyruvate is intrinsically enolizable because it is a ketone, and its enolizability is enhanced relative to other ketones by the presence of the carboxyl group and its capacity for metal complexation. Therefore, the enolization of pyruvate does not present a difficult problem at the active site of pyruvate carboxylase. The dehydration of bicarbonate, however, requires ATP to generate a substantial amount of carbon dioxide at the active site. ATP in effect absorbs the water of dehydration by using it to cleave a phosphodiester bond in a tightly controlled manner involving the formation of the specific intermediate carboxy phosphate:

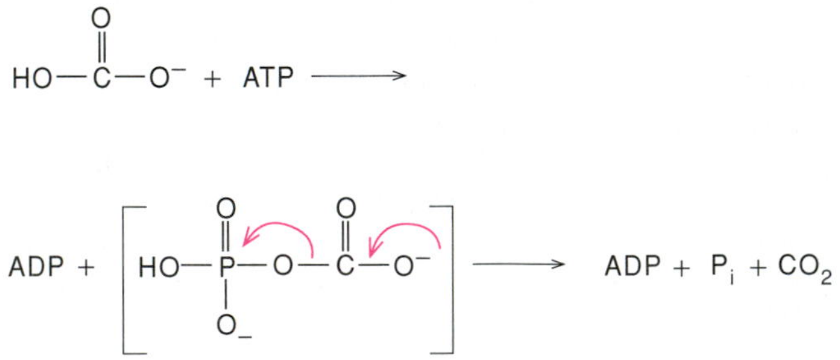

$$HO{-}\overset{O}{\overset{\|}{C}}{-}O^- + ATP \longrightarrow$$

$$ADP + \left[HO{-}\overset{O}{\overset{\|}{P}}{-}O{-}\overset{O}{\overset{\|}{C}}{-}O^-\right] \longrightarrow ADP + P_i + CO_2$$

(In the bracketed intermediate, the phosphorus also carries an O^-.)

The carbon dioxide generated at the active site of pyruvate carboxylase in this reaction is highly mobile and reactive and could easily escape into solution, where it would quickly be quenched by water to form bicarbonate. This is prevented from occurring by the presence of biotin at the active site. Biotin reacts reversibly with carbon dioxide and stabilizes it as N^1-carboxybiotin (see chapter 11).

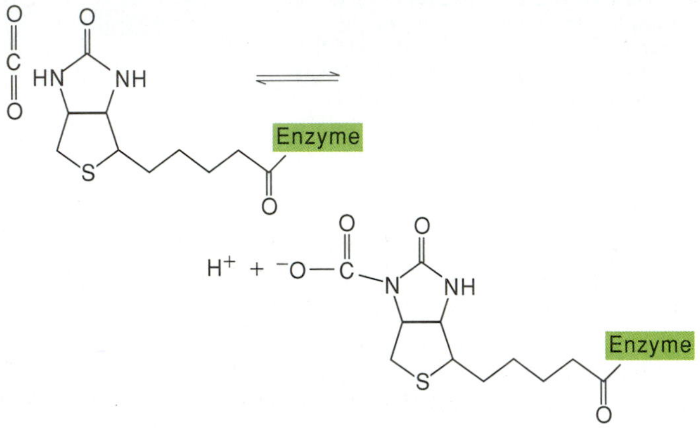

$$O{=}C{=}O + \text{Biotinyl-Enzyme} \rightleftharpoons H^+ + N^1\text{-Carboxybiotinyl-Enzyme}$$

In the form of N^1-carboxybiotin the carbon dioxide is potentially available for carboxylating enolpyruvate, but it is stabilized by biotin against side reactions and against escape from the enzyme. In its stabilized form as N^1-carboxybiotin, which is covalently bonded to the enzyme, the carboxyl group can be translocated from the ATP-dependent carboxylation site to the site at which pyruvate is enolized. At the pyruvate enolization site carbon dioxide is regenerated from N^1-carboxybiotin and reacts with enolpyruvate to form oxaloacetate.

The structural mobility of the biotinyl and N^1-carboxybiotinyl group in the enzyme is ensured by the covalent linkage of biotin to the ε-amino group of a lysine residue. Between the four methylene groups of biotin and the four of lysine, there is a high degree of bond rotational freedom potentially available to facilitate translocation of the biotin ring between sites. The torsionally free bonds are indicated in color in the structure of the N^1-carboxybiotinyl-enzyme.

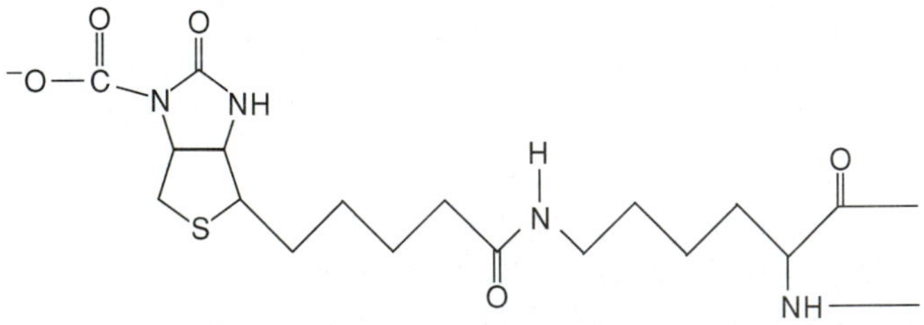

N^1–Carboxybiotinyl-enzyme

The carboxylation of pyruvate supplies thermodynamic activation for the next step in the sequence. The standard free energy change for decarboxylation of β-keto carboxylic acids is negative and rather large. Thus the decarboxylation will make the formation of phosphoenolpyruvate more thermodynamically favorable than it would otherwise be. Another important example of this interesting strategy of activation by carboxylation is seen in fatty acid synthesis, in which carboxylation of acetyl-coenzyme A serves a similar function (chapter 17).

Oxaloacetate is converted to phosphoenolpyruvate (PEP) in a reaction catalyzed by phosphoenolpyruvate carboxykinase. In many species, including mammals, this reaction involves a GTP-to-GDP conversion, rather than ATP-to-ADP:

$$\text{Oxaloacetate} + \text{GTP} \rightarrow \text{PEP} + \text{GDP} + CO_2 \quad \textbf{(12)}$$

The use of GTP, UTP, or CTP in a metabolic reaction or sequence is energetically equivalent to the use of ATP, because the nucleoside diphosphate that is produced is rephosphorylated at the expense of ATP in the reaction catalyzed by nucleoside diphosphate kinase:

$$\text{ATP} + \text{NDP} \rightleftharpoons \text{ADP} + \text{NTP} \quad \textbf{(13)}$$

NDP may be CDP, GDP, or UDP, and NTP the corresponding triphosphate. The enzyme is unusual in its lack of specificity for the acceptor nucleoside diphosphate.

Carboxylation is mechanistically, as well as energetically, an appropriate means of activating pyruvate for conversion to phosphoenolpyruvate, as seen in figure 13.15. When a β-keto acid loses CO_2, electrons move into the α-β bond and displace a pair of electrons from the carbonyl double bond. The usual result is that a proton is picked up from the surroundings, yielding as the initial product the enol form of the ketone. The enol form will tautomerize almost quantitatively to the keto form. However, in the case of the phosphoenolpyruvate carboxykinase reaction, the enzyme positions one GTP molecule appropriately so that the carbonyl oxygen atom of oxaloacetate, on becoming negatively charged, attacks the terminal phosphorus atom of the cofactor, displacing GDP. Thus the product is locked in the high-energy enolic form. The negative free energy changes that accompany β decarboxylation and the hydrolysis of GTP are sufficiently large to allow the conversion of oxaloacetate to phosphoenolpyruvate to proceed with a standard free energy change of about -4 kcal/mole and an equilibrium constant of about 800.

If we add the equations for the reactions catalyzed by pyruvate carboxylase, phosphoenolpyruvate carboxykinase, and nucleoside diphosphate kinase, we obtain the overall reaction for conversion of pyruvate to phosphoenolpyruvate:

$$H_2O + \text{pyruvate} + 2\ \text{ATP} \rightarrow \text{PEP} + 2\ \text{ADP} + P_i + 2\ H^+ \quad \textbf{(14)}$$

Thus two molecules of ATP are used to reverse the conversion that yields one molecule of ATP in the glycolytic direction, and both directions of conversion are thermodynamically favorable.

Other Routes from Pyruvate to Phosphoenolpyruvate

Different pathways are used by bacteria and green plants for the conversion of pyruvate to phosphoenolpyruvate. Many bacterial species contain an enzyme called phosphoenolpyruvate synthase, which catalyzes the direct conversion of pyruvate to phosphoenolpyruvate, with the production of AMP and inorganic phosphate ion from ATP. Since a second molecule of ATP must be used to convert the product AMP to ADP, the conversion is actually coupled to two ATP-to-ADP conversions, and the overall equilibrium constant is identical to that of the conversion by way of oxaloacetate:

$$H_2O + \text{pyruvate} + \text{ATP} \rightarrow \text{PEP} + \text{AMP} + P_i + 2\ H^+ \quad \textbf{(15)}$$

$$\text{ATP} + \text{AMP} \rightleftharpoons 2\ \text{ADP} \quad \textbf{(16)}$$

$$\text{Sum: } H_2O + \text{pyruvate} + 2\ \text{ATP} \rightarrow \text{PEP} + 2\ \text{ADP} + P_i + 2\ H^+ \quad \textbf{(14)}$$

Some species of green plants (those that carry out "C-4" photosynthesis) fix CO_2 into oxaloacetate as part of a transport mechanism that increases the concentration of CO_2 at the photosynthetic centers (chapter 16). Since phosphoenolpyruvate is the CO_2 acceptor, these plants must carry out a massive conversion of pyruvate to phosphoenolpyruvate. This conversion is effected by an enzyme called pyruvate phosphate dikinase, which catalyzes the transfer of one phosphate group to pyruvate and another to inorganic phosphate ion. Hydrolysis of the resulting pyrophosphate and conversion of AMP to ADP by the action of adenylate kinase result again in the same stoichiometry for conversion of pyruvate to phosphoenolpyruvate as in the other cases:

$$\text{Pyruvate} + P_i + \text{ATP} \rightarrow \text{PEP} + \text{AMP} + PP_i + 2H^+ \quad \textbf{(17)}$$

$$PP_i + H_2O \rightarrow 2\ P_i + H^+ \quad \textbf{(18)}$$

$$\text{ATP} + \text{AMP} \rightleftharpoons 2\ \text{ADP} \quad \textbf{(16)}$$

$$\text{Sum: } H_2O + \text{pyruvate} + 2\ \text{ATP} \rightarrow \text{PEP} + 2\ \text{ADP} + P_i + 3\ H^+ \quad \textbf{(14)}$$

Conversion of Phosphoenolpyruvate to Fructose-1,6-Bisphosphate Uses the Same Enzymes as Glycolysis

The interconversion of the phosphorylated three-carbon acids phosphoenolpyruvate, glycerate-2-phosphate, and glycerate-3-phosphate has already been discussed in this chapter. This pool is linked to the fructose-1,6-bisphosphate/triose phosphate pool by the reactions that convert glycerate-3-phosphate to glyceraldehyde-3-phosphate. The linkage here, however, is organized differently from the reactions that link other pools. In this case the same reactions, catalyzed by the same enzymes, function in both directions (see fig. 13.14). That is possible only because the free energy difference between the two pools is approximately equal to the free energy of hydrolysis of ATP, so that material can move from one pool to the other in either direction at reasonable physiological concentrations of the components of the two pools. That is, the one molecule of ATP per molecule of substrate that is generated in the pathway from the fructose-1,6-bisphosphate/triose phosphate pool to the phosphorylated acid pool is enough to make the reverse conversion thermodynamically feasible. In our water-flow analogy, the amount of electrical energy generated when water flows downhill (from fructose-1,6-bisphosphate/triose phosphate to phosphorylated acids) is equal to the amount of energy needed to pump water uphill in the reverse direction, so that no additional energy input is necessary.

From figure 13.14 it can be seen that there is no pair of oppositely directed reactions with different ATP stoichiometries (no pseudocycle) in the whole segment of glycolysis and gluconeogenesis between phosphoenolpyruvate and fructose-1,6-bisphosphate. Thus the stoichiometry for conversions in the two directions must be the same. Both directions of conversion do, in fact, occur (from fructose-1,6-bisphosphate to phosphoenolpyruvate in glycolysis and from phosphoenolpyruvate to fructose-1,6-bisphosphate in gluconeogenesis). It follows that the components of both pools, as well as glycerate-1,3-bisphosphate, which lies between them, must be so close to equilibrium in the cell that the direction of flux will depend on mass action.

On first thought, it seems unlikely that material could flow freely between these two pools without the pumping effect of a pseudocycle. The standard free energy change for the phosphate-glycerate kinase reaction is large and negative, and we would not expect that this reaction, which leads to the formation of glycerate-3-phosphate in the glycolytic direction, could be reversed under physiological conditions. The equilibrium constants of neighboring reactions, however, are such as to bring the equilibrium constant for the overall interconversion between fructose-1,6-bisphosphate and phosphoenolpyruvate into a range where reversal is feasible. During glycolysis the value of ΔG for the conversion of fructose-1,6-bisphosphate to phosphoenolpyruvate is negative. A rather small shift in the concentrations of those compounds can cause the value of ΔG to change sign, so that the conversion of phosphoenolpyruvate to fructose-1,6-bisphosphate becomes thermodynamically favorable. This change may be enhanced by a change in the $NAD^+/NADH$ ratio.

Fructose-Bisphosphate Phosphatase Converts Fructose-1,6-Bisphosphate to Fructose-6-Phosphate

Fructose-1,6-bisphosphate is converted to fructose-6-phosphate by hydrolysis of the phosphoryl ester bond at C-1, a reaction catalyzed by fructose-bisphosphate phosphatase. This reaction and that catalyzed by phosphofructokinase thus make up a metabolic pseudocycle. The standard free energy change for the hydrolysis is about −4 kcal/mole, corresponding to an equilibrium constant of nearly 1,000. Thus, as in any such pseudocycle the two conversions—the phosphorylation of fructose-6-phosphate to form fructose-1,6-bisphosphate with ATP as the phosphate donor, and the hydrolysis of fructose-bisphosphate to form fructose-6-phosphate—are both highly favorable thermodynamically under any conditions that will arise in a living cell.

Summary of Gluconeogenesis

Gluconeogenesis is the reverse of glycolysis, being the process by which sugars are produced from simpler nonsugar compounds such as lactate or amino acids. The two sequences together constitute a giant pseudocycle. However, in contrast to most pseudocycles, many of the reactions in the sequence are common to both directions. Thus most of the enzymes used and most of the intermediates produced are the same in both sequences. Only at three points, all outside the pools, are reactions of glycolysis and gluconeogenesis distinct and catalyzed by different enzymes.

The pools are the hexose phosphate pool (glucose-1-phosphate, glucose-6-phosphate, and fructose-6-phosphate), the aldolase pool (fructose-1,6-bisphosphate, glyceraldehyde-3-phosphate, and dihydroxyacetone phosphate), and the three-carbon pool (glycerate-3-phosphate, glycerate-2-phosphate, and phosphoenolpyruvate). Within each pool the direction of reaction is determined by simple mass-action considerations. The direction of overall conversion (glycolysis or gluconeogenesis) depends on the rates at which the connecting reactions remove materials from some pools and supply materials to others.

The connecting reactions (except for those catalyzed by glyceraldehyde-3-phosphate dehydrogenase and glycerate-3-phosphate kinase, which connect the aldolase pool and the three-carbon pool) have large free energies of reaction and are far from equilibrium in the cell. In order to achieve a favorable equilibrium in both directions the conversions must involve different ATP stoichiometries. Overall we find that gluconeogenesis consumes ATP, whereas glycolysis produces ATP.

The differences in ATP consumption for the connecting reactions are illustrated in figure 13.14. In going from the hexose monophosphate pool to the aldolase pool, one ATP is required. In the reverse direction, there is no net gain or loss of ATP. In going from the three-carbon pool to pyruvate, there is a net gain of one ATP. In the opposite direction, essentially two ATPs are required.

Conversion of Hexose Phosphates to Polysaccharides

Before turning to the regulation of glycolysis and gluconeogenesis, let's briefly consider the conversion of sugars to polysaccharides, as these are the main energy-storage forms of carbohydrates. The subject of polysaccharide synthesis is taken up again in chapter 21, where we will elaborate on the synthesis of complex polysaccharides.

When fructose-6-phosphate is generated by photosynthesis (see chapter 16) or gluconeogenesis, an equimolar amount of glucose-1-phosphate will usually be removed from the hexose monophosphate pool by conversion to storage polysaccharide (glycogen in animals and many kinds of microorganisms, starch in green plants).

The first step in the conversion of glucose-1-phosphate to starch or glycogen is activation at the expense of a nucleoside triphosphate. In green plants the activated form is ADP-glucose, formed by transfer of an AMP moiety from ATP to the

Figure 13.16

The elongation step in starch synthesis. (*a*) The activated form of glucose in starch synthesis is ADP-glucose. This is formed from glucose and ATP as shown. (*b*) The ADP-glucose does not react directly with the elongating starch molecule. Rather, the starch synthase produces an oxonium ion intermediate which is attacked by the terminal C-4 hydroxyl on the growing polymer.

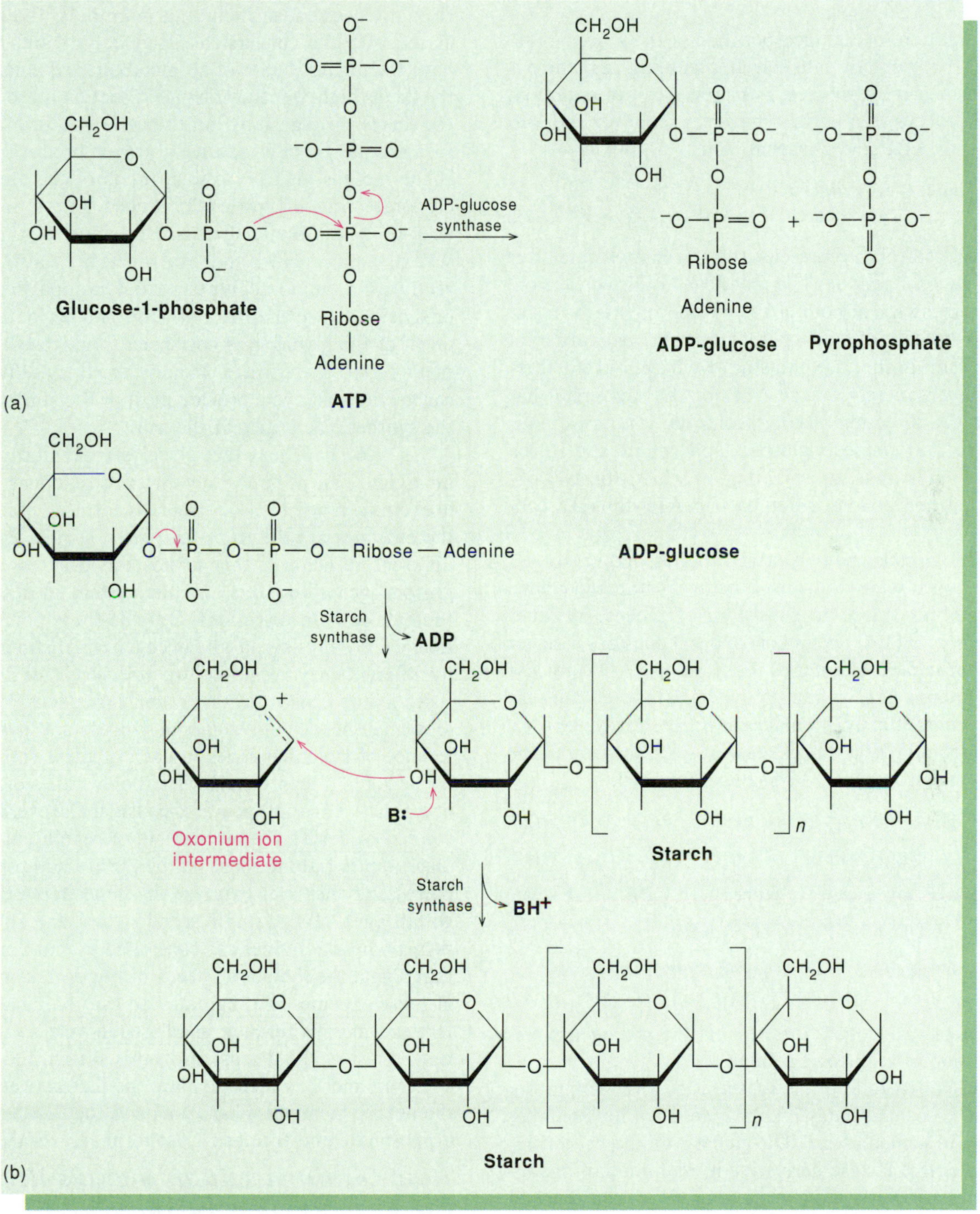

phosphate of glucose-1-phosphate by the enzyme ADP-glucose synthase. An oxygen atom of the phosphoryl group of glucose-1-phosphate attacks the α phosphorus atom of ATP, displacing pyrophosphate (fig. 13.16). Since the bond broken and the bond formed are both pyrophosphate bonds, the change in standard free energy for this reaction is close to zero and the equilibrium constant is near 1. However, if the pyrophosphate is hydrolyzed by the enzyme pyrophosphatase, as probably occurs under most conditions in living cells, the standard free energy change for the two reactions combined is large and negative, about -7.5 kcal/mole.

In the reaction catalyzed by starch synthase, the oxygen on C-4 of the terminal glucosyl residue of a starch chain attacks the C-1 of a glucoxonium ion, resulting in the addition of a glucosyl unit to the polymeric starch molecule (see fig. 13.16*b*). The standard free energy for the elongation step in starch synthesis is about -3 kcal/mole.

The equation for the incorporation of the glucosyl residue of glucose-1-phosphate into starch, assuming that the pyrophosphate formed is hydrolyzed, is the reverse of the equation for the phosphorolysis of a polysaccharide, except for the conversion of ATP to ADP and phosphate ion:

$$\text{Glucose-1-phosphate} + \text{ATP} + \text{starch} \rightarrow (\text{starch}^{+}) + \text{ADP} + 2\,\text{P}_i + \text{H}^{+} \quad \textbf{(19)}$$

In this equation, (starch^{+}) represents the starch molecule after addition of a glucosyl residue. The reactions in this conversion, which includes cleavage of both of the pyrophosphate bonds of ATP and the formation of a new pyrophosphate bond, are a bit more complex than in the case of a simple kinase, but the thermodynamic effect is merely that an ATP-to-ADP conversion has been added in the direction of polysaccharide synthesis. Thus the pseudocycle that connects glucose-1-phosphate and starch is energetically equivalent to any other in which the two oppositely directed conversions differ by one ATP-to-ADP conversion.

In most animals, many bacterial species, and yeasts and other fungi, glycogen serves the same function as starch in plants. Glycogen resembles starch in consisting of glucose residues linked primarily by $\alpha(1,4)$ acetal bonds, but it contains a larger number of 1,6 branches (chapter 6). Bacteria use ADP-glucose as the glucosyl donor in glycogen synthesis, but vertebrates use UDP-glucose. Energetically this makes no difference. Since UTP is regenerated from UDP at the expense of ATP, the sum of the reactions is the same as when ADP-glucose is used:

$$\text{H}^{+} + \text{G-1-P} + \text{UTP} \rightleftharpoons \text{UDP-glucose} + \text{PP}_i + \text{H}_2\text{O} \quad \textbf{(20)}$$

$$\text{H}_2\text{O} + \text{PP}_i \rightarrow 2\,\text{P}_i + \text{H}^{+} \quad \textbf{(18)}$$

$$\text{UDP-glucose} + \text{glycogen} \rightarrow (\text{glycogen}^{+}) + \text{UDP} + \text{H}^{+} \quad \textbf{(21)}$$

$$\text{UDP} + \text{ATP} \rightleftharpoons \text{UTP} + \text{ADP} \quad \textbf{(22)}$$

$$\text{Sum: Glucose-1-phosphate} + \text{ATP} + \text{glycogen} \rightarrow (\text{glycogen}^{+}) + \text{ADP} + 2\,\text{P}_i + \text{H}^{+} \quad \textbf{(23)}$$

In these equations, (glycogen^{+}) represents the glycogen molecule after addition of a glucosyl residue.

Because of its roles in the synthesis of glycogen, in isomerization of hexose phosphates, and as a precursor for a number of biosynthetic intermediates, UDP-glucose might be considered the most central hexose derivative in mammalian metabolism, aside from glycolysis and gluconeogenesis. In bacteria and plants, those roles are divided between ADP-glucose (production of storage polysaccharide) and UDP-glucose (sugar interconversions and biosynthesis).

Regulation of Glycolysis and Gluconeogenesis

Several features of the glycolysis-gluconeogenesis system indicate that its regulation is of special importance to the organism. In many organisms, including mammals, the sequences associated with the conversions of glycolysis and gluconeogenesis have the highest fluxes of all metabolic sequences. Not only are the fluxes high, but they change direction more frequently than the fluxes through most other pathways. In mammalian liver, for example, there is massive glycogen production after a meal. If the meal is high in protein, much of the glucose that is incorporated into glycogen is produced from pyruvate and oxaloacetate. These metabolites are products of the catabolism of several amino acids, and oxaloacetate is also produced indirectly from amino acids that are degraded to α-ketoglutarate or succinate (see chapter 19). During a period of fasting, much or all of the glycogen is converted to glucose-1-phosphate and metabolized to pyruvate. In addition, this pathway and the tricarboxylic acid cycle provide most of the starting materials for the synthetic activities of the cell.

We have seen that glycolysis and gluconeogenesis make up a giant pair of oppositely directed reactions, that is, a pseudocycle (see fig. 13.14). Whereas in most pseudocycles all of the enzymes are different for paired sequences functioning in different directions, this is not the case for the glycolysis-gluconeogenesis system. In this system most of the enzymes function in both directions. Only at three points are the reactions of glycolysis and gluconeogenesis distinct and catalyzed by different enzymes in the two directions (see fig. 13.14). These three points constitute three small pseudocycles in the paired sequences between glycogen or glucose and pyruvate. The difference of stoichiometries in each of these small pseudocycles is one ATP. When considering the overall stoichiometries and energetics, it is appropriate to consider the overall pathway; but when considering the mechanism of regulation of the flux it is more appropriate to focus on the three small pseudocycles. The paired enzymes that catalyze the oppositely directed reactions within each of the small pseudocycles are all regulatory enzymes whose activities are controlled by small-molecule factors that signal the metabolic needs of the organism. The activities of these enzymes are regulated so that only one of the regulatory enzymes within each small pseudocycle is active at any given time. Most of the discussion in this section focuses on pseudocycles Ia and II, which regulate the flux between glycogen and the hexose monophosphate pool and the flux between fructose-6-phosphate and fructose-1,6-phosphate, respectively.

How Do Intracellular Signals Regulate the Key Enzymes Involved in Glycolysis and Gluconeogenesis?

Most of the intracellular signals for the pathways we are discussing consist of small molecules that are a direct indication of the energy charge: ATP, AMP, and citrate. An example of

Table 13.2
Activities of Some of the Enzymes that Control the Fluxes in the Glycolysis-Gluconeogenesis Pathways

	Pseudocycle I		Pseudocycle II	
	Glycolysis	*Gluconeogenesis*	*Glycolysis*	*Gluconeogenesis*
Intracellular controls	Glycogen phosphorylase	Glycogen synthase	Phosphofructokinase	Fructose bisphosphate phosphatase
High energy charge	↓	↑	↓	↑
Low energy charge	↑	↓	↑	↓
Hormonal controls				
Low blood glucose	↑	↓	↓	↑

the way in which these signals operate is the well-studied case of phosphofructokinase (see chapter 10). This enzyme has a site (sometimes more than one) for binding regulatory factors that is separate from the active site. The binding of these small molecules changes the structure of the enzyme in a way that affects the active site. In the presence of high levels of ATP, phosphofructokinase requires high substrate concentrations before it becomes very active. By contrast, at high AMP concentrations (low ATP) the enzyme is quite active at very low substrate concentrations (see fig. 10.6). A high citrate level, like a high ATP level, favors a lowering in the activity of the enzyme. The small molecules ATP, AMP, and citrate function as allosteric effectors that bind to the regulatory locus and thereby affect the conformation at the active site of the enzyme. Appropriately, ATP and citrate, which are an indication of a high energy charge, encourage the low-activity form of the enzyme, and AMP, an indication of a low energy charge, favors the high-activity form.

Phosphofructokinase is a member of pseudocycle II. The other enzyme that is a member of pseudocycle II, fructose bisphosphate phosphatase, is affected in just the opposite way by energy charge and the small molecules that reflect the energy charge. The net result is that a low energy charge favors the high-activity form of phosphofructokinase and the low-activity form of fructose bisphosphate phosphatase. This result encourages glycolysis and inhibits gluconeogenesis. A high energy charge has just the opposite effects on these two paired enzymes. A summary of how energy charge effects the regulatory enzymes in pseudocycles I and II is given in table 13.2. In all cases the effects are consistent with the energy needs of the cell. Thus a low energy charge favors glycolysis and discourages gluconeogenesis, whereas a high energy charge favors gluconeogenesis and discourages glycolysis.

Hormonal Controls Override the Usual Intracellular Controls

In certain cells, such as liver cells, the response to energy charge, or to signals that reflect the energy charge, can be quite different, since such cells have a responsibility to assure the energy needs of the entire organism. In particular, the liver senses the blood glucose level. A low blood glucose level indicates an organism in need of glucose or energy. This condition triggers an extracellular signal (to be discussed next) that elevates the activities of both glycogen phosphorylase and fructose bisphosphate phosphatase and lowers the activity of both glycogen synthase and phosphofructose kinase.

Since the liver's metabolic functions are not dictated by local needs so much as by the general need for controlling the blood glucose level, additional regulatory inputs are needed. As in many other cases in which the activities of different tissues and organs are correlated for the good of the whole organism, hormones are involved in this additional level of control.

When the level of glucose in the blood falls slightly, small clusters of cells in the pancreas (A or α cells of the islets of Langerhans) sense the decrease. They respond by releasing the peptide hormone glucagon into the blood. On reaching the liver, glucagon causes the conversion of the glycogen phosphorylase from its normal form (phosphorylase *b*) to its hormone-activated form (phosphorylase *a*). The activity of the enzyme increases and it becomes independent of AMP control. Thus the hormone, reflecting a systemic (organismwide) need, overrides the AMP effect that provides control in response to local needs. Such override of local control to satisfy systemic needs is a common feature of hormone action. Epinephrine, a small hormone derived from tyrosine (see chapter 24), affects phosphorylase in the same way as does glucagon. Epinephrine is made in the adrenal glands and is liberated into the blood in response to signals from the central nervous system. The two hormones have a similar effect on liver cells even though they originate from different sources, and as we will see, they do not have the same effect on all cell types.

The Action of Glucagon and Epinephrine Is Mediated by Cyclic AMP

In the 1950s Earl Sutherland demonstrated activation of phosphorylase by epinephrine and glucagon in cell-free systems. He found that when he centrifuged a broken cell extract, the

Figure 13.17

Synthesis of cyclic AMP. A catalytic site on adenyl cyclase (:B) removes a proton from the C-3 oxygen, which then attacks the α-phosphate and displaces the pyrophosphate group.

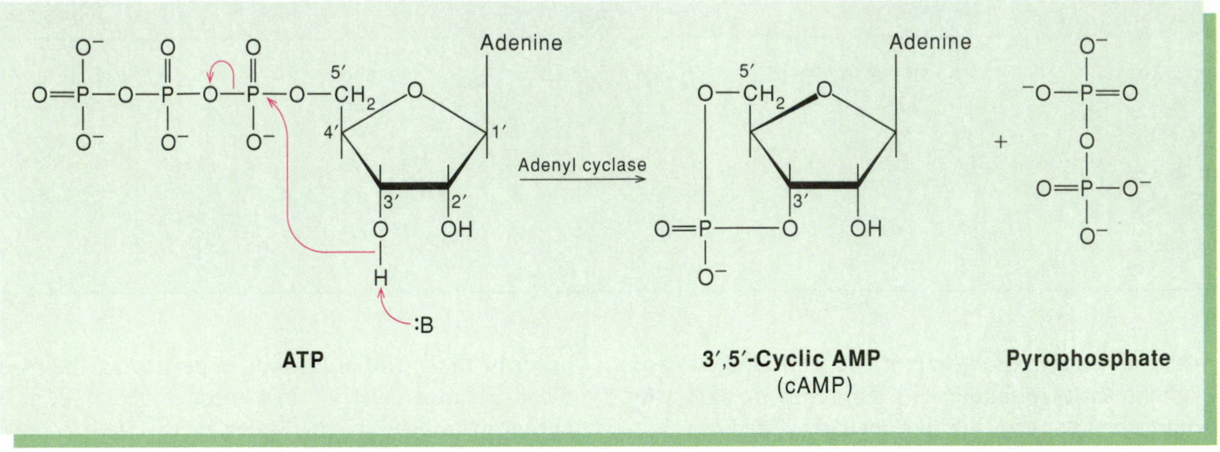

membrane-containing fraction, if supplied with ATP and either epinephrine or glucagon, produced a low-molecular-weight factor that could be isolated and would stimulate the activity of glycogen phosphorylase in the supernatant fraction. Further analyses showed that these hormones cause a membrane-bound enzyme in the particulate fraction to convert ATP to cyclic AMP (fig. 13.17). The cell membrane contains receptors for glucagon and epinephrine on its outer surface. When either hormone binds to its specific receptor, the enzyme adenylate cyclase, on the inner surface of the membrane, is stimulated to catalyze the production of cyclic AMP. The detailed mechanism for triggering of cAMP synthesis will be discussed in chapter 24.

Cyclic AMP triggers a cascade of reactions that ultimately leads to glycogen breakdown. The immediate action of cAMP is to activate a protein kinase, which phosphorylates a number of proteins including phosphorylase kinase. The phosphorylation of phosphorylase kinase converts it from an inactive to an active form. Phosphorylase kinase in the active form catalyzes the conversion of phosphorylase *b* to phosphorylase *a* by phosphorylation of a specific serine residue (see chapter 10). As we noted previously, this conversion eliminates the dependence of glycogen phosphorylase on AMP. The entire cascade of effects triggered by the hormones is shown in figure 13.18.

In its normal or base state, glycogen synthase is relatively active. The same cyclic-AMP-dependent protein kinase that phosphorylates phosphorylase kinase also catalyzes the phosphorylation of glycogen synthase. Whereas phosphorylation of phosphorylase kinase leads to increased activity of glycogen phosphorylase, the phosphorylation of glycogen synthase decreases its activity. The phosphorylated enzyme is much less active. The result is that when glycogen breakdown is stimulated in response to epinephrine or glucagon, the synthesis of glycogen is prevented. In this way the simultaneous operation of both arms of the glycogen-hexose-monophosphate pseudocycle is prevented.

Regulatory responses must necessarily be reversible if regulation is to be sensitive to changing metabolic needs. All of the effects we have seen are promptly reversed when the need for them has ended. Cyclic AMP is hydrolyzed to ordinary 5′-AMP by an enzyme called cyclic AMP phosphodiesterase, and there are protein phosphatases that catalyze the hydrolytic removal of the phosphoryl groups that had been attached to proteins by protein kinases.

The Effects of Glucagon and Epinephrine Are Not the Same on All Tissues

Both glucagon and epinephrine affect liver cells in the same way through the production of cAMP. Nevertheless, the physiological significance of the two hormones is quite different. Glucagon is part of the system that is responsible for stabilization of the concentration of glucose in the blood. In contrast to the homeostatic function of glucagon (its support of the status quo), epinephrine is an alarm signal that leads to deviation from the homeostatic status quo. Its release is controlled by the central nervous system; specifically, it is released in response to fear or anger. Its function is to prepare the body for a sudden extreme energy output if that becomes necessary; hence it is often termed the "fight or flight" hormone. In addition to many physiological effects, such as dilation of the air passages leading to the lungs and redirection of the circulatory system to favor skeletal muscles, epinephrine stimulates the conversion of phosphorylase *b* to *a* in the skeletal muscle. The result is to build up the levels of glycolytic intermediates, and presumably to eliminate a time lag in energy mobilization when and if the organism begins either to fight or to flee.

Epinephrine causes the conversion of phosphorylase *b* to phosphorylase *a* both in liver and in skeletal muscle, since it is desirable to increase the concentrations of intermediates both locally (in the muscle) and in reserve (in the liver). In contrast, glucagon, which has the same effect on liver phosphorylase as

Figure 13.18

Molecular basis for the stimulation of glycogen breakdown by the hormones epinephrine and glucagon. The epinephrine or the glucagon activates adenyl cyclase, which activates protein kinase, which activates phosphorylase kinase. Phosphorylase kinase in turn activates phosphorylase, which cleaves glucose from glycogen. This cascade of regulatory interactions results in a very large amplification, so that a small amount of hormone can have a large effect on the rate of breakdown of glycogen.

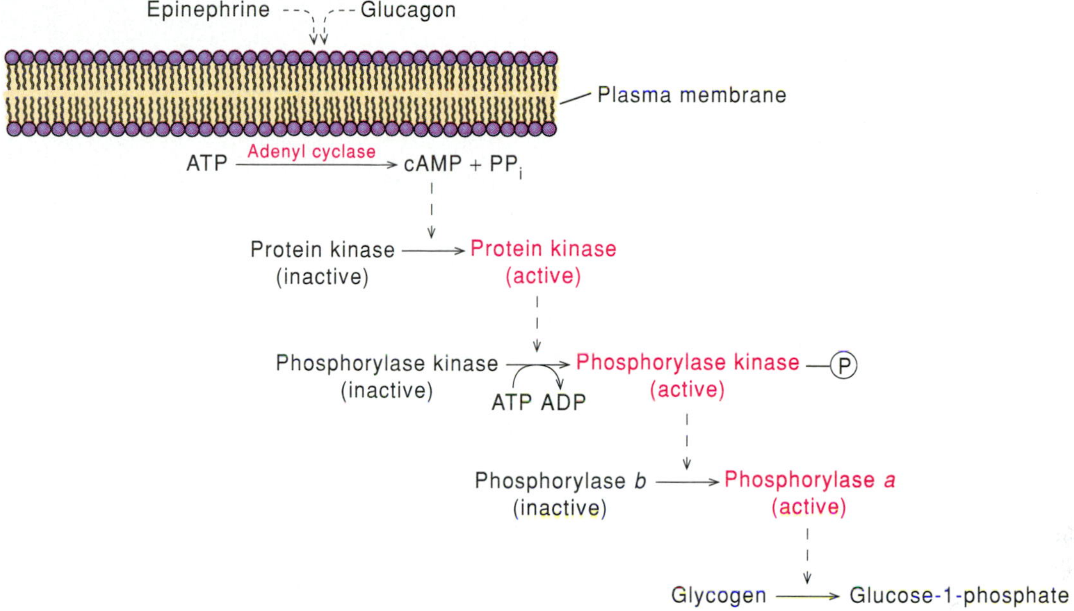

epinephrine, does not affect muscle phosphorylase. This is because muscle cells lack glucagon receptors. Muscle is strictly a consumer tissue and does not contribute to stabilization of the level of glucose in blood (except under starvation conditions, see chapter 24); thus it should not, and does not, break down its glycogen reserves in response to low blood glucose.

The Hormonal Regulation of the Flux between Fructose-6-Phosphate and Fructose-1,6-Bisphosphate Is Mediated by Fructose-2,6-Bisphosphate

The same protein kinase that phosphorylates glycogen phosphorylase and glycogen synthase does not phosphorylate the enzymes of pseudocycle II, but the enzymes of both pseudocycles are nevertheless regulated by the same hormonal signal, glucagon. In this case the enzyme that gets phosphorylated by the protein kinase is one responsible for the synthesis of fructose-2,6-bisphosphate. When the synthesizing enzyme is phosphorylated it becomes inactive, and the level of the fructose-2,6-bisphosphate compound drops as a result. More than this, the synthesizing enzyme becomes active in the breakdown of the fructose-2,6-bisphosphate compound. Fructose-2,6-bisphosphate is a potent activator of phosphofructokinase and a potent inhibitor of fructose bisphosphate phosphatase. Thus the negative effect of glucagon on phosphofructokinase activity arises because the hormone leads to a lowering in the level of a potent activator of the enzyme. Since the same compound is a potent inhibitor of fructose bisphosphate phosphatase, the effect of glucagon on the latter enzyme is exactly the opposite.

The effect of energy charge and the small molecules reflecting the energy charge, as well as the effect of hormones, on the key enzymes of pseudocycles I and II are reviewed in table 13.2 and figure 13.19.

Summary of the Regulation of Glycolysis and Gluconeogenesis

The connecting reactions between the three pools associated with glycolysis and gluconeogenesis can be thought of as being organized into three small pseudocycles (see fig. 13.14). The ATP stoichiometries are different for going in the two directions in these pseudocycles. This difference assures that the reactions are thermodynamically feasible in either direction, but it does not assure the direction of the reaction under different metabolic circumstances. The actual direction taken is determined by pairs of regulatory enzymes that catalyze oppositely directed reactions within each pseudocycle. These enzymes are sensitive to regulatory factors that prevent the conversions within the pseudocycles from proceeding in both directions simultaneously. In the discussion on regulation of these pathways we have focused on the regulatory elements of pseudocycle I, involving the interconversion of glycogen and glucose-1-phosphate, and pseudocycle II, involving the interconversion of fructose-1-phosphate and fructose-1,6-bisphosphate.

In all unicellular organisms and in most cells of multicellular organisms the enzymes that regulate the flux across pseudocycles I and II are regulatory enzymes sensitive to inhibition and activation by small molecules that bind covalently to their surfaces and thereby change their conformations. These

Figure 13.19

Some major points of regulation in the glycolytic-gluconeogenic pathways. Normal cellular controls favor glycogen breakdown and glycolysis when the energy charge is low. When the blood glucose level is low, hormonal controls on the liver cells favor glycogen breakdown and gluconeogenesis. The target sites for control factors are indicated by red arrows, with a plus or a minus at the arrowhead to indicate activation and inhibition, respectively.

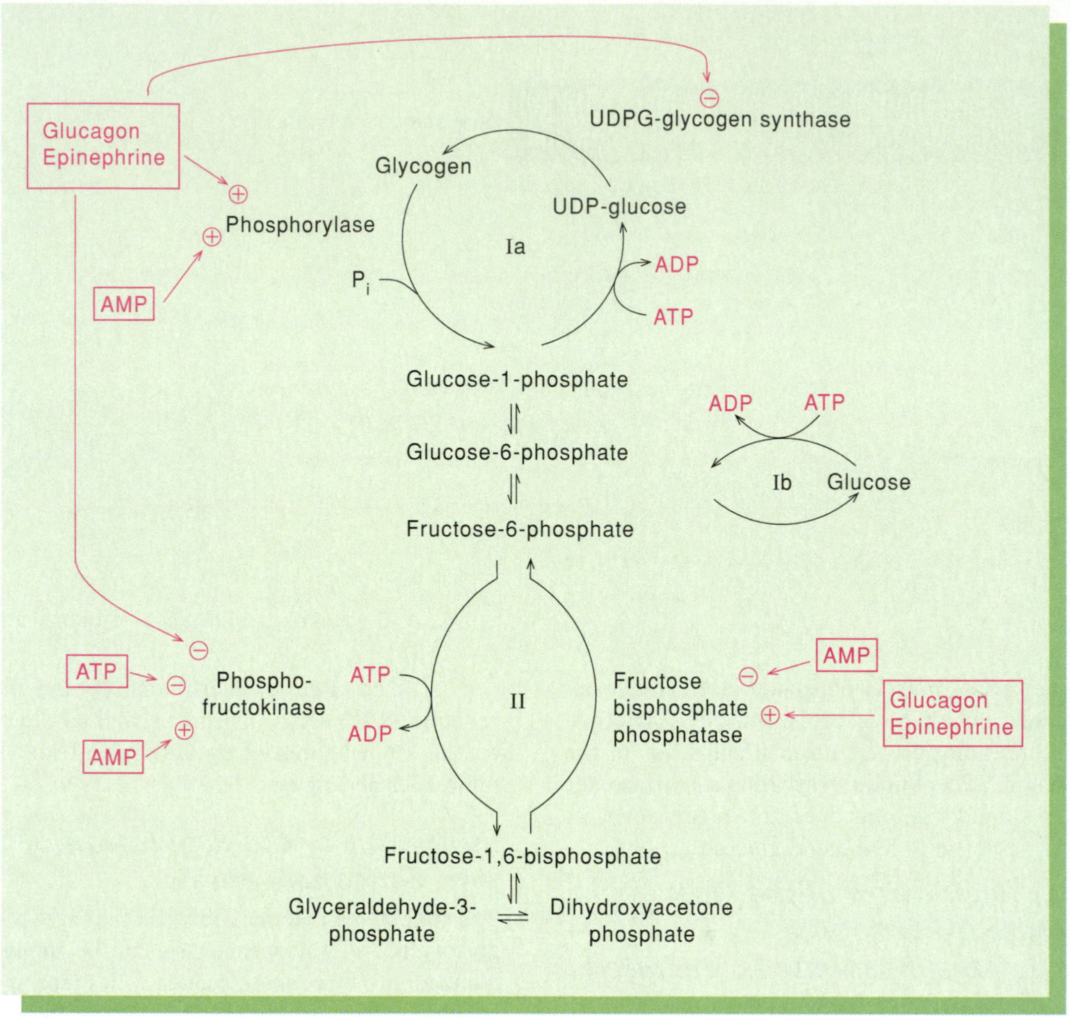

small molecules are a direct reflection of the energy charge of the cells; they include molecules such as ATP, AMP, and citrate. The effects that they have on the activities of the relevant enzymes are indicated in table 13.2 and figure 13.19. The net effect is that the small-molecule effectors that indicate a high energy charge inhibit degradation and stimulate synthesis, whereas the small molecules that indicate a low energy charge stimulate degradation and inhibit synthesis.

In certain cells of very special organs such as the liver a different situation exists with respect to the behavior of these enzymes. This difference is related to the fact that certain cells have a dual responsibility, to nourish themselves and to nourish other tissues. The cells of the liver have the obligation to maintain the blood glucose at close to a constant level as long as that is possible. The regulatory mechanisms that are involved in the organismwide responsibilities of the liver involve hormones that sense the blood glucose level. For example, if the blood glucose level is low, then glucagon is secreted by certain cells in the pancreas only to bind to the surface of certain hormone-responsive liver cells. Upon binding of glucagon, an enzyme on the inner membrane of these cells will begin to synthesize cyclic AMP. This small molecule activates a protein kinase, which in turn activates another protein kinase that phosphorylates a number of enzymes involved in carbohydrate metabolism. The effect of the phosphorylation reactions is to send a new, overriding signal to the enzymes involved in regulating the flux over the glycolysis-gluconeogenesis pathways. Therefore the enzymes respond to a low blood glucose level by taking corrective action. The signal sometimes enhances the effects of normal intracellular signals and sometimes results in opposite effects.

Figure 13.20

Stage 1 of the pentose phosphate pathway: the oxidation of glucose-6-phosphate to ribulose-5-phosphate and CO_2 and the reduction of $NADP^+$. Net Reaction: Glucose-6-phosphate + 2 $NADP^+$ → Ribulose-5-phosphate + CO_2 + 2 NADPH + 2 H^+.

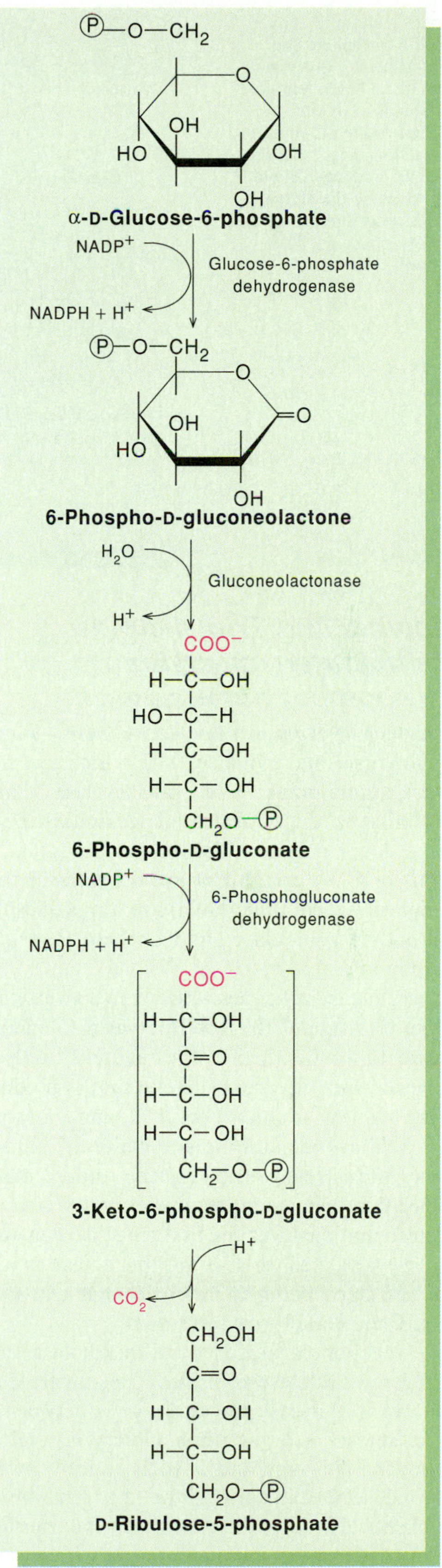

The Pentose Phosphate Pathway

Many kinds of organisms and some mammalian organs, notably liver, contain an alternative pathway for the oxidation of hexoses. Rather than producing a cleavage into trioses, this pentose phosphate pathway consists of oxidation of glucose-6-phosphate to carbon dioxide and a five-carbon sugar phosphate. The carbon atoms of the pentose are then shuffled around in various ways by two enzymes that catalyze transfers of two-carbon and three-carbon pieces between molecules. The final products can contain from three to seven carbon atoms. They are triose phosphate, which can feed into the glycolytic pathway; erythrose-4-phosphate, which is used in the synthesis of aromatic amino acids; ribose-5-phosphate, which is required in the synthesis of nucleic acids (chapter 26); or regenerated hexose phosphate. The major functions of this pathway are generation of NADPH and supply of ribose-5-phosphate.

NADPH Is Generated by Oxidation of Glucose-6-Phosphate

NADPH is required for many biosynthetic sequences. It is generated in different kinds of cells by a variety of reactions, including an $NADP^+$-linked oxidation of malate to pyruvate and CO_2 and transfer of hydride ion from NADH to $NADP^+$ in a mitochondrial reaction that is driven by metabolic energy. However, in some cases, including the mammalian liver, a major part of the NADPH requirement is met by oxidation of glucose-6-phosphate to ribulose-5-phosphate and CO_2. The four electrons that are released by the oxidation are transferred to $NADP^+$.

The reactions in this conversion are shown in figure 13.20. The first oxidation, catalyzed by glucose-6-phosphate dehydrogenase, is at C-1, converting the hemiacetal derivative of the aldehyde group to the lactone of the corresponding acid, 6-phosphogluconic acid. After hydrolysis of the lactone, the second oxidation is at C-3, converting the secondary alcohol to a ketone. The expected product, 3-keto-6-phospho-gluconic acid, is decarboxylated to yield ribulose-5-phosphate, which is the product released by the enzyme.

At this point the metabolic function of the sequence, when it is serving to supply electrons for biosynthesis, has been fulfilled. Two molecules of NADPH have been generated for each molecule of glucose-6-phosphate that was oxidized. It is only necessary to convert ribulose-5-phosphate, the end product of the oxidative sequence, to compounds in the mainstream of metabolism.

Figure 13.21

The transaldolase-catalyzed conversion of fructose-6-phosphate and erythrose-4-phosphate to glyceraldehyde-3-phosphate and sedoheptulose-7-phosphate. This is a two-step conversion. The first step is similar to the aldolase reaction except that the dihydroxyacetone produced is held at the catalytic site, while the aldose product diffuses away and is replaced by another aldose molecule. The second step involves an aldol condensation.

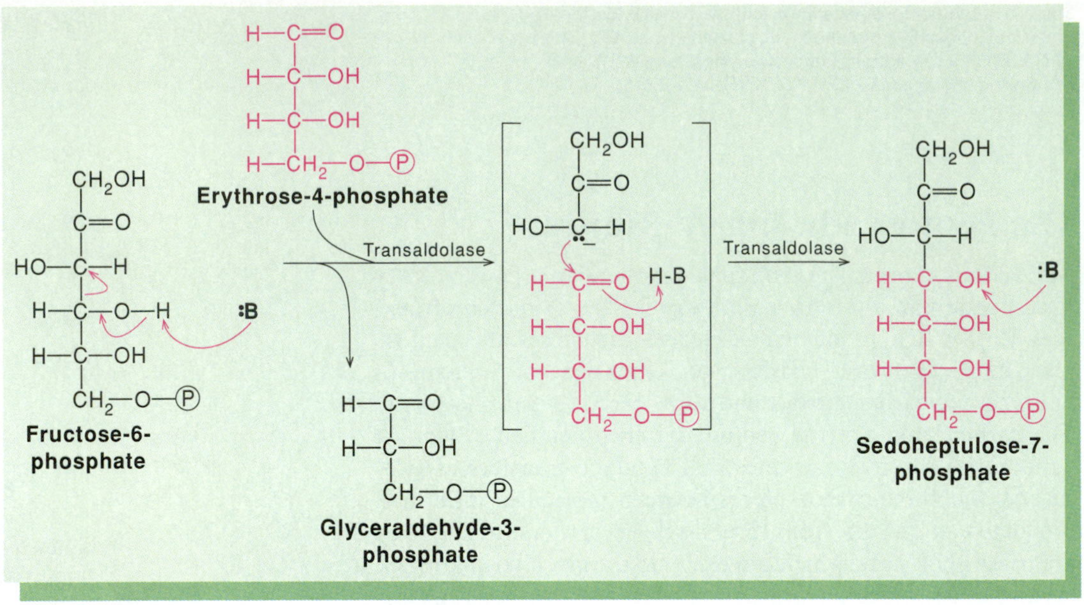

Transaldolase and Transketolase Catalyze the Interconversion of Many Phosphorylated Sugars

The key enzymes involved in these conversions—in addition to phosphofructokinase and aldolase, with which you are already familiar—are transaldolase and transketolase. The two enzymes are similar in their substrate specificities. Both require a ketose as a donor and an aldose as an acceptor. The steric requirements for C-1 through C-4 are the same as the requirements of aldolase, except that aldolase requires phosphorylation at C-1, and both transaldolase and transketolase require a free hydroxyl group at C-1.

Transaldolase catalyzes a two-step conversion. The first step, an aldol cleavage of the bond between C-3 and C-4 of a ketose, is essentially identical to the reaction catalyzed by aldolase. However, the dihydroxyacetone that is produced in the transaldolase reaction from carbons 1, 2, and 3 is not released. Rather, it is held at the catalytic site while the aldose product diffuses away and is replaced by another aldose molecule. An aldol condensation then generates the second product of the reaction, a ketose that contains the first three carbon atoms of the original ketose attached to C-1 of the acceptor aldose (fig. 13.21). The two steps, being an aldol cleavage and an aldol condensation, have no cofactor requirement.

The reaction catalyzed by transketolase superficially resembles the transaldolase reaction. The substrate specificity is identical, but in this case the cleavage is between carbons 2 and 3 of the ketose. A two-carbon moiety is retained on the enzyme following cleavage and is subsequently transferred to an acceptor aldose. But in spite of the apparent similarity, this reaction is chemically quite different from the transaldolase reaction. Aldol condensations and cleavages, which occur readily in mildly alkaline solution, are feasible because protons on a carbon atom adjacent to a carbonyl carbon are moderately acidic. The carbanion formed on dissociation of such a proton can participate in a nucleophilic addition to the carbonyl carbon of another molecule of aldehyde or ketone, as in the transaldolase reaction. A ketol condensation, in contrast, would involve the carbonyl carbon as the nucleophile. That is not energetically feasible because the polarity of the carbonyl bond precludes a negative charge on carbon.

Transketolase is one of several enzymes that catalyze reactions that involve intermediates with a negative charge on what was initially a carbonyl carbon atom. All such enzymes require thiamin pyrophosphate (TPP) as a cofactor (chapter 11). The transketolase reaction is initiated by addition of the thiamin pyrophosphate anion to the carbonyl of a ketose phosphate, for example xylulose-5-phosphate (fig. 13.22). The adduct next undergoes an aldol-like cleavage. Carbons 1 and 2 are retained on the enzyme surface in the form of the glycolaldehyde derivative of TPP. This intermediate can exist as a resonance-stabilized carbanion, and thus can condense with the carbonyl of another aldolase. If the reactants are xylulose-5-phosphate and ribose-5-phosphate, the products will be 3-phosphate-glyceraldehyde and the seven-carbon ketose sedoheptulose-7-phosphate (see fig. 13.23).

Production of Ribose-5-Phosphate and Xylulose-5-Phosphate

Both transaldolase and transketolase require as substrates a ketose phosphate as the donor molecule and an aldose phosphate as the acceptor. Further, both enzymes require at carbons 3 and 4 the same steric configuration as is found in glucose and fructose. Ribulose-5-phosphate, the first pentose phosphate to be formed in the pentose phosphate pathway, does not have the correct configuration to serve as a substrate for transaldolase

Figure 13.22

The transketolase-catalyzed conversion of xylulose-5-phosphate and ribose-5-phosphate to glyceraldehyde-3-phosphate and sedoheptulose-7-phosphate. Although the aldolase and ketolase reactions superficially resemble each other, they proceed by very different mechanisms. This is because in the aldolase reaction the carbon adjacent to a carbonyl acts as a nucleophilic agent, whereas the ketolase reaction involves an intermediate with a negative charge on what was originally a carbonyl carbon. The latter type of reaction is more complex and requires thiamin pyrophosphate. This mechanism starts in a very similar way to the one shown in figure 13.12, but the difference in substrates results in quite different products.

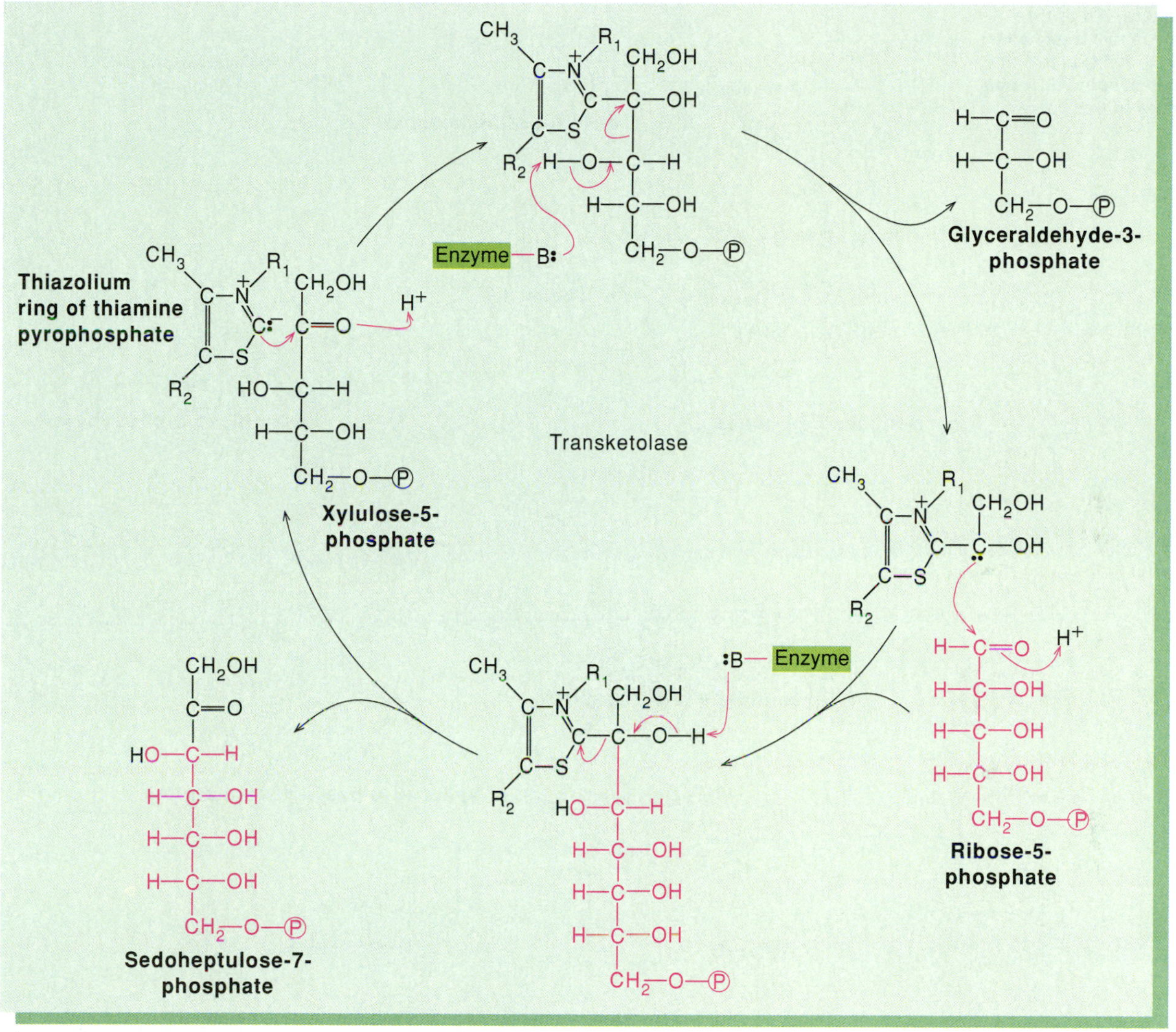

and transketolase. Both a suitable donor ketose and an acceptor aldose can be made by isomerizations of ribulose-5-phosphate, however, and enzymes that catalyze those isomerizations are found in cells that can carry out the pentose phosphate pathway (fig. 13.23).

Conversion of ribulose-5-phosphate to ribose-5-phosphate is a simple ketose-aldose isomerization similar to the interconversion of glucose-6-phosphate and fructose-6-phosphate. Conversion of ribulose-5-phosphate to xylulose-5-phosphate involves epimerization at C-3, and an enediol intermediate seems likely. Once a donor molecule, xylulose-5-phosphate, and an acceptor, ribose-5-phosphate, have been generated, conversion to triose phosphate and hexose phosphate is straightforward. Because of the alternative pathways that are possible, and that probably occur simultaneously, no single set of reactions will describe these conversions uniquely.

One possible set of pathways is shown in figure 13.23. If the triose phosphate formed is converted to hexose phosphate, the overall pathway can be seen as regenerating five molecules of hexose phosphate for each six used initially.

$$\text{6 glucose-6-phosphate} + 12\ NADP^+ \rightarrow 6\ CO_2 + \text{5 fructose-6-phosphate} + 12\ NADPH + 10\ H^+ \quad \textbf{(24)}$$

The pathway was first formulated in this form, hence the name pentose phosphate cycle.

Alternatively, it is possible to write a sequence of reactions, including action of phosphofructokinase and aldolase on seven-carbon intermediates, in which the carbon of ribulose-5-phosphate is converted mainly to glyceraldehyde-3-phosphate. Such a pathway, with the triose phosphate entering the glycolytic sequence, would amount to a bypass or shunt

Figure 13.23

Stage 2 of the pentose phosphate pathway. The groups with color background are those transferred in transketolase-catalyzed reactions. The groups in bold type are transferred in the transaldose-catalyzed reactions. All of the reaction arrows in this figure are double-headed to indicate that the reactions can go in either direction with little change in free energy.

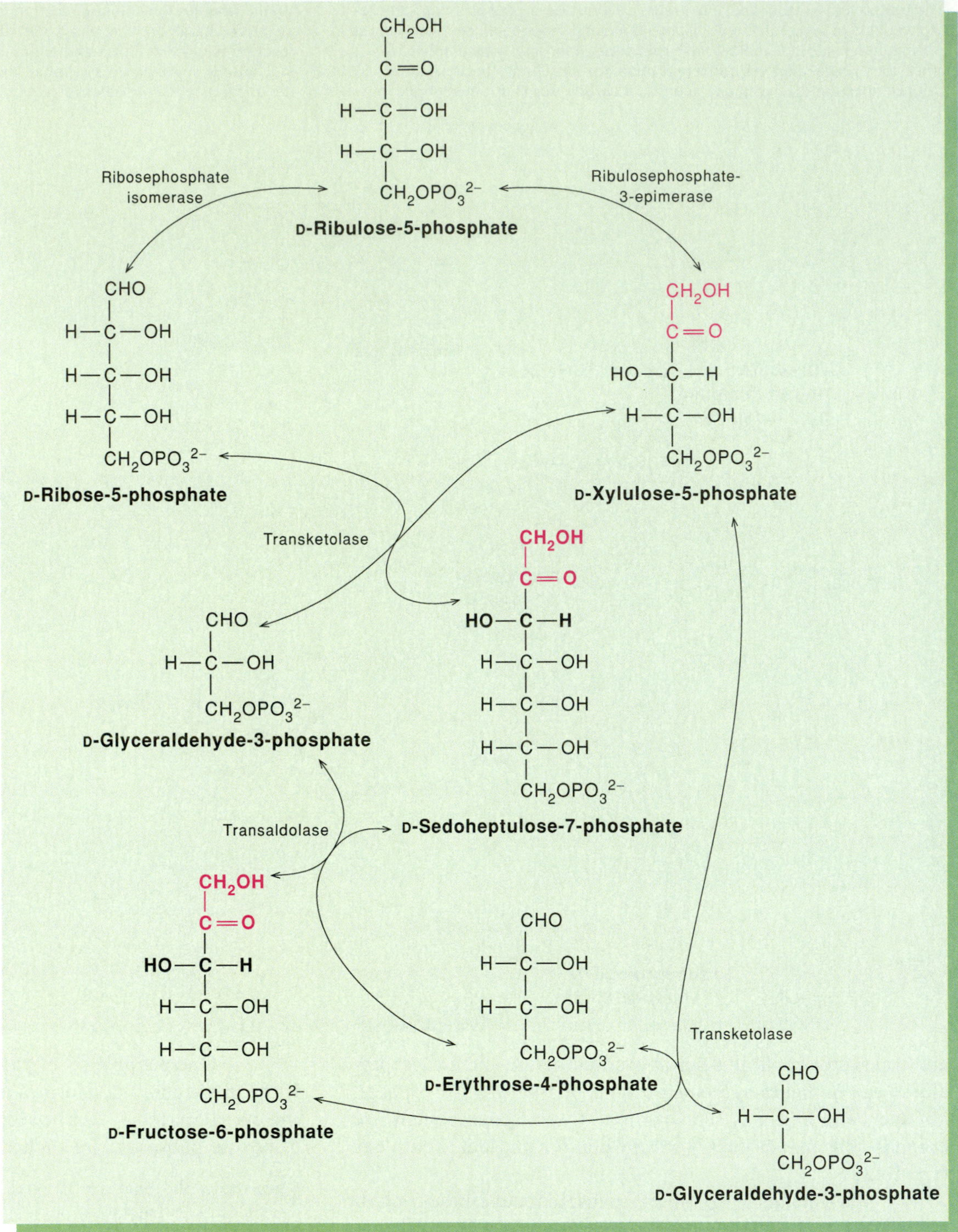

around the first reactions of glycolysis, and the name hexose monophosphate shunt is sometimes used. It will be evident that any amount of ribose-5-phosphate or erythrose-4-phosphate that may be needed for biosynthetic sequences can also be obtained from this oxidative pentose phosphate pathway.

Since the transaldolase and transketolase reactions are symmetrical with respect to types of bonds cleaved and formed, their equilibrium constants are near 1. The pool of sugar phosphates is thus near equilibrium in cells that contain these enzymes. In a water-flow analogy, the sugar phosphates and the reactions that interconvert them resemble a large swamp, with ill-defined flows along many interconnecting channels. Water may be fed in from any direction, and may leave the swamp in any direction.

The swamp analogy is quite realistic for the reactions catalyzed by transaldolase and transketolase and the associated enzymes. The physiological function of these enzymes is to facilitate the conversion of one kind of sugar to another, and depending on the type of cell involved and on metabolic conditions, any of the component sugar phosphates may be added to the pool, and any others removed.

In addition to the regeneration of NADPH and the synthesis of ribose-5-phosphate, the reactions of the pentose phosphate pathway are used also in photosynthesis. For each molecule of hexose phosphate that is produced in photosynthesis there are ten molecules of glyceraldehyde-3-phosphate that must be converted to six molecules of ribulose-5-phosphate, which after phosphorylation serves as the acceptor for CO_2 in continuing photosynthesis (chapter 16). In our water analogy, we may say that in the pentose phosphate pathway in liver, for example, pentose phosphate flows into the swamp at one point and triose phosphate or hexose phosphate flows out at another. In photosynthesis, triose phosphate is dumped into the swamp at one point and pentose phosphate flows out at another. The same enzymes are involved in both cases.

In keeping with their function in regeneration of NADPH, the enzymes of the pentose phosphate pathway are not found in muscle, a tissue in which carbohydrates are utilized almost exclusively for generation of mechanical energy and there is little biosynthetic activity. Liver, in which biosyntheses are important, contains considerable amounts of transaldolase and transketolase, as well as the enzymes that catalyze the oxidation of glucose-6-phosphate to ribulose-5-phosphate and the isomerization of the pentose phosphates.

Regulation of the Pentose Phosphate Pathway

Regulation of the pentose phosphate pathway has not been studied extensively. Both of the oxidative steps are regulated by the NADPH/$NADP^+$ ratio, and stimulation of glucose-6-phosphate dehydrogenase as the energy charge falls slightly has been observed. Clearly, in cells that contain the enzymes of the pentose phosphate pathway, the ratio of rates at which ATP and NADPH are regenerated will be determined largely by partitioning of hexose phosphate between the reactions catalyzed by phosphofructokinase and by glucose-6-phosphate dehydrogenase. Since both glycolysis and the pentose phosphate pathway lead to regeneration of ATP, it seems likely that the most important input into regulation at this branchpoint is the NADPH/$NADP^+$ ratio.

Summary

The reactions that are discussed in this and the next chapter make up the central pathways of metabolism. They are closely related to all aspects of the biochemistry of the cell. They supply the starting materials for biosynthesis of the components of the cell and the reducing power needed for those biosyntheses, and also the starting materials and reducing power for long-term storage of energy as fat.

Quantitatively, the most important role of these pathways lies in the generation of energy to power the activities of the cell or organism, and in the short-term storage of energy as polysaccharides (starch and glycogen). The parts of the pathway that do not require oxygen and that are found in nearly all cells, aerobic or anaerobic, are discussed in this chapter. Our discussion emphasizes the energetic aspects of glycolysis (the conversion of hexoses or storage polysaccharides to three-carbon compounds) and of gluconeogenesis (the conversion of three-carbon compounds to hexoses or storage polysaccharides). The following points are most important.

1. The intermediates of glycolysis and gluconeogenesis are phosphorylated. The breakdown of storage polysaccharide yields glucose-1-phosphate, and the first step in the metabolism of free glucose is phosphorylation to glucose-6-phosphate. Those two compounds are freely interconvertible with fructose-6-phosphate.
2. The three sugar phosphates just named make up the hexose phosphate pool, which is a major crossroads of energy metabolism. Glucose-1-phosphate is produced from storage polysaccharides and is used in their synthesis; glucose-6-phosphate is produced from glucose and is the precursor for blood glucose in mammals and the starting material for the pentose phosphate pathway; fructose-6-phosphate is produced by gluconeogenesis and photosynthesis and is the starting material for glycolysis.
3. Glycolysis and gluconeogenesis depend on pools of intermediates that are close to equilibrium. Most of the reactions that connect the pools are far from equilibrium in the metabolizing cell. The pools are the hexose phosphate pool (glucose-1-phosphate), the aldolase pool (fructose-1,6-phosphate, glyceraldehyde-3-phosphate, and dihydroxyacetone phosphate) and the three-carbon pool (glycerate-3-phosphate, glycerate-2-phosphate, and phosphoenolpyruvate).
4. Within each pool of intermediates, the direction of reaction is determined by simple mass-action considerations. The direction of overall conversion (glycolysis or gluconeogenesis) depends on the rates at which the connecting reactions remove materials from some pools and supply materials to others.
5. The connecting reactions (except for those catalyzed by glyceraldehyde-3-phosphate dehydrogenase and glycerate-3-phosphate kinase, which connect the aldolase pool and the three-carbon pool) have large free energies of reaction and are far from equilibrium in the cell. They are subject to regulatory signals that indicate the metabolic needs of the cell or organism. As a result of those regulatory interactions, polysaccharide is stored when nutrients are available in excess of current needs and storage compounds are broken down to supply energy when necessary.

6. In mammals, the liver is a special case. It is a major site of storage of glycogen, and it regulates the level of glucose in the blood. When the blood glucose is low, the liver replenishes it by hydrolysis of glucose-6-phosphate, which is made simultaneously by breakdown of glycogen and by gluconeogenesis. When the liver's supply of glycogen is depleted, blood glucose is replenished by gluconeogenesis, and body proteins are broken down to obtain amino acids as starting materials.
7. The pentose phosphate pathway serves a variety of functions: (a) the production of NADPH for biosynthesis; (b) the production of ribose, required mainly for nucleic acid synthesis, and (c) the interconversion of a variety of phosphorylated sugars.

Selected Readings

Atkinson, D. E., *Cellular Energy Metabolism and Its Regulation.* New York: Academic Press, 1977. A general discussion, covering some topics in this and the preceding and following chapters in somewhat greater depth than the treatment in this book.

Beitner, R., *Regulation of Carbohydrate Metabolism.* Boca Raton, Fla.: CRC Press, 1985. Two-volume discussion, at a rather advanced level, of many aspects of the regulation of mammalian carbohydrate metabolism with frequent discussions of clinical conditions. Many references.

Dawes, E. A., *Microbial Energetics.* Glasgow and London: Blackie [New York: Chapman and Hall], 1986. Discussion of bacterial metabolism, with emphasis on the central pathways and the generation and use of ATP. A good source for learning something of the diversity of metabolic adaptations among different groups of bacteria.

Hers, H. G., and L. Hue, Gluconeogenesis and related aspects of glycolysis. *Ann. Rev. Biochem.* 52:617–653, 1983.

Hochachka, P. W., and G. N. Somera, *Biochemical Adaptation.* Princeton: Princeton University Press, 1984. An excellent and extensive discussion of how biochemical processes, including many discussed in this book, are adapted by various types of organisms in fitting themselves for survival under specific and often difficult conditions.

Hoffman, E., Phosphofructokinase—a favorite of enzymologists and students of metabolic regulation. *Trends Biochem. Sci.* 3:145–147, 1978.

Katz, J., and R. Rognstad, Futile cycling in glucose metabolism. *Trends Biochem. Sci.* 3:171–174, 1978.

Pilkis, S. J., M. R. El-Maghrabi, and T. H. Claus, Hormonal regulation of hepatic gluconeogenesis and glycolysis. *Ann. Rev. Biochem.* 57:755–784, 1988.

Roehrig, K. L., *Carbohydrate Biochemistry and Metabolism.* Westport, Ct.: Avi Publishing Co., 1984. Carbohydrate metabolism discussed at a rather elementary level. Covers several topics that are not included in this book, including disorders of carbohydrate metabolism with brief discussions of many types of human genetic diseases in which carbohydrate metabolism is impaired.

Srivastava, D. K., and S. A. Bernhard, Metabolite transfer via enzyme-enzyme complexes. *Science* 234:1,081–1,087, 1986. Unusual mode of transfer observed for intermediates in the glycolytic pathway.

Van Schaftingen, E., Fructose-2,6-bisphosphate. *Adv. Enzymol.* 59:315–395, 1987. An allosteric effector that regulates the flow between fructose-6-phosphate and fructose-1,6-bisphosphate.

Problems

1. Suppose that you have isolated a facultative microorganism that you are growing anaerobically in a medium containing a carbohydrate. Explain, on the basis of your knowledge of metabolism, why each of the following statements about the fermentation is false.
   (a) The culture must be growing on glucose because bacteria ferment few other compounds.
   (b) The products of the fermentation must be more highly oxidized than the substrates, otherwise no energy is conserved.
   (c) The culture cannot be producing any CO_2.
   (d) Addition of fluoride to the anaerobic culture will not alter the ratio of 2-phosphoglycerate to phosphoenolpyruvate.
2. A yeast culture is fermenting glucose to ethanol. In order to ensure that the CO_2 released during fermentation is radio-labeled, what carbon(s) of glucose must be labeled with ^{14}C?
3. Assume that a mutant form of glyceraldehyde-3-phosphate dehydrogenase was found to hydrolyze the oxidized enzyme-bound intermediate with water rather than phosphate.
   (a) Write a chemical reaction that describes the hydrolysis, showing the structures of the products.
   (b) What would be the effect, if any, on the ATP yield from glycolysis of glucose to lactate?
   (c) What would be the effect of the mutation on an obligately aerobic microorganism?
4. Suppose that you are seeking bacterial mutants with altered triose phosphate isomerase (TPI). The organism of interest is known to use the glycolytic pathway with the production of lactate.
   (a) Explain why the absence of TPI would be lethal to an organism fermenting glucose exclusively through the glycolytic pathway.
   (b) Suppose that you have an organism that uses glycolysis and an oxidative pathway as energy sources. Mutants of that organism having only 10% of the TPI activity present in the wild-type cells grew slowly on glucose under anaerobiosis but grew faster aerobically. Explain the metabolic basis of the observation.
   (c) You have constructed a plasmid that would direct the synthesis of dihydroxyacetone phosphate phosphatase and have introduced the plasmid into the mutant organism described in part (b). Predict whether the plasmid-bearing organism with an active DHAP

phosphatase could grow on glucose or glycerol either anaerobically or aerobically. What are the metabolic considerations you used to make your predictions?

5. There are two sites of ADP phosphorylation in glycolysis. These processes are called substrate level phosphorylations. Arsenate (AsO_4^{3-}), an analog of phosphate, uncouples ATP formation resulting from glyceraldehyde-3-phosphate oxidation but not that resulting from dehydration of glycerate-2-phosphate. Explain.
6. The disaccharide sucrose can be cleaved by either of two methods:

$$\text{Sucrose} + H_2O \xrightarrow{\text{Invertase}} \text{glucose} + \text{fructose}$$

or

$$\text{Sucrose} + P \xrightarrow{\text{Sucrose phosphorylase}} \text{glucose-1-phosphate} + \text{fructose}$$

   (a) Given that the $\Delta G°'$ value for the invertase reaction is -7.0 kcal/mole, calculate the $\Delta G°'$ value for sucrose phosphorylase. Assume that the $G°'$ for hydrolysis of glucose-1-phosphate is -5.0 kcal/mole. From the calculated value of $\Delta G°'$, calculate the equilibrium constant for sucrose phosphorylase at 25° C.
   (b) Explain the metabolic advantage to the cell of cleaving sucrose with phosphorylase rather than with invertase.
7. What concentration of glucose would be in equilibrium with 1.0-mM glucose-6-phosphate, assuming that hexokinase is present and the concentration ratio of ATP to ADP is 5? Is it reasonable to expect an actively metabolizing cell to maintain the concentration of glucose necessary to sustain the concentration of glucose-6-phosphate at 1mM?
8. 2-Phosphoglycerate and phosphoenolpyruvate differ only by dehydration between C-2 and C-3, yet the difference in $\Delta G°'$ of hydrolysis is about -12 kcal/mole. How does dehydration "trap" so much chemical energy?
9. Explain the metabolic role of alcohol dehydrogenase in yeasts that are (a) fermenting glucose to ethanol and (b) aerobically oxidizing ethanol.
10. Elevated pyruvate concentration inhibits the heart muscle lactate dehydrogenase (LDH) isoenzyme but not the skeletal muscle LDH isoenzyme.
    (a) What would be the consequence of having only the heart isoenzyme form in skeletal muscle?
    (b) Would there necessarily be a negative consequence if the skeletal muscle isoenzyme were the only LDH in heart muscle?
11. Beginning with pyruvate, show which reactions of gluconeogenesis introduce the four chiral centers into glucose.
12. We have stated that the equilibrium constant for the pyruvate kinase reaction is 10^6. Assume that the steady-state concentration of ATP is 2 mM and of ADP is 0.2 mM. Calculate the concentration ratio of pyruvate to PEP under these conditions. Does your calculation support or refute the assertion that pyruvate kinase reaction is metabolically irreversible?
13. In gluconeogenesis, the thermodynamic barrier imposed by pyruvate kinase is overcome by coupling two separate reactions for the synthesis of PEP from pyruvate.
    (a) Write the two chemical reactions used to bypass the pyruvate kinase reaction.
    (b) Calculate the overall $\Delta G°'$ of the two reactions you wrote in part (a). (Assume that GTP is the thermodynamic equivalent of ATP.) What can you now surmise about the feasibility of PEP formation from pyruvate by this route?
14. Write a chemical reaction for the $NADP^+$-dependent oxidation of 6-phosphogluconate to ribulose-5-phosphate.
15. Fructose-2,6-bisphosphate is a potent activator of the liver phosphofructokinase (PFK-1) and a potent inhibitor of liver fructose-1,6-bisphosphate phosphatase (FBPase-1). Fructose-2,6-bisphosphate is the product of a second phosphofructokinase (PFK-2) and is hydrolyzed to fructose-6-phosphate by FBPase-2. The activities of PFK-2 and FBPase-2 reside on a single, bifunctional protein in liver. The bifunctional protein is under glucagon control imposed via cAMP. (Reference: H.-G. Hers, *Arch. Biol. Med. Exp.* 18:243–251, 1985.)
    (a) Under what metabolic conditions would PFK-2 be active? FBPase-2?
    (b) Gluconeogenesis in liver is stimulated by the hormone glucagon. The activity of PFK-2/FBPase-2 bifunctional enzyme under glucagon regulation shifts from an active PFK-2 to an inactive PFK-2. Inactivation of the PFK-2 alone would still not be adequate to stimulate gluconeogenesis sufficiently for the organism. Explain.
    (c) cAMP-dependent phosphorylation of PFK-2/FBPase-2 not only inhibits PFK-2 but stimulates FBPase-2. Under these conditions, gluconeogenesis is sufficiently rapid to meet cellular demand. Explain.
    (d) What would you predict as the relative activities of the following enzymes in the liver of a rat made diabetic through chemical means (administration of alloxan or streptozotocin)? PFK-2, FBPase-2, PFK-1, FBPase-1, pyruvate carboxylase, PEP carboxykinase.
    (e) If the diabetic animal were treated with insulin, what changes would you predict in the activities of the liver enzymes cited in part (d)? In the concentration of cAMP?
16. Muscle pyruvate kinase (PK) responds hyperbolically to its substrate, PEP, but the liver form of the enzyme responds sigmoidally. Fructose-1,6-bisphosphate is an allosteric activator of liver pyruvate kinase, but it apparently has no effect on the muscle enzyme.
    (a) If liver PK responded hyperbolically to PEP and were otherwise unregulated, how might gluconeogenesis be affected?
    (b) What is the metabolic advantage of having the liver PK activated by fructose-1,6-bisphosphate?

14 CHAPTER

The Tricarboxylic Acid Cycle

In chapter 13 we saw that only a small fraction of the total free energy content of glucose or any other oxidizable substrate is released under anaerobic conditions. The reason is the limited extent to which the oxidation of an organic substrate can occur in the absence of oxygen. Essentially, catabolism under anaerobic conditions means that every oxidative event, in which electrons are removed from an organic compound, must be accompanied by a reductive event, in which electrons are returned to another organic compound, often closely related to the first compound. Since the difference in free energy in such coupled oxidation-reduction reactions is small, relatively little energy can be obtained from the starting substrate under these conditions. Thus a cell carrying out fermentation of glucose to lactate transfers electrons from glyceraldehyde-3-phosphate by means of NADH to pyruvate and has access to only about 7% of the total free energy content of the glucose molecule. The cell must accordingly content itself with the generation of only two ATP molecules per molecule of glucose fermented. Most of the energy of the glucose molecule remains untapped in the lactate molecule.

Given access to oxygen, however, the cell can do much more with the oxidizable organic molecules available to it, and the energy yield increases dramatically. We are now ready to explore the lower half of the central metabolic pathways requiring oxygen (fig. 14.1). With oxygen available as the electron acceptor, the carbon atoms of glucose (or another substrate) can be oxidized fully to CO_2, and all the electrons that are removed during the multiple oxidation events are transferred ultimately to oxygen. In the process, the ATP yield per glucose is almost twenty times greater than that possible under anaerobic conditions. Therein lies the advantage of the aerobic way of life.

It is not surprising, then, that aerobic processes capable of extracting further energy from pyruvate have come to play so prominent a role in energy metabolism. The overall process

Figure 14.1

Outline of the main reactions involved in carbohydrate metabolism that will be considered in this chapter. Most of these reactions belong to the tricarboxylic acid cycle, which provides a means for catabolizing three carbon units all the way to CO_2. This process, which requires oxygen, yields considerably more energy and reducing power than simple glycolysis. The reactions shown in black are taken from the master diagram of figure 12.5. Additional reactions, not shown in figure 12.5, are drawn in color. Two of these reactions are part of the glyoxylate bypass, a means of synthesizing four-carbon and six-carbon units from the two-carbon level of acetyl-CoA. The reactions of the glyoxylate bypass are found in plants, fungi, and microorganisms but not in vertebrates. Citrate formed in the mitochondrion can be transferred to the cytosol, where it leads to the formation of acetyl-CoA and NADPH (not shown) used in biosynthesis.

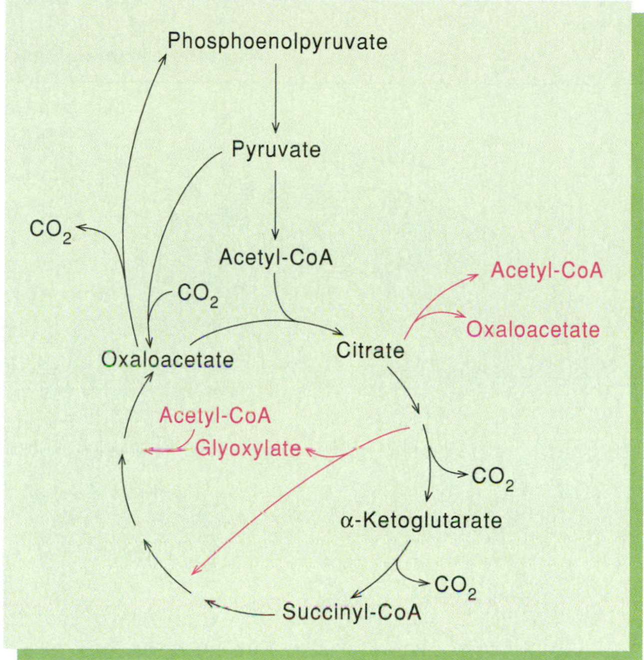

is aerobic respiratory metabolism, and its distinguishing characteristics are (1) the use of oxygen as the ultimate electron acceptor, (2) the complete oxidation of organic substrates to CO_2 and water, and (3) the conservation of much of the free energy as ATP.

The oxygen required for aerobic metabolism actually serves as the terminal electron acceptor only, providing for the continuous reoxidation of reduced coenzyme molecules (most prominent of which are NADH and $FADH_2$). It is these coenzyme molecules (in their oxidized form) that actually carry out the stepwise oxidation of organic intermediates derived from pyruvate. Aerobic respiratory metabolism can therefore be thought of in terms of two separate but intimately linked processes: the actual oxidative metabolism, in which electrons are removed from organic substrates and transferred to coenzyme carriers, and the concomitant reoxidation of the reduced coenzymes by transfer of electrons to oxygen, accompanied indirectly by the generation of ATP.

Under aerobic conditions, the glycolytic pathway becomes the initial phase of glucose catabolism. As shown in figure 14.2, the other three components of respiratory metabolism are the tricarboxylic acid (TCA) cycle, which is responsible for further oxidation of pyruvate, the electron-transport chain, which is required for the reoxidation of coenzyme molecules at the expense of molecular oxygen, and the oxidative phosphorylation of ADP to ATP, which is driven by a "proton gradient" generated in the process of electron transport. Overall, this leads to the potential formation of 36 to 38 molecules of ATP per molecule of glucose in the typical eukaryotic cell.

In this chapter, discussion will focus on the TCA cycle and its central role in the aerobic catabolism of carbohydrates. The following chapter will then explain how the free energy present in the reduced coenzymes that are generated by glycolysis and the TCA cycle is conserved as ATP during the companion processes of electron transport and oxidative phosphorylation.

Discovery of the TCA Cycle

The discovery of the TCA cycle began with a series of biochemical experiments performed in the early 1900s on anaerobic suspensions of minced animal tissues. The experiments established that the suspensions contained enzymes that could transfer hydrogen atoms from various low-molecular-weight organic acids to other reducible compounds, such as the dye methylene blue. (Methylene blue was a convenient indicator in these experiments because it is converted from a blue to a colorless form by reduction.) Only a few organic acids were active in the reduction: succinate, fumarate, malate, and citrate. It was later observed that in the presence of oxygen the same suspensions oxidized these acids to CO_2 and water.

Albert Szent-Gyorgyi found that when small amounts of these organic acids were added to a tissue suspension, considerably more oxygen was consumed than was required to oxidize the added acid. He concluded that in the presence of oxygen the organic acids had a catalytic effect on the oxidation of glucose or other carbohydrates in the tissue slices. Szent-Gyorgyi also observed that a specific inhibitor of succinate dehydrogenase, malonate, blocked the utilization of oxygen (respiration) by muscle suspensions. This finding suggested that oxidation of succinate is an indispensable reaction in muscle oxidative metabolism. When the same response was found in many other tissues, the phenomenon seemed to have fundamental and perhaps universal significance.

Next, Hans Krebs studied the interrelationships between the oxidative metabolism of different organic acids. For his experiments he used slices of pigeon flight muscle, which are particularly active in oxidative metabolism. He found a select group of organic acids that was oxidized very rapidly by extracts from the muscle. His list included the acids that Szent-Gyorgyi had found plus oxaloacetate, α-ketoglutarate, isocitrate, and *cis*-aconitate. Like Szent-Gyorgyi, Krebs also found that catalytic amounts of the same organic acids stimulated the oxidation of pyruvate or carbohydrate, and that their catalytic

Figure 14.2

The components of respiratory metabolism include glycolysis, the tricarboxylic acid (TCA) cycle, the electron-transport chain, and the oxidative phosphorylation of ADP to ATP. Glycolysis converts glucose to pyruvate; the TCA cycle fully oxidizes the pyruvate (by means of acetyl-coenzyme A) to CO_2 by transferring electrons stepwise to coenzymes; the electron-transport chain reoxidizes the coenzymes at the expense of molecular oxygen; and the energy of coenzyme oxidation is conserved in the form of a transmembrane proton gradient that is then used to generate ATP.

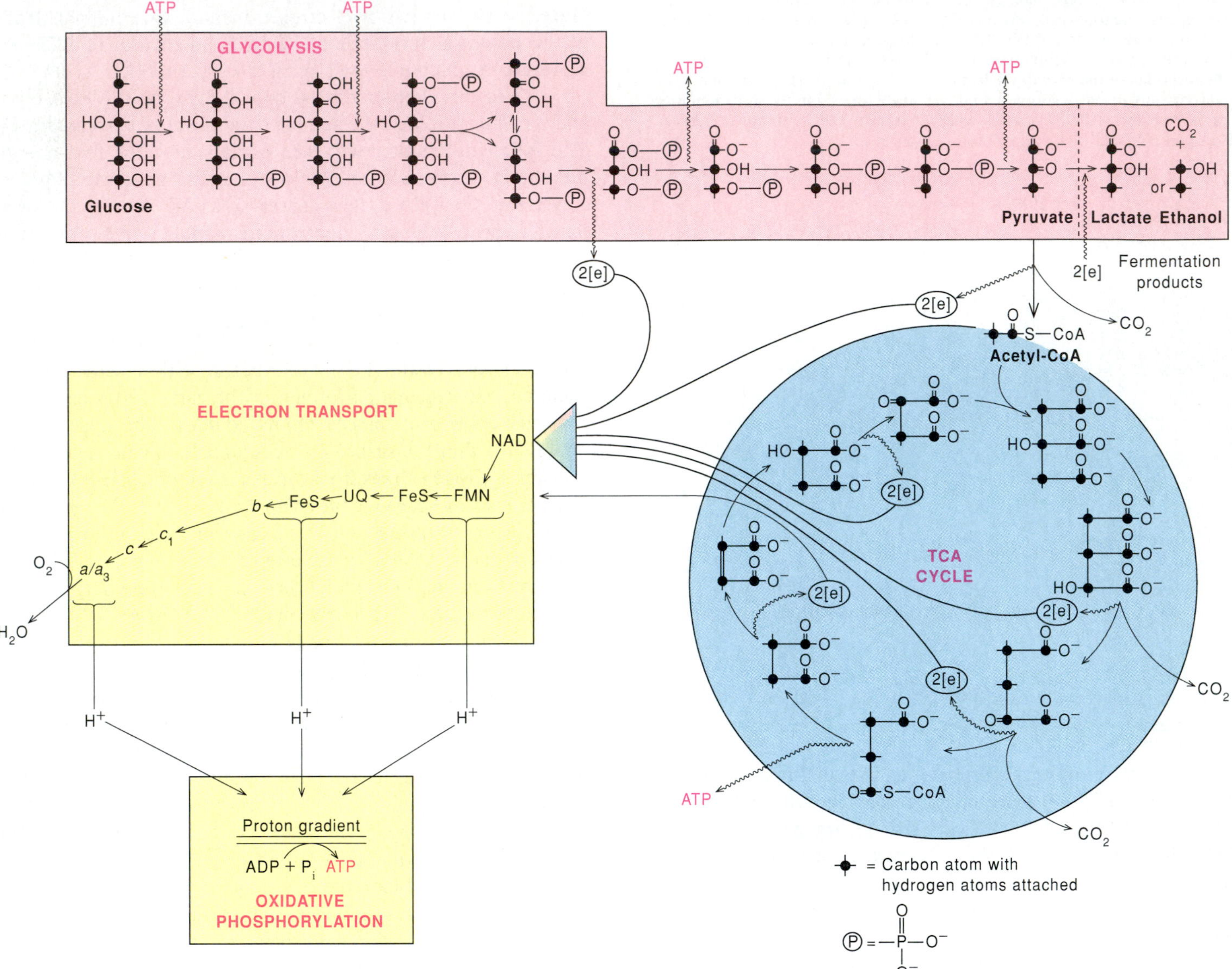

effect was blocked by malonate. Since malonate specifically inhibited the conversion of succinate to fumarate by succinic acid dehydrogenase, he concluded that this reaction was one of a series of steps essential for the complete oxidization of pyruvate or carbohydrate. Additional steps in the process were presumed to involve the other organic acids that had been found to catalyze the process. The organic acids with catalytic potential could be arranged in a chain related by biochemical conversions:

Citrate → *cis*-aconitate → isocitrate → α-ketoglutarate → succinate → fumarate → malate → oxaloacetate

But if this chain of reactions were to act catalytically, then there must be some way of regenerating all of the intermediates of the chain from a single intermediate. A cyclical process would be the simplest way of doing this. For the chain to operate as a cycle, one or more additional reactions needed to be discovered. Krebs found that in the absence of oxygen, small amounts of citrate could be formed from added oxaloacetate and pyruvate. Could this reaction be the missing link in the cyclical process? Krebs thought so, and proposed that the first step in pyruvate oxidation involved its condensation with

Figure 14.3

Original tricarboxylic acid (TCA) cycle proposed by Krebs. This cycle is also called the citrate cycle or the Krebs cycle. To start the cycle in operation, pyruvate loses one of its carbons and condenses with a four-carbon dicarboxylic acid, oxaloacetic acid, to form a six-carbon tricarboxylic acid, citrate. In one turning of the cycle, two carbons are lost as CO_2, thus returning the citrate to oxaloacetate. The conversion blocked by malonate is indicated by a colored bar.

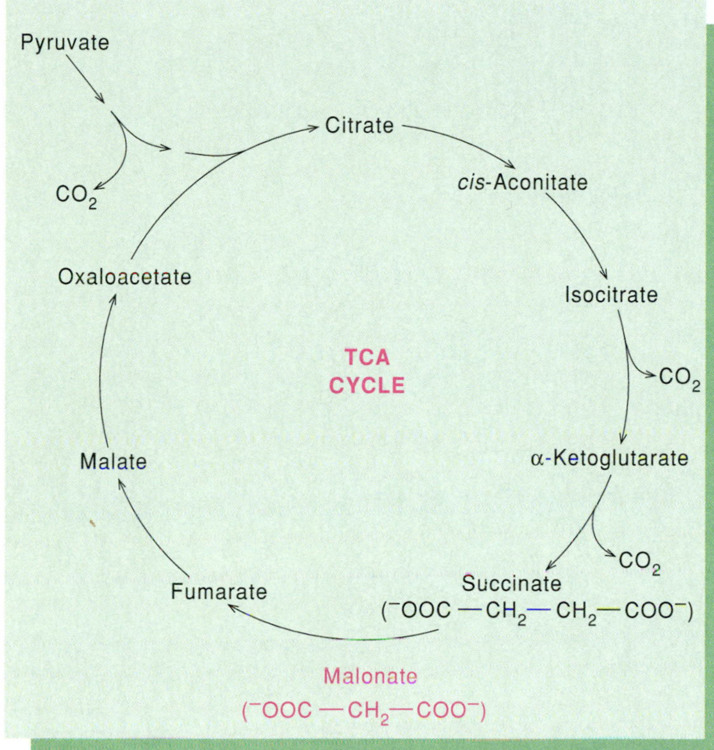

oxaloacetate to form citrate and CO_2. The oxidation of more complex carbohydrates then could be explained by their prior conversion to pyruvate through the glycolytic pathway.

Further support for the cyclical nature of the chain came from observations of malonate-inhibited muscle slices. In such preparations the addition of any of the intermediates of the chain led to accumulation of succinate. Even the addition of fumarate, the immediate product of succinate oxidation, led to accumulation of succinate.

When the oxidation of substantial amounts of pyruvate was blocked by malonate, stoichiometric amounts of pyruvate would react if either oxaloacetate or its precursors malate or fumarate was added. None of the precursors to succinate in the chain was effective in this regard. This strongly supported Krebs's hypothesis that oxidation of pyruvate involved its condensation with oxaloacetate and the subsequent series of conversions in the chain described by Krebs. The Krebs cycle in its originally proposed form is shown in figure 14.3. It is also known as the tricarboxylic acid (TCA) cycle or the citrate cycle.

Steps in the TCA Cycle

The tricarboxylic acid cycle (fig. 14.4) begins with acetyl-coenzyme A, which is obtained either by oxidative decarboxylation of pyruvate available from glycolysis or by oxidative cleavage of fatty acids, as described in chapter 17. Regardless of source, acetyl-CoA transfers its acetyl group to a four-carbon acceptor (oxaloacetate), thereby generating citrate, the six-carbon compound for which the cycle was originally named. In a cyclic series of reactions, the citrate is subjected to two successive decarboxylations and several oxidative events, leaving a four-carbon compound from which the starting oxaloacetate is eventually regenerated. Each turn of the cycle involves the entry of two carbons from acetyl-CoA and the release of two carbons as CO_2, thus providing for regeneration of the oxaloacetate with which the cycle began. As a result, the cycle is balanced with respect to carbon flow and functions without net consumption or buildup of oxaloacetate (or any other intermediate), unless side reactions occur that either feed carbon into the cycle or drain it off into alternative pathways.

The Oxidative Decarboxylation of Pyruvate Leads to Acetyl-CoA

Although the acetyl-CoA with which the cycle begins may be derived catabolically from fatty acids or amino acids, the major source in most cells is the pyruvate available from the glycolytic breakdown of carbohydrate. The gap from the glycolytic pathway to the TCA cycle is bridged by the oxidative decarboxylation of pyruvate (with which glycolysis ends) to yield acetate in the form of acetyl-CoA (with which the TCA cycle commences). The decarboxylation of pyruvate is achieved by a complicated sequence of events that is catalyzed by a cluster of three enzymes called the pyruvate dehydrogenase complex. In eukaryotic cells, this enzyme complex is located in the mitochondria, as are all the other reactions of aerobic energy metabolism beyond pyruvate. Since the glycolytic pathway occurs in the cytosol, it is as pyruvate that the carbon derived from glucose (or other carbohydrate substrates) enters the mitochondria.

Decarboxylation of an α-keto acid like pyruvate is a difficult reaction for the same reason as are the ketol condensations or cleavages that were discussed in the preceding chapter; both kinds of reactions require the participation of an intermediate in which the carbonyl carbon carries a negative charge. In all such reactions that occur in metabolism, the intermediate is stabilized by prior condensation of the carbonyl group with thiamin pyrophosphate. In figure 14.5 thiamin pyrophosphate and its hydroxyethyl derivative are written in the doubly ionized form (see chapter 11) because those are the forms that actually participate in the reactions. The un-ionized forms will greatly predominate, however, and the reactive carbanions are shown only for clarity.

Figure 14.4

The TCA cycle. In each turn of the cycle, acetyl-CoA from the glycolytic pathway or from ß oxidation of fatty acids enters and two fully oxidized carbon atoms leave (as CO_2). ATP is generated at one point in the cycle and coenzyme molecules are reduced. The two CO_2 molecules lost in each cycle originate from the oxaloacetate of the previous cycle rather than from incoming acetyl from acetyl-CoA. This point is emphasized by the use of color. Since the two ends of the succinate molecule are identical, each carbon atom of succinate, fumarate, malate, and oxaloacetate has a 50% chance of being derived from the last acetyl group to enter the cycle.

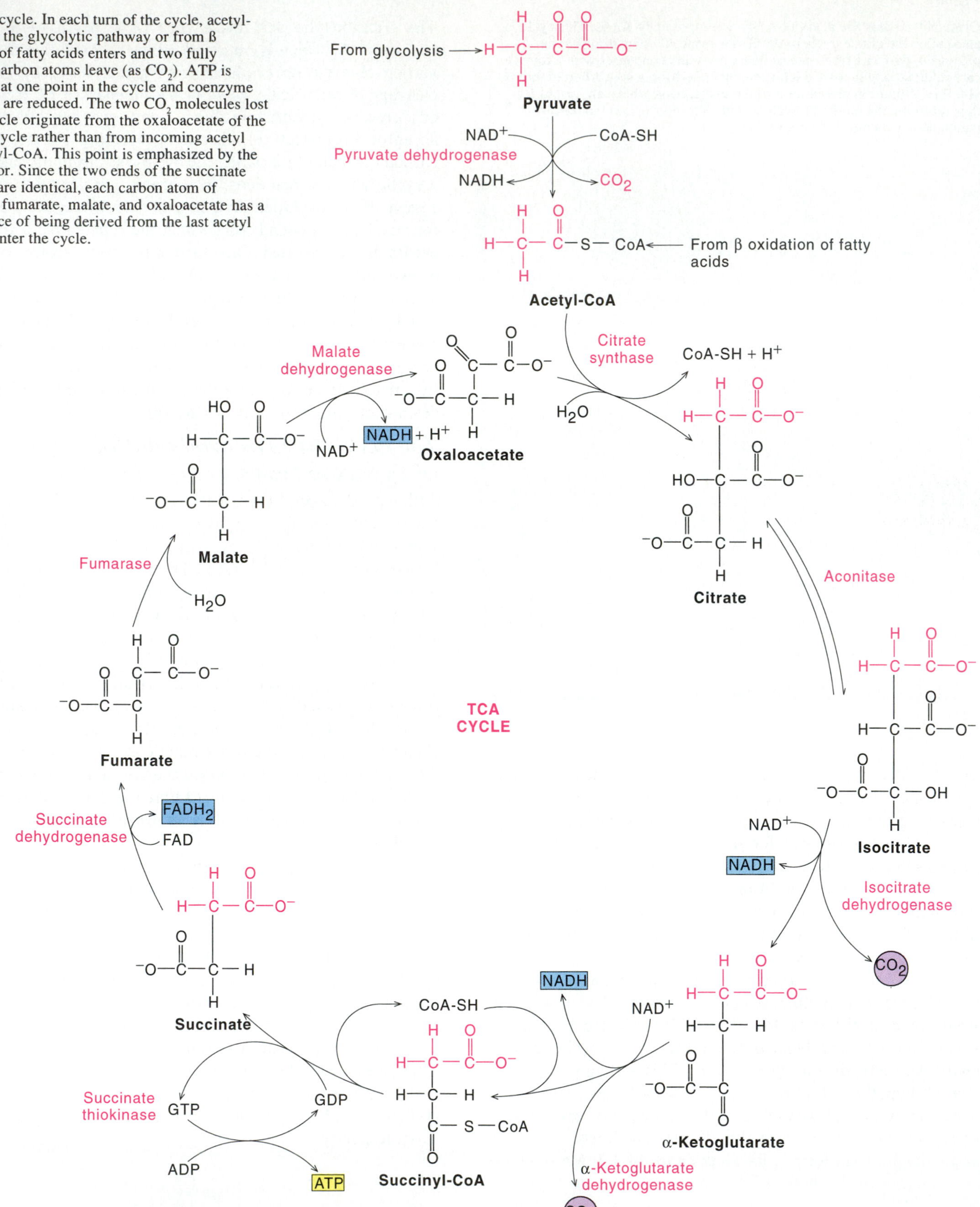

Figure 14.5

The conversion of pyruvate to acetyl-CoA. The reactions are catalyzed by the enzymes of the pyruvate dehydrogenase complex. This complex has three enzymes: pyruvate decarboxylase, dihydrolipoyl transacetylase, and dihydrolipoyl dehydrogenase. In addition, three coenzymes are required: thiamin pyrophosphate, lipoic acid, and NAD^+. Lipoic acid is covalently attached to the transacetylase component of the complex by an amide bond between the carboxyl group of lipoic acid and the terminal amino group of a lysine residue of the enzyme. In fact one or more lipoic acids are involved in each transfer reaction for reasons indicated in chapter 11 (see fig. 11.14). Reactants are products shown in color.

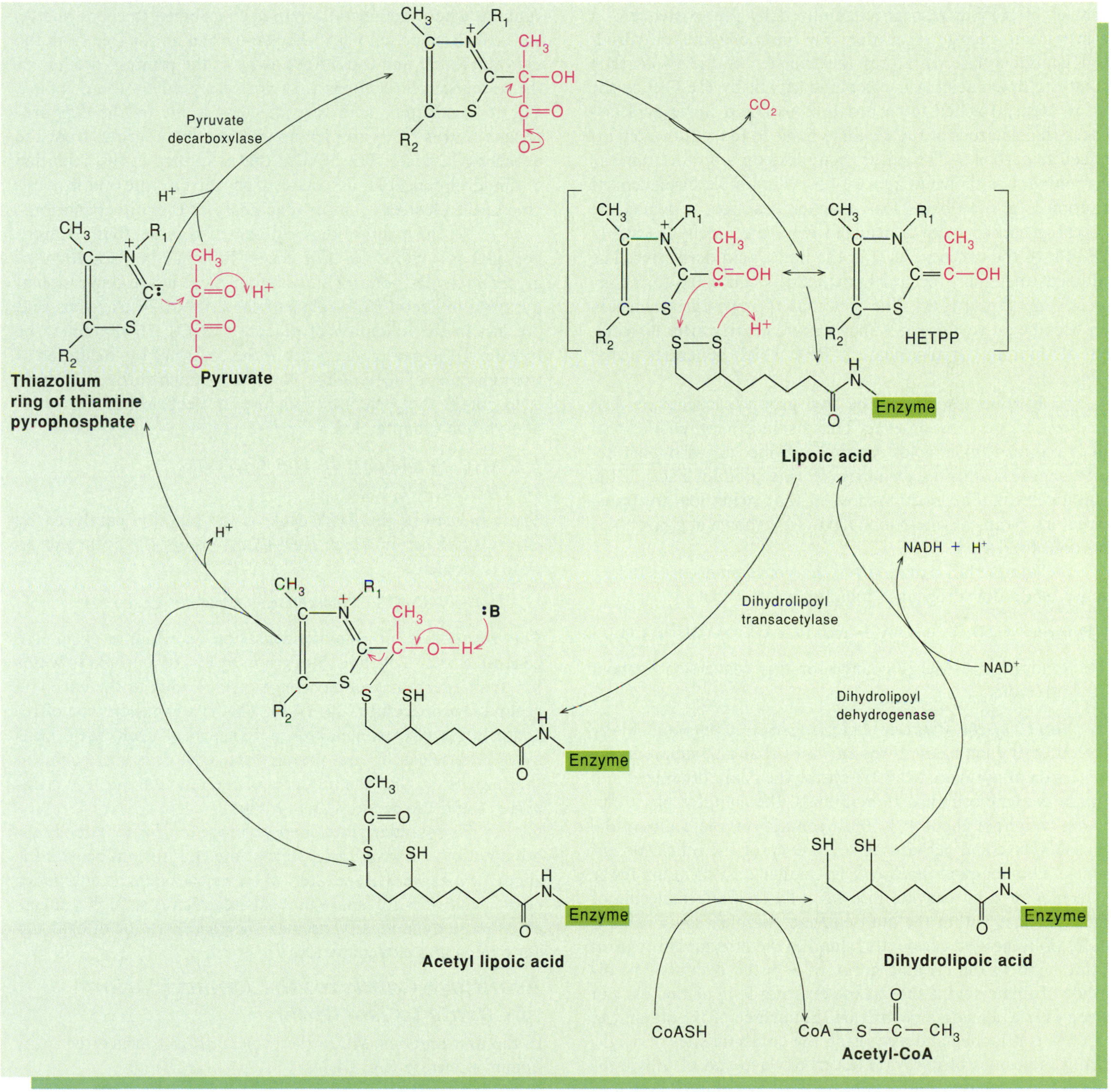

In the first step of the conversion, catalyzed by pyruvate decarboxylase, a carbon atom from thiamin pyrophosphate adds to the carbonyl carbon of pyruvate. Decarboxylation produces the key reactive intermediate, hydroxyethyl thiamin pyrophosphate (HETPP), which we saw as an intermediate in the conversion of pyruvate to acetaldehyde in the alcoholic fermentation (see chapter 13). As shown in figure 14.5, the ionized form of HETPP is resonance-stabilized by the existence of a form without charge separation. The next enzyme, dihydrolipoyl transacetylase, catalyzes the transfer of the two-carbon moiety to lipoic acid. A nucleophilic attack by HETPP on the sulfur atom attached to carbon 8 of oxidized lipoic acid displaces the electrons of the disulfide bond to the sulfur atom attached to carbon 6. The sulfur then picks up a proton from the environment as shown in figure 14.5. This simple displacement reaction is also an oxidation-reduction reaction, in which the attacking carbon atom is oxidized from the aldehyde level in HETPP to the carboxyl level in the lipoic acid derivative. The oxidized (disulfide) form of lipoic acid is converted to the reduced (mercapto) form. The fact that the two-carbon moiety has become an acyl group is shown more clearly after dissociation of thiamin pyrophosphate (TPP), which generates acetyl lipoic acid.

Further transfer of the acyl group to coenzyme A is catalyzed by the same enzyme. This displacement reaction produces reduced lipoic acid. A third enzyme, dihydrolipoyl dehydrogenase, catalyzes oxidation of this product back to the disulfide form. The electrons lost in that oxidation are transferred first to an enzyme-bound flavin (not shown in figure 14.5) and then to NAD^+.

The overall equation for the conversion catalyzed by the pyruvate dehydrogenase complex is

$$\text{Pyruvate} + NAD^+ + \text{CoA} \rightarrow \text{acetyl-CoA} + NADH + CO_2$$

The standard free energy change for this conversion is about -8 kcal/mole.

The Nature of the Pyruvate Dehydrogenase Complex Many molecules of each of the three enzymes that participate in the conversion of pyruvate to acetyl-coenzyme A are organized into a giant enzyme complex. In mammals, the complex has a molecular weight of about 9×10^6; it contains 60 molecules of the transacetylase and perhaps 20 to 30 each of the other two enzymes. The complex in *E. coli* is smaller and contains fewer molecules of each enzyme (see fig. 11.14). Each molecule of transacetylase within the multienzyme complex contains molecules of lipoic acid covalently bound to the enzyme through an amide bond to the ϵ-amino group of a lysine residue. The disulfide of lipoic acid is thus at the end of a long chain, and can sweep over a considerable area on the surface of the subunit to which it is attached and also reach the catalytic sites of neighboring molecules of pyruvate decarboxylase and of dihydrolipoyl dehydrogenase. Thus it appears that the catalytic activities of the complex depend on the ability of the sulfur atoms of the tethered lipoic acid to successively visit the three types of catalytic sites that are contained in the complex (see fig. 11.14). Pyruvate binds at the decarboxylase site and is converted to HETPP as discussed earlier; then the disulfide group of oxidized lipoic acid picks up the two-carbon moiety, oxidizing it to an acetyl group, and carries it to a site on the transacetylase subunit, where the acetyl group is transferred to coenzyme A. In the form of acetyl-CoA, the two-carbon unit is free from the enzymic tether and can diffuse away as the primary product of the sequence of reactions that are catalyzed by the complex. The reduced lipoic acid is oxidized at a site on the dehydrogenase, and is ready to pick up another acetyl group from the decarboxylase site. The NADH that is formed in the oxidation of dihydrolipoic acid dissociates from the enzyme and is available to the electron-transfer machinery of the mitochondrion.

At this point in the oxidation of glucose, four electrons per glucose molecule have been lost in the oxidation of glyceraldehyde-3-phosphate and four more in the conversion of pyruvate to acetyl-CoA. Thus of the total of 24 electrons that are lost in the oxidation of glucose to CO_2, 16 remain to be transferred to oxidizing agents in the course of the oxidation of two molecules of acetyl-CoA. A major function of the TCA cycle is to obtain these electrons for use in electron-transfer phosphorylation (chapter 15).

Citrate Synthase Is the Gateway to the TCA Cycle

The first step of the TCA cycle is the reaction catalyzed by citrate synthase, in which acetyl-CoA enters the cycle and citrate is formed:

$$\text{Acetyl-CoA} + \text{oxaloacetate} \rightarrow \text{citrate} + \text{CoA} + H^+$$

This reaction is an aldol condensation, in which a carbanion generated at C-2 of the acetyl group by loss of a proton to water (or to an acceptor group on the enzyme) adds to the carbonyl group of oxaloacetate (fig. 14.6). This is a common type of reaction in acetyl-CoA metabolism. In nearly all such cases, coenzyme A is released from the product while it is still bound to the enzyme, so that the products of the reaction are free coenzyme A and the condensation product.

The standard free energy change for this reaction is approximately -8 kcal/mole, so that the equilibrium constant is about 10^6. As we will see later, when we consider the energetics of the cycle, it is fortunate that this reaction is so highly favorable thermodynamically, because the concentration of oxaloacetate in cells is extremely low.

Aconitase Catalyzes the Isomerization of Citrate to Isocitrate

In the first reaction within the cycle, citrate is converted to its isomer isocitrate (fig. 14.7):

$$\text{Citrate} \rightleftharpoons \textit{cis}\text{-aconitate} \rightleftharpoons \text{isocitrate}$$

Figure 14.6

Formation of citrate, catalyzed by citrate synthase. A carbanion of acetyl-CoA, generated by loss of a proton to water (or to an acceptor group on the enzyme), adds to the carbonyl group of oxaloacetate. The immediate product of the condensation is probably citryl-coenzyme A, but the products that dissociate from the catalytic site of the enzyme are citrate and free coenzyme A.

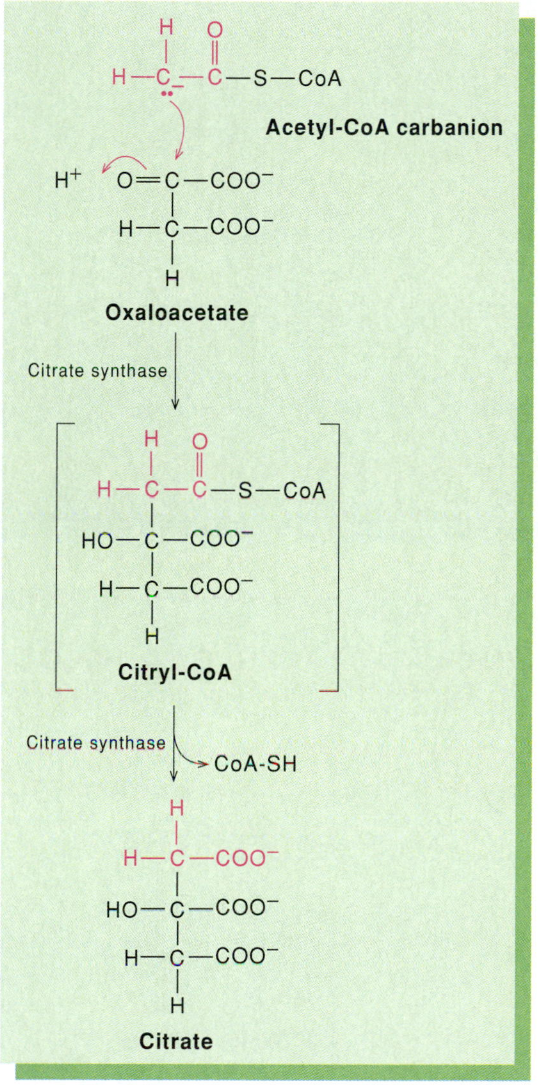

Acetyl-CoA carbanion

Oxaloacetate

Citrate synthase

Citryl-CoA

Citrate synthase

CoA-SH

Citrate

Figure 14.7

Interconversion of citrate, *cis*-aconitate, and isocitrate, catalyzed by aconitase. At equilibrium the relative concentrations of these compounds are about 90, 6, and 4, respectively.

Citrate

Aconitase H_2O H_2O

cis-Aconitate

Aconitase H_2O H_2O

Isocitrate

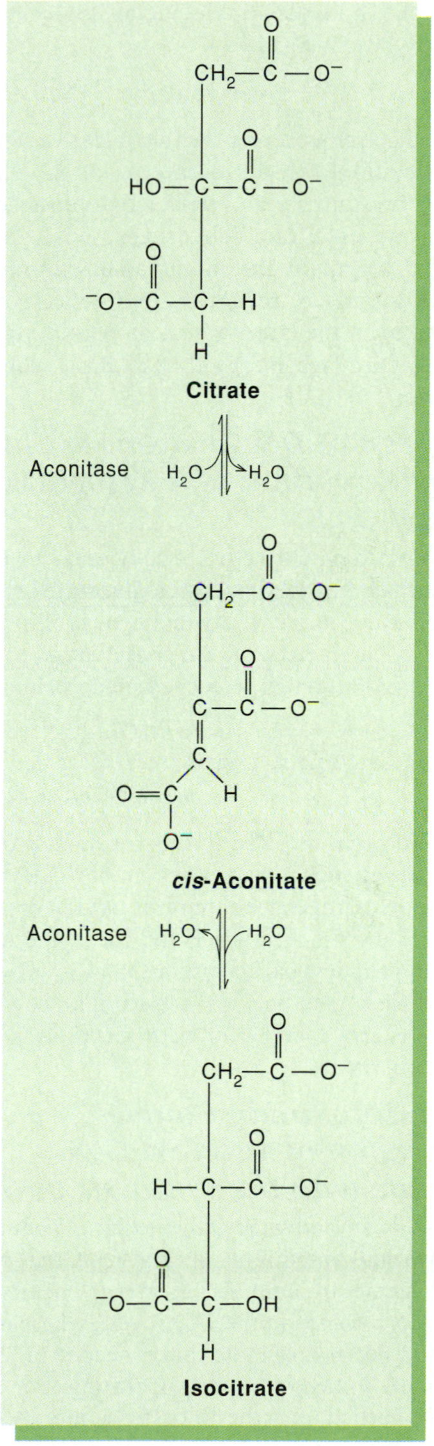

Aconitase, the enzyme that catalyzes this isomerization, is named for the fact that the unsaturated compound corresponding to citrate and isocitrate, *cis*-aconitate, can also serve as substrate or product. In any case, aconitase catalyzes the attainment of equilibrium between citrate, isocitrate, and *cis*-aconitate. At equilibrium the relative concentrations are about 90, 6, and 4 respectively. Aconitate is not known to have any function in metabolism, and it can be disregarded in the consideration of the TCA cycle.

Isocitrate Dehydrogenase Catalyzes the First Oxidation in the TCA Cycle

The first oxidative conversion in the TCA cycle is catalyzed by isocitrate dehydrogenase. Since the product is α-ketoglutarate, the conversion evidently involves two steps: oxidation of the secondary alcohol to a ketone, producing oxalosuccinate, followed by β decarboxylation (fig. 14.8):

$$\text{Isocitrate} + NAD^+ \rightarrow \alpha\text{-ketoglutarate} + NADH + CO_2$$

NAD^+ is the electron acceptor for the oxidative step, and Mg^{2+} or Mn^{2+} is required for the decarboxylation. The oxalosuccinate that is presumably an intermediate does not dissociate from the enzyme. Isocitrate dehydrogenase is a large enzyme; the molecular weight of the mammalian enzyme is about 4×10^5. It has a range of regulatory properties, some of which will be discussed in the later section on regulation of the TCA cycle. The standard free energy change of the reaction is about -2 kcal/mole.

α-Ketoglutarate Dehydrogenase Catalyzes the Decarboxylation of α-Ketoglutarate to Succinyl-CoA

The second of two oxidative decarboxylation reactions of the TCA cycle is that catalyzed by α-ketoglutarate dehydrogenase. The reaction sequence is entirely analogous to that of pyruvate dehydrogenase, complete with the conservation of some of the energy of the oxidation in a coenzyme-containing derivative succinyl-CoA:

$$\alpha\text{-Ketoglutarate} + NAD^+ + \text{CoA} \rightarrow \text{succinyl-CoA} + NADH + CO_2$$

The enzyme complex involved in this reaction also is very similar to the pyruvate dehydrogenase complex. Indeed, the same dihydrolipoyl dehydrogenase subunit is used in both complexes. The product of such conversions is the coenzyme A ester of the acid containing one less carbon atom than the substrate, in this case succinyl-CoA (see fig. 14.4). As might be expected, the standard free energy change for both reactions is similar also, about -8 kcal/mole.

Succinate Thiokinase Couples the Conversion of Succinyl-CoA to Succinate with the Synthesis of GTP

Succinyl-CoA is an activated intermediate. That activation is useful for the small amount of succinyl-CoA that is used in the synthesis of heme in animals. However, nearly all of the succinyl-CoA is retained in the TCA cycle, where it leads to the regeneration of the oxaloacetate that is needed for condensation with acetyl-CoA to keep the cycle operating. For that use, activation is not needed, and the thioester is merely converted to succinate:

$$\text{Succinyl-CoA} + \text{GDP} \rightarrow \text{succinate} + \text{GTP}$$

The reaction is catalyzed by an enzyme that may have originally evolved for the reverse reaction (if succinyl-CoA was required in early anaerobic organisms). In any case, in the direction

Figure 14.8

The oxidative decarboxylation of isocitrate to α-ketoglutarate, catalyzed by mitochondrial isocitrate dehydrogenase. The intermediate, oxalosuccinate, is not released from the enzyme. :B represents a catalytic side chain from the enzyme.

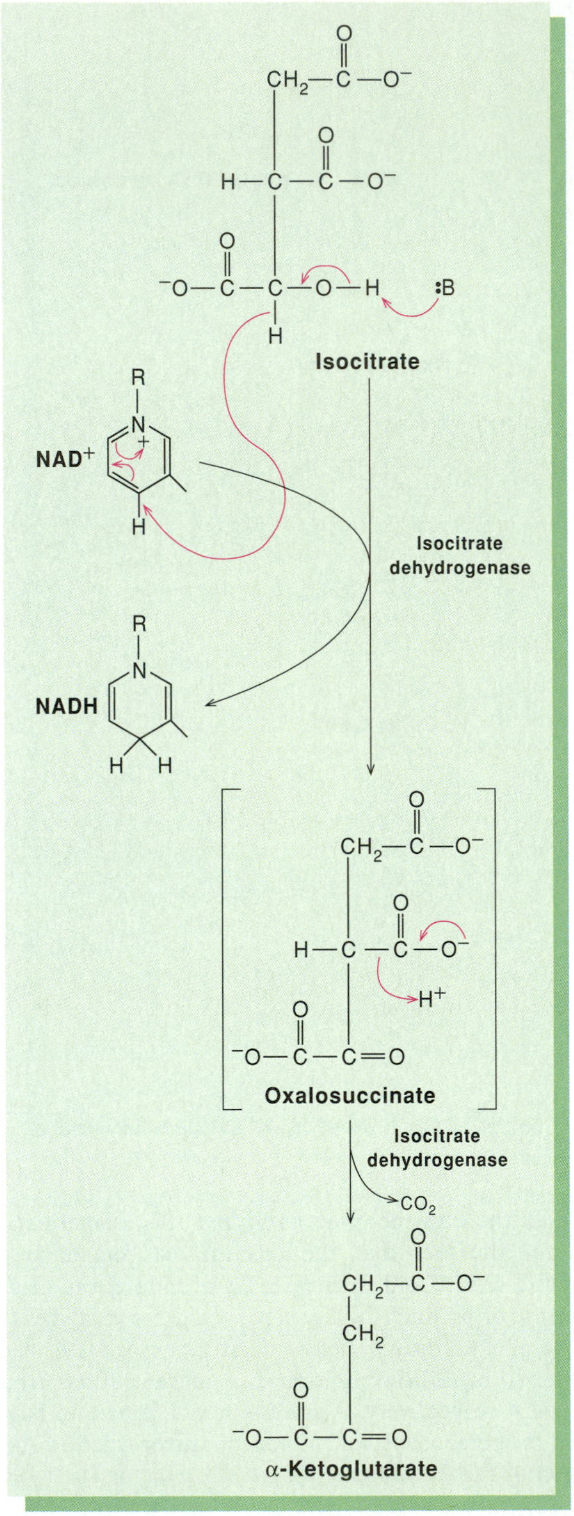

in which it occurs in the TCA cycle, the reaction produces one molecule of GTP in mammals (ATP in some other kinds of organisms) for each molecule of succinyl-CoA converted. The reaction pathway is complex, and involves an intermediate in which a phosphate group is attached to a histidine residue of the enzyme. Probably CoA is first displaced by inorganic phosphate, forming succinyl phosphate. A nitrogen atom of a specific histidine residue then attacks phosphorus, displacing succinate and forming an *N*-phosphoryl derivative. The final step is an attack by GDP (or ADP) on the phosphorus atom of that derivative, forming GTP (or ATP). The standard free energies of hydrolysis of pyrophosphates, such as ATP and GTP, and of CoA esters are very similar; consequently the equilibrium constant for the overall conversion is small, about 3.

Succinate Dehydrogenase Catalyzes the Oxidation of Succinate to Fumarate

In terms of carbon atoms, we might say that since two carbons entered the cycle as the acetyl group and two have left as CO_2, the functional part of the cycle is finished at this stage, and that thus the four carbons of succinate must merely be converted to oxaloacetate to serve again as an acetyl acceptor. However, that assumption would be wrong. When the TCA cycle serves as a means of oxidizing acetyl groups, that is, when it is not producing biosynthetic starting materials at a significant rate, its function is to supply electrons to the electron-transfer phosphorylation system. When we reach succinate, four electrons have been transferred to NAD^+. Thus, of the eight electrons that must be lost in the oxidation of acetate to CO_2, four remain in succinate. In terms of its oxidative function, the cycle is only half finished at this point.

The next step in the TCA cycle, the oxidation of succinate to fumarate, involves insertion of a double bond into a saturated hydrocarbon chain:

$$\text{Succinate} + \text{FAD} \rightarrow \text{fumarate} + \text{FADH}_2$$

This is not an easy or common reaction in organic chemistry. It is, however, a very important type of reaction in metabolic chemistry, and is an integral step in the oxidation of carbohydrates, fats, and several amino acids.

Because of the nature of the reaction, a strong oxidizing agent is required. The electron acceptor that is used in most oxidative steps of catabolism, NAD^+, is not a strong enough oxidizing agent to allow a reasonable equilibrium constant. Flavoproteins are stronger oxidizing agents than NAD^+, and succinate dehydrogenase, which catalyzes the oxidation of succinate to fumarate, is a flavoprotein enzyme. The oxidation of succinate by the electron acceptor of succinate dehydrogenase is about 16 kcal (65 kJ) more favorable than it would be if the electron acceptor were NAD^+. This makes the equilibrium constant more favorable by a factor of about 10^{11}. In this example we see how important the choice of cofactors can be. In general FAD is a better oxidizing agent than NAD^+, and NADH is a better reducing agent than $FADH_2$.

The succinate dehydrogenase of mammalian mitochondria consists of two subunits with molecular weights of about 70,000 and 27,000. Each subunit contains iron-sulfide centers (chapter 15) that appear to participate in electron transfer. The enzyme is integrally attached to the inner mitochondrial membrane (see chapter 15). Because the $FADH_2$ that is produced in the reaction is covalently attached to the enzyme (in contrast to the NADH produced by other dehydrogenases of the cycle, which can diffuse from one catalytic site to another), succinate dehydrogenase presumably must be fixed in position relative to enzymes of the electron-transfer pathway, in order to facilitate direct transfer of electrons. Since the FAD/$FADH_2$ couple probably transfers electrons directly to other carriers, this reaction is probably closely coupled to further electron-transfer reactions.

Fumarase Catalyzes the Addition of Water to Fumarate to Form Malate

Fumarate is converted to L-malate by stereospecific addition of water across the double bond:

$$\text{Fumarate} + H_2O \rightarrow \text{L-malate}$$

The enzyme that catalyzes this reaction, named fumarase, is a tetramer of identical subunits, each with a molecular weight of about 48,000. The equilibrium constant for the reaction is about 4.

Malate Dehydrogenase Catalyzes the Oxidation of Malate to Oxaloacetate

The final oxidation, and final step, of the cycle is the conversion of malate to oxaloacetate:

$$\text{L-malate} + \text{NAD}^+ \rightarrow \text{oxaloacetate} + \text{NADH} + \text{H}^+$$

Although oxidation of an alcohol to a carbonyl is not easy, this enzyme, rather surprisingly, uses NAD^+ as the oxidizing agent. The standard free energy change is about 7 kcal/mole, and the equilibrium constant is near 10^{-5}. Consequently the steady-state concentration of oxaloacetate is very low. This potential problem will be discussed further in the section on energetics of the cycle. The enzyme consists of two identical subunits, each with a molecular weight of about 66,000.

Stereochemical Aspects of TCA Cycle Reactions

Some early applications of isotopic tracer techniques to the TCA cycle led to temporary confusion, but in the end they led to a new generalization concerning the stereochemistry of interaction between enzymes and a certain type of substrates.

In the early 1940s, before the discovery of ^{14}C and when acetyl-coenzyme A was unknown, two research groups used the stable isotope ^{13}C and home-made mass spectrometers (the technique was so new that instruments were not available commercially) to study carbon flow in the TCA cycle. Pyruvate and

$^{13}CO_2$ were added to pigeon liver preparations to form carboxy-labeled oxaloacetate. Malonate was added to stop the TCA cycle at succinate. The expected result was that half of the ^{13}C would be found in succinate and the other half in CO_2 (fig. 14.9).

To the surprise of the investigators and the biochemical community, it was found that although pyruvate was consumed and succinate was produced, there was no ^{13}C in the succinate. Further experiments showed that the isotope had been incorporated, but was all lost in the oxidative decarboxylation of α-ketoglutarate to succinate. Since the citrate molecule has a plane of symmetry, it was taken for granted that the two $-CH_2-COO^-$ arms of the molecule must be chemically equivalent and that the $-OH$ group would have an equal chance of migrating into either arm in the aconitase reaction. In that case, the label would have an equal chance of being lost or retained in the decarboxylation of α-ketoglutarate. For a few years this experimental result was thought to prove that citrate as such could not be an intermediate of the cycle, and that some derivative or analog of citrate must be the immediate product of the condensation of oxaloacetate with the unknown active two-carbon intermediate.

Figure 14.9

Stereochemical relationships in the synthesis and metabolism of citrate. When oxaloacetate labeled with ^{13}C in the carbonyl group ß to the keto group (*) was used as substrate, the researchers expected that half of the label would be found in succinate and half in CO_2. That prediction was based on the assumption that the two $-CH_2-COO^-$ arms of citrate must be equivalent in every way. In fact, all of the label was found in CO_2. Thus only the intermediates shown on the left were produced. This result shows that both the condensation of acetyl-CoA with oxaloacetate and the isomerization of citrate are stereospecific reactions. The carbon atoms supplied by acetyl-CoA are shown in color. Clearly, neither of those atoms is lost in the first turn of the cycle after their entry.

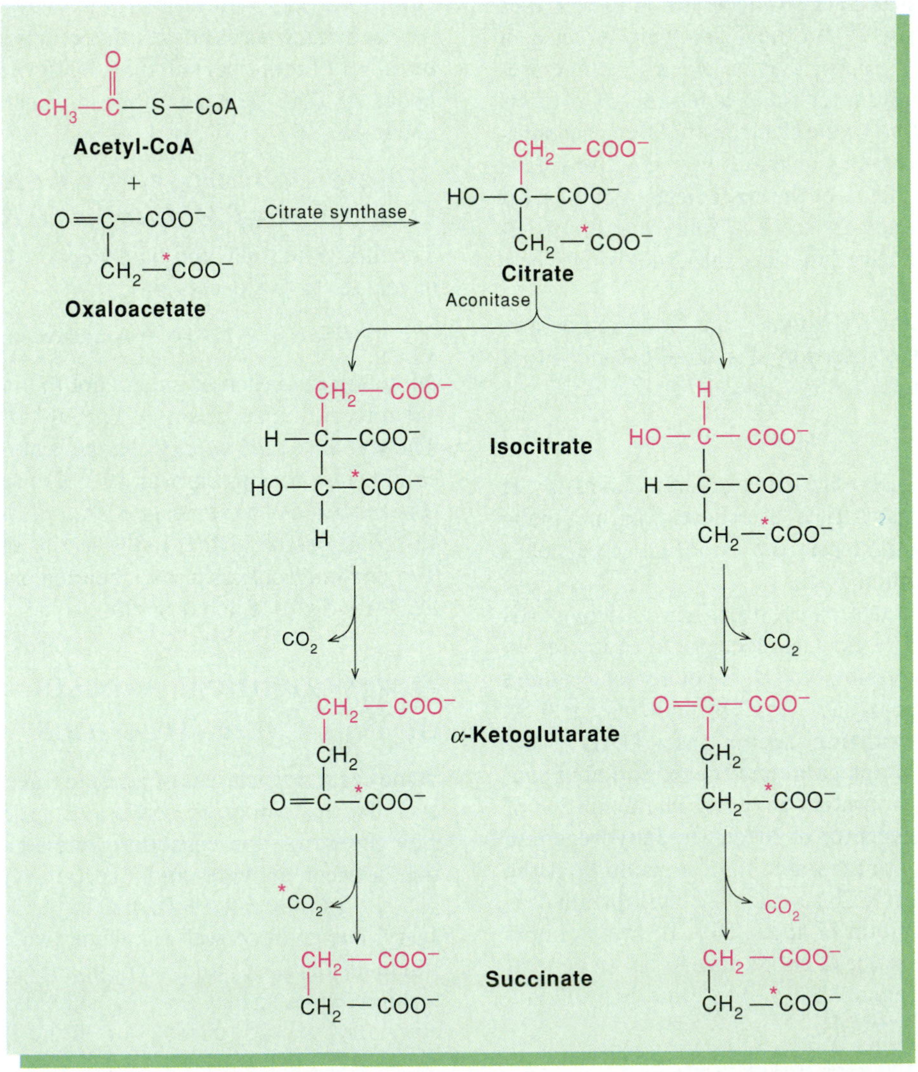

Then, in 1948, Ogston suggested that citrate was not necessarily excluded by the isotopic evidence, because the two $—CH_2—COO^-$ arms might actually not be equivalent when citrate was the substrate for an enzymic reaction. He pointed out that if the substrate were attached to the enzyme at three points, its orientation would be fixed by those attachments and it would be impossible for the two identical arms to exchange positions. Thus only one of them could occupy the position that allowed it to participate in the reaction (fig. 14.10; see also box 14A).

Ogston's concept is a valid and important generalization, but the suggestion of a three-point attachment is too restrictive. Three constraints are necessary to fix an object in three-dimensional space, but they need not all be points of attachment. In principle the two *a* groups of a molecule of the type Ca_2bd could be distinguished by an enzyme if the three constraints were one point of attachment, one pocket, into which *b* could fit but *d* could not, and the position of the reactive groups of the catalytic site. Westheimer and Vennesland showed a few years later that ethanol dehydrogenase distinguishes with 100% specificity between the two hydrogen atoms on C-1 of ethanol (see chapter 11 and fig. 11.7). Three sterically distinct points of attachment between an enzyme and a substrate as small as ethanol seem unlikely.

Ogston's contribution led to an interesting extension of concepts concerning stereochemistry of enzyme action. Compounds of the type Ca_2bd are termed prochiral, and it is recognized that an enzyme that either synthesizes such a compound or uses it as a substrate will nearly always do so stereospecifically. Once the point has been made, it seems very reasonable. In the case of citrate synthase, for example, it is inherently likely that the planar carbonyl carbon of oxaloacetate will lie flat on an enzyme surface, and that only one side of the atom will be available for attack by acetyl-coenzyme A.

Figure 14.10

Citrate is shown with three of the substituents from the central carbon atom making contact with the enzyme surface. Binding in this way makes the two $—CH_2—COO^-$ groups nonequivalent, as they must be in the TCA cycle.

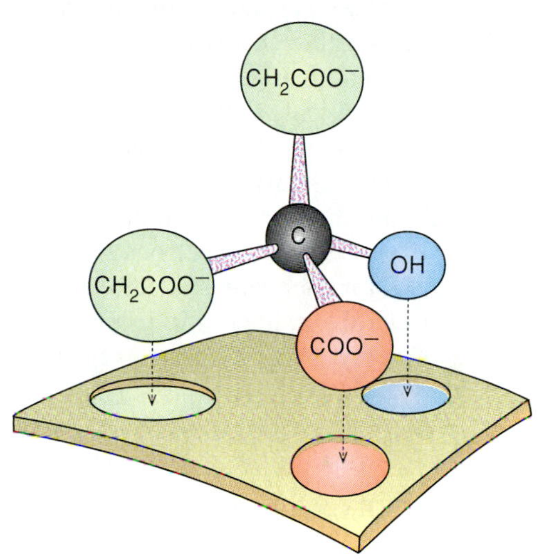

Potter's Demonstration of Prochirality

Ogston pointed out that an enzyme might discriminate between the two carboxymethyl groups of citrate, and thus that the evidence did not exclude citrate as an intermediate in the TCA cycle. Only further experimental evidence could show whether, in fact, citrate synthase and aconitase act with the kind of steric consequences that Ogston had suggested might be possible. What was necessary was to isolate citrate that had been produced from pyruvate and labeled oxaloacetate and then to use that citrate as the substrate for aconitase. V. R. Potter performed the critical experiment. An inhibitor of aconitase was used in the first incubation, so that citrate would accumulate and could be isolated. When this citrate was then used as substrate for a liver preparation poisoned with malonate, as in the previous experiment, the same result was obtained: all of the ^{13}C was in CO_2, and none was in succinate. This result proved that citrate synthase introduces the labeled carboxymethyl group into only one of the two possible steric positions in citrate, and that aconitase distinguishes between the two positions in its action. Ogston's suggestion was confirmed, and citrate was shown to be an intermediate in the cycle.

This episode is an interesting specific illustration of the interplay between hypothesis (or reasoning) and experiment on which scientific advance is based. First, unexpected experimental results seemed to prove that citrate could not be an intermediate in the TCA cycle. Then it was pointed out that the accepted concepts were not necessarily valid, so that the results did not actually argue either for or against the participation of citrate as an intermediate. Finally, experiments that were performed because of the new hypothesis proved that citrate is actually formed by liver extracts and is converted to succinate by the same extracts, so that it is an intermediate. The results of the new experiments also showed that the novel suggestions of the new hypothesis (discrimination between identical groups in a nonchiral molecule) were valid. The experiments would not have been performed without the stimulus of the hypothesis; on the other hand, the hypothesis was only a suggestion until the experiments demonstrated its validity.

ATP Stoichiometry of the TCA Cycle

By summing the component reactions of the TCA cycle, we arrive at the following overall summary reaction:

$$\text{Acetyl-CoA} + 2\,H_2O + 3\,NAD^+ + FAD + ADP + P_i \rightarrow 2\,CO_2 + 3\,NADH + 2\,H^+ + FADH_2 + CoA + ATP$$

Looking at this summary reaction, you may wonder why it doesn't seem to reflect the substantially greater ATP yield that is supposed to be characteristic of aerobic metabolism. The answer is that the potential for making ATP is stored in the reduced coenzyme molecules on the right-hand side of the reaction. Reoxidation of these compounds liberates a large amount of free energy. To see how that energy is released, we must wait until we examine the final phases of chemotrophic energy metabolism in chapter 16. It is only as the electrons are transferred stepwise from the high-energy coenzyme carriers to molecular oxygen that the coupled generation of ATP occurs. We will see that the oxidative phosphorylation system regenerates three molecules of ATP for each pair of electrons from NADH, and two molecules of ATP for each pair of electrons from $FADH_2$. Thus in the normal aerobic mitochondrial metabolism, oxidation of one mole of acetyl groups leads to regeneration of 12 moles of ATP.

If we take pyruvate rather than acetyl-CoA as the starting point, the equation for the TCA cycle is

$$\text{Pyruvate} + 4\,NAD^+ + FAD + ADP\ (\text{or GDP}) + P_i + 2\,H_2O \rightarrow 3\,CO_2 + 4\,NADH + 2\,H^+ + FAD + ATP\ (\text{or GTP})$$

and we see that the mitochondrial part of carbohydrate metabolism yields 15 moles of ATP per mole of pyruvate, or 30 moles per mole of glucose.

Thermodynamics of the TCA Cycle

In the metabolic pathway for the oxidation of pyruvate there are four reactions (those catalyzed by pyruvate dehydrogenase, citrate synthase, isocitrate dehydrogenase, and α-ketoglutarate dehydrogenase) for which the equilibrium constants are large. For four others (those catalyzed by aconitase, succinate thiokinase, succinate dehydrogenase, and fumarase), the equilibrium constant is fairly close to 1, and for one reaction (catalyzed by malate dehydrogenase), it is very small.

Since the value of the equilibrium constant for the malate dehydrogenase reaction is about 10^{-5} and if the physiological ratio of $NAD^+/NADH$ is about 10, the ratio of oxaloacetate to malate at equilibrium should be about 10^{-4}. Thus if the concentration of malate is around 1 mM, the concentration of oxaloacetate would be about 0.1 μM, or 10^{-7} M, at equilibrium. Estimates of the concentration of oxaloacetate in the mitochondrial matrix range from around 10^{-8} M to slightly over 10^{-7} M, which is in good agreement with the thermodynamic expectations.

Clearly, the equilibrium constant for the malate dehydrogenase reaction is most unfavorable for operation of the TCA cycle. It seems strange that such an unfavorable reaction should occur in a sequence (the TCA cycle) that has one of the largest flux rates in metabolism. Probably this is possible only because the following reaction, catalyzed by citrate synthase, is so thermodynamically favorable. The equilibrium constant for this reaction is about 5×10^5. If the concentrations of acetyl-coenzyme A and of free coenzyme A are approximately equal, at equilibrium the ratio of citrate concentration to oxaloacetate concentration will be equal to the equilibrium constant. Then if [oxaloacetate]/[malate] is about 10^{-4} and [citrate]/[oxaloacetate] is about 5×10^5, the ratio [citrate]/[malate] could be as high as 50. That value is only approximate. The actual steady-state ratio would be expected to be significantly below the equilibrium ratio. Further, the values of the equilibrium constants are not precise; effects of ionic strength have not been taken into account, and the assumptions as to the [AcCoA]/[CoASH] and [NAD^+]/[NADH] ratios are tenuous. But crude as they must be, these calculations show that the very small equilibrium constant for the malate dehydrogenase reaction does not pose a serious thermodynamic problem for the operation of the TCA cycle.

The Amphibolic Nature of the TCA Cycle

The sole purpose of the TCA cycle, when it is functioning as such, is the oxidation of acetate to CO_2, with concomitant conservation of the energy of oxidation as reduced coenzymes and eventually as ATP. Strictly speaking, then, the TCA cycle has but a single substrate, acetyl-CoA. In most cells, however, there is considerable flux of four-, five-, and six-carbon intermediates into and out of the cycle, which occurs in addition to the primary function of the cycle. Such side reactions serve two main purposes: (1) to provide for the synthesis of compounds derived from any of several intermediates of the cycle, and (2) to replenish and augment the supply of intermediates in the cycle as needed. Because the TCA cycle can function both in a catabolic mode and as a source of precursors for anabolic pathways, it is often called an amphibolic pathway (from the Greek, *amphi* = "both"). Even under anaerobic conditions, the TCA cycle enzymes are used to supply biosynthetic intermediates (box 14B). Some of the main biosynthetic pathways that begin with intermediates in the TCA cycle, as well as the ways in which the supply of intermediates in the cycle is replenished, are indicated in figure 14.11. We will discuss these pathways only briefly here, as they will be discussed in greater detail in subsequent chapters. (For fats, see chapters 17, 22, and 23; for amino acids, see chapters 18 and 19.)

Oxaloacetate and α-ketoglutarate are used in the synthesis of several amino acids; succinyl-CoA is used in heme synthesis; and citrate is the source of the acetyl-CoA in the cytosol, which is used for the synthesis of fats and other lipids and some amino acids. These are some of the major drains on the TCA cycle. Now we will see how these drains are compensated for.

Reactions that replenish the intermediates in the TCA cycle are termed anaplerotic, from a Greek root that means "filling up." It is not necessary to replenish the intermediate

Figure 14.11

The TCA cycle, showing some of the branchpoint pathways (in color) that either drain or replenish the TCA intermediates.

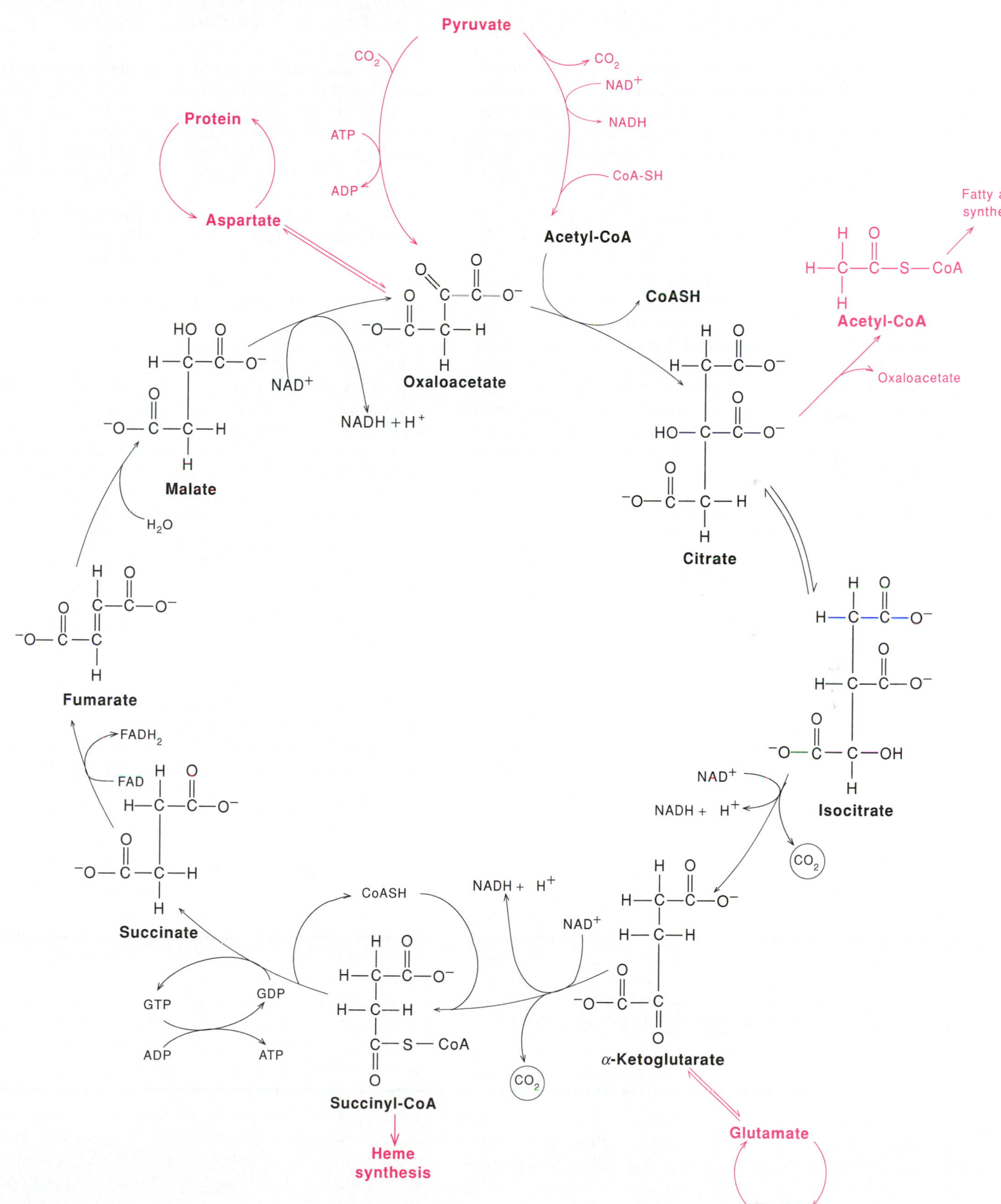

14B

BOX

Generation of Biosynthetic Intermediates Under Anaerobic Conditions

Many of the metabolites that serve as starting materials for biosynthesis are intermediates in glycolysis or the pentose phosphate pathway and so are available when those sequences are active. The others, acetyl-coenzyme A, oxaloacetate, α-ketoglutarate, and succinyl-coenzyme A, can be produced from pyruvate, the end product of glycolysis. The individual reactions are considered in this chapter.

Acetyl-coenzyme A is produced from pyruvate by decarboxylation, followed by oxidation of the carbonyl carbon to the level of carboxylate and activation by formation of the thiol ester. Two electrons are lost in this conversion. Acetyl-coenzyme A is one of the most widely used biosynthetic starting materials. It supplies all of the carbon atoms of fats and several other classes of lipids, and also contributes some of the carbon atoms of several amino acids and other metabolites.

Oxaloacetate is produced by carboxylation of pyruvate. This reaction requires an ATP-to-ADP conversion, but no oxidation or reduction is involved. Oxaloacetate is required in the synthesis of aspartate, asparagine, threonine, isoleucine, methionine, and, in some species, lysine.

Oxaloacetate is converted to α-ketoglutarate by a sequence of reactions that is used frequently in metabolism for the conversion of an α-keto acid to the next higher homolog (the corresponding compound containing one more $-CH_2-$ group). This sequence includes condensation with acetyl-CoA, isomerization by apparent migration of a hydroxyl group, oxidation, and β decarboxylation. α-Ketoglutarate is used in the syntheses of glutamate, glutamine, proline, arginine, and, in some species, lysine.

The production of acetyl-CoA and of α-ketoglutarate from pyruvate involves oxidation, and under anaerobic conditions this causes problems. Two electrons must be disposed of for each molecule of acetyl-CoA that is produced, and four electrons for each molecule of α-ketoglutarate. In the absence of oxygen or an alternative exogenous electron acceptor, those electrons must be fed back into metabolism. That is, they must be used in the reduction of some metabolite, just as the electrons from the 3-phosphoglyceraldehyde dehydrogenase are used in the reduction of pyruvate to lactate or ethanol in anaerobic glycolysis.

It would be possible to merely convert pyruvate to lactate in order to allow synthesis of α-ketoglutarate, but the common laboratory bacterium *E. coli,* and probably many other kinds of bacteria, when growing anaerobically use an alternative set of reactions that dispose of four electrons, rather than two, for each molecule of pyruvate. The pathway is the reduction of oxaloacetate to succinate. Two electrons are used in the first reduction, to malate. This product is dehydrated to fumarate, which is further reduced to succinate.

Thus the reduction of one molecule of oxaloacetate to succinate balances the oxidation that is involved in the production of two molecules of acetyl-CoA or one molecule of α-ketoglutarate. Since NAD^+ is the electron acceptor in the oxidation of pyruvate to acetyl-CoA and on to α-ketoglutarate, it is NADH that must be used in the reduction of oxaloacetate to malate and of fumarate to succinate.

Succinyl-coenzyme A is a starting material in the synthesis of heme. In bacteria, heme is used mainly in the production of cytochromes, which may not be needed in the absence of oxygen. If it is required, succinyl-CoA can be made under anaerobic conditions from succinate with the expenditure of one molecule of ATP or GTP.

Thus by use of the metabolic pathways considered in the preceding chapter and the conversions outlined here, all of the starting materials needed for the biosynthetic sequences of a cell can be made anaerobically from carbohydrates. These pathways are used by at least some bacterial species today in the absence of oxygen, and it is reasonable to assume that they were used in the necessarily anaerobic metabolism of organisms before oxygen was added to the atmosphere by the photosynthetic activity of cyanobacteria about three billion years ago. This example illustrates the most probable way that metabolic cycles have evolved. The reactions of preexisting linear sequences are put to new use as a consequence of the development of one or a few reactions that convert the sequences into a cycle. In this case, only α-ketoglutarate dehydrogenase (and perhaps succinate thiokinase) were needed to convert the two branches of the biosynthetic pathway, which were balanced with respect to oxidation-reduction, to a cyclic oxidative pathway. Because of the difficulty of oxidizing succinate to fumarate (see text), the NAD-linked enzyme was not suitable, and a new FAD-linked enzyme (succinate dehydrogenase) was evolved. Even today, *E. coli* and probably many other organisms make the NAD-linked fumarate reductase under anaerobic conditions and FAD-linked succinate dehydrogenase in the presence of air. Under anaerobic conditions, the conversion of oxaloacetate to succinate consumes four electrons and thus permits necessary oxidative steps such as the conversion of pyruvate to α-ketoglutarate, whereas under aerobic conditions the conversion of succinate to oxaloacetate liberates four electrons for use in oxidative phosphorylation. The cyclic sequence still supplies all of the biosynthetic starting materials that were provided by the linear branched pathways.

that is used in a biosynthetic pathway directly, as the replenishment of any intermediate will occur by a feeding-in process at any point in the cyclical pathway.

When carbohydrates are being metabolized, TCA cycle intermediates are replenished by production of oxaloacetate from pyruvate. In mammals, this reaction is catalyzed by pyruvate carboxylase, and one ATP-to-ADP conversion is associated with the carboxylation. Other properties of the reaction will be discussed later in this chapter, in connection with regulation of the TCA cycle and related metabolic sequences.

In prokaryotic organisms and some eukaryotes, oxaloacetate is fed into the cycle by carboxylation of phosphoenolpyruvate. Energetically the carboxylation of phosphoenolpyruvate, catalyzed by phosphoenolpyruvate carboxylase, is equivalent to the sum of the pyruvate kinase and pyruvate carboxylase reactions, which are used by mammals:

$$H^+ + \text{phosphoenolpyruvate} + \text{ADP} \rightarrow \text{pyruvate} + \text{ATP} + H_2O$$

$$H_2O + \text{pyruvate} + \text{ATP} + CO_2 \rightarrow \text{oxaloacetate} + \text{ADP} + P_i + H^+$$

$$\text{Sum: Phosphoenolpyruvate} + CO_2 \rightarrow \text{oxaloacetate} + P_i$$

The Glyoxylate Cycle

Usually condensation of acetyl-coenzyme A with oxaloacetate to form citrate is a signal that the metabolic fate of the acetyl carbons is sealed; the inevitable result, by means of the TCA cycle, is their oxidation and eventual release as CO_2. However, the glyoxylate cycle, shown in figure 14.12, represents an alternative pathway that also begins with citrate formation, but results in anabolism to the four-carbon level rather than catabolism to the one-carbon level. Comparison of the glyoxylate cycle with the TCA cycle (see fig. 14.12) reveals that two of the five reactions of the glyoxylate cycle are unique to this pathway, while the other three are also part of the TCA cycle. Specifically, the glyoxylate cycle effectively bypasses the two steps of the TCA cycle in which CO_2 is released. Furthermore, two molecules of acetyl-CoA are taken in per turn of the cycle rather than just one, as in the TCA cycle. The net result is the conversion of two molecules of two carbons each (i.e., the acetate of acetyl-CoA) into one four-carbon compound, succinate.

The glyoxylate cycle is an indispensable metabolic capability for those species of bacteria, protozoans, fungi, and algae that grow on a two-carbon substrate such as acetate or ethanol. It is also an essential reaction sequence for seedlings of fat-storing plant species that must effect net synthesis of sugars and other cellular components from the acetyl-CoA produced by oxidation of storage triglycerides. In such plant seedlings, and many other eukaryotic organisms that possess this capability, the enzymes of the glyoxylate cycle (and those of related metabolic pathways to be discussed later) are compartmentalized together in specialized organelles called glyoxysomes.

Species capable of growth on two-carbon substrates need, in addition to the glyoxylate cycle, some preparatory sequence for converting the substrate into acetyl-CoA. If the substrate is acetate, activation requires only formation of the CoA derivative, catalyzed by acetate thiokinase, with ATP hydrolysis as the driving force:

$$CH_3\overset{\overset{\large O}{\|}}{C}O^- + \text{CoASH} + \text{ATP} \rightarrow CH_3\overset{\overset{\large O}{\|}}{C}\,\text{SCoA} + \text{AMP} + PP_i$$

If the substrate is ethanol, it must first be oxidized in two steps to the level of acetate:

$$\underset{\text{Ethanol}}{CH_3CH_2OH} \xrightarrow{NAD^+ \; NADH + H^+} \underset{\text{Acetaldehyde}}{CH_3CHO} \xrightarrow{NAD^+ \; NADH + H^+} \underset{\text{Acetate}}{CH_3COO^-}$$

Other two-carbon substrates are also possible; each requires specific processing to convert it to the acetyl-CoA with which the glyoxylate cycle itself begins.

Both the initial formation of citrate from oxaloacetate and acetyl-CoA and its subsequent conversions by means of aconitate to isocitrate are already familiar from the TCA cycle. The only difference is that in eukaryotic cells the citrate synthase and aconitase enzymes that carry out these reactions in the glyoxysomes are organelle-specific isozymes, differing in physical and enzymatic properties from the enzymes responsible for the same reactions that occur in the mitochondria as part of the TCA cycle.

The two reactions unique to the glyoxylate cycle are those responsible for generation and subsequent utilization of glyoxylate, the two-carbon compound from which the cycle (and the organelle) derives its name. Isocitrate is split into two molecules rather than being oxidatively decarboxylated, as would occur in the TCA cycle (see fig. 14.12). The products are glyoxylate and succinate, with two and four carbons, respectively. The reaction is catalyzed by isocitrate lyase, an enzyme found only in those microbial and plant species that are able to carry out net growth (and hence synthesis of higher-carbon compounds) from the two-carbon level.

The succinate arises from the "upper" four carbon atoms of isocitrate (see fig. 14.12), while the glyoxylate corresponds to the "lower" two carbons. The succinate represents the immediate product of the glyoxylate cycle and becomes in turn the starting point for synthesis of other compounds needful to the cell or organism. The glyoxylate, however, becomes the acceptor for the acetate group from the second acetyl-CoA molecule that enters the cycle. The enzyme responsible for this reaction is malate synthase. Like its companion enzyme isocitrate lyase, malate synthase is found only in species capable of net growth from the two-carbon level.

Figure 14.12

Comparison of the TCA cycle and the glyoxylate cycle. In the TCA cycle one molecule of acetyl-CoA is oxidized to two molecules of CO_2. In the glyoxylate cycle (color) two molecules of acetyl-CoA are converted to one molecule of oxaloacetate. As indicated, the glyoxylate cycle uses some of the enzymes of the TCA cycle. Only enzymes operative in the glyoxylate cycle are shown. In plant cells, enzymes of the glyoxylate cycle are located in specialized organelles called glyoxysomes. In yeasts and other eukaryotic organisms, these enzymes are located in the cytosol and are structurally distinct molecules called isozymes.

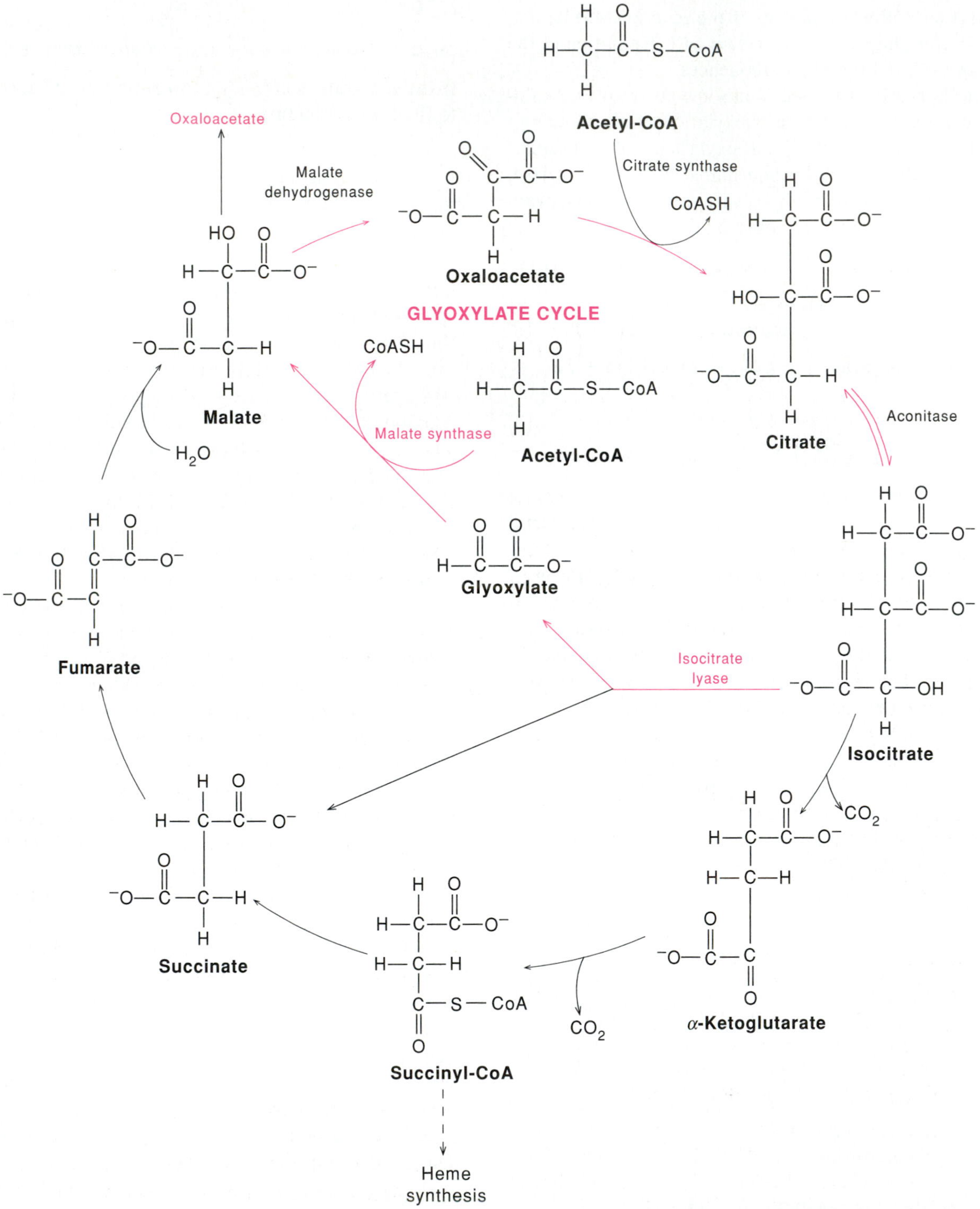

Figure 14.13

Role of the glyoxylate cycle in gluconeogenesis from (*a*) two-carbon compounds or (*b*) fatty acids. The pathway involves reaction sequences that in plant cells are localized within lipid bodies, glyoxysomes, mitochondria, and the cytoplasm.

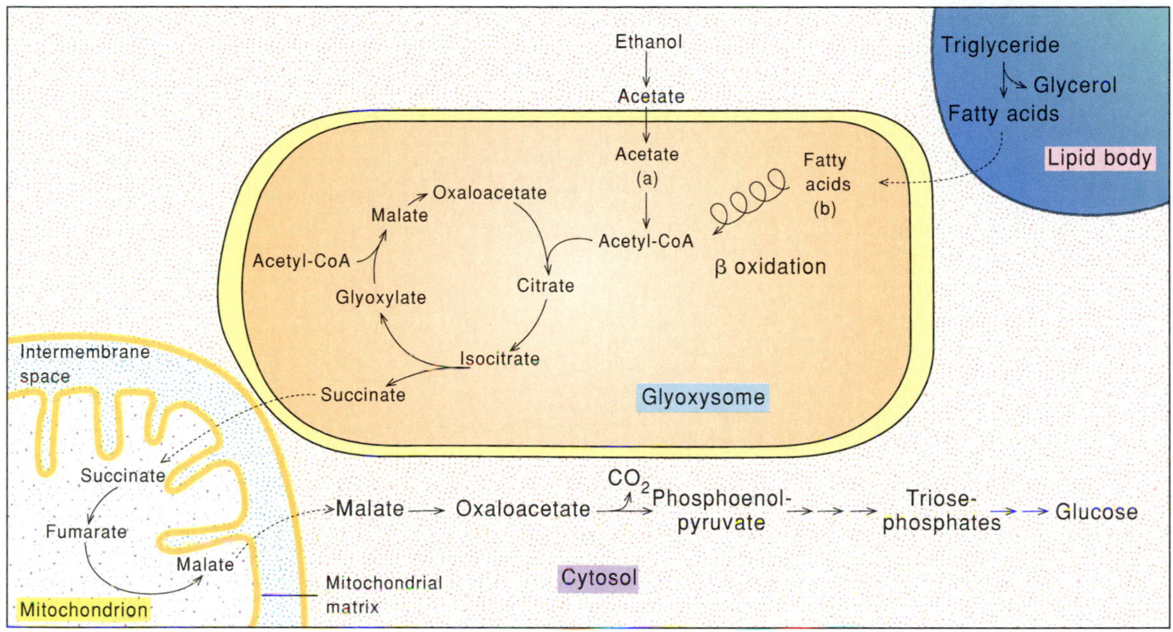

The glyoxylate cycle is completed by oxidative conversion of malate to oxaloacetate, a reaction catalyzed here, as in the TCA cycle, by malate dehydrogenase with NAD^+ as the electron acceptor. In fat-storing plant species, the isozyme of malate dehydrogenase that is involved in the glyoxylate cycle is specific for the glyoxysomes in which the process is localized.

The glyoxylate cycle itself can be summarized by the following overall reaction:

$$2\ \text{Acetyl-CoA} + NAD^+ \rightarrow \text{succinate} + NADH + 3\ H^+ + 2\ \text{CoASH}$$

Utilization of the Succinate Requires Passage from the Glyoxysome to the Mitochondria

Since the glyoxylate cycle is the mechanism by which biosynthesis of more complex molecules from the two-carbon level is initiated, and succinate is the immediate product of the cycle, let's examine the pathways by which other compounds can be synthesized from succinate. As you are already aware, succinate is also an intermediate in the TCA cycle, so that we are in effect dealing with an aspect of the amphibolic nature of the TCA cycle.

Of greatest significance to an organism that depends on a two-carbon substrate for all its carbon needs is the gluconeogenic (sugar-synthesizing) route from succinate to the hexose level, since access to the six-carbon level essentially guarantees access to all other biosynthetic pathways in the cell. Gluconeogenesis in organisms that do not possess the enzymes specific to the glyoxylate cycle usually proceeds from molecules with at least three carbon atoms, such as pyruvate or lactate (see chapter 13).

Synthesis of hexoses from succinate proceeds by means of phosphoenolpyruvate by the pathway shown in figure 14.13. The succinate arising from glyoxylate cycle activity is converted by TCA enzymes located in the mitochondria, first to fumarate and then to malate, which can diffuse from the mitochondria into the cytosol for the further conversion steps. This transfer of succinate to the mitochondria is necessary for further processing because the glyoxysome does not possess the necessary TCA cycle enzymes. The steps from oxaloacetate to glucose were discussed in chapter 13 (see fig. 13.14).

Finally, note that the glyoxylate bypass permits the synthesis of sugars from fatty acid degradation, since acetyl-CoA is the ultimate product of this pathway (see fig. 14.13). The subject of fatty acid degradation is treated in chapter 17.

Oxidation of Other Substrates by the TCA Cycle

The TCA cycle, strictly speaking, has only one input fuel, acetyl-CoA. Catabolism of carbohydrates and fats leads to the production of acetyl-CoA, so the TCA cycle is ideally suited to serve as the major oxidative sequence in the catabolism of those types of compounds. However, degradation of the amino acids that result from the hydrolysis of protein produces a number of intermediates, among which are α-ketoglutarate, succinyl-CoA,

Figure 14.14

The conversion of oxaloacetate to pyruvate. The nucleotide triphosphate used in the first step is regenerated in the second step. Thus ATP or GTP functions as a true coenzyme in this reaction.

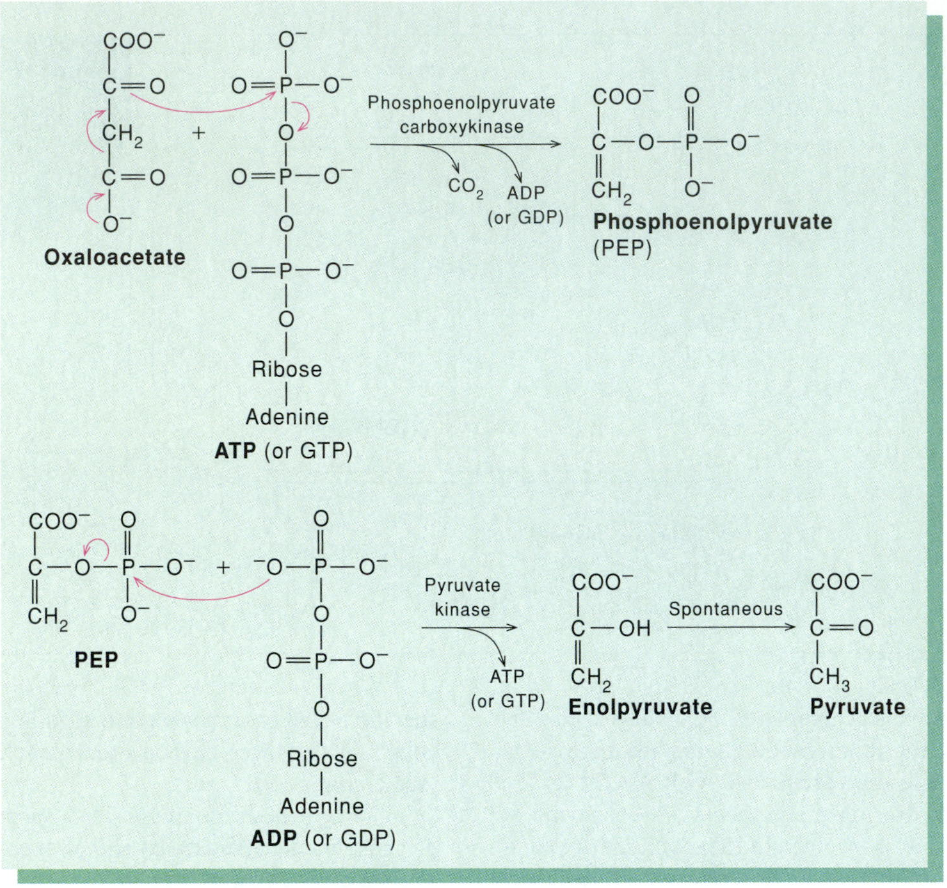

Oxaloacetate + **ATP** (or GTP) —Phosphoenolpyruvate carboxykinase→ **Phosphoenolpyruvate** (PEP) + CO_2 + ADP (or GDP)

PEP + **ADP** (or GDP) —Pyruvate kinase→ **Enolpyruvate** + ATP (or GTP) —Spontaneous→ **Pyruvate**

fumarate, and oxaloacetate (chapter 19). α-Ketoglutarate, succinyl-CoA, and fumarate can be oxidized to oxaloacetate, but the cycle as such cannot oxidize oxaloacetate further. In the presence of acetyl-CoA, each molecule of oxaloacetate used in the synthesis of citrate is regenerated in the cycle; thus there is no net oxidation of oxaloacetate in that case either.

The problem of how to oxidize oxaloacetate is solved by the action of phosphoenolpyruvate carboxykinase, which we discussed in connection with gluconeogenesis in the preceding chapter. This enzyme catalyzes the conversion of oxaloacetate to phosphoenolpyruvate with the help of ATP or GTP, and thus permits the total oxidation of oxaloacetate to CO_2 by the enzymes of the TCA cycle. The mechanism for the conversion of oxaloacetate to pyruvate is shown in figure 14.14:

$$\text{Oxaloacetate} + \text{ATP} \rightarrow \text{phosphoenolpyruvate} + \text{ADP} + CO_2$$

$$H^+ + \text{phosphoenolpyruvate} + \text{ADP} \rightarrow \text{pyruvate} + \text{ATP}$$

$$\text{CoA} + \text{pyruvate} + \text{NAD}^+ \rightarrow \text{acetyl-CoA} + CO_2 + \text{NADH}$$

$$H_2O + \text{oxaloacetate} + \text{acetyl-CoA} \rightarrow \text{citrate} + \text{CoA} + H^+$$

Sum:

$$H_2O + 2\ \text{Oxaloacetate} + \text{NAD}^+ \rightarrow \text{citrate} + 2\ CO_2 + \text{NADH}$$

If we add the equation for the oxidation of citrate to oxaloacetate in one turn of the cycle,

$$P_i + \text{citrate} + 3\ \text{NAD}^+ + \text{FAD} + \text{ADP} \rightarrow \text{oxaloacetate} + 3\ \text{NADH} + \text{FADH}_2 + \text{ATP} + 2\ CO_2 + H^+$$

we obtain the equation for the total oxidation of oxaloacetate:

$$H_2O + P_i + \text{oxaloacetate} + 4\ \text{NAD}^+ + \text{FAD} + \text{ADP} \rightarrow 4\ CO_2 + 4\ \text{NADH} + \text{FADH}_2 + \text{ATP} + H^+$$

NADH and $FADH_2$ will feed electrons into the oxidative phosphorylation system, so the oxidation of a molecule of oxaloacetate will cause the regeneration of 15 molecules of ATP, which is the same as the yield from one molecule of pyruvate.

In addition to the amino acids that are converted to intermediates of the TCA cycle, others are converted to pyruvate or acetyl-CoA and thus enter the cycle in the usual way. In fact, all of the twenty protein amino acids are metabolized by way of the TCA cycle (see fig. 19.1). Thus, although it is often thought of as part of the pathway for carbohydrate metabolism, the TCA cycle is actually the central oxidative sequence for all three of the major types of carbon and energy sources: carbohydrates, fats, and proteins.

Regulation of TCA Cycle Activity

The TCA cycle is carefully regulated to ensure that its level of activity corresponds closely to cellular needs. The cycle serves two functions: (1) furnishing reducing equivalents (such as NADH and to a lesser extent as $FADH_2$) to the electron-transport chain, and (2) by means of side reactions, providing substrates for biosynthetic reactions. Both of these functions are reflected in the regulation to which the cycle is subject.

In its primary role as a means of oxidizing acetyl groups to CO_2 and water, the TCA cycle is sensitive both to the availability of its substrate, acetyl-CoA, and to the accumulated levels of its principal end products, NADH and ATP. Actually the ratio $NADH/NAD^+$ and the energy charge or the ATP/ADP ratio are more important than the individual concentrations. Other regulatory parameters to which the TCA cycle is sensitive include the ratios of acetyl-CoA to free CoA, acetyl-CoA to succinyl-CoA, and citrate to oxaloacetate. The major known sites for regulation are shown in figure 14.15. These include two enzymes outside the TCA cycle (pyruvate dehydrogenase and pyruvate carboxylase) and three enzymes inside the TCA cycle (citrate synthase, isocitrate dehydrogenase, and α-ketoglutarate dehydrogenase). As might be suspected *a priori,* each of these sites of regulation represents an important metabolic branchpoint.

The Pyruvate Branchpoint Partitions Pyruvate between Acetyl-CoA and Oxaloacetate

Acetyl-CoA is the only compound that can enter the TCA cycle when it is operating purely oxidatively, but one molecule of oxaloacetate must enter for each molecule of citrate, α-ketoglutarate, or succinyl-CoA that is removed for use in biosynthesis. It follows that pyruvate is the most important branchpoint in the metabolism of a cell that is living on carbohydrate. The partitioning of pyruvate between decarboxylation to form acetyl-CoA and carboxylation to form oxaloacetate is in effect partitioning between the two major metabolic uses of pyruvate: oxidation of carbon for regeneration of ATP and conversion to starting materials for biosynthesis.

Some of the regulatory interactions that affect partitioning at pyruvate are shown in figure 14.15. We will focus first on the effects of acetyl-CoA, which is a negative modifier for pyruvate dehydrogenase and a very strong positive modifier for pyruvate carboxylase. To illustrate the regulatory roles of those effects, we may consider a mitochondrion in which the TCA cycle is functioning only oxidatively, so that no input of oxaloacetate is required. If conditions change so that α-ketoglutarate begins to be removed for biosynthesis, the rate at which oxaloacetate is regenerated by the cycle will be reduced by the rate of α-ketoglutarate removal. The citrate synthase reaction can proceed no more rapidly than the rate at which oxaloacetate, one of its substrates, is supplied, so the rate of citrate synthesis, too, will decrease. Consequently acetyl-CoA will be used more slowly, which will cause its concentration to increase slightly. The increase in acetyl-CoA concentration will lead to a decrease in the activity of the pyruvate dehydrogenase complex and an increase in the activity of pyruvate carboxylase. As a result of those effects, the partitioning of pyruvate is changed so as to produce more oxaloacetate, thus replenishing the cycle and allowing it to continue to function in spite of the removal of α-ketoglutarate.

Since the system responds to the concentration of acetyl-CoA and thus indirectly to the concentration of oxaloacetate, it will adjust automatically to maintain a functional concentration of oxaloacetate regardless of whether the intermediate removed from the cycle is citrate, α-ketoglutarate, succinyl-CoA, oxaloacetate itself, or any combination of these. The same effects will cause the rate of conversion of pyruvate to oxaloacetate to decrease when the rate of removal of biosynthetic precursors decreases.

In addition to the negative effects of acetyl-CoA, pyruvate dehydrogenase is also negatively regulated by ATP and NADH. These effects are in the correct direction to cause the rate of acetyl-CoA synthesis to vary with the need for electrons and for regeneration of NADH and ATP.

In addition to the negative effect of these small molecule modifiers, the pyruvate dehydrogenase complex, at least in mammals, is subject to regulation by covalent modification. Each of the giant complexes contains a few molecules of a protein kinase and a protein phosphorylase. The kinase catalyzes phosphorylation of specific serine hydroxyl groups of the pyruvate decarboxylase portion of the complex, and the phosphorylase catalyzes hydrolytic removal of those phosphoryl groups. The phosphorylated enzyme is relatively inactive. Thus the phosphorylation-dephosphorylation system also contributes to regulation of the rate of conversion of pyruvate to acetyl-CoA. The action of the kinase, and the resultant decrease in the activity of the pyruvate complex, is favored by high ATP/ADP and $NADH/NAD^+$ ratios and by high acetyl-CoA. These effects act in the same direction as the direct effects of the modifier metabolites on the enzymes of the complex, and the types of regulation must therefore reinforce each other.

Citrate Synthase Is Negatively Regulated by NADH and the Energy Charge

Thus far we have discussed the two most important enzymes rimming the TCA cycle that are involved in its regulation. This leaves us with the three enzymes, all within the cycle, that are most important in regulating the activity of the cycle. The first of these, citrate synthase, catalyzes the formation of citrate from acetyl-CoA and oxaloacetate. Regulation, by energy charge and other parameters, of the rate of glycolysis and of the pyruvate dehydrogenase reaction play important roles in controlling the rate of citrate synthesis. For example, yeast citrate synthase has been shown to respond sensitively to variation in the value of

Figure 14.15

Major regulatory sites of the TCA cycle, with activators (⊕) and inhibitors (⊖) of specific reactions shown in color.

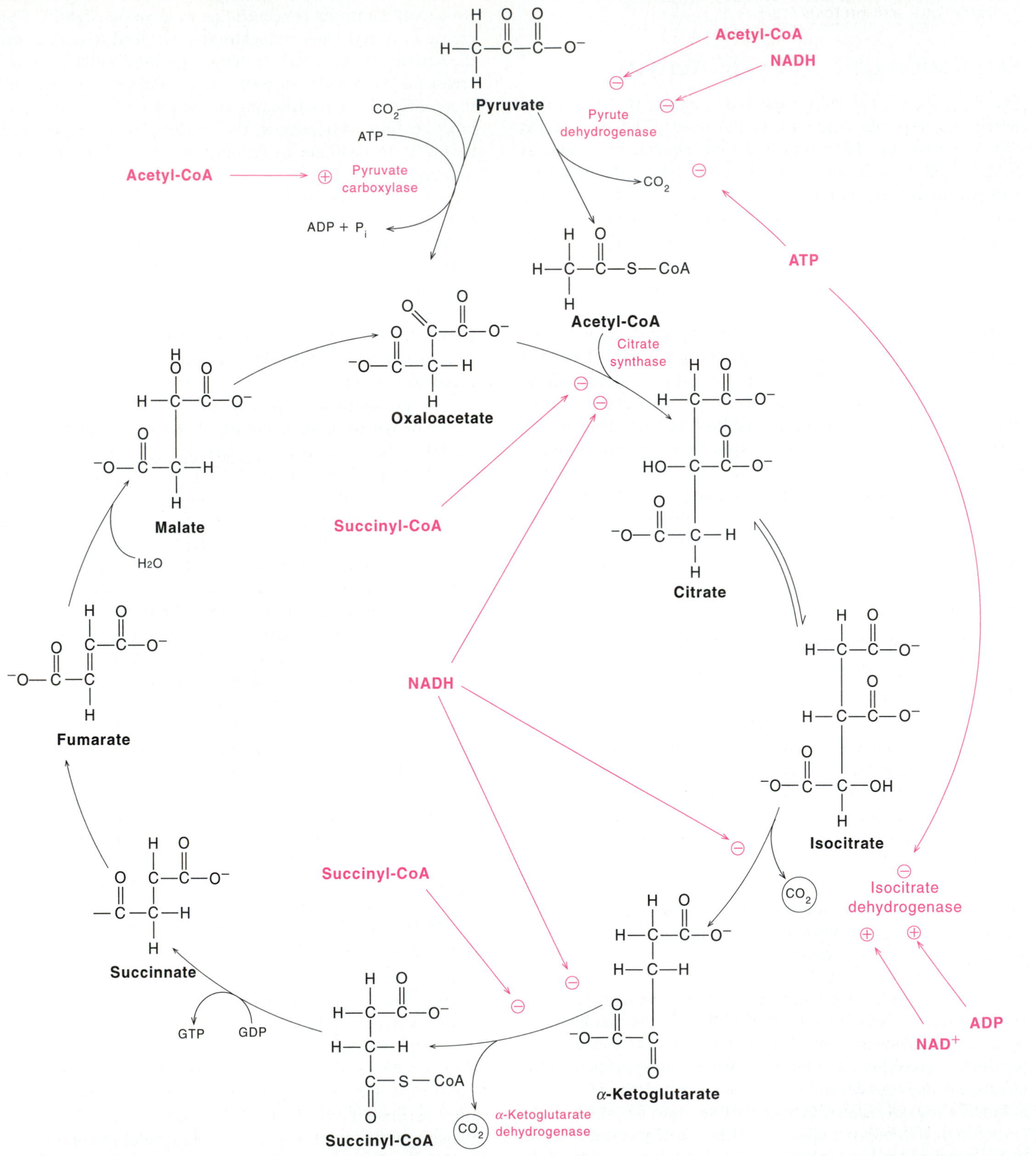

Figure 14.16

Cycling between mitochondria and cytosol in the supply of acetyl-CoA for use in biosynthetic sequences in the cytosol. Citrate moves from the interior of the mitochondria to the cytosol. In the cytosol it is cleaved to acetyl-CoA and oxaloacetate by citrate lyase. The equilibrium constant for this reaction is favorable because an ATP-to-ADP conversion is involved. Most of the oxaloacetate is reduced to malate. The malate may be taken up by mitochondria or oxidized to pyruvate and CO_2, generating NADPH for use in biosynthetic sequences in the cytosol. The pyruvate enters the mitochondria, where it may be converted to oxaloacetate or an acetyl-CoA by the usual routes (see fig. 14.15).

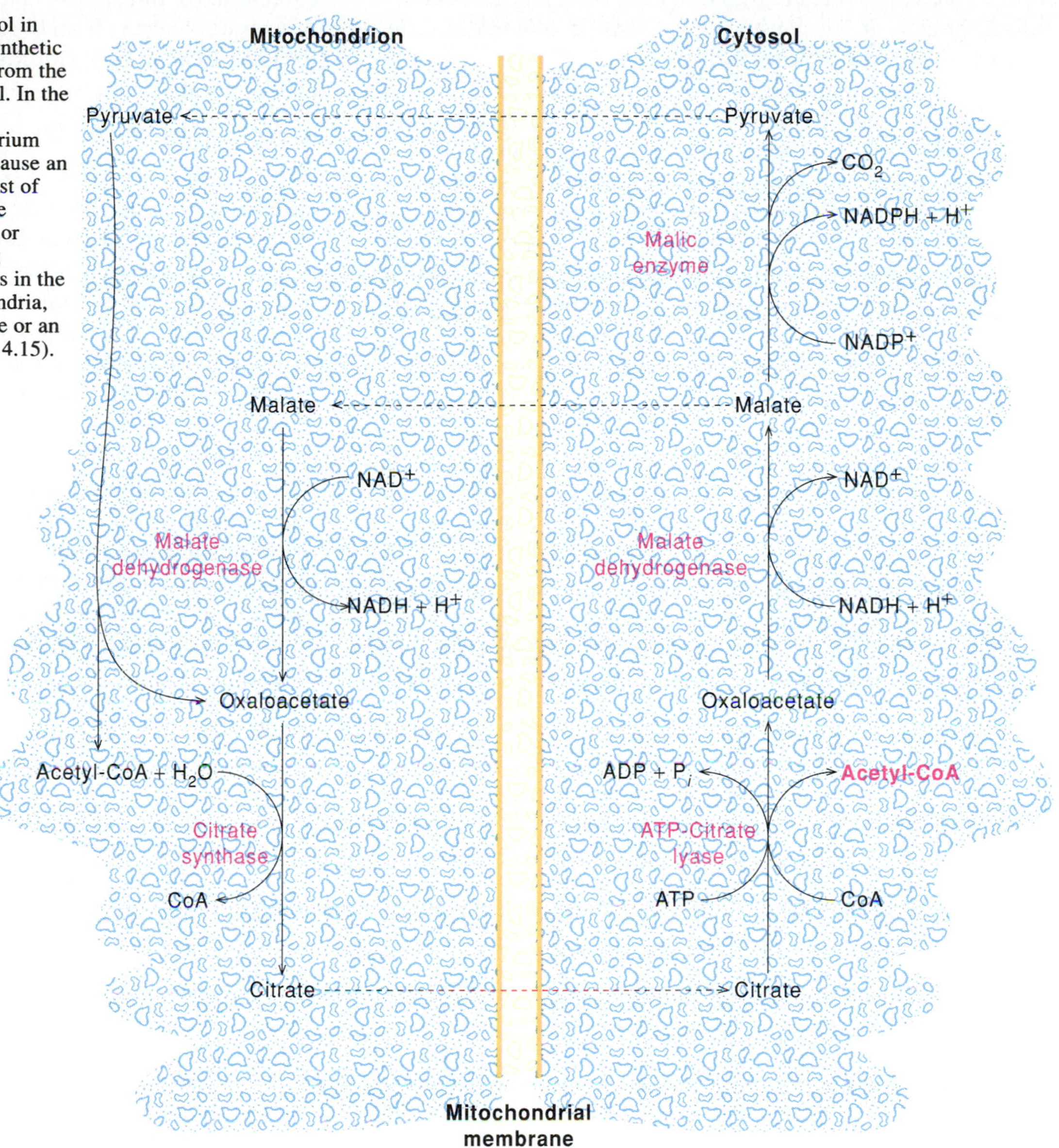

the energy charge. Synthesis of citrate is also favored by a low $NADH/NAD^+$ ratio. This is clearly metabolically desirable, since the operation of the cycle should favor an increase in this ratio or the reduced form of the coenzyme.

Isocitrate Dehydrogenase Is Regulated by the NADH/NAD⁺ Ratio and the Energy Charge

The equilibrium constant for the conversion of citrate to isocitrate is small so these two intermediates make up a metabolic pool. Since regulation is not expected within a metabolic pool for reasons discussed in chapters 12 and 13, the next possible regulatory site as we go around the TCA cycle is the conversion of isocitrate to α-ketoglutarate. This reaction is catalyzed by isocitrate dehydrogenase, which is a highly regulated enzyme. In yeast the activity of isocitrate dehydrogenase is positively regulated by the fourth order of the isocitrate concentration and the second order of the NAD^+ and AMP concentrations. It is negatively regulated by NADH making it very sensitive to the $NADH/NAD^+$ ratio. The properties of the mammalian enzyme are similar to those of the yeast enzyme, with the notable exception that ADP replaces AMP as a positive modifier.

In order to understand why isocitrate dehydrogenase is so intensely regulated we must consider reactions beyond the TCA cycle, and indeed beyond the mitochondrion (fig. 14.16). Of the two compounds citrate and isocitrate, only citrate is permeable to the barrier imposed by the mitochondrial membrane. Citrate, which passes from the mitochondrion to the cytosol, plays a major role in biosynthesis both because of its immediate regulatory properties and because of the chain of covalent reactions that it initiates. In the cytosol, citrate undergoes a cleavage reaction in which acetyl-CoA is produced. The other cleavage product, oxaloacetate, can be utilized directly in various biosynthetic reactions or it can be converted to malate. Malate can be returned to the mitochondrion by diffusion, or it can be converted in the cytosol to pyruvate, which also results in the reduction of $NADP^+$ to NADPH. The pyruvate is either utilized directly in biosynthetic processes, or like malate, can return to the mitochondrion by diffusion.

The acetyl-CoA produced in the cytosol from citrate breakdown is used in a number of biosynthetic reactions including the synthesis of lipids, some amino acids, vitamins, cofactors, and pigments. This is the main source of acetyl-CoA in the cytosol as the acetyl-CoA produced in the mitochondrion cannot diffuse across the mitochondrial membrane. The NADPH produced indirectly in the cytosol from the citrate is also of great importance in a number of biosynthetic reactions.

In addition to its importance in providing cytosolic acetyl-CoA and NADPH, citrate also serves as a major regulator of the rate of fatty acid synthesis. As we shall see (see chapter 17) citrate is a very strong positive modifier of the first reaction in fatty acid synthesis. It should be remembered (see chapter 13) that citrate is a negative modifier of phosphofructokinase thereby exerting a negative effect on glycolysis, which also occurs in the cytosol.

The effect of small molecule modifiers on isocitrate dehydrogenase is appropriate in that an excess of citrate is an indication of a high-energy charge. As a result, conditions that favor inhibition of isocitrate dehydrogenase will favor an accumulation of mitochondrial citrate leading to an increased diffusion rate of citrate from the mitochondrion to the cytosol where the citrate can exert its multiple positive effects on biosynthesis and its negative effects on glycolysis.

α-Ketoglutarate Dehydrogenase Is Negatively Regulated by NADH

α-Ketoglutarate is also a branchpoint metabolite, since it can be transaminated to form glutamate (chapter 19). Glutamate is needed in protein synthesis directly, and also is a precursor of a number of other amino acids. Thus it is consumed in the cytosol in rather large amounts in a cell that is synthesizing protein rapidly. Therefore we would expect that the reaction catalyzed by a α-ketoglutarate dehydrogenase would be regulated so as to retain carbon in the cycle when energy is in short supply, and to allow the concentration of α-ketoglutarate to rise, facilitating its transamination to glutamate and exit from the cycle, when the energy supply is high. When NADH is low, the enzyme will compete more strongly for α-ketoglutarate and thus maximize the regeneration of ATP while tending to decrease biosynthetic activities that would use ATP. No adequate test of the effect of adenine nucleotides on this enzyme has been reported.

Summary

This chapter is mainly concerned with the contribution of the tricarboxylic acid cycle to carbohydrate metabolism. Our discussion has centered on the following points.

1. The TCA cycle is the main source of electrons for oxidative phosphorylation, and thus the major energetic sequence in the metabolism of aerobic cells or organisms. It serves as the main distribution center of metabolism, receiving carbon from the degradation of carbohydrates, fats, and proteins and, when appropriate, supplying carbon for the synthesis of carbohydrates, fats, or proteins. Every aspect of the metabolism of an aerobic organism is directly dependent on the TCA cycle.
2. The major capabilities of the enzymes of the TCA cycle include the following:
   (a) Oxidation of acetyl-CoA, yielding electrons for regeneration of ATP by oxidative phosphorylation.
   (b) With the help of phosphoenolpyruvate carboxykinase, pyruvate kinase, and pyruvate decarboxylase, oxidation of oxaloacetate or any other intermediate of the cycle, yielding electrons for use in the oxidative phosphorylation system.
   (c) Conversion of oxaloacetate and acetyl-CoA to citrate, α-ketoglutarate, or succinyl-coenzyme A for use in biosynthesis.
   (d) When intermediates of the cycle are supplied—for example, from protein degradation—production of phosphoenolpyruvate as a starting material for gluconeogenesis.
   (e) With the help of isocitrate lyase and malate synthase, conversion of acetyl-CoA to oxaloacetate or any other intermediate of the cycle for use in biosynthesis or gluconeogenesis.
3. The rate of entry of carbon into the cycle is regulated by many factors, including the $NADH/NAD^+$ ratio, the energy charge, and the need for biosynthetic intermediates. Regulatory effects may be exerted on the behavior of pyruvate dehydrogenase and on the behavior of citrate synthase.
4. Partitioning between the oxidative and biosynthetic functions of the cycle occurs mainly at pyruvate in mammals. The amount of pyruvate that is carboxylated to form oxaloacetate is determined by the rate at which intermediates of the cycle are removed for use in biosyntheses. In many other kinds of organisms a similar partitioning occurs at phosphoenolpyruvate.
5. The glyoxylate cycle provides an alternative fate for acetyl-CoA. Reactions of the glyoxylate cycle permit the synthesis of four- or six-carbon compounds from two-carbon compounds. This is accomplished with four enzymes from the TCA cycle and two that are unique to the glyoxylate cycle. The latter two enzymes are possessed by certain plants and microorganisms.

Selected Readings

Atkinson, D. E., *Cellular Energy Metabolism and Its Regulation.* New York: Academic Press, 1977. A solid, detailed discussion of energy metabolism by an author who is very much a part of the scene.

Broda, E., *The Evolution of Bioenergetic Processes.* New York: Pergamon Press, 1975. An excellent discussion of cellular energetics from an evolutionary perspective.

Gest, Howard, Evolutionary roots of the citric acid cycle in prokaryotes. *Biochem. Soc. Symp.* 54:3–16, 1987. A fascinating account of how the evolution of the cycle is traced by studying the way in which enzymes of the cycle are used in current-day microorganisms.

Krebs, H. A., The history of the tricarboxylic acid cycle. *Perspect. Biol. Med.* 14:154, 1970. An engaging historical account by the man who masterminded much of it.

Mehlman, M., and R. W. Hanson, *Energy Metabolism and the Regulation of Metabolic Processes in Mitochondria.* New York: Academic Press, 1972. The title speaks for itself—a good reference.

Williamson, J. R., and R. V. Cooper, Regulation of the citric acid cycle in mammalian systems. *FEBS Lett.* 117(Suppl.):K73, 1980. A well-rounded, current review of TCA cycle regulation as a contemporary research theme, from a symposium dedicated to Hans Krebs.

Problems

1. Using your knowledge of metabolism, determine whether the following statements are true or false and explain the reasoning behind your decision.
   - (a) Dihydrolipoamide dehydrogenase catalyzes the only oxidation-reduction reaction in the pyruvate dehydrogenase complex.
   - (b) Hydrolysis of the thioester bond of acetyl-CoA yields insufficient energy to drive phosphorylation of ADP.
   - (c) The methyl group of each acetyl-CoA molecule entering the TCA cycle is derived from the methyl group of pyruvate.
   - (d) Even if aconitase were unable to discriminate between the two ends of the citrate molecule, the CO_2 released would still come from the oxaloacetate rather than the acetyl-CoA substrate of the citrate synthase reaction.
   - (e) Malate cannot be converted to fumarate because the TCA cycle is unidirectional.
2. Assume that you have a buffered solution containing pyruvate dehydrogenase and all the enzymes of the TCA cycle but none of the cycle intermediates.
   - (a) If you add 3 μmoles each of pyruvate, CoASH, NAD^+, GDP, and P_i, how much CO_2 will be evolved? What other products are formed?
   - (b) In addition to the reagents in (a), you add 3 μmoles each of the TCA cycle intermediates. How much CO_2 is evolved? Explain.
   - (c) If you were to add to the system described in (a) an electron acceptor that reoxidized NADH, would there be increased CO_2 evolution? Why or why not?
   - (d) Explain the effect on CO_2 evolution of adding the NADH-reoxidizing system to the system described in (b), assuming that you also added excess GDP and P_i.
3. Suppose that you supply $[1-{}^{14}C]$-labeled glucose to an aerobic bacterial culture and rapidly isolate the intermediates of the glycolytic and TCA pathways. Indicate which of the carbons will be initially labeled in the following intermediates, and explain the reasoning behind your answer. Assume that the activity of pyruvate carboxylase can be ignored in formulating your answer.
   - (a) Glyceraldehyde-3-phosphate
   - (b) Acetyl-CoA
   - (c) Citrate
   - (d) α-Ketoglutarate
   - (e) Succinate
   - (f) Malate
4. What would you expect to be the metabolic consequences of the following mutations in yeast?
   - (a) Inability to synthesize malate synthase.
   - (b) Pyruvate carboxylase that is not activated by acetyl-CoA.
   - (c) Pyruvate dehydrogenase that is inhibited by acetyl-CoA more strongly than is the wild-type enzyme.
5. The substrate hydroxypyruvate

$$HO-CH_2-\overset{\overset{O}{\|}}{C}-\overset{\overset{O}{\|}}{C}-O^-$$

is metabolized to pyruvate in a five-step process requiring the four intermediates whose structures are shown below.

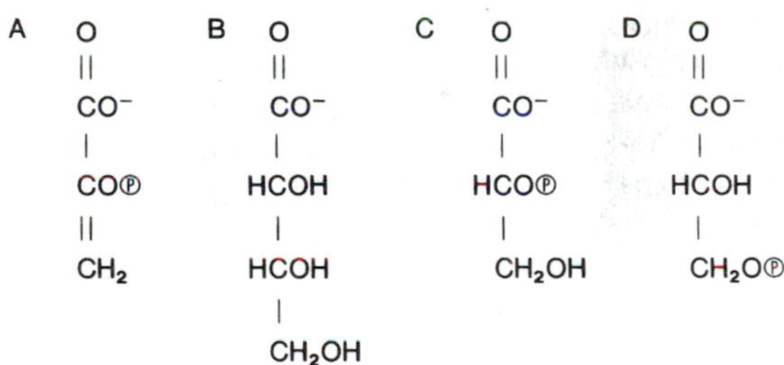

The letter designating the intermediate does not necessarily reflect the order in which it is formed. The metabolic conversion requires NADH and catalytic quantities of both ATP and ADP. Assume that the pathway begins with an NADH-mediated reaction.
   - (a) Designate the order in which the intermediates are formed in the metabolism of hydroxypyruvate to pyruvate.
   - (b) Write an overall equation for the metabolism of hydroxypyruvate to pyruvate.
   - (c) Name each of the intermediates (A–D) and indicate which are intermediates in the glycolytic pathway.
   - (d) Explain why only catalytic rather than stoichiometric amounts of ADP and ATP are required in the pathway.

6. Under anaerobic conditions, *E. coli* synthesizes an NADH-dependent fumarate reductase rather than succinate dehydrogenase, the flavoprotein that oxidizes succinate to fumarate.
   (a) Write an equation for the reaction catalyzed by fumarate reductase.
   (b) NADH produced by the glyceraldehyde-3-phosphate dehydrogenase reaction is reoxidized by reducing an organic intermediate. Rather than reduce pyruvate to lactate, anaerobic *E. coli* utilize the fumarate reductase. However, under anaerobiosis, the activity of α-ketoglutarate dehydrogenase is virtually nonexistent. Show how fumarate is formed, using reactions beginning with PEP and including the necessary TCA cycle enzymes. (Spiro, S., and J. R. Guest, *TIBS* 16:310–314(1991).)
   (c) What is the metabolic advantage to anaerobic *E. coli* in using the fumarate reductase pathway rather than lactate dehydrogenase to reoxidize NADH?
7. Cite some metabolic advantages in having lipoic acid covalently bound to the acyl transfer enzymes of pyruvate dehydrogenase and α-ketoglutarate dehydrogenase complexes.
8. Consider the glyceraldehyde-3-phosphate dehydrogenase-phosphoglycerokinase enzymes of glycolysis and the succinate thiokinase of the TCA cycle. Compare the mechanisms of incorporation of inorganic phosphate into the respective nucleoside triphosphates.
9. Explain why a genetic deficiency of dihydrolipoamide dehydrogenase decreases the oxidation rate of pyruvate and α-ketoglutarate in mitochondria. Assume that each of the dehydrogenase complexes were assayed with their respective substrates. What effect would the defect have on ATP production in the mitochondria?
10. The pyruvate dehydrogenase complex may have been regulated by phosphorylation of any one of the three different enzymes in the complex, yet regulation occurs on the first enzyme of the complex. How is regulation of the complex consistent with the regulation observed in metabolic pathways whose enzymes are not physically associated?
11. *E. coli* growing on acetate as the sole carbon source partitions isocitrate between the TCA cycle and glyoxalate bypass by reversible phosphorylation of isocitrate dehydrogenase (ICDH). Phosphorylated ICDH is catalytically inactive. Explain how decreasing the ICDH activity effectively shifts isocitrate to the glyoxalate bypass. (Reference: H. G. Nimmo, *TIBS.* 9:475–478, (1984).)
12. A bifunctional regulatory enzyme (ICDH kinase/phosphatase) catalyzes the phosphorylation/dephosphorylation of ICDH and is under metabolic control. Predict the relative activities of the kinase and phosphatase function of the bifunctional enzyme when *E. coli* is using acetate as the sole carbon source, under each of the following conditions.
   (a) Oxaloacetate is being diverted to an anabolic pathway.
   (b) AMP levels are high.
   (c) The culture is shifted to a glucose-containing medium.
13. Plants effectively accomplish the glyoxalate bypass without resorting to phosphorylation of the mitochondrial ICDH. How is partitioning of the glyoxalate bypass substrates and products accomplished in plants? (Tolbert, N. E., *Ann. Rev. Biochem.* 50:133–157(1981).)
14. Net movement of acetyl-CoA from the mitochondria to the cytosol is an ATP-consuming process. When the acetyl-CoA transfer is accompanied by formation of NADPH at the expense of NADH, the reaction stoichiometry is as follows:

$$2\ \text{ATP} + \text{NADH} + \text{NADP}^+ + \text{acetyl-CoA}_{\text{mit}} \rightarrow \text{NAD}^+ + \text{NADPH} + \text{acetyl-CoA}_{\text{cyt}} + 2\ \text{ADP} + 2\ \text{P}_i$$

   The stoichiometry belies the number of reactions involved in the separate metabolic compartments.
   (a) Construct a metabolite flow diagram that illustrates the net movement of acetyl-CoA from the mitochondria to the cytosol and which accommodates all the intermediate steps.
   (b) For each of the enzyme-catalyzed reactions, indicate which cofactors or coenzymes (if any) are required.
   (c) Explain why there is no net consumption of TCA cycle intermediates in the reaction sequence you derived in part (a).
   (d) What are some sources of cytoplasmic NADPH other than the NADP$^+$-malic enzyme?
15. Although there is no net synthesis of glucose from acetyl-CoA in mammals, acetyl-CoA has two major functions in gluconeogenesis. Explain the functions of acetyl-CoA in the synthesis of glucose from lactate in mammalian liver.

15

CHAPTER

Electron Transport and Oxidative Phosphorylation

In the last two chapters we have followed the catabolism of glucose to pyruvate and then to CO_2 and H_2O. These are oxidative processes. When glyceraldehyde-3-phosphate is oxidized to 1,3-diphosphoglycerate in glycolysis, two electrons stripped from the substrate are transferred to NAD^+. Electrons also move to NAD^+ when pyruvate is oxidized to acetyl-CoA, and at three steps in the TCA cycle (see fig. 15.1). At another point in the TCA cycle, electrons move from succinate to a flavin, FAD. Clearly, these reactions can continue only if the reduced coenzymes, NADH and $FADH_2$, are somehow reoxidized to NAD^+ and FAD.

As we saw in chapter 13, some types of cells can live anaerobically by transferring electrons from NADH back to an organic metabolite. If they are deprived of O_2, yeast reduce acetaldehyde to ethanol and muscle cells reduce pyruvate to lactate. Some bacteria survive by using inorganic oxidants such as nitrate. Before photosynthetic organisms began to release O_2 into the earth's atmosphere, virtually all heterotrophic organisms must have depended on processes similar to these. But the amount of ATP that can be obtained from these reactions is relatively small. For each molecule of glucose that it converts to ethanol a yeast cell gleans only two ATPs; bacteria that transfer electrons to nitrite, nitrate, or sulfate do not fare any better.

Photosynthetic cyanobacteria developed the ability to produce O_2 approximately 2.4 billion years ago. The subsequent evolution of cells that could use O_2 as an electron acceptor for the reoxidation of NADH and $FADH_2$ led to an enormous increase in the capacity to make ATP. In this chapter we will explore the series of oxidation-reduction reactions by which electrons move from the reduced coenzymes to O_2, and discuss how cells couple these reactions to the synthesis of ATP.

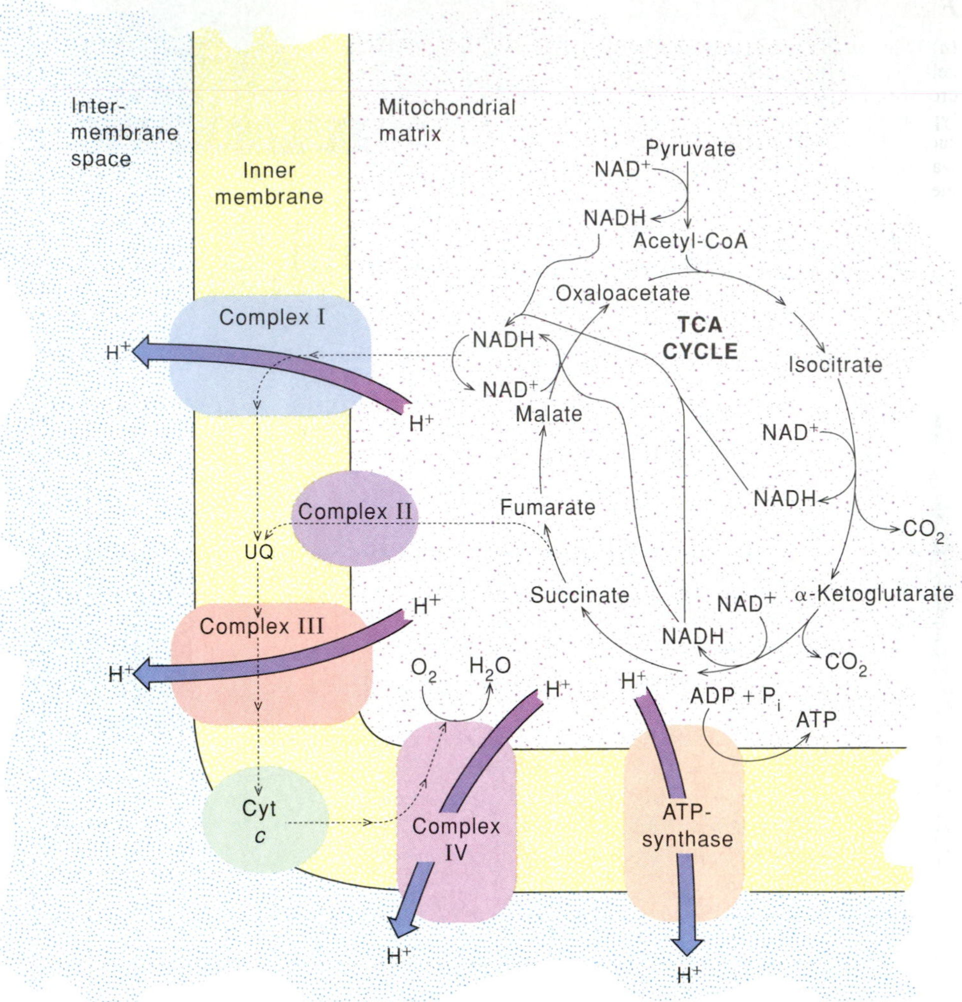

Figure 15.1

Overview of ATP production in mitochondria. The TCA cycle enzymes in the mitochondrial matrix oxidize acetyl-CoA to CO_2, transferring electrons to NAD^+ and to the bound FAD of succinate dehydrogenase. NAD^+ also is reduced in the conversion of pyruvate to acetyl-CoA. The NADH and bound $FADH_2$ created by these reactions are reoxidized by multiprotein electron-transfer complexes in the mitochondrial inner membrane. These complexes pass electrons stepwise to molecular oxygen. NADH dehydrogenase (complex I) moves electrons from NADH to a mobile carrier, ubiquinone (UQ). Succinate dehydrogenase (complex II) also transfers electrons to UQ. Complex III reoxidizes the reduced UQ and transfers electrons to cytochrome *c* (cyt *c*). Finally, complex IV moves electrons from cytochrome *c* to O_2. The oxidation-reduction reactions are coupled indirectly, but tightly, to the formation of ATP. As electrons pass through complexes I, III, and IV, protons are taken up from the solution inside the mitochondrion and released on the outside, setting up a gradient of pH and electrical potential across the inner membrane. Protons flow back into the mitochondrion through an ATP-synthase enzyme, which uses the free energy liberated from the proton influx to drive the synthesis of ATP. The parts of this figure shown in black are found in figure 12.5.

The Mitochondrial Electron-Transport Chain

In eukaryotic cells, the TCA cycle and most of the associated reactions of aerobic energy metabolism occur in mitochondria, cylindrical organelles that typically are several μm long and about 0.5 μm in diameter. Figure 15.2*a* shows an electron micrograph of a kidney cell mitochondrion. Mitochondria have two separate membranes, a smooth outer membrane and an inner membrane with extensive infoldings known as cristae (see fig. 15.2*b*). The inner membrane separates the mitochondrion into two distinct spaces, the internal matrix region and the intermembrane space between the inner and outer membranes.

The mitochondrial outer membrane has only a few known enzymatic activities, and is permeable to molecules with molecular weights up to about 5,000. The inner membrane is very different. It has an unusually high ratio of protein to phospholipid (about 3:1 by weight), and is impermeable to most ions and polar molecules. This permeability barrier prevents protons or cofactors such as NADH from moving freely between the mitochondrial matrix and the intermembrane space. The proteins of the inner membrane include the enzymes that are responsible for oxygen consumption and the formation of ATP. One of the enzymes of the TCA cycle, succinate dehydrogenase, also is embedded in the inner membrane. The other enzymes of the TCA cycle are in the matrix. The matrix also contains DNA and ribosomes, which are responsible for the synthesis of a few of the proteins of the mitochondrion. (Most mitochondrial proteins are coded in nuclear DNA and are imported into the mitochondrion after synthesis on cytosolic ribosomes.) Several enzymes that participate in ATP utilization, including creatine kinase and adenylate kinase, reside in the intermembrane space.

A role of mitochondria in O_2 uptake, or respiration, was suggested in histological studies by Michaelis in the early 1900s. In 1913, Otto Warburg found that oxygen consumption was associated with subcellular particles that he separated from soluble cellular components by filtration. In the late 1940s, Eugene Kennedy and Albert Lehninger showed that isolated mitochondria are able to carry out all of the reactions of the TCA cycle and the transport of electrons from substrates such as succinate to O_2, and that this flow of electrons is coupled to the formation of ATP.

Figure 15.2

(*a*) Thin-section electron micrograph of a mitochondrion in a frog kidney cell. (Courtesy of Dr. J. Luft, University of Washington.) (*b*) Schematic cross-sectional view of a mitochondrion. The morphology illustrated here is typical of mitochondria in tissues that have an active aerobic metabolism, such as kidney. The cristae resemble flattened sacks extending most of the way across the mitochondrion. They are all extensions of the inner membrane, which separates the matrix from the intermembrane space. The space within a crista connects to the intermembrane space through five or six tubular channels around the edges. (In the electron micrograph, some of the cristae appear not to be connected to the inner membrane, but this is because the cross section does not pass through the connecting regions.) The intermembrane space communicates with the cytosol through pores in the outer membrane.

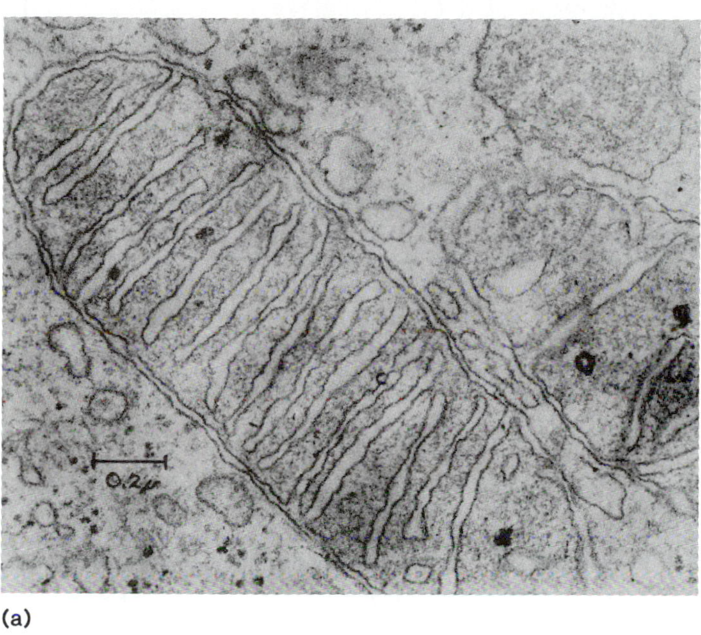

(a)

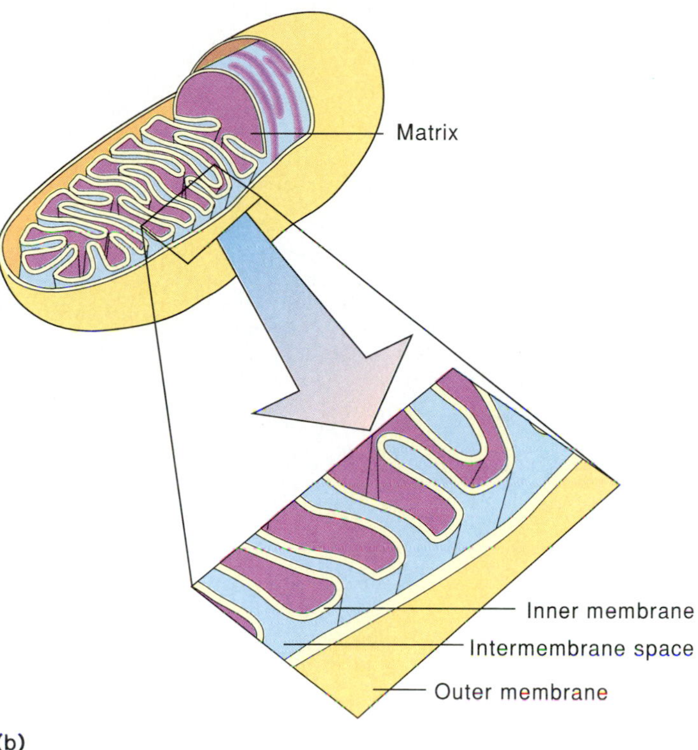

(b)

A Series of Cytochromes Carry Electrons toward O_2

Although many types of cells consume O_2 briskly, early biochemists noted that isolated metabolites such as sugars or organic acids do not react directly with O_2 at a significant rate. This fact suggested that an enzyme activates O_2 in some way to increase its reactivity. In the early 1920s, Warburg found that membranous fractions prepared from a variety of cells contained such an enzyme, and he concluded that iron played a central role in its function. The evidence for his conclusion was indirect. Warburg had discovered a number of inhibitors of respiration, including CO, N_3^-, and CN^-. Of these, CO was particularly interesting because the inhibition by CO could be reversed by light. CO was known to form complexes with Fe that had the property of dissociating when they were illuminated.

By measuring the effectiveness of light of different wavelengths at reversing the inhibition of respiration by CO, Warburg was able to obtain the absorption spectrum of the enzyme-CO complex. The spectrum indicated that the Fe is bound to the enzyme in the form of a heme. Hemes are iron-porphyrins that form the prosthetic groups of hemoglobin and myoglobin (see chapter 5) and of a large group of proteins called cytochromes. Warburg's "respiratory enzyme" proved to be a cytochrome.

The wider participation of cytochromes in respiration was recognized by David Keilin in the 1930s. Keilin found that aerobic cells contain three types of cytochromes, which he named *a, b,* and *c.* The three groups are distinguished by different substituents on the periphery of the porphyrin ring and by different modes of attachment of the porphyrin to the protein. The *b* cytochromes contain Fe-protoporphyrin IX, as do hemoglobin and

Figure 15.3

Iron protoporphyrin IX (heme) is found in the *b*-type cytochromes, and in hemoglobin and myoglobin. In heme *c*, cysteine residues of the protein (R) are attached covalently by thioether links to the two vinyl (—CH=CH_2) groups of protoporphyrin IX. Heme *c* is found in the *c* cytochromes. In heme A, which is found in the *a* cytochromes, a 15-carbon isoprenoid side chain is attached to one of the vinyls and a formyl group replaces one of the methyls.

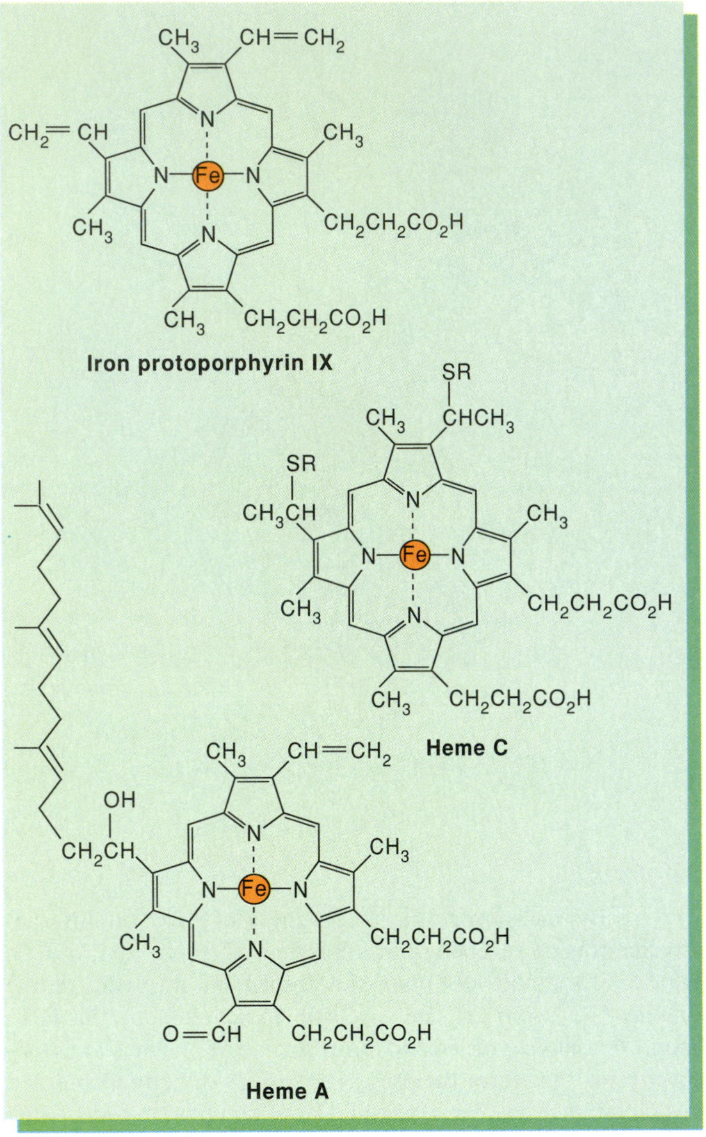

Iron protoporphyrin IX

Heme C

Heme A

Figure 15.4

Optical absorption spectra of a 10-μM solution of cytochrome *c* in the reduced and oxidized states. The three characteristic bands in the spectrum of the reduced cytochrome are called the α, β, and γ bands; the γ band is also referred to as the Soret band.

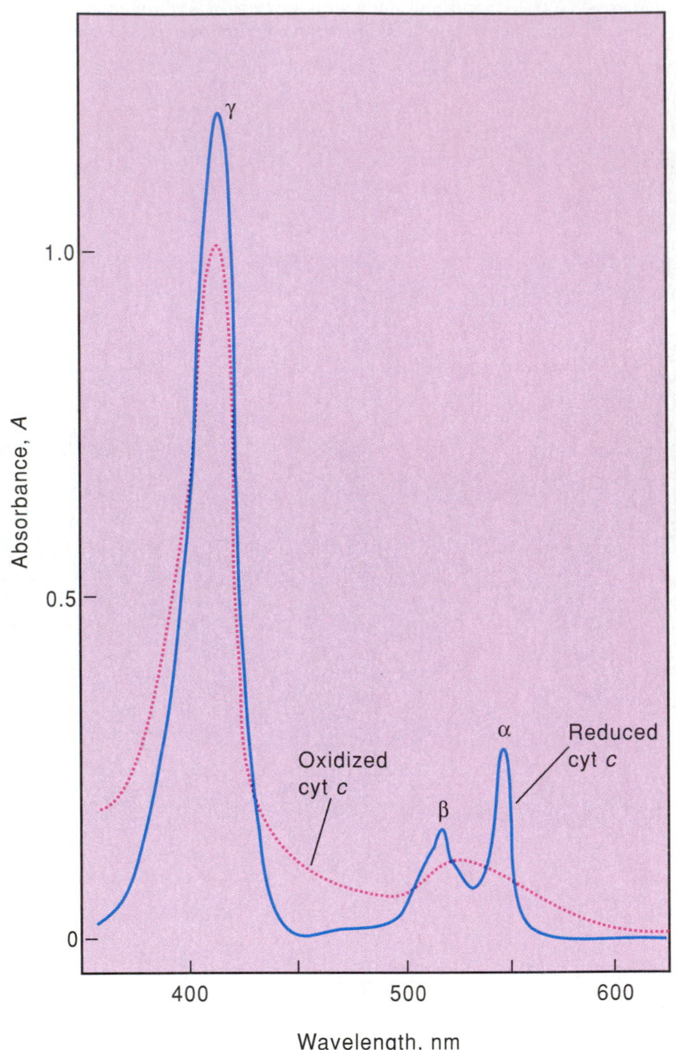

myoglobin; the *a* cytochromes contain a modified prosthetic group, heme A (fig. 15.3). The *c* cytochromes contain heme C, in which the vinyl groups of protoporphyrin IX are attached covalently to the protein by thioether links to cysteine residues. In the *a* and *b* cytochromes and the O_2-carrying proteins, the heme binds noncovalently to the protein. The three types of cytochromes also have characteristic absorption spectra: Their long-wavelength absorption band (the α band) is near 600 nm in the *a* cytochromes, 560 nm in the *b* cytochromes, and 550 nm in the *c* cytochromes (fig. 15.4).

One of the mitochondrial *c*-type cytochromes is a small, water-soluble protein that is associated loosely with the inner membrane. This cytochrome is usually called simply cytochrome *c*. Two *b* cytochromes (cytochromes b_L and b_H), two *a* cytochromes (*a* and a_3), and another *c* cytochrome (c_1) are embedded in the same membrane as parts of large complexes that we will describe in a later section.

The compact structure of cytochrome *c* has been determined by x-ray crystallography (fig. 15.5). The planar heme group is packed in the center of the molecule, surrounded mainly by hydrophobic amino acids. Near one edge of the heme there is a cluster of eight lysine residues. We will see that this positively charged region plays a role in the activity of the cytochrome. The Fe of the heme attaches on one side to the sulfur

Figure 15.5

The structure of cytochrome *c* from heart muscle. The α-carbon chain is represented in blue, and the heme in yellow with its iron atom in red. The side chains of the lysine residues clustered near one edge of the heme are shown in green. The figure is based on the crystal structure described by T. Takano and R. E. Dickerson.

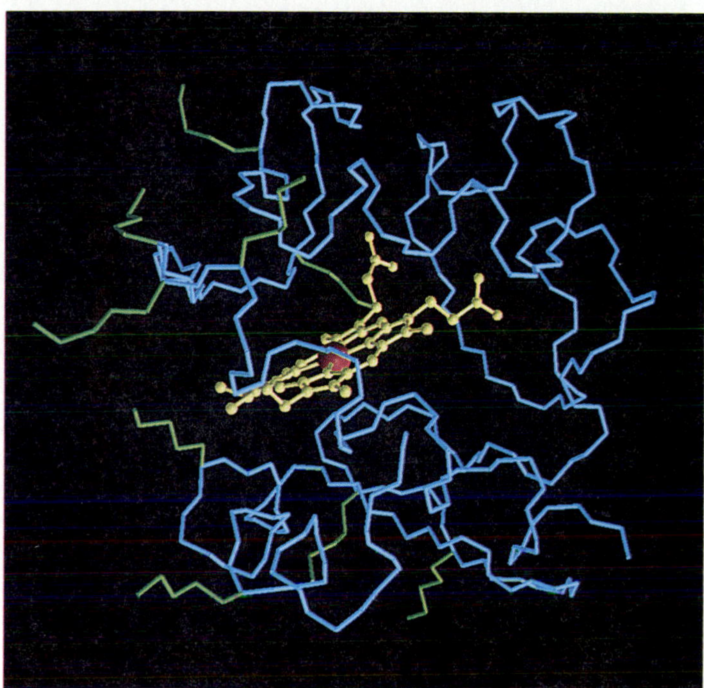

atom of a methionine residue and on the other to a histidyl nitrogen (fig. 15.6). By contrast, in hemoglobin and myoglobin, the methionine is replaced by a second, more distant histidine residue, leaving space for O_2, H_2O, or CO to bind to the Fe (see chapter 5).

The Fe atoms of the cytochromes undergo oxidation and reduction during respiration, cycling between the ferrous (Fe^{2+}) and ferric (Fe^{3+}) oxidation states. The absorption spectra of the oxidized and reduced forms differ (see fig. 15.4), and Keilin used this property to measure changes in the oxidation-reduction state of the cytochromes in living cells. Under anaerobic conditions, the cytochromes rapidly became reduced; in the presence of O_2, they became oxidized. If one of Warburg's inhibitors of respiration (CO, N_3^-, or CN^-) was added, it blocked the oxidation. Other types of inhibitors (amytal, rotenone, and malonate) blocked the reduction of the cytochromes. But Keilin found that purified cytochrome *c* was not oxidized directly by O_2. These observations suggested that the transfer of electrons from a substrate such as succinate (AH_2) to cytochrome *c* was catalyzed by a dehydrogenase, and the transfer of electrons from cytochrome *c* to O_2 by another component, which Keilin called cytochrome oxidase:

AH_2 → (Dehydrogenase) → $A + 2\,H^+$; 2 cytochrome *c* (Fe^{3+}) → 2 cytochrome *c* (Fe^{2+}) ; 2 cytochrome *c* (Fe^{2+}) → 2 cytochrome *c* (Fe^{3+}) ; $\frac{1}{2}O_2 + 2\,H^+$ → (Cytochrome oxidase) → H_2O

Figure 15.6

Edge-on views of the heme groups in cytochrome *c* and hemoglobin. The axial ligands in cytochrome *c* are histidyl and methionyl residues of the protein. In hemoglobin, there are two histidyl residues at different distances from the Fe, leaving room for O_2 to bind to the Fe on one side.

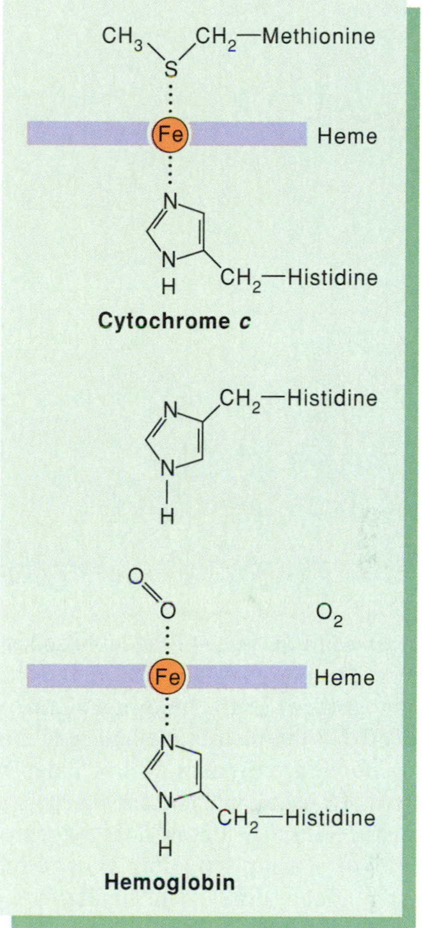

By 1940, it was clear that Keilin's cytochrome oxidase was identical with Warburg's respiratory enzyme, and that it involved the two *a*-type cytochromes, *a* and a_3. Cytochrome *c* appeared to pass electrons to cytochrome *a*, which in turn reduced cytochrome a_3. Cytochrome a_3 evidently reacts with O_2 and is the site of inhibition by CO and CN^-:

cytochrome *c* (Fe^{2+}) → cytochrome *c* (Fe^{3+}) ; cytochrome *a* (Fe^{3+}) → cytochrome *a* (Fe^{2+}) ; cytochrome *a* (Fe^{2+}) → cytochrome *a* (Fe^{3+}) ; cytochrome a_3 (Fe^{3+}) → cytochrome a_3 (Fe^{2+}) ; cytochrome a_3 (Fe^{2+}) → cytochrome a_3 (Fe^{3+}) ; $\frac{1}{4}O_2 + H^+$ → $\frac{1}{2}H_2O$

The *b* cytochromes and cytochrome c_1 were fitted into the scheme between the reducing substrates and cytochrome *c*. The idea thus developed that the respiratory apparatus includes a chain of electron carriers that operate in a defined sequence:

AH_2 → cytochromes b,c_1 → cytochrome *c* → cytochrome *a* → cytochrome a_3 → O_2

Figure 15.7

Structures of flavin mononucleotide (FMN) and flavin adenine dinucleotide (FAD).

Flavin mononucleotide (FMN) **Flavin adenine dinucleotide** (FAD)

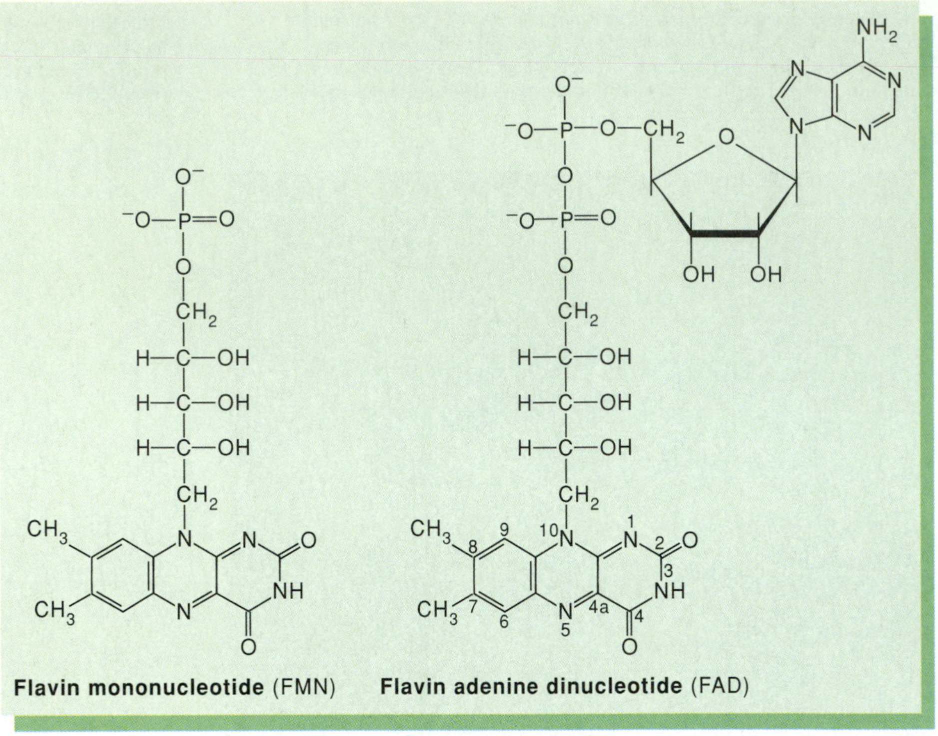

The next point that must be established is how the links in the respiratory chain are organized. Are the electron carriers actually bound together like the links of a chain, or do they diffuse independently in the membrane and react only when they collide? If the individual carriers diffuse independently, what prevents electrons from taking shortcuts such as jumping from a *b* cytochrome directly to a_3? How does cytochrome a_3 reduce O_2 to H_2O, instead of simply binding O_2 and releasing it unmodified, as hemoglobin does? And finally, how is the movement of electrons through the respiratory chain coupled to the formation of ATP? To pursue these questions we need to consider other electron carriers that participate in the system along with the cytochromes.

Other Electron Carriers in the Respiratory Chain

In addition to cytochromes, the mitochondrial electron-transport system contains at least four other types of electron carriers: flavins, quinones, iron-sulfur complexes, and copper atoms.

Flavins The dehydrogenases that remove electrons from succinate or NADH contain flavins as prosthetic groups. NADH dehydrogenase contains flavin mononucleotide (FMN); succinate dehydrogenase contains covalently bound flavin adenine dinucleotide (FAD). The structures of these coenzymes are shown in figure 15.7. Flavins can undergo one-electron reduction to semiquinone forms or two-electron reduction to the dihydroflavins, $FMNH_2$ and $FADH_2$ (see fig. 11.10). The reduction can be measured spectrophotometrically by the disappearance of a broad absorption band at 450 nm.

The inner membrane of the mitochondrion also has a flavoprotein, glycerol-3-phosphate dehydrogenase, that oxidizes glycerol-3-phosphate to dihydroxyacetone phosphate, reducing the bound FAD:

$$HOCH_2CHOHCH_2OP + FAD \rightarrow HOCH_2COCH_2OP + FADH_2$$

In the mitochondrial matrix, there are at least eight other flavoprotein dehydrogenases, including the acyl-CoA dehydrogenases that participate in the oxidation of fatty acids (chapter 17). All of these dehydrogenases in the matrix transfer electrons to another membrane-bound flavoprotein, the electron-transfer flavoprotein, which reduces still another flavoprotein, the electron-transfer-flavoprotein-ubiquinone oxidoreductase. Like NADH dehydrogenase, succinate dehydrogenase, and glycerol-3-phosphate dehydrogenase, this enzyme passes electrons to the cytochromes by way of iron-sulfur complexes and a quinone.

Iron-Sulfur Centers NADH dehydrogenase, succinate dehydrogenase, and the electron-transfer-flavoprotein-ubiquinone oxidoreductase contain iron atoms that are bound by the sulfur atoms of cysteine residues of the protein, in association with additional, inorganic sulfide atoms. Proteins that contain such iron-sulfur complexes are termed nonheme iron proteins. The iron-sulfur centers of most of the nonheme iron proteins contain either two iron atoms and two sulfide atoms bound to four cysteines, or four iron atoms and four sulfides, again bound to four cysteines. Structures of these complexes are shown in figure 15.8. Succinate dehydrogenase has three iron-sulfur centers, one with a [2 Fe-2 S] structure, one with [4 Fe-4 S], and one with a cluster

Figure 15.8

Structures of 2 Fe-2 S (*a*) and 4 Fe-4 S (*b*) iron-sulfur centers. In both structures, the arrangements of the sulfur ligands around the iron atoms are approximately tetrahedral. The cysteine residues are part of the polypeptide chain.

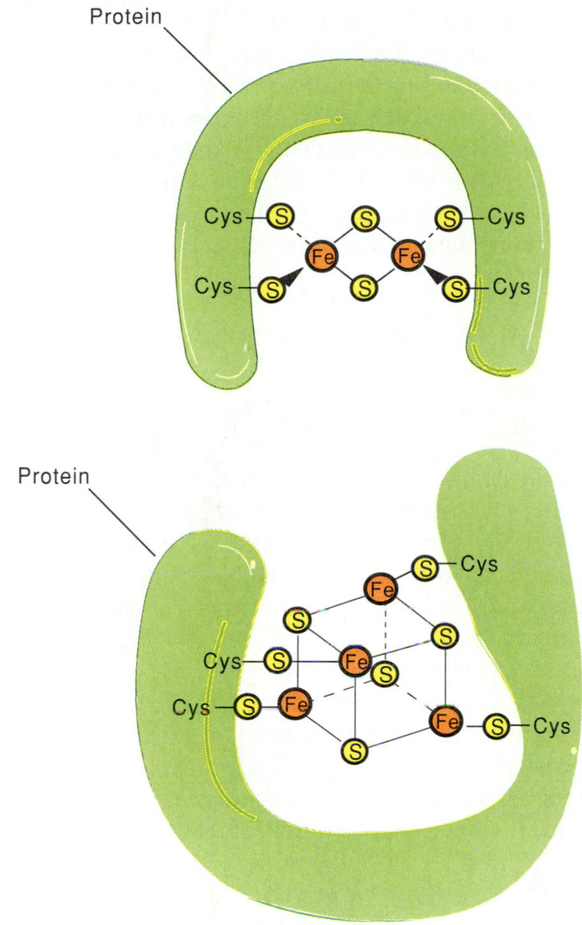

Figure 15.9

Structures of ubiquinone in its oxidized state (UQ) and in the fully reduced dihydroquinone or quinol state (UQH_2). The UQ found in eukaryotes has 50 carbons (10 prenyl units) in the side chain; shorter side chains are found in some bacteria. The side chain makes ubiquinone soluble in the phospholipid bilayer of the mitochondrial inner membrane.

Ubiquinone (UQ)

$2e^- + 2H^+$

Dihydroubiquinone (UQH_2)

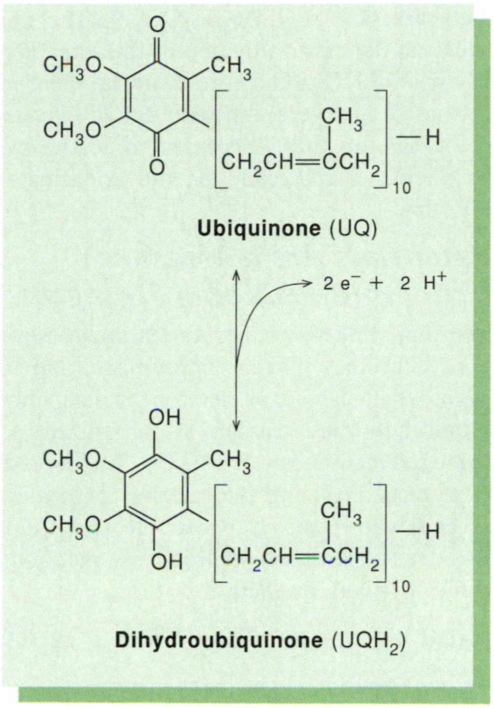

containing three iron atoms and three (or possibly four) sulfides. Another type of iron-sulfur protein found in the mitochondrial inner membrane, the Rieske iron-sulfur protein, appears to have its two iron atoms bound to two cysteines and two histidine residues, in addition to two sulfide atoms. Iron-sulfur centers undergo one-electron oxidation-reduction reactions, which can be viewed as involving a transition of one of the iron atoms between its Fe^{2+} and Fe^{3+} states. However, the actual molecular orbitals of the iron-sulfur centers are more complex than this description suggests, and the changes in electron density are distributed over more than one iron atom.

Ubiquinone Ubiquinone (UQ) is a benzoquinone with a long side chain (fig. 15.9). Because of its discovery by two groups of investigators, it still goes under two different names, ubiquinone and coenzyme Q. The former name indicates the ubiquitous nature of the quinone in aerobic cells ranging from bacteria to plants and mammals. The length of the side chain varies from six prenyl (five-carbon) units in some bacteria to ten in mammals, but the different ubiquinones are functionally interchangeable. Some bacteria contain a naphthoquinone with a similar side chain (menaquinone, or vitamin K_2) in place of, or in addition to UQ. With either UQ or the menaquinones, the hydrocarbon tail makes the molecule strongly hydrophobic. In bacteria, the UQ is found in the cytoplasmic membrane; in eukaryotes it is found mainly in the mitochondrial inner membrane. Compared with the concentrations of the cytochromes in these membranes, the concentration of UQ is relatively large. In heart mitochondria, for example, the concentration of UQ is about seven times that of cytochrome a_3. Most of the UQ probably is not bound tightly to proteins, but rather moves freely in the phospholipid bilayer of the membrane.

Ubiquinone undergoes a two-electron reduction to the dihydroquinone or quinol, UQH_2 (see fig. 15.9). It also can accept a single electron and stop at the semiquinone, which can be either anionic ($UQ^{\cdot -}$) or neutral ($UQH^{\cdot}$), depending on the pH and on the nature of the binding site when the semiquinone is bound to a protein. The · in $UQ^{\cdot -}$ or $UQH^{\cdot}$ indicates that the semiquinone contains an unpaired electron.

The reduction of UQ can be measured by the disappearance of an absorption band at 275 nm. By this technique it was shown that the addition of a substrate such as succinate caused a rapid reduction of essentially all the UQ present in the mitochondrial membrane, and that the UQH_2 could be reoxidized by the cytochrome system in the presence of O_2. To determine whether UQ is a necessary participant in electron transport from succinate to O_2, the quinone was removed from the mitochondria by extraction with an organic solvent. The depleted mitochondria were incapable of respiration but recovered this activity when reconstituted with UQ. Similar results were obtained with electron donors other than succinate, including glycerol-3-phosphate and substrates that transfer their electrons initially to NADH, such as malate, α-ketoglutarate, and β-hydroxybutyrate ($CH_3CHOHCH_2CO_2^-$). The ability of a variety of electron donors to function in this capacity suggests that the large pool of UQ in the mitochondrial inner membrane accepts electrons from several different dehydrogenases, and is responsible for passing these electrons on to the cytochrome system. The kinetics of UQ reduction and oxidation are consistent with this view.

Redox Potentials Are a Measure of Oxidizing and Reducing Strengths

We have seen that the respiratory system includes a variety of molecules—cytochromes, flavins, ubiquinone, and iron-sulfur proteins—all of which can act as electron carriers. Our goal now is to understand how these carriers are organized to transport electrons from reduced substrates to O_2. To proceed, we first need a way of characterizing the relative tendencies of these different molecules to release or accept electrons.

Consider the lactate dehydrogenase reaction, in which NADH reduces pyruvate to lactate:

$$\text{NADH} + \text{H}^+ + \text{pyruvate} \rightarrow \text{NAD}^+ + \text{lactate}$$

Conceptually, we can imagine the reaction occurring in two steps. In the first step, NADH undergoes oxidation to NAD^+, releasing two electrons and a proton; in the second, pyruvate undergoes reduction by accepting two electrons and two protons:

$$\text{NADH} \rightarrow \text{NAD}^+ + 2\,\text{e}^- + \text{H}^+ \quad (1)$$

$$\text{Pyruvate} + 2\,\text{e}^- + 2\,\text{H}^+ \rightarrow \text{lactate} \quad (2)$$

Actually, it is a hydride ion (H^-) that is transferred from NADH to pyruvate (see chapter 11), but the details of the mechanism do not concern us for the moment. Each of the steps involves a redox couple of two molecules that are interconvertible by the transfer of electrons and protons.

The standard free energy change ($\Delta G^{\circ\prime}$) for the overall reaction between NADH and pyruvate to form NAD^+ and lactate is −6.22 kcal/mole. The negative sign indicates that, given equimolar concentrations of the reactants and products, the reaction proceeds spontaneously in the direction of lactate. This means that NADH is a stronger electron-donor than lactate. Now how does NADH compare with $FMNH_2$ or with cytochrome *c*? To deal quantitatively with questions like these, it is useful to define a parameter called the standard redox potential, E°, which expresses the relative tendency of a pair of molecules like the NAD^+/NADH couple to release or accept electrons. Like the electrical potentials that govern the direction of electron flow from one physical region to another, redox potentials are specified in units of volts. Because electron transfer reactions frequently involve protons also, an additional symbol is used to indicate that the E° value applies to a particular pH; an $E^{\circ\prime}$ thus refers to an E° at pH 7. (People who work in bioenergetics often refer to E° as "the midpoint" potential and use the symbols E_m and $E_{m,7}$ in place of E° and $E^{\circ\prime}$.)

Suppose we have two redox couples D_{ox}/D_{red} and A_{ox}/A_{red}, where the subscripts ox and red indicate the oxidized and reduced forms of the molecules. Consider the reaction in which D_{red} acts as an electron donor (reductant) and A_{ox} as the electron acceptor (oxidant). We take the difference between the E° values of two couples (ΔE°) to be proportional to the standard free energy change for the reaction:

$$D_{red} + A_{ox} \rightarrow D_{ox} + A_{red} \quad (3)$$

$$\Delta E^\circ = E^\circ_A - E^\circ_D = -\Delta G^\circ/nF \quad (4)$$

or

$$\Delta G^\circ = -nF\,\Delta E^\circ \quad (5)$$

Here *n* is the number of electrons transferred from the donor to the acceptor (two, in the case of the lactate dehydrogenase reaction), and the proportionality constant *F* is the Faraday constant (23,060 cal·volt^{-1} mole^{-1} or 96.5 kJ·volt^{-1} mole^{-1}). Note that these expressions specify a convention concerning the signs of E° values: The overall reaction is spontaneous (ΔG° negative) if ΔE° is positive, that is, if electrons are transferred from a couple with a more negative E° value to one with a more positive value. In other words, relatively negative E° values are associated with strong reductants, and relatively positive values with strong oxidants. With $n = 2$, a difference of 0.1 V between the E° values corresponds to a ΔG° of −4.61 kcal/mole (19.3 kJ/mole).

The relationship between ΔG° and ΔE° is the same as the expression that describes electron flow between two physical regions that are at different electrical potentials, such as the two terminals of a battery. Electrons tend to flow spontaneously in the direction of more positive potential. In fact, we can measure ΔE° values by separating the two redox couples physically into different compartments connected by a salt bridge and a voltmeter so that electrons can flow from one solution to the other through the meter (fig. 15.10). If the solutions in the two compartments are both under standard conditions, the meter will sense a voltage difference equal to the ΔE°.

So far, we have defined only the difference between two E° values. To set the individual values, it is necessary to choose a particular redox couple as a reference. The reference that is used most commonly by biochemists is the standard hydrogen half-cell, in which protons at 1 M [H^+] are reduced to H_2 at a pressure of 1 atm ($2\,H^+ + 2\,e^- \rightarrow H_2$). This half-cell is arbitrarily assigned an E° value of zero. The choice of pH 0 as the standard condition for this half-cell does not restrict our choice of conditions for the cell in the other compartment of the apparatus. Relative to the standard hydrogen half-cell, the NAD^+/NADH couple at pH 7 has an $E^{\circ\prime}$ of −0.32 V, and the pyruvate/lactate couple has an $E^{\circ\prime}$ of −0.19 V.

Figure 15.10

Apparatus for measuring the difference between the $E°$ values of two redox couples. The cell on the left contains equimolar concentrations of NADH and NAD^+; that on the right, equimolar concentrations of $FMNH_2$ and FMN. If both solutions are at pH 7, the voltmeter will sense the difference between the two $E°'$ values (0.10 V, negative on the left, in this example). To determine the $E°'$ of one of the redox couples, the other couple is replaced by a standard redox couple such as the standard hydrogen half-cell (H_2/H^+ at pH 0 and 1 atm H_2). The dependence of one of the $E°$ values on pH can be measured by changing the pH in the experimental cell and holding that in the reference cell constant. Experimentally, the calomel redox couple ($Hg/HgCl_2$) is often used as a working reference instead of the hydrogen half-cell, and the results are corrected by subtracting the $E°$ of the calomel couple (0.24 V). It may be necessary to add a catalytic amount of another organic redox couple as a mediator to facilitate the transfer of electrons to and from the metallic electrodes.

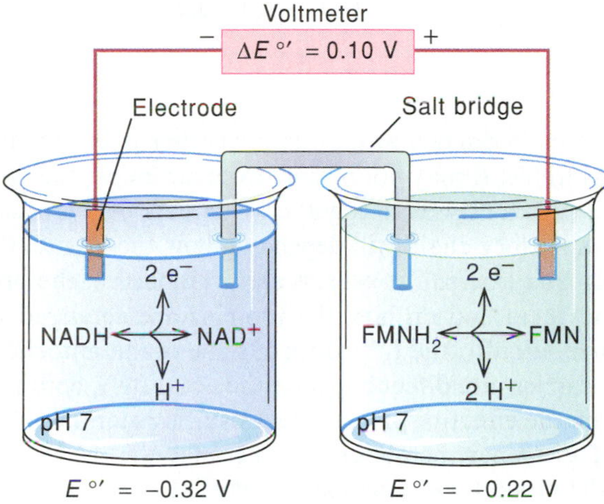

Table 15.1 gives the $E°'$ values of a number of biochemical redox couples, including some of the components of the respiratory chain. These are listed in order of increasing $E°'$, which means that under standard conditions a given couple will reduce any of the couples below it in the table.

As we mentioned earlier, quinones and flavins can undergo one-electron reduction to semiquinone forms or can be fully reduced by the addition of two electrons. The $E°$ values for the individual one-electron steps can be very different. In the case of UQ, the one-electron redox couples UQ/UQH· and UQH·/UQH_2 have $E°'$ values of approximately 0.03 and 0.19 V. Because the $E°'$ of the second couple is more positive than that of the first, the second electron is thermodynamically easier to add than the first. The $E°'$ for the overall two-electron reduction of UQ to UQH_2 is the mean of the $E°'$ values for the two one-electron steps, 0.11 V.

$E°$ values are called "standard" redox potentials because they refer to standard conditions, which usually are chosen to be 1-M concentrations of all reactants and products except for H^+, OH^-, and H_2O. By the principle of mass action, a solution containing a redox couple D_{ox}/D_{red} can be made more oxidizing by increasing the concentration ratio $[D_{ox}]/[D_{red}]$, or more reducing by decreasing this ratio. The voltage measured with the apparatus shown in figure 15.10 will reflect such changes. Suppose that the couple in one of the compartments is the standard hydrogen half-cell, and that the concentrations here are held constant while $[D_{ox}]$ or $[D_{red}]$ is varied in the other compartment. The voltage difference measured between the two compartments can be defined as the redox potential, E, in the solution containing D_{ox} and D_{red}. The relationship between E and $E°$ is the same as that between the free energy change in a reaction (ΔG) and the standard free energy change ($\Delta G°$).

Table 15.1
$E°'$ Values of Biochemical Redox Couples

Redox Couple	$E°'$ (V)	n[a]
Succinate + CO_2 + 2 H^+ + 2 e^- ⇌ α-ketoglutarate + H_2O	−0.67	2
Glycerate-3-phosphate + 2 H^+ + 2 e^- ⇌ glyceraldehyde-3-phosphate + H_2O	−0.55	2
α-Ketoglutarate + CO_2 + 2 H^+ + 2 e^- ⇌ isocitrate	−0.38	2
NAD^+ + H^+ + 2 e^- ⇌ NADH	−0.32	2
Glycerate-1,3-bisphosphate + 2 H^+ + 2 e^- ⇌ glyceraldehyde-3-phosphate + P_i	−0.29	2
Lipoic acid + 2 H^+ + 2 e^- ⇌ dihydrolipoic acid	−0.29	2
FMN + 2 H^+ + 2 e^- ⇌ $FMNH_2$	−0.22[b]	2
FAD + 2 H^+ + 2 e^- ⇌ $FADH_2$	−0.22[b]	2
Acetaldehyde + 2 H^+ + 2 e^- ⇌ ethanol	−0.20	2
Pyruvate + 2 H^+ + 2 e^- ⇌ lactate	−0.19	2
Oxaloacetate + 2 H^+ + 2 e^- ⇌ malate	−0.17	2
Fumarate + 2 H^+ + 2 e^- ⇌ succinate	−0.03	2
Cytochrome $b_L(Fe^{3+})$ + e^- ⇌ cytochrome $b_L(Fe^{2+})$	−0.03	1
UQ + H^+ + e^- ⇌ UQH·	+0.03[c]	1
Cytochrome $b_H(Fe^{3+})$ + e^- ⇌ cytochrome $b_H(Fe^{2+})$	+0.05	1
UQ + 2 H^+ + 2 e^- ⇌ UQH_2	+0.11[c]	2
UQH· + H^+ + e^- ⇌ UQH_2	+0.19[c]	1
Rieske Fe-S (Fe^{3+}) + e^- ⇌ Fe-S (Fe^{2+})	+0.28	1
Cytochrome $c_1(Fe^{3+})$ + e^- ⇌ cytochrome $c_1(Fe^{2+})$	+0.23	1
Cytochrome c (Fe^{3+}) + e^- ⇌ cytochrome c (Fe^{2+})	+0.24	1
Cytochrome a (Fe^{3+}) + e^- ⇌ cytochrome a (Fe^{2+})	+0.28	1
Cytochrome a_3 (Fe^{3+}) + e^- ⇌ cytochrome $a_3(Fe^{2+})$	+0.35	1
O_2 + 4 H^+ + 4 e^- ⇌ 2 H_2O	+0.82	4

[a] n is the number of electrons transferred.

[b] This value is for the free coenzyme. $E°'$ values for flavoproteins range from −0.3 to 0 V.

[c] For UQ in aqueous ethanol.

Considering how ΔG depends on $\Delta G°$ and on the concentrations of the reactants and products (see chapter 2), we can see that E must depend on the concentration ratio as follows:

$$E = E° + (2.303\ RT/nF) \log \{[D_{ox}]/[D_{red}]\} \quad (6)$$

This is one of two expressions that go under the name of the Nernst equation. If $[D_{ox}]/[D_{red}]$ is increased by a factor of 10, E increases by an amount equal to 2.303 RT/nF. At 25° C, 2.303 RT/nF is 0.060 V for $n = 1$, and 0.030 V for $n = 2$.

The Nernst equation is analogous to the Henderson-Hasselbach equation, which describes how the ratio of base and acid concentrations depends on the pH and the pK_a of an acid. Figure 15.11 shows how the $[D_{ox}]/[D_{red}]$ ratio depends on E for several redox couples with different values of $E°$. Note that, because of the way n enters into equation (6), the curves shown in figure 15.11 are steeper if $n = 2$ than if $n = 1$. $[D_{ox}]/[D_{red}]$ increases by a factor of 10 for every 0.06 V increase in E when $n = 1$, and for every 0.03 V when $n = 2$.

Let us now return to the point that $E°$ values can depend on pH. When UQ, for example, is reduced to UQH_2, two protons are added to the molecule along with each pair of electrons. By mass action, increasing the proton concentration should favor the reduction. Lowering the pH thus must make UQ a stronger oxidant. This relation can be expressed quantitatively by including the proton concentration along with the concentrations of UQ and UQH_2 in the equation for the free energy change associated with the oxidation-reduction reaction. The result is that, if the pH is decreased by one pH unit, the $E°$ value becomes more positive by $(2.303RT/F)\ (n_{H^+}/n)$, where n_{H^+} is the number of protons taken up in the reaction (two in the case of the UQ/UQH_2 couple) and n again is the number of electrons transferred (two in this case). At 25° C, $(2.303RT/F)\ (n_{H^+}/n)$ is 0.060 V per pH unit when $n_{H^+} = n$.

E° Values of Electron Carriers in the Respiratory Chain Given equal concentrations of reactants and products, electrons will tend to flow spontaneously from a carrier with a more negative $E°$ value to one with a more positive value. The sequence of carriers in the respiratory chain should, therefore, be generally consistent with the relative $E°$ values of the carriers. The linear sequence of *b*-, *c*-, and *a*-type cytochromes that was proposed above agrees with this expectation. We can verify this by plotting the $E°'$ values as a function of the carriers' suggested positions in the chain (fig. 15.12). However, the $E°$ values do not allow us to determine the exact organization of the respiratory chain, particularly in regions where there are numerous carriers with similar values of $E°$. There are several reasons for this. First, it is not really $\Delta E°$ that determines the direction of electron flow, but the difference between the operating redox potentials, ΔE. Electrons can move in the direction of more negative $E°$ if the reactants (the reduced form of the electron donor and the oxidized form of the electron acceptor) are present at sufficiently high concentrations relative to the products (the oxidized donor and the reduced acceptor). This is simply an application of the law of mass action. Further, even if electron transfer from a particular carrier to another is thermodynamically favorable, it may not be able to occur unless the two carriers are sufficiently close together. Contacts between carriers in the respiratory chain will depend on how the hemes, flavins, quinones, and iron-sulfur centers are positioned in the proteins that bind them and on how the proteins are arranged in the membrane. Finally, the $E°$ values of some of the mitochondrial electron carriers are difficult to measure accurately, because they depend on the effective pH and electrostatic potential in the interior of the inner membrane and because the measurements can be disturbed by interactions among the carriers.

Figure 15.11

Theoretical titration curves for three different redox couples, showing how the portion of a couple that is in the oxidized form increases as the redox potential in the solution is made more positive. Curve A is for a redox couple that undergoes a 1-electron reaction ($n = 1$) with an $E°$ value of −0.10 V. Curve B is for an $E°$ of +0.05 V and $n = 2$. Curve C is for an $E°$ of +0.25 V and $n = 1$. The ordinate shows $100 \times [D_{ox}]/([D_{ox}]+[D_{red}])$, as calculated from the Nernst equation [equation (6)]. When $E = E°$, 50% of a redox couple is in the oxidized form. The dashed lines indicate $E°$ on the abscissa.

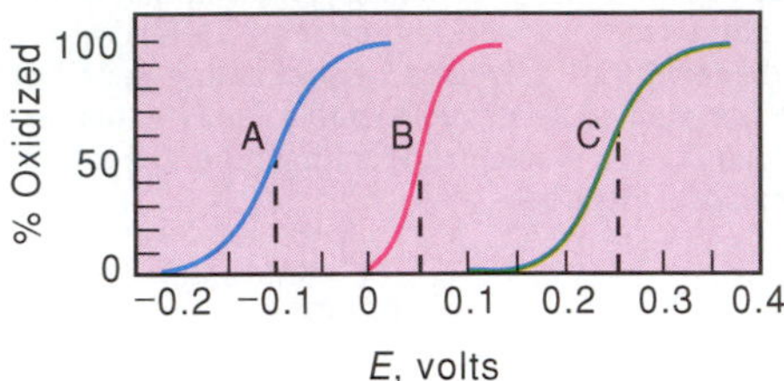

The Sequence of Carriers Can Be Deduced from Kinetic Studies and from the Effects of Specific Inhibitors

Because we cannot deduce the sequence of electron carriers in the respiratory chain simply on the basis of the $E°$ values, it is necessary to bring additional experimental techniques to bear on the problem. Spectrophotometric techniques have been particularly helpful, because they make it possible to examine the rates at which the different electron carriers accept and release electrons in intact, functioning mitochondria. Other approaches that have played key roles are the use of specific inhibitors and the isolation of protein complexes that represent small sections of the respiratory chain.

Difference Spectra Extremely sensitive and rapid spectrophotometric techniques for examining the redox states of the respiratory carriers in intact mitochondria were developed by Britton Chance in the 1950s. Chance measured the differences between the optical absorption spectra of two samples of mitochondria that were under different conditions. Figure 15.13 shows such a difference spectrum. For this measurement, the electron carriers in the control or reference sample were predominantly in their oxidized states because the solution contained excess O_2. The other sample was provided with an excess of a reducing substrate and was allowed to use up all its O_2, so

Figure 15.12

Approximate $E^{\circ\prime}$ values of some of the electron carriers in the respiratory chain, plotted as a function of the approximate positions of the carriers in the chain. The diagram includes components of the cytochrome bc_1 complex that are discussed later in the chapter. For simplicity, it shows a single $E^{\circ\prime}$ value for ubiquinone, although the UQH_2/UQH• and UQH•/UQ redox couples have different values. The $E^{\circ\prime}$ for the FMN associated with NADH dehydrogenase is not known accurately; the value shown is for free FMN. The $E^{\circ\prime}$ values and positions for the iron-sulfur centers associated with NADH dehydrogenase also are uncertain; only two of the six to eight iron-sulfur centers in this region of the respiratory chain are shown. The right-hand scale gives the standard free energy change ($\Delta G^{\circ\prime}$) for transferring two equivalents of electrons from a carrier with the corresponding $E^{\circ\prime}$ value to O_2.

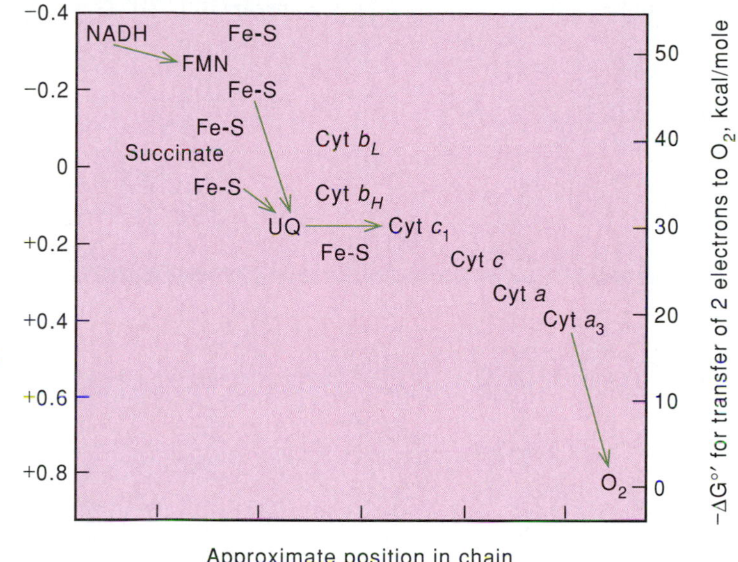

Figure 15.13

Reduced-minus-oxidized difference spectrum of the electron carriers in rat liver mitochondria. The plot shows the difference (ΔA) between the absorption spectrum of a sample of mitochondria that had used up all the O_2 in the solution and the spectrum of a reference sample that contained O_2. Both samples were provided with ß-hydroxybutyrate, which can enter mitochondria and serve as a reductant for the internal NAD^+. The peak near 340 nm reflects reduction of NAD^+ to NADH; the sharp positive peaks at longer wavelengths, reduction of cytochromes. The peaks near 420 and 440 nm, which are due mainly to the reduction of *c* and *a* cytochromes, are superimposed on a broad, negative band that results partly from the reduction of flavins. (For the individual absorption spectra of oxidized and reduced cytochrome *c*, see figure 15.4.) Absorbance changes due to the *b* cytochromes are not well resolved from those due to the *c* and *a* cytochromes in this spectrum. A difference spectrum can be sensitive to small changes in one of the samples even if both samples have a high absorbance. (For details, see B. Chance and G. R. Williams, Respiratory enzymes in oxidative phosphorylation, II. Difference spectra. *J. Biol. Chem.* 217:395–407, 1955.)

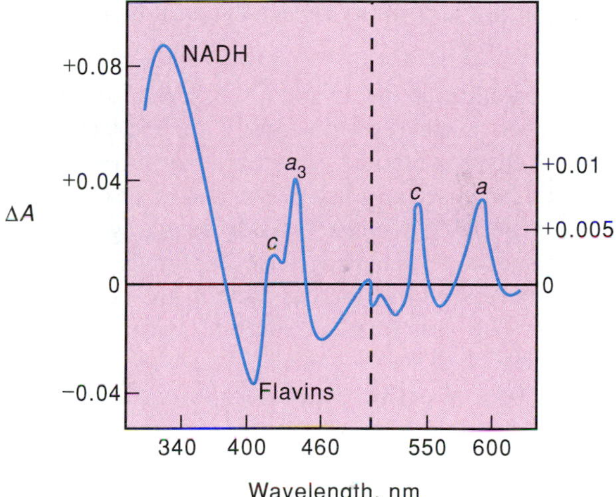

that the electron carriers became largely reduced. The spectrum is therefore a reduced-minus-oxidized difference spectrum. The positive peak near 340 nm is due to the reduction of NAD^+ to NADH; the sharp positive peaks at longer wavelengths, to reduction of the cytochromes (compare the curves for reduced and oxidized cytochrome *c* in figure 15.4). Reduction of flavins causes a broad, negative absorbance change in the region from 400 to 480 nm, underlying the sharp absorbance increases due to the cytochromes.

The reducing substrate used in the experiment shown in figure 15.13 was β-hydroxybutyrate. A dehydrogenase in the mitochondrial matrix oxidizes β-hydroxybutyrate to β-ketobutyrate, transferring a pair of electrons to NADH:

$$CH_3CHOHCH_2CO_2^- + NAD^+ \rightarrow CH_3COCH_2CO_2^- + NADH + H^+$$

The NADH dehydrogenase in the inner membrane then transfers electrons from NADH to the respiratory chain. Malate, α-ketoglutarate, or any other substrate that is capable of reducing mitochondrial NAD^+ to NADH gives a spectrum similar to that shown in the figure. Succinate also gives a similar spectrum, except that the peak near 340 nm is not seen. Recall that succinic dehydrogenase is a flavin enzyme that does not involve NAD^+. This raises the question of whether succinate dehydrogenase reduces one set of cytochromes and NADH dehydrogenase reduces another, independent set, or whether both enzymes reduce the same cytochromes. Chance found that if a strong chemical reductant such as $Na_2S_2O_4$ was added to a mitochondrial sample that had been reduced with either β-hydroxybutyrate or succinate, there was little additional reduction of cytochromes. Any of the normal respiratory substrates evidently is capable of reducing virtually all of the cytochromes in the membrane. As mentioned above, either succinate or NADH-linked substrates also can reduce essentially all of the UQ. The flavins, however, are reduced only partially by succinate, but almost completely by a combination of succinate and malate. These observations fit the idea that all of the cytochromes and UQ are part of a common network, to which various flavoprotein dehydrogenases can feed electrons.

Using similar techniques, Britton Chance, Quentin Gibson, and others also measured the kinetics with which the different electron carriers became oxidized after the addition of

O_2 to an anaerobic suspension of mitochondria, or became reduced after addition of substrate. When O_2 was added, cytochrome a_3 became oxidized first, followed by cytochrome *a*, the *c* cytochromes, the *b* cytochromes, and flavins, in that order. Later work by Klingenberg showed that UQ became oxidized at about the same time as the *b* cytochromes. These observations suggest that the electron carriers are arranged in the sequence:

$$\text{Succinate} \rightarrow \text{FAD NADH} \rightarrow \text{FMN} \rightarrow [\text{UQ, cyt } b] \rightarrow [\text{cyt } c_1, \text{cyt } c] \rightarrow \text{cyt } a \rightarrow \text{cyt } a_3 \rightarrow O_2$$

This is consistent with the sequence shown in figure 15.12. Such kinetic measurements have to be interpreted cautiously, however. The observed rate of oxidation or reduction of a carrier depends on the difference between the rates of the reactions in which the carrier accepts electrons from components upstream and gives electrons to components downstream. In principle, the net rate might not accurately reflect the position of the carrier in the chain.

Inhibitors In addition to CO and CN^-, several specific inhibitors have been useful experimentally for exploring the sequence of the electron carriers. Amytal (a barbiturate), rotenone (a plant toxin long used as a fish poison and an insecticide), and piericidin-A specifically block NADH dehydrogenase. Malonate blocks succinate dehydrogenase. Antimycin, a toxin obtained from *Streptomyces griseus,* acts in the region of the *b* cytochromes. The structures of some of these inhibitors are shown in figure 15.14.

The sites of action of the respiratory inhibitors were studied by measuring difference spectra between samples that were treated with the inhibitors and untreated controls. For these experiments, both of the mitochondrial suspensions contained a reducing substrate in addition to O_2, so that the electron carriers were in a steady state of partial reduction. If CO or CN^- was added to one of the samples, all of the electron carriers in the sample became more reduced. This result is consistent with the idea that these agents block the functioning of cytochrome a_3 at the downstream end of the respiratory chain, preventing the removal of electrons from the chain by O_2. If antimycin was added, NAD^+, flavins, and the *b* cytochromes became more reduced, but cytochromes *c*, c_1, *a*, and a_3 all became more oxidized. The situation here is analogous to the construction of a dam across a stream: When the gates are closed, the water level increases upstream from the dam and decreases downstream. The results with antimycin suggest that this inhibitor blocks electron flow between the *b* cytochromes and cytochrome c_1. Although this conclusion has needed modification in the light of more recent work that we will discuss later, antimycin was very helpful for establishing the general organization of the respiratory chain. The observation that antimycin did not inhibit the reduction of UQ, for example, showed that the quinone fits into the chain upstream of cytochromes *c*, c_1, *a*, and a_3.

Figure 15.14

Structures of several inhibitors of electron transport. Amytal, rotenone, and piericidin-A block NADH dehydrogenase; antimycin blocks in the region of the *b* cytochromes (see fig. 15.18).

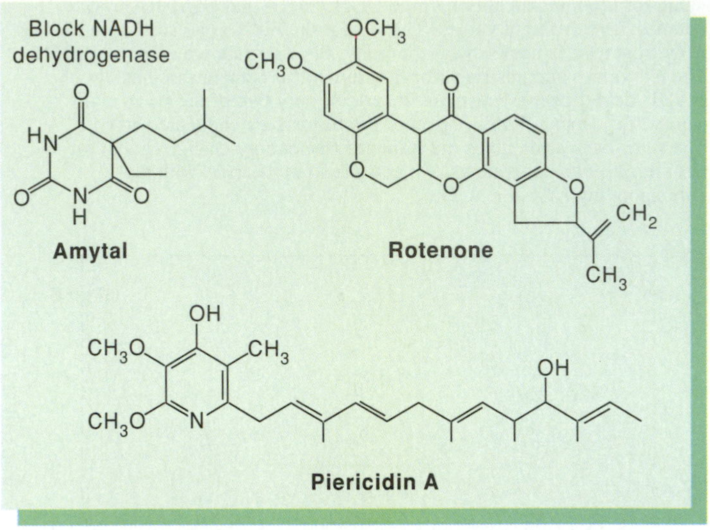

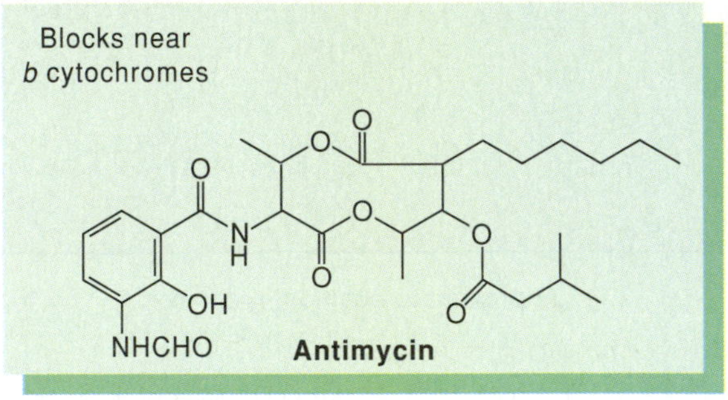

Most of the Electron Carriers Occur in Large Complexes

Another way to explore the organization of the respiratory chain is to isolate fragments of the chain, after partial disruption of the mitochondrial inner membrane with detergents such as sodium deoxycholate. Following this approach, Yousef Hatefi, David Green and others found that all of the major electron carriers except for cytochrome *c* and UQ occur in the form of four large complexes (fig. 15.15). The isolated NADH dehydrogenase complex (NADH-ubiquinone oxidoreductase, or complex I) catalyzes the reduction of UQ by NADH; the succinate dehydrogenase complex (succinate-ubiquinone oxidoreductase, or complex II), catalyzes reduction of UQ by succinate. The cytochrome bc_1 complex (ubiquinone-cytochrome *c* oxidoreductase, or complex III) transfers electrons from reduced ubiquinone (UQH_2) to cytochrome *c*. And finally cytochrome oxidase (complex IV) transfers electrons from cytochrome *c* to O_2. Each of these complexes contains a large number of polypeptide subunits, including some that have no electron-carrying prosthetic groups (table 15.2).

Figure 15.15

The multisubunit complexes of the respiratory chain. Complexes I (NADH dehydrogenase) and II (succinate dehydrogenase) transfer electrons from NADH and succinate to UQ. Complex III (the cytochrome bc_1 complex) transfers electrons from UQH_2 to cytochrome *c*, and complex IV (cytochrome oxidase), from cytochrome *c* to O_2. The arrows represent paths of electron flow. NADH and succinate provide electrons from the matrix side of the inner membrane, and O_2 removes electrons on this side. Cytochrome *c* is reduced and oxidized on the opposite side of the membrane, in the lumen of a crista or in the intermembrane space. The side of the inner membrane that faces the intermembrane space is generally referred to as the cytosolic side, because small molecules can move freely between the solution here and the cytosol.

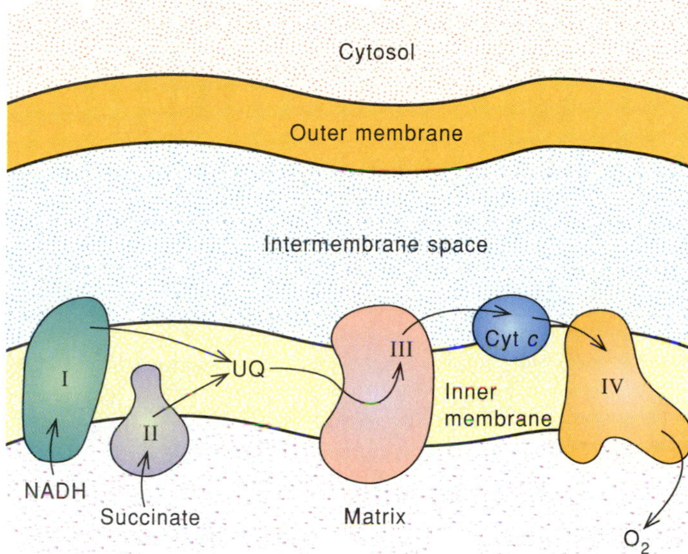

Complex I: The NADH Dehydrogenase Complex The NADH dehydrogenase complex is the largest complex in the mitochondrial inner membrane. It has at least 26 different polypeptides, with a total molecular weight of about 10^6. In addition to FMN, it contains approximately seven iron-sulfur centers, which add up to about 23 iron atoms per FMN. Although the structure of the complex is not known, a flavoprotein subunit and most of the iron-sulfur proteins appear to be clustered in the center, surrounded by a shell of hydrophobic proteins. The iron-sulfur centers include both 2-Fe and 4-Fe complexes. They fall in two groups, one with $E°'$ values near that of NAD^+/NADH (−0.32 V), and the other with $E°'$ values near that of UQ/UQH_2 (approximately +0.10 V). NADH probably reacts with the FMN, reducing it to $FMNH_2$, and the reduced flavin then transfers electrons to an iron-sulfur center (fig. 15.16). Electrons then move from one iron-sulfur center to another, and eventually to UQ. In the presence of the inhibitors amytal, rotenone, or piericidin-A, NADH is still able to reduce the flavin and most, if not all, of the iron-sulfur centers, but the release of electrons to UQ is blocked.

NADH dehydrogenase is oriented in the inner membrane so that its binding site for NADH faces inward toward the matrix space (see fig. 15.15). This orientation is appropriate for oxidizing the NADH that is generated in the matrix by enzymes of the TCA cycle. The UQH_2 that is formed by complex

Table 15.2

Components of the Mitochondrial Electron-Transport Complexes

Complex	Approx. mol. wt.[a]	Polypeptides[b]	Prosthetic Groups
I NADH dehydrogenase	1×10^6	At least 26	FMN, Fe-S centers
II Succinate dehydrogenase	1×10^5	2 Fe-S proteins, 2 others	FAD, Fe-S centers
III Cytochrome bc_1 complex	2×10^5	2 cytochromes, 1 Fe-S protein, 6–8 others	*b*- and *c*-type hemes (cyt b_L, b_H, c_1), Fe-S centers
IV Cytochrome oxidase	2×10^5	Up to 13	*a*-type hemes (cyt a, a_3), 2 Cu

[a]Complex III occurs as a dimer in the membrane. The molecular weight listed is for a monomer. The same is true of complex IV.

[b]The contents of some of the smaller polypeptides in purified preparations of the complexes vary, depending on the sources of the enzymes. It is not yet clear whether all of these polypeptides are integral components of the complexes.

Figure 15.16

Electron transfer through complex I. At the top, the reduction of FMN to $FMNH_2$ by NADH requires the uptake of one proton from the matrix. $FMNH_2$ subsequently transfers a pair of electrons to Fe-S centers, releasing protons to the solution on the cytosolic side of the membrane. Electrons then move to UQ via additional Fe-S centers. The detailed path of electrons through the six to eight different Fe-S centers in the complex is not known. When UQ is reduced to UQH_2, two more protons are taken up from the matrix. This scheme does not account for two additional protons that are transferred across the membrane.

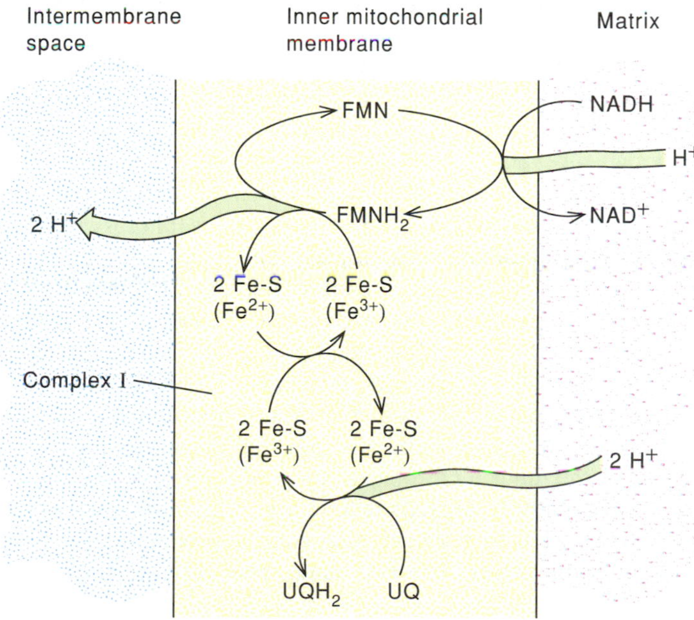

I diffuses to complex III in the central, hydrophobic region of the membrane's phospholipid bilayer (see fig. 15.15), but the protons that are taken up when the quinone undergoes reduction come from the solution on the matrix side of the membrane (see fig. 15.16). In the course of the electron transfer reactions, protons also are released to the solution on the cytosolic side. We will see the significance of these proton movements when we discuss the formation of ATP.

Complex II: The Succinate Dehydrogenase Complex Succinate dehydrogenase is the only enzyme of the TCA cycle that is embedded in the mitochondrial inner membrane. It contains two iron-sulfur proteins, with molecular weights of about 70,000 and 27,000. The substrate oxidation site is on the larger of these subunits, and as with NADH dehydrogenase, is on the matrix side of the membrane. The large subunit has a 4-Fe center and an FAD that is bound covalently through its C-8a methyl group to a histidine residue of the protein. The role of the 4-Fe center is unclear, because the $E°'$ of the center (−0.26 V) is substantially more negative than that of the succinate/fumarate couple (−0.03 V). Succinate could not reduce this iron-sulfur center to any appreciable extent unless the [succinate]/[fumarate] concentration ratio were extremely high. The smaller iron-sulfur subunit has a 2-Fe center with an $E°'$ near 0.0 V, and a 3-Fe center with a more positive $E°'$. Complex II also has two other small polypeptides, which may participate in the interaction with UQ.

Complex III: The Cytochrome bc_1 Complex The cytochrome bc_1 complex contains two *b*-type cytochromes. One of these, cytochrome b_L (or b_{566}), has an $E°'$ of about −0.03 V. The other, cytochrome b_H (or b_{562}), has an $E°'$ of about +0.05 V. (The subscripts L and H denote "low-potential" and "high-potential.") Both of the *b* cytochromes are on a single 30,000-Da polypeptide. The complex also contains cytochrome c_1 ($E°' \approx 0.23$ V), a 2-Fe iron-sulfur protein ($E°' \approx 0.28$ V), several subunits that have been implicated in interactions with UQ, and between four and six additional subunits without known prosthetic groups. The iron-sulfur protein is frequently called the Rieske iron-sulfur protein after its discoverer, John Rieske. It is unusual in that the iron atom is bound to two histidine residues, in addition to two cysteines and two sulfide atoms. The entire cytochrome bc_1 complex can be isolated from heart mitochondria as a dimer with a total molecular weight of about 450,000. Similar complexes are found in the photosynthetic electron-transfer chains of chloroplasts and photosynthetic bacteria (chapter 16). In mitochondria, the site at which cytochrome *c* undergoes reduction is on the side of the inner membrane that faces the intermembrane space (see fig. 15.15).

Many of the subunits of complex III have been sequenced and found to contain stretches of hydrophobic amino acids that could form transmembrane α helices stretching from one side of the inner membrane to the other. The cytochrome *b* subunit, for example, appears to have nine such transmembrane helices, connected by segments that contain predominantly hydrophilic amino acids. The two hemes probably are bound to pairs of histidyl residues in adjacent α helices, as shown in figure 15.17. An electron jumping between the hemes could move across the membrane.

Figure 15.17

The hemes of cytochromes b_L and b_H probably are sandwiched between histidyl residues of two parallel α helices in the same subunit of the cytochrome bc_1 complex (complex III). Helices extending across the phospholipid bilayer are predicted to form in nine regions of the protein where there are stretches of approximately 20 predominantly hydrophobic amino acids. The positions of histidines in two of these regions are highly conserved in the amino acid sequences of the *b* cytochromes obtained from different species. The propionate side chains of the hemes (see fig. 15.3) probably form salt bridges to arginyl groups at the ends of the helices.

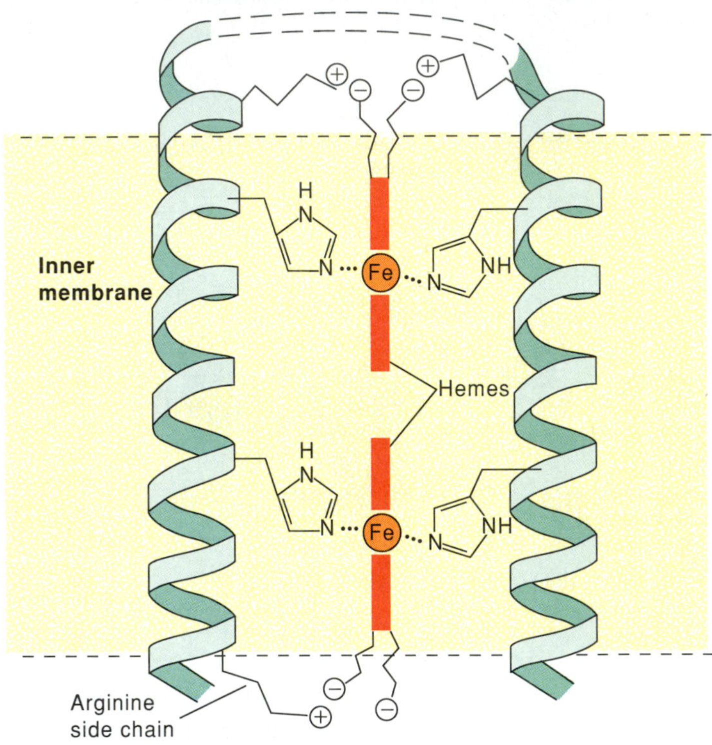

The Path of Electrons Through Complex III: The Q Cycle The path of electrons from UQ to cytochrome *c* is circuitous. The first indication of this was the surprising finding that, under certain conditions, addition of an oxidant such as O_2 causes the *b* cytochromes to become more reduced, instead of more oxidized. To see this effect, we first treat mitochondria with antimycin, which blocks electron transfer in the region of the *b* cytochromes as noted above. (The exact site at which antimycin acts will be discussed shortly.) We then reduce cytochromes *c* and c_1 and the Rieske iron-sulfur protein by adding an electron donor that reacts directly with cytochrome *c*, such as sodium ascorbate. The *b* cytochromes are not reduced under these conditions because their $E°'$ values are more negative than the $E°'$ of ascorbate. However, if we add succinate the *b* cytochromes still do not become reduced. This is odd, because the $E°'$ of succinate (−0.03 V) is considerably more negative than that of cytochrome b_H, and about the same as that of cytochrome b_L.

Figure 15.18

Electron transport through the cytochrome bc_1 complex: the Q cycle. In this figure, the thin black arrows represent the path of electron flow; the large arrows represent the paths of UQ in its various oxidation states and of protons. UQH_2 is oxidized to UQ in two steps at an enzyme site on the cytosolic side of the inner membrane (*top part of the figure*), transferring the first electron to the Fe-S protein and the second electron to cytochrome b_L. The stoichiometries indicated are for two molecules of UQH_2 undergoing these reactions. One of the two molecules of UQ that are produced diffuses to a site on the matrix side of the membrane (*bottom*), where it is reduced back to UQH_2. The two electrons required for the reduction come through the two *b* cytochromes. Antimycin inhibits this reaction. The UQH_2 that is generated at the reduction site diffuses back to the oxidation site, where it joins the pool of UQH_2 coming from complexes I and II. Electrons leaving the UQH_2 oxidation site reduce cytochrome *c* (*top*), which leaves the complex on the cytosolic side of the membrane to go to cytochrome oxidase. Protons also are released to the solution on this side of the membrane. Protons are taken up from the solution on the matrix side of the membrane at the UQ reduction site. The reduction of UQ to UQH_2 involves an intermediate semiquinone (UQH•) similar to that formed in the oxidation of UQH_2 to UQ, but the details of the reactions at the reduction site are not yet clear.

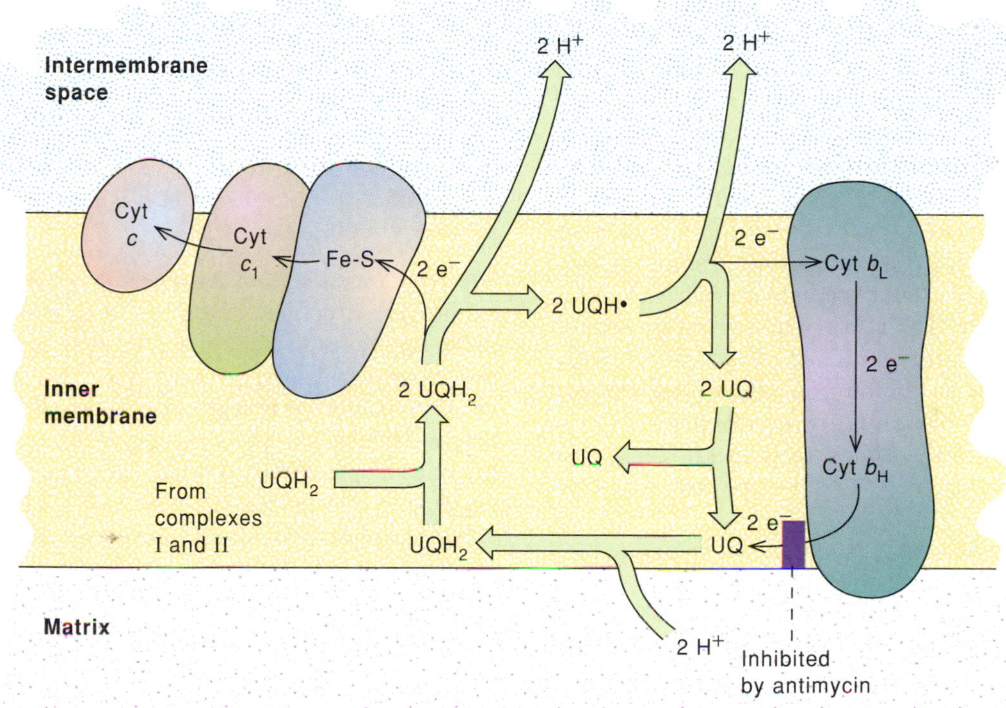

The succinate does reduce the UQ in the membrane. Finally, when we add O_2, cytochromes *c* and c_1 become oxidized and the *b* cytochromes become reduced. Electrons evidently cannot be transferred from UQH_2 to the *b* cytochromes, unless electrons are also withdrawn from the cytochrome bc_1 complex.

These puzzling observations can be explained by the following scheme, in which UQH_2 first transfers one of its two electrons to the Rieske iron-sulfur protein, and then gives the other electron to cytochrome b_L:

$$UQH_2 \xrightarrow[\text{Fe-S}(Fe^{3+}) \;\rightarrow\; \text{Fe-S}(Fe^{2+})]{H^+} UQH\cdot \xrightarrow[\text{cyt } b_L(Fe^{3+}) \;\rightarrow\; \text{cyt } b_L(Fe^{2+})]{H^+} UQ$$

In this scheme, oxidation of UQH_2 by the iron-sulfur protein generates the semiquinone UQH·, which then serves as the reductant for cytochrome b_L. Remember that the UQ/UQH· redox couple is a stronger reductant then the UQH·/UQH_2 couple (see table 15.1).

The reduced iron-sulfur protein subsequently transfers an electron to cytochrome c_1 and on to cytochrome *c*. Further experiments suggest that the reduced cytochrome b_L transfers its electron across the membrane to the other *b* cytochrome, cytochrome b_H, which then contributes the electron for the reduction of another molecule of UQ at a second site in the complex:

$$\begin{matrix} \text{cyt } b_L(Fe^{2+}) \\ \text{cyt } b_L(Fe^{3+}) \end{matrix} \Bigg) \Bigg(\begin{matrix} \text{cyt } b_H(Fe^{3+}) \\ \text{cyt } b_H(Fe^{2+}) \end{matrix} \Bigg) \Bigg(\begin{matrix} \tfrac{1}{2}UQH_2 \\ \tfrac{1}{2}UQ + H^+ \end{matrix}$$

If each of these steps occurs twice, so that two electrons pass through the *b* cytochromes, a molecule of UQ could be reduced fully to UQH_2. At this point, we may seem to be back where we started. Note, however, that for each UQH_2 that is regenerated, two molecules of UQH_2 were oxidized. There is a net oxidation of one UQH_2 to UQ, and two electrons will proceed to cytochrome *c* and on their way to O_2 (fig. 15.18).

This scheme for electron transport through the cytochrome bc_1 complex is known as the Q cycle. As shown in figure 15.18, the sites at which UQH_2 and $UQ^{\cdot}$ undergo oxidation face the intermembrane space, whereas the UQ reduction site is on the matrix side. UQ and UQH_2 evidently diffuse through the membrane from one site to the other.

In addition to providing distinct sites for the oxidation and reduction of UQ, the cytochrome bc_1 complex must impose constraints on the reactions of the quinone with the oxidant or reductant at each site. The iron-sulfur protein, for example, must be prevented from taking two electrons from UQH_2. If the iron-sulfur protein removed a second electron, oxidizing $UQH^{\cdot}$ to UQ, reduction of the *b* cytochromes would be thwarted. In addition, the semiquinones $UQ^{\cdot}$ and $UQH^{\cdot}$ must react only at the sites where they are generated, rather than diffusing away to react on the opposite side of the membrane.

Antimycin blocks electron transfer between cytochrome b_H and UQ at the reduction site (see fig. 15.18). Using antimycin to inhibit the reactions of cytochrome b_H at this site makes it possible to focus experimentally on the reduction of the *b* cytochromes at the UQH_2 oxidation site. Inhibitors that act at other points have been found, and also have been helpful for elucidating the scheme. The role of the iron-sulfur protein has been explored by selective extraction of this protein from the cytochrome bc_1 complex. If the iron-sulfur protein is removed, electron transfer from UQH_2 to cytochrome c_1 is prevented. This shows that the iron-sulfur protein operates upstream of cytochrome c_1, even though its $E°$ is more positive than that of the cytochrome (see fig. 15.12).

As indicated in figure 15.18, the oxidation and reduction of UQ by the cytochrome bc_1 complex is coupled to the uptake of protons from the solution on the matrix side of the inner membrane and the release of protons on the cytosolic side. Two protons are taken up on the matrix side and four protons are released on the cytosolic side for each pair of electrons that proceed to cytochrome *c*. We will return to this point a little later.

Complex IV: Cytochrome Oxidase Cytochrome oxidase contains two atoms of copper, in addition to the hemes of cytochromes *a* and a_3. The copper atoms undergo one-electron oxidation-reduction reactions between the cuprous (Cu^+) and cupric (Cu^{2+}) states. One of the copper atoms (Cu_B) is close to the iron atom cytochrome a_3 (fig. 15.19). The other (Cu_A) is associated with cytochrome *a*, but not so intimately. Cytochrome *a* and Cu_A both have effective $E°'$ values of about 0.28 V, somewhat more positive than that of cytochrome *c* (0.22 V). Cytochrome a_3 and Cu_B have $E°'$ values near 0.35 V in the absence of O_2, but these values become much more positive when the complex binds O_2.

Reduction of O_2 to 2 H_2O requires the addition of four electrons. Addition of one electron to O_2 to produce the superoxide radical ($O_2^{\cdot -}$) is thermodynamically unfavorable, and has to be pulled along by a further reduction to the level of peroxide (H_2O_2) or H_2O. (Although the $E°'$ for the overall reduction of O_2 to H_2O is $+0.82$ V, the redox couple $O_2/O_2^{\cdot -}$ has an $E°'$ of -0.33 V.) This is one of the reasons that O_2 is unreactive in the absence of the enzyme. Cytochrome oxidase appears to get around this thermodynamic obstacle by binding O_2 to Cu_B as well as to the Fe of cytochrome a_3, and then transferring two electrons at a time (see fig. 15.19).

Cytochrome oxidase isolated from mammalian mitochondria has between six and thirteen different subunits, ranging in molecular weight from 5,000 to 50,000. The total molecular weight of the complex is approximately 200,000. The three largest subunits are synthesized in the mitochondria; the smaller ones are made in the cytosol. The hemes and Cu atoms all appear to be located in two of the large subunits. Several of the subunits span the phospholipid bilayer of the inner membrane, and a large domain of the protein protrudes from the membrane's cytosolic surface. Cytochrome *c* transfers electrons to cytochrome *a* and Cu_A on the cytosolic side of the membrane, but the reduction of O_2 by cytochrome a_3 and Cu_B occurs on the matrix side. To get from one of these sites to the other, electrons must traverse the membrane.

Electron Transfer from One Complex to the Next Having described the structures of the four multisubunit complexes that make up the respiratory chain, we now turn to the question of how electrons move between the complexes. The isolated electron-carrier complexes can be combined to reconstitute longer stretches of the respiratory chain by mixing them in the presence of phospholipids and a detergent such as sodium deoxycholate. When the detergent is removed by dilution or dialysis, the phospholipids assemble into vesicles and the proteins are incorporated into the phospholipid bilayers. If UQ is added, vesicles that contain the NADH dehydrogenase and cytochrome bc_1 complexes will catalyze electron transfer from NADH to cytochrome *c*. In the presence of cytochrome *c*, vesicles containing the cytochrome bc_1 and cytochrome oxidase complexes will transfer electrons from UQH_2 to O_2.

Within each of the individual complexes, the electron carriers occur in a fixed molar ratio. In a reconstitution experiment, however, the complexes can be mixed in any ratio. The kinetics of electron transfer in the mixture suggest that the complexes do not bind to each other directly. Instead, the movement of electrons from NADH dehydrogenase to the cytochrome bc_1 complex appears to be mediated by the diffusion of UQH_2 from one complex to the other within the phospholipid bilayer. Similarly, the movement of electrons from the cytochrome bc_1 complex to cytochrome oxidase appears to occur by the diffusion of reduced cytochrome *c* along the surface of the bilayer (see fig. 15.15). Remember that cytochrome *c* differs from the other cytochromes in being a water-soluble protein. It is attached loosely to the membrane surface by electrostatic interactions that are relatively weak at physiological salt concentrations.

There is additional evidence that the electron-transfer complexes are not connected in fixed chains in the mitochondrial membrane. First, the different complexes are not present

Figure 15.19

A schematic drawing of heme a_3 and Cu_B in cytochrome oxidase (complex IV). The other two electron carriers (heme *a* and Cu_A) are not shown. (*a*) The oxidized enzyme, in which heme a_3 is in the ferric (+3) state, and Cu_B is cupric (+2). The heme is viewed edge-on, as in figure 15.6. One of its axial ligands probably is a histidine; the other axial ligand is shared with Cu_B, and is probably OH^- as indicated here, but could be Cl^- or O. (*b*) When heme a_3 and Cu_B are reduced to the ferrous and cuprous states by two electrons provided by heme *a* and Cu_A, the bridging ligand is replaced by a molecule of O_2. (*c*) Two electrons now can be transferred to the O_2, one from the Fe and one from the Cu. (*d*) The transfer of two more electrons causes the bridging O—O bond to split, with one of the oxygen atoms remaining on the Fe and the other on the Cu. A water molecule now is released, and the enzyme returns to its original state. The protons that are picked up in the course of these reactions come from the matrix side of the membrane. This scheme does not account for the additional protons that the enzyme pumps across the membrane.

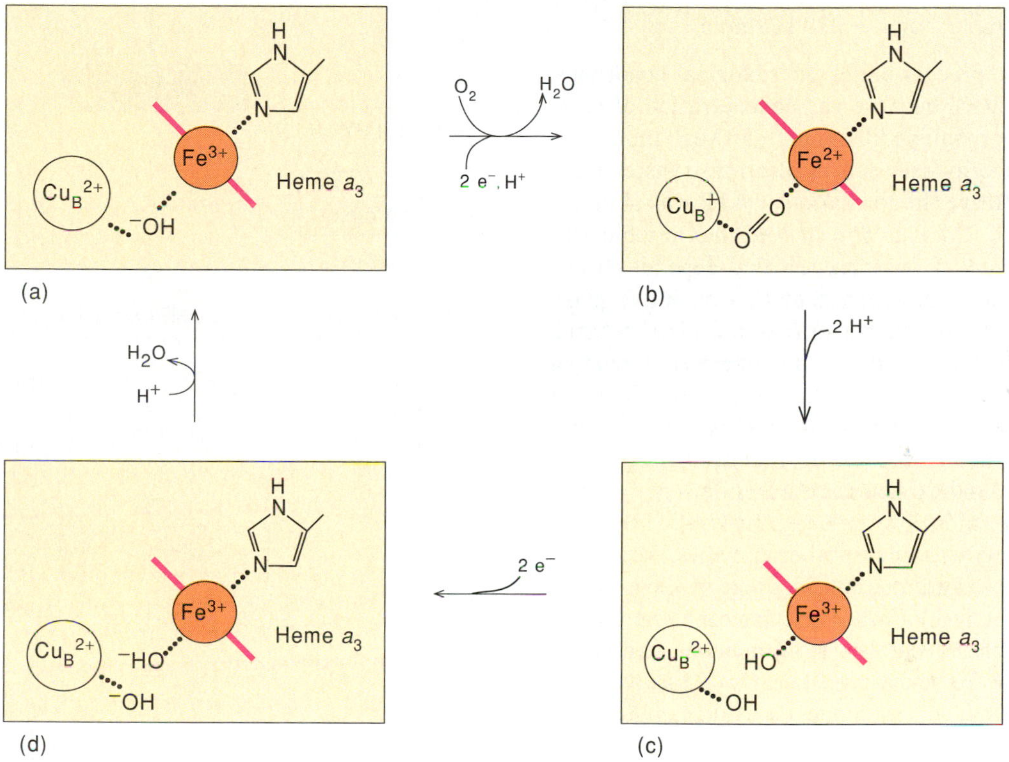

in equimolar concentrations. In rat liver mitochondria, for example, the concentration of cytochrome oxidase is about twice that of the cytochrome bc_1 complex and more than six times that of NADH dehydrogenase. If the respiratory complexes were connected in discrete chains, each of which ended with a single cytochrome oxidase complex, only one-sixth of the chains would have NADH dehydrogenase. This would make it hard to explain how an NADH-linked substrate such as malate can rapidly reduce all of the cytochromes in the membrane. Second, if most of the cytochrome oxidase complexes in the membrane are inhibited with CO, the few molecules that remain uninhibited are still able to catalyze the oxidation of all the cytochrome *c* by O_2. This observation suggests that cytochrome *c* can diffuse rapidly from one cytochrome oxidase complex to another, rather than remaining bound to an individual complex. Direct measurements of the diffusion indicate that cytochrome *c*, UQ, and the complexes themselves diffuse laterally in the membrane at different rates, which means that they cannot all stay stuck together. The diffusion rates are such that random collisions between a complex and its reaction partner (reduced or oxidized UQ or cytochrome *c*) could occur at relatively high frequencies. Finally, the region of the cytochrome *c* molecule that interacts with the cytochrome bc_1 complex appears to be the same region that interacts with cytochrome oxidase. This fact indicates that cytochrome *c* could not stay fixed in position attached to one of these complexes and transfer an electron to the other; it must dissociate and at least turn around. The site on cytochrome *c* that interacts with both of the complexes is the cluster of lysine residues near one edge of the heme (see fig. 15.5). It seems likely that the positively charged lysines also interact with the negatively charged phospholipid headgroups of the membrane, orienting the cytochrome *c* so that its reactive site faces the surface.

Oxidative Phosphorylation

From the preceding discussion, it should be clear that the movement of a pair of electrons from NADH through the respiratory chain to O_2 involves a large, negative $\Delta G^{\circ\prime}$:

$$\Delta E^{\circ\prime} = E^{\circ\prime}_{O_2/H_2O} - E^{\circ\prime}_{NADH/NAD^+}$$

$$= +0.82 \text{ V} - (-0.32 \text{ V}) = 1.14 \text{ V}$$

$$\Delta G^{\circ\prime} = -nF\,\Delta E^{\circ\prime}$$

$$= -(2 \text{ electrons/NADH})(23 \text{ kcal V}^{-1} \text{ mole}^{-1})(1.14 \text{ V})$$

$$= -52.6 \text{ kcal/mole} \quad \text{or} \quad -220 \text{ kJ/mole}$$

The actual ΔG for the process under physiological conditions, though smaller than $\Delta G^{\circ\prime}$ because the concentration of O_2 is only about 200 μM, is probably still about −50 kcal/mole. Evidence that the free energy released in electron-transfer reactions could be used to drive the formation of ATP was obtained about 1940 by Herman Kalckar, who showed that aerobic cells make ATP from ADP and P_i by a process that depends on respiration. It was not clear, however, that this oxidative phosphorylation occurred in mitochondria, or that it involved NADH. When NADH was added to a crude homogenate of liver or muscle tissue, it was oxidized, but no ATP was made. We know now that the NADH dehydrogenase of animal mitochondria can oxidize only NADH that is in the mitochondrial matrix (see figs. 15.15 and 15.16), and exogenous NADH does not get into the mitochondria. The oxidation of exogenous NADH that was observed experimentally was due to other enzymes. Resolution of the problem had to await the development of methods for preparing mitochondria free of other cellular constituents, and of presenting NADH on the matrix side of the inner membrane. When these methods were devised in the early 1950s, Kennedy and Lehninger found that mitochondria do oxidize endogenous NADH, and that between two and three ATPs are generated per NADH oxidized.

How is it possible to couple such different types of reactions as electron transfer and the formation of a phosphate anhydride bond? We will explore this question in the following sections.

The P/O Ratio Measures the Efficiency of Oxidative Phosphorylation

The number of molecules of ATP formed per pair of electrons transferred down the respiratory chain is termed the P/2e⁻ ratio. If the electrons travel all the way to O_2, the P/2e⁻ ratio is usually called the P/O ratio because two electrons are used for each atom of oxygen that is reduced to H_2O. When NADH is used as the reducing substrate, the measured P/O ratio is about 2.5; with succinate it is about 1.5. These observations suggest that there are three "coupling sites" for ATP synthesis somewhere in the electron-transport chain between NADH dehydrogenase and O_2, and two coupling sites between succinate dehydrogenase and O_2. Because the electron-transport chains for NADH and succinate converge at UQ, there would appear to be one coupling site between NADH and UQ, and two between UQH_2 and O_2 (fig. 15.20; also see fig. 15.15). The locations of these last two sites can be identified by using a substrate that feeds electrons directly to cytochrome *c*, such as ascorbic acid, in the presence of antimycin to block electron flow through complex III. This gives a P/O of about 1.0, indicating that there is one coupling site between cytochrome *c* and O_2 (see fig. 15.20). Complementary experiments can be done by adding oxidants that remove electrons from the chain at various points. For example, the last coupling site can be bypassed by adding additional cytochrome *c* as an electron acceptor, along with CN^- to block cytochrome oxidase. Electron transfer from succinate to cytochrome *c* gives a P/2e⁻ ratio of about 0.5.

There thus appears to be one coupling site in each of complexes I, III, and IV, but none in complex II. Efraim Racker and his co-workers verified the capacity of the individual complexes I, III, and IV to support the formation of ATP, by incorporating purified preparations of the electron-transport complexes into phospholipid vesicles along with the mitochondrial ATP-synthase enzyme that we will describe a little later. By referring to figure 15.12, you can see that the flow of two electrons through each of complexes I, III, and IV involves a sufficiently large negative $\Delta G^{\circ\prime}$ to support the phosphorylation

Figure 15.20

Approximately 2.5 molecules of ADP can be phosphorylated to ATP for each pair of electrons that traverse the electron-transport chain from NADH to O_2. About 1.5 molecules of ATP are formed for a pair of electrons that enter the chain via succinate dehydrogenase or other flavoproteins such as glycerol-3-phosphate dehydrogenase. Approximately 1.0 ATP is formed for a pair of electrons that enters via cytochrome *c*. It therefore appears that electron flow through each of complexes I, III, and IV is coupled to phosphorylation, but that the stoichiometry of ATP formation is lower in complex III than in complexes I and IV.

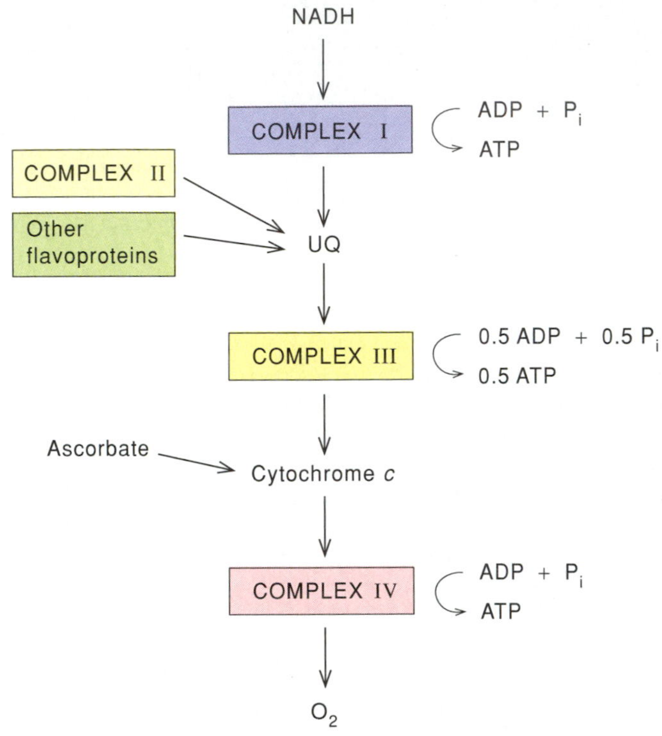

Figure 15.21

The rate of respiration by a suspension of mitochondria can be increased dramatically by the addition of a small amount of ADP. An oxidizable substrate (succinate) and P_i first are added in excess at the times indicated by the vertical arrows, but the respiratory rate remains low until ADP is added. The P/O ratio can be determined by dividing the amount of ADP added by the amount of oxygen taken up during the period of rapid respiration. Since another period of rapid respiration is obtained upon the addition of more ADP, we conclude that essentially all of the ADP gets used up.

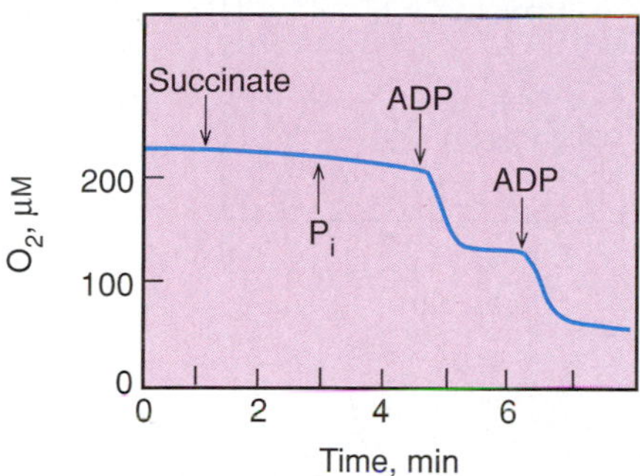

Figure 15.22

Structures of two uncouplers of oxidative phosphorylation. 2,4-Dinitrophenol (DNP) was one of the first uncouplers to be discovered; carbonylcyanide-*p*-trifluoromethoxyphenylhydrazone (FCCP) works at much lower concentrations (on the order of 1μM). DNP has a weakly dissociable proton on the oxygen atom; FCCP has a similar proton on one of the nitrogens.

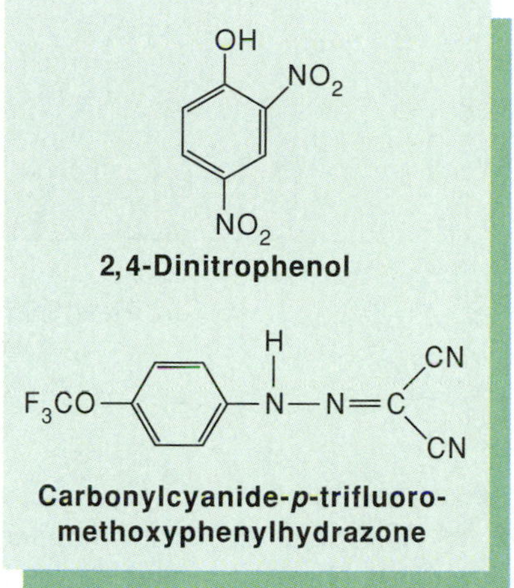

of ADP to ATP. In the presence of 10-mM Mg^{2+}, $\Delta G°'$ for the reaction $ADP + P_i \rightarrow ATP + H_2O$ is about 7.5 kcal/mole, which is equivalent to a $\Delta E°'$ of about 0.16 V for $n = 2$ electrons/molecule. However, the actual ΔG for ATP formation in a suspension of respiring mitochondria can be as large as 15 kcal/mole because of a high $[ATP]/[ADP][P_i]$ ratio.

The linkage between respiration and phosphorylation is so tight that the electron-transfer reactions are strongly inhibited if phosphorylation is blocked. For example, respiration slows down greatly if mitochondria run out of ADP. The experiment shown in figure 15.21 illustrates this point. The trace shows a measurement of the concentration of O_2 in a suspension of mitochondria as a function of time. When succinate is added, O_2 is consumed at a very low rate, indicating that something more than an electron donor is needed in order for respiration to occur. The addition of a small amount of ADP in the presence of excess P_i causes a brief period of brisk respiration, as reflected by a rapidly decreasing O_2 concentration. O_2 uptake evidently stops when all the ADP is converted to ATP, because it can be started again by the addition of more ADP.

Experiments like that shown in figure 15.21 provide one way of measuring P/O ratios. The P/O ratio is proportional to the amount of ADP that must be added in order to cause a certain amount of O_2 uptake.

The regulation of the rate of electron transport by ADP is called respiratory control. Chance and Williams, who first studied the phenomenon, noted that the rate of respiration can be limited by a variety of factors, including the availability of oxidizable substrates, ADP, and P_i, and the intrinsic cycling rates of the electron carriers. They used the term "state 4" for the condition in which ADP is limiting, and "state 3" for the condition in which ADP, P_i, and substrate are present in excess and the mitochondria are respiring at the maximal rate sustainable by the electron carriers. With succinate as substrate, the respiration rate in state 3 is typically about six times greater than that in state 4. Although it makes physiological sense for a cell to avoid oxidizing substrates wastefully when the concentration of ADP is low and the cell's energy charge is high, there is no indication that respiratory control involves an allosteric regulatory mechanism. It simply reflects the fact that ADP and ATP are, in essence, substrates and products of the respiratory apparatus.

Uncouplers Release Electron Transport from Phosphorylation

The coupling of respiration and phosphorylation is lost if the integrity of the mitochondrial inner membrane is disrupted. For example, if mitochondria are placed in a hypotonic solution so that they swell, their rate of respiration can increase even though no ADP is added. The P/O ratio simultaneously decreases to zero. These effects can be related to an increase in the leakage of protons across the inner membrane.

There also is a large group of molecules that cause an uncoupling of electron transfer from the phosphorylation of ADP. Representative of these uncouplers are carbonylcyanide-*p*-trifluoromethoxyphenylhydrazone and 2, 4-dinitrophenol (fig. 15.22). Studies of the actions of uncouplers have played a pivotal role in attempts to understand the mechanism of oxidative phosphorylation. If an uncoupler is added to a suspension of mitochondria in the presence of an oxidizable substrate, O_2 uptake commences immediately and continues until essentially all of the O_2 in the solution is used up (fig. 15.23*a*). This happens even in the absence of added ADP or P_i. In leaky mitochondria, or in the presence of an uncoupler, the free energy that is released in the electron-transport reactions is not captured in the form of ATP. Though structurally diverse, uncouplers generally are lipophilic weak acids. We will return to this point shortly.

Figure 15.23

(*a*) Addition of an uncoupler to a suspension of mitochondria causes brisk O_2 consumption, which continues until all the O_2 in the solution is used up. An oxidizable substrate (succinate) is added before the uncoupler, but P_i is not required. (*b*) Oligomycin, an inhibitor of the ATP-synthase, blocks the stimulation of respiration caused by ADP. It does not block the stimulation caused by an uncoupler.

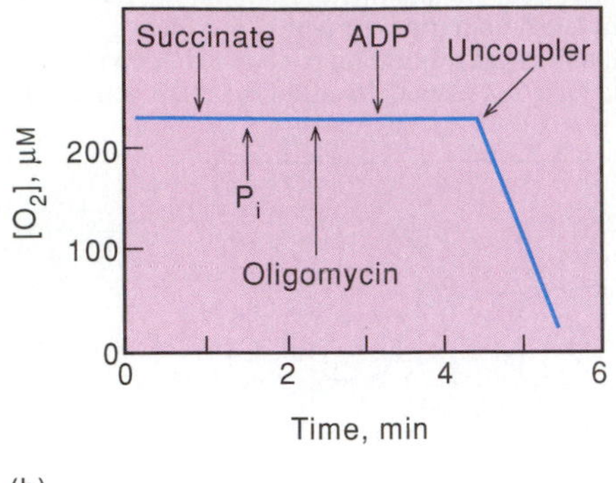

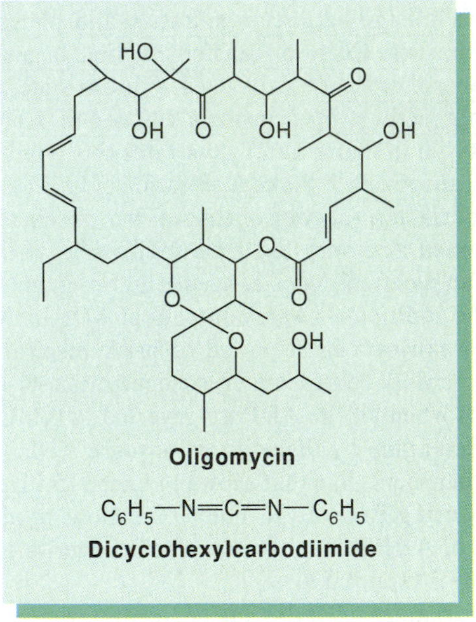

Other agents are found to interfere with the phosphorylation of ADP to ATP in a different manner. These true inhibitors, which include oligomycin and dicyclohexylcarbodiimide (DCCD) (fig. 15.24), prevent ADP from increasing the rate of respiration (see fig. 15.23*b*). They do not block the stimulation of respiration that is caused by an uncoupler. This fact suggests that the inhibitors act directly on the ATP-synthase that converts ADP and P_i to ATP.

Figure 15.24

Structure of oligomycin and dicyclohexylcarbodiimide (DCCD), two inhibitors of the ATP-synthase. Both molecules also inhibit the ATPase activity of the enzyme. Oligomycin, like the electron-transport inhibitor antimycin and many other antibiotics, is obtained from species of *Streptomyces*.

When experiments like those of figures 15.21 and 15.23 were first done, it was reasonable to suggest that the mechanism of oxidative phosphorylation might resemble that of the glycolytic enzyme glyceraldehyde-3-phosphate dehydrogenase. As we discussed in chapter 13, the oxidation of glyceraldehyde-3-phosphate by NAD^+ results in the formation of an enzyme-bound thioester of 3-phosphoglyceric acid, which then reacts with P_i to form glycerate-1,3-bisphosphate (see fig. 13.10). The essence of the mechanism is that the substrate that undergoes oxidation is converted to a labile intermediate that later is broken down to yield ATP. If the intermediate is not broken down (for example, because no P_i is added), the oxidation-reduction reactions will come to a halt. We can generalize the mechanism as follows:

$$AH_2 + B \rightarrow {\sim}A + BH_2$$

$${\sim}A + ADP + P_i \rightarrow A + ATP \quad \text{(in several steps)}$$

where AH_2 and B are the reductant and oxidant substrates, A and BH_2 are the oxidized and reduced products, and ~A is a labile, intermediate form of one of the products. In the case of glyceraldehyde-3-phosphate dehydrogenase, the "squiggle" (~) represents a thioester bond to the enzyme; in other cases, it could represent a phosphoric anhydride bond, or any other structure with a large, negative $\Delta G^{\circ\prime}$ of hydrolysis.

The effects of uncouplers of oxidative phosphorylation at first seem to fit in with this scheme. If the uncoupler somehow discharged ~A, for example by causing hydrolysis of a bond symbolized by the ~, the electron carrier A would be set free and electron transport to O_2 could continue. However, there evidently is something different about oxidative phosphorylation, as compared with the phosphorylation catalyzed by glyceraldehyde-3-phosphate dehydrogenase, because uncouplers have no effect on the latter reaction. Nor do they affect any of the other soluble enzymes that make or use ATP. On the other hand, a molecule that acts as an uncoupler at any one of the three coupling sites of oxidative phosphorylation invariably has a similar effect at the other two sites. This observation suggests that uncouplers act on some species or state that is formed at all three sites. If we continue to denote this species or state as ~, the reactions of oxidative phosphorylation would appear to need rewriting in the following form:

$$A_{red} + B_{ox} \rightleftharpoons A_{ox} + B_{red} + \sim \quad (7)$$

$$\sim + ADP + P_i \rightleftharpoons ATP + H_2O \quad \text{(blocked by oligomycin)} \quad (8)$$

$$\sim \xrightarrow{\text{Uncoupler}} \text{dissipated state} \quad (9)$$

where A_{red}, A_{ox}, B_{red}, and B_{ox} are the reduced and oxidized forms of electron carriers that participate at any one of the coupling sites. Although the nature of the ~ has suddenly become rather vague, we can say at this point that the ~ does not seem to involve phosphate or ADP, because the stimulation of respiration by uncouplers does not require the addition of either of these materials (see fig. 15.23*a*).

Note that the reactions that generate the hypothetical ~ and use it to make ATP have been written as being reversible. This reversibility can be demonstrated experimentally. In box 15A we show that the ~ that drives the formation of ATP can be used to push electron transfer backwards. Here, let's consider a reversal of just the process that converts the ~ to ATP. Suppose that ATP is added to a suspension of mitochondrial inner membrane vesicles in the presence of a battery of inhibitors that block electron transport through the regions of all three coupling sites (amytal, antimycin, and CN^-, for example). A limited amount of ATP is hydrolyzed to ADP and P_i but then the hydrolysis stops. If an uncoupler is added, hydrolysis of ATP continues. Uncouplers thus can stimulate an ATPase activity in mitochondria. This observation is consistent with reactions (8) and (9). As ATP is hydrolyzed, ~ builds up, slowing down the hydrolysis. Removal of ~ by an uncoupler (reaction 9) pulls reaction (8) to the left. The ATPase activity is blocked by oligomycin, just as the forward reaction that synthesizes ATP is, in agreement with the view that the ATPase and the ATP-synthase are the same enzyme.

The Chemiosmotic Theory Proposes that Phosphorylation Is Driven by Proton Movements

The effects of uncouplers suggest that the mechanism by which the respiratory chain couples phosphorylation to electron transfer is fundamentally different from the mechanisms of energy coupling in soluble enzymes. In 1961, Peter Mitchell suggested a radically different theory. He proposed that the ~ is not a compound, or a labile chemical bond, but rather an electrochemical potential gradient for protons across the mitochondrial inner membrane. Figure 15.25 illustrates the basic idea. Electron transport down the respiratory chain results in the movement of protons across the membrane, from the mitochondrial matrix to the intermembrane space and the cytosol. The removal of protons from the matrix causes the pH in this region to rise.

Figure 15.25

According to the chemiosmotic theory, the flow of electrons through the electron-transport complexes is coupled to the movement of protons across the inner membrane from the matrix to the intermembrane space. This raises the pH in the matrix, and leaves the matrix negatively charged with respect to the intermembrane space and the cytosol. Protons flow passively back into the matrix through a channel in the ATP-synthase, and this flow is coupled to the formation of ATP.

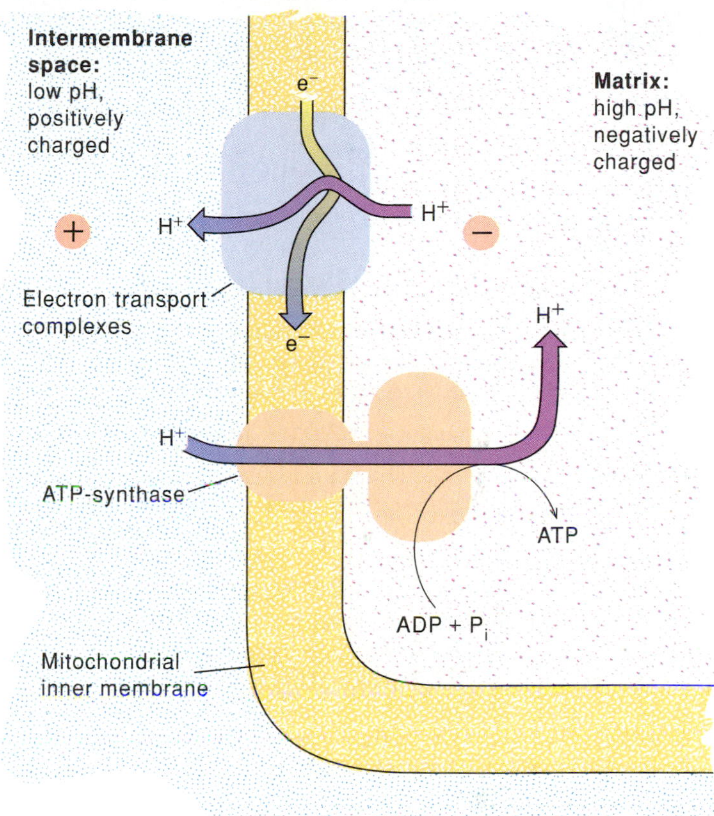

Because protons are positively charged, the matrix also becomes negatively charged with respect to the cytosol. The differences in pH and electrical potential across the membrane provide a driving force that tends to pull protons from the cytosol back into the matrix. Part of the free energy decrease associated with the electron-transfer reactions thus is exchanged for an electrochemical potential difference for protons between the solutions on the two sides of the inner membrane. The membrane is relatively impermeable to protons, except for special channels that are part of an ATP-synthase. Mitchell suggested that the ATP-synthase couples the inward movement of protons through these channels, down the electrochemical potential gradient, to the formation of ATP. This proposal is called the chemiosmotic theory because it emphasizes that chemical reactions can drive, or be driven by, movements of molecules or ions between osmotically distinct spaces separated by membranes. Let's examine some of the evidence supporting the theory.

15A
BOX

Reversal of Oxidative Phosphorylation

Suppose that the respiratory chain is inhibited in complex III by antimycin, but that electron transport through complexes I and II is not blocked. The following reactions then might occur:

$$\text{NADH} + \text{UQ} \underset{}{\overset{\text{Complex I}}{\rightleftharpoons}} \text{NAD}^+ + \text{UQH}_2 + \sim$$

$$\text{Fumarate} + \text{UQH}_2 \overset{\text{Complex II}}{\rightleftharpoons} \text{UQ} + \text{succinate}$$

If we neglect the ~ formed in complex I, the equilibrium of the net reaction

$$\text{NADH} + \text{fumarate} \rightleftharpoons \text{NAD}^+ + \text{succinate}$$

would lie far to the right. (The $E°'$ value of the NADH/NAD$^+$ couple is −0.32 V; that of the succinate/fumarate couple, −0.03 V.) But the actual reaction includes the formation of ~:

$$\text{NADH} + \text{fumarate} \rightleftharpoons \text{NAD}^+ + \text{succinate} + \sim$$

We thus should drive the electron-transfer reactions to the left if we add ATP to generate ~. This can be demonstrated experimentally as shown in trace *a* of figure 1. The reversal of electron transport is prevented by oligomycin, which blocks the formation of ~ from ATP, or by an uncoupler, which dissipates the ~ as rapidly as it is formed (traces *b* and *c*).

A significant extension of these experiments is to generate the ~, not by the addition of ATP, but by electron transport through one of the other coupling sites. For example, electron flow through complex IV will occur if ascorbate is added to reduce cytochrome *c;* antimycin can be added to block flow through complex III. We now can ask whether the ~ that is generated in complex IV can be used in complex I to drive electrons from succinate to NADH. The answer is yes (trace *d*). Again, this process is prevented by adding an uncoupler to dissipate the ~ as it is formed. It is not blocked by oligomycin, and it does not require the presence of P_i or ADP, because it does not involve the phosphorylation reactions directly.

Experiments of the type just described indicate that the ~ that is generated at any of the three coupling sites can be used to drive electron transfer backwards in either complex I or complex III. This agrees with the observation that uncouplers can dissipate the ~ that is formed at any of the sites. It also implies that if the ~ is a compound, it must be able to diffuse relatively freely from one site to another. Alternatively, if the ~ is a state of some sort, the same state must be expressed at multiple sites. We will see that this second interpretation appears to be the correct one.

Figure 1

Reversal of electron transport through complex I can be driven by ATP. The traces show the absorbance at 340 nm of suspensions of membrane vesicles made from the mitochondrial inner membrane. An increase in the absorbance reflects the reduction of NAD$^+$ to NADH (see fig. 15.13). Antimycin is present to block electron flow through complex III. Under these conditions, succinate alone cannot reduce NAD$^+$, but it can if ATP is added (trace *a*). The reduction of NAD$^+$ driven by ATP is blocked by oligomycin or uncouplers (traces *b* and *c*). In trace *d,* ascorbate is used to reduce cytochrome *c,* so that electrons can flow through complex IV, below the block imposed by antimycin. The free energy decrease associated with electron transfer in complex IV can be used to drive electrons from succinate to NAD$^+$ in the absence of ATP, ADP, and P_i. Oligomycin now has no effect, because the ATP-synthase does not participate in the reactions (trace *e*), but the process is still sensitive to uncouplers (trace *f*). Experiments like these are easiest to interpret if they are done in membrane vesicles, because in intact mitochondria, succinate can be converted to fumarate and then to malate, which can reduce NAD$^+$ directly.

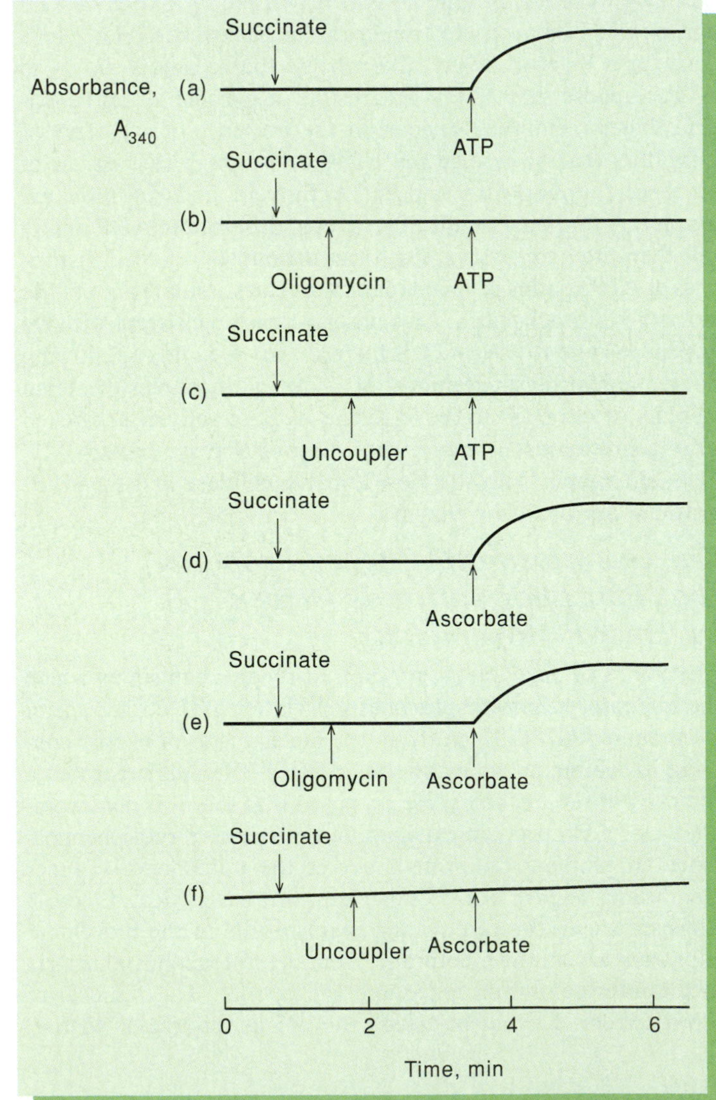

Figure 15.26

Respiring mitochondria extrude protons. A weakly buffered suspension of mitochondria is provided with an excess of an oxidizable substrate such as ß-hydroxybutyrate and allowed to use up all the O_2 in the solution. The traces show measurements of the pH of the suspension, with pH decreases plotted upward. A small amount of O_2 is added at the upward arrow, allowing respiration to occur for a few seconds. The pH of the solution decreases suddenly at this point (experiment A). The pH change persists for a period of several minutes, long after the O_2 has been used up. If an uncoupler is added, the pH returns abruptly to nearly its original level. Experiment B was done in the presence of the detergent triton X-100, which disrupts the inner membrane's phospholipid bilayer and makes the membrane permeable to protons. Respiration still occurs under these conditions, but the pH change is not seen. In experiment A, the number of protons that move across the membrane can be estimated from the pH change, and can be related to the number of oxygen atoms consumed. In this experiment, about six protons appear to be translocated per oxygen atom consumed. (Since ß-hydroxybutyrate reduces the mitochondrial NAD^+, a pair of electrons move through all three coupling sites for each oxygen atom that is consumed.) However, this is an underestimate of the amount of proton translocation, because the phosphate OH^- transport system dissipates part of the pH gradient. Larger $\Delta H^+/O$ ratios are measured if the phosphate transporter is inhibited. (For more details see P. Mitchell and J. Moyle, *Nature* 208:147-151, 1965, and *Biochem. J.* 105:1147-1162, 1967; also A. Alexandre et al., *J. Biol. Chem.* 255:10721-30, 1980.)

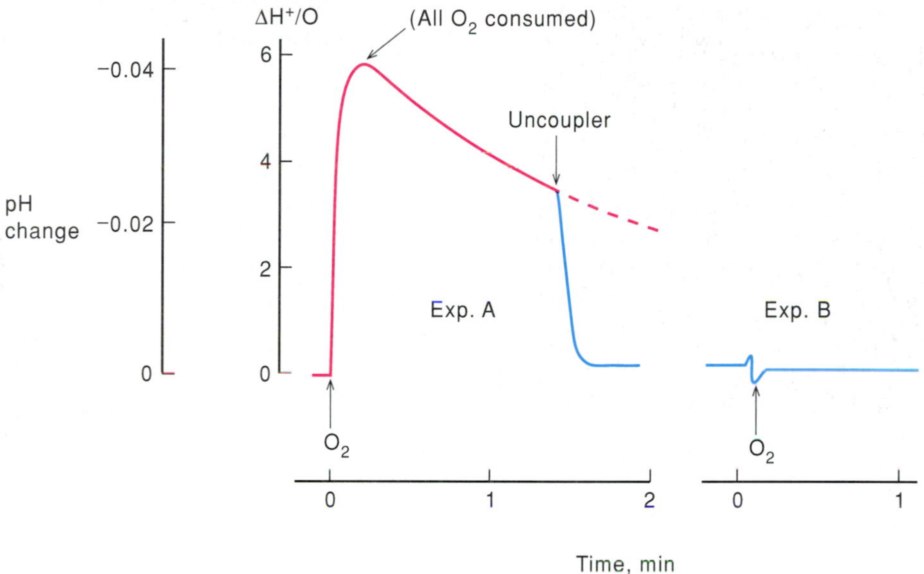

Electron Transport Creates an Electrochemical Potential Gradient for Protons across the Inner Membrane

Peter Mitchell and Jenifer Moyle showed that protons are pumped out of mitochondria during respiration. When O_2 is added to a weakly buffered suspension of mitochondria, the pH measured by a glass electrode immersed in the suspension drops, indicating that the proton concentration outside the mitochondria increases (fig. 15.26). The pH change is not seen if the integrity of the mitochondrial membrane has been broken by the addition of a detergent. This fact indicates that the protons probably emerge from the matrix space inside the mitochondria, rather than simply coming from a process that causes a net production of free protons, such as the hydrolysis of a phosphate ester. By making the membrane leaky to protons, the detergent allows any protons that are pumped out to go right back in.

Mitchell and Moyle calculated that mitochondria depleted of ADP can build up a pH gradient of about 0.05 pH units across the inner membrane, depending somewhat on the experimental conditions. More recent measurements indicate that approximately ten protons are pumped out of the mitochondrion for each pair of electrons that move down the full respiratory chain from NADH to O_2, and approximately six protons for a pair of electrons coming from succinate. Between two and four protons thus appear to be extruded per pair of electrons that move through each of the three coupling sites.

The pH gradient that is created by respiration collapses abruptly if an uncoupler is added (see fig. 15.26). This observation supports the proposal that the intermediate state (~) that couples electron transport to phosphorylation is an electrochemical potential gradient for protons across the membrane. In the absence of an uncoupler, the pH change caused by a brief period of respiration decays slowly, in agreement with the view that the mitochondrial inner membrane is not very permeable to protons. Mitchell and Moyle reinforced this conclusion by experiments in which they added a small amount of HCl to a suspension of mitochondria and measured the rate at which protons leaked through the membrane. The leakage occurs on a time scale of minutes, which is much slower than the milliseconds-to-seconds time scale of oxidative phosphorylation.

The idea that the membrane is largely impermeable to protons fits well with what we know about the structure of biological membranes (see chapters 7 and 32). A charged species such as a hydronium ion (H_3O^+) cannot pass readily through the hydrocarbon region of the phospholipid bilayer. In the absence of an uncoupler, the effective conductance of the phospholipid bilayer to protons is approximately 10^6 times lower than that of the aqueous phases on either side of the membrane.

The ability of uncouplers to collapse a pH gradient across the membrane is understandable when we recall that uncouplers are lipophilic weak acids. Its lipophilic character allows an uncoupler to diffuse relatively freely through the phospholipid bilayer. Because it is a weak acid, it can pick up a proton from the solution on one side of the membrane, carry the proton across the bilayer, and release it on the opposite side. The unprotonated molecule then can diffuse back across the membrane to pick up another proton. Uncoupling also is caused by gramicidin A, a hydrophobic peptide antibiotic that can form a tubular channel extending across the membrane (see chapter 32). Protons and Na^+ or K^+ ions move rapidly in either direction through the gramicidin channel, short-circuiting the membrane.

The pumping of protons across the membrane by the respiratory chain creates a transmembrane electrical potential gradient, in addition to a pH gradient. This has been shown by examining the movements of lipophilic anions and cations. If the matrix space becomes negatively charged with respect to the region outside of the mitochondrion, there will be a driving force that tends to pull any positively charged ion into the matrix, and to push any negatively charged ion out. Whether or not a particular ion actually moves across the membrane will depend on whether the ion can pass through the phospholipid bilayer. As we have just discussed, hydronium ions cannot traverse the bilayer readily in the absence of an uncoupler, but some lipophilic ions can. For example, in the triphenylmethylphosphonium and tetraphenylarsonium ions, $(C_6H_5)_3CH_3P^+$ and $(C_6H_5)_4As^+$, the positive charge is sufficiently buried by phenyl and methyl groups that the ion can move across the membrane relatively freely. Other lipophilic cations that pass rapidly through phospholipid bilayers are the complexes of the ionophore valinomycin with K^+ or Rb^+ (fig. 15.27). Respiring mitochondria are found to take up lipophilic cations, such as the triphenylmethylphosphonium ion, or K^+ in the presence of valinomycin. They extrude lipophilic anions such as thiocyanate (CNS^-). These observations are consistent with the idea that the efflux of protons driven by electron transport makes the interior of the mitochondrion negatively charged relative to the solution outside.

By measuring the concentrations of a lipophilic ion inside and outside, we can estimate the electrical potential difference across the mitochondrial membrane. The calculation proceeds as follows. The free energy change for the movement of one mole of a cation C^+ into the mitochondrion is the difference between the molar free energies for the ion on the two sides of the membrane. This difference depends both on the ratio of the concentrations of the ion on the two sides and on the difference between electrical potentials:

$$\Delta G_{C^+} = G_{C^+(in)} - G_{C^+(out)} = RT \ln \{[C^+]_{in}/[C^+]_{out}\} + zF\,\Delta\psi \quad \mathbf{(10)}$$

where $[C^+]_{in}$ and $[C^+]_{out}$ are the concentrations of the cation inside and outside, z is the valency of the ion (+1 for a univalent cation), F is the Faraday constant, and $\Delta\psi$ is the difference in electrical potential across the membrane (the electrical potential inside minus that outside). This is the second of the two Nernst equations. (Equation 6 was the first.) If the ion can pass freely across the membrane, the concentrations on the two sides will come to equilibrium, but the two concentrations generally will not be equal. Equilibrium means that $\Delta G_{C^+} = 0$, which occurs when $\ln \{[C^+]_{in}/[C^+]_{out}\} = -zF\,\Delta\psi/RT$, that is, when the difference in chemical potential exactly balances the difference in electrical potential. Measurements of ion concentration ratios have indicated that the electrical potential ($\Delta\psi$) across the inner membrane of respiring mitochondria is typically on the order of −0.15 V, inside negative.

How Do the Electron-Transfer Reactions Push Protons across the Membrane?

We noted that NADH, flavins, and quinones bind and release protons in addition to electrons when they undergo reduction and oxidation. Suppose that an electron-transfer reaction in which protons are taken up from the solution occurs on the matrix side of the inner membrane, and a reaction that releases protons to the solution occurs on the cytosolic side. If protons are transferred along with electrons between the two reaction sites, a flow of electrons through the system will result in a movement of protons across the membrane (fig. 15.28). The key point here is that, unlike reactions in free solution, enzymatic reactions in organized structures such as membranes can have a vectorial, or directional, character.

This basic idea accounts well for the proton translocation that occurs in complex III, the cytochrome bc_1 complex. As explained above (see fig. 15.18), electron transport through complex III involves the oxidation of UQH_2 to UQ on the cytosolic side of the inner membrane and reduction of UQ to UQH_2 on the matrix side. Hydrogen atoms are carried across the membrane by the diffusion of UQH_2 from one of these catalytic sites to the other. The net transfer of two electrons from UQH_2 through the complex to cytochrome *c* would be coupled to the uptake of two protons from the mitochondrial matrix and release of four protons to the cytosol, which agrees well with the measured stoichiometry of proton translocation.

The mechanism of proton translocation in complexes I and IV is not so clear. In the scheme that we presented for electron transfer through complex I (see fig. 15.16), two protons are released to the cytosol for each pair of electrons that move from NADH to UQ. This scheme fails to explain how FMN and $FMNH_2$, charged molecules that would not diffuse freely through the phospholipid bilayer, could have access to two enzymatic sites on opposite sides of the membrane. In addition, measurements indicate that four protons are transported per pair of electrons, two protons more than the scheme predicts. In complex IV, two protons are taken up from the solution on the matrix side for each $\frac{1}{2}O_2$ that is reduced to H_2O, or for each pair of electrons transferred. Because cytochrome *c* is oxidized on the cytosolic side of the membrane, while O_2 is reduced on the matrix side (see fig. 15.15), there is a net transfer of electrons across the membrane from the cytosolic side to the matrix. This inward movement of negative charge accounts for part of the contribution of complex IV to the electrical potential difference across the membrane. However, electron transfer from

Figure 15.27

Valinomycin, another antibiotic obtained from *Streptomyces* species, is one of a group of compounds termed ionophores, which facilitate the movement of ions across phospholipid membranes. Valinomycin has the cyclic structure [-D-valine-lactate-L-valine-hydroxyvalerate-]$_3$. (Hydroxyvaleric acid is $(CH_3)_2CHCHOH\text{-}CO_2H$.) The links between D-valine and lactate and between L-valine and hydroxyvalerate are ester bonds; the links between lactate and L-valine and between hydroxyvalerate and D-valine are peptide bonds. Molecules with alternating ester and peptide linkages of this sort are called depsipeptides. This figure shows the crystal structure of the complex formed between valinomycin and K^+. The K^+ ion binds to the six ester oxygen atoms, which point in to the center of the ring. The amide oxygens are all hydrogen-bonded to nitrogens, and the methyl and isopropyl groups all point outward. The complex is soluble in organic solvents, and can pass through the phospholipid bilayer of the mitochondrial inner membrane. K^+ dissociates reversibly from the complex in the aqueous solution on either side of the membrane. The transport of K^+ across membranes by valinomycin is discussed further in chapter 32.

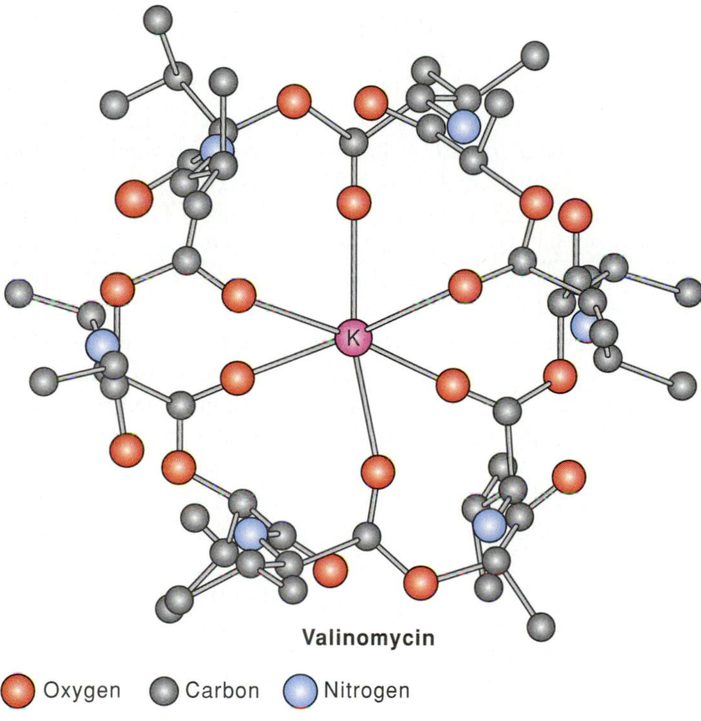

Figure 15.28

A general scheme showing how electron transport can result in proton translocation. A, B, D, and E are electron carriers that bind protons when they are reduced. A and D undergo reduction to AH_2 and DH_2 on the matrix side of the membrane, picking up protons from the solution on this side. They then transfer protons along with electrons to B and E. BH_2 and EH_2 are reoxidized on the cytosolic side, releasing protons here. Electrons move without protons from BH_2 to D. A set of reactions that move protons and electrons in one direction across a membrane, and that move electrons alone back in the other direction, is frequently called a loop. The scheme shown here has one and a half loops. The scheme in figure 15.16 has one loop.

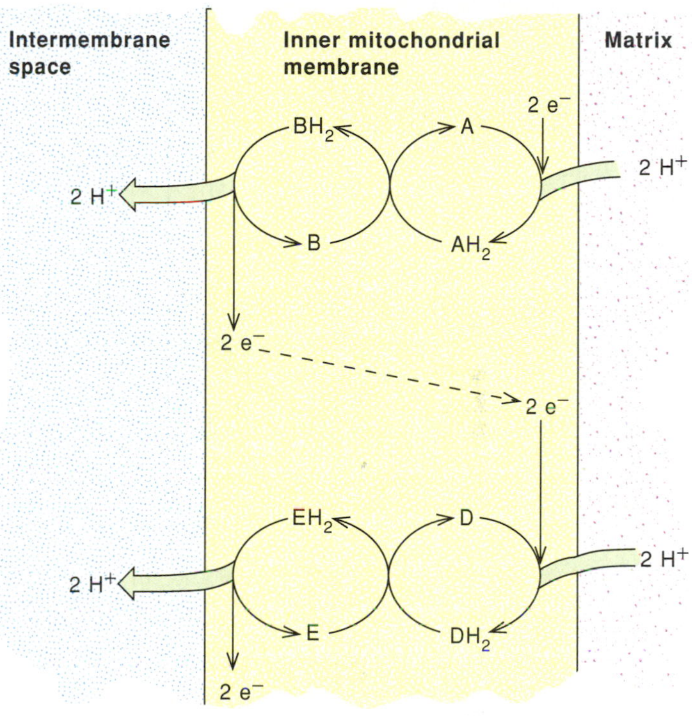

cytochrome *c* to O_2 also appears to result in the outward movement of two additional protons per electron. How these protons are pumped is not known.

The Movement of Protons Back into the Matrix Drives the Formation of ATP

We have seen thus far that the flow of electrons from reducing substrates to O_2 causes protons to move out of the matrix space, setting up a pH gradient and an electrical potential difference across the inner membrane. The chemiosmotic theory postulates that protons moving back into the matrix via an ATP-synthase drive the formation of ATP. Evidence for this proposal is that an electrochemical potential gradient for protons can support the formation of ATP in the absence of any electron-transfer reactions. A transient pH gradient that will tend to pull protons into the matrix can be set up by first incubating mitochondria at pH 9, so that the inside becomes alkaline, and then suddenly lowering the pH of the suspension medium to 7 (fig. 15.29). By using K^+ and valinomycin, it also is possible to set up an electrical potential gradient that adds to the force pulling protons into the matrix. The mitochondria first are loaded with K^+ in the absence of valinomycin, placed in a medium without K^+, and then suddenly treated with valinomycin. The valinomycin allows K^+ to flow rapidly out of the mitochondria, leaving a net negative charge inside. Generation of ATP in experiments of this sort was first achieved with chloroplast membranes by Andre Jagendorf (see chapter 16). Similar results were later obtained by investigators studying vesicles made from the mitochondrial inner membrane.

Additional evidence that proton movements are coupled to ATP synthesis comes from the observation that the hydrolysis of ATP that occurs when the ATP-synthase runs backwards results in the movement of protons *out* of the mitochondria. Remember that the breakdown of ATP apparently generates the same ~ that is created by respiration. In experiments analogous to those shown in figure 15.26, hydrolysis of

Figure 15.29

An electrochemical potential gradient for protons can be set up across the mitochondrial inner membrane in the absence of electron transfer reactions. In step 1, mitochondria are incubated at high pH in the presence of KCl to reduce the proton concentration inside, and to load the inside with K^+ and Cl^-. In step 2, the pH and the KCl concentration outside the mitochondria are decreased. The ionophore valinomycin (Val) is added so that K^+ can come out. Because the counter-ion Cl^- cannot move rapidly across the membrane, the efflux of K^+ down its concentration gradient leaves the inside of the mitochondrion with a net negative charge, relative to the outside. This difference in electrical potential, together with the higher pH inside, favors the influx of protons. Protons moving inward through the ATP-synthase can drive the formation of ATP.

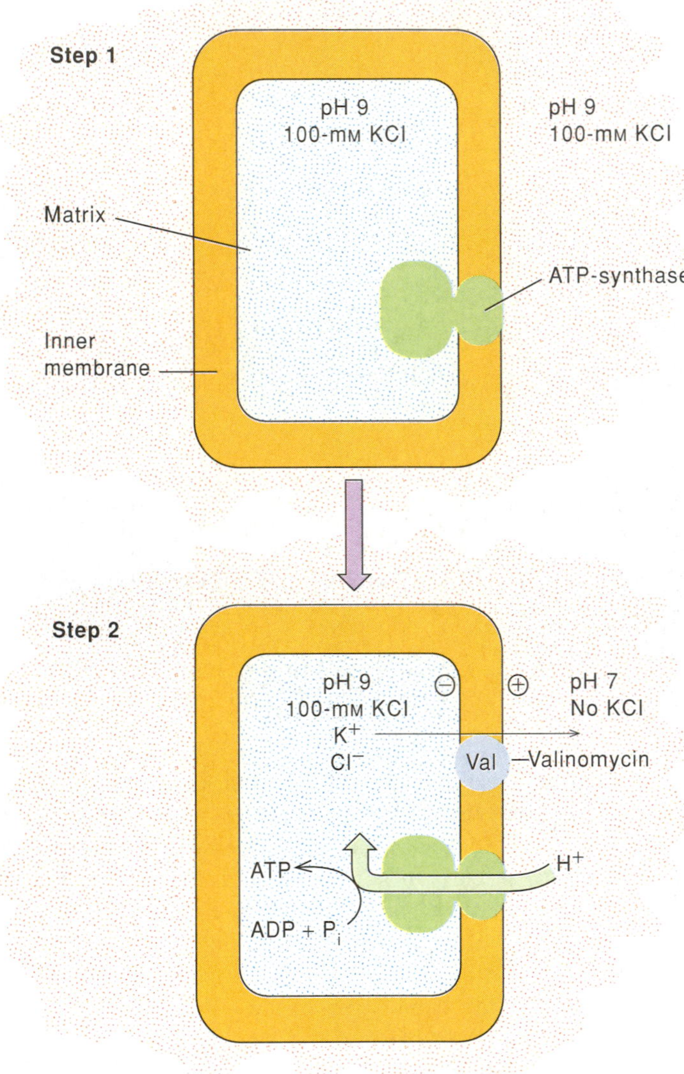

ATP by mitochondria results in the extrusion of protons and the uptake of lipophilic cations. These processes are blocked by oligomycin, the inhibitor of the ATP-synthase/ATPase. (As you might predict, oligomycin does not block proton extrusion when the extrusion is driven by electron transport.)

Current estimates are that three protons move into the matrix through the ATP-synthase for each ATP that is synthesized. We will see soon that one additional proton moves into the mitochondrion in connection with the uptake of ADP and P_i and the export of ATP, giving a total of four protons per ATP. How large a free energy change is associated with moving these protons? Rewriting equation (10) for the movement of one mole of protons, we have

$$\Delta G_{H^+} = G_{H^+(in)} - G_{H^+(out)} = RT \ln\{[H^+]_{in}/[H^+]_{out}\} + F\,\Delta\psi$$
$$= -2.3RT\,\Delta pH + F\,\Delta\psi \qquad \textbf{(11)}$$

where $\Delta pH = pH_{in} - pH_{out}$. If the pH in the matrix space is 0.05 pH units above that in the cytosol, the term $-2.3RT\,\Delta pH$ amounts to -0.07 kcal/mole (-0.3 kJ/mole). If the electrical potential difference across the membrane is -0.15 V, $F\,\Delta\psi$ is -3.46 kcal/mole (-14.3 kJ/mole). This gives a total ΔG_{H^+} of -3.53 kcal/mole (-14.6 kJ/mole), with the term $F\,\Delta\psi$ making the dominant contribution. If four protons move across the membrane for each molecule of ATP that is synthesized, the free energy change associated with the proton translocation would be $4 \times (-3.53)$, or -14.1 kcal per mole of ATP (58.4 kJ/mole), which is about twice as large as the $\Delta G°'$ of ATP hydrolysis (-7.5 kcal/mole or -31.4 kJ/mole). The free energy decrease associated with proton movement thus would be large enough to account for the high [ATP]/[ADP][P_i] ratio that is maintained under physiological conditions. The thermodynamic driving force for proton translocation is frequently termed the proton-motive force (Δp) and is expressed quantitatively as $\Delta G_{H^+}/F$ in units of volts.

Deferring for the moment the question of how the ATP-synthase is able to make ATP at the expense of the proton-motive force, let's see how the stoichiometry of proton movements relate to the P/O ratios we discussed earlier. When mitochondria respire and form ATP at a constant rate, protons must be moving inward at a rate that just balances the rate at which the electron-transport reactions pump protons out. Suppose that ten protons are pumped out for each pair of electrons that traverse the respiratory chain from NADH to O_2, and that four protons move back in for each ATP that is synthesized. Because the rates of proton efflux and influx must balance, 2.5 molecules of ATP (10/4) will be formed for each pair of electrons that go to O_2. The P/O ratio thus is given by the ratio of the proton stoichiometries. If the oxidation of succinate drives the extrusion of six protons per pair of electrons, the P/O ratio for this substrate would be 6/4, or 1.5. These ratios agree with the measured P/O ratios for the two substrates. Note that in the chemiosmotic theory the P/O ratio does not need to have an integer value. Also, if some of the protons pumped out leak back across the membrane without passing through the ATP-synthase, the P/O ratios will be lower than these values.

The Proton-Conducting ATP-Synthase or ATPase: F_1 and F_o

Let us now turn to the ATP-synthase. What is its structure, and how is it able to use a proton-motive force to drive the formation of ATP? Studies of the ATP-synthase began in the early 1960s, when Efraim Racker and his colleagues identified several enzymatic components that appeared to participate in oxidative

Figure 15.30

(*a*) Negatively stained electron micrograph of part of a mitochondrion, showing head-pieces of the F_1 ATP-synthase lining the inner membrane. The wormlike white area is a crista; the large dark areas are part of the matrix. The F_1 head-pieces show up as white spots projecting from the membrane of the crista into the matrix. (From B. Chance and D. Parsons, "Cytochrome Function in Relation to Inner Membrane Structure of Mitochondria," *Science* 142:1176, 1963, © 1963 by the AAAS.) (*b*) Negatively stained electron micrograph of mitochondrial membrane vesicles. These vesicles are capable of oxidative phosphorylation. F_1 head-pieces line the outer surface of the membranes. (*c*) Electron micrograph of the purified F_1 ATPase. (*d*) Mitochondrial membrane vesicles that were depleted of F_1 by treatment with urea and trypsin. These membranes contain a functional electron-transfer chain, but do not form or hydrolyze ATP. Note that the surfaces of the membranes appear smooth. (Micrographs *b, c,* and *d* courtesy of Dr. E. Racker.)

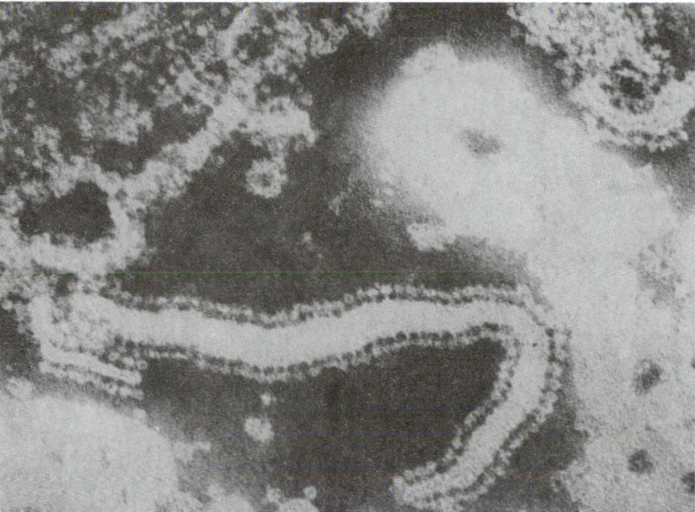

(a)

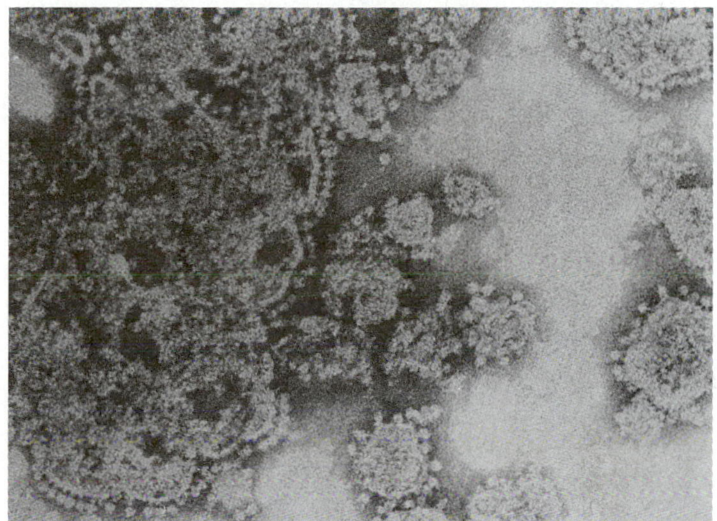

(b)

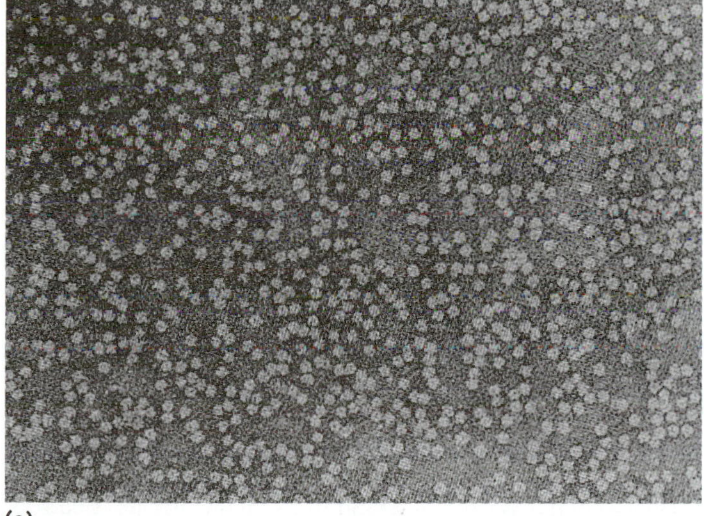

(c)

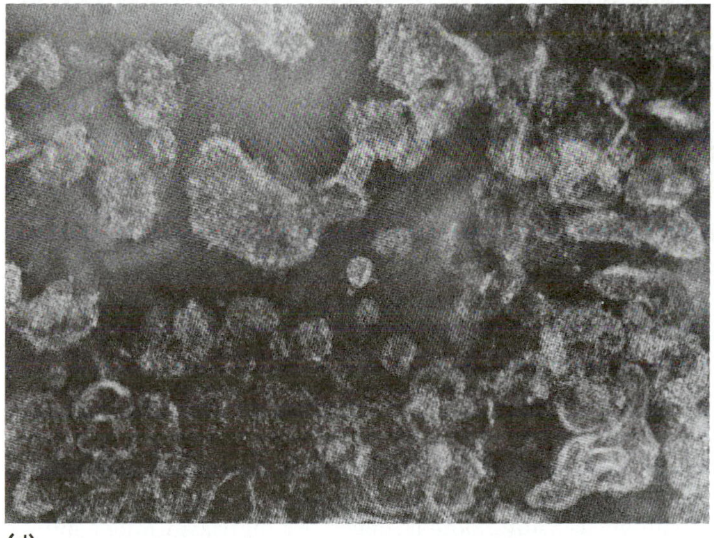

(d)

phosphorylation. The experimental strategy was to disrupt mitochondria and fractionate the proteins released from the membranes. In some cases, the depleted membranes lost their ability to synthesize ATP, although the electron-transfer reactions were undisturbed. Recombining the solubilized proteins, or coupling factors, with the membranes restored oxidative phosphorylation.

The first coupling factor to be purified, F_1, was an active ATPase when it was removed from the mitochondria. Unlike the membrane-bound ATPase in intact mitochondria, the solubilized F_1 ATPase was not inhibited by oligomycin. However, it recovered its sensitivity to oligomycin if it was reattached to the membranes in combination with another coupling factor, F_o. (The subscript "o" stands for oligomycin.)

F_o and F_1 both turned out to be parts of the proton-conducting ATP-synthase. Together they form a multiprotein complex that, like the electron-transfer complexes, is partially embedded in the mitochondrial inner membrane. F_1, which contains the catalytic sites for ATP formation or hydrolysis, consists of five different polypeptide subunits (α, β, γ, δ, and ϵ) with the stoichiometry $\alpha_3\beta_3\gamma\delta\epsilon$. Its total molecular weight is about 360,000. In electron micrographs of mitochondria, the F_1 complexes can be seen as spheres with diameters of about 85 Å, protruding from the surface of the inner membrane (fig. 15.30). F_o, a complex of approximately five hydrophobic polypeptides, is an integral component of the inner membrane and acts as a base-piece and stalk that holds F_1 to the membrane (fig. 15.31).

Figure 15.31

Components of the proton-conducting ATP-synthase. The F_1 head-piece includes three α and three β subunits and one copy each of three other subunits (γ, δ, and ε). F_0 includes a cluster of 9 to 12 copies of a small peptide, which appears to form a transmembrane channel for protons, and several additional subunits.

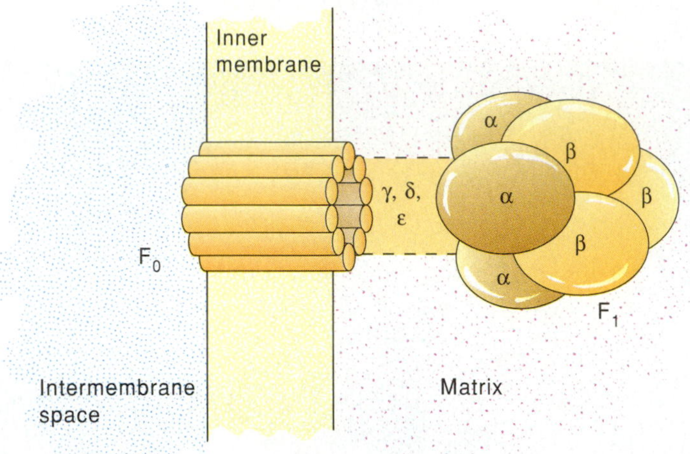

Complexes similar to the mitochondrial F_0 and F_1 have been isolated from the chloroplast thylakoid membrane (see chapter 16) and from bacteria. In the enzyme from *E. coli,* which has been particularly well studied, F_0 has three subunits (*a, b,* and *c*) with the stoichiometry $a_1b_2c_{10}$.

The F_1-F_0 ATPase complex is a useful market for distinguishing the two surfaces of the mitochondrial inner membrane. In intact mitochondria, the spheres line the side of the membrane facing the matrix (see fig. 15.30*a*). In preparations of submitochondrial vesicles, made by breaking mitochondria by sonic oscillations, the ATPase is on the outside (see fig. 15.30*b*). When the mitochondria are broken, the saclike cristae evidently pinch off and seal to form vesicles so that the side of the membrane that originally faced the matrix becomes the outer surface of the vesicle. You can see how this might occur by examining figure 15.2. The idea that the matrix side of the membrane faces outward in the isolated vesicles is consistent with observations on the orientations of the electron carriers. NADH dehydrogenase, for example, faces the matrix in intact mitochondria but reacts with added NADH in submitochondrial particles; cytochrome *c* attaches to the cytosolic side of the membrane in mitochondria but to the inner surface in the particles. Further, the electron-transport chain pumps protons out of intact mitochondria, but *into* the isolated vesicles.

The Mechanism of Action of the ATP-Synthase

If the purified F_1-F_0 complex is incorporated into a phospholipid vesicle and an electrochemical potential gradient for protons is set up across the membrane, the complex can synthesize ATP.

Figure 15.32

ATP synthesis can be obtained in an artificial system in which the mitochondrial ATP-synthase is incorporated into membrane vesicles together with bacteriorhodopsin, a membrane protein obtained from *Halobacterium halobium.* When bacteriorhodopsin is excited with light, it pumps protons across the membrane (see chapters 32 and 36). The solution inside the vesicle thus becomes acidic and positively charged relative to the external solution. Protons can move back out through the ATP-synthase, and this movement is coupled to the formation of ATP. The formation of ATP is blocked by uncouplers or oligomycin. Note that the F_1 headpiece of the ATP-synthase, which is found on the inner surface of the mitochondrial inner membrane, is on the outside of the liposome in this artificial system. The ATP-synthase also faces outward in submitochondrial particles made by breaking the inner membrane with sonic oscillations. In all cases, ATP synthesis is associated with the movement of protons through F_0 in the direction toward F_1. The hydrolysis of ATP results in proton movements in the opposite direction.

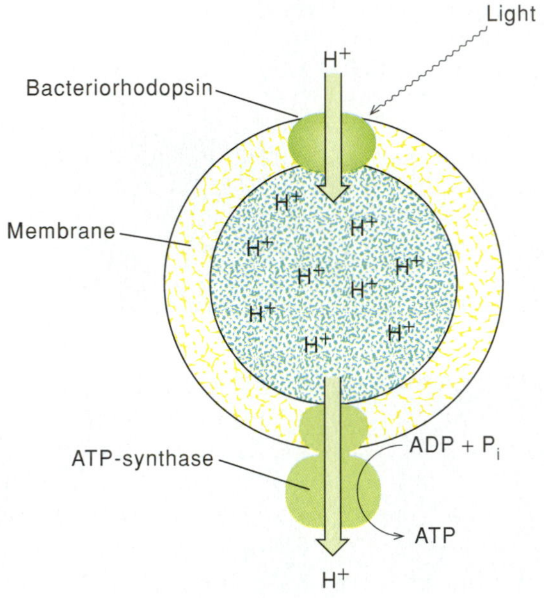

To show this, Racker and Stoeckenius generated the proton gradient by incorporating a light-driven proton pump, bacteriorhodopsin from *Halobacterium halobium* (see chapters 7 and 36), into membrane vesicles along with the F_1-F_0 complex. When the vesicles were illuminated, ATP was formed (fig. 15.32). This experiment provided a dramatic demonstration that the ATP-synthase does not have to be attached directly to the electron-transport complexes of the respiratory chain, for no electron carriers were included in the liposomes. It added strong support to the idea that the coupling of phosphorylation to electron transport is mediated by a proton-motive force that can be generated at one site on the membrane and used at relatively distant sites.

F_0 appears to contain the proton-conducting channel of the ATP-synthase. If F_0 alone is incorporated into liposomes, it makes the membranes leaky specifically to protons. The leaks can be blocked by either of the phosphorylation inhibitors oligomycin or dicyclohexylcarbodiimide (DCCD). Oligomycin reacts reversibly and noncovalently; DCCD reacts irreversibly

15B

BOX

Natural Uncouplers: Thermogenesis in Brown Fat and Skunk Cabbage

Several unusual types of cells use the respiratory chain for thermogenesis, the generation of heat, rather than the formation of ATP. Newborn mammals, and also the adults of some mammals that are adapted to live in cold climates, do this in a specialized adipose tissue called brown fat. The brown color is due largely to the cytochromes in the tissue's abundant mitochondria. Heat produced by brown-fat mitochondria is important for maintaining body temperature in the newborn and for the arousal of hibernating animals. Brown fat generates heat at a rate of about 400 W/kg, far above the rate of about 1 W/kg that is typical of other resting mammalian tissues.

The inner membrane of brown-fat mitochondria contains a protein, thermogenin, that acts as a channel for anions. The protein allows OH^- or Cl^- ions to pass rapidly across the membrane. Because a movement of OH^- from the mitochondrial matrix to the cytosol has the same consequences as a movement of H^+ in the opposite direction, thermogenin acts as an uncoupler. The free energy released in the electron-transfer reactions is stored transiently as an electrochemical potential gradient for protons, but then is degraded largely to heat.

The uncoupling activity of thermogenin is regulated by ATP, ADP, GTP, and GDP, which bind tightly to the protein and inhibit anion transport. Very low concentrations of fatty acids increase anion permeability. Fatty acids released from triacylglycerol stores in the tissue could be important in the hormonal control of thermogenesis. The amount of thermogenin in the brown-fat mitochondria also changes in response to physiological conditions. In animals that live at low temperatures, it can represent as much as 15% of the protein in the inner membrane.

Skunk cabbage uses a different strategy for thermogenesis. It oxidizes UQH_2 by an alternative pathway that bypasses the cytochrome bc_1 complex and cytochrome oxidase. Electron transport by this pathway does not result in proton translocation. Heat produced by respiration may help the plant thrive early in spring, when the climate is cool.

---

to form a covalent bond with a critical aspartate residue in one of the subunits. In the enzyme from *E. coli,* the site of inhibition is in subunit *c,* which is a small, hydrophobic polypeptide with a molecular weight of about 8,000. Analysis of the amino acid sequence suggests that each of the ten copies of the *c* subunit in the F_1-F_0 complex is folded into a hairpinlike structure with two transmembrane α helices. It has been suggested that the multiple copies of the subunit assemble to form a tubular channel across the membrane at the base of F_1 (see fig. 15.31).

How the movement of protons through F_0 drives the formation of ATP in F_1 is not known. However, studies by Paul Boyer, Harvey Penefsky, and others have shown that the enzyme binds ADP and ATP very tightly. The reaction

$$ADP + P_i \rightleftharpoons ATP + H_2O$$

occurs between the bound nucleotides on the enzyme, even in the absence of a proton-motive force. P_i evidently reacts directly with ADP to give ATP in a single step, without the formation of a phosphorylated enzyme intermediate. Remarkably, the equilibrium constant for the formation of ATP is close to 1 when the nucleotides are bound to the enzyme. When the nucleotides are free in solution, the equilibrium constant is about 10^{-5}.

Soluble enzymes offer several precedents for a substantial difference in the equilibrium constant of a reaction that forms ATP, depending on whether the reactants and products are bound to the enzyme or are free in solution. With both pyruvate kinase and 3-phosphoglycerate kinase, the equilibrium constant for the formation of bound ATP on the enzyme is much more favorable than the equilibrium constant for the reaction of the free materials. This difference could reflect the fact that the $\Delta G°'$ for ATP formation in solution includes an entropy decrease associated with the ordering of water molecules around the polyphosphate chain. Such an entropy decrease can be much smaller if the ADP, P_i, and ATP are bound on an enzyme.

The proton-motive force does not have much effect on the equilibrium constant for the formation of ATP on the ATP synthase. Instead, it alters the binding and release of ADP, P_i, and ATP from the enzyme. The nucleotides evidently are bound in an environment that favors the formation of ATP, and a movement of protons in response to the protein-motive force can change the enzyme's conformation in such a way that the ATP is released.

Although there is much evidence that movement of protons through the ATP-synthase drives the formation of ATP, changes in the [ATP]/[ADP][P_i] ratio have been measured in mitochondria under conditions when there is little change in the proton-motive force. One way to explain such observations is to postulate that some of the protons that are pumped by the electron-transport reactions are not released into the aqueous solution at the surface of the membrane, but rather are conducted more directly to the ATP-synthase. The proton-motive force between the solutions on the two sides of the membrane thus might reflect only a part of the driving force that is expressed at the ATP-synthase.

In some unusual cell types, natural uncouplers exist whose function is to channel the energy of the respiratory chain to heat production rather than ATP formation (box 15B).

Transport of Substrates, P_i, ADP, and ATP into and out of Mitochondria

One of the principles underlying the chemiosmotic theory is that the mitochondrial inner membrane is basically impermeable to charged and highly polar molecules. As we have discussed, NADH, NAD^+, and H_3O^+ cannot pass freely across the membrane. How, then, can P_i, adenine nucleotides, and substrates such as pyruvate and citrate move into and out of mitochondria? The answer is that the inner membrane has a set of transport systems that specifically catalyze the movements of these materials.

Uptake of P_i and Oxidizable Substrates Is Coupled to the Release of OH^- Ions

The transport of materials across biological membranes is discussed in detail in chapter 32, where we will develop the idea that the movement of one substance across a membrane can be coupled to movement of another substance in the same or the opposite direction. The uptake of β-galactosides by some bacteria, for example, is linked to the movement of protons into the cell. Like the proton-conducting ATP-synthase, transport systems in the cell membrane are able to harness the free energy decrease associated with moving protons down an electrochemical potential gradient. They use this free energy to move sugars and other nutrients thermodynamically uphill, against the concentration gradient of the nutrient. This concept was first advanced by Mitchell as an extension of the chemiosmotic theory, and it has turned out to apply to the movements of many materials into and out of mitochondria.

The uptake of P_i by mitochondria is coupled to an outward movement of OH^-. If the phosphate moves in the form of $H_2PO_4^-$, carrying one negative charge, an exchange of P_i for OH^- will be electrically neutral: There is no net movement of charge in either direction. The transport of P_i thus will be relatively insensitive to the electrical potential gradient $\Delta\psi$ that the respiratory chain sets up across the inner membrane. It will, however, be sensitive to the pH gradient. An outward movement of OH^- is essentially equivalent to an inward movement of H^+, and is thermodynamically downhill in respiring mitochondria because the pH is higher in the matrix than in the cytosol (fig. 15.33*a*). Mitochondria thus are capable of taking up P_i from the cytosol even when the concentration of P_i inside exceeds that outside.

Pyruvate, which is generated in the cytosol by glycolysis but consumed in the mitochondria by the TCA cycle, also is transported into the mitochondria in exchange for OH^-. Succinate and malate are taken up by a transport system that can exchange either of these dicarboxylic acids for P_i (fig. 15.33*b*). The same transport system also can exchange one of the dicarboxylic acids for another. A separate system exchanges the tricarboxylic acids citrate and isocitrate for each other or for a dicarboxylic acid. Mitochondria thus can achieve the uptake of any of these substrates by a series of exchanges of one carboxylic acid for another, of an acid for P_i, and ultimately of P_i for OH^-.

Figure 15.33

(*a*) The phosphate/hydroxide exchange protein carries a phosphate ion ($H_2PO_4^-$) in one direction across the mitochondrial inner membrane in exchange for an OH^- ion moving in the opposite direction. These movements are reversible, but a net efflux of OH^- is thermodynamically downhill because the respiratory chain pumps protons out of the matrix and raises the pH there. Because OH^- efflux is linked to $H_2PO_4^-$ uptake, phosphate is concentrated in the matrix. (*b*) The phosphate/dicarboxylic acid exchange protein exchanges succinate (or some other dicarboxylic acids) for phosphate (HPO_4^{2-}). Phosphate efflux is thermodynamically favorable because the phosphate concentration in the matrix exceeds that in the cytosol, as a result of the action of the phosphate/hydroxide exchange protein. (*c*) The ADP/ATP exchange protein exchanges ADP^{3-} for ATP^{4-}. An outward movement of ATP removes one negative charge from the matrix, and is favored because proton pumping by the respiratory chain gives the matrix a negative charge relative to the cytosol.

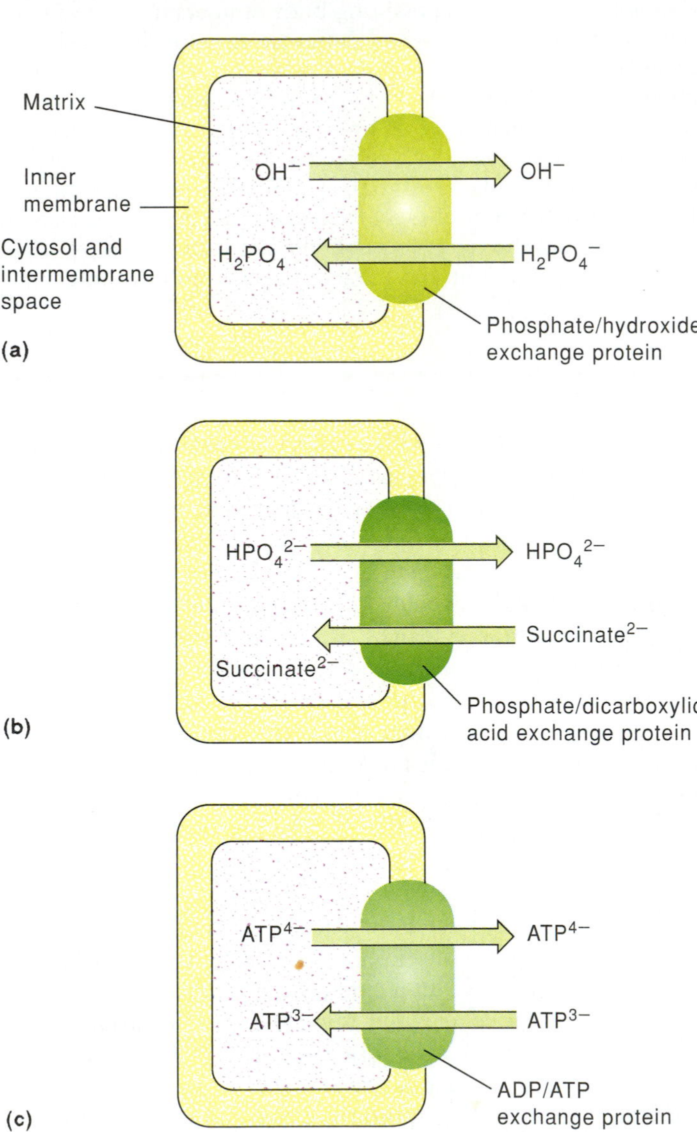

Figure 15.34

Structures of two inhibitors of the ATP/ADP exchange protein. Atractyloside is obtained from a species of thistle; bongkrekic acid, from a fungus found in decaying coconuts (*bongkrek* in Indonesian). As we will see in chapter 32, the ATP/ADP exchange protein has two distinct conformational states. In one of these states it binds ADP and atractyloside; in the other, it binds ATP and bongkrekic acid.

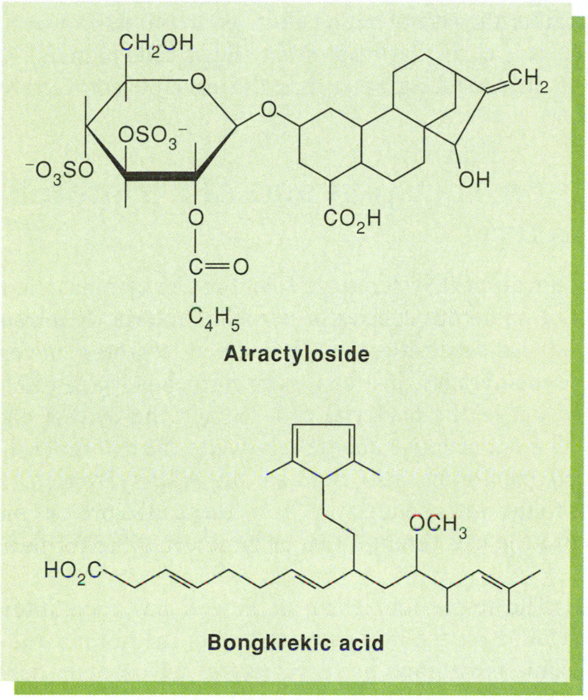

Atractyloside

Bongkrekic acid

Export of ATP Is Coupled to ADP Uptake

ATP, which is produced in the mitochondria largely for use elsewhere in the cell, is exported by a transport system that exchanges ATP for ADP. This exchange is not electrically neutral. At pH 7, the net charge on ATP is approximately -4 whereas that on ADP is about -3. The outward movement of a molecule of ATP in exchange for uptake of an ADP removes one negative charge from the matrix, and is driven by the electrical gradient, $\Delta\psi$ (see fig. 15.33*c*). The exchange is said to be electrogenic. The ATP/ADP exchange protein, or adenine nucleotide translocator, is actually the most abundant protein in the mitochondrial inner membrane. It is sensitive to several specific inhibitors, atractyloside, and bongkrekic acid (fig. 15.34). The mechanism of the ATP/ADP exchange is discussed in chapter 32.

The combined effect of exchanging extramitochondrial ADP^{3-} and $H_2PO_4^-$ for mitochondrial ATP^{4-} and OH^- is to move one proton into the mitochondrial matrix for every molecule of ATP that the mitochondria release into the cytosol. This proton translocation must be considered, along with the movement of protons through the ATP synthase, in order to account for the P/O ratio of oxidative phosphorylation. If three protons pass through the ATP synthase, and the adenine nucleotide and P_i transport systems move one additional proton, then in total, four protons move into the matrix for each ATP provided to the cytosol. As we explained earlier, translocation of four protons per ATP is consistent with the relationship between the proton-motive force and the measured ΔG for ATP formation under physiological conditions.

Electrons from Cytosolic NADH Are Imported by Shuttle Systems

The mitochondrial inner membrane does not contain a transport system for NAD^+ or NADH. In animal cells, most of the NADH that must be oxidized by the respiratory chain is generated in the mitochondrial matrix by the TCA cycle or by the oxidation of fatty acids. However, NADH also can be generated by glycolysis in the cytosol, and this NADH must be reoxidized to NAD^+ in some manner. If O_2 is available, it clearly is advantageous to reoxidize the NADH by the respiratory chain, rather than by the formation of lactate or ethanol. Approximately 2.5 molecules of ATP can be formed for each NADH that is oxidized in the mitochondria, whereas no ATP is made when NADH is oxidized by the cytosolic lactate dehydrogenase or alcohol dehydrogenase.

Plant mitochondria have a second NADH dehydrogenase that is distinct from complex I and can oxidize cytosolic NADH, but this enzyme is not found in animals. If NADH itself cannot enter their mitochondria, animal cells must have a shuttle system that transfers electrons from cytosolic NADH to the respiratory chain. There are several such systems. The simplest involves the reduction of dihydroxyacetone phosphate to glycerol-3-phosphate in the cytosol, followed by reoxidation of the glycerol-3-phosphate by the mitochondrial glycerol-3-phosphate dehydrogenase:

NADH + H^+ → NAD^+ (Cytosolic dehydrogenase); Dihydroxyacetone phosphate ⇌ Glycerol-3-phosphate; FAD → $FADH_2$ (Mitochondrial dehydrogenase)

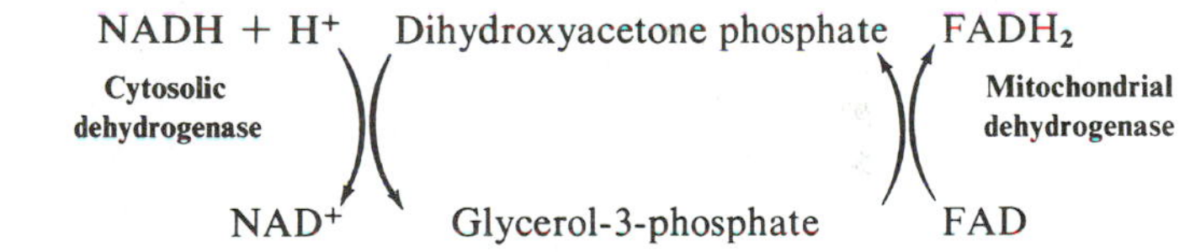

The catalytic site of the mitochondrial glycerol-3-phosphate dehydrogenase is on the cytosolic surface of the inner membrane, so the glycerol-3-phosphate does not have to pass through this membrane in order to be reoxidized.

The dihydroxyacetone phosphate/glycerol-3-phosphate shuttle has the shortcoming that oxidation of glycerol-3-phosphate by the respiratory chain generates only 1.5 ATP, instead of the 2.5 that can be generated by the oxidation of mitochondrial NADH (see fig. 15.20). The mitochondrial glycerol-3-phosphate dehydrogenase passes electrons to UQ, below the coupling site in complex I. There is another, more complicated shuttle system that allows the electrons to pass through complex I. In this scheme, cytosolic NADH reduces oxaloacetate to malate, which is carried across the inner membrane by a specific transporter. Inside the mitochondria, the malate is reoxidized to oxaloacetate, reducing mitochondrial NAD^+ to NADH:

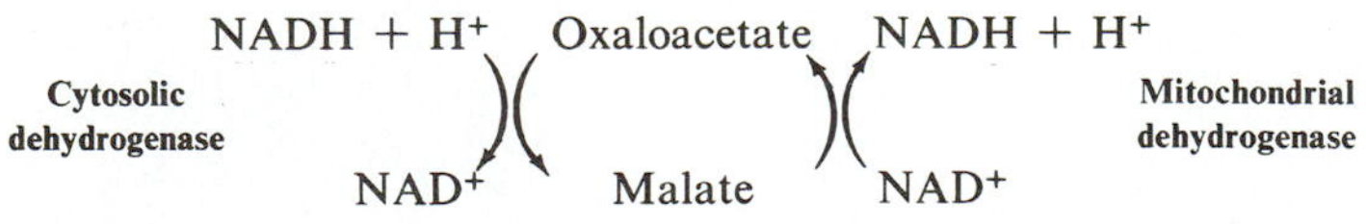

NADH + H^+ → NAD^+ (Cytosolic dehydrogenase); Oxaloacetate ⇌ Malate; NAD^+ → NADH + H^+ (Mitochondrial dehydrogenase)

The internal NADH then can be oxidized by the respiratory chain.

To continue the oxidation of cytosolic NADH by this shuttle system, oxaloacetate must return from the mitochondria to the cytosol. Oxaloacetate itself cannot cross the mitochondrial inner membrane; there is no transporter for it. However, oxaloacetate can react with the amino acid glutamate to form aspartate and α-ketoglutarate (see chapter 18). The inner membrane has transporters that couple exchange of α-ketoglutarate for malate and exchange of glutamate for aspartate. Working together, these enzymes can complete the shuttle. Because the glutamate/aspartate transporter also moves a proton into the mitochondrion, decreasing the proton-motive force across the membrane, oxidation of cytosolic NADH by the oxaloacetate/malate shuttle still does not generate as much ATP as can be obtained from mitochondrial NADH; the net yield is probably about 2.25 ATP per NADH.

Complete Oxidation of Glucose Yields about 30 Molecules of ATP

The shuttle systems that operate between the cytosol and mitochondria must be taken into account when we calculate the total yield of ATP from a molecule of glucose that is oxidized to CO_2 and H_2O. To make such a calculation, recall first that two molecules of ATP are formed from each glucose in glycolysis, and two more are formed via GTP in the TCA cycle. At the same time, two molecules of NADH are produced in the cytosol by glycolysis, and eight molecules of NADH are generated in the mitochondrial matrix by the pyruvate dehydrogenase complex and the TCA cycle. The two molecules of succinate proceeding through the succinate dehydrogenase step of the TCA cycle reduce two molecules of FAD to $FADH_2$. Reoxidation of the eight mitochondrial NADH molecules and two $FADH_2$ molecules by the respiratory chain can generate approximately 23 molecules of ATP (2.5 for each NADH and 1.5 for each $FADH_2$). The two NADH molecules that are formed in the cytosol could contribute their reducing equivalents to the mitochondria in the form of either malate or glycerol-3-phosphate. Operation of the glycerol-3-phosphate shuttle would provide two pairs of electrons at the level of UQH_2, yielding three molecules of ATP (2×1.5). The malate shuttle allows electrons to pass through complex I also, leading to the formation of about 4.5 ATP (2×2.25). Thus, depending on which shuttle system is in operation, a total of either 26 or 27.5 ATPs can be formed by oxidative phosphorylation in addition to the four ATPs formed in glycolysis and the TCA cycle. Exporting the two molecules of ATP formed in the TCA cycle would require the uptake of two protons by the mitochondria, which would reduce the total amount of ATP by about 0.5. (Remember that the adenine nucleotide transporter and the ATP-synthase together move four protons for each ATP they provide to the cytosol.) This leaves a grand total of from 29.5 to 31 ATP, depending on the shuttle.

The actual yields of ATP under physiological conditions are probably less than these limiting values. Some of the free energy that is stored transiently in the form of the proton-motive force may be lost as a result of nonspecific leakage of protons or other ions through the membrane, or may be used to drive reactions other than the formation of ATP. In addition, NADH that is formed in the cytosol may be used there for reductive biosynthetic reactions such as the formation of fatty acids, rather than contributing electrons to the respiratory chain. Even so, it is clear that respiration allows cells to make ATP in amounts that are far greater than the amounts provided by glycolysis.

Electron Transport and ATP Synthesis in Bacteria

Electron transport systems are found in the cytoplasmic membranes of numerous species of aerobic bacteria. Electron flow from reduced substrates to O_2 is coupled to proton movements in these membranes, just as it is in mitochondria. Protons are pumped out of the bacterial cell, leaving the cytosol alkaline and with a net negative charge relative to the external solution. The cell membrane also contains an ATP-synthase/ATPase similar to the mitochondrial ATP-synthase, and protons moving back into the cell through this enzyme drive the formation of ATP.

The respiratory chain of *E. coli* has been studied in some detail. Like the mitochondrial inner membrane, the cell's cytoplasmic membrane has a variety of flavoprotein dehydrogenases that transfer electrons to UQ. However, instead of cytochrome *c* and complexes II and IV, there are two different multienzyme complexes, either of which can catalyze electron transfer from UQH_2 all the way to O_2. One of these complexes, the cytochrome *o* complex, contains two modified *a* hemes and one copper atom. Several of its polypeptides appear to be homologous to subunits of the eukaryotic cytochrome oxidase. The alternative cytochrome *d* complex contains two *b* hemes and two copies of the unusual heme *d*, but no copper or iron-sulfur centers. The cells make varying amounts of the two complexes, depending on the availability of O_2. Electron flow through either complex is coupled to proton translocation.

A proton-translocating ATPase also is found in anaerobic bacteria that do not have electron-transport systems. The role of the ATPase in these species is to pump protons out of the cell, at the expense of ATP made by fermentative pathways such as glycolysis. The proton efflux creates an electrochemical potential gradient, which the bacteria use to drive the uptake of sugars and other nutrients. The enzymes that catalyze this transport of nutrients are discussed further in chapter 32.

Summary

In this chapter we have focused on the oxidation-reduction reactions that are coupled to the synthesis of ATP. The following points are highlights of our discussion.

1. In eukaryotes, the TCA cycle and most of the reactions of aerobic energy metabolism occur in mitochondria. An inner membrane separates the mitochondrion into two distinct spaces, the internal matrix space and the intermembrane space. All but one of the enzymes of the TCA cycle are in the matrix space; succinate dehydrogenase is in the inner membrane.
2. In the TCA cycle, electrons are passed to NAD^+ and FAD. An electron-transport system in the inner membrane reoxidizes the reduced coenzymes (NADH and $FADH_2$) at the expense of molecular oxygen. Flow of electrons to O_2 releases the free energy that supports the synthesis of most of the ATP formed during aerobic metabolism.
3. Electron transfer to O_2 occurs stepwise, via a series of flavoproteins, cytochromes (heme-proteins), iron-sulfur proteins, quinones, and copper atoms. Most of the electron carriers are collected in four large complexes, which communicate via two mobile carriers, ubiquinone (UQ) and cytochrome *c*.
4. Complex I transfers electrons from NADH to UQ, and complex II transfers electrons from succinate to UQ. Both of these complexes contain flavins and numerous iron-sulfur centers.
5. Complex III, which contains three cytochromes (cytochromes b_L, b_H, and c_1) and one iron-sulfur protein, passes electrons from reduced ubiquinone (UQH_2) to cytochrome *c*. Complex IV contains two cytochromes (*a* and a_3) and two copper atoms, and transfers electrons from cytochrome *c* to O_2. The transfer of electrons through complex III involves a cyclic series of reactions (the Q cycle), in which UQH_2 and UQ undergo oxidation and reduction at two distinct sites.
6. As electrons move through complexes I, III, and IV toward O_2, protons are taken up from the solution on the matrix side of the membrane and released on the cytosolic side. This raises the pH of the matrix slightly above that of the cytosol and leaves the matrix negatively charged relative to the cytosol, creating an electrochemical potential difference or proton-motive force that tends to pull protons from the cytosol back into the matrix.
7. Proton movements through the F_o base-piece of the ATP-synthase in the inner membrane drive the formation of ATP by causing the release of bound ATP from the catalytic site on the F_1 head-piece of the enzyme.
8. Respiration is tightly coupled to the formation of ATP. Approximately 2.5 molecules of ATP are synthesized for each pair of electrons that pass down the electron-transport chain from NADH to O_2. Uncouplers, which are lipophilic weak acids, can dissipate the proton-motive force by carrying protons across the membrane. Respiration then occurs rapidly even in the absence of the phosphorylation reactions.
9. The proton-motive force also drives the uptake of P_i and ADP into the mitochondrial matrix and the export of ATP to the cytosol. By exchanging P_i for an organic acid, and exchanging one organic acid for another, mitochondria concentrate pyruvate and the other substrates that provide electrons to the electron-transport chain.

Selected Readings

Boyer, P., A perspective of the binding change mechanism for ATP synthesis. *FASEB Journal* 3:2164, 1989. A review of work on the ATP-synthase.

Chan, S. I., and P. M. Li, Cytochrome *c* oxidase: understanding nature's design of a proton pump. *Biochem.* 29:1, 1990. A review of the operation of complex IV.

Chance, B., and G. R. Williams, Respiratory enzymes in oxidative phosphorylation II. Difference spectra. *J. Biol. Chem.* 217:395, 1955. One of a series of papers developing kinetic techniques for elucidating the sequence of electron carriers in the respiratory chain.

Ernster, L. (ed.), *Bioenergetics.* Elsevier: Amsterdam, 1984. A collection of reviews covering electron transport, the ATP-synthase, translocation of ions across the mitochondrial inner membrane, thermogenesis in brown fat, and other topics in bioenergetics.

Hinkle, P. C., A. Kumar, A. Resetar, and D. L. Harris, Mechanistic stoichiometry of mitochondrial oxidative phosphorylation. *Biochem.* 30:3576, 1991. Measurements of P/O ratios and a discussion of the amount of ATP synthesized during the oxidation of glucose.

Mitchell, P., and J. Moyle, Stoichiometry of proton translocation through the respiratory chain and adenosine triphosphatase systems of rat liver mitochondria. *Nature* 208:147, 1965. The initial observations that electron transport moves protons outward across the mitochondrial inner membrane and that ATP hydrolysis does the same.

Reynafarje, B., and A. L. Lehninger, The K^+/site and H^+/site stoichiometry of mitochondrial electron transport. *J. Biol. Chem.* 254:6331, 1978. Valinomycin is used to allow K^+ to move inward across the inner membrane, in response to the electrical potential difference created by H^+ efflux. The number of protons pumped out is found to be larger than previously estimated.

Trumpower, B., The proton motive Q cycle. *J. Biol. Chem.* 265: 11409, 1990. A review of the structure and operation of complex III and the Q cycle.

Problems

1. Compare iron-sulfur proteins, flavoproteins, and quinones with respect to the following.
   (a) Chemical nature of the functional group that undergoes oxidation-reduction.
   (b) Number of reducing equivalents per redox center involved in electron donor/acceptor reactions of physiological importance. If semiquinones are formed, so indicate and include them in the reduction scheme.
   (c) Stoichiometry of protons taken up per electron.
2. (a) Describe how heme is bound to the protein portion of the a/a_3-, *b*-, and *c*-type cytochromes.
   (b) Although there are three types of cytochromes in rat liver mitochondria, CO and CN^- inhibit electron transfer only at the cytochrome a/a_3 complex. Why do these inhibitors interact with cytochrome a/a_3 but not with cytochrome *b* or cytochrome *c*?
3. Calculate the standard redox potential change ($\Delta E°'$) and the standard free energy change ($\Delta G°'$) for the following reactions at pH 7.0. Write a balanced equation for each reaction.
   (a) Cyt $c(Fe^{2+})$ + cyt $a_3(Fe^{3+}) \rightarrow$ cyt $c(Fe^{3+})$ + cyt $a_3(Fe^{2+})$
   (b) 4 cyt $c(Fe^{2+}) + O_2 + 4\ H^+ \rightarrow$ 4 cyt $c(Fe^{3+})$ + 2 HOH
   (c) Oxidation of succinate by succinate: cytochrome *c* reductase.
   (d) Reduction of extramitochondrial NAD^+ by dihydroubiquinone, via the α-glycerolphosphate shuttle.
4. (a) In biological oxidation-reduction reactions, does the stoichiometry of electron transfer (reducing equivalents/mole) differ among the 1 Fe, 2 Fe-2 S, and 4 Fe-4 S centers?
   (b) 4 Fe-4 S centers function in electron transport over a wide range of reduction potentials. There is nothing inherent in the iron-sulfur cluster to suggest this range of reduction potentials. Therefore, what other component(s) must dictate reduction potential?
5. What percent of cytochrome *c* will be in the oxidized form in a solution held at +0.30 V and pH 7.0?
6. Given the standard reduction potentials for cytochrome *c* and ubiquinone at pH 7.0 (see text), calculate the corresponding values at pH 6.0 and 8.0.
7. (a) Acetylation of one or more lysines near the edge of the heme in cytochrome *c* decreases both the rate of electron transfer to the cytochrome *c* from complex III and transfer of electrons from the reduced cytochrome to cytochrome oxidase. What does this suggest concerning operation of the respiratory chain?
   (b) What types of amino acid residues on complex III and complex IV would you expect to interact with cytochrome *c*?
8. Rotenone, which blocks the transfer of electrons from $FMNH_2$ of the NADH dehydrogenase to ubiquinone, is a potent insecticide and fish poison.
   (a) Explain why rotenone is lethal to insects and fish.
   (b) Would you expect the use of rotenone as an insecticide to be potentially hazardous to other animals (e.g., humans)? Why or why not?
   (c) If isolated mitochondria were respiring with succinate as substrate, would you expect a change in O_2 consumption upon addition of rotenone? If β-hydroxybutyrate were the respiratory substrate?
9. We can estimate the overall ATP yield for the oxidation of specific metabolic intermediates by considering both oxidative and substrate level phosphorylation. In principle, total molar yields of "high-energy phosphate" (ATP or equivalent) from cytosolic and mitochondrial processes divided by the molar consumption of O in the mitochondria yield theoretical P/O (or ADP/O) ratios. Using the value of P/O = 2.5 for NADH oxidation and P/O = 1.5 for succinate oxidation in the mitochondria, calculate theoretical P/O ratios for the oxidations given below. Assume that all required enzymes and cofactors are present and that extramitochondrial NADH is oxidized via the α-glycerolphosphate shuttle.
   (a) Oxidation of lactate to CO_2 and HOH.
   (b) Oxidation as in part (a), with 2,4-dinitrophenol present.
   (c) Oxidation of dihydroxyacetone phosphate to CO_2 and HOH.
   (d) Oxidation as in part (c), with 2,4-dinitrophenol present.
10. (a) Explain the necessity of having ubiquinone (UQ) concentration in excess of other mitochondrial electron-transfer components.
    (b) Suppose that you are examining muscle tissue mitochondria in which the UQ content is well below normal. You find that concentrations of all other electron-transfer components are within the normal range. Predict the effect of UQ deficiency on oxidation of: (i) NADH-producing substrates, (ii) succinate, (iii) ascorbate plus a redox mediator.
    (c) Explain why UQ-deficient muscle tissue has greater than normal concentrations of lactate. (Source: Ogasahara, Engel, Frens, and Mack, *PNAS* 86:2379–2382, 1989.)
11. (a) Explain what is meant by "tightly coupled" mitochondria. How can we determine whether mitochondria are tightly coupled?
    (b) What is the importance of "respiratory control" in oxidation of metabolites?
    (c) In what metabolic circumstance is it advantageous to the organism to have mitochondria uncoupled?

12. The uncoupling reagent, 2, 4-dinitrophenol (2,4-DNP), is highly toxic to humans, causing marked increase in metabolism, body temperature, profuse sweating, and in many instances, collapse and death. For a brief period in the 1940s, however, doses of 2,4-DNP presumed to be sublethal were given as a means of weight reduction in humans.
   (a) Explain why administration of 2,4-DNP results in increased metabolic rate as evidenced by increased O_2 consumption.
   (b) How are the metabolic events that occur upon administration of 2,4-DNP pertinent to regulation of glycolysis and the TCA cycle?
   (c) Why would consumption of 2,4-DNP lead to hyperthermia and profuse sweating?
   (d) Explain how 2,4-DNP uncouples oxidative phosphorylation.
13. A suspension of mitochondria is incubated with pyruvate, malate, and ^{14}C-labeled triphenylmethylphosphonium [TPP] chloride under aerobic conditions. The mitochondria are rapidly collected by centrifugation and the amount of ^{14}C that they contain is measured. In a separate experiment, the volume of the mitochondrial matrix space has been determined so that the concentration of TPP cation in the matrix can be calculated. The internal concentration is found to be 1,000 times greater than that in the external solution.
   (a) What is the apparent electrical potential difference ($\Delta\psi$) across the inner membrane? Express your answer in the appropriate units and indicate which side of the membrane is positive.
   (b) Qualitatively, how might $\Delta\psi$ be affected by the addition of an uncoupler?
14. Differentiate between electrogenic and neutral transport systems in mitochondria. How is electrogenic transport influenced by the membrane potential? What is the effect of neutral transport on the pH gradient?
15. (a) Suppose you prepare submitochondrial particles (SMP) that can oxidize succinate and catalyze oxidative phosphorylation. Protons are transported to the interior of the formed particles. How does the orientation of the vectorial H^+ transport in SMP compare with the orientation in mitochondria?
   (b) Uncouplers do not prevent succinate oxidation but the particle interior is no longer acidified. Explain.
   (c) Atractyloside, an inhibitor of adenine nucleotide translocator, blocks oxidative phosphorylation in intact mitochondria. Would atractyloside inhibit oxidative phosphorylation in the SMPs? Would you expect the SMPs to oxidize added NADH? Explain.

16

CHAPTER

Photosynthesis and Other Processes Involving Light

In chapter 15 we saw that mitochondria convert the chemical free energy of electron-transfer reactions into a transmembrane electrochemical potential gradient for protons. The vectorial aspects of metabolism presented new ideas that biochemists initially found difficult to grasp. It was necessary to consider free energy changes that accompany movements of ions between different compartments of a cell, in addition to the free energy changes associated with the electron-transfer reactions themselves. In this chapter we come upon a transfer of energy that initially may seem even more esoteric, the transfer of energy between light and matter. We will examine this phenomenon in three different biological situations: (1) the conversion of light into chemical energy in photosynthesis, (2) the effects of light on phytochrome, a pigment that regulates circadian and seasonal cycles of growth and flowering in plants, and (3) the production of light by chemical reactions in bioluminescence. We begin with photosynthesis, which will be the major concern of the chapter.

Photosynthesis

Heterotrophic organisms, including animals, fungi, and most types of bacteria, live by degrading complex molecules provided by other organisms. Life on earth obviously could not continue indefinitely in this manner without an independent mechanism for synthesizing complex molecules from simple ones. The energy that sustains this synthesis comes almost entirely from the sun, and is captured in the process of photosynthesis (fig. 16.1). Plants and other photosynthetic organisms convert, or "fix," about 10^{11} tons of carbon from CO_2 into organic compounds annually. In spite of the enormous magnitude of this conversion, the total amount of fixed carbon on earth is decreasing as a result of consumption. As our reserves of energy diminish, it becomes increasingly important that we understand how photosynthesis works and how our activities affect it.

When most of us think of photosynthesis, we think of green trees or perhaps of fields of grain. Actually, about one-third of the carbon fixation that occurs on earth takes place in the oceans and is carried out by microorganisms. In addition to plants, several groups of bacteria are capable of photosynthesis.

Figure 16.1

Photosynthesis harnesses the energy of sunlight for biosynthesis. In plants, the reactions of photosynthesis occur in the thylakoid membranes and the surrounding stroma of chloroplasts. The thylakoid membranes contain pigment-protein complexes that serve as antennas, absorbing light and passing its energy to the reaction centers of two distinct photosystems. In the reaction centers, excited chlorophyll molecules (P680 in photosystem II, P700 in photosystem I) give off electrons, reducing a series of nearby electron acceptors. Photosystem I pushes electrons to $NADP^+$ by way of iron-sulfur proteins and a stromal flavoprotein (FP). Photosystem II reduces plastoquinone (PQ). Electrons flow spontaneously from the reduced PQ to photosystem I via the cytochrome b_6f complex and a mobile protein (plastocyanin, PC). The oxidized species generated in photosystem II strip electrons from H_2O, releasing O_2. The electron-transfer reactions release protons in the thylakoid lumen and take up protons from the stroma, setting up a transmembrane pH gradient. An ATP-synthase couples the formation of ATP to the flow of protons back to the stroma. Thus the electron-transfer reactions driven by light provide the stroma with both NADPH and ATP.

Enzymes in the stroma use these materials to convert CO_2 to carbohydrates. CO_2 first combines with ribulose-1, 5-bisphosphate to form 2 molecules of glycerate-3-phosphate. ATP and NADPH are used to reduce the glycerate-3-phosphate to glyceraldehyde-3-phosphate. ATP also is used to recycle some of the glyceraldehyde-3-phosphate to ribulose-1, 5-bisphosphate for further CO_2 fixation. The remaining glyceraldehyde-3-phosphate can be exported from the chloroplast or converted to other carbohydrates.

Many of the photosynthetic reactions resemble reactions we have seen in earlier chapters: the formation of hexoses and pentoses from trioses resembles the reactions of gluconeogenesis and the pentose phosphate pathway (see chapter 13); the cytochrome-b_6f complex is similar to the mitochondrial cytochrome bc_1, and the proton-conducting ATP-synthase is similar to the mitochondrial ATP-synthase (see chapter 15). The major difference between the operation of mitochondria and chloroplasts is the direction of electron flow. In mitochondria, electrons flow from reduced organic compounds to O_2; photosynthetic organisms use the energy of light to push electrons from H_2O to fix CO_2 to form carbohydrates.

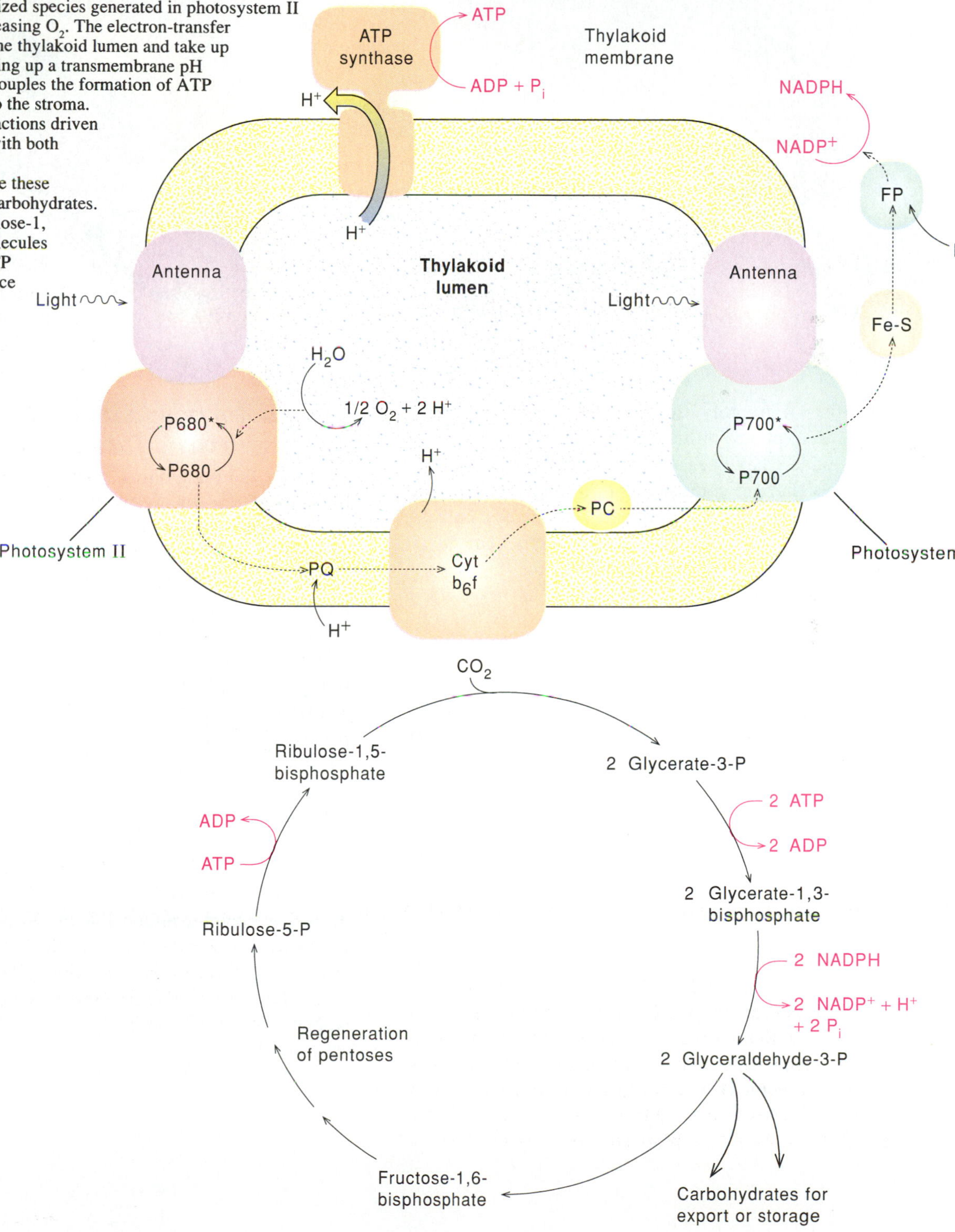

Figure 16.2

(*a*) Electron micrograph of a chloroplast in a lettuce leaf. The organelle is shaped like a flattened sausage with a width of about 10 μM and a thickness of about 3 μM. It is surrounded by a double outer membrane or envelope (E). The stroma (S) contains DNA, ribosomes, and the soluble enzymes of CO_2 fixation. Extending throughout the stroma is the thylakoid membrane, which is differentiated into stacked regions or grana (G) and unstacked stromal lamellae (SL). (Courtesy of Dr. Charles Arntzen.) (*b*) A schematic drawing of the chloroplast membrane systems. The highly folded thylakoid membrane separates the thylakoid lumen from the stroma.

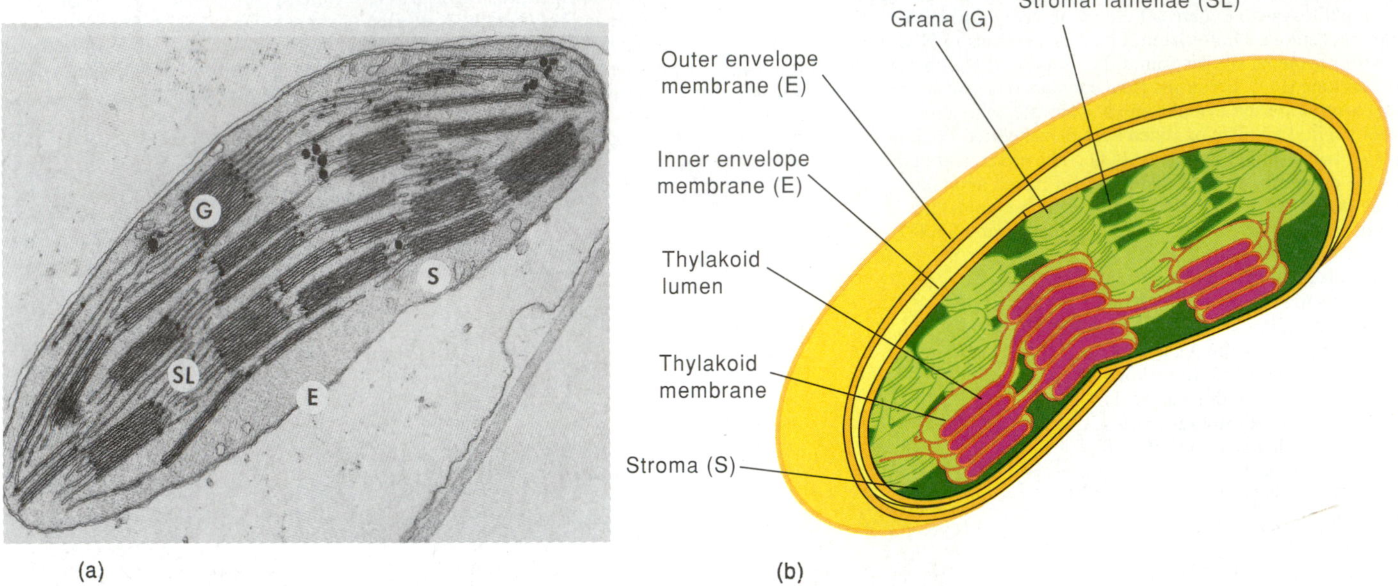

Bacteria have been extremely useful in the study of photosynthesis because their photosynthetic apparatus is simpler than that of plants, and it has been easier to purify the pigment-protein complexes that are components of this apparatus in the bacteria. Although there are important differences between photosynthesis in bacteria and plants, the photochemical reactions that capture the energy of light are basically the same.

An overall equation for CO_2 fixation as it occurs in plants is

$$6\ CO_2 + 6\ H_2O + \text{light} \rightarrow C_6H_{12}O_6 + 6\ O_2 \quad (1)$$

or, more generally,

$$CO_2 + H_2O + \text{light} \rightarrow (CH_2O) + O_2 \quad (2)$$

where (CH_2O) represents part of a carbohydrate molecule. Electrons and protons are removed from H_2O, O_2 is evolved, and CO_2 is reduced to the level of a carbohydrate. One group of photosynthetic bacteria, the cyanobacteria, carry out the same process. Other types of photosynthetic bacteria carry out similar overall processes except that they do not evolve O_2 because they use materials other than H_2O as a source of electrons.

If we leave out the light, the equilibrium for the synthesis of glucose from CO_2 and H_2O lies vanishingly far to the left. The equilibrium constant at 27° C is 10^{-496}! Our goal is to explore how photosynthetic organisms use light to drive the reaction in the direction of carbohydrates, against this enormous thermodynamic gradient. How can light do chemistry?

The Photochemical Reactions of Photosynthesis Take Place in Membranes

In plants, the reactions of photosynthesis take place in specialized subcellular organelles, the chloroplasts. Figure 16.2*a* shows an electron micrograph of a chloroplast from a lettuce leaf. Chloroplasts are bounded by an envelope of two membranes, and they have an extensive internal membrane called the thylakoid membrane. In electron micrographs of thin-sectioned chloroplasts, the thylakoid membrane gives the appearance of a large number of separate sheets or flattened vesicles. It actually is a single membrane that is highly folded and encloses a distinct compartment, the thylakoid lumen (fig. 16.2*b*). In places, the folded membrane is tightly stacked into disklike structures called grana. The chlorophyll found in chloroplasts is bound to proteins that are integral constituents of the thylakoid membrane, and it is here that the initial conversion of light into chemical energy occurs. The thylakoid membrane also contains a collection of electron carriers and an ATP-synthase similar to the proton-translocating ATP-synthase of mitochondria. The enzymes responsible for the actual fixation of CO_2 and the synthesis of carbohydrates are soluble proteins and reside in the stroma that surrounds the thylakoid membrane (see fig. 16.2). The stroma also contains DNA and ribosomes, which are responsible for synthesizing some of the proteins found in the chloroplast.

Figure 16.3

Cross-sectional view of a cell of *Rhodospirillum rubrum*, a purple photosynthetic bacterium. A double membrane system surrounds the cell. The inner membrane is extensively invaginated into tubules (arrows). These look circular when they are cut in cross section. (Courtesy of Dr. Gerald Peters.)

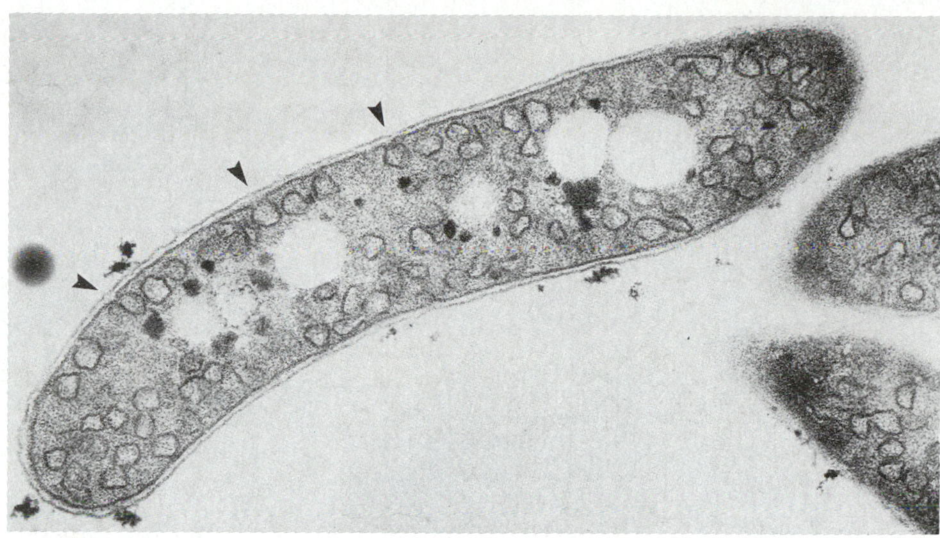

Table 16.1
Properties of Photosynthetic Organisms

Group	Evolve O_2	Contain Chloroplasts	Type of Chlorophyll	Number of Photosystems
Algae and higher plants	Yes	Yes	Chlorophyll *a* and *b*	2
Cyanobacteria	Yes	No	Chlorophyll *a* and *b*	2
Purple bacteria	No	No	Bacteriochlorophyll *a* or *b*	1

Algae are members of the plant kingdom and contain chloroplasts similar to those of higher plants. Prokaryotic photosynthetic organisms, which include the cyanobacteria and several groups of purple or green bacteria, do not have chloroplasts. In prokaryotes, the photochemical reactions of photosynthesis take place in the membrane that encloses the cell. This membrane has extensive invaginations resembling the cristae of the mitochondrial inner membrane (fig. 16.3). Table 16.1 summarizes the distinctions between some of the major groups of photosynthetic organisms.

Photosynthesis Depends on the Photochemical Reactivity of Chlorophyll

With the exception of certain halophilic bacteria (see chapter 36), all known photosynthetic organisms take advantage of the photochemical reactivity of one or another type of chlorophyll. Figure 16.4 shows the structures of several of the different types of chlorophyll that occur in nature. Chlorophylls resemble hemes, the prosthetic groups of the cytochromes and hemoglobin, and they are derived biosynthetically from protoporphyrin IX. However, they differ from hemes in four major respects: (1) Chlorophylls have an additional ring (ring V) with carbonyl and carboxylic ester substituents. (2) The central metal atom is magnesium rather than iron. (3) In the case of chlorophyll *a* and chlorophyll *b*, the two major chlorophylls in plants and cyanobacteria, one of the pyrrole rings (ring IV) is reduced by the addition of two hydrogens. In bacteriochlorophyll *a* and bacteriochlorophyll *b*, which occur in the purple and green bacteria, two of the rings are reduced (rings II and IV). Finally (4), the propionyl side chain of ring IV is esterified with a long-chain isoprenoid alcohol. Chlorophylls *a* and *b* contain the alcohol phytol; bacteriochlorophylls *a* and *b* have either phytol or geranylgeraniol, depending on the species of bacteria. Photosynthetic organisms also contain small amounts of pheophytins

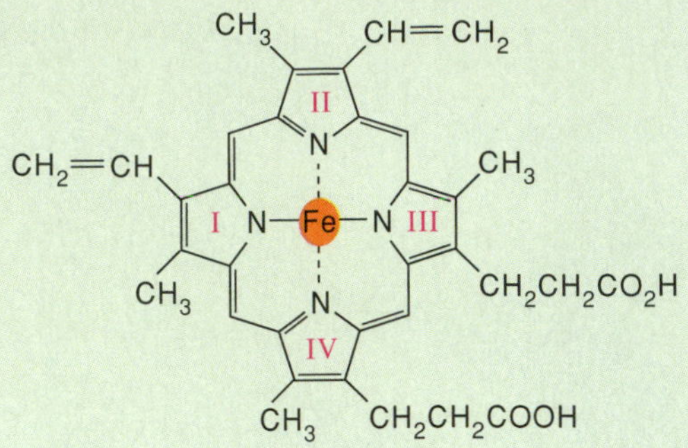

CH_3 $CH{=}CH_2$ $CH_2{=}CH$ CH_3 I II III IV N Fe CH_3 $CH_2CH_2CO_2H$ CH_3 CH_2CH_2COOH

Protoporphyrin IX

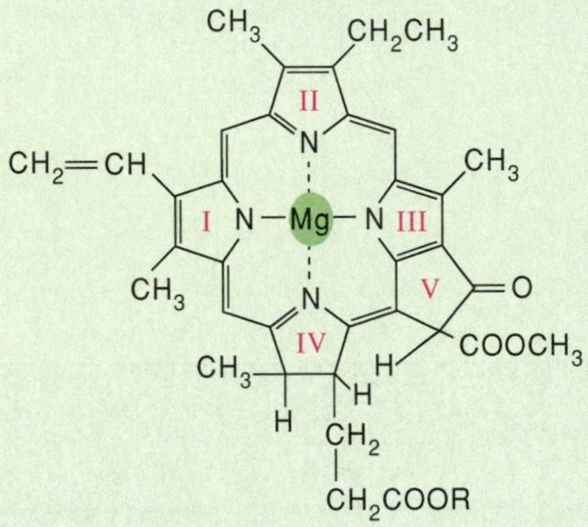

CH_3 CH_2CH_3 $CH_2{=}CH$ CH_3 I II III IV V N Mg CH_3 O $COOCH_3$ CH_3 H H CH_2 CH_2COOR

Chlorophyll *a*

CH_3 H H CH_2CH_3 O CH_3C CH_3 I II III IV V N Mg CH_3 O $COOCH_3$ CH_3 H H H CH_2 CH_2COOR or R'

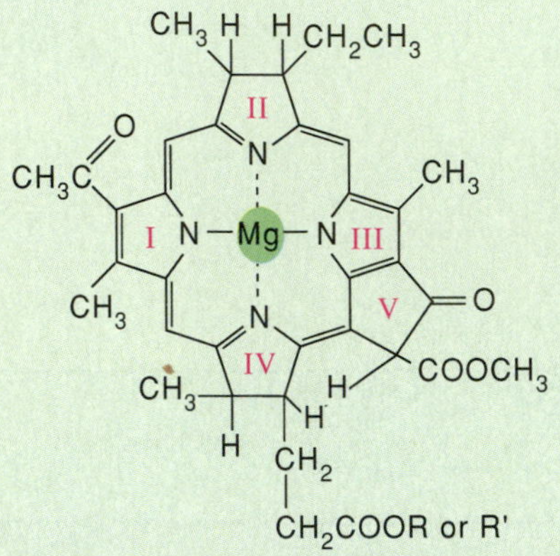

Bacteriochlorophyll *a*

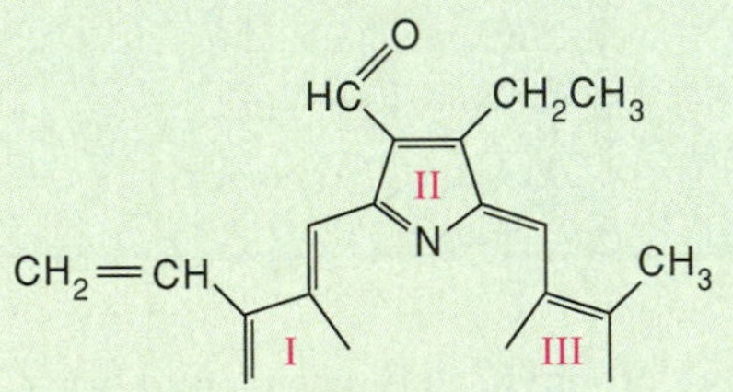

O HC CH_2CH_3 $CH_2{=}CH$ I II III N CH_3

Chlorophyll *b*

CH_3 H $CHCH_3$ O CH_3C I II III N CH_3

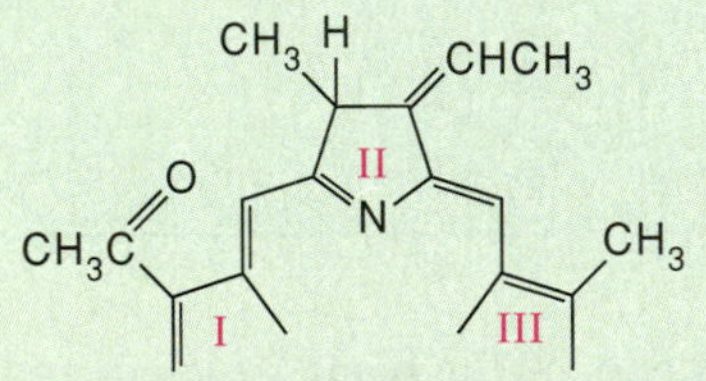

Bacteriochlorophyll *b*

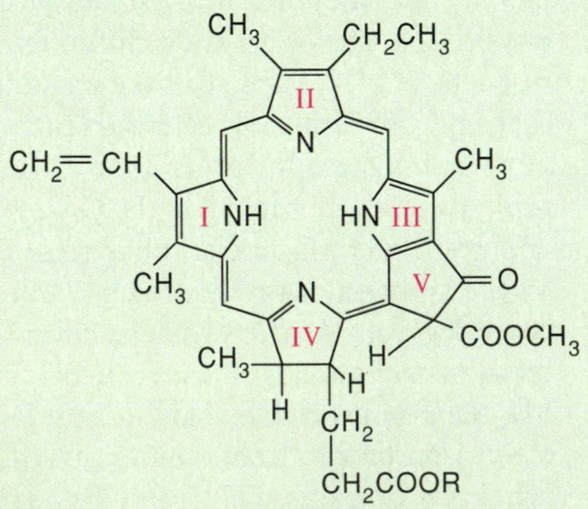

CH_3 CH_2CH_3 $CH_2{=}CH$ CH_3 I NH II III HN IV V N CH_3 O $COOCH_3$ CH_3 H H H CH_2 CH_2COOR

Pheophytin *a*

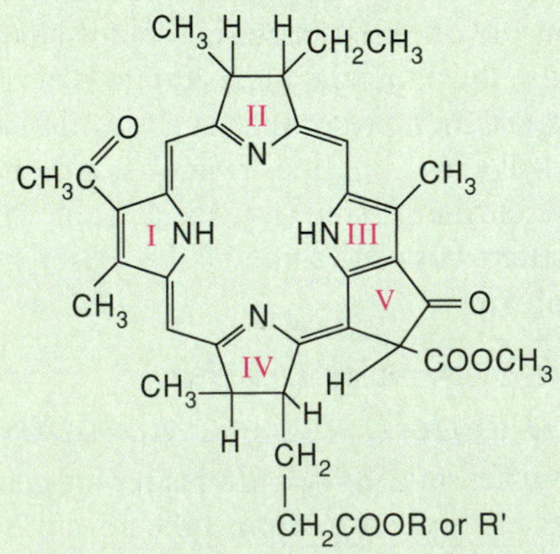

CH_3 H H CH_2CH_3 O CH_3C CH_3 I NH II HN III IV V N CH_3 O $COOCH_3$ CH_3 H H H CH_2 CH_2COOR or R'

Bacteriopheophytin *a*

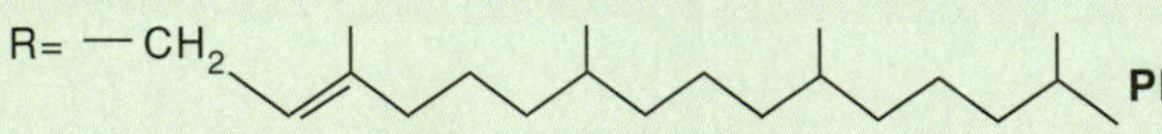

R= $-CH_2$ **Phytyl side chain**

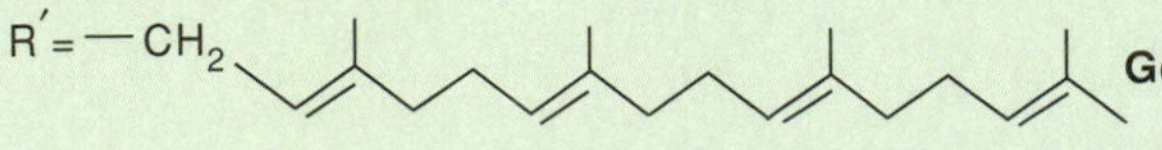

R′= $-CH_2$ **Geranylgeranyl side chain**

Figure 16.4

Structures of protoporphyrin IX (the prosthetic group of hemoglobin, myoglobin, and the *c*-type cytochromes), and several types of chlorophyll and bacteriochlorophyll. Chlorophyll *a* and chlorophyll *b* are the main types of chlorophyll in plants and the cyanobacteria. Purple photosynthetic bacteria contain either bacteriochlorophyll *a* or bacteriochlorophyll *b*, depending on the bacterial species. (The green photosynthetic bacteria contain still another form of bacteriochlorophyll.) Chlorophyll *b* and bacteriochlorophyll *b* are the same as chlorophyll *a* and bacteriochlorophyll *a* except for the substituents on ring II. Pheophytins and bacteriopheophytins are the same as the corresponding chlorophylls or bacteriochlorophylls except that two hydrogen atoms replace Mg.

or bacteriopheophytins, which are the same as the corresponding chlorophylls or bacteriochlorophylls except that two hydrogens replace the magnesium (see fig. 16.4). We will see that the pheophytins and bacteriopheophytins play special roles as electron carriers in photosynthesis.

The reduction of ring IV in chlorophyll *a* or *b* makes the conjugated aromatic ring system decidedly asymmetrical. This changes the optical absorption spectrum of the molecule dramatically. Whereas the long-wavelength absorption band of a cytochrome (the α band) is relatively weak (see fig. 15.4), chlorophyll *a* has an intense absorption band at 676 nm (fig. 16.5). Chlorophyll *b* has a similar band at 642 nm. Bacteriochlorophylls *a* and *b*, in which the asymmetry of the conjugated system is even more pronounced, have extremely strong absorption bands in the region of 770 nm (see fig. 16.5). All of the chlorophylls thus absorb light very well, particularly at relatively long wavelengths.

A Physical Definition of Light Before we discuss how chlorophylls can transform light energy into chemical energy, we must review some of the basic properties of light. Light is an electromagnetic field that oscillates sinusoidally in space and time (fig. 16.6). It interacts with matter in packets, or quanta, called photons, each of which contains a definite amount of energy. A mole of photons is called an einstein. The relationship between the energy ϵ of a photon and the frequency ν of the oscillating electromagnetic field is given by

$$\epsilon = h\nu \qquad (3)$$

where h is Planck's constant (6.63×10^{-27} erg · s or 4.12×10^{-15} eV · s). The energy per einstein is

$$E = N\epsilon = Nh\nu \qquad (4)$$

where N is Avogadro's number (6.02×10^{23} photons/einstein). The frequency ν is the number of oscillations per second at a given point in space. The wavelength λ of the oscillations (the distance between successive peaks in the amplitude of the field) depends on both ν and the velocity c at which the peaks move through space:

$$\lambda = c/\nu \qquad (5)$$

Figure 16.5

Absorption spectra of chlorophyll *a* and bacteriochlorophyll *a* in ether. Note that the long-wavelength absorption bands are much stronger than the α band of reduced cytochrome *c* (see fig. 15.4). In the chlorophyll-protein complexes found in photosynthetic organisms, the long-wavelength absorption band generally is shifted to even longer wavelengths. This probably reflects interactions between neighboring chlorophylls, which are bound to the proteins as dimers or larger groups.

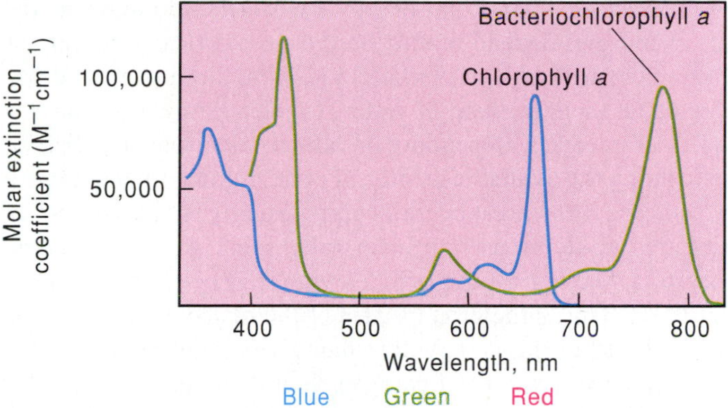

Figure 16.6

Light is an oscillating electromagnetic field. The lengths of the arrows in this diagram represent the strength of the electric field at a particular time, as a function of position in a ray of light proceeding along the *y* axis. A similar diagram could represent the field as a function of time at a particular point in space. The polarization of the light is defined by the orientation of the electric field: the beam shown here is polarized parallel to the *z* axis. The magnetic field, which is not shown here, is in phase with the electric field but is parallel to the *x* axis.

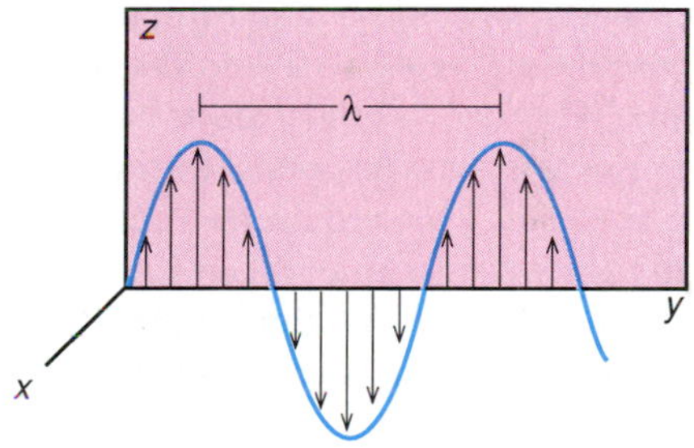

Light travels with a velocity of 3×10^{10} cm/s in a vacuum. In a dense medium such as water the velocity is less, depending inversely on the refractive index of the medium. Photon energies usually are stated in units of electron volts (eV), where 1 eV is the energy associated with moving one electron across a potential difference of 1 V and is equivalent to 23,060 kcal/mole (96,480 kJ/mole). Blue light, with a wavelength in the region of 450 nm ($\nu = 6.7 \times 10^{14}$ s^{-1}), has an energy of 2.75 eV, or 64 kcal/einstein, and far-red light (700 nm) has an energy of 1.77 eV (41 kcal/einstein).

Radiation with wavelengths much below 400 nm or above 750 nm is invisible to the human eye, and some authors prefer not to call it "light." However, such radiation can be important biologically. Many photosynthetic bacteria are adapted to use radiation in the region between 800 and 900 nm, and some species do well with even longer wavelengths.

How Light Interacts with Molecules The electrons in a molecule are held in a set of molecular orbitals, each of which is associated with a particular energy. In an isolated molecule, the orbital energies depend mainly on the electrostatic interactions of the electrons with each other and with the nuclei, and are more or less independent of time. A molecule thus can have a variety of energies, depending on how its electrons are distributed among the available orbitals. For an organic molecule with $2n$ electrons, the lowest overall energy usually is obtained when there are two electrons with antiparallel spins in each of the first n orbitals, leaving all the orbitals with higher energies empty (fig. 16.7). This is the ground state of the molecule. A molecule that is placed in this state will remain there indefinitely, as long as the electronic potential energies do not change.

When light interacts with a molecule, the oscillating electric field of the light makes the electronic potential energies strongly time-dependent. The result of this can be that a photon is absorbed and an electron moves from one of the occupied molecular orbitals to an unoccupied orbital with a higher energy (see fig. 16.7). Two requirements must be met in order for this to occur. First, the difference between the energies of the two orbitals must be the same as the photon's energy, $h\nu$. This is why a given type of molecule absorbs light of some wavelengths and not of others. The second, more complex requirement has to do with the shapes of the two orbitals and the orientation (polarization) of the oscillating field relative to the disposition of the orbitals in space. The two orbitals must have different geometrical symmetries and must be oriented in an appropriate way with respect to the field. This requirement explains why some absorption bands are stronger than others.

If chlorophyll (or any other molecule) absorbs a photon, it is excited to a state that lies above the ground state in energy. This usually occurs with no change in electronic spin (see fig. 16.7). As long as the spins of the two unpaired electrons remain antiparallel, so that the molecule has no net electronic spin, the molecule is said to be in an excited *singlet* state.

A molecule in an excited singlet state can decay back to the ground state by releasing energy in several different ways (see fig. 16.7). One possibility is simply to emit a photon; this is fluorescence. The wavelength of the fluorescence is generally longer than that of the light that was originally absorbed, because readjustments of the molecular geometry decrease the energy of the excited molecule somewhat before the molecule fluoresces. The extra energy is given off to the environment as heat. If the molecule has several absorption bands, as the chlorophylls do, the wavelength of the fluorescence is usually slightly longer than that of the longest-wavelength absorption band. The molecule relaxes to the lowest, or "first," excited

Figure 16.7

When a molecule absorbs light, an electron is excited to a molecular orbital with higher energy. The horizontal bars in this diagram represent molecular orbitals for electrons. Each orbital can hold two electrons with antiparallel spins (arrows pointing upward or downward). Only the top few of the filled orbitals are shown here. Absorption of light raises an electron from one of these orbitals to an orbital that is normally unoccupied. For this to occur, the energy of the photon must match the difference between the energies of the two orbitals. The choice of an upward or downward arrow is arbitrary, but there is no change of spin during the excitation. As long as the spins of the two unpaired electrons remain antiparallel, so that the molecule has no net electronic spin, the molecule is said to be in an excited "singlet" state. An excited molecule can return directly to the ground state by giving off energy as fluorescence or heat, or by transferring energy to another nearby molecule, or it can transfer an electron to another molecule (see fig. 16.8).

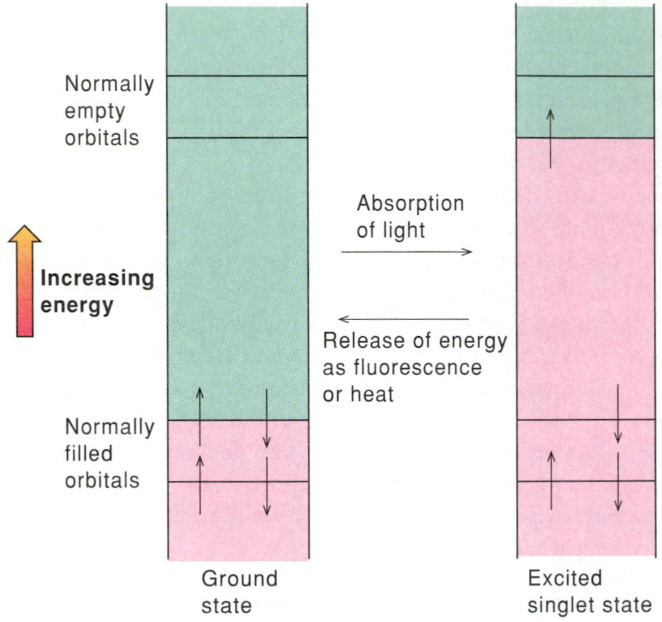

singlet state before the emission occurs. In some cases, the molecule can decay all the way to the ground state by radiationless processes, converting the excitation energy entirely into heat. Another decay mechanism is to transfer the energy to a neighboring molecule by the process of resonance energy transfer. This phenomenon plays an important role in photosynthesis, and we will return to it later. A fourth possibility is for the excited molecule to transfer an electron to a neighboring molecule.

Light Causes an Electron-Transfer Reaction Electron transfer can be a favorable path for the decay of an excited molecule, because an electron in the upper, normally unoccupied orbital is bound less tightly than one in a lower, normally filled orbital. If the absorption of a photon increases the energy of the molecule by ϵ electron volts, where $\epsilon \approx h\nu$, it will lower the standard redox potential for removing an electron from the excited molecule by ϵ volts, compared with the $E°$ for the molecule in the ground state. In the case of chlorophyll *a*, the $E°$ for oxidation in the ground state is approximately +0.5 V, and $h\nu$ for the long-wavelength absorption band is about 1.7 eV. The $E°$ of the excited molecule is therefore about −1.2 V. This means that in

Figure 16.8

The photochemical process that initiates photosynthesis is an electron-transfer reaction. The horizontal bars and vertical arrows represent molecular orbitals and electrons, as in figure 16.7. Absorption of light increases the free energy of a chlorophyll complex (Chl) by $h\nu$, making the transfer of an electron to an acceptor (A) thermodynamically favorable. The oxidized chlorophyll complex (Chl^+) extracts an electron from a donor (D).

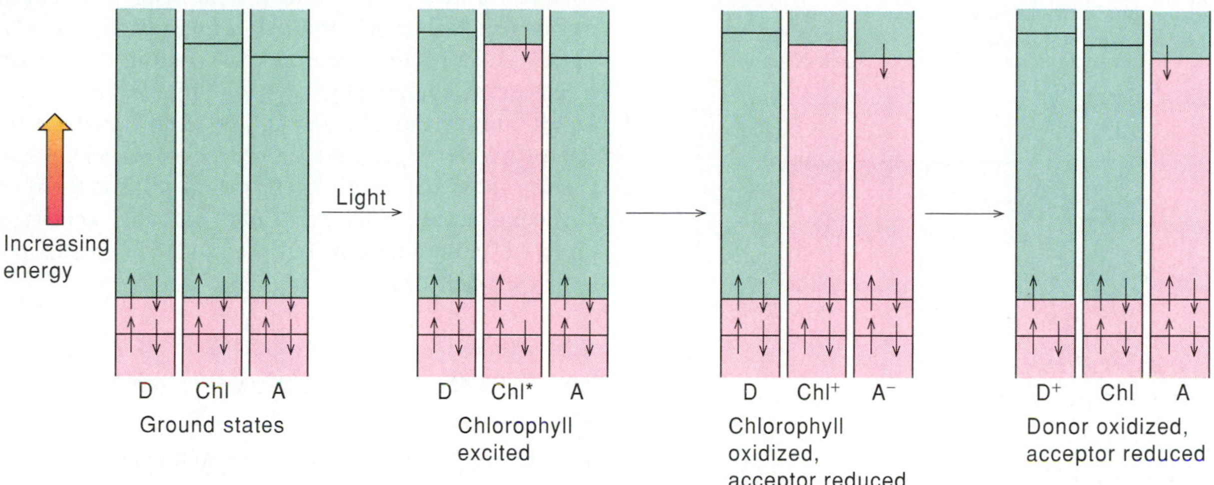

the excited state chlorophyll *a* is an extremely strong reductant. For comparison, recall that $NAD^+/NADH$ has an $E°'$ of only −0.32 V. The basic principle underlying the photochemistry of photosynthesis is that excitation causes a molecule of chlorophyll or bacteriochlorophyll (or a complex of several such molecules) to release an electron. The chlorophyll complex is oxidized, and another molecule becomes reduced (fig. 16.8). The oxidized chlorophyll species that is formed is a relatively strong oxidant and can extract an electron from a third molecule.

The idea that light drives the formation of oxidants and reductants was first advanced by C. B. van Niel in the 1920s. It was known at the time that the purple photosynthetic bacteria thrive only if they are provided with a reduced substrate such as an organic acid. Some species grow well on a reduced inorganic material such as H_2S. Van Niel noticed that although the bacteria do not evolve O_2, the reactions they carry out have a formal resemblance to the process that occurs in plants. If we let H_2B represent a reduced substrate, and let B represent an oxidized product, we can write the process that occurs in the purple bacteria as

$$CO_2 + 2\,H_2B \xrightarrow{\text{Light}} (CH_2O) + H_2O + 2\,B \qquad (6)$$

The equation we presented earlier for CO_2 fixation in plants and cyanobacteria (equation 2) can be put in a similar form by replacing H_2B and 2 B with H_2O and O_2:

$$CO_2 + 2\,H_2O \xrightarrow{\text{Light}} (CH_2O) + H_2O + O_2 \qquad (7)$$

To van Niel, this suggested that the essence of photosynthesis in both plants and bacteria is the photochemical separation of oxidizing and reducing power. The substance that is reduced in the photochemical reaction could be used to reduce CO_2 to carbohydrates in enzyme-catalyzed reactions that do not require light. The material that is oxidized photochemically could be discharged by the oxidation of H_2O to O_2 in plants, or by the oxidation of some other material in bacteria (fig. 16.9).

Figure 16.9

Van Niel proposed that the reductant generated in a photochemical electron-transfer reaction (A^-) is used to reduce CO_2 to carbohydrate. He suggested that the oxidant (D^+) oxidizes to H_2O to O_2 in plants, and oxidizes some other material (H_2B) in the purple bacteria. Only the initial charge separation requires light. (Chl = chlorophyll.)

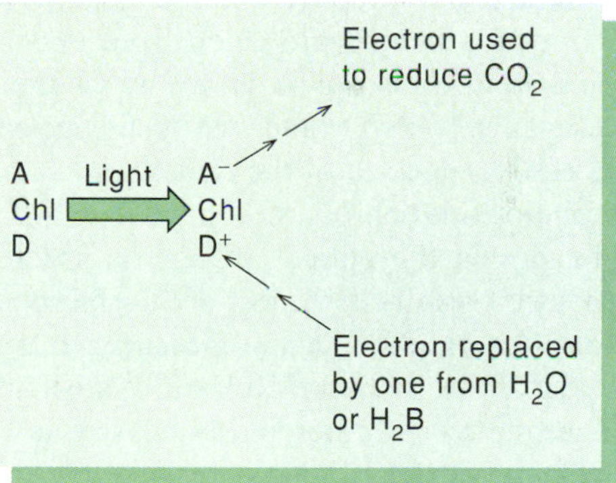

Support for these ideas came from experiments done by Robin Hill in 1939. Hill discovered that isolated chloroplasts would evolve O_2 if they were illuminated in the presence of an added nonphysiological electron acceptor such as ferricyanide [$Fe(CN)_6^{3-}$]. The electron acceptor became reduced in the process. Because no fixation of CO_2 occurred under these conditions, it was clear that the photochemical reactions of photosynthesis can be separated from the reactions that involve CO_2.

Figure 16.10

When a suspension of chloroplasts is illuminated, its optical absorbance decreases in the regions around 430 and 700 nm. The absorbance changes reflect the oxidation of a special chlorophyll complex (P700). Addition of a chemical oxidant such as potassium ferricyanide causes similar absorbance changes. (Source: B. Kok, "Partial purification and determination of oxidation reduction potential of the photosynthetic chlorophyll complex absorbing at 700 mμ," in *Biochimica et Biophysica Acta* 48:527, 1961. Copyright © 1961 Elsevier Science Publishers, Amsterdam, The Netherlands.)

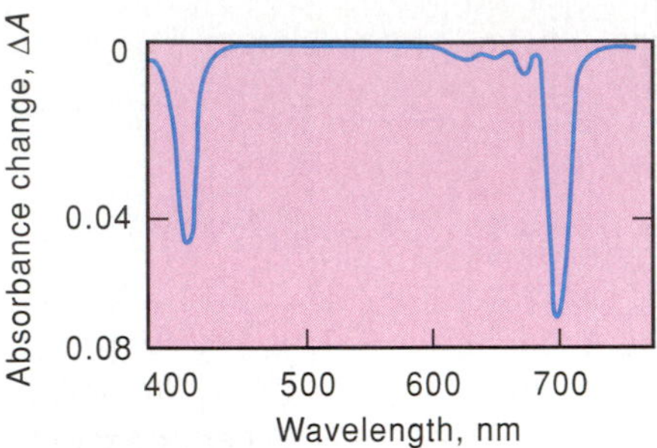

Photooxidation of Chlorophyll Generates a Cationic Free Radical

When chlorophyll or bacteriochlorophyll is oxidized, the product is a positively charged free radical. The situation differs subtly from the oxidation of a cytochrome. When a cytochrome is oxidized, the electron that is removed comes from the iron, which changes from ferrous (Fe^{2+}) to ferric (Fe^{3+}). In chlorophyll, the electron is removed not from the magnesium, but rather from the aromatic π-electron system of the molecule. The positive charge of the oxidized chlorophyll and the spin of the unpaired electron that remains behind are delocalized extensively over the π-electron system. This can be shown by studying the electron spin resonance (ESR) and electron nuclear double-resonance (ENDOR) spectra of the radical.

The photooxidation of chlorophyll can be detected by measuring changes in the optical absorption spectrum of the molecule. Oxidation results in the loss of the chlorophyll's characteristic absorption bands. The first measurements of this sort were made in the 1950s by Bessel Kok and Louis Duysens. Kok found that illumination of chloroplasts caused an absorbance decrease at 700 nm (fig. 16.10). He suggested that this reflected the photooxidation of a reactive chlorophyll complex, which he called P700. (P stood for "pigment," and 700 for the wavelength at which the unoxidized complex had its main absorption band.) Duysens made similar observations on *Rhodospirillum rubrum,* a purple photosynthetic bacteria. Here the reactive bacteriochlorophyll complex absorbed at 870 nm, and Duysens called it P870. A second type of reactive complex in chloroplasts, P680, was discovered subsequently by H. Witt and his colleagues. We will see below that P700 and P680 are parts of two distinct photochemical systems, photosystem I and photosystem II. P700 and P680 also are found in the cyanobacteria.

The photooxidation of P870, P700, or P680 generates a cationic radical ($P870^+$, $P700^+$, or $P680^+$) in which the charge and the unpaired electron are delocalized over the macrocyclic ring system. In fact, the ESR and ENDOR spectra of the $P870^+$ radical indicate that the unpaired electron is delocalized over *two* bacteriochlorophyll *a* molecules, which form a closely interacting pair. The strong interaction between the two bacteriochlorophylls probably explains why the absorption band of the complex is at 870 nm, whereas the long-wavelength band of monomeric bacteriochlorophyll *a* in solution is at 770 nm (see fig. 16.5). In bacterial species that contain bacteriochlorophyll *b* instead of bacteriochlorophyll *a,* the reactive dimers absorb at 960 nm instead of 870, and are often called P960. For simplicity, we will neglect the differences in wavelength among the various species of purple bacteria and will use the term P870 in a general sense. P700 and P680 probably consist of similar dimers of chlorophyll *a,* but the evidence on this point is not entirely conclusive.

The Reactive Chlorophyll Is Bound to Proteins in Complexes Called Reaction Centers

The chlorophyll or bacteriochlorophyll that undergoes photooxidation is bound to a protein in a complex called a reaction center. Reaction centers have been purified by disrupting chloroplasts or bacterial membranes with detergents. This was first achieved with purple photosynthetic bacteria, particularly by Roderick Clayton, and the structure of the bacterial complex is still much better understood than are the structures of the plant reaction centers. When they are excited with light, purified bacterial reaction centers can carry out the initial photochemical transfer of an electron from P870 to a series of electron acceptors.

Reaction centers of the purple bacteria generally contain three polypeptides with a total molecular weight of about 100,000. Bound noncovalently to the protein are four molecules of bacteriochlorophyll and two bacteriopheophytins. The reaction center also contains two quinones and one nonheme iron atom. In some bacterial species, both quinones are ubiquinone. In others, one of the quinones is menaquinone (vitamin K_2), a naphthoquinone that resembles ubiquinone in having a long isoprenoid side chain (fig. 16.11). Reaction centers from some purple bacteria, such as *Rhodopseudomonas viridis,* also have a cytochrome subunit with four *c*-type hemes.

The crystal structure of reaction centers from *R. viridis* was solved by Hartmut Michel, Johann Deisenhofer, Robert Huber, and their colleagues in 1984. This was a landmark achievement because it was the first high-resolution crystal structure to be obtained of an integral membrane protein. Reaction centers from another species, *Rhodobacter sphaeroides,* subsequently proved to have essentially the same structure except that they lack the bound cytochrome. In both species, the bacteriochlorophyll and bacteriopheophytin, the nonheme iron atom and the quinones are all bound to two of the polypeptides, which are folded into a series of α helices that pass back and forth across the cell membrane (fig. 16.12*a*). The third polypeptide resides largely on the cytoplasmic side of the membrane, but it also has one transmembrane α helix. The cytochrome subunit in *R. viridis* sits on the external (periplasmic) surface of the membrane.

Figure 16.11

The structures of ubiquinone, menaquinone (vitamin K_2), plastoquinone, and phylloquinone (vitamin K_1). Purple photosynthetic bacteria contain ubiquinone, menaquinone, or both, depending on the bacterial species; chloroplasts contain plastoquinone and phylloquinone. The number of isoprenoid units in the ubiquinone side chain (*n*) varies from 8 to 10.

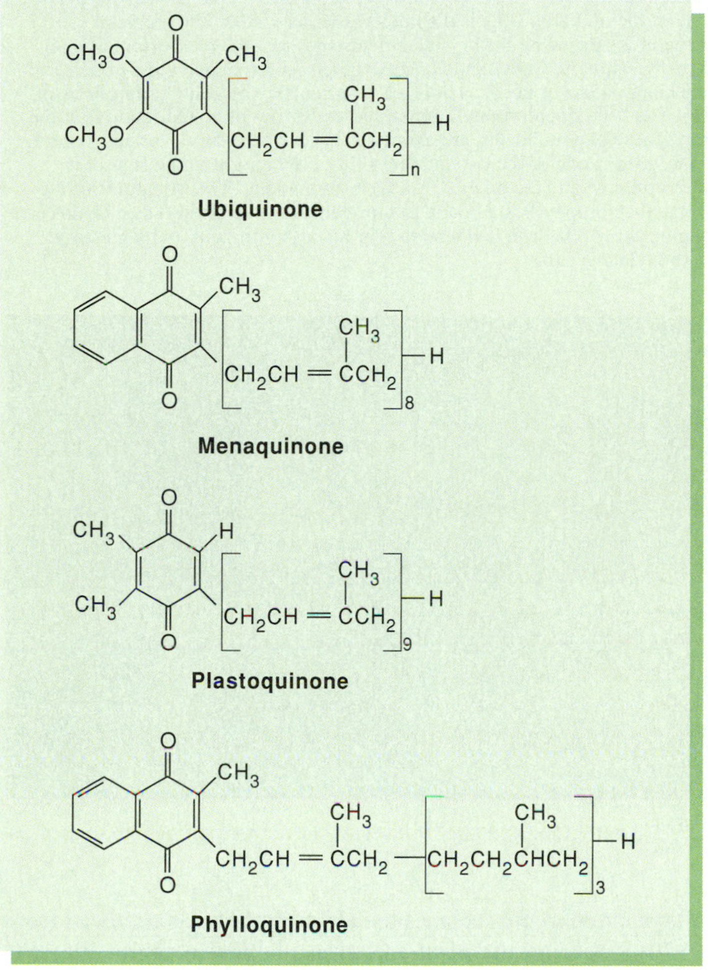

Ubiquinone

Menaquinone

Plastoquinone

Phylloquinone

Figure 16.12*b* shows the arrangement of the pigments in greater detail. Two of the four bacteriochlorophylls are packed closely together. Studies of the reaction center's optical absorption spectrum indicate that this special pair of bacteriochlorophylls is P870, the reactive complex that releases an electron when the reaction center is excited with light.

Although the structures of the plant reaction centers are not yet known in detail, photosystem II reaction centers resemble the reaction centers of purple photosynthetic bacteria both functionally and structurally. The amino acid sequences of their two major polypeptides are homologous to those of the two polypeptides that hold the pigments in the bacterial reaction center. Also, the reaction centers of photosystem II contain a nonheme iron atom and two molecules of plastoquinone, a quinone that is closely related to ubiquinone (see fig. 16.11), and they contain one or more molecules of pheophytin *a* and several molecules of chlorophyll *a* in addition to the two that probably make up P680.

The reaction center of photosystem I is larger and more complex. It contains two polypeptides with molecular weights of about 80,000 and at least seven other polypeptides ranging in size from 8,500 to 21,000. The reactive chlorophyll *a* complex P700 resides on the two large polypeptides, along with about 60 additional molecules of chlorophyll *a*, two quinones, and an iron-sulfur center. Some types of green photosynthetic bacteria have reaction centers that resemble those of photosystem I.

In Purple Bacterial Reaction Centers, Electrons Move from P870 to Bacteriopheophytin and Then to Quinones

Let's now consider the acceptors that extract an electron from P870 in the purple bacterial reaction center. The first acceptor whose reduction can be well resolved kinetically is a bacteriopheophytin. Note that the reaction center contains two bacteriopheophytins and two additional bacteriochlorophylls, in addition to the special pair of bacteriochlorophylls that make up P870 (see fig. 16.12*b*). When isolated reaction centers are excited with a short flash of light, an excited singlet state of the reactive bacteriochlorophylls (P870*) is formed essentially instantaneously. This state decays in about 3×10^{-12} s, and as it does, a $P870^+BPh^-$ radical-pair is created. Here $P870^+$ is the cationic radical formed by removing an electron from P870, and BPh^- is the anionic radical formed by adding an electron to a bacteriopheophytin (fig. 16.13). The creation of the radical-pair state can be detected spectrophotometrically by the disappearance of absorption bands of P870 and the bacteriopheophytin, and the formation of new absorption bands attributable to the two radicals. (The bacteriopheophytin that undergoes reduction is the one on the right in figures 16.12*a* and *b*. The two bacteriopheophytins are bound to the polypeptides in slightly different ways, and are distinguishable by their different absorption spectra.) It is likely that the electron that moves from P870 to the BPh passes through the intervening molecule of bacteriochlorophyll, but if a $P870^+BChl^-$ intermediate is formed, this radical-pair lives for less than 1×10^{-12} s before the electron moves on to the BPh.

The $P870^+BPh^-$ radical-pair lasts for about 2×10^{-10} s, decaying by the movement of an electron from BPh^- to one of the quinones (Q_A). This leaves the reaction center in the state $P870^+Q_A^-$, where Q_A^- is the anionic semiquinone radical (see fig. 16.13).

The electron carriers that participate in these first few steps are all fixed in position close together in the reaction center, and have little freedom of motion. One indication of this fact is that the electron-transfer reactions, in addition to being phenomenally fast, are almost independent of temperature. They actually increase slightly in speed with decreasing temperature. This behavior indicates that the reactants do not need to diffuse together in order for an electron to move from one molecule to another, and they do not have to acquire any additional thermal

Figure 16.12

(a) Structure of the reaction center of *Rhodopseudomonas viridis*. The α-carbon chains of the two main polypeptides are shown in yellow and gold. Each of these subunits has five α-helical regions that pass back and forth across the phospholipid bilayer of the cell membrane. A third subunit, shown in light yellow, sits on the cytosolic side of the membrane but also has one transmembrane α helix. The cytochrome subunit, which resides on the outer (periplasmic) surface of the membrane, is shown in purple at the top, with its four heme groups in red. The four bacteriochlorophylls are represented in shades of blue and green, the two bacteriopheophytins in white, the iron atom in red, the two quinones in purple, and a carotenoid molecule in orange. The phytyl side chains of the bacteriochlorophylls and bacteriopheophytins have been truncated for clarity; the isoprenoid tail of one of the quinones (Q_B) is disordered and not visible in the crystal structure. (Based on the crystal structure described by J. Deisenhofer, O. Epp, K. Miki, R. Huber and H. Michel.) (b) An expanded view of some of the components of the *R. viridis* reaction center. The color coding is as in (a). The bacteriochlorophyll dimer that undergoes photooxidation is at the top. These two molecules are about 3 Å apart where they overlap in ring I. In *R. viridis*, one of the two quinones (Q_A, on the right in the figure) is menaquinone and the other (Q_B, left) is ubiquinone. The Mg atoms of the bacteriochlorophylls are bound to nitrogen atoms of histidyl side chains (not shown here). The iron is bound to four histidyl nitrogens and a glutamyl carboxylate group.

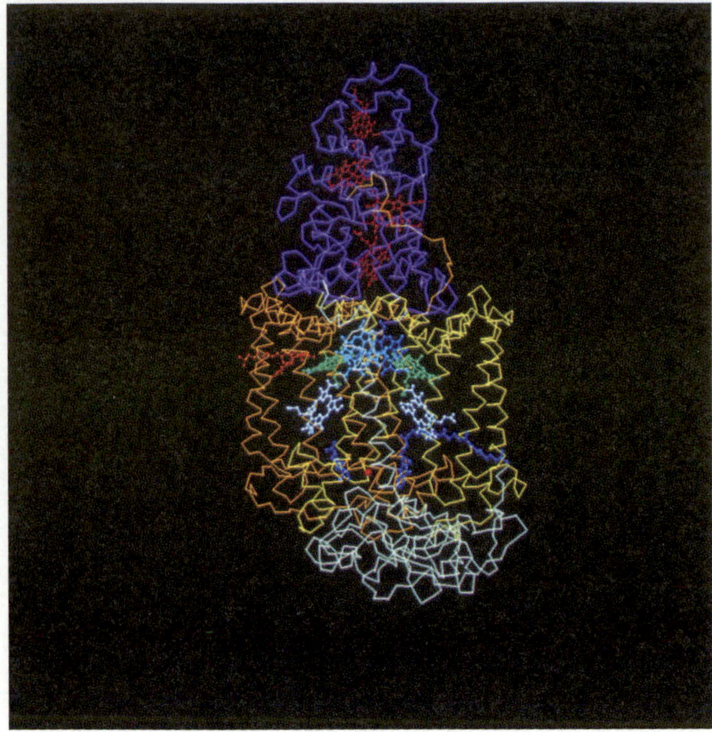

(a)

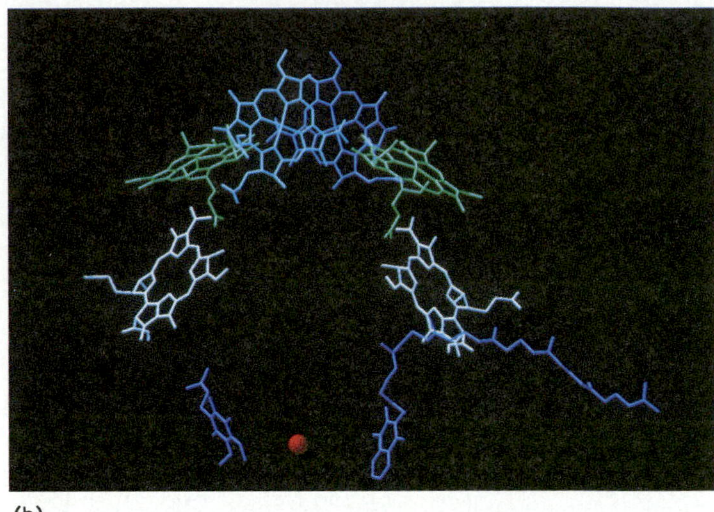

(b)

energy from their surroundings. The initial electron-transfer steps are in a sense "solid-state" processes. Further, the reactions are amazingly efficient. Their efficiency can be expressed in terms of the quantum yield of $P870^+Q_A^-$, which is the number of moles of $P870^+Q_A^-$ formed per einstein of light absorbed. The measured quantum yield in purified reaction centers is 1.02 ± 0.04. Essentially every time the reaction center is excited, an electron moves from P870 to Q_A.

A Cyclic Electron-Transport Chain Returns Electrons to P870 and Moves Protons Outward across the Membrane; Flow of Protons Back into the Cell Drives the Formation of ATP

From Q_A^-, an electron moves to the second quinone that is bound to the reaction center (Q_B in figure 16.14). This step takes about 10^{-4} s, considerably longer than the earlier steps, and it becomes even slower with decreasing temperature. In the meantime, a *c*-type cytochrome replaces the electron that was removed from P870, preparing the reaction center to operate again. Electron transfer between the cytochrome and $P870^+$ takes between 10^{-7} and 10^{-3} s, depending on the bacterial species. In *R. viridis,* the electron donor is the bound cytochrome with four hemes shown at the top of figure 16.12*a*. In other species, it often is a soluble *c*-type cytochrome with a single heme, more like the mitochondrial cytochrome *c*.

When the reaction center is excited a second time, a second electron is pumped from P870 to the bacteriopheophytin and on to Q_A and Q_B. This change places Q_B in the fully reduced form, Q_B^{2-}. The uptake of two protons transforms the reduced quinone to the uncharged quinol, QH_2, which dissociates from the reaction center into the phospholipid bilayer of the cell membrane. In intact bacteria, the protons that are taken up in the formation of QH_2 come from the cytosol of the cell (see fig. 16.14). Like the mitochondrial inner membrane, the bacterial membrane contains a relatively high concentration of quinone, which again can be either ubiquinone or menaquinone, depending on the species of bacteria.

Figure 16.13

The initial electron-transfer steps in reaction centers of purple photosynthetic bacteria. (P870* = the first excited singlet state of P870; BPh = bacteriopheophytin; Q_A = menaquinone or ubiquinone, depending on the bacterial species.) In this scheme, various states of the photosynthetic apparatus are positioned vertically according to their free energies, with the states that have the highest free energies at the top.

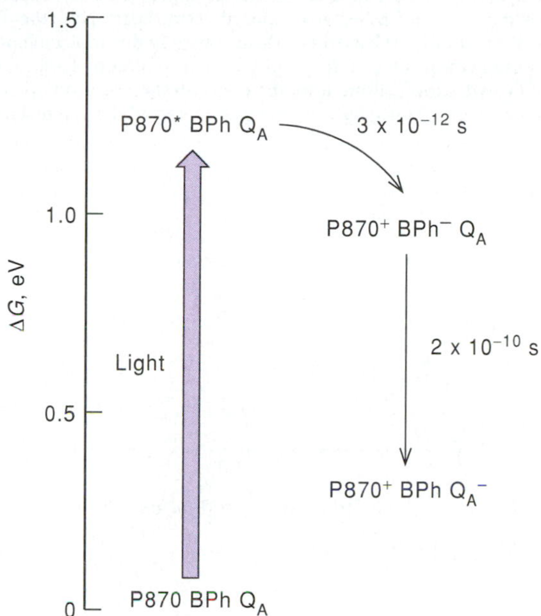

Figure 16.14

In purple photosynthetic bacteria, electrons return to P870⁺ from the quinones Q_A and Q_B via a cyclic pathway. When Q_B is reduced with two electrons, it picks up protons from the cytosol and diffuses to the cytochrome-bc_1 complex. Here it transfers one electron to an iron-sulfur protein and the other to a *b*-type cytochrome, and releases protons to the extracellular medium. The electron-transfer steps catalyzed by the cytochrome-bc_1 complex probably include a Q cycle similar to that catalyzed by complex III of the mitochondrial respiratory chain (see fig. 15.18). The *c*-type cytochrome that is reduced by the iron-sulfur protein in the cytochrome-bc_1 complex diffuses to the reaction center, where it either reduces P870⁺ directly or provides an electron to a bound cytochrome that reacts with P870⁺. In the Q cycle, four protons probably are pumped out of the cell for every two electrons that return to P870. This proton translocation creates an electrochemical potential gradient across the membrane. Protons move back into the cell through an ATP-synthase, driving the formation of ATP.

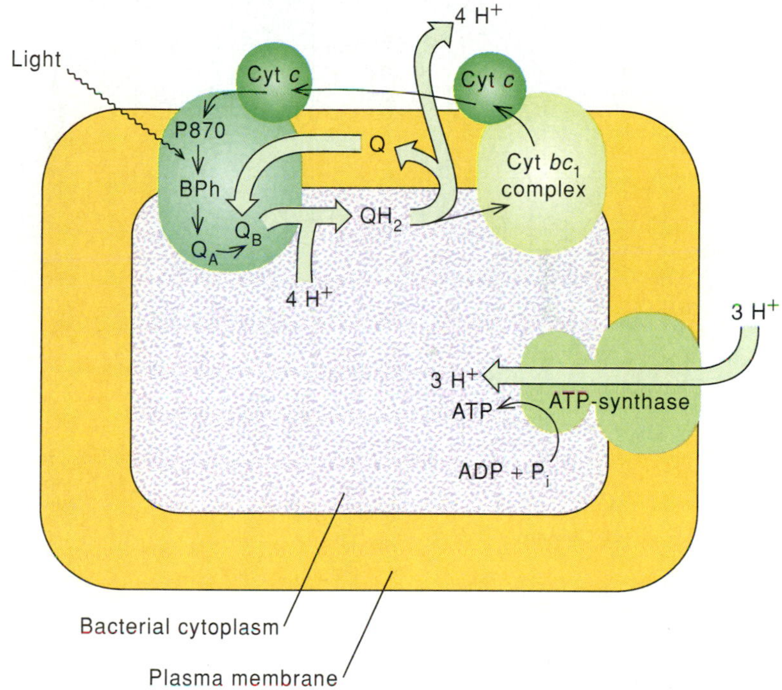

QH_2 is reoxidized by a cytochrome-bc_1 complex that is very similar to complex III of the mitochondrial respiratory chain. Electrons move through the cytochrome-bc_1 complex to a *c*-type cytochrome, which then diffuses to the reaction center and provides an electron for the reduction of P870⁺ (see fig. 16.14). As it does in mitochondria, the movement of electrons through the cytochrome-bc_1 complex results in the release of protons on the extracellular surface of the membrane. The proton extrusion can be explained well by the Q cycle that we discussed in connection with the mitochondrial complex (see fig. 15.18). The flow of electrons from QH_2 back to P870 thus drives the movement of protons outward across the cell membrane, generating a transmembrane electrochemical potential gradient for protons. The inside of the cell becomes negatively charged relative to the external medium, and the pH of the cytosol becomes higher than the external pH.

The flow of protons back into the bacterial cell, down the electrochemical potential gradient, is mediated by an ATP-synthase resembling the proton-conducting ATP-synthase of the mitochondrial inner membrane (see chapter 15). As in mitochondria, the movement of protons through a channel in the F_0 base-piece of the enzyme is linked to the formation of ATP (see fig. 16.14).

Note that the electron-transport system of purple photosynthetic bacteria is cyclic. Excitation of the reaction center with light creates a strong reductant (BPh⁻) and a strong oxidant (P870⁺), and electrons return from BPh⁻ to P870⁺ via quinones, the cytochrome-bc_1 complex, and *c*-type cytochromes. The cyclic flow of electrons results in the formation of ATP, but no net oxidation or reduction. How, then, can we explain van Niel's observation that the bacteria carry out a net transfer of electrons from organic acids to carbohydrates (see fig. 16.9)? The biosynthesis of carbohydrates requires a reductant such as NADPH, in addition to ATP. The answer to this puzzle is that the bacteria use ATP to support the transfer of electrons from succinate or other substrates to NADP⁺. These reactions are carried out by dehydrogenase complexes in the cell membrane. As we discussed in chapter 15 (see box 15A), a similar reduction of NAD⁺ by succinate can occur in mitochondria if ATP is added.

An Antenna System Transfers Energy to the Reaction Centers

The reactive chlorophyll or bacteriochlorophyll molecules of P700, P680, or P870 account for only a small fraction of the total pigment in photosynthetic membranes. Chloroplasts contain on the order of 300 chlorophylls per P700 and P680, and

Figure 16.15

(*a*) A molecule in an excited state (Chl_D^*) can transfer its energy to another molecule (Chl_A). The donor molecule returns to its ground state (Chl_D) and the acceptor is elevated to an excited state (Chl_A^*). This requires that the difference in energy between Chl_D^* and Chl_D (the excitation energy of Chl_D) be the same as the difference between Chl_A^* and Chl_A (the excitation energy of Chl_A). When the energies match in this way, there is a resonance between the two states $\{Chl_D^*\ Chl_A\}$ and $\{Chl_D\ Chl_A^*\}$. (*b*) The energy of light absorbed by pigment-protein complexes in the antenna system hops rapidly from complex to complex by resonance energy transfer until it is trapped in an electron-transfer reaction in a reaction center.

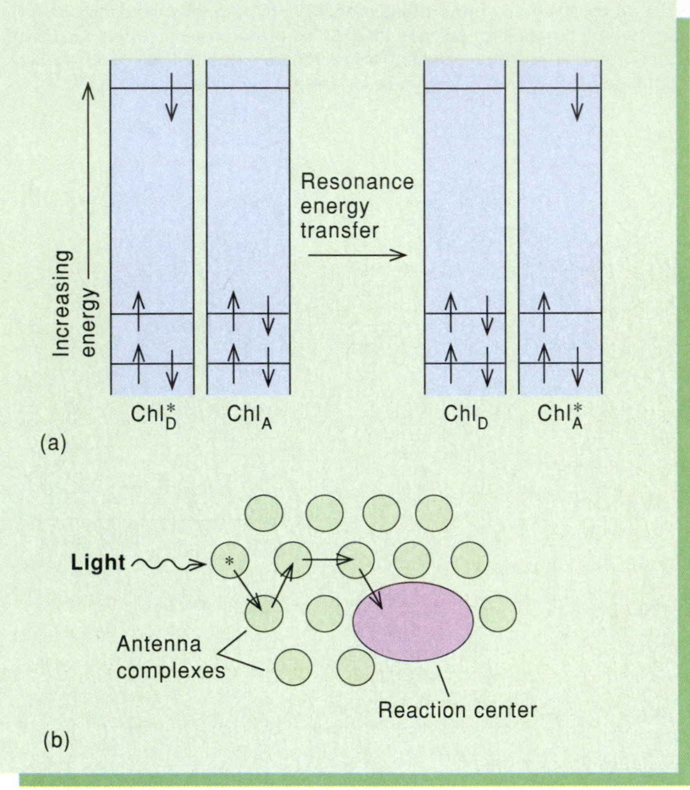

Figure 16.16

The amount of O_2 released when a suspension of algae (*Chlorella pyrenoidosa*) was excited with a train of short flashes of light, as a function of (*a*) the time between the flashes and (*b*) the intensity of the flashes. Each flash lasted about 10 μs. The total amount of O_2 released was measured after several thousand flashes and was divided by the number of flashes and by the amount of chlorophyll in the suspension. (What happens on the first few flashes will be discussed later.) The measurements in (*a*) were made with flashes comparable to the strongest flashes used in (*b*). For the measurements shown in (*b*), the flashes were spaced 20 ms apart. Note that the maximum amount of O_2 released per flash was only one molecule of O_2 per several thousand molecules of chlorophyll. With saturating flashes, the amount of O_2 evolution is limited by the concentration of reaction centers, whereas most of the chlorophyll in the algae is part of the antenna system.

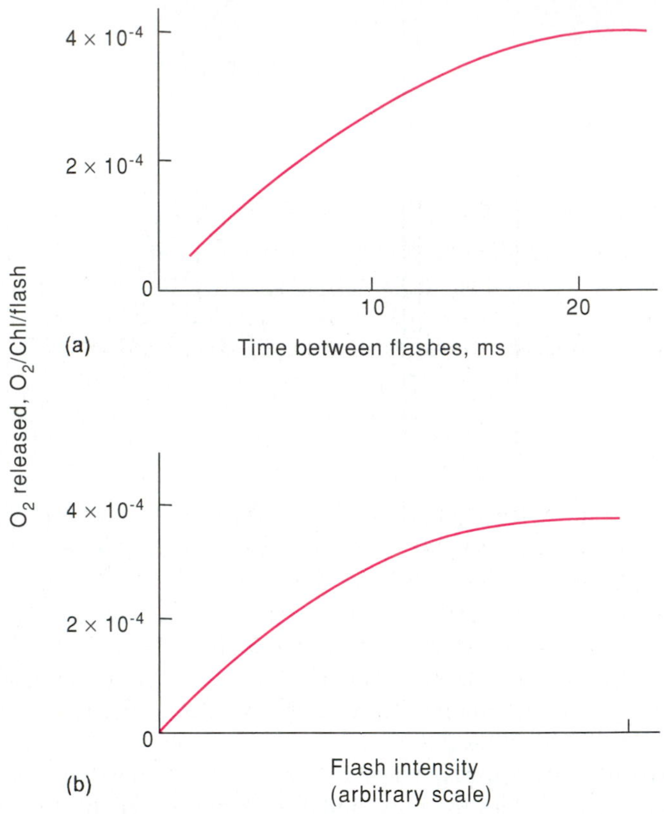

the cell membranes of purple photosynthetic bacteria have from 25 to several hundred bacteriochlorophylls per P870, depending on the species. Most of the chlorophyll or bacteriochlorophyll is not photochemically active. Instead, it serves as an antenna. When one of the molecules in the antenna system is excited with light, it can transfer its energy to a neighboring molecule by resonance energy transfer. Energy absorbed anywhere in the antenna migrates rapidly from molecule to molecule until it is trapped by an electron-transfer reaction in a reaction center (fig. 16.15). Measurements of the lifetime of fluorescence from the antenna system indicate that, after the antenna absorbs a photon, the energy is trapped in a reaction center within about 10^{-10} s in purple bacteria and within about 5×10^{-10} s in chloroplasts.

The distinction between the antenna system and the reaction centers grew out of experiments done by Robert Emerson and William Arnold in the 1930s. Emerson and Arnold measured the amount of O_2 evolution that occurred when they excited suspensions of green algae with a train of short flashes of light. To obtain the maximum O_2 per flash, they found that they had to allow a period of about 2×10^{-2} s of darkness between successive flashes. The amount of O_2 evolved per flash decreased if the flashes were spaced more closely together (fig. 16.16*a*). Each flash evidently generated a product that had to be consumed before the photosynthetic apparatus was ready to work again. If the next flash arrived too soon, its energy was wasted, mainly as heat and fluorescence. We know now that the chlorophyll complexes that undergo oxidation in the reaction centers (P700 and P680 in algae) have to be returned to their reduced states before they can react again. The acceptors that remove electrons from the chlorophyll complexes also have to be reoxidized.

Emerson and Arnold found that the amount of O_2 evolved per flash increased as they raised the strength of the flashes (i.e., as they excited the algae with a larger and larger number of photons on each flash), but reached a plateau when

Figure 16.17

Structures of β-carotene, a major carotenoid in many types of plants, and spheroidene, a common carotenoid of photosynthetic bacteria. There are about 350 different natural carotenoids. These vary widely in color, depending principally on the number of conjugated double bonds.

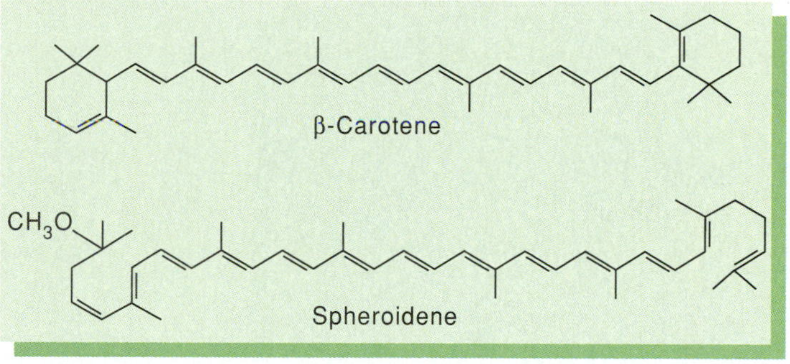

β-Carotene

CH_3O

Spheroidene

they made the flashes sufficiently strong (see fig. 16.16*b*). Even with flashes of saturating intensity and optimal timing, the amount of O_2 was very small relative to the chlorophyll content of the algae. The cells contained about 2,500 molecules of chlorophyll for each molecule of O_2 that they evolved. The low yield of O_2 was extremely puzzling at the time because it was generally accepted that chlorophyll was intimately involved in the photochemistry of CO_2 fixation and O_2 evolution. Now it is clear that most of the chlorophyll is part of the antenna system. When the flash intensity is high, the amount of O_2 evolution that can occur on each flash is limited by the concentration of reaction centers, and this is much smaller than the total concentration of chlorophyll. The apparent discrepancy between the number of chlorophyll molecules per P700 or P680 (about 300) and the number of chlorophylls per O_2 (several thousand) will make more sense after we have discussed the fact that chloroplasts have to absorb about eight photons for each molecule of O_2 that is evolved.

The chlorophyll or bacteriochlorophyll molecules that make up the antenna are bound noncovalently to integral membrane proteins with molecular weights that vary from 6,000 to 40,000, depending on the organism. Each polypeptide typically carries only two or three pigment molecules. These small complexes aggregate into larger arrays that allow excitations to hop rapidly from complex to complex.

Along with bacteriochlorophyll or chlorophyll, the antenna systems contain a variety of carotenoids and other pigments. These *accessory pigments* fill in the antenna's absorption spectrum in regions where chlorophylls do not absorb well. As shown in figure 16.5, chlorophyll *a* absorbs red or blue light well, but not green. Carotenoids, which are long, linear polyenes (fig. 16.17), have absorption bands in the green region. The energy they absorb is transferred to the chlorophyll molecules of the antenna, and from there to the reaction centers. This transfer is thermodynamically downhill because the carotenoid's excited singlet state has a higher energy than the lowest excited singlet state of chlorophyll.

Figure 16.18

The Z scheme for the photosynthetic apparatus of plants. Two photochemical reactions are required to drive electrons from H_2O to NADP. In this scheme, the solid arrows represent paths of electron flow. The electron carriers are positioned vertically according to their $E^{\circ\prime}$ values, with the strongest reductants (most negative $E^{\circ\prime}$ values) at the top. Electron flow downward is thermodynamically spontaneous.

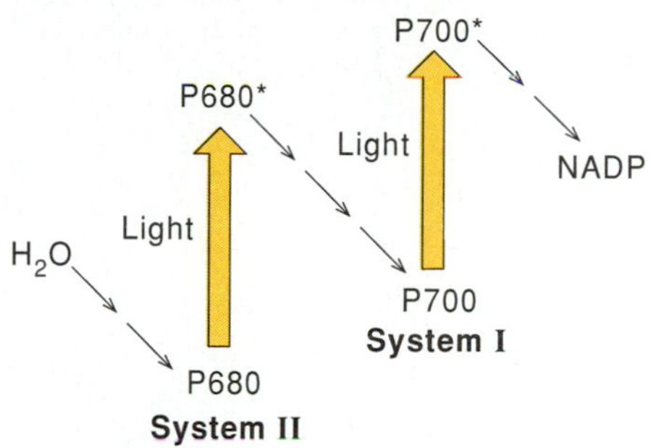

Chloroplasts Have Two Photosystems Linked in Series

We mentioned earlier that chloroplasts contain two types of reactive chlorophyll complexes, P700 and P680. Together with their antennae and their initial electron acceptors and donors, the reaction centers that contain P700 or P680 form two distinct assemblies called *photosystem I* and *photosystem II*. Much evidence indicates that photosystems I and II are connected in series, as shown in figure 16.18. Excitation of P700 (photosystem I) generates a strong reductant, which transfers electrons to NADP by way of several secondary electron carriers. Excitation of P680 (photosystem II) generates a strong oxidant, which oxidizes H_2O to O_2. The reductant formed in photosystem II injects electrons into a chain of carriers that connect the two photosystems. This scheme, which is called the Z scheme, was first suggested by R. Hill and F. Bendall in 1960. Note that the Z scheme differs from the bacterial electron transport chain that we discussed above in being linear rather than cyclic.

Figure 16.19 gives a more detailed picture of the Z scheme. The electron acceptors on the reducing side of photosystem II resemble those of purple bacterial reaction centers. The acceptor that removes an electron from P680 appears to be a molecule of pheophytin *a*. The second and third acceptors are molecules of plastoquinone, whose structure was shown in figure 16.11. As in the bacterial reaction center, electrons move one at a time from the first plastoquinone to the second. When the second plastoquinone becomes doubly reduced, it picks up protons from the stromal side of the thylakoid membrane and dissociates from the reaction center to join a pool of plastoquinone molecules in the membrane.

The chain of carriers between the two photosystems includes the cytochrome-b_6f complex and a copper protein called plastocyanin. The cytochrome-b_6f complex is very similar to the mitochondrial and bacterial cytochrome-bc_1 complexes. It contains a cytochrome with two *b*-type hemes (cytochrome b_6), an

Figure 16.19

A more detailed version of the Z scheme. [Mn complex = a complex of four Mn atoms bound to the reaction center of photosystem II; Y_Z = tyrosine side chain; Phe *a* = pheophytin *a*; Q_A and Q_B = two molecules of plastoquinone; Fe-S = an iron-sulfur protein similar to the Reiske iron-sulfur protein of the mitochondrial cytochrome bc_1 complex; Cyt *b/f* = cytochrome b_6f; PC = plastocyanin; Chl *a* = chlorophyll *a*; Q = phylloquinone (vitamin K_1); Fe-S_X, Fe-S_A, and Fe-S_B = iron-sulfur centers in the reaction center of photosystem I; FD = ferredoxin; FP = flavoprotein (ferredoxin-NADP oxidoreductase).] The sequence of electron transfer through Fe-S_A and Fe-S_B is not yet clear.

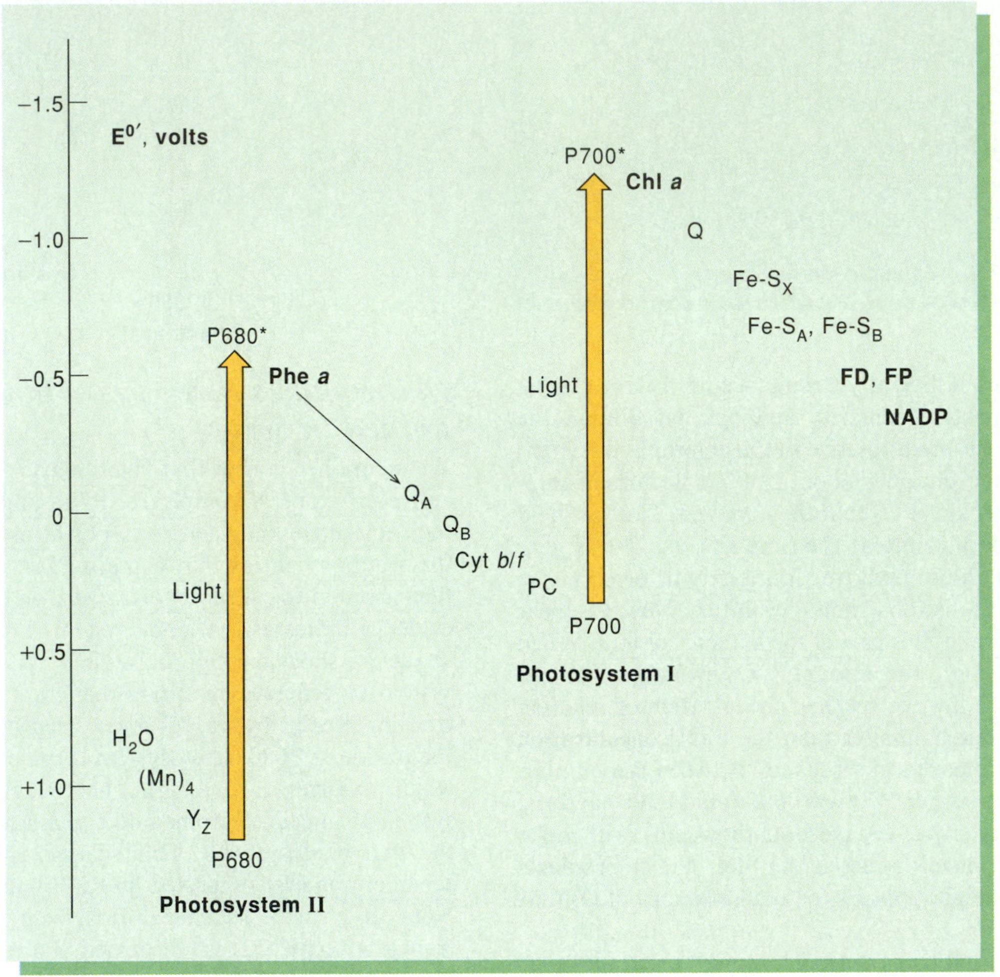

Figure 16.20

The quantum yield of O_2 evolution from a suspension of algae (*Chlorella pyrenoidosa*) as a function of the excitation wavelength. Measurements were made without any supplementary light (lower curve) and with supplementary blue-green light (upper curve). The quantum yield was calculated as moles of O_2 evolved per einstein of light incident on the sample; O_2 evolution caused by the supplementary light alone was subtracted. Without the supplementary light the quantum yield falls off precipitously in the far-red region above 680 nm. The antenna of photosystem I absorbs light well in this region, but that of photosystem II does not. Supplementary blue-green light, which is absorbed well by photosystem II, increases the quantum yield with which the far-red light can be used. (Source: R. Emerson et al., "Some factors influencing the long-wavelength limit of photosynthesis," in *Proceedings National Academy of Sciences USA* 43:133, 1957. Copyright © 1957 National Academy of Sciences, Washington, D.C.)

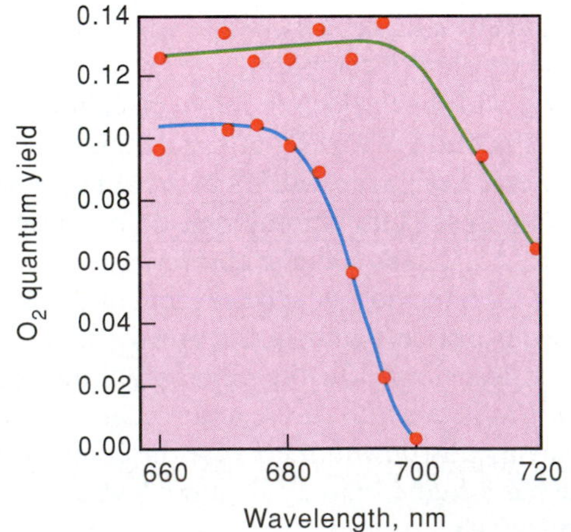

16A
BOX

Carotenoids Protect Cells against Damage by O_2

In addition to transferring excitation energy to the chlorophylls, carotenoids play an important role in protecting the cell against damage by O_2 at high light intensities. If the antenna is flooded with light too rapidly for the reaction centers to keep pace, the antenna chlorophylls discharge most of the extra energy as fluorescence and heat. However, the excited molecules also have an opportunity to evolve into an excited triplet state, in which the spins of the two unpaired electrons are parallel (see the illustration, part *a*). Excited triplet states are relatively long-lived. They cannot decay to the ground (singlet) state unless the electronic spin changes again, and this does not happen readily. One way that they do decay is by reacting with molecular O_2, which has a triplet ground state. The reaction returns the chlorophyll to its ground state and promotes the O_2 to an excited singlet state (illustration, part *b*). This change can have lethal consequences for the cell because singlet O_2 is extremely toxic. It reacts irreversibly with a variety of groups in proteins, nucleic acids, and lipids.

Carotenoids intervene to prevent these destructive side reactions by quenching the excited triplet chlorophyll before it has a chance to react with O_2 (illustration, part *c*). In this process, the chlorophyll returns immediately to its ground state, and the carotenoid is elevated to an excited triplet state. The carotenoid triplet state cannot generate singlet O_2, because it lies below singlet O_2 in energy. Instead, it decays harmlessly to the ground state. Carotenoids also can quench singlet O_2 itself.

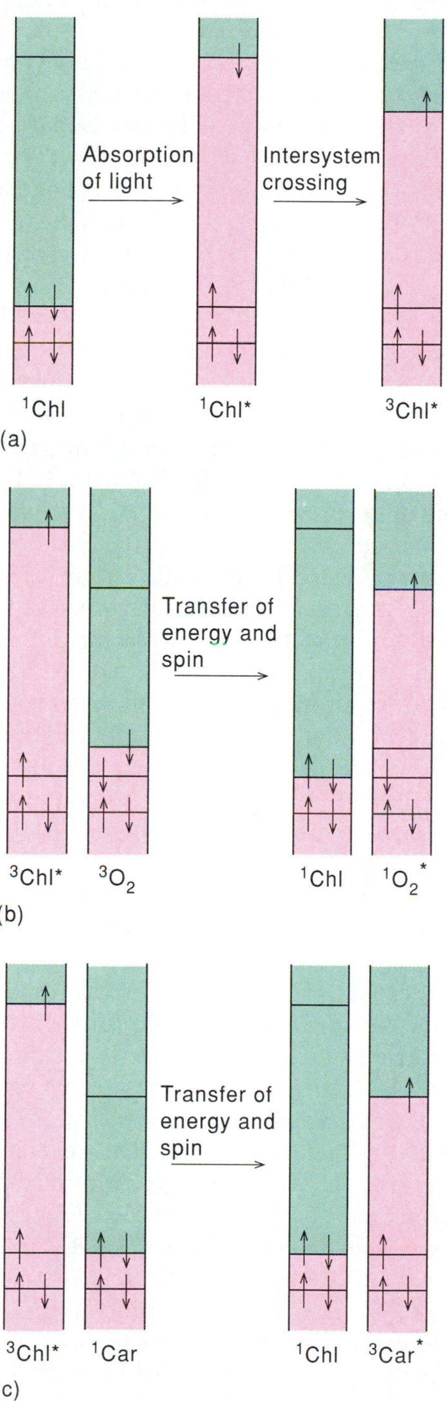

iron-sulfur protein, and the *c*-type cytochrome *f*. As electrons move through the cytochrome-b_6f complex from reduced plastoquinone to cytochrome *f*, the plastoquinone probably executes a Q cycle similar to the cycle that was presented for ubiquinone in the mitochondrial complex III and in the bacterial photosynthetic apparatus (see fig. 15.18). The cytochrome-b_6f complex provides electrons to plastocyanin, which transfers them to $P700^+$ in the reaction center of photosystem I. Additional details on the electron carriers between P700 and NADP and between H_2O and P680 will be presented later, after we discuss some of the evidence supporting the Z scheme.

Some of the earliest observations that led to the Z scheme came from measurements of how the quantum yield of O_2 evolution in algae depends on the excitation wavelength. The quantum yield of O_2 evolution is the molar ratio of O_2 evolved to photons absorbed. In green algae, the quantum yield is relatively independent of wavelength between 400 and 675 nm, but falls off drastically in the far-red region near 700 nm (fig. 16.20). This seems odd, because most of the absorbance in the far-red region is due to chlorophyll *a*. How could light absorbed by chlorophyll *a* be used less efficiently than light absorbed by carotenoids or other accessory pigments? A clue came from R.

Emerson's finding in 1956 that 700-nm light is used more efficiently if it is superimposed on a background of weak blue-green light (see fig. 16.20). It was found subsequently that the blue-green light actually does not have to be presented simultaneously with the red light. Illumination with blue-green light improves the utilization of far-red light even if the blue-green light is turned off several seconds before the red light is turned on.

These observations can be explained by the Z scheme, if the absorption spectra of the antennae associated with photosystems I and II are somewhat different. Because the two photosystems must operate in series, light will be used most efficiently when the flux of electrons through photosystem II is equal to that through photosystem I. If light of some wavelengths excites one of the photosystems more frequently than the other, some of the light will be wasted.

The effectiveness with which light of different wavelengths excites the two photosystems can be explored by blocking one of the systems. For example, herbicides such as 3-(3,4-dichlorophenyl)-1,1-dimethylurea (DCMU) block electron flow between the two molecules of plastoquinone in the reaction center of photosystem II. Photosystem I continues to function well in the presence of DCMU if a reductant is added to provide electrons to the carriers in the chain connecting the two photosystems. Experiments with various excitation wavelengths showed that photosystem I can absorb and use virtually all wavelengths of light up to about 740 nm. Similar studies of photosystem II showed that it is not excited well by far-red light. The pool of electron carriers between the two photosystems will therefore be drained of electrons if algae are illuminated with far-red light in the absence of added reductants. P700 will remain oxidized and will be unable to respond to the light. Photosystem II does absorb blue-green light well, so illumination with green light will replenish the pool. Electrons remain in the pool when the blue-green light is turned off, and that is why far-red light can be used effectively for a time even after a period of darkness.

Subsequent experiments by Louis Duysens provided strong support for the idea that photosystem I withdraws electrons from carriers situated between the two photosystems while photosystem II feeds electrons to these components. Duysens and his colleagues measured the redox states of several of the carriers directly. The easiest component to measure is cytochrome *f*, because its absorption spectrum changes markedly when the cytochrome undergoes oxidation. (Refer to figure 15.5 to see the absorption spectra of a *c*-type cytochrome in the oxidized and reduced forms, and to figure 16.19 for the position of cytochrome *f* in the Z scheme.) Illumination of algae with far-red light causes essentially all of the cytochrome *f* in the cells to become oxidized, as we would expect if far-red light drives photosystem I but not photosystem II (fig. 16.21). When a supplementary green light is turned on, some of the cytochrome returns quickly to the reduced state. This makes sense if green light can drive photosystem II. During the illumination, individual cytochrome molecules will cycle repeatedly between

Figure 16.21

Illumination of a suspension of red algae (*Porphyridium cruentum*) with far-red light (680 nm) causes an oxidation of cytochrome *f*, as measured by an absorbance change at 422 nm. (The box below the trace indicates the period when the far-red light was on.) Green (562-nm) light superimposed on top of the far-red light causes the cytochrome to become more reduced. Additional measurements showed that green light also could cause cytochrome oxidation, if it was turned on in the absence of the far-red light, when the cytochrome was initially reduced. This result indicates that the green light can drive either photosystem I or photosystem II. Far-red light can cause only cytochrome oxidation, because it is absorbed only by photosystem I.

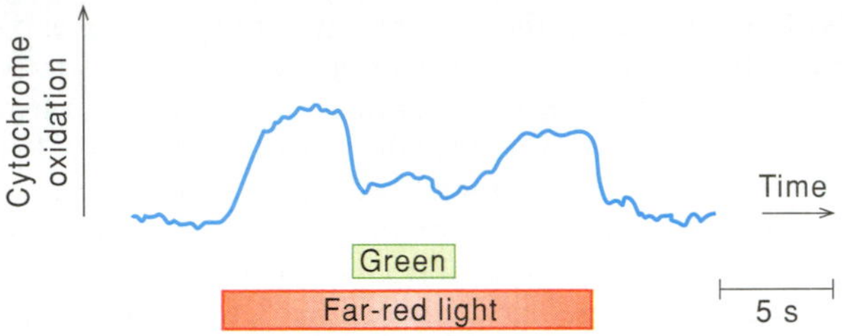

the oxidized and reduced states, so that the cytochrome population as a whole is in a steady state of partial oxidation. Green light actually can cause either a net oxidation or a net reduction of cytochrome *f*, depending on the initial conditions, so it must be able to excite either photosystem II or photosystem I. Far-red light, however, can cause only oxidation of the cytochrome.

The electron carriers in photosystem II are located mainly in the stacked, granal regions of the thylakoid membrane, whereas those of photosystem I occur mainly in the unstacked stromal lamellae (fig. 16.22). The cytochrome-b_6f complexes that participate in the transport of electrons between the two photosystems are found in both regions of the membrane. The distribution of photons between the two photosystems depends partly on the relative numbers of antenna complexes in the two regions of the membrane, and can change as a result of movements of antenna complexes from one region to another. Movements of one type of antenna complex appear to be regulated by phosphorylation of the protein. When chloroplasts are illuminated with light that is absorbed better by photosystem I than by photosystem II, the mobile antenna complex moves into the stacked regions, and a larger fraction of the excitations then are passed to photosystem II.

If the reaction centers of photosystem I and photosystem II are segregated into separate regions of the thylakoid membrane, how can electrons move readily from photosystem I to photosystem II? The answer appears to be that the plastoquinone that is reduced in the reaction center of photosystem II can diffuse rapidly in the plane of the membrane, just as ubiquinone does in the mitochondrial inner membrane. Plastoquinone thus carries electrons from photosystem II to the cytochrome-b_6f complex. Similarly, plastocyanin is a small protein that can diffuse in the thylakoid lumen. Plastocyanin thus can act as a mobile electron carrier from the cytochrome-b_6f complex to a reaction center of photosystem I, just as cyto-

Figure 16.22

The photosystem II complexes are located mainly in the stacked (granal) regions of the thylakoid membrane, whereas photosystem I complexes are most abundant in the unstacked stromal lamellae. The cytochrome b_6f complex is located in both regions of the membrane. Some types of antenna complexes also reside in both regions and can shift from one region to the other in response to a change in conditions. The ATP-synthase occurs mainly in the unstacked regions. These conclusions came from studies in which thylakoid membranes were fragmented gently by ultrasound and then separated by density-gradient centrifugation into fractions representing stacked and unstacked membranes.

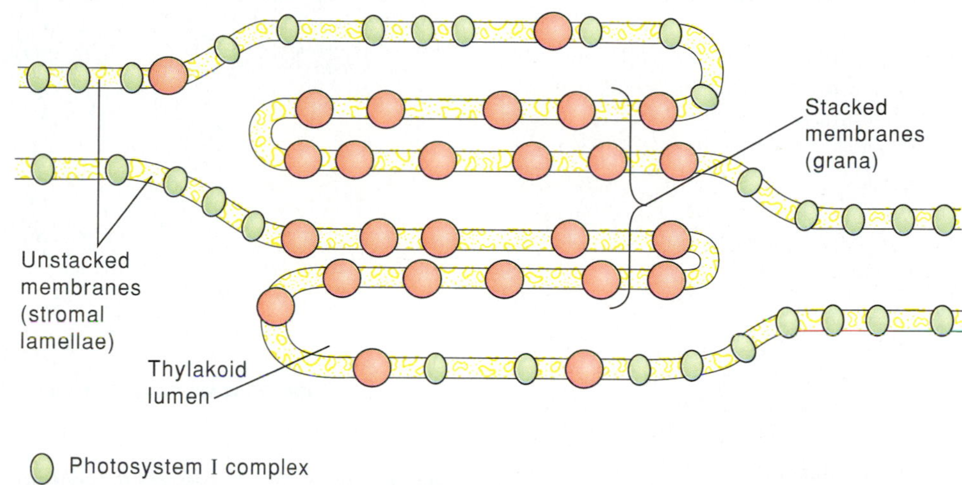

chrome *c* carries electrons from the mitochondrial cytochrome-bc_1 complex to cytochrome oxidase and as a *c*-type cytochrome provides electrons to the reaction centers of purple bacteria (see figure 16.14).

Photosystem I Reduces NADP⁺ by Way of Iron-Sulfur Proteins

On the reducing side of photosystem I, the earliest electron acceptor appears to be a molecule of chlorophyll *a* (see fig. 16.19). The second acceptor appears to be a quinone, phylloquinone (vitamin K_1) (fig. 16.11). In these respects, photosystem I resembles photosystem II and the purple photosynthetic bacteria, which use pheophytin *a* and bacteriopheophytin *a* as initial electron acceptors and a quinone as the secondary acceptor. From this point on, however, photosystem I is different. Its next set of electron carriers consists of iron-sulfur proteins instead of additional quinones.

The first of the iron-sulfur proteins to be characterized was ferredoxin, a small, soluble protein found in the chloroplast stroma. Ferredoxin is designated as FD in figure 16.19. It contains two iron atoms and two atoms of inorganic sulfide, which are bound to the sulfur atoms of four cysteine residues in the type of complex that was illustrated in figure 15.8*a*. Photosystem I also contains three additional iron-sulfur clusters that are more firmly associated with the reaction center. These are designated Fe-S_X, Fe-S_A, and Fe-S_B in figure 16.19. Fe-S_X, Fe-S_A, and Fe-S_B each probably has four iron atoms and four inorganic sulfides held by four cysteines in the cubic structure shown in figure 15.8*b*. The cysteines of Fe-S_X are provided by the two main polypeptides of the reaction center, which also bind P700 and the chlorophyll and quinone that act as the initial electron acceptors. Each polypeptide evidently contributes two cysteines to the Fe-S_X cluster. Fe-S_A and Fe-S_B are both bound to a separate, small polypeptide.

As we discussed in chapter 15, iron-sulfur proteins undergo one-electron oxidation-reduction reactions. The $E°'$ values for these transitions are about -0.7 V for Fe-S_X, -0.59 for Fe-S_B, -0.54 for Fe-S_A, and -0.40 V for FD. The quinone that is reduced in photosystem I probably transfers an electron to Fe-S_X. Fe-S_X reduces Fe-S_A and Fe-S_B, which in turn reduce FD (see fig. 16.19). From FD, electrons move to a flavoprotein ferredoxin-NADP oxidoreductase and then to $NADP^+$.

O_2 Evolution Requires the Accumulation of Four Oxidizing Equivalents in the Reaction Center of Photosystem II

The oxidation of H_2O to O_2 requires the removal of four electrons for each O_2 produced. According to the Z scheme each electron must traverse the photochemical reactions of both photosystem I and photosystem II, so at least 8 photons (2×4) have to be absorbed for each O_2 that is released. Currently accepted experimental measurements of the number of photons that are needed, the quantum requirement, are indeed on the order of 8 to 12. (The quantum requirement is the reciprocal of the quantum yield of O_2 evolution.) The quantum requirement will exceed 8 if some of the photons that are absorbed are lost as heat or fluorescence from the antenna, or if some of the electrons pumped through the photosystems are not removed to NADP, but instead cycle back into the electron-transport chain between the two photosystems.

If the photosystem II reaction center can transfer only one electron at a time, how does the photosynthetic apparatus assemble the four oxidizing equivalents that are needed for the

Figure 16.23

O_2 evolution from a suspension of chloroplasts that was excited with a series of 24 flashes after having been kept in the dark for several minutes. Little or no O_2 is evolved on the first two flashes. O_2 evolution peaks on the third flash and on every fourth flash thereafter. The oscillations in O_2 evolution are damped, and after many flashes the yields converge on the level indicated by the horizontal line. This level is the same as the average yield measured in experiments like those illustrated in figure 16.16.

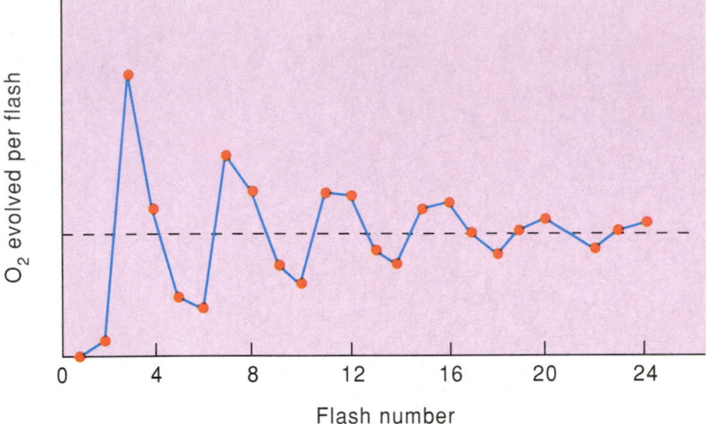

Figure 16.24

The five oxidation states of the O_2-evolving apparatus. One electron (e^-) is removed photochemically in each of the transitions between states S_0 and S_4. S_4 decays spontaneously, releasing O_2. Protons are released at several steps of the cycle.

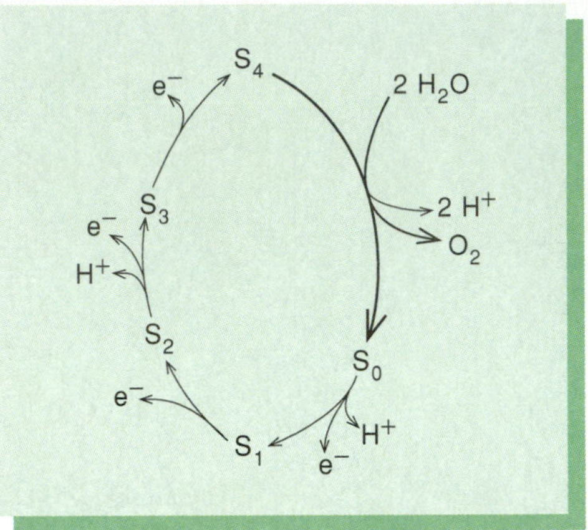

oxidization of H_2O to O_2? One possibility would be that several different photosystem II reaction centers cooperate, but this seems not to happen. Instead, each reaction center progresses independently through a series of oxidation states, advancing to the next state each time it absorbs a photon. O_2 evolution occurs only when a reaction center has accumulated four oxidizing equivalents. This conclusion comes principally from measurements of the amount of O_2 that is evolved on each flash when algae or chloroplasts are excited with a series of short flashes after a period of darkness. Pierre Joliot found that essentially no O_2 is released on the first or second flashes (fig. 16.23). On the third flash, however, there is a burst of O_2. After this, the amount of O_2 released on each flash oscillates, going through a maximum every fourth flash.

Kok pointed out that this pattern can be explained if the photosystem II reaction center cycles through five different oxidation states, S_0 through S_4, as shown in figure 16.24. When the system reaches state S_4, O_2 is given off and the reaction center returns to state S_0. The fact that the first burst of O_2 comes on the third flash instead of the fourth can be explained if the reaction centers relax mainly into S_1 rather than into S_0 during the dark period before the flashes. Added reductants can in fact convert the reaction centers from S_1 to S_0, so that the first peak of O_2 occurs on the fourth flash. The gradual damping of the oscillations in figure 16.23 is due to reaction centers that get out of phase, either because they miss being excited on one of the flashes or because they are excited twice and advance two steps.

The component that undergoes oxidation as photosystem II progresses from one of the S states to the next is a complex of four atoms of manganese that probably is bound to one or both of the two central polypeptides of the reaction center. $P680^+$ draws electrons from the manganese complex by way of a tyrosine in one of these polypeptides (Y_z in figure 16.19). In the course of this reaction, the phenolic side chain of the tyrosine is oxidized transiently to a free radical. Although the structure of the manganese complex is not yet known, the transitions of the complex from state S_0 to S_1, from S_1 to S_2, and from S_3 to S_4 probably represent sequential oxidations of three of the manganese atoms from the Mn(III) level to Mn(IV). The transition from S_2 to S_3 may involve oxidation of a histidine side chain of the polypeptide.

The $E°$ of the $P680^+/P680$ redox couple must be greater than $+1$ V, because the oxidized manganese complex that $P680^+$ generates is powerful enough to oxidize water. At pH 5, which is approximately the pH of the thylakoid lumen, the O_2/H_2O couple has an $E°$ of $+0.94$ V. $P700^+/P700$, by contrast, has an $E°'$ of only about $+0.50$ V. It is puzzling that $P680^+$ is so much stronger an oxidant than $P700^+$, if they both consist of chlorophyll *a*. However, the redox potential of the chlorophyll radical is likely to depend strongly on the polarity of the surroundings in the protein.

Flow of Electrons from H_2O to $NADP^+$ Drives Proton Transport into the Thylakoid Lumen; Protons Return to the Stroma through an ATP-Synthase

As in the mitochondrial inner membrane and in the cell membrane of purple photosynthetic bacteria, the flow of electrons through the chloroplast's cytochrome-b_6f complex results in the translocation of protons across the thylakoid membrane. When plastoquinone undergoes reduction in photosystem II and in the cytochrome-b_6f complex, protons are taken up from the stromal side of the membrane. When the plastoquinone is reoxidized,

Figure 16.25

Transport of two electrons from photosystem II through the cytochrome b_6f complex to photosystem I results in the movement of four protons from the chloroplast stroma to the thylakoid lumen. The proton translocation probably occurs in a Q cycle resembling that illustrated in figure 15.18. Two more protons are released in the lumen for each molecule of H_2O that is oxidized to O_2, and one additional proton is removed from the stroma for each molecule of $NADP^+$ reduced to NADPH. (Two protons are taken up when the flavoprotein is reduced by photosystem I; one ends up on NADPH, and the other is returned to the solution.) The flow of protons from the thylakoid lumen back to the stroma through an ATP-synthase (CF_0-CF_1) drives the formation of ATP. The abbreviations used in this figure are the same as in figure 16.19.

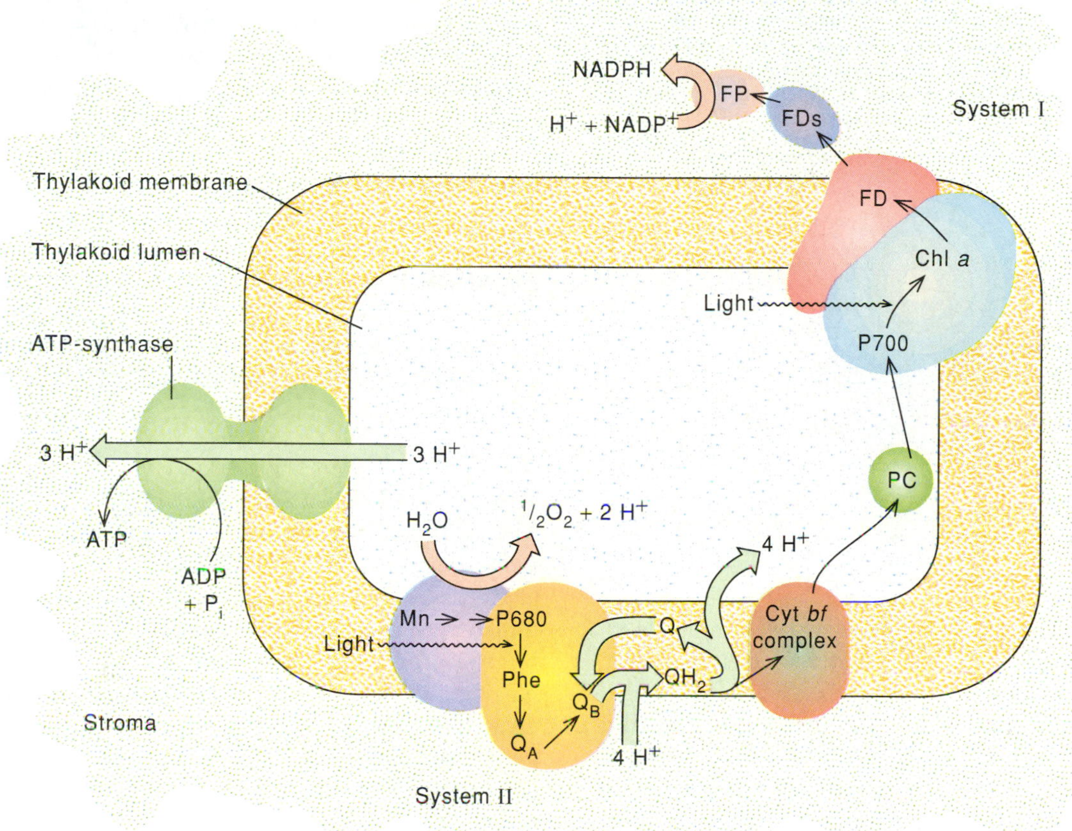

protons are released into the thylakoid lumen. This lowers the pH in the lumen, and makes the inside positively charged with respect to the stroma. In current schemes of the Q cycle catalyzed by the cytochrome-b_6f complex, four protons move across the membrane for each pair of electrons that proceed to photosystem I. Two more protons are released in the lumen for each molecule of H_2O that is oxidized to $\frac{1}{2}$ O_2, and one proton is taken up from the stroma for each $NADP^+$ that is reduced to NADPH (fig. 16.25).

Proton movement back out through the thylakoid membrane is conducted by an ATP-synthase, and this movement drives the formation of ATP. The chloroplast ATP-synthase is structurally very similar to the mitochondrial ATP-synthase (see fig. 15.31). Its head-piece, which contains the catalytic sites, is generally called CF_1. Its hydrophobic base-piece, CF_0, includes the proton-conducting channel. As in the case of the mitochondrial enzyme, CF_1 acts as an ATPase if it is removed from CF_0.

Observations on chloroplasts played a key role in the development of the chemiosmotic theory of oxidative phosphorylation, which we discussed in chapter 15. Andre Jagendorf and his colleagues discovered that if they illuminated chloroplasts in the absence of ADP, the chloroplasts developed the capacity to form ATP when ADP was added later, after the light was turned off. The amount of ATP that could be synthesized was much greater than the number of electron-transport assemblies in the thylakoid membranes, so the energy to drive the phosphorylation could not have been stored in an energized form of one of the electron carriers. Protons were taken up from the solution outside the thylakoid membranes during the illumination, and the ability to form ATP was correlated with the magnitude of the pH gradient across the membrane. If a pH gradient in the right direction was set up across the thylakoid membrane by suddenly raising the pH on the outside, chloroplasts could make ATP without any illumination at all (fig. 16.26).

Figure 16.26

Chloroplasts can use an electrochemical potential gradient for protons to form ATP in the dark. In a two-step experiment, spinach chloroplasts first were equilibrated at an acidic pH (pH 4.0). The pH of the solution then was raised quickly to a value between 6.7 and 8.8, as indicated on the abscissa of the graph, and ADP and P_i were added. Because the pH in the thylakoid lumen was buffered at pH 4, the sudden increase in the external pH created a pH gradient of between 2.7 and 4.8 pH units across the thylakoid membrane. Proton efflux through the ATP-synthase can drive the formation of ATP. Under optimal conditions, one molecule of ATP was formed for approximately every five chlorophylls in the sample (see the ordinate scale on the graph). For comparison, chloroplasts contain only about one P700 and one P680 per 400 chlorophylls. (Source: A. Jagendorf and E. Uribe, "ATP formation caused by acid-base transition of spinach chloroplasts," in *Proceedings National Academy of Sciences USA* 55:197, 1966. Copyright © 1966 National Academy of Sciences, Washington, D.C.)

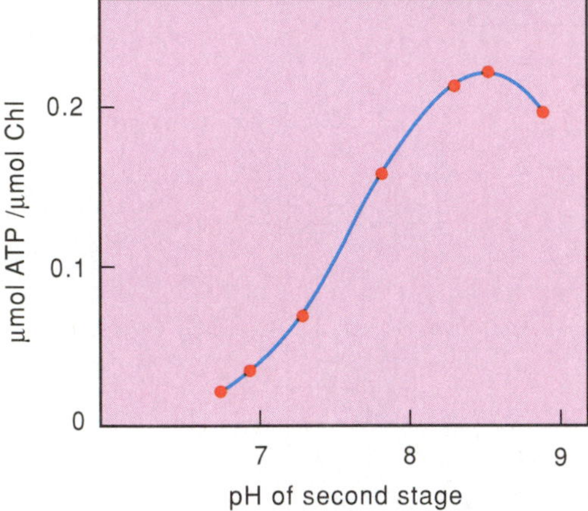

The formation of ATP by photosynthetic systems is often called photophosphorylation to distinguish it from the process that is coupled to respiration in mitochondria. Although the two processes are very similar, they do differ in a few details. The respiratory chain pumps protons outward across the inner membrane, from the mitochondrial matrix to the intermembrane space (see chapter 15). The chloroplast electron-transport chain pumps protons into the thylakoid lumen. The electrochemical potential gradient for protons across the mitochondrial membrane consists mainly of an electrical potential; in chloroplasts it consists mainly of a pH gradient, which can be greater than three pH units during strong illumination. In both cases, however, the F_1 or CF_1 head-piece of the proton-conducting ATP-synthase is on the side of the membrane that becomes more alkaline. The relative importance of the pH gradient or the electrical potential probably depends on the permeability of the membrane to other ions such as Cl^-. Movement of Cl^- into the thylakoid lumen reduces the electrical potential difference across the membrane, allowing the buildup of a larger pH gradient.

In the Z scheme, photosystem II, the cytochrome-b_6f complex, and photosystem I operate in series to move electrons from H_2O to $NADP^+$ and to create an electrochemical potential gradient for protons across the thylakoid membrane. In addition to this linear pathway, chloroplasts in some plant species may use a cyclic electron-transfer scheme that includes photosystem I and the cytochrome-b_6f complex, but not photosystem II. If the concentration of $NADP^+$ is low, electrons

Figure 16.27

The reductive pentose cycle, or Calvin cycle. The number of arrows drawn at each step in the diagram indicates the number of molecules proceeding through that step for every three molecules of CO_2 that enter the cycle. The entry of three molecules of CO_2 results in the formation of one molecule of glyceraldehyde-3-phosphate (box on right), and requires the oxidation of six molecules of NADPH to $NADP^+$ and the breakdown of nine molecules of ATP to ADP.

ejected by the reaction center of photosystem I can be passed from one of the iron-sulfur proteins FD_A and FD_B to a *b*-type cytochrome (cytochrome b_{563}), and from there to the plastoquinone pool between the two photosystems (see fig. 16.19). From plastoquinone, electrons can return to P700 via the cytochrome-b_6f complex and plastocyanin. The Q cycle that is included within this cyclic electron pathway results in proton translocation across the thylakoid membrane, and thus could contribute to the transmembrane electrochemical potential gradient that drives the ATP-synthase. Cyclic electron transport probably does not occur to a significant extent under physiological conditions in cyanobacteria and terrestrial plants, but may be important in some algae.

Carbon Fixation: The Reductive Pentose Cycle

The ATP and NADPH that are generated by the photosynthetic electron-transfer reactions are used to drive the fixation of CO_2. Reactions in which CO_2 is incorporated into carbohydrates were discovered in the early 1950s by Melvin Calvin and his co-workers in some of the first biochemical studies employing radioactive tracers. A key experiment was to expose algae to a brief period of illumination in the presence of $^{14}CO_2$ and then to disrupt the cells quickly and search for organic molecules that had become labeled with ^{14}C. The material that turned out to be labeled most rapidly was glycerate-3-phosphate, and almost all of its ^{14}C was found to be in the carboxyl group. To identify the precursor of the glycerate-3-phosphate, Calvin's group looked for changes in the steady-state concentrations of other compounds when they turned a continuous light on or off or when they suddenly raised or lowered the concentration of CO_2. These experiments showed that glycerate-3-phosphate is formed from ribulose-1,5-bisphosphate. A molecule of CO_2 is incorporated in the process, so that one molecule of the five-carbon sugar ribulose-1,5-bisphosphate gives rise to two molecules of the three-carbon glycerate-3-phosphate (fig. 16.27). The reaction is catalyzed by ribulose bisphosphate carboxylase, which is found in the chloroplast stroma.

Further studies revealed that other enzymes in the stroma can convert glycerate-3-phosphate back to ribulose-1,5-bisphosphate. The reactions of this reductive pentose cycle, or Calvin cycle, are shown in figure 16.27. The 3-phosphoglycerate is first phosphorylated to glycerate-1,3-bisphosphate at the expense of ATP and then reduced to glyceraldehyde-3-phosphate by NADPH. (Note that the nucleotide specificity in the reductive step differs from that of the cytosolic glyceraldehyde-3-phosphate dehydrogenase, which uses NAD^+ and NADH.)

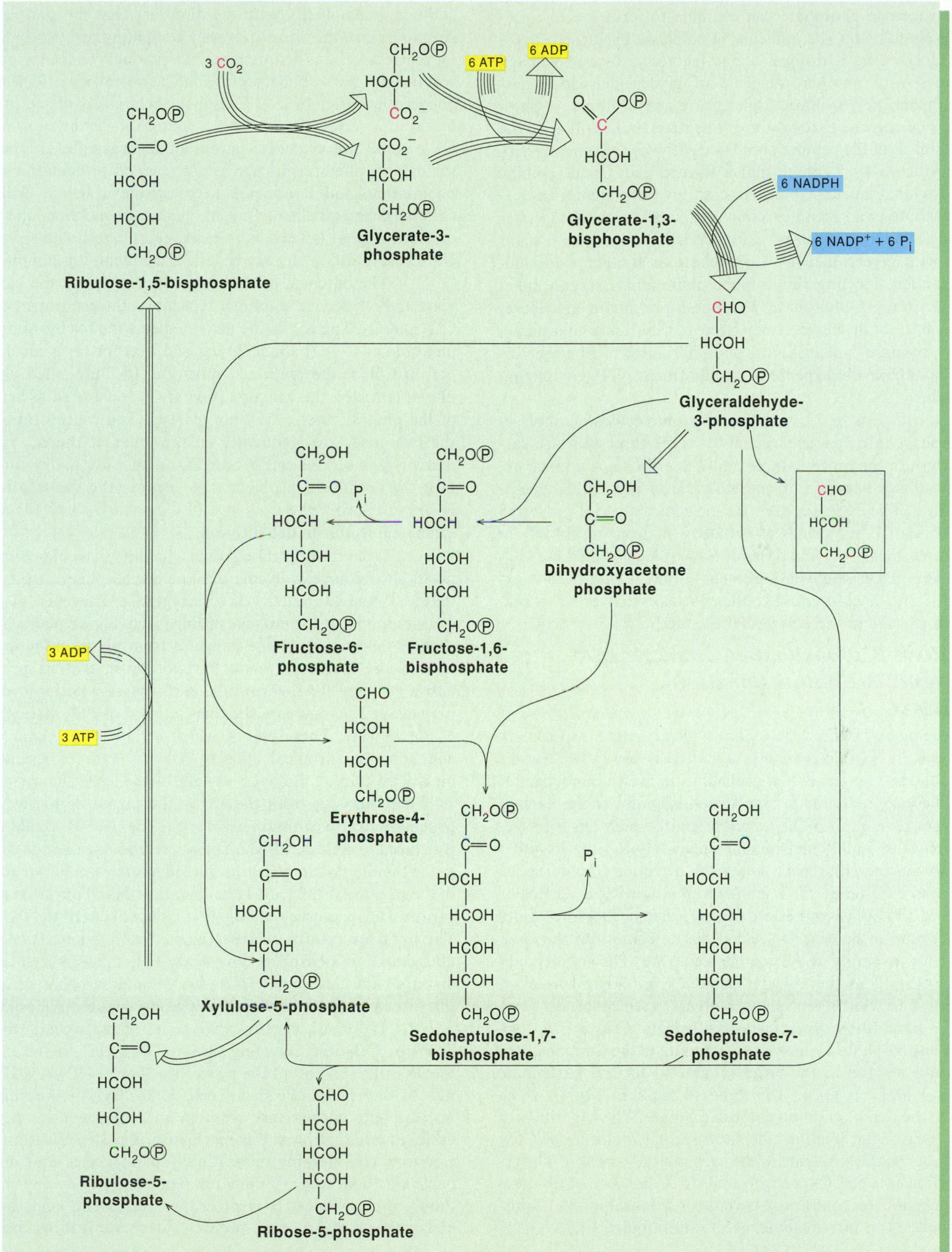

3 CO_2
6 ATP
6 ADP
Ribulose-1,5-bisphosphate
Glycerate-3-phosphate
Glycerate-1,3-bisphosphate
6 NADPH
6 $NADP^+$ + 6 P_i
Glyceraldehyde-3-phosphate
Dihydroxyacetone phosphate
Fructose-1,6-bisphosphate
P_i
Fructose-6-phosphate
3 ADP
3 ATP
Erythrose-4-phosphate
Xylulose-5-phosphate
Sedoheptulose-1,7-bisphosphate
P_i
Sedoheptulose-7-phosphate
Ribulose-5-phosphate
Ribose-5-phosphate

Glyceraldehyde-3-phosphate and its isomerization product dihydroxyacetone phosphate can combine to form fructose-1,6-bisphosphate under the influence of aldolase. Fructose-6-phosphate, formed by hydrolysis of the fructose-1,6-bisphosphate, combines with another molecule of glyceraldehyde-3-phosphate, generating xylulose-5-phosphate and erythrose-4-phosphate. The enzyme that catalyzes this reaction is similar to the transketolase of the pentose phosphate pathway (see chapter 13). The erythrose-4-phosphate that is formed goes on to combine with dihydroxyacetone phosphate to give sedoheptulose-1,7-bisphosphate in a second reaction catalyzed by aldolase. After hydrolysis to sedoheptulose-7-phosphate, the seven-carbon sugar reacts with glyceraldehyde-3-phosphate in another transketolase reaction, forming ribose-5-phosphate and a second molecule of xylulose-5-phosphate. Xylulose-5-phosphate and ribose-5-phosphate both can be isomerized to ribulose-5-phosphate. Finally, ribulose-5-phosphate is phosphorylated to ribulose-1,5-bisphosphate at the expense of an additional ATP, completing the cycle.

In figure 16.27, note that three molecules of ribulose-1,5-bisphosphate are regenerated for every three that are carboxylated. In the process, three molecules of CO_2 are taken up, and there is a net gain of one molecule of glyceraldehyde-3-phosphate. The expenses are nine molecules of ATP converted to ADP and six molecules of NADPH oxidized to NADP, or three molecules of ATP and two of NADPH consumed per CO_2 fixed. The glyceraldehyde-3-phosphate that is produced is exported from the chloroplast to the cytosol or converted to hexoses for storage in the chloroplast as starch.

Ribulose Bisphosphate Carboxylase/ Oxygenase, Photorespiration, and the C_4 Cycle

Ribulose bisphosphate carboxylase typically accounts for more than half the soluble protein in a leaf. It is surely the world's most abundant enzyme, and probably the most abundant protein. The enzyme found in plants contains eight copies each of two types of subunits. The larger subunit, which has a molecular weight of 56,000 and contains the catalytic site, is synthesized within the chloroplast under the direction of chloroplast DNA and ribosomes. The smaller subunit, with a molecular weight of 14,000, is synthesized on cytoplasmic ribosomes under the direction of nuclear DNA and has to come into the chloroplast for assembly of the complete enzyme. The smaller subunit is lacking in the enzyme isolated from some species of bacteria, and what role it plays in plants is still unclear.

In addition to serving as a substrate, CO_2 activates ribulose bisphosphate carboxylase by binding to the ε-amino group of a lysyl residue in the large subunit to form a carbamate (lysine — NH — CO_2^-). This process requires Mg^{2+}, which binds to the carboxyl group of the carbamate. The Mg^{2+} in turn forms part of the binding side for a second molecule of CO_2, which acts as the substrate in the carboxylase reaction. The reactive molecule of CO_2 probably adds to the enolate of ribulose-1,5-bisphosphate to form 2-carboxy-3-ketoarabinitol-1,5-bisphosphate as an intermediate, as shown in figure 16.28.

Studies of ribulose bisphosphate carboxylase took on additional complexity with the discovery that the same active site on the enzyme also catalyzes a competing reaction in which O_2 replaces CO_2 as a substrate. The products of the oxygenase reaction are 3-phosphoglycerate and a two-carbon acid, 2-phosphoglycolate (fig. 16.29). Phosphoglycolate is oxidized to CO_2 by O_2 in a series of additional reactions involving enzymes in the cytosol, mitochondria, and another organelle, the peroxisome. This photorespiration is not coupled to oxidative phosphorylation, and it appears to constitute a severe drain on chloroplast metabolism. In some plants, as much as one-third of the CO_2 that is fixed is released again by photorespiration. If photorespiration has any benefit to the plant, it is not obvious.

The outcome of the competition between the carboxylase and oxygenase reactions depends on the concentrations of CO_2 and O_2. The K_m of the activated enzyme for the substrate molecule of CO_2 is about 20 μM and that for O_2 is about 200 μM, so CO_2 is the preferred substrate. In illuminated chloroplasts, however, the concentration of O_2 is elevated as a result of the photosynthetic splitting of H_2O. The concentrations of the two gases are frequently on the order of the K_m values, making O_2 a serious competitor. Plants that live in dry climates face the additional problem that opening the stomata in the leaves to improve the exchange of CO_2 and O_2 with the atmosphere can result in dehydration.

If photorespiration is only a liability, it would seem that plants should have evolved a ribulose bisphosphate carboxylase that had minimal activity as an oxygenase. However, the oxygenase activity of the enzyme purified from higher plants is only slightly below that of the enzymes from photosynthetic bacteria. Some species of plants have, however, evolved an alternative strategy for favoring the carboxylase reaction over the oxygenase: They increase the concentration of CO_2 in the region of the enzyme. These plants, which include corn, sugar cane, and numerous tropical species, have a layer of specialized mesophyll cells at the outer surface of the leaf. The mesophyll cells take up CO_2 from the air by the carboxylation of phosphoenolpyruvate to obtain oxalacetate (fig. 16.30). Oxalacetate is reduced to malate, which then moves from the mesophyll cells to the bundle sheath cells that surround the vascular structures in the interior of the leaf. Here malate is decarboxylated to pyruvate in an oxidative reaction that reduces NADP to NADPH. The pyruvate returns to the mesophyll cell, where it is phosphorylated to phosphoenolpyruvate. The phosphorylation is driven by the splitting of ATP to AMP and pyrophosphate and the subsequent hydrolysis of the pyrophosphate to phosphate.

The result of this cyclic series of reactions is the delivery of CO_2 and reducing power (NADPH) to the bundle sheath cells, at a cost of the hydrolysis of one ATP to AMP (see fig. 16.30). The bundle sheath cells fix the CO_2 by the ribulose-bisphosphate carboxylase reaction and the reductive pentose cycle. Plants that have this mechanism for concentrating CO_2 can grow considerably more rapidly than species that do not, particularly in hot, dry climates. The auxiliary cycle is called the C_4 cycle because it involves the four-carbon acids malate and oxalacetate. In some species, oxalacetate is transaminated

Figure 16.28

The reaction catalyzed by ribulose bisphosphate carboxylase involves 2-carboxy-3-ketoarabinitol-1, 5-bisphosphate as an enzyme-bound intermediate. The intermediate probably forms by the addition of CO_2 to the enolate of ribulose-1, 5-bisphosphate. The substrate is known to be CO_2 rather than bicarbonate.

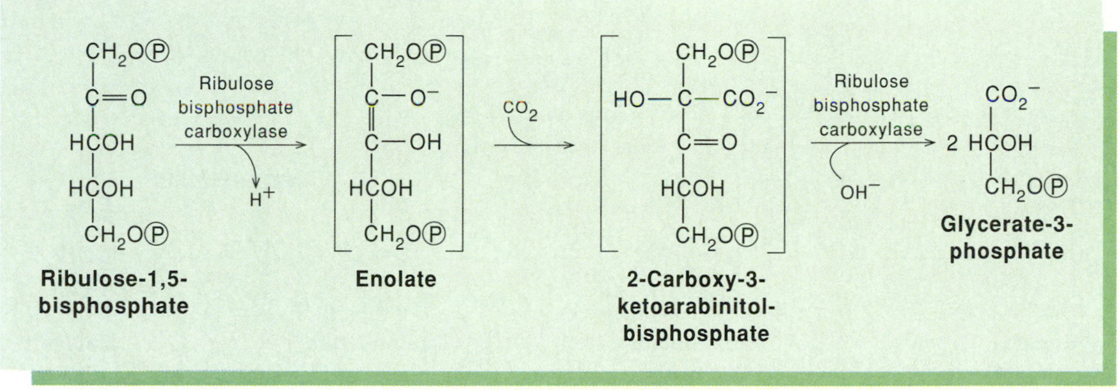

Figure 16.29

Photorespiration results from the oxygenation reaction catalyzed by ribulose-bisphosphate carboxylase/oxygenase. 2-Phosphoglycolate generated by the reaction moves from the chloroplast to the cytosol, where other enzymes break it down to CO_2, H_2O, and P_i. The oxygenation reaction, like carboxylation, does not require light directly. It occurs mainly during illumination, however, because the formation of the substrate, ribulose-1, 5-bisphosphate, requires ATP and NADPH (see fig. 16.27).

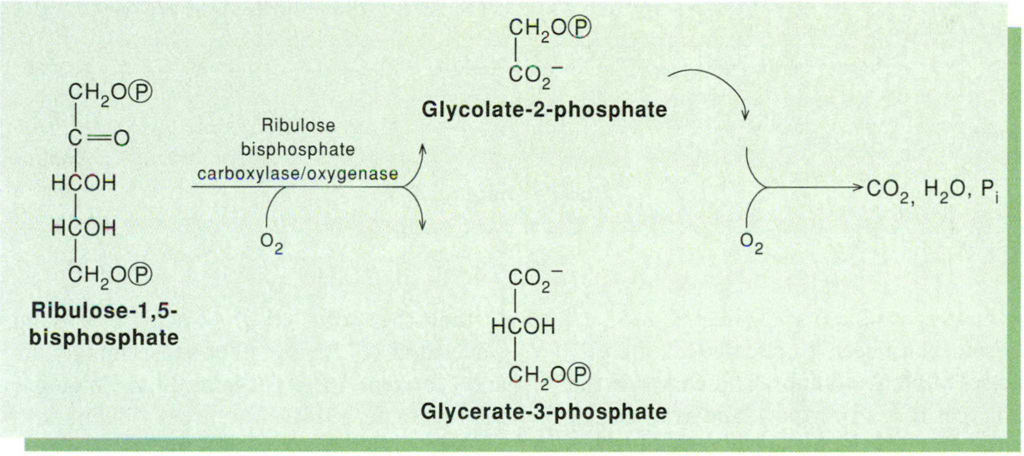

to aspartate rather than being reduced to malate, and it is aspartate that moves to the bundle sheath cells. There is currently interest in the possibility of increasing agricultural productivity by using genetic engineering techniques to introduce the C_4 pathway into plant species that lack it.

Other Biochemical Processes Involving Light

Living things depend on light as a source of energy. They also have found a number of other uses for light. Vision is one of these and is taken up separately in chapter 36, where it can be treated in the necessary detail. In the remainder of this chapter, we will consider bioluminescence and the role of light in the synchronization of circadian and seasonal rhythms.

Phytochrome Synchronizes Circadian and Seasonal Rhythms in Plants

Most eukaryotic organisms have endogenous rhythms of metabolic activity with periods of about 24 hours. The circadian, or diurnal, rhythms are triggered by light, and under natural conditions they are locked in phase with the 24-h cycle of day and night. Many organisms also have seasonal cycles that are set by the length of the day. In mammals, the synchronization of circadian and seasonal rhythms depends mainly on visual responses to light, but organisms that lack nervous systems depend on other types of photoreceptors. In some unicellular organisms such as paramecia, the photoreceptive pigment appears to be a flavin or flavoprotein. In higher plants, the major receptor is a pigment-protein complex called phytochrome.

Figure 16.30

The C-4 pathway requires the cooperation of two types of cells. Mesophyll cells (*top*) take up CO_2 from the air and export malate to the bundle sheath cells (*bottom*). The bundle sheath cells return pyruvate to the mesophyll cells and fix the CO_2 using ribulose bisphosphate carboxylase and the reductive pentose cycle.

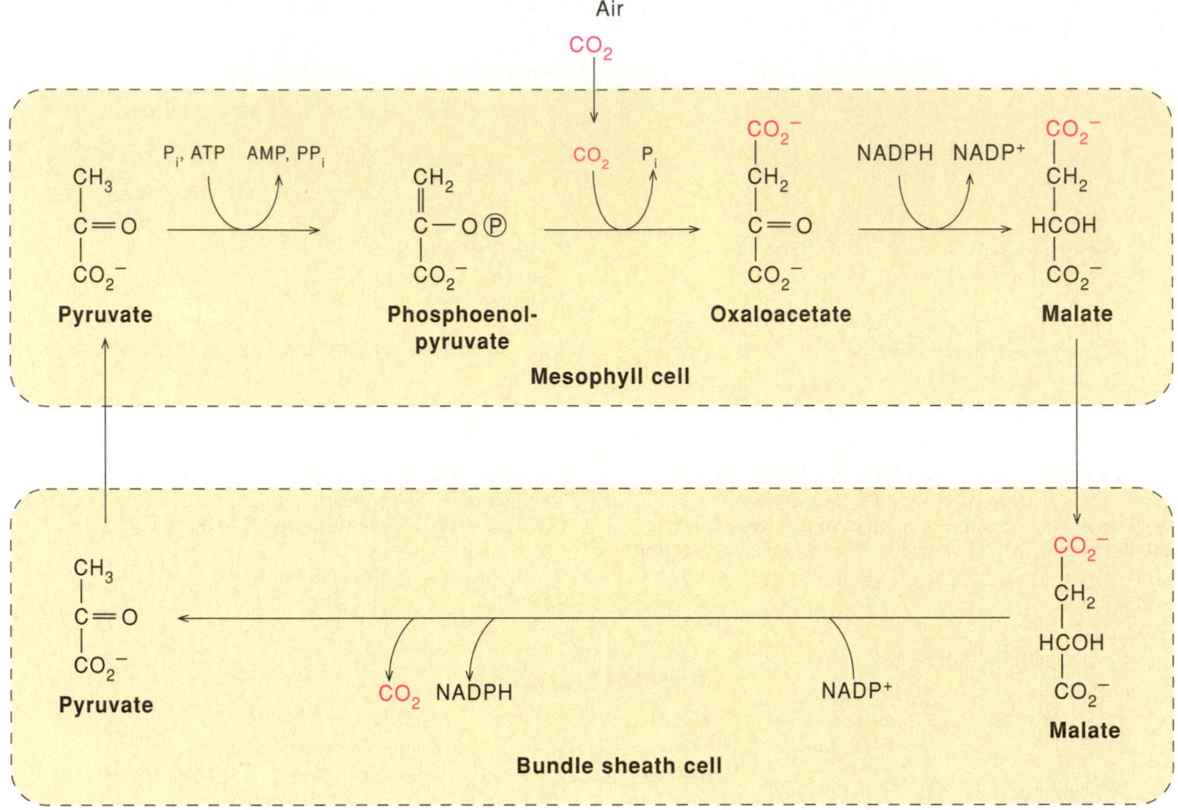

In addition to the synchronization of diurnal and seasonal rhythms, phytochrome has been implicated in literally hundreds of other responses of plants to light. These include the germination of seeds, changes in ion transport, and synthesis of numerous enzymes. Some of its effects involve changes in gene expression, including an increase in the rate of transcription of the gene for the small subunit of ribulose bisphosphate carboxylase. Other effects, such as changes in the movement of Ca^{2+} across the cytoplasmic membrane, seem too rapid for this, and may reflect a more immediate site of action. Some of the effects appear to be mediated by the Ca^{2+}-binding protein calmodulin.

The phytochrome protein is a dimer of subunits with molecular weights of about 120,000. The pigment is an open-chain tetrapyrrole (fig. 16.31). Phytochrome has two interconvertible forms (fig. 16.32). When the molecule is in one form, called P_R, phytochrome has an optical absorption maximum near 670 nm. The absorption of light converts phytochrome from P_R to the other form (P_{FR}), which absorbs maximally at longer wavelengths (720 nm). (The subscripts R and FR stand for "red" and "far red.") P_{FR} is evidently the biologically active form of phytochrome, but its structure is still unknown, and exactly what it does is not clear.

Phytochrome decays slowly from P_{FR} back to P_R in the dark. It also can be returned to P_R immediately by exciting it with light (see fig. 16.32). Because P_R absorbs maximally at 670 nm and P_{FR} at 720 nm, phytochrome can be cycled back and forth between the active and inactive forms by excitation flashes that alternate between the two wavelengths. Continuous illumination sets up a steady-state mixture, in which the amount of P_{FR} depends on the color of the light. Because chlorophyll absorbs red light better than it does far-red light, the color of the light reaching the leaf depends on whether the leaf is shaded by other leaves. The information collected by phytochrome can therefore be useful in controlling the direction and rate of growth of the plant.

Figure 16.31

The prosthetic group of phytochrome is an open-chain tetrapyrrole, which is bound to the protein as a cysteine thioether.

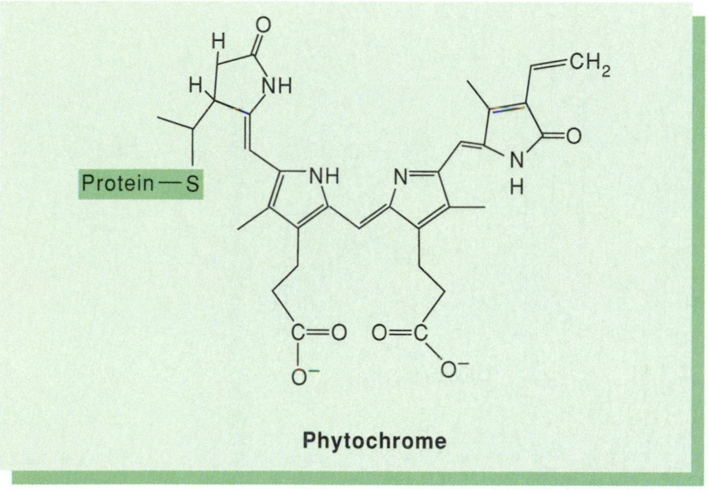

Figure 16.32

Phytochrome exists in two interconvertible forms, P_R and P_{FR}, which have different absorption spectra. The form that is synthesized initially, P_R, has an absorption band at 670 nm. When it is excited with light, P_R is converted to the biologically active species, P_{FR}, which absorbs at 720 nm. P_{FR} decays back to P_R in the dark, and it can be converted to P_R by excitation with light. P_{FR} also is degraded relatively quickly by proteolysis. P_R and P_{FR} may be conformational isomers that share a common excited state.

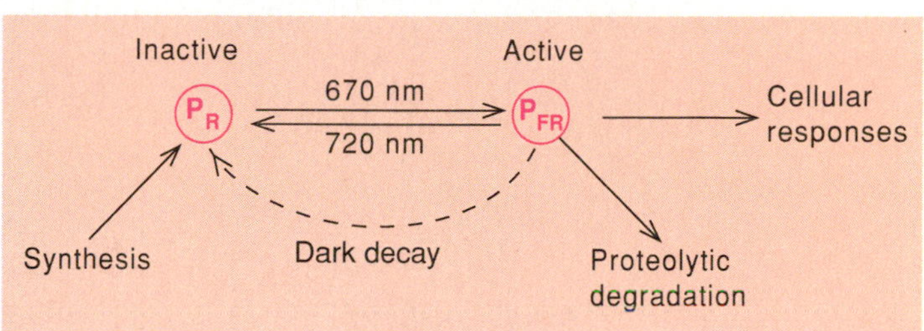

Bioluminescence

Conceptually, bioluminescence is the reverse of photosynthesis. A reduced substrate reacts with O_2 and is converted to an oxidized product in an excited electronic state. The excited molecule then decays to the ground state, emitting light. The process occurs in several groups of bacteria and fungi; in marine invertebrates such as sponges, shrimp, and jellyfish; and in a variety of terrestrial creatures, including earthworms, centipedes, and insects. The bacteria that emit light generally live symbiotically with fish in special luminous organs. In some cases, the evolutionary benefits of bioluminescence seem clear: Fireflies use it for communication; fish, to attract or locate prey, or to confuse predators. In other cases, such as in fungi, the benefits are not obvious, and one is struck by the expense of the process. The energy that is released in the oxidation-reduction reaction could, in principle, be directed into ATP production instead of being emitted as light.

The substrates that undergo oxidation with the emission of light are all called luciferins, although their structures vary among different bioluminescent species. The oxidized products are termed oxyluciferins, and the enzymes that catalyze the oxidations are called luciferases. Figures 16.33 and 16.34 show the structures of the luciferins and oxyluciferins used by two species, the sea pansy (*Renilla reniformis*) and the firefly (*Photinus*). The figures also indicate likely intermediates in the oxidation reactions. Although the luciferins have rather different structures in the two species, the chemistry appears to be much the same. O_2 reacts with a heterocyclic ring, forming a peroxide at a carbon that is bound to a nitrogen and to a carboxylic amide or ester. The peroxide cyclizes to give a cyclic peroxide, or dioxetanone, with the release of the amine or alcoholic group that was bound to the carboxyl. The cyclic peroxide then decomposes with the release of CO_2, forming the oxyluciferin in an excited electronic state.

The decomposition of the dioxetanone intermediate may involve an internal oxidation-reduction reaction in which an electron is transferred to the peroxide from another part of the molecule. Such a reaction could generate a linked pair of free radicals ($D^{+\cdot}A^{-\cdot}$) analogous to the species that are formed in the photochemical reactions of photosynthesis. After CO_2 splits off, an electron could return from the reduced component to the oxidized component, putting the product in an excited state.

One difference between fireflies and *Renilla* is that the luciferase of fireflies activates the carboxyl group of the luciferin by forming an acyladenylate (see fig. 16.34). This step resembles the activation of amino acids for protein synthesis (chapter 29), but it is unique to fireflies among the bioluminescent systems that have been studied. Since the activation requires ATP, measurements of the light that is emitted by firefly oxyluciferin provide a sensitive assay for ATP.

Luminescent bacteria use a long-chain aliphatic aldehyde such as decanal as a luciferin, oxidizing it to the carboxylic acid. $FMNH_2$ is oxidized simultaneously to FMN. Again, a peroxide appears to be an intermediate in the oxidation. The excited product is probably a hydroxy derivative of FMN, which subsequently loses H_2O to form FMN.

In bioluminescent organisms, the molecule that actually emits the light is sometimes not the oxyluciferin itself, but a pigment on another protein, to which the excited oxyluciferin transfers its energy. This is conceptually like a reverse operation of the light-harvesting antenna systems of photosynthesis.

Figure 16.33

Bioluminescent reactions in the sea pansy *Renilla reniformis*. The structures in brackets are plausible enzyme-bound intermediates, but have not been identified conclusively. They are based partly on analogous nonenzymatic reactions of dioxetanones and on the observation that one of the O atoms of the CO_2 that is released comes from the O_2.

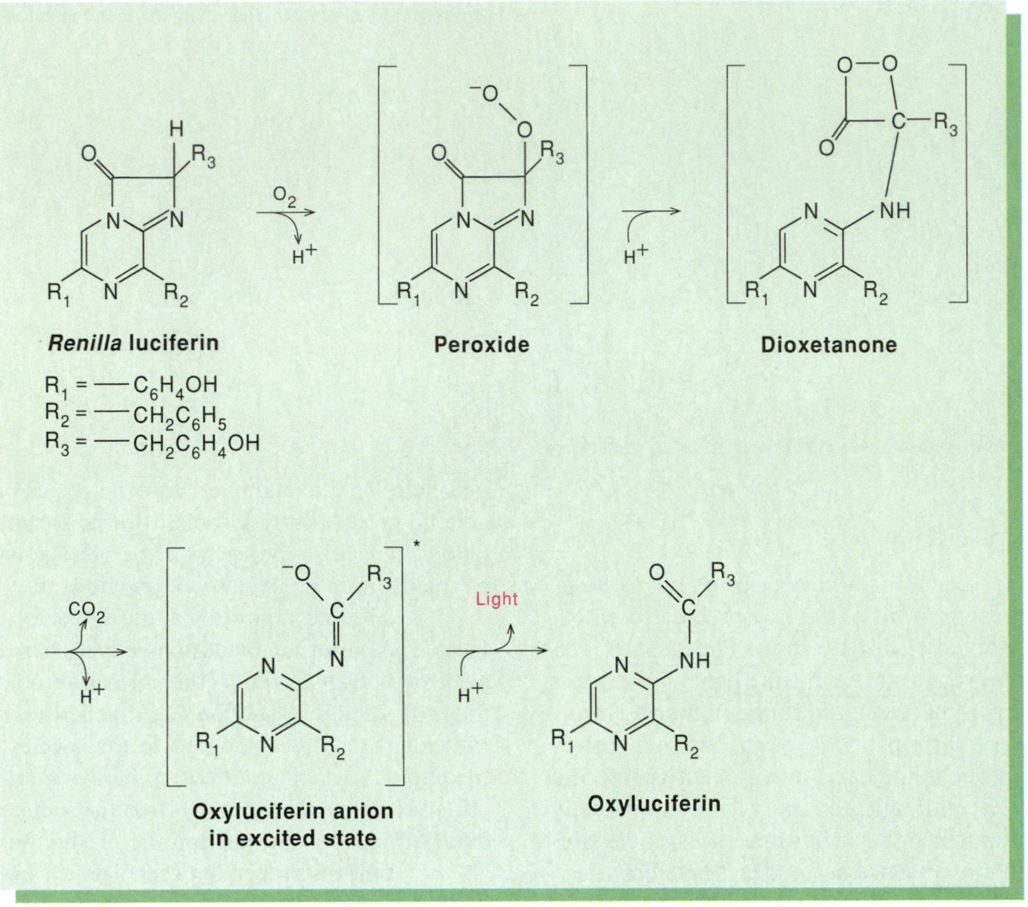

***Renilla* luciferin** → (O_2, H^+) → **Peroxide** → (H^+) → **Dioxetanone** → (CO_2, H^+) → **Oxyluciferin anion in excited state** → (Light, H^+) → **Oxyluciferin**

$R_1 = —C_6H_4OH$
$R_2 = —CH_2C_6H_5$
$R_3 = —CH_2C_6H_4OH$

Summary

In this chapter we have examined photochemical reactions, primarily those that occur in photosynthesis. The chief points made in the chapter are as follows.

1. The primary source of energy for biosynthetic reactions is sunlight. Photosynthesis, the process of capturing sunlight and converting it into chemical energy, occurs in numerous species of bacteria and algae, in addition to higher plants. In plants, the photochemical reactions of photosynthesis take place in the chloroplast thylakoid membrane. In bacteria, they take place in the cell membrane.
2. Almost all photosynthetic organisms take advantage of the fact that chlorophyll or bacteriochlorophyll becomes a strong reductant when it is excited with light. Photooxidation of chlorophyll or bacteriochlorophyll occurs in pigment-protein complexes, the reaction centers. Chloroplasts contain two types of reaction centers: photosystem I, which contains a reactive chlorophyll complex called P700, and photosystem II, which contains P680. The reactive complex in purple photosynthetic bacteria (P870) is a bacteriochlorophyll dimer.
3. The thylakoid membrane and the bacterial cell membrane also contain molecules that serve as antennae. The antenna systems consist of small pigment-protein complexes assembled into large arrays that can contain hundreds of pigment molecules per reaction center. When the antenna absorbs a photon, the excitation energy moves rapidly from complex to complex by resonance energy transfer until the energy is trapped in a reaction center.
4. The bacterial reaction center contains two molecules of bacteriopheophytin and two of bacteriochlorophyll in addition to the two bacteriochlorophylls of P870. When P870 is excited, it transfers an electron to one of the bacteriopheophytins. The bacteriopheophytin then reduces a quinone, Q_A, which in turn reduces another quinone, Q_B.

Figure 16.34

Bioluminescent reactins of fireflies. The excited oxyluciferin that emits light is believed to be a dianion, in which the phenolic and enolic oxygens are both ionized. The ionization state can be deduced by comparing the emission spectrum of the luminescence with those of model compounds.

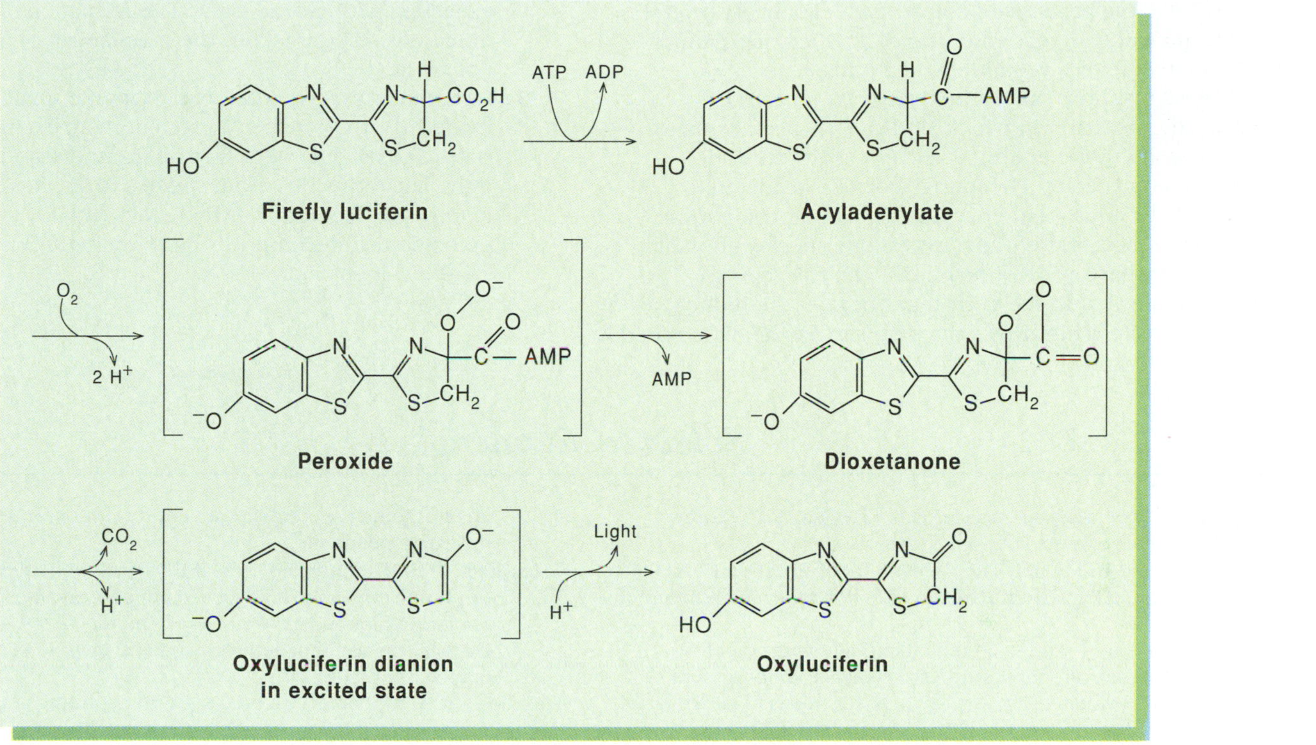

Firefly luciferin → (ATP → ADP) → Acyladenylate → (O_2, → 2 H^+) → [Peroxide] → (→ AMP) → [Dioxetanone] → (→ CO_2, → H^+) → [Oxyluciferin dianion in excited state] → (Light, H^+) → Oxyluciferin

In some bacteria, Q_A and Q_B are both ubiquinone; in other species one or both of the quinones is menaquinone. When P870 is excited a second time, a second electron is sent through the same carriers to Q_B. The electron released by P870 is replaced by one from a *c*-type cytochrome. The doubly reduced Q_B picks up protons from the solution on the inside of the cell. Electrons move from the quinol (QH_2) to the *c*-type cytochrome via a cytochrome-bc_1 complex, and protons are released to the solution outside the cell. This movement of protons across the membrane creates an electrochemical potential gradient across the membrane. Proton movement back into the cell is mediated by an ATP-synthase and drives the formation of ATP.

5. Photosystems I and II of chloroplasts operate in series. The photooxidation of P680 in photosystem II generates a strong oxidant that can oxidize H_2O to O_2. In this process, a manganese complex bound to a protein progresses through five different oxidation states (S_0 to S_4), advancing to the next higher state each time P680 undergoes photooxidation. When the complex reaches state S_4, O_2 is given off. The electron released by P680 goes to a pheophytin, then to a molecule of plastoquinone, and from there to a second plastoquinone. When the second plastoquinone has been doubly reduced, it picks up protons from the chloroplast stromal space and dissociates from the reaction center. Electrons move from the reduced plastoquinone to P700 of photosystem I by means of a cytochrome-b_6f complex and a copper protein (plastocyanin).
6. Photooxidation of P700 in photosystem I appears to reduce a chlorophyll, which transfers electrons to a series of membrane-bound iron-sulfur proteins, probably by way of a quinone. From the bound iron-sulfur proteins, electrons move to the soluble iron-sulfur protein ferredoxin, and then to a flavoprotein that reduces NADP. There also may be some cyclic electron flow, in which electrons go from the bound iron-sulfur proteins to the carriers between the two photosystems.
7. Electron movement through the cytochrome-b_6f complex between the chloroplast photosystems results in proton translocation from the stroma to the thylakoid lumen. In addition, protons are released in the lumen when H_2O is oxidized, and are taken up in the stromal space when NADP is reduced. Protons move from the thylakoid lumen back to the stroma through an ATP-synthase (CF_0-CF_1), driving the formation of ATP.

8. ATP and NADPH are used for the incorporation of CO_2 into carbohydrates. CO_2 reacts with ribulose-1,5-bisphosphate to give two molecules of 3-phosphoglycerate, under the influence of ribulose bisphosphate carboxylase/oxygenase. 3-Phosphoglycerate can then be converted back to ribulose-1,5-bisphosphate at the expense of ATP and NADPH in the reductive pentose cycle. For every three molecules of CO_2 that enter the cycle, there is a gain of one molecule of glyceraldehyde-3-phosphate.
9. Ribulose bisphosphate carboxylase/oxygenase also catalyzes a reaction in which ribulose-1,5-bisphosphate reacts with O_2, generating 2-phosphoglycolate and 3-phosphoglycerate. Phosphoglycolate is oxidized to CO_2 and H_2O by additional enzymes. This photorespiration appears to be a wasteful process. Some species of plants use an auxiliary cycle of reactions (the C_4 cycle) to concentrate CO_2 in the cells that carry out the reductive pentose cycle. This allows the plants to fix CO_2 at elevated rates.
10. Biological systems have found other uses for light. In plants, light regulates a broad range of physiological processes through phytochrome, a complex of a protein and an open-chain tetrapyrrole. Phytochrome has two interconvertible forms, P_R and P_{FR}. P_R is converted to P_{FR} when it is excited with red light. P_{FR} reverts to P_R upon excitation with far-red light. The formation of P_{FR} initiates numerous responses, but the mechanism of its action is unknown.
11. In bioluminescence, the oxidation of a small molecule (a luciferin) by O_2 creates a product in an excited electronic state. The product decays to its ground state by emitting light. Luciferins vary from species to species, but many are heterocyclic compounds that react with O_2 to form a cyclic peroxide that decomposes by releasing CO_2.

Selected Readings

Amesz, J. (ed.), *Photosynthesis.* Amsterdam: Elsevier, 1987. Review articles on many aspects of photosynthesis.

Deisenhofer, J., and H. Michel, The Photosynthetic Reaction Center from the Purple Bacterium *Rhodopseudomonas viridis. Science* 245:1463, 1989.

Duysens, L. N. M., and J. Amesz, Function and identification of two functional systems in photosynthesis. *Biochim. Biophys. Acta* 64:243, 1962. Evidence supporting the Z scheme is obtained by measuring oxidation and reduction of cytochrome *f* in algae illuminated with light of various colors.

Forbush, B., B. Kok, and M. McGloin, Cooperation of charges in photosynthetic O_2 evolution: II. Damping of flash yield, oscillation, deactivation. *Photochem. Photobiol.* 14:307, 1971. Oscillations in O_2 yield during a train of flashes provide evidence for the four S states.

Jagendorf, A. T., and E. Uribe, ATP formation caused by acid-base transition of spinach chloroplasts. *Proc. Natl. Acad. Sci. USA* 55:197, 1966. Chloroplasts can form ATP in the dark if an electrochemical potential gradient for protons is set up across the thylakoid membrane.

Lorimer, G. H., The carboxylation and oxygenation of ribulose-1,5-bisphosphate: The primary events in photosynthesis and photorespiration. *Ann. Rev. Plant Physiol.* 32:349, 1981. Mechanisms and biochemical significance of the carboxylase and oxygenase reactions.

Rutherford, A. W., Photosystem II, the water-splitting enzyme. *Trends Biochem. Sci.* (TIBS) 14:227, 1989. A readable account of current results and speculations on the oxygen-evolving complex.

Scheer, H. (ed.), *Chlorophylls.* Boca Raton, FL: CRC Press, 1991. A collection of review articles.

Woodbury, N. W., M. Becker, D. Middendorf, and W. W. Parson, Picosecond kinetics of the initial photochemical electron transfer reaction in bacterial photosynthetic reaction centers. *Biochem.* 24:7516, 1985. Fast spectrophotometric techniques are used to follow the initial steps in reaction centers purified from photosynthetic bacteria.

Problems

1. The lowest-energy absorption band of photosystem P870 occurs at 870 nm.
   (a) Calculate the energy of an einstein of light at this wavelength.
   (b) Estimate the effective standard redox potential ($E°'$) of P870 in its first excited singlet state, given that the $E°'$ for oxidation in the ground state is +0.45 V.
2. The traces at right show measurements of optical absorbance changes at 870 and 550 nm when a suspension of membrane vesicles from photosynthetic bacteria was excited with a short flash. Downward deflection of the traces represent absorption decreases. Explain the observations. (Absorption spectra of a *c*-type cytochrome in its reduced and oxidized forms are described in the previous chapter.)

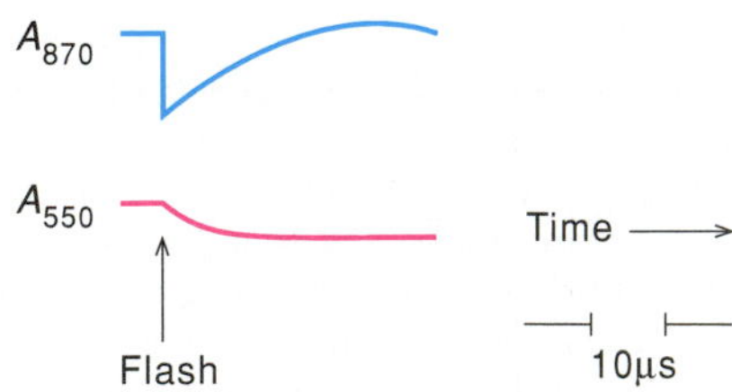

3. You add a nonphysiological electron donor to a suspension of chloroplasts. When you illuminate the chloroplasts, the donor becomes oxidized. How could you determine whether this process involves both photosystems I and II? (In principle, the donor could transfer electrons either to some component on the O_2 side of photosystem II or to a component between the two photosystems.)
4. Ubiquinone has an absorbance band at 275 nm. This band bleaches when the quinone is reduced to either the semiquinone or the dihydroquinone. The anionic semiquinone has an absorption band at 450 nm, but neither the quinone nor the dihydroquinone absorbs at this wavelength. A suspension of purified bacterial reaction centers was supplemented with extra ubiquinone and reduced cytochrome *c* and was then illuminated with a series of short flashes of light. The absorbance at 275 nm decreased on the odd-numbered flashes as shown on the first curve below. The absorbance at 450 nm increased on the odd-numbered flashes but returned to the original level on the even-numbered flashes as shown in the second curve.

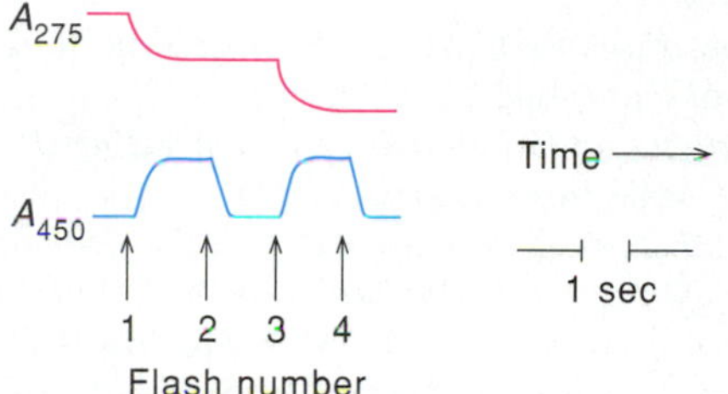

   (a) Explain the patterns of absorbance changes at the two wavelengths.
   (b) Why is it necessary to have reduced cytochrome *c* present in order to see these effects?
5. When chloroplasts are illuminated with a weak continuous white light, the intensity of fluorescence from the antenna chlorophylls is relatively low. (About 2% of the photons absorbed are reemitted as fluorescence.) The intensity of the fluorescence increases severalfold if the redox potential is decreased to approximately 0 mV. (Assume that the plastoquinone pool is reduced.) Explain the increase in fluorescence.
6. When chloroplasts are illuminated with a strong continuous light, the intensity of fluorescence from the antenna chlorophyll increases with time as shown in curve A.

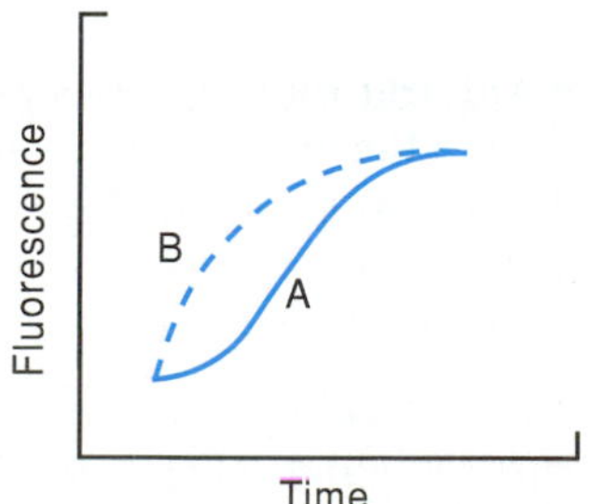

   (a) Explain the increase in fluorescence.
   (b) The herbicide DCMU (3-(3,4-dichlorophenyl)-1,1-dimethylurea) blocks electron transfer between the two plastoquinones in the reaction center of photosystem II. In the presence of DCMU, the fluorescence increases more abruptly (curve B). Explain.
   (c) What effect would DCMU have on $NADP^+$ reduction? On O_2 evolution?
7. Explain why, in green plants, the quantum efficiency of photosynthesis drops sharply at wavelengths longer than 680 nm.
8. Define photophosphorylation. Differentiate between cyclic and noncyclic photophosphorylation in green plants and describe how photosystems I and II are involved in each process.
9. Would you predict that plant cells lacking carotenoids would be more or less susceptible to damage by photooxidation? Explain.
10. O_2 is an alternative substrate for ribulose bisphosphate carboxylase/oxygenase and is also a competitive inhibitor with respect to CO_2 fixation. Explain.
11. Compare and contrast the initial process of CO_2 fixation in C_3 and C_4 plants. In each case, indicate which stable organic molecule initially becomes labeled with ^{14}C-labeled CO_2.
12. C_4 plants ultimately use ribulose bisphosphate carboxylase/oxygenase, yet have lower rates of photorespiration than do C_3 plants. Explain.
13. Reduction of three moles of CO_2 to form one mole of triose phosphate requires nine moles ATP and six moles NADPH.
   (a) What is the source of NADPH used in the reduction of CO_2?
   (b) Account for the ATP consumed in the formation of triose phosphate.
   (c) Assume that the CO_2 initially is added to PEP in the C_4 pathway. What is the additional cost in ATP per mole of CO_2 added?
   (d) Is additional NADPH required for CO_2 fixation in the C_4 plant?

17 CHAPTER

Metabolism of Fatty Acids

Fatty acids, like carbohydrates, are important structural elements and important sources of energy. In this chapter we will focus on the role of fats, in particular the fatty acids, in energy metabolism. We will see that long-chain fatty acids are assembled together in blocks of two carbon atoms and similarly that they are degraded in blocks of two carbon atoms (fig. 17.1). Despite this obvious similarity, the processes of synthesis and breakdown are very different. In contrast to what happens with carbohydrates, for fatty acids the two processes use completely different sets of enzymes, and they occur in different cellular compartments. These differences make it easy to provide regulatory mechanisms for ensuring that the two processes do not occur simultaneously.

Fatty Acid Degradation

In this section we will deal with reactions associated with fatty acid breakdown, and in the following section we will turn to the mechanisms for fatty acid synthesis. Finally, we will consider the regulatory mechanisms that determine the conditions under which either of these processes occurs.

Fatty Acids Originate from Three Sources: Diet, Adipocytes, and de novo *Synthesis*

Fatty acids are mainly derived from the diet. In fact, 30–40% of the calories ingested each day in the average human diet are provided by the fatty acid components of complex lipids (e.g., triacylglycerols, phospholipids). These lipids are degraded by

Figure 17.1

Synthesis and degradation of fatty acids. The master diagram (fig. 12.5) indicates a double arrow connecting acetyl-CoA and fatty acids. This is a gross simplification. Both synthesis and degradation of fatty acids occur by multistep pathways involving totally different enzymes located in different parts of the cell. Both pathways are similar in that two carbon atoms are either added (in synthesis) or deleted (in degradation) in each round of a repetitious, cyclical process.

The acetyl-CoA used as substrate for fatty acids in the cytosol is synthesized in the cytosol from citrate, which originates from the mitochondrion (see fig. 14.16 or 17.28). This acetyl-CoA condenses with CO_2 to form malonyl-CoA, which eliminates a CO_2 after an initial condensation reaction. After three more steps a two-carbon unit is added to a growing fatty acid chain. This cycle repeats itself many times (most commonly six) until a fatty acid chain with many carbon atoms (most commonly 16 carbons) has been synthesized. Variations of this pathway give rise to unsaturated and polyunsaturated fatty acids.

The degradation of fatty acids also involves a repetitious process yielding a two-carbon unit (acetyl-CoA) in every cycle. This acetyl-CoA can be fed into the TCA cycle or, alternatively, it can be used to make ketone bodies.

Elaborate controls prevent synthesis and degradation of fatty acids from occurring simultaneously. The synthesis and degradation of ketone bodies occurs in the mitochondria; frequently ketone bodies formed in liver mitochondria are catabolized in the mitochondria of other tissues, where they serve as a source of energy.

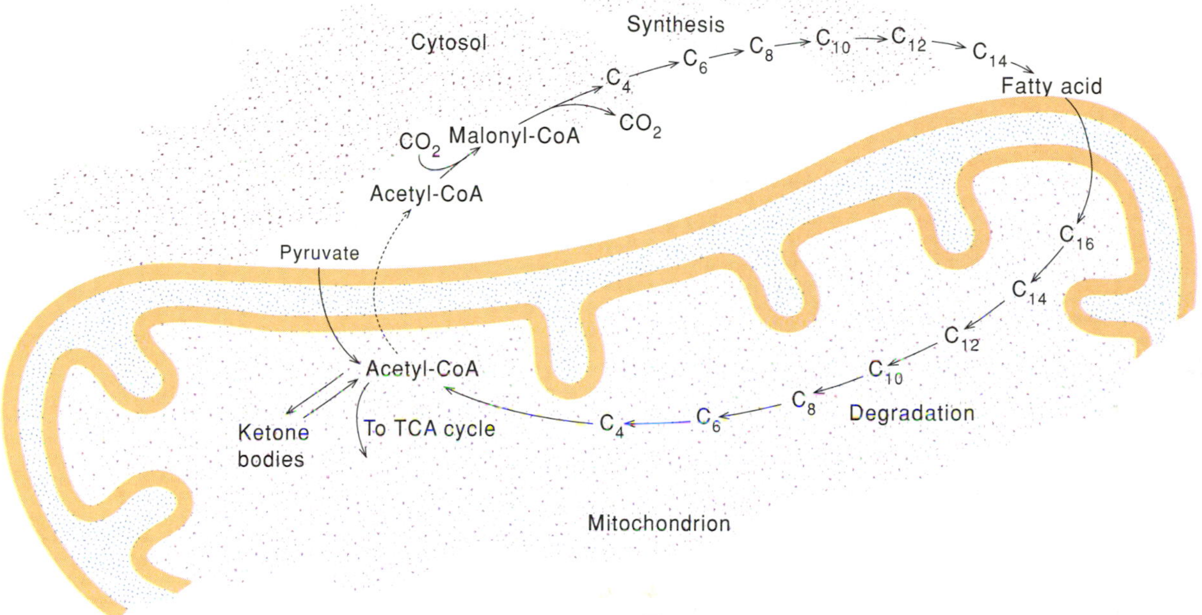

lipases and phospholipases, which are secreted into the intestinal lumen. The fatty acids released by this process are absorbed in the intestine, and the triacylglycerols are resynthesized in cells of the intestinal wall (fig. 17.2). Here the triacylglycerols are packaged into spherical lipoproteins known as chylomicrons, which are secreted by the intestine into lymphatic vessels. Once in the circulatory system the triacylglycerol components of the chylomicrons are degraded to fatty acids and glycerols by lipases that are attached to cells of various tissues that line the capillary vessels (fig. 17.3). The released fatty acids are transported into these tissues for generation of energy, synthesis of structural phospholipids, or storage of energy as triacylglycerols.

If the diet does not provide sufficient lipid to satisfy immediate needs, the lipids stored in the adipocytes may be called upon (see fig. 7.3). If this source is inadequate, then lipids may be supplied directly as they are synthesized in most cells from carbohydrates or to some extent from amino acids.

Fatty Acid Breakdown Occurs in Blocks of Two Carbon Atoms

Investigations into the mechanism of fatty acid breakdown began around the turn of the century. In 1905, Fritz Knoop reported a series of experiments indicating that fatty acids were oxidized by removal of two carbons at a time. He fed rabbits fatty acids in which the methyl group had been replaced with a phenyl group. If the altered fatty acid contained an even number of carbons (for example, $C_6H_5—CH_2(CH_2)_2COOH$), the primary metabolite was phenylacetic acid ($C_6H_5—CH_2COOH$), which was excreted as a glycine conjugate, phenylaceturic acid ($C_6H_5—CH_2CO—NHCH_2COOH$). When he fed the rabbits a phenyl derivative of a fatty acid with an odd number of carbons (for example, $C_6H_5—CH_2(CH_2)_3COOH$), he found benzoic acid (C_6H_5COOH), which was excreted in the urine as its glycine conjugate, hippuric acid ($C_6H_5CO—NH—CH_2COOH$). These results led Knoop to propose that fatty acids are oxidized at the β carbon (hence β oxidation) and degraded to acetic acid and a fatty acid with two fewer carbons:

$$RCH_2CH_2COOH \rightarrow R\overset{O}{\overset{\|}{C}}CH_2COOH \rightarrow R\overset{O}{\overset{\|}{C}}OH + CH_3COOH$$

Figure 17.2

Triacylglycerols are digested in the intestine by pancreatic lipase. The fatty acids and monoacylglycerol are absorbed into the intestinal wall. The cells in the wall, enterocytes, resynthesize triacylglycerol and package the lipid with proteins to give lipoprotein particles called chylomicrons, which are secreted from the enterocytes into the lymph.

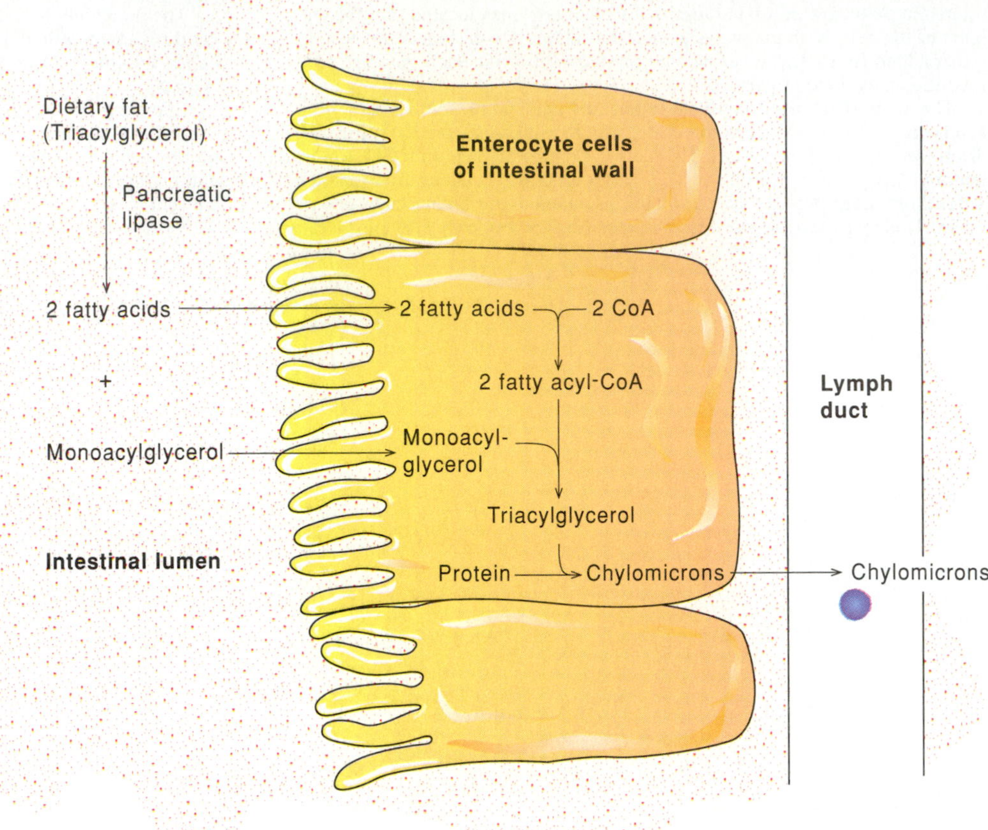

Figure 17.3

The chylomicrons are delivered to other tissues in the body via the bloodstream. In the tissues the chylomicrons bind to the endothelial cells of the capillaries and are degraded to fatty acids and glycerol by lipoprotein lipase. This enzyme is secreted by cells in the tissue and is bound to the endothelial cells.

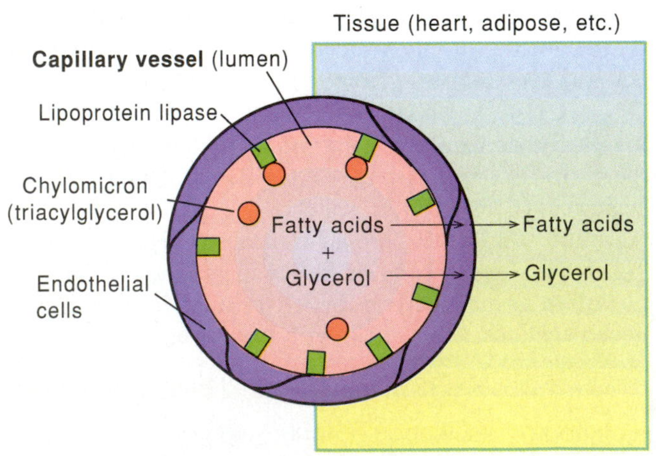

The next major experimental step was the demonstration in 1943 by J. M. Munoz and Luis Leloir that fatty acids could be oxidized in a cell-free system. This was followed by Albert Lehninger's demonstration that fatty acid oxidation occurred in liver mitochondria and, apparently, involved an "active acetate." Experiments by Fritz Lipmann proved that coenzyme A was involved in the formation of "active acetate":

$$\text{Acetate} + \text{ATP} + \text{CoA} \rightarrow \text{"active acetate"}$$

Subsequently, in 1951, Feodor Lynen, working with yeast, demonstrated that "active acetate" was acetyl-CoA. At this stage, several laboratories conceived of the idea that CoA might play a role in the activation of fatty acids for β oxidation, and by 1954 the basic outline of β oxidation as we know it today was developed.

The Oxidation of Saturated Fatty Acids Occurs in Mitochondria

The outline for β oxidation of a saturated fatty acid is shown in figure 17.4. In the first reaction, the acyl-CoA is dehydrogenated to yield the α-β (or 2–3)-trans-enoyl-CoA and $FADH_2$. This enoyl-CoA is subsequently hydrated stereospecifically to yield the 3-L-hydroxyacyl-CoA. The hydroxyl group is oxidized

Figure 17.4

(*a*) The β oxidation of fatty acids consists of four reactions in the matrix of the mitochondrion. Each cycle of reactions results in the formation of acetyl-CoA and an acyl-CoA with two fewer carbons. (*b*) The cyclic nature of the β oxidation cycle. In some illustrations the —SH group of CoA is indicated for emphasis. Likewise in some illustrations the —S— group in an acyl-CoA structure is indicated.

$RCH_2CH_2-C(=O)-CoA$
Acyl-CoA

Acyl-CoA dehydrogenase ($FAD \rightarrow FADH_2$)

$RC(H)=C(H)-C(=O)-CoA$
Enoyl-CoA

Enoyl-CoA hydrase (H_2O)

$RC(OH)(H)-CH_2-C(=O)-CoA$
3-Hydroxyacyl-CoA

3-L-Hydroxyacyl-CoA dehydrogenase ($NAD^+ \rightarrow NADH + H^+$)

$RC(=O)-CH_2-C(=O)-CoA$
β-Ketoacyl-CoA

Thiolase (CoA)

$RC(=O)-CoA + CH_3-C(=O)-CoA$
Acyl-CoA **Acetyl-CoA**

(a)

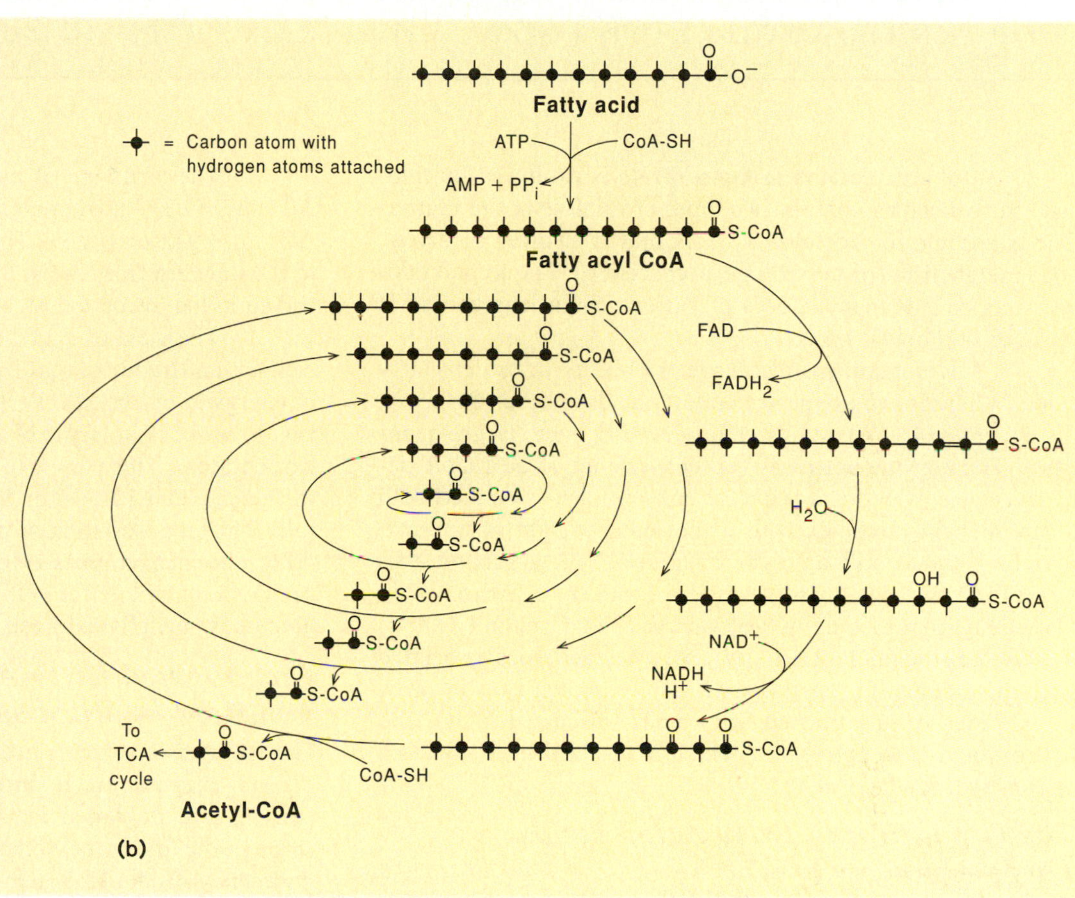

by NAD^+ and a dehydrogenase to yield β-ketoacyl-CoA and NADH. The final step in the sequence involves a thiolytic cleavage to form acetyl-CoA and an acyl-CoA that is two carbons shorter than the initial substrate for β oxidation. This acyl-CoA can undergo another round of β oxidation to yield $FADH_2$, NADH, acetyl-CoA, and acyl-CoA. The enzymatic steps are repeated until, in the last sequence of reactions, butyryl-CoA ($CH_3CH_2CH_2CO-CoA$) is degraded to two acetyl-CoA's.

The first reaction in β oxidation (fig. 17.4*a*) is catalyzed by acyl-CoA dehydrogenases, which have FAD tightly, but noncovalently, bound. There are three different soluble matrix enzymes with similar molecular weights (180,000) that show short-, medium-, or long-chain acyl-CoA specificity. In each case the products are the *trans* α, β-enoyl-CoA and enzyme-bound $FADH_2$. The electrons from $FADH_2$ are channeled into the electron-transport chain via ubiquinone (see chapter 15).

The second reaction (fig. 17.4*a*) is catalyzed by enoyl-CoA hydrase. Two different mitochondrial enzymes have hydrase activity. One is crotonase, a soluble matrix enzyme (M_r = 165,000) that prefers short-chain acyl-CoA's. The other prefers long-chain acyl-CoA's and is probably membrane-associated. 3-L-Hydroxyacyl-CoA dehydrogenase is a soluble matrix enzyme that catalyzes the oxidation of the β-hydroxyl group with NAD^+ as a specific cofactor (fig. 17.4*a*).

Table 17.1
Equations for the Complete Oxidation of Palmitoyl-CoA to CO_2 and H_2O

$CH_3(CH_2)_{14}CO—CoA + 7\ FAD + 7\ H_2O + 7\ NAD^+ + 7\ CoA \rightarrow 8\ CH_3COSCoA + 7\ FADH_2 + 7\ NADH + 7\ H^+$	(1)
$14\ H^+ + 7\ FADH_2 + 14\ P_i + 14\ ADP + 3\frac{1}{2}\ O_2 \rightarrow 7\ FAD + 21\ H_2O + 14\ ATP$	(2)
$7\ NADH + 28\ H^+ + 21\ P_i + 21\ ADP + 3\frac{1}{2}\ O_2 \rightarrow 7\ NAD^+ + 28\ H_2O + 21\ ATP$	(3)
$96\ H^+ + 8\ Acetyl\text{-}CoA + 16\ O_2 + 96\ P_i + 96\ ADP \rightarrow 8\ CoA + 104\ H_2O + 16\ CO_2 + 96\ ATP$	(4)
$131\ H^+ + CH_3(CH_2)_{14}CO—CoA + 131\ P_i + 131\ ADP + 23\ O_2 \rightarrow 131\ ATP + 16\ CO_2 + 146\ H_2O + CoA$	(5)

The last reaction in figure 17.4*a* is catalyzed by thiolase. Mitochondria contain two types of thiolase in the matrix. One is specific for acetoacetyl-CoA and is involved in ketone body metabolism (discussed in a later section). The second thiolase acts on 3-ketoacyl-CoA's of various chain lengths and is involved in β oxidation.

Microorganisms also have the ability to oxidize fatty acids. When *E. coli* are grown on fatty acids instead of glucose, the enzymes of β oxidation are produced in much larger quantities. The enzyme activities (except for that of acyl-CoA dehydrogenase, which is a separate enzyme) are associated with a multienzyme complex with a molecular weight of approximately 250,000 and an $\alpha_2\beta_2$ structure. The α subunit (M_r = 78,000) has the enoyl-CoA hydrase and the 3-hydroxyacyl-CoA dehydrogenase activities, whereas the β subunit (M_r = 42,000) has the thiolase activity. Two other activities associated with the α subunit are enoyl-CoA isomerase, which is involved in oxidation of unsaturated fatty acids (discussed later), and hydroxyacyl-CoA epimerase, which is involved in β oxidation of D-β-hydroxy fatty acids.

The Oxidation of Fatty Acids Yields Large Amounts of ATP

The equations for the complete oxidation of palmitoyl-CoA are shown in table 17.1. Equation (1) in the table shows the oxidation of palmitoyl-CoA by the enzymes of β oxidation. Each of the products of equation (1) is further oxidized by the respiratory chain, equations (2) and (3), or by the tricarboxylic acid cycle and the respiratory chain, equation (4). When the reactions of equations (1) and (4) are summed, the result is equation (5). Hence one molecule of palmitoyl-CoA can be oxidized to yield 131 ATP + 16 CO_2 + 146 H_2O and CoA. If the starting material were palmitic acid, two fewer ATPs and one fewer H_2O would be produced. This is because the formation of the CoA derivative of the fatty acid requires two ATP equivalents and one molecule of H_2O. Water seems almost incidental to this process, but it is interesting to note that the water supplied by oxidation of fatty acids is used as a major source of H_2O by the killer whale. This animal lives in the sea but cannot drink the sea water.

The oxidation of palmitate to CO_2 and H_2O yields a $\Delta G°'$ of −2,340 kcal/mole. The formation of 129 ATPs has a $\Delta G°'$ of +942 kcal/mole. Thus approximately 40% (942/2,340) of the standard free energy for the oxidation of palmitate is conserved in the formation of ATP.

The yield of ATP from the oxidation of palmitic acid can be compared with that from glucose. Both are major sources of energy in our bodies. Perhaps the most meaningful comparison to make is in terms of the number of ATPs produced per carbon atom. This comes to 36/6 or 6 ATPs per carbon atom for glucose and 129/16 or 8 ATPs per carbon atom for palmitate. Thus the oxidation of palmitate yields substantially more ATP. The chemical reason for this difference is that most of the carbons in palmitate are in the completely reduced state, whereas glucose is partially oxidized, with six oxygens in the molecule.

Unsaturated Fatty Acids Are Also Oxidized in Mitochondria

Unsaturated fatty acids are degraded by β oxidation, but additional enzymes are required. As seen in figure 17.5, oleoyl-CoA can be degraded in the same manner as stearoyl-CoA through the first three cycles of β oxidation. The result, Δ^3-*cis*-dodecenoyl-CoA (12:1 Δ^3), however, is not a substrate for the acyl-CoA dehydrogenase. This step is bypassed by an isomerization of the double bond by enoyl-CoA isomerase to the Δ^2-*trans*-dodecenoyl-CoA, which is a normal substrate for enoyl-CoA hydrase, and the normal route for β oxidation resumes. Since the dehydrogenase is not used in one round of β oxidation of oleoyl-CoA, one fewer $FADH_2$ and two fewer ATPs are made than in the β oxidation of stearoyl-CoA (18:0). Enoyl-CoA isomerase has been purified from liver mitochondria (M_r = 90,000) and is active with Δ^3-*cis* or Δ^3-*trans* acyl-CoAs that contain from 6 to 16 carbons.

Polyunsaturated fatty acids also are degraded by β oxidation, but the process requires enoyl-CoA isomerase and an additional enzyme, 2,4-dienoyl-CoA reductase (fig. 17.6). For example, the degradation of linoleoyl-CoA begins, like that of oleoyl-CoA, with three rounds of β oxidation and results in a Δ^3-*cis* unsaturated fatty acid that is not a substrate for the acyl-CoA dehydrogenase. Isomerization of the double bond to the

Figure 17.5

Unsaturated fatty acids such as oleoyl-CoA can also be degraded to acetyl-CoA in mitochondria. Three cycles of ß oxidation result in an acyl-CoA (Δ^3-*cis*-dodecenoyl-CoA) that is not a substrate for acyl-CoA dehydrogenase. This problem is circumvented by isomerization of the double bond to the Δ^2-*trans* position by enoyl-CoA isomerase. Complete oxidation of the remainder of the molecule by the enzymes of ß oxidation is now possible.

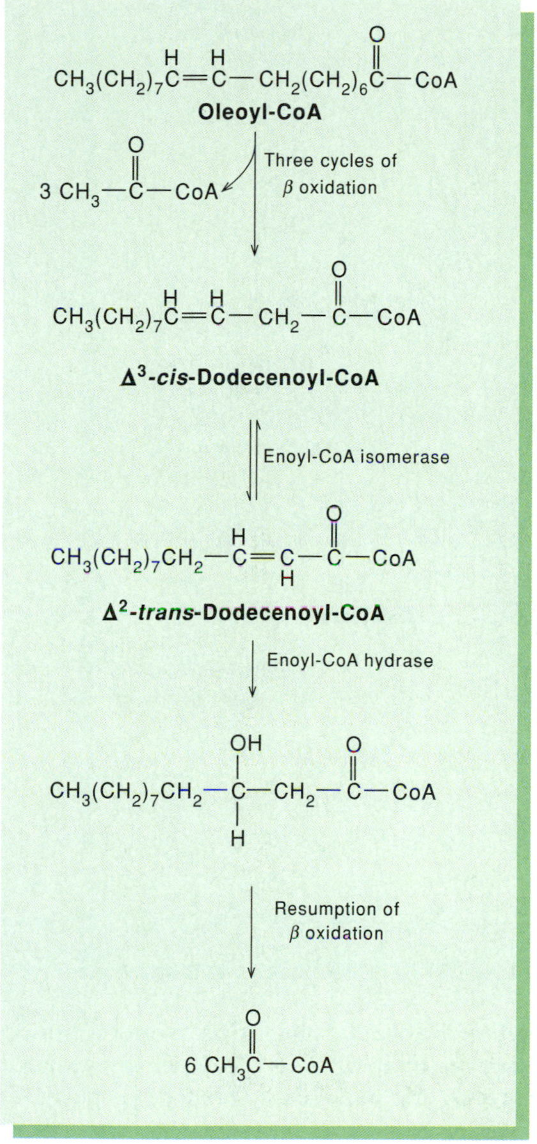

$CH_3(CH_2)_7CH{=}CH{-}CH_2(CH_2)_6C({=}O){-}CoA$

Oleoyl-CoA

↓ Three cycles of β oxidation (→ $3\ CH_3{-}C({=}O){-}CoA$)

$CH_3(CH_2)_7CH{=}CH{-}CH_2{-}C({=}O){-}CoA$

Δ^3-*cis*-Dodecenoyl-CoA

⇅ Enoyl-CoA isomerase

$CH_3(CH_2)_7CH_2{-}CH{=}CH{-}C({=}O){-}CoA$

Δ^2-*trans*-Dodecenoyl-CoA

↓ Enoyl-CoA hydrase

$CH_3(CH_2)_7CH_2{-}CH(OH){-}CH_2{-}C({=}O){-}CoA$

↓ Resumption of β oxidation

$6\ CH_3C({=}O){-}CoA$

Figure 17.6

Polyunsaturated fatty acids such as linoleoyl-CoA are also oxidized in the mitochondria. In addition to the enzymes of ß oxidation and enoyl-CoA isomerase, the process requires 2,4-dienoyl-CoA reductase.

$CH_3(CH_2)_4CH{=}CH{-}CH_2{-}CH{=}CH{-}CH_2(CH_2)_6C({=}O){-}CoA$

Δ^9-*cis*, Δ^{12}-*cis*

↓ Three cycles of β oxidation

$CH_3(CH_2)_4CH{=}CH{-}CH_2{-}CH{=}CH{-}CH_2{-}C({=}O){-}CoA + 3\ CH_3{-}C({=}O){-}CoA$

Δ^3-*cis*, Δ^6-*cis*

↓ Enoyl-CoA isomerase

$CH_3(CH_2)_4CH{=}CH{-}CH_2{-}CH_2{-}CH{=}CH{-}C({=}O){-}CoA$

Δ^2-*trans*, Δ^6-*cis*

↓ One cycle of β oxidation

$CH_3(CH_2)_4CH{=}CH{-}CH_2{-}CH_2{-}C({=}O){-}CoA + CH_3{-}C({=}O){-}CoA$

Δ^4-*cis*

↓ Acyl-CoA dehydrogenase

$CH_3(CH_2)_4CH{=}CH{-}CH{=}CH{-}C({=}O){-}CoA$

Δ^2-*trans*, Δ^4-*cis*

↓ 2,4-Dienoyl-CoA reductase ($NADPH + H^+$ → $NADP^+$)

$CH_3(CH_2)_4CH_2{-}CH{=}CH{-}CH_2{-}C({=}O){-}CoA$

Δ^3-*trans*

↓ Enoyl-CoA isomerase

$CH_3(CH_2)_4CH_2{-}CH_2{-}CH{=}CH{-}C({=}O){-}CoA$

Δ^2-*trans*

↓ Four cycles of β oxidation

$5\ CH_3{-}C({=}O){-}CoA$

Acetyl-CoA

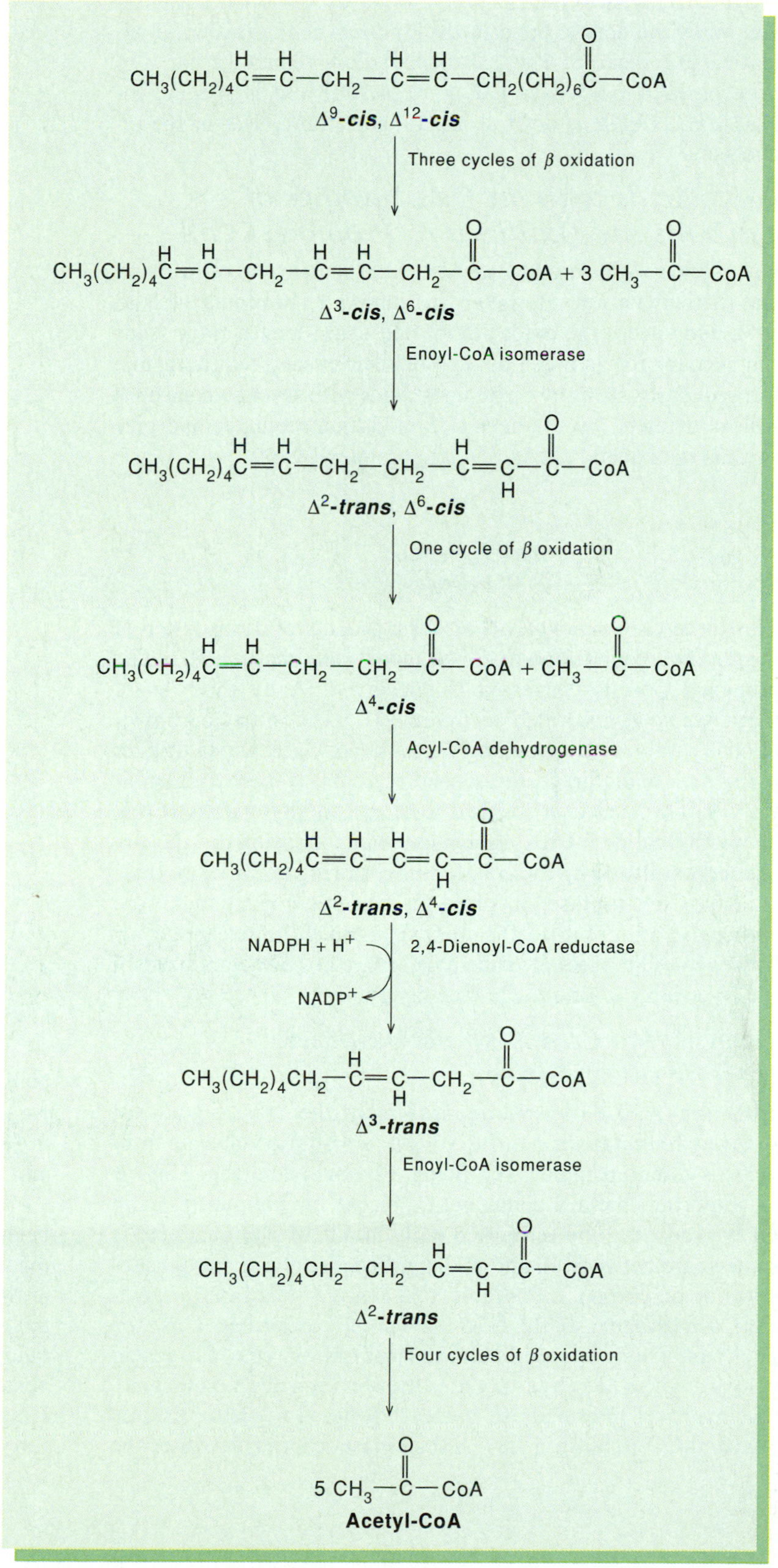

Δ^2-*trans* position by enoyl-CoA isomerase allows the resumption of one cycle of β oxidation and generation of a Δ^4-*cis*-enoyl-CoA. Subsequently, acyl-CoA dehydrogenase produces Δ^2-*trans*, Δ^4-*cis*-dienoyl-CoA (see fig. 17.6), which is converted by an NADPH-dependent 2,4-dienoyl-CoA reductase to Δ^3-*trans* enoyl-CoA, then to the Δ^2-*trans* isomer by enoyl-CoA isomerase. With the double bond in the Δ^2-*trans* configuration, β oxidation can resume. The 2,4-dienoyl-CoA reductase has also been purified from liver (M_r = 124,000) and is specific for NADPH. NADH is not a substrate nor an inhibitor of the reductase.

Fatty Acids with an Odd Number of Carbons Are Oxidized to Propionyl-CoA

Fatty acids with an odd number of carbons are rare in many mammalian tissues. However, in ruminant mammals such as cows and sheep, the oxidation of odd-chain-length fatty acids can account for as much as 25% of their energy requirements. Consequently, straight-chain fatty acids with seventeen carbons will be oxidized by the normal β oxidation sequence and give rise to seven acetyl-CoAs and one propionyl-CoA:

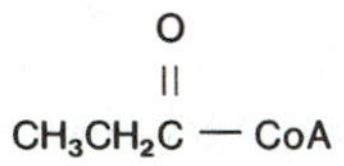

$$CH_3CH_2\overset{\overset{\displaystyle O}{\|}}{C}-CoA$$

Propionyl-CoA

This three-carbon acyl-CoA also is a product of degradation of the amino acids valine and isoleucine (see chapter 19). The propionyl-CoA is converted to succinyl-CoA by three enzymatic steps, as indicated in figure 17.7. The initial carboxylation is catalyzed by propionyl-CoA carboxylase, which utilizes biotin as a cofactor. In the second reaction, D-methylmalonyl-CoA is converted to its optical isomer, L-methylmalonyl-CoA, by methylmalonyl-CoA racemase. The last step in the sequence, catalyzed by the cobalamin-requiring enzyme, methylmalonyl-CoA mutase, involves an unusual migration of the carbonyl-CoA group to the methyl group in an exchange for hydrogen. The product, succinyl-CoA, can be metabolized in the tricarboxylic acid cycle (see fig. 14.4).

Figure 17.7

Propionyl-CoA generated by the ß oxidation of odd-chain fatty acids is converted to succinyl-CoA, an intermediate of the tricarboxylic acid cycle.

$$CH_3CH_2\overset{\overset{\displaystyle O}{\|}}{C}-CoA + ATP + CO_2 + H_2O$$

Propionyl-CoA

↓ Propionyl-CoA carboxylase

$$H-\overset{\overset{\displaystyle COO^-}{|}}{\underset{\underset{\underset{\displaystyle O}{\|}}{\displaystyle C-CoA}}{\underset{|}{C}}}-CH_3$$

D-Methylmalonyl-CoA

⇅ Methylmalonyl-CoA racemase

$$H_3C-\overset{\overset{\displaystyle COO^-}{|}}{\underset{\underset{\underset{\displaystyle O}{\|}}{\displaystyle C-CoA}}{\underset{|}{C}}}-H$$

L-Methylmalonyl-CoA

↓ Methylmalonyl-CoA mutase

$$H_2\underset{\underset{\underset{\underset{\displaystyle O}{\|}}{\displaystyle C-CoA}}{|}}{C}-\overset{\overset{\displaystyle COO^-}{|}}{CH_2}$$

Succinyl-CoA

Fatty Acids Can Also Be Oxidized by α or ω Oxidation

Although β oxidation is quantitatively the most significant pathway for catabolism of fatty acids, α oxidation of some fatty acids is essential to our well-being. Phytanic acid (fig. 17.8) is an important dietary component, present in ruminant fat and dairy products. The estimated daily intake of phytanic acid is somewhere between 50 and 100 mg. Because of the methyl substitution on carbon 3, phytanic acid is not a substrate for acyl-CoA dehydrogenase, the first enzyme in β oxidation. This step is circumvented by another mitochondrial enzyme, fatty acid α-hydroxylase, that hydroxylates the α carbon of phytanic acid. The hydroxyl intermediate is decarboxylated to yield pristanic acid and CO_2 (see fig. 17.8). A thiokinase reaction activates the acid to its CoA derivative. Pristanyl-CoA is unsubstituted at carbon 3 and can be oxidized by acyl-CoA dehydrogenase and the normal enzymes of β oxidation to produce propionyl-CoA and an acyl-CoA.

A minor pathway for the oxidation of fatty acids has been observed in rat liver microsomes. This pathway involves oxidation of the terminal methyl, called the ω (omega) carbon, or the adjacent methylene carbon of fatty acids by NADPH and molecular oxygen (fig. 17.9). The pathway is probably not quantitatively significant for the oxidation of long-chain fatty acids, and the enzymes involved have not been characterized. However, ω oxidation may be important for the metabolism of fatty acids with 6 to 10 carbons.

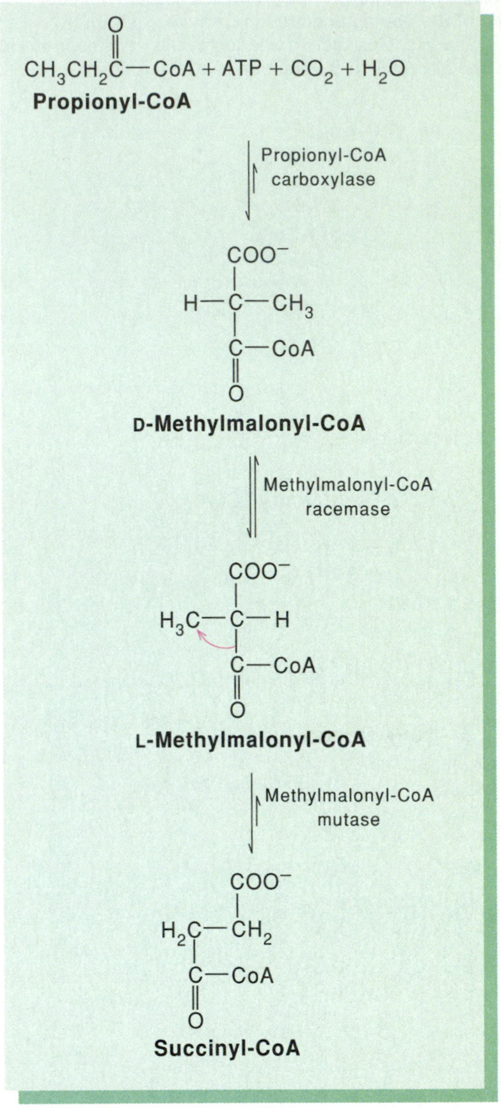

Figure 17.8

The oxidation of phytanic acid involves oxidation of the α carbon and decarboxylation. The product, pristanic acid, can be degraded by β oxidation after conversion to the CoA derivative.

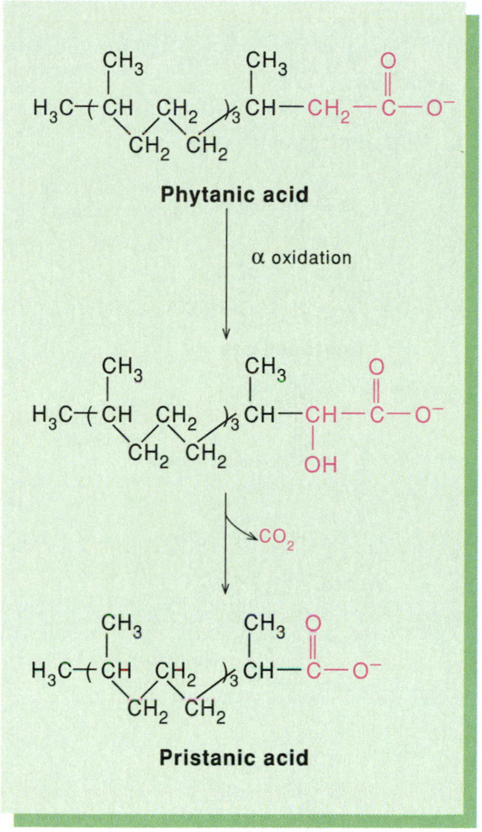

Figure 17.9

The oxidation of the methyl group (ω oxidation) of certain fatty acids is known to occur. However, the enzymes involved have not been well described.

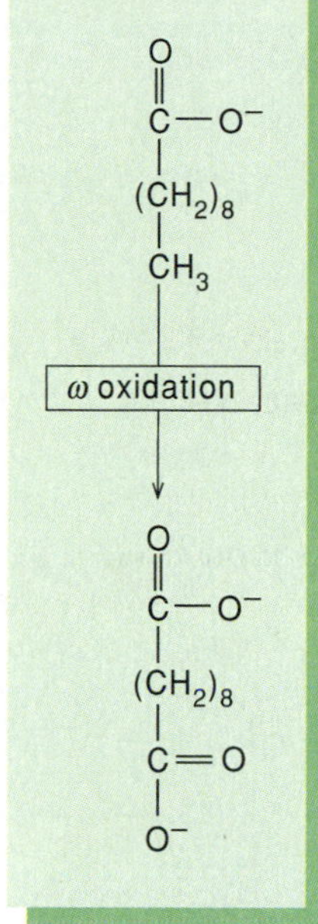

Ketone Bodies Formed in the Liver Are Used for Energy in Other Tissues

Once fatty acids are degraded in the mitochondria of liver, the acetyl-CoA can undergo a number of metabolic fates. Of central importance, as we learned in chapter 14, is utilization of acetyl-CoA by the tricarboxylic acid cycle. An alternative fate is the synthesis of ketone bodies (acetoacetate, β-hydroxybutyrate, and acetone), which takes place only in the mitochondrial matrix, as depicted in figure 17.10. In the first reaction, catalyzed by acetoacetyl-CoA thiolase, two acetyl-CoAs condense to form acetoacetyl-CoA. This reaction is thermodynamically unfavorable in that the equilibrium favors thiolytic cleavage of acetoacetyl-CoA. Hence this compound is formed when the levels of acetyl-CoA rise. A third molecule of acetyl-CoA reacts with acetoacetyl-CoA to yield β-hydroxy-β-methylglutaryl-CoA (HMG-CoA) in a reaction catalyzed by HMG-CoA synthase. As we will see in chapter 23, the same two reactions are the first steps in cholesterol biosynthesis, which takes place in the cytosol. In the formation of ketone bodies, the next reaction is catalyzed by HMG-CoA lyase and yields acetoacetate and acetyl-CoA. The acetoacetate can be reduced to β-hydroxybutyrate by an enzyme on the inner membrane of the mitochondrion, β-hydroxybutyrate dehydrogenase. Although the acetoacetate also can be decarboxylated to form acetone, this is normally of minor importance. However, patients with uncontrolled diabetes have high levels of ketone bodies in their plasma, and their breath smells of acetone.

Ketone body synthesis is primarily a liver function, since HMG-CoA synthase is present in large quantities only in this tissue. Acetoacetate and β-hydroxybutyrate diffuse into the blood, where they are carried to other tissues and converted into acetyl-CoA, as described in figure 17.11. The reactions catalyzed by β-hydroxybutyrate dehydrogenase and thiolase are common to both the synthesis and degradation of the ketone bodies. However, the second enzyme in the sequence (see fig. 17.11), β-oxoacid-CoA-transferase, is present in all tissues but liver. Hence the ketone bodies are largely made in the liver and metabolized to CO_2 and energy in nonhepatic tissues. Ketone bodies supply acetyl-CoA for fatty acid and cholesterol biosynthesis and are an important source of energy for the brain during prolonged starvation.

In untreated type I diabetes, there is a reduced supply of glucose to most tissues, and fatty acid levels are greatly elevated. The result is an increase in acetyl-CoA and an excess of ketone bodies. A large rise in the serum levels of ketone bodies lowers the pH because of the increased concentration of these

Figure 17.10

The biosynthesis of ketone bodies (acetoacetate, hydroxybutyrate, and acetone) occurs in the mitochondria of liver.

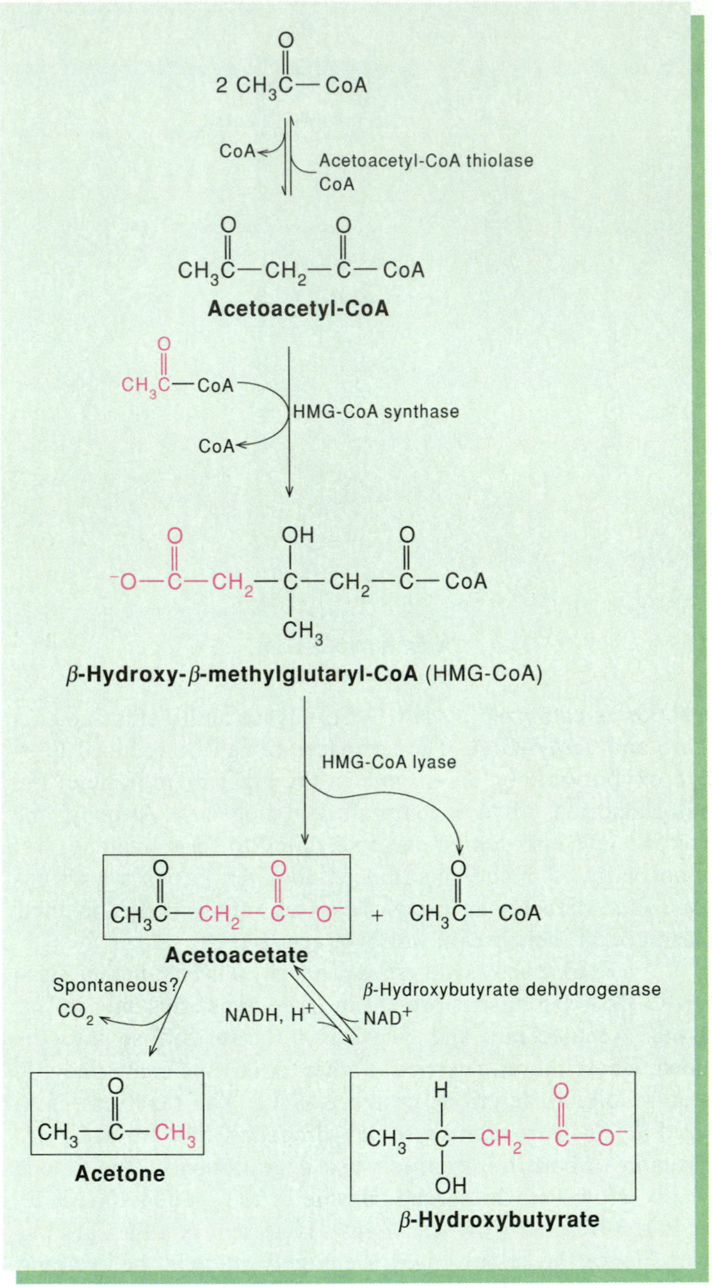

Figure 17.11

Ketone bodies are converted back to acetyl-CoA in the mitochondria of nonhepatic tissues and used as a source of energy in these tissues.

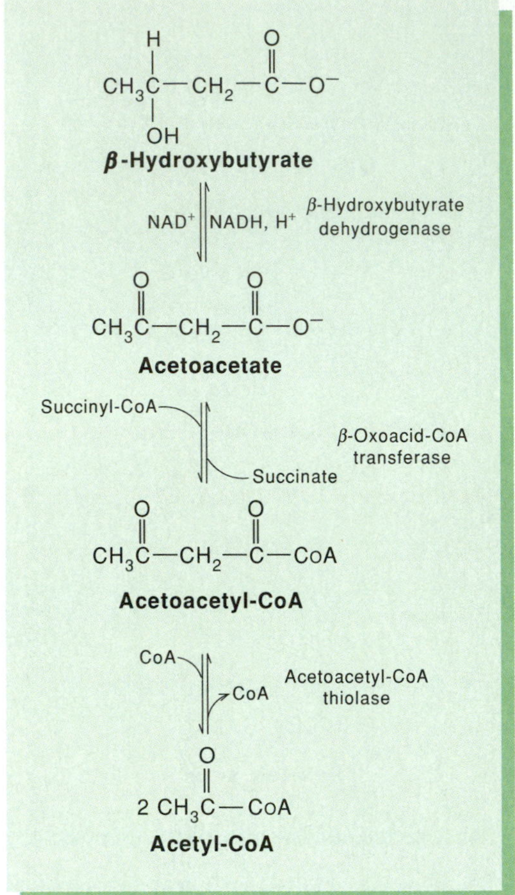

acids. This condition can be hazardous because of the effect of lower pH on the oxygen binding to hemoglobin and for other reasons.

β Oxidation Also Occurs in Peroxisomes

Peroxisomes are subcellular organelles whose main function is to process substances for elimination (see chapter 5). In 1976, Paul Lazarow and Christian de Duve showed that peroxisomes from rat liver contained a β-oxidation system. The reactions of β oxidation in peroxisomes are similar to those in mitochondria (see fig. 17.4), with the notable exception of the initial dehydrogenation. This reaction is catalyzed by the FAD-containing enzyme acyl-CoA oxidase:

$$\text{(1)}\quad E—FAD + RCH_2CH_2\overset{\overset{\Large O}{\|}}{C}—CoA \rightarrow RC\overset{H}{=}\underset{H}{C}—\overset{\overset{\Large O}{\|}}{C}—CoA + E—FADH_2$$

$$\text{(2)}\quad E—FADH_2 + O_2 \rightarrow H_2O_2 + E—FAD$$

Oxidized flavin is regenerated by reaction with oxygen, which is reduced to H_2O_2, and catalase reduces the H_2O_2 to H_2O. Since the electrons from the β oxidation are not shuttled into the respiratory chain, the peroxisomal pathway is not involved in energy production. Acyl-CoA oxidase is inactive with low-molecular-weight acyl-CoAs such as hexanoyl-CoA or butyryl-CoA. It is suspected that these acyl-CoAs are further catabolized in the mitochondria.

In view of all these limitations you may well wonder what function β oxidation serves in the peroxisomes. It appears that peroxisomes are uniquely involved in the oxidation of very-long-chain fatty acids such as 24:0 and rare fatty acids found in the brain (C_{26}–C_{38} polyenoic fatty acids). The importance of this fact is underscored by the occurrence of rare metabolic disorders where there is an accumulation of very-long-chain fatty acids. In all such cases these disorders have been correlated with defects in the peroxisomal oxidation system.

Summary of Fatty Acid Degradation

In this section we have seen that fatty acids are oxidized in units of two carbon atoms. The immediate end products of this oxidation are reduced coenzymes of FAD and NAD^+, which supply energy, through the respiratory chain and acetyl-CoA, that can be used in various ways. For lipids that do not have a straight-chain saturated structure, there are auxiliary enzymes to keep the oxidation process flowing smoothly. Finally, for fatty acids that are apparently too big to enter the mitochondria there is an auxiliary oxidation system in the peroxisomes, whose main purpose is to insure against an accumulation of these giant fatty acids.

Biosynthesis of Saturated Fatty Acids

Research on the mechanism of fatty acid synthesis goes back to the very early years of biochemical investigations. The finding that most fatty acids contain an even number of carbon atoms, between four and twenty, led Rapier in 1907 to postulate that they were produced by condensation of an activated two-carbon compound. After isotopes became fundamental tools for the biochemist in the late 1930s and early 1940s, experiments were performed that clearly implicated an acetate derivative as the two-carbon compound. In a series of experiments done in 1944 and 1945, David Rittenberg and Konrad Bloch fed mice acetate that had been labeled with both deuterium and ^{13}C ($C^2H_3{}^{13}COOH$) and found that both isotopes were incorporated into fatty acids. Subsequently, Feodor Lynen discovered a central role for acetyl-CoA in fatty acid biosynthesis. Precisely how acetyl-CoA was converted into fatty acids eluded workers until the late 1950s, when the involvement of malonyl-CoA was discovered. From this point on, progress was rapid and the scheme for fatty acid biosynthesis as we know it today was elucidated.

The synthesis of saturated fatty acids is very similar in most organisms. In eukaryotes, synthesis occurs in the cytosol. The overall reaction for the formation of the commonly synthesized 16-carbon palmitic acid is

$$\underset{\text{Acetyl-CoA}}{CH_3\overset{\overset{O}{\|}}{C}\,CoA} + 7\ \underset{\text{Malonyl-CoA}}{{}^-O\overset{\overset{O}{\|}}{C}\,CH_2\,\overset{\overset{O}{\|}}{C}\,CoA} + 14\ NADPH + 20\ H^+ \longrightarrow \underset{\text{Palmitic acid}}{CH_3(CH_2)_{14}COO^-} + 7\ CO_2 + 8\ CoA + 14\ NADP^+ + 7\ H_2O$$

Comparing this equation with the equation for the complete oxidation of palmitoyl-CoA (see table 17.1, equation 1), we find two major differences in carriers and intermediates. The principal electron carrier system in the anabolic pathway is the NADPH-$NADP^+$ system; the principal electron carriers in the catabolic pathway are FAD-$FADH_2$ and NAD^+-NADH. The second striking difference between the two pathways is that malonyl-CoA is the principal substrate in the anabolic pathway, but plays no role in the catabolic pathway. These differences reflect the fact that the two pathways do not use any enzymes in common. Indeed, in animal cells the reactions even occur in separate cell compartments; biosynthesis takes place in the cytosol, whereas degradation occurs in the mitochondria and peroxisomes.

Most Reactions for Fatty Acid Synthesis Are Organized in a Multienzyme Complex

There are eight enzyme-catalyzed reactions involved in the conversion of acetyl-CoA into fatty acid (fig. 17.12). Except for the first reaction, involving the synthesis of malonyl-CoA from acetyl-CoA, in animal cells all of the reactions of fatty acid synthesis occur on a multienzyme complex known as fatty acid synthase (fig. 17.13). This complex contains a single polypeptide chain; normally such multienzyme complexes form dimers. Each monomer also contains an acyl carrier protein (ACP), which carries all the intermediates during fatty acid synthesis. The main function of the ACP is served by a covalently linked 4′-phosphopantetheine coenzyme that is bound to the ACP by an ester linkage between the phosphate of the coenzyme and a serine hydroxyl side chain of the protein (fig. 17.14). The sulfhydryl group of the phosphopantetheine is the attachment site for the intermediates during fatty acid synthesis. Apparently, the long flexible arm of phosphopantetheine permits the attached substrate to visit successively all of the active sites during the course of the reactions. The use of a multienzyme complex, together with ACP, increases the efficiency of fatty acid synthesis. For each step in the pathway the next enzyme is always near at hand, and the loss of intermediates is minimized. Thus it is not surprising that similar multienzyme complexes are believed to exist for species ranging from bacteria to higher eukaryotes.

In *E. coli* and plants, each of the enzyme activities exists on a separate protein, which can be isolated from the others by conventional methods of purification. Table 17.2 summarizes some of the main properties of the *E. coli* enzymes. These enzymes probably exist inside the cell as a complex that disaggregates when the cells are homogenized. However, such an arrangement remains to be demonstrated.

Figure 17.12

Outline of the reactions for fatty acid synthesis. Fatty acids grow in steps of two-carbon units. To make a 16-carbon fatty acid requires many repeats of the same cycle of reactions. In animal cells the palmitate is released as the free acid. In *E. coli* the ACP derivative is used directly for the biosynthesis of phospholipids (discussed in chapter 22). Most of the reactions of fatty acid synthesis are believed to take place on a multiprotein complex.

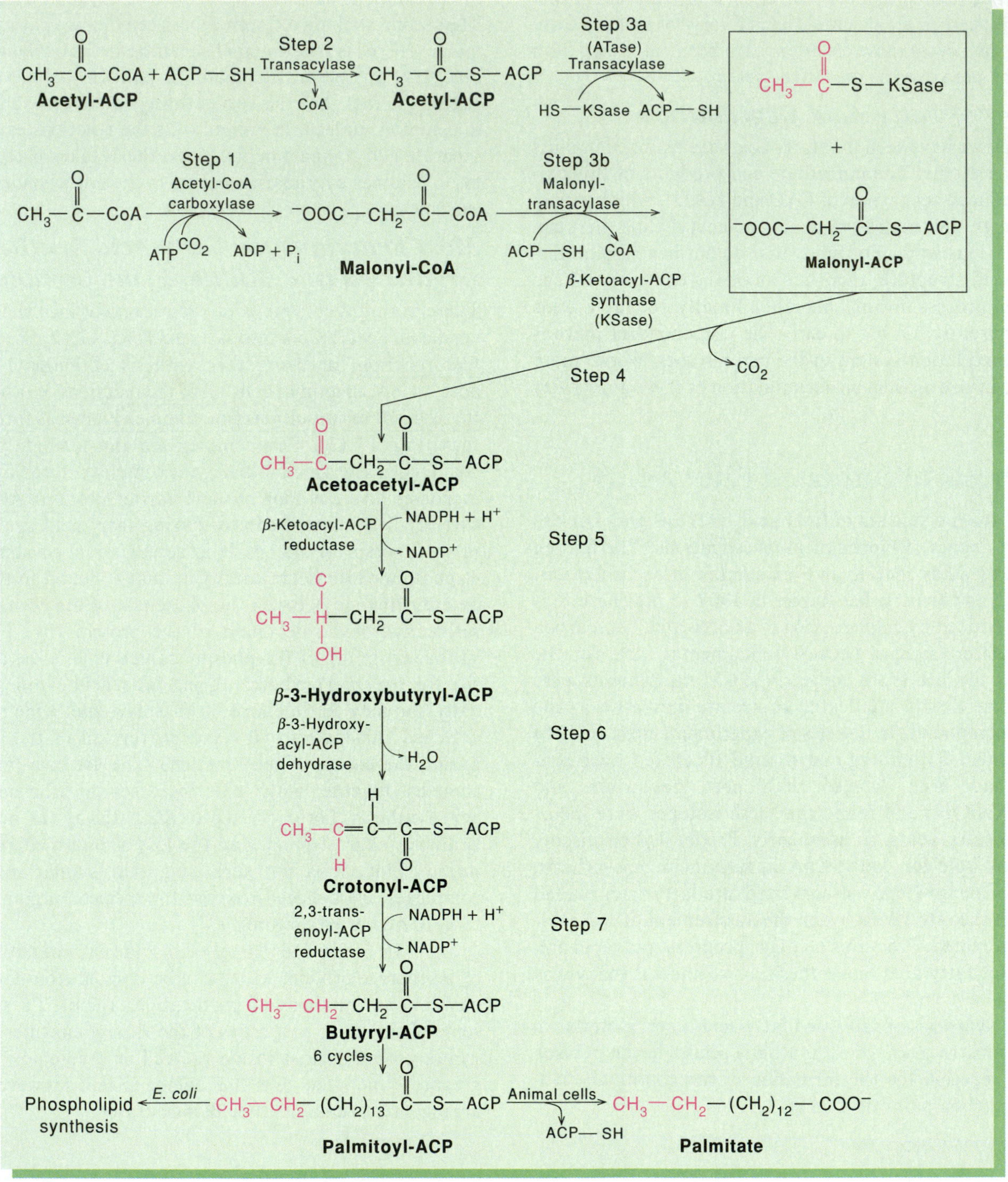

Figure 17.13

A proposed organization of the enzymatic activities of fatty acid synthase from animal liver. The multienzyme complex exists as a dimer of two giant identical polypeptides (M_r = 250,000). The two peptides are organized in a head-to-tail configuration in such a way that it is theoretically possible for the complex to make two fatty acids at the same time. The substrates, acetyl-CoA and malonyl-CoA, are transferred onto the enzyme by acetyl-CoA-ACP transacylase (ATase) and malonyl-CoA-ACP transacylase (MTase). Subsequently, ß-ketobutyryl-ACP and CO_2 are formed in a condensation reaction catalyzed by ß-ketoacyl-ACP synthase (KSase). The polypeptide with this condensing activity has a cysteine residue (Cys—SH) at the active site. The ß-ketobutyryl-ACP is reduced to the ß-hydroxy derivative by ß-ketoacyl-ACP reductase (KRase), dehydrated to enoyl-ACP by ß-hydroxyacyl-ACP dehydrase (DHase) and finally reduced to butyryl-ACP by enoyl-ACP reductase (ERase). The butyryl group is then transferred to the cysteine residue of KSase and another malonyl group is transferred to ACP by MTase. Another condensation occurs and the ß-ketohexanoyl group is converted to hexanoyl-ACP by the enzymes KRase, DHase, and ERase. This enzymatic process continues to recycle until palmitoyl-ACP is made. The palmitate is released from the complex by a thioesterase (TEase). This model is similar to the one proposed by Dr. S. Wakil and co-workers (*Journal of Biological Chemistry* 258:15312, 1983).

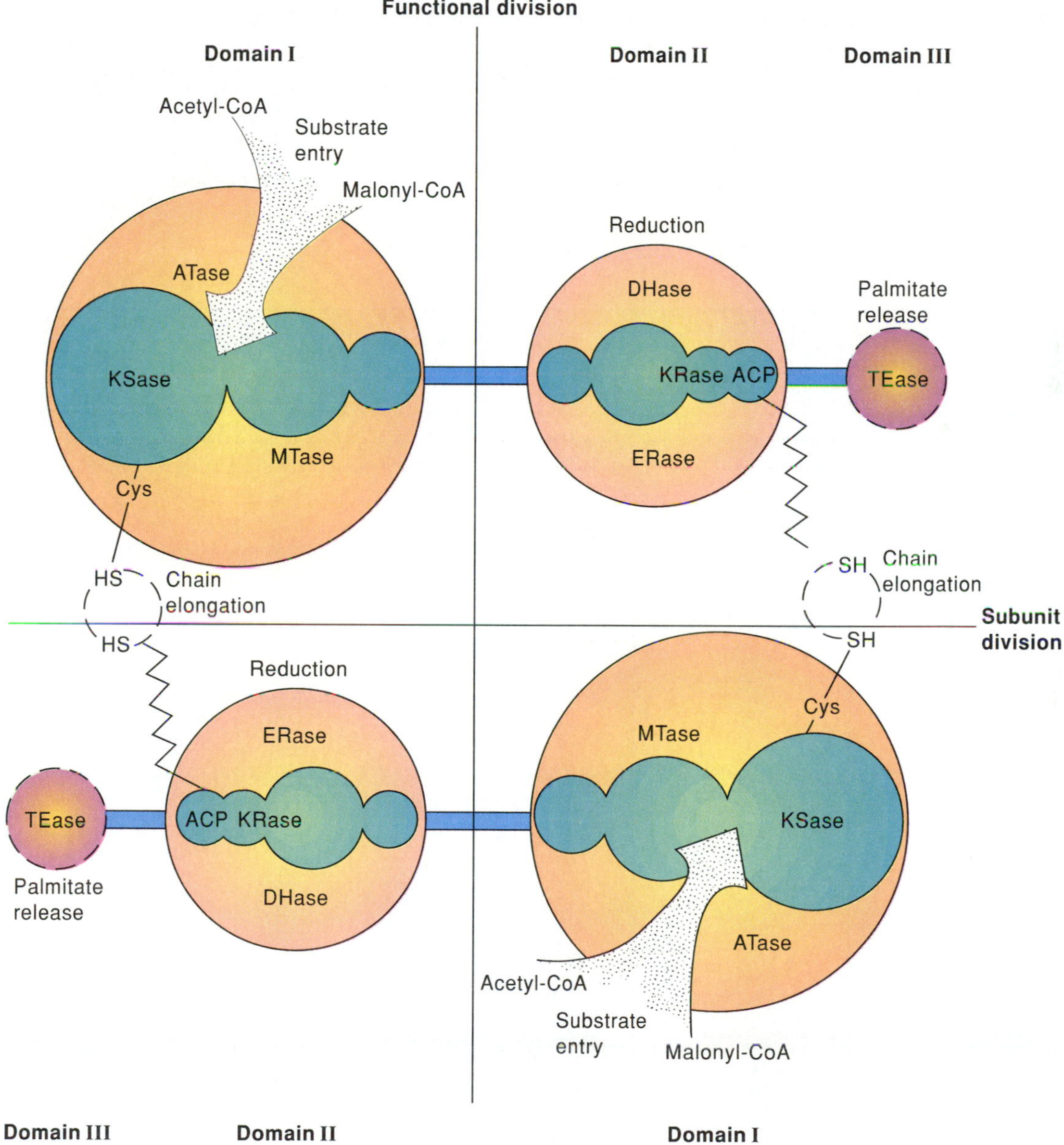

Figure 17.14

The phosphopantetheine group in acyl carrier protein (ACP) and in CoA.

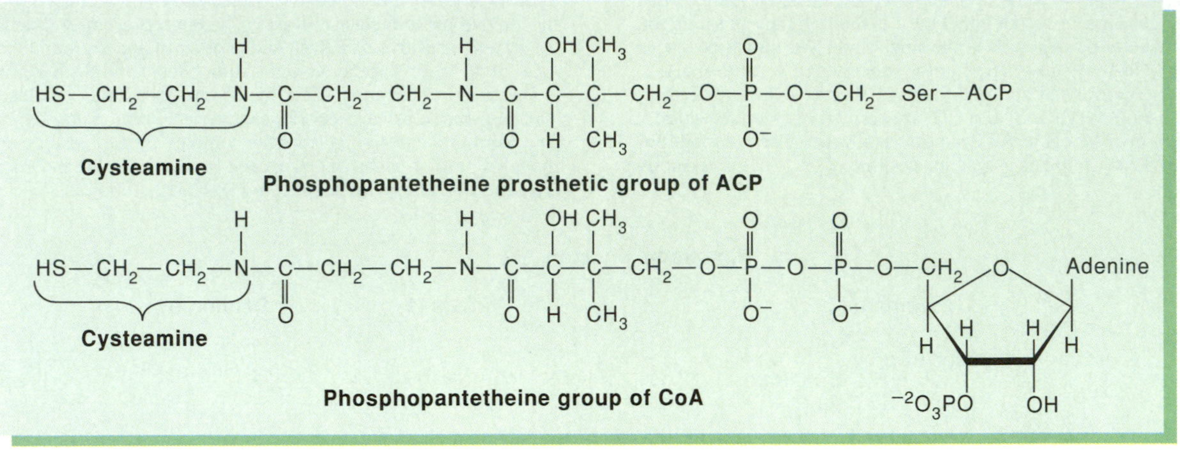

HS—CH$_2$—CH$_2$—N(H)—C(=O)—CH$_2$—CH$_2$—N(H)—C(=O)—C(OH)(H)—C(CH$_3$)(CH$_3$)—CH$_2$—O—P(=O)(O$^-$)—O—CH$_2$—Ser—ACP

Cysteamine

Phosphopantetheine prosthetic group of ACP

HS—CH$_2$—CH$_2$—N(H)—C(=O)—CH$_2$—CH$_2$—N(H)—C(=O)—C(OH)(H)—C(CH$_3$)(CH$_3$)—CH$_2$—O—P(=O)(O$^-$)—O—P(=O)(O$^-$)—O—CH$_2$ (ribose: Adenine, H, H, H, H, $^{-2}O_3PO$, OH)

Cysteamine

Phosphopantetheine group of CoA

Table 17.2

The Enzymes of Fatty Acid Synthase from *E. coli*

Enzyme	M_r	Subunits	Specificity	Miscellaneous
Acetyl-CoA-ACP transacylase	29,000	2 × 29,000	Acetyl-CoA 100% Butyryl-CoA 10% Hexanoyl-CoA 4.5%	Acetyl-CoA + enz. ⇌ acetyl-enz. + CoA Acetyl-enz. + ACP ⇌ acetyl-ACP + enz.
Malonyl-CoA-ACP transacylase	36,700	Single	Acetyl-CoA is a competitive inhibitor; K_I = 115 μM	Malonyl-CoA + enz. ⇌ malonyl-enz. + CoA Malonyl-enz. + ACP ⇌ malonyl-ACP + enz. Malonate is esterified to serine on the enzyme
β-Ketoacyl-ACP synthase I	80,000	2 × 40,000, apparently identical	Active with C_2–C_{14} ACP, but inactive with C_{16} ACP; inactive with CoA derivatives	Acetyl-ACP + enz. ⇌ acetyl-enz. + ACP Acetyl-enz. + malonyl-ACP ⇌ acetoacetyl-ACP + CO_2 + enz.
β-Ketoacyl-ACP synthase II	88,000	2 × 44,000	Active with C_2–C_{14} ACP and 16:1$^{\Delta 9}$-ACP	
β-Ketoacyl-ACP reductase	—	—	Specific for NADPH; active with C_4–C_{16} β-ketoacyl-ACP	The product has the D configuration
β-Hydroxyacyl-ACP dehydrase	—	—	Specific for D isomer; inactive with L-β-hydroxyacyl-ACP; active with C_4–C_{16} derivatives; lowest activity with C_{10} substrate; inactive with CoA derivatives	The product is *trans*
Enoyl-ACP reductase	—	—	There may be two enzymes, one specific for NADH and one for NADPH	

The First Step in Fatty Acid Synthesis Does Not Take Place on the Multienzyme Synthase

The first reaction in fatty acid synthesis, the formation of malonyl-CoA from acetyl-CoA, is catalyzed by acetyl-CoA carboxylase, and, as we have noted, it is the only reaction in animal fatty acid synthesis that does not take place on the multienzyme synthase. The properties of acetyl-CoA carboxylase are quite similar to those of pyruvate carboxylase, which is important in the gluconeogenesis pathway (see chapter 13, equation 11). We will see that the activity of acetyl-CoA carboxylase plays an important role in the control of fatty acid biosynthesis in most organisms. This is a typical situation for the first enzyme in a biosynthetic pathway. The enzyme contains the biotin enzyme covalently linked by means of the ϵ group of a lysine in the protein (fig. 17.15).

Figure 17.15

Structure of biotin linked to biotin carboxyl carrier protein (BCCP). BCCP is one of the components of acetyl-CoA carboxylase isolated from *E. coli.*

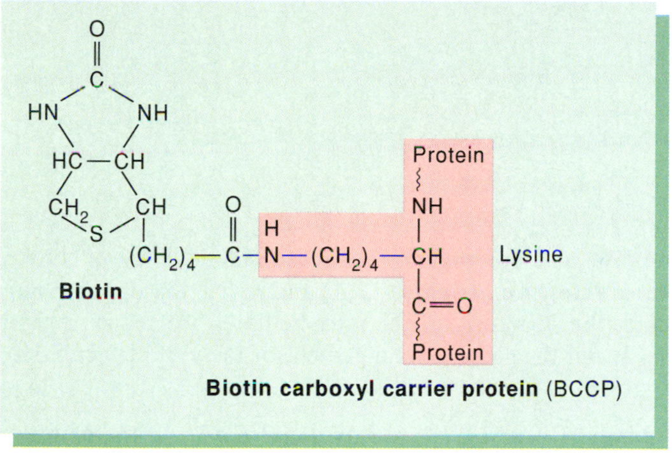

The acetyl-CoA carboxylase of *E. coli* consists of three protein components: biotin carboxyl carrier protein (BCCP) ($M_r = 22{,}500$), biotin carboxylase ($M_r = 98{,}000$; two subunits of 49,000), and carboxyltransferase ($M_r = 130{,}000$). The carboxyltransferase component has an A_2B_2 structure, and the molecular weights of the two types of subunits are 35,000 and 30,000. The reaction sequence (fig. 17.16) involves an initial carboxylation of BCCP, catalyzed by biotin carboxylase. Subsequently, the carboxyltransferase transfers the CO_2 from BCCP to acetyl-CoA. The probable mechanism for the carboxyltransferase reaction involves a carbanion attack on the biotin-linked carboxyl group, as shown in figure 17.17.

A distinctly different form of acetyl-CoA carboxylase is found in the cytosol of animal tissues. The rat liver enzyme is a dimer composed of two subunits ($M_r = 265{,}000$), with one biotin per subunit. In contrast to the situation in *E. coli,* the three functional parts of acetyl-CoA carboxylase in rat liver occur as a single multifunctional polypeptide. The dimeric form of the enzyme has very low activity, but incubation with citrate results in activation of the enzyme (fig. 17.18) and subsequent aggregation of the dimer to a polymer with a molecular weight between 4 and 8 million. The enzyme is deactivated and depolymerized when incubated with malonyl-CoA or palmitoyl-CoA. The significance of palmitoyl-CoA on the rate of fatty acid biosynthesis will be discussed when we look at the regulation of fatty acid metabolism, later in the chapter. Incidentally, citrate does not activate or polymerize the *E. coli* enzyme.

Seven Reactions Are Catalyzed by the Fatty Acid Synthase

Following malonyl-CoA synthesis, the remaining steps in fatty acid synthesis take place on the fatty acid synthase multienzyme complex (see fig. 17.13). This complex contains seven distinct enzyme activities (ATase, MTase, KSase, DHase, ERase, KRase, and TEase) in three distinct domains (domains I, II,

Figure 17.16

Reactions catalyzed by acetyl-CoA carboxylase. In *E. coli,* BCCP and the two enzymatic activities (biotin carboxylase and carboxyltransferase) can be separated from each other. In contrast, in liver all three components exist on a single multifunctional polypeptide.

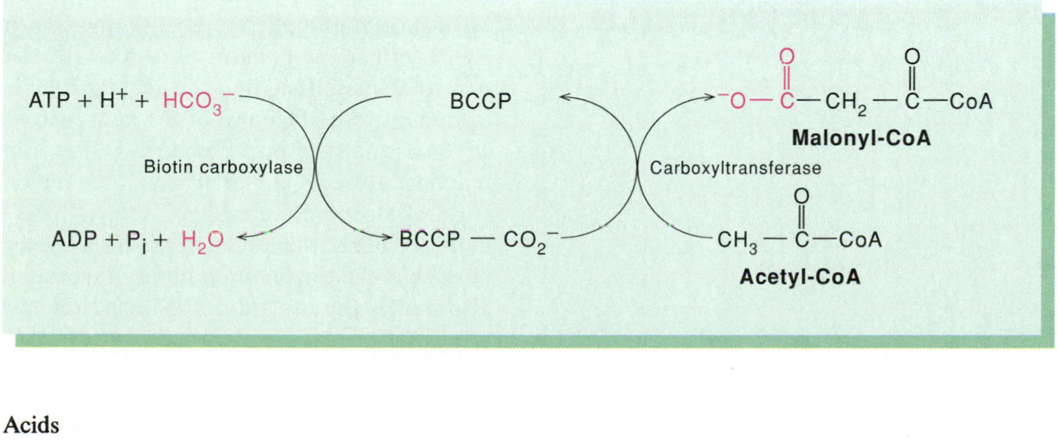

Figure 17.17

Mechanism for the carboxylation of acetyl-CoA. Acetyl-CoA forms a carbanion by proton loss. This carbanion attacks the carbon in carboxybiotin to give malonyl-CoA and biotinate. The biotinate anion is returned to the neutral form by addition of a proton. The R group attached to carboxybiotin is the carboxyl carrier protein.

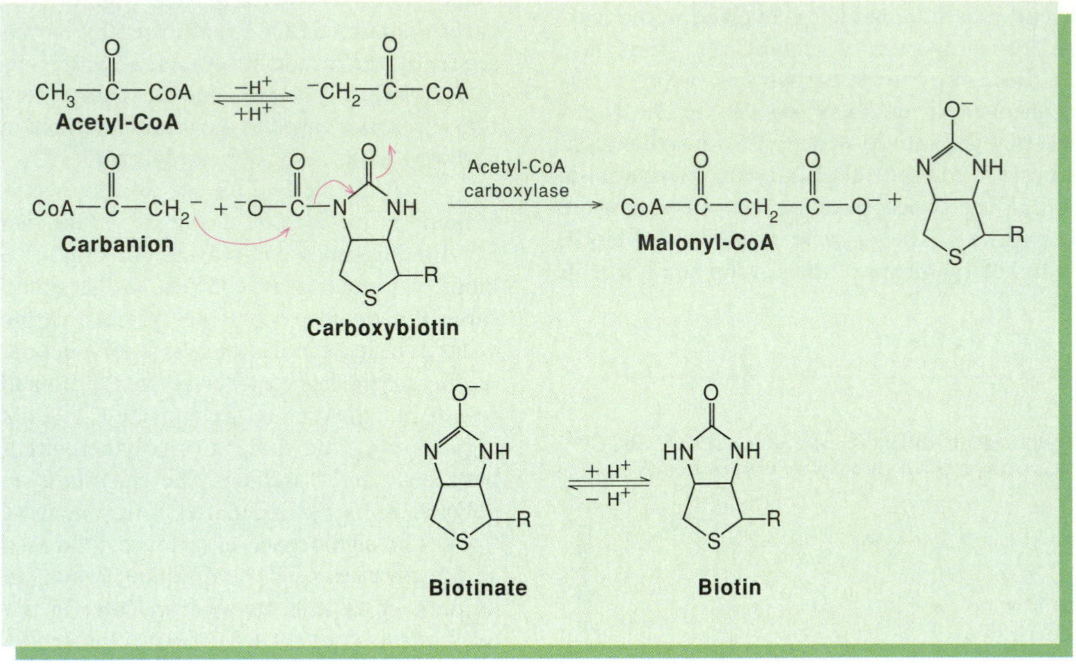

$$CH_3{-}\overset{O}{\overset{\|}{C}}{-}CoA \underset{+H^+}{\overset{-H^+}{\rightleftharpoons}} {}^{-}CH_2{-}\overset{O}{\overset{\|}{C}}{-}CoA$$

Acetyl-CoA

$$CoA{-}\overset{O}{\overset{\|}{C}}{-}CH_2^- + {}^{-}O{-}\overset{O}{\overset{\|}{C}}{-}N\ldots NH \xrightarrow{\text{Acetyl-CoA carboxylase}} CoA{-}\overset{O}{\overset{\|}{C}}{-}CH_2{-}\overset{O}{\overset{\|}{C}}{-}O^- + \text{biotinate}$$

Carbanion — **Carboxybiotin** — **Malonyl-CoA**

$$\text{Biotinate} \underset{-H^+}{\overset{+H^+}{\rightleftharpoons}} \text{Biotin}$$

Biotinate — **Biotin**

Figure 17.18

Activation of acetyl-CoA carboxylase from rat liver *in vitro*. Citrate activates the enzyme and converts it to a polymer. Either palmitoyl-CoA or malonyl-CoA can reverse this process. Citrate does not activate or polymerize the *E. coli* enzyme.

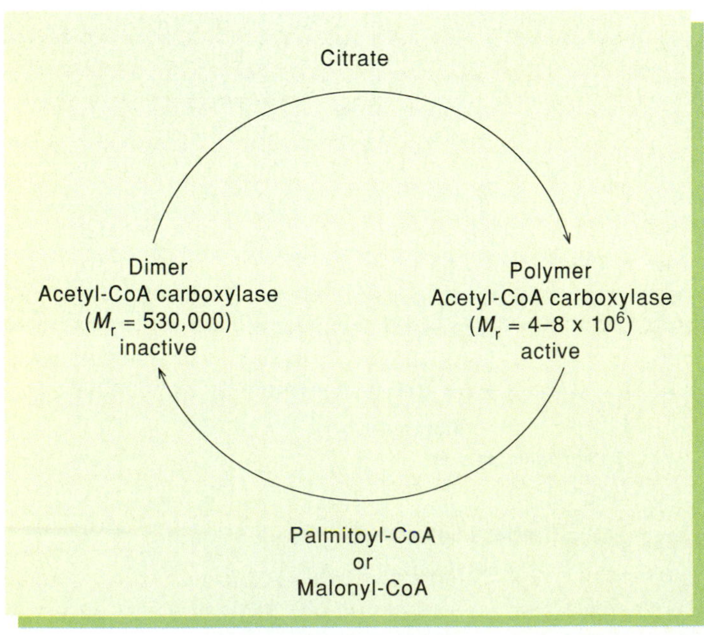

and III). The best way to follow the individual steps is by repeated reference to figures 17.12 and 17.13; the steps are numbered, the structures of the intermediates are shown in figure 17.12, and the general arrangement of the enzyme activities is suggested in figure 17.13.

Binding of Acetyl-CoA and Malonyl-CoA In the first reaction to occur on the fatty acid synthase, the acetyl transacylase (ATase) catalyzes the transfer of the acetyl group from acetyl-CoA to the cysteine-SH group (step 2a) on the β-ketoacyl-ACP synthase (KSase of domain I, figure 17.13) and the ACP (domain II) is charged with a malonyl group from malonyl-CoA in a reaction catalyzed by malonyl transacylase (MTase), which also resides on domain I (step 2b).

The Condensation Reaction In the condensation reaction (step 3) the acetyl group from the cysteine-SH group on β-ketoacyl-ACP synthase of domain I reacts with the malonyl group on ACP of domain II so that the acetyl group becomes the methyl-terminal two-carbon unit of the new acetoacetyl group. The reaction is catalyzed by β-ketoacyl-ACP synthase (KSase). In this reaction we see a major reason why malonyl-CoA is used instead of a second molecule of acetyl-CoA. The release of CO_2 in this condensation reaction provides the extra thermodynamic push to make the reaction highly favorable (i.e., exergonic). It also makes the central carbon a better nucleophilic agent for attacking the carbonyl carbon of the acetyl group (fig. 17.19).

Figure 17.19

Formation of acetoacetyl-ACP, catalyzed by ß-ketoacyl-ACP synthase (KSase). The acetyl group is bound to the cysteine residue of KSase (labeled enzyme in the figure). The carbonyl group of the enzyme-bound acetyl is attacked by the central carbon on the malonyl bound to ACP. Finally the C—S linkage is broken, resulting in acetoacetyl-ACP.

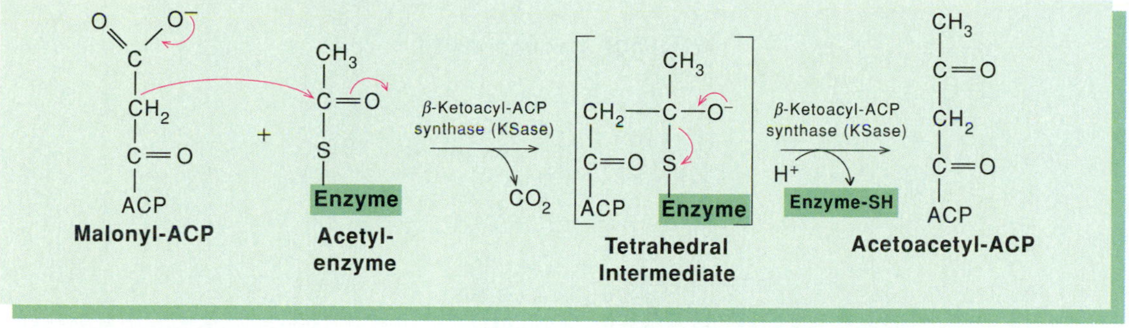

Malonyl-ACP + Acetyl-enzyme →(β-Ketoacyl-ACP synthase (KSase); CO_2)→ [Tetrahedral Intermediate] →(β-Ketoacyl-ACP synthase (KSase); H^+; Enzyme-SH)→ Acetoacetyl-ACP

Because of the way in which the functional groups are arranged (see fig. 17.13), the condensation reaction involves functional groups from different protein subunits.

The Reduction Reactions The object of the next three reactions (steps 4–6) in the fatty acid synthesis cycle is to reduce the β-carbonyl group. This is accomplished in three steps. The carbonyl is first reduced to a hydroxyl. This is dehydrated to produce a *trans* double bond, which is further reduced to give a fully saturated derivative. These reactions take place on domain II of the fatty acid synthase (see fig. 17.13). Chemically, these reactions are nearly the same as the reverse of three steps in the β-oxidation pathway except that the hydroxy group is in the D configuration for fatty acid synthesis and the L configuration for β oxidation. Also, different cofactors and different enzymes are used for synthesis and catabolism of fatty acids.

To return to the pathway, we find that β-ketoacyl-ACP reductase (KRase) catalyzes the first reduction of acetoacetyl-ACP to β-hydroxybutyryl-ACP. Mechanistically this reduction is similar to the reduction of pyruvate to lactate (see chapter 13). Following this reduction, the elements of water are removed by β-hydroxyacyl-ACP dehydrase (DHase) to yield crotonyl-ACP. The double bond resulting from the dehydration is further reduced by 2,3-*trans*-enoyl-ACP reductase (ERase), again using NADPH.

Continuation Reactions At this point we have seen one full round of reactions on the fatty acid synthase. Each enzyme activity of the complex (except the TEase on domain III) has been used precisely once. The resulting butyryl group, formed in the last step, is transferred from the pantetheine —SH group to a cysteine —SH group on the β-ketoacyl-ACP synthase, as was the acetyl group in the first cycle. This process is now ready to repeat itself. It continues, usually for six more cycles, until the palmitoyl group is formed; this group is released from the enzyme by a thioesterase (TEase), located on domain III (see fig. 17.13).

The head-to-tail arrangement of the dimeric synthase (see fig. 17.13) gives rise to an unusual situation in which the functional division of the dimeric synthase cuts across the subunit division. Thus domains II and III of one subunit work together with domain I of the other subunit. It seems likely that two fatty acids could be synthesized simultaneously, one on each of the functional synthases.

We will now examine some further modifications of the palmitoyl group. Its incorporation into more complex lipids is described in chapter 22. Box 17A describes an assay for the fatty acid synthase of liver.

Biosynthesis of Monounsaturated Fatty Acids Follows Two Routes

Two chemically distinct pathways exist for the introduction of a *cis* double bond into saturated fatty acids—the anaerobic pathway, as typified in *E. coli,* and the aerobic pathway, found in many eukaryotes and studied mainly in mammalian liver.

As the name "anaerobic" implies, the double bond of the fatty acid is inserted in the absence of oxygen. Biosynthesis of monounsaturated fatty acids follows the pathway described previously for saturated fatty acids until the intermediate β-hydroxydecanoyl-ACP is reached (fig. 17.20). At this point, a new enzyme, β-hydroxydecanoyl-ACP dehydrase, becomes involved. This dehydrase can form the α-β *trans* double bond, and saturated fatty acid synthesis can occur as previously discussed. But in addition, this dehydrase is capable of isomerization of the double bond to a *cis* double bond as shown in figure 17.20. The β-γ unsaturated fatty acyl-ACP is subsequently elongated by the *E. coli* fatty acid synthase to yield palmitoleoyl-ACP ($16{:}1^{\Delta 9}$). The conversion of this compound to the major unsaturated fatty acid of *E. coli, cis*-vaccenic acid ($18{:}1^{\Delta 11}$), appears to involve β-ketoacyl-ACP synthase II, which shows a preference for palmitoleoyl-ACP as a substrate (see table 17.2). The subsequent conversion of β-keto-*cis*-vaccenyl ACP to *cis*-vaccenyl-ACP is catalyzed by the usual enzymes of fatty acid biosynthesis.

In contrast to the anaerobic pathway found in *E. coli,* the aerobic pathway in eukaryotic cells introduces double bonds after the C_{16} or C_{18} saturated fatty acid has been synthesized. In rat liver and other eukaryotic cells, an enzyme complex associated with the endoplasmic reticulum desaturates stearoyl-CoA (18:0) to oleoyl-CoA ($18{:}1^{\Delta 9}$). This reaction requires NADH and O_2 and results in the remarkable formation of a

Figure 17.20

Anaerobic pathway for biosynthesis of monounsaturated fatty acids in *E. coli*. Synthesis of monounsaturated fatty acids follows the pathway described previously for saturated fatty acids until the intermediate ß-hydroxydecanoyl-ACP is reached. At this point there is an apparent competition between the enzymes involved in saturated and unsaturated fatty acid synthesis.

Acetyl-ACP + 4 Malonyl-ACP

Normal pathway for saturated fatty acid synthesis

$CH_3(CH_2)_5-CH_2-CH(OH)-CH_2-C(=O)-$ACP

β-Hydroxydecanoyl-ACP

β-Hydroxydecanoyl-ACP dehydrase

H_2O

$CH_3(CH_2)_5-CH=CH-CH_2-C(=O)-$ACP ($\gamma$, β)

β-Hydroxy-decanoyl-ACP dehydrase

$CH_3(CH_2)_5CH_2-CH=CH-C(=O)-$ACP ($\beta$, α)

Saturated fatty acid synthesis

$16{:}1^{\Delta 9}$ —ACP

Saturated fatty acid synthesis

16:0 —ACP

β-Ketoacyl-ACP synthase II and saturated fatty acid synthesis

$18{:}1^{\Delta 11}$ —ACP

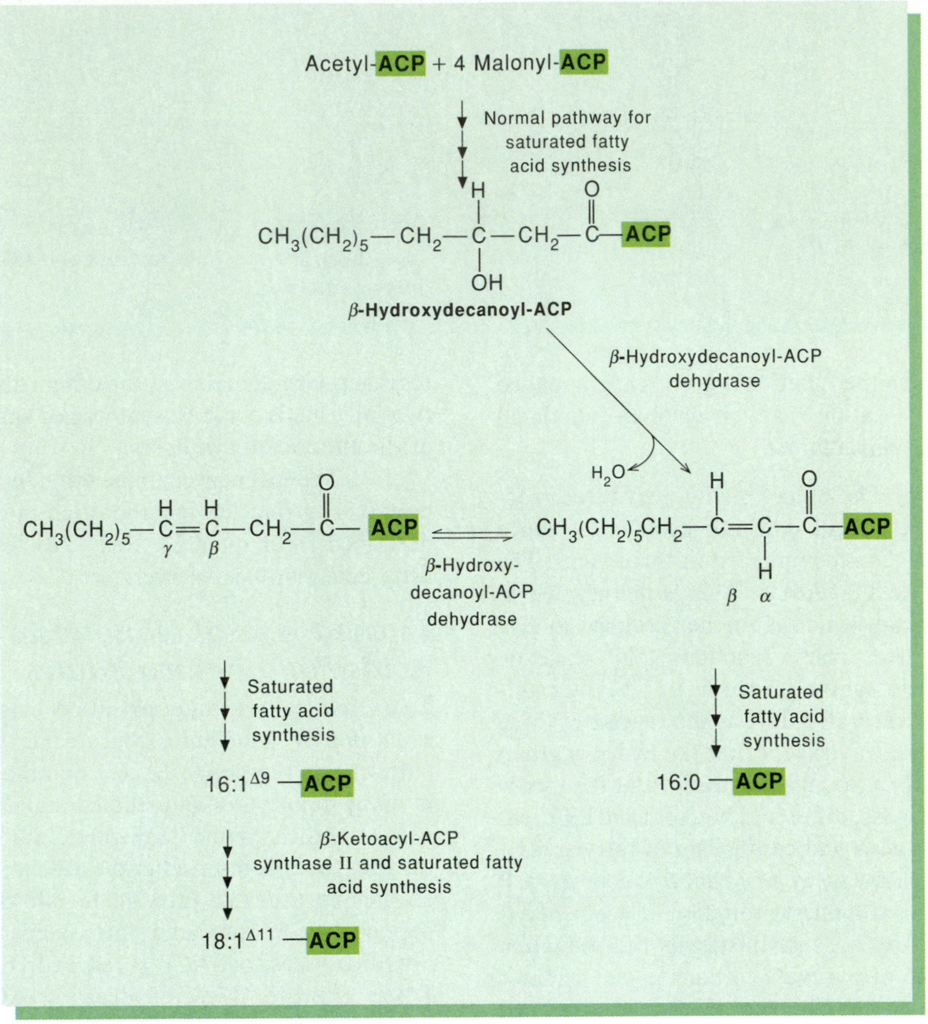

double bond in the middle of an acyl chain with no activating groups nearby. Although many elegant experiments have been performed, the chemical mechanism for desaturation of long-chain acyl-CoA's remains unclear.

Desaturation requires the cooperative action of two enzymes, cytochrome b_5 reductase and stearoyl-CoA desaturase, and the action of cytochrome b_5. A scheme for this set of reactions is shown in figure 17.21. Cytochrome b_5 reductase (M_r = 43,000) is a flavoprotein that transfers electrons from NADH by means of flavin (F) to cytochrome b_5, a heme-containing protein (M_r = 16,700) in which Fe^{3+} is reduced to Fe^{2+}. Both cytochrome b_5 and the reductase are amphipathic proteins; that is, each has a hydrophobic peptide tail that anchors the protein into the membrane of the endoplasmic reticulum and a hydrophilic portion that is outside the membrane surface (see chapter 7 for details about membrane-bound proteins). Stearoyl-CoA desaturase utilizes two electrons from cytochrome b_5 coupled with an atom of oxygen to form a *cis* double bond in the Δ^9 position of stearoyl-CoA. The desaturase (M_r = 53,000) has 62% nonpolar amino acids, which is probably the main reason it is tightly embedded in the membrane. There is also one atom of nonheme iron per molecule of enzyme.

Biosynthesis of Polyunsaturated Fatty Acids Occurs Mainly in Eukaryotes

E. coli does not have polyunsaturated fatty acids, whereas eukaryotes produce a large variety of polyunsaturated fatty acids. Mammals cannot desaturate between the Δ^9 position and the methyl end of an acyl chain, whereas plants have the enzymes to desaturate at positions Δ^{12} and Δ^{15}. Thus mammals have a dietary requirement for linoleic acid ($18{:}2^{\Delta 9,12}$) and linolenic acid ($18{:}3^{\Delta 9,12,15}$), which are essential components of phospholipids

17A BOX

Assaying the Activity of Fatty Acid Synthase

The activity of fatty acid synthase can be assayed by the incorporation of [2-^{14}C]malonyl-CoA or [^{3}H]acetyl-CoA into fatty acid. This scheme illustrates a common principle often utilized in the assay of lipid biosynthetic enzymes. The radioactive substrate is a water-soluble molecule that can easily be separated from the lipid product by extraction of the reaction mixture with an organic solvent such as petroleum ether.

Step 1. Prepare the incubation mixture, which contains:

0.3 mM NADPH
0.1 M phosphate buffer, pH 6.8
5 μM mercaptoethanol
3 μM EDTA
50 μM malonyl-CoA
12 μM[^{3}H]acetyl-CoA (specific radioactivity = 2.0×10^6 dpm μmole)
Enough distilled H_2O to bring the final volume to 1 ml

Mercaptoethanol is used to keep the SH residues of the enzyme in the reduced state. EDTA chelates divalent cations such as Mg^{2+}, which might inhibit the reaction. All of the materials can be purchased from companies that sell chemical compounds. Radioactive compounds (e.g., [^{3}H]acetyl-CoA) are sold by companies that specialize in the manufacture of radioisotopes.

Step 2. Equilibrate the mixture for 5 min at 37°C in a shaking water bath.

Step 3. Add enzyme (e.g., 2 mg protein), mix thoroughly, and incubate at 37°C for 5 min.

Step 4. Stop the reaction by the addition of 0.1 ml 18% perchloric acid. The perchloric acid lowers the pH to ~1, and thus denatures and inactivates the enzyme.

Step 5. Add 1 ml ethanol and 2 ml petroleum ether, mix thoroughly, and allow the phases to separate. Transfer the upper ether layer to a tube and extract the lower aqueous phase two more times with petroleum ether. The ethanol and water separate from the ether layer, which floats on the aqueous layer. Because the pH is approximately 1, the fatty acid product is protonated (RCOOH) and can therefore be easily extracted into the ether phase.

Step 6. Evaporate the combined petroleum ether extracts, add liquid scintillation fluid, and determine the radioactivity by liquid scintillation spectrophotometry. Liquid scintillation counters are sophisticated instruments that are obtainable from companies that specialize in their manufacture and sale. Their cost ranges between \$15,000 and \$30,000.

Step 7. Calculate the specific activity of the enzyme. To do this, divide the dpm incorporated into the fatty acid by the specific radioactivity of the acetyl-CoA, the time of the incubation, and the milligrams of protein added to the assay. For example:

$$\text{Specific activity} = \frac{\text{dpm in fatty acid}}{\text{specific radioactivity of acetyl-CoA} \cdot \text{min} \cdot \text{mg protein}}$$
$$= \frac{50{,}000 \text{ dpm}}{2 \times 10^6 \text{ dpm}/\mu\text{mole} \cdot 5 \text{ min} \cdot 2 \text{ mg protein}}$$
$$= 2.5 \times 10^{-3}\ \mu\text{mole fatty acid formed/min} \cdot \text{mg protein}$$

The specific activity is a measure of the activity of an enzyme as a function of time and amount of protein. When fatty acid synthesis is reduced, for example during a fast, the specific activity will be much lower (about tenfold) than after a carbohydrate-rich meal. The specific activity can also be a measure of the purity of an enzyme preparation. Thus fatty acid synthase in rat liver cytosol might have an activity of 1×10^{-3} μmole fatty acid formed per minute per milligram of protein. By contrast, a pure enzyme, free of all other cytosolic proteins, might have 1000-fold higher specific activity (1 μmole fatty acid formed per minute per milligram of protein).

When the fatty acid synthase is highly purified, it also can be assayed by a spectrophotometric method in which the oxidation of NADPH is followed. As $NADP^+$ is formed, there is a decrease in the absorbance at 340 nm.

---

in membranes, and precursors to eicosanoids (see chapter 22). However, enzyme complexes occur in the endoplasmic reticulum of animal cells that desaturate at Δ^5 if there is a double bond at the Δ^8 position, or at Δ^6 if there is a double bond at the Δ^9 position. These enzymes are different from the Δ^9-desaturase, but they do appear to utilize cytochrome b_5 reductase and cytochrome b_5.

The major polyunsaturates of mammals are derived either from diet or from desaturation and elongation of $18{:}2^{\Delta 9,12}$ or $18{:}3^{\Delta 9,12,15}$. A scheme for the synthesis of arachidonic acid ($20{:}4^{\Delta 5,8,11,14}$) from linoleic acid is shown in figure 17.22. This example illustrates the principle by which polyunsaturated fatty acids are made in animals. The elongation step is catalyzed by a series of membrane-bound enzymes that are present in the endoplasmic reticulum. These enzymes use malonyl-CoA as the donor for the two-carbon unit, and the chemical mechanism seems to be similar to that described earlier for fatty acid synthesis (see fig. 17.13). The liver enzymes also will elongate other polyunsaturated fatty acyl-CoA's. In addition, endoplasmic reticulum enzymes will elongate C_{16} and C_{18} CoA's to produce the C_{22} and C_{24} CoA's characteristic of sphingolipids (see chapter 22). The latter elongation enzymes are most active in brain tissue during the synthesis of myelin.

Figure 17.21

The aerobic pathway for formation of oleoyl-CoA in eukaryotes. In eukaryotes the double bonds are introduced after the C_{16} and C_{18} saturated fatty acid has been synthesized.

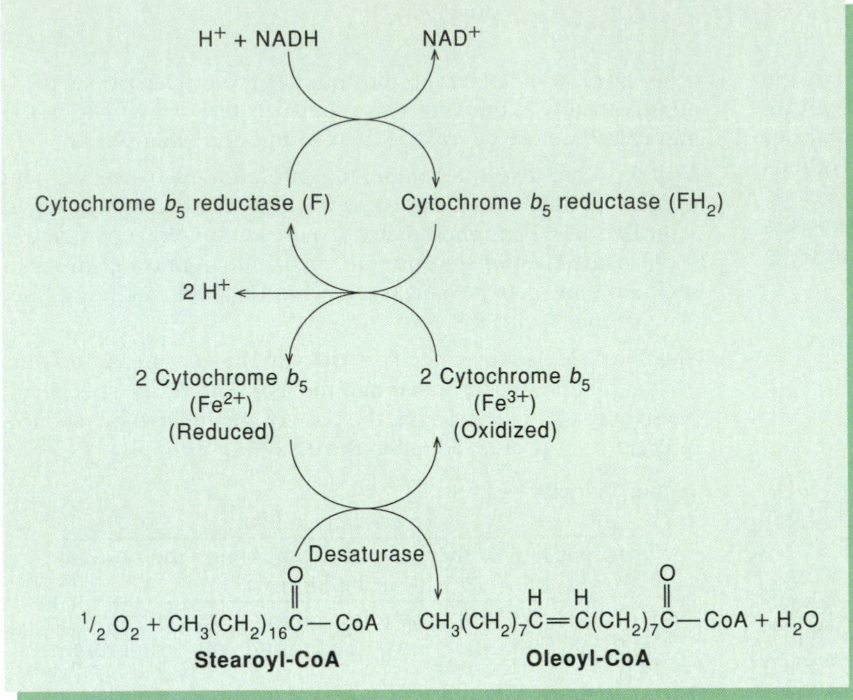

Figure 17.22

Synthesis by mammalian tissues of arachidonic acid from linoleic acid. The Δ^5 and Δ^6 desaturases are separate enzymes and are also different from the Δ^9 desaturase (fig. 17.21). The mechanisms, however, seem to be the same, involving cytochrome b_5 and cytochrome b_5 reductase. The enzymes for elongation of unsaturated fatty acid such as 18:3 to 20:3 occur on the endoplasmic reticulum.

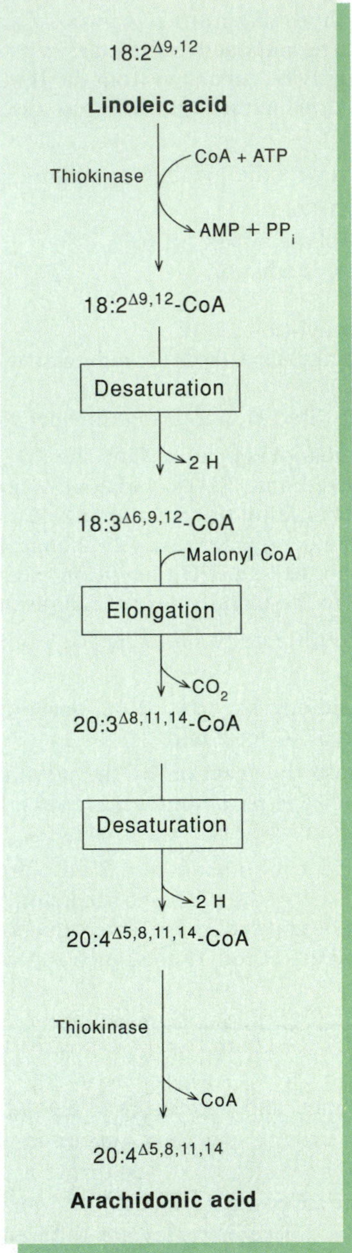

Summary of the Pathways for Synthesis and Degradation

Before discussing the specific aspects of regulation of fatty acid metabolism, let us review the main steps in the processes of synthesis and degradation. Figure 17.23 illustrates these processes in a way that emphasizes the parallels. In both cases two-carbon units are involved. Despite this superficial similarity, each step in the two processes involves different enzymes and different coenzymes. Furthermore, the processes take place in different parts of the cell; degradation takes place in the mitochondria and synthesis takes place in the cytosol. The differences in location of the two processes and in the enzymes used make it possible to regulate the two processes independently.

Regulation of Fatty Acid Metabolism

Given the principles of pathway regulation discussed in chapter 12, you can anticipate that there must be controls to ensure that fatty acid synthesis and breakdown do not both occur at the same time, and that the pathways that are active at any given time will be the ones that best suit the needs of the organism. It seems likely that, when the organism has satisfied its immediate energy needs and the limited storage space for glycogen has been filled, a switch will be turned to direct nutrients to fatty acid synthesis. Since fatty acids are useful for both energy storage and as precursors for a variety of membrane lipids, it also seems likely that the vital structural need for fatty acids will be given a high priority.

Figure 17.23

Parallels between synthesis and degradation of fatty acids. Both processes involve two carbons at a time and very similar intermediates, even though they go in opposite directions. CoA is also heavily involved in both processes. Here the similarities end. The enzymes used in the two processes are totally different and the remaining coenzymes are different. In the degradative direction FAD and NAD^+ are used, whereas in the synthetic direction the coenzyme NADPH is used. Degradation occurs in the mitochondrial matrix and synthesis occurs in the cytosol.

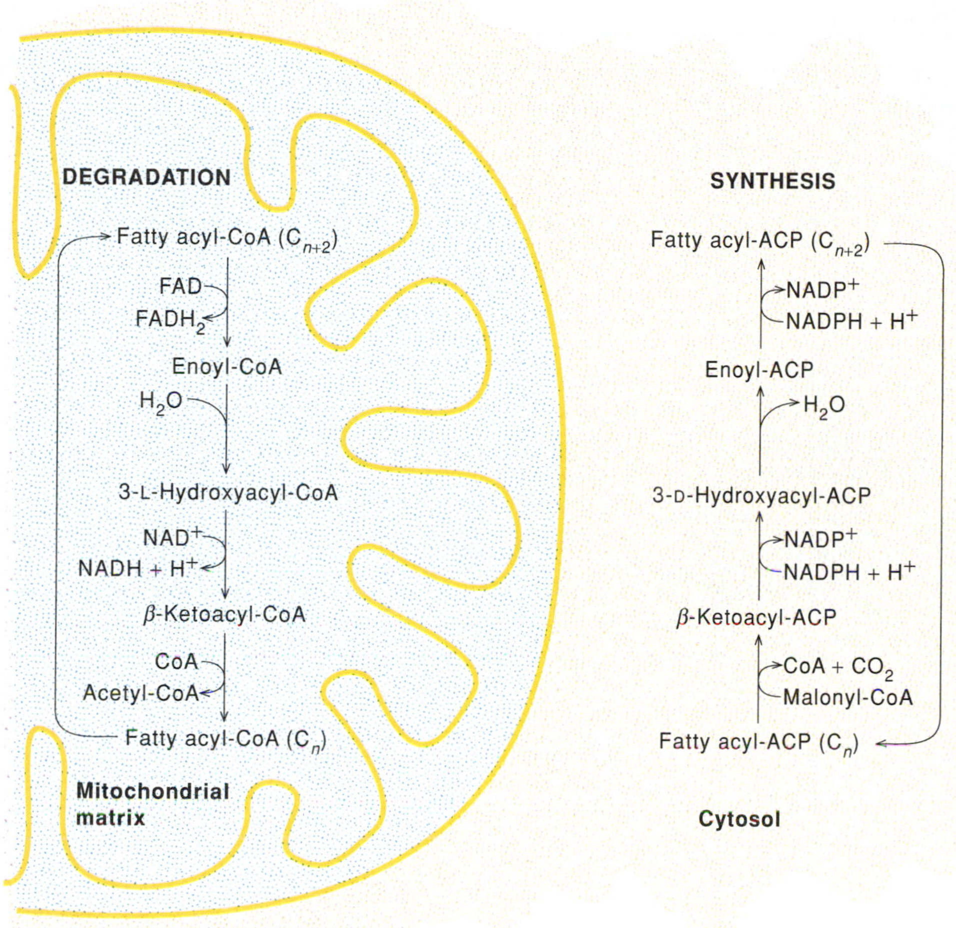

Lipid Utilization Is Regulated by Controlling the Release of Fatty Acids from Adipose Tissue

Although triacylglycerols are found in the liver and intestine, they are primarily found in adipose (fat) tissue, which functions as the main storage depot for lipid. The specialized cell in this tissue is called the adipocyte (see fig. 7.3). The cytoplasm of the adipocyte is packed with vesicles that are rich in triacylglycerols and that serve as the long-term energy reserves in mammals. When the energy supply from the diet becomes limited, the animal responds to the deficiency with a hormonal signal that is transmitted to the target tissue by the release of hormones, especially epinephrine and glucagon. The way in which the hormones work on the adipocytes is similar to the way they work on cells containing glycogen storage granules. First the hormones bind to the plasma membrane of the target cell, which stimulates the synthesis of cyclic AMP (cAMP). As shown in figure 17.24 (also see fig. 13.18), the cAMP activates a protein kinase that phosphorylates a key enzyme; in the case of the adipocytes the key enzyme is triacylglycerol lipase. The lipase hydrolyzes the triacylglycerol to diacylglycerol with release of a fatty acid from carbon 1 or 3 of the glycerol backbone. This reaction is thought to be the rate-limiting step in the complete hydrolysis of the triacylglycerols. The diacylglycerols and monoacylglycerols are rapidly hydrolyzed to fatty acids and glycerol.

The unesterified fatty acids move through the plasma membranes of the adipocytes and endothelial cells of the blood capillaries into the bloodstream, where they become bound to plasma proteins, in particular serum albumin. Passive diffusion appears to account for the movement of fatty acids across the adipocyte membrane into the blood plasma. Hence the rate of

Figure 17.24

When certain hormones (e.g., epinephrine) bind to their receptors in adipose tissue, adenylate cyclase is activated. The cAMP that is formed activates protein kinase A, which phosphorylates triacylglycerol lipase. The phosphorylated form of this enzyme is the active species, and triacylglycerols are degraded to fatty acids. The fatty acids are released into the bloodstream, bound by albumin, and delivered to energy-deprived tissues.

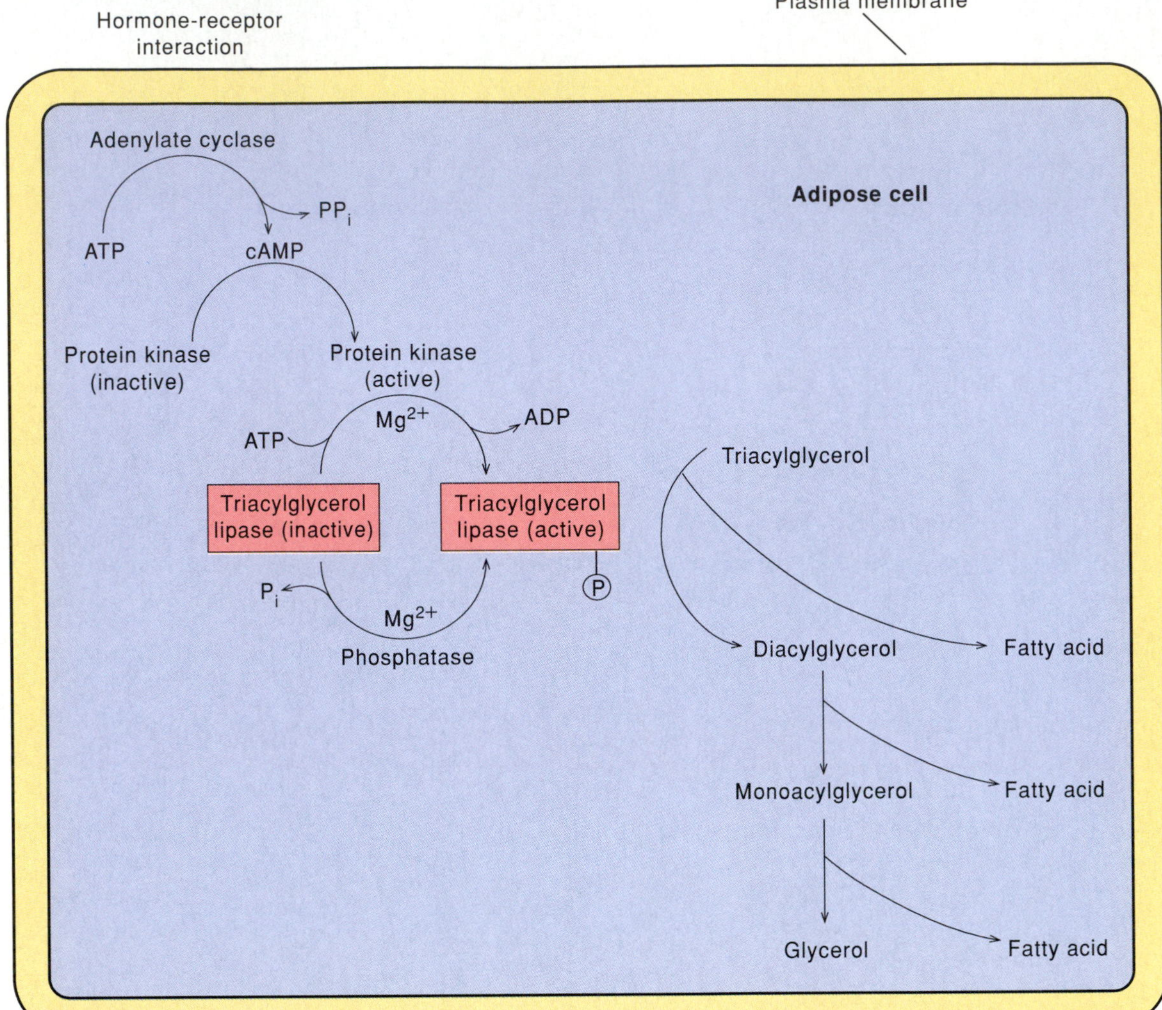

transfer depends on the concentrations of fatty acids, both in the adipocytes and in the plasma. The glycerol also can be released into the plasma and removed by the liver for glucose production.

Albumin carries the fatty acids to energy-deficient tissues, where fatty acids move from the plasma to the tissues. Until recently it was thought that uptake of fatty acids also occurs by passive diffusion. It now appears that transport of fatty acids into cells is mediated by a specific plasma membrane transport system. Fatty acid uptake is coupled to sodium transport and is mediated by a fatty acid binding protein in the membrane (more information on coupled transport systems can be found in chapter 32). The amount of fatty acid removed by a tissue depends on the relative concentrations of fatty acids, both in the plasma and in the cells of the tissues. For example, cardiac muscle utilizes fatty acids as the major oxidative source of energy for ATP synthesis and therefore removes large amounts from the circulation. At the other extreme, the brain does not utilize fatty acids for energy production but depends almost exclusively on carbohydrates.

Fatty Acid Binding Proteins and Acyl-CoA Binding Protein May Be Important in the Intracellular Traffic of Fatty Acids

Because of their hydrophobic tail, fatty acids preferentially associate with membrane lipids. How do the fatty acids move from the plasma membrane to the mitochondria and other organelles in the cell? The answer is not known, but fatty acid binding proteins are major candidates for this job. These are small proteins of 127–132 amino acids that bind long-chain

Figure 17.25

Acyl-CoA is not transported across the inner membrane of the mitochondrion. Instead, the acyl-CoA reacts with carnitine to yield the acyl carnitine derivative. This reaction is catalyzed by carnitine acyltransferase I, which is located on the outer mitochondrial membrane. The acyl carnitine is transported across the inner membrane by a specific carrier protein. Once inside the matrix of the mitochondrion, the acyl carnitine is converted back to its acyl-CoA derivative, the substrate for the start of β oxidation. This reaction is catalyzed by carnitine acyltransferase II, which is located on the mitochondrial inner membrane. Note that acyltransferases I and II are oriented in their respective membranes so that the reactions they catalyze occur in intermembrane space and the mitochondrial matrix, respectively. The carnitine itself can pass between the various cellular compartments.

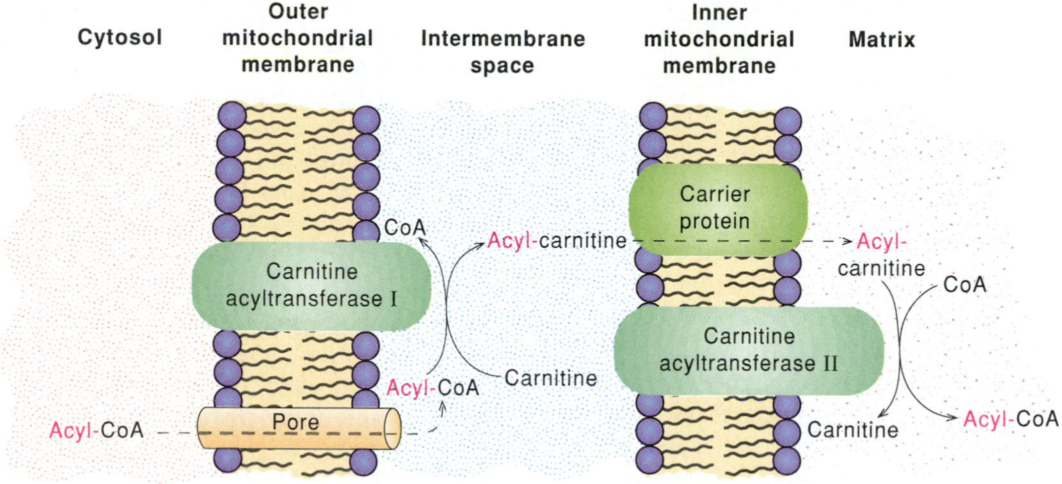

(16–20 carbons) saturated and unsaturated fatty acids to a single binding site with a K_d of 1–4 μM. The rat liver fatty acid binding protein (M_r = 14,273) is an abundant cytosolic protein and accounts for 3–5% of the mass of cytosolic proteins. Approximately 60% of the long-chain fatty acids found in liver cytosol are bound to this protein.

Although the fatty acid binding proteins also bind acyl-CoA's, another cytosolic protein has recently been identified that binds acyl-CoA's with greater avidity. The acyl-CoA binding protein (M_r = 9,938) does not bind unesterified fatty acids. This protein is present at lower concentrations (approx. 0.5% of cytosolic protein) than fatty acid binding protein and is a candidate for the transfer of acyl-CoA's between organelles in cells.

Transport of Fatty Acids into Mitochondria Is Regulated

The level of fatty acids in the bloodstream favors utilization of the fatty acids, but further controls are needed within the cell because the needs of different cell types vary widely. Within the cell the catabolism of fatty acids is regulated by controlling the flow of fatty acids into the mitochondria, where the oxidation apparatus is located.

Fatty acids taken into cells are first activated in the cytosol by reaction with coenzyme A and ATP to yield fatty acyl-CoA in a reaction catalyzed by acyl-CoA ligase (also known as thiokinase):

$$\text{RCOO}^- + \text{ATP} + \text{CoA} \xrightarrow{Mg^{2+}} \text{RCO—CoA} + \text{PP}_i + \text{AMP}$$

Regulation of acyl-CoA flow into the mitochondria results from the fact that acyl-CoA derivatives cannot pass directly into the mitochondria. First they must be converted to their acyl carnitine derivatives, which can cross the inner membrane of the mitochondria:

$$\text{RCO—CoA} + (CH_3)_3N^+\text{—}CH_2\underset{\underset{\text{OH}}{|}}{\text{CH}}CH_2COO^- \rightleftharpoons$$

Carnitine

$$(CH_3)_3N^+\text{—}CH_2\underset{\underset{\underset{\text{RC=O}}{|}}{\underset{\text{O}}{|}}}{\text{CH}}CH_2COO^- + \text{CoA}$$

Acyl carnitine

This reaction is catalyzed by carnitine acyltransferase. There are at least three acyltransferases associated with mitochondria: one specific for short-chain fatty acids (carnitine acetyltransferase), and two specific for the longer-chain fatty acids (carnitine acyltransferases I and II).

There is a protein carrier in the inner mitochondrial membrane that can transport carnitine, acetyl carnitine, and short- and long-chain acyl carnitine derivatives across the membrane. The transfer of fatty acyl carnitine into the mitochondria involves an exchange with free carnitine, as illustrated in figure 17.25. Once inside the mitochondria, the reaction is reversed by carnitine acyltransferase II to yield a fatty acyl-CoA. Thus there are at least two distinct pools of acyl-CoA in the cell, one in the cytosol and the other in the mitochondria. Acyltransferases I and II occupy different locations. As shown in figure 17.25, acyltransferase I resides on the outer membrane of the mitochondria, whereas acyltransferase II is on the inner membrane, facing the mitochondrial matrix.

This elaborate chain of reactions provides a number of possible points for regulating the supply of acyl-CoA's for oxidation. The major control point is carnitine acyltransferase I, which is strongly inhibited by malonyl-CoA. Recall that malonyl-CoA is a substrate for fatty acid synthesis. Thus high levels of malonyl-CoA, which indicate that fatty acid synthesis is in progress, also prevent fatty acid catabolism. We will have more to say later about factors that regulate the concentration of malonyl-CoA.

Fatty Acid Synthesis Is Limited by Substrate Supply

Just as fatty acid oxidation is limited by substrate supply, so is fatty acid synthesis limited by substrate supply. The overall equation for fatty acid synthesis indicates a need for acetyl-CoA, malonyl-CoA, and NADPH. As we have seen, the malonyl-CoA is derived from acetyl-CoA, so we can think of acetyl-CoA as the main substrate for fatty acid synthesis. The supply of acetyl-CoA and NADPH necessary for fatty acid synthesis is generated from the substrates that originate in the glycolytic pathway and the TCA cycle. Thus glycolysis generates pyruvate, and in the mitochondria the pyruvate is converted to acetyl-CoA as well as oxaloacetate. In animals the excess acetyl-CoA cannot be directly utilized for fatty acid synthesis. This is because acetyl-CoA cannot cross the mitochondrial membrane to the site of fatty acid synthesis in the cytosol. The way in which the acetyl-CoA in the mitochondria leads to the synthesis of acetyl-CoA and NADPH in the cytosol is an interesting example of selective permeability, which we discussed in chapter 14 (see fig. 14.14). Two other reactions, one catalyzed by glucose-6-phosphate dehydrogenase and one catalyzed by 6-phosphogluconate dehydrogenase, also are important contributors of NADPH for fatty acid synthesis (see chapter 13).

Fatty Acid Synthesis Is Regulated by the First Step in the Pathway

An abundance of NADPH and acetyl-CoA creates the possibility for fatty acid synthesis, but since these substrates can be used in a variety of ways, it is essential to have another, more specific control that regulates the synthesis of fatty acids. In fact, the first reaction in the pathway catalyzed by acetyl-CoA carboxylase is regulated by more than one mechanism.

We have already seen that citrate activates the liver acetyl-CoA carboxylase and converts it to a high-molecular-weight polymer (see fig. 17.18). The activation by citrate is appropriate because excess citrate is a good indication that the energy needs of the cell are satisfied. Furthermore, citrate is the most important source of cytosolic acetyl-CoA, which is supplied by a mechanism previously described (see fig. 14.14). In this connection, recall from chapter 14 that isocitrate dehydrogenase is strongly inhibited by a high $NADH/NAD^+$ ratio, another good indication that the energy status of the cell is high. The inhibition of isocitrate dehydrogenase blocks the TCA cycle and leads to a buildup of citrate, which is a precursor of the cytosolic acetyl-CoA.

The acetyl-CoA carboxylase is inhibited by palmitoyl-CoA. Palmitoyl-CoA also inhibits glucose-6-phosphate degradation by the pentose phosphate pathway, a major supplier of NADPH. This control can be thought of as a case of end-product inhibition, since palmitoyl-CoA is a major end product of fatty acid synthesis. Palmitoyl-CoA likewise inhibits the fatty acid synthase, but to a lesser extent than it inhibits the carboxylase.

The acetyl-CoA carboxylase is also inhibited through a chain of reactions initiated by the hormones glucagon and epinephrine. Remember that high levels of these hormones serve as a general signal that energy is in short supply or soon may be needed in large amounts (see chapter 13). At such a time it would be inappropriate to divert energy to the synthesis of fatty acids. These hormones bind to receptors on the plasma membrane and stimulate a kinase that phosphorylates acetyl-CoA carboxylase (fig. 17.26). Recall that hormones are active only on cells that carry the specific hormone receptors. Both liver and adipose tissue respond to glucagon and epinephrine. The hormone insulin, which often counteracts the action of glucagon, has the opposite effect on the carboxylase; the mechanism of insulin action is under study.

Regulation of Fatty Acid Oxidation in the Heart Occurs Later in the Cycle

In most tissues, regulation of β oxidation is a complex topic that relates to lipid biosynthesis and carbohydrate metabolism. In the heart there is very little fatty acid biosynthesis, so the regulation picture is much simpler. Oxidation of fatty acids is the major source of energy for the heart. If energy use by the heart is decreased, the reduced activity of the tricarboxylic acid cycle and oxidative phosphorylation will cause an accumulation of acetyl-CoA and NADH. An increase in acetyl-CoA in the mitochondria inhibits thiolase and thus inhibits β oxidation (fig. 17.27). The increase in NADH and the lack of NAD^+ for the 3-hydroxyacyl-CoA dehydrogenase may also be important in retarding oxidation. It is noteworthy that the regulation of β oxidation in heart tissue seems to occur at the later enzymes in the cycle and not at the initial reactions, as is usually the case in regulation of a metabolic pathway.

The Controls for Fatty Acid Metabolism Discourage Simultaneous Synthesis and Breakdown

The controls for fatty acid metabolism are designed so that they satisfy most of our expectations for what control systems should do. One of the most important aspects of control is that it should discourage simultaneous synthesis and breakdown, which would result in a loss of energy for no purpose. At the level of hormones we see that the two major hormones implicated in fatty acid metabolism, glucagon and epinephrine, encourage fatty acid breakdown. They encourage mobilization of fatty acids from storage sites, and simultaneously discourage synthesis, by inhibiting the formation of malonyl-CoA from acetyl-CoA (fig. 17.28). So the main effect of the hormones is to stimulate breakdown while inhibiting synthesis.

Figure 17.26

Regulation of acetyl-CoA carboxylase by phosphorylation and dephosphorylation. Glucagon is known to activate cAMP-dependent protein kinase; this kinase phosphorylates both serine 77 and serine 1200 of rat acetyl-CoA carboxylase, which inactivates the enzyme. However, there is also an AMP-dependent kinase that will phosphorylate serine 79 and serine 1200 and inactivate the rat acetyl-CoA carboxylase. The relative importance of these two kinases in regulating the carboxylase *in vivo* is still unclear. Likewise, the phosphorylated enzyme is a substrate for several different protein phosphate phosphatases and the physiologically relevant phospha-tases are not known. Epinephrine may inhibit the carboxylase via a Ca^{2+}-dependent protein kinase.

The dephosphorylated form of the carboxylase does not require citrate for activity, but the phosphorylated form of the enzyme can be activated by citrate *in vitro*. This reaction is reminiscent of the effect of glucose-6-phosphate on glycogen synthase as discussed in chapter 13. The active, dephosphorylated form of glycogen synthase has only a small requirement for glucose-6-phosphate, whereas high concentrations of this activator are required to activate the phosphorylated form of glycogen synthase.

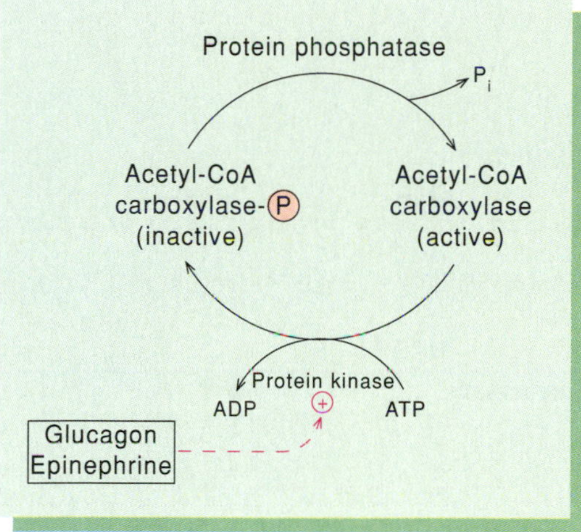

Similarly, factors that stimulate the first enzyme in the pathway for fatty acid synthesis discourage its breakdown. This is because the first enzyme in the pathway leads to the formation of malonyl-CoA, which is a potent inhibitor of carnitine acyltransferase I, an enzyme required to transport fatty acids across the inner mitochondrial membrane. In the absence of such transport there is no way in which the fatty acids can reach the active site for breakdown.

Another point to note about the controls of fatty acid metabolism is that they are designed to meet the metabolic needs of the organism. Thus an excess of palmitoyl-CoA in the cytosol indicates a lack of need for more of the same, and such a condition serves to shut off synthesis of fatty acids by inhibiting the enzymes of fatty acid synthesis, as well as to inhibit reactions that result in the production of NADPH. Similarly, an abundance of citrate indicates that the energy needs of the organism are being met, and that this would therefore be a good time to synthesize fatty acids for energy storage. Consistent with this picture, we see that citrate is a key compound involved in the synthesis of the substrates for fatty acid synthesis, NADPH and acetyl-CoA. Moreover, citrate is a specific activator of acetyl-CoA carboxylase, which carries out the first reaction in the biosynthetic pathway.

Figure 17.27

The rate of ß oxidation in the heart appears to be regulated by the feedback inhibition of thiolase by acetyl-CoA. The colored arrows indicate inhibition. The product inhibition of 3-hydroxyacyl-CoA dehydrogenase by NADH may also be an important regulatory mechanism.

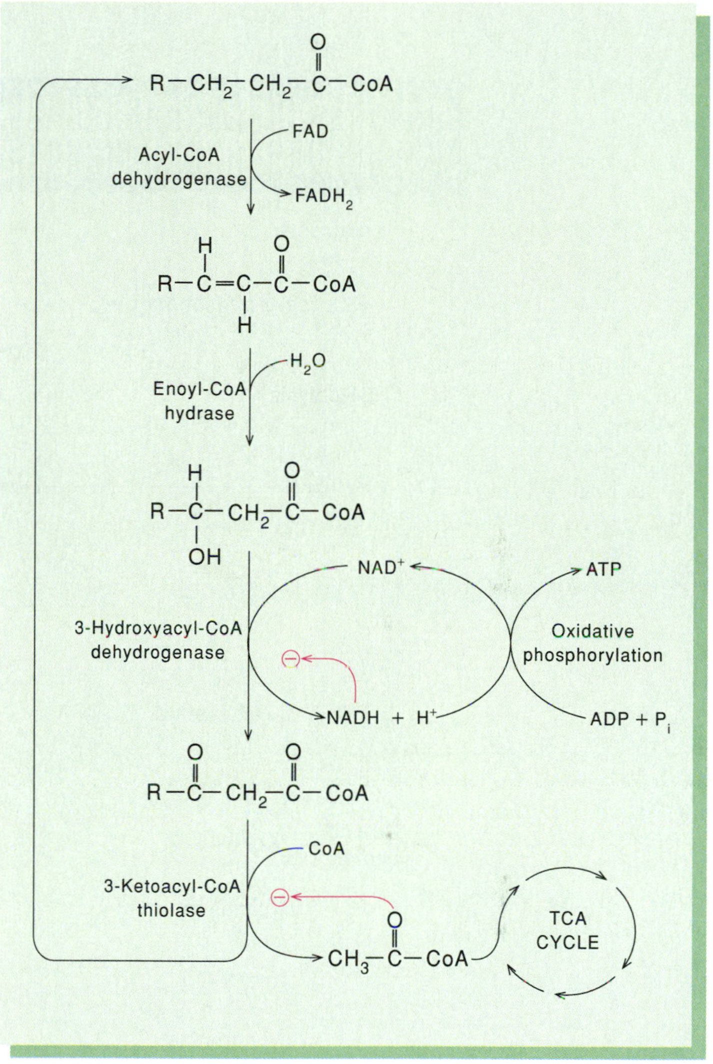

R—CH₂—CH₂—C(=O)—CoA

Acyl-CoA dehydrogenase (FAD → $FADH_2$)

R—CH=CH—C(=O)—CoA

Enoyl-CoA hydrase (H_2O)

R—CH(OH)—CH₂—C(=O)—CoA

3-Hydroxyacyl-CoA dehydrogenase (NAD^+ → NADH + H^+); Oxidative phosphorylation (ADP + P_i → ATP)

R—C(=O)—CH₂—C(=O)—CoA

3-Ketoacyl-CoA thiolase (CoA)

CH₃—C(=O)—CoA → TCA CYCLE

In broad outline, the mechanisms of synthesis and degradation of fatty acids are very similar in mammalian liver and *E. coli*. However, the control mechanisms are quite different. *E. coli* does not use fatty acids for storage of energy. Hence there is no need for elaborate controls to prevent simultaneous synthesis and degradation. In fact, just the opposite need might

Figure 17.28

Overview of the conversion of carbohydrate to lipid in rat liver cells and its regulation. Colored arrows with pluses and minuses indicate points of activation and inhibition, respectively. Glucagon and epinephrine are the main hormones involved in regulation. Citrate and palmitoyl-CoA are the main substrates involved in regulation.

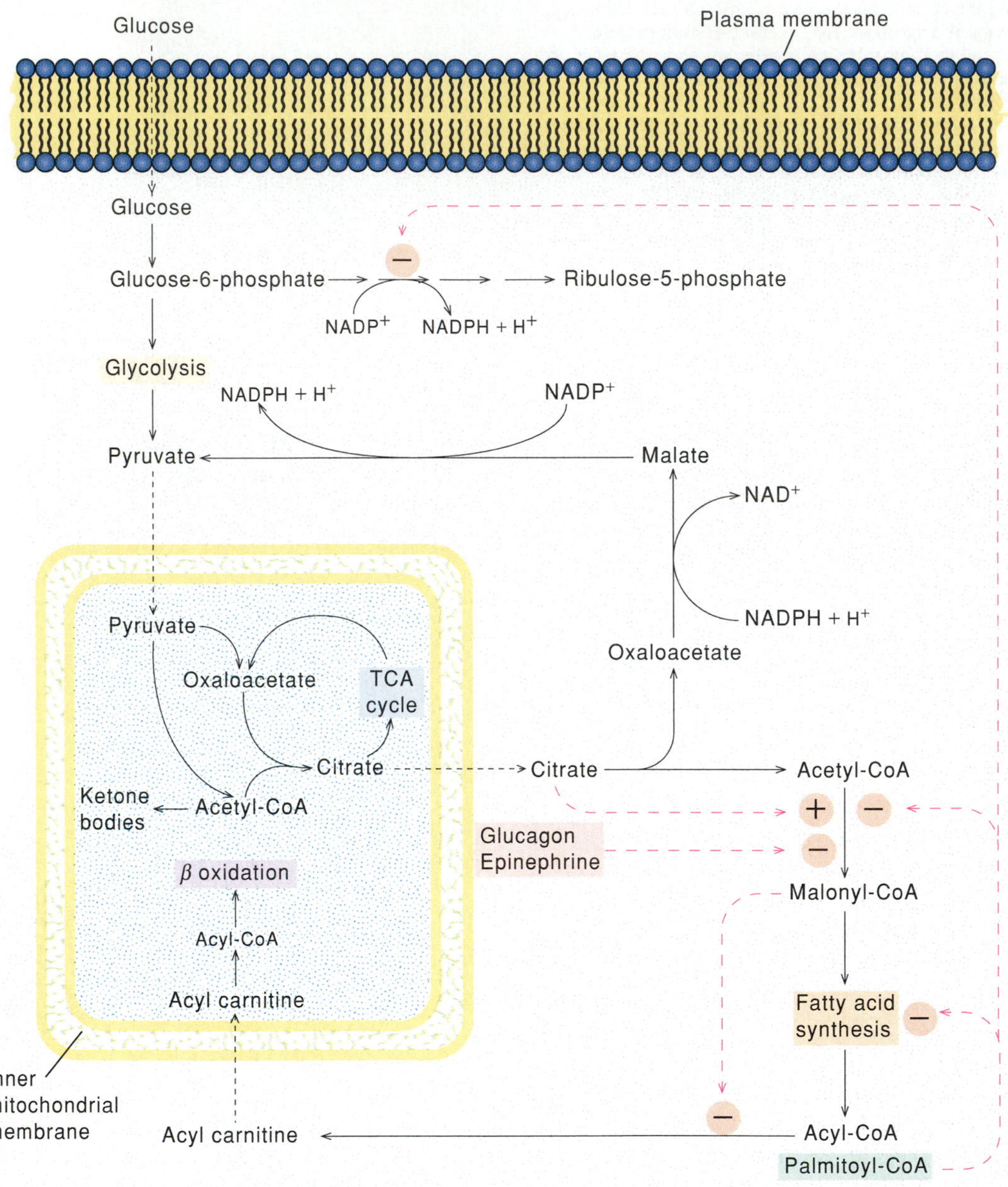

occur under circumstances where lipid was the main source of energy. Under such conditions there would be a need for lipid oxidation to supply energy and simultaneously there would be a need for lipid synthesis to satisfy structural needs. Thus it seems likely that *E. coli* has an entirely different set of controls for regulating synthesis and degradation according to its special needs. A simple end-product mechanism could regulate the synthetic pathways quite satisfactorily. As far as the degradation pathways are concerned, *E. coli* has an elaborate system (see chapter 30) that determines which carbon source will be used to supply energy. Glucose is the favored source. Lipid would not be utilized for such purposes unless the supply of glucose and other more readily metabolizable carbon sources was exhausted. Control of lipid utilization would almost certainly occur at the plasma membrane, where selective permeases would be involved.

Chain Length and Ratio of Unsaturated to Saturated Fatty Acids Are Regulated in E. coli

The chain length of fatty acids in *E. coli* is regulated by the activities and specificities of several enzymes. First, neither of the β-ketoacyl-ACP synthases of *E. coli* (see table 17.2) uses eighteen-carbon substrates effectively. Second, β-ketoacyl-ACP synthase II is required for elongation of palmitoleate ($16{:}1^{\Delta 9}$) to *cis*-vaccenate ($18{:}1^{\Delta 11}$) and is therefore responsible for *E. coli* having unsaturated fatty acids with eighteen carbons. Third, the lengths of fatty acids are also determined by the rate of utilization of acyl-ACPs for phospholipid synthesis. Thus there appears to be a competition between the rate of elongation of acyl-ACPs and that of utilization of acyl-ACPs for phospholipid synthesis. This competition appears to be the main reason why palmitate, and not stearate, is the most plentiful saturated fatty acid in *E. coli*.

The proportion of unsaturated and saturated fatty acids synthesized is regulated also. This is a sensitive function of temperature; the ratio of monounsaturated to saturated fatty acids increases as temperature decreases. It seems likely that this correlation serves the purpose of keeping the membrane in a fluid state. Remember that unsaturated fatty acids do not pack as well in membrane structures as saturated fatty acids (see chapter 7). The major fatty acid made at lower temperatures is *cis*-vaccenate, and its concentration decreases when *E. coli* is grown at higher temperatures. In contrast, the level of palmitoleate does not change with temperature. From this observation, coupled with studies on an *E. coli* mutant defective in β-ketoacyl-ACP synthase II, we may conclude that this enzyme is solely responsible for thermal regulation of the fatty acid composition of *E. coli*. How the lack of synthase II inhibits the initial introduction of a *cis* double bond by β-hydroxydecanoyl-ACP dehydrase (see fig. 17.20) is not understood.

Long-Term Dietary Changes Lead to Adjustments in the Level of Enzymes

Before closing this section on regulation, we should point out that, over an extended period, dietary conditions can result in appreciable shifts in the levels of enzymes concerned with fatty acid metabolism. For example, the concentrations of fatty acid synthase and acetyl-CoA carboxylase in rat liver are reduced four- to fivefold after fasting. When the rat is fed a fat-free diet, the concentration of fatty acid synthase is 14-fold higher than in a rat maintained on a normal rat chow. Current evidence indicates that the levels of these enzymes are governed by the rate of enzyme synthesis, not degradation. It appears that synthesis of mRNA, in turn, is controlled by the rate of transcription of DNA. A question of current interest is how this transcription of DNA is regulated.

Summary

In this chapter we have focused on the synthesis and degradation of long-chain fatty acids and on how these processes are regulated. Most of our discussion was concerned with how these reactions take place in the mammalian liver, although occasionally we referred to other animal tissues and to *E. coli* and plants. The following points are the most important.

1. Fatty acids originate from three sources: diet, adipocytes, and *de novo* synthesis.
2. The degradation of fatty acids occurs by an oxidation process in the mitochondria. The breakdown of the 16-carbon saturated fatty acid palmitate occurs in blocks of two-carbon atoms by a cyclical process. The active substrate is the acyl-CoA derivative of the fatty acid. Each cycle involves four discrete enzymatic steps. In the process of oxidation the energy is sequestered in the form of reduced coenzymes of FAD and NAD^+. These reduced coenzymes lead to ATP production through the respiratory chain. The oxidation of fatty acids yields more energy per carbon than the oxidation of glucose. This is because saturated fatty acids are in the most reduced state.
3. Unsaturated fatty acids are also oxidized in mitochondria with the help of certain additional enzymes that facilitate a continuous flow of the oxidation process. Some oxidation of fatty acids also takes place in peroxisomes, especially that involving long-chain fatty acids.
4. The main end product of fatty acid oxidation is acetyl-CoA. The acetyl-CoA can be used by the tricarboxylic acid cycle, or alternatively, ketone bodies may be formed from condensation of acetyl-CoA's. Ketone bodies are largely made in the liver and subsequently diffuse into the blood to be carried to other tissues, where they are converted into acetyl-CoA for various metabolic purposes.
5. Biosynthesis also occurs in steps of two-carbon atoms. Biosynthesis takes place in the cytosol. In addition to occurring in a different cellular compartment, biosynthesis involves totally different enzymes and different coenzymes.
6. Fatty acid synthesis takes place in seven steps. All except the first step take place on a macroenzyme complex. The intermediates on this complex are carried by attachment of the acid end in thioester linkage to phosphopantetheine coenzyme. The multienzyme complex greatly increases the efficiency of fatty acid synthesis, because for each step in the pathway the next enzyme is always near at hand, and the loss of intermediates is minimized.
7. Regulation of fatty acid metabolism takes place in such a way that synthesis and degradation never occur simultaneously, which would be wasteful. Control factors assure that synthesis occurs only when there is an energy excess, and degradation only when there is an energy need. The hormones epinephrine and glucagon stimulate degradation and inhibit synthesis. They stimulate degradation by facilitating release of fatty acids stored in the adipocytes. They inhibit synthesis by inactivating the first enzyme in the pathway for synthesis, acetyl-CoA carboxylase. Most other control factors interfere with the supply of substrate for either of the two processes.

Selected Readings

Cook, H. W., Chapter 5, Fatty acid desaturation and chain elongation in eucaryotes. In D. E. Vance and J. E. Vance (eds.), *Biochemistry of Lipids, Lipoproteins and Membranes.* Amsterdam: Elsevier Science Publishers, 1991. Provides an advanced and current treatment of fatty acid desaturation and its regulation, also cites other key references to this field.

Deuel, H. J., *The Lipids: Biochemistry,* vol. 3. New York: Interscience, 1957. A comprehensive and classical treatise on the biochemistry of lipids until the mid-1950s.

Goodridge, A. G., Chapter 4, Fatty acid synthesis in eucaryotes. In D. E. Vance and J. E. Vance (eds.), *Biochemistry of Lipids, Lipoproteins and Membranes.* Amsterdam: Elsevier Science Publishers, 1991. Provides an advanced treatment of regulation of fatty acid synthesis and cites other key references related to this topic.

Jackowski, S., J. E. Cronan, and C. O. Rock, Chapter 2, Lipid metabolism in procaryotes. In D. E. Vance and J. E. Vance (eds.), *Biochemistry of Lipids, Lipoproteins and Membranes.* Amsterdam: Elsevier Science Publishers, 1991. Contains current and advanced information on the metabolism of fatty acids in *E. coli* and other prokaryotes.

McGarry, J. E., and D. W. Foster, Regulation of hepatic fatty acid oxidation and ketone body production. *Ann. Rev. Biochem.* 49:395, 1980. A now classic review article that summarizes the evidence for the regulation of β oxidation by malonyl-CoA.

Najjar, V. A., *Fat Metabolism.* Baltimore: The Johns Hopkins Press, 1954. A good summary of the early work on fatty acid metabolism.

Schulz, H., Chapter 3, Oxidation of fatty acids. In D. E. Vance and J. E. Vance (eds.), *Biochemistry of Lipids, Lipoproteins and Membranes.* Amsterdam: Elsevier Science Publishers, 1991. Provides an advanced and current summary of fatty acid oxidation in prokaryotes and eukaryotes.

Sweetser, D. A., R. O. Heuckeroth, and J. I. Gordon, The metabolic significance of mammalian fatty acid binding proteins: abundant proteins in search of a function. *Ann. Rev. Nutr.* 7:337, 1987. Summarizes at an advanced level recent developments on understanding the structure and function of fatty acid binding proteins.

Wakil, S. J., Fatty acid synthase, a proficient multifunctional enzyme. *Biochemistry* 28:4523, 1989. Reviews the evidence for the current model of the mammalian fatty acid synthase as depicted in figure 17.13.

Problems

1. (a) Consider the complete oxidation of glucose (M_r = 180) via glycolysis and the TCA cycle and calculate the moles ATP generated during the oxidation of 1 mole of glucose to CO_2 and H_2O. Assume that the free energy of ATP hydrolysis under physiological conditions is −11 kcal/mole, and assume mitochondrial P/O ratios of 2.5 for NADH oxidation and 1.5 for succinate (or equivalent) oxidation. Estimate the free energy conserved as ATP during the oxidation. Calculate the free energy conserved per gram of glucose oxidized.
   (b) Repeat the calculations for ATP formation but consider the complete oxidation of 1 mole of palmitic acid (M_r = 256) via β oxidation and TCA cycle. Estimate the free energy conserved as ATP energy from palmitate oxidation. Calculate the free energy conserved per gram of palmitic acid oxidized.
   (c) Bearing in mind that respiratory metabolism is an oxidative process, how might you explain the difference in energy content of fat and carbohydrate on a weight basis?
   (d) Explain the rationale behind the use of fat rather than carbohydrate as energy reserve in plants and animals.
2. Explain the role of carnitine acyltransferases in fatty acid oxidation.
3. Carnitine deficiency in liver is correlated with hypoglycemia. Suggest a plausible explanation for hypoglycemia in the carnitine-deficient human.
4. Explain why α oxidation is an obligatory step in the oxidation of phytanic acid.
5. How might a deficiency of vitamin B_{12} affect oxidation of propionyl-CoA formed during β oxidation of odd-chain-length fatty acids or of pristanic acid?
6. Oxidation of reduced fatty acyl-CoA dehydrogenase requires electron-transfer flavoprotein (ETF) and the enzyme ETF-ubiquinone oxidoreductase.
   (a) Write the reaction catalyzed by the ETF-ubiquinone oxidoreductase. Why may we consider the enzyme essentially irreversible in the mitochondria under physiological conditions?
   (b) What effect would you expect a nonreducible structural analog of ubiquinone to have on ETF-ubiquinone oxidoreductase activity? On the fatty acyl-CoA dehydrogenase activity of mitochondria?
7. (a) Liver mitochondria convert long-chain fatty acids to ketone bodies (acetoacetate and β-hydroxybutyrate) that are subsequently transported in the plasma to nonhepatic tissues. Suggest some metabolic advantages of supplying ketone bodies to nonhepatic tissues.
   (b) In what way is β-hydroxybutyrate a better energy source than acetoacetate for nonhepatic tissues?
   (c) Outline the oxidation of β-hydroxybutyrate to acetyl-CoA in heart mitochondria.
8. Predict the effect on oxidation of ketone bodies and of glucose in nonhepatic tissue of individuals with markedly diminished β-oxyacid-CoA-transferase activity. Predict the effect if the activity were absent.
9. What is the role of the NADPH-dependent 2,4-dienoyl-CoA reductase in the oxidation of linolenoyl-CoA?

10. Except for malonyl-CoA formation, all the individual reactions for palmitate synthesis reside on a single multifunctional protein (fatty acid synthase) in animal cells. It has been shown that a dimer of the multifunctional protein is required to catalyze palmitate synthesis. Explain the molecular basis of this observation.

11. (a) For an *in vitro* synthesis of fatty acids with purified fatty acid synthase, the acetyl-CoA was supplied as the ^{14}C-labeled derivative

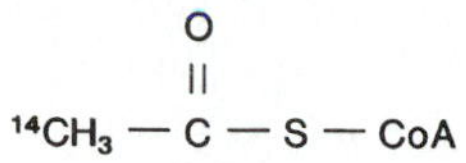

$$^{14}CH_3-\overset{\overset{\displaystyle O}{\|}}{C}-S-CoA$$

The other reactants, including the malonyl-CoA were not radioactive. Where would the ^{14}C-label be found in palmitic acid?

(b) If the malonyl-CoA were supplied as the only labeled compound deuterated as shown below, how many deuterium atoms would be incorporated in palmitate? On which carbon(s) would these deuterium atoms reside?

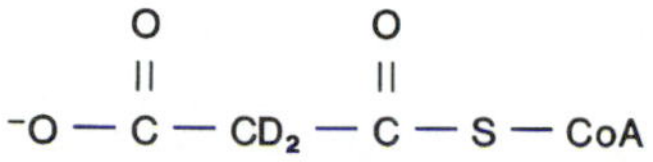

$$^{-}O-\overset{\overset{\displaystyle O}{\|}}{C}-CD_2-\overset{\overset{\displaystyle O}{\|}}{C}-S-CoA$$

(c) If [3-^{14}C]malonyl-CoA (shown below) were used in the reaction, which atoms in palmitate would be labeled? Why?

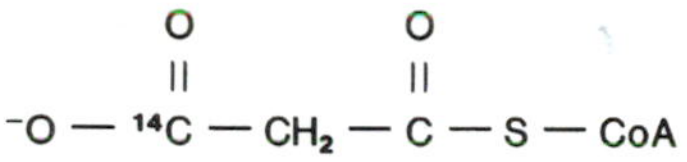

$$^{-}O-\overset{\overset{\displaystyle O}{\|}}{^{14}C}-CH_2-\overset{\overset{\displaystyle O}{\|}}{C}-S-CoA$$

12. Citrate is both a lipogenic substrate and a regulatory molecule in mammalian fatty acid synthesis.
(a) Explain each function of citrate in fatty acid synthesis.
(b) Write reactions (including structures) outlining the role of citrate as a lipogenic substrate.

13. What are the metabolic sources of NADPH used in fatty acid biosynthesis? How many moles of NADPH are required for the synthesis of 1 mole of palmitic acid from acetyl CoA?

14. (a) Why is the location of biosynthesis and β oxidation of fatty acids in separate metabolic compartments essential to regulation of fatty acid metabolism in the hepatocyte?
(b) Would you expect an inhibitor of the extramitochondrial carnitine acyltransferase to mimic the effect of malonyl-CoA on β oxidation? (Assume that the inhibitor can penetrate the cell membrane.) Explain the rationale for your answer.

15. Which catalytic activity of the mammalian fatty acid synthase determines the chain length of the fatty acid product?

Biosynthesis of the Building Blocks

5

PART

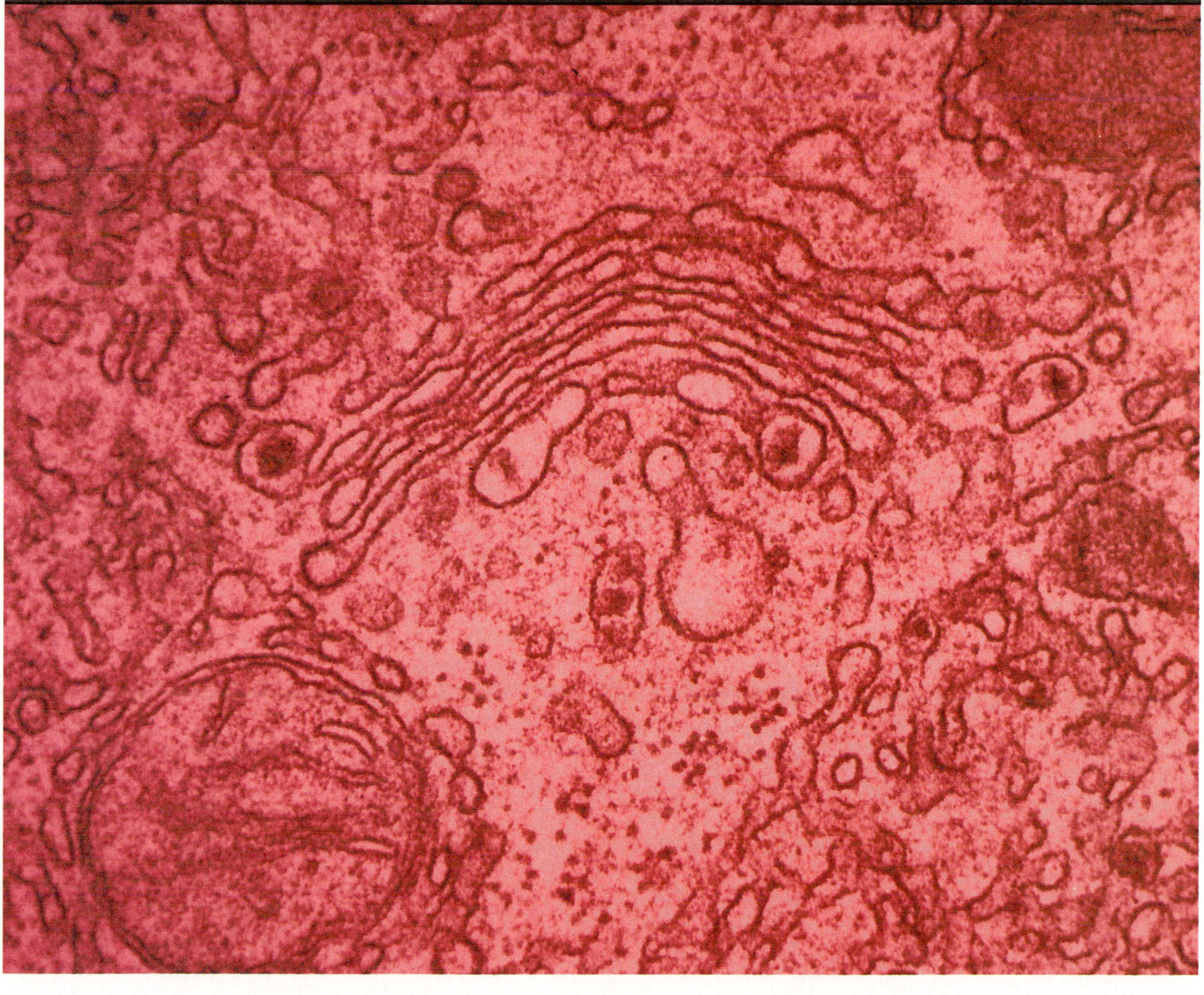

Electron micrograph of a thin section from a human pancreatic cell showing the Golgi and smooth endoplasmic reticulum. (© Robert Bollender/Visuals Unlimited.)

Proteins, carbohydrates, and lipids are complex molecules that are constructed from far simpler molecular building blocks. For each of these species of biomolecules the building blocks are similar in size and possess the same functional groups for polymerization. In autotrophs—the most self-sufficient organisms—the synthesis of these building blocks starts from N_2, H_2O, and CO_2. In heterotrophs, it normally starts from the carbon skeletons of low-molecular-weight carbohydrates and a nitrogen source, usually ammonia or amino acids. The biochemical energy and reducing power derived from catabolism are used to drive the anabolic processes of the cell. Following synthesis, the building blocks are activated through the formation of high-energy esters so that they can be readily condensed into larger molecules.

In part 5 we will limit ourselves to discussing the biosynthesis of amino acids, nucleotides, carbohydrates, and lipids; the synthesis of nucleic acids and of proteins will be treated in part 6. We begin part 5 by examining amino acid synthesis and utilization (chapters 18 and 19). Then we devote one chapter each to nucleotides (chapter 20) and carbohydrates (chapter 21) and two chapters to lipids (chapters 22 and 23). In conclusion, in chapter 24 we discuss the integration of metabolism. Because of the dominant role of chemical messengers in the integration of metabolism in multicellular eukaryotes, we also include a detailed discussion of hormones (chapter 24).

Biosynthesis of Amino Acids

mino acids are best known as the building blocks of protein, and indeed that is a main function of the twenty L-amino acids most commonly found in proteins (see chapter 3). However, not all amino acids are found in proteins. Many nonprotein amino acids are toxins, precursors of the compounds that form cell walls, or components of a variety of biologic peptides. Moreover, even those amino acids that can serve as the building blocks of proteins have other functions. They are precursors to many important small-molecule compounds, including nucleotides (see chapter 20) and porphyrins (see chapter 19); they form parts of lipid molecules (see chapter 22); and they are precursors for several coenzymes. Amino acids may also be deaminated and their carbon skeletons used to synthesize other molecules or catabolized to release energy (see chapter 19).

Figure 18.1

Outline of the biosynthesis of the twenty amino acids found in proteins. The *de novo* biosynthesis of amino acids starts with carbon compounds found in the central metabolic pathways. The central metabolic pathways are drawn in black and the additional pathways are drawn in color. Some key intermediates are illustrated and the number of steps in each pathway is indicated alongside the conversion arrow. All amino acids are emphasized by boxes. Dashed arrows from pyruvate to both diaminopimelate and isoleucine reflect the fact that pyruvate contributes some of the side-chain carbon atoms for each of these amino acids. Note that lysine is unique in that two completely different pathways exist for its biosynthesis.

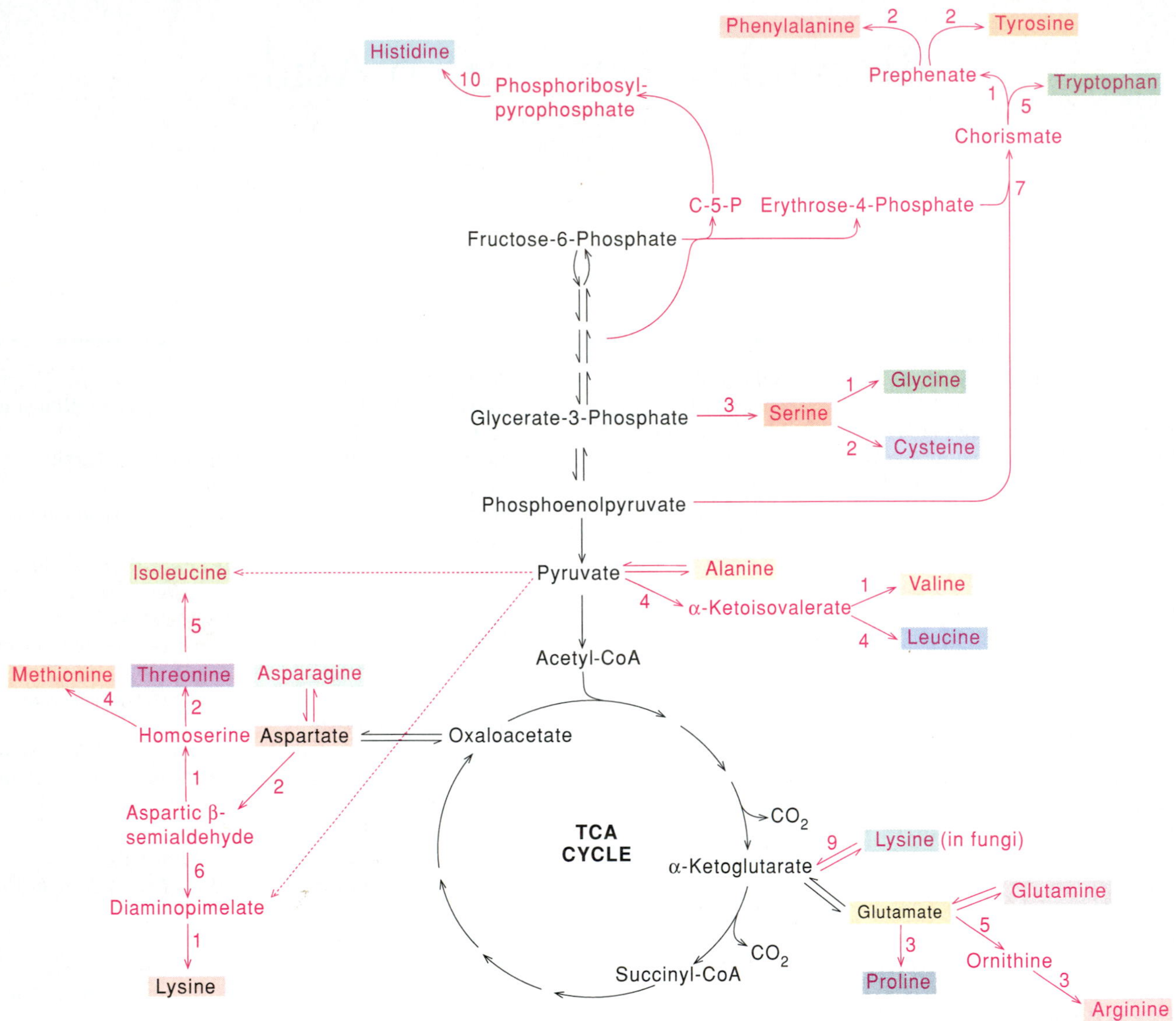

In this chapter we will focus on the biosynthesis of amino acids (fig. 18.1) and their role in bringing inorganic nitrogen and sulfur into the biological world. In the following chapter we will consider other aspects of amino acid metabolism, their catabolism, and their role as precursors of other biomolecules.

We will begin by considering the biosynthesis of amino acids in microorganisms. Then we will consider the amino acid pathways that occur in mammals. Late in the chapter we will briefly discuss nonprotein amino acids, and finally, we will examine the problem of nitrogen fixation in biological systems.

Using Microorganisms for the Study of Amino Acid Biosynthesis

It is not surprising that *E. coli* has served as the organism of choice for examining the pathways for amino acid biosynthesis. *E. coli* has the capacity for synthesizing all twenty amino acids commonly found in proteins and it also is an ideal organism for combining genetic and biochemical techniques to analyze these pathways. The principles involved are similar to those we discussed in connection with analysis of multistep pathways in chapter 12.

18A

BOX

Genetic Complementation and the Analysis of Biochemical Pathways

One of the main goals in genetics research is to determine the number of genes involved in conferring a particular trait. For this purpose, the single most powerful technique is complementation analysis. To apply this technique, we first isolate a large number of mutants possessing the abnormal phenotype of interest. Then we mate the mutants, ideally in all possible pairwise combinations, to see whether they complement, i.e., to see whether they compensate for each other's defects. Finally, we use the phenotypes of the offspring of these mated pairs to classify the mutants into the same or different complementation groups (i.e., complementing pairs), according to whether the offspring show mutant-type or wild-type phenotypes.

For example, let us consider the trait for the ability to synthesize tryptophan. The enzymes needed in this pathway are encoded by a set of genes that we designate *trp*. In *E. coli*, the wild-type phenotype is TRP$^+$, whereas mutants that have defects in one or more of the *trp* genes are TRP$^-$. Although bacteria do not mate in the conventional manner, there are several ways of constructing merodiploids, i.e., cells that contain partial gene duplications, and for present purposes we can treat clones of these cells as the equivalent of the F_1 generation in a standard genetic cross. For a single defect, if two sets of the *trp* genes are present in the same cell, two phenotypes are possible. Either the same gene will be defective in both sets of genes and the cell will remain TRP$^-$, or different genes will be defective in the paired sets and the cell will be TRP$^+$. The TRP$^+$ phenotype is easy to score merely by growing the mated pairs on agar plates containing minimal medium (see fig. 12.14). When we do this, we find that the *trp*$^-$ mutants fall into five complementation groups (see illustration). By further analysis we can show that each of these complementation groups corresponds to a gene encoding one of the five proteins or protein subunits involved in tryptophan biosynthesis.

It is important to appreciate what complementation analysis tells us about a biochemical pathway like tryptophan biosynthesis and what questions it leaves open. In the present example, the complementation analysis tells us that there are five distinct proteins, one associated with each complementation group, but it does not tell us what functions are served by each protein, nor does it tell us the order in which they function. These are questions that need to be investigated through biochemical analysis of the kind described in the text.

Figure 1

Complementation analysis of *trp*$^-$ mutants of *E. coli*. A large number of *trp*$^-$ mutants are isolated. These mutants are crossed in pairwise fashion. Crosses result either in no clones that grow on minimal medium (−) or some clones that grow on minimal medium (+). On the basis of this simple ± test, mutants can be classified into five complementation groups. These are labeled *E, D, C, B,* or *A*. Members within the same complementation group do not produce *trp*$^+$ progeny. Those from different complementation groups usually do.

Step 1. Isolate a large number of *trp* mutants

trp_1
trp_2
trp_3
trp_4
⋮
trp_n

Step 2. Carry out crosses between various mutants and examine phenotypes of resulting crosses; for example.

$trp_1 \times trp_2 \rightarrow$ TRP$^-$
$trp_1 \times trp_3 \rightarrow$ TRP$^-$
$trp_1 \times trp_4 \rightarrow$ TRP$^+$

Step 3. Classify mutants. Mutants that give wild-type phenotype in a cross contain mutations in different complementation groups. Mutants that give mutant phenotype in a cross contain mutations in the same complementation group. In this way five different complementation groups are found.They are labeled A, B, C, D, E.

Step 4. Crosses between mutants in different complementation groups should give results shown in table.

	A	B	C	D	E
A	−	+	+	+	+
B	+	−	+	+	+
C	+	+	−	+	+
D	+	+	+	−	+
E	+	+	+	+	−

Typically, the investigation of an amino acid biosynthetic pathway in *E. coli* begins with the accumulation of mutants that are deficient in only one amino acid needed for growth (auxotrophs). Such a deficiency arises from a mutation in a single gene that encodes an enzyme required for a single step in the biosynthesis of the amino acid. By complementation analysis (box 18A), we should be able to determine how many steps there are in the synthesis of a particular amino acid. Complementation analysis involves combining the relevant genes from two mutants into one cell. If two mutants do not complement (auxotrophic condition), they must have a defect in the same gene. If the two mutants complement (prototrophic condition), so that the cell regains its capacity to synthesize the amino acid, the defects must be in different genes. As a rule, mutations in

Figure 18.2

Immediate consequences of a point mutation in a biosynthetic pathway. When the mutation leads to an inactive enzyme, the chain of reactions leading to the end product in the pathway is broken, and frequently large amounts of intermediate are produced, accumulate in the cell, and may leak to the environment.

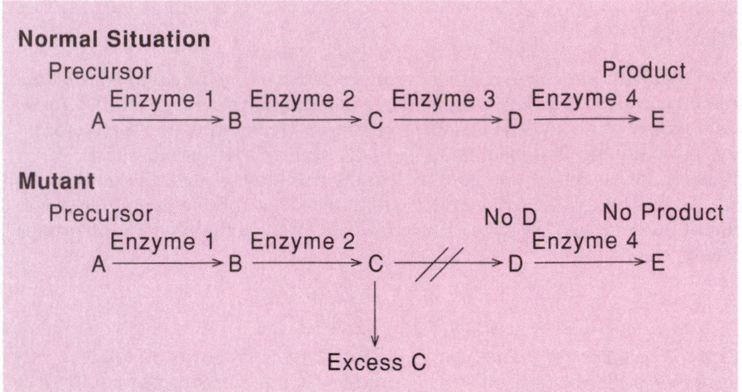

the same gene do not complement; such mutations are said to be in the same complementation group. Conversely, mutations in different genes usually do complement. If there are five complementation groups that affect the synthesis of a particular amino acid, it usually means that there are five genes and, correlated with them, five enzymes involved in the synthesis of the amino acid.

By revealing the number of enzymes involved, complementation group analysis sets the stage for biochemical analysis. We know that a mutation in a single gene encoding an enzyme required for a single step in the pathway has two major consequences (fig. 18.2; see also fig. 12.15): (1) no amino acid is produced, and (2) the substrate for the defective enzyme is produced in large excess. The abundance of substrate is a clue that helps us to isolate an intermediate or a closely related derivative from mutant cells. By determining the structure of intermediates isolated from different mutants and piecing this information together, we can build up an approximate picture of the pathway.

In the next phase of analysis, we isolate the relevant enzymes from whole cells. To this end, we first make extracts from normal (wild type) or mutant cells. Then we test these extracts in conjunction with different intermediates to see whether the extracts can carry out the conversion of one intermediate to the next in the pathway. By progressive subfractionation and testing of such an extract, we reach a point at which the isolated enzyme is reasonably pure. The process of isolating all the enzymes for a particular pathway usually takes many years of investigation, but it is absolutely necessary if we wish to get a complete picture of the pathway. Then, in the final phase of analysis, we characterize the purified enzymes by their structure and the reactions they catalyze.

The Pathways to Amino Acids Arise as Branchpoints from a Few Key Intermediates in the Central Metabolic Pathways

A detailed analysis of the amino acid biosynthetic pathways shows that all amino acids arise from a few intermediates in the central metabolic pathways (see fig. 18.1). Amino acids that arise from a common intermediate are said to be in the same amino acid family. In the next six sections, our discussion of pathways is organized under the headings of different amino acid families.

As we will see, most of the carbon flow from the central metabolic routes into amino acids is irreversible. However, in nearly all cases the flow is an orderly one that provides the cell with amino acids in the amounts needed for maintenance and growth. The orderly flow is achieved by relatively simple, yet almost faultless regulatory mechanisms.

The Biosynthesis of Amino Acids of the Glutamate Family: L-Glutamate, L-Glutamine, L-Proline, and L-Arginine

A common element in all amino acids is the α-amino group. Directly or indirectly, the α-amino groups are all derived from ammonia by way of the amino groups of L-glutamate. The other amino acids of the glutamate family are glutamine, proline, and arginine. In fungi, lysine is also included in this family. Here we will elaborate only on the pathways for glutamine, proline, and arginine.

The Direct Amination of α-Ketoglutarate Leads to Glutamate

The simplest route to glutamate (and therefore, amino group formation) is that exhibited by many bacteria when grown in a medium containing an ammonium salt as the sole nitrogen source. The reaction is a reductive amination catalyzed by glutamate dehydrogenase (fig. 18.3). In organisms such as *E. coli,* the enzyme is specific for NADPH as the hydrogen donor, as we might expect of a biosynthetic reaction involving a reductive step.

Some organisms have both an NAD^+/NADH- and an $NADP^+$/NADPH-dependent glutamate dehydrogenase. In such cases, the NAD^+-dependent enzyme is considered to play a catabolic role that converts glutamate back to α-ketoglutarate. Both enzymes, however, catalyze reversible reactions, and assignment to anabolic or catabolic functions is largely inferential. On the other hand, in some organisms (e.g., certain water molds), the presumed catabolic enzyme is inhibited by ATP, CTP, and fructose bisphosphate, indicators of a high energy charge, and is stimulated by AMP, an indicator of a low energy charge.

Figure 18.3

The conversion of ammonia into the α-amino group of glutamate and into the amide group of glutamine. The direct amination of α-ketoglutarate by NH_4^+ occurs only under conditions of high NH_4^+ concentrations, which are rarely found in nature.

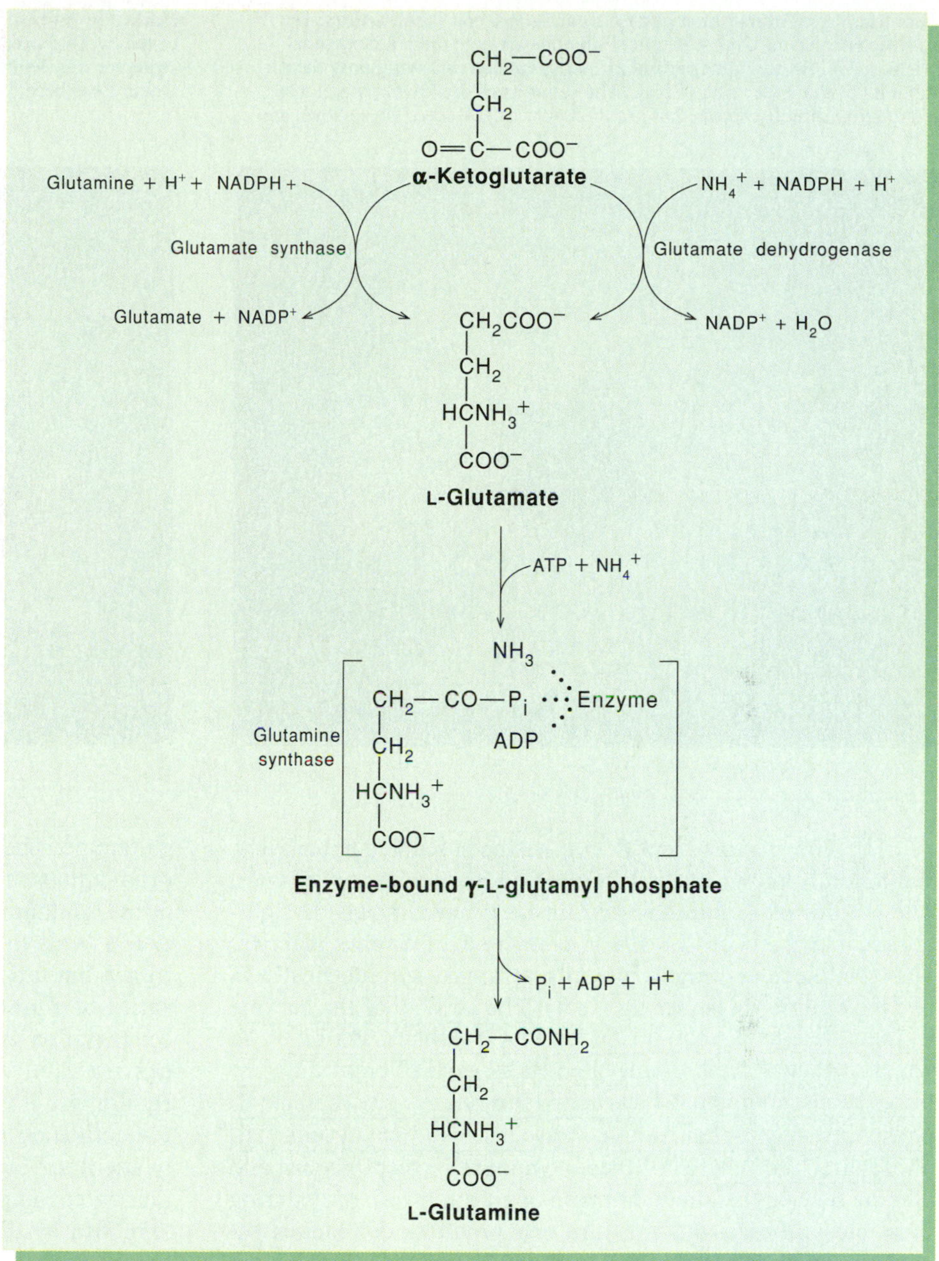

Studies on the glutamate dehydrogenases from green plants revealed that the plant enzymes can use either NADH or NADPH as the hydrogen donor in the amination reaction. However, for reasons given a little later, it is unlikely that the glutamate dehydrogenase of plants plays a significant role in glutamate biosynthesis. It is more likely that in green plants, as in many bacteria and fungi, and in probably all bacteria when they are not being grown under NH_4^+ excess, the formation of the amino group of glutamate occurs not from NH_4^+, but from the amide group of glutamine (see the following discussion). In such cases, the primary conversion from NH_4^+ to organic nitrogen is catalyzed by glutamine synthase (fig. 18.4).

Amidation of Glutamate to Glutamine Is a Highly Regulated Process

The reaction catalyzed by glutamine synthase is one that involves activation of the γ-carboxyl group of glutamate to yield a γ-glutamyl enzyme complex and the cleavage of ATP to ADP and P_i (see fig. 18.3). In a second step, γ-glutamyl transfer to NH_4^+ occurs.

Glutamine synthase is a key enzyme in the flow of nitrogen to organic compounds, and its activity is subject to elaborate controls that sense the cell's need for nitrogen-containing compounds (fig. 18.5).

Figure 18.4

The structure of glutamine synthase from *Salmonella typhimurium*. The enzyme consists of twelve identical subunits arranged like a hexagonal prism. (*a*) View down the sixfold axis of symmetry, showing only the six subunits of the upper ring in pink. The subunits of the lower ring, in blue are roughly directly below those of the upper ring. Active sites shown are marked by pairs of Mn^{2+} ions (white spheres). (*b*) Side view along one of the twofold axes, showing only the six nearest subunits. This structure was determined by x-ray crystallography. (Courtesy of David Eisenberg, UCLA.)

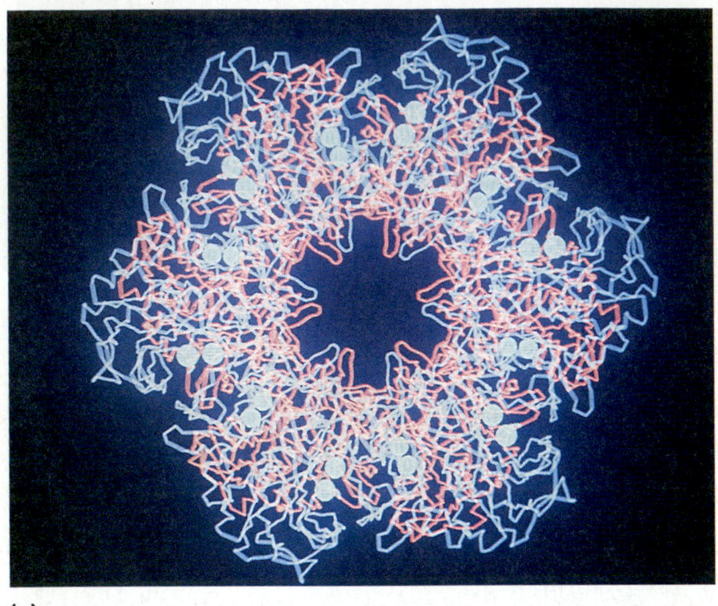

(a)

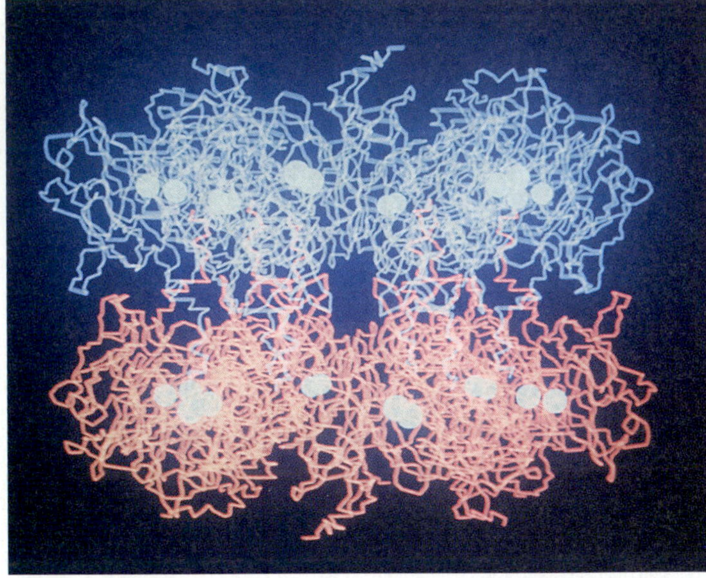

(b)

The enzyme from *E. coli* has been studied extensively, and much is known of its structure and its regulatory behavior. The studies of Stadtman and Ginsberg have shown that glutamine synthase of *E. coli* is composed of twelve identical 50,000-molecular-weight subunits arranged symmetrically in two hexameric rings (see fig. 18.4). The activity of the enzyme is regulated in a complex pattern of feedback inhibition in which a partial inhibition is effected by each of eight different nitrogenous compounds: carbamylphosphate, glucosamine-6-phosphate, tryptophan, alanine, glycine, histidine, cytidine triphosphate, and AMP. All these compounds except glycine and alanine receive the amide nitrogen of glutamine directly during their biosynthesis, and thus are end products of reaction sequences leading from glutamine. Glycine and alanine, although not direct end products, could be looked upon as "indicators" of the sufficiency of the nitrogen supply of the cell.

Even more important in the regulation of *E. coli* glutamine synthase activity is the reversible ATP-dependent adenylylation of a specific tyrosyl residue on each subunit. As the enzyme becomes progressively more adenylylated (up to the fully adenylylated form of twelve AMP groups per enzyme molecule), the enzyme becomes progressively less active.

The adenylylation reaction and its reversal by a phosphorolytic deadenylylation are regulated by the nitrogen supply in the cell. The immediate small-molecule effectors of this regulatory system are glutamine and α-ketoglutarate. A high glutamine concentration or a high glutamine/α-ketoglutarate ratio signals nitrogen excess. Conversely, a high α-ketoglutarate/glutamine ratio signals nitrogen limitation. Nitrogen excess leads to inactivation of glutamine synthase, just as nitrogen limitation leads to activation (see fig. 18.5). The activation of glutamine synthesis favors fixation of ammonia in a condensation with glutamate to form glutamine. Although the enzyme catalyzes the reverse reaction (like any enzyme), the equilibrium is very much in favor of glutamine formation. Both the activation (deadenylylation) and the inactivation (adenylylation) are controlled by a cascade of regulatory interactions illustrated in figure 18.5. As a rule, a metabolic signal is greatly amplified by the cascade arrangement. The cascades involved here are unusual in that the same two regulatory proteins function in both cascades. Both regulatory proteins have two enzymatically distinct sites arranged so that only one site is active. Which site is active in the two regulatory proteins depends on the relative concentrations of glutamine and α-ketoglutarate, as already indicated.

Let us first consider the situation under conditions of nitrogen excess (see fig. 18.5). The first regulatory protein in the cascade at high glutamine is converted into a uridylyl-removing enzyme. This enzyme hydrolyzes UMP from a PII · 4UMP protein that acts in concert with adenylyltransferase, at a high glutamine/α-ketoglutarate ratio, to adenylate glutamine synthase. The resulting adenylated enzyme is inactive.

Figure 18.5

Regulation of glutamine synthase in *E. coli*. The activity of glutamine synthase is inhibited by adenylylation. Both adenylylation and deadenylylation are regulated by a cascade of controls that are responsive to the concentrations of glutamine and α-ketoglutarate. The concentrations of these two compounds are an indication of the nitrogen supply. Thus under conditions of nitrogen excess (top part of figure) the concentration of glutamine is high and the ratio of glutamine to α-ketoglutarate is high. This leads to inactivation of the glutamine synthase. Under conditions of nitrogen limitation (bottom part of figure) the α-ketoglutarate concentration is high and the ratio of α-ketoglutarate to glutamine is high, so the glutamine synthase is activated.

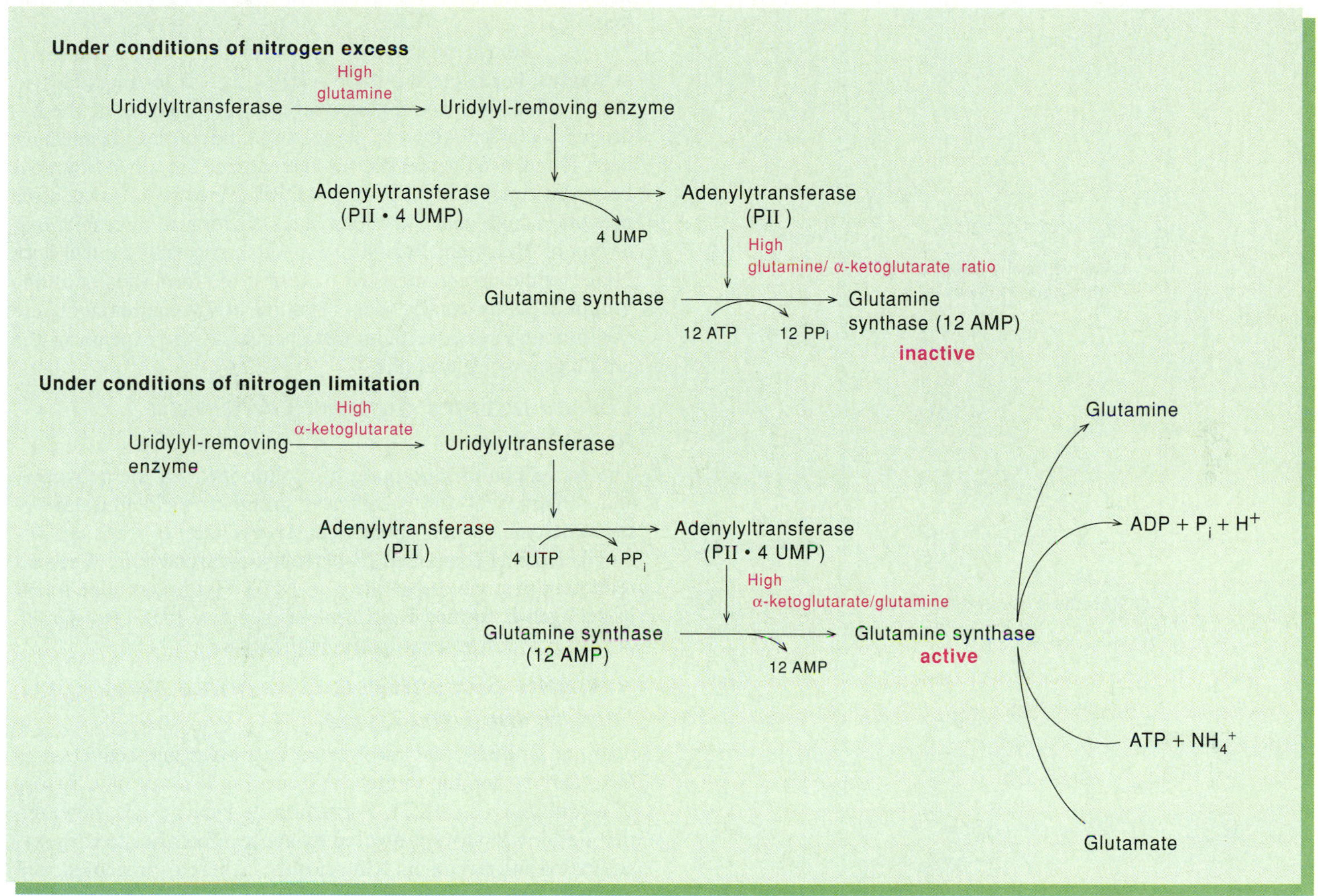

Under conditions of nitrogen limitation, the first regulatory enzyme in the cascade is converted into a uridylyltransferase at high α-ketoglutarate concentrations. This enzyme uridylates the PII protein. The PII · 4UMP protein and adenylyltransferase deadenylates the glutamine synthase · 12AMP when the α-ketoglutarate/glutamine ratio is high.

The control of glutamine synthase by this pattern of adenylylation and deadenylylation may be limited to some of the Gram negative bacteria. There may, however, be different physiologically analogous modifications of other glutamine synthases. For example, at least some of the glutamine synthases of eukaryotic cells are octameric structures that respond to regulation by dissociation to a tetrameric form under conditions of restricted NH_4^+ supply.

The Amination of α-Ketoglutarate by the Amide Group of L-Glutamine Also Leads to Glutamate

The direct reductive amination of α-ketoglutarate with ammonia described earlier (see fig. 18.3) is probably an exception in nature. Only a few forms of life make glutamate in this way, and they do so only when they are grown in the presence of a high concentration of NH_4^+. The prevalent reaction is the amination of α-ketoglutarate by glutamate synthase, a reductive reaction in which the source of the amino group is not ammonia itself but an "activated" form, the amide group of glutamine, which must be amidated again to sustain the continued synthesis of amino groups (see fig. 18.3). Although this mode of

Figure 18.6

The biosynthesis of proline. Proline is synthesized in three steps from L-glutamate. The middle step involves a cyclization that occurs spontaneously.

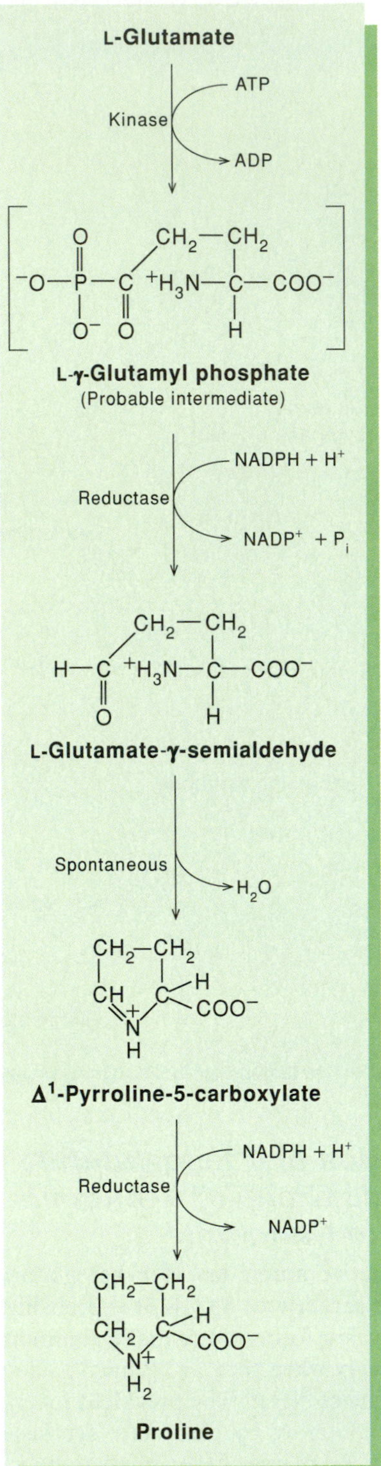

amino group formation requires a high-energy phosphate for the formation of the amide group, there is considerable advantage, since amide group formation can proceed at a much lower concentration of NH_4^+ than can amination by glutamate dehydrogenase. For example, the K_m for NH_4^+ of the *E. coli* glutamate dehydrogenase is 1.1 mM, whereas that of the *E. coli* glutamine synthase is about 0.2 mM. (Furthermore, the equilibrium of the direct amination reaction strongly favors deamination.)

The hydrogen donor varies for the various glutamate synthases. For bacteria, it is NADPH, as it is for most biosynthetic reductive reactions. For several fungi, however, the hydrogen donor is NADH. In plants, two kinds of the enzyme have been found. One is specific for ferredoxin; the other functions *in vitro* with either NADH or NADPH. Since NH_4^+ is not often present in high concentrations in soil (owing to microbial oxidation of NH_4^+ and regulatory mechanisms that control both NO_3^- reduction and nitrogen fixation), the formation of amino groups in plants usually occurs in a low NH_4^+ environment, and the amination of α-ketoglutarate occurs at the expense of the amide group of glutamine.

Three Enzymes Convert Glutamate to Proline

The conversion of glutamate to proline involves the activation and reduction of the γ-carboxyl group to yield glutamic-γ-semialdehyde, which spontaneously cyclizes to yield a five-membered ring compound Δ¹-pyrroline-5-carboxylate. A second reduction then yields proline (fig. 18.6). Hydroxyproline found in collagen is formed from proline (see box 18B; also see fig. 29.23) as a posttranslational modification.

Arginine Biosynthesis Uses Some Reactions Seen in the Urea Cycle

Another pathway that involves an activation and reduction of the γ carboxyl of glutamate is that leading to L-arginine by way of L-ornithine (fig. 18.7). For ornithine biosynthesis, however, the α-amino group is protected by acetylation prior to carboxyl activation and reduction. Ring closure is thereby prevented, and glutamyl residues destined for arginine biosynthesis are effectively sequestered from those destined for proline biosynthesis.

Ornithine is converted to arginine by means of a series of reactions that we will encounter when we discuss urea formation in chapter 19.

The Biosynthesis of Amino Acids of the Serine Family (L-Serine, Glycine, and L-Cysteine) and the Fixation of Sulfur

The diversion of 3-phosphoglycerate from the glycolytic pathway into the serine biosynthetic pathway is important, not only for the formation of L-serine, L-cysteine, and glycine needed for incorporation into protein, but also for other functions these amino

Hydroxyproline Is Formed After the Collagen Polypeptide Chain Is Made

Twenty amino acids are incorporated into polypeptide chains at the time of polypeptide synthesis. Following polypeptide synthesis, changes occur in a number of the amino acid side chains, greatly increasing the variety of structures that proteins can form. Notable is the formation of hydroxyproline in collagen. Collagen contains a high percentage of 4-hydroxyproline (see chapter 5), which is formed from proline after the proline is incorporated into the collagen polypeptide chain.

Prolyl hydroxylase

Ascorbate, Fe^{2+}

This can be shown by using radioactively labeled precursors. If ^{14}C-labeled 4-hydroxyproline is administered to rats, the collagen synthesized is nonradioactive. In contrast, if ^{14}C-labeled proline is administered, both the proline and the hydroxyproline become labeled. The formation of hydroxyproline from proline is catalyzed by the enzyme prolyl hydroxylase, which requires ascorbic acid (vitamin C) for activity. A deficiency of ascorbic acid in the diet leads to the disease known as scurvy, which is associated with skin lesions, blood vessel fragility, and poor wound healing. Scurvy appears to be the result of a weakening in the structure of the collagen fibers.

L-Ascorbic acid (vitamin C)

acids serve. For example, the conversion of serine to glycine serves to generate one-carbon units that can be used in purine, thymine, and methionine biosynthesis as well as to replenish the methyl group transferred from methionine in many methylation reactions. The carbons of glycine contribute to purine and heme-containing compounds, to glutathione (as do those of cysteine), and in animals, to certain detoxification products. Oxidation of glycine provides an additional source of one-carbon units from the α carbon. In many plants and microorganisms, sulfur in the form of sulfide is first incorporated into cysteine and then later is transferred to methionine and other sulfur-containing compounds. In the process, the carbons of cysteine are returned to the glycolytic pathway in the form of pyruvate. Serine itself is incorporated directly into phospholipids and into tryptophan, which, therefore, also might be considered a member of the serine family of amino acids. However, it will be more appropriate to consider tryptophan biosynthesis later, along with the formation of the other aromatic amino acids.

Three Enzymes Convert 3-Phospho-D-Glycerate (Glycerate-3-Phosphate) to Serine

The initial reaction in L-serine biosynthesis is the oxidation of 3-phosphoglycerate by an NAD^+-linked dehydrogenase (fig. 18.8). Even though the reaction is freely reversible, the enzyme activity is regulated by the end product of the biosynthetic sequence, serine. The product of the reaction, 3-phosphohydroxypyruvate, is converted to *O*-phosphoserine by a specific phosphoserine-glutamate transaminase. Removal of the phosphate group by a specific phosphoserine phosphatase yields serine. Phosphoserine is also found as a protein constituent, but it is almost always derived from the posttranslational phosphorylation of a seryl residue by a specific protein kinase. Such kinases serve an important function in regulating the activity of some enzymes (e.g., see chapters 10 and 13).

Two More Enzymes Convert L-Serine to Glycine

The pathway from serine to glycine consists of a single complex step catalyzed by serine hydroxymethyltransferase (fig. 18.9). The reaction is a pyridoxal-phosphate-dependent aldol cleavage to yield glycine and an "active" formaldehyde that is transferred to a tetrahydrofolate cofactor. This reaction is an important one in supplying one-carbon units for other biosynthetic reactions. It has been calculated, however, that the amount of methylene tetrahydrofolate generated during the biosynthesis of the glycine in bacterial protein is not quite sufficient to account for the synthesis of those cellular constituents that require one-transfers of carbon for their formation (you may wish to review the reactions involving the folic acid coenzymes in chapter 11). The most likely source of these required extra one-carbon units is the α carbon of glycine; glycine is converted to NH_3, CO_2, and a one-carbon unit as 5,10-methylene tetrahydrofolate by a four-protein complex, the glycine cleavage enzyme. This system has been found in bacteria, plants, and animals.

Figure 18.7

The biosynthesis of arginine. This synthesis begins by the N-acetylation of the α-amino group. The γ-carboxyl group is phosphorylated before reduction to an aldehyde that is subsequently transaminated. In some organisms the acetyl group, whose purpose is to protect the α-amino group from reacting during intermediate steps, is removed as shown. In others, it is preserved by transfer to another glutamate as the initial step in arginine biosynthesis. In the latter, the acetyl-CoA-dependent formation of *N*-acetylglutamate serves only an anaplerotic role. The ornithine formed after the deacetylation is converted to arginine by means of a series of reactions that are encountered in urea formation (see fig. 19.6).

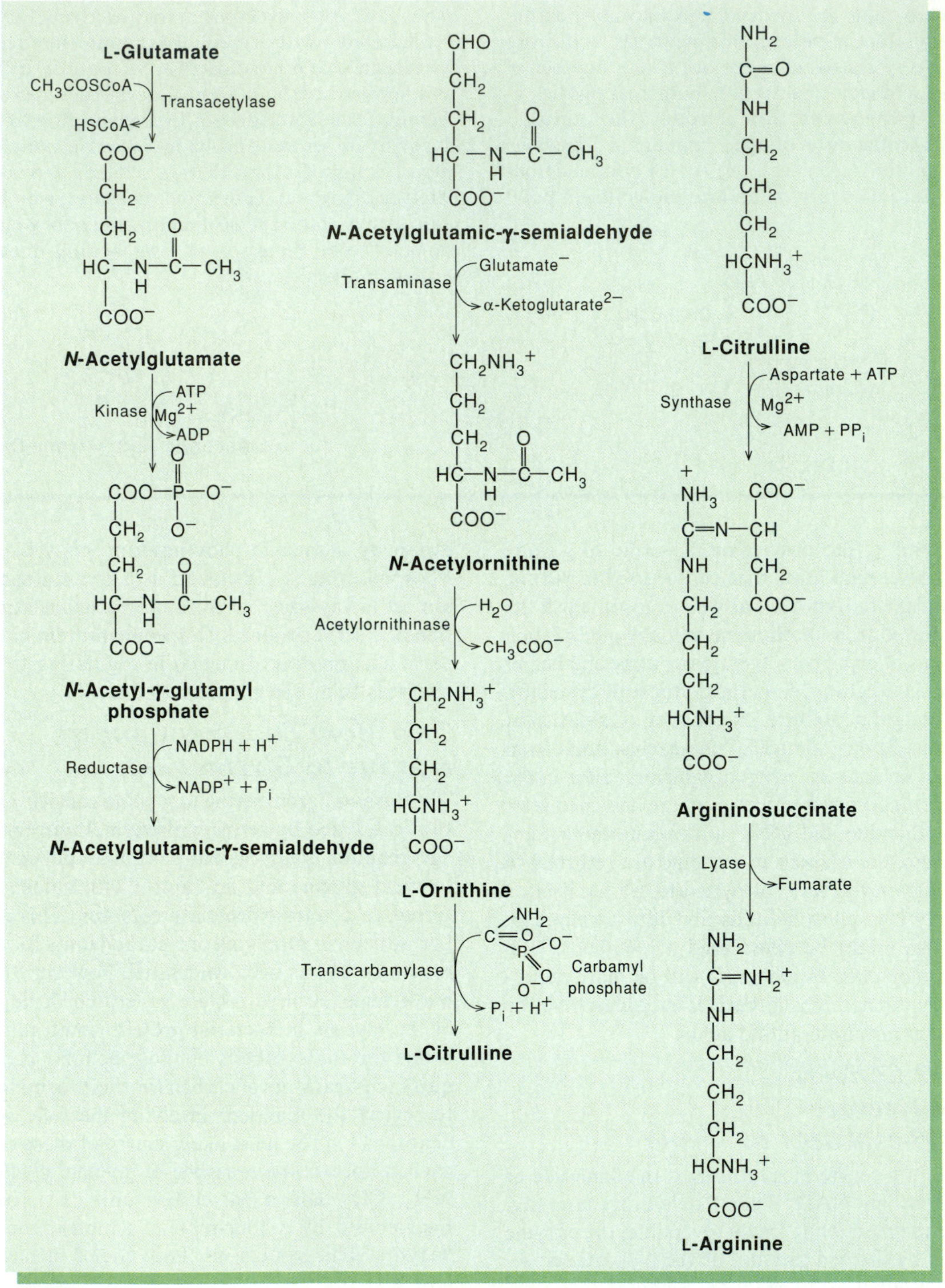

Figure 18.8

The biosynthesis of serine. The end product, serine, inhibits 3-phosphoglycerate dehydrogenase, the first enzyme in the pathway.

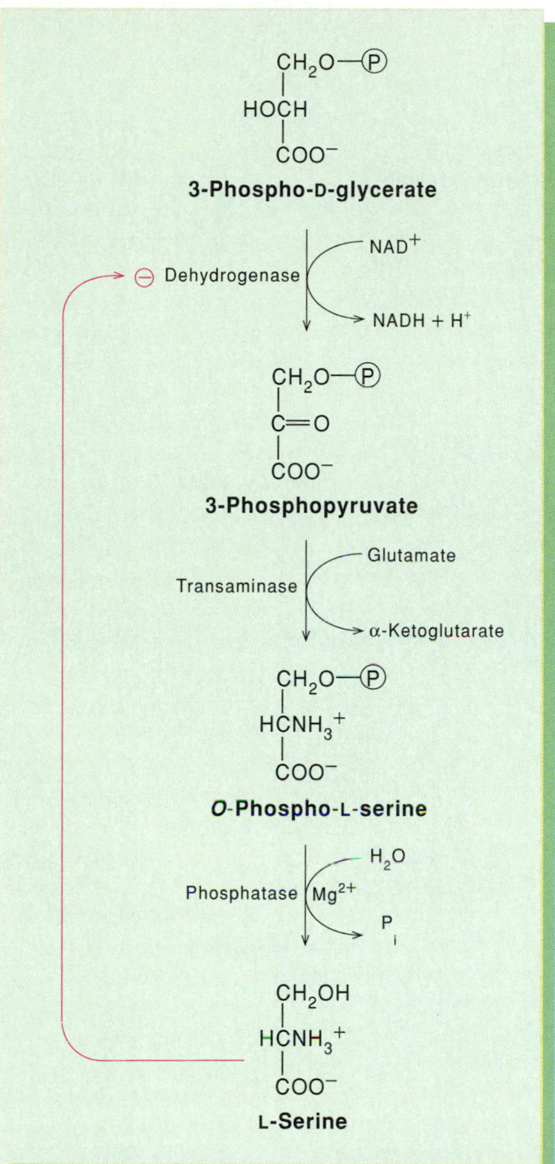

Figure 18.9

Biosynthesis and oxidation of glycine. Both of these reactions are important for supplying one-carbon units for metabolism by way of the tetrahydrofolate coenzyme.

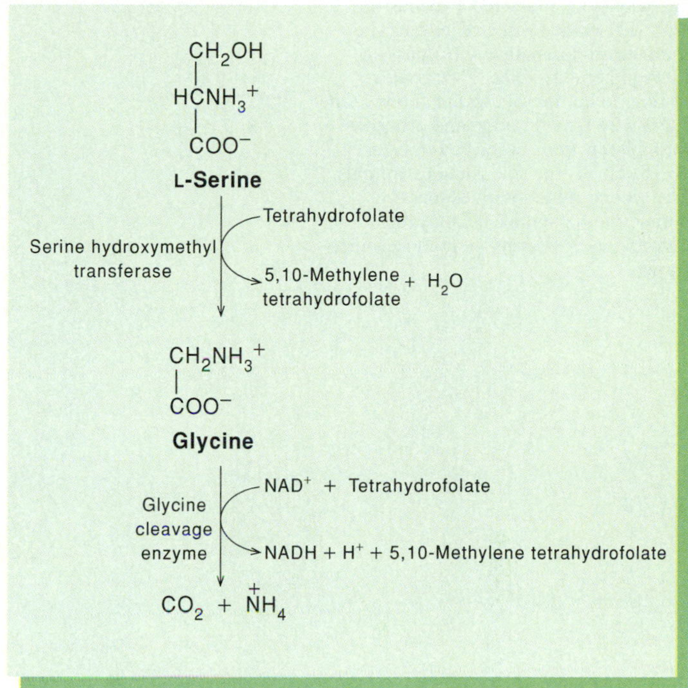

Cysteine Biosynthesis Involves Sulfhydryl Transfer to Activated Serine

The biosynthesis of L-cysteine results from a sulfhydryl transfer to an activated form of serine. The form of the sulfhydryl group used in cells is unclear. For most plants and microorganisms, the sulfur source is sulfate, which must be reduced to the level of sulfide. This reduction is a complex process that we will discuss briefly in the following subsection. For some organisms, it appears that the initial sulfhydryl transfer is made to an activated form of homoserine, a methionine precursor. Subsequently, transfer of the sulfur to cysteine occurs by a transsulfuration pathway.

The direct sulfhydrylation pathway to L-cysteine has been most thoroughly studied in *E. coli* and the related *Salmonella typhimurium* and in the eukaryotic green alga *Chlorella.* The initial step, which is inhibited by cysteine, is a transfer of the acetyl group of acetyl-CoA to serine, to yield *O*-acetylserine (fig. 18.10*a*). The reaction is catalyzed by serine transacetylase. The formation of cysteine itself is catalyzed by O-acetylserine sulfhydrylase, a reaction in which *O*-acetylserine serves as a β-alanyl donor (alanine activated at the β carbon) and H_2S as the β-alanyl acceptor.

The direct sulfhydrylation of *O*-acetylserine to yield cysteine is the most common mechanism for sulfide incorporation. Nevertheless, in some forms the major, if not the sole, mechanism for sulfur incorporation is by means of homocysteine synthase, an enzyme considered later, when we discuss L-methionine biosynthesis. Under such conditions, cysteine formation occurs by transsulfuration, with the intermediate formation of L,L-cystathionine (fig. 18.10*b*). Cystathionine biosynthesis from serine and homocysteine is catalyzed as a simple condensation by cystathionine-β-synthase. The cleavage of cystathionine to yield cysteine, α-ketobutyrate, and NH_4^+ is catalyzed by γ-cystathionase, a pyridoxal-phosphate-containing enzyme. This transsulfuration pathway from methionine also serves as a route for methionine catabolism (see fig. 19.30).

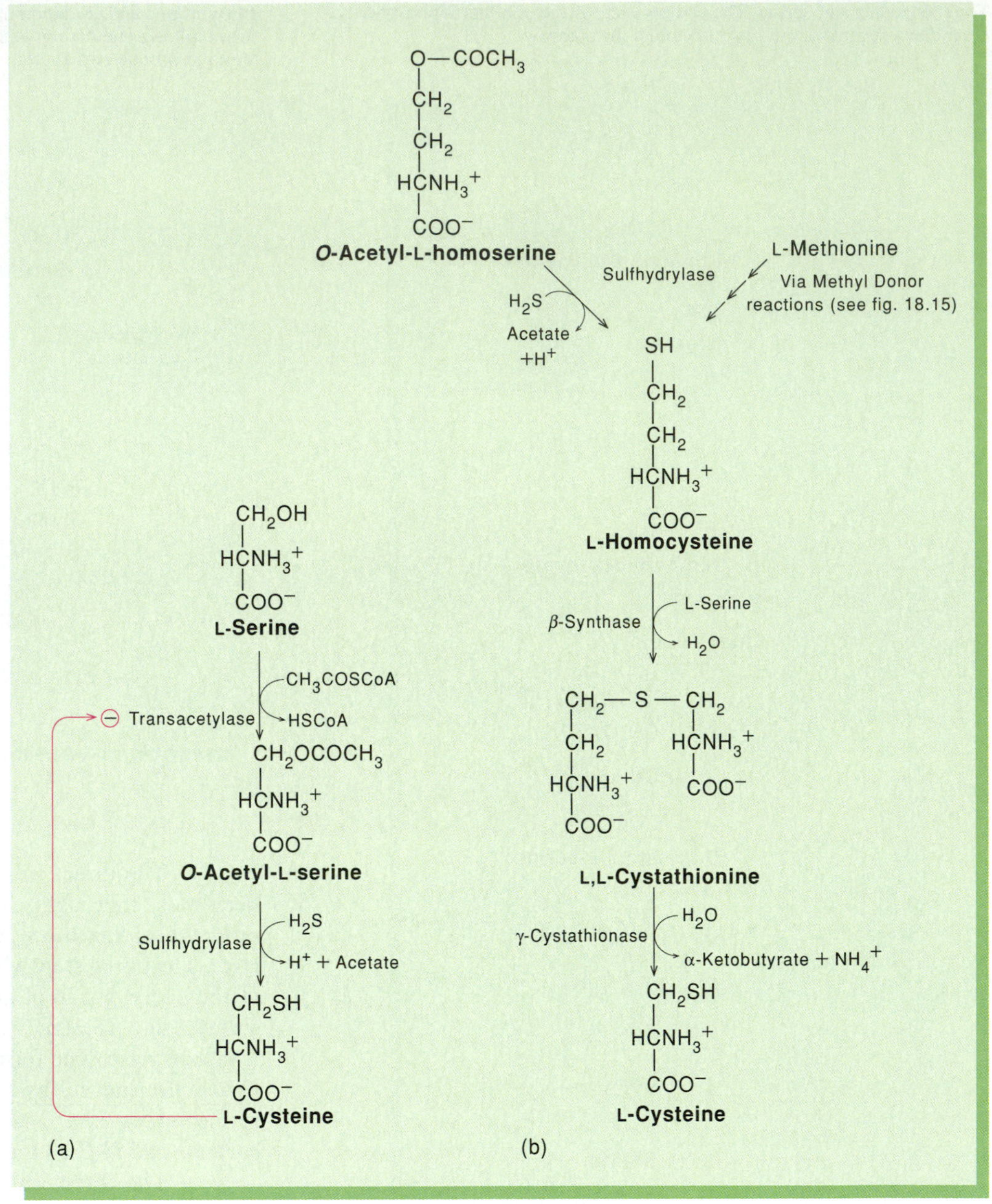

Figure 18.10

The biosynthesis of cysteine by direct sulfhydrylation and by a transsulfuration route in which the sulfur is derived from homocysteine. The direct sulfhydrylation pathway (*a*) is indicated as occurring with H_2S as the source of sulfur. The transsulfuration pathway (*b*) passes through homocysteine. Two routes leading to homocysteine are shown. One starts with L-methionine and proceeds through reactions described in figure 18.15. This is the sole route in animals. The second route involves the homocysteine-synthase-catalyzed conversion of *O*-acetyl-L-homoserine as shown.

Sulfate Must Be Reduced to Sulfide before Incorporation into Amino Acids

Most sulfur exists in the inorganic form of sulfate ion, SO_4^{2-}. In order to be incorporated into amino acids it must first be reduced to H_2S. This change requires a system for reduction found only in plants and microorganisms. The eight-electron reduction of SO_4^{2-} to H_2S occurs in two stages. The first stage requires activation of sulfate (fig. 18.11) catalyzed by adenylylsulfate pyrophosphorylase, which yields adenosine-5′-phosphosulfate (APS). Further activation is required in *E. coli* and certain other bacteria. This is achieved by phosphorylation of the 3′-OH by APS kinase to yield 3′-phosphoadenosine-5′-phosphosulfate (PAPS) with ATP as the phosphate donor. The reduction of the sulfonyl moiety of APS (in yeast and plants) or of PAPS (in *E. coli*) to sulfite occurs by transfer of the sulfonyl group to a thiol acceptor, such as thioredoxin (a vicinal dithiol protein), to yield an $—S—SO_3^-$ derivative. Reaction with a second thiol group (on thioredoxin or on a second molecule of glutathione) yields sulfite and an oxidized acceptor.

The six-electron reduction of sulfite to sulfide is catalyzed by sulfite reductase without the release of any free intermediates. Sulfite reductase is a multisubunit complex composed of a flavoprotein and a protein containing both heme and iron-sulfur centers.

Figure 18.11

Formation of 3′-phosphoadenosine-5′-phosphosulfate, an active intermediate involved in sulfate reduction. The eight-electron reduction of SO_4^{2-} to H_2S is poorly understood except for the initial steps in the activation of sulfate. Reduction in yeast and plants involves the APS derivative shown. In *E. coli* a PAPS derivative is used.

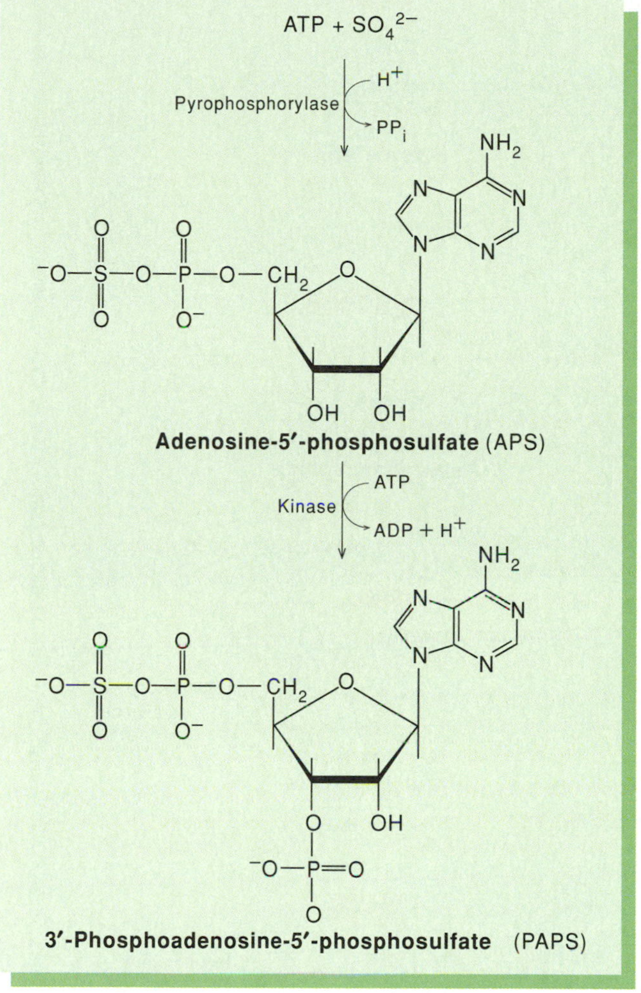

ATP + SO_4^{2-}

Pyrophosphorylase (H^+ → PP_i)

Adenosine-5′-phosphosulfate (APS)

Kinase (ATP → ADP + H^+)

3′-Phosphoadenosine-5′-phosphosulfate (PAPS)

Figure 18.12

The biosynthesis of aspartate and asparagine. The more prevalent transamination reaction is shown. In a limited number of microorganisms an ammonia-dependent asparagine synthase has been found. The asparagine synthase involved in the typical transamination reaction shown in the figure also catalyzes a pyrophosphate cleavage of ATP.

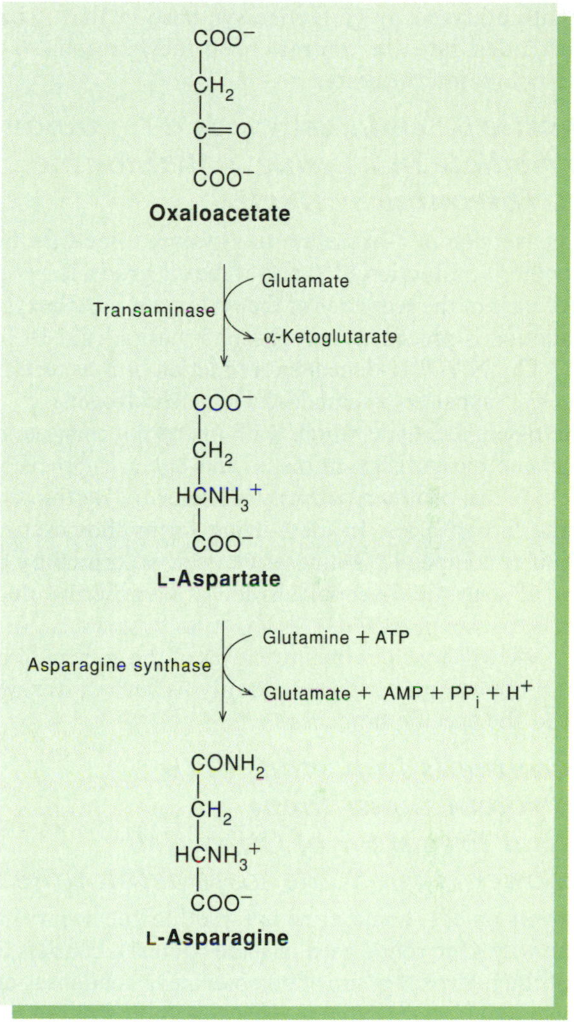

COO^-–CH_2–C=O–COO^-

Oxaloacetate

Transaminase (Glutamate → α-Ketoglutarate)

COO^-–CH_2–$HCNH_3^+$–COO^-

L-Aspartate

Asparagine synthase (Glutamine + ATP → Glutamate + AMP + PP_i + H^+)

$CONH_2$–CH_2–$HCNH_3^+$–COO^-

L-Asparagine

Control of the conversion of sulfate to sulfide may occur in *E. coli* by cysteine inhibition of the active transport of sulfate into the cell.

The Biosynthesis of Some Amino Acids of the Aspartate Family: L-Aspartate, L-Asparagine, L-Methionine, and L-Threonine

The formation of the aspartate family of amino acids (asparagine, methionine, threonine, isoleucine, and, in plants and bacteria, lysine) and the conversion of aspartate to the pyrimidine nucleotides account for an even more significant drain of carbon from the tricarboxylic acid cycle in organisms such as *E. coli* than does that of the α-ketoglutarate family. In addition, the nitrogen of aspartate is used both in the formation of inosinate and its conversion to adenylate and in the conversion of citrulline to arginine.

Aspartate Is Formed from Oxaloacetate in a Transamination Reaction

L-Aspartate is formed from oxaloacetate in a transamination reaction with glutamate as the amino donor (fig. 18.12). In *E. coli,* the major protein exhibiting aspartate-glutamate transaminase activity is transaminase A, an enzyme that also exhibits activity with the aromatic acids but not with leucine.

Asparagine Biosynthesis Requires ATP and Glutamine

In most organisms it is likely that the formation of L-asparagine occurs by an ATP-dependent transfer of the amide group of glutamine to the β carboxyl of aspartate by asparagine synthase (see fig. 18.12). The basic mechanism of amidation is different from that catalyzed by glutamine synthase in that it involves β-aspartyladenylate (in contrast to an acylphosphate) as an enzyme-bound intermediate.

L-Aspartic-β-Semialdehyde Is a Common Intermediate in L-Lysine, L-Methionine, and L-Threonine Synthesis

In the conversion of L-aspartate to L-lysine, L-methionine, and L-threonine, a reduction of the β-carboxyl group is necessary. As in the case of the reduction of the glutamate α-carboxyl group the reduction is preceded by a phosphorylation (fig. 18.13).

The NADPH-dependent reduction of β-aspartyl phosphate by aspartic-β-semialdehyde dehydrogenase yields aspartic-β-semialdehyde, which is a branchpoint compound from which lysine biosynthesis in plants and bacteria proceeds (see fig. 18.13). The common pathway is longer for methionine and threonine biosynthesis. In most living forms, homoserine is a branchpoint compound. Homoserine itself is formed by the reduction of aspartic-β-semialdehyde by homoserine dehydrogenase.

We will consider the intricacies of the control of carbon flow over the common aspartate family pathway after we have described the specific branches.

Methionine Is Important Both as a Protein Constituent and as a Precursor of Other Cell Components via S-Adenosylmethionine

The conversion of L-homoserine to L-methionine occurs in more than one way. One route, used in some bacteria, consists of acylation of the hydroxyl group of homoserine, a condensation with cysteine to yield cystathionine, cleavage to homocysteine, and methylation to yield methionine (fig. 18.14). The second route, found in yeast and fungi, was referred to earlier and involves a sulfhydryl transfer to homoserine activated by an acyl group to yield homocysteine directly. The reaction, catalyzed by homocysteine synthase, is shown in figures 18.10 and 18.14.

Although methionine is itself an end product of the pathway and is incorporated into protein, another important biosynthetic intermediate is *S*-adenosylmethionine (SAM). This intermediate serves as a methyl donor in many reactions and as a precursor of the propylamine groups in spermidine and spermine (see fig. 19.25). In some organisms SAM participates in the control of the pathway leading from homoserine to methionine.

Figure 18.13

The common aspartate family pathway. In some green plants, *O*-phosphohomoserine rather than homoserine itself is the point at which methionine biosynthesis and threonine biosynthesis diverge.

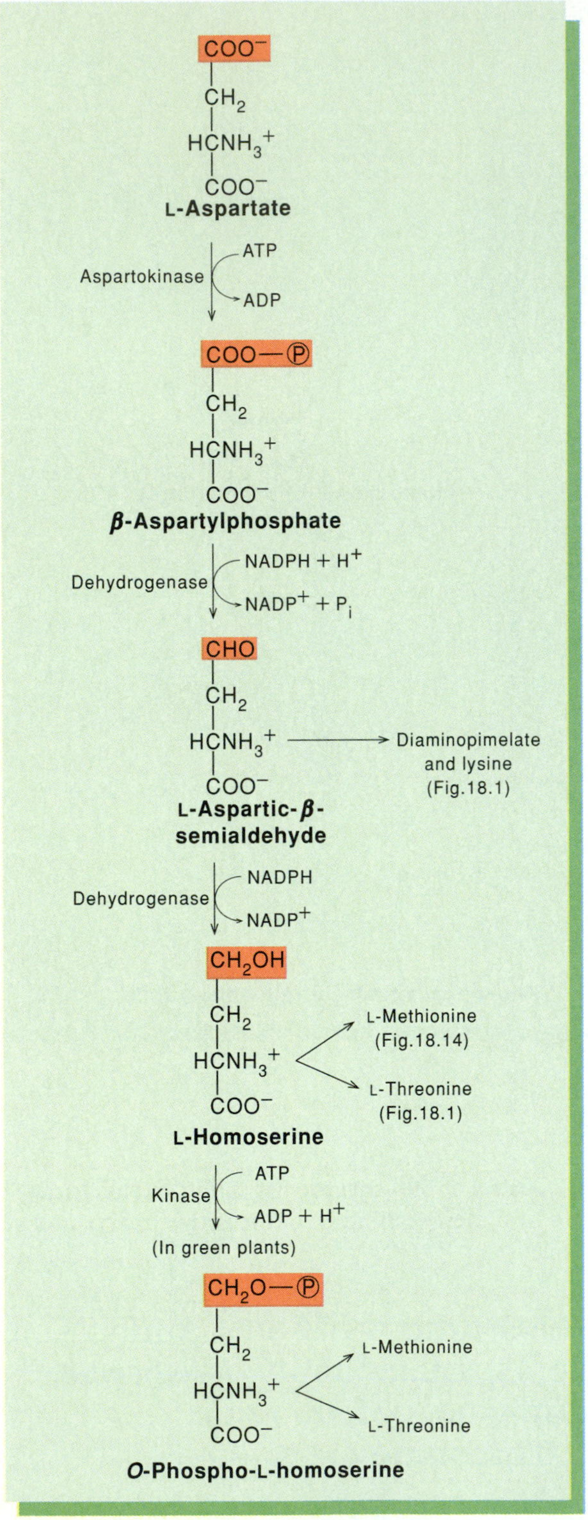

COO^- / CH_2 / $HCNH_3^+$ / COO^-

L-Aspartate

Aspartokinase (ATP → ADP)

$COO—Ⓟ$ / CH_2 / $HCNH_3^+$ / COO^-

β-Aspartylphosphate

Dehydrogenase ($NADPH + H^+$ → $NADP^+ + P_i$)

CHO / CH_2 / $HCNH_3^+$ / COO^- → Diaminopimelate and lysine (Fig.18.1)

L-Aspartic-β-semialdehyde

Dehydrogenase (NADPH → $NADP^+$)

CH_2OH / CH_2 / $HCNH_3^+$ / COO^- → L-Methionine (Fig.18.14); L-Threonine (Fig.18.1)

L-Homoserine

Kinase (ATP → $ADP + H^+$)

(In green plants)

$CH_2O—Ⓟ$ / CH_2 / $HCNH_3^+$ / COO^- → L-Methionine; L-Threonine

***O*-Phospho-L-homoserine**

Figure 18.14

The biosynthesis of methionine. The acyl group employed to activate homoserine varies among different organisms. When a phosphoryl group is used as in green plants, the intermediate, *O*-phosphohomoserine, is a branchpoint compound that is converted to either threonine or methionine. The direct sulfhydrylation of activated homoserine (by either H_2S or carrier-bound sulfide) that is found in some forms is indicated by a broken line. Finally, the methylation of homocysteine may occur by either a tetrahydrofolate-dependent route (shown) or a cobalamin-dependent route (not shown).

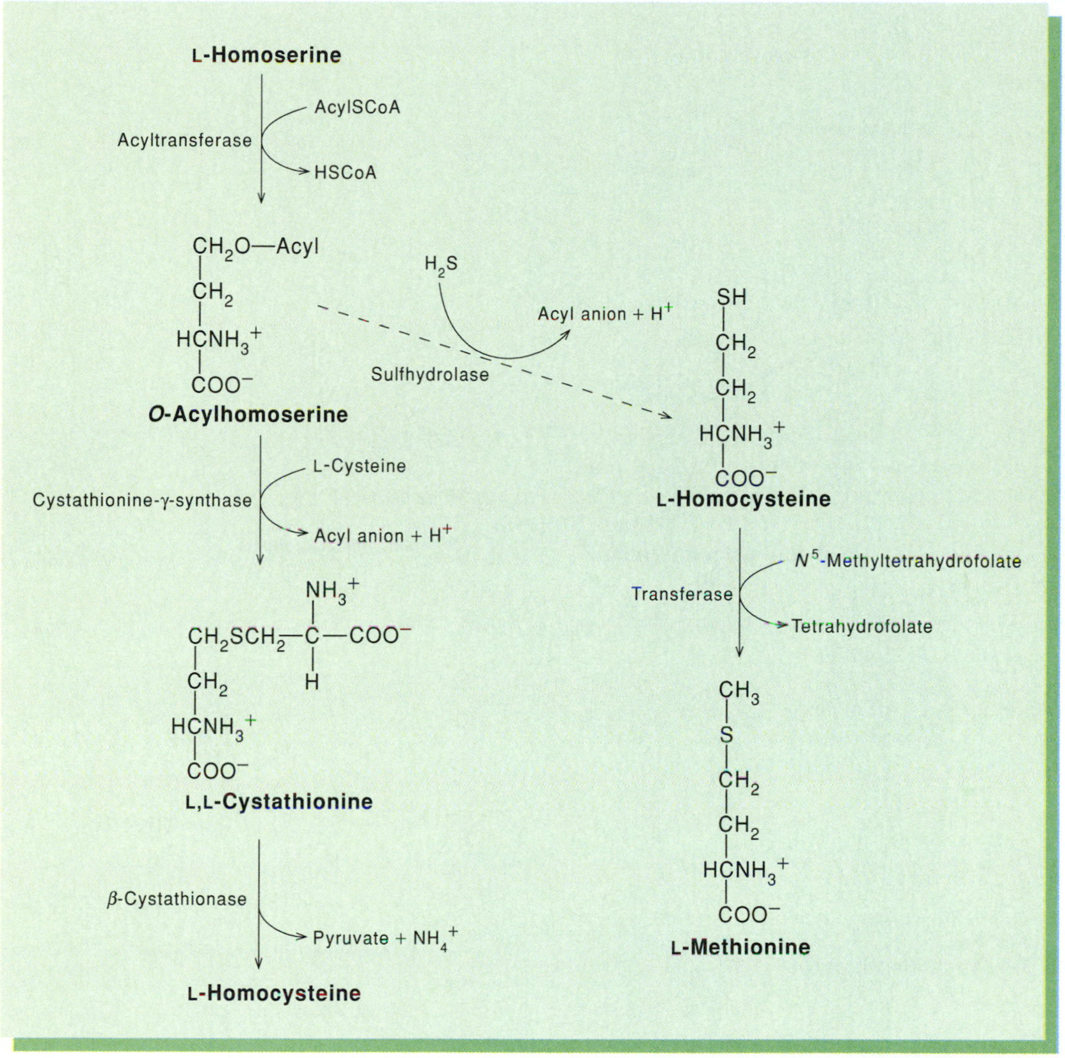

An adenosylation of methionine by SAM synthase activates methionine for either methyl group or, with a decarboxylase, propylamine transfers. As figure 18.15 shows, the activation is unusual in that all three phosphate groups of ATP are cleaved in the activation reaction. In the methyl donor reaction, *S*-adenosylhomocysteine is liberated by one of several methyltransferases and then cleaved by *S*-adenosylhomocysteine hydrolase to homocysteine and adenosine, which can be regenerated to methionine and ATP, respectively. In the reactions in which a propylamine group is transferred to putrescine or to spermidine, methylthioadenosine is liberated (see fig. 19.25). The latter is converted to free adenine and 5′-methylthioribose-1-phosphate by either of two mechanisms. One, demonstrated in rat liver homogenates, is the direct conversion via a phosphorylase. The second, demonstrated in plants, protozoans, and bacteria, is via a nucleosidase (hydrolytic) to yield the nonphosphorylated thiomethyl sugar, which is phosphorylated in a subsequent kinase reaction. The adenine can be returned to the adenylate pool by one of the pyrophosphorylases described in chapter 20, and 5′-methylthioribose-1-phosphate is converted to methionine and formate by a series of reactions in which carbons 2 to 3 become carbons 1 to 4 of methionine.

Figure 18.15

The formation of "active" methionine, or *S*-adenosylmethionine (SAM), and some of its reactions. *S*-adenosylmethionine is synthesized in one step from L-methionine and ATP. The SAM so formed can serve as a propylamine donor or a methyl donor as shown. In the methyl donor reaction, the *S*-adenosylhomocysteine formed is cleaved to homocysteine and adenosine, which can be used to regenerate methionine and ATP, respectively. The methylthioadenosine liberated in propylamine donor reactions is converted to 5′-methylthioribose-1-phosphate, which is converted to methionine and formate by a series of reactions.

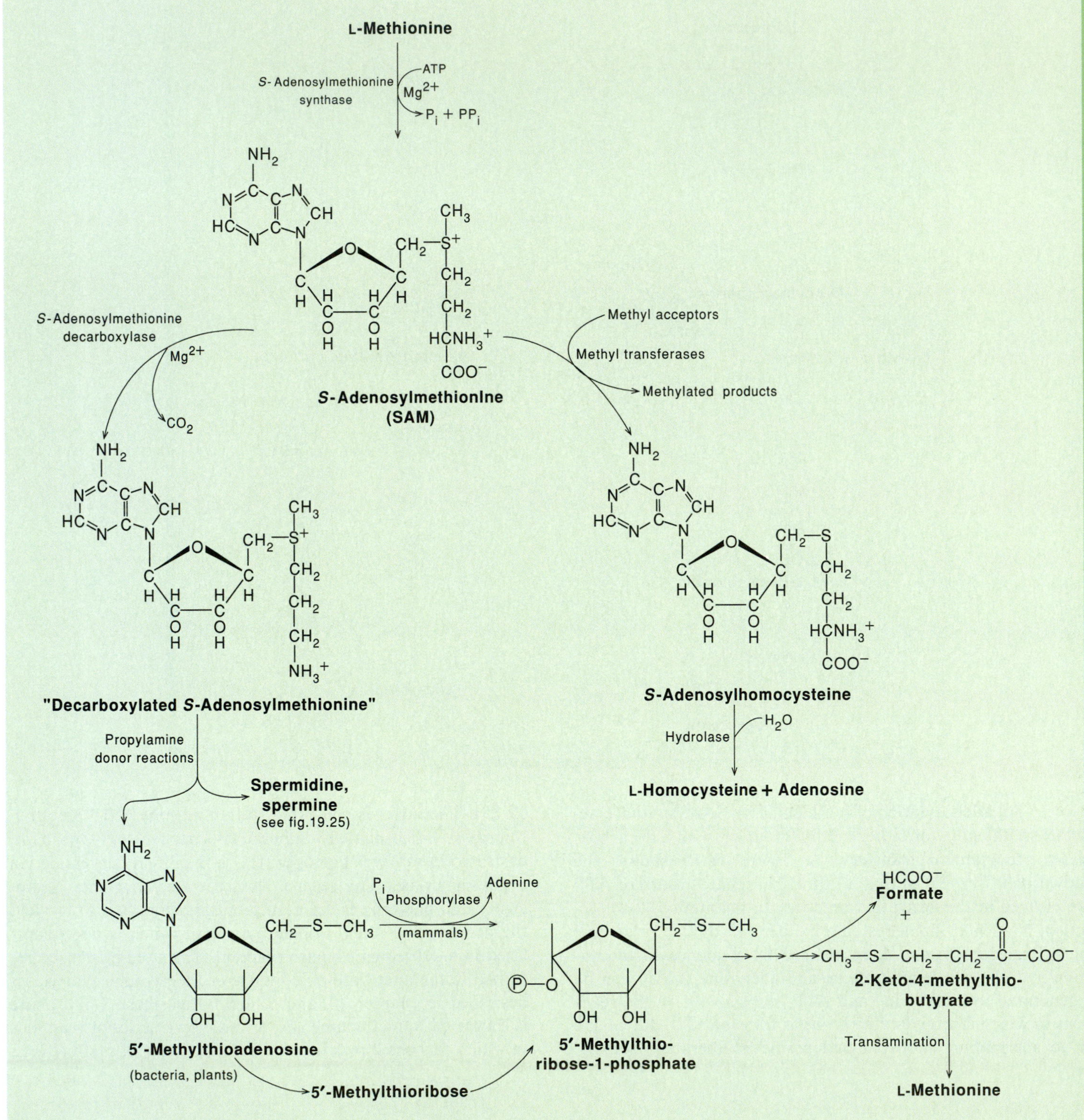

Figure 18.16

Regulation of the aspartokinases of *E. coli*. The activity of these enzymes is regulated at the level of enzyme synthesis and at the level of enzyme activity. In *E. coli* there are three aspartokinases that catalyze the conversion of L-aspartate to ß-aspartylphosphate. The formation of these aspartokinases is repressed by different amino acids that are end products of this highly branched pathway. In addition, the activities of aspartokinase I and homoserine dehydrogenase I are inhibited by threonine and that of aspartokinase III is inhibited by lysine (arrows for inhibited reactions are not shown).

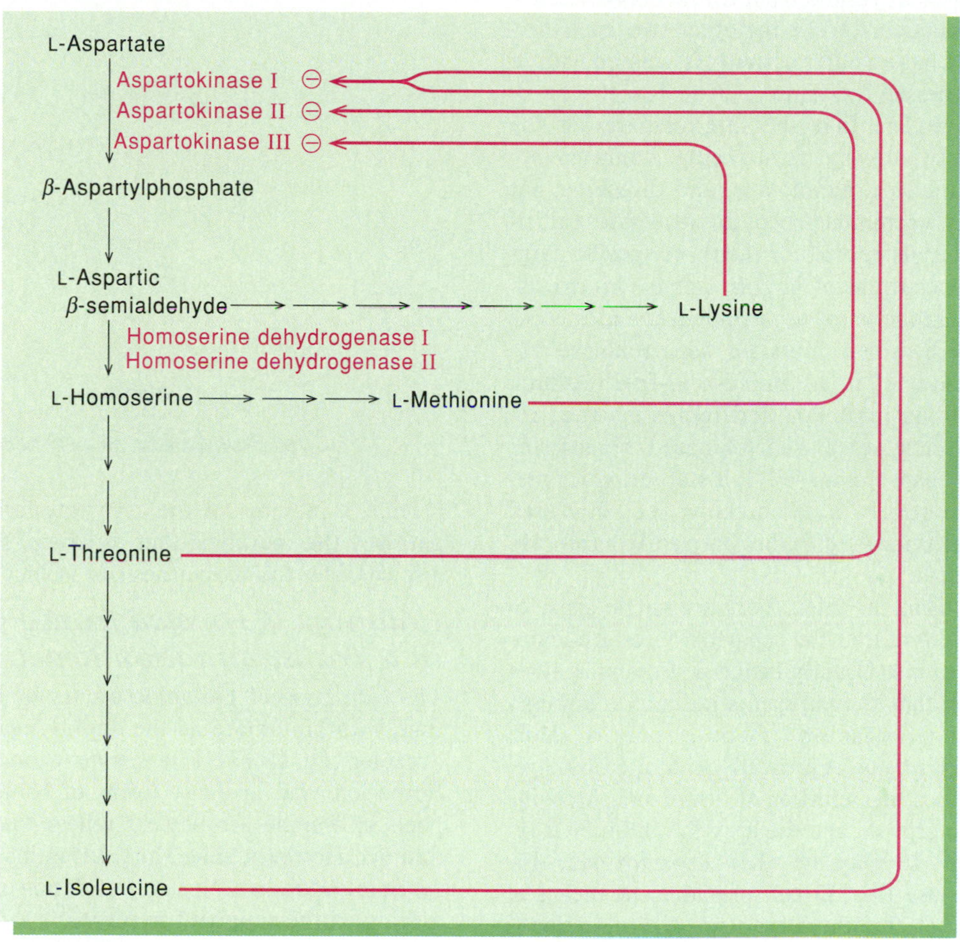

The Carbon Flow in the Aspartate Family Is Regulated at the Aspartokinase Step

The flow of carbon into the common aspartate family pathway must be regulated in a way that provides ample amounts of the branchpoint compounds aspartic-β-semialdehyde (except in fungi) and homoserine (or *O*-phosphohomoserine in most green plants) but does not lead to oversynthesis of either. In most organisms there appear to be negligible pools of either branchpoint compound—perhaps only those amounts bound to product or substrate sites of the respective enzymes. Therefore feedback control cannot be exerted by these intermediates. Rather, in one way or another, control must be exerted by the end products (lysine, methionine, threonine, and isoleucine), which accumulate in measurable amounts.

Although an essentially common pathway has evolved for formation of the aspartate family of precursors, the patterns of feedback control vary considerably. They are of two general kinds. In one, there are single aspartokinases, but the control of the enzyme is multivalent. By multivalent, we mean that more than one of the amino acid end products is required to inhibit the enzyme. The other basic pattern is one of multiple aspartokinases, each of which is controlled differently.

The most extensively studied pattern is that of the enteric bacteria, exemplified by *E. coli* (fig. 18.16). *E. coli* possesses three different proteins with aspartokinase activity. One enzyme is inhibited by threonine. This enzyme is part of a protein that also exhibits homoserine dehydrogenase activity, which is also sensitive to inhibition by threonine. The enzyme is called aspartokinase I–homoserine dehydrogenase I. Another distinctly different protein, aspartokinase II–homoserine dehydrogenase II, is not inhibited by any of the multiple end products of the aspartate family pathway, but its synthesis is repressed by methionine (as are other methionine biosynthetic enzymes).

The synthesis of aspartokinase I–homoserine dehydrogenase I is repressed by a combination of two end products, threonine and isoleucine (multivalent repression).

The third aspartokinase in *E. coli,* aspartokinase III, is inhibited by lysine. The inhibition is enhanced synergistically by phenylalanine, leucine, or, to a lesser extent, methionine. Any physiological advantage that results from this synergistic inhibition has not been explained. Unlike the other two aspartokinases, aspartokinase III has no other activity associated with it. Its synthesis is also repressed by lysine.

We must emphasize that the intermediates of the common aspartate family pathway are not being channeled into the branches leading to lysine, methionine, and threonine, but rather, there is probably a common pool of intermediates (albeit small) from which materials needed for the three specific pathways are drawn. In some strains of *E. coli,* lysine-sensitive aspartokinase III is the predominant aspartokinase, and there is very little of the methionine-repressible aspartokinase II–homoserine dehydrogenase II in an amino-acid-free medium. Under conditions of strong inhibition and repression (i.e., repressed synthesis of the enzyme) of aspartokinase I–homoserine dehydrogenase I and of aspartokinase III, a starvation for methionine would be prevented by a derepression (i.e., increased synthesis of the enzyme) of the aspartokinase II–homoserine dehydrogenase II.

In those organisms in which there are single aspartokinases, it is common to find that the aspartokinases are inhibited by lysine and threonine. Usually lysine or threonine alone is weakly inhibitory, so that the pattern is actually a strongly synergistic one. In such organisms there is only a single homoserine dehydrogenase, and it is usually inhibited by threonine alone. In some cases, the addition of lysine and threonine is strongly inhibitory to growth, and the growth inhibition is reversed by methionine. It may be that aspartokinase II–homoserine dehydrogenase II of *E. coli* provides one means of avoiding this complication. In other organisms the inhibition is less severe, and the proteins are relatively insensitive to the feedback inhibitors after growth in the presence of the inhibitory amino acids. The physical basis of the desensitization is unknown.

The precise patterns of regulation of synthesis of amino acids of the aspartate family vary widely, and in other microorganisms and in plants the studies have not been as thorough as they have in *E. coli* and related organisms, which appear to have a rather unique pattern of regulation.

The Biosynthesis of Amino Acids of the Pyruvate Family: L-Alanine, L-Valine, and L-Leucine

The pyruvate family of amino acids consists of L-alanine, L-valine, and L-leucine. In addition, pyruvate contributes two carbons to isoleucine and, on the average, two and one-half carbons to lysine. As we mentioned earlier, the biosynthesis of isoleucine, a member of the aspartate family, is carried on by a pathway that parallels that of valine. For this reason, we will also consider the biosynthesis of isoleucine here.

Figure 18.17

The biosynthesis of alanine. This reaction is not regulated. It is readily reversible and therefore constitutes no major drain on the pyruvate supply.

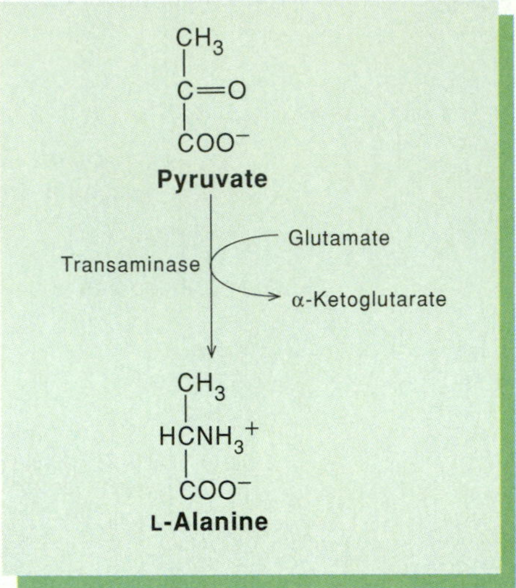

L-Alanine Is Formed from Pyruvate in a Transamination Reaction

The formation of L-alanine occurs by a transamination reaction, with glutamate as the amino donor and pyruvate as the acceptor (fig. 18.17). There is no feedback control over alanine formation, and in many forms of bacteria large intracellular pools of alanine are present unless the nitrogen supply is restricted. However, since the transaminases catalyze completely reversible reactions, this accumulation of alanine does not effect a drain on the supply of pyruvate.

Isoleucine and Valine Biosynthesis Share Four Enzymes

The four enzymes required for valine biosynthesis are also required for the last four, parallel steps in isoleucine biosynthesis (fig. 18.18). The first step in valine biosynthesis is a condensation between pyruvate and "active" acetaldehyde (probably hydroxyethyl thiamine pyrophosphate) to yield α-acetolactate. The enzyme usually has a requirement for FAD, which, in contrast to most flavoproteins, is rather loosely bound to the protein. The same enzyme transfers the acetaldehyde group to α-ketobutyrate, yielding α-aceto-α-hydroxybutyrate, the isoleucine precursor. The α-ketobutyrate, unlike pyruvate, is not one of the key intermediates in many of the central metabolic routes; therefore a specific pathway to α-ketobutyrate must be present.

For nearly all plants, fungi, and bacteria, the normal route to α-ketobutyrate is that from aspartate by way of threonine, which in turn is deaminated to α-ketobutyrate. The

Figure 18.18

The biosynthesis of isoleucine and valine. The reactions leading to valine are catalyzed by the same enzymes that catalyze the corresponding reactions in isoleucine biosynthesis.

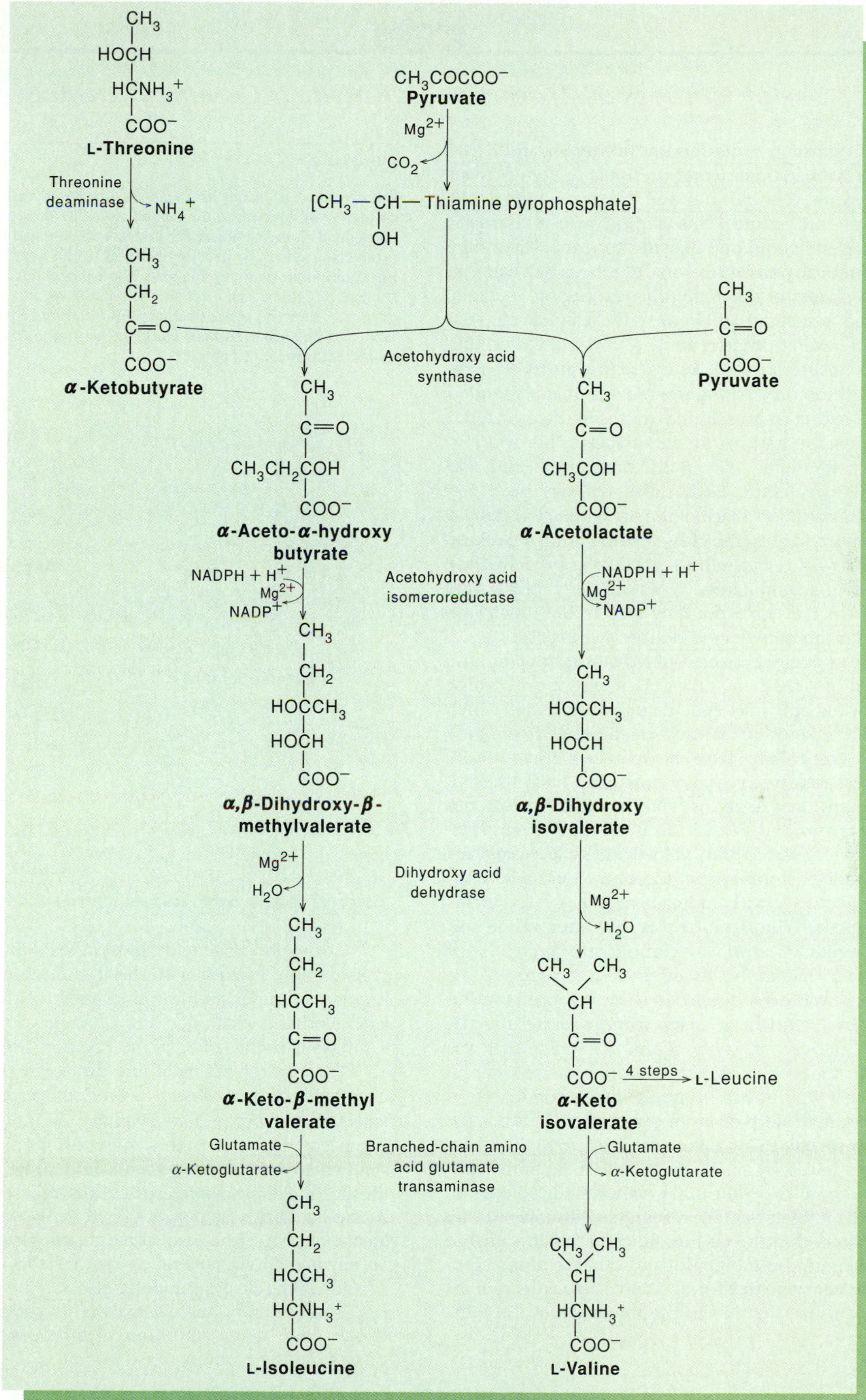

Use of Isotopes as Tracers to Delineate a Complex Pathway

Early studies with mutants of both *Neurospora* and *E. coli* revealed that certain mutants, presumably altered in but a single gene, required not one, but two amino acids, isoleucine and valine. This finding was an apparent contradiction to the one gene, one enzyme concept. When they were examined genetically and nutritionally, it became clear that there were several classes of these doubly auxotrophic mutants. One class was found to accumulate, in the culture fluids, material that fed mutants of several other classes. Analysis revealed that the active material consisted of the α-keto acid precursors of valine and isoleucine. Furthermore, among the mutants that responded to the keto acids, one class was found to accumulate material that fed other isoleucine and valine auxotrophs. This accumulated material was identified as α,β-dihydroxyisovalerate and α,β-dihydroxy-β-methylvalerate. Both these findings were followed by the demonstration of a lack of the branched-chain amino acid-glutamate transaminase in the class accumulating the α-keto acids and the lack of what is now called dihydroxy acid dehydrase in the dihydroxy acid accumulators. It was this loss of a single enzyme that catalyzes the corresponding step in both pathways which accounted for the unexpected double auxotrophy.

The valine and isoleucine auxotrophs that responded to the dihydroxy acids did not feed any other class except one, which appeared to be blocked only in isoleucine biosynthesis and which responded as well to α-ketobutyrate as to α-aminobutyrate. These single blocked auxotrophs were later shown to lack threonine deaminase, the enzyme required only for isoleucine biosynthesis.

While these nutritional analyses were in progress, isotopic studies with *Neurospora* and yeast indicated that pyruvate carbons were being incorporated into valine and that exogenous nonradioactive aspartate, homoserine, threonine, and α-ketobutyrate competed effectively with radioactive glucose for incorporation into isoleucine. The paradox was that when valine and isoleucine were degraded carbon by carbon, after growth with either labeled lactate (metabolically equivalent to pyruvate) or labeled acetate, there was no sequence of three carbons in valine (see illustration) that could have arisen from pyruvate directly, nor was there any four-carbon sequence in isoleucine that was labeled with acetate carbon, as aspartate and threonine were labeled. Strassman and Weinhouse proposed that a condensation of a two-carbon fragment derived from pyruvate with either pyruvate (to yield acetolactate) or α-ketobutyrate (to yield acetohydroxybutyrate), followed by an intramolecular migration, could account for the isotopic distribution in valine and isoleucine. Shortly after this proposal appeared, researchers discovered that acetolactate accumulated in the culture fluids of mutants unable to reduce and rearrange the molecule, and other workers identified the condensing enzyme that formed both acetohydroxy acids and the isomeroreductase that led to the formation of the dihydroxy acids.

Figure 1

Incorporation of lactate and acetate carbon into valine and isoleucine. The colored symbols next to the carbon atoms in lactate, acetate, and the other compounds indicate where the carbons are situated in the precursors and products. Radioactive labeling patterns of this type provide important information for pathway analysis. The isotope studies indicate that no sequence of three carbons in valine could have arisen from lactate (or pyruvate) directly, nor is there any four-carbon sequence in isoleucine that is labeled with acetate carbon. Thus the incorporation of these carbons into the amino acids must be complex.

$CH_3—CHOH—COOH$
Lactate

$CH_3—COOH$
Acetate

$CH_3—CH(CH_3)—CHNH_2—COOH$
Valine

$COOH—CH_2—CH(NH_2)—COOH$
Aspartic acid

$CH_3—C(=O)—C(CH_3)OH—COOH$
Acetolactic acid

$CH_3—CH(CH_2—CH_3)—CH(NH_2)—COOH$
Isoleucine

$CH_3—C(=O)—C(CH_2—CH_3)OH—COOH$
Acetohydroxybutyric acid

Valine and isoleucine biosynthesis in plants has become of considerable interest with the finding that three classes of herbicides are potent inhibitors of most of the plant acetohydroxy acid synthases examined and of some of the acetohydroxy acid synthase isozymes of bacteria (see box 18E).

The control of metabolite flow over the pathways to valine and isoleucine is subject to one complication. In many organisms, acetohydroxy acid synthase is inhibited by valine. The complete inhibition of this enzyme not only would prevent oversynthesis of valine, but also would prevent isoleucine biosynthesis, for which acetohydroxy acid synthase is the second enzyme in the pathway. It is perhaps for this reason that many organisms also contain a valine-insensitive acetohydroxy acid synthase. Such an enzyme allows isoleucine synthesis to occur in the presence of valine, but also potentially allows an oversynthesis of valine under some conditions. Control of the synthesis of isoleucine itself is achieved by an inhibition of threonine deaminase, the first enzyme in the pathway to isoleucine.

enzyme threonine deaminase contains pyridoxal phosphate and functions as a dehydratase (see chapter 11, equation 6), presumably liberating α-aminocrotonate, which upon rearrangement to α-iminobutyrate spontaneously yields α-ketobutyrate and ammonia. There are two additional ways to form α-ketobutyrate, but we will not consider these in this section.

Conversion of the acetohydroxy acids to the β-dihydroxy acid precursors of valine and isoleucine is a complex reaction catalyzed by acetohydroxy acid isomeroreductase. The α,β-dihydroxy acids are both converted to the α-keto acid precursors of valine and isoleucine by a dihydroxy acid dehydrase. Finally, the two amino acids are formed in transamination reactions in which glutamate is the amino donor (branched-chain amino acid-glutamate transaminase).

The pathways to isoleucine and valine illustrate well the way studies with nutritionally deficient mutants (auxotrophs), isotope incorporation experiments, and enzymatic analysis have been used to decipher the biosynthetic pathways to the amino acids (box 18C).

L-Leucine Is Formed from α-Ketoisovalerate in Four Steps

The biosynthesis of L-leucine involves the lengthening of the carbon chain of α-ketoisovalerate, an intermediate in valine biosynthesis (fig. 18.19).

In the Biosynthesis of the Aromatic Family of Amino Acids (L-Tryptophan, L-Phenylalanine, and L-Tyrosine) Chorismate Is a Key Intermediate

The aromatic amino acids phenylalanine, tyrosine, and tryptophan are all formed by means of the shikimate pathway to build the benzene ring. This pathway is also important for the formation of the aromatic nuclei in or the aromatic precursor of vitamins E and K, folic acid, ubiquinone, and plastoquinone and certain metal chelators, such as enterochelin. The branchpoint compound for all these diverse products is chorismate, which has a prearomatic cyclohexadiene nucleus (fig. 18.20).

The overall route of chorismate synthesis, beginning with the condensation of two intermediates from the central metabolic routes, phosphoenolpyruvate and erythrose-4-phosphate, is illustrated in figure 18.20. Chorismate is the final product of the common pathway to all aromatic compounds, including phenylalanine, tyrosine, and tryptophan.

Prephenate Is a Common Intermediate in L-Phenylalanine and L-Tyrosine Synthesis

In some organisms, the pathways from chorismate to L-phenylalanine and L-tyrosine, shown in figure 18.21, are truly separate pathways, even though the first step from chorismate is the same. For example, in *E. coli* and related organisms, one protein, chorismate mutase P-prephenate dehydratase, converts chorismate to phenylpyruvate, with the intermediate, prephenate, being enzyme-bound. A second protein, the NAD-dependent chorismate mutase T-prephenate dehydrogenase, converts

Figure 18.19

The biosynthesis of leucine from α-ketoisovalerate, the branchpoint intermediate also used in L-valine synthesis. In the first step the carbon chain of α-ketoisovalerate is lengthened by two carbon atoms. The steps to leucine are completed by specific isomerase, dehydrogenase, and transaminase reactions.

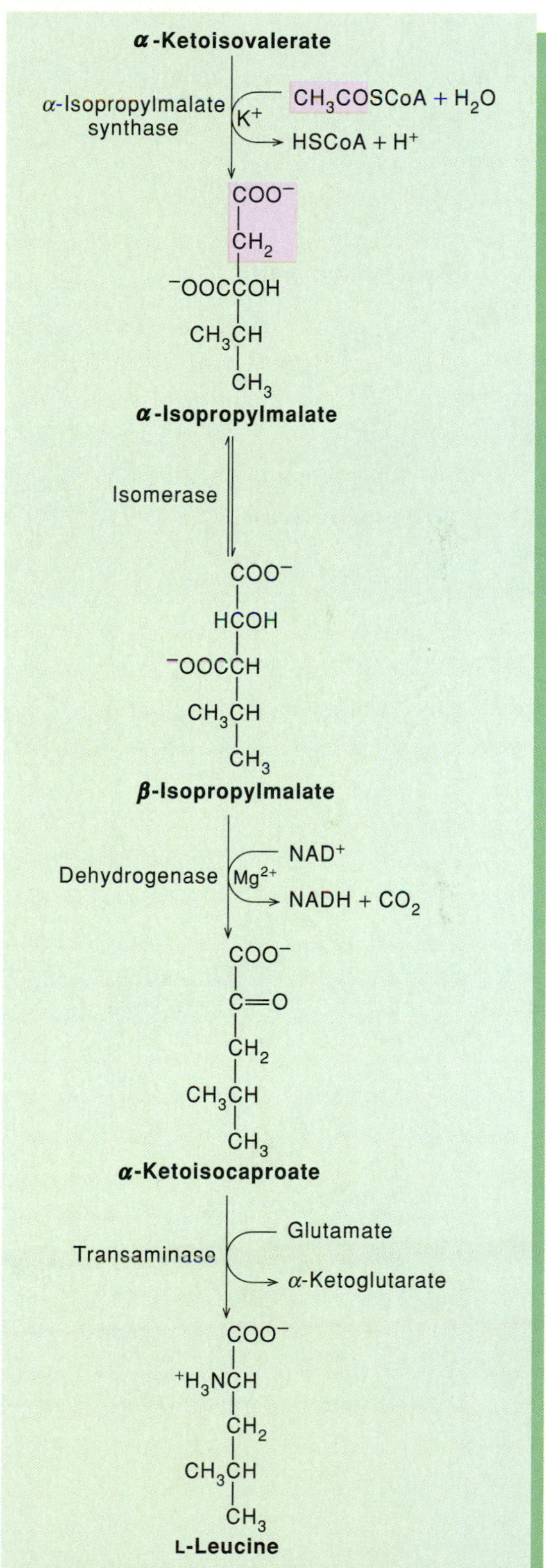

Figure 18.20

The common aromatic (shikimate) pathway leading to chorismate biosynthesis. Note that NAD^+ is required for the conversion of 3-deoxy-D-*arabino*-heptulosonate-7-phosphate to 3-dehydroquinate, but there is no net change in the redox state during the conversion of substrate to product. This was also seen in the epimerization of fructose (see chapter 21) and is indicative of an oxidized intermediate. Chorismate formed by this series of reactions is a common intermediate for phenylalanine tyrosine and tryptophan biosynthesis.

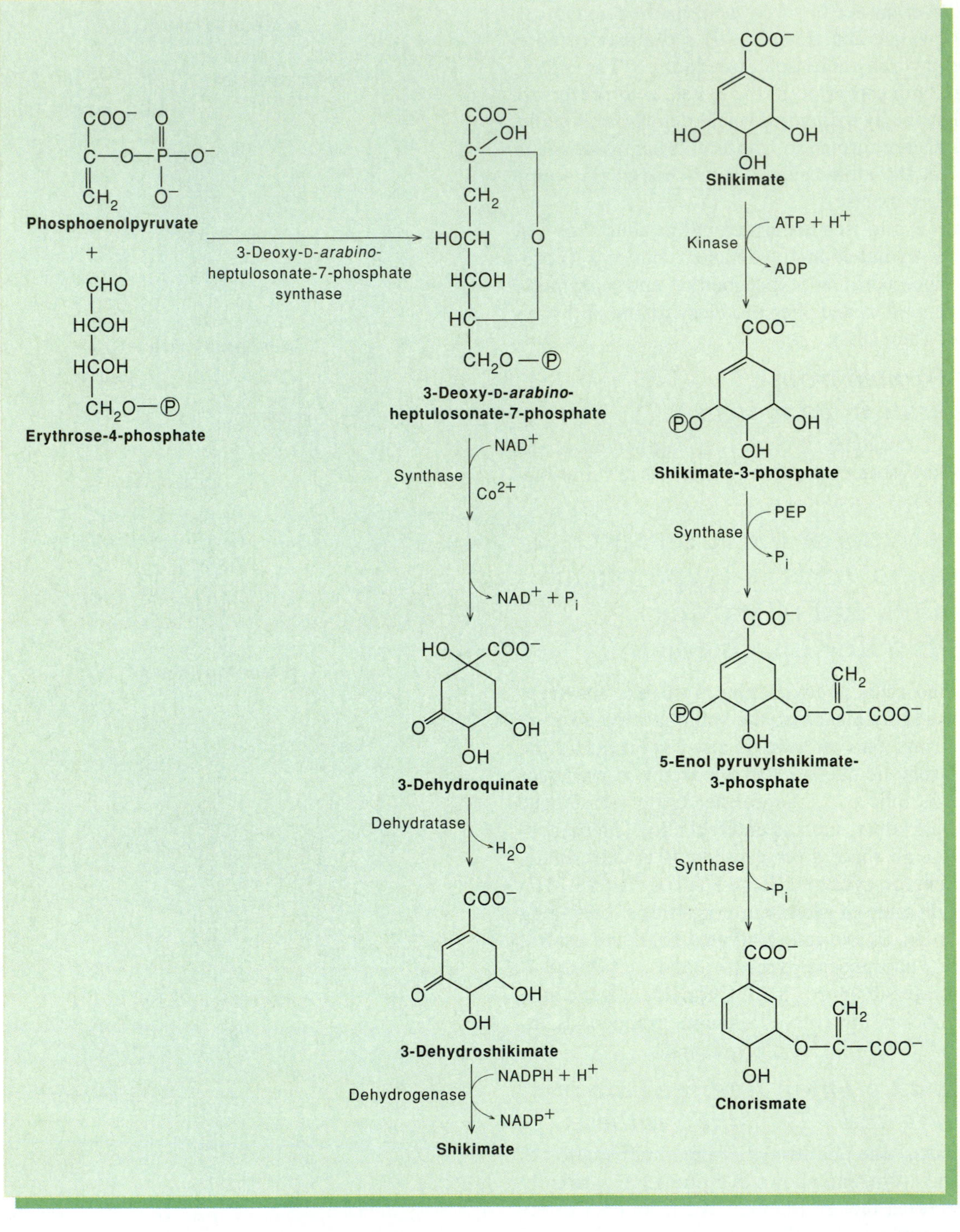

Figure 18.21

The biosynthesis of phenylalanine and tyrosine from the branchpoint compound chorismate. The lines between the mutase and the dehydratase and between the mutase and the dehydrogenase indicate that in both pathways we are dealing with bifunctional enzymes that are exclusive to the indicated pathways. Thus, although prephenate is the first intermediate in both pathways, it is not a branchpoint intermediate in the usual sense.

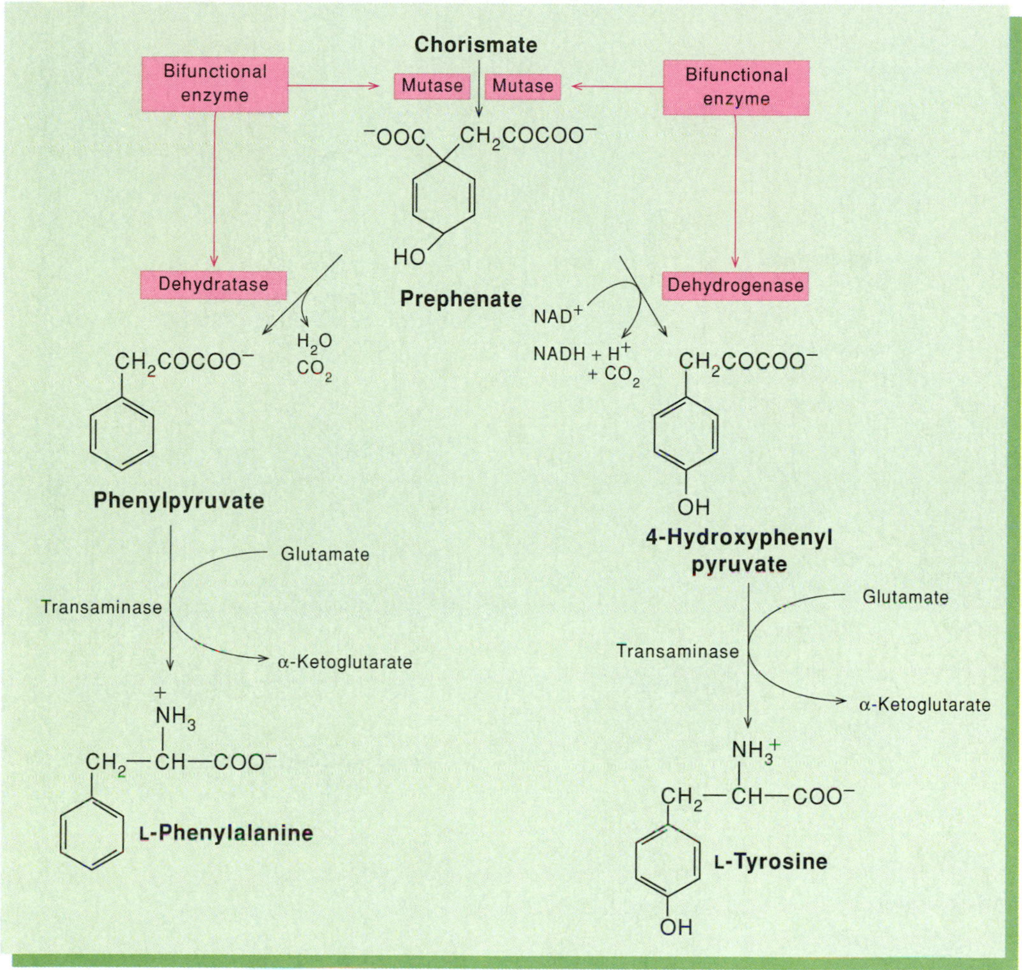

chorismate to 4-hydroxyphenylpyruvate, again with the formation of prephenate occurring as an enzyme-bound intermediate. Both enzymes utilize prephenate, the phenylalanine biosynthetic enzyme yielding phenylpyruvate, and the tyrosine biosynthetic enzyme yielding 4-hydroxyphenylpyruvate. The final aromatization step, the removal of water from prephenate in phenylalanine biosynthesis, or the removal of hydrogen in tyrosine biosynthesis, is accompanied by the loss of the ring carboxyl as CO_2.

Tryptophan Is Synthesized in Five Steps from Chorismate

The pathway leading from chorismate to L-tryptophan (fig. 18.22) is the most thoroughly studied of any biosynthetic pathway. In *E. coli,* the details of the enzymatic steps, the correlation between DNA sequence and the protein products, and the factors controlling the transcription of the structural genes far exceed those known for any other set of related genes. Although this has not been the work of any one group, the extensive gene-enzyme analysis of Charles Yanofsky laid the foundation for others to explore details of some of the enzymatic steps by physical and kinetic approaches. Comparative studies in other bacteria and in fungi have revealed a variation upon the themes found in *E. coli,* particularly with respect to the distribution on one protein or another of the sequence of enzyme activities, which are identical in all forms. In addition, these studies have also revealed differences in the way the genes are arranged in the DNA and in the way expression of those genes is controlled.

The first specific step in tryptophan biosynthesis is the glutamine-dependent conversion of chorismate to the simple aromatic compound anthranilate. Like most other glutamine-dependent reactions, the reaction can also occur with ammonia

Figure 18.22

The biosynthesis of tryptophan from the branchpoint compound, chorismate in *E. coli*. The first step involves the conversion of chorismate to the aromatic compound anthranilate. The anthranilate is transferred to a ribose phosphate chain. The product is cyclized to indoleglycerol phosphate by the removal of water and loss of the ring carboxyl by indoleglycerol phosphate synthase. Finally, in a replacement reaction catalyzed by tryptophan synthase, glyceraldehyde-3-phosphate is removed from indoleglycerol phosphate and the enzyme-bound indole is condensed with serine. The structure of phosphoribosyl pyrophosphate is shown in figure 18.23.

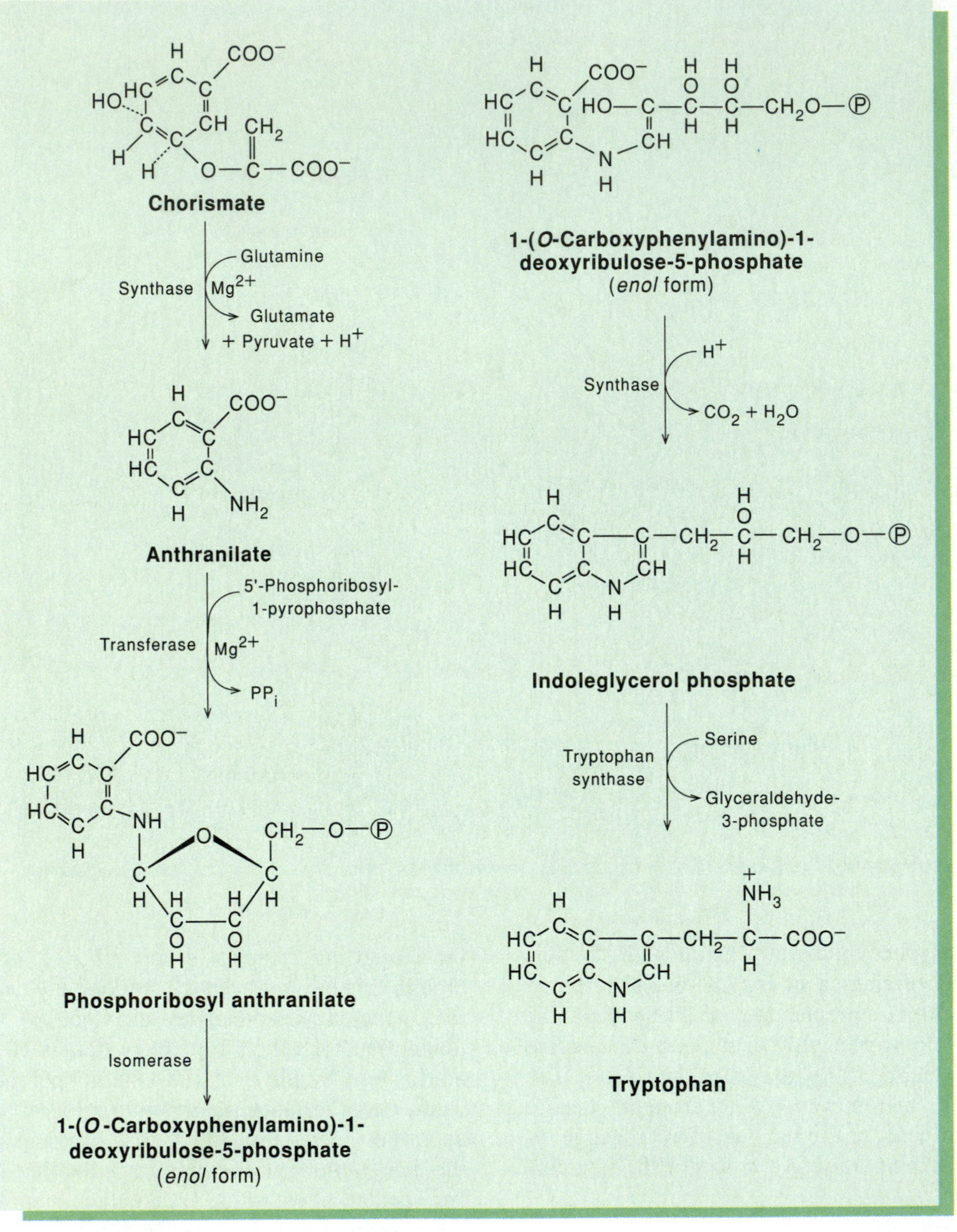

as the source of the amino group. However, high concentrations of ammonia are required. Thus far, almost all the anthranilate synthases examined have the glutamine amidotransferase activity (component II) and the chorismate-to-anthranilate activity (component I) on separate proteins. Component I actually catalyzes two reactions with 2-amino-4-deoxychorismate being an enzyme-bound intermediate.

Anthranilate is transferred to a ribose phosphate chain in a phosphoribosyl-pyrophosphate-dependent reaction catalyzed by anthranilate phosphoribosyltransferase. An isomerase catalyzes an Amadori rearrangement, in which the ribosyl moiety of phosphoribosylanthranilate becomes a ribulosyl moiety. The product, 1-(*o*-carboxyphenylamino)-1-deoxyribulose-5′-phosphate, is cyclized to indoleglycerol phosphate by the

Table 18.1
Distribution of Tryptophan Biosynthetic Enzyme Activities on Different Proteins in Bacteria and Fungi

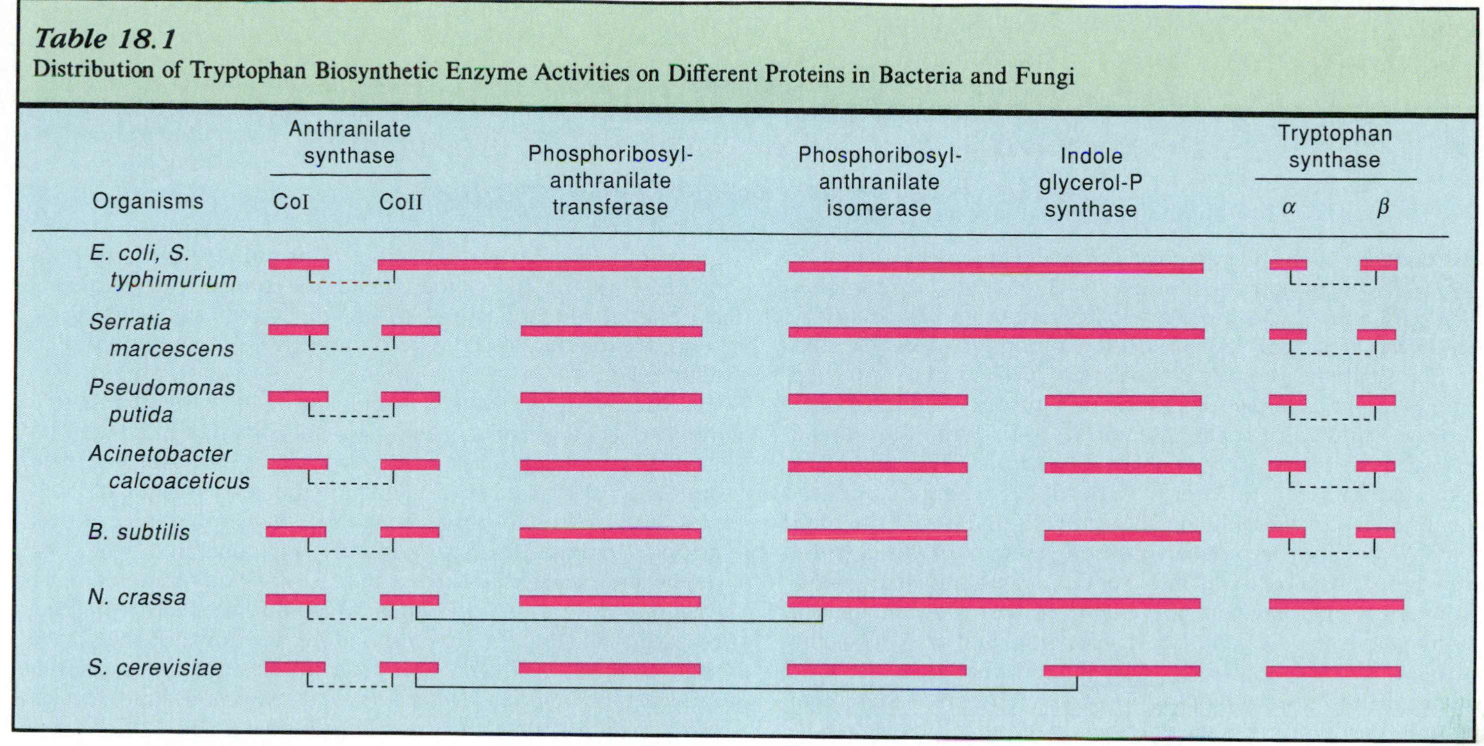

Anthranilate synthase
Organisms
CoI
CoII
Phosphoribosyl-anthranilate transferase
Phosphoribosyl-anthranilate isomerase
Indole glycerol-P synthase
Tryptophan synthase
α
β
E. coli, S. typhimurium
Serratia marcescens
Pseudomonas putida
Acinetobacter calcoaceticus
B. subtilis
N. crassa
S. cerevisiae

Note: └──┘ = covalent linkage; └‐‐‐‐‐‐‐┘ = obligatory association required for full activity. A single band covering two activities indicates a single polypeptide.

removal of water and loss of the ring carboxyl by indoleglycerol phosphate synthase. The final step in tryptophan biosynthesis is a replacement reaction, catalyzed by tryptophan synthase, in which glyceraldehyde-3-phosphate is removed from indoleglycerol phosphate and the enzyme-bound indole so formed condensed with serine.

At one time it was thought that indole itself was an intermediate in tryptophan synthesis, but this notion was dispelled by a combination of biochemical and genetic studies (box 18D).

Among different organisms these five enzyme activities are distributed on different proteins (table 18.1). For example, in *E. coli,* indoleglycerol phosphate synthase catalyzes both the isomerization of phosphoribosylanthranilate and the cyclization step. Of particular interest is the occurrence on a single protein of the catalytic activities for nonconsecutive reactions in some cases. If in such cases the proteins were separate from each other in the cell, this arrangement, for example, in *Neurospora,* would necessitate the product of one reaction leaving the product site of one enzyme to be acted upon by another enzyme and then returning to the substrate site of a third enzyme on the same protein that exhibited the first enzyme activity. The persistence of this arrangement during evolution makes attractive the idea that all the tryptophan biosynthetic enzymes exist in the cell as a single multienzyme (and multiprotein) complex. However, if so, the complex must be quite labile, since individual gene products are so readily separated.

The evolution of the gene fusion that resulted in the conversion of the α and β peptides of tryptophan synthase as found in *E. coli* to the single peptide with β and α domains as found in fungi and yeast has been studied experimentally. Fusion of the *trpB* and *trpA* genes in the order they occur on the chromosome to yield a single β-α chain results in a protein of only limited activity. In contrast, when the gene fusion is done in the reverse of the genetic order (i.e., *A* to *B*), the resulting α-β chain yields a protein that is much more active. In addition, a short interdomain peptide region is required for full activity in a fusion protein.

Carbon Flow in the Biosynthesis of Aromatic Amino Acids Is Regulated at Branchpoints

Metabolite flow to tryptophan is controlled by inhibition of anthranilate synthase by tryptophan. Regulation of metabolite flow in phenylalanine and tyrosine biosynthesis varies from organism to organism, owing to the variety of enzyme patterns in the conversion of chorismate to the two amino acids. In *E. coli* and related organisms, phenylalanine inhibits both activities of chorismate mutase P-prephenate dehydratase, whereas tyrosine inhibits only the mutase activity of chorismate mutase T-prephenate dehydrogenase.

There are two general patterns of control over the common aromatic pathway. One is that found in *E. coli* and related organisms. The pattern is similar to that of the common aspartate family pathway of the same organism in that there are three isozymic deoxy-*arabino*-heptulosonate-7-phosphate synthases. Each is inhibited by one of the three aromatic amino acids. (There is, in addition, a tryptophan-specific repression of the tryptophan-sensitive enzyme, a tyrosine-specific repression

18D BOX

Demonstration that Indole Is Not a True Intermediate in the Tryptophan Biosynthetic Pathway

The tryptophan pathway provides another example in which nutritional studies with mutants of *Neurospora* and *E. coli,* isotope incorporation studies, and enzymatic analyses have been exploited to reveal the steps in a biosynthetic pathway. For example, early studies with tryptophan-requiring organisms found in nature revealed that some could use indole (a compound known to be formed by the microbial degradation of tryptophan; see chapter 19), while others could use anthranilate. Later, after Beadle and Tatum introduced the approach of studying metabolism with mutants of the bread mold *Neurospora,* tryptophan-requiring mutants of this organism were found that could use indole or either anthranilate or indole. Still later, similar mutants of *E. coli* were found, and mutants of both organisms were described that accumulated one or the other of these compounds. Clearly, those mutants that grew on anthranilate or indole were blocked in some step before these compounds, and those that accumulated them were blocked in the step after them.

Incorporation studies with isotopes showed that when anthranilate was converted to tryptophan, the carboxyl group of anthranilate was lost as carbon dioxide, but the nitrogen was retained. Because the enzymes in the tryptophan biosynthetic pathway have only a limited specificity, it was possible to substitute 4-methylanthranilate in *E. coli* extracts that could convert anthranilate to indole. This "nonisotope" label was conserved during the conversion to yield 6-methyl indole:

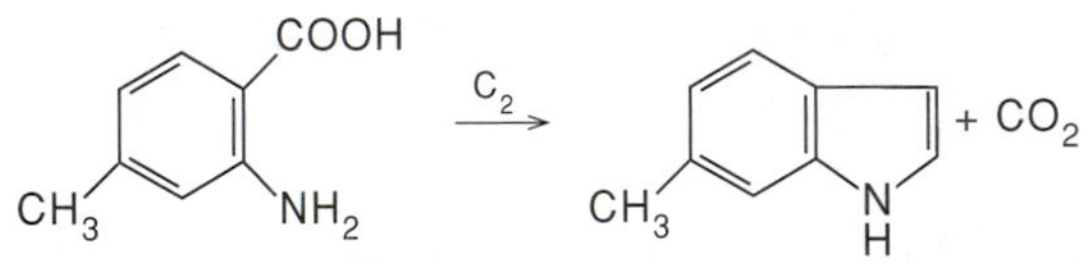

4-Methylanthranilate **6-Methyl indole**

It was thus clear that some two-carbon unit replaced the carboxyl carbon of anthranilate. Further studies with such *E. coli* extracts indicated that phosphoribosyl pyrophosphate was a good cosubstrate for the formation of indole from anthranilate. Fractionation of these extracts, as well as examination of mutants blocked between anthranilate and indole, revealed that an intermediate in this conversion was indole-3-glycerol phosphate. Extracts from one group of such mutants could not form indole-3-glycerol phosphate, while the other group could not convert it to indole and glyceraldehyde-3-phosphate. The latter group was found to accumulate the dephosphorylated derivative, indole-3-glycerol, in culture fluids.

The two intermediates in the conversion of anthranilate to indole-3-glycerol phosphate, phosphoribosylanthranilate and 1-(*o*-carboxyphenylamino)-1-deoxyribulose-5′-phosphate, were originally postulated to account for the involvement of phosphoribosyl pyrophosphate in indole-3-glycerol phosphate formation. Support for the postulate was obtained when the dephosphorylated derivative of the second of these intermediates was found in the culture fluids of certain bacterial tryptophan-requiring mutants. The corresponding derivative of the first intermediate has not been found, probably because of its instability. Indeed, this compound, when formed in extracts, is rapidly broken down to anthranilate and ribose-5-phosphate.

For several years, indole, which was accumulated by some mutants and used to satisfy the tryptophan requirement by others, was considered an intermediate in tryptophan biosynthesis. Such a role for indole would have been of interest, since it appeared to be an exception to the generalization that biosynthetic intermediates had to bear a charge. It was found that extracts of cells that utilized indole did indeed catalyze the condensation of indole with serine, and extracts of cells that accumulated indole catalyzed the cleavage of indole-3-glycerol phosphate to indole and glyceraldehyde-3-phosphate. Furthermore, *E. coli* mutants of these two classes clearly were affected in separate genes. However, the two products of these genes catalyzed their corresponding reactions faster when they were associated in a complex of the form $\alpha_1\beta_2$, where α and β stand for different protein subunits. The complex itself catalyzes the overall reaction

$$\text{Indole-3-glycerol phosphate} + \text{serine} \rightarrow \text{tryptophan} + \text{glyceraldehyde-3-phosphate}$$

faster than either of the separate reactions. The same was found with extracts of *Neurospora* in which the two partial reactions were catalyzed by the same protein and with which no evidence for indole as a free intermediate could be found. Thus it became clear that indole, although historically important in deciphering the pathway to tryptophan, occurs only as a bound intermediate.

---

of the tyrosine-sensitive enzyme, and a tryptophan plus phenylalanine-specific multivalent repression of the phenylalanine-sensitive enzyme.) As in the synthesis of the intermediates in the aspartate family common pathway, the three enzymes contribute to a common pool of deoxy-*arabino*-heptulosonate-7-phosphate that is drawn upon for all the compounds formed from chorismate. Indeed, in some strains, the phenylalanine-sensitive enzyme is predominant, whereas in others, the tyrosine-sensitive enzyme is predominant.

Another pattern is found in *Bacillus subtilis*. The single deoxy-*arabino*-heptulosonate-7-phosphate synthase is carried on the same protein that exhibits chorismate mutase activity.

The protein is complexed with another protein that exhibits shikimate kinase activity. Both the deoxy-*arabino*-heptulosonate-7-phosphate synthase activity and the shikimate kinase activity are inhibited by chorismate and prephenate, which may inhibit by virtue of binding to the substrate and product sites of the chorismate mutase. The chorismate mutase activity is inhibited by prephenate. Prephenate dehydratase is inhibited by phenylalanine, whereas prephenate dehydrogenase is inhibited by tyrosine.

The Biosynthesis of Histidine

The pathway to histidine in all plants and bacteria involves the transfer of N-1 and C-2 of the adenine moiety of ATP to the ribose phosphate moiety of phosphoribosyl pyrophosphate (PRPP) as shown in figure 18.23. This transfer is initiated by a condensation reaction between the two parent compounds. After a series of additional reactions involving ring opening, isomerization, and an amino-group transfer, the residue from the ATP molecule is released as 5-aminoimidazole-4-carboxamide ribotide. The latter, an intermediate in the purine nucleotide biosynthetic pathway (see chapter 20) is then "recycled." In this cyclic process, the 6-amino group of the adenine ring which had been derived from aspartate becomes the amide group of the 5-aminoimidazole-4-carboxamide ribotide. During purine biosynthesis, this amide group is also derived from aspartate and becomes N-1 of the purine ring. Thus there is no way to distinguish an adenosine derivative that has been formed directly by the *de novo* pathway from one that has been recycled from the histidine pathway.

Another interesting feature of the histidine pathway is the conversion of imidazoleglycerol phosphate to a carbonyl derivative that can undergo transamination. Following the incorporation of the α-amino group, which provides a positively charged molecule, the phosphate group is removed, thus illustrating the importance of charged groups on biosynthetic intermediates. We have previously noted, in chapter 16, the importance of charged intermediates in catabolism.

Histidine biosynthesis has been studied primarily in a few bacteria, *Neurospora,* and yeast. In several bacteria, the phosphatase and the dehydratase activities are carried on a single bifunctional protein. In yeast there is a single protein that exhibits the pyrophosphatase, cyclohydrolase, and dehydrogenase activities (the second, third, and tenth steps). These findings, along with the fact that significant "pools" of the intermediates are not found, raise the question of whether all the enzymes of histidine biosynthesis might not be arranged in a single complex, as has been suggested for the enzymes involved in tryptophan biosynthesis.

Amino Acids and Nutritional Needs in Mammals

In view of the central importance of amino acids in proteins, we might expect that all organisms would possess the necessary enzymes to synthesize the protein amino acids. Quite surprisingly, we find that this is not the case. In mammals, for example, fewer than half of the protein amino acids can be synthesized by *de novo* pathways. The remainder must be supplied by nutrients. Frequently this requirement leads to nutritional problems, since many diets that might contain sufficient carbohydrates and other essential components are deficient in one or more of the required amino acids.

Why did the evolutionary process not favor the preservation of pathways for synthesizing all the amino acids needed in higher animals? Can this be a case of too great a genetic load? That seems unlikely, since only 84 genes are required to encode all of the enzymes needed to make the twenty amino acids found in proteins. Mammals are believed to have in excess of 50,000 genes. Another possible reason for the loss of competency is that some of the intermediates in amino biosynthesis may have toxic effects on other sophisticated biochemical processes occurring in higher eukaryotes. There is no proof for this explanation, but it is at least a tenable hypothesis. We do know that amino acid excess and certain degradative intermediates can lead to metabolic problems (see chapter 19).

Plants and bacteria, which possess the ability to synthesize all protein amino acids, obviously require more enzymes for amino acid biosynthesis than animals, which can synthesize only a limited number of amino acids. Recognizing this fact, certain chemical firms have been busy attempting to synthesize toxins (herbicides) that are selective against plants. These herbicides are designed so that they interact specifically with enzymes in the amino acid pathways uniquely possessed by plants (box 18E).

An Essential Amino Acid Is One that Must Be Supplied in the Diet

The inability of mammals to synthesize all of the amino acids they require has led to the classification of amino acids as essential and nonessential. An "essential" amino acid, in this classification, means one that must be supplied in the diet if the organism is to maintain a positive nitrogen balance. As we will see, the absence of a *de novo* pathway for the biosynthesis of an amino acid does not automatically mean that the amino acid must be obtained from the diet. For one thing, it is frequently possible for the organism to make one amino acid from another amino acid. Furthermore, an alternative pathway for synthesis may supply sufficient amino acid for maintenance, if not for growth.

The classical differentiation between essential and nonessential amino acids was made for rats by W. C. Rose. He based his identification on the weight gain of growing white rats that were fed diets containing nineteen of the twenty amino acids found in proteins. His results appear in the first two columns of table 18.2. For humans, there have been only a few studies. These were based not on weight gain or loss, but on short-term maintenance of a positive nitrogen balance. If there was a negative nitrogen balance (total nitrogen excretion exceeding total nitrogen intake) during the period when a single amino acid was excluded from the diet, that was taken to mean that tissue protein was being degraded and used to supply the missing amino

Figure 18.23

The biosynthesis of histidine. The 5-aminoimidazole-4-carboxamide ribotide formed during the course of histidine biosynthesis is also an intermediate in purine nucleotide biosynthesis. Therefore it can be readily regenerated to an ATP, thus replenishing the ATP consumed in the first step in the histidine biosynthetic pathway (see fig. 20.14).

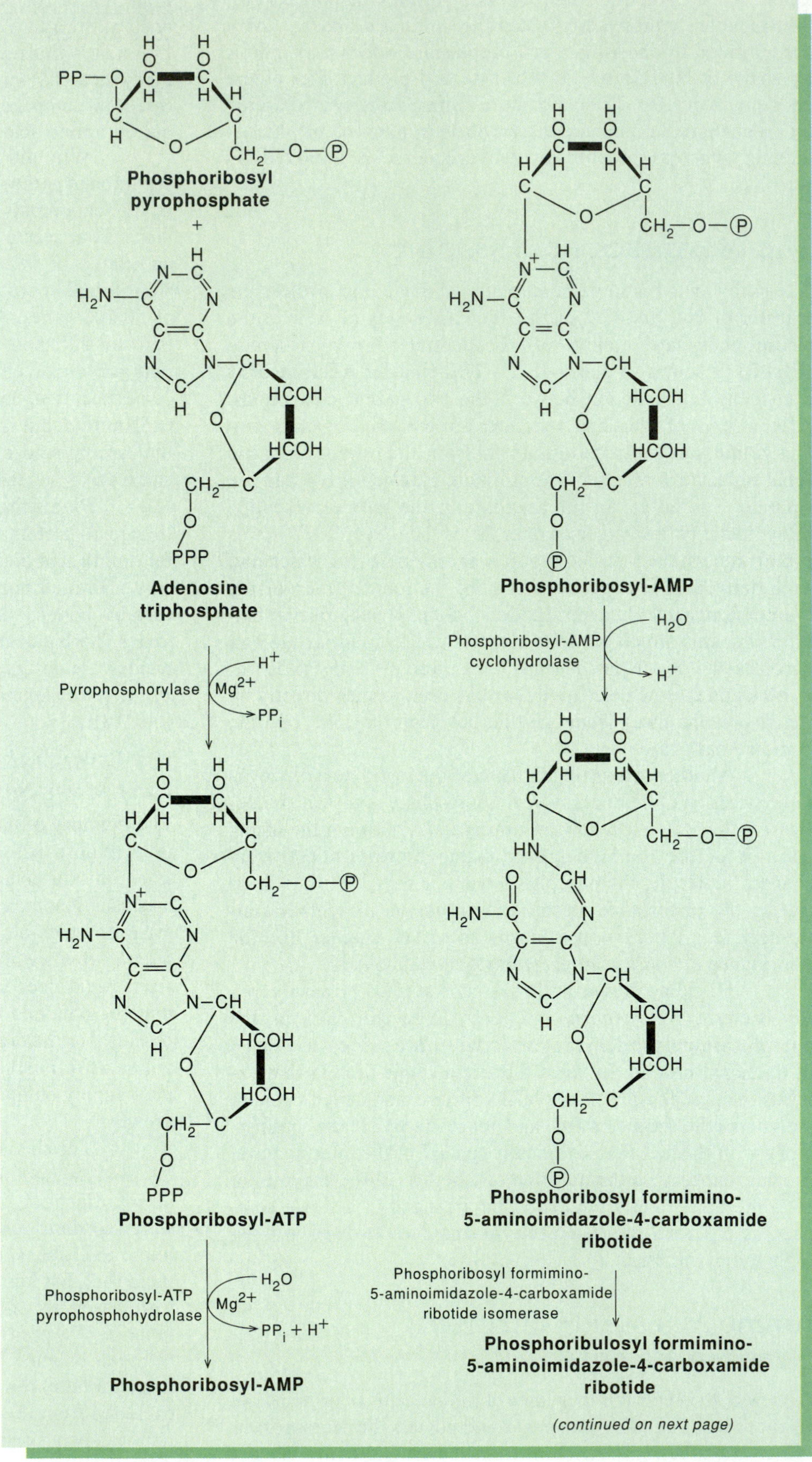

(continued on next page)

Phosphoribulosyl formimino-5-aminoimidazole-4-carboxamide ribotide

Amidocyclase

Glutamine → Glutamate

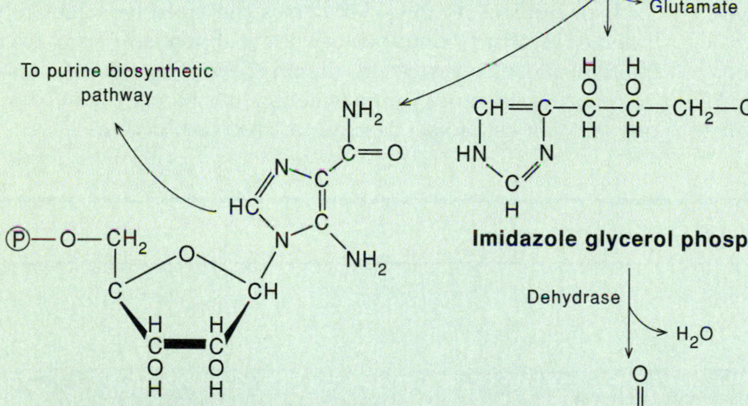

5-Aminoimidazole-4-carboxamide ribotide

Imidazole glycerol phosphate

Dehydrase

H_2O

Imidazole acetol phosphate

Transaminase

Glutamate → α-Ketoglutarate

Histidinol phosphate

Histidinol phosphate

Phosphatase

H_2O → P_i

L-Histidinol

Dehydrogenase

NAD^+ → $NADH + H^+$

$NAD^+ + H_2O$ → $NADH + 2H^+$

L-Histidine

18E
BOX

Amino Acid Pathways Absent in Mammals Offer Targets for Safe Herbicides

In recent years, agribusiness firms have empirically developed several compounds that inhibit essential steps in the biosynthesis of amino acids found in plants but missing in animals. One of these compounds, glyphosate, is a highly specific inhibitor of 5-enol-pyruvylshikimate-3-phosphate synthase (an enzyme needed for aromatic amino acid biosynthesis; see fig. 18.20). Glyphosate is the active ingredient in the widely used herbicide "Roundup."

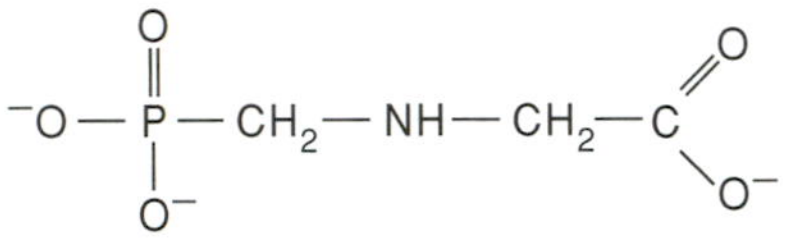

$$^{-}O-\overset{\overset{O}{\|}}{\underset{\underset{O^-}{|}}{P}}-CH_2-NH-CH_2-C\overset{\nearrow O}{\searrow O^-}$$

Glyphosate

Three other classes of compounds, although quite different from each other, are all inhibitors of acetohydroxy acid synthase (an enzyme required for branched-chain amino acid biosynthesis; see fig. 18.18). These three classes are sulfonylureas, imidazolinones, and triazolopyrimidines, which are the active ingredients in, respectively, "Oust," "Sceptor," and a third commercial herbicide still under development.

Because animals do not synthesize either the aromatic or branched-chain amino acids, these materials can be applied to kill unwanted vegetation without causing harm to domestic animals or humans. Certain derivatives can often be selectively applied to combat noxious plants without appreciable harm to crops. More promising, however, is the prospect of using biotechnology to incorporate genes for enzymes specifically resistant to one of the herbicides into seeds used for crop production.

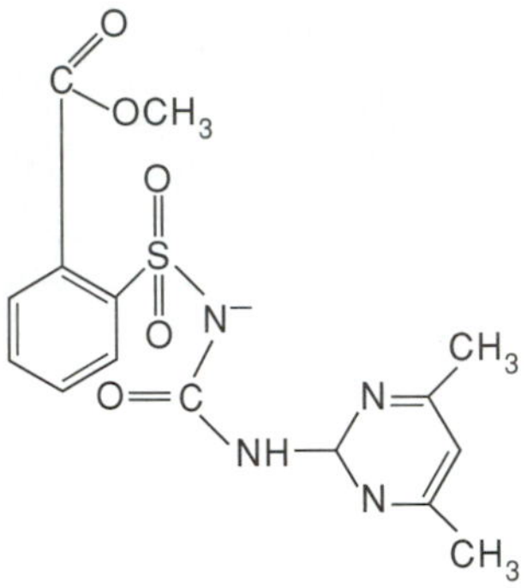

Sulfometuron methyl

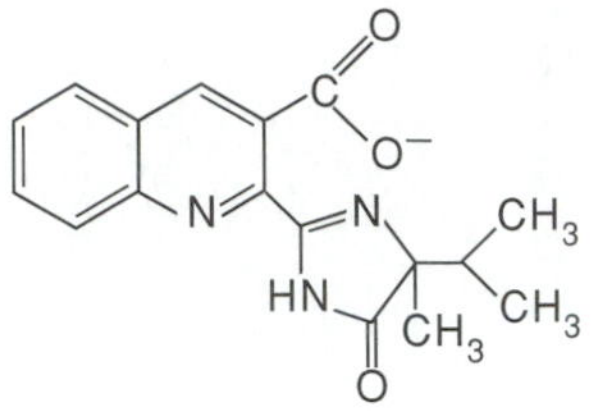

Imazaquin

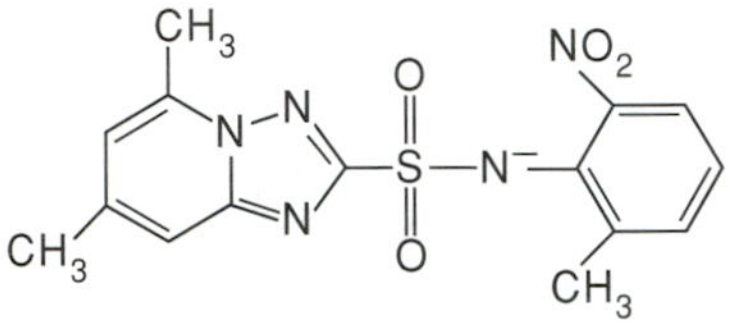

1,2,4-Triazolo-(1,5-*a*)-2,4-dimethyl-3-(*N*-sulfonyl-)2-nitro-6-methyl-sulfonanilide

Table 18.2
The Essential Amino Acids

For Weight Gain in Protein-Starved Adult Rats		For Positive Nitrogen Balance in Adult Humans		For Mouse L Cells in Culture	
Essential	**Nonessential**	**Essential**	**Nonessential**	**Essential[a]**	**Nonessential**
Histidine	Alanine	Isoleucine	Alanine	Arginine	Alanine
Isoleucine	Arginine[b]	Leucine	Arginine	Cysteine	Asparagine
Leucine	Asparagine	Lysine	Asparagine	Glutamine	Aspartate
Lysine	Aspartate	Methionine	Aspartate	Histidine	Glutamate
Methionine	Cysteine	Phenylalanine	Cysteine	Isoleucine	Glycine
Phenylalanine	Glutamate	Threonine	Glutamate	Leucine	Proline
Threonine	Glutamine	Tryptophan	Glutamine	Lysine	Serine
Tryptophan	Glycine	Valine	Glycine	Methionine	
Valine	Proline		Histidine[c]	Phenylalanine	
	Serine		Proline	Threonine	
	Tyrosine		Serine	Tryptophan	
			Tyrosine	Tyrosine	
				Valine	

[a]The medium also contained 0.25–1% dialyzed horse serum.

[b]Arginine is required in the diet of young rats.

[c]Histidine is required in infant humans.

Figure 18.24

Amino acids that are synthesized *de novo* in mammals. All such amino acids are related by a small number of steps to glycolysis or TCA cycle intermediates. (Also see fig. 18.1.)

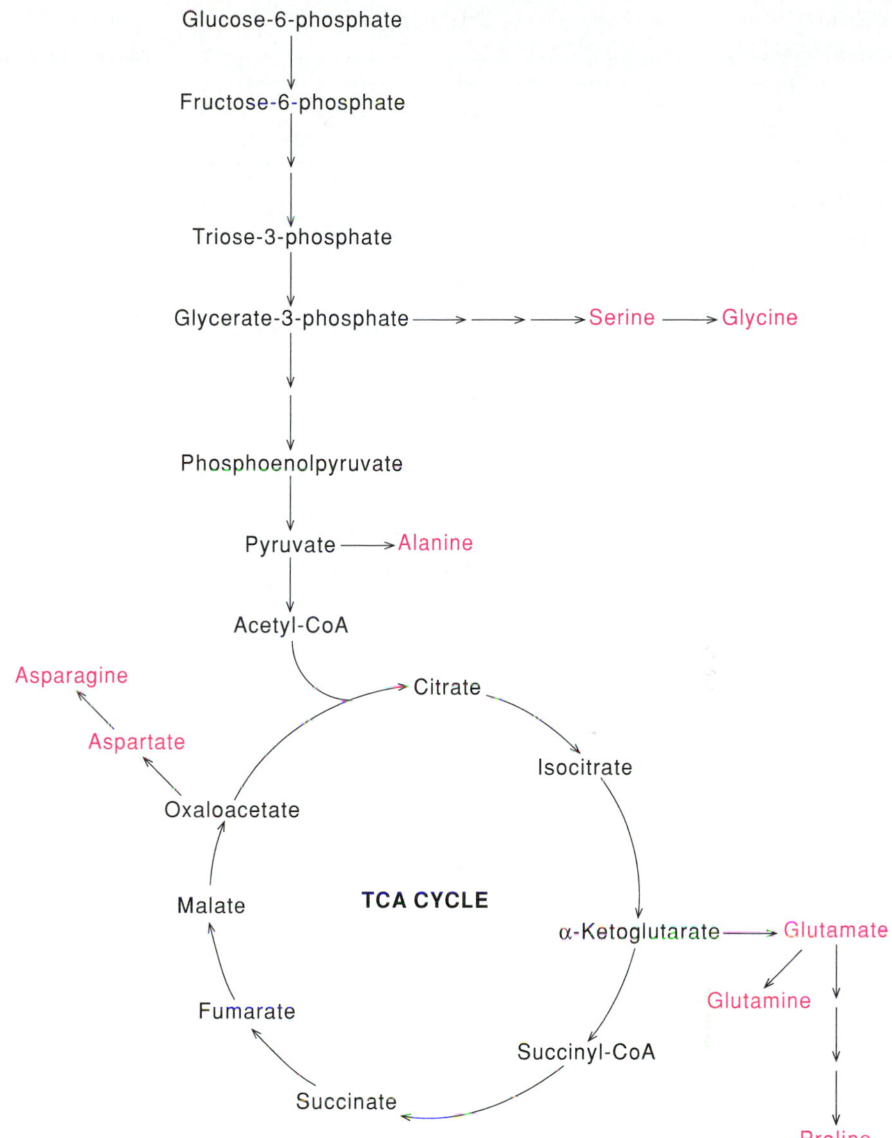

acid. Table 18.2 lists the findings. For comparison, the amino acids found to be essential and nonessential for a strain (L) of mouse fibroblasts grown in cell culture are also included in table 18.2.

The results in table 18.2 do not mean, for example, that rats have all the enzymes they need for converting the several central metabolites to amino acids listed in column 2, or that humans have the enzymes needed to form those in column 4. Rather, they indicate that given nineteen other amino acids, the particular amino acid can be formed either by the *de novo* pathway used by plants or by a "salvage" pathway at the expense of another amino acid.

It is true that many animals, including humans, do have the enzymes of several of the "short" biosynthetic pathways—those to alanine, glutamate, glutamine, proline, aspartate, asparagine, serine, and glycine (fig. 18.24). The enzymes needed for proline biosynthesis (a kinase and two reductases) are present in animal tissues. In addition, animals contain a proline oxidase that yields Δ^1-pyrroline-5-carboxylic acid, which because of the equilibrium with glutamic-γ-semialdehyde (see fig. 19.24) and a transaminase, allows for a nonacetylated route to ornithine. Ornithine, in turn, can be converted to arginine by means of four of the five enzymes that are essential for urea formation (see fig. 19.6). Thus, although the pathway by which bacteria and plants form arginine is not present in animal tissue, arginine can be formed, but, in some organisms (e.g., rats), not at a rate sufficient for growth. Two other amino acids normally present in the diet also can be formed from other amino acids

Table 18.3
"Salvage" Pathways for Formation of Certain Nonessential Amino Acids from Other Amino Acids

Amino Acid Formed	Formed From	Enzymes Required
Arginine	Proline	Proline oxidase Ornithine-glutamate transaminase Ornithine transcarbamoylase Argininosuccinate synthase Argininosuccinate lyase
Cysteine	Methionine	*S*-adenosylmethionine synthase α-Methyltransferase *S*-adenosylhomocysteinase Cystathionine-β-synthase Cystathionine-γ-lyase
Tyrosine	Phenylalanine	Phenylalanine-4-monooxygenase

by unidirectional routes. For example, cysteine can be formed from dietary methionine (see fig. 19.30), and tyrosine can be formed by hydroxylation of phenylalanine by a pteridine-dependent monooxygenase (see fig. 19.16). These routes and the corresponding enzymes are listed in table 18.3.

Nonprotein Amino Acids

In addition to the twenty amino acids most frequently found in protein, there are many amino acids in plants, bacteria, and animals that are not found in proteins. Some are found in peptide linkages in compounds that are important as cell wall or capsular structures in bacteria or as antibiotic substances produced by bacteria and fungi. Others are found as free amino acids in seed and other plant structures. Some amino acids are never found in proteins. These nonprotein amino acids, numbering into the hundreds, include precursors of normal amino acids, such as homoserine and diaminopimelate, intermediates in catabolic pathways, such as pipecolic acid, D-enantiomers of "normal" amino acids, and amino acid analogs, such as azetidine-2-carboxylic acid and canavanine, that might be formed by unique pathways or by modification of normal amino acid biosynthetic pathways.

A Wide Variety of D-Amino Acids Are Found in Microbes

Certain D-amino acids along with some L-enantiomers are commonly found both in microbial cell walls and in many peptide antibiotics. For example, the peptidoglycans of bacteria contain both D-alanine and D-glutamate. The latter is present in a γ-glutamyl linkage. In some forms, the α carboxyl of the D-glutamyl residue is either amidated or in peptide linkage with glycine. D-Lysine or D-ornithine is found in the glycopeptide of some Gram positive organisms. The capsule of the anthrax bacillus is composed of a nearly pure homopolymer of D-glutamate in γ linkage. Other bacilli also produce γ-linked polyglutamates, some of which form separate D-glutamate and L-glutamate chains, while others form a copolymer of D- and L-glutamate. A wide variety of D-amino acids have been found in antibiotics, and some of these are listed in table 18.4.

Table 18.4
Some D-Amino Acids Found in Peptide Antibiotics

Antibiotic	D-Amino Acids Present	Produced by
Actinomycin C_1	D-Valine	*Streptomyces parralus* and others
Bacitracin A	D-Asparagine D-glutamate, D-ornithine, D-phenylalanine	*Bacillus subtilis*
Circulin A	D-Leucine	*Bacillus circulans*
Fungisporin	D-Phenylalanine, D-valine	*Penicillium* species
Gramicidin S	D-Phenylalanine	*Bacillus brevis*
Malformin A_1, C	D-Cysteine, D-leucine	*Aspergillus niger*
Mycobacillin	D-Aspartate, D-glutamate	*Bacillus subtilis*
Polymixin B_1	D-Phenylalanine	*Bacillus polymyxa*
Tyrocidine A, B	D-Phenylalanine	*Bacillus brevis*
Valinomycin	D-Valine	*Streptomyces fulrissimus*

At least some of the D-amino acids are formed by racemases as free intermediates and are subsequently incorporated into peptide bonds. An example is D-alanine, found in the bacterial cell wall peptidoglycan. L-Alanine is converted to D-alanine by a racemase that contains pyridoxal phosphate as a cofactor. The racemization is followed by the formation of a D-alanyl-D-alanine dipeptide, which is accompanied by the conversion of ATP to ADP. The dipeptide is subsequently incorporated into the glycopeptide.

In most cases of formation of D-amino acid–containing peptides studied thus far, the L form of the amino acid is the substrate for the incorporating enzyme. In contrast, the free D-amino acid is ordinarily a poor substrate for the incorporation reaction. Whether the racemization occurs on the enzyme or inversion occurs afterwards remains to be determined in most cases. In the case of the D-valyl residue formed in penicillin, a tripeptide derivative containing L-valine is an intermediate, and conversion is thought to occur by way of an α-β-dehydro form of the valyl residue.

Table 18.5
Some Naturally Occurring Nonprotein Amino Acids

Compound	Occurrence	Remarks
Branched-chain and cyclopropane amino acids		
$CH_3-CH_2-CH(CH_3)-CH_2-CHNH_2-COOH$ 2-Amino-4-methylcaproic acid (homoisoleucine)	California buckeye	Leucine antagonizes toxicity
$(CH_3)_2-NCH_2-CHNH_2-COOH$ 2-Amino-3-dimethylaminopropionic acid (azaleucine)	*Streptomyces neocaliberis*	Leucine antagonizes toxicity
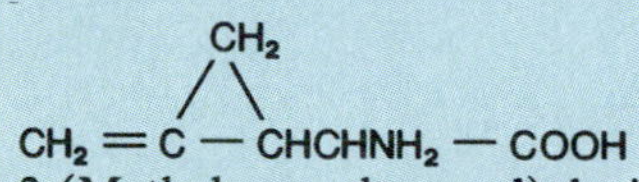 CH_2 (ring) $CH_2=C-CHCHNH_2-COOH$ 2-(Methylenecyclopropyl)glycine	Lychee seeds	Leucine antagonizes toxicity
Sulfur-containing amino acids		
$CH_3-SCH_2-CHNH_2-COOH$ *S*-Methylcysteine	Broad bean (*Phaseolus vulgaris*)	—
$CH_3-CH=CH-S-CH_2-CHNH_2-COOH$ *S*-(Prop-1-enyl)cysteine	Garlic	—
Aromatic and heterocyclic amino acids		
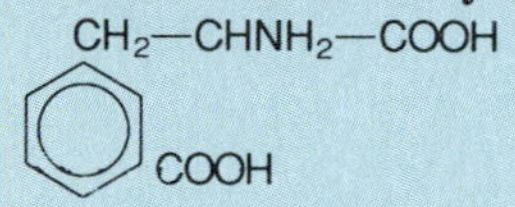 CH_2-CHNH_2-COOH COOH 3-(3-Carboxyphenyl)alanine	Iris (*Iris pseudacoras*)	—
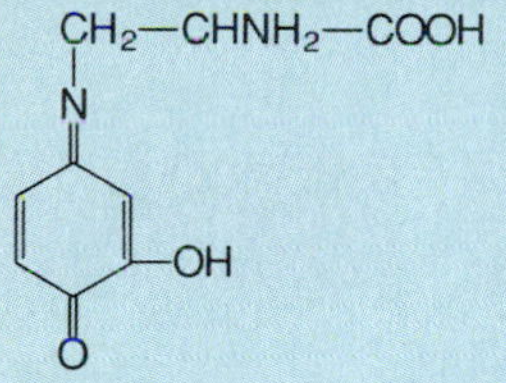 CH_2-CHNH_2-COOH N OH O *β*-*N*-(3-Hydroxy-4-pyridone)alanine (mimosine)	Mimosa tree	Tyrosine; toxic to nonruminants
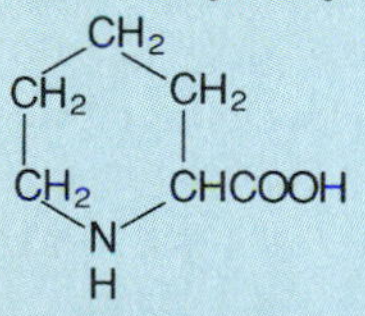 CH_2 CH_2 CH_2 CH_2 CHCOOH N H Pipecolic acid	Widely distributed in plants	Probably not toxic; an intermediate in lysine catabolism
Acidic amino acids		
$HOOC-CH(CH_3)-CH_2-CHNH_2-COOH$ 4-Methylglutamic acid	Sweet pea	—
Basic amino acids		
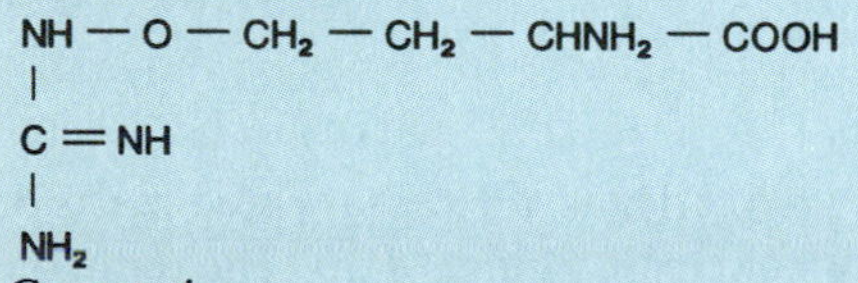 $NH-O-CH_2-CH_2-CHNH_2-COOH$ $C=NH$ NH_2 Canavanine	Jack bean and other legumes	Arginine antagonizes toxicity
$NH_2-CH_2-CHNH_2-COOH$ 2,3-Diaminopropionic acid	Seeds of acacia and mimosa	As the oxalyl derivative, acts as a neurotoxin

There Are Hundreds of Naturally Occurring Amino Acid Analogs

Among the hundreds of nonprotein amino acids found in nature are many that might be considered naturally occurring amino acid analogs and many that are toxic and antagonistic to the usual twenty amino acids found in proteins. Some are found as components of antibiotics; others have been identified as antibiotic substances themselves. The frequent occurrence of toxic amino acids in the seeds of plants suggests that they might play a role in the protection of the seeds from insects or other predators.

Some typical examples of these naturally occurring analogs are given in table 18.5 along with their sources and the antagonistic L-amino acid, where such an antagonism is known. It should be pointed out, however, that not all the "analogs" are toxic. For example, pipecolic acid, the next higher homolog of proline and an intermediate in lysine degradation, does not interfere in any demonstrable way with proline metabolism.

Sources of Nitrogen for Amino Acid Biosynthesis: The Nitrogen Cycle

Nitrogen passes through various forms and valence states as a result of its interactions with different living forms. Collectively the transitions constitute a cycle that is illustrated in figure 18.25. The valence states range from +5 in nitrates to −3 in ammonia or organic materials. In the 0 valence state, nitrogen is a gas. In all other valence states, it usually exists as a solid. The passage of nitrogen from one form to another involves a chain of widely distributed organisms, as indicated in figure 18.25.

NH_3 is the form in which nitrogen is incorporated into organic materials, but it is less often available to plants or bacteria for biosynthesis than other forms of nitrogen. When present for any length of time in nature, NH_3 will either be assimilated into organic materials or be oxidized by nitrifying bacteria (such as *Nitrosomonas* and *Nitrobacter*) to nitrite and nitrate. Reduction of nitrate by plants or by bacteria seldom yields NH_3 in excess. Nitrogen fixation, by which nitrogen of the atmosphere is reduced to NH_3, occurs in a very limited number of microorganisms and plants. NH_3 is rarely produced in excess by this process either. Thus most plants must assimilate NH_3 at the low levels that are generated by their own nitrate and nitrite reductases or from the reduction of nitrogen by microorganisms.

Nitrogen Fixation Involves an Enzyme Complex Called Nitrogenase

The biological fixation of nitrogen by both free-living and symbiotic nitrogen-fixing bacteria involves an enzyme complex called nitrogenase (fig. 18.26). Nitrogenases consist of two proteins. One is an iron-containing protein called component II, and the other, containing molybdenum and iron (or vanadium and iron in some nitrogenases), is called component I (see fig. 18.26). The Fe protein from *Clostridium pasteurianum* is a dimer of two identical subunits, each with a molecular weight of 29,000, surrounding an iron-sulfur center that contains four iron atoms and four acid-labile sulfur atoms. The Mo-Fe protein is considerably more complex, and in *C. pasteurianum* is a tetramer of molecular weight 220,000, with two molybdenum atoms, about 32 iron atoms, and about as many acid-labile sulfur atoms. The four subunits are of two kinds, two with a molecular weight of 50,000 and two with a molecular weight of 60,000. There are probably four iron-sulfur centers, each containing four iron atoms and four labile sulfur atoms, and two extractable iron-molybdenum-sulfur "cofactors" that contain eight iron atoms, six labile sulfur atoms, and one molybdenum atom. Although it might be expected that only six reducing equivalents would be required for each mole of N_2 reduced to NH_3, in actual fact, a necessary part of the reaction catalyzed by nitrogenase obligatorily converts at least two protons (H^+) to hydrogen (H_2) for each molecule of N_2 reduced. The overall reaction for the reaction is:

$$N_2 + 8\ H^+ + 8\ e^- + 16\ ATP \rightarrow 2\ NH_3 + H_2 + 16\ ADP + 16\ P_i$$

Figure 18.25

The nitrogen cycle depicts the flow of nitrogen in the biological world. The proteins of animals and plants are cleaved by many microorganisms to free amino acids from which ammonia (or ammonium ion) is released by deamination. Urea, the main nitrogen excretion product of animals, is hydrolyzed to NH_3 and CO_2. *Nitrosomonas* soil bacteria obtain their energy by oxidizing NH_3 to nitrite, NO_2^-. *Nitrobacter* obtain their energy by oxidizing nitrite, NO_2^-, to nitrate, NO_3^-. Plants and many microorganisms reduce nitrate for incorporation into amino acids, completing the cycle. Other microorganisms reduce nitrate partly to NH_3 and partly to N_2, which is lost to the atmosphere. Atmospheric nitrogen, N_2, can be recaptured, reduced, and converted into organic substances by a limited number of nitrogen-fixing bacteria and algae.

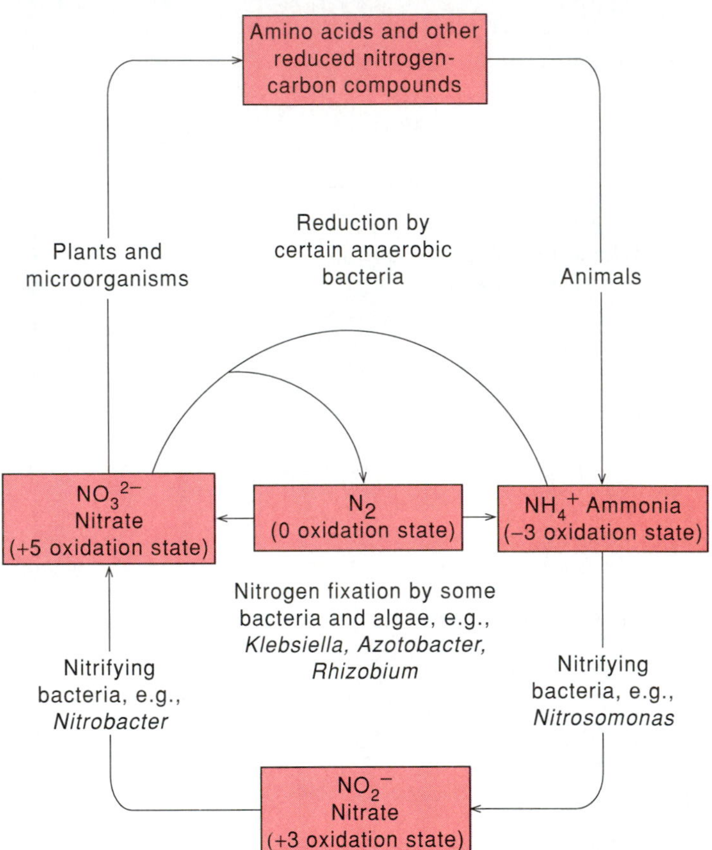

Why hydrogen production is a necessary step in the reduction of dinitrogen can be seen in figure 18.26. The reduction of nitrogen by nitrogenase can be viewed as a cycle of eight electron transfers, each of which occurs in the cyclic transfer of a single electron from reduced ferredoxin or flavodoxin to the Fe protein (component II) and the regeneration of the reduced electron donor by cell metabolism (e.g., pyruvate oxidation). Upon reduction, the Fe protein binds two molecules of MgATP and then associates with the Fe-Mo protein component (component I). The transfer of the electron to the Fe-Mo protein is accompanied by the hydrolysis of the two MgATP molecules,

Figure 18.26

The eight-step nitrogenase cycle. Nitrogenase is composed of two multiprotein complexes known as components I and II. Component II (the Fe protein) binds to component I (in most nitrogenase complexes an Fe-Mo protein) after picking up a single electron from reduced ferredoxin (Fd) or flavodoxin (Fld) and binding two ATP molecules. Upon cleavage of the two ATP molecules, the electron is transferred from component II to component I, and component II is released to begin a new Fe-protein cycle. After three single-electron (e^-) transfers, a H_2 molecule is released from component II and a N_2 molecule becomes bound. After eight turns of the Fe-protein cycle, the N_2 molecule is fully reduced and two NH_3 molecules are released, allowing nitrogenase to begin a new cycle. The dinitrogen is probably reduced in stages as discussed in the text. Wherever there is an electron (e^-) there should also be a proton (H^+) and wherever there is an ATP or an ADP there should also be a magnesium cation (Mg^{2+}). The protons and magnesium cations have been left out of the diagram to avoid crowding. The pathway followed by the electron transfer is indicated by colored arrows. The eight-step cyclical process is arranged in a giant circle. It should be understood that a functional unit of nitrogenase contains only one molecule of components I and II.

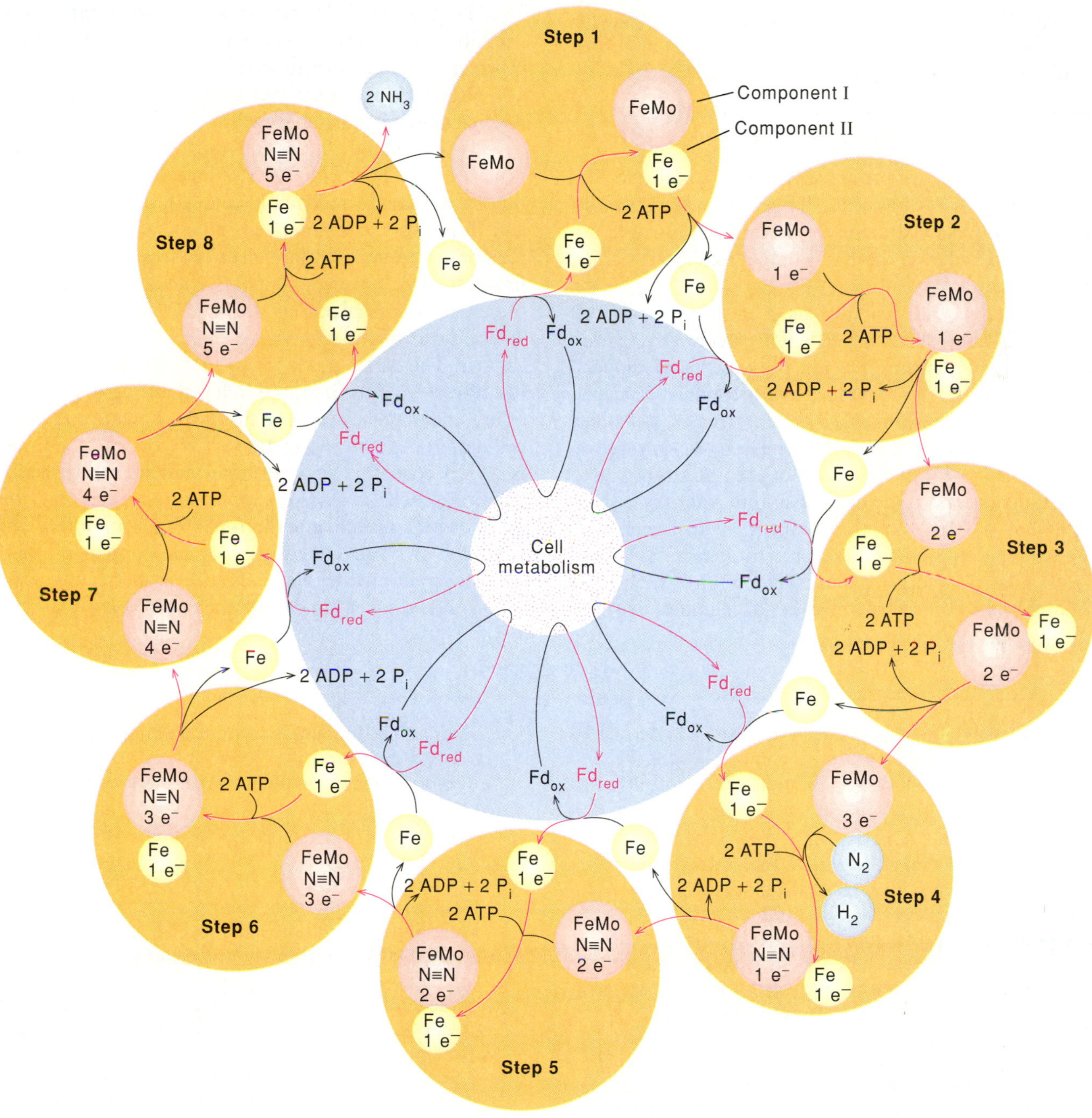

and the disassociation of the two components. Thus the sequential functioning of eight Fe protein reduction-oxidation cycles is required for one Fe-Mo protein cycle. As the diagram shows, after three such transfers, one H_2 molecule is given off to allow for the binding of the dinitrogen to the active site.

Under most conditions, however, nitrogen fixation can be more costly than the overall reaction that we have shown implies. Unless reduced ferredoxin and ATP are supplied to component II at a very high rate, the partially reduced component I can release H_2 and "slip back" to a less reduced state, thus initiating a futile cycle. Although the sequence of reactions leading from N_2 to 2 NH_3 would in effect be the sequential formation of diimide, hydrazine, and finally ammonia,

$$N_2 \xrightarrow{2H^+ + 2e^-} HN{=}NH \xrightarrow{2H^+ + 2e^-} H_2N{-}NH_2 \xrightarrow{2H^+ + 2e^-} 2NH_3$$

the actual intermediates are presumably bound to the Mo-Fe-S centers in a state as yet unknown. Thus far, only in *Azotobacter chroococcum* has hydrazine formation by a nitrogenase been demonstrated and even there it is only a minor but significant product.

Nitrogenases can function only in the absence of oxygen. For strict anaerobes, this does not present any special problems, since the organism grows in the absence of oxygen. Some nitrogen fixers are strict aerobes and there must be a special means to protect the nitrogenase system from oxygen. One genus, *Azotobacter,* protects its nitrogenase by a very active electron-transport system that removes oxygen at a rapid rate. At least part of that rapid rate of oxygen removal may be due to the ability of the Fe-protein component of nitrogenase to reduce O_2, forming H_2O_2 (which in turn is broken down by catalase or peroxidase). Another genus, *Rhizobium,* which fixes nitrogen symbiotically in the root nodules of legumes, depends on leghemoglobin, a special hemoglobin made by plants that absorb O_2 and transports it to the respiring bacteria at a partial pressure of oxygen low enough to allow nitrogenase to function.

Nitrate and Nitrite Reduction Play Two Physiological Roles

Nitrate and nitrite reduction play two physiological roles. One, exhibited primarily by bacteria, is a dissimilatory role in which nitrate and nitrite serve as terminal electron acceptors. In other words, the reduction serves to oxidize reducing equivalents, e.g., NADH, generated during oxidation of substrates. The other, exhibited by bacteria, fungi, and plants, is an assimilatory role in which nitrite is formed from nitrate and is reduced to NH_3 at a rate no greater than that required for synthesis of nitrogenous compounds during growth.

In general, the assimilatory nitrate and nitrite reductases are soluble enzymes that utilize reduced pyridine nucleotides or reduced ferredoxin. In contrast, the dissimilatory nitrate and less well characterized nitrate reductases are often membrane-bound terminal electron acceptors that are tightly linked to cytochrome b_1 pigments. Such complexes allow one or more sites of energy conservation (ATP generation) coupled with electron transport.

The NH_4^+ produced by fermentative bacteria that utilize nitrite as an oxidant, or produced during the decomposition of organic materials, can be utilized by plants and bacteria for cell material. However, under suitably aerobic conditions, NH_3 is rapidly converted again to nitrite and nitrate by the nitrifying bacteria, such as *Nitrosomonas europaea* and *Nitrobacter agilis,* which are chemolithotrophs that gain energy from the oxidation of NH_3. The ubiquity of such organisms assures the persistence of nitrate, which, is utilized better by green plants than is NH_3, except under alkaline conditions.

Summary

In this chapter we have discussed the biosynthesis of amino acids and the roles that certain amino acids play in bringing inorganic nitrogen and sulfur into bioorganic compounds. The following points are the highlights of this discussion.

1. Amino acid biosynthesis is best studied in microorganisms such as *E. coli,* in which all twenty of the amino acids found in proteins are synthesized. Microorganisms are also ideal for such studies because both genetic and biochemical techniques can be harnessed to analyze the pathways. Typically, research begins by isolating mutants defective in single steps in the pathway for a particular amino acid and analyzing the consequences of the mutation.
2. The pathways to amino acids arise as branchpoints from a few key intermediates in the central metabolic pathways.
3. Inorganic nitrogen for amino acid biosynthesis must be derived from nitrate, nitrite, or ammonia in the environment. However, it is incorporated into organic form only as ammonia. For most cells and under most conditions, this conversion occurs by means of the amidation of glutamate to yield glutamine. This amide group is then used as a donor of "active" ammonia in numerous reactions, including the reductive amination of α-ketoglutarate. The amino group of glutamate thus formed then serves as the source of amino groups in the biosynthetic pathways of all the amino acids found in proteins. In only some cell types, growing in the presence of a high concentration of ammonia, does the amination of α-ketoglutarate and the amidation of aspartate (to yield asparagine) occur directly by free ammonia.
4. The pathway for synthesizing the α-ketoglutarate family of amino acids has two main branches. One of these involves a chain-lengthening process that yields lysine. Glutamate is also a precursor of proline and arginine. Both routes require activation and reduction of a carboxyl group. In the route to arginine, protection of the α-amino group is required to prevent the cyclization reaction essential for proline biosynthesis. The formation of the guanidine group of arginine requires carbamoyl phosphate, which is also a precursor of the pyrimidines.

5. The serine family includes L-serine, glycine, and L-cysteine. Three enzymes convert 3-phospho-D-glycerate to serine and two more enzymes convert L-serine to glycine. Cysteine biosynthesis involves sulfhydryl transfer to activated serine. Inorganic sulfate must be reduced to sulfide before it is incorporated into amino acids.
6. The biosynthetic route for the aspartate family is a highly branched pathway leading first to its amide, asparagine, then to lysine following condensation with pyruvate, then to methionine following sulfur and methyl group transfer, and then to threonine. In these pathway branches, the α-amino group of aspartate is preserved. The amino group is lost, however, in the route to isoleucine, by which threonine contributes four carbons to isoleucine. Carbon flow in the aspartate family is regulated in the aspartokinase step.
7. The pyruvate family consists of its α-amino analog, alanine, valine (derived from the condensation of two pyruvate molecules), and leucine (made by lengthening the carbon skeleton of valine, much like the chain-lengthening reaction in the fungal lysine pathway, which we have not shown). The steps to valine are paralleled by those to isoleucine and, indeed, are catalyzed by the same enzymes—the difference being that α-ketobutyrate rather than pyruvate is the acceptor of the two-carbon fragment in the condensation step in isoleucine biosynthesis.
8. The aromatic amino acids tyrosine, phenylalanine, and tryptophan derive their aromatic rings from the shikimate pathway, with final aromatization of the ring occurring only in the specific branch pathways leading to the final products. The ring itself arises from a condensation between erythrose-4-phosphate and phosphoenolpyruvate, followed by cyclization. The α amino group of tryptophan arises only indirectly by transamination with glutamate, since there is an exchange of serine for three carbons that had originated from ribose-5-phosphate.
9. The histidine pathway is a complex one in which a —C—N— unit of adenine serves as a nucleus for condensation with a ribosylphosphate moiety and another nitrogen derived from glutamine. The residue from adenine is, in fact, an intermediate in the purine nucleotide biosynthetic pathway and can thus be recycled by replenishing the lost —C—N— unit. The phosphate group is retained until after the α-amino group is incorporated, an example of the principle that metabolic intermediates bear charged groups.
10. Only eight of the *de novo* pathways for amino acid biosynthesis can be found in humans. These amino acids are all related by a small number of steps to glycolysis or TCA cycle intermediates. A number of additional amino acids can be formed from these amino acids. Essential amino acids are those that must be supplied in the diet.
11. In addition to the twenty amino acids commonly found in proteins, there are many amino acids and amino acid analogs that serve other functions. (Some nonprotein compounds formed from amino acids are discussed in chapter 19.)
12. NH_3 is the form in which nitrogen is incorporated into organic materials. Nitrogen exists in the -3 valence state in NH_3. Nitrogen itself actually passes through various forms and valence states as a result of its interactions with different living forms. The valence states range from $+5$ in nitrates to -3 in ammonia or organic materials. In the 0 valence state, nitrogen is a gas. The passage of nitrogen from one form to another involves a chain of widely distributed organisms. The biological fixation of gaseous nitrogen by both free-living and symbiotic nitrogen-fixing bacteria involves an enzyme complex called nitrogenase.

Selected Readings

Battersby, A. R., C. J. R. Fookes, G. W. J. Matcham, and E. McDonald, Biosynthesis of the pigments of life: Formation of the macrocycle. *Nature* 285:17, 1980. This paper discusses the steps in tetrapyrrole biosynthesis and the pathways diverting this nucleus to chlorophylls, hemes, cytochromes, and other macrocyclic pigments.

Bender, D. A., *Amino Acid Metabolism* Wiley, 1985.

Fowden, L., P. J. Lea, and E. A. Bell, The nonprotein amino acids of plants. *Adv. Enzymol.* 50:117, 1979. A discussion of the occurrence and biosynthesis of naturally occurring amino acid analogs in plants.

Katz, E., and A. L. Demain, The peptide antibiotics of *Bacillus:* Chemistry, biogenesis and possible functions. *Bacteriol. Rev.* 41:449, 1977. A description of several peptide antibiotics showing the distribution of D-amino acid in these compounds.

Kishore, G. M., and D. M. Shah, Amino acid biosynthesis inhibitors as herbicides. *Ann. Rev. Biochem.* 57:627–663, 1988. The focus is on the biosynthesis of essential amino acids.

Meister, A., *Biochemistry of the Amino Acids,* vols. 1 and 2. New York: Academic Press, 1965. The two-volume classic provides a thorough discussion of amino acid literature, occurrence, properties, and metabolism of amino acids up to that time.

Miflin, B. J. (ed.), *The Biochemistry of Plants: A Comprehensive Treatise,* vol. 5, *Amino Acids and Derivatives.* New York: Academic Press, 1980. This volume contains ten chapters by several authors detailing amino acid biosynthesis pathways in plants.

Neidhardt, F. C., J. L. Ingraham, K. B. Low, B. Magasanik, M. Schaechter, and H. E. Umbarger (eds.), *Escherichia coli and Salmonella typhimurium: Cellular and Molecular Biology, Vol. 1.* Washington: American Society for Microbiology, 1987. This volume contains seven chapters by several authors describing in detail the pathways of amino acid biosynthesis in bacteria with particular emphasis on enzymatic and genetic control mechanisms.

Torchinsky, Yu M., Transamination, its discovery, biological and chemical aspects (1937–1987). *Trends Biochem. Sci.* 12:115–117, 1987.

Yamada, K., S. Kinoshita, T. Tsunoda, and K. Aida (eds.), *The Microbial Production of Amino Acids.* New York: John Wiley and Sons, 1972. A collection of essays describing microbial processes used in Japanese industry for the production of amino acids. Includes examples in which the regulatory mechanisms functioning in most cells have been modified or bypassed.

Problems

1. Molecules with structures as diverse as carbamoylphosphate, tryptophan, and cytidine triphosphate are feedback inhibitors of the *E. coli* glutamine synthase. The feedback inhibition is cumulative, with each metabolite exerting a partial inhibition of the enzyme. Why would complete inhibition of the glutamine synthase by a single metabolite be metabolically unsound?
2. Given the structural diversity of the compounds that feedback-inhibit glutamine synthase, would you predict that they interact at a common regulatory site? Why or why not?
3. A mutant of *E. coli* is discovered with a defect in serine hydroxymethyltransferase. What amino acid supplement(s) would you expect the mutant to require? Explain. Although the mutant does not require L-methionine, its growth is stimulated by addition of this amino acid to the medium. Suggest a reason for this observation.
4. Aspartate, asparagine, glutamate, glutamine, and proline are among the amino acids that are not essential for humans. Glucose may be considered the precursor for each of these amino acids. Explain.
5. How does increased synthesis of aspartate and glutamate affect the TCA cycle? How does the cell accommodate this effect?
6. Contrast the mechanisms of amidation of glutamate to glutamine and aspartate to asparagine.
7. For each carbon in serine (α carboxyl; α carbon; β carbon) indicate which carbons of glucose-6-phosphate are incorporated. Assume that glucose is supplied solely from the medium and is not being synthesized from precursors.
8. A genetic defect in cystathionine-β-synthase leads to homocystinuria in humans. Explain.
9. The biosynthesis of L-proline from glutamate involves an internal amination rather than a transamination. Explain what is meant by an "internal amination" in the context of proline biosynthesis.
10. Write a scheme illustrating the catalytic role of homocysteine in methyl transfer reactions involving *S*-adenosylmethionine and the pool of one-carbon metabolites.
11. In what sense may indole be viewed as an "intermediate" in L-tryptophan biosynthesis?
12. The accumulation of biosynthetic intermediates, or of metabolites derived from these intermediates, has proven to be valuable in the analysis of biosynthetic pathways in microorganisms. It was found that these accumulations occurred only after the required amino acid had been consumed and growth had stopped. How might you account for this observation?
13. When ^{14}C-labeled 4-hydroxyproline was administered to rats, the 4-hydroxyproline in newly synthesized collagen was not radio-labeled. Explain.
14. Although phenylalanine is an amino acid essential to humans, tyrosine is not. Explain why this is true.

19

CHAPTER

The Metabolic Fate of Amino Acids

Although amino acids serve as important components or precursors of many biological compounds, they all ultimately undergo degradation reactions. Each amino acid is degraded by a specific pathway (fig. 19.1). In complex eukaryotes the main purpose of degradation is sometimes merely to remove excess amino acids, which can be toxic to the organism. Degradative reactions can also serve to supply the organism with nitrogen, carbon, or energy. Indeed, amino acids constitute vital carbon and energy sources for some cells, either because the cells cannot utilize fatty acids or carbohydrates or because these sources of energy are not available. Our primary object in this chapter will be to describe some of the main degradation pathways.

We will also consider a few biosynthetic processes that are initiated with preformed amino acids. These include processes leading to the formation of the porphyrin nucleus found in many oxygen- and electron-carrying proteins, of biologically active amines, of glutathione, of peptide antibiotics, and of several other important metabolites. In later chapters we will discuss the reactions by which amino acids are incorporated into proteins (chapter 29), cell walls (chapter 21), and pyrimidines and purines (chapter 20).

Protein Degradation

Amino acids that are catabolized come primarily from three different sources: dietary proteins, storage proteins, and metabolic turnover of endogenous proteins. The catabolism of dietary proteins and amino acids is a characteristic of higher animals, whereas the catabolism of storage protein is best illustrated by the germination of protein-storing seeds such as beans or peas. In addition, all cells exhibit metabolic turnover of many proteins; in this process protein-containing structures, and the amino acids to which they are degraded, can be recycled into other proteins or derivatives that involve amino acids as precursors.

Protein catabolism begins with hydrolysis of the covalent peptide bonds that link successive amino acid residues together in a polypeptide chain (fig. 19.2). This process is termed proteolysis, and the enzymes responsible for the action are called proteases. For proteins ingested by higher animals, proteolysis occurs in the gastrointestinal tract and depends on proteases secreted by the stomach, pancreas, and small intestine. For endogenous protein or proteins ingested by unicellular animals, proteolytic digestion occurs within the cell.

Figure 19.1

Pathways for the degradation of the twenty amino acids found in proteins. The strategy followed for amino acid degradation in gross respects is similar (except for direction) to the strategy for amino acid biosynthesis. Thus the α-amino groups are usually removed at an early stage in degradation and the carbon skeletons filter into the central metabolic pathways. However, there are enormous differences in the specific pathways used in the two processes. Only a handful of amino acids involve similar sequences in both directions, and most of these involve highly reversible reactions, indicated by double arrows. The differences can be seen by comparing figures 19.1 and 18.1. In degradation pathways the carbon skeletons are funneled almost exclusively to intermediates in the TCA cycle. The dashed arrows in this figure associated with tyrosine and isoleucine reflect the fact that the carbon skeletons of these amino acids are split into two components, which are separately processed.

Amino acid degradation serves three purposes: (1) supplying energy, (2) supplying intermediates for the synthesis of other compounds, and (3) removing harmful excesses of certain amino acids. In this diagram those parts represented in the master diagram of figure 12.5 are shown in black and the remainder of the diagram is shown in color. The number of steps in each pathway is indicated by a number alongside the conversion arrow.

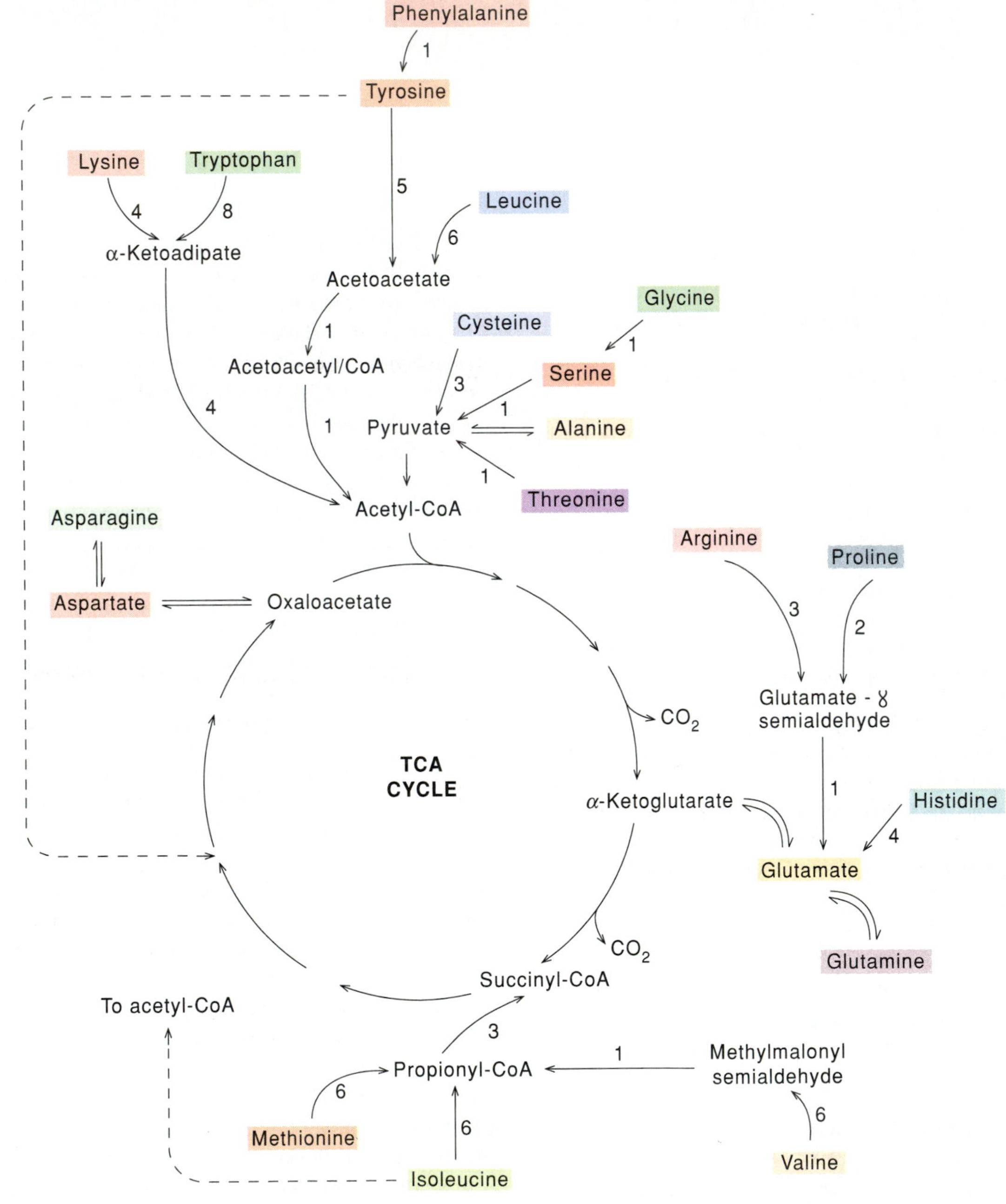

Figure 19.2

A protease hydrolyzes a peptide bond. Proteases have varying degrees of specificity, depending on the chemical nature of the R group and the location of the peptide linkage. Exopeptidases attack one or both ends of a polypeptide chain, while endopeptidases attack interior linkages.

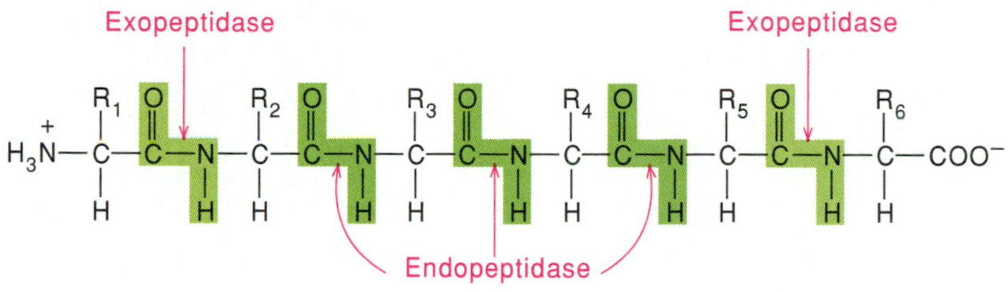

Figure 19.3

Transamination and deamination. Glutamate transaminase catalyzes the transfer of the α-amino group of an amino acid to α-ketoglutarate. The reaction is highly reversible, since the reacting functional groups of the products are identical to those of the reactants. Transamination is not deamination. Transamination yields ammonia only if it is linked to another type of deamination process. Here net deamination results from the combined action of glutamate transaminase and glutamate dehydrogenase. In this process the α-ketoglutarate is recycled.

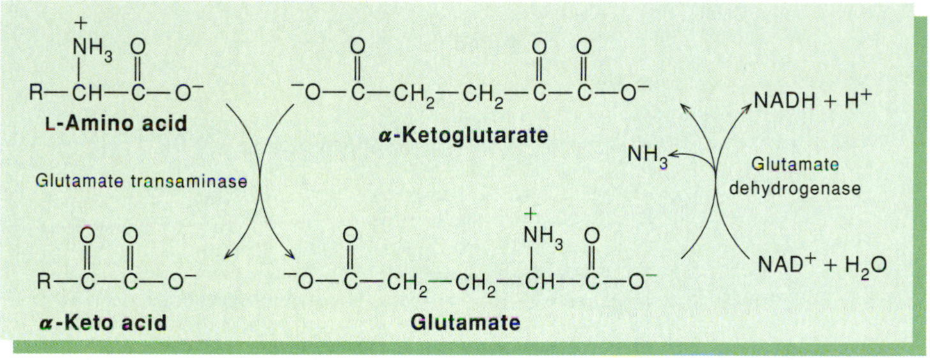

$R{-}CH(\overset{+}{N}H_3){-}C({=}O){-}O^-$ **L-Amino acid**

$^-O{-}C({=}O){-}CH_2{-}CH_2{-}C({=}O){-}C({=}O){-}O^-$ **α-Ketoglutarate**

Glutamate transaminase

$R{-}C({=}O){-}C({=}O){-}O^-$ **α-Keto acid**

$^-O{-}C({=}O){-}CH_2{-}CH_2{-}CH(\overset{+}{N}H_3){-}C({=}O){-}O^-$ **Glutamate**

NH_3 Glutamate dehydrogenase $NADH + H^+$ $NAD^+ + H_2O$

The products of proteolytic digestion are free amino acids and small peptides. Further digestion of peptides then depends on peptidases, which are characteristic of the intestinal mucosa. Peptidases act on their substrate either by hydrolyzing internal peptide bonds (in the case of endopeptidases) or by removing successive amino acids from the end of the peptide (exopeptidases). Exopeptidases are referred to as aminopeptidases or carboxypeptidases, depending on the end of the peptide from which digestion proceeds. Many single-cell microorganisms make proteases and peptidases, which they secrete into the surrounding medium so as to break down potential nutrient proteins to a size suitable for absorption. Finally, each amino acid is broken down by a specific pathway, and many of those pathways converge to yield intermediates in the common metabolic routes (see fig. 19.1).

Nitrogen Removal from Amino Acids

Degradative pathways for most amino acids begin by removal of the α-amino nitrogen. There are two major routes of deamination: transamination and oxidative deamination. Both of these processes are of major importance.

Transamination Is the Most Widespread Form of Nitrogen Removal

The process of transamination is illustrated in figure 19.3 for an undesignated amino acid donating its amino group to the TCA cycle intermediate α-ketoglutarate. The reaction leads to an α-keto acid and glutamate.

Nearly all transaminases contain pyridoxal-5′-phosphate as the coenzyme. The mechanism for transamination involving pyridoxal phosphate was discussed in chapter 11 (see fig. 11.3*a*). First the amino acid forms a Schiff's base adduct with the coenzyme. A proton is removed from the α carbon of the amino acid and donated to carbon 4′ of pyridoxal-5′-phosphate (an intramolecular transaldimination). Upon hydrolysis of the isomerized adduct, the corresponding α-keto acid is released, and the enzyme is in its pyridoxamine-5′-phosphate form. Most transaminases involved in amino acid breakdown exhibit a fairly broad specificity for the α-amino acid with which their pyridoxal forms can react or the corresponding α-keto acid with which their pyridoxamine forms can react. It is commonly found that one member of the amino donor-acceptor pair is glutamate-α-ketoglutarate. Because glutamate can be deaminated directly (see the following discussion), the general reaction shown in figure 19.3 is a common one.

Figure 19.4

Mechanism of the glutamate-dehydrogenase-catalyzed reaction. The reaction involves hydride transfer from glutamate to NAD^+, followed by transimidation to an α-amino group of a lysyl side chain of the enzyme, and finally by hydrolysis to α-ketoglutarate.

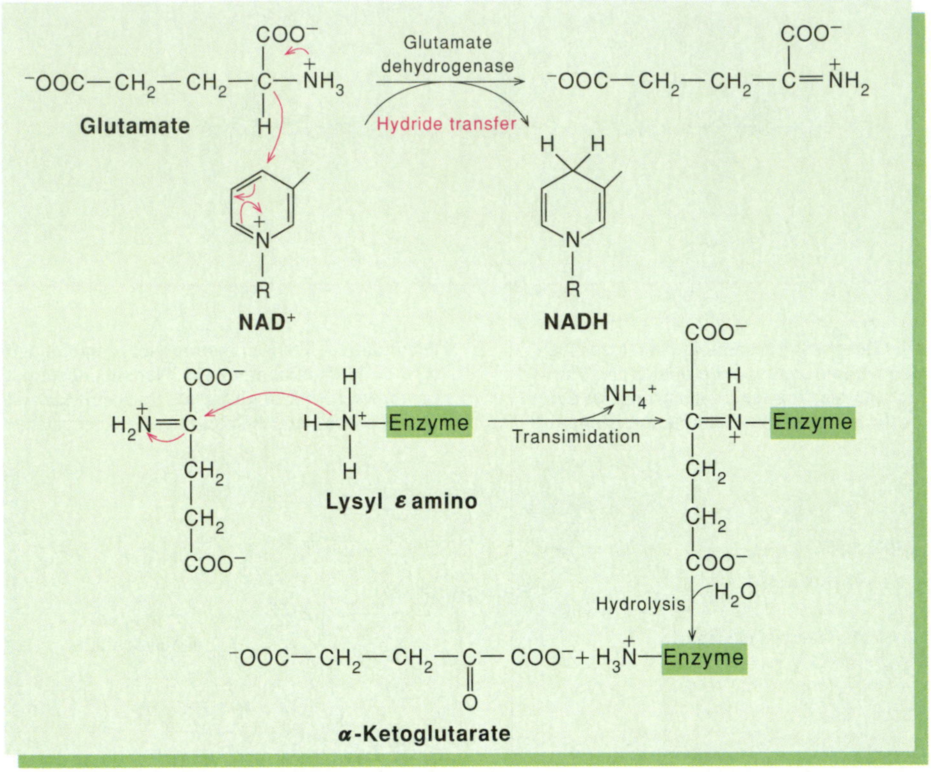

Oxidative Deamination Is Required for Net Deamination

Transamination does not result in any net deamination, since one amino acid is replaced by another amino acid. The main function of transamination is to funnel the amino nitrogen into one or a few amino acids. For glutamate to play a role in the net conversion of amino groups to ammonia, a mechanism for glutamate deamination is needed so that α-ketoglutarate can be regenerated for further transamination. The regeneration is accomplished by the oxidative deamination of glutamate, a reaction catalyzed by an NAD^+-linked enzyme, glutamate dehydrogenase. This broadly distributed enzyme is located in the mitochondria of eukaryotic cells. It catalyzes release of the α-amino group of glutamate, leading to the regeneration of α-ketoglutarate.

$$\text{Glutamate} + NAD^+ + H_2O \rightarrow \alpha\text{-ketoglutarate} + NH_4^+ + NADH$$

The overall process of transamination of α-ketoglutarate and regeneration of the α-ketoglutarate is shown in figure 19.3.

The catalysis by glutamate dehydrogenase involves covalent bond formation between an intermediate and the enzyme (fig. 19.4). In the first step of the reaction there is a hydride transfer from the α carbon of the amino acid to NAD^+. The resulting electron-deficient imino carbon is attacked by a lysyl side chain of the enzyme, leading to the displacement of ammonia and formation of an imino linkage with the enzyme. In the last step a hydrolysis restores the enzyme to its original state and α-ketoglutarate is released. Like glutamate, some amino acids that can undergo transamination can be deaminated more directly by oxidative reactions, either by a flavoprotein or by an NAD^+-linked enzyme.

Whether transamination or direct deamination is more important as an initial step in amino acid breakdown depends on the organism or tissue under investigation. However, where two mechanisms are available for one amino acid in a given cell type, it may well be that both mechanisms are employed.

The Fate of Nitrogen Derived from Amino Acid Breakdown

The NH_3 resulting from deamination of amino acids is converted to ammonia either directly or indirectly (e.g., by means of a transamination to yield a readily deaminated product such as glutamate). In microorganisms using a single amino acid as a nitrogen source, the ammonia so liberated is assimilated and used to form other nitrogen-containing cell components. When the amino acid is a carbon source, much more ammonia is liberated than is needed for biosynthesis, and it is disposed of by excretion to the medium. This simple disposal mechanism is adequate for free-living microorganisms, since the ammonia is carried away in the surrounding medium or escapes into the atmosphere.

Ammonia is also the major nitrogenous end product in some of the simpler aquatic and marine animal forms, such as protozoans, nematodes, and even bony fishes, aquatic amphibians, and amphibian larvae. Such animals are called ammonotelic. But in many animals, NH_3 is toxic, and its removal by simple diffusion would be difficult. Thus, in terrestrial snails and amphibians, as well as in other animals with environments in which water is limited, urea is the principal end product (fig. 19.5). Urea formation also helps in maintaining osmotic balance with seawater in the cartilagenous fishes. In such animals, most of the urea secreted by the kidney glomerulus is reabsorbed by the tubules. Indeed, the amount of nitrogen excreted by the kidneys of fishes is small compared with that excreted by the gills, and in most fishes, ammonia is the major form of excreted nitrogen.

Another form of "detoxified" ammonia that is used in nitrogen excretion is uric acid. Uric acid is the predominant nitrogen excretory product in birds and terrestrial reptiles (turtles excrete urea, whereas alligators excrete ammonia unless they are dehydrated, when they, too, excrete uric acid). Uric acid formed as a product of amino acid catabolism involves the *de novo* pathway of purine biosynthesis; therefore, its formation from NH_3 liberated upon amino acid catabolism will not be described here. In mammals, uric acid is exclusively an intermediate in purine catabolism, and in most mammals (primates excluded), it is further converted by uricase to allantoin.

Urea Formation in the Animal Liver Detoxifies NH_3

The formation of urea in the liver involves the multistep conversion of ornithine to arginine. Urea itself is formed from arginine by the action of arginase, which regenerates ornithine. The overall cyclic pathway was deduced by Krebs and Henseleit in 1932 from their biochemical investigations on rat liver slices. The details, shown in figure 19.6, have been developed through the efforts of many workers, particularly P. P. Cohen, S. Grisolia, and S. Ratner. Here we will discuss the urea cycle as it occurs in the mammalian liver.

The complete urea cycle uses five enzymes, argininosuccinate synthase, arginase and argininosuccinate lyase, which function in the cytosol, and ornithine transcarbamoylase and carbamoyl phosphate synthase, which function in the mitochondria. Additional specific transport proteins are required for the mitochondrial uptake of L-ornithine, NH_4^+, and HCO_3^- and for the release of L-citrulline.

The free ammonia formed by oxidative deamination of glutamate is converted into carbamoyl phosphate in a reaction requiring two ATP molecules:

$$NH_4^+ + HCO_3^- + H_2O + 2\,ATP \rightarrow \text{carbamoyl phosphate} + HPO_4^{2-} + 2\,ADP$$

Figure 19.5

Excretory forms of nitrogen in different organisms. NH_3 is the most common end product of nitrogen metabolism. In many organisms NH_3 is toxic. To prevent the harmful excess of ammonia it is converted to urea or uric acid before excretion.

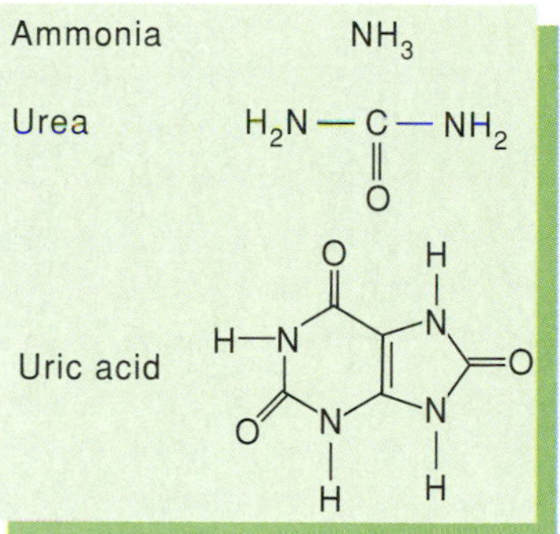

The reaction involves three steps, all of which take place on the same enzyme (fig. 19.7). In the first step the bicarbonate ion is activated and prepares the carbon for a nucleophilic attack by ammonia, which leads to the intermediate carbamate (step 2). In a reaction closely related to step 1 a second phosphoryl group is transferred to carbamate to form carbamoyl phosphate (step 3).

The carbamoyl group of carbamoyl phosphate has a high group transfer potential, which is displayed by its transfer to the terminal amino group of ornithine to form L-citrulline (see fig. 19.6). In the process inorganic phosphate is released. Before a further reaction can occur the citrulline must be transported across the mitochondrial membrane to the cytosol, where the remaining reactions leading to urea formation occur. Citrulline reacts with L-aspartate in an ATP-dependent reaction to form argininosuccinate, AMP, and PP_i. The PP_i is subsequently hydrolyzed to inorganic phosphate, so in effect the cost of this step is two ATP molecules. Argininosuccinate is cleaved to fumarate and L-arginine. The fumarate returns to the pool of TCA cycle intermediates, whereas the arginine becomes hydrolyzed to urea and ornithine. The ornithine is reutilized in further rounds of the urea cycle. Urea diffuses through the bloodstream and is ultimately eliminated through the kidney in the urine. The stoichiometry for the urea cycle is

$$CO_2 + NH_4^+ + 3\,ATP + \text{aspartate} + 2\,H_2O \rightarrow \text{urea} + 2\,ADP + 2\,P_i + AMP + PP_i + \text{fumarate} + 5\,H^+$$

In each turning of the urea cycle two nitrogens are eliminated, one originating from the oxidative deamination of glutamate and the other coming from the α-amino group of aspartate. Since the PP_i is subsequently hydrolyzed, it takes four high-energy phosphates to form a single molecule of urea. Thus the cost of this form of detoxification of ammonia is surprisingly high.

Figure 19.6

The urea cycle is a mechanism for removing unwanted nitrogen. Sources of nitrogens involved in urea formation are shown in color. Five enzymes are used in the urea cycle. Three of these function in the cytosol and two, as shown, function in the mitochondrial matrix. Specific carriers in the inner mitochondrial membrane transport ornithine, citrulline, ammonium ion, and HCO_3^- (CO_2) into and out of the mitochondrial matrix.

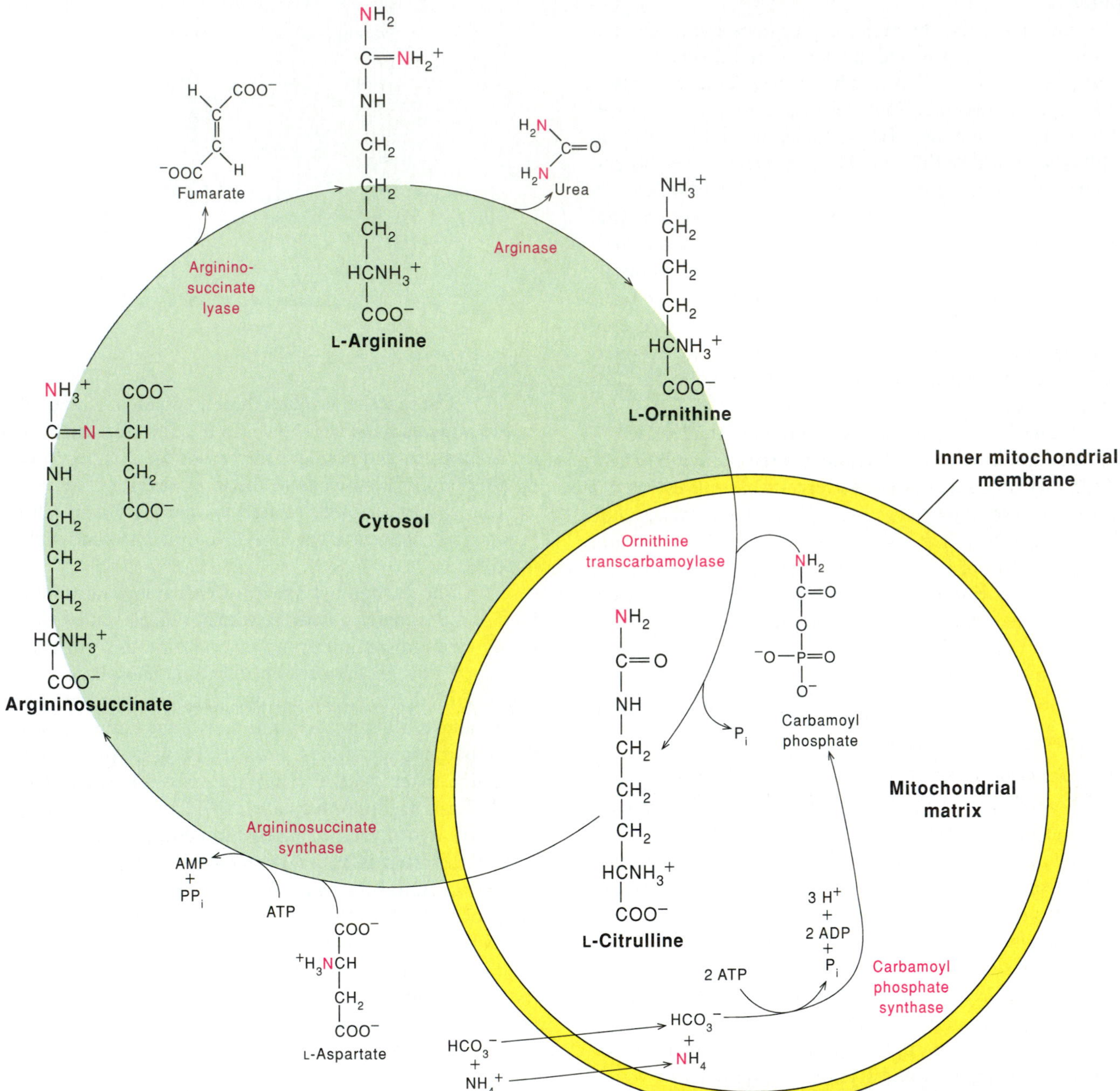

The Urea Cycle and the TCA Cycle Are Linked

In the urea cycle the carbon skeleton of the aspartate is released as fumarate. This product links the urea cycle with the TCA cycle. Fumarate is hydrated to malate, which is oxidized to oxaloacetate. The carbons of oxaloacetate can stay in the TCA cycle by condensation with acetyl-CoA to form citrate, or they can leave the TCA cycle either by gluconeogenesis to form glucose or by transamination to regenerate the aspartate as shown in figure 19.8. Since Krebs was involved in the discoveries of both the urea cycle and the TCA cycle, the interaction between the two cycles shown in figure 19.8 is sometimes referred to as the Krebs bicycle.

Figure 19.7

The mechanism of formation of carbamoyl phosphate. The reaction involves three steps, all of which take place on the same enzyme, carbamoyl phosphate synthase.

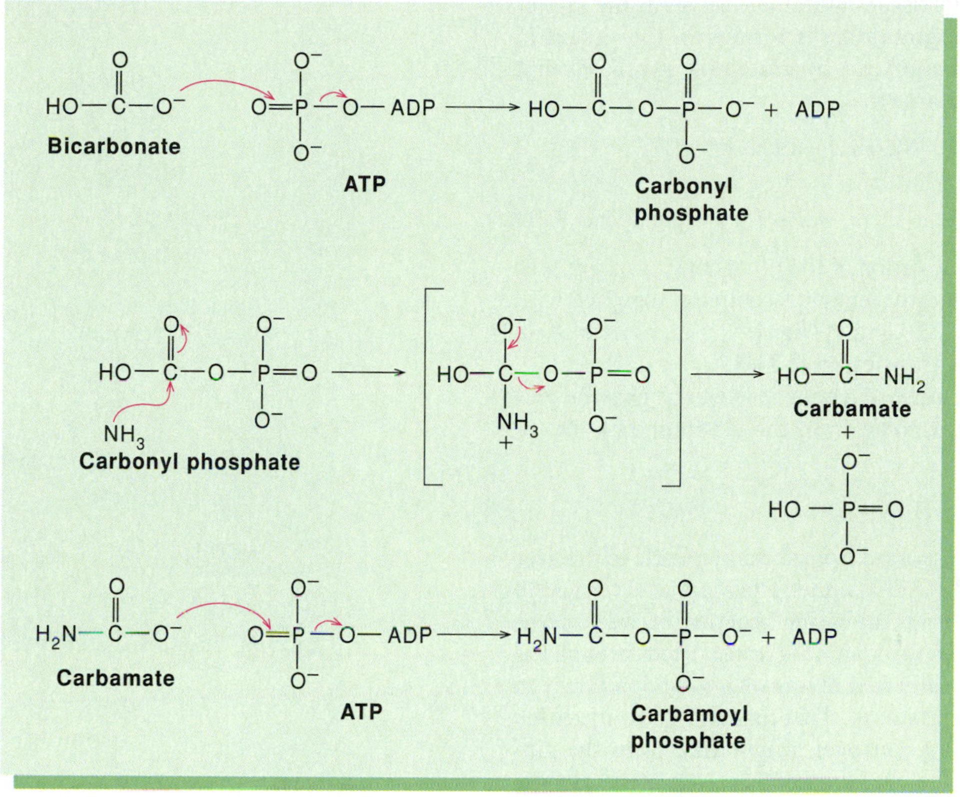

$$\underset{\textbf{Bicarbonate}}{HO{-}C({=}O){-}O^-} + \underset{\textbf{ATP}}{O{=}P(O^-)_2{-}O{-}ADP} \longrightarrow \underset{\textbf{Carbonyl phosphate}}{HO{-}C({=}O){-}O{-}P({=}O)(O^-){-}O^-} + ADP$$

$$\underset{\textbf{Carbonyl phosphate}}{HO{-}C({=}O){-}O{-}P(O^-)_2{=}O} + NH_3 \longrightarrow \left[HO{-}C(O^-)(\overset{+}{N}H_3){-}O{-}P(O^-)_2{=}O \right] \longrightarrow \underset{\textbf{Carbamate}}{HO{-}C({=}O){-}NH_2} + HO{-}P(O^-)_2{=}O$$

$$\underset{\textbf{Carbamate}}{H_2N{-}C({=}O){-}O^-} + \underset{\textbf{ATP}}{O{=}P(O^-)_2{-}O{-}ADP} \longrightarrow \underset{\textbf{Carbamoyl phosphate}}{H_2N{-}C({=}O){-}O{-}P({=}O)(O^-){-}O^-} + ADP$$

Figure 19.8

The "Krebs bicycle" involves interaction between components of the TCA cycle (on the left) and the urea cycle (on the right). This interaction explains the origin of the amino group contributed by aspartate to urea formation. The amino group originates from a transamination reaction involving oxaloacetate. The resulting aspartate is deaminated to fumarate, which can be recycled to oxaloacetate.

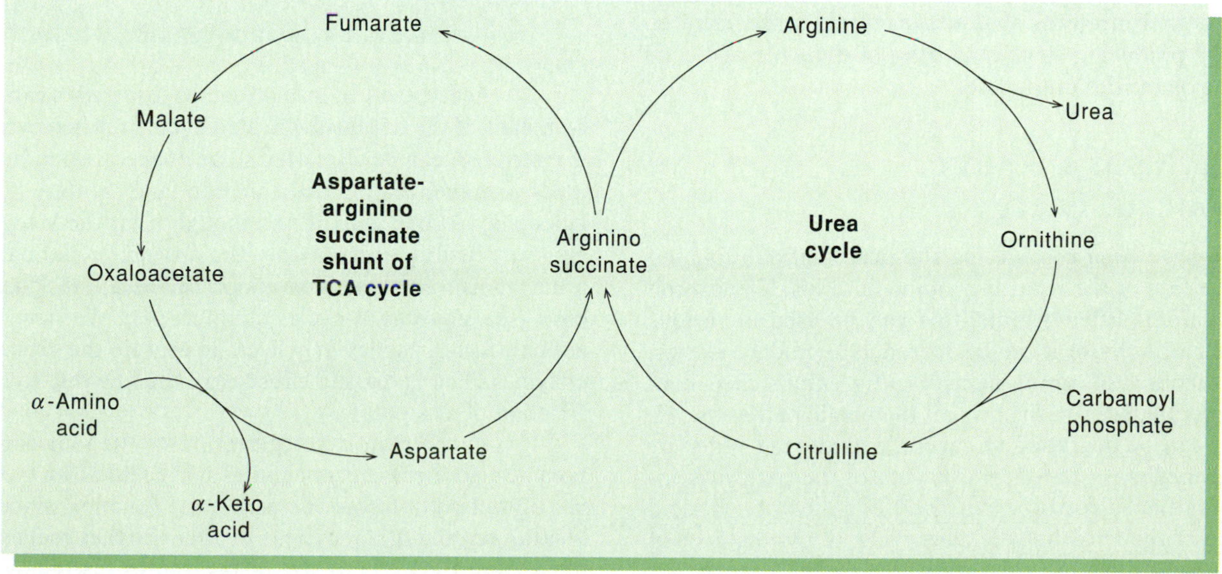

Different Carriers Transport Ammonia to the Liver

In mammals the urea cycle is a unique function of the liver. Ammonia formed in other tissues must be carried in a nontoxic form to the liver. In many tissues glutamine serves as the carrier of excess nitrogen. The glutamine is formed in the tissues in question in a reaction, catalyzed by glutamine synthase, that combines NH_3 with glutamate:

$$\text{ATP} + NH_4^+ + \text{glutamate} \xrightleftharpoons{\textbf{Glutamine synthase}} \text{ADP} + P_i + \text{glutamine} + H^+$$

This reaction involves activation of the γ-carboxyl group of glutamate to yield a γ-glutamyl enzyme complex, together with the cleavage of ATP to ADP and P_i (fig. 19.9). In a second step, the γ-glutamyl group is transferred to NH_4^+.

After the glutamine reaches the liver, the enzyme glutaminase releases the ammonia from the glutamine by the reaction

$$\text{Glutamine} + H_2O \rightarrow \text{glutamate} + NH_4^+$$

NH_3 is also transported from skeletal muscle to the liver in the form of the amino acid alanine. The alanine is formed in the muscle tissue by a transamination reaction between pyruvate and glutamate. Then the alanine is transported through the bloodstream to the liver, where it reacts with α-ketoglutarate to re-form pyruvate and glutamate. This reaction is catalyzed by alanine transaminase. The nitrogen originating from the glutamate is processed by the urea cycle. When the blood glucose concentration is low, the pyruvate resulting from alanine transamination is used to make glucose via the gluconeogenesis pathway. The glucose can be returned to the skeletal muscle to supply quick energy. Thus the transport of alanine from muscle to liver results in a reciprocal transfer of glucose to muscle. The entire cyclical process is referred to as the glucose-alanine cycle (fig. 19.10). Its importance is proportional to the muscular activity of the organism. Recall that active muscle tissue operates anaerobically, producing large quantities of pyruvate and consuming large quantities of glucose.

Amino Acids as a Source of Carbon and Energy

Thus far we have been considering the deamination of amino acids and the fate of the resulting ammonium ion. The carbon skeleton remaining after deamination can be used in various biosynthetic pathways, or it can be degraded to produce energy.

Catabolism of amino acids usually entails their conversion to intermediates in the central metabolic pathways. All amino acids can be degraded to carbon dioxide and water by appropriate enzyme systems. In every case, the pathways involve the formation, directly or indirectly, of a dicarboxylic acid intermediate of the tricarboxylic acid cycle, of pyruvate, or of acetyl-CoA (see fig. 19.1).

Figure 19.9

Glutamine synthase catalyzes the synthesis of glutamine in an ATP-dependent reaction. Free ammonia is used as the amino group donor, and the active intermediate is an enzyme-bound γ-L-glutamyl phosphate.

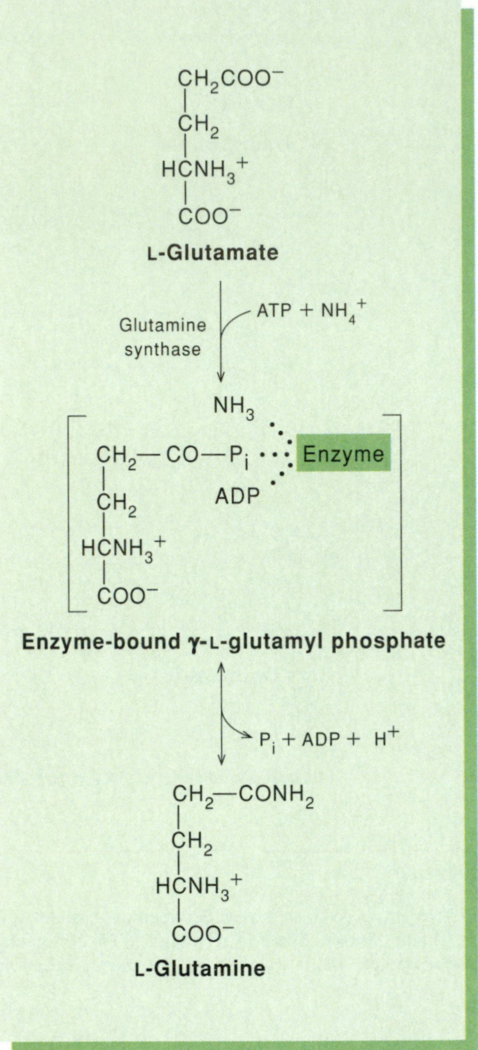

Acetyl-CoA so formed can be oxidized to carbon dioxide by means of the tricarboxylic acid cycle or, when cycle function is restricted, can be converted to acetoacetate and lipid. Amino acids metabolized to acetoacetate and acetate are termed ketogenic. At one time, it was thought that the ketogenic property was readily explained by the absence in animal tissues of a mechanism for a net conversion of acetate residues into glucose (see glyoxylate cycle, chapter 14). We now know that animal tissues, particularly liver, do contain the glyoxylate cycle enzymes. The ketogenic effect may thus be due to the limited function of this pathway.

In contrast, α-ketoglutarate or the four-carbon dicarboxylic acids derived from amino acid breakdown (see fig. 19.1) can stimulate tricarboxylic acid cycle function, since they play a catalytic role in the cycle. For their further metabolism they must leave the cycle by one of two routes (see chapter 14). By one route, the conversion of oxaloacetate to phosphoenolpyruvate results in gluconeogenesis when carbohydrate utilization is

Figure 19.10

The glucose-alanine cycle. Active muscle functions anaerobically and synthesizes alanine by a transamination reaction between glutamate and pyruvate. The alanine is transported to the liver, where the pyruvate is regenerated and converted to glucose by gluconeogenesis. The glucose then is transported back to the muscle tissue, where it is used for energy production in glycolysis.

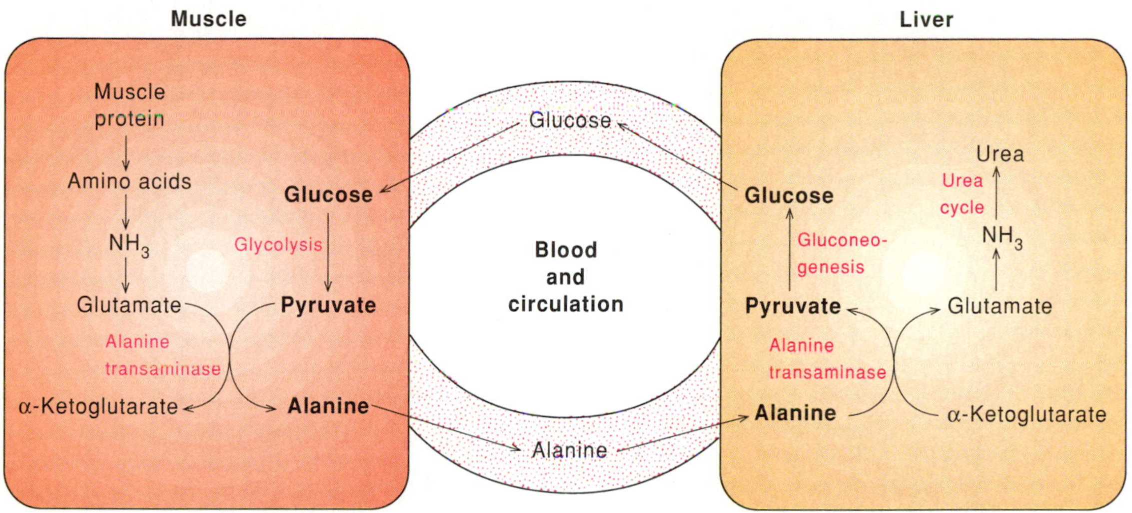

restricted. For this reason, such amino acids are considered glycogenic. By the other route, pyruvate is formed and, after conversion of the latter to acetyl-CoA, can be oxidized completely to carbon dioxide and water provided there is ample tricarboxylic acid cycle function.

Our discussion of the catabolic pathways is organized according to groups of amino acids that give rise to the same main pathway intermediates.

Five Amino Acids Are Degraded to Acetyl-CoA by Way of Pyruvate

The carbon skeletons of ten amino acids yield acetyl-CoA. Five of these—alanine, glycine, serine, cysteine, and, indirectly, threonine—are degraded to acetyl-CoA by way of pyruvate (fig. 19.11). Another five—phenylalanine, tyrosine, tryptophan, lysine, and leucine—go by other routes to acetyl-CoA. We will discuss first the amino acids that are converted to pyruvate.

Alanine Alanine undergoes a reversible transamination directly to pyruvate. Recall that this reaction is part of the glucose-alanine cycle described earlier.

Threonine Threonine is degraded in more than one way. In the pathway shown in figure 19.11, threonine is converted to glycine and acetyl-CoA via threonine dehydrogenase and α-amino-β-ketobutyrate lyase. The glycine so formed can give rise to a second acetyl-CoA via pyruvate as described above.

Glycine and Serine Glycine itself is degraded in two ways, only one of which leads to pyruvate. The pathway to pyruvate involves the conversion of glycine to serine by addition of a hydroxymethyl group carried by N^5,N^{10}-methylenetetrahydrofolate (see fig. 11.16). Subsequently the serine is converted to pyruvate by a specific serine dehydratase, unless, of course, there is a shortage of serine for biosynthesis.

The major pathway for the catabolism of glycine involves the oxidative cleavage of glycine to CO_2,NH_4^+, and a methylene group ($—CH_2—$), which is accepted by tetrahydrofolate in a reversible reaction catalyzed by glycine synthase (also called glycine cleavage enzyme):

$$\text{Glycine} + FH_4 + NAD^+ \rightleftharpoons N^5,N^{10}\text{-methylene } FH_4 + CO_2 + NADH + NH_4^+$$

Thus even though glycine does not enter the TCA cycle by this mode of degradation, its degradation products are not wasted: The methyl group donated to the coenzyme is used in one-carbon metabolism, and the NADH also produced in this process can be used directly to yield energy via the electron-transport system (since the glycine cleavage system is mitochondrial) or indirectly via a transhydrogenase to yield reducing power for biosynthesis. As we have emphasized in our previous discussions of biosynthetic pathways, most pathways have a general requirement for reducing power that is dependent on NADPH rather than NADH.

Cysteine There are several pathways for the catabolism of cysteine. All of these ultimately lead to the formation of pyruvate. The main pathway in animal cells occurs in the three steps, shown in figure 19.12. Cysteinesulfinate, an intermediate in this pathway, is also a biosynthetic intermediate; upon decarboxylation and oxidation it produces taurine (2-aminoethanesulfonate), a component of certain bile acids (see chapter 23). Frequently two cysteines are disulfide-linked into a cystine. On such occasions the cystine is first reduced by an NADH-linked cystine reductase.

Figure 19.11

Outline of the catabolism of threonine, serine, cysteine, and glycine to acetyl-CoA by way of pyruvate. When a single enzyme is involved in the transition, the enzyme name is indicated next to the reaction arrow. Otherwise the number of steps is indicated.

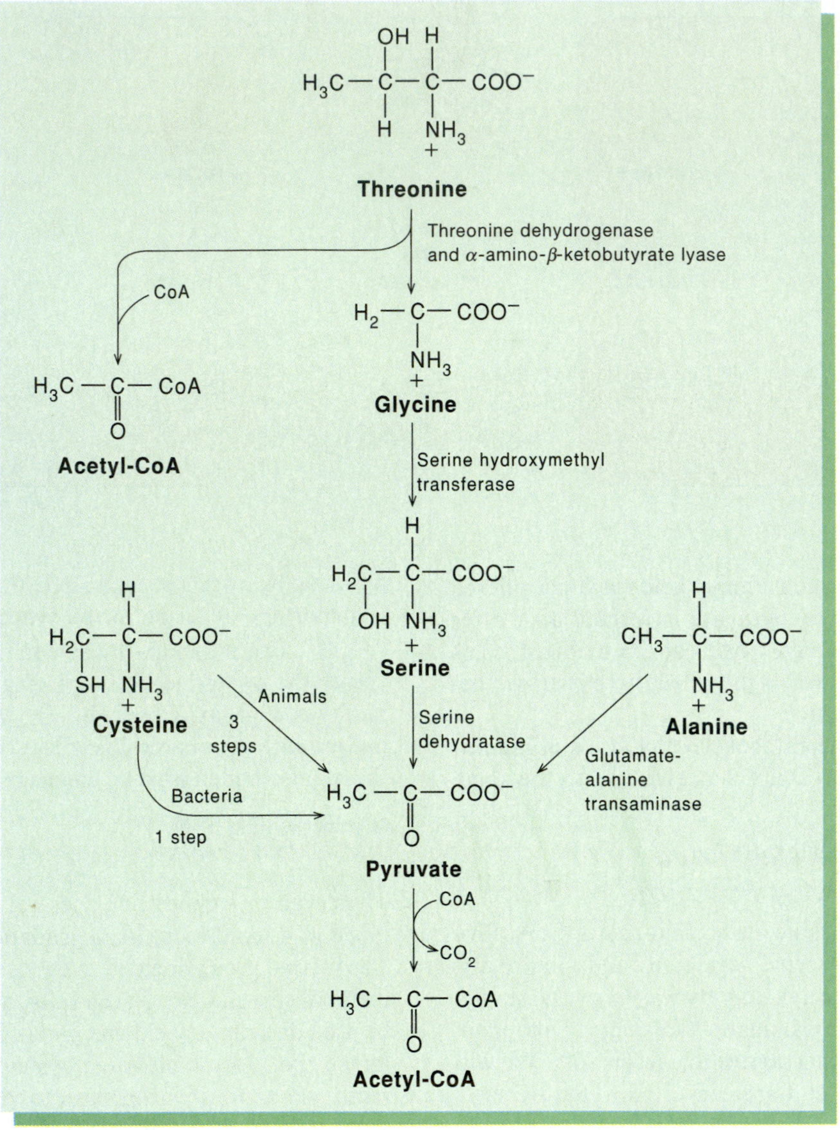

Seven Amino Acids Are Degraded to Acetyl-CoA without Forming Pyruvate

The pathways for the degradation of phenylalanine, tyrosine, tryptophan, lysine, isoleucine, and leucine also lead to acetyl-CoA, but they do not go by way of pyruvate (fig. 19.13). All these pathways contain many steps, as indicated by the numbers next to the reaction arrows. As shown in figure 19.11, threonine can yield two acetyl-CoA's, one via pyruvate and one not via pyruvate.

Lysine and Leucine The pathways for leucine, lysine, and tryptophan are similar in the final steps and resemble the steps in the β oxidation of fatty acids (see chapter 17).

Tyrosine The pathways for the aromatic amino acids phenylalanine, tyrosine, and tryptophan are especially noteworthy not only because their side chains create special complications for degradation, but because they provide numerous intermediates for the biosynthesis of useful compounds.

The oxidation of tyrosine (and phenylalanine) by the liver proceeds by way of acetoacetate and the dicarboxylic acid fumarate. Thus tyrosine and phenylalanine are both ketogenic and glycogeni(

The first step in tyrosine catabolism involves its conversion to 4-hydroxyphenylpyruvate by a tyrosine-glutamate transaminase of rather broad specificity (fig. 19.14). The next

Figure 19.12

The main pathway for the conversion of cysteine to pyruvate in animals takes place in three steps. The first intermediate in this pathway, cysteinesulfinate, is a branchpoint that can also lead to taurine. Taurine is a component of certain bile acids (see chapter 23).

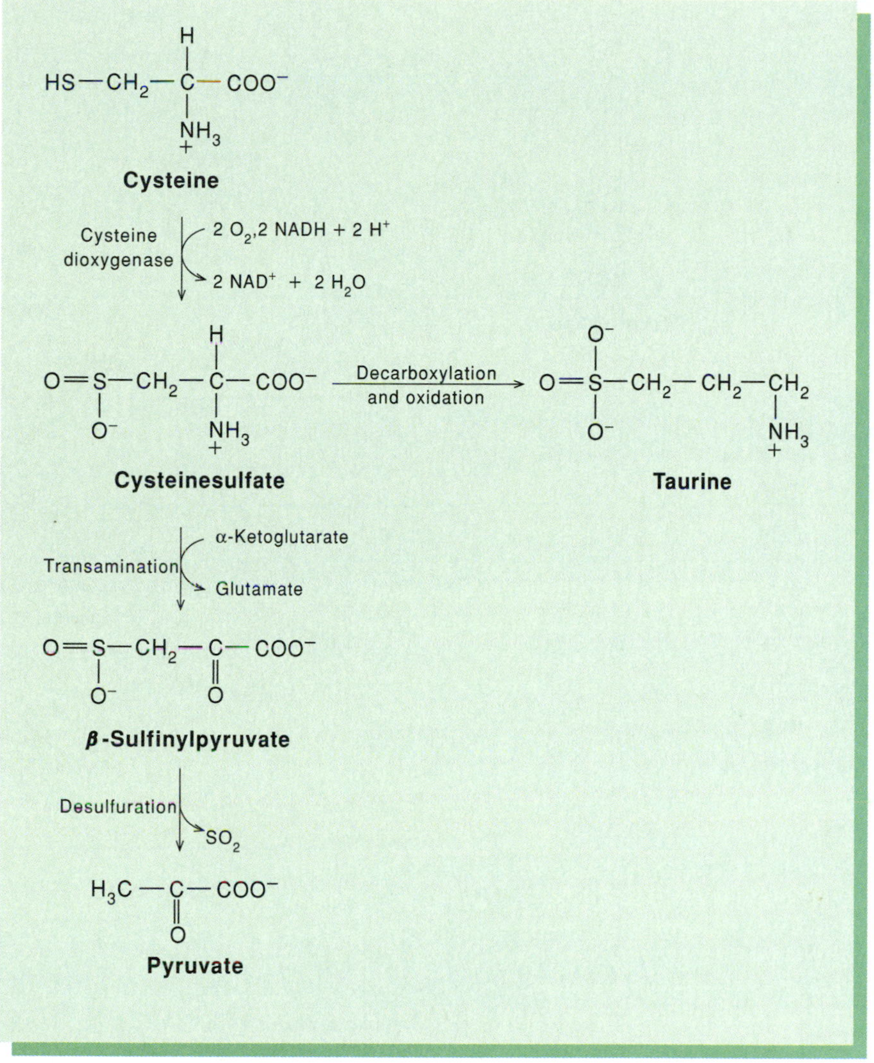

$$HS{-}CH_2{-}CH(NH_3^+){-}COO^-$$

Cysteine

Cysteine dioxygenase: $2\ O_2, 2\ NADH + 2\ H^+ \rightarrow 2\ NAD^+ + 2\ H_2O$

$$O{=}S(O^-){-}CH_2{-}CH(NH_3^+){-}COO^- \xrightarrow{\text{Decarboxylation and oxidation}} O{=}S(O^-)(O^-){-}CH_2{-}CH_2{-}CH_2{-}NH_3^+$$

Cysteinesulfate **Taurine**

Transamination: α-Ketoglutarate → Glutamate

$$O{=}S(O^-){-}CH_2{-}C({=}O){-}COO^-$$

***β*-Sulfinylpyruvate**

Desulfuration: $\rightarrow SO_2$

$$H_3C{-}C({=}O){-}COO^-$$

Pyruvate

step is catalyzed by 4-hydroxyphenylpyruvate dioxygenase, a copper-containing enzyme that is stimulated by ascorbate. The enzyme is called a dioxygenase because both atoms of the oxygen become incorporated into the product. The product, homogentisate, results from oxidation of the aromatic ring and an oxidative decarboxylation and migration of the side chain. The aromatic ring is further oxidized and cleaved by homogentisate-1,2-dioxygenase to 4-maleylacetoacetate. As is apparent from the name, this enzyme is also a dioxygenase. Nearly all cleavages of aromatic rings in biological systems are catalyzed by dioxygenases. The enzyme requires ferrous iron and is also stimulated by ascorbate. An isomerase, maleylacetoacetate isomerase, yields the *trans* compound 4-fumarylacetoacetate, which is hydrolytically cleaved to fumarate and acetoacetate by fumarylacetoacetate hydrolase.

Another route of tyrosine metabolism is that leading to melanin, which results from a two-stage attack on tyrosine by tyrosinase (fig. 19.15), yielding first dihydroxyphenylalanine (dopa). The latter is oxidized as a cosubstrate by tyrosinase to yield the 3,4-quinone. The quinone is unstable and undergoes a series of spontaneous reactions that ultimately lead to melanin, a black pigment. In animals tyrosinase is found only in the organelles known as melanosomes, which are present in specialized pigment-producing melanocytes found in the epidermis and certain other tissues.

Dopa is also synthesized from tyrosine by tyrosine hydroxylase as a precursor to certain neurohormones (norepinephrine and epinephrine). This reaction occurs exclusively in the adrenal glands (see chapter 24). Between tyrosinase and tyrosine hydroxylase we have an example of two enzymes that carry out the same reaction but for totally different purposes.

Figure 19.13

Outline of the catabolism of lysine, tryptophan, phenylalanine, tyrosine, and leucine. The numbers of enzyme-catalyzed steps are indicated next to the reaction arrows.

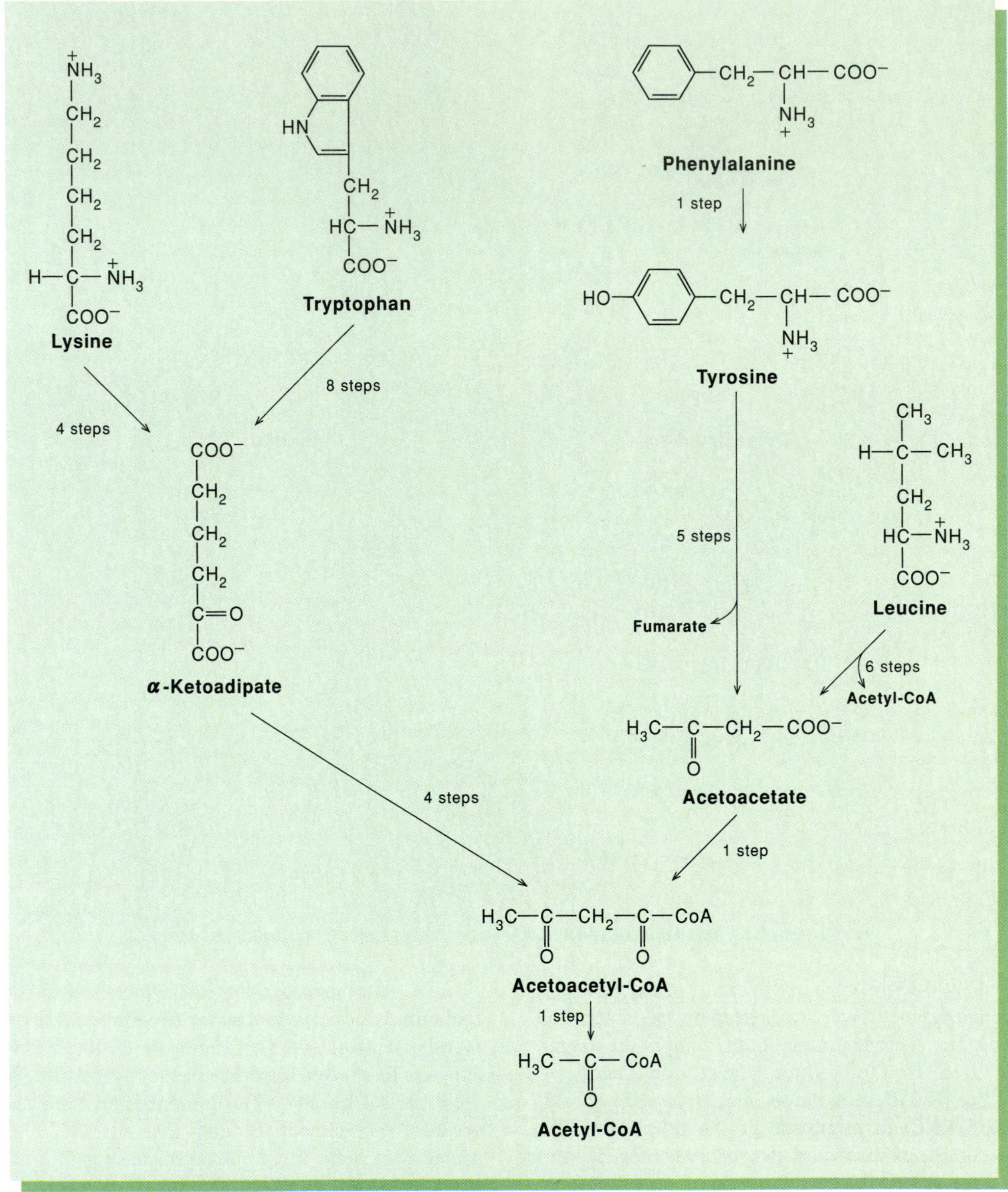

Figure 19.14

The conversion of tyrosine to fumarate and acetoacetate.

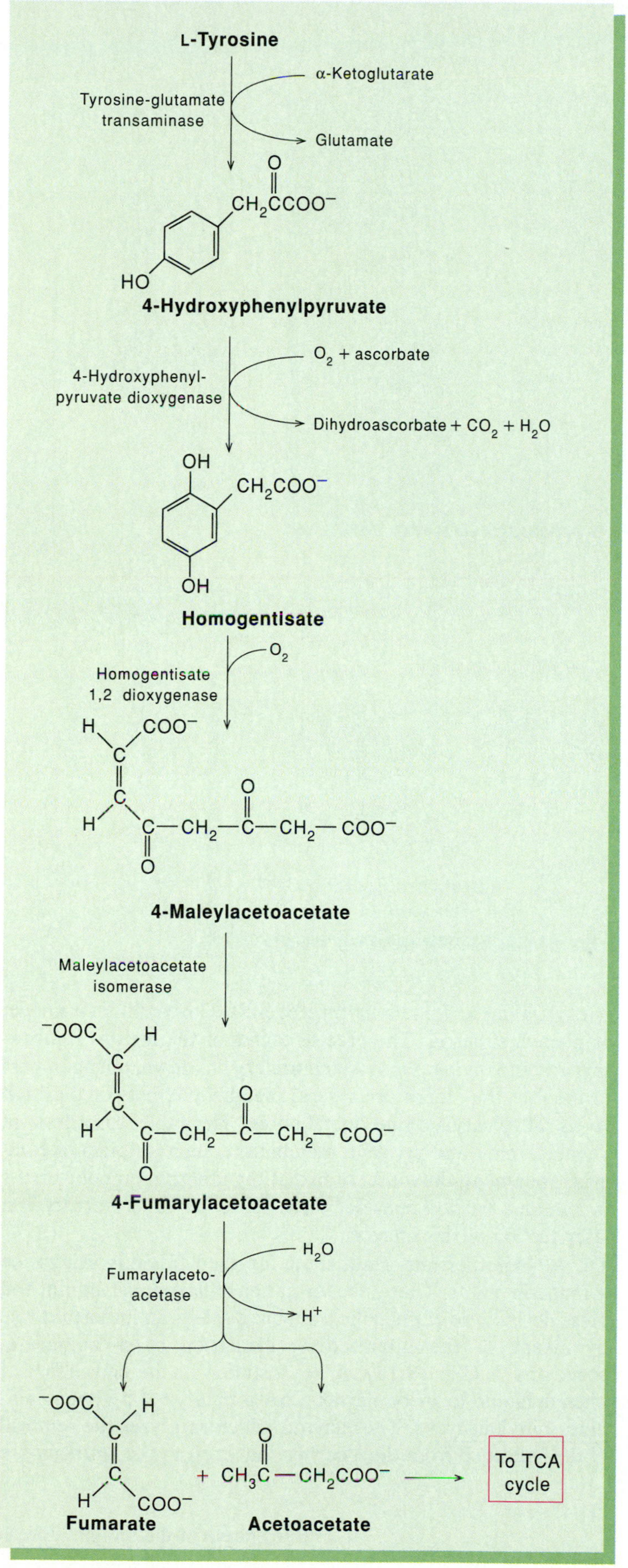

L-Tyrosine

Tyrosine-glutamate transaminase

α-Ketoglutarate → Glutamate

4-Hydroxyphenylpyruvate

4-Hydroxyphenylpyruvate dioxygenase

O_2 + ascorbate → Dihydroascorbate + CO_2 + H_2O

Homogentisate

Homogentisate 1,2 dioxygenase

O_2

4-Maleylacetoacetate

Maleylacetoacetate isomerase

4-Fumarylacetoacetate

Fumarylacetoacetase

H_2O → H^+

Fumarate + **Acetoacetate** → To TCA cycle

Figure 19.15

Melanin formation from tyrosine. Melanin is the black pigment found in hair and skin.

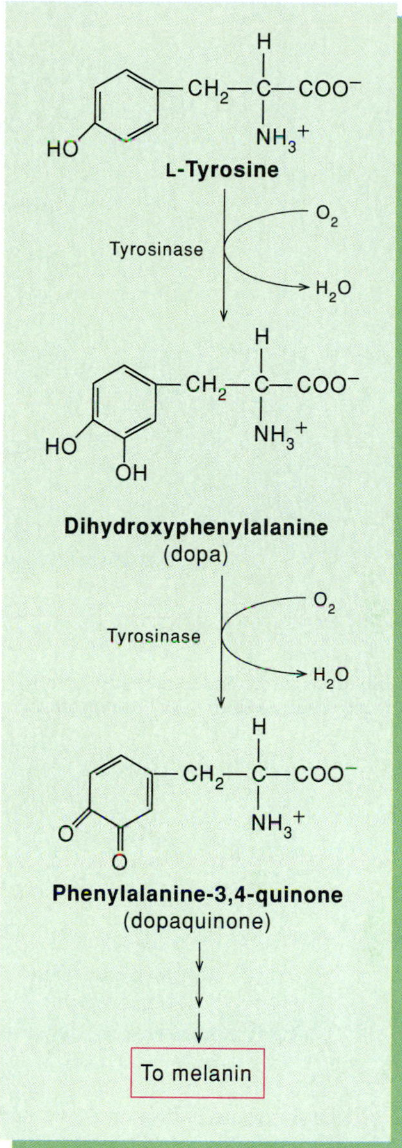

Figure 19.16

The formation of tyrosine from phenylalanine. This reaction occurs in one step. The enzyme requires tetrahydrobiopterin, a folic-acidlike compound, as a cosubstrate. Tetrahydrobiopterin is occasionally used as an electron carrier coenzyme.

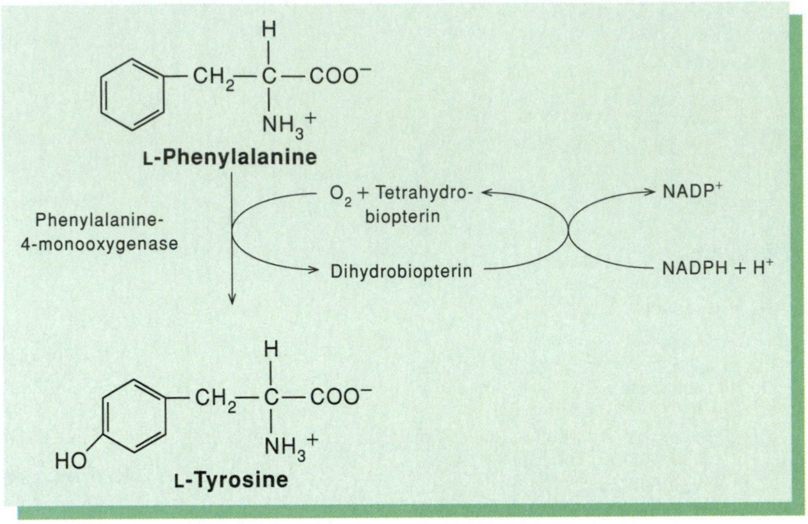

Figure 19.17

The structures of dihydrobiopterin and tetrahydrobiopterin. Dihydrobiopterin is the oxidized form of the coenzyme.

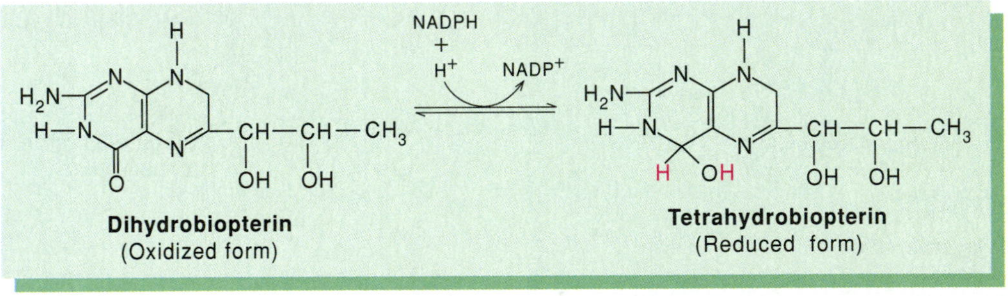

Phenylalanine Phenylalanine is broken down normally by way of tyrosine through the action of phenylalanine-4-monooxygenase, as indicated in figure 19.16. The enzyme requires tetrahydrobiopterin, a folic-acidlike compound, as a cosubstrate. Tetrahydrobiopterin is an infrequently used electron carrier coenzyme (fig. 19.17). Dihydrobiopterin is its oxidized form. The biopterin is kept in the reduced form by NADPH, the ultimate hydrogen donor in the hydroxylation reaction. The presence of phenylalanine-4-monooxygenase accounts for the fact that tyrosine is not an essential amino acid in mammals, provided the dietary supply of phenylalanine is sufficient.

Minor pathways for phenylalanine breakdown in animals involve transamination to yield phenylpyruvate. Although phenylpyruvate can be reduced to phenyllactate, and metabolized to other phenyl derivatives, the disposal of dietary phenylalanine by these routes is insufficient, so that in the inherited absence of the hydroxylation to tyrosine, high blood levels of phenylalanine and phenylpyruvate result. The condition is known as phenylketonuria. The precise causes of the mental retardation accompanying phenylketonuria are unknown, owing in part to the fact that there are several metabolic effects of the high levels of phenylalanine metabolites. Heritable disorders in phenylalanine and tyrosine metabolism, such as phenylketonuria, are among the most studied of the inborn metabolic errors in humans. We will consider these and other inborn errors in a later section of this chapter.

Another important route for phenylalanine utilization is found in plants. There, the formation of flavonoids, lignin, and other derivatives of phenolic compounds plays an important role by means of the intermediate formation of *p*-coumarate-coenzyme A (fig. 19.18). A key reaction in the conversion of phenylalanine to *p*-coumarate-CoA is catalyzed by phenylalanine ammonia-lyase. The enzyme, which catalyzes the removal of the hydrogen from the α carbon that is *cis* to the amino group

Figure 19.18

Conversion of phenylalanine and tyrosine to *p*-coumaryl-CoA, the precursor to flavonoids, lignin, and other compounds in plants. The phenylalanine ammonia-lyase enzyme uses phenylalanine or tyrosine as a substrate.

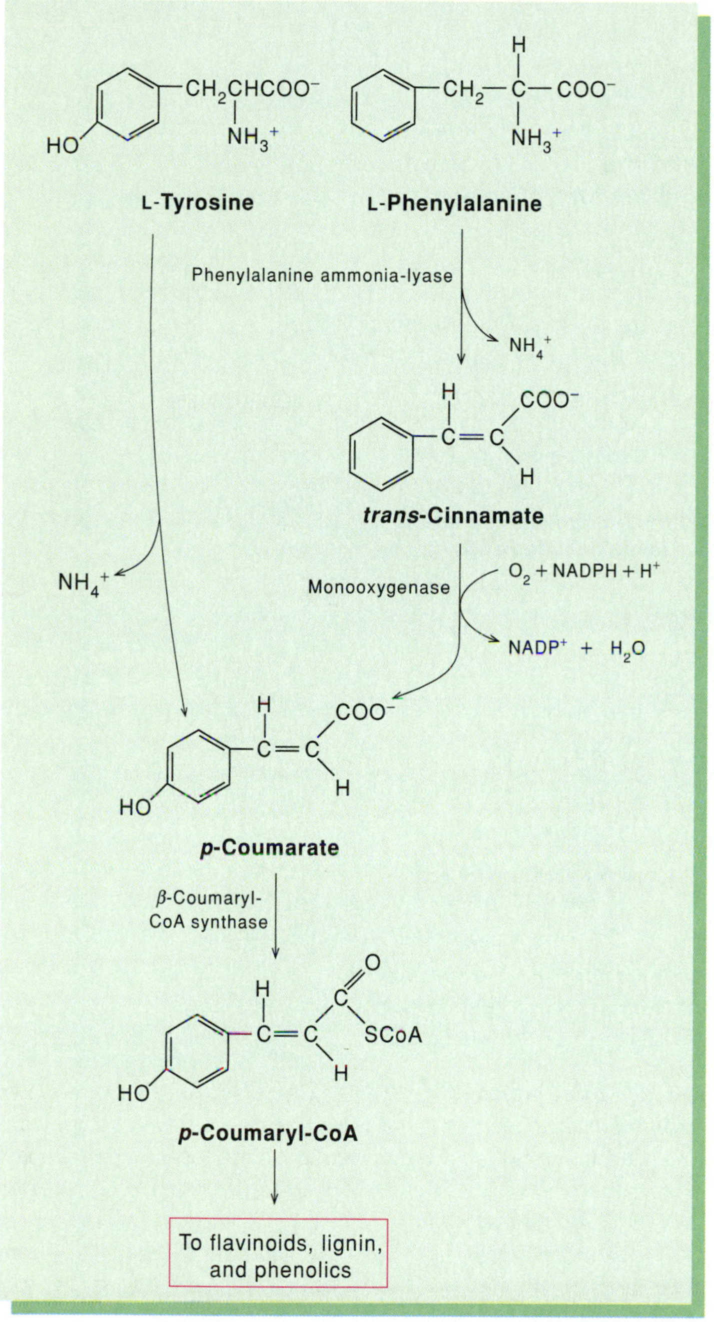

L-Tyrosine → (NH_4^+) → *p*-Coumarate

L-Phenylalanine → Phenylalanine ammonia-lyase (NH_4^+) → *trans*-Cinnamate → Monooxygenase ($O_2 + NADPH + H^+$ → $NADP^+ + H_2O$) → *p*-Coumarate

p-Coumarate → β-Coumaryl-CoA synthase → *p*-Coumaryl-CoA → To flavinoids, lignin, and phenolics

Figure 19.19

The structure of dehydroalanine. The enzyme phenylalanine ammonia-lyase contains at its N-terminal end a dehydroalanine residue that directly participates in the catalysis.

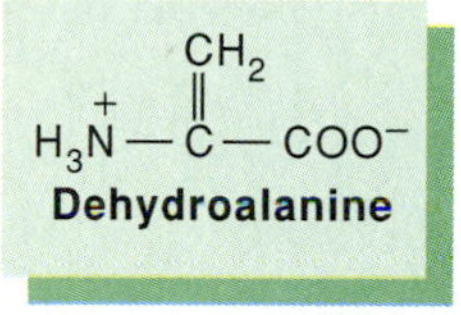

$$H_3\overset{+}{N}-\underset{}{C}(=CH_2)-COO^-$$

Dehydroalanine

to yield *trans*-cinnamate, contains a dehydroalanine residue at the N-terminal end of the peptide chain (fig. 19.19). The amino group of the dehydroalanine is thought to be in an imine linkage with some other group on the protein that provides the electron sink required for elimination of the amino group from the α carbon. The *trans*-cinnamate is, in turn, oxidized to 4-hydroxycinnamate (*p*-coumarate) by *trans*-cinnamate-4-monooxygenase. The enzyme requires FAD and NADPH, which is oxidized in the hydroxylation process. *p*-Coumarate can also be formed directly from tyrosine through the action of phenylalanine ammonia-lyase on tyrosine. *p*-Coumarate is then converted to its coenzyme A derivative by *p*-coumaryl-CoA synthase. It is the coenzyme A derivative that is the branchpoint for a variety of biosynthetic routes found in plants.

Tryptophan The major pathways for tryptophan catabolism in the mammalian liver and for many microorganisms proceed by way of kynurenine (fig. 19.20). Kynurenine itself can be metabolized in liver by way of α-ketoadipate, which is also an intermediate in lysine degradation. An interesting variant of the kynurenine pathway allows for the synthesis of the vitamin nicotinamide.

The first step in the breakdown of tryptophan is catalyzed by tryptophan oxygenase, which yields *N*-formylkynurenine (see fig. 19.20). The enzyme is a dioxygenase and cleaves the indole ring by incorporating an oxygen atom on both C-2 and C-3 of the indole ring. Kynurenine itself is formed by the liberation of formate by kynurenine formamidase.

Kynurenine is converted to 3-hydroxykynurenine by the NADPH-dependent kynurenine-3-monooxygenase (see fig. 19.20). Kynureninase, a pyridoxal phosphate enzyme, catalyzes a hydrolytic cleavage of the alanine side chain to yield 3-hydroxyanthranilate. The aromatic ring is cleaved to 2-amino-3-carboxymuconate-6-semialdehyde (ACS) by 3-hydroxyanthranilate oxygenase. Again, this enzyme is a dioxygenase, and oxygen atoms are incorporated on both the carbons at the site of ring cleavage. Ferrous ions are required by the enzyme. ACS can be spontaneously cyclized to quinolinate with the liberation of a molecule of H_2O. Quinolinate is an intermediate in the biosynthesis of nicotinamide, a synthesis that many animals have a limited capacity to perform. The ACS is decarboxylated by a specific decarboxylase to yield 2-aminomuconate-6-semialdehyde. 2-Aminomuconate-6-semialdehyde is oxidized by an NAD-dependent aminomuconate semialdehyde dehydrogenase. The resulting 2-aminomuconate is reduced to α-ketoadipate by an NAD(P)H-dependent reductase.

The further catabolism of α-ketoadipate results in the liberation of two molecules of CO_2 and two of acetyl-CoA (fig. 19.21). The first step is a coenzyme-A-dependent oxidative decarboxylation by an enzyme probably identical to α-ketoglutarate dehydrogenase. The product, glutaryl-CoA, is oxidized by a flavin-linked dehydrogenase to an intermediate common to

Figure 19.20

The main route of tryptophan degradation in mammals leads to α-ketoadipate. The intermediate 2-amino-3-carboxymuconate-6-semialdehyde (ACS) is a branchpoint metabolite that can also lead to nicotinamide via quinolinate as indicated.

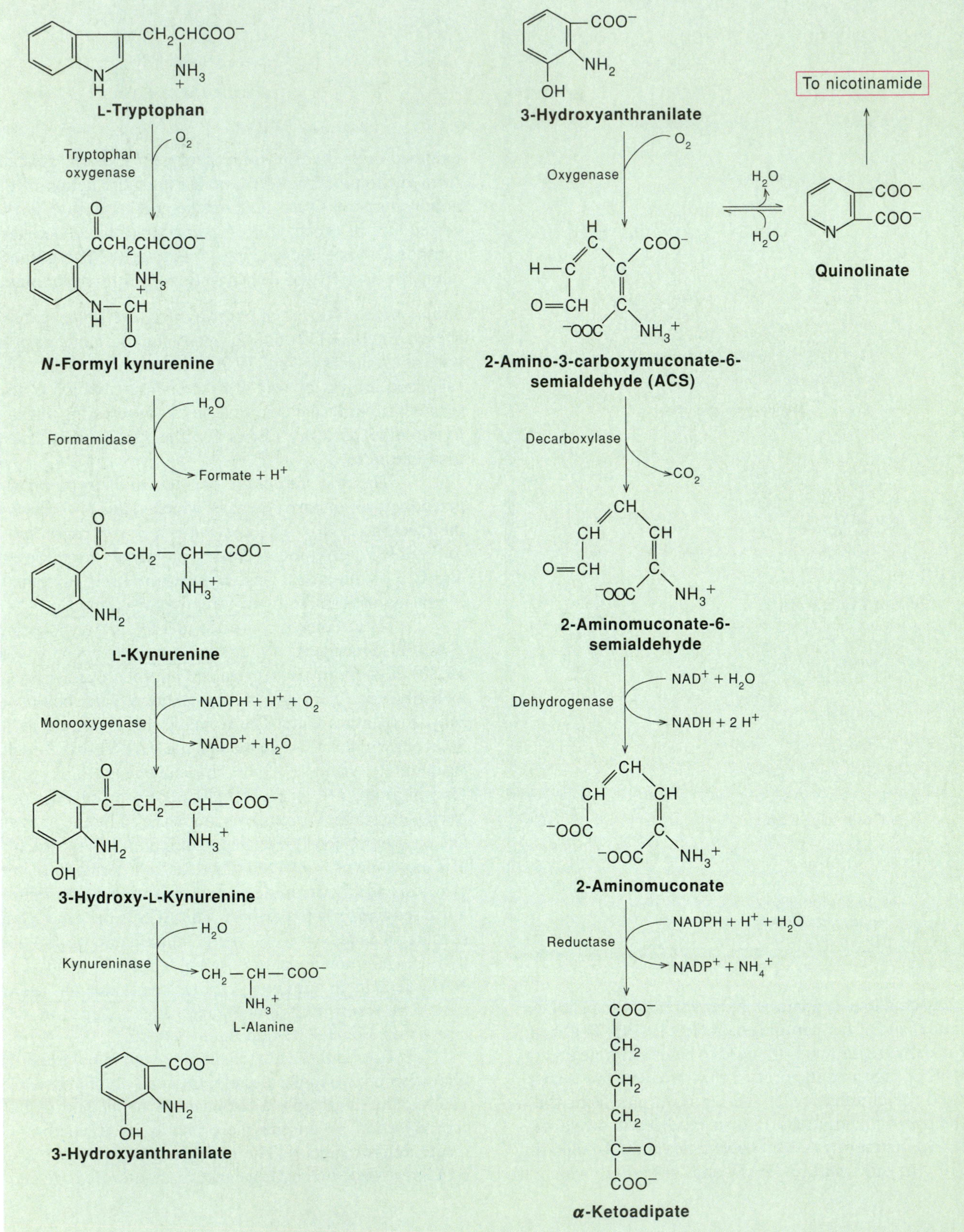

Figure 19.21

The conversion of α-ketoadipate to acetyl-CoA. The sulfur in the CoA-containing compounds is indicated. α-Ketoadipate is also formed in the liver by the breakdown of lysine (see fig. 19.13). Thus the degradative pathways for tryptophan and lysine converge at the level of α-ketoadipate. Two molecules of CO_2 are released and two molecules of NADH are produced in the process of converting α-ketoadipate into two molecules of acetyl-CoA.

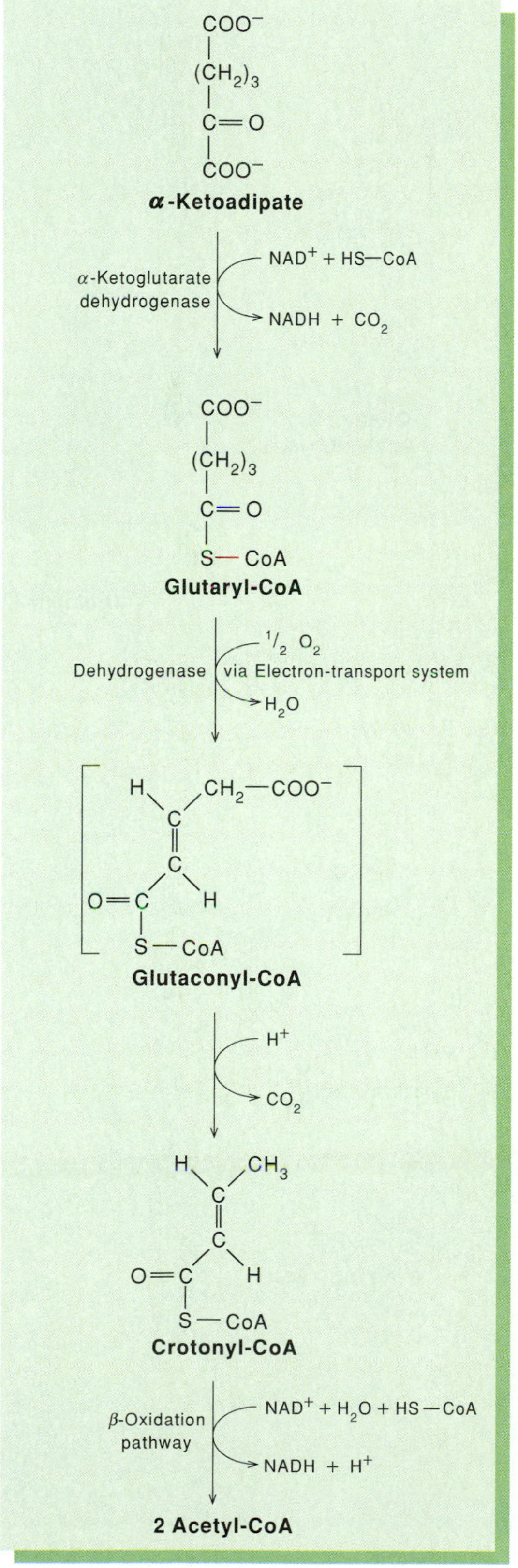

the oxidation of fatty acids, crotonyl-CoA. (The α-keto derivative glutaconyl coenzyme is probably an enzyme-bound intermediate in this reaction.) Finally, two molecules of acetyl-CoA are formed by the action of the fatty acid oxidizing enzymes (see chapter 17). Recall that α-ketoadipate is also formed in the liver by the breakdown of lysine (see fig. 19.13). Thus the degradative pathways for lysine and tryptophan converge at the level of α-ketoadipate.

Five Amino Acids Are Degraded to α-Ketoglutarate

α-Ketoglutarate is the endpoint for degradation of five amino acids, arginine, histidine, proline, glutamic acid, and glutamine (fig. 19.22). As we saw (chapter 18), it is also the starting point for the synthesis of these five amino acids. Hence these are reversible conversions, like those we saw in the glycolytic-gluconeogenic interconversions (see chapter 13), but fewer enzymes are used in common.

Glutamine and Glutamate Reactions involving the interconversions of glutamine, glutamate, and α-ketoglutarate were discussed earlier in this chapter when we were considering deamination.

Proline Proline is converted into Δ′-pyrroline-5-carboxylate in a reaction catalyzed by proline oxidase. This compound is in equilibrium with glutamate-γ-semialdehyde, which is also an intermediate in arginine catabolism (fig. 19.23).

Arginine Arginine is one of the few amino acids for which there is no transaminase reaction. In addition to serving as a source of carbon and energy, arginine is a precursor of various essential polyamines and, as we have seen, is an intermediate or a component in the urea cycle.

Arginine is converted to ornithine by more than one route (fig. 19.24; also see fig. 19.26). We have discussed one, which provides a mechanism for urea formation and is important in the nitrogen metabolism in many animal species. Another route is one that provides a source of energy for many microorganisms. It is called the arginine dihydrolase pathway. The first enzyme, arginine deiminase, converts arginine to citrulline, with the liberation of NH_3 (see fig. 19.24). Citrulline can be cleaved by a degradative ornithine transcarbamoylase to yield ornithine and carbamoyl phosphate. The carbamoyl phosphate so formed serves as a high-energy phosphate donor for ATP formation in a reaction catalyzed by carbamate kinase. In some animal tissues it appears that the same reaction can be catalyzed by an acetate kinase.

Ornithine, whether formed by the arginase of the urea cycle or arginine dihydrolase pathway, is broken down in most organisms by a transaminase to yield glutamate-γ-semialdehyde or its cyclized derivative, Δ′-pyrroline-5-carboxylate (see fig. 19.24). This compound can be further oxidized to glutamate by Δ′-pyrroline-5-carboxylate dehydrogenase, an NAD^+-linked enzyme, or it can be reduced to proline by the normal proline biosynthetic enzyme Δ′-pyrroline-5-carboxylate reductase in an NADPH-requiring reaction. This route to proline from ornithine accounts for the interconvertibility of ornithine and proline that is seen in many cells and tissues.

Figure 19.22

Outline of the catabolism of histidine, arginine, proline, and glutamine to glutamate and then to the TCA component α-ketoglutarate.

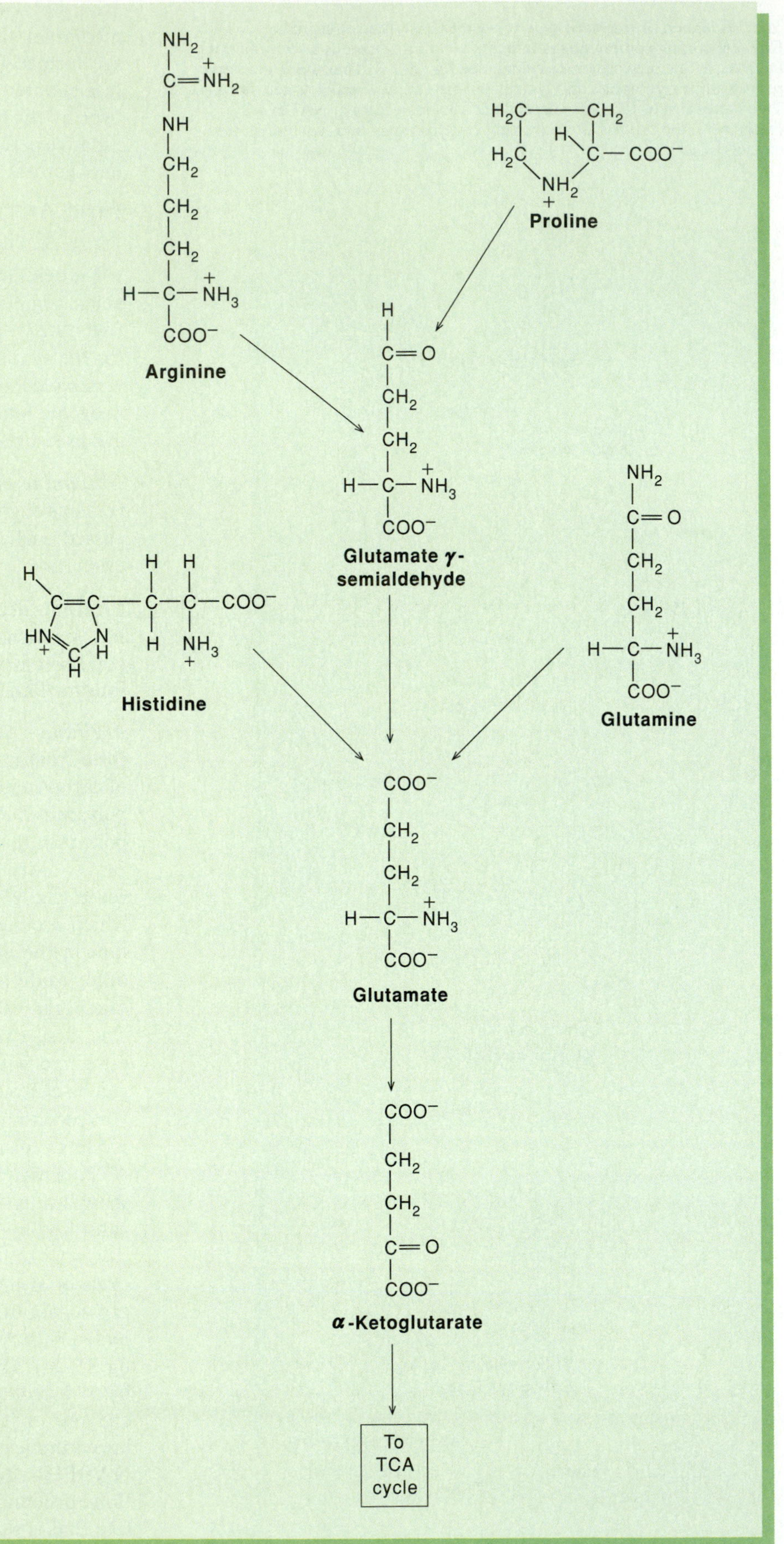

Figure 19.23

Degradation of proline to glutamate-γ-semialdehyde. This can be thought of as a two-step reaction. The first step is enzyme-catalyzed. The second step is spontaneous and reversible.

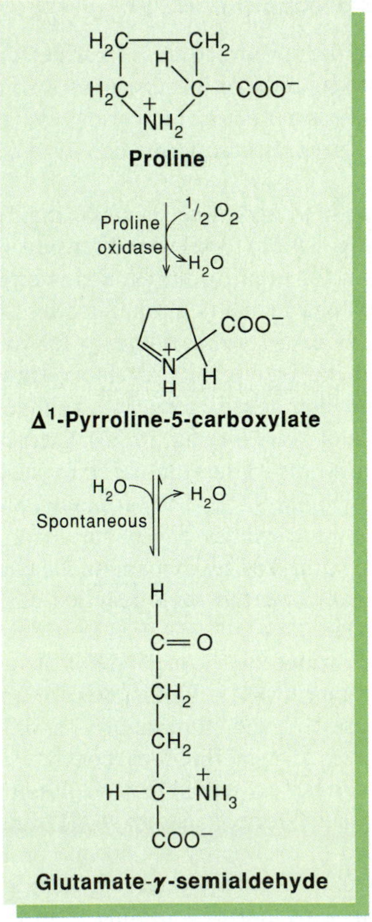

Proline

Proline oxidase ($^1/_2\ O_2$, H_2O)

Δ^1-Pyrroline-5-carboxylate

Spontaneous (H_2O)

Glutamate-γ-semialdehyde

Figure 19.24

Some of the reactions involved in arginine catabolism. Arginine is converted to ornithine by two different routes. One is involved in urea formation (see fig. 19.6). The other route, known as the dihydrolase pathway, provides a source of energy for many microorganisms. The first enzyme in this pathway converts arginine to citrulline. The citrulline is cleaved to ornithine and carbamoyl phosphate. The carbamoyl phosphate serves as a high-energy phosphate donor for ATP formation. Ornithine can be converted to proline, glutamate, and several polyamines, including putrescine, spermidine, and spermine.

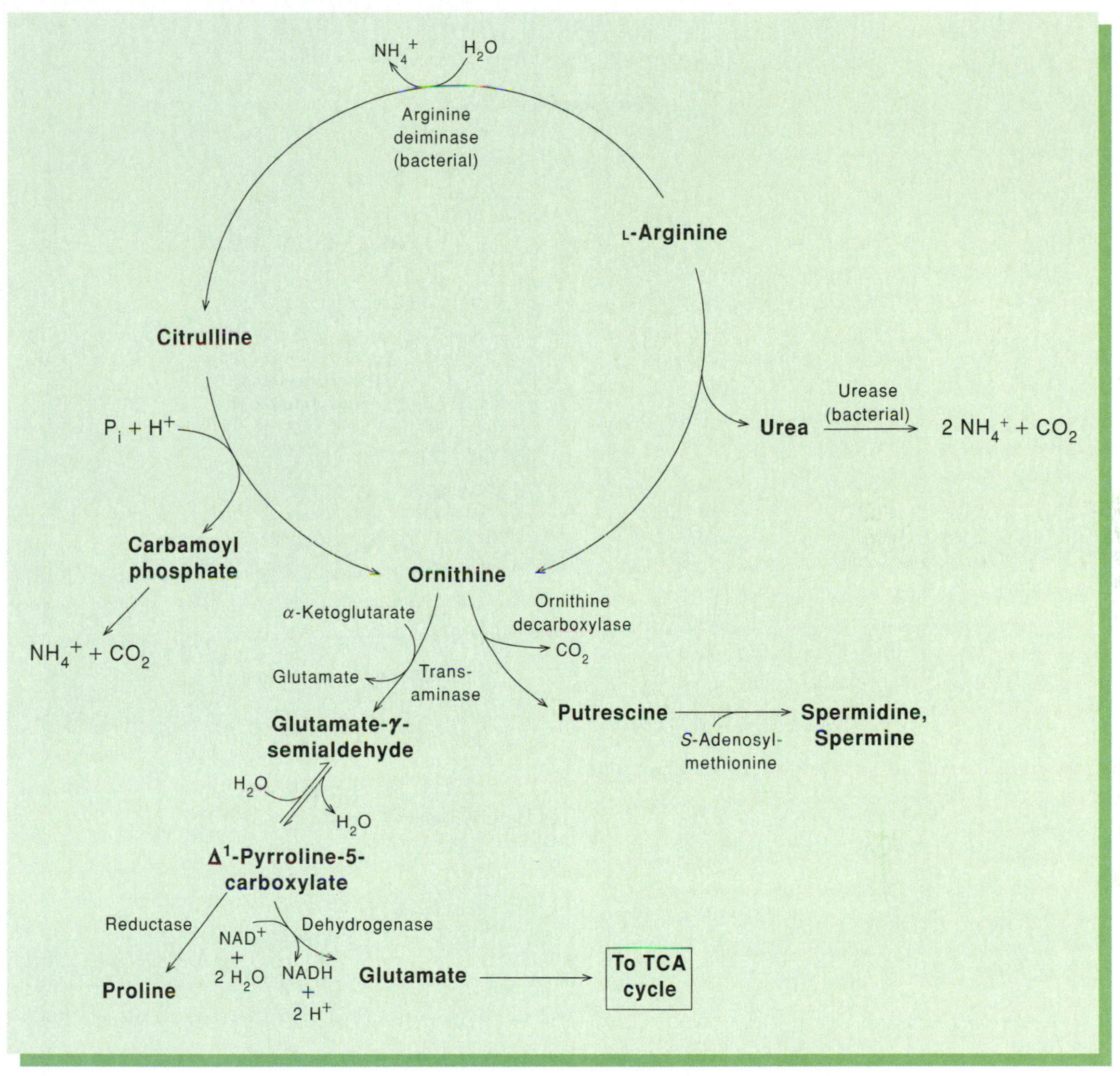

Amines produced from arginine. Putrescine, the decarboxylated product of ornithine, is important as an intermediate in spermidine and spermine formation. These two polyamines are generally found in association with nucleic acids. The decarboxylation is catalyzed by ornithine decarboxylase, and in most bacteria this route is the sole or primary route to putrescine formation (see fig. 19.24).

The synthesis of the omnipresent polyamines spermidine and spermine requires not only the formation of putrescine, but the generation of a propylamine group. The propylamine donor is a product derived from *S*-adenosylmethionine by a specific decarboxylase (fig. 19.25). The same enzyme transfers a propylamine group to spermidine to yield spermine.

The first step in creatine phosphate formation involves a transfer of the amidino group of arginine to glycine. The amidino group of arginine is transferred to a number of substances by a simple displacement reaction. For example, creatine is synthesized by means of a pair of reactions in which the amidino group of arginine is transferred to glycine to yield ornithine and guanidinoacetate, and a methyl group is transferred to the α nitrogen (fig. 19.26). The reversible phosphorylation of the terminal amino group results in the high-energy storage compound creatine phosphate.

Creatine phosphate is an important reservoir of high-energy phosphate groups in skeletal muscle. Its major asset is that it can be rapidly mobilized by transferring its phosphate

Figure 19.25

The utilization of putrescine and *S*-adenosylmethionine for the formation of spermidine and spermine. Polyamines are thought to have many functions; they are invariably found complexed with nucleic acids, both DNAs and RNAs.

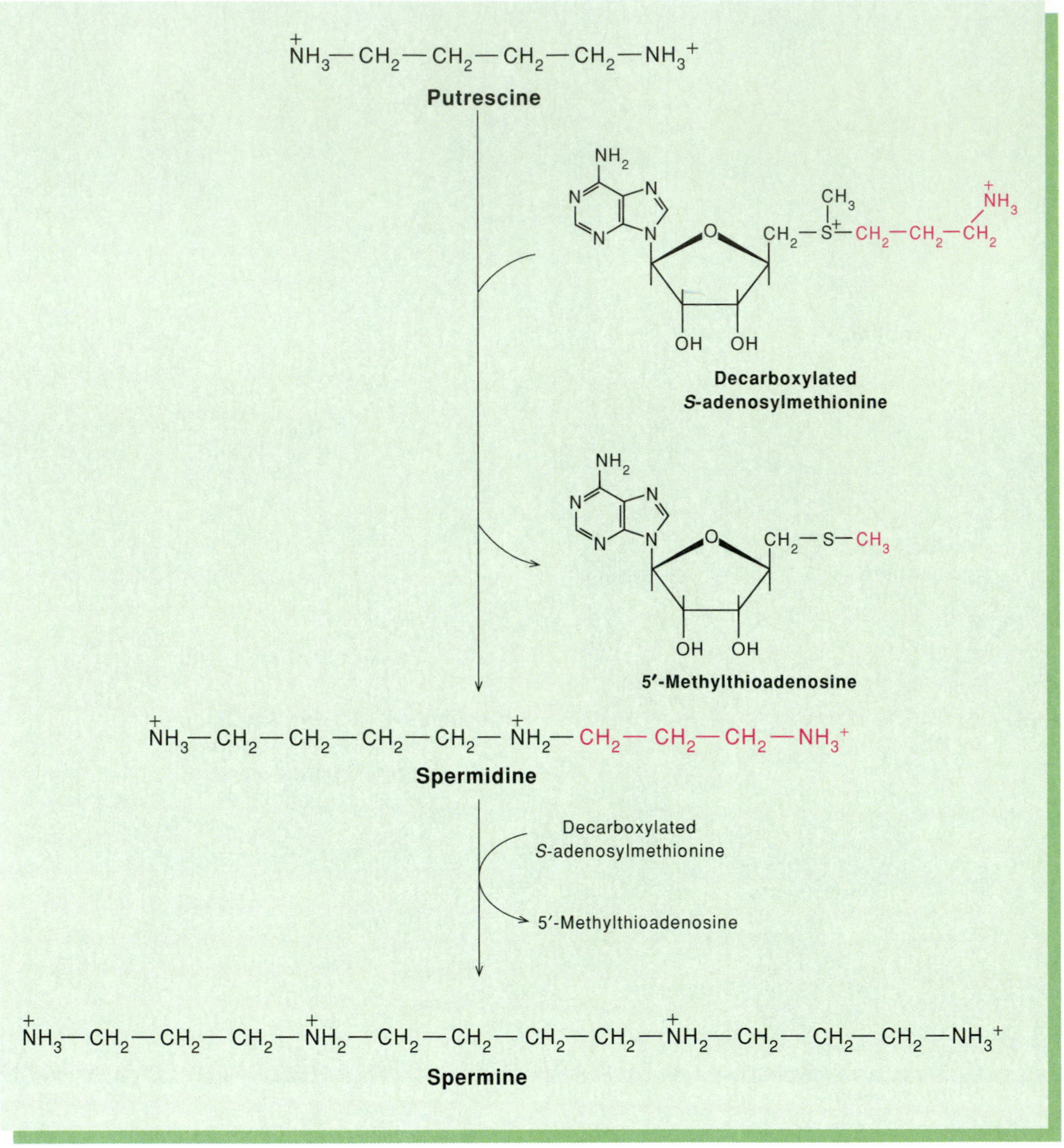

$\overset{+}{N}H_3-CH_2-CH_2-CH_2-CH_2-NH_3^+$

Putrescine

Decarboxylated *S*-adenosylmethionine

5′-Methylthioadenosine

$\overset{+}{N}H_3-CH_2-CH_2-CH_2-CH_2-\overset{+}{N}H_2-CH_2-CH_2-CH_2-NH_3^+$

Spermidine

Decarboxylated *S*-adenosylmethionine

5′-Methylthioadenosine

$\overset{+}{N}H_3-CH_2-CH_2-CH_2-\overset{+}{N}H_2-CH_2-CH_2-CH_2-CH_2-\overset{+}{N}H_2-CH_2-CH_2-CH_2-NH_3^+$

Spermine

to ADP with the help of creatine kinase. There is about five times as much creatine phosphate in skeletal muscle as there is ATP. In many invertebrates arginine phosphate plays a similar role.

Histidine The major route for histidine catabolism in mammals involves the conversion of histidine to glutamate. In the process of breakdown, histidine contributes a carbon atom to one-carbon metabolism (fig. 19.27). In the first step histidase catalyzes the removal of NH_3 with the formation of urocanate. This step is followed by an internal oxidation and reduction involving addition of the elements of water in a reaction catalyzed by urocanase. The resulting intermediate, 4-imidazolone-3-propionate, contains an imidazolone ring, which is opened by a hydrolytic reaction. Cleavage of the product *N*-formimino-L-glutamate leads to formimino group transfer to the N^5 position of tetrahydrofolate and free glutamate. Animals that are deficient in folic acid excrete large amounts of formiminoglutamate.

Figure 19.26

Formation of creatine and creatine phosphate. In step 1 the amidino group of arginine is transferred to the α nitrogen of glycine leading to the formation of guanidinoacetate. The ornithine formed in step 1 can be reutilized via the urea cycle. In step 2 the same α nitrogen is methylated leading to the formation of creatine which is reversibly phosphorylated to creatine phosphate.

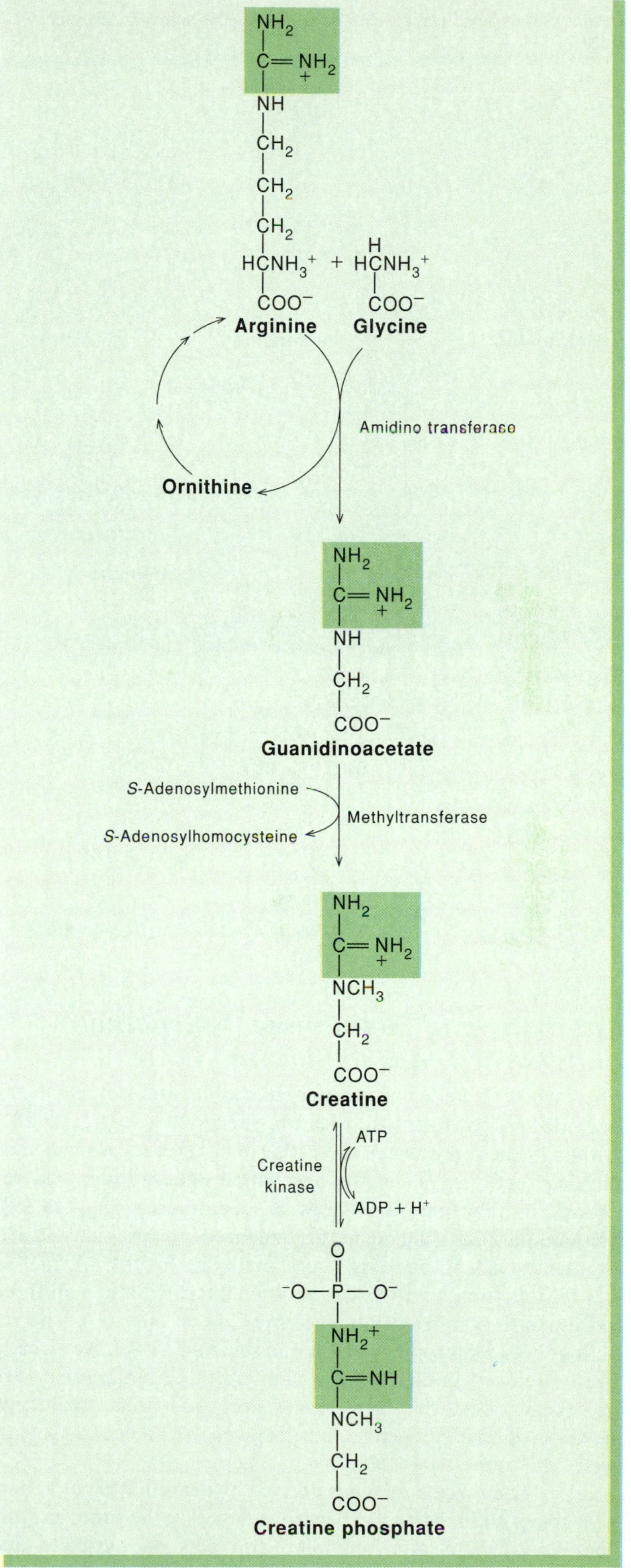

Figure 19.27

The catabolism of histidine. In the final step of histidine breakdown to glutamate, tetrahydrofolate is converted to N^5-formiminotetrahydrofolate. In this way histidine breakdown contributes a carbon atom to C-1 metabolism.

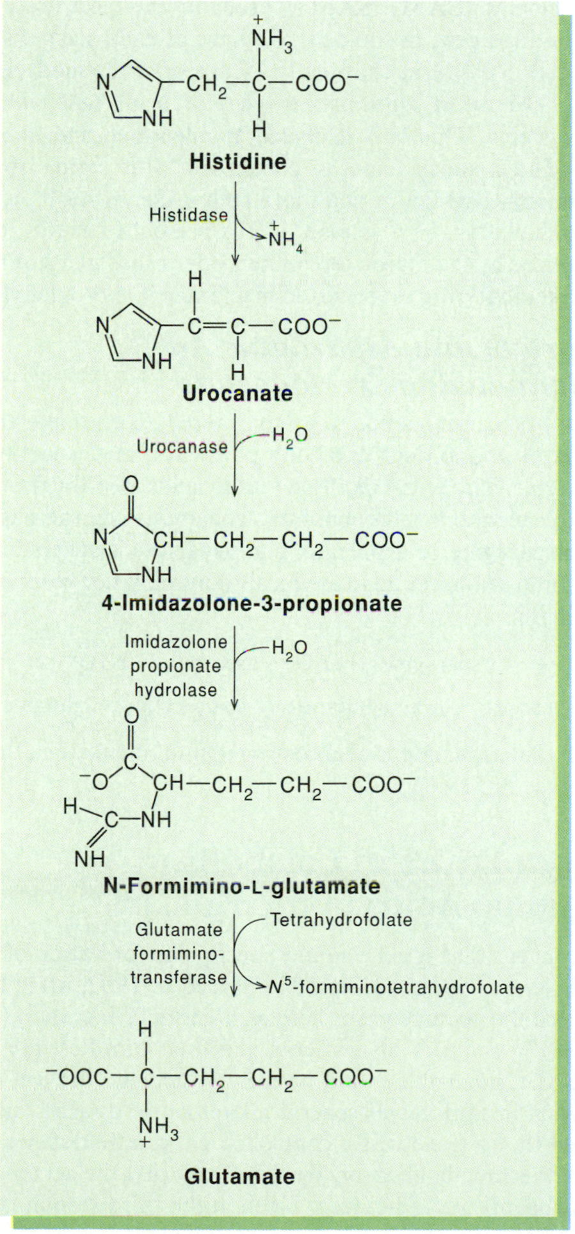

Histidine

Histidase → NH_4^+

Urocanate

Urocanase ← H_2O

4-Imidazolone-3-propionate

Imidazolone propionate hydrolase ← H_2O

N-Formimino-L-glutamate

Glutamate formiminotransferase: Tetrahydrofolate → N^5-formiminotetrahydrofolate

Glutamate

Figure 19.28

Outline of the catabolism of methionine, isoleucine, and valine to succinyl-CoA.

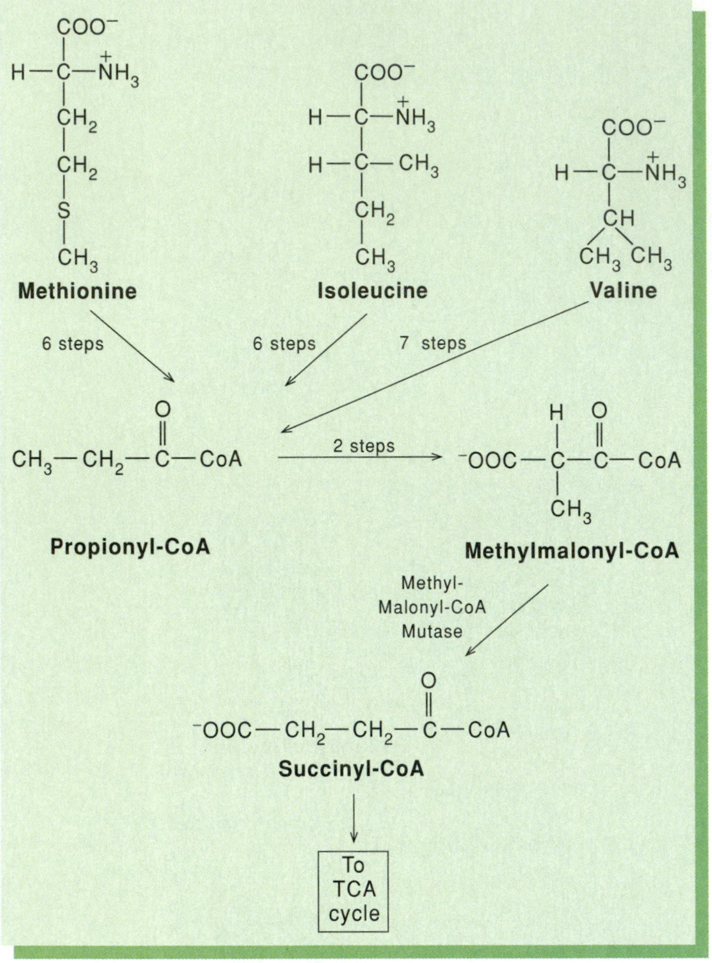

Catabolism of Methionine, Isoleucine, and Valine Leads to Succinyl-CoA

The carbon skeletons of methionine, valine, and isoleucine are degraded by pathways that lead to succinyl-CoA (fig. 19.28). Although these pathways are rather long, there are some simplifying features: Isoleucine and valine undergo identical reactions in the first four steps of degradation (fig. 19.29); methionine and isoleucine are reduced to propionyl-CoA. Propionyl-CoA is converted into methylmalonyl-CoA in two steps. Methylmalonyl-CoA, a common intermediate in all three pathways, is converted into succinyl-CoA in one step. The reactions between propionyl-CoA and succinyl-CoA have already been discussed in chapter 17. Thus all three amino acids are glycogenic, since they give rise to propionyl-CoA. Isoleucine, however, is also ketogenic, since it gives rise to acetyl-CoA as well.

The three keto acids derived by deamination of valine, isoleucine, and leucine are decarboxylated by the same enzyme complex. This enzyme complex also acts on pyruvate and α-ketobutyrate, which are products of both threonine and methionine metabolism, respectively. Some persons are genetically defective for the presence of this enzyme complex. They accumulate substantial amounts of branched-chain α-keto acids in the urine and suffer from a variety of disorders (see table 19.1). Strict control of the diet, allowing only low amounts of these three amino acids, alleviates the immediate symptoms of this disease, but unless begun early after birth cannot prevent or reverse the mental retardation.

Methionine The catabolism of methionine involves nine steps leading to succinyl-CoA. Figure 19.30 illustrates the first six steps of this pathway; the last three steps, from propionyl-CoA, have already been discussed in chapter 17. We will focus on the first six steps, which are unique to methionine catabolism. These reactions illustrate some interesting one-carbon and sulfur metabolic processes.

In the first step methionine is adenylated to *S*-adenosylmethionine (SAM). SAM is probably the most used transmethylating agent in the cell. We have already seen examples of the use of SAM in chapter 18. Transfer of the methyl group from SAM to an appropriate receptor leads to *S*-adenosylhomocysteine. This is hydrolyzed to adenosine and homocysteine. The homocysteine is condensed with serine to yield cystathionine, which in one more step is converted to cysteine and α-ketobutyrate. Cysteine, if it is present in excess, can be catabolized by the three-step pathway described in figure 19.12. The α-ketobutyrate is converted in one step into propionyl-CoA.

Aspartate and Asparagine Are Deaminated to Oxaloacetate

The last two amino acids to be considered, to complete our discussion of amino acid catabolic pathways, are aspartate and asparagine. The entry of these amino acids into the TCA pool via oxaloacetate involves only two enzymes. Asparaginase converts asparagine to aspartate, and aspartate is reversibly converted into oxaloacetate in a typical transamination reaction with glutamate:

$$\text{Asparagine} + H_2O \rightarrow \text{aspartate} + NH_4^+$$

$$\text{Aspartate} + \alpha\text{-ketoglutarate} \rightleftharpoons \text{oxaloacetate} + \text{glutamate}$$

Recall that aspartate is also converted into fumarate in the urea cycle (see fig. 19.6).

Inborn Errors in Catabolism of Amino Acids in the Human

The concept that a gene might specify the formation of a specific enzyme was introduced by Garrod in 1902, after he had analyzed the occurrence of homogentisic acid excretion (alkaptonuria) in some of his patients and their families. He recognized the inheritable and therefore genetic nature of the condition in families of several alkaptonuric patients and postulated that a genetically controlled enzymatic deficiency underlay this metabolic error. We now know that the accumulation of homogentisate indicates a block in the third step in tyrosine

Figure 19.29

Isoleucine and valine undergo identical reactions in the first four steps of degradation.

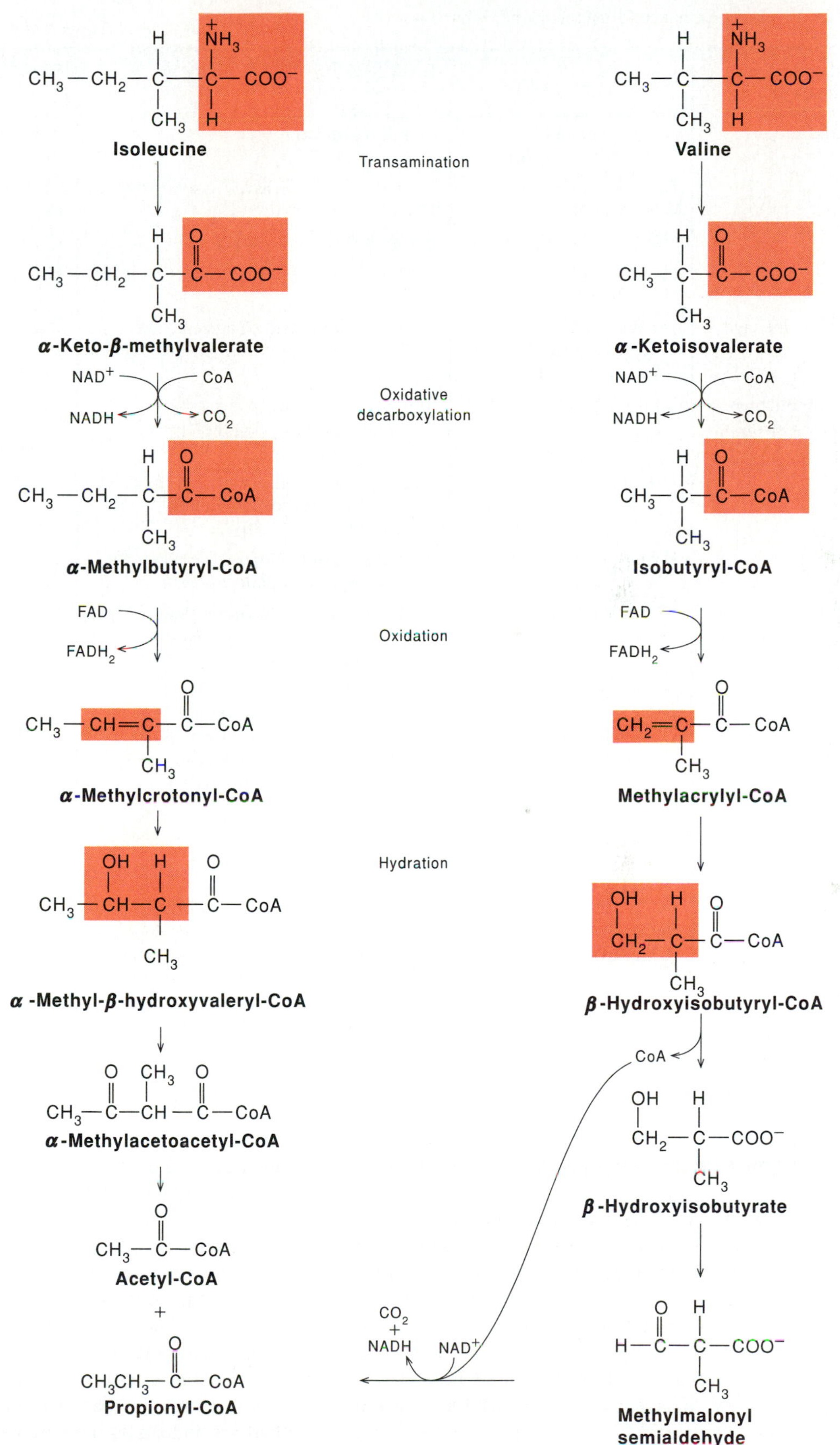

Table 19.1
Some Inborn Errors of Amino Acid Metabolism in Humans

Amino Acid Catabolic Pathway Involved	Condition	Distinctive Clinical Manifestation	Enzymatic Block or Deficiency
Arginine and the urea cycle	Argininemia and hyperammonemia	Mental retardation	Arginase
	Hyperammonemia	Neonatal death, lethargy, convulsions	Carbamoyl phosphate synthase
	Ornithinemia	Mental retardation	Ornithine decarboxylase
Glycine	Hyperglycinemia	Severe mental retardation	Glycine-cleavage system
Histidine	Histidinemia	Speech defects; mental retardation in some cases	Histidase
Isoleucine, leucine, and valine	Branched-chain ketoaciduria ("maple syrup urine disease")	Neonatal vomiting, convulsions, and death; mental retardation in survivors	Branched-chain keto acid dehydrogenase complex
Isoleucine, methionine, threonine, and valine	Methylmalonic acidemia	Similar to preceding except that methylmalonate accumulates	Methylmalonyl-CoA mutase (some patients respond to vitamin B_{12} therapy)
Leucine	Isovaleric acidemia	Neonatal vomiting, acidosis, lethargy, and coma; survivors mentally retarded	Isovaleryl-CoA dehydrogenase
Lysine	Hyperlysinemia	Mental retardation and some noncentral nervous system abnormalities	Lysine-ketoglutarate reductase
Methionine	Homocystinuria	Mental retardation common; several eye diseases and thromboembolism common; osteoporosis and faulty bone structures	Cystathionine-β-synthase
Phenylalanine	Phenylketonuria and hyperphenyl-alaninemia	Vomiting is an early neonatal symptom, but mental retardation and other neurologic disorders develop in the absence of dietary treatment	Phenylalanine L-monoxygenase
Proline	Hyperprolinemia, type I	Probably not etiologically associated with any disease; proline excreted	Proline oxidase
Tyrosine	Alkaptonuria	Homogentisic acid in urine darkens on standing; in adult years, pigment deposits cause darkening of skin, cartilage; arthritis develops	Homogentisic acid oxidase
	Albinism	The most common type, oculocutaneous albinism, results in white hair, pink skin, and an extreme photophobia owing to lack of pigment in the eye	Tyrosinase of the melanocyte is absent

breakdown. The condition results from a defective gene for the dioxygenase enzyme that catalyzes this reaction (see fig. 19.14 and table 19.1).

Since 1902, researchers have described many metabolic diseases that are due to the inability of the affected individuals to dispose of certain dietary components. The diseases may be difficult to treat in the cases of errors in amino acid catabolism, since the culprit amino acid is one of the normal constituents of protein and is required for growth and development as well as for replacement of those body proteins that undergo rapid turnover. Because the affected fetus is usually carried by a mother heterozygous for the deficiency (carrying one normal and one defective gene) and whose own metabolism is essentially normal, the development of the fetus is essentially normal. Thus management of these diseases is possible, but it is dependent on a diagnosis soon after birth. Treatment consists in the use of a low-protein diet, carefully selected to supply enough of the culprit amino acid for protein formation but not enough to allow high plasma levels of it or of the offending metabolites. Supplements of nonoffending amino acids, prepared by synthesis or by fermentation processes, could be employed to compensate in part for the low-protein diet. Indeed, the chemical industries have made such preparations available.

Figure 19.30

Degradation of methionine to propionyl-CoA. The first step in methionine breakdown leads to the formation of *S*-adenosylmethionine (SAM), which is used in many transmethylating reactions. Another useful product formed during methionine breakdown is cysteine.

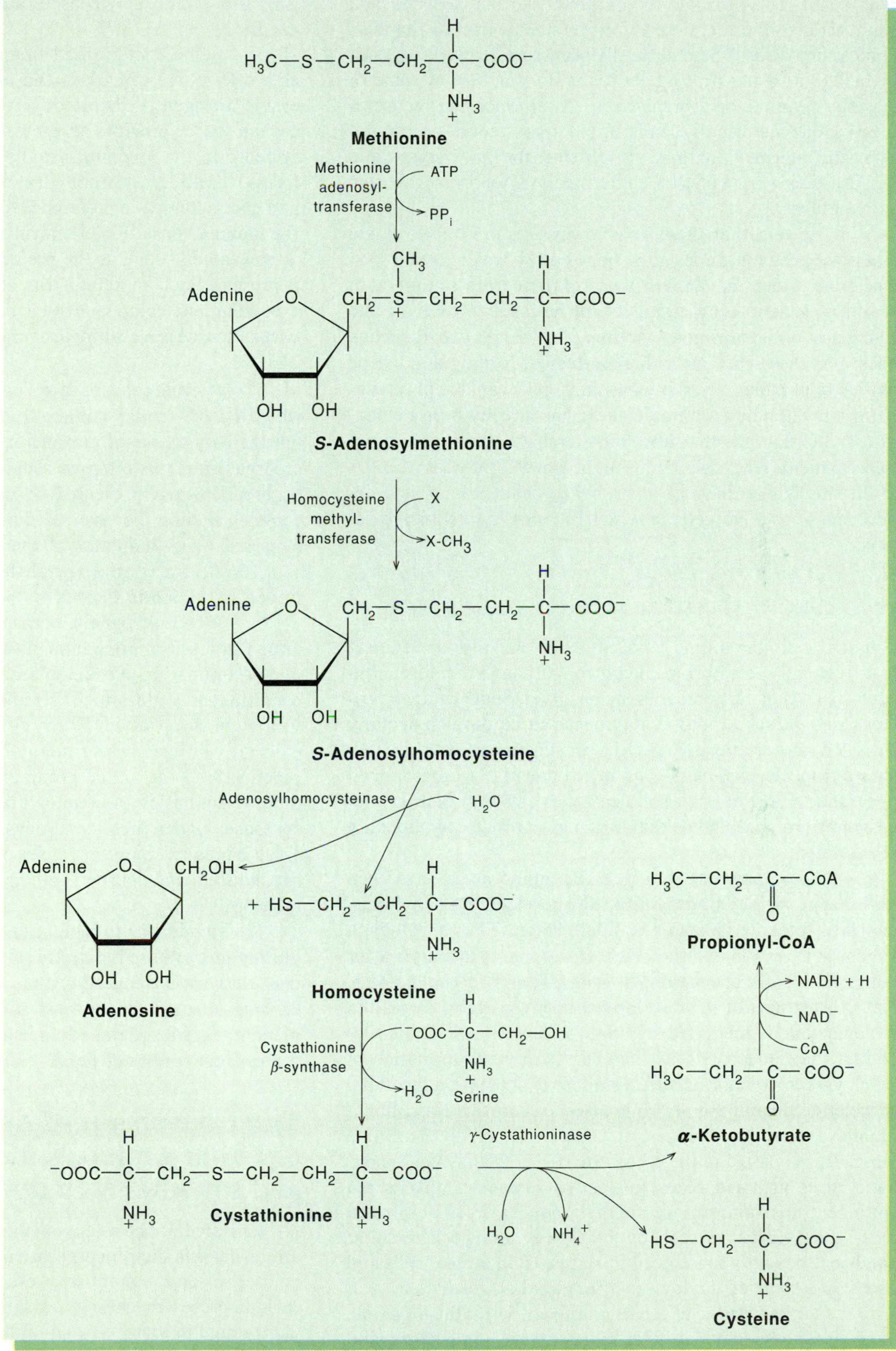

Some of the diseases of amino acid catabolism are listed in table 19.1. To appreciate the nature of the disease listed, you should refer to the text describing the metabolism of the individual amino acids. These naturally occurring defects have been invaluable in demonstrating the obligatory nature of some of the steps in amino acid breakdown. A mutation that causes a defective enzyme usually leads to (1) a substantial accumulation of the intermediate that is a substrate for that enzyme, and (2) a drastic lowering of all the intermediates following that step in the pathway.

The fact that these genetic diseases are due to single (recessive) gene mutations gives hope on two fronts. It will soon be possible to identify carriers of any of these traits so that, with the aid of genetic counseling, it will be possible to avoid the occurrence of homozygous victims that carry two defective genes. For those rare individuals affected, it may one day be possible to provide, through transplant, cells capable of metabolizing the offending amino acids. It has already been demonstrated, for example, that fibroblasts from human maple syrup disease patients (see table 19.1) can be transfected with a cDNA for the missing component of branched-chain α-keto acid dehydrogenase to yield cells in which the missing activity is restored.

Regulation of Amino Acid Breakdown

In animal cells the amino acid catabolic enzymes are subject most frequently to a hormonal control, although diet also influences the level of certain catabolic enzymes. Some catabolic enzymes that attack amino acids appear to be developmentally controlled and are formed only in certain tissues or at certain times during development. The developmentally programmed appearance of certain catabolic enzymes could, in fact, be mediated by hormonal signals that are themselves developmentally programmed.

A few enzymes that degrade amino acids have been measured at various stages during the development of animals, from late fetal periods to the adult stage. Thus tryptophan oxygenase is very low in the newborn rat, but it increases after about twelve days, concomitantly with a rise in adrenal activity. That at least part of the developmentally controlled formation of the enzyme is mediated by adrenal activity is indicated by the fact that glucocorticoids induce the *de novo* formation of the enzyme in young rats and stimulate its formation in adults. Liver serine deaminase also increases around this time, but it exhibits a transient increase in activity at birth. Ornithine transcarbamoylase and the other urea-forming enzymes appear shortly after birth and, like serine and threonine deaminases and ornithine transaminase, are further induced by high-protein diets. Table 19.2 lists several amino-acid-degrading enzymes that have been shown to be induced or repressed in animal cells and tissues, and some of the factors affecting their formation.

Our knowledge of the regulation of formation of amino acid degradation enzymes is far more extensive in bacteria and fungi. Toxic affects of amino acid excesses are far less a problem in unicellular organisms. As a result, regulation in microorganisms is dominated by considerations of nutritional and energy needs.

Some bacteria and fungi can use certain amino acids as a sole source of carbon and energy. In the process, considerable nitrogen is liberated, so that the amino acid used as a carbon source provides an excess of nitrogen, which is then excreted into the medium, usually as ammonia. However, most forms will utilize a carbohydrate or an organic acid as a carbon and energy source in preference to an amino acid. In such forms, the induced formation of enzymes metabolizing the amino acid is prevented as long as the preferred carbon and energy source is present. Such an antagonism of the induction of enzymes catabolizing one compound by a preferred or more readily used carbon source is an example of catabolite repression (see chapter 30).

Microorganisms also can occur in environments in which a good (readily metabolizable) carbon source is available, but the only source of nitrogen is an amino acid. For some microorganisms, this nitrogen source would be unavailable, since its breakdown is prevented by catabolite repression. A fairly common finding, however, is that catabolite repression can be bypassed by an induction of the particular catabolic pathway as a result of a control signal that is conditioned through nitrogen starvation in the cell.

When nitrogen is in excess, the predominant mode of amino-group formation in many bacteria is by means of the reductive amination of α-ketoglutarate by free NH_3. Under these conditions less glutamine is required and glutamine synthase formation is repressed. When nitrogen is limited, NH_3 is more efficiently utilized by the energy-dependent glutamine synthase reaction, and the amide group of glutamine is used in the reductive amination of α-ketoglutarate catalyzed by glutamate synthase. Under these conditions, formation of glutamine synthase is induced. Some examples of microorganism enzymes that are induced when carbon and energy limit growth are given in table 19.3.

In addition to regulation of its synthesis, the activity of glutamine synthase is directly subject to an elaborate set of controls that sense the need of the cell for nitrogen. Although the enzyme is regulated in most animals and plants in a similar manner, our most detailed understanding of how it is regulated comes from studies of the *E. coli* enzyme (see fig. 18.5).

The Conversion of Amino Acids to Other Amino Acids and to Other Metabolites

As we stated at the outset of this chapter, the primary fate of amino acids is their incorporation into protein. However, amino acids also serve a number of other important functions. These include processes leading to the formation of the porphyrin nucleus found in many oxygen- and electron-carrying proteins, of biologically active amines, of glutathione, of peptide antibiotics, and of several other important metabolites. We will consider these topics in the remainder of this chapter.

Table 19.2
Regulation of Some Typical Amino Acid Degradative Enzymes in Animal Cells and Tissue

Amino Acid	Enzyme	Cell or Tissue	Factors Affecting Formation
Arginine and its precursor, ornithine	Ornithine-glutamate transaminase	Liver	High-protein diet; glucagon or cyclic AMP stimulate formation *in vivo* and in cell culture; corticosteroids and glucose repress *in vivo*
	Urea cycle enzymes	Liver	High-protein diet stimulates
Serine	Serine deaminase	Liver	High-protein diet stimulates
Tryptophan	Tryptophan oxygenase	Liver	Glucocorticoids stimulate formation
Tyrosine	Tyrosine-glutamate transaminase	Liver	Glucocorticoids stimulate formation
Threonine	Threonine deaminase	Liver	High-protein diet stimulates formation

Table 19.3
The Regulation of Some Typical Amino Acid Degradative Pathways in Bacteria and Fungi

Amino Acid	Organism	Inducer	Key Enzymes in Degradation
Alanine	*E. coli*	D- or L-alanine	Alanine racemase, D-alanine dehydrogenase
Arginine	*S. cerevisiae*	Arginine	Arginase, ornithine transaminase
Glutamate	Fungi	Glutamate	NAD-linked glutamate dehydrogenase
Histidine	*K. aerogenes*	Urocanate	Histidase, urocanase
Proline	*E. coli*	Proline	Proline oxidase
D-Serine	*E. coli*	D-Serine	D-Serine deaminase
Tryptophan	*E. coli*	Tryptophan	Tryptophanase

Porphyrin Biosynthesis Starts with the Condensation of Glycine and Succinyl-CoA

The early isotope tracer experiments of David Shemin led to discovery of the way in which the immediate precursor of the porphyrin needed for the cytochromes and for hemoglobin is formed. These studies indicated that the glycine methylene carbon and nitrogen were incorporated along with both carbons of acetate. Subsequent enzymic studies in both bacteria and animals revealed that a condensation reaction occurs between succinyl-CoA and glycine to yield δ-aminolevulinate and CO_2 (presumably by way of an enzyme-bound β-keto acid, α-amino-β-ketoadipate) (fig. 19.31).

Aminolevulinate is also the precursor to porphobilinogen in plants, blue-green algae, and most eubacteria, but it is not formed by a condensation of glycine and succinyl-CoA. Rather it is formed from glutamate by reduction of the α-carboxyl group to yield α-glutamyl semialdehyde. As we emphasized in describing reduction of other carboxyl groups, an "activation" of the carboxyl group is required. The reaction in the δ-aminolevulinate pathway is unique in that the glutamate is transferred to a tRNA acceptor. Hence the reaction requires the expenditure of two high-energy phosphate bonds. The glutamyl tRNA is the substrate for a specific reductase. The reduced product, α-glutamyl semialdehyde, undergoes an unusual intramolecular transamination reaction in which the amino group on C-2 of the semialdehyde is transferred to the C-1 position (which becomes C-5 in δ-aminolevulinate).

The pyrrole monomer porphobilinogen arises from the condensation of two molecules of δ-aminolevulinate with the loss of two water molecules. The reaction is catalyzed by δ-aminolevulinate dehydrase. The condensation of four porphobilinogen molecules to yield the branchpoint compound in tetrapyrrole synthesis, uroporphyrinogen III, is a complex reaction requiring two enzymes, uroporphyrinogen I synthase, which catalyzes a head-to-tail condensation of four porphobilinogen molecules (box 19A), and uroporphyrinogen III cosynthase, which inverts one of the units and closes the ring.

Glutathione Is γ-Glutamylcysteinylglycine

The tripeptide γ-glutamylcysteinylglycine, or glutathione, is found in nearly all cells and plays a variety of roles. The tripeptide is formed in two steps catalyzed by ATP-requiring reactions. The first step is the condensation of glutamate with cysteine:

$$\text{Glutamate} + \text{cysteine} + \text{ATP} \xrightarrow{\textbf{γ-Glutamylcysteine synthase}} \text{γ-glutamylcysteine} + \text{ADP} + P_i$$

The second step is the condensation of the dipeptide with glycine:

$$\text{γ-Glutamylcysteine} + \text{glycine} + \text{ATP} \xrightarrow{\textbf{Glutathione synthase}} \text{glutathione} + \text{ADP} + P_i$$

Figure 19.31

Tetrapyrrole biosynthesis. The sequence by which four porphobilinogen residues are converted to uroporphyrinogen III is the sequential head-to-tail condensation of the four residues by uroporphyrinogen I synthase (porphobilinogen deaminase) to yield the unrearranged hydroxymethylbilane. This unstable intermediate is rearranged and cyclized by uroporphyrinogen II cosynthase to yield uroporphyrinogen III. Hydroxymethylbilane in the absence of cosynthase will spontaneously cyclize to yield the unrearranged urobilinogen I, which is not an intermediate in the pathway (hence the name uroporphyrinogen I synthase for the deaminase).

19A
BOX

Mechanism of Polymerization in a Linear Tetrapyrrole

Four molecules of porphobilinogen undergo a head-to-tail condensation catalyzed by uroporphyrinogen I synthase to yield a tetrapyrrole. Asterisks indicate nitrogen and carbon atoms derived from glycine; the others are derived from succinyl-CoA.

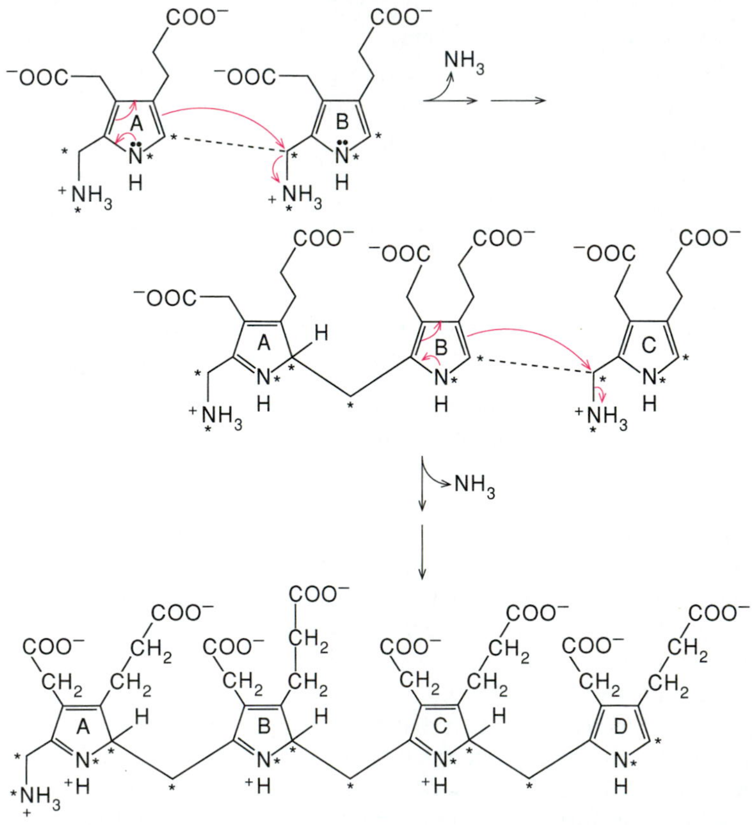

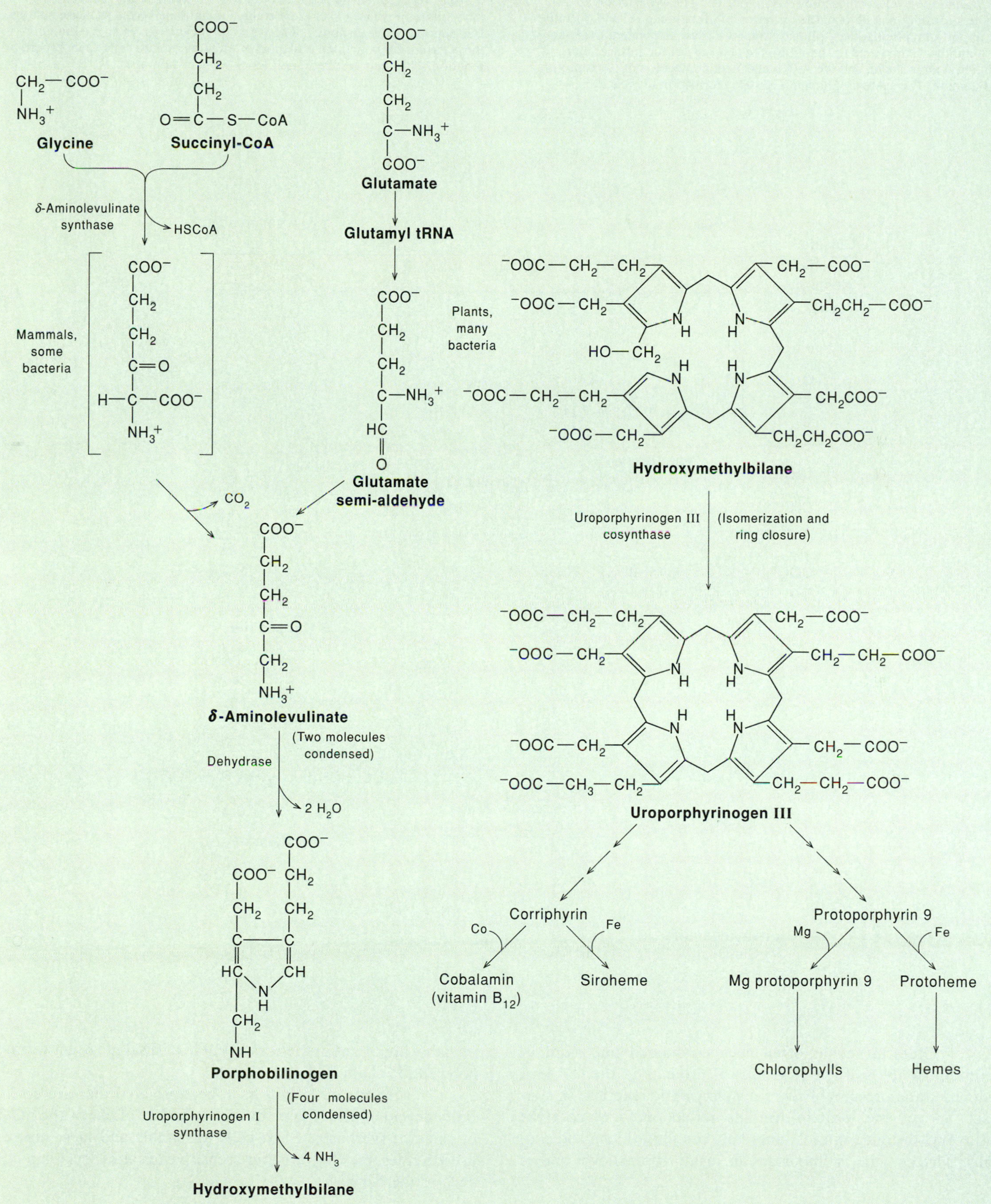
Glycine
Succinyl-CoA
Glutamate
δ-Aminolevulinate synthase
HSCoA
Glutamyl tRNA
Mammals, some bacteria
Plants, many bacteria
Glutamate semi-aldehyde
CO_2
δ-Aminolevulinate
Dehydrase
(Two molecules condensed)
2 H_2O
Porphobilinogen
Uroporphyrinogen I synthase
(Four molecules condensed)
4 NH_3
Hydroxymethylbilane
Hydroxymethylbilane
Uroporphyrinogen III cosynthase
(Isomerization and ring closure)
Uroporphyrinogen III
Corriphyrin
Co
Fe
Cobalamin (vitamin B_{12})
Siroheme
Protoporphyrin 9
Mg
Fe
Mg protoporphyrin 9
Protoheme
Chlorophylls
Hemes

Figure 19.32

The γ-glutamyl cycle proposed by Alton Meister. The cycle involves enzymes forming glutathione, the excretion of glutathione (1) and, in cells that import glutathione, the γ-glutamyltranspeptidase-dependent transport (2) of another amino acid (or peptide), the cleavage of the intracellular γ-glutamyl amino acid, and the ATP-dependent conversion of 5-oxoproline to glutamate. The cysteinylglycine (Cys-Gly) formed in the reaction is transported by an uncharacterized transport system (3) and cleaved by an intracellular protease (4) or cleaved by a membrane-bound protease and the free amino acids transported. The γ-glutamyl cycle enzymes are found in those tissues for which the transport of glutathione into cells is an important function. This includes liver and kidney cells in animals.

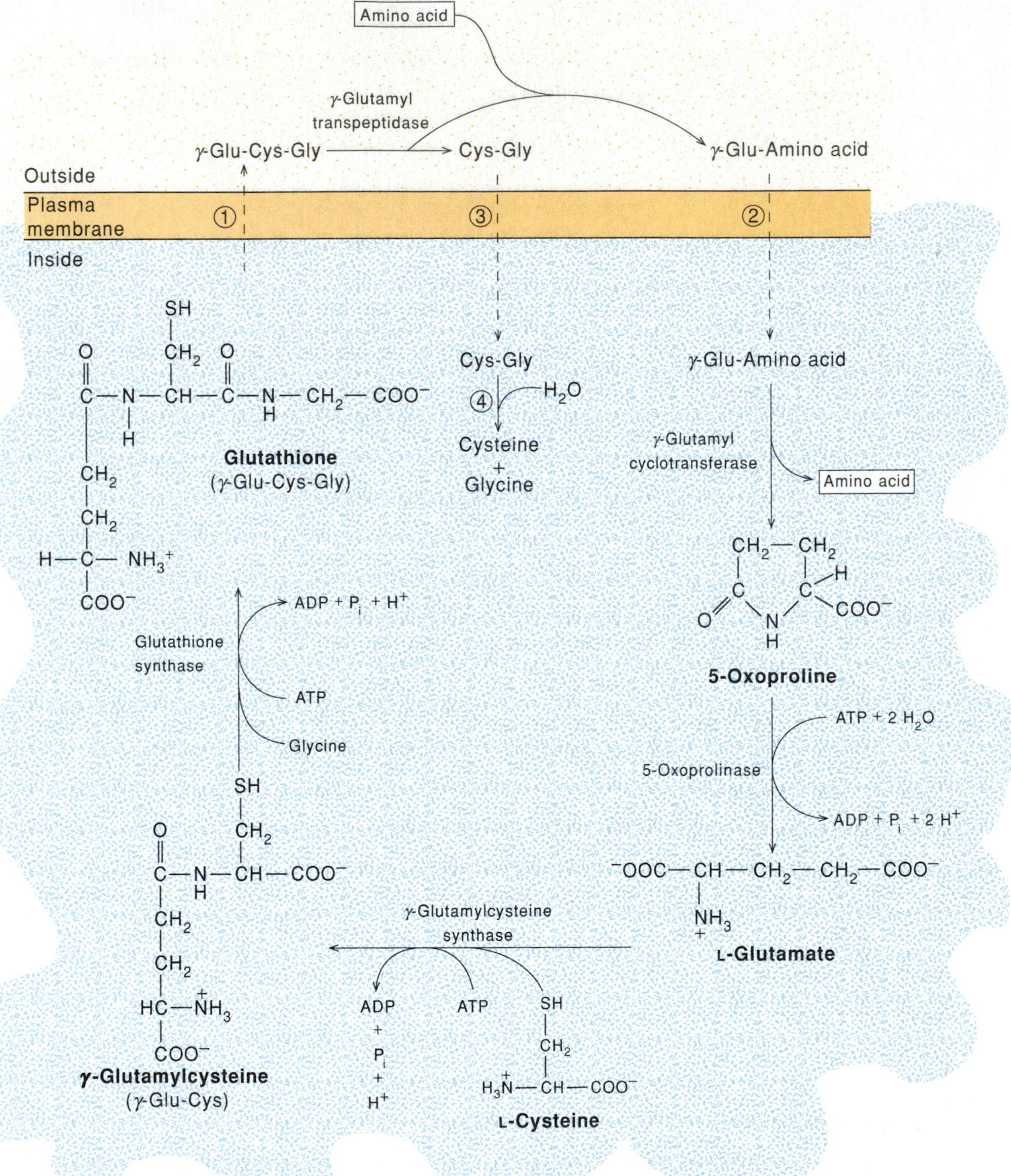

Glutathione has often been considered important in maintaining the sulfhydryl groups of proteins in the cell in a reduced state, presumably by a nonenzymatic reaction. In contrast, an enzyme, protein-disulfide reductase, does catalyze sulfhydryl-disulfide interchanges between glutathione and proteins. The enzyme is important in insulin breakdown and is probably important in the reassortment of disulfide bonds during polypeptide chain folding.

Mutants of *E. coli* have been isolated that are essentially devoid of glutathione owing to the loss of one or the other of the two synthases. Such cells are viable and have normal growth rates, but they are more sensitive to sulfhydryl reagents, such as mercurials.

Glutathione also plays a role as a reduced carrier for the reduction of glutaredoxin, which, like thioredoxin, is a hydrogen donor for nucleotide reductase (see chapter 20) and for the reduction of activated sulfate to sulfite (see chapter 18). Glutathione is also important in maintaining the iron of hemoglobin in the ferrous state. In all these roles, the reduction of oxidized glutathione by the NADPH-dependent glutathione reductase provides for the regeneration of reduced glutathione.

A role for glutathione that is independent of its reducing property is one as a γ-glutamyl donor in the γ-glutamyl cycle (fig. 19.32). Of particular significance is the fact that cells exhibiting the activities of the cycle contain substantial amounts of γ-glutamyl transpeptidase activity on the outer surface of their cell membranes. This enzyme is thought to transfer the γ-glutamyl group of extracellular glutathione to an extracellular amino acid. The γ-glutamylamino acid is transported into the cell, where, as a substrate for γ-glutamyl cyclotransferase, the amino acid and 5-oxoproline are released. 5-Oxoprolinase is converted to glutamate in an ATP-dependent cleavage catalyzed by 5-oxoprolinase. The glutathione is then regenerated from glutamate and the cysteine and glycine that were released by cysteinylglycine dipeptidase.

The γ-glutamyl cycle enzymes are found in those tissues for which the transport of glutathione into cells is an important function. Whereas glutathione is exported by most cells, it is efficiently transported only into cells that contain the membrane-bound γ-glutamyl transpeptidase. Thus the γ-glutamyl transpeptidase appears to facilitate salvage of glutathione secreted by some tissues into the bloodstream as well as to permit an energy-driven transport system for amino acids. Either oxidized or reduced glutathione is a substrate for the transpeptidase. The enzyme also catalyzes the hydrolysis of glutathione to glutamate and cysteinylglycine and glutamine to glutamate and ammonia. Many amino acids can serve as acceptors for the γ-glutamyl group, including γ-glutamylamino acids (to yield γ-glutamyl-γ-glutamylamino acids) and even glutathione itself. Such a transport across cell membranes is probably especially important for cysteine and methionine, as well as for glutathione.

That this cycle does play a role in transport is shown by the fact that incorporating glutamyl transpeptidase isolated from kidney into erythrocyte membranes stimulates the uptake of glutamate and alanine supplied along with glutathione to such preparations.

There is another class of enzymatic conversions involving glutathione that occurs in the cytosol of the liver. The enzymes are called glutathione S-transferases. They are similar enzymes, few in number, that exhibit a broad range of activities owing to their ability to bind many hydrophobic substances, such as bilirubin, steroids, and polycyclic aromatic hydrocarbons. Near this hydrophobic binding site is a second site specific for glutathione. Whether the hydrophobic ligand is also a substrate depends on whether it has an electrophilic atom that can undergo nucleophilic attack by glutathione, resulting in conjugation with glutathione and release of the electrophilic atom:

$$RX + GSH \rightarrow RSG + HX$$

Figure 19.33

The structure of the antibiotic gramicidin S. Gramicidin S is a cyclic decapeptide composed of a repeated sequence of five amino acids.

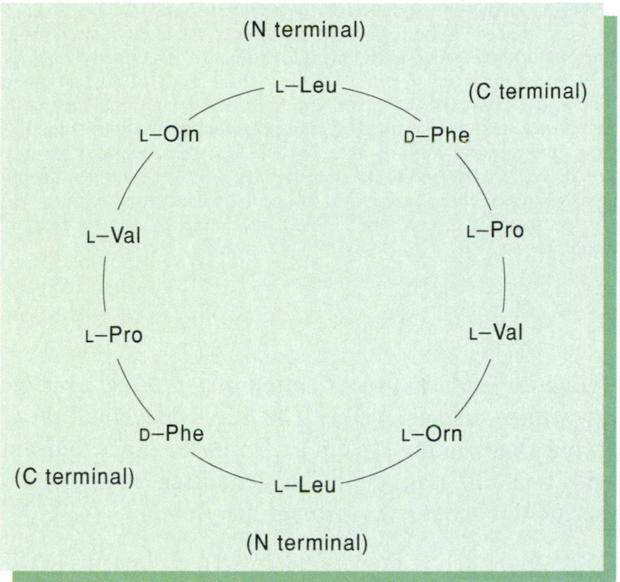

The glutathione conjugation product may be further metabolized or may simply be excreted. The glutathione S-transferases thus serve to solubilize, detoxify, and initiate the catabolism of a wide variety of hydrophobic substances.

Gramicidin Is a Cyclic Decapeptide Synthesized on a Protein Template

The antibiotic gramicidin S produced by the bacterium *Bacillus brevis* is a cyclic decapeptide (fig. 19.33) composed of a repeated sequence of five amino acids (—D-Phe—L-Pro—L-Val—L-Orn—L-Leu—)$_2$. Whereas in glutathione biosynthesis the order of amino acids is determined by a specific enzyme for each peptide linkage made, in gramicidin the ordering is mainly a function of the attachment points of amino acids on the enzyme surface. Thus one of the two enzymes involved in gramicidin synthesis appears to be serving a dual role of enzyme and template. The synthesis of more complex polypeptides requires the intricate biochemical machinery used in protein synthesis. Gramicidin synthesis is divided into five phases.

1. *Activation, thioesterification, and racemization of L-phenylalanine.* The light enzyme of the gramicidin-forming system activates L-phenylalanine as the aminoacyl adenylate and transfers the phenylalanyl residue to a thiol group on the enzyme. Racemization of L-phenylalanine occurs at this thioester stage:

$$\text{ESH} + \text{ATP} + \text{L-Phe} \rightarrow (\text{L-Phe} \sim \text{AMP})\text{ESH} + \text{PP}_i \rightarrow \text{ES} \sim \text{L-Phe} + \text{AMP} \rightarrow \text{ES} \sim \text{D-Phe}$$

Figure 19.34

The formation of gramicidin S on a protein template. (*a*) The activated amino acids are held in thioester linkage on the light and heavy enzymes. The phenylalanyl residue has undergone racemization and is being transferred to the pantotheine arm on the heavy enzyme. (*b*) The first peptide bond is about to be formed by transfer of the phenylalanyl group from the pantotheine group to the prolyl residue. (*c*) The phenylalanylprolyl residue is about to be transferred to the free thiol group of the pantotheine arm. The light enzyme has accepted another phenylalanyl residue that has already undergone racemization. (*d*) The phenylalanylprolyl residue is about to be transferred to the valyl residue. (*e*) The first pentapeptidyl group after being made was transferred to the waiting site 6. The second pentapeptidyl group has just been completed and is about to be condensed with the first to yield the decapeptide gramicidin S. The pantotheine arm is now free to repeat the process.

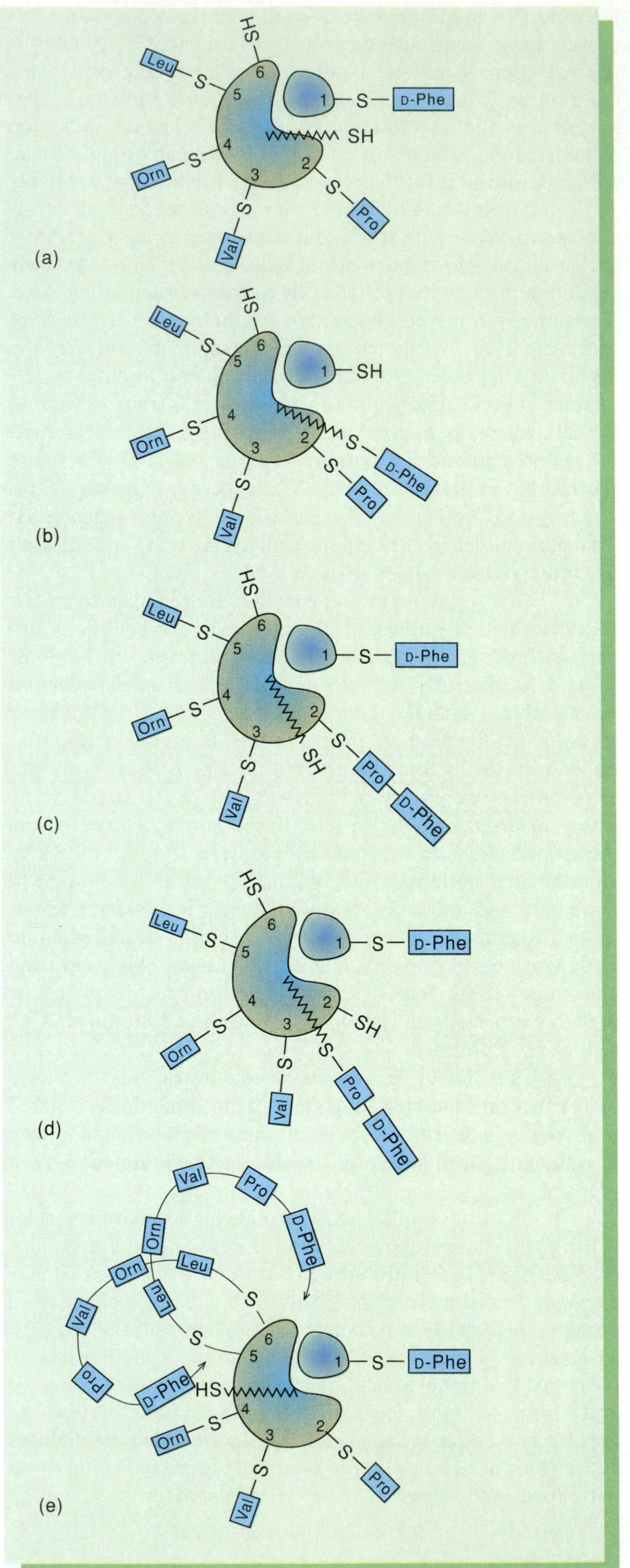

2. *Activation and thioesterification of* L-*proline,* L-*valine,* L-*ornithine, and* L-*leucine.* The heavy enzyme of the gramicidin-forming system activates the other four amino acids found in gramicidin S and transfers each to a specific thiol-containing site on the protein:

$$\underset{\text{SH}\quad\text{SH}\quad\text{SH}\quad\text{SH}}{\underline{\qquad E \qquad}} + \text{Pro} + \text{Val} + \text{Orn} + \text{Leu} + 4\ \text{ATP} \rightarrow$$

$$\underset{\text{S}\sim\text{Pro}\quad\text{S}\sim\text{Val}\quad\text{S}\sim\text{Orn}\quad\text{S}\sim\text{Leu}}{\underline{\qquad E \qquad}} + 4\ \text{AMP} + 4\ \text{PP}_i$$

3. *Transfer of* D-*phenylalanine to the heavy enzyme and initiation of peptide formation.* The heavy enzyme contains a covalently linked 2-nm-long pantotheine arm that is thought to serve as a carrier of the growing peptide chain. It is to the —SH group of this pantotheine arm on the heavy enzyme that the D-phenylalanyl group is probably transferred from the light enzyme (transthiolation). The first peptide bond would then be formed by a transpeptidation reaction that liberates the pantotheine thiol group.
4. *Elongation.* The liberated pantotheine arm is now free to undergo transthiolation with the newly formed phenylalanylprolyl residue and to move to the valyl-thiol site to repeat the transpeptidation step. This step is repeated at the ornithinyl and leucyl sites.
5. *Cyclization.* After the pentapeptide is formed, it is cyclized with an identical peptide in a head-to-tail fashion. One possible mechanism, implied in figure 19.34, involves the transfer of the first pentapeptide to a thiol "waiting" site, and when a second pentapeptide has been completed, cyclization occurs by two additional transpeptidation reactions (see fig. 19.34). Another possibility is that the cyclization occurs by an *inter*molecular transpeptidation involving two heavy enzymes, each containing one completed pentapeptidyl residue at the terminal thiol site (not shown).

Summary

Amino acids can serve as a source of energy, carbon, or nitrogen. In addition to these uses amino acids are frequently catabolized simply because they are present in potentially harmful excess. In our discussion of the metabolic fate of amino acids, we have made the following points.

1. For most amino acids the α-amino group is removed at an early stage in catabolism, usually in the first step. Transaminases are specific for different amino acids. Frequently α-ketoglutarate is the acceptor for the amino group, in which case it is converted into glutamate. The α-ketoglutarate can be regenerated from the glutamate by oxidative deamination.
2. A great deal of excess NH_3 frequently results from amino acid catabolism. This excess ammonia must be eliminated. In bacteria and lower eukaryotes the ammonia can usually be removed by simple diffusion, but in higher eukaryotes this is not feasible. Since the ammonia is frequently quite toxic, before removal it is detoxified by conversion to uric acid or urea. An intricate pathway resulting in the conversion of ammonia into urea involves five enzymes—three located in the cytoplasm and the remaining two in the mitochondrial matrix.
3. All amino acids can be degraded to CO_2 and water via the TCA cycle by the appropriate enzymes; the pathways are often complex and contain branchpoints to useful biosynthetic products. In every case, the pathways involve the formation of a dicarboxylic acid intermediate of the TCA cycle, of pyruvate, or of acetyl-CoA.
4. Our discussion of amino acid catabolism is organized according to the common intermediates formed during degradation. Alanine, glycine, threonine, serine, and cysteine are degraded to acetyl-CoA by way of pyruvate. Phenylalanine, tyrosine, tryptophan, lysine, isoleucine, and leucine also lead to acetyl-CoA, but not via pyruvate. Threonine falls into this category also. Arginine, histidine, proline, glutamic acid, and glutamine are all degraded to α-ketoglutarate. Catabolism of methionine, valine, and isoleucine leads to succinyl-CoA. Aspartate and asparagine are converted to oxaloacetate on degradation.
5. The importance of catabolic pathways is underscored by a broad spectrum of human metabolic diseases, in each of which one enzyme for normal amino acid catabolism is either missing or defective.
6. The formation of amino acid degradation enzymes is frequently regulated by hormones or nutritional factors. In microorganisms the formation of degradative enzymes is most often dependent on whether or not the amino acid is needed to supply energy, carbon, or nitrogen for growth.
7. Many biologically important routes of amino acid utilization—other than those leading to incorporation into proteins—are known. Some of these routes are distinctly anabolic pathways in which the amino acids, whether supplied from exogenous sources or synthesized *de novo* by the organism, serve as an initial substrate in an independent biosynthetic pathway. Some simple pathways involve the conversion of one amino acid to another, such as the formation of tyrosine from phenylalanine or ornithine from proline. These pathways are not usually found in organisms that form the amino acids *de novo*. On the other hand, the utilization of glycine in the formation of porphyrin derivatives occurs by very complex and highly branched pathways. Some of the other biologically important pathways lead to the biosynthesis of small peptides, including glutathione and the antibiotic gramicidin.

Selected Readings

Barker, H. A., Amino acid degradation by anaerobic bacteria. *Ann. Rev. Biochem.* 50:23, 1981. A review of an important group of fermentation pathways of amino acid breakdown that occur in nature and could not be covered in this chapter.

Christen, P., and D. E. Metzler (eds.), *Transaminases.* New York: John Wiley and Sons, 1985. A series of review chapters describing in detail the scope and mechanisms of transamination reactions.

Ledley, F. D., H. E. Grenett, M. McGinnis-Shelnutt, and S. L. C. Woo, Retroviral-mediated gene transfer of human phenylalanine hydroxylase into NIH 3T3 and hepatoma cells. *Proc. Natl. Acad. Sci. USA* 83:409, 1986.

Mazelis, M., Amino acid catabolism. In B. J. Mifflin (ed.), *The Biochemistry of Plants,* vol. 5. New York: Academic Press, 1980. A survey of some of the amino acid catabolic pathways that have been found in plants.

Meister, A., *Biochemistry of the Amino Acids,* vol. 2, 2d ed. New York: Academic Press, 1965. A very complete survey of amino acid catabolic pathways as they were known up to that time.

Meister, A., Glutathione metabolism and its selective modification. *J. Biol. Chem.* 263:17205, 1988. A mini-review describing the many important metabolic roles for glutathione.

Warren, M. J., and A. I. Scott, Tetrapyrrole assembly and modification into the ligands of biologically functional cofactors. *Trends Biol. Sci.* 51:486–491, 1990.

Wellner, D., and A. Meister, A survey of inborn errors of metabolism and transport in man. *Ann. Rev. Biochem.* 50:911, 1981. This review documents the importance of the pathways that break down amino acids in humans.

Also see the readings at the end of chapter 18.

Problems

1. What consequences would a severe deficiency in pyridoxal phosphate have on amino acid metabolism?
2. Differentiate between transamination and oxidative deamination as mechanisms to deaminate amino acids.
3. Compare the mechanisms for release of ammonia from the amine and amide groups of glutamine used in urea formation.
4. Would you anticipate elevated arginase activity in the liver of an untreated diabetic animal? Why or why not?
5. Explain the central role of pyruvate in amino acid catabolism in oxygen-limited muscle tissue.
6. Glutamic acid, valine, and aspartic acid are called glycogenic amino acids, whereas lysine and leucine are called ketogenic amino acids. Explain. Why are tyrosine and phenylalanine considered both ketogenic and glycogenic amino acids?
7. Glutamate and aspartate are deaminated to α-keto acids that are TCA cycle intermediates, yet cannot be completely oxidized by the TCA cycle alone. Explain. Propose a pathway for the complete oxidation of the deamination products of glutamate and aspartate.
8. Predict the effect of glycine cleavage enzyme deficiency on (a) 1-carbon pool and (b) energy production from glycine catabolism.
9. In what way is arginine catabolism uniquely associated with the "high-energy phosphate" reservoir in the skeletal muscle?
10. (a) The branched-chain α-keto acid dehydrogenase catalyzes an oxidative decarboxylation that requires thiamine pyrophosphate, lipoamide, and FAD as cofactors. Review the chemistry of the pyruvate dehydrogenase (an enzyme that also catalyzes an oxidative decarboxylation) and suggest a mechanism for the branched-chain α-keto acid dehydrogenase.
    (b) In principle, why is it possible for individuals with decreased activity of dihydrolipoamide dehydrogenase to also have increased serum levels of the branched-chain α-keto acids?
11. Isoleucine, valine, and methionine are all glycogenic amino acids. Genetic defects that affect catabolism of these amino acids include deficiency in propionyl-CoA carboxylase and methylmalonyl-CoA mutase activities. Given this information, propose a common pathway to glucose-6-phosphate from these amino acids.
12. Some patients with the deficiencies described in problem 11 accumulate propionate, whereas methylmalonic acidemia is observed in other patients. In each case, explain why the accumulated metabolite is predicted to result from the genetic deficiency.
13. A patient is found to have a defect in the enzyme 4-hydroxyphenylpyruvate dioxygenase. What diet would you recommend as treatment for this condition? What is the basis for your recommendation?
14. Concentration of phenylalanine in the blood of neonates is used to screen for phenylketonuria. Explain the biochemical basis for the correlation of elevated blood phenylalanine concentration and PKU. Explain why restriction of dietary phenylalanine is critically important for youngsters who are deficient in phenylalanine hydroxylase.
15. (a) Unlike proteins, L-glutathione is not a primary gene product. What "information" is used to direct the synthesis of L-glutathione?
    (b) Predict the effect of a glutathione synthase inhibitor on cells exposed to oxidative stress.

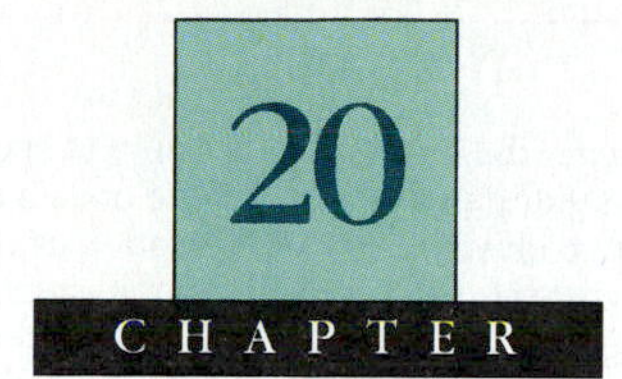

Nucleotides

This chapter deals with the biosynthesis of ribonucleotides and deoxyribonucleotides, their role in metabolic processes, and the pathways for their degradation (fig. 20.1). The biosynthesis of nucleotides is a vital process, since these compounds are indispensable precursors for the synthesis of both RNA and DNA. Without RNA synthesis, protein synthesis is halted; and unless cells can synthesize DNA, they cannot divide. Nucleotides are also necessary for constant repair of DNA, a process necessary for cell survival.

It is not surprising that inhibitors of nucleotide biosynthesis are very toxic to cells. As we will see, their toxicity has been used to advantage in the treatment of cancer as well as in the treatment of certain diseases resulting from infections by viruses, bacteria, or protozoans.

Nucleotides play important roles in all major aspects of metabolism. ATP, an adenine nucleotide, is the major substance used by all organisms for the transfer of chemical energy from energy-yielding reactions to energy-requiring reactions such as biosynthesis. Other nucleotides are activated intermediates in the synthesis of carbohydrates, lipids, proteins, and nucleic acids. Adenine nucleotides are components of many major coenzymes, such as NAD^+, $NADP^+$, FAD, and CoA. The critical role played by nucleotides as regulators of metabolism in both prokaryotic and eukaryotic organisms is described in other chapters.

Pathways for the metabolic degradation of nucleotides are also very important to the organism, as demonstrated by the fact that several genetic defects causing blocks in these pathways have serious consequences for the health of the organism.

The physiological importance of nucleotides is reflected in the careful regulation of their intracellular levels as well as intra- and extracellular levels of nucleosides and nucleobases. Levels of ATP, ADP, and AMP are tightly controlled in a variety of cells, and intracellular levels of dATP, dCTP, dGTP, and dTTP are also carefully regulated under normal

Figure 20.1

Synthesis and breakdown of nucleotides. Reactions from the master diagram of figure 12.5 are shown in black; the remaining reactions discussed in this chapter are shown in color. Dashed arrows are used for catabolic pathways.

It is remarkable how few substrates are involved in the biosynthesis of nucleotides. The sugar is ribose, derived from the pentose phosphate pathway, or deoxyribose, derived by reduction of ribonucleotides. The ribose is used in the activated form, phosphoribosylpyrophosphate (PRPP). In pyrimidine nucleotide synthesis, the pyrimidine ring is synthesized first and then condensed with a ribose. In purine nucleotide synthesis the purine base is built on the sugar. Most of the nitrogen and carbon components of both pyrimidine and purine rings are contributed one at a time in single steps. In addition, aspartate is a major contributor to both the purine and pyrimidine rings, whereas glycine contributes to the purine ring. Once the nucleotide monophosphates have been constructed, they are converted to triphosphates in two steps. However, other conversions usually are carried out only at specific levels of phosphorylation. For example, the reduction of ribose occurs at the diphosphate level for the uridine nucleotide and the formation of thymine nucleotide from uridine nucleotide occurs at the monophosphate level. The skeleton of glutamine is not used in purine or pyrimidine construction, but glutamine is frequently used in place of NH_3 as the amine group donor in various biosynthetic steps. Single carbons are most frequently contributed by formyl group donors.

Degradation of nucleotides starts with complete removal of phosphate. This is followed by cleavage of the ribose as a ribose-1-phosphate leaving the purine or pyrimidine base intact. These breakdown products can be reutilized for new nucleotide construction by so-called salvage pathways (not shown), or they may be further degraded to excretion products. Unlike the catabolism of carbohydrates, fats or proteins, nucleotide degradation, except for the recovery of the ribose, makes no contribution to the central metabolic pathways.

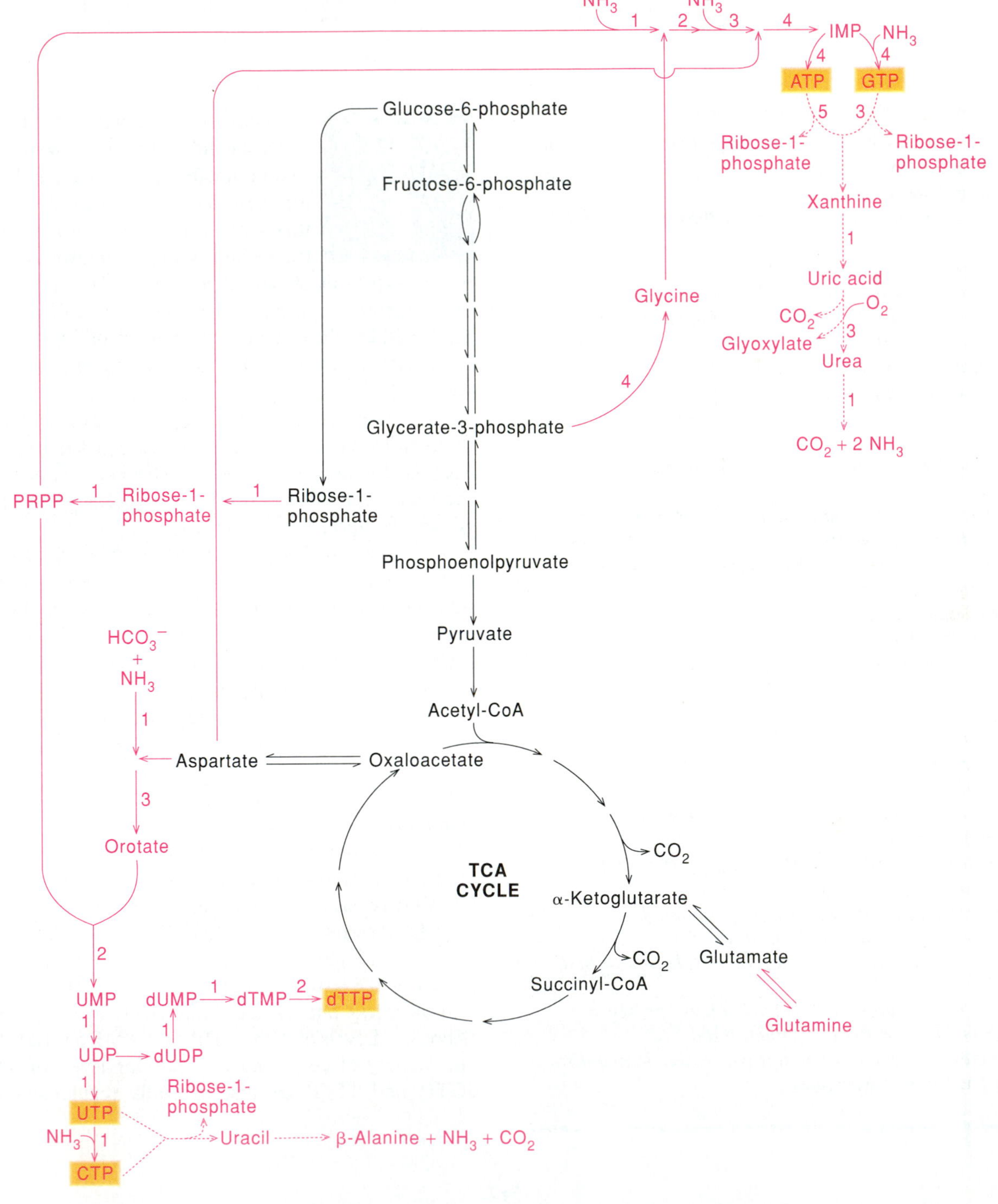

Figure 20.2

Structure of a nucleotide. A nucleotide has three components: a phosphoryl group, a pentose, and a nitrogenous base. The carbon atoms of the pentose (ribose in the example shown) are given prime designations. The deoxyribonucleotides have deoxyribose as the pentose. This sugar lacks a hydroxyl group at C-2′ of the pentose. The phosphoryl group is attached as an ester of one of the pentose hydroxyls, most commonly at C-5′. The base (adenine in the example shown) is attached by a glycosidic bond from a ring nitrogen to C-1′ of the pentose. When the base is a purine (e.g., adenine, guanine, or hypoxanthine) and the pentose is deoxyribose, this glycosidic bond is acid-labile. A nucleoside has no phosphoryl group, only a hydroxyl group at C-5′.

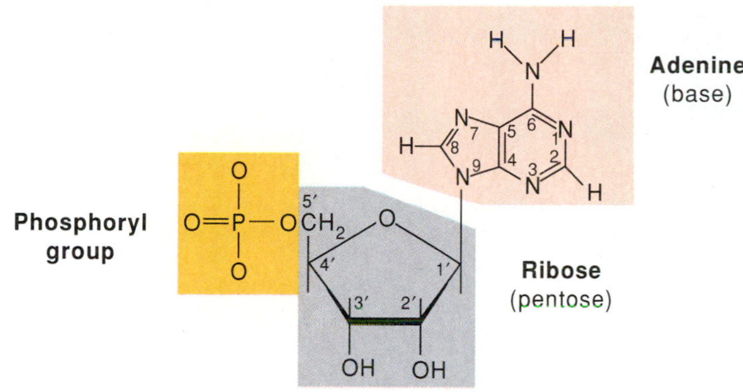

conditions. This regulation is determined by the concentration and location of the enzymes of nucleotide metabolism and by levels of substrates, products, and effectors.

Nucleotide Components: A Phosphoryl Group, a Pentose, and a Base

Each nucleotide is composed of three parts: (1) a heterocyclic nitrogenous base, (2) a pentose, and (3) a phosphoryl group (fig. 20.2). Nucleotides differ from one another through differences in each of these components.

Nucleotide bases (or nucleobases) belong to two classes: pyrimidines and purines (table 20.1). The former have a single six-membered ring containing two nitrogen atoms. In pyrimidine nucleotides, N-1 of the pyrimidine base is attached to the pentose C-1′. The common pyrimidine bases in nucleotides are uracil, cytosine, and thymine (fig. 20.3). Purine bases have a six-membered pyrimidine ring fused to a five-membered imidazole ring, the fused system containing four nitrogen atoms. In purine nucleotides, the glycosidic linkage is between N-9 of the purine and C-1′ of the pentose. The common purine bases in nucleotides are adenine, guanine, and hypoxanthine (see fig. 20.3). Nucleotide catabolism leads to the purines xanthine and uric acid (see fig. 20.27), and xanthosine monophosphate is a nucleotide intermediate of metabolism (see fig. 20.16).

Table 20.1
Names of Common Bases, Nucleosides, and Nucleotides

Base	Nucleoside[a]	Nucleotide
Purines		
Adenine	Adenosine (A)	AMP[b] or adenylate
Guanine	Guanosine (G)	GMP or guanylate
Hypoxanthine	Inosine (I)	IMP or inosinate
Pyrimidines		
Uracil	Uridine (U)	UMP or uridylate
Thymine	Thymidine (T)	TMP or thymidylate
Cytosine	Cytidine (C)	CMP or cytidylate

[a]With the exception of thymidine, these are the names of the ribonucleosides, and deoxyribonucleosides are indicated by the prefix *deoxy* in front of the nucleoside name; thus deoxyadenosine contains deoxyribose instead of ribose. Thymidine indicates the deoxyribose derivative of thymine. Standard one-letter abbreviations are shown in parentheses.

[b]AMP is the abbreviation for the most commonly used name for this nucleotide, adenosine-5′-monophosphate or adenosine-5′-phosphate. Similarly, ADP and ATP designate the 5′-diphosphate and -triphosphate of adenosine, respectively (see fig. 20.4). Adenylate (or adenylic acid) is an alternative nomenclature.

Figure 20.3

Structures of three common ribonucleotides (*a*) and four common deoxyribonucleotides (*b*). See table 20.1 for alternative names and for names of the corresponding bases and nucleosides.

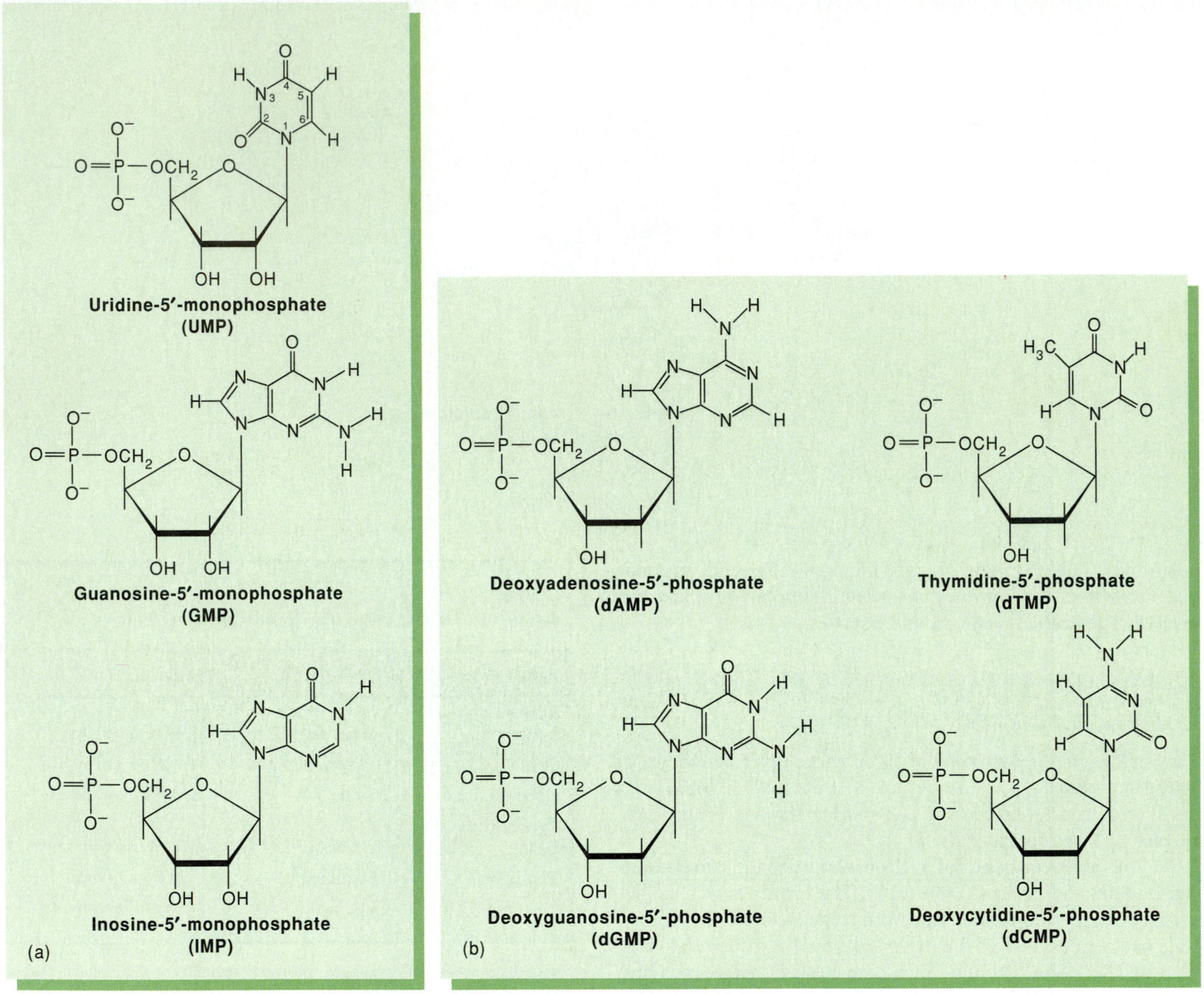

(a) Uridine-5′-monophosphate (UMP); Guanosine-5′-monophosphate (GMP); Inosine-5′-monophosphate (IMP)

(b) Deoxyadenosine-5′-phosphate (dAMP); Thymidine-5′-phosphate (dTMP); Deoxyguanosine-5′-phosphate (dGMP); Deoxycytidine-5′-phosphate (dCMP)

The pentose component of naturally occurring nucleotides is ribose or 2-deoxyribose (i.e., ribose with a hydrogen instead of a C-2′ —OH). In nucleotides the purine or pyrimidine is attached to C-1′ of the pentose in the β configuration. This means that the base is *cis* relative to C-5′ and *trans* relative to the C-3′ —OH. The major function of deoxyribonucleotides (those that have 2-deoxyribose as the pentose) is to serve as building blocks for DNA. Although ribonucleotides similarly serve as the units for RNA synthesis, they also have a multitude of other functions in cell metabolism. In some synthetic nucleosides with therapeutic properties, other pentose components, such as arabinose, are present. (See fig. 6.2 for the structure of arabinose.)

The phosphoryl group of nucleotides is most commonly substituted on the C-5′ —OH of the pentose. However, in cyclic nucleotides a single phosphoryl group is esterified to both the

Figure 20.4

The general structure of a nucleoside monophosphate, diphosphate, and triphosphate.

Figure 20.5

Tautomeric equilibrium of guanine and adenine. The keto and amino forms are strongly favored for guanine and adenine, respectively. Comparable tautomeric equilibria exist for thymine, uracil, and cytosine.

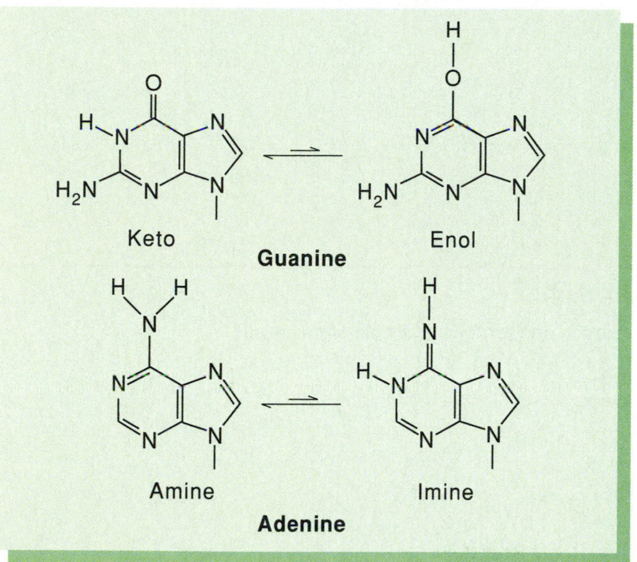

Table 20.2

Ionization Constants of the Ribonucleotides (Presented as p*K* Values)

	Base	Secondary Phosphate	Primary Phosphate
Adenosine-5′-phosphate (AMP)	3.8	6.1	0.9
Uridine-5′-phosphate (UMP)	9.5	6.4	1.0
Cytidine-5′-phosphate (CMP)	4.5	6.3	0.8
Guanosine-5′-phosphate (GMP)	2.4, 9.4	6.1	0.7

C-5′ —OH and the C-3′ —OH (e.g., see fig. 13.17). In nucleic acids each nucleotide unit has one phosphoryl group esterified to the C-5′ —OH and another at the C-3′ —OH, and these phosphoryl groups link the nucleotide units.

A nucleoside has no phosphoryl group; it consists of a purine or pyrimidine linked to ribose or deoxyribose. The simplest nucleotides are therefore nucleoside-5′-monophosphates. However, nucleoside-5′-diphosphates and nucleoside-5′-triphosphates (fig. 20.4) are nucleotide forms that are also extremely important in cell metabolism. In diphosphates a second phosphoryl group is added to the first in acid anhydride linkage (also called pyrophosphate linkage), and in triphosphates a second anhydride bond links another phosphoryl group. The nucleoside diphosphates (NDPs) and the nucleoside triphosphates (NTPs) dissociate three and four protons, respectively, from their phosphate groups. In the triphosphates, the phosphate immediately attached in ester linkage to the 5′-carbon is designated α; the middle phosphate, in pyrophosphate linkage, is called β; and the terminal phosphate, also in pyrophosphate linkage, is called γ. The NDPs and NTPs can form complexes with Mg^{2+} or Ca^{2+} and probably exist in these complexes in the cell. The NDPs and NTPs have a number of important metabolic functions in the cell: they serve as energy-carrying enzyme cofactors (e.g., see chapters 13–16) and as substrates for the biosynthesis of nucleic acids (see chapters 26 and 28).

All of the commonly occurring bases in nucleotides are capable of existing in two tautomeric forms, which differ by the placement of a proton and some electrons. For example, guanosine can undergo a change from a keto form to an enol form as shown in figure 20.5. The keto form is so strongly favored that it is difficult to detect even trace amounts of the enol form at equilibrium. Similarly, the keto forms of thymidine or uridine are strongly preferred. Adenosine and cytidine can isomerize to imino forms, but the amino forms are strongly preferred (see fig. 20.5). Even though the unusual tautomers are present in very small amounts, it is conceivable that when present in DNA they contribute to the mutation process.

Some nucleotides undergo protonation in acid and some undergo deprotonation in base; the relative p*K* values are listed in table 20.2. At neutrality there is no charge on any of the bases. Three of the bases, A, C, and G, undergo protonation as the pH

Figure 20.6

Uncharged and protonated forms of adenosine. The charged base resonates between the two structures shown on the right. Cytosine protonates in a similar way.

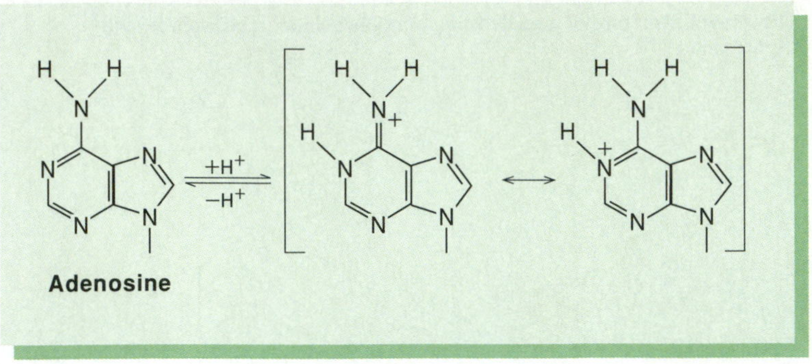

Adenosine

Figure 20.7

Uncharged and protonated forms of guanosine.

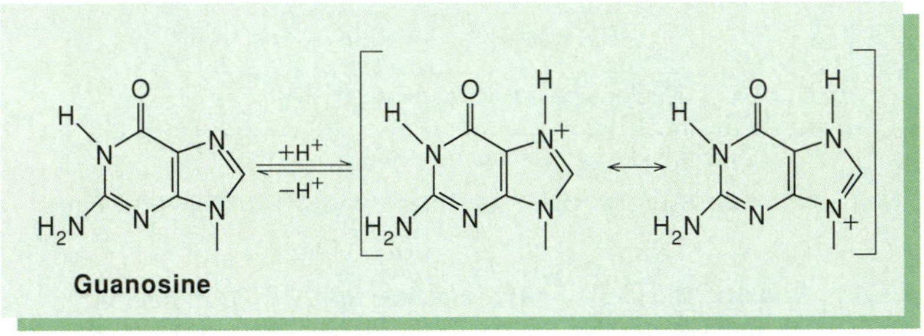

Guanosine

Figure 20.8

Absorption spectra of the common ribonucleoside-5′-monophosphates in protonated and unprotonated forms. The lower pH gives absorbance for the protonated form. Absorbance values are relative to the value at the maximum for each compound. Molar absorbances (that is, the absorbance of a 1-M solution) at the wavelength and pH giving the maximum absorbance are as follows: AMP, 15.4×10^3; UMP, 10.0×10^3; CMP, 13.2×10^3; GMP, 13.7×10^3. The comparable deoxyribonucleotides and the nucleoside di- and triphosphates have similar absorbance curves and molar absorbances.

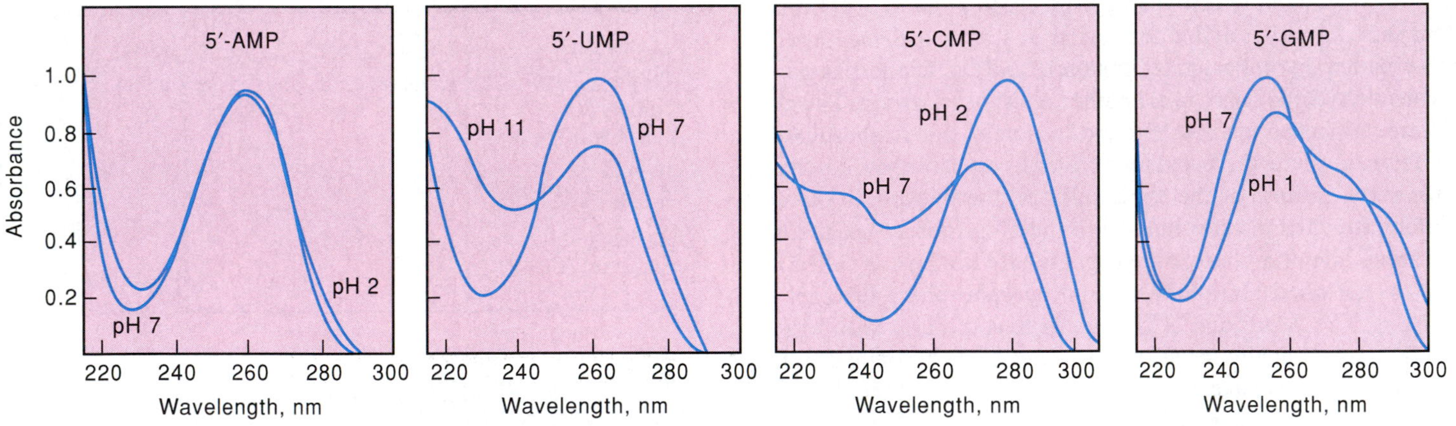

is lowered. The adenine moiety in AMP (adenosine monophosphate) protonates on the N-1 position of the purine rather than on the amino group (fig. 20.6). The charged form is stabilized by the resonance hybrids shown. In CMP the proton adds to the comparable N-3 ring nitrogen. In guanosine a proton adds to N-7 rather than the amino group (fig. 20.7), again indicating the unusually low basicity of the amino groups on the nucleotides compared with primary aliphatic amines. On the basic side of neutrality both UMP and GMP lose a proton from the imino nitrogens at positions 3 and 1, respectively. As you might expect, the ionization constants for the primary and secondary dissociations of the phosphate group do not differ appreciably for the various nucleosides (see table 20.2).

Owing to the large number of conjugated double bonds in their nitrogen bases, all nucleotides show absorption maxima in the near-ultraviolet range (fig. 20.8). The spectrum is pH-dependent since protonation or deprotonation changes the electronic distribution in the base rings. The ultraviolet absorption of the nucleotides has been useful in many ways for the study of mononucleotides and polynucleotides. Indeed, it has been most useful in studying nucleic acid conformation (chapter 25).

Figure 20.9

Pathways of purine metabolism. Double-headed arrows indicate reversible enzymatic reactions. Separate arrows in opposite directions between metabolites indicate a different enzyme participating in each direction. The diagram is arranged in tiers: purines at the bottom, nucleosides at the next level, then nucleoside mono-, di-, and triphosphates (in ascending order).

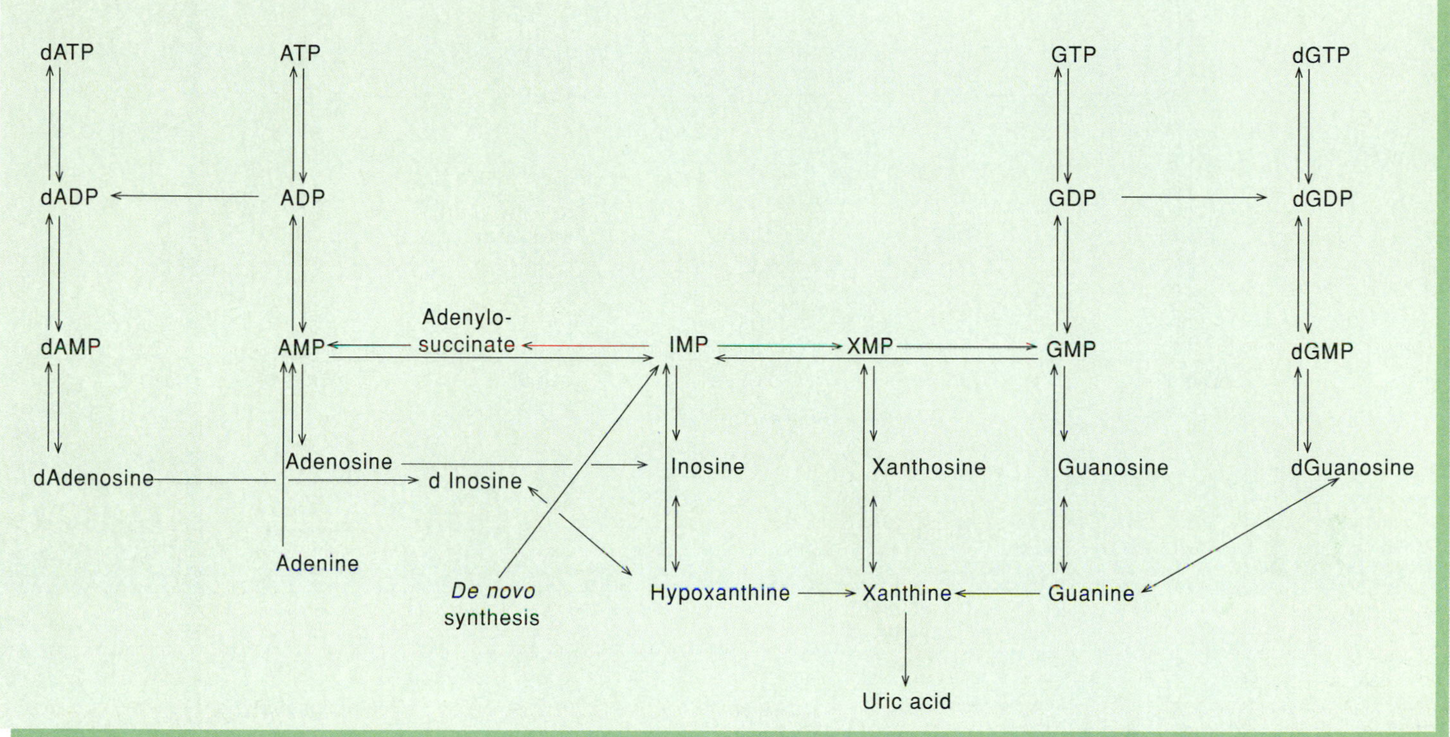

Overview of Nucleotide Metabolism

The pathways by which cells synthesize, interconvert, and catabolize various purine and pyrimidine nucleotides are summarized schematically in figures 20.9 and 20.10. Cells of different types, or even the same cells in different stages of development, differ greatly in their ability to carry out some of the reactions involved, with some cells favoring one set of reactions and others another. In the rest of the chapter we will deal with the details of these pathways.

Synthesis of Purine Ribonucleotides *de Novo*

Purine nucleotides can be synthesized in three ways: by *de novo* synthesis, by reconstruction from purine bases through the addition of a ribose phosphate moiety, or by phosphorylation of nucleosides. The first two pathways are the more important quantitatively, and in both of them phosphoribosylpyrophosphate (PRPP) is an essential precursor. It has a similar important role in pathways for pyrimidine nucleotide biosynthesis. PRPP is synthesized from ribose-5-phosphate, which cells synthesize either from glucose-6-phosphate by an oxidative pathway or from intermediates of glycolysis by a nonoxidative pathway (see chapter 13).

The formation of PRPP from ribose-5-phosphate and ATP is catalyzed by ribose-5-phosphate pyrophosphokinase (fig. 20.11). This is an unusual kinase because the pyrophosphoryl group is transferred rather than the phosphoryl group. As might be anticipated for an enzyme with a key position in several biosynthetic pathways, its activity is regulated by a number of metabolites. Inorganic phosphate is an activator, and the Mg^{2+} ion is both cofactor and activator. Inhibitors include ADP and glycerate-2,3-bisphosphate, which are competitive with respect to ribose-5-phosphate. By contrast, AMP and GDP are noncompetitive inhibitors. At any specific time, the concentration of these metabolites as well as those of the substrates will determine the activity of the kinase. In particular, the nucleotide inhibitors will curtail synthesis of PRPP when the energy stores of the cell are low.

The ultimate precursors of the purine ring were established by administering isotopically labeled compounds to pigeons and tracing the incorporation of labeled atoms into the purine ring of uric acid. Birds were used in these experiments because they excrete waste nitrogen largely as uric acid, a purine derivative that is easily isolated in pure form. Chemical degradation of the uric acid revealed the origins of the atoms as depicted in figure 20.12. A flow scheme showing the successive incorporation of atoms from these precursors is shown in figure 20.13.

Figure 20.10

Pathways of pyrimidine metabolism. Arrows and layout have a similar significance as in figure 20.9.

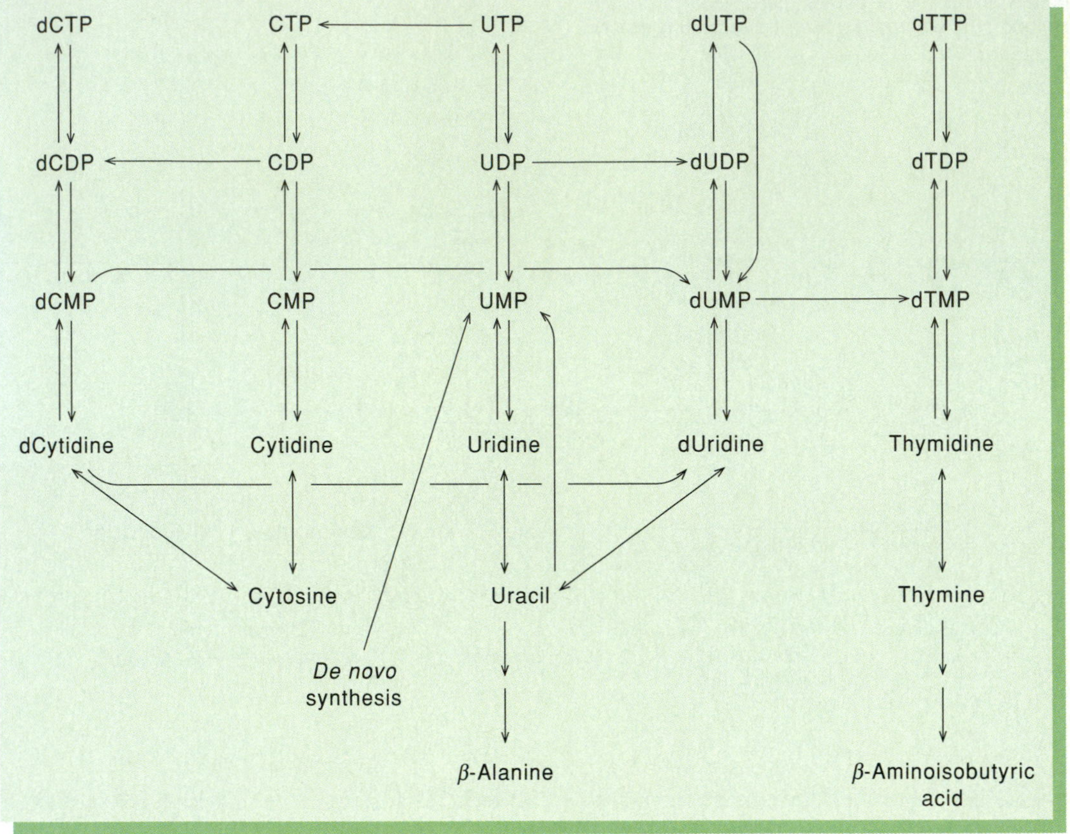

Figure 20.11

Synthesis of phosphoribosylpyrophosphate (PRPP). This is an unusual kinase-catalyzed reaction because the group transferred is the pyrophosphate group rather than the phosphate group.

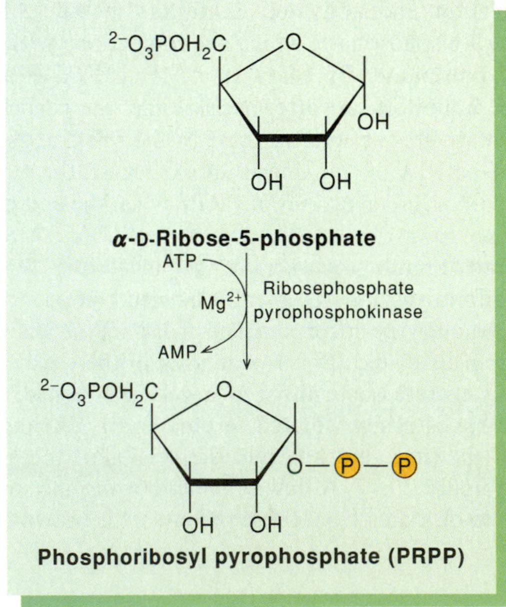

Figure 20.12

Precursors of the purine ring of uric acid in pigeons as determined by isotope labeling experiments. The indicated precursor substances were administered one at a time to pigeons. Each precursor was labeled with isotopic nitrogen or carbon, and in each case the excreted uric acid was purified and degraded chemically. The isotope content of the various degradation products indicated that precursors contributed the specific atoms indicated.

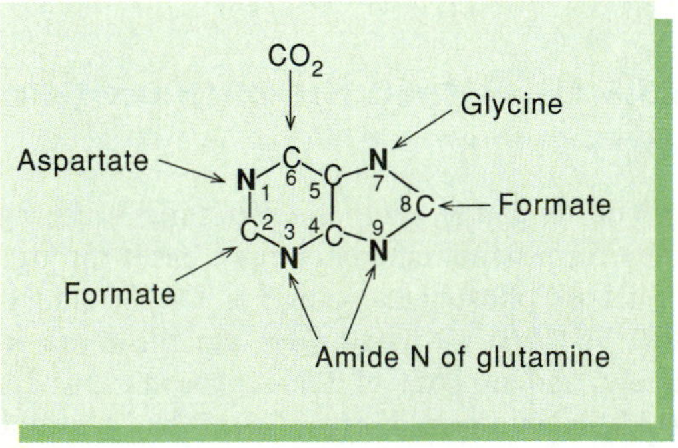

Figure 20.13

Summary of incorporation of precursors into the purine ring of IMP. Formate as well as other indirect donors of one carbon, such as serine, donate their atoms by means of the formyl group of a folate derivative in steps 3 and 9.

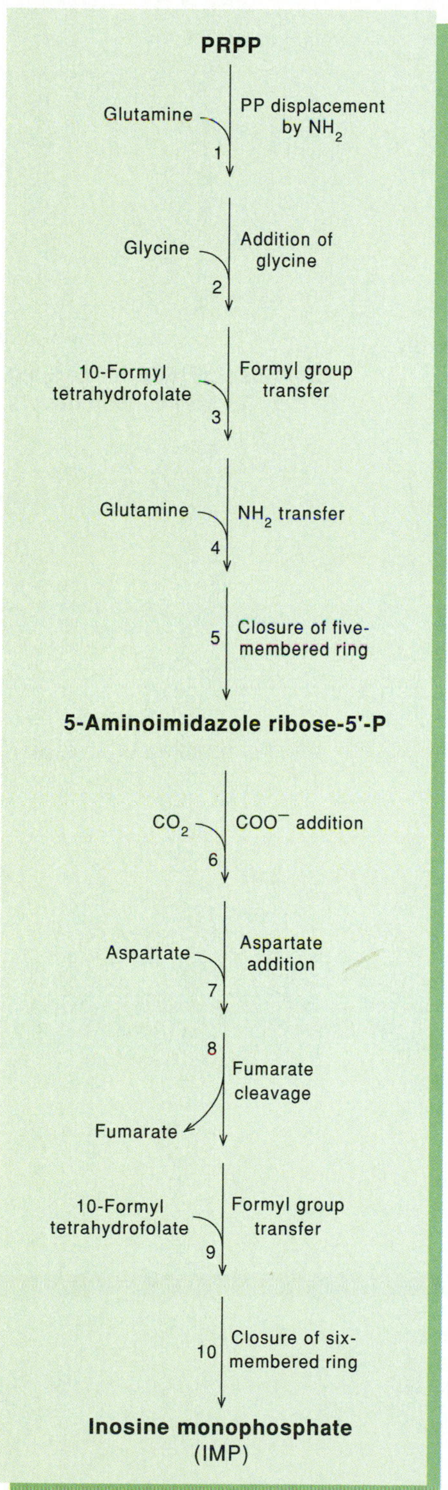

Inosine Monophosphate (IMP) Is the First Purine Nucleotide Formed

The pathway from PRPP to the first complete purine nucleotide, inosine monophosphate, involves ten steps and is shown in figure 20.14. It would seem logical that the purine should be built up first, followed by addition of ribose-5-phosphate, but this is not the case. The starting point is PRPP, to which the imidazole ring is added; the six-membered ring is built up afterward.

The first step, in which phosphoribosylamine is formed, is catalyzed by glutamine phosphoribosylpyrophosphate amidotransferase, an enzyme containing nonheme iron. The reaction involves inversion of the configuration at C-1 of the ribose and leads to the β configuration that is characteristic of naturally occurring nucleotides. This step involves commitment of PRPP to the purine biosynthetic pathway and, as you might expect, is subject to important feedback inhibitory effects by purine nucleotides. We will examine these effects a little later.

In step 2, an amide bond is formed by the synthase between the carboxyl group of glycine and the amino group of phosphoribosylamine, with ATP supplying energy and being hydrolyzed to ADP and inorganic phosphate. After these two steps have introduced atoms 4, 5, 7, and 9 of the purine ring, the remaining atoms are introduced one by one (steps 3, 4, 6, 7, and 9).

In step 3, carbon 8 is introduced as a formyl group that is transferred from 10-formyltetrahydrofolate (10-formyl-H_4 folate). (To review the way tetrahydrofolate derivatives accept a one-carbon unit from donors such as serine, glycine, or formate and transfer it to a suitable acceptor in biosynthetic reactions, see chapters 11 and 18; in chapters 11 and 18 10-formyl H_4 folate is written as N^{10}-formyl H_4 folate, an alternative terminology that is also still in use.) Now the five components that will constitute the imidazole part of the purine are present, but before ring closure occurs, N-3 of the purine ring is introduced (step 4) by transfer of another amino group from glutamine to phosphoribosylformylglycinamide. ATP provides energy for the amido group transfer, being itself hydrolyzed to ADP and phosphate.

The imidazole ring is closed in an essentially irreversible cyclization requiring the presence of Mg^{2+} and K^+ (step 5). Then C-6 of the purine ring is introduced by addition of bicarbonate in the presence of a specific carboxylase (step 6). This carboxylation is unusual in that it does not seem to involve biotin and is not coupled with any energy-yielding process such as ATP hydrolysis. The equilibrium is unfavorable for the formation of the carboxylate. *In vivo* the reaction proceeds at physiological concentrations of bicarbonate because of coupling with subsequent steps that are thermodynamically favorable.

Next, in steps 7 and 8, N-1 of the purine ring is contributed by aspartate. Aspartate forms an amide with the 4-carboxyl group, and the succinocarboxamide so formed is then cleaved with release of fumarate. Energy for carboxamide formation is provided by ATP hydrolysis to ADP and phosphate.

5-Phospho-α-D-ribosyl-1-pyrophosphate

Glutamine + H_2O → Glutamate + PP_i (① Mg^{2+}; Glutamine PRPP amidotransferase)

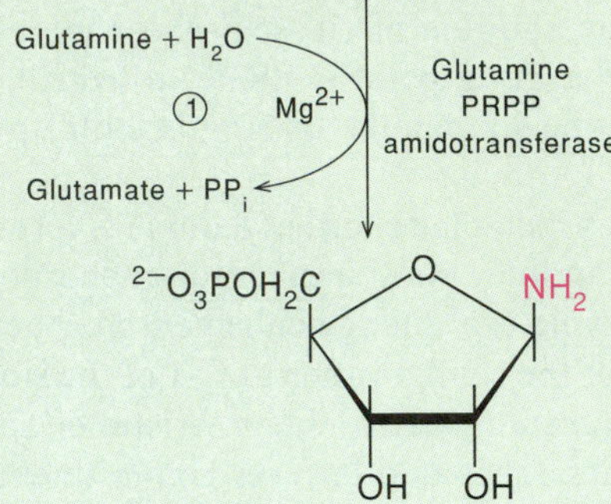

5-Phospho-β-D-ribosylamine

Glycine + ATP → ADP + P_i (② Synthase)

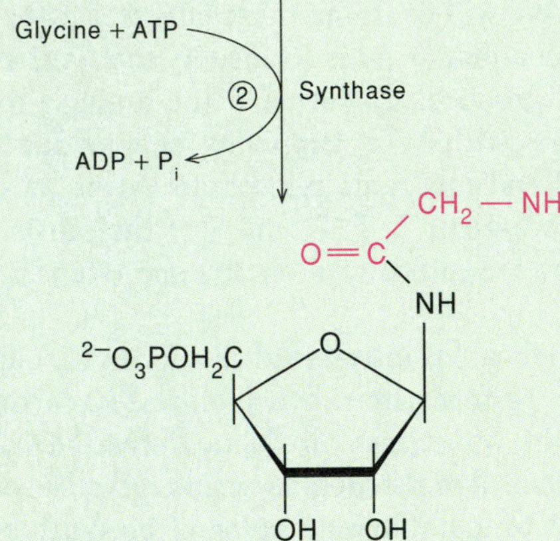

5′-Phosphoribosylglycinamide

10-Formyl tetrahydrofolate → Tetrahydrofolate (③ Formyltransferase)

CH_2—NH—CHO; O═C; NH; Ribose-5-phosphate

5′-Phosphoribosyl-N-formylglycinamide

ATP + Gln + H_2O → ADP + Glu + P_i (④ Synthase)

CH_2—NH—CHO; HN═C; NH; Ribose-5-phosphate

5′-Phosphoribosyl-N-formylglycinamidine

ATP → ADP + P_i (⑤ Mg^{2+}, K^+; Synthase)

H_2N; N; N; Ribose-5-phosphate

5′-Phosphoribosyl-5-aminoimidazole

CO_2 (⑥ Carboxylase)

^-OOC; H_2N; N; N; Ribose-5-phosphate

5′-Phosphoribosyl-5-aminoimidazole-4-carboxylate

ATP + Asp → ADP + P_i (⑦ Mn^{2+}; Synthase)

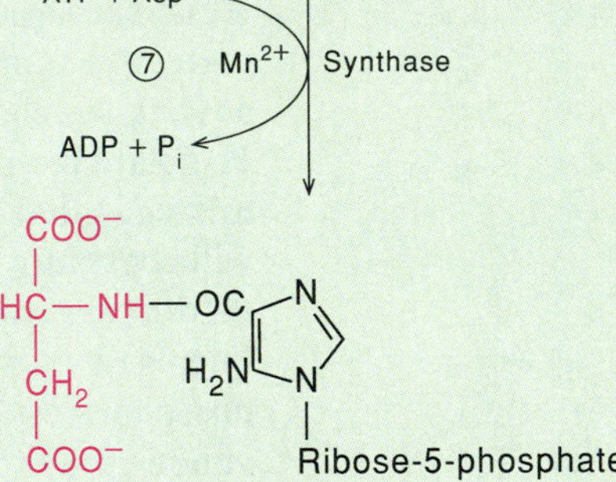

5′-Phosphoribosyl-4-(N-succinocarboxamide)-5-aminoimidazole

→ Fumarate (⑧ Adenylosuccinate lyase)

NH_2—OC; H_2N; N; N; Ribose-5-phosphate

5′-Phosphoribosyl-4-carboxamide 5-aminoimidazole

5′-Phosphoribosyl-4-carboxamide-5-aminoimidazole

10-Formyl-tetrahydrofolate → Tetrahydrofolate (⑨ Formyltransferase)

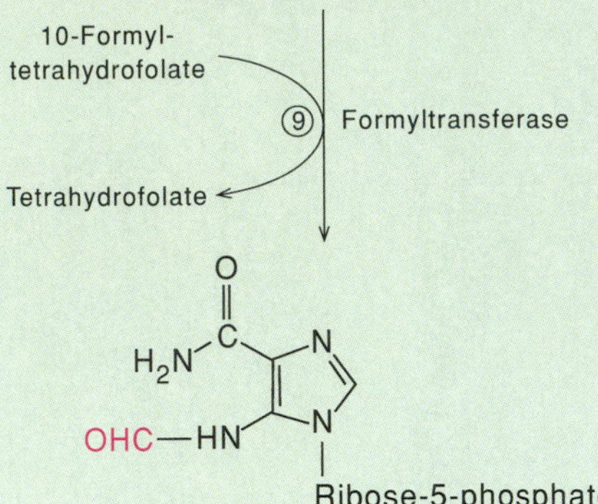

5′-Phosphoribosyl-4-carboxamide-5-formamidoimidazole

→ H_2O (⑩ IMP Cyclohydrolase)

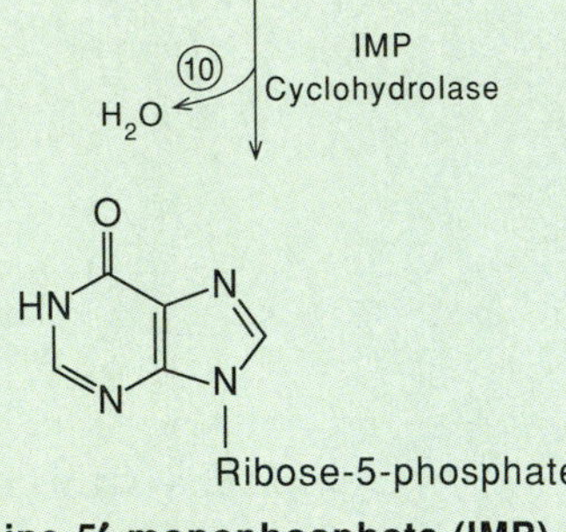

Inosine-5′-monophosphate (IMP)

Figure 20.14

Biosynthetic pathway to inosine monophosphate. Color indicates the group or atom introduced at each step. The first step, in which phosphoribosylamine is formed, involves inversion of the configuration at C-1 of the ribose. In step 2 an amide bond is formed by the synthase between the carboxyl group of glycine and the amino group of phosphoribosylamine. An additional formyl group is added to the glycine amino group in the next step. An additional amide group is donated by a glutamine, and next the five-membered imidazole ring is formed. Further groups are introduced through successive attachments to the imidazole, and finally the purine ring is completed in inosine-5′-monophosphate.

Figure 20.15

Utilization of fumarate formed in reactions in which aspartate acts as donor of an amino group to generate ATP and regenerate aspartate (ETC = electron-transport chain).

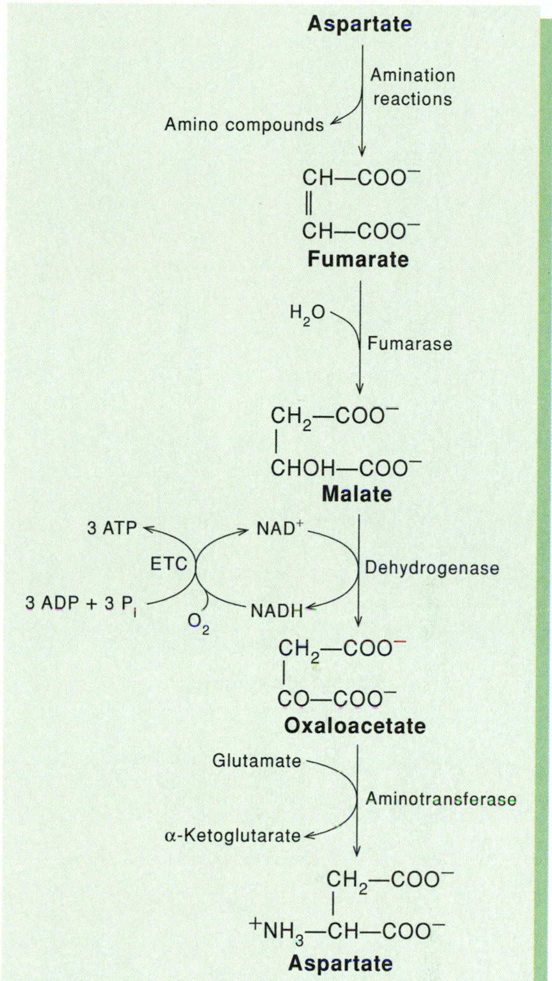

These reactions resemble the conversion of citrulline to arginine in the urea cycle (chapter 19) and the conversion of IMP to AMP (see fig. 20.16). It is worth noting that fumarate released in all these synthetic pathways can be converted to oxaloacetate by fumarase and malate dehydrogenase, and in the process, NAD^+ is reduced to NADH (by malate dehydrogenase). Reoxidation of the NADH by the electron-transport chain then generates three ATPs from ADP and P_i. The ATPs supply some of the energy required for purine synthesis, and the oxaloacetate formed can be used to replenish the supply of aspartate by transamination with glutamate (fig. 20.15).

The final atom of the purine ring is provided in step 9 by donation of a formyl group from 10-formyltetrahydrofolate to the 5-amino group of the almost completed ribonucleotide. In the final step, ring closure is effected by elimination of water to form IMP (inosine monophosphate), the first product with a complete purine ring. Although this final ring closure does not require energy from ATP, closure of the imidazole ring does, and the synthesis of IMP from ribose-5-phosphate requires a total of six high-energy phosphate groups from ATP (assuming hydrolysis of pyrophosphate released during phosphoribosylamine synthesis, step 1 of figure 20.14).

IMP Is Converted into AMP and GMP

IMP does not accumulate in the cell, but is converted to AMP, GMP, and the corresponding diphosphates and triphosphates. The two steps of the pathway from IMP to AMP (fig. 20.16) are typical reactions by which the amino group from aspartate is introduced into a product. The 6-hydroxyl group of IMP (tautomeric with the 6-keto group) is first displaced by the amino of aspartate to give adenylosuccinate, and the latter is then cleaved nonhydrolytically by adenylosuccinate lyase to yield fumarate and AMP. In the condensation of aspartate with IMP, cleavage of GTP to GDP and phosphate provides energy to drive the reaction.

Conversion of IMP to GMP also proceeds by a two-step pathway (fig. 20.16): first, dehydrogenation of IMP to xanthosine-5′-phosphate (XMP), and second, transfer of an amino group from glutamine to C-2 of the xanthine ring to yield GMP. The second reaction also involves the cleavage of ATP to AMP and inorganic pyrophosphate. The latter is in turn hydrolyzed to inorganic phosphate by the ubiquitous inorganic pyrophosphatase in a reaction with a very favorable equilibrium. This hydrolysis is coupled with the GMP synthase reaction because pyrophosphate is a product of the latter and a substrate of the pyrophosphatase. The net result is that the release of two high-energy phosphate groups is used to drive the GMP synthase reaction to completion.

Synthesis of Pyrimidine Ribonucleotides *de Novo*

Like purine nucleotides, pyrimidine nucleotides can be synthesized either *de novo* or by the "salvage pathways" from nucleobases or nucleosides (see fig. 20.10). However, salvage is less efficient because, except in the case of utilization of uracil, by bacteria, and to some exent by mammalian cells, nucleobases are not converted to nucleotides directly but only via nucleosides.

Figure 20.16

Conversion of IMP to AMP and GMP. In both cases two steps are required. Note that the formation of AMP requires GTP and the formation of GMP requires ATP. This tends to balance the flow of the IMP down the two pathways.

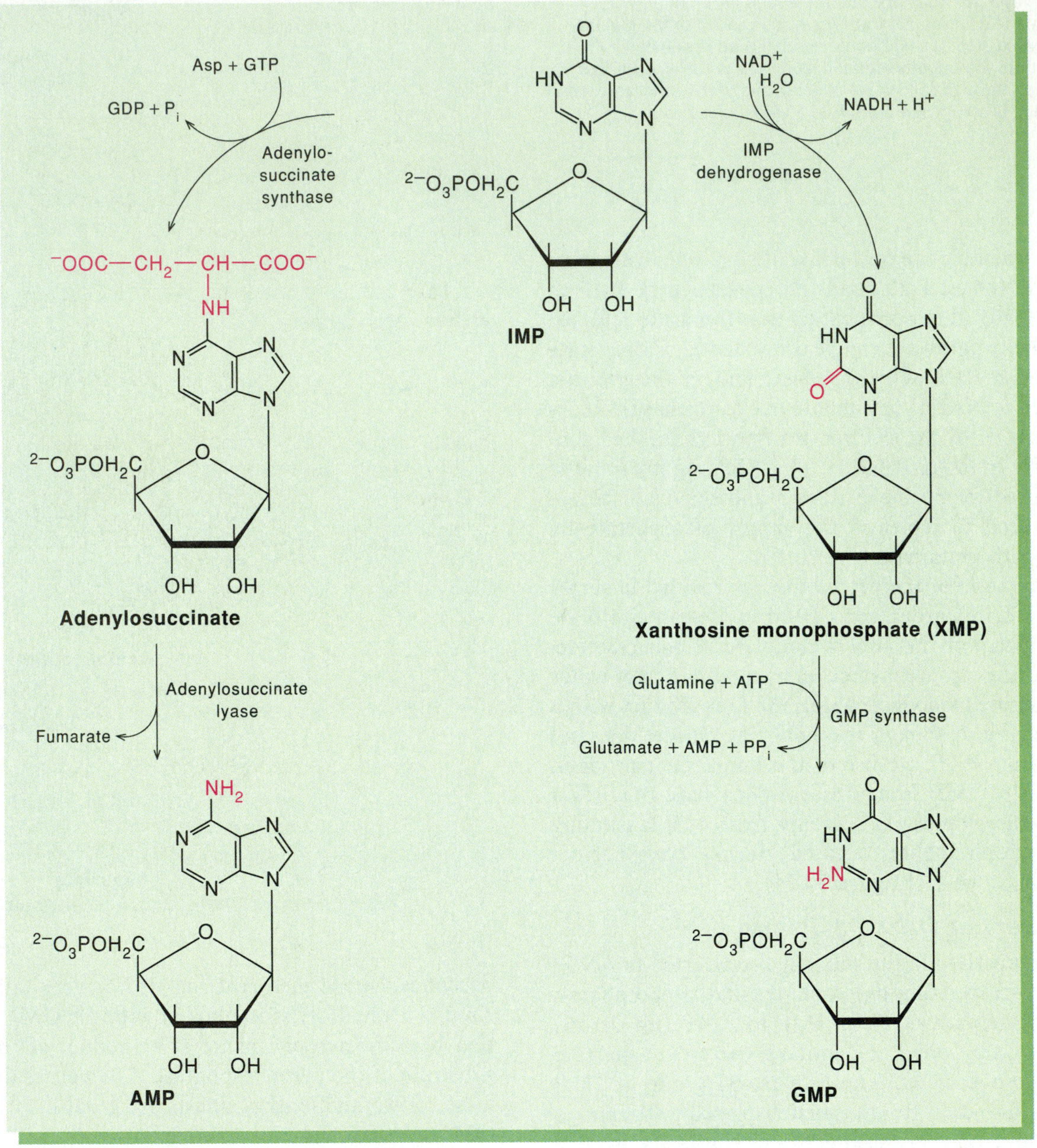

The biosynthetic pathway to pyrimidine nucleotides is simpler than that for purine nucleotides, reflecting the simpler structure of the base. In contrast to the biosynthetic pathway for purine nucleotides, in the pyrimidine pathway the pyrimidine ring is constructed before ribose-5-phosphate is incorporated into the nucleotide. The first pyrimidine mononucleotide to be synthesized is orotidine-5′-monophosphate (OMP), and from this compound, pathways lead to nucleotides of uracil, cytosine, and thymine. OMP thus occupies a central role in pyrimidine nucleotide biosynthesis, somewhat analogous to the position of IMP in purine nucleotide biosynthesis. Like IMP, OMP is found only in low concentrations in cells and is not a constituent of RNA.

Early clues to the nature of the pyrimidine pathway were provided by the observations that orotic acid (6-carboxyuracil, fig. 20.17) can satisfy the growth requirement of mutants of the fungus *Neurospora* that are unable to make pyrimidines, and that isotopically labeled orotate is an immediate precursor of pyrimidines in *Neurospora* and a number of bacteria.

Figure 20.17

The structure of orotic acid. Loss of proton leads to orotate. Deposits of sodium orotate cause a painful condition.

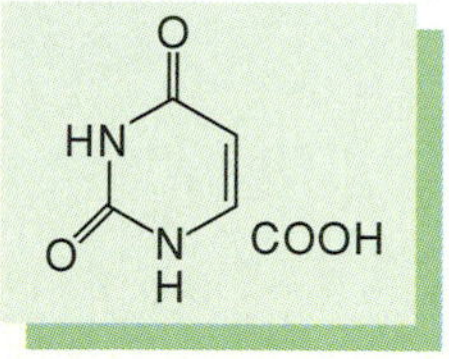

UMP Is a Precursor of Other Pyrimidine Mononucleotides

The pathway for UMP synthesis is shown in figure 20.18. It starts with the synthesis of carbamoyl phosphate, catalyzed by carbamoyl phosphate synthase. This enzyme is present in microorganisms and in the cytosol of all eukaryotic cells that are capable of forming pyrimidine nucleotides. Eukaryotes also have another carbamoyl phosphate synthase, a distinct enzyme that uses ammonia as a substrate instead of glutamine. It is associated with citrulline formation in the pathway for arginine biosynthesis, and in mammals it is a mitochondrial enzyme present predominantly in the liver, where it catalyzes a step in the urea cycle (chapter 19).

Carbamoyl phosphate synthase does not contain biotin and is not activated by it. The enzyme product, carbamoyl phosphate, next reacts with aspartate to form carbamoyl aspartate. The reaction is catalyzed by aspartate carbamoyltransferase (aspartate transcarbamoylase or aspartate transcarbamylase), and the equilibrium greatly favors carbamoyl aspartate synthesis.

In the third step, the pyrimidine ring is closed by dihydroorotase to form L-dihydroorotate. Dihydroorotate is then oxidized to orotate by dihydroorotate dehydrogenase. This flavoprotein in some organisms contains FMN and in others both FMN and FAD. It also contains nonheme iron and sulfur. In eukaryotes it is a lipoprotein associated with the inner membrane of the mitochondria. In the final two steps of the pathway, orotate phosphoribosyltransferase yields orotidine-5′-phosphate (OMP), and a specific decarboxylase then produces UMP.

Low activities of orotidine phosphate decarboxylase and (usually) orotate phosphoribosyltransferase are associated with a genetic disease in children that is characterized by abnormal growth, megaloblastic anemia, and the excretion of large amounts of orotate. When affected children are fed a pyrimidine nucleoside, usually uridine, the anemia decreases and the excretion of orotate diminishes. A likely explanation for the improvement is that the ingested uridine is phosphorylated to UMP, which is then converted to other pyrimidine nucleotides so that nucleic acid and protein synthesis can resume. In addition, the increased intracellular concentrations of pyrimidine nucleotides inhibit carbamoyl phosphate synthase, the first enzyme in the pathway of orotate synthesis.

CTP Is Formed from UTP

After UMP is synthesized, it is phosphorylated to UTP. Then CTP is formed by reaction of UTP with glutamine in a reaction driven by the concomitant hydrolysis of ATP to ADP and inorganic phosphate (fig. 20.19). Both the mammalian and bacterial cytidine triphosphate synthases can use ammonia as a donor in place of glutamine, but this reaction is of no physiological significance because the K_m for ammonia is very high and the reaction rate low. With glutamine as the amino donor, GTP is an allosteric activator for CTP synthase from *E. coli* and probably for the mammalian enzyme as well (fig. 20.20). However, there is no stimulation of CTP synthesis from ammonia, a result interpreted to mean that the allosteric effect of GTP is specifically on the release of ammonia from glutamine. This ammonia is then channeled into reaction with UTP in the enzyme active site.

Biosynthesis of Deoxyribonucleotides

Tracer studies with isotopically labeled precursors have shown that both in mammalian tissues and in microorganisms, deoxyribonucleotides are formed from corresponding ribonucleotides by replacement of the 2′—OH group with hydrogen.

There are three types of ribonucleotide reductase that catalyze this reduction of the ribose ring. The first is widely distributed in nature, occurring in mammalian and plant cells, in yeast, and in some prokaryotes. This type of reductase contains a tyrosyl radical closely associated with nonheme iron, and the prototype is the reductase from *E. coli*. The latter consists of two nonidentical subunits, both contributing to the active site, and is specific for the reduction of diphosphates (ADP, GDP, CDP, and UDP).

The second type of ribonucleotide reductase is restricted to certain microorganisms, including some bacteria, several species of algae, and at least one fungus. Ribonucleotide reductase of this type, for which the enzyme from *Lactobacillus leichmannii* is the prototype, requires adenosyl cobalamin (coenzyme B_{12}) as an obligatory coenzyme. It does not contain nonheme iron; it uses either nucleoside diphosphates or triphosphates, depending on the source of the enzyme; and it consists of only one type of polypeptide chain (although this may form oligomers in some cases). A third type of reductase present in some prokaryotes contains manganese.

An unusual feature of ribonucleotide reductase is that the reaction it catalyzes involves a radical mechanism. The *E. coli* type of reductase initiates this reaction by the tyrosyl radical-nonheme iron, whereas the *L. leichmannii* reductase initiates radical formation by homolytic cleavage of the C–Co bond of the adenosylcobalamin coenzyme (coenzyme B_{12}). Hydroxyurea and related compounds inhibit the mammalian reductase by abolishing the radical state of the tyrosine residue. Inhibition of DNA synthesis by such compounds is secondary to this effect.

Figure 20.18

Biosynthesis of UMP. Color indicates the parts of the intermediates derived from aspartate. Bold type indicates atoms derived from carbamoyl phosphate. In contrast to purine nucleotide synthesis, where ring formation starts on the sugar, in pyrimidine biosynthesis the pyrimidine ring is completed before being attached to the ribose.

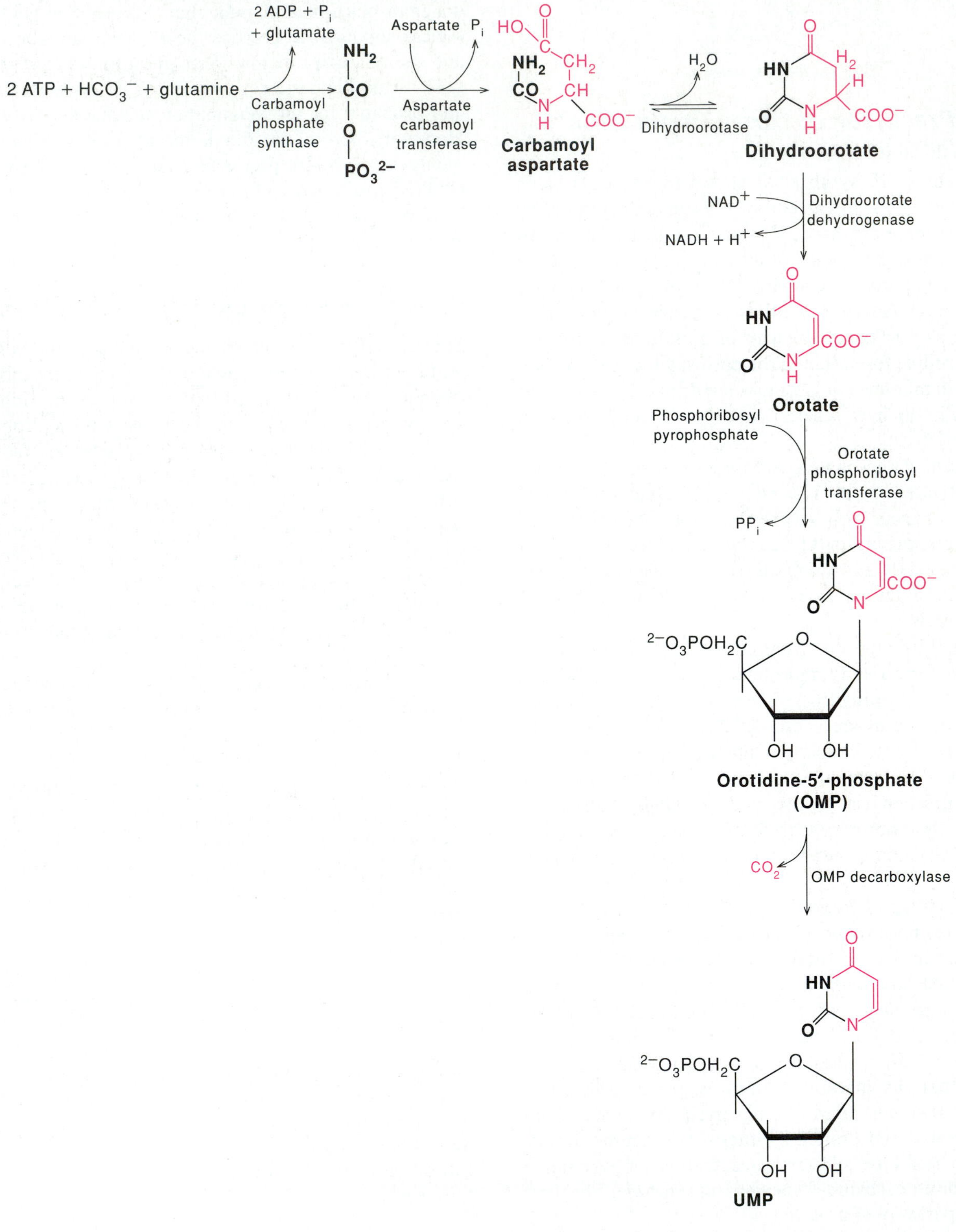

Figure 20.19

The production of CTP by amination of UTP. The conversion of the pyrimidine ring of uracil to cytosine occurs at the level of the nucleotide triphosphate. The enzyme responsible for this conversion is known as cytidine triphosphate synthase.

UTP + ATP + Gln → CTP + ADP + P_i + Glu

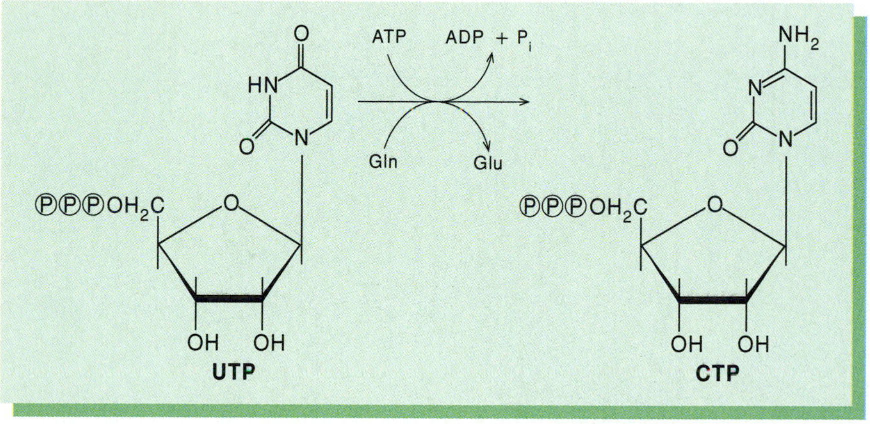

Figure 20.20

Cytidine triphosphate synthase is activated by GTP.

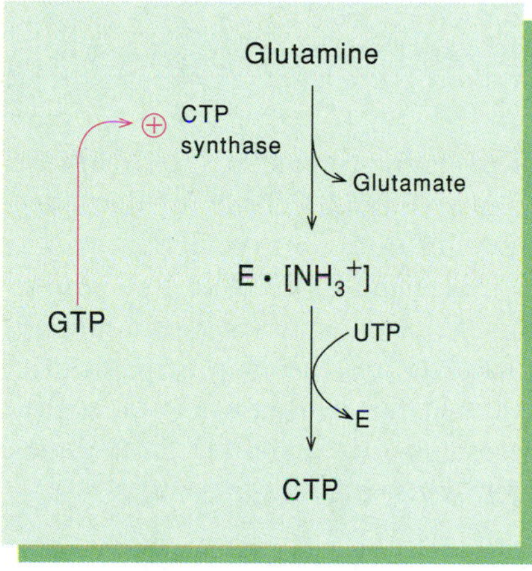

For both major types of reductase, the physiological reducing substrate is a low-molecular-weight (13,000) electron-transport protein, thioredoxin. Thioredoxin has two half-cystine residues that are separated in the polypeptide chain by two other residues. The oxidized form of thioredoxin, with a disulfide bridge between the half-cystines, is reduced by NADPH in the presence of a flavoprotein, thioredoxin reductase. The reduced form of thioredoxin, with two cysteine residues present, is the reducing substrate for ribonucleotide reduction. The flow of electrons from NADPH to ribose is shown in figure 20.21.

An alternative electron-transport system for ribonucleotide reduction has been discovered in *E. coli.* In this case, the ultimate source of electrons is again NADPH, but they are passed to glutathione (in a reaction catalyzed by glutathione reductase), and the reduced glutathione, in turn, reduces a small protein called glutaredoxin. It is the reduced glutaredoxin that acts as the reducing substrate in ribonucleotide reduction. The distribution of glutaredoxin in nature remains to be determined. Neither thioredoxin nor glutaredoxin is essential to *E. coli,* but the bacteria cannot survive when both proteins are lost.

Thymidylate Is Formed from dUMP

Since DNA contains thymine (5-methyluracil) as a major base instead of uracil, the synthesis of thymidine monophosphate (dTMP or thymidylate) is essential to provide dTTP (thymidine triphosphate), needed for DNA replication together with dATP, dGTP, and dCTP.

Thymidylate is synthesized from dUMP, and there are two pathways by which the latter may be formed in cells. The major precursor of dUMP is dCMP, which is converted to dUMP by deoxycytidylate deaminase:

$$\text{dCMP} + H_2O \rightarrow \text{dUMP} + NH_3$$

This enzyme is widely distributed in animal tissues, and the enzyme produced by yeast, by bacteriophage T2-infected *E. coli,* and by animal cells has been studied extensively. Deamination of 5-methyldeoxycytidylate and 5-hydroxymethyldeoxycytidylate is also catalyzed by this enzyme.

Another route to dUMP is the reduction of UDP to dUDP, followed by phosphorylation of dUDP to dUTP (or direct reduction of UTP to dUTP in some microorganisms). The dUTP is then hydrolyzed to dUMP. This circuitous route to dUMP is dictated by two considerations. First, the ribonucleotide reductase in most cells acts only on ribonucleoside diphosphates, probably because this permits better regulation of its activity. Second, cells contain a powerful deoxyuridine triphosphate diphosphohydrolase (dUTPase). It prevents the incorporation of dUTP into DNA by keeping intracellular levels of dUTP low by means of the reaction

$$\text{dUTP} + H_2O \rightarrow \text{dUMP} + PP_i$$

Figure 20.21

Ribonucleotide reductase and the thioredoxin system. In *Lactobacillus leichmannii*, vitamin B_{12} is involved in the reduction of ribonucleotide diphosphate to deoxyribonucleotide.

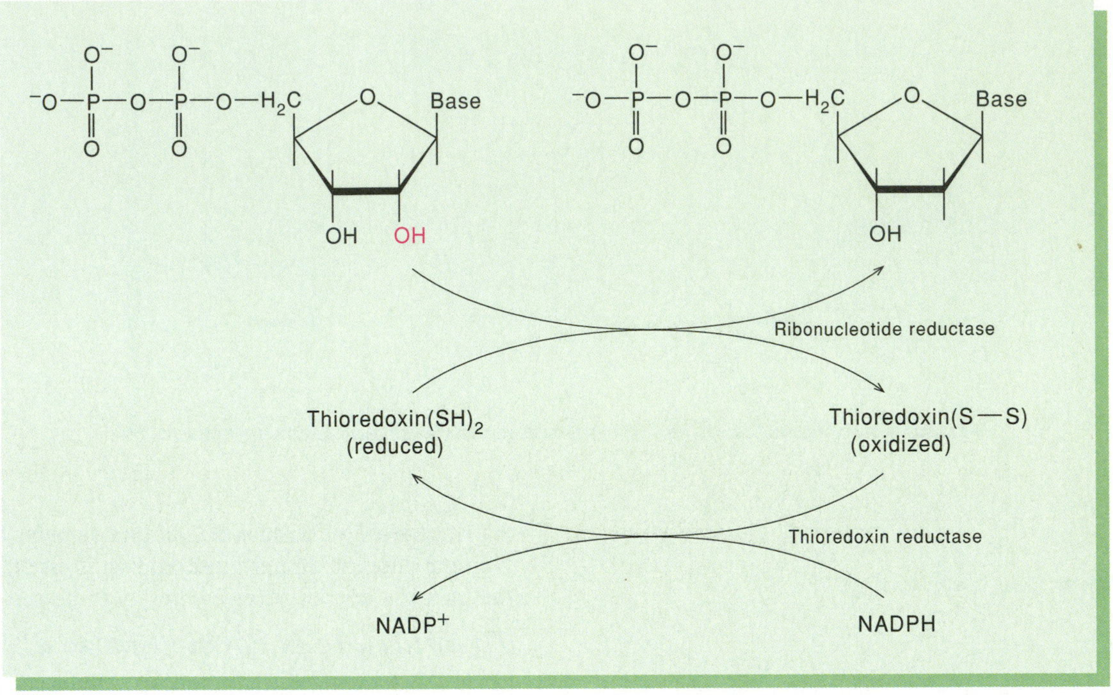

Methylation of dUMP to give thymidylate is catalyzed by thymidylate synthase and utilizes 5,10-methylenetetrahydrofolate as the source of the methyl group. This reaction is unique in the metabolism of folate derivatives in that the folate derivative acts both as a donor of the one-carbon group and also as its reductant, using the reduced pteridine ring as the source of reducing potential. Consequently, in this reaction, unlike any other in folate metabolism, dihydrofolate is a product (fig. 20.22). Since folate derivatives are present in cells at very low concentrations, continued synthesis of thymidylate requires regeneration of 5,10-methylenetetrahydrofolate from dihydrofolate. As shown in figure 20.22, this occurs in two steps catalyzed by the enzymes dihydrofolate reductase, an $NADP^+$-linked dehydrogenase, and serine hydroxymethyltransferase. Thymidylate synthesis can be interrupted, and consequently the synthesis of DNA arrested, by the inhibition of either thymidylate synthase or dihydrofolate reductase. Many potent inhibitors are known for each of these enzymes, the best known being 5-fluoro-dUMP for thymidylate synthase and methotrexate, trimethoprim, and related compounds for dihydrofolate reductase. We will discuss these and other inhibitors of nucleotide synthesis in a later section.

You should note that reduced folates are present in cells as polyglutamate forms, with up to five additional glutamate residues attached to the terminal carboxyl of the folates. The glutamate residues are attached to each other in γ-peptide linkage. These polyglutamate forms of methylenetetrahydrofolate are the true substrates for thymidylate synthase; they have much lower K_m values and give higher maximum velocity than the monoglutamate. Analogous polyglutamate forms of 10-formyltetrahydrofolate are the true cofactors for purine synthesis. Human thymidylate synthase, as well as the enzyme from bacteria, has been crystallized and the three-dimensional structure deduced from x-ray diffraction results.

Formation of Nucleotides from Bases and Nucleosides (Salvage Pathways)

In addition to the pathways for synthesis *de novo*, mammalian cells and microorganisms can readily form mononucleotides from purine and pyrimidine bases and their nucleosides. In this way bases and nucleosides formed by constant breakdown of mRNA and other nucleic acids can be reconverted to useful nucleotides, and the energy expended by the cell in synthesizing the bases is retained.

Transporter systems are present in the membranes of mammalian cells and of some microorganisms for the efficient uptake of bases and nucleosides. The carriers for bases are quite specific and mediate facilitated diffusion (chapter 32). Two types

Figure 20.22

Thymidylate biosynthesis. (R = *p*-aminobenzoyl-L-glutamate.) Colored symbols indicate atoms that are precursors of the methyl group of thymidylate. The dTMP is formed from dUMP. 5,10-methylenetetrahydrofolate donates the methyl group in this reaction. This methyl group donor is regenerated by a two-step process involving first the reduction of 7, 8-dihydrofolate to tetrahydrofolate and thence to 5,10-methylenetetrahydrofolate, in which serine supplies the methylene group.

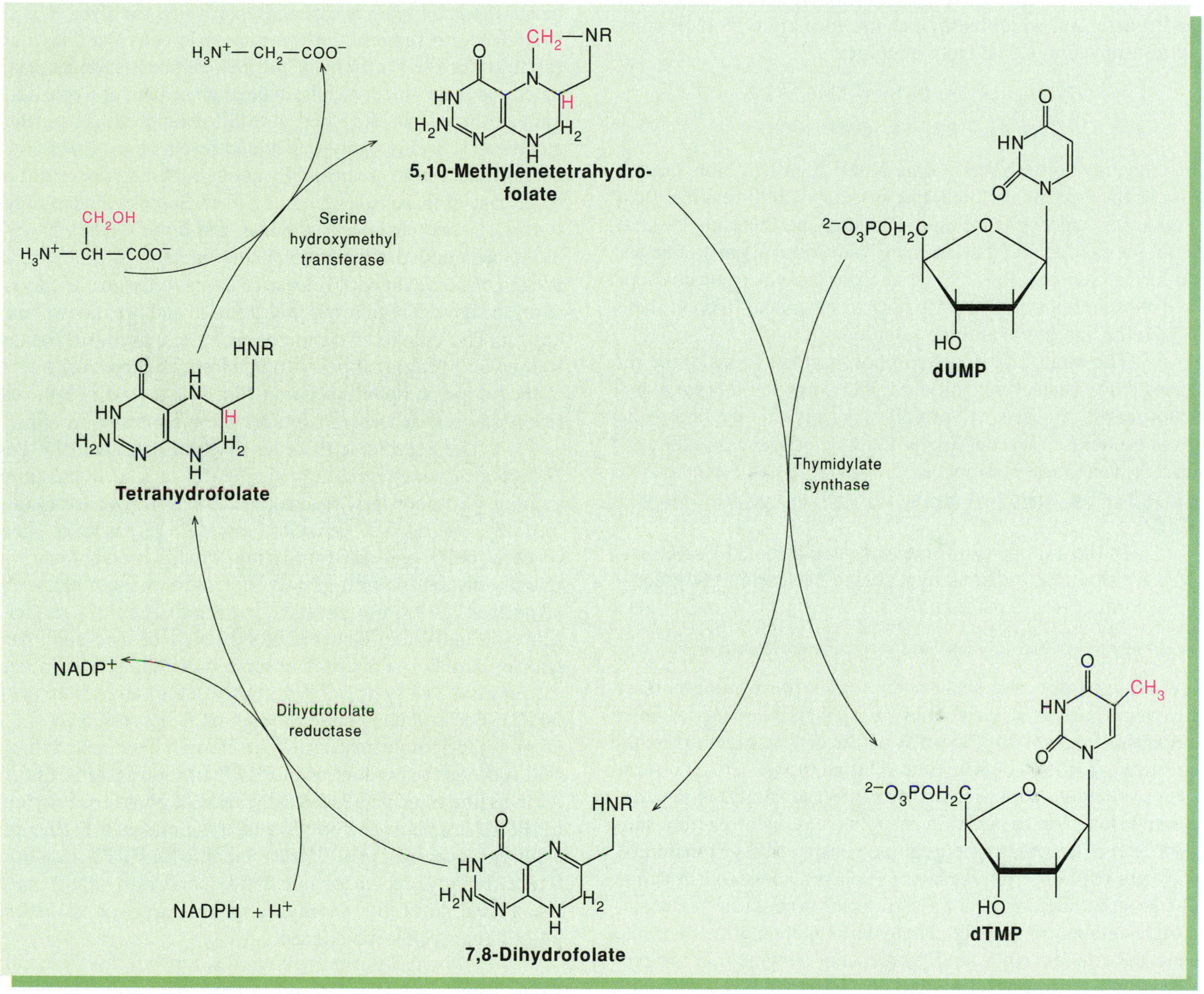

of carrier for facilitated diffusion of nucleosides are widely distributed in membranes of mammalian cells, and concentrative uptake of nucleosides occurs in cells of the intestinal epithelium and of the kidney tubule. Once inside the cell, the bases and nucleosides are converted to nucleotides by enzymes that are widely distributed in mammalian tissues and in microorganisms. The same transport systems and salvage enzymes are responsible for converting base or nucleoside analogs to therapeutically active nucleotide forms.

An ecto-5′-nucleotidase, which is attached to the outer surface of the plasma membrane of many types of mammalian cell, dephosphorylates purine and pyrimidine ribo- and deoxyribonucleoside monophosphates to the corresponding nucleosides. An important function of the ecto-nucleotidase may be the assimilation of nucleotides arising from the dissolution of dying cells. The nucleosides formed by the ecto-5′-nucleotidase are then transported into the cell and are reconverted to nucleotides by nucleoside kinases, as described in the following section.

Purine Phosphoribosyltransferases Convert Purines to Nucleotides

In mammals specific enzymes for converting purine bases to nucleotides are present in many organs, and in heart muscle this may be the main source of purine nucleotides. The most important of these enzymes is hypoxanthine-guanine phosphoribosyltransferase, which catalyzes the formation of IMP from hypoxanthine and GMP from guanine:

$$\text{Hypoxanthine} + \text{PRPP} \rightleftharpoons \text{IMP} + \text{PP}_i$$

$$\text{Guanine} + \text{PRPP} \rightleftharpoons \text{GMP} + \text{PP}_i$$

The enzyme may also form some XMP from xanthine, but affinity of the enzyme for the latter is nearly 1,000-fold less than for guanine and hypoxanthine. This enzyme therefore returns the major products of purine nucleotide catabolism to nucleotide forms (see fig. 20.9). Some microorganisms contain a separate phosphoribosyltransferase for xanthine in addition to those for adenine and hypoxanthine-guanine.

The equilibria in these phosphoribosyltransferase reactions favor nucleotide synthesis, and since the inorganic pyrophosphate released is rapidly hydrolyzed by inorganic pyrophosphatase, the coupling of these reactions makes the synthesis of nucleotide irreversible. However, the efficiency of salvage is heavily dependent on the intracellular concentration of PRPP.

In theory, hypoxanthine and guanine could also be salvaged by the widely distributed purine nucleoside phosphorylase according to the reaction:

$$\text{Purine} + \text{ribose-1-phosphate} \rightleftharpoons \text{purine nucleoside} + \text{P}_i$$

However, this does not seem to be a quantitatively important salvage pathway and the role of phosphorylase seems to be in catabolism, presumably because of the low intracellular concentration of ribose-1-phosphate relative to P_i.

Adenine phosphoribosyltransferase catalyzes the conversion of adenine to AMP in many tissues, by a reaction similar to that of hypoxanthine-guanine phosphoribosyltransferase, but is quite distinct from the latter. It plays a minor role in purine salvage since adenine is not a significant product of purine nucleotide catabolism (see fig. 20.9). The function of this enzyme seems to be to scavenge small amounts of adenine that are produced during intestinal digestion of nucleic acids or in the metabolism of 5′-deoxy-5′-methylthioadenosine, a product of polyamine synthesis.

The role of hypoxanthine-guanine phosphoribosyltransferase in purine salvage has been confirmed by the abnormally high excretion of purines (as uric acid) in humans who lack hypoxanthine-guanine phosphoribosyltransferase. Studies of purine metabolism in cultures of cells from patients with this hereditary disorder also support this conclusion. Another study focused on patients who lack the enzyme xanthine oxidase (to be discussed later). The amounts of xanthine and hypoxanthine excreted by these patients were compared with estimates of the amounts synthesized daily. The results indicated that about 90% of the purines formed were reutilized. The high recovery rate is presumably possible because of the restricted distribution, in tissues of vertebrates, of catabolic enzymes acting on free purines.

Besides this salvage role, however, hypoxanthine-guanine phosphoribosyltransferase is probably important also for the transfer of purines from liver to other tissues. Purine biosynthesis *de novo* is especially active in the liver, and there is evidence to suggest that extrahepatic cells that have a low capacity for the synthesis of purines *de novo,* such as erythrocytes and bone marrow cells, depend on uptake of hypoxanthine and xanthine from the blood to fulfill their needs for purine nucleotides. It seems likely that blood levels of xanthine and hypoxanthine, which are normally about 0.04 mM, are maintained by release of these bases from the liver. Some evidence suggests that the bases released by the liver are largely taken up by red blood cells and converted to purine nucleotides that are later broken down again with release of bases to tissues. If this is the case, the factors regulating their release and breakdown are unknown. The uptake of purine bases by extrahepatic tissues, as well as by bacteria, appears to be closely linked to the activity of the purine phosphoribosyltransferases. At least in some cases, this is so because the transferases are membrane proteins.

The neurologic disorder of children called the Lesch-Nyhan syndrome is due to a congenital lack of hypoxanthine-guanine phosphoribosyltransferase. The disorder is characterized by aggressive behavior, mental retardation, spastic cerebral palsy, and self-mutilation. Purine metabolism is profoundly disturbed, with greatly increased *de novo* biosynthesis of purines (200 times normal), overproduction of uric acid (6 times normal), and elevated blood levels of uric acid. Severe gout is caused by the latter in some individuals. The increased rate of purine biosynthesis is probably due to several factors (fig. 20.23). Because nucleotide production is depressed by the lack of phosphoribosyltransferase, the normal feedback inhibition that controls the production of PRPP amidotransferase is lifted. At the same time, decreased utilization of phosphoribosylamine by PRPP by phosphoribosyltransferase makes a higher intracellular concentration of PRPP available for PRPP amidotransferase. Finally, decreased nucleotide pools also would increase the PRPP level by deregulating the activity of ribose-5-phosphate pyrophosphokinase.

Although purine nucleosides are intermediates in the catabolism of nucleotides and nucleic acids in higher animals and humans, these nucleosides do not accumulate and are normally present in blood and tissues only in trace amounts. Nevertheless, cells of many vertebrate tissues contain kinases capable of converting purine nucleosides to nucleotides. Typical of these is adenosine kinase, which catalyzes the reaction

$$\text{Adenosine} + \text{ATP} \rightarrow \text{AMP} + \text{ADP}$$

2′-Deoxyadenosine is also a substrate, though a relatively poor one. In some mammalian cells deoxycytidine kinase also phosphorylates deoxyadenosine and deoxyguanosine (see next section). In addition, indirect evidence suggests the existence of a specific deoxyadenosine kinase in human cells. Deoxyadenosine has received special attention as a substrate for kinases because

Figure 20.23

Mechanism of overproduction of purine nucleotides in the congenital deficiency of hypoxanthine-guanine phosphoribosyltransferase. The loss of the transferase prevents the recycling of hypoxanthine and guanine. This increases uric acid production as well as the *de novo* synthesis of purine nucleotides.

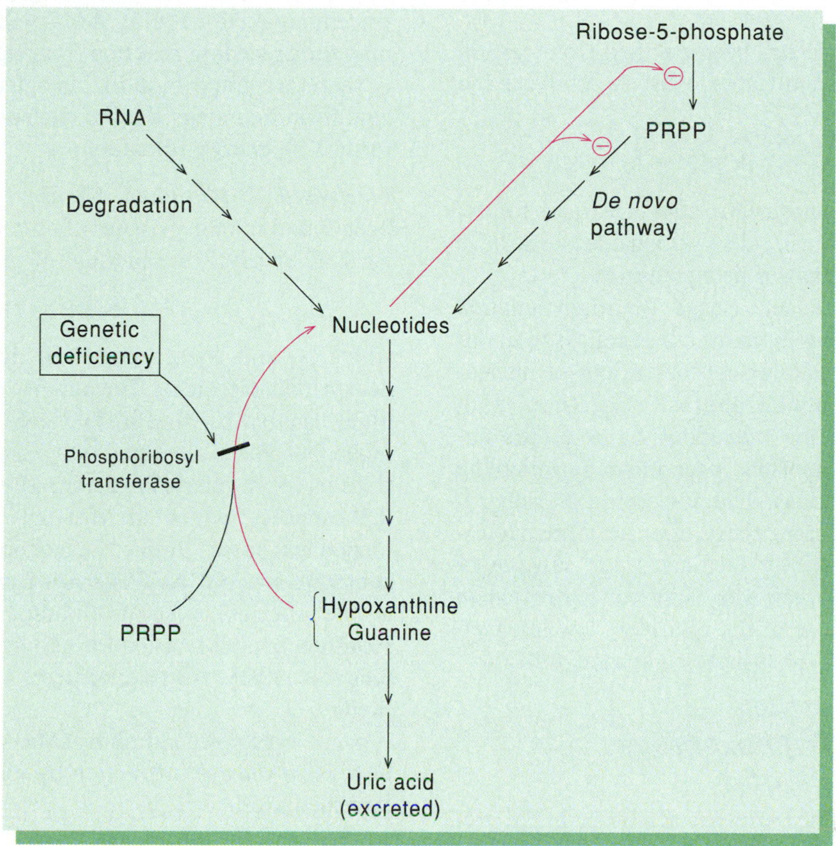

congenital inability to catabolize this nucleoside results in immune deficiency disease. A mitochondrial kinase specific for deoxyguanosine has been reported. None of these kinases has been studied extensively, and their role in various tissues and organs is uncertain.

Salvage of Pyrimidines Is Less Important for Mammals and Goes through Mononucleotides

In contrast to the extensive salvage and reutilization of purine bases, pyrimidine bases are poorly utilized by mammalian tissues. Many bacteria take up pyrimidine bases efficiently for nucleotide synthesis, however, and a phosphoribosyltransferase for uracil has been identified in a few species of bacteria.

An alternative route for conversion of uracil to UMP, both in bacteria and in higher animals, is by successive reactions catalyzed by uridine phosphorylase and uridine kinase, respectively:

$$\text{Uracil} + \text{ribose-1-phosphate} \rightleftharpoons \text{uridine} + P_i$$

$$\text{Uridine} + \text{ATP} \xrightarrow{Mg^{2+}} \text{UMP} + \text{ADP}$$

Mammalian uridine phosphorylase will also accept deoxyuridine as a substrate, and thymidine is slowly attacked, but other nucleosides are not substrates. Uridine kinase activity has been demonstrated in a variety of bacteria and animal cells, including tumors, and is especially high in cells of high growth rate. Cytidine is the only other physiological nucleoside to act as a substrate for bacterial uridine kinase, but a variety of nucleoside triphosphates can act as phosphoryl donors.

Thymidine is similarly converted to dTMP according to the following reactions:

$$\text{Thymine} + \text{deoxyribose-1-phosphate} \rightleftharpoons \text{thymidine} + P_i$$

$$\text{Thymidine} + \text{ATP} \rightarrow \text{dTMP} + \text{ADP}$$

Thymine phosphorylase has been purified from bacteria and mammalian tissues. Thymidine is the preferred substrate, but deoxyuridine and 5-substituted deoxyuridines are also cleaved. Thymidine kinase is widely distributed and the cytosolic enzyme from mammalian tissues is the most extensively studied of the deoxyribonucleoside kinases. It also accepts deoxyuridine as substrate, is inhibited by dTTP, and has complex kinetics. The activity of thymidine kinase in cells increases dramatically during rapid growth and DNA synthesis. Infection of cells with any of several viruses also induces thymidine kinase activity, together with certain other enzymes concerned with deoxyribonucleotide synthesis. The viral thymidine kinase is quite different from the host enzyme. Significantly, the viral enzymes are subject to none of the allosteric controls of the host enzymes.

A deoxycytidine kinase has been purified from certain bacteria, from thymus, and from tumor cells. It catalyzes the reaction

$$\text{Deoxycytidine} + \text{ATP} \rightleftharpoons \text{dCMP} + \text{ADP}$$

Deoxyadenosine and deoxyguanosine are also substrates for deoxycytidine kinase, but the purine deoxyribonucleosides have much higher K_m values than that of deoxycytidine.

As in the case of adenosine kinase, the main function of pyrimidine nucleoside kinases in mammalian cells is to maintain a balance between intracellular concentrations of nucleosides and nucleoside monophosphates. Together with nucleotidases that hydrolyze the monophosphates to the nucleosides, they constitute an important mechanism by which the cell regulates levels of nucleotides. The nucleoside phosphorylase–nucleoside kinase route for synthesis of pyrimidine nucleoside monophosphates is relatively inefficient for salvage of pyrimidine bases because of their very low concentration in plasma and tissues, and because of the relatively low intracellular concentration of ribose-1-phosphate compared with P_i.

Conversion of Nucleoside Monophosphates to Triphosphates Goes through Diphosphates

The products of the biosynthetic pathways discussed in the preceding sections are, in most cases, mononucleotides. In cells, a series of kinases (phosphotransferases) converts these mononucleotides to their metabolically active diphosphate and triphosphate forms.

Nucleoside Monophosphate Kinases Bacteria and other microorganisms, as well as animal cells, contain a variety of kinases that catalyze reactions of the general type

$$\text{(d)NMP} + \text{ATP} \rightleftharpoons \text{(d)NDP} + \text{ADP}$$

Four types of nucleoside monophosphate kinases are known. These catalyze, respectively, the phosphorylation of (1) GMP and dGMP, (2) AMP and dAMP, (3) dCMP, CMP, and UMP, and (4) dTMP.

The second of the kinases mentioned, which uses AMP as substrate, is referred to as adenylate kinase. Its activity is high in tissues where the turnover of energy from adenine nucleotides is great, e.g., in liver and muscle, and its activity is also high in mitochondria. In these tissues its function is to make more energy available as ATP bond energy. When ADP is formed from ATP in energy-consuming reactions, more ATP can be formed from ADP according to the reaction

$$2\ \text{ADP} \rightleftharpoons \text{AMP} + \text{ATP}$$

Under conditions where energy-generating reactions convert intracellular ADP to ATP, AMP will be phosphorylated by running the preceding reaction from right to left. Adenylate kinase is therefore important in biological systems for maintaining equilibrium among adenine nucleotides as they are depleted or formed by energy transfers.

Nucleoside Diphosphate Kinases Enzymes of this type have been found in many tissues of animals, plants, and microorganisms. They catalyze reactions of the following general type:

$$\text{N}_1\text{TP} + \text{N}_2\text{DP} \rightleftharpoons \text{N}_1\text{DP} + \text{N}_2\text{TP}$$

where N_1 and N_2 are purine or pyrimidine ribonucleosides or deoxyribonucleosides. The activity of NDP kinases is relatively high, usually 10- to 100-fold greater than the activity of the monophosphate kinases. As a result, intracellular concentrations of triphosphates are normally much higher than those of diphosphates, which in turn are often higher than those of monophosphates. Unlike the monophosphate kinases, which are substrate-specific, NDP kinases from all sources are active with a wide range of nucleoside diphosphates and triphosphates. They require a divalent cation for activity, and although many metal ions can satisfy this requirement, Mg^{2+} is the physiological cofactor.

Nucleoside diphosphate kinases from many sources have been shown to function by forming a phosphoryl-enzyme intermediate:

$$\text{E} + \text{N}_1\text{TP} \rightleftharpoons \text{E} \sim \text{P} + \text{N}_1\text{DP}$$

$$\text{E} \sim \text{P} + \text{N}_2\text{DP} \rightleftharpoons \text{E} + \text{N}_2\text{TP}$$

In several cases it has been shown that the phosphoryl group is attached to N-1 of a histidine side chain.

In addition to nucleoside diphosphokinases, a number of other cellular enzymes are capable of converting nucleoside diphosphates to triphosphates. Thus, within many cells, GDP and dGDP are converted to their triphosphates by phosphoglycerate kinase and pyruvate kinase at rates comparable to that caused by nucleoside diphosphate kinase.

Inhibitors of Nucleotide Synthesis

There are several distinct types of inhibitors of nucleotide biosynthesis, each type acting at different points in the pathways to purine or pyrimidine nucleotides. All these inhibitors are very toxic to cells, especially rapidly growing cells such as those of tumors or bacteria, because interruption of the supply of nucleotides seriously limits the cell's capacity to synthesize the nucleic acids necessary for protein synthesis and cell replication.

In some cases, the toxic effect of such inhibitors makes them useful in cancer chemotherapy or in the treatment of bacterial infections. However, these agents can also damage the replicating cells of the intestinal tract and bone marrow. This danger imposes limits on the doses that can be used safely.

6-Mercaptopurine (table 20.3) and related thiopurines are potent inhibitors of purine nucleotide biosynthesis, but they are inactive until they are converted to the corresponding ribonucleoside 5′-phosphates. 6-Mercaptopurine is converted by the action of hypoxanthine-guanine phosphoribosyltransferase to the nucleotide 6-thioinosine-5′-monophosphate (T-IMP) (fig. 20.24). The latter inhibits several enzymes of purine biosynthesis, and it is uncertain which effect is primarily responsible for the toxicity. It blocks conversion of IMP to adenylosuccinate and to XMP, key reactions in the formation of AMP and GMP (see fig. 20.16). T-IMP is also capable of "pseudofeedback" inhibition of glutamine PRPP amidotransferase (see fig. 20.14), the first committed step in the purine nucleotide pathway. This enzyme is highly responsive to intracellular concentrations of both normal ribonucleoside 5′-monophosphates and analogs. 6-Mercaptopurine is used clinically in the treatment of leukemia.

Tiazofurin is a thiazole-C-nucleoside with anticancer activity. It is metabolized to tiazofurin-5′-monophosphate and then to thiazole-4-carboxamide adenine dinucleotide (TAD), an analog of NAD. TAD specifically inhibits IMP dehydrogenase (see fig. 20.16), with consequent depletion of the pools of GMP, GDP, and GTP. The corresponding analog with selenium in place of sulfur is also under trial as an anticancer agent.

Two inhibitors are shown in table 20.3 that specifically interfere with steps in pyrimidine nucleotide biosynthesis. *N*-(Phosphonacetyl)-L-aspartate (PALA) is a powerful inhibitor of the carbamoyl transferase reaction (see fig. 20.18). PALA was synthesized to act as an analog of the transition state intermediate (fig. 20.25) postulated to be formed in the aspartate carbamoyltransferase reaction, and its tight binding to the enzyme is probably due to its resemblance to the intermediate.

Pyrazofurin is a fermentation product. It is converted by kinase action to the 5′-phosphate (see fig. 20.24), which appears to mimic the substrate of orotidylate decarboxylase, for which it is a powerful inhibitor. All three compounds are under trial as anticancer agents.

5-Fluorouracil, an agent used in treating solid tumors, interferes with thymidylate synthesis (see fig. 20.22). Its major inhibitory effect occurs after its conversion in the cell to 5-fluoro-2′-deoxyuridine-5′-monophosphate (see fig. 20.24). The latter acts as an analog of dUMP and binds very tightly to thymidylate synthase (see fig. 20.22) in the presence of methylenetetrahydrofolate, forming a covalent complex with the enzyme that is unable to undergo the normal catalytic reaction. 5-Fluorouracil may also exert a cytotoxic effect through incorporation of 5-fluorouridine phosphate into RNA.

Among inhibitors that interfere with the synthesis of both pyrimidine and purine nucleotides are hydroxyurea and α-(*N*)-heterocyclic carboxaldehyde thiosemicarbazone (see table 20.3). These compounds interfere with the synthesis of deoxyribonucleotides by inhibiting ribonucleotide reductase of mammalian cells, an enzyme that is crucial and probably rate-limiting in the biosynthesis of DNA. They probably act by disrupting the iron-tyrosyl radical structure at the active site of the reductase, and hydroxyurea is in clinical use as an anticancer agent.

Reactions involving glutamine as a substrate are inhibited by the glutamine analogs azaserine and 6-diaza-5-oxo-1-aminohexanoic acid (see table 20.3). Inhibition is irreversible, with formation of a covalent bond between the inhibitor and an amino acid side chain at the catalytic site. The specific reactions inhibited are those catalyzed by glutamine PRPP amidotransferase and phosphoribosyl-*N*-formylglycinamidine synthase in the *de novo* purine pathway and carbamoyl phosphate synthase and CTP synthase in the pyrimidine pathway.

Acivicin is an analog with a more distant resemblance to glutamine. It is, nevertheless, a potent inhibitor of several steps in purine nucleotide biosynthesis that utilize glutamine. The enzymes it inhibits are glutamine PRPP amidotransferase (step 1, fig. 20.14), phosphoribosyl-*N*-formylglycinamidine synthase (step 4, fig. 20.14), and GMP synthase (see fig. 20.16). In pyrimidine nucleotide biosynthesis the enzymes inhibited are carbamoyl synthase (step 1, fig. 20.8) and CTP synthase (see fig. 20.9). Acivicin is under trial for the treatment of some forms of cancer.

Another group of inhibitors prevents nucleotide biosynthesis indirectly by depleting the level of intracellular tetrahydrofolate derivatives. Sulfonamides are structural analogs of *p*-aminobenzoic acid (fig. 20.26) and they competitively inhibit the bacterial biosynthesis of folic acid at a step in which *p*-aminobenzoic acid is incorporated into folic acid. Sulfonamides are widely used in medicine because they inhibit growth of many bacteria. When cultures of susceptible bacteria are treated with sulfonamides, they accumulate 4-carboxamide-5-aminoimidazole in the medium, because of a lack of 10-formyltetrahydrofolate for the penultimate step in the pathway to IMP (see fig. 20.14). Methotrexate, trimethoprim, and a number of related compounds inhibit the reduction of dihydrofolate to tetrahydrofolate, a reaction catalyzed by dihydrofolate reductase. These inhibitors are structural analogs of folic acid (see fig. 20.26) and bind at the catalytic site of dihydrofolate reductase, an enzyme catalyzing one of the steps in the cycle of reactions involved in thymidylate synthesis (see fig. 20.22). These inhibitors therefore prevent synthesis of thymidylate in replicating cells, and as a secondary effect synthesis of purine nucleotides is also decreased because of 10-formyltetrahydrofolate depletion. Methotrexate is used as an anticancer drug, and trimethoprim, which specifically inhibits bacterial dihydrofolate reductase, is used for treating certain bacterial infections.

Table 20.3
Some Inhibitors of Nucleotide Metabolism Together with Enzymes Inhibited

Inhibitor	Enzymes Inhibited	Inhibitor	Enzymes Inhibited
Inhibitors of purine nucleotide biosynthesis		Inhibitors of both purine and pyrimidine nucleotide biosynthesis	
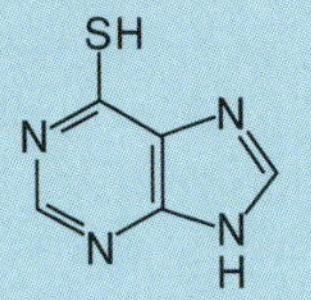 **6-Mercaptopurine**	Figure 20.16, adenylosuccinate synthase, IMP dehydrogenase; Figure 20.14, reaction 1	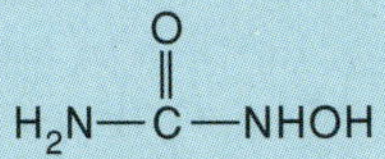 $H_2N—C(=O)—NHOH$ **Hydroxyurea**	Figure 20.21, ribonucleotid reductase
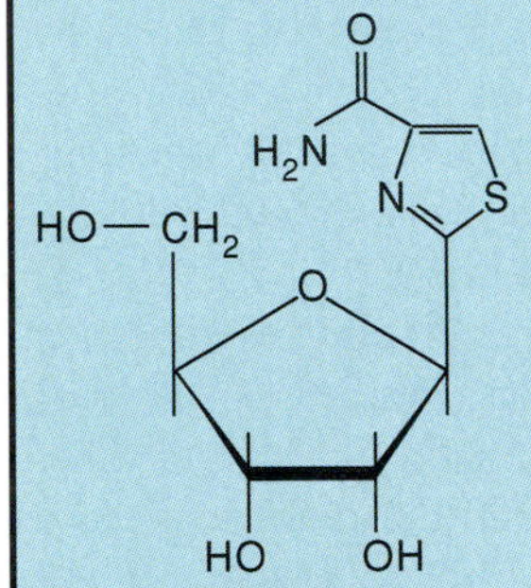 **Tiazofurin**	Figure 20.16, IMP dehydrogenase	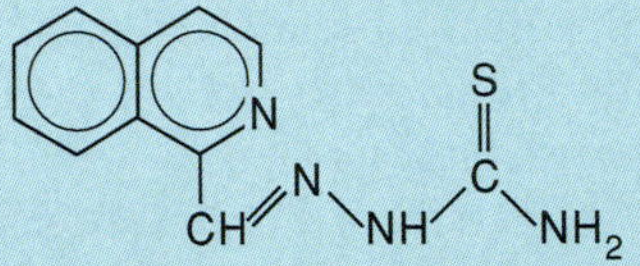 **α-(*N*)-Heterocyclic carboxaldehyde thiosemicarbazone**	
Inhibitors of pyrimidine nucleotide biosynthesis			
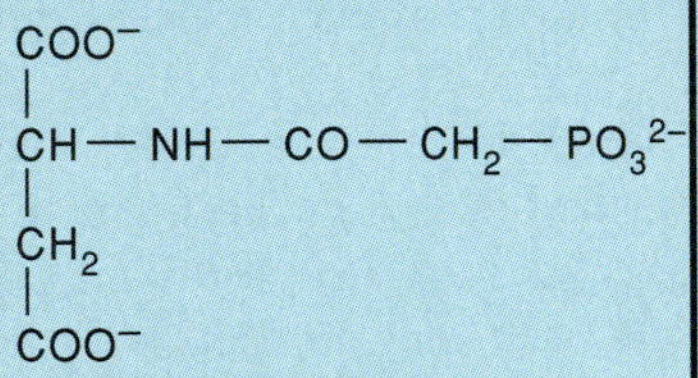 $^{-}OOC—CH(CH_2—COO^{-})—NH—CO—CH_2—PO_3^{2-}$ ***N*-(Phosphonacetyl)-L-aspartate (PALA)**	Figure 20.18, aspartate carbamoyltransferase	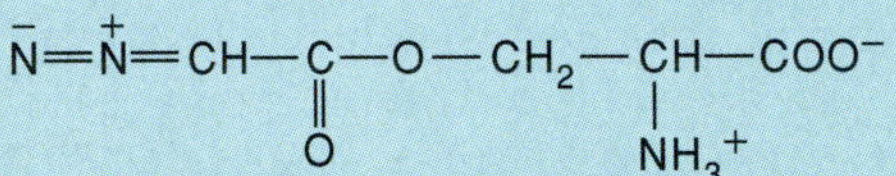 $\bar{N}=\overset{+}{N}=CH—C(=O)—O—CH_2—CH(NH_3^+)—COO^-$ **Azaserine(*O*-diazoacetyl-L-serine)**	Figure 20.14, reactions 1 a 4; Figure 20.18, reaction 1 Figure 20.20
		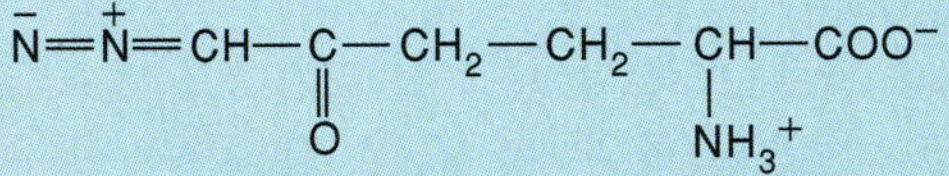 $\bar{N}=\overset{+}{N}=CH—C(=O)—CH_2—CH_2—CH(NH_3^+)—COO^-$ **6-Diazo-5-oxo-L-2-aminohexanoic acid (6-diazo-5-oxo-norleucine, DON)**	
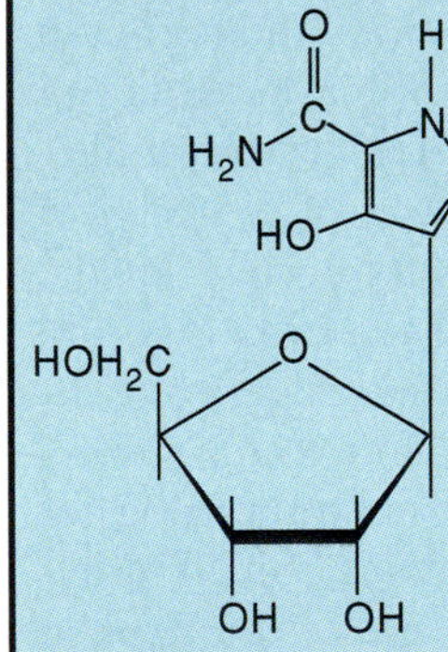 **Pyrazofurin (pyrazomycin)**	Figure 20.18, orotidylate decarboxylase	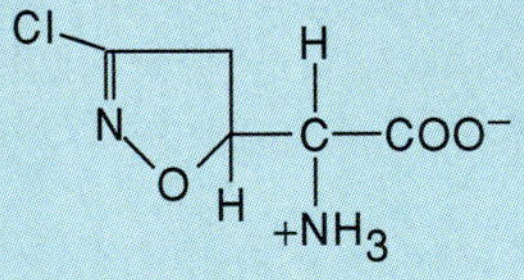 **Acivicin**	Figure 20.14, reactions 1 a 4; Figure 20.16, GMP synthase; Figure 20.18, reaction 1
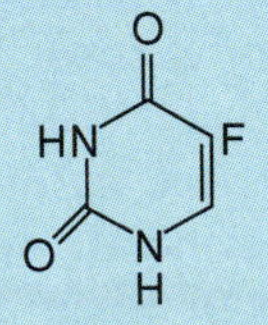 **5-Fluorouracil**	Figure 20.22, thymidylate synthase		

Inhibitor	Enzymes Inhibited
Indirect inhibitors of nucleotide biosynthesis	
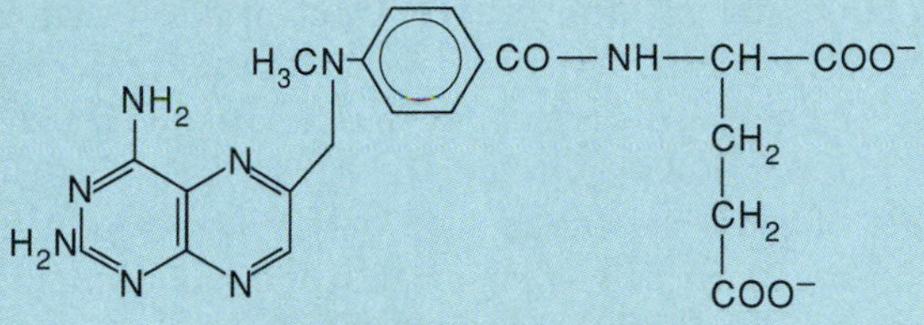 **Methotrexate (amethopterin)**	
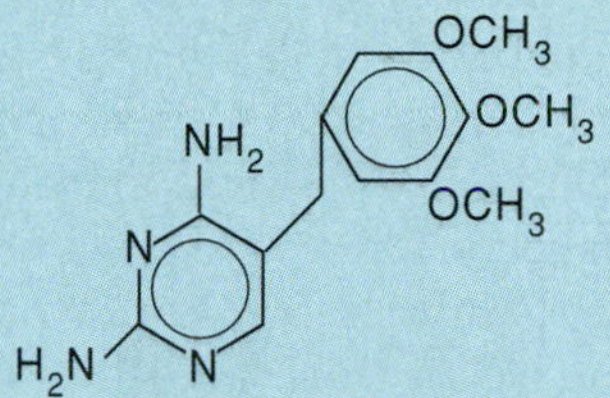 **Trimethoprim**	Figure 20.22, dihydrofolate reductase
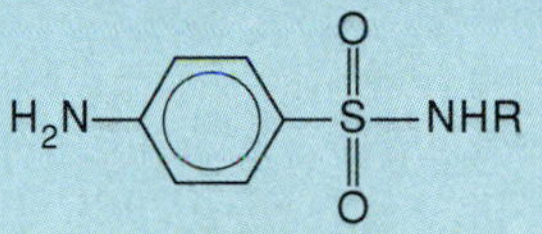 **Sulfonamide**	Dihydropteroate synthase

Figure 20.24

Activated forms of some inhibitors of nucleotide metabolism.

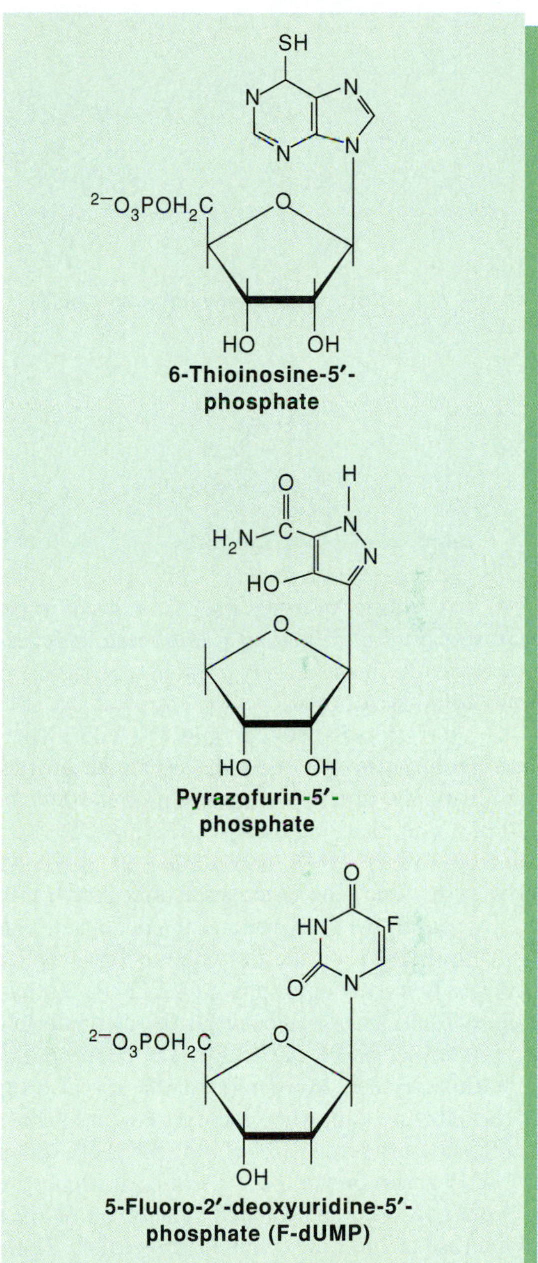

Figure 20.25

Postulated transition state intermediate in the aspartate carbamoyltransferase (fig. 20.18).

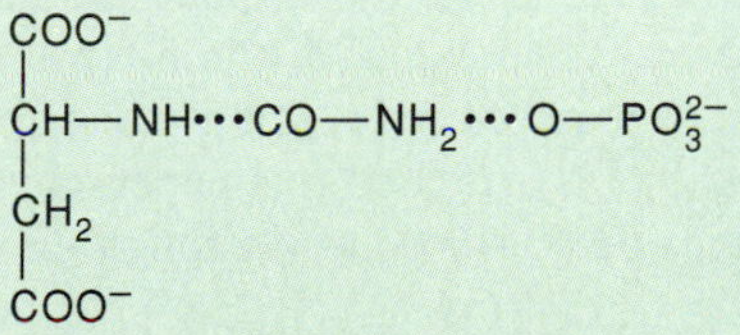

COO^-
|
$CH-NH\cdots CO-NH_2\cdots O-PO_3^{2-}$
|
CH_2
|
COO^-

Figure 20.26

Normal metabolites with which indirect inhibitors of nucleotide metabolism compete. Methotrexate and trimethoprim compete with folate and, more importantly, with dihydrofolate (fig. 20.22); sulfonamides (sulfa drugs) compete with *p*-aminobenzoate.

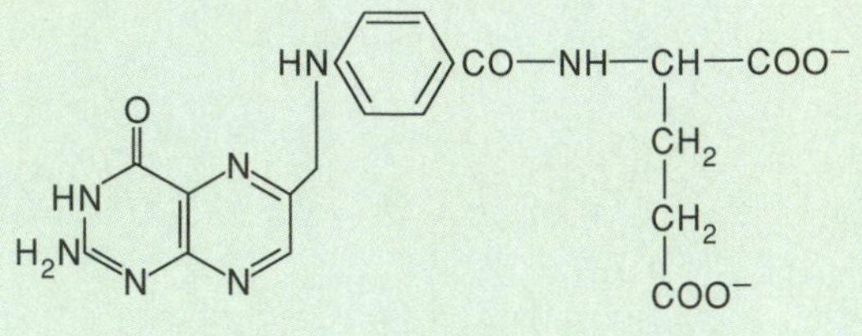

Folic acid (pteroylglutamic acid)

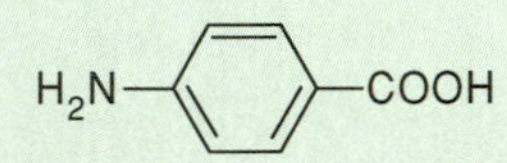

p-Aminobenzoic acid

A number of drugs that are of great importance for the treatment of viral infections and of certain types of cancer are nucleosides or have closely related structures. Examples are shown in table 20.4.

3′-Azido-3′-deoxythymidine (AZT) has received much attention because it is the only drug with proven effectiveness in arresting the course of acquired immune deficiency syndrome (AIDS). The virus that causes this disease, human immunodeficiency virus (HIV), contains an RNA-directed DNA polymerase. Once the virus enters a host cell, this reverse transcriptase catalyzes the synthesis of a double-stranded DNA copy of the viral RNA, as the first step in virus replication. AZT is converted to its monophosphate (AZTMP) by thymidine kinase; thymidylate kinase then converts the monophosphate to the triphosphate (AZTTP). AZTTP is a powerful inhibitor of the reverse transcriptase. More importantly, it is also a good substrate for this enzyme and, when incorporated into the growing DNA chain, promptly terminates chain extension, because there is no 3′ —OH group on the AZT residue to which the next nucleotide unit can become attached. The effect of AZTTP on DNA polymerase of the host cell is much weaker. The effect of AZT is further increased by the accumulation of considerable amounts of AZTMP in cells. This monophosphate inhibits thymidylate kinase, with a consequent decrease in the concentration of dTTP in the cell. With less competition from dTTP there is more efficient incorporation of AZTTP into viral DNA by the reverse transcriptase. 2′,3′-dideoxyinosine (ddI) is under trial for use against AIDS and is similarly incorporated into viral DNA by the reverse transcriptase with chain termination.

Acyclovir (ACV) is not a true nucleoside, because the guanine residue is attached to an open-chain structure, but the latter mimics deoxyribose well enough for the compound to be accepted as a substrate by a thymidine kinase specified by certain herpes-type viruses. The normal thymidine kinase in mammalian cells will not recognize ACV as a substrate, however, so only virus-infected cells convert ACV to its monophosphate. Once the first phosphate has been added, the second phosphate is added by cellular guanylate kinase, while several other cellular kinases can add the third phosphate. The triphosphate is a more potent inhibitor of the viral DNA polymerases than of cellular DNA polymerases, and also inactivates the former but not the latter. The net result is that ACV has been an effective treatment of, and prophylaxis for, genital herpes, can result in dramatic relief of pain associated with "shingles" caused by reactivation of latent varicella-zoster virus, and has been successful in many patients with herpes encephalitis.

Gancyclovir (GCV), like ACV, is an open-chain analog of deoxyguanosine. It is more effective than ACV in the treatment of infections by human cytomegalovirus (HCMV). These infections can cause birth defects in newborns, and can be fatal or cause blindness in immunosuppressed individuals such as transplant recipients, cancer patients receiving chemotherapy, or AIDS patients. HCMV-infected cells are more efficient at forming the triphosphate of the drug, though the mechanism is unclear. Furthermore, the viral DNA polymerase is much more sensitive to inhibition (and chain termination) by the triphosphate than are cellular DNA polymerases.

Cytosine arabinose (araC) is taken up by cells and converted to its triphosphate (araCTP), which is a substrate for cell DNA polymerases, and is a relative chain terminator. This means that the cell DNA polymerases have difficulty in adding the next nucleotide after araCMP has been added to the chain. Incorporation of two or more successive araCMP residues causes chain termination. araC is a component of a combination of drugs with proven effectiveness against some forms of leukemia.

Catabolism of Nucleotides

Dietary nucleic acids are unaffected by gastric enzymes, but in the small intestine, ribonuclease and deoxyribonuclease I, which are secreted in the pancreatic juice, hydrolyze nucleic acids mainly to oligonucleotides. The oligonucleotides are further hydrolyzed by phosphodiesterases, also secreted by the pancreas, to yield 5′- and 3′-mononucleotides. Most of the mononucleotides are then hydrolyzed to nucleosides by various group-specific nucleotidases or by a variety of nonspecific phosphatases. The resulting nucleosides may be absorbed intact by the intestinal mucosa, or they may undergo phosphorolysis by nucleoside phosphorylases and by nucleosidases to free bases:

$$\text{Nucleoside} + P_i \rightleftharpoons \text{base} + \text{ribose-1-phosphate}$$

$$\text{Nucleoside} + H_2O \rightleftharpoons \text{base} + \text{ribose}$$

Little is known about these enzymes and their specificity. The mucosa of the small intestine are rich in nucleoside phosphorylase, and most of the remaining nucleoside is probably hydrolyzed to bases in this tissue.

Table 20.4
Other Nucleoside Analogs Used in Therapy

Analog	Enzyme Inhibited by Active Form	Clinical Condition Treated
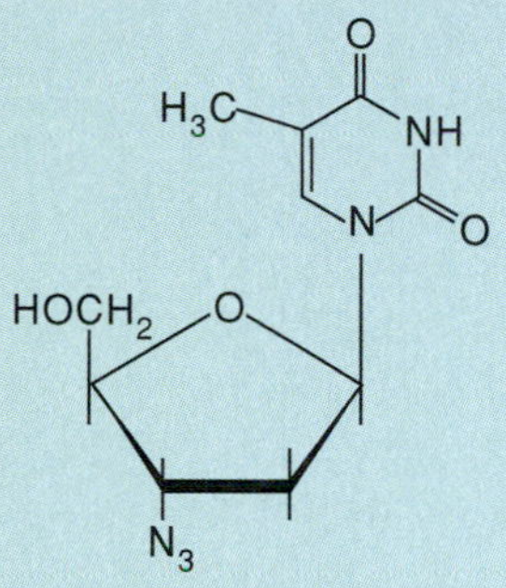 **Azidothymidine (AZT)**	Viral DNA polymerase	AIDS (Human immuno-deficiency virus infection)
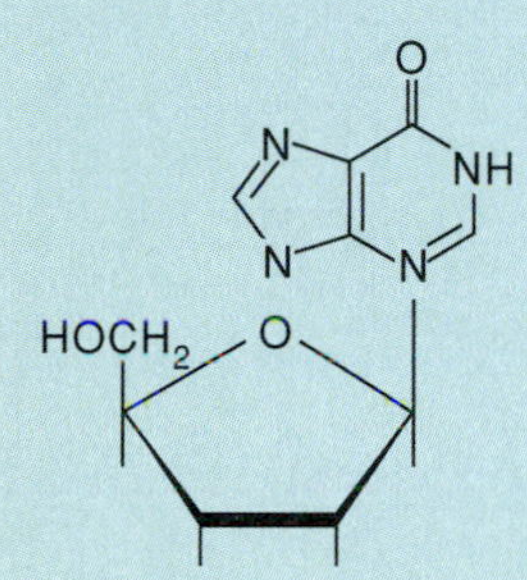 **2′,3′-Dideoxyinosine** (ddI)	Viral DNA polymerase	AIDS (Human immuno-deficiency virus infection)
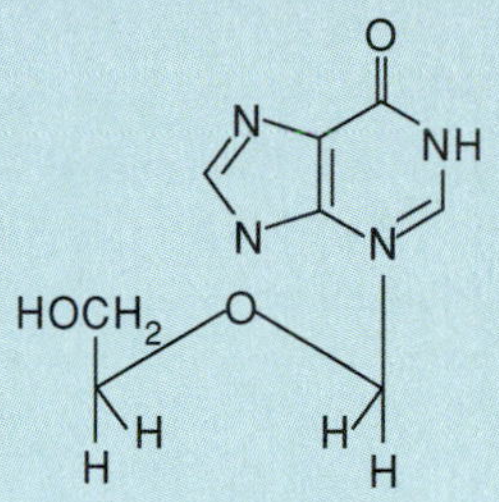 **Acyclovir (ACV)**	Viral DNA polymerase	Herpes simplex virus infections
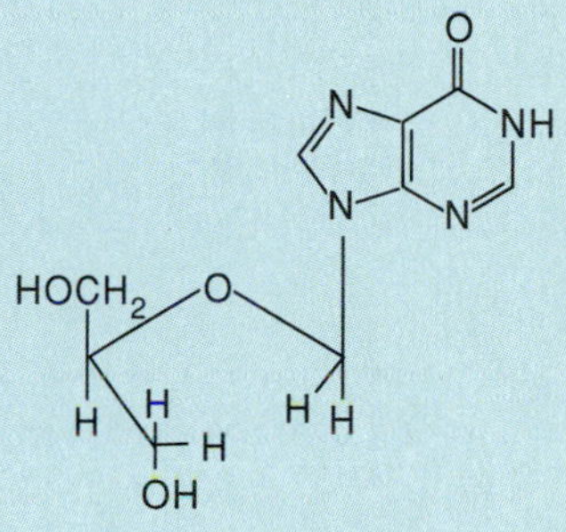 **Gancyclovir (GCV)**	Viral DNA polymerase	Human cytomegalovirus infections
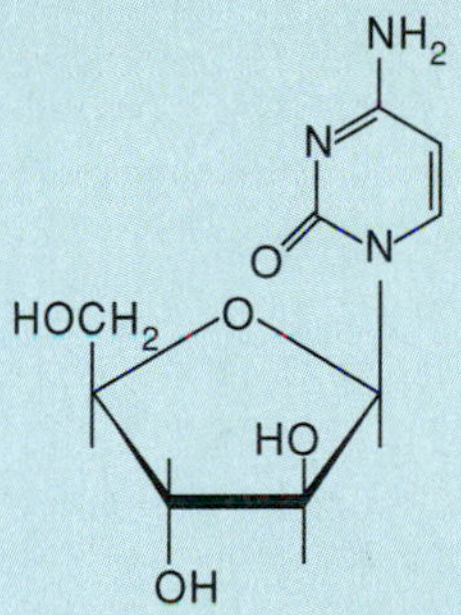 **Cytosine arabinoside (araC)**	DNA polymerase	Leukemia

Experiments with labeled nucleic acid indicate that purines and pyrimidines of ingested nucleic acids are used only to a small extent for synthesis of tissue nucleic acids, and in the case of purines, most of the bases were shown to be catabolized. This finding is consistent with the presence in intestinal mucosa of a high level of xanthine oxidase, a catabolic enzyme.

Intracellular Catabolism of Nucleotides Is Highly Regulated

Nucleotides are also catabolized within cells by several types of intracellular nucleotidase that hydrolyze the phosphate ester groups to release inorganic phosphate. One of the 5′-nucleotidases that act on monophosphates in mammalian cells is probably located in lysosomes. It has a pH optimum of 5.0 and a broad substrate specificity. In lymphocytes there are two cytoplasmic purine nucleotidases, one acting on deoxyribonucleotides, the other on ribonucleotides. In these cells three types of cytoplasmic nucleotidase act on pyrimidine nucleotides; one acts on ribonucleotides, another on deoxyribonucleotides, and a third is specific for thymidylate.

Although allosteric regulation of 5′-nucleotidases has not been demonstrated, attack of these enzymes on nucleotides is limited under normal circumstances, either by their intracellular localization or by the effects of nucleotide concentrations. Nevertheless, they appear to be involved in cycling of nucleotides along the pathways shown in figures 20.9 and 20.10. The balance between the activity of nucleoside kinases and 5′-nucleotidases serves to regulate intracellular nucleoside and

Figure 20.27

Major pathways of purine degradation in animals. Primates excrete uric acid. Mammals other than primates catabolize uric acid to other end products (see fig. 25.29). In contrast to the catabolism of carbohydrates, lipids or amino acids, the catabolism of nucleotides results in no energy production in the form of ATP. In both GMP and AMP catabolism ribose-1-phosphate is released.

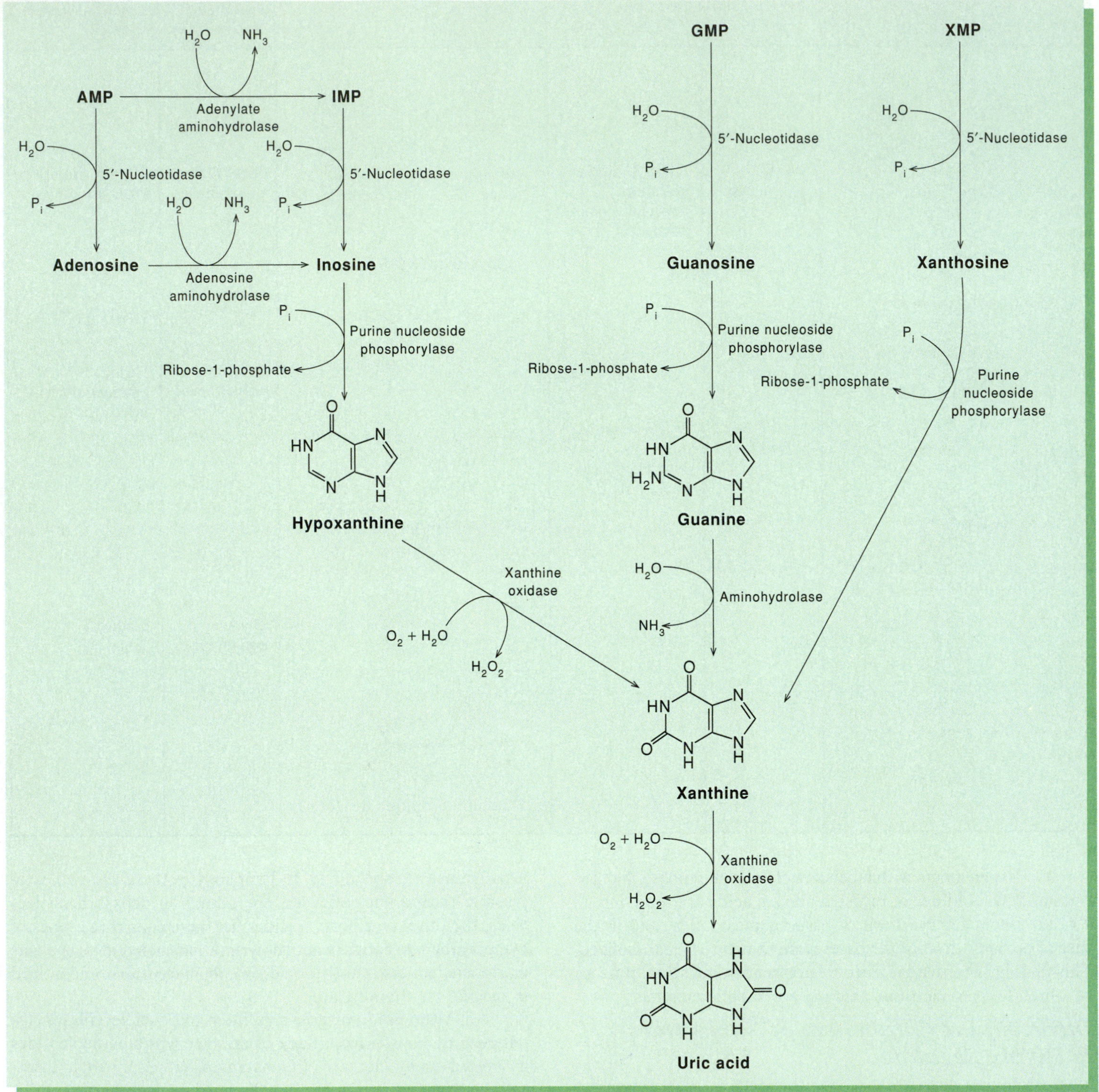

nucleotide levels. An increase in nucleotide levels will result in increased nucleotidase-catalyzed hydrolysis of nucleotides to nucleosides, which are then transported out of the cell or broken down by the nucleoside phosphylases that we described in connection with salvage pathways. The nucleobases formed can exit from the cell with the help of transport proteins. Low intracellular nucleotide levels will result in net formation of nucleotides by kinase action on nucleosides, and by phosphoribosyltransferase action on purine bases, as well as by *de novo* synthesis. This is evident from results of experiments with inhibitors of some of the enzymes involved in these pathways, such as adenosine deaminase and purine nucleoside phosphorylase. In certain tissues specialized nucleotidases have a specific role. For example, an acid phosphatase that serves to provide inorganic phosphate in bone is an iron-stimulated nucleotidase acting on di- and triphosphates.

Purines Are Catabolized to Uric Acid and Then to Other Products

After purine nucleotides have been converted to the corresponding nucleosides by 5′-nucleotidases and by phosphatases, inosine and guanosine are readily cleaved to the nucleobase and ribose-1-phosphate by the widely distributed purine nucleoside phosphorylase. The corresponding deoxynucleosides yield deoxyribose-1-phosphate and base with the phosphorylase from most sources. Adenosine and deoxyadenosine are not attacked by the phosphorylase of mammalian tissue, but much AMP is converted to IMP by an aminohydrolase (deaminase), which is very active in muscle and other tissues (fig. 20.27). It has recently been discovered that inherited deficiency of purine nucleoside phosphorylase is associated with a deficiency in the cellular type of immunity, but not in humoral immunity.

An adenosine aminohydrolase (deaminase) is also present in many mammalian tissues. This enzyme is of interest because hereditary deficiency of the enzyme is linked to a severe (usually fatal) defect in the immune system, marked by a serious deficiency in lymphocytes and consequent inability to combat infections.

Inosine formed by either route is then phosphorolyzed to yield hypoxanthine. Although, as we have previously seen, much of the hypoxanthine and guanine produced in the mammalian body is converted to IMP and GMP by a phosphoribosyltransferase, about 10% is catabolized. Xanthine oxidase, an enzyme present in large amounts in liver and intestinal mucosa and in traces in other tissues, oxidizes hypoxanthine to xanthine, and xanthine to uric acid (see fig. 20.27). Xanthine oxidase contains FAD, molybdenum, iron, and acid-labile sulfur in the ratio 1:1:4:4, and in addition to forming hydrogen peroxide, it is also a strong producer of the superoxide anion $\cdot O_2$, a very reactive species. The enzyme oxidizes a wide variety of purines, aldehydes, and pteridines.

Guanine aminohydrolase (guanine deaminase or guanase), present in liver, brain, and other mammalian tissues, provides another pathway to xanthine, this time from guanine. Subsequent oxidation of xanthine to uric acid then occurs.

Figure 20.28

The structure of allopurinol, an analog of hypoxanthine. Allopurinol inhibits xanthine oxidase and is used in the treatment of gout.

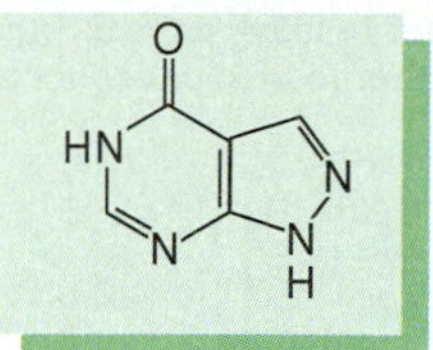

Gout is a relatively common (≈3 per 1,000 persons) derangement of purine metabolism that is associated with elevated plasma levels of uric acid. The excessive uric acid leads to painful deposits of monosodium urate in the cartilage of joints, especially of the big toe. Uric acid deposits also may occur as calculi in the kidney, with resultant renal damage. The genetics are complex and incompletely understood. Individuals suffering from gout and other metabolic disorders producing elevation of serum uric acid may be treated with the xanthine oxidase inhibitor allopurinol (fig. 20.28), an analog of hypoxanthine. Allopurinol is also a substrate of xanthine oxidase, but the product, oxypurinol, binds very tightly to the reduced form of xanthine oxidase and inactivates the enzyme. Allopurinol is nontoxic and administration causes a marked decrease in the uric acid concentration in serum, and in urinary excretion of uric acid. During allopurinol treatment, serum levels of hypoxanthine and xanthine are prevented from accumulating by operation of the salvage pathway, and because the consequent accumulation of IMP, GMP, and AMP causes feedback inhibition of purine biosynthesis (see fig. 20.31).

Mammals other than primates further oxidize urate by a liver enzyme, urate oxidase, which is a copper protein. The product, allantoin, is excreted. Humans and other primates, as well as birds, lack urate oxidase and hence excrete uric acid as the final product of purine catabolism. In many animals other than mammals, allantoin is metabolized further to other products that are excreted: allantoic acid (some teleost fish), urea (most fishes, amphibians, some mollusks), and ammonia (some marine invertebrates, crustaceans, etc.). This pathway of further purine breakdown is shown in figure 20.29.

Pyrimidines Are Catabolized to β-Alanine, NH_3, and CO_2

A number of deaminases present in many cells are able to deaminate cytosine or its nucleosides or nucleotides to the corresponding uracil derivatives. Cytosine aminohydrolase (deaminase) appears to occur only in microorganisms (yeast and bacteria), but cytidine aminohydrolase is widely distributed in bacteria, plants, and mammalian tissues. A distinct deoxycytidine aminohydrolase is present in various mammalian tissues and tumors, in plants, and in bacteria. A deoxycytidylate aminohydrolase that is similarly distributed produces dUMP, which is susceptible to attack by 5′-nucleotidase to give deoxyuridine. Although the physiological function of these aminohydrolases is not completely understood, the uridine and deoxyuridine

Figure 20.29

Degradation of uric acid to excretory products. Mammals other than primates oxidize uric acid further to allantoin. Humans and other primates as well as birds lack urate oxidase and hence excrete uric acid as the final product of purine catabolism. In many animals other than mammals, allantoin is metabolized further to urea or ammonia and CO_2 as shown.

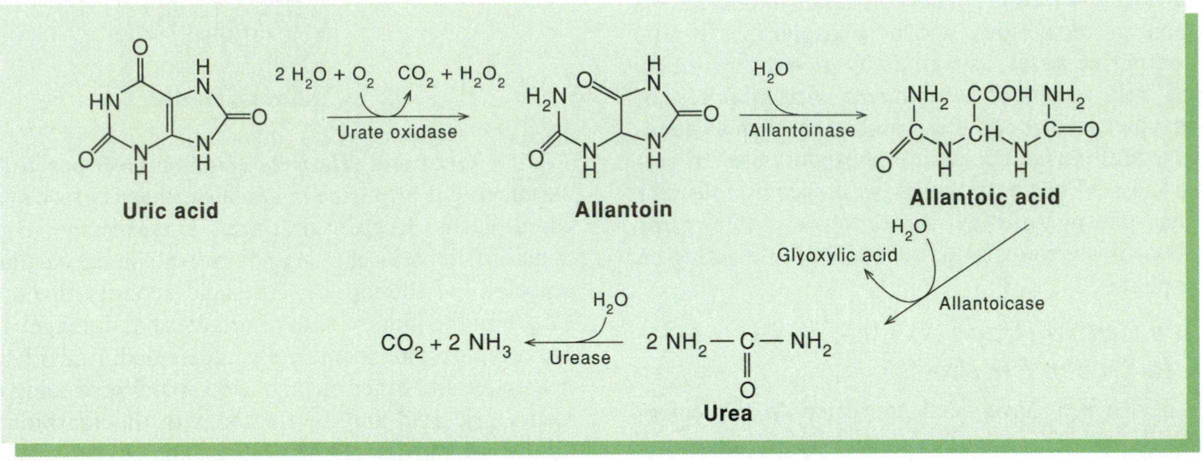

formed can be further degraded by uridine phosphorylase to uracil as previously discussed, so that these reactions provide a pathway for converting nucleotides of uracil and cytosine to uracil and ribose-1-phosphate or deoxyribose-1-phosphate (fig. 20.30). Similarly, thymine nucleosides and nucleotides can be converted by 5′-nucleotidase and phosphorylase to thymine.

Enzymes present in mammalian liver are capable of the catabolism of both uracil and thymine. The first reduces uracil and thymine to the corresponding 5,6-dihydro derivatives. This hepatic enzyme uses NADPH as the reductant, whereas a similar bacterial enzyme is specific for NADH. Similar enzymes are apparently present in yeast and plants. Hydropyrimidine hydrase then opens the reduced pyrimidine ring, and finally the carbamoyl group is hydrolyzed off from the product to yield β-alanine or β-aminoisobutyric acid, respectively, from uracil and thymine (see fig. 20.30).

Regulation of Nucleotide Metabolism

Among the reaction pathways that we have described, there exist many possibilities for futile cycles, in which nucleotides built up in the biosynthetic pathways are broken down in catabolic pathways to products closely related to the starting materials. As an example, AMP synthesized from IMP by adenylosuccinate synthase and adenylosuccinate lyase may be hydrolyzed back to IMP by adenylate aminohydrolase (see fig. 20.27). The net result is the conversion of aspartate to fumarate and ammonia and the hydrolysis of GTP to GDP and P_i. To avoid such futile cycles, both biosynthetic and catabolic processes are under tight regulatory controls. The efficiency of these controls is demonstrated by the increased activity of many enzymes involved in nucleotide biosynthesis when cells are proliferating. Evidently, regulatory mechanisms increase nucleotide biosynthesis as intracellular nucleotides are used for the synthesis of RNA and DNA. As we have seen, drastic consequences can attend impairment of the control machinery, as in the Lesch-Nyhan syndrome or intervention with drugs such as 6-mercaptopurine.

Although much remains to be discovered about the details of the regulation of nucleotide metabolism, a number of important control points are rather well understood. We will discuss these here, together with their known effects on intracellular nucleotide pools.

Purine Biosynthesis Is Regulated at Two Levels

Many lines of evidence indicate that the first committed step in *de novo* purine nucleotide biosynthesis, production of glutamine PRPP amidotransferase, is rate-limiting for the entire sequence. Consequently, regulation of this enzyme is probably the most important factor in control of purine synthesis *de novo* (fig. 20.31). The enzyme is inhibited by purine-5′-nucleotides, but the nucleotides that are most inhibitory vary with the source of the enzyme. Inhibition constants (K_I) are usually in the range 10^{-3} to 10^{-5} M. The maximum effect of this end-product inhibition is produced by certain combinations of nucleotides (e.g., AMP and GMP) in optimum concentrations and ratios, indicating two kinds of inhibitor binding sites. This is an example of a concerted feedback inhibition.

The rate of the amidotransferase reaction is also governed by intracellular concentrations of the substrates L-glutamine and PRPP. Competing metabolic reactions or drugs that alter the supply of these substrates also affect the rate of IMP synthesis.

Figure 20.30

Degradation of pyrimidine bases. Parts of this pathway are widely distributed in nature. The entire pathway is found in mammalian liver. As in purine nucleotide catabolism, no ATP results from catabolism, and the ribose-1-phosphate is released during catabolism before destruction of the base.

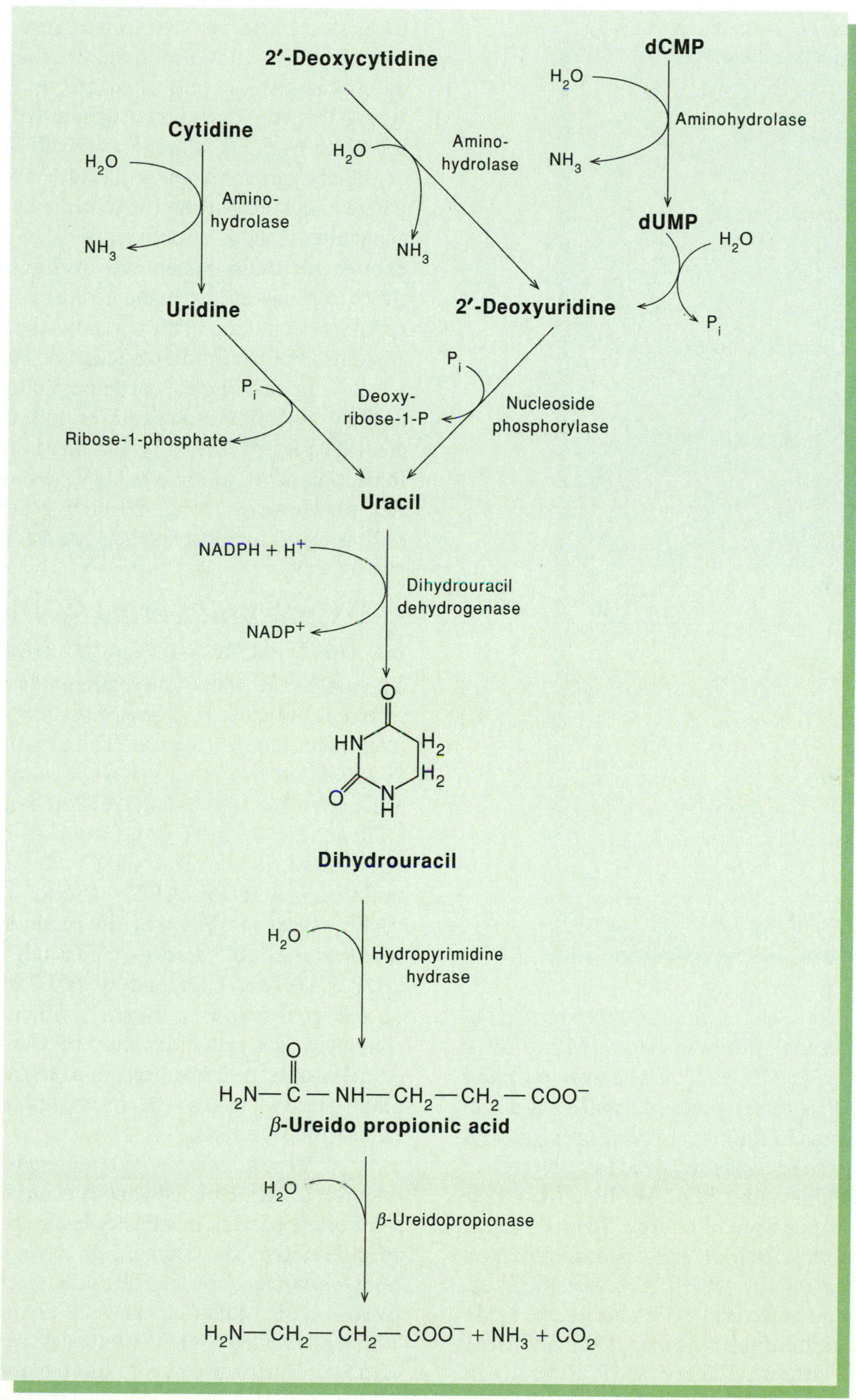

Figure 20.31

Regulations of purine biosynthesis. Red arrows show points of feedback inhibition. In addition to the feedback inhibition, GTP stimulates ATP synthesis and ATP stimulates GTP synthesis. This helps to assure a balance between the pools of the two nucleoside triphosphates. The full biosynthetic pathways are shown in figures 20.14 and 20.16.

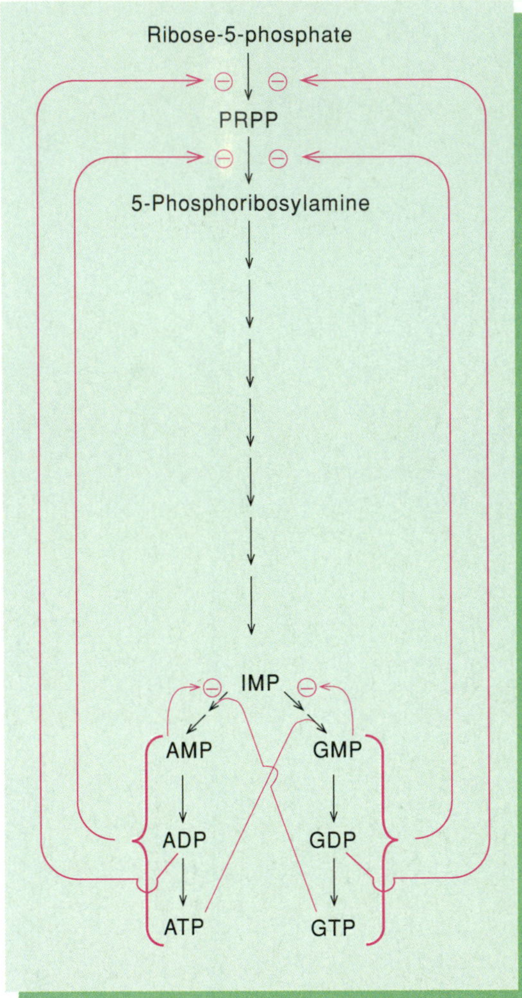

The second important level of regulation of purine nucleotide synthesis is in the branch pathways from IMP to AMP and to GMP (see fig. 20.16). The first of the two reactions leading from IMP to AMP is the irreversible synthesis of adenylosuccinate. This requires GTP as a source of energy and is inhibited by IMP. Of the two reactions required to convert IMP to GMP, the first is irreversible and is inhibited by GMP, while the second requires ATP as a source of energy. Thus there are two types of regulation at this level of purine nucleotide synthesis: (1) a "forward" control, by which increased GTP accelerates AMP synthesis and increased ATP accelerates GMP synthesis, and (2) feedback inhibition, by which AMP and GMP each regulate their own synthesis. Excess AMP also may be converted to IMP by adenylate aminohydrolase and thus can serve as a source of GMP. Adenylate aminohydrolase is activated by ATP and inhibited by GTP, which may serve to control this potential conversion of adenine nucleotides to guanine nucleotides. Finally, when the energy reserves of the cell are low, feedback inhibition of ribose-5-phosphate pyrophosphokinase by ADP and GDP restricts the synthesis of PRPP.

Pyrimidine Biosynthesis Is Regulated at the Level of Carbamoyl Aspartate Formation

In bacteria, the first committed step in pyrimidine nucleotide biosynthesis is the formation of carbamoyl aspartate from carbamoyl phosphate and aspartate. In *E. coli,* the enzyme catalyzing this step, aspartate carbamoyltransferase, is powerfully inhibited by CTP, which acts chiefly by decreasing the affinity of the enzyme for aspartate (see chapter 10). ATP has the opposite effect, activating the enzyme by increasing its affinity for aspartate. Concentrations of ATP and CTP in *E. coli* are high enough for these nucleotides to influence the intracellular activity of aspartate carbamoyltransferase. However, this is not a regulatory enzyme in all bacterial species and is not involved in regulation of pyrimidine nucleotide synthesis in animal cells.

In eukaryotes, carbamoyl phosphate synthase is inhibited by pyrimidine nucleotides and stimulated by purine nucleotides; it appears to be the most important site of feedback inhibition of pyrimidine nucleotide biosynthesis in mammalian tissues. However, it has been suggested that under some conditions, orotate phosphoribosyltransferase may be a regulatory site as well.

Deoxyribonucleotide Synthesis Is Regulated by Both Activators and Inhibitors

The manner in which the reduction of ribonucleotides to deoxyribonucleotides is regulated has been studied with reductases from relatively few species. The enzymes from *E. coli* and from Novikoff rat liver tumor have a complex pattern of inhibition and activation (fig. 20.32). ATP activates the reduction of both CDP and UDP. As dTTP is formed by metabolism of both dCDP and dUDP, it activates GDP reduction, and as dGTP accumulates, it activates ADP reduction. Finally, accumulation of dATP causes inhibition of the reduction of all substrates. This regulation is reinforced by dGTP inhibition of the reduction of GDP, UDP, and CDP and by dTTP inhibition of the reduction of the pyrimidine substrates. Since there is evidence that ribonucleotide reductase may be the rate-limiting step in deoxyribonucleotide synthesis in at least some animal cells, these allosteric effects may be important in controlling deoxyribonucleotide synthesis.

The adenosylcobalamin-requiring ribonucleotide reductases from lactobacilli and certain other microorganisms have a different pattern of allosteric effects, the principal one being specific activation effects. For example, in the case of the *L. leichmannii* enzyme, dGTP activates ATP reduction, dATP activates CTP reduction, and dCTP activates UTP reduction. These effects may serve to adjust the relative rates of reduction of the various substrates to more equal values. In addition, the synthesis of *L. leichmannii* enzyme is repressed by the presence in the growth medium of an excess of vitamin B_{12} (cyanocobalamin) or of a deoxyribonucleoside such as thymidine. The

Figure 20.32

Proposed scheme for the regulation of deoxyribonucleotide synthesis in *E. coli* and mammalian cells. Red arrows indicate points of activation and inhibition, respectively. (Source: L. Thelander and P. Reichard, "Reduction of ribonucleotides," in *Annual Review of Biochemistry* 48:133, 1979. Copyright © 1979 Annual Reviews Inc. Palo Alto, Calif.)

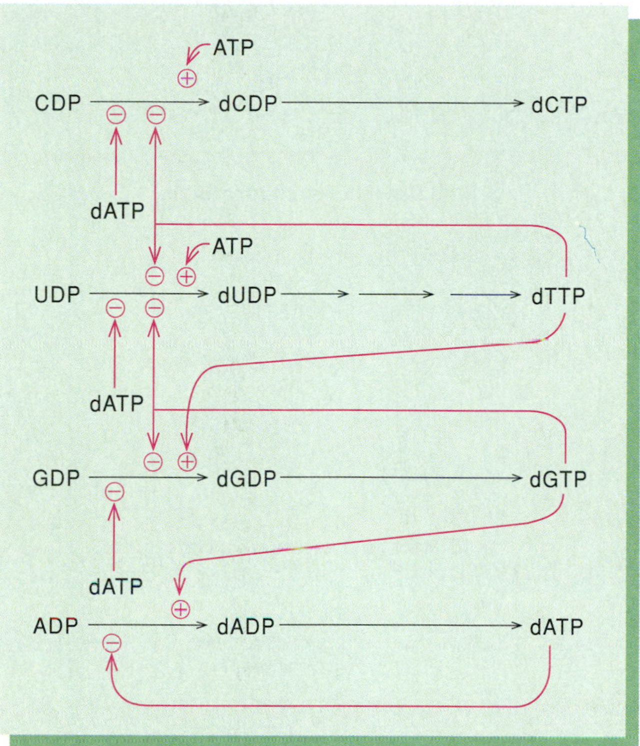

Figure 20.33

Phases in the life cycle of a typical eukaryotic cell. The cell cycle is divided into the resting stage (G_0), the prereplication stage (G_1), the synthesis or replication stage (S), the postreplication stage (G_2) and the mitotic stage (M).

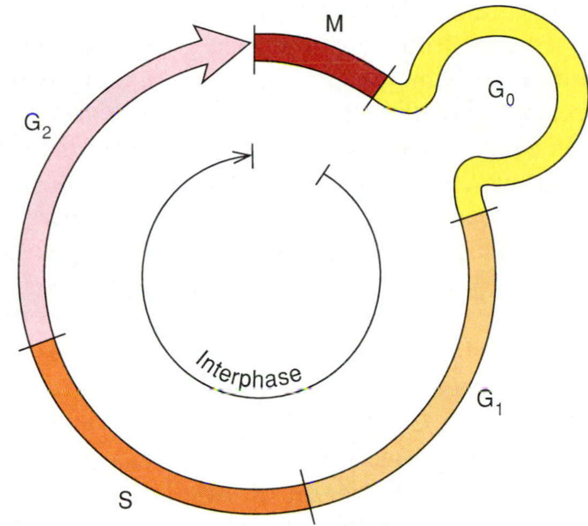

repressor for enzyme synthesis is probably dTTP or a closely related nucleotide, which accumulates in the cell when rapid ribonucleotide reduction occurs as a result of an ample cobalamin supply or when deoxynucleoside is supplied. Further deoxyribonucleotide synthesis is then slowed by the decreased rate of reductase synthesis.

A second enzyme on the pathway to dTTP that is subject to allosteric control is deoxycytidylate deaminase, which supplies dUMP for thymidylate synthesis. The enzyme has been studied in depth in mammalian cells, yeast, and bacteriophage T2-infected *E. coli*. In each case the enzyme is allosterically activated by dCTP (hydroxymethyl dCTP for the phage enzyme) and inhibited by dTTP.

Enzyme Synthesis Also Contributes to Regulation of Deoxyribonucleotides during the Cell Cycle

Many of the enzymes participating in *de novo* synthesis of deoxyribonucleotide triphosphates, as well as those responsible for interconversion of deoxyribonucleotides, increase in activity when cells prepare for DNA synthesis. The need for increased DNA synthesis occurs under three circumstances: (1) when the cell proceeds from the G_0 or resting stage of the cell cycle to the S or synthetic or replication stage (fig. 20.33); (2) when it performs repair after extensive DNA damage; and (3) after infection of quiescent cells with virus. When cells leave G_0, for example, enzymes like thymidylate synthase and ribonucleotide reductase increase as well as the corresponding mRNAs. These increases in enzyme amount supplement allosteric controls that increase the activity of each enzyme molecule. Corresponding decreases in amounts of these enzymes and their mRNAs occur when DNA synthesis is completed.

Metabolites Are Channeled Along the Nucleotide Biosynthesis Pathways

In addition to the regulatory controls described in the preceding sections, which are mainly allosteric feedback mechanisms, evidence is accumulating that nucleotide biosynthetic pathways are closely controlled through the phenomenon of channeling. This involves an assembly of enzymes catalyzing successive steps in the biosynthetic pathway so that metabolic intermediates pass directly from one enzyme to the next. Thus the metabolites are channeled along the metabolic pathway, with restricted opportunity for their diffusion into the medium or entry into the general metabolic pool in the cell.

The most clearly established way in which channeling of nucleotide precursors is achieved involves multifunctional enzymes, in which several catalytic activities occur on a single polypeptide chain. Another possibility is the noncovalent association of pathway enzymes in complexes that may sometimes be concentrated in a particular intracellular location.

Examples of multifunctional enzymes are provided by the pyrimidine biosynthetic pathway. Although in most prokaryotes, six structural genes code for the six enzymes involved in the *de novo* synthesis of UMP, in eukaryotes the number of genes is reduced because of the production of multifunctional proteins. In the fungus *Neurospora,* a single protein has both carbamoyl phosphate synthase activity and aspartate carbamoyltransferase activity, but in mammalian cells a single protein not only has both these activities, but also has dihydroorotase activity. The latter protein is an oligomer (probably a trimer) of a large polypeptide (M_r = 200,000), and there is evidence to indicate that the multifunctional polypeptide is a single gene product. This multifunctional enzyme channels carbamoyl phosphate and carbamoyl aspartate, provided dihydroorotate is rapidly removed, which is the case in the normal cell.

In mammalian cells, the last two steps of the pathway to UMP (see fig. 20.18) are catalyzed by a bifunctional protein that is the product of a single gene. This protein therefore has both orotate phosphoribosyltransferase and OMP decarboxylase activities. Although added OMP is accepted as substrate for the decarboxylase, OMP formed from orotate and PRPP is not released, but is preferentially utilized at the decarboxylase site for UMP formation.

In the purine pathway, a trifunctional enzyme catalyzes three reactions concerned with generation of the formyl donors for steps 3 and 9 (see fig. 20.14). The enzymatic reactions catalyzed by this protein are shown in figure 20.34. Since 10-formyltetrahydrofolate inhibits the cyclohydrolase, this may serve to regulate the amount of the formyl donor that is available. Channeling is further enhanced by noncovalent association of other enzymes of the pathway with the dehydrogenase-cyclohydrolase-synthase enzyme. Evidence has been obtained that this loose complex contains serine hydroxymethyltransferase (which generates methylenetetrahydrofolate), the transformylases catalyzing steps 3 and 9 of the purine pathway, and probably all the other enzymes of the pathway. These enzyme activities largely remain associated through certain mild purification procedures. This association of pathway enzymes greatly increases channeling and permits the efficient generation and use of the unstable 10-formyltetrahydrofolate.

Intracellular Concentrations of Ribonucleotides Are Much Higher than Those of Deoxyribonucleotides

Methods are available for analysis of nucleotides in eukaryotic cells and in mammalian tissues. Concentrations are frequently expressed in terms of picomoles per 10^6 cells or picomoles per microgram of DNA, since it is easier to express analyses on this basis than as intracellular molar concentration. Some estimates in molar terms are available, but the values depend on a number of factors. The presence of nucleosides or bases in the extracellular fluid not only causes increases in the intracellular concentration of the corresponding nucleotides, but through allosteric effects may also affect the size of other nucleotide pools. Transfer of a cell culture to fresh medium may also cause perturbation of intracellular nucleotide levels. Finally, as the cell

Figure 20.34

Formation of formyl group donors for purine biosynthesis (R = *p*-aminobenzoyl-L-glutamate). A trifunctional enzyme catalyzes the three reactions shown, which generate the formyl donors for steps in purine nucleotide synthesis (see steps 3 and 9 in fig. 20.14).

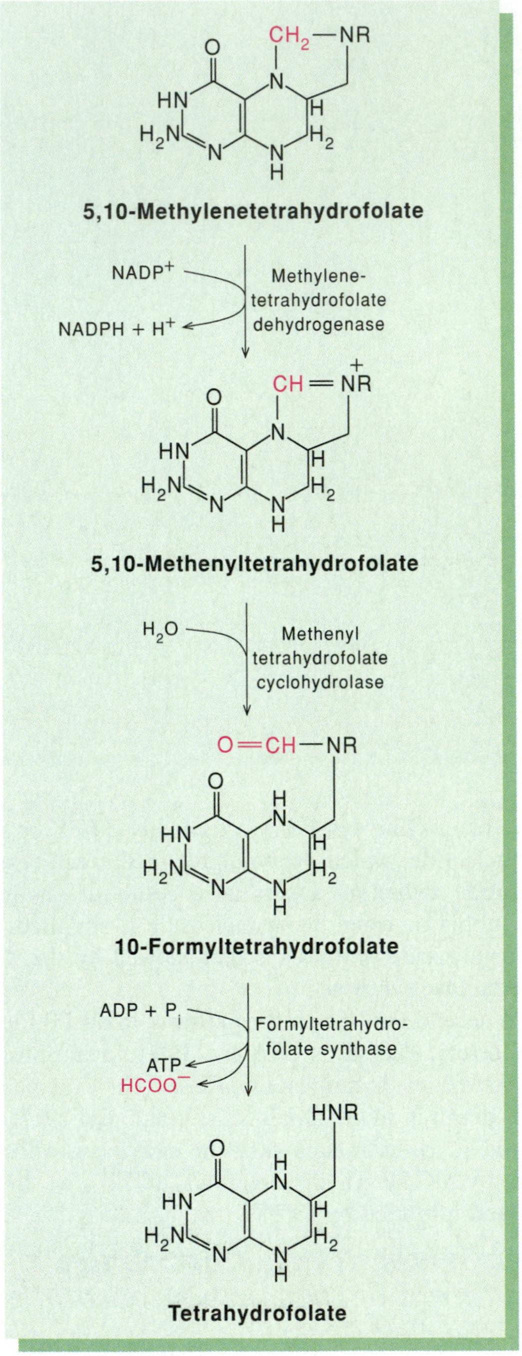

progresses through its life cycle (fig. 20.33) there are marked changes in concentration of some nucleotides, particularly an increase of the deoxyribonucleoside triphosphates as the cell enters S phase. Conditions that damage cells, such as starvation, lack of oxygen, or presence of toxic materials, also affect nucleotide levels. Consequently, there are no "correct" values of nucleotide levels, even for a specific type of cell. Some general statements are possible, however. ATP is generally present in normal cells at a concentration of 2–10 millimolar, but other ribonucleoside triphosphates are at lower concentrations (0.05 to 2 mM), depending on conditions and the cell type. The ribonucleoside mono- and diphosphates have lower concentrations than the corresponding triphosphates, but ADP is typically comparable with or even higher than GTP, CTP, or UTP. Deoxyribonucleoside triphosphates (dNTPs) are present in much lower concentrations, 2–60 μM. The concentration of the dNTP present at lowest concentration, dCTP in some cells and dGTP in others, is sufficient for only a few minutes of DNA synthesis, so that during DNA replication, synthesis of dNTPs must keep pace with utilization for DNA synthesis.

In prokaryotes, the levels of deoxyribonucleotides vary from undetectable to 200 picomoles per 10^6 cells (compared with 3 to 30 picomoles of dATP, dCTP, and dGTP per 10^6 eukaryotic cells). As in eukaryotic cells, dTTP is usually present at higher concentrations than the other deoxyribonucleoside triphosphates.

The arrest of cell growth by many agents is associated with depletion of one or more of the deoxyribonucleotide pools. Thus thymidine at millimolar concentrations arrests cell growth and decreases the concentration of dCTP dramatically, whereas the concentrations of dATP, dGTP, and especially dTTP increase. This effect is considered to be mediated by the allosteric inhibition of ribonucleotide reductase, referred to earlier. Hydroxyurea, another agent that arrests cell growth by blocking DNA synthesis, depletes the pools of dATP and dGTP, and in some cells it is the effect on the latter, also brought about by ribonucleotide reductase inhibition, that is probably critical.

T4 Bacteriophage Infection Stimulates Nucleotide Metabolism

Infection of *E. coli* by T4 phage results in the induction of nearly 30 proteins. Many of these are enzymes that ensure an abundant supply of deoxynucleotides above those produced by the host. As a result, deoxynucleotide concentrations are increased many times. Phage-coded enzymes and related proteins include thioredoxin, ribonucleotide reductase, dihydrofolate reductase, dCMP deaminase, thymidylate synthase, and deoxyribonucleotide kinase. However, in some instances the phage relies completely on host enzymes that are normally present at high levels. Examples of such enzymes are adenylate kinase and nucleoside diphosphate kinase. Completely novel enzymes coded by the phage are endonucleases II and IV, which supply nucleotide directly by degrading host DNA.

The phage DNA contains no cytosine; instead hydroxymethylcytosine is incorporated. To accomplish this the phage induces enzymes that hydrolyze dCTP and dCDP, synthesize 5-hydroxymethyl dCMP from dCMP and methylenetetrahydrofolate, and phosphorylate hydroxymethyl dCMP. All these enzymes help to ensure rapid and specific synthesis of phage DNA while preventing synthesis of host DNA.

Biosynthesis of Nucleotide Coenzymes

Many of the nucleotides considered thus far play important roles in metabolism and are discussed in other chapters. However, nucleotide coenzymes such as flavin nucleotides, NAD^+, $NADP^+$, and coenzyme A are also extremely important in metabolism. In this section, we will discuss the pathway for the completion of each coenzyme.

Riboflavin, i.e., 7,8-dimethyl-10-(1′-D-ribityl)isoalloxazine, is synthesized by microorganisms such as the fungus *Eremothecium* and mutants of the yeast *Saccharomyces* in a pathway that starts from GTP. Riboflavin is an essential dietary constituent for mammals and is converted in the body to the mononucleotide or dinucleotide forms that function as the prosthetic groups of many enzymes. Riboflavin is converted to riboflavin-5′-phosphate, more commonly called flavin mononucleotide (FMN), by flavokinase (ATP:riboflavin phosphotransferase), as shown in figure 20.35. The enzyme has been purified from yeast, plants, and liver. It is also present in a variety of other animal tissues (kidney, brain, spleen, and heart).

The other nucleotide form of riboflavin, flavin adenine dinucleotide (FAD), is formed from FMN in a reversible reaction catalyzed by flavin nucleotide pyrophosphorylase (see fig. 20.35). This enzyme is also widely distributed in nature and has been observed in plants, yeast, lactobacilli, and many animal tissues.

The nicotinamide moiety of the coenzymes nicotinamide adenine dinucleotide (NAD^+) and nicotinamide adenine dinucleotide phosphate ($NADP^+$) is synthesized by several routes. In liver and other animal tissues, tryptophan degradation forms, among other products, quinolinic acid (chapter 19), which is converted to nicotinate mononucleotide (deamidonicotinamide mononucleotide, deamido-NMN) by quinolinate phosphoribosyltransferase (fig. 20.36). In the cytosol of cells of many mammalian tissues, and in yeast and other microorganisms, there is present a nicotinate phosphoribosyltransferase that also forms deamido-NMN (see fig. 20.36). A very similar phosphoribosyltransferase present in the cytosol of all animal tissues investigated acts on nicotinamide. These transferases are responsible for utilization of nicotinate and nicotinamide in the diet. The role of ATP in these reactions is unclear. Some transferases do not require it, for others it seems to be an allosteric regulator, and in yet other cases ATP seems to be hydrolyzed to yield ADP and P_i in equimolar amounts with deamido-NMN formation.

Figure 20.35

Biosynthesis of flavin mononucleotide (FMN) and flavin adenine dinucleotide (FAD) from riboflavin. In the first reaction a kinase transfers a single phosphate to the terminal hydroxyl of the ribose. In the second reaction the AMP moiety is transferred to the phosphate.

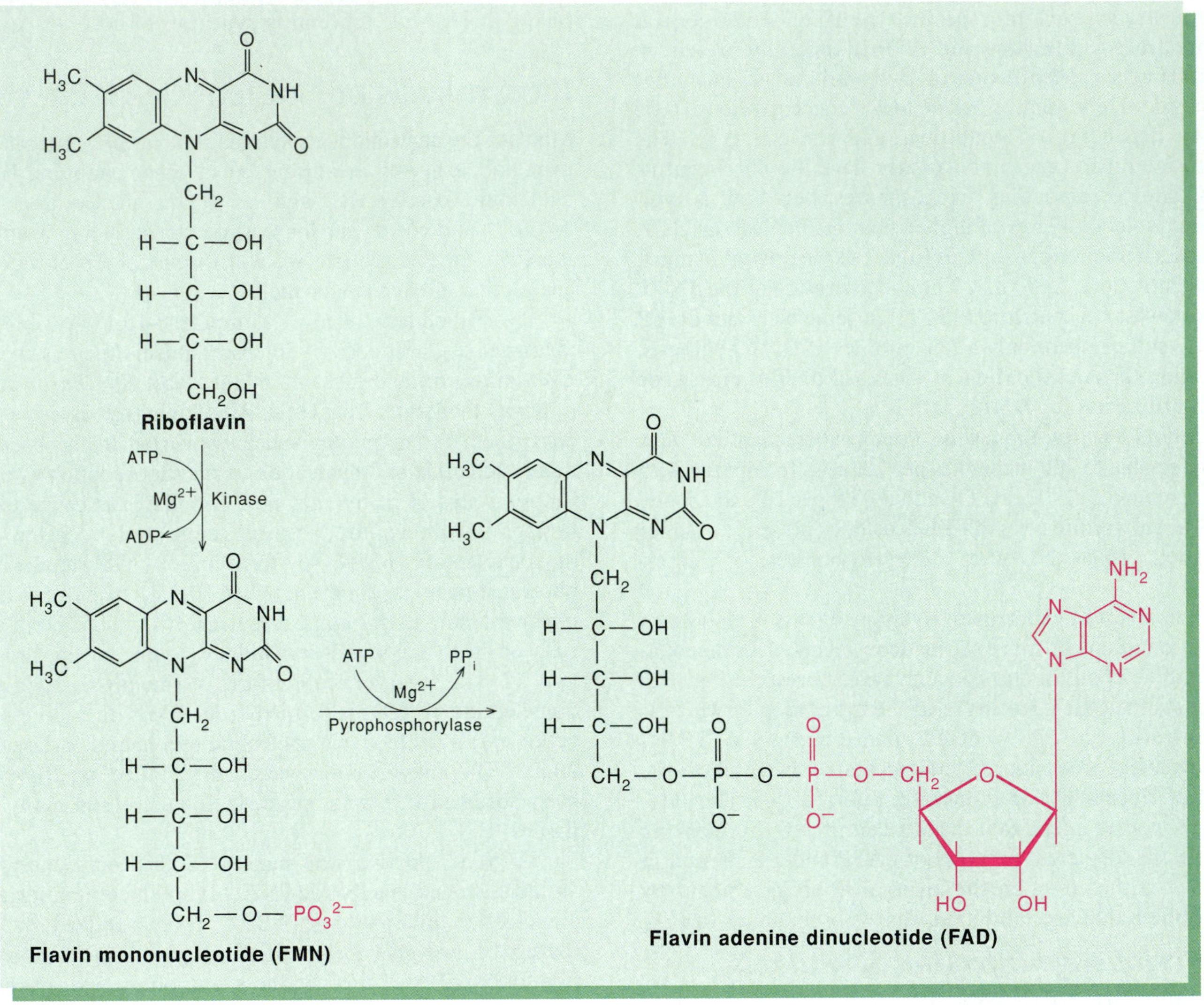

Riboflavin

ATP, Mg^{2+}, Kinase, ADP

Flavin mononucleotide (FMN)

ATP, Mg^{2+}, PP_i, Pyrophosphorylase

Flavin adenine dinucleotide (FAD)

The mononucleotides so formed are converted to the corresponding dinucleotides by NMN adenyltransferase (see fig. 20.36). In mammalian cells it appears to be a single enzyme that catalyzes both reactions, but an adenyltransferase acting only on NMN has been isolated from some bacteria (*Lactobacillus fructosus*). A cytoplasmic NAD^+ synthase present in yeast, liver, and other tissues transfers the amino group from glutamine at the expense of ATP hydrolysis (see fig. 20.36).

A cytoplasmic kinase present in liver, mammary gland, and brain is responsible for the formation of $NADP^+$ from NAD^+:

$$NAD^+ + ATP \rightarrow NADP^+ + ADP$$

NADH is not a substrate and inhibits competitively with respect to NAD^+.

Coenzyme A is synthesized in the mammalian liver from pantothenic acid (pantoyl-β-alanine), which is required in the mammalian diet. The five steps in the synthesis are shown in figure 20.37. In the last step, a specific kinase transfers a phosphoryl group to the 3′-hydroxyl of the adenylate portion of the molecule.

Figure 20.36

Biosynthesis of NAD^+. In animal tissues tryptophan degradation leads to quinolinate, which is converted to deamidonicotinamide mononucleotide (deamido-NMN). Deamido-NMN can also be formed from nicotinate in some mammalian tissues and in many microorganisms. The deamido-NMN is subsequently converted into NAD^+. Another route to NAD^+ starts from nicotinamide. The phosphoribosyltransferase required for this pathway is found in the cytosol of animal tissues. These transferases are responsible for utilization of nicotinate and nicotinamide in the diet.

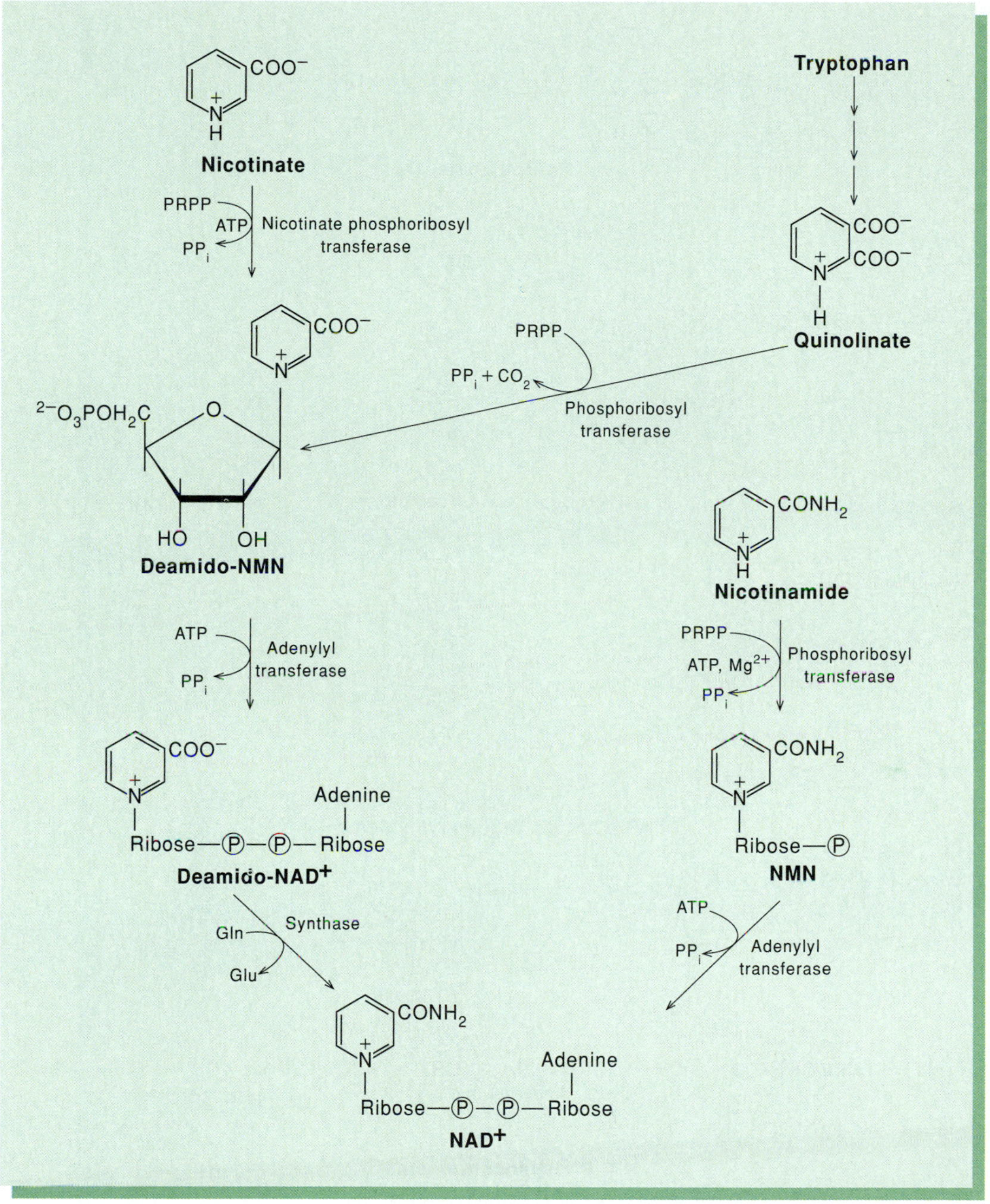

Figure 20.37

Biosynthesis of coenzyme A from pantothenate. This synthesis occurs in the mammalian liver. Pantothenate must be supplied in the diet. Color indicates the groups introduced at the kinase and synthase steps.

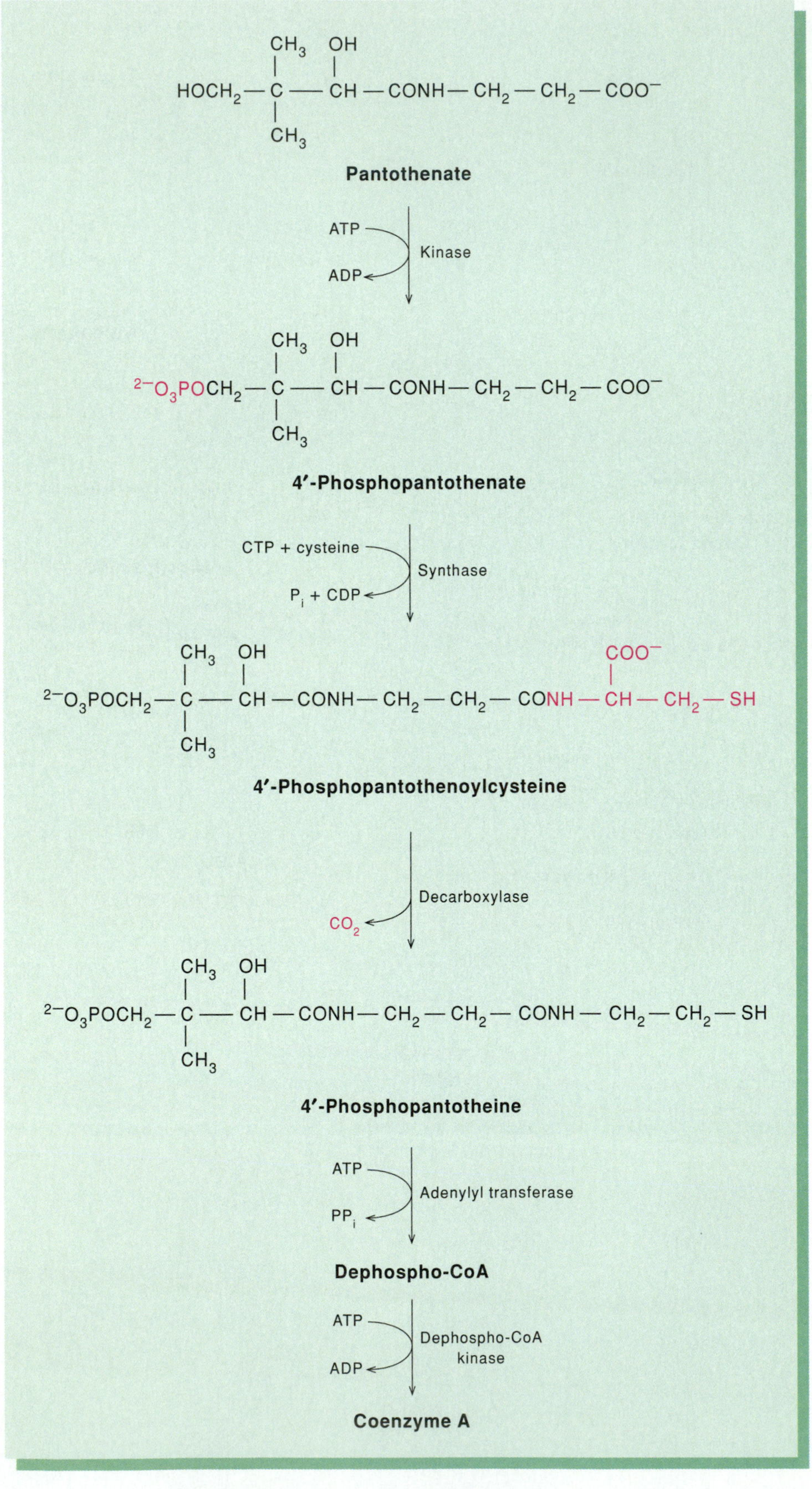

$$HOCH_2-C(CH_3)_2-CH(OH)-CONH-CH_2-CH_2-COO^-$$

Pantothenate

ATP → ADP (Kinase)

$$^{2-}O_3POCH_2-C(CH_3)_2-CH(OH)-CONH-CH_2-CH_2-COO^-$$

4′-Phosphopantothenate

CTP + cysteine → P_i + CDP (Synthase)

$$^{2-}O_3POCH_2-C(CH_3)_2-CH(OH)-CONH-CH_2-CH_2-CONH-CH(COO^-)-CH_2-SH$$

4′-Phosphopantothenoylcysteine

Decarboxylase → CO_2

$$^{2-}O_3POCH_2-C(CH_3)_2-CH(OH)-CONH-CH_2-CH_2-CONH-CH_2-CH_2-SH$$

4′-Phosphopantotheine

ATP → PP_i (Adenylyl transferase)

Dephospho-CoA

ATP → ADP (Dephospho-CoA kinase)

Coenzyme A

Summary

Nucleotides are the building blocks for nucleic acids; they are also involved in a wide variety of metabolic processes. They serve as the carriers of high-energy phosphate and as the precursors of several coenzymes and regulatory small molecules. Nucleotides can be synthesized *de novo* from small-molecule precursors or, through salvage pathways, from the partial breakdown products of nucleic acids. The highlights of our discussion in this chapter are as follows.

1. The ribose for nucleotide synthesis comes from glucose, either by means of the pentose phosphate pathway or from glycolytic intermediates through transketolase-transaldolase reactions. Ribose-5-phosphate is converted to phosphoribosylpyrophosphate (PRPP), the starting point for purine synthesis. This pathway also incorporates into purines atoms from glycine, aspartate, glutamate, CO_2, and one-carbon fragments carried by folates. IMP synthesized by this route is converted by two-step pathways to AMP and GMP, respectively.
2. The biosynthetic pathway to UMP starts from carbamoyl phosphate and results in the synthesis of the pyrimidine orotate, to which ribose phosphate is subsequently attached. CTP is subsequently formed from UTP. Deoxyribonucleotides are formed by reduction of ribonucleotides (diphosphates in most cells). Thymidylate is formed from dUMP.
3. All biosynthetic pathways are under regulatory control by key allosteric enzymes that are influenced by the end products of the pathways. For example, the first step in the pathway for purine biosynthesis is inhibited in a concerted fashion by nucleotides of either adenine or guanine. In addition, the nucleoside monophosphate of each of these bases inhibits its own formation from inosine monophosphate (IMP). On the other hand, adenine nucleotides stimulate the conversion of IMP into GMP, and GTP is needed for AMP formation.
4. Several inhibitors of nucleotide biosynthesis are known. Each is extremely toxic, especially to rapidly growing cells, where the need for nucleic acid synthesis is greatest. In limited amounts, some of the inhibitors have chemotherapeutic value in the treatment of cancer and other illnesses. Some analogs of normal nucleosides are proving to be useful in the treatment of AIDS and certain other viral infections.
5. Nucleic acids and nucleotides are degraded to nucleosides or free bases before they are ingested. Nucleotides or their partial degradation products may be reutilized for nucleic acid synthesis, or they may be further catabolized for excretion or for use in the synthesis of other products. Purine nucleotides are degraded via guanine, hypoxanthine, and xanthine to uric acid, which in some species is degraded further before excretion. Inherited deficiencies in some of the enzymes involved in nucleotide degradation and salvage cause severe impairment of health, a fact testifying to the importance of the degradative pathways. Nucleotides of uracil and cytosine are degraded via uridine and uracil to simpler substances such as β-alanine.
6. In addition to the nucleotides used as substrates in nucleic acid synthesis, there are a number of other nucleotide-containing molecules that serve various purposes in the cell. These include the coenzymes NAD^+, $NADP^+$, FAD, and CoA.

Selected Readings

Foster, J. W., and A. G. Moat, Nicotinamide adenine dinucleotide biosynthesis and pyrimidine nucleotide cycle metabolism in microbial systems. *Microbial Rev.* 44:83–105, 1980. Comprehensive review.

Hoffee, P. A., and M. E. Jones (eds.), *Methods in Enzymology,* vol. 51. *Pyrimidine Nucleotide Metabolism.* New York: Academic Press, Inc., 1978. Contains short summaries of information about most enzymes of the pathways, with references and details of preparation and assay.

Jones, M. E., Pyrimidine nucleotide biosynthesis in animals: Genes, enzymes and regulation of UMP biosynthesis. *Ann. Rev. Biochem.* 49:253–279, 1980. Authoritative outline of the regulatory properties of the two multifunctional proteins responsible for pyrimidine nucleotide synthesis in animals.

Kornberg, A., *DNA Replication,* 2d ed. San Francisco: Freeman, 1989. See especially chapter 1.

Manfredi, J. P., and E. W. Holmes, Purine salvage pathways in myocardium. *Ann. Rev. Physiol.* 47:691–705, 1985. Although this review applies specifically to salvage in heart muscle, it is an excellent summary of purine salvage pathways and the metabolism of the purine nucleotides formed.

Mathews, C. K., L. K. Moen, and R. G. Sargent, Enzyme interactions in deoxyribonucleotide synthesis. *Trends Biochem. Sci.* 13:394–397, 1988.

Nordlund, P., B–M. Sjoberg, and H. Eklund, Three-dimensional structure of the free radical protein of ribonucleotide reductase. *Nature* 345:593–598, 1990.

Plageman, P. G. W., R. M. Woehlheuter, and C. Woffendin, Nucleoside and nucleobase transport in animal cells. *Biochem. Biophys. Acta* 947:405–443, 1988. Reviews the field in detail.

Reichard, P., Interactions between deoxyribonucleotide and DNA synthesis. *Ann. Rev. Biochem.* 57:349–374, 1988. Comprehensive review of deoxyribonucleotide synthesis and its relation to DNA synthesis.

Stadel, J. M., A. D. Lean, and R. J. Lefkowitz, Molecular mechanisms of coupling in hormone receptor-adenylate cyclase systems. *Adv. Enzymol.* 53:1–43, 1982. A current account by a major contributing group, emphasizing the more biochemical aspects of this important system.

Weber, G., Enzyme pattern-targeted chemotherapy. *Adv. Enzyme Regul.* 24:118, 1985. This volume contains several chapters about inhibitors of nucleotide metabolism and their mechanism of actions.

Problems

1. When phosphoribosylpyrophosphate (PRPP) is incubated in alkali, 5-phosphoribose-1,2-cyclic phosphate is formed with the release of phosphate. Draw a chemical reaction mechanism for this reaction. (Hint: The reaction is similar to the mechanism for the base hydrolysis of RNA.)
2. Using the information in problem 1, design an experiment to prove that ribose-5-phosphate pyrophosphokinase (rib-5-P + ATP → PRPP + AMP) transfers a pyrophosphate group making it an unusual kinase. (In most cases a pyrophosphate group is constructed in two steps. For example, in the formation of mevalonate pyrophosphate two phosphates are *separately* transferred to form the pyrophosphate group.) Also, could you prove that the pyrophosphate group in PRPP is in the α position? (Hint: Use [γ-^{32}P]ATP and [α-^{32}P]ATP.) Draw a chemical reaction mechanism for this pyrophosphokinase.
3. In bacteria, pyrimidine biosynthesis is regulated at aspartate carbamoyltransferase, while most of the regulation in humans is at carbamoyl phosphate synthase. Why does this observation make biochemical sense?
4. If mammalian cells in tissue culture are treated with increasing concentrations of *N*-(phosphonacetyl)-L-aspartate (PALA)—a transition-state analog inhibitor of aspartate carbamoyltransferase—cells resistant to this toxin can be isolated. When the enzyme aspartate carbamoyltransferase was assayed, its activity was elevated 100-fold in such cells. Also, activities of carbamoyl phosphate synthase and dihydroorotase were elevated about 100-fold. No other pyrimidine pathway enzyme activities were elevated. Explain these observations.
5. Genetic defects in adenosine deaminase (ADA) in humans leads to severe defects in the immune system. This disease has been treated by injecting the enzyme adenosine deaminase and more recently by gene therapy. Discuss how this defect could lead to toxic effects on lymphocytes. (Hint: Deoxyadenosine is metabolized by ADA.)
6. What would happen if you gave allopurinol to a chicken? (Give a biochemical explanation.)
7. Explain why Lesch-Nyhan patients suffer from severe gout. Although these patients can be treated with allopurinol to relieve the symptoms of gout, this treatment has no effect on the severe mental retardation. Suggest a possible explanation.
8. Propose a chemical reaction mechanism for the second enzyme in purine biosynthesis, GAR synthase (phosphoribosylamine + ATP + glycine → GAR + ADP + P_i).
9. Explain how antifolates like methotrexate selectively kill cancer cells. Why do these patients when treated lose their hair, the intestinal mucosa, cells of the immune system, etc.?
10. Which atoms of nucleotide bases isolated from a hydrolysate of DNA would be labeled by the following precursors?
    (a) [3-^{14}C] serine.
    (b) [2-^{14}C] glucose.
    (c) [^{14}C] CO_2.
11. Compare the number of high-energy phosphate bonds required for the synthesis of GTP and the synthesis of CTP. Assume that PRPP and folate one-carbon derivatives are available.
12. The pathway for the *de novo* synthesis of dTTP is more complex than the pathway for synthesis of the other deoxyribonucleotides. Illustrate this difference by reference to known inhibitors of DNA synthesis.
13. Allopurinol administered together with 6-mercaptopurine under certain conditions enhances the anticancer effectiveness of the latter. How can the known site of action of allopurinol explain this effect?
14. The growth of many bacteria is inhibited by sulfanilamide and other sulfa drugs. The toxic effect of sulfanilamide can be reversed by *p*-aminobenzoate. How do sulfa drugs work and why are they not very toxic to people?

21

CHAPTER

Biosynthesis of Complex Carbohydrates

We have examined the reactions of carbohydrates associated with energy metabolism (chapters 13 and 16) and the involvement of carbohydrates in amino acid and nucleotide metabolism (chapters 18–20). In this chapter we will deal with the remaining major aspect of carbohydrate metabolism, that of complex carbohydrates, i.e., carbohydrates that contain more than one type of monomeric unit.

Since the main building block of all polysaccharides is the hexose, we will begin this chapter by considering the synthesis of hexoses. Following this, we will consider some aspects of the synthesis of disaccharides and, briefly, the synthesis of some homopolymeric polysaccharides. In the remainder of the chapter we will focus on the synthesis of complex oligosaccharides and polysaccharides. In that discussion we will give special emphasis to glycoproteins, because there have been major advances in this all-important subject in recent years.

Figure 21.1 gives an overview of the main topics that we will be discussing in this chapter.

Monosaccharide Biosynthesis

We discussed the biosynthesis of glucose from simpler starting materials in chapter 13. The biosynthesis of other hexoses is linked to glucose by a complex network of single-step reactions (fig. 21.2). In fact, glucose serves as the precursor for the synthesis of many other hexoses without any rearrangement of the central carbon atoms.

Although the list of hexoses of biologic origin is large, we can make certain generalizations about the network of pathways by which they are metabolically connected. Neutral hexoses are never interconverted; some interconversions occur at the level of the monophosphorylated hexose, but the majority of interconversions occur at the level of the nucleotide sugars. Finally, hexose modification (e.g., addition of an *N*-acetyl group, dehydrogenation) generally takes place before polymerization to form a nucleotide sugar.

The Hexose Monophosphate Pool Includes Mannose as Well as Glucose and Fructose

Derivatives of glucose, mannose, fructose, and galactose are the most common sugars found in oligo- and polysaccharides. Three of these hexoses—glucose, mannose, and fructose—belong to

Figure 21.1

Outline of the synthesis of complex carbohydrates. Simple carbohydrates include homopolymers of glucose that serve as the building blocks for starch and cellulose. Complex carbohydrates involve more than one type of hexose building block, and the building blocks are usually covalently modified. The starting point for the synthesis of complex carbohydrates is the hexose monophosphate pool. Those members of the pool that were represented in the master diagram (fig. 12.5) are drawn in black; the remainder of the diagram is drawn in color. Mannose is also a member of the hexose monophosphate pool that is found in many complex carbohydrates. The most common hexose that is not a member of the pool is galactose, which can interact with the pool only via its UDP-activated derivative. In addition to being activated by formation of nucleotide diphosphate derivatives, the hexoses that form complex carbohydrates frequently undergo covalent modification. Thus glucose can form an amine by interaction with glutamine and an *N*-acetyl amine by further reaction with acetyl-CoA. Many other types of modification occur (see text). The simplest complex carbohydrates are the straight-chain copolymers such as hyaluronic acid, which contains a strict alternating sequence of glucuronic acid and *N*-acetylglucosamine. Bacterial cell walls are formed from a complex building block that attaches itself to a long-chain phospholipid. This lipid serves two functions: (1) it activates the building block and (2) it transfers the building block from inside the plasma membrane to a region outside of the plasma membrane, where it can add to the growing three-dimensional structure that constitutes the cell wall.

In eukaryotes there is a large class of protein-carbohydrate conjugates called glycoproteins that contain straight- or branched-chain carbohydrates attached to amino acid side chains of the protein. In O-linked glycoproteins the carbohydrate moiety is attached to an oxygen in the protein; in N-linked glycoproteins the carbohydrate is attached to a nitrogen in the protein. All proteins destined to form glycoproteins are synthesized in close association with the endoplasmic reticulum, so that when protein synthesis is completed the protein finds itself in the lumen of the endoplasmic reticulum. The addition of the carbohydrate moiety follows quite different paths for the N- and O-linked glycoproteins. Formation of an N-linked glycoprotein begins in the endoplasmic reticulum and continues in the Golgi. First in the endoplasmic reticulum, a branched-chain oligosaccharide is formed on the outer surface of the endoplasmic reticulum, which is attached to a phospholipid similar to the one that functions as a carrier in bacterial cell wall synthesis. Once formed, this glycolipid inverts so that the carbohydrate is transferred to the lumen of the endoplasmic reticulum. A specific glycosyltransferase transfers the carbohydrate moiety from the lipid carrier to a nitrogen group of a protein. The nascent glycoprotein may undergo further modifications of its carbohydrate component in the endoplasmic reticulum or later, when the nascent glycoprotein is transferred en masse to the Golgi. O-linked glycoproteins are formed without the assistance of the phospholipid carrier and they are formed mostly if not exclusively in the Golgi. Both types of glycoproteins are destined for different parts of the cell or the extracellular milieu of the organism.

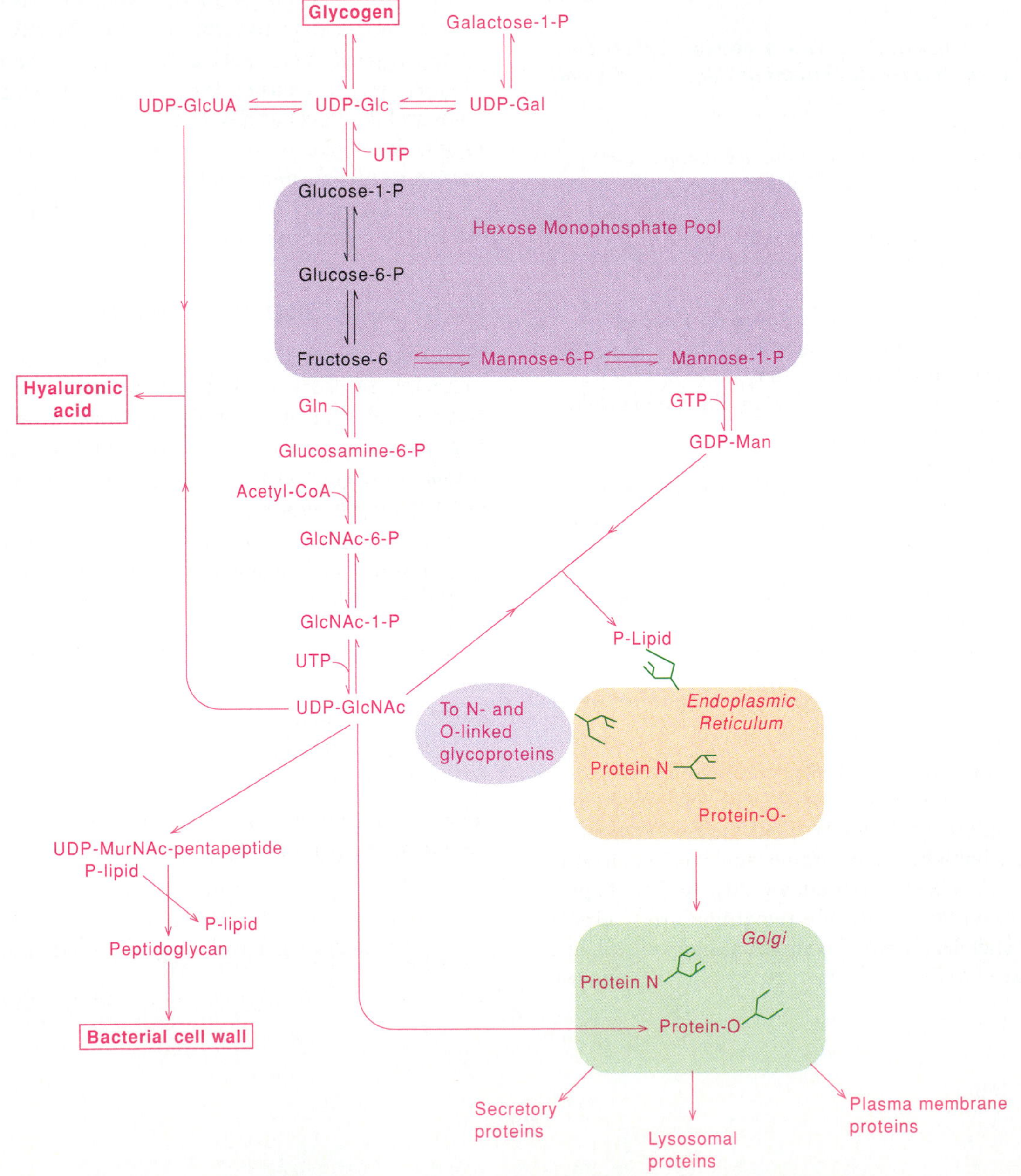

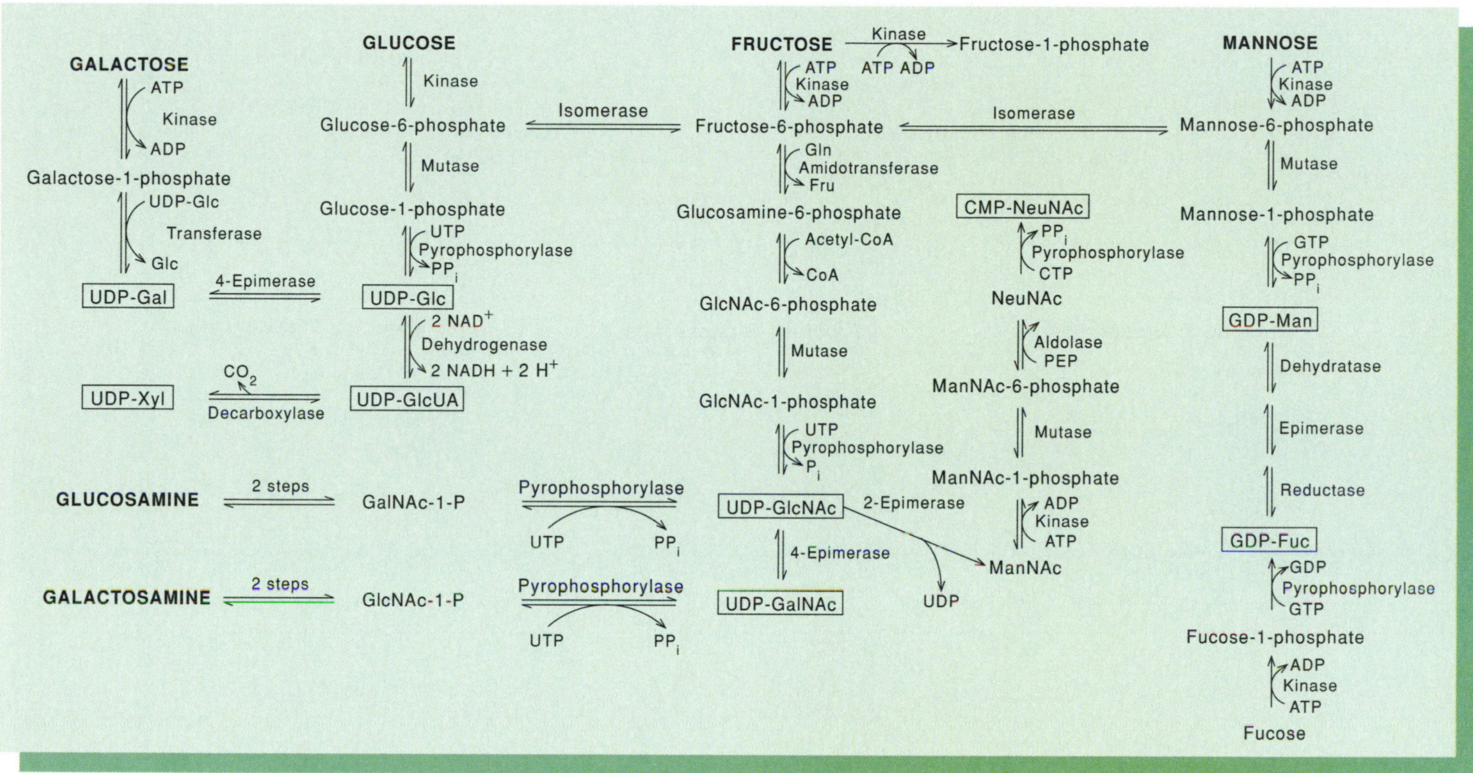

Figure 21.2

Monosaccharide interconversions. The following abbreviations are used here and throughout the chapter: Gal = galactose, Xyl = xylose, Glc = glucose, GlcUA = glucuronic acid, GlcNAc = *N*-acetylglucosamine, GalNAc = *N*-acetylgalactosamine, Fuc = fucose, Man = mannose, ManNAc = *N*-acetylmannosamine, NeuNAc = *N*-acetylneuraminic acid (equivalent to Sia = sialic acid). Actually, sialic acid is a more general term including N- and O-substituted derivatives of neuramic acid. Compounds in capital boldface type are the unmodified sugars or their amine derivatives. The nucleotide sugars are boxed.

the same hexose monophosphate pool (fig. 21.3). They are readily interconverted, with little or no difference in the free energy of the different compounds involved.

We discussed the interconversion of glucose and fructose phosphates in chapter 13, but we omitted mannose from that discussion because it is not centrally involved in either glycolysis or gluconeogenesis. Mannose is, however, a major component of many complex carbohydrates. Mannose-6-phosphate, the 2-epimer of glucose-6-phosphate, can be made directly from fructose-6-phosphate in a reaction analogous to the interconversion of fructose-6-phosphate and glucose-6-phosphate (box 21A). The phosphomannoisomerase enzyme that catalyzes this isomerization holds the substrate so that a proton can be added to the planar C-2 of the intermediate enediol on the side opposite that on which addition occurs in the phosphoglucoisomerase reaction (see fig. 21.2). Subsequently mannose-6-phosphate can be converted to mannose-1-phosphate by a mutase specific for phosphomannose. Mannose and mannose derivatives are incorporated into complex polysaccharides by way of GDP-mannose, which is made from mannose-1-phosphate and GTP in reactions analogous to the activation reactions that we have seen previously (see fig. 21.2).

Galactose Is Not a Member of the Hexose Monophosphate Pool

Galactose, the other main hexose found in structural polysaccharides, is not a member of the central hexose monophosphate pool. The only route from the main pool and galactose goes through UDP-activated derivatives of glucose (see fig. 21.2). By the simplest route galactose, the 4-epimer of glucose (see fig. 6.2), is produced by way of UDP-glucose. The reactions are

$$\text{Glucose-1-phosphate} + \text{UTP} \rightarrow \text{UDP-glucose} + \text{PP}_i$$

$$\text{UDP-glucose} \rightleftharpoons \text{UDP-galactose}$$

UDP-glucose 4-epimerase, which catalyzes the isomerization of UDP-glucose to UDP-galactose, contains a tightly bound molecule of NAD^+ or $NADP^+$, even though no net oxidation is involved. It seems likely that the —OH group at C-4 is transiently oxidized to a carbonyl, which can then be reduced by addition of a hydride ion to either side, thus producing either UDP-glucose or UDP-galactose. Two monosaccharides are considered <u>epimers</u> (e.g., galactose is the 4-epimer of glucose) when they differ only by the configuration of a single hydroxyl group other than at C-1; <u>anomers</u> differ in configuration at C-1 (α or β).

Figure 21.3

Members of the hexose monophosphate pool. The sugars shown are all freely interconvertible, with little change in free energy involved in the conversions.

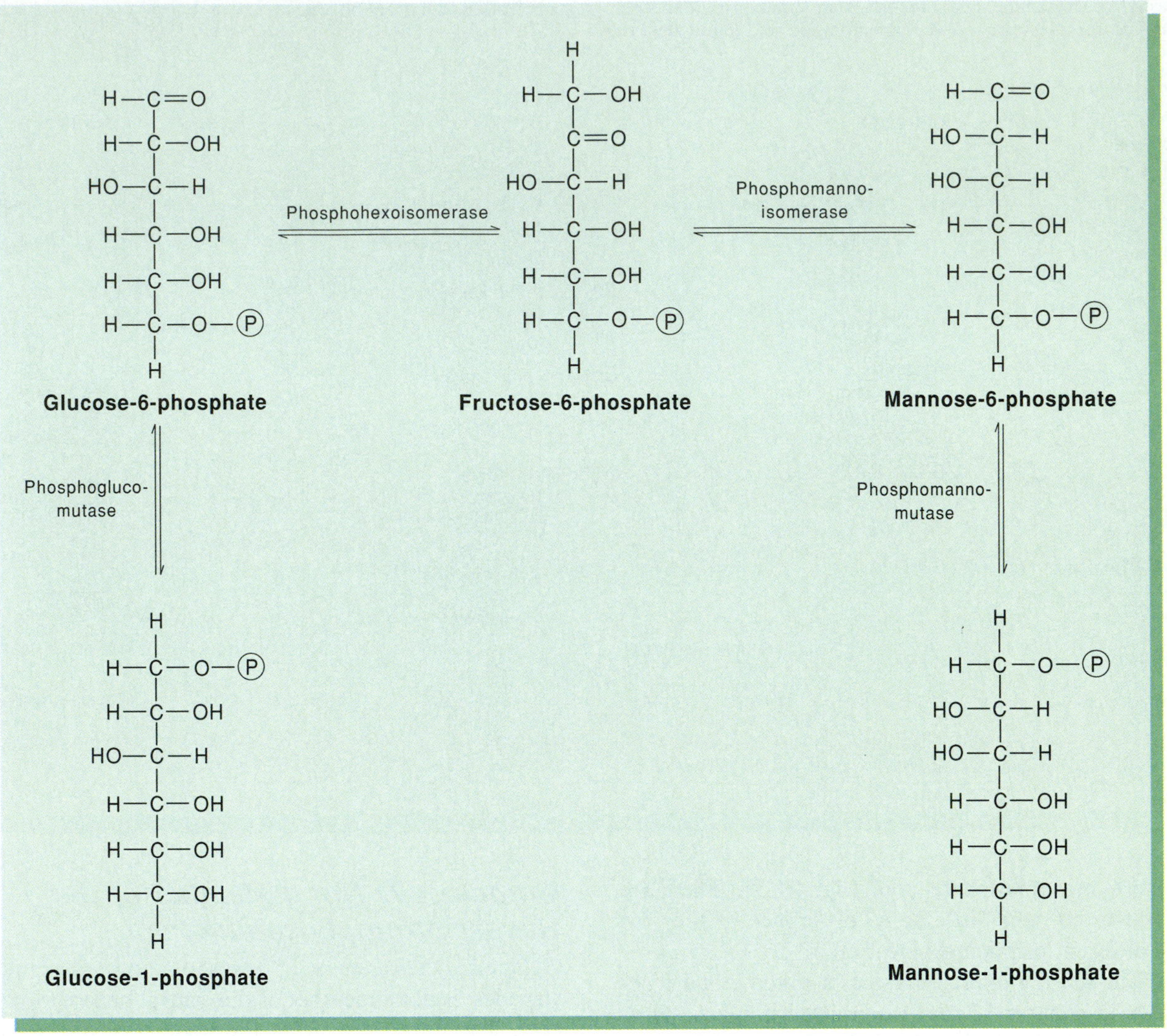

Hexose Modifications Involve Alterations or Additions of Small Substituents

Thus far we have considered only the isomerization of hexoses. Many modifications of hexoses involve the alteration of existing groups or the addition of small groups to the hexose moiety. Oxidation of carbon 6 to a carboxylate group produces a uronic acid. The dehydrogenases that catalyze those oxidations use NAD^+ as the oxidizing agent. For example, UDP-glucose (UDP-Glc) is converted to UDP-glucuronic acid (UDP-GlcUA) in this way.

Replacement of the —OH group at carbon 2 by —NH_2 yields an amino sugar, which is usually modified further. For example, fructose-6-phosphate is the monosaccharide precursor of several important derivatives (fig. 21.4). The process begins with glucosamine formation, a one-step reaction in which the amide nitrogen of glutamine is transferred to the C-2 carbon of fructose. This is the first step in a biosynthetic pathway leading to several different sugar monomers. The amine is acetylated to *N*-acetylglucosamine-6-phosphate, which is activated by reaction with UTP to form UDP-*N*-acetylglucosamine. This derivative may be directly incorporated into a polymer or may be converted to other polymer precursors. In one step UDP-*N*-acetylglucosamine can be epimerized to UDP-*N*-acetylgalactosamine, or in six steps it can be converted to CMP-*N*-acetylneuraminic acid, also called CMP-sialic acid (see fig. 6.11 for the structure of this sialic acid). The CMP-sialic acids are the only nucleotide sugars that occur as nucleoside monophosphate derivatives.

Little is known about the regulation of synthesis of various nucleotide sugar derivatives. If that regulation follows the general scheme suggested in chapter 12 for a biosynthetic

21A
BOX

Mechanism of the Interconversion of Fructose-6-Phosphate and Mannose-6-Phosphate

The interconversion of fructose-6-phosphate and mannose-6-phosphate (figure 1) is very similar to that observed for glucose-6-phosphate and fructose-6-phosphate, involving the same enediol intermediate (see fig. 13.7). In the illustration, the A and B groups refer to catalytic sites on the enzyme.

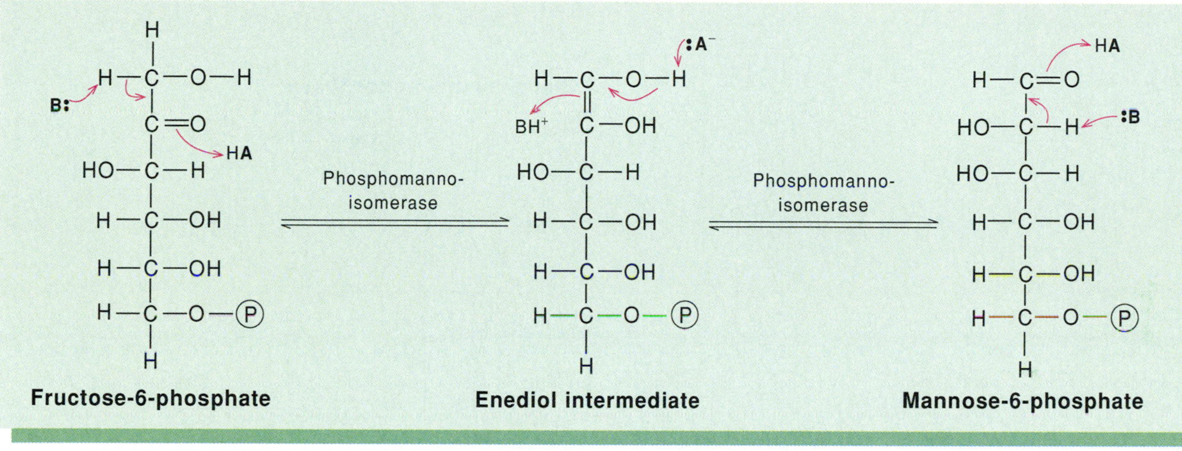

Fructose-6-phosphate ⇌ (Phosphomanno-isomerase) Enediol intermediate ⇌ (Phosphomanno-isomerase) Mannose-6-phosphate

pathway, then the first enzyme in a sequence should be negatively regulated by the end product of the sequence. In agreement with this principle, it has been found in rat liver that UDP-*N*-acetylglucosamine regulates its own synthesis by inhibiting the amidotransferase that catalyzes the conversion of fructose-6-phosphate to glucosamine-6-phosphate, the first step that is specific to this pathway (see fig. 21.4).

Disaccharide Biosynthesis

In disaccharide synthesis only one of the participating hexoses is activated. The disaccharide lactose is formed in the mammary gland from D-glucose and UDP-galactose by the action of lactose synthase (fig. 21.5). The reaction involves nucleophilic displacement of UDP from UDP-galactose by the C-4 hydroxyl group of a free glucose.

Formation of lactose involves an unusual mechanism for controlling enzyme specificity. Lactose synthase is actually a complex of two proteins: (1) galactosyltransferase is found not only in the mammary gland, but in all body tissues. It catalyzes the reaction

UDP-galactose + *N*-acetyl-D-glucosamine →
UDP + *N*-acetyllactosamine.

(2) α-Lactalbumin of milk has no catalytic activity of its own. Rather, it alters the specificity of galactosyltransferase so that the latter will utilize D-glucose instead of *N*-acetyl-D-glucosamine as the galactose acceptor. As a result, α-lactalbumin makes lactose instead of *N*-acetyllactosamine:

UDP-galactose + D-glucose → UDP + lactose

Sucrose is synthesized from glucose and fructose. First, fructose-6-phosphate is produced from glucose-6-phosphate. The latter is also converted via glucose-1-phosphate to uridine diphosphate glucose (UDP-glucose), which then reacts with fructose-6-phosphate to give UDP and sucrose-6-phosphate. The phosphate is removed by a single enzymatically catalyzed hydrolysis to yield sucrose and inorganic phosphate. In some plants, sucrose is formed simply by the reaction of UDP-glucose with fructose.

Energy-Storage Polysaccharides Are Simple Homopolymers

Most polysaccharides that are used for energy storage are simple homopolymers of glucose linked by $\alpha(1, 4)$-glycosidic bonds. Some aspects of the synthesis of these polyglucose molecules were described in chapter 13. In all cases the C-1 atom of the

Figure 21.4

Some derivatives formed from fructose-6-phosphate. This figure elaborates on one branch of figure 21.2.

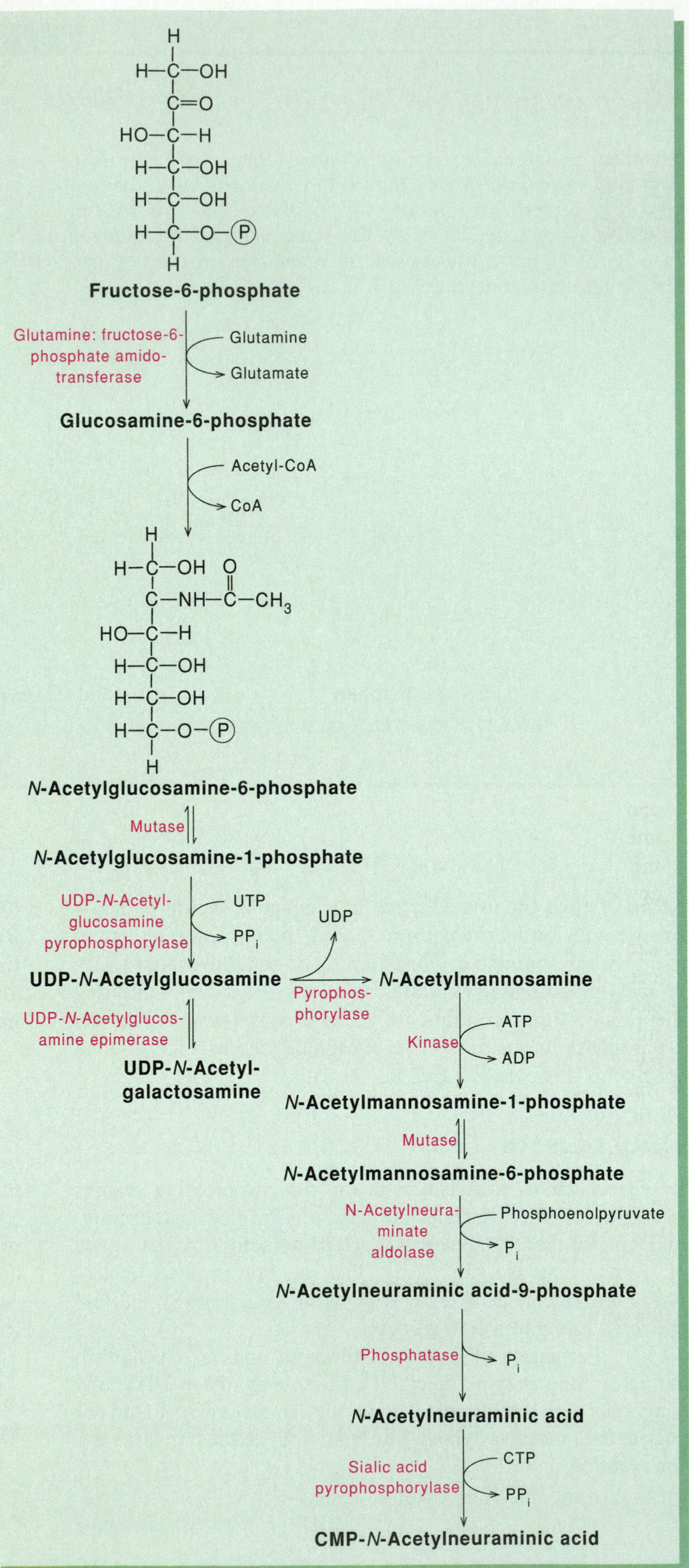

Figure 21.5

The mechanism of lactose formation. The more realistic chair forms are shown for the hexoses to make it easier to appreciate the stereochemistry of the reaction. Note how the configuration at C-1 becomes inverted in this reaction. Lactose synthase is two proteins: β(1,4)-galactosyltransferase and α-lactalbumin. The mechanism for this reaction differs from that found in polysaccharide synthesis or degradation, neither of which shows inversions; these latter reactions involve an oxonium ion intermediate.

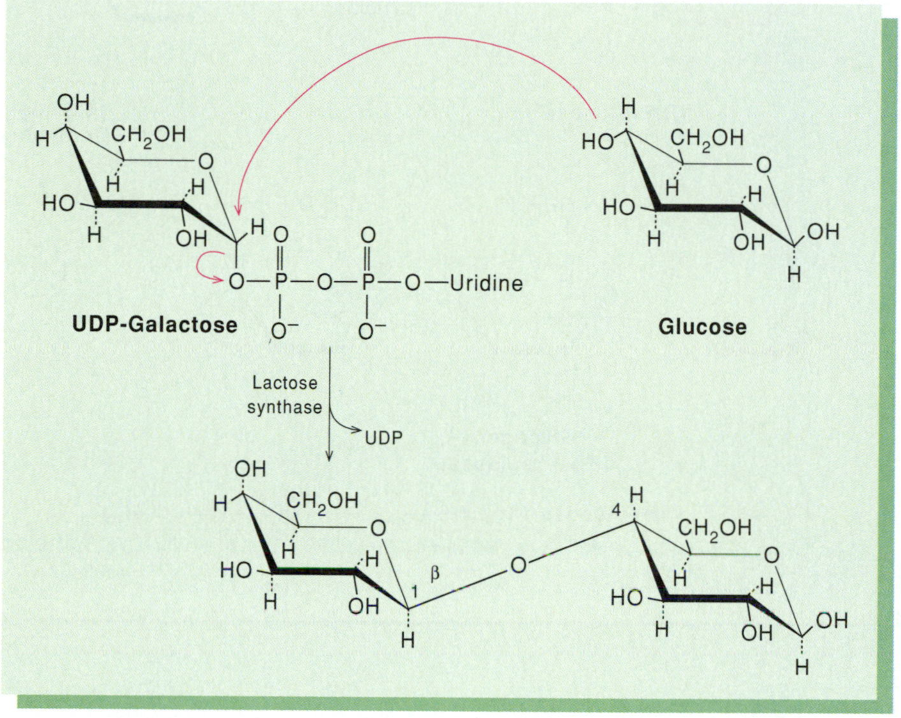

monomer is activated as in disaccharide synthesis (fig. 21.6). Sometimes this activation is supplied by a UDP-derivative, sometimes by an ADP-derivative (table 21.1). The $\Delta G°$ of this reaction is about -3.2 kcal/mole.

The enzyme glycogen synthase requires a primer oligosaccharide with at least four glucose residues, to which it adds successive glucosyl groups. In addition to $\alpha(1,4)$ bonds, glycogen contains $\alpha(1,6)$ bonds. The latter bonds are made by the branching enzyme amylo-(1,4-1,6)-*trans*-glycosylase. This enzyme transfers a terminal oligosaccharide fragment of six or seven glucosyl residues from the end of the main glycogen chain to the 6-hydroxyl group of a glucose residue somewhere in a glycogen chain. The reaction produces a branched-chain polymer from a straight-chain polymer, as shown in figure 21.7. As you might expect, the free energy change in this reaction is very small, since very similar chemical linkages are involved. In plant tissues, starch synthesis occurs by an analogous pathway that is catalyzed by amylose synthase. ADP-glucose is the preferred glucose donor.

Since the function of glycogen is to provide a readily accessible storage form of energy, its breakdown is just as important as its synthesis. Glycogen breakdown proceeds by a different route, involving the action of inorganic phosphate and the enzyme glycogen phosphorylase on the polymer (see chapter 13). One of the most interesting aspects of glycogen metabolism has to do with the intricate mechanism that controls synthesis and breakdown, which is related to the organism's energy requirements (see chapter 13).

Structural Polysaccharides Include Homopolymers and Heteropolymers

The most abundant structural polysaccharide is plant cellulose, a straight-chain homopolymer of glucose with a $\beta(1,4)$ linkage. Cellulose is formed by the same general mechanism as glycogen, using nucleoside diphosphate sugars. In addition, chitin in insects, which is a $\beta(1,4)$ homopolymer of *N*-acetylglucosamine (GlcNAc), is formed in a similar reaction from UDP-*N*-acetylglucosamine (UDP-GlcNAc).

The animal polysaccharide hyaluronic acid presents a variation characteristic of glycosaminoglycans (see table 6.1). Like many linear polysaccharides found in protein or lipid conjugates, hyaluronic acid consists of a strictly alternating sequence of two different hexoses (see table 6.1). The monomers of hyaluronic acid are *N*-acetylglucosamine and glucuronic acid (GlcUA). Hyaluronic acid is formed by successive addition of UDP-glucuronic acid and UDP-*N*-acetylglucosamine to the ends of a growing chain. This process involves two enzymes, one specific for the addition of each monomer.

Whereas most polysaccharide syntheses use nucleoside diphosphate sugars as substrates, in the formation of dextran the disaccharide sucrose serves as the substrate. The energy of

Figure 21.6

Elongation step in glycogen synthesis. The gross chemistry of this reaction is quite similar to that of lactose synthesis except that there is no inversion about C-1, since it proceeds by a different mechanism, probably similar to that used by starch synthase (see fig. 13.16).

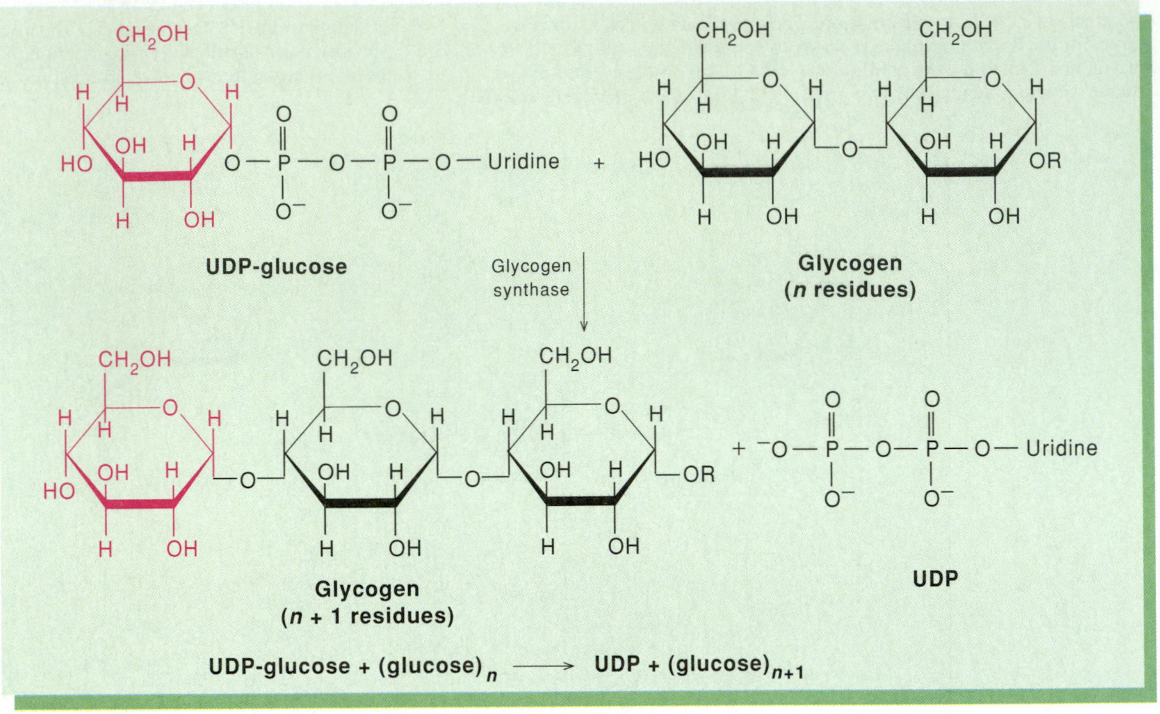

Table 21.1 Some Storage Polysaccharides				
Source	**Polysaccharide**	**Monosaccharide Component(s)**	**Glycosyl Donor**	**Polymer Structure**
Primarily muscle and liver cells of animals	Glycogen	Glucose	UDP-glucose	$\alpha(1,4)$ with $\alpha(1,6)$ branchpoints
Bacterial glycogen	Glycogen	Glucose	ADP-glucose	$\alpha(1,4)$ with $\alpha(1,6)$ branchpoints
Green algae	Amylose	Glucose	ADP-glucose	Linear $\alpha(1,4)$
Leaves, stem, roots, and seeds of higher plants	Amylopectin	Glucose	ADP-glucose	Linear $\alpha(1,4)$ with $\alpha(1,6)$ branchpoints
Some bacteria	Dextran	Glucose	Sucrose	Linear $\alpha(1,6)$ with $\alpha(1,2)$, $\alpha(1,3)$ or $\alpha(1,4)$ branchpoints

the glycosidic bond between glucose and fructose in this disaccharide drives the formation of an $\alpha(1,6)$ polymer of glucose according to the following reaction:

$$n \text{ Sucrose} \xrightarrow{\textbf{Dextran sucrase}} \text{dextran} + n \text{ fructose}$$

Dextrans formed by bacteria growing on the surface of teeth are an important component of dental plaque.

Oligosaccharide Biosynthesis in Higher Animals

One of the most exciting areas in contemporary biological research has to do with identifying the functions of the oligosaccharide moieties covalently attached to proteins and lipids. As we discussed in chapters 6 and 7, the carbohydrates of glycoproteins and glycolipids have a wide variety of structures that are based on a common theme. The variations generally involve

Figure 21.7

Schematic diagram showing the action of the "branching enzyme" in glycogen formation. A terminal hexasaccharide fragment is shown as being transferred from a 1,4 straight-chain linkage to a 1,6 branchpoint. No activation is involved because the energy change on reaction is very small.

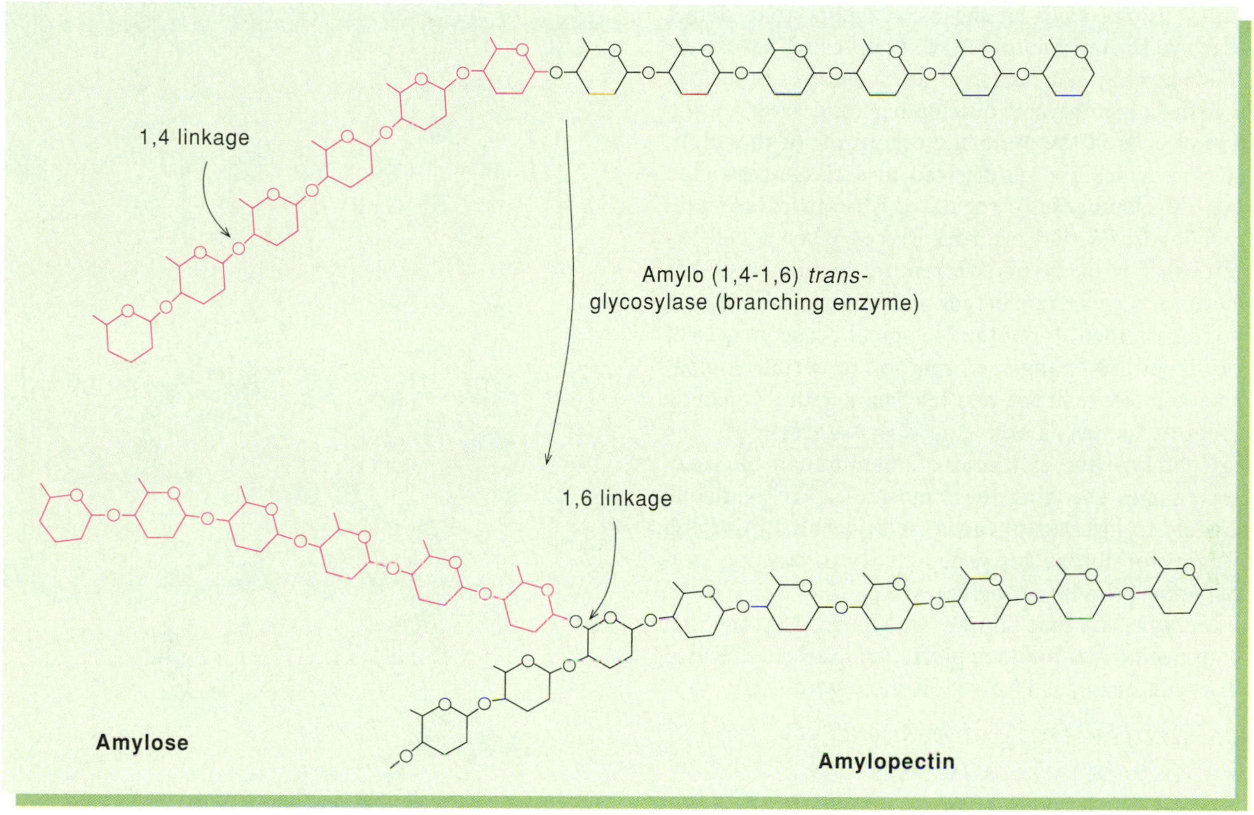

the number of different sugars, the ways in which they are linked together, their branching patterns, and the differences in the sugar combinations that terminate each branch. We know from nuclear magnetic resonance spectra of carbohydrates (see chapter 6), and also from the existence of antibodies that recognize specific combinations of sugars, that a change in the linkage of only one sugar of a complex branched structure may significantly change the conformation of an oligosaccharide. Thus the potential for carbohydrates to act as recognition groups (markers) for carbohydrate-binding proteins is enormous. In fact, there are several instances in which biologically important carbohydrate-protein recognition occurs. For example, influenza virus has a glycoprotein termed hemagglutinin (because it causes agglutination or clumping of red blood cells), which binds to cells via the sialic acid residues of cell surface glycoconjugates (glycoproteins or glycolipids). Some hemagglutinins bind only to sialic acid linked $\alpha(2,3)$ to galactose, while others bind only to sialic acid linked $\alpha(2,6)$ to galactose. These binding specificities are due to the presence of a particular amino acid at a single position in the sialic acid binding site of the hemagglutinin. This exquisite specificity shows the potential for the functional consequences of changes in carbohydrate structure. Other important roles of carbohydrate as recognition markers include the binding of cholera toxin to a specific glycolipid, GM_1, the binding of lysosomal enzymes via phosphorylated, N-linked oligomannosyl carbohydrates by the mannose-6-phosphate receptor, the binding of terminal sialyted, fucosylated lactosamine sugar sequences by cellular adhesion molecules, and the recognition of mammalian eggs by sperm. It is likely that the species specificity of egg fertilization is due to the inability of sperm from one species to recognize the oligosaccharides of glycoproteins in the zona pellucida that surrounds the eggs of another species.

The carbohydrate-binding proteins of mammalian cells are termed vertebrate lectins. Like the plant lectins described in chapter 6, a vertebrate lectin is a protein of nonimmune origin that binds carbohydrates and may cause agglutination of cells or precipitation of glycoproteins. Some vertebrate lectins are integral membrane proteins, while others are soluble glycoproteins that are found in tissue fluids. They have been classified into two major groups, depending on whether they require calcium for binding (C-type) or whether they depend on a thiol group (—SH) for binding (S-type). Correlative evidence implies several biological functions for vertebrate lectins, including clearance of aged red blood cells from the circulation, tissue-specific localization of lymphocytes (box 21B), transport of lysosomal enzymes to lysosomes, and specific connections between nerve cells.

Studies using plant lectins as reagents that detect changes in carbohydrate structures (see chapter 6) have been instrumental in showing that carbohydrate structures at the cell surface change during embryonic development and, later, during differentiation of cells in the maturing organism. In the adult, certain oligosaccharides are synthesized in a tissue-restricted manner. Such developmentally regulated structures are precisely the carbohydrates that are reexpressed when a cell becomes cancerous. It is likely that vertebrate lectins exist that recognize these carbohydrate changes and initiate a biological response. Progress in identifying the biological consequences of carbohydrate structural changes, in relation to certain biological states, should improve in the next few years with the cloning of many vertebrate lectins as well as glycosyltransferases—the enzymes that catalyze the synthesis of mammalian oligosaccharides. Once clones encoding these molecules are available, it will be possible to investigate causal relationships by trying various strategies for abrogating gene activity or causing molecules to be inappropriately expressed.

In the discussion that follows, we will see that the rich variety of oligosaccharides found in glycoconjugates is reflected by a rich variety of enzymes involved in their synthesis.

Oligosaccharides Are Synthesized by Specific Glycosyltransferases

All oligosaccharide structures of glycoconjugates are synthesized by glycosyltransferases according to the following general reaction:

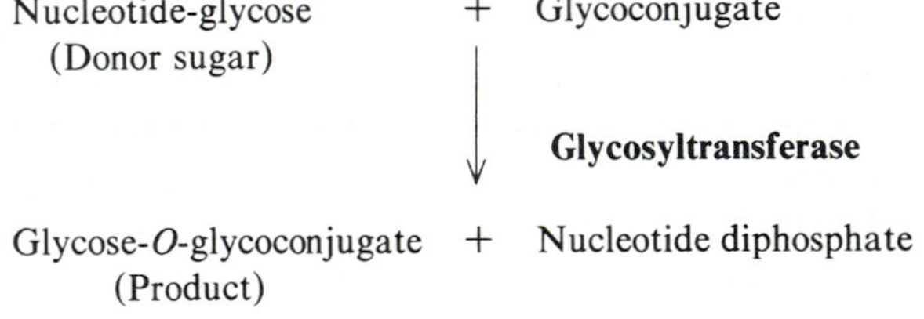

The reaction is driven towards the formation of product by the hydrolysis of the released nucleotide diphosphate. Most glycosyltransferase enzymes require a divalent cation (usually manganese) and most are highly specific for the acceptor substrate, nucleotide-sugar, and linkage by which the new sugar is attached to its acceptor. For example, there are two different enzymes that attach GlcNAc in $\beta(1,2)$ linkage to mannose for the synthesis of a biantennary N-linked carbohydrate (fig. 21.8). This is because the carbohydrate acceptor to which the GlcNAc is added differs markedly in each case. The specificity of glycosyltransferases is generally such that there exists virtually one unique enzyme for each and every type of glycosidic bond known.

Figure 21.8

Glycosyltransferases are specific for the acceptor, the sugar transferred, and the linkage whose formation they catalyze. The *N*-acetylglucosaminyltransferases termed GlcNAc-TI and GlcNAc-TII both transfer *N*-acetylglucosamine to mannose in ß(1,2) linkage, but their acceptor substrates are completely different. This sequence of biosynthetic steps was deduced from analyses of a mammalian cell mutant that lacks GlcNAc-TI activity.

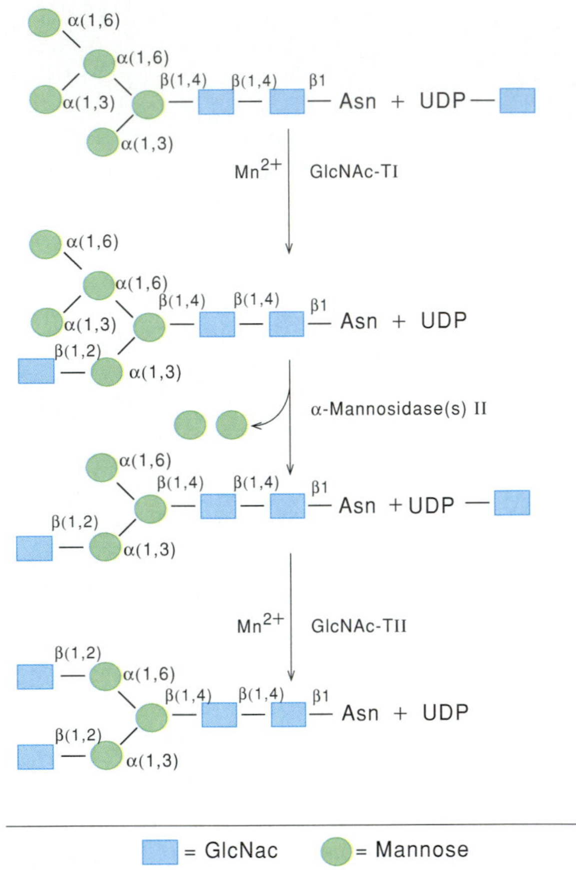

Oligosaccharides Are Synthesized in a Concerted Fashion

In the biosynthesis of oligosaccharides, each sugar is added in a stepwise and orderly fashion to the previous sugar. For example, in the case of O-glycosidically linked carbohydrates attached to Ser or Thr, the first sugar is added directly to the protein by a specific *N*-acetylgalactosaminyl glycosyltransferase. The second sugar is subsequently added by a different glycosyltransferase and so on until a mature structure is obtained. The final structure of an oligosaccharide will depend on the different glycosyltransferases present in a cell, the conformation of the protein during sugar addition, and the speed with which the protein traverses the secretory pathway. Thus O-glycosidically linked carbohydrates can be as simple as one sugar (GalNAc or GlcNAc) or can contain many sugars in a variety of linkages and branching patterns.

21B
BOX

ELAM-1: A Cell Adhesion Molecule with a Lectin Domain that Functions to Bind Leukocytes

Since many different molecules with lectin domains have now been identified, it appears that there is enormous potential available to the cell for utilizing carbohydrate–protein interactions for physiological recognition events. For example, consider the well-characterized endothelial cell LEC-CAM (cell adhesion molecule with a lectin domain) known as ELAM-1 (endothelial leukocyte adhesion molecule 1). This glycoprotein is synthesized after endothelial cells are activated by a cytokine and is expressed within 4–6 h as a transmembrane molecule on the cell surface. Incidentally, LEC-CAM molecules are now called selectin.

ELAM-1 functions to bind leukocytes and facilitate their transport from the bloodstream across the endothelial cell layer of postcapillary venules to sites of inflammation or tissue damage. The means by which ELAM-1 recognizes the correct subset of leukocytes from all the cells in the bloodstream is at least in part via their carbohydrates. The structure that appears to be necessary and sufficient for ELAM-1 binding *in vitro* is NeuNAcα(2,3) Galβ(1,4)Fucα(1,3)GlcNAc(β1)(see illustration). However, there is evidence that ELAM-1 ligands are shed from the leukocyte cell surface; if so, the physiological ligand may be a particular glycoconjugate that carries this carbohydrate structure. This carbohydrate is an example of a developmentally regulated structure that is expressed at different stages of ontogeny and is often reexpressed in cancer cells. Its availability at the cell surface may be controlled by the regulated expression of a specific α(1,3)fucosyltransferase gene.

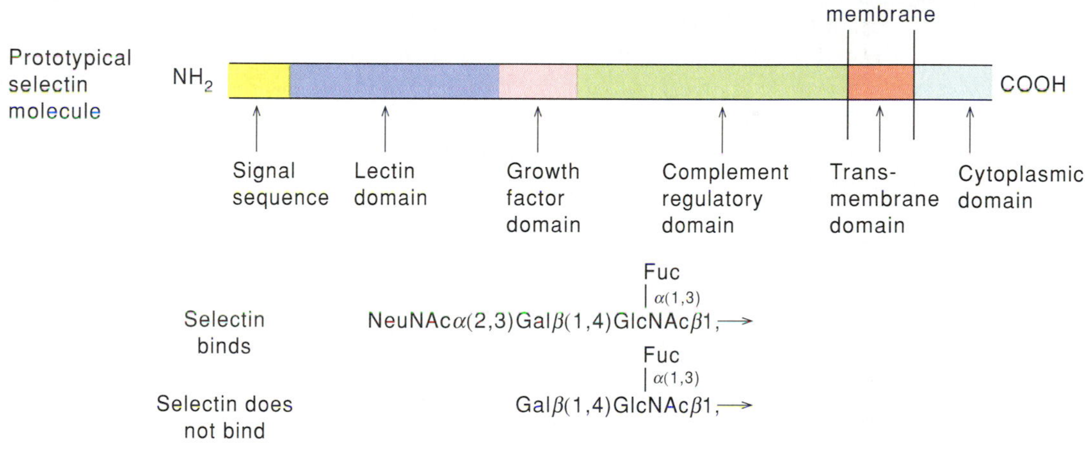

In addition to high specificity of glycosyltransferases, another major reason for the ordered addition of oligosaccharides is that the glycosyltransferases that catalyze sugar addition are compartmentalized within the cell. Therefore, during biosynthesis glycoproteins come in contact with particular glycosyltransferases when they traverse different cellular compartments. Nearly all glycoproteins (with the exception of cytoplasmic glycoproteins carrying *O*-GlcNAc or *O*-mannose residues) have their oligosaccharide portions synthesized in the lumen of membrane-bound compartments.

The early stages of glycoprotein synthesis are further complicated by the fact that oligosaccharides are added and often modified before protein synthesis is completed. We will not discuss the mechanism of protein synthesis in any detail until chapter 29, but there are certain facts that must be presented here if we are to understand how oligosaccharide synthesis is associated with glycoprotein synthesis.

In a eukaryotic cell, protein synthesis takes place on ribosomes (see chapters 1 and 5). All the ribosomes in a cell are identical except that they become transiently associated with different messenger RNA molecules, which determine the specific protein that will be synthesized. Ribosomes in the process of protein synthesis exist as independent bodies in the cell cytosol or as membrane-complexed bodies on the endoplasmic reticulum. The endoplasmic reticulum is a membrane-bounded organelle that is found in the cytoplasm (see fig. 1.1). Ribosomes bind to the endoplasmic reticulum by a complex sequence of events. All proteins destined to be N-glycosylated and most that are O-glycosylated are sequestered from the cytoplasm by translocation during synthesis into the lumen of the endoplasmic reticulum (fig. 21.9). These proteins are marked by the presence of a short (~20) hydrophobic amino acid stretch termed a signal sequence at their extreme NH_2 terminus. During

Figure 21.9

Diagram of events involved in the targeting and translocation of proteins destined to be completed in the endoplasmic reticulum. Proteins that include a signal sequence of approximately 20 hydrophobic amino acids at their NH_2 terminus are recognized by a signal recognition particle (SRP). This complex of proteins and 7S RNA binds to the signal sequence of the nascent protein and the large ribosomal subunit. SRP binding causes translation arrest and the targeting of the translocation complex to the endoplasmic reticulum (ER) where it is bound by the SRP receptor as well as a ribosome receptor protein. As protein synthesis resumes, SRP and its receptor are released from the ribosome, and the growing chain traverses the membrane of the endoplasmic reticulum and enters the lumen. Signal peptidase degrades the signal sequence, creating a new, NH_2 terminus. When protein synthesis is complete, soluble ER proteins are released into the lumen of the ER. Proteins with hydrophobic regions elsewhere in the protein may span the ER membrane so that their C terminus or N terminus or both are on the cytoplasmic face of the ER membrane.

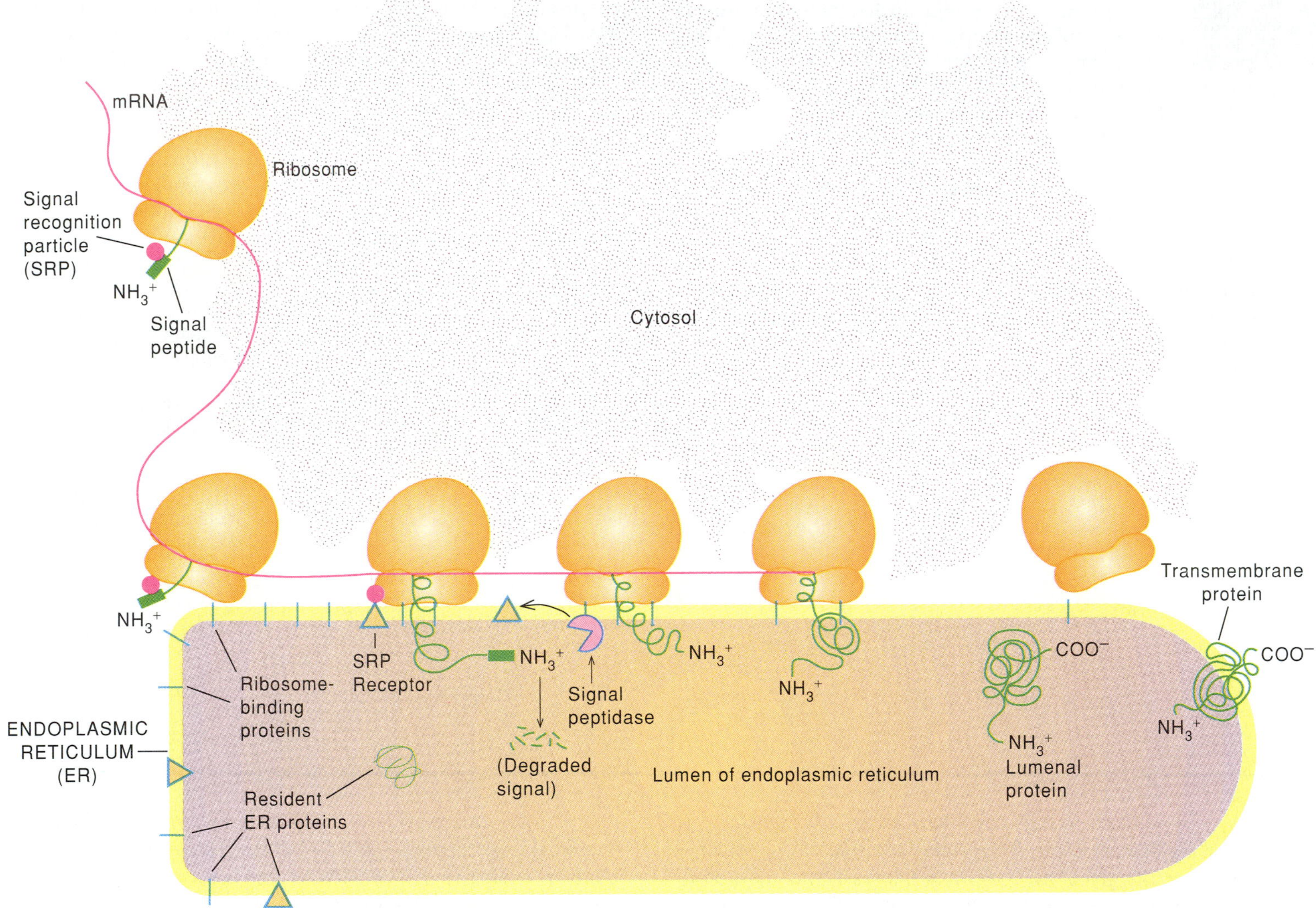

translation, the signal sequence emerges first from the large ribosomal subunit. The signal sequence and ribosome are recognized by a complex termed the signal recognition particle (SRP). The binding of SRP blocks further translation until binding between SRP and the SRP receptor of endoplasmic reticulum membranes is achieved. Once the ribosome complex is bound via receptor proteins that recognize ribosomes and the SRP receptor, translation resumes and the protein being synthesized is extruded into the endoplasmic reticulum.

The biosynthesis of O- and N-linked oligosaccharides follows two different routes. N-linked oligosaccharides are added to certain Asn-X-Ser/Thr sequences as a complex of sugars originally synthesized on a lipid carrier. There may be some glycoproteins that receive O-linked *N*-acetylgalactosamine in a compartment just beyond the endoplasmic reticulum, but most O-linked glycosylation occurs in the Golgi complex. This is a membranous organelle to which all nonresident proteins of the endoplasmic reticulum are transferred by a budding process diagrammed schematically in figure 21.10. In the Golgi complex, oligosaccharide biosynthesis continues with the sequential addition (and removal) of sugar residues. After exiting the ***trans*** Golgi network by a budding process, glycoproteins are delivered to the plasma membrane as resident integral membrane proteins, or they are secreted from the cell or sent to lysosomes.

Figure 21.10

Schematic diagram showing the relative locations of nucleus, endoplasmic reticulum (ER), Golgi complex, *trans* Golgi network, and plasma membrane. Glycoproteins synthesized in the lumen of the ER pass to the *cis* cisterna of the Golgi complex by a sequential membrane budding and fusion mechanism. The Golgi cisternae are classified into *cis*, medial, and *trans* in the order of increasing distance from the nucleus. Mature glycoproteins exit from the *trans* Golgi network in membrane-bound secretory vesicles. Depending on the nature of the membranes, the vesicles have one of several different fates. Following fusion with other membranes, their contents are delivered outside the cell, to the plasma membrane, or to other organelles inside the cell, such as lysosomes (see fig. 21.14).

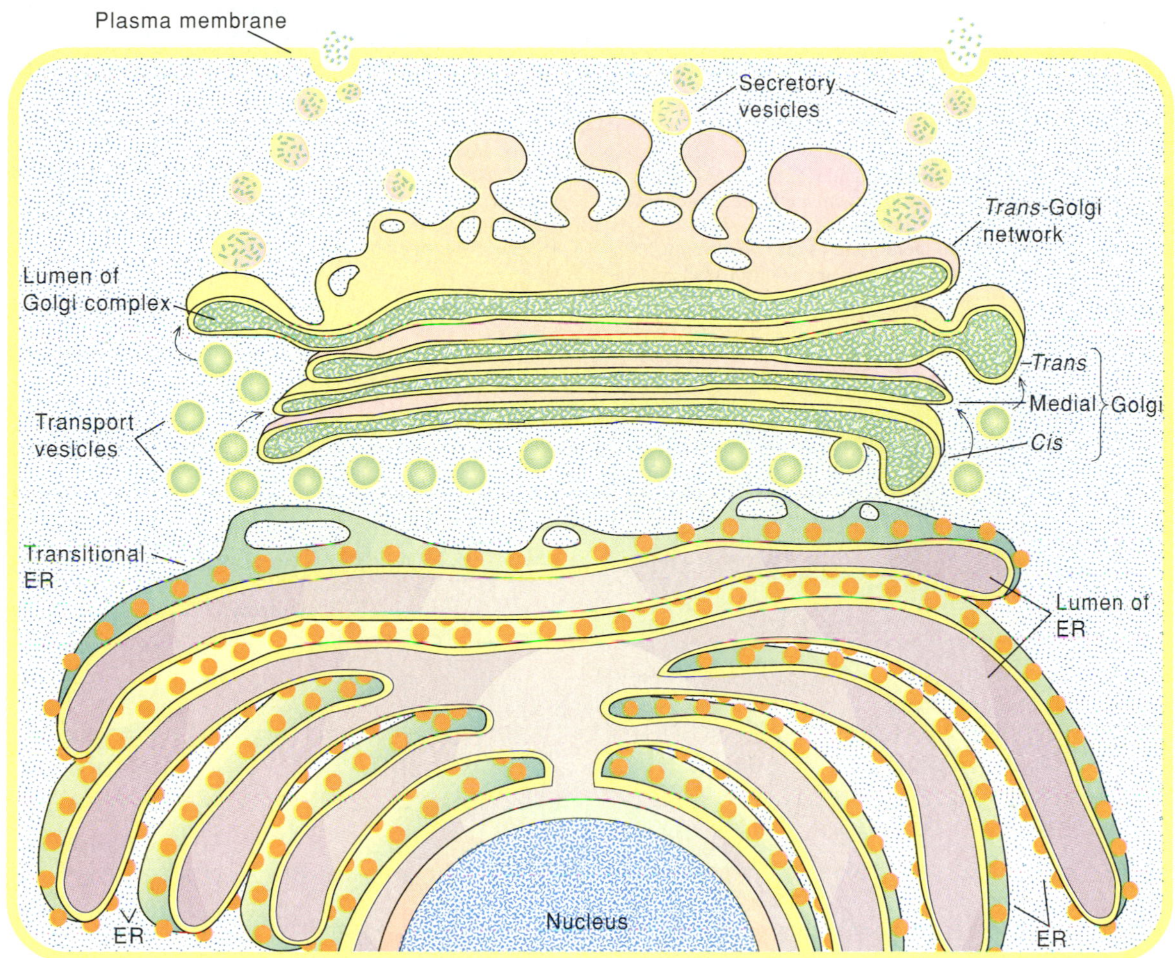

Biosynthesis of N-Linked Oligosaccharides

The N-linked oligosaccharides are divided into three major classes: complex or lactosamine-containing (lactosamine is the disaccharide Galβ(1,4)GlcNAc), hybrid, and oligomannosyl. An example of a structure in each class is given in chapter 6 (see fig. 6.23). All structures include a common core of five sugars linked to Asn: Manα(1,3)Manα(1,6)Manβ(1,4)GlcNAcβ(1,4) GlcNAc(β1),Asn. The reason for this is that all the structures are initially synthesized by a common pathway that appears to exist in all eukaryotes, including yeast. This pathway involves the sequential addition of sugars to dolichol phosphate, a membrane-associated polyprenol lipid (fig. 21.11).

Dolichol phosphate (Dol-P) serves as the substrate for a glycosyltransferase that adds phospho-*N*-acetylglucosamine from UDP-GlcNAc to form Dol-P-P-GlcNAc. The first GlcNAc transfer and the subsequent addition of four sugar residues occur on the cytoplasmic side of the endoplasmic reticulum membrane as shown in figure 21.12. The first seven sugars are transferred to the growing oligosaccharide from their appropriate nucleotide-sugar conjugates (UDP-GlcNAc or GDP-Man), which are present in the cytoplasm. At this stage the Dol-P-oligosaccharide portion is translocated to the lumenal side of the endoplasmic reticulum membrane. The next sugars are added from Dol-P-Man and Dol-P-Glc precursors rather than the corresponding nucleotide sugar derivatives. The Dol-P sugars themselves are synthesized by the transfer of mannose from GDP-mannose or of glucose from UDP-glucose to dolichol phosphate to form Dol-P-Man and Dol-P-Glc, respectively. These reactions probably occur on the cytoplasmic face of the endoplasmic reticulum, where Dol-P has access to both nucleotide sugars. The final structure assembled on dolichol contains two *N*-acetylglucosamines, nine mannoses, and three glucoses linked in the manner shown in figure 21.12. This oligosaccharide moiety is transferred in a block by the enzyme oligosaccharyltransferase to the Asn residues in certain Asn-X-Ser(Thr) sequences of growing peptide chains. The transfer and the subsequent biosynthesis of N-linked carbohydrates are shown schematically in figure 21.13.

Figure 21.11

The structure of dolichol phosphate. The hydrocarbon portion of the molecule has a high affinity for membrane structures. The phosphate end forms an activated complex with oligosaccharide intermediates.

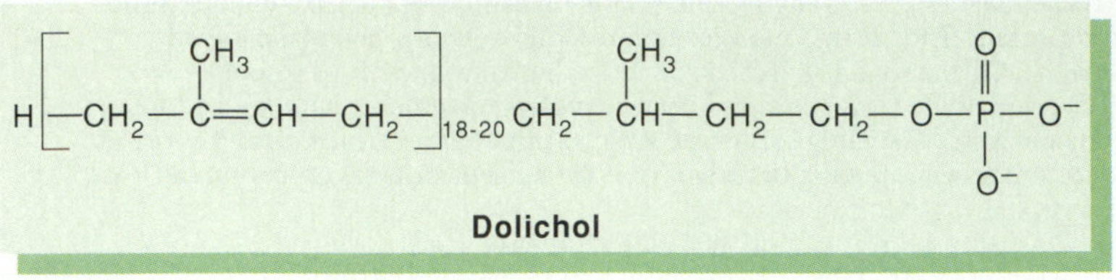

$$H\left[-CH_2-\underset{}{\overset{CH_3}{\overset{|}{C}}}=CH-CH_2-\right]_{18\text{-}20}CH_2-\overset{CH_3}{\overset{|}{CH}}-CH_2-CH_2-O-\underset{O^-}{\underset{|}{\overset{O}{\overset{\|}{P}}}}-O^-$$

Dolichol

Figure 21.12

Synthesis of the dolichol-linked oligosaccharide. The reaction starts on the cytoplasmic side of the endoplasmic reticulum. Seven single-step reactions lead to the heptasaccharide that is linked to dolichol phosphate. The heptasaccharide is translocated across the membrane of the endoplasmic reticulum to the lumen, where seven additional one-step additions take place. The activated sugars for these additions are formed on the cytosolic side and translocated to the lumen as hexose-P-X-dolichol complexes. The initial step in the pathway is inhibited by tunicamycin. The synthesis of Dol-P-Man is inhibited by amphomycin.

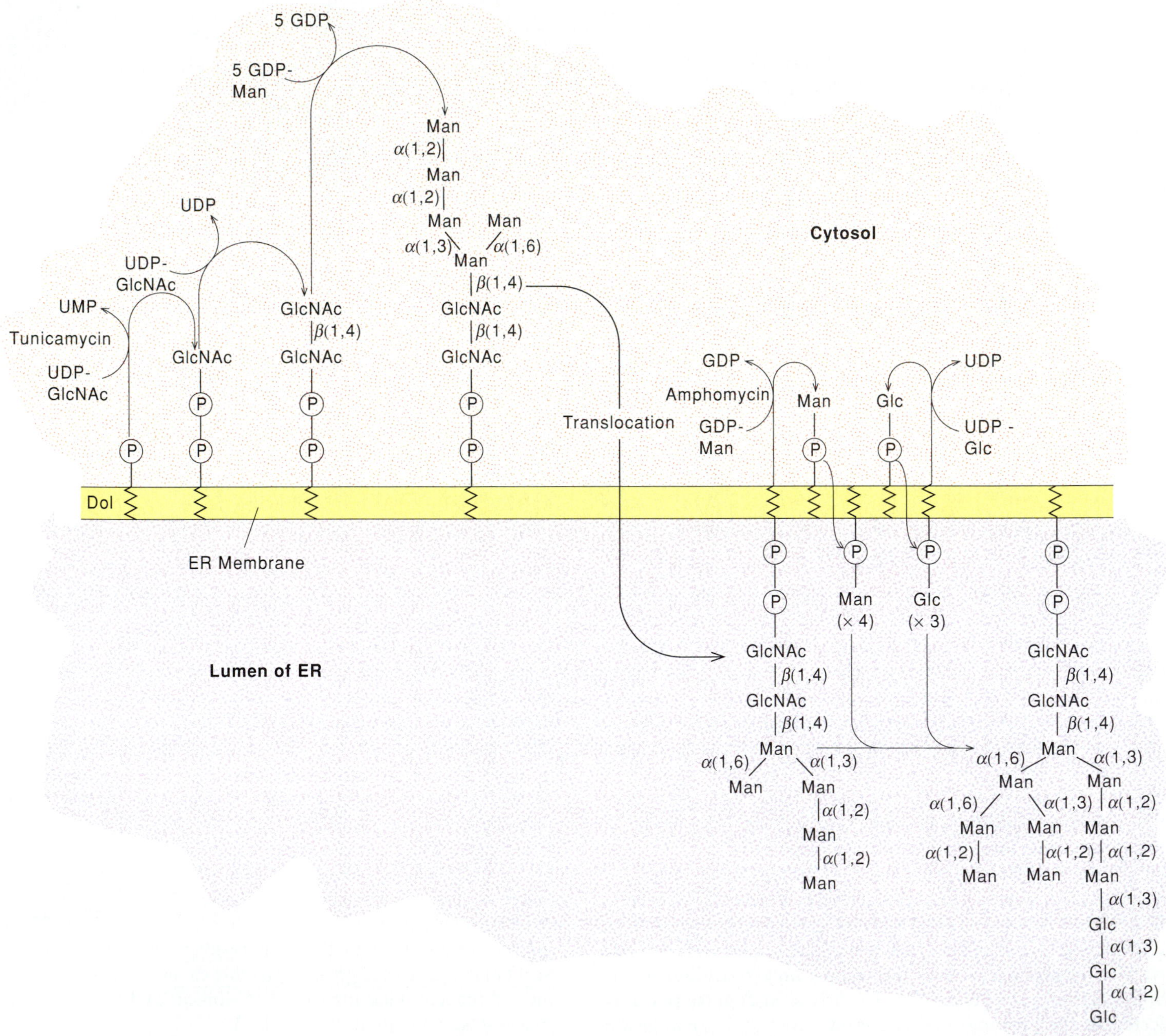

Figure 21.13

Schematic diagram illustrating the transfer of oligosaccharide to protein in the endoplasmic reticulum and the subsequent processing and maturation of oligosaccharides in the Golgi complex. In the endoplasmic reticulum the oligosaccharide synthesized on dolichol-P is transferred to the Asn(N) of an Asn-X-Ser(Thr) sequence in the growing peptide chain. Subsequent trimming of the oligosaccharide occurs by specific exoglycosidases that sequentially remove terminal glucoses and a specific mannose residue. When synthesis of the protein portion is complete, the glycoprotein is transferred to the lumen of the *cis* membranes of the Golgi complex. For most glycoproteins further processing of the $Man_8GlcNAc_2$ oligosaccharide occurs in this compartment. However, soluble lysosomal enzymes are recognized by a phospho-*N*-acetylglucosaminyltransferase, which transfers P-GlcNAc to the C-6 of mannose residues. Subsequently an α-*N*-acetylglucosaminidase removes the GlcNAc, exposing Man-6-P which is recognized by the Man-6-P receptor. Lysosomal enzymes may have several N-linked carbohydrates, some of which are converted to typical complex carbohydrates in the later compartments of the Golgi complex.

In the medial Golgi the synthesis of a biantennary structure terminating in GlcNAc is achieved (see fig. 21.8). Also in this compartment additional GlcNAc residues may be added to the core mannose residues to give as many as seven branches. Fucose, galactose, and sialic acid residues are added in the *trans* Golgi and *trans* Golgi network compartments.

The membranes of the Golgi compartments contain nucleotide-sugar translocases specific for transferring each nucleotide-sugar across the membrane.

Inhibitors can block the pathway at specific steps, causing the accumulation of biosynthetic intermediates. Castanospermine inhibits the α-glucosidase that removes the first glucose; deoxymannojirimycin inhibits the α-mannosidase I that removes mannose from the Man_8 intermediate, and swainsonine blocks the α-mannosidase II that removes mannose from the Man_5 intermediate. In the presence of swainsonine, hybrid N-linked carbohydrates are synthesized.

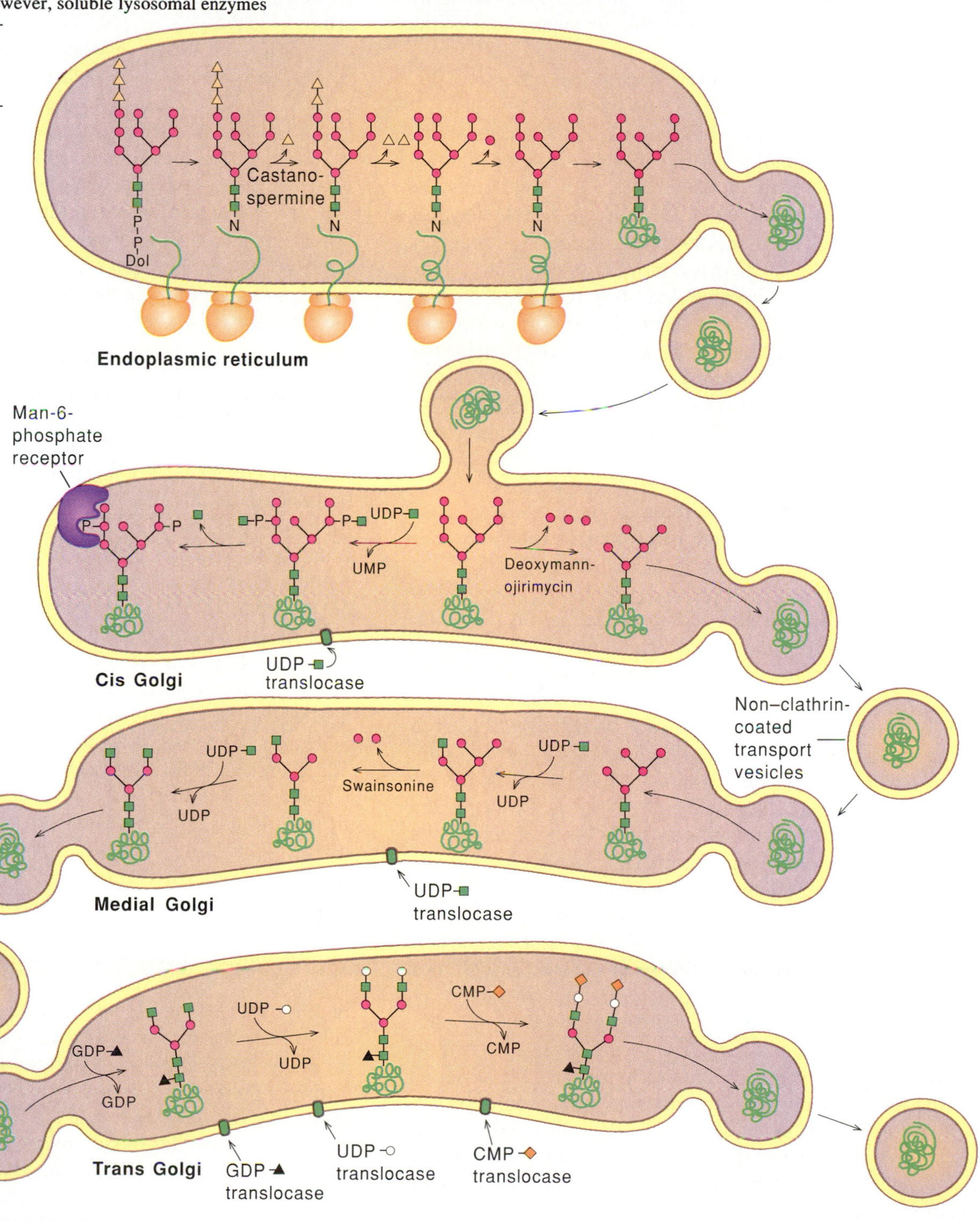

Once attached to the protein, the oligosaccharide is processed (or trimmed) by specific exoglycosidases that sequentially remove terminal sugars (see fig. 21.13). The glucoses are removed by at least two different α-glucosidases, α-glucosidase I, which removes the first glucose, and α-glucosidase II, which removes the next two glucoses. Following this, the first mannose is removed by a specific endoplasmic reticulum α-mannosidase. Yeast and mammalian cells appear to follow exactly the same pathway to the generation of the Man_8 intermediate. At this point many glycoproteins are translocated by vesicular transport to the lumen of the *cis* membranes of the Golgi complex.

In the *cis* Golgi, either the Man_8 oligosaccharide is further processed or, if it is attached to a lysosomal hydrolase, it is modified by the addition of P-GlcNAc residues that are transferred from UDP-GlcNAc by an α-phospho-*N*-acetylglucosaminyltransferase. This transferase acts only on the oligomannosyl carbohydrates of certain lysosomal enzymes, which it recognizes by a specific protein–protein interaction. Once transferred, the GlcNAc is removed by an α-*N*-acetylglucosaminidase, thereby exposing Man-6-P residues for interaction with the Man-6-P receptor. Lysosomal hydrolases subsequently traverse all compartments of the Golgi and therefore may also contain N-linked carbohydrates of the complex type at other glycosylation sites. In yeast, the mannose oligosaccharide is elongated by the addition of many mannose residues in the Golgi complex.

The α-mannosidase I in the *cis* Golgi removes mannose residues from the Man_8 intermediate in mammalian cells. Altogether three mannoses are removed to generate the Man_5 substrate for *N*-acetylglucosaminyltransferase I (GlcNAc-TI), whose action initiates the conversion of oligomannosyl carbohydrates to hybrid or complex carbohydrates (see fig. 21.8). The addition of the first branch GlcNAc residue (which occurs in medial Golgi cisternae), generates the substrate for the last processing glycosidase, α-mannosidase II. The second, branch GlcNAc residue is added by GlcNAc-TII (see fig. 21.8), generating an intermediate that can be acted on by several other GlcNAc branching transferases and by a fucosyltransferase. The addition of all GlcNAc residues probably occurs in the medial Golgi, while the fucose is probably added in the *trans* Golgi. Also in the *trans* Golgi, branch GlcNAc residues are substituted with galactose, which is in turn substituted with sialic acid. Many other types of terminal sugar additions have been described, including the presence of long polylactosamine (Galβ(1,4)GlcNAc) sequences, fucose on galactose and GlcNAc residues, GalNAc and galactose on galactose residues and so on. The structure of mature complex N-linked carbohydrates may vary enormously as the result of a variety of factors. In addition to a wide range of different sugar combinations, the presence of sulfated and phosphorylated residues has also been described. Most of these latter modifications are thought to occur in the *trans* Golgi or *trans* Golgi network. These compartments are variable in size and potentially also in function in different cell types.

The specific location of the glycosylation reactions of N-linked carbohydrate biosynthesis is a major factor determining final oligosaccharide structure. Glycoproteins that reside in a particular compartment bear oligosaccharides that reflect the pathway to that point. Similarly, if a glycoprotein accumulates in a certain compartment because of a mutation that prevents normal transit (e.g., causes misfolding), it will bear the oligosaccharides typical of its final location (box 21C). It is therefore possible to tell in which compartment a glycoprotein resides by determining the structure of its N-linked carbohydrates.

Not all of the vesicles that bud from the *trans* Golgi network are destined for the plasma membrane (fig. 21.14). Lysosomal hydrolases bound to the Man-6-P receptor are sorted into vesicles coated with clathrin that fuse with an acidic prelysosome. In this environment of acidic pH (≤ 5), the binding between Man-6-P receptors and lysosomal hydrolases is broken and they subsequently are sorted independently. Vesicles carrying lysosomal hydrolases ultimately deliver them to a mature lysosome, while the Man-6-P receptor (which is an integral membrane protein) cycles between the plasma membrane, Golgi membranes, and endosomes.

Biosynthesis of O-Linked Oligosaccharides

Glycoproteins in the secretory pathway receive GalNAc in O-glycosidic linkage from UDP-GalNAc via a transferase that acts directly on Ser or Thr residues of proteins. No dolichol-linked intermediates are involved. Most kinetic labeling studies indicate that GalNAc is added in Golgi membranes, although there are a few exceptions that hint at the existence of a GalNAc transferase in a compartment just beyond the endoplasmic reticulum and before the *cis* Golgi. The second sugar in O-linked oligosaccharides is usually galactose, and that is added in the *trans* Golgi compartment. The remainder of the O-linked oligosaccharide is synthesized in the Golgi by sequential sugar additions from nucleotide-sugars catalyzed by specific glycosyltransferases. There is a certain order to the sequence of sugars added, but no established core sequence after the GalNAc is added to the protein. Cytoplasmic and nuclear proteins with *O*-GlcNAc residues that are not further modified are synthesized by a cytoplasmic glycosyltransferase.

The O-linked oligosaccharides of membrane glycoproteins are usually not very large although they may exhibit branching and various terminal sugar sequences. However, in body fluids very large O-linked structures are found on proteins termed mucins. These oligosaccharides often terminate with sugars that are highly immunogenic. These sugars form the basis of the human blood type, and are therefore known as blood group substances. Because blood-group-specific sugars are expressed on many O-linked oligosaccharides of cells and fluids, and are highly immunogenic, individuals will mount an immune response against blood that is of a different type. For this reason, individuals of similar blood groups can accept blood from each other, but individuals of different types frequently cannot.

Assay for Transit Between Golgi Membranes

The compartmentalization of glycosylation reactions allows us to assay the transit between Golgi membranes. Donor Golgi membranes lacking the GlcNAc-TI enzyme cannot further modify the oligosaccharides of a marker glycoprotein even if donor-donor membrane transport occurs. However, when a marker glycoprotein is transported into an acceptor Golgi membrane compartment prepared from wild-type cells that have GlcNAc-TI activity, GlcNAc is added to the oligosaccharides of the marker glycoprotein. With the help of radio-labeled UDP-GlcNAc we can therefore measure transport between the *cis* and medial Golgi compartments. Similar assays using different mutants can be used to follow transport between ER and Golgi or between different medial and *trans* compartments. Such assays have been important in identifying molecules involved in vesicular transport reactions.

Figure 21.14

Sorting of proteins for their final destination occurs in the *trans* Golgi network (TGN). At 20° C secreted proteins accumulate in the TGN. By raising the temperature to 37° C, the sorting and secretion of synchronized populations of marker proteins can be studied. Two major routes to the plasma membrane have been identified—the constitutive pathway and the regulated pathway. Proteins in the latter pathway (including many prohormones) accumulate in secretory granules that fuse with the plasma membrane upon receiving a biochemical stimulatory signal. As the name suggests, passage of integral membrane proteins and soluble proteins through the constitutive pathway occurs constantly, with no requirement for a stimulatory signal. Soluble lysosomal hydrolases bound to the Man-6-P receptor exit from the TGN in clathrin-coated vesicles. These vesicles fuse with an acidic, prelysosomal compartment where the lysosomal hydrolases are released from the Man-6-P receptor and are ultimately delivered to lysosomes. The Man-6-P receptor does not go to lysosomes but recycles between the plasma membrane, Golgi membranes, and prelysosomes.

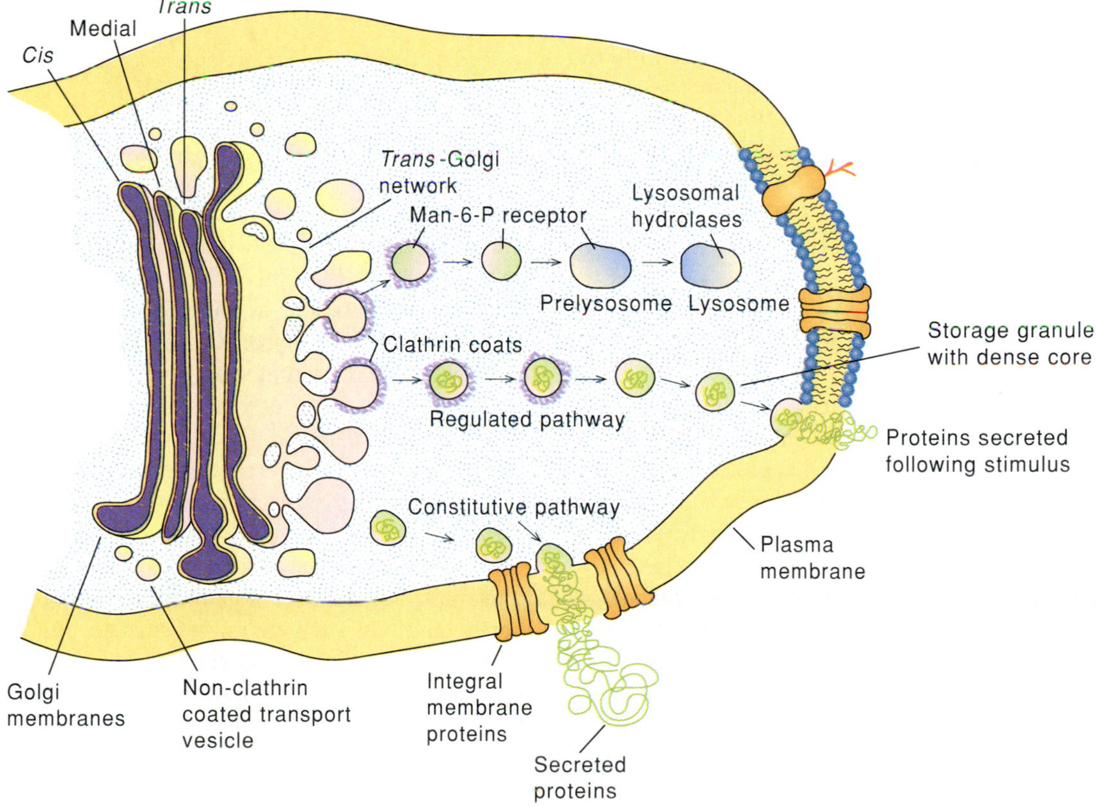

Table 21.2 The Human ABO Blood Group Scheme	
Gene Type	**Antigen**
AA	A
AB	A,B
BB	B
AO	A
BO	B
OO	—

One of the better-known blood-grouping schemes is the ABO scheme. Blood is considered to be of type A, B, or O. All individuals contain two genes for blood type. The possible combinations give rise to six different gene types and four different combinations of blood group antigens in different individuals (table 21.2). Cells of these individuals carry A antigens (in gene types *AA* or *AO*), B antigens (in gene types *BB* or *BO*), a mixture of A and B antigens (in gene type *AB*), or neither of these antigens (in gene type *OO*). The explanation of these correlations between gene types and blood group antigens is that the *A* gene and the *B* gene encode different glycosyltransferases. The *A* gene encodes a glycosyltransferase that catalyzes the addition of a terminal *N*-acetylgalactosamine (GalNAc) residue onto a core oligosaccharide, and the *B* gene encodes a similar enzyme that adds a galactose (Gal) residue to the same site (fig. 21.15). When *A* and *B* genes are present, both structures are found, but when only *O* genes are present, the site on the oligosaccharide is left unsubstituted. The presence or absence of these glycosyltransferases is readily detectable in the milk of the lactating female.

Specific Inhibitors and Mutants Are Used to Explore the Roles of Glycoprotein Carbohydrates

The functions of the carbohydrate moieties of glycoproteins are difficult to study because of the intimate physical association between oligosaccharides and the protein backbone. However, there are several ways to produce glycoproteins with carbohydrates that are severely truncated or missing entirely, so that we can deduce the function of the affected carbohydrates.

For glycoproteins that contain N-linked carbohydrates, we can completely inhibit oligosaccharide addition by synthesizing the glycoprotein in the presence of tunicamycin (see fig. 21.12). The protein portion of many glycoproteins is synthesized and translocated through the secretory pathway essentially normally in the presence of tunicamycin and may be studied by functional assays. Alternatively, we can eliminate each glycosylation sequence by site-directed mutagenesis of the cloned glycoprotein. Following transfection, the protein portion devoid of carbohydrates will be produced. This approach has the advantage that the altered protein is synthesized in a normal cell in contrast to a drug-treated cell, in which all glycoproteins are affected, as happens with tunicamycin treatment. An analogous approach is to introduce cloned glycoprotein genes into *E. coli,* which has none of the tranferases required for eukaryotic N- or O-linked glycosylation. However, *E. coli* has other modifying enzymes that may give rise to a protein with modifications not usually associated with it in mammalian cells. Therefore, in structure/function studies we usually try to synthesize mammalian proteins in a mammalian host.

Figure 21.15

The structure and reactions at the oligosaccharide termini of the ABO human blood group antigens. Two enzymes add different hexoses to the termini of O-linked glycoproteins. Individuals may carry both or neither or only one of these enzymes. Individual differences are reflected in the structures of their blood group antigens. These differences are genetically inherited. The *A*, *B*, and *O* genes have been cloned and it seems they are all derived from a single gene. Only four nucleotide differences were detected between *A* and *B* genes, and these must be responsible for the different specificities of the A and B transferases. The *O* gene produces a nonfunctional protein.

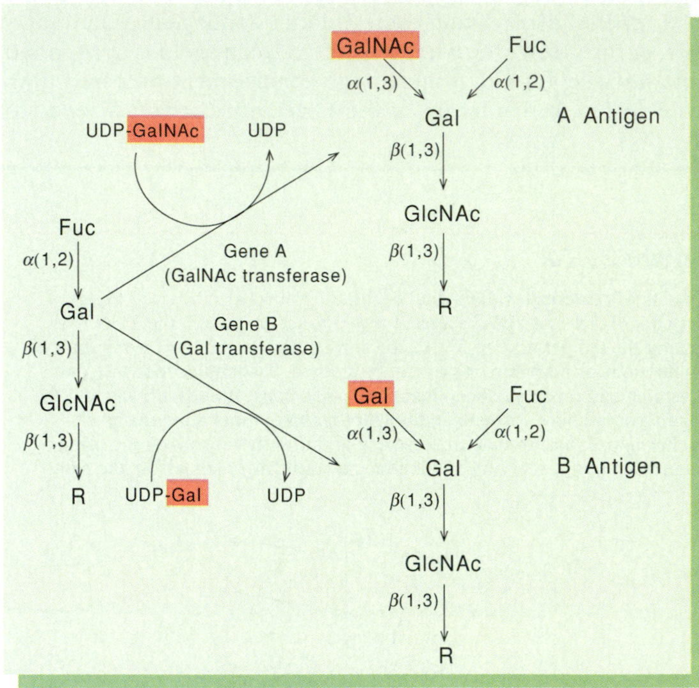

Although several glycoproteins have been efficiently produced without their oligosaccharides, many glycoproteins cannot be synthesized devoid of carbohydrates because they require oligosaccharides for proper folding and translocation competence. In such cases, we instead aim to produce glycoproteins with truncated carbohydrates, to see the effect of the new structures on the biological activity of the glycoprotein. One way to achieve this aim is to use inhibitors that stop oligosaccharide processing, such as castanospermine, deoxymannojirimycin, or swainsonine (table 21.3). The glycoproteins thus produced have immature N-linked carbohydrates that are quite distinct in structure from mature complex oligosaccharides. A complementary approach, which has the advantage of producing glycoproteins with truncated oligosaccharides that are

Table 21.3
Effects of Certain Inhibitors on Glycoprotein Synthesis

Glycosylation Inhibitor	Major N-linked Carbohydrates Found on Glycoproteins in Presence of Inhibitor
Castanospermine	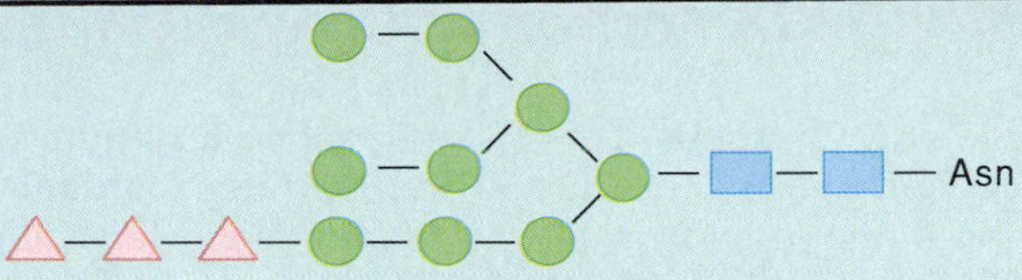 Asn ($Glc_3Man_9GlcNAc_2Asn$)
Deoxymannojirimycin	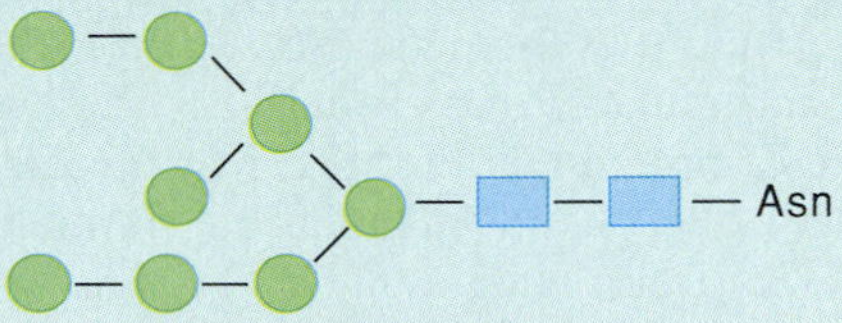 Asn ($Man_8GlcNAc_2Asn$)
Swainsonine	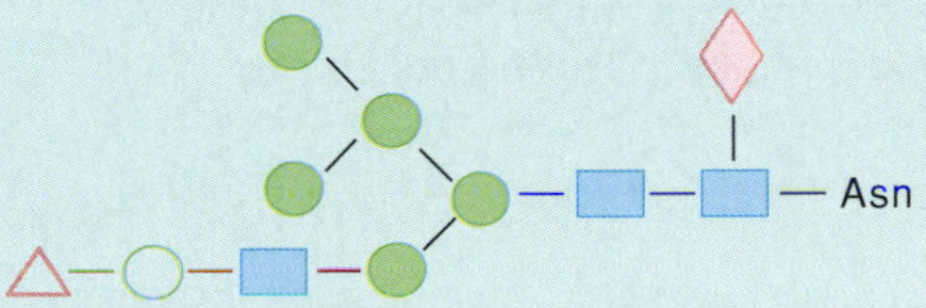 Asn (Hybrid)

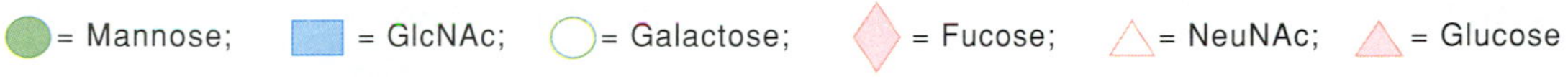

= Mannose; = GlcNAc; = Galactose; = Fucose; = NeuNAc; = Glucose

quite homogeneous, is to use mutants of mammalian cells or yeast that are unable to complete the synthesis of mature N-linked carbohydrates. When cloned glycoproteins are transfected into such mutants, the carbohydrates produced will be specifically lacking in particular structural features.

There are many glycosylation mutants of cultured mammalian cells and yeast. They have been selected as rare survivors of treatments that kill cells that express a particular carbohydrate or glycoprotein at the cell surface. For example, many plant lectins are toxic to mammalian cells and can be used to select for mutants that no longer bind the lectin at the cell surface. We can then isolate glycosylation mutants if the glycoprotein in question requires its carbohydrates for stable cell-surface expression. Some glycoproteins require a cluster of O-linked oligosaccharides close to the membrane to protect them from being attacked by a protease. Glycoproteins with a glycosylphosphatidyl inositol (GPI) anchor must have Dol-P-Man for their synthesis (box 21D). In mutants lacking Dol-P-Man synthase, a complete GPI anchor is not synthesized and glycoproteins that are normally localized to the plasma membrane are secreted from the cell. Loss of Dol-P-Man synthase also affects N-linked carbohydrate biosynthesis by prohibiting the addition of the last four mannose residues to the dolichol oligosaccharide (see fig. 21.13). However, the cell compensates by adding three glucose residues to the $Man_5GlcNAc_2PPDol$ intermediate and transfers that to protein. The protein is processed and extended to form the usual range of complex N-linked carbohydrates by an "alternative" pathway, which requires that GlcNAc-TI transfer GlcNAc to $Man_3GlcNAc_2Asn$ instead of $Man_5GlcNAc_2Asn$ (see figs. 21.8 and 21.13). Thus although glycoproteins synthesized in this mutant may carry normal complex N-linked carbohydrates, they will have only oligomannosyl residues up to $Man_5GlcNAc_2Asn$. This $Man_5GlcNAc_2Asn$ intermediate is distinct from the processing intermediate of the medial Golgi and is not susceptible to cleavage by endo H.

Glycosylation mutants that are particularly useful for producing glycoproteins with modified carbohydrates are summarized in table 21.4. In some mutants both N- and O-linked carbohydrates are modified, while in others only N-linked carbohydrates are affected. We could obtain glycoproteins with altered carbohydrates by treating the glycoproteins with glycosidases to remove particular sugars, but sugars and oligosaccharides are often sterically protected by the protein, and therefore complete deglycosylation would require treatment of the unfolded (often denatured) protein. For this reason, the use of glycosylation mutants frequently provides a superior approach for making altered carbohydrates.

Synthesis of the Glycan Portion of Glycosylphosphatidylinositol

The endoplasmic reticulum is the site of synthesis of the glycan portion of the glycosylphosphatidylinositol (GPI) anchor. The first reaction in the pathway is the transfer of GlcNAc from UDP-GlcNAc to C-6 of *myo*-inositol in a phosphatidylinositol lipid. The GlcNAc is subsequently deacetylated to become $GlcNH_2$ (glucosamine) and three mannoses are added sequentially. One or more of the mannose residues are donated from Dol-P-Man, and mutants that cannot synthesize Dol-P-Man cannot synthesize complete GPI anchors. After three mannose residues are added, phosphatidylethanolamine is incorporated, but the donor of this group remains to be identified. Most glycan moieties have other sugars such as galactose or GalNAc added to the mannose core residues, and these are probably added in the Golgi. However, the immature GPI is transferred to protein in the endoplasmic reticulum. This occurs very soon after synthesis in a transamidation reaction between the carboxyl-terminal amino acid of the protein and the amino group of phosphatidylethanolamine. Recognition of proteins to receive a GPI anchor involves a hydrophobic stretch of amino acids (not unlike a signal sequence) in the C-terminal region, which is removed by cleavage at an internal amino acid at a certain distance from the hydrophobic domain. The cleavage enzyme prefers amino acids with small functional groups. Thus glypiation of a protein involves removal of C-terminal amino acids and transfer of a preformed GPI anchor in the endoplasmic reticulum almost immediately after translation of the mRNA is completed. The C-terminal peptide removed in this reaction is not easy to isolate, but its sequence can be deduced from the cDNA that encodes a glypiated protein.

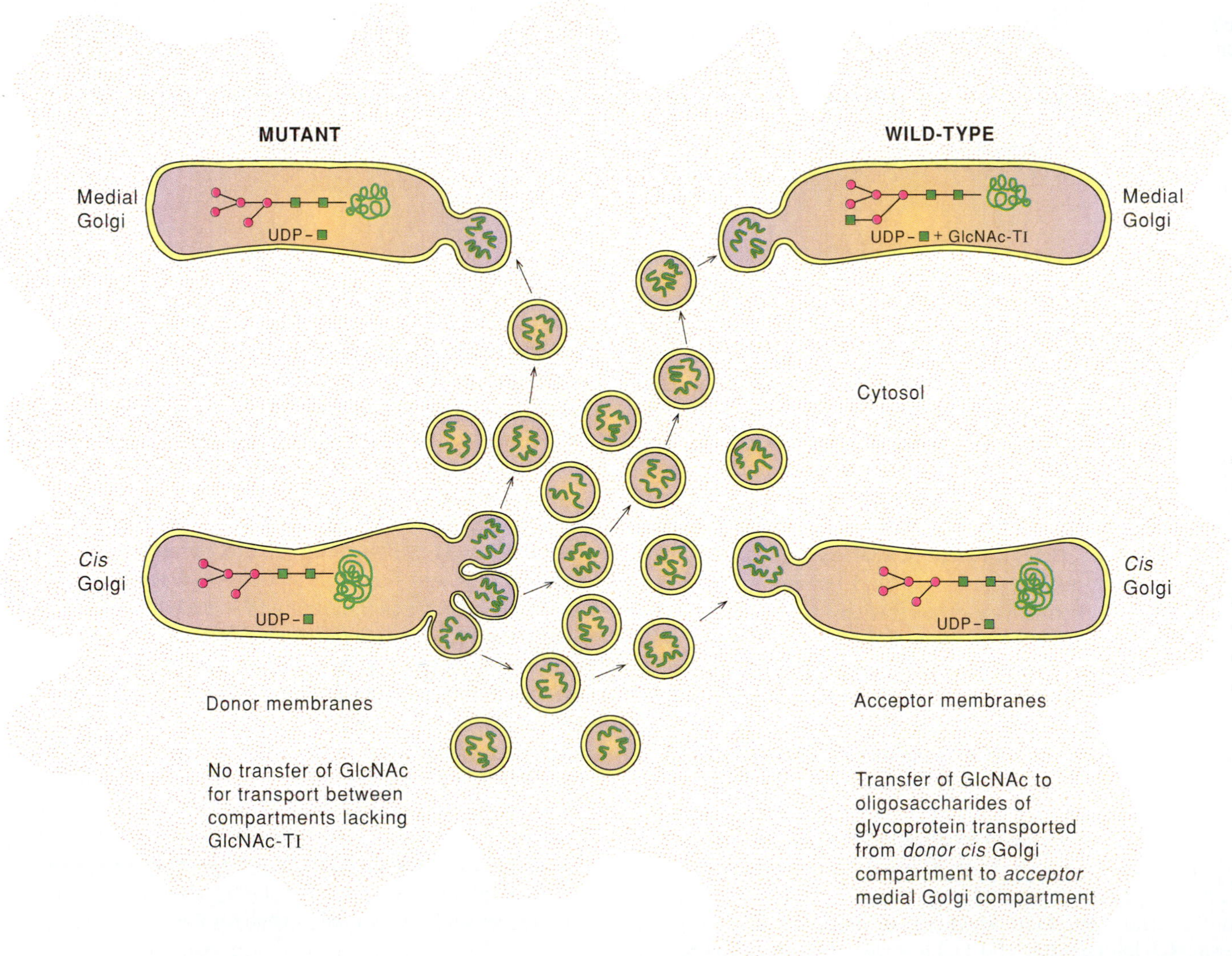

MUTANT

WILD-TYPE

Medial Golgi

UDP-■

UDP-■ + GlcNAc-TI

Medial Golgi

Cytosol

Cis Golgi

UDP-■

UDP-■

Cis Golgi

Donor membranes

Acceptor membranes

No transfer of GlcNAc for transport between compartments lacking GlcNAc-TI

Transfer of GlcNAc to oligosaccharides of glycoprotein transported from *donor cis* Golgi compartment to *acceptor* medial Golgi compartment

Table 21.4

Glycosylation Mutants that Are Useful for Producing Glycoproteins with Modified Carbohydrates

Mutant Cell Line[a]	**Glycosylation Enzyme Deficiency**	**Carbohydrates on Glycoproteins** *N-Linked*	*O-Linked*
Lec1 Clone 15B	GlcNAc-TI	Asn	Ser
ldlD	UDP-Glc-4-epimerase	Asn	Ser
Lec8 Clone 13	UDP-Gal Golgi translocase	Asn	Ser
Lec 2 Clone 021	CMP-NeuNAc Golgi translocase	Asn	Ser

= Mannose; = GlcNAc; = Galactose; = GalNAc; = NeuNAc

[a] All mutants were derived from Chinese hamster ovary (CHO) cells.

Bacterial Cell Wall Biosynthesis

A segment of bacterial cell wall has the structure of a two-dimensional network containing a parallel array of linear heteropolysaccharide strands extended in one direction, which are cross-linked with strands of an oligopeptide in the perpendicular direction (see fig. 6.29). Biosynthesis of the cell wall is unusual in two respects: (1) It is an example of the synthesis of a regularly cross-linked polymer, and (2) part of the synthesis takes place inside the plasma membrane and part takes place outside the plasma membrane. For descriptive purposes, the synthesis of peptidoglycan can be conveniently broken into three stages, which occur at different locations in the cell: (1) synthesis of UDP-*N*-acetylmuramyl-pentapeptide, (2) polymerization of *N*-acetylglucosamine and *N*-acetylmuramyl-pentapeptide to form the linear peptidoglycan strands, and (3) cross-linking of the peptidoglycan strands.

Synthesis of the UDP-N-Acetylmuramyl-Pentapeptide Monomer Occurs in the Cytoplasm

The first stage in cell wall synthesis (fig. 21.16) involves the synthesis of UDP-*N*-acetylmuramyl-pentapeptide. First, the condensation of *N*-acetylglucosamine-1-phosphate with UTP leads to the formation of UDP-*N*-acetylglucosamine. A specific transferase catalyzes a reaction with phosphoenolpyruvate to give the 3-enolpyruvylether of UDP-*N*-acetylglucosamine. The pyruvyl group is then reduced to lactyl by an NADPH-linked reductase, thus forming the 3-*O*-D-lactylether of *N*-acetylglucosamine. This compound is known as UDP-*N*-acetylmuramic acid (see fig. 21.16 for its detailed structure).

Conversion of UDP-*N*-acetylmuramic acid to its pentapeptide form occurs by the sequential addition of the necessary amino acids. Each step requires ATP and a specific enzyme that ensures the addition of amino acids in the proper sequence; L-alanine is added first, followed by D-glutamic acid, L-lysine (attached by its α-amino group to the γ-carboxyl group of the glutamic acid), and finally the dipeptide D-alanyl-D-alanine is added as a unit. The latter dipeptide is formed by two enzymatic reactions: conversion of L-alanine to D-alanine by a racemase, followed by the linking of the two alanine residues in an ATP-requiring reaction to form D-alanyl-D-alanine. All of these reactions occur in the cytoplasm of the bacterial cell.

Formation of Linear Polymers of the Peptidoglycan Is Membrane-Associated

The most complex stage in peptidoglycan synthesis takes place on the plasma membrane. It may be divided into five steps, which are illustrated in figure 21.17. This stage involves the polymerization of *N*-acetylglucosamine and *N*-acetylmuramyl-pentapeptide-containing residues into peptidoglycan strands.

Figure 21.16

The first stage of cell wall synthesis: formation of UDP-*N*-acetylmuramyl-pentapeptide (full structure shown at bottom). Points of inhibition by the antibiotic penicillin and phosphonomycin are indicated.

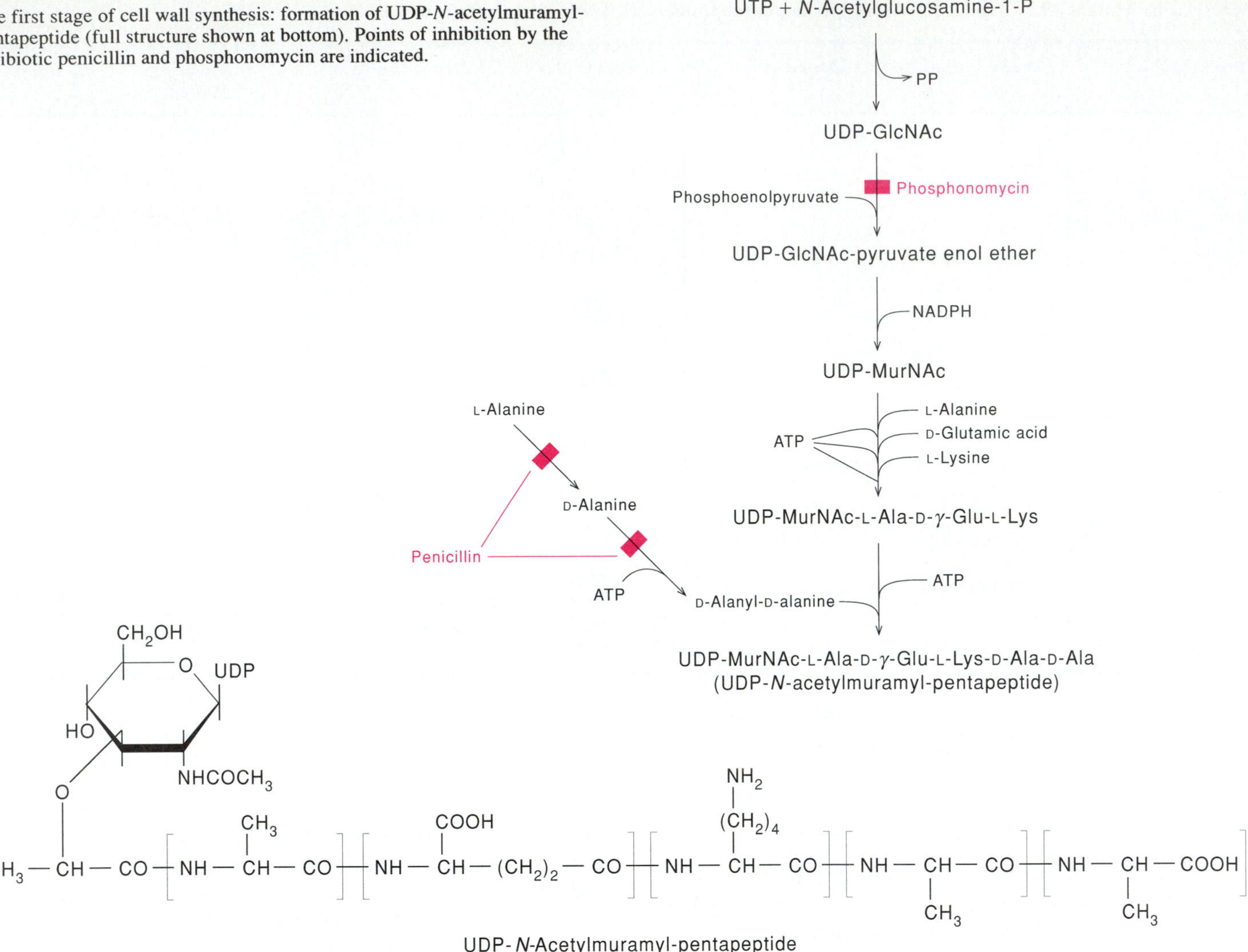

In step 1 (see fig. 21.17) UDP-*N*-acetylmuramyl-pentapeptide reacts with a 55-carbon isoprenyl alcohol known as undecaprenol phosphate. This lipid is similar in structure (fig. 21.18) and function to the dolichol phosphates involved in glycoprotein synthesis in eukaryotes. A pyrophosphate linkage is formed with the lipid and UMP is released.

In step 2 (see fig. 21.17), *N*-acetylglucosamine is added to the lipid intermediate by means of a typical transglycosylation from UDP-*N*-acetylglucosamine, and UDP is released. In step 3, five glycine residues are sequentially added to the ϵ-amino group of lysine. Curiously, these glycines are activated by ester formation to a transfer RNA molecule. This form of amino acid activation is rarely seen except in protein synthesis (see chapter 29). Clearly the primary function of transfer RNAs is for protein synthesis. The involvement of transfer RNA in cell wall synthesis seems opportunistic.

In step 4, the disaccharide-oligopeptide unit is transferred from the lipid intermediate to the growing peptidoglycan, and lipid pyrophosphate is generated. It is in this step that we see the dual function of the undecaprenol lipid. The lipid not only serves to activate the monomer for addition to polymer, but it transports the monomer from the cytoplasmic side of the membrane to the extracellular side of the membrane, where cell wall assembly must take place. Little is known about the details of this transport process.

In the fifth and final step, one phosphate is hydrolyzed to regenerate the phospholipid, which then can react once again with UDP-*N*-acetylmuramyl-pentapeptide and participate in another cycle, resulting in the addition of a new unit to the growing peptidoglycan strand. The antibiotic bacitracin is a specific inhibitor of the dephosphorylation of the pyrophosphate form of the lipid.

Figure 21.17

The second stage of cell wall synthesis. An ATP-requiring amidation of glutamic acid that occurs between steps 2 and 3 has been omitted. Points of action of the antibiotic inhibitors bacitracin and vancomycin are indicated.

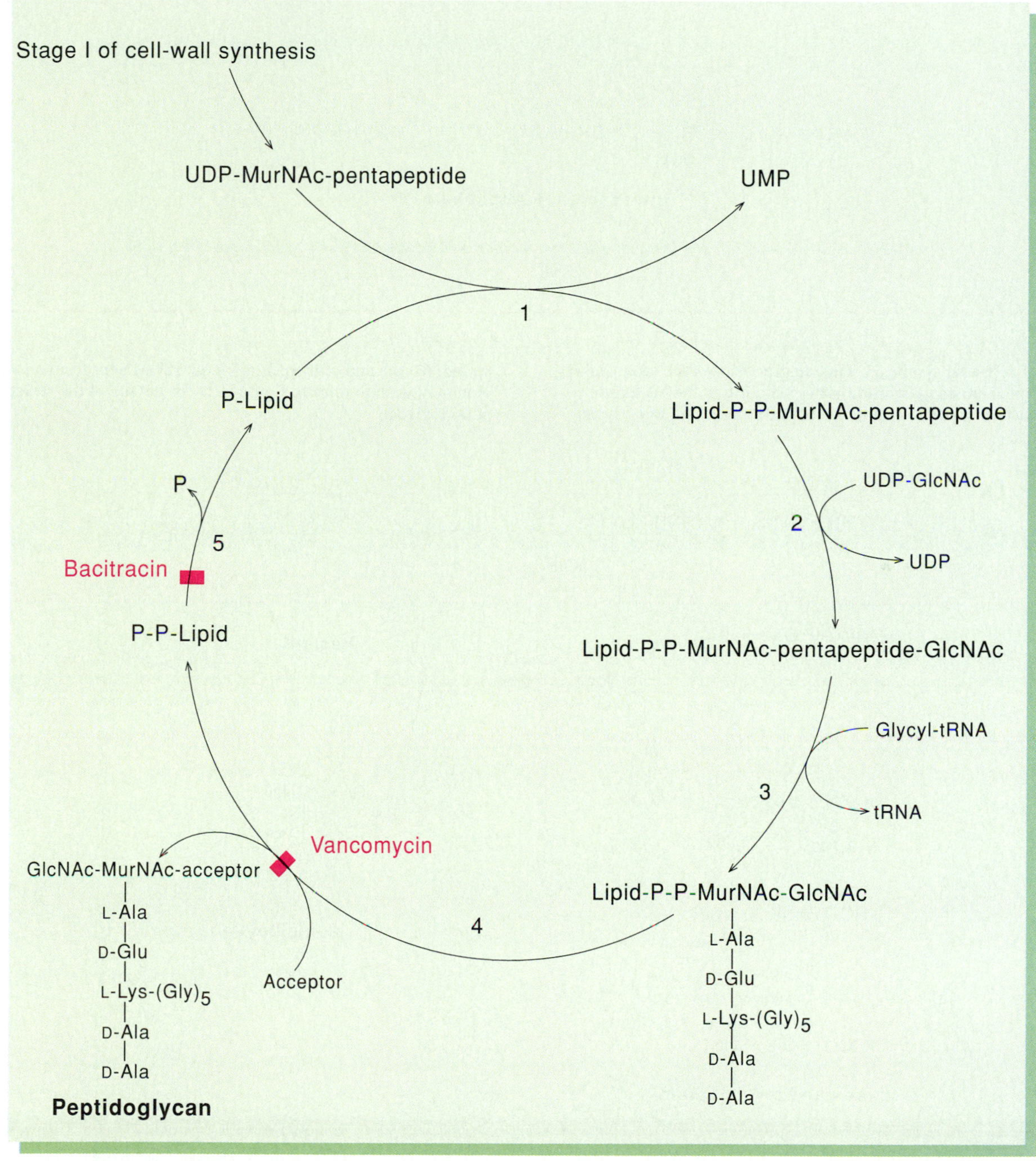

Cross-Linking of the Peptidoglycan Strands Occurs on the Noncytoplasmic Side of the Plasma Membrane

The cross-linking of peptidoglycan strands takes place outside the cell membrane, at the site of the preexisting wall. Since there is no ATP or other obvious energy source available there, a mechanism independent of any external energy source has evolved for this reaction. The reaction is a transpeptidation in which the terminal amino end of an open cross-bridge attacks the terminal peptide bond in an adjacent strand to form a cross-link. The terminal alanine residue from the strand that becomes cross-linked is thus eliminated (fig. 21.19).

Figure 21.18

The structure of undecaprenol phosphate. Isoprene phosphates such as undecaprenol phosphate are important carriers and activators in the synthesis of oligosaccharides. The synthesis of undecaprenol phosphate is described in chapter 22. Note the similarities to dolichol phosphate.

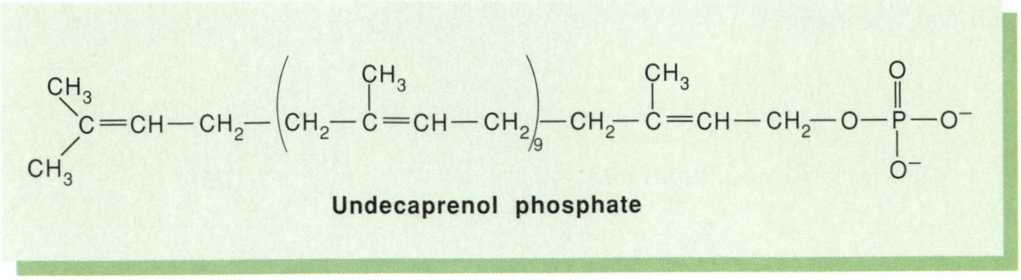

$$(CH_3)_2C{=}CH{-}CH_2{-}\left(CH_2{-}\overset{CH_3}{\overset{|}{C}}{=}CH{-}CH_2\right)_9{-}CH_2{-}\overset{CH_3}{\overset{|}{C}}{=}CH{-}CH_2{-}O{-}\overset{O}{\overset{\|}{P}}(O^-){-}O^-$$

Undecaprenol phosphate

Figure 21.19

The third stage of cell wall synthesis. This diagram shows the cross-linking reaction and the mechanism of inhibition by penicillin in the bacterium *Staphylococcus aureus*. (*a*) The end of the peptide side chain of a glycan strand; (*b*) the end of the pentaglycine substituent from an adjacent strand. You may wish to refer to figure 6.29 for details of the structure of the peptidoglycan.

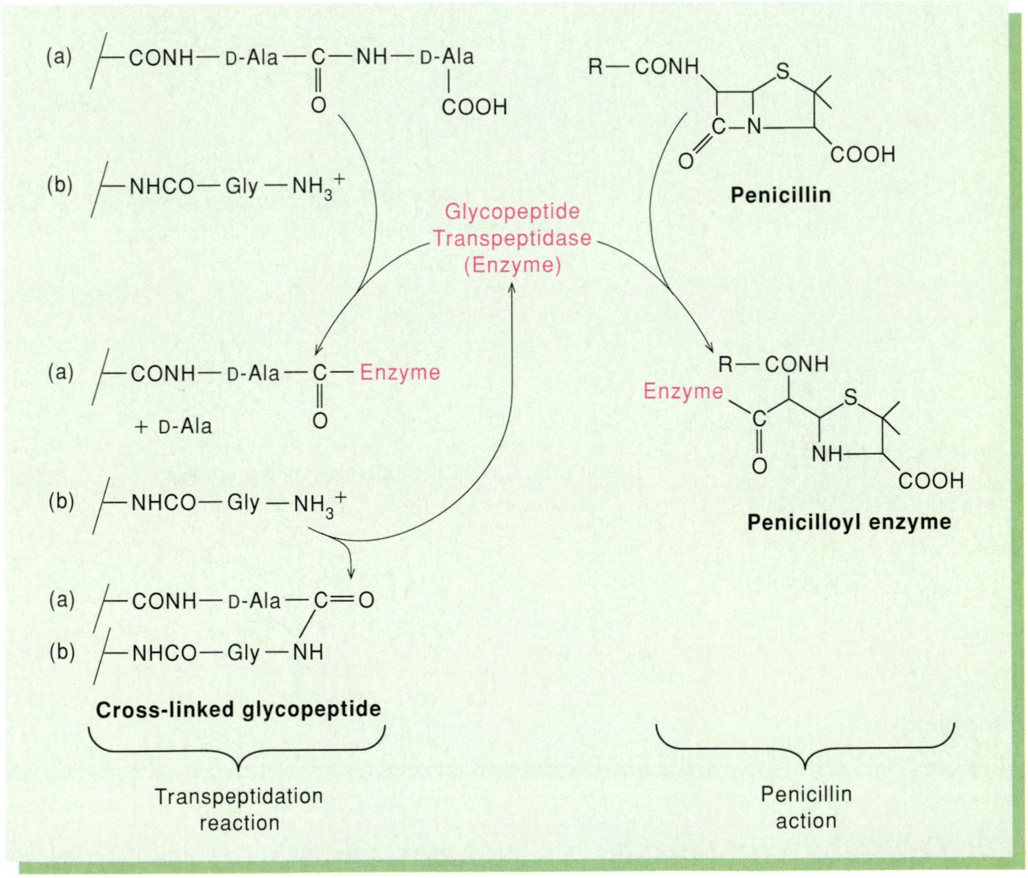

Figure 21.20

Stereomodels of penicillin (middle, left) and of the D-alanyl-D-alanine end of the peptidoglycan strand (middle, right). Arrows indicate the position of the CO—N bond in the ß-lactam ring of penicillin and of the CO—N bond in D-alanyl-D-alanine at the end of the peptidoglycan strand.

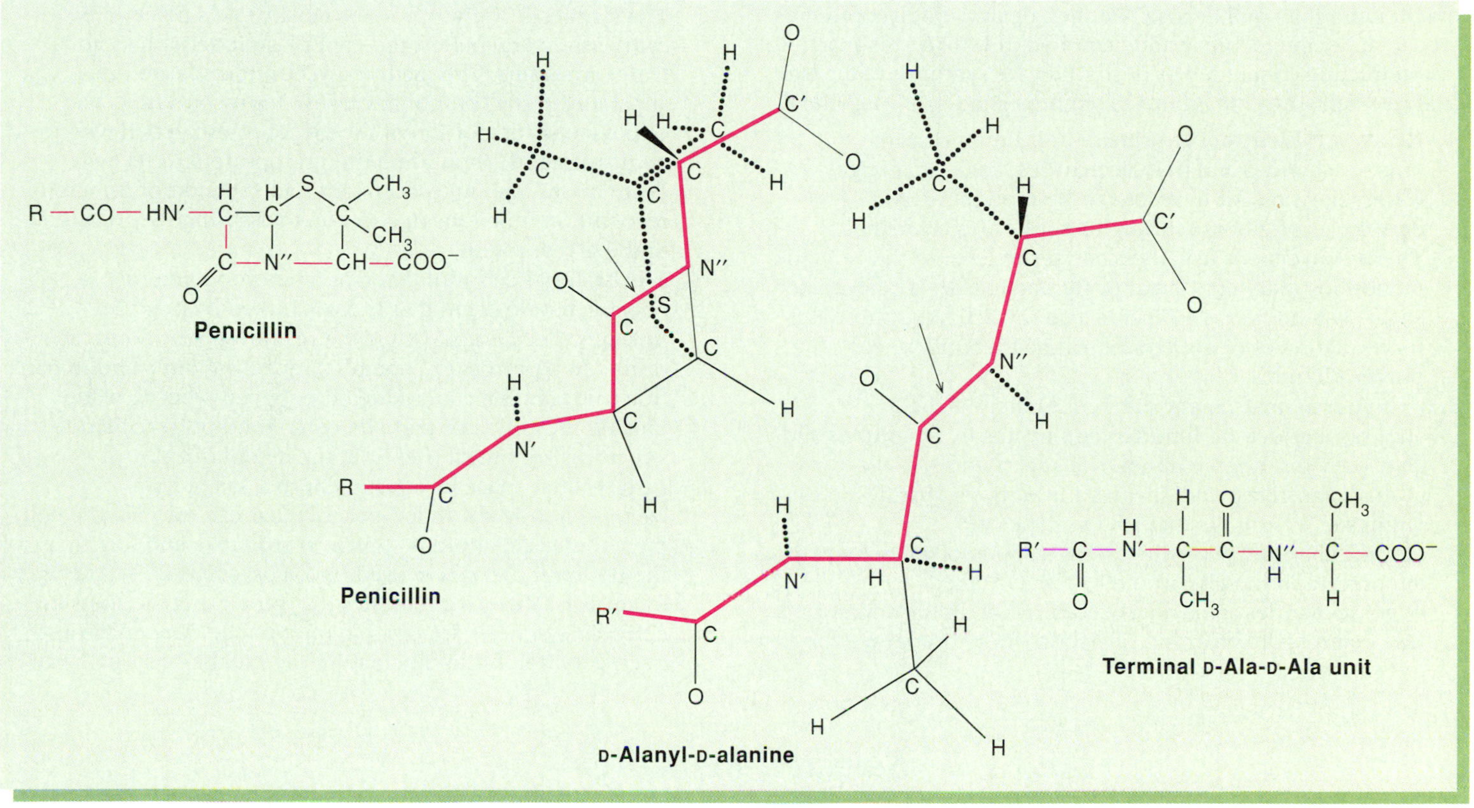

Penicillin Inhibits the Transpeptidation Reaction

Penicillin has been unequaled for usefulness in combatting bacterial diseases and infections. During the 50 years since Fleming brought penicillin to the attention of microbiologists, many biochemists and pharmacologists have been interested in the mechanism by which this potent antibiotic kills bacteria. It is known that penicillin inhibits the cross-linking reaction by acting as a structural analog of the terminal D-alanyl-D-alanine residue of the peptidoglycan strand. The similarity between the conformation of the penicillin molecule and one of the conformations of the dipeptide D-alanyl-D-alanine is shown in figure 21.20.

It is thought that the transpeptidase (TPase) first reacts with the substrate to form an acyl enzyme intermediate, with the elimination of D-alanine, and that this active intermediate then reacts with another strand to form the cross-link and regenerate the enzyme. Because penicillin is an analog of alanyl-alanine, it should fit the substrate binding site, with the highly reactive CO — N bond in the β-lactam ring in the same position as the bond involved in the transpeptidation. It thus has the potential to acylate the enzyme, forming a penicilloyl enzyme, and thereby inactivate it (see fig. 21.19). In support of this view is the fact that penicilloyl is the piece of the antibiotic found in inhibited enzymes. An acyl group derived from the substrate is used to form an acyl enzyme intermediate. Most important for verifying the proposed mechanism of penicillin action, the antibiotic-derived penicilloyl moiety and the substrate-derived acyl moiety are substituted on the same site in the penicillin-sensitive enzymes from a variety of genera of bacteria.

A number of other antibiotics in addition to penicillin (phosphonomycin, bacitracin, and vancomycin) block other stages in cell wall synthesis. Their points of action are indicated in figures 21.16 and 21.17. In addition to their biological and medical importance, these antibiotics have been extremely useful in elucidating the biosynthetic pathway. This is because they permit accumulation of the product before the blocked step; the product can then frequently be isolated and confirmed as a genuine intermediate in the pathway.

Summary

In this chapter we have focused on the synthesis of complex carbohydrates. We began by examining the hexoses that are the building blocks of complex carbohydrates, then moved to some aspects of the synthesis of disaccharides and simple homopolysaccharides, and to a brief consideration of heteropolymers that contain more than one hexose. We then dealt with glycoproteins that contain complex linear and branched carbohydrates attached to proteins, and concluded by describing the synthesis of the bacterial cell wall. The chief points in our presentation are as follows.

1. Hexoses, which are the primary building blocks of oligosaccharides and polysaccharides, come in a large variety of types. All hexoses can be thought of as derivatives of glucose through a series of conversions. These conversions usually occur at the level of the monophosphorylated sugar or the nucleoside diphosphate sugar. The nucleoside sugar is also the activated substrate for formation of disaccharides, oligosaccharides, and polysaccharides.
2. In higher animals, many different branched-chain oligosaccharides are found as conjugates in glycolipids and glycoproteins. A large number of sugars and specific glycosyltransferases are involved in oligosaccharide synthesis. We can distinguish two types of oligosaccharides, according to their mode of attachment to the protein in the glycoprotein. The O-linked oligosaccharides are synthesized directly on the amino acid side chain hydroxyl group of a serine or a threonine. The N-linked oligosaccharides are synthesized first on a long-chain dolichol phosphate and then transferred to the asparagine side chain of a receptor protein. The protein-attached oligosaccharide is processed by removal of certain sugars and addition of others.
3. The synthesis of glycoproteins mostly takes place in two cytoplasmic organelles, the endoplasmic reticulum and the Golgi apparatus. The mature glycoproteins leave the Golgi apparatus in the form of microvesicles by budding. The oligosaccharide portion of the glycoprotein is believed to be instrumental in guiding the glycoprotein to its final destination. The oligosaccharide can serve other important recognition functions in addition to assuring the location of the glycoprotein.
4. The bacterial cell wall contains a heteropolymeric polysaccharide chain that is cross-linked by peptide linkages. The complexity of the resulting peptidoglycan results in part from the complex repeating units and in part from the fact that a cross-linked polymer is being made outside the cell. The partially completed polysaccharide structures are transferred from the cytoplasm to extracellular space by attachment to a long-chain bactoprenol lipid (undecaprenol) that can traverse the cell membrane. This lipid is similar in structure and function to the dolichol phosphate used in oligosaccharide synthesis in animals. Various antibiotics that block specific steps in cell wall synthesis are of great importance. They have also been very helpful in elucidating the biochemical pathway.

Selected Readings

Albeijon, C., and C. B. Hirschberg, Topography of glycosylation reactions in the endoplasmic reticulum. *Trends in Biochem. Sciences* 17:32–36, 1992.

Doering, T. L., W. J. Masterson, G. W. Hart, and P. T. Englund, Biosynthesis of glycosylphosphatidylinositol membrane anchors. *J. Biol. Chem.* 265:611–614, 1990.

Edelman, G. M., Cell adhesion and the molecular processes of morphogenesis. *Ann. Rev. Biochem.* 54:135–170, 1985.

Elbein, A. D., Inhibitors of the biosynthesis and processing of N-linked oligosaccharides. *CRC Crit. Rev. Biochem.* 16:21–49, 1984.

Elbein, A. D., Inhibitors of the biosynthesis and processing of N-linked oligosaccharide chains. *Ann. Rev. Biochem.* 56:497–534, 1987.

Fukuda, M. N., Hempas disease: Genetic defect of glycosylation. *Glycobiology* 1:9–15, 1990.

Fukuda, M. N., K. A. Masri, A. Dell, L. Luzzatto, and K. W. Moremen, Incomplete synthesis of N-glycans in congenital dyserythropoietic anemia type II caused by a defect in the gene encoding α-mannosidaseII. *Proc. Natl. Acad. Sci. USA* 87:7443–7447, 1990.

Hirschberg, C. B., and M. D. Snider, Topography of glycosylation in the rough endoplasmic reticulum and the Golgi apparatus. *Ann. Rev. Biochem.* 56:63–87, 1987.

Kochetkov, N. K., and V. N. Shibaev, Glycosyl esters of nucleoside pyrophosphates. *Adv. Carbohydr. Chem. Biochem.* 28:307–325, 1973. A concise review of the chemistry and biochemistry of nucleoside pyrophosphate sugars and derivatives.

Kornfeld, R., and S. Kornfeld, Assembly of asparagine-linked oligosaccharides. *Ann. Rev. Biochem.* 54:631–664, 1985.

Krieger, M., P. Reddy, K. Kozarsky, D. Kingsley, and M. Penman, Analysis of the synthesis, intracellular sorting and function of glycoproteins using a mammalian cell mutant with reversible glycosylation defects. *Methods in Cell Biol.* 32:57–84, 1989.

Kukowsaka-Latallo, J. F., R. D. Larsen, R. P. Nair, and J. B. Lowe, A cloned human cDNA determines expression of a mouse stage-specific embryonic antigen and the Lewis blood group α(1,3/1,4)fucosyltransferase. *Genes Dev.* 4:1288–1303, 1990.

Kukuruzinska, M. A., M. L. E. Bergh, and B. J. Jackson, Protein glycosylation in yeast. *Ann. Rev. Biochem.* 56:915–944, 1987.

Larsen, R. D., L. K. Ernst, R. P. Nair, and J. B. Lowe, Molecular cloning, sequence and expression of a human GDP-L-fucose: β-D-galactosidase 2-α-L-fucosyltransferase cDNA that can form the H blood group antigen. *Proc. Natl. Acad. Sci. USA* 87:6674–6678, 1990.

Paulson, J. C., and K. J. Colley, Glycosyltransferases: Structure, localization, and control of cell type-specific glycosylation. *J. Biol. Chem.* 264:17615–17618, 1989.

Pfeffer, S. R., and J. E. Rothman, Biosynthetic protein transport and sorting by the endoplasmic reticulum and Golgi. *Ann. Rev. Biochem.* 56:829–852, 1987.

Phillips, M. L., E. Nudelman, F. C. A. Gaeta, M. Perez, A. K. Singhal, S. Hakomori, and J. C. Paulson, ELAM-1 mediates cell adhesion by recognition of a carbohydrate ligand, sialosyl-Le^x. *Science* 250:1130–1132, 1990.

Rothman, J. E., The compartmental organization of the Golgi apparatus. *Sci. Am.* 253(3):74–89, 1985.

Stanley, P., Glycosylation mutants and the functions of mammalian carbohydrates. *Trends in Genetics* 3:77–81, 1987.

Stanley, P., Glycosylation mutants of animal cells. *Ann. Rev. Genet.* 18:525–552, 1984.

Yamamoto, F., H. Clausen, T. White, J. Marken, and S. Hakomori, Molecular genetic basis of the histo-blood group ABO system. *Nature* 345:229–233, 1990.

Yamamoto, F., and S. Hakomori, Sugar-nucleotide donor specificity of histo-blood group A and B transferases is based on amino acid substitutions. *J. Biol. Chem.* 265:19257–19262, 1990.

Problems

1. How is lactose made in mother's milk? What is unusual about the subunit structure of lactose synthase?
2. The synthesis of dextrans does not use nucleotide sugars as seen in the synthesis of other glucose polymers. What is the driving force for this reaction, i.e., how would you analyze the thermodynamics?
3. How can lectins be used to identify complex carbohydrate structures on glycolipids or glycoproteins? (Hint: You could label the lectins with radioactive iodine [^{125}I] or use lectins bound to a column matrix.)
4. A great variety of different oligosaccharides result from a limited number of sugars. Explain.
5. How can two different genes for different glycosyltransferases determine ABO blood group types in humans? Explain why blood group O is considered a universal donor. If you have AB type blood, why can you accept any blood type? A small number of people lack the H antigen (the glycosyltransferase that adds Fuc α1, 2) and have Bombay type blood. What blood type could you give to a person with Bombay type blood and why?
6. There is a lot of interest in expressing human genes in bacteria to produce a useful human protein in large amounts. Why would this approach not be expected to work with some human proteins?
7. A mutant cell line that does not synthesize Dol-P-Man does not express glypiated proteins on its cell surface. Why?
8. Mutants with a glycosylation defect are used to follow transport between cellular compartments. Using one of these mutants, design an assay to detect transit between Golgi compartments that lie beyond the medial Golgi.
9. How would you provide biochemical evidence that a particular glycoprotein is a resident protein of the endoplasmic reticulum?
10. Explain why fibroblasts from a patient with I-cell disease (discussed in chapter 6) will secrete lysosomal enzymes when grown in tissue culture.
11. Why does penicillin kill susceptible bacteria only when they are growing?
12. What complications have to be overcome to synthesize complex carbohydrates (such as cell wall components and O-antigens) outside the cell?

22

CHAPTER

Biosynthesis of Membrane Lipids and Related Substances

hus far we have been concerned with the metabolism of fatty acids as it relates to the storage and release of energy (see chapter 17). In this chapter we will focus on the metabolism of lipids that serve other roles (fig. 22.1). Most fatty acids that are not utilized for energy storage perform important structural roles as integral components of membranes. A select group of fatty acids are used to make lipophilic compounds with physiologic activity.

Phospholipids

Phospholipids are ideal compounds for making membranes because of their amphipathic nature (see chapter 7). The polar head groups of phospholipids prefer an aqueous environment, whereas the nonpolar acyl substituents do not. As a result, phospholipids spontaneously form bilayer structures (see fig. 7.17). Bilayers are the dominant structural feature of most membranes in which proteins are embedded. The amphipathic nature of phospholipids has a great influence on the mode of biosynthesis of the phospholipids. Thus most of the reactions involved in lipid synthesis occur on the surface of membrane structures with enzymes that themselves are amphipathic.

In E. coli, *Phospholipid Synthesis Leads to Phosphatidylethanolamine, Phosphatidylglycerol, and Diphosphatidylglycerol*

E. coli contains three important classes of phospholipids: phosphatidylethanolamine (75–85%), phosphatidylglycerol (10–20%), and diphosphatidylglycerol (5–15%). All three of these phospholipids share the same biosynthetic pathway up to the formation of CDP-diacylglycerol (fig. 22.2), after which the pathways branch.

Most of the enzymes for phospholipid synthesis are located on the inner plasma membrane of *E. coli*. Glycerol-3-phosphate acyltransferase, the first enzyme in the pathway, preferentially utilizes saturated fatty acyl derivatives (palmitoyl-CoA or palmitoyl-ACP) for the initial acylation of glycerol-3-phosphate (the mechanism for acyltransferase-catalyzed reactions is described in box 22A). The second enzyme (see fig. 22.2),

Figure 22.1

Outline of pathways for the synthesis of complex lipids. Most of the metabolism of complex lipids occurs on membrane surfaces because of the insoluble nature of the substrates. Lipids play three major roles: (1) they act as storehouses of chemical energy, as with triacylglycerols; (2) they are structural components of membranes (the boxed compounds); and (3) they act as regulatory compounds (underlined), either as eicosanoids, which function as local hormones, or as phosphorylated inositols and diacylglycerols, which function as second messengers. In this figure the pathways shown in black are taken from the master diagram of figure 12.5; the remainder of the pathways discussed in this chapter are shown in color. Some of these pathways are uniquely found in either eukaryotes or prokaryotes; other pathways are common to both prokaryotes and eukaryotes (consult text).

Synthesis of most lipids starts from glycerol-3-phosphate, which is formed in one step from the central metabolic pathways, and acyl-CoA, which arises in one step from activation of a fatty acid. In two acylation steps the key compound phosphatidic acid is formed. This can be converted to many other lipid compounds as well as to CDP diacylglycerol, which is a key branchpoint intermediate that can be converted to many other lipid compounds. Two routes to phosphatidylethanolamine and phosphatidylcholine are shown: (1) the pathway found in prokaryotes, which involves CDP-diacylglycerol as an intermediate, and (2) the pathway found in eukaryotes, which starts with transport across the plasma membrane of ethanolamine and/or choline. The modified derivatives of these compounds are directly condensed with diacylglycerol to form the corresponding membrane lipids. Modification of head groups or tail groups on preformed lipids is a common reaction. For example, the ethanol part of the head group in phosphatidylethanolamine can be replaced in one step by either serine or choline. Similarly, the complex polyunsaturated fatty acid arachidonic acid may replace other, simpler fatty acids in a preformed phospholipid.

Sphingomyelin and glycosphingolipids are exceptional, since they do not form through the usual intermediates.

Phospholipids containing arachidonic acid serve as the precursor, through liberation of arachidonic acid, of a wide variety of eicosanoids. Similarly, the membrane lipid phosphatidylinositol can undergo hydrolysis to inositol and diacylglycerol. Both of these hydrolysis products are regulatory molecules that serve as second messengers (so called as they are formed in response to hormone binding to the plasma membrane).

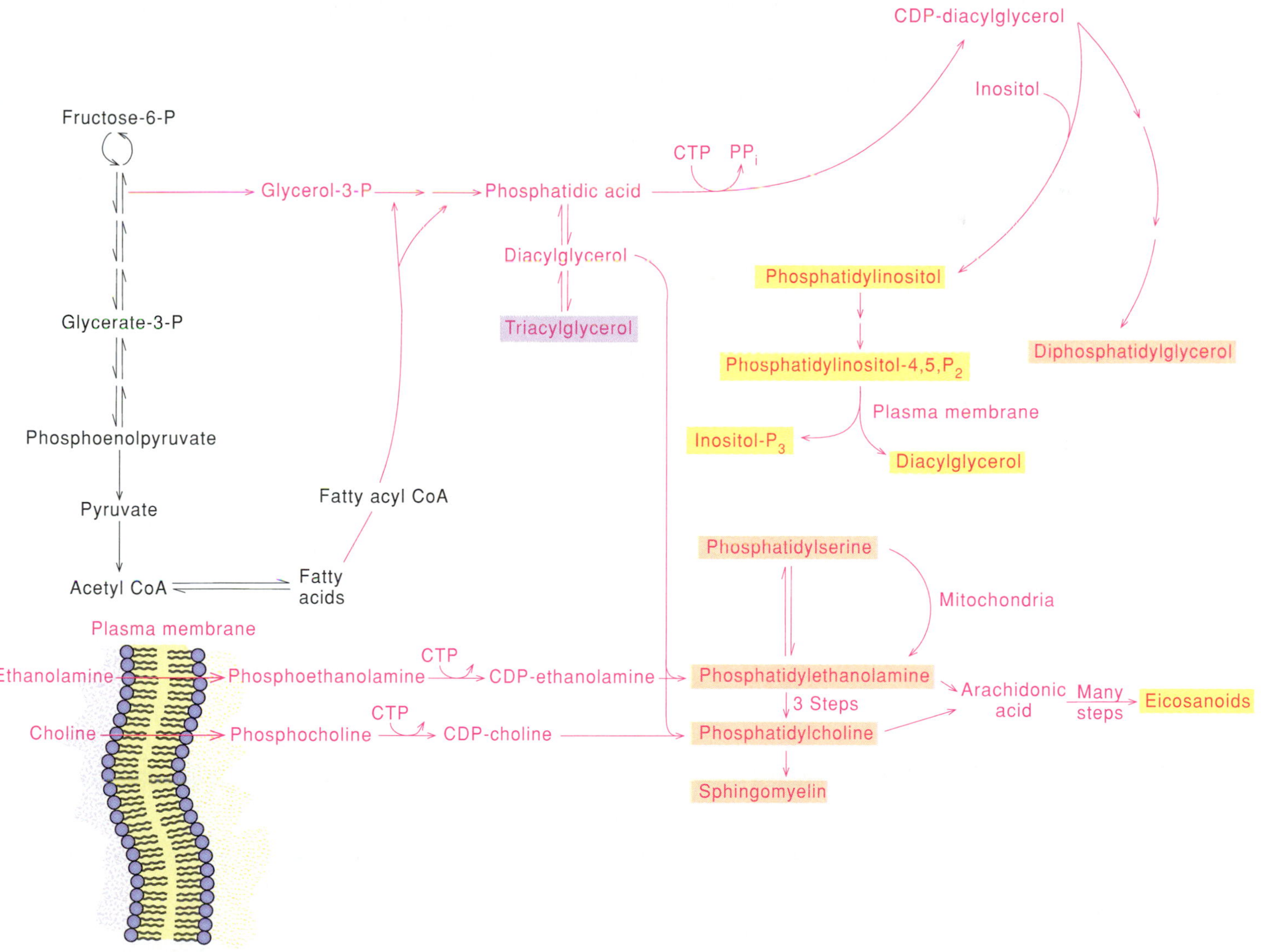

Figure 22.2

The first phase of phospholipid synthesis in *E. coli* and eukaryotes. Additional routes to and from phosphatidic acid, found exclusively in eukaryotes, are shown in brackets.

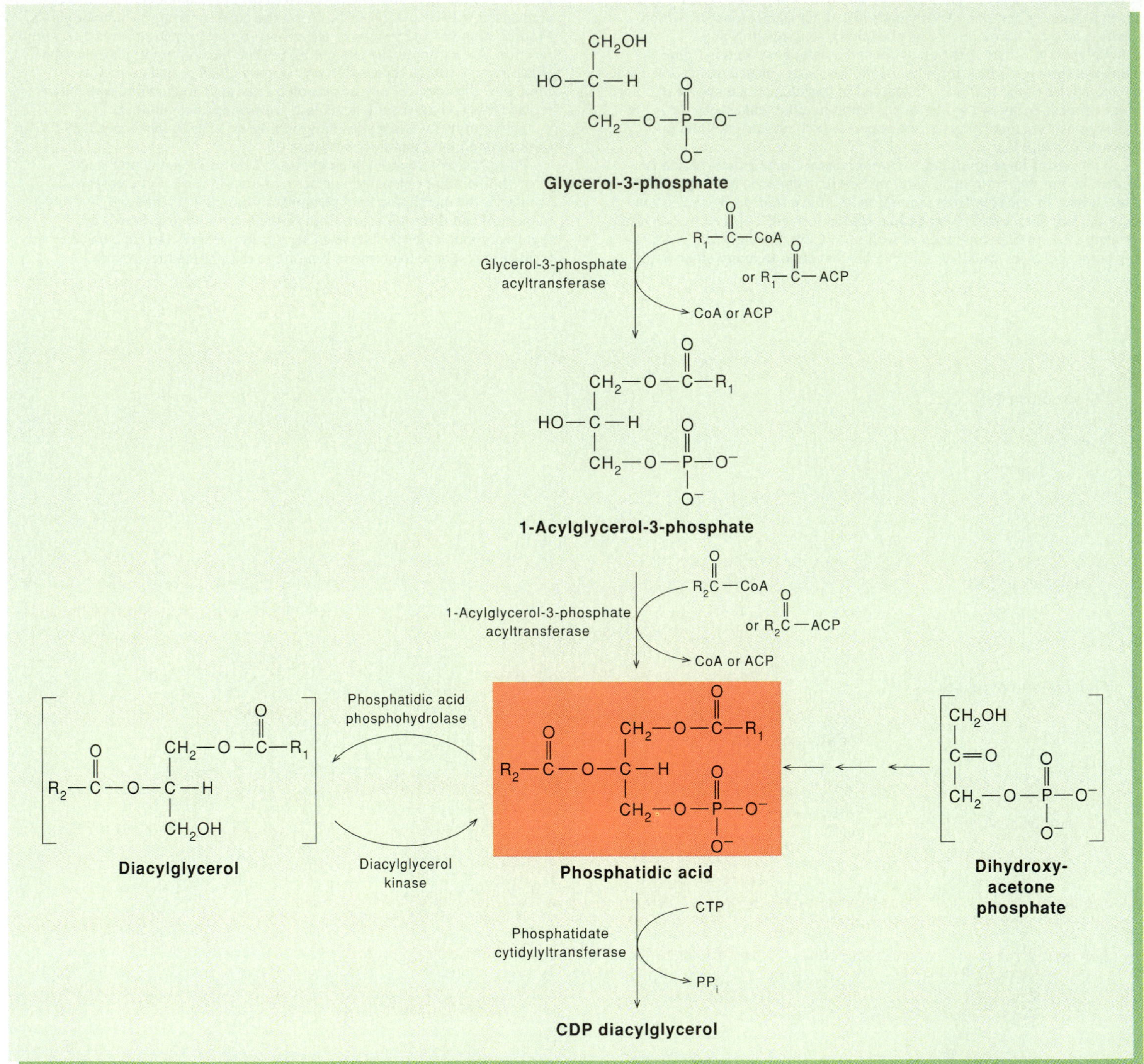

22A BOX

Mechanism of Action of Acyltransferases

$$\begin{matrix}CH_2OH\\|\\HO-C-H\\|\\CH_2\\|\\O-Ⓟ\end{matrix} + \begin{matrix}R\\|\\C=O\\|\\S\\|\\CoA\end{matrix} \xrightarrow[-H^+]{} \left[\begin{matrix}CH_2-O-\overset{\overset{R}{|}}{C}-O^-\\ | \quad\quad\quad |\\HO-C-H \quad S \quad H^+\\|\quad\quad\quad |\\CH_2 \quad\quad CoA\\|\\O-Ⓟ\end{matrix}\right] \xrightarrow{CoA} \begin{matrix}CH_2-O-\overset{\overset{O}{\|}}{C}-R\\|\\HO-C-H\\|\\CH_2-O-Ⓟ\end{matrix}$$

Nucleophilic addition to the neutral activated acyl group is a favored process and coenzyme A is a good leaving group from the tetrahedral intermediate. This point was discussed in chapter 11 in connection with the formation of acetoacetyl-CoA, and we see it again here in the acyltransferase reactions involving glycerol or glycerol derivatives and acyl-CoA compounds. The reaction is initiated by a nucleophilic attack of acyl-CoA by a hydroxyl, giving rise to a tetrahedral intermediate. This is followed by expulsion of the CoA.

1-acylglycerol-3-phosphate acyltransferase, catalyzes phosphatidic acid formation; it will use either acyl-CoA or acyl-ACP as substrate and prefers acyl residues with a double bond. The substrate specificity of these two acyltransferases accounts for the fact that saturated fatty acids are usually found in the SN-1 position and unsaturated fatty acids are usually found in the SN-2 position. A third transferase reaction converts phosphatidic acid to CDP-diacylglycerol with CTP as cosubstrate. CDP-diacylglycerol is a branchpoint intermediate. It either gets converted to phosphatidylserine and ultimately phosphatidylethanolamine or gets converted to mono- or diphosphatidylglycerol (fig. 22.3).

Phospholipid Synthesis in Eukaryotes Is More Complex

Phospholipid synthesis in eukaryotes is more complex than in *E. coli*. This is because eukaryotes have many more membrane structures, each with its own specific function, and also because eukaryotes store fatty acids as triacylglycerols.

In the first phase of phospholipid synthesis from glycerol-3-phosphate to phosphatidic acid, the pathways in *E. coli* and eukaryotes are very similar (see fig. 22.2). The major difference is that there are two additional pathways (see fig. 22.2): phosphatidic acid is made from dihydroxyacetone phosphate (DHAP) and diacylglycerol is a major product of phosphatidic acid catabolism. Once the phosphatidic acid is made, it is rapidly converted to diacylglycerol or CDP-diacylglycerol, which are metabolized in various ways.

Diacylglycerol Is the Key Intermediate in the Biosynthesis of Phosphatidylcholine and Phosphatidylethanolamine

Phosphatidylcholine and phosphatidylethanolamine, which are quantitatively the most important phospholipids in eukaryotic cells (see table 7.8), are derived from diacylglycerol as shown in figure 22.4. The biosynthesis of phosphatidylcholine begins with the transport of choline into the cell. Choline is an essential ingredient in the human diet and cannot be made by animals except indirectly, as described in the next paragraph. Once inside the cell, the choline is rapidly phosphorylated to phosphocholine by the action of choline (ethanolamine) kinase, a cytosolic enzyme. Phosphocholine and CTP form CDP-choline in a reaction catalyzed by CTP:phosphocholine cytidylyltransferase

Figure 22.3

The second phase of phospholipid synthesis in *E. coli*, from CDP-diacylglycerol to the end products.

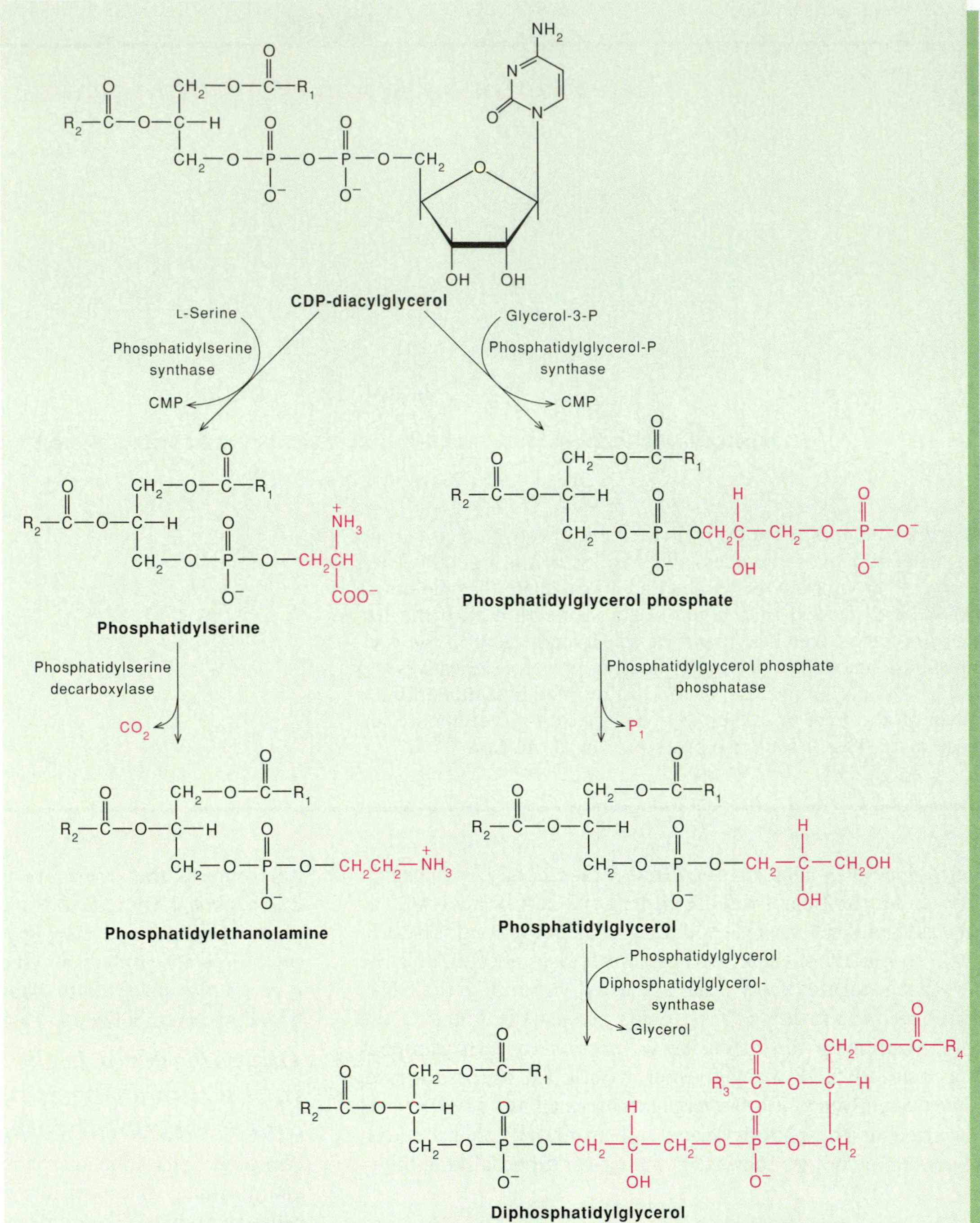

(see fig. 22.4). The activity of this enzyme is usually rate-limiting for phosphatidylcholine biosynthesis and is activated by translocation from its inactive form in the cytosol to the endoplasmic reticulum, where it is activated by the phospholipids in this membrane (discussed a little later). The CDP-choline immediately reacts with diacylglycerol, a reaction catalyzed by an enzyme localized on the endoplasmic reticulum. The biosynthesis of phosphatidylethanolamine (see fig. 22.4) proceeds from ethanolamine in a comparable series of reactions.

In liver, yeast, and the bacterium *Pseudomonas,* there is a pathway for the conversion of phosphatidylethanolamine to phosphatidylcholine in which methyl groups are transferred from

Figure 22.4

The second phase of phospholipid synthesis in eukaryotes. Choline or ethanolamine enters the cell via active transport mechanisms and is immediately phosphorylated by the same enzyme, choline (ethanolamine) kinase. The phosphorylated derivatives of choline and ethanolamine are activated to their CDP derivatives by separate enzymes. The last reaction occurs on the endoplasmic reticulum. The diacylglycerol used as a substrate in this reaction may alternatively be converted to storage lipid (triacylglycerol).

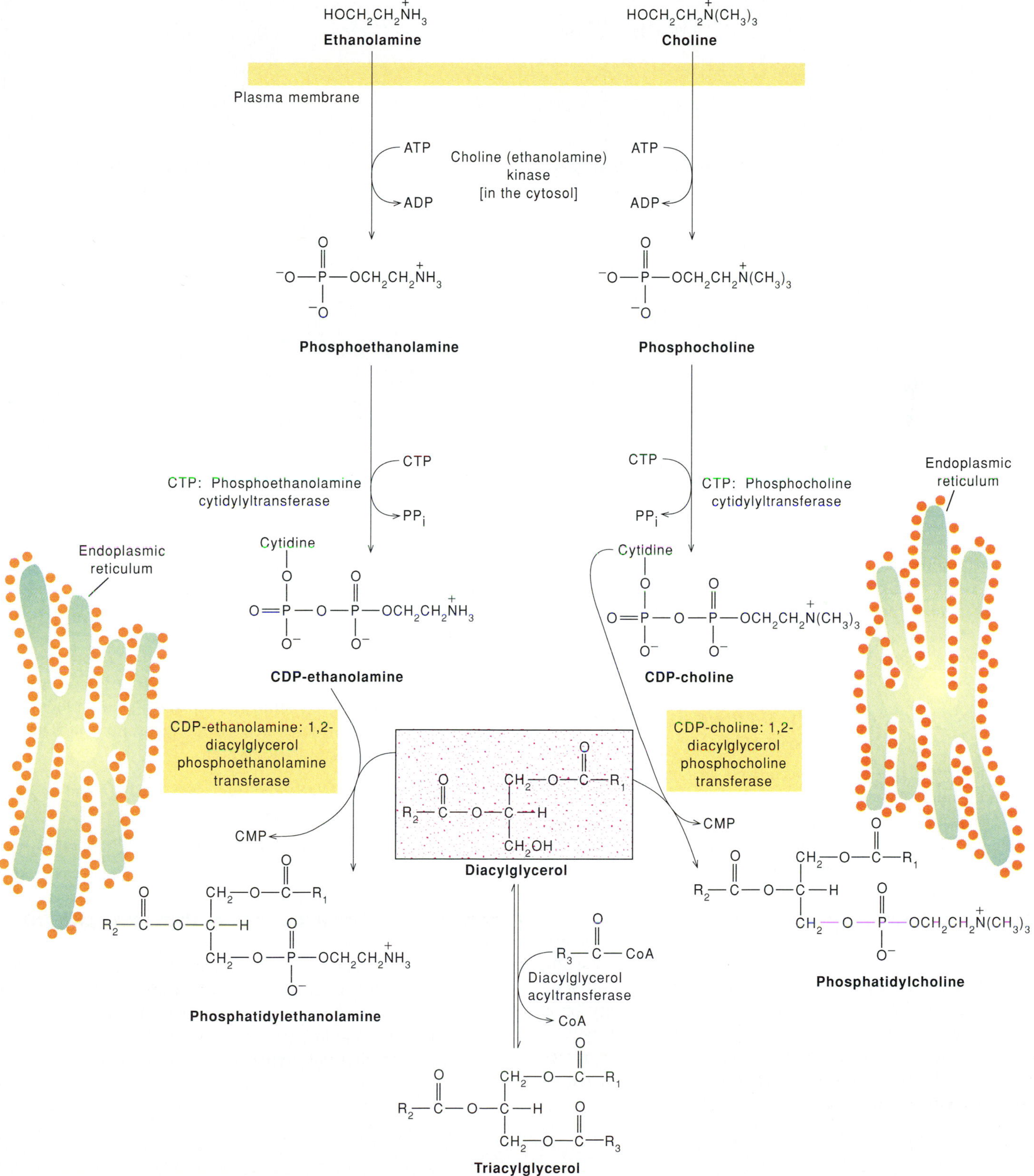

Figure 22.5

Conversion of phosphatidylethanolamine to phosphatidylcholine by phosphatidylethanolamine-*N*-methyltransferase. (AdoMet is a standard abbreviation for *S*-adenosyl-L-methionine. Recall that SAM may also be used. AdoHcy is a standard abbreviation for *S*-adenosyl-L-homocysteine.) The structures of AdoMet and AdoHcy and the general mechanism of the methylation reaction are presented in box 22B.

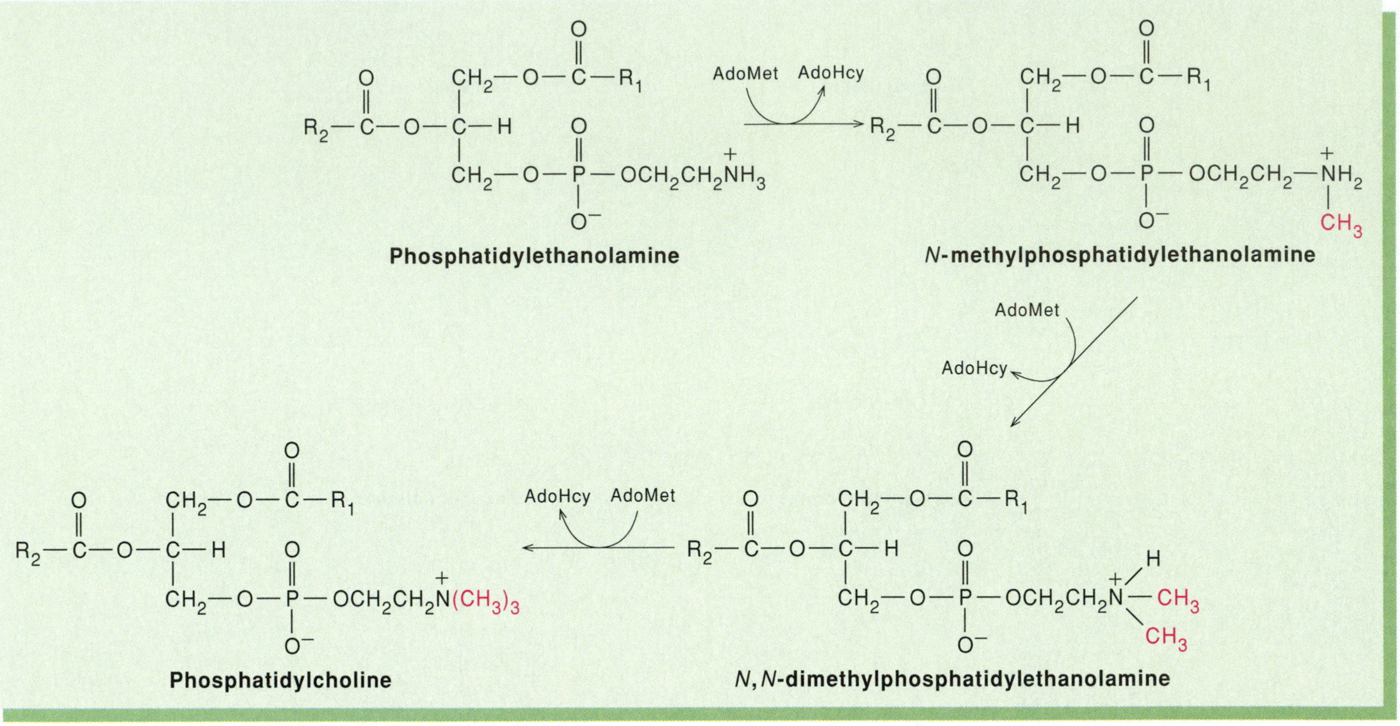

S-adenosylmethionine (SAM) to phosphatidylethanolamine in three consecutive reactions (fig. 22.5). The methylation of phosphatidylethanolamine, together with the subsequent degradation of phosphatidylcholine, is the only known mechanism by which liver can produce choline. The pathway for methylation by SAM is discussed in box 22B.

Fatty Acid Substituents May Be Switched at the SN-1 and SN-2 Positions

Lung tissue manufactures a specialized species of phosphatidylcholine, dipalmitoylphosphatidylcholine, in which palmitic acid is the fatty acyl substituent on both the SN-1 and SN-2 positions of the glycerol backbone. This species of phosphatidylcholine is the major component of lung surfactant, which maintains surface tension in the lung alveoli so that they do not collapse when air is expelled.

The probable pathway for the synthesis of dipalmitoylphosphatidylcholine is illustrated in figure 22.6. The starting species of phosphatidylcholine is made by the CDP-choline pathway (see fig. 22.4). The fatty acid at the SN-2 position is hydrolyzed by phospholipase A_2 and the lysophosphatidylcholine is reacylated with palmitoyl-CoA. This modification allows alteration of the properties of the phospholipid without resynthesis of the entire molecule.

The synthesis of dipalmitoylphosphatidylcholine in lung tissue is an example of the modulation of the fatty acid composition of a phospholipid by "remodeling." Deacylation-reacylation of phosphatidylcholine occurs in other tissues and provides an important route for alteration of the fatty acid substituents at both the SN-1 and SN-2 positions. For example, fatty acids at the SN-2 position are replaced by arachidonic acid, which is stored there until needed for eicosanoid biosynthesis, to be discussed later in this chapter.

Phosphatidylinositol-4,5-Bisphosphate, a Mediator of Hormone Action, Is Synthesized via CDP-Diacylglycerol

The derivatives of CDP-diacylglycerol, which are key intermediates in prokaryotic phospholipid synthesis, are also synthesized in eukaryotes, where they can give rise to phosphatidylglycerol and diphosphatidylglycerol (see fig. 22.3) in the mitochondria. CDP-diacylglycerol also gives rise to phosphatidylinositol (fig. 22.7), which accounts for about 5% of the lipids present in membranes (see table 7.8). Also present, at much lower concentrations, are phosphatidylinositol-4-phosphate and phosphatidylinositol-4,5-bisphosphate. The latter

Figure 22.6

Biosynthesis of dipalmitoylphosphatidylcholine. R_2 is usually an unsaturated fatty acid. Thus this two-step reaction results in the replacement of an unsaturated by a saturated fatty acid at the C-2 position on the glycerol backbone. Dipalmitoylphosphatidylcholine is the major component in lung surfactant, a substance that maintains surface tension in the lung alveoli so that they do not collapse when air is expelled.

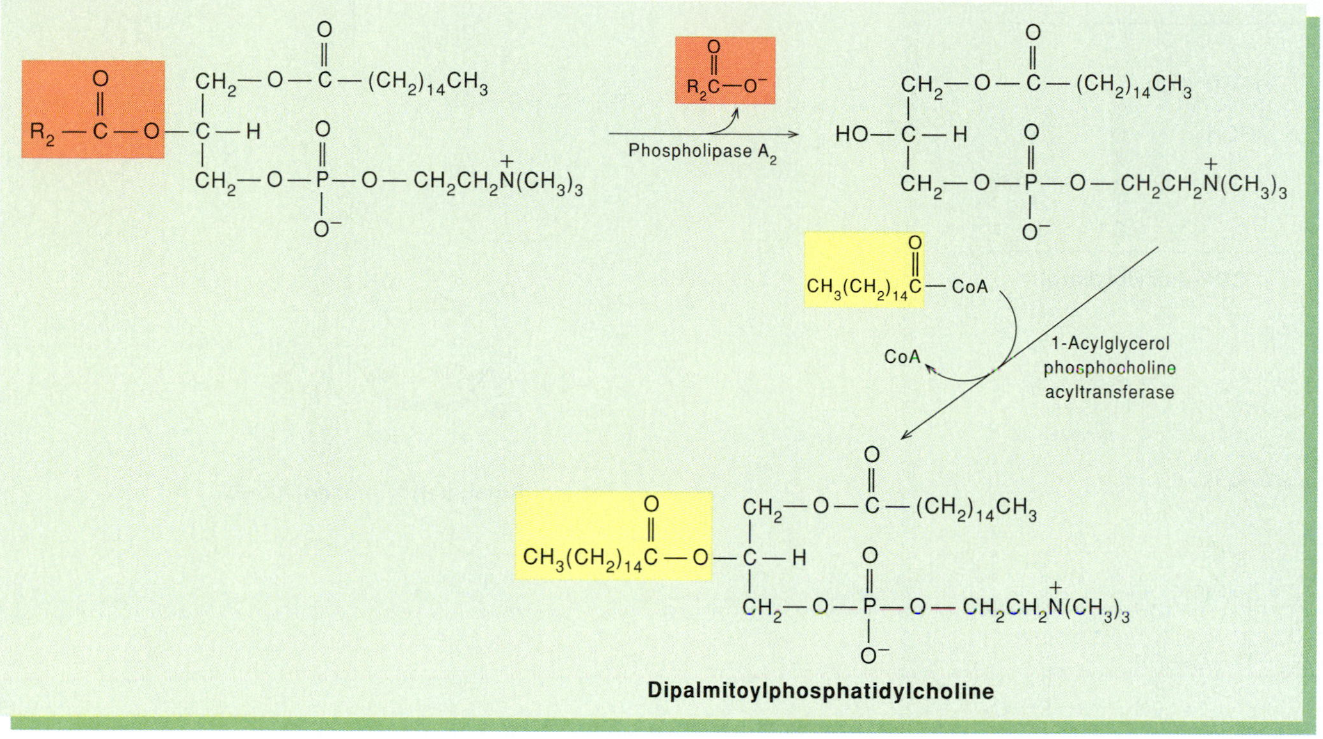

Phospholipase A_2

$CH_3(CH_2)_{14}C$— CoA

CoA

1-Acylglycerol phosphocholine acyltransferase

Dipalmitoylphosphatidylcholine

compound is degraded to inositol-1,4,5-P_3 and diacylglycerol by phospholipase C (fig. 22.8). Each of these degradation products has important regulatory functions (see chapter 24). The inositol-1,4,5-P_3 is involved in mobilization of calcium from intracellular stores (endoplasmic reticulum) and the rise in cytosolic calcium activates various enzymes (see chapter 24). The other product of the phospholipase C reaction, diacylglycerol, activates an enzyme called protein kinase C. This enzyme also requires calcium and phosphatidylserine for activity. Protein kinase C plays a major regulatory role as phosphorylating agent of an extremely diverse group of proteins (see chapter 24).

The Metabolism of Phosphatidylserine and Phosphatidylethanolamine Is Closely Linked

In prokaryotes, phosphatidylserine is made from CDP-diacylglycerol (see fig. 22.3). The enzyme for carrying out this conversion is absent in animal cells, which rely on a base exchange reaction in which serine and ethanolamine are interchangeable (fig. 22.9). Phosphatidylserine formed by the base exchange reaction can be decarboxylated to phosphatidylethanolamine in the mitochondria by phosphatidylserine decarboxylase (see fig. 22.9). These two reactions establish a cycle that has the net effect of converting serine to ethanolamine. This is a major mechanism for the biosynthesis of ethanolamine in eukaryotic cells. The primary fate of this ethanolamine is use in the synthesis of phosphatidylethanolamine via the CDP-ethanolamine pathway (see fig. 22.4). Thus phosphatidylserine is converted to phosphatidylethanolamine and ethanolamine but is also itself an important membrane lipid and is an activator of protein kinase C (not mentioned earlier).

Biosynthesis of Alkyl and Alkenyl Ethers

Some phospholipids contain either an *O*-alkyl or an *O*-alkenyl ether species at the SN-1 position, instead of the more common acylester linkage (see chapter 7). The biosynthetic pathway for the alkyl ether species of phosphatidic acid in eukaryotes is indicated in figure 22.10. The initial step involves the acylation of dihydroxyacetone phosphate (DHAP), a reaction discussed in connection with phosphatidic acid biosynthesis. Subsequently, an exchange reaction replaces the 1-acyl group with an alkyl group derived from an alcohol. This alcohol is formed by reduction of an acyl-CoA by NADPH or NADH (fig. 22.11). Following the exchange reaction, reduction of the ketone and acylation of the 2-hydroxyl group occurs. Once formed, the 1-alkyl ether derivative of phosphatidic acid is used for the synthesis of other phospholipids.

Certain alkyl ether species of phosphatidylcholine possess potent biological activity. For example, 1-alkyl-2-acetylglycerophosphocholine (fig. 22.12), also known as platelet-activating factor, reduces blood pressure in hypertensive rats and causes blood platelets to aggregate at very low hormone levels (10^{-10} M).

Figure 22.7

Reaction that converts CDP-diacylglycerol to phosphatidylinositol in eukaryotic cells.

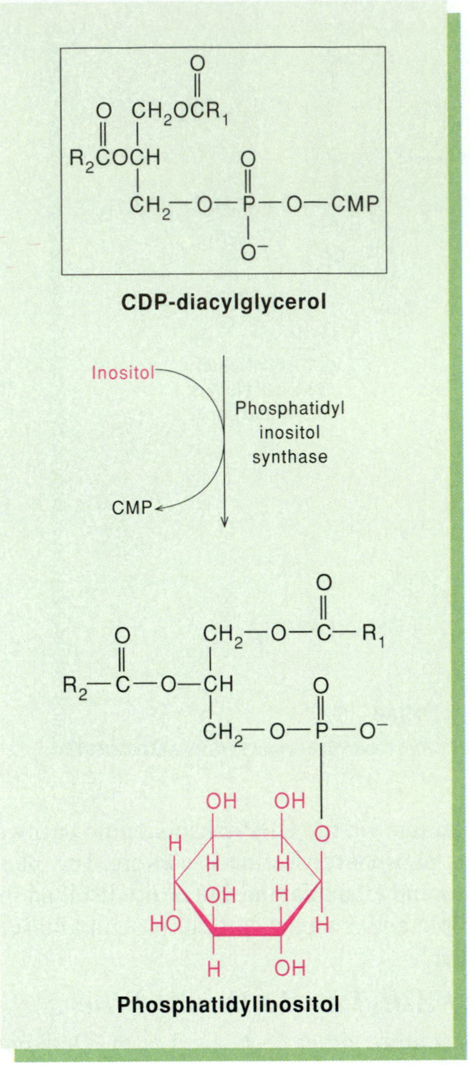

Figure 22.8

Phospholipase C degradation of phosphatidylinositol-4,5-P_2.

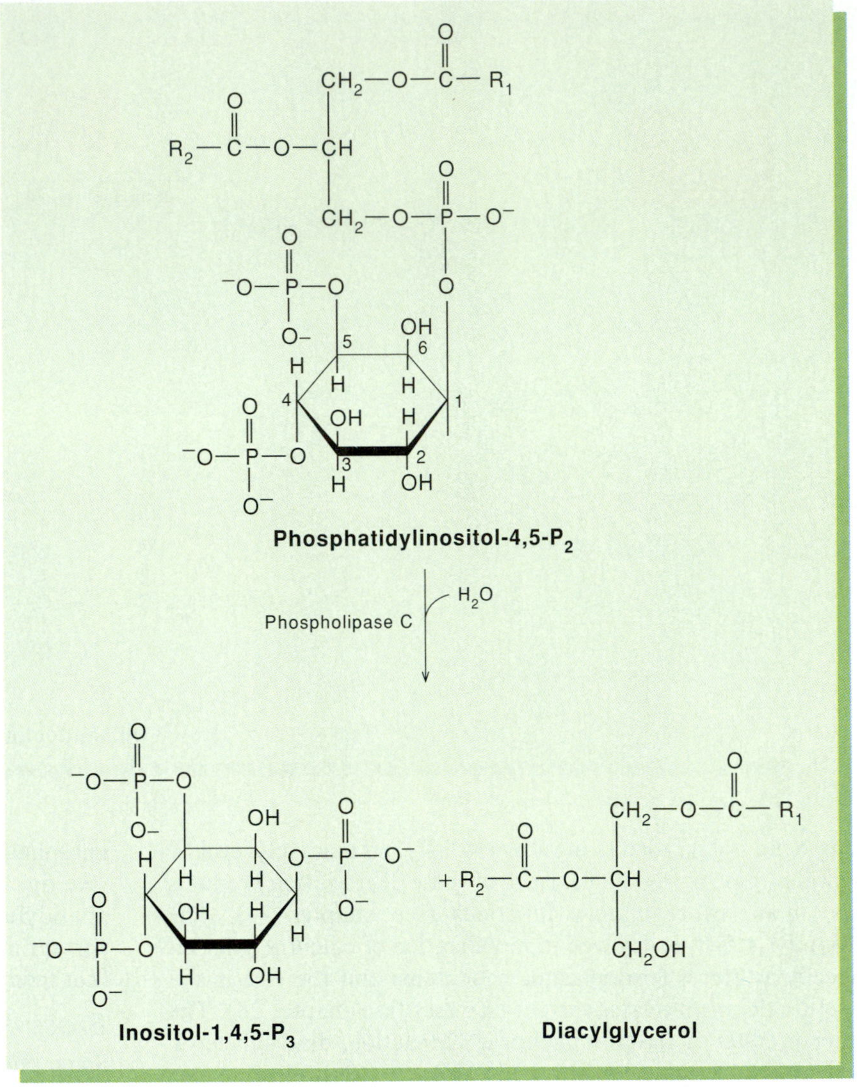

In many tissues, 1-alkyl-2-acylphosphatidylethanolamine can be desaturated by an endoplasmic reticulum enzyme, 1-alkyl-2-acylglycerophosphoethanolamine desaturase, to yield the corresponding unsaturated derivative called a plasmalogen (fig. 22.13). This enzyme requires O_2, NADH, and cytochrome b_5, the same cofactors required for the desaturation of stearoyl-CoA (see fig. 17.21). In many tissues plasmalogens are minor constituents, but in heart tissue nearly 50% of phosphatidylethanolamine contains the alkenyl ether at position SN-1. Alkenyl-ether-containing phospholipids can protect cells against the deleterious effects of singlet oxygen, which at high concentrations can kill cells.

Ether lipids are also found in microorganisms, notably the eukaryotic protozoans and archaebacteria. Alkyl ether bonds are more stable to hydrolysis than alkyl ester bonds. This greater stability probably accounts for the omnipresence of ether lipids in the membranes of archaebacteria, which frequently experience extremes of pH, salt, and temperature.

In the Liver, Regulation Gives Priority to Structural Lipids Over Energy-Storage Lipids

It is not clear what regulates the flux of diacylglycerol to phosphatidylcholine, phosphatidylethanolamine, or triacylglycerol (see fig. 22.4). However, at least in liver, the requirements for the synthesis of the essential membrane components, phosphatidylcholine and phosphatidylethanolamine, are met before an appreciable amount of energy-storage lipid (triacylglycerol) is made. The regulation of the synthesis of these lipids, as well as the synthesis of phosphatidylserine, phosphatidylglycerol, diphosphatidylglycerol, and the inositol phospholipids, is

Figure 22.9

Phosphatidylserine biosynthesis in animals is catalyzed by a base exchange enzyme on the endoplasmic reticulum. Decarboxylation of phosphatidylserine occurs in mitochondria. The cyclic process of phosphatidylserine formation from phosphatidylethanolamine and the reformation of phosphatidylethanolamine by decarboxylation has the net effect of converting serine to ethanolamine. This is a major mechanism for the synthesis of ethanolamine in many eukaryotes.

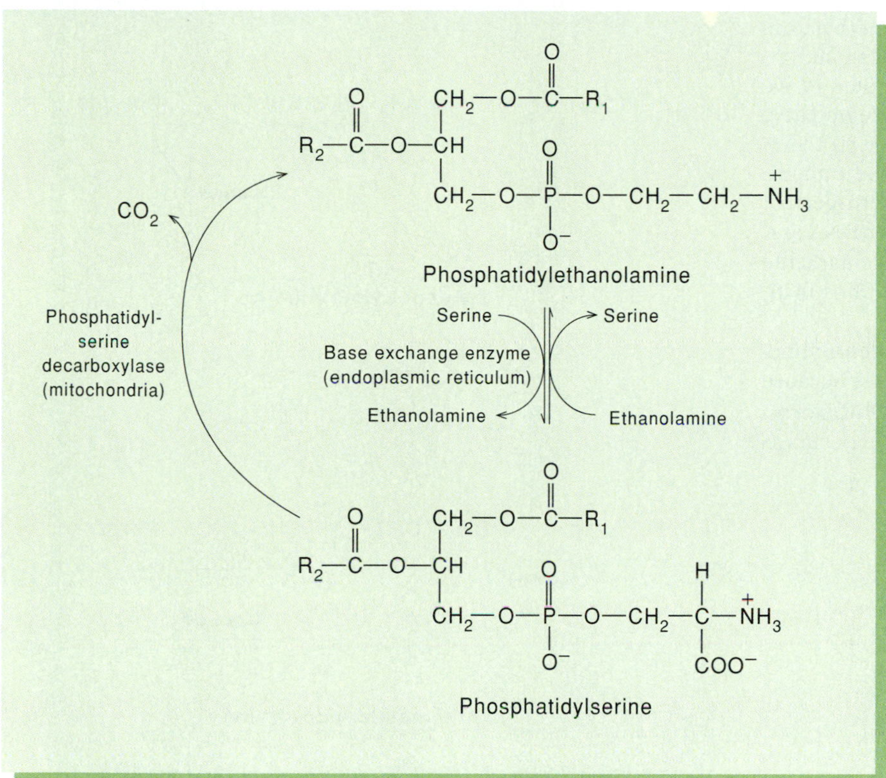

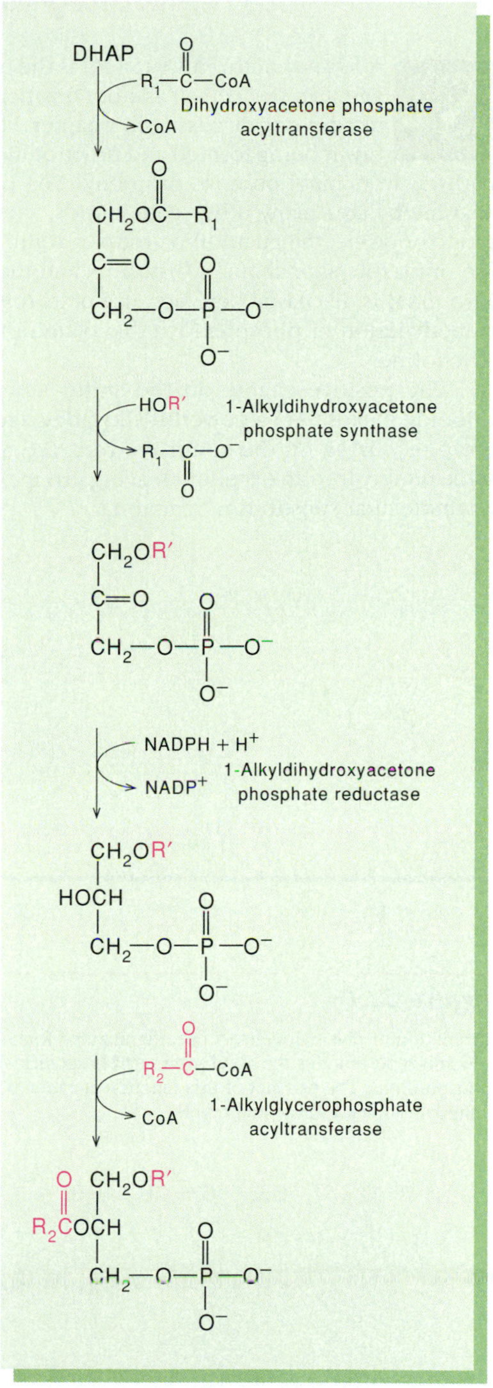

Figure 22.11

Acyl-CoA reductase catalyzes the formation of long-chain alcohols from fatty acyl-CoA.

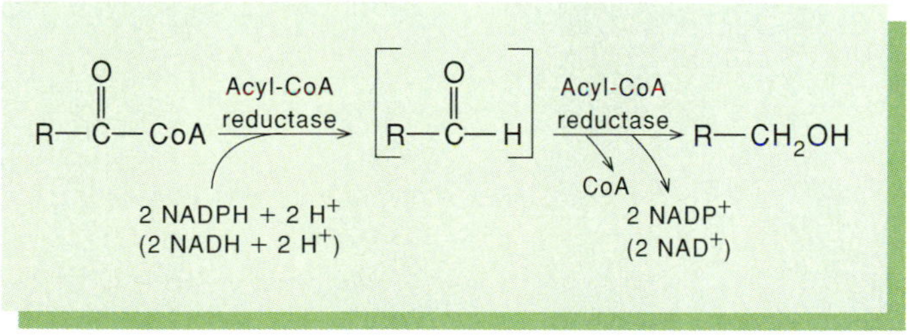

$$R-\overset{O}{\overset{\|}{C}}-CoA \xrightarrow[\text{2 NADPH + 2 H}^+ \text{ (2 NADH + 2 H}^+)]{\text{Acyl-CoA reductase}} \left[R-\overset{O}{\overset{\|}{C}}-H\right] \xrightarrow[\text{CoA, 2 NADP}^+ \text{ (2 NAD}^+)]{\text{Acyl-CoA reductase}} R-CH_2OH$$

Figure 22.12

Structure of 1-alkyl-2-acetylglycerophosphocholine, a potent phospholipid that will reduce hypertension in rats and cause platelets to aggregate.

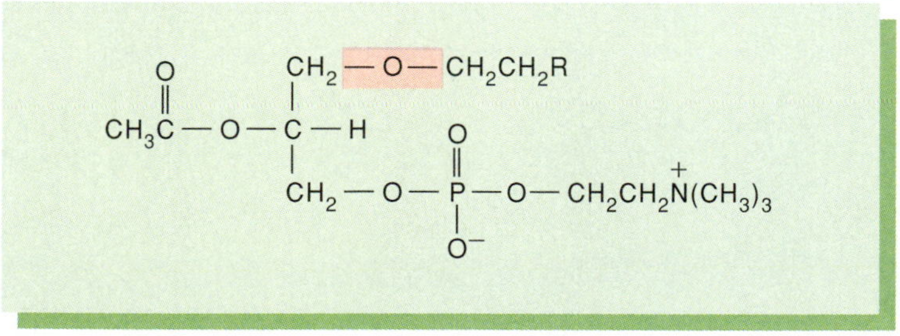

$$CH_3\overset{O}{\overset{\|}{C}}-O-\underset{CH_2-O-\overset{O}{\overset{\|}{P}}(O^-)-O-CH_2CH_2\overset{+}{N}(CH_3)_3}{\overset{CH_2-O-CH_2CH_2R}{C}}-H$$

Figure 22.10

Biosynthesis of the alkyl ether species of phosphatidic acid. The first two enzymes in this sequence are restricted to peroxisomes.

DHAP + $R_1-\overset{O}{\overset{\|}{C}}-CoA$ → CoA (Dihydroxyacetone phosphate acyltransferase)

$CH_2O\overset{O}{\overset{\|}{C}}-R_1$ / $C{=}O$ / $CH_2-O-PO_3^{2-}$

+ HOR' → $R_1-\overset{O}{\overset{\|}{C}}-O^-$ (1-Alkyldihydroxyacetone phosphate synthase)

CH_2OR' / $C{=}O$ / $CH_2-O-PO_3^{2-}$

+ NADPH + H^+ → $NADP^+$ (1-Alkyldihydroxyacetone phosphate reductase)

CH_2OR' / $HOCH$ / $CH_2-O-PO_3^{2-}$

+ $R_2-\overset{O}{\overset{\|}{C}}-CoA$ → CoA (1-Alkylglycerophosphate acyltransferase)

CH_2OR' / $R_2\overset{O}{\overset{\|}{C}}OCH$ / $CH_2-O-PO_3^{2-}$

22B
BOX

Pathway for Methylation by S-Adenosylmethionine

S-Adenosylmethionine (SAM) is the primary methylating agent in the cell. It is not considered a coenzyme, so we did not discuss it in chapter 11. In chapter 19 we saw it being formed and functioning in the degradative pathway from methionine to propionyl-CoA (see fig. 19.30). SAM also methylates many other compounds, including the ϵ-amino group of lysine, the guanidino group of arginine, the nitrogens of the imidazole side chain of histidine, and the bases of DNA and RNA. SAM is discussed here because of its role in the remarkable trimethylation of phosphatidylethanolamine to form phosphatidylcholine.

The positive charge on the sulfur atom of *S*-adenosylmethionine makes SAM a powerful alkylating agent. This is because the plus charge on the sulfur converts the *S*-adenosylhomocysteine moiety into an excellent leaving group. Alkylation occurs as a bimolecular substitution reaction.

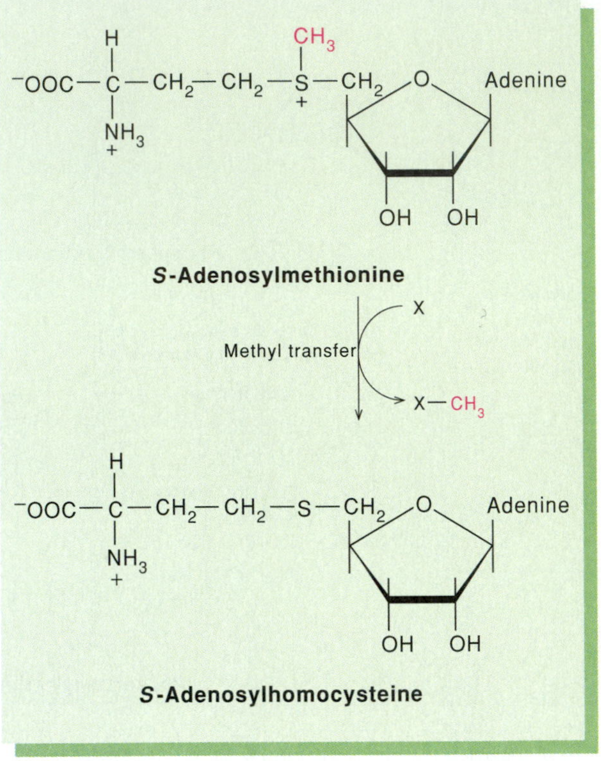

S-Adenosylmethionine

Methyl transfer: X → $X-CH_3$

S-Adenosylhomocysteine

Figure 22.13

Formation of plasmalogens. A specific enzyme found in the endoplasmic reticulum desaturates the alkyl group and is specific for the ethanolamine phospholipid. The product of this reaction is called plasmalogen and is abundant in heart tissue and myelin.

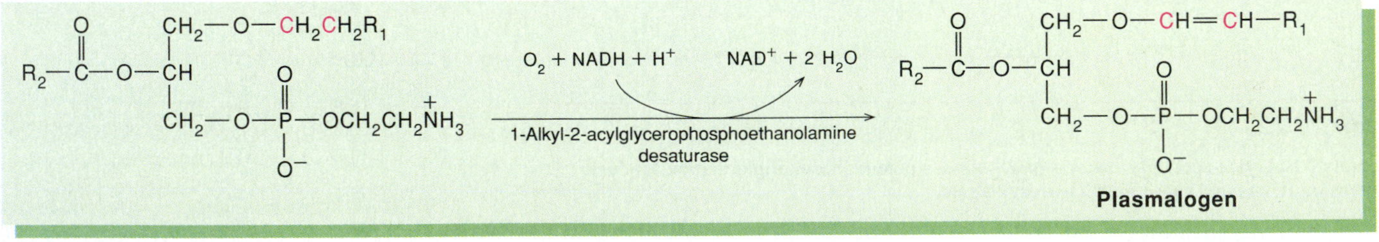

$O_2 + NADH + H^+$ → $NAD^+ + 2\ H_2O$

1-Alkyl-2-acylglycerophosphoethanolamine desaturase

Plasmalogen

Figure 22.14

Proposed mechanisms for regulation of CTP:phosphocholine cytidylyltransferase (CT) in rat liver. The enzyme is found in an inactive form in the cytosol and in an active form bound to the endoplasmic reticulum. The distribution between these two subcellular fractions appears to be primarily regulated by (1) the concentration of phosphatidylcholine, (2) the state of phosphorylation of the enzyme, (3) the concentration of fatty acids, and (4) the concentration of diacylglycerol. See the text for a more complete discussion of this enzyme.

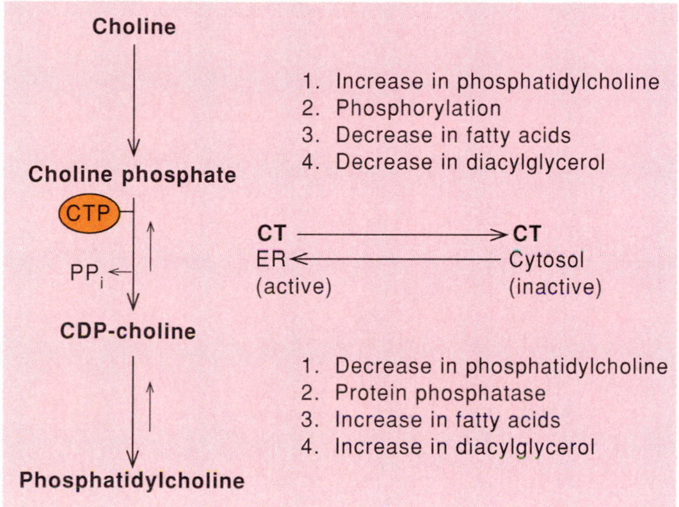

intimately related to the maintenance, proliferation, and functions of the plasma membrane and internal cellular membranes such as those of the endoplasmic reticulum and the Golgi.

The rate of phosphatidylcholine biosynthesis appears to be regulated principally by the activity of CTP:phosphocholine cytidylyltransferase, a dimer of identical subunits with a molecular weight of 41,720. The enzyme exists in both the cytosolic and endoplasmic reticulum fractions, but only the membrane-bound enzyme is active (fig. 22.14). Several mechanisms have been defined that govern the distribution of cytidylyltransferase between the cytosol and endoplasmic reticulum. (1) Perhaps the most important is feedback regulation by the amount of phosphatidylcholine in the endoplasmic reticulum. When the concentrations of phosphatidylcholine decrease in this membrane, there is an increased binding of enzyme to the endoplasmic reticulum, where it is activated, thus promoting the synthesis of phosphatidylcholine. As the levels of phosphatidylcholine return to normal values, the cytidylyltransferase is released into the cytosol, where it is inactive. (2) A protein kinase phosphorylates the enzyme and causes its release from the membrane (see fig. 22.14), a process that can be reversed by a phosphatase (see fig. 22.14). The nature of this protein kinase is not currently known, but it does not appear to be cAMP-dependent protein kinase or protein kinase C. (3) Fatty acids promote binding of the cytidylyltransferase to the endoplasmic reticulum, while removal of the fatty acid releases the enzyme into the cytosol (see fig. 22.14). (4) Finally, diacylglycerol, a substrate for phosphatidylcholine biosynthesis (see fig. 22.4), will also cause an increase in binding of cytidylyltransferase to the endoplasmic reticulum. When fatty acids are present in abundance, the diacylglycerol levels increase and this will not only provide substrate for phosphatidylcholine biosynthesis, but will also activate the cytidylyltransferase. These novel mechanisms for control of enzyme activity allow the liver to modulate phosphatidylcholine biosynthesis in a rapid and reversible fashion.

Cyclic AMP-mediated reactions inhibit fatty acid biosynthesis, which in turn decreases the synthesis of diacylglycerol. Hence the supply of diacylglycerol can limit the biosynthesis of phosphatidylcholine, phosphatidylethanolamine, and triacylglycerol. When sufficient diacylglycerol is present, the requirements for the synthesis of the essential membrane components, phosphatidylcholine and phosphatidylethanolamine, are met before an appreciable amount of energy-storage lipid (triacylglycerol) is made. The synthesis of phosphatidylcholine is regulated by the supply of CDP-choline through the cytidylyltransferase reaction as discussed in the previous paragraph. Similarly, the supply of CDP-ethanolamine will regulate the biosynthesis of phosphatidylethanolamine. When the cell's needs for these phospholipids are met, any additional diacylglycerol and fatty acyl-CoA is channeled into triacylglycerol.

The Final Reactions for the Biosynthesis of Phospholipids Occur on the Cytosolic Surface of the Endoplasmic Reticulum

The final reactions for the biosynthesis of phosphatidylcholine, phosphatidylethanolamine, phosphatidylserine, and phosphatidylinositol all occur on the cytosolic surface of the endoplasmic reticulum and the Golgi membranes in rat liver (fig. 22.15). By contrast, phosphatidylglycerol and diphosphatidylglycerol are synthesized in and remain largely in mitochondria. Two questions concerning the distribution of lipids in membranes remain unanswered: (1) Since phospholipid synthesis occurs on the cytosolic side of the endoplasmic reticulum and phospholipids are found on both leaflets of the bilayer, how do these lipids reach the inner leaflet of the bilayer? (2) How does the cell sort and transport phospholipids from the site of synthesis to other membranes in the cell?

The first question was addressed in chapter 7. In answer to the second, one possibility is that phospholipids are translocated within the cell as part of membranes. In this view, membrane vesicles with specific membrane proteins bud from the endoplasmic reticulum and move through the cytoplasm, eventually fusing with the membrane of a particular organelle. Specific membrane proteins present on the surface of the vesicle may direct the vesicle to a specific organelle or permit fusing upon collision.

Another possible mechanism for intracellular lipid transfer is provided by phospholipid exchange proteins found in the cytosol of eukaryotic cells. These proteins catalyze the exchange of phospholipid molecules between two membranes. For

Figure 22.15

Glycerolipid synthesis on the endoplasmic reticulum from rat liver. Fatty acids are inserted into the cytoplasmic surface of the endoplasmic reticulum (1), and activated to form acyl-CoA thioesters (2). The acyl chains may be elongated and/or desaturated (3). Glycerol-phosphate undergoes acyl-CoA-dependent esterification to form phosphatidic acid (4). The action of phosphatidic acid phosphatase (5) forms diacylglycerols that are converted to phosphatidylcholine and phosphatidylethanolamine by acquisition of phosphocholine and phosphoethanolamine polar head groups (6). Phosphatidylserine synthesis occurs by base exchange (7). Triacylglycerol synthesis occurs by esterification of diacylglycerol (8). CDP-diacylglycerol is an intermediate in the synthesis of phosphatidylinositol (9). Once formed, the glycerolipids may move to the lumenal surface of the endoplasmic reticulum (10). (C = choline; E = ethanolamine; I = inositol; S = serine; X = polar head group C, E, I, or S; P = PO_4.) (Source: R. M. Bell et al., "Lipid topogenesis" in *Journal of Lipid Research* 22:391, 1981. Copyright © 1981 Federation of American Societies for Experimental Biology, Bethesda, Md.)

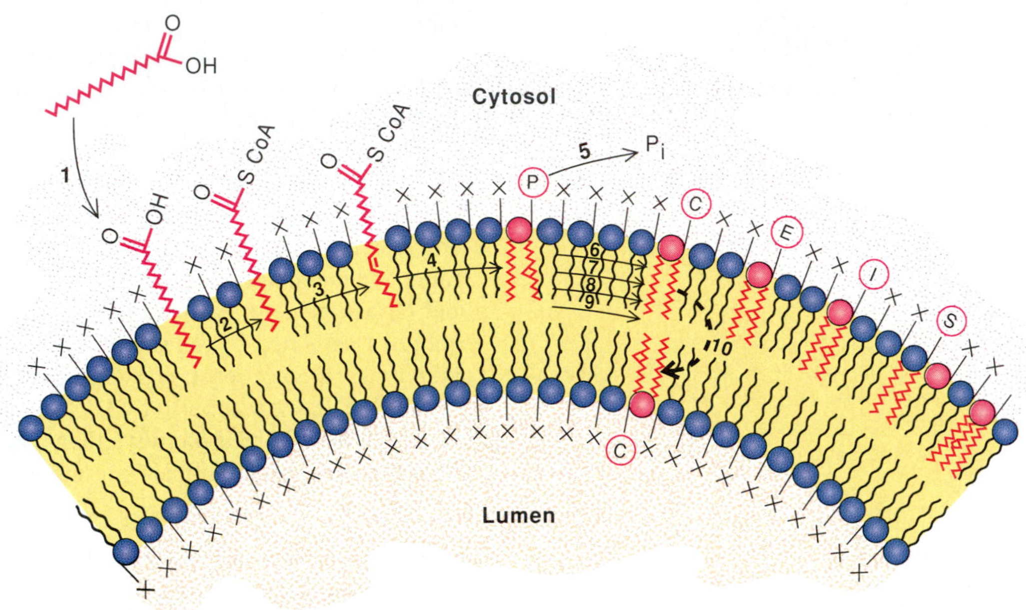

example, a phosphatidylcholine molecule on the endoplasmic reticulum exchanges with a molecule of phosphatidylcholine bound to the exchange protein. The protein moves to a mitochondrion, where another exchange occurs. In this way a phosphatidylcholine molecule can be moved from the endoplasmic reticulum to a mitochondrion. Some isolated exchange proteins show no specificity for the polar head group, while other proteins are highly specific. However, for the membranes to grow, a net transfer of lipid is required, and this effect has been difficult to demonstrate with these proteins. At a minimum, the proteins could function in renewal of the phospholipids in various membranes by exchange, since the phospholipids in eukaryotic cells do turn over.

Phospholipases Degrade Phospholipids

Enzymes that degrade phospholipids are called phospholipases. They are classified according to the bond cleaved in a phospholipid (fig. 22.16). Phospholipases A_1 and A_2 selectively remove fatty acids from the SN-1 and SN-2 positions, respectively. Phospholipase C cleaves between glycerol and the phosphate moieties; phospholipase D hydrolyzes the head group moiety, X, from the phospholipid. The lysophospholipids are degraded by lysophospholipases. Phospholipases are found in all types of cells and in various subcellular locations within eukaryotic cells. Some of these enzymes show specificity for particular polar head groups; others are nonspecific.

In some cases the functions of phospholipases are purely degradative, preparing phospholipids for elimination or catabolism. But in many cases they play important roles in synthesis and regulation. For example, phospholipase A_2 is involved in the first step in the biosynthesis of dipalmitoylphosphatidylcholine (see fig. 22.6). As we have mentioned, this is a typical "remodeling reaction," which involves the replacement of one acyl group by another. Phospholipase A_2 also plays an important regulatory role in the release of arachidonic acid from the SN-2 position. This molecule is a precursor of hormones known as eicosanoids (see fig. 22.22).

Phospholipase C degrades phosphatidylinositol-4,5-P_2 to inositol-1,4,5-P_3 and diacylglycerol (see fig. 22.8 and fig. 24.16). Both of these hydrolysis products are major regulatory molecules.

Sphingolipids

The common structural feature of all sphingolipids is a long-chain hydroxylated secondary amine (see table 7.4). These lipids are components of all mammalian cells and are abundant in myelin sheath, a multilayered membranous structure that protects and insulates nerve fibers. They are also found as components of plasma lipoproteins (see chapter 23).

Figure 22.16

Reactions catalyzed by phospholipases. X can be any of the head groups: choline, ethanolamine, serine, glycerol, or inositol.

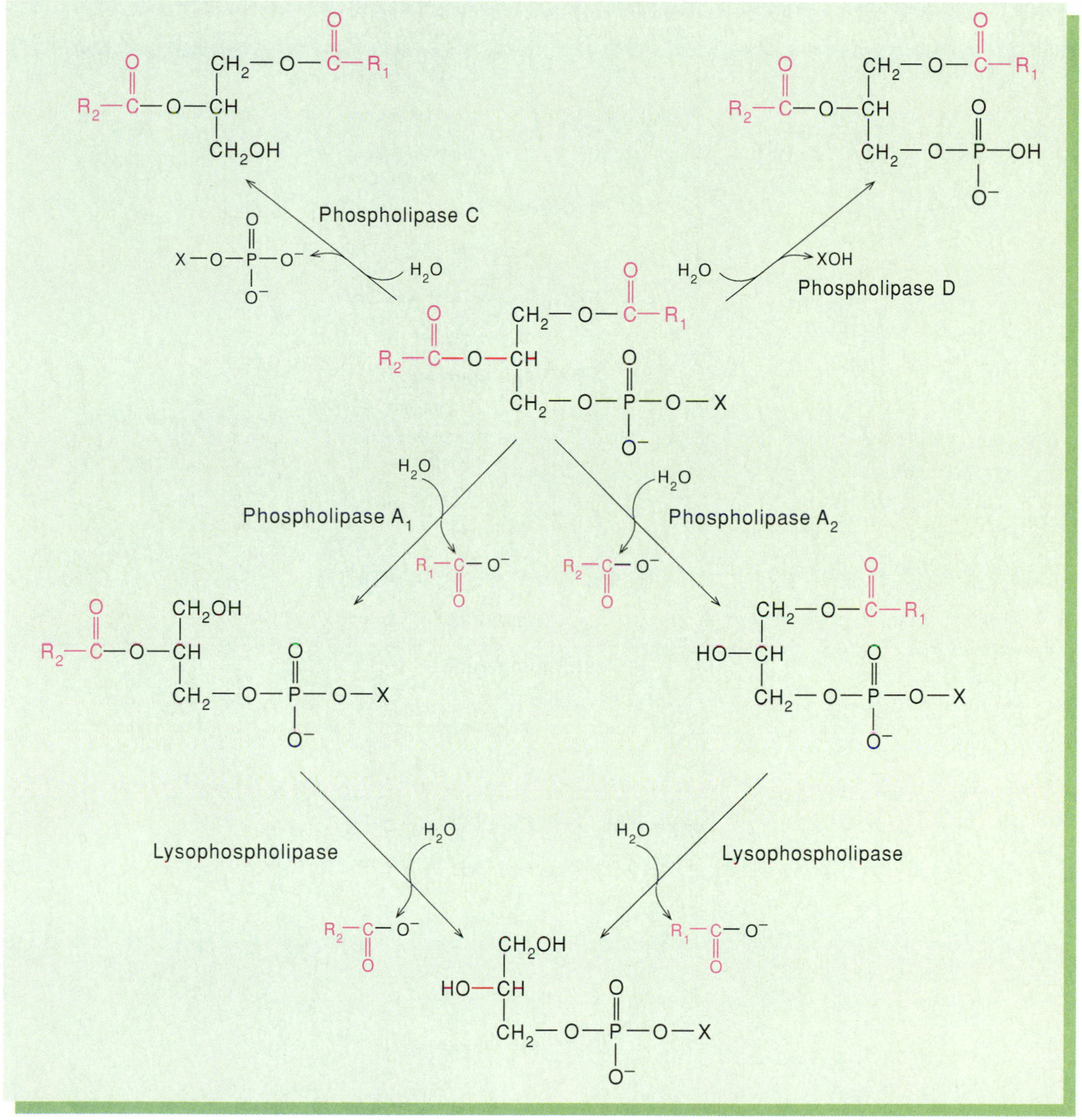

Sphingomyelin Is Formed Directly from Ceramide

Sphingenine is the major long-chain base present in sphingolipids. Sphingenine is derived from sphinganine by a desaturation reaction. The biosynthesis of sphinganine occurs on the endoplasmic reticulum and involves a condensation of palmitoyl-CoA with serine and a subsequent reduction of the 3-ketone (fig. 22.17). (Note: The sphingolipids are built on a serine-derived core rather than a glycerol core as is the case with other phospholipids.) Once the long-chain base is formed, it reacts with the acyl-CoA to form ceramide (see fig. 22.17).

Sphingomyelin is synthesized by the transfer of the phosphocholine moiety of phosphatidylcholine to ceramide (see fig. 22.17). The enzyme is located on the lumenal side of the Golgi membrane.

Glycosphingolipid Synthesis Also Starts from Ceramide

The synthesis of glycosphingolipids shows many similarities to the synthesis of the O-linked glycoproteins (see chapter 21). Thus the carbohydrates are added to the acceptor lipid by transfer from a sugar nucleotide. The enzymes involved in these reactions are called glycosyltransferases and are thought to be

Figure 22.17

The pathways to sphingomyelin and glycosphingolipids. Figure 22.18 elaborates on some pathways from ceramide to glycosphingolipids.

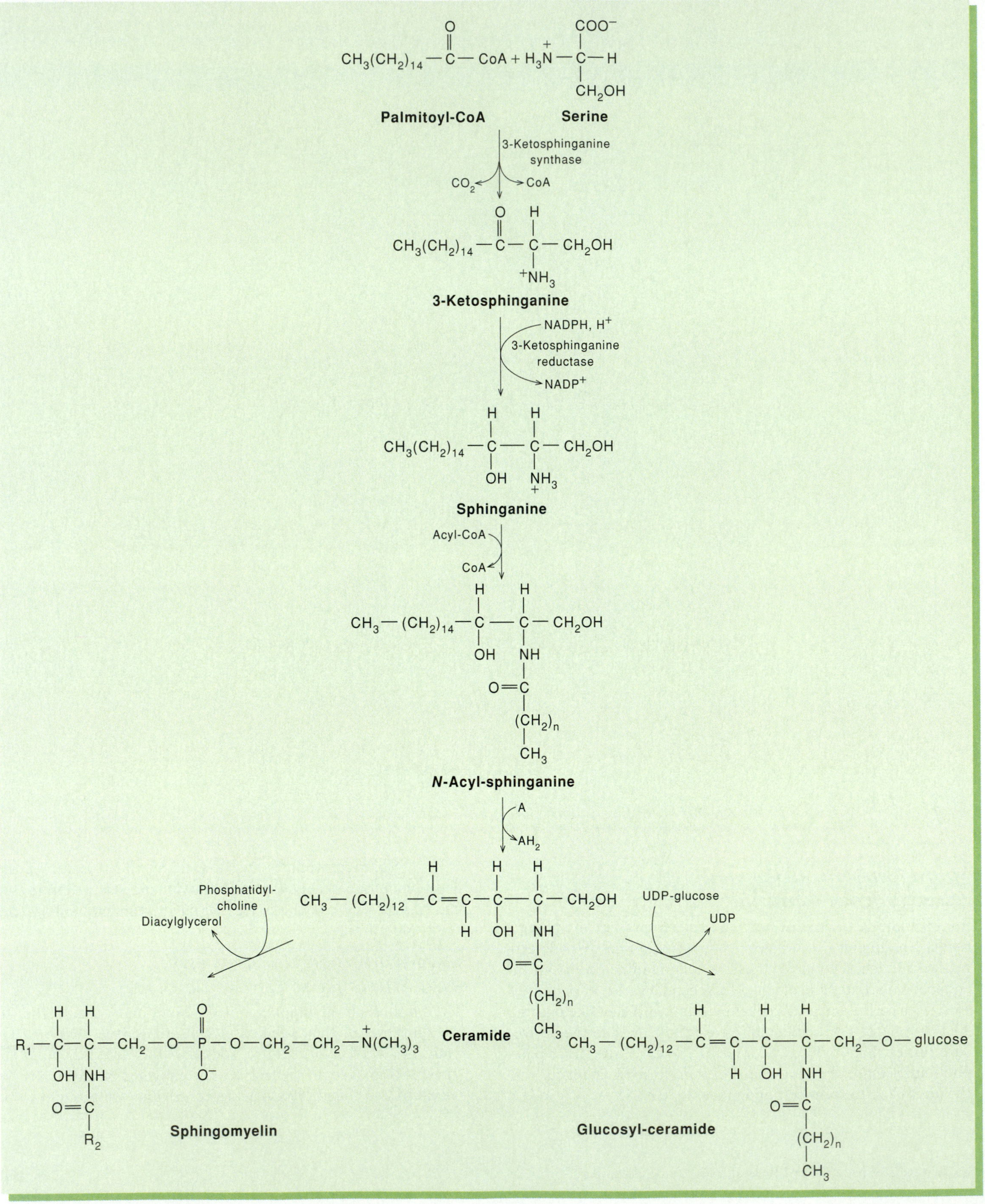

Figure 22.18

Outline of biosynthesis of some glycosphingolipids. (Glc = glucose, Gal = galactose, GalNAc = *N*-acetylgalactosamine, NeuNAc = *N*-acetylneuraminic acid, PAPS = 3′-phosphoadenosine, 5′-phosphosulfate.) PAPS is a sulfate analog of 3′-phospho-ADP. It has a high transfer potential for sulfate. The formation of PAPS is discussed in figure 18.11.

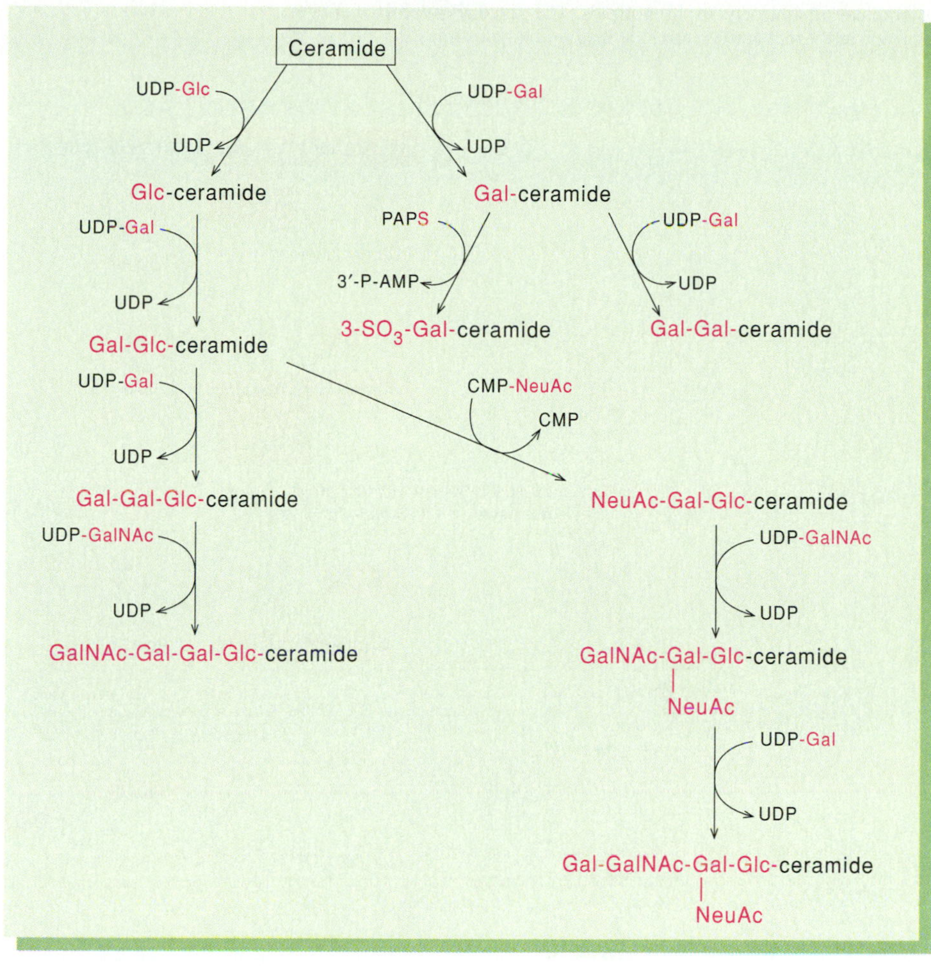

specific for each reaction. Most of the glycosyltransferases involved in the biosynthesis of the glycosphingolipids are located on the lumenal side of the Golgi apparatus. Finally, the nucleotide sugars used in biosynthesis are transferred across the membrane into the lumen of the Golgi by a transporter located on the Golgi membrane. The biosynthesis of certain key glycosphingolipids is shown in figure 22.18.

Glycosphingolipids Function as Structural Components and as Specific Cell Receptors

In all instances examined, the glycolipids are oriented asymmetrically in the bilayer of the plasma membrane, facing the outside of the cell. Many observations indicate that the glycosphingolipids are more than just structural lipids and have specific cell-surface recognition properties that are similar to plasma-membrane-bound glycoproteins (see chapters 6 and 21). This conclusion is consistent with the fact that many cell-surface recognition properties are due to the protruding carbohydrate moieties, which may be bound by either lipids or proteins. For example, glycosphingolipids are blood group antigens and a person who diplays the B blood type has comparable B antigens on both its membrane-bound glycosphingolipids and glycoproteins. Similarly, the carbohydrate moieties displayed by the glycosphingolipids differ according to stage of development of the organism. Cells transformed by tumor viruses also display altered glycosphingolipids, just as is the case with membrane-bound glycoproteins.

Defects in Sphingolipid Catabolism Are Associated with Metabolic Diseases

The degradation of glycosphingolipids occurs in a stepwise fashion in lysosomes (fig. 22.19). Investigations of the degradation process have been stimulated by the occurrence of a number of human genetic diseases, the sphingolipidoses, each of which results from the accumulation of one of these lipids. In all but a few cases, the diseases are caused by a deficiency in the activity of one of the enzymes involved in the catabolism of the sphingolipids. The principal diseases in which defects in the catabolism of a sphingolipid have been detected are shown in table 22.1. Here we will describe two of the disorders.

Fabry's disease was noted in 1898 by J. Fabry and independently by W. Anderson. Some characteristic symptoms are skin rash, pain in the extremities, and renal impairment, which is accompanied by hypertension. Patients lead a reasonably normal life until approximately their fourth decade, when the kidneys usually fail. Little progress was made in understanding

Figure 22.19

Catabolism of some glycosphingolipids. The glycosyl hydrolase enzymes involved in these reactions are localized in the lysosomes.

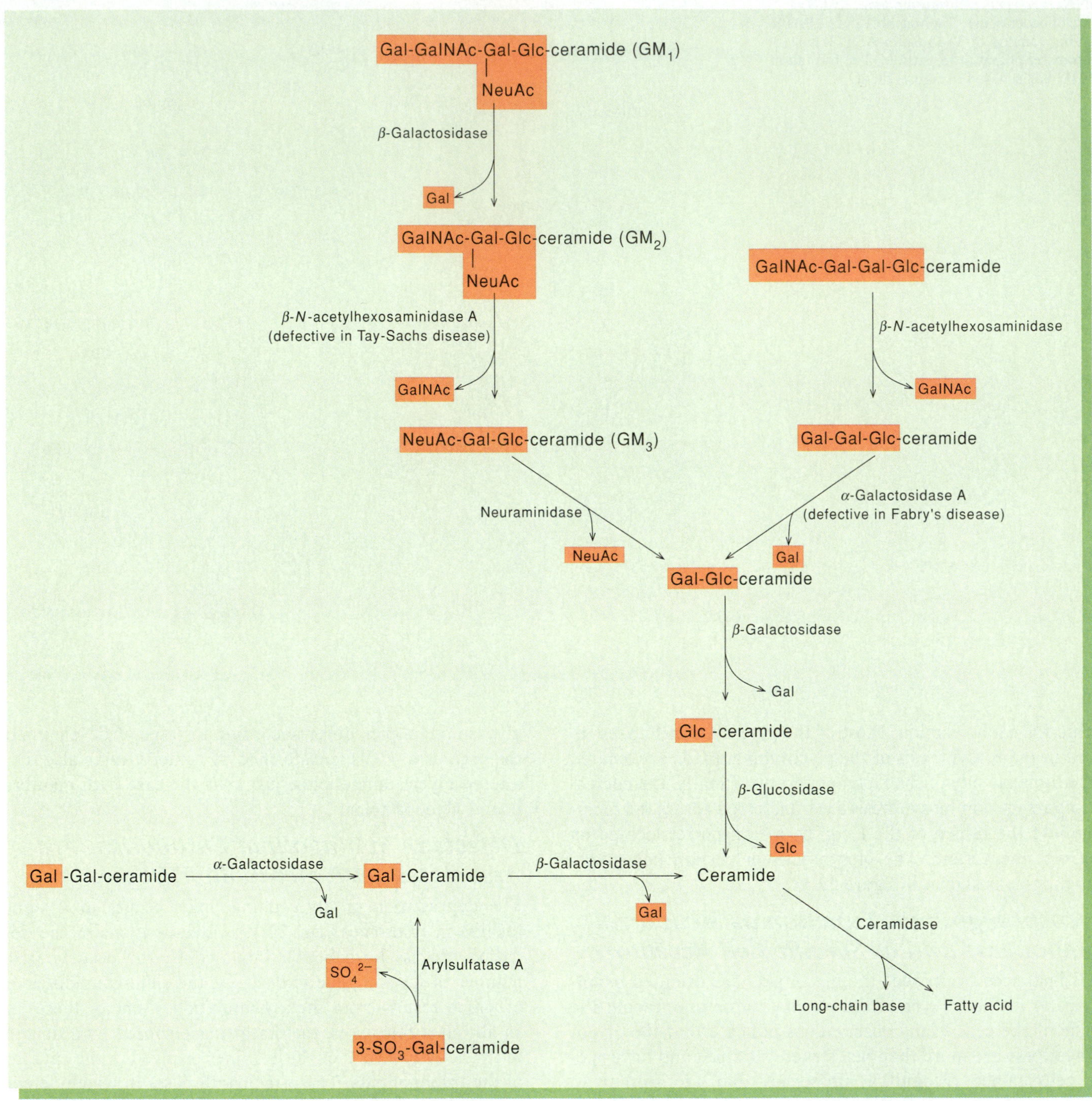

the cause of this disease until Charles Sweeley and Bernard Klionsky described, in 1963, the structure of Gal-Gal-Glc-ceramide as the major lipid that accumulates in the Fabry kidneys. In 1967 Roscoe Brady demonstrated that the enzymatic defect was a deficiency of the enzyme that degrades trihexosylceramide, and this enzyme was later shown by J. A. Kint to be α-galactosidase A (see fig. 22.19). It is now possible to diagnose the disease, which is X-chromosome-linked, with biochemical tests based on these discoveries. For example, in prenatal diagnosis, cells are obtained from the amniotic fluid by amniocentesis and cultured; then the activity of α-galactosidase A is determined.

Table 22.1
Inherited Diseases of Sphingolipid Catabolism

Disease	Enzyme Activity that Is Deficient	Reaction
1. Ceramidase deficiency: Farber's lipogranulomatosis	Ceramidase	Ceramide ⟶ fatty acid + long-chain base
2. Sphingomyelin lipidosis: Niemann-Pick disease	Sphingomyelinase	Sphingomyelin ⟶ ceramide + phosphocholine
3. Glucosylceramide lipidosis: Gaucher's disease	β-Glucosidase	Glc-ceramide ⟶ Glc + ceramide
4. Galactosylceramide lipidosis: globoid cell leukodystrophy	β-Galactosidase	Gal-ceramide ⟶ Gal + ceramide
5. Sulfatide lipidosis: metachromatic leukodystrophy	Arylsulfatase A	3′-SO_3^--Gal-ceramide ⟶ Gal-ceramide + SO_4^{2-}
6. Fabry's disease	α-Galactosidase A	Gal-Gal-Glc-ceramide ⟶ Gal + Gal-Glc-ceramide
7. GM_1[a] gangliosidosis	GM_1-β-galactosidase	GM_1 ⟶ Gal + GM_2
8. Tay-Sachs disease (GM_2 gangliosidosis)	Hexosaminidase A	GM_2 ⟶ GM_3 + GalNAc
9. Sandhoff's disease	Hexosaminidases A + B	GM_2 ⟶ GM_3 + GalNAc

[a]GM is the general abbreviation for a ganglioside. Subscripts are added to distinguish different members of this group of the glycosphingolipids. The structure of GM_2 is shown in figure 7.12.

There is presently no effective treatment for Fabry's disease. Preliminary studies have suggested that patients might respond to infusions of purified α-galactosidase A, but the studies have had limited success because of the short lifetime of the infused enzyme in the blood. Fortunately, Fabry's disease is rare, with only a few hundred cases reported throughout the world.

The disease described by Warren Tay in 1881, now known as Tay-Sachs disease, is quantitatively more important. It is estimated that 30 to 50 children with Tay-Sachs disease are conceived each year in the United States, largely by Jewish parents. This disease is devastating; the children usually do not survive beyond the age of 3. In Tay-Sachs disease there is an accumulation of ganglioside GM_2, especially in the brain, as a result of a deficiency of hexosaminidase A.

There is no way of treating Tay-Sachs disease. Enzyme replacement is not considered a likely remedy because infused enzyme would not penetrate the blood-brain barrier. However, the incidence of the disease has been dramatically reduced by prenatal diagnosis. Tay-Sachs disease is an autosomal recessive disease and so can arise only if both parents are carriers, i.e., if each parent carries a single defective gene for the hexosaminidase A enzyme. In that case there is a 25% chance that an offspring will have the disease.

Eicosanoids: Hormones Derived from Membrane-Bound Arachidonic Acid

Eicosanoids are a diverse group of hormones most of which are derived from the C_{20} polyunsaturated fatty acid, arachidonic acid (see fig. 17.22). Most prominent among the group are a series of cyclopentanoic acids known as prostaglandins (fig. 22.20). Closely related to the prostaglandins are the oxygenated eicosanoids known as thromboxanes (TX) (fig. 22.21). They differ from prostaglandins in the ring structure, which for thromboxanes is a cyclic ether (oxane ring).

The arachidonic acid precursor is in effect stored in the membrane as part of a phospholipid complex. Upon demand, the arachidonic acid is released from the phospholipid by a specific phospholipase A_2. Alternatively, the phospholipid may be degraded by a phosphatidylinositol-specific phospholipase C to yield a diacylglycerol, which is subsequently cleaved by diacylglycerol lipase to yield arachidonic acid. The release of arachidonic acid from the phospholipid is believed to be the rate-limiting step for the synthesis of eicosanoids.

The initial steps for the synthesis of prostaglandins and thromboxanes are the oxidation and cyclization of arachidonic acid to yield PGG_2 and the PGH_2 in reactions catalyzed by a single bifunctional enzyme, prostaglandin endoperoxide synthase, contained on the endoplasmic reticulum. The formation of PGG_2 is catalyzed by the cyclooxygenase component of the enzyme (fig. 22.22), and the subsequent formation of PGH_2 is catalyzed by the peroxidase component. The cyclooxygenase activity has the unusual property of catalyzing its own destruction. Approximately once in every 400 substrate turnovers, the cyclooxygenase activity is irreversibly inactivated. This "suicide reaction" occurs both *in vivo* and *in vitro,* but the chemical mechanism has not been described.

The anti-inflammatory effect of aspirin appears to be due to the acetylation of a serine at the active site of the cyclooxygenase, which irreversibly inactivates the cyclooxygenase activity of the enzyme (fig. 22.23). Aspirin has no effect on the peroxidase activity of the enzyme.

Figure 22.20

Structure of prostaglandin PGE_2. All prostaglandins are derived from C_{20} fatty acids. The names and numbering system are based on the hypothetical parent compound, prostanoic acid, shown at the top. The PG stands for prostaglandin. The E specifies the type, and the subscript refers to the number of double bonds. From the structure it can be seen that the double bonds of PGE_2 are *trans*Δ^{13} and *cis*Δ^{5}.

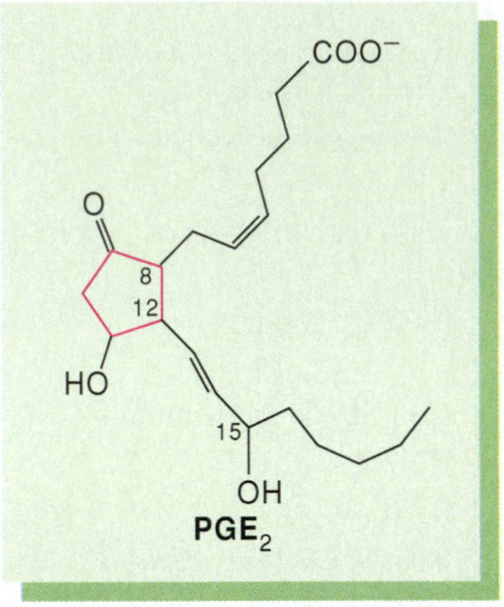

Figure 22.21

The structure of the thromboxane TXA_2. Thromboxanes differ from prostaglandins in having a cyclic ether (oxane) ring structure.

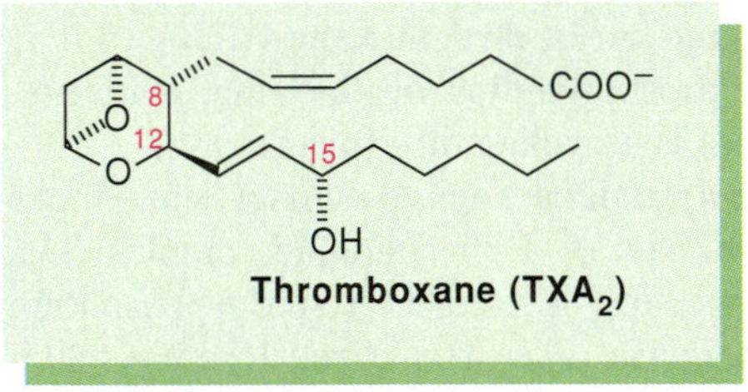

PGH₂ is the precursor of various prostaglandins and thromboxanes. Each step is catalyzed by a separate tissue-specific enzyme as indicated in figure 22.24. The tissue localization of the enzyme relates to the role of the eicosanoid; for example TXA_2 (made in platelets) causes platelet aggregation, whereas PGI_2 (made in arterial walls) inhibits this process.

In addition to the formation of prostaglandins and thromboxanes, arachidonic acid is metabolized to several other compounds. The hydroxyeicosatetraenoic acids, or HETEs (fig. 22.25), are formed from arachidonic acid by the enzyme lipoxygenase. Mammalian lipoxygenases found in white blood cells catalyze the insertion of oxygen in the 5, 12, or 15 position of various eicosanoic acids. The reaction mechanism for these insertions involves addition of oxygen to a double bond with the formation of a conjugated *cis-trans* diene (fig. 22.26). Subsequently the hydroperoxy group is reduced to an alcohol to form the corresponding hydroxy-eicosanoic acid. The physiological functions of hydroxy-eicosanoic acids are not well understood.

Figure 22.22

Reaction catalyzed by prostaglandin endoperoxide synthase. This enzyme has two catalytic activities, as indicated next to the reaction arrows.

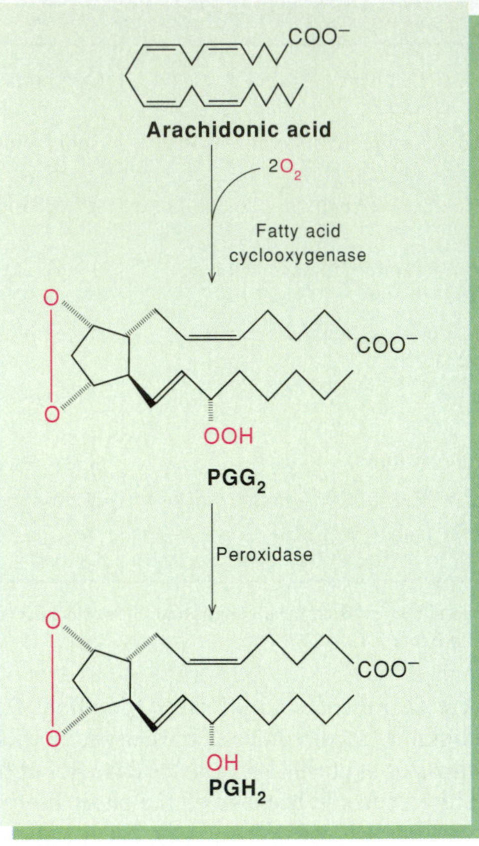

Figure 22.23

Aspirin inactivates cyclooxygenase by acetylation of a serine, probably at or near the active site.

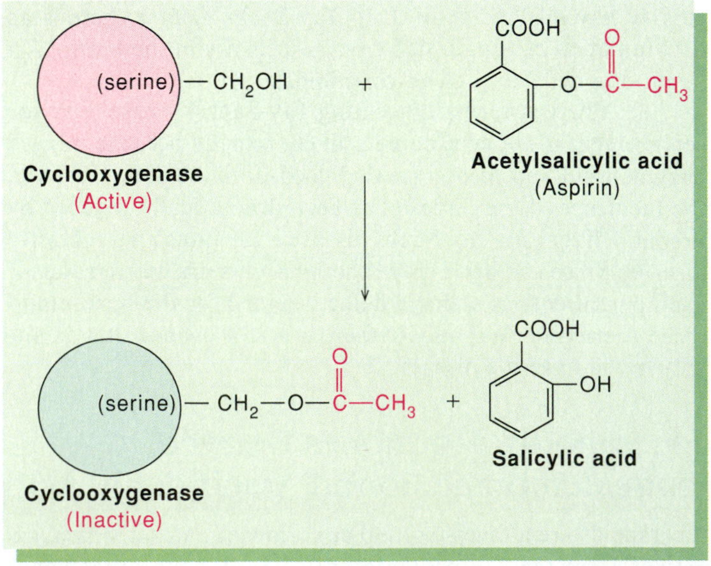

Figure 22.24

Formation of prostaglandins and thromboxanes.

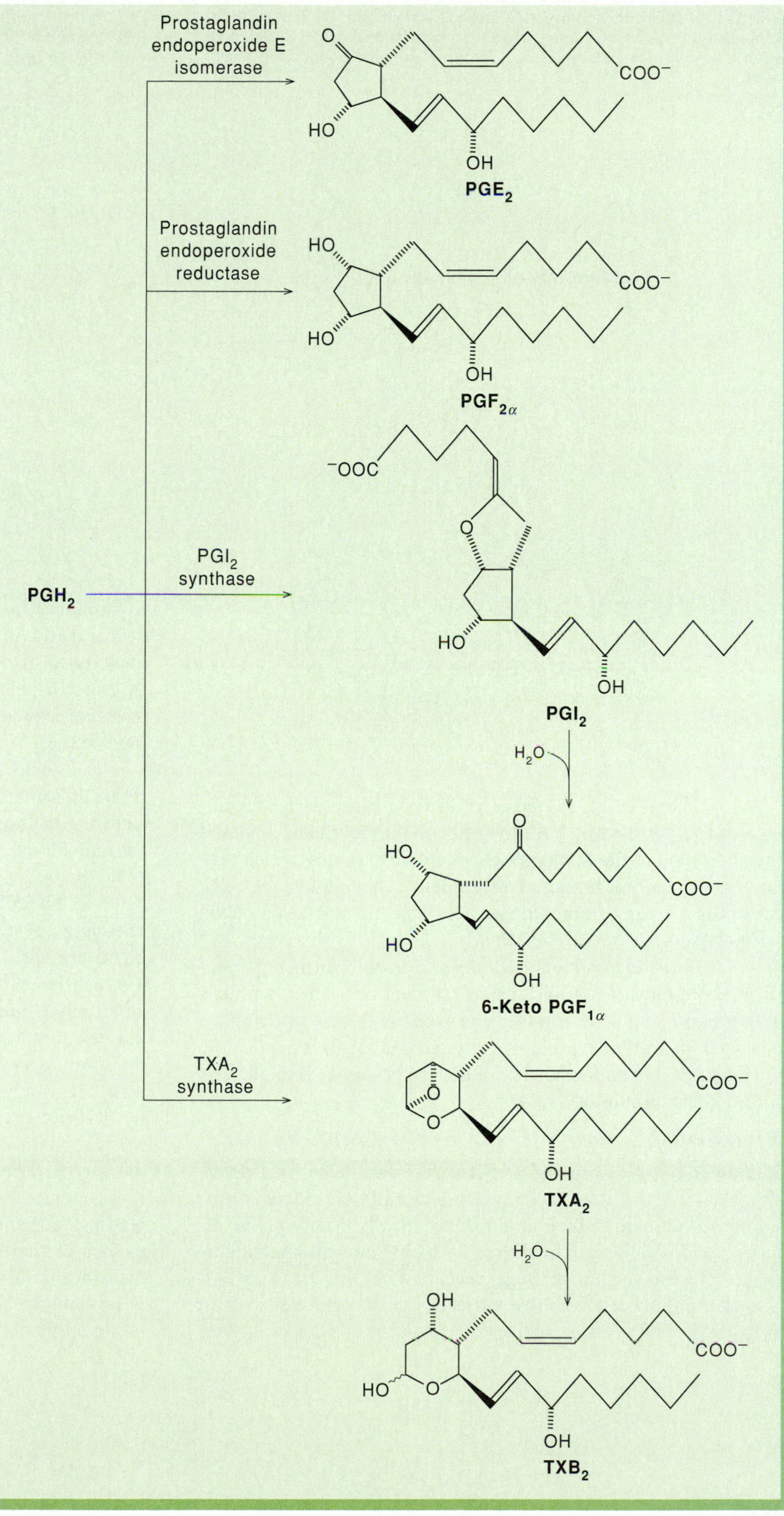

Figure 22.25

Structures of three common hydroxyeicosatetraenoic acids (HETEs). These compounds are all formed directly from arachidonic acid by the action of a specific lipoxygenase, which inserts a hydroxyl group at the indicated location.

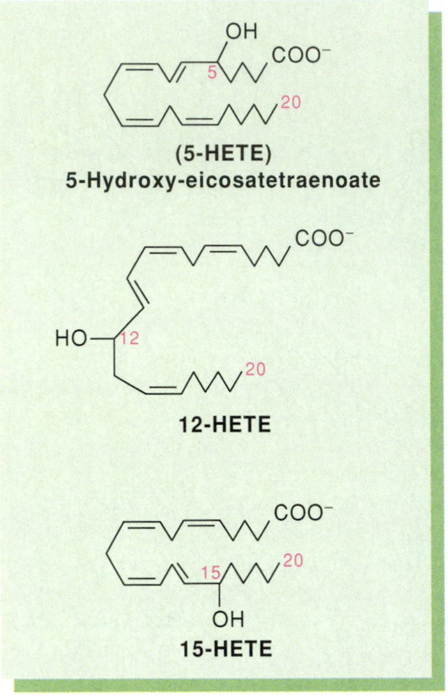

OH
COO^-
5
20
(5-HETE)
5-Hydroxy-eicosatetraenoate

COO^-
HO
12
20
12-HETE

COO^-
15
20
OH
15-HETE

Figure 22.26

The mechanism of the reaction catalyzed by lipoxygenases. This is a simple dioxygenase reaction; there is no net oxidation-reduction of either the fatty acid or the oxygen. A *cis-trans* conjugated diene is formed in the reaction.

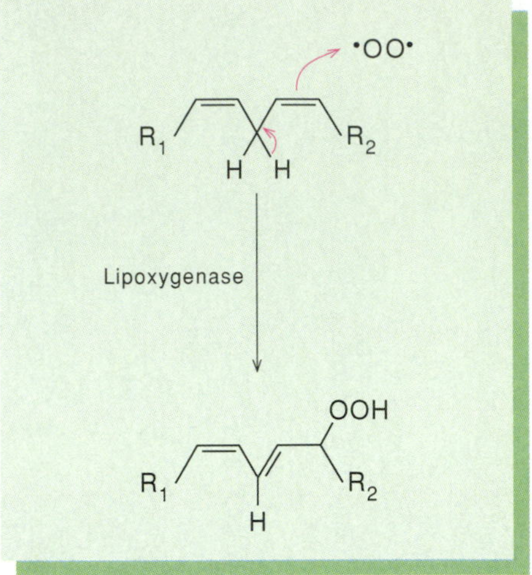

The rate of synthesis of lipoxygenase-derived compounds is controlled by both the release of arachidonic acid and the activation of the lipoxygenase. Certain peptides, which promote chemically stimulated movement (chemotaxis) of cells, cause the release of arachidonic acid and the activation of the 5-lipoxygenase.

If the eicosanoids are to serve useful hormonal functions, there must be a rapid means for their removal when their action is no longer desirable. Consistent with this principle, it has been found that eicosanoids have short half-lives *in vivo;* the PGE and PGF compounds do not normally survive a single pass through the circulatory system.

Eicosanoids Exert Their Action Locally

The diversity of eicosanoid structures is paralleled by a diversity of biological effects. They are generally considered to be hormones that exert their main effect locally. Many of the effects of eicosanoids are mediated by cAMP, a well-known hormone effector substance (e.g., see chapters 13, 17, and 24). Specific binding of PGE_2 to a variety of cells correlates with an activation of adenylate cyclase and the accumulation of cAMP. In human adipocytes, PGE_2 causes a 15-fold increase in the concentration of cAMP. In platelets PGI_2, rather than PGE_2, appears to mediate the increase in cAMP. There is evidence that specific, high-affinity receptors exist for prostaglandins just as is the case for other hormones that bind to the outer plasma membrane.

The eicosanoids have been implicated as mediators of tissue inflammation since aspirin's anti-inflammatory effect was shown to be the result of its inactivation of cyclooxygenase. How eicosanoids cause tissue inflammation is the focus of much current research.

The role of prostaglandins in blood clotting is also generating a great deal of research and speculation. PGI_2 lowers blood pressure, relaxes coronary arteries, and inhibits platelet aggregation. TXA_2 has the opposite effects. PGI_2 is made in the endothelial lining of blood vessel walls. It appears to inhibit platelet aggregation by binding to a receptor on the plasma membrane, an action that leads to increase of cAMP in platelets. TXA_2 appears to suppress the increase in cAMP caused by PGI_2. It has been speculated that the synthesis of PGI_2 may prevent platelets from binding to arterial walls. In damaged areas of arteries, the synthesis of PGI_2 may be decreased and the presence of TXA_2 would cause the platelets to aggregate, leading to the formation of a blood clot. The physiological relevance of these observations is unclear. Finally, the fact that aspirin in mild doses is recommended to prevent heart attacks is probably related in some way to its inhibitory effect on eicosanoid formation, but this point needs further study.

Summary

Fatty acids are components of a rich variety of more complex lipid molecules that play structural roles in membranes. Fatty acids are also the precursors of compounds with hormonelike activity. In this chapter we have focused on the biosynthesis of these compounds, with some mention of their functions. The following points are the highlights of our discussion.

1. Lipid synthesis is unique in that it is almost exclusively localized to the surface of membrane structures. The reason for this restriction is the amphipathic nature of the lipid molecules. In bacteria, where there are no organelles, lipid synthesis is localized to the inner plasma membrane. In eukaryotes, various organelles are involved in lipid synthesis, most notably the endoplasmic reticulum, the mitochondria, and the Golgi apparatus.
2. Phospholipids are biosynthesized by acylation of either glycerol-3-phosphate or dihydroxyacetone phosphate to form phosphatidic acid. This central intermediate can be converted into phospholipids by two different pathways. In one of these, phosphatidic acid reacts with CTP to yield CDP-diacylglycerol, which in bacteria is converted to phosphatidylserine, phosphatidylglycerol, or diphosphatidylglycerol. In *E. coli,* the major phospholipid, phosphatidylethanolamine, is synthesized by means of this route though the decarboxylation of phosphatidylserine. In the second pathway, found in eukaryotes, phosphatidic acid is hydrolyzed to diacylglycerol, which reacts with CDP-ethanolamine or CDP-choline to yield phosphatidylethanolamine or phosphatidylcholine, respectively. Alternatively, the diacylglycerol may react with acyl-CoA to form triacylglycerol.
3. There are numerous reactions by which the acyl groups or polar head groups of phospholipids might be modified or exchanged. Phospholipids are degraded by specific phospholipases. The transfer of phospholipids among membranes is facilitated by phospholipid exchange proteins. Budding and fusion also play a role in the distribution of lipids between the different membranes.
4. In *E. coli,* lipids are used almost exclusively for phospholipid synthesis, and regulation is known to occur at an early stage in fatty acid synthesis. In the mammalian liver, fatty acids are important precursors of both the structural phospholipids and also of the energy-storage lipid, triacylglycerol. The need for structural lipids is satisfied before fatty acids are shunted into energy storage. The rate of triacylglycerol synthesis is regulated by the availability of fatty acid, from both diet and biosynthesis. That availability is primarily a function of nutrient excess. The regulation of phosphatidylcholine biosynthesis occurs primarily at the reaction catalyzed by CTP:phosphocholine cytidylyltransferase. This enzyme is activated by translocation from the cytosol, where it is inactive, to the endoplasmic reticulum, where it is activated. The regulation of the biosynthesis of other phospholipids in normal cells is not well understood.
5. The sphingolipids are important structural lipids found in eukaryotic membranes. The acylation of sphinganine produces ceramide, which reacts with phosphatidylcholine to give sphingomyelin. Ceramide also can react with activated carbohydrates (e.g., UDP-glucose) to form the glycosphingolipids. Studies on the catabolism of the sphingolipids have revolved around many inherited diseases that result from a defect in the enzymatic degradation of single sphingolipids.
6. Prostaglandins and thromboxanes are related compounds known as eicosanoids, which have a large variety of biological activities. Prostaglandins and thromboxanes are biosynthesized from C_{20} polyunsaturated fatty acids, primarily arachidonic acid. In the initial reaction, arachidonic acid is converted to PGH_2 by prostaglandin endoperoxide synthase. PGH_2 is then converted to PGE_2, PGF_{2a}, TXA_2, or PGI_2. Alternatively, arachidonic acid can be converted to hydroxyeicosatetraenoic acids. The eicosanoids are local hormones that affect the function of a cell by binding to a receptor on the cell surface. This receptor in turn interacts with a GTP binding protein that alters the synthesis of a second messenger such as cAMP.

Selected Readings

Brindley, D. N., Metabolism of triacylglycerols. Chapter 6 in D. E. Vance and J. E. Vance (eds.), *Biochemistry of Lipids, Lipoproteins and Membranes.* Amsterdam: Elsevier Science Publishers, 1991. This chapter provides an advanced discussion of phosphatidic acid and triacylglycerol metabolism.

Dennis, E. A., Phospholipases. In *Enzymes XVI.* New York: Academic Press, 1983. This article provides a thorough review of phospholipases.

Jackowski, S., J. E. Cronan, and C.O. Rock, Lipid metabolism in procaryotes. Chapter 2 in D. E. Vance and J. E. Vance (eds.), *Biochemistry of Lipids, Lipoproteins and Membranes.* Amsterdam: Elsevier Science Publishers, 1991. This chapter provides an advanced treatment of the genetics and metabolism of phospholipids in *E. coli.*

Scriver, C. R., A. L. Beaudet, W. S. Sly, and D. Valle (eds.), *The Metabolic Basis of Inherited Disease,* 6th ed., vol. II New York: McGraw-Hill, 1989. This book contains in-depth chapters on sphingolipid metabolism and many of the sphingolipidoses.

Smith, W. L., P. Borgeat, and F. A. Fitzpatrick, The eicosanoids: cyclooxygenase, lipoxygenase and epoxygenase pathways. Chapter 10 in D. E. Vance and J. E. Vance (eds.), *Biochemistry of Lipids, Lipoproteins and Membranes.* Amsterdam: Elsevier Science Publishers, 1991. This chapter provides advanced information on the biochemistry, metabolism, and functions of the eicosanoids.

Snyder, F., Metabolism, regulation and function of ether-linked glycerolipids and their bioactive species. Chapter 8 in D. E. Vance and J. E. Vance (eds.), *Biochemistry of Lipids, Lipoproteins and Membranes.* Amsterdam: Elsevier Science Publishers, 1991. This chapter contains advanced information on the biochemistry of the lipids that contain an alkyl ether or an alkenyl ether and platelet activating factor.

Sweeley, C. C., Sphingolipids. Chapter 11 in D. E. Vance and J. E. Vance (eds.), *Biochemistry of Lipids, Lipoproteins and Membranes.* Amsterdam: Elsevier Science Publishers, 1991. This is an advanced chapter on the chemistry, metabolism, and function of the sphingolipids.

Vance, D. E. (ed.), *Phosphatidylcholine Metabolism,* Boca Raton, Fla.: CRC Press, 1989. This advanced monograph contains 13 chapters on all aspects of phosphatidylcholine biosynthesis and catabolism.

Vance, D. E., Phospholipid metabolism and cell signalling in eucaryotes. Chapter 7 in D. E. Vance and J. E. Vance (eds.), *Biochemistry of Lipids, Lipoproteins and Membranes.* Amsterdam: Elsevier Science Publishers, 1991. This chapter covers phospholipid metabolism at an advanced level and discusses the role of phosphatidylinositol and other phospholipids in the generation of second messengers in the cell.

Waite, M., Phospholipases. Chapter 9 in D. E. Vance and J. E. Vance (eds.), *Biochemistry of Lipids, Lipoproteins and Membranes.* Amsterdam: Elsevier Science Publishers, 1991. This chapter provides advanced knowledge about phospholipases and the hydrolysis of phospholipids.

Waite, M., *The Phospholipases.* New York: Plenum Press, 1987. This book provides a complete coverage of phospholipases from the technicalities of assay of phospholipases to the proposed mechanism of catalysis.

Weissmann, G., Aspirin. *Sci. Am.* 264:84–90, 1991. An interesting article that discusses the mechanism of action of aspirin in connection with eicosanoid metabolism.

Problems

1. Why are human genetic defects in phospholipid biosynthesis not observed?
2. Phosphatidylcholine biosynthesis appears to be regulated principally at the step catalyzed by CTP:phosphocholine cytidylyltransferase. Does the type of regulation observed make biochemical sense? Draw a chemical reaction mechanism for this enzyme.
3. The majority of phospholipids contain an unsaturated fatty acid at the C-2 position. Give an example of a phospholipid that has a saturated fatty acid in this position. How is it synthesized?
4. Explain why the reactions from choline to phosphatidylcholine in eukaryotic cells are thermodynamically feasible.
5. Where are ether-linked phospholipids found and what are some of their functions?
6. A patient accumulated the lipid Gal $\beta(1,4)$-Glc $\beta(1,4)$-ceramide. What enzymatic reaction might be defective?
7. Arachidonic acid is the major precursor of prostaglandins and thromboxanes. If a person were unable to absorb arachidonic acid from the diet but could absorb linoleic acid, could that person still make PGE_2?
8. Low doses of aspirin (one aspirin every other day) are recommended to prevent heart attacks and strokes. Why would three or four tablets per day not work better? (Hint Remember that TXA_2 is made in platelets and that PGI_2 is made in the arterial walls.)
9. Snake venoms contain many types of lipases, including phospholipase A_2. Why would small amounts of this enzyme contribute to some of the toxic effects of snake venom? (Bee venom contains a protein that stimulates phospholipase A_2.)

23

CHAPTER

Metabolism of Cholesterol

Steroids are tetracyclic hydrocarbons that can be considered derivatives of perhydrocyclopentanophenanthrene (see fig. 7.14). They differ from one another in the degree of saturation of each of the four hydrocarbon rings and in the side-chain substituents attached to these rings.

Steroids are much more common in eukaryotes than in prokaryotes. Cholesterol, the most prominent member of the steroid family, is an important component of many eukaryotic membranes (see chapter 7). In addition, it is the precursor of the other two major classes of steroids: the steroid hormones and the bile acids (fig. 23.1).

Steroid hormones play a key role in the regulation of metabolism. These hormones come in a rich variety, each interacting in a highly specific manner with a receptor protein to effect gene expression in the appropriate target tissue (see chapter 24). Bile acids are the primary degradation product of cholesterol. The bile acids are made in the liver, stored in the gall bladder, and secreted into the small intestine. There they aid in the solubilization of lipids, facilitating their digestion by intestinal lipases.

All of these biologic roles of the steroids figure prominently in human well-being. Defects in cholesterol metabolism are a major cause of cardiovascular disease. It is no wonder that steroids are a central concern in medical biochemistry. In this chapter we will discuss the metabolism of these complex lipids and the plasma lipoproteins in which they and other complex lipids are transported to various tissues.

Biosynthesis of Cholesterol

Early in the 1930s, the structure of cholesterol was finally determined, an achievement that concluded a brilliant chapter in structural organic chemistry. However, the solution to that problem led to the formulation of many new ones. In particular, it was not clear how such a complex structure could be assembled from small molecules.

Work on the biosynthesis of cholesterol began in earnest after Rudolf Schoenheimer and David Rittenberg, at Columbia University, developed isotopic tracer techniques for the analysis of biochemical pathways. In 1941, Rittenberg and Konrad Bloch were able to show that deuterium-labeled acetate ($C^2H_3COO^-$) was a precursor of cholesterol in rats and mice. Subsequently, in collaboration with Edward Tatum and others, Bloch proved that the carbon skeleton of the sterol ergosterol of *Neurospora crassa* was entirely derived from acetate. In 1949,

Figure 23.1

Formation of cholesterol and some of its derivatives. All of the carbon atoms of cholesterol are derived from acetyl-CoA by way of mevalonate in a pathway with 33 reaction steps (shown in black, as these steps are derived from the master diagram of figure 12.5). From cholesterol a wide variety of steroids and bile acids and bile salts are formed (shown in color, as these reactions are specifically discussed in this chapter). Since many of the reactions leading to cholesterol derivatives are organ-specific, the reactions occurring in specific organs are encircled and the organ indicated by a label. Most of the biosynthetic processes occur on the surface of the endoplasmic reticulum within the cells of the specified organ. A few reactions occur in the mitochondria and are so labeled. Numbers associated with the arrows indicate the approximate number of steps in the sequence, where that is relevant. Most of the reactions are unidirectional. Excess cholesterol is disposed of by conversion to bile acids and excreted as bile salts (as shown). Excess steroid hormones are inactivated by covalent modification, linked to glucuronic acid, and then excreted in the urine (not shown).

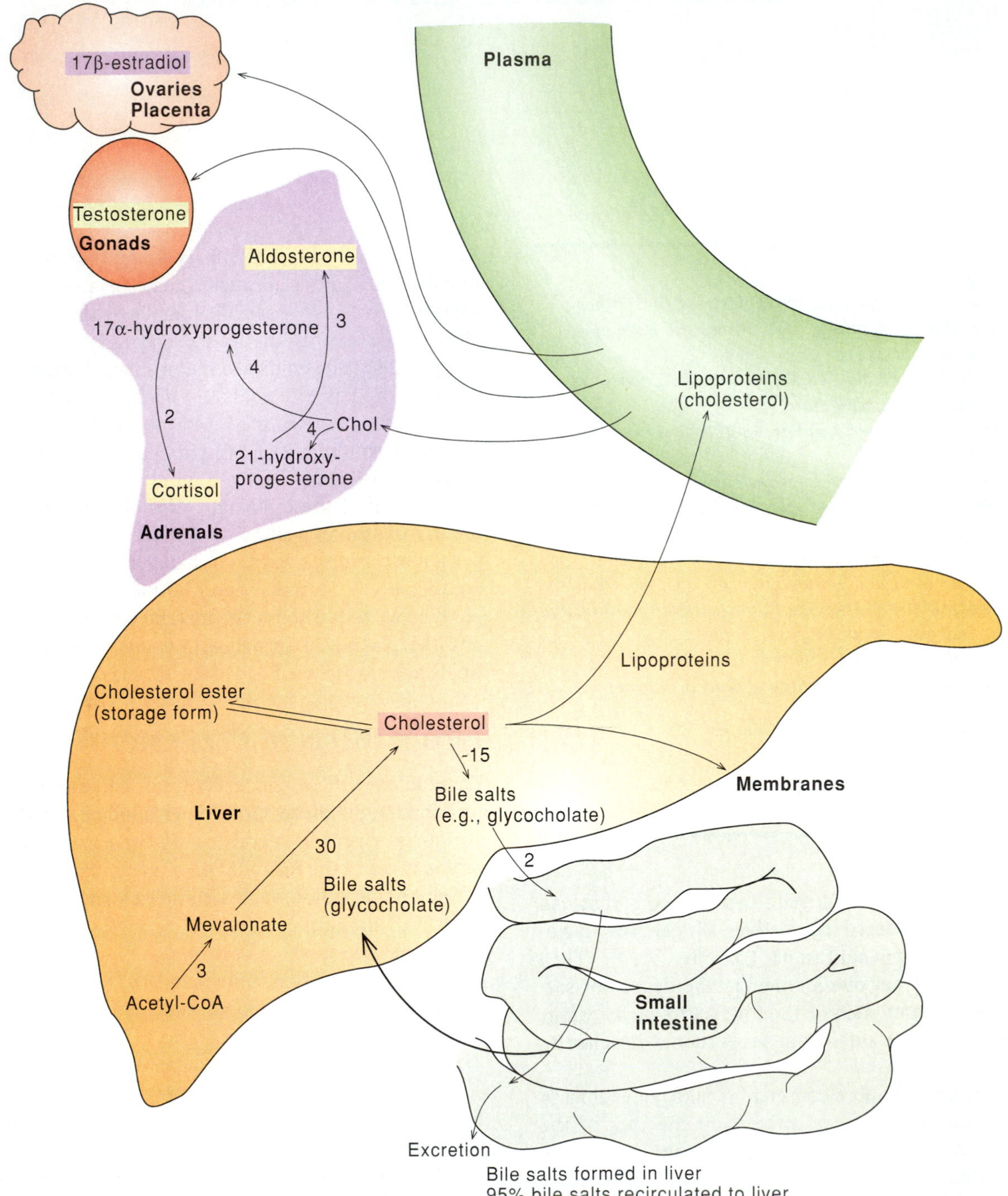

Figure 23.2

Basic scheme for cholesterol biosynthesis proposed by Bloch in 1952 (left) and scheme proposed for the cyclization of squalene by Woodward and Bloch in 1953.

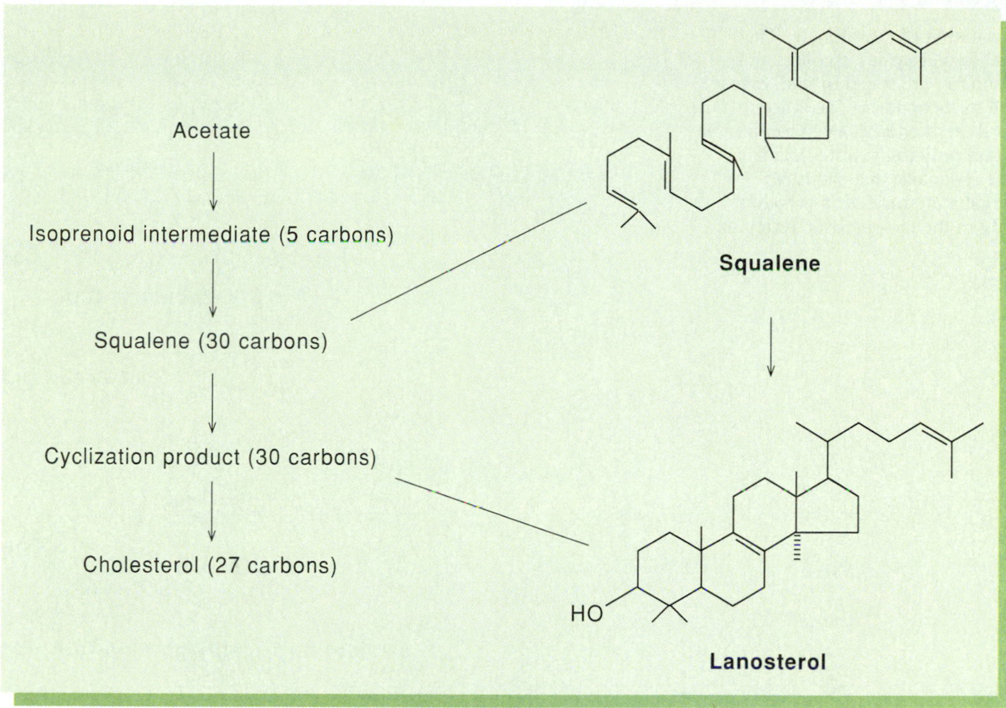

James Bonner and Barbarin Arreguin postulated that three acetates could combine to form a single five-carbon unit called isoprene:

$$CH_2=C(CH_3)-C(H)=CH_2$$

This proposal agreed with an earlier prediction of Sir Robert Robinson, that cholesterol was a cyclization product of squalene, a 30-carbon polymer of isoprene units. Thus Bloch postulated a scheme for the biosynthesis of cholesterol as shown in figure 23.2. And in 1952, Bloch and Robert Langdon demonstrated that squalene could readily be converted into cholesterol.

Two difficult problems remained. What was the structure of the isoprenoid intermediate, and how did squalene cyclize to form cholesterol? In 1953, R. B. Woodward and Bloch postulated a cyclization scheme for squalene (fig. 23.2) that was later shown to be correct. In 1956, the unknown isoprenoid precursor was identified as mevalonic acid by Karl Folkers and others at Merck, Sharpe and Dohme Laboratories. The discovery of mevalonate provided the missing link in the basic outline of cholesterol biosynthesis. Since that time, the sequence and the stereochemical course for the biosynthesis of cholesterol have been defined in detail.

Mevalonate Is a Key Intermediate in Cholesterol Biosynthesis

The sequence of cholesterol biosynthesis begins with a condensation in the cytosol of two molecules of acetyl-CoA, a reaction catalyzed by *thiolase* (fig. 23.3). The next step requires the enzyme β-hydroxy-β-methylglutaryl-CoA (HMG-CoA) synthase. This enzyme catalyzes the condensation of a third acetyl-CoA with β-ketobutyryl-CoA to yield HMG-CoA. HMG-CoA is then reduced to mevalonate by HMG-CoA reductase. The activity of this reductase is primarily responsible for control of the rate of cholesterol biosynthesis.

HMG-CoA is an important intermediate for the biosynthesis of both cholesterol and ketone bodies (see chapter 17). The biosynthesis of cholesterol is catalyzed by enzymes in the cytosol and enzymes bound to the endoplasmic reticulum. The synthesis of ketone bodies, however, is restricted to the mitochondrial matrix. Thus thiolase and HMG-CoA synthase are found in both mitochondria and cytosol of rat liver. In contrast, HMG-CoA lyase, which cleaves HMG-CoA to ketone bodies (see chapter 17), is located only in mitochondria. HMG-CoA reductase is bound to the endoplasmic reticulum.

The Rate of Mevalonate Synthesis Determines the Rate of Cholesterol Biosynthesis

The thiolase and HMG-CoA synthase exhibit some regulatory properties in rat liver (cholesterol feeding causes a decrease in these enzyme activities in the cytosol, but not in the mitochondria). However, primary regulation of cholesterol biosynthesis appears to be centered on the HMG-CoA reductase reaction. HMG-CoA reductase is found on the endoplasmic reticulum, has a molecular weight of 97,092, and consists of 887 amino acids in a single polypeptide chain. The structure of the enzyme was deduced by Michael Brown and Joseph Goldstein from the sequence of a piece of DNA (cDNA) derived from mRNA that codes for the reductase. The enzyme has two domains (fig. 23.4).

Figure 23.3

Formation of mevalonate. The first two enzymes, thiolase and synthase, are found in both cytosol and mitochondria. The lyase that catalyzes ketone body formation is found only in the mitochondria. The reductase that catalyzes mevalonate formation is found only in the endoplasmic reticulum.

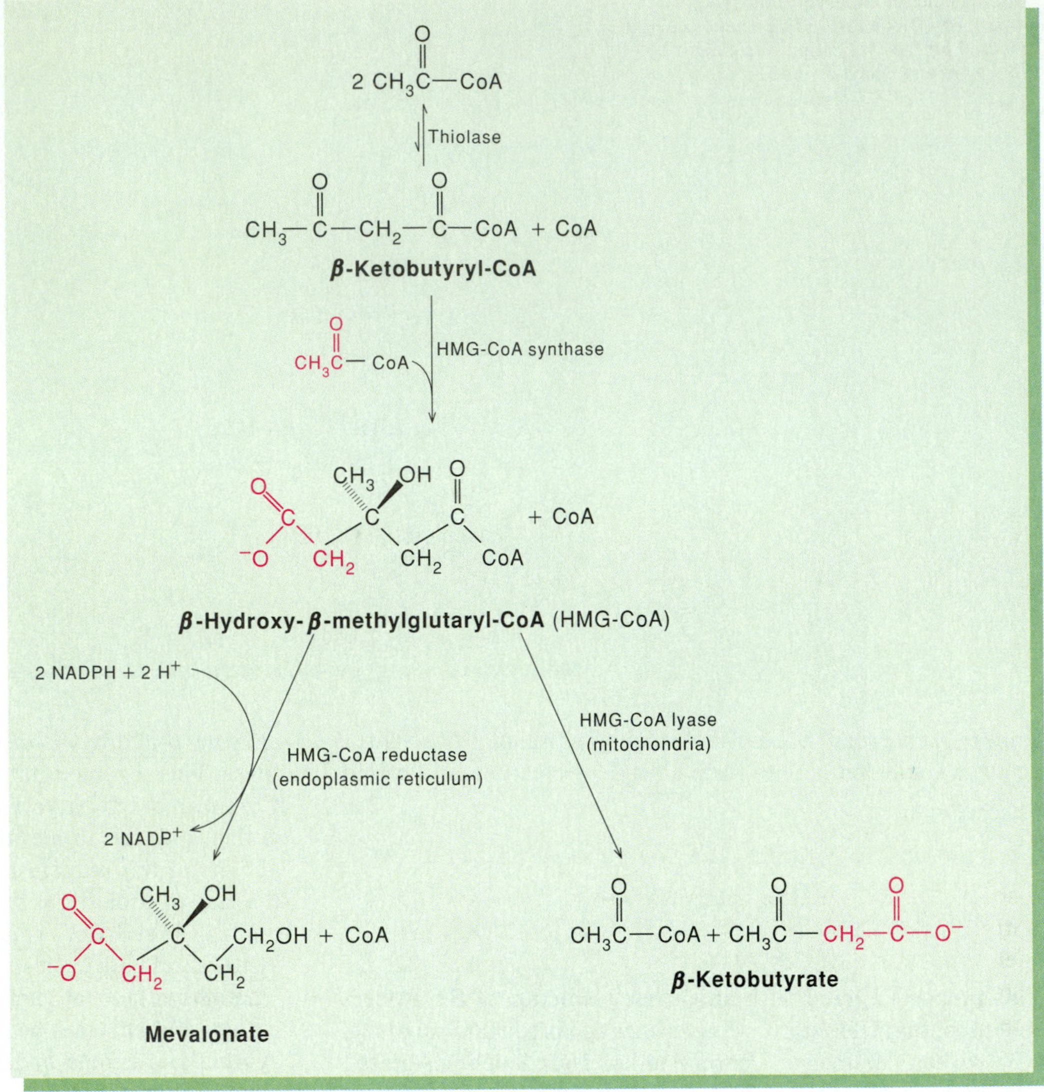

Figure 23.4

Proposed structure of HMG-CoA reductase, derived from studies of recombinant DNA that codes for the enzyme. The enzyme is attached to the endoplasmic reticulum membrane and consists of two domains: the hydrophobic domain, embedded in the membrane, and the catalytic domain, which protrudes into the cytosol. (Source: L. Liscum et al., "Domain structure of 3-hydroxy-3-methylglutaryl coenzyme A reductase, a glycoprotein of the endoplasmic reticulum," in *Journal of Biological Chemistry*, 260:522-530, 1985. Copyright © 1985 American Society for Biochemistry and Molecular Biology Inc., Bethesda, Md.)

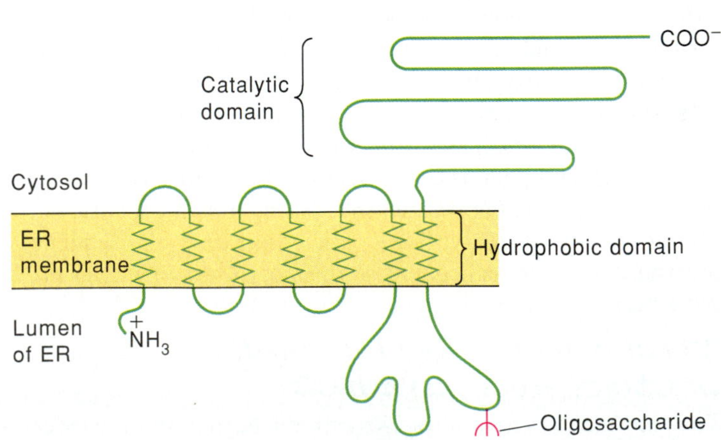

Figure 23.5

Scheme for modulation of the activity of HMG-CoA reductase by phosphorylation/dephosphorylation. The kinases and phosphatases shown in this figure are believed to mediate short-term changes in the activity of HMG-CoA reductase. The kinase is the same enzyme that catalyzes the phosphorylation of acetyl-CoA carboxylase (see fig. 17.26).

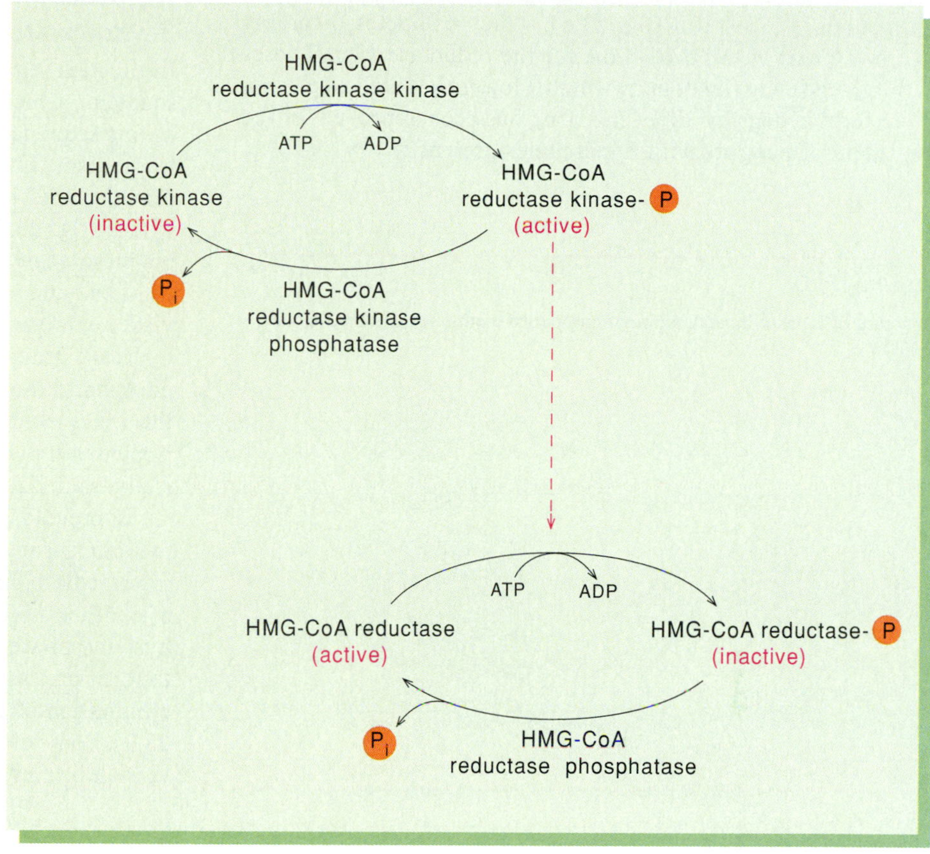

The amino-terminal domain has a molecular weight of 35,000, with seven hydrophobic segments that are thought to cross the membrane as shown in figure 23.4. The carboxyl-terminal domain has a molecular weight of 62,000, contains the catalytic site of the enzyme, and is thought to protrude into the cytosol.

The activity of the reductase is regulated by three distinct mechanisms. The first control point is at the level of gene expression. The amount of mRNA produced is modulated by the supply of cholesterol. When cholesterol is in excess, the amount of mRNA for HMG-CoA reductase is reduced, hence less enzyme is made. Depletion of cholesterol enhances the synthesis of reductase mRNA.

The second regulatory mechanism involves the rate of degradation of HMG-CoA reductase. As stated in chapter 17, the amount of an enzyme in a cell is determined by both its rate of synthesis and its rate of degradation. It has been known for some time that the half-life for HMG-CoA reductase is between 2 and 4 h, about 10-fold lower than that of many other proteins on the endoplasmic reticulum. In other words, HMG-CoA reductase is rapidly degraded within the cell. The rate of degradation of the reductase appears to be modulated by the supply of cholesterol. Thus when cholesterol is abundant, the rate of enzyme degradation is twice as fast as when there is a limited supply of cholesterol. The effect of cholesterol on enzyme degradation is mediated by the membrane domain of the enzyme. In support of this point, a mutant enzyme lacking the membrane domain was found to be active in the synthesis of mevalonic acid even though free in the cytosol, and the mutant enzyme had an extended half-life of five times the normal. Moreover, the supply of cholesterol did not affect the half-life. More work needs to be done to determine the exact mechanism of regulating the half-life of this enzyme.

The third regulatory mechanism is phosphorylation/dephosphorylation of the reductase, which causes inactivation and activation as outlined in figure 23.5. HMG-CoA reductase kinase kinase catalyzes the phosphorylation and activation of HMG-CoA reductase kinase. This kinase phosphorylates HMG-CoA reductase, which results in an inactivation of the reductase. Both kinases are cAMP-independent. Their effects are reversed by the action of protein phosphatases. These kinases and phosphatases mediate short-term changes in the activity of HMG-CoA reductase.

Cholesterol biosynthesis is also controlled by plasma low-density lipoproteins, which we will discuss in the context of lipoprotein metabolism later in this chapter.

It has long been recognized that a correlation exists between a high level of serum cholesterol and cardiovascular disease (e.g., most heart diseases, stroke). Most serum cholesterol originates from the liver; thus a drug that would specifically reduce cholesterol biosynthesis has been sought. It would be logical for such a drug to inactivate HMG-CoA reductase, since this enzyme catalyzes the key regulatory step in the pathway.

Several fungal metabolites have been isolated that are competitive inhibitors of HMG-CoA reductase. One of the most active compounds is lovastatin (fig. 23.6), which competes favorably ($K_I = 0.6$ nM) with HMG-CoA for the reductase. Small doses of this drug (8 mg/kg of body weight) lower the levels of plasma cholesterol in dogs by 30%. This drug has been approved for the treatment of patients with hypercholesterolemia.

Figure 23.6

Structure of lovastatin acid, a potent competitive inhibitor of HMG-CoA reductase.

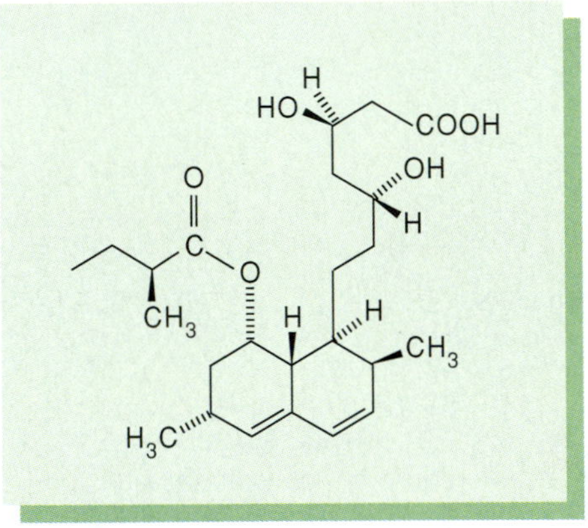

It Takes Six Mevalonates and Ten Steps to Make Lanosterol, the First Tetracyclic Intermediate

In the next segment of the pathway, mevalonate is converted to squalene, which is cyclized to form lanosterol. The first stage in this sequence of reactions is the synthesis of the five-carbon isoprenoid intermediates isopentenyl pyrophosphate and dimethylallyl pyrophosphate. The synthesis involves four enzymes (fig. 23.7). The action of mevalonate kinase and phosphomevalonate kinase produces 5-pyrophosphomevalonate. The third enzyme (see fig. 23.7), pyrophosphomevalonate decarboxylase, catalyzes a decarboxylation and elimination of the 3-hydroxyl group. The decarboxylase probably acts by an initial phosphorylation of the 3-hydroxyl group with ATP, followed by the *trans* elimination of the carboxyl and phosphate to give isopentenyl pyrophosphate. This intermediate can be enzymatically isomerized to 3,3-dimethylallyl pyrophosphate. These two C_5 isoprenoid pyrophosphates react to produce the C_{10} intermediate geranyl pyrophosphate (fig. 23.8). Subsequently, a C_{15} intermediate, farnesyl pyrophosphate, is formed by the reaction of another molecule of isopentenyl pyrophosphate with geranyl pyrophosphate. Two molecules of farnesyl pyrophosphate react to form presqualene pyrophosphate, which rearranges with the elimination of PP_i to yield the C_{30} intermediate squalene (fig. 23.9). Some of the complex stereochemistry involved in the conversion of mevalonate to squalene is discussed in box 23A.

Figure 23.7

The conversion of mevalonate to isopentenyl pyrophosphate and dimethylallyl pyrophosphate. Mevalonate is converted to isopentenyl pyrophosphate in three steps. Each of these steps requires one ATP cleavage. The conversion of isopentenyl pyrophosphate to dimethylallyl pyrophosphate is readily reversible and does not require any further expenditure of ATP.

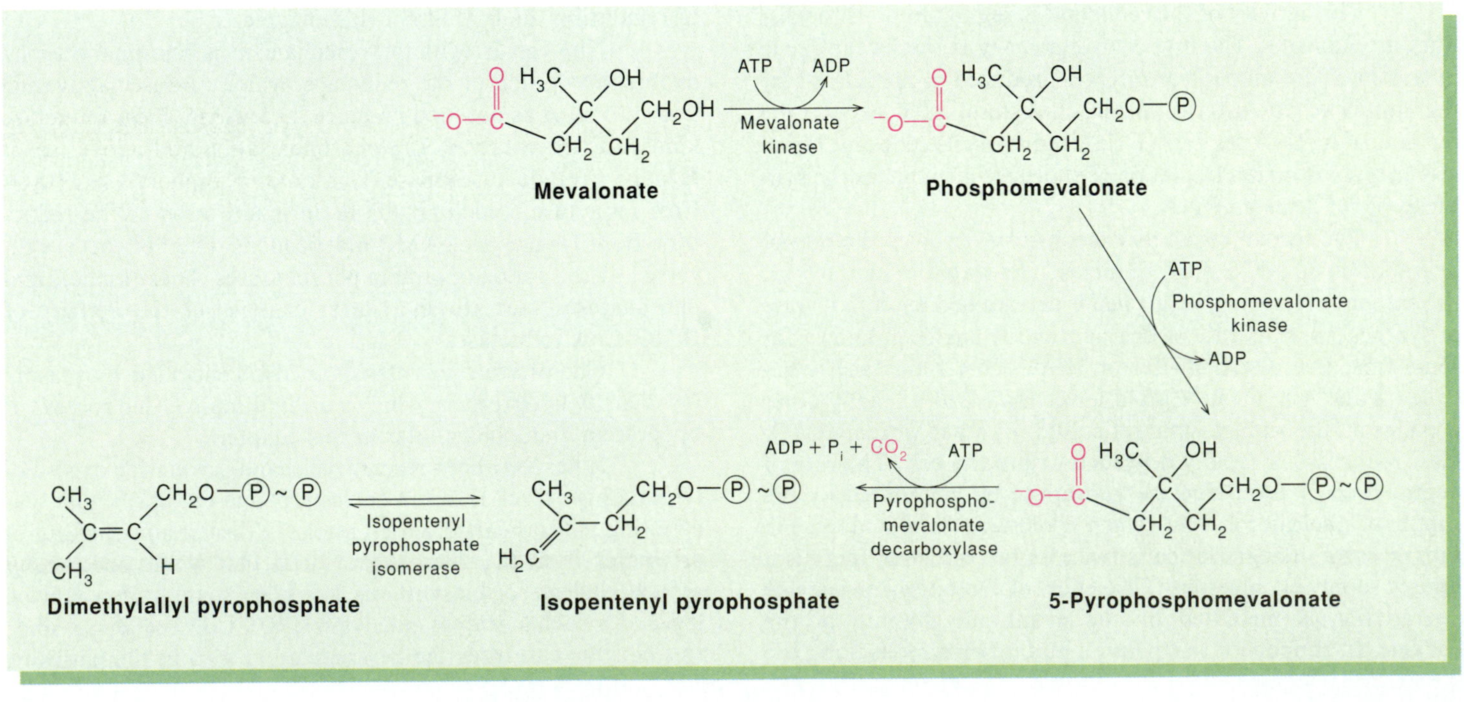

The enzymes that convert mevalonate to farnesyl pyrophosphate are probably cytosolic, whereas farnesyl transferase (squalene synthase) is tightly associated with the endoplasmic reticulum (see fig. 23.9). Even though such membrane-bound enzymes are difficult to solubilize, this enzyme has been isolated from yeast. In a polymeric form and in the presence of NADPH, farnesyl transferase will convert two farnesyl pyrophosphates to squalene. When dissociated into its depolymerized form, the farnesyl transferase catalyzes only the formation of presqualene pyrophosphate.

Figure 23.8

Biosynthesis of farnesyl pyrophosphate. Farnesyl pyrophosphate is a C_{15} intermediate containing three C_5 isoprenoid subunits. The two transferase reactions involved in the formation of farnesyl pyrophosphate occur by virtually identical mechanisms as shown.

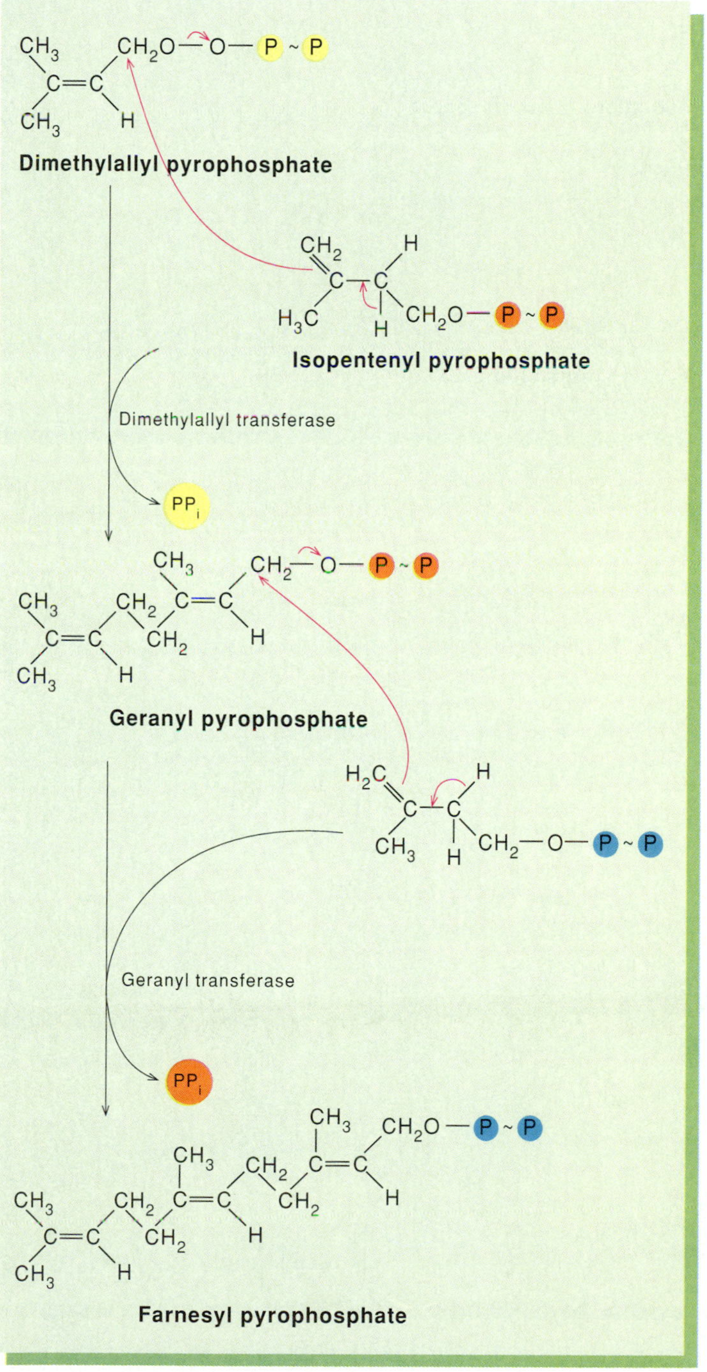

Dimethylallyl pyrophosphate

Isopentenyl pyrophosphate

Dimethylallyl transferase

PP_i

Geranyl pyrophosphate

Geranyl transferase

PP_i

Farnesyl pyrophosphate

The two remaining reactions in the biosynthesis of lanosterol are shown in figure 23.10. In the first of these reactions, squalene-2,3-oxide is formed from squalene by squalene monooxygenase, an endoplasmic-reticulum-bound enzyme, in a reaction that requires O_2, NADPH, FAD, phospholipid, and a cytosolic protein. As can be seen in figure 23.9, squalene is a symmetrical molecule, hence the formation of squalene oxide can be initiated from either end of the molecule. The oxide is converted into lanosterol by another endoplasmic-reticulum-bound enzyme, 2,3-oxidosqualene lanosterol cyclase. The reaction can be formulated as proceeding by means of a protonated intermediate that undergoes a concerted series of *trans*-1,2 shifts of methyl groups and hydride ions to produce lanosterol (see fig. 23.10).

From Lanosterol to Cholesterol Takes Another Twenty Steps

The last sequence of reactions in the biosynthesis of cholesterol involves approximately twenty enzymatic steps, starting with lanosterol. In mammals the major route comprises a series of double-bond reductions and demethylations (fig. 23.11). The exact position in the scheme for the reduction of the Δ^{24} double bond is not established. Otherwise, the sequence of reactions involves the oxidation and removal of the 14α-methyl group followed by the oxidation and removal of the two methyl groups at position 4 in the sterol. The final reaction is a reduction of the Δ^7 double bond in 7-dehydrocholesterol.

An alternative pathway from lanosterol to cholesterol (see fig. 23.11) initially involves three demethylations to give zymosterol and then isomerization of the Δ^8 double bond to the Δ^5 position to produce desmosterol (see fig. 23.11). The final reaction in this pathway is the reduction of the Δ^{24} double bond.

The enzymes involved in the transformation of lanosterol to cholesterol are all located on the endoplasmic reticulum. In addition to these enzymes, two cytosolic proteins have been found that stimulate several of the membrane-associated reactions that convert squalene to cholesterol. How these soluble proteins actually function in cholesterol biosynthesis is a problem of current interest.

Lipoprotein Metabolism

Unesterified fatty acids are carried in plasma by albumin (chapter 17). The plasma also transports more complex lipids among the various tissues as components of lipoproteins (particles composed of lipids and proteins). In this section we will be concerned with the structure and metabolism of these lipoproteins.

Figure 23.9

Formation of squalene from farnesyl pyrophosphate. Farnesyl transferase is tightly complexed to the endoplasmic reticulum. The enzyme is active in both protomeric and polymeric forms. In the protomeric form the enzyme catalyzes only to the intermediate presqualene pyrophosphate. Only in the polymeric form does the reaction proceed beyond the first step to squalene.

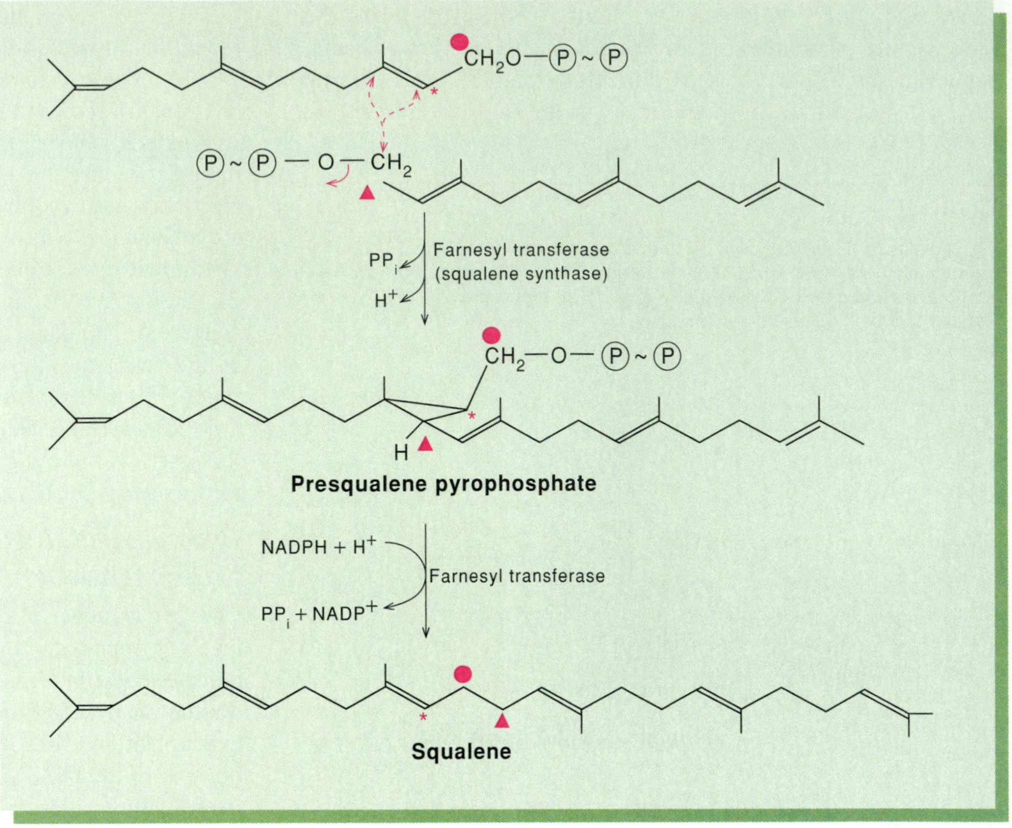

Figure 23.10

The transformation of squalene into lanosterol. The squalene monooxygenase reaction requires O_2, NADPH, FAD, phospholipid, and a cytosolic protein. The cyclase reaction has no cofactor requirements. The reaction proceeds by means of a protonated intermediate that undergoes a concerted series of *trans*-1,2 shifts of methyl groups and hydride ions to produce lanosterol.

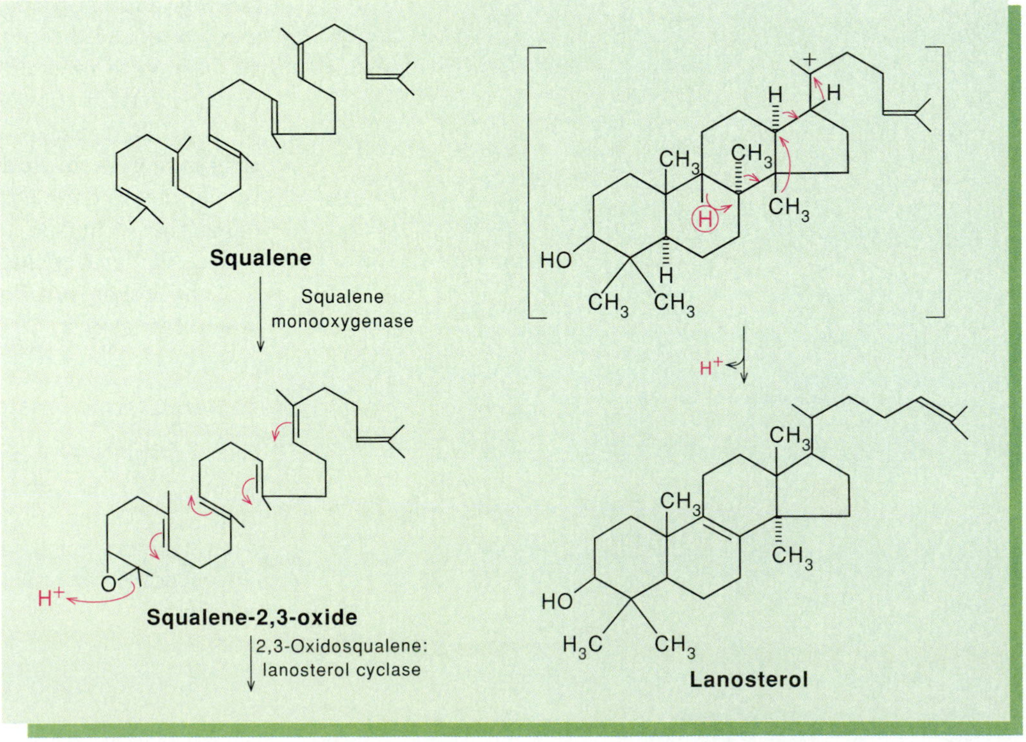

23A BOX

Stereochemistry of the Conversion of Mevalonate to Squalene

As we have seen, the biosynthesis of cholesterol from acetyl-CoA is accomplished by means of a complicated route that involves more than 30 different enzymes, numerous cofactors, and at least two cytosolic proteins. Although this certainly represents enough complexity for most people, George Popjak and John Cornforth, in Britain, recognized that it was possible to define the conversion of mevalonate to squalene in a more precise and elegant manner. These two scientists observed in the 1960s that there were 14 "stereochemical ambiguities" in the conversion of pyrophosphomevalonate to squalene. In other words, there were 2^{14}, or 16,384, theoretically possible stereochemical pathways by which mevalonate could be transformed into squalene. This in itself was a remarkable observation. More remarkably, these two men and their collaborators subsequently were able to define precisely which one of these 16,384 possible stereochemical pathways actually occurred.

It is beyond the scope of this introductory text for us to examine each of the 14 stereochemical ambiguities and their resolution. Two examples should demonstrate the principles involved. The reaction catalyzed by pyrophosphomevalonate decarboxylase could involve either a *cis* or a *trans* elimination of the carboxyl and hydroxyl groups to produce isopentenyl pyrophosphate (fig. 23.7). Cornforth et al. solved this stereochemical ambiguity by the synthesis of a stereospecifically deuterium-labeled pyrophosphomevalonate (figure 1) that was incubated with the decarboxylase. The product of the reaction was isolated, and after several chemical transformations, Cornforth and coworkers were able to distinguish which of the two possible isomers of isopentenyl pyrophosphate was formed. The product was solely the result of a *trans* elimination (illustration 1). Thus the first stereochemical ambiguity was resolved.

Another stereochemical problem was to determine which of the two hydrogens from C-2 of isopentenyl pyrophosphate was lost in its isomerization to dimethylallyl pyrophosphate (figure 2). Isopentenyl pyrophosphate was chemically labeled with deuterium on the 2-R or 2-S position (the use of the RS terminology is explained in box 11A). Incubation of this substrate with the isomerase and subsequent characterization of the product demonstrated that the pro R hydrogen (H_R) was specifically removed during the isomerization reaction (illustration 2). Further information on the 14 stereochemical ambiguities and their resolution can be found in the book by Ronald Bentley (see Selected Readings).

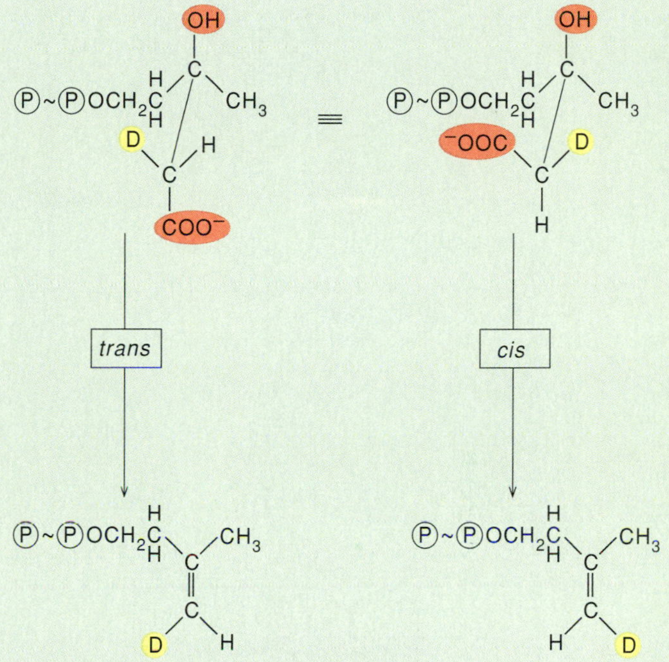

Figure 1

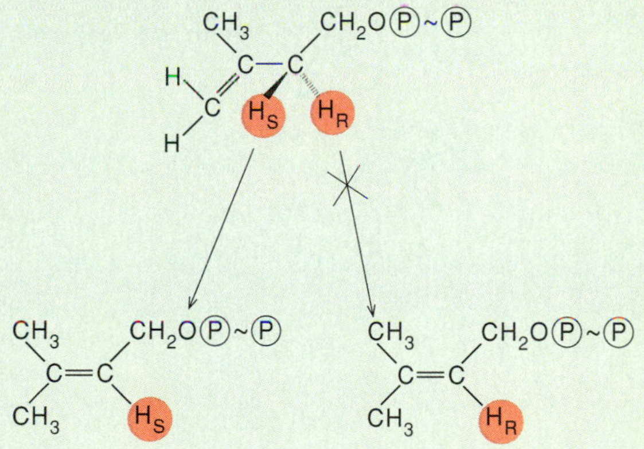

Figure 2

Figure 23.11

Two pathways for the conversion of lanosterol to cholesterol. The major route in mammals proceeds through 7-dehydrocholesterol.

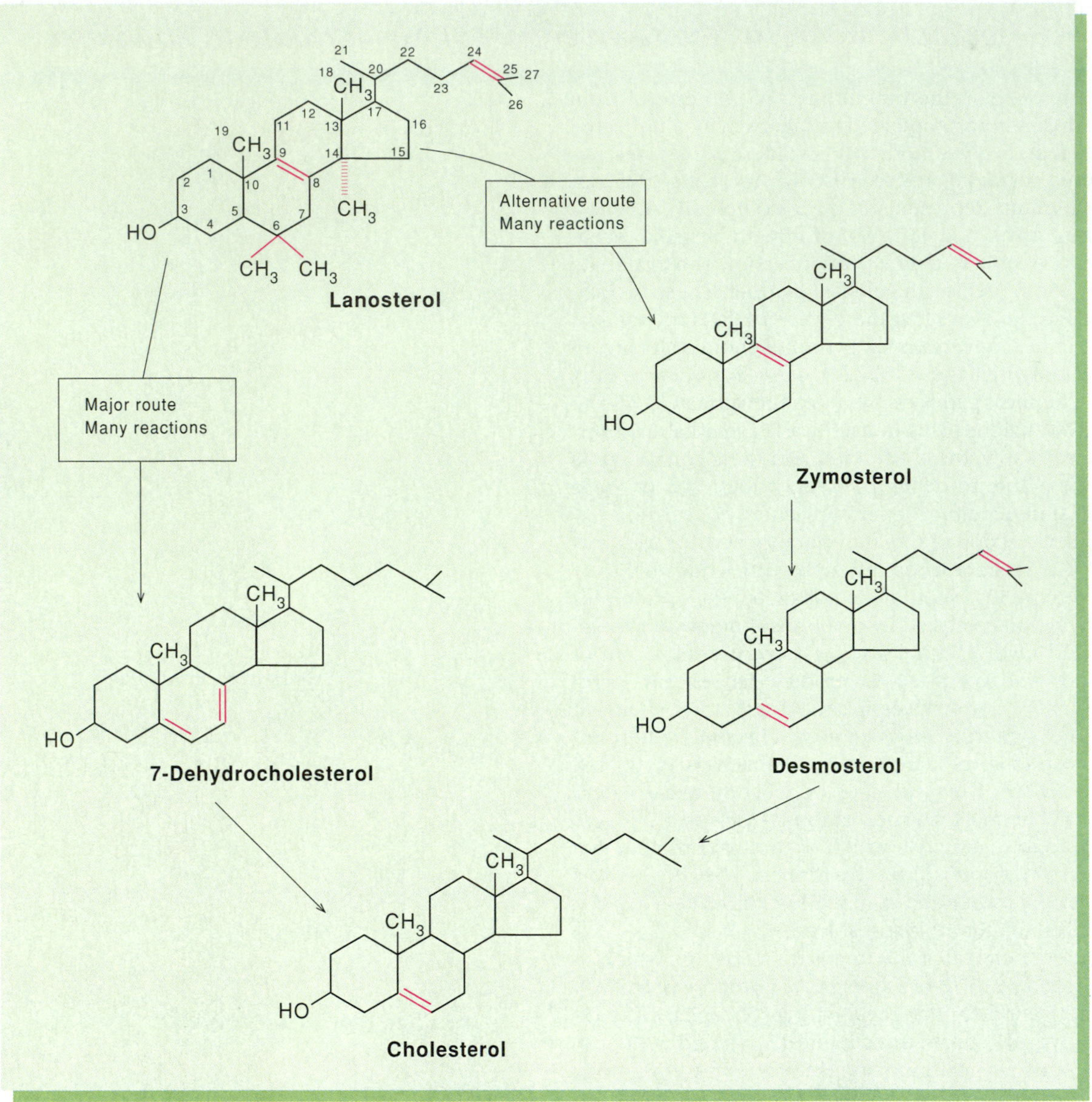

Table 23.1 Normal Concentrations of the Major Lipid Classes in Plasma in Humans	
Lipid	**Concentration (g/l)**
Total lipid	3.6–6.8
Cholesterol and cholesterol ester	1.3–2.6
Triacylglycerol	0.8–2.4
Phospholipid	1.5–2.5

There Are Five Types of Lipoproteins in Human Plasma

The amounts and types of lipids found in human plasma fluctuate according to the dietary habits and metabolic states of the individual. The normal ranges for the lipid levels in plasma are shown in table 23.1. In plasma these lipids are associated with proteins in the form of lipoproteins, which are classified into five major types on the basis of their density (table 23.2 and fig. 23.12). The lipoproteins of lowest density, the chylomicrons, are the largest in size and contain the most lipid and the smallest percentage of protein. At the other extreme are the high-density

Table 23.2
Composition and Density of Human Lipoproteins

	Chylomicron	VLDL	IDL	LDL	HDL
Density (g/ml)	<0.95	0.95–1.006	1.006–1.019	1.019–1.063	1.063–1.210
Diameter (nm)	75–1,200	30–80	25–35	18–25	5–12
Components (% dry weight)					
Protein	1–2	10	18	25	33
Triacylglycerol	83	50	31	10	8
Cholesterol and cholesterol esters	8	22	29	46	30
Phospholipids	7	18	22	22	29
Apoprotein composition	A-I, A-II B-48 C-I, C-II, C-III	B-100 C-I, C-II, C-III E	B-100 C-I, C-II, C-III E	B-100	A-I, A-II C-I, C-II, C-III D E
Classification by electrophoresis	Omega	Pre-beta	Between beta and pre-beta	Beta	Alpha

Figure 23.12

Electron micrographs of low-density lipoproteins (LDL), very-low-density lipoproteins (VLDL), chylomicrons (Chylo), and high-density lipoproteins (HDL). Lipids are transported in plasma as components of these particles. The larger particles contain a higher percentage of lipid and therefore are less dense, whereas the smaller particles have less lipid, a higher percentage of protein, and are more dense. (Photograph courtesy of Dr. Robert Hamilton, University of California, San Francisco.)

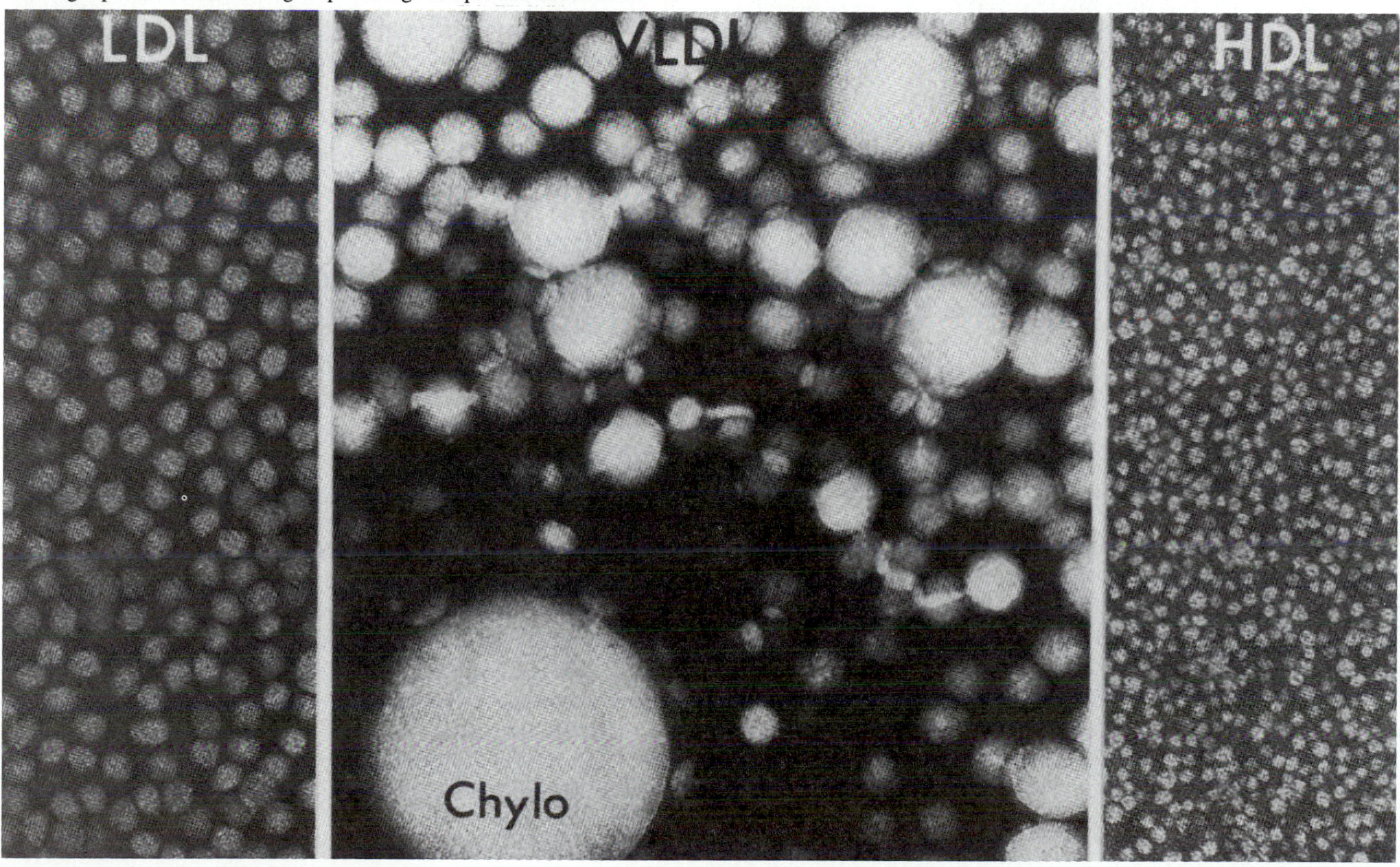

Figure 23.13

Generalized structure of human lipoproteins. The lipoproteins are spherical and vary in diameter from 10 nm to as much as 1,000 nm, depending on the particular proteins and lipids. Each of the lipoprotein classes contains a neutral lipid core composed of triacylglycerol and/or cholesterol ester. Around the core is a layer of protein, phospholipid, and cholesterol that is oriented with the polar portions exposed to the surface of the lipoprotein.

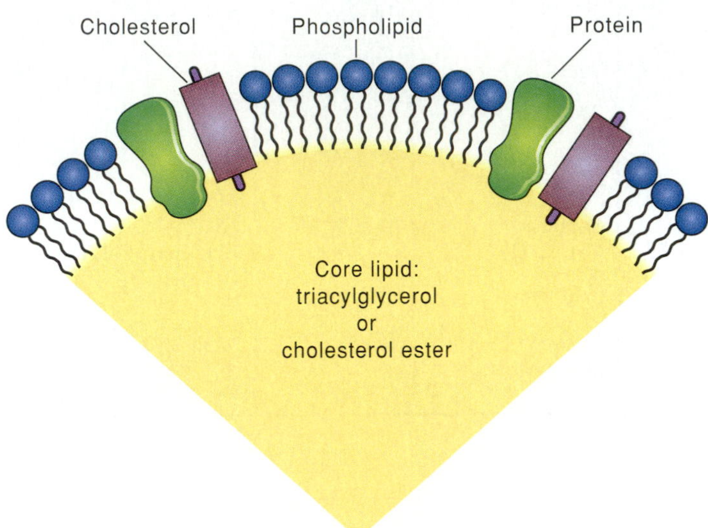

lipoproteins (HDL), which are the smallest particles and contain the highest percentage by weight of protein and lowest percentage of lipid. Between these two classes, in both size and composition, are the low-density lipoproteins (LDL), the intermediate-density lipoproteins (IDL), and the very-low-density lipoproteins (VLDL).

The structures of the various lipoproteins appear to be similar (fig. 23.13). Each of the lipoprotein classes contains a neutral lipid core composed of triacylglycerol and/or cholesterol ester. Around this core is a layer of protein, phospholipid, and cholesterol oriented with the polar portions exposed to the surface of the lipoprotein.

There are at least nine apoproteins* (see table 23.2) associated with the lipoproteins, as well as several enzymes and a cholesterol ester transfer protein. The structure and function of these apoproteins has been intensely studied in the past decade, and some of the properties of these apoproteins are summarized in table 23.3. Most of the apoproteins have been sequenced and contain regions that are rich in hydrophobic amino acids, which facilitate binding of phospholipid.

Lipoproteins Are Made in the Endoplasmic Reticulum of the Liver and Intestine

Of the various lipid components of the lipoproteins, only the biosynthesis of cholesterol esters has not been mentioned. Cholesterol ester is the storage form of cholesterol in cells. It is synthesized from cholesterol and acyl-CoA by acyl-CoA:cholesterol acyltransferase (ACAT) (fig. 23.14), which is located on the cytosolic surface of liver endoplasmic reticulum. The synthesis of the apoproteins takes place on ribosomes that are bound to the endoplasmic reticulum. As we mentioned previously, the biosynthesis of cholesterol, triacylglycerols, and phospholipids also occurs on the endoplasmic reticulum.

How the various components of the lipoproteins are assembled and secreted into the plasma is not known. Current ideas for this process suggest the transfer of the components from the endoplasmic reticulum to the Golgi apparatus, where secretory (membrane-encapsulated) vesicles are formed. These vesicles would subsequently fuse with the plasma membrane and release their lipoprotein contents into plasma (fig. 23.15).

The plasma lipoproteins appear to be made mainly in the liver and intestine. In the rat, approximately 80% of the plasma apoproteins originate from the liver; the rest come from the intestine. Most of the components of chylomicrons, including apoprotein A, apoprotein B-48, phospholipid, cholesterol, cholesterol ester, and triacylglycerols, are products of the intestinal cells. The chylomicrons are secreted into lymphatic capillaries, which eventually enter the bloodstream at the large subclavian vein and therefore bypass the liver. The liver appears to be the major source of VLDL and HDL, which include apo A-I, A-II, B-100, C-I, C-II, C-III, and E and the lipid components of these lipoproteins. Low-density lipoprotein is produced from VLDL, as we will see in a moment.

Chylomicrons and Very-Low-Density Lipoproteins (VLDL) Transport Cholesterol and Triacylglycerol to Other Tissues

Chylomicrons serve as the mode of transport of triacylglycerol and cholesterol ester from the intestine to other tissues in the body. Very-low-density lipoprotein functions in a similar manner for the transport of lipid from the liver to other tissues. These two types of triacylglycerol-rich particles are initially degraded by the action of lipoprotein lipase, an extracellular enzyme that is most active within the capillaries of adipose tissue, cardiac and skeletal muscle, and lactating mammary gland. Lipoprotein lipase catalyzes the hydrolysis of triacylglycerols (see fig. 17.2). The enzyme is specifically activated by apoprotein C-II, which is associated with chylomicrons and VLDL (see table 23.2). As a result, this lipase supplies the heart and adipose tissue with fatty acids, derived from these lipoproteins in the plasma. In both the heart and adipose tissue, the fatty acids produced by lipoprotein lipase can be used for energy or stored as a component of triacylglycerols. Alternatively, the fatty acids can be bound by albumin and transported to other tissues.

As the lipoproteins are depleted of triacylglycerol, the particles become smaller. Some of the surface molecules (apoproteins, phospholipids) are transferred to HDL. In the rat, "remnants" that result from chylomicrons and VLDL catabolism are taken up by the liver. In humans, the uptake of remnant VLDL also occurs, but much of the triacylglycerol is further degraded by lipoprotein lipase to give the intermediate-density lipoprotein IDL. This particle is converted into LDL via the action of lipoprotein lipase and enriched in cholesterol ester via

*The apoprotein is the protein part of the lipoprotein.

Table 23.3
Properties of the Apoproteins of the Major Human Lipoprotein Classes

Apoprotein	M_r	Plasma Concentration (mg/100 ml)	Miscellaneous
A-I	29,016	90–120	Major protein in HDL (64%); contains 245 amino acids and no carbohydrate; activates LCAT*
A-II	17,400	30–50	Mainly in HDL (20% of dry mass); two identical chains with 77 amino acids each, joined by a disulfide at residue 6
B-100	513,000	80–100	Major protein in LDL; very difficult to solubilize in detergents, is made in the liver and binds to the LDL receptor (fig. 23.18)
B-48	241,000	<5	A protein found exclusively in chylomicrons
C-I	7,000	4–7	Contains 57 amino acids
C-II	9,000	3–8	Contains 80–85 amino acids; activates lipoprotein lipase
C-III	9,300	8–15	Contains 79 amino acids and inhibits lipoprotein lipase
D	19,000	8–10	Associated with HDL; function unknown
E	33,000	3–6	Found on VLDL and chylomicrons and binds to the LDL receptor (fig. 23.18)

*LCAT = Lecithin: cholesterol acyltransferase.

Figure 23.14

Biosynthesis of cholesterol esters. The acyl-CoA: cholesterol acyltransferase involved in cholesterol ester synthesis is located on the cytosolic surface of liver endoplasmic reticulum.

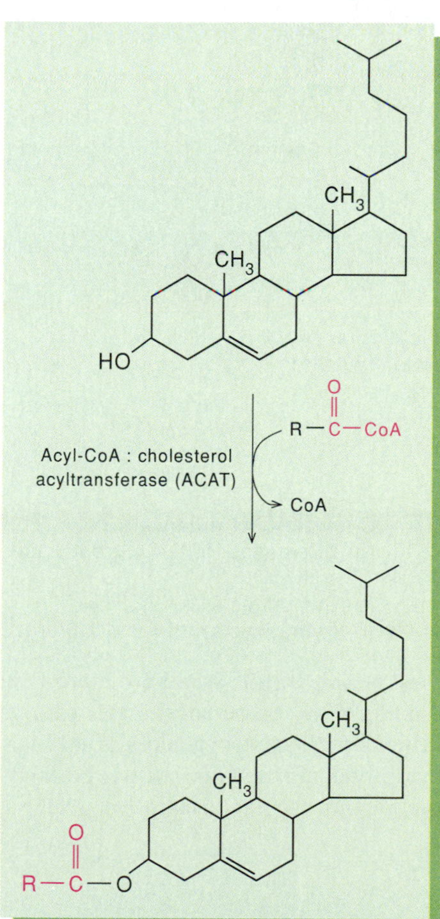

Figure 23.15

Postulated scheme for the synthesis, assembly, and secretion of VLDL by a hepatocyte (liver cell). (1) Synthesis: the apoproteins, phospholipid, triacylglycerol, cholesterol, and cholesterol esters are synthesized in the endoplasmic reticulum. (2) Assembly: these components are assembled into a prelipoprotein particle in the lumen of the endoplasmic reticulum. (3) Processing: the particle moves to the Golgi, where additional phospholipids and perhaps also cholesterol and cholesterol ester are added. (4) Vesicle formation: a secretory vesicle containing the lipoprotein particles is formed and fuses with the plasma membrane. (5) Secretion: the VLDL is released into the circulation.

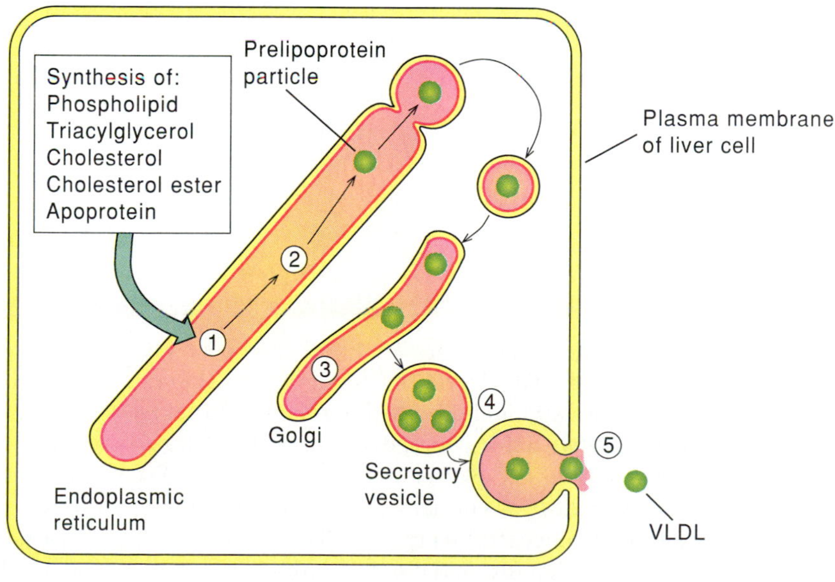

transfer from HDL by the cholesterol ester transfer protein. The half-life for clearance of chylomicrons and remnants from plasma of humans is 4 to 5 min. The clearance value for VLDL is 1 to 3 h.

Low-Density Lipoproteins (LDL) Are Removed from the Plasma by the Liver, Adrenals, and Adipose Tissue

Each day approximately 45% of the plasma pool of low-density lipoprotein is removed from human plasma by both the liver and extrahepatic tissues (particularly the adrenals and adipose tissue). The mechanism for the uptake of LDL in extrahepatic tissue has been extensively described by Michael Brown and Joseph Goldstein. They studied the uptake of LDL by human skin fibroblasts grown in cultures in Petri dishes. LDL particles bind to the cell surface by specific receptors that congregate in areas of the plasma membrane called coated regions (figs. 23.16 and 23.17). These areas of plasma membranes engulf the LDL particles (in a process called endocytosis) to form "coated vesicles," which are somehow directed toward and fuse with lysosomes. The LDL particles are degraded within the lysosomes by the action of proteases and lysosomal acid lipases (lipid degradative enzymes). The cholesterol, or a derivative of cholesterol, diffuses from the lysosomes, suppresses the activity of HMG-CoA reductase, and stimulates the activity of acyl-CoA:cholesterol acyltransferase (ACAT). ACAT catalyzes the synthesis of cholesterol esters (see fig. 23.14), which are then stored within the cell. The cholesterol (or its derivative) also suppresses the synthesis of the LDL receptors and thereby limits the uptake of LDL.

The LDL receptor is a glycoprotein with 839 amino acids and consists of five domains (fig. 23.18). Domain 1 is the binding site for lipoproteins that contain apo B-100 and apo E. Domain 2 is 35% homologous to part of the extracellular domain of the *precursor* to epidermal growth factor (growth factors are discussed in chapter 24). The function of this domain is not clear. The cytosolic portion (domain 5) is required for congregation of the LDL receptors in the coated regions of the plasma membrane (see fig. 23.16). The number of LDL receptors per cell can vary between 15,000 and 70,000, depending on the cell's requirement for cholesterol. Once an LDL molecule is bound to the receptor, both are rapidly internalized by endocytosis ($t_{1/2}$ = 3 min).

Serious Diseases Result from Cholesterol Deposits

Brown and Goldstein were able to deduce the pathway for the uptake and catabolism of LDL largely as a result of their studies on the inherited disease familial hypercholesterolemia. Patients with the homozygous (two defective genes) form of this disease have grossly elevated levels of plasma cholesterol (650–1,000 mg/100 ml) (see table 23.1 for normal values), which is largely carried by an elevated concentration of LDL. One result is the formation of cholesterol deposits in the skin (xanthomas) in various areas of the body. Of greater consequence is the deposit of

Figure 23.16

Receptor-mediated uptake of LDL by human skin fibroblasts. Specific LDL receptors are located in coated regions of the plasma membrane. LDL binding results in uptake by endocytosis and formation of a coated vesicle. This vesicle fuses with a lysosome containing many hydrolytic enzymes that degrade the lipoprotein, releasing cholesterol.

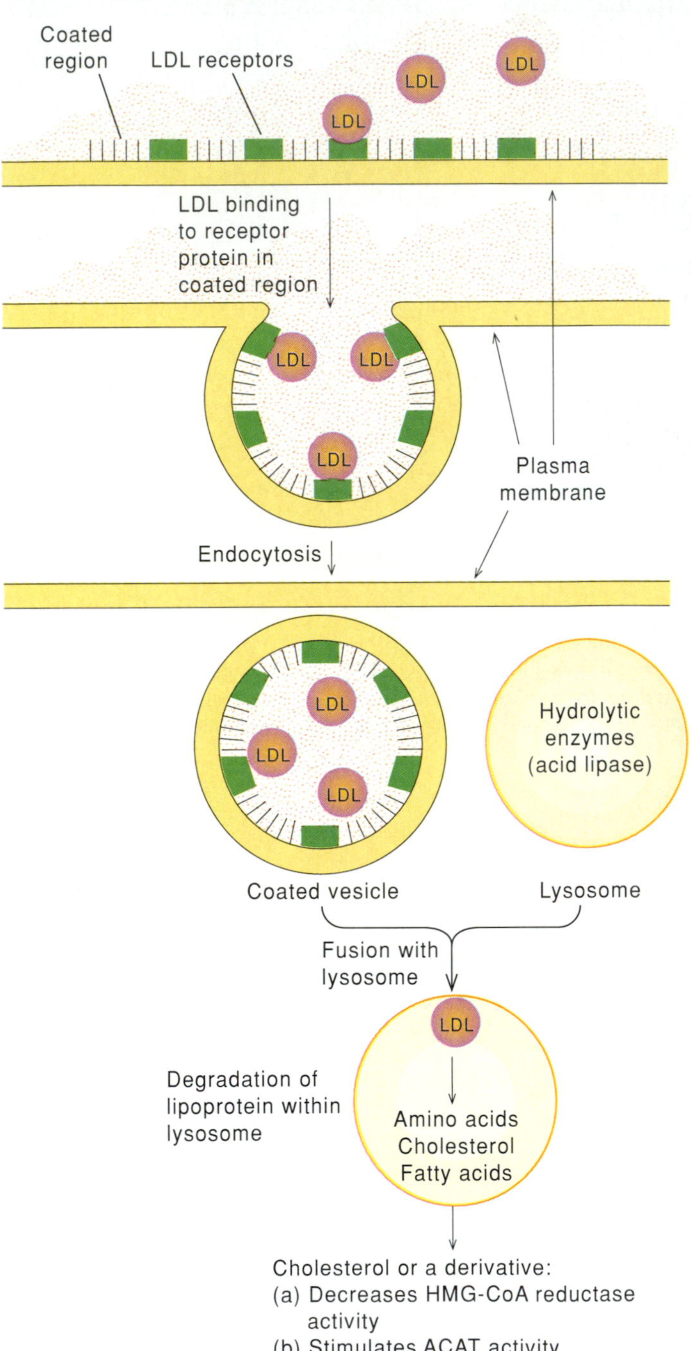

cholesterol in arteries, which results in atherosclerosis, a condition that is the underlying cause of most cardiovascular diseases. In fact, patients with homozygous familial hypercholesterolemia have symptoms of heart disease by the early teens and usually die from cardiovascular disease before the age of

Figure 23.17

Electron micrograph of LDL particles (made electron dense with covalently bound ferritin) bound to coated regions of a human skin fibroblast (97,000×). (From R. G. W. Anderson, M. S. Brown, and J. L. Goldstein, "Role of the coated and endocytic vesicle in the uptake of receptor-bound low density lipoprotein in human fibroblasts." *Cell* 10:351, 1977. © Cell Press.)

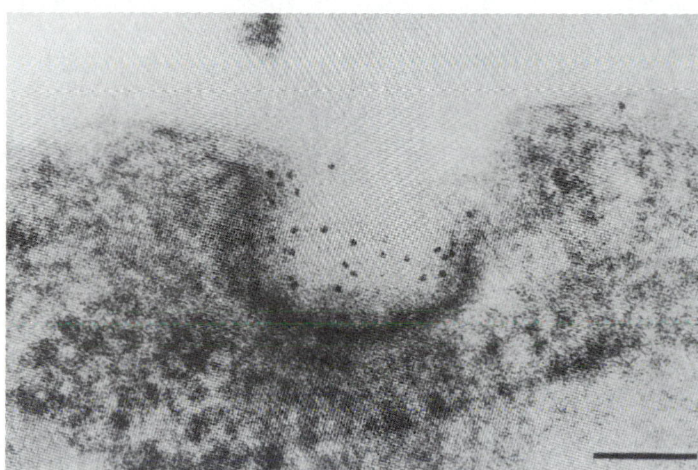

20. The heterozygotes (individuals with one normal and one defective gene) manifest similar but less severe symptoms. Their plasma cholesterol is in the range of 250–550 mg/100 ml of plasma, and they generally do not have a heart attack before the age of 40. The frequency of the heterozygous form of familial hypercholesterolemia has been estimated at 1 in 500, and that of the homozygous form is, in all likelihood, about 1 in 1 million.

Four different classes of biochemical mutations have been shown to cause familial hypercholesterolemia (fig. 23.19). The most common defect (class 1) is in the synthesis of the receptor. The other classes of mutations are defects in the transport of the receptor to the Golgi (class 2), defects in the binding of LDL (class 3), and inability of the receptors to cluster in coated pits (class 4).

A related disorder, Wolman's disease, has provided further evidence for the receptor-mediated pathway of LDL uptake (see fig. 23.16). Wolman's disease is a very rare inborn error of metabolism (approximately 25 cases diagnosed since 1956) that is characterized by the accumulation of cholesterol esters and triacylglycerols in various tissues. The disease can be diagnosed within several weeks of birth but is fatal, usually within 6 months. It is caused by a complete lack of a lysosomal acid lipase, which is responsible for the normal catabolism of cholesterol esters and triacylglycerols in lysosomes. Cholesterol ester storage disease is a related disorder, caused by a substantial reduction in the activity of lysosomal acid lipase (1 to 20% of normal). The symptoms are far less severe than in Wolman's disease, and patients have survived to the age of 40.

The importance of the work done on the LDL receptor and familial hypercholesterolemia was recognized in 1985 when Joseph Goldstein and Michael Brown were awarded the Nobel Prize in Physiology or Medicine.

Figure 23.18

The LDL receptor: a single protein with five domains. The cytosolic portion (domain 5) is required for congregation of the LDL receptors in the coated regions of the plasma membrane (see fig. 23.16). Once an LDL molecule is bound to the receptor, both are rapidly internalized by endocytosis.

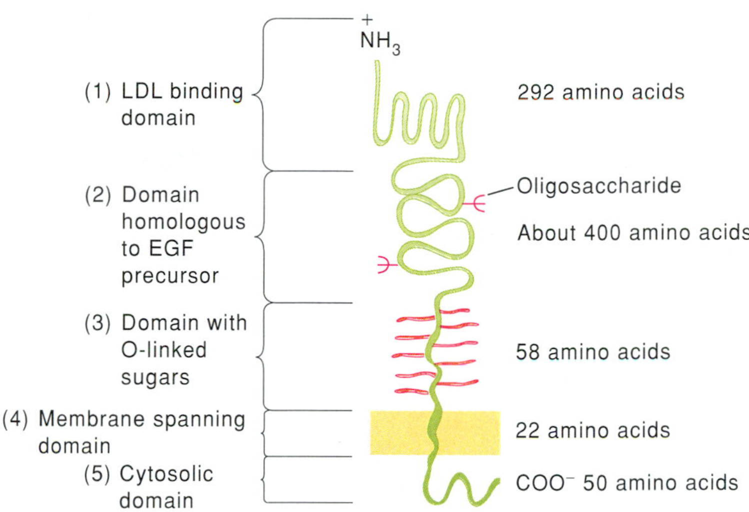

Figure 23.19

Four classes of mutations that disrupt the structure and function of the LDL receptor. Each class of mutation interferes with a different step in the process by which the receptor is synthesized. Class 1 mutations result in defective receptor synthesis in the endoplasmic reticulum. Class 2 mutations lead to defective receptor processing in the Golgi. Class 3 mutations result in defective receptor binding sites for LDL particles. Finally, class 4 mutations result in the inability of a receptor to cluster in coated pits.

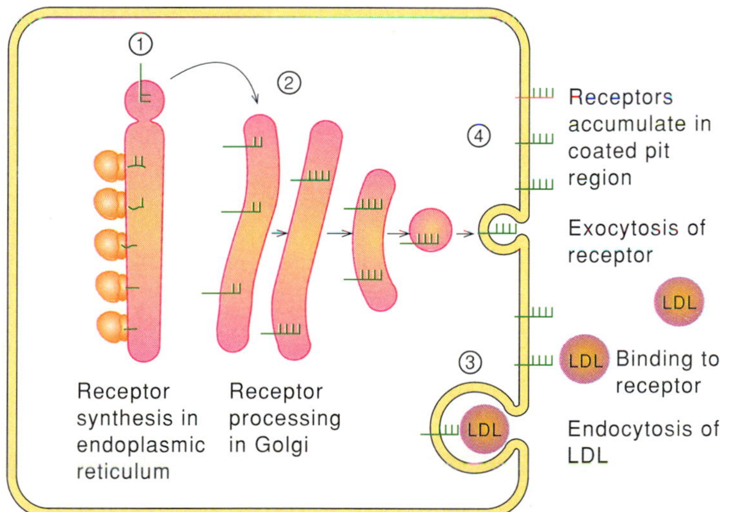

High-Density Lipoproteins (HDL) May Reduce Cholesterol Deposits

The catabolism of high-density lipoproteins is a complex process that is currently under investigation. The half-life of HDL in human plasma (5 to 6 days) is much longer than for the other lipoproteins. When HDL is secreted into plasma from liver, it has a discoid shape and is almost devoid of cholesterol ester.

Figure 23.20

Reaction catalyzed by lecithin: cholesterol acyltransferase (LCAT). The resulting cholesterol ester is transferred to VLDL and LDL particles by a lipid transfer protein.

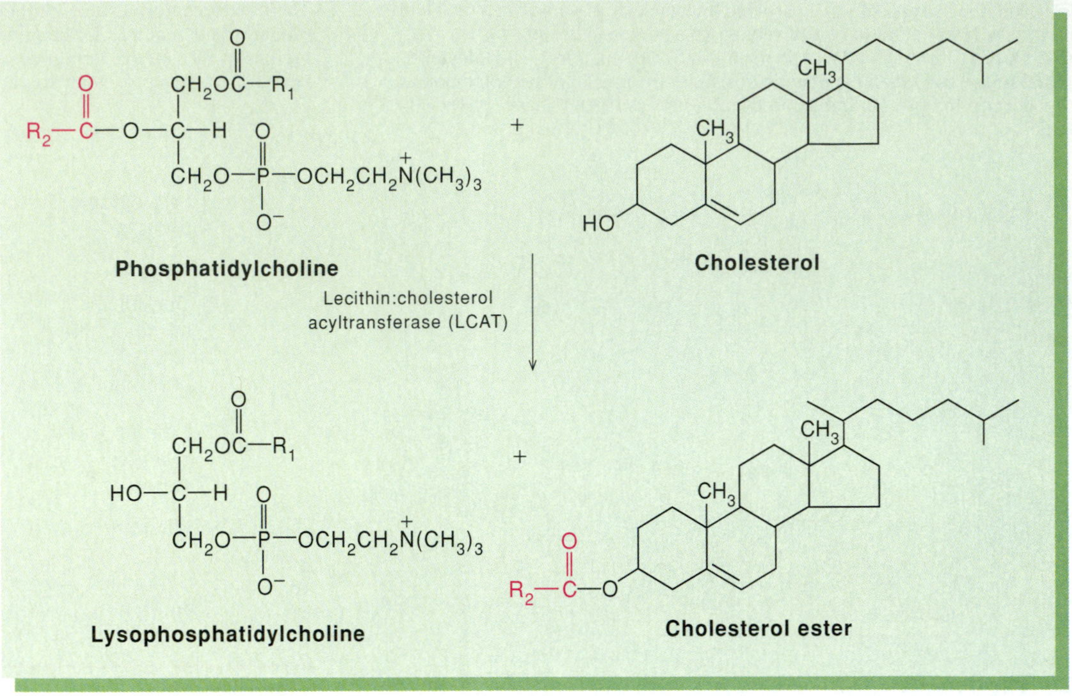

Phosphatidylcholine + Cholesterol

Lecithin:cholesterol acyltransferase (LCAT)

Lysophosphatidylcholine + Cholesterol ester

Figure 23.21

Formation of 7-hydroxycholesterol. The committed and rate-limiting reaction for bile acid synthesis is catalyzed by the endoplasmic reticulum enzyme 7α-hydroxylase.

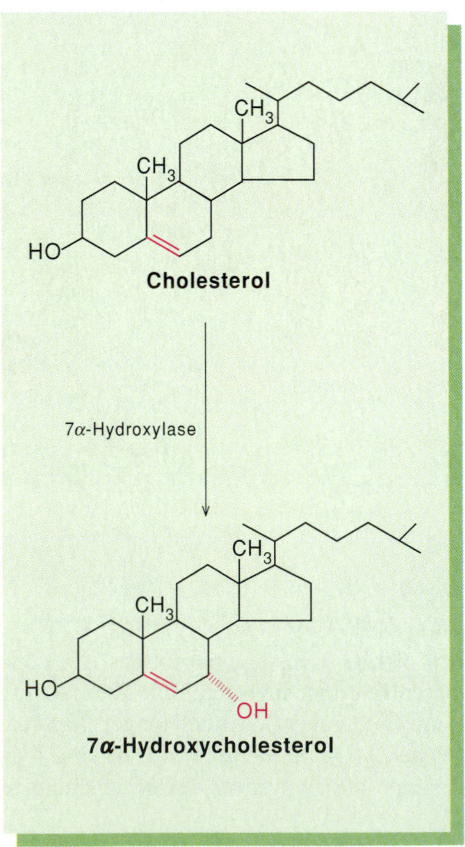

Figure 23.22

The reaction for the mixed function oxidase activity of 7α-hydroxylase.

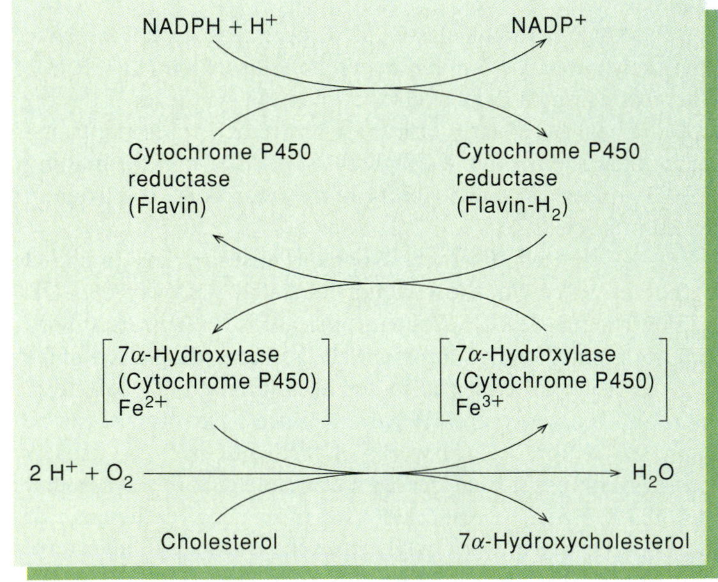

Figure 23.23

Conversion of 7α-hydroxycholesterol to cholic acid. The increase in the number of polar groups in the conversion to cholic acid improves the water solubility. Bile acids possess a 5 hydrogen: thus the A and B rings are no longer coplanar but have an A/B *cis* configuration as discussed in chapter 7. This configuration improves the detergent properties of the bile acids so that they are better able to solubilize lipids in the intestine and aid digestion.

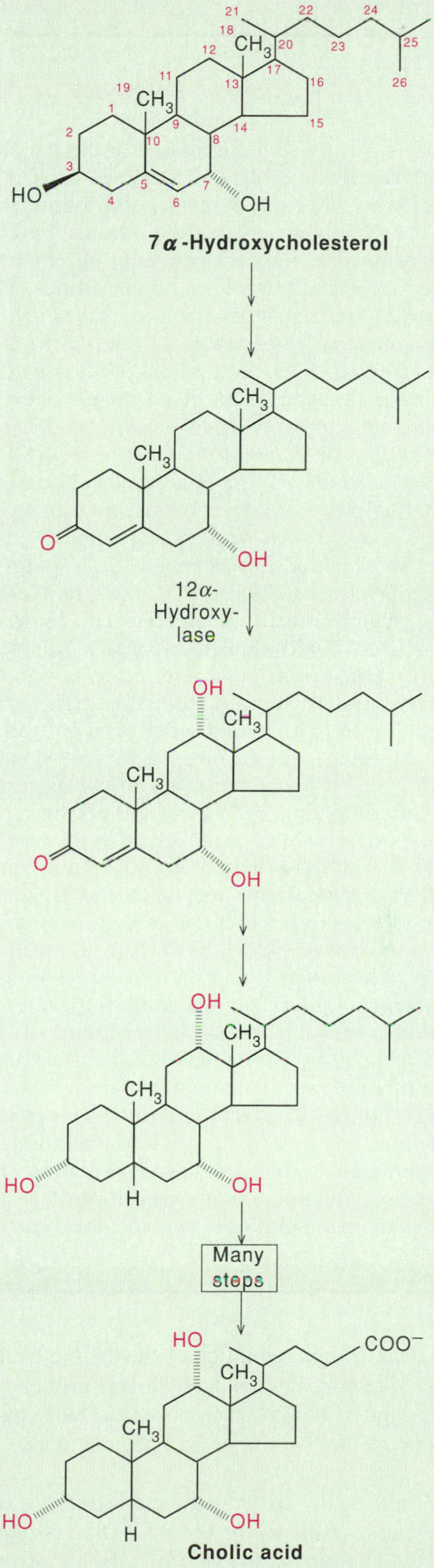

These newly formed HDL particles are converted into spherical particles by the accumulation of cholesterol ester. The cholesterol ester is derived from cholesterol and phosphatidylcholine on the surface of the HDL particle in a reaction catalyzed by lecithin:cholesterol acyltransferase (LCAT) (fig. 23.20). LCAT is a glycoprotein (24% carbohydrate by weight) with a molecular weight of 59,000. This enzyme is associated with HDL in plasma and is activated by apoprotein A-I, a component of HDL (see table 23.3). Associated with the LCAT-HDL complex is cholesterol ester transfer protein, which catalyzes the transfer of cholesterol esters from HDL to VLDL or LDL. In the steady state, cholesterol esters that are synthesized by LCAT would be transferred to these other lipoproteins and catabolized as noted earlier. The HDL particles themselves turn over, but how they are degraded is not firmly established.

Although elevated levels of cholesterol and LDL in human plasma are linked with an increased incidence of cardiovascular disease, recent data have shown that an increase in concentration of HDL in plasma is correlated with a lowered risk of coronary artery disease. Why does an elevated HDL level in plasma appear to protect against cardiovascular disease, whereas an elevated LDL level seems to cause this disease? The answer to this question is not known. An explanation currently favored is that HDL functions in the return of cholesterol to the liver, where it is metabolized and secreted. The net effect would be a decrease in the amount of plasma cholesterol available for deposit in arteries (see box 23B for more information on cholesterol and heart disease).

Bile Acid Metabolism

The conversion of cholesterol to bile acids is quantitatively the most important mechanism for degradation of cholesterol. In a normal human adult approximately 0.5 g of cholesterol is converted to bile acids each day. The regulation of this process operates at the initial biosynthetic step catalyzed by the endoplasmic reticulum enzyme 7α-hydroxylase (fig. 23.21). The 7α-hydroxylase is one of a group of enzymes called mixed function oxidases, which are involved in the hydroxylation of the sterol molecule at numerous specific sites. A mixed function oxidase is an enzyme complex that catalyzes hydroxylation of a substrate and production of H_2O from a single molecule of O_2. The 7α-hydroxylase is one of several enzymes referred to as cytochrome P450. The hydroxylation of cholesterol also requires NADPH: cytochrome P450 reductase (fig. 23.22).

The subsequent conversion of 7-hydroxycholesterol to cholic acid is outlined in figure 23.23. These reactions involve oxidation of the 3β-hydroxyl group, isomerization of the double

23B

BOX

Cholesterol Metabolism and Heart Disease

There is considerable discussion in the lay press about cholesterol and its link to cardiovascular disease because there is a direct relationship between elevated levels of cholesterol in the plasma and the incidence of heart disease. Experts generally agree that people with levels of total cholesterol in plasma above 240 mg/dl (6.2 millimoles/l) for many years are at increased risk of having a heart attack compared with people whose plasma cholesterol level is below 200 mg/dl (5.2 millimoles/l). It is generally recommended that adults should endeavor to achieve levels of total cholesterol (includes both free cholesterol and cholesterol ester) in plasma of 200 mg/dl (5.2 millimoles/l) or less. As discussed in the text, plasma cholesterol is largely carried in LDL as cholesterol ester. The cholesterol ester carried by LDL is sometimes referred to in the lay press as "bad cholesterol."

However, the relationship between cholesterol levels in plasma and cardiovascular disease is not so simple, since high levels of HDL (which contains 30% cholesterol by weight—see table 23.2) appear to protect against heart attack. Not surprisingly, the cholesterol carried by HDL is referred to as "good cholesterol" in the lay press. In addition, there are many other factors that have been linked to increased incidences of heart attack by epidemiological studies. Among these factors are smoking, high blood pressure, genetic background (recall our discussion of familial hypercholesterolemia), diabetes, and obesity.

How can people achieve total cholesterol levels of 200 mg/dl or less? If they were lucky and chose their parents carefully, they will have genes that protect them from high cholesterol levels. Such people can eat a diet that is high in fat and cholesterol and still remain well below the 200 mg/dl threshold. Most adults are not in this category and therefore have to use dietary or drug treatments to reduce plasma cholesterol levels. Cholesterol is enriched in animal meats and dairy products and is absent in vegetables. Thus many people can reduce their plasma cholesterol levels by eating smaller amounts of meat and dairy products. Another complication is that the serum levels of cholesterol are affected by the amount of saturated and polyunsaturated fats in the diet. Saturated fats tend to increase plasma cholesterol levels, whereas polyunsaturated fats tend to protect against increased cholesterol in the plasma. Again, meat and dairy products are enriched in saturated fatty acids. In contrast, many fish have high levels of polyunsaturated fatty acids, particularly $C_{22:5}$ and $C_{22:6}$. People in countries such as Japan, who have diets enriched in fish, have lower cholesterol levels in plasma and a reduced incidence of heart disease compared with people in North America and many European countries. Thus, reducing the intake of fat from 40 to 30% of dietary calories by eating more carbohydrate, vegetables, and fish will often result in a 15 to 20% decrease in serum cholesterol. However, there is a wide variation among individuals in how successful the dietary approach is, since genetic factors again come into play.

Recently a new drug called lovastatin (see fig. 23.6) has become available by prescription for treatment of hypercholesterolemia. This compound, a competitive inhibitor of HMG-CoA reductase, sharply reduces the rate of cholesterol biosynthesis. Equally important, in response to the reduction of cellular cholesterol levels there is increased expression of LDL receptors, which in turn allows more LDL to be cleared from the bloodstream, particularly by the liver. Thus the combination of this drug plus dietary restriction can result in striking reductions in plasma cholesterol levels (e.g., from 300 to 200 mg/dl). The drug is used effectively by heterozygotes with familial hypercholesterolemia. Interestingly, lovastatin has no significant effect on the serum cholesterol levels in homozygotes with familial hypercholesterolemia, since these patients do not have functional LDL receptors.

High levels of plasma cholesterol do not directly cause heart attacks. Rather, high levels of serum cholesterol over long periods are somehow involved in the development of a disease of the arteries called atherosclerosis. Atherosclerotic plaques are complex lesions in arterial walls that contain abnormal deposits of cholesterol esters. Precisely how high cholesterol levels in the plasma relate to the development of atherosclerosis is not understood and is a major frontier of medical research today. The actual sudden onset of heart attack appears to result from the adherence of platelets to the atherosclerotic lesion and the subsequent formation of a clot which occludes the blood flow in a coronary artery. The heart tissue is therefore deprived of blood supply, a condition that leads to tissue death.

---

bond, 12α-hydroxylation, reduction of the double bond, and reduction of the 3-keto group to a 3α-hydroxyl group. Additional hydroxylations and oxidation reactions on the side chain lead to cholic acid, one of the two major human bile acids (see chapter 7).

The bile acids are mostly converted to the corresponding bile salts, as shown for the formation of glycocholate in figure 23.24. The bile salts are critically important for the solubilization of lipids in the intestine, as we saw in chapter 7.

Metabolism of Steroid Hormones

Information on the hormonal functions of steroids and the mechanism by which steroids work is presented in the chapter on hormone action (chapter 24). Here we will focus on the enzymatic processes by which cholesterol is converted to the major steroid hormones.

The initial reaction in steroid hormone biosynthesis is catalyzed by desmolase (side-chain cleavage complex), which is found in the mitochondria of steroid-producing tissues (e.g.,

Figure 23.24

Conversion of a bile acid into a bile salt in the formation of glycocholate. Glycocholate is an example of a bile salt. The bile salts solubilize lipids in the small intestine so that they can be degraded by lipases.

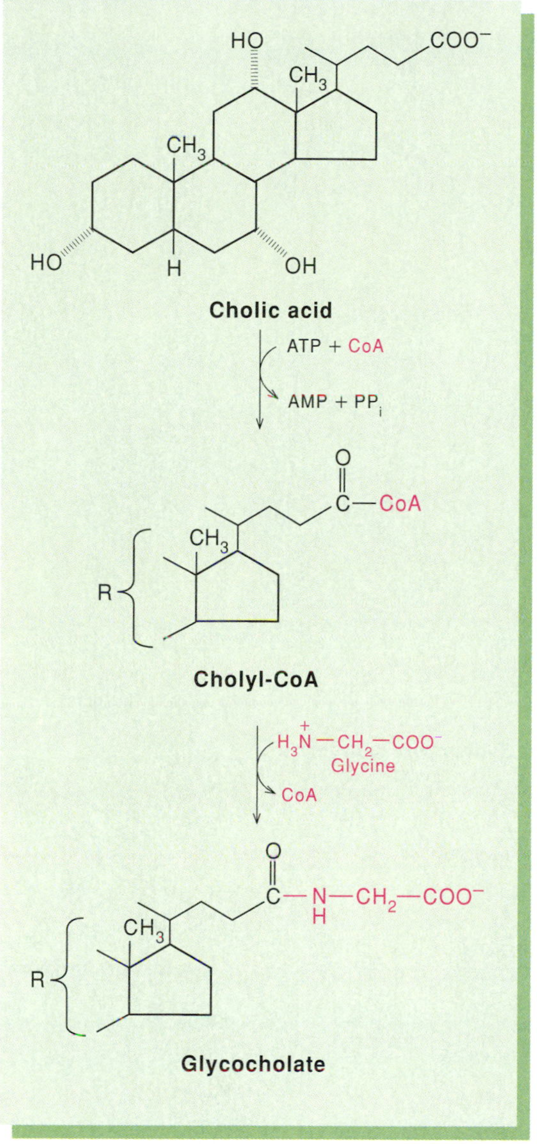

Figure 23.25

Biosynthesis of progesterone. Desmolase is found in the mitochondria of steroid-producing tissues. It is converted to pregnenolone in the mitochondria and then transferred to the endoplasmic reticulum, where it is converted to progesterone. Progesterone is a hormone as well as the precursor of several other hormones.

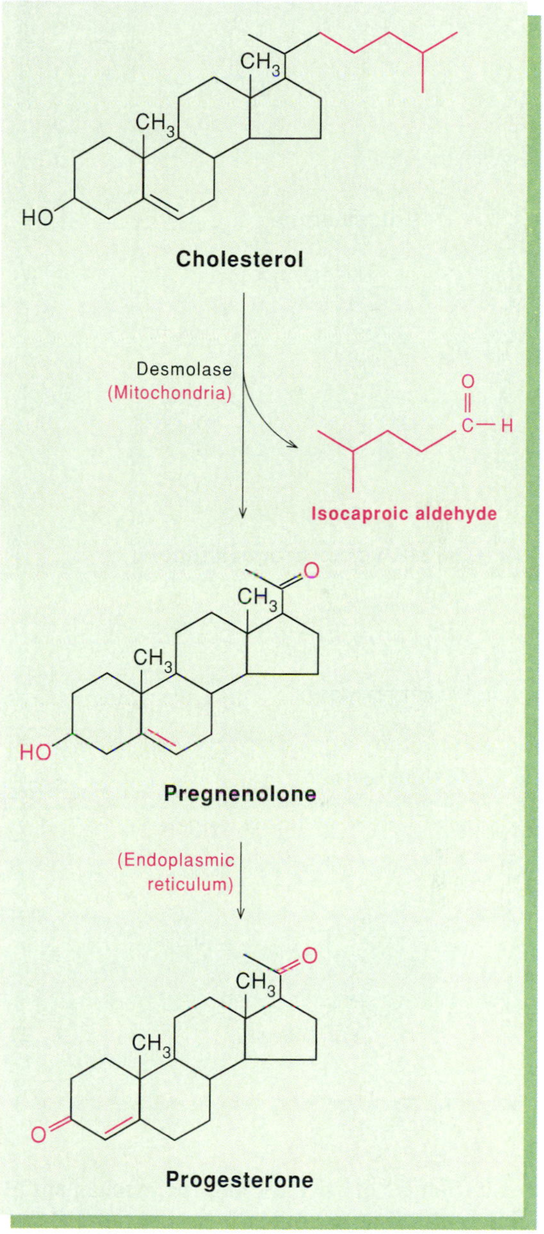

adrenals, gonads). The reaction is shown in figure 23.25. Desmolase appears to consist of two hydroxylases containing cytochrome P450, and a lyase. The product, pregnenolone, is subsequently transferred to the endoplasmic reticulum, where an oxidation of the hydroxyl group and isomerization of the double bond produces progesterone (see fig. 23.25). Progesterone is a steroid hormone, and it or pregnenolone is the biosynthetic precursor of all other steroid hormones.

The initial step in the conversion of progesterone to aldosterone (fig. 23.26) is catalyzed by a 21-hydroxylase, present on endoplasmic reticulum from the adrenal cortex but absent in gonads and placenta. This enzyme (M_r = 47,000) is also a cytochrome P450 protein and has been purified from adrenocortical endoplasmic reticulum. It is distinct from the 11-β-hydroxylase and 18-hydroxylase, two mitochondrial enzymes also involved in aldosterone synthesis (see fig. 23.26). It is curious that the biosynthesis of aldosterone from cholesterol begins in mitochondria with desmolase. Then the next reactions are catalyzed by enzymes on the endoplasmic reticulum, and finally the 21-hydroxy-progesterone is carried back to the mitochondria for the last enzymatic steps. The reason for this subcellular

Figure 23.26

Conversion of progesterone to aldosterone. The initial reaction, involving the 21-hydroxylase enzyme, occurs on the endoplasmic reticulum of the adrenal cortex. Two mitochondrial enzymes are involved in the next step, leading to aldosterone.

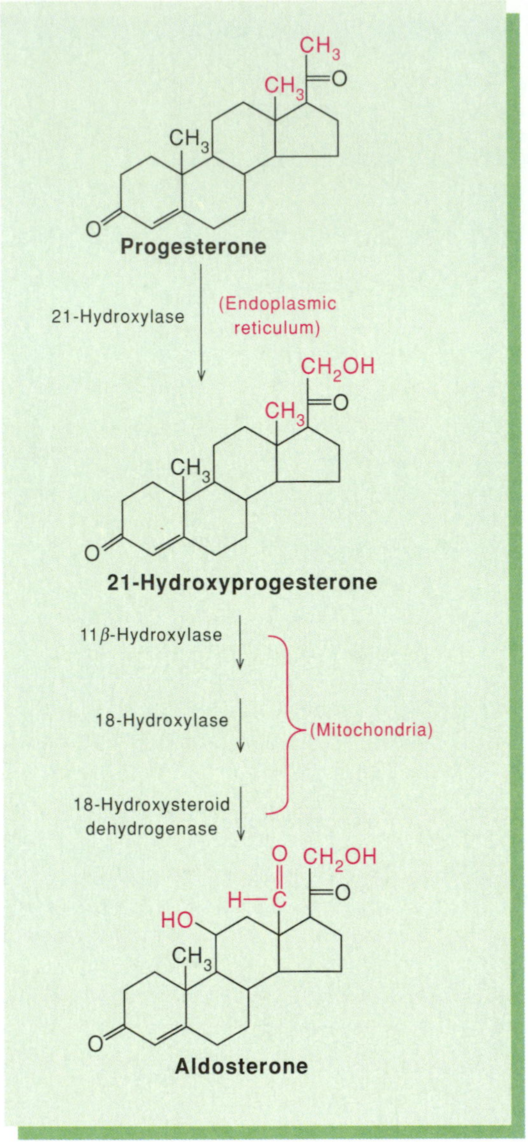

Figure 23.27

Two pathways for the conversion of progesterone to other steroid hormones. The first reaction, catalyzed by 17α-hydroxylase, is found in all steroid-secreting organs. Subsequent conversions to cortisol and testosterone are organ-specific as shown.

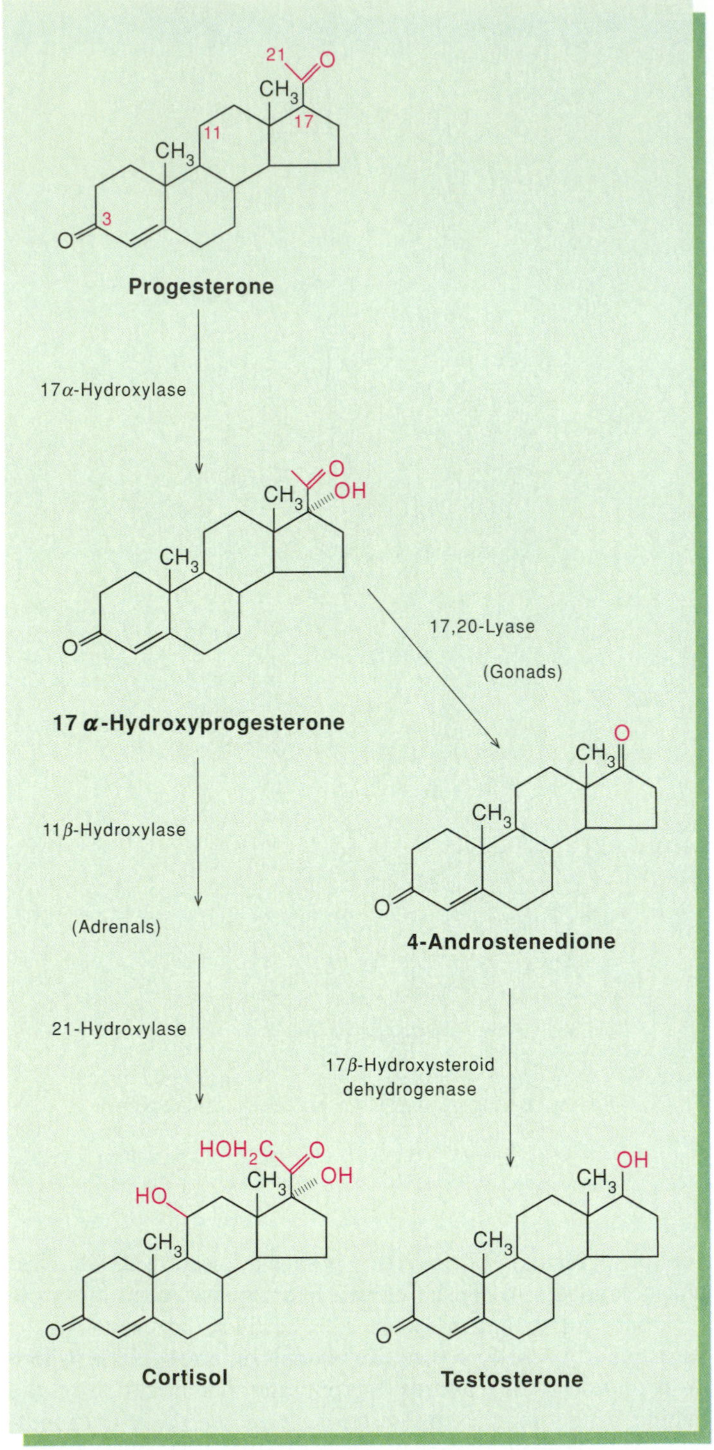

compartmentation is not obvious, and the mechanism by which the cell directs the intermediates from one subcellular site to another is not known.

The 17α-hydroxylase (fig. 23.27) is a mixed function oxidase found on the endoplasmic reticulum in all steroid-secreting organs. It is the key enzyme for directing steroids into the synthesis of glucocorticoids, androgens, and estrogens. Hydroxylations at the 11β-position and the 21-position direct the 17-OH progesterone into cortisol, a glucocorticoid made in the adrenals (see fig. 23.27). Alternatively, the gonads will direct

Figure 23.28

The conversion of testosterone into estradiol. These reactions are catalyzed by a complex of enzymes called the aromatase system, which consists of three enzyme activities located on the endoplasmic reticulum. The aromatase system is found in the ovaries and the placenta.

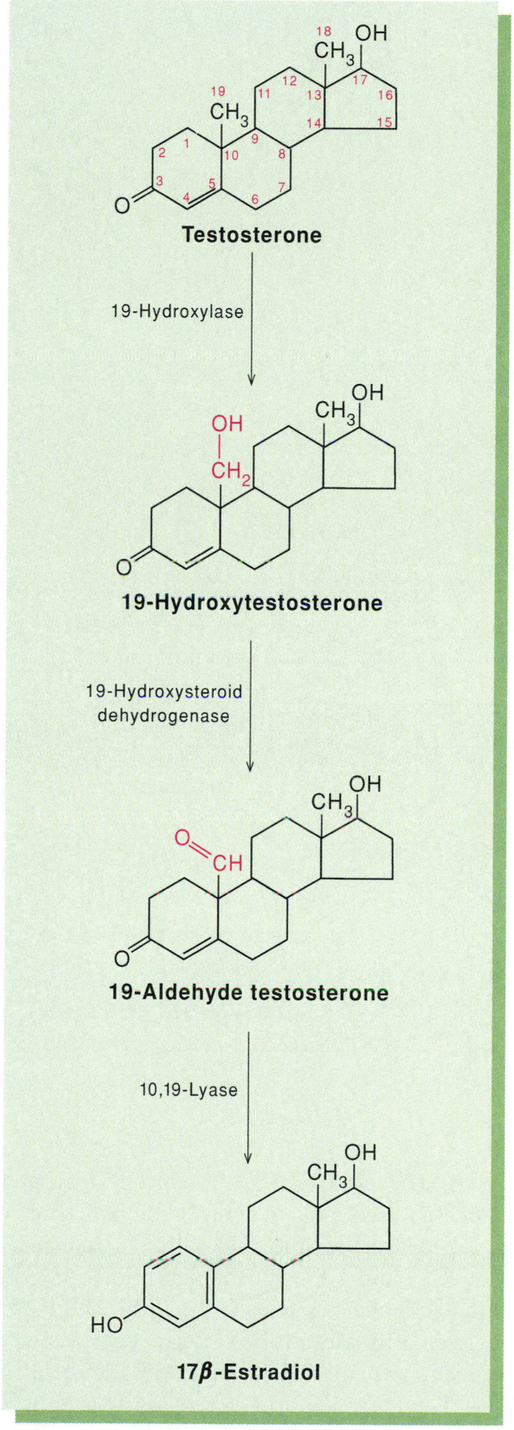

Testosterone

19-Hydroxylase

19-Hydroxytestosterone

19-Hydroxysteroid dehydrogenase

19-Aldehyde testosterone

10,19-Lyase

17β-Estradiol

Figure 23.29

Inactivated steroid hormones are conjugated to glucuronic acid prior to excretion in the urine.

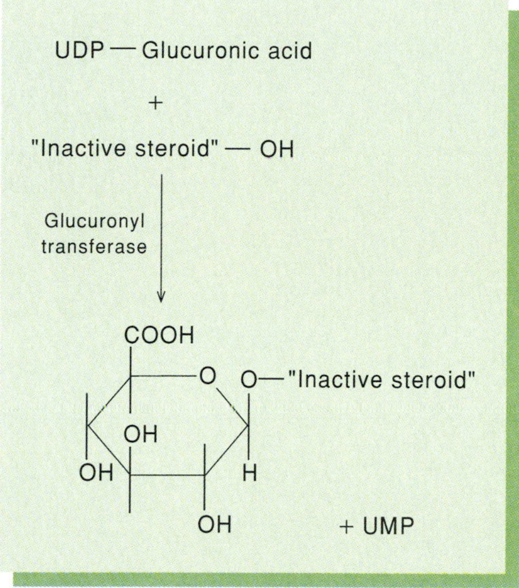

17-OH progesterone to the synthesis of *testosterone* as the result of the action of 17,20-lyase and a reduction of the 17-keto group (see fig. 23.27). Testosterone can be converted to dihydrotestosterone as discussed in chapter 24.

The female sex hormones arise from testosterone via the 19-hydroxylated intermediate (fig. 23.28). The enzyme complex responsible for this conversion is called the aromatase system and is one of the few reactions by which mammals can make an aromatic ring. This complex is associated with the endoplasmic reticulum of cells in the ovary and placenta.

Thus the second major degradative fate of cholesterol in mammals is the formation of steroid hormones. The enzymes involved are found in mitochondria and endoplasmic reticulum of steroid-producing tissues. The hydroxylases are in each case mixed function oxidases that have cytochrome P450 at the active site. We will discuss the remarkable effects of these potent hormones in chapter 24.

There is no known pathway in mammals by which the steroid ring nucleus can be degraded to smaller molecules such as acetate. The carbon atoms of the ring system cannot, therefore, be used as a source of metabolic energy. However, many reactions occur, particularly in the liver, for the partial catabolism and inactivation of the steroid hormones. Frequently, these reactions involve reduction of ketone groups or double bonds. These inactive steroids are conjugated to glucuronic acid (fig. 23.29) or sulfate. The excretion of such derivatives results in a very large number of steroid metabolites in the urine. For example, twenty different metabolites of estrogen, conjugated to sulfate or glucuronic acid, have been identified in human urine.

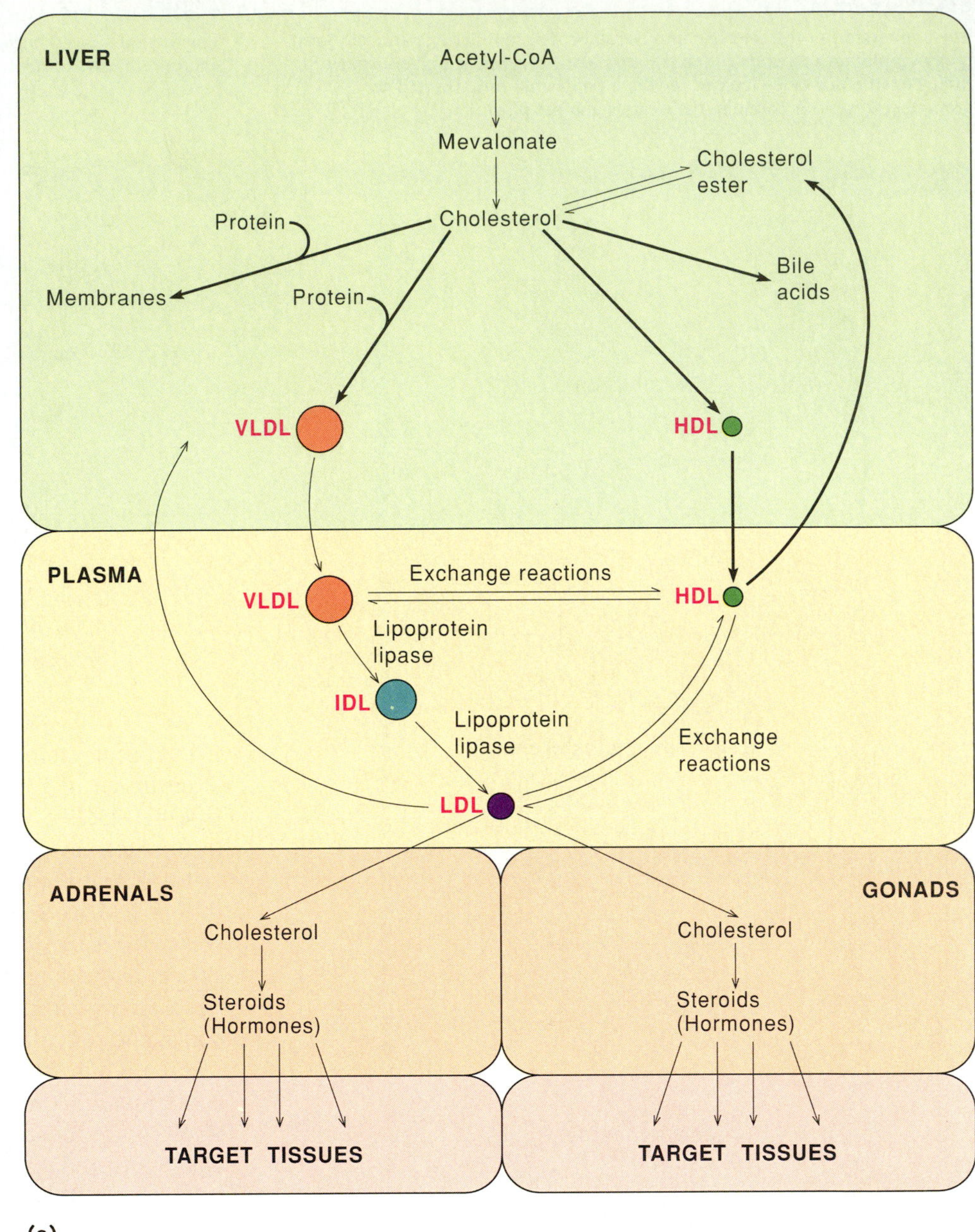

Figure 23.30

Fate of cholesterol. (*a*) Cholesterol biosynthesized in the liver has several alternative fates. (*b*) Cholesterol obtained from the diet can enter the plasma and subsequently the liver.

Overview of Mammalian Cholesterol Metabolism

The metabolism of cholesterol in mammals is extremely complex. A summary sketch (fig. 23.30) helps to draw the major metabolic interrelationships together. Cholesterol is biosynthesized largely in the liver (fig. 23.30*a*) or taken in through the diet (fig. 23.30*b*). From the intestine, cholesterol is secreted into the plasma mainly as a component of chylomicrons. These particles are quickly degraded by lipoprotein lipase and the remnants are removed by the liver. Apoproteins and lipid components of the chylomicrons and remnants appear to exchange with HDL. Cholesterol made in the liver (fig. 23.30*a*) has several alternative fates. It can be (1) secreted into plasma as a component of HDL and VLDL, (2) stored in droplets as cholesterol ester, (3) used as a structural component of cell membranes, or (4) converted into bile salts. In plasma, VLDL is degraded to IDL and LDL by the action of lipoprotein lipase and through exchange reactions with HDL. The LDL serves as a major carrier of cholesterol to extrahepatic cells, which include the adrenals and gonads. LDL and HDL are also returned to the liver. The steroid hormones made in the adrenals and gonads are delivered to various target tissues and promote a wide range of metabolic effects. The steroid hormones are eventually excreted as glycosyl conjugates in the urine. The bile salts made in the liver are delivered to the upper intestine, where they aid in the solubilization of dietary lipid. Most of the bile salts are resorbed in the lower intestine and returned to the liver by the portal vein.

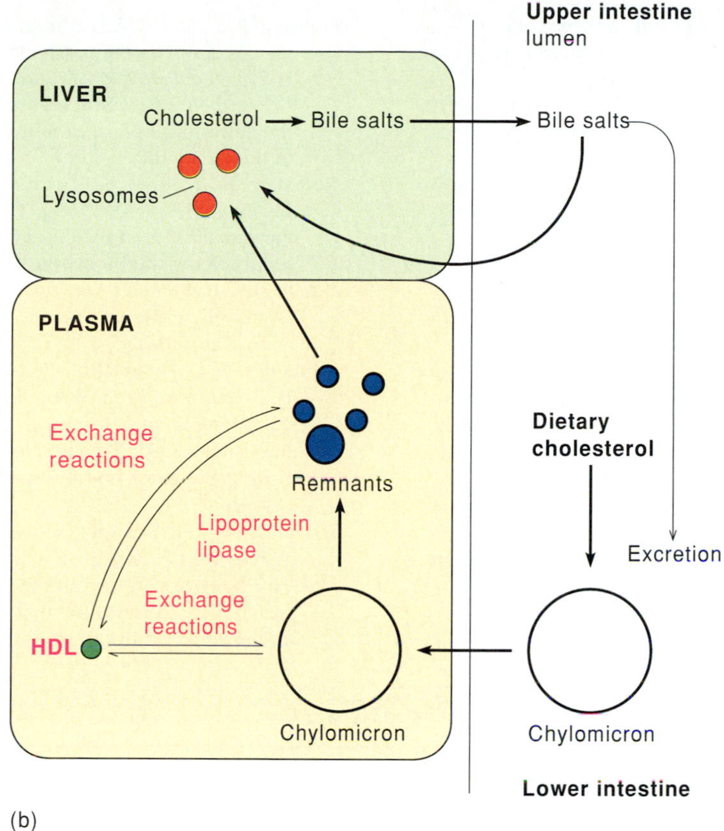

(b)

Summary

In this chapter we have dealt primarily with the metabolism of cholesterol, the most prominent member of the steroid family of lipids, and with the associated plasma lipoproteins. The chief points in our discussion are as follows.

1. Steroids are derivatives of the tetracyclic hydrocarbon perhydrocyclopentanophenanthrene. The biosynthesis of steroids begins with the conversion of three molecules of acetyl-CoA into mevalonate, the decarboxylation of mevalonate, and its conversion to isopentenyl pyrophosphate. Six molecules of isopentenyl pyrophosphate are polymerized into squalene, which is cyclized to yield lanosterol. Lanosterol is converted to cholesterol, which is the precursor of bile acids and steroid hormones.
2. The rate of cholesterol biosynthesis appears to be regulated primarily by the activity of HMG-CoA reductase. This key enzyme is controlled by the rate of enzyme synthesis and degradation and by phosphorylation/dephosphorylation reactions, and synthesis of the reductase is inhibited by cholesterol delivered to cells by means of low-density lipoproteins (LDL).
3. Cholesterol and phospholipids are carried in plasma by lipoproteins, which are synthesized and secreted by the intestine and liver. The major lipoproteins are chylomicrons, very-low-density lipoproteins (VLDL), low-density lipoproteins (LDL), and high-density lipoproteins (HDL). The triacylglycerols in chylomicrons and VLDLs are degraded in plasma by lipoprotein lipase, and the fatty acids and monoacylglycerols are absorbed primarily by heart, skeletal, and adipose tissue. LDLs are removed from plasma by an endocytotic process after binding to specific LDL receptors on the plasma membrane. The LDLs are enzymatically degraded in the lysosomes. In familial hypercholesterolemia, the specific receptors for LDL uptake are inactive. High levels of LDL are associated with an increased risk of cardiovascular disease, whereas high levels of HDL seem to protect against this disease.
4. Bile acids are C_{24} carboxylic acids that are biosynthetically derived from cholesterol. The 7α-hydroxylation of cholesterol is the committed and rate-limiting reaction in the synthesis of bile acids. Salts formed from the bile acids are secreted into the small intestine and aid the solubilization and digestion of lipids.
5. Steroid hormones are biosynthesized from cholesterol in the adrenal cortex, gonads, and placenta. These steroids are important hormones for many specific physiological processes.

Selected Readings

Bentley, R., *Molecular Asymmetry in Biology,* vol. 2. New York: Academic Press, 1970. This book contains a very lucid explanation of the stereochemistry of cholesterol biosynthesis.

Bloch, K., Cholesterol: evolution of structure and function. Chapter 12 in D. E. Vance and J. E. Vance (eds.), *Biochemistry of Lipids, Lipoproteins and Membranes.* Amsterdam: Elsevier Science Publishers, 1991. This article provides an interesting view of how the structure of cholesterol evolved to optimize its function in cells.

Brown, M. S., and J. L. Goldstein, A receptor-mediated pathway for cholesterol homeostasis. *Science* 232:34–47, 1986. An article describing their Nobel Prize winning research on the LDL receptor and familial hypercholesterolemia.

Davis, R., Lipoprotein structure and function. Chapter 14 in D. E. Vance and J. E. Vance (eds.), *Biochemistry of Lipids, Lipoproteins and Membranes.* Amsterdam: Elsevier Science Publishers, 1991. This chapter provides an advanced discussion on the assembly and secretion of very-low-density lipoproteins.

Edwards, P. A., Regulation of sterol biosynthesis and isoprenylation of proteins. Chapter 13 in D. E. Vance and J. E. Vance (eds.), *Biochemistry of Lipids, Lipoproteins and Membranes.* Amsterdam: Elsevier Science Publishers, 1991. The complexities of the regulation of cholesterol biosynthesis are explained in this chapter.

Fielding, P. E., and C. J. Fielding, Dynamics of lipoprotein transport in the circulatory system. Chapter 15 in D. E. Vance and J. E. Vance (eds.), *Biochemistry of Lipids, Lipoproteins and Membranes.* Amsterdam: Elsevier Science Publishers, 1991. This article reviews the current literature on the intricacies of lipoprotein metabolism in the circulatory system.

Goldstein, J. L., and M. S. Brown, Regulation of the mevalonate pathway. *Nature* 343:425–430, 1990. This article describes regulatory mechanisms for cholesterol biosynthesis within the context of the regulation of the biosynthesis of other isoprenoid derivatives.

Makin, H. L. J., *Biochemistry of Steroid Hormones,* 2d ed. Oxford: Blackwell, 1984. An advanced and comprehensive treatment of steroid hormones.

Schneider, W. J., Removal of lipoproteins from plasma. Chapter 16 in D. E. Vance and J. E. Vance (eds.), *Biochemistry of Lipids, Lipoproteins and Membranes.* Amsterdam: Elsevier Science Publishers, 1991. This article provides a clear explanation of the current literature on the uptake of lipoproteins into cells and tissues.

Scriver, C. R., A. L. Beaudet, W. S. Sly, and D. Valle, *The Metabolic Basis of Inherited Disease,* 6th ed., vol. I. New York: McGraw-Hill, 1989. This book has an introductory chapter on lipoprotein structure and metabolism followed by many excellent chapters on disorders of cholesterol and lipoprotein metabolism.

Problems

1. How can elevated levels of cholesterol in the liver lower cholesterol biosynthesis? What effect will elevated cholesterol have on LDL receptors?
2. During routine investigations, the plasma from a family of rats (group 1) was found to have very low concentrations of cholesterol. When the microsomal HMG-CoA reductase from liver was assayed, extremely low activities were found. When the cytosol from normal rats (group 2) was added to the microsomal fraction from group 1 rats, the HMG-CoA reductase activity was gradually restored to normal values. What enzyme activity (or activities) might be deficient in the group 1 rats?
3. A person with diabetes is found to show no signs of ketone bodies in plasma even when in diabetic shock. Which enzyme(s) of ketone body synthesis might be deficient? If cholesterol synthesis were normal, would this be a clue as to which enzyme(s) might be deficient?
4. A patient homozygous for familial hypercholesterolemia (FH) was treated with lovastatin to lower LDL levels in the blood. This treatment did not have any effect on LDL levels. Why? After a number of heart attacks, a heart and liver transplant were done and LDL levels were dramatically lowered. Why were both organs replaced?
5. An FH heterozygote had plasma LDL-cholesterol levels about twice the normal levels. How could this patient be treated to lower LDL levels? What is the biochemical rationale for these treatments?
6. Liver cells in culture are given 2-[^{14}C]-acetate. Where would this label appear in HMG-CoA?
7. A deficiency of apoprotein C-II results in the disease hyperlipoproteinemia type I, in which there is a massive increase in the concentration of plasma triacylglycerol. Provide an explanation for this clinical finding.
8. What is the hereditary defect in Wolman's disease? Would you expect HMG-CoA reductase activity to be high or low in skin fibroblasts cultured from patients with this disorder? Would the number of LDL receptors be high or low in these fibroblasts?
9. There is an inherited disease in which lecithin:cholesterol acyltransferase (LCAT) is deficient. What effect would you expect this deficiency to have on the composition of HDL and other lipoproteins in plasma?

24

CHAPTER

Integration of Metabolism and Hormone Action

In chapter 12 we described strategies used to organize biochemical conversions so that all essential conversions are thermodynamically feasible and kinetically regulated. In the eleven chapters that followed, we witnessed countless examples employing these basic strategies. This is an appropriate time to see how various organizational strategies are integrated in the metabolism of a multicellular organism. The integration process is complicated by the fact that specific tissues assume unique roles in the overall metabolism of the organism. The best-understood multicellular organisms are vertebrates, and the aspect of their metabolism that is best understood is that dealing with energy production and consumption. Consequently, we will examine the integration of energy metabolism for several vertebrate tissues. First we will consider how energy metabolism is manipulated between these tissues. Then we will explore the underlying mechanisms that permit a smooth flow between energy-supplying and energy-consuming tissues.

The Three Major Forms of Energy Storage

In addition to relying on nutrition to supply the chemical fuel for their energy needs, vertebrates maintain fuel reserves in various tissues. These reserves are of three types: glycogen, triacylglycerols, and proteins. Greater than 90% of the fuel reserves in an average human exist in the form of triacylglycerols stored in the adipose tissue (table 24.1). In obese individuals this fuel reserve can be severalfold higher. Despite their abundance, triacylglycerols are held in check until the more readily utilizable glycogen reserves located in liver and muscle tissues are close to exhaustion. Generally, glycogen reserves of the liver play the most widely useful role despite the greater abundance of glycogen often found in muscle tissue. This is because, of the two

Table 24.1
Fuel reserves in a normal 70-kg human[a] (kcal)

Organ	Glucose or Glycogen	Triacylglycerols	Mobilizable Proteins
Liver	400	450	400
Brain	8	0	0
Muscle	1,200	450	24,000
Adipose tissue	80	135,000	40

[a]Data from G. T. Cahill, *Clin. Endocrinol. Metab.* 5:598, 1976.

tissues, only the liver is capable of degrading glycogen to (dephosphorylated) glucose that can be secreted into the bloodstream. Once the glycogen reserves have been depleted from the liver, the energy-hungry organism turns to the lipid (triacylglycerol) reserves sequestered in adipose tissue. The first step in mobilizing lipid for energy consumption involves the hydrolysis of triacylglycerols to fatty acids and glycerol. The fatty acids so produced are transported to other tissues, where they are degraded to the activated two-carbon unit acetyl-CoA in a repetitious multistep process (see chapter 17). A third major fuel reserve, protein, mostly from muscle tissues, can be mobilized by breakdown to amino acids that are transported to the liver. In the liver the amino acids are deaminated and converted into TCA cycle intermediates. The use of muscle tissue proteins to satisfy energy needs is bound to physically weaken the organism, so it comes as no surprise that muscle tissue is used for energy purposes only as a last resort. The best-known exception to this rule occurs during periods of prolonged muscle inactivity, as in the case of a temporarily incapacitated limb.

There has to be a way in which the energy-requiring tissues can send signals to the energy-producing tissues about their needs. The signals are supplied by circulating hormones, which regulate the metabolic activities of different tissues in the organism (fig. 24.1).

Energy Demands and Contributions of Different Tissues

Some tissues are mainly energy suppliers, while others are mainly energy consumers. Still other tissues are important both as consumers and as suppliers. In this section we will survey energy metabolism in five well-characterized vertebrate tissues: liver, adipocytes, striated muscle, smooth heart muscle, and brain.

Brain Tissue Makes No Contributions to the Fuel Needs of the Organism

Brain tissue does not contribute to the energy needs of other tissues. Rather, it makes major demands on the glucose supply, accounting for upwards of 50% of the glucose consumption in the resting human. To satisfy its needs, the brain normally absorbs glucose from the bloodstream, which it metabolizes to CO_2 and H_2O by a combination of glycolysis and the TCA cycle. Energy consumption by the brain is approximately constant: A sleeping brain and a wide-awake, mentally alert brain burn about the same amount of glucose. Because of its vital importance and delicate constitution, brain tissue is somehow given priority in meeting demands on the available supplies of blood glucose in times of starvation or other forms of stress, when the glucose supply is limited. Under conditions of prolonged starvation the blood glucose level drops, but even then the energy needs of brain tissue can be met by ketone bodies, which are metabolized via the TCA cycle.

Heart Muscle Utilizes Fatty Acids in Preference to Glucose to Fulfill Its Energy Needs

Continuous operation is an obvious necessity for the heart muscle. Its energy needs vary considerably, depending on the physical activity of the organism. For reasons that are unclear, the heart relies mainly on fatty acids supplied from the bloodstream to fulfill its energy needs. Perhaps this is because the fatty acid supply is more reliable than the fluctuating carbohydrate supply. The fatty acids are metabolized by β oxidation in the densely packed mitochondria of the heart muscle cells. Most organisms have a very extensive supply of fatty acids; thus the functioning of the heart muscle is assured. Under starvation conditions the heart muscle can switch to using ketone bodies supplied from the liver via the bloodstream. The ketone bodies are degraded in the mitochondria of the heart muscle by way of the TCA cycle. Like the brain, heart muscle makes no energy contribution to other tissues of the organism.

Skeletal Muscle Can Function Aerobically or Anaerobically

Whereas the heart muscle operates under strictly aerobic conditions, skeletal muscle can function either aerobically or anaerobically. The energy consumption of skeletal muscle varies enormously with the extent of muscle activity. During periods of mild exertion, muscle tissue requires moderate levels of glucose, which can be supplied by the blood glucose or the breakdown of the glycogen reserves present in the muscle tissue. This glucose is metabolized aerobically all the way to CO_2 and H_2O. During times of great exertion, muscle tissue uses oxygen faster

Figure 24.1

Some aspects of the regulation of the metabolism of cells in a typical mammal. Each cell type serves a particular function and has specific requirements for maintenance and growth. Some cells are primarily energy producers and others are primarily energy consumers. The activities of different cell types are regulated by an intricate hierarchy of hormone-secreting cells. In order to be susceptible to a specific hormone, a cell must possess a specific hormone receptor. On contact with the receptor the hormone causes a structural change in the hormone receptor, which in turn triggers a chain of reactions in the cell.

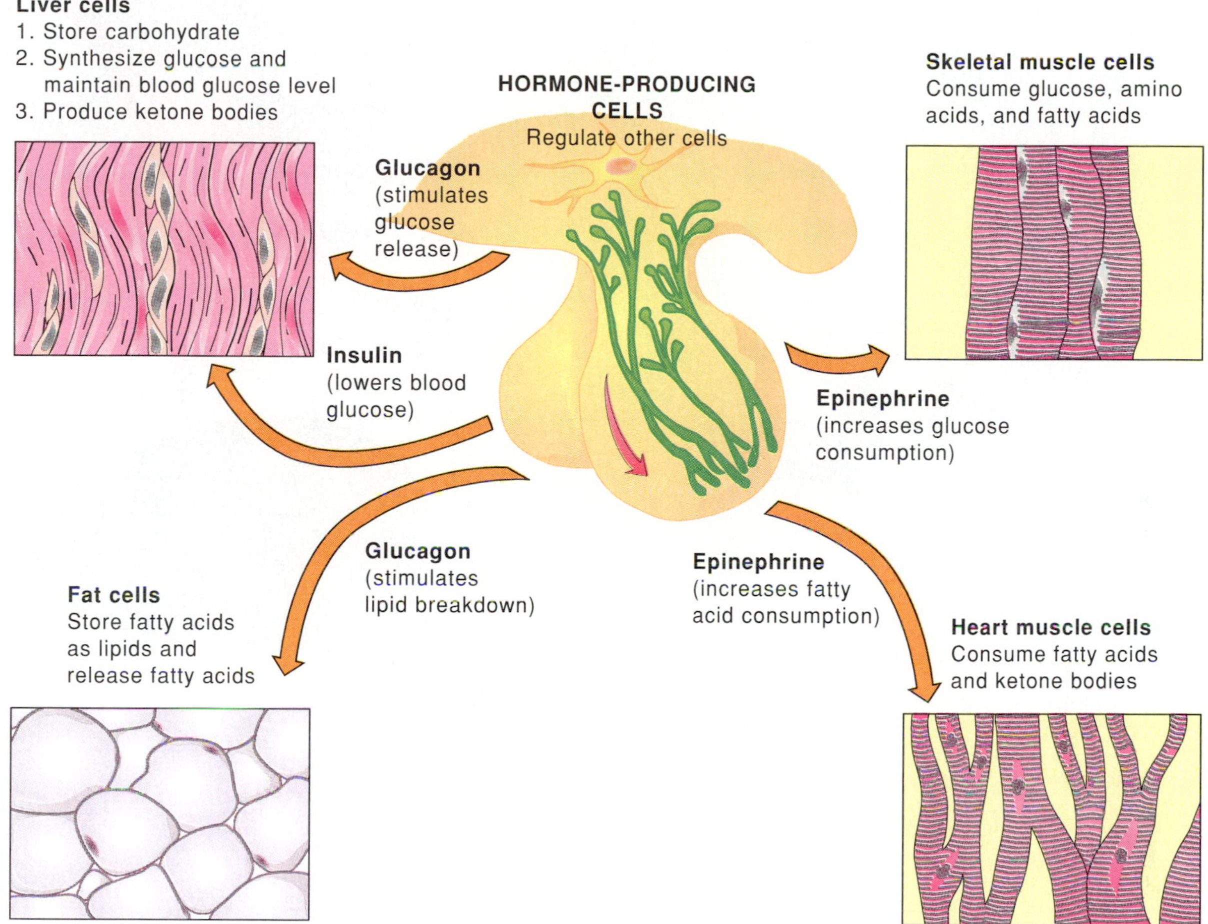

than it can be supplied by the bloodstream, so that the muscle must operate anaerobically. Under these conditions, glucose breakdown stops at the 3-carbon acid, lactate. The lactate so produced is secreted into the bloodstream and picked up by the liver, which converts the lactate back to glucose. This glucose can be returned to the muscle for further glycolysis. The cycling of glucose and lactate between skeletal muscle and liver is known as the Cori cycle (fig. 24.2). Under conditions where the blood glucose level is low, muscle tissue can utilize fatty acids as an alternative supply of energy. The fatty acids are mobilized via the β oxidation pathway in the muscle mitochondria. Muscle lactate can also be harnessed by heart muscle, which catabolizes it for energy purposes (box 24A).

In times of starvation, muscle tissue is usually quite inactive, so its energy demands are low. In fact, under such conditions the muscles tend to supply energy to the rest of the organism. The mechanism for doing so is the degradation of muscle protein to amino acids, which are transported to the liver tissue. In the liver the amino acids are converted to glucose or used for biosynthesis of other proteins essential for maintenance of the organism's vital processes.

Adipose Tissue Maintains Vast Fuel Reserves in the Form of Triacylglycerols

Under conditions of nutritional excess, fatty acids are absorbed by adipose tissue where they are converted to storage lipids in the form of triacylglycerols. The triacylglycerols can be mobilized at a later time, when the carbohydrate energy reserves are low. The process of mobilization involves the conversion of triacylglycerols to fatty acids, which are transported by the bloodstream to the liver for further processing (fig. 24.3).

Figure 24.2

The cycling of lactate and glucose between muscle and liver in the Cori cycle. Under conditions of intense activity the muscle operates anaerobically so that the end product of glucose breakdown is lactate. This product can pass through the bloodstream to the liver and be converted to glucose, which can be returned to the muscle.

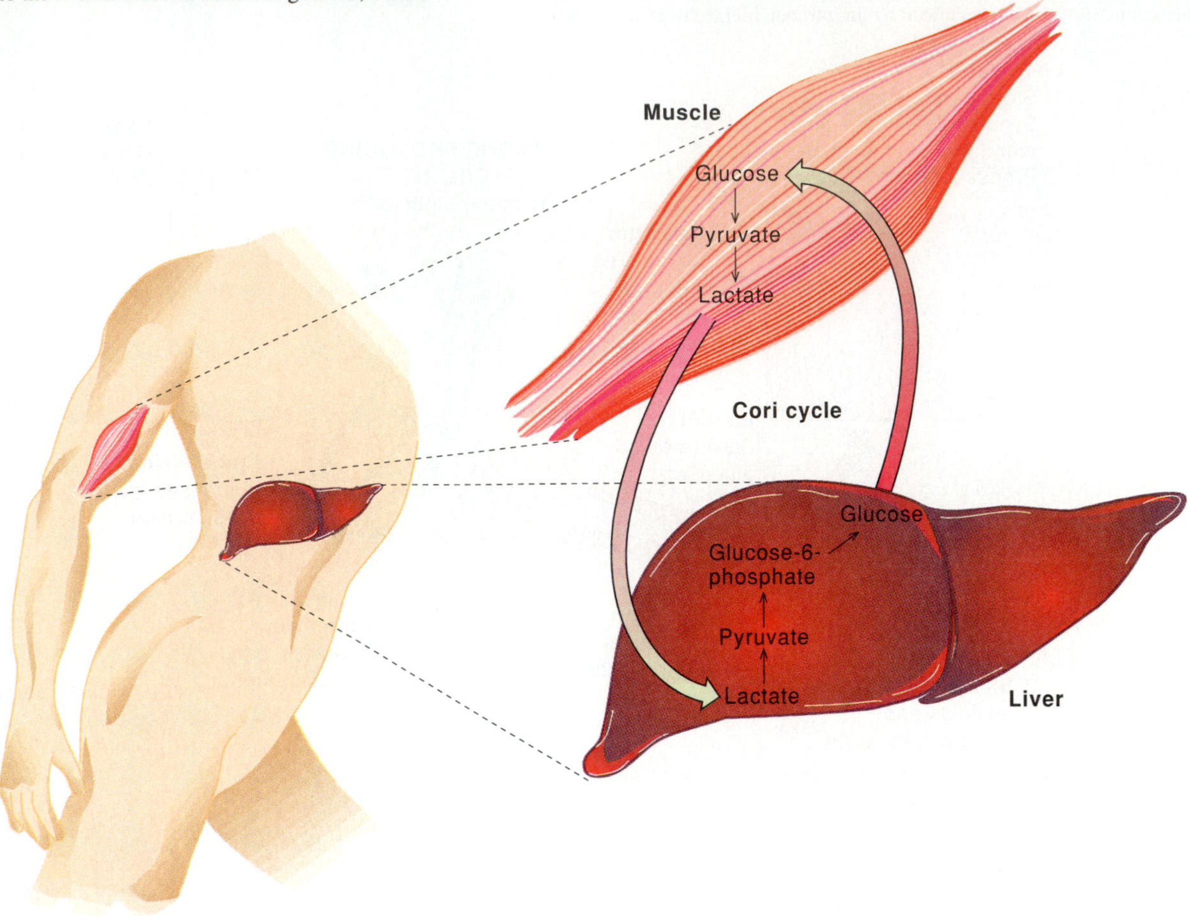

Figure 24.3

Synthesis and degradation of triacylglycerols in adipose tissue. Fatty acids are delivered to adipose tissue. In times of energy excess these are converted to triacylglycerols and stored until needed, at which point the triacylglycerols are converted back to fatty acids.

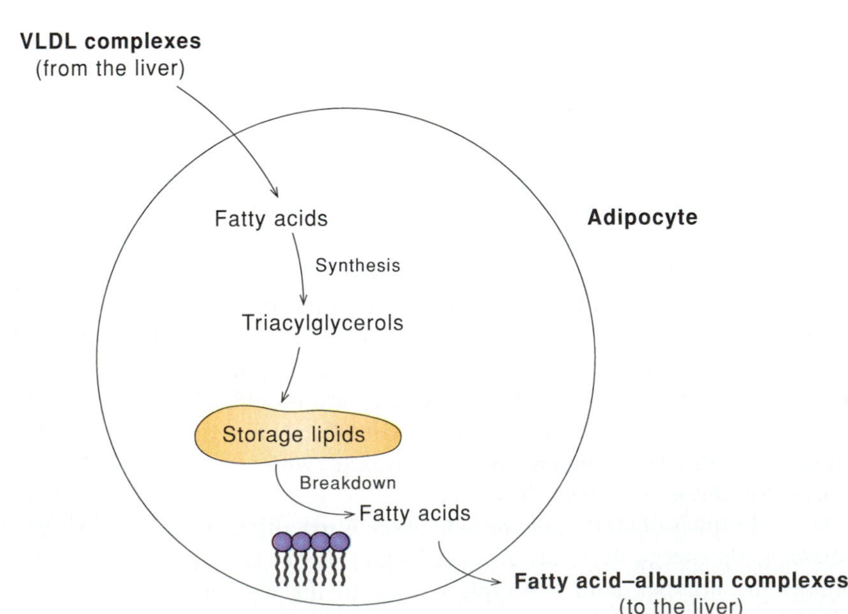

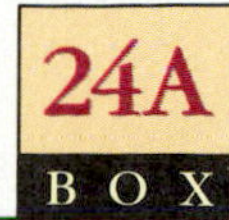

The Lactate Dehydrogenase of Heart Muscle

Most vertebrates possess at least two genes for lactate dehydrogenase that make similar but nonidentical polypeptides called M and H. In embryonic tissue, both genes are equally active, resulting in equimolar amounts of the two gene products and a statistical array of tetramers (M_4, M_3H, M_2H_2, H_3M_1, and H_4 in the ratios of 1:4:6:4:1). These so-called isoenzymes, or isozymes, can usually be detected by differing electrophoretic mobilities. As embryonic tissue multiplies and differentiates, the relative amounts of the M and H forms change. In pure heart tissue, the H_4 tetramer predominates. In skeletal muscle, which functions anaerobically under stress, the M_4 isozyme predominates. It seems likely that the M and H forms were designed to serve different functions. A clue to these functions may be revealed by the inhibiting effect of pyruvate on the dehydrogenase. Pyruvate can form a covalent complex with NAD^+ at the active site of the enzyme according to the following reaction:

$$\text{ADPR}-N^+\text{(pyridine ring, }CONH_2\text{)} + CH_3COCOO^- \longrightarrow \text{ADPR}-N\text{(dihydropyridine ring, H, }CH_2COCOO^-, CONH_2\text{)}$$

The H_4 tetramer shows a much greater inhibition by this compound than the M_4 tetramer. Active muscle tissue is anaerobic and produces a good deal of pyruvate. Inhibition of lactate dehydrogenase under anaerobic conditions would shrink the supply of NAD^+ and shut down the glycolytic pathway (see chapter 13) with disastrous consequences. In fact, active muscle tissue has augmented levels of pyruvate but converts this readily to lactate. This conversion is possible because the predominant form of lactate dehydrogenase in muscle is M_4, which is only poorly inhibited by excess pyruvate. The function of LDH in aerobic tissue is less clear. Heart muscle is aerobic tissue, and consequently, most of its pyruvate is funneled into the Krebs cycle for greater energy production (see chapter 14). The lactate dehydrogenase of heart muscle, H_4, might be inhibited by pyruvate to prevent waste of this potential high-energy carbon source or excessive buildup of pyruvate resulting from the conversion of incoming lactate to pyruvate. It seems likely that the H_4 enzyme is used in such tissues to convert absorbed lactate into pyruvate.

The Liver Is the Central Clearing House for all Energy-Related Metabolism

Glucose is the most readily utilizable energy source of the organism, and it is a primary function of the liver to maintain blood glucose at a reasonable level for absorption into most tissues. Thus in times of glucose excess the liver absorbs glucose and converts it to glycogen, which it stores for future energy needs. At other times, when the glucose concentration in the bloodstream drops, the liver converts glycogen into glucose, which it secretes into the bloodstream. When the blood glucose level falls and the liver's glycogen reserves are also exhausted, the liver still has the capacity to synthesize glucose via gluconeogenesis from amino acids that are supplied from protein breakdown. Under starvation conditions the liver forms increasing amounts of ketone bodies (see fig. 17.10). This is due to elevated concentrations of acetyl-CoA, which favor the formation of ketone bodies. The ketone bodies are secreted and used as a source of energy by other tissues.

The major events in fuel storage and energy utilization are summarized in table 24.2. In this table, energy metabolism is considered under conditions of glucose excess, such as just after a meal; under conditions where the glucose supply is scant, as after prolonged periods between meals; and under starvation conditions, when the glucose supply is severely limited.

The metabolic state of the organism is usually reflected by small molecules present in the plasma. Figure 24.4 illustrates the plasma levels of glucose, ketone bodies, and fatty acids as a function of the number of days of starvation. As you can see, the glucose level drops about 30% as the period of starvation becomes prolonged; the fatty acid level rises about twofold, while the level of ketone bodies rises severalfold.

Pancreatic Hormones Play a Major Role in Maintaining Blood Glucose Levels

Thus far we have argued that it is the liver's job to maintain normal blood glucose levels. But the liver cannot by itself sense the blood glucose levels. It relies on hormonal signals transmitted by the pancreas (fig. 24.5). The pancreas sends two quite different signals. When the blood glucose level is high, the pancreas secretes insulin, which binds to specific receptors located on the outer plasma membranes of insulin-responsive cells (target cells). The effect of insulin on liver cells is to stimulate uptake of excess glucose, which is converted to glycogen. When the blood glucose level is low, the pancreas secretes the protein hormone, glucagon, which binds to different receptors on liver cells. This hormone stimulates the pathway leading to glycogen breakdown and subsequent secretion of glucose by the liver.

Table 24.2
Main Energy-Related Reactions in Five Vertebrate Tissues

	High Blood Glucose	Low Blood Glucose	Starvation
Brain	Glucose ↓ $CO_2 + H_2O$	Glucose ↓ $CO_2 + H_2O$	Ketone bodies ↓ $CO_2 + H_2O$
Heart muscle	Fatty acids ↓ $CO_2 + H_2O$	Fatty acids ↓ $CO_2 + H_2O$	Ketone bodies ↓ $CO_2 + H_2O$
Skeletal muscle	Glucose ↓ $CO_2 + H_2O$ Lactate	Fatty acids ↓ Ketone bodies	Fatty acids ↓ Ketone bodies Proteins ↓ Amino acids
Adipose tissue	Fatty acids ↓ Triacylglycerols	Triacylglycerols ↓ Fatty acids	Triacylglycerols ↓ Fatty acids
Liver	Glucose ↙ ↘ Glycogen Fatty acids	Glycogen ↓ Glucose Fatty acids ↓ $CO_2 + H_2O$	Fatty acids ↓ Ketone bodies Amino acids ↓ Glucose

Figure 24.4

The plasma levels of fatty acids and ketone bodies increase during starvation, whereas the glucose levels decrease.

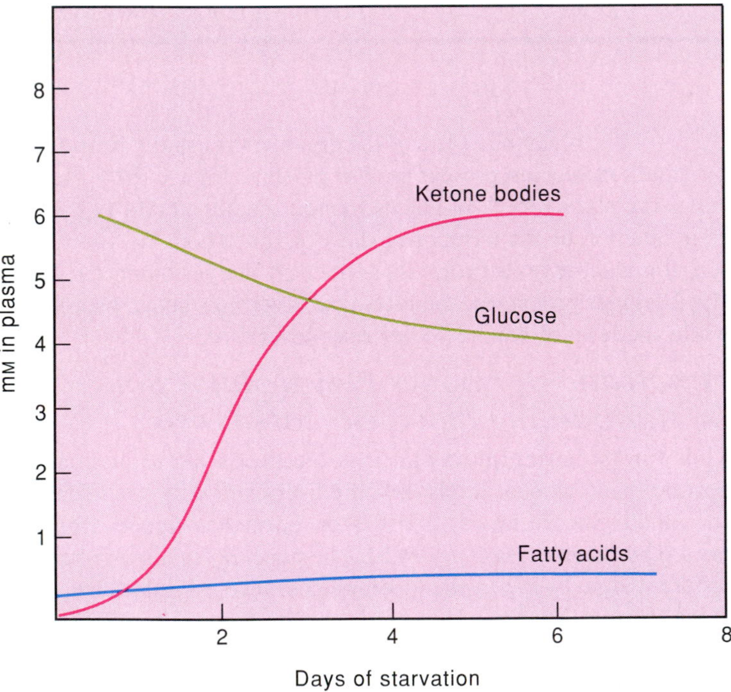

Glucagon and insulin bind to specific receptors on the outer plasma membrane of a target cell. In the case of glucagon, this binding stimulates the enzyme adenylate cyclase, on the inner surface of the membrane, to catalyze the production of cyclic AMP. Depending on the cell type, the cAMP exerts different effects inside the cell. In liver cells cAMP sets off a chain of reactions that results in the breakdown of glycogen to glucose-1-phosphate and subsequent glycolysis (see fig. 13.18). In adipocytes the cAMP also leads to glycogen breakdown, but the main effect in the case of adipocytes is to activate a key enzyme in the pathway for lipid breakdown called triacylglycerol lipase. This enzyme hydrolyzes the triacylglycerol to diacylglycerol with release of fatty acid, the rate-limiting step in the complete hydrolysis of the triacylglycerols. Fatty acids are transported to other tissues (see chapter 23), where they can be metabolized as a source of energy.

Glucagon and insulin are only two of the many hormones that regulate vertebrate metabolism, and cAMP is only one of several so-called second messengers that transduce the hormone signal to the inside of the cell. In the remainder of this chapter we will take a comprehensive look at the hormone systems that provide the main mechanisms for regulating energy metabolism and other metabolic activities as well.

Hormones: Major Vehicles for Intercellular Communication

Hormones, such as glucagon and insulin, override the normal cellular controls.* Whatever the cells were doing before they received their hormonal instruction, their activities are redirected. The result is that hormone-responsive cells do things under the influence of hormones that they would not ordinarily do if they were responding only to their own needs. Usually the action triggered by the hormone brings no immediate gain to the cell acted on, but since it is beneficial for the organism, it ultimately benefits all cells of the organism.

*One of the clearest examples of how a hormonal signal can override an intracellular signal is presented for glycogen phosphorylase in figure 10.20.

Figure 24.5

The opposing effects of insulin and glucagon on the blood glucose level. Insulin and glucagon are both secreted by the pancreas. In many cases they act on the same tissues, but their effects are opposite. Insulin promotes the storage of glucose as glycogen, the use of glucose as an energy source, and the synthesis of proteins and fats. As a result, insulin tends to lower the blood glucose level. Glucagon acts in the opposite direction and its action therefore tends to raise the blood glucose level. The secretion of these hormones is regulated by the blood glucose level. A low blood glucose level favors the secretion of glucagon, and a high blood glucose level favors the

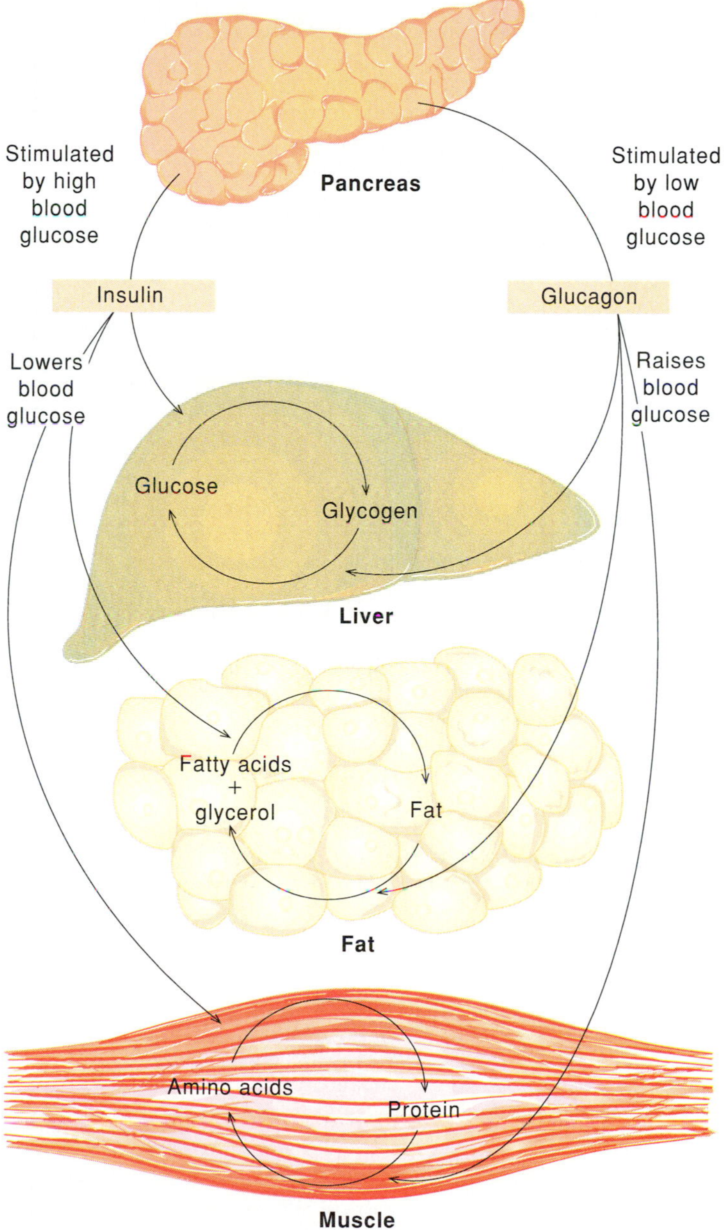

The two hormones that we have just discussed are one part of a larger picture, which includes many hormones that directly affect most tissues in the vertebrate. Each hormone is synthesized in a specialized endocrine gland (fig. 24.6 and table 24.3) and is secreted under special circumstances. Once secreted, the hormone diffuses throughout the entire organism,

Figure 24.6

Location of major endocrine glands in humans. The hypothalamus regulates the anterior pituitary, which regulates the hormonal secretions of the thyroid, adrenals, and gonads (ovary in the female and testis in the male).

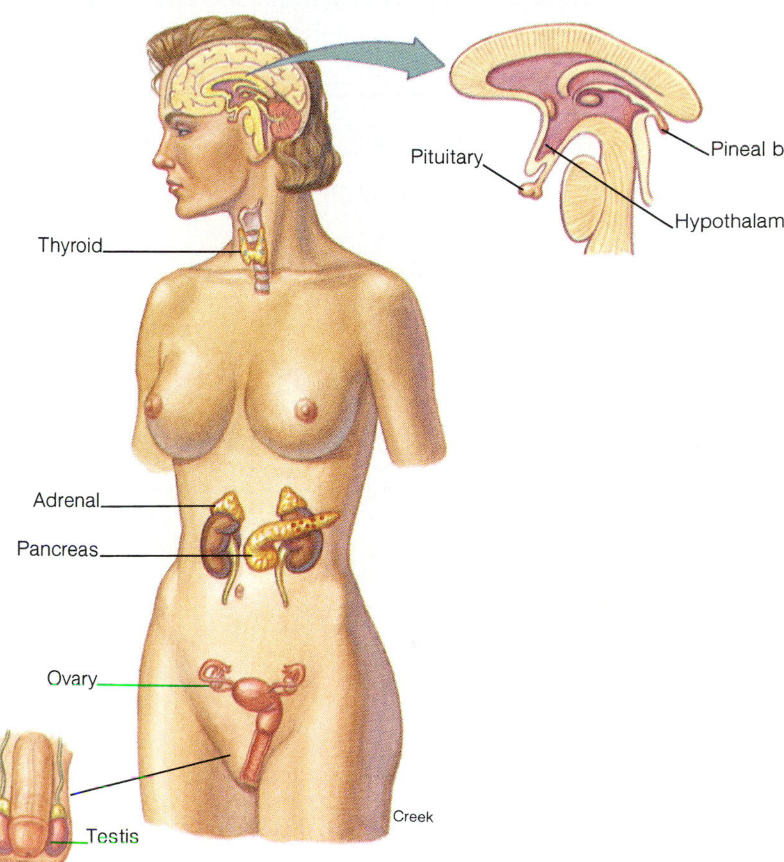

but it triggers reactions only in those cells that carry specific receptors for the hormone.

In the following sections we will discuss how hormones are synthesized, transported, and degraded. Then we will explore basic aspects of hormone–receptor interaction and the more direct biochemical consequences of these interactions. We will also indicate some of the ways in which the hormonal circuits themselves are regulated.

Synthesis and Secretion of Hormones by the Endocrine Glands

Table 24.3 lists many of the better-known hormones of vertebrates. Although many different classes of compounds are used as hormones by animals and plants, most vertebrate hormones fall into one of three classes: polypeptides, amino acid derivatives, and steroids.

Polypeptide Hormones Are Stored in Secretory Granules after Synthesis

Insulin, the first polypeptide hormone to be identified, was discovered by Banting and Best in 1922. They found that this substance, which they isolated from the pancreas, would restore

Table 24.3
Vertebrate Hormones[a]

Hormone	Structure	Function
Pineal		
Melatonin	*N*-Acetyl-5-methoxytryptamine	Regulates circadian rhythms
Hypothalamus[b]		
Corticotropin-releasing factor (CRF or CRH)	Polypeptide (41 residues)	Stimulates ACTH and β-endorphin secretion
Gonadotropin-releasing factor (GnRF) or (GnRH)	Polypeptide (10 residues)	Stimulates LH and FSH secretion
Prolactin-releasing factor (PRF)	(May be TRH)	Stimulates prolactin secretion
Prolactin-release inhibiting factor (PIF)	(May be 56-residue peptide from GnRH precursor)	Inhibits prolactin secretion
Growth hormone-releasing factor (GRF or GRH)	Polypeptide (40 and 44 residues)	Stimulates GH secretion
Somatostatin (Growth hormone-release inhibiting factor, SIF)	Polypeptide (14 and 28 residues)	Inhibits GH and TSH secretion
Thyrotropin-releasing factor (TRF or TRH)	Polypeptide (3 residues)	Stimulates TSH and prolactin secretion
Pituitary		
Oxytocin (ocytocin)	Polypeptide (9 residues)	Uterine contraction, milk ejection
Vasopressin (antidiuretic hormone, ADH)	Polypeptide (9 residues)	Blood pressure, water balance
Melanocyte-stimulating hormones (MSH)	α Polypeptide (13 residues) β Polypeptide (18 residues) γ Polypeptide (12 residues)	Pigmentation
Lipotropin (LPH)	β Polypeptide (93 residues) γ Polypeptide (60 residues)	Fatty acid release from adipocytes
Corticotropin (adrenocorticotropic hormone, ACTH)	Polypeptide (39 residues)	Stimulates adrenal steroid synthesis
Thyrotropin (thyroid-stimulating hormone, TSH)	2 Polypeptides (α, 96 residues; β, 112 residues)	Stimulates thyroid hormone synthesis
Growth hormone (GH)	Polypeptide (191 residues)	General anabolic effects; stimulates release of insulinlike growth factor-I
Prolactin	Polypeptide (197 residues)	Stimulates milk synthesis
Luteinizing hormone (LH)	2 Polypeptides (α, 96 residues; β, 121 residues)	Ovary: luteinization, progesterone synthesis; testis: interstitial cell development, androgen synthesis
Follicle-stimulating hormone (FSH)	2 Polypeptides (α, 96 residues; β, 120 residues)	Ovary: follicle development, ovulation, estrogen synthesis; testis: spermatogenesis
Thyroid		
Thyroxine and triiodothyronine	Iodinated dityrosine derivatives (see fig. 24.9)	General stimulation of many cellular reactions
Calcitonin	Polypeptide (32 residues)	Ca^{2+} and P_i metabolism
Calcitonin gene related peptide (CGRP)	Polypeptide (37 residues)	Vasodilator
Parathyroid		
Parathyroid hormone (PTH)	Polypeptide (84 residues)	Ca^{2+} and P_i metabolism

[a]Only the more common hormones of known structure are listed.
[b]Most of the hypothalamic releasing factors are also called hypothalamic regulatory hormones.

Hormone	Structure	Function
Alimentary tract[c]		
Gastrin	Polypeptide (17 residues)	Stimulates acid secretion from stomach and pancreatic secretion
Secretin	Polypeptide (27 residues)	Regulates pancreas secretion of water and bicarbonate
Cholecystokinin	Polypeptide (33 residues)	Secretion of digestive enzymes
Motilin	Polypeptide (22 residues)	Controls gastrointestinal muscles
Vasoactive intestinal peptide (VIP)	Polypeptide (28 residues)	Gastrointestinal relaxation; inhibits acid and pepsin secretion
Gastric inhibitory peptide (GIP)	Polypeptide (43 residues)	Inhibits gastrin secretion
Somatostatin	Polypeptide (14 residues)	Inhibits gastrin secretion; inhibits glucagon secretion
Heart		
Atrial natriuretic peptide (ANP)	Several active peptides cleaved from precursor polypeptide of 126 residues	Smooth muscle relaxation; diuretic activity
Pancreas		
Insulin	2 Polypeptides (21 and 30 residues)	Glucose uptake, lipogenesis, general anabolic effects
Glucagon	Polypeptide (29 residues)	Glycogenolysis, release of lipid
Pancreatic polypeptide	Polypeptide (36 residues)	Glycogenolysis, gastrointestinal regulation
Somatostatin	Polypeptide (14 residues)	Inhibition of somatotropin and glucagon release
Adrenal cortex		
Glucocorticoids	Steroids (cortisol, corticosterone)	Many diverse effects on protein synthesis and inflammation
Mineralocorticoids	Steroids (aldosterone)	Maintains salt balance
Adrenal medulla		
Epinephrine	Tyrosine derivative (see fig. 24.10)	Smooth-muscle contraction, heart function, glycogenolysis, lipid release
Norepinephrine	Tyrosine derivative (see fig. 24.10)	Arteriole contraction, lipid release
Gonads		
Estrogens (ovary)	Steroids (estradiol, estrone)	Maturation and function of secondary sex organs
Progestins (ovary)	Steroids (progesterone)	Ovum implantation, maintenance of pregnancy
Androgens (testes)	Steroids (testosterone)	Maturation and function of secondary sex organs
Inhibins A and B	1 Polypeptide (α, 134 residues; β, 115 and 116 residues)	Inhibit FSH secretion
Placenta		
Estrogens Progestins	Steroids	Maintenance of pregnancy
Choriogonadotropin	2 Polypeptides (α, 96 residues; β, 147 residues)	Similar to LH
Placental lactogen	Polypeptide (191 residues)	Similar to prolactin
Relaxin	2 Polypeptides (22 and 32 residues)	Muscle tone
Liver		
Angiotensin[d]	Polypeptide (8 residues)	Responsible for essential hypertension
Kidney		
1,25-dihydroxyvitamin D_3	Steroid	Calcium uptake, bone formation

[c]Many of these peptides are also found in the brain, where they may modulate neural activity.

[d]The liver secretes α_2-globulin, which is cleaved by renin, a kidney enzyme, to give a decapeptide, proangiotensin, from which the carboxyl-terminal dipeptide is removed to give angiotensin.

Figure 24.7

Processing pathway of preproopiomelanocortin. This precursor polypeptide is cleaved into a variety of active peptides. With the exception of the signal peptidase cleavage site, the cleavage sites are generally pairs of basic amino acids, although one site contains four. Which active peptides are produced depends on the processing pathway, which varies in different cell types. Thus in the anterior and intermediate lobes of the pituitary gland, the precursor polypeptide is cleaved to yield corticotropin and ß-lipoprotein. In the intermediate lobe only, these polypeptide hormones are further cleaved to yield γ-MSH, α-MSH, γ-lipotropin, and ß-endorphin.

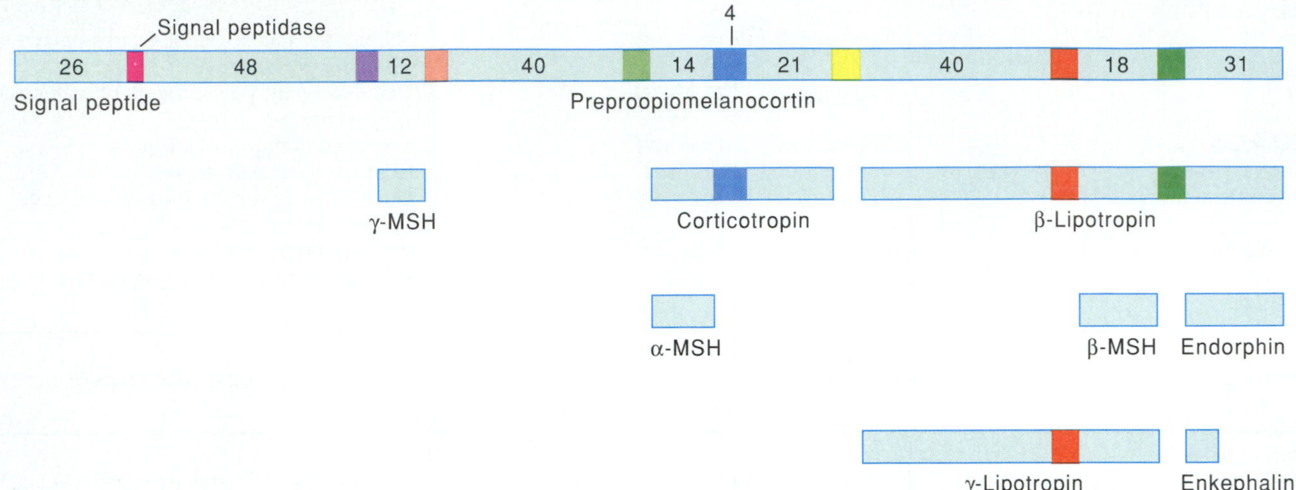

normal glucose utilization in experimental animals lacking a pancreas. Insulin was also the first protein to be sequenced, a landmark accomplishment achieved by Fred Sanger in 1955. About twenty years later, Steiner discovered that the two polypeptide chains of insulin are synthesized as a single polypeptide, proinsulin, which folds and is cross-linked by disulfide bonds (see fig. 29.21). An internal peptide is then removed by the concerted action of specific proteases. All of these events occur within the pancreatic β cells that synthesize insulin.

With the advent of techniques to isolate and translate mRNA in cell-free systems, it was discovered that the primary translation product of insulin mRNA—called preproinsulin—is even larger than proinsulin (see chapter 29). Like all secreted polypeptide hormones, proinsulin is synthesized with a hydrophobic signal sequence at the amino terminal that directs the nascent polypeptide into the endoplasmic reticulum and is then removed. Surprisingly, in some cases several different peptide hormones can be liberated from the same precursor by proteolytic processing. A striking example is preproopiomelanocortin (fig. 24.7), which is a precursor for corticotropin (ACTH), β-lipotropin, three melanocyte-stimulating hormones (MSH), endorphin, and an enkephalin. When several different polypeptide hormones are cleaved from a common precursor, the cleavage pattern can vary to yield a different spectrum of peptides, depending on the cell type. Proteolytic processing of these precursors often occurs at the site of dibasic amino acids.

All polypeptide hormones are synthesized from mRNA as precursors, which contain signal peptides that direct them into the lumen of the endoplasmic reticulum. In a few cases (e.g., growth hormone and prolactin) there is no further processing. However, in most cases further processing is required. The peptides synthesized by neurosecretory cells of the hypothalamus, for example, are often much larger than the final hormones. Oxytocin and vasopressin, each nine residues long, represent the amino terminals of precursors called proneurophysins, which are 160 and 215 amino acids long, respectively. After oxytocin or vasopressin is cleaved from proneurophysin in the hypothalamus, where it is synthesized, the hormone remains associated with the neurophysin as it passes down the axons to the posterior pituitary, where it is secreted. The neurophysin may serve to protect the hormone from degradation prior to secretion. Somatostatin, another hypothalamic hormone, is a 14-amino-acid product cleaved from the carboxyl terminal of a precursor that contains 121 amino acids. Thyrotropin regulatory hormone (TRH) is the smallest known polypeptide hormone (pyroGlu-His-ProNH_2). In the synthesis of TRH, the glutamyl residue is cyclized and the carboxyl-terminal amide originates from an adjacent glycine. The precursor polypeptide (255 amino acids) contains five copies of the sequence Lys-Arg-Gln-His-Pro-Gly-Arg-Arg within it (fig. 24.8).

Polypeptide hormones are usually stored in secretory granules after their passage through the endoplasmic reticulum and Golgi. Release of these hormones into the bloodstream is accomplished by fusing the secretory granule membranes with the plasma membrane. This event is often regulated by other hormones.

Thyroid Hormones and Epinephrine Are Amino Acid Derivatives

Thyroxine (T_4) and the more potent triiodothyronine (T_3) are cleaved from a large precursor protein called thyroglobulin. Thyroglobulin exists as a dimer of two identical polypeptides

Figure 24.8

Biosynthesis of thyrotropin-releasing hormone (TRH). Five copies of TRH are contained within a 255-amino-acid precursor polypeptide (pre-pro-TRH). The precursor has a hydrophobic signal peptide (dark green) and five copies of the sequence Lys-Arg-Gln-His-Pro-Gly-Arg-Arg (light green). The dibasic amino acids are recognized by specific proteases to liberate Gln-His-Pro-Gly, which is subsequently converted into TRH. The amino-terminal glutamine is converted into pyroglutamine, and the carboxyl-terminal amide is derived from the neighboring glycine, which is removed by a specific enzyme.

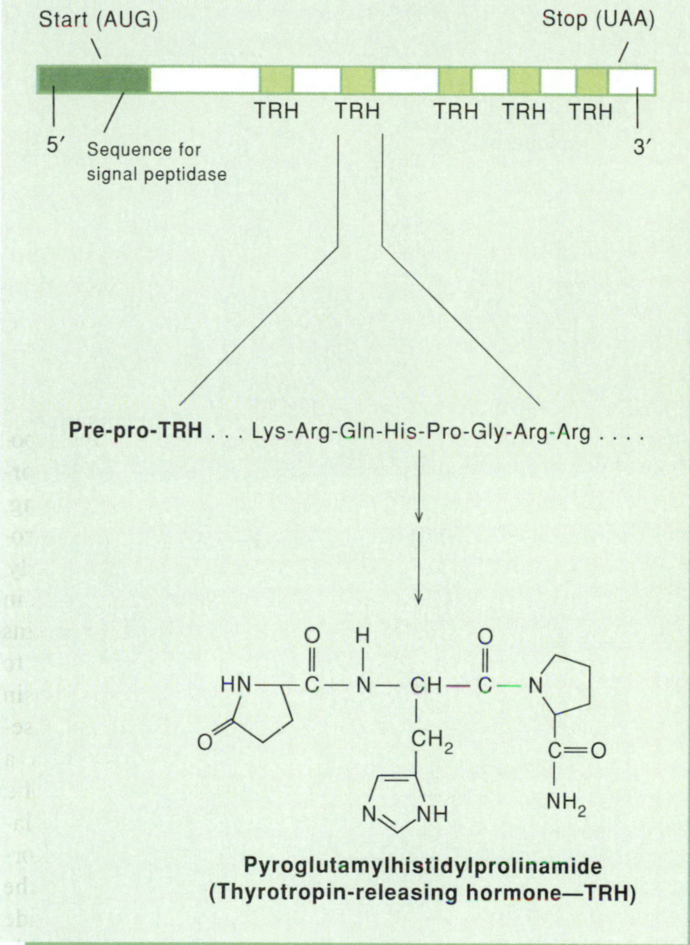

($M_r \approx$ 330,000). It is a storage protein for iodine and can be considered a prohormone of the circulating thyroid hormones. Thyroglobulin is secreted into the lumen of the thyroid gland, where several of the tyrosine residues are iodinated in one or two positions by a special peroxidase; then two iodinated residues condense as shown in figure 24.9.

The secretion of thyroid hormones starts with endocytosis of the modified thyroglobulin, followed by fusion of the endocytotic vesicles with lysosomes. The lysosomal enzymes then degrade the thyroglobulin, liberating triiodothyronine and thyroxine into the circulation. Only about five molecules of T_3 and T_4 are generated from each molecule of thyroglobulin. Thyroid hormone secretion is stimulated by thyrotropin (TSH), a pituitary hormone that activates adenylate cyclase in its target cells.

Epinephrine, originally called adrenalin, was the first hormone to be isolated, characterized, and synthesized. Epinephrine and its precursor, norepinephrine, are synthesized from tyrosine in the chromaffin cells of the adrenal medulla. They are also synthesized by neurons of the central and peripheral nervous system. The biosynthetic pathway is shown in figure 24.10. The first step, which involves oxidation of tyrosine to 3,4-dihydroxyphenylalanine (dopa), is catalyzed by tyrosine hydroxylase. This is the rate-limiting enzyme in the pathway. The amount and activity of tyrosine hydroxylase are regulated by cAMP-dependent mechanisms that are responsive to the neurotransmitter acetylcholine. The latter is liberated by special neurons that impinge upon the chromaffin cells. Second, dopa is decarboxylated to dopamine, which is then β-hydroxylated to produce norepinephrine. Finally, the N-methylation of norepinephrine (*S*-adenosylmethionine is the methyl donor) produces epinephrine. Both of these catecholamines (norepinephrine and epinephrine) are stored in chromaffin granules, where they are complexed with ATP and proteins called chromogranins. Neural stimulation of the medulla is mediated by acetylcholine, which binds to receptors on the membranes of medullary cells (see chapter 35). This event leads to a local depolarization and an influx of calcium. As a result, the chromaffin granules fuse with the cell membrane, and a packet of catecholamines, ATP, and protein is extruded into the extracellular fluid.

Steroid Hormones Are Derived from Cholesterol

Steroid hormones are derived from cholesterol by a stepwise removal of carbon atoms and hydroxylation. The steroid hormones are synthesized by cells of the adrenal cortex (in the case of glucocorticoids and mineralocorticoids) and the gonads (in the case of estrogens, progestins, and androgens). The hormone names just mentioned are generic names for entire classes of compounds that interact with specific receptors; for example, estrogen is the generic name for a family that incudes 17β-estradiol, estrone, and diethylstilbestrol (DES), a synthetic nonsteroidal estrogen.

The step-by-step synthesis of steroid hormones from cholesterol (C_{27}) was presented in chapter 23 (see figs. 23.25 through 23.28). Note that pregnenolone (C_{21}) and progesterone (C_{21}) are intermediates in the biosynthesis of all of the major adrenal steroids, including cortisol (C_{21}), corticosterone (C_{21}), and aldosterone (C_{21}). The same two compounds are intermediates in the synthesis of the gonadal steroid hormones, testosterone (C_{19}) and 17β-estradiol (C_{18}). Because the synthesis of all these hormones follows a common pathway, a defect in the activity or amount of an enzyme along that pathway can lead to both a deficiency in the hormones beyond the affected step and an excess of the hormones, or metabolites, prior to that step.

Figure 24.9

Pathway of thyroxine (T_4) and triiodothyronine (T_3) synthesis. Thyroid cells actively transport iodine (I-), which is incorporated into a few tyrosine residues of thyroglobulin by the enzyme iodoperoxidase. After condensation of iodinated tyrosine residues, the thyroglobulin is proteolytically degraded, liberating thyroxine and triiodothyronine.

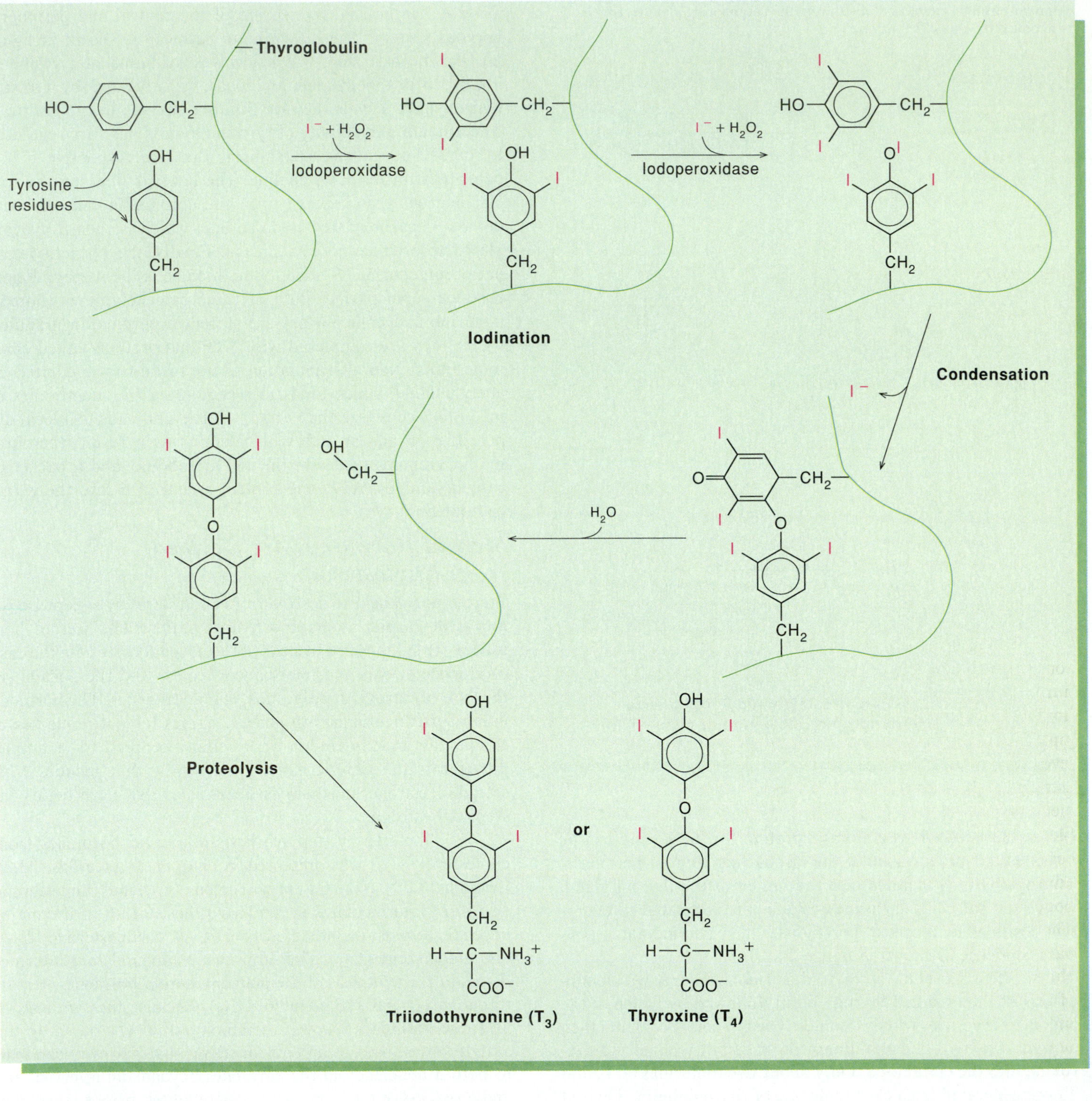

Figure 24.10

Pathway of epinephrine synthesis. Epinephrine and its precursor, norepinephrine, are synthesized from tyrosine. The synthesis occurs in the chromaffin cells of the adrenal medulla and in neurons of the central and peripheral nervous system. The first step, which is catalyzed by tyrosine hydroxylase, is the rate-limiting step in the pathway.

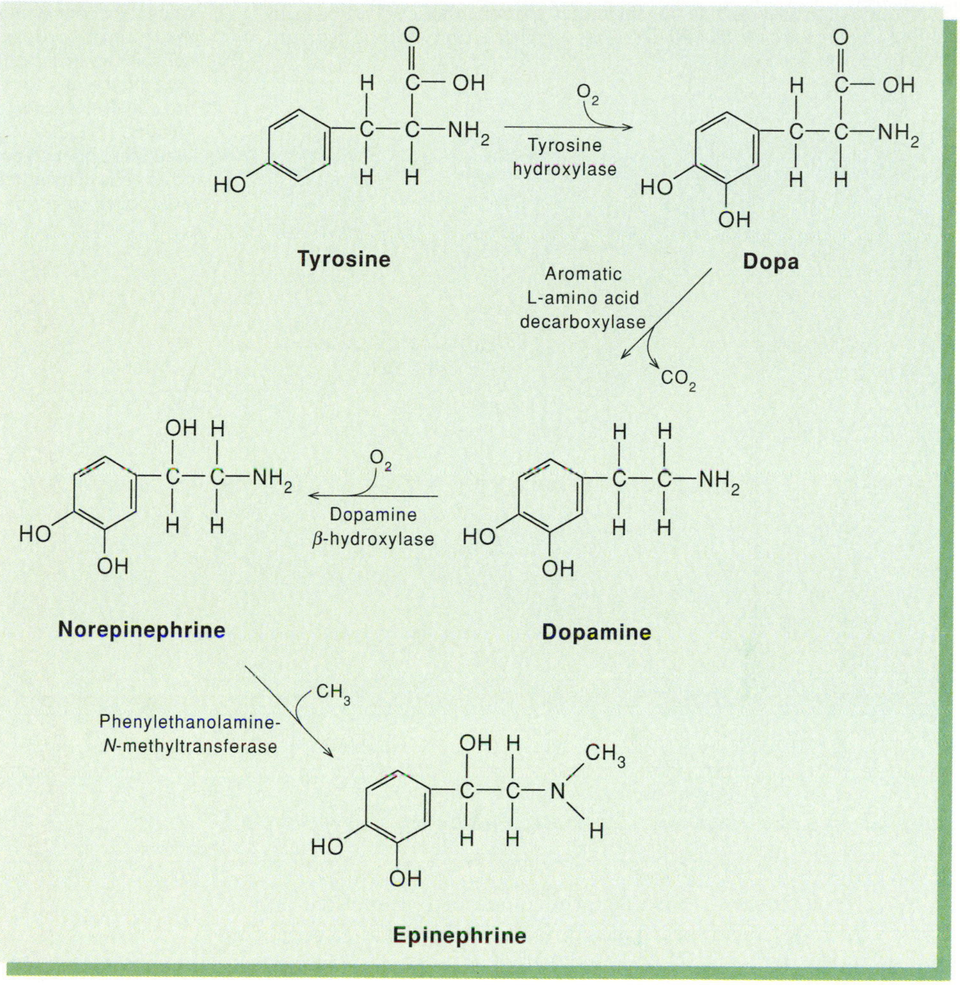

Deficiencies in each of the six enzymes involved in the conversion of cholesterol to aldosterone have been observed in humans. Each deficiency gives rise to a characteristic steroid hormone imbalance, with telling clinical consequences. For example, a deficiency in 17-hydroxylase gives rise to inadequate levels of cortisol as well as inadequate levels of androgens and estrogens, with severe effects on sexual maturation. A deficiency in the next enzyme along the pathway, 21-hydroxylase, blocks the synthesis of adrenal glucocorticoids and mineralocorticoids and leads to an overproduction of testosterone by the adrenals. This overproduction of testosterone is due to metabolic shunting of progesterone into the sex steroid pathway. The synthesis of androgens is exacerbated by the lack of feedback inhibition of cortisol on the hypothalamus; the result is chronic production of CRF and ACTH and perpetual activation of adrenal steroid biosynthesis. In females, excessive androgen production, due to a defect in 21-hydroxylase, leads to masculinization or, to use the clinical term, female pseudohermaphrodism—that is, a genetic female with male appearance. This reversal of sexual phenotype is explained by the fact that during embryonic development of mammals, the external genitalia develop from common precursor cells. In the absence of hormonal stimulation, they will develop into female structures, but androgens will direct their development into male structures.

Testosterone is both a hormone and a prohormone. The high levels of testosterone normally produced in the male by the testes play a major role in the growth and function of many tissues in addition to reproductive organs. Essentially all the sexual differences in nonreproductive tissues, such as muscle, liver, and brain, are a consequence of androgen action. Although testosterone is the major circulating androgen, many target cells reduce this steroid to 5α-dihydrotestosterone, a steroid that binds to the androgen receptor with higher affinity than testosterone (fig. 24.11). When 5α-reductase is defective, the androgen receptors are only partially activated and a full androgen response is not obtained. A deficiency in this enzyme therefore leads to abnormal development of male genitalia, i.e., they are of female phenotype (a clinical condition referred to as male pseudohermaphrodism, type 2). In this example, testosterone can be considered a prohormone of a more active androgen.

Figure 24.11

Conversion of testosterone to 5α-dihydrotestosterone (5α-DHT). Receptor-binding studies indicate that 5α-DHT has a higher affinity for the androgen receptor than testosterone.

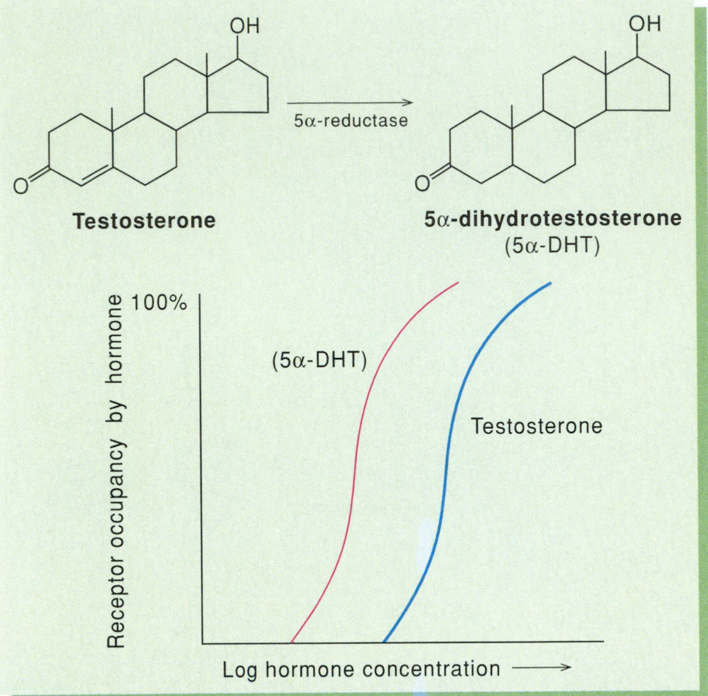

Testosterone is also a prohormone of metabolites that bind to different receptors. For example, some of the effects of androgens on the production of red blood cells (erythropoiesis) are due to the reduction of testosterone to 5β-dihydrotestosterone within precursor cells. 5β-Dihydrotestosterone binds to a receptor that is distinct from the one that binds testosterone and 5α-dihydrotestosterone (see fig. 24.10). Testosterone also influences erythropoiesis by stimulating the kidney to produce a specific growth factor, erythropoietin. Another striking example of testosterone as a prohormone occurs in the brain, where testosterone influences neural development and activity (e.g., male-specific mating behavior and bird songs) by being converted into 17β-estradiol, which interacts with estrogen receptors. Testosterone is metabolized to 17β-estradiol by aromatase, the same enzyme involved in 17β-estradiol synthesis in the ovary (fig. 24.12).

Cell specificity of mineralocorticoid action is achieved in a different manner. Aldosterone, cortisol, and corticosterone bind with similar affinities to mineralocorticoid and glucocorticoid receptors. However, aldosterone activates only its own receptor in target tissues, e.g., kidney, because of an enzyme, 11β-hydroxysteroid dehydrogenase, that converts the prevalent glucocorticoids into inactive 11-keto derivatives but does not affect aldosterone.

Figure 24.12

Metabolic conversion of testosterone by target cells. Testosterone (T) is the predominant androgen in the bloodstream. When testosterone enters target cells, it can be metabolized in a variety of different ways. It can either (*a*) bind directly to androgen receptors (R^a) or (*b*) be reduced to 5α-dihydrotestosterone (5α-DHT), which then binds to R^a with higher affinity. (*c*) Other target cells reduce testosterone to 5ß-dihydrotestosterone (and other 5ß metabolites), which bind to a distinct receptor ($R^ß$). (*d*) Yet other cells convert testosterone into an estrogen, 17ß-estradiol, which binds to estrogen receptors (R^e).

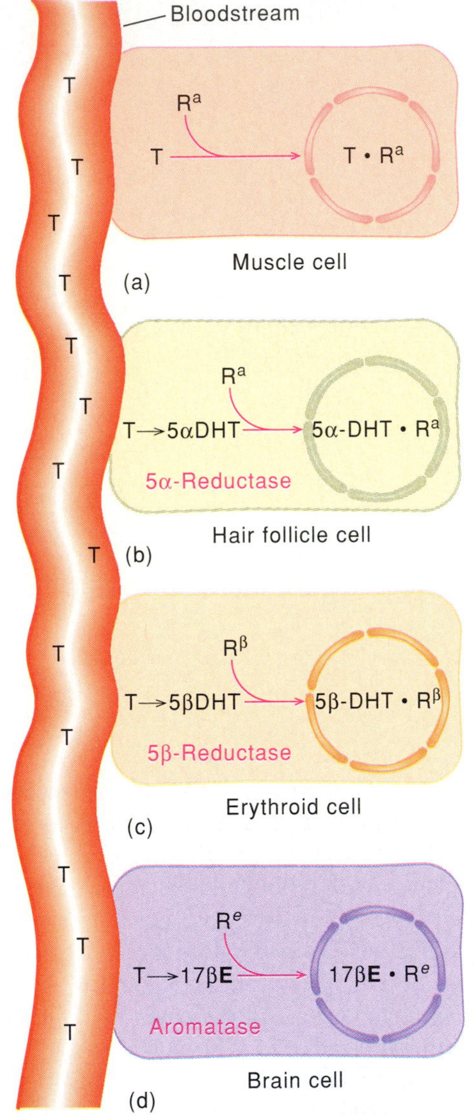

Vitamin D_3 is a precursor of the hormone 1,25-dihydroxyvitamin D_3. Vitamin D_3 is essential for normal calcium and phosphorus metabolism. It is formed from 7-dehydrocholesterol by ultraviolet photolysis in the skin. Insufficient exposure to sunlight and absence of vitamin D_3 in the diet leads to rickets, a condition characterized by weak, malformed bones. Vitamin D_3 is inactive, but it is converted into an active compound by two hydroxylation reactions that occur in different

Figure 24.13

The conversion of vitamin D_3 to an active compound. Vitamin D_3 is formed from 7-dehydrocholesterol by ultraviolet photolysis in the skin. Vitamin D_3 is inactive, but it is converted into an active compound by two hydroxylation reactions that occur in different organs. The first reaction occurs in the liver and results in 25-hydroxyvitamin D_3. The second hydroxylation occurs in the kidney and results in 1,25-dihydroxyvitamin D_3.

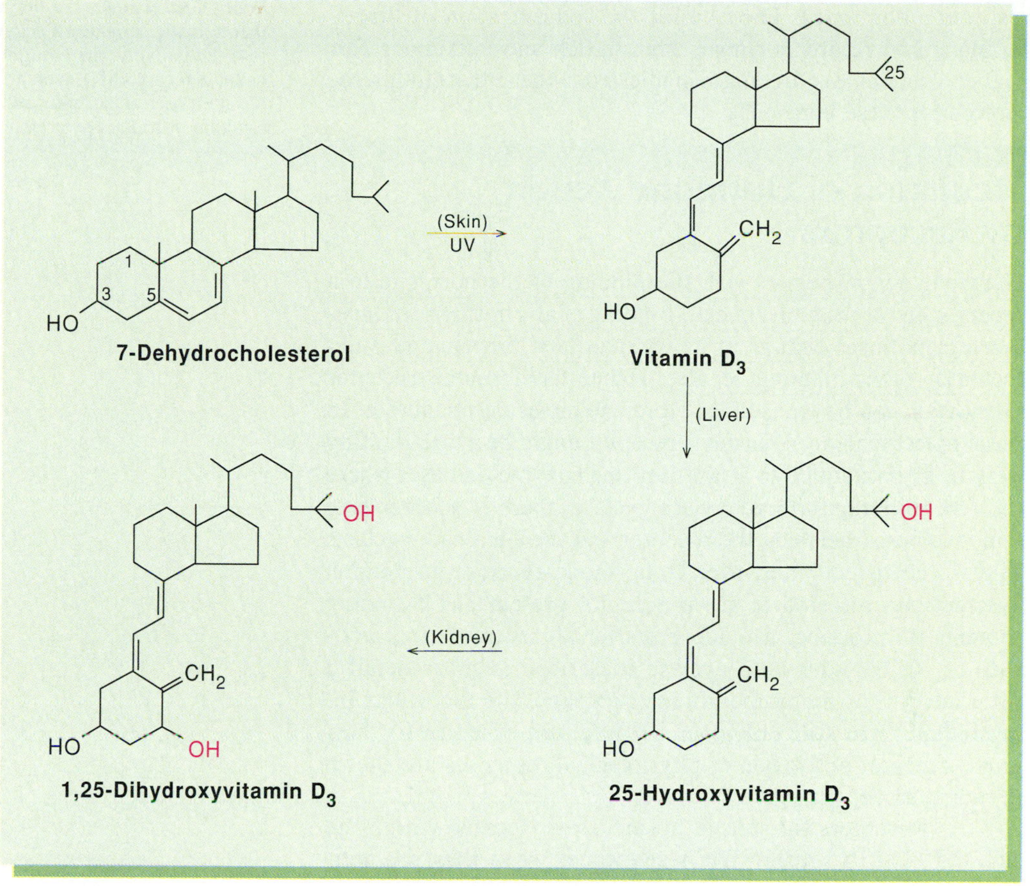

organs. The first hydroxylation occurs in the liver, which produces 25-hydroxyvitamin D_3, abbreviated $25(OH)D_3$; the second hydroxylation occurs in the kidney and gives rise to the active product 1,25-dihydroxyvitamin D_3 $24,25(OH)_2D_3$ (fig. 24.13). The hydroxylation at the 1 position that occurs in the kidney is stimulated by parathyroid hormone (PTH), which is secreted from the parathyroid gland in response to low circulating levels of calcium. In the presence of adequate calcium, $25(OH)D_3$ is converted into an inactive metabolite, $24,25(OH)_2D_3$. The active derivative of vitamin D_3 is considered a hormone because it is transported from the kidneys to target cells, where it binds to nuclear receptors that are analogous to those of typical steroid hormones. $1,25(OH)_2D_3$ stimulates calcium transport by intestinal cells and increases calcium uptake by osteoblasts (precursors of bone cells).

Regulation of the Circulating Hormone Concentration

The occupancy of hormone receptors can fluctuate greatly and is ultimately determined by the concentration of "free" hormone in the blood. The major determinants of hormone concentrations are (1) the rate of hormone secretion from endocrine cells and (2) the rate of hormone removal by clearance or metabolic inactivation. As we have seen, most hormones (with the exception of steroids) are stored in secretory granules. When the hormone is needed, the granule membranes fuse with the plasma membrane to liberate their contents into the bloodstream. This event is triggered by signals from other hormones or by neural signals. Stimulation of hormonal secretion is usually coupled with an increase of hormone synthesis, so that hormonal stores are replenished.

Most hormones have a half-life in the blood of only a few minutes because they are cleared or metabolized very rapidly. The rapid degradation of hormones allows target cells to respond transiently. Polypeptide hormones are removed from the circulation by serum and cell-surface proteases, by endocytosis followed by lysosomal degradation, and by glomerular filtration in the kidney. Steroid hormones are taken up by the liver and metabolized to inactive forms, which are excreted into the bile duct or back into the blood for removal by the kidneys. Catecholamines are metabolically inactivated by O-methylation, by deamination, and by conjugation with sulfate or glucuronic acid.

Thyroid hormones and most steroid hormones are associated with carrier proteins in the serum. The carrier proteins are called, appropriately, thyroxine-binding globulin, transcortin (for cortisol), and sex-steroid-binding protein. These proteins have a high affinity ($K_d \sim 10^{-9}$ to 10^{-8} M) for their respective hormones. They buffer the concentration of "free" hormone and retard hormone degradation and excretion. The carrier proteins are distinguishable from the intracellular receptors for these hormones.

Mediation of Hormone Action by Receptors

Hormone action begins with the binding of the hormone to a receptor on (or in) a target cell. Binding of any hormone induces a conformational change in its receptor, and this change is detected by other macromolecules. Hence the hormone–receptor interaction can be transduced from one molecule to another. In the simplest scheme, a hormone receptor might be a rate-limiting enzyme or be coupled to a rate-limiting enzyme. In cases where the receptors regulate adenylate cyclase, there is another protein interposed between the receptor and the adenylate cyclase. The protein directly activated by hormone–receptor interaction is sometimes referred to as an acceptor protein. GTP-binding proteins (G proteins) are acceptors for all receptors that activate or inhibit adenylate cyclase (described in more detail a little later). For many membrane receptors, the molecular intermediates are still unknown. Generalized schemes for hormone–receptor activation of physiological responses are shown in figure 24.14.

Receptors for steroid hormones are located within the cell and bind to specific DNA sequences when they are activated by the appropriate hormones. The binding of these receptors to DNA in the promoter region of a gene activates transcription, in most cases by helping to assemble an efficient initiation complex. The acceptor proteins in these cases may be other transcription factors.

The important point is that all hormones act by binding to receptors that are located either on the cell membrane or inside the cell. Binding of the hormone induces a conformational change in the receptor that is transmitted to its active site (if it is an enzyme) or to other macromolecules. In either case, a chain of events is elicited that ultimately affects a vast array of metabolic processes, ranging from alterations in enzyme activities to changes in gene expression. With time, these changes may lead to profound alterations in cell growth, morphology, and function.

Most Membrane Receptors Generate a Diffusible Intracellular Signal

Activation of most membrane-associated hormone receptors generates a diffusible intracellular signal called a second messenger. Many different receptors generate the same second messenger. Five intracellular messengers are currently known: cyclic AMP, cyclic GMP, inositol triphosphate, diacylglycerol, and calcium. There are a few membrane receptors for which second messengers do not exist or have not yet been identified (table 24.4).

Figure 24.14

Possible mechanisms of hormone action. (*a*) The hormone (H) could theoretically activate an enzyme (E) directly as an allosteric effector. (*b*) Alternatively, there might be a separate binding protein for the hormone, called a receptor (R), which then activates an enzyme. (*c*) Another possibility interposes an acceptor protein (A) between the receptor and the enzyme. Each interaction is reversible.

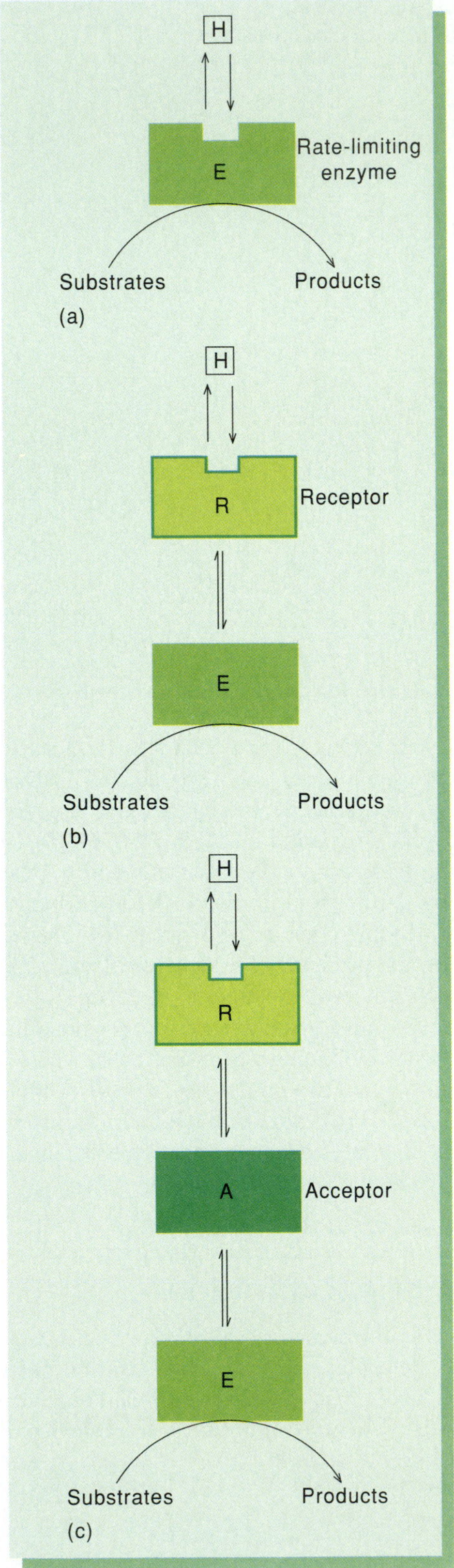

Table 24.4
Hormones for Which Second Messengers Have Not Been Found

Catecholamines (acting on α_1 receptors)
Placental lactogen
Growth hormone
Insulin
Oxytocin
Prolactin
Somatostatin

Table 24.5
Hormones that Activate or Inhibit Adenylate Cyclase

Activators
Corticotropin (ACTH)
Calcitonin
Catecholamines (acting on β_1 and β_2 receptors)
Choriogonadotropin
Follicle-stimulating hormone (FSH)
Glucagon
Gonadotropin-releasing hormone (GnRH)
Growth hormone-releasing hormone (GRH)
Luteinizing hormone (LH)
Lipotropin (LPH)
Melanocyte-stimulating hormones (MSH)
Parathormone (PTH)
Secretin
Thyrotropin regulatory hormone (TRH)
Thyrotropin (TSH)
Vasoactive intestinal peptide (VIP)
Vasopressin
Inhibitors
Angiotensin
Catecholamines (acting on α_2 receptors)

The Adenylate Cyclase Pathway Is Triggered by a Membrane-Bound Receptor

Many hormones bind to receptors that act through an intermediary protein to either activate or inhibit adenylate cyclase (table 24.5). Cyclic AMP (cAMP) is formed from ATP by adenylate cyclase, which is bound to the inside of the cell membrane.

As we have indicated, the interaction between the hormone receptor and the adenylate cyclase is mediated by a G protein. Most G proteins are heterotrimers of α, β, and γ subunits that are associated with the membrane and can make contact with both the membrane receptor and an effector molecule such as adenylate cyclase as shown in figure 24.15. There are

Figure 24.15

The adenylate cyclase pathway of hormone receptor action. When the receptor is unoccupied, the G protein α subunit has GDP bound and it is complexed with the subunits; in this form it cannot activate adenylate cyclase. Binding of hormone activates the receptor, which leads to replacement of GDP by GTP and the activation subunit then interacts productively with adenylate cyclase to stimulate the synthesis of cAMP. The intrinsic GTPase activity of the α subunit leads to GDP production and the inactivation of the G protein. If hormone remains bound, the G protein can be activated again. Meanwhile the cAMP produced binds to the R subunits of cAMP-dependent protein kinase, leading to the dissociation of two catalytic subunits, which can then phosphorylate critical proteins; in this case the activation of phosphorylase kinase is shown.

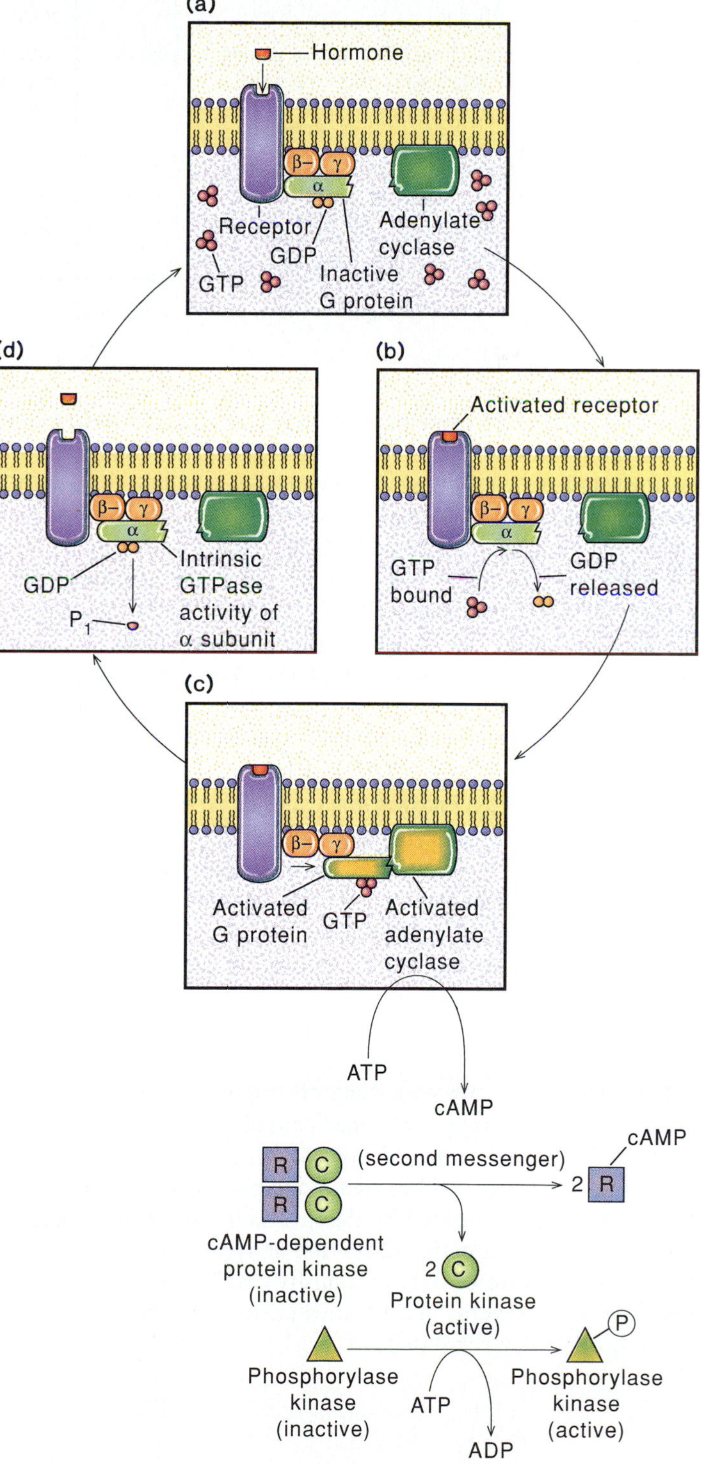

Table 24.6
G Protein Receptors and Effectors[a]

G Protein	Receptors for	Effectors	Signaling pathways
G_s	Adrenaline, noradrenaline, histamine, glucagon, ACTH, luteinizing hormone, follicle-stimulating hormone, thyroid-stimulating hormone, and others	Adenylyl cyclase Ca^{2+} channels	↑ cAMP ↑ Ca^{2+} influx
G_{olf}	Odorants	Adenylyl cyclase	↑ cAMP (olfaction)
G_{t1} (rods)	Photons	cGMP phosphodiesterase	↓ cGMP (vision)
G_{t2} (cones)	Photons	cGMP phosphodiesterase	↓ cGMP (color vision)
G_{i1}, G_{i2}, G_{i3}	Noradrenaline, prostaglandins, opiates, angiotensin, many peptides	Adenylyl cyclase Phospholipase C Phospholipase A_2 K^+ channels	↓ cAMP ↑ Inositol triphosphate, diacylglycerol, Ca^{2+} Arachidonate release Membrane polarization
G_0	Probably many, but not yet defined	Phospholipase C Ca^{2+} channels	↑ Inositol triphosphate, diacylglycerol, Ca^{2+} ↓ Ca^{2+} influx
Gq	Noradrenalin, vasopressin	Phospholipase C	↑ Inositol bisphosphate, diacylglycerol, Ca^{2+}

[a]H. R. Bourne et al., "The GTPase superfamily: A conserved switch of diverse cell functions" in *Nature,* 348:126, 1990. Copyright © 1990 Macmillan Magazines Ltd., London, England.

several classes of G proteins with different specificities (table 24.6). In the absence of hormone activation, the α subunit of the G protein binds GDP at an allosteric site. Upon binding of hormone, the conformation of the α subunit changes so that the GDP is displaced and replaced by GTP. The GTP-bound α subunit possibly dissociates from the β, γ subunits allowing it to activate an effector molecule. In the case of receptors coupled with G_s, the α_s-GTP complex activates adenylate cyclase, which proceeds to synthesize a burst of cAMP. The period of activation is brief, as the α subunit also contains a GTPase activity that hydrolyzes a phosphate residue from the GTP, thereby returning the α subunit to its original inactive state. Further cAMP synthesis is contingent on the same chain of reactions that began with hormone binding to the G protein.

The cAMP formed on the inner side of the plasma membrane diffuses into the cytoplasm and binds to regulatory subunits of cAMP-dependent protein kinase. This enzyme is composed of four subunits: two catalytic subunits, C; and two regulatory subunits, R. Binding of cAMP to the regulatory subunits promotes the dissociation of the two catalytic subunits, which then become active (see fig. 24.14). The catalytic subunits phosphorylate a wide variety of proteins on serine or threonine residues, inducing conformational changes that alter their function. Many substrates of the cAMP-dependent kinase are rate-limiting enzymes in key metabolic pathways. The activity of the kinase is determined by the intracellular cAMP concentration, which is a function of its rate of synthesis by adenylate cyclase and its rate of degradation by cAMP phosphodiesterases. There are several phosphodiesterases and their activities can be hormonally regulated.

Another similar class of receptors are coupled to G_i proteins. When these G proteins are activated, they in turn activate a variety of effectors as indicated in table 24.6. One of these effectors is adenylate cyclase again, but in this case the α_i-GTP subunit inhibits enzyme activity thus lowering cAMP levels.

The Guanylate Cyclase Pathway

Another second messenger system involves receptors that either are coupled to guanylate cyclase or guanylate cyclase itself. One class of receptors, those that are activated by atrial, cardiac or brain natriuretic factors, is bifunctional in that the external domain of these membrane receptors binds the peptide hormones resulting in the activation of an internal cytoplasmic domain that has guanylate cyclase activity and produces cGMP from GTP. This is the best example of the simplest receptor-effector coupling scheme illustrated in figure 24.14. Another class of receptors activates intracellular guanylate cyclases by an unknown mechanism. The cGMP that is produced can either act directly on target molecules, e.g., ion channels, or it can activate a cGMP protein kinase that can phosphorylate many proteins resulting in a variety of metabolic changes, depending on the cell type. For example, a cAMP phosphodiesterase can be activated by cGMP-dependent phosphorylation, resulting in lowering of cAMP levels; this effect helps explain how cAMP and cGMP can have opposite effects on cells. The cGMP kinases differ from the cAMP kinases described previously, in that the regulatory and catalytic domains are part of the same molecule rather than being encoded by separate genes.

Figure 24.16

Phosphatidylinositol-4,5-bisphosphate (PIP_2) and the two second messengers, diacylglycerol and inositol triphosphate, that are derived from it.

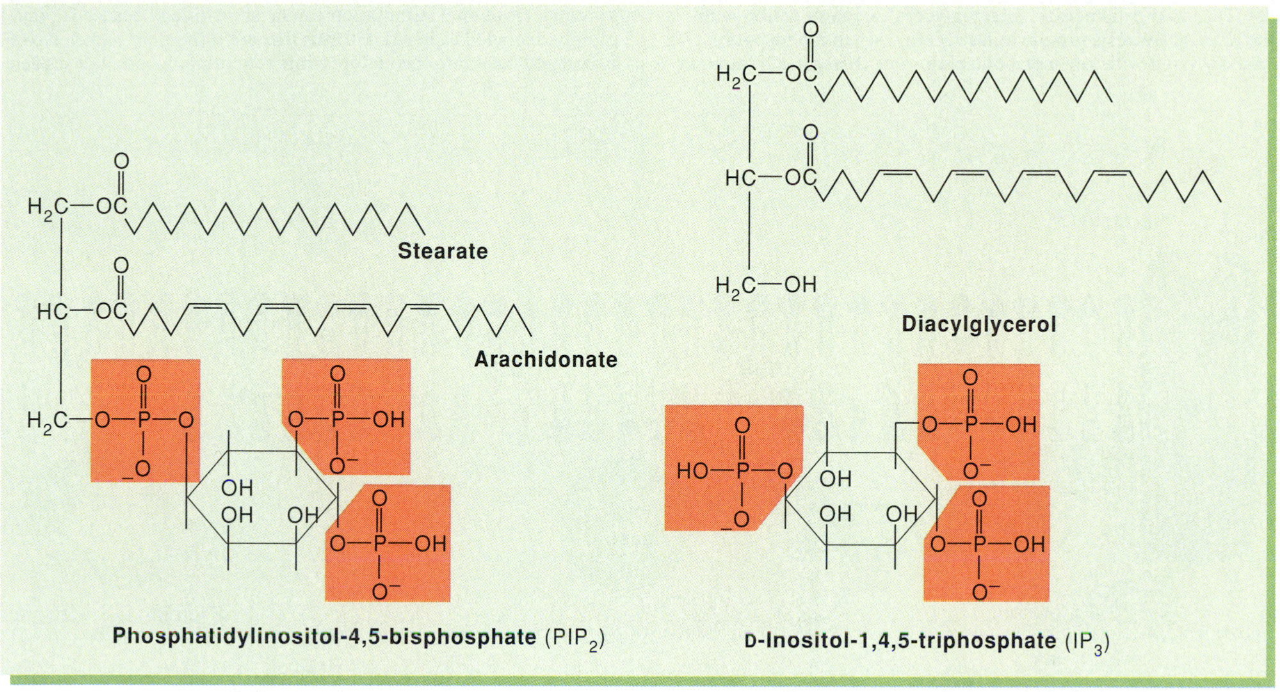

Stearate

Arachidonate

Phosphatidylinositol-4,5-bisphosphate (PIP_2)

Diacylglycerol

D-Inositol-1,4,5-triphosphate (IP_3)

Calcium and the Inositol Triphosphate Pathway

The outline of another important second messenger system has been elucidated during the last few years. Chemical messengers that act via this system include a variety of hormones (e.g., catecholamines, vasopressin, and angiotensin as well as some neurotransmitters (e.g., acetylcholine acting on pancreatic acinar cells to stimulate secretion of digestive enzymes, or acting on pancreatic β cells to stimulate insulin secretion). The receptors for these hormones and neurotransmitters are membrane-bound proteins with their binding sites facing the outside of the cell. Their activation by an appropriate signal is transmitted by a G protein (G_0, or Gq, table 24.6), which then activates a phosphodiesterase (phospholipase C) that cleaves the polar inositol triphosphate (IP_3) from phosphatidylinositol-4,5-bisphosphate (PIP_2). As a result, IP_3 can enter the cytoplasm, while the diacylglycerol moiety remains in the membrane (fig. 24.16). Both breakdown products of PIP_2 play important second messenger roles (fig. 24.17). IP_3 stimulates the release of calcium from intracellular stores (residing in the endoplasmic reticulum) into the cytoplasm. The calcium is bound by the protein calmodulin, which then activates one or more calcium-dependent protein kinases. Meanwhile, the diacylglycerol, along with phosphatidylserine, activates a membrane-associated protein kinase (C-kinase) that phosphorylates serine and threonine residues (see fig. 24.17). Thus several kinases are activated by this complex system and they, like the cAMP-dependent protein kinases, modify the function of rate-limiting enzymes and regulatory proteins involved in a variety of metabolic pathways. In addition to intracellular messengers, the breakdown of PIP_2 also stimulates the production of extracellular modulators of hormone activity. Thus, following the breakdown of PIP_2, arachidonic acid (one of the main fatty acids found in the diacylglycerol moiety of PIP_2) is metabolically converted into eicosanoids of different sorts. (Recall that we discussed the formation of eicosanoids, as well as their function as local hormones, in chapter 22.)

IP_3 is rapidly degraded to inactive IP_2 and then on to inositol. Meanwhile, diacylglycerol is phosphorylated and then converted to CDP-diacylglycerol, which combines with inositol to form phosphatidylinositol. The latter is subsequently phosphorylated in two steps to PIP_2. The degradation and resynthesis of PIP_2 completes the so-called phosphatidylinositol cycle.

Steroid Receptors Modulate the Rate of Transcription

The receptors for all classes of steroid hormones, including the receptors for 1,25-dihydroxyvitamin D_3, and thyroid hormones are intracellular proteins that are not very abundant, usually only 10^2 to 10^5 molecules per cell.

Figure 24.17

The phosphatidylinositol (PIP_2) pathway. Binding of certain hormones and some other ligands to a variety of membrane receptors (R) leads to activation of GTP-binding proteins, which then activate a phospholipase C that cleaves the inositol triphosphate (IP_3) moiety from PIP_2 and leaves diacylglycerol (DG) in the membrane. Diacylglycerol, in conjunction with phosphoserine (PS), activates a protein kinase (C-kinase) that phosphorylates enzymes associated with key metabolic pathways, thereby activating or inactivating them. Meanwhile, the polar IP_3 moiety binds to intracellular receptors on the endoplasmic reticulum, resulting in the liberation of calcium into the cytosol. The calcium binds to calmodulin, which then activates another group of protein kinases (the Ca^{2+}/CaM-dependent protein kinases). Hormonal stimulation can be short-lived because IP_3 and DG are rapidly degraded to inactive forms that are ultimately recycled to PIP_2; Ca^{2+} is pumped back into the endoplasmic reticulum, where it is sequestered.

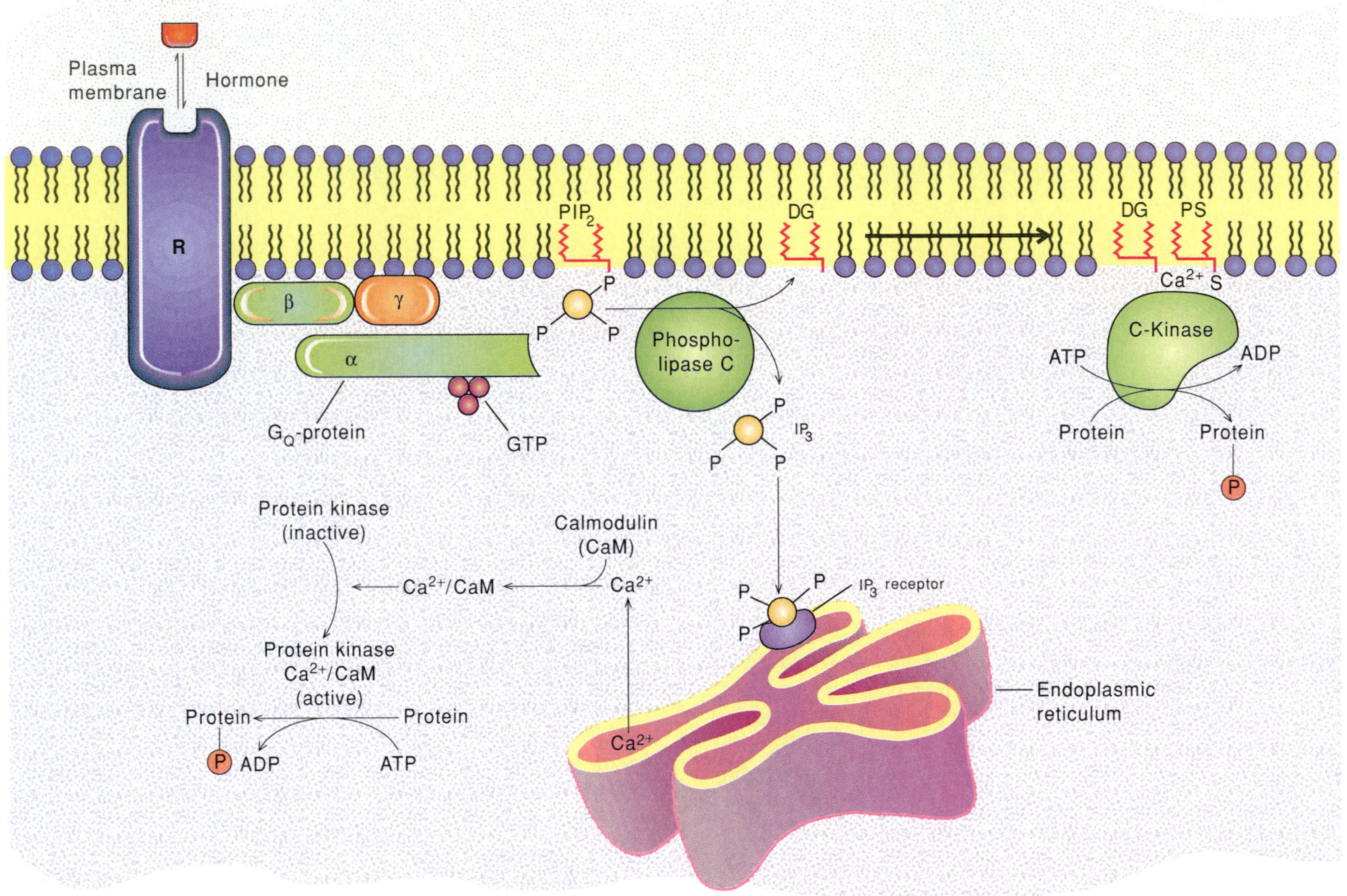

All of these classes of receptors have a similar structure composed of several functional domains including regions that bind the hormone, bind to DNA, activate transcription and allow dimerization. In the absence of hormone these receptors are usually sequestered by other proteins located either in the cytoplasm or nucleus, which precludes undesirable activity. For example, glucocorticoid receptors are sequestered by hsp90 in the cytoplasm. Binding of hormone, e.g., corticosterone, to the COOH-terminal domain of the receptor causes a conformational change that liberates the receptor and allows it to interact with another activated receptor, forming a homodimer (fig. 24.18). The receptor dimers recognize specific DNA sequences that are located in the promoter region of the genes that these receptors regulate (table 24.7). The positioning of receptors at the promoter allows their activation domains to stimulate transcription. Transcriptional regulation is discussed in detail in chapter 31.

Thyroid hormone receptors differ from steroid hormone receptors in that they are bound to DNA all the time. In the absence of thyroid hormones, these receptors inhibit the expression of genes to which they bind; the addition of hormone converts them into transcriptional activators.

Hormones Are Organized into a Hierarchy

The synthesis of many hormones is regulated by a cascade of hormones (fig. 24.19). Frequently, a hypothalamic hormone impinges on the pituitary to stimulate synthesis of a hormone that activates hormone synthesis in yet another organ. The end products of these cascades generally feedback-inhibit the production of hormones at the beginning of the cascade (fig. 24.20).

Another common mechanism for modulating hormonal response involves two (or more) hormonal inputs with both positive and negative effects. The hypothalamic peptides somatostatin and GRF have opposite effects on GH synthesis

Figure 24.18

Activation of steroid hormone receptors by the hormone. In the absence of the hormone, the steroid receptors are complexed through the hormone-binding domain to another protein known as heat shock protein 90 (hsp90). Both the hormone-binding domain and the hsp90 prevent functional interaction of the receptor with DNA. Binding of the hormone frees the receptor from hsp90 and promotes dimerization of the receptor, which can then bind to the palindromic hormone response element (HRE) and activate transcription.

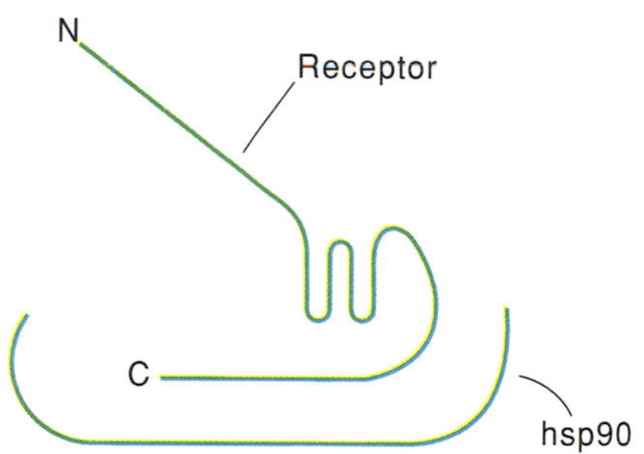

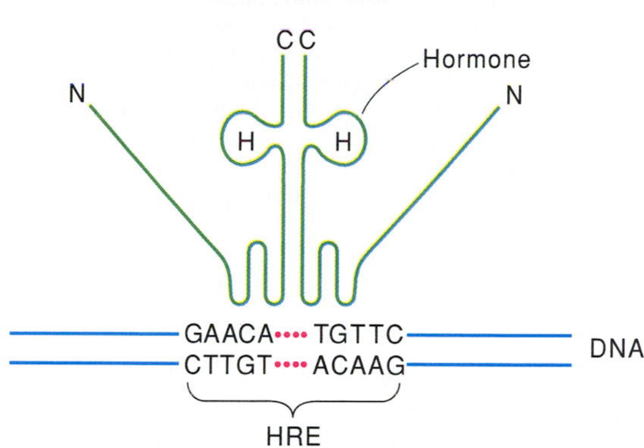

Table 24.7
Regulation of Specific Genes by Steroid and Thyroid Hormones

Glucocorticoids	
Tyrosine aminotransferase	Liver
Tryptophan oxygenase	Liver
Glutamine synthase	Liver, retina
Phosphoenolpyruvate carboxykinase	Kidney
Ovalbumin	Oviduct
Conalbumin	Oviduct (liver)[a]
α-Fetoprotein (↓)	Liver
α_2-Globulin	Liver
Metallothionein	Liver
Proopiomelanocortin (↓)	Pituitary
Mammary tumor virus	Mammary gland
Estrogens	
Ovalbumin	Oviduct
Conalbumin	Oviduct (liver)
Ovomucoid	Oviduct
Lysozyme	Oviduct
Vitellogenin	Liver
apo-VLDL	Liver
Glucose-6-*P*-dehydrogenase	Uterus
Progestins	
Avidin	Oviduct
Ovalbumin	Oviduct
Conalbumin	Oviduct
Uteroglobin	Uterus
Androgens	
β-Glucuronidase	Kidney
Aldolase	Prostate
Prostate-binding proteins	Prostate
Ovomucoid	Oviduct
Ovalbumin	Oviduct
1,25-Dihydroxyvitamin D_3	
Calcium-binding protein	Intestine
Ostercalcin	Bone
Ecdysone	
Dopa-decarboxylase	Epidermis
Vitellogenin	Fat body
Larval serum protein I	Fat body
Thyroid Hormones	
Carbamyl phosphate synthase	Liver
Growth hormone	Pituitary
Prolactin (↓)	Pituitary
α-Glycerophosphate dehydrogenase	Liver (mitochondria)
Malic enzyme	Liver

Figure 24.19

Growth hormone cascade. Neurosecretory cells in the hypothalamic region of the brain are activated by neurotransmitters from other neurons to secrete growth hormone releasing factor (GRF), a 44-amino-acid peptide that travels through the portal circulation to the anterior pituitary, where it binds to membrane receptors on somatotroph cells. GRF binding stimulates the production of cAMP, which activates growth hormone (GH) synthesis and secretion. GH, a 191-amino-acid polypeptide, passes into the bloodstream and travels to the liver, where it binds to membrane receptors that probably produce a second messenger (as yet unknown) that activates transcription of insulinlike growth factor-I (IGF-I) gene, which codes for a 71-amino-acid peptide. IGF-I is secreted from the liver into the bloodstream and ultimately binds to membrane receptors (which are tyrosine kinases similar to the insulin receptor) located on many peripheral cells. Activation of these receptors leads to cellular proliferation under appropriate conditions.

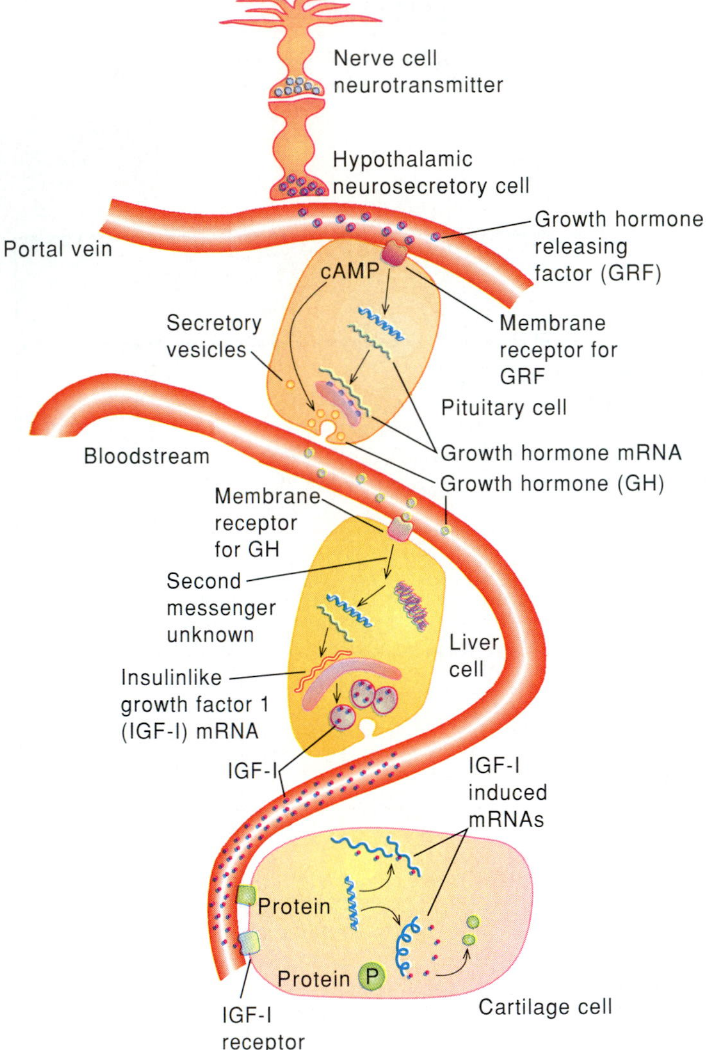

Figure 24.20

Control of hormone synthesis and secretion in the anterior pituitary. Neurosecretory neurons in the hypothalamus liberate polypeptides that either stimulate ⊕ or inhibit ⊖ hormone synthesis by specialized pituitary cells containing the appropriate receptors. For example, GnRH stimulates gonadotroph cells in the pituitary to synthesize and secrete LH and FSH. These pituitary hormones then impinge upon target cells, typically stimulating them to make other low-molecular-weight hormones. The end products of these cascades feedback-inhibit hormone production at either or both hypothalamic and pituitary levels.

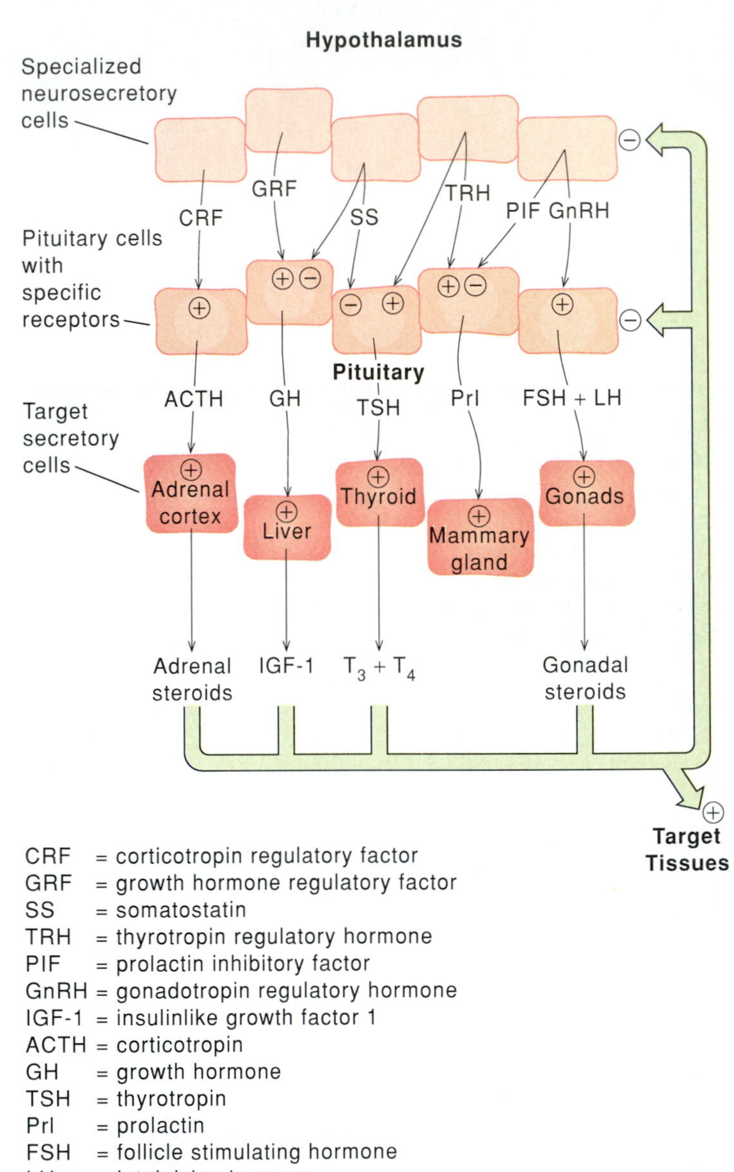

and secretion. Similarly, glucagon and insulin have opposite effects on gluconeogenesis in the liver (see the discussion earlier in this chapter) and some of the effects of ecdysone on gene expression in insects are blocked by juvenile hormone (a terpene derivative; fig. 24.21).

There is also considerable "cross-talk" between different hormones at the level of receptor function. For instance, the diacylglycerol-activated protein kinase C phosphorylates and thereby inhibits the activity of insulin and epidermal growth factor receptor kinases (epidermal growth factor is discussed in a later section). Likewise, when cAMP-dependent protein kinases are active they can inhibit β-adrenergic receptors by phosphorylating them.

Cells also have mechanisms that tend to prevent chronic stimulation. Exposure of cells to epinephrine leads to an initial sharp rise in cAMP levels; however, cAMP levels fall nearly to

Figure 24.21

Structure of ß-ecdysone and juvenile hormone. These hormones play major roles in the growth and maturation of insects by controlling the timing for molting of the insect exoskeleton.

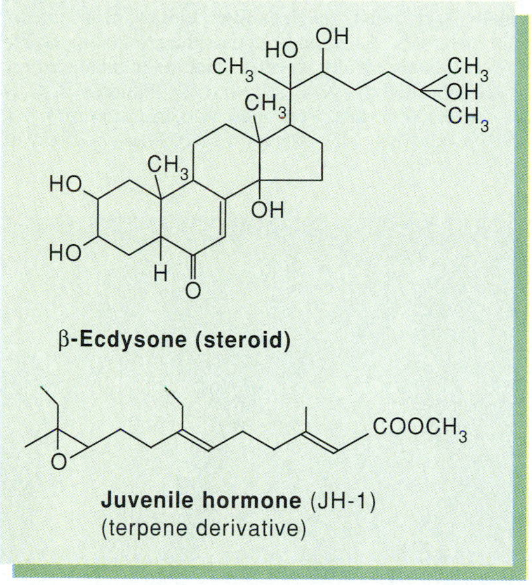

β-Ecdysone (steroid)

Juvenile hormone (JH-1)
(terpene derivative)

basal levels within an hour or so, despite the continuous presence of saturating amounts of epinephrine. Furthermore, if the hormone is removed and the cells are challenged within a few hours, the secondary response is lower. This phenomenon, called desensitization, is associated with both an uncoupling of receptors from adenylate cyclase activation and a decrease in the number of receptors accessible to hormone binding. It occurs only after productive receptor function. The main consequence of desensitization is that exposure to hormone results in transient activation of cellular events, rather than chronic activation.

The response to many polypeptide hormones also diminishes with chronic stimulation, but the time course is much longer (many hours to days) and the mechanism is different from that involved in desensitization. Binding of insulin, calcitonin, LH, TRH, and EGF to their respective receptors promotes a physiological response but ultimately leads to the clearance of these receptors from the surface and a blunting of that response. The hormone-mediated loss of receptors is often referred to as down-regulation. In this case, the receptors are internalized and degraded. Hormone binding leads to a clustering of receptors, as if they were cross-linked. These clusters, or patches, aggregate in membrane structures called coated pits (the intracellular side of the pits is coated with a scaffolding protein called clathrin). Small endosome vesicles that engulf the receptors and their ligands bud off from the coated pits, migrate within the cell, and associate with other membranous structures known collectively as GERL (Golgi-endoplasmic-reticulum-lysosomes). After a few hours they fuse with lysosomes, at which point the lysosomal enzymes degrade both the receptor and the hormone (fig. 24.22). Replacement of the hormone receptors requires protein synthesis.

Figure 24.22

The down-regulation of receptors by endocytosis. The hormone-mediated loss of receptors is often referred to as down-regulation. Continuous activation of receptors by hormone often leads to patching, a clustering of receptors as if they were cross-linked. Endocytosis of the patches removes them from the cell surface. The endocytotic vesicles, sometimes called receptosomes, fuse with lysosomes, where the contents are degraded by the lysosomal enzymes. Receptosomes also may allow entry of receptors into other cell compartments, such as the nucleus, by fusing with these organelles.

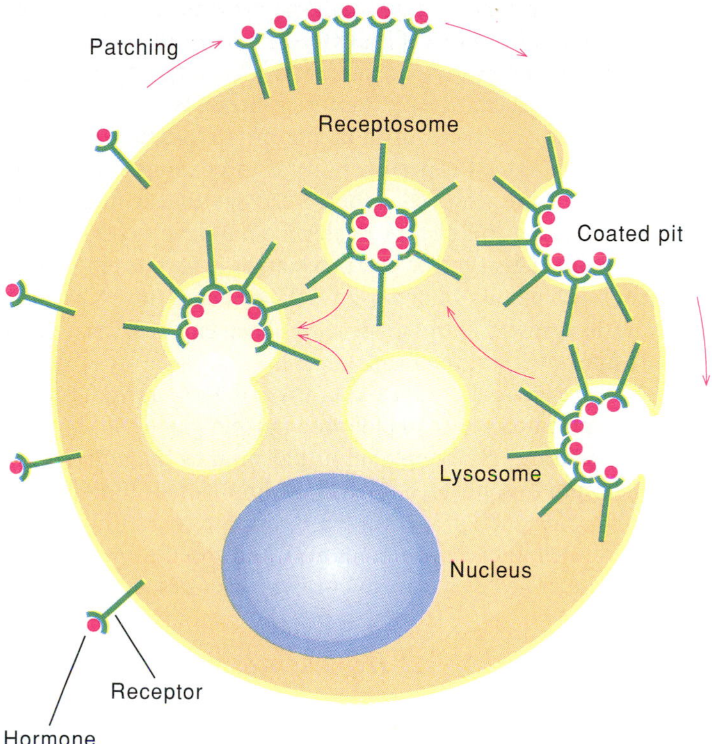

Most hormones are released in a cyclic manner (examples include insulin, glucagon, growth hormone, and many of the hypothalamic releasing hormones). In some cases the cycles appear to be autonomously regulated, but in others they are clearly entrained by neural and hormonal signals that may vary, depending on the developmental stage or physiological condition. The periodic release of peptide hormone can promote distinctly different responses than would be achieved by chronic release.

Diseases Associated with the Endocrine System

Human diseases related to endocrine dysfunction can be broadly grouped into (1) overproduction of a particular hormone, (2) underproduction of a hormone, and (3) target-cell insensitivity to a hormone.

Overproduction of Hormones Is Commonly Caused by Tumor Formation

Most cases of hormonal overproduction are associated with enlargement of the normal endocrine organ, frequently owing to a tumor. Pituitary neoplasms usually affect the production of only one pituitary hormone as a result of the cancerous proliferation of the cell type that normally synthesizes that hormone. Examples include giantism (acromegaly), which is associated with proliferation of the somatotroph cells that synthesize growth hormone, and Cushing's syndrome, which is usually due to overproduction of ACTH. Adrenal and parathyroid tumors that lead to the overproduction of various adrenal steroids and parathyroid hormones have also been described.

Occasionally, tumors of organs that do not usually produce a given hormone (ectopic tumors) synthesize and secrete peptide hormones, as in ACTH synthesis by certain lung tumors and GRH production from a pancreatic islet tumor. Expression of hormones by these ectopic tumors represents a curious activation of gene expression in an inappropriate tissue. The same result can be obtained experimentally by producing animals in which a structural gene codes for a hormone that is under the transcriptional control of a promoter from a gene expressed in another tissue. This type of overproduction is illustrated in experiments with metallothionein-growth hormone (fig. 24.23). In this situation growth hormone gene has been engineered so that it is not under normal feedback control.

The excessive production of thyroid hormone in Graves' disease is associated with an enlarged thyroid gland, but in this case a circulating immunoglobulin that mimics the activity of thyroid stimulating hormone (TSH) is implicated. Finally, we saw in an earlier section how an inappropriate steroid hormone may be secreted in excessive amounts when specific enzymes in the adrenal steroid biosynthetic pathway are present at inadequate levels.

Figure 24.23

A transgenic mouse that synthesizes rat growth hormone ectopically, and a normal littermate. A chimeric gene with the mouse metallothionein promoter fused to the structural gene of rat growth hormone (rGH) was introduced into the germline of mice by microinjection of cloned DNA into fertilized eggs. Because of the metallothionein promoter, these genes were expressed in many large organs (such as liver, kidney, heart, intestine) that do not normally make GH. As a consequence, the circulating GH level was elevated several hundredfold, an effect that led to increased growth of those mice that inherited the gene. (From R. D. Palmiter, R. L. Brinster, R. E. Hammer, and R. M. Evans, Reprinted by permission from *Nature* 300, 611-615: © 1982 MacMillan Magazines Limited. Photo courtesy of Ralph Brinster.)

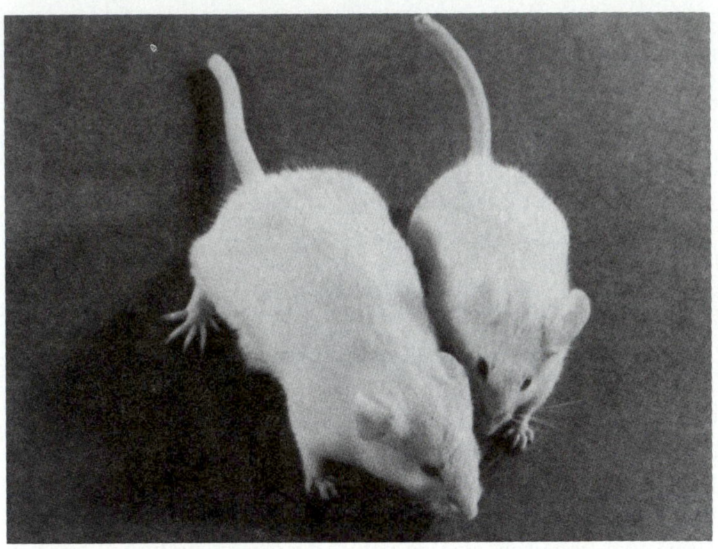

Underproduction of Hormones Has Multiple Causes

A wide variety of defects could lead to inadequate production of hormones. In some conditions, there are not enough cells producing a given hormone. For example, in juvenile-onset diabetes there is a decrease in the number of pancreatic islet cells that synthesize insulin. In other, more extreme cases, the gene coding for the hormone may be missing or defective; for example, some forms of dwarfism are due to a lack of the growth hormone gene. In most cases, however, the gene coding for the hormone is present but inadequate amounts of active hormone are produced. Such defects can have many possible explanations. Some of them are genetic. For example, a mutation may affect the rate of synthesis of mRNA coding for the hormone precursor, or a mutation may affect the processing of the mRNA, or an amino acid substitution may decrease the activity or processing of the mRNA, or an amino acid substitution may decrease the activity or processing of a polypeptide hormone. The techniques of molecular biology are being used to discover the causes of many of these genetic disorders. Other defects in hormone synthesis are the consequence of an inadequate supply of precursors. For example, iodine is essential for thyroid hormone synthesis, and its absence leads to goiter. This condition, involving enlargement of the thyroid, is caused by high concentrations of TSH that result from lack of normal feedback by T_3 and T_4 on the hypothalamus. Likewise, synthesis of vitamin D requires ultraviolet irradiation of 7-dehydrocholesterol, without which the formation of $1,25(OH)_2D_3$ cannot occur, so that rickets ensues. A rare cause of rickets involves a defect in the enzyme that converts $25(OH)D_3$ to $1,25(OH)_2D_3$, the active form of vitamin D. Similarly, defects in enzymes involved in adrenal steroid biosynthesis lead to Addison's disease, and a defect in 5α-reductase leads to one form of testicular feminization, the result of inadequate production of dihydrotestosterone.

Target-Cell Insensitivity Results from a Lack of Functional Receptors

The most dramatic examples of target-cell insensitivity are those in which the correct receptors are lacking. In complete testicular feminization, androgen receptors are missing from all cells. The consequence is phenotypic expression of female characteristics in genotypic males. A rare form of dwarfism (Laron dwarfs) is associated with high plasma levels of growth hormone (GH) but low levels of insulinlike growth factor-I (IGF-I); these individuals have a defect in the GH receptor. Occasionally, antibodies are directed against receptors and interfere with normal hormone binding, e.g., in one form of adult-onset diabetes. Remarkably, the antibody itself promotes insulin effects.

Table 24.8
Vertebrate Growth Factors

Factor[a]	M_r	Cell Types Affected
Epidermal growth factor (EGF)[b]	6,400	Epithelial and mesodermal cells
Tumor growth factor-α (TGF-α)	7,000	Same
Insulinlike growth factor-I (IGF-I)	7,000	Same
Insulinlike growth factor-II (IGF-II)	7,000	Same
Fibroblast growth factor (FGF)[c]	13,000	Same
Platelet-derived growth factor (PDGF)[b]	31,000	Same
Nerve growth factor (NGF)[c]	13,000	Sensory and sympathetic neurons
Erythropoietin	23,000	Erythroid cell precursors
Macrophage colony stimulating factor (M-CSF or CSF-I)[b]	70,000 (dimer)	Macrophage precursors
Granulocyte colony stimulating factor (G-CSF)	25,000	Granulocyte precursors
Granulocyte-macrophage colony stimulating factor (GM-CSF)	23,000	Granulocytes and macrophage precursors
Multicolony stimulating factor (Multi-CSF) or Interleukin-3 (IL-3)	25,000	Precursors of most hematopoietic cells
Interleukin-2 (IL-2)[c]	13,000	T lymphocytes

[a]The genes for most of these growth factors and many of their receptors have been cloned.
[b]The EGF receptor is homologous to the *erbB* and *neu* oncogenes, and the M-CSF receptor is homologous to the *fms* oncogene. PDGF is homologous to the *sis* oncogene.
[c]There are many members of the FGF, NGF, and interleukin families of growth factors.

Growth Factors

In addition to the many hormones that have been discussed, there exist a large number of growth factors that have hormonelike activities. As their name implies, growth factors often stimulate the proliferation of particular cells. Unlike hormones, most growth factors are synthesized by a variety of cell types rather than in a specialized endocrine gland. However, the mechanisms of action of growth factors and hormones are likely to be very similar.

All of the growth factors that have been characterized so far are proteins (table 24.8). Many of them are cleaved from larger precursors. The 53-amino-acid epidermal growth factor (EGF) is cleaved from a precursor of 1,168 amino acids. This precursor is a membrane-spanning protein, with the EGF moiety and nine related sequences in the extracellular domain. EGF is homologous to several other growth factors, including α-tumor growth factor and a protein secreted by vaccinia virus. These factors bind to membrane receptors and are thought to trigger intracellular events in much the same way as hormones.

The EGF receptor, for example, shows striking similarity to the insulin receptor in overall organization and amino acid sequence; however, instead of separate α and β chains, the EGF receptor is composed of a single polypeptide that appears to be a composite of the α and β chains of the insulin receptor. As is true with the insulin receptor, the intracellular domain of the EGF receptor has tyrosine kinase activity.

The receptor for insulinlike growth factor-I(IGF-I) is also similar to the insulin receptor in structure and tyrosine kinase activity. Thus in this case it is likely that the hormones (insulin and IGF-I) and their receptors have each evolved from common ancestral genes by duplication and divergence.

The receptors for one of the colony-stimulating factors (M-CSF) and for platelet-derived growth factor (PDGF) are also membrane proteins with tyrosine kinase activity. Although the mechanisms by which these receptors regulate cellular activities are not clear yet, a major hypothesis is that they phosphorylate (and thereby either activate or inhibit) regulatory enzymes, including some serine/threonine protein kinases. PDGF also appears to activate the inositol triphosphate/diacylglycerol pathway by phosphorylating phospholipase C on tyrosine residues (see fig. 24.16). This would also lead to activation of several serine/threonine kinases.

Normal cells require growth factors (mitogens) for proliferation. In the absence of these factors they reversibly withdraw from the cell cycle and become arrested. Transformed cells (tumor cells) have a relaxed cell cycle control and can traverse the cell cycle in the absence of added growth factors. Cell transformation can be achieved by activation or inappropriate expression of a variety of cellular or viral genes. The isolation and characterization of many of these transforming genes (oncogenes) has revealed striking homology with growth factors or their receptors. For example, the oncogene product *sis* resembles PDGF in its sequence and its affinity for specific receptors, and the EGF receptor is homologous to the *erbB* oncogene. Thus a prevalent idea is that many oncogenes lead to abnormal cell proliferation by directly or indirectly providing a rate-limiting factor, be it a growth factor, its receptor, or an intracellular intermediate (see chapter 34).

Plant Hormones

Plant hormones (also called growth regulators) coordinate the growth and development of plants. The major hormones discovered to date fall into six classes: auxins, cytokinins, gibberellins, abscisic acid, ethylene and oligosaccharides. The structures of representative members of each class are shown in figure 24.24. All of these hormones are low-molecular-weight compounds; indeed, one hormone, ethylene, is a gas. The first polypeptide hormone of higher plants, systemin, was recently described. It is an 18-amino-acid peptide that is released upon wounding and activates the plant defense system. Polypeptide and steroid hormones have not been described for higher plants, although some yeasts and fungi use these compounds as mating factors. Each class of compounds elicits many diverse responses, and there is considerable interaction among different plant hormones in the control of physiological processes. Furthermore, the same process is controlled by different hormones (or combinations of hormones) in different species. These considerations make it difficult to analyze, and generalize about, the mechanism of plant hormone action.

Auxins are synthesized in the apical buds of growing shoots. They stimulate growth of the main shoot, but inhibit the development of lateral shoots; this effect has led to the horticultural practice of pinching off the apical buds to stimulate the formation of bushy plants. The curvature of plants toward the light (phototropism) is thought to be due to transport of auxins away from the light, which stimulates more rapid growth of cells on the darker side of the shoot. Auxins bind to specific membrane proteins. The affinity of auxins for these proteins, coupled with their location and abundance on target cells, supports the view that the membrane proteins may be the receptors that mediate auxin action.

Auxin stimulation of growth occurs in two phases. The earliest response (called the rapid response) is an increase in proton transport out of the cell, which occurs after a lag of a few minutes. This hydrogen-ion pump is thought to be coupled with a membrane ATPase, but it is not clear whether the receptor interacts directly with the ATPase or whether other intermediates are involved. It is thought that lowering the extracellular pH activates enzymes that partially degrade the cell wall, thereby loosening it and allowing for cell expansion. A subsequent effect of auxin (the slow response) is to increase the synthesis of proteins and nucleic acids, resulting in sustained growth. Auxin has been shown, for example, to increase the amount of cellulase mRNA during pea cell expansion. A 20-kDa membrane protein that binds auxins has been characterized that appears to be the auxin receptor in that an antibody against it blocks auxin-stimulated events.

Cytokinins are adenine derivatives that are produced in the roots and that promote growth and differentiation of numerous tissues. Cytokinins can overcome the auxin-mediated inhibition of lateral shoots; in other tissues both hormones act synergistically. Plant cells can be grown in culture if auxins and cytokinins are provided. With relatively balanced concentrations of both hormones, cells will proliferate but remain unorganized. If the ratio of auxins to cytokinins is high, shoots will develop; if the ratio is low, root development will be favored. Intermediate concentrations of both hormones will promote undifferentiated growth, with neither roots nor shoots. As these results show, plant cells can display a wide range of physiological responses. They are much more plastic in their developmental potential than animal cells. Indeed, normal plants can be grown from a single tissue culture cell. While this diversity of responses is fascinating, it has also made it difficult to define the mechanism of action of plant hormones.

Gibberellins also promote a shoot elongation, and frequently they act synergistically with auxins. Gibberellins stimulate the accumulation of specific mRNAs, e.g., amylase mRNA in germinating seeds. This fact suggests that their receptors (or a second messenger) act at the genetic level.

Abscisic acid is antagonistic to many other plant hormones. It inhibits germination of seeds and shoot growth while promoting resting bud formation and leaf senescence. Wilting stimulates the synthesis of abscisic acid in the chloroplasts within mesophyll cells of the leaves. The abscisic acid in this case is a stress signal that stimulates the guard cells to close and thus minimize water loss through the stomata. Abscisic acid stimulates K^+ efflux from guard cells and into adjacent cells; thus it may act by means of membrane receptors to modulate ion pumps, as was suggested for auxins.

Oligosaccharin is one member of complex oligosaccharides that function in defense against disease, in control of plant growth, and in differentiation. They are released from the cell wall by specific degradative enzymes. For example, when bacteria or fungi infect plants, they produce enzymes that degrade the cell wall, liberating oligosaccharides. Some of these compounds stimulate the plant to make antibiotics or proteases that help the plant defend against the pathogen or insect. There they stimulate the plant cells to produce antibiotics that inhibit the growth of the pathogen. When certain plant cells are damaged they can produce the enzymes required to liberate oligosaccharides; this allows a hormonal response even when the pathogen does not produce the appropriate enzymes. Auxins also stimulate the production of oligosaccharides, which then counteract the auxin-stimulated growth; thus these oligosaccharides serve as feedback inhibitors. The inhibition of lateral shoot growth that has been attributed to auxin may actually be due to the production of oligosaccharides.

Addition of oligosaccharides to combinations of auxin plus cytokinins also influences the differentiation of shoots and roots in tissue culture. Indeed, many of the pleiotropic effects originally attributed to auxins and other plant hormones may actually be mediated by oligosaccharins.

Figure 24.24

Some common plant hormones. All of these hormones are low-molecular-weight compounds. One hormone, ethylene, is a gas.

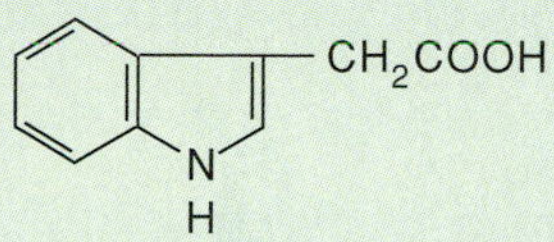

CH_2COOH

Auxin
(indole acetic acid)

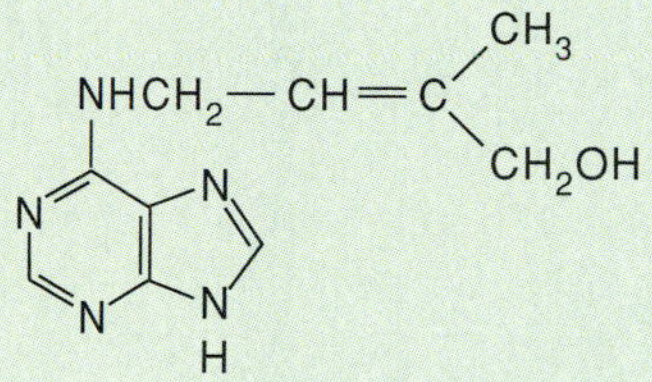

$NHCH_2—CH=C(CH_3)(CH_2OH)$

Cytokinin
(zeatin)

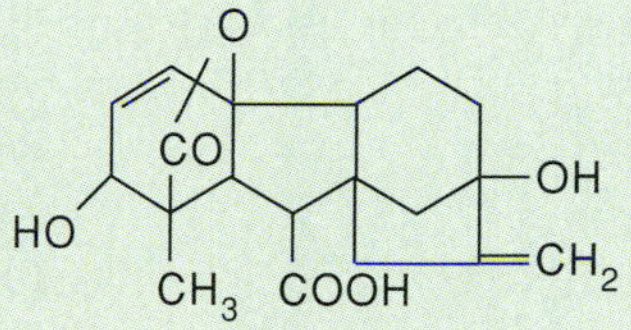

Gibberellin
(gibberellic acid)

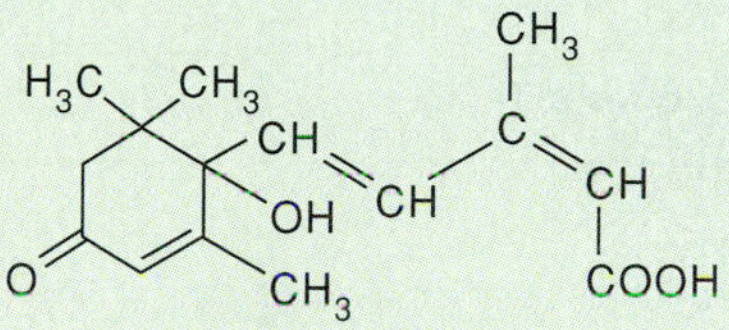

Abscisic acid

$H_2C=CH_2$

Ethylene

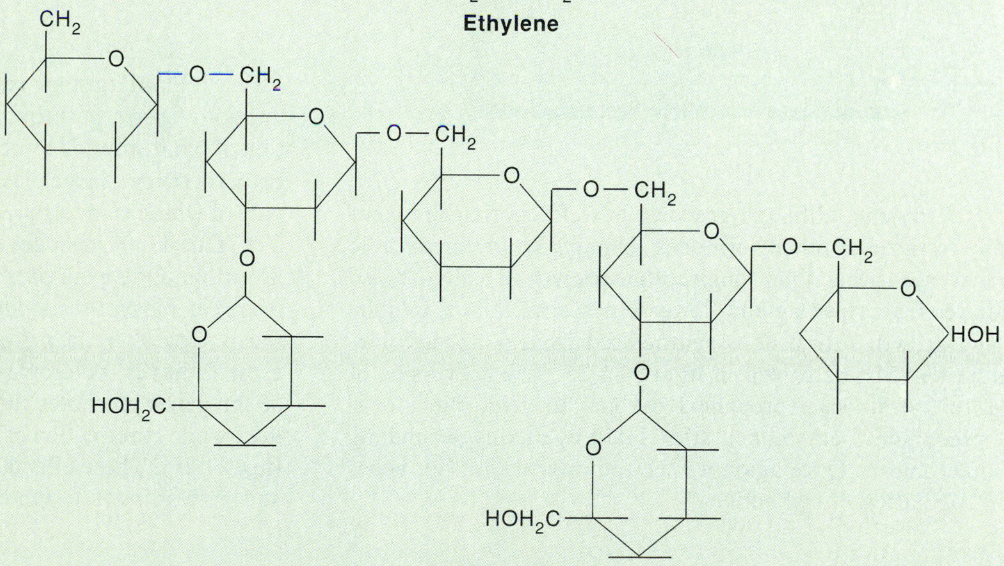

Oligosaccharin
(a heptaglucoside)

Figure 24.25

Synthesis of ethylene, a gaseous plant hormone involved in fruit ripening and flower senescence.

$$CH_3-S-CH_2-CH_2-\overset{NH_3^+}{\overset{|}{C}H}-COO^-$$

Methionine

$\downarrow$ ATP $\rightarrow$ $PP_i + P_i$

$$CH_3-\underset{\text{Adenosine}}{\underset{|}{S}}-CH_2-CH_2-\overset{NH_3^+}{\overset{|}{C}H}-COO^-$$

***S*-Adenosylmethionine**

$\downarrow$

$$CH_3-\underset{\text{Adenosine}}{\underset{|}{S}}$$

$+$

$$\begin{matrix} CH_2 & & NH_3^+ \\ | & C & \\ CH_2 & & COO^- \end{matrix}$$

1-Aminocyclopropane-1-carboxylic acid (ACC)

$\downarrow$ O_2 $\rightarrow$ $HCOOH + NH_3 + CO_2$

$$H_2C{=}CH_2$$

Ethylene

Figure 24.26

Crown gall tumors of plants are caused by certain strains of the bacterium *Agrobacterium* that can infect plant cells and introduce new genetic information coding for plant hormones. The transforming DNA of these bacterial strains is carried on a large plasmid; it contains genes coding for rate-limiting enzymes in hormone biosynthesis. (*a*) The transforming DNA of wild-type bacterial plasmid codes for rate-limiting enzymes involved in both auxin and cytokinin biosynthesis, resulting in undifferentiated growth (called callus). (*b*) Mutations that disrupt the gene involved in auxin biosynthesis give rise to tumors that produce only cytokinins, which leads to a proliferation of leaves and shoots. (*c*) Alternatively, mutations that disrupt the gene involved in cytokinin synthesis give rise to tumors that produce only auxin. In these tumors, only roots differentiate.

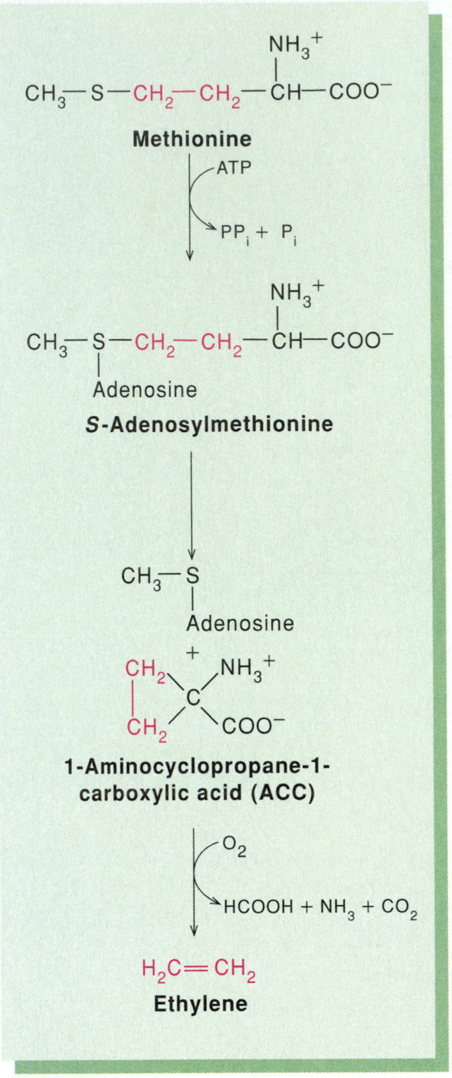

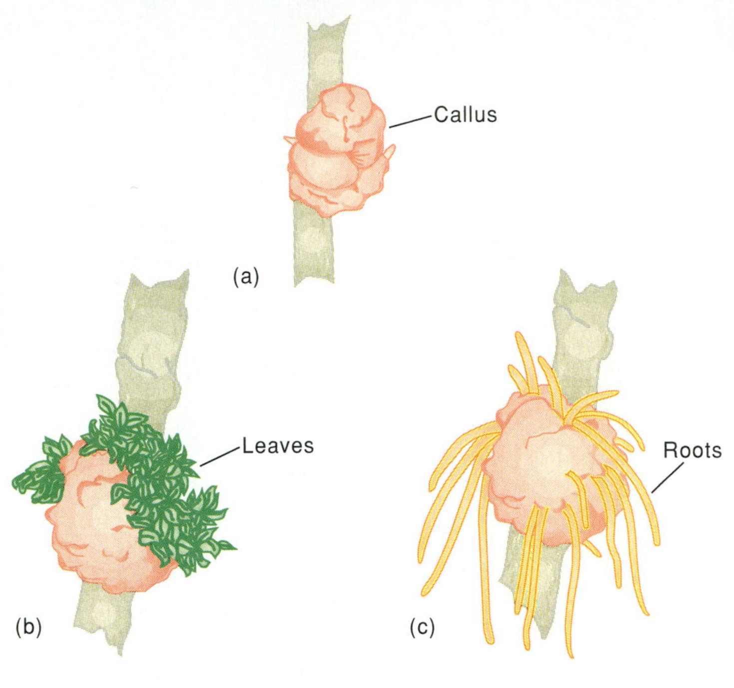

Ethylene, although gaseous, has effects that are comparable to those of other hormones. It plays an important role in transverse rather than longitudinal growth of cells. It also stimulates fruit ripening and flower senescence, and it inhibits seedling growth. Ethylene is synthesized from *S*-adenosylmethionine (SAM), as shown in figure 24.25. The conversion of SAM to 1-aminocyclopropane-1-carboxylic acid, the immediate precursor of ethylene, is stimulated by auxins, wounding, and anaerobiosis. Once again we see the interplay of hormones in the regulation of cell activity.

Plant tumors result from uncontrolled hormone production. Crown gall tumors, for example, are due to the infection of plant wounds by certain strains of *Agrobacterium*. These bacteria carry a large plasmid, the tumor-inducing or T_i plasmid, part of which is incorporated into the plant genome (see chapter 27). This DNA encodes several genes that stimulate cell proliferation. One gene product is a rate-limiting enzyme involved in auxin biosynthesis, and another controls cytokinin biosynthesis. Together they promote the rapid, but undifferentiated, growth that is the crown gall. Interestingly, inactivation of one of the genes promotes the growth of shoots at the site of infection, while inactivation of the other promotes the growth of roots (fig. 24.26). These effects are reminiscent of the action of auxins and cytokinins in tissue culture.

Summary

In this chapter we have focused on the ways in which various metabolic activities are integrated, with special attention to the nature and functioning of hormones. The following points are the highlights of our discussion.

1. Tissues store biochemically useful energy in three major forms: carbohydrates, lipids, and proteins. Each tissue makes characteristic demands on and contributions to the energy supply of the organism.
2. Hormones are chemical messengers formed in specific tissues. They circulate between tissues of multicellular organisms and serve to coordinate metabolic activities, maintain homeostasis of essential nutrients, and prepare the organism for reproduction.
3. Most hormones fall into three classes: polypeptides, steroids, and amino acid derivatives. Polypeptide hormones are synthesized from large precursors. Steroid hormones are derivatives of cholesterol. Thyroid hormones and epinephrine are amino acid derivatives.
4. A number of factors—synthesis, rate of release, and rate of elimination—determine the concentration of circulating hormone.
5. Hormones act by reversibly binding to proteins called receptors, an event that results in a conformational change that is detected by other macromolecules (acceptors) and that eventually leads to activation of rate-limiting enzymes. Each class of hormones binds to specific receptors that activate other membrane proteins.
6. Most membrane receptors generate a diffusible intracellular signal called a second messenger. Five intracellular messengers are currently known: cyclic AMP, cyclic GMP, inositol triphosphate, diacylglycerol, and calcium. Second messengers usually activate or inhibit the action of one or more enzymes.
7. Steroid hormones penetrate the cell and bind to receptors in the nucleus, and activate (or sometimes repress) transcription of specific genes. Thyroid hormones act similarly.
8. A large number of diseases are due to either overproduction or underproduction of hormones, or to insensitivity of target tissues to circulating hormones. Knowledge of hormone biosynthesis, secretion, and interaction with target cells is essential to an understanding of the biochemical basis of these disorders.
9. In addition to classical hormones, there are other chemical messengers called growth factors that serve to coordinate growth of tissues during development.
10. Plants make several hormones that regulate growth and differentiation.

Selected Readings

Annual Reviews of Biochemistry and *Annual Reviews of Physiology.* Over 40 volumes in each series with many relevant reviews in the area of hormone receptors and hormone action. Provides good access to primary literature.

Baxter, J. D., and K. M. MacLoed, Molecular basis for hormone action. In P. K. Bondy and L. E. Rosenberg (eds.), *Metabolic Control and Disease.* Philadelphia: Saunders, 1980. A useful summary.

Berridge, M. J., Inositol triphosphate and diacylglycerol: two interacting second messengers. *Ann. Rev. Biochem.* 56:159–193, 1987.

Carafoli, E., and J. T. Penniston, The calcium signal. *Sci. Amer.* 253(5):70–78, 1985.

Carpenter, G., Receptors for epidermal growth factor and other polypeptide mitogens. *Ann. Rev. Biochem.* 56:881–914, 1987.

Collins, S., M. G. Caron, and R. J. Lefkowitz, From ligand binding to gene expression: new insights into the regulation of G-protein-coupled receptors. *Trends Biochem. Sci.* 17:37–39, 1992.

Czech, M. P., J. K. Klarlund, K. A. Yagaloff, A. P. Bradford, and R. E. Lewis, Insulin receptor signaling. *J. Biol. Chem.* 263:11017–11020, 1988.

deGroot, L. J., et al. (eds.), *Endocrinology,* 3 vols. New York: Grune and Stratton, 1979. Over 2,000 pages of comprehensive treatment; primarily from a medical point of view.

DeVos, A. M., M. Ultsch, and A. A. Kossiakoff, Human growth hormone and extracellular domain of its receptor: crystal structure of the complex. *Science* 255:306–312, 1992.

Gerisch, G., Cyclic AMP and other signals controlling cell development and differentiation in *Dictyostelium. Ann. Rev. Biochem.* 56: 853–879, 1987.

Gilman, A. G., G proteins: Transducers of receptor-generated signals. *Ann. Rev. Biochem.* 56:615–650, 1987.

Guillemin, R., Peptides in the brain: The new endocrinology of the neuron. *Science* 202:390, 1978. Nobel laureate speech.

Kikkawa, U., A. Kishimoto, and Y. Nishizuku, The protein kinase C family: heterogeneity and its implications. *Ann. Rev. Biochem.* 58:31–44, 1989.

Napier, R. M., and M. A. Venis, From auxin-binding protein to plant hormone receptor? *Trends Biochem. Sci.* 16:72–75, 1991.

O'Malley, B. W., and L. Birnbaumer, *Receptors and Hormone Action,* 3 vols. New York: Academic Press, 1977. Review articles by many authors cover most aspects of hormone action.

Pelech, S. L., and D. E. Vance, Signal transduction via phosphatidylcholine cycles. *Trends Biochem. Sci.* 14:28–30, 1989.

Pilkis, S. J., M. R. El-Maghrabi, and T. H. Claus, Hormonal regulation of hepatic gluconeogenesis and glycolysis. *Ann. Rev. Biochem.* 57:755–784, 1987.

Rasmussen, H., The cycling of calcium as an intracellular messenger. *Sci. Am.* 261(4):66–73, 1989.

Recent Progress in Hormone Research. New York: Academic Press. An annual publication with nearly 40 volumes. A good place to find a recent summary.

Rhee, S. G., P. -G. Suh, S. -H. Ryu, and S. Y. Lee, Studies of linositol phospholipid-specific phospholipase C. *Science* 244:546–550, 1989.

Riddiford, L. M., and J. W. Truman, Biochemistry of insect hormones and insect growth regulators. In *Biochemistry of Insects.* New York: Academic Press, 1978.

Simon, M. I., P. Strathmann, and N. Gautam, Diversity of G proteins in signal transduction. *Science* 252:802–808, 1991.

Simpson, I. A., and S. W. Cushman, Hormonal regulation of mammalian glucose transport. *Ann. Rev. Biochem.* 55:1059–1089, 1986.

Sutherland, E. W., Studies on the mechanism of hormone action. *Science* 177:401, 1972. Nobel laureate speech related to discovery of cAMP as second messenger.

Yarden, Y., and A. Ullrich, Growth factor receptor tyrosine kinases. *Ann. Rev. Biochem.* 57:443–478, 1988.

Problems

1. Most metabolic conversions can occur in either direction by using different pathways. Eukaryotic cells frequently take advantage of subcellular compartments to separate oppositely directed pathways. Use fatty acid synthesis and degradation as an example and discuss the design of the paths from the point of view that they are thermodynamically favorable and kinetically regulated (be sure to consider subcellular compartments). Discuss the design of glycolysis and gluconeogenesis in the liver.
2. If a starving person (one who has gone a number of weeks with no food) is given a shot of insulin, what will happen? Explain your answer.
3. Why are all known hormone receptors proteins? Could other macromolecules serve as receptors?
4. The contraceptive pill contains synthetic progestin or progestin plus estrogen. How do you suppose the pill works?
5. List several reasons why polypeptide hormones are synthesized as precursors.
6. A patient has a hypothyroid condition. He has low serum T_3 and T_4 levels and elevated serum TSH. Upon injection of TRH his serum TSH goes even higher. Is his defect primary (thyroid), secondary (pituitary), or tertiary (hypothalamus)?
7. Activation of most membrane associated hormone receptors generates a second messenger. What is a second messenger and what are the five second messengers currently known? What role do G proteins play in second messenger formation?
8. If cells are repeatedly exposed to hormones, the secondary response is lower or does not occur. Explain the phenomenon of desensitization (how does it happen) and why is this useful to the cell?
9. Is vitamin D a hormone or a vitamin? Explain your answer.
10. Inhibitors of protein synthesis have been shown to block both the rapid and slow auxin-mediated growth responses. How could you explain these observations?

Storage and Utilization of Genetic Information

6

PART

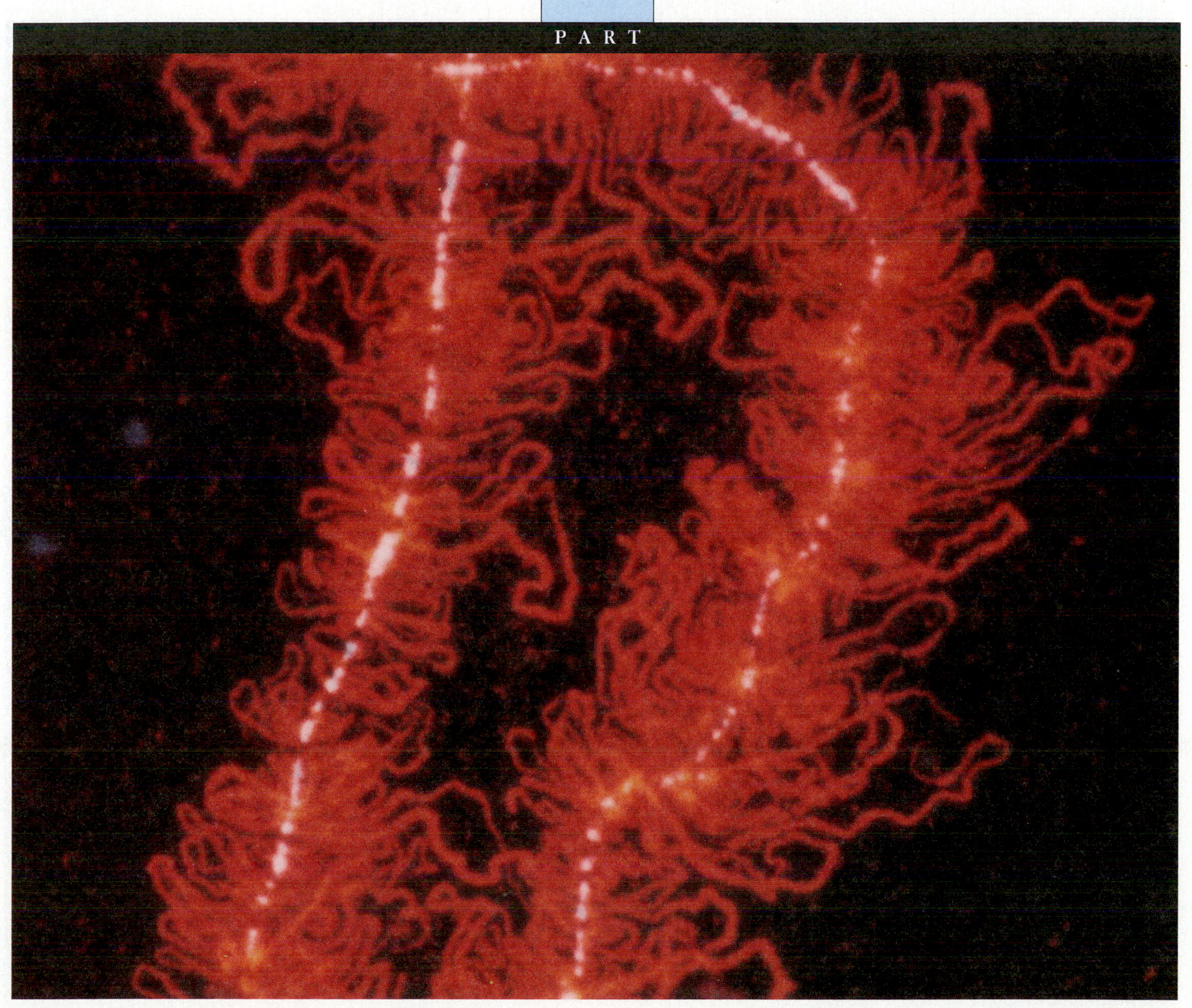

Portion of a lampbrush chromosome from an oocyte of the newt *Notophthalmus viridescens* (1,200X). The axis fluoresces white after staining with the DNA-specific dye DAPI. The lateral transcription loops fluoresce red after indirect immunofluorescent staining with a monoclonal antibody. (From M.B. Roth and J.G. Gall, Reproduced from the *Journal of Cell Biology,* 1987, 105, 1047–1054 by Copyright permission of the Rockefeller University Press.)

In the seven chapters in this part, we examine the means by which genetic information is expressed within an organism and passed on to its descendants.

Genetic information is stored in the chromosome as a sequence of nucleotides in a DNA molecule (chapter 25). This information is parceled into packets called genes, each of which usually carries the information for ordering the amino acids in a single polypeptide chain. Although we think of DNA as carrying a fixed sequence of nucleotides that is precisely replicated prior to cell division (chapter 26), there are both natural and artificial ways in which these sequences can be altered. In nature, sequences change by the processes of recombination and mutagenesis. In the laboratory, a new discipline called recombinant DNA technology (chapter 27) permits us to isolate, amplify, and systematically alter selected segments of chromosomes.

Between cell divisions, genes are transcribed into RNAs (chapter 28), which are then translated into proteins (chapter 29). As you would by now expect, transcription and translation are regulated so that gene products are synthesized in amounts meeting the requirements of the cell (chapters 30 and 31).

25

CHAPTER

Structures of Nucleic Acids and Nucleoproteins

Among the biopolymers, nucleic acids occupy a unique position. Not only are they involved in many important reactions, but they contain all biochemically useful information in the form of a genetic code. This information must be faithfully passed by replication from one cell generation to the next; it must be translated into proteins; and it should be capable of change by mutation to produce the variability on which evolutionary selection processes feed. In the following five chapters our focus will be on the biochemistry of DNA, RNA, and protein, with occasional reminders of the genetic significance of these processes.

Appropriately, we will begin this part of the text by considering certain experiments that demonstrated the genetic significance of nucleic acids. Following this, we will turn to the structural properties of nucleic acids (fig. 25.1). Next, we will discuss reactions involving DNA—first purely biochemical reactions (chapter 26), and then *in vitro* and *in vivo* reactions involving DNA manipulation (chapter 27). In chapter 28 we will deal with the process of information transfer from DNA to RNA, and finally, in chapter 29 we will consider the mechanism of information transfer from RNA to protein.

The Genetic Significance of Nucleic Acids

After the discovery, around the turn of the century, that genes are carried by chromosomes, a great deal of effort went into characterizing the sizes and shapes of chromosomes. But it was not until much later that significant progress was made in elucidating the chemical nature of the gene. As early as 1900, biologists were aware that chromosomes are composed of both nucleic acids and proteins. However, the seemingly simple chemical composition of nucleic acids misled early investigators into believing that nucleic acids were a purely structural component of the chromosome. They believed then that the arrangement of specific proteins along the chromosome accounted for gene specificity. This notion was dispelled in 1944 when Oswald T. Avery and his colleagues at Rockefeller Institute (now University) demonstrated that purified deoxyribonucleic acid (DNA) contains the genetic determinants of the bacterium *Diplococcus pneumoniae* (now Streptococcus). In this section we will discuss Avery's results, as well as some other historically important observations on the genetic significance of nucleic acids.

Figure 25.1

DNA is a long linear polymer of nucleotides. The nucleotides are very similar but they differ in several minor respects. All nucleotides are composed of a purine or a pyrimidine linked to a deoxyribose on one side. On the other side the deoxyribose is attached to a phosphate group. There are two purines, adenine and guanine, and two pyrimidines, cytosine and thymine. The polymeric structure is held together by phosphodiester linkages formed between the 5′ phosphate group of one nucleotide and the 3′OH group of the adjacent nucleotide. This gives the polynucleotide strands a directional sense (5′ to 3′). Two chains come together to form a duplex structure with hydrogen bonds formed between purines and pyrimidines on opposing chains (*a*). The two-chain structure twists to form a right-handed duplex structure (*b*). In eukaryotes the DNA wraps around a histone protein octamer, forming a hollow core (*c*). This structure forms a 200-nm-wide (2,000-Å) fiber by further coiling (*d*). The chromosome is a coiled-coil structure formed from very long 200-nm filaments that extend over the entire length of the chromosome (*e*).

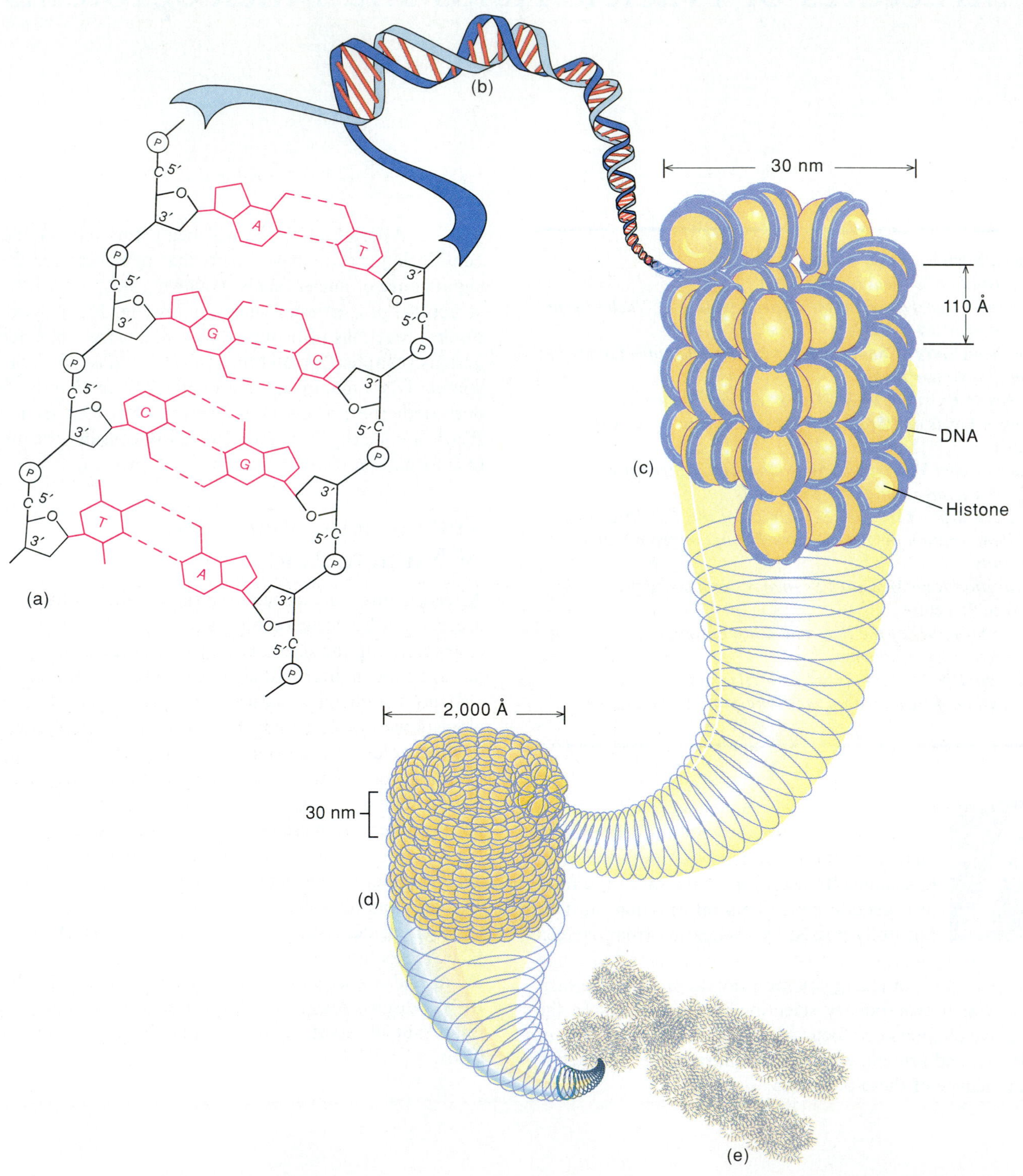

Figure 25.2

In vivo and *in vitro* evidence that DNA causes transformation of *D. pneumoniae*. (*a*) Transformation experiment by Griffith. R bacteria are nonvirulent. S bacteria are virulent. A mixture of R bacteria and heat-killed S bacteria is also virulent if transformation has occurred. (*b*) Transformation experiment by Avery and co-workers. When bacteria from a liquid culture are spread on a semisolid medium, each cell adheres to the medium at random. As time passes, the cells and their offspring grow and divide, leading to visible clones or colonies, each of which arose from a single cell. R and S cells each have a distinct clonal morphology. While R cells produce small rough clones, S colonies are smooth. R cells exposed to DNA from S cells produce a mixture of both types of clones: untransformed R colonies and transformed S colonies.

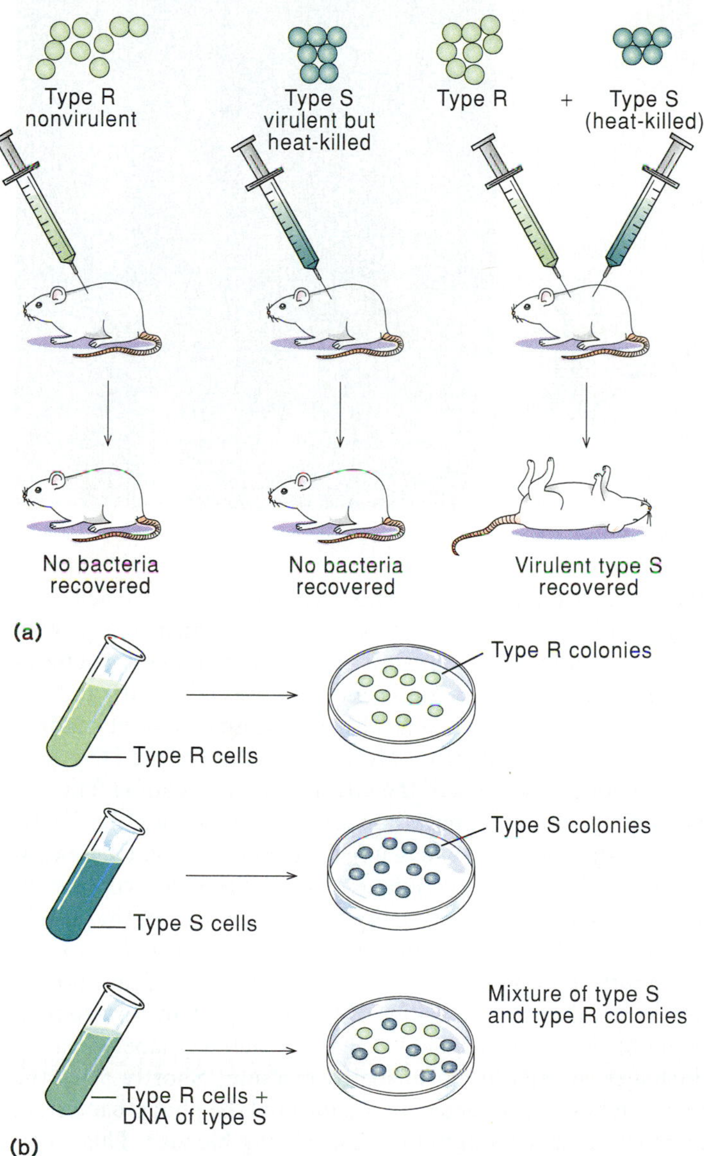

Transformation Is Mediated by DNA

In 1928 Fred Griffith was experimenting with two different strains of pneumococcus. Type S bacteria (S for smooth, from the appearance of bacterial colonies on agar plates) are encapsulated by polysaccharide. The capsules protect them from the host immune system, making the S bacteria pathogenic; even when small numbers of S bacteria are injected into mice, death results. By contrast, R bacteria (R for rough colonies) are nonencapsulated; they are readily attacked by the mouse's immune system and consequently are nonpathogenic. Although heat-killed S bacteria by themselves are nonvirulent, Griffith found that when heat-killed S bacteria were mixed with live R bacteria and introduced into a susceptible laboratory mouse, death of the animal frequently occurred (fig. 25.2*a*). S bacteria could then be detected in the blood. Apparently the genetic factor required for encapsulation was transferred from killed S cells to live R cells.

Griffith's result was duplicated *in vitro* by Dawson and Sia in 1930. The two bacterial strains, when grown in a liquid growth medium, can be distinguished by the different appearance of the R and S colonies on plates (see fig. 25.2*b*). Dawson and Sia found that, in parallel with the results that Griffith had obtained in the mouse, heat-treated extracts of S cells transformed R cells into S cells. Then Alloway, in Avery's laboratory, succeeded in isolating a stable, alcohol-precipitable transforming agent from killed smooth cultures. The *in vitro* results represented an important advance in experimental techniques. The ability to isolate the active agent (or "transforming principle," as it was then called) from S cells allowed investigators to use a quantitative approach similar to that used by biochemists to purify an enzyme.

More than ten years later, Avery provided convincing proof that the active agent in the S cell extracts was the cellular DNA. He and his co-workers did this by purifying the DNA and showing that it was extremely active in transforming R cells into S cells *in vitro* (see fig. 25.2*b*). The transforming activity in the purified extract was destroyed if the extract was first incubated with the enzyme DNase, an endodeoxyribonuclease that specifically degrades DNA. The activity was not affected by RNase, proteases, or enzymes that degrade capsular polysaccharides. Later studies showed that DNA could be used to transfer many other genetic traits between the appropriate pairs of donor and recipient bacterial strains. For example, resistance to the antibiotics streptomycin and penicillin could be transferred, with the DNA of resistant cells, to sensitive cells. In these studies, it appeared that the active genetic material being transferred faithfully reflected the mutational history and hence the genetic patterns of the donor strains. The alteration of the genetic composition of one cell strain as a result of exposure to DNA of another strain is called transformation.

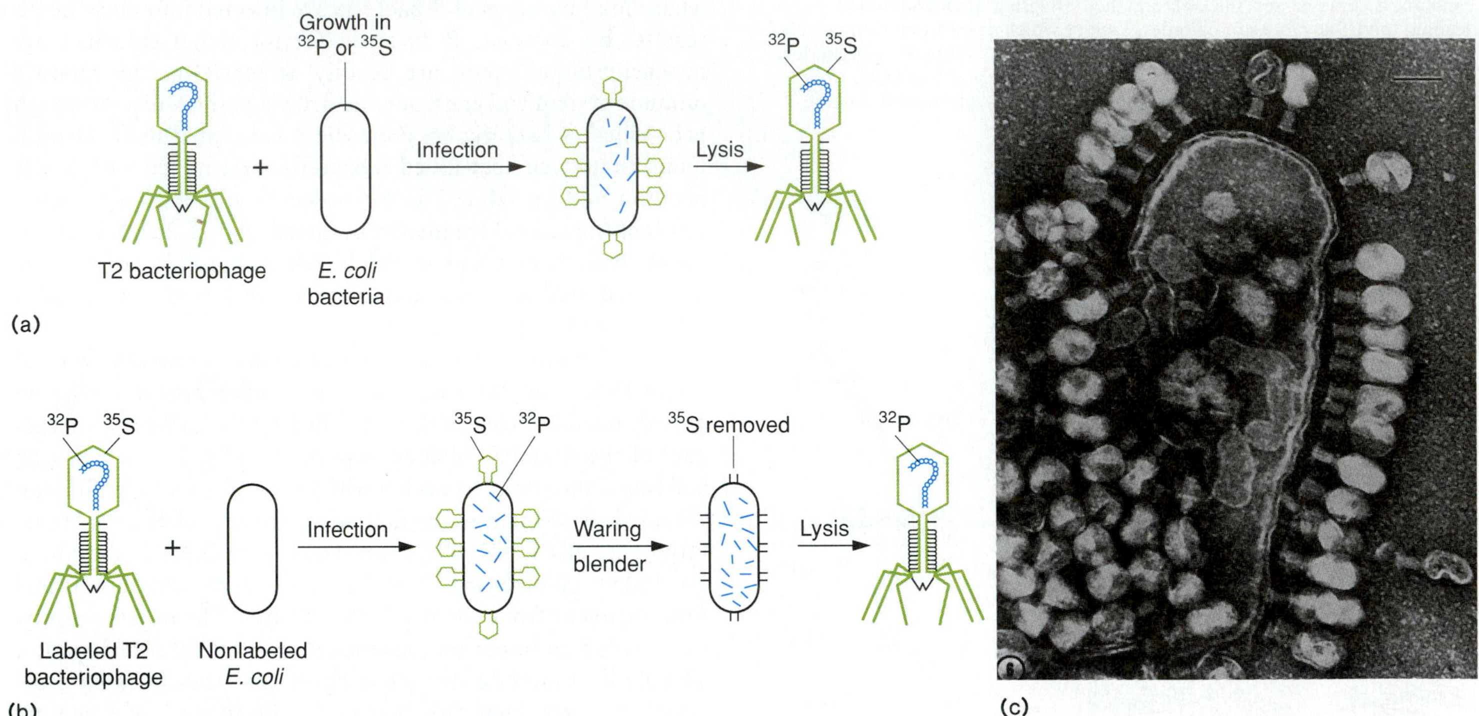

Figure 25.3

Hershey-Chase experiment demonstrating that the DNA but not the protein of the T2 bacteriophage is passed from parent to progeny phage particles. (*a*) Viruses with ^{32}P-labeled DNA or ^{35}S-labeled protein were prepared by growing the viruses on bacteria in growth medium containing $^{32}P\text{-}PO_4^{3-}$ or $^{35}S\text{-}SO_4^{2-}$, respectively. (*b*) When labeled phage particles infect an unlabeled *E. coli* cell, only the ^{32}P label appears in the progeny phage particles, showing that the DNA structure but not the protein structure is preserved during phage replication. (*c*) T4 phage (closely related to phage T2) lysing an *E. coli* cell. (Courtesy of Lee D. Simon, Wakesman Institute for Microbiology, Rutgers University.)

Transformation occurs naturally in only a small number of bacterial species other than *Streptococcus pneumoniae.* In organisms that are not transformed naturally, transformation can be induced by first making the cell envelope (e.g., of *E. coli*) partially permeable so that it can take up DNA. This technique is called artificial transformation. Conventional methods of inducing transformation include Ca^{2+} treatment and temperature shock, but for most organisms a more efficient method is electroporation, or electrical shock. For instance, by exposing yeast to electric field pulses it is possible to obtain transformation efficiencies that are 10^4 greater than that obtained by other methods. Transformation is an essential step in the cloning of genes. Transformation with self-replicating plasmids that may be carrying different genes has made it possible to select cells carrying the gene(s) of interest from large populations of cells.

Studies on Viruses Confirm the Genetic Nature of Nucleic Acid

For some time, doubts lingered about whether DNA, protein, or capsular polysaccharide was the transforming factor. During this period, other investigators performed some elegant experiments showing that nucleic acid was the genetic substance of bacterial viruses (bacteriophages) and plant viruses.

In 1952, A. D. Hershey and Martha Chase demonstrated the independent functions of viral protein and viral nucleic acid in the *E. coli* bacteriophage T2. This bacterial virus is composed of about equal weights of DNA and protein. The linear, duplex DNA is encompassed by a protein shell in a polyhedral "head" or capsid, which is connected to a protein tail that facilitates infection of cells. When a single bacteriophage particle infects a bacterium, it causes cell death and cell disruption (lysis), accompanied by the release of several hundred viruses within 30 min after infection. The fate of the DNA and the protein components of the infecting viruses can be followed by labeling with an appropriate radioactive isotope (fig. 25.3).

In an initial round of infection, Hershey and Chase used inorganic ^{32}P-labeled phosphate to label the DNA component of the virus and ^{35}S-labeled inorganic sulfate to label the protein component. The phosphorus became incorporated into the phosphoryl groups of the viral DNA and the sulfate into the sulfur-containing amino acids of the viral protein. In a second round of infection, they added the radioactively labeled bacteriophages to cells in an unlabeled medium. Shortly after the bacterial host was infected, they subjected the phage-bacterium complex to vigorous agitation in a Waring blender. This treatment sheared the phage from the attachment sites on the bacterial cell wall. The bacteria were then sedimented by centrifugation, leaving the DNA-free viral protein in the supernatant. Analysis for radioactivity showed that most of the ^{32}P-labeled viral DNA sedimented with the bacteria, whereas most of the ^{35}S-labeled viral protein remained in the supernatant. That result demonstrated that most of the viral DNA entered the cells (see fig. 25.3*b*) during infection. Quantitative isotopic analysis of the progeny phage particles indicated that they retained less than 1% of the original infecting viral protein but about two-thirds of the original infecting DNA. In other words, most of the original DNA that entered the cell on infection also became an integral part of the progeny phage, but the protein did not. Since the genetic material should survive replication, the results strongly implicated DNA as the genetic component of bacteriophage T2. Long after these experiments were performed, it was shown that purified, deproteinized phage DNA could itself produce phage after being taken up by host cells. Infection of bacteria with phage-derived DNA is referred to as transfection.

Early experiments on plant viruses provided even more convincing evidence identifying nucleic acid as the genetic substance of a virus. This work was done on tobacco mosaic virus (TMV), the first virus to be crystallized. TMV is a plant virus with a molecular weight of 40×10^6. The virion contains one molecule of single-stranded RNA with a molecular weight of 2×10^6 and 2,130 identical protein subunits, each with a molecular weight of 18,000. In 1956 Gierer and Schramm showed that the deproteinized viral RNA produced lesions on tobacco plant leaves similar in appearance to the lesions produced by the intact virus.

Shortly thereafter (1957) Fraenkel-Conrat and Singer obtained similar evidence that RNA was the genetic substance of TMV (fig. 25.4). Using two different strains of the virus, HR and TMV, each with readily identifiable protein components, they generated reconstituted viruses *in vitro* in all possible combinations:

HR-protein + HR-RNA
HR-protein + TMV-RNA
TMV-protein + HR-RNA
TMV-protein + TMV-RNA

The reconstituted virus made from purified RNA and protein components had infectious activities comparable to those of the normal viruses. After infection the protein components were examined. For all combinations the type of viral protein found in progeny virus was determined by the type of RNA in the reconstituted particles, a very elegant proof that the nucleic acid carries the genetic determinants of the viral protein.

The studies that we have described in this section show that the genetic information required for replication of a cell (e.g., pneumococcus) or a virus (T2, TMV) is carried by nucleic acids. In the case of a cell the genetic information is always carried by DNA. In the case of viruses the genetic information can be carried by either single- or double-stranded DNA or by RNA, depending on the virus. Sometimes the bacterial or viral DNAs are linear overall; sometimes they are circular. The only case in which circular RNAs have been encountered is in viroids—low-molecular-weight, single-stranded RNAs that cause several important diseases in cultivated plants but are not encapsidated. Viroids are the smallest known agents that cause infectious disease.

Figure 25.4

Reconstitution experiment with tobacco mosaic virus. The RNA and protein components of TMV may be separated and reconstituted. If two strains are used and if the RNA and protein parts from different viruses are used in the reconstitution experiment, the progeny virus will always reflect the parental type from which the RNA was taken.

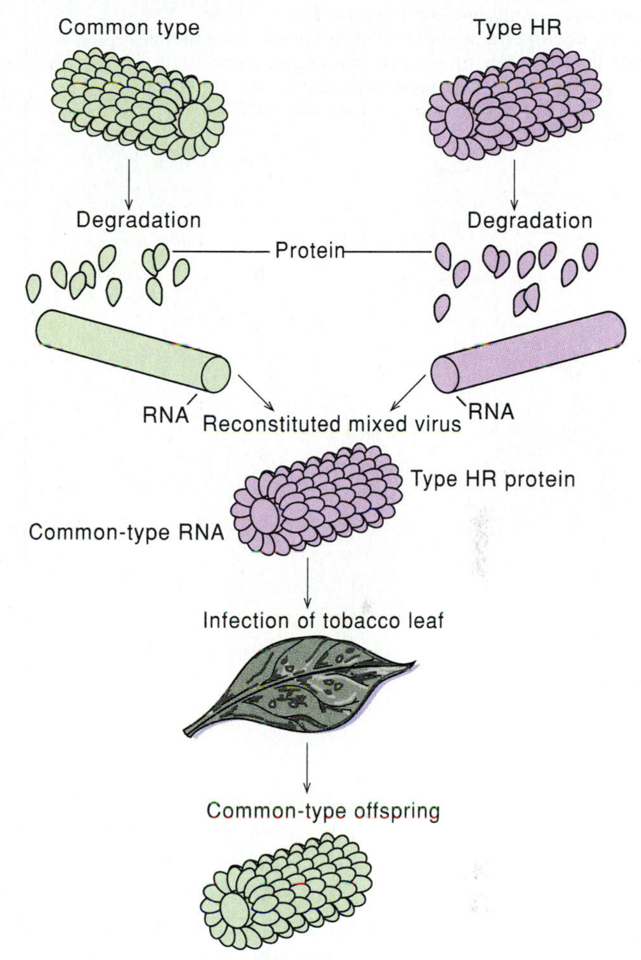

Structural Properties of DNA

The early work equating the genetic material with certain nucleic acids plunged genetics into an entirely new vocabulary of chemical terms. The genetic consequences of these early studies could be understood, but the chemistry was new. The remainder of this chapter focuses on the chemical and structural properties of nucleic acids.

As we have seen, genes can be composed of either DNA or RNA. However, RNA is the genetic component only in some viruses. Otherwise, DNA provides the genetic information, and

Figure 25.5

Genome size in different cells, viruses, and plasmids. Plasmids are small circular DNA molecules that replicate autonomously in cells harboring them. Unlike viruses, they do not form any complex nucleoprotein structures. In the case of plasmids, most viruses, and bacteria, the genome size is equivalent to the size of the chromosomal DNA because there is only one chromosome. For all of the remaining eukaryotic organisms that are listed, the genome is subdivided into two or more chromosomes. Some organisms contain over one hundred chromosomes. All chromosomes are believed to contain a single DNA molecule.

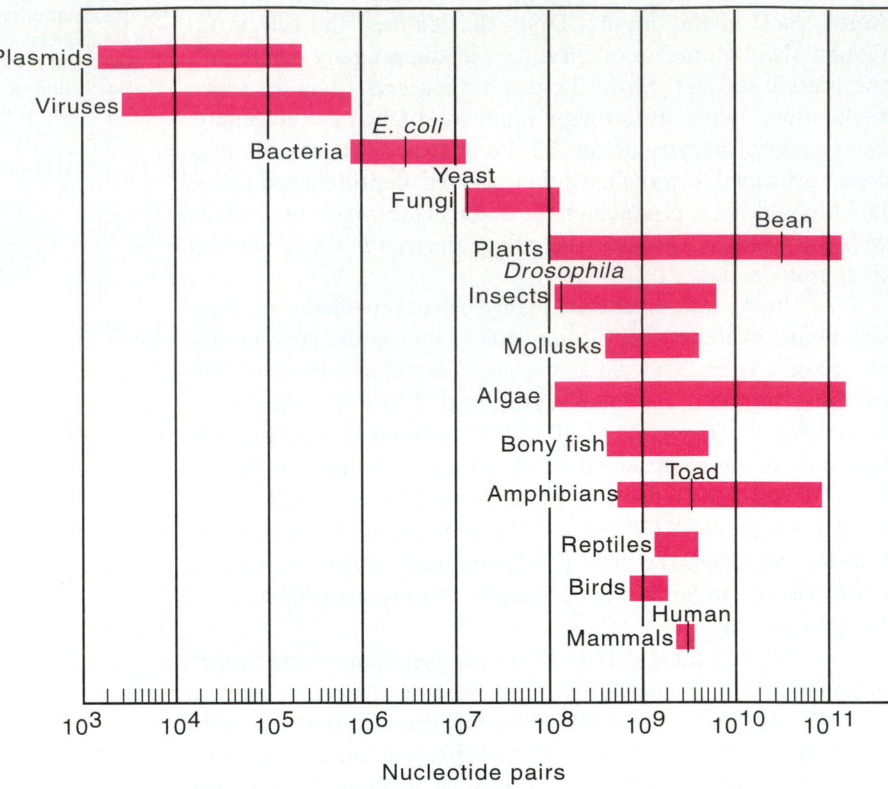

this is transcribed into RNA, which, in most cases, is subsequently used to direct protein synthesis. The different functions of the two cellular nucleic acids are reflected by their different locations in the cell; DNA is found almost exclusively in the cell nucleus, whereas the mature forms of RNA are located mainly in the cytoplasm, where protein synthesis takes place.

The amount of DNA per cell differs widely among different organisms (fig. 25.5). Mammalian cells contain about 1,000 times as much DNA as bacterial cells. Bacterial viruses such as the T-type bacteriophages that infect *E. coli* contain one-tenth to one-twentieth as much DNA as the bacterial host chromosome. The DNA of the smallest viruses is, in turn, about one-tenth of the amount found in the smallest T-phage DNA, containing barely enough genetic material to accommodate about ten genes. This finding is consistent with the fact that viruses do not contain sufficient genetic information for independent growth, but can only grow parasitically in the host cells they infect. On the other hand, the amount of DNA per cell is not always directly proportional to the amount of genetic information that an organism carries. This is because complex eukaryotes contain in their chromosomes a great deal of noninformational DNA, whose function is frequently unclear.

In addition to the main body of DNA associated with the cell or the virus, there is informational DNA found in special organelles such as chloroplasts and mitochondria. This DNA carries genes for some ribosomal RNAs and transfer RNAs and for proteins associated with the respective organelles. The simplest type of system using informational DNA is the plasmid. Plasmids vary in size and exist as naked DNA molecules that carry anywhere from 5 to 100 genes. They are very widespread in bacteria. They are active in both prokaryotes and eukaryotes and in the past ten years they have proved to be extremely useful for DNA manipulation (see chapter 27).

The Polynucleotide Chain Contains Mononucleotides Linked by Phosphodiester Bonds

Nucleotides are the building blocks of nucleic acids; we discussed nucleotide structures and biochemistry in chapter 20, which you may wish to review at this point or as the need for relevant details arises. When a 5′-phosphomononucleotide is joined by a phosphodiester bond to the 3′-OH group of another mononucleotide, a dinucleotide is formed. The 3′-5′-linked phosphodiester internucleotide structure of nucleic acids was firmly established in 1951 by Alexander Todd, who in 1957 was awarded a Nobel Prize for Chemistry as a result. Repetition of the phosphodiester linkage leads to the formation of polydeoxyribonucleotides in DNA or polyribonucleotides in RNA. The structure of a short polydeoxyribonucleotide is shown in figure 25.6. The polymeric structure consists of a sugar phosphate diester backbone with bases attached as distinctive side chains to the sugars.

Figure 25.6

The structure of a deoxyribonucleotide. Drawn in abbreviated form at lower left. The illustrated structure is written pTpApCpG.

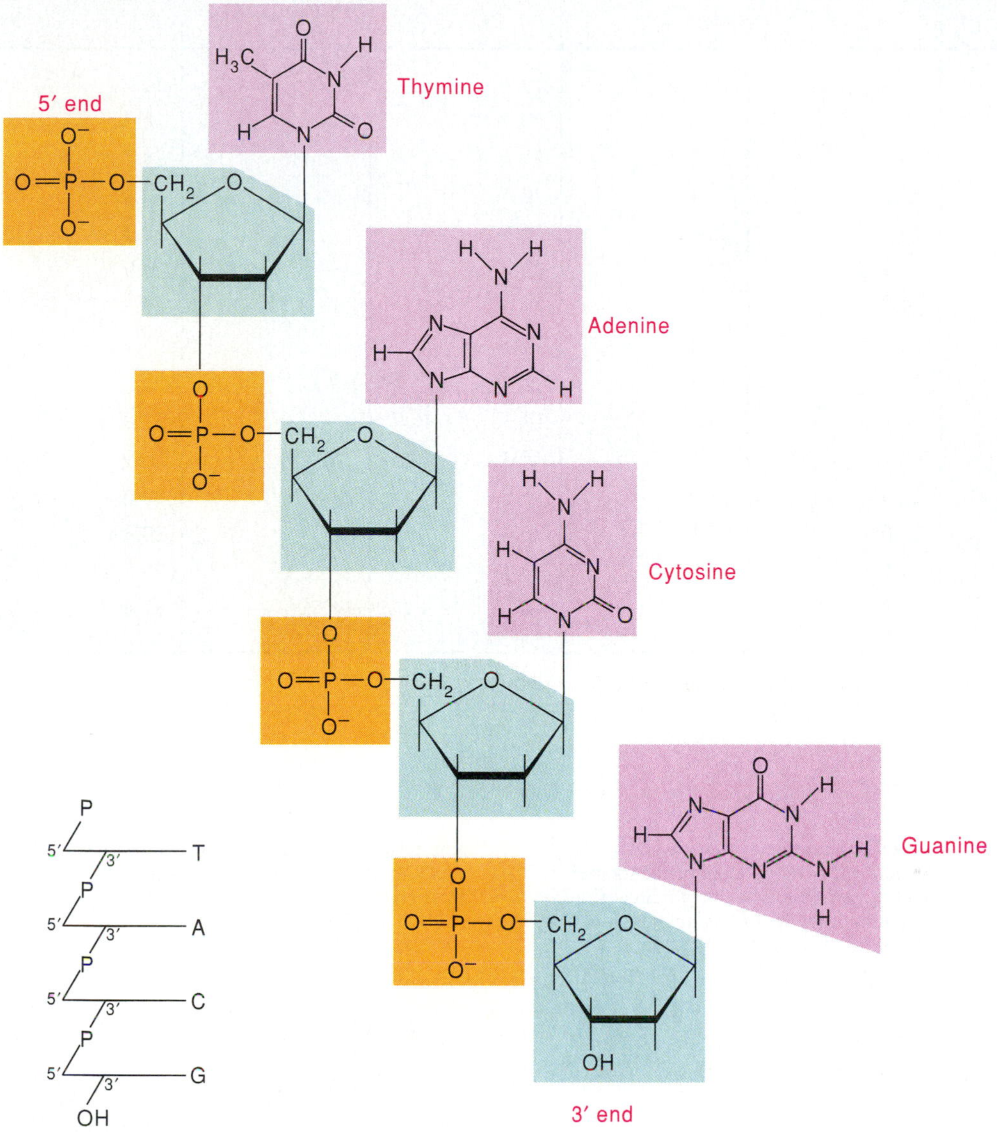

The polynucleotide chain has a directional sense with a 5′ and a 3′ end. Either of these ends may contain a free hydroxyl group or a phosphorylated hydroxyl group. The structure shown in figure 24.6 contains a phosphate group on the 5′ end but none on the 3′ end. By convention, we write a nucleic acid sequence from the 5′ to the 3′ end. Thus we would write the structure in figure 24.6 pTpApCpG. If it had no phosphate on the 5′ end, we would write the structure TpApCpG; alternatively, if the terminal phosphate were on the 3′ end rather than the 5′ end, we would write the structure TpApCpGp. When the phosphates are not indicated, we write the oligonucleotide TACG. If the oligonucleotide is part of a larger polynucleotide, we indicate this fact by dashes on either end: -TACG-. The letter "d" or "r" sometimes precedes the capital letter of the nucleotide to indicate a deoxyribo- or a ribo-derivative.

Most DNAs Exist as Double-Helix (Duplex) Structures

Like most other types of biological macromolecules, nucleic acids adopt highly organized three-dimensional structures. The dominant factors that determine nucleic acid conformation are the limitations imposed by the stereochemistry of the polynucleotide chains, the high negative charge resulting from the regularly repeating phosphate groups, and the noncovalent affinities between purine and pyrimidine bases.

A body of chemical information that proved vital to understanding DNA structure came from Erwin Chargaff's analyses of the nucleotide composition of duplex DNAs from various sources (table 25.1). Although the base compositions varied over a wide range, Chargaff found that within the DNA

Table 25.1

Base Composition of DNAs from Different Sources

	(A) Adenine	(G) Guanine	(C) Cytosine	(5-MC) 5-Methylcytosine	(T) Thymine	$\frac{A + T}{G + C + 5\text{-MC}}$
Human	30.4	19.6	19.9	0.7	30.1	1.53
Sheep	29.3	21.1	20.9	1.0	28.7	1.38
Ox	29.0	21.2	21.2	1.3	28.7	1.36
Rat	28.6	21.4	20.4	1.1	28.4	1.33
Hen	28.0	22.0	21.6		28.4	1.29
Turtle	28.7	22.0	21.3		27.9	1.31
Trout	29.7	22.2	20.5		27.5	1.34
Salmon	28.9	22.4	21.6		27.1	1.27
Locust	29.3	20.5	20.7	0.2	29.3	1.41
Sea urchin	28.4	19.5	19.3		32.8	1.58
Carrot	26.7	23.1	17.3	5.9	26.9	1.16
Clover	29.9	21.0	15.6	4.8	28.6	1.41
Neurospora crassa	23.0	27.1	26.6		23.3	0.86
Escherichia coli	24.7	26.0	25.7		23.6	0.93
T4 bacteriophage	32.3	17.6		16.7[a]	33.4	1.91

[a]In T bacteriophage all of the cytosine exists in the 5-hydroxymethyl form 5-HMC.

Figure 25.7

Dimensions and hydrogen bonding of (*a*) thymine to adenine and (*b*) cytosine to guanine. Note that there are two hydrogen bonds formed in the A-T base pair and three in the G-C base pair. The overall dimensions of the base pairs are the same. Consequently they will fit at any position in an otherwise regular polymeric structure (Source: S. Arnott et al., "Fourier synthesis studies of lithium DNA, part III: Hoogsteen models" in *Journal of Molecular Biology* 11:391, 1965. Copyright © 1965 Academic Press Ltd., London, England.)

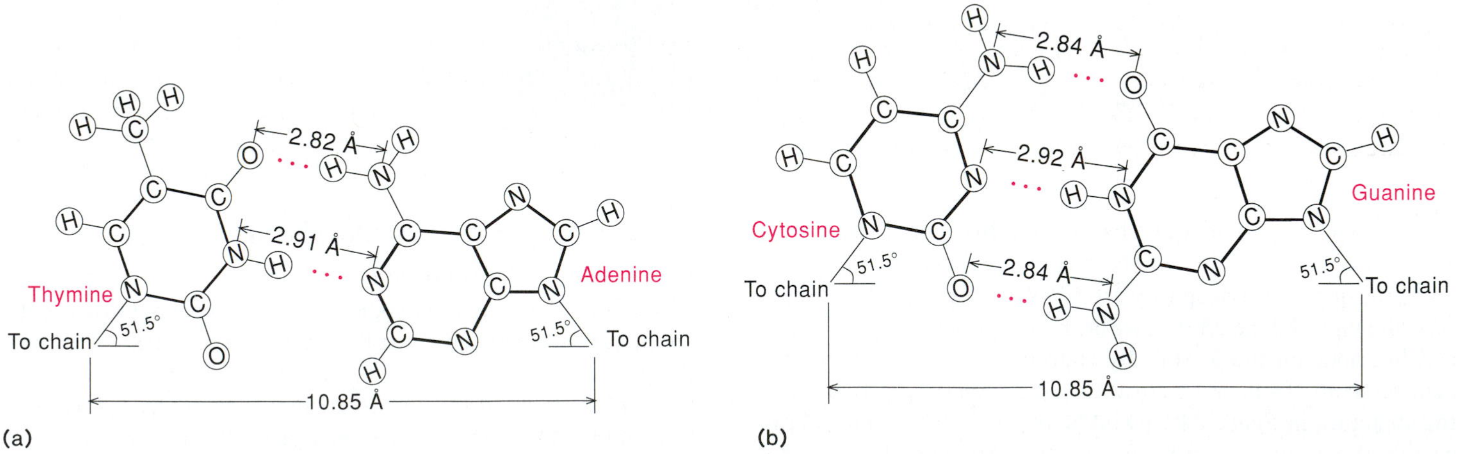

of each source that he examined, the amount of adenine (A) was very nearly equal to the amount of thymine (T), and the amount of guanine (G) was very nearly equal to the amount of cytosine (C). The cytosine is present as both unmodified cytosine and, to a lesser extent, 5-methylcytosine, which results from postreplicative modification. The two equalities were the first indication that regular complexes occur between A and T and between G and C in DNA.

While searching for the meaning of these equalities, James Watson discovered, with the help of molecular models, that between A and T and between G and C it was possible to form hydrogen-bonded base-paired structures that have the same overall dimensions (fig. 25.7). Two hydrogen bonds are formed in the A-T base-paired structure and three in the G-C base-paired structure. The hydrogen-bonded pairs are formed between bases of opposing strands and can arise only if the directional senses of the two interacting chains are opposite or antiparallel (fig. 25.8). Watson brought this information to the attention of his crystallographer colleague Francis Crick, who realized that the x-ray diffraction pattern produced by DNA could be interpreted in terms of a helix (see box 25A) composed of two polynucleotide strands. In this structure the planes of the base pairs are perpendicular to the helix axis, and the distance between adjacent pairs along the helix axis is 3.4 Å, bringing them into close contact (fig. 25.9). The structure repeats itself after about ten residues, or once every 34 Å along the helix axis; the repeating distance is referred to as the pitch length or just the pitch. An average-size bacterial gene is about 1,000 base pairs (bp) in length, equivalent to 100 helical turns. The complementary structure of duplex DNA suggests how the genetic material is faithfully replicated as well as how it is expressed.

An important feature of the helical structure is the grooved nature of the surface resulting from the helical twist. Alternating wide (major) and narrow (minor) grooves are displayed in a side view of the helix structure (see fig. 25.9*a* and *b*). Different sections of the purine and pyrimidine bases are exposed in these two grooves as indicated in figure 25.9*c*. Many different proteins interact with DNA; most of the interactions occur with the phosphoryl groups on the outer surface of the structure and with the purine and pyrimidine bases in the wide groove because of its greater accessibility. We will consider specific instances of DNA–protein interactions later (chapters 28, 30, and 31).

Hydrogen Bonds and Stacking Forces Stabilize the Double Helix

Several factors account for the stability of the double-helix structure. The negatively charged phosphoryl groups are all located on the outer surface, where they have a minimum effect on one another. The repulsive electrostatic interactions generated by these charged groups are often partly neutralized by interaction with cations such as Mg^{2+}, basic polyamines (such as putrescine and spermidine), and the positively charged side chains of chromosomal proteins. The core of the helix is composed of the base pairs, held together by the specific hydrogen bonds and also by favorable stacking interactions between the planes of adjacent base pairs. These stacking interactions are complex, involving dipole–dipole interactions and van der Waals forces; they result in a stacking energy comparable in magnitude to the stabilizing energy generated by the hydrogen bonds between the base pairs. The result is that stacking is maximized in most nucleic acid structures. In this connection, it is noteworthy that two fully extended polynucleotide strands can form a hydrogen-bonded base-paired complex that leads to a stepladderlike structure (fig. 25.10*b*). In this structure the chains do not form a helix but lie straight, with a distance of 6.8 Å between identical residues in the direction of the long axis. This 6.8-Å distance between adjacent base pairs produces a gap that would presumably have to be filled by water. Such a conformation does not result in an energetically favorable structure, however, since the planes of the bases would prefer to be in closer contact with each other than with water. The stepladder structure can be converted into the helix structure by a simple right-handed twist (see fig. 25.10*a*). When this is done, the distance between base pairs decreases until they are in close contact, with a spacing of 3.4 Å.

Figure 25.8

Segment of DNA, drawn to emphasize the hydrogen bonds formed between opposing chains. Each type of base is represented by a different color, with the sugar-phosphate backbones in black. Note the three hydrogen bonds in the G-C pairs and the two in the A-T pairs (A, red; T, green; G, yellow; C, blue). The two strands are antiparallel: one strand (left side) runs 5′ to 3′ from top to bottom and the other strand (right side) runs 5′ to 3′ from bottom to top. The planes of the base pairs are turned 90° to show the hydrogen bonds between the base pairs. (Adapted from "The Synthesis of DNA" by Arthur Kornberg. Copyright © 1968 by Scientific American, Inc.)

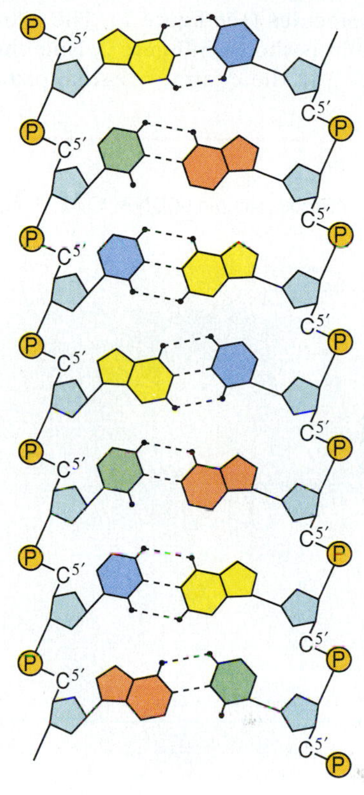

25A BOX

X-Ray Diffraction of DNA

X-ray diffraction of DNA is performed in a manner similar to x-ray diffraction of fibrous proteins (see box 4A), except that the stretched fiber in this case contains many DNA molecules (see figure 1). The diffraction pattern recorded on the film is shown in figure 2. Note the strong 3.4-Å and 34-Å spacings and the central crosslike pattern, which reflects a helix structure. Watson and Crick were the first to appreciate the significance of these features. They concluded that the diffraction pattern was consistent with a hydrogen-bonded antiparallel double-helix structure—the model that is now accepted as a correct description of the structure of DNA.

Figure 1

Diffraction pattern of a fibrous sample of DNA. © M. H. F. Wilkins.

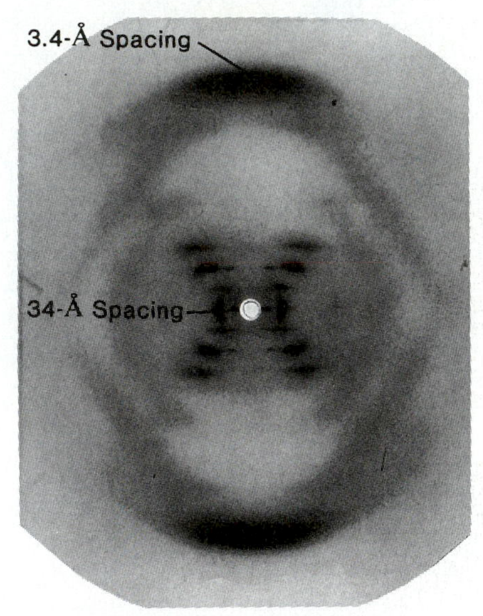

Figure 2

Camera setup for obtaining DNA diffraction pattern.

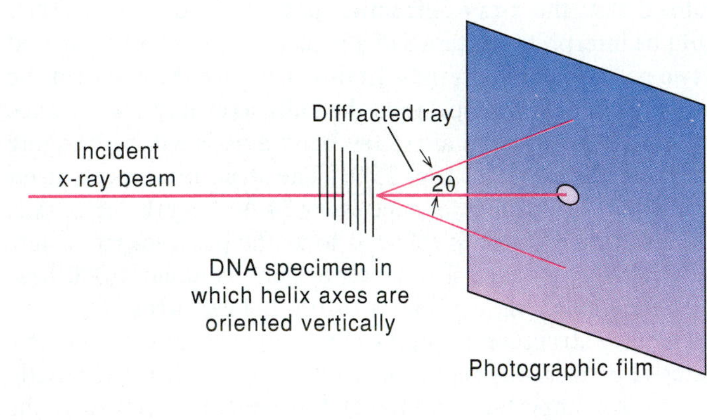

Conformational Variants of the Double-Helix Structure

The same base-pairing arrangement is found in all naturally occurring double-helix structures. However, the inherent flexibility in the furanose ring of the sugar and the degrees of freedom generated by several rotatable single bonds per residue—six in the sugar phosphate backbone and one in the C-1′ N-glycosidic linkage (fig. 25.11)—lead to considerable variation in the conformations adopted by double-helix structures. Four puckered conformations for the sugar, with small differences in stability, are shown in figure 25.12. In the double-helix structure shown in figure 25.8, the furanose rings are in the C-2′-endo conformation. This is believed to be the major conformation adopted by the sugars in DNA, both when free in aqueous solution and in chromosomes. It is known as the B form of DNA. When some of the water is removed from the hydrated DNA fibers, the double helices push even closer together and the structure changes to the so-called A form, which has about eleven bases per turn and base pairs that are tilted about 20° with respect to the helix axis (fig. 25.13*b*). In the A form, the furanose rings have changed their pucker to the C-3′-endo conformation. The furanose rings in RNA have a stronger preference for the C-3′-endo conformation, with the result that RNA double helices adopt a structure similar to the DNA A form, even at high degrees of hydration.

The most striking conformational variant observed for a DNA double helix with Watson-Crick base pairing is referred to as the Z form. This structure was first detected by

Figure 25.9

The most common form of the double-helix DNA. The base-paired structure shown in figure 25.8 forms the helix structure shown in (*a*) and (*b*) by a right-handed twist. The two strands are antiparallel as indicated by the curved arrows in (*a*). In (*b*) a space-filling model depicts the sugar-phosphate backbones as strings of mostly gray, red, white, and yellow spheres, while the base pairs are rendered as horizontal flat plates composed of dark blue spheres. In (*c*) the orientation of the groups in the base pairs with respect to the wide and narrow grooves is indicated. The wide and narrow grooves formed in the right-handed duplex structure are sometimes referred to as the major and minor grooves.

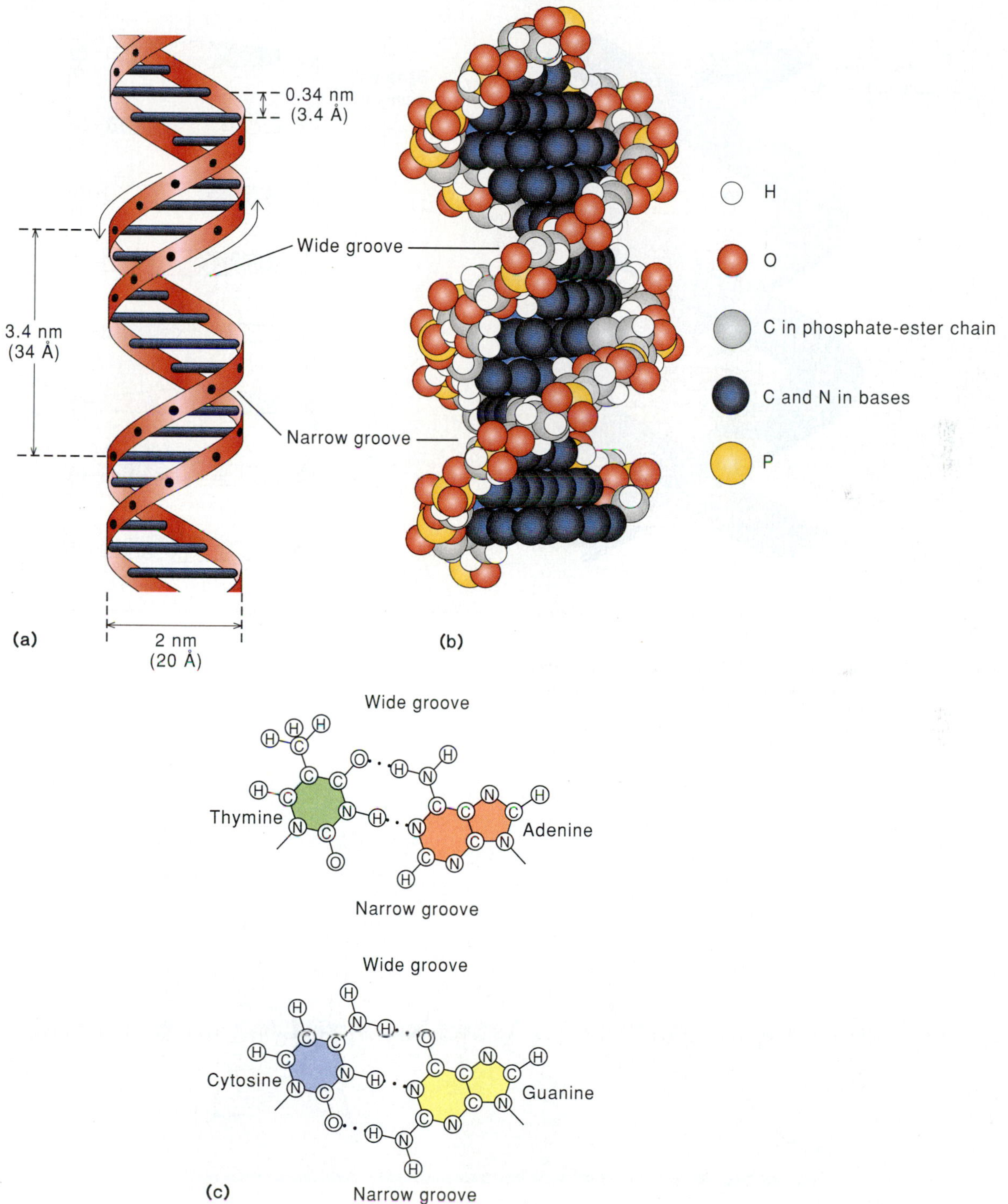

Figure 25.10

Different conformations of base-paired DNA: (*a*) the normal spiral ladder, (*b*) the untwisted straight ladder. The second structure is unstable; it can be converted into a spiral ladder by a right-handed twist, a change that permits the planes of the base pairs to come into close contact.

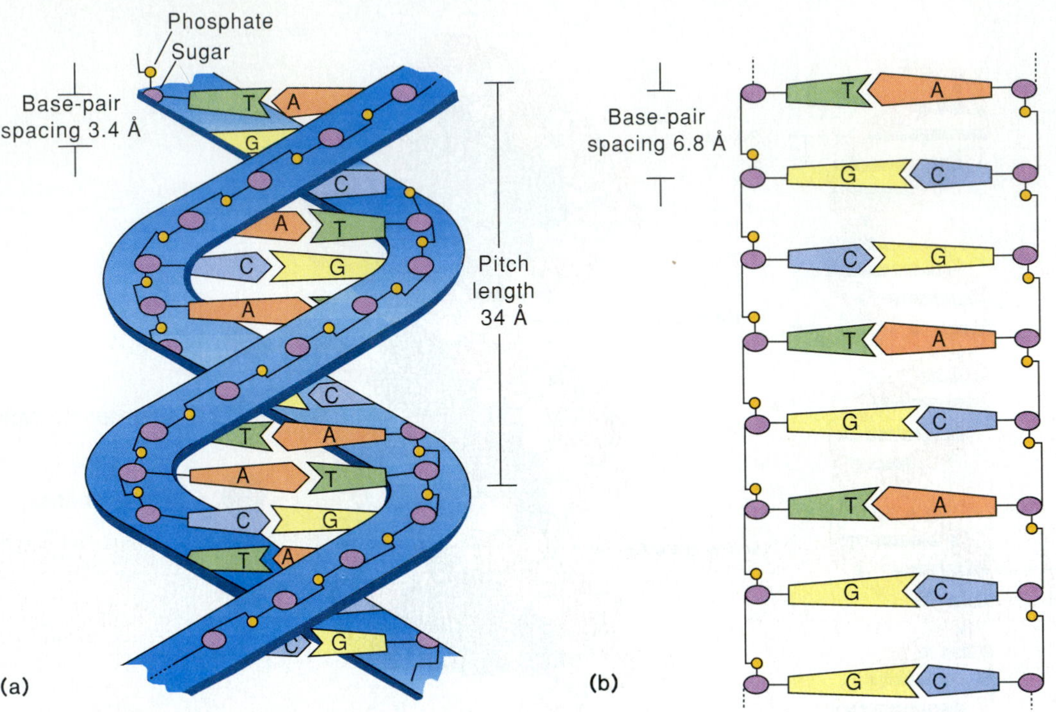

Figure 25.11

A segment of polynucleotide chain with rotatable bonds indicated by curved arrows. The presence of so many rotatable bonds permits the backbone of a polynucleotide chain to adopt many different conformations. (Source: W. K. Olson and P. J. Flory, "Different conformations of deoxyribose in DNA" in *Biopolymers*, 11:1, 1972. Copyright © 1972 John Wiley & Sons, Inc., New York, N.Y.)

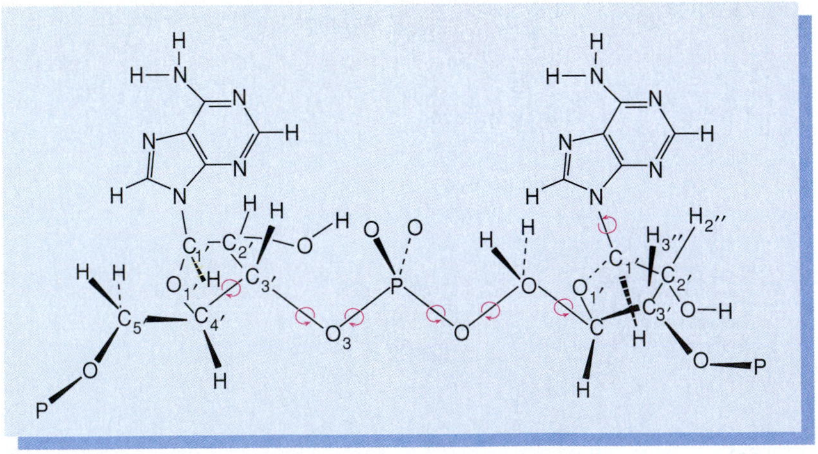

Figure 25.12

Four pucker conformations of the furanose rings of ribose and deoxyribose that are deemed energetically feasible. The flexibility of these furanose rings contributes to the possible variations in the conformation of the double helix.

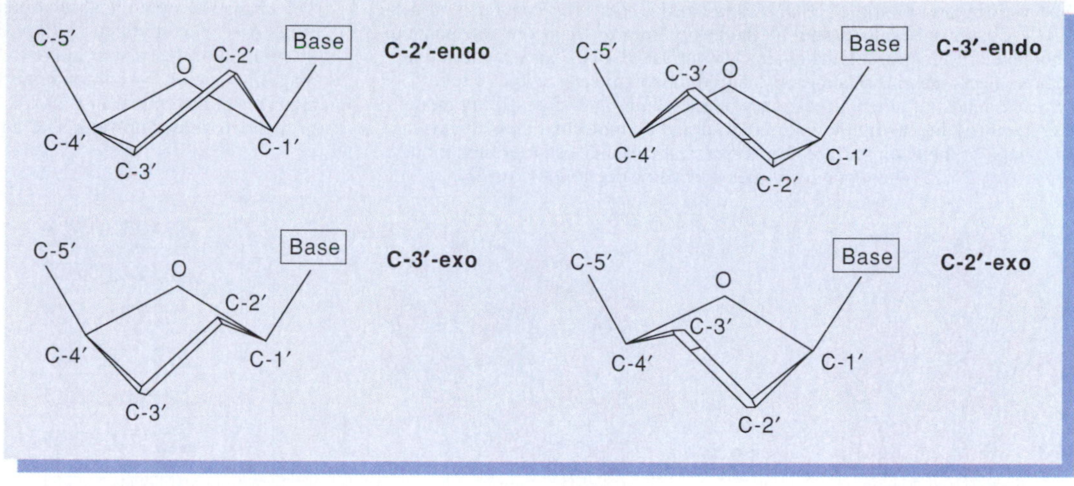

Figure 25.13

The A and B forms of DNA. (*Top*) A view perpendicular to the helix axis. (*Bottom*) A view of two adjacent base pairs, looking down the helix axis. A single turning is shown for each duplex. DNA usually is found in the B form. The A form has been observed in microcrystals of DNA from which a significant amount of the water has been removed. Dimensions of the two structures are compared in table 25.2.

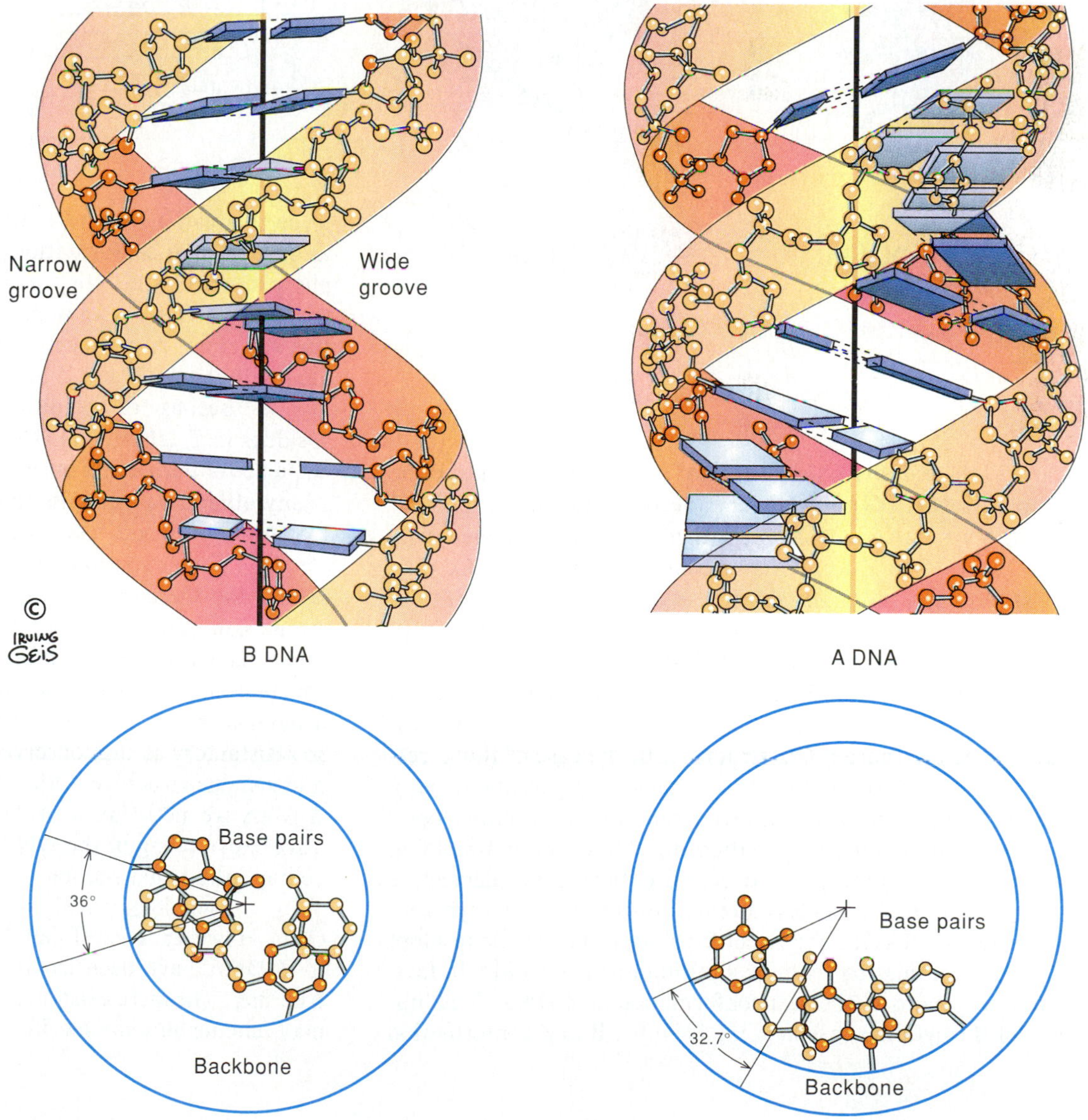

Narrow groove

Wide groove

© Irving Geis

B DNA

A DNA

36°

Base pairs

Backbone

Base pairs

32.7°

Backbone

Figure 25.14

Space-filling models of (*a*) Z DNA and (*b*) B DNA. The irregularity of the Z DNA backbone is illustrated by the heavy lines that go from phosphate to phosphate residues along the chain. In contrast, B DNA has a smooth line that connects the phosphate groups and the two grooves, neither one of which extends into the helix axis of the molecule. The space-filling model is excellent for displaying the volume occupied by molecular constituents and the shape of the outer surface. By contrast, the framework representations of DNA (fig. 25.13) show connectivities and allow one to look inside.

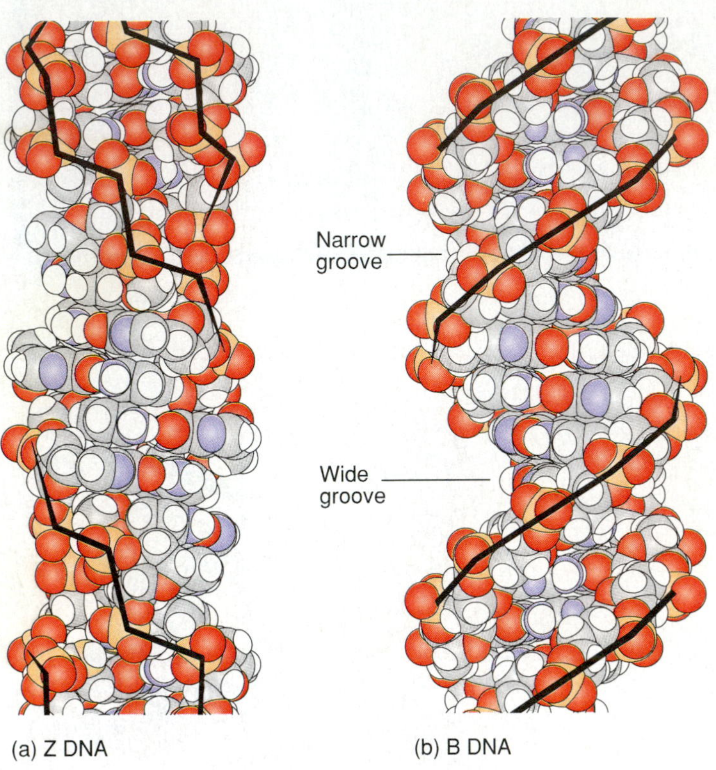

Figure 25.15

The change in topological relationship if a four-base-pair segment of B DNA is converted into Z DNA. Such a conversion could be accomplished by rotation of the bases relative to those in B DNA. This rotation is shown diagrammatically by coloring one surface of the bases. All of the colored areas are at the bottom in B DNA. In the segment of Z DNA, however, four of them are turned upward. The turning is indicated by the curved arrows.

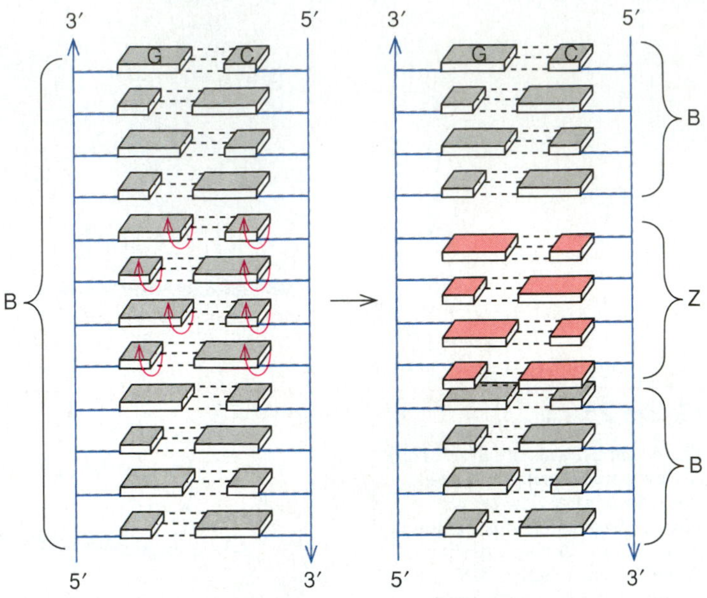

Alex Rich and his co-workers for the deoxyoligonucleotide d(CpGpCpGpCpG), which crystallizes into an antiparallel double helix with a left-handed rather than a right-handed twist (fig. 25.14). The Z form is a considerably slimmer helix than the B form and contains twelve base pairs per turn rather than ten. In the Z form, the planes of the base pairs are rotated approximately 180° with respect to the helix axis from their orientation in the B form (fig. 25.15). The flipping of the base pairs involves different conformational changes in the G and C residues in the alternating GC structure. In the case of the G residues the base is rotated by 180° about the glycosidic bond, a change resulting in a transition from the anti conformation found in B DNA to the syn conformation (fig. 25.16). Model-building studies indicate that the anti conformation of the nucleotide (found in B DNA) has less steric crowding than the syn conformation but that it is far easier for a purine nucleotide to adopt the syn conformation than for a pyrimidine nucleotide. In fact, cytidine remains in the anti conformation in Z DNA. The flipping of the cytidine base in going from the B to Z conformation involves rotation of the entire cytidine residue while maintaining the anti conformation. The effect is to make the sugar-phosphate backbone follow a zigzag course (see fig. 25.14). Thus the name Z DNA is an appropriate descriptive designation for this structure. A number of the structural parameters associated with the A, B, and Z helices are summarized in table 25.2.

Because of the different orientations of the G and C residues in Z DNA, this DNA conformation requires that the sequence of purine and pyrimidine bases be strictly alternating. Many other arrangements that involve alternating purine and pyrimidine residues can adopt the Z conformation. For example, a duplex containing alternating T and G residues in one strand and the complementary A and C residues in the other strand can adopt a Z conformation. An alternating A-and-T DNA sequence has never been observed in a Z conformation. The reason is believed to be the way that water molecules orient around an A-T base pair in the Z helix. The arrangement is not so satisfactory as that observed for the G-C base pair, creating a less stable structure. Other factors that favor the stability of Z DNA are methylation of the 5 position of C residues and negative supercoiling of the DNA. We will discuss negative supercoiling in the next section.

The biological significance of Z DNA is currently unclear. However, several cellular proteins that bind specifically to Z DNA have been isolated from the nuclei of *Drosophila* fruit flies. The mere existence of such proteins suggests that they may function in some specific role when they encounter stretches

Figure 25.16

The syn and anti conformations of deoxyadenosine and deoxycytidine. Purine nucleosides can readily adapt to either conformation by a simple rotation about the C-1—N-9 glycosidic bond. Pyrimidine nucleosides are considerably less stable in the syn conformation because of steric hindrance between the sugar and the C-2 carbonyl group.

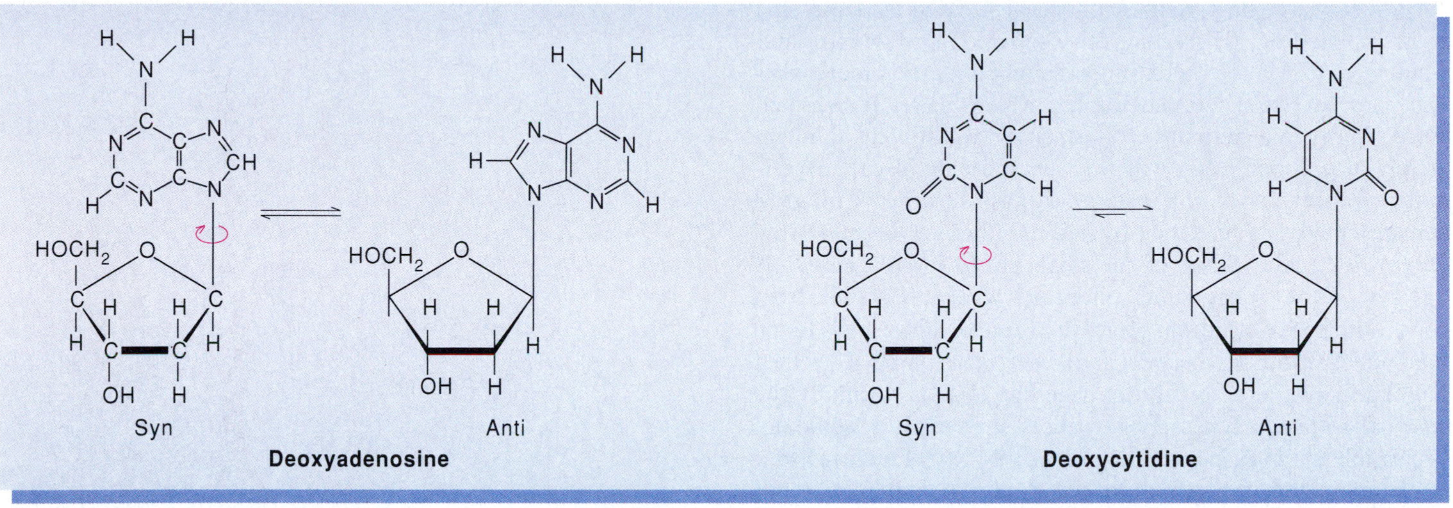

Syn Anti
Deoxyadenosine
Syn Anti
Deoxycytidine

Table 25.2 Some Structural Parameters of A, B, and Z DNA	A DNA	B DNA	Z DNA
Helix sense	Right-handed	Right-handed	Left-handed
Residues per turn	11	10	12 (6 dimers)
Rise per residue	2.55 Å	3.4 Å	3.7 Å
Helix pitch	28 Å	34 Å	45 Å
Base pair tilt	20°	6°	7°
Rotation per residue	33°	36°	−60° (per dimer)
Glycosidic conformation			
Deoxycytidine	Anti	Anti	Anti
Deoxyguanosine	Anti	Anti	Syn
Sugar pucker			
Deoxycytidine	C-3′-endo	C-2′-endo	C-2′-endo
Deoxyguanosine	C-3′-endo	C-2′-endo	C-3′-endo

of DNA that can adopt a Z conformation. A mutation-genetic analysis of such Z-DNA-binding proteins would be helpful to ascertain their role(s). Current speculation on the biological significance of the Z conformation centers on the general notion that the conformation plays a regulatory role in gene expression and possible genetic recombination.

In addition to the A, B, and Z helices, many other conformations have been observed for DNA. All, however, preserve Watson-Crick hydrogen bonding. Synthetic polyribonucleotide complexes have been made that simulate the normal duplex structure. For example, poly(A) and poly(U) in 1:1 base-pair ratios form a duplex resembling the A form of DNA. In 1:2 ratios of one A chain to two U chains they form a triplex that displays additional kinds of hydrogen bonding. Two chains of A make a double helix and four chains of G can form a quadruple helix. A great variety of structures and hydrogen-bonding patterns is possible with polymers that have special sequences. The unusual hydrogen-bonded structures are difficult to detect in nature. Nevertheless, triple helices have been detected *in vivo*. To this, we would add that unusual pairings occur in transfer RNA (see chapter 28).

Duplex Structures Can Form Supercoils

The detection of different conformations of DNA underscores the inherent flexibility built into the DNA duplex. All the conformations discussed thus far involve regular linear duplexes. Energetically favorable interactions with other molecules, particularly proteins, can induce additional conformations that do not result in major changes in either base pairing or stacking. Several conformations are believed to play important roles in different situations. Bends are known to be important in structures formed by chromosomes (discussed later in this chapter). Cruciforms, in which a single chain folds back on itself into a hairpinlike duplex, are important as intermediates in DNA and RNA synthesis (chapters 26 and 28). Supercoiled DNA is a very common type of tertiary structure in which the double-helix segments twist around each other. Supercoiled DNA is topologically constrained by being covalently closed and circular, or by being complexed to proteins so that the ends of the DNA cannot rotate freely.

DNA can form right-handed (negatively supercoiled) or left-handed (positively supercoiled) supercoils (fig. 25.17). Negative supercoiling imparts a torsional stress to the DNA that favors unwinding, whereas positive supercoiling favors tighter winding of the double helix. Supercoiling imparts a more compact structure to a circular duplex, which makes it sediment more rapidly in a centrifuge. Thus, either a positively or a negatively supercoiled DNA will sediment more rapidly than a circular duplex with no supercoiling. Jerome Vinograd demonstrated this tendency by adding ethidium bromide (fig. 25.18) to circular DNA of the small phage PM2. This DNA has about 40 negatively supercoiled turns when it is isolated from cells. Addition of ethidium bromide to duplex DNA leads to the binding of ethidium between adjacent base pairs, a type of binding known as intercalation (see fig. 25.18). Normally the base pairs are nearly closely packed, so such binding would not be possible without an alteration in the DNA structure. In order to accommodate the ethidium, the duplex unwinds by about $-27°$ per base pair. This unwinding reduces the negative supercoiling in naturally occurring circular duplexes such as PM2 DNA. As more ethidium is added, enough unwinding takes place to eliminate all the supercoiling. At this point the DNA is in the most extended state and thus has its slowest sedimentation rate (fig. 25.19). Addition of more ethidium causes the DNA to adopt a positively supercoiled form that again increases its sedimentation rate.

Supercoiling of circular duplex DNA is quantitatively considered in terms of the linking number (L), an integer that specifies the number of complete turns made by one strand around the other. The linking number can change only if a covalent linkage in the DNA backbone is broken. If a molecule of DNA is projected onto a two-dimensional surface, the linking number is defined as the excess of right-handed over left-handed crossings of one strand over the other. Linear duplex DNA with free ends adopts a conformation in solution close to the B form, with about ten base pairs per turn. Therefore a closed circular duplex with this extent of twist is presumed to be under no torsional strain and is said to be relaxed. Because B DNA is a right-handed helix, the linking number is normally positive by the sign convention we have adopted. The values of the linking number of relaxed DNA, $L°$, will be distributed over a narrow range of integral values centered around 1 per 10 base pairs. DNA with a mean linking number smaller than this is termed negatively supercoiled, or underwound; DNA with a larger linking number is termed positively supercoiled or overwound. The deviation of the linking number from its relaxed value, $\Delta L = L - L°$, can be partitioned between twist (altered double helix coiling) and supercoiling:

$$L = \text{twist } (T) + \text{supercoiling } (S)^*$$

*It is common to use writhe (W) instead of supercoiling (S) in this equation. Writhe includes conformations other than supercoiled that are topologically equivalent as far as the linking number is concerned.

Figure 25.17

Topology of negative and positive supercoil.

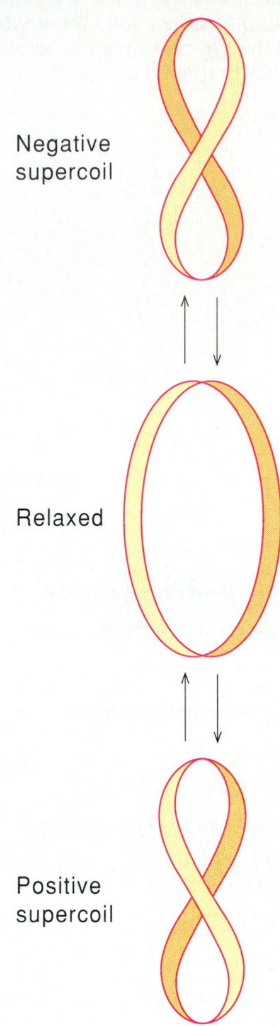

At the present time we do not know precisely how L will partition between twist and supercoiling for helices under torsional stress. For example, consider the hypothetical situation illustrated in figure 25.20. A 360-base-pair structure in the circular relaxed form ($L = +36$, $T = +36$, $S = 0$) is indicated to the left. Exposure to the bacterial enzyme DNA gyrase will introduce negative supercoils into such a structure in a reaction that requires ATP (see chapter 26). If four negative supertwists are introduced, L will be reduced to $+32$. Barring other changes, T will remain fixed and S will become -4. In fact, the torsional strain introduced by the four negative supertwists will tend to reduce T, causing a partial unwinding of the duplex. At one extreme, this effect could lead to the unwinding of four helical turns, in which case the supercoiling would disappear entirely ($L = +32$, $T = +32$, $S = 0$). The actual situation would probably lead to a reduction in the negative value of S and a concomitant reduction in T that would be spread over the entire duplex without any localized total unwinding of the double helix as pictured. Note that the linkage number L changes only when covalent bonds are broken, as in the case of gyrase treatment.

Figure 25.18

Ethidium bromide and DNA intercalated with ethidium bromide. Molecules such as ethidium bromide can intercalate DNA because they are flat rings of the same thickness as DNA base pairs. In order to accommodate the ethidium bromide molecule the duplex must untwist in the region of intercalation. This increases the separation between the planes of the base pairs. We saw the effect of complete untwisting in figure 25.10.

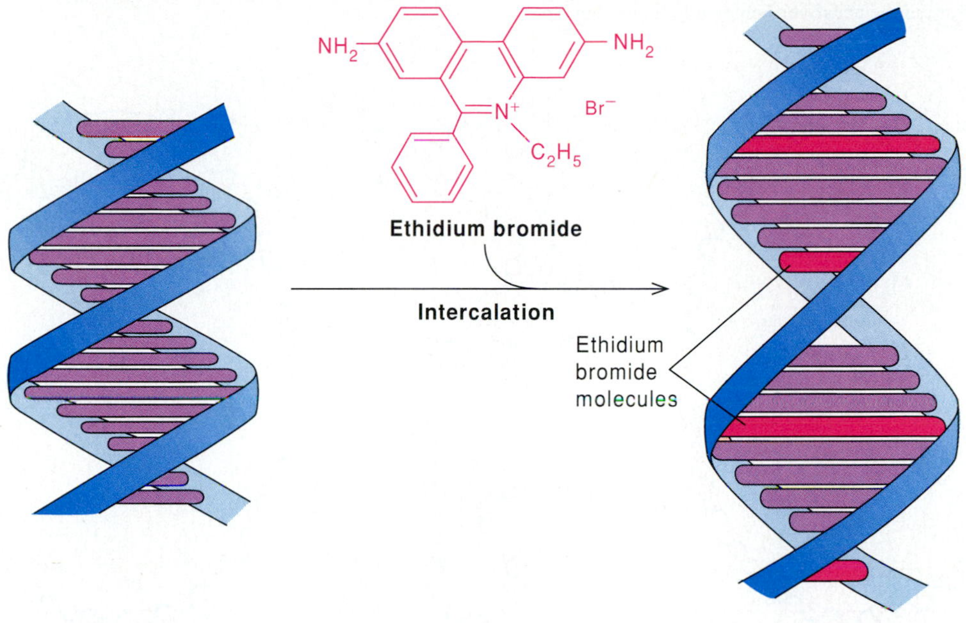

For DNA with a molecular weight of less than 10^7, agarose gel electrophoresis is a most effective method for assessing the extent of supercoiling. (For analysis of higher-molecular-weight DNAs, see boxes 25B and 25C). DNA isomers differing by 1 in linking number form separate bands in the gel (fig. 25.21). The more highly supercoiled molecule will migrate more rapidly through the gel as a result of its more compact structure.

In chapter 26 we will discuss enzymes called topoisomerases, which relax positively or negatively supercoiled DNA, as well as topoisomerases such as DNA gyrase (mentioned earlier), which only generate negatively supercoiled DNA from relaxed DNA. Gyrases that catalyze the formation of negatively supercoiled DNA have been found only in bacteria. Consistent with this fact, all double-helix DNA found in bacteria that is topologically constrained by not having free ends, such as circular DNA, is negatively supercoiled. Circular DNA isolated from virus-infected eukaryotic cells (for example, simian virus 40, or SV40 DNA) is frequently found to be negatively supercoiled, but only after deproteinization. In such instances, the DNA is not supercoiled in cells. Instead, the supercoiling results from removal of the proteins that normally are bound to the DNA and cause it to be underwound in its native state.

The biological importance of supercoiling has been clearly established only for DNA in bacteria. Drugs, such as novobiocin, that specifically inhibit DNA gyrase have a lethal effect except in strains that have a novobiocin-resistant DNA gyrase. On the other hand, mutants containing an altered topoisomerase, relaxing enzyme, which relaxes negatively supercoiled DNA, have been isolated from *E. coli*. Careful analysis of such mutants has shown that they are double mutants;

Figure 25.19

The sedimentation coefficients of closed circular phage PM2 DNA in 2.85-M CsCl containing varying amounts of ethidium bromide. The more compact the structure, the faster it will sediment. Supercoiled DNA is more compact than relaxed circular DNA. Thus when ethidium bromide is added to negatively supercoiled DNA the sedimentation constant decreases, indicating that the DNA is becoming more relaxed. A minimum sedimentation constant is reached when the fully relaxed structure is obtained. Addition of ethidium bromide beyond this point results in an increase in the sedimentation constant again, as the DNA becomes positively supercoiled. (Source: B. M. J. Revet et al., "Direct determination of the superhelix density of closed circular DNA by isometric titration" in *Nature (New Biol.)* 229:10, 1971. Copyright © 1971 Macmillan Magazines Ltd., London, England.)

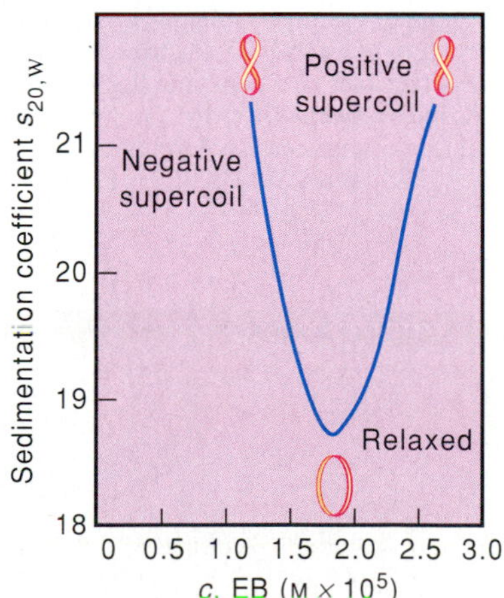

Figure 25.20

A circular duplex molecule in different topological states. (Adapted from a diagram supplied by M. Gellert.) The linking number can be changed only by breakage and re-formation of the phosphodiester linkages, as shown in the conversion of the relaxed circular form (1) to the strained negatively supercoiled form (2). The strain in the negatively supercoiled form can be partitioned in different ways between twist (T) and supercoiling (S) as shown in the interconversion between (2) and (3). No phosphodiester linkages are broken in making this interconversion and consequently there is no change in the linking number (L) as indicated.

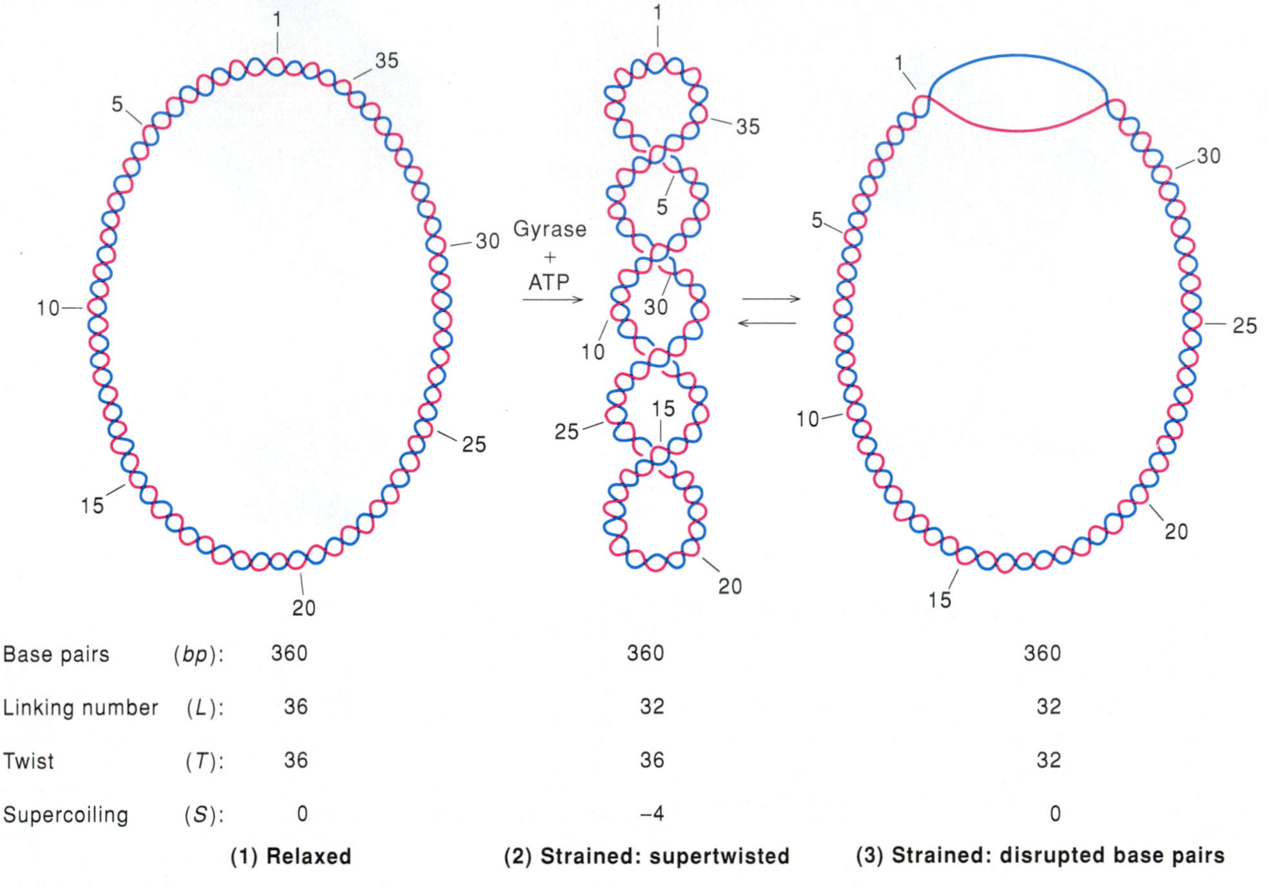

	(1) Relaxed	(2) Strained: supertwisted	(3) Strained: disrupted base pairs
Base pairs (bp):	360	360	360
Linking number (L):	36	32	32
Twist (T):	36	36	32
Supercoiling (S):	0	−4	0

Figure 25.21

Electrophoretic patterns of highly supercoiled or partially supercoiled DNA. Strip A represents a sample of circular duplex DNA obtained by deproteinization of the animal virus SV40. In strips B and C the DNA has been exposed for increasing times to an enzyme (topoisomerase) that catalyzes relaxation. Adjacent bands differ by 1 in linking number. (From W. Keller, Characterization of purified DNA-relaxing enzyme from human tissue culture cells, *Proc. Nat. Acad. Sci.* 72:2553, 1975.)

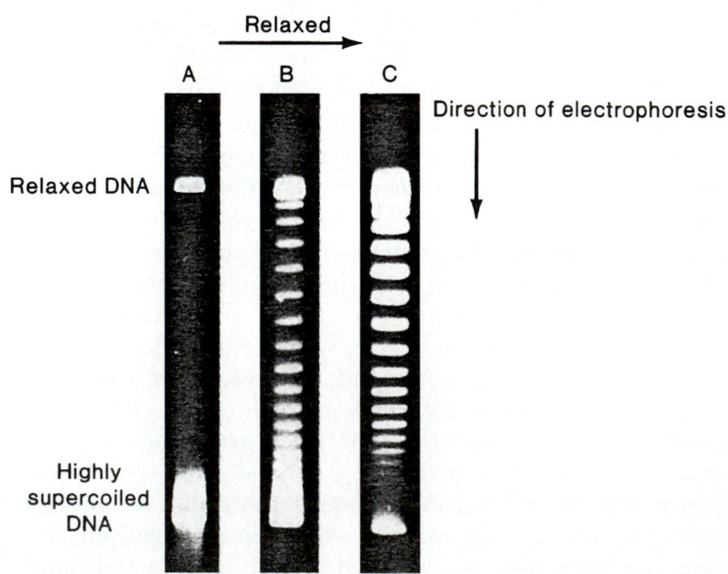

25B
BOX

Equilibrium-Density-Gradient Centrifugation

In equilibrium-density-gradient centrifugation, macromolecules such as nucleic acids or proteins are dissolved in a concentration of CsCl whose density is similar to that of the macromolecule. At high speeds, a nearly linear gradient of the salt is established. The macromolecules will form a narrow band (whose width is inversely related to their molecular weight) at a position where their density is equal to the density of the solution. If a mixture of DNA, RNA, and protein is centrifuged, the protein will band at a buoyant density of about 1.25 g/cc; DNA (depending on the percent G + C) will band at 1.710 g/cc; and RNA will settle to the bottom of the gradient, since its density is unusually high, about 1.9 g/cc. The greater the G + C content of DNA, the higher its buoyant density; the density in CsCl can in fact be used to determine the G + C content of native DNA. In contrast to velocity sedimentation, position of DNA in a density gradient is independent of linear duplex DNA size; the band width, however, is reciprocally related to size.

CsCl gradients can also be used to distinguish ^{15}N-labeled DNA from ^{14}N- or $^{14}N/^{15}N$-labeled DNA, or DNA containing 5-bromouracil from unsubstituted DNA. This feature has proved useful to establish the semiconservative replication of DNA and to separate replicated from unreplicated DNA (see chapter 26).

Finally, density-gradient centrifugation is very useful in the preparative separation of covalently closed circular (e.g., plasmid) DNA from linear (e.g., fragmented genomic) DNA in the presence of intercalating agents such as ethidium bromide. Because of its topological constraints, covalently closed circular DNA binds less of the dye. Since binding makes DNA lighter, its density is greater than that of linear (or nicked circular DNA) bound to ethidium bromide.

that is, they have an altered DNA gyrase that is less active, in addition to the defective relaxing enzyme. This observation suggests that in a normal bacterium the extent of negative supercoiling is carefully adjusted by the relative activities of the gyrase and relaxing enzymes. Negative supercoiling imparts a torsional stress or tension to the DNA structure that can be relieved to some extent by unwinding of the double helix. Both the initiation of DNA synthesis (chapter 26) and the transcription of certain DNA sequences or genes (chapter 28) are strongly dependent on negative supercoiling, probably because of the necessity to unwind at least one turn of the double helix during the initiation process. This same torsional stress explains why Z helix formation is encouraged by negative supercoiling; since in Z DNA the twists are left-handed, the transition from the B to Z form will tend to relax negatively supercoiled DNA.

DNA Denaturation Involves Separation of the Two Strands

The process of separating the polynucleotide strands of duplex nucleic acid structures is called denaturation. Denaturation disrupts the secondary binding forces that hold the strands together. Recall that the secondary binding forces include the edge-to-edge hydrogen bonds between the base pairs of opposing strands and the face-to-face stacking forces between the planes of adjacent base pairs. Individually, these secondary forces are weak, but when they act cooperatively they give rise to a DNA duplex that is highly stable in aqueous solution. The conditions required to denature DNA provide us with a measure of the strength of these interactions.

One of the simplest ways to denature DNA is by heating. The extent of denaturation at any temperature can be readily measured by the change in ultraviolet absorbance of a solution of DNA. A substantial rise in absorbance (hyperchromic shift) accompanies the transition from the native to the denatured state. When a solution of native DNA is slowly heated, the absorption at the ultraviolet maximum (at 260 nm) remains constant until an elevated temperature is reached, at which point the absorbance increases by about 40% over a narrow temperature range. This rise in absorbance coincides with the disruption of the regular base-paired structure, and the two polynucleotide strands separate from one another. The sharpness of the disruption of the regularly hydrogen-bonded base-paired native structure may be likened to the melting of a pure organic compound. It is customary to refer to this ultraviolet absorption temperature profile as a melting curve (curve *a* in figure 25.22).

The melting temperature, T_m, of DNA is defined as the temperature at the midpoint of the absorption increase. This is about 85° C for the example shown in figure 25.22. Rapid cooling of the denatured DNA solution leads to re-formation of intrastrand hydrogen bonds, but in a nonspecific, irregular manner. The absorbance decreases, but only by about three-fourths of the total original increase, and the decrease occurs over a much broader range of temperatures, as shown by curve *b* in figure 25.22. On subsequent reheating and cooling, the absorbance follows this cooling curve in the appropriate direction, a result indicating that denaturation produces an irreversible change.

25C

BOX

Application of Gel Electrophoresis for Chromosome Analysis

A variety of methods, both absolute (light scattering) and empirical (sedimentation, viscosity, gel electrophoresis), exist to determine the size of nucleic acids. The most accurate method is to determine the base sequence of the nucleic acid. Then, sequenced restriction fragments can be used as standards for comparison to samples whose molecular weight is being studied.

To analyze heterogeneous DNA populations, gel electrophoresis offers the best approach. However, very large DNA molecules—in the range of 150 kb or greater—tend to migrate with size-independent mobilities. (The abbreviation kb stands for "kilobase pair," or 1,000 base pairs.) A greatly improved technique can resolve large DNA molecules, the size of the *E. coli* genome and intact, individual yeast chromosomal DNA, on agarose gels. The gently isolated DNA samples are subjected alternately to two approximately orthogonal electric fields. First one field is applied and turned off. Then the other field is applied and turned off. This process is repeated with a pulse time of many seconds over a period of several hours. The precise pulse time depends on the size of the molecules being separated, being longer for larger molecules.

When large DNA molecules enter a gel in response to an electric field, the molecule must elongate parallel to the field. When the field is shut off and a new field is applied, perpendicular to the long axis of the DNA, the molecule must reorient. The reorientation time should be quite sensitive to the size or molecular weight of the DNA. Since the length of the stretched-out DNA molecule is generally larger than the pores in an agarose gel, this reorientation is absolutely essential if the DNA is to undergo any net migration in response to the new field. Therefore, the smaller molecules have a mobility advantage over the larger molecules, which results in a size-based separation. (See figure.)

Another development is the electrophoretic analysis of partially denatured DNA, a technique that is very useful in the detection of single base substitutions of cloned DNA. Fragments of DNA that are wild type or that contain a single base mismatch migrate into a polyacrylamide gel containing an ascending denaturing gradient of urea and formamide in the same direction as the electrical field. At a critical depth the mobility of partially denatured DNA slows down abruptly. The distance moved is a measure of the stability of the sequence, rather than the overall base composition or size of the remaining unmelted duplex. Differences in the partial denaturation of wild-type and mutant DNA molecules allow one to separate and detect them. This method, developed by L. Lerman, provides a quick method to detect differences, even a single base-pair change, in cloned homologous fragments from wild-type and mutant sources.

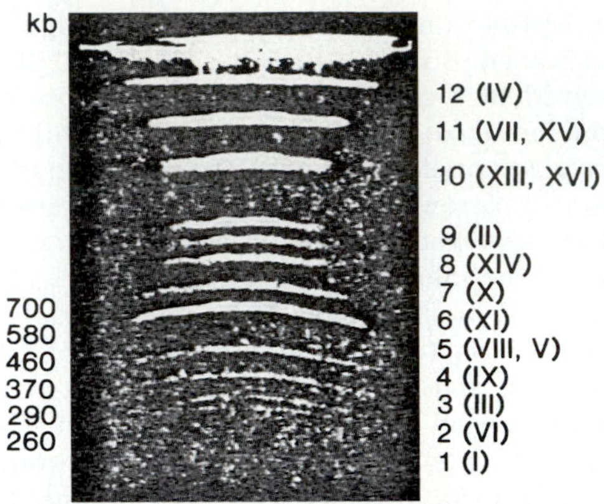

Figure 1

Separation of intact DNA from the yeast chromosome. The chromosome assignments are on the right; size markers, in kb, are on the left. The DNA bands were detected by staining with ethidium bromide. (From G. F. Carle and M. V. Olson, "An electrophoretic karyotype for yeast," *Proc. Natl. Acad. Sci.* 82, 3756–3760, 1985.)

Some DNAs do not occur naturally in the double-helix form. The melting curve is an excellent tool for detecting DNAs in the single-stranded conformation. For example, in certain bacteriophages that infect *E. coli,* such as ϕX174, fd, or M13, the DNA exists as a single, circular strand. The ultraviolet absorption temperature curve for the DNA of these phages is similar in shape to that observed for denatured DNA (curve *b* in figure 25.22). Broad melting curves also are characteristic of most RNAs, which rarely have regions of regular base pairing that extend for more than ten or twenty residues.

Melting curves also have provided evidence that the stability of the double-helix structure is a function of its base composition. The midpoint of denaturation (T_m) of naturally occurring DNAs is precisely correlated with the average base composition of the DNA: the higher the mole percent of G-C base pairs, the higher the T_m (fig. 25.23). This seems reasonable, since the G-C base pair contains three hydrogen bonds, whereas the A-T base pair contains only two (see fig. 25.7); thus we would expect DNA with a greater G-C content to be more stable. As we indicated earlier, base stacking is also believed to contribute to the stability of the duplex structure. In general, the interaction energy gained by stacking between adjacent G-C base pairs is greater than that gained by interaction between A-T base pairs. An excellent source of natural DNAs that differ in their content of G-C base pairs is bacteria. Members of the genus *Clostridium* have only 25% G-C in their DNA,

Figure 25.22

Effect of temperature on the relative absorbance of native, renatured, and denatured DNA. When native DNA is heated in aqueous solution, its absorbance does not change until a temperature of about 80° C is reached, after which the absorbance rises sharply, by about 40% (curve *a*). On cooling the absorbance falls, but along a different curve, and it does not return to its original value (curve *b*). Renatured DNA, in which the two strands have been brought back into perfect register, shows a sharp melting curve similar to native DNA (curve *c*). Renatured DNA is prepared from denatured DNA by holding the temperature at about 25° C below the denaturation temperature for an extended time. This subject is discussed in detail later in the text. The temperature at which the native DNA is half denatured is labeled T_m.

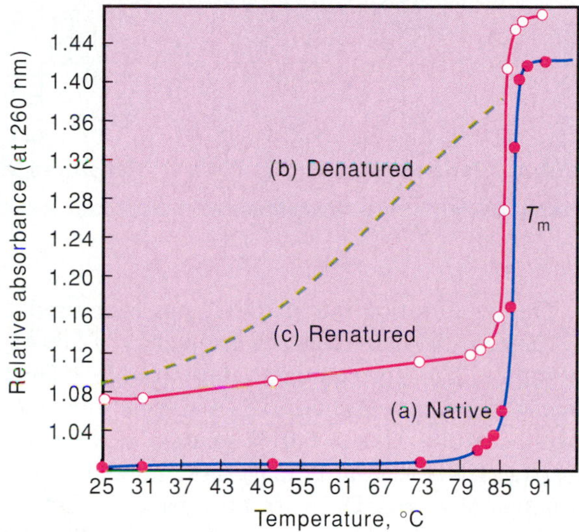

Figure 25.23

Dependence of the temperature midpoint (T_m) of DNA on the content of guanine and cytosine. As the percentage of G + C increases, the T_m increases. Two curves are shown to illustrate the point that the denaturation temperature is shifted to lower values when the ionic strength is lowered.

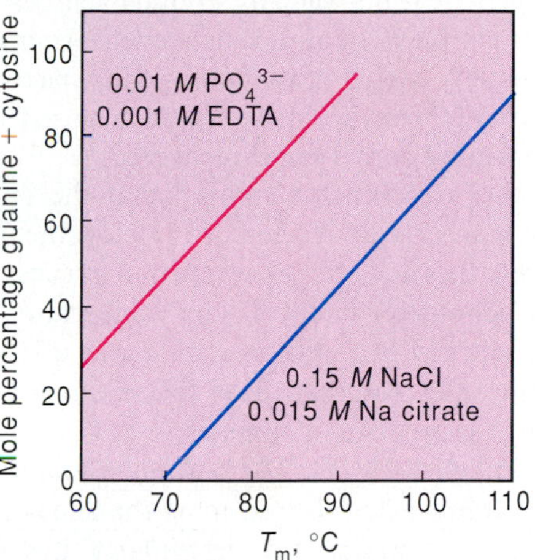

whereas some *Micrococcus* species have DNA with 72% G-C. *E. coli* DNA is about 50% G-C. Whereas the base composition of the DNA of bacteria along the genome is uniform, reflected in a narrow, symmetrical peak in a CsCl density gradient, the DNA of higher eukaryotic cells is not so uniform, with a minor fraction of DNA displaying a G-C content different from the major fraction of DNA. This minor fraction can be detected in a CsCl density gradient as a small band separate from the main band (see box 25B). The smaller band is termed satellite DNA.

Other factors present in aqueous solution can affect the stability of the double-helix structure in a positive or a negative way. For example, salt has a stabilizing effect, which is mainly due to the repulsive electrostatic interactions between the negatively charged phosphate groups. Salt shields this charge interaction and therefore stabilizes the duplex structure. Thus DNA in 0.15-M NaCl denatures at a T_m about 20° C higher than DNA in 0.01-M phosphate (see fig. 25.23). In pure water (no salt present) DNA denatures at room temperature. Extremes of pH also have a destabilizing effect on the double-helix structure. When the pH is above 11.5 or below 2.3, there is extensive deprotonization or protonization, respectively, of the hydrogen-bonding groups of the bases, which in turn disrupts the hydrogen-bonded structure. Alkali is an excellent DNA denaturant; it permits rapid separation of the strands without degradation. Alkali both denatures and degrades RNA to 2′(3′) mononucleotides.

Many solutes that can form hydrogen bonds also lower the melting temperature (decrease the stability) of double-helix structures. The organic compounds formamide and urea are frequently used to lower the denaturation temperature as well as to prevent reaggregation of denatured strands on subsequent cooling. This is important in DNA manipulations, where the wish is to avoid nonspecific aggregation. Reagents that increase the solubility of the DNA bases (e.g., methanol) or disrupt the water shell around them (e.g., trifluoracetate) reduce the hydrophobic interactions between the bases and lower the T_m. Most proteins that bind to DNA inhibit denaturation. However, some DNA-binding proteins destabilize the native state. Proteins of this class usually bind preferentially to single-stranded DNA, thereby favoring separation of the double strands. For example, gene 32 of bacteriophage T4 encodes a DNA-binding protein that is essential for bacteriophage T4 DNA replication. This protein induces the local unwinding of the DNA duplex, a process that facilitates its replication (chapter 26). Addition of appropriate amounts of the gene-32-encoded protein leads to complete denaturation of individual DNA molecules. This complete denaturation occurs because the gene-32 protein exhibits cooperative binding; that is, the binding of one gene-32 protein molecule to DNA facilitates binding of a second gene-32-encoded molecule at a neighboring site and continued binding at adjacent sites. A number of DNA-binding proteins similar to the gene-32 protein have been found in both prokaryotic and eukaryotic cells. They are believed to play an important role in DNA replication, recombination, and repair.

DNA Renaturation Involves Duplex Formation from Single Strands

We have pointed out that when a solution of heat-denatured DNA is allowed to cool rapidly, the regularly hydrogen-bonded structure does not re-form. However, reassembly of the two separated polynucleotide strands into the native structure, called renaturation, is possible under certain specialized conditions.

The first indication that renaturation was possible came from studies of Julius Marmur and Paul Doty, using transforming DNA in a biological assay. The basis of the assay was that denatured pneumococcus DNA is inactive in DNA-mediated transformation, whereas native DNA is active in transferring genetic traits, such as streptomycin resistance, from a donor to a recipient strain. When transforming DNA was heated and rapidly cooled it was biologically inactive; however, when denatured DNA was slowly cooled, a large fraction of the initial transforming activity was recovered.

Further experiments showed that the optimum temperature for renaturation is about 25° below the T_m. As is the case with denaturation, the renaturation process can be followed spectrophotometrically. If the temperature of a denatured DNA solution is maintained at $T_m - 25°$ C for a long period of time, the absorbance of the solution gradually decreases until it approaches a value close to that of native DNA (see curve *c* in figure 24.22). The optimum temperature for renaturation is frequently referred to as the annealing temperature. At such a temperature, irregularly hydrogen-bonded structures are unstable but regularly hydrogen-bonded structures are stable. Consequently, prolonged exposure of denatured DNA at the annealing temperature allows the DNA bases to explore various configurations until complementary regions of pairing are formed between otherwise separated strands.

Spectrophotometric methods and biological assays are not the only means of monitoring renaturation. Other methods are sometimes used, especially when it is desirable to separate the denatured and renatured fractions or when small, radioactive amounts of nucleic acids are being analyzed. One of these methods involves digestion of the nucleic acid with enzymes that preferentially degrade single-stranded DNAs. For this purpose, the S1 nuclease derived from the mold *Aspergillus* is commonly used. The amount of DNA resistant to digestion by this nuclease gives an accurate measure of the amount of duplex structure in a sample of partially renatured DNA. A more commonly used method involves binding to a column made from a calcium phosphate gel known as hydroxyapatite. For reasons that are unclear, duplex DNA binds to hydroxyapatite at salt concentrations where single-stranded DNA does not bind. This method must be used with caution as a quantitative measure of renaturation, because a DNA molecule that is partially duplex and partially single-stranded will also bind to hydroxyapatite. Hydroxyapatite chromatography is also useful as a preparative technique for separating rapidly reassociating fragments from slowly reassociating fragments. The significance of the rate of reassociation is explained later.

Renaturation Rate Measures Sequence Complexity

Marmur and Doty and their co-workers' discovery of the phenomenon of renaturation met with considerable interest. Many applications have grown out of the technique of renaturation or annealing (see box 25D).

Figure 25.24

Steps in denaturation and renaturation of a DNA duplex. In step 1 the temperature is raised to the point where the two strands of the duplex separate. If denatured DNA is slowly cooled, the events depicted as steps 2 and 3 follow. In step 2 there is a second-order reaction in which two complementary strands of DNA must collide and form interstrand hydrogen bonds over a limited region. Step 3 is a first-order reaction in which additional hydrogen bonds form between the complementary strands that are partially hydrogen-bonded (zippering). Once complementary strands are partially bonded, the zippering reaction will occur rapidly. In the overall process, step 2 is rate-limiting.

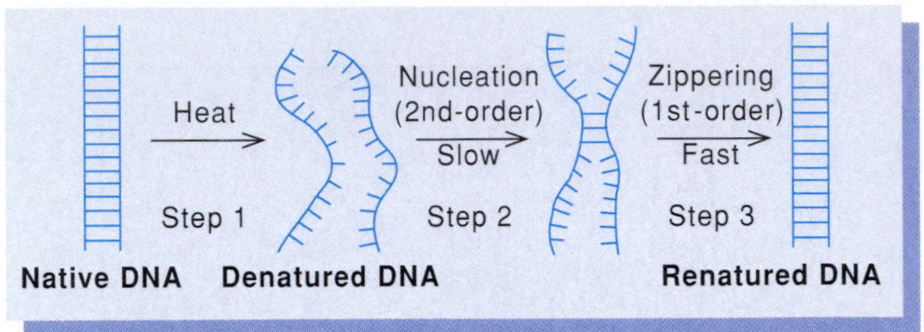

The renaturation rate of DNA has been used extensively to compare the base sequence complexity of DNAs from different sources. For a given weight concentration of DNA, the more homogeneous in sequence a DNA sample, the more rapidly it will renature. Thus a 0.01% solution of denatured T4 bacteriophage DNA renatures much more rapidly than does a 0.01% solution of *E. coli* DNA. In turn, bacterial DNA renatures much more rapidly than mammalian DNA. The steps involved in denaturation and renaturation are indicated in figure 25.24. Renaturation is a two-step process. In the first step, called nucleation, contact is made between two complementary regions of DNA. Nucleation is followed by a relatively rapid zippering up of adjoining base residues into a duplex structure. Thus nucleation is the rate-limiting step, and it is this step that must be examined to determine the time course of renaturation. Because nucleation involves interaction between two molecules, it should occur at a rate proportional to the square of the concentration of single strands; that is, it is a second-order reaction.

If *c* is the concentration of single-stranded DNA at time *t*, then the second-order rate equation for loss of single-stranded DNA is

$$-\frac{dc}{dt} = k_2c^2$$

where k_2 is the second-order rate constant. Given a starting concentration c_0 of completely denatured DNA, the amount of single-stranded DNA left after renaturation for time *t* is given by

$$\frac{c}{c_0} = \frac{1}{1 + k_2c_0t}$$

At time $t_{1/2}$, when half of the DNA is renatured, $c/c_0 = 0.5$ and $t = t_{1/2}$, from which it follows that

$$c_0t_{1/2} = \frac{1}{k_2}$$

Techniques Using Nucleic Acid Renaturation

Nucleic acid hybridization takes place between any two complementary single-stranded nucleic acids that are annealed at $T_m - 25°$ C for prolonged periods of time. Since the rate of duplex formation depends on the concentration of the interacting complementary strands, the method can be used to measure the abundance of a specific nucleic acid in a mixture. Thus the number of *Drosophila* ribosomal RNA genes can be estimated by titrating a fixed amount of denatured genomic DNA with increasing amounts of rRNA. Specific DNA-RNA hybrids are detected by first immobilizing denatured DNA onto nitrocellulose filters, then by adding labeled RNA to serve as a probe. The labeled RNA will be retained by the nitrocellulose only if it is hybridized by complementarity to the DNA. The method can be modified for use in isolating specific mRNA sequences; RNA hybridized to immobilized, denatured DNA can be eluted following denaturation of the hybrids.

Hybridization can be used to detect specific sequences in the nucleus or chromosomes using a labeled probe and cells fixed to a microscope slide. Such *in situ* experiments serve to locate genes on chromosomes and were useful in demonstrating that highly repeated sequences in eukaryotic DNA are found in heterochromatin. In the Southern blotting technique, individual DNA fragments generated by restriction endonuclease digestion are first separated by gel electrophoresis; they are denatured and then transferred to a nitrocellulose support. The DNA fragments can then be probed for specific sequences using a suitable labeled DNA or RNA probe. In Northern blot hybridization, immobilized RNA separated by agarose gel electrophoresis under denaturing conditions is transferred to nitrocellulose and is detected by a labeled, complementary single-strand probe. These blotting techniques can be used to determine the size and abundance of the immobilized nucleic acid. The blotting and filter hybridization techniques are very useful in studying the level of expression of various genes and the effect of inducers, of repression, and of mutations on the level of expression.

Hybridization has also been exploited to detect specific sequences in recombinant clones. A labeled probe is applied to bacterial colonies (or plaques) that have been immobilized and the DNA denatured. Hybridization allows a worker to detect transformants (or transfectants) harboring recombinant vectors that carry the sequence of interest.

As a rule, c/c_0 is plotted as a function of c_0t. The resulting curve is referred to as a "cot" curve (fig. 25.25). Data are shown for DNAs and RNAs with varying nucleotide complexity N, which is defined as the number of nucleotides in a nonrepeating sequence. If there are no repeating sequences in the cellular DNA, then N is equal to the number of nucleotides in the genome. As you can see, $c_0t_{1/2}$ is proportional to N for these samples. This family of curves has been used to calibrate more complex situations.

For a given nucleotide complexity, the renaturation rate is also a function of the length L of the nucleic acid, i.e., the actual number of bases per single-stranded nucleic acid molecule present in a renaturation mixture. It can be shown that

$$k_2 = \frac{L^{0.5}}{N}$$

and at $c_0t_{1/2}$ should be proportional to $N/L^{0.5}$. When DNA samples of different initial lengths are compared, the effect of length is usually eliminated by shearing the DNAs of the different samples to a more or less uniform length, so that $c_0t_{1/2}$ can be used directly as a measure of N. This is usually the parameter of interest in a renaturation rate experiment.

Most prokaryotic DNAs yield simple monophasic cot curves with one inflection point (see fig. 25.25). In such instances, N is directly proportional to the amount of DNA in the chromosome. However, for the total DNA from a complex eukaryotic organism such as a human, the curves are more complex, containing more than one inflection point. A precise interpretation of the cot curve for human DNA, shown in figure 25.26, is not possible, but the following explanation is considered most likely. About 2% of the DNA (the foldback portion) renatures very rapidly. This behavior is due mostly to single-stranded regions of DNA that fold back on themselves, thereby forming hairpin duplex (cruciform) structures. Structures giving rise to cruciforms must have possessed inverted repeating sequences called palindromes. The next class of sequences to renature (designated the fast segment in the figure) accounts for about 5% of the total DNA of higher eukaryotes and has a $c_0t_{1/2}$ value of about 10^{-2}. The small amount of DNA, together with the rapid reassociation kinetics, suggests that about 5% of the nuclear DNA (fast) is present in a very large number of copies per genome (highly repetitive DNA). In the same figure, a region labeled intermediate can be seen that accounts for an additional 20% of the DNA (middle repetitive DNA). Finally, about 70% of the DNA has reassociation kinetics in the range expected for single-copy or unique DNA, i.e., DNA whose sequence is present as a single copy per haploid genome.

This type of renaturation kinetic analysis has been used for the gross characterization of DNA in many cell types as shown in figure 25.27. These data show that bacterial DNA

Figure 25.25

Reassociation of double-stranded nucleic acids from various sources. The genome size is indicated by arrows near the upper nomographic scale. Over a factor of 10^9, this value is proportional to the c_0t (the "cot") required for half-reaction. All DNAs were sheared so that they have approximately the same fragment size (about 400 nucleotides, single-stranded). Correction has been made to give the rate that would be observed at 0.18 M sodium ion concentration. No correction for temperature has been applied, since it was approximately optimum in all cases. The labels for the different DNAs should not concern the average reader. MS2 is the RNA sequence obtained from a bacterial virus. Mouse satellite and calf (nonrepetitive fraction) are fractions of the genome obtained from the indicated animals. (Source: R. J. Britten and D. E. Kohne, "Repeated sequences in DNA" in *Science,* 161:529, 1968. Copyright © 1968 American Association for the Advancement of Science, Washington, D.C.)

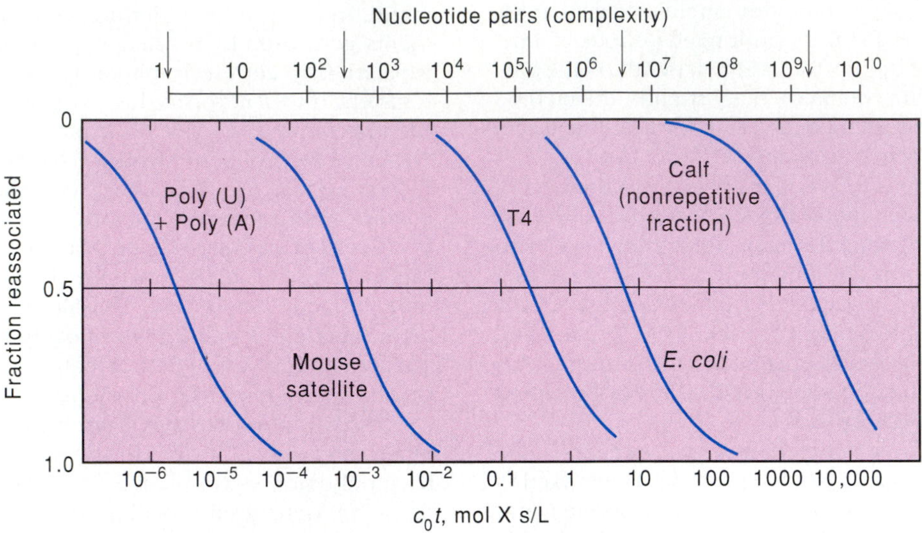

Figure 25.26

The cot curves for total human nuclear DNA. The total human DNA (and the DNA from other complex eukaryotic organisms) does not give a simple monophasic plot, but shows more than a single inflection point. Labels indicate the type of DNA that renatures at different cot values. (Source: Unpublished data of A. R. Mitchell.)

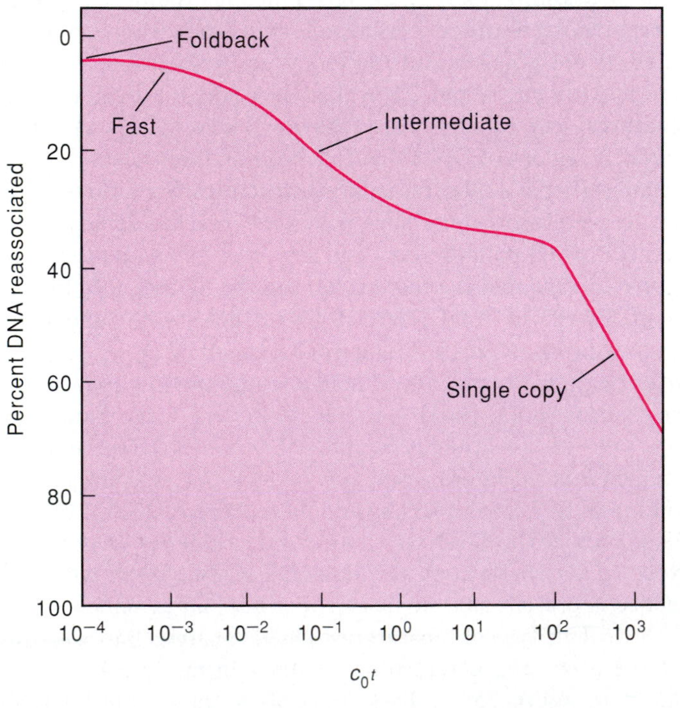

Figure 25.27

Distribution of single-copy DNA and repetitive-sequence DNA in various organisms. The width of bands and the number below the bands indicate the fraction of total cellular DNA in each class. For example, in calf about 55% of the DNA sequences are single copy, about 38% are present in somewhat less than 10 copies, and about 3% are present in somewhat less than 10 copies. (Adapted from "Repeated segments of DNA" by Roy J. Britten and David E. Kohne. Copyright © 1970 by Scientific American, Inc. All rights reserved.)

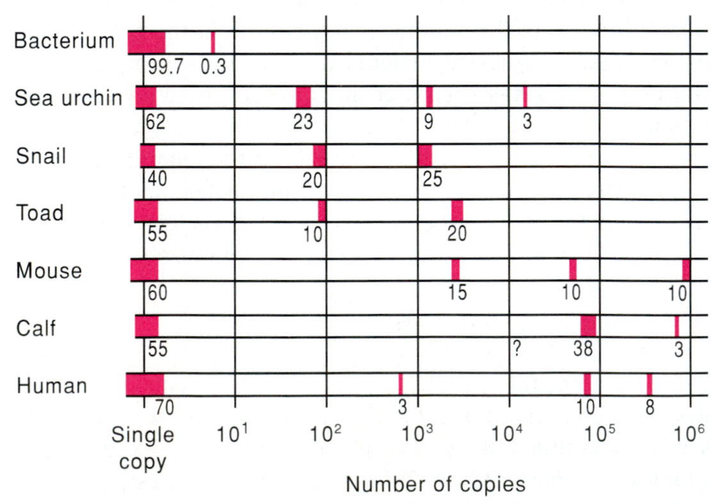

(*E. coli*) contains very little repetitive DNA (0.3%). The amount that is found is mainly accounted for by the eight genes for *E. coli* ribosomal RNA that have nearly identical sequences. In eukaryotes, single-copy DNA (i.e., nonrepetitive DNA) accounts for 40 to 70% of the DNA, most of the remainder being roughly divided between middle repetitive ($<10^4$ copies per genome) and highly repetitive ($>5 \times 10^4$ copies per genome). Further analyses of eukaryotic gene structure by a variety of other techniques have shown that most genes encode proteins that belong to the unique (single-copy) class. The middle repetitive class includes transfer RNA genes and ribosomal RNA genes that are involved in protein synthesis (chapter 29) as well as the genes encoding the nuclear proteins called histones. Some highly repetitive sequences occur in tandem; and still other repetitive DNA elements are distributed at random throughout the genome. It has also been argued that some families of repetitive DNA, referred to as "selfish DNA," represent "parasitic" sequences that replicate together with the genome without conferring any positive or negative characteristics to the host cells that harbor these sequences. Satellite DNA detected in CsCl density-gradient centrifugation is usually enriched in highly repetitive DNA.

Chromosome Structure

All types of nucleic acids interact with proteins. Chromosomal DNA forms stable nonspecific complexes with structural proteins that stabilize their tertiary structure; it also forms transient complexes with enzymes and regulatory proteins that are involved in DNA and RNA metabolism.

We have discussed nucleoprotein complexes at various points in this text where they are relevant to metabolism. Let us now look at the structure of DNA in *E. coli* and in eukaryotic nucleohistone and chromatin.

Physical Structure of the Bacterial Chromosome

The single chromosome of *E. coli* contains about 2×10^9 daltons or about 3×10^6 base pairs of DNA. If all of this DNA were in a duplex structure stretched end-to-end it would be 1 mm long, which is about 50 cell diameters. In fact, the chromosome is circular, centrally located in the cell, and highly folded, so that it is only about 2 μm across. By contrast to what happens in eukaryotes, no dramatic change in chromosome morphology is seen prior to cell division. Clearly, the degree of compaction observed throughout the cell cycle does not interfere with transcription to any great extent, since all the genes in *E. coli* are readily expressible.

Electron micrographs of the *E. coli* chromosome suggest a folded circular structure containing about 40 to 100 supercoiled loops (diagrammatically indicated in figure 25.28). It is believed that the folded structure is held together by an RNA-protein core, although the manner in which this is done is not well understood. The structure is further stabilized because the core forms a complex with positively charged polyamines and

Figure 25.28

The *E. coli* chromosome exists as a circular, folded, supercoiled duplex (*a*). This can be converted to a partially unfolded structure by brief treatment with RNase (*c*). There are 50 to 100 loops in the structure; supercoiling may be selectively eliminated from individual loops by single-strand nicking of the DNA within the loop (*b*).

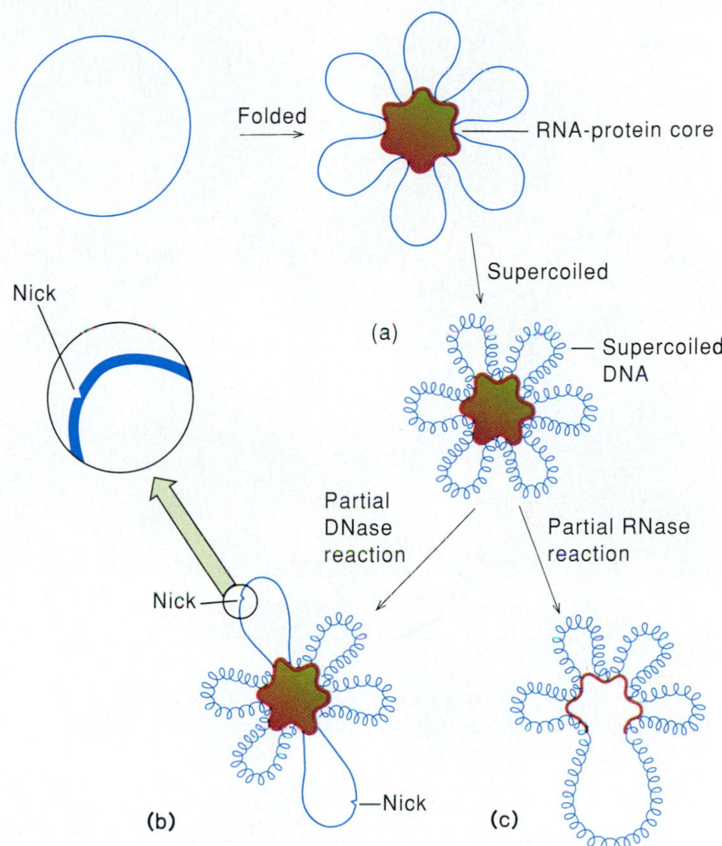

certain basic proteins. Pettijohn and his co-workers have provided evidence of such a core. They first showed that the individual supercoiled loops maintain their supercoiling independently of one another. Thus if a single nick is introduced into one of the loops by limited DNase action, that loop adopts an expanded relaxed conformation, but supercoiling in the other loops is maintained (see fig. 25.28). Limited RNase or protease treatment causes the partial breakdown of the looped structures without interfering with the supercoiling (see fig. 25.28). These results have led to the proposal that each of the loops is a domain whose lateral motion is restricted by an RNA-protein core complex.

The Genetic Map of E. coli

The circular *E. coli* chromosome contains enough base pairs to make about 3,000 average-size genes (1,000 base pairs each). The relative positions of over half of these genes are known. In regions where our understanding is reasonably complete, we gain the impression of an efficiently organized genome. Coding regions are interspersed with regulatory regions; there is no evidence for significant stretches of DNA with no function.

Figure 25.29

Swollen fibers of chromatin from the nucleus of the chicken red blood cell. The electron micrograph is enlarged about 325,000X and negatively stained with uranyl acetate. (Micrograph courtesy of A. L. Olins and D. E. Olins.)

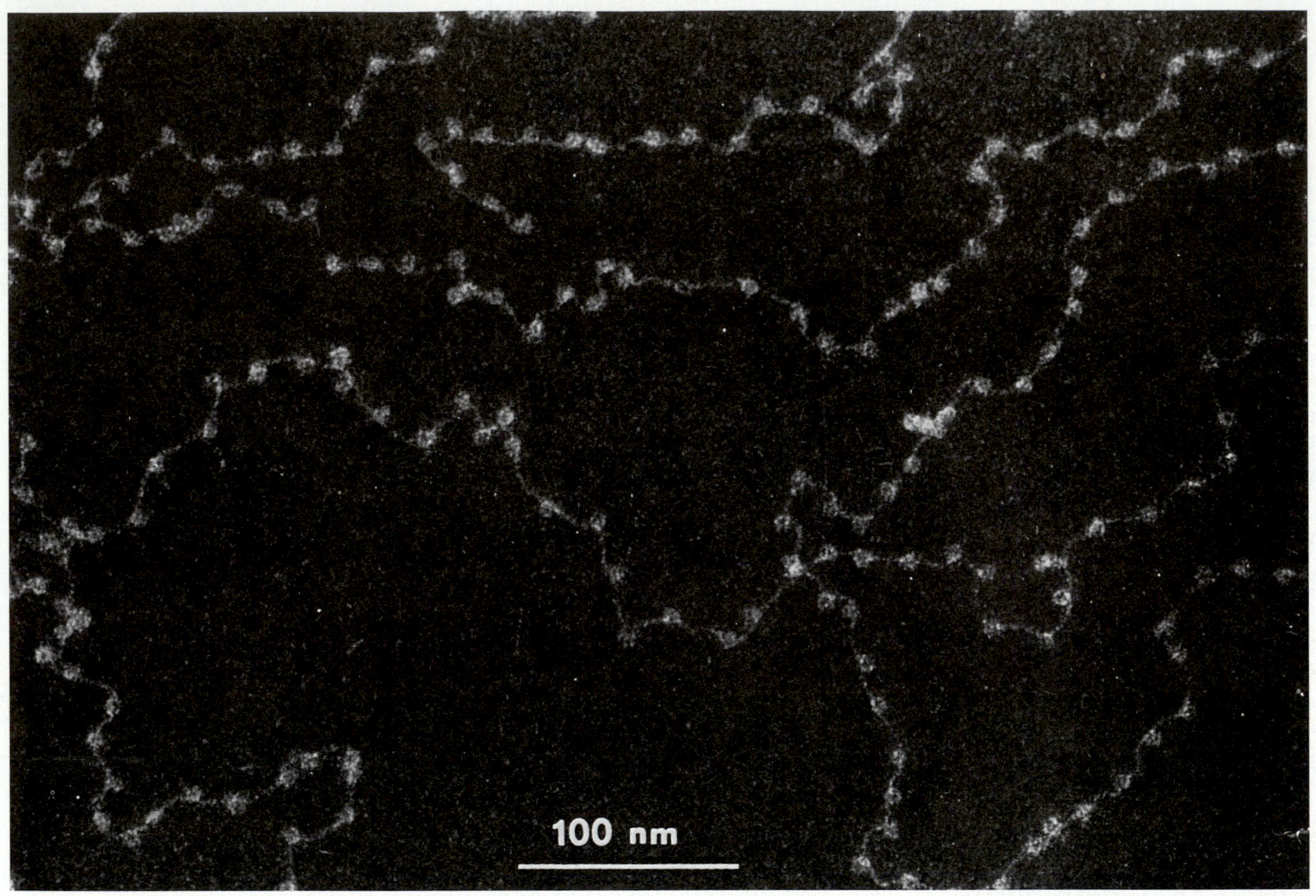

Frequently genes with a related function are tightly clustered. These clustered genes are usually transcribed into single expression units (messenger RNAs) containing the information for the synthesis of several functionally related proteins.

Eukaryotic DNA Is Complexed with Histones

DNA in eukaryotic chromosomes exists in a highly compacted form known as nucleohistone, a complex of DNA with approximately equal weights of five proteins known collectively as histones. Since the same kind of compaction occurs with duplex DNA of almost any sequence, it is presumed that a repeating aspect of the structure is recognized by the protein.

Electron microscopic and x-ray diffraction studies on compacted chromatin suggest that chromatin is a coiled coil. Data are scant, and some of the x-ray spacings observed still require explanation. A breakthrough came when D. E. Olins and A. L. Olins observed that chromatin viewed after sudden swelling in water showed a beaded structure (fig. 25.29). The beads, called nucleosomes, contain most of the histone; they are about 10 nm in diameter, and the spacing between the beads is about 14 nm. Brief enzymatic digestion of chromatin with micrococcal nuclease or some other endonucleases fragments this structure. The DNA from this partial digestion gives rise to a banded pattern on agarose gel electrophoresis that indicates nucleoprotein structures containing 200 base pairs of DNA or multiples thereof (400, 600, 800 bp, etc.). When the products of partial micrococcal nuclease digestion were fractionated by ultracentrifugation, the fractions examined in the electron microscope showed a direct correlation between the size of the DNA estimated on gels and the number of nucleosomes. Thus the most rapidly moving DNA band seen on gels was derived from a structure containing one nucleosome, and the second-fastest migrating species contained nucleosome dimers, etc. Evidently, the brief treatment with endonuclease preferentially cleaves DNA in the internucleosomal region, where the DNA

Table 25.3 Characteristics of Histones			
Name	**Ratio of Lysine to Arginine**	M_r	**Copies per Nucleosome**
Histone H1[a]	20	21,000	1 (not in bead)
Histone H2a	1.2	14,500	2 (in bead)
Histone H2b	2.5	13,700	2 (in bead)
Histone H3	0.7	15,300	2 (in bead)
Histone H4	0.8	11,300	2 (in bead)

[a]Not found in lower eukaryotes such as yeast.

is least likely to be protected from enzyme attack. More exhaustive nuclease treatment gives rise to a single band on gels that contain a single nucleosome and 140 base pairs of DNA. The effect of the more exhaustive nuclease treatment is to degrade all the DNA that is not in direct contact with the nucleosome. Combined electrophoresis and electron microscopy results have led to a model for the beaded structure in which the nucleosome core particles contain clumps of histone complexed to 140 base pairs of DNA duplex, with about 60 base pairs of duplex serving as linkers between the core particles.

The histones present in chromatin have been extensively characterized, and their sequences are known. The classes of histones and their molecular weights are listed in table 25.3. The lysine-rich histone H1 is not present in the nucleosome core particle, as evidenced by its release on extensive nuclease treatment. Consistent with this conclusion is the finding that H1 is the only histone that readily exchanges between free and chromatin-bound histone. The other eight histones, two each from the other four histone classes, form a tightly complexed core particle that is conserved when chromosomes duplicate.

Many fine points about the structure of chromatin remain to be determined. An illustration of the probable DNA coiled-coil structure of the nucleosome core particle is presented in figure 25.30. There are 140 base pairs of DNA in a nuclease-resistant nucleosome core that make about 1.75 superhelical turns around a histone octamer. An additional 60 base pairs of spacer DNA connect the core particles. Electron microscopic observations indicate that the amount of linker DNA is actually different in different species and in different tissues of the same species, varying from about 20 to 95 base pairs.

Nuclear magnetic resonance spectroscopy reveals that nucleosomal DNA has a secondary structure quite similar to that of B DNA. The DNA is wound around the histone complex. Treatment with pancreatic DNase I (instead of with micrococcal nuclease) produces single-stranded fragments that are multiples of about ten nucleotides. The current interpretation of this observation is that the digestion by DNase I is confined to the exposed side of the DNA in the nucleosome core (see fig. 25.30*a*), the approach of the enzyme to the other side being hindered by the presence of the histones.

Salt bridges between positively charged basic amino acid side chains of histones and the DNA phosphates play an important role in stabilizing the DNA-histone complex.

Figure 25.30

(*a*) Path of DNA that can account for the bipartite structure of the nucleosome core is a super helix with an external diameter of 110 Å and a pitch of 27 Å; the turns of the 20-Å-wide DNA helix are nearly in contact. There are about 80 nucleotide pairs of DNA per turn; the nucleosome core, an enzymatically reduced form of the nucleosome consisting of some 140 nucleotide pairs, has about one and three-quarter turns wrapped on it. The histone octamer complex, containing two each of histones H2a, H2b, H3, and H4, is packed on the inside of the DNA coiled-coil structure. (From "The Nucleosome" by Roger D. Kornberg and Aaron Klug. Copyright © 1981 Scientific American Inc. All rights reserved.) In (*b*) we see this histone octamer inserted into the nucleosome core. The H3-H4 tetramer is shown in yellow and an H2a-H2b dimer is shown at each end in purple.

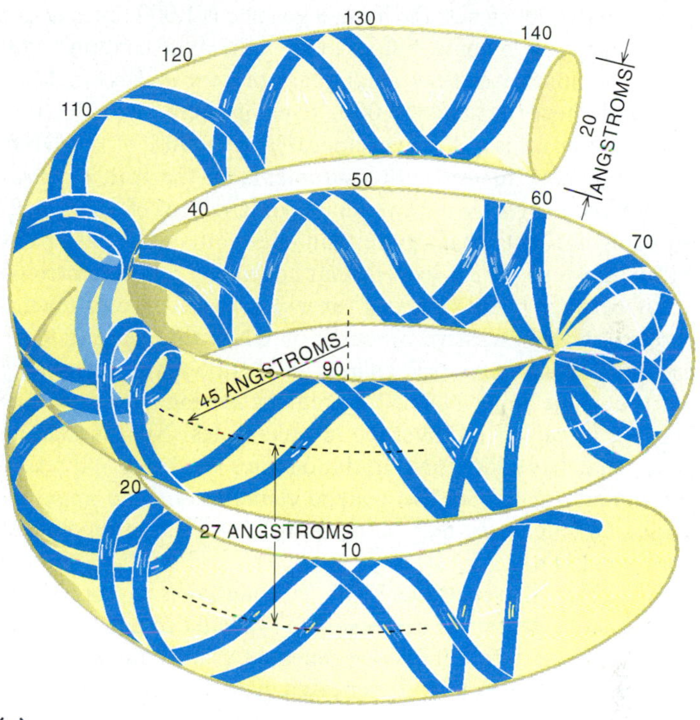

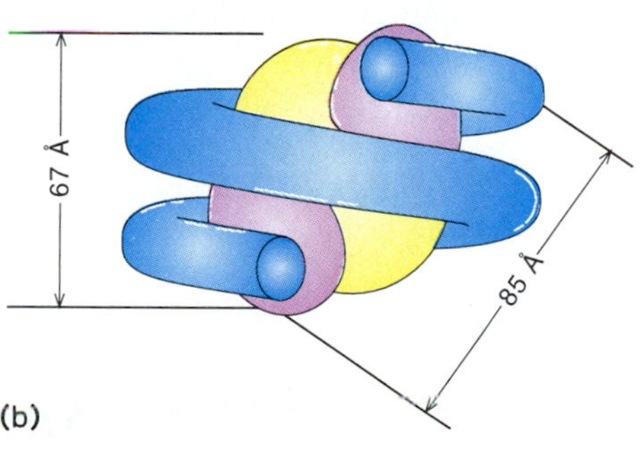

Higher-order structures beyond that of the nucleosome itself have been inferred from light, x-ray, and neutron-scattering microscopy and visualized by electron microscopy. Thus by electron microscopy two types of fibers have been seen: a 10-nm fiber and a 30-nm fiber. The nucleosome can be arranged in a zigzag fashion to produce a fibril that is 10 nm wide, in which the nucleosomes are packed edge to edge rather than face to face. When the ionic strength is raised, the fibrils reversibly condense into an irregular supercoiled fiber about 30 nm in diameter. The nucleosome particles in this fiber are thought to have their cylindrical axes approximately perpendicular to the long axis of the fiber with six to seven nucleosomes per turn (fig. 25.31). Further coiling of these structures is necessary to explain the structures seen in mitotic chromosomes (e.g., see fig. 25.1), but evidence on the precise nature of these structures is lacking.

Organization of Genes within the Eukaryotic Chromosome

The organization of genes within the eukaryotic chromosome is far more complex and less well understood than in prokaryotes. It is clear that a much lower percentage of the DNA is used for coding in complex eukaryotes than in prokaryotes. *E. coli* contains about 3,000 genes; the human genome is 1,000 times larger than the *E. coli* genome, but it is very unlikely to contain 1,000 times the number of functional genes present in *E. coli*. Most estimates hover around 50,000. Even if there were 200,000 human genes, this still leaves approximately 90% of the DNA with no coding function. Measurements on the fruit fly *Drosophila melanogaster* indicate that coding regions of some genes probably account for no more than one-tenth of the base pairs within the gene. The discovery that noncoding regions (introns) occur between coding regions (exons) goes a long way toward explaining the excessive amounts of DNA that appear to be present in eukaryotes (see chapter 28), although it is probably not the whole story. Noncoding regions serving as control loci may be larger on the average in eukaryotes. Nonfunctioning genes that have lost their initiation sites for being expressed (promoters) and repetitive genetic elements with no apparent coding function also help to account for the large amount of noncoding DNA.

Not all repetitive DNA in eukaryotes is noncoding. Thus histone genes, for example, are typically reiterated many times. The five different histone genes are usually clustered, and this cluster is then tandemly repeated many (up to 100 or more) times. Ribosomal RNA genes are also tandemly clustered. Other nonidentical but functionally related genes that show clustering include the globin genes and the immunoglobulin genes.

Figure 25.31

Helical superstructures might be formed with increasing salt concentration (*bottom to top*) as is suggested here. The zigzag pattern of nucleosome (1, 2, 3, 4) closes up, eventually to form a solenoid, a helix with about six nucleosomes per turn. (The helix is probably more irregular than it is in this drawing.) Cross-linking data indicate that H1 molecules on adjacent nucleosomes make contact. Extrapolation from the zigzag form to the solenoid suggests (but does not prove) that the aggregation of H1 at higher ionic strengths gives rise to a helical H1 polymer (not shown) running down the center of the solenoid. In the absence of H1 (*bottom*) no ordered structures are formed. The details of H1 associations are not known at this time; the drawing is meant to indicate only that H1 molecules contact one another and linker DNA. (From "The Nucleosome" by Roger D. Kornberg and Aaron Klug. Copyright © 1981 Scientific American Inc. All rights reserved.)

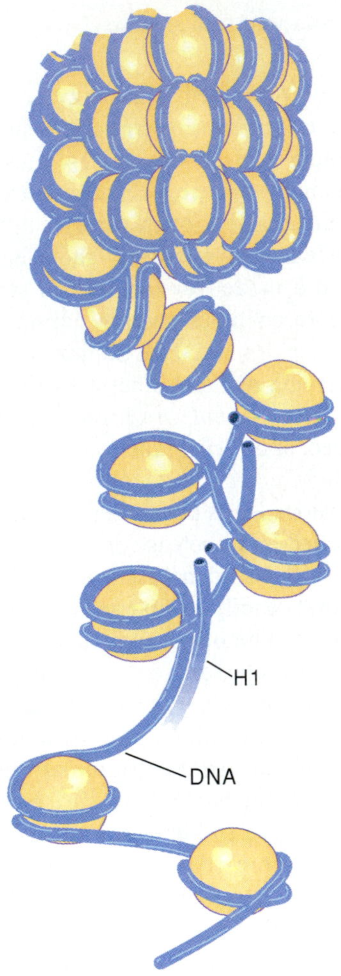

Summary

In this chapter we have been concerned with the nature and structure of DNA, RNA, and associated nucleoproteins. The highlights of the discussion are as follows.

1. The genetic material of cells and viruses consists of DNA or RNA. That DNA bears genetic information was first shown when the heritable transfer of various traits from one bacterial strain to another was found to be mediated by purified DNA.
2. All nucleic acids consist of covalently linked nucleotides. Each nucleotide has three characteristic components: a purine or pyrimidine base, a pentose, and a phosphate group. The purine or pyrimidine bases are linked to the C-1′ carbon of a deoxyribose sugar in DNA or a ribose sugar in RNA. The phosphate groups are linked to the sugar at the C-5′ and C-3′ positions. The purine bases in both DNA and RNA are always adenine (A) and guanine (G). The pyrimidine bases in DNA are thymine (T) and cytosine (C); in RNA they are uracil (U) and cytosine. The bases may be postreplicatively or posttranscriptionally modified by methylation or other reactions in certain circumstances.
3. DNA exists most typically as a double-stranded molecule, but in rare instances it exists (in some phages and viruses) in a single-stranded form. The continuity of the strands is maintained by repeating 3′,5′-phosphodiester linkages formed between the sugar and the phosphate groups; they constitute the covalent backbone of the macromolecule.

The side chains of the covalent backbone consist of the purine or pyrimidine bases. In double-stranded, or duplex, DNA the two chains are held together in an antiparallel arrangement.

4. The base composition of DNA varies characteristically from one species to another in the range of 25 to 75% guanine plus cytosine. Specific pairing occurs between bases on one strand and bases on the other strand. The complementary base pairs are either A and T, which can form two hydrogen bonds, or G and C, which can form three hydrogen bonds. The duplex is stabilized by the edge-to-edge hydrogen bonds formed between these planar base pairs and face-to-face interactions (stacking) between adjacent base pairs. Twisting of the duplex structure into a helix makes stacking interactions possible.
5. The right-handed helical structure, known as B DNA, is the most commonly occurring conformation of linear duplex DNA in nature. In this structure, the distance between stacked base pairs is 3.4 Å, with approximately ten base pairs per helical turn. The inherent flexibility of the structure, however, makes a variety of conformations possible under different conditions. In some instances, nucleotide sequence and degree of hydration dictate which conformations will be favored. DNA interacts with a variety of proteins inside the cell, and these proteins can also have a significant influence on its secondary and tertiary structure.
6. Circular DNA molecules, which are topologically confined so that their ends are not free to rotate, can form supercoils that are either right-handed (negative) or left-handed (positive). Negative supercoiling exerts a torsional tension favoring the untwisting of the primary right-handed double helix, whereas positive supercoiling has the opposite effect. Negatively supercoiled DNAs are most commonly observed in prokaryotes, which contain an enzyme that generates the supercoiled structure.
7. When duplex DNA (or RNA) is heated, it dissociates (denatures) into single strands. The temperature at which denaturation occurs (the melting temperature) is a measure of the stability of the duplex and is a function of the G-C content of the DNA.
8. A preparation of denatured DNA may be renatured by maintaining the temperature about 25° C below the melting temperature. The rate of renaturation is a measure of the sequence complexity of the DNA. In prokaryotes, which consist predominantly of unique sequences, the complexity (the number of base pairs) is approximately equal to the genome size. However, complex eukaryotic cells contain DNAs of varying sequence complexity that renature at quite different rates. The fastest-renaturing fractions are present in many copies per nucleus (usually detected as satellites in CsCl density gradients), whereas the slowest-renaturing fractions are present in single copies. Analyses by other techniques have shown that some of the repetitive DNA sequences exist as tandemly repeated structures, while other types of repetitive sequences are dispersed throughout the genome.
9. In the chromatin of eukaryotic cells, DNA forms a coiled-coil structure with approximately equal weights of five basic proteins known as histones. Four of these histones in pairs form an octamer around which the DNA duplex is wound in a left-handed helix. The DNA octamer complex is called a nucleosome.
10. Each nucleosome contains about 140 base pairs of DNA in a nuclease-resistant nucleosome core and approximately 60 base pairs of spacer between core particles. Histone H1 binds to the chromatin independently of the octamer and is the first histone to dissociate from the chromatin when the ionic strength is raised. Beyond the nucleosome, the higher-order structure of the chromosome involves coiled-coil structures with varying degrees of regularity.

Selected Readings

Avery, O. T., C. M. MacLeod, and C. McCarthy, Studies on the chemical nature of the substance inducing transformation of pneumococcal types. *J. Exp. Med.* 79:137–158, 1944.

Cantor, C. R., C. L. Smith, and M. K. Mathew, Pulsed-field gel electrophoresis of very large molecules. *Ann. Rev. Biophys. Chem.* 17:287–304, 1988.

Dickerson, R. E., The DNA helix and how it is read. *Sci. Am.* 249(6)d:94–111, 1983.

Felsenfeld, G., DNA. *Sci. Am.* 253(4):58–66, 1985.

Hershey, A. D., and M. Chase, Independent functions of viral proteins and nucleic acid in growth of bacteriophage. *J. Gen. Physiol.* 36:39–56, 1952.

Hillary, C. M., J. T. Finch, B. F. Luisi, and A. Klug, The structure of an oligo(dA),oligo(dT) tract and its biological implications. *Nature* 330:221–236, 1987.

Kim, S. H., Three-dimensional structure of transfer RNA. *Prog. Nuc. Acid Res. Mol. Biol.* 17:181–216, 1973.

Kornberg, R. D., and A. Klug, The nucleosome. *Sci. Am.* 244(2):52–64, 1981.

Lerman, L. S., S. G. Fischer, I. Hurley, K. Silverstein, and N. Lumelsky, Sequence-determined DNA separations. *Ann Rev. Biophys. Bioeng.* 13:399–423, 1983.

Morse, R. H., and R. T. Simpson, DNA in the nucleosome. *Cell* 54: 285–287, 1988.

Nadeau, J. G., and D. M. Crothers, Structural basis for DNA bending. *Proc. Natl. Acad. Sci.* 86:2622–2626, 1989.

Noller, H. F., Structure of ribosomal RNA. *Ann. Rev. Biochem.* 53: 119–162, 1984.

Rich, A., A. Nordheim, and A. H.-J. Wang, The chemistry and biology of left-handed Z DNA. *Ann. Rev. Biochem.* 53:791–846, 1984.

Richmond, T. J., J. T. Finch, B. Rushton, D. Rhoades, and A. Klug, Structure of the nucleosome core particle at 7 Å resolution. *Nature* 311:532–537, 1984.

Saenger, W., *Principles of Nucleic Acid Structure.* New York: Springer-Verlag, 1984.

Schmid, M. B. Structure and function of the bacterial chromosome. *Trends Biochem. Sci.* 13:131–135, 1988.

Schwartz, D. C., and C. R. Cantor, Separation of yeast chromosome-sized DNAs by pulsed field gradient gel electrophoresis. *Cell* 37:67–75, 1984.

Strobel, S. A., L. A. Doucette-Stamm, L. Riba, D. E. Housman, P. B. Dervan. Site-specific cleavage of human chromosome 4 mediated by triple-helix formation. *Science* 254: 1639–1642, 1991.

Structures of DNA, *Cold Spring Harbor Symp. Quant. Biol.* 47, 1983.
van Holde, K. E., *Chromatin.* New York: Springer-Verlag, 1988.
Watson, J. D., and F. H. C. Crick, Molecular structure of nucleic acids. *Nature* 171:737–738, 1953.
Wells, R. D., D. A. Collier, J. C. Hanvey, M. Shimizu, and F. Wohlrab, The chemistry and biology of unusual DNA structures adopted by oligopurine · oligopyrimidine sequences. *Faseb J.* 2:2939–2949, 1988.

Problems

1. Briefly describe how Avery was able to show that DNA is the genetic material in cells.
2. Summarize the evidence that RNA is the genetic material in tobacco mosaic virus (TMV).
3. Describe two physical methods that could be used to estimate the base composition of DNA. What would the data look like with two DNA samples, one with high G-C content and another with high A-T content? (Assume that the concentration of the samples is equal.)
4. Why is DNA denatured at either low pH (pH 2) or high pH (pH 11) and why is DNA stable at pH 7? (Hint: See p*K* values in table 20.2.)
5. Standard conditions for hydrolyzing RNA to nucleotides are 0.3 *N* NaOH, 37°, for 16 h. Draw a chemical reaction mechanism for this hydrolysis. Why is DNA not hydrolyzed under these conditions?
6. What effect would the following reagents have on the T_m of duplex DNA: 7-M urea, 90% formamide, higher concentrations of NaCl, pure water, and T4 gene 32-encoded protein? Explain how these chemicals act to affect DNA duplex stability.
7. Why can't RNA duplexes or RNA-DNA hybrids adopt the B conformation?
8. There are DNA-binding proteins that specifically bind to Z DNA. How could these proteins help stabilize DNA in the Z configuration? (Hint: How do single-stranded DNA-binding proteins destabilize duplex structures?)
9. Linear duplex DNA can bind more ethidium bromide than covalently closed circular DNA of the same molecular weight. Why? How could ethidium bromide be used to separate these two forms of DNA on CsCl gradients? (Hint: The ethidium cation has a density less than that of water.)
10. What is the structure of the nucleic acid called A, given the clues listed below?
    (a) Nucleic acid A sediments as a single species in a neutral sucrose density gradient.
    (b) In an alkaline sucrose gradient, one-half (structure B) of the mass of A sediments down the tube, the other half remains at the meniscus (top of tube).
    (c) The buoyant density of A is much higher than that of duplex DNA of the same G + C content.
    (d) The thermal transition profile of B is broad when A is melted.
    (e) When A is heated and rapidly cooled it gives rise to B and C. The buoyant density of C is heavier than that of B.
    (Hint: Single-stranded DNA is more dense than duplex DNA, and RNA is more dense than DNA.)
11. Renaturation of randomly sheared denatured DNA can be measured by the hypochromic shift at 260 nm, by S1-nuclease resistance, or by retention on hydroxyapatite. Which method is likely to give an overestimate for the extent of renaturation?
12. Give the relative times for 50% renaturation of the following pairs of denatured DNAs, starting with the same initial DNA concentrations.
    (a) T4 DNA and *E. coli* DNA, each sheared to an average single-strand length of 400 nucleotides.
    (b) Unsheared T4 DNA and sheared T4 DNA.
13. When histone proteins are isolated from chromatin their mass is equal to the DNA, and the ratio of four of the histones is 1:1:1:1 (H2a:H2b:H3:H4), while H1 is found in half the yield (0.5). Discuss whether or not these data fit the bead-and-string model for nucleosomes.
14. You are given a sample of nucleic acid. How would you determine (a) whether it is DNA or RNA and (b) whether it is single- or double-stranded?

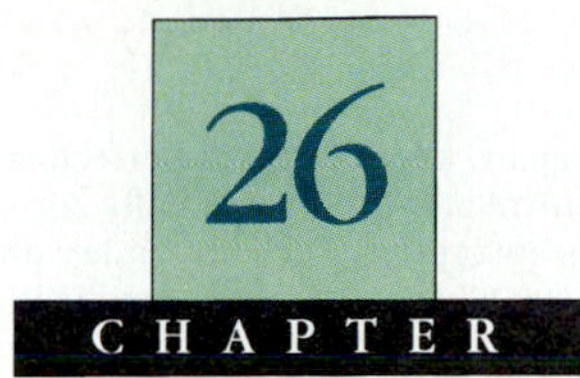

DNA Replication, Repair, and Recombination

Since the genetic information is contained in the sequence of bases in the DNA molecule, it is crucial that DNA be replicated with a very low error frequency. Otherwise the information transfer system would be destabilized. DNA must also resist damage and, if damage occurs, it must have a repair system that restores the original sequence. An elaborate enzymatic machinery has evolved to do just that. Recombination of segments of DNA between different chromosomes is sometimes used to facilitate repair and sometimes is used to facilitate orderly change. Many enzymes are involved in recombination processes. Our focus in this chapter is on the biochemistry of replication. We will also consider critical features of the repair and recombination processes (fig. 26.1).

The Semiconservative Replication of Duplex DNA

Double-helix DNA is relatively inert. Unwinding the double helix is essential to replication and transcription. When the DNA duplex replicates, each chain serves as a template for the synthesis of a complementary chain (fig. 26.2). The rules of Watson-Crick base pairing are strictly obeyed in the polymerization of monomers binding to a DNA template. Although semiconservative replication seems like a predictable mode of replication for a DNA molecule, it required a great deal of ingenuity to devise an experiment to demonstrate it the first time.

Matthew Meselson and Franklin Stahl conceived of a way of demonstrating the semiconservative mode of replication of the entire *E. coli* genome. Their approach made use of isotopes that would result in DNAs of altered densities after replication. For this purpose they grew *E. coli* cells for several generations on a defined medium in which all the nitrogen was the heavy ^{15}N isotope (normal nitrogen is ^{14}N). As a result, the DNA in the progeny cells had a greater than normal density, since the ^{15}N became incorporated into the bases of the DNA during replication. Next, they transferred the bacteria to growth medium containing normal ^{14}N-nitrogen, and allowed the cells

Figure 26.1

Major phases of metabolism associated with DNA: replication, repair, and recombination. *Replication* occurs by a semiconservative mechanism at specified origins of replication. Synthesis of new chains involves absorption of complementary nucleotides at the growth points, followed by covalent bond formation to the growing chain. Growth is always in the $5' \rightarrow 3'$ direction (arrowhead in the figure represents the 3′ ends of the polymers). Since the two chains in a mature DNA duplex are oriented in an antiparallel manner, continuous growth is possible on one chain only. Growth on the opposing chain occurs in short spurts in a direction opposite to the general direction of replication. Replication is semiconservative; that is, after duplication each duplex contains one old chain from the "parent" duplex and one newly formed chain.

Repair is divided into two phases: recognition of an imperfection followed by removal of the imperfection, and repair synthesis. Some imperfections occur during the replication process; these imperfections are usually corrected immediately by the polymerase, which carries the necessary proofreading functions. Other imperfections occur at postreplication stages and are corrected by an auxiliary repair system that recognizes imperfections in the already formed DNA. There are a wide range of degradative enzymes that serve different functions; some of these nucleases operate in conjunction with repair or recombination systems and some function merely to break down the DNA. Nucleases can be broadly classified into exonucleases, which attack the DNA at free ends (3′ or 5′), and endonucleases, which attack the DNA somewhere in the interior.

Recombination processes exist for various purposes. The most general recombination system serves to recombine segments of DNA from homologous chromosomes. This type of recombination involves breakage, rejoining, and repair.

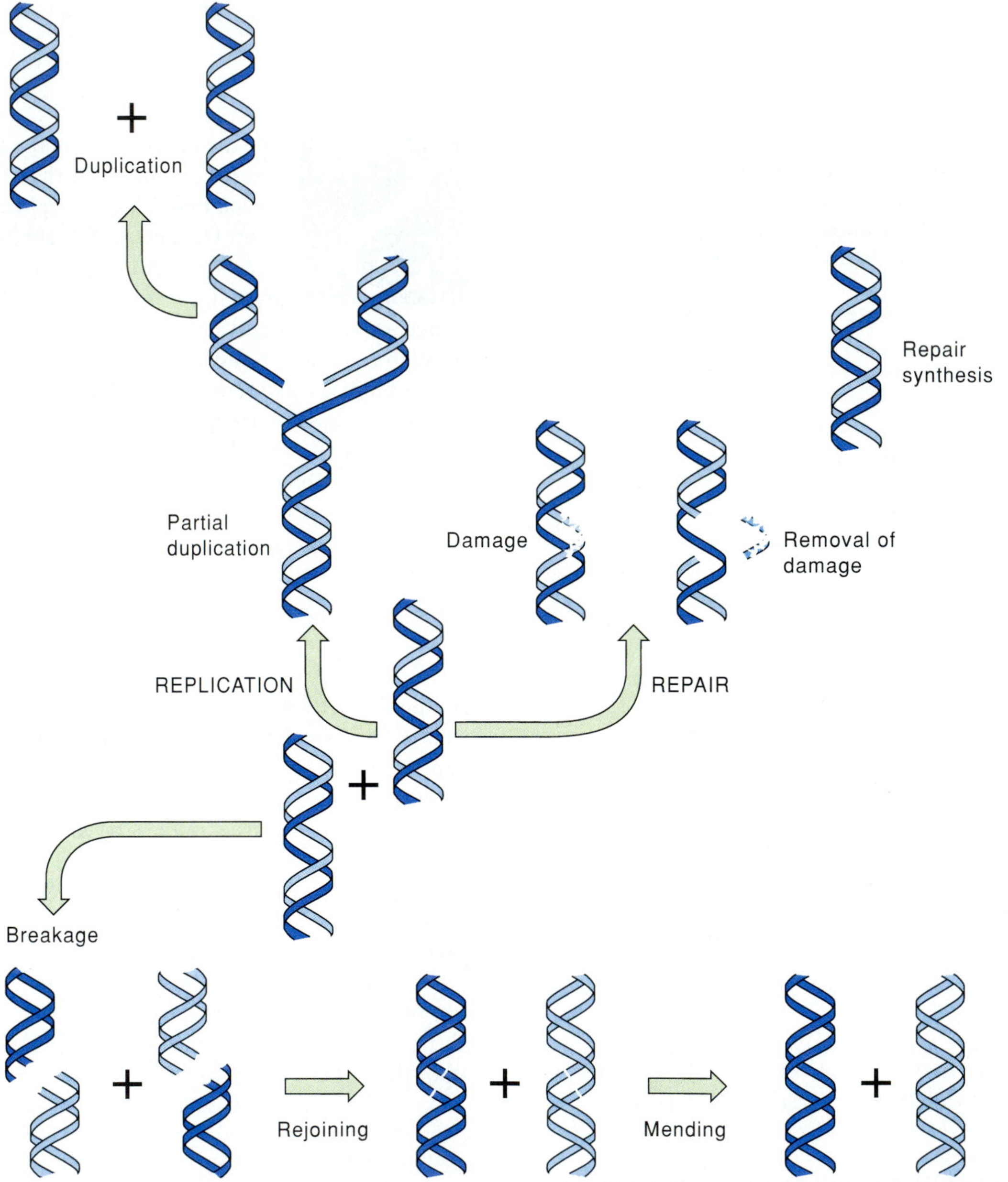

Figure 26.2

Watson-Crick model for DNA replication. The double helix unwinds at one end. New strand synthesis begins by absorption of mononucleotides to complementary bases on the old strands. These ordered nucleotides are then covalently linked into a polynucleotide chain, a process resulting ultimately in two daughter DNA duplexes.

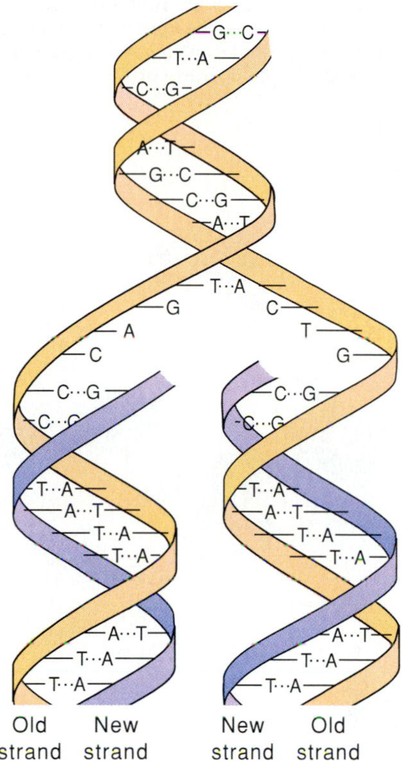

Figure 26.3

The Meselson-Stahl experiment demonstrating semiconservative replication for *E. coli* chromosomal DNA. CsCl density-gradient centrifugation is used to discriminate between DNAs of different densities. *E. coli* DNA has different densities when cells are grown in ^{14}N or ^{15}N medium (frames 1 and 2). When cells containing pure heavy DNA (^{15}N-^{15}N DNA) are grown in ^{14}N medium for one generation, all of the DNA is of intermediate density (^{14}N-^{15}N). After two generations of growth in ^{14}N medium, the cells contain equal amounts of light (^{14}N-^{14}N) DNA and intermediate density DNA (frame 5). In subsequent generations the hybrid DNA reappears in constant amounts, but the amount of light DNA increases. These results support the model of a semiconservative mode of DNA replication.

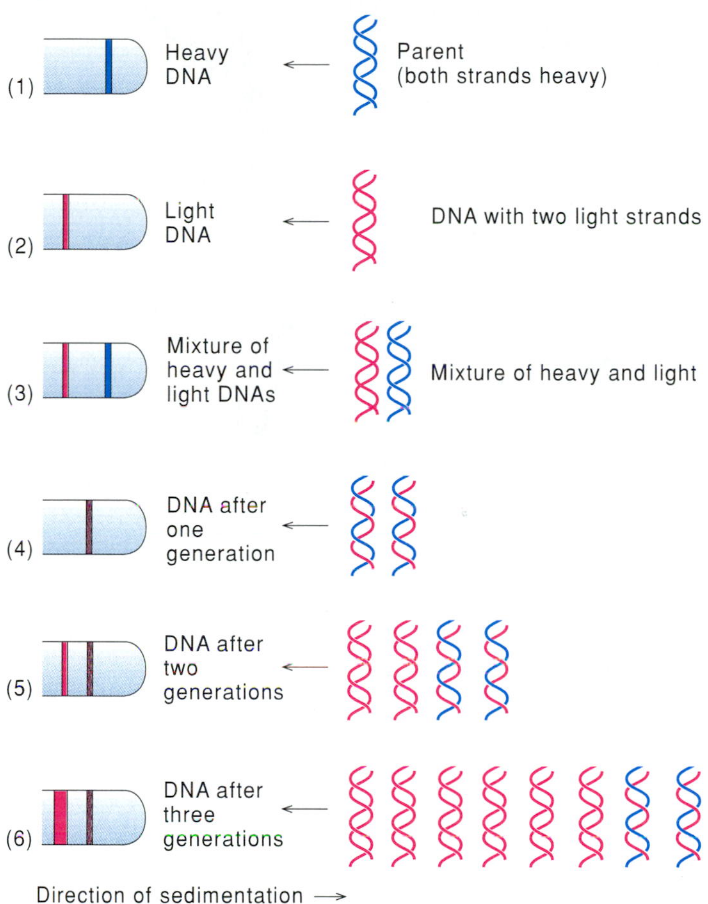

to go through one or more doublings. Then they isolated the DNA from these cells and analyzed it by CsCl density-gradient centrifugation (see box 25B). Pure ^{15}N-DNA produces a single band of DNA (fig. 26.3, frame 1). The same is true for ^{14}N-DNA (fig. 26.3, frame 2). The only difference is that denser DNA produces a band farther down the centrifuge tube. Thus the location of the DNA in the tube made it possible to monitor the density of the DNA. When cells containing pure ^{15}N-DNA were allowed to grow in ^{14}N medium for precisely one generation time, the only band visible in the isolated DNA was that corresponding to ^{15}N-^{14}N hybrid DNA (fig. 26.3, frame 4). This observation argued in favor of a semiconservative mode of replication, and the results in subsequent generations agreed with this view. Thus the next generation grown in light label showed equal amounts of hybrid-density and light-density DNA (fig. 26.3, frame 5), and the following generation showed a ratio of 3:1 of light to hybrid (fig. 26.3, frame 6). Besides demonstrating that *E. coli* DNA was replicated semiconservatively, the density transfer experiments also showed that all the chromosomal DNA replicated once, before it initiated a new round of DNA synthesis.

Similar experiments have been performed on mammalian cells grown in tissue culture, using bromouracil as a density label. Bromouracil contains a bromine atom instead of a methyl group on the 5 position of thymine. Bromouracil, when incorporated into DNA, can substitute for most of the thymidine, leading to DNA with a substantially higher than normal density. The results with eukaryotic DNA were found to parallel those in *E. coli*. Thus it appears that semiconservative replication of cellular DNA is a general phenomenon.

Taylor, Woods, and Hughes devised another method for labeling DNA to follow its replication. They selectively labeled chromosomal DNA with tritiated (^3H-labeled) thymidine and then visualized the distribution of the radioactive label directly in autoradiographs of chromosomes (fig. 26.4*a*); the same pattern of labeling was visible after the cells subsequently replicated in the absence of tritiated thymidine (fig. 26.4*b*). Bean seedlings were used for those experiments because cell division is very rapid in their growth tips. A sufficient period of time was allowed for some of the cells in the seedlings to undergo one round of DNA duplication (incorporating the radioactive label)

Figure 26.4

Autoradiographs of *Vicia faba* chromosomes labeled with [^{3}H]thymidine. The labeled thymidine becomes incorporated into the chromosomal DNA. A suitably labeled preparation is flattened and subjected to film exposure. Small dots indicate radioactive disintegration in the exposed film. (*a*) The first metaphase after replication in the presence of [^{3}H]thymidine. (*b*) The second metaphase after an additional replication in nonradioactive medium. (*c*) A diagrammatic interpretation of the results shown in (*a*) and (*b*). Radioactive single strands of DNA are shown in color. Radioactive chromatids at metaphase are also indicated in color. Colchicine has been used to inhibit spindle fiber formation and thus the anaphase separation of sister chromatids. Under these "C-metaphase" conditions, separation of sister chromatids is delayed. In (*a*) both sister chromatids are labeled uniformly. In (*b*) the sister chromatids are not labeled uniformly. The large chromosome at the top has one chromatid labeled and one virtually unlabeled. The homolog to its right has two exchanges (a labeled segment moved into the lower chromatid). The two small chromosomes to the lower left of it are lying one on top of the other. Both have one sister chromatid exchange. The small chromosome to the upper left has one sister chromatid exchange and the one at the lower left is lightly labeled but probably has two exchanges. (Autoradiographs courtesy of J. H. Taylor.)

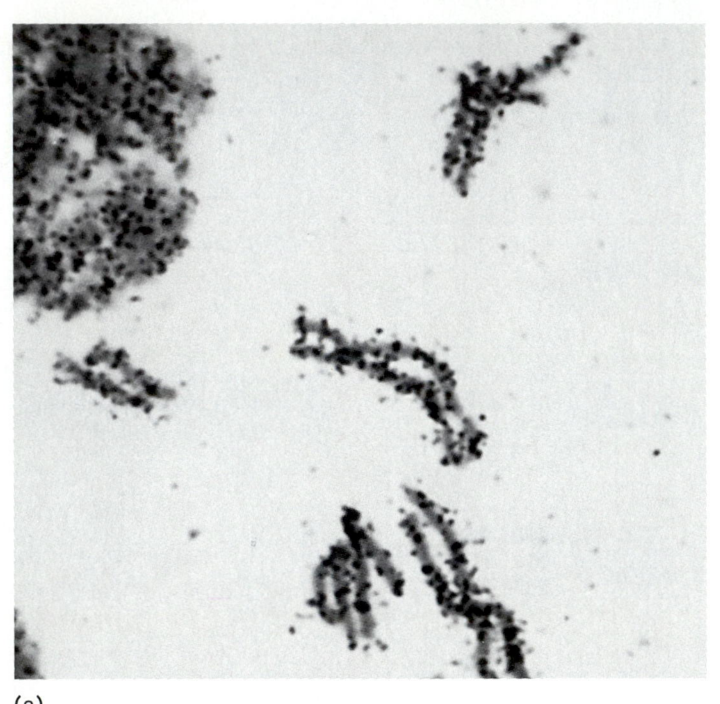

(a)

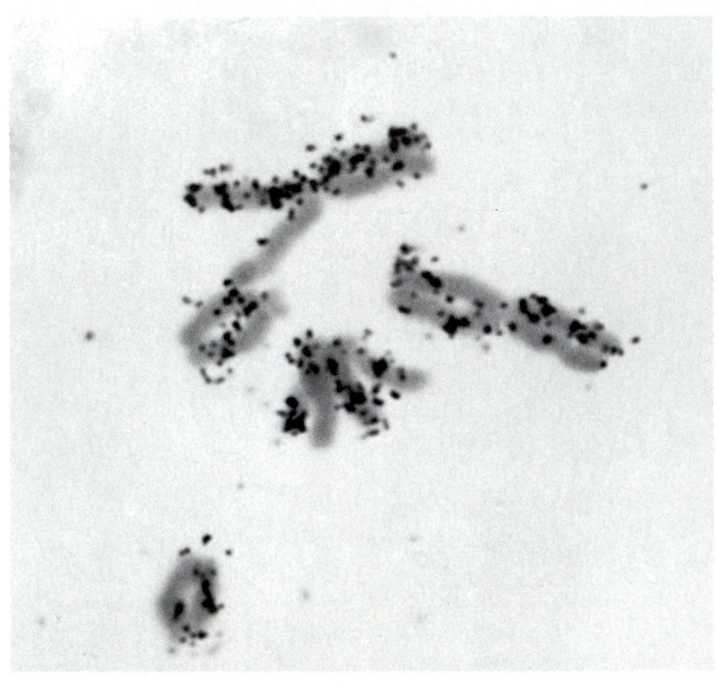

(b)

In presence of colchicine

Duplication with labeled thymidine

First C-metaphase after labeling; (a) above

Duplication without labeled thymidine

Second C-metaphase after labeling; (b) above

(c)

and cell division. After this, the seedlings were transferred to a fresh solution containing colchicine with the ^{3}H label. Colchicine is a plant alkaloid that inhibits normal mitosis by interacting with microtubule proteins necessary for mitotic spindle formation; it does not inhibit DNA synthesis nor the replication of chromosomes but delays the formation of daughter cells by inhibiting chromatid segregation. The advantage of colchicine treatment in this experiment is that in tissue so treated many cells become arrested in the state where the sister chromatids are paired. Cells that were examined shortly after the transfer from the [^{3}H]thymidine medium were examined at mitosis, and all the chromosomes appeared to be labeled uniformly (see fig. 26.4*a*). When the cells were allowed to duplicate their chromosomes in unlabeled medium, only one of the two chromatids in each chromosome pair was labeled (see fig. 26.4*b*). That is exactly the result that would be expected if each chromatid is composed of a single linear duplex of DNA that replicates semiconservatively (see fig. 26.4*c*). Similar results have been obtained for other eukaryotic cell types.

The universality of the semiconservative mode of DNA replication follows from the complementary nature of the DNA duplex. As the duplex unwinds, it presents two templates for the binding of complementary nucleotides. Subsequently, polymerization results in two duplexes, each with one old strand and one new strand. Thus the genetic information contained in the base sequence is directly transferred from one generation of DNA to the next by the capacity of DNA single strands to serve as templates for the assembly of complementary mononucleotides.

DNA Synthesis in Prokaryotes

More is known about the synthesis of genomic DNA in *E. coli* than in any other cell. The *E. coli* bacterium contains a single circular chromosome with about 4.5×10^{6} base pairs. Most other bacteria have chromosomes of about the same size and, in some cases, the chromosomes have been found to be circular as well. The number of genes required for DNA synthesis is probably between 20 and 50. In figure 26.5 the locations of some of these genes are indicated, together with the locations of unique initiation (*oriC*) and termination points for DNA replication. Under optimal growth conditions the *E. coli* chromosome is replicated in 20–30 min.

The Circular Bacterial Chromosome Replicates Bidirectionally

Replication of the *E. coli* chromosome can be visualized by autoradiography of intact ^{13}H-labeled chromosomes, using a gentle isolation technique developed by Cairns and Davern. After completing one round of replication in labeled medium, chromosomes appear as uniformly labeled circles (fig. 26.6, top). Initiation of a second round of replication leads to the formation of a replication eye (fig. 26.6, middle). As synthesis proceeds in labeled medium, the size of the replication eye increases; the replicating chromosome at this stage is referred to as a theta structure because it resembles the Greek letter θ (fig. 26.6, bottom). Semiconservative replication is consistent with the density of the autoradiographic tracks observed on parts of the chromosome after one and two rounds of replication in [^{3}H]thymidine.

It is reasonable to conclude that the replication eye contains two partially separated parental DNA strands that are base-paired with strands of newly synthesized DNA. Not resolved by this type of observation is the question of whether replication occurs in one direction or both directions about the origin of replication. If growth is unidirectional, we would expect one growth point (fig. 26.7*a*), called a growth fork (or replication fork); if growth is bidirectional, we would expect two growth points or growth forks (fig. 26.7*b*). Although examples of both types of replication are observed, it has been shown in several cases that bacterial chromosomes, including that of *E. coli*, replicate bidirectionally.

Convincing evidence of bidirectional replication of *E. coli* was obtained by measuring gene frequency during replication. Shortly after initiation, those genes near the origin of replication that have been duplicated must be present at twice the number of copies as those genes far removed from the origin.

Figure 26.5

A diagram of *E. coli* chromosome. The origin and approximate region of termination of replication are indicated. Locations of some genes involved in DNA replication are also indicated. The functions of many of these genes are described in tables 26.1 and 26.2.

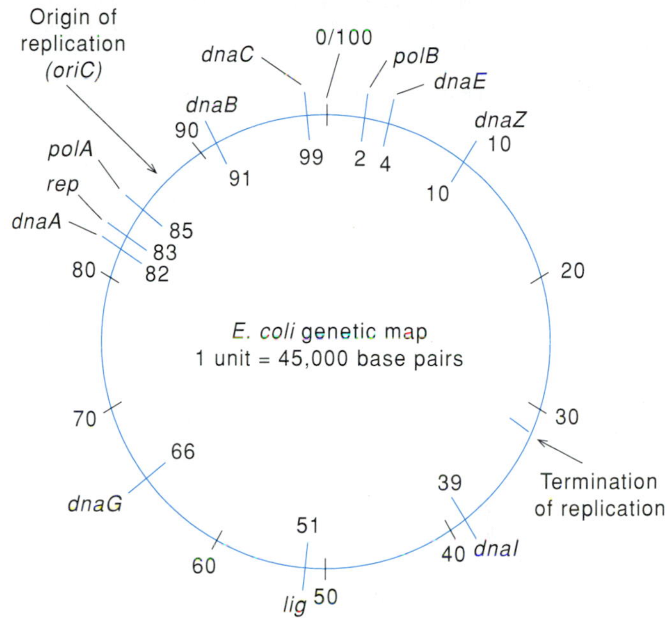

Figure 26.6

Simulated autoradiographs of the *E. coli* chromosome after one or more replications in the presence of [^{3}H]thymidine. After one round of replication the autoradiograph shows a circular structure that is uniformly labeled. The second round of replication begins with the formation of a replication eye. One branch in the replication eye is twice as strongly labeled as the remainder of the chromosome, indicating that this branch contains two labeled strands. This structure is consistent with semiconservative replication for the *E. coli* chromosome.

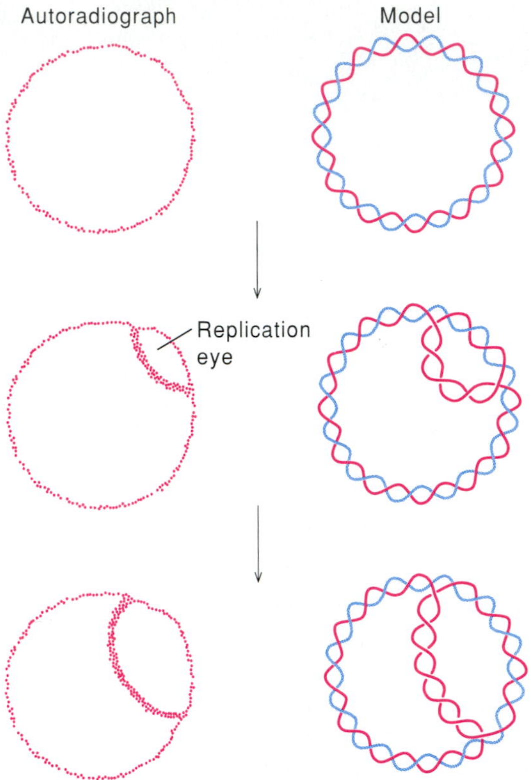

Figure 26.7

Schematic diagrams of two different modes of DNA synthesis at the growth fork(s). In unidirectional replication (*a*) there is one growth fork; in bidirectional replication (*b*) there are two. Color indicates regions containing newly synthesized DNA. Measurements of gene frequency make it possible to distinguish between unidirectional and bidirectional synthesis. (*c*) In unidirectional replication, the gene frequencies should be highest for genes located on one side of the origin. (*d*) In bidirectional replication, the gene frequencies should be equally high for regions symmetrically disposed about the origin.

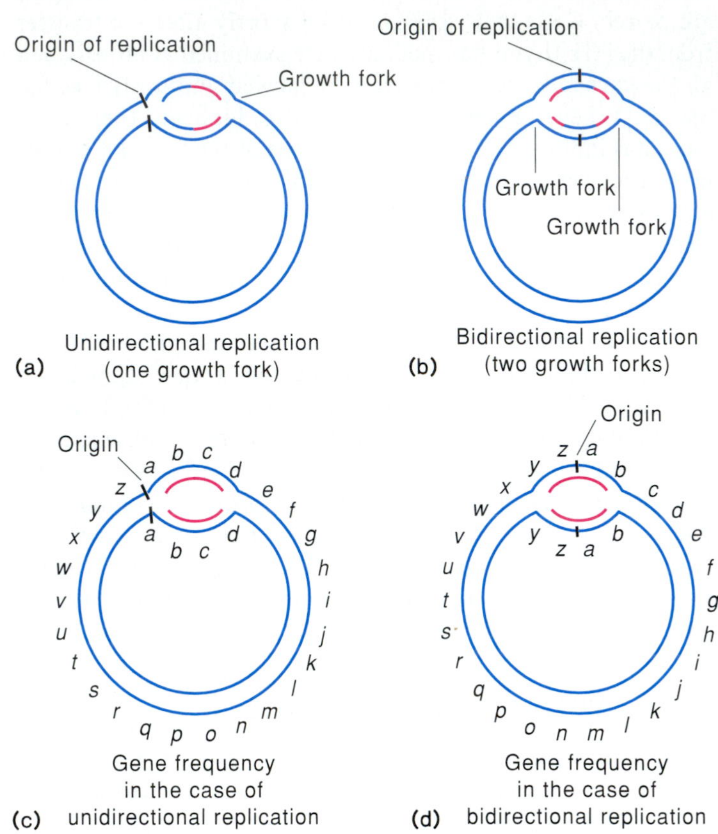

The numbers of genes can be assayed biologically by transformation (chapter 25) or transduction. If replication is unidirectional, a steady increase in gene number should occur in one direction along the circular chromosome (see fig. 26.7*c*). On the other hand, if replication is bidirectional, the increase in gene number should proceed in both directions from a common point, the origin (see fig. 26.7*d*). Detailed measurements indicated that the latter situation is true for the chromosomes of *E. coli* and *Bacillus subtilis*. These measurements also made it possible to locate the initiation and termination points on the chromosomes (see fig. 26.5).

The genetic evidence for bidirectional replication of the *E. coli* chromosome was supported by direct cytologic evidence. Bacteria were grown for a very short time in the presence of radioactive thymidine, and replicating chromosomes were examined by autoradiography; both of the forks in the replicating structures were intensely labeled (as in the colored portions of figure 26.7*b*). This result shows that both forks must be active during replication, a finding consistent with bidirectional growth. Replication was also shown to be bidirectional in bacteriophage λ DNA, through electron microscopy. By analyzing circular λ molecules at different stages of their replication, it was found that both ends of the replication eye moved relative to a single, unique cut made by a restriction endonuclease (we will consider the properties of restriction endonucleases later in this chapter and more fully in chapter 27). The experiments on bidirectionality of replication of the *E. coli* genome have been possible because there is a unique origin of replication. Once initiated, each replication fork travels about the same distance, each fork terminating at a point 180° removed from *oriC*.

Growth in Each Direction Is Discontinuous on at Least One Strand

Continuous synthesis on both strands of a replication fork would require synthesis in the 5′→3′ direction on one strand and in the 3′→5′ direction on the other strand because of the antiparallel nature of duplex DNA (fig. 26.8*a*). That possibility seems unlikely because the only known enzymes that catalyze DNA synthesis do so in the 5′→3′ direction. For this reason replication was postulated to be discontinuous on one of the branches at the replication fork (fig. 26.8*b*). Careful electron microscopic examination of replication forks in bacterial viruses has in fact

Figure 26.8

Models for synthesis at the replication fork. (*a*) Continuous synthesis on both strands. Note that both growth arrows are pointing in the same direction, which would require growth in the 5′ → 3′ direction on one strand and in the 3′ → 5′ direction on the other strand. If growth occurs only in the 5′ → 3′ direction, synthesis would have to be discontinuous on one strand, as in (*b*). Alternatively, it could be discontinuous on both strands (*c*).

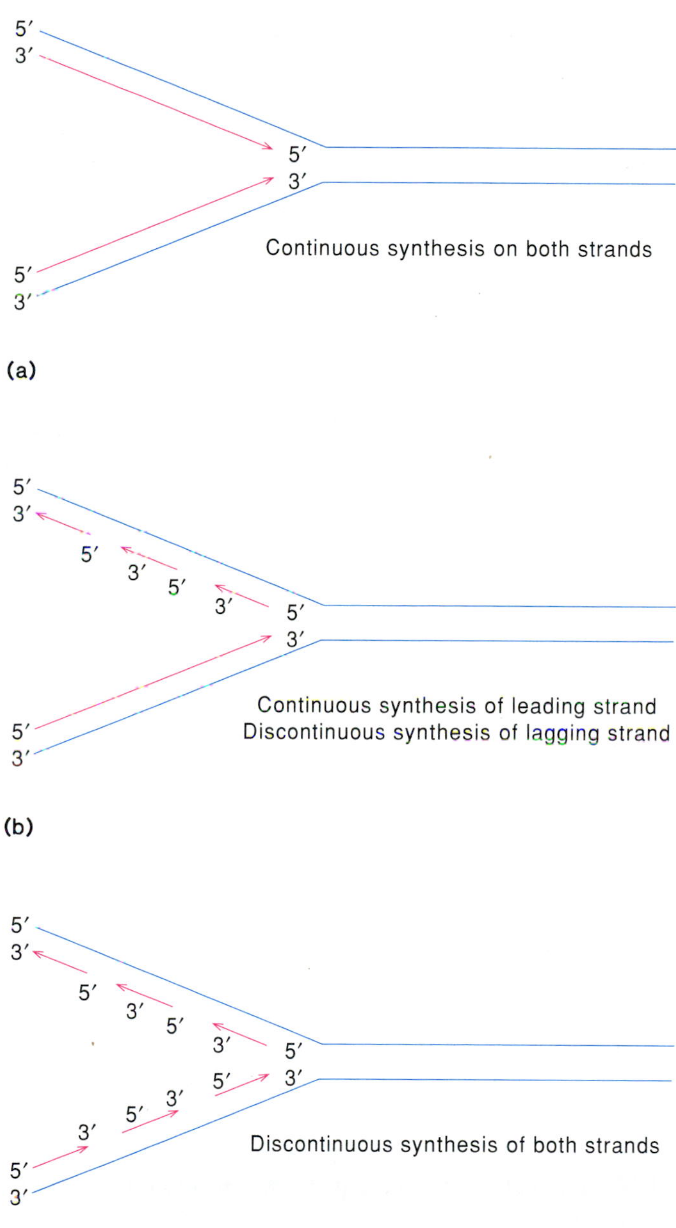

Figure 26.9

Sedimentation analysis of *E. coli* DNA from cells labeled with [^{3}H]thymidine for different lengths of time. Sedimentation analysis is done on an alkaline sucrose gradient. The alkali denatures the DNA so that it becomes single-stranded. (*a*) Short-term labeling (2 to 10 s) preferentially labels the most slowly sedimenting DNA, in the size range of 1,000–2,000 base pairs. In long-term labeling (1 to 2 min), most of the labeled DNA is of much higher molecular weight. (*b*) In a pulse labeling experiment, the short-term labeling is done in the same way as in (*a*). The long-term labeling includes the addition of a large amount of nonradioactive substrate after a short time. This added substrate is called a chase, as it lowers the specific radioactivity of the labeled substrate so that no further incorporation of radioactivity is detectable. Long-term labeling is done in order to follow the fate of the short-term-labeled DNA.

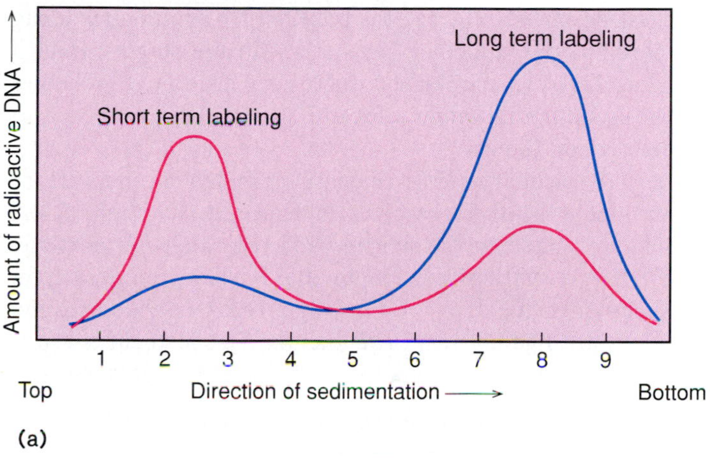

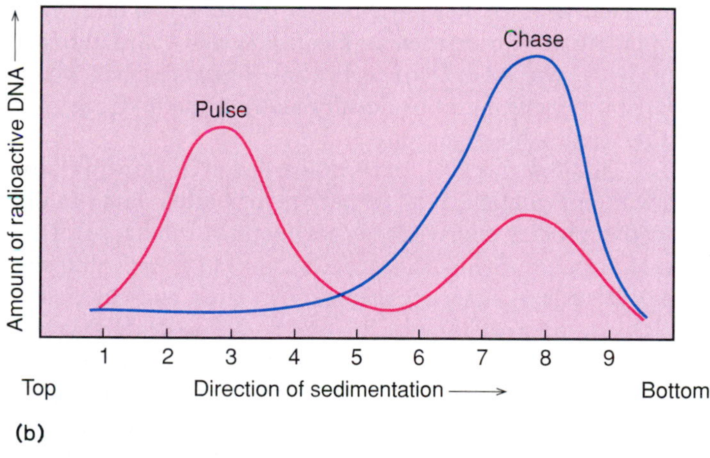

shown that transient gaps sometimes are apparent on one of the daughter DNA strands, close to the replication fork. Observations such as this led to the notion of a leading strand and a lagging strand (fig. 26.8*c*). In this model, 5′→3′ synthesis of the leading strand could occur continuously in the same direction as the unwinding of the replication fork. Synthesis of the lagging strand in the 5′→3′ direction could occur in discontinuous spurts in a direction opposite to DNA unwinding.

This concept of discontinuous synthesis was supported by Okazaki, who found that at least half the newly synthesized DNA is first made as small pieces (Okazaki fragments) that later become incorporated into large segments of DNA. Small replication fragments were detected by exposing growing cells to tritiated thymidine for a very short time (2 to 10 s) (a pulse), followed by rapid isolation of the radioactively labeled DNA. If replication was allowed to continue in unlabeled medium for several minutes (the chase period) most of the labeled DNA was found in much larger pieces of DNA (fig. 26.9). Sedimentation analysis in alkali (the alkali denatures the duplex DNA into single strands) provided an estimated length of 1,000–2,000 bases for the bacterial Okazaki fragments. The researchers concluded that the replicating polymerase must operate by synthesizing short fragments on the lagging strand as new points for the initiation of replication are presented by the progressive unwinding of the double helix at the growth fork.

A closer examination of the Okazaki fragments revealed short stretches of ribonucleotides at their 5′ ends. From this and many other observations made with different *in vitro* systems, it appears that a new DNA chain can be initiated only by attaching the first deoxynucleotide through its 5′ phosphate to the 3′-OH of a short RNA chain. An RNA oligonucleotide that functions in this capacity is called a primer. Primers are synthesized *de novo* at various secondary priming sites along the chromosome, and they base-pair with the single-stranded template DNA in the regions where they are found. Whether these secondary initiation sites are specific or are chosen at random is not known.

A summary of the currently accepted mechanism for discontinuous synthesis is given in figure 26.10. First, RNA primers are made on the single-strand region of the template; then DNA is synthesized. Finally, the RNA is removed from the fragments, the gaps are filled by a DNA polymerase, and the nicks are ligated. In *E. coli* the *dnaG* gene encodes a primase that generates primers from ribonucleotide precursors.

Figure 26.10

A model for discontinuous DNA synthesis. Synthesis occurs in a region that has been partially single-stranded. First, RNA primers are formed at various points on the single-stranded region (1). The DNA synthesis starts at the 3′ ends of the primers (2). The primers are removed (3). Gaps between DNA fragments are filled in by further DNA synthesis (4). The fragments are ligated to make one long continuous piece of DNA (5). Newly synthesized RNA and DNA are indicated in red.

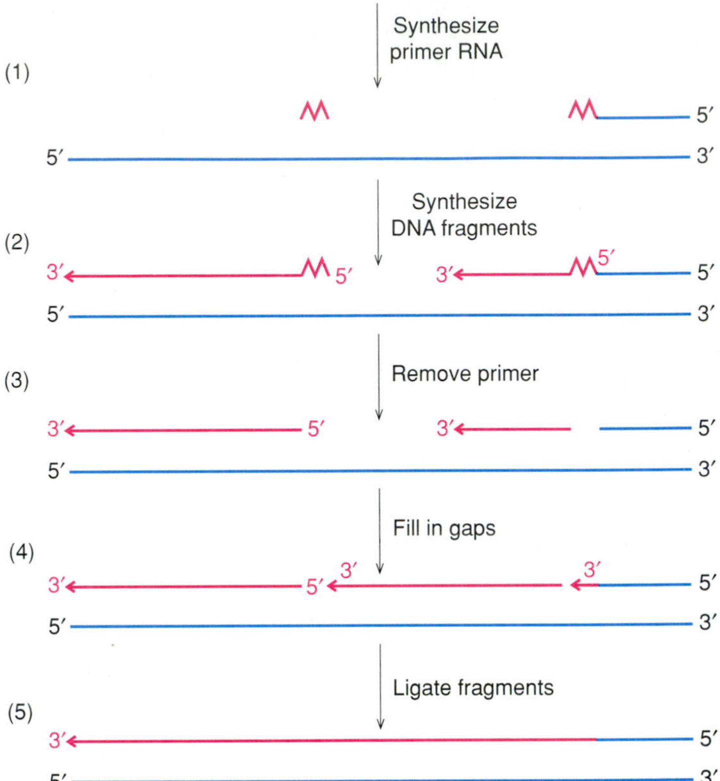

An Assemblage of Proteins Is Involved in the Replication Process

Many proteins are required for DNA replication in addition to the polymerizing enzyme. The process of discovering these enzymes has combined many approaches. Historically, three general methods have been used for identifying and characterizing the proteins involved in DNA replication. They are purification, reconstitution, and mutation. Ideally, all three methods are used together in suitable organisms.

The first method, purification, involves isolating proteins with enzymatic activities that are logically related to the replication process, such as DNA polymerases and ligases. This is the classical biochemical approach and can be applied to any biological system. After an enzyme is isolated and characterized, several approaches may be used to demonstrate that the purified enzyme is active in the replication process *in vivo*. Sometimes this can be done by using inhibitors that act on both the purified protein extracts and at the cellular level. The concentration of inhibitor required to inhibit the purified enzyme *in vivo* should be approximately the same as that required to inhibit the purified enzyme *in vitro*. Inhibitors are particularly useful in comparing and characterizing various polymerases that are present in the same cell and in comparing DNA polymerases extracted and purified from different organisms by monitoring their relatedness. Thus yeast DNA polymerase I is probably related to mammalian DNA polymerase α, since these are both resistant to the inhibitor butylphenyl-dGTP. In prokaryotes, mutations in specific genes have been very useful for confirming the functions of isolated proteins in the replication process. The induction of a new enzyme activity associated with a biological process, such as virus infection or cell proliferation, also provides useful evidence for the involvement of the enzyme.

A second method used to identify proteins needed for replication is reconstitution. Whole-cell lysates containing all of the components necessary for replication are fractionated and the DNA replication system is then reconstituted with various combinations of the purified or partially purified proteins. Components of the replication system are recognized on the basis of their ability to restore overall activity *in vitro*. This procedure can be applied to any organism, even when relevant genetic mutants are not available. However, mutant studies are usually required for final confirmation that a protein carries out a particular function *in vivo*. We will see an example of such confirmation a little later, when we discuss the role of the DNA polymerase I.

The third method of analysis uses mutants as the primary tool. This approach is most effective with organisms having well-characterized means of carrying out genetic analysis and exchange. It requires the isolation of conditionally lethal mutants, i.e., mutants that grow normally under one set of conditions but fail to do so under another set of conditions. Most commonly, temperature-sensitive DNA replication mutants are used. Such mutants have been isolated for *E. coli;* they grow normally at a low temperature (25–30° C), called the permissive temperature, but poorly or not at all at a high temperature (41° C), referred to as the nonpermissive temperature. Preliminary analysis of the temperature-sensitive step provides clues to the stage of replication affected. For example, the length of time required for DNA synthesis to stop, after cells have been shifted from permissive to nonpermissive temperatures, can indicate whether the mutation occurs in a protein involved in the initiation or the elongation reactions of DNA synthesis. If the mutation is in the gene for a protein required for elongation, DNA synthesis will stop almost immediately at the nonpermissive temperature. If the mutation affects a protein required only for initiation of replication, DNA synthesis will continue for some time and stop when the round of replication in progress is completed.

Once the stage of replication has been identified, *in vitro* assays are used to aid in purifying the corresponding proteins from fully competent wild-type cells. Extracts from cells with the temperature-sensitive defect will not synthesize DNA at the elevated temperature, but activity can be restored by adding the corresponding protein from wild-type cells. This complementation test can be used as an assay for the purification of particular replication proteins. To prove that the correct protein has been purified from wild-type cells, the proteins from the temperature-sensitive mutant also must be purified and shown to be abnormal. Frequently they exhibit unusual thermolability.

A Polymerase with the Expected Characteristics Is Discovered by Kornberg

The first DNA-polymerizing enzyme to be discovered was found in cell-free extracts of *E. coli*. We will examine this polymerase in some detail as an example of how biochemical and genetic studies can be used to determine the properties of an enzyme and its normal *in vivo* role. We will then consider the other major classes of proteins involved in *E. coli* DNA replication.

The Watson-Crick proposal of a complementary duplex structure for DNA stimulated a search for a DNA polymerase enzyme with certain implied properties. The enzyme should require an intact DNA chain to serve as a template for the binding of complementary bases, and the *de novo* synthesized DNA should be complementary to the template DNA chain. Kornberg and his co-workers isolated from cells of *E. coli* a DNA-synthesizing enzyme that satisfied these requirements. They gave it the name DNA polymerase; it is now known as DNA polymerase I (Pol I) or the Kornberg enzyme.

In its purified state, the Kornberg enzyme requires a DNA template, the four commonly occurring deoxynucleoside triphosphates, and Mg^{2+} ions for synthesizing DNA. The enzyme catalyzes the addition of mononucleotides to the 3′-OH end of a growing chain (fig. 26.11*a*). At the same time, the linkage between the α and β phosphates is broken, releasing an inorganic pyrophosphate group (PP_i). In fact, the energy provided by this cleavage drives the linking of the mononucleotide to the growing DNA chain. The bases added to the growing chain are determined by the sequence of bases in the DNA template. As nucleotides complementary to those on the template are added, the single-stranded DNA template is progressively converted to a double helix.

Physiochemical studies have shown that the enzyme has a complex surface, with specific attachment sites for the template chain, the growing chain, and monomer nucleoside triphosphate. The enzyme is highly selective, since it links to the growing chain only those nucleotides that form Watson-Crick base pairs with the template strand. As the new chain is lengthened by synthesis, the enzyme moves along the template one base at a time. That the newly synthesized DNA is a replica of the template strand has been shown in a number of ways—by gross base-composition analysis, by melting-curve profile of the hybrid formed between the template strand and the newly synthesized strand (see chapter 25 for an explanation of nucleic acid melting curves), and finally, through a series of manipulations, by the *de novo* synthesis of genetically active DNA. A unique property of DNA and RNA polymerases is that they require a template to exhibit their activity.

Phosphodiester bond formation probably occurs as a nucleophilic attack, by the 3′ hydroxyl group of the terminal mononucleotide residue at the growing end of the chain, on the α phosphorus atom of the entering nucleoside-5′-triphosphate. The attack causes displacement of the pyrophosphate group and formation of the internucleotide linkage (fig. 26.11*b*).

In addition to catalyzing the characteristic template-directed polymerization, DNA polymerase I is associated with an activity that can remove mononucleotides from the 3′ end of a polynucleotide strand. This second function is an exonuclease activity; it leads to hydrolytic cleavage of the bond between the 5′ phosphate of the terminal residue and the 3′-OH group of the penultimate residue. Since this is the same linkage made during synthesis, the degradation reaction may be thought of as a reversal of polymerization. For net chain propagation to occur beyond the 3′-OH end of the DNA strand, the polymerization rate must exceed the depolymerization rate. Polymerization is much faster than depolymerization if the correct base-paired nucleotide is inserted in the growing chain. If a mismatched base is accidentally inserted, the opposite is true and the mismatched base is usually removed. The combination of polymerization-depolymerization reactions serves as an "editing" or "proofreading" error-reducing mechanism in DNA synthesis. If the editing function is defective because of a mutation, a generalized increase in mutations occurs in cells harboring the altered nuclease.

Figure 26.11

Template and growing strands of DNA. (*a*) Nucleotides are added one at a time to the 3′-OH end of the growing chain. Only residues that form Watson-Crick H-bonded base pairs with the template strand are added.

(*b*) Covalent bond formation between the 3′-OH end of the growing chain and the 5′ phosphate of the mononucleotide is accompanied by removal of the two terminal phosphates from the substrate nucleoside triphosphate.

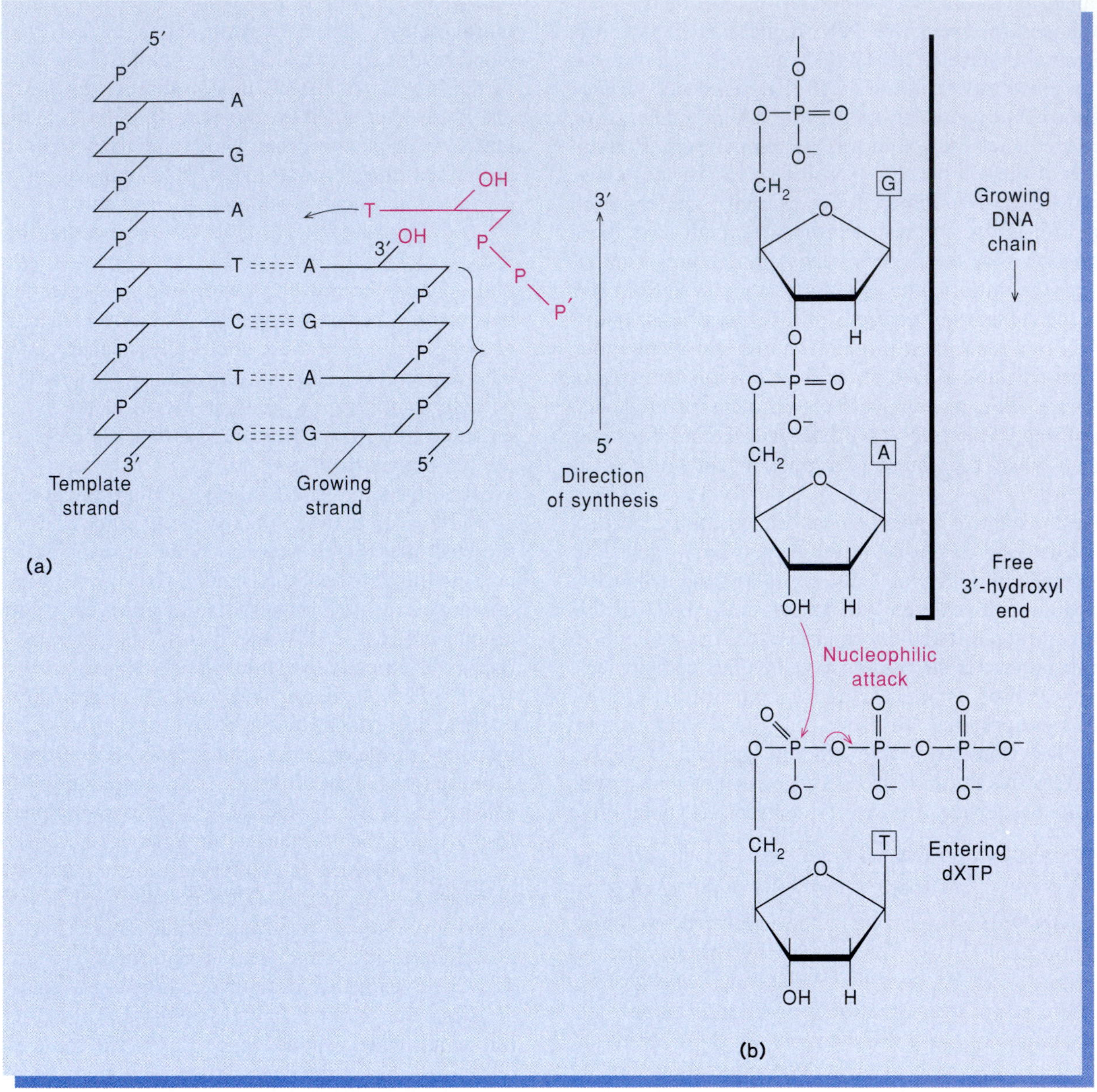

Pol I has another associated activity, catalyzing the 5′→3′ degradation of DNA. Whereas the 3′→5′ nuclease activity of the enzyme is much more effective on unpaired or mispaired bases, the 5′→3′ activity preferentially cleaves base-paired regions. The ability of DNA Pol I to degrade DNA (or RNA) in the 5′→3′ direction, as well as to carry out a polymerization reaction, suggests a possible function in removing the RNA primer prior to gap filling (see fig. 26.10). In fact, RNA primer removal and gap filling are probably catalyzed in rapid succession by the same DNA polymerase I molecule.

Crystallography Has Combined with Genetics to Produce a Detailed Picture of DNA Pol I Function

The next question that arises is whether the three activities of Pol I just described all originate from a single active site or more than one active site. Cleavage of Pol I by the protease subtilisin leads to a small fragment (M_r = 30,000) with 5′→3′ nuclease activity and a large fragment (M_r = 70,000), called the Klenow fragment, exhibiting the polymerization and 3′→5′ depolymerization activities.

Figure 26.12

Schematic drawing of the large proteolytic (Klenow) fragment of *E. coli* DNA polymerase I. The domain catalyzing the 3′→5′ exonuclease reaction includes residues 324–520, while the polymerase domain includes residues 521 to the C terminus and contains a large cleft. (From D. L. Ollis, P. Brick, R. Hamlin, N. G. Xuong, and T. A. Steitz, Structure of large fragment of *Escherichia coli* DNA polymerase I complexed with dTMP, *Nature* 313:762–766, 1958.)

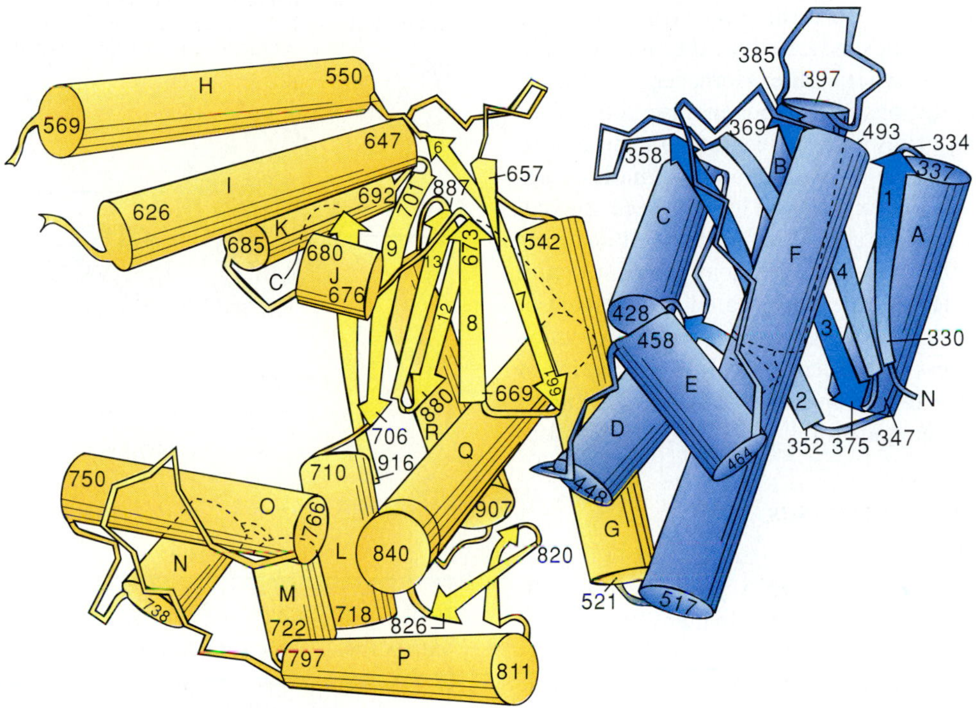

A series of investigations by Tom Steitz and his coworkers has led to a detailed understanding of the molecular anatomy of Klenow peptide fragment. Crystallography shows that the 605-amino-acid polypeptide is folded into two structural domains containing about 200 and 400 amino acids (fig. 26.12). The N-terminal third of the Klenow fragment (residues 324–517 in the Pol I sequence) forms the smaller of the two domains. This domain has a core of β-pleated sheet with α helices on both sides. In the crystal, this domain can bind two divalent metal ions and a molecule of dNMP. The larger domain (residues 521–928) forms a structure with a deep cleft, about 20–24 Å wide and 25–35 Å deep. A six-stranded antiparallel β sheet forms the bottom of the cleft and large protrusions of α helices form its sides. Steitz points out that this structure is similar in shape to a right hand grasping a rod. Thus one side of the cleft forms a wall, 50 Å long, that can be compared with the curled fingers of a right hand. The other side of the cleft is formed primarily by two long α helices, H and I, projecting from the protein like a thumb. Located at the tip of the thumblike protrusion are 50 amino acid residues connecting helices H and I that are partially disordered, indicating that this region is flexibly attached to the rest of the molecule.

The cleft in the large domain can accommodate double-stranded B DNA. The J and K α helices are placed partially into a major groove and may function like the thread of a nut to fix the exact position of the DNA major groove relative to the protein. Thus during DNA synthesis the polymerase would follow a spiral path along the DNA. This Klenow fragment-DNA complex is consistent with calculations that indicate that virtually all of the positive electrostatic charge of the Klenow fragment lies within the cleft. Some experimental evidence for the assignment of the DNA binding site is provided by genetics. The mutation polA6, which weakens the enzyme's interaction with DNA, is an Arg → His change within helix K.

When duplex DNA is bound, the flexible 50-amino-acid subdomain connecting helices H and I could fold across the top of the binding cleft, allowing the protein to surround the DNA template completely. This firm grip on the template should encourage polymerase to stay bound to the template so that it could catalyze the formation of many phosphodiester linkages in swift succession. The alternative to an enzyme that operates in this way (processively) would be a polymerase that dissociates every time it catalyzes the formation of a single phosphodiester linkage. It seems likely that a processive enzyme would be much more efficient, since it would not have to go through the time-consuming events of dissociation and reassociation with the template every time a nucleotide is added to the growing polymer. All DNA polymerases are processive enzymes, some more than others, as we will see when we discuss the mammalian DNA polymerases.

It is known that deoxynucleoside monophosphates (dNMPs) inhibit the 3′,5′-exonuclease reaction but not the polymerase reaction. Thus it is likely that the dNMP binding site corresponds to the position of the DNA 3′ terminus in the exonuclease active site. The dNMP phosphate would therefore mark the position of the phosphodiester bond to be cleaved in the exonuclease reaction. Crystal structure studies show that dNMP binds to the small domain of Klenow fragment. Furthermore, mutant studies show that discrete amino acid changes that eliminate the binding of dNMP also eliminate the 3′, 5′-exonuclease activity without perturbing the structure of the Klenow fragment. Such studies indicate that the small domain contains the 3′,5′-exonuclease active site. Moreover, the absence of an effect of these mutations on the polymerase activity indicates that the two active sites must be separate to some degree.

Although the precise structure for the smaller fragment cleaved in the subtilisin experiment has not been studied, it appears highly likely that the structure of intact Pol I consists of three domains with the approximate layout shown in figure 26.13. The 5′,3′-exonuclease activity is located downstream of the other two activities, a position consistent with its function of removing RNA primers on nascent strands. The location of the 3′,5′ activity relative to the polymerase activity is known more precisely; the two enzyme sites are located eight bases apart.

Two More DNA Polymerases Are Discovered in E. coli

Kornberg's early work was done on DNA polymerase in cell-free systems. The fact that the enzyme had so many of the properties expected for a DNA-replicating enzyme led most observers to believe that it was indeed the DNA-replicating enzyme of *E. coli*. But to obtain final proof that an enzyme functions in a given capacity *in vivo,* it is essential to isolate mutants in which the enzyme's behavior has come about by correlating *in vitro* biochemical behavior with the behavior of mutants in which DNA replication and repair are affected.

From the time of the discovery of DNA polymerase I, about twenty years were needed to clarify its physiological function. A major step in that direction was taken by Cairns and De Lucia, who laboriously scanned several thousand colonies of heavily mutagenized *E. coli* and found one that contained almost no DNA polymerase I polymerizing activity (1 to 2% of normal). That mutant grew well under normal conditions, a fact suggesting that the DNA-polymerizing activity of the Kornberg enzyme was not needed for replication. However, the mutant was more sensitive to ultraviolet irradiation than its wild-type parent. Ultraviolet irradiation was known to damage DNA. Therefore the increased sensitivity of the mutant suggested that, although DNA polymerase I was not the chromosome-replicating polymerase, it might be involved in repairing chromosome damage caused by ultraviolet light.

In cell-free extracts from a mutant *E. coli* strain that did not contain DNA polymerase I polymerizing activity, two additional DNA-polymerizing enzymes were subsequently detected. These were named DNA polymerases II and III. The behavior of conditional lethal mutants of DNA polymerase III led to the conclusion that DNA polymerase III is the main replication enzyme. Some of the properties of the three enzymes are summarized in table 26.1.

Figure 26.13

A schematic drawing of the apparent domain structure of *E. coli* DNA polymerase I. The solid line represents the experimentally determined Klenow fragment structure; the dashed lines indicate a possible location for the small fragment produced by proteolytic cleavage of Pol I. The orientation of the nicked DNA substrate is such that the 5′→3′ exonuclease domain would be able to interact with the DNA 5′ downstream of the primer terminus. (From C. M. Joyce and T. A. Steitz, "DNA polymerase I: From crystal structure to function via genetics," *Trends Biochem. Sci.* 12:288-292, 1987.)

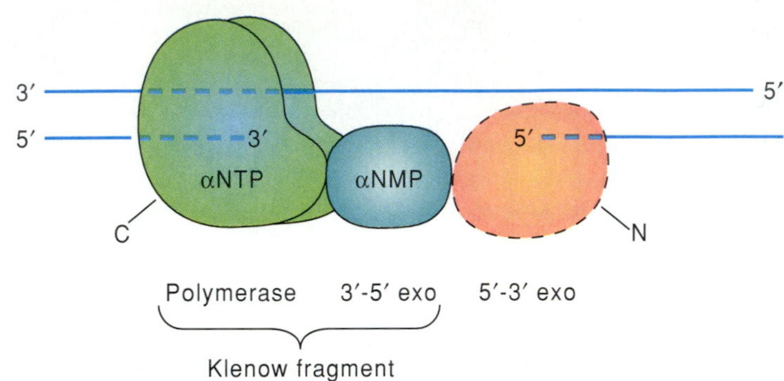

Table 26.1
Properties of Polymerases I, II, and III of *E. coli*

	Pol I	Pol II	Pol III
Molecules per cell	400	—	15
Turnover number[a]	600	30	9,000
Structural gene[b]	*polA*	*polB*	*polC* (*dnaE*)
5′→3′ polymerizing activity	+	+	+
3′→5′ exonuclease activity	+	+	+
5′→3′ exonuclease activity	+	−	−
Mutant loci	*polA*	*polB*	*polC, dnaN, dnaX, dnaZ, dnaQ*
Lethality	Viability reduced only when 5′→3′ exonuclease affected	No effect	Conditional lethality

[a]Nucleotides polymerized/min/molecule of enzyme at 37° C.
[b]Only the structural gene for the largest protein subunit in the enzyme is recorded.

For a few years following the observations of Cairns and De Lucia, it was assumed that DNA polymerase I was not important in replication but only in repair. However, further genetic and biochemical studies proved otherwise. The original mutated DNA polymerase I was defective only in its polymerizing function; its two degradation functions were intact. Subsequently, workers isolated conditional lethal mutants in which the 5′→3′ degradation function of Pol I was affected; the mutants cannot elongate DNA under nonpermissive conditions. Thus it was demonstrated that the degradation activity of DNA polymerase I is indispensable for chromosome replication.

DNA polymerase III, which functions as the main replicating enzyme, is a complex of at least seven different proteins (table 26.2). Genetic loci for some of the protein subunits found in the polymerase III enzyme have been identified (see table 26.2). Mutants in which θ is affected have not yet been isolated, so the physiological significance of θ remains unknown.

Table 26.2
Subunits and Subassemblies of Pol III Holoenzyme

Subunit	kDa	Gene	Subassembly			
α	130	*dnaE*	pol III (core)	pol III′	pol III*	holoenzyme
ϵ	27.5	*dnaQ*	pol III (core)	pol III′	pol III*	holoenzyme
θ	10		pol III (core)	pol III′	pol III*	holoenzyme
τ	71	*dnaX*		pol III′	pol III*	holoenzyme
γ	47.5	*dnaX*	γ complex		pol III*	holoenzyme
δ	35		γ complex		pol III*	holoenzyme
δ'	33		γ complex		pol III*	holoenzyme
χ	15		γ complex		pol III*	holoenzyme
ψ	12		γ complex		pol III*	holoenzyme
β	40.6	*dnaN*				holoenzyme

Polynucleotide Ligase Links Long Chains Together

Discontinuous DNA synthesis requires an enzyme for joining the newly synthesized segments (see fig. 26.10). Initially, two enzymes of this type were discovered, and they have been given the general name polynucleotide ligase. In the ligase-catalyzed reaction, a single phosphodiester bond forms between long runs of discontinuous chains held in proper juxtaposition by a template chain (fig. 26.14). Since the joining reaction is vital to DNA replication, it is not surprising to find that ligases are ubiquitous in living cells. They differ in the source of activation energy used for phosphodiester bond formation. The *E. coli* bacterial ligase (encoded by the *lig* gene, table 26.3) uses nicotinamide adenine dinucleotide (NAD) as the source of energy: the T4 bacteriophage and mammalian DNA ligases use ATP. Ligase is important in recombinant DNA techniques because it can seal fragments that have single-stranded complementary termini. Such fragments can be generated by restriction endonucleases. If the concentration of DNA fragments is high enough, ligase will also covalently link blunt-ended fragments that lack single-stranded complementary termini.

Topoisomerases Catalyze Supercoiling

Supercoiled helices are frequently found in closed circular DNA molecules or in otherwise constrained DNA segments (see chapter 25). With the help of supercoiled DNA and advances in the technology of DNA analysis by agarose gel electrophoresis (see fig. 25.21), researchers have discovered several kinds

Figure 26.14

Steps in the sealing of a DNA nick, catalyzed by DNA ligase. The bacterial ligase uses NAD⁺ to make an enzyme-AMP intermediate. Mammalian DNA ligases and bacteriophage T4 ligase use ATP for the same purpose.

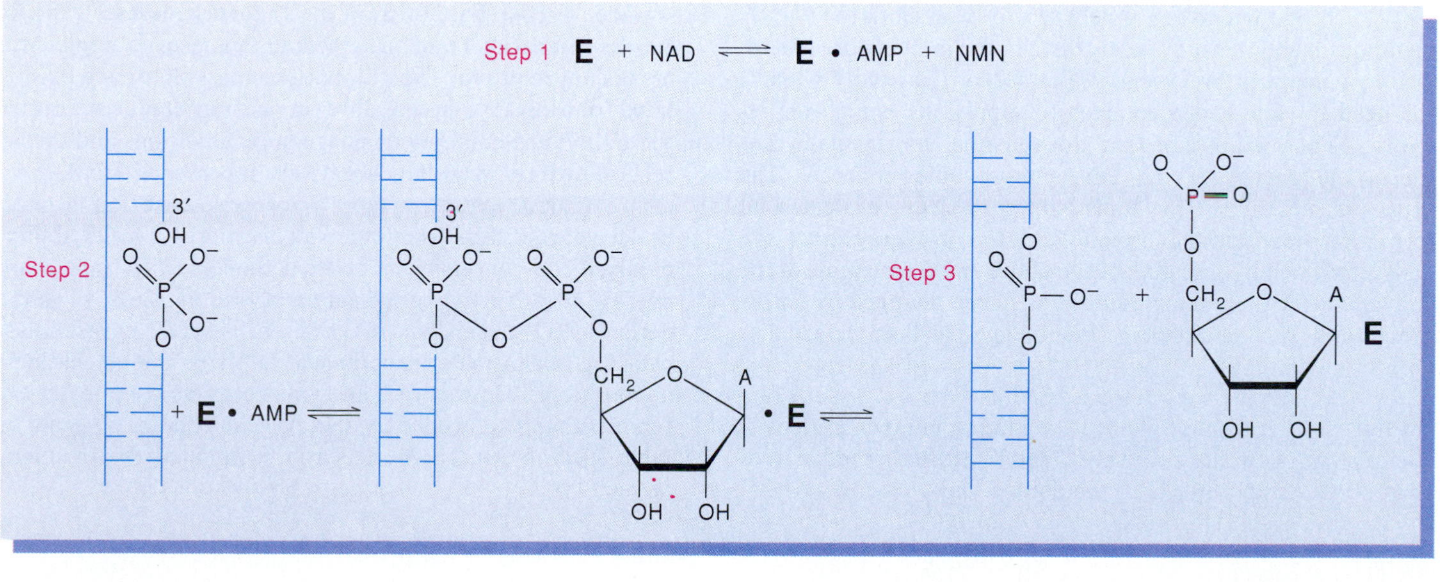

Table 26.3
Other Proteins Involved in DNA Synthesis in *E. coli*

Gene	Function of Encoded Protein	Size (Daltons × 10^3)
polA	In DNA polymerase I, extending Okazaki fragments and removing primers	109
dnaB	Interacts with primase (functions as hexamer)	50 monomer
dnaC	Stimulates priming	25 monomer
dnaG	Primase (acts as dimer)	60 monomer
dnaY	—	—
lig	Ligase	75
rep	Helicase, ATP-dependent	66
ssb	Single-strand DNA-binding protein	74
GyrA	Subunit of topoisomerase II encoding GyrA (sensitive to nalidixic acid)	100 monomer
GyrB	Subunit of topoisomerase II encoding GyrB (sensitive to coumermycin and novobiocin)	95 monomer

The collection of the *dnaB-*, *dnaC-*, and *dnaG*-encoded proteins, together with i, n, n″ and Y, constitute the primosome. Factor Y first recognizes a primosome assembly site, stimulating its DNA-dependent ATPase. The n protein then binds to the factor Y:DNA complex, with the subsequent addition of protein i, n″, and *dnaB-* and *dnaC*-encoded proteins, which together constitute the *prepriming complex*. It can move along DNA in the 5′→3′ direction when primase is bound (forming the *primosome*); primers are synthesized at various locations on the DNA template.

of enzymes that can either reduce or increase the linking number (L) of a supercoiled helix. Those enzymes, called topoisomerases, were first identified as activities capable of relaxing negatively supercoiled DNA in *E. coli* and in mouse embryo cells.

Topoisomerases are believed to be essential for DNA metabolism and they are widely distributed in living organisms. The enzymes can break and rejoin DNA repeatedly without the need for any added cofactor to supply the energy for rejoining. Thus it appears that the reaction intermediate conserves the energy of the DNA phosphodiester bond. The proposed mechanism for supercoiling involves covalent-bond formation between the protein and the broken end of the DNA, followed by resealing of the DNA and dissociation of the enzyme, after the linking number has been changed by one or more units. A 3′-phosphotyrosine linkage has been detected as an intermediate.

Topoisomerases are classified as type I or type II, according to whether they change the linking number in steps of one or steps of two, respectively. Type I enzymes produce transient single-strand breaks in the double helix, whereas type II enzymes produce transient double-strand breaks (fig. 26.15). Type I topoisomerase of *E. coli,* also known as omega protein, is the best-understood enzyme of this class. Its molecular weight is 110,000 and the native protein exists as a monomer in solution. The enzyme shows a preference for highly negatively supercoiled DNA and is inactive on positively supercoiled DNA. This enzyme can also catalyze the catenation (interlocking) of double-stranded circular DNA, or the separation of catenanes into simple circles provided at least one circle contains a single-stranded break (fig. 26.16). Type II topoisomerases can carry out the catenation reaction as well; in this case, no single-strand breaks are required. Type I topoisomerases that have been isolated from eukaryotic cells differ in two important respects from the *E. coli* protein: they do not require Mg^{2+} ions and they can relax positively as well as negatively supercoiled DNA.

DNA gyrase, a type II topoisomerase found only in prokaryotes, differs from other topoisomerases in its ability to catalyze the conversion of relaxed duplex DNA into a high-energy, negatively superhelical form (see fig. 26.15). That reaction is ATP-dependent. In the absence of ATP, gyrase can still catalyze relaxation of superhelical DNA or the catenation reactions described earlier. The gyrase enzyme contains two different subunits, encoded by the *gyrA* and the *gyrB* genes (see table 26.3). Normally it exists as a tetramer with two subunits of each type.

Figure 26.15

Type I and type II topoisomerases relax negatively supercoiled DNA in steps of one and steps of two, respectively. Type II topoisomerases can also add additional negative supercoils, as indicated by the double reaction arrow. The latter reaction requires energy input, which is encoded by ATP cleavage.

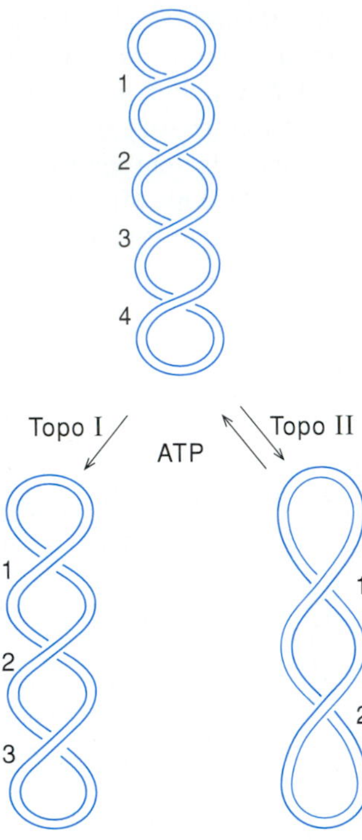

Figure 26.16

Catenation by topoisomerases. (*a*) Two circular DNAs can be catenated by type I topoisomerase only if one of the DNAs is nicked. This is not necessary when using a type II topoisomerase. (*b*) Electron micrographs of catenated DNA before (i) and after (ii) incubation with DNA gyrase. The catenane contains one large circular DNA and one small circular pBNP66 plasmid DNA. (Source: M. Gellert et al., "DNA gyrase: Site-specific interactions and transient double-strand breakage of DNA" in *Cold Spring Harbor Symposium on Quantitative Biology*, 45:301, 1981. Copyright © 1981 Cold Spring Harbor Laboratory Press, Cold Spring Harbor, N.Y.)

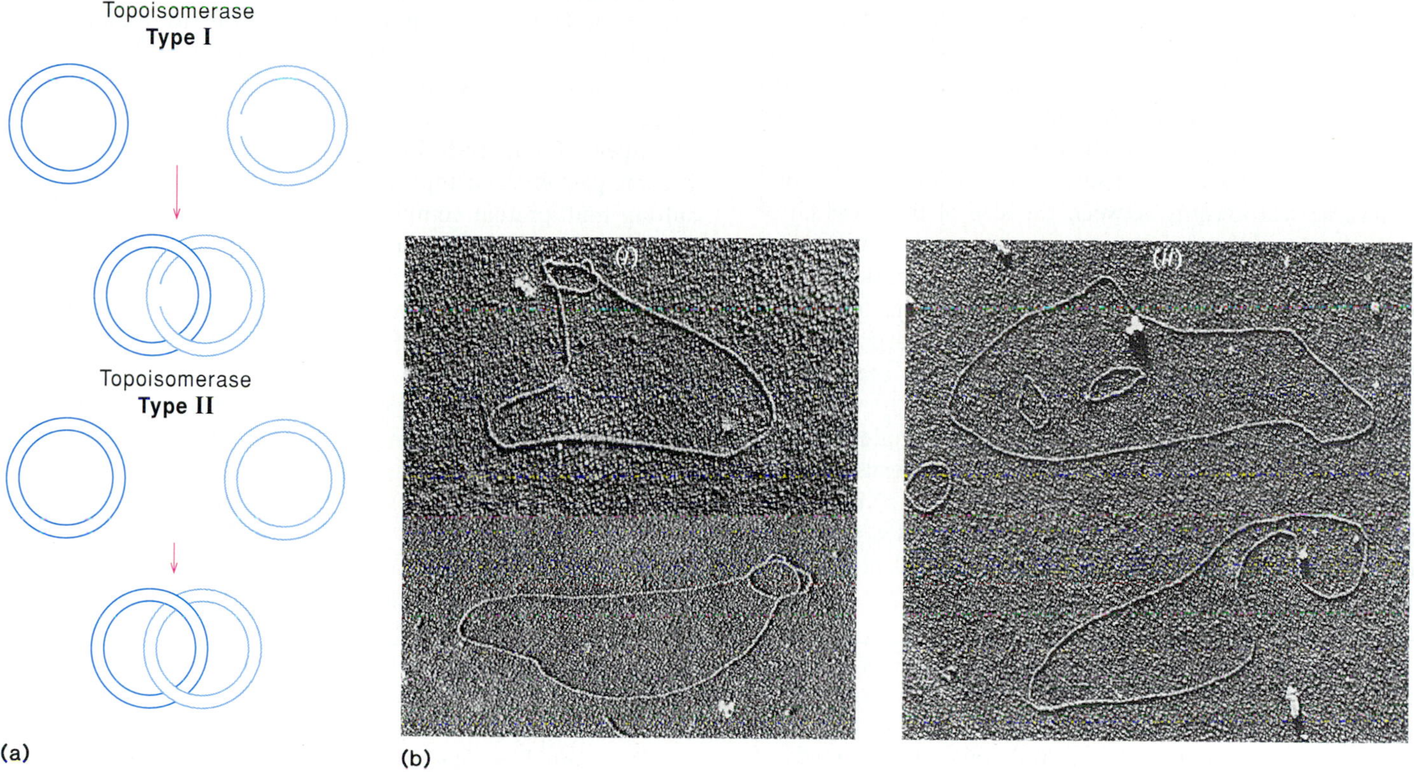

Topoisomerases must serve vital functions, since mutants carrying defective topoisomerases grow very poorly. We do not know all the functions of topoisomerases, but probably they include the elimination of tangled DNA by decatenation reactions, as well as the elimination of the extreme supercoiling that would otherwise result from the unwinding of the duplex structure during DNA replication (see the following subsection). Topoisomerase is likely to act in the final stages of the separation of replicating circular DNA.

Helicases Catalyze Duplex Unwinding

The parental double helix must unwind at the growth fork so that it can present additional single-stranded regions to serve as templates for continued replication. Enzymes that catalyze unwinding are called helicases. The rep protein of *E. coli* is believed to be a helicase, since DNA elongation is slower when the rep protein is not present. During the unwinding of duplex DNA by helicase, ATP is consumed, indicative that the reaction requires energy. Helicase is one of the components of the prepriming *E. coli* replication complex, probably facilitating both priming and replication fork movement.

The unwinding of a circular duplex should tend to generate positively supercoiled DNA. As unwinding progresses, the increasing degree of positive supercoiling would lead to considerable strain in the duplex structure. It seems likely that those tensions are relieved in bacterial systems by the counteracting tendency of gyrase to cause negative supercoiling. Type I bacterial topoisomerases cannot relax positively supercoiled DNA and therefore would be of little value in this regard. By contrast, in eukaryotes containing no gyraselike enzyme, the type I topoisomerases are competent in relaxing positively supercoiled DNA.

Single-Strand Binding Protein Stabilizes Single Strands

Single-strand DNA-binding protein (SSB) is found in abundance in *E. coli*. It has several functions, including the presentation of the template to the polymerase and productive interactions with several replication proteins. This and other evidence (to be described later) indicate that SSB plays a role in DNA replication, and probably in repair and recombination as well. It is believed to act in concert with the rep protein to facilitate unwinding at the growth fork. Since it binds preferentially to single-stranded DNA, it probably facilitates unwinding by inhibiting rewinding. Possibly, also, it protects the single-stranded DNA from attack by endogenous nucleases.

Single-Stranded DNA Viruses Replicate through a Duplex Intermediate

Our understanding of prokaryotic DNA replication has proceeded most rapidly through investigations on viruses because of their relative simplicity. Bacteriophages that infect *E. coli* vary considerably in size and structure. Each type replicates its chromosome in a unique manner and relies on the host enzymatic machinery in a way that suits its needs. In general, there is an inverse relationship between the size of the virus chromosome and the degree of dependency on host enzymes. Large viruses, such as bacteriophage T4, encode several genome-replicating functions that have properties analogous to those of closely related host protein, but that for various reasons are preferred by the virus (sometimes only because they are made in larger amounts).

The bacteriophage ϕX174 has a small, single-stranded circular DNA chromosome (5,386 bases) that depends on many *E. coli* proteins for replication. During replication, ϕX174 DNA passes from a single-stranded to a double-stranded form and then back to a single-stranded form. The replication process is particularly well understood, and findings with ϕX174 not only reveal the replication process of the viral DNA but also give insights concerning certain host proteins used in host chromosome replication. The ease with which ϕX174 DNA can be isolated intact and assayed biologically, through its ability to infect *E. coli* spheroplasts, makes it an ideal template for the study of the fidelity of its replication *in vitro.*

When ϕX174 infects *E. coli* it loses its protein coat, so that only the circular viral DNA enters the cytoplasm. During the first phase of the replication process, the single-stranded viral DNA must be converted to a double-stranded form. This double-stranded form serves two purposes, as an intermediate in replication (replicative form) and as a template for synthesis of viral RNA. Since no viral RNA or proteins can be made until after the viral DNA is converted to the double-stranded form, it follows that the single-to-double-strand transition must be carried out by preexisting host cellular proteins. This point has been verified by experiments showing that the double-stranded form of the virus is made in the presence of chloramphenicol, which inhibits protein synthesis, whereas further steps in viral DNA replication cannot occur in the presence of this antibiotic.

Kornberg, his co-workers, and Hurwitz and his colleagues have constructed a cell-free system containing ϕX174 DNA and purified proteins that can carry out many of the steps believed to be involved in ϕX174 DNA replication *in vivo.* Before DNA synthesis can be initiated on the single-stranded circular DNA, a primer must be synthesized by the host primase. Two types of complexes have been made in which primase is active. The first complex is formed by adding primase to a preformed complex of DNA and dnaB-encoded protein. Primer formation occurs in this system at random sites on the DNA. In the second complex, which is sequence-specific, several additional proteins are required before dnaB protein or primase can bind (fig. 26.17). These include n, n′, n″, dnaC, and single-strand binding (SSB) protein. When the DNA is uniformly coated with SSB, random primer formation is prevented because the dnaB protein cannot displace SSB from the DNA to form a complex. Under these conditions n′ makes an initial complex at a specific site on the DNA (step I in fig. 26.17). This site is adjacent to a 44-base region on the single-stranded DNA that forms a hairpin double helix to which SSB does not bind. Other proteins—n, n″—join the initial complex, resulting in the displacement of some SSB. Finally, the dnaB, dnaC, and i proteins become part of the complex, as more SSB is displaced; the resulting multiprotein complex is known as the preprimosome. Addition of primase to this complex completes assembly of the primosome (step II, fig. 26.17). The primosome can migrate on the single-stranded DNA in the 5′→3′ direction. In this process, still more SSB is removed from the single-stranded DNA (step III, fig. 26.17).

This movement of the primosome is an energy-dependent reaction requiring ATP. At various points during its journey the primosome pauses to synthesize a short stretch of primer (step IV, fig. 26.17). DNA polymerase III holoenzyme uses these primers as initiation sites for DNA synthesis (step V, fig. 26.17). Although primase initiates limited stretches of RNA synthesis at several locations on the template, the necessary primosome can be assembled only at a unique site around base 2,300 (step I, fig. 26.17). The primosome complex is very stable and can survive many rounds of replication. The Okazaki fragments formed by primase and DNA polymerase III (step V, fig. 26.17) require DNA polymerase I and ligase for completion (step VI, fig. 26.17). DNA polymerase I closes the gaps as it removes the RNA on the 5′ ends of the fragments. Ligase joins the nicked fragments of DNA together. The circular duplex so formed is referred to as replicative form DNA (RF). If gyrase is present, this molecule will become negatively supercoiled. The RF is used as a template to synthesize more RF and viral messenger RNA, which, as we will see, is transcribed entirely from the minus strand of the RF. The viral messenger RNA serves in turn as the template for the synthesis of viral proteins that are essential in subsequent steps in viral DNA replication.

After single-stranded viral DNA (SS DNA) has been converted to the replicative form duplex (SS→RF), the parental RF is duplicated to produce multiple copies of RF (RF→RF). In this reaction the viral DNA, known as the plus strand, and the complementary DNA, known as the minus strand, are replicated by distinct mechanisms. Synthesis of the viral plus strand begins when the viral gene A protein induces the cleavage of the viral plus strand in the RF at position 4,305 (fig. 26.18). The bifunctional A protein not only cleaves the plus strand at this point, but also becomes covalently attached to the 5′ end of the interrupted strand (see fig. 26.18). More copies of the replicative duplex form are initiated from this point. Nucleotides are added continuously to the 3′-OH end of the plus strand. The new plus strand displaces the old plus strand during this phase of the reaction. After one viral single-stranded plus

Figure 26.17

Proposed mechanism for the conversion of single-stranded φX174 DNA to the duplex replicative form (SS → RF reaction). (Source: K. Arai et al., "Mechanism of *dnaB* protein action V. Association of *dnaB* protein, protein *n′* and other prepriming proteins in the primosome of DNA replication" in *Journal of Biological Chemistry*, 256:5280, 1981. Copyright © 1981 American Society for Biochemistry and Molecular Biology, Inc., Bethesda, Md.)

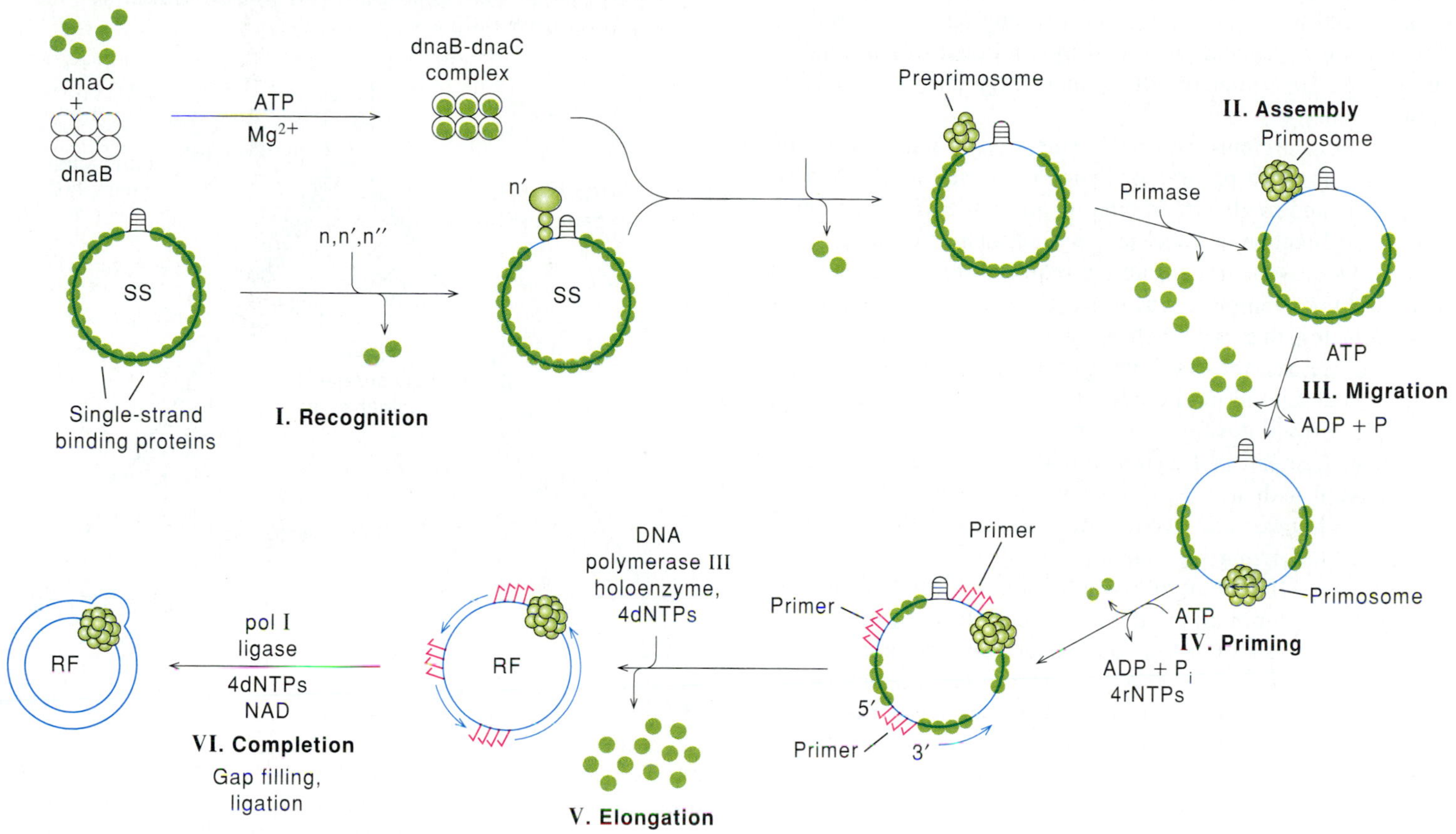

dnaC
+
dnaB
ATP
Mg^{2+}
dnaB-dnaC complex
Preprimosome
II. Assembly
Primosome
Primase
SS
Single-strand binding proteins
n,n′,n″
I. Recognition
n′
SS
ATP
III. Migration
ADP + P
Primosome
ATP
IV. Priming
ADP + P_i
4rNTPs
Primer
Primer
5′
Primer
3′
DNA polymerase III holoenzyme, 4dNTPs
V. Elongation
RF
pol I
ligase
4dNTPs
NAD
VI. Completion
Gap filling, ligation
RF

Figure 26.18

Scheme for φX174 RF replication in two stages. Continuous replication initiated by gene A protein cleavage generates viral (+) circles, and discontinuous replication of the viral circles by the SS → RF system produces RF. In the presence of phage-encoded maturation and capsid proteins, viral circles (RF → SS) are encapsulated rather than replicated. (Source: A. Kornberg, *DNA Replication*. Copyright © 1980, W. H. Freeman, New York, N.Y.)

Gene A protein
Replication origin
(+) Strand synthesis
3′
5′
−
RF DNA
SSB protein
rep protein
ATP
SSB protein
3′
−
rep protein
RF
Gene A protein
Mg^{2+}
pol III holoenzyme
ATP, dNTPs
SSB
(−) Strand synthesis
SS
+
Gene A protein
+
−
Virus proteins
+
Phage
Protein coat

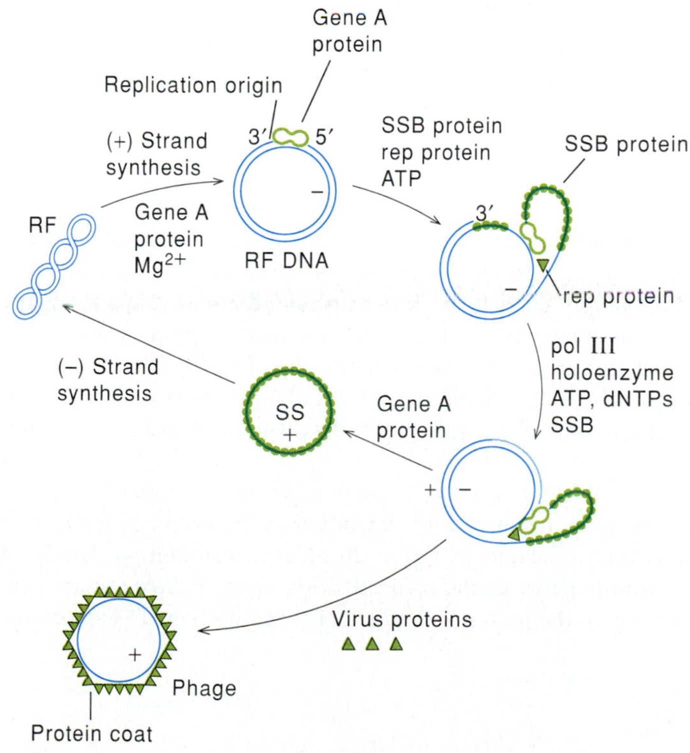

strand has been displaced, it becomes covalently closed through the action of the viral gene A protein (acting as a specific ligase) and can either serve as the template for synthesis of more RF or become a component (in the genome) of the mature virus. During the synthesis of plus strand it can be seen that the minus strand is being used as the template in a continuous fashion: its action could be likened to that of a rolling circle. The rolling-circle mode of replication is used by a number of bacterial viruses (e.g., bacteriophage λ) at some stage during their replication cycle.

Host proteins required during plus-strand synthesis include the host rep protein, or helicase, as well as DNA polymerase III and single-strand binding protein (SSB). As we have noted, the function of helicase is to facilitate unwinding of the duplex. Helicase is not absolutely required for DNA synthesis using a duplex template, but it greatly accelerates the process. SSB acts together with helicase by complexing with single-stranded regions, thus inhibiting rewinding. The viral gene A protein is crucial in maintaining the integrity of the displaced virus plus strand during synthesis. It does this by holding the otherwise free end and ensuring that the strand closes at the proper point without releasing itself from the template.

What we know about additional steps in the replication process comes primarily from an analysis of the events occurring *in vivo*. After about 60 copies of the replicative form have been made (20 min after infection), the infected cell switches to making single-stranded circles exclusively. Presumably some regulatory device is involved to ensure that further single-stranded circles do not get converted into duplex forms at this stage. The protein encoded by the phage genes *C* and *J* are thought to play a key role in diverting the single-stranded DNA into mature viral particles. In that process, the plus-strand circles get packaged into phage protein shells as they are being synthesized, in preparation for making completed viruses. The packaging step concludes the replication cycle for ϕX174 DNA.

Initiation of Chromosomal Replication in E. coli *Occurs at a Unique Location*

Replication of the *E. coli* chromosome has been more difficult to study than replication of phage DNA, in part because of its large size. As we stated earlier, genetic evidence has indicated that host proteins involved in ϕX174 replication might also be involved in similar ways in *E. coli* DNA replication. In particular, the movement of the primosome complex suggests a mechanism for discontinuous DNA synthesis on the lagging strand of the replication fork. The primosome could migrate on the lagging strand as the template unwinds, stopping periodically to allow primase to form primer (fig. 26.19). The elongation steps, so clearly elucidated in ϕX174 minus-strand synthesis, could account for lagging-strand synthesis at the *E. coli* replication fork.

That leaves two aspects of *E. coli* chromosome replication to be accounted for: initiation at the origin of replication (location indicated in figure 26.5) and termination. Study of termination awaits the reconstitution of an *in vitro* system containing all the proteins necessary for replication. To explore the mechanism of initiation, Kornberg and his co-workers developed a cell-free system for bidirectional replication using small circular DNA molecules, known as hybrid plasmids, that contain the origin of the *E. coli* chromosome (*oriC*) together with purified proteins from *E. coli*. These ongoing studies are beginning to show how bacterial chromosome synthesis is initiated.

Figure 26.19

Role of the primosome at the bacterial replication fork. The primosome, assembled at or near the DNA replication origin, migrates continuously on the lagging strand as the replication fork moves. (Source: K. Arai et al., "Movement and site selection for priming by the primosome in phage ϕX174 DNA replication," in *Proceedings of the National Academy of Sciences USA* 78:711, 1981. Copyright © 1981 National Academy of Sciences, Washington, D.C.)

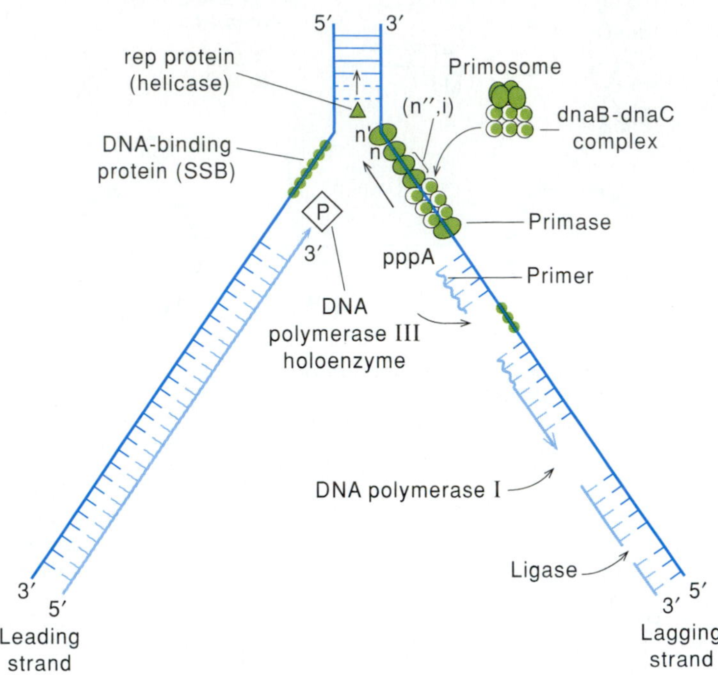

The use of plasmids is crucial in such studies because they greatly facilitate detection of initiation on a much smaller and more tractable piece of DNA. Plasmids contain about 245 base pairs (bp) of the minimal origin sequence from the *oriC* region of the *E. coli* chromosome. The 245-bp *oriC* region has a considerable number of GATC sites in which the cytosine residues are specifically methylated by an *E. coli* enzyme encoded by the *dam* gene. Methylation may be important in the control of initiation of replication at *oriC*.

The initiation of replication at *oriC* can be divided into three steps:

1. Generation of a template (requiring DNA gyrase, topoisomerase I, and a histonelike DNA-binding protein called HU).
2. Transcription and priming (requiring RNA polymerase and the dnaA and dnaB proteins).
3. Primer processing (requiring RNase H, an enzyme that degrades RNA when it is base-paired to DNA or RNA, and some yet unidentified factor).

Figure 26.20

Hypothetical asymmetric dimeric structure suggested by Kornberg for the DNA polymerase III holoenzyme of *E. coli.* The holoenzyme appears to be organized as an asymmetric dimer; a pair of core subassemblies, each with a potential for polymerase action, has an asymmetric distribution of auxiliary subunits that may endow each one with different properties, one suited to the continuous synthesis of one strand and the other to the discontinuous synthesis of the other strand. The α subunit contains the polymerase activity. The ε subunit contains the 3′ to 5′ exonuclease proofreading activity, which is active only on attachment to the α subunit. Several of the subunits are essential to processivity. Among them, the β subunit is held loosely in the holoenzyme and dissociates during the cycling of the polymerase from one template to another. (Source: A. Kornberg, "DNA replication" in *Journal of Biological Chemistry* 263:1-4, 1988. Copyright © 1988 American Society for Biochemistry and Molecular Biology Inc., Bethesda, Md.)

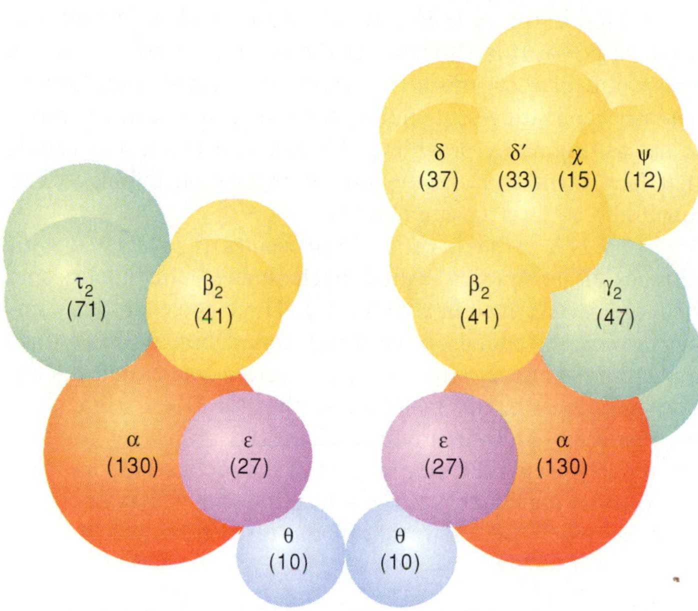

Figure 26.21

Hypothetical scheme for concurrent replication of leading and lagging strands by an asymmetric, dimeric polymerase associated with a primosome and a helicase in a replisome. In this hypothetical scheme for concurrent replication, looping of the lagging strand by 180° gives it the same orientation as the leading strand at the fork. A primer generated by primase is extended by polymerase as the lagging strand template is drawn past it. When synthesis reaches the 5′ end of the previous nascent fragment, the lagging strand template is released and unlooped. Helicase action and continuous synthesis of the leading strand periodically expose lengths of template for priming of nascent fragments. (Source: A. Kornberg, "DNA replication" in *Journal of Biological Chemistry,* 263:1-4, 1988. Copyright © 1988 American Society for Biochemistry and Molecular Biology Inc., Bethesda, Md.)

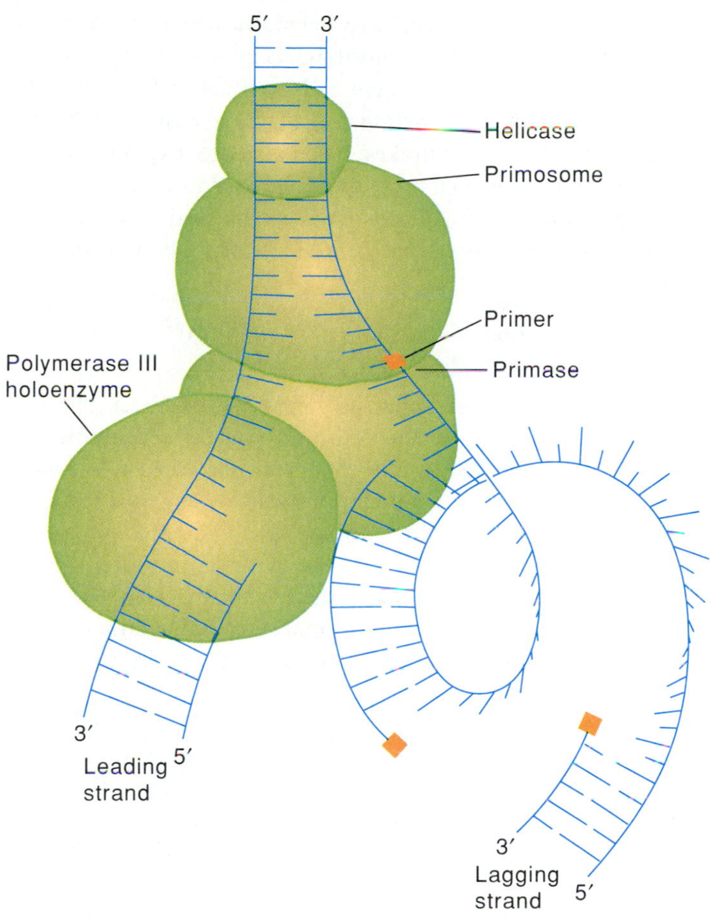

In addition to the primosomal proteins and the Pol III holoenzyme, other, as yet unidentified proteins are required for *oriC*-dependent DNA replication. The requirements for RNA polymerase and DNA gyrase have been shown by their specific sensitivities to the inhibitors rifampicin and nalidixic acid (and novobiocin), respectively. Rifampicin is known specifically to inhibit RNA polymerase (see chapter 28), and nalidixic acid and novobiocin specifically inhibit the action of the subunits of DNA gyrase (see chapter 25). The characterization of the proteins required in this system is under intensive investigation. It is noteworthy that RNA polymerase catalyzes primer synthesis at *oriC* but that primase is used at other sites along the *E. coli* chromosome.

Synthesis May Take Place Concurrently on Both Strands

Concurrent replication of both strands, rather than the jerky sequence of synthesis of one strand and then the other, might be achieved if priming of nascent fragments of the lagging strand were integrated with continuous synthesis of the leading strand. Concurrent replication would also require a more complex holoenzyme possessing primase and twin active sites for polymerization (fig. 26.20). Evidence for such a structure includes (1) twin polymerase subassemblies in the Pol III holoenzyme and (2) complexing of primase by some polymerases. In this hypothetical scheme for concurrent replication, looping of the lagging strand by 180° gives it the same orientation as the leading strand at the replication fork (fig. 26.21). A primer generated by primase is extended by polymerase as the lagging strand template is drawn past it. When synthesis reaches the 5′ end of the previous nascent fragment, the lagging strand template is released. Helicase action and continuous synthesis of the leading strand periodically expose lengths of template for priming of nascent fragments.

DNA Synthesis and Chromosomal Replication in Eukaryotes

The most productive approach to understanding DNA replication in eukaryotes has been to isolate proteins that exhibit activities logically related to DNA replication and to isolate replicative intermediates. Different forms of DNA polymerase, DNA ligases that require ATP, topoisomerases, single-stranded DNA-binding proteins, unwinding enzymes, and degradation enzymes have all been isolated from eukaryotic cells. We can presume that many of these have vital functions *in vivo;* some of the functions have been verified in yeast with the help of mutants.

Two major systems are currently being exploited in studying eukaryotic DNA synthesis. They are animal viruses and yeast. The yeast cell system and its extensively known genetics makes it possible to isolate mutants by conventional means. The use of yeast also makes it possible to exploit reverse genetics—that is, purified proteins can be used to clone the genes encoding them, then the cloned genes can be used to disrupt, by homologous integrative transformation, their genomically located counterparts.

Eukaryotic Cells Also Have Several DNA Polymerases

Four nuclear species of DNA polymerases have been characterized in eukaryotes, designated α, β, δ and ϵ. An additional polymerase γ is located in mitochondria. The amount of DNA polymerase α is relatively high in cells actively engaged in DNA synthesis. This and its key role in the replication of SV40 DNA (see below) implicate the α polymerase in host chromosomal DNA synthesis. The δ enzyme has an important function in the DNA replication of certain animal viruses including SV40; it seems likely that it has a comparable function in host chromosomal DNA synthesis.

DNA polymerases β and γ are present in eukaryotic cells in resting and differentiated states. The levels of β and γ enzymes do not change markedly in proliferating cells. The function of the β enzyme is unclear, although its localization in the nucleus has led to the speculation that it may be involved in repair or replication of nuclear DNA. DNA polymerase γ is usually localized in mitochondria, and is believed to be involved in replication of the organelle's DNA.

Eukaryotic Chromosomes Contain Several Origins of Replication

Although the general features of DNA replication in eukaryotes are thought to be similar to those in prokaryotes, there are some interesting differences. The chromosomes of higher eukaryotic organisms are quite large, in some cases a thousand times larger than their bacterial counterparts. In order for such large DNA molecules to replicate in a reasonable length of time, they must have multiple origins of replication. The simultaneous synthesis of DNA at several points along the chromosome has been demonstrated by incorporating radioactive nucleotides into the replicating chromosomes for a short period and then observing the distribution of radioactive DNA by autoradiography (fig. 26.22). Multiple regions (or replicons) of incorporated label are observed, with replication proceeding bidirectionally from each of these regions. Termination of replication occurs at the point where the growth forks from two adjacent replication units meet (see fig. 26.22). DNA on the lagging strand of a fork is made discontinuously. The Okazaki fragments are much shorter than those found in prokaryotes, averaging between 100 and 200 nucleotides in length. Synthesis of these DNA fragments is initiated on RNA primers that are found covalently attached to the 5′ ends of newly synthesized fragments. Chromosomal replication occurs only during the S phase of the cell cycle.

Are eukaryotic replication origins located at unique sites? This is the case in animal viruses such as SV40 and polyoma DNA. It is also true for the yeast *S. cerevisiae;* in this eukaryotic microorganism the origins are termed ARS (autonomously replicating sequences). ARS sequences can be shown to act as origins by providing this function to yeast plasmids that lack them and are the site of replication bubbles when monitored by electron microscopy.

DNA in eukaryotic chromosomes is associated with histones in complexes called nucleosomes (the nucleosome structure is described in chapter 25). These nucleoprotein complexes serve to condense exceedingly long DNA molecules into much more compact structures, but during replication the nucleosomes must be disassembled so that the DNA strands can be separated. The disassembly of the DNA-histone complex presumably occurs directly in front of the replication fork. Two observations suggest that nucleosome disassembly may be a rate-limiting step in the migration of the replication fork through chromatin:

1. The rate of migration of replication forks is slower in eukaryotes (with nucleosomes) than in prokaryotes (lacking nucleosomes).
2. The length of the Okazaki replication fragments is similar to the length of DNA between adjacent nucleosomes (about 200 base pairs).

Before the newly replicated DNA is reassembled into nucleosomes, the RNA primers must be removed by an as yet unknown enzyme, the gaps must be filled in by DNA polymerase, and the replication fragments must be linked together by DNA ligase. Replication origins are activated in a specific temporal order, the overall number of origins depending on the time required to replicate the cell's genome. Those genes that are actively transcribed are replicated in early S phase; if an actively transcribed gene is transposed to a late-replicating region, it is still replicated early in S phase.

The yeast genome has 17 chromosomes, with a combined size of 1.35×10^4 kb. There are about 400 ARS sequences, where DNA replication is believed to be initiated. DNA-binding proteins that bind specifically to ARS sequences have been identified; they could be analogous to the SV40 viral T antigen, to be described later. Many mutant genes that affect

Figure 26.22

Multiple-origin model for eukaryotic chromosomal DNA replication. (*a*) Autoradiograph of short-term labeling of a eukaryotic chromosome during replication and its interpretation. (*b*) Overall replication scheme for a eukaryotic chromosome. Only a short region of the chromosome is shown. It is believed that replication origins are relatively free of proteins.

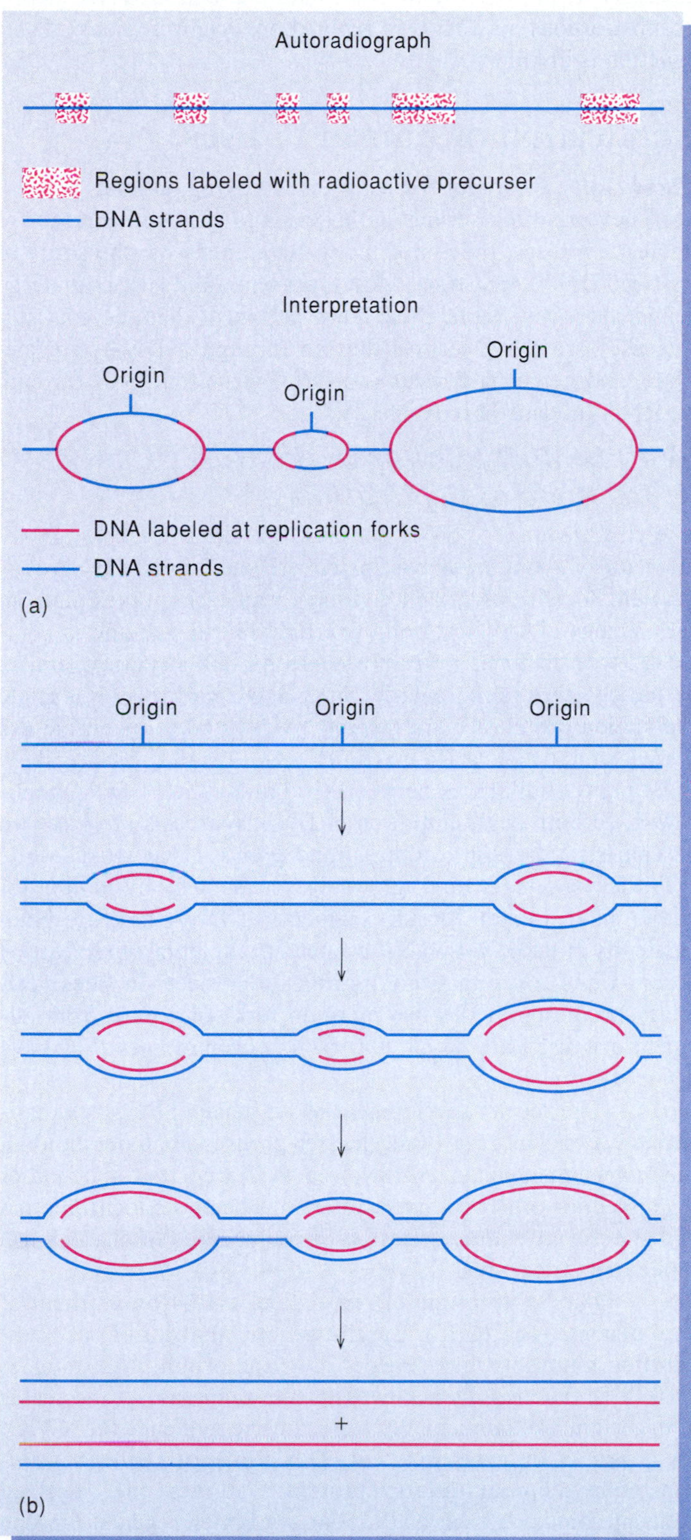

replication in yeast have been mapped and their effects have been characterized. The products of some of these genes are known. Most notably it is known that yeast DNA polymerase I, encoded by the CDC 17 gene, is similar to mammalian DNA polymerase α while yeast polymerase III, encoded by the CDC 2 gene, is most like the mammalian DNA polymerase δ.

Telomerase Facilitates Replication at the Ends of a Eukaryotic Chromosome

The linear structure of the eukaryotic chromosome creates a problem in replication as DNA chains always start from a primer. If an RNA primer was to form as a complement on the 3′ end of a DNA strand there would be no obvious way of replacing the primer with DNA. As a result, a segment of DNA from the ends of the chromosomal DNA would be lost every time it goes through a round of replication. This problem is overcome by a special DNA structure at the chromosome ends and a unique enzyme, DNA telomerase.

The ends of linear eukaryotic chromosomes are terminated in a loop or hairpin by a covalent linkage formed between the 3′ and 5′ ends of opposing chains. Short sequences tandemly repeated many times (telomeres) adjoin the chromosome ends. In humans the repeating sequence is TTAGGG at the 3′ ends and complementary sequences at the opposing 5′ ends. Prior to replication, the hairpin end is scissored and the protruding 3′ end is extended by a unique reverse transcriptase that is partly RNA and partly protein. The RNA of this enzyme associates via hydrogen bonds with complementary sequences on the 3′ ends of the DNA and serves as a template for extension of the 3′ end of the DNA. This 3′ extension of the DNA provides the repetitive sequences occurring in the mature chromosome as well as a region on which an RNA primer can be formed, which need not be replaced by DNA during the chromosome replication process.

Mitochondrial DNA Replicates Continuously on Both Strands

Mitochondria in mammalian cells contain circular, supercoiled DNA (mtDNA) about 5 μ in size ($M_r = 10 \times 10^6$; about 15,000 bp). It appears to be replicated by DNA polymerase γ, since this is the only DNA polymerase present in mammalian mitochondria. Furthermore, mtDNA replicates even in cells where DNA polymerase α is absent or inhibited by aphidicolin, a highly specific inhibitor of α polymerase (fig. 26.23).

The replication of animal mtDNA has been studied most extensively in mouse L cells (fig. 26.24), which have two origins of replication, one for each strand (referred to as H and L). DNA replication is initiated at the first site on the H strand (O_H) to form a displacement loop (D loop). It is a triple-stranded structure that includes the newly synthesized short daughter H strand. Replication continues unidirectionally until completion. When H strand synthesis is two-thirds complete, L strand synthesis is initiated (O_L) and elongated in the direction opposite to that of H strand synthesis. The daughter molecules segregate and synthesis proceeds to completion prior to closure with the

Figure 26.23

Structure of the DNA polymerase α inhibitor aphidicolin.

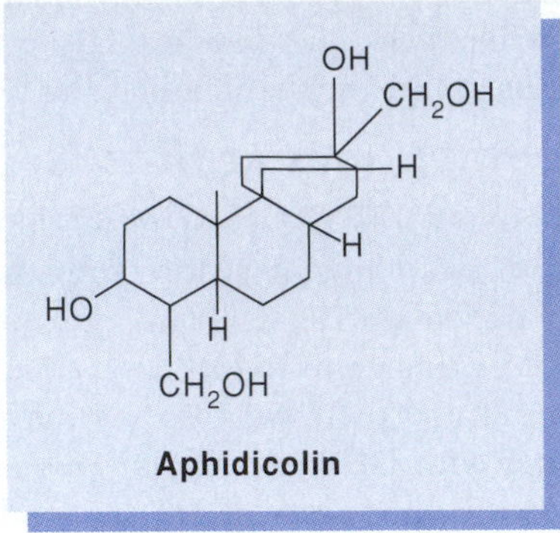

Figure 26.24

Replication model for mouse mitochondrial DNA. (Blue solid line, parental heavy (H) strands. Black solid lines, parental light (L) strands. Blue dashed lines, daughter H strands. Black dashed lines, daughter L strands.) The order of replication is clockwise, starting at D mtDNA. O_H and O_L are the origins of H- and L-strand synthesis, respectively. The double arrows reflect the metabolic instability of D-loop strands and consequent equilibrium between D mtDNA and C mtDNA. Expanded D-loop replicative intermediates are termed Exp-D prior to initiation of L-strand synthesis and Exp-D(l) after initiation of L-strand synthesis. The caret marks the interruption of at least one phosphodiester bond in the H strand of the daughter molecule. (ß Gpc = gapped circular daughter molecule.) Each replicative form is discussed in order in the text.

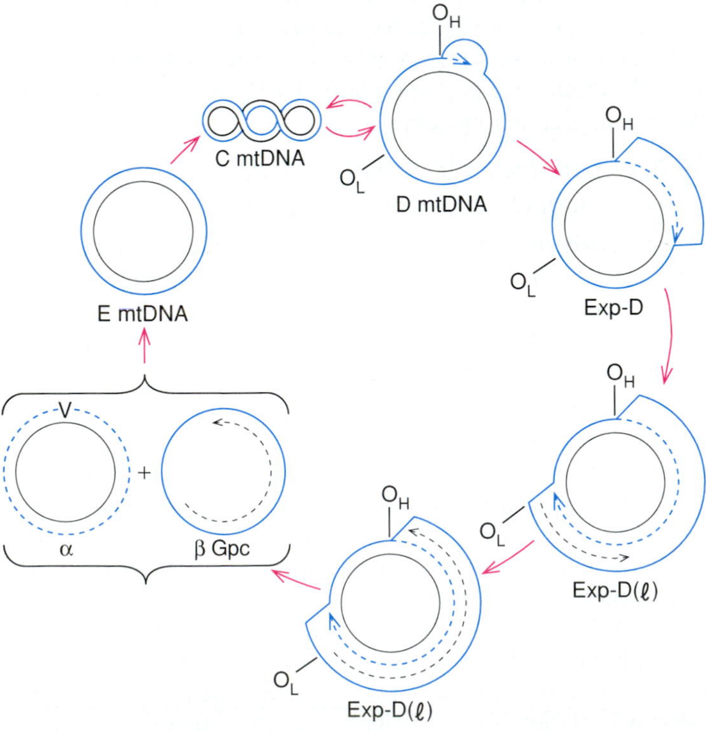

introduction of about 100 negative superhelical turns. The overall rate of mtDNA synthesis is about 270 nucleotides/min, about 0.5% the rate at which *E. coli* DNA replicates.

An interesting observation is that mammalian mtDNA contains a small percentage of ribonucleotides, which have been monitored by measuring susceptibility to alkali and to ribonuclease H. It has been found that mouse L-cell mtDNA has ribosubstitutions in the two replication origin regions. Their function is unknown.

Replication of Animal Viruses

There are two major reasons for studying animal viruses: (1) They are often serious pathogens and (2) as in the case of bacterial viruses, they offer many advantages for the study of host-cell DNA replication. Most types of animal viruses undergo semiconservative replication; however, two of them do not. Retroviruses are RNA viruses that go through a DNA intermediate, and hepatitis B virus is a DNA virus that goes through an RNA intermediate (table 26.4).

SV40 Is Most Similar to the Host in Its Mode of Replication

The viral genome of SV40 consists of a circular duplex DNA molecule of about 5,000 base pairs with one origin of DNA replication. As with most DNA viruses, replication takes place in the nucleus of the host cell. The SV40 viral genome is complexed with histones to form a nucleoprotein structure similar to that observed for chromatin. Since SV40 encodes only a single replication protein (T antigen), the virus makes extensive use of the cellular replication machinery. As a result there are likely to be many similarities between viral and cellular DNA replication. In both cases initiation of DNA synthesis involves two nascent strands. The leading strand grows continuously while the other, lagging strand, grows discontinuously by joining together small (about 200 bp) segments of DNA that are independently initiated with RNA primers. Completion of replication occurs when two oppositely moving forks meet. In linear cellular chromosomes the two merging forks originate from adjacent origins, while in circular SV40 chromosomes they have a single origin.

The recent development of an efficient cell-free replication system has greatly accelerated progress in understanding the molecular mechanisms involved. An important dividend of the dissection of the cell-free system has been the identification and functional characterization of components of the cellular replication apparatus.

The SV40 origin of replication is a 64-bp segment of viral genome (fig. 26.25). Careful genetic analysis of base substitution mutations has revealed that the origin consists of at least three functionally distinct sequence domains. At the center of the origin are four copies of a pentameric sequence (GAGGC) organized as an inverted repeat. This sequence element is recognized by the viral initiation protein, T antigen; the T antigen molecule binds to each of the four pentamer repeats. On one

Table 26.4
Animal Viruses Requiring DNA Synthesis

Family	Examples	Genome Size (kb)	Virion
1. dsDNA Viruses Papovavirus	Polyoma, SV40, wart viruses	5–8	Naked, icosahedral
Adenovirus	Many human and animal adenoviruses	35–40	Naked, icosahedral
Herpesvirus	Herpes simplex I, II, varicella-zoster (chickenpox), Epstein-Barr virus	120–200	Enveloped, icosahedral
2. ssDNA Viruses Parvovirus	Adeno-associated virus, minute virus of mice, canine, feline, and human parvovirus	4–5	Naked, icosahedral
3. RNA-DNA Viruses Retrovirus	Rous sarcoma, avian, feline, and murine leukemia, mouse mammary tumor virus, HTLV	7–10 (diploid)	Enveloped, icosahedral
4. DNA-RNA Viruses Hepadnavirus	Hepatitis B virus of humans, birds, and rodents	3	Enveloped, icosahedral

Figure 26.25

The SV40 origin of replication. The origin consists of at least three functionally distinct sequence domains. At the center of the origin are four copies of a pentameric sequence (GAAGC) organized as an inverted repeat. This sequence element is recognized by the viral initiation protein, T antigen. On one side of the T antigen-binding site is a 17-bp segment containing A-T base pairs. On the other side of the T antigen-binding site is a 15-bp imperfect palindrome of unknown function. All three sequence domains of the origin are required for SV40 DNA replication.

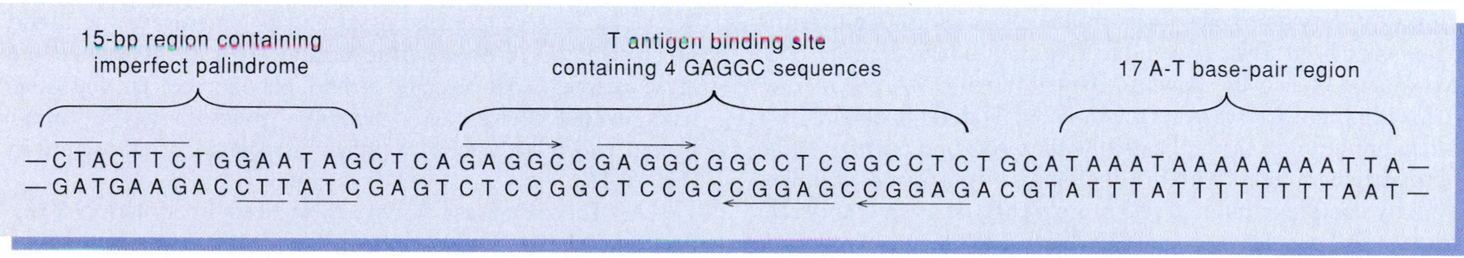

side of the T antigen-binding site is a 17-bp segment containing A-T base pairs. It is suspected that this is the initial site of strand opening during initiation of SV40 DNA replication. On the other side of the T antigen-binding site is a 15-bp imperfect palindrome of unknown function. All three sequence domains of the origin are required for SV40 DNA replication.

In addition to its specific binding activity, the 82 kDa T antigen has an intrinsic helicase activity. Once it is bound to the origin, T antigen is capable of catalyzing the ATP-dependent unwinding of the two DNA strands. Unwinding appears to be a critical step that establishes the replication forks and generates the substrate that is required for the priming and elongation of nascent strands.

In addition to specific nucleotide sequence elements, the T antigen-mediated unwinding reaction requires accessory proteins contributed by the host cell. For example, a single-stranded DNA-binding protein is required to prevent reassociation of the single strands exposed during unwinding. It seems likely that this function is normally fulfilled by a cellular protein designated replication protein RP-A. This protein is absolutely required for SV40 DNA replication in the reconstituted cell-free system. The protein consists of three tightly associated subunits of 70 kDa, 32 kDa, and 14 kDa. The largest subunit binds specifically to single-stranded DNA. Heterologous single-stranded DNA binding proteins, such as *E. coli* SSB, will substitute for RP-A in the unwinding reaction; however, *E. coli* SSB cannot replace RP-A in the complete DNA replication reaction, a fact indicating that RP-A must play other roles in the replication process.

Of the four distinguishable nuclear DNA polymerase activities α, β, δ, and ϵ, it appears that α and δ are required for SV40 DNA replication. DNA polymerase α has long been considered to be the major replicative polymerase in animal cells. The enzyme is composed of four distinct subunits. The largest subunit (180 kDa) contains the polymerase active site while the smallest subunit (50 kDa) contains a primase capable of synthesizing short RNA transcripts that can serve as primers for subsequent DNA chain elongation by the catalytic subunit. The mammalian DNA polymerase α is not a highly processive enzyme, as fewer than 100 nucleotides are polymerized per binding event under the usual assay conditions.

There is evidence that DNA polymerase α can form a specific complex with the SV40 T antigen. Thus antibodies that bind to T antigen (or DNA polymerase α) will usually coprecipitate the two proteins from crude cell extracts. T antigen is also associated with a cellular function referred to as p53. Murine p53 has been shown to compete with DNA polymerase α for binding to T antigen (see table 34.3) and appears to be a strong inhibitor of the replication both *in vivo* and *in vitro*. These observations coupled with the finding that transformed cells frequently contain a mutated p53 protein suggest a regulatory role for p53 in host chromosomal DNA synthesis.

DNA polymerase δ has no primase activity. A 37-kDa protein called proliferating cell nuclear antigen (PCNA) greatly augments the activity of DNA polymerase δ with template/primers containing long single-stranded regions. In the presence of PCNA, DNA polymerase δ is a highly processive enzyme capable of catalyzing the polymerization of at least 1,000 nucleotides per binding event. PCNA has no effect on the activity or processivity of DNA polymerase α. It has been demonstrated by direct reconstitution that PCNA is required for efficient SV40 DNA replication in cell-free systems and is probably involved in DNA chain elongation. In the absence of PCNA, initiation of DNA synthesis at the origin occurs, but only short nascent strands, containing a maximum of a few hundred nucleotides, are synthesized.

It has been suggested that DNA polymerase δ serves as the leading-strand polymerase and DNA polymerase α serves as the lagging-strand polymerase. Such a model is consistent with the known biochemical properties of the two enzymes. We would expect the leading-strand polymerase to be highly processive and to derive little benefit from an associated primase activity. The lagging-strand polymerase, on the other hand, would require only moderate processivity, but would benefit enormously from a tightly associated primase activity (fig. 26.26).

Topoisomerase activity is required for SV40 DNA replication. It has two distinct roles. One role is to act as a swivel to relieve superhelical tension that would otherwise hinder the unwinding of the parental strands as the replication forks advance. Either of the two known mammalian DNA topoisomerases, topoisomerase I or topoisomerase II, could provide this

Figure 26.26

Hypothetical scheme for the concurrent replication of leading and lagging strands at the replication fork on the SV40 chromosome. The scheme follows the same general notion as the one proposed by Kornberg for *E. coli* chromosomal replication (see fig. 26.21). In this scheme polymerase δ functions on the leading strand and polymerase α functions in conjunction with primase on the lagging strand. The 37-kDa protein PCNA greatly augments the activity of polymerase δ but has no effect on polymerase α. (Source: B. Stillman, "Initiation of eukaryotic DNA replication *in vitro*" in *BioEssays* 9:56, 1988.)

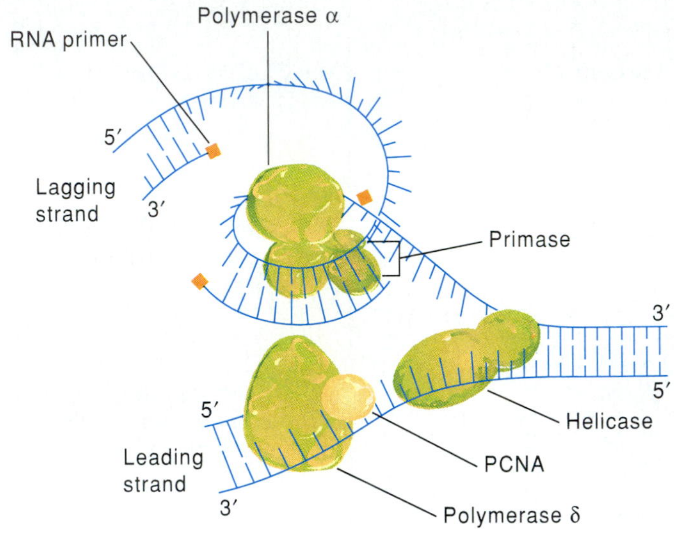

Figure 26.27

(*a*) A model of synthesis of adenoviral DNA in which synthesis of a new strand displaces the parental strand. If the sequences of the inverted terminal repetition base-pair form a panhandle intermediate, then the double-stranded panhandle has the same terminal structure as parental viral DNA and will presumably be recognized by the enzyme complex responsible for initiation of viral DNA synthesis. Although not shown here, *r* strands displaced during the first step in synthesis could obviously form an analogous panhandle intermediate, with the same terminal double-strand sequence. (*b*) A similar displacement mechanism in which the possible role of the 5′ terminal protein in initiation of adenoviral DNA synthesis is illustrated. (From J. Tooze, *The Molecular Biology of DNA Tumor Viruses,* Cold Spring Harbor Press, New York, 1980. Reprinted with permission.)

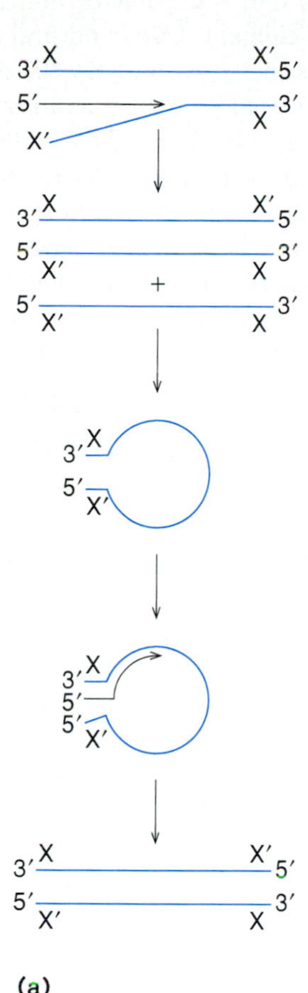

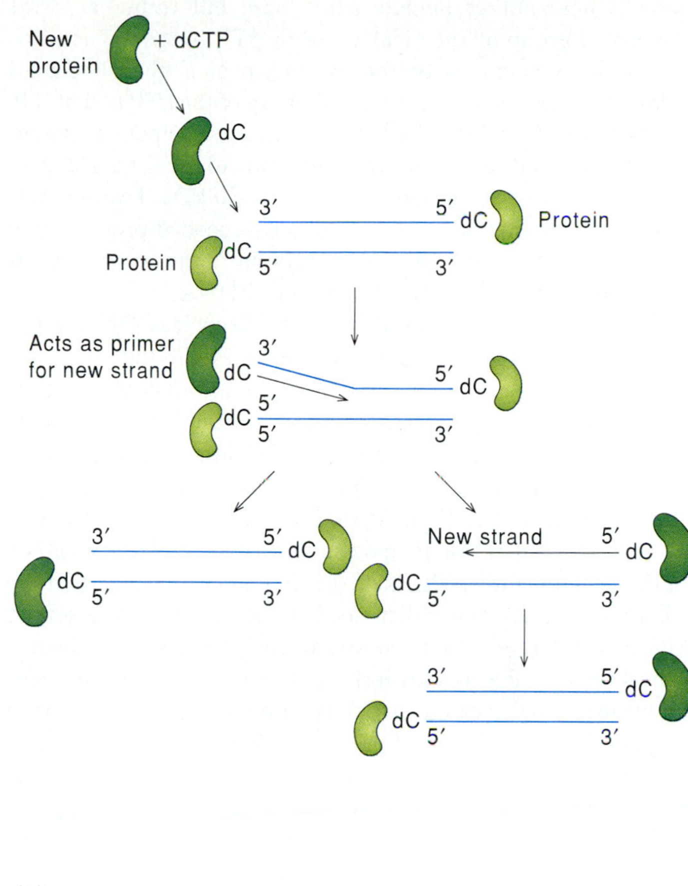

function. The second role for topoisomerase in SV40 DNA replication is to mediate the separation of the newly synthesized daughter duplexes at the completion of DNA replication. In general, the links between the parental strands are not completely removed prior to termination of SV40 DNA synthesis, so the immediate products of replication consist of two circular DNA molecules that are multiply intertwined. The final act of replication is the segregation of these intertwined daughter molecules into two separate unlinked molecules. This reaction is catalyzed by topoisomerase II. Genetic experiments in yeast indicate that the topoisomerase probably plays these same two roles during the replication and segregation of cellular chromosomes.

Adenovirus Uses a Protein Primer

Adenovirus DNA replication occurs by a process that is significantly different from chromosomal DNA replication. Replication initiates by a novel protein priming mechanism, and all daughter strands are elongated by a continuous mode of synthesis such as occurs at the leading strand of a chromosomal replication fork. The biochemical dissection of a soluble *in vitro* system capable of faithfully replicating adenovirus DNA has led to the identification of a number of the proteins involved. Adenovirus DNA replication requires the participation both of virus-encoded replication proteins and of host-cell-encoded factors.

The adenovirus genome encodes three proteins that play central roles in viral DNA replication: the 80-kDa terminal protein precursor (pTP), the 140-kDa adenovirus DNA polymerase (Ad pol), and a single-stranded DNA binding protein (DBP).

The genomes of the human adenoviruses are double-stranded linear DNA molecules containing approximately 35,000 bp. The 5′ terminus of each strand of the viral genome is covalently attached to a virus-encoded protein (TP) with a molecular weight of about 45,000. In addition, the nucleotide sequences at the extreme ends of the genome are identical, but in a reverse orientation (when adenovirus DNA is denatured, the single-stranded DNA can form a panhandle structure; see fig. 26.27). Both of these structural features play important roles in the initiation of viral DNA replication.

Synthesis of adenovirus DNA begins about 8 h after cellular infection by the virus and reaches a maximum about 10 h later. Initiation of DNA synthesis begins asynchronously at one end. DNA synthesis invariably requires both a primer and a template. In the case of adenovirus, the primer for DNA synthesis is not another nucleic acid chain but rather a serine β-hydroxyl group of the viral-encoded pTP protein. The initiation process begins with the formation of a phosphodiester linkage between the serine hydroxyl group of the pTP and dCTP. Formation of the pTP-dCMP complex requires factors present in uninfected cells as well as the 140-kDa Ad pol. The pTP protein active during initiation has a size of 80 kDa. During maturation this protein is cleaved by a virus-encoded protease to a protein of 55 kDa and this latter protein is the one found attached to the 5′ ends of the mature viral DNA.

In vivo studies indicate that adenovirus DNA replication takes place in two stages (see fig. 26.27). In the first stage, DNA synthesis is initiated at either terminus but does not occur simultaneously at the two DNA ends of the duplex viral genome by the protein priming mechanism. The initiation process results in formation of a replication fork that moves from one end of the genome to the other. At each replication fork only one of the two parental DNA strands serves as a template for DNA synthesis. Thus the products of the first stage of replication are a daughter duplex and a displaced single strand. In the second stage of DNA replication, the strand complementary to the displaced single strand is synthesized. The initial step in this process probably involves the circularization of the single-stranded template by annealing of its self-complementary termini. The resulting duplex panhandle has the same structure as the terminus of the duplex adenovirus genome and is presumably recognized by the same initiation machinery that operates in the first stage of replication. Following a second initiation event, complementary strand synthesis proceeds from one end of the template to the other, generating a second daughter duplex. In both stages of adenovirus DNA replication there is only one priming event per nascent daughter strand, so all viral strands are synthesized in a continuous fashion from their 5′ termini to their 3′ termini.

Two other animal viruses, polio virus and encephalomyocarditis virus, have proteins attached to their terminal nucleotide that are probably involved in a protein-priming mechanism of replication initiation.

Herpes Virus Encodes Many of the Proteins Required for the Replication Process

In contrast to SV40, herpes simplex virus (HSV), with its much larger genome, encodes many, if not all, of the proteins that are involved in its DNA replication. The complete set of viral genes necessary has been identified. Several of these purified proteins have functions expected of replication proteins, including a DNA polymerase, a helicase, a primase, a single-stranded DNA-binding protein, and an origin recognition protein.

Bovine Papilloma Virus Replicates Once per Cell Division

Adenovirus, SV40, and HSV are all examples of viruses that normally multiply by productive cytocidal infection. In all of these cases, viral DNA replication begins soon after infection and continues at a high rate until the death of the host cell. In contrast, bovine papilloma virus (BPV) represents an example of a virus that is capable of multiplying as a stable extrachromosomal element. Under normal circumstances the host is not killed. As in the case of SV40, the BPV genome is relatively small and encodes only a small number of proteins involved in DNA replication; viral DNA synthesis strongly depends on host-cell replication proteins. Genetic analysis has provided evidence for a negative control system that apparently ensures that each viral genome is replicated once and only once during each cell cycle. This result suggests that bovine papilloma virus represents an excellent model system for investigating the mechanisms involved in regulating DNA replication.

Retroviruses Are RNA Viruses that Replicate through a DNA Intermediate

RNA viruses that cause tumors (oncogenic RNA viruses) contain an unusual DNA polymerase called reverse transcriptase. This enzyme carries out template-directed polymerization and shares many features with the cellular DNA polymerization reaction. The main difference is that RNA is the preferred template for reverse transcriptase, which catalyzes the polymerization of deoxynucleoside triphosphates only (fig. 26.28). If a primer is added, for example poly(dT), which will base-pair with the poly(A) at the 3′ end of mRNA, then an RNA-DNA hybrid is the initial product of the reaction. A second round of DNA synthesis on the hybrid produces a double-stranded DNA. In this manner, viral RNA sequences can be transcribed into double-stranded DNA sequences called cDNA. The duplex DNA copy (cDNA) of the viral RNA is subsequently integrated into the host chromosome, where it functions in the synthesis of further viral RNA, often causing malignant transformation of the cells. In the laboratory, reverse transcriptase is used extensively for conversion of mRNA species into DNA sequences (cDNA) for cloning and sequencing studies (see chapter 27).

Interestingly, reverse transcriptase has now been demonstrated in bacteria; however, its role there is unknown. G. Fink has made an interesting proposal for one possible role of the reverse transcriptase encoded by the Ty transposable elements in *S. cerevisiae,* namely, that the enzyme could reverse-transcribe mRNA to generate cDNA *in vivo.* Then by homologous recombination, known to occur in yeast, the cDNA would integrate into the genome, replacing any introns at the site of recombination. This proposal could explain the rare presence of introns in the *S. cerevisiae* genome.

Hepatitis B Virus Is a DNA Virus that Replicates through an RNA Intermediate

Hepatitis B is an enveloped virus with a small DNA genome. The genome is circular, but both of its DNA strands are linear. A gap in one strand is bridged by an incomplete complementary strand. The longer strand has a protein bound at its 5′ end. A DNA polymerase present in the mature virus particle can elongate the 3′ ends of the incomplete strands. The unique feature of this virus is that replication involves an RNA intermediate.

Figure 26.28

Synthesis of viral DNA from the viral RNA genome, as shown in a model of the generation of double-stranded DNA carrying two copies of a long terminal repeat (LTR = [3′ ■ 5′]). The sequence X is a marker for the plus strand: the complementary sequence X′ occurs on the minus strand. (*a-c*) Synthesis of minus-strand DNA from the genomic RNA template using a tRNA primer (represented by ⌡ in step (*a*). This initial DNA synthesis of minus strand reaches a pause site at the end of the RNA template as shown in (*c*). (*d*) Degradation of the RNA portion of the resulting DNA-RNA hybrid by RNase H partially exposes the newly synthesized DNA. (*e*) Bridge formation between the newly synthesized segment of minus strand and repeated sequences at the 3′ end of genomic RNA permits continued synthesis of the minus-strand DNA. This synthesis of minus-strand DNA continues until it reaches the end of the RNA template. (*f*) At the same time, DNA synthesis of the plus-strand DNA begins in the region of the LTR on the minus-strand DNA using the plus-strand RNA as a primer. (*g*) This synthesis pauses after 300 bases, that is, when it reaches the end of the template. The tRNA primer is degraded by RNase H (*h*). Continued synthesis of plus-strand DNA requires bridge formation between sequences repeated in a completed minus-strand DNA and in the 300-nucleotide fragment. (*i*) Completion of synthesis and tidying up results in a duplex with copies of the LTR at both ends. The DNA duplex is now ready for the next series of reactions, which results in the integration of the viral DNA sequences into the host genome.

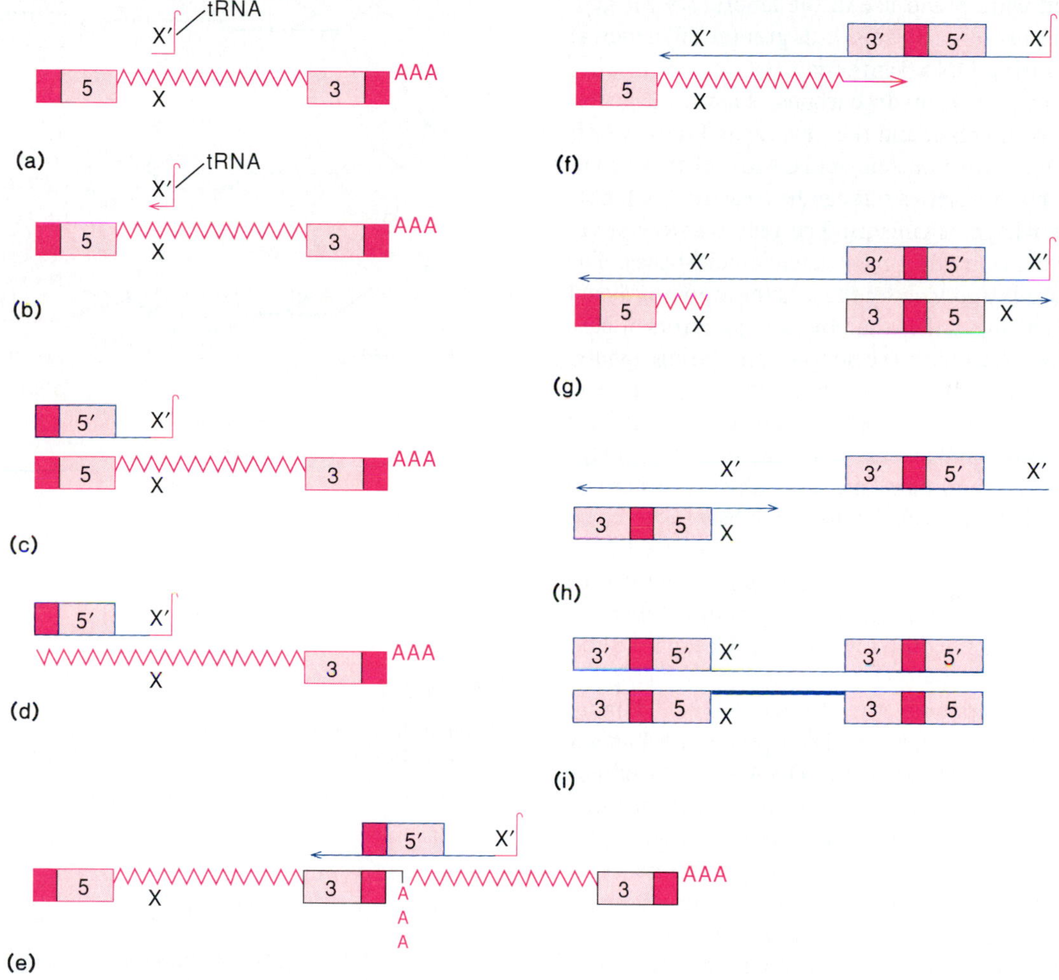

Cellular Systems for DNA Repair

We have seen that chromosomes are usually formed by a single DNA molecule regardless of their size. Such a large molecule makes a very easy target to damage and, in fact, limited damage occurs quite frequently to these large intricate polymers. A single lesion in the DNA molecule could prevent the whole chromosome from replicating and could thus cause the death of the cell. Clearly, it is highly advantageous for a living system to repair the damage. The proofreading functions of the polymerizing enzymes have already been described as a means of correcting some errors (mainly mismatches) made during synthesis. For correcting damage to existing DNA, more elaborate systems exist. DNA repair systems have received increasing attention as it has become clear that most types of repairable damage to DNA are both mutagenic and carcinogenic.

DNA damage is caused by a variety of physical, chemical, and biological agents. Physical agents include UV and ionizing radiation. Damage can be in the form of a missing, incorrect, or modified base or an alteration in the structural integrity of the DNA strands by breaks, cross-links, or dimerization of bases, usually pyrimidines.

Some chemicals that modify the bases in nucleic acids are very specific for one base. Hydroxylamine (NH_2OH) forms a specific adduct with cytosine that can then base-pair with adenine. Nitrous acid, however, is a general reagent used to deaminate bases, converting adenine to hypoxanthine, guanine to xanthine, and cytosine to uracil. Other compounds that cause DNA damage alkylate the bases or form other covalent adducts. Nitrosamine derivatives usually must be converted to more reactive species by biological oxidation before becoming active as base-alkylating reagents. The formation of alkyl adducts may alter the pairing characteristics of the bases. *N*-methyl-*N*-nitrosoguanidine is a particularly potent nitrosamine that has found widespread use in the laboratory for generating mutations in bacteria. It acts with greater efficiency at the growing point during DNA synthesis.

Aromatic polycyclic hydrocarbons, such as 2-acetamidofluorene, benzo(a)pyrene, and the mycotoxin known as aflatoxin B, form adducts with nucleic acids. Most of these compounds are inactive by themselves but can be converted to highly reactive derivatives within organisms. The reactive derivatives are nonselective and can modify most nucleic acid bases. The bulky groups that are introduced by the adducts apparently act by preventing base pairing rather than by causing errors of base pairing, such as those seen with the simpler alkylating agents.

Adjacent pyrimidine bases in a DNA strand form dimers with high efficiency after absorption of ultraviolet light (fig. 26.29). By contrast, purines are quite resistant to damage by ultraviolet. Pyrimidine dimers formed within an otherwise intact DNA duplex have provided a useful substrate to assay for DNA repair. These dimers can be repaired directly by enzymatic photoreactivation (fig. 26.30). The photoreactivation enzyme binds to the DNA containing the pyrimidine dimer and uses visible light to cleave the dimer without breaking any phosphodiester bonds.

Pyrimidine dimers and other forms of DNA damage can also be removed by a general excision repair mechanism. In this form of repair the lesion in the DNA is removed and repaired. The well-defined genetic system of *E. coli* has promoted an understanding of the enzymatic steps in excision repair, particularly in response to UV-induced damage. The genes affecting DNA excision repair fall into three categories: those exclusively involved in repair processes, those involved in both repair and recombination processes, and those also involved in other aspects of DNA replication (table 26.5).

The removal of ultraviolet-induced pyrimidine dimers in bacterial cells is catalyzed by two mechanisms (fig. 26.31). Special glycosylases that recognize abnormal or incorrectly paired bases can cleave N-glycosidic bonds to generate an apurinic or apyrimidinic (AP) site (fig. 26.31). Alternatively, a double incision is made in the strand that carries the lesion by the enzymes of the *uvrA*, *uvrB*, and *uvrC* system (fig. 26.31*b*). In the case of pyrimidine dimers, one incision is made seven nucleotides 5′ to the pyrimidine dimer, and then a second cut is made three or four nucleotides 3′ to the same dimer. The *uvrD* gene product (shown to be helicase II), together with DNA polymerase I and possibly a single-strand DNA-binding protein, releases the 12–13-nucleotide oligomer generated by the incision of the uvrABC complex. After the release of the oligomer, DNA polymerase I and ligase are required to resynthesize the excised region and ligate the nicks.

Figure 26.29

Structure of a thymine dimer formed in DNA by exposure to short-wavelength ultraviolet light.

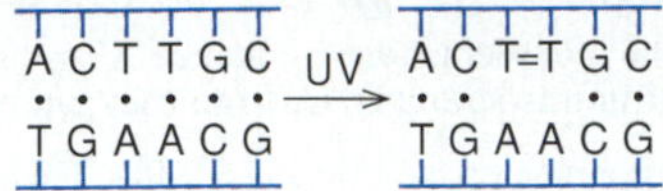

(a)

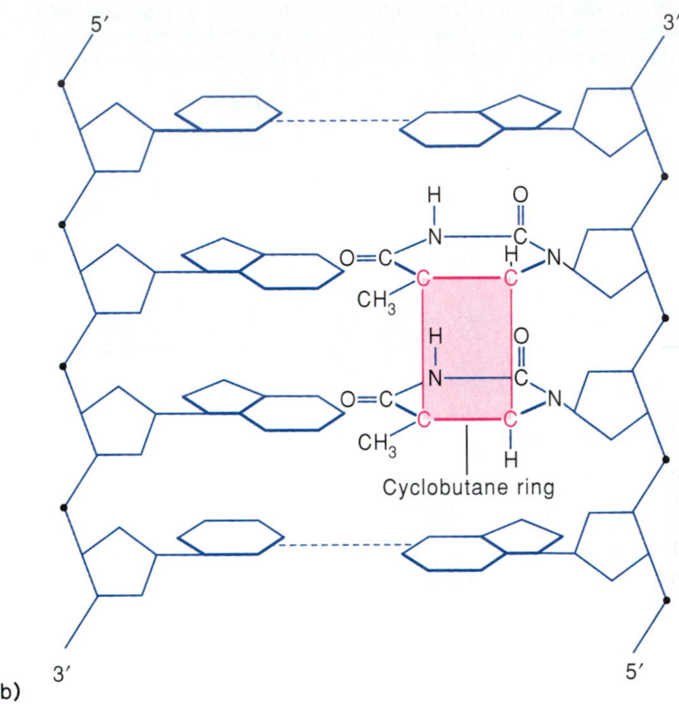

(b)

Figure 26.30

Thymine dimers may be monomerized from DNA by enzymatic photoreactivation. In this case no nucleotides are removed in the repair reaction.

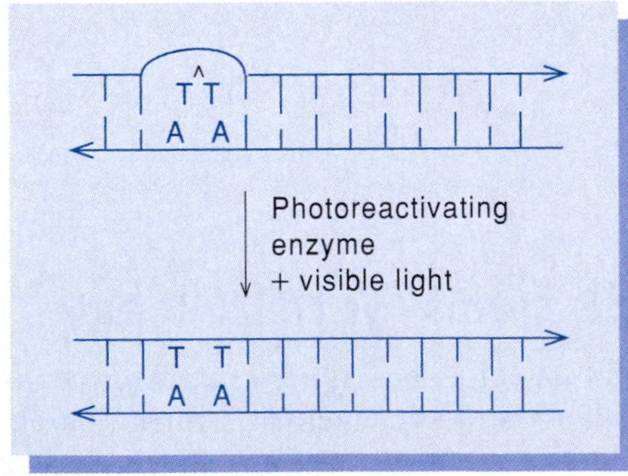

Table 26.5
Some *E. coli* Genes that Affect Responses to DNA Damage

Gene	Map Location	Function
uvr Genes		
uvrA	91	Gene products work together to make initial incision at or near the site of DNA damage.
uvrB	17	
uvrC	42	
uvrD		Helicase
rec Genes		
recA	58	Structural gene for recA protein. *RecA* carries a highly specific protease that is activated by DNA damage.
recB	60	Encodes two subunits of exonuclease V. Believed to make the initial nick required to initiate DNA recombination.
recC	60	
Other Genes		
lexA	90	Controlling gene for *recA* and other SOS functions
polA	85	Structural gene for DNA polymerase I
polC	4	Structural gene for DNA polymerase III (same as *dnaE*)
lig	51	Structural gene for DNA ligase

Figure 26.31

Pyrimidine dimers and other forms of DNA damage can be removed by a general excision repair mechanism. The first reaction in this form of repair involves forming nicks about the damaged region of the DNA. In (*a*) we see the mode of incision of UV-irradiated DNA by the pyrimidine-dimer-specific glycosylase and AP endonuclease activities of *M. luteus* and bacteriophage T4. In (*b*) we see the mode of incision of the uvrABC endonuclease of *E. coli.* (Source: G. Walker, "Inducible DNA repair systems" in *Annual Review of Biochemistry,* 54:425, 1985. Copyright © 1985 Annual Reviews Inc., Palo Alto, Calif.)

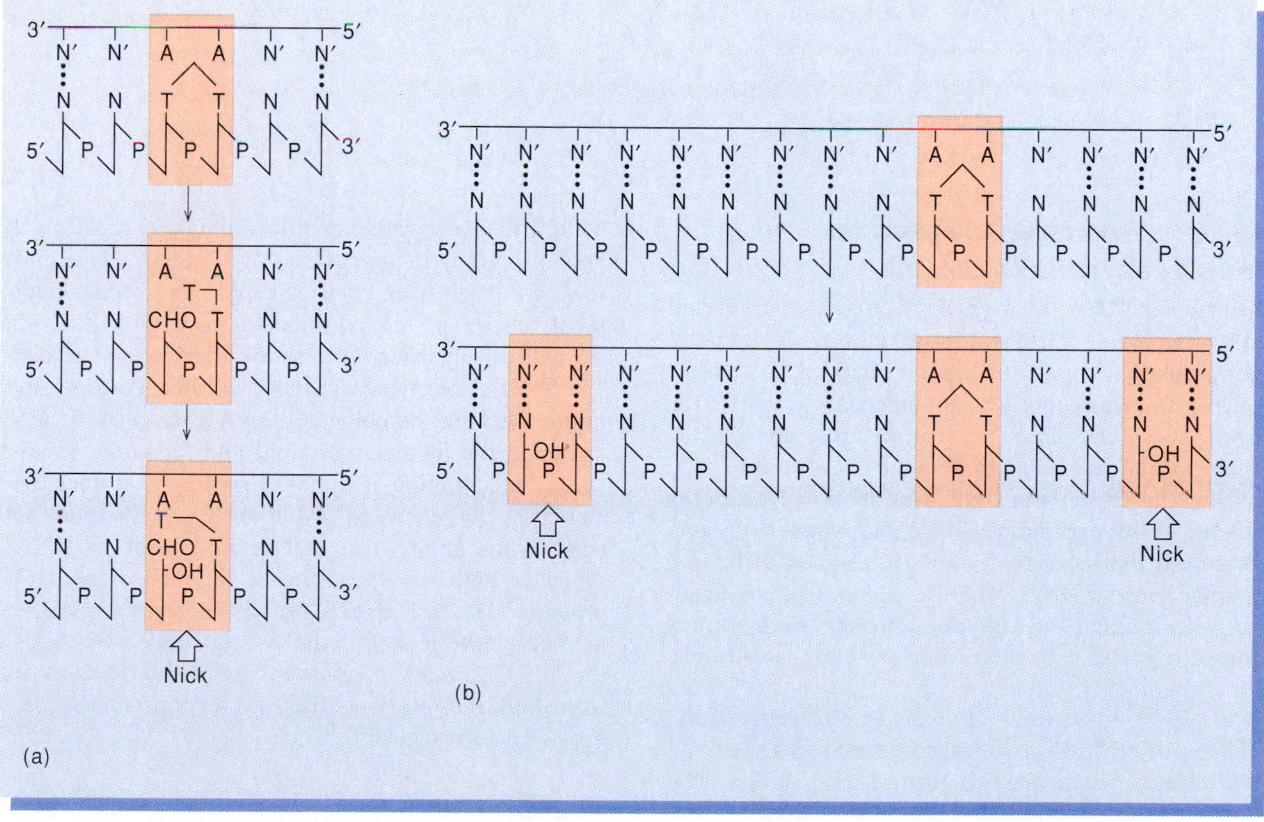

Figure 26.32

Model for the SOS regulatory system. In normally growing cells the SOS functions associated with DNA repair are not expressed. This is because lexA repressor inhibits their transcription. LexA repressor also inhibits its own expression and that of recA. SOS functions are turned on by a series of reactions that starts with DNA damage. DNA damage results in an inducing signal that activates the protease function of recA. This protease cleaves lexA protein, so that all genes that were formerly inhibited by lexA can be expressed. Once the damage is repaired, the level of lexA repressor builds up again and the SOS genes return to their usual repressed state.

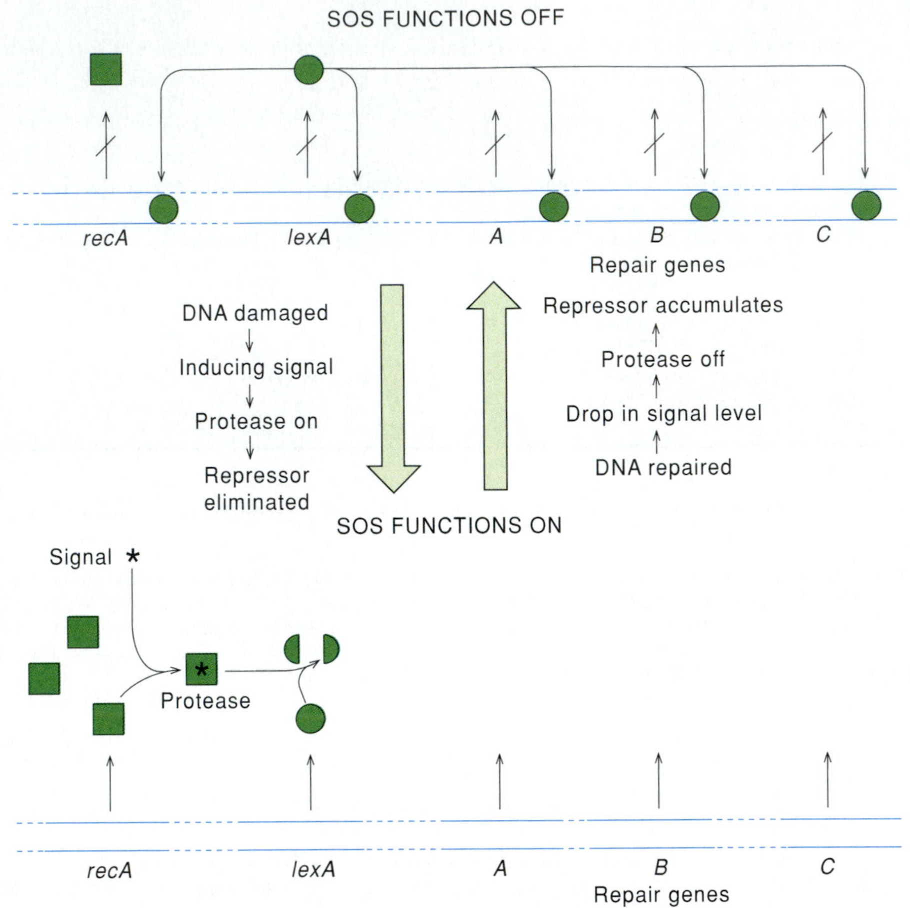

A Special Glycohydrolase Removes Uracil Residues from DNA

Uracil appearing in DNA, as a result of misincorporation of dUTP or deamination of cytosine, can be removed by a uracil-DNA glycohydrolase. The resulting gap is filled with a cytosine residue. Since cytosines deaminate spontaneously at a low rate, this glycohydrolase is considered to be quite important over the course of evolutionary time for preserving the G-C pairs in DNA. This may be a major reason why DNA uses thymine rather than uracil. Both of these pyrimidines have the same hydrogen-bonding potential, but an enzyme that is designed to remove misplaced uracils in the DNA would have less trouble distinguishing between a uracil and a thymine than between a uracil that was correctly paired with an adenine and one that was mispaired with a guanine.

In *E. coli* the synthesis of many of the enzymes involved in the repair process is regulated by the SOS system. At the heart of the SOS system are two genes, *lexA* and *recA* (fig. 26.32). Under normal conditions the lexA protein acts as a repressor binding to approximately 17 genes whose encoded proteins are involved in excision repair, gap repair, double-strand break repair, and methyl-directed mismatch repair. These genes are collectively referred to as *din* (damage-inducible) genes.

The recA protein has two roles. It catalyzes recombination between homologous DNA molecules in *E. coli* (discussed in the following section) and regulates, in an activated state, the induction of the SOS responses. It does the latter by mediating the cleavage of the lexA protein at an—AlaGly—bond. Thus an insult to DNA that leads to DNA damage somehow activates the protease function of recA. When the protease cleaves the lexA protein, it results in a greatly augmented synthesis of proteins associated with DNA repair. Once the DNA damage is removed, the activated recA function is turned off and newly synthesized lexA protein again represses the DNA repair genes.

In humans, the inability to repair DNA damage is associated with rare genetic syndromes. The best known is xeroderma pigmentosum. There are several variations of this condition. People with the disease are unable to repair the DNA damage caused by exposure to ultraviolet light and some chemicals. Individuals with different maladies associated with defective DNA repair mechanisms show different sensitivities to damaging agents. This fact suggests that there are several enzyme systems for DNA repair in humans. Hypersensitivity to DNA-damaging agents may be caused by a defect in a single enzyme in any one of the different pathways used for DNA repair. The multiplicity of defects in repair in different patients has been confirmed by complementation studies carried out with cells grown in culture; there are about six to eight complementation groups, suggesting at least an equal number of loci that encode enzymes involved in repair of DNA damage in human cells.

The Three Types of DNA Recombination

DNA recombination is the process of rearranging DNA segments or chromosomes. In nature, DNA recombination is very widespread; it serves different functions and it occurs in many different ways. There are three basically different types of recombination. The first and most common is called general recombination. This form of recombination takes place between identical or nearly identical segments of chromosomes such as homologously paired eukaryotic chromosomes during meiosis (box 26A), or bacteriophage λ chromosomes during multiplication. The second type of recombination, site-specific recombination, is limited to highly select regions of the genome and to very specific functions. One of the best-understood examples of site-specific recombination is the integration of bacteriophage λ DNA into the *E. coli* host chromosome. The third type of recombination, nonspecific recombination, occurs between nonhomologous regions of chromosomes, requiring little or no site specificity. The main function of this type of recombination might be to bring about a controlled level of mutation.

The details of recombination are fascinating, but they are more appropriately dealt with in a genetics text or a molecular biology text. We will limit ourselves in this section to a few comments about general recombination.

In the 1960s Meselson proposed a three-step model for general recombination: (1) formation of staggered breaks in two parental DNAs; (2) base pairing between single-stranded regions of the two parental types; and (3) repair synthesis (fig. 26.33). Meselson's model explained many experimental results. In fact, it went beyond that. The proposal that base pairing takes place between single-stranded regions of the different parental DNAs provided an explanation for why chromosomes tend to recombine in homologous regions.

The main approach to the enzymology of recombination has been to correlate mutations that affect recombination with actual enzyme activities. The ultimate goal of these studies is to reconstitute a cell-free system with purified enzymes that can carry out recombination. Most of this work with bacteria has taken place in *E. coli*. A considerable amount of work is now being carried out with the yeast *S. cerevisiae*.

Figure 26.33

Meselson model for phage recombination (circa 1964). Phage form staggered breaks. Base pairing leads to an annealed complex with gaps. Gaps are mended by repair synthesis.

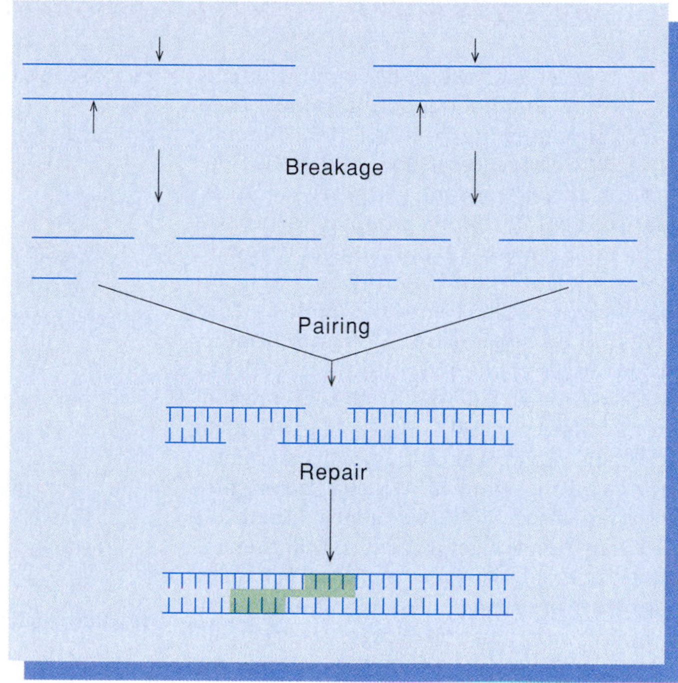

Considering the proposed steps in the recombination process, it seems likely that several activities would be required. These include (1) an endonuclease activity to make the initial incision or incisions in the DNA, (2) an activity to catalyze single-strand invasion of a duplex structure and subsequent unwinding and rewinding of base-paired structures, (3) enzymes for repair, synthesis, and ligation of DNA, (4) an enzyme that removes mismatched bases, and (5) another endonuclease to make scissions that resolve the complex into separate chromosomes.

The search for enzyme activity associated with recombination has been based on analysis of organisms carrying mutations that affect general recombination. Such mutants, known as *rec* mutants, were first discovered more than fifteen years ago, and new ones are still being found. *E. coli rec* mutations, *recA, recB, recC,* and *recD* reduce recombination efficiency to different extents. The *recA* mutants are totally deficient in homologous recombination, while *recB* or *recC* mutants can recombine only about 10^{-4} times as well as wild-type cells. In addition, mutations in two other genes will reverse the defect of *recB* or *recC* mutants; no comparable second-site suppressors of *recA* have been found. This lack suggests that *recA* is absolutely required for homologous recombination but that there may be alternate pathways to recombination that do not require *recB* and *recC*.

Recombination between Homologous Chromosomes Occurs during Meiosis

Most eukaryotic cells contain two sets of homologous chromosomes, each originating from one of the parents. Before meiosis begins there is a duplication of chromosome mass as in premitosis (see fig. 1.21). Meiosis is divided into two stages called meiosis I and meiosis II (figure 1*a* and *b*). Each cell entering meiosis ultimately gives rise to four cells. Meiosis I involves the separation of homologous chromosomes of the parent diploid cell and the consequent halving in the number of chromosomes per cell. Meiosis II involves an equational separation of identical sister chromatids. During meiosis I homologous chromosomes pair, and frequently there is recombination between them so that they exchange homologous regions (figure 1*c*). Since most genes are represented by two or more alleles, the recombination event gives rise to different combinations of alleles, which are destined for different germ cells.

Recombination is extremely important to the species because it permits the testing of different combinations of alleles, combinations that may prove to be more or less beneficial to the organism than the original ones. Meiotic recombination has also been used by geneticists for gene mapping. Classical mapping of genes is based on the principle that the recombination frequency between alleles of different genes is a function of the distance between them. Thus the physical chromosome length is related to the "genetic length" measured by recombination. The unit of genetic length, defined as the length of chromosome over which there occurs a 1% recombination frequency for a given mating pair, is called a centimorgan (cM).

Figure 1

Meiosis. The process of meiosis involves two cell divisions with two segregation cycles, meiosis I and meiosis II. These cycles are pictured for a hypothetical cell containing four chromosomes. (*a*) During the first cycle homologous chromosomes pair and then segregate. (*b*) During the second cycle the sister chromatids from each chromosome segregate. The second cycle is very much like mitosis (see fig. 1.21). Each diploid cell that enters meiosis ultimately yields four haploid cells. These haploid cells are the sex cells for the next round of mating. (*c*) Recombination between homologous chromosomes during meiosis I leads to different combinations of alleles. In this illustration alleles for the same gene are represented by corresponding capital or lower-case letters.

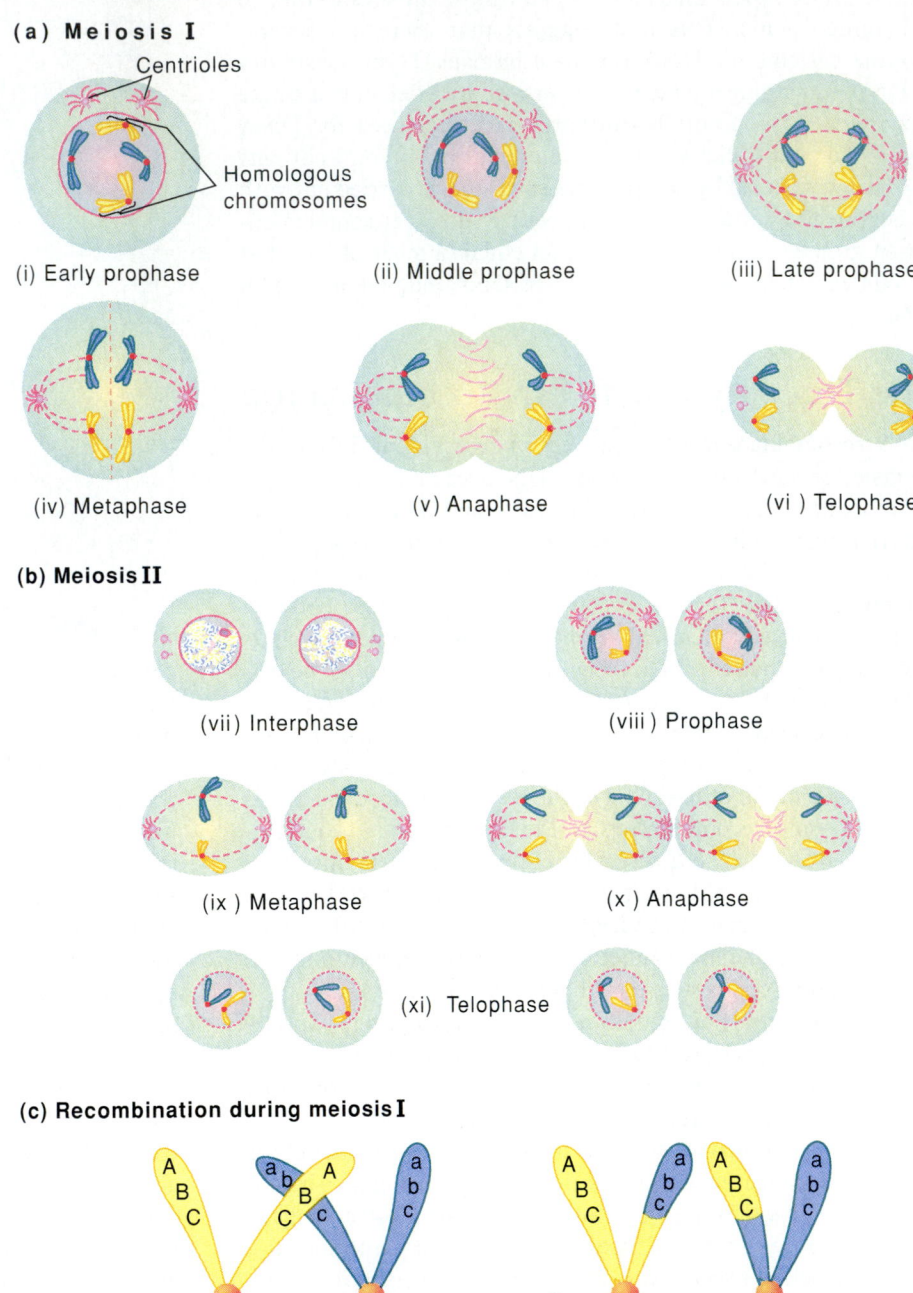

Figure 26.34

Reactions catalyzed by purified recA protein *in vitro*. RecA catalyzes a number of different reactions between DNA strands, all of them involving the unwinding and winding of base-paired structures. (*a*) D-loop formation by interaction between supercoiled circular duplex DNA and single-stranded DNA. (*b*) Partial circular duplex formation by displacement of one strand from a duplex structure. (*c*) Strand exchange between a gapped circular duplex structure and a linear duplex structure. (*d*) Complex formation between two helices, one of which is gapped.

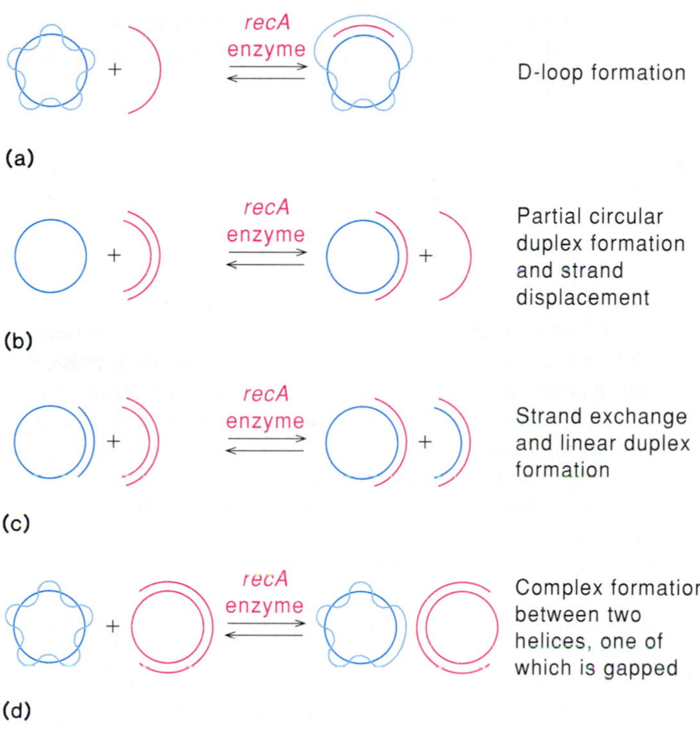

Figure 26.35

DNA unwinding by exoV enzymes. This is best observed *in vitro* in the presence of ATP and Ca^{2+}. The Ca^{2+} inhibits the exonucleolytic activity of the enzyme. The ATP provides the energy that drives the enzyme in a concerted manner into the duplex structure. The duplex unwinds in front of the path of the enzyme and rewinds in back of the enzyme.

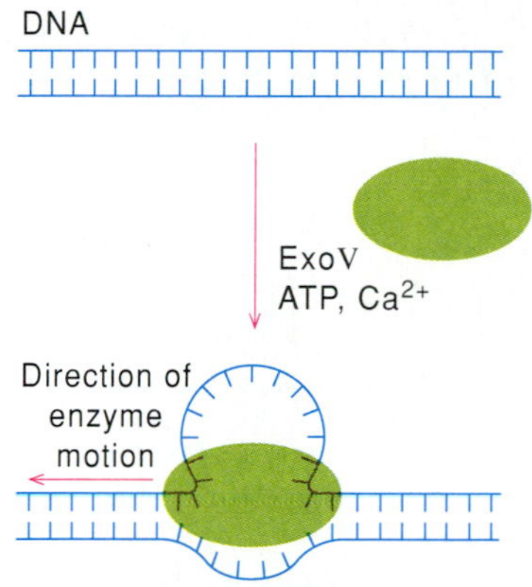

We have discussed the repair role played by the protease function of *recA* (see fig. 26.31). The second class of reactions catalyzed by recA protein indicates a prominent role in the initiation of homologous pairing during genetic recombination. Thus the purified recA protein catalyzes various forms of complex formation exchange between duplex and single-stranded DNAs in reactions that entail the cleavage of ATP (fig. 26.34). D-loop formation can result when the recipient is a circular duplex and the donor is a single strand (see fig. 26.34*a*). Partial circular duplex formation can result from the exchange between a circular single strand and a linear duplex (see fig. 26.34*b*). Strand exchange can occur where a single-stranded fragment is removed in the presence of a double-stranded fragment (see fig. 26.34*c*). Finally, recA protein will also cause a complex to form between two circular helices, provided that at least one input helix is gapped on one strand (see fig. 26.34*d*). Interestingly, while the two helices must contain homologous sequences in order to form a four-stranded structure, the gap may lie in a nonhomologous region. The fact that mutations in *recA* completely eliminate recombination, as well as the finding that the recA-related protein catalyzes strand exchange reactions like those required in all three of the recombination models we have been discussing, makes it highly likely that recA plays a key role in these processes *in vivo*. If this is so, then recA protein is clearly situated at the hub of activities in the recombination complex.

The recA enzyme will not catalyze any recombination event unless at least one free end of a single strand or a DNA duplex is available. Thus it seems unlikely that the recA protein could be involved in making the initial incision(s) in the duplex structure that are required to initiate recombination. There are reasons for believing that the *recB* and *recC* genes are involved in this capacity. First, a mutation in either of these genes reduces recombination by a factor of about 10^{-4}. Second, these two genes together with *recD* encode the subunits of a 330 kDa enzyme exonuclease V (exoV). Surprisingly *recD* mutants, which result in an exoV complex lacking all known enzyme activities, possess an increased rate of recombination *in vivo*. Normally exoV is both a helicase and a nuclease. The helicase activity is best demonstrated *in vitro* under conditions where the nuclease activity is blocked; this can be done by adding Ca^{2+}. When linear duplex DNA is incubated with exoV in the presence of ATP and Ca^{2+}, it begins to unwind at one end. As the unwinding progresses, the single-stranded regions collapse back on each other to re-form a duplex. A double-loop structure is maintained in the vicinity of the migrating enzyme (fig. 26.35).

What is the point of this migrating behavior of exoV? The exoV enzyme may scan the duplex structure to find a site where it would be appropriate to make the strand scission required to initiate recombination. These indications come from detection of an octanucleotide sequence known as *chi*, 5′-G-C-T-G-G-T-G-G-3′, which is associated with regions highly active for recombination. The *chi* sequence does not stimulate recombination in special *recBC* mutants where exoV retains some recombination proficiency but lacks detectable nuclease activity. This observation has led to the suggestion that exoV normally scans the duplex and makes strand incisions in or near regions containing the *chi* sequence.

The recombination process triggered by the initial scissions made by the recBCD complex and mediated by the recA enzyme are not sufficient for recombination between interacting chromosomes. A third enzyme, resolvase, is required for efficient separation of the recombining chromosomes in the recA-chromosome complex. This protein is encoded by the *ruvC* gene in *E. coli,* and is absolutely required for homologous recombination in the bacterium. Its effectiveness has also been demonstrated in a cell-free *in vitro* system.

Incidentally the fact that the recA enzyme carries two enzyme activities, one for homologous recombination and one for regulating the concentration of DNA repair genes including itself (the recA protease, which cleaves the lexA repressor), leads to the suspicion that recA is also more directly involved in DNA repair. Recombination between two homologous chromosomes badly damaged in different regions could lead to one normal chromosome. Although, technically a haploid organism, *E. coli* carries two copies of the bacterial chromosome for a considerable period of time before cell division. Thus the necessary identical pairs of chromosomes are present for homologous recombination.

Summary

This chapter deals with reactions involved in DNA synthesis, degradation, repair, and recombination. The chief points to remember are as follows.

1. DNA replication proceeds by the synthesis of one new strand on each of the parental strands. This mode of replication is called semiconservative and it appears to be universal. DNA synthesis initiates from a primer at a unique point on a prokaryotic template such as the *E. coli* chromosome. From the initiation point, DNA synthesis proceeds bidirectionally on the circular bacterial chromosome. The bidirectional mode of synthesis is not followed by all chromosomes. For some chromosomes, usually small in size, replication is unidirectional.
2. In eukaryotic systems, replication can start at several points (still not well defined) along the chromosome. Replication is usually bidirectional about each initiation site. The termination points of replication are interspersed between initiation sites. In most cases of unidirectional or bidirectional replication, synthesis occurs nearly (but not exactly) simultaneously on both strands of the parent DNA template. Since synthesis can occur only in the $5' \rightarrow 3'$ direction on the growing chain, and since the two strands in the parent duplex are oriented in opposite directions, this means that synthesis can occur continuously on only the leading strand. On the other (lagging) strand it must pause for the template to unwind. Synthesis on the lagging strand does not occur continuously but rather in small discontinuous spurts, generating Okazaki fragments.
3. Many proteins are required for DNA synthesis and chromosomal replication. These include polymerases; helicases, which unwind the parental duplex; enzymes that fill in the gaps and join the ends in the case of lagging-strand synthesis; enzymes that synthesize RNA primers at various points along the DNA template; topoisomerases, which permit rotation and supercoiling; and single-strand DNA-binding proteins, which stabilize single-stranded regions that are transiently formed during replication. Most of these proteins have been isolated from whole cells and studied in cell-free systems.
4. In *E. coli,* mutations have been isolated in the genes encoding a number of these enzymes. Many of these mutations are conditional, since the functional enzymes involved are required for DNA synthesis and cell viability. Mutants carrying mutationally altered proteins have been important in confirming their roles predicted from cell-free studies.
5. In eukaryotes, most of the work on enzymes involved in DNA synthesis has been done without mutants. However, mutants are currently being isolated in *S. cerevisiae* that prove as useful as those of *E. coli* to help characterize the mechanism of DNA replication. Considerable progress has been made in studying the *in vitro* replication of animal viruses such as SV40 and adenovirus. The importance ascribed to the enzymes that have been characterized is largely based on a comparison of their properties with similar prokaryotic enzymes whose functions are better understood.
6. Many enzymes that act on DNA are involved in processes other than DNA synthesis. They include DNA repair enzymes, DNA degradation enzymes, and DNA recombination enzymes.

Selected Readings

Bauer, W. R., F. H. C. Crick, and J. H. White, Supercoiled DNA. *Sci. Am.* 243(4):118–133, 1980.

Beese, L. S., and T. A. Steitz, Structural basis for the 3′–5′-exonuclease activity of *E. coli* DNA polymerase I: a two metal ion mechanism. *EMBO J.* 10:25–33, 1991.

Bell, S. P., and B. Stillman, ATP-dependent recognition of eukaryotic origins of DNA replication by a multiprotein complex. *Nature* 357:128–134, 1992.

Blackburn, E. H., Structure and function of telomeres. *Nature* 350: 569–573, 1991.

Bohr, V. A., and K. Wasserman, DNA repair at the level of the gene. *Trends Biochem. Sci.* 13:429–432, 1988.

Bramhill, D., and A. Kornberg, A model for initiation at origins of DNA replication. *Cell* 54:915–918, 1988.

Campbell, J. L., Eukaryotic DNA replication. *Ann. Rev. Biochem.* 55:733, 1986.

Cedar, H., DNA methylation and gene activity. *Cell* 53:3–4, 1988.

Challberg, M. D., and T. J. Kelly, Animal viruses and DNA replication. *Ann. Rev. Biochem.* 58:671–717, 1989.

Cox, M. M., and I. R. Lehman, Enzymes of general recombination. *Ann. Rev. Biochem.* 56:229–262, 1987.

Demple, B., and P. Karran, Death of an enzyme: suicide repair of DNA. *Trends Biochem. Sci.* 8:137–139, 1983.

Dunderdale, H. J., F. E. Benson, C. A. Parsons, G. J. Sharples, R. G. Hoyd, and S. C. West, Formation and resolution of recombination intermediates by *E. coli.* RecA and RunC protein. *Nature* 354: 506–510, 1991.

Fangman, W. F., and B. J. Brewer, Activation of replication origins with yeast chromosomes. *Ann. Rev. Cell. Biol.* 7:375–402, 1991.

Fink, G. R., J. D. Boeke, and D. J. Garfinkel, The mechanisms and consequences of retrotransposition. *Trends Genet.* 2:118–123, 1986.

Holliday, R., A different kind of inheritance. *Sci. Am.* 260(4):60–73, 1989. On DNA methylation.

Itoh, T., and J. Tomizawa, Antisense RNA. *Ann. Rev. Biochem.* 60: 631–652, 1991. Includes an excellent description of the initiation of DNA synthesis for Co1 E1 plasmid.

Kohlstaedt, L. A., J. Wang, J. M. Friedman, P. A. Rice, and T. A. Steitz, Crystal structure at 3.5 Å resolution of HIV-1 reverse transcriptase complexed with an inhibitor. *Science* 256:1783–1790, 1992.

Kong X-P, R. Onrust, M. O'Donnell, and J. Kuriyan, Three-dimensional structure of the β subunit of *E. coli* DNA polymerase III holoenzyme: a sliding clamp. *Cell* 69:425–437, 1992.

Kornberg, A., and T. A. Baker, *DNA replication,* 2d ed. Freeman, New York, 1991. A magnificent up-to-date, clearly written and thorough treatment by the chairman of the board.

Landy, A., Dynamic, structural and regulatory aspects of site-specific recombination. *Ann. Rev. Biochem.* 58:913–950, 1989.

Linn, D., and W. Maas, Reverse transcriptase-dependent synthesis of a covalently linked branched DNA-RNA compound in *E. coli* B. *Cell* 56:891–904, 1989.

Lohman, T. M., W. Bujalowski, and L. B. Overman, *E. coli* single strand binding protein. *Trends Biochem. Sci.* 13:250–255, 1988.

Maxwell, A., and M. Gellert, Mechanistic aspects of DNA topoisomerases. *Adv. Prot. Chem.* 38:69–107, 1986.

McHenry, C. S., DNA polymerase III holoenzyme of *E. coli. Ann. Rev. Biochem.* 57:519–550, 1988.

Meselson, M., and F. W. Stahl, The replication of DNA in *Escherichia coli. Proc. Natl. Acad. Sci.* 44:671–682, 1958. A classic paper.

Messer, W., and W. Noyer-Weidner, Timing and targeting: the biological function of Dam methylation in *E. coli. Cell* 54:734–737, 1988.

Modrich, P., DNA mismatch correction. *Ann. Rev. Biochem.* 56:435–466, 1987.

Newlin, C. S., Yeast chromosome replication and segregation. *Microbiological Reviews* 52:568–601, 1988.

Ogawa, T., and T. Okazaki, Discontinuous DNA replication. *Ann. Rev. Biochem.* 57:519–550, 1988.

Radman, M. and R. Wagner, The high fidelity of DNA replication. *Sci. Am.* 259(2):40–46, 1988.

Sancar, A., and G. B. Sancar, DNA repair enzymes. *Ann. Rev. Biochem.* 57:29–67, 1988.

Sancar, G. B., and A. Sancar, Structure and function of DNA photolyases. *Trends Biochem. Sci.* 12:259–261, 1987.

Shapiro, J. A., *Mobile Genetic Elements.* New York: Academic Press, 1983.

Smith, G. R., Homologous recombination in *E. coli:* Multiple pathways for multiple reasons. *Cell* 58:807–809, 1989.

Stahl, F. W., Genetic recombination. *Sci. Am.* 256(2):90–101, 1987.

Varmus, H., Reverse transcription. *Sci. Am.* 257(3):56–64, 1987.

Wang, E. H., P. N. Friedman, and C. Prives, The murine p53 protein blocks replication of SV40 DNA *in vitro* by inhibiting the initiation functions of SV40 large T antigen. *Cell* 57:379–392, 1989.

Wang, J. C., DNA topoisomerases. *Ann. Rev. Biochem.* 54:665–697, 1985.

Wang, T. S. F., Eukaryotic DNA polymerases. *Ann. Rev. Biochem.* 60:513–553, 1991.

Problems

1. Explain how Meselson and Stahl were able to demonstrate that *E. coli* replicates its DNA in a semiconservative mode. How would the data appear for the first and second doublings if the mode of replication were dispersive? Does the dispersive mode of DNA synthesis occur in cells? (Explain your answer.)

2. The genetic map below represents the distribution of eight genes (*a–h*) as they occur on a bacterial chromosome. The plot is a graphic representation of the average number of copies for each of the eight genes, for a cell that grows with a doubling time equal to the time required for a complete round of DNA replication.
   (a) Estimate the location of the origin of replication.
   (b) Infer whether replication is bidirectional or unidirectional.

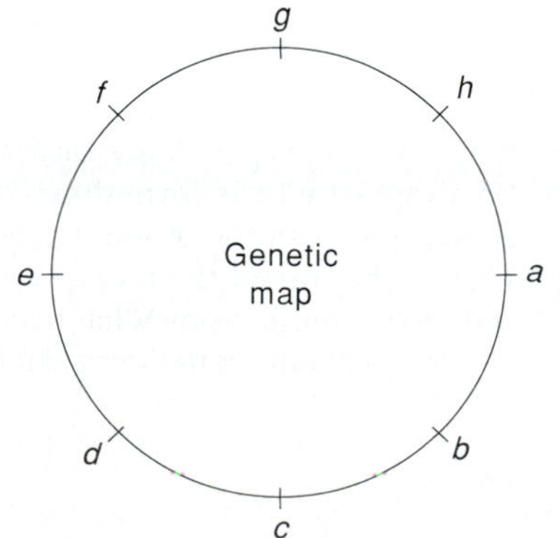

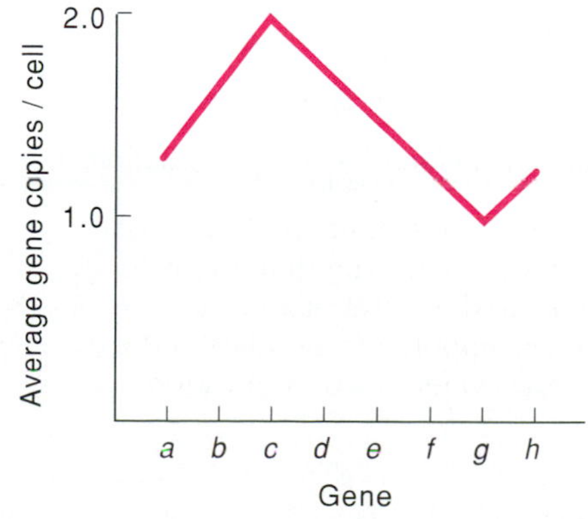

3. Draw the chemical reaction mechanism for the formation of a phosphodiester linkage during DNA synthesis. Discuss the significance of the pyrophosphate product that is formed. What is the significance of the Mg^{2+} requirement?
4. Would you expect the replication of either ϕX174 DNA or the *E. coli* chromosome to be sensitive to inhibition by the antibiotic rifampicin? If inhibition occurs, at what stage during replication does it occur?
5. Draw the chemical reaction mechanism for DNA ligase in *E. coli* (uses NAD as the source of energy and forms a covalent intermediate with an ϵ-amino group of lysine). Why are ligation reactions that require ATP more thermodynamically favorable?
6. *E. coli* has a genomic complexity of about 4×10^6 base pairs (bp) and each replication fork can move at about 10^3 bp per second. How long would it take to replicate the *E. coli* chromosome? With an ample carbon source and ideal growth conditions, cells of *E. coli* can divide in about 20 min. How can this shorter division time occur if the rate of fork migration remains constant at 10^3 bp per second?
7. Humans have about 3×10^9 base pairs of DNA in their genome and the replication forks migrate much slower than in bacteria (about 30 bp per second). How long would it take to replicate the entire genome if it was a single continuous piece of DNA (one chromosome)? How many replication origins would be required to replicate this DNA in an hour?
8. In the graph below, *E. coli* was labeled with radioactive thymidine for a short pulse (10 s) followed by a chase with an excess of nonradioactive thymidine. The DNA was extracted and centrifuged in alkaline sucrose gradients (under high pH conditions the DNA denatures). Explain what these data imply, and interpret these results in light of our current model for DNA replication.

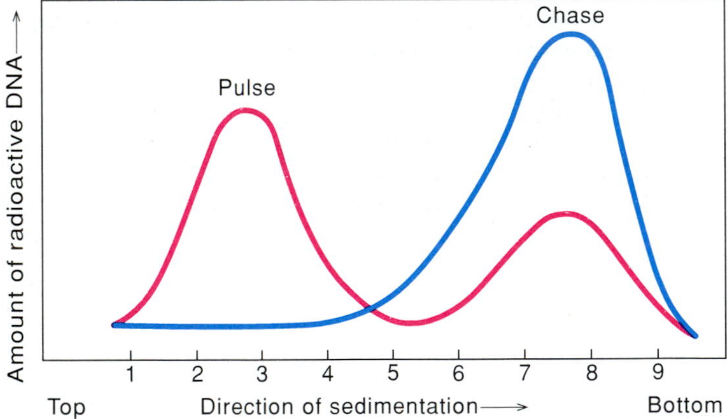

9. Cairns and De Lucia isolated a mutant strain of *E. coli* that had only about 1% of the DNA Pol I activity found in wild-type cells, yet the strain replicated its DNA at a normal rate. Explain how this discovery was important in understanding the role of the different DNA polymerases in replication and repair.
10. All enzymes that make DNA in a template-dependent fashion require a primer. Why does a primer increase the fidelity of DNA synthesis and why is this primer usually RNA?
11. What is the novel priming system for adenovirus DNA replication? Why is this system an advantage to the virus? What other viruses have similar priming systems?
12. Outline in general terms how a retrovirus replicates its RNA. What unique viral protein is used in this replication process? What serves as the primer for retrovirus replication?
13. Eukaryotic DNA is replicated at a slower rate than prokaryotic DNA. One reason may be the requirement for the deposition of histone proteins on DNA (histone synthesis and DNA replication are coupled). Describe a model for the replication of eukaryotic DNA and nucleosome formation.
14. Normal human fibroblasts were grown in culture and then exposed to UV light. A short time later the DNA was extracted and applied to an alkaline sucrose gradient, and the data in graph (*a*) were observed. Another sample of cells was also exposed to UV light but about 12 h were allowed to pass before the DNA was extracted and applied to an alkaline sucrose gradient (graph *b*). Explain these data from what you know about DNA repair. Another sample of fibroblast cells, isolated from a patient with xeroderma pigmentosum, was exposed to UV light and then applied to a gradient after a short time. Would you expect the data to resemble those in graph (*a*) or (*b*)? Why?

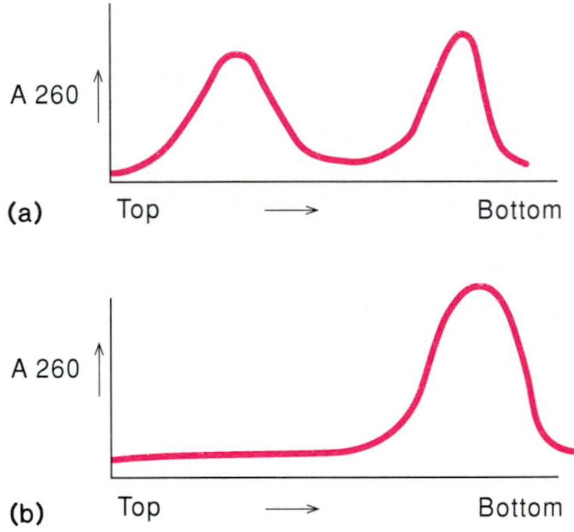

15. Why is the uracil-DNA glycohydrolase very important in DNA repair? Why is thymidine-containing and not uracil-containing DNA the modern product of evolution?
16. What is the role of the recA protein in *E. coli* DNA repair and in recombination? What use are *recA* mutants in biochemical and genetic research?

27

CHAPTER

DNA Manipulation and Its Applications

In recent years, researchers have developed a broad array of techniques to investigate the fine structure of DNA (fig. 27.1). The techniques of DNA manipulation have also proved to be of enormous value in related fields. In particular, they have made it possible for us to map and isolate genes from complex eukaryotes—feats that, for technical reasons, could never have been accomplished by classical genetic methods. Beyond this, they have also enabled us to redesign genes to achieve a variety of practical goals. These newly designed genes may be inserted into the same species from which the original unmodified genes came, or into other species.

In this chapter we will first consider DNA sequencing and the synthesis of DNAs with a predesignated sequence. Then we will explore the different approaches for isolating specific genes or gene segments. Following this, we will examine the methods that are currently available for restructuring existing DNA sequences. Finally, we will look at some of the major advances in our understanding of gene structure and location that have resulted from the new technology. This part of the discussion will focus on two examples: the mapping of the human globin gene family and the mapping of the gene responsible for the genetically inherited disease cystic fibrosis.

Sequencing of DNA

Initial efforts at sequencing nucleic acids were confined to RNA molecules that could be readily isolated in pure form. The first sequence to be determined was that for tyrosine tRNA from yeast. From roughly 100 pounds of yeast, Robert Holley was able to isolate enough of the tyrosine tRNA to carry out a sequence analysis. This historically significant effort required a number of enzymes and chromatographic techniques. We will not, however, elaborate on this accomplishment, because sequencing of RNA is no longer done directly. In fact, the sequencing of RNA has been replaced by the sequencing of cDNA, which results from the transcription of RNA into DNA.

Figure 27.1

Procedures for manipulation of DNA. DNA manipulation is divided into two steps: (1) isolating the selected region(s) of the chromosome that is under study and (2) studying the properties of the isolated DNA. Most of the procedures that are discussed in this chapter are indicated in this figure. First the nuclear (genomic) DNA is isolated. This DNA is degraded into smaller, more manageable fragments by treatment with a restriction enzyme that cleaves at defined sequences. Small, manageable fragments of DNA may also be isolated by reverse-transcribing the messenger RNA to obtain cDNA. Either of these DNAs, genomic or cDNA, is then inserted into a vector for purposes of amplification (either a plasmid or a virus may be used for this purpose). The plasmid mixture containing a mixture of DNAs is then used to transfect cells, which are plated to form clones. Each clone contains many copies of a vector linked to an inserted DNA segment. The mixture of clones is called a cDNA or a genomic DNA library, depending on the source of the inserted DNA. Selective detection procedures are used to find the clones that carry the desired DNA. The plasmids are isolated from these clones, and cells are transfected again with the purified plasmid to obtain a large amount of DNA. DNA isolated by this means can be used in a number of ways: (1) Segments of the desired inserted DNA may be selectively amplified by the polymerase chain reactions (PCR); (2) the DNA may be sequenced; (3) cells may be transfected with the purified plasmid to make protein, to study the effect of the encoded protein, or to change the properties of the cells. Alternatively, the DNA may be changed by directed mutagenesis and the properties of the changes may be investigated. The DNA in conjunction with an appropriate library also provides a useful tool for locating the gene in the chromosome by chromosome walking (and/or jumping) and for isolation of DNA fragments in the adjacent regions of the chromosome.

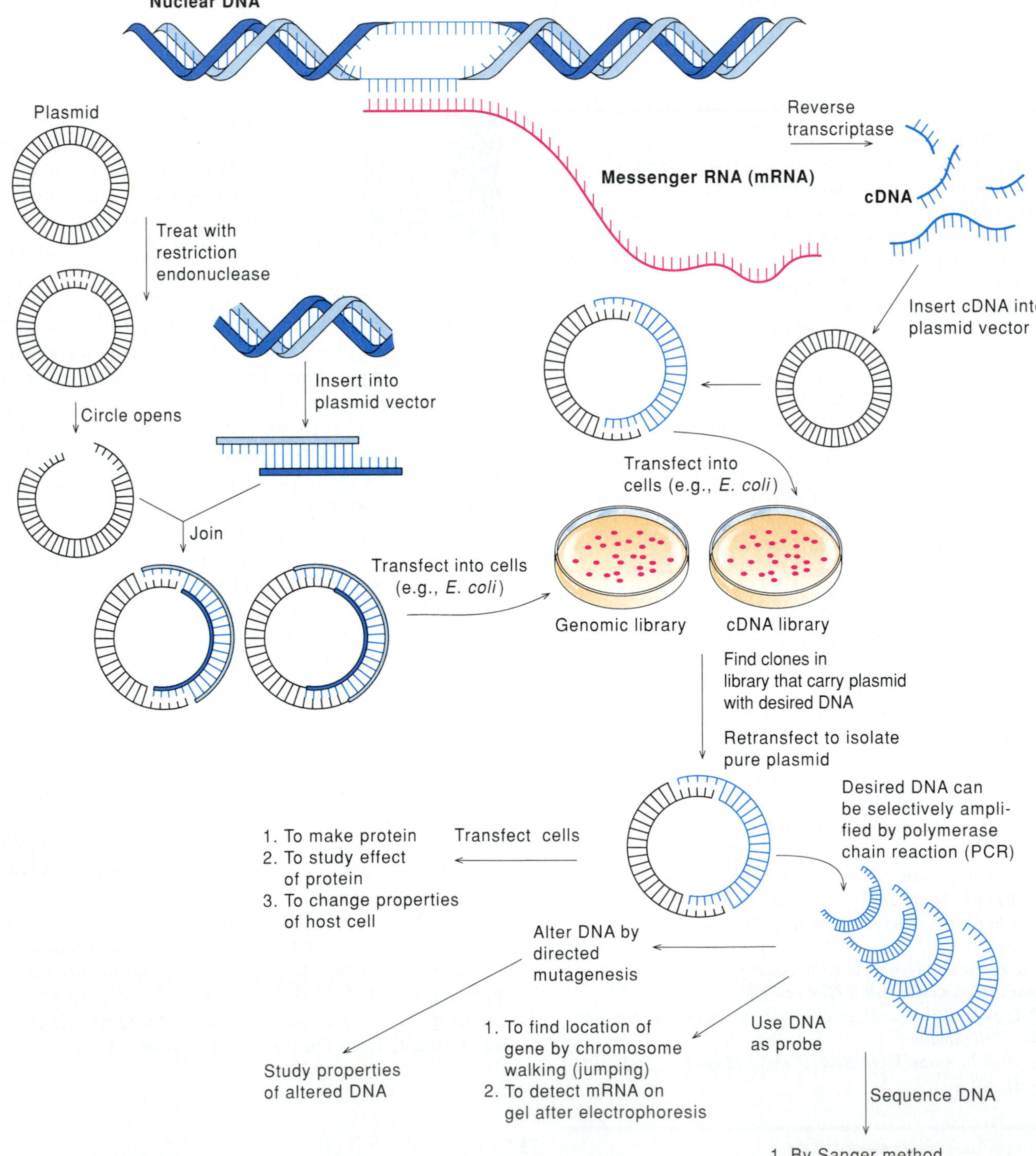

Figure 27.2

Sequencing end-labeled DNA by limited, base-specific chemical cleavage. (*a*) A sequence of three reactions leading to strand cleavage at a guanine. If the entire DNA is subjected to this reaction sequence for a limited time, the result is (*b*), a family of labeled fragments that will be cleaved at different guanine sites.

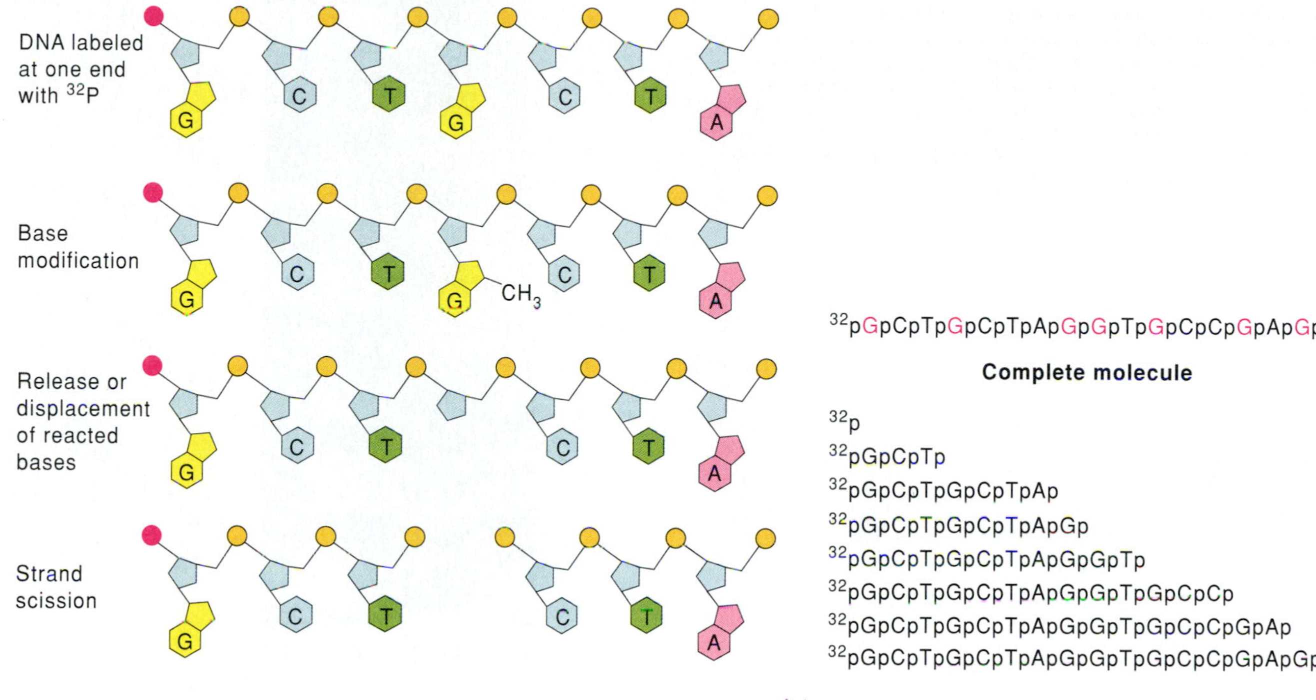

Two quite different methods have been developed for sequencing DNA. One method, involving cleavage of preexisting DNA, uses a chemical approach. This was developed by Walter Gilbert and Alan Maxam. The second method, involving premature termination of newly synthesized DNA, uses an enzymatic approach and was developed by Fred Sanger, the same person who received the Nobel Prize for sequencing the first protein, insulin. Both Gilbert and Sanger received Nobel Prizes for their work on DNA sequencing. Both methods merit description here, as they are both useful in different ways.

The Maxam-Gilbert Sequencing Procedure Uses Chemical Methods for Cleavage of Preexisting DNAs

The Maxam-Gilbert sequencing procedure is important both for sequencing and because it provides a technique for determining where proteins bind to the DNA (see chapter 30). For initial sequencing of DNA, either single- or double-stranded molecules can be used.

In the first step a specific fragment of DNA is ^{32}P-labeled at its 5′ end with polynucleotide kinase. This enzyme transfers the $^{32}PO_4^-$ group from $\gamma^{32}P$-ATP to the 5′ hydroxyl end of the deoxyribonucleotide chain. If double-stranded DNA is used, both 5′ ends will become labeled, so such a molecule cannot be sequenced directly. Instead the complementary strands or the ends of a doubly labeled DNA must first be separated. This can be done by denaturation followed by gel electrophoretic separation of the individual chains, or by treatment of the DNA with an enzyme (usually a restriction enzyme) that cleaves the molecule into two segments that can be separated by electrophoresis.

Next, the DNA is treated with a chemical reagent that specifically reacts with one of the four bases. Reaction is carried out for a limited period of time, so that on the average only a few residues in the polynucleotide chain react with the reagent. The modified base introduces a linkage amenable to backbone cleavage by subsequent chemical treatment (fig. 27.2). To a first approximation, all bases of a given type in a chain are equally susceptible to modification. The net result is a family of products labeled at the 5′ end with ^{32}P and terminating at the point of cleavage (see fig. 27.2).

Four chemical reactions are used that cleave DNA preferentially at guanines (G $>$ A), adenines (A $>$ G), cytosine alone (C), and cytosines and thymines equally (C + T). When the products of the four reactions are resolved according to size by gel electrophoresis, the DNA sequence can be "read" from the pattern of radioactive bands as an autoradiogram (fig. 27.3), which is obtained by placing a sheet of film over the gel for a suitable exposure time.

Figure 27.3

Autoradiogram of a sequencing gel according to Maxam and Gilbert. Only a portion of the autoradiogram is shown. To obtain this autoradiogram four different reaction mixtures were used. Each reaction mixture started with the same labeled DNA, then was subjected to a different set of limited digestion conditions. In column 1, the digestion conditions led to cleavage at guanines. In column 2, the conditions were adjusted so that fragmentation occurred at both guanines and adenines. In column 3, digestion conditions that resulted in cleavage at cytosines were used, and finally, in column 4, digestion conditions that resulted in approximately equal extents of cleavage at the cytosines and thymines were used. The different digestion products were electrophoresed (from top to bottom in the figure). The smaller the chain size, the farther the product migrated in the gel. It is possible to "read" the sequence by tracing the individual bands one step at a time, starting at the bottom of the gel. The sequence of the DNA for this gel pattern is given in the center of the autoradiogram.

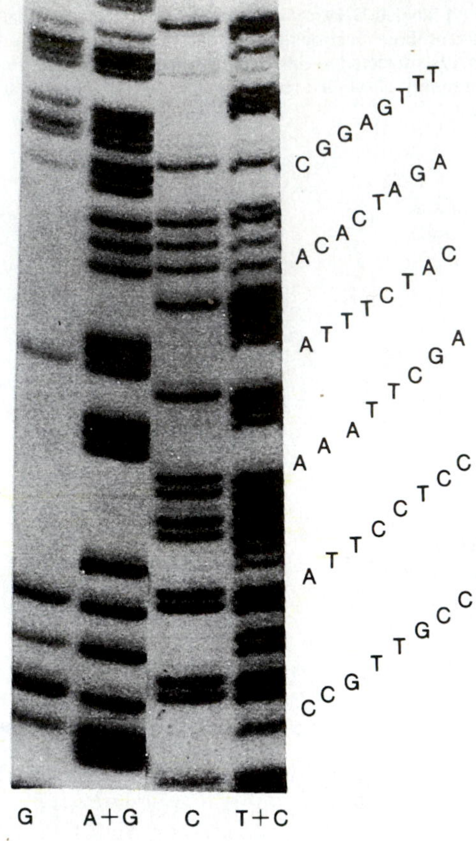

Figure 27.4

Treatment of DNA with dimethylsulfate, resulting in methylation at N-3 in adenine (*a*) and at N-7 in guanine (*b*). Guanine is about five times more reactive than adenine. After reaction for a suitable period of time with the dimethylsulfate, the pH is adjusted to pH 7.0 to permit removal of the modified base and subsequent cleavage of the polynucleotide chain.

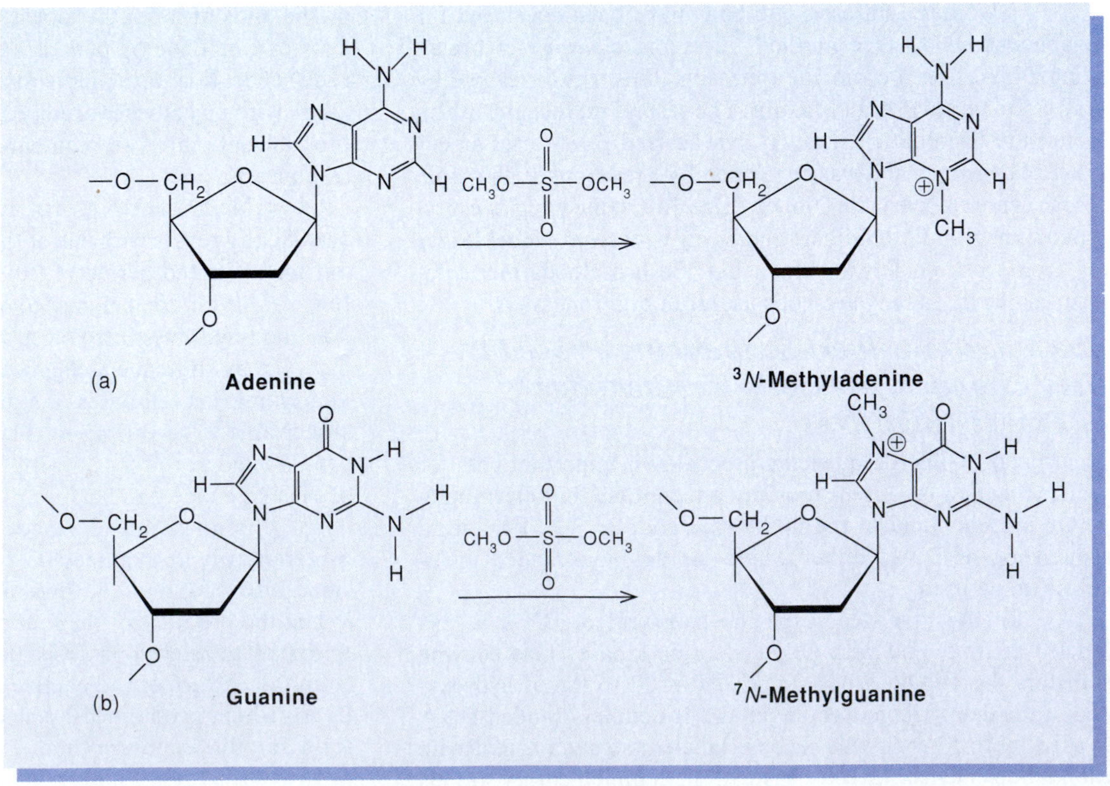

$CH_3O-SO_2-OCH_3$

(a) Adenine → ^3N-Methyladenine

(b) Guanine → ^7N-Methylguanine

Figure 27.5

Reaction of thymine (*a*) and cytosine (*b*) bases with hydrazine. The products of the reaction are identical. Preferential reaction of cytosine occurs in high salt. In low salt the two bases react at approximately equal rates.

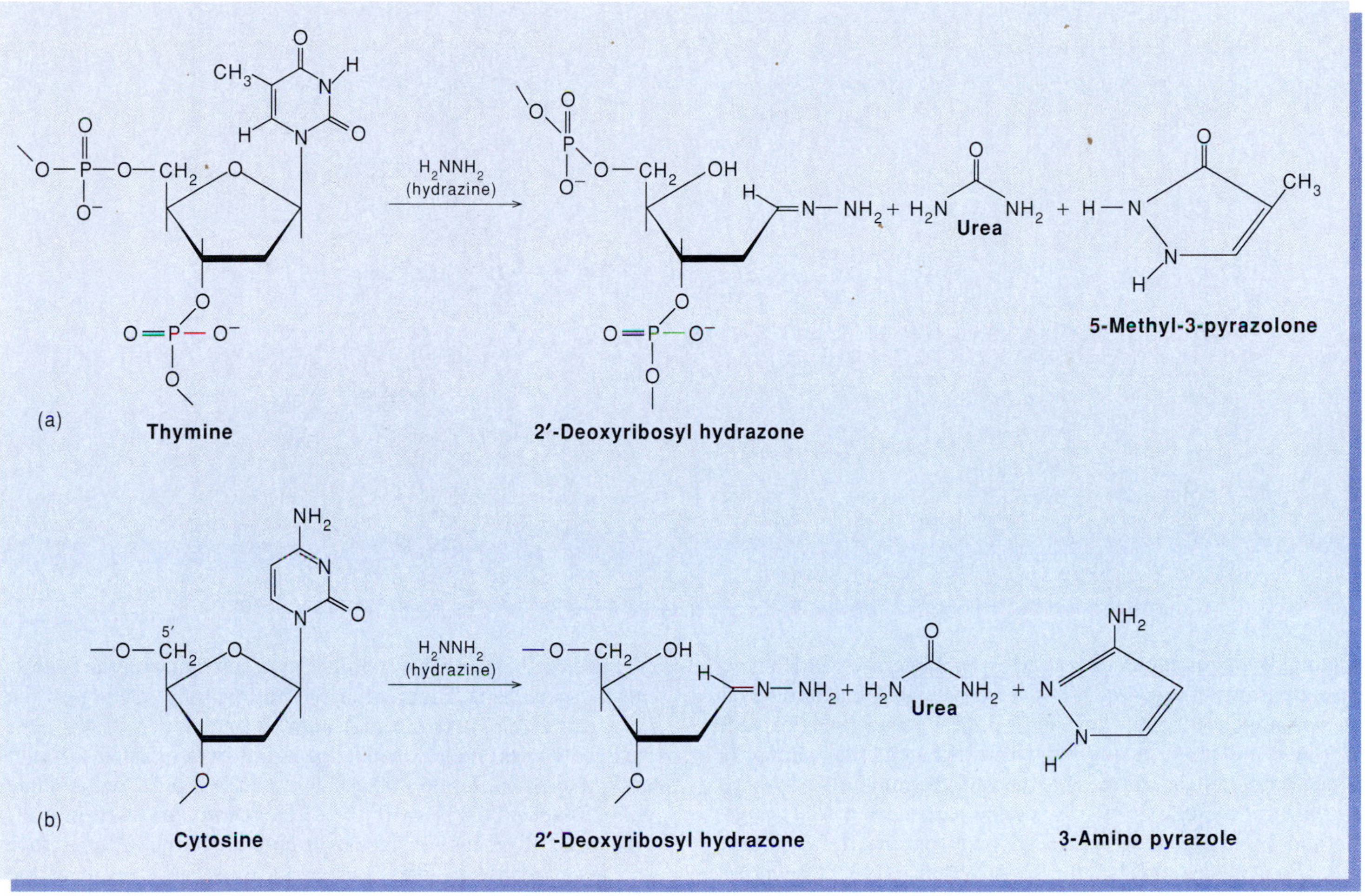

For the purine-specific reaction, an aliquot of the DNA is treated with dimethylsulfate, which methylates the guanines in DNA at the N-7 position and the adenines at the N-3 position (fig. 27.4). The glycosidic bond of a methylated purine is cleaved on heating at neutral pH, leaving the sugar free. Alkali at 90° C will then cleave the sugar from the neighboring phosphate groups. When the resulting end-labeled fragments are resolved on a gel, the autoradiogram contains a pattern of dark and light bands (G > A lane). An adenine-enhanced cleavage can be obtained by treating the methylated DNA with acid, which releases methylated adenine preferentially (A > G lane).

Other aliquots are reacted with hydrazine, which attacks cytosine and thymine bases (fig. 27.5). After a partial reaction in aqueous hydrazine, the phosphate backbone of the DNA is cleaved with 0.5-M piperidine (fig. 27.6). The final gel pattern contains bands of similar intensity owing to the cleavages at cytosines and thymines (C + T lane). However, if 2-M NaCl is included in the hydrazine reaction, the reaction of the thymine residues is suppressed. Then the piperidine breakage produces bands only from cytosines (C lane). Consequently, if the results with and without added hydrazine are compared, C's and T's in the sequence can be distinguished.

Many sequences have been determined by the Maxam-Gilbert method; it is still the tool of choice in cases where the goal is to determine the site of binding of a protein (see chapter 30).

The Sanger Method for Sequencing Uses Newly Synthesized DNAs that Are Randomly Terminated

For pure sequencing the Sanger method is more popular and easier to automate. The method employs chain-terminating dideoxynucleoside triphosphates to produce a continuous series of

Figure 27.6

Cleavage of the hydrazine derivative of the cytosine and the thymine in the presence of piperidine.

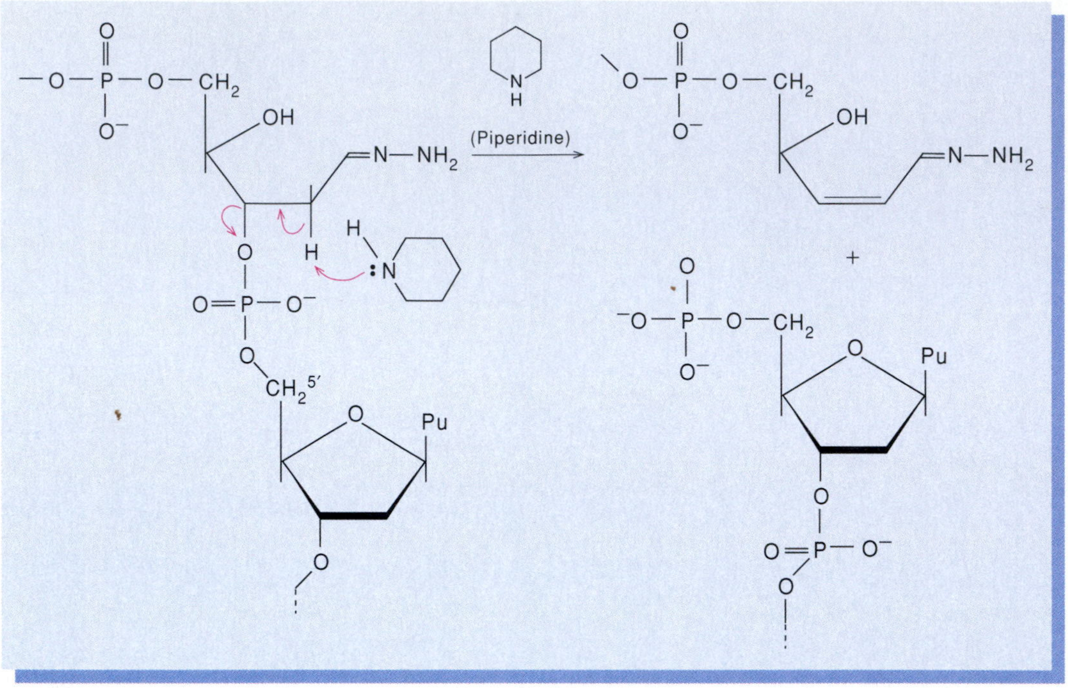

fragments in reactions catalyzed by polymerase. Dideoxynucleoside triphosphates (ddXTPs) resemble deoxynucleoside triphosphates except that they lack a 3′-OH group. They can add to a growing chain during polymerization, but they cannot be added onto and therefore serve as chain terminators.

Either DNA or RNA may be sequenced by the Sanger method. For RNA to be sequenced, reverse transcriptase is used to make a DNA copy of the RNA. When DNA is being sequenced, DNA polymerase is used to make a complementary copy of a primed single-stranded DNA fragment. By choosing an appropriate primer, the region of the nucleic acid that is copied can be predetermined (fig. 27.7).

Synthetic reaction mixtures are then set up, each containing one or more radioactive deoxyribonucleoside triphosphates to label the fragments for detection by autoradiography. Each mixture contains a single, limiting amount of dideoxynucleoside triphosphate to randomly terminate the synthesized fragments at one of the four nucleotides. The products of four separate reaction mixtures, each containing a different dideoxynucleoside triphosphate, are analyzed (see fig. 27.7). Reaction 2, using dideoxyadenosine triphosphate (ddATP), contains all fragments with an A terminus; reaction 2, using ddCTP, contains all C terminations; and so on. Following synthesis, the reaction products are separated from the template by denaturation and fractionated by electrophoresis on polyacrylamide gels. After electrophoresis the positions of the fragments on the gel are detected by autoradiography. The sequence is read directly from the autoradiogram, starting with the fastest-moving (smallest) fragment at the bottom, and moving up the gel (see fig. 27.7). If the first band is in reaction 3, it is a G residue; the next highest band, appearing from reaction 4, would be T; and so on. Up to 800 residues can be read from a single gel. ^{35}S-labeled dXTPs (fig. 27.8) frequently are used in preference to ^{32}P because of the longer half-life of the sulfur isotope and also because it gives sharper patterns, as a result of the shorter pathlength of its β-particle emission.

One of the latest adaptations of the Sanger method uses colored fluorescent derivatives attached to the dideoxynucleotide terminators. The advantage of this approach is that each reaction corresponding to an unknown DNA sequence can be carried out in a single test tube and loaded onto a single column for electrophoresis.

Oligonucleotides of the Desired Sequence Can Be Made in Vitro

Short DNA chains (oligodeoxynucleotides) are essential to make primers in the Sanger sequencing procedure; they are also valuable for other procedures that we will discuss shortly. Chemical methods for making synthetic DNA originated in the laboratory of H. G. Khorana. Khorana's goal was to chemically synthesize an entire gene and then to see whether it would function properly when reinserted into the organism. He wisely chose a small tRNA gene for this study, the gene for the 76-nucleotide tRNA for alanine. His success in the total synthesis of this RNA was a historic achievement and led others to refine the methods and apply them in other areas.

Figure 27.7

The Sanger dideoxynucleoside method of sequencing DNA. (*a*) A suitable template is chosen, and the primer is chosen so that DNA synthesis begins at the point of interest. In addition to the template-primer complex the reaction mixture contains all four radioactive deoxyribonucleoside triphosphates and small amounts of a single dideoxynucleoside triphosphate. The dideoxy compound serves as a chain terminator. (*b*) After synthesis in the presence of DNA polymerase I, the products of the reaction mixture are separated by gel electrophoresis and analyzed by autoradiography. For a given dideoxy compound all fragments terminating with that particular base should give rise to bands on the gel. The interpretation of the gel pattern is given in (*c*). The smallest labeled fragment moves the fastest and appears at the bottom of the gel. (*d*) A typical sequencing film. The sequence begins CAAAAAACGG. (Courtesy of GIBCO-BRL, Life Technologies, Inc., Gaithersburg, Md.)

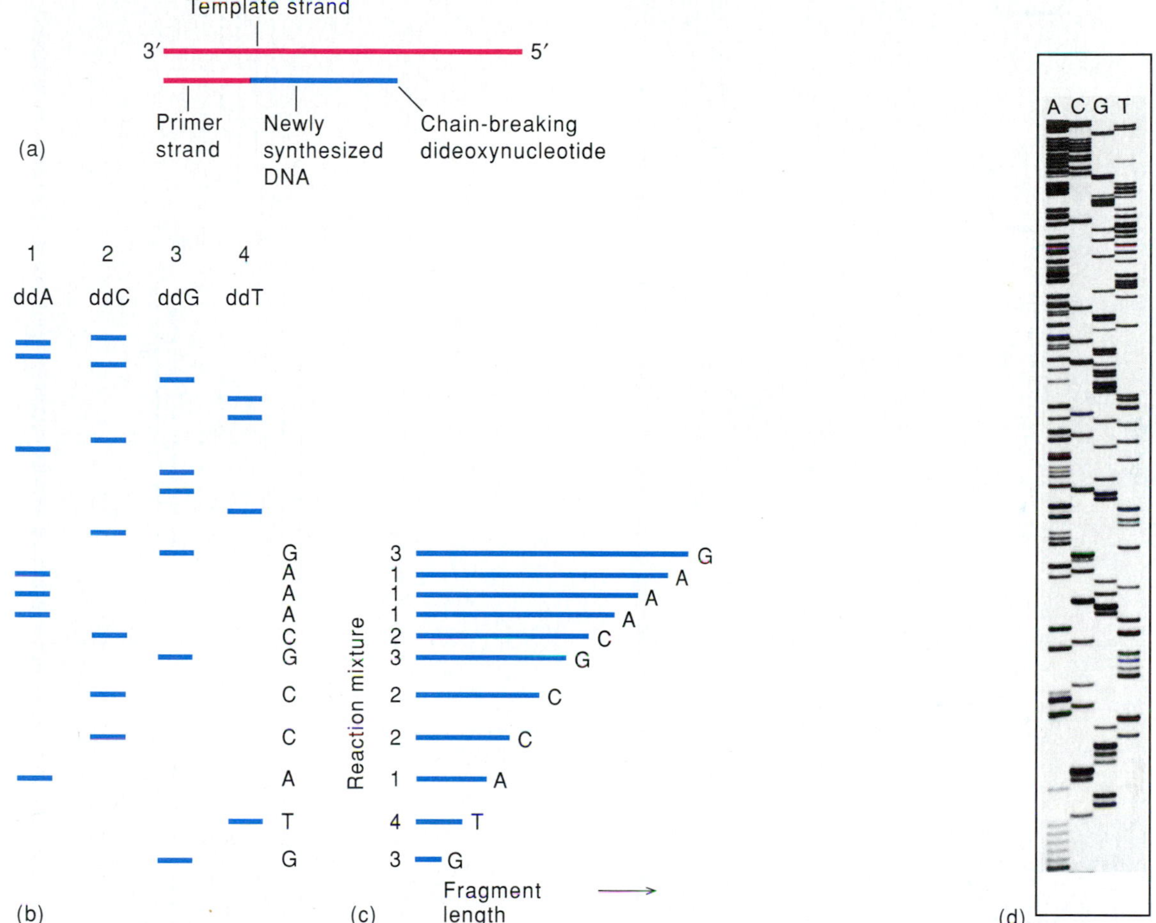

Various chemical methods of synthesis have evolved over the years, but they all have two features in common: (1) 3′ activation is involved in the synthesis instead of 5′ activation as in biosynthesis, and (2) protecting groups are required in intermediate steps to prevent other groups from reacting. Currently, synthesis centers around use of highly reactive phosphite groups, which involve phosphorus in the +3 valence; this is oxidized to +5 phosphorus after incorporation by iodine. The chemistry of the phosphite-triester scheme for synthesizing oligonucleotides is shown in figure 27.9. This method involves sequential treatment of a 5′-O-protected nucleoside with (a) $R'OPCl_2$, (b) a 3′-O-protected nucleoside, and (c) iodine-water. The chain is extended to the desired length by successive repeats of this cycle. After construction of the desired sequence, all phosphoryl protecting groups and the 3′-O-terminal protecting groups are removed (d) to give a synthetic oligonucleotide.

Figure 27.8

Sulfur-labeled substrate is preferred to phosphorus-labeled substrate because of its longer half-life and the shorter pathlength of its ß emission, which gives sharper banded patterns after autoradiography.

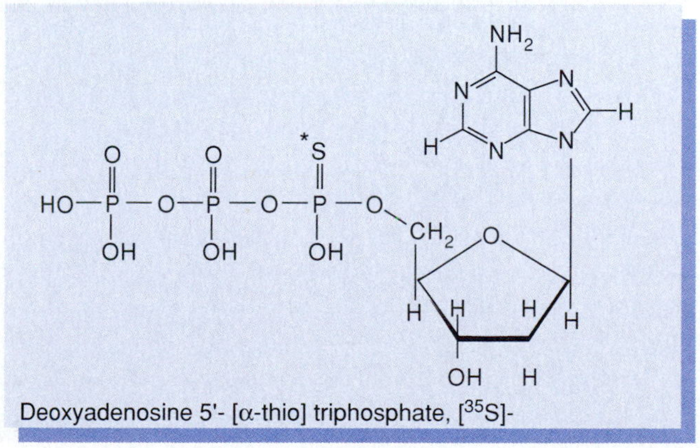

NH_2, N, N, H, H, N, N, O, O, *S, HO—P—O—P—O—P—O, OH, OH, OH, CH_2, O, H, H, H, H, OH, H

Deoxyadenosine 5′- [α-thio] triphosphate, [^{35}S]-

Figure 27.9

Phosphite-triester method for the synthesis of oligonucleotides. The procedure for making a dinucleotide is shown. This is extended to making oligonucleotides (up to approximately 100 residues) by repeating steps (*a*) through (*c*) in a cyclic fashion.

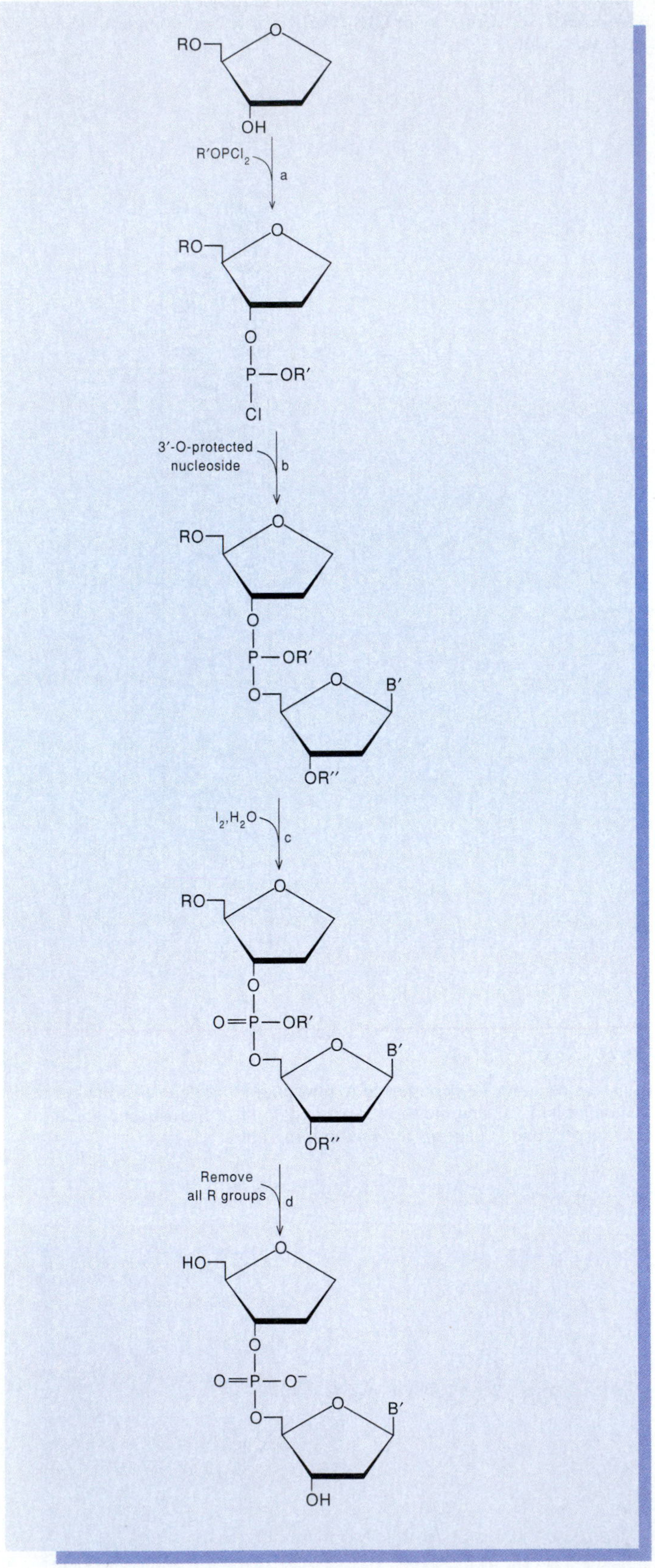

Figure 27.10

Schematic diagram showing how the tRNA gene for alanine tRNA was synthesized. First, small oligonucleotides representing different fragments of the gene were synthesized. Then these were annealed and covalently linked with the help of the enzyme DNA ligase. The process of annealing was repeated until the full gene was assembled.

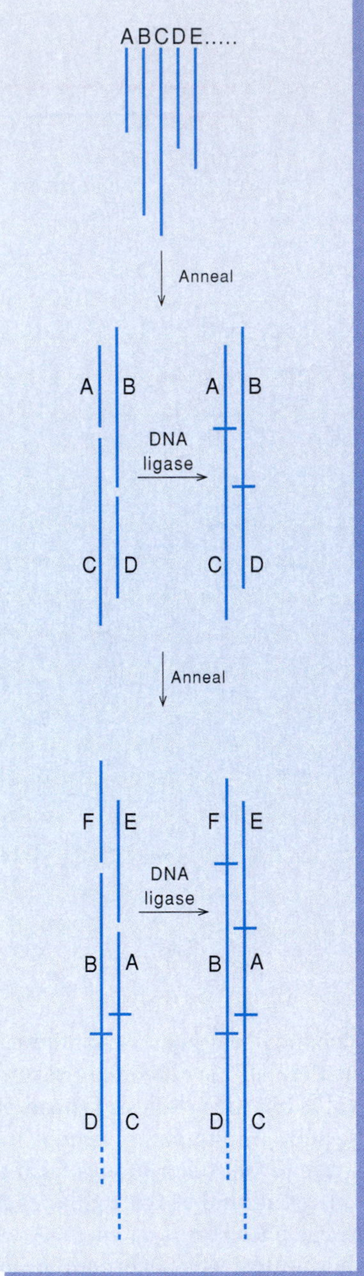

Figure 27.11

Steps in the polymerase chain reaction (PCR). The DNA to be amplified is denatured and annealed with two oligonucleotides that flank the region of interest. These oligonucleotides (or primers) are extended. Extension continues to the ends of the DNA strands. The products are again denatured and annealed to primers for a second round of extension. This process of denaturation, annealing, and primer extension is repeated many times. The primary product of the reaction is duplex DNA, bounded by the sequences of the primers.

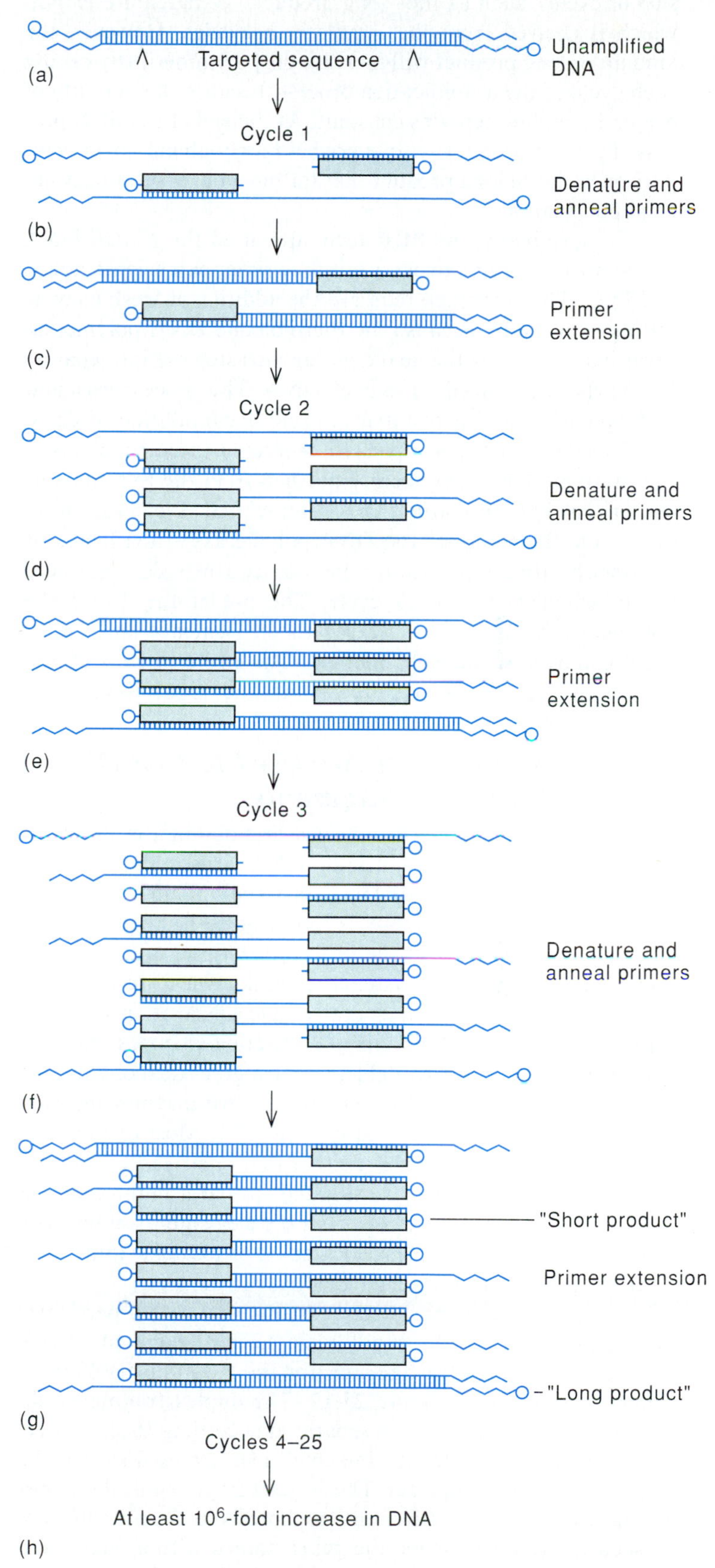

Drawing on the technology developed by Merrifield for the solid-phase synthesis of polypeptides (mentioned in chapter 3), procedures were developed for coupling the 5′-OH group of the initial nucleotide to a resin support, to facilitate the stepwise addition of nucleotides. All mixing and washing steps were thereby greatly simplified and the procedure has become amenable to automation. In fact, it is now possible to program the synthesis of a 100-mer by merely pushing a button for the order of nucleotides.

On occasion, when larger oligonucleotides are desired they can be assembled by synthesizing overlapping, complementary oligonucleotides, which are then annealed and ligated (fig. 27.10).

Sequences of Any Type Can Be Amplified in Vitro *by the Polymerase Chain Reaction*

There are two widely used methods for amplifying defined segments of DNA *in vitro*. One is DNA cloning, which we will describe in a later section. A simpler and more recently discovered procedure is the preferred method where only a limited quantity of the amplified DNA is needed. This method, called the polymerase chain reaction (PCR), entails the use of two oligonucleotide primers, which flank the DNA segment to be amplified (fig. 27.11*a*). The primers must complement opposite strands so that, after annealing, their 3′ ends in effect face each other (fig. 27.11*b*).

There are three steps in the PCR procedure, which are usually repeated many times in a cyclical manner:

1. Denaturation of the original double-stranded DNA sample at high temperature.
2. Annealing of the oligonucleotide primers to the DNA template at low temperature (37° C).
3. Extension of the primers using DNA polymerase.

These steps are illustrated in figure 27.11. Each set of three steps comprises a cycle. The extension products of one primer provide a template for the other primer in a subsequent cycle so that each successive cycle essentially doubles the amount of DNA synthesized in the previous cycle. The result is an exponential accumulation of the specific target fragment to approximately 2^n, where n is the number of cycles. The specific target fragment is also referred to as the "short product" and is defined as the region between the 5′ ends of the extension primers. Each primer is physically incorporated into one strand of the short product.

Other products are also synthesized during the succession of cycles, such as the "long product," of indefinite length, which is derived from the template molecules. However, the amount of long product will increase only arithmetically during each cycle of the amplification process, because the quantity of original template remains constant. At the end of the PCR process the short product is so overwhelmingly abundant in comparison with the long product that for most purposes it does not need purification.

Originally, the PCR technique used the *E. coli* DNA polymerase I enzyme to carry out the primer extension reaction at 37° C. This approach required the addition of fresh enzyme during each cycle, because the thermolabile DNA polymerase I was inactivated by the heat denaturation step used to separate the newly synthesized strands of DNA. The process was slow and expensive, and it resulted in rapid accumulation of denatured enzyme in the sample. To overcome these problems, a thermostable DNA polymerase purified from the thermophilic bacterium *Thermus aquaticus* (abbreviated Taq) was introduced into the reaction. Taq DNA polymerase is unaffected by the denaturation temperature and consequently does not need to be replenished at each cycle. The use of the Taq DNA polymerase has greatly increased the convenience and with it the popularity of the PCR method. The entire process of amplification requires about three hours and requires very few manipulations.

Restriction Enzymes Are Used to Cut DNA into Well-Defined Fragments

In order to sequence DNA, we must first obtain a well-defined piece of DNA. In the case of small DNA viruses, the virus provides a unique polymer of DNA in large amounts. However, even in such cases the DNA is too large to be sequenced directly, and therefore must be broken down into many smaller, well-defined fragments. Systematic breakdown of duplex DNA requires restriction enzymes that cleave DNA at specific recognition sequences. Hundreds of restriction enzymes with different specificities are available, giving a great deal of choice in where to cut the genome. Some representative enzymes and their recognition sites are indicated in table 27.1. Most of these enzymes recognize a sequence of either four or six successive base pairs. The cleavage sites are situated so that a blunt-ended or staggered-ended DNA results from the cleavage reaction. As a rule, the recognition sites are located on an axis of symmetry so that the freshly cut ends have identical structures.

A viral genome cleaved exhaustively with a particular restriction enzyme will usually yield several fragments. Some restriction enzyme cleavage sites for the 5,300-bp SV40 virus genome are shown in figure 27.12. The duplex fragments obtained after cleavage can be separated according to size by gel electrophoresis. Nondenaturing conditions are used so that the duplex strands stay together. The larger a fragment is, the slower it migrates on the gel. After electrophoresis for a time sufficient to separate the fragments, the gel is stained with a fluorescent dye such as ethidium bromide and is viewed under long-wavelength ultraviolet light (long-wavelength UV is used because it does not damage the DNA). Individual fragments may be extracted from the gel for sequencing, PCR amplification, or cloning (described later).

Table 27.1
Recognition Sequences and Cutting Sites of Selected Restriction Enzymes

Enzyme	Recognition Sequence
AluI	A G↓C T
BamHI	G↓G A T C C
BglII	A↓G A T C T
ClaI	A T↓C G A T
EcoRI	G↓A A T T C
HaeIII	G G↓C C
HindII	G T Py↓Pu A C
HindIII	A↓A G C T T
HpaII	C↓C G G
KpnI	G G T A C↓C
MboI	↓G A T C
PstI	C T G C A↓G
PvuI	C G A T↓C G
SalI	G↓T C G A C
SmaI	C C C↓G G G
XmaI	C↓C C G G G

The problem of determining how a set of restriction fragments is normally interconnected may be systematically resolved by determining the sequences of another set of fragments cut with a different restriction enzyme. The overlapping information thereby obtained should permit a determination of the complete sequence of the intact genome. Overlapping information can be obtained also by comparing the digestion products from a DNA sample that has been partially digested with one that has been allowed to go to complete digestion with the same enzyme. These two approaches are used to different extents with different chromosomes.

DNA Cloning

While restriction enzymes permit DNAs to be cut into well-defined fragments, other enzymes permit DNAs from different sources to be linked together to form recombinant DNAs (table 27.2). DNA to be amplified by cloning is linked to a plasmid or a virus that can be extensively replicated. After amplification the DNA of interest is cut from the plasmid or virus with the

Figure 27.12

Cleavage map of the SV40 genome. The zero point of the map is the unique *Eco*RI site. For clarity, the circular genome is shown opened at the R1 site, and the cleavage sites (and resulting fragments) for each restriction enzyme are indicated on a separate line.

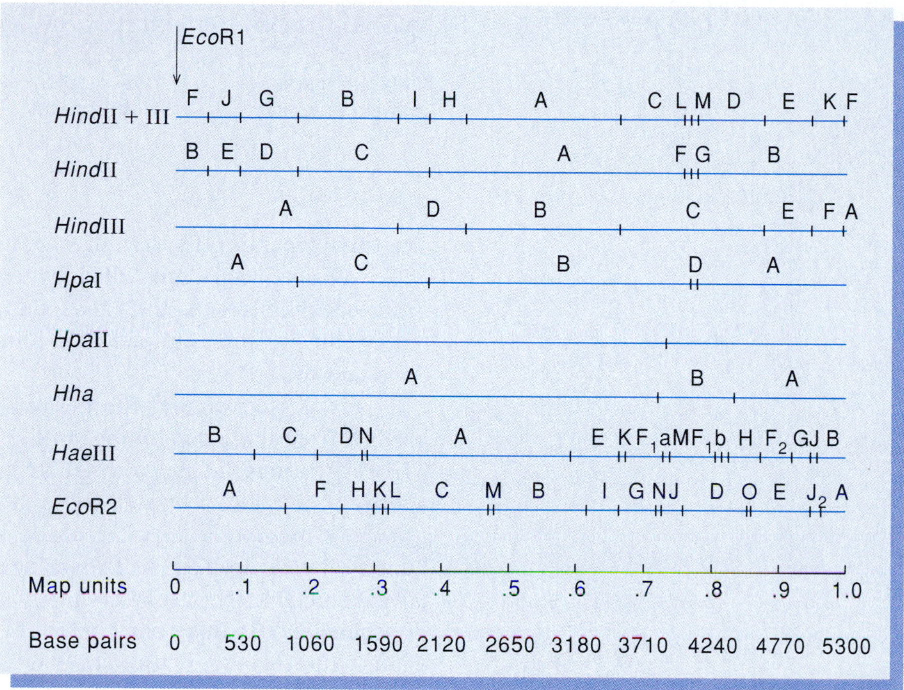

appropriate restriction enzyme and reisolated by electrophoresis. This method of amplification is more work but also opens up more possibilities than the PCR method that we described earlier.

Plasmids Are Used to Clone Small Pieces of DNA

In the basic procedure for DNA cloning, an autonomously replicating plasmid and a DNA insert are cut with a restriction enzyme and then the pieces are annealed and covalently joined by the action of DNA ligase. The resulting recombinant molecules are then transfected into *E. coli,* where they replicate. When plasmid vectors are used, a population of permeabilized cells is bathed in the plasmid DNA containing the inserted DNA. Since only a small number of cells become transfected by this procedure, there needs to be a way to select cells that carry the desired hybrid plasmids. A particularly useful plasmid vector for selecting transfected cells, called pBR322, is itself a hybrid plasmid (fig. 27.13). This plasmid contains two genes, amp^r and tet^r, which confer resistance to penicillin and tetracycline, respectively.

To begin, *Pst*I restriction fragments of foreign DNA may be inserted into the unique *Pst*I restriction site on pBR322 (see fig. 27.13). This is done by digesting pBR322 with *Pst*I, mixing the product of this step with the restriction fragments

Table 27.2
Some Enzymes Used in DNA Recombinant Methodology

Enzyme	Function
Polynucleotide ligase	Joining of two DNA molecules
Polynucleotide kinase	Adding a phosphate to the 5′-OH on a polynucleotide
Terminal transferase	Adding homopolymer tails to the 3′-OH ends of a linear duplex
DNA polymerase I	Repair synthesis
λ exonuclease	Stepwise removal of nucleotides from 5′ ends of a duplex to expose the 3′ ends
Type II restriction endonucleases	Specific cleaving of DNAs to produce staggered ends or blunt ends
Alkaline phosphatase	Removing terminal phosphates from either the 5′ or 3′ end or both
Reverse transcriptase	Synthesizing a DNA copy of an RNA molecule
S1 endonuclease	Removing unpaired single-strand regions

Figure 27.13

Structure of the pBR322 plasmid (*a*) and construction of a hybrid plasmid containing the pBR322 vector and a segment of foreign DNA (*b*). For pBR322 the unique sites for various restriction enzymes are indicated. Also indicated are the locations of the tetracycline (*tet*ʳ) and the ampicillin (*amp*ʳ) resistance genes and the origin for DNA replication. The hybrid plasmid is constructed by treating the plasmid and the foreign DNA with the *Pst*I restriction enzyme and mixing the two DNAs together in the presence of DNA ligase.

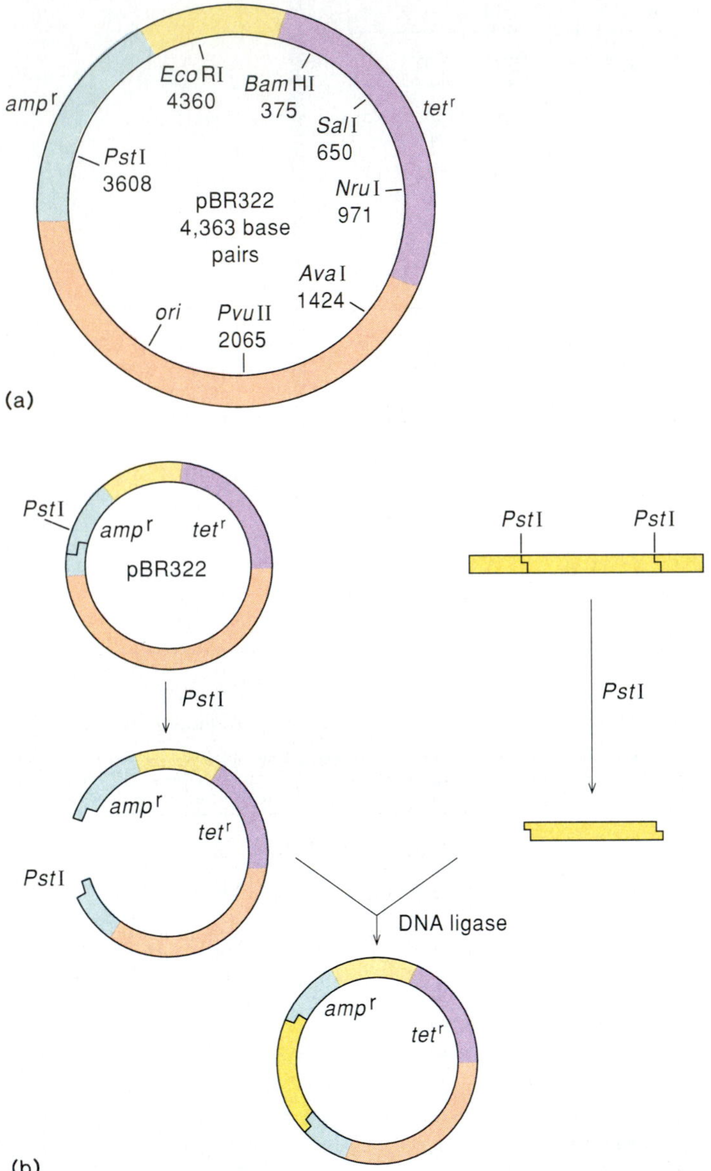

Figure 27.14

Replica plating technique for detecting hybrid plasmid-containing cells.

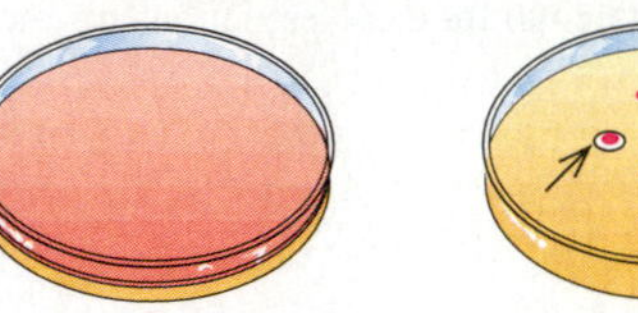
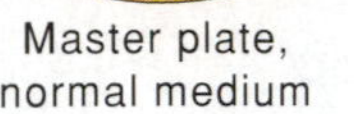
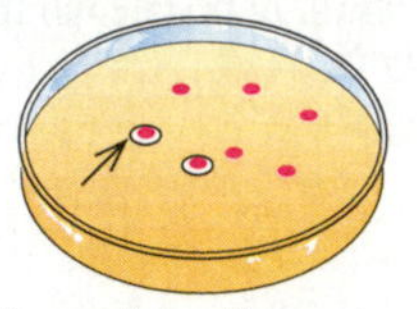
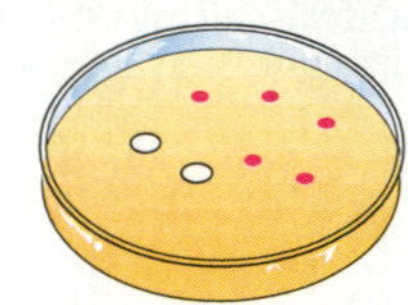

to be cloned at low temperatures, to permit annealing to take place between the two DNAs, and finally ligating the annealed fragments with DNA ligase. The end product will contain some of the original pBR322 and some pBR322 with the inserted foreign DNA. When this mixture is used in transfection, most cells will not be transfected, some will be transfected with pBR322, and some will be transfected with the desired hybrid plasmid.

The three types of cells will differ in their drug resistance properties. Normal cells are killed by tetracycline or penicillin. Transfected cells with the DNA inserted in the plasmid are tetracycline-resistant but penicillin-sensitive, since the insert has disrupted the *tet*ʳ gene.

Cells containing the desired plasmids can be distinguished from those containing pBR322 by replica plating (fig. 27.14). On semisolid agarose plates containing growth medium, a large population of treated bacteria is spread and allowed to grow overnight (see fig. 27.14). A seemingly homogeneous "lawn" of cells develops on the surface of the gel. Actually, the lawn results from the growth of many microcolonies to the point of confluency. At this point a piece of velvet is lightly pressed against the surface of the plate and this impression is transferred to other agarose plates containing growth medium with tetracycline or penicillin plus tetracycline. Only the transfected cells will produce colonies on the plates containing antibiotics, and because of their small number, each of these will give rise to readily detectable clones. The clones that are present on the tetracycline-containing plates but missing on the penicillin plus tetracycline plates are the ones most likely to contain the desired hybrid plasmids (see fig. 27.14). These clones are usually plucked from the tetracycline plates and retested to eliminate any uncertainty about the original drug testing.

Once their identity is confirmed, the appropriate plasmid-containing cells are grown in liquid culture. After a moderate density of growth is achieved, the plasmid DNA is selectively amplified by overnight growth. Plasmid DNA replication continues for several hours, until each cell contains 1,000 to 2,000 copies of the small circular plasmid DNA. This DNA is readily separable from the host DNA and can be characterized by its rate of migration on gel electrophoresis (fig. 27.15), or other more specific tests, to see whether it contains the inserted DNA sequence. If desired, the inserted sequence may be removed from the plasmid vector by digestion with *Bam*HI, the restriction enzyme used in the initial construction of the hybrid plasmid. The cleaved fragments can be readily separated by gel electrophoresis.

Plasmids such as pBR322 are excellent, easy-to-use vectors, but other types of vectors offer specific advantages. Let us now look at some of these alternatives.

Figure 27.15

Electrophoretogram of restriction enzyme digests of pBR322 and pBR322 with a DNA insert at the *Pst*I site. The insert is assumed to have no internal *Bam*HI restriction sites. In channels A and B the pBR322 is predigested with *Pst*I and *Bam*HI, respectively. The resulting DNA migrates with the same mobility because the plasmid has one site for each of these enzymes and therefore has the same molecular weight. In C and D the hybrid plasmid containing a DNA insert is treated with *Bam*HI and *Pst*I, respectively. In C the hybrid plasmid has been linearized by one cut at the *Bam*HI site in the *amp*^{r} gene. It runs more slowly than the pBR322 as it is larger because of the DNA insert. In D the plasmid cuts at two *Pst*I sites located between the pBR322 sequences and the insert sequences. Consequently one segment migrates at the rate of a linearized pBR322 plasmid. The other segment, also linearized, migrates at a rate characteristic of the size of the DNA insert. The electrophoresis is run from left to right; fragments are stained with ethidium bromide and photographed with UV light.

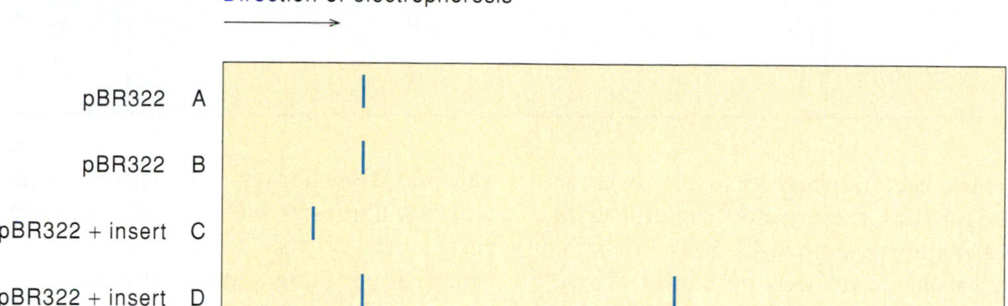

Bacteriophage λ Vectors Are Useful for Cloning Larger Segments of DNA

Bacteriophage λ possesses a number of advantages as a cloning vector. DNA fragments as large as 24 kb can be propagated using this vector. The primary pool of clones can be amplified by limited phage growth as plaques, and the entire collection of phage clones (recognized as clear plaques) can be stored for long periods in a small volume (fig. 27.16).

Since λ phage will not accommodate molecules of DNA that are much longer than the viral genome, this phage cannot be used as a vector for cloning substantial DNA fragments unless a significant portion of the viral DNA is removed beforehand. Fortunately, the central third of the genome contains genes that are not essential for plaque production, and it can therefore be deleted. We now have available a number of bacteriophage λ vectors that accommodate foreign DNA fragments generated by a variety of restriction endonucleases. The recombinant DNA molecules that incorporate some of these vectors can be introduced directly into *E. coli* by transfection. Alternatively, recombinant DNA molecules can be packaged into phage particles and subsequently infected into suitable host cells.

Figure 27.16

The nutrient agar plate contains a continuous lawn of *E. coli* bacteria except for circular clearings that represent phage plaques. Each plaque was originally derived from a single phage particle infecting a single *E. coli* bacterium.

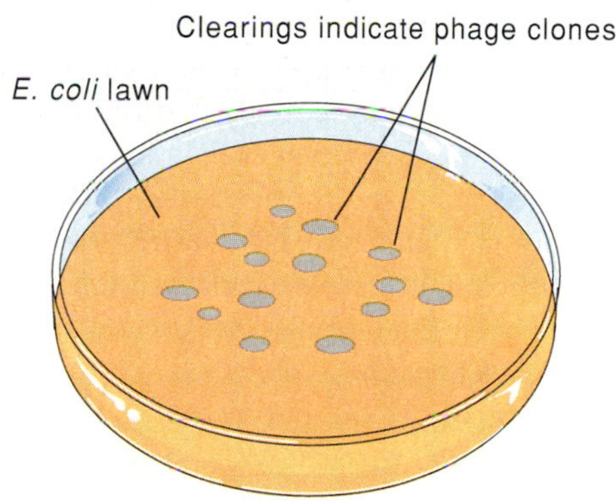

Cosmids Are Used to Clone the Largest Segments of DNA

While plasmids and bacteriophage λ are both useful vectors, the size of the DNA fragments that can be cloned in them is limited. With plasmids, the larger the fragment of foreign DNA inserted, the lower the efficiency of ligation and transfection, and thus the cloning of DNA fragments larger than 15 kb is experimentally difficult. Even in λ vectors, the length of the nonessential region of λ DNA limits fragment size to 24 kb or less. Also, the original λ vectors do not allow propagation of viable bacterial cells that carry the inserted DNA fragment; the insert is propagated as part of a virus that lyses the cell.

For cloning large DNA fragments, the vectors needed are cosmids. The first part of their name, *cos,* comes from the fact that cosmids contain the cohesive ends or *cos* sites of normal λ. These ends are essential for packaging the DNA into λ phage heads. The last part of their name, *mid,* indicates that cosmids carry a plasmid origin of replication like the one found in the pBR322 plasmid.

Cosmids can be used for cloning in the same way as any other plasmid vector. However, because cosmids also contain the *cos* sites, cosmid DNA along with an inserted DNA fragment can be packaged as a λ phage. The result after packaging is a defective but nevertheless infectious phage particle. Once the cosmid and the inserted DNA fragment are introduced by infection into a λ-sensitive cell, the plasmid replicates.

Table 27.3
Theoretical Number of Clones Required to Fully Represent the Entire Genome of Various Organisms

Size of Cloned DNA Fragment (bp)	Genome Size (bp)		
	2×10^6 (e.g., bacteria)	2×10^7 (e.g., fungi)	3×10^9 (e.g., mammals)
5×10^3	400	4,000	600,000
10×10^3	200	2,000	300,000
20×10^3	100	1,000	150,000
40×10^3	50	500	75,000

Since cosmids lack the entire bacteriophage genome except for the region adjacent to the *cos* sites, these vectors can propagate exogenously derived DNA fragments of up to 40–50 kb in length.

Several cosmid vectors have been developed. These differ in size (and therefore cloning capacity), useful cloning sites, and selectable properties, or markers. For example, the pJB8 vector is a 5.1-kb plasmid containing several possible cloning sites and the amp^r selectable marker. Its cloning capacity is limited to DNA fragments between 30 and 47 kb. A feasible protocol for using this vector is described in box 27A.

Shuttle Vectors Can Be Cloned into Cells of Different Species

Vectors that include replication systems derived from more than one host species are known as shuttle vectors. Such vectors commonly include a replication system able to function in *E. coli* and one that works in a second host, which may be bacterial or eukaryotic. Initial cloning and amplification of the DNA segment to be studied is often carried out in *E. coli* because it is easier to make large quantities in a culture of *E. coli*. The recombinant DNA molecule—consisting of the "bifunctional vector" plus the cloned segment of DNA—is then introduced into the second host, where the purpose is usually to measure the expression of the genes carried by the vector. Shuttle vectors that can replicate in both *E. coli* and yeast are the most common.

Constructing a DNA Library

Cloning can produce a single vector-linked DNA fragment or a whole collection of independently isolated vector-linked DNA fragments derived from a single organism. Such a collection, termed a library, can serve as a good source of well-defined sequences from a given organism, since each clone of a library harbors a purified sequence from that organism. Within the entire library some sequences may be repeated but other sequences may be missing. The ideal library, which can only be approached, represents all of the sequences with the smallest possible number of clones.

As we saw at the outset of this chapter (see fig. 27.1), a library can be prepared in two ways. The genome may be fragmented and ligated to the appropriate vector to produce a genomic DNA library. An alternative approach is to construct a cDNA library, in which the DNA fragments are obtained by reverse transcription from the cellular RNA. Each type of library has advantages and disadvantages, and for a specific purpose one library is usually preferred over the other. In either case, the vast majority of DNA within the library is uncharacterized. Therefore the task of finding the desired genes or sequences within a library is often harder than the task of constructing the library. We will describe ways of selecting the correct clones after we have discussed the two types of libraries in more detail.

A Genomic DNA Library Contains Clones with Different Genomic Fragments

A major problem in constructing a genomic DNA library is to maximize the probability that all segments of the genome will be represented. If the genomic DNA is prepared by cutting with a restriction enzyme, an added problem is the possibility that the enzyme will cleave genes of interest at one or more sites.

To increase the likelihood of isolating desired genes in one piece, different restriction enzymes can be used on different parallel preparations. But even if the genes of interest are not cut by the enzyme(s) chosen, the DNA fragments produced may be inconveniently small to work with. An enzyme that recognizes a sequence of six bases (a 6-cutter) gives an average fragment size of 4,096 bp*, which is a reasonable size for making a plasmid library but much smaller than the size desirable for cloning in λ or a cosmid vector. Therefore, when large randomly generated fragments are desired, the method of choice usually entails making an incomplete digest with a 4-cutter restriction enzyme, which produces overlapping ends that can be readily cloned into the chosen vector as described earlier. The extent of digestion is controlled so that cleavage occurs at only some of the restriction enzyme recognition sites and the average size of the fragments produced is in the desired range. The conditions used thus depend on whether the product is going to be cloned in a plasmid, a λ phage, or a cosmid vector.

Table 27.3 gives the minimum number of clones (that is, the size of the library) required to fully represent the entire genome in a genomic DNA library, as a function of the average

$*(1/4)^6 = 1/4096$

27A
BOX

Construction of a Cosmid

The pJB8 vector is a 5.1-kb plasmid containing a λ cos site, an *amp*ʳ selectable marker, an origin of replication, and several possible cloning sites. To construct it, first one of the staggered-end ligation sites of the plasmid was opened with the appropriate restriction enzyme (step 1). In this case the *Bam*HI site was chosen. The linearized vector was then treated with alkaline phosphatase (step 2) to prevent recircularization of the vector in the subsequent ligation. The remaining procedures were dictated by the goal of this protocol, which was to produce very large cloned segments of DNA. Thus ligation had to yield large DNA fragments that were produced by random cutting (and thus were representative of the entire genome) and that could be inserted into the *Bam*HI site of the cosmid vector. On a random basis, any very large DNA fragment is likely to be cut several times by even a 6-cutter restriction enzyme under conditions of complete digestion. In order to avoid this problem and to obtain maximum randomness of fragments of the appropriate size, the *Mbo*I restriction enzyme was used under conditions of very limited digestion. This enzyme recognizes the four-base sequence of GATC, whereas *Bam*HI recognizes a six-base sequence, GGATCC. The two enzymes produce identical overlapping fragments (see upper right of illustration). Thus the eukaryotic fragments should ligate efficiently with the *Bam*HI restricted cosmid vector (step 3). Because limited digestion conditions were used, only some of the *Mbo*I (and *Bam*HI) sites were cleaved, and these varied from molecule to molecule on a random basis. Thus some molecules in the population were likely to contain an uncleaved enzyme recognition site at any given position. After ligation, the resulting concatamers were packaged *in vitro* into λ particles and introduced into a suitable *E. coli* strain (step 4). Transformants were then selected with ampicillin.

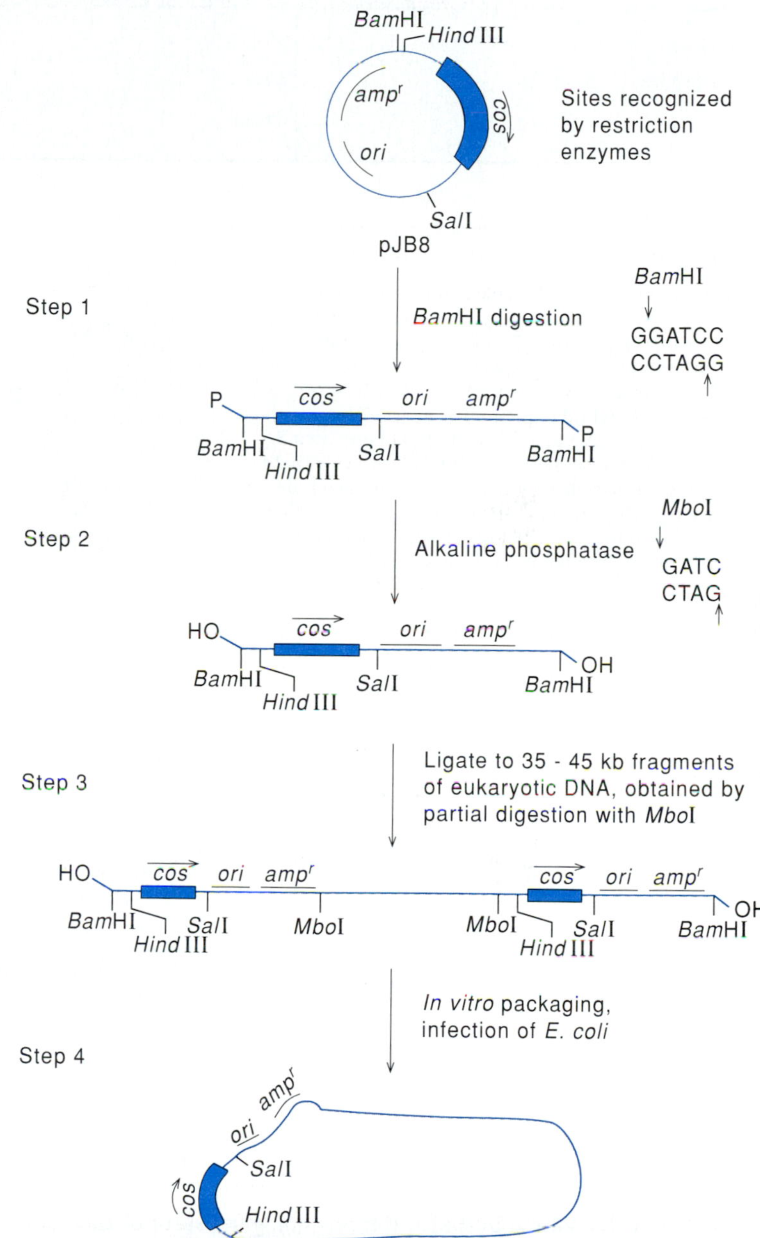

Figure 1

A plasmid containing a λ *cos* site is linearized by treatment with the BamHI restriction enzyme (step 1). The linear structure is treated with alkaline phosphatase to remove the 5′ terminal phosphates (step 2). Eukaryotic restriction fragments are ligated to the linearized plasmid in the presence of a large excess of the latter (step 3). The concatamer shown, which contains two *cos* sites, packages sufficiently into a λ phage that can be introduced by infection into an *E. coli* cell. The transfected structure readily circularizes (step 4). (From Geoffrey Zubay, *Genetics*. Copyright © 1987 Benjamin/Cummings Publishing Company Inc., Menlo Park, Calif. Reprinted by permission of the author.)

size of the cloned fragments and the size of the genome. Since DNA fragments in a population are cloned on a random basis, there is a 50% chance of finding a given single-copy gene in a library of the indicated size. A clone bank should be three to ten times the minimum size to give a high probability that a particular segment will be represented.

A cDNA Library Contains Clones Reflecting the mRNA Sequences

A cDNA library consists of a collection of clones that contain DNA copies of the cellular or organismic RNA. If the RNA is obtained from a differentiated multicellular organism, then the library varies in composition according to the type of cell used

Table 27.4
Distribution of Population of mRNA Molecules of Typical Eukaryotic Cells

Abundance Class	Fraction of the Total mRNA Population in Each Abundance Class	Number of Different mRNA Sequences in Each Abundance Class	Number of Copies of a Unique mRNA Sequence Present per Cell
High	22%	30	3,500
Medium	49%	1,090	230
Low	29%	10,670	14

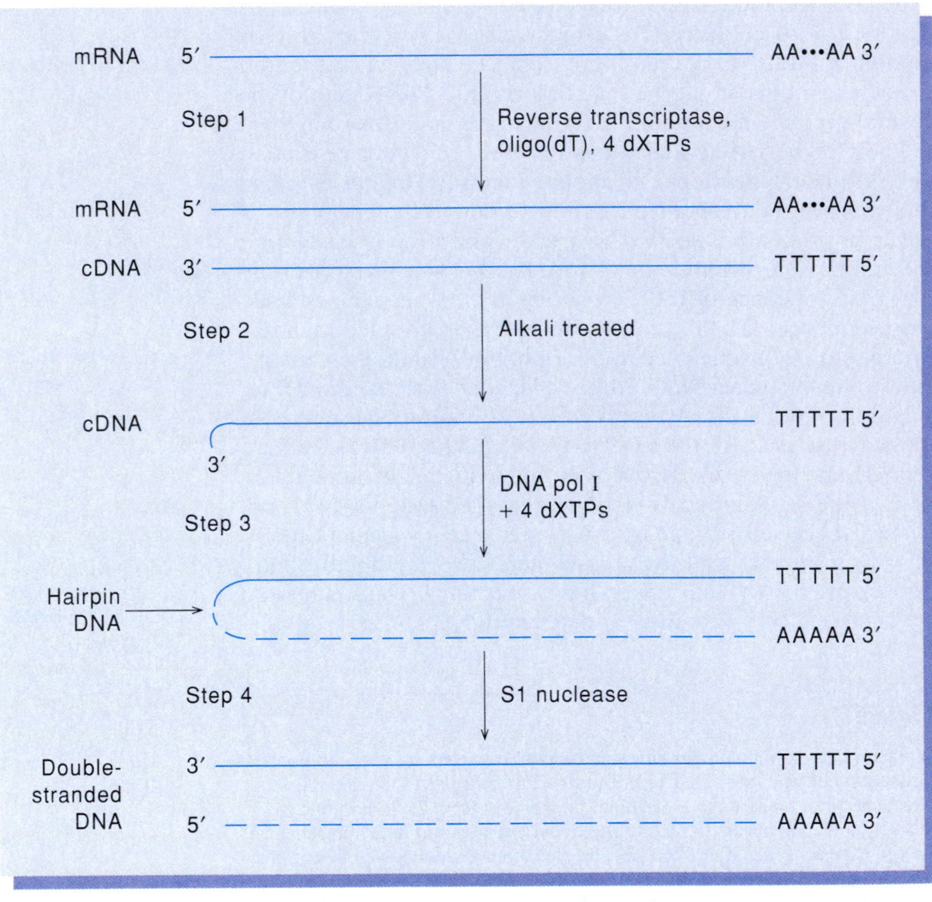

Figure 27.17

Formation of a cDNA duplex from a poly(A) mRNA. Step 1: To the poly(A) mRNA are added reverse transcriptase, oligo(dT), and the four-deoxynucleotide triphosphates. In favorable circumstances this step results in a full-length DNA-RNA hybrid duplex. Step 2: The RNA is removed and degraded by alkali treatment. Step 3: The remaining single-stranded DNA serves as both template and primer for the synthesis of a complementary DNA strand. The newly synthesized DNA is linked to the primer template DNA by a hairpin turn. Step 4: The hairpin is cleaved with S1 nuclease.

as the RNA source and also the physiological state of the cell. This variation is a reflection of the relative abundances of particular messenger RNAs made by different cell types. If a cDNA species corresponding to a particular gene product is desired, it is often possible to select a cell type known to synthesize a large amount of the corresponding mRNA. Thus pituitary cells can be used if cDNA encoding growth hormone is desired, and liver cells can be used if a serum albumin cDNA is the goal. mRNAs present in low amounts will clearly require the screening of a larger library than mRNAs present in medium or high abundance (table 27.4).

Sometimes some prepurification of the crude RNA is possible to increase the chances of finding a particular cDNA in the library. A technique commonly used with RNA from eukaryotic cells is to purify the crude RNA by chromatography over a cellulose column containing oligo(dT) chains. This procedure enriches for mRNA because most eukaryotic mRNAs contain a 3′-poly(A) tail that sticks to the oligo(dT) chains on the column.

The poly(A) tail not only aids in the fractionation step, but it also provides an important feature for the subsequent synthesis of the cDNA (fig. 27.17). The crude poly(A)-mRNA

Figure 27.18

Insertion of cDNA into pBR322 plasmid by the homopolymer tailing method. The cDNA duplex is tailed with terminal transferase and dCTP. pBR322 is opened with the *Pst*I enzyme and tailed with G residues. The two DNAs spontaneously anneal when mixed. The complex is transfected directly allowing for the repair synthesis and ligation to take place *in vivo*.

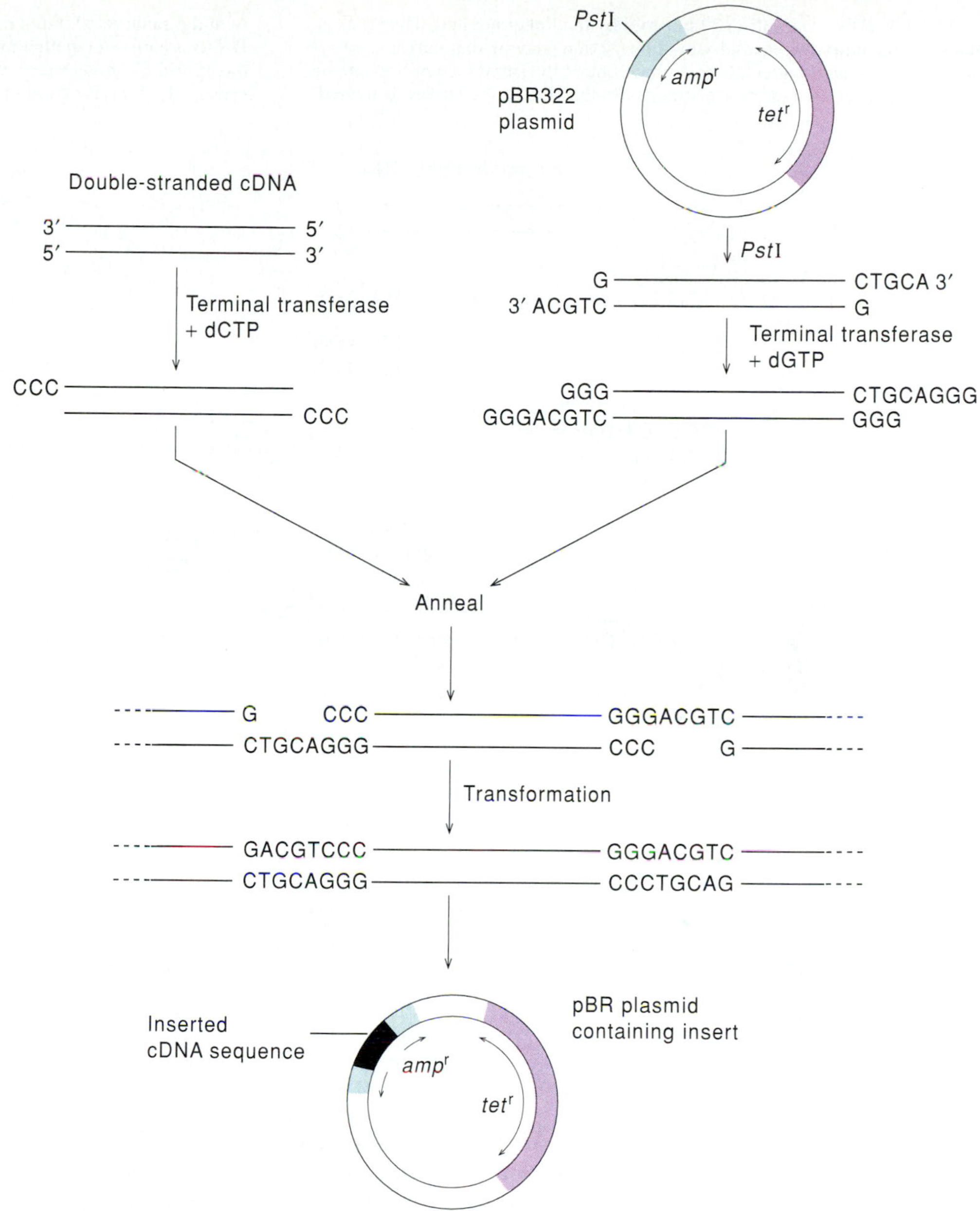

fraction serves as a template for the synthesis of the first strand of the cDNA, using the enzyme reverse transcriptase (step 1, fig. 27.17). When poly(A)-mRNA is the template, a short poly(dT) chain makes an ideal primer as it complexes with the poly(A)-containing portion at the 3′ end of the RNA. Synthesis of the polynucleotide chain then proceeds along the mRNA template in the 5′ direction. The product of this reaction is an RNA-DNA hybrid duplex, which, in favorable cases, extends the full length of the mRNA. The RNA strand is removed from the duplex by alkali treatment (step 2, fig. 27.17) and the reisolated single strand of enzymatically synthesized DNA is used as a template for the synthesis of a complementary second strand of DNA by DNA polymerase I (step 3, fig. 27.17). This enzyme also has a requirement for primer; commonly the 3′ end of the first strand of DNA seems to be able to loop around and serve as a primer and template for its own synthesis, leading to the formation of a duplex DNA molecule having its strands joined together at one end like a hairpin. The next step (step 4, fig. 27.17) in the preparation of a cDNA duplex involves scission of the hairpin. This can be done by brief treatment with S1 nuclease, which acts as an endonuclease on single-stranded regions of the DNA.

After S1 cleavage, appropriate ends are attached to the newly synthesized duplex to make it suitable for insertion into a plasmid or other vector. There are two ways of doing this. One is called homopolymer tailing. The double-stranded cDNA is tailed with one base, and a suitable plasmid vector such as pBR322 is tailed after opening with the complementary base. A detailed protocol for this is shown in figure 27.18. The double-stranded cDNA is 3′-tailed with C residues. The plasmid is

Figure 27.19

Insertion of cDNA into pBR322 plasmid by the linker method. The strategy here is to open up the plasmid with a restriction enzyme that makes staggered cuts and to attach linkers that contain the same recognition site to the cDNA. After the linkers are attached to the cDNA, the duplex is treated with the same restriction enzyme (*Pst*I) to expose the overhangs. The two DNAs are mixed together and ligated. After transfection, cells containing the hybrid plasmids are recognized by tetracycline resistance and ampicillin sensitivity. Identification of the insert is discussed in the text.

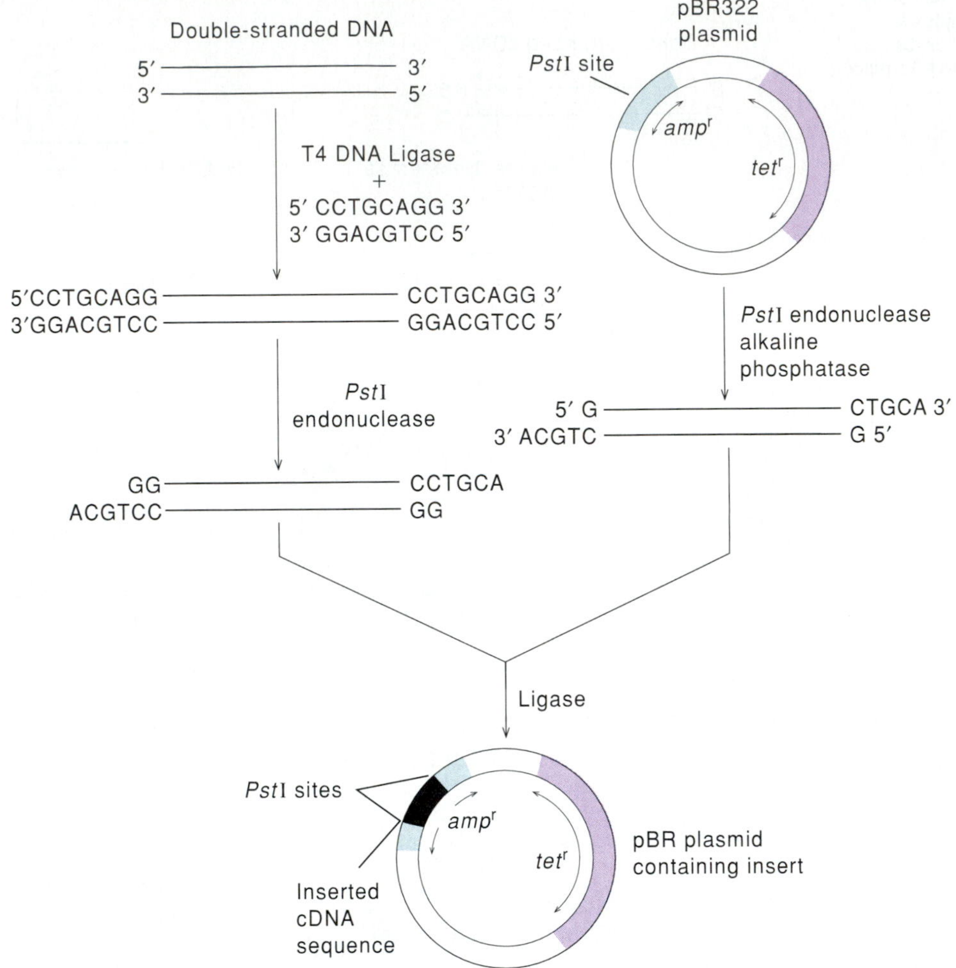

linearized by treatment with *Pst*I restriction endonuclease at a unique site in the *amp*ʳ gene. This treatment results in a linear duplex with a four-base 3′-OH overhang, TGCA. This is 3′-tailed with G residues. The two-tailed duplexes are mixed together, whereupon they spontaneously anneal. The annealed complex can be directly transfected, leaving the final folding-in to take place in the bacterial cytoplasm.

Another way of inserting the cDNA into a vector is to tail it with linkers. Linkers are synthetic single-stranded oligonucleotide segments (6, 8, 10, or 12 bases in length) that self-associate to form symmetrical blunt-ended double-stranded molecules containing the recognition sequence for a particular restriction enzyme. Figure 27.19 shows an eight-base linker (CCTGCAGG) containing a *Pst*I recognition site. This linker self-associates to produce an eight-base blunt-ended duplex structure that adds to the double-stranded cDNA in the presence of T4 ligase. The resulting product is treated with *Pst*I to produce the characteristic 3′ overhang. The plasmid, linearized with *Pst*I, and the two DNAs are mixed and reacted with ligase to produce plasmid with the insert. Considerable plasmid will reclose without incorporating the insert, but this material can be distinguished from plasmid with the insert because the latter has lost its ampicillin resistance. Alternatively, the linearized plasmid DNA can be treated with alkaline phosphatase to remove the 5′ phosphates from its termini. Phosphate removal prevents the plasmid from recircularizing. When the terminally dephosphorylated duplex is reacted with the cDNA insert in the presence of ligase, a circular structure with two nicks is formed. Upon introduction into a bacterial cell, these nicks are mended.

Although a cDNA library requires more work to prepare than a genomic DNA library, it has major advantages for certain purposes:

1. Since only a small fraction of the genomic DNA of eukaryotes is expressed at any one time in any one cell type, the variety of clones is greatly reduced. If the desired genetic sequence is among those cloned, screening is much easier.

2. cDNA contains only a small number of repetitive sequences, whose presence can complicate screening when cloned DNA is used as a probe. A probe with repetitive sequences is likely to anneal to any structure that has similar repetitive sequences, whereas the goal is usually to use a probe that will anneal to a unique sequence in the genome.

3. Because intervening sequences or introns* present in eukaryotic DNA are not present in the RNA (and consequently in the cDNA), the cloned cDNA frequently can be used in *E. coli* or other microorganisms to obtain the proteins encoded by the eukaryotic gene. A DNA containing introns is not properly expressed in *E. coli*.

4. A single cDNA clone is pure in the sense that it contains only the sequences corresponding to a single mRNA. For this reason the results obtained when using the cDNA probe are often easier to interpret than data from probing experiments that employ cloned genomic DNA segments, which sometimes may contain two or more different genetic sequences.

There Are Numerous Approaches to Picking the Correct Clone from a Library

A library can contain thousands or even tens of thousands of different kinds of clones (see table 27.3), and the screening needed to isolate a clone that has the DNA of interest can take a lot of time. Cloned DNAs do not usually include easily selectable genetic markers, such as a gene conferring resistance to a specific antibiotic.

Most currently used procedures for screening large numbers of colonies for plasmids or phage that contain specific DNA inserts are variants of the colony hybridization method developed by Grunstein and Hogness. This procedure makes use of a specific radioactive probe that contains some sequences complementary to those in the DNA of interest. The colonies to be screened are first grown on agar Petri plates (fig. 27.20). A replica of each plate is made on another agar plate, which is stored for reference. A replica is also made on a nitrocellulose filter. The colonies formed on the filter are lysed and the contents denatured simultaneously by treatment with sodium hydroxide. After heating, the denatured DNA is fixed on the filter at each site where a colony was located. The DNA on the filter is then hybridized with a radioactively labeled nucleic acid probe complementary to the specific DNA sequence to be selected. The presence of hybridized probe at sites occupied by DNA derived from colonies that include the DNA fragment of interest is detected by autoradiography on film. The colony whose DNA hybridizes with the nucleic acid probe can then be picked from the reference plate, which contains a viable bacterial colony at a corresponding location.

A parallel procedure developed by Benton and Davis makes it possible to screen bacteriophage λ recombinant DNA clones by hybridization to single plaques *in situ*. The number of phage plaques that can be placed on a single Petri plate is much larger than the number of individual bacterial colonies that can be placed on the same size plate (about 10,000 phage plaques versus about 200 bacterial colonies); therefore the method is especially useful for screening large libraries of eukaryotic DNA for genes that may be present at a low frequency (i.e., single-copy genes).

Figure 27.20

Colony hybridization procedure used to identify bacterial clones harboring a plasmid containing a specific DNA. Step 1: Replica-plate the colonies containing plasmids onto nitrocellulose paper. Step 2: Lyse cells with NaOH and fix denatured DNA to paper. Step 3: Hybridize to ^{32}P-labeled DNA carrying desired sequence and autoradiograph the product. Locations of desired DNA should be emphasized in autoradiograph. Clones carrying desired plasmids (circled) may then be isolated from a corresponding agar replica plate carrying untreated colonies.

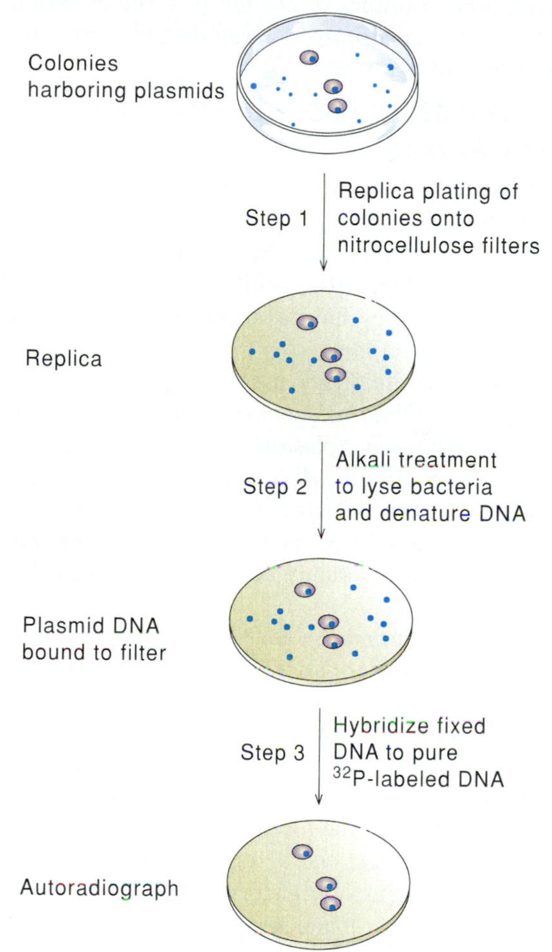

In cases where no nucleic acid probes are available, antibodies directed against the gene-encoded protein have proved valuable in the primary screening of cDNA expression libraries. The technical details of this approach are discussed in references given at the end of this chapter.

Cloning in Systems Other Than *E. coli*

Despite the success and broad applications of *E. coli* cloning systems, gene products cannot always be made in this bacterium. Sometimes they are not synthesized in their entirety, or they are rapidly broken down after synthesis. In addition, for

*Introns are noncoding regions located between coding regions. They are much more common in eukaryotes than prokaryotes (see chapter 28).

the study of certain processes indigenous to other species (e.g., photosynthesis, antibiotic production) it is often necessary to use a host bacterial species that carries out the process naturally, rather than *E. coli*.

Effective gene cloning systems are available for a variety of bacterial hosts, including *Bacillus subtilis, Streptomyces* species, and *Agrobacter tumefaciens.* Cloning systems have also been developed for eukaryotic hosts. In this section we will consider examples of cloning in three eukaryotic systems—the yeast *Saccharomyces cerevisiae,* mammalian cells in tissue culture, and plant cells.

Yeast Is the Most Popular Eukaryotic Cell for Cloning

Simple and generally useful techniques exist for isolating and amplifying virtually any yeast gene in *E. coli.* Moreover, by transformation these genes can be returned to the yeast cell. Such procedures have greatly facilitated the molecular dissection of yeast genes; they have also provided a variety of new approaches to gene mapping.

Yeast gene isolations usually begin by taking fragments of total yeast DNA and cloning them into an *E. coli* vector, either λ bacteriophage or a plasmid. After transfection or transformation, respectively, *E. coli* cells containing specific regions of the yeast DNA are selected. Cells containing the desired fragments are then amplified, purified, and sometimes modified, after which the fragments may be returned to yeast by transfection.

The first isolation of a gene encoding a protein in yeast took advantage of the fact that many yeast genes function in *E. coli.* For example, a plasmid containing an origin for replication in *E. coli* and the yeast *LEU2* gene was selected from a yeast plasmid library by its ability to restore to a leu^- mutant of *E. coli* the ability to grow in the absence of added leucine. The purified clone containing the *LEU2* gene could then be returned by transfection to a yeast cell defective in *LEU2.* Subsequent analysis showed that stable transformants contained the *LEU2* gene, integrated at the homologous site in the yeast genome. Thus the yeast recombination system, like the *E. coli* recombination system, appears to have the capacity to guide a gene introduced by transfection to a site of homology. Selection methods now exist for isolating mammalian cells showing homologous recombination of transfected sequences (see Selected Readings article by Mansour et al.).

The transformation frequency in yeast, which is usually quite low, may be greatly increased by incorporating special yeast DNA sequences into the plasmid used for transfection. The best-known of these special sequences derives from the so-called 2μ plasmid found in most *Saccharomyces* strains. Addition of a fragment of the 2μ plasmid into a plasmid containing a selectable gene, such as *LEU2,* increases the transformation frequency a thousandfold or more. High-frequency transformation using all or part of the 2μ plasmid is usually due to autonomous replication of the plasmid, which is maintained in multiple copies. Other types of high-frequency transformation have been observed when certain yeast sequences, referred to as autonomously replicating sequences (ARS), are an integral part of the transforming plasmid. These probably contain a yeast chromosomal origin for DNA synthesis (see the discussion of ARS sequences in chapter 26).

Integrative transformation with specially constructed plasmids has opened the door to many forms of gene mapping that would be much more difficult to do by conventional mapping techniques. In addition, it has permitted the construction of unusual gene arrangements. For example, if a yeast-integrating plasmid contains a gene to be transposed and a segment of yeast DNA homologous to another location on the yeast gene, then it should be possible to transpose the selectable gene to another position on the genome.

Studies on Cloned Genes in Mammals Start with Tissue Culture Cells

Mammalian cells from various sources can be adapted for growth as single cells in liquid culture or on plates. Such cells can be formally treated like bacteria or yeast; single cells give rise to genetically homogeneous colonies. Cultured cells have proved to be effective recipients of cloned DNAs.

The simplest procedure for transfection of mammalian cells utilizes purified DNA. The DNA is mixed with calcium chloride and sodium phosphate to give a finely divided calcium phosphate–DNA precipitate. Treated cells appear to take up the DNA by endocytosis. Transfection efficiencies, measured as the fraction of treated cells that become transfected, vary from 10^{-4} to several percent by this method.

Cultured mammalian cells permit many types of studies that would be impractical to carry out on whole animals. However, only a few mutants provide readily selectable genetic markers; this limitation creates a problem in selecting for transformed colonies. One of the best systems available for selection involves the thymidine kinase (*tk*) gene of herpes simplex virus. Mutants that are tk^- have been isolated for a number of cell types, including mouse, rat, and human. Such cells cannot grow on the selective medium known as HAT. The tk^- cells can be transformed to tk^+ cells by plasmid DNA that contains the herpes simplex virus *tk* gene. Other genes of interest can be inserted into the same plasmid vector so that selection of tk^+ cells after transformation leads to a high probability of coselection of adjacent genes.

One of the most versatile experimental systems using DNA transfection is the mouse teratocarcinoma cell (TCC) system. With this system it is possible to transfect cells in tissue culture as in other tissue culture systems and then to implant the cells into a growing embryo so that the cells become part of the animal. By this procedure we can study the effects of cells altered by transfection in any organ of the whole mammalian organism.

The teratoma is a unique type of tumor found in many kinds of mammals. It is composed both of neoplastic cells, such as occur in other tumors, and also of many kinds of differentiated cells. A typical teratoma may include nerve cells, muscle cells, blood cells, skin cells, and other differentiated cells all mixed together with neoplastic stem cells. Only the stem cells are neoplastic, producing more stem cells or more differentiated cells, usually both. In many ways the stem cells behave like embryonic cells. They can be dissociated into single cells and cultured *in vitro* like bacterial cells. Cells grown from the culture can be reintroduced into the animal by subcutaneous implantation, in which case they produce a tumor. Most remarkably, they can be introduced into an early embryo (blastocyst) to produce a hybrid chimera where, in favorable cases, all of the tissues possess some cells from the tumor parent (fig. 27.21). Even egg cells have been isolated that are derived from the tumor parent. These egg cells, when fertilized by normal sperm, result in progeny that could truly be said to have a tumor for a mother. This example shows that the neoplastic stem cells are capable of reverting to completely normal behavior when subjected to the embryonic environment.

Taking advantage of this quite remarkable result, Bea Mintz and her co-workers have shown that teratocarcinoma (TCC) stem cells can function as vehicles for the introduction of specific recombinant genes into mice. For this purpose, TCC cells were first grown in single-cell tissue culture, and cells with a thymidine kinase deficiency were selected and treated with DNA containing the human β-globin gene and the thymidine kinase (*tk*) gene. Then tk^{+}-transformed cells were selected with the help of HAT medium as explained earlier. Further tests revealed that the majority of the transformants also had copies of the human β-globin gene. Thus, although there is no facile selection procedure for the human β-globin gene, it can be successfully cotransfected by linking it to the *tk* gene. Such altered TCC cells can then be introduced into early embryos to engineer new mouse strains carrying the human β-globin gene. This experiment shows that virtually any gene can be transfected into the whole animal by this two-step procedure.*

Oncogenes Can Be Selected from a Genomic Library by Subculture Cloning

Sometimes a gene can be identified when it confers an obvious phenotype on a mammalian cell. However, even in such cases indirect selection procedures often must be used because of the limitations in the cloning vehicles currently available for mammalian cells. In a strategy called subculture cloning, a genomic library of mammalian DNA is made in *E. coli* and then clones from the library are tested in subgroups in mammalian cells until one exhibiting the desired properties is identified. The DNA of interest is then propagated in *E. coli*. This selection method is illustrated in figure 27.22 for the isolation of an oncogene from a chicken lymphoma. A library containing 200,000 phage clones was divided into sublibraries, and the DNA of the clones in each sublibrary was introduced by transfection into mouse fibroblast cells growing in tissue culture. Cellular transformation was assayed by the appearance of a clone or clones of rapidly dividing cells (foci) superimposed on a background of slowly dividing, nontransformed cells. The sublibrary that assayed positively was further subdivided and retested. This procedure was repeated until a clone was identified that conferred the rapid growth properties associated with oncogene transformation.

Subculture cloning depends on the sensitivity of the assay used for detecting a small number of positive clones in a heterogeneous population. Alternatives to subculture cloning

Figure 27.21

Manipulation of mouse teratoma cells. The cells from a teratocarcinoma may be dispersed and grown in tissue culture. These cells can be injected into an embryo (A), in which case the resulting animal is a chimera in which some cells come from the original parents and others arise from the cells injected into the blastocyst. (B) Alternatively, these cells may be implanted subcutaneously, in which case the animal develops a tumor at the site of implantation.

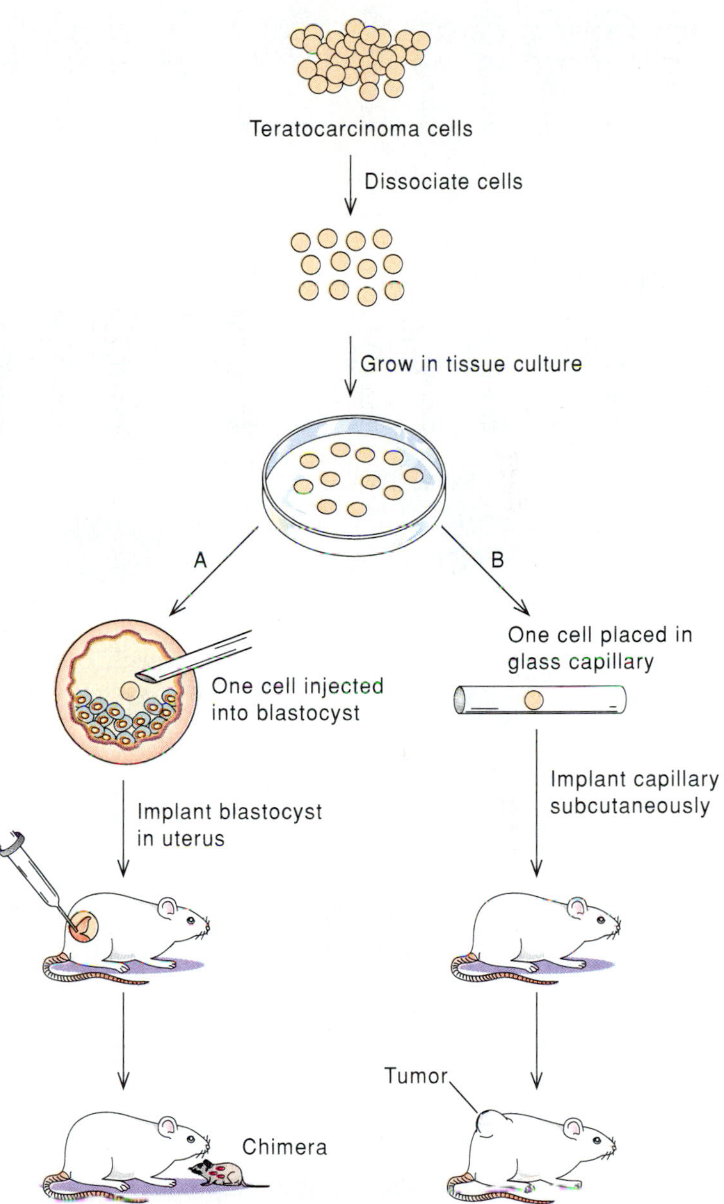

*Embryonal stem cells (ES cells) are frequently used in place of TCC cells.

Figure 27.22

Subculture cloning procedure. Subculture cloning is used in successive steps to isolate the clone that contains the gene of interest. In this case the gene of interest was an oncogene originating from a chicken lymphoma and identified by the rapid growth characteristics it confers on transformed cells. A λ library was made with the chicken lymphoma nuclear DNA. The library was subdivided into ten approximately equal lots (sublibraries). From each lot the DNA was isolated and used in a transformation assay. Lot 2 scored positively. This lot was further subdivided and the procedure was repeated again and again until a pure plaque carrying the oncogene was isolated.

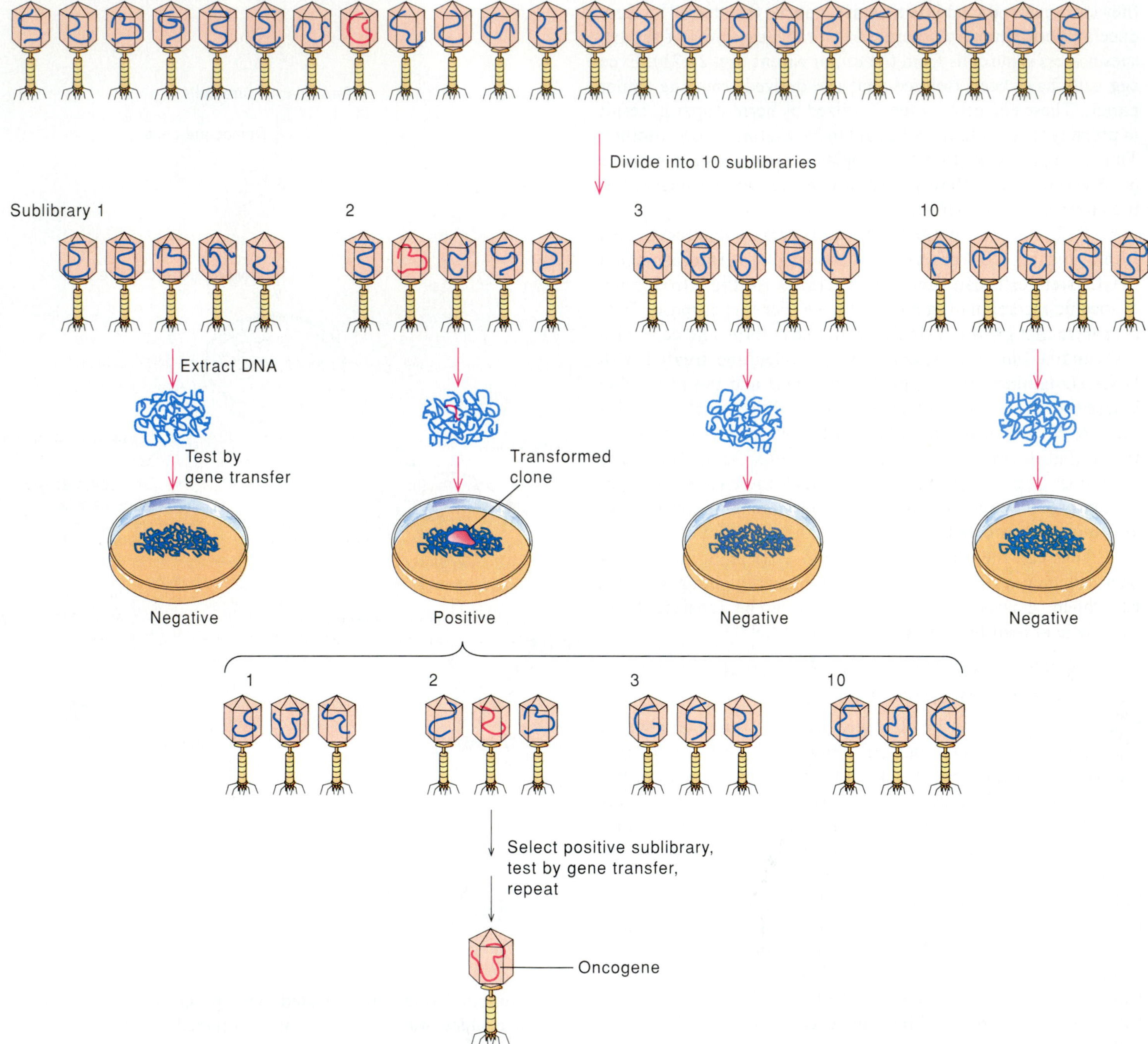

have been devised, but these methods still clone genes in bacteria and then detect them in mammalian cells. However, papilloma viruses, which can function as true plasmid vectors in mammalian cells, may permit the direct cloning and selection of genes in mammalian cells.

Cloning in Plants Has Been Accomplished with a Bacterial Plasmid

Tissue culture systems for plants are hard to manipulate and grow for extended periods of time. The difficulties of cloning DNA in plants are compounded by the fact that there is no plasmid that can be introduced directly into plant cells. Much plant DNA cloning uses a naturally occurring host-vector system involving *Agrobacter tumefaciens,* a pathogenic soil bacterium that causes crown gall tumors in a wide variety of dicotyledonous plants. (It is noteworthy that this is the only known case of naturally occurring nucleic acid transfer between a prokaryote and a eukaryote.) These plant tumors are caused by tumor-inducing (T1) plasmids in some agrobacteria; related bacteria that lack the plasmid are not tumorigenic.

The T1 plasmids are large circular DNA duplexes ranging in size from 160 to 240 kb. Bacterial infection of the plant leads to development of a tumor, usually near the junction of root and stem, as a consequence of a plasmid segment called T-DNA. T-DNA is integrated into the nuclear genome of the plant cells, where it alters the metabolism of the tumor cells so that small molecules known as opines are synthesized in large quantities. Different T1 plasmids encode genes that lead to the synthesis of different types of opines. While these compounds are of no apparent use to the plant's cells, they provide a growth substance for the agrobacteria that contain the T1 plasmid. Other genes carried by the T1 plasmid confer upon the host bacteria the ability to metabolize the T-DNA-encoded opine and use it as a substrate to promote bacterial growth.

It is impractical to use the entire T1 plasmid as a cloning vector because of its size. However, foreign genes can be transferred into plants by introducing them into the T-DNA segment of T1, reinserting the T-DNA into T1, and then letting the natural biological process of T-DNA transfer to the plants cells do the rest (see box 27B). Some plant cells carrying part or all of the T1 plasmid can regenerate whole plants, so that it is possible to measure the effects of transfected genes on any tissues of the whole plant.

Site-Directed Mutagenesis

By combining different procedures of DNA manipulation, we now can make discrete changes in a given gene. This technique, called site-directed mutagenesis, is one of the most important manipulations in modern genetics. The first site-directed mutagenesis studies were carried out by Shortle and Nathans in 1978 with the help of the mutagen sodium bisulfite, which deaminates C residues and thus converts them into U residues. Directed mutagenesis as it is practiced today is based on the chemical synthesis of a deoxyoligonucleotide that contains discrete changes in its sequence from that normally observed in the genome under investigation. These changes may be single base or multibase, and they may involve base changes, base deletions, or base additions.

There are many variations of site-directed mutagenesis. We can start out with a circular single-stranded DNA and anneal it to a synthetic primer DNA carrying the desired changes (fig. 27.23). After extending the primer, we transfect the resulting product. Finally we select clones of cells containing the plasmid with the desired changes.

For most site-directed mutagenesis, however, the polymerase chain reaction (PCR) will probably be the method of choice in the future. The use of PCR for this purpose requires

Figure 27.23

Scheme for oligonucleotide-directed mutagenesis of double-stranded circular plasmid DNA. Supercoiled plasmid circles are nicked in one strand and rendered partially single-stranded by treatment with exonuclease. The gapped circles are hybridized with a homologous oligodeoxynucleotide carrying, by design, some mismatches. *In vitro* DNA synthesis, primed in part by the oligodeoxynucleotide, leads to heteroduplex plasmid circles. (Source: G. Dalbadie-McFarland et al., "Oligonucleotide-directed mutagenesis as a general and powerful method for studies of protein function" in *Proceedings of the National Academy of Sciences USA* 79:6408-6412, 1982. Copyright © 1982 National Academy of Sciences, Washington, D.C.)

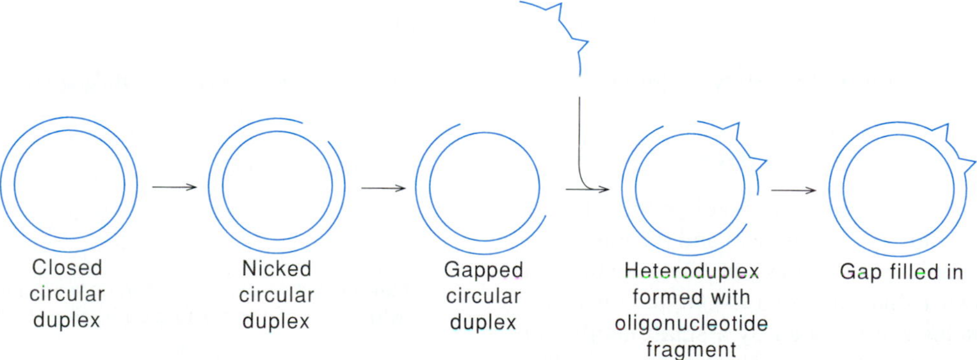

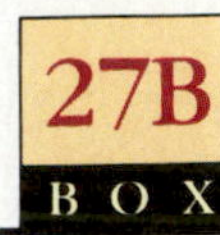

Construction of a T1 Plasmid Derivative Suitable for Cloning

The following procedure is quite complex, but it introduces some new and very worthwhile principles of cloning. First, a segment of T-DNA that included a preselected restriction enzyme cleavage site was cloned into pBR322 for replication in *E. coli* (step 1). The plant gene to be cloned was covalently linked to a selective genetic marker, the kanamycin resistance (*kan*^{r}) gene (step 2). The segment containing both the *kan*^{r} gene and the plant gene was spliced into a T-DNA segment cloned in the pBR322 plasmid (step 3), which was used for amplification of the plant gene in *E. coli.* However, since pBR322 does not replicate in agrobacteria, the segment containing the T-DNA and the *kan*^{r} gene was inserted into the plasmid pRK290, which replicates in either bacterial host (step 4). The modified T-DNA fragment, now attached to pRK290, was introduced by transformation into an agrobacterium that contained a T1 plasmid (step 5).

Transformed cells retained both the T1 plasmid and several copies of the pRK290 plasmid. The T-DNA segment in pRK290 recombines *in vivo* with the homologous T-DNA region in the T1 plasmid, thus inserting the modified T-DNA segment (step 6). But this event occurs spontaneously at a very low frequency, so it was necessary to devise a way of screening for those cells that contained the T1 plasmids with a modified T-DNA segment. A strategy referred to as plasmid eviction was used to accomplish this. It was reasoned that if pRK290 could be evicted under conditions requiring kanamycin resistance, the most likely survivors should be T1 plasmid recombinants containing the modified T-DNA segments with the *kan*^{r} gene. It is known that when two closely related plasmids cohabit the same cell, incompatibility between them results in the loss of one. R751MG2, a plasmid closely related to pRK290 but carrying a gentamycin resistance gene

Figure 27.24

Illustration of a general method of mutagenesis using PCR. Primers are represented as short lines with arrowheads pointing in the 3′ direction. —^— indicates a mismatched base.

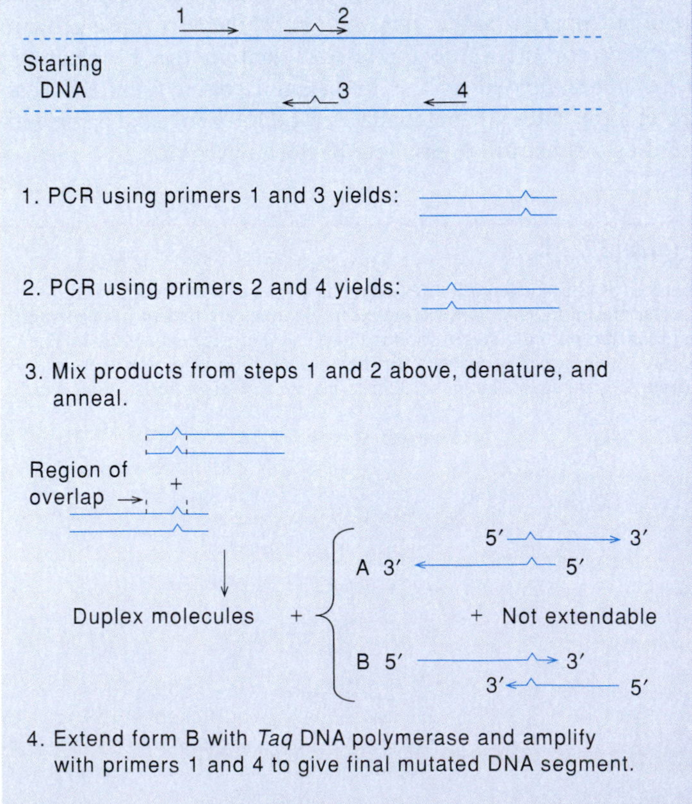

a piece of duplex starting DNA, two outside flanking primers that are perfect complements to segments of opposing strands, and two complementary primers with the desired changes in their sequence as diagrammed in figure 27.24. The steps followed parallel the steps of PCR amplification.

PCR amplification can be coupled with classical cloning methods using cloning vectors. For this purpose the PCR product should contain restriction sites at its ends that are suitable for cloning. Thus PCR amplification and cloning need not be thought of as alternatives for certain purposes but as complementing one another to give a greater variety of approaches.

Recombinant DNA Techniques

The human globin family of proteins is a paradigm for studying differential gene activity during development and the molecular basis of genetic disorders in gene expression. Hemoglobin is a tetramer containing two α-like and two β-like subunits (see chapter 5). These proteins are encoded by a small number of genes that are expressed sequentially during development. The

($gent^r$) and lacking a T-DNA segment, was introduced by conjugation with *E. coli* containing pRK290 and the T1 plasmid. To direct the process so that pRK290 was evicted instead of R751pMG2, the conjugation was done in the presence of both gentamycin and kanamycin (step 7). Only cells that carried both of the drug resistance genes could survive; these cells had lost the pRK290 plasmid, but they retained the kan^r gene in T1 plasmids that had acquired the T-DNA/kan^r gene segment by recombination. The R751MG2 plasmid was also present. The T1 plasmid in the agrobacterium could then be used for transfer of the desired gene to a plant cell system.

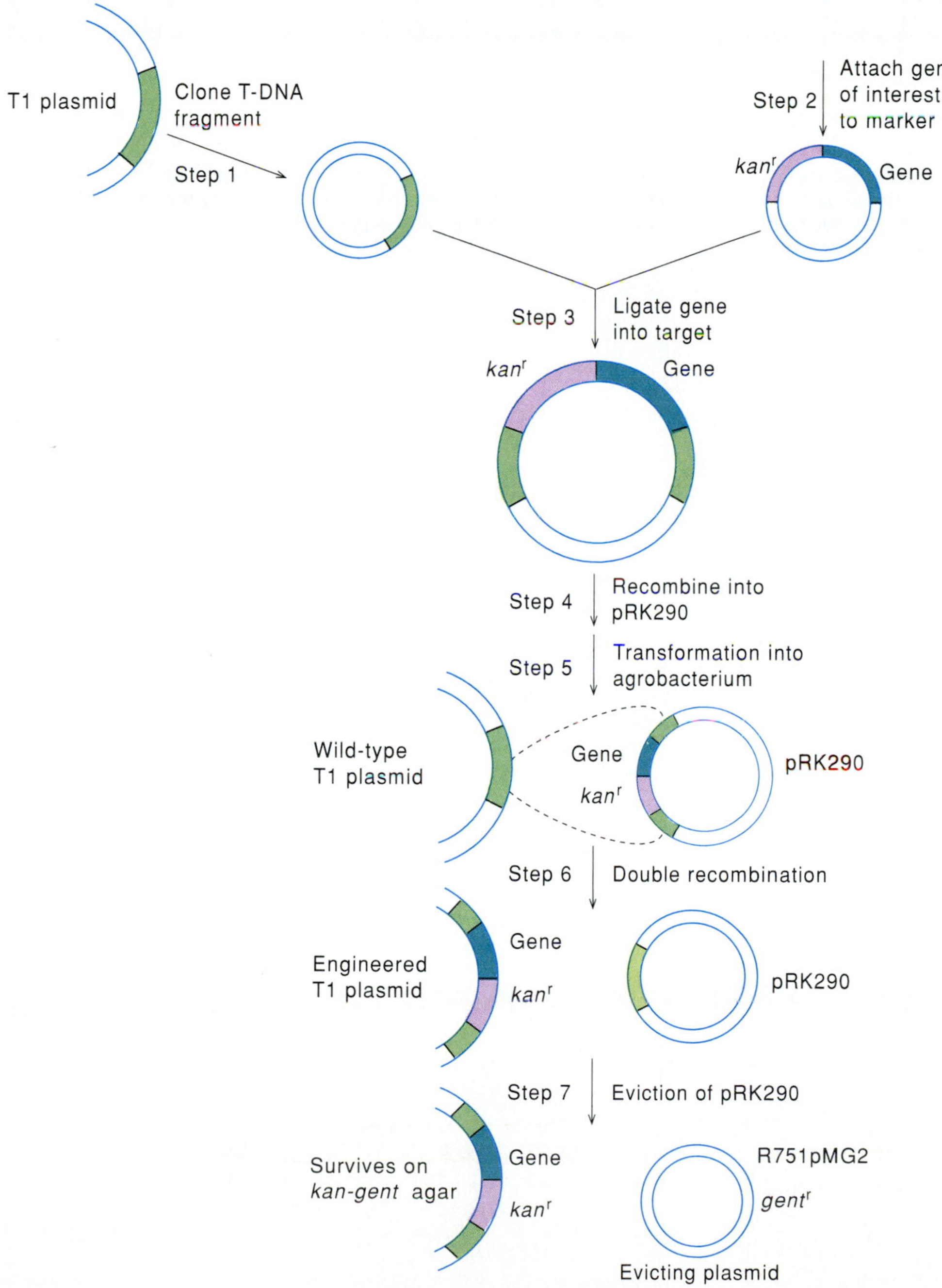

Figure 1

Construction of T1 plasmid for cloning. (From Geoffrey Zubay, *Genetics*. Copyright © 1987 Benjamin/Cummings Publishing Company Inc., Menlo Park, Calif. Reprinted by permission of the author.)

Figure 27.25

Changes in types of hemoglobin observed in early development. A single switch in gene expression is observed for α-like chains. Two switches in gene expression are observed for β-like chains. The corresponding tetrameric hemoglobin molecules observed at different stages in development are also indicated.

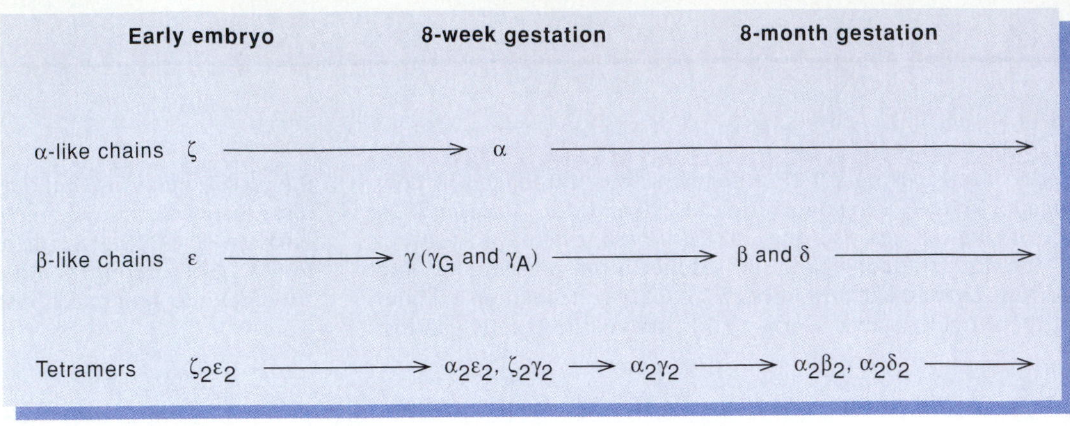

Figure 27.26

The chromosomal localization and genomic organization of the human globin genes. The α- and β-globin gene complexes are positioned on chromosomes 16 and 11, respectively. For each complex, the arrangement of genes on the chromosome is depicted above, and the general structure of the major gene is shown below, together with the location of the intervening sequences or introns (IVS) and codon numbers. Coding regions are shown by solid boxes and IVS regions by open boxes.

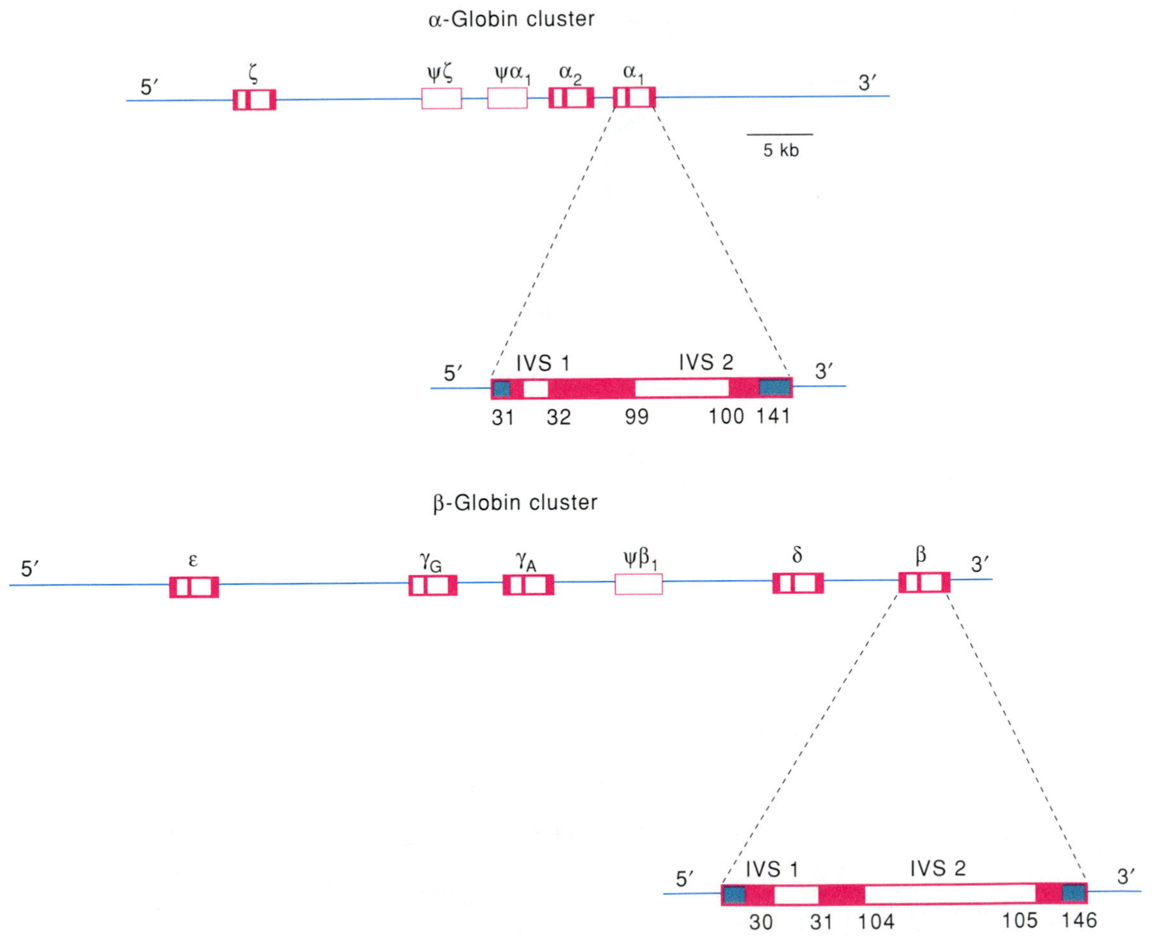

information summarized in figure 27.25 indicates that the α-like and β-like globin gene families have coordinated programs for expression: Two switches in gene expression (embryonic to fetal to adult) are observed for the β-like genes, while a single switch results in activation of adult α-globin production early in fetal life. A combination of classical and recombinant DNA techniques has been used to show that the α-like genes are located in a single cluster on chromosome 16, while the β genes are located in a single cluster on chromosome 11 (fig. 27.26).

cDNA Probes Were Used for Detecting the Genomic Globin Sequences

The first time that the technique of mapping with recombinant DNA probes was applied, it was to the human globin genes. The procedure used was that of DNA hybridization in solution. Radioactive human α- and β-globin cDNA probes reanneal to form stable duplexes with the corresponding human genomic DNA sequences, and do not cross-react with each other or with mouse globin genes. The presence of duplex material in a hybridization reaction containing human globin cDNA and hybrid cell genomic DNA thus indicates the presence of the human globin gene DNA in the hybrid. However, the liquid hybridization procedure has two drawbacks. First, it requires large amounts of DNA. More important, in many cases the donor gene probe does cross-react with the homologous recipient cell gene, so that interpreting the assay is difficult. This disadvantage limits the range of genes for which this mapping procedure is useful.

The Southern blotting procedure avoids both difficulties (fig. 27.27). In this procedure, genomic DNA isolated from somatic cell hybrids is digested with restriction endonucleases to yield specific DNA fragments. These are fractionated according to size by means of agarose gel electrophoresis. The fractions are then denatured, transferred to nitrocellulose sheets, and hybridized with isotopically labeled DNA probes. Noncoding sequences, flanking and intervening, in a structural gene diverge rapidly, so that even genes highly conserved between two species can usually be resolved.

Isolation of the human β-globin gene was greatly simplified because it was possible to obtain the mRNA in essentially pure form from reticulocytes, the cells that carry vast quantities of hemoglobin. From this mRNA a cDNA probe was made and this probe was used to isolate the gene from a genomic library. All members of the human genomic library that anneal to the radioactive cDNA probe were isolated and each of these was cross-hybridized. Restriction-site mapping to each of these DNAs, followed by sequence analysis, resulted in a complete description of the β-globin gene. This analysis showed that the β-globin gene was appreciably larger than the β-globin messenger RNA; in addition to containing regions that are present in the final mRNA, the gene contains two intervening regions not represented in the mRNA sequences (see fig. 27.27). We will have more to say about intervening sequences in the next chapter.

Figure 27.27

The steps involved in assaying by Southern blotting. The DNA to be analyzed is digested with a restriction enzyme (1). The resulting fragments are electrophoresed on an agarose gel (2). The DNA fragments on the gel are transferred to a cellulose nitrate sheet by placing the cellulose nitrate sheet next to the gel and passing solvent through the gel into the sheet. Flow of the solvent is maintained by blotting the far side of the cellulose nitrate sheet with paper towels. The DNA, first denatured with alkali, flows with the solvent but gets stuck in the sheet (3). The sheet is hybridized to radioactively labeled DNA containing the gene sequence of interest (4). The hybridized sheet is autoradiographed to determine the location of the labeled restriction fragment on the gel (5).

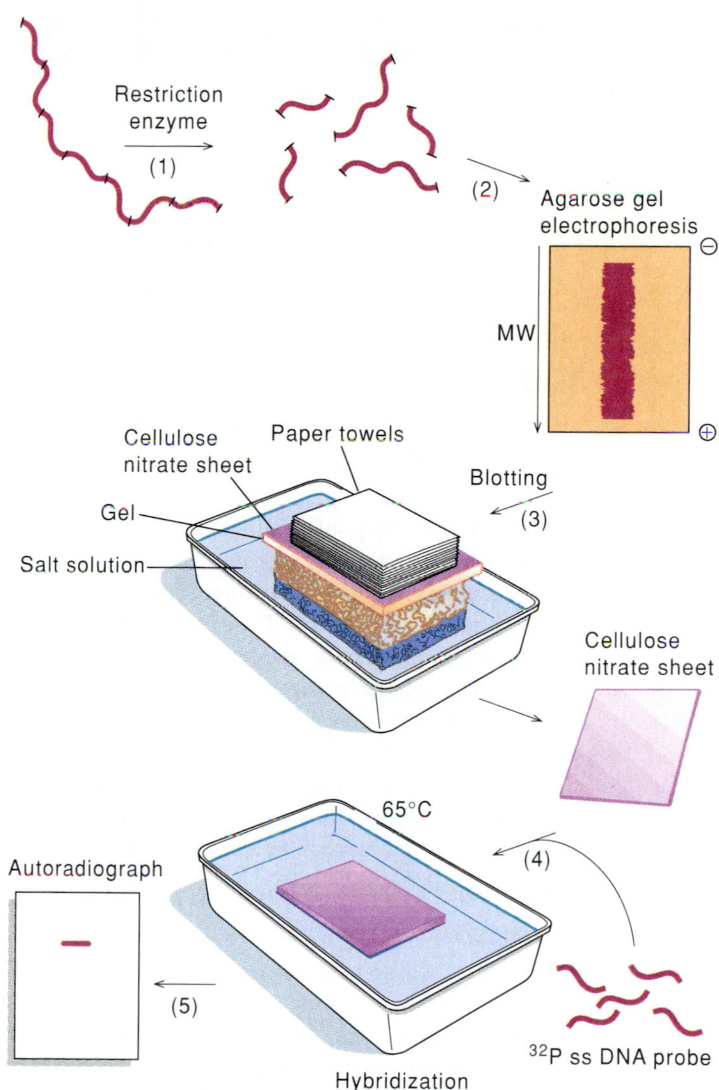

Chromosome Walking Was Used to Identify and Isolate the Regions Around the Adult Globin Gene

Further studies of the β-globin gene resulted in an analysis of the region surrounding the gene. The original cDNA probe could detect only the members of the genomic library that contained sequences homologous to those present in the probe. In order to explore the region flanking the β-globin gene, members of the

Figure 27.28

The linkage map of the human ß-globin gene locus as shown by the structural analysis of overlapping λ genomic clones. Both λHßG1 and λHßG3 clones contained the entire ß-globin gene. Other clones detected by "walking" led to the discovery of other ß-globin-like genes. These included four genes that are expressed and two pseudogenes that are not expressed. The genomic segments of the clones isolated are shown together with the cleavage sites for the enzyme *Eco*R1. The numbers on the top line indicate the size of the fragments in kilobase pairs.

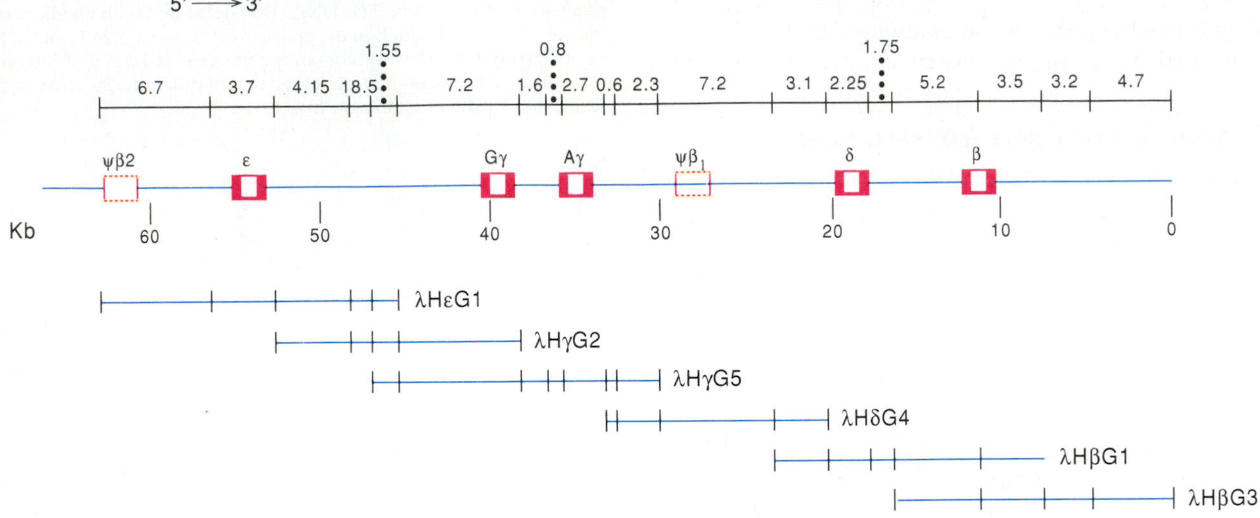

genomic library that hybridized with the original cDNA were themselves converted into radioactive probes, and these were used to locate additional members of the library that contained sequences flanking the β-globin gene. This process led to the detection of sequences bordering the sequences of the β-globin gene. A cyclic repetition of the process resulted in a gradual extension of sequence information in and around the β-globin gene. Using the library in this cyclic manner to extend the map is known as chromosome walking (fig. 27.28). In the case of the β-globin gene the walk was quite rewarding; it was discovered that all other β-globinlike genes were present nearby.

DNA Polymorphism Has Been Used for Prenatal Diagnosis of Sickle-Cell Anemia

A gene represented by two or more different alleles within a population is said to be polymorphic. Allelic differences are caused by sequence differences; in other words, the DNA itself is polymorphic. A DNA polymorphism might be in the coding (protein-specifying) region or in a portion of the gene that regulates the amount of gene product or elsewhere. In fact, DNA polymorphism appears to be a general phenomenon that results in sequence differences between homologous regions of DNA in different individuals. These differences can be detected even when no other differences can be found.

Operationally, then, DNA polymorphism should be recognizable by sequence differences. But rather than directly sequencing a gene to detect polymorphisms, differences called restriction fragment length polymorphisms (RFLPs) can be used to highlight rapidly the regions of sequence differences between individuals. The technique depends on the presence of a polymorphic marker—a particular restriction enzyme recognition site present in some individuals but not in others. That site will cause the length of fragments cut by a given restriction enzyme to be smaller than it is in individuals not possessing the marker.

Kan and Dozy in 1978 were the first to discover allele-specific linked polymorphism in the globin genes. Using the *Hpa*I restriction enzyme, they showed that the normal β-globin gene was contained within a 7.6-kb *Hpa*I fragment, but that the β-globin gene of sickle-cell anemia, Hb^S, was contained within a 13-kb fragment (fig. 27.29). Further analysis showed that the RFLP in this case resulted from a *Hpa*I restriction site, 5 kb to the 3′ side of the β-globin gene, that was present in the normal case and absent in Hb^S. A possible explanation for why this RFLP was associated with the sickle-cell trait is that the population, after a mutation, diverged into the 7.6-kb type and a few 13-kb types before introduction of the sickle-cell gene. After the Hb^S mutation was introduced into the 13-kb type, it became greatly expanded because of the selective advantage of the sickle-cell trait in certain populations.

Knowledge of this polymorphism could be used for pre- or postnatal diagnosis for the sickle-cell gene. Such information has been of practical value in genetic counseling. Incidentally, it is now possible to diagnose sickle-cell disease with greater precision because the point mutation leading to the Hb^S protein itself produces a recognizable RFLP.

Figure 27.29

Inheritance pattern of an RFLP associated with sickle-cell disease. Humans carry two alleles for the same gene, and each offspring inherits one allele from each of its parents in an entirely random fashion. Normal individuals are homozygous for normal *Hb* alleles; individuals with sickle-cell trait are heterozygous, with the one normal *Hb* allele and one Hb^S allele; and individuals with sickle-cell disease are homozygous for the Hb^S allele. At the top (*a*) we see a three-generation pedigree analysis for a family that carries both the normal and the sickle-cell gene for ß-globin. Males are represented by squares and females by circles. An open pink circle or square indicates an individual who is homozygous normal. A half-filled circle or square (pink/purple) indicates a heterozygous individual with sickle-cell trait. A filled circle or square (purple) indicates a homozygous individual with sickle-cell disease. In (*a*) both sets of grandparents produce a heterozygous individual with sickle-cell trait. Since one of the grandparents is homozygous normal and the other is heterozygous, there is a 50% chance that the grandparent mating will give rise to a heterozygous offspring as shown and also a 50% chance that they will have normal offspring (not shown). The two heterozygous parents have an increased chance of having abnormal offspring because in this mating each parent carries one abnormal gene or allele. There is a 25% chance of a homozygous sickle-cell anemic offspring, a 50% chance of an abnormal offspring with sickle-cell trait, and a 25% chance of a normal offspring. Below the pedigree chart is the electrophoretic pattern of a *HpaI* digest probed with ß-globin cDNA (*b*). At the bottom we see an interpretation of the normal and abnormal DNAs (*c*).

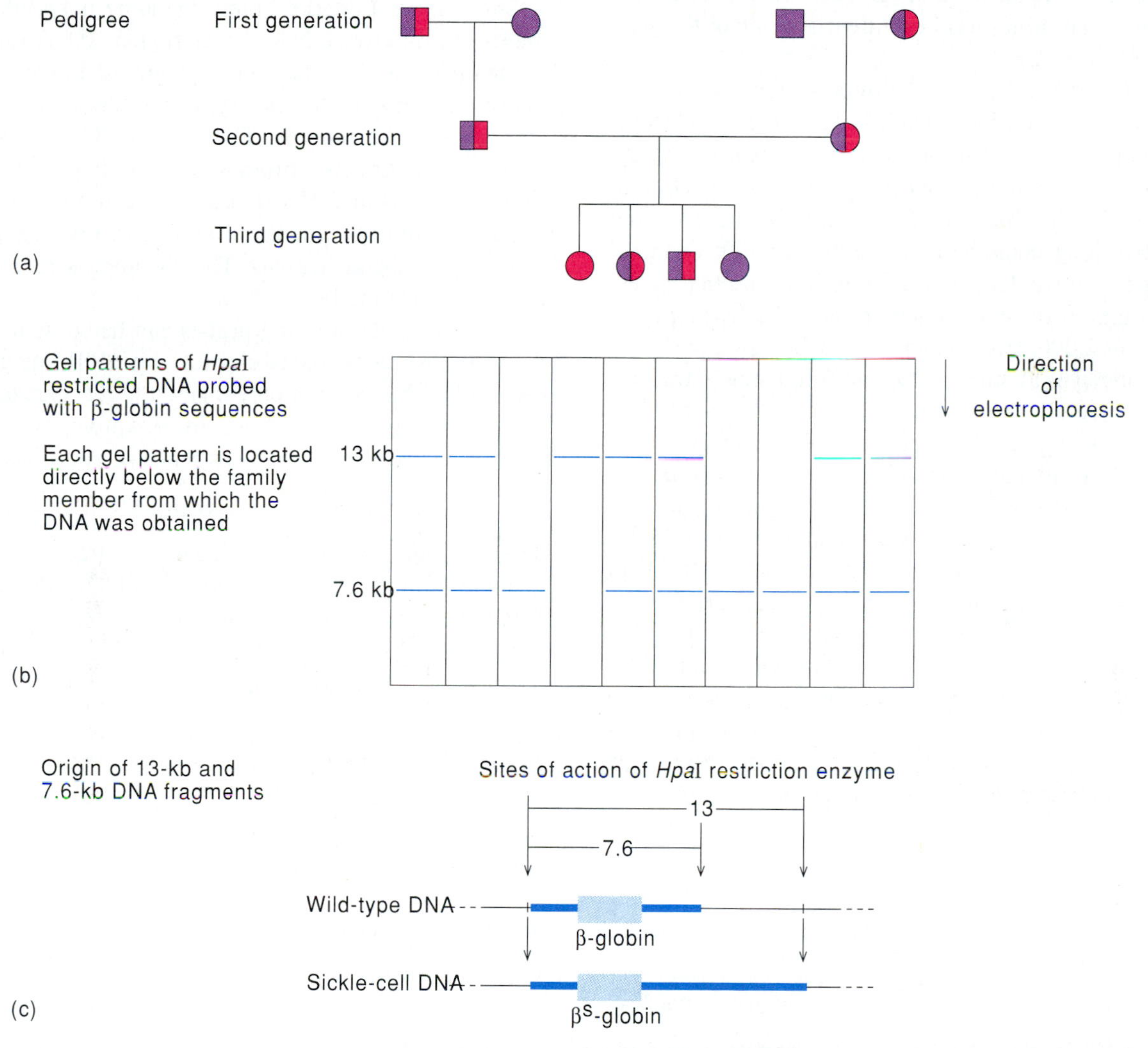

Walking and Jumping Were Both Used to Map the Cystic Fibrosis Gene

By combining the linkage information obtained from RFLP mapping with other DNA manipulation techniques, it has been possible to locate genes causing serious genetic disorders even when these genes are known only from their inheritance patterns. The list of serious disorders that can be linked to single genetic loci is growing. It includes Huntington's disease, Duchenne's muscular dystrophy, polycystic kidney disease, cystic fibrosis, chronic granulomatous disease, peripheral neurofibromatosis, central neurofibromatosis, familial polyposis coli, and multiple endocrine neoplasia. One of the most spectacular achievements in recent times has been identification of the gene causing cystic fibrosis (CF).

Cystic fibrosis is the most common serious genetic disorder in Caucasian populations. The major clinical symptoms include chronic pulmonary disease, pancreatic exocrine insufficiency, and an increase in the concentration of sweat electrolyte. Bearers of this disease, who are readily diagnosed, often die of congestive lung complications before age 30. Pedigree analysis shows that a single gene inherited in autosomal recessive fashion results in the disease syndrome. The frequency of the disease is 1 in 2,000, from which it may be calculated that the carrier frequency is about 1 in 20 (the frequency of the heterozygote for a rare allele is twice the square root of the frequency of the homozygote). By classical genetic analysis (see box 26A), the *CF* locus has been assigned to the long arm of chromosome 7 near the *Met* locus. A map of the region containing the *Met* locus and the *CF* gene is shown in figure 27.30.

In many genetic disorders the analysis must begin before the responsible gene and its protein product are known. Thus it is not possible to locate the gene directly, as in the case of the β-globin gene, and then determine its approximate location in subsequent analysis. For genes such as cystic fibrosis the approximate location is determined by conventional recombination analysis, and then researchers attempt to close in on the gene by the process called reverse genetics.

Detailed genetic mapping by recombination frequency is not practical with human genes that are separated by map distances of less than 1 centimorgan (cM)* because of the small number of test recombinant crosses that are ordinarily available for observation. Unfortunately, 1 cM on the human genome is still equivalent to a physical distance of about 10^6 (1 million) bp. One obstacle to the use of chromosome walking over such long distances is the size limitation of probes. Even cosmids, which provide the largest probes, cannot harbor probes larger than 40 kb, so it would take 25 cosmids end-to-end to span 10^6 base pairs. A walk involving this many probes would take a very long time, if it were possible at all. Another hindrance to such a long walk is that the human genome is sprinkled with segments of repetitive DNA. Those regions interrupt the walk because upon encountering such a region the probe will anneal to many members of the library, which could be situated almost anywhere in the genome. Therefore walking must be replaced by a procedure called chromosome jumping. Probes for jumping carry segments of DNA that are not ordinarily next to one another and that will therefore anneal to two regions that are a considerable distance apart. This distance is determined by how the probes are made.

Suitable jumping probes can be constructed in several ways. In one particularly elegant approach, the genome is first digested with a restriction enzyme that recognizes rare sites in the DNA (fig. 27.31). *Not*I, for example, is a "rare cutting enzyme" that recognizes a sequence of eight bases,

GC↓GGCCGC

The fragments produced by complete digestion of the human genome with *Not*I are, on average, about 500 kb long. These fragments are circularized in the presence of a small marker DNA, so that thereafter the marker DNA is always located between the ends of the original large fragments. The circular

*Genetic loci 1 cM apart recombine 1% of the time (see box 26A).

Figure 27.30

Map of restriction fragment length polymorphisms (RFLPs) closely linked to the cystic fibrosis (*CF*) gene. The inverted triangle near the right-hand end indicates the location of the ΔF_{508} mutation characteristic of most persons with cystic fibrosis disease. (Source: B-S. Kerem et al., "Identification of the cystic fibrosis gene: Genetic analysis" in *Science* 245:1075, 1989. Copyright © 1989 American Association for the Advancement of Science, Washington, D.C.)

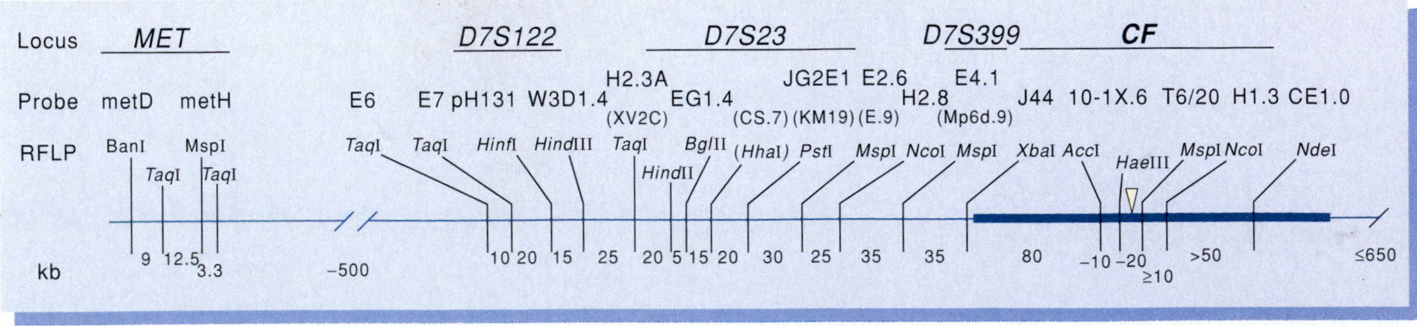

Figure 27.31

Construction of a jumping library (left) and a complementary linking library (right). The jumping library is constructed by exhaustive digestion of total nuclear DNA with a rare cutting *Not*I restriction enzyme. The cut fragments are circularized in the presence of a small gene required for amplification in a special *E. coli* host. The circularized molecules are recut with *Bam*HI restriction enzyme and ligated into a λ vector. The linking library is constructed from the same starting DNA. In this case the DNA is partially digested with *Sau*3A restriction enzyme. The resulting fragments are circularized in the presence of the same small gene required for amplification and the circularized fragments are recut with *Not*I. The linear fragments are ligated into a suitable λ vector and amplified in the special *E. coli* host.

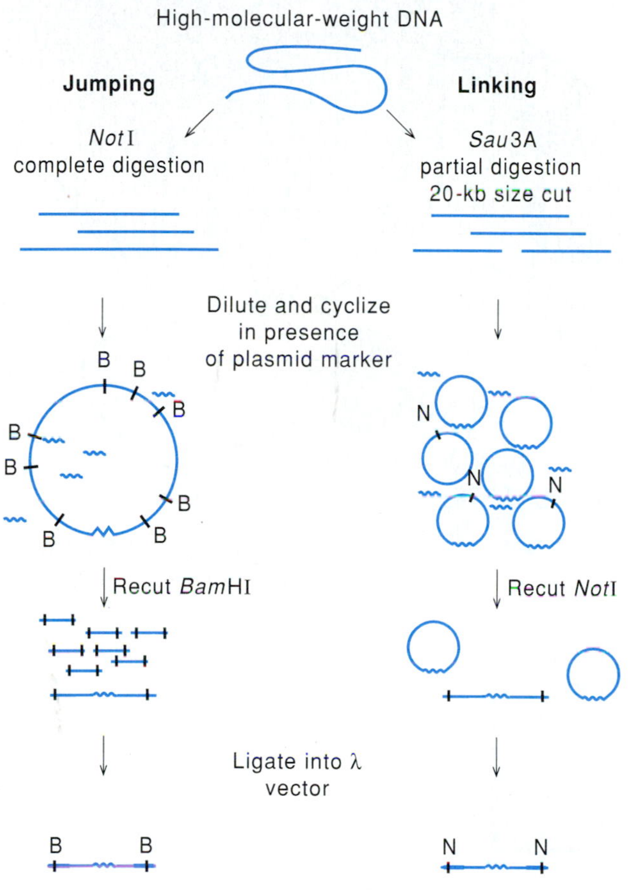

DNA is recut with a restriction enzyme that produces fragments suitable in size for cloning in λ. The marker DNA used contains a gene essential for λ multiplication on a suitable *E. coli* host. In this way the only λ clones that develop are those that carry the marker DNA and hence the ends of the original restriction fragments.

A jumping library is of very limited value on its own because it permits only one jump if used by itself. This is because the members of such a library would not be expected to cross-anneal. Therefore, a jumping library is always used in conjunction with a complementary linking library (fig. 27.31). Starting from the same DNA sample from which the jumping library was made, a partial digest is made with a restriction enzyme that gives considerably smaller restriction fragments, about 20 kb in size. These fragments are circularized in the presence of the same marker DNA that was used to construct the jumping library, and the resulting circular pieces are recut by complete digestion with the *Not*I enzyme. Only the circular pieces containing a *Not*I site are linearized by this treatment, so only these fragments become part of the linking library. The linearization step is followed by ligation of the linear fragments into the λ vector and selection on the same *E. coli* strain that requires the marker DNA for λ replication.

Each member of the linking library carries sequences on both sides of *Not*I restriction sites in the original DNA. In contrast, in the jumping library each member carries sequences from one side of two adjacent *Not*I sites. The jumping library and the complementary linking library are used in strict alternation as illustrated in figure 27.32. Since there are two directions in which jumping can occur initially, the direction is determined by the first annealing. After the first annealing the direction of future jumps is fixed. By using other markers it is possible to determine whether the direction of the first jump was the one of interest. Sometimes several jumps have to be made before the direction can be known.

In the cystic fibrosis investigation, starting from each *Not*I site determined by the jumping-linking library analysis, the sequences surrounding the *Not*I sites were scrutinized with

Figure 27.32

Directional jumping by alternating between a jumping library and a complementary linking library. Because of the way in which these two libraries were constructed (see fig. 27.31) each member of the linking library should contain sequences on both sides of a single *Not*I site, whereas each member of the jumping library should contain sequences on one side of adjacent *Not*I sites. By alternating between the two libraries it should be possible to traverse large regions on the genome relatively quickly and in a unique direction. The process starts with a member of either library that originates from the general region of the cystic fibrosis gene. If the search starts with a member of the jumping library as suggested in the figure, then this member is used to probe the linking library. The member of the linking library found in this way is then used to probe the jumping library, and the process is repeated as many times as necessary.

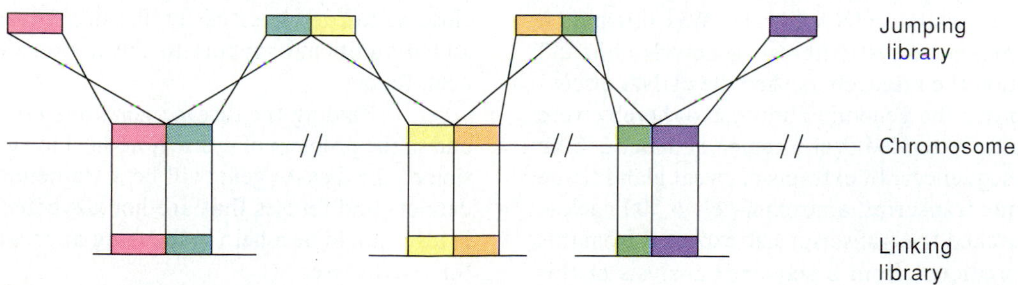

Figure 27.33

Jumping and walking to find the cystic fibrosis gene. Following each jump, the locus defined by the jump was used as a starting point for a chromosome walk. Each DNA segment so found was hybridized to a sweat gland cDNA library until a match was found. It seemed likely that the sweat gland cDNA library would have a good representation of the cystic fibrosis gene transcript because the disease shows sweat gland involvement.

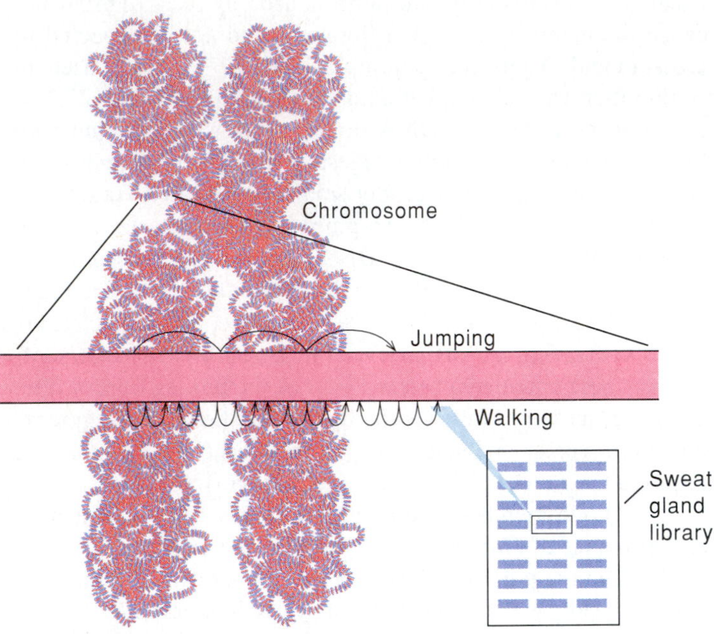

a walking library made from the same starting DNA. The purpose of these local walks was to get as much information as possible about the region, in the hope that by trial and error one of the probes would be the lucky one that overlapped the cystic fibrosis gene. The joint procedure of jumping and walking is illustrated in figure 27.33.

All of this analysis would have been in vain in the absence of some criterion for knowing when the goal of finding the cystic fibrosis gene had been reached. This is where clues from the physiological nature of the condition became useful. In a brilliant strategy a cDNA library was prepared from the messenger RNA fraction of sweat gland tissue. Recall that in cystic fibrosis there is sweat gland malfunction. Therefore it seemed that the mRNA for the cystic fibrosis gene might be well represented in the messenger RNA fraction of the sweat gland cells.

While the walking and jumping process was in progress, each new segment-mapping in the general region of interest was tested against the sweat gland cDNA library. Finally, a member of the walking library was found that annealed with a member of the sweat gland cDNA library. Was this match fortuitous or did it mean the cystic fibrosis gene was in hand? To answer this question the researchers used the cDNA discovered in this way to probe the genomic library, and thereby were able to map a gene that extended over a region of about 250 kb with 23 intervening sequences. In extracts of sweat gland tissue they detected a unique transcript, approximately 6,500 nucleotides in size, that matched the transcript size expected from this gene. The protein predicted from a sequence analysis of this

Figure 27.34

Predicted structure of the CFTR protein. Cylinders represent membrane-spanning helical segments. The cytoplasmically oriented NBFs are shown as blue spheres with slots to indicate the points of entry of nucleotides. R represents the large polar domain, which is linked to two halves of the protein molecule. Charged amino acids are shown as small circles with the charge sign. Net charges on the internal and external loops joining the membrane cylinders and on regions of the NBFs are contained in open squares. Potential sites for phosphorylation by protein kinases A or C (PKA or PKC) and N-glycosylation (N-linked CHO) are indicated. (K = Lys; R = Arg; H = His; D = Asp; E = Glu.)

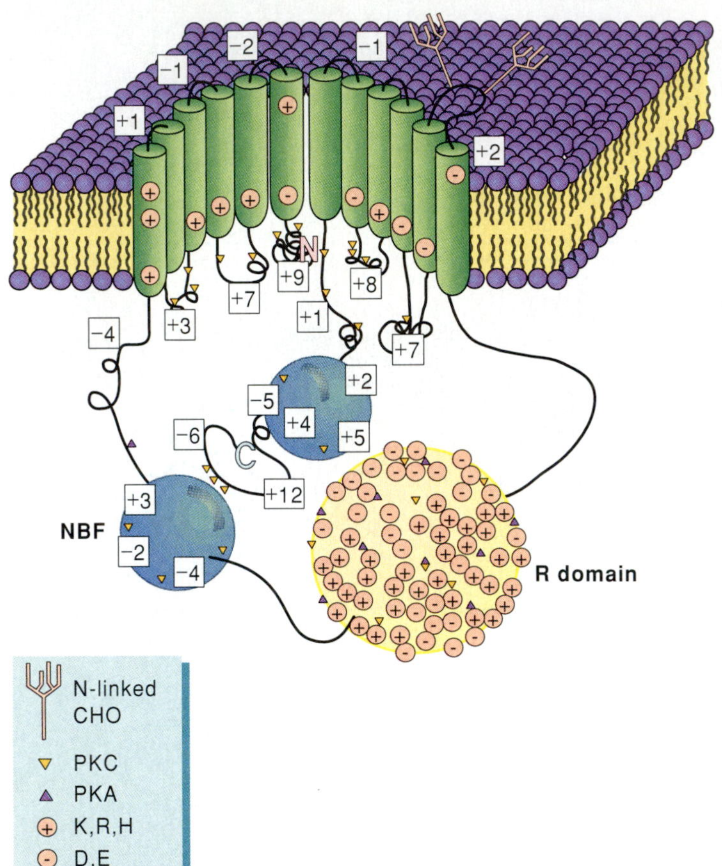

transcript consists of two similar motifs, each with (1) a domain having properties consistent with membrane association and (2) a domain believed to be involved in ATP binding (fig. 27.34). Finally the researchers discovered that most CF patients carry a three-base deletion in this transcript, eliminating a phenylalanine residue from the protein. Thus they reasoned that the protein in cystic fibrosis patients should bear a defect. This defect correlated with the notion that CF patients have a faulty membrane protein that leads to the secretion problems characteristic of the disease. The fact that the abnormal gene is located as close as can be detected by classical genetics to the CF locus added additional support to the notion that the CF gene had been found.

Finding the disease gene has not, of course, meant an end to the problem of cystic fibrosis. However, the characterization of the disease gene will be a tremendous aid in diagnosing carriers and fetuses that are homozygous for the disease gene. It also should be a help in focusing approaches to finding a cure for the disease.

Summary

In this chapter we have described methods for sequencing, synthesizing, cloning, and restructuring DNA, and we have seen how recombinant DNA technology is being used to advance our knowledge of gene structure and function. Our discussion has focused on the following points.

1. To sequence DNA, researchers use chemical methods to cleave specific bases in preexisting DNA, or they carry out DNA synthesis under conditions where the synthesis is interrupted at specific bases.
2. By chemical methods it is possible to synthesize oligodeoxynucleotides up to a length of 100 monomers.
3. A specific segment of DNA can be synthesized *in vitro* by a polymerase chain reaction (PCR), in which short segments of DNA bordering the segment of interest are added to a mixture containing the segment of interest, a special heat-resistant DNA polymerase, and the deoxyribotriphosphate substrates. The DNAs are first denatured, then annealed and then synthesized. This cycle is repeated twenty or more times by raising the temperature to stop synthesis and then lowering the temperature to permit annealing and further synthesis. The outcome is a mixture in which the vast majority of the DNA is newly synthesized DNA bounded by the sequences of the added primers.
4. Another procedure for amplification cuts DNA containing the segment of interest into small pieces with a restriction enzyme. The cut pieces are incorporated into a plasmid or virus vector to be amplified in a suitable host. After growth, the mixture is plated to produce a mixture of bacterial or viral clones. The clones of interest are often identified by hybridizing replica plates of clones with a radioactive probe, then subjecting them to autoradiography.
5. Most cloning has been done in *E. coli.* Yeast is the most used eukaryotic host. Cloning is also possible in a number of plant and animal cells.
6. Mapping with recombinant DNA probes was first applied to the human globin genes. Starting probes were obtained by isolating the globin messenger from reticulocytes and converting it into a cDNA, which was then used to scan a human genomic library for cross-hybridizing members. Once detected and purified, these cross-hybridizing members carrying globin messenger sequences were themselves converted to radioactive probes and used to further scan the genomic library for nearby sequences. Repeating this cycle several times, a process known as a chromosome walk, revealed a region around the adult hemoglobin gene that contained several closely related genes associated with hemoglobin.
7. Frequently alleles of the same gene can be distinguished by restriction site differences in the genes themselves or in nearby locations. Alleles identified in this way are said to show restriction fragment length polymorphism (RFLP). The allele responsible for sickle-cell disease was identified in this way.
8. The cystic fibrosis gene was mapped by chromosome walking and jumping, a newer approach in which the relevant probes contain segments of the genome that are normally located about 500 kb from one another. A cDNA library was made from normal sweat gland tissue, chosen because of the disease's association with abnormal release of sweat salt, which suggested that the sweat gland would contain an abundance of the messenger associated with the gene. By hybridizing the genomic DNA probes with the cDNA sweat gland library, a segment of genome was identified as a candidate for the cystic fibrosis gene. This gene was characterized in detail and found to encode a complex transmembrane protein that carries a specific amino acid change in most persons with cystic fibrosis. This correlation provided overwhelming support for the conclusion that the gene responsible for cystic fibrosis had been mapped and characterized.

Selected Readings

Caruthers, M. H., A. D. Barone, S. L. Beaucage, D. R. Dodds, E. F. Fisher, L. J. McBride, M. Matteucci, Z. Stabinsky, and J. Y. Tang, Chemical synthesis of deoxyoligonucleotides. *Methods Enzymol.* 154: 287–313, 1987.

Cohen, S. N., A. Change, H. Boyer, and R. Helling, Construction of biologically functional bacterial plasmids *in vitro. Proc. Natl. Acad. Sci.* 70:3240–3244, 1973.

Gusella, J. F., DNA polymorphism and human disease. *Ann. Rev. Biochem.* 55:831–854, 1986.

Hunkapiller, T., R. J. Kaiser, B. F. Koop, and L. Hood, Large-scale and automated DNA sequence determination. *Science* 254:59–67, 1991. State of the art on the mammoth project to sequence the human genome.

Jackson, D. A., R. H. Symons, and P. Berg, Biochemical method for inserting new genetic information into DNA of Simian Virus 40: Circular SV40 DNA molecules containing lambda phage genes and the galactose operon of *E. coli. Proc. Natl. Acad. Sci.* 69: 2904–2909, 1972.

Kerem, B., J. M. Rommens, J. A. Buchanan, D. Markiewicz, T. K. Cox, A. Chakravarti, M. Buchwald, and L. C. Tsui, Identification of the cystic fibrosis gene: Genetic analysis. *Science* 245:1073–1079, 1989.

Mansour, S. L., K. R. Thomas and M. R. Capecchi, Disruption of the proto-oncogene *int-2* in mouse embryo-derived stem cells: a general strategy for targeting mutations to non-selectable genes. *Nature* 336:348–352, 1988.

Mullis, K. B., The unusual origin of the polymerase chain reaction. *Sci. Am.* 262:56–65, 1990.

Riordan, J. R., J. M. Rommens, B. S. Kerem, N. Alon, R. Rozmahel, Z. Grezelczak, J. Zielenski, S. Lok, N. Plasvsic, J.-L. Chou, M. T. Drumm, M. C. Iannuzzi, F. S. Collins, and L-C. Tsui, Identification of the cystic fibrosis gene: Cloning and characterization of complementary DNA. *Science* 245:1066–1073, 1989.

Sambrook, J., E. F. Fritsch, and T. Maniatis, *Molecular Cloning: A Laboratory Manual,* 2d ed. A three-volume collection that is thorough and up-to-date.

Sanger, F., Sequences, sequences, and sequences. *Ann. Rev. Biochem.* 57:1–28, 1988. A scientific memoir.

Problems

1. What are the advantages and disadvantages of the two methods of sequencing DNA (Gilbert's chemical method and Sanger's chain termination method)?
2. What are the limitations on sequencing RNA by the indirect method of sequencing the cDNA obtained from the reverse transcription of RNA?
3. What are the major advantages of the polymerase chain reaction (PCR) method for amplifying defined segments of DNA as opposed to the use of conventional cloning methods? How might the PCR method be used to test for infection with the AIDS virus and how would this be an improvement over the antibody test currently used? (The current ELISA test is an indirect test for the presence of antibodies against the HIV proteins.)
4. You have just isolated a novel recombinant clone and purified the desired insert (a 10-kb linear duplex DNA) from the vector. Now you wish to map the recognition sequences for restriction endonucleases A and B. You cleave the DNA with these enzymes and fractionate the digestion products according to size by gel electrophoresis. You observe the following:
   (a) Digestion with A alone gives two fragments, of lengths 3 and 7 kb.
   (b) Digestion with B alone gives three fragments, of lengths 0.5, 1, and 8.5 kb.
   (c) Digestion with A plus B generates four fragments, of lengths 0.5, 1, 2, and 6.5 kb.

   Draw a restriction map for the insert, showing the relative positions of the cleavage sites with respect to one another.
5. Describe the procedure you would use to clone a DNA fragment into the *Bam*HI site of pBR322.
6. A small circular duplex viral DNA with about 5,000 bp contains a 72-bp direct repeat in tandem. Each 72-bp segment contains a cleavage site for the *Eco*RI restriction enzyme that is not present in the rest of the virus. How could you produce a modified viral chromosome with only one copy of the 72-bp segment?
7. How large a genomic library should you construct in order to detect and isolate a 15-kb gene out of a genome containing 3×10^9 bp?
8. Why isn't a primer required to promote second-strand synthesis using the single-stranded DNA produced by reverse transcriptase from an RNA template?
9. If you were interested in isolating a cDNA for human serum albumin, why would you use a cDNA library established from mRNA isolated from liver? If you wanted to isolate the gene for albumin, why could you use a genomic library established from any human tissue?
10. The analysis of DNA regulatory sequences can be undertaken via the technique of DNA-mediated gene transfer, using a gene that provides a functional assay ("reporter gene") for the effect of the putative regulatory sequence on gene expression. How could this be done?
11. Explain how the thymidine kinase gene (*tk*) can be used as a selectable genetic marker for isolating transformed mammalian cells containing DNA of interest.
12. Site-directed mutagenesis is one of the most powerful tools available to the biochemist. What are some of the applications of this technique? How can the PCR method be used to do site-directed mutagenesis, and what is the advantage of this method?
13. Briefly outline how you could map the globin gene family using recombinant DNA techniques.
14. What is the procedure called chromosome jumping? How was this procedure used to map the cystic fibrosis gene?

28

CHAPTER

RNA Synthesis and Processing

In the early 1950s George Gamov suggested that DNA might serve as a template for protein synthesis. However, biochemical investigations showed that in eukaryotes the processes of RNA and protein synthesis are compartmentalized in different parts of the cell—the nucleus and the cytoplasm, respectively. For this reason, it seemed unlikely that DNA could serve directly as a template for protein synthesis.

Base composition studies by Erwin Chargaff, which had provided information that was vital in deducing DNA structure and in suggesting a mechanism for DNA replication, were less helpful in regard to the structure and synthesis of RNA. Total cellular RNA did not reflect the base composition of the DNA. However, Volkin and Astrachan showed that when a DNA bacterial virus, T2 bacteriophage, infects an *E. coli* cell, the newly synthesized RNA (which is exclusively transcribed from the phage DNA) reflects the base composition of the DNA. From experiments such as this, and from an understanding of the DNA duplex structure, there emerged the hypothesis that DNA, by unwinding, serves as a template for the synthesis of RNA.

In the early 1950s Crick proposed what he called the "central dogma" of molecular biology, which states that genetic information flows from DNA to RNA and thence to protein:

$$\text{DNA} \rightarrow \text{RNA} \rightarrow \text{protein}$$

This dogma provided researchers with the theoretical framework for experiments designed to demonstrate the existence of a transcribing enzyme that copies the DNA sequence into an RNA molecule.

The first enzyme discovered that could catalyze polynucleotide synthesis was polynucleotide phosphorylase. This enzyme, isolated by Severo Ochoa and Marianne Grunberg-Manago in 1955, could make long chains of $5' \rightarrow 3'$ linked polyribonucleotides starting from nucleoside diphosphates. However, no template dependence could be found for this synthesis, and the sequence was only controllable in a crude way by adjusting the relative concentrations of different nucleotides in the starting materials.

Sam Weiss was the first to obtain evidence for a true transcribing activity in cell-free extracts of rat liver (1959). His experimental design was influenced by the theoretical framework provided by Crick and by Kornberg's discovery that *in vitro* DNA synthesis required nucleoside triphosphates for substrates. With crude liver extracts, Weiss was able to demonstrate a capacity for RNA synthesis that was severely inhibited by DNase. This was the beginning of systematic investigations of the biochemistry of RNA metabolism. A continuous expansion of research effort in related areas over the past 30 years has provided us with a wealth of understanding about the transcription process and other aspects of RNA metabolism.

We considered various aspects of nucleotide metabolism in chapter 20. In this chapter we will focus on RNA synthesis and the reactions that follow synthesis, whereby the initially transcribed RNA is modified into a mature functional form. These reactions are summarized in figure 28.1.

Evidence that RNA Sequences Are Derived from DNA Sequences

The base composition studies of Volkin and Astrachan had suggested a template-product relationship between DNA and RNA, and Sol Spiegelman and Benjamin Hall pursued this result a step further. Using ^{3}H-labeled T2 DNA and ^{32}P-labeled RNA that was synthesized after T2 infection of *E. coli* cells, they were able to demonstrate that the newly synthesized RNA forms a specific complex with the viral DNA (fig. 28.2). To show that this RNA contains sequences homologous to those in DNA, advantage was taken of the annealing technique developed by Doty and Marmur. They reasoned that it should be possible to make a DNA-RNA duplex in the same way as a DNA-DNA duplex, by annealing. Since such a complex should contain one strand of DNA and one strand of RNA, it is referred to as a hybrid duplex, and the process of making the duplex is referred to as hybridization. Following hybridization, the product was subjected to CsCl density-gradient centrifugation. (See chapter 26 for a general explanation of this technique and for its application to the study of DNA replication.) After equilibrium was reached, they analyzed various fractions from the centrifuge tube for ^{32}P and ^{3}H, using isotope-counting methods that make it possible to distinguish between ^{32}P and ^{3}H in the same sample. Thus they could determine the locations of DNA and RNA by measuring ^{32}P and ^{3}H in the various fractions. They found that the ^{3}H gave a sharp peak near the middle of the tube, while the ^{32}P gave a broad peak with a maximum near the bottom. RNA is denser than DNA, which accounts for its lower average position in the centrifuge tube. The excitement in this experiment centered around a sharp peak of ^{32}P coincident with the ^{3}H label (see fig. 28.2). This peak indicated the presence of a complex that involved both DNA and RNA. Proof that the complex was specific was obtained by carrying out the annealing experiment with other DNAs. Only when T2 DNA was used did the ^{32}P peak appear. Furthermore, the peak appeared only if the T2

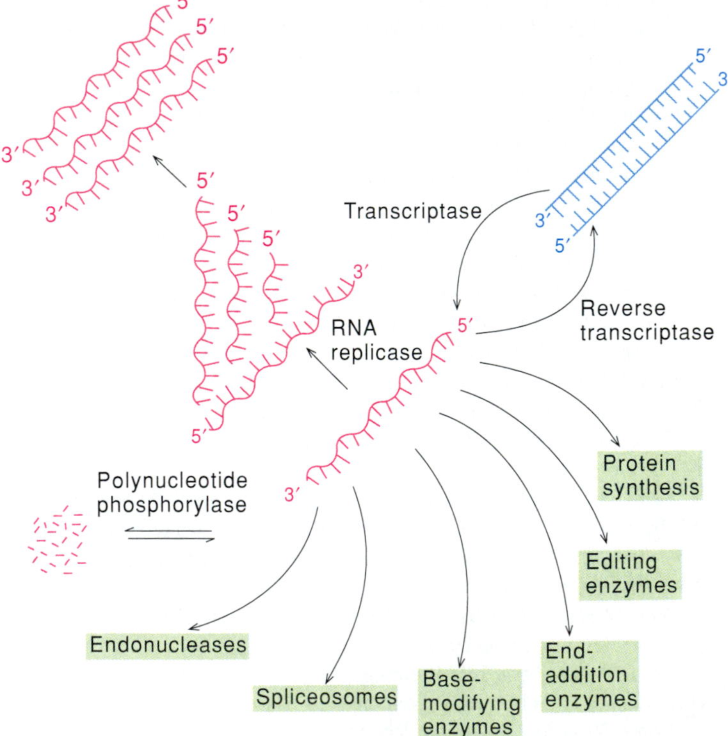

Figure 28.1

Different types of enzyme-catalyzed reactions involving RNA. RNA is best known as the intermediary *messenger RNA* that carries the information from DNA for the synthesis of protein and for other polynucleotides that are directly involved in protein synthesis, *ribosomal RNA* and *transfer RNA*. Certain RNAs possess catalytic activity leading to the evolutionary argument that RNA as template and catalyst preceded protein. The first reaction to be discovered whereby polynucleotides are synthesized was catalyzed by the bacterial enzyme *polynucleotide phosphorylase*. Polynucleotide phosphorylase uses nucleotide diphosphates as substrates and has no template requirement. All other *RNA polymerases* use nucleotide triphosphates as substrates. Double helix DNA is the template that is used in the synthesis of most cellular RNAs. The DNA carries specific sequences of bases called *promoters* that are recognized as RNA polymerase binding sites and initiation points for RNA synthesis. Other sites on the DNA are recognized as termination points for RNA synthesis. A gene always has a promoter and a termination point and the information in between that is transcribed by the RNA polymerase. Most transcripts carry the information for making protein. The enzyme *reverse transcriptase*, which is very widely distributed, is capable of using RNA as a template for the synthesis of DNA. This is a minor reaction in most cells but it may play a highly significant evolutionary role and it is indispensable to the life cycle of certain viruses as we saw in the previous chapter. Another RNA polymerizing enzyme known as *RNA replicase* is used by certain RNA viruses to replicate viral RNAs.

Many enzymes are involved in RNA processing. These enzymes have been named according to the types of reactions they catalyze. *Endonucleases* and *exonucleases* hydrolyze phosphodiester linkages in preformed RNAs. *Splicing enzymes* cleave internal segments and knit the remaining polymers back together. *Base modifying enzymes* alter or replace the original bases present in the polymer at select points. *End addition enzymes* add additional nucleotides at either the 3′ or 5′ ends of the ribopolymer. Recently discovered *editing enzymes* remove certain nucleotides and add additional nucleotides.

Figure 28.2

Demonstration that phage RNA has sequences complementary to phage DNA. Spiegelman and Hall used ^{3}H-labeled T2 DNA and ^{32}P-labeled RNA, the latter made after T2 infection of *E. coli* cells. The two nucleic acids were mixed together, annealed, and then centrifuged to equilibrium in a CsCl density gradient. In (*a*) the DNA was first heat-denatured and then mixed with RNA and annealed at 65° C prior to centrifugation. In (*b*) the DNA denaturation step was left out. Most of the RNA goes to the bottom of the centrifuge tube because of its high density. The DNA bands about one-third of the way from the top. In (*a*) some RNA also bands at approximately the same location as the DNA, but in (*b*) this is not the case. The comigration of a fraction of the RNA with the DNA is believed to be due to the formation of a specific DNA-RNA hybrid duplex. The RNA is much smaller than the DNA, so the RNA in the hybrid duplex migrates at the density of the DNA. In (*b*) no hybrid duplex forms because the DNA was not denatured before carrying out the annealing process.

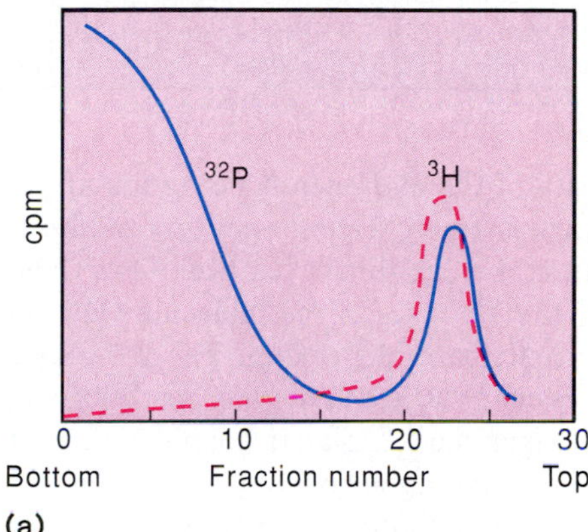

(a)

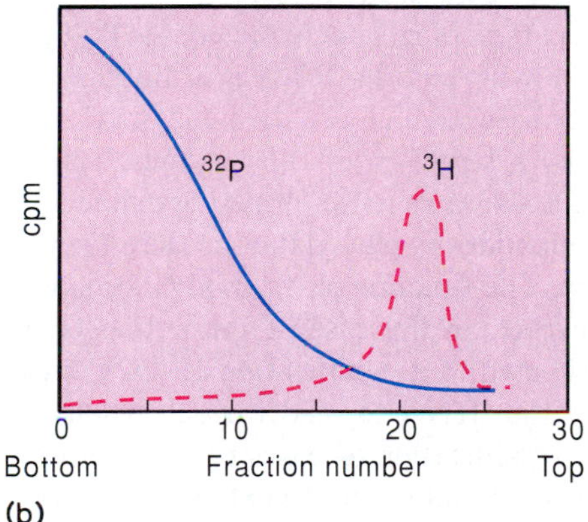

(b)

DNA was denatured before the annealing step. This result indicated that the two DNA strands must be separated to make the DNA-RNA complex, as would be expected in an annealing operation.

The hybridization technique developed in Spiegelman's laboratory was to take its place, alongside the DNA-DNA annealing technique developed by Doty and Marmur, among the techniques most often used in molecular biology. It has since been used to characterize RNAs and their location of synthesis on the genome. Historically, that first hybridization experiment gave credence to the proposal that RNA was a direct transcript of a DNA chain.

The Three Major Classes of RNA

Although most RNAs are synthesized as a result of transcription from a DNA template, different strategies are used in their synthesis and in their posttranscriptional modification. The strategy used in a given case is strongly related to the function of the RNA. For this reason, it is important that we discuss the different classes of RNA found in the cell before we consider their mode of synthesis.

There are three major types of cellular RNA involved in protein synthesis: messenger RNA (mRNA), ribosomal RNA (rRNA) and transfer RNA (tRNA). In addition, there are other RNAs, some of which function as primers for DNA synthesis, as component parts of ribonucleases, and in the posttranscriptional modification of eukaryotic mRNA precursors. The properties of the RNAs found in *E. coli* are summarized in table 28.1. You will find it helpful to refer to this table in the course of the following discussion.

Messenger RNA Carries the Information for Polypeptide Synthesis

Messenger RNA is transcribed from DNA by RNA polymerase, and its sequence contains the information for the sequence of amino acids in the protein product. Specific mRNAs are synthesized by the cell in response to the conditions under which it finds itself, and the cell can thereby control the kinds and amounts of proteins it produces. Prokaryotic cells have unstable mRNAs that turn over rapidly, with an average half-life of 1 to 3 min, whereas eukaryotic cells, which do not have to be able to respond as rapidly to changing conditions, often have more stable mRNAs.

The mRNA fraction is heterogeneous in size, ranging from 500 to 6,000 nucleotides in *E. coli*. This range reflects not only the heterogeneity in size of the proteins of the cell, but also the fact that some prokaryotic mRNAs contain the information to encode more than one protein. Such mRNAs are referred to as polycistronic mRNAs, each cistron containing the information for the synthesis of a single polypeptide chain.

In prokaryotes, the messenger is engaged in protein synthesis even before the entire mRNA molecule has been synthesized. This rapid utilization of nascent transcript minimizes the opportunity for processing of the transcript. In contrast, in eukaryotes, where the processes of transcription and translation are sharply divided, the nascent transcript undergoes an elaborate regimen of processing, usually at both ends and frequently internally, before it is transported to the cytoplasm for use as mRNA.

Transfer RNA Carries Amino Acids to the Template for Protein Synthesis

The second major type of RNA is transfer RNA. Its function is to carry amino acids to the mRNA template, where the amino acids will be linked in a specific order. The tRNAs contain both a site for the attachment of an amino acid and a site, the anticodon, that recognizes the corresponding three-base codon on

Table 28.1
Types of RNA in *E. coli*

Type	Function	Number of Different Kinds	Number of Nucleotides	Percent of Synthesis	Percent of Total RNA in Cell	Stability
mRNA	Messenger	Thousands	500–6,000	40–50	3	Unstable ($t_{1/2}$ = 1 to 3 min)
rRNA	Structure and function of ribosomes	3 { 23S, 16S, 5S	2,800 1,540 120	50	90	Stable
tRNA	Adapter	50–60	75–90	3	7	Stable
RNA primers	DNA replication	?	<50	<1	<1	Unstable
RNA component of RNase P	Catalytic	1	377	<1	<1	?

the mRNA (see chapter 29). There are about 50 different kinds of tRNAs in a bacterial cell. Each amino acid is enzymatically attached to the 3′ end of one or more tRNAs by a specific aminoacyl-tRNA synthase that recognizes both the amino acid and the tRNA. This two-step process is discussed in chapter 29.

A great deal is known about the structure of tRNA because the entire nucleotide sequences of a large number of tRNAs have been determined, and the three-dimensional structures of some of them have been obtained by x-ray crystallography. The complete three-dimensional structure of phenylalanine tRNA of yeast, $tRNA^{Phe}$, is shown in figure 28.3. This molecule contains 76 nucleotides in a single chain with four loops and a stem.

Most of the bases are involved in a complex secondary and tertiary structure. The four loops in the molecule are referred to as the TψC loop, the anticodon loop, the D loop, and the variable loop. The variable loop is so named because it contains a different number of bases in different tRNAs. The remainder of the molecule (with the exception of the D loop in some instances) contains the same number of residues in all tRNAs, even though there is considerable sequence variability. The $tRNA^{Phe}$ molecule contains 20 base pairs that are hydrogen-bonded in Watson-Crick fashion. It contains an additional 40 or so hydrogen bonds, most of which are not of the Watson-Crick type. These additional hydrogen bonds and the accompanying base stacking stabilize the tRNA in the complex folded structure shown in figure 28.3. Some of the unusual hydrogen-bonded interactions involved in stabilizing the complex tertiary structure are shown in figure 28.4. These structures involve two to four residues from different regions of the RNA. Not only are the hydrogen-bonding patterns unusual, but in some cases phosphate and the C-2′-OH group participate in the hydrogen bonding. Tertiary structures of most tRNAs are very similar except for the variable loop, even though tRNAs show considerable sequence variability.

Earlier (chapter 25) we stressed the importance of base stacking in stabilizing duplex nucleic acid structures. Despite the irregularity of the hydrogen bonding found in tRNA, base stacking, as assessed by a close scrutiny of the three-dimensional structure, is a prominent feature of the tRNA structure. Only four of the 76 bases in the molecule (D^{16}, D^{17}, U^{47}, and G^{20}) do not participate in stacking. The D^{16} and D^{17} dihydrouracil residues cannot stack under any conditions, because these residues no longer have a planar configuration. It is believed that stacking and hydrogen bonding contribute about equally to the conformational stability of the tRNA.

Another significant feature of the tRNA structure is the high percentage of ribonucleotide bases that differ from the usual four. We will see that tRNA is initially synthesized from the usual four bases, which are then enzymatically modified in many cases by a large variety of enzymes. Some of the more than 30 such unusual bases so formed are shown in figure 28.5. Of the 76 nucleotides in phenylalanine tRNA, twelve are of the modified type. The function of the modification is not known in general. However, in the case of the anticodon region of the tRNAs it is known that modification of the 5′ base can change the ability of the tRNA to read various synonym codons (see chapter 29). Modification of bases in other types of RNA has also been observed, but not to the extent that it occurs in tRNA.

The detailed structural investigations of tRNA greatly broadened our expectations in considering the less-well-understood structures formed by other nuclear and cytoplasmic RNAs. This single example illustrates the fact that nucleic acids with a properly adjusted primary sequence can adopt complex secondary and tertiary structures. *A propos* of this point, Francis Crick once said that transfer RNA is "an RNA trying to look like a protein." The same could be said of the ribosomal RNAs.

Ribosomal RNA Is an Integral Part of the Ribosome

Ribosomal RNA is a structural and functional component of ribosomes, the large ribonucleoprotein bodies that house the machinery for most of the reactions associated with protein synthesis. Ribosomes in prokaryotes are referred to as 70S ribo-

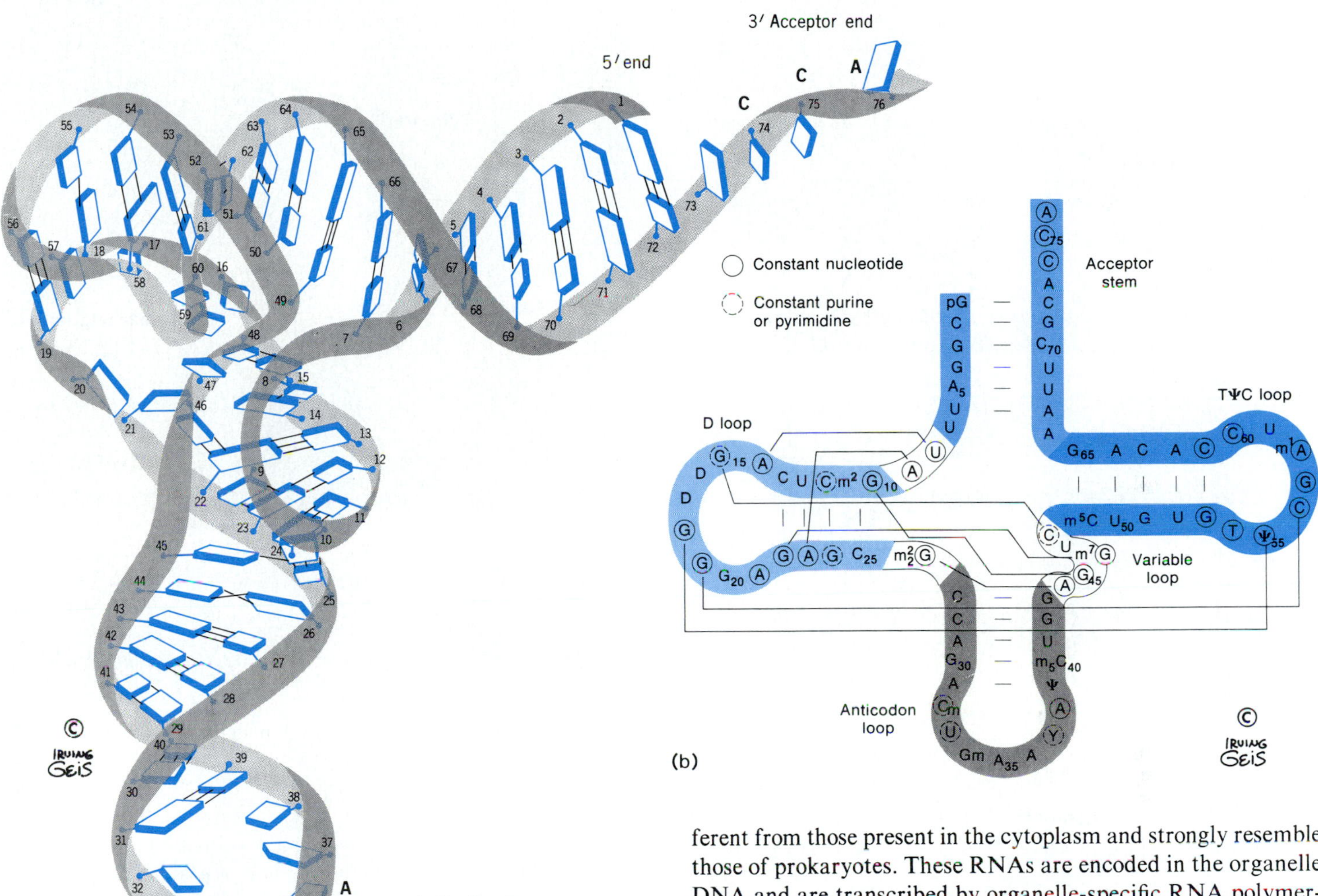

Figure 28.3

The tertiary structure of yeast phenylalanine tRNA. (*a*) The full tertiary structure. Purines are shown as rectangular slabs, pyrimidines as square slabs, and hydrogen bonds as lines between slabs. (Source: G. J. Quigley and A. Rich, "Structural domains of transfer RNA molecules" in *Science*, 194:796, 1976. Copyright © 1976 American Association for the Advancement of Science, Washington, D.C.) (*b*) Nucleotide sequence. Residues that appear in most of the yeast tRNAs and residues that appear to be constantly a purine or a pyrimidine are indicated. Residues involved in tertiary base pairing are shown connected by solid lines. Several of the nucleotides are methylated. These are indicated by a small m.

somes, a measure of their rate of sedimentation in a centrifuge and hence their size (S refers to Svedberg units, which are defined in chapter 5). A 70S ribosome consists of two subunits, the 50S subunit and the 30S subunit, each of which is made up of RNA and protein. The 50S subunit contains 23S and 5S rRNAs. The 30S subunit contains a single 16S rRNA (fig. 28.6). Eukaryotic ribosomes are similar in structure, although they are somewhat larger (80S) and contain mostly larger RNAs (25–28S, 18S, 5S, and an additional 5.5 to 5.8S; this additional rRNA corresponds in sequence to the first 150 nucleotides of the prokaryotic large rRNA subunit). Chloroplasts and mitochondria have ribosomes and rRNAs that are distinctly different from those present in the cytoplasm and strongly resemble those of prokaryotes. These RNAs are encoded in the organelle DNA and are transcribed by organelle-specific RNA polymerases.

Ribosomal RNA Secondary Structure Is Highly Conserved

The primary sequences of many rRNAs and their genes have now been determined, and the results have done much to shape our understanding of rRNA structure and function. Sequence analysis has revealed that conservation of secondary structure is the major paradigm of ribosomal RNA evolution. In general, rRNA has evolved by the introduction of complementary base changes that conserve regions of secondary structure. Much evidence indicates that regions that can vary in primary sequence, without the necessity of compensating changes in a complementary region turn out to be single-stranded parts of the rRNA. There are few regions in rRNA where primary sequence is conserved, and these are generally located in single-stranded regions of the molecule. The secondary structures of rRNAs that have been deduced by this method of phylogenetic sequence comparison are shown in figure 28.7.

Figure 28.4

Some of the tertiary hydrogen-bonded interactions found in yeast phenylalanine tRNA. Superscripts refer to base number in the tRNA molecule. It can be seen that there are many hydrogen-bonding arrangements in a tRNA structure that are not found in a duplex DNA structure. (Source: G. J. Quigley and A. Rich, "Structural domains of transfer RNA molecules" in *Science,* 194:796, 1976. Copyright © 1976 American Association for the Advancement of Science, Washington, D.C.)

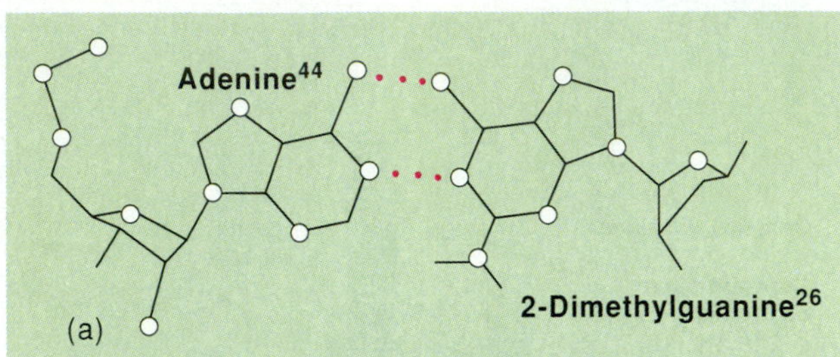

Figure 28.5

The structures of some modified bases found in tRNA. The parent ribonucleosides are shown with the numbering of the atoms in the purine and pyrimidine rings.

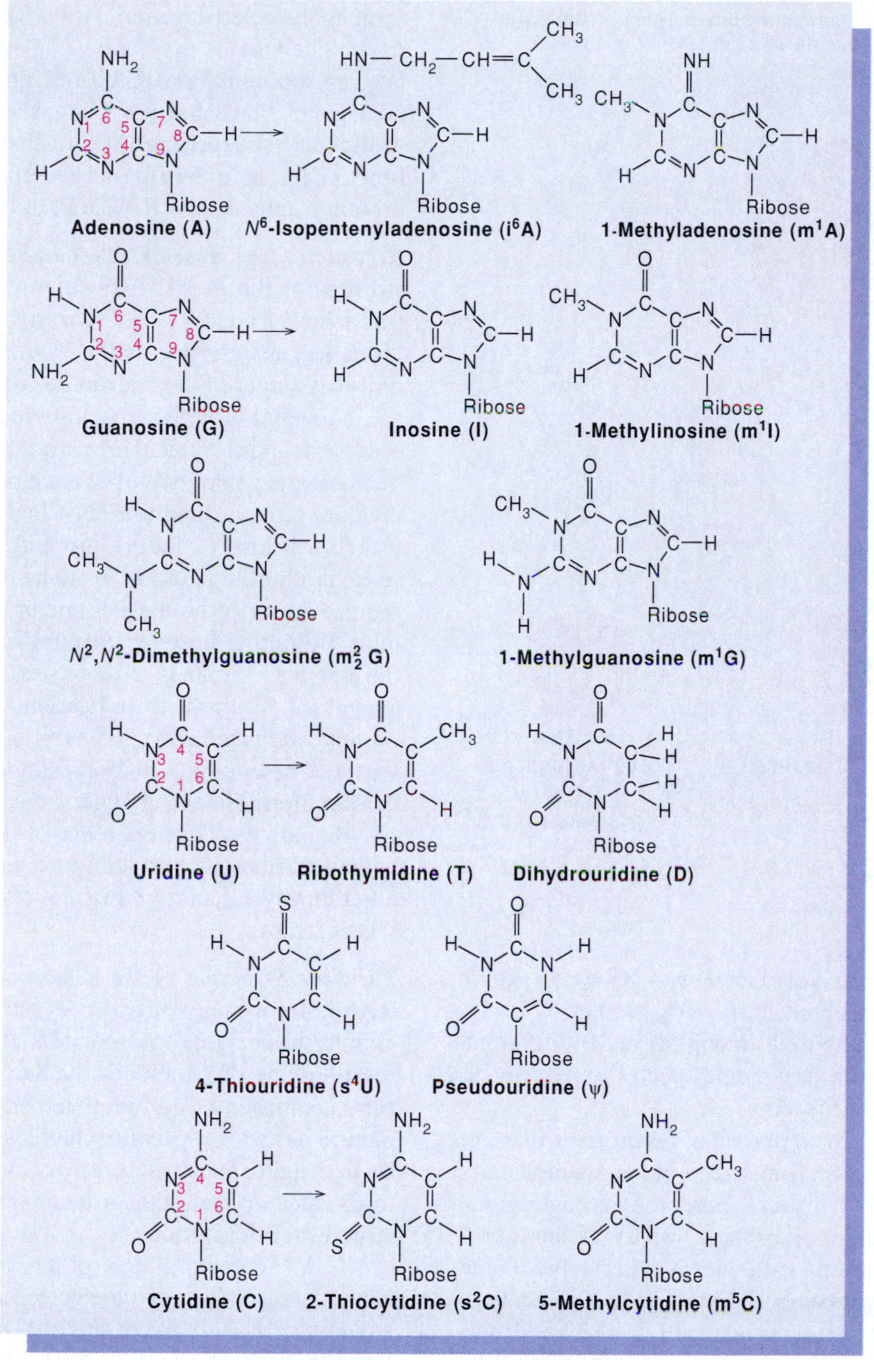

Adenosine (A) → **N^6-Isopentenyladenosine (i^6A)**; **1-Methyladenosine (m^1A)**

Guanosine (G) → **Inosine (I)**; **1-Methylinosine (m^1I)**

N^2,N^2-Dimethylguanosine (m_2^2G); **1-Methylguanosine (m^1G)**

Uridine (U) → **Ribothymidine (T)**; **Dihydrouridine (D)**

4-Thiouridine (s^4U); **Pseudouridine (ψ)**

Cytidine (C) → **2-Thiocytidine (s^2C)**; **5-Methylcytidine (m^5C)**

For years, the pattern of rRNA structure observed in common bacteria and in many eukaryotes—the presence of discrete pieces of defined sizes—has been considered an integral feature of ribosome structure. This view is now changing with the discovery that the number and size of the individual pieces of rRNA can vary widely in different organisms. Ribosomal RNAs in some organisms have been found to occur in as many as a dozen separate pieces as the result of transcription from independent genetic units. It now appears that rRNA consists of discrete and functionally essential domains that need not be joined together. Much of the difference in the overall length that distinguishes rRNA from different species appears to arise by insertions or deletions between these individual segments. Thus rRNA seems to be composed of discrete functional domains that can be independently inherited and separated by segments of variable length.

Figure 28.6

Composition of the *E. coli* ribosomes. The 70S ribosome can dissociate into a 50S and a 30S subunit. *In vitro* this can be done by lowering the Mg ion concentration. The individual subunits can be dissociated into their constituent RNAs and proteins by exposure to urea denaturant. Molecular weights are given for the subunits and the proteins, and the numbers of nucleotides are given for the RNAs.

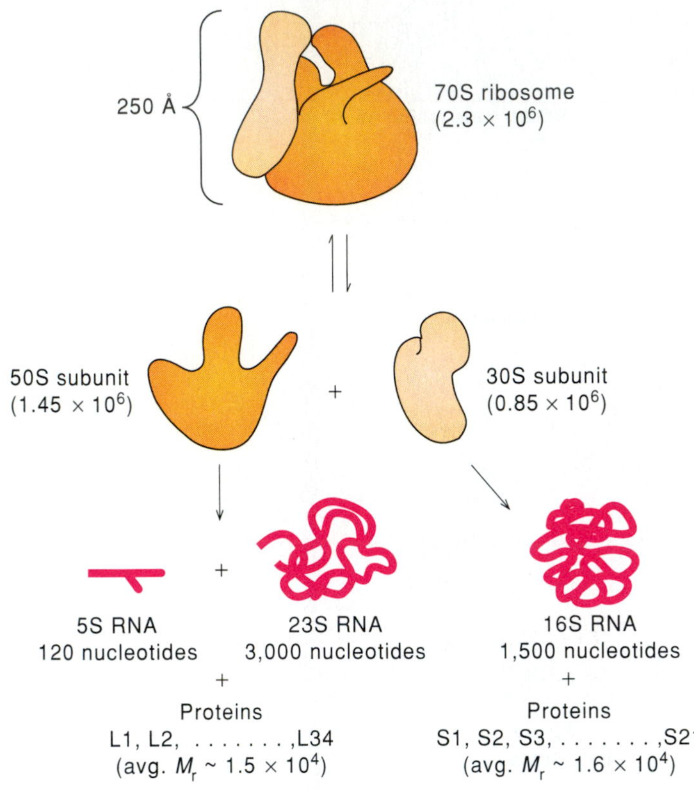

Ribosomes Contain a Discrete Set of Proteins Ribosomes from different species vary from about 30 to 70% protein by mass. This mass is generally distributed among 50 to 100 different proteins, but ribosomes from any single species appear to be composed of a unique set of proteins.

E. coli ribosomes are composed of 56 proteins that are mostly basic in charge and small in size, with an average mass of about 20,000 daltons. Fifty-three of these proteins have unique sequences and occur once per ribosome. Twenty of these proteins belong to the 30S subunit with one copy per subunit (see box 28A). One protein occurs on both the small (S20) and large (L26) subunits. Another protein, designated L7/L12, yields two electrophoretic spots because it is partially acylated at its amino terminal. This protein is present in four copies per 50S as a highly elongated complex that appears to form a stalk on the subunit.

For many years it was generally assumed that the ribosomal proteins are the functionally important components of the ribosomes. However, despite a great deal of effort, the function of individual ribosomal proteins has consistently eluded discovery. The location of some proteins near functional binding sites has been shown by chemical cross-linking of bound ligands. Mutational alterations in others have been shown to cause resistance to antibiotic inhibitors of protein synthesis. Remarkably as many as half of the individual *E. coli* ribosomal proteins, as well as at least some eukaryotic ribosomal proteins, can be completely deleted genetically without inactivating the ribosome. The view is developing that the primary function of ribosomal proteins is to bind and thus assemble the individual domains of ribosomal RNA into their appropriate locations. It would not be surprising to learn that ribosomal proteins play a functional role in the ribosome similar to the role played by the protein component of RNase P, to be discussed later.

Ribosomes Can Assemble Spontaneously The *in vitro* reconstitution of the *E. coli* 30S subunit from purified components was achieved about twenty years ago by Masayasu Nomura and his colleagues. Subsequently, *in vitro* assembly of the 50S subunit was achieved. In addition to the structural components of the individual subunits, assembly required the appropriate ionic conditions and elevated temperature. Nomura and his co-workers demonstrated that assembly is an ordered process in which individual proteins bind to rRNA in a prescribed sequence. Some proteins, primary binding proteins, bind to rRNA in the absence of all others. Other proteins, secondary binding proteins, require the prior binding of one or more primary binding proteins. Still other proteins, tertiary binding proteins, cannot join the assembly until one or more secondary binding proteins have assembled. From such indications Nomura developed an assembly pathway for the 30S subunit (fig. 28.8). He also showed that the assembling ribosome undergoes temperature-dependent conformational changes at specific points along the assembly pathway. The existence of the pathway implies that the individual ribosomal proteins interact with both rRNA and each other during assembly, which in turn implies underlying spatial relationships.

The Fine Structure of the Ribosome is Beginning to Emerge Results from many different experimental approaches are beginning to coalesce to produce a three-dimensional picture of the ribosome that includes the location of its individual structural components and functional sites. The development of this picture has been especially challenging because the ribosome is large, fragile, and structurally complex and has resisted efforts to produce crystals that are capable of giving high-resolution structural information.

The current view of the overall morphology of ribosomes is based largely on electron micrographic studies of the subunits of *E. coli*. From this analysis it appears that both subunits are asymmetric (fig. 28.9). The large subunit is visualized as having several protuberances at one end. A large protuberance in the center is flanked by smaller ones on either side. The small subunit is visualized as having a head, neck, and platform, all at one end of the structure.

The relative location of individual ribosomal proteins within the two subunits has been carefully examined by two different techniques. One method involves determining which ribosomal proteins can be chemically cross-linked to each other, and has yielded an elaborate grid of spatial relationships based

Figure 28.7

Comparison of secondary structures of 16S-like rRNAs from (*a*) eubacteria (*E. coli*), (*b*) archaebacteria (*H. volcanii*), (*c*) eukaryotes (*S. cerevisiae*), and (*d*) a structure showing features common to all sequenced 16S-like rRNAs, including mitochondrial types.

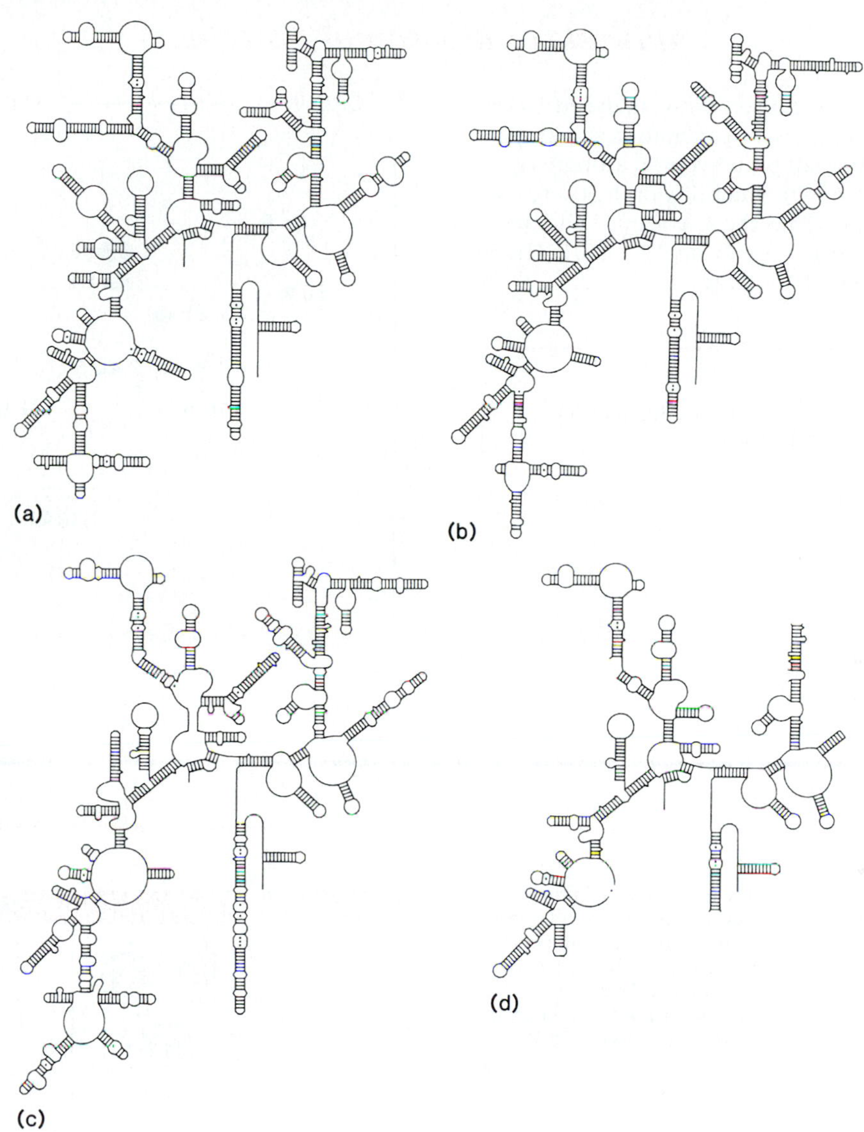

on the frequency of cross-linking. The other method relies on neutron diffraction. This method employs individually deuterated ribosomal proteins and locates their relative centers of mass within the subunit. The two methods of determining the location of ribosomal proteins have yielded a relatively consistent spatial picture.

Recently, information concerning protein locations within the 30S subunit has been combined with the secondary structure model of 16S rRNA and the location of protein binding sites in rRNA to generate an initial, partial three-dimensional picture of how the rRNA and proteins interact in this ribosomal subunit (fig. 28.10).

Other Cellular RNAs Are Involved in DNA Synthesis and RNA Processing

In addition to the three major classes of RNA involved in protein synthesis, chromosomal DNA also carries the information for the synthesis of other, minor cellular RNAs.

DNA synthesis requires a primer—an oligonucleotide that is base-paired with the template strand and provides a 3′ hydroxyl group upon which to start polymerizing a new DNA strand (see chapter 26). The cell uses small RNAs as primers for DNA replication. In some cases in prokaryotic cells these primers are synthesized by RNA polymerase (the main tran-

28A
BOX

Analysis of Ribosomal Proteins

Ribosomal proteins are conveniently analyzed by two-dimensional electrophoresis. Generally, a single system is sufficient to resolve all proteins from a single type of ribosomal subunit, and the mobility of the proteins on the gel is used to identify them. Thus the *E. coli* 30S subunit contains 21 proteins that are designated S (for small) 1 to 21. The 50S subunit proteins are designated L for large.

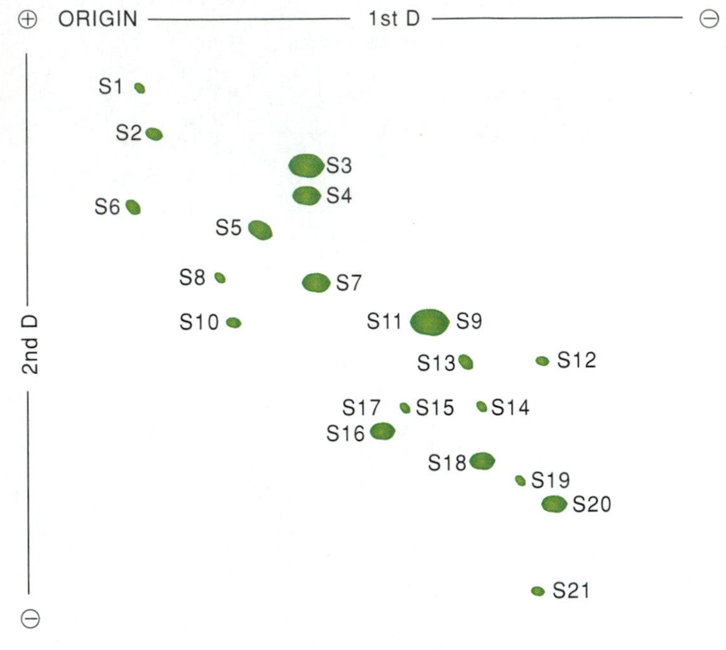

Figure 1

A two-dimensional gel electrophoretogram of *E. coli* 30S proteins. (Courtesy of Bishwajit Nag and Robert R. Traut.)

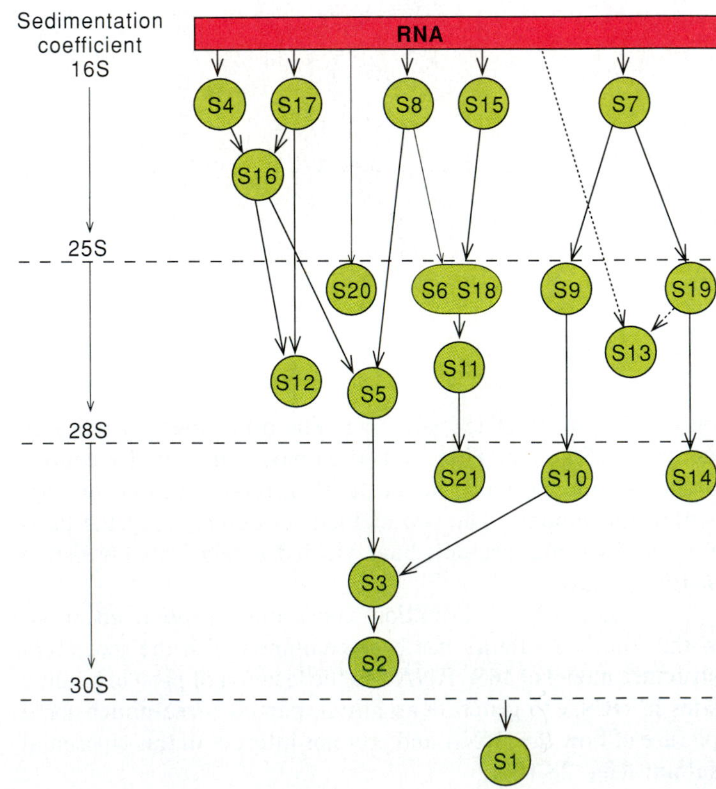

Figure 28.8

Assembly map of the 30S subunit. Initially only five ribosomal proteins associate with the rRNA. The binding of further proteins occurs after these initial proteins have bound, and these proteins make contacts with both the ribosomal RNA and the already bound proteins. The assembly process can be monitored by the changing sedimentation constant of the ribosome. The folding of the rRNA in the completed structure, together with the location of the individual ribosomal protein subunits, is illustrated in figure 28.10.

Figure 28.9

Gross shapes of the *E. coli* ribosomal subunits and the ribosome.

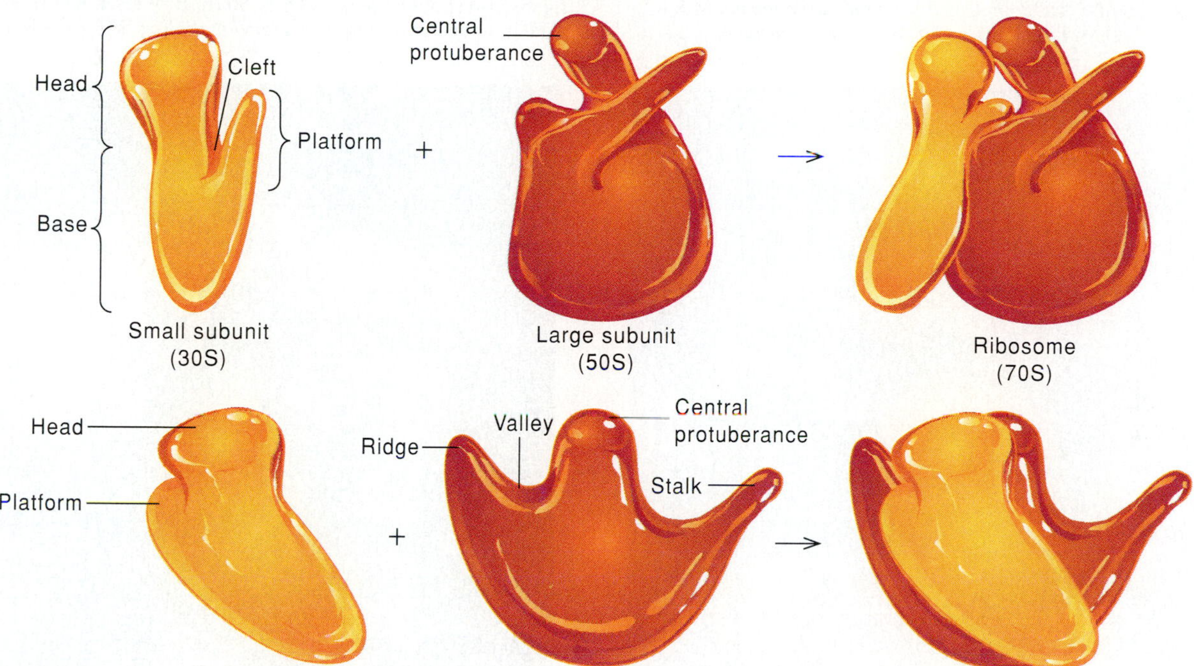

scribing enzyme), but most primers in *E. coli* are made by a special enzyme, called DNA primase, that is encoded by the *dnaG* gene.

Eukaryotic cells contain in their nuclei a unique collection of small RNAs, called small nuclear RNAs (snRNAs), which are complexed with certain proteins to form small nuclear ribonucleoprotein particles (snRNPs). These RNAs have been named U1, U2, U3, . . . , U13 RNA and range in size from 100 to 220 bases. One, U3, is found in the nucleolus, the site of rRNA synthesis. All are very abundant and there may be as many as 1 million copies per nucleus. Their base sequences are highly conserved among organisms, and all seem to contain unusual trimethylguanosine structures at their 5′ ends. Each snRNP contains an snRNA and six to twelve proteins, some of which are common to all snRNPs. The role of all of the snRNPs is not fully known, but they seem to be involved in processing and maturation of RNA precursors including splicing (U1, U2, U4, U5, U6), polyadenylation of pre-mRNA (U11), formation of 3′ ends of histones (U7), and maturation of rRNA (U3).

Transcription in Prokaryotes

There are many differences in synthesis and processing of RNAs in prokaryotes and eukaryotes. We will first deal with the reactions as they occur in *E. coli,* the best studied prokaryote. All DNA-dependent RNA polymerases carry out the following reaction:

$$\text{NTP} + (\text{NMP})_n \xrightarrow[\text{DNA}]{\text{Mg}^{2+}} (\text{NMP})_{n+1} + \text{PP}_i$$

The DNA template strand determines which base will be added to the growing RNA molecule, by base pairing similar to that used to direct the semiconservative replication of DNA. For example, a cytosine in the template strand of DNA means that a complementary guanine will be incorporated into the corresponding position of the RNA. Synthesis proceeds in a 5′ → 3′ direction, with each new nucleotide being added onto the 3′-OH end of the growing RNA chain.

E. coli *RNA Polymerase Transcribes the Three Major Classes of RNA*

Prokaryotic polymerases are used to transcribe all major genes of the prokaryotic genome. They do this in a discriminatory way so that the organism is supplied with just the amounts of appropriate mRNAs, tRNAs, and rRNAs that the organism requires.

The Bacterial RNA Polymerase Contains Five Subunits at the Time of Initiation The purification of substantial amounts of RNA polymerase from *E. coli* has made it possible to study the structure of the enzyme. Richard Burgess, Andrew Travers, John Dunn, and Ekehard Bautz showed that the active enzyme molecule is a pentamer containing four different polypeptide chains with a total molecular weight of about 500,000. The subunits of the enzyme can be separated by electrophoresis on polyacrylamide gels. The four different polypeptide chains, termed β', β, σ^{70}, and α, have molecular weights of 155,000, 151,000, 70,000, and 36,500, respectively. Additional σ-like proteins have

Figure 28.10

Arrangement of components in the *E. coli* 30S ribosomal particle. In (*a*) the relative locations of the ribosomal proteins, numbered 1-21, are shown. (Illustration prepared by Dr. Malcolm Capel from data described in M.S. Capel, M. Kjeldguard, D.M. Engelman, and P.B. Moore, *Journal of Molecular Biology* 200:66-87, 1988.) In (*b*) the conformation of the rRNA is shown. The location of the proteins is indicated by numbers given in the figure. (Illustration prepared by S. Stern, B. Weiser, and H. F. Noller from data described in *Journal of Molecular Biology* 204:447-481, 1988.)

(b)

also been identified, and we will discuss them later. Some of the properties of the polymerase subunits are summarized in table 28.2.

A complex with the subunit structure $\alpha_2\beta\beta'\sigma^{70}$ can carry out the functions necessary for correct and efficient synthesis of RNA and is referred to as holoenzyme. Holoenzyme can be reversibly separated into two components by chromatography on a phosphocellulose column:

$$\underset{\text{Holoenzyme}}{\alpha_2\beta\beta'\sigma^{70}} \rightleftarrows \underset{\text{Core polymerase}}{\alpha_2\beta\beta'} + \underset{\text{Sigma-70}}{\sigma^{70}}$$

One component, called core polymerase ($\alpha_2\beta\beta'$), retains the capability to synthesize RNA, but it is defective in the ability to bind and initiate transcription at the appropriate initiation sites (promoters) on the DNA. The other component, σ^{70}, has no RNA synthetic activity, but when added back to core polymerase, it re-forms holoenzyme with its ability to bind tightly and selectively at promoters and initiate RNA chains efficiently. Because of its role in binding and initiation, σ^{70} is often referred to as an initiation factor.

Although the precise functions of the various subunits of core are not known, β' is a basic (positively charged) polypeptide thought to be involved in DNA binding. The β subunit is the site of binding of several inhibitors of transcription and is thought to contain most or all of the active sites for phosphodiester bond formation. The α subunit is necessary in order to reconstitute active enzyme from separated subunits and appears to be the site of contact with a number of transcription factors, such as CAP.

Table 28.2
E. coli RNA Polymerase Subunits and Regulatory Factors

Subunit or Protein	Gene Name	Map Position (min)	Polypeptide M_r	No. in Enzyme	Function	Properties
β' (beta')	*rpoC*	90	155,000	1	DNA binding?	Basic
β (beta)	*rpoB*	90	151,000	1	Active site	Acidic
σ^{70} (sigma-70)	*rpoD*	67	70,000	1	Promoter recognition, initiation	Acidic
α (alpha)	*rpoA*	72	36,500	2	Site of activator contact	Acidic
CAP	*crp*	74	23,000	2	Activation	Basic
ρ (rho)	*rho*	85	46,000	6	Termination	Basic
nusA	*nusA*	69	55,000	1	Termination	Acidic

Genetics of the Subunits Since RNA polymerase is an essential cellular enzyme, most mutations in the polymerase genes are lethal. Those that can be studied are the ones that affect the enzyme activity only under certain conditions, e.g., at higher than normal temperatures. As we saw in chapter 26, mutants that can grow normally at one temperature but not at another are called temperature-sensitive mutants. Another class of mutants is composed of those in which the enzyme is resistant to an inhibitor of RNA synthesis, which acts by binding to the enzyme. We will give examples of such inhibitors later in this chapter.

Strains of *E. coli* have been obtained with mutations in the α, β, β', and σ^{70} subunits of RNA polymerase. These mutants have made it possible to map the genes for the corresponding subunits. The circular map of the *E. coli* genome is shown in figure 28.11; on it we show the positions of the genes known to be involved in RNA synthesis, modification, processing, and degradation. It is evident that these genes are scattered throughout the genome. At first it may seem surprising that the genes encoding the subunits of RNA polymerase are not all clustered together. Their dispersion is probably due to the fact that only β and β' are required in equimolar amounts. The genes for β and β' are in the same region at 90 min that contains the genes for two ribosomal proteins. The gene for α is found far from those for β and β', at 73 min on the map, and also is in a region containing genes for several ribosomal proteins. The gene for the σ^{70} subunit is found at 67 min, in a region containing the genes for DNA primase and a ribosomal protein.

Figure 28.11

Genetic map of *E. coli* showing the location of some genes involved in RNA metabolism. The map is divided into 100 minutes. Three types of genes are indicated: (1) genes involved in RNA synthesis, (2) genes involved in regulation, and (3) genes involved in processing and degradation. (Source: B. J. Bachmann, "Linkage map of *E. coli* K12, edition 7" in *Microbiological Reviews,* 47:180, 1983. Copyright © 1983 American Society for Microbiology, Washington, D.C.)

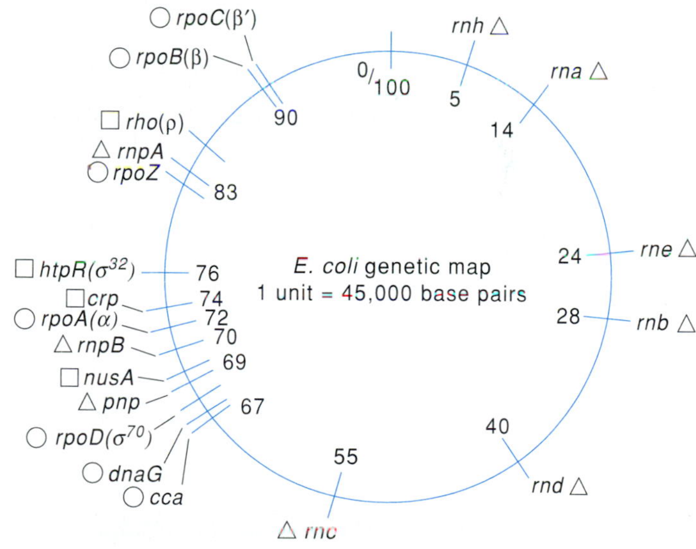

Transcription Involves Initiation, Elongation, and Termination

The overall transcription cycle, involving binding, initiation, elongation, and termination, is shown in figure 28.12. Let us now look at this cycle in some detail.

Transcription Units and Promoter Signals In addition to the sequences that code for proteins, precise sequences along the DNA signal RNA polymerase start and stop sites. The promoter region contains the information that tells the RNA

Figure 28.12

Schematic of the overall transcription cycle.

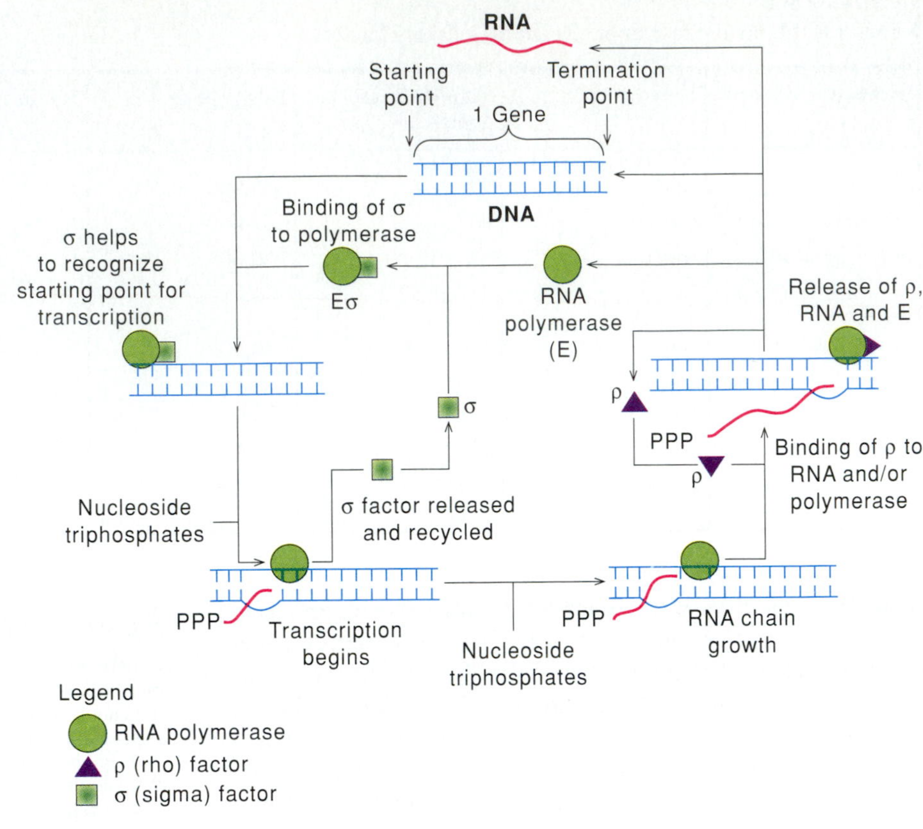

Figure 28.13

Important features of a typical transcription unit. DNA is shown with promoter and terminator regions expanded below. RNA is transcribed starting in the promoter region at +1 and ending after the stem and loop of the terminator. The protein resulting from translation of this RNA is shown above with its N and C termini indicated.

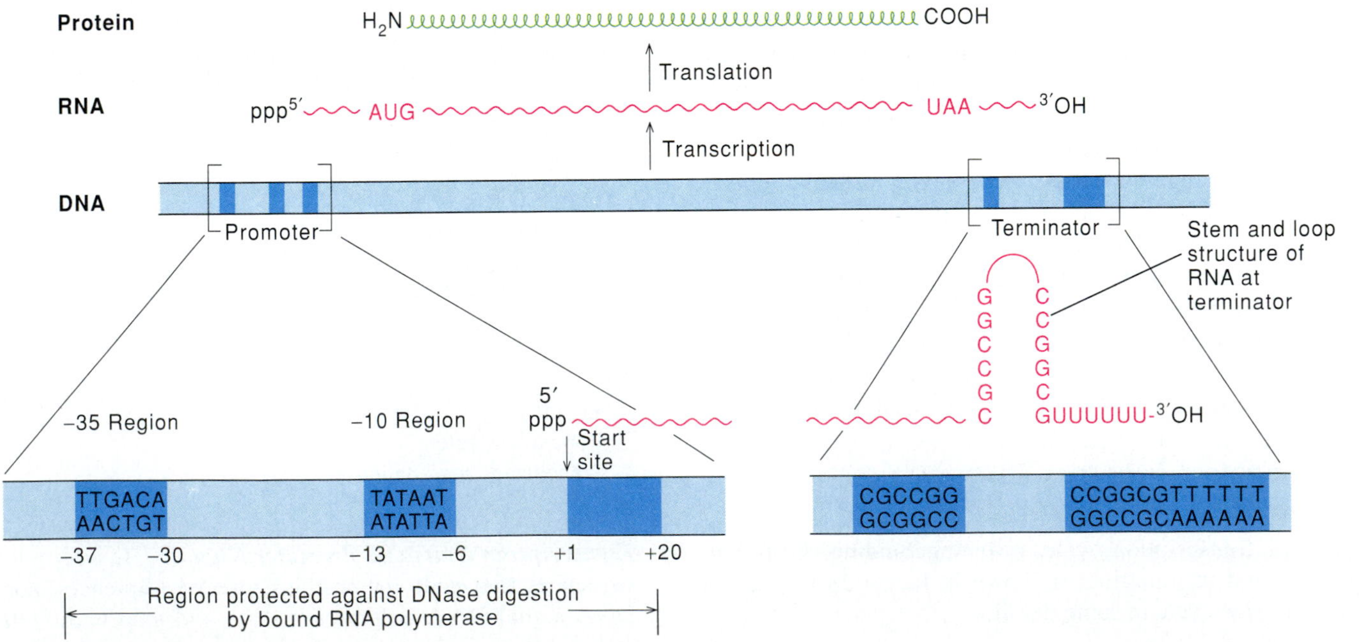

polymerase where to bind, how tightly to bind, and how frequently to initiate an RNA chain. It also often contains sites at which additional regulatory proteins bind; these proteins influence binding and initiation by RNA polymerase. The terminator region contains DNA sequences that cause the RNA polymerase to stop transcribing. Then, either spontaneously or with the aid of a termination factor, the RNA polymerase and RNA are released from the template. The transcribed region, including the start and stop signals, is called a transcription unit or operon. It may include the structural genes for several proteins whose syntheses are controlled coordinately. An example of a transcription unit in *E. coli* is shown in figure 28.13.

A large number of transcription units have been studied, and many promoter and terminator regions have been sequenced. Common features have been identified in the sequence of over 100 different promoters, and in the case of several promoters, further information about DNA bases important for binding has been obtained. Promoter features are summarized in figure 28.13.

RNA polymerase binding at the promoter protects about 60 base pairs of DNA from digestion by DNase. The protected region runs from −40 (40 bp before the RNA chain starting site) to +20 (20 bp after the starting site). Two sites within this 60-bp region, one centered at −35 and one centered at −10, contain specific recognition sequences that are common to most but not all bacterial promoters. Mutations in these two sites affect the ability of RNA polymerase to bind and initiate.

The DNA base sequence of each promoter differs somewhat from the "average" promoter sequence. This is to be expected because promoters differ tremendously in their "strength" or frequency of RNA initiation. Some promoters are very "weak"; for example, the promoter for the lac repressor* produces an mRNA only once in 20 to 40 min. This frequency is estimated as follows: Each *E. coli* contains about twelve molecules of lac repressor. Since each repressor is composed of four identical subunits, this represents about 48 molecules of repressor polypeptide. If each mRNA is translated by about 50 ribosomes before it decays, then only one mRNA is needed per generation of 20 to 40 min. In contrast, the promoters for rRNA genes must be utilized once every second or two in order to produce over 10,000 rRNAs per generation from the seven copies of the rRNA gene. This 2,000-fold difference in promoter strength is encoded in the base sequence of the promoter.

Binding at Promoters In the cell, RNA polymerase transcribes only selected regions of the DNA and makes RNA complementary to only one of the DNA strands in any particular region. This selectivity is possible because holoenzyme is able to recognize and form a stable complex with DNA at specific promoter regions. RNA polymerase is able to form unstable nonspecific complexes at any place on the template, mainly by an ionic interaction with the DNA phosphates, but it either rapidly dissociates and rebinds, or slides along the DNA until it reaches a promoter region. It then forms a moderately stable complex with the promoter DNA, most probably interacting stereospecifically with particular nucleotides in the −35 region of the promoter. These complexes can form at 0° C, where the DNA remains double helical or unmelted, and are referred to as closed promoter complexes (fig. 28.14). The next step involves a conformational change in the polymerase and a temperature-dependent melting of about ten base pairs of DNA from positions −9 to +2. The resulting very stable complex is termed the open promoter complex.

*The lac repressor will be discussed in chapter 30. It is not important at this point that we know anything about the lac repressor except what is stated here.

Figure 28.14

Various types of RNA polymerase-DNA binding complexes. A nonspecific complex is formed at any point along the DNA. The closed promoter complex is formed at a polymerase binding site. Following the formation of the closed promoter complex, an open promoter complex is formed at the same site.

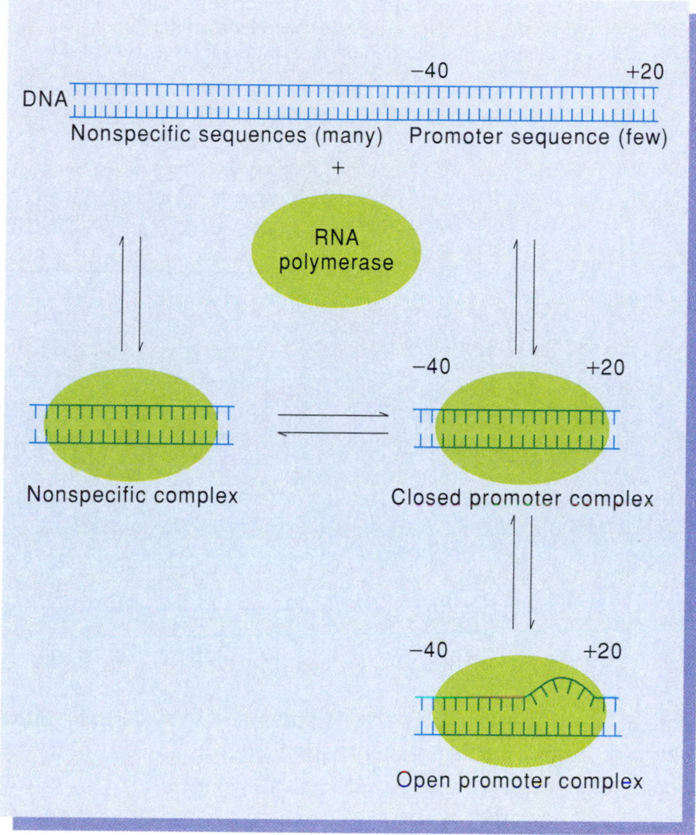

The existence of consensus sequences in both the −35 and −10 regions of the promoter carries with it the implication that these regions are important in forming the open promoter complex with RNA polymerase. One possibility is that both regions are complexed with polymerase in the open promoter complex. Another possibility is that one region serves for initial recognition of the polymerase, but that in the final complex the polymerase binds only to the other region.

Several approaches to establish more precisely the regions of contact (or close approach) between polymerase and DNA were developed in Walter Gilbert's laboratory. One

Figure 28.15

(*Upper*) Sequence of the T7 A3 promoter, showing the stronger contacts to the RNA polymerase. (▼ = phosphate contacts; ○ and ∧ = purines that the polymerase protects from methylation or whose susceptibility to this methylation is enhanced, respectively; * = methylated purines that interfere with polymerase binding.) The most probable bases for the −10 region and the −35 region are shown above the corresponding regions in the A3 promoter sequence; +1 represents the start of transcription. The minimal region unwound by the polymerase is represented by a separation of the strands. (*Lower*) Planar representation of the cylindrical projection of the DNA molecule (10.5 base pairs per turn), with contacts to the polymerase marked. (● = phosphate contacts ⊛ = methylated purines that interfere with polymerase binding, and other symbols are as in the upper drawing.) Contact regions, strands, front view, back view, and initiation site are indicated. Regions likely to interact with polymerase are shaded with vertical lines. (Source: U. Siebenlist and W. Gilbert, "Contacts between *E. coli* RNA polymerase and an early promoter of phage T7" in *Proceedings. National Academy of Sciences USA* 77:122, 1980. Copyright © 1980 National Academy of Sciences, Washington, D.C.)

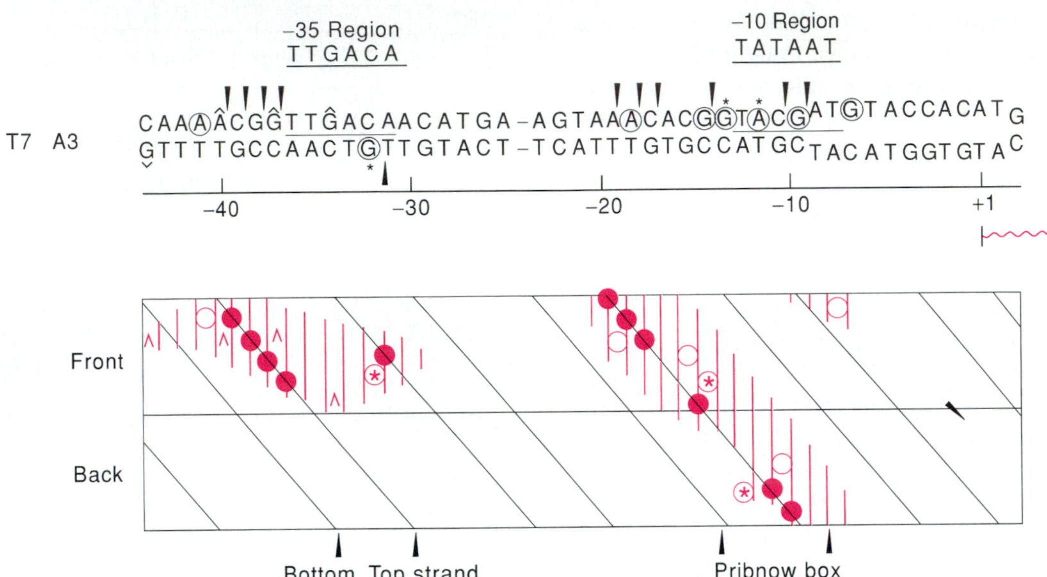

strategy used was to make the polymerase-DNA complex and then determine which regions showed altered susceptibility to specific chemical probes. Another strategy was to carry out a partial reaction with a specific chemical probe and see which sites interfered with polymerase complex formation. For these experiments, two chemical probes have been most useful. The first, dimethylsulfate, methylates double-stranded DNA at the N-7 position of guanine in the major groove and the N-3 position of adenine in the minor groove. The methylated purine will depurinate on heating; alkali then can cause a β-elimination reaction and create a series of breaks in the DNA chain. The second probe, ethylnitrosourea, preferentially ethylates the DNA phosphates. The resulting phosphotriesters serve as cleavage sites when the DNA is exposed to alkali.

The results of these experiments are depicted in figure 28.15. Domains of interaction are focused on the −35 region and the −10 region. This result provides strong support for the conclusion that polymerase binds simultaneously to both of these regions.

From the experiments with dimethylsulfate, evidence emerged that the polymerase produces an unwinding of a segment of the DNA duplex from −9 to +2. Dimethylsulfate, which reacts with the N-1 of adenine and the N-3 of cytosine only in single-stranded DNA, reacted with this region in the presence of polymerase. In the double helix these nitrogens are normally unavailable for reaction.

Another type of experiment was used to establish the thymine contacts with polymerase. For this purpose, thymine was replaced in the DNA by 5-bromouracil. Ultraviolet light normally cleaves DNA at the bromouracils. A nearby bound protein can either protect the DNA from such cleavage or become cross-linked to the DNA at position 5′ of the bromouracil. This approach was particularly valuable in suggesting which of the polymerase subunits are in direct contact with the DNA. Observations for the *lac*UV5 promoter show that the σ^{70} subunit cross-links at position −3 of the sense strand (the sense strand carries the same sequence as the transcribed RNA) and that the β subunit cross-links at position +3 of the sense strand.

The chemical-probe approach used to establish points of close contact between polymerase and DNA also provided useful information on regulatory protein interaction with DNA (see the discussion on CAP and *lac* repressor in chapter 30).

Initiation at Promoters Once the polymerase binds to the promoter and strand separation occurs, initiation usually proceeds rapidly (1 to 2 s). The first, or initiating, nucleoside-5′-triphosphate, which is usually ATP or GTP, binds to the enzyme. The binding is directed by the complementary base in the DNA template strand at position +1, the start site. A second nucleoside-5′-triphosphate (NTP) binds, and initiation occurs upon formation of the first phosphodiester bond by a nucleophilic attack of the 3′ hydroxyl group of the initiating NTP on the α phosphorus atom of the second NTP. Inorganic pyrophosphate derived from the second NTP is a product of the reaction. This process is illustrated in figure 28.16.

The initiation process can be followed during an *in vitro* transcription reaction in several ways. NTP radioactively labeled with ^{32}P in the β or γ phosphate positions can be used in transcription reactions. Since the initiating NTP retains its 5′ triphosphate the beginning of the RNA chain, or the 5′ triphosphate end, will be exclusively labeled. Another way is to add ^{32}P-labeled pyrophosphate ($\overset{**}{pp}_i$) to an otherwise unlabeled reaction and measure initiation by a reaction called pyrophosphate exchange. The labeled pyrophosphate participates in the reverse of the phosphodiester-bond formation reaction, allowing the 3′-terminal nucleotide to be removed and the label to be incorporated into its β and γ positions. Pyrophosphate exchange also can occur during chain elongation:

$$\text{pppApU} + {}^{32}\overset{**}{pp}_i \underset{\text{Bond formation}}{\overset{\text{Pyrophosphorolysis}}{\rightleftharpoons}} \overset{**}{pp}\text{pA} + \text{pppU}$$

A third method for measuring initiation depends on the finding that in the absence of the nucleotide needed for the third position on the RNA chain, the dinucleotide formed, pppXpY, dissociates from the active site and the initiation process must start again. This phenomenon is called abortive initiation. For example, if only ATP and [α-^{32}P]UTP are added to a reaction mixture where an RNA chain that begins with the sequences pppApUpCp is being synthesized, the reaction produces the dinucleotide pppA^{32}pU. The production of this dinucleotide can be quantified to measure initiation of this chain. Some promoters *in vitro* undergo considerable abortive initiation, even in the presence of all four nucleoside triphosphates, releasing oligonucleotides of two to eight residues several times before finally succeeding in producing a long RNA chain. Therefore, promoter strength should be equated not strictly with the frequency of initiation at a promoter, but rather with the frequency with which long RNA chains are produced.

Elongation of RNA After initiation has occurred, chain elongation proceeds by the successive binding of the nucleoside triphosphate complementary to the next base in the template strand, bond formation with pyrophosphate release, and translocation of the polymerase one base farther along the template strand. Transcription proceeds in the 5′ → 3′ direction, antiparallel to the 3′ → 5′ strand of the template DNA. Once elongation has produced an RNA chain about ten bases long, the σ^{70} subunit dissociates from the holoenzyme to yield core polymerase, which continues the elongation reaction until a termination signal is reached. The released σ^{70} is available to bind to a free core polymerase and re-form holoenzyme capable of binding at a promoter and initiating a new RNA chain (see fig. 28.12).

In vivo, mRNA chains grow at a rate of about 45 nucleotides per second, nicely matched with the rate of translation of 15 amino acids per second. rRNA is synthesized about twice as fast as mRNA.

Figure 28.16

Details of phosphodiester bond formation. The α, β, and γ phosphates are indicated on the initiating NTP, which in this case is ATP. The colored ovals represent NTP binding sites on the RNA polymerase. The biochemistry of bond formation in RNA synthesis is very similar to that in DNA synthesis.

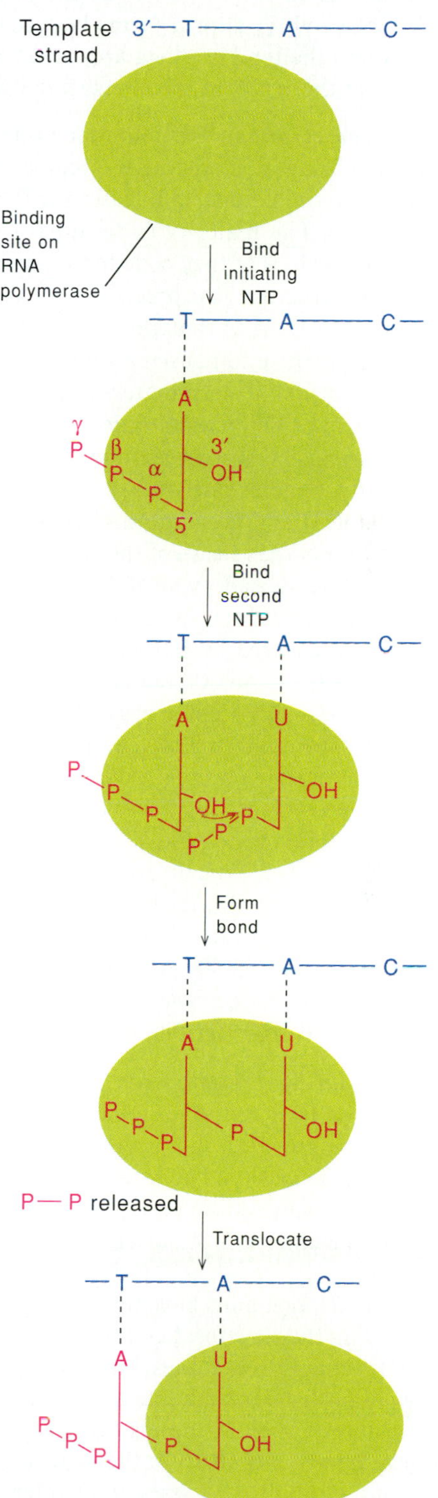

As the polymerase travels along the DNA, it must continually cause an opening or strand separation of the DNA so that a single DNA template strand is available at the active site of the enzyme. For the transcription reaction to be energetically feasible, one base pair must re-form behind the active site for every base pair opened in front of it. It is likely that a short transient RNA-DNA hybrid duplex forms between the newly synthesized RNA and the ten-base-long unpaired region of the DNA and helps hold the RNA to the elongating complex.

Termination of Transcription Termination of transcription involves stopping the elongation process at a region on the DNA template that signals termination and releases the RNA product and RNA polymerase. The majority of terminators that have been studied are similar in that they code for a double-stranded RNA stem-and-loop structure just preceding the 3′ end of the transcript (see fig. 28.13). It is thought that such a structure causes the RNA polymerase to pause or stop elongating.

Two main types of terminators have been distinguished. The first is capable of termination with no accessory factors and contains about six uridine residues following the stem and loop. Apparently, a particularly weak interaction between the 3′ terminal oligo(U) and the oligo(dA) region in the template DNA strand allows the release of the RNA from the transcription complex. The second type of terminator lacks the oligo(U) region and requires termination factor rho for RNA chain release. Rho, discovered by Jeffery Roberts in 1969, is a hexamer of subunits (M_r = 46,000 per subunit). Rho does not bind tightly to DNA or to RNA polymerase. It does bind tightly to RNA, especially cytosine-rich RNA, and in its presence rho hydrolyzes ribonucleoside triphosphates to nucleoside diphosphates. A current theory is that rho binds to sites on RNA and, by hydrolyzing nucleoside triphosphates, either moves along the RNA or winds the RNA around it until it reaches an RNA polymerase that has stopped at a terminator, where rho causes release. Another termination factor, nusA, is an acidic protein with a molecular weight of 55,000 that is able to bind to core polymerase after sigma is released and aid in the release of RNA and RNA polymerase at some terminators.

Other Factors Regulating Transcription As more transcription units are studied in detail, it becomes apparent that the basic processes just described can be modulated in a great number of ways, both by including additional signals in the DNA sequence and by supplying additional regulatory factors. Some of these regulatory mechanisms are discussed in more detail in chapters 30 and 31.

Many transcription units have terminator signals called attenuators preceding the structural gene or between two structural genes. Attenuators sometimes cause termination and sometimes allow read-through to the structural part of the gene. The efficiency of termination at these sites is variable and can be regulated in response to changes in the growth conditions of the cell. Attenuators are also discussed in chapter 30.

Protein factors can inhibit transcription (negative control) or stimulate it (positive control). The binding of a repressor molecule to a specific site (operator) overlapping the promoter region prevents transcription by sterically interfering with RNA polymerase binding. This is an example of negative control. An example of positive control is the stimulation of transcription caused by the binding of the catabolite activator protein (CAP) in the presence of cyclic AMP (cAMP) to sites just adjacent to the promoters of a number of genes coding for enzymes involved in the catabolism of a variety of sugars. In *in vitro* transcription reactions, these promoters bind RNA polymerase holoenzyme only weakly and are greatly activated by the addition of CAP and cAMP.

Although most transcription in *E. coli* appears to utilize σ^{70} as an initiation factor, several additional initiation factors have been discovered that bind to core polymerase to form holo-like enzymes that recognize different promoters. One such factor is the product of the *htpR* gene, which is involved in heat shock regulation. This factor is called sigma-32 (σ^{32}) because its molecular weight is 32,000. It becomes bound to core polymerase in *E. coli* cells that have been subjected to heat shock and directs the polymerase to bind to and initiate at a class of promoters (the heat shock promoters) responsible for high-level expression of a class of a dozen or so heat shock proteins. Another such factor, σ^{54}, is a protein (M_r = 54,000) encoded by the *ntrA* gene that is needed for expression of certain genes implicated in nitrogen metabolism. Recently three additional sigmas have been identified. Sigma-E is thermostable and recognizes the promoter for σ^{32}. Sigma-F (M_r of about 28,000) is involved in transcription of several genes coding for flagellar proteins and enzymes involved in chemotaxis. Sigma S (M_r = 41,500), a product of the *katF* gene, is involved in regulating *katE* and other genes turned on during carbon starvation. The consensus sequences associated with different classes of promoters are given in table 28.3.

Multiple sigmas have also been found in the bacterium *Bacillus subtilis.* Several different σ-like subunits are present in growing cells, and additional ones appear during sporulation to allow expression of sporulation-specific genes. Furthermore, the infection of *B. subtilis* with certain bacteriophages, such as SP01 or SP82, results in virus-coded σ-like factors that bind to core polymerase in place of the major sigma, σ^{43}, and direct RNA polymerase to viral promoters.

Differences between Eukaryotic and Prokaryotic Transcription

Although the basic mechanism by which RNA is synthesized is quite similar in prokaryotes and eukaryotes, there are several important differences. The DNA in eukaryotic cells is complexed to form chromatin (see chapter 25). The chromatin, in turn, is condensed to form chromosomes. Only a small fraction

Table 28.3
Summary of *E. coli* Sigma Factors

Sigma Factor	Gene	Consensus Sequence −35 Region	−10 Region	Genes Recognized
σ^{70}	*rpoD*	TTGACA	TATAAT	Most genes
σ^{32}	*rpoH (htpR)*	CCCTTGAA	CCCCAT-TA	Heat-shock regulated
σ^{54}	*rpoN (ntrA)*	$CTGGCACN_5TTGCA$		Nitrogen-regulated
σ^{E}	Not identified	GAACTT	TCTGA	*rpoH, htrA*
σ^{F}	fliA	TAAA	GCCGATAA	Flagellar, chemotaxis
σ^{S}	*rpoS (katF)*	Unknown		*katE,* Starvation regulated

of the chromosome is actively being transcribed at any given time. In transcriptionally active regions, the DNA must be partially exposed and be more accessible to RNA polymerase. The DNA in these exposed regions is found to be sensitive to cleavage by mild treatment with bovine pancreatic DNase I and appears to contain bound RNA polymerase, additional nonhistone proteins, and modified histones. In addition, DNA in active regions in some species is undermethylated when compared with DNA from regions not active in transcription. Although prokaryotic DNA is not complexed with histones, it seems to be coated with small basic proteins; however, most parts of the bacterial DNA are accessible to RNA polymerase binding and transcription.

As we stated earlier, transcription and translation occur simultaneously in prokaryotes. RNA polymerase moves along a gene generating an mRNA chain, and as soon as a bit of mRNA is synthesized, ribosomes bind to it and start translating it. The situation is very different in eukaryotes, where transcription and translation occur in separate compartments of the cell. The nucleus, where the chromosomes are located, is the site of DNA-dependent RNA synthesis. Large RNA precursors to mRNA are synthesized in the nucleus, become complexed with proteins to form ribonucleoprotein particles (RNPs), and then are modified, and processed to form smaller mRNAs that are transported across the nuclear membrane to the cytoplasm, where they are translated into protein. These differences lead to a complex process of modification and processing in the eukaryotic RNAs.

Eukaryotes Have Three Nuclear RNA Polymerases

Unlike prokaryotes, in which all major types of RNA are synthesized by one RNA polymerase holoenzyme, eukaryotic cells have become more specialized in their transcription capabilities. They contain different nuclear DNA-dependent RNA polymerases, each responsible for synthesizing different classes of RNA.

Nuclear extracts can be fractionated by chromatography on DEAE-cellulose to give three peaks of RNA polymerase activity. (The use of column chromatography is explained in chapter 5.) These three peaks correspond to three different RNA polymerases (I, II, and III), which differ in relative amount, cellular location, type of RNA synthesized, subunit structure, response to salt and divalent cation concentration, and sensitivity to the mushroom-derived toxin α-amanitin. The three polymerases and some of their properties are summarized in table 28.4.

RNA polymerase I is located in the nucleolus and synthesizes a large precursor (fig. 28.17) that is later processed to form rRNA. It is resistant to inhibition by α-amanitin at concentrations even greater than 1,000 μg/ml. RNA polymerase II is located in the nucleoplasm and synthesizes large precursor RNAs (sometimes called heterogeneous nuclear RNA or hnRNA) that are processed to form cytoplasmic mRNAs. It is also responsible for the synthesis of most viral RNA in virus-infected cells. It is very sensitive to α-amanitin, being 50% inhibited by about 0.05 μg/ml. RNA polymerase III is also located in the nucleoplasm and synthesizes small RNAs such as 5S RNA and the precursors to tRNAs. This enzyme is somewhat resistant to α-amanitin, requiring about 5 μg/ml to reach 50% inhibition.

All three of these enzymes have been purified extensively and have complex subunit structures. All contain two subunits with molecular weights larger than 120,000 and six to ten smaller subunits. Although each has some unique subunits, some subunits are common to two or three. At present, very little is known about the functions of the individual subunits. The subunit structure for RNA polymerase II is quite similar for enzymes purified from a variety of eukaryotes, including human, calf, mouse, wheat, cauliflower, acanthamoeba, yeast, and slime molds.

The genes for the largest subunits of RNA polymerases I and II from yeast and of RNA polymerase II from *Drosophila* have been cloned and sequenced. Similar amino acid sequences are present in several regions of these three subunits and, surprisingly, are also found in the β' subunit of *E. coli.*

Crude enzyme preparations of RNA polymerases I, II, and III have been shown to be capable of selective transcription of defined DNA templates, initiating at sites known in some

Table 28.4
Comparison of Eukaryotic DNA-Dependent RNA Polymerases

Type	Location	RNAs Synthesized	Sensitivity to α-Amanitin
RNA polymerase I	Nucleolus	Pre-rRNA	Resistant
RNA polymerase II	Nucleoplasm	hnRNA, mRNA	Sensitive
RNA polymerase III	Nucleoplasm	Pre-tRNA, 5S RNA	Sensitive to very high levels
Mitochondrial	Mitochondria	Mitochondrial	Resistant
Chloroplast	Chloroplasts	Chloroplast	Resistant

Figure 28.17

Electron micrograph of regions of the nucleolus of the newt *Triturus* that are believed to be active in transcribing ribosomal RNA. Clarifying diagram below. (Magnification 18,000 X.) (From O. L. Miller, Jr., and B. R. Beatty, "Visualization of nuclear genes," *Science* 164:955-957, 1969; cover photo of *Science* 23 May 1969. Copyright 1969 by AAAS.)

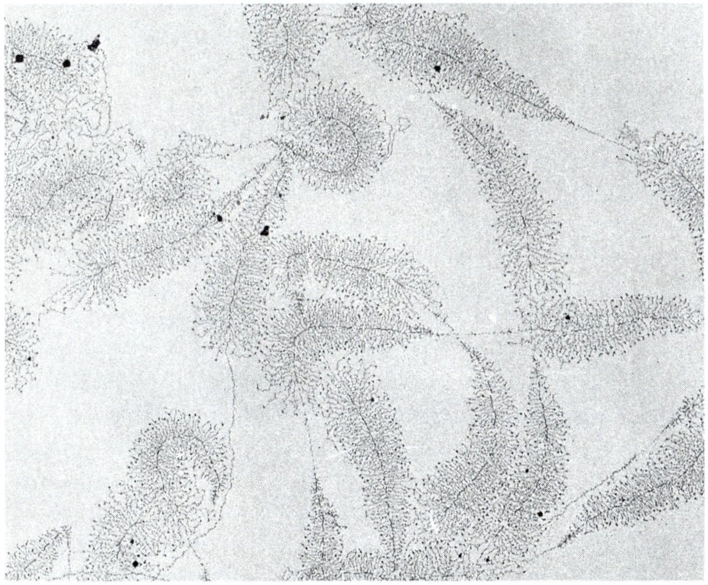

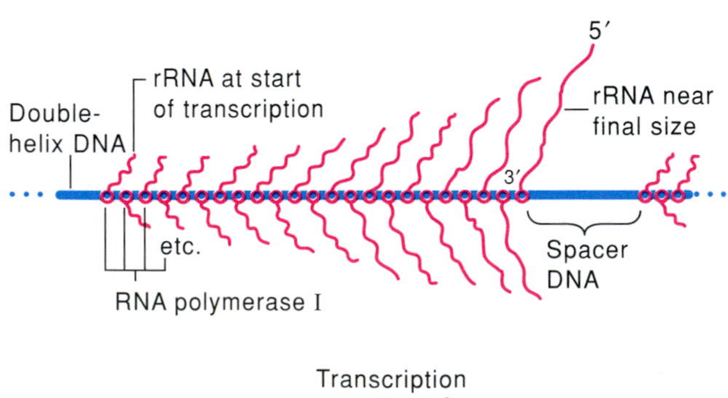

cases to be utilized *in vivo.* Fractionation of these extracts has revealed additional proteins that are required for *in vitro* transcription. These additional proteins are called transcription factors (TFs). With all three polymerases, two or more transcription factors are required, and some or most of these factors must bind to the promoter before the RNA polymerase can bind (fig. 28.18).

Because many eukaryotic genes have been cloned during the last few years, it has become possible to compare the DNA sequences preceding the genes that may act as promoterlike signals for RNA polymerase II. One feature that stands out is a common sequence, TATAAA, often called a TATA box, usually found 25 to 30 bases before the transcription start site in many but not all genes. Certain proteins selectively bind to the TATA box and these proteins appear to be essential for the formation of the initiation complex.

Phil Sharp and Leonard Guarente have shown that four transcription factors are required in addition to polymerase II for initiation from the major late promoter of adenovirus. *In vitro* studies indicate that these factors assemble in an orderly fashion (see fig. 28.18*b*). Thus complexes were generated by sequential binding of TFIID, TFIIA, TFIIB, RNA polymerase II, and TFIIE. TFIIA derived from yeast or mammalian cells forms a complex with yeast TFIID and the TATA element. TFIIB binds to this complex and probably acts as a bridge to the polymerase and the initiation site. Finally TFIIE binds. It is believed that this complex may function for a large number of eukaryotic promoters that contain TATA boxes. Progress in our understanding in this area promises to be accelerated by the observation that many transcriptional factors are interchangeable between yeast and mammals. For investigatory purposes yeast has distinct advantages to mammals, as it is more manipulable both genetically and biochemically.

Several additional promoter regions have been identified that are needed for elevated levels of transcription. These additional regions tend to be gene-specific. Both yeast and vertebrate (fig. 28.19) genes often have regions from about −200 to −40 bases upstream from the transcription start site where upstream activator elements (UASs) are located. These elements bind additional transcription activation factors, which

Figure 28.18

Formation of the initiation complex for transcription for the three major classes of eukaryotic RNA polymerase, Pol I, Pol II, and Pol III. Each polymerase consists of many subunits (not shown). In addition to the firmly bound subunits there are a number of protein factors called transcription factors (TFs) which only associate with the polymerases at the initiation site for transcription. For all three polymerases some of these transcription factors must bind to the promoter before the polymerase can bind. In each case we see an orderly progression of binding of factors and polymerase. In the case of Pol III, three transcription factors are involved for 5S rRNA genes and only two are involved for tRNA genes.

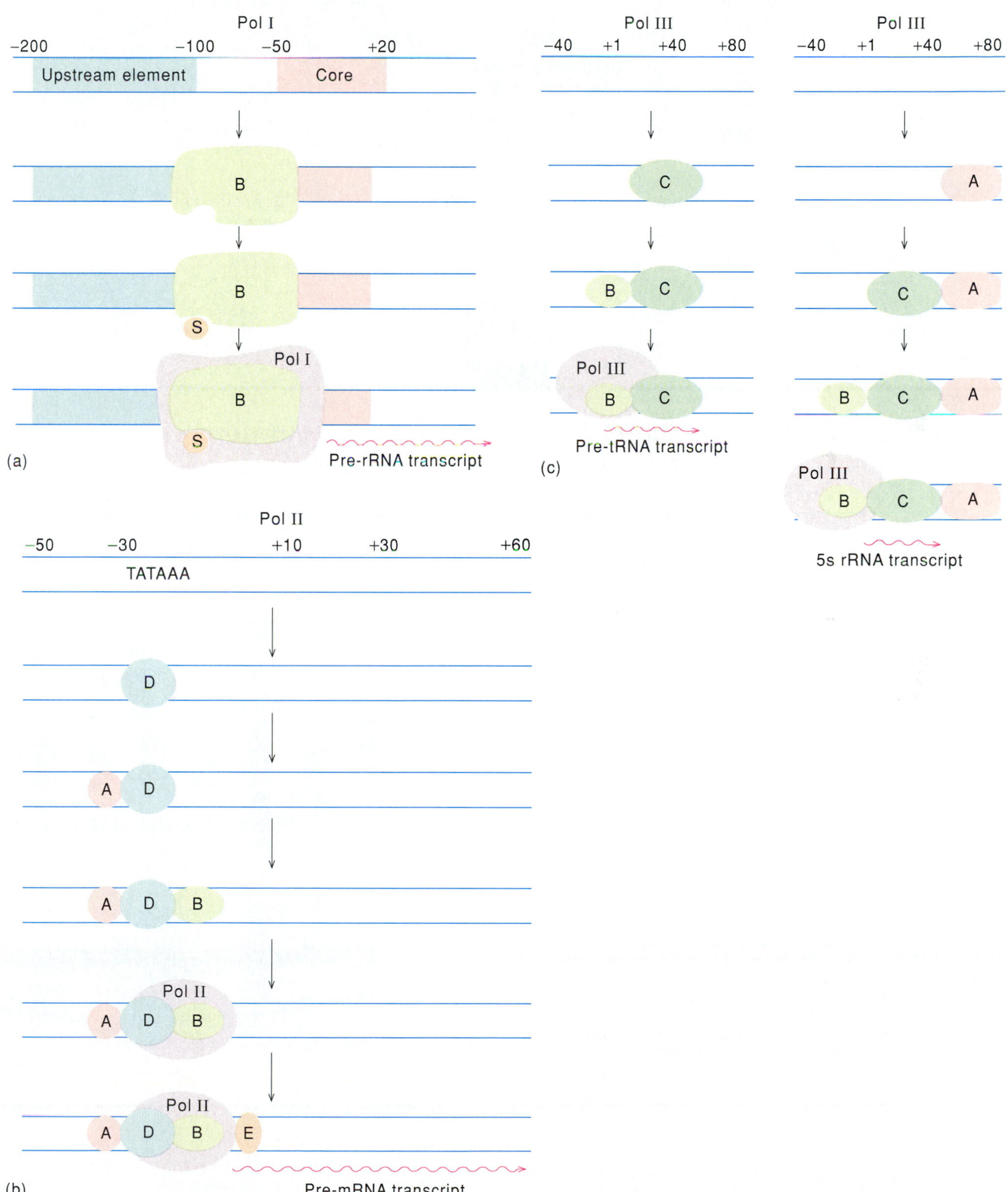

Figure 28.19

Cis elements involved in transcription in yeast (*a*) and in vertebrates (*b*). Upstream activator sequences in yeast are similar to upstream enhancers in vertebrates. Yeast has no parallel to downstream enhancers found in vertebrates.

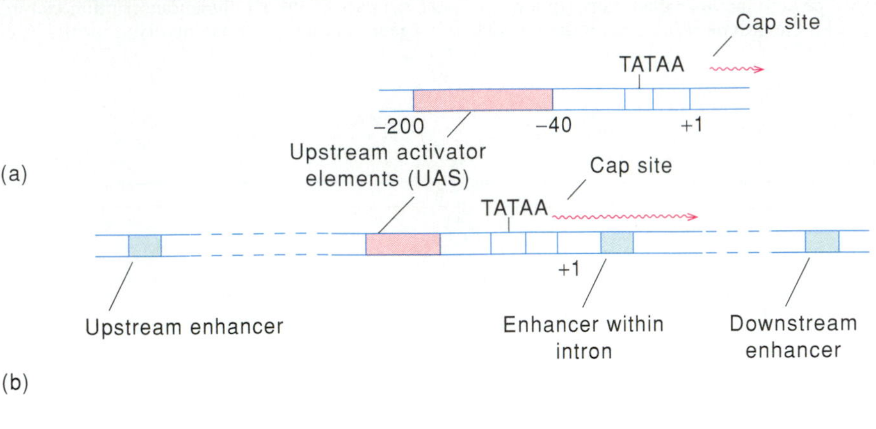

28B BOX

Use of the Footprinting Technique to Determine the Binding Site of a DNA-Binding Protein

In order to confirm the site of action of TFIIIA, the approximate location of its binding site on a 5S gene of *Xenopus borealis* was determined by the so-called footprinting technique. DNA fragments containing the 5S gene were 5′ end labeled with ^{32}P, mixed with the protein, and digested with DNase I. The resulting DNA fragments were electrophoresed on a polyacrylamide gel and autoradiographed. The region of the DNA that was protected from DNase attack by TFIIIA binding appears as a series of blank spots ("footprint") on the autoradiogram. The footprint shows that the protected region is situated between the 45th and the 90th base pairs.

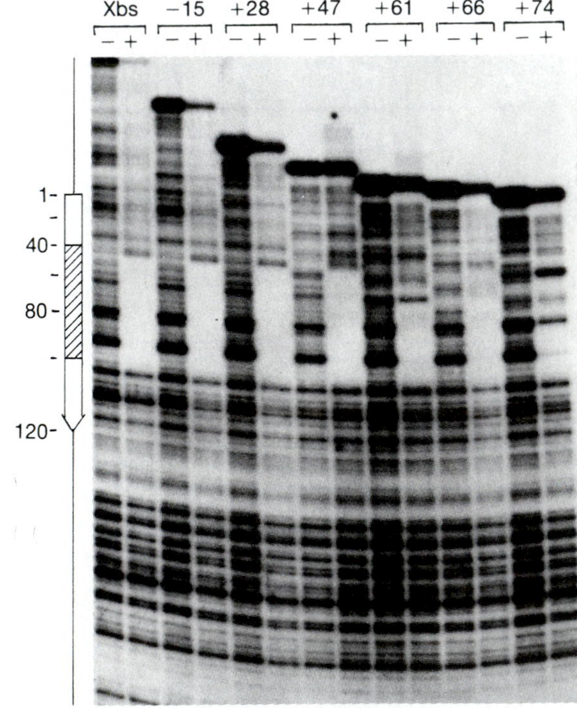

Figure 1

DNase I protection (footprinting) experiment on 5S DNA of *Xenopus*. The diagram on the left indicates the region on the gel that corresponds to the 5S RNA gene. Arrow points in the direction of transcription. Cross-hatched area indicates the region that binds transcription factor protein. Column labeled Xbs refers to an intact gene containing 160 bp of the 5′ flanking sequence, the 5S RNA gene (120 bp), and the 3′ flanking sequence (138 bp). The various deleted 5S DNAs are preceded by 74 bp of the plasmid pBR322 sequence. Numbers in other columns refer to portions of the 5S gene that have been deleted. All samples were subjected to partial digestion with DNase I, then were electrophoresed and autoradiographed. In + columns, transcription factor protein was added before DNase I treatment. (Courtesy Donald D. Brown of the Carnegie Institute of Washington.)

Table 28.5
RNA-Synthesizing Enzymes

	Template	Primer	Molecular Weight of Subunit(s)	Gene Name	Substrate	Inhibition by Rifampicin
Template-dependent						
Enzymes from bacteria						
Holoenzyme (*E. coli*)	DNA	—	155,000 151,000 70,000 36,500	*rpoC* *rpoB* *rpoD* *rpoA*	4 NTPs	Yes
DNA primase	DNA	—	65,000	*dnaG*	4 NTP, 4 dNTP	No
Enzymes from phage or phage-infected bacteria						
T7 RNA polymerase	T7 DNA	—	99,000	T7 *gene1*	4 NTPs	No
N4 RNA polymerase	N4 DNA	?	350,000	Viral	4 NTPs	No
Qβ replicase	Qβ RNA	—	65,000 55,000 43,000 35,000	*rpsA* Viral *tuf* *tsf*	4 NTPs	No
Template-independent						
CCA enzyme	—	3′ end tRNA	45,000	*cca*	CTP ATP	No
Poly(A) polymerase (eukaryotic)	—	3′ end mRNA			ATP	No
Polynucleotide phosphorylase	—	3′ end RNA	86,000 48,000	*pnp*	4 NDPs	No

interact with those that bind at the TATA box (fig. 28.19). In addition, vertebrate DNA contains enhancer elements, which activate specific genes at some distance. They can be found upstream, downstream, and even in the middle of genes. Enhancer elements may influence transcription as far as 10 kb from the gene. As in the case of the UASs, the enhancer elements are binding sites for additional transcription activators. We will have considerably more to say about these additional transcription activator elements in chapter 31, where we focus on the regulation of gene expression.

The promoter region for RNA polymerase I also involves the cooperative binding of additional transcription factors. In this case the transcription factors, of which there are two, bind upstream of the transcription start site (see fig. 28.18*a*).

In contrast to the initiation complex for RNA polymerases I and II, the region necessary for selective transcription of 5S RNA by RNA polymerase III is located in a region 40 to 80 bases downstream of the transcription start site (see fig. 28.18*c*). One of the additional proteins needed for selective transcription of the 5S gene (TFIIIA) binds to this site and in conjunction with two other protein factors (TFIIIB and TFIIIC), directs RNA polymerase III to bind and initiate transcription.

The binding site for factor TFIIIA was determined by the DNA "footprinting" technique (box 28B), using a similar approach to that used to determine the location of binding of the *E. coli* polymerase in the T7 promoter (see fig. 28.15).

The Pol III polymerase is also used to transcribe tRNA genes. In this case TFIIIA is not used, but the other two transcription factors are still used (see fig. 28.18*c*).

Chloroplasts Have Their Own RNA Polymerases

In addition to nuclear RNA polymerases, certain organelles encode their own RNA polymerases. The RNA polymerase found in chloroplasts is very similar in structure to the bacterial RNA polymerase, a fact supporting the notion that these organelles are bacterial in origin. By contrast, RNA polymerases found in mitochondria often contain only one protein subunit.

Other RNA Synthesis

Although most RNA in cells is synthesized by cellular DNA-dependent RNA polymerases, there are several other enzymes that are capable of forming phosphodiester bonds and of synthesizing additional RNA in normal cells or virus-infected cells. Some properties of these enzymes are summarized in table 28.5.

DNA Primase Makes Primer RNA for DNA Synthesis

RNA primers used for the initiation of DNA synthesis during replication are made by DNA primase, the protein product of the *dnaG* gene of *E. coli* (also see chapter 25). It usually binds to DNA in association with another protein, the product of the *dnaB* gene, although it can initiate primer synthesis by itself at certain hairpin structures in single-stranded DNA. The primase can use either NTPs or dNTPs as substrates *in vitro* and synthesizes primers 10 to 50 nucleotides long that are complementary to the DNA template. Synthesis is in the $5' \rightarrow 3'$ direction. As we mentioned earlier, RNA polymerase is also capable of synthesizing primers. The use of primase or RNA polymerase to make primer is controlled by the DNA. One or the other is used exclusively in specific instances.

Many Viruses Make Their Own RNA Polymerases

There are three strategies used by different DNA viruses to accomplish transcription of viral DNA. The first type utilizes the host RNA polymerase, in some cases modifying it or synthesizing new promoter-specificity factors to direct it to read the viral promoters. Examples of such viruses are bacteriophage ϕX174, λ, and T4 (of *E. coli*), and SP01 and SP82 (of *B. subtilis*), as well as animal viruses SV40 and adenovirus.

The second type of virus utilizes the host RNA polymerase to transcribe some early expressed viral genes. One of these "early" genes codes for a new RNA polymerase that transcribes exclusively the remaining "late" viral genes. *E. coli* bacteriophages T7, T3, and SP6 are the best-known examples of this type. T7 RNA polymerase is a single polypeptide with a molecular weight of 99,000. It is not inhibited by the antibiotics rifampicin and streptolydigin, which inhibit host RNA polymerase. It recognizes specifically the T7 late promoters, all of which contain a nearly identical sequence of 18 to 22 nucleotides just before the 5′ triphosphate terminal GTP start site. T7 RNA polymerase is also capable of termination at specific points on the template. Thus it seems to be able to carry out the basic polymerization reaction and specific initiation and termination with a much simpler subunit structure than the host holoenzyme. However, with simplicity it loses versatility. Unlike the host polymerase, it is not able to recognize a wide variety of related but nonidentical promoter sequences, nor is it able to be regulated by positive and negative control factors that act at sites of initiation and termination.

A third type of virus, exemplified by bacteriophage N4, carries a viral RNA polymerase within its virion. This enzyme enters the cell along with the viral DNA and transcribes some early viral genes. Some of these genes code for specificity factors that direct the host RNA polymerase to transcribe late genes. Vaccinia virus is another example of a virus that contains a virion-encapsulated RNA polymerase. It shares some of the properties of the N4 enzyme.

RNA-Dependent RNA Polymerases of RNA Viruses The RNA genomes of single-stranded RNA bacterial viruses (such as Qβ, MS2, R17, and f2) are themselves mRNAs. Bacteriophage Qβ codes for a polypeptide ($M_r \approx 55{,}000$) that combines with three host proteins to form an RNA-dependent RNA polymerase (replicase). The three host proteins are ribosomal protein S1 (*rpsA* gene) and two elongation factors for protein synthesis, EF-Tu (*tuf* gene) and EF-Ts (*tsf* gene) (see table 28.5). The Qβ replicase can use the Qβ RNA plus strand as a template to make a complementary RNA transcript (minus strand); following this the minus strand is used as a template to make more viral RNA plus strands (see fig. 28.1). Like the DNA-dependent RNA polymerases, the replicase utilizes ribonucleoside-5′-triphosphates and transcribes in a $5' \rightarrow 3'$ direction. The phage RNA must first act as an mRNA to direct the synthesis of a component of the replicase, since uninfected cells do not have an RNA-dependent RNA polymerase or replicase.

RNA tumor viruses (retroviruses) that infect animal cells exhibit a different replication strategy. They carry in their virion an enzyme that can use the RNA viral genome as a template to synthesize a DNA copy. Since this process is the reverse of transcription, the enzyme is called reverse transcriptase. The result is a DNA-RNA hybrid. A second DNA strand is then synthesized, displacing the RNA strand. Once a double-stranded DNA copy is made, it is integrated into the host genome, and additional virus RNA is synthesized by the host RNA polymerase II in a normal DNA-dependent fashion.

Reverse transcriptase activity has been found in virus-free cultures of yeast and fruit flies, a fact suggesting that this enzyme may be involved in normal cellular metabolism. There is evidence that certain segments of the genomic DNA can jump from one location to another with the combined assistance of RNA polymerase and reverse transcriptase. This capability could provide a major mechanism for evolutionary change. Finally, reverse transcriptase has been found in bacteria and is associated with the synthesis of a molecule containing both RNA and DNA sequences.

3′-End Addition Enzymes Add Nucleotides to tRNA

Three enzymes are known that add ribonucleotides posttranscriptionally to the 3′ hydroxyl end of RNA. None of them are DNA-dependent. One adds the CCA sequence that all tRNAs have at their 3′ ends. The 3′ terminal adenine is the base to which the amino acid is covalently attached by the aminoacyl-tRNA synthase (see chapter 29). The 3′-CCA is relatively unstable and is continually being added when needed by this enzyme, called the CCA enzyme, or tRNA nucleotidyltransferase.

In eukaryotes, 100 to 200 adenosine residues are added to the 3′ end of most mRNAs by a poly(A) polymerase. This addition occurs in the nucleus before the mRNA is processed and transported to the cytoplasm, as we will describe later. The third enzyme capable of adding nucleotides posttranscriptionally to the end of RNA is polynucleotide phosphorylase.

Polynucleotide Phosphorylase Makes Polynucleotides from Nucleotide Diphosphates

Polynucleotide phosphorylase was the first enzyme found that could synthesize long polynucleotide chains *in vitro*. Two forms of the enzyme have been isolated, a form with three identical subunits (M_r = 86,000) and a form with these subunits and two additional subunits (M_r = 48,000). Unlike all the other enzymes we have discussed, polynucleotide phosphorylase utilizes ribonucleoside-5′-diphosphates instead of triphosphates as substrates for RNA synthesis. It catalyzes the reaction

$$\text{NDP} + (\text{NMP})_n \xrightleftharpoons{\text{Mg}^{2+}} (\text{NMP})_{n+1} + \text{P}_i$$

Polynucleotide phosphorylase does not require a template and randomly incorporates bases into RNA, depending on the relative concentration of the four NDPs in the reaction medium. The enzyme takes advantage of a primer if one is available to provide a 3′ hydroxyl end and synthesizes RNA in the 5′ → 3′ direction. The reaction is readily reversible, and we do not know whether this enzyme plays a role primarily in degradation or in synthesis of RNA in the cell.

RNA Ligase of Bacteriophage T4 Links RNAs Together

Perhaps the most unusual enzyme capable of synthesizing RNA is the bacteriophage T4 RNA ligase. It can link together, in an ATP-dependent reaction, a 3′-OH terminus on an "acceptor" and a 5′-PO_4 terminus on a "donor," as shown below:

$$\text{ATP} + \cdots + \underset{\text{Acceptor}}{\text{XpY-3'-OH}} + \underset{\text{Donor}}{\text{PO}_4\text{-5'Zp}\cdots} \rightarrow \underset{\text{Ligated product}}{\cdots\text{XpYpZp} + \cdots} + \text{ADP} + \text{PP}_i$$

A template strand is not required to align the ends of the reactants. The donor and acceptor end can be on the same RNA molecule, in which case a circular RNA product is produced. The enzyme has been used to end-label the 3′-OH end of RNA molecules by the addition of the short donor [5′-^{32}P]Cp.

This enzyme can be used to synthesize defined sequences of RNA. The ligase also can accept deoxyribose polymers as donors. We do not know whether this enzyme carries out the ligation reaction *in vivo* after T4 infection of *E. coli*.

Posttranscriptional Alterations of Transcripts

Most major types of RNA synthesized by cellular DNA-dependent RNA polymerases undergo changes before they can carry out their functions (fig. 28.20). The two exceptions are prokaryotic messengers and eukaryotic 5S ribosomal RNA. Two types of changes are sometimes distinguished. Modification involves additions to or alterations of existing bases or sugars. In some cases it involves addition of one or more nucleotides. Processing, strictly speaking, involves phosphodiester bond cleavage and loss of certain nucleotides. These changes are summarized in table 28.6. The types of alterations that pre-rRNA and pre-tRNA transcripts undergo are very similar in prokaryotes and eukaryotes. We will focus on the processing of these two classes of transcripts in *E. coli* because they are better understood. The complex processing of eukaryotic messengers will also be discussed. We will see that splicing of eukaryotic messengers is a unique process, although occasional splicing also occurs for other types of transcripts.

Processing and Modification of tRNA Requires Several Enzymes

Transfer RNAs are processed from larger precursors in both prokaryotic and eukaryotic cells. This processing involves two types of enzymes that can cleave phosphodiester bonds in RNA. Endoribonucleases cleave at internal sites in the RNA, resulting in two smaller RNAs. Exonucleases sequentially remove single nucleotides from one end of the RNA. Some of the processing ribonucleases of *E. coli* are shown in table 28.7.

As an example, the processing of the *E. coli* tyrosine $tRNA_1$ is diagrammed in figure 28.21. The initial transcript has, in addition to the 85 nucleotides of the final product, 41 nucleotide residues at the 5′ end and 225 residues at the 3′ end. The initial transcript probably folds to form the typical cloverleaf tRNA structure. Processing begins when a specific endonuclease, called RNase F, cleaves the precursor at a site three nucleotides beyond what will be the 3′ end of the mature tRNA. Another endonuclease, RNase P, then cleaves the remaining RNA to produce the mature 5′ end. At the 3′ end, exonuclease RNase D sequentially removes the additional nucleotides and stops, leaving the 3′ terminal CCA sequence. (Some tRNAs encode the CCA terminal sequence; some do not.) Individual bases on the tRNA molecule are then modified by a variety of enzymes, including methylases, deaminases, thiolases, pseudouridylating enzymes, and transglycosylases. Some of the modified bases are presented in figure 28.5. In addition, some eukaryotic tRNAs are processed to remove internal RNA sequences, as we will see later.

Processing of Ribosomal RNA Precursor Leads to Three RNAs

Both eukaryotic and prokaryotic cells synthesize large precursors to rRNA that are processed to produce the mature rRNAs. The most detailed studies of the numerous enzymatic steps involved in this process have been carried out in *E. coli*. This processing scheme is summarized in figure 28.22.

The initial transcript is over 5,500 nucleotides long and includes, reading from the 5′ end of the RNA: the 16S rRNA, a spacer region with one or two tRNAs, the 23S rRNA, the 5S rRNA, and in some cases one or two more tRNAs. Extra bases are found preceding and following each of these RNAs. Primary processing events include the endonucleolytic action by RNase III to produce pre-16S and pre-23S rRNAs and then

Figure 28.20

Comparative processing of major transcripts in prokaryotes and eukaryotes.

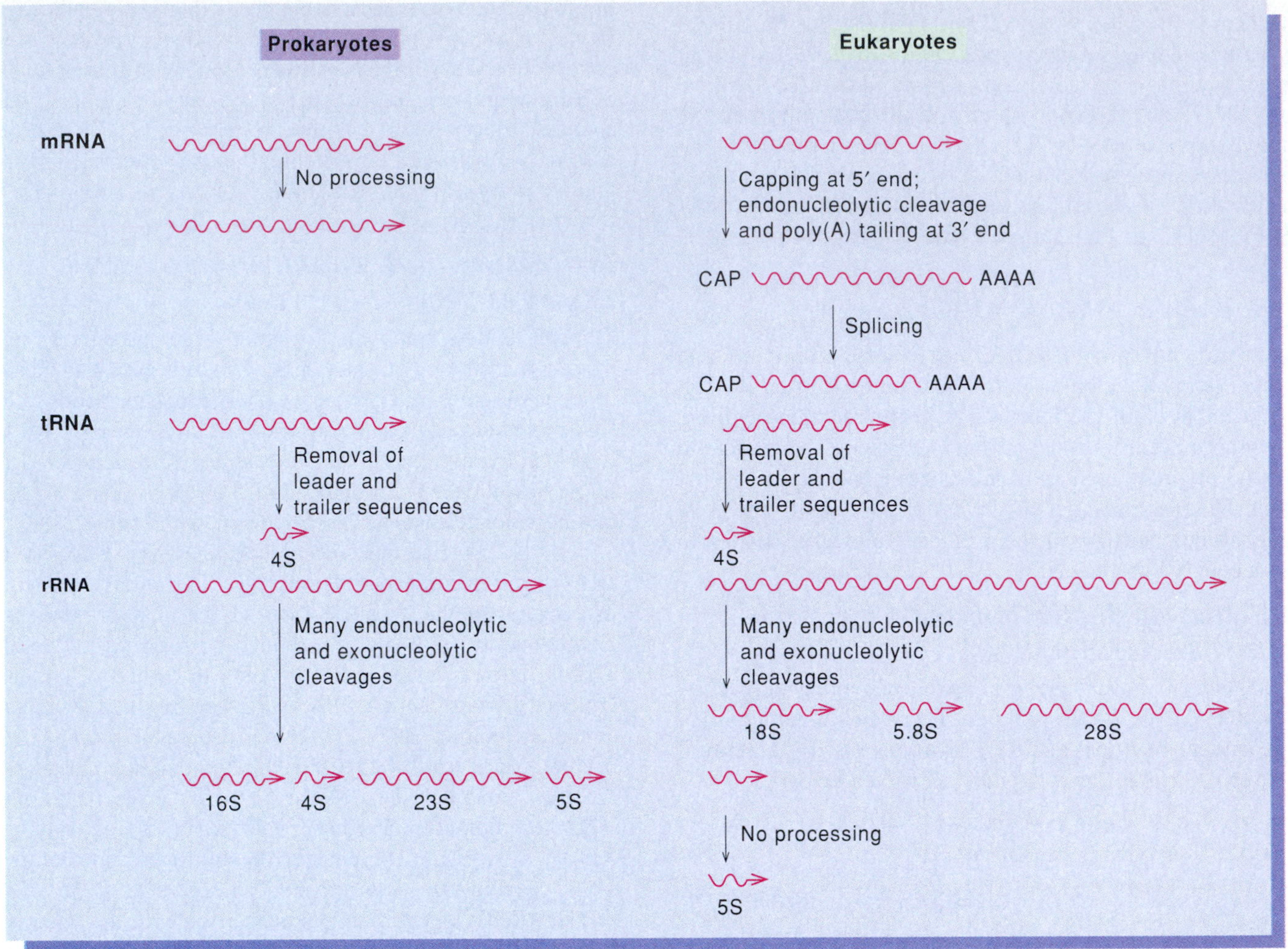

Table 28.6

Summary of RNA Modification and Processing

RNA	Precursor	Modification	Processing	Products
mRNA				
Prokaryotic	None	None? Polyadenylation?	In some cases specific cleavage by endoribonucleases	mRNAs
Eukaryotic	hnRNA	Capping, methylation, polyadenylation	In many cases splicing out introns	mRNAs
rRNA	Pre-rRNA	Methylation	Specific cleavage	16S, 23S, 5S, spacer tRNA (18S, 28S, and 5.8S in eukaryotes)
tRNA	Pre-tRNA	Many modified bases	Specific cleavage by endonucleases, trimming by exonucleases, CCA addition, removal of intervening sequences in eukaryotes	Mature tRNAs

Table 28.7
Representative Enzymes Involved in Processing and Degradation

Enzyme	For *E. coli* Genes Gene Name	Map Position	Type	Product	Specificity
Processing					
RNase III	*rnc*	55′	endo	3′-OH, 5′-PO_4	Specific, long, double-stranded RNA
RNase D	*rnd*	40′	3′ → 5′ exo	5′-NMPs	Nonspecific, but stops at CCA
RNase E	*rne*	24′	endo		Specific
RNase F	*rnf*	?	endo		Specifically cuts 3′ to tRNA-like structures
RNase P	{ *rnpA* *rnpB*	83′ 70′	endo	3′-OH, 5′-PO_4	Specifically cuts 5′ to tRNA-like structures
RNase M16			endo(s)		Specific, cuts pre-16S to 16S RNA
RNase M23			endo(s)		Specific, cuts pre-23S to 23S RNA
RNase M5			endo		Specific, cuts pre-5S to 5S RNA
Degradation					
RNase I	*rna*	14′	endo	3′-PO_4 oligos	Nonspecific
RNase II	*rnb*	28′	3′ → 5′ exo	5′-NMP	Nonspecific
Polynucleoside phosphorylase	*pnp*	69′	3′ → 5′ exo	5′-NDP	Nonspecific
RNase H	*rnh*	5′	endo		Nonspecific, digests RNA out of RNA-DNA duplex
Bovine pancreatic RNase A			endo	Py-3′-PO_4	Specific, cuts 3′ to pyrimidines
Aspergillus RNase T1			endo	G-3′-PO_4	Specific, cuts 3′ to guanine
Aspergillus S1 nuclease			endo	5′-NMP	Nonspecific, cuts single-stranded RNA or DNA
Bovine spleen phosphodiesterase			5′ → 3′ exo	3′-NMP	Nonspecific
Snake venom phosphodiesterase			3′ → 5′ exo	5′-NMP	Nonspecific

the action of specific ribonucleases to produce the tRNAs and pre-5S rRNA. Secondary processing by endonucleases M16, M23, and M5 results in mature 16S, 23S, and 5S RNAs, respectively. These three endonuclease activities, as well as that of RNase F, have not been completely purified and characterized, and each may represent more than one enzyme. Extra bases on the 3′ end of the tRNAs are removed by exonuclease RNase D. This processing scheme has been deduced by observing the accumulation of intermediates in mutant strains defective in one or more of the nucleases and by cleaving the intermediates *in vitro* with purified or partially purified nucleases.

Processing of Messenger RNA in Eukaryotes Involves Changes at Both Ends and Often Removal of Noncoding Segments in the Middle of Genes

In prokaryotes, many mRNAs function in translation with no prior processing. There is evidence, however, that some are processed by specific endonucleolytic cleavage, often cutting polycistronic messengers into smaller units. In eukaryotes, a much more complex process occurs to produce functional mature

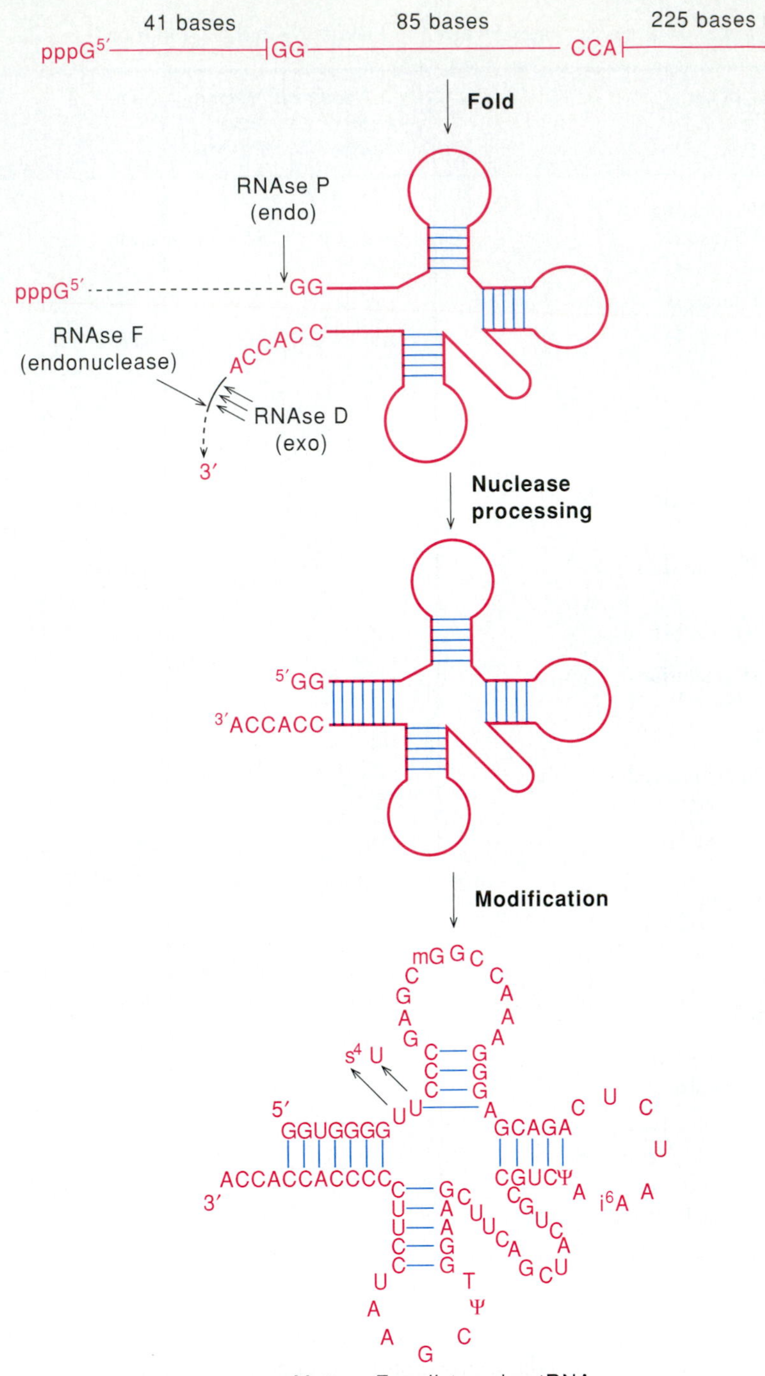

Figure 28.21

Processing and modification of *E. coli* tyrosine tRNA. (T = ribothymidine; ψ = pseudouridine; i6A = isopentyladenosine; mG = methylguanosine; s4U = thiouridine.) See figure 28.5 for the structures of these modified bases.

mRNA. Changes occur in sequence. First there is a modification at the 5′ end of the message called capping. Then the message is cleaved by an endonuclease and a poly(A) tail is added to the 3′ end. In many cases, especially in higher eukaryotes, noncoding regions called introns are removed from the middle of genes before they can be used in translation.

Capping Occurs before the Nascent Transcript is Completed

In 1975, Aaron Shatkin and co-workers found that most viral and cellular mRNAs contain an unusual methylated nucleotide at the 5′ terminus. This entire methylated terminal oligonucleotide is called a cap structure (fig. 28.23). The cap structure is formed in the nucleus by a series of enzymatic reactions (fig. 28.24). First, a triphosphatase converts a 5′-triphosphate to a

Figure 28.22

Processing of *E. coli* ribosomal RNA. The ribosomal RNA is transcribed as one long RNA molecule, which contains the sequences for the three ribosomal RNAs and one or two tRNA molecules. There are many processing sites and many different enzymes involved in the processing, as indicated by the vertical arrows and the symbols associated with these arrows. The various nucleases are described in table 28.7. (Source: D. Apirion and P. Gegenheimer, "Processing of bacterial RNA" in *FEBS Letters,* 125:1, 1981. Copyright © 1981 Elsevier Science Publishers B.V., Amsterdam, The Netherlands.)

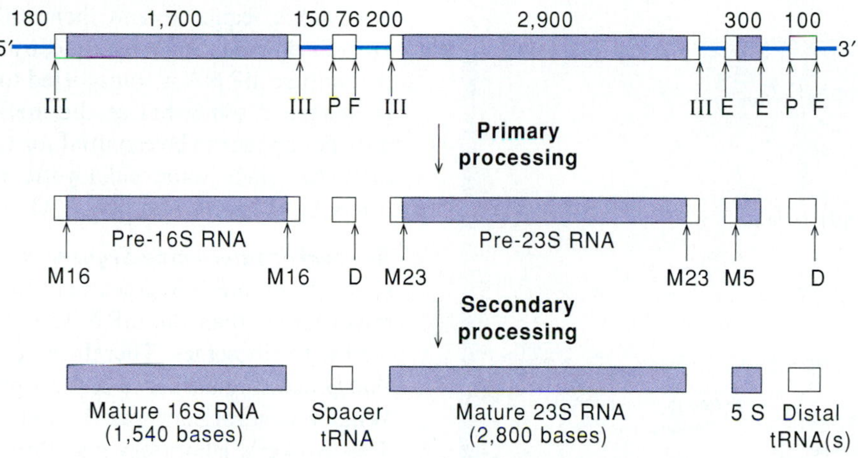

Figure 28.23

Structure of the 5′ methylated cap of eukaryotic mRNA. A 7-methylguanosine (in color) is attached through a triphosphate linkage formed between its 5′-OH and the 5′-OH of the terminal residue in the initial transcript. Note that the 2′-OH groups on the last two bases of the initial transcript have also been modified by methylation (in color). N_1, N_2, and N_3 can be any purine or pyrimidine bases.

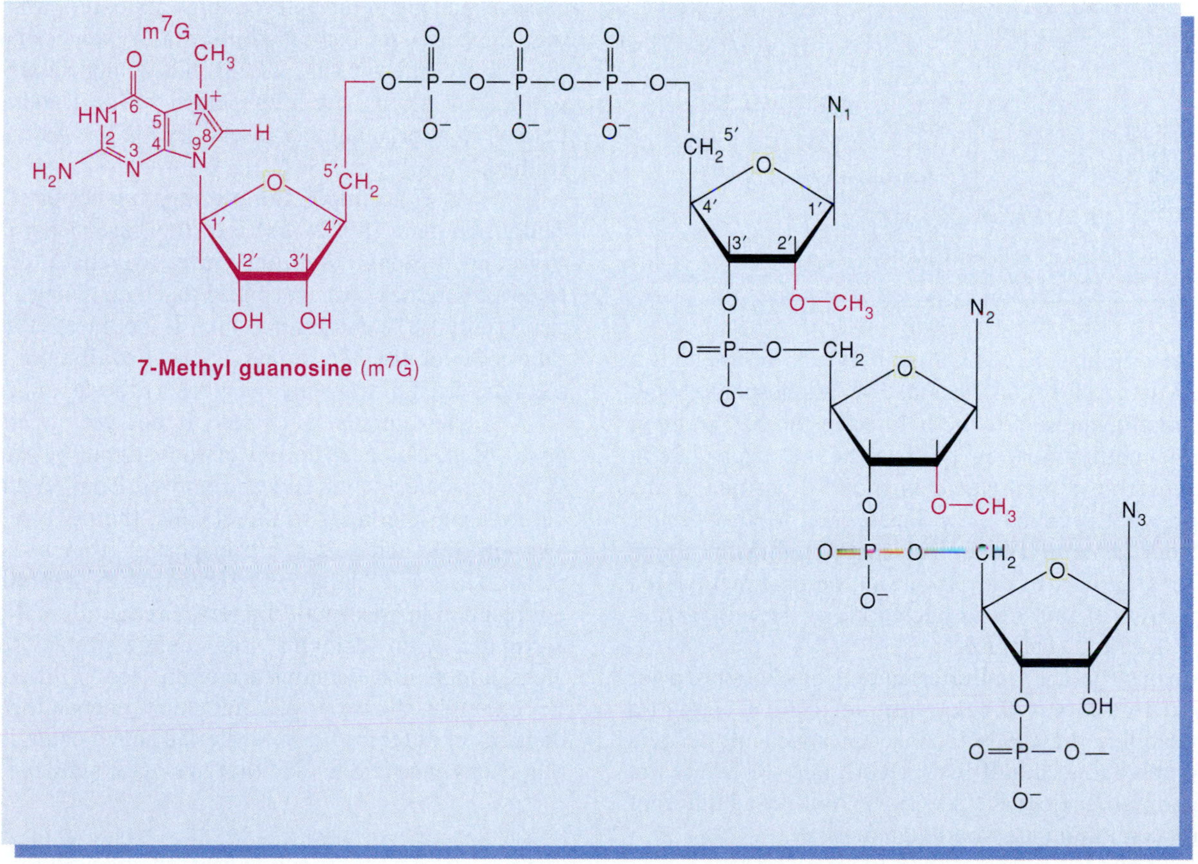

Figure 28.24

Proposed reaction sequence for cap formation in HeLa cells. The enzyme catalyzing each reaction is shown on the left. (Source: S. Venkatesan and B. Moss, "Donor and acceptor specificities of HeLa cell mRNA guanylyltransferase" in *Journal of Biological Chemistry* 255:2835, 1980. Copyright © 1980 American Society for Biochemistry and Molecular Biology Inc., Bethesda, Md.)

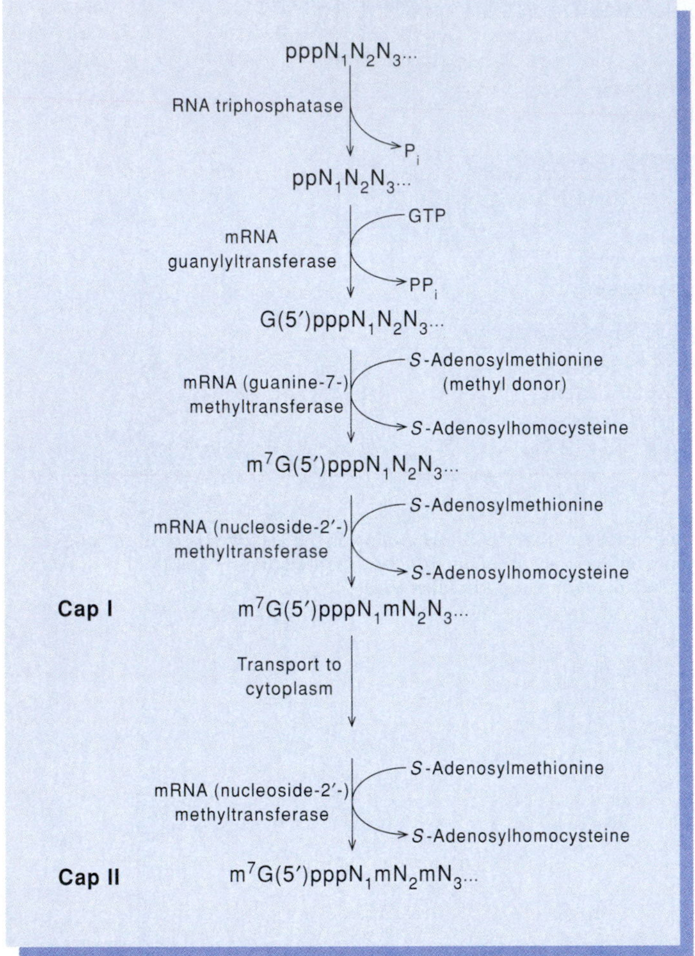

5′-diphosphate terminus. Second, a guanylyltransferase adds a GMP to the 5′ end to form an unusual 5′-5′ triphosphate bond, and this terminal guanosine is methylated in the N-7 position by a guanine-7-methyltransferase. Then the first nucleotide in the initial transcript is methylated in the 2′-0 position of the ribose to form what is called a cap I structure. Methyl groups are derived from the methyl donor *S*-adenosylmethionine. Some mRNAs, after transport to the cytoplasm, become methylated in the 2′-0 position of the second nucleotide of the initial transcript to form a cap II structure.

The cap structure facilitates binding of ribosomes prior to initiation of translation of eukaryotic mRNAs; it may also function to stabilize the mRNA, since uncapped messengers have considerably reduced half-lives. rRNA and tRNA are not capped. Most small nuclear RNAs (snRNAs) have a different cap structure, containing a trimethylguanosine.

Polyadenylation of the 3′ End Most eukaryotic mRNAs found associated with ribosomes contain 50 to 150 adenine nucleotides on their 3′ ends. This poly(A) is not coded by the DNA template but is added to the mRNA before it leaves the nucleus. In several cases studied, the process occurs in at least two steps. First, the RNA is cleaved about twelve nucleotides past an AAUAAA sequence, and then 200 to 250 adenylate residues are added, one residue at a time, by a poly(A) polymerase. After the mature mRNA is transported to the cytoplasm, the poly(A) is shortened somewhat as the mRNA ages. Poly(A) addition does not appear to be essential for transport or translation of all mRNAs, since some eukaryotic mRNAs, including histone mRNAs, do not contain poly(A).

Removal of Intervening Sequences It had long been known that some of the mRNA precursors in the nucleus (hnRNAs) are much larger than the mRNAs found in the cytoplasm associated with ribosomes. Therefore, it came as no surprise to learn that processing occurs to remove parts of the precursor RNAs. What was surprising was the finding made simultaneously by Tom Broker's laboratory and Phil Sharp's laboratory in 1977 that the parts that are removed are not at the ends of the molecules but are interspersed with the coding regions. The phenomenon was first observed for adenovirus transcripts.

One of the early demonstrations of sequence removal resulted from finding that mouse β-globin precursor mRNA did not form a perfect hybrid with DNA complementary to mature mRNA. When the hybrid was observed with an electron microscope, a loop in the heteroduplex appeared, which suggested that the precursor contained internal sequences not present in the mature mRNA (fig. 28.25). These noncoding intervening sequences (introns) are interspersed with the coding sequences (exons) in a large number of eukaryotic RNAs that have been studied.

As an example, the primary transcript of the chicken ovalbumin gene (fig. 28.26) is 7,700 bases long and has 7 introns and 8 exons. The intervening sequences, or introns, are removed and the exons are spliced together, giving a final product that is only 1,872 nucleotides long. It includes 1,158 nucleotides that code for the 386 amino acids of ovalbumin and 714 nucleotides from untranslated regions at the 5′ and 3′ ends.

The function of introns is not yet understood. Some genes (e.g., histones) do not contain them, yet function well. When a particular intron was removed from SV40 virus DNA before using the latter to infect cells, the mRNA that formed was unstable and was not transported from nucleus to cytoplasm. However, similar experiments with other mRNAs have not resulted in abnormal function. Frequently, splice points are correlated with "domains" that define protein structural regions, and similar domains are often seen in different proteins. For example, the exons of hemoglobin encode three structural domains of different types, while the heavy-chain immunoglobulin exons encode domains that are quite similar in structure.

Figure 28.25

Electron micrograph showing mouse ß-globin precursor mRNA (nascent transcript) hybridized with DNA (cDNA) complementary to mature mRNA (upper photo). A control experiment (lower photo) shows that mature mRNA forms a perfect hybrid with DNA complementary to mature mRNA (cDNA) as expected. The intron region in the nascent transcript is indicated by a loop in the upper figure. Schematics indicating the RNA and cDNA are shown on the right. A second small intron is present in the precursor mRNA near the 5′ end and is the reason that the 5′ end of the RNA is not hybridized to the cDNA in the upper photo. (From A. Kinniburgh, J. Mertz, and J. Ross, "The precursor of mouse ß-globin contains two intervening sequences," *Cell* 14:681, 1978. © Cell Press.)

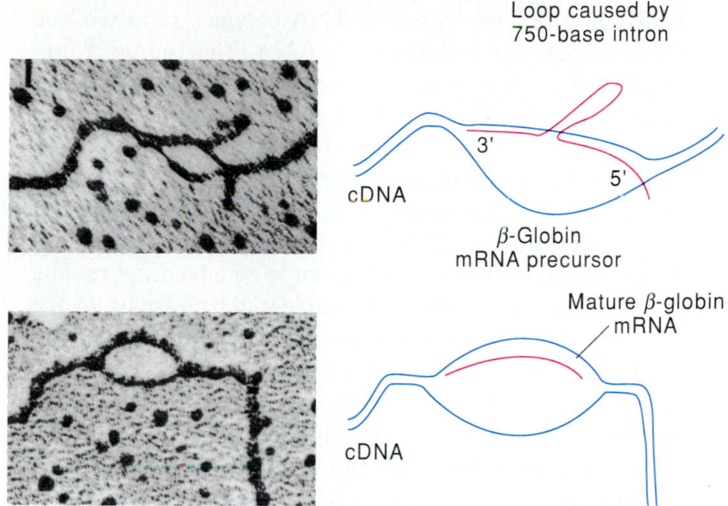

Figure 28.26

Maturation of ovalbumin mRNA. First, the entire ovalbumin gene is transcribed into a precursor RNA, the primary transcript. The transcript is capped at the 5′ end and the poly(A) tail is added at the 3′ end. Then the transcripts of the introns are excised and the adjacent exon transcripts are ligated in a series of splicing steps; an intermediate, from which five of the seven intron transcripts have been eliminated, is illustrated. These steps are accomplished in the cell nucleus. After splicing, mature messenger is transferred to cytoplasm. *L* indicates the 5′ leader region, which is part of the mature RNA that is not translated. (From "Split genes" by Pierre Chambon. Copyright © 1981 Scientific American Inc. All rights reserved.)

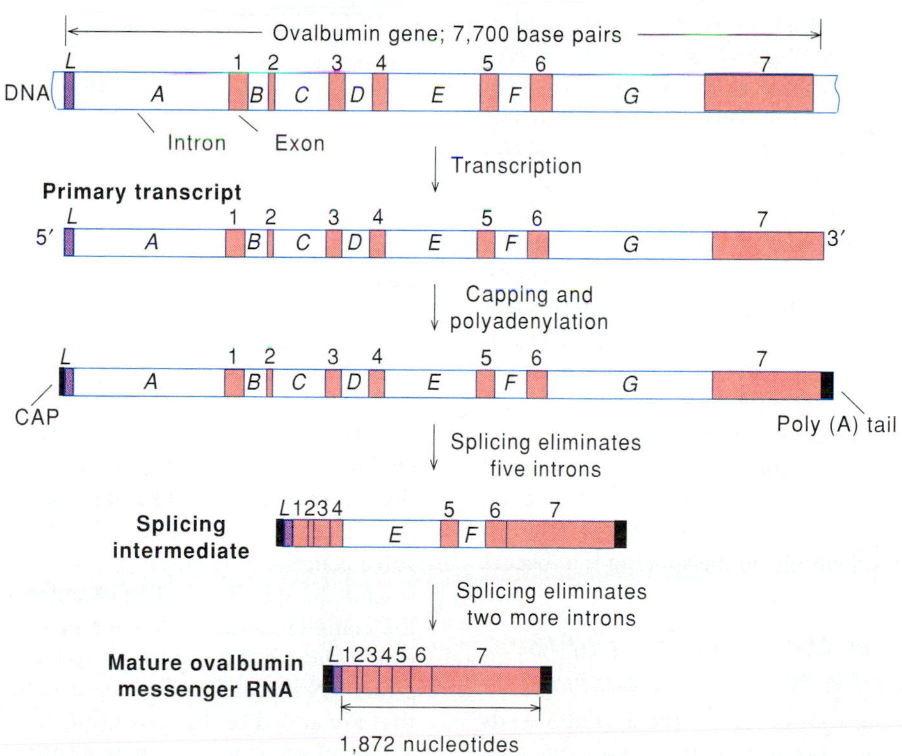

By attaching the promoter for the bacteriophage SP6 RNA polymerase to the β-globin gene, it is possible to transcribe the gene *in vitro* with SP6 RNA polymerase to produce abundant amounts of β-globin pre-mRNA. This source of precursor mRNA has been used to determine the steps and factors involved in splicing. The process (fig. 28.27) involves cleavage of the pre-mRNA at the 5′ splice site to generate the proximal exon and an RNA species containing the intron in a "lariat" configuration connected to the distal exon (fig. 28.27, step 1). The lariat is formed via a 2′-5′ phosphodiester bond, which joins the 5′ terminal guanosine of the intron to an adenosine residue within the intron at a spot 18–40 nucleotides upstream of the 3′ splice site. These two RNA species are probably held together in a noncovalent complex until the next step (fig. 28.27, step 2) in the reaction, which is cleavage at the 3′ splice site to generate the free intron RNA and ligation of the two exons via a 3′-5′ phosphodiester bond. U1 and U2 snRNAs as well as several additional protein factors appear to be necessary for the reactions to occur *in vitro*. These various factors are sometimes seen as a large 40S–60S complex, termed a spliceosome.

Abnormal mRNA splicing appears to be one cause of the human disease β^+-thalassemia. The inefficient splicing of β-globin mRNA precursors in affected individuals seems to be due to mutations in the intron and leads to very low levels of mature mRNA and thus to a β-globin deficiency.

Removal of intervening sequences in eukaryotes is not restricted to mRNA processing. It also occurs in the processing of rRNA and some tRNAs. A temperature-sensitive mutant of yeast has been isolated that accumulates certain tRNA precursors at the nonpermissive temperature. One of these is a tyrosine tRNA precursor that has a 14-base intervening sequence that can be removed by cell extracts *in vitro*. The reaction occurs in two steps and is shown in figure 28.28.

First, a 14-base sequence is removed by an ATP-independent endonucleolytic cleavage to produce two half tRNA molecules. These cleavages produce 3′-PO_4 and 5′-OH termini, unlike the 3′-OH and 5′-PO_4 termini produced by most RNA processing enzymes, such as RNase III or RNase P. The two termini are then ligated in a second reaction that requires ATP.

Some yeast tRNAs have intervening sequences that are up to 60 bases long, but they seem to be processed by the same endonuclease, which appears to recognize a structural feature in the precursor rather than a particular sequence. The function of the yeast intervening sequences is not known, nor is it known whether their splicing is at all similar to the splicing that occurs in mRNA.

RNA Editing Involves the Processing of the Coding Sequences of a Nascent Transcript

In addition to the processing that occurs at the 5′ and 3′ ends of pre-mRNAs and the removal of noncoding regions (introns) from coding regions, there are an increasing number of examples where the coding sequences within the nascent transcripts are changed. In cases where this occurs, the changes, known as RNA editing, are crucial since they often convert a nontranslatable transcript into one that can function as a message. A case in point is the mitochondrial message for cytochrome B oxidase in the flagellated protozoan *Leishmania tarentolae*. In this case several uracil residues are added at different points in the nascent transcript (fig. 28.29). These editing changes involve complex formation between the nascent transcript and a so-called guide RNA. The changes in sequence are dictated by the complementary sequence in the guide RNA in the region where the editing occurs. A string of U's at the 3′ end of the guide RNA is believed to serve as the source for additional U's that are added to the transcript. In addition to the guide RNA, a special enzyme system is required, which is capable of inserting and removing bases at various points in the nascent transcript. Why make a message that requires such complex changes before it can function in translation? One possibility is that this is part of a regulatory mechanism that controls the amount of cytochrome B oxidase.

Figure 28.27

Splicing scheme for pre-mRNA. In step 1 the 2′-OH on an adenosine attacks a phosphate that is 5′-linked to a guanine residue. This leads to a lariat configuration connected to the distal exon. The lariat is formed via a 2′-5′ phosphodiester bond, which joins the 5′ terminal guanosine of the intron to an adenosine residue within the intron, 18-40 nucleotides upstream of the 3′ splice site. In the next step there is a cleavage at the 3′ splice site to generate the free intron RNA and a ligation of the two exons via a 3′-5′ phosphodiester bond. Bases usually found in the region of the splice sites are indicated. The spliceosome is presumed to hold the reacting components in the proper juxtaposition to facilitate the two-step reaction.

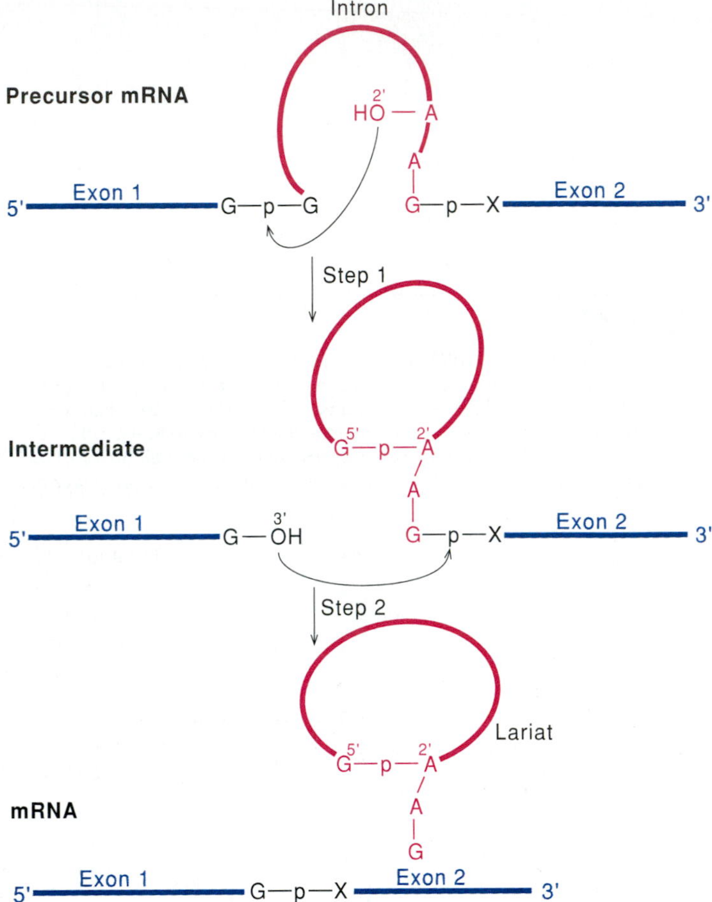

Figure 28.28

Processing of yeast tyrosine tRNA to remove intervening sequence. First, a 14-base sequence is removed by an ATP-independent endonucleolytic cleavage. The two newly produced termini are then ligated in a second reaction, which requires ATP.

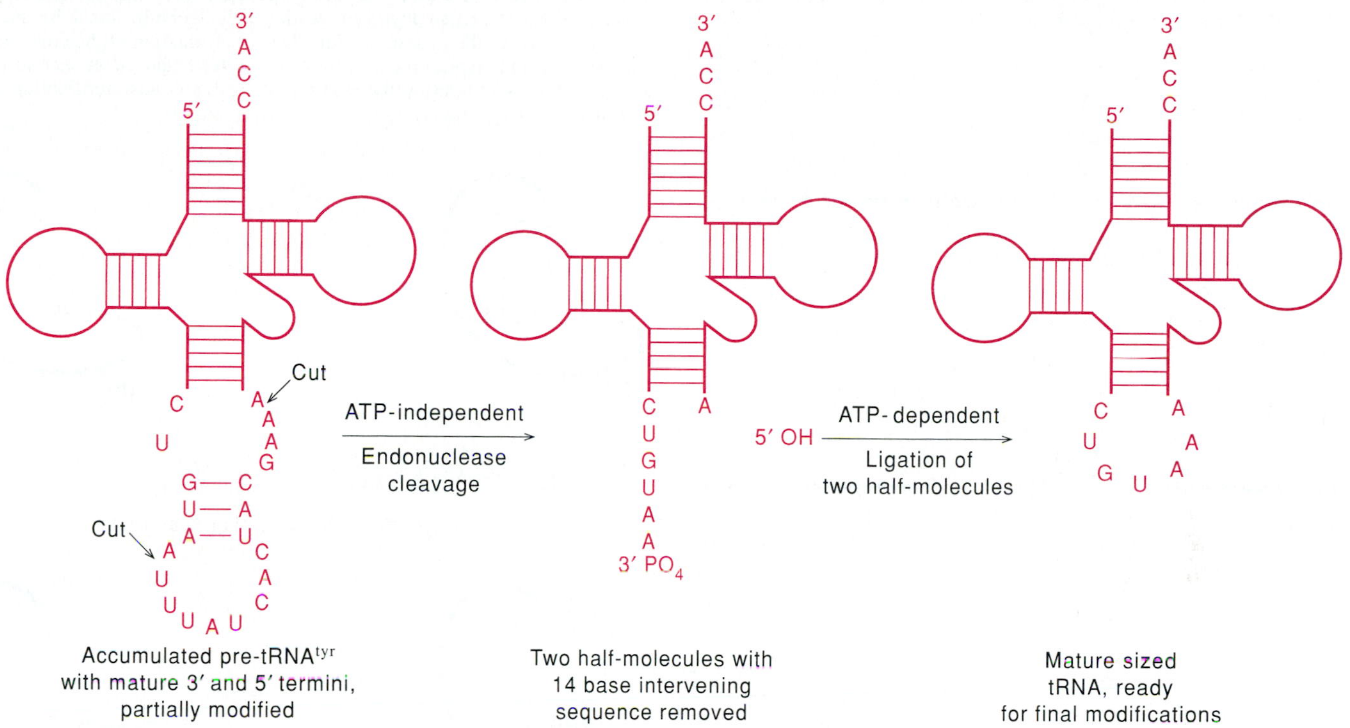

Figure 28.29

Editing of the cytochrome B message in the flagellated protozoan *Leishmania tarentolae*. Editing, which involves the insertion of eleven U residues, creates a translatable RNA where no translation was possible before. Structure shown for mRNA is after editing.

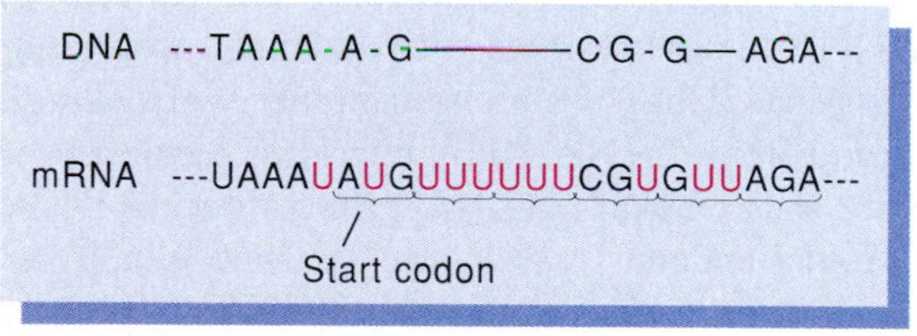

DNA ---TAAA-A-G———CG-G—AGA---

mRNA ---UAAAUAUGUUUUUUCGUGUUAGA---

Start codon

Some RNAs Are Self-Splicing

In 1982, the splicing of rRNA from the protozoan *Tetrahymena* was shown by Tom Cech to involve a startling mechanism. When the pre-rRNA was incubated under the appropriate conditions of salt, Mg^{2+}, and guanosine nucleotide, splicing of the RNA occurred without the help of a protein enzyme! This ability of RNA to act as a catalyst for its own splicing has given rise to the term "ribozyme" and has broadened our definition of enzymes.

The mechanism of the self-splicing reaction is shown in figure 28.30. The first step is a transesterification reaction in which the 3′ hydroxyl group of the guanosine cofactor attacks the phosphodiester bond at the 5′ splice site. The second step is another transesterification reaction in which the 3′ hydroxyl group of the upstream exon attacks the phosphodiester bond at the 3′ splice site and displaces the 3′ hydroxyl group of the intron. The final reaction products are the joined exons and the excised intron.

Since its initial discovery, self-splicing has been found to occur for RNAs from a wide variety of organisms. However, the fraction of unprocessed RNAs that can be shown to undergo self-splicing *in vitro* is quite small in all cases. Certain precursor RNAs that exhibit self-splicing produce intron lariats, just like those seen in the commonly observed splicing reactions that take place in most precursor RNAs with introns. In this case the guanosine nucleotide is not required. This fact suggests that originally RNAs may have been self-splicing, and that the spliceosome components have evolved to improve catalytic efficiency and perhaps to regulate the process.

Self-splicing of RNA shows that in certain cases RNA has enzymelike activity. However, it does not by itself demonstrate true enzyme behavior. Recall that an enzyme is a substance that acts as a catalyst—it accelerates a reaction without itself being consumed in the reaction. More recently Zaug and Cech have shown that a fragment of the self-splicing ribosomal RNA intervening sequence of *Tetrahymena thermophila* can act as an enzyme *in vitro*. This RNA fragment has been shown to catalyze the breakage and rejoining of oligonucleotide substrates in a sequence-dependent manner, with a $K_m = 42\ \mu M$

Figure 28.30

Self-splicing of pre-rRNA from the protozoan *Tetrahymena.* The first step is a transesterification reaction in which the 3′ hydroxyl group of a guanosine attacks the phosphodiester bond at the 5′ splice site. The second step involves another transesterification reaction in which the 3′ hydroxyl group of the upstream exon attacks the phosphodiester bond at the 3′ splice site and displaces the 3′ hydroxyl group of the intron.

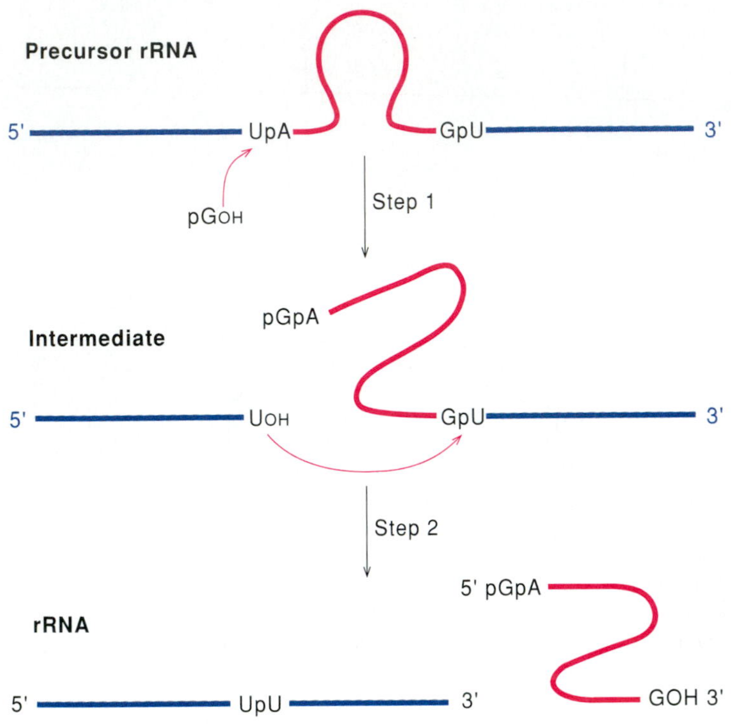

Figure 28.31

Model for the enzymatic mechanism of the 19-base oligonucleotide intervening sequence isolated by Cech from a *Tetrahymena* precursor rRNA. (*a*) The oligonucleotide enzyme (ribozyme) is shown with the oligopyrimidine binding site (RRRRRR), a sequence of six purines near its 5′ end, and a guanine with a 3′ hydroxyl group at its 3′ end. (*b*) The enzyme binds its substrate, a pentacytidylic acid. (*c*) Nucleophilic attack by the terminal 3′-OH of the guanine residue leads to formation of the covalent intermediate and displacement of the tetracytidylic acid. (*d*) A second pentacytidylic acid binds at the enzyme site and, by transesterification, a hexanucleotide is formed.

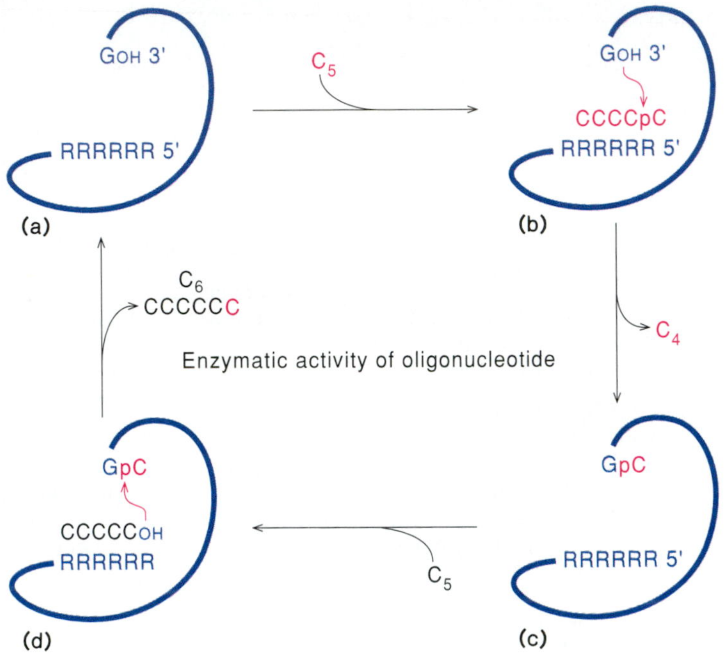

and a $k_{cat} = 2\ min^{-1}$. With pentacytidylic acid as the substrate, successive cleavage and rejoining reactions lead to the synthesis of higher polymers of polycytidylic acid (fig. 28.31).

The discovery of Cech and his colleagues that RNA can function as an enzyme raises many possibilities both for biochemical reactions as they are currently understood and for reactions that may have played a major role in prebiotic evolution (see later discussion).

Degradation of RNA by Ribonucleases

Although a large number of nonspecific exonucleases and endonucleases have been identified in many organisms, the role of most of them is not understood. Some are extracellular or secreted enzymes that presumably function in breaking down RNA to recycle the purine and pyrimidine bases. Intracellular ribonucleases also may be involved in recycling, as well as in some aspects of processing, as described earlier. Some of the better-known enzymes are listed in table 28.7.

E. coli RNase I is an endonuclease that is located in the periplasmic space between the cell membrane and the cell wall. RNase II is an intracellular 3′ exonuclease that rapidly degrades RNA fragments. Mutants defective in these enzymes were very important in the development of *in vitro* translation systems, since extracts from cells defective in these enzymes allowed mRNA to be translated without being rapidly degraded.

RNase H has the unusual property of degrading the RNA strand from a RNA-DNA hybrid molecule and may be involved in the removal of RNA primers during DNA replication. Some endonucleases have marked preferences for cleavage after certain bases. Pancreatic RNase A cleaves at the 3′ side of pyrimidines, while RNase T1 cleaves only after guanosine residues. Both leave 3′ phosphate termini. Because of their specificity, these enzymes have proven very useful in the analysis of RNA sequences. S1 nuclease digests single-stranded DNA or RNA.

Finally, there are exonucleases that digest in the 3′ → 5′ direction and others that digest in the 5′ → 3′ direction. All known intracellular exonucleases in *E. coli* degrade from 3′ → 5′ and produce 5′-NMPs or 5′ NDPs, which can be readily recycled for use in RNA synthesis.

It is likely that degradation of mRNA proceeds by numerous endonucleolytic cleavages and then 3′ → 5′ exonucleolytic digestion of the resulting fragments. Differences in decay rates of bacterial mRNAs result from differences in mRNA sequence and structure. RNase II and PNPase, the major bacterial exonucleases involved in mRNA turnover, rapidly degrade single-stranded RNA from the 3′ end, but are impeded by 3′ stem-and-loop structures.

Some Ribonucleases Are RNAs

RNase P, which we mentioned earlier, was the first nucleic-acid-containing ribonuclease to be discovered. The enzyme is a ribonucleoprotein composed of both a 20-kDa protein and a 377-nucleotide RNA molecule. The RNA component is catalytically active, whereas the protein component helps to maintain the three-dimensional structure of the RNA.

Certain circular and low-molecular-weight single-stranded plant pathogenic RNAs called viroids can self-cleave at specific sites in the presence of Mg^{2+}. This property is considered important in their replication *in vivo* by a rolling-circle mechanism. All RNAs produced by the rolling-circle mechanism are polymers of indefinite length. The self-cleavage reaction results in linear monomeric molecules that can subsequently be circularized. The self-cleavage site occurs at a specific locus in a secondary structure that has been described as a hammerhead structure (fig. 28.32). Short RNA molecules that anneal to form a hammerhead structure with a consensus sequence undergo self-cleavage as shown in figure 28.32*b* and *c*.

Catalytic RNA May Have Great Evolutionary Significance

For many years prior to the discovery of catalysis by RNA, evolutionists concerned with the origin of life struggled with the question of which came first, nucleic acid or protein. Since protein makes an excellent catalyst but a poor template for transferring information, it seemed likely that nucleic acid and protein must have coevolved. The trouble with this idea is that protein synthesis requires a great number of enzymes. If protein is the only substance that can function as an enzyme, then we clearly have a dilemma. However, if RNA can function as an enzyme, we have a way out. The first systems for protein synthesis could have employed RNA enzymes.

In this same evolutionary context, recall that coenzymes frequently can function as catalysts in the absence of the enzymes with which they are normally associated, although their catalytic activity is usually greatly diminished. In the preprotein period of evolution, it seems likely that coenzymes in conjunction with RNA molecules may have played a far greater role than they do today.

Figure 28.32

Hammerhead structures that undergo self-cleavage. Cleavage occurs at site indicated by arrow. (*a*) The structure formed by a sequence found in a viroid or a virusoid. (*b*) A short circular RNA molecule with the "consensus" sequence that also undergoes cleavage. (*c*) A hybrid molecule, formed by two RNAs, that undergoes a comparable cleavage reaction. In the presence of excess "substrate" the 19-mer functions as an enzyme, since the same 19-mer can cleave many substrates.

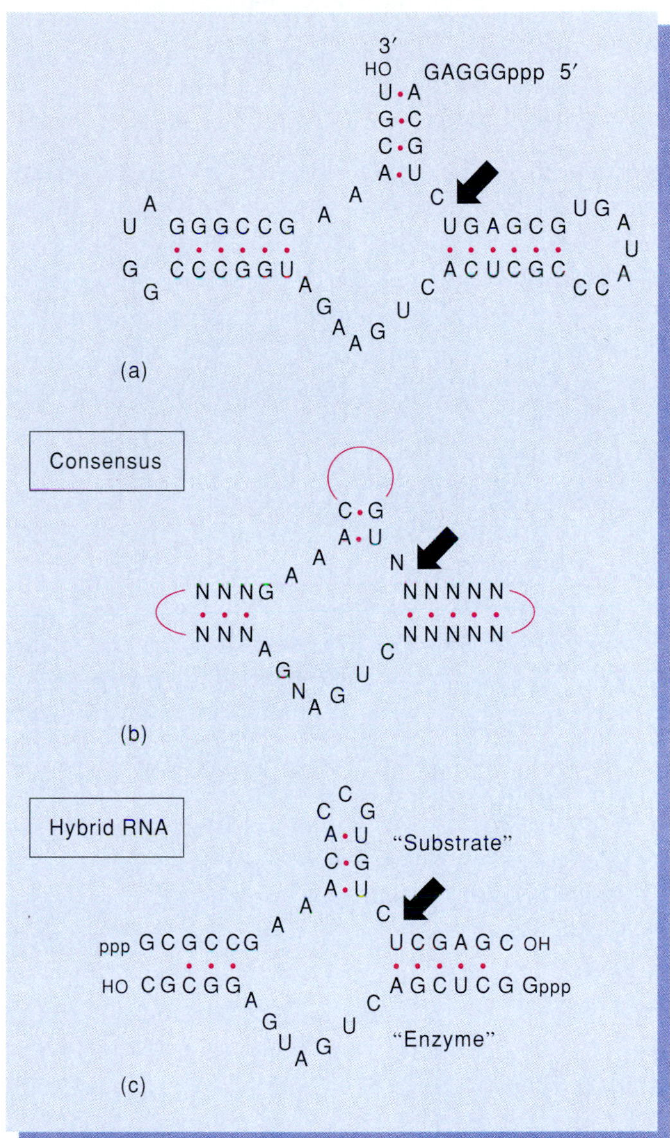

Inhibitors of RNA Metabolism

A large variety of inhibitors of RNA synthesis have been identified. These inhibitors have proved useful in elucidating transcription mechanisms, and some have allowed the isolation of mutant strains with enzymes that are resistant to their inhibition. The inhibitors (fig. 28.33) fall into several classes.

Some Inhibitors Act by Binding to DNA

The best-known example of an inhibitor that binds to DNA is actinomycin D, an antibiotic produced by *Streptomyces antibioticus*. The inhibition of RNA synthesis is caused by the insertion (intercalation) of its phenoxazone ring between two G-C base pairs, with the side chains projecting into the minor

Figure 28.33

Inhibitors of RNA synthesis.

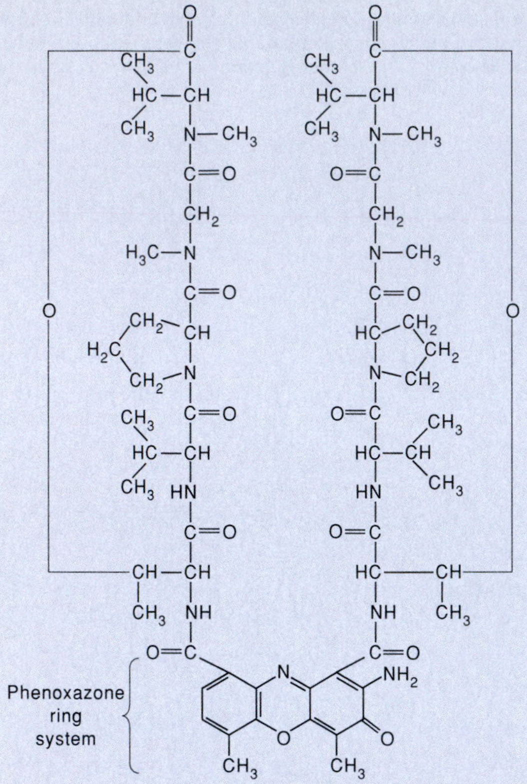

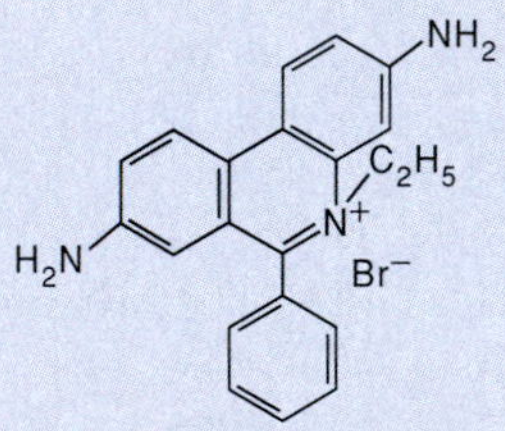

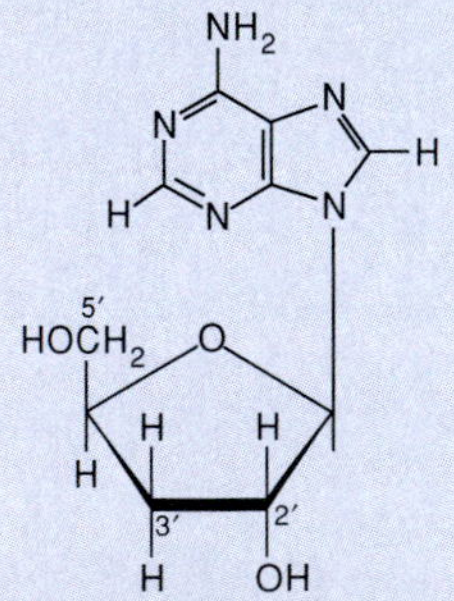

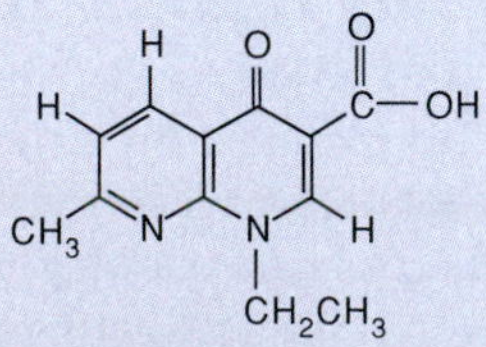

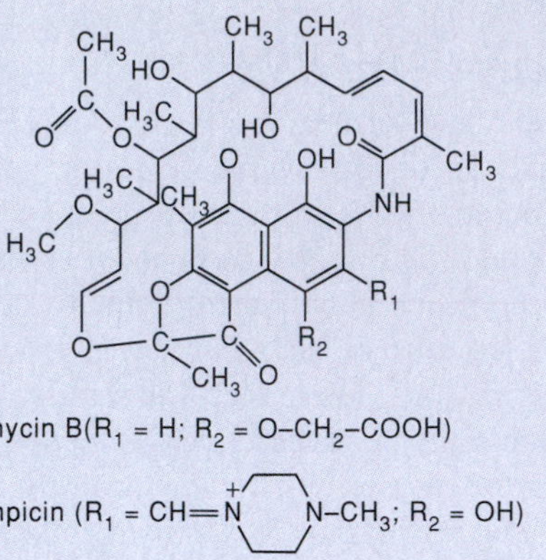

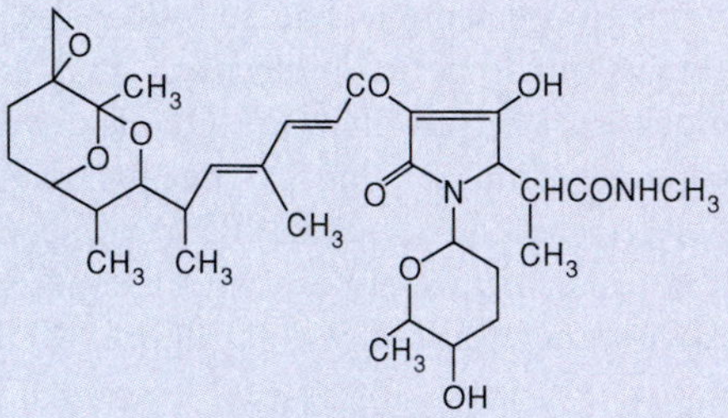

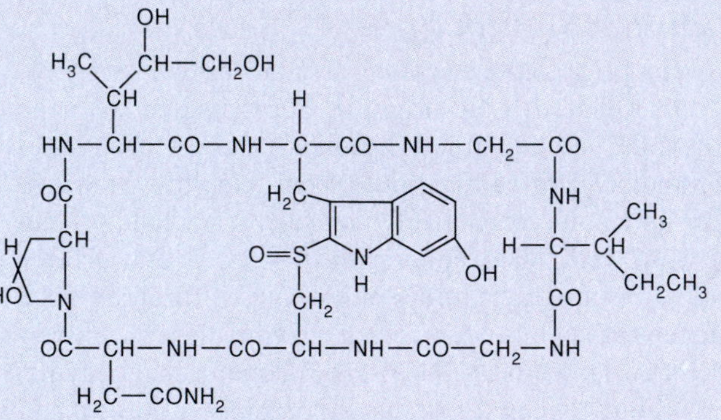

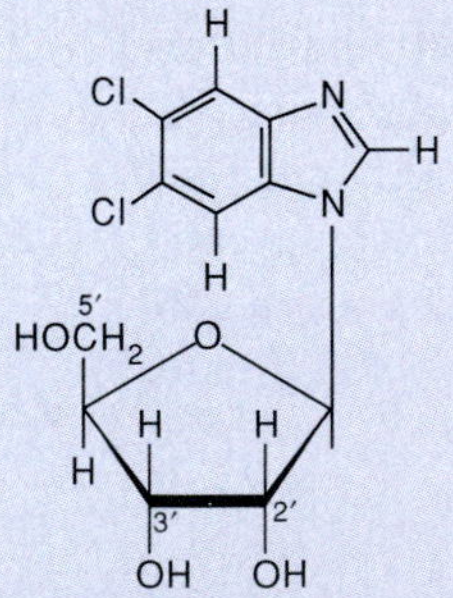

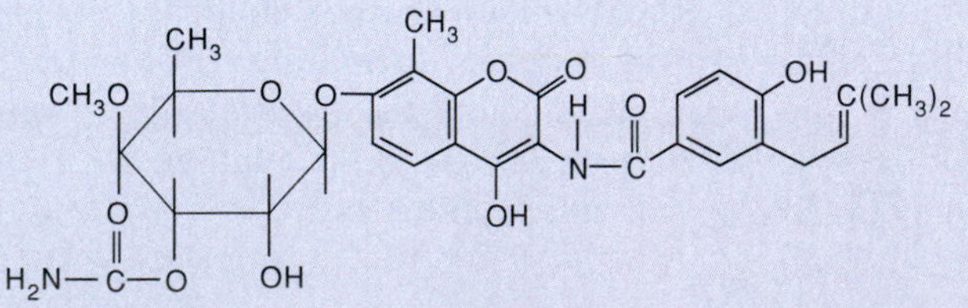

groove of the double helix, hydrogen-bonded to guanine residues. RNA polymerase binding to DNA that contains actinomycin D is only slightly impaired, but RNA chain elongation in both eukaryotes and prokaryotes is blocked. Ethidium bromide also intercalates into DNA and at low concentrations preferentially binds to negatively supercoiled DNA (see chapter 25). It has been used to selectively inhibit transcription in mitochondria, which contain supercoiled DNA.

Some Inhibitors of Transcription Bind to DNA Gyrase

DNA gyrase is an enzyme in *E. coli* that is needed to maintain the proper degree of supercoiling in DNA. Transcription of some genes seems to be affected by the supercoiled state of the template, and transcription of these genes can be selectively inhibited or stimulated by compounds that interfere with the action of DNA gyrase. Examples of compounds that inhibit DNA gyrase are nalidixic acid, which binds to one of the two gyrase subunits, and coumermycin and novobiocin, which bind to the other (see chapters 25 and 26).

Some Inhibitors Bind to RNA Polymerase

Rifampicin is a synthetic derivative of a naturally occurring antibiotic, rifamycin, that inhibits bacterial DNA-dependent RNA polymerase but not T7 RNA polymerase or eukaryotic RNA polymerases. It binds tightly to the β subunit. While it does not prevent promoter binding or formation of the first phosphodiester bond, it effectively prevents synthesis of longer RNA chains. It does not inhibit elongation when added after initiation has occurred. Another antibiotic, streptolydigin, also binds to the β subunit, but it is able to prevent all bond formation, whether involved in initiation or elongation.

The most useful inhibitor of eukaryotic transcription has been α-amanitin, a major toxic substance in the poisonous mushroom *Amanita phalloides*. The toxin preferentially binds to and inhibits RNA polymerase II. At high concentrations it also can inhibit RNA polymerase III, but not RNA polymerase I or bacterial, mitochondrial, or chloroplast RNA polymerase.

Cordycepin in its 5′-triphosphorylated form is a substrate analog and can be incorporated into growing RNA chains by most RNA polymerases. Cordycepin is a 3′-deoxyadenosine. It causes chain termination, since it does not contain the 3′ hydroxyl group necessary for the formation of the next phosphodiester bond.

Summary

In this chaper we have described the synthesis, transcription, and posttranscriptional modification of the three major classes of RNA. The main points we have covered are as follows.

1. RNA is synthesized in the $5' \rightarrow 3'$ direction by the formation of 3′, 5′-phosphodiester linkages between the four ribonucleoside triphosphate substrates, analogous to the process of DNA synthesis. The sequence of bases in RNA transcripts catalyzed by DNA-dependent RNA polymerases is specified by the complementary DNA template strand.
2. Some newly synthesized RNA transcripts are the functional species, while others must be modified and/or processed into the mature functional species. Modifying enzymes add nucleotides to the 5′ or 3′ ends or alter bases within the RNA, such as by methylation of specific residues. Specific processing enzymes cleave RNA internally, splice together noncontiguous regions of a transcript, or remove nucleotides from the 5′ or 3′ ends.
3. The major classes of RNA in both prokaryotes and eukaryotes are messenger RNA, ribosomal RNA, and transfer RNA. These distinct classes of RNA play specific functional or structural roles in the translation of genetic information into proteins. Other classes of RNAs transcribed by DNA-dependent RNA polymerases include primers for DNA synthesis, small RNAs that may be directly involved in processing of larger RNA precursors, and still others that may play as yet unidentified regulatory or structural roles in cells.
4. DNA-dependent synthesis of RNA in *E. coli* is catalyzed by one enzyme, consisting of five polypeptide subunits. The complete holoenzyme is composed of four polypeptides (the core enzyme) and an additional polypeptide that confers specificity for initiation at promoter sequences in the DNA template.
5. The steps involved in transcription—binding of polymerase, initiation, elongation, and termination—have been studied in great detail. Elucidation of these steps, as well as the identification and function of the proteins and DNA sequences involved at each step, has been greatly facilitated by the use of specific inhibitors of RNA synthesis, mutants, and a variety of *in vitro* techniques.
6. In eukaryotes, transcription is catalyzed by at least four distinct polymerases, each specializing in the synthesis of a different class of RNA. RNA polymerases I, II, and III are responsible for the synthesis of rRNA, mRNA, and small RNA (such as tRNA) transcripts, respectively. Specific RNA polymerases are responsible for transcription in mitochondria and chloroplasts. Although the basic mechanism of RNA synthesis by these enzymes is similar to that of prokaryotic RNA polymerase, many details of the steps, proteins, and DNA regulatory sequences involved in eukaryotic transcription are less well understood than in prokaryotes.
7. Various other RNA polymerases have been identified and studied. These include DNA primase, specific viral-induced or viral-encoded polymerases, and polymerases that are not dependent on DNA templates for RNA synthesis. Other enzymes involved in RNA metabolism include processing nucleases, degradation nucleases, ligases, and modifying enzymes. Many of these enzymes have been at least partially purified and characterized, and some have proved useful for the *in vitro* manipulation and analysis of RNAs.

8. One of the more interesting processing reactions of nascent transcripts involves the removal of intervening sequences. This type of processing is referred to as splicing. Most splicing reactions appear to require host proteins. However, it is clear that certain splicing reactions require only RNA, a fact demonstrating that the RNA is capable of functioning like an enzyme in the making and breaking of phosphodiester linkages in polyribonucleotides.

Selected Readings

Altman, S., M. Baer, C. Guerrier-Takada, and A. Vioque, Enzymatic cleavage of RNA by RNA. *Trends Biochem. Sci.* 11:515–518, 1986.

Bass, B. L., Splicing: the new edition. *Nature* 352:283–284, 1991.

Bear, D. G., and D. W. Peabody, The *E. coli* rho protein: An ATPase that terminates transcription. *Trends Biochem. Sci.* 13:343–348, 1988.

Bjork, G. R., J. U. Ericson, C. E. D. Gustafsson, T. G. Hdagervall, Y. H. Josson, and P. M. Wikstrom, Transfer RNA modification. *Ann. Rev. Biochem.* 56:263–287, 1987.

Buratowski, S., S. Hahn, L. Guarente, and P. A. Sharp, Five intermediate complexes in transcription initiation by RNA polymerase II. *Cell* 550–561, 1988.

Burtis, K. C., and B. X. Baker, *Drosophila* double sex gene controls somatic sexual differentiation by producing alternatively spliced mRNAs encoding related sex-specific polypeptides. *Cell* 56: 997–1010, 1989.

Cattaneo, R., RNA editing: in chloroplast and brain. *Trends Biochem. Sci.* 17:4–6, 1992. A recent review dealing with RNA editing.

Cech, T., RNA editing: world's smallest introns? *Cell* 64:667–669, 1991.

Cech, T. R., RNA as an enzyme. *Sci. Am.* 225(5):64–75, 1986.

Cech, T. R., and B. L. Bass, Biological catalysis by RNA. *Ann. Rev. Biochem.* 55:599–630, 1986.

Chambon, P., Split genes. *Sci. Am.* 244:60–66, 1981.

Chowrira, B. M., A. Berzal-Herranz, and J. M. Burke, Novel guanosine requirement for catalysis by the hairpin ribozyme. *Nature* 354: 320–323, 1991. In some cases the guanine amino group serves a catalytic role in self-splicing.

Crick, F., Central dogma of molecular biology. *Nature* 227:561–563, 1970.

Dahlberg, A. E., The functional role of ribosomal RNA in protein synthesis. *Cell* 57:525–529, 1989.

Darnell, J. E., Jr., RNA. *Sci. Am.* 253(4):68–78, 1985.

Deutscher, M. P., The metabolic role of RNases. *Trends Biochem. Sci.* 13:136–139, 1988.

Dorit, R. L., L. Schoenbach, and W. Gilbert, How big is the universe of exons? *Science* 250:1377–1382, 1990.

Forster, A. C., A. C. Jeffries, C. C. Sheldon, and R. H. Symons, Structural and ionic requirements for self-cleavage of virusoid RNAs and trans self-cleavage of viroid RNA. *Cold Spring Harbor Symposia on Quantitative Biology* 52:249–259, 1987.

Futcher, B., Supercoiling and transcription, or vice versa? *Trends Genet.* 4:271–272, 1988.

Geiduschek, E. P., and G. P. Tocchini-Valentini, Transcription by RNA polymerase III. *Ann. Rev. Biochem.* 57:873–914, 1988.

Haas, E. S., D. P. Morse, J. W. Brown, F. J. Schmidt, and N. R. Pace, Long-range structure in ribonuclease P RNA. *Science* 254:853–856, 1991.

Hall, B. D., and S. Spiegelman, Sequence complementarity of T2-DNA and T2 specific RNA. *Proc. Natl. Acad. Sci.* 47:137–146, 1964. The first use of RNA-DNA hybridization.

Helmann, J. D., and M. J. Chamberlin, Structure and function of bacterial sigma factors. *Ann. Rev. Biochem.* 57:839–872, 1988.

Hou, Y.-M., and P. Schimmel, A simple structural feature is a major determinant of the identity of a transfer RNA. *Nature* 333:144–145, 1988.

Khoury, G., and P. Gruss, Enhancer elements. *Cell* 33:313–314, 1983.

Leff, S. D., M. G. Rosenfeld, and R. M. Evans, Complex transcriptional units: Diversity in gene expression by alternative RNA processing. *Ann. Rev. Biochem.* 55:1091–1117, 1986.

Lührmann, R., B. Kastner, and M. Bach, Structure of spliceosomal snRNPs and their role in pre-mRNA splicing. *Biochemica et Biophysica Acta* 1087:265–292, 1990.

Nikoliu, D. B., S-H. Hu, J. Lin, A. Gasch, A. Hoffman, M. Horikoshi, N-H. Chua, R. G. Roeder, and S. K. Burly, Crystal structure of TF IID TATA-box binding protein. *Nature* 360:40–45, 1992.

Padgett, R. A., P. J. Grabowski, M. M. Komarska, S. Seller, and P. A. Sharp, Splicing of messenger RNA precursors. *Ann. Rev. Biochem.* 55:1119–1150, 1988.

Patzelt, E., K. L. Perry, and N. Agabian, Mapping of branch sites in trans-spliced pre-mRNAs of *Trypanosoma brucei. Molecular and Cellular Biology* 9:4291–4297, 1989. Cases where the splicing event involves transcripts from different chromosomes.

Pestka, S., J. A. Langer, K. C. Zoon, and C. E. Samuel, Interferons and their actions. *Ann. Rev. Biochem.* 56:757–777, 1987.

Peterson, M. G., J. Inostroza, M. E. Maxon, F. Osvaldo, A. Admon, D. Reinberg, and R. Tjian, Structure and functional properties of human general transcription factor IIt. *Nature* 354:369–373, 1991.

Proudfoot, N. J., How RNA polymerase II terminates transcription in higher eukaryotes. *Trends Biochem. Sci.* 114:105–110, 1989.

Steitz, J. A., "Snurps." *Sci. Am.* 258(6):56–63, 1988.

Stuart, K., RNA editing in mitochondrial mRNA of trypanosomatids. *Trends Biochem. Sci.* 16:68–72, 1991. RNA editing is processing that involves the removal, addition, and modification of nucleotides in the coding regions of nascent transcripts.

Problems

1. One strand of DNA is completely transcribed into RNA by RNA polymerase. The base composition of the DNA template strand is: G = 20%, C = 25%, A = 15%, T = 40%. What would you expect the base composition of the newly synthesized RNA to be?

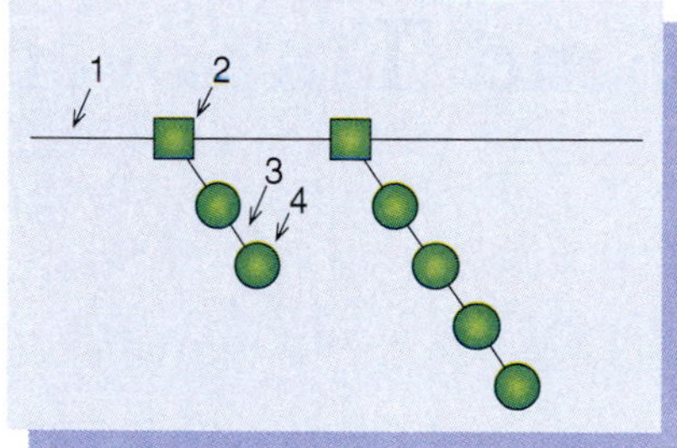

2. The illustration is a schematic drawing representing a portion of Miller's electron microscope picture of transcription and translation in progress in *E. coli.*
   (a) Identify 1, 2, 3, and 4.
   (b) Indicate the 3′ and 5′ ends of the sense (template) strand of DNA.
   (c) Indicate the 3′ and 5′ ends of the mRNA.
   (d) Draw four peptides in the process of being synthesized, indicating relative lengths.
   (e) Indicate the N- and C-terminal ends of the longest peptide.
   (f) Indicate with an arrow the direction in which RNA polymerase is moving.
   (g) What parts of this diagram would be different in eukaryotes?
3. Although 40 to 50% of the RNA being synthesized in *E. coli* at any given time is mRNA, only about 3% of the total RNA in the cell is mRNA. Explain.
4. In *E. coli,* the mRNA fraction is heterogeneous in size, ranging from 500 to 6,000 nucleotides. The largest mRNAs have many more nucleotides than needed to make the largest proteins. Why are some of these mRNAs so large in bacteria?
5. When yeast phenylalanine tRNA is digested with a small amount of RNase (partial digest), such as T2 RNase, almost all of the cleavage sites are found in the anticodon loop. Explain this observation.
6. Speculate on the advantage of having three rRNAs (16S, 23S, and 5S) as part of the same RNA precursor such as found in *E. coli.*
7. In *E. coli* the precise spacing between the −35 and −10 conserved promoter elements has been found to be a critical determinant of promoter strength. What does this suggest about the interaction between RNA polymerase and these elements? What sorts of evidence could you obtain about this interaction by doing "footprint" experiments? (Also explain how you would do these experiments.)
8. What is the maximum theoretical rate of initiation at an *E. coli* promoter, assuming that the diameter of its RNA polymerase is about 200 Å and the rate of RNA chain growth is 45 nucleotides per second?
9. Cordycepin-5′-triphosphate, 3′-deoxy ATP, is a ribonucleoside triphosphate analog that can be bound as a substrate by RNA polymerase. What effect does this analog have on transcription and how could this information be used to determine the direction of transcription?
10. The following hypothetical RNA is a precursor made in a eukaryotic nucleus and its synthesis is inhibited by low levels of α-amanitin:

pppAUUAUGCCGAUAAGGUAAGUA-(N_{100})-
AUCUCCCUGCAGGGCGUAACCAAUAAACGACGACGACGUCACC

Indicate the final processed RNA found in the cytoplasm and point out important features.

11. In early research with intact eukaryotic mRNA, the RNA appeared to have two 3′ ends and no 5′ terminus. Explain these observations.
12. What is unique about the promoter for RNA polymerase III? Diagram how this promoter works.
13. T7 and SP6 RNA polymerases are used to make RNA transcripts *in vitro* as important tools in modern molecular biology. The addition of inorganic pyrophosphatase to these reactions can increase the yield of the RNA product. Explain the increased yield.
14. Draw the chemical reaction mechanism for the formation of a "lariat" in the processing of introns from eukaryotic mRNA.
15. A particular eukaryotic DNA virus was found to code for two mRNA transcripts, one shorter than the other, from the same region on the DNA. Analysis of the translation products revealed that the two polypeptides shared the same amino acid sequence at their amino-terminal ends, but were different at their carboxyl-terminal ends. The longer polypeptide was coded by the shorter mRNA! Suggest an explanation.
16. You are a graduate student working on an *in vitro* processing system for intron removal with a precursor rRNA isolated from *Tetrahymena.* The purified RNA will process by itself without any proteins being added to the reaction. How could you prove that this RNA was autocatalytic and that the reaction was not just due to some protein contamination in your incubation?
17. Draw a chemical reaction mechanism for the first step in the self-splicing reaction of group I introns. (Group I intron splicing requires a guanosine co-factor.)

29

CHAPTER

Protein Synthesis, Targeting, and Turnover

Proteins are informational macromolecules, the ultimate heirs of the genetic information encoded in the sequence of nucleotide bases within the chromosomes. Each protein is composed of one or more polypeptide chains and each peptide chain is a linear polymer of amino acids. The order of the amino acids (see table 3.1) commonly found in the polypeptide chain is determined by the order of nucleotides in the corresponding messenger RNA template. In this chapter we will examine four aspects of protein metabolism (fig. 29.1): (1) the process whereby amino acids are ordered and polymerized into polypeptide chains; (2) posttranslational alterations in polypeptides, which occur after they are assembled on the ribosome; (3) the targeting process whereby proteins move from their site of synthesis to their sites of function; and (4) the proteolytic reactions that result in the return of proteins to their starting material, amino acids.

Figure 29.1

Overview of reactions in protein synthesis. (aa_1, aa_2, aa_3 = amino acids 1, 2, 3.) Protein synthesis requires transfer RNAs for each amino acid, ribosomes, messenger RNA, and a number of dissociable protein factors in addition to ATP, GTP, and divalent cations. First the transfer RNAs become charged with amino acids, then the initiation complex is formed. Peptide synthesis does not start until the second aminoacyl tRNA becomes bound to the ribosome. Elongation reactions involve peptide bond formation, dissociation of the discharged tRNA, and translocation. The elongation process is repeated many times until the termination codon is reached, an event that signals the end of polypeptide synthesis. Termination is marked by the dissociation of the messenger RNA from the ribosome and the dissociation of the two ribosomal subunits. The polypeptide chain sometimes folds into its final form without further modifications. Frequently the folded polypeptide chain is modified by removal of part of the polypeptide chain or by addition of various groups to specific amino acid side chains. The completed polypeptide chain will migrate to different locations according to its structure. It may remain in the cytosol or it may be transported into one of the cellular organelles or across the plasma membrane into extracellular space. Proteins have different lifetimes. Eventually a protein will be degraded, usually down to its component amino acids.

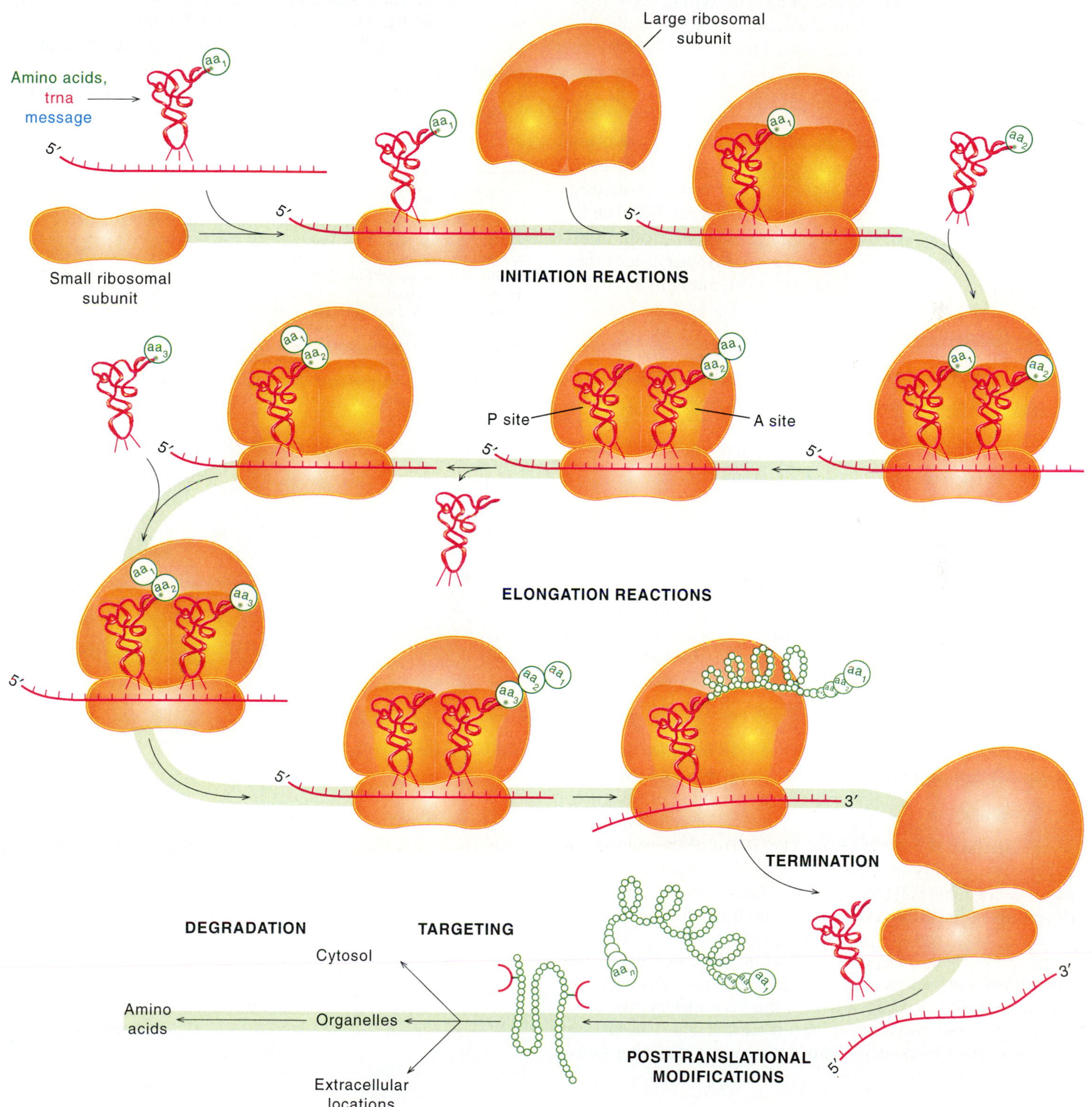

The Cellular Machinery of Protein Synthesis

Amino acids are assembled into polypeptides on ribosomes. Prior to their interaction with messenger RNA (mRNA), amino acids are covalently attached to transfer RNA (tRNA) to form aminoacyl-tRNAs. The aminoacyl-tRNAs attach to specific sites on the mRNA. mRNA contains the instructions for translation in the form of the genetic code that specifies the amino acid sequence of the polypeptide to be synthesized. Each ribosome binds to and moves along the messenger RNA while producing a single polypeptide. The direction or polarity of translation is from the 5′ to the 3′ terminus on the message, while the polypeptide is synthesized from the amino to the carboxyl terminus. As a rule, the information in a single mRNA molecule is translated simultaneously by a number of ribosomes that form a structure called a polysome. In prokaryotic cells, translation of mRNA begins while it is still being transcribed so that assemblies of nascent polypeptides and mRNAs can be seen on chromosomal DNA (fig. 29.2). In eukaryotic cells, transcription and translation are separate and polysomes occur either free in the cytoplasm or bound to membranes in the endoplasmic reticulum (fig. 29.3).

RNA molecules play crucial roles in protein synthesis. In addition to mRNA and tRNA, the ribosome also contains ribosomal RNA (rRNA). The tRNAs and the mRNAs are independent entities or are found in complexes with other molecules. Ribosomal RNAs, on the other hand, exist solely as complexes with proteins in the form of ribosomes.

Messenger RNA Is the Template for Protein Synthesis

The mRNA molecule carries the genetic message in the form of a sequence of nucleotides that determines the order of amino acids in the polypeptide chain. Each amino acid is represented in the mRNA by a sequence of three nucleotides called codons. Codons are arranged in a contiguous reading frame, which is flanked on either side by bases that are not translated. These untranslated regions frequently have roles in regulating the processing and expression of the message. The 5′ end of the reading frame begins with a start codon, usually consisting of the nucleotides AUG, a sequence that codes for the amino acid methionine. Methionine is always used to initiate translation. The 3′ end of the reading frame contains one or more of three stop codons, UAA, UAG, or UGA. Stop codons serve as signals to terminate the polypeptide chain. The 3′ end of the message in eukaryotes usually contains a posttranscriptionally added poly(A) tail. This poly(A) tail has no known role in translation except that it probably increases the lifetime of the message.

The 5′ end of the mRNA plays a special role in the selection of the start codon. This selection process occurs in fundamentally different ways in prokaryotes and eukaryotes. In prokaryotes (fig. 29.4) the mRNA contains a specific ribosome-binding site upstream of the initiating start codon. This binding site allows the ribosome to identify the correct initiating AUG. As we mentioned in chapter 28, prokaryotic mRNAs are frequently polycistronic, which means that they encode more than one polypeptide chain. Internal ribosome binding sites allow for independent initiation at internal reading frames. Eukaryotic mRNAs, on the other hand (see fig. 29.4), are usually monocistronic, encoding only one polypeptide chain. In keeping with this architecture, eukaryotic ribosomes interact with a ribosome entry site at the 5′ terminus of mRNA and move along it by a scanning mechanism to find the start codon. Identification of the ribosome entry site may be partly facilitated by the cap structure (see fig. 28.23) at the 5′ end of the mRNA. Initiation in eukaryotes usually occurs at the first AUG sequence downstream from the ribosome entry site. In multicellular eukaryotes the initiating AUG sequence is selected by a preferred initiation context of adjacent bases.

Figure 29.2

(*a*) Electron micrograph of *E. coli* polysomes. Ribosomes are the dark structures connected by the faintly visible mRNA strand. A DNA strand connecting the polysomes from which the mRNA is being transcribed is visible as a horizontal line. (From O. L. Miller, B. A. Hankalo, and C. A. Thomas, "Visualization of bacterial genes in action," *Science* 169:392, 1970, © 1970 by the AAAS.) (*b*) Line drawing for clarification. In bacteria, translation usually begins before transcription is completed. This results in a DNA-mRNA-ribosome complex as shown. As the mRNA grows, the number of ribosomes associated with it increases. Here the mRNA appears to be growing from left to right. It is not possible to see the growing polypeptide chains on the ribosome because of their small size.

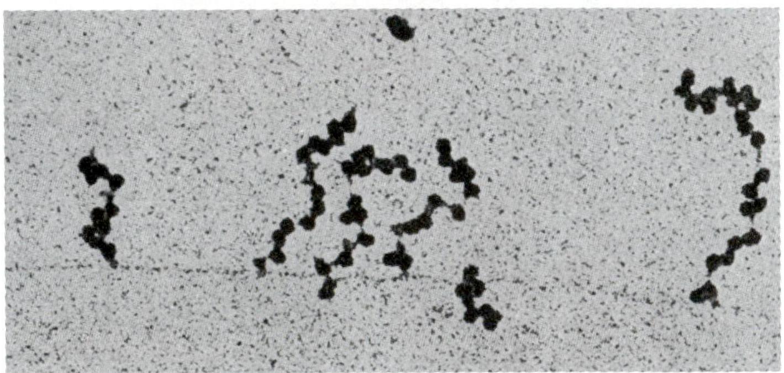

(a)

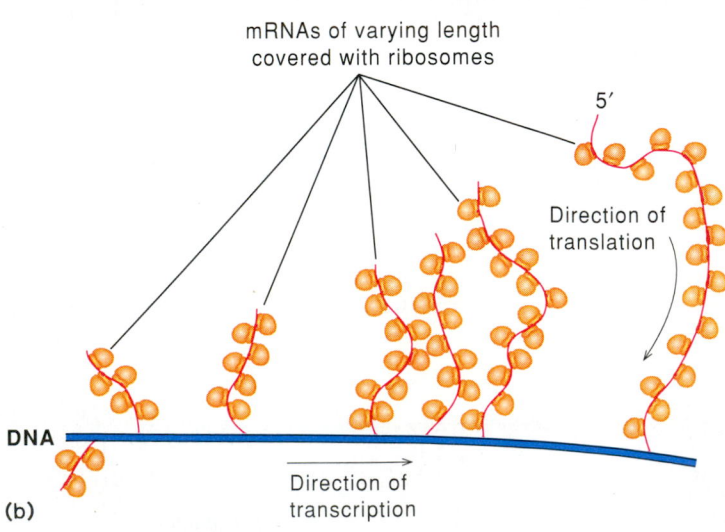

mRNAs of varying length covered with ribosomes

5′

Direction of translation

DNA

Direction of transcription

(b)

Transfer RNAs Order Activated Amino Acids on the mRNA Template

Transfer RNAs contain two crucial functional sites—a site for attachment of the amino acid, and a site that interacts with the mRNA during translation on the ribosome. At least one type of tRNA corresponds to each of the twenty types of amino acids that are incorporated into proteins. For accurate translation to occur, tRNAs must be distinguishable from each other by the molecules that recognize these specific types while still being recognizable by the molecules that interact with all tRNAs. A major challenge has been to understand how the similarities and differences in the structures of different tRNAs are related to the functions that they play in protein synthesis. Here we will summarize the structural features that are found in all tRNAs, allowing them to perform their common functions.

tRNAs that correspond to a single amino acid type are known as cognate tRNAs. In writing, the different cognate tRNAs are designated by a superscript. For example, $tRNA^{Phe}$ and $tRNA^{Ser}$ designate the tRNAs that correspond to the amino acids phenylalanine and serine, respectively. Some amino acids have several different cognate tRNAs, which are called isoacceptor tRNAs. Isoacceptor tRNAs are distinguished from each other by subscripts. A single cell typically contains a total of 50 or more different tRNAs.

$tRNA^{Phe}$ was the first tRNA to be crystallized, and its three-dimensional structure is one of the best understood (see fig. 28.3). The other tRNAs are built on a structural theme similar to that of $tRNA^{Phe}$, so that a single figure can describe the common features of their primary, secondary, and tertiary structures (fig. 29.5). All tRNAs are composed of a single polynucleotide chain of from 70 to 95 residues. This chain folds back on itself to form a cloverleaf structure composed of four double-stranded stems and four single-stranded loops. The overall folding is such that the 5′ and 3′ ends of the molecule are brought together to create a stem with seven regular Watson-Crick base pairs. This stem is known as the acceptor stem because during protein synthesis the amino acid is transiently attached to the ribose at this terminus. The 3′ end of the molecule

Figure 29.3

(*a*) Electron micrograph of mammalian rough endoplasmic reticulum. Continuous sheets of membrane create a compartment distinct from the surrounding cytosol. (From S. L. Wolfe, *Biology of the Cell*, 2d ed., Wadsworth, Belmont, Calif., 1981.) (*b*) Clarifying line drawing shows expanded region of endoplasmic reticulum with ribosomes attached to the surface of the membrane facing the cytosol. The term rough endoplasmic reticulum arose because at low magnification the attached ribosomes give the endoplasmic reticulum a rough appearance.

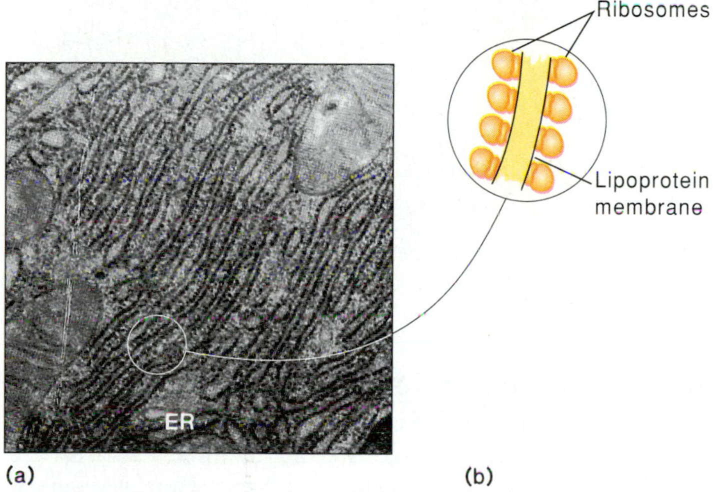

Figure 29.4

Simplified diagram of mRNA structure. (*a*) Typical eukaryotic mRNA. An AUG start codon is located near the 5′ end of the mRNA. The single reading frame ends with one of the three trinucleotide sequences that represents a stop codon. Frequently, but not always, the 5′ end of the mRNA is capped and the 3′ end contains a poly(A) tail. The cap structure is described in figure 28.23. (*b*) Typical bacterial mRNA. Some bacterial mRNAs contain more than one reading frame; some contain only one. Each reading frame contains a start and a stop codon. Recognition of the start codons is facilitated by the presence of a ribosome recognition sequence (Shine-Dalgarno). The space between any two reading frames in the same transcript varies from one transcript to the next.

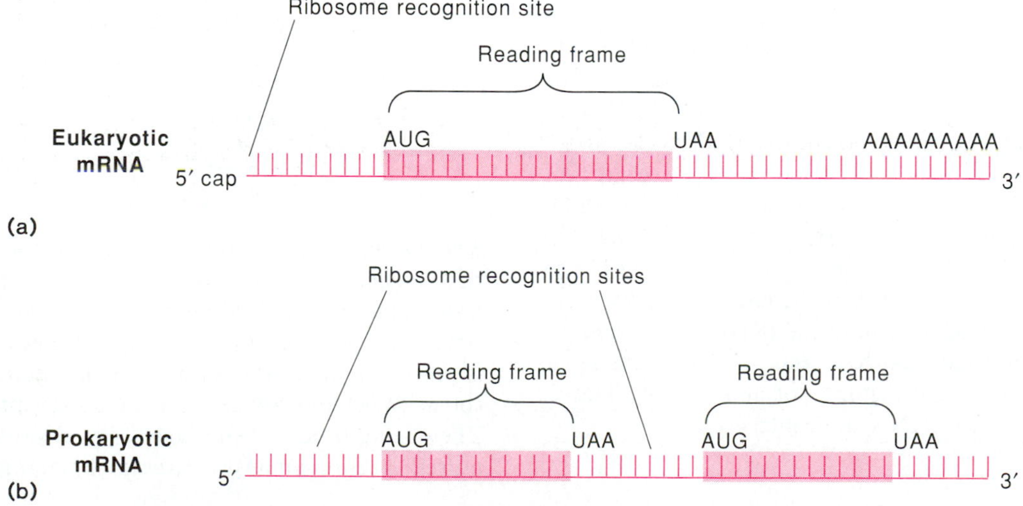

Figure 29.5

The general structure of a tRNA molecule. (*a*) Representation in the form of a cloverleaf which is the clearest way of observing the secondary structure. (*b*) A more realistic drawing of the three-dimensional folded structure. Color coding shows how the various loops in the cloverleaf structure correspond to the parts of the folded structure. For a more detailed drawing showing the types of hydrogen bonds and the structures of the individual bases, see figure 28.3.

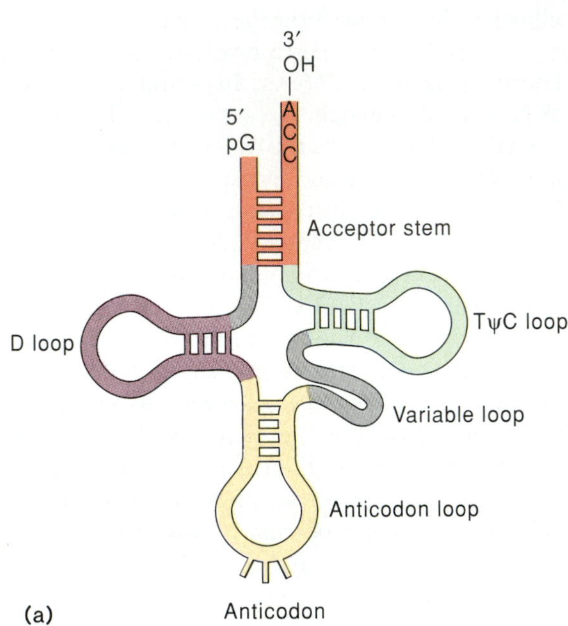

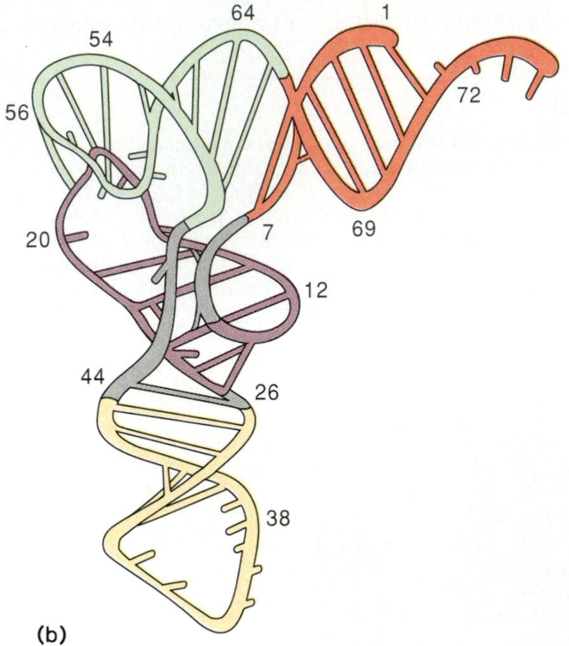

is always terminated with the sequence CCA, which identifies the tRNA to the components that interact with all of these molecules during protein synthesis.

The unpaired loops of tRNA are named according to their unique structural features. Loop I varies in size from 7 to 11 unpaired bases and frequently contains the unusual base dihydrouracil; it is designated the D loop. Loop II contains the three bases known as the anticodon and is therefore designated the anticodon loop. This portion of tRNA plays a key role in translation by pairing with the complementary codon in mRNA and thus serves to align amino acids in their appropriate sequence. Loop III, the variable loop, may contain as few as 3 or as many as 21 bases, making it the major site of size variation in tRNA. Loop IV contains the unusual ribothymidine and pseudouridine bases as part of an invariant sequence; for this reason it is known as the TψC loop.

The secondary cloverleaf structures of all tRNAs in solution fold into a three-dimensional L-shaped structure like that of $tRNA^{Phe}$ (fig. 29.5*b*). This structure is composed of two helical arms joined at right angles. The ribose moiety to which the amino acid is joined is at the end of one arm, identifying it as the acceptor arm. The anticodon is at the end of the other arm, identifying it as the anticodon arm. In the tRNA structures thus far solved, the 3′-terminal ribosome and anticodon are separated by a constant diagonal distance of about 75 Å. This L shape is the hallmark of a tRNA molecule.

Ribosomes Are the Site of Protein Synthesis

We considered the structure of the ribosome in some detail in the previous chapter without referring to those sites that are functionally important in protein synthesis. Here we can localize some of the functional sites on the two ribosomal subunits (fig. 29.6). The mRNA binds to the smaller ribosomal subunit. The peptidyl transferase is an integral part of the 50S subunit,* and the elongation factor EF-G binds to the 50S subunit. The nascent polypeptide chain exits through a channel in the 50S subunit. There are two functional sites on the 50S subunits, called the P site and the A site, where adjacent tRNAs are bound to the messenger. These sites cannot be seen in figure 29.6.

Before we go into a detailed discussion of the reactions that occur on the ribosome, it would be useful to consider the crucial interaction between the anticodon on the tRNAs and the codons on the mRNAs.

The Genetic Code

The concept of a genetic code grew out of the realization that both nucleic acids and proteins are linear polymers made of a limited number of building blocks, plus the knowledge that the structure of proteins is genetically determined. The genetic code is simply the sequence relationship between nucleotides in genes (or mRNA) and the amino acids in the proteins they encode. The coding ratio is the number of nucleotides in an mRNA required to represent an amino acid. Since there are four different

*Recently it has been shown that the peptidyl transferase is part of the rRNA of the 50S subunit.

Figure 29.6

Functional sites on the prokaryotic ribosome. This figure shows a ribosome in the process of elongating a polypeptide chain. The mRNA and the EF-Tu-aminoacyl-tRNA complex are more closely associated with the 30S subunit. The peptidyl transferase and the EF-G elongation factor are associated with the 50S subunit, as is the nascent polypeptide chain.

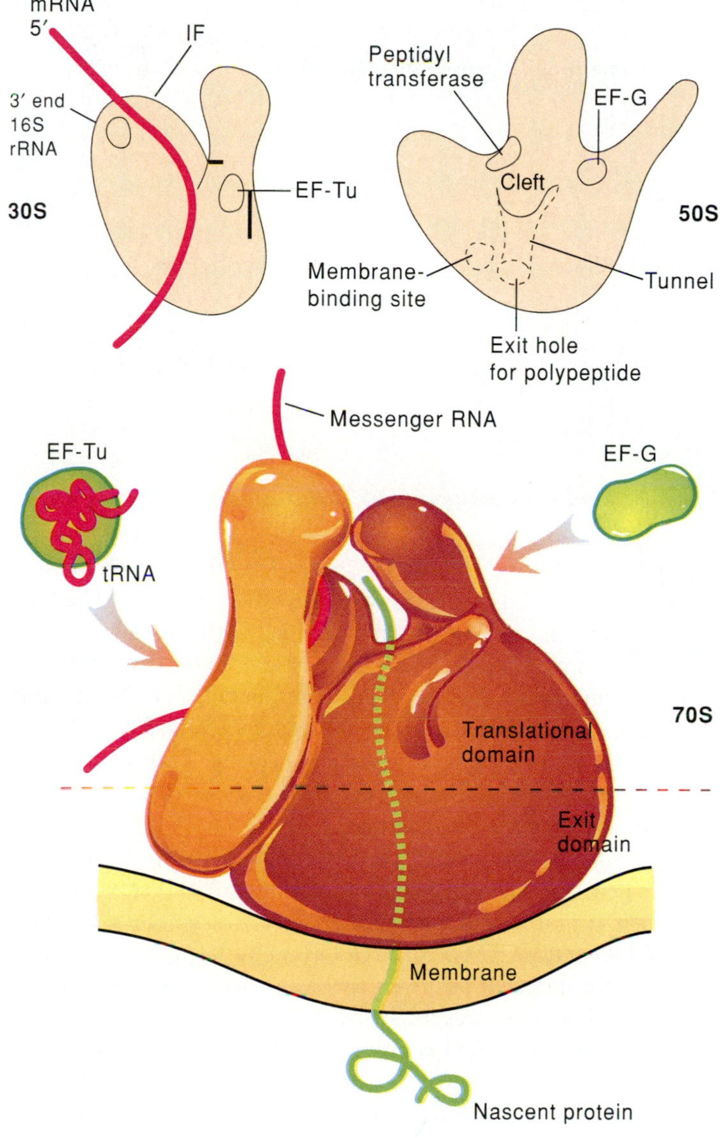

nucleotides in RNA and 20 different amino acids in proteins, the minimum acceptable coding ratio is 3: A singlet code, in which one nucleotide represents one amino acid, could code for only four different amino acids because there are only four different nucleotides in DNA or RNA. A doublet code, in which two nucleotides code for a single amino acid, could code for 16 amino acids since there are 16 possible doublet sequences in RNA. A triplet code, made from four nucleotides taken three at a time, generates a total of 64 possible triplet sequences. This would be more than enough to represent the 20 amino acids found in proteins.

In fact, a triplet code creates a problem of a different sort. What is the use of the extra trinucleotide sequences? Allowing for the three triplets that represent stop signals, we are still left with 61 possible triplets to code for just 20 amino acids. A nondegenerate code, in which a unique triplet codes for each amino acid, would imply that 41 triplet sequences are never found in translation reading frames. On the other hand, a degenerate code, in which all triplets are used, would require that each amino acid be represented, on the average, by more than one triplet sequence. We will return to this point later. For now, we may simply note that in most cases, point mutations in which one base pair in the DNA is replaced by another result either in no amino acid change or in a change to another amino acid. This is the result that we would expect if most of the triplet sequences do in fact represent amino acids.

Another question relates to the location of codons in the translation reading frame. Codons could be arranged in a close-packed side-by-side arrangement. In this event each nucleotide within the translation reading frame would represent a code letter within one, and only one, codon. Alternatively, codons could be separated by one or more "spacer nucleotides." In this event some nucleotides within the reading frame would represent code letters within codons, while others would serve as spacer nucleotides. Finally, we can imagine an overlapping codon arrangement, in which all nucleotides represent code letters and some nucleotides represent code letters in more than one codon. However, biochemical experiments that we will describe later in this chapter favor a nonoverlapping triplet code without spacers. Further genetic experiments in support of this type of code (not elaborated on here) indicate that when a single base pair is added or deleted, a change occurs in the reading frame of the message downstream of the change. Such an alteration in the DNA is known as a frameshift mutation. A frameshift mutation will produce a new termination point because the frameshift changes the reading of the original stop codon and produces a new stop codon.

The Code Was Deciphered with the Help of Synthetic Messengers

In 1961, Marshall Nirenberg and Heinrich Mattaei were attempting to synthesize proteins in cell-free extracts of *E. coli*. These extracts, containing the essential cellular components, were supplemented with nucleotides, salts, and radioactive amino acids, which were used so that the expected small amounts of protein could be detected. For mRNA they chose to use viral RNA from tobacco mosaic virus (TMV), because their goal was to make viral protein. As a control for these first "incorporation" experiments, they needed an mRNA that would not be expected to code for a protein. For this purpose their choice was an artificial polynucleotide produced by Marianne Grunberg-Manago and Severo Ochoa with a newly discovered enzyme, polynucleotide phosphorylase. Initially, Grunberg-Manago and Ochoa had thought that this enzyme might be responsible for cellular RNA synthesis, but they later came to realize that its more likely role was to degrade RNA by a phosphorolysis reaction:

$$RNA_n + nP_i \rightarrow (\text{nucleoside diphosphates})_n$$

As a result of that realization, polynucleotide phosphorylase was relegated to the synthesis of unusual polynucleotides as laboratory curiosities. One such polynucleotide, poly(U), contained only uridine residues. When Nirenberg and Mattaei added this control RNA to their cell-free system, substantial incorporation was evident. They quickly showed that this synthetic messenger specifies the synthesis of a polypeptide containing only phenylalanine residues (polyphenylalanine).

A wave of excitement was set in motion by this discovery. It was rapidly demonstrated that poly(A) promotes polylysine synthesis and poly(C) promotes polyproline synthesis. From these observations it seemed clear that the code words UUU, AAA, and CCC correspond to the amino acids phenylalanine, lysine, and proline, respectively.

Polynucleotide phosphorylase, running in the reverse direction, produces an "RNA" of random sequence whose composition reflects the mixture of nucleoside diphosphates in the reaction mixture. Mixed polynucleotides containing two bases were used in the incorporating system and shown to incorporate a pattern of amino acids that was consistent with a triplet code, but the observed incorporation could not define the code sequence.

The chemical synthesis of short oligonucleotides of known sequence provided a way out of the dilemma. First, Philip Leder and Nirenberg showed that nucleotide triplets cause the specific binding of aminoacyl-tRNA to ribosomes. They cleverly observed that this binding could be readily measured by simply passing a solution containing radioactive aminoacyl-tRNA, ribosomes, and the trinucleotide through a nitrocellulose filter. Ribosomes, along with bound ligands, were adsorbed to the filter and the coding specificity of some triplets was quickly established. Thus UUC, UCU, and CUU were shown to specify phenylalanine, serine, and leucine, respectively. However, many triplets did not give clear binding signals and there was concern that the binding reaction might not display the correct specificity.

The code was ultimately defined by the translation of polynucleotides of repeating sequences. These long-repeat-sequence polynucleotides were produced by chemically synthesizing short, defined-sequence oligomers and then amplifying them enzymatically. When repeating dinucleotides were translated (fig. 29.7), they yielded repeating dipeptides, and when repeating trinucleotides (not shown) were translated, they yielded up to three individual peptides, each containing only a single amino acid. Eventually the sequence of bases from natural messages was correlated with the sequence of bases in the proteins they encoded.

The Code Is Highly Degenerate

A triplet code, one made from four nucleotides taken three at a time, generates a total of 64 different triplet sequences or codons. Three of these codons, as we will see, are utilized to terminate translation and are not generally used to specify amino acids. The remaining 61 codons and the 20 amino acids can be neatly summarized by grouping codons with the same first and second bases into a grid (table 29.1). The four horizontal sections are composed of codons with the same first base. The four vertical sections are composed of codons with the same second base. The boxes representing the vertical/horizontal intersections contain codon families whose members differ only in their 3′-terminal base. For example, the codons UCU, UCC, UCA, and UCG comprise a family encoding serine. Thus the genetic code is degenerate, i.e., one amino acid is generally specified by multiple or synonymous codons.

Figure 29.7

Demonstration that poly(UC) codes for a Ser-Leu repeating peptide. When poly(UC) containing a strict alternating sequence of U and C is used as an mRNA, only serine and leucine are incorporated. Analysis of the polypeptide product indicates that it contains an alternating sequence of serine and leucine.

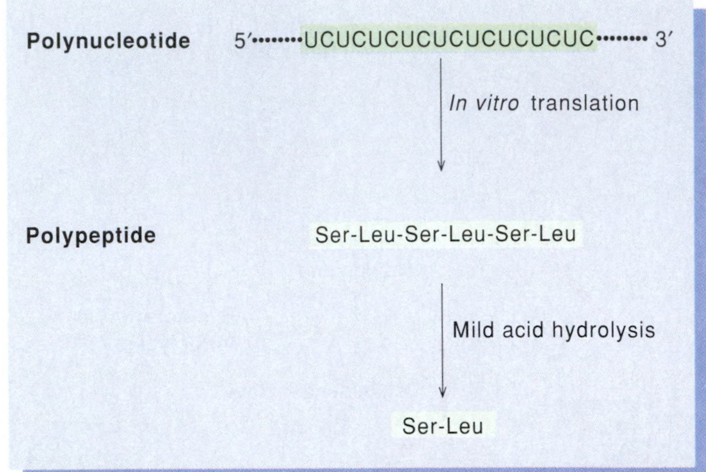

As we have indicated, the codon AUG is the only one that is generally used to specify methionine, but it serves a dual function in that it is also used to initiate translation. Occasionally, GUG and UUG are also read as an initiating codon in bacteria, but in internal positions these codons are always read as valine and leucine, respectively. In eukaryotes, initiation at codons other than AUG is much less frequent than in prokaryotes. Weak initiation occasionally occurs at GUG, CUG, and ACG codons in eukaryotic systems. The UGA triplet also serves a dual function; it is usually recognized as a stop, but on occasion it serves as a codon for selenocysteine (box 29A).

Organisms differ in the frequency with which they utilize synonymous codons. Some synonymous codons are used frequently and some are used infrequently. For *E. coli* and yeast this frequency of use, known as codon usage, correlates with the abundance in the organism of the tRNAs that recognize particular synonymous codons. Also, for the same two species, proteins that occur in the greatest abundance employ high-usage codons more frequently and low-usage codons less frequently than the average proteins. It is not clear that this same trend is followed in multicellular eukaryotes such as *Drosophila* and humans.

Table 29.1
The Genetic Code

First position (5′ end)	Second position: U	Second position: C	Second position: A	Second position: G	Third position (3′ end)
U	UUU } Phe UUC UUA } Leu UUG	UCU UCC UCA } Ser UCG	UAU } Tyr UAC UAA } STOP UAG	UGU } Cys UGC UGA STOP UGG Trp	U C A G
C	CUU CUC CUA } Leu CUG	CCU CCC CCA } Pro CCG	CAU } His CAC CAA } Gln CAG	CGU CGC CGA } Arg CGG	U C A G
A	AUU AUC } Ile AUA AUG Met	ACU ACC ACA } Thr ACG	AAU } Asn AAC AAA } Lys AAG	AGU } Ser AGC AGA } Arg AGG	U C A G
G	GUU GUC GUA } Val GUG	GCU GCC GCA } Ala GCG	GAU } Asp GAC GAA } Glu GAG	GGU GGC GGA } Gly GGG	U C A G

Figure 29.8

Examples of standard (*a*) and wobble (*b* and *c*) base pairs formed between the first base in the anticodon and the third base in the codon.

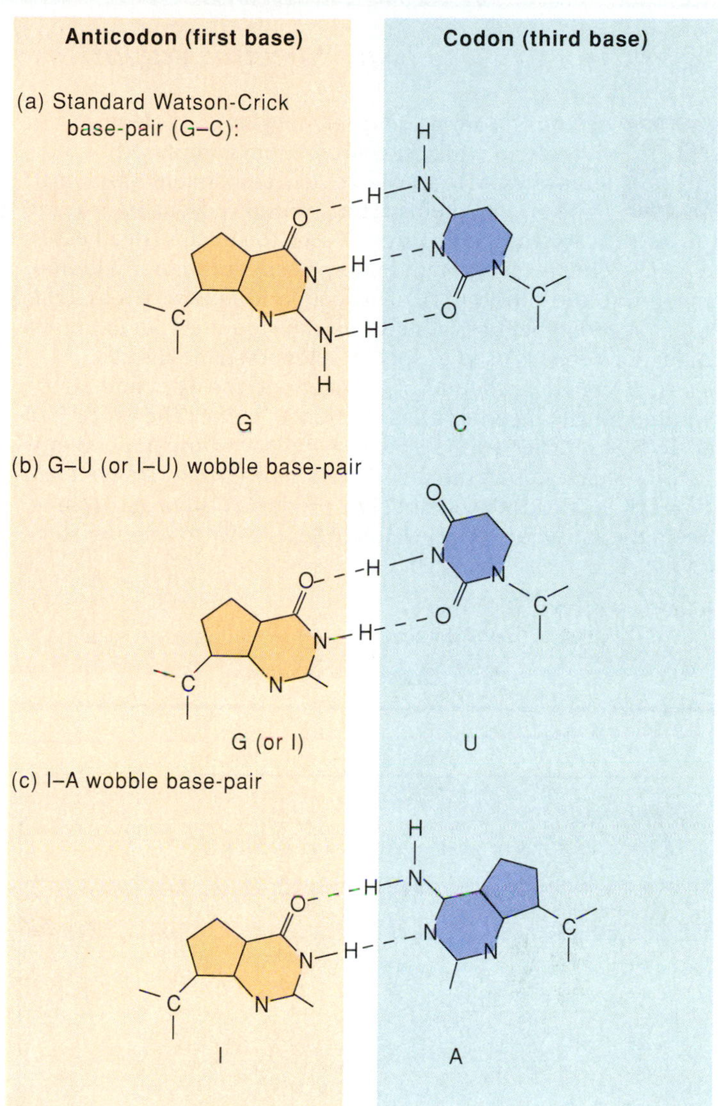

Wobble Introduces Ambiguity into Codon–Anticodon Interactions

In order for 61 triplets to act as codons, tRNAs must interact specifically with each triplet. Strict Watson-Crick base-pairing between codon and anticodon would require 61 different anticodons and, correspondingly, 61 different tRNAs. As the characterization of tRNAs progressed it became clear that in many cases individual tRNAs could recognize more than one codon. In all cases the different codons recognized by the same tRNA were found to contain identical nucleotides in the first two positions and a different nucleotide in the third position (in the 3′ position of the codon). This relationship is the reason that the genetic code can be so neatly arranged as shown in table 29.1, by codon families that differ in their third base.

The 3′-terminal redundancy of the genetic code and its mechanistic basis were first appreciated by Francis Crick in 1966. He proposed that codons and anticodons interact in an antiparallel manner on the ribosome in such a way as to require strict Watson-Crick pairing (that is, A-U and G-C) in the first two positions of the codon but to allow other pairings in its 3′-terminal position. Nonstandard base-pairing between the 3′-terminal position of the codon and the 5′-terminal position of the anticodon alters the geometry between the paired bases; Crick's proposal, labeled the wobble hypothesis, is now viewed as correctly describing the codon–anticodon interactions that underlie the translation of the genetic code.

By inspecting the geometry that would result from different wobble pairings (fig. 29.8) and recognizing that inosine, the deaminated form of adenine, frequently occurs in tRNAs in the 5′ position of the anticodon, Crick grasped the relationship between codon and anticodon. According to this relationship, table 29.2, when C or A occurs in the 5′ position of an

29A BOX

Site-Specific Variation in Translation Elongation

The unusual amino acid selenocysteine (a derivative of cysteine in which the sulfur atom is replaced by a selenium atom) is an essential component in a small number of proteins. These proteins occur in prokaryotes and eukaryotes ranging from *E. coli* to humans. In all cases, selenocysteine is incorporated into protein during translation in response to the codon UGA. This codon usually serves as a termination codon but occasionally, in some required but unknown context of bases, is used to specify selenocysteine instead.

In *E. coli,* the products of four genes (*selA, B, C,* and *D*) are required for the incorporation of selenocysteine. The product of the *selC* gene is $tRNA^{Ser}$ (a suppressor tRNA whose anticodon is UCA). The first step in the incorporation of selenocysteine is catalyzed by seryl-tRNA synthase. The products of *selA* and *D* function in the subsequent conversion of Ser-$tRNA^{Ser}$ to selenocysteyl-$tRNA^{Ser}$. The probable pathway is

$$tRNA^{Ser} \rightarrow \text{seryl-}tRNA^{Ser} \rightarrow \text{phosphoseryl-}tRNA^{Ser} \rightarrow \text{selenocysteyl-}tRNA^{Ser}$$

Incorporation of selenocysteyl-$tRNA^{Ser}$ into protein in response to the UGA codon requires SELB (the protein product of the *selB* gene in *E. coli*). SELB is homologous in sequence to EF-Tu and probably replaces it in translation by specifically recognizing selenocysteyl-tRNA and UGA in the appropriate sequence context. Selenocysteyl-$tRNA^{Ser}$, in combination with SELB, must be capable of competing with termination factors for the translation of the termination codon when it occurs in the "right" context of bases. This process is known as *site-specific variation* in translation elongation.

Other proteins of unknown function that are homologous to EF-Tu are known to exist. Conceivably, these proteins might participate in the incorporation of other rare amino acids into unique positions in proteins.

Table 29.2
The Wobble Rules of Codon–Anticodon Pairing

5′ Base of Anticodon	3′ Base of Codon
C	G
A	U
U	A or G
G	C or U
I	U, C, or A

anticodon, it can pair only with G or U, respectively, in the 3′ position of a codon. tRNAs containing either G or U in the 5′ or wobble position of the anticodon can each pair with two different codons, while an inosine (I) in this position produces a tRNA that can pair with three codons differing in the 3′ base. Subsequent sequence analysis of many tRNAs has proved this hypothesis as a feature of the translation of the "universal" genetic code (but see the following discussion).

A careful comparison of the wobble rules with the genetic code indicates that the minimum number of tRNAs required to translate all 61 codons is 31. With the addition of $tRNA_i^{Met}$ the total comes to 32. Most cells contain many more than this minimum number of tRNA types.

The Code Is Not Quite Universal

It was originally believed that exactly the same genetic code is utilized by all genetic systems. Initial experiments supported this conclusion, since it was found that some mammalian mRNAs could be faithfully translated in cell-free bacterial systems. We now know, however, that significant variations in the meaning of specific code words occur in many genetic systems. The exact scope and nature of these variations are still being discovered, but the current view is that the variations in genetic meaning all represent divergences from the standard or "universal" genetic code described earlier. In other words, there is no indication that variations in the genetic code resulted from independent origins of the genetic code. Much of the variation in genetic meaning derives from alterations in the pairing of codon and anticodon on the ribosome.

Many of the known variations in the genetic code are found in genes of mitochondria and chloroplasts. It is easy to see why these genetic systems might be more plastic, since they frequently encode only ten to twenty proteins. The remainder of the organellar proteins are derived by importing nuclear gene products.

The tRNAs used to translate mitochondrial mRNAs are entirely derived from mitochondrial chromosomes. The first clue that something was unusual about the mitochondrial genetic code was that only 24 types of tRNA could be found. Recall that according to Crick's rules of wobble, 32 tRNAs minimally are required for the translation of all 61 codons. One possible

Table 29.3
The Genetic Code of Yeast Mitochondria

First position (5′ end)	Second position: U	Second position: C	Second position: A	Second position: G	Third position (3′ end)
U	UUU, UUC } Phe AAG UUA, UUG } Leu AAU*	UCU, UCC, UCA, UCG } Ser AGU	UAU, UAC } Tyr AUG UAA, UAG } STOP	UGU, UGC } Cys ACG UGA, UGG } Trp ACU*	U C A G
C	CUU, CUC, CUA, CUG } Thr GAU	CCU, CCC, CCA, CCG } Pro GGU	CAU, CAC } His GUG CAA, CAG } Gln GUU*	CGU, CGC, CGA, CGG } Arg GCA[b]	U C A G
A	AUU, AUC } Ile UAG AUA, AUG } Met UAC[a]	ACU, ACC, ACA, ACG } Thr UGU	AAU, AAC } Asn UUG AAA, AAG } Lys UUU*	AGU, AGC } Ser UCG AGA, AGG } Arg UCU*	U C A G
G	GUU, GUC, GUA, GUG } Val CAU	GCU, GCC, GCA, GCG } Ala CGU	GAU, GAC } Asp CUG GAA, GAG } Glu CUU*	GGU, GGC, GGA, GGG } Gly CCU	U C A G

Source: S. G. Bonitz et al., "Codon recognition rules in yeast mitochondria," in *Proceedings National Academy of Sciences USA* 77:3167, 1980. Copyright © 1980, National Academy of Sciences, Washington, D.C.

The codons (5′ → 3′) are at the left and the anticodons (3′ → 5′) are at the right in each box. (* designates U in the 5′ position of the anticodon that carries the —$CH_2NH_2CH_2COOH$ grouping on the 5′ position of the pyrimidine.)

[a]Two tRNAs for methionine have been found. One is used in initiation and one is used for internal methionines.

[b]Although an Arg tRNA has been found in yeast mitochondria, the extent to which the CGN codons are used is not clear.

solution to this conundrum was that mitochondrial genes do not utilize all 61 codons. Another possibility was that the wobble rules might be different for mitochondria. In fact, the latter is the case. Crick's original wobble rules stated that at least two different tRNAs are needed to translate four codon families. In all of these cases (with the exception of the codon family specifying Arg; see the footnote to table 29.3), single tRNAs have been found responsible for specifying all four code words and these tRNAs all contain a U in the "wobble" position of their anticodons. It appears that the mitochondrial ribosome allows these tRNAs to pair with all four members of the codon family. The six mitochondrial tRNAs that pair with the normal two codons contain an altered U in the wobble position and this modification causes them to conform to the normal "wobble" rules.

Another peculiarity of the mitochondrial code emerged from a study of yeast codon usage. By comparing tables 29.3 and 29.1, you will see that the mitochondrial code has several differences in code word meaning. The codons beginning with CU represent Thr instead of Leu, the AUA codon represents Met instead of Ile, and the UGA codon represents Trp rather than a stop signal.

The Rules Regarding Codon–Anticodon Pairing Are Species-Specific

The genetic code differs very little between species. By contrast, there are considerable differences between species in the anticodon translation system of tRNA, as evidenced by the mitochondrial tRNA system. In all systems the bases in the anticodon-codon complex run antiparallel, as in standard double-helix pairing, and in all cases only Watson-Crick-like base pairing occurs between the first two bases in the codon and the opposing bases in the anticodon segment of the tRNA. However, for the 3′ base in the codon, the rules for pairing vary with the species and with the base in question. These rules, summarized in table 29.4, are as follows.

When the 5′ base in the anticodon is a G it can pair with either a U or a C, and this is true in all organisms. When the 5′ base in the anticodon is an A it can pair only with a U in the codon. However, it is rare that an A is found in this position of the anticodon. An A in this position in eukaryotes is usually deaminated to an inosine (I) base, which has an expanded capacity for pairing. Base A can pair only with U, but I can pair with U, C, or A. In eubacteria, deamination is limited to the conversion of the ACG sequence to an ICG anticodon. When C

Table 29.4
Anticodon–Codon Pairing

Anticodon First Base	Codon Third Base	Examples
U	U, C, A, G	Mitochondrial code in family boxes
*U	A, G	Mitochondrial code in two-codon sets
†U	A	Eukaryotes
‡U	U, A, G	Eubacteria in family boxes
C	G	All codes
*C	A	Bacteria, isoleucine codon AUA
G	U, C	All codes
A	U	Rare
I	U, C, A	Eukaryotes, ICG in eubacteria

*, †, ‡ = Various modifications of U (Yokoyama et al., 1985)
*C = modified C

Source: From Thomas Jukes, et al., *Cold Spring Harbor Symposium on Quantitative Biology* 32:775, 1987. Copyright © 1987 Cold Spring Harbor Laboratory Press, Cold Spring Harbor, N.Y.

is the 5′ base in the anticodon it pairs with G only. The only known exception to this rule is found in eubacteria, where the C is covalently modified in the tRNA that recognizes the AUA codon. Thus in this one instance the modified C can pair with a 3′ A in the codon. A U base in the 5′ position of the anticodon shows the greatest variability. An unmodified U can pair with any of the four bases in the 3′ position of the codon. This situation is reflected in the U-family box in mitochondria (table 29.4).

U can also be modified in various ways. In mitochondria one type of modification permits a U to pair with either an A or a G in the two-codon sets. In eukaryotes, another type of U modification limits U to pairing with an A in the codon, and in eubacteria, a third type of U modification permits U pairing with U, A or G but not C in the family boxes.

The Steps in Translation

Protein synthesis is one of the most complex biochemical processes known, involving more than 100 different proteins and more than 30 kinds of RNA molecules. The process begins by the attachment of amino acids to specific tRNA molecules. Subsequent steps take place on the ribosome; amino acids are transported to the ribosome on their tRNA carriers and they do not leave the ribosome until they have become an integral part of a completed polypeptide chain.

Synthases Attach Amino Acids to tRNAs

A unique class of enzymes, called aminoacyl-tRNA synthases, attach amino acids to their cognate tRNAs. This attachment serves two functions: (1) The linkage between the amino acid and the tRNA activates the amino acid, making the subsequent formation of a peptide bond energetically favorable. (2) The tRNA directs the amino acid to a designated location on a messenger RNA so that the amino acid will be incorporated at the appropriate location in the polypeptide chain. Thus aminoacyl-tRNA synthases provide both energy and specificity for protein synthesis.

All synthases act by a common reaction mechanism. This reaction proceeds in two separate steps (fig. 29.9). In the first step the synthase recognizes its corresponding amino acid and its second substrate, ATP, and forms a mixed anhydride bond between the carboxyl group of the amino acid and the phosphate of AMP with the release of PP_i:

$$\text{Amino acid} + \text{ATP} \rightarrow \text{aminoacyl-AMP} + \text{PP}_i \qquad \textbf{(1)}$$

The equilibrium constant for this reaction is about 1.0 so that the energy derived from the cleavage of the phosphate anhydride of ATP is conserved in the mixed anhydride. Aminoacyl-AMP remains tightly bound to the enzyme and, as we will soon show, this fact has allowed workers to crystallize this important complex and analyze its structure.

The second reaction catalyzed by the aminoacyl-tRNA synthases results in the attachment of the amino acid through an ester linkage to the 3′-terminal ribose of tRNA:

$$\text{Aminoacyl-AMP} + \text{tRNA} \rightarrow \text{aminoacyl-tRNA} + \text{AMP} \qquad \textbf{(2)}$$

Synthases differ with respect to their site of attachment to tRNA. Some synthases specifically form the 2′-ester, some specifically form the 3′-ester, and still others produce a mixture of the two. The specificity of the synthases was determined by analyzing their ability to act on tRNA derivatives lacking one or the other terminal hydroxyl group. Once esterified to the terminal ribose, the aminoacyl group can migrate between the vicinal 2′ and 3′ hydroxyl groups. Thus, in cells, aminoacyl-tRNAs are mixtures of 2′ and 3′ esters. Only the 3′ derivative is a substrate for the subsequent transpeptidation reaction catalyzed by the ribosome.

The sum of the two reactions catalyzed by aminoacyl-tRNA synthases is:

$$\text{Amino acid} + \text{ATP} + \text{tRNA} \rightarrow \text{aminoacyl-tRNA} + \text{AMP} + \text{PP}_i \qquad \textbf{(3)}$$

This overall reaction is reversible but is driven to completion by the subsequent hydrolysis of PP_i to two equivalents of P_i through the action of ubiquitous pyrophosphatases. Thus the formation of aminoacyl-tRNA consumes two equivalents of ATP. The energy that ultimately drives the formation of the peptide linkage during protein synthesis is derived from the ester linkage that joins amino acids to tRNA.

Figure 29.9

Formation of aminoacyl-tRNA. This is a two-step process involving a single enzyme that links a specific amino acid to a specific tRNA molecule. In the first step (1) the amino acid is activated by the formation of an aminoacyl-AMP complex. This complex then reacts with a tRNA molecule to form an aminoacyl-tRNA complex (2).

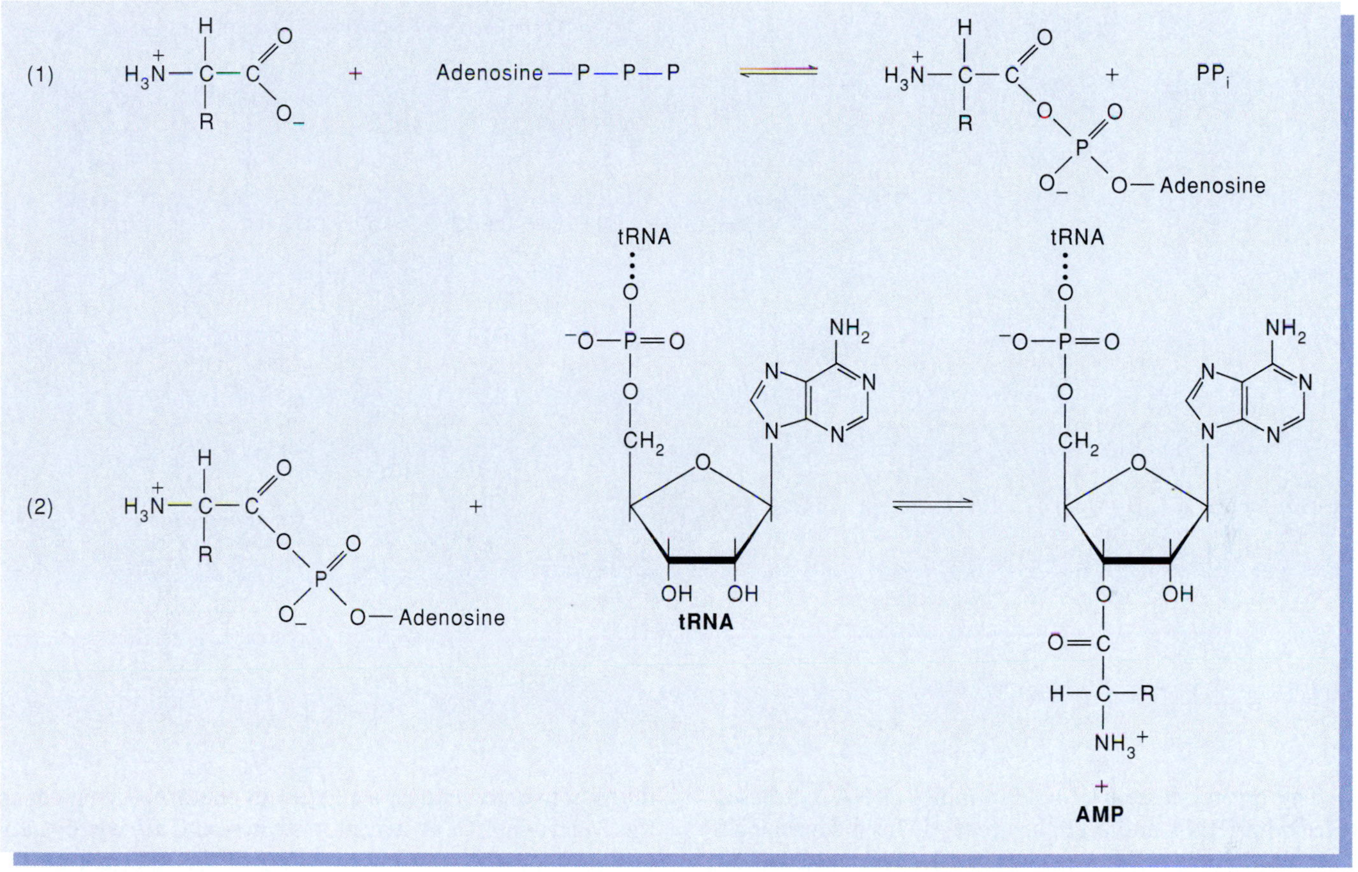

Each Synthase Recognizes a Specific Amino Acid and Specific Regions on Its Cognate tRNA

Even though synthases differ widely in size and subunit complexity, they all appear to be built on a common structural theme. Our understanding of the chemistry of the synthase reaction and the types of the active sites involved in these reactions was advanced substantially by the crystallization and structural solution of tyrosyl-tRNA synthase complexed with the reaction intermediate tyrosyl-adenylate (fig. 29.10). The reaction intermediate is bound in a deep cleft in the enzyme and interacts with it through eleven hydrogen bonds. Six of these bonds are with the AMP moiety and five are with the tyrosyl moiety of the intermediate. The amino acid selectivity of tyrosyl-tRNA synthase is thus determined primarily by the formation of specific hydrogen bonds with the amino acid.

Most cells contain only one synthase for each of the twenty amino acids specified by the genetic code. Each enzyme must be capable not only of recognizing its unique amino acid but also of recognizing the one or more corresponding cognate tRNAs and of discriminating between them and structurally similar noncognate tRNAs. Solving the puzzle of how synthases recognize tRNAs has been one of the major challenges in understanding the nature of the translation of the genetic code itself. The identity relationship between the tRNA and its synthase has, in fact, been called the "second genetic code."

Surprisingly, synthases fall into two categories with respect to the importance of the anticodon as a specificity element that they recognize. Some synthases recognize the anticodon and some do not. This distinction has been demonstrated in several ways. First, some tRNAs can be genetically altered in their anticodon without changing the specificity of their recognition by the cognate synthase. Some suppressor tRNAs (to be discussed later) are formed in this way. In a similar manner, some synthases are capable of recognizing their cognate tRNA even though the anticodon is chemically altered or removed. Other synthases apparently require an unaltered anticodon in order to recognize their cognate tRNAs. Four examples are shown in figure 29.11 (also see box 29B).

Figure 29.10

Recognition of an adenylylated amino acid by the proper synthetase. Shown is adenyl tyrosine bound to the tyrosyl synthetase. A network of H bonds not only stabilizes the reaction but also serves to discriminate between different amino acid residues. MC designates main chain (backbone) carbonyl or amino groups participating in hydrogen bonding.

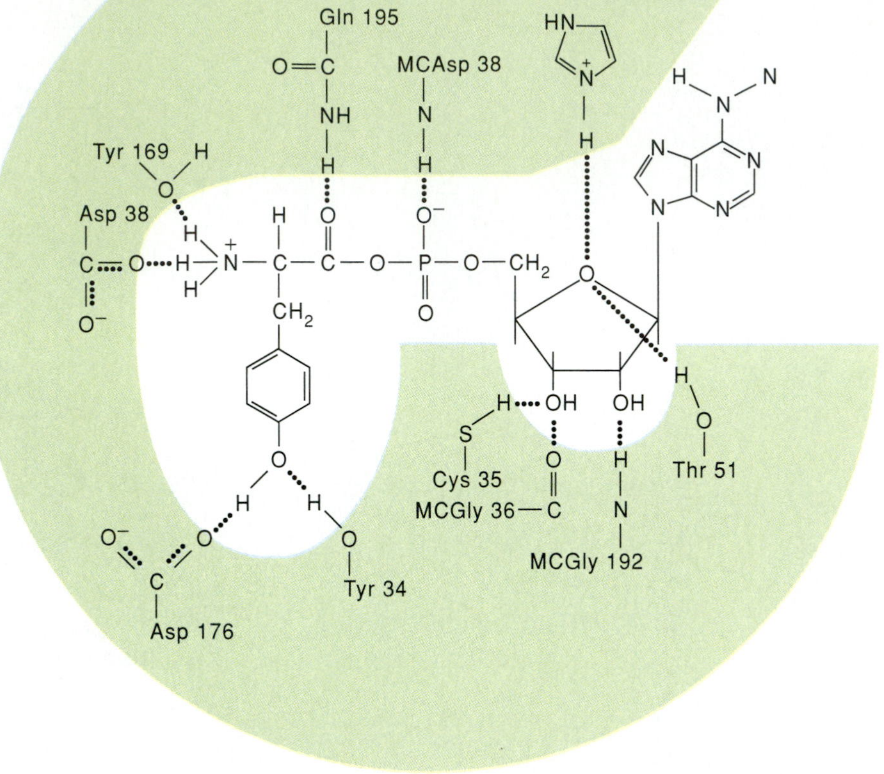

The crystal structure of glutaminyl-tRNA synthase, complexed with tRNA and ATP, has recently been determined by Tom Steitz and his colleagues (fig. 29.12). This accomplishment provides the first structure of a tRNA-protein complex and thus offers important insight into the general nature of the recognition of tRNAs by proteins.

Glutaminyl-tRNA synthase is known to require the integrity of the anticodon loop of tRNA for recognition, and is thus presumed to interact with both ends of the tRNA. In keeping with this interpretation, the synthase is asymmetric and longer (~100 Å) than the tRNA. Moreover, the crystal structure demonstrates the occurrence of significant contact between the protein and bases in the anticodon loop. The additional contacts between the synthase and tRNA appear to occur along the inside of the L-shaped structure. The recognition of the tRNA appears to arise from direct hydrogen bonding between amino acid residues of the protein and bases of the tRNA over a wide region of the tRNA structure. At this stage we do not know which of these protein-nucleic acid contacts are crucial for recognition.

The acceptor stem of the bound tRNA substrate plunges deep into the active site pocket created by the dinucleotide fold, also the site of ATP binding. Surprisingly, binding in the active site induces a significant conformational change in the 3′-terminal CCA acceptor sequence. This conformational change appears to cause the melting of the A-U pair at the end of the acceptor stem in a manner that allows these bases to interact with amino acid side chains of the protein. At this point in our understanding, it seems likely that the interactions that allow synthases to identify their cognate tRNAs are both diverse and complex.

Aminoacyl-tRNA Synthases Can Correct Acylation Errors

Many aminoacyl-tRNA synthases appear to contain a second catalytic site that serves to correct errors by hydrolyzing incorrectly matched amino acids and tRNAs. These proofreading hydrolysis reactions are best understood in the case of isoleucyl-tRNA synthase. The amino acids isoleucine and valine, which differ by only a single methylene group, are among the most difficult to discriminate. Isoleucyl-tRNA synthase is capable of distinguishing these two amino acids at its aminoacylation site but occasionally and mistakenly forms valyl-tRNA^{Ile}. When isoleucyl-tRNA synthase encounters valyl-tRNA^{Ile} the hydrolytic active site specifically degrades it:

$$\text{Valyl-tRNA}^{Ile} + H_2O \rightarrow \text{valine} + \text{tRNA}^{Ile}$$

Figure 29.11

Identity elements in four tRNAs. Each circle represents one nucleotide. Filled circles indicate nucleotides that serve as recognition elements to the appropriate aminoacyl-tRNA synthase. It is possible that there are other identity elements in these structures still to be discovered.

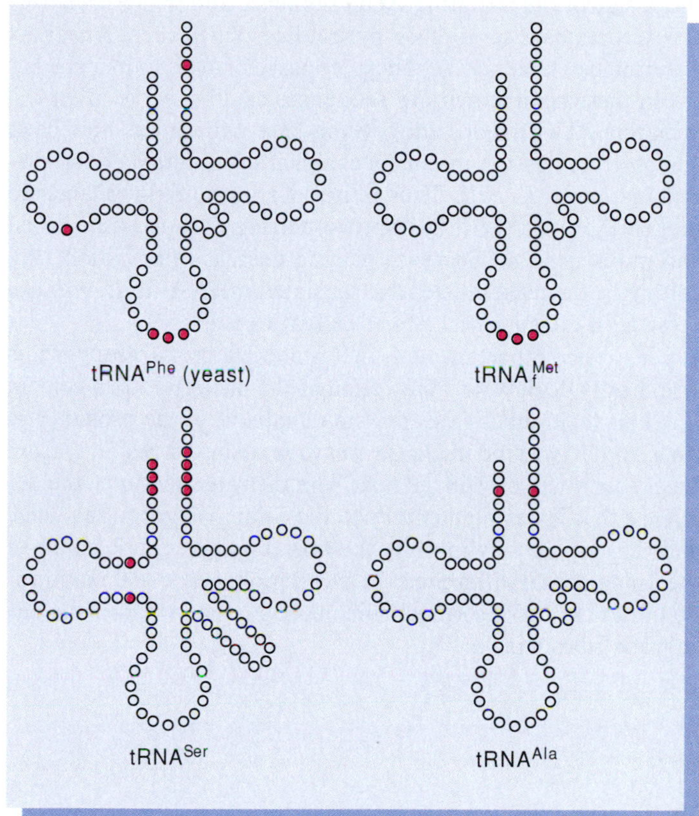

In this way the erroneous incorporation of valine in the place of isoleucine is avoided. The sequential action of aminoacylation and proofreading sites contributes to an overall error frequency of translation of less than 1 in 10,000.

A Unique tRNA Initiates Protein Synthesis

The translation of every protein begins with the incorporation of the amino acid methionine. A unique initiator tRNA, which we will designate $tRNA_i^{Met}$, is responsible for the incorporation of this initiating methionine in all protein-synthesizing systems and it also plays an important role in selecting the appropriate translation start site in mRNA. There are generally only two tRNAs in cells that specify methionine. We will designate the one that is responsible for the incorporation of internal methionines simply as $tRNA^{Met}$. A single methionyl-tRNA synthase is responsible for activating both methionine isoacceptor tRNAs.

Figure 29.12

Solvent-accessible surface representation of the GlnRS enzyme complexed with tRNA and ATP. The region of contact between tRNA and protein extends across one side of the entire enzyme surface and includes interactions from all four protein domains. The acceptor end of the tRNA and the ATP are seen in the bottom of the deep cleft. Protein is inserted between the 5′ and 3′ ends of the tRNA and disrupts the expected base pair between U1 and A72. (From M. G. Rould, J. J. Persona, D. Söll, and T. Steitz, "Structure of *E. coli* glutamyl-tRNA synthetics complexed with $tRNA^{Gln}$ and ATP at 2.8-Å resolution, implications for tRNA discrimination," *Science* 246:1135-1142, 1989, © 1989 by the AAAS.)

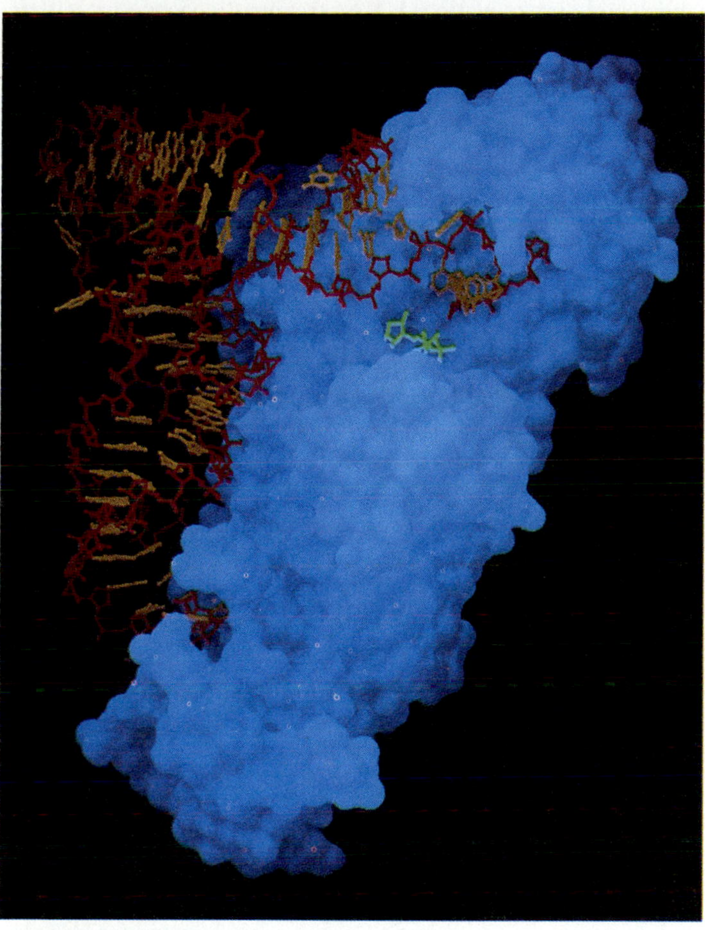

The functional discrimination between the two isoacceptor tRNAs is accomplished by protein initiation and elongation factors. Initiation factors recognize $tRNA_i^{Met}$ and elongation factors recognize $tRNA^{Met}$. Clearly, $tRNA_f^{Met}$ and $tRNA^{Met}$ must possess structural features that allow them to be distinguished from all other tRNAs by the single methionyl-tRNA synthase, but they must be sufficiently different that they can be discriminated by the factors.

In prokaryotic systems, a specific formylating enzyme exists that can recognize $tRNA_f^{Met}$ and formylate its amino terminus utilizing N^{10}-formyltetrahydrofolate as the formyl donor. This reaction serves to ensure that the initiator does not participate in the elongation reactions. This recognition step was lost during evolution and is not possessed by eukaryotic systems.

Translation Begins with the Binding of mRNA to the Ribosome

One of the major demands of protein synthesis is to select the appropriate initiating codon, generally AUG, for translation. This is accomplished at the level of the ribosome by the binding of the small ribosomal subunit to mRNA. The recognition of the appropriate start codon is accomplished in different ways in prokaryotes and eukaryotes. This fundamental difference in initiating mechanism is responsible for the fact, noted earlier, that each eukaryotic mRNA generally encodes only one protein, while single prokaryotic mRNAs frequently encode more than one.

In eukaryotes, the AUG sequence closest to the 5′ end of the mRNA usually serves as the start codon for the single protein encoded by each mRNA. The small ribosomal subunit binds at the 5′ end of the mRNA and moves along the strand until it encounters an AUG sequence that is recognized by base pairing with the anticodon of Met-$tRNA_i^{Met}$. In higher eukaryotes, but not in yeast, the recognition of this initiating AUG is facilitated by an initiation context of flanking bases. The preferred initiation context is GCCGCCpurCC$\overline{\text{AUG}}$G. How this sequence is recognized is unknown, but when the initiation context of the first AUG departs from this sequence the 40S subunit can bypass it and initiate at the next AUG downstream.

In prokaryotic systems, the initiating AUG codon may occur at any point within the mRNA and there may be more than one start site within the same mRNA. How do prokaryotic ribosomes select the appropriate initiating codon from the much more abundant internal AUG sequences? The answer was suggested when Shine and Dalgarno, in the early 1970s, noticed that bacterial mRNAs generally contain a complementary purine-rich region (which has become known as the Shine-Dalgarno sequence) centered approximately ten bases toward the 5′ side of the initiating AUG sequence and that *E. coli* 16S RNA contains a seven-base pyrimidine-rich sequence near its 3′ terminus (table 29.5). They proposed that base pairing between these complementary sequences could serve to align the initiating AUG for decoding. Such base pairing has now been demonstrated as the major mechanism of initiation codon discrimination in *E. coli*. Thus mutations in the Shine-Dalgarno sequence of mRNA that improve pairing enhance translation and mutations that decrease pairing decrease translation. Accordingly, the most efficiently translated mRNAs in *E. coli* are those with the strongest Shine-Dalgarno pairing.

The importance of the Shine-Dalgarno sequence is graphically depicted by the action of the bacterial toxin colecin E3. This toxin inactivates the small subunit of the prokaryotic ribosome by specific nuclease cleavage about 50 residues from the 3′ terminus of 16S rRNA. The cleavage disrupts the sequence that is complementary to the Shine-Dalgarno sequence and thus specifically inhibits the initiation process. Because of the fundamental differences between prokaryotic and eukaryotic initiation just described, colecin E3 does not inhibit the eukaryotic ribosome.

29B
BOX

Discrimination of $tRNA^{Ala}$ by Its Synthase

The recognition of $tRNA^{Ala}$ by its synthase has been extensively studied by Paul Schimmel and his colleagues. These investigators have systematically altered the structure of the substrate RNA and created ever-smaller versions of it in order to determine the critical features of its structure that allow the enzyme to recognize it. The features identified in this way have turned out to be surprisingly simple. A single G3:U70 pair in the acceptor stem appears to be a major determinate in the discrimination of this tRNA by its synthase. This relationship was demonstrated in several ways. First, many alterations, except in this single pair, had little or no effect on the recognition of the tRNA by the synthase. Second, substituting this single pair into $tRNA^{Cys}$, which differs in sequence from $tRNA^{Ala}$ at forty other locations, allowed its recognition by alanyl-tRNA synthase. Finally, the synthase was able to acylate a microhelix (a stem and loop containing only seven base pairs patterned after the sequence of the acceptor stem of $tRNA^{Ala}$), provided that it contained the crucial G:U pair. The structure of this synthase-tRNA complex has not been determined, but the crucial recognition, in contrast to that of $tRNA^{Gln}$, appears simple and limited to only a small region of the acceptor stem.

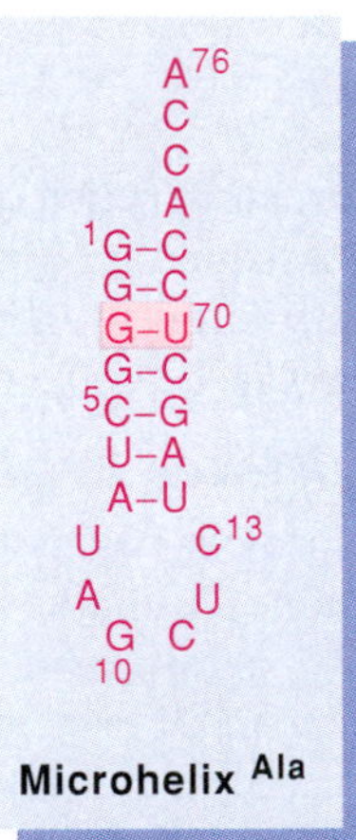

Dissociable Protein Factors Play Key Roles at the Different Stages in Protein Synthesis on the Ribosome

At each stage in protein synthesis on the ribosome—initiation, elongation and termination—a different set of protein factors is engaged by the ribosome. Why do such protein factors, which are crucial to the translation, exist separate from the ribosome?

Table 29.5
The Sequence of the 3′ End of *E. coli* 16S RNA and Some Shine-Dalgarno Sequences at the 5′ End of Bacterial mRNAs

	The pyrimidine-rich complement to the Shine-Dalgarno sequence
16S rRNA	3′ · · · HOAUUCCUCCACUA · · · 5′
lacZ mRNA	5′ · · · ACACAGGAAACAGCUAUG · · · 3′
trpA mRNA	5′ · · · ACGAGGGGAAAUCUGAUG · · · 3′
RNA polymerase β mRNA	5′ · · · GAGCUGAGGAACCCUAUG · · · 3′
r-Protein L10 mRNA	5′ · · · CCAGGAGCAAAGCUAAUG · · · 3′
	The purine-rich Shine-Dalgarno sequence; The initiation codon

Why must they cycle on and off the ribosome, and why are their functions not performed by firmly bound ribosomal proteins? The answers to these questions appear to lie in the economy of structure that is afforded by the cycling strategy. Since factors cannot interact simultaneously with the ribosome, it makes sense for the ribosome to interact with them sequentially at a single site.

Protein Factors Aid Initiation

Even though specific differences distinguish the initiation process in eukaryotes and prokaryotes, three things must be accomplished to initiate protein synthesis in all systems: (1) the small ribosomal subunit must bind the initiator tRNA, (2) the appropriate initiating codon on mRNA must be located, and (3) the large ribosomal subunit must associate with the complex of the small subunit, the initiating tRNA and mRNA. Nonribosomal proteins, known as initiation factors (IFs), participate in each of these three processes. IFs interact transiently with a ribosome during initiation and thus differ from ribosomal proteins, which remain continuously associated with the same ribosome.

E. coli has three initiation factors (fig. 29.13) bound to a small pool of 30S ribosomal subunits. One of these factors, IF-3, serves to hold the 30S and 50S subunits apart after termination of a previous round of protein synthesis. The other two factors, IF-1 and IF-2, promote the binding of both fMet-$tRNA_i^{Met}$ and mRNA to the 30S subunit. As we noted before,

Figure 29.13

Formation of the initiation complex for protein synthesis in prokaryotes. *E. coli* has three initiation factors bound to a pool of 30S ribosomal subunits. One of these factors, IF-3, holds the 30S and 50S subunits apart after termination of a previous round of protein synthesis. The other two factors, IF-1 and IF-2, promote the binding of both fMet-$tRNA^{fMet}$ and mRNA to the 30S subunit. The binding of mRNA occurs so that its Shine-Dalgarno sequence pairs with 16S rRNA and the initiating AUG sequence with the anticodon of the initiator tRNA. The 30S subunit and its associated factors can bind fMet-$tRNA^{fMet}$ and mRNA in either order. Once these ligands are bound, IF-3 dissociates from the 30S subunit, permitting the 50S subunit to join the complex. This releases the remaining initiation factors and hydrolyzes the GTP, which is bound to IF-2.

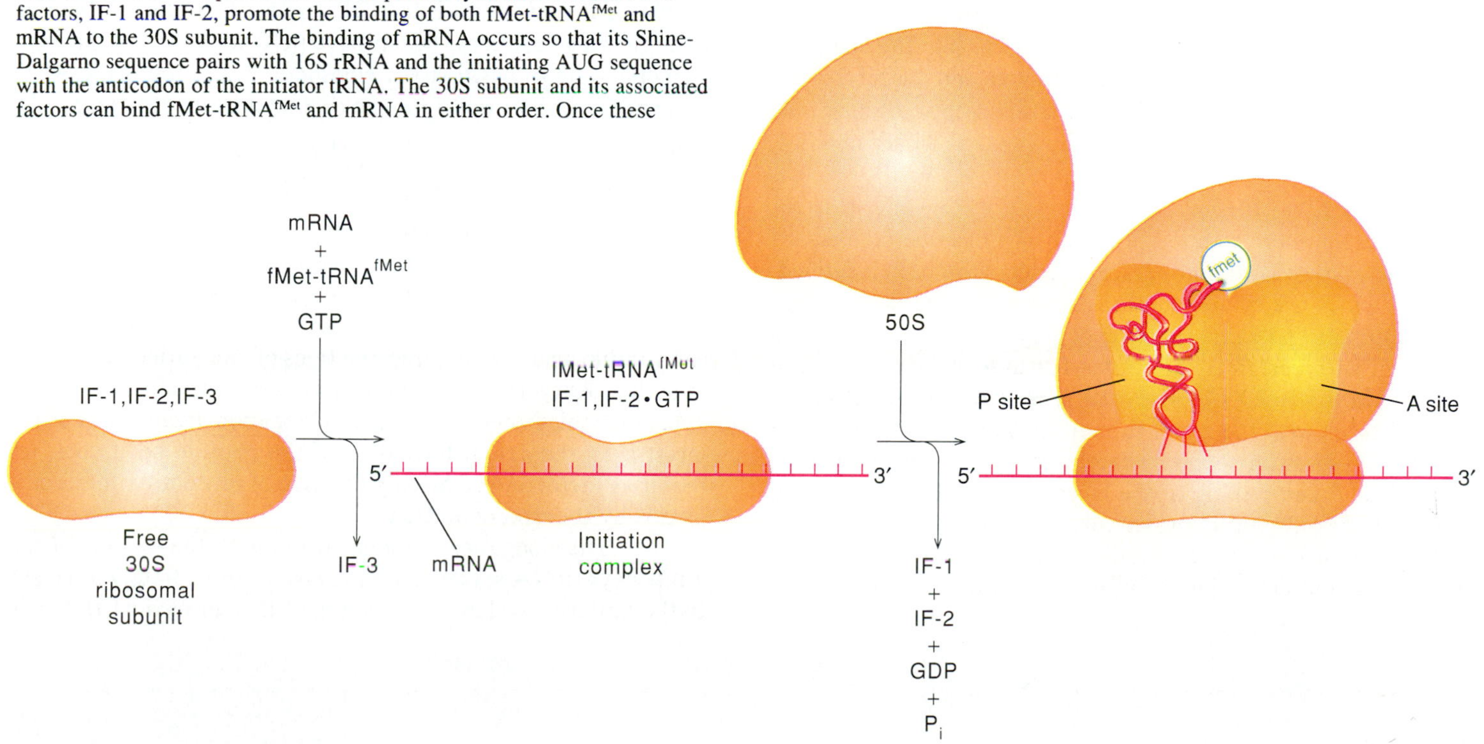

Figure 29.14

Formation of the initiation complex for protein synthesis in eukaryotes. The reaction begins with the small subunit held apart from the large subunit by an antiassociation factor and ends with the hydrolysis of GTP and joining of the large subunit as in prokaryotes. The intervening reactions are different. A much more complex spectrum of initiation factors (eIFs) is involved, and the exact function of only a few of these factors is known with certainty. The Met-tRNA^{fMet} first binds to the small subunit as a ternary complex with the eIF-2 and GTP. This ternary complex then binds to the 5′ end of mRNA with the aid of several factors, one of which, eIF-4F, contains a subunit that specifically binds to the terminal cap structure of the mRNA. Binding to mRNA is followed by a scanning reaction that moves the small subunit along the mRNA, usually to the first AUG, in a reaction driven by the hydrolysis of ATP to ADP and P_i.

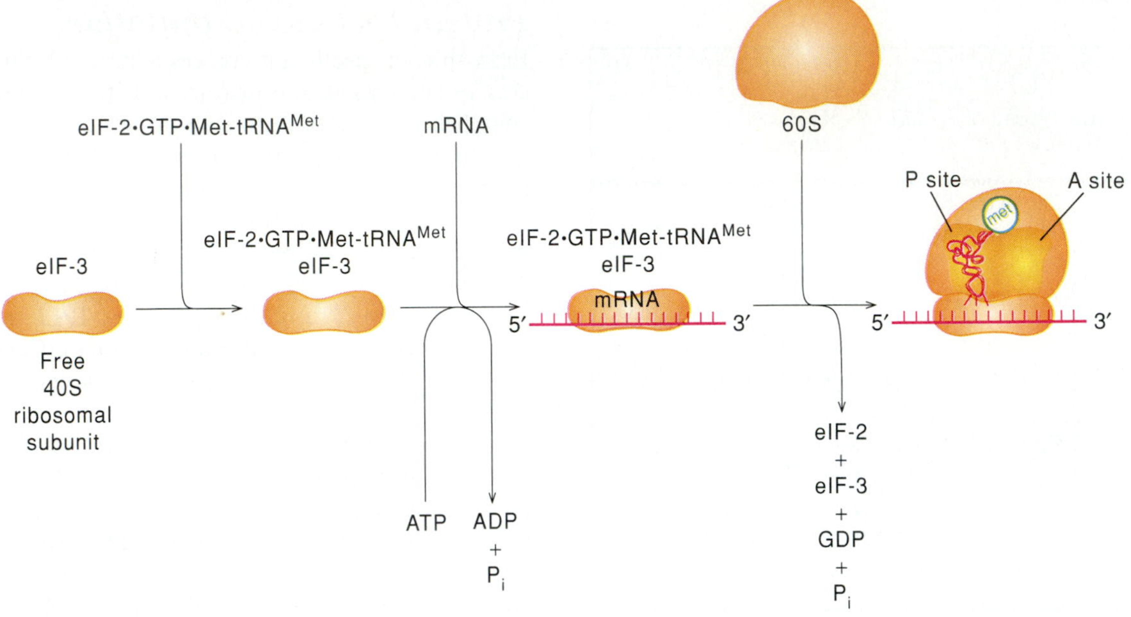

the binding of mRNA occurs so that its Shine-Dalgarno sequence pairs with 16S RNA and the initiating AUG sequence with the anticodon of the initiator tRNA, although this latter interaction has not been rigorously proven. The 30S subunit and its associated factors can apparently bind fMet-tRNA$_i^{Met}$ and mRNA in either order. Once these ligands are bound, IF-3 dissociates from the 30S, permitting the 50S to join the complex. This releases the remaining initiation factors and hydrolyzes the GTP that is bound to IF-2. The initiation step in prokaryotes requires the hydrolysis of one equivalent of GTP to GDP and P_i.

Initiation of protein synthesis in eukaryotic cells requires a much more complex spectrum of initiation factors, abbreviated eIF. At least nine separate factors have been identified, some of which are composed of as many as eleven different peptide subunits. The exact function of only a few of these factors is known with certainty. The main features of the process are outlined in figure 29.14. As in the prokaryotic system, the reaction begins with the small subunit held apart from the large subunit by an antiassociation factor and ends with the hydrolysis of GTP and joining of the large subunit. The intervening reactions are different in prokaryotes and eukaryotes. The Met-tRNA$_i^{Met}$ first binds to the small subunit as a ternary complex with eIF-2 and GTP. The resulting preinitiation complex then binds to the 5′ end of mRNA with the aid of several factors, one of which, eIF-4F, contains a subunit that specifically binds to the terminal cap structure of mRNA. Binding to mRNA is followed by a scanning reaction that moves the small subunit along the mRNA, usually to the first AUG, in a reaction driven by the hydrolysis of ATP to ADP and P_i. The mRNA binding and scanning reactions, which have no counterpart in the prokaryotic system, position the ribosome on the initiating AUG sequence in the appropriate flanking nucleotide sequence.

Three Elongation Reactions Are Repeated with the Incorporation of Each Amino Acid

At the conclusion of the initiation process, the ribosome is poised to translate the reading frame associated with the initiation codon. The translation of the contiguous codons in mRNA is accomplished by the sequential repetition of three reactions with each amino acid. These three reactions of elongation are similar in both prokaryotic and eukaryotic systems; two of them require nonribosomal proteins known as elongation factors (EF). Interestingly, the actual formation of the peptide bond does not require a factor and is the only reaction of protein synthesis catalyzed by the ribosome itself.

The elongation reactions begin with the binding of the aminoacyl-tRNA specified by the codon immediately adjacent to the initiating codon. The binding of this aminoacyl-tRNA is

Figure 29.15

Addition of the second aminoacyl-tRNA to the ribosome complex and the accompanying EF-Tu, EF-Ts cycle in *E. coli*. The purpose of the cycle is to regenerate another protein aminoacyl-tRNA complex suitable for transferring further aminoacyl-tRNAs to the A site on the ribosome.

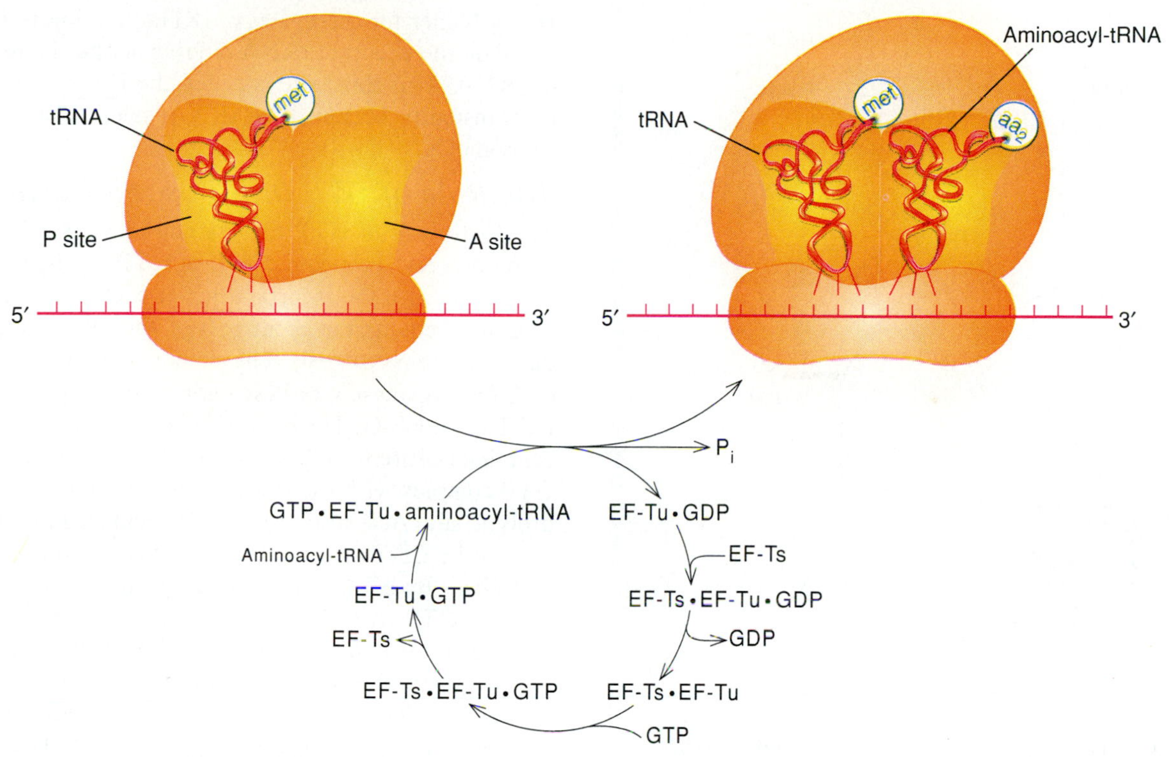

catalyzed by an aminoacyl-tRNA binding factor, designated EF-Tu in bacteria and EF-1 in eukaryotic systems. The factor interacts with the ribosome as a ternary complex bound to both aminoacyl-tRNA and GTP. The binding of this complex to the ribosome is coupled to GTP hydrolysis. A productive complex forms if, and only if, the anticodon of the tRNA in the complex is complementary to the codon bound to the A site on the ribosome. Following the binding of aminoacyl-tRNA, EF-Tu is released from the ribosome as a complex with GDP. A second elongation factor, EF-Ts, catalyzes the regeneration of the EF-Tu-GTP complex so that it can again bind aminoacyl-tRNA. The series of reactions involved in this regeneration is depicted in figure 29.15. EF-1 is a multisubunit protein that combines the functional properties of EF-Tu and EF-Ts.

Peptide bond formation occurs immediately following the dissociation of the binding factor from the ribosome. This reaction is known as transpeptidation and the enzymatic center that catalyzes it is known as peptidyl transferase, although it promotes the conversion of an ester to a peptide bond. Transpeptidation is generally assumed to proceed by nucleophilic attack by the amino group of the incoming aminoacyl-tRNA on the carbonyl of the ester of peptidyl-tRNA with the formation of a tetrahedral intermediate (fig. 29.16).*

*H. F. Noller, V. Hossarth, and L. Zimniak, Ribosomes stripped of all their protein can still catalyze transpeptidation. *Science* 256:14–16, 1992.

The final step in elongation is known as translocation (fig. 29.17). This reaction, like aminoacyl-tRNA binding, is catalyzed by a factor (the translocation factor, known as EF-G in prokaryotic systems and EF-2 in eukaryotic systems) that cycles on and off the ribosome and hydrolyzes GTP in the process. The overall purpose of translocation is to physically move the ribosome along the mRNA to expose the next codon for translation. The mechanical basis of this movement reaction has long eluded understanding. Recent evidence suggests that translocation occurs in two discrete steps that may involve the relative movement of the two ribosomal subunits.

The structural differences between bacterial and eukaryotic elongation factors are highlighted by the selective action that diphtheria has on eukaryotic systems (box 29C).

The Ribosome Binds tRNA at Three Sites

The original model of ribosome function, proposed over 25 years ago by James D. Watson, pictured the ribosome as having two sites of interaction with tRNA. One site was called the P site, because it was visualized as holding the growing peptide chain. The second site was called the A site, because it was visualized as holding the aminoacyl-tRNA prior to peptide bond formation. This two-site model is the minimal stochastic model of ribosome function compatible with the stepwise addition of a single

Figure 29.16

Formation of the first peptide linkage. The formylmethionine group is transferred from its tRNA at the P site to the amino group of the second aminoacyl-tRNA at the A site on the ribosome. This involves nucleophilic attack by the amino group of the second amino acid on the carboxyl carbon of the methionine. The resulting bond formation attaches both amino acids to the tRNA at the A site.

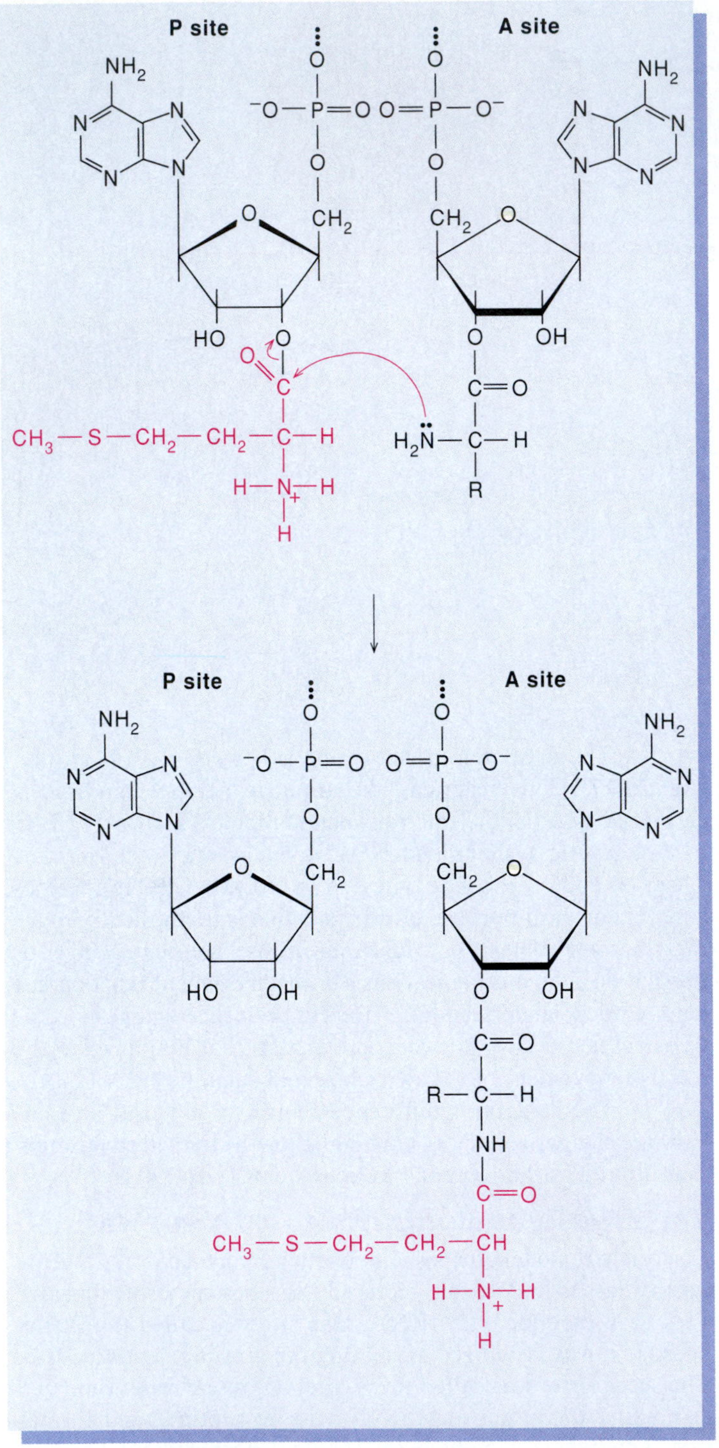

amino acid from a tRNA to a growing chain that is also attached to a tRNA. The model has been vigorously debated over the ensuing years but has resisted attack, largely because of the difficulty of precisely measuring the nature and location of tRNA binding to the ribosome. Recently the two-site model has given way to compelling evidence obtained by Danesh Moazed and Harry Noller for three sites of tRNA binding to the ribosome, based on the effects of tRNA binding on the chemical reactivity of rRNA. This third site is called the E site and is believed to be transiently occupied by tRNA prior to its exit from the ribosome.

The Role of GTP in Ribosomal Reactions

The hydrolysis of GTP plays a conspicuous role in the translation process. Two equivalents of GTP are hydrolyzed during elongation with the incorporation of each amino acid. This hydrolysis accounts for about half of the total energy consumed during protein synthesis. The chemical and functional purposes of GTP hydrolysis are best understood in the case of *E. coli* EF-Tu and EF-G. The sites of GTP binding and hydrolysis are contained entirely on these factors. The interaction of the factor-GTP complex with the ribosome is believed to activate the hydrolytic site. Activation can also be achieved in other ways; hydrolysis by EF-Tu is stimulated by the antibiotic kirromycin and hydrolysis by EF-G is promoted by various organic solvents.

GTP hydrolysis leads to inversion of configuration of the released P_i and is believed to occur by direct attack of water on the γ phosphorus of GTP through an in-line displacement mechanism. There are no covalent intermediates in this reaction. Rather, the change of bound ligand to GDP that results from hydrolysis and release of P_i is believed to change the conformation of the factor. The factor when bound to GTP is thought to be in the "on" configuration, able to bind to the ribosome and through binding to cause either aminoacyl-tRNA binding or translocation. The binding of GDP to the factor puts it in the "off" configuration and causes it to dissociate from the ribosome. Hydrolysis itself is not required for these changes to occur. This point is demonstrated by the fact that the nonhydrolyzable analog of GTP, GMPPCP, containing a methylene bridge instead of an oxygen between the β and γ phosphorus, can substitute for it in the reactions catalyzed by the elongation factors (see fig. 29.18). The use of this analog slows these reactions by delaying the dissociation of the factors from the ribosome.

The translation factors that interact with GTP are members of the G protein superfamily, which includes the signal-transducing G proteins that link membrane receptors with their intracellular targets (see chapter 24) and the Ras proteins that function as growth regulators (see chapter 34). All G proteins bind and hydrolyze GTP and it is believed that they all act by the same general mechanism we have just described. Thus when bound to GTP these proteins are in their "active" configuration and when bound to GDP as a result of hydrolysis they are converted to their "inactive" configuration.

Figure 29.17

The translocation reaction in *E. coli*. The translocation reaction occurs immediately after peptide synthesis. It involves displacement of the discharged tRNA from the P site and concerted movement of the peptidyl-tRNA and mRNA so that the peptidyl-tRNA is bound to the P site and the same three nucleotides in the mRNA. The A site is vacated and ready for the addition of another aminoacyl-tRNA. Translocation in eukaryotes is similar except that the EF-2 factor is involved instead of the EF-G factor.

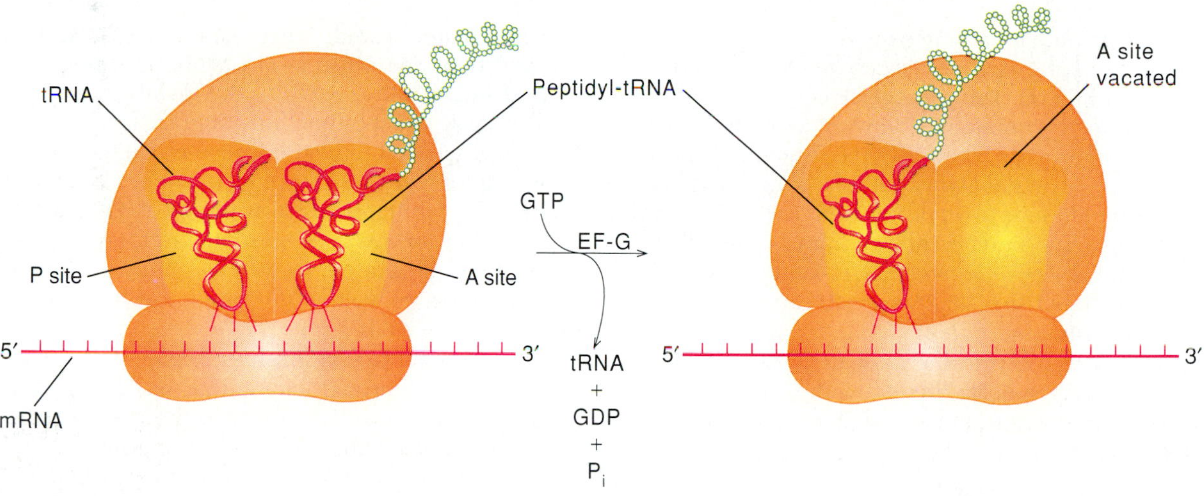

Figure 29.18

The structure of guanylyl methylene diphosphonate (GMPPCP).

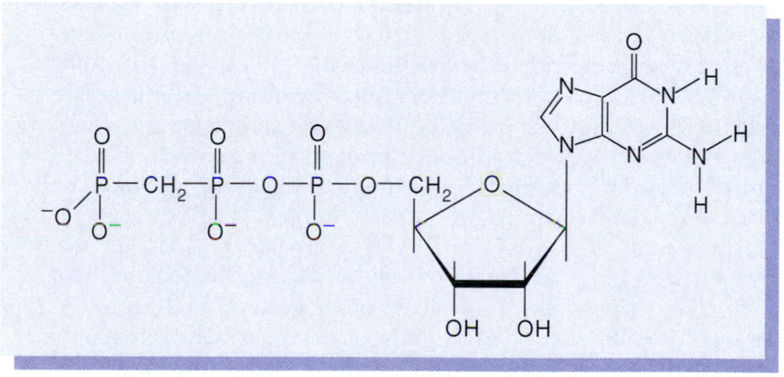

Termination of Translation Requires Release Factors and Termination Codons

The last step in translation involves the cleavage of the ester bond that joins the now complete peptide chain to the tRNA corresponding to its C-terminal amino acid (fig. 29.19). This process of termination, in addition to the termination codon, requires release factors (RFs). The freeing of the ribosome from mRNA during this step requires the participation of a protein called ribosome releasing factor (RRF).

Cells usually do not contain tRNAs that can recognize the three termination codons. In *E. coli,* when these codons arrive in the A site on the ribosome they are recognized by one of three release factors. RF-1 recognizes UAA and UAG, while RF-2 recognizes UAA and UGA. The third release factor, RF-3, does not itself recognize termination codons but stimulates the activity of the other two factors.

The consequence of release factor recognition of a termination codon in the A site is to alter the peptidyl transferase center on the large ribosomal subunit so that it can accept water as the attacking nucleophile rather than requiring the normal substrate, aminoacyl-tRNA (fig. 29.20). In other words, the termination reaction serves to convert the peptidyl transferase into an esterase. This feature of the termination reaction is clearly seen in the simple *in vitro* reaction that occurs when *E. coli* ribosomes are combined with fMet-$tRNA_i^{Met}$, RF-1, and two separate nucleotide triplets, AUG and UAA. Formylmethionine is produced by hydrolysis and this reaction is specifically inhibited by antibiotics, such as sparsomycin, that inhibit peptidyl transferase.

29C
BOX

Mechanisms of Damage Produced by Certain Toxins

A single exotoxin, diphtheria toxin, is responsible for the pathogenesis of *Corynebacterium diphtheriae* and the disease diphtheria. The pathogenic consequences can be prevented by immunization with toxoid, an inactivated form of the purified toxin. Curiously, the structural gene for the toxin is carried by a bacterial virus, called β phage, that must infect the bacterium to induce toxin production. The widespread immunization against diphtheria employed in the United States has caused β phage, but not *C. diphtheriae,* to largely disappear. A catalytically identical but structurally very different exotoxin is produced by *Pseudomonas aeruginosa.*

In the cytoplasm of the cell, the catalytic portion of diphtheria toxin acts as a very specific protein-modifying enzyme. It catalyzes the ADP-ribosylation and consequent inactivation of EF-2 by the following reaction:

$$\text{EF-2} + \text{NAD}^+ \rightarrow \text{ADP-ribosyl-EF-2} + \text{nicotinamide} + \text{H}^+$$

This reaction is reversible when conducted *in vitro,* but under the conditions of pH and nicotinamide concentration that exist in the cell, it is irreversible. Thus diphtheria toxin kills cells by irreversibly destroying the ability of EF-2 to participate in the translocation step of protein synthesis elongation. A number of other protein toxins have subsequently been found to ADP-ribosylate and inactivate cellular proteins involved in other essential cellular pathways. For example, cholera and pertussis toxins ADP-ribosylate and inactivate proteins important to cAMP metabolism.

The enzymatic specificity of diphtheria toxin deserves special comment. The toxin will ADP-ribosylate EF-2 in all eukaryotic cells *in vitro* whether or not they are sensitive to the toxin *in vivo,* but it will not modify any other protein, including the bacterial counterpart of EF-2. This narrow enzymatic specificity has called attention to an unusual posttranslational derivative of histidine, diphthamide, that occurs in EF-2 at the site of ADP-ribosylation (see illustration). While the unique occurrence of diphthamide in EF-2 explains the specificity of the toxin, it raises questions about the functional significance of this modification in translocation. Interestingly, some mutants of eukaryotic cells selected for toxin resistance lack one of several enzymes necessary for the posttranslational synthesis of diphthamide in EF-2 that is necessary for toxin recognition, but these cells seem perfectly competent in protein synthesis. Thus the *raison d'être* of diphthamide, as well as the biological origin of the toxin that modifies it, remains a mystery.

Some other toxins inhibit protein synthesis by inactivating the ribosome through alterations of rRNA. A fungal toxin, α-sarcin, inactivates the ribosome by specific nuclease cleavage of a single phosphodiester bond in a purine-rich region found in the 23–28S RNA of all ribosomes. A second group of toxins, known as the ribosome-inactivating proteins, is abundant in plants of many species. The best-known example of this type of toxin is ricin, the toxic agent in the castor bean. These proteins all act by the curious mechanism of removing a single adenine residue, by N-glycolytic cleavage, from the site in rRNA adjacent to the one that is cleaved by α-sarcin. Both of these modifications of rRNA alter the ribosome's ability to interact with factors, a fact suggesting that this region of rRNA is part of the factor interaction site on the large subunit. One of the ribosome-inactivating proteins, trichosanthin, was originally isolated from a plant tuber as the active principle of an Oriental folk medicine used to induce abortions. More recently, trichosanthin has been found to selectively inhibit viral protein synthesis in HIV-infected T-cells, and it is now undergoing clinical trials as a therapeutic agent for the treatment for AIDS. The biological basis of this selective inhibition is unknown.

Ribosomes Can Change Reading Frame and Jump during Translation

Normally, ribosomes march in lock step down the mRNA, one codon at a time, beginning at the initiating AUG and ending at a termination codon. In this way translation is rigidly maintained in a single reading frame. In the last few years, a number of specific exceptions to this seemingly rigid rule have been discovered in both prokaryotic and eukaryotic systems. In these cases, the correct expression of specific messages requires the ribosome to violate the rule of lock-step translation. One way that this occurs is known as translational frameshifting. Frameshifting generally occurs only one base at a time, but this change or slipping can be in either the +1 or −1 direction. Another violation of the rule of lock-step translation involves the failure of the ribosome to translate a substantial portion of mRNA, by skipping as many as 50 bases downstream in mRNA. This process is known as translational jumping. From an informational point of view, translational jumping is analogous to splicing in mRNA except that the information is not removed but is simply ignored by the ribosome. Both translational frameshifting and

Figure 1

Inactivation of the EF-2 factor by diphtheria toxin through the ADP-ribosylation of a modified histidine side chain.

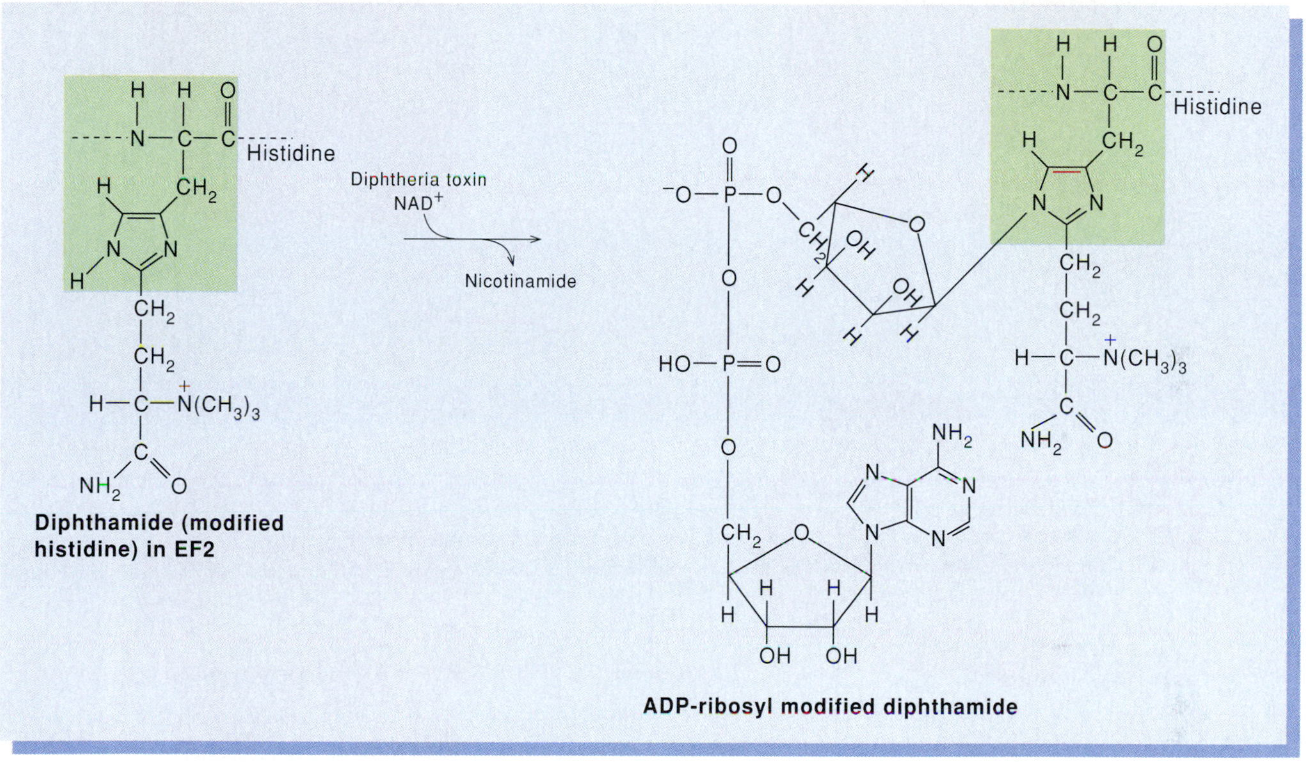

translational jumping occur at specific sites in mRNA and in some cases are known to require the presence of specific sequences and specific structures in mRNA. While the nature of these sequences and structures have been identified in several cases, the mechanisms by which they are read have not been defined.

A notable instance of frameshifting occurs in the gene for the release factor RF-2, which recognizes the UGA stop codon in *E. coli*. This gene does not make RF-2 unless a frameshift takes place at an in-frame UGA sequence. When there is an abundance of the release factor, the frameshift is unlikely to occur because release occurs when the ribosome reaches this point on the messenger. However, when the release factor is in short supply, the ribosome is likely to pause at this site and then undergo a frameshift to make functional release factor. In this example, frameshifting results in a negative feedback mechanism for regulating the amount of release factor that is synthesized.

Translational fidelity of bacterial ribosomes is influenced in a more general way by the binding of antibiotics to the ribosome (box 29D).

Figure 29.19

The release reaction in *E. coli*. The release reaction occurs when the codon adjacent to the anticodon-codon complex is one of the stop codons, e.g., UAA. The stop is recognized by release factor proteins that cause the peptidyl transferase to transfer the nascent polypeptide to water, forming a free polypeptide. Following the release of the polypeptide, the final tRNA and the mRNA dissociate from the ribosome, and the ribosome dissociates into its constituent subunits.

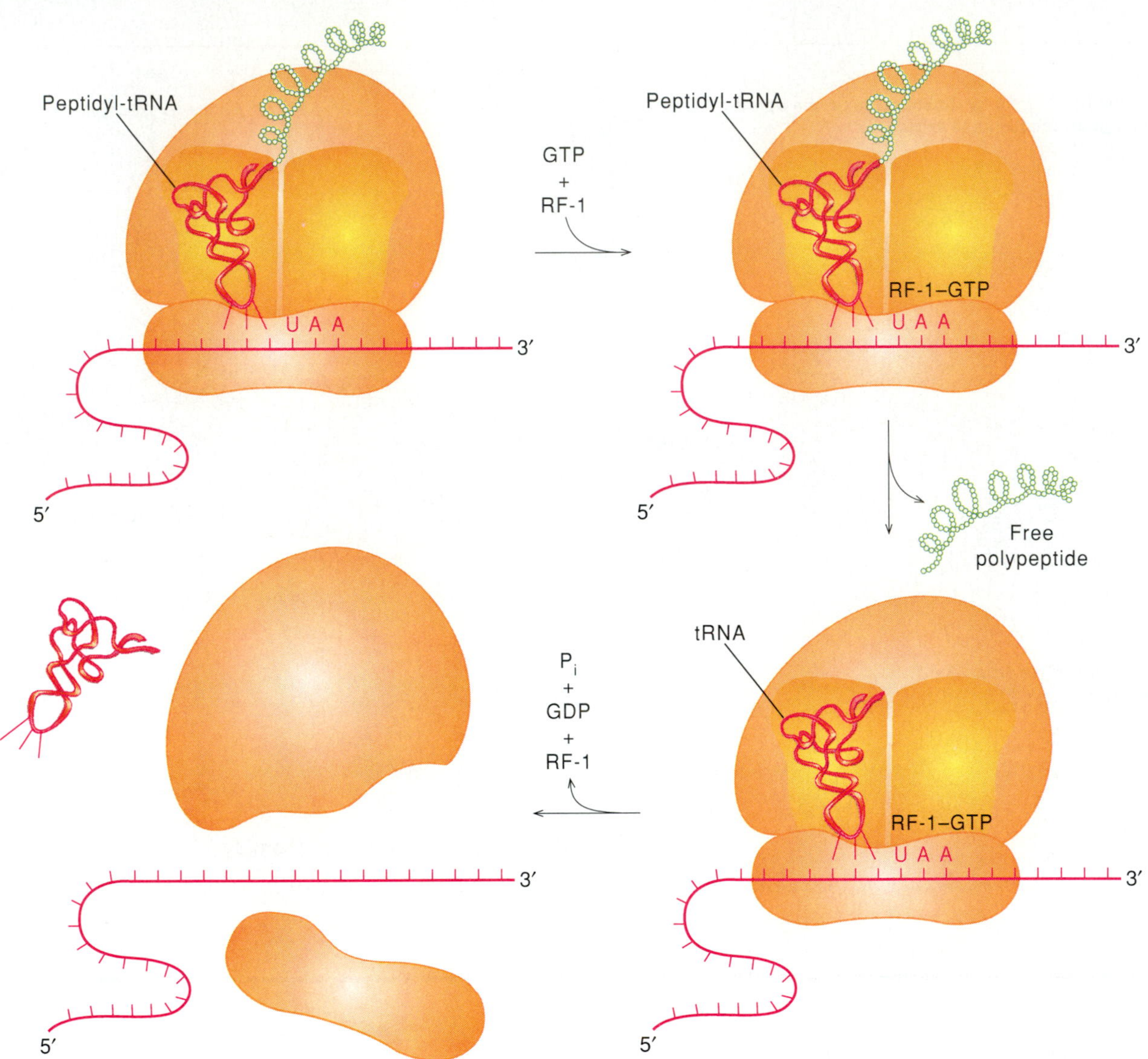

Targeting and Posttranslational Modification of Proteins

Despite the fact that only twenty amino acids (plus selenocysteine and formylmethionine in prokaryotic systems) are known to be directly specified by the genetic code, chemical analysis of mature proteins has revealed hundreds of different amino acids, all of them structural variants on the original twenty. This structural diversity, which greatly expands the chemical lexicon of proteins, results from posttranslational modification of the primary products of translation. Our knowledge of the nature and significance of enzymatic reactions that bring about these important alterations is still very incomplete.

In addition to the modification of amino acid side chains there are many cases where parts of the originally synthesized polypeptide chain are removed during the process of maturation. The types of modification and processing that polypeptide chains undergo is strongly related to the site of protein synthesis and to the mechanisms that are involved in targeting the polypeptide chain to its final destination. In fact, processing frequently begins during polypeptide synthesis and continues for some time thereafter. A major division involving the types of processing reactions that polypeptide chains undergo is related to the site of protein synthesis and the final destination of the protein. Recall (see fig. 5.1) that proteins synthesized on free polysomes either remain in the cytosol or are targeted to the

Figure 29.20

An *in vitro* assay for release factors.

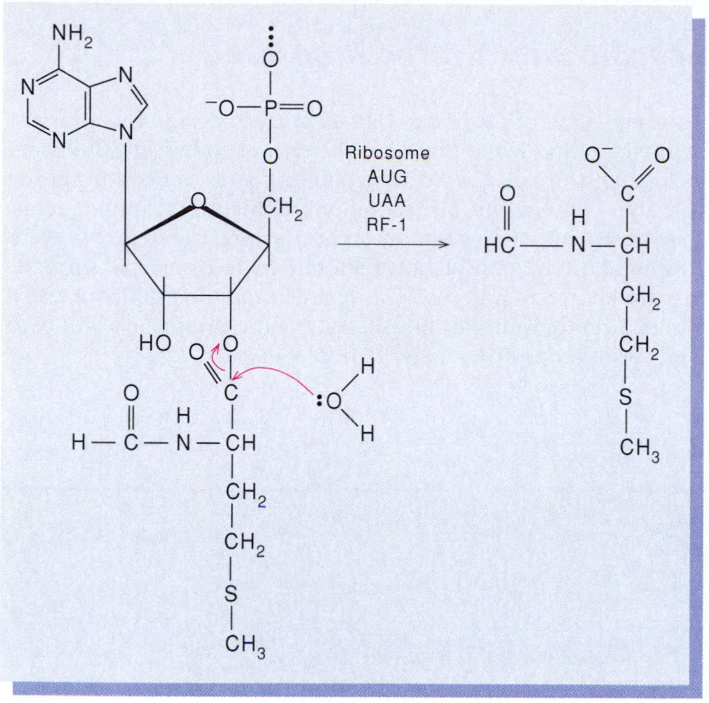

Figure 29.21

Biosynthesis of insulin. Insulin is synthesized by membrane-bound polysomes in the ß cells of the pancreas. The primary translation product is preproinsulin, which contains a 24-residue signal peptide preceding the 81-residue proinsulin molecule. The signal peptide is removed by signal peptidase, cutting between Ala (−1) and Phe (+1), as the nascent chain is transported into the lumen of the endoplasmic reticulum. Proinsulin folds and two disulfide bonds cross-link the ends of the molecule as shown. Before secretion, a trypsin-like enzyme cleaves after a pair of basic residues 31, 32 and 59, 60; then a carboxypeptidase B-like enzyme removes these basic residues to generate the mature form of insulin.

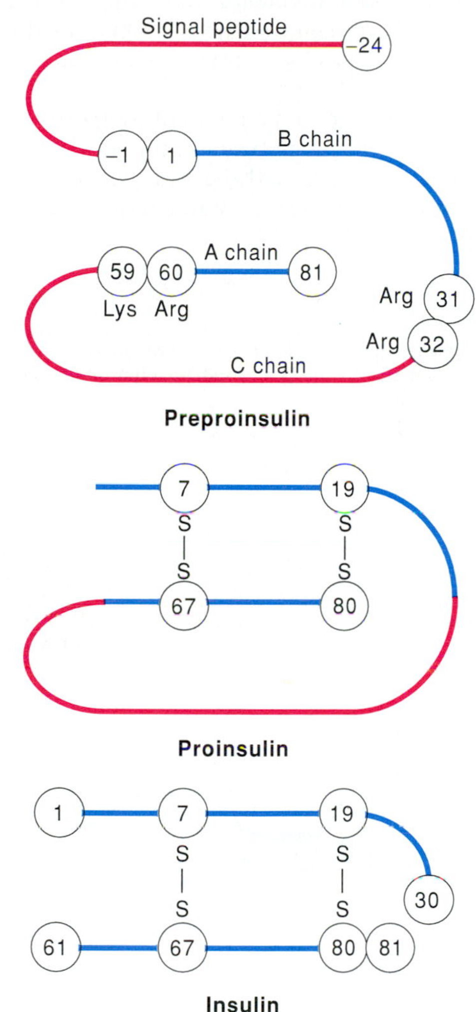

mitochondria, the chloroplasts, or the nucleus. Those that will remain in the cytosol are often ready to function as soon as they are released from the ribosome. Those that must be transported to another site usually undergo substantial modification during the transport process. Many proteins, however, are synthesized on membrane-bound polysomes rather than on free polysomes. In bacteria such proteins are synthesized on polysomes associated with the inner plasma membrane; in eukaryotes, they are synthesized on polysomes associated with the endoplasmic reticulum. Proteins that are synthesized on ribosomes bound to the endoplasmic reticulum either remain in the ER or are targeted to the Golgi, the secretory granules, the plasma membrane, or the lysosomes. In general, proteins that are synthesized on membrane-bound polysomes undergo more extensive modification before they reach their final destination and become fully functional (see chapter 21).

Proteins Are Targeted to Their Destination by Signal Sequences

Despite the apparent complexity of the problem, protein transport in all systems is accomplished by a single, rather simple underlying mechanism; each polypeptide destined for transport contains an amino acid sequence known as a signal or leader sequence that identifies the polypeptide to the appropriate transporting system. This generality was first recognized in the middle 1970s by Gunter Blobel, who articulated the underlying signal hypothesis (see fig. 21.9). Frequently, the signal sequence is cleaved from the parent polypeptide during the transport process. A protein containing a signal sequence that is cleaved upon transport is known as a preprotein. A protein containing a peptide sequence that must be removed for the protein to be active is known as a proprotein, and a protein that contains both of these sequences is known as a preproprotein.

Many protein hormones are synthesized in a preproprotein form. For example, insulin mRNA is translated into a single polypeptide chain, preproinsulin, that contains 84 residues of proinsulin plus a 23-residue signal peptide (fig. 29.21). Within the islets of Langerhans of the pancreas, the N-terminal sequence is removed cotranslationally in targeting proinsulin to the Golgi apparatus, and the two disulfides joining the ends of the molecule are formed. Following this, the

29D

BOX

Antibiotics Inhibit by Binding to Specific Sites on the Ribosome

Many antibiotics (generally, small organic compounds with therapeutic utility) prevent bacterial growth by inhibiting translation. This is not surprising, since translation is both a complex and metabolically essential process. Also, the bacterial ribosome is structurally distinct from the ribosome in the eukaryotic cytoplasm and thus specific bacterial inhibitors can be found.

A great many antibiotic inhibitors of ribosome function belong to the class known as aminoglycoside antibiotics. Of these, streptomycin is the best known and the best investigated. Streptomycin binding produces a variety of functional alterations in the ribosome. One of the first to be recognized was the loss of translational fidelity. When bound to the small subunit, streptomycin distorts its structure so as to allow altered codon–anticodon pairing and the consequent incorporation of incorrect amino acids. Indeed, mutations to streptomycin resistance frequently prevent antibiotic binding and involve alterations in ribosomal proteins that are known to play a role in maintaining the fidelity of translation. Streptomycin binding also alters the ribosome's ability to participate properly in the initiation reactions.

Figure 1

The structures of some antibiotic inhibitors of protein synthesis. All of the inhibitors shown function by binding to specific sites on the ribosome.

$R=CH_3NH$

Streptomycin

Chloramphenicol

Tetracycline

Erythromycin

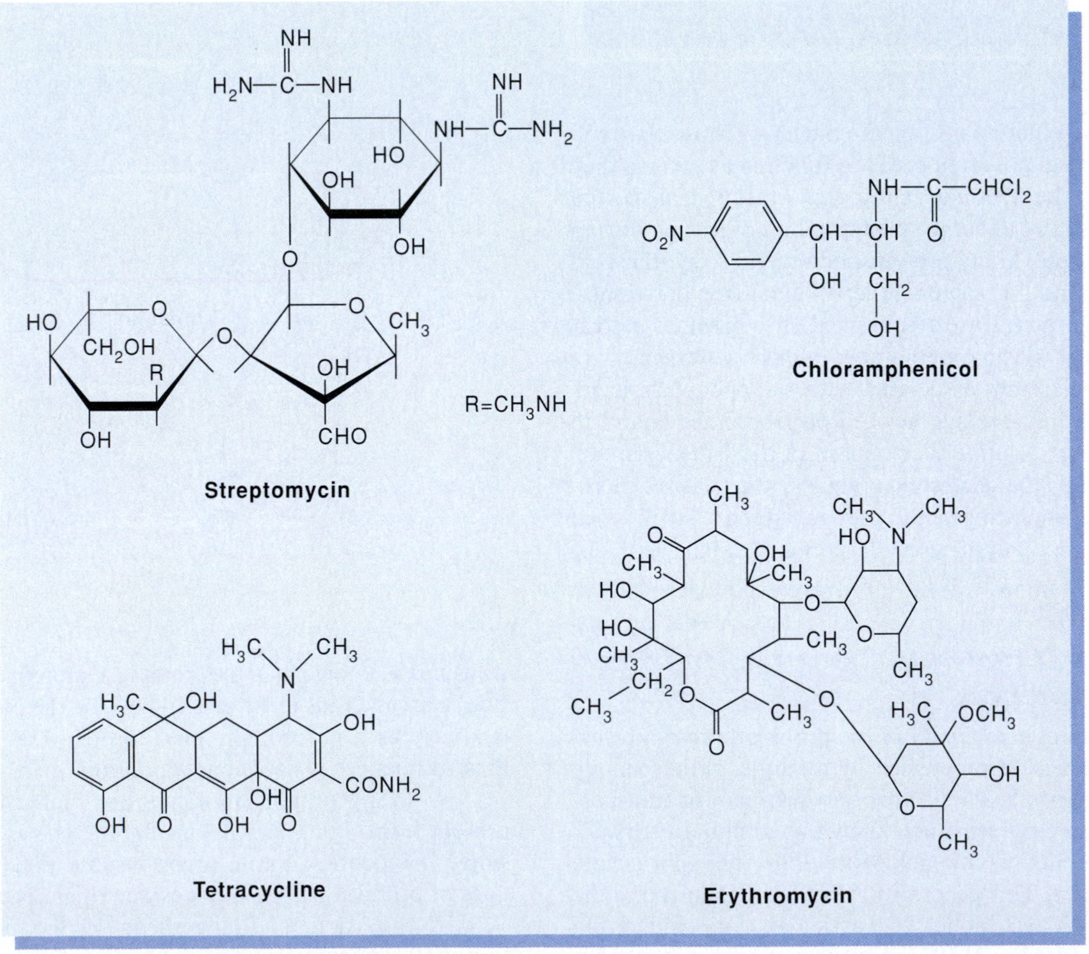

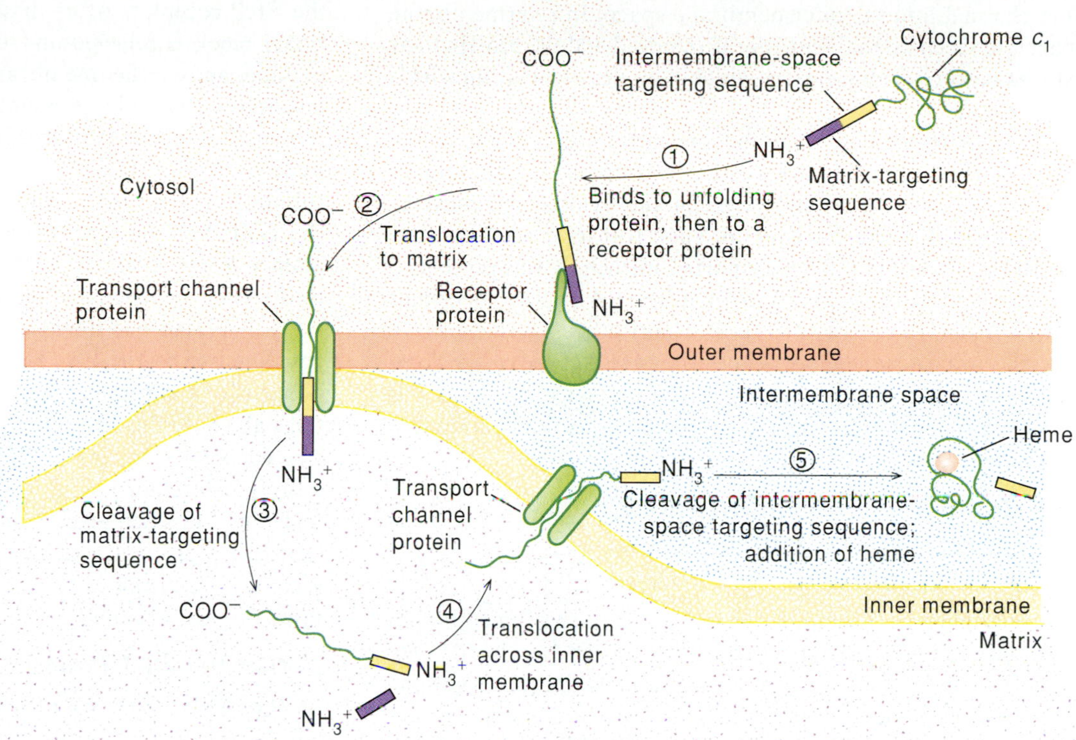

Figure 29.22

Two successive translocations are required to target proteins such as cytochrome c_1 to the intermembrane space. The precursors of cytochrome c_1 have two uptake-targeting sequences at its N terminus. The first targets the polypeptide to the matrix. In the matrix the first target sequence is cleaved by a specific protease. The second target sequence is thereby exposed and targets the polypeptide to intermembrane space, where the second target sequence is removed by another protease. The molecule folds and adds heme to become fully functional. (Source: F. U. Hartl et al., in *Cell* 51:1021-1027, 1987. Copyright © 1987 Cell Press, Cambridge, Mass.; and E. C. Hurt and A. P. G. M. van Loon, "How proteins find mitochondria and intramitochondrial compartments" in *Trends in Biochemical Sciences* 11:204-207, 1986. Copyright © 1986 Elsevier Trends Journals, Cambridge, England.)

C-peptide region of proinsulin is removed to yield the circulating form of insulin with 51 residues in two disulfide-linked peptides.

Signal sequences usually occur at the amino terminus of polypeptides to be transported. These sequences vary in length from 10 to 40 amino acid residues and are generally characterized by the occurrence of a block of hydrophobic residues. The presence of hydrophobic residues reflects the fact that protein transport is essentially a problem of the movement of proteins across lipid membranes. The hydrophobic residues both target the protein to the appropriate compartment and initiate penetration into the membrane.

The sequence rules that identify signal sequences to individual cellular compartments are not completely clear. The nature of the recognition, however, is phylogenetically conserved because signal sequences generally target proteins to the appropriate compartment in different species. Also, the specificity of the signal is independent of the protein to be transported because cytoplasmic proteins, for example, can be targeted to membrane locations by the addition of appropriate signal sequences.

Some Mitochondrial Proteins Are Transported after Translation

Some proteins, especially those destined for the eukaryotic mitochondria and chloroplasts, are transported after their synthesis on free polysomes is complete. Such transport is known as posttranslational transport. In the case of posttranslational transport it is believed that the polypeptide to be transported must be unfolded from its native folded configuration by a system of polypeptide chain binding proteins (PCBs) before it can pass through the membrane. Posttranslational transport into the mitochondrion requires both ATP and a proton gradient. Presumably the energy from one or both of these sources is used to unfold the protein or separate it from the PCB system in order to allow it to pass through the membrane.

The transport of proteins from the cytoplasm to the mitochondrion is further complicated by the fact that mitochondria themselves are compartmentalized. They have both an outer membrane and an inner membrane (see chapter 15). Some proteins are targeted to either of these membranes, some are targeted to the fluid layer enclosed by the inner membrane, called the matrix, and others such as cytochrome c_1 are targeted to the fluid layer bounded by the inner membrane and the outer membrane, known as the intermembrane space (fig. 29.22). Cytochrome c_1 has two signal sequences located in series at the amino-terminal end of the preprotein. The first is recognized by a specific receptor protein attached to the outer membrane of the mitochondria. This receptor protein guides the polypeptide chain to a transport channel, across which it is transported into the matrix. Once in the matrix, the first signal sequence is removed by a specific protease. A second signal sequence, which is exposed by this cleavage reaction, results in the transport of the polypeptide chain by a similar mechanism across a transport channel into the intermembrane space. In intermembrane space the second signal sequence is removed and the protein folds into its mature configuration, which includes associating with a heme group.

As in the case of mitochondrial proteins, some of the proteins of chloroplasts are synthesized directly in the organelle, while others are synthesized in the cytosol and must be transported. The mechanisms for such transport are very similar to those observed for transported mitochondrial proteins.

All nuclear proteins are synthesized on free polysomes in the cytosol. In contrast to the transport processes that involve specific amino-terminal sequences that are cleaved, more complex structural features are recognized in nuclear proteins and they are somehow selectively transported in an intact state into the nucleus.

Eukaryotic Proteins Targeted for Secretion Are Synthesized in the Endoplasmic Reticulum

The endoplasmic reticulum (ER) is the largest membrane-bounded organelle in a typical eukaryotic cell (see fig. 21.10). The ER is a continuous network of tubules and cisternae extending throughout the cytoplasm, with a total surface area many times that of the plasma membrane. Most of the ER is studded with ribosomes to form the rough endoplasmic reticulum (RER). The ribosomes of the RER are the site of synthesis of membrane and secretory proteins and are the starting point for the protein secretory pathway. The membrane and lumen of the ER contain a characteristic set of proteins that function to process secretory and membrane proteins. After only a short time in the RER, secretory proteins are transported, by a process of vesicle budding and fusion, to the Golgi apparatus and from there to the cell surface, secretory granules, or lysosomes. As you will recall, the Golgi apparatus is a flattened stack of membranes that is the primary site of protein targeting as well as a principal site of carbohydrate addition to form glycoproteins (see chapter 21).

Let us review the way in which proteins that are destined for assembly into membranes, for secretion, or for targeting to other areas of the cell enter the lumen of the ER (see fig. 21.9). The initial targeting of nascent polypeptides to the ER membrane results from the cotranslational recognition of a signal sequence by a ribonucleoprotein complex called the signal recognition particle or SRP. The SRP is a complex that consists of six different proteins and a single 300-residue RNA molecule designated 7S RNA. Exactly how this complex functions is not clear, but it appears to be a structure especially adapted to recognize and bind to the signal sequence when the total nascent chain has reached a length of about 90 amino acid residues. Upon binding, and by an unknown mechanism, the SRP arrests the translation of the nascent chain while it searches for a specific receptor known as the SRP receptor or "docking protein." The SRP receptor is an integral membrane protein that protrudes on the inner face of the ER membrane. When docking is complete, the SRP dissociates from the ribosome, which is now bound by the signal sequence of its nascent polypeptide to the SRP receptor. After dissociation of the SRP, the translational block is relieved and translocation of the nascent polypeptide across the membrane begins. It has recently been discovered that GTP is essential for the docking maneuver and that its hydrolysis is probably essential for the subsequent release of SRP. Furthermore, sequencing of one of the protein subunits of SRP, SRP54, has revealed that it is homologous to the GTP binding domain of the G proteins. Thus signal sequence recognition by SRP may operate in a manner analogous to the GTP-dependent functioning of the G protein family.

Numerous nonmembranous proteins are retained by and function in the ER. Retention of these proteins is accomplished by a surprisingly simple mechanism. Soluble ER proteins all share the common C-terminal sequence, Lys-Asp-Glu-Leu (or KDEL in single-letter language). There appears to be a protein receptor bound to the ER membrane that recognizes and binds this sequence. The passive nature of this retention mechanism is illustrated by the fact that if the four C-terminal residues are removed from an ER protein, it is no longer retained by the organelle.

Proteins that Pass through the Golgi Become Glycosylated

Glycosylation is a conspicuous modification of proteins as they pass through the Golgi apparatus (see chapter 21). Many of the cell-surface and secretory proteins produced here are glycoproteins. Because of the diversity of the glycosylation reactions that occur in the Golgi and the fact that this organelle is the principal site of protein transport, it seems likely that the carbohydrate moieties attached to proteins are responsible for targeting them to their destinations.

One clear-cut example supporting this conclusion is the role of mannose-6-phosphate residues in targeting proteins to the lysosome. Digestive enzymes destined for the lysosome are identified by the presence of mannose-6-phosphate by binding to a specific mannose-6-phosphate receptor protein. The critical step in targeting glycoproteins from the Golgi to the lysosome is their recognition by the enzymes that catalyze the two-step phosphorylation of terminal mannose residues in their N-linked oligosaccharide side chains. Failure to phosphorylate the mannose residues results in the secretion of lysosomal enzymes. This malfunction, you may remember, is associated with the syndrome known as I-cell disease, a condition that leads to the crowding of lysosomes with damaged proteins that it cannot degrade (see chapters 6 and 21).

Processing of Collagen Does Not End Until after Secretion

Recall that collagen is an extracellular matrix protein that serves as a major constituent of many connective tissues (see figs. 4.10 to 4.13). Collagen fibrils have a distinctive banded pattern with a periodicity of 680 Å. Individual fibrils are composed of three polypeptide chains wound around one another in a right-handed helix with a total length of 3,000 Å. Each of the polypeptide chains in the triple helix has a repetitious tripeptide sequence,

Figure 29.23

Major events in the posttranslational processing of collagen.

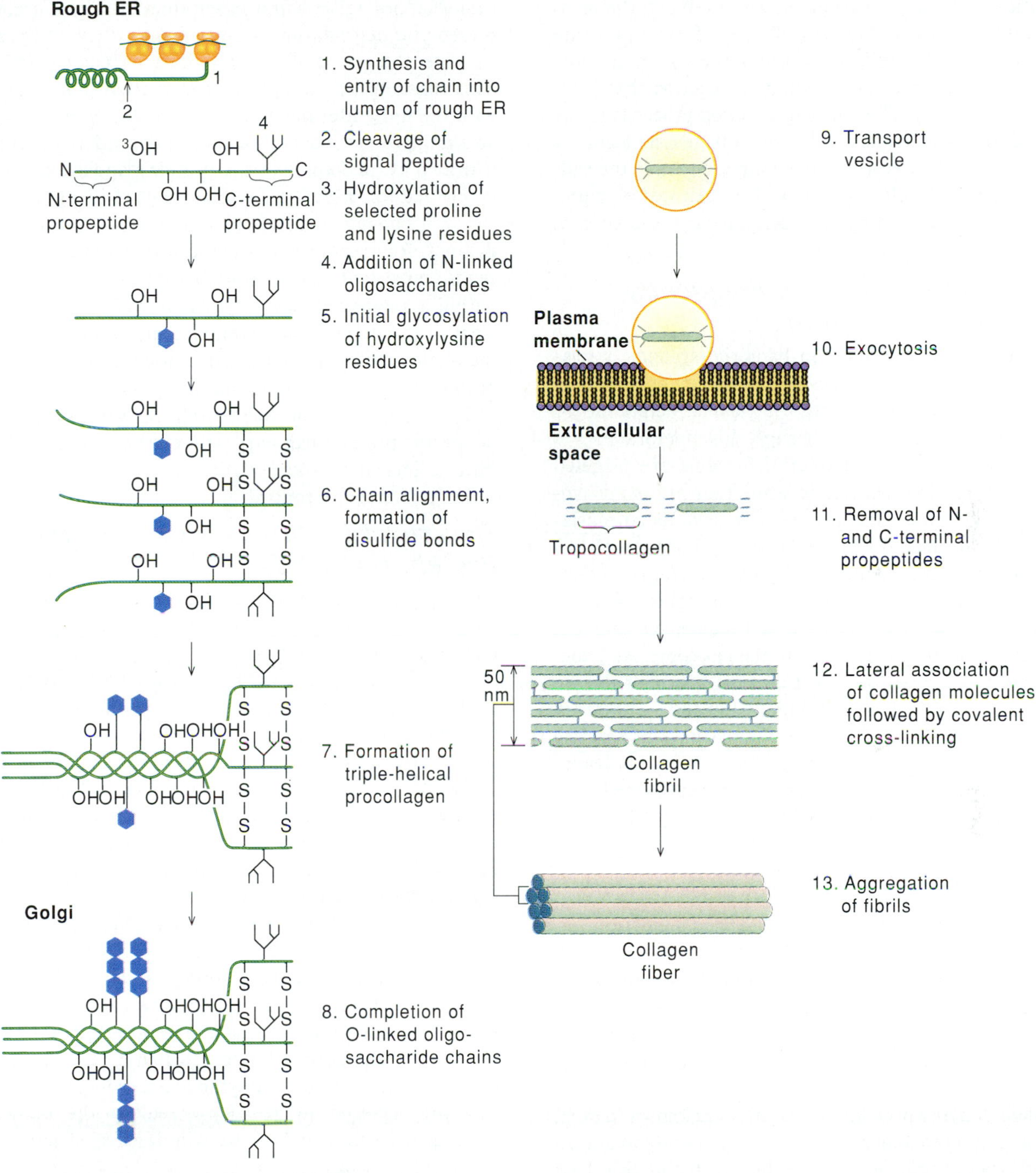

Gly-X-Y, where X is frequently a proline and Y is frequently a hydroxyproline. The latter amino acid is not one of the twenty that are specified genetically, so it must be formed posttranslationally by a modification of some of the prolines.

Since collagen is a secreted protein, we know that it must follow the route of synthesis that starts on a ribosome bound to the endoplasmic reticulum (fig. 29.23). This is where the fun begins, as we find that the nascent collagen polypeptide has extensive N and C termini (150 and 250 amino acids, respectively, for type I collagen) that are not found in the mature collagen. The function of these extensions appears to be to facilitate the initial interaction of chains in triplets and to stabilize the triple helices once they have been formed. The first modification reaction to take place in the endoplasmic reticulum is hydroxylation of specific proline residues and some lysine residues by two different enzyme systems. Glycosylation begins soon thereafter with O-glycosylation of certain hydroxylysine residues and N-glycosylation of certain asparagine residues. Next, the modified polypeptide chains, in clusters of three, form interchain disulfides near the C termini. This brings the chains in close proximity, thereby facilitating the winding reaction that leads to triple helix formation. The winding reaction proceeds in the C to N direction. The triplex helix procollagen molecule is packaged into a secretory vesicle. Following exocytosis, the ends of the procollagen molecule are removed by extracellular proteases. Collagen fibrils form by the spontaneous association of the mature collagen molecules.

Bacterial Protein Transport Frequently Occurs during Translation

The problem of protein transport in bacterial systems is relatively simple. Polypeptides synthesized in the cytoplasm may function there, may be inserted into the plasma membrane, or may be secreted by being passed through this membrane.

Most noncytoplasmic bacterial proteins are targeted into or across the plasma membrane while they are being synthesized on the ribosome. This process is known as cotranslational transport. Cotranslational secretion of proteins in *E. coli* involves a group of proteins encoded by the *sec* (for secretion) genes. Some of these proteins are membrane proteins that function by recognizing and binding the signal sequence of nascent polypeptide chains as they emerge from the ribosome. As a consequence of cotranslational transport and the action of the sec proteins, most bacterial ribosomes engaged in the synthesis of secreted proteins are tethered to the cytoplasmic face of the plasma membrane by their nascent peptide chains. A leader peptidase that is an integral membrane protein cleaves the leader sequence from the secreted polypeptide.

Protein Turnover

Cellular proteins are continuously being formed and degraded. At first glance, continuous degradation appears to be wasteful. However, protein degradation is of major biological importance in regulating protein levels, in protecting against the accumulation of abnormal proteins, in controlling growth and development, and in allowing adaptation to changing environmental conditions.

Although many proteolytic enzymes are known to exist, we are just beginning to understand how they operate to govern the levels and types of proteins within cells—to distinguish those proteins that must be degraded from those that must be preserved.

The Lifetimes of Proteins Differ

The level of a protein within a cell is determined by the balance between its rates of synthesis and degradation. As a consequence, changes in protein levels can be brought about by changes either in synthetic or degradative rates. Moreover, a rapid rate of degradation ensures that the concentration of a protein rises or falls rapidly when its synthetic rate is changed.

The degradation of cellular proteins, like the decay of radioisotopes, follows first-order kinetics. Therefore, the rate of proteolytic degradation is conveniently defined by a protein's half-life, the period of time required for it to be 50% degraded. This characteristic of protein degradation should be contrasted with the finite lifetime of an erythrocyte, which is the result of an aging process. For the purpose of degradation, each molecule is drawn at random from the protein population and therefore is selected for proteolysis independent of its age.

The half-lives of eukaryotic proteins are different and distinct. Representative examples are listed in table 29.6. It is readily apparent that degradative rates of individual proteins within a single cell can vary over a wide range. Generally, at least in the mammalian liver, enzymes that occupy important metabolic control points are degraded most rapidly, while especially long-lived proteins are rarely the sites of metabolic control. In addition, it has been established that degradative rates of specific proteins can vary with changes in physiological conditions. Thus the protein degradative mechanism is in some way tuned to metabolic control.

Structural Features Can Determine Protein Half-Lives

How are individual proteins selected for hydrolysis by the proteolytic machinery? Two sequence characteristics have been identified as correlating with rates of protein degradation.

Several years ago it was observed through sequence analysis that rapidly degraded liver proteins, those with half-lives less than 2 h, nearly all contain regions of their sequence that are rich in the amino acids proline, glutamate, serine, and threonine. These regions, involving from 10 to 60 amino acid residues, have been designated PEST sequences on the basis of the single-letter designations of their constituent amino acids. Presumably, the PEST sequences create structural domains that are recognized by proteolytic enzymes.

A second structural feature that has been correlated with protein degradative rates is the N-terminal residue of the mature form of cytoplasmic proteins. This relationship, summarized in table 29.7, has become known as the N-terminal rule. Here the presumption is that the amino-terminal residue is at least partly responsible for recognition by the degradative machinery. Proteins that are degraded according to the N-terminal rule are believed to be recognized by the ubiquitin-ATP-dependent pathway, which we will describe shortly.

Abnormal Proteins Are Selectively Degraded

A very important function of protein degradation is to protect the organism against the consequences of intracellular accumulation of abnormal proteins. Bacterial and eukaryotic cells do not accumulate structurally altered proteins because such proteins are rapidly degraded. Thus it appears that essentially all cells possess a degradative system that is capable of recognizing "abnormal" proteins.

Table 29.6
Half-Lives of Some Proteins in Mammalian Cells

Enzyme	Half-Life (h)
Rapidly degraded	
1. c-myc, c-fos, p53 oncogenes	0.5
2. Ornithine decarboxylase	0.5
3. δ-Aminolevulinate synthase	1.1
4. RNA polymerase I	1.3
5. Tyrosine aminotransferase	2.0
6. Tryptophan oxygenase	2.0
7. β-Hydroxylβ-methylglutaryl coenzyme A reductase	2.0
8. Deoxythymidine kinase	2.6
9. Phosphoenolpyruvate carboxykinase	5.0
Slowly degraded	
1. Arginase	96
2. Aldolase	118
3. Cytochrome b_5	122
4. Glyceraldehyde-3-phosphate dehydrogenase	130
5. Cytochrome *b*	130
6. Lactic dehydrogenase (isoenzyme 5)	144
7. Cytochrome *c*	150

Table 29.7
Correlation between Half-Lives of Cytosolic Proteins and Amino Acid Residue at the N Terminal

Amino-Terminal Residue	Half-Life
Stabilizing	
Methionine	>20 hours
Glycine	
Alanine	
Serine	
Threonine	
Valine	
Destabilizing	
Isoleucine	~30 minutes
Glutamate	
Tyrosine	~10 minutes
Glutamine	
Proline	~7 minutes
Highly destabilizing	
Leucine	~3 minutes
Phenylalanine	
Aspartate	
Lysine	
Arginine	~2 minutes

Source: A. Bachmjuir et al., "In vivo half-life of a protein is a function of its amino-terminal residue," in *Science* 234:179, 1986. Copyright © 1986 American Association for the Advancement of Science, Washington, D.C.

The features of this degradative system are best seen in *E. coli,* where altered intracellular proteins can be readily manipulated. For example, it has been known for many years that incomplete chains of β-galactosidase are rapidly degraded even though the completed protein is entirely stable. In a similar way, many protein alterations that result from miscoding induced by streptomycin, or by the incorporation of amino acid analogs, also fail to accumulate because the protein products that contain them are rapidly degraded. This system presumably limits expression of heterologous proteins in bacteria that are degraded because they cannot form their native structure.

Another manifestation of the defense against abnormal proteins is seen in the heat shock response. This is a defense reaction, involving programmed changes in gene transcription and translation, that is exhibited by essentially all cells under stressful conditions. The cells reduce their overall rates of gene transcription and translation and for a brief time produce a small repertoire of proteins called heat shock proteins (hsp's). As noted in chapter 4, some hsp's play a role in the folding and transport of proteins during normal cellular function. In the heat shock response it is believed that these proteins protect against the presence of damaged proteins by binding to them and promoting either their refolding or their proteolytic degradation.

Proteolytic Hydrolysis Occurs in Mammalian Lysosomes

The classic studies of Christian DeDuve in the 1960s established that mammalian cells contain a degradative organelle, the lysosome, that is produced in the Golgi and contains a large number of proteases and other hydrolytic enzymes. A similar vacuole containing hydrolytic enzymes is also present in yeast and higher plants. Together, the hydrolytic enzymes in these organelles are capable of completely degrading many macromolecules and delivering their monomeric units to the cytoplasm for further metabolism. The metabolic importance of the lysosome is demonstrated by the existence of various lysosomal storage diseases, in which one or more hydrolytic enzymes is missing. An accumulation of undegraded molecules results from these genetic diseases, frequently with devastating consequences.

A primary function of the lysosome is to digest protein-containing particles derived from the extracellular space. One mechanism of delivery is the process of endocytosis. Endocytosis is the invagination of a group of occupied receptors on the plasma membrane. Most mammalian cells can also engulf large extracellular particles by the less specific processes of pinocytosis and phagocytosis. The endocytic vesicles formed by these processes fuse with lysosomes to form secondary lysosomes where hydrolysis occurs. Lysosomes also function to degrade intracellular proteins, especially under conditions of nutritional deprivation. Under these circumstances, lysosomes can engulf cytoplasmic contents to form autophagic vacuoles and thus recycle cellular proteins.

Lysosomal proteases are called cathepsins, a name derived from the Greek term meaning "to digest." The interior of the lysosome is acidic and the cathepsins, like all lysosomal hydrolases, possess acidic pH optima and exhibit little enzymatic activity at neutral pH. This characteristic protects the cell from autolytic breakdown that might result from leakage of lysosomal contents into the neutral cytoplasm.

Ubiquitin Tags Proteins for Proteolysis

A number of different proteolytic systems are thought to be responsible for the degradation of soluble proteins in the cytoplasm of eukaryotic cells. One of the best understood is that which involves ATP and the protein ubiquitin. Ubiquitin is a small protein of only 76 residues. It occurs universally in eukaryotic cells and is highly conserved in sequence; only three residues distinguish the ubiquitin in yeast and humans. The covalent attachment of ubiquitin to proteins is thought to "tag" them for subsequent hydrolysis by cellular proteases.

Three enzymes participate in the ATP-ubiquitin system that prepares proteins for proteolysis (fig. 29.24). In the first step of this series of reactions, the C-terminal glycine residue of ubiquitin is activated by forming a thiol ester with a specific activating enzyme designated E1. This reaction is driven by the hydrolysis of ATP to AMP and PP_i. Ubiquitin is then transferred to a sulfhydryl group of a second protein (E2) that is a substrate for a group of ubiquitin-targeting proteins designated E3. The E3 proteins are responsible for identifying proteins for degradation (some on the basis of the identity of their N-terminal residue and some because they are "abnormal") and catalyzing the covalent transfer of ubiquitin from E2 to these proteins. This attachment is via isopeptide bonds that join the carboxyl group of the previously activated C-terminal glycine residue of ubiquitin with ϵ-amino groups of lysine side chains of the targeted proteins. Proteins tagged in this way are then recognized and degraded by specific proteases that release ubiquitin so that it can recycle.

ATP plays a conspicuous and rather surprising role in the degradation of proteins. Clearly, the hydrolysis of peptide bonds is a reaction that in and of itself does not require the input of energy. Nonetheless, ATP is required for the action of many proteolytic enzymes, independent of its role in ubiquitination that we have just seen.

One particularly well-understood ATP-dependent protease is the La protease that is the product of the *lon* gene of *E. coli*. This protease, like many ATP-dependent proteases, is a large protein and its ability to hydrolyze proteins is tightly coupled to its ability to hydrolyze ATP. Approximately two ATPs are hydrolyzed to ADP and P_i for each peptide bond that is hydrolyzed. It appears that the hydrolysis of ATP is required to activate the proteolytic active site of La. Other proteases seem to require ATP as an allosteric effecter to activate their hydrolytic sites, but this ATP is not hydrolyzed.

Thus ATP serves at least three independent roles in intracellular proteolysis: (1) it functions in the tagging of proteins through covalent attachment of ubiquitin, (2) it activates proteases such as La through hydrolysis, and (3) it serves as a positive allosteric effecter of other proteases without being hydrolyzed. The presumed function of this energetic requirement is to ensure the fidelity of protein degradation. It is, after all, nearly as important to cellular function to degrade proteins correctly as it is to synthesize them correctly.

Figure 29.24

The ubiquitin marking system targets certain proteins for degradation. At least three enzymes, E1, E2, and E3, are involved in addition to ubiquitin-specific proteases.

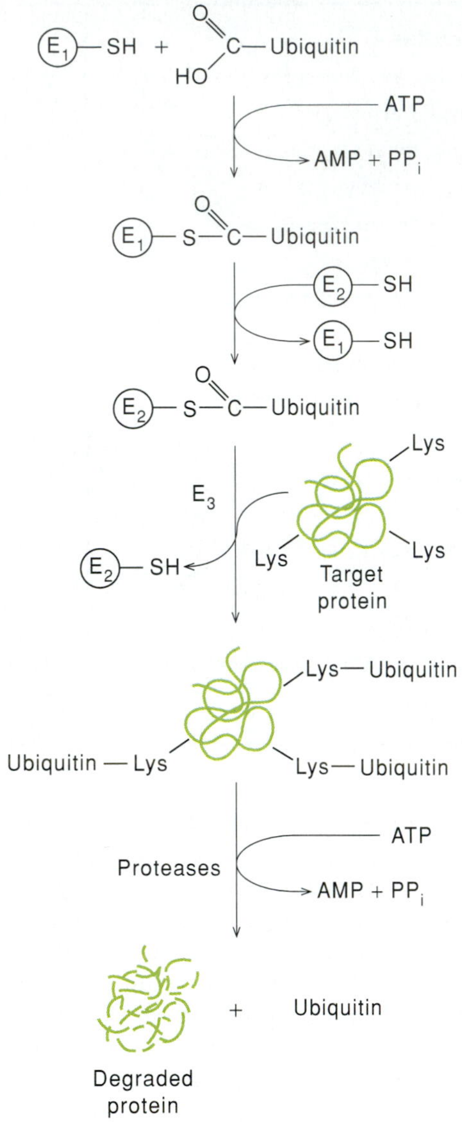

Summary

We have focused in this chapter on the complex mechanisms of protein synthesis. The following points are central to this subject.

1. Three types of RNA carry out protein synthesis: ribosomal RNA, transfer RNA, and messenger RNA. Ribosomal RNA is invariably complexed with many proteins to form ribosomes, on which amino acids are assembled into polypeptides. The amino acids are brought to the ribosomes attached to transfer RNAs. The messenger RNA contains the instructions for translation in the form of the genetic code. Messenger RNAs form transient complexes with ribosomes. Individual aminoacyl-tRNAs bind to specific sites on the messenger RNAs. The interacting site on the messenger is the codon; the interacting site on the tRNA is the anticodon.
2. The part of the messenger that is translated is the reading frame. Eukaryotic messages carry only one reading frame, whereas prokaryotic messengers may carry more than one. In prokaryotes the initiation codons are recognized by a ribosome binding site upstream of the start codon.
3. Most transfer RNAs have common parts and uncommon parts. The common parts facilitate binding of the aminoacyl-tRNAs to common sites on the ribosome. The uncommon sites permit specific reactions with charging enzymes that covalently attach the correct amino acids to the correct tRNA. Another uncommon site on the tRNAs is the anticodon, which leads to specific complex formation with the complementary codon site on the messenger.
4. Attachment of the amino acid to the tRNA is catalyzed by a specific aminoacyl synthase, which recognizes all the cognate tRNAs for a specific amino acid.
5. A unique methionyl-tRNA binds to the initiation codon on all messages.
6. The genetic code is the sequence relationship between nucleotides in the messenger RNA and amino acids in the proteins they encode. Triplet codons are arranged on the messenger in a nonoverlapping manner without spacers.
7. The code was deciphered with the help of synthetic messengers with a defined sequence, by analyzing the types of polypeptide chains that were made when these messengers were used in an *in vitro* protein-synthesizing system.
8. The genetic code is highly degenerate, with most amino acids represented by more than one codon. In many cases the 3′ base in the codon may be altered without changing the amino acid that is encoded.
9. The codon–anticodon interaction is limited to Watson-Crick pairing for the first two bases in the codon but is considerably more flexible in the third position.
10. Translation begins with the binding of the ribosome to mRNA. A number of protein factors transiently associate with the ribosome during different phases of translation: initiation factors, elongation factors, and termination factors.
11. Initiation factors contribute to the ribosome complex with the messenger RNA and the initiator methionyl-tRNA. Elongation factors assist the binding of all the other tRNAs and the translocation reaction that must occur after each peptide bond is made. Termination factors recognize a stop signal and lead to the termination of polypeptide synthesis and the release of the polypeptide chain and the messenger from the ribosome.
12. A large number of antibiotics have been characterized that inhibit protein synthesis. These antibiotics are usually made by a particular microorganism and they inhibit protein synthesis in a broad family of other organisms, mostly bacterial.
13. Specific enzymes catalyze folding after polypeptide synthesis.
14. Proteins are targeted to their destination by signal sequences built into the polypeptide chain. These signals are usually located at the N-terminal end of the protein and are generally cleaved during protein maturation.
15. Posttranslational modifications include many covalent alterations: polypeptide processing, attachment of carbohydrate or lipid groups to specific side chains, and addition of many other low-molecular-weight ligands to side chains.
16. Intracellular protein degradation is not random. Different proteins have quite different half-lives, which are related to specific structural features. Imperfectly folded proteins and polypeptide fragments are frequently degraded most rapidly. In eukaryotes, lysosomes play a major role in protein degradation.

Selected Readings

Bachmjuir, A., and A. Varshavsky, The degradation signal is a short-lived protein. *Cell* 56:1019–1032, 1989.

Beasley, E. M., and G. Schatz, Import of proteins into mitochondria. *Chemtracts* 2:305–317, 1991.

Bjork, G. R., J. U. Ericson, C. E. D. Gustafsson, T. G. Hagervall, Y. H. Jonsson, and P. M. Wilkstrom, Transfer RNA modification. *Ann. Rev. Biochem.* 56:263–288, 1987.

Böck, A., K. Forchhammer, J. Heider, and C. Baron, Selenoprotein synthesis: an expansion of the genetic code. *Trends Biochem. Sci.* 16:463–467, 1991.

Bond, J. S., and P. E. Butler, Intracellular proteases. *Ann. Rev. Biochem.* 56:333, 1987. An overview of the types of proteolytic enzymes that are found in cells and how they may function in biologically important cleavages of proteins.

Brunori, M., M. C. Silvestrini, and M. Pocchiari, The scrapie agent and the prion hypothesis. *Trends Biochem. Sci.* 13:309–313, 1988.

Burgess, T. L., and R. B. Kelly, Constitutive and regulated secretion of proteins. *Ann. Rev. Cell. Biol.* 3:243–294, 1987.

Crick, F. H. C., Codon-anticodon pairing; the wobble hypothesis. *J. Mol. Biol.* 19:548–555, 1966.

Ellis, R. J., and S. M. van der Vies, Molecular chaperones. *Ann. Rev. Biochem.* 60:321–348, 1991.

Englesberg-Kulka, H., and R. Schoulakerk-Schwarz, A flexible genetic code, or why does selenocysteine have no unique codon? *Trends Biochem. Sci.* 13(11):419–421, 1988.

Ferguson, M. A. J., and A. F. Williams, Lipids as membrane tethers for proteins. *Ann. Rev. Biochem.* 57:285, 1988.

Fessler, J. H., and L. I. Fessler, Biosynthesis of procollagen. *Ann. Rev. Biochem.* 47:129–162, 1978.

Finley, D., and A. Varshavsky, The ubiquitin system functions and mechanisms. *Trends Biochem. Sci.* 10:343–347, 1985.

Fox, T. D., Natural variation in the genetic code. *Ann. Rev. Gen.* 21:67, 1987. A review of the exceptions to the "universal" genetic code.

Gold, L., Posttranslational regulatory mechanisms in *E. coli. Ann. Rev. Biochem.* 57:199, 1988. A summary of the mechanisms of initiation and regulation of translation in prokaryotic systems.

Hardesty, B., and G. Kramer, *Structure, Function and Genetics of Ribosomes.* New York: Springer-Verlag, 1986. The third book in the Springer Series in Molecular Biology, describing current research on protein synthesis.

Hershey, J. W. B., Protein phosphorylation controls translation rates. *J. Biol. Chem.* 264:20823, 1989. Describes how protein kinases are thought to regulate translation in eukaryotic systems.

Hershko, A., Ubiquitin-mediated protein degradation. *J. Biol. Chem.* 263:15237–15240, 1988.

Hill, W. E., A. Dahlberg, R. A. Garrett, and J. R. Warner, The ribosome: structure function and evolution. American Society of Microbiology, Washington, D.C., 1990. The fourth book in a series describing current research on protein synthesis.

Hoagland, M. B., M. L. Stephenson, J. F. Scott, L. I. Hecht, and P. Zamecnik, A soluble ribonucleic acid intermediate in protein synthesis. *J. Biol. Chem.* 231:241–257, 1958. Describes the pioneering tracer studies that chart the course of amino acid into polypeptide chain.

Kartles, J. R., and A. I. Hubbard, Plasma membrane protein sorting in epithelial cells; do secretory pathways hold the key? *Trends Biochem. Sci.* 13:181–184, 1988.

Kleinkauf, H., and H. Dohren, Nonribosomal polypeptide formation on multifunctional proteins. *Trends Biochem. Sci.* 8:281–283, 1983.

Kozak, M., The scanning model for translation: An update. *Mol. Cell. Biol.* 8:2737, 1989. The current view of the way in which the eukaryotic ribosome selects the initiating codon, by the originator of the scanning model.

Moore, P. B., The ribosome returns. *Nature* 33:223–227, 1988.

Nirenberg, M. W., and J. H. Mattaei, The dependence of cell-free protein synthesis in *E. coli* upon naturally occurring or synthetic polyribonucleotides. *Proc. Natl. Acad. Sci.* 47:1588–1602, 1961. The landmark paper reporting the finding that poly(U) stimulates the synthesis of polyphenylalanine.

Noller, H. F., V. Hoffarth, and L. Zimniak, Resistance of peptidyl transferase to protein extraction procedures. *Science* 256:1416–1418, 1992. Demonstration that peptidyl transferase is an RNA.

Normanly, J., and J. Abelson, tRNA identity. *Ann. Rev. Biochem.* 58:1029, 1989. A summary of the chemical features of tRNA.

Pain, V. M., Initiation of protein synthesis in mammalian cells. *Biochem. J.* 235:625, 1986. A comprehensive review of the complex process of translational initiation in mammalian systems, emphasizing the mechanism of the process and its regulation.

Parker, J., Errors and alternatives in reading the universal genetic code. *Microbiol. Rev.* 53:273, 1989. A summary of the current knowledge of alternative mechanisms of translation.

Pelham, H. R. B., Control of protein exit from the endoplasmic reticulum. *Ann. Rev. Cell. Biol.* 5:1, 1989. Describes the current picture of how proteins are sorted and transported through the endoplasmic reticulum.

Pfeffer, S. R., and J. E. Rothman, Biosynthetic protein transport and sorting by the endoplasmic reticulum and Golgi. *Ann. Rev. Biochem.* 56:829, 1987. An excellent overview of the major features of protein targeting in eukaryotic cells.

Proud, C. G., Guanine nucleotides, protein phosphorylation and control of translation. *Trends Biochem. Sci.* 11:73–77, 1986.

Rechsteiner, M., S. Rogers, and K. Rote, Protein structure and intracellular stability. *Trends Biochem. Sci.* 12:390–394, 1987.

Rich, A., and S. H. Kim, The three-dimensional structure of transfer RNA. *Sci. Am.* 238(1):52–62, 1978.

Rothman, J. E., Polypeptide chain binding proteins: Catalysts of protein folding and related processes in cells. *Cell* 59:591, 1989. A description of the proteins that are thought to be involved promoting the formation of three-dimensional structure in proteins.

Rould, M. A., J. J. Perona, and T. A. Steitz, Structural basis of anticodon loop recognition by glutaminyl-tRNA synthetase. *Nature* 352: 213–218, 1991.

Schimmel, P., Parameters for the molecular recognition of transfer RNAs. *Biochemistry* 28:2747, 1989. A summary of the relationship between the structure of tRNAs and their recognition by synthases.

Schimmel, P. R., Aminoacyl tRNA synthetases: general scheme of structure-function relationships in the polypeptides and recognition of transfer RNAs. *Ann. Rev. Biochem.* 56:125–158, 1987.

Siegel, V., and P. Walter, Each of the activities of signal recognition particle (SRP) is contained within a distinct domain. *Cell* 52:39–49, 1988.

Thompson, R. C., EFTu provides an internal kinetic standard for translational accuracy. *Trends Biochem. Sci.* 13:91–93, 1988.

Tobias, J. W., T. E. Shrader, G. Rocap, and A. Varshavsky, The N-end rule in bacteria. *Science* 154:1374–1377, 1991.

Von Figura, K., and A. Hasilik, Lysosomal enzymes and their receptors. *Ann. Rev. Biochem.* 55:167–193, 1986.

Webb, R., and L. A. Sherman, Chaperones classified. *Nature* 359: 458–486, 1992.

Problems

1. Compare the translation initiation signals in prokaryotic and eukaryotic systems, and describe those features of each type of mRNA that determine the frequency with which a particular message is translated. What consequences do these differences have for gene organization in the two systems?
2. Transfer RNA molecules are rather large, considering the fact that the anticodon is a trinucleotide. Why is this the case?
3. Draw the chemical reaction mechanism for the formation of an aminoacyl-tRNA.
4. The relationship between tRNAs and their synthases is sometimes called the "second genetic code." Explain.
5. Explain this statement: "The universal genetic code is not quite universal."
6. How much energy is required to synthesize a single peptide bond in protein synthesis? How does this compare with the free energy of formation of the peptide linkage, which is about 5 kcal/mole?
7. Bromouracil is a base replacement mutagen, while acridine causes single-base insertions or deletions. Mutations caused by acridine frequently can be compensated for by secondary mutations several nucleotides distant from the first, while compensation for those caused by bromouracil require changes within the same codon. Explain.
8. Assume that you have a copolymer with a random sequence containing equimolar amounts of A and U. What amino acids would be incorporated and in what ratio, when this copolymer is used as an mRNA?

9. Assuming that translation begins at the first codon, deduce the amino acid sequence of the polypeptide encoded by the following mRNA template:

AUGGUCGAAAUUCGGGACACCCAUUUGAAGAAACAGAUAGCUUUCUAGUAA

10. The effect of single-point mutations on the amino acid sequence of a protein can provide precise identification of the codon used to specify a particular residue. Assuming a single base change for each step, deduce the wild-type codon in each of the following cases.
    (a) Gln ⟶ Arg ⟶ Trp
    (b) Glu ⟶ Lys ⟶ Ile

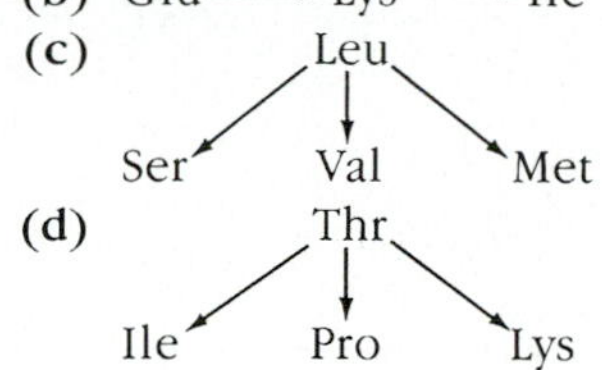

11. Give examples of several proteins that are not functional without posttranslational modification or processing.
12. Why are PEST sequences more likely to be found in regulatory proteins than in structural proteins?
13. Scientists have tried to isolate the peptidyl transferase from ribosomes for many years without success. It is now thought that this activity is part of the ribosome (large subunit). Discuss this point, in view of what you know about other catalytic RNP complexes (RNA-protein complexes).
14. Explain why toxins such as ricin and α-sarcin are such potent toxins, i.e., act at such low concentrations (one molecule per cell). What is unique about this group of toxic enzymes?
15. The signal recognition particle (SRP) has sometimes been called the "third ribosomal subunit." Do you think this is a valid comment? Explain.
16. Explain why normal hemoglobin is very stable, while the same protein containing the valine analog aminochlorobutyrate is very unstable.

Regulation of Gene Expression in Prokaryotes

In chapter 12 we made the point that a unique aspect of biochemistry is the functional relationship between different reactions. Similarly, in this chapter and the next, we will see that in the regulation of gene expression, the most characteristic aspect of the processes involved is that RNAs and proteins are made and broken down in a way that best suits the needs of the organism. Thus, for example, the special characteristics of prokaryotes dictate the forms of regulation that are most used. Posttranscriptional regulation of messages is rare because the processes of transcription and translation take place in the same cell compartment. Moreover, short messenger lifetimes usually eliminate the need for elaborate mechanisms of translational control.

An overview of the points where gene expression are regulated in prokaryotes is presented in figure 30.1. In this chapter our emphasis will be on major modes of transcriptional regulation in *E. coli* and the bacteriophage λ. We will deal first, and at greater length, with the regulation of gene expression for bacterial genes in *E. coli,* and then with the regulation of gene expression in bacterial viruses, especially λ.

Regulation of Gene Expression in *E. coli*

E. coli contains a single chromosome with about 4.5×10^6 base pairs; this is sufficient to encode about 3,000 genes. Under conditions of active growth, only about 5% of the genome is actively transcribed at any given time; the remainder is either silent or transcribed at a very low rate. When growth conditions change, some active genes are turned off and some inactive genes are turned on. The cell always retains its totipotency (its potential to express any of its genes), so that within a short time—seconds to minutes in most cases—and given appropriate circumstances, any gene can be turned fully on or off.

The level of transcription for any particular gene usually results from a complex series of control elements organized into a hierarchy that coordinates the metabolic activities of the cell. For example, when the rRNA genes are highly active, so are the genes for ribosomal proteins, and the latter are regulated in such a way that stoichiometric amounts of most of the ribosomal proteins are produced. When glucose is abundant, most genes involved in utilizing more complex carbon sources are turned off in a process called catabolite repression. If the glucose supply is depleted and lactose is present, then the genes involved in lactose catabolism are expressed.

Figure 30.1

Gene expression in prokaryotes is controlled at several levels. At the DNA level the number of genes can be controlled by selective DNA amplification or rearrangement. Gross gene amplification is particularly important in the case of viruses. Rearrangement is a rare mechanism of control, which is observed in cases where there is a need to switch between closely related alternative genes for the same function. Regulation of transcription is the most important mode of regulating gene expression in both bacteria and bacteriophages. This form of regulation usually involves the initiation site for transcription, RNA polymerase and regulatory proteins. Translation control mechanisms usually operate at the initiation point for translation. Degradation mechanisms for controlling the amounts of particular RNAs and proteins are also an important type of control of gene expression.

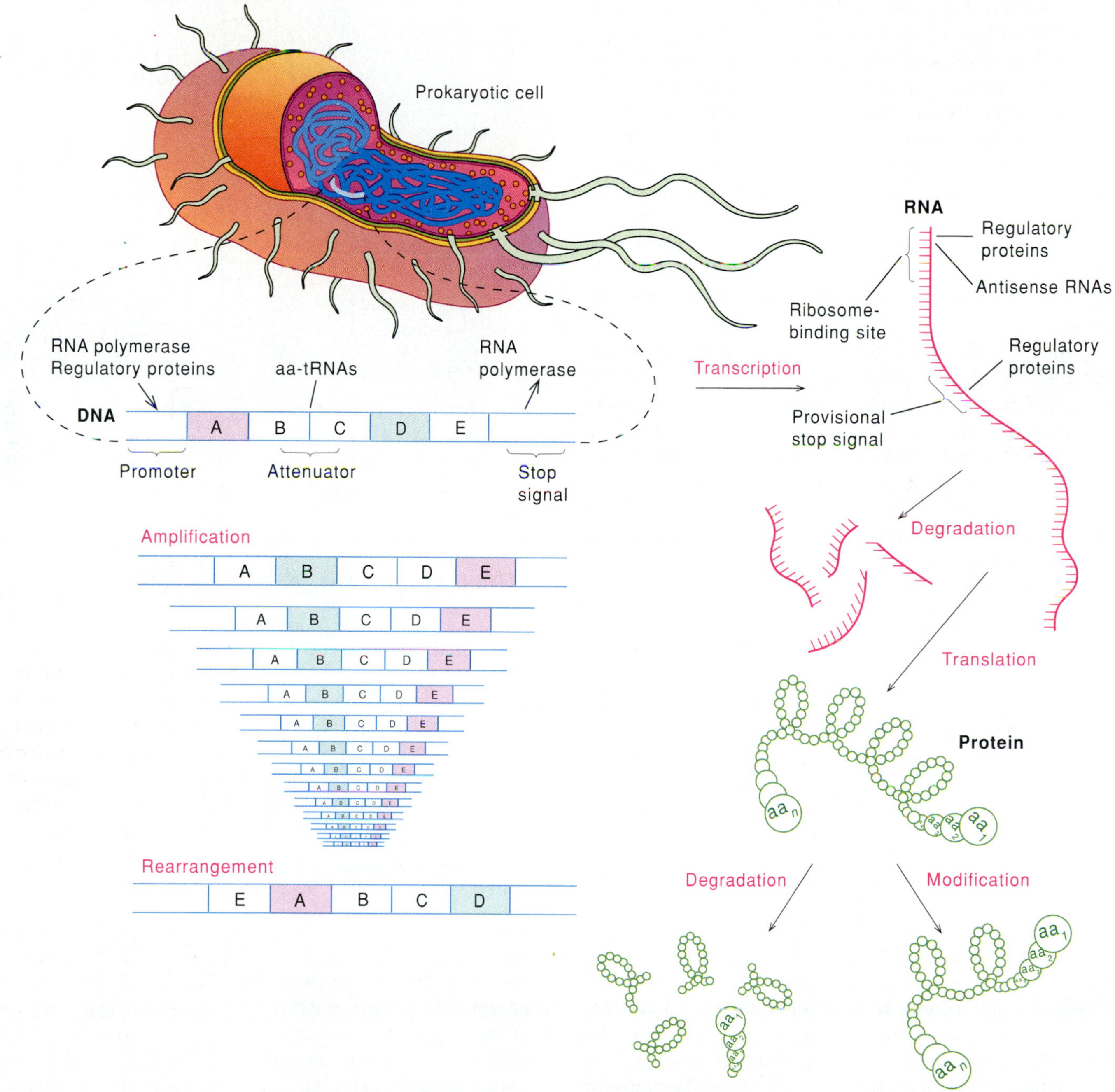

In *E. coli,* the producton of most RNAs and protein is regulated at the transcription level. Rapid response to changing conditions is ensured partly by a short mRNA lifetime—on the order of 1 to 3 min for most mRNAs. Some mRNAs have appreciably longer lifetimes (10 min or longer) and thus the potential for much higher levels of protein synthesis per mRNA molecule. These atypical mRNAs, at least in some instances, also may be subject to translational control.

The Initiation Point for Transcription Is a Major Site for Regulating Gene Expression

The most common way of regulating transcription in bacteria is by controlling the rate of initiation. We described the basic mechanics of transcription in chapter 28. Recall that prior to initiation of transcription, the RNA polymerase holoenzyme attaches to a 35- to 40-nucleotide segment of the DNA called the promoter. The affinity between DNA and RNA polymerase is controlled by a sequence of bases in the DNA. Two main areas of contact are recognized, one centered in the −35 region of the promoter, with a favored sequence of TTGACA, and one centered in the −10 region, with the favored sequence TATAAT. These two hexanucleotide regions are referred to as polymerase binding site 1 (PBS1) and polymerase binding site 2 (PBS2), respectively.

The favored sequences, or consensus sequences, as they are commonly called, were identified by comparing the sequences found in about 400 different promoters. Each base in the consensus sequence represents the base most commonly found in that position. Naturally occurring promoters have never been found with the consensus sequences at both sites. However, artificially constructed promoters bearing the consensus sequence are highly active in transcription. Furthermore, the activity of naturally occurring promoters is frequently proportional to the concordance between their promoter sequences and the consensus sequence. Future research may show that other regions are also important in promoter-RNA polymerase interactions.

From the time it first makes contact with the promoter to the time that it achieves the proper orientation for initiation, the promoter-polymerase complex may go through several metastable states. The final conformation of this complex, adopted immediately before initiation, is referred to as the rapid-start complex or open promoter complex. In this state, the polymerase is in contact with PBS1 and PBS2, and about eleven base pairs—from the −9 to the +1 positions at the origin of transcription—are unpaired (fig. 30.2). Once the rapid-start complex has been formed, initiation of RNA synthesis is rapid, taking only a fraction of a second in the presence of ribonucleotide triphosphates.

The rate of initiation of transcription can be regulated in several ways, all of which influence the rate of formation of the rapid-start complex. The primary sequence of nucleotides in the promoter region is the first factor that has an influence. The closer this sequence is to the consensus sequence, the greater is the affinity of the polymerase for the promoter. Second, for some promoters, negative supercoiling of the DNA serves as an appreciable stimulus to transcription, probably because negative supercoiling facilitates unwinding and unpairing of the double helix (see chapter 25), such as is observed in the −9 to +1 region of the promoter in the rapid-start complex.*

The rate of initiation of transcription also can be altered by changes in the RNA polymerase structure (see chapter 28). These changes include subunit replacement, subunit covalent modification, and small-molecule-induced allosteric transition. When the temperature rises from 30 to 42° C, the usual σ subunit (σ^{70}) is partially replaced by an alternative σ factor (σ^{32}), a shift that changes the types of promoters recognized by the polymerase. In bacteriophage T4 infection, the

Figure 30.2

Schematic diagram of DNA conformation in the rapid-start complex. Two regions, PBS1 and PBS2, most important in polymerase binding, are lettered with most favored sequences. Transcription starts at the +1 base pair.

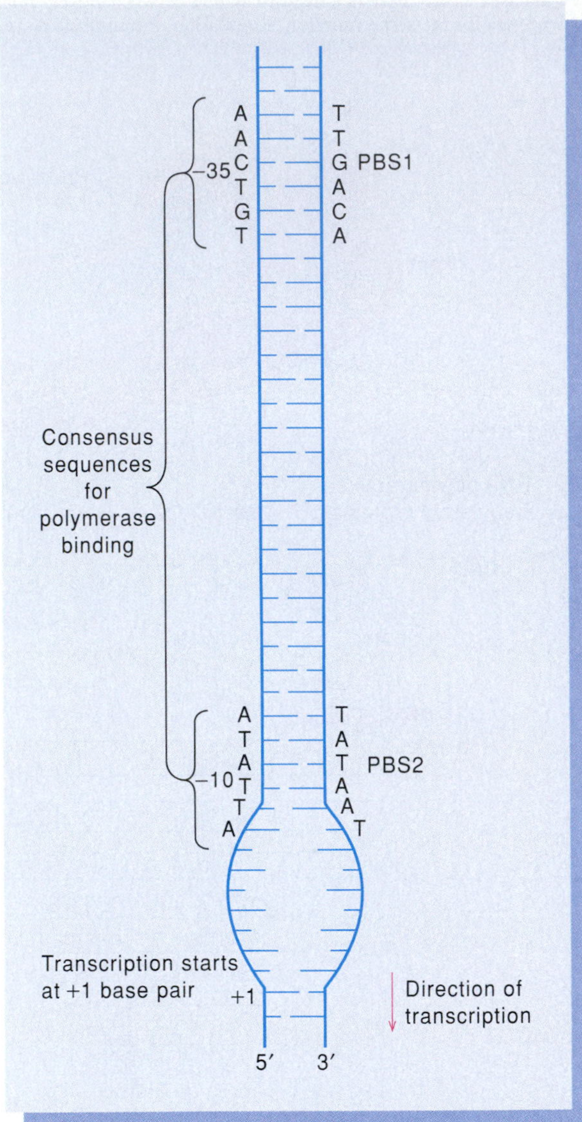

subunits of the polymerase become ribose-adenylated, lowering the affinity of polymerase for bacterial promoters and raising the affinity for phage promoters. Binding of guanosine tetraphosphate (ppGpp) to RNA polymerase changes the structure of the polymerase so that it has a greatly lowered affinity for rRNA, tRNA, and ribosomal protein promoters and at the same time a somewhat greater affinity for some other promoters.

Finally, the rate of initiation of RNA synthesis can be controlled by auxiliary regulatory proteins that affect the rate of formation of the rapid-start complex in either a positive or a negative way; such regulatory proteins are known as activators or repressors, respectively. Each promoter has a natural affinity for RNA polymerase that is determined by its promoter sequence. Activators increase this affinity, thereby increasing the

*Negative supercoiling does not always stimulate transcription. Some genes are unaffected by the extent of supercoiling. The gene for DNA gyrase is actually much more active when the DNA is relaxed, with no supercoiling. Since DNA gyrase is the enzyme that catalyzes negative supercoiling, this finding indicates that relaxed DNA serves as a signal for the selective synthesis of DNA gyrase.

rate of transcription. Repressors have the opposite effect. For example, the lac repressor inhibits productive polymerase binding to the *lac* operon promoter. The CAP protein is an activator that stimulates polymerase binding to the same promoter. Strictly speaking, CAP should be referred to as an apoactivator rather than an activator, because it promotes transcription only when it is complexed to the small molecule coactivator 3′5′-cyclic AMP (cAMP). Let us now look at the ways in which CAP and the lac repressor function.

Regulation of the Three-Gene Cluster Containing the Gene for β-Galactosidase Occurs at the Transcription Level

The stretch of chromosome in *E. coli* that is known as the *lac* operon contains three genes associated with the metabolism of the dissacharide lactose. The story of research into the *lac* operon is one that we will present here in some detail, because it can help give you a sense of the close interplay between genetics and biochemistry and of how important various techniques have been to advances in our understanding of gene expression. Improvement in technical skills and advances in our understanding of the *lac* operon have gone hand in hand. Progress has depended not only on the development of various genetic techniques, but also on elucidation of the basic mechanisms of DNA, RNA, and protein synthesis. More recently, it has depended on cell-free synthesis techniques, use of restriction enzymes for isolation and cloning of small discrete segments of DNA, methods for determining nucleotide sequences, and x-ray crystallography.

The *lac* operon has a molecular weight of about 4×10^6 and thus constitutes about 0.1% of the *E. coli* chromosome. Its DNA is separated into two functional portions: the controlling elements and the structural genes, which code for the three proteins specified by the *lac* operon—β-galactosidase, lactose permease, and thiogalactoside transacetylase (fig. 30.3).

β-Galactosidase (*z* gene) hydrolyzes β-galactosides, in particular lactose, to produce the monosaccharides glucose and galactose. The permease protein (*y* gene) is associated with the β-galactoside active transport system. High concentrations of permease lead to the concentration of galactosides by as much as 100-fold over their concentration in the external medium. The *a* gene codes for thiogalactoside transacetylase. Unlike defects in the other structural genes of the operon, defects in the *a* gene do not affect the ability of the bacteria to grow on lactose. Transacetylase is known to catalyze the transfer of an acetyl group from acetyl-CoA to a thiogalactoside to form an acetylthiogalactoside. Surprisingly, the physiological role of this enzyme is not clear despite all the work that has been done on the *lac* operon.

The elements controlling the operon consist of a promoter locus *p*, an operator locus *o*, and a repressor gene *i*. The operator is the site on the chromosome where the lac repressor binds. The promoter contains a site for RNA polymerase binding and an adjacent site for binding the apoactivator protein CAP. The repressor gene is located near the operon and encodes the lac repressor protein.

Figure 30.3

Different genetic elements of the *lac* operon. The operon contains a control region, the promoter-operator region, and three structural genes, *z*, *y*, and *a*. The *i* gene, a repressor, is also shown. It is not part of the operon, but it is located at an adjacent site on the genome.

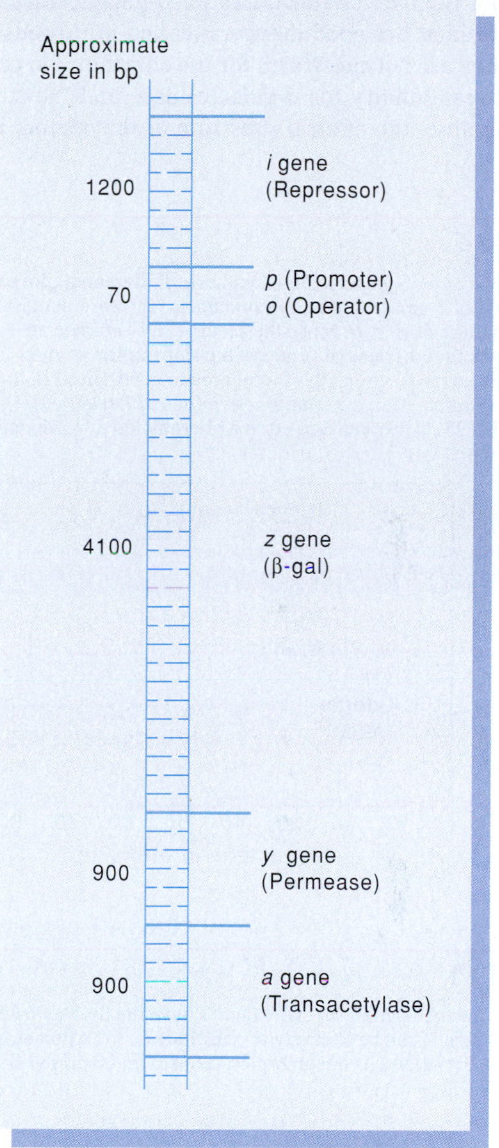

β-Galactosidase Synthesis Is Augmented by a Small Molecule Inducer Wild-type *E. coli* cells grown in the absence of a galactoside contain an average of 0.5 to 5.0 molecules of β-galactosidase per cell, whereas bacteria grown in the presence of an excess of lactose or certain lactose analogs contain 1,000 to 10,000 molecules per cell. Radioactive amino acid has been used as a tracer to show that the increase in enzyme activity observed on induction results from *de novo* protein synthesis. When excess β-galactoside inducer is added, enzyme activity increases at a rate proportional to the increase in total protein within the culture (fig. 30.4). Enzyme formation reaches its maximum rate within 3 min after inducer is added. Removal of inducer leads to cessation of enzyme synthesis in about the same amount of time.

Many compounds have been tested for their capacity to induce β-galactosidase. All inducers contain an intact, unsubstituted galactosidic residue (fig. 30.5). Many compounds that are not themselves substrates for β-galactosidase, such as thiogalactosides, are good inducers (called gratuitous inducers because they are not substrates for the enzyme). No correlation exists between affinity for β-galactosidase and the capacity to induce. Lactose, the natural substrate of the operon, is not the inducer *in vivo*. Rather, allolactose, which is formed as an intermediate in lactose metabolism in the presence of the very limited amount of β-galactosidase that occurs in the uninducer cells, is believed to be the natural inducer (fig. 30.6). The three proteins of the *lac* operon are coordinately induced, i.e., they are induced to the same extent by the same inducer. These results suggest that the receptor molecule for inducer is distinct from the structural components of the operon and that the inducer acts at one site.

A Gene Was Discovered That Leads to Repression of Synthesis in the Absence of Inducer Two distinct types of mutations have been observed in the lactose system. One class of mutations includes structural gene mutations: (1) β-galactosidase mutations ($z^+ \rightarrow z^-$), expressed as the loss of the capacity to synthesize active β-galactosidase, (2) permease mutations ($y^+ \rightarrow y^-$), expressed as the loss of the capacity to concentrate lactose, and (3) transacetylase mutations ($a^+ \rightarrow a^-$), expressed as the loss of the capacity to form thiogalactoside transacetylase. The other class of mutations involves controlling elements of the operon, such as *i* gene mutations ($i^+ \rightarrow i^-$), expressed as the capacity to synthesize large amounts of β-galactosidase even in the absence of inducer. This type of i^- mutation is called a constitutive mutation. Structural gene mutations usually affect only the enzyme in whose gene the alteration occurs.* In contrast, constitutive mutations in the *i* gene invariably affect the amounts of β-galactosidase, permease, and transacetylase synthesized, but not their structures (at this point you may wish to refer to the review of genetic mutation and concepts in box 30A).

*This is strictly true only when synthetic inducer is used. As we explained earlier, a small amount of β-galactosidase is required to convert lactose to allolactose, the natural inducer.

Figure 30.4

Effect of inducer on ß-galactosidase synthesis. Differential plot expressing accumulation of ß-galactosidase as a function of increase in mass of cells in a growing culture of *E. coli*. Since the abscissa and ordinate are expressed in the same units (micrograms of protein), the slope of the straight line gives galactosidase as the fraction (*P*) of total protein synthesized in the presence of inducer. (Source: Melvin Cohn, *Bacteriological Reviews* 21:140, 1957. Copyright © 1957 American Society for Microbiology, Washington, D.C.)

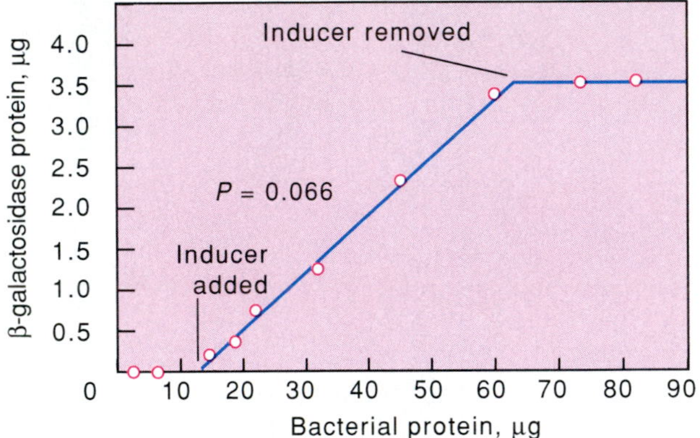

Figure 30.5

Inducers of the *lac* operon. (*a*) All inducers have the ß-galactoside structure shown, in which R can be a variety of substituents. (*b*) Allolactose is the natural inducer when cells are grown on lactose. (*c*) Isopropyl-ß-D-thiogalactoside (IPTG) is a synthetic inducer useful in the laboratory; the ß oxygen is replaced by a ß sulfur atom. This change prevents hydrolysis by ß-galactosidase. For this reason IPTG is called a gratuitous inducer.

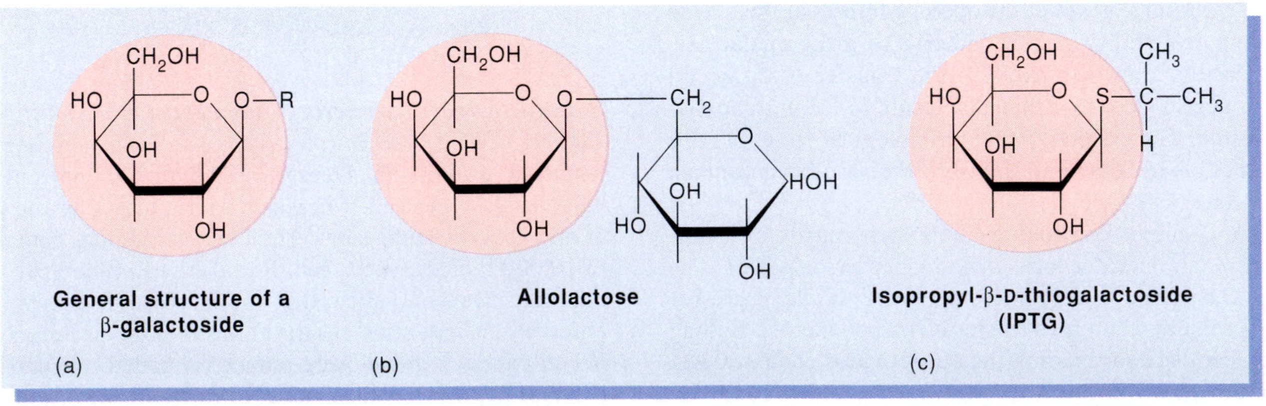

General structure of a β-galactoside (a)

Allolactose (b)

Isopropyl-β-D-thiogalactoside (IPTG) (c)

Some of the most informative early studies were performed with partial diploids (merodiploids) that contain the relevant genes on both the cellular chromosome and an F-factor plasmid. Merodiploids of the type $z^+y^-a^-/Fz^-y^+a^+$ or $z^-y^+a^+/Fz^+y^-a^-$ are wild type, that is, they behave normally. They metabolize lactose and form normal amounts of both β-galactosidase and transacetylase. Furthermore, it does not matter which *lac* operon is on the *E. coli* chromosome and which is on the F-factor chromosome. This efficient complementation between structural gene mutations indicates that they belong to independent genes or cistrons. The most significant feature of i^- mutations is that they simultaneously affect all three gene-product proteins, each independently determined by its different structural genes, *z, y,* or *a*.

The study of merodiploids of the types i^+z^-/Fi^-z^+ or i^-a^+/Fi^+a^- demonstrated that the i^+ inducible allele is dominant to the i^- constitutive allele, and that it is active on the same chromosome (*cis*), or on a different chromosome (*trans*) with respect to both *a* and *z* (table 30.1). The fact that it is effective in the *trans* position as well as the *cis* position shows that *i* gene mutations belong to an independent gene, governing the expression of the *z, y,* and *a* genes through production of a diffusible cytoplasmic component. The dominance of the inducible to the constitutive allele suggests that the former corresponds to the active form of the *i* gene.

Further understanding of *i* gene function has come from study of a mutant gene designated i^s. This mutant has lost its capacity to synthesize all structural gene products of the *lac* operon. In merodiploids of the constitution i^s/i^+, the i^s is dominant; that is, the merodiploids cannot synthesize either β-galactosidase or transacetylase even when inducer is present (see table 30.1). The most reasonable explanation for the i^s mutant is that it is an allele of *i* in which the structure of the repressor is changed so that it can no longer be antagonized by the inducer.

A Locus Adjacent to the Operon Is Found to be Required for Repressor Action In the *lac* system, rare dominant constitutive mutants (o^c) were isolated by selecting for constitutivity in cells diploid for the *lac* region, including the *i* gene, thus virtually eliminating the much more frequently occurring recessive (i^-) constitutive mutants. (Since two copies of *i* gene are present in such cells, both *i* genes would have to mutate simultaneously to i^- to give constitutive behavior.) If the probability of an $i^+ \rightarrow i^-$ mutation is 10^{-6}, the probability of two such simultaneous events in the same cell is 10^{-12}. By recombination studies, the o^c mutations were mapped in the *lac* region between the *i* and *z* loci, generating the gene order shown in figure 30.3. Genetic and biochemical evidence has shown that the *o* locus is

Figure 30.6

Conversion of lactose to allolactose, the natural inducer of the *lac* operon. Ultimately lactose is broken down to its constituent monosaccharides, galactose and glucose.

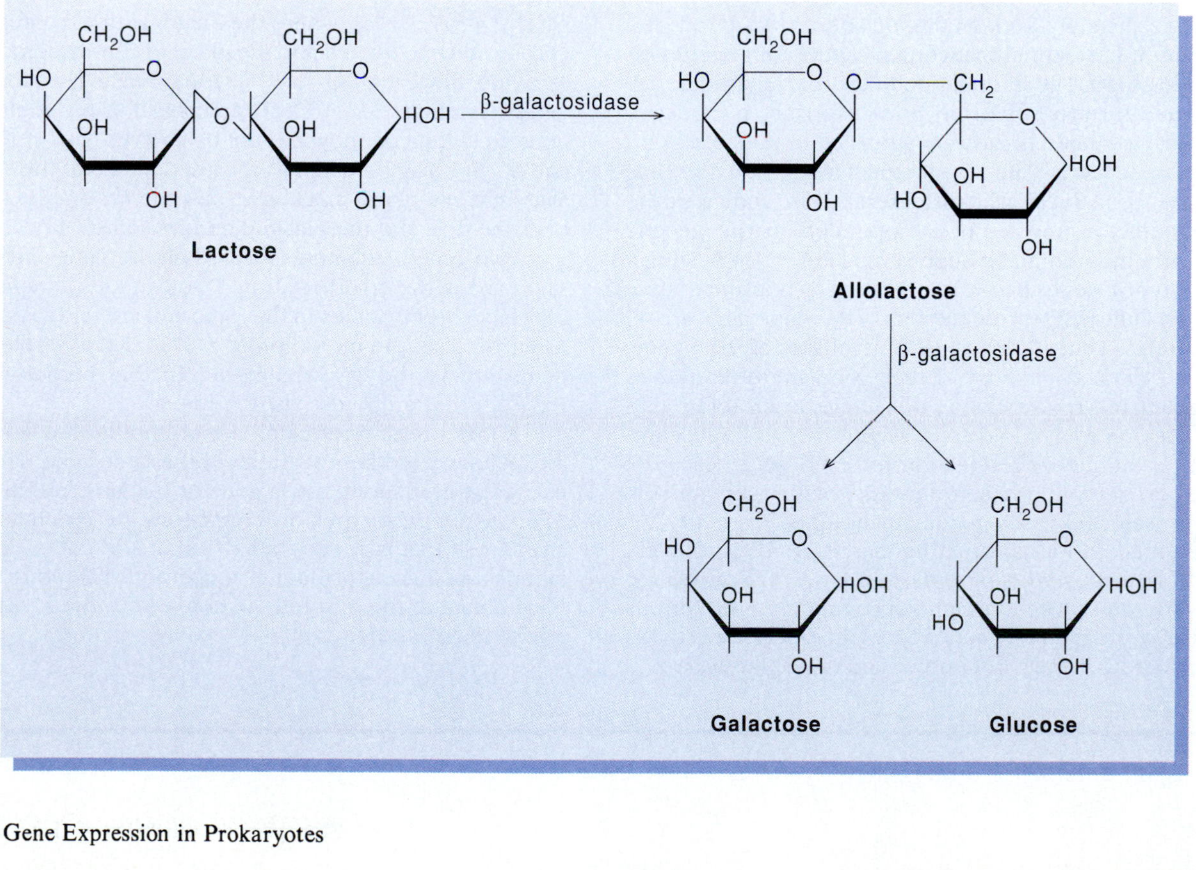

Table 30.1

Synthesis of β-Galactosidase and Galactoside Transacetylase by Haploids and Partial Diploids of *E. coli* Regulator Mutants

Strain No.	Genotype	β-Galactosidase		Galactoside-transacetylase	
		Noninduced	Induced	Noninduced	Induced
1	$i^+z^+y^+$	<0 · 1	100	<1	100
2	$i_6^-z^+y^+$	100	100	90	90
3	$i_3^-z^+y^+$	140	130	130	120
4	$i^+z_1^-y^+/\mathrm{F}i_3^-z^+y^+$	<1	240	1	270
5	$i_3^-z_1^-y^+/\mathrm{F}i^+z^+y_U^-$	<1	280	<1	120
6	$i_3z_1^-y^+/\mathrm{F}i^-z^+y^+$	195	190	200	180
7	$\Delta_{izy}/\mathrm{F}i^-z^+y^+$	130	150	150	170
8	$i^sz^+y^+$	<0 · 1	<1	<1	<1
9	$i^sz^+y^+/\mathrm{F}i^+z^+y^+$	<0 · 1	2	<1	3

Bacteria are grown in glycerol as carbon source and induced, when stated, by isopropyl-β-D-thiogalactoside (IPTG), 10^{-4} M. Values are given as a percentage of those observed with induced wild type. Δ_{izy} refers to a deletion of the whole *lac* region. It should be noted that organisms carrying the wild allele of one of the structural genes (*z* or *y*) on the F factor form more of the corresponding enzyme than the haploid. This is presumably due to the fact that one to two copies of the F*lac* plasmid are present per cell. In i^+/i^- heterozygotes, values observed with uninduced cells are sometimes higher than in the haploid control. This is due to the presence of a significant fraction of i^-/i^- homozygous recombinants in the population. Subscripts in column labeled genotype refer to different mutants. (Data obtained from J. Monod.)

Genetic Concepts and Genetic Notation

Much of the early work on the *lac* operon was purely genetic. It is essential that certain aspects of genetics be understood. The information in this box should be adequate for those with no prior exposure to genetics except for what they have already encountered in this text.

Genes are specified by one or more small letters in italic. Thus *z* indicates the gene for β-galactosidase and *lac* indicates the operon. Frequently a superscript is appended to the genetic symbol. The two most common superscripts are +, indicating a normal (wild-type) gene, and −, indicating a nonfunctioning (mutant) gene. Different representations of the same gene are referred to as alleles. Thus z^+ and z^- are both alleles of the *z* gene.

Cells that carry a single copy of each gene are referred to as haploids. Cells that carry two copies of each gene are referred to as diploids. Bacteria are haploid cells, as they carry a single chromosome with a unique representation for each gene. Bacterial cells that are partial diploids (merodiploids) may occur naturally or they may be selected for by genetic techniques.

A favored method for constructing merodiploids is to infect the bacterial cell with a virus or a plasmid DNA that carries the extra genes of interest. The F plasmid is commonly used for this purpose. In strict usage, the genetic representation for a cell carrying the *lac* operon on the chromosome and the F plasmid would be $z^+y^+a^+//\mathrm{F}z^+y^+a^+$, where the diagonal lines represent the host chromosome to the left and the plasmid chromosome to the right. As a rule, however, only one diagonal line is used to separate the genetic symbols. Also, for convenience, if all the alleles for a given gene are wild type they may not be shown. Thus $z^+y^+a^+/\mathrm{F}z^-y^+a^+$ and $z^+/\mathrm{F}z^-$ may be taken as representations of the same genetic state in cases where it is known that the *lac* operon is present on both the host and the plasmid chromosome.

Two genetic elements located on the same chromosome are said to be in the *cis* orientation. Two genetic elements located on different chromosomes in the same cell are said to be in the *trans* orientation. In the merodiploid $z^-y^-a^+/\mathrm{F}z^+y^+a^+$, the two mutant genes are in the *cis* orientation. In the merodiploid $z^-y^+a^+/\mathrm{F}z^+y^-a^+$, they are in the *trans* orientation.

A major reason for using merodiploids is to study the interaction between different alleles of the same gene. This often tells us a great deal about how a gene or the gene product functions. The two simplest types of interactions are dominant and recessive. A cell that is $z^+/\mathrm{F}z^-$ behaves like a z^+ cell as far as the metabolism of β-galactosidase is concerned. Therefore the z^+ allele is dominant to the z^- allele, or conversely, the z^- allele is recessive to the z^+ allele.

adjacent to but distinct from the *z* gene. Thus constitutive o^c mutations affect the quantity of β-galactosidase synthesized but not its structure.

In merodiploids of the type o^c/o^+, β-galactosidase and transacetylase are constitutively synthesized, a result showing that the o^c mutation is dominant to o^+. In merodiploids of the type o^cz^+/o^+z^-, the o^c mutation is dominant; but in o^cz^-/o^+z^+, it is recessive for galactosidase expression. Thus the o^c mutation is dominant only in the *cis* position, i.e., when it is adjacent to a wild-type z^+ gene. From this evidence, François Jacob and Jacques Monod inferred that the $o^+ \rightarrow o^c$ mutations correspond to a modification of the specific repressor-accepting structure of the operator. This modification identifies the operator locus, i.e., the genetic segment responsible for the structure of the operator, but not necessarily the operator itself.

Genetic Studies on the Repressor Gene and the Operator Locus Lead to a Model for Repressor Action The behavior of the various mutations we have just discussed led Jacob and Monod to propose a model for the regulation of protein synthesis. The genetic elements of this model consist of a structural gene or genes, a regulator gene, and an operator locus (fig. 30.7).

1. The structural gene produces a messenger RNA molecule that serves as a template for protein synthesis.
2. The regulator gene (not itself part of the operon) produces a repressor that can interact with the operator locus.
3. The operator is always adjacent to the structural genes it controls.
4. The operator and its associated structural genes are referred to as the operon.

The repressor molecule combines with the operator locus to prevent the structural gene(s) from synthesizing messenger RNA. In induction, the inducer (or antirepressor, a more descriptive term) combines with the repressor to prevent its interaction with the operator; this region of the genome is then free to combine with RNA polymerase. The Jacob-Monod operon hypothesis has provided a tremendous stimulus for investigations directed toward understanding not only the *lac* system, but other genetic regulatory systems as well.

Biochemical Studies Verify the Genetic Predictions Regarding Repressor Action Genetic studies were most important in arriving at a hypothesis to explain how the repressor worked, but biochemical studies were necessary to demonstrate that the hypothesis was correct. To do the biochemical experiments, it was first necessary to isolate repressor.

According to the operon hypothesis, antirepressor is supposed to combine with repressor. In order to isolate repressor, Walter Gilbert and Benno Muller-Hill used ^{14}C-labeled isopropyl-β-D-thiogalactoside (IPTG), one of the strongest known antirepressors (see fig. 30.5), to monitor repressor purification from a crude cell extract. Unlike the natural inducer allolactose, IPTG not only binds strongly to repressor but is completely stable in *E. coli* cells or crude extracts. This characteristic makes it most useful for experimental purposes. A crude cell-free extract was fractionated by standard protein purification procedures, and the fraction containing the repressor that binds IPTG was found. (See box 30B for details of how binding is measured.)

Figure 30.7

Schematic model illustrating the operon hypothesis. This diagram is modified from the original proposed by Jacob and Monod, who thought *i* gene repressor was an RNA rather than a protein. (*a*) The *i* gene encodes a repressor that binds tightly to the operator *o* locus, thereby preventing transcription of the mRNA from the *z*, *y*, and *a* structural genes. (*b*) When inducer is present, it combines with repressor, changing its structure so it can no longer bind to the operator locus. Inducer also can remove repressor already complexed with the *o* locus.

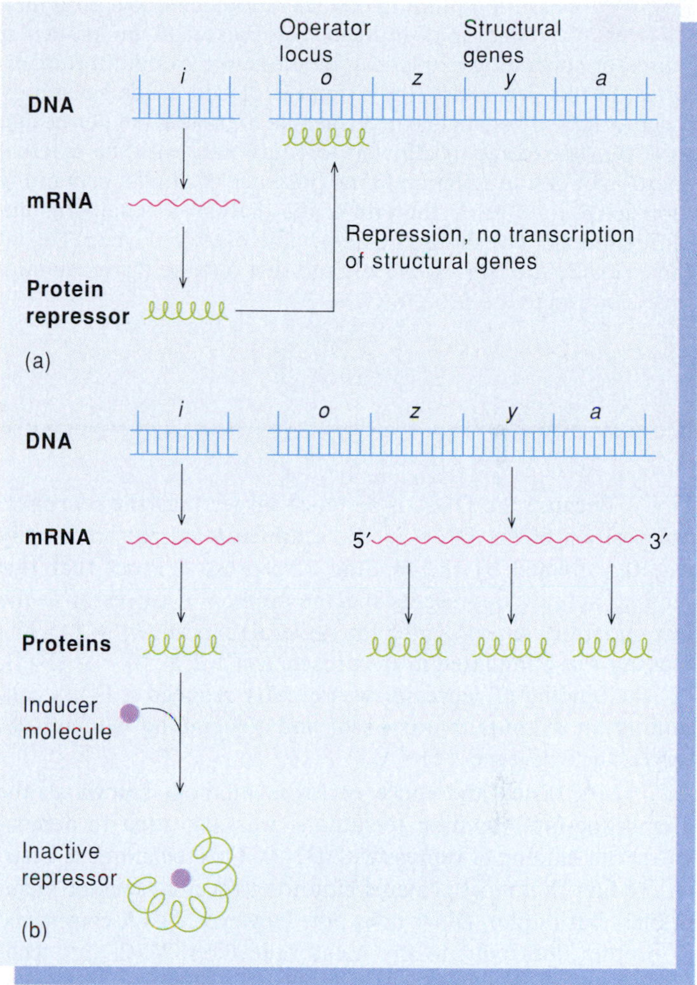

With purified repressor available, the Jacob-Monod hypothesis about binding of repressor to operator could be directly tested. The binding of purified repressor to DNA was studied in two ways. First, ^{35}S-labeled repressor was mixed with λ *lac* DNA (i.e., bacteriophage λ DNA containing the *lac* operon) and centrifuged through a sucrose gradient for a limited time. In the absence of binding, the repressor, detected by its radioactivity, sediments much more slowly than the DNA. In the absence of inducer, some repressor comigrates with the

30B

BOX

Measuring the Binding of Inducer to Repressor

The theory for equilibrium binding is discussed in chapter 10. Here we describe a popular technique for measuring binding of a small molecule to a protein.

Binding of inducer to repressor in any particular repressor-containing extract can be measured by equilibrium dialysis. In this procedure, the extract is placed inside a semipermeable dialysis bag that is impermeable to protein but permeable to IPTG. The extract is allowed to equilibrate with the external solution by gentle agitation in the presence of a buffer containing radioactive inducer. At the end of the dialysis, a certain amount of free inducer (I_f) should be detectable outside the bag. The inducer inside the bag should exceed that outside by the amount that is bound to the protein (I_b).

The strength of binding of repressor to IPTG can also be determined by equilibrium dialysis. When the ratio I_b:I_f is plotted versus I_b, the slope should equal the negative of the formation constant for binding, K_f, where K_f is defined by the equation

$$K_f = \frac{(\text{repressor} - \text{IPTG})}{(\text{repressor})(\text{IPTG})}$$

The formation constant for this complex, using wild-type repressor, is approximately 10^6 M^{-1}, or 10^6 moles^{-1} liters; the K_f for allolactose binding to repressor is about 100 times higher. Variants of wild-type repressor purified to the same degree have also been examined. The mutant repressor i^t, which was isolated from strains that are easier to induce, has about double this affinity. The repressor encoded by i^s, isolated from noninducible strains, shows no affinity for IPTG, as predicted (see main text).

DNA. Because the DNA is so much larger than the repressor, the position of the DNA in the sedimentation pattern is only slightly affected by the binding of repressor. Proof that this binding actually represents specific binding of repressor to the *lac* operator was provided by several experiments: (1) The binding was eliminated in the presence of 1.2×10^{-4}-M IPTG; (2) the binding of repressor was greatly reduced if DNA containing an o^c mutant was used; and (3) binding was not observed for wild-type λ DNA.

A faster and more versatile method, known as the membrane-filter binding technique, was also used to demonstrate the binding of repressor to DNA. This technique is based on the fact that most proteins bind to nitrocellulose filter membranes, but duplex DNA does not. However, DNA complexed to protein does bind to the membrane filter. With this technique, actively labeled DNA is used in conjunction with unlabeled purified repressor. A solution containing the labeled DNA, with or without added repressor, is passed through the membrane filter with the help of mild suction. Only when repressor is present is the label retained by the filter. Other variables were tested, and the same conclusions were reached as when the centrifugation technique was used.

The operator binding site for repressor was further characterized by sequence analysis. This analysis revealed a region of about 36 bases, extending from about −8 to +28, that shows extensive symmetry (fig. 30.8). Twenty-eight of the 36 bases in this region are arranged so that if the double helix is rotated 180° the same sequence is observed. This type of symmetry in DNA is referred to as dyad symmetry. The lac repressor is composed of an even number of identical subunits (four) and most proteins so composed also contain axes of dyad symmetry. Long before the sequences of any operator binding site had been determined or the structure of repressors was known in any detail, Zubay and Chambers had predicted that dyad symmetry would play a major role in the interaction between regulatory protein and DNA. The repressor binding site in this case consists of two symmetrically disposed binding sites on the DNA surface that simultaneously bind to complementary sites on the protein. As we will see, this symmetrical aspect of repressor-operator interaction appears to be common to many DNA–regulatory protein interactions.

Several pieces of evidence add support to the notion that the 36-base region includes the repressor binding site: (1) All eight o^c mutations shown involve base replacements near the center of this region. (2) Repressor binding substantially decreases the reactivity toward dimethylsulfate of several purine bases in this region (see fig. 30.8). Dimethylsulfate reacts with the N-7 position on guanine and the N-3 position on adenine in the double helix (see chapter 27 for a discussion of these chemical reactions). Finally, if all the thymines in the DNA are replaced by bromouracil, a number of the bromouracil bases become cross-linked in the repressor in the presence of ultraviolet light. Taking into account the normal twist of the double helix, all the groups shown by chemical methods to be in the vicinity of the repressor are situated on one side of the double helix. This finding strongly supports the notion that repressor binds on one side of the double helix, over the 36-base region covered by the symmetry axis.

The RNA polymerase and repressor binding sites (promoter and operator) overlap (fig. 30.9), so that the binding of one protein would preclude the binding of the other. Although there are other DNA sequences where the repressor might bind

Figure 30.8

The operator locus (the presumptive repressor binding site). Bases are numbered +1 for the first base transcribed and −1 for the base before that. Regions showing dyad symmetry are underlined and overlined. Arrows indicate point mutations leading to the constitutive phenotype (o^c). Circled bases are those groups that are strongly protected against reaction with dimethylsulfoxide when lac repressor is bound. Shaded circles indicate those groups that become cross-linked to repressor in the presence of ultraviolet light when thymine in the DNA is replaced by 5-bromouracil.

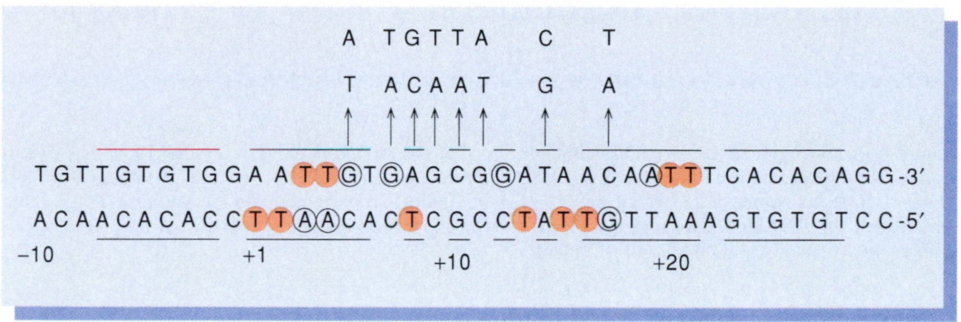

to prevent transcription, most repressors bind so that they overlap the promoter region, thereby preventing the initial formation of the RNA polymerase-DNA complex.

An Activator Protein Is Discovered that Augments Operon Expression In the absence of active repressor or when the repressor is combined with inducer, RNA polymerase can bind to the *lac* operon and transcribe *lac* mRNA. The extent to which it does this depends on other factors in the growth medium, particularly on the presence of sugars other than lactose. For example, when *E. coli* cells are grown in the presence of glucose and lactose as the sole carbon sources, the cells will first utilize glucose and afterwards utilize the lactose. This phenomenon is referred to as diauxic growth. Careful analysis of cells growing under such conditions has shown that during the glucose phase of growth, the *lac* operon is poorly expressed and induction of the operon must occur before the lactose can be utilized. Repression of *lac* operon expression by glucose or its closely related derivatives, such as glucose-6-phosphate or fructose, is known by the general term catabolite repression (or glucose repression). Enzymes of the glycolytic pathway that directly utilize glucose are always present in the cell. The enzymes of the *lac* operon are not needed as long as this more directly utilizable carbon source is available. Many genes in *E. coli* associated with catabolism are subject to this type of repression for the same reason.

A turning point in our understanding of catabolite repression was provided by Sutherland's finding that the level of intracellular cAMP in *E. coli* is drastically lowered in the presence of glucose, from about 10^{-4} to 10^{-7} M. Subsequently I. Pastan and R. Perlman showed that large quantities of cAMP in the growth medium could partially reverse the glucose catabolite repression effect on the *lac* operon. The fact that cAMP has a stimulatory effect on the expression of genes subject to catabolite repression was demonstrated directly in a cell-free system in Zubay's laboratory. The system contained λ *lac* DNA, a cytoplasmic extract of *E. coli*, and the substrates and other low-molecular-weight components necessary for transcription of the DNA and translation of the resulting mRNA. In this system 10^{-3}-M cAMP stimulated β-galactosidase synthesis as much as 30-fold.

Subsequently, Beckwith and his colleagues isolated mutants of *E. coli* that were permanently catabolite-repressed. Such mutants proved to be incapable of expressing a number of genes that required cAMP for activation. The mutants fell into two categories, those that would be phenotypically corrected by growing in the presence of 5×10^{-3}-M cAMP and those that could not. The first class of mutants was defective in the synthesis of cAMP, and the second was defective in the protein(s) with which cAMP interacts to bring about stimulation of β-galactosidase synthesis (as well as that of other catabolite-repressed proteins). Further work showed that these two types of mutants originate in the genes now referred to as *cya* and *crp*, respectively. When the cell-free system was prepared from extracts of crp^- cells, it produced a low level of β-galactosidase that was not stimulated by adding cAMP. The defect could be corrected by adding soluble protein from crp^+ cells. By making use of this *in vitro* complementation assay, Zubay was able to fractionate the crp^+ cells and isolate a single protein that was responsible for the activity.

The protein encoded by the *crp* gene, the apoactivator CAP, is a dimer composed of identical subunits with a molecular weight of 22,000. CAP binds to DNA, and this binding is greatly stimulated in the presence of cAMP. The structural alterations produced in CAP by cAMP binding apparently alter its conformation so that it can form a strong complex with DNA. CAP binds preferentially to the *lac* promoter region, although the selectivity for preferential binding to the specific site is substantially less than is the case for lac repressor.

A series of genetic deletions demonstrate that the site necessary for CAP stimulation of the *lac* operon is in the −50 to −80 base-pair region. Here there exists a 14-pair segment between −53 and −68 that shows dyad symmetry for 12 of the 14 base pairs (fig. 30.10).

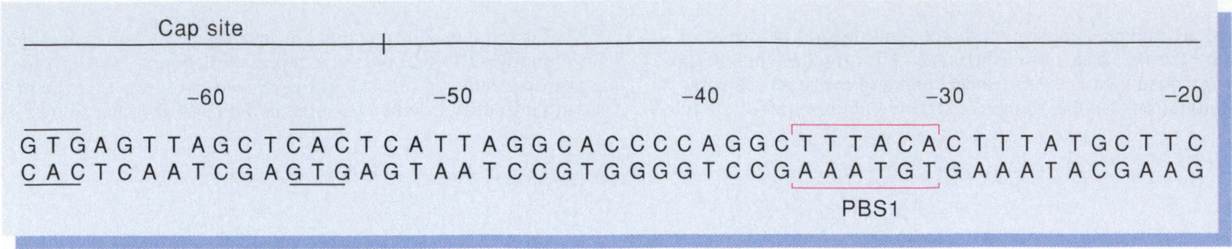

Figure 30.9

DNA sequence in the *lac* promoter-operator region. Bases are numbered +1 for the first base transcribed and −1 for the base before that. CAP and repressor-binding sites are indicated. Regions within these two sites showing dyad symmetry are underlined and overlined. Two regions, PBS1 and PBS2, where polymerase binds most strongly in a number of promoters, are bracketed in color. Note that PBS1 has five of six bases in the idealized sequence (TT-ACA), whereas PBS2 has four out of six (TAT-T). Arrows indicate replacements in the mutant *UV5* promoter that lead to high-level expression in the absence of cAMP-CAP.

As in the case with lac repressor, chemical-probe experiments have been carried out to characterize the region binding to CAP. In these experiments, a small DNA segment containing the CAP-binding site was used. Dimethylsulfate (DMS), which reacts with the adenine N-3 and the guanine N-7 positions, and ethylnitrosourea (EtNu), which reacts with the phosphates, were used. Experiments with the DMS probe showed that CAP binding protects the two outer guanines on each side of the presumptive binding site from methylation, and also that if the DNA is first premethylated, binding of CAP is strongly inhibited. The pattern of protection demonstrates symmetry and interaction in two adjacent, large grooves of the DNA. Experiments with EtNu confirmed the importance of the symmetry and also showed that, like lac repressor, CAP interacts with only one side of the DNA helix, over a region covered by the symmetry axis.

In the absence of bound repressor, the full sequence of reactions in the initiation of transcription is summarized by the following set of equations.

$$\text{cAMP} + \text{CAP} \rightleftharpoons \text{cAMP-CAP}$$

$$\text{cAMP-CAP} + \text{DNA} \rightleftharpoons \text{cAMP-CAP-DNA}$$

$$\text{cAMP-CAP-DNA} + \text{polymerase} \rightleftharpoons \text{cAMP-CAP-DNA-polymerase}$$

First, the coactivator cAMP combines with the apoactivator CAP, which then binds in the −60 region of the promoter. This complex stimulates the binding of RNA polymerase to an adjacent site on the promoter. CAP stimulates binding of RNA polymerase by an affinity between CAP and polymerase. The transcription rate of the *lac* operon in the presence of cAMP-CAP is enhanced 20- to 50-fold, as a direct result of the increased affinity of RNA polymerase holoenzyme for the promoter in the presence of CAP and cAMP.

The overall strategy in creating a promoter sequence responsive to cAMP-CAP activation could be summarized as follows. The nucleotide sequence in the RNA polymerase binding site is adjusted so that polymerase by itself produces a low level of transcription. An adjacent site for binding CAP is created so that when CAP is bound, the additional affinity contributed by favorable contacts between the CAP and the polymerase converts this sequence into a high-level promoter. A mutant (*UV5*) containing two base replacements in the *lac* promoter produces high-level transcription in the absence of cAMP-CAP and no further increase when the latter complex is present. The base replacements in the *UV5* mutant in the PBS2 site (see fig. 30.8) bring the promoter sequence closer to the consensus sequence and therefore should lead to a stronger polymerase binding site in the absence of the cAMP-CAP complex, as is in fact observed.

Figure 30.10

CAP-binding region in the *lac* promoter. Regions showing dyad symmetry that are believed to interact strongly with CAP are overlined and underlined. Circled bases are those that are strongly protected from reaction with dimethylsulfate when CAP is bound. Point mutations L8 and L29, which produce a promoter that is not stimulated in transcription by cAMP, are indicated. Colored dots indicate those phosphate positions which, if ethylated by ethylnitrosourea, block CAP binding. If the area of interest exists as a normal DNA double helix when binding CAP, then most of the groups implicated in CAP binding would appear on one side of the DNA.

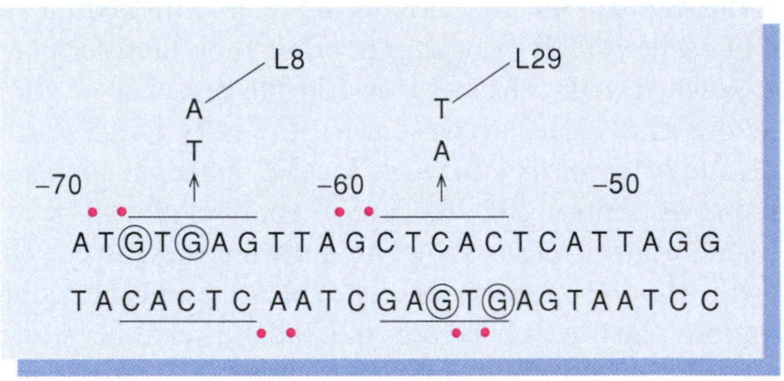

Many Genes Important in Catabolism Are Stimulated by CAP The cAMP-CAP activator is essential for regulating a wide variety of genes involved in catabolism. This complex usually functions in series with a gene-specific regulator so that any particular gene subject to cAMP-CAP activation will be fully expressed only if both control switches are in the "on" position. The galactose (*gal*) operon, which encodes enzymes for galactose catabolism, has a promoter regulated by cAMP-CAP as well as by a specific repressor that is antagonized by galactose.

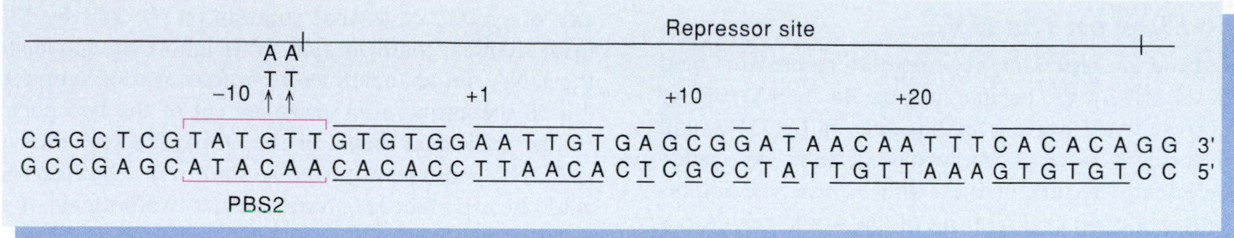

The *gal* operon appears to have a second start signal close by (five bases before the first start signal) that is not subject to cAMP-CAP control; the function of this second promoter is not yet understood. It may be related to the fact that in *E. coli* galactose acts both as a carbon source (catabolite) and as a substrate for cell-wall synthesis (anabolite). The enzymes of the *gal* operon convert galactose to an activated complex, uridine diphosphogalactose (UDPG), that is an intermediate in both the catabolic and the anabolic pathways. Thus the *gal* operon should be responsive to two types of controls that reflect different needs.

The arabinose (*ara*) operon also uses the pleiotropic cAMP-CAP activator and a specific regulator, but here the similarity with *lac* and *gal* ends. The specific regulatory protein that is encoded by the *araC* gene regulates its own synthesis by functioning as a repressor. The sugar L-arabinose serves as both antirepressor and coactivator. When it complexes with the araC protein, the complex dissociates from the repressor binding site and becomes converted into an activator complex that binds at a different site, adjacent to the polymerase binding site for the operon. In this case the binding of cAMP-CAP stimulates the binding of the L-arabinose-araC complex to the adjacent site, which in turn stimulates the nearby binding of RNA polymerase.

The studies on *lac* and *ara* have led to the impression that activators stimulate expression by an energetically favorable interaction between the protein surface of the activator and the protein surface of another activator or the polymerase bound to an adjacent site on the DNA.

It is clear that the repressor proteins discussed thus far interfere with the transcription process in one of two ways: (1) through binding at a site that overlaps the polymerase binding site at the promoter (as in the *lac* operon) or (2) through binding at a site that interferes indirectly with the polymerase binding at the promoter by interfering with CAP binding (as in the *ara* operon). By contrast, the activator complex cAMP-CAP acts by binding at a site adjacent to the polymerase binding site, thereby augmenting polymerase–promoter interaction. In some cases it may be that the most important difference between activators and repressors is not how they bind but where they bind to the DNA.

The Synthesis of the CAP Protein Is Negatively Regulated at the Transcription Level The synthesis of most regulatory genes is itself regulated. We have seen that the synthesis of *araC* is negatively regulated by the araC protein, acting as a repressor for the synthesis of its own messenger. The same is true for the trp repressor that we will discuss a little later. In the case of CAP, synthesis is also negatively regulated, but in this case the mechanism is unusual in that it involves RNA–RNA interaction. A second promoter in the region of the *crp* gene is strongly activated by cAMP-CAP. Transcription from this promoter is initiated twelve nucleotides downstream and on the opposite strand from the initiation site of the *crp* gene. The initial nucleotides of the divergent RNA are complementary to ten of the first eleven nucleotides of the *crp* mRNA (fig. 30.11). The next eleven nucleotides of *crp* mRNA and the 5′ end of *crp* mRNA could produce a structure similar to a ρ-independent terminator (see chapter 28), a change leading to inhibition of *crp* transcription.

Figure 30.11

Model for the inhibition of *crp* transcription by divergent RNA. The proposed RNA-RNA hybrid between the 5′ end of *crp* mRNA and the initial 14 nucleotides of the divergent RNA is shown. The complete structure of the proposed ρ-independent terminator is shown below. The structure consists of the RNA-RNA duplex followed by an A•U-rich RNA•DNA hybrid of 11 bp. Okamoto and Freundlich proposed that the RNA•RNA hybrid causes the polymerase to pause, and as suggested for ρ-independent terminators, the instability of the A•U-rich RNA•DNA hybrid leads to the release of the transcript and termination of transcription. (Source: K. Okamoto and M. Freundlich, "Mechanism for the autogenous control of the *crp* operon: transcriptional inhibition by a divergent RNA transcript," in *Proceedings. National Academy of Sciences USA* 83:5000–5004, 1986. Copyright © 1986 National Academy of Sciences, Washington, D.C.)

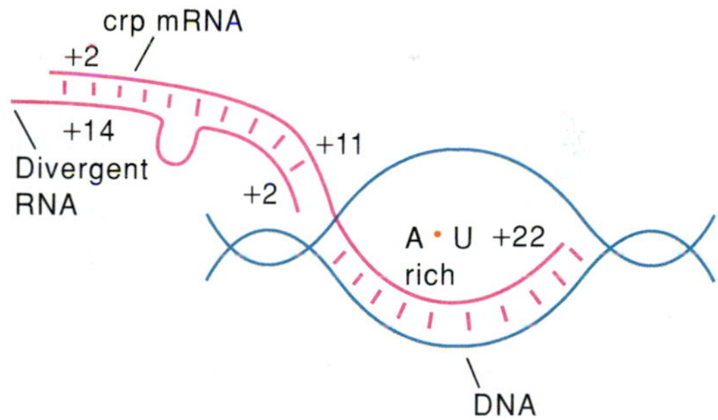

Gene Regulatory Proteins Bind Strongly to Specific Sites on the DNA

Like CAP and the lac repressor, gene regulatory proteins bind strongly and selectively to specific sites on the DNA. Before precise structures for the mutual binding sites on the DNA and the regulatory proteins were known, two predictions were made regarding the general nature of these sites. These predictions were based on certain general features of DNA and regulatory protein structure.

The first feature has to do with symmetry. Many regulatory proteins are composed of an even number of identical subunits. Structures of this composition usually are highly symmetrical; in particular, they usually have at least one twofold axis of symmetry, so that rotation on the axis by 180° results in an identical conformation. If we ignore the sequence of bases, the DNA duplex also has a twofold axis of symmetry that is due to the antiparallel arrangement of the two polynucleotide chains. In a local region of the DNA, the sequence of bases in the DNA could be adjusted so that over a limited region there could be a perfect or almost perfect twofold axis of symmetry. Such a situation would generate two identical sites on the DNA. If one of these sites were used for binding to a regulatory protein, then, given the appropriate dimensions and symmetry, the other site might also be used for binding to the same protein. A hypothetical two-site model of this type is illustrated for the interaction between CAP and DNA in figure 30.12.

Figure 30.12

Schematic diagram of the cAMP-CAP-DNA complex. CAP is a dimer composed of identical subunits. The binding site on the DNA is arranged so that rotation of either the DNA or the CAP dimer by 180° on the dyad axis should produce an identical structure. Results from the laboratory of T. Steitz suggest that CAP binding induces a pronounced bend (90° or more) in the DNA.

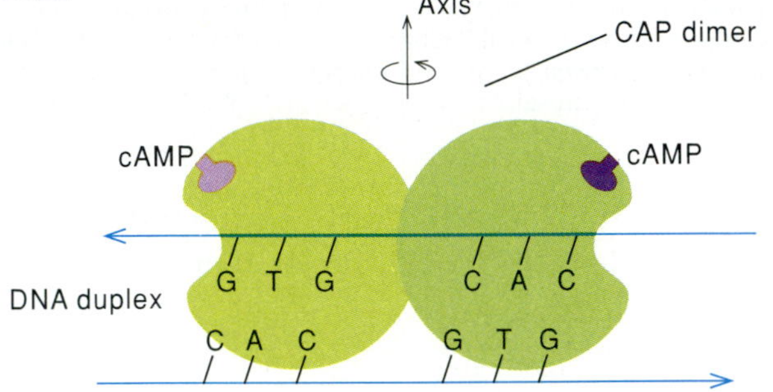

The two-site model received its first direct support from sequence data on the lac repressor and CAP DNA binding sites. Both of these sequences showed a significant amount of the required symmetry (see figs. 30.8 and 30.10). Final confirmation of the two-site model resulted from x-ray crystallographic studies on regulatory proteins and their complexes with DNA. We now know that, in many cases, two nearly identical sites on the DNA do indeed interact with two nearly identical sites on the regulatory protein on a twofold axis of symmetry. The clear advantage of the two-site interaction is that it approximately doubles the strength of the interaction between the DNA and the protein.

Another and possibly even more general aspect of the complex formed between DNA and regulatory protein has to do with the location at which the interaction takes place. The formation of a specific complex must involve interaction be-

Figure 30.13

The structures of three regulatory proteins. They all possess twofold axes of symmetry, and the protruding helical cylinders that interact with adjacent major grooves on the DNA (red) are spaced about 34 Å apart. N and C labels indicate the N and C termini of the polypeptide chains.

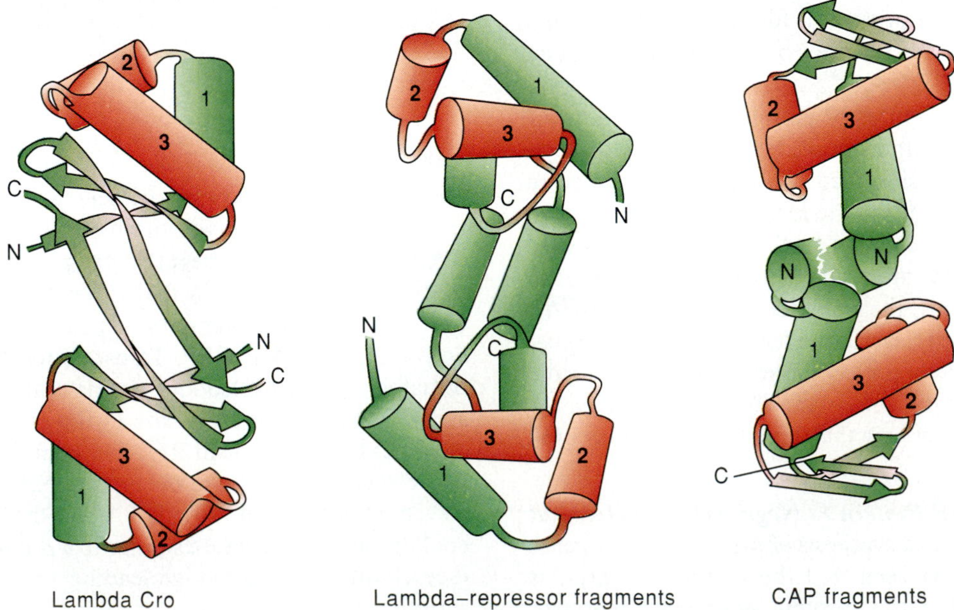

tween specific sites on the protein and specific sites on the DNA. On the DNA it seems likely that the interaction involves sites on base pairs exposed in the major groove. This is because the major groove is much more accessible for interaction with a large molecule than the minor groove. In fact, the major groove is just the right size to accommodate an α helix, whereas the minor groove could not do so. Thus we might expect interaction between DNA and a regulatory protein to involve interaction between a protruding α-helix segment of the regulatory protein and the major groove of the DNA.

Progress beyond these predictions to an understanding of the specific interactions between DNA and protein took more than a quarter of a century of intensive effort by a few highly skilled groups of x-ray crystallographers. We now know the detailed molecular structures of a number of regulatory proteins.

Figure 30.14

Model for location of α_3 and α_2 segments when cro binds to DNA.

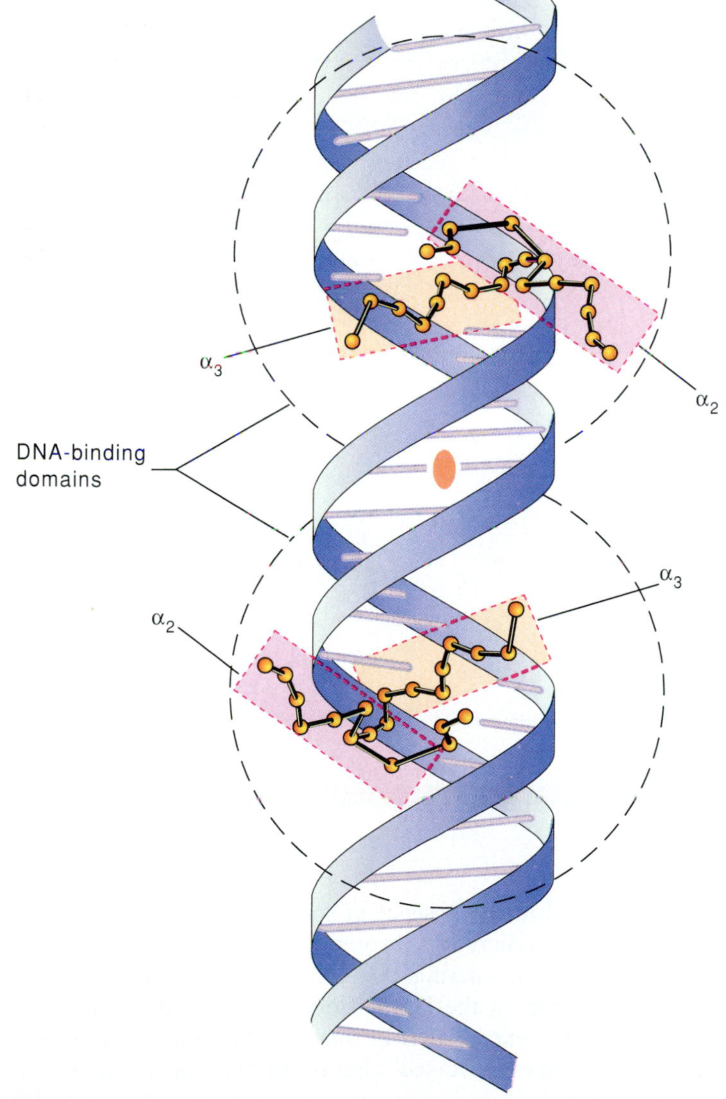

This list includes CAP, the λ repressor, the phage 434 repressor, the 434 cro protein, and the tryptophan repressor. The structures of these regulatory proteins show surprisingly similar features (fig. 30.13). All of the regulatory proteins possess two-fold axes of symmetry and two protruding cylinders of α helix spaced 34 Å apart, which is just the distance necessary for them to interact with adjacent major grooves of the DNA. All regulatory proteins of this type also possess at least one additional segment of α helix; the overall structure is referred to as the helix-turn-helix motif because of the arrangement between the helix segment that interacts with the DNA (the recognition helix) and the additional segment of helix. Although the recognition helices in the protein dimer are spaced at the ideal distance for interaction with adjacent major grooves, their orientation is different for each protein. Only in the case of cro would the recognition helix run parallel to the major groove (fig. 30.14). Despite this lack of perfect orientation of the helix segments in the other regulatory proteins, the amino acid side chains are sufficiently long that interaction between the bases in the major groove and the amino acid side chains should be possible.

Although the helix-turn-helix motif is the most common one thus far found for specific DNA-binding proteins in prokaryotes, it is not the only one. Another radically different motif that has been observed for the met and arc repressors involves a dimeric molecule in which two antiparallel polypeptide chains in the extended β configuration are tucked into the large groove where they interact with the DNA. We will see (see chapter 31) that additional DNA-binding motifs have also been discovered in eukaryotes, which are quite different from both of these.

Final proof of the manner in which the regulatory protein interacts with the DNA was obtained by cocrystal studies on the regulatory protein and the relevant region of the DNA made synthetically. Such studies confirmed the primary interaction with the major groove and permitted a detailed analysis of the specific interactions between the DNA and the protein.

In most of these complexes with protein, the DNA maintains a conformation that is close to the B form. The regulatory proteins interact with the DNA on a two-fold axis of symmetry so that the protein binds on one side of the DNA in two adjacent major grooves. The affinity between the protein and the DNA results partly from nonspecific contacts between amino acid side chains on the protein and the sugar-phosphate backbone of the DNA and partly from specific contacts made between the amino acid side chains and the base pairs of the DNA. In addition to its size, the major groove has one further advantage for interaction. The hydrogen-bonding groups on the base pairs that are exposed in the major groove present a much greater number of possibilities for interaction with complementary hydrogen-bonding groups of the amino acid side chains of the regulatory proteins. (Figure 25.9*c* shows which groups are exposed in the major and minor grooves.) The specific contacts between the amino acid side chains of the λ repressor are illustrated in figure 30.15. In all cases the specific contacts involve hydrogen-bonding groups on the purines in the major grooves

Figure 30.15

The specific interactions between λ repressor and one half of its operator binding site. Identical contacts are made in the other half of the binding site. The numbers associated with the amino acid side chains refer to the distance of the amino acids from the amino-terminal end of the protein. Nucleotides are numbered from the central dyad at the operator, which passes through base pair O. N_6 and N_7 positions on adenines and N_7 and O_6 positions on guanines involved in H bond formation are labeled. Numbering of positions in purine rings is described in figures 1.6, 20.2, and 20.12. (Source: T. Steitz, *Quarterly Reviews of Biophysics* 23:236, 1990. Copyright © 1990 Cambridge University Press, Cambridge, England.)

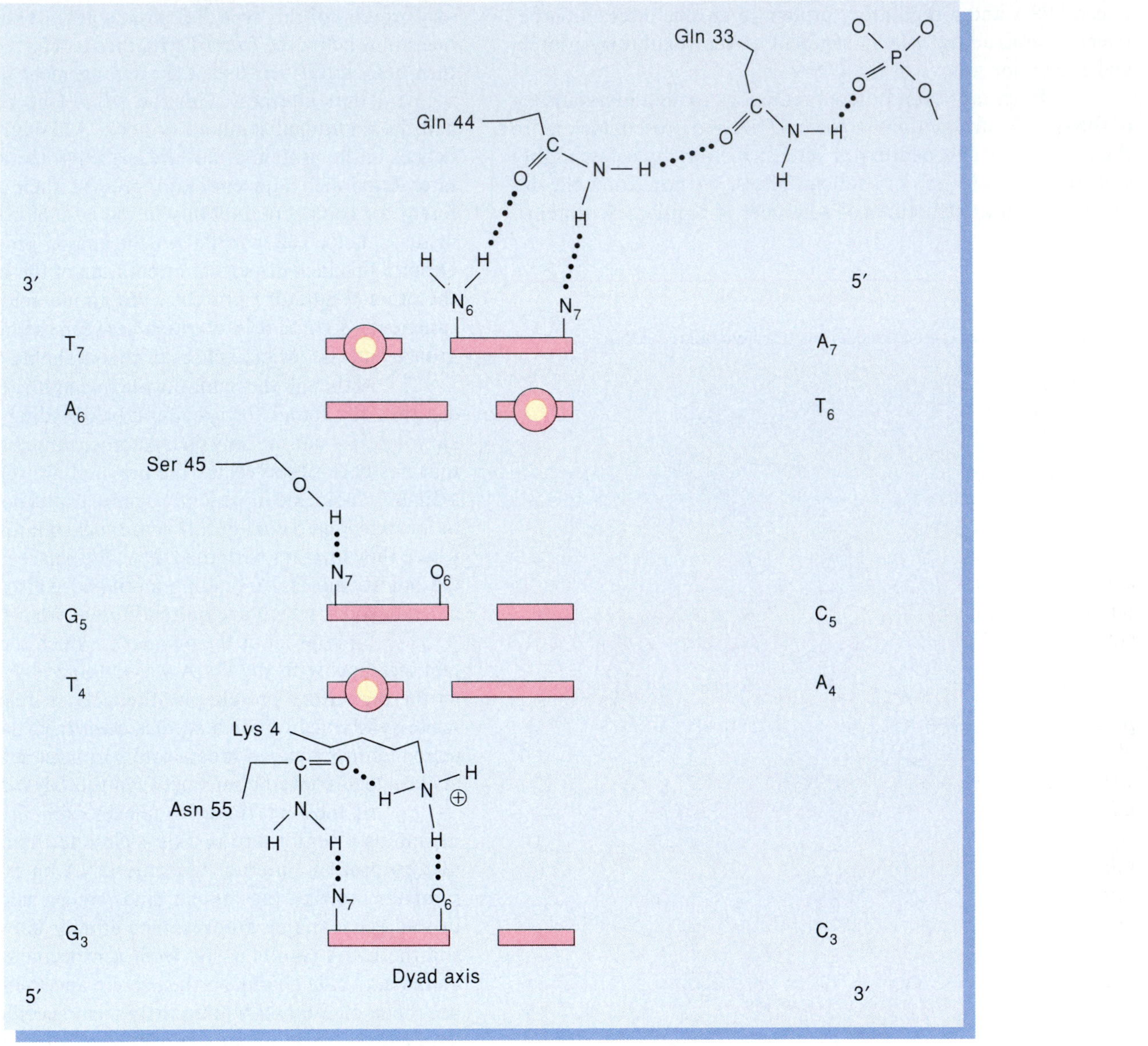

of the DNA with polar amino acid side chains emanating from the interacting α-helix segment. The amino acid side chains make hydrogen bonds directly with the hydrogen-bonding groups on specific purines.

A common feature shared by many but not all regulatory proteins is their sensitivity to small-molecule effectors. For example, the lac repressor adopts two conformations, one that is favored in the absence of small-molecule effectors like allolactose, and one that is favored when binding to allolactose. In the former conformation the repressor has a high affinity for the *lac* operator; in the latter it does not. A similar situation exists for CAP, except that in this case the role of the small-molecule effector is reversed. That is, the binding of cAMP converts the protein from a structure with a low affinity to one with

Figure 30.16

The tryptophan operon, indicating the size of the different genes, the polypeptide chains, the resulting enzyme complexes, and the reactions catalyzed by the enzyme complexes.

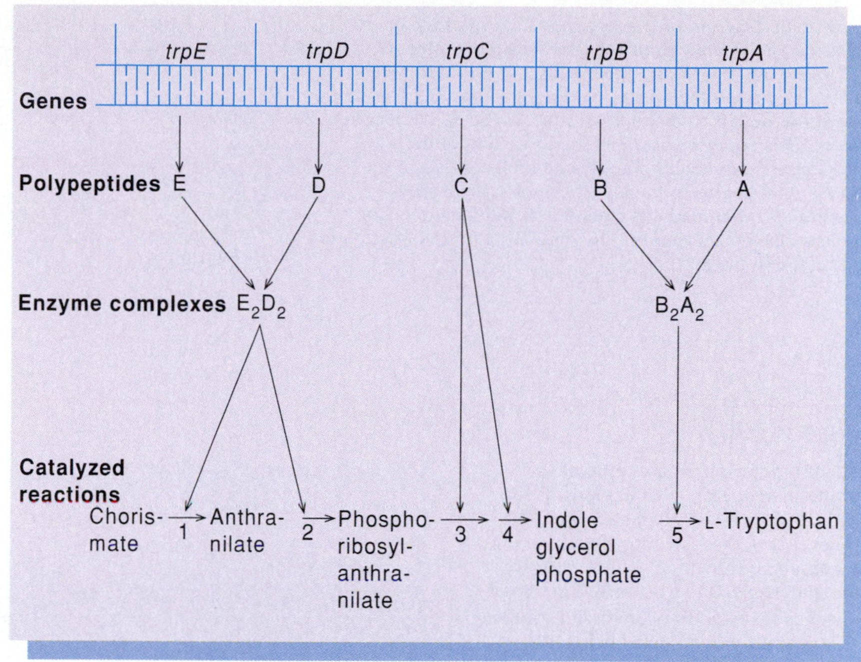

a high affinity for a specific site on the DNA. Although the binding sites for cAMP in CAP are known from structural studies, we still have no idea why the binding of cAMP influences the binding of CAP to the DNA in such a profound way.

Before leaving this section on the binding of regulatory proteins, it is important to point out that sometimes the binding of a regulatory protein can have a major effect on the DNA structure. For example, in the case of the binding of CAP to DNA the alteration in DNA structure is quite large. The first evidence for this came from studies of the electrophoretic mobility of complexes between CAP and segments of DNA containing the CAP-binding site. They migrated in an electric field at an anomalously fast rate as though the DNA was effectively smaller after complex formation than before. It was proposed that this behavior is due to the formation of a bend in the DNA binding to CAP. Seven years later, the crystal structure of this complex has confirmed that a bend is formed of greater than 90 degrees. It seems likely that this major change in the DNA tertiary structure must be important in the action of CAP on gene expression, but it is too early to say precisely what this effect would be.

Enzymes that Catalyze Amino Acid Biosynthesis Are Regulated at the Level of Transcription Initiation

Of all the genes concerned with anabolic processes, the genes uniquely involved in synthesis of the amino acid tryptophan are the best understood. Our understanding is mostly a result of the efforts of Charles Yanofsky and his colleagues, who have used a wide variety of genetic and biochemical techniques to probe the complexities of this system. Expression of the genes involved in tryptophan biosynthesis is regulated at the level of transcription. The way in which transcription is regulated in a biosynthetic operon is quite different from the way in which it is regulated for the catabolic genes. These differences reflect different strategies for assuring that the cell has sufficient amounts of a particular substrate required for protein biosynthesis.

Wild-type *E. coli* can synthesize all twenty types of amino acids from simpler substrates, but for most of them, it does so only when they are not available in adequate amounts from the external growth medium. For example, production of enzymes for tryptophan synthesis is sharply reduced when the external tryptophan supply is high. Lowering the available L-tryptophan selectively stimulates synthesis of the messenger RNA and the five polypeptide chains that form the enzymes associated with the biosynthesis of tryptophan from chorismate (fig. 30.16). The five contiguous structural genes are transcribed as a single polygenic messenger RNA approximately seven kilobases in length. Initiation of transcription is regulated in part by the interaction of the tryptophan aporepressor, the protein product of the *trpR* gene, with its target site on the DNA, the *trp* operator, *trpO* (fig. 30.17). Binding of L-tryptophan to the aporepressor causes a structural alteration essential for strong specific binding to the *trpO* locus. Like lac repressor, the trp repressor binds at a site that overlaps the RNA polymerase binding site, and the region where the repressor binds contains a number of bases arranged with dyad symmetry (fig. 30.18). It seems likely that the bases on the dyad-symmetry axis are

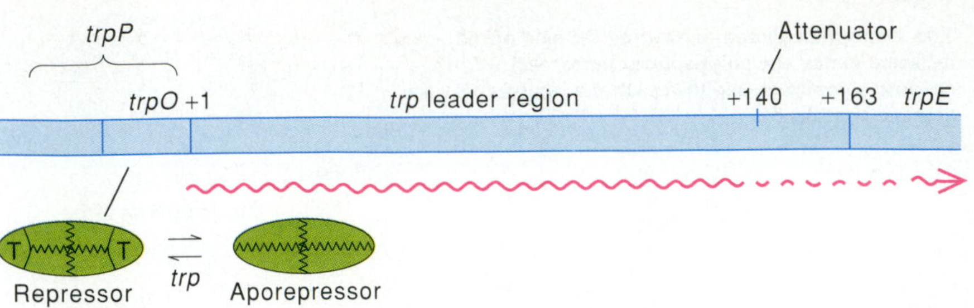

Figure 30.17

Schematic diagram of the repressor control of *trp* operon expression. The *trp* promoter (*P*) and *trp* operator (*O*) regions overlap. The *trp* aporepressor is encoded by a distantly located *trpR* gene. L-tryptophan binding converts the aporepressor to the repressor that binds at the operator locus. This complex prevents the formation of the polymerase-promoter complex and transcription of the operon that begins in the leader region (*trpL*). Only a fraction of the transcripts extend beyond the attenuator locus in the leader region. The regulation of this fraction is discussed in the text.

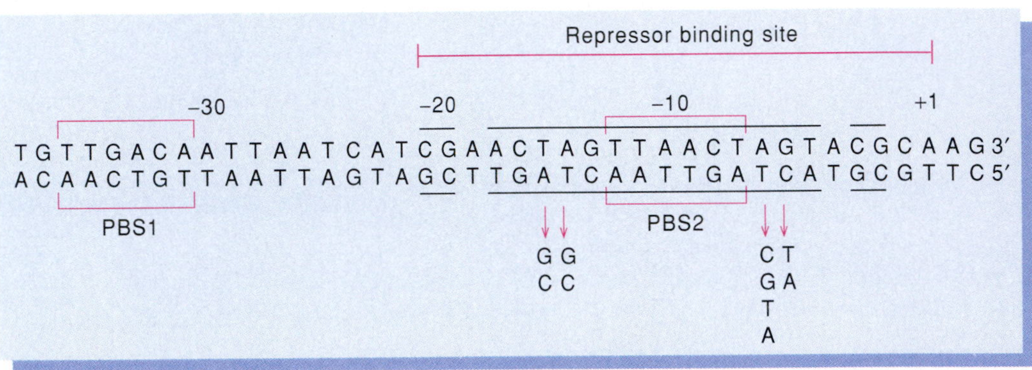

Figure 30.18

The promoter-operator region of the tryptophan operon. Two regions, PBS1 and PBS2, where polymerase binds are bracketed. Regions within the repressor-binding site showing dyadic symmetry are underlined and overlined. Single-base changes that lead to operator constitutive (o^c) mutants are indicated below the duplex.

strongly involved in the binding site for repressor and probably reflect similar symmetry in the repressor itself. The repressor is composed of four identical subunits, each with a molecular weight of 12,400.

The most significant difference between the action of the trp and lac repressors relates to the function of the small-molecule effector. In the case of lac, the effector molecule allolactose acts as an antirepressor, causing release of repressor from the operator; in the case of trp, the effector molecule L-tryptophan acts as a corepressor, stimulating the binding of repressor to the operator. It should be obvious that the difference in action of these small-molecule effectors, whose concentrations dictate the level of operon activity, is well suited to the different metabolic needs of the cell satisfied by the two operons.

In Many Cases Transcription of Amino Acid Biosynthetic Operons Is Also Regulated after the Initiation Point by Attenuators

From this point on, the close parallel between the regulation of the *trp* operon and the *lac* operon ends. The *trp* operon has no positive control system like cAMP-CAP, but it regulates transcription in another way. This is a provisional stop signal called an attenuator, located 141 bases downstream from the initiation site for transcription. Only a fraction of the messenger RNAs (0.1 to 0.9, depending on metabolic conditions) that are initiated transcribe through this provisional stop signal to the end of the operon. The existence of the stop signal was first suspected when it was discovered that a genetic deletion of some of the bases between the initiation site for transcription and the first structural gene (*trpE*) raised the level of expression of the operon 8- to 10-fold. This was true even in strains with a defective repressor gene (*trpR*$^-$ strains). How could this be, and what could be the mechanism of action? An important clue was provided by sequence analysis of the 162 bases in the *trp* leader region, that is, the region between the initiation site for transcription and the initiation site for translation of the first structural gene (fig. 30.19). This leader region contains a potential initiation codon (bases 27 to 29), two tandem *trp* codons (bases 54 to 59), and a terminator codon (bases 69 to 71). A so-called leader peptide of 14 amino acids would result from translation of this region.

There are numerous reasons for believing that translation of the leader peptide up to or through the *trp* codons regulates attentuation of mRNA transcription. First of all, selective starvation of cells for tryptophan relieves attenuation and permits most RNA polymerase molecules to read through the leader region. The only other amino acid that relieves attenuation of the *trp* operon when it is lacking is arginine; arginine starvation is about 80% as effective as tryptophan starvation. Notice that an *arg* codon is located adjacent to the two *trp* codons in the leader region. Most telling of all is the finding that a leader mu-

Figure 30.19

The leader region for the tryptophan operon. The region of the leader RNA containing the hypothesized leader polypeptide is shown. The translation start of the trpE protein is also shown.

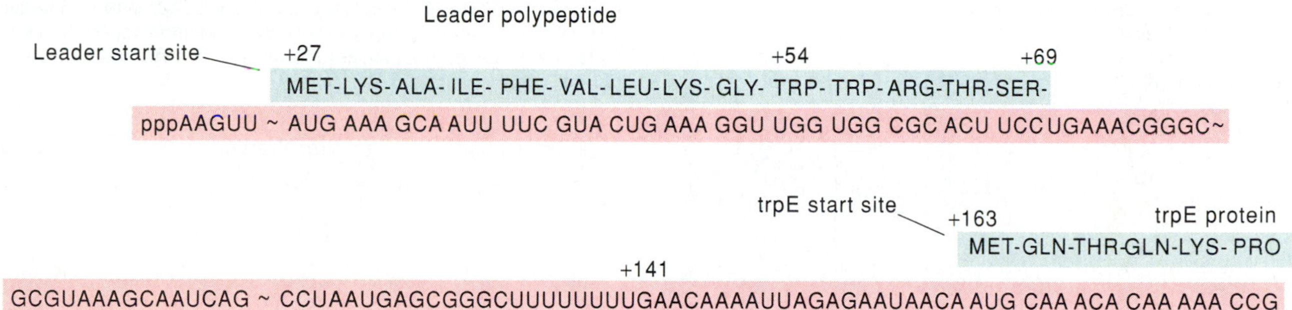

tation resulting in the replacement of the AUG start codon by AUA, which should eliminate translation of the leader peptide, also prevents transcription beyond the attenuator.

Other experiments indicated that the fraction of $tRNA^{Trp}$ that is charged with an amino acid is a crucial factor in the attenuation response. This has been examined *in vivo* by comparing the *trp* operon enzyme levels in *trpR*⁻ strains that are otherwise normal with strains that are defective in some respect in charged $tRNA^{Trp}$ (table 30.2). Such structural defects in $tRNA^{Trp}$ or in the charging enzyme elevates expression, probably by permitting polymerase to transcribe through the attenuator.

These results support the hypothesis that transcription read-through requires partial translation of the leader sequence. However, only if the translation pauses or stops in the region where the *trp* or *arg* codons occur is read-through favored. A careful examination of the secondary-structure possibilities in the attenuator region suggests why this is so. The leader region RNA between bases 50 and 141 has the potential to form a variety of base-paired conformations. Figure 30.20 illustrates the most likely secondary structures that form in terminated *trp* leader RNA. These are based on analysis of regions of the transcript that show resistance to RNase T1 digestion under mild conditions and the base pairing established by studies of defined oligonucleotides. Four regions of base pairing that can form three stem-and-loop structures have been proposed. Region 1, which includes the tandem *trp* codons and the leader peptide translation stop codon (bases 54 to 68), can base-pair with region 2 (bases 76 to 91). Although region 2 (bases 74 to 85) also should be able to base-pair with region 3 (bases 108 to 119), stem-and-loop 2 · 3 has not been observed *in vitro,* presumably because stem-and-loop 3 · 4 and stem-and-loop 1 · 2 form preferentially. Region 3 (bases 114 to 121) can base-pair with region 4 (bases 126 to 134). The existence of this stem-and-loop is inferred from the resistance of the GC-rich region from residue 107 to the 3′ end of the transcript to RNase T1 digestion. The stem-and-loop structure formed between regions 3 and 4, followed by a sequence of U residues, is a common structure in transcription termination (see chapter 28). Hence we would expect that conditions under which this structure is preserved would favor transcription termination. In support of this conjecture, a number of single-base replacement mutations have been isolated that lead to mispairing in the 3 · 4 stem, and all of these lower the level of transcription termination in the leader to some extent.

The absence of translation of the leader would not perturb the 1 · 2 and 3 · 4 structures (fig. 30.21*a*). Consistent with this fact, changing the initiator codon of the leader peptide by a single base prevents read-through, as we discussed earlier. On

Table 30.2
Attenuation of the *trp* Operon Is Affected by *trpT, trpX,* and *trpS* mutations in *trpR*⁻ Strains

Strain (*trpR* Derivative)	*trp* Enzyme Levels (Normalized)	
	Wild Type (*trpa*⁺)	*trp*Δ*LD102 (trpa*⁻*)*
Wild type	1.0	7.1
$trpT_{ts}$	7.1	—
trpX	4.5	6.7
trpS9969	2.7	7.5

All strains were *trpR*⁻ (derepressed) and grown in the presence of excess tryptophan; *trp*Δ*LD102* deletes the attenuator, $trpT_{ts}^-$ is a temperature-sensitive lesion in the $tRNA^{Trp}$ gene. *trpX*⁻ leads to a defect in the modification of the adenine adjacent to the anticodon sequence of the tRNA. *trpS9969* is a mutant in the tryptophenyl-tRNA synthase gene; the resulting enzyme has one-hundredth the affinity for tryptophan shown by the wild-type enzyme, resulting in much lower levels of charged $tRNA^{Trp}$ than normal.

Figure 30.20

Proposed secondary structures in the leader RNA. Four regions, labeled 1, 2, 3, and 4 at the left, can base-pair to form three stem-and-loop structures (1.2, 2.3, and 3.4). The arrows in the main figure mark the RNAse T1 cleavage sites.

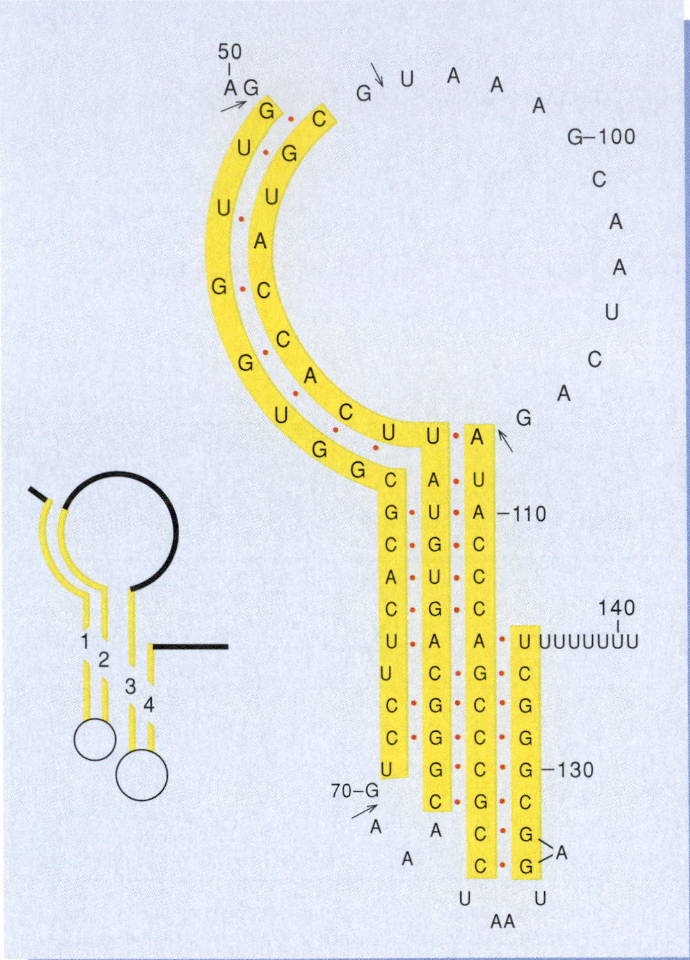

Figure 30.21

Model for attenuation in the *trp* operon, showing ribosome and leader RNA. (*a*) Where no translation occurs, as when the leader AUG codon is replaced by an AUA codon, stem-and-loop 3.4 is intact and termination in the leader is favored. (*b*) Cells are selectively starved for tryptophan so that the ribosome stops prematurely at the tandem *trp* codons. Under these conditions, stem-and-loop 2.3 can form, and this is believed to lead to the disruption of stem-and-loop 3.4, permitting attenuation. (*c*) All amino acids, including excess tryptophan, are present so that stem-and-loop 3.4 is present.

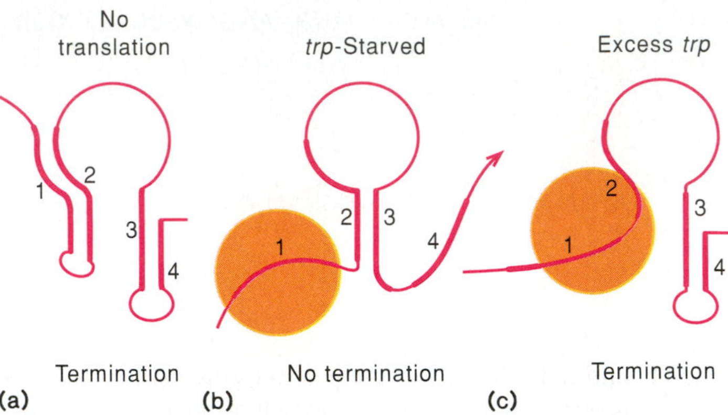

the other hand, selective starvation resulting from either tryptophan or arginine deprivation stimulates transcription read-through. Probably this is because the ribosome stalls in the region of the *trp* or *arg* codons (bases 54 to 62). The resulting rupture of the base-pairing 1 · 2 structure would make region 2 available for pairing with region 3. This would encourage disruption of the stem-and-loop 3 · 4 structure and thus permit transcription read-through (fig. 30.21*b*). In the presence of an adequate supply of all amino acids, translation would proceed beyond this critical region, so that region 2 would not be available for base pairing and the 3 · 4 loop would be maintained, an arrangement favoring transcription termination at the attenuator (fig. 30.21*c*).

This attenuator mechanism of control is amazingly simple, since it requires no proteins other than those normally used for transcription and translation. We would therefore expect the same mechanism to be used repeatedly for other operons involved in amino acid biosynthesis. An indication that the attenuator mechanism might be functioning would be the finding that the biosynthetic enzymes for a particular amino acid were increased when charging of the cognate tRNA was reduced. Indeed, for the several other amino acid biosynthetic pathways in *E. coli* for which tRNA charging is involved in regulation, attenuator mechanisms have been found. Analysis of the leader region preceding the structural genes for the biosynthesis of three such amino acids in *E. coli,* phenylalanine, histidine, and leucine, reveals structures with the features of this type of control mechanism: (1) leader sequences containing several consecutive codons for that amino acid followed by (2) sequences that can be organized into stem-and-loop structures that could signal termination. Even the multivalently controlled* threonine operon (*thr*) and isoleucine and valine operons (*ilv*) of *E. coli* exhibit this mechanism of expression. The leader sequence of the *thr* operon specifies a 21-amino-acid peptide containing 8 threonine codons and 4 isoleucine codons, and that of the *ilv* operon specifies a 32-amino-acid peptide, of which 14 are leucine, isoleucine, or valine. Comparable studies of one of the genes for arginine biosynthesis (*argF*), however, failed to show attenuation. This gene, which is required for arginine biosynthesis, appears to be exclusively regulated by a repressor mechanism. By contrast, the *his* operon has no repressor and appears to be regulated exclusively with an attenuator.

In view of these findings, why are both types of transcription controls used in the *trp* and certain other operons? Two conceivable advantages have been suggested. First, for a system

*The *thr* operon is repressed in the presence of excess threonine and isoleucine. Repression of the *ilv* operon requires leucine, isoleucine, and valine in excess.

with two metabolic switches, the amplification factor between fully on and fully off should be the product of what could be achieved with either switch alone. Second, under some conditions, for example, during very rapid protein synthesis, it is possible that a higher level of amino acid is required to keep the tRNA at an adequately charged level. Finally, we should note that the *trpR* protein has other roles; it is also a repressor for the *trpR* gene and for the *aroH* gene, which encodes one of three isozymes that catalyze the first reaction in aromatic biosynthesis.

Genes for RNA Polymerase and Ribosomes Are Coordinately Regulated

Over 100 genes in *E. coli* participate in the synthesis of the RNAs and proteins that comprise the enzymatic machinery for transcription and translation. The relevant gene products make up between 20 and 40% of the dry cell mass. In rapidly growing cells, about 85% of the RNA is ribosomal, 10% is tRNA, and most of the remainder is mRNA. The various RNAs and proteins are produced according to their need. For ribosomes and tRNAs, the synthesis rate is roughly proportional to the cell growth rate; the relative amounts of the three rRNAs (16S, 23S, and 5S), the 60 or so tRNAs, and the 50 ribosomal proteins is consistent with the stoichiometric needs for making ribosomes. RNA polymerase, which contains four different subunits (see chapter 28), is produced in equimolar amounts of β and β', approximately equal in sum to the number of α subunits. The dissociable σ subunit of RNA polymerase is produced in somewhat less than stoichiometric amounts. The synthesis of RNA polymerase subunits is loosely coupled to the synthesis of ribosomal protein subunits. Demands for large amounts of all these RNAs and proteins and precise regulation of their relative amounts have led to complex gene arrangements and controls at both the transcription and translation levels.

Control of rRNA and tRNA Synthesis by the* rel *Gene Under conditions of rapid growth, *E. coli* cells contain about 10^4 ribosomes. The maximum rate of reinitiation at the ribosomal gene promoter is about one per second. In rapid growth, *E. coli* can duplicate once every 20 min, which would allow for the synthesis of only about 1,200 molecules of rRNA if there were only one gene for ribosomal RNA. In fact, there are seven copies for ribosomal RNA operons in the bacterial chromosome, which makes it possible for rRNA synthesis to maintain the necessary pace under conditions of rapid growth. These operons are dispersed at seven locations around the circular *E. coli* chromosome. Each operon is transcribed into one long transcript that is processed into one each of 16S, 23S, 5S, and usually at least one 4S transfer RNA (fig. 30.22). The order of the genes in the operon, starting from the initiation site for transcription, is 16S, 4S, 23S, and 5S. In some of the operons, additional 4S genes for tRNA are located downstream from the 5S gene. All the known ribosomal RNA operons appear to have two strong promoters located in tandem; this arrangement probably permits a more rapid rate of reinitiation for the genes than would otherwise be possible.

Figure 30.22

A typical rRNA (*rrn*) operon contains two promoters and genes for 16S, 23S, and 5S rRNA and a single 4S tRNA gene. The four fully processed RNAs are derived from a single intact 30S primary transcript.

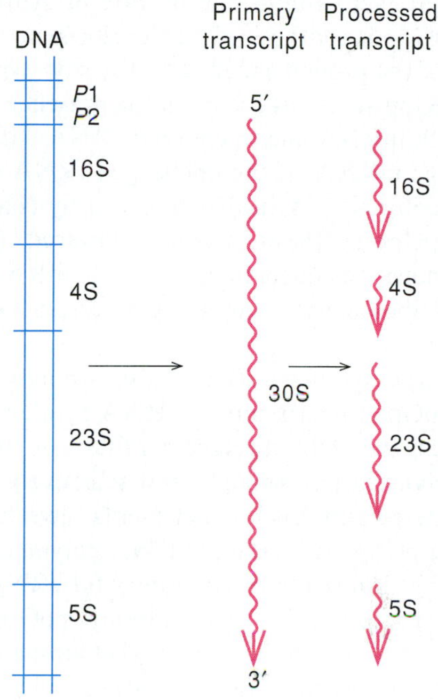

As we stated earlier, rRNA synthesis is usually maintained at a rate that is proportional to the gross rate of protein synthesis. In a normal wild-type cell, when protein synthesis is limited (e.g., by amino acid availability), Michael Cashel and John Gallant showed that there is a rapid rise in ppGpp concentration from about 50 to 500 μM (fig. 30.23). Concomitantly, there is an abrupt cessation of rRNA synthesis. This is part of the phenotype known as the stringent response: First, amino acid deprivation or other factors that slow down protein synthesis provoke an increased rate of accumulation of ppGpp; this in turn leads to an inhibition of rRNA synthesis.

Observations on different mutants indicate that the concentration of ppGpp is regulated by a careful balance between its rate of synthesis (controlled by the *rel* gene) and rate of breakdown (controlled by the *spoT* gene). Correlated observations indicate that the synthesis of rRNA is inversely proportional to the ppGpp concentration. In a *relA* mutant cell, neither the rapid rise in ppGpp concentration nor the cessation of rRNA synthesis is seen when amino acids are removed. The *relA* gene encodes a protein that is required for ppGpp synthesis. If amino acids are reintroduced into the growth medium of a wild-type culture, the ppGpp concentration falls rapidly (half-life about 20 s), and the rate of rRNA synthesis rises rapidly. In a mutant called *spoT*, the normal rise in ppGpp level is observed on amino acid starvation, but the level of ppGpp falls much more slowly on readdition of amino acids to the growth medium. Correlated with this behavior, the rate of rRNA synthesis in a *spoT* mutant also increases very slowly on readdition of amino acids.

The synthesis of ppGpp has been studied both in crude cell-free extracts of *E. coli* and in a partially purified system to determine what factors influence its rate of synthesis. It was found that ppGpp is synthesized on the ribosome from GTP in the presence of the protein encoded by the wild-type *relA* gene. Maximum ppGpp synthesis occurs in the presence of ribosomes associated with mRNA and uncharged tRNA with anticodons specified by the mRNA. If the uncharged tRNA bound to the ribosome acceptor site (A site) is replaced by charged tRNA, the rate of ppGpp synthesis is greatly lowered. If uncharged tRNA anticodons are not complementary to mRNA codons exposed on the ribosome for protein synthesis, ppGpp synthesis does not occur.

Cell-free synthesis studies also strongly support the notion that ppGpp directly inhibits RNA synthesis. Thus in a cell-free system the DNA-directed synthesis of rRNA with *E. coli* RNA polymerase is strongly and selectively inhibited by 100 to 200 μM ppGpp. Such experiments have led to the hypothesis that ppGpp, by binding to RNA polymerase, alters its structure so that it has a lowered affinity for rRNA promoters.

The *in vivo* and *in vitro* studies on ppGpp and rRNA have resulted in a model for how the level of amino acid charging of tRNA controls the rate of rRNA synthesis (fig. 30.24). First, uncharged tRNA that is codon-specific for the exposed codons on the mRNA becomes bound to the ribosome acceptor site, creating a situation unfavorable for protein synthesis but favorable for ppGpp formation. Second, the ppGpp diffuses and binds to RNA polymerase, thereby lowering its affinity for rRNA promoters.

Figure 30.23

Guanosine tetraphosphate (ppGpp) concentration under normal conditions, after amino acid starvation, and after readdition of amino acids (• = wild-type cells; ▲ = *relA* cells; and ■ = *spoT* cells).

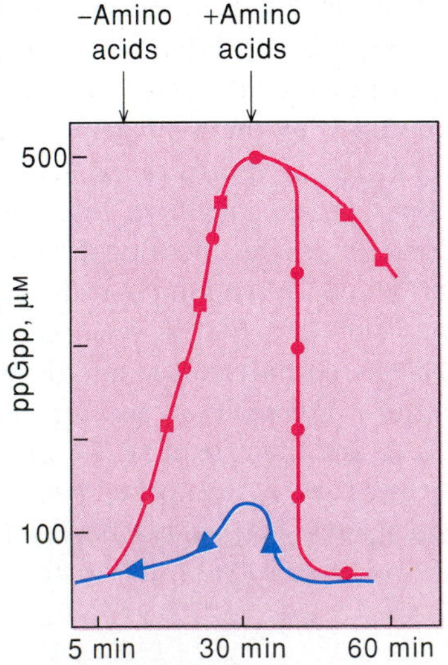

This alteration of the RNA polymerase by ppGpp affects the ability of RNA polymerase to interact with promoters in a differential way. For the promoters of the rRNA operons, the polymerase-promoter interaction is strongly inhibited by ppGpp. For certain other promoters the effect of ppGpp is actually stimulating. As first observed in the cell-free system, ppGpp stimulates expression from the *lac* and *trp* operons. In much more recent studies, Cashel has found that the biosynthetic operons associated with arg, gly, his, leu, lys, phe, ser, thr, and val biosynthetic enzymes have an absolute requirement for ppGpp. The role of ppGpp in these examples seems clear. An absence of amino acid leads to ppGpp buildup which then stimulates the genes required for biosynthesis of the amino acid.

The observations on ppGpp's role in rRNA synthesis show that this small molecule is an important control factor regulating rRNA synthesis, but it does not eliminate the possibility that other factors affect the level of rRNA. *In vivo* and *in vitro* evidence indicates that the inhibitory effect of ppGpp on transcription extends to most tRNA and ribosomal protein genes. Ribosomal protein gene expression also appears to be regulated at the translational level.

Figure 30.24

Schematic diagram of ppGpp synthesis and the hypothesized mechanism for its action. ppGpp is synthesized on the ribosome when there is a peptidyl-tRNA on the P site of the ribosome and an uncharged tRNA on the A site. The ppGpp probably inhibits rRNA synthesis by complexing with the RNA polymerase.

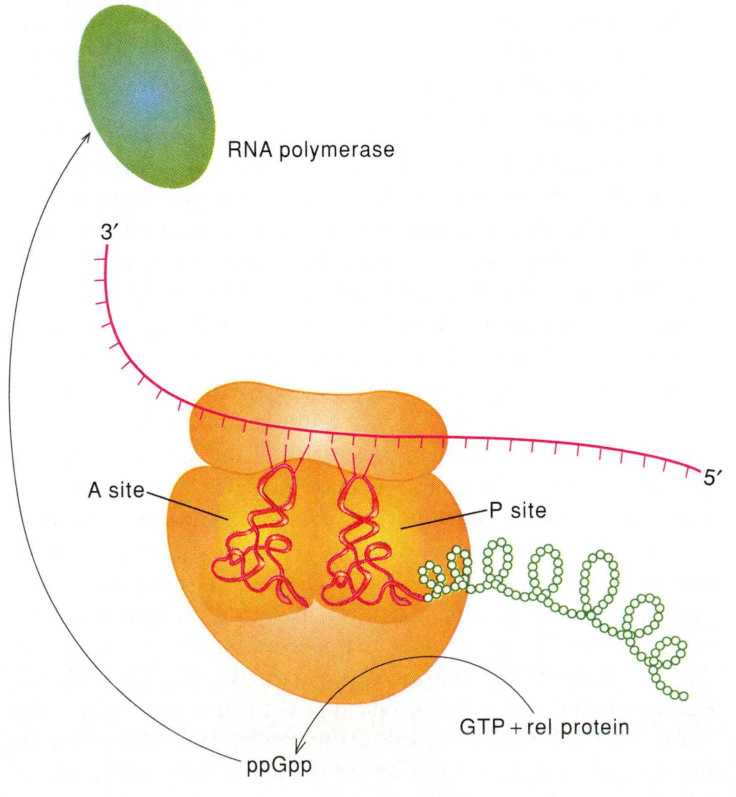

Translational Control of Ribosomal Protein Synthesis In exponentially growing *E. coli* cells, synthesis rates of all ribosomal proteins (except L7, L12, and S6) are identical and coordinately regulated. Nomura and his co-workers have suggested that free ribosomal proteins inhibit the translation of their own mRNA and that as long as the assembly of ribosomes removes ribosomal proteins, the corresponding mRNA escapes this feedback inhibition. This hypothesis has been tested *in vitro* using a protein-synthesizing system with various template DNA molecules carrying ribosomal protein genes; it has been tested *in vivo* by examining the effect of overproduction of certain ribosomal proteins on the synthesis of other ribosomal proteins using various recombinant plasmids. By these means it was found that certain ribosomal proteins selectively inhibit the synthesis of other ribosomal proteins whose genes are part of the same operon as their own; this autogenous or self-imposed inhibition occurs at the level of the translation of the mRNA rather than at the level of transcription of mRNA. Figure 30.25 illustrates how this scheme works for the regulatory protein L1. L1 and L11 are encoded by the P_{L11} operon. L1 can form a complex with either the 5′ end of its own mRNA or with 23S rRNA. It binds more strongly to the 23S rRNA. However, if there is an excess of L1 over the available 23S rRNA, then L1 will bind to the 5′ end of the mRNA, thereby inhibiting the synthesis of both L1 and L11. In this way the amounts of L1 and L11 protein synthesized are kept in register with the amounts of rRNA synthesized. Other operons carrying ribosomal protein genes are believed to be regulated in a similar manner.

A meaningful overall pattern of control appears to be emerging that is consistent with the requirement to balance the relative amounts of rRNA and ribosomal proteins synthesized. Most of the ribosomal protein genes occur in clusters or operons with promoters at one end (e.g., fig. 30.26). In some cases, individual operons are regulated by one of the encoded ribosomal proteins; in other cases, operons appear to be subdivided into "units of regulation," and individual units of regulation are regulated by one of the gene products (e.g., the P_{SPC} operon in fig. 30.26).

Regulation of RNA Polymerase Synthesis Is Coordinated with Ribosomal Protein Synthesis The genes for the RNA polymerase subunits β and β' (*rpoB* and *rpoC*, respectively) and L7/L12 and L10 (*rplJ* and *rplL*, respectively) belong to a single transcription unit (fig. 30.27). The gene for the α subunit (*rpoA*) is cotranscribed with four ribosomal protein genes coding for the proteins S13, S11, S4, and L17 (see fig. 30.26). This suggests that the regulatory system for RNA polymerase subunit synthesis and that for ribosomal protein synthesis are shared. Even more fascinating is the finding that the gene for the σ subunit of RNA polymerase (*rpoD*) is located on the same operon as the gene for DNA primase (*dnaG*) and the small ribosomal subunit protein S21 (*rpsU*). This operon, known as the σ operon, clearly encodes proteins essential to DNA, RNA, and protein synthesis, which suggests that regulation of the operon may be important to balance all three of these biosynthetic processes.

Despite the use of a common promoter, the synthesis of the RNA polymerase proteins and certain nearby ribosomal proteins is not always coordinately regulated. We will discuss only the β operon, because it is better understood. For brief times, synthesis of L7/L12 and L10 is inhibited *in vivo* under conditions of amino acid deprivation, while the synthesis of β and β' is not. Furthermore, the RNA polymerase inhibitor rifampicin causes a transient stimulation of β and β' synthesis, but not of L10 and L7/L12. This evidence for lack of coordinate regulation implies the existence of a separate regulatory device for

Figure 30.25

Schematic diagram explaining autogenous inhibition by L1. L1 can form a complex either with the 5′ end of its own mRNA or with 23S rRNA. If there is an excess of L1 over the available 23S rRNA, then L1 will bind to the 5′ end of the mRNA and inhibit both L1 and L11 synthesis. The inhibition of L1 translation by this binding depends on the fact that the translation of L1 and L11 is somehow coupled.

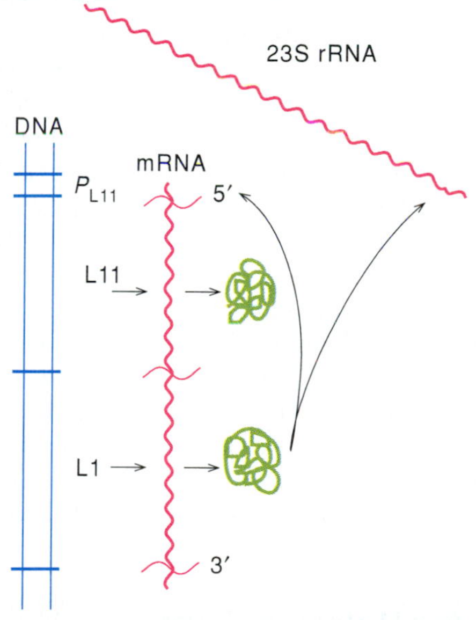

Figure 30.26

A giant cluster containing 25 r-protein genes and the gene for RNA polymerase. This cluster is divided into three operons with promoters P_α, P_{SPC}, and P_{S10}. The corresponding messages are subject to autogenous inhibition at the translation level by S4, S8, and L4 protein, respectively. S8 regulates only genes downstream from L5; these include S14, S8, L6, L18, S5, L30, and L24. Translation is also autogenously regulated by one or both of these proteins. Parentheses around S19 and L22 indicate that the order of genes is uncertain.

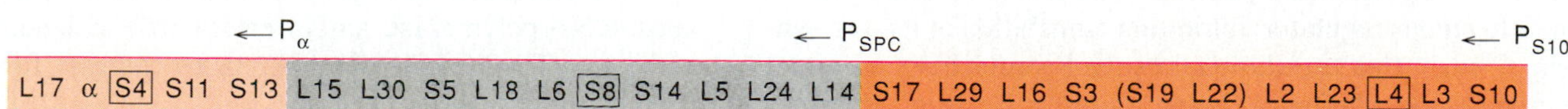

Figure 30.27

DNA and transcripts in the β operon. Only a small number of transcripts read through the attenuator (ATT) located 69 nucleotides beyond the 3′ end of the *rplL* gene. The read-through transcripts are cleaved by RNase III about 200 nucleotides beyond the end of the *rplL* gene. L10 ribosomal protein and RNA polymerase are believed to inhibit translation of the promoter-proximal and promoter-distal messages, respectively. L10 may only inhibit its own synthesis. Normally about five times as much L7/L12 is synthesized at L10. Ribosomes can initiate independently at the initiation site for L7/L12, and it is possible that L7/L12 may regulate this initiation. Bruckner and Matzura have shown that another promoter, two genes upstream from the P_{L10} promoter, may be the true initiation point for this operon *in vivo*.

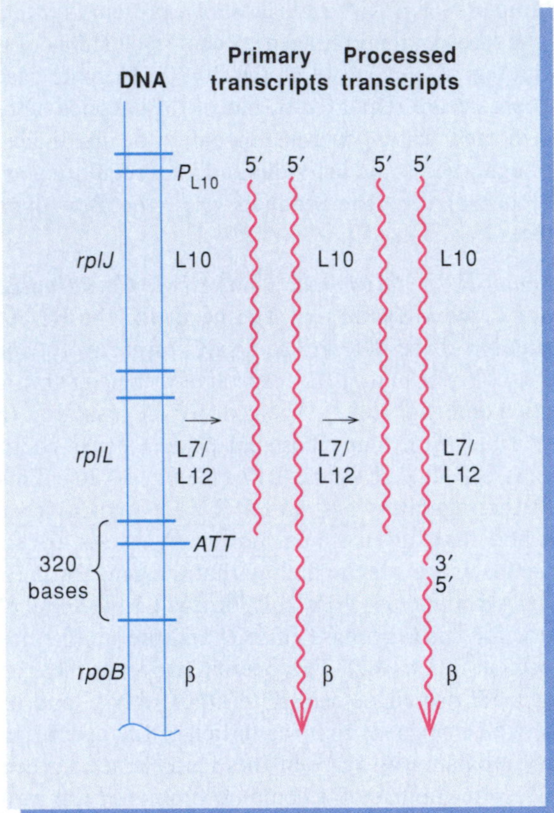

controlling L10 and L7/L12 and for controlling β and β′ subunit synthesis. There are 320 untranslated bases between the *rplL* gene and the downstream *rpoB* gene (see fig. 30.27); this is a suspiciously large segment for an intercistronic region, suggesting that it functions as more than just a spacer between cistrons. Under normal growth conditions, only 20% of the transcripts that initiate at the L10 promoter P_{L10} read through the entire operon. The remainder terminate 69 nucleotides beyond the 3′ end of the *rplL* gene. The nonattenuated transcript is normally cleaved by RNase III in the intercistronic region at a point about 200 nucleotides beyond the end of the *rplL* gene. This cleavage divides the mRNA for the operon into two segments, one encoding the ribosomal proteins and the other the two RNA polymerase proteins. The L10 ribosomal protein is an autogenous regulator, inhibiting translation of its own synthesis. Clearly this translational control would have no direct effect on the translation of the message for β and β′. Recently it has been discovered that RNA polymerase holoenzyme inhibits the translation of β and β′. Thus both segments of the operon appear to be regulated at the translational level by different specific autogenous regulatory proteins.

The finding that RNA polymerase inhibits the synthesis of its own subunits immediately suggests an explanation for the differential *in vivo* effect of amino acid deprivation (increased ppGpp) and rifampicin on expression of the ribosomal protein genes and the polymerase protein genes. Whereas ppGpp or rifampicin should inhibit initiation of transcription at the P_{L10} promoter, inhibiting both ribosomal protein and polymerase subunit synthesis, the binding of either rifampicin or ppGpp to RNA polymerase could interfere with its ability to act as an autogenous inhibitor. This interference should stimulate β and β′ synthesis and affect all the messages for β and β′ already synthesized, as well as any that are in the process of being made.

Overview of Gene Regulation in E. coli

We have seen that in most cases gene expression in *E. coli* is regulated at the level of transcription. However, there are notable exceptions. For example, the case of the ribosomal protein genes, where autogenous regulation plays an important controlling role at the translational level. Controls for the expression of any particular gene are frequently multiple and organized into a hierarchy. In the case of the *lac* operon, for example, transcription is positively regulated by the cAMP-CAP complex that also regulates a large number of catabolic genes in a similar manner; the *lac* operon is also negatively regulated in a highly specific way by the *lac* repressor and lactose.

Bacterial regulatory systems are geared toward a rapid response to changing environmental conditions. Since environmental conditions may change in either direction, the regulatory responses must also be of a highly reversible type. Bacterial systems achieve this flexibility by maintaining low levels of regulatory proteins, which are controlled by the concentration of small-molecule effectors. The concentration of the effector molecules can vary over wide limits.

For the remainder of this chapter we will turn our attention to bacteriophage control systems. The patterned responses of phage systems have led to very different strategies for control of gene expression. By and large, phage systems are regulated by the concentrations of the regulatory proteins themselves; these regulatory proteins are not subject to the influence of small-molecule effectors.

Host-Virus Relationships

Before leaving the subject of bacterial gene regulation, let us review what we know about some of the parasites of *E. coli*. Most bacterial parasites can be classified as plasmids or viruses. Plasmids are DNA molecules that have no extracellular form; they are the ultimate parasites because they are totally dependent on host enzymes for their survival. Therefore it is not surprising that plasmids replicate in moderation, using the normal host RNA polymerase, and interfere little with normal host metabolism. The lifestyle of viruses is, by contrast, far more varied and complex.

Table 30.3
Sizes of Some *E. coli* Phages and Genomes

Phage	Particle Shape	$M_r(\times 10^6)$	DNA Length (bp × 10³)	Structure of Chromosome	Special Properties of Genome
ϕX174	Icosahedron	1.8	5.4[a]	Single strand, circular	Duplex replicative form
M13	Filament	2.1	6.4[a]	Single strand, circular	Duplex replicative form
T7	Head, short tail	26.5	40.0	Duplex, linear	Terminal redundancy
λ	Head, tail	32.0	48.6	Duplex, linear	Cohesive ends form replicative circles; lysogenic
P1	Head, tail	58.6	88.0	Duplex, linear	Terminal redundancy; circularly permuted
T4	Head, tail	110.0	166.0	Duplex, linear	Terminal redundancy; circularly permuted
R17	Icosahedron	1.0	3.0[a]	Single strand, linear	RNA (not DNA)
Mu		25.00	38.00	Duplex, linear	Replicates in the integrated state
T5	Head, tail	75.0	113.0	Duplex, linear	One strand interrupted

[a]For single-stranded phages, length is for bases (b) rather than base pairs (bp).

Viruses fall into three categories: lytic, nonlytic, and temperate. When lytic viruses infect a cell, they invariably redirect its metabolism to suit their own needs. Upon completion of a life cycle that produces mature virus particles, a virus-encoded enzyme disrupts the cell wall, the virus particles escape, and the cell dies. Lytic viruses include bacteriophage T4 and bacteriophage T7 (table 30.3), which redirect host cell metabolism in different ways. Infection by T4 modifies the host RNA polymerase in a manner that favors T4 phage transcription over host transcription. By contrast, when T7 infects a cell, the first few phage genes to be transcribed use the host RNA polymerase unchanged. Then the remainder of the T7 genes are transcribed by a T7-encoded RNA polymerase. This polymerase is so active that the host polymerase cannot compete effectively with it for substrate. Hence all genes requiring the host polymerase for expression are turned off by default. Like T4, the T7 virus produces a lysozyme that disrupts the cell wall, permitting the viruses to escape.

Nonlytic viruses are relatively rare (see M13 in table 30.3). These viruses are able to replicate in moderation and exit from the cell without disruption of the cell wall or cell membrane.

The most interesting type of behavior is shown by temperate viruses such as λ (fig. 30.28). Upon infecting a sensitive host, λ can adopt either an active lytic state or a dormant lysogenic state. In the active lytic state λ behaves very much like T4. The dormant state is regulated in a unique manner.

Figure 30.28

Electron micrograph of bacteriophage lambda. (Courtesy of Roger Hendrix.)

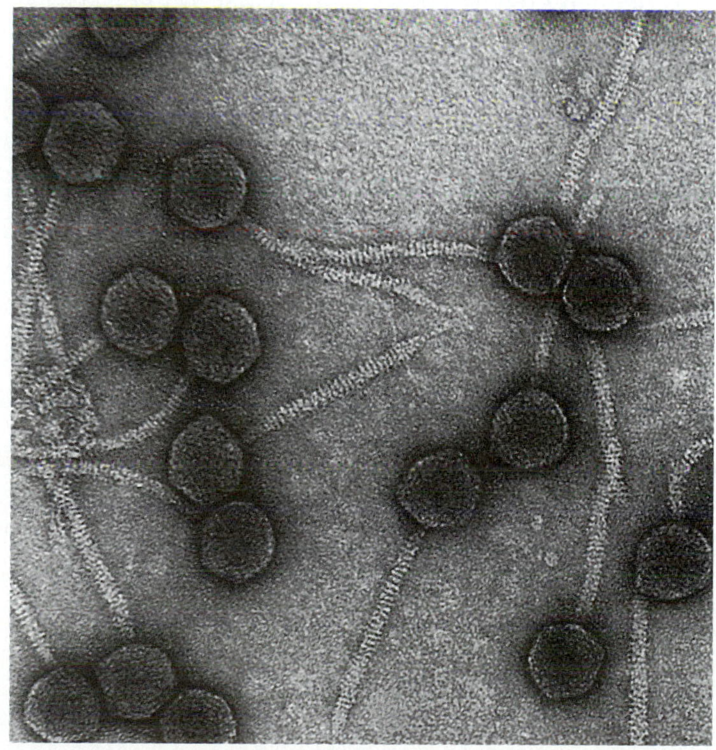

Figure 30.29

A segment of the λ genome containing the P_{RM} and nearby P_L and P_R promoters. Transcription from P_{RM} results in cI synthesis. cI binds near the promoters. An expanded view is shown of the three cI binding sites flanking the P_{RM} and P_R promoters. Transcripts that can originate from the three promoters are shown by the wavy lines. The O_{R1}, O_{R2}, and O_{R3} sites are binding sites for the cI repressor.

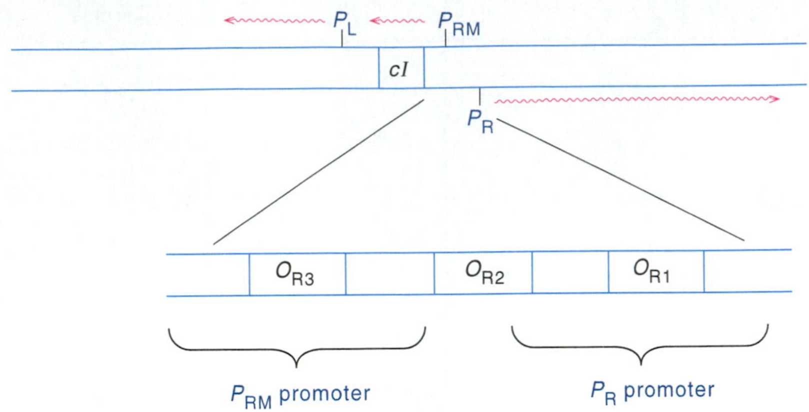

The Dormant State of λ Is Maintained by a Repressor

About half the time, when λ infects a cell it adopts a dormant lysogenic state, in which the virus is linearly integrated into the host genome. The lysogenic state is maintained by a moderate amount of the λ-encoded cI repressor. This protein binds as a dimer at two locations on the virus genome, near the promoter for early leftward transcription, P_L, and near the promoter for early rightward transcription, P_R (fig. 30.29). The cI repressor binding site near P_R overlaps the polymerase binding site at the P_R promoter. This binding site is a complex consisting of three adjacent binding sites for the cI repressor dimer, O_{R1}, O_{R2}, and O_{R3}. At moderate concentrations (about 100 copies of cI repressor per cell) O_{R1} and O_{R2} are occupied. At higher concentrations of cI the O_{R3} site is also occupied. Binding at O_{R3} prevents overproduction of cI, which would be wasteful and possibly make it difficult to induce the active lytic form of λ under the appropriate conditions. Binding of repressor at O_{R1} and O_{R2} has two effects: It inhibits early rightward transcription and it stimulates the promoter for repressor maintenance, P_{RM}. Thus at moderate concentrations, the cI repressor functions simultaneously as a repressor and an activator. It represses rightward transcription because it binds at a site that overlaps the polymerase binding site for this promoter. On the other hand, it activates the P_{RM} promoter because it binds at a site that makes favorable contact with the polymerase at the P_{RM} promoter.

As long as a moderate concentration of the cI repressor is present, λ remains dormant. Exposure of lysogenic cells to ultraviolet light or DNA-damaging drugs such as mitomycin C leads to activation of the virus. This is because DNA damage results in activation of the host recA protease, which cleaves the cI repressor and destroys its capacity to function as a repressor. The activated recA protease cleaves the cI repressor in much the same way that it cleaves the lexA repressor (see fig. 26.32).

Events That Follow Infection of E. coli *by Bacteriophage λ*

The infection process by λ starts by attachment of the mature virus particle to the bacterial membrane and injection of the viral chromosome (fig. 30.30). This is followed by circularization catalyzed by the cellular DNA ligase enzyme. Only two genes are expressed at this time, the regulatory genes *N* and *cro* from the P_L and the P_R promoters, respectively. The N protein combines with the host RNA polymerase, permitting extension of the early right and early left transcripts. This extension leads to expression of other regulatory genes—*cII, cIII,* and *Q*—as well as the genes for DNA replication and recombination. At this point the infectious process can proceed in either of two mutually exclusive directions, the lytic direction or the lysogenic direction. If the lysogenic route is followed, most early genes are turned off and only the regulatory gene *cI* and the integration gene *int* are turned on. If lysis proceeds, all early genes are turned off and the head, tail, and lysis genes are turned on.

In all, λ has six regulatory genes—*cI, cII, cIII, cro, N,* and *Q*. The developmental processes accompanying λ infection correlate with the roles of the proteins encoded by these six regulatory genes (table 30.4) and the factors that govern the level of their expression.

Very Early Expression Results in Two Short Transcripts, One from P_R *and One from* P_L

Six promoters have been identified in λ—P_L, P_R, P_{RE}, P_{int}, P_{RM}, and $P_{R'}$. Only two of of these promoters, P_L and P_R, are recognized by the unmodified host RNA polymerase (fig. 30.31). These promoters lead to the expression of the N and cro regulatory proteins. The transcript originating from the P_L promoter is terminated at the t_{L1} terminator just on the other side of the *N* gene. The transcripts originating from the P_R promoter are

Figure 30.30

Overview of λ development.

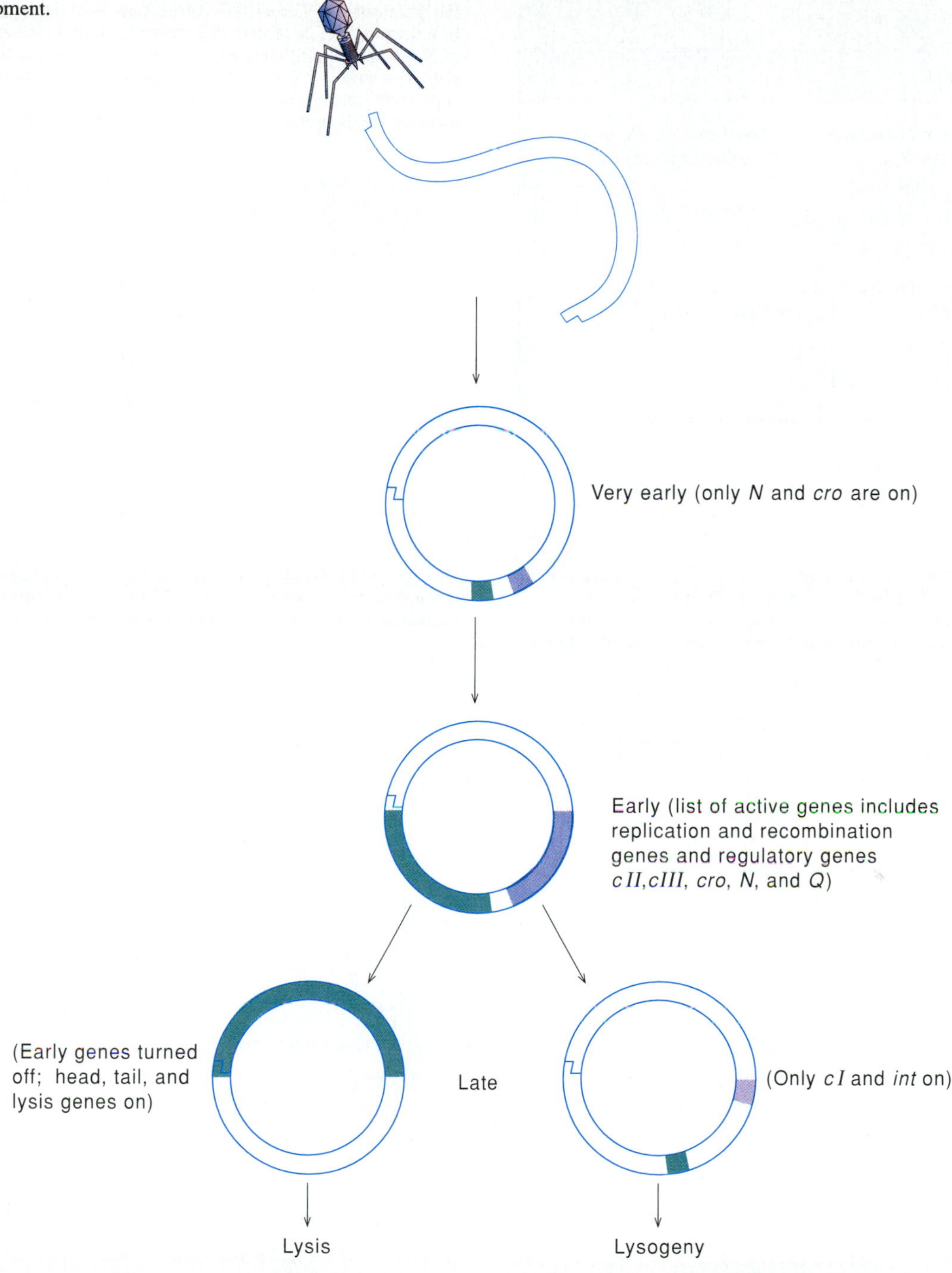

mostly terminated at t_{R1}, just on the other side of *cro*. Some rightward transcripts are not terminated at t_{R1}; these are terminated at t_{R2}.

Early Expression Leads to Extension of the Very Early Transcripts

The N protein resulting from very early expression is an antiterminator. The presence of N protein leads to extension of the transcripts originating from the P_L and the P_R promoters and the resulting expression of a number of early genes. From leftward transcription the gene for regulatory protein cIII and the recombination genes are expressed. From rightward transcription the DNA replication genes and genes for the regulatory proteins cro, cII, and Q are expressed.

The utilization of the N protein is fascinating (fig. 30.32). In the early right and early left transcripts there are sites where the RNA polymerase pauses during the process of elongating the transcripts. At these points, known as *nut* sites (for N utilization), the N protein combines with the RNA polymerase, altering its properties so that it no longer recognizes the early termination sites.

Table 30.4
Regulatory Proteins of λ

Protein	Function
cI	At low concentrations represses P_R and P_L and activates P_{RM}; at high concentrations represses P_R, P_L, and P_{RM}
cII	Activator for P_{int} and P_{RE}
cIII	Stabilizes cII
cro	At low concentrations represses P_{RM}; at high concentrations represses P_{RM} and P_R
N	An antiterminator at t_{L1}, t_{R1}, and t_{R2}
Q	An antiterminator at t_{6S}

Figure 30.32

Early transcripts. The early left transcript is initiated from the P_L promoter. In the absence of N protein this transcript terminates at t_{L1}. In the presence of N protein the polymerase picks up an N protein at the N utilization site, nut_L. This makes it possible for the polymerase to transcribe through t_{L1}. The early right transcript tends to terminate at t_{R1} unless N protein is present. In this case the N protein becomes bound to the polymerase at the nut_R site.

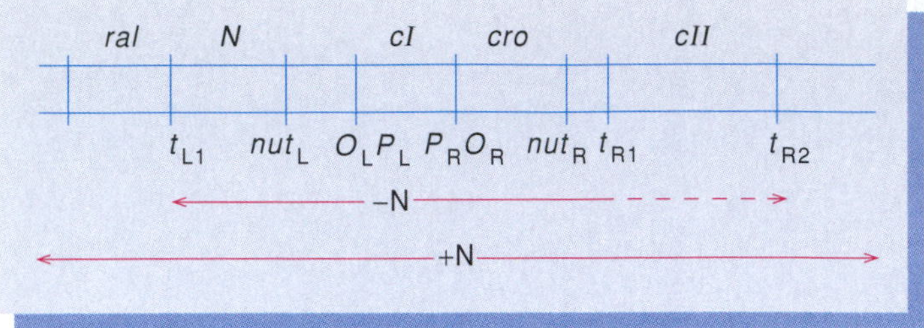

Figure 30.31

Bacteriophage lambda DNA. *Top:* Circular map indicating locations of main control genes and early and late functions. *Bottom:* Expanded region containing control genes and main promoters and terminators. Arrows indicate main transcripts and conditions under which they are active. In the absence of N protein, about half the transcription beginning at P_R reads through t_{R1} to t_{R2} (indicated by dashed portion of arrow). More information on regulatory proteins and promoters is given in table 30.4.

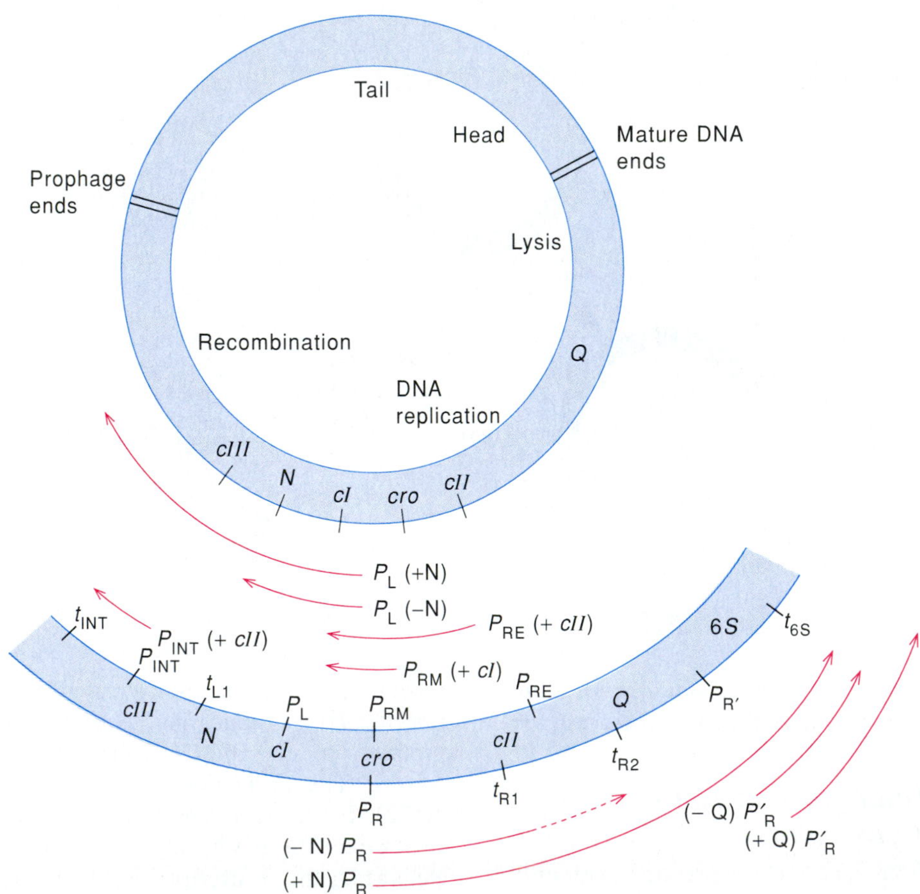

Figure 30.33

A segment of the λ genome containing the P_{RM} and nearby P_L, P_R, and P_{RE} promoters. Transcripts that can originate from the three promoters are shown by the wavy lines. The P_{RE} and P_{RM} require different activators for expression. The transcript from P_{RE} is ten times more effective in cI expression because it has a Shine-Dalgarno sequence for ribosome attachment.

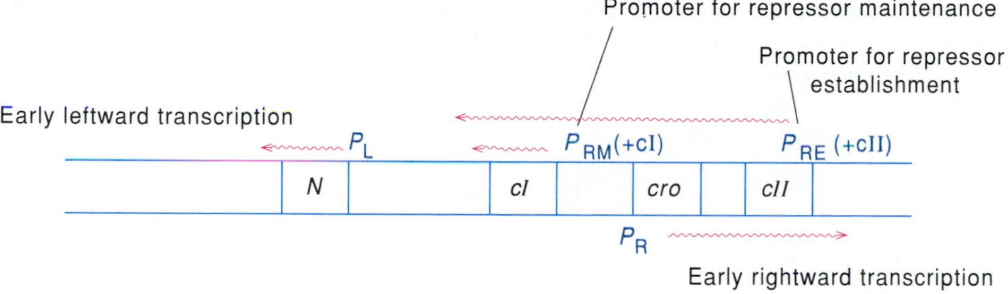

Late Expression Along the Lysogenic Pathway Results from Elevated Expression of the cII Regulatory Protein

Late expression is the phase during which commitment to the lytic or the lysogenic pathway occurs. The choice depends on the level of cII protein that develops shortly after infection. At high levels of cII the lysogenic pathway is followed: at low levels of cII the lytic pathway is followed. Precisely what controls the level of cII expression relative to other events is unclear. To some extent the level of cII expression depends on cIII expression, since the main function of cIII is to stabilize the otherwise quite unstable cII protein. The crucial importance of cII is easier to understand. The cII protein is a positive activator for the P_{RE} and the P_{int} promoters (fig. 30.33). The P_{RE} promoter leads to synthesis of the cI repressor (RE stands for repressor establishment). The cI protein is a negative regulator of the P_L and the P_R promoters. Thus the presence of low to moderate levels of cI shuts off both of the main early promoters. The P_{int} promoter, as the name implies, is responsible for synthesis of the int protein, which is required for λ integration. Integration completes the lysogeny process, and thereafter the cI protein keeps both the P_L and the P_R promoters in the "off" position. Eventually this repression silences the P_{RE} promoter also.

Since the P_{RE} promoter was responsible for the synthesis of cI in the first place, we may well ask what happens to the level of cI protein in the lysogenic cell. Under these conditions the cI gene is expressed exclusively from the less efficient transcript* originating from the P_{RM} promoter. This promoter is inactive unless cI is already present. As we have already indicated, cI at moderate levels activates the P_{RM} promoter. At very high levels, cI inhibits further cI production. In this way a level of about 100 copies of cI protein per cell is maintained. Only a traumatic incident can disturb the dormant lysogenic state. This is exactly what happens when the cell is exposed to DNA-damaging conditions. Under these conditions the pool of cI proteins is completely destroyed and the dormant bacteriophage enters the lytic cycle.

*This transcript is less efficient because the start codon is so close to the 5′ end that there is no Shine-Dalgarno sequence to favor ribosome binding.

Late Expression Along the Lytic Pathway Occurs When the Buildup of cI Repressor Is Slow

The lytic pathway is followed when the buildup of cI repressor does not occur fast enough to shut down early transcription. The extension of the rightward transcript, stimulated by the N protein, leads to expression of the *Q* gene, producing a key regulatory protein for late transcription. Like N, the Q protein is an antiterminator. Its action, which takes place at the t_{6S} terminator, leads to transcription of all of the late genes including the lysis gene and the genes for the head and tail proteins of the mature phage. At late times other factors encourage rightward transcription from the $P_{R'}$ promoter.

Cro Protein Is Essential to the Lytic Cycle

The cro protein and the cI protein bind to the same sites. Despite this fact, the cro protein is required for lysis, whereas the cI repressor is required for lysogeny. These requirements can be shown with mutants. A *cI* mutant invariably undergoes lysis, whereas a *cro* mutant can lysogenize but cannot complete the lytic cycle. This remarkable difference in the behavior of cro and cI is due to the fact that they bind to the same sites but with different relative affinities. Cro binds preferentially to O_{R3} and less strongly to O_{R1} and O_{R2}. The binding of cro to O_{R3} turns off cI expression from the P_{RM} promoter. At high concentrations cro binds to one or more of the other sites, turning off the P_R promoter. The P_L promoter is also turned off at high concentrations of cro. The most important physiological effect of cro is the turning off of the P_{RM} promoter. In the absence of cro, cI expression increases because of the increase in the number of cI gene copies resulting from early replication of viral DNA during the lytic cycle. This buildup will turn off early rightward and early leftward transcription before sufficient transcripts have been made to ensure a start on the lytic pathway. The cro protein prevents the buildup of cI protein that would otherwise result from the increase in gene copies. The binding properties of cI and cro to the tripartite operator shared by P_{RM} and P_R is shown in figure 30.34.

Figure 30.34

Segment of the λ genome showing the three operators O_{R1}, O_{R2}, and O_{R3} around the P_{RM} and the P_R promoters. The cI and cro regulatory proteins bind to these operators with different relative affinities. The net result of these differing affinities is that cI is required for lysogeny and cro is required for the lytic cycle.

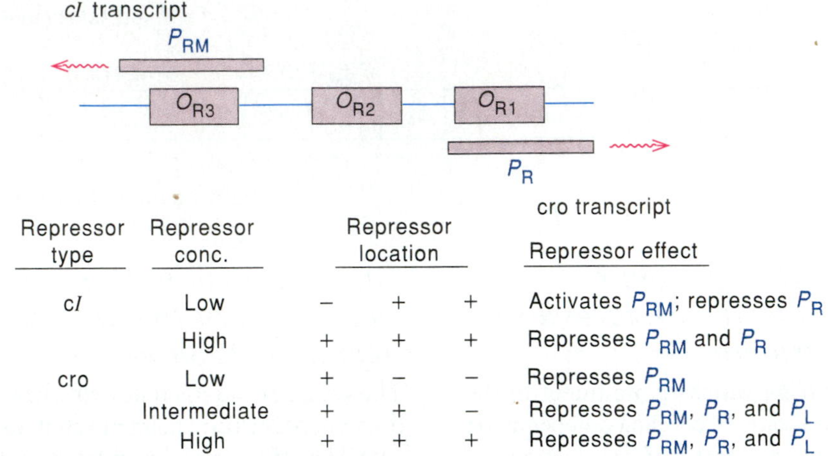

Repressor type	Repressor conc.	Repressor location			Repressor effect
cI	Low	−	+	+	Activates P_{RM}; represses P_R
	High	+	+	+	Represses P_{RM} and P_R
cro	Low	+	−	−	Represses P_{RM}
	Intermediate	+	+	−	Represses P_{RM}, P_R, and P_L
	High	+	+	+	Represses P_{RM}, P_R, and P_L

Summary

In this chapter we have discussed the regulatory systems of the *E. coli* bacterium and the λ bacteriophage. The main points in our presentation are as follows.

1. *E. coli* carries about 4,400 genes. Only a small fraction of the genome is actively transcribed at any given time. But all of the genes are in a state where they can be readily turned on or turned off in a reversible fashion. The level of transcription is regulated by a complex hierarchy of control elements.
2. In the most common form of control, expression is regulated at the initiation site of transcription. There are several ways of doing this, all of them revolving around protein or small-molecule factors that influence the binding of RNA polymerase at the transcription start site.
3. The *lac* operon, a cluster of three genes involved in the catabolism of lactose, exemplifies both positive and negative forms of control that influence the rate of initiation of transcription. Jacob and Monod identified the repressor as a negative control element that is *trans* dominant and the operator as a *cis* dominant site for binding the repressor. Transcription is initiated by RNA polymerase at the promoter, which overlaps the operator site on one side of the three structural genes of the *lac* operon. The tight complex between repressor and operator prevents initiation and it is broken when lactose is present. The lactose is readily converted to allolactose, which binds to the lac repressor. This changes the structure of the repressor so that it dissociates from the DNA.
4. Initiation of transcription proceeds at a greatly increased rate when cyclic AMP is present. This is because cyclic AMP forms a complex with the CAP activator protein, which then binds at a site adjacent to the polymerase binding site. The CAP protein enhances polymerase binding at the adjacent site by cooperative binding.
5. Lactose is the substrate of the enzymes of the *lac* operon. In the absence of lactose, there is no use for enzymes of the *lac* operon.
6. CAP and cAMP activate a large number of genes in *E. coli* that are concerned with catabolism. When glucose is present, the cAMP is greatly lowered and the *lac* operon is expressed at a very low level, even when lactose is present. This is because glucose is a more readily metabolizable carbon source than lactose.
7. The *trp* operon contains a cluster of five structural genes associated with tryptophan biosynthesis. Initiation of transcription of the *trp* operon is regulated by a repressor protein that functions similarly to the lac repressor. The main difference is that the trp repressor action is subject to control by the small-molecule effector tryptophan. When tryptophan binds the repressor, the repressor binds to the *trp* operator. Thus the effect of the small-molecule effector here is opposite to its effect on the *lac* operon. When tryptophan is present, there is no need for the enzymes that synthesize tryptophan.
8. The *trp* operon has a control locus called an attenuator about 150 bases after the transcription initiation site. The attenuator is regulated by the level of charged tryptophan tRNA, so that between 10 and 90% of the elongating RNA polymerases transcribe through this site to the end of the operon. Low levels of *trp* tRNA encourage transcription through the attenuator.
9. Ribosomal RNA and protein synthesis are both controlled at the level of initiation of transcription. This is a result of the direct binding of guanosine tetraphosphate, ppGpp, to the RNA polymerase. This binding decreases the affinity of RNA polymerase for the initiation sites of transcription. Guanosine tetraphosphate is synthesized when the general level of amino-acid-charged tRNA is low.

10. The synthesis of ribosomal proteins is regulated at the level of translation. Certain ribosomal proteins bind to specific sites on the ribosomal RNAs or their own mRNAs. In the absence of the ribosomal RNAs, they bind to their own mRNAs, which inhibits their translation. This form of translational control seems to regulate the rate of synthesis of ribosomal proteins so that it does not exceed the rate of ribosomal RNA synthesis.
11. Viruses borrow heavily on the host enzymatic machinery to obtain energy for synthesis, as well as for replication, transcription, and translation. The virus infective cycle is strongly irreversible. Virus infection is followed by the gradual turning on of viral genes. Viral enzymes are the first viral gene products; in late infection, the virus structural proteins are favored. The irreversible lytic cycle of the virus is directed by a cascade of controls.
12. In λ the host RNA polymerase is used throughout. Regulation is achieved through a series of repressors and activators, as well as two viral proteins that bind directly to the RNA polymerase. The viral proteins that bind to the polymerase modify it so that it can transcribe through provisional stop signals.
13. When λ phage infects an *E. coli* cell, it does not always produce viral progeny. Sometimes it integrates its genome into the host genome and replicates only as the host genome replicates. This so-called lysogenic state can be disrupted by DNA-damaging conditions such as exposure to UV radiation. Under these conditions the dormant viral genome enters the active replication cycle.
14. Bacterial regulatory proteins are controlled by small-molecule effectors; viral regulatory proteins are not. Bacterial genes are regulated in a highly reversible manner; viral genes are usually turned on only once.

Selected Readings

Anderson, J. E., M. Ptashne, and S. C. Harrison, Structure of the repressor-operator complex of bacteriophage 434. *Nature* 306: 846–852, 1987.

Brennan, R. G., and B. W. Matthews, The helix-turn-helix DNA binding motif. *J. Biol. Chem.* 264:1903–1906, 1989. A mini review.

Gilbert, W., and B. Muller-Hill, Isolation of the lac repressor. *Proc. Natl. Acad. Sci.* 56:1891–1898, 1966. First isolation of a repressor protein.

Gold, L., Posttranscriptional regulatory mechanisms in *Escherichia coli. Ann. Rev. Biochem.* 56:199–234, 1988.

Goodrich, J. A., and W. R. McClure, Competing promoters in prokaryotic transcription. *Trends Biochem. Sci.* 16:394–396, 1991. Two or more bacterial promoters are often found in close proximity, and may compete for the binding of RNA polymerase.

Green, P. J., O. Pines, and M. Inouye, The role of antisense RNA in gene regulation. *Ann. Rev. Biochem.* 55:569–597, 1986.

Helman, J. D., and M. J. Chamberlain, Structure and function of bacterial sigma factors. *Ann. Rev. Biochem.* 57:839–872, 1988.

Jacob, F., and J. Monod, Genetic regulatory mechanisms in the synthesis of proteins. *J. Mol. Biol.* 3:318–356, 1961. A classic paper.

Kustu, S., A. K. North, and D. S. Weiss, Prokaryotic transcriptional enhancers and enhancer-binding proteins. *Trends Biochem. Sci.* 16:397–401, 1991. First discovered in eukaryotes, enhancers have now been found to exist for a number of prokaryotic genes.

Nomura, M., R. Gourse, and G. Baughman, Regulation of the synthesis of ribosomes and ribosomal components. *Ann. Rev. Biochem.* 53:75–117, 1984.

Ptashne, M., *A Genetic Switch: Gene Control and Phage* λ. Cambridge, Mass.: Cell Press, and Palo Alto, Calif.: Blackwell Scientific, 1987.

Ptashne, M., A. D. Johnson, and C. O. Pabo, A genetic switch in a bacterial virus. *Sci. Amer.* 247(5):128–140, 1982.

Simons, R. W., and N. Kleckner, Biological regulation by antisense RNA in prokaryotes. *Ann. Rev. Genet.* 22:87–600, 1988.

Steitz, T. A., Structural studies of protein-nucleic acid interaction: the sources of sequence-specific binding. *Quarterly Reviews of Biophysics* 23:205–280, 1990. A very readable and very thorough review of the subject with excellent illustrations, written by one of the pioneers in the field.

Storz, G., L. A. Tartaglia, and B. N. Ames, Transcriptional regulator of oxidative stress-inducible genes: Direct activation by oxidation. *Science* 248:189–194, 1990.

Stwinowski, Z., R. W. Schevitz, R.-G. Zhang, C. L. Lawson, A. Joachimiak, R. Q. Marmorstein, B. F. Luisi, and P. B. Sigler, Crystal structure of trp repressor/operator complex at atomic resolution. *Nature* 335:321–329, 1988.

Weintraub, H., Antisense RNA and DNA. *Sci. Am.* 262(1):40–46, 1990.

Wolberger, C., Y. Dong, M. Ptashne, and S. C. Harrison, Structure of phage 434 Cro/DNA complex. *Nature* 335:789–795, 1988.

Yanofsky, C., Operon-specific control by transcription attenuation. *Trends Genet.* 3:356–360, 1987.

Zubay, G., M. Lederman, and J. DeVries, DNA-directed peptide synthesis III. Repression of β-galactosidase synthesis and inhibition of repressor by inducer in a cell-free system. *Proc. Natl. Acad. Sci.* 58:1669–1675, 1967. Showing that repressor works in a cell-free system.

Zubay, G., D. Schwartz, and J. Beckwith, Mechanism of activation of catabolite-sensitive genes: A positive control system. *Proc. Natl. Acad. Sci.* 66:104–110, 1970. First isolation of an activator protein.

Problems

1. What set of data originally led Jacob and Monod to suggest the existence of a repressor in *lac* operon regulation?
2. In a cell that is z^-, what would be the relative thiogalactoside transacetylase concentration, compared with wild type, under the following conditions?
   (a) After no treatment.
   (b) After addition of lactose.
   (c) After addition of IPTG.

3. Consider a negatively controlled operon with two structural genes (*A* and *B*, for enzymes A and B), an operator gene (*O*), and a regulatory gene (*R*). The first line of data below gives the enzyme levels in the wild-type strain after growth in the absence or presence of the inducer. Complete the table for the other cultures.

	Uninduced		Induced	
Strains	Enz A	Enz B	Enz A	Enz B
Haploid strains:				
(1) $R^+O^+A^+B^+$	1	1	100	100
(2) $R^+O^cA^+B^+$				
(3) $R^-O^+A^+B^+$				
Diploid strains:				
(4) $R^+O^+A^+B^+/R^+O^+A^+B^+$				
(5) $R^+O^cA^+B^+/R^+O^+A^+B^+$				
(6) $R^+O^+A^-B^+/R^+O^+A^+B^+$				
(7) $R^-O^+A^+B^+/R^+O^+A^+B^+$				

4. In a diploid situation would you expect a *crp*$^+$ to be dominant to a *crp*$^-$? Referring to *lac* expression, describe the phenotypes of *crp*$^-$, *crp*$^+$, and *crp*$^+$/*crp*$^-$.
5. Although *E. coli* promoters generally conform to a rather well-defined consensus sequence, no perfect match to this consensus has ever been observed in a naturally occurring promoter. Suggest an explanation.
6. What is the advantage of using IPTG as an inducer of the *lac* operon?
7. The lac repressor has an "on" rate constant for the binding of the *lac* operator (when cloned into λ) of about 5×10^{10} $\text{M}^{-1}\ \text{s}^{-1}$. This value is much greater than the calculated diffusion-controlled process, which is about 10^8 $\text{M}^{-1}\ \text{s}^{-1}$ for a molecule the size of the lac repressor. Explain why this repressor binding works better than expected.
8. Explain how histidine biosynthesis is controlled in *E. coli.*
9. A mutation in the *trp* leader region is found to result in a reduction in the level of *trp* operon expression when the mutant is grown in rich medium. However, when the mutant is grown in a medium lacking glycine, a stimulation in the level of trp enzymes is observed. Explain these observations. What would you anticipate would be the effect of growing the mutant in a medium lacking both glycine and tryptophan?
10. In *E. coli* there are no pools of free rRNAs or ribosomal proteins floating around in the cell even when the bacteria are grown at different growth rates. Explain how *E. coli* coordinates the biosynthesis of the ribosome.
11. Explain why rRNA synthesis is slow to resume when amino acids are added back to a *spoT* mutant in *E. coli.*
12. Describe the principal differences between patterns of control of gene expression used by bacterial host and bacteriophage systems.
13. In the infection of the host cell by λ, cI repressor favors lysogeny, while cro repressor favors the lytic cycle, even though both repressors bind to the same sites. Explain the basis of their different effects.
14. How is the synthesis of the CAP protein regulated? What is unusual about this regulation?
15. Gene regulatory proteins in bacteria were predicted (before their precise structure was known) to interact in the major groove of DNA by a two-site model of binding. Describe the data that showed this model to be correct.

31

CHAPTER

Regulation of Gene Expression in Eukaryotes

All eukaryotes carry their DNA in a membrane-bounded organelle, the nucleus. This creates a fundamental difference between the way eukaryotes and prokaryotes carry out transcription and translation. In prokaryotes the processes are strongly coupled and occur concurrently; in eukaryotes transcription is completed before the start of translation, and the nascent transcripts are subject to an elaborate regimen of posttranscriptional controls. There are other differences that affect the number and types of regulatory mechanisms. Eukaryotic cells usually contain many more genes, than prokaryotes distributed on several chromosomes. Multicellular eukaryotes contain cells of different types, which are limited in the types of genes they can express. These differences result in increasingly complex mechanisms for gene regulation in eukaryotes (fig. 31.1).

In this chapter we will emphasize modes of gene regulation that are unique to eukaryotes. First we will consider a unicellular eukaryote; then we will take a look at some of the complex problems associated with multicellular eukaryotes that show extensive cellular differentiation.

Gene Regulation in Yeast, a Unicellular Eukaryote

A cell of the budding yeast *Saccharomyces cerevisiae* occupies more than ten times the volume of an *E. coli* cell. Its genome is about four times the size of that of *E. coli,* and is distributed among seventeen chromosomes. Yeast cells are nearly spherical in shape, divide by budding, and are bounded by a cytoplasmic membrane that is surrounded by a thick polysaccharide wall (fig. 31.2). The cells contain a nucleus, an endoplasmic reticulum, ribosomes, a Golgi apparatus with secretory vesicles, and several types of granular and vesicular inclusions. The amount of endoplasmic reticulum and the number of mitochondria and ribosomes fluctuate with growth conditions suggesting that regulatory devices are at work.

The separation of transcription from translation permits elaborate schemes for the processing of nascent transcripts prior to their translation. Nascent transcripts destined to become mRNA are modified by capping the 5′ end and are tailed by 3′-terminal polyadenylic acid, (see chapter 28). Splicing occurs for some transcripts, mainly ribosomal protein messages and certain tRNAs. In mature messages the 5′ proximal initiator

Figure 31.1

Overview of some of the unique features associated with regulation of gene expression in higher organisms. Eukaryotes may be unicellular, as in the case of yeast, or multicellular, as in the case of vertebrates. Regulation of gene expression in all eukaryotes bears a close resemblance to regulation of gene expression in simple unicellular prokaryotes. Yet there are some striking differences. Thus multicellular eukaryotes are composed of differentiated cells that can express only a limited number of the total genes in their chromatin. And most of the chromatin in any particular cell type is highly condensed (heterochromatic), but a small fraction of the chromatin is expanded or swollen (euchromatic). It is the latter portion of the chromatin that has the potential to be expressed. The regions of the chromatin that are euchromatic are different in different cell types. To convert a region of the genome from the potentially active to the active state requires the binding of regulatory proteins to the promoter. The role of these proteins is to create an environment favorable for the binding of RNA polymerase. Here we come to one of the most striking differences between prokaryotes and eukaryotes. In prokaryotes the regulatory proteins usually bind in the immediate vicinity of the RNA polymerase. In eukaryotes the regulatory proteins may be bound to the DNA at several places, ranging from a point adjacent to the polymerase binding site to points a thousand or more bases away. Although the explanation for how distantly bound regulatory proteins can influence gene expression is not totally clear, the consensus is that chromosome folding is the most likely way of bringing such proteins into the immediate proximity of the RNA polymerase binding site. The function of all of the regulatory proteins binding to a single promoter of the DNA is to promote formation of the transcription initiation complex.

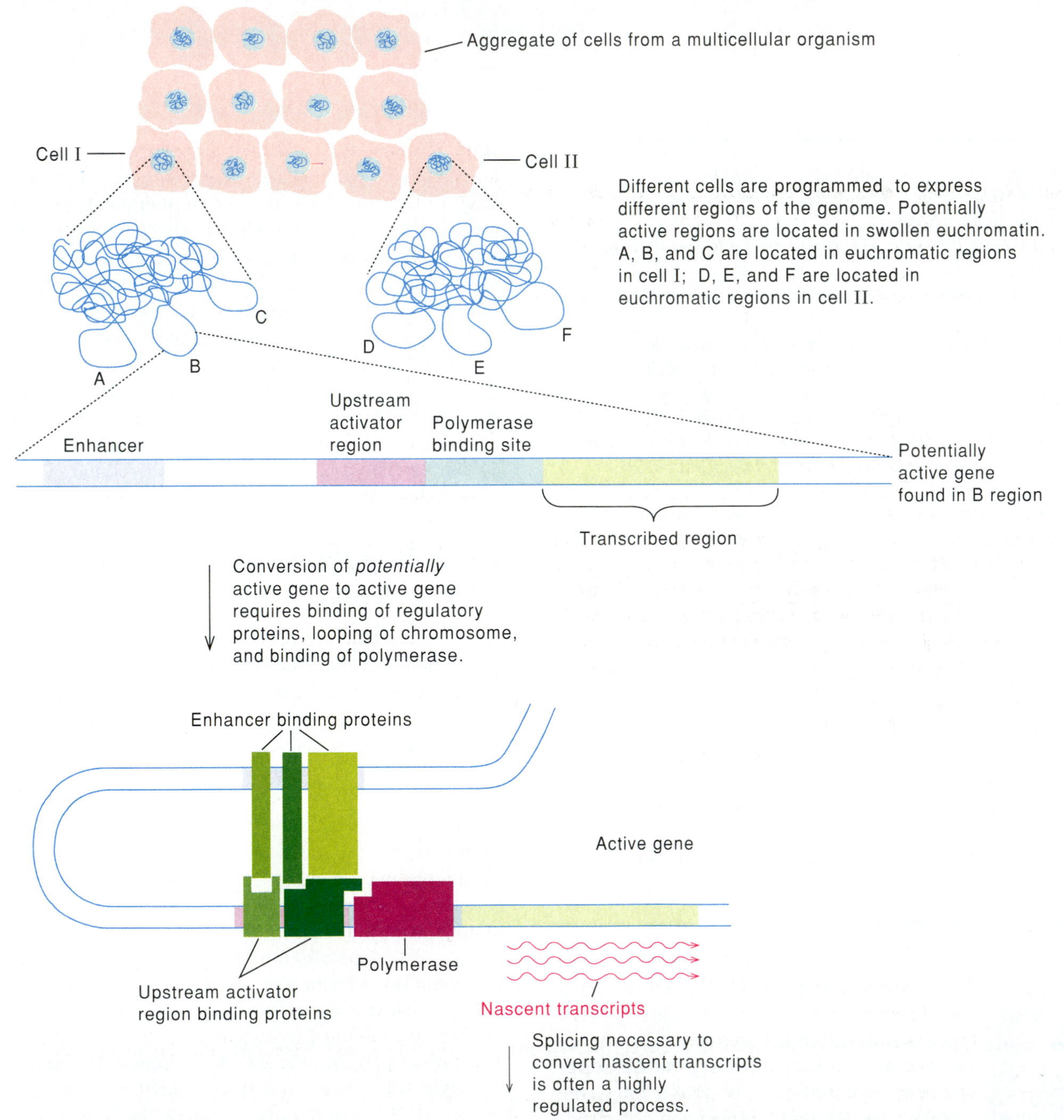
Aggregate of cells from a multicellular organism
Cell I
Cell II
Different cells are programmed to express different regions of the genome. Potentially active regions are located in swollen euchromatin. A, B, and C are located in euchromatic regions in cell I; D, E, and F are located in euchromatic regions in cell II.
A
B
C
D
E
F
Enhancer
Upstream activator region
Polymerase binding site
Potentially active gene found in B region
Transcribed region
Conversion of *potentially* active gene to active gene requires binding of regulatory proteins, looping of chromosome, and binding of polymerase.
Enhancer binding proteins
Active gene
Upstream activator region binding proteins
Polymerase
Nascent transcripts
Splicing necessary to convert nascent transcripts is often a highly regulated process.

Figure 31.2

Diagram of a haploid yeast cell. A yeast cell contains many of the organelles characteristic of a typical eukaryotic cell. Duplication occurs by a budding process. The bud gradually grows until, just before pinching off, it contains a nuclear equivalent of chromosomes, as well as some mitochondria and other elements present in the cytoplasm. Mechanical strength of the yeast cell is guaranteed by a thick cell wall.

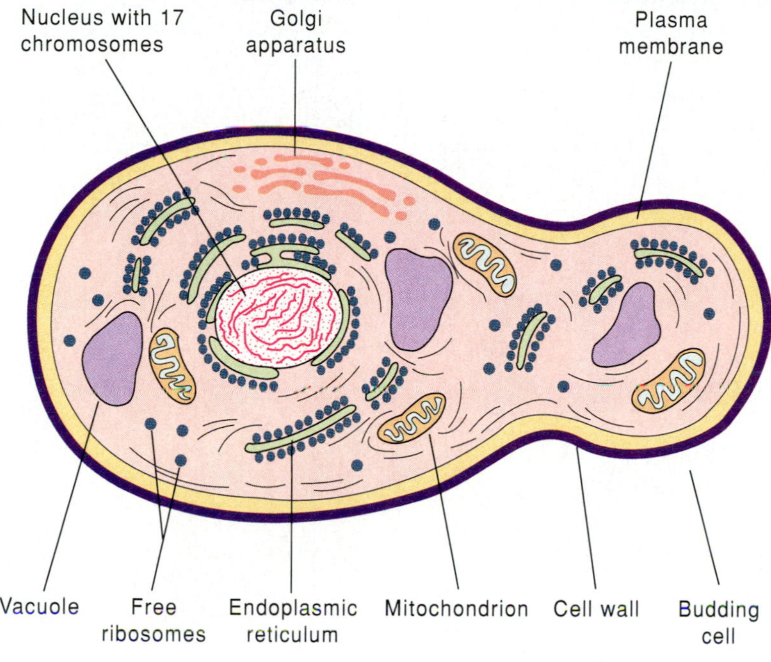

codon (AUG) is usually recognized as a translation start (see chapter 29). This restriction eliminates the value of polycistronic mRNAs and probably accounts for the greater dispersal of genes serving common pathways that are found in yeast compared with bacteria.

A comparison of the gene arrangements and messages for histidine biosynthesis in *E. coli* and *S. cerevisiae* illustrates some of the contrasting modes of gene arrangement and gene expression in these two organisms. In *E. coli,* all ten histidine biosynthetic enzymes are encoded by a single polygenic mRNA, used for the synthesis of the corresponding individual polypeptide chains. In yeast, genes for three of these activities (steps 2, 3, and 10 in the biosynthesis) are clustered at the *HIS4* locus, yielding a single transcript that specifies a single polypeptide chain of molecular weight 9.5×10^4. This polypeptide chain folds into a multifunctional enzyme that carries out three enzymatic functions (fig. 31.3) that in *E. coli* are executed by separate proteins. It is not uncommon to find multifunctional polypeptide chains in eukaryotes for functions that are carried out by two or more polypeptide chains in *E. coli.*

Galactose Metabolism Is Regulated by Specific and General Control Factors in Yeast

In yeast four enzyme activities are required for galactose utilization, each of them encoded by a unique gene. Three of these genes, *GAL7, GAL10,* and *GAL1,* are clustered on chromosome

Figure 31.3

The *HIS4* gene encodes three enzyme activities. It produces a single transcript that results in a single enzyme. The enzyme is multifunctional, carrying out steps 2 (B), 3 (A), and 10 (C) in histidine biosynthesis.

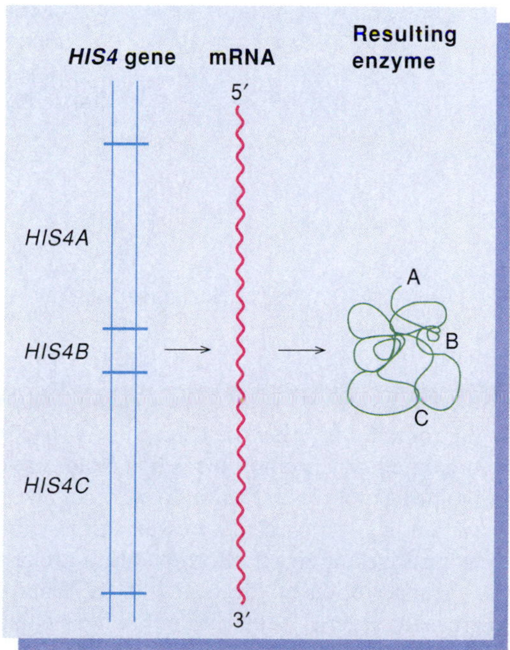

Figure 31.4

Model for the regulation of enzymes of the *GAL* system. (*a*) Three structural genes synthesize distinct mRNAs and enzymes (transferase, epimerase, and kinase). Arrows next to the genes indicate the direction of transcription. Synthesis requires the GAL4 protein. However, it is inactive in the absence of inducer because of complex formation with the GAL80 protein. The GAL80 protein does not prevent the GAL4 protein from binding to specific sites on the DNA, but it does prevent GAL4 protein from activating transcription once bound. (*b*) GAL4 protein becomes active when the GAL80 protein is removed by adding inducer, which binds to the GAL80 protein.

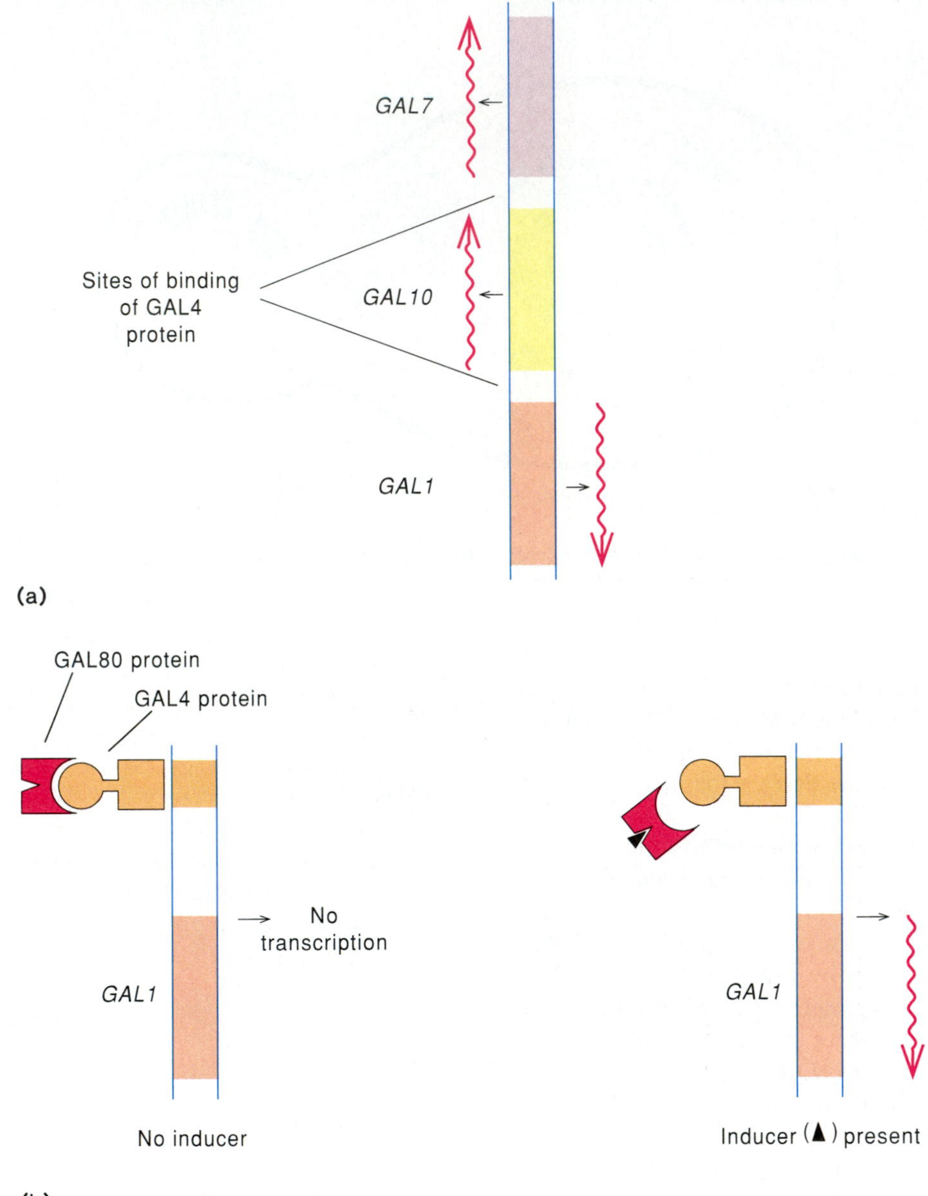

II, whereas the fourth, for galactose transport, is specified by a gene (*GAL2*) located on chromosome XII. (Note: in yeast the normal wild-type genotype is capitalized and the mutant genotype is written in lower case.) Expression of these four genes is regulated by positive and negative controls that are of a highly specific type. Transcription of the *GAL1*, *GAL7*, and *GAL10* genes is increased several thousandfold when galactose is present.

Specific Controls Affect Expression of the* GAL *Genes Despite their clustering and their coordinate induction by galactose, the *GAL7*, *GAL10*, and *GAL1* genes are not transcribed into a single mRNA molecule (fig. 31.4). The *GAL7* and *GAL10* genes are transcribed from the same DNA strand. However, the *GAL1* gene, approximately 600 base pairs from *GAL10*, is transcribed from the complementary DNA strand. Discrete transcripts are generated for the synthesis of three distinct proteins that contain the different enzyme activities: transferase (GAL7), epimerase (GAL10), and kinase (GAL1).

The yeast system is ideally suited for genetic studies as the organism can be grown in either the haploid or the diploid state. Genetic analysis of the *GAL* system indicates two *trans*-acting gene products that regulate expression, GAL4 and GAL80. Recall from our observations on prokaryotes that *trans*-acting gene products usually are proteinaceous. Most *gal4* mutants are uninducible for the *GAL* structural genes in the haploid state, but inducible in the diploid state when paired with a *GAL4* wild type. This suggests that the active GAL4 product is probably a positive control protein like CAP in the *lac* operon. Rare *GAL4*[c] mutants result in constitutive expression of the *GAL* genes. This suggests that the structure of the GAL4 protein is modified so that its function is no longer repressible. Most *gal80* mutants also give rise to constitutive expression of the *GAL* genes. This suggests that the active form GAL80 is a repressor of *GAL* gene expression. Rare *GAL80*[s] mutants are uninducible in the haploid state, or in the diploid state where they are paired with wild-type *GAL80*. It is believed that wild-type GAL80 protein binds the inducer galactose and that this binding converts the protein to an inactive form that does not inhibit the action of the GAL4 protein. The GAL80[s] protein appears to have lost the site for galactose binding. The results of the different mutants for *GAL1* and *GAL10* gene expression are summarized in table 31.1.

Based on the genetic studies with the different regulatory gene mutations, a model has been proposed for the mechanisms of regulation of the *GAL* system (see fig. 31.4*b*). In this model the GAL4 protein binds to a site upstream of the gene and promotes RNA polymerase II-dependent transcription of the target genes, and the GAL80 protein represses transcription by binding to the GAL4 protein. Induction entails dissociation of the GAL4-GAL80 protein complex, permitting the GAL4 protein to function as an activator. Induction results from the binding of galactose to the GAL80 protein, altering its structure so that it has a lowered affinity for the GAL4 protein.

The proposed model requires the presence of *cis*-acting sequences upstream of the target genes. These GAL4 protein-binding sequences are termed upstream activating sequences (UASs). The GAL4 protein binds to four related 17-bp dyad symmetric sequences upstream from *GAL1* and *GAL10*, activating their (divergent) transcription. The specific DNA sequences to which the GAL4 protein binds have been determined by reacting the DNA with dimethylsulfate and then following the methylation pattern by nucleotide sequencing to determine which guanine bases are protected by the regulatory protein. The sequences that showed protection in a wild-type *GAL4* strain were not protected in a *gal4* mutant derivative. Further manipulations of the promoter region showed that only one of the 17-bp dyad sequences is needed for normal regulation.

The model for how GAL4 protein functions depicts a complex protein, with binding sites for DNA and GAL80 protein as well as a site involved in the activation of transcription. Deletion studies were performed to monitor whether these different sites could be allocated to different parts of the 881-amino-acid GAL4 protein. These studies showed that a mutant protein containing only the first 98 N-terminal amino acids binds DNA but cannot bind the GAL80 protein or activate transcription. Additional mutant studies indicate that amino acids 148–196 and 768–881 are required to activate transcription and that amino acids 851–881 are required for GAL80 protein binding.

Table 31.1
Phenotypes of Haploid and Diploid Yeast with Regulatory Gene Mutations

Genotype	GAL10 (epimerase)[a] Noninduced	Induced	GAL1 (kinase) Noninduced	Induced
Wild type	−	+	−	+
gal4	−	−	−	−
gal4/GAL4	−	+	−	+
gal80	+	+	+	+
GAL4[c]	+	+	+	+
GAL80[s]	−	−	−	−
GAL80[s]/*GAL80*	−	−	−	−

[a]In the induced state, galactose is added to the growth medium. It should be noted that GAL10 and GAL1 respond in identical fashion to the different mutations.

Further studies of the GAL4 protein showed that different parts of the GAL4 protein and the lexA bacterial repressor protein are functionally interchangeable. To demonstrate this, Mark Ptashne and his co-workers carried out a domain-swap experiment in which the DNA-binding domain of the lexA regulatory protein from *E. coli* was inserted in place of the GAL4 DNA-binding domain. As you might expect, this protein was ineffective at inactivating the wild-type *GAL1* gene for transcription (fig. 31.5). However, if the DNA-binding site for the GAL4 protein was replaced by the DNA-binding site for the lexA protein, then this hybrid protein functioned in transcription of the *GAL1* gene in the normal way.

The DNA-binding domain of GAL4 protein was once thought to have the structure of a simple zinc finger, which we will discuss later (see fig. 31.20). Now, however, it appears that the DNA-binding domain of this protein is more complex—that it involves two zinc ions and six cysteines (fig. 31.6).

Catabolite Gene Products Are Regulated by Repression and Inactivation

A complicating factor in the regulation of the *GAL* genes emerges from two observations regarding yeast that is grown on glucose: (1) The expression of the *GAL* genes is strongly inhibited even in the presence of galactose, and (2) the GAL4 protein does not protect its target DNA sequences from dimethylsulfate. Your first reaction may be to think of the catabolite repression effect that glucose has on *lac* operon expression. However, closer inspection shows that there are significant differences.

Figure 31.5

Ptashne's domain-swap experiment. The GAL4 protein normally binds at a site upstream from the *GAL1* gene. In the absence of GAL80 this activates transcription from the *GAL1* gene. If the DNA-binding domain of GAL4 is replaced by the DNA-binding domain of the lexA protein, there is no transcription from the *GAL1* gene because the hybrid regulatory protein does not bind at the appropriate site. However, if the DNA-binding site is also changed to that of the DNA-binding site for the lexA protein, then the *GAL1* gene becomes active again.

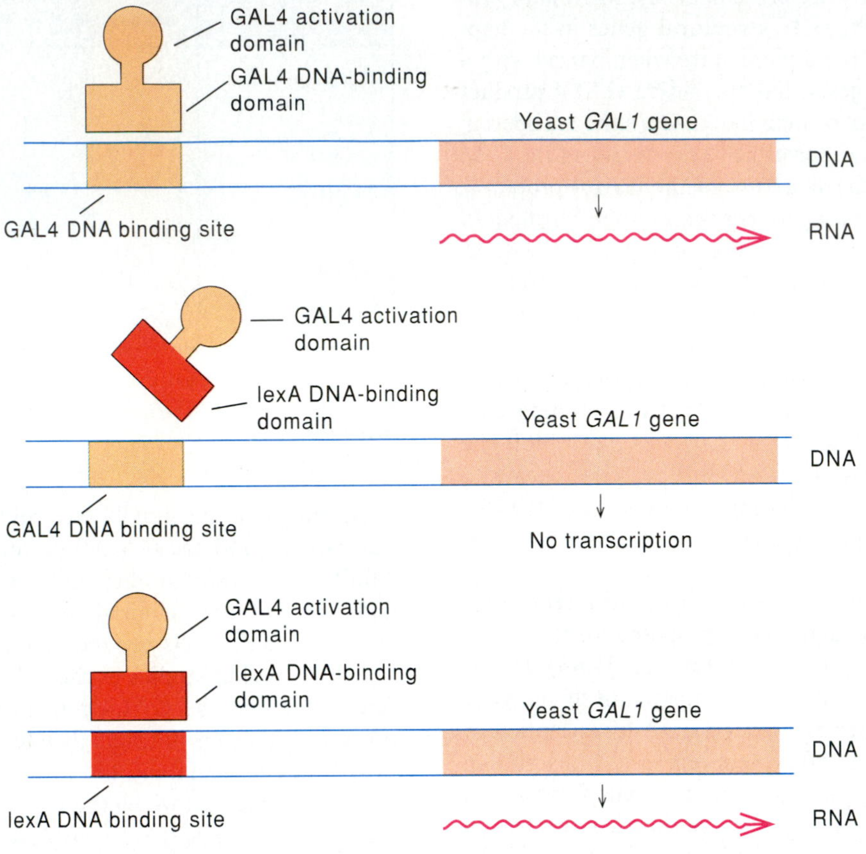

Figure 31.6

Proposed structure for the zinc thiolate cluster of GAL4. (Source: B. L. Vallee and D. S. Auld, "Zinc coordination, function, and structure of zinc enzymes and other proteins," in *Biochemistry* 29:5647, 1990. Copyright © 1990 American Chemical Society, Washington, D.C.)

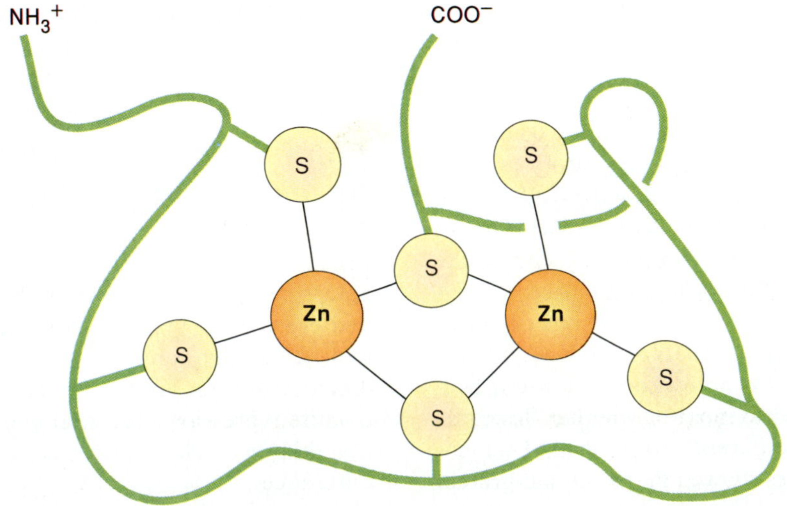

First of all, it is clear that in the yeast system cAMP is not involved, because the effects are the same even in yeast strains that are defective in cAMP synthesis. Second, the presence of glucose not only inhibits the synthesis of the GAL structural proteins, it inhibits the activity of the GAL proteins that are already present. The former effect is called catabolite repression; the latter effect is called catabolite inactivation. The mechanisms underlying these effects of glucose are being intensively investigated.

Positive and Negative Controls of Transcription Are Common in Yeast

In addition to the *GAL* system, the *cis*-acting regulatory regions of many other yeast genes have been investigated. The gene encoding iso-1-cytochrome *c* (*CYC1*) is under positive control; the gene *HAP1* regulates *CYC1* dependence on heme. HAP1 protein also regulates the gene *CYC7*, closely related to *CYC1*, as well as the genes *CYT1* (cytochrome c_1) and *CTT1* (catalase). Another *cis*-acting element, *UAS2*, is regulated by three proteins, HAP2, HAP3, and HAP4, that form a heteromeric duplex to the sequence CCAAT. The transcription of *HIS3* and *HIS4* (involved in histidine biosynthesis) is derepressed by starvation for histidine or other amino acids (general amino acid control). Starvation of yeast for any one of ten amino acids derepressed the synthesis of more than 30 enzymes in nine different amino acid biosynthetic pathways. These coregulated genes have two to four copies of a *cis*-active DNA sequence, 5′-TGACTC-3′, in far upstream regions. Several *trans*-acting genes function in a regulatory hierarchy in the starvation response. The gene *GCN4* functions most directly as a positive regulator of transcription by interaction of the GCN4 protein with the TGACTC sequence in the promoter regions of the genes regulated by this general control protein. The synthesis of GCN4 itself is regulated in a novel way at the translational level. The GCN4 mRNA has four AUG codons at different points in its extensive leader sequence. Starvation conditions, and *trans*-acting factors interact to regulate translation from the fourth translation initiation codon that leads to GCN4 protein synthesis.

Figure 31.7 depicts the *cis*-acting sites of several yeast genes. In all cases these include TATA boxes (T). The open boxes indicate the sites that mediate positive control of transcription. For example, the site that interacts with the GAL4 protein that regulates *GAL10* gene expression is located more than 100 base pairs upstream of its TATA box. The *CYC1* gene has two homologous sites, at about −275 and −225, that mediate activation by heme. UAS control sites can be located at considerable distances from the genes they influence. Furthermore, they can be oriented in either direction relative to these genes. Hence the control sites appear to differ from most controls found in prokaryotes, but are quite similar to enhancer control elements found in multicellular eukaryotes (see chapter 13 and the discussion later in this chapter).

How different are the controlling regions in yeast from bactiera? Yeast genes often have several controlling sites that are variable in their spacing. The multiple UASs for a given gene permit a variety of regulatory transcriptional responses, yet the TATA box still determines the site for transcription initiation. In yeast the canonical TATAAA sequence can be replaced by closely related sequences such as CATTTATT.

Figure 31.7

Location of control elements for a select group of yeast genes. The genes have all been found to contain upstream activation sites (UASs) or upstream sites of repression. Locations of TATA boxes and initiation sites are also indicated. In some cases there is more than one initiation site.

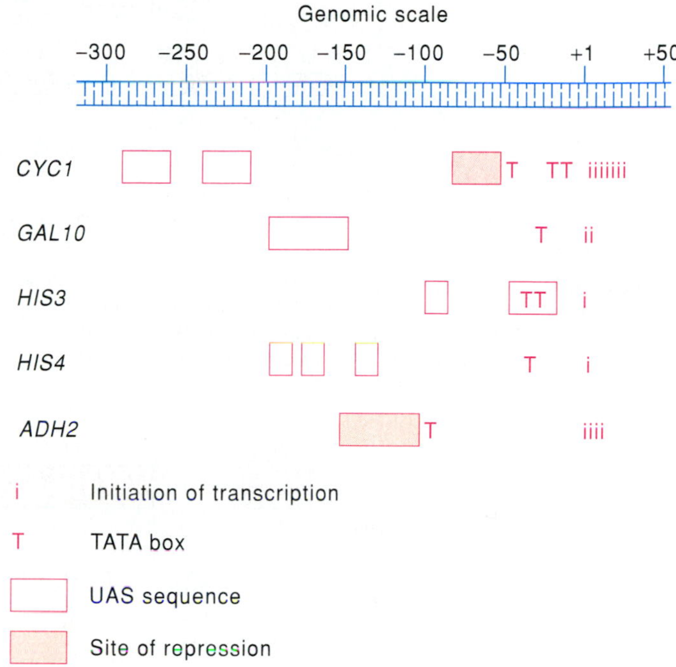

Specific negative regulators that bind directly to DNA are relatively rare in yeast. An example is the protein encoded by one of the genes of the *MATα* locus discussed in the next section.

Mating Type Is Determined by Transposable Elements

Yeast has two haploid cell mating types, *MATα* (or simply *α*) and *MATa* (or *a*) that can fuse to form a single diploid (*a*/*α*) cell. Diploid cells can grow indefinitely as diploid cells or they can undergo sporulation and by meiosis convert back into an equal number of *a* and *α* cells. The haploid mating type of *S. cerevisiae* is determined by the *MAT* locus. Haploid cells with the homothallism (*HO*) gene are able, at almost every cell division, to switch between the *a* and *α* mating types. This efficient switching generates progeny of opposite mating types from the progeny of a single cell, which readily fuse to produce *a*/*α* diploids that do not switch or mate, but can sporulate when starved. Haploid heterothallic (*ho*) strains interchange their mating types rarely, on the order of 10^{-6} events/generation. During interconversion of the mating type, the DNA sequences of *MAT* are replaced by either the *MATα* or *MATa* sequence copied from one of the two silent storage copies *HMLα* or *HMRa* (fig. 31.8). When the sequences are transposed from *HMLα* to *MAT*, the *α* mating type is expressed; transposition from *HMRa* to *MAT* leads to expression of the opposite, or *a*, mating type.

Figure 31.8

Yeast chromosome III, showing the *HML*, *MAT*, and *HMR* loci. *W* is a region common to *MAT* and *HML*, but not *HMR* (~750 bp). *X* is a region found at *MAT*, *HML*α, and *HMRa* (~700 bp). Y_a is a specific substitution found at *MATa* and *HMRa* (~600 bp). Y_α is a specific substitution found at *HML*α and *MAT*α (750 bp). Z_1 is a region found at *MAT*, *HML*α, and *HMRa* (~250 bp). Z_2 is a region found at *MAT* and *HML*α (~70 bp).

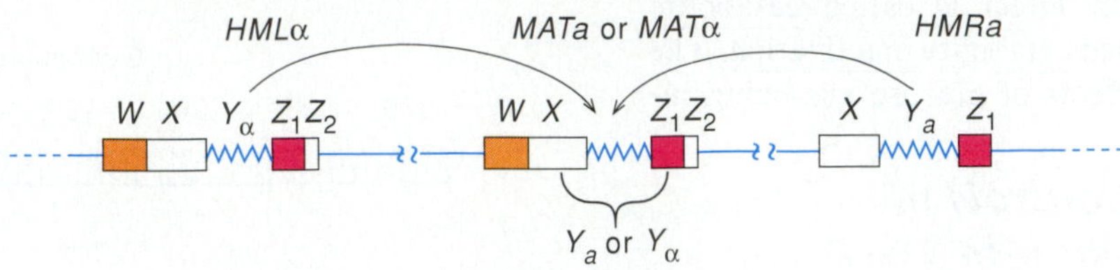

Figure 31.9

Structure of mating loci determinants in yeast. The genetic regions *HML*α and *HMRa* are normally silent, whereas *MATa* or *MAT*α are active. *MATa*, which contains the Y_a segment, expresses transcript *a*1. *MAT*α, which contains the Y_α segment, expresses two transcripts, α_1 and α_2. The structures in and around the Y_a and Y_α segments are the same at the storage locus and the expression locus. The inactivity at the storage loci results from far upstream *cis*-acting elements not present at *MAT*. These elements, in conjunction with the four *SIR* gene products, negatively regulate expression of mating type genes at the storage loci.

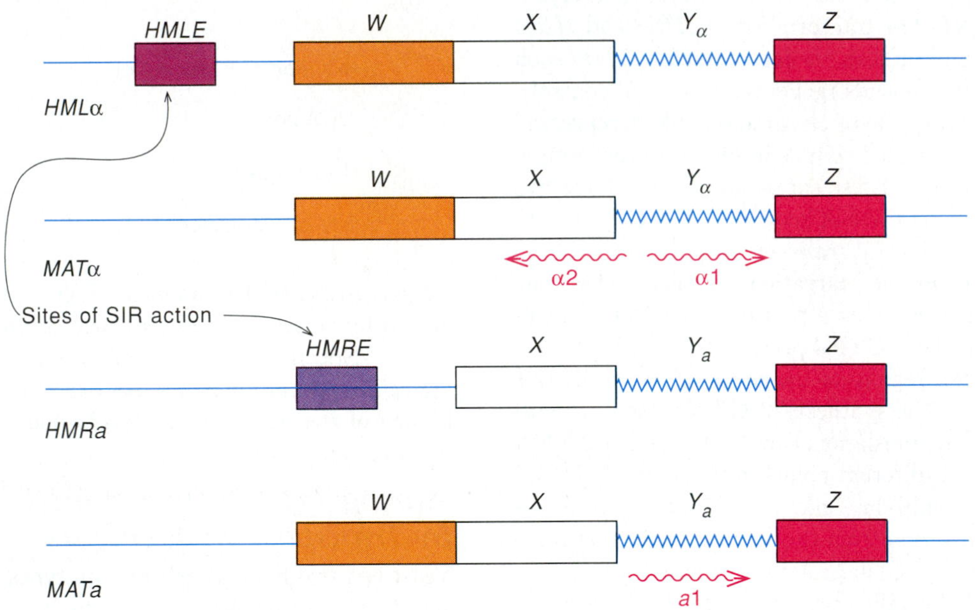

Two immediately interesting questions arise about regulation associated with mating-type switching of haploid yeast strains: (1) What regulates mating-type switching? And (2) why are the α and *a* sequences expressed only when they are present at the *MAT* locus? Partial answers are available to both of these questions. In both cases it appears that *trans*-acting control elements are involved.

The *MATa* and the *MAT*α sequences that are stored at the *HML* and *HMR* loci are expressed only when they are present at the *MAT* locus. The explanation for this behavior was revealed by the finding of four unlinked genes, *SIR1–SIR4*. In all *sir* mutant strains the *MATa* and *MAT*α sequences at *HML* and *HMR* that are normally silent express as they do at *MAT*. The *trans*-acting regulatory factors, encoded by the *SIR* genes, act as repressors at "silencer" sites (*HMLE* and *HMRE*) adjacent to *HMR* and *HML*. Remarkably, these silencer sites are more than a kilobase away from the regions of transcription initiation (fig. 31.9). The silencer sites are not found at or near to the mating type locus (*MAT*), which is why the *a* and α coding sequences are expressed when transposed to *MAT*. The *cis*-acting *HMRE* sequence is able to repress in either orientation and over distances 2,600 bp away from the transcription initiation site.

The important question remaining is, How does a single genetic locus determine the haploid yeast-cell mating type? Haploid cells that carry *MAT*α behave as α cells; cells that carry *MATa* behave as *a* cells; and, finally, diploid cells that carry both *MAT*α and *MATa* do not express any haploid specific genes. To explain the key role of *MAT* it was proposed that the *MAT* locus encodes regulatory proteins that control unlinked genes

Figure 31.10

Regulation of *a*-specific genes (*a*sg) and α-specific genes (αsg) in α and *a* haploid cells. In an α cell, α2 inhibits *a*sg and α1 activates αsg. In an *a* haploid cell, the *a*1 transcript made at *MATa* does not appear to exert any regulatory function. (⊕ = positively regulated; ⊖ = negatively regulated.)

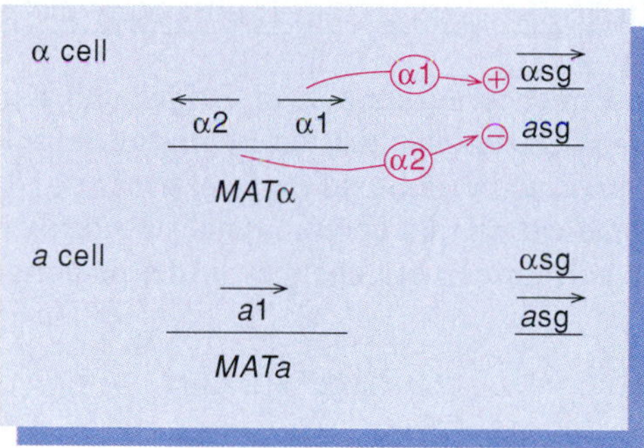

Figure 31.11

Regulation of meiosis in yeast. Haploid cells are not able to initiate meiosis because they express *RME1*, which is a negative regulator of meiosis. Diploid cells are able to initiate meiosis because they make a complex of α2 and *a*1 that inhibits the synthesis of the RME1 product.

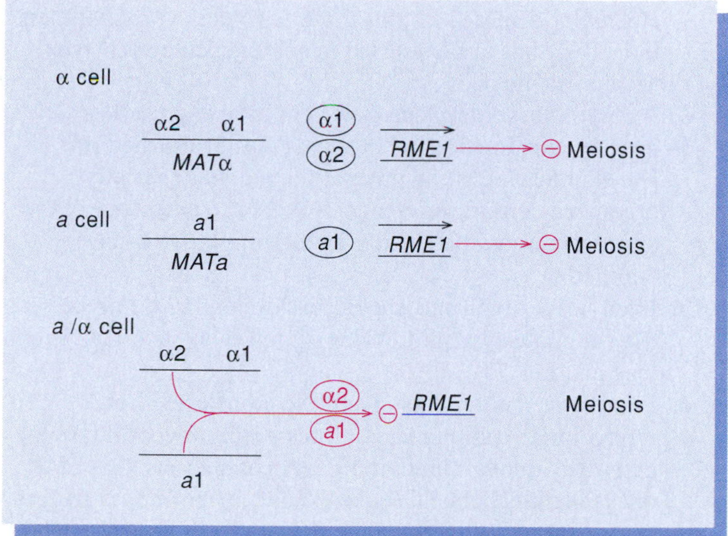

that determine cell type. Subsequently it was shown that the *MATα* locus contains two genes, *MATα1* and *MATα2*, that encode the α1 and α2 regulators, respectively. According to Ira Herskowitz and his co-workers, α1 turns on expression of the α-specific genes and α2 turns off expression of the *a*-specific genes (fig. 31.10). A number of recent observations support this model. For example, the mRNA of *STE6* (an *a*-specific gene) is synthesized in cells that lack the α2 product. It has also been shown that α2 is a sequence-specific DNA-binding protein that recognizes a 32-bp operator sequence upstream of the *a*-specific genes *STE2, STE6, MFA1, MFA2,* and *BAR1*. When this operator is placed between the *CYC1* UAS and its transcription initiation site, α2 brings about repression, or negative control *in vivo.* Even when the operator is placed upstream of the UAS sequence it still represses expression of *CYC1,* but to a lesser degree. Thus the operator need not overlap with essential promoter sequences (as in prokaryotes) to permit repression by α2.

In addition to beginning to understand the regulatory mechanisms that give the haploid cells their specific properties, researchers have made considerable progress in determining the mechanism that regulates meiosis in diploid cells. Diploid *a*/α cells do not mate but can undergo sporulation and meiosis when the diploid cells are starved for nitrogen and have a poor carbon source. It is believed that the reactions leading to meiosis are triggered by a complex of the α2 and *a*1 proteins, which represses the haploid-specific gene *RME1* (regulator of meiosis), allowing the cells to initiate meiosis (see fig. 31.11).

The control of mating type in yeast illustrates one use of mobile genetic elements. Mobile genetic elements are commonly found in both prokaryotes and eukaryotes, but they rarely insert at specific sites as they must in this case. Because of their random insertion behavior, most mobile genetic elements are more apt to be vehicles for evolutionary change than mechanisms for regulating specific genes as in yeast mating-type switching. In higher organisms, mating type (sex) is controlled in part by gene dosage. For example, in humans, males carry one X chromosome whereas females carry two X chromosomes. The mechanism used by yeast and other fungi is more easily influenced by factors in the external environment and may be more suitable for a unicellular organism, which is more exposed to its surroundings.

This concludes our discussion of the regulatory behavior of yeast. There are many ways in which *E. coli* and yeast are quite similar. All of the regulatory processes are highly reversible, so that each cell retains its totipotency, that is, its potential to express any of its genes. This situation does not hold true in more complex eukaryotes. The model developed by Jacob and Monod to explain the regulation of the *lac* operon of *E. coli* at the genetic level (that is, the interaction of a regulatory protein with a specific region of DNA) remains one of the most important conceptual developments in gene expression and development.

General Patterns of Gene Regulation in Multicellular Eukaryotes

In terms of the types of regulatory mechanisms used, yeast is probably typical of the relatively simple, undifferentiated eukaryote. Even in yeast we see the beginning signs of cell differentiation processes and hormonal systems for intercellular communication that typify more complex multicellular eukaryotes. Yeast differentiates into three cell types—the *a* and α haploid types and *a*/α diploids. Furthermore, haploid yeast cells secrete peptides known as pheromones that affect cells of the opposite mating type and change their pattern of gene expression so as to favor cell-to-cell contact and mating. Such similarities between yeast and more complex eukaryotes have made yeast valuable as a model system for studying cellular differentiation and hormonal control, but they do not eliminate the need to study these processes more directly in higher forms.

A number of unique regulatory problems must be confronted in higher eukaryotes:

1. The arrangement of DNA in the nucleus must be efficiently disposed so that those few genes (probably less than 1%) that are expressed in any particular cell type are accessible.
2. As the embryo develops, new genes are expressed and already expressed genes in some cases are turned off. Development must be precisely timed and spatially organized so that different cell types are produced in the appropriate numbers and orientation relative to one another.
3. Even adult organisms engage in developmental processes. The immune system provides examples of such activities (see chapter 33).
4. Finally, in the developing and the adult organism, intercellular communication is essential to coordinate the activities among different cells. A complex battery of diffusible biochemical signals known as hormones trigger responses at the cell membrane in some cases and at the nucleus in others (see chapter 24).

In this chapter we will consider some aspects of these problems, and we will treat additional aspects in chapters 33 and 34.

Nuclear Differentiation Starts in Early Development

In the mature adult state, a multicellular eukaryote contains many cells, each with a potential for the constitutive or inducible expression of only a small, select fraction of the total genome. Since most differentiated cells in the adult contain the normal amount of genomic DNA, the expression of only a fraction of the genome in any particular cell type suggests that the average nucleus may have lost its totipotency.

The first direct evidence that nuclei irreversibly differentiate during development came from Briggs and King in 1952. Working with frog eggs, they showed that the original nucleus from an unfertilized egg could be replaced by the diploid nucleus from a developing animal (fig. 31.12). If the inserted nucleus was taken from the donor very early in its development (at the blastula stage), a mature frog developed. If, however, the nucleus came from a later stage of development, no growth or only limited growth ensued. These nuclear transplantation experiments demonstrated that during development the nucleus assumes a pattern of expression that is difficult or impossible to reverse even when the nucleus is returned to the environment of the egg cytoplasm.

In 1962 Gurdon took this type of experiment one step further by showing that adult frogs could be developed by injecting enucleated eggs with nuclei from the intestinal epithelium of tadpoles. The success frequency was much lower than when nuclei from earlier stages of development were used. Nevertheless, such results showed the influence of the cytoplasm on nuclear expression; they also demonstrated that in certain cases nuclei already tentatively committed to a specific pathway of development can be reprogrammed by placing them in a different environment. The most important lesson to be gained from such experiments is that the nucleus tends to assume a stable pattern of expression that is ultimately dictated by its surroundings.

These pioneering studies of Briggs and King and of Gurdon, demonstrating nuclear differentiation and dedifferentiation, are supported by a broad range of genetic and biochemical findings indicating that chromosomal structural differences often can be correlated with changes in the potential for gene expression.

Chromosome Structure Varies with Gene Activity

In interphase cells, chromosomes exist are composed of chromatin. Chromatin can assume either a compact form known as heterochromatin or a swollen form known as euchromatin. Most nuclei contain both types of chromatin. The heterochromatin is generally concentrated around the nucleolus and the inside of the nuclear envelope. Some chromosomes or parts of chromosomes are always heterochromatic (constitutive heterochromatin); others are heterochromatic only during certain times of the cell cycle or in certain cell types (facultative heterochromatin). The ratio of euchromatin to heterochromatin increases with increasing protein synthetic activity. Many studies

Figure 31.12

Nuclear transplantation technique. The nucleus of an unfertilized egg is mechanically removed with a micropipette (left), and a nucleus from a blastula cell is removed and microinjected into the enucleated egg.

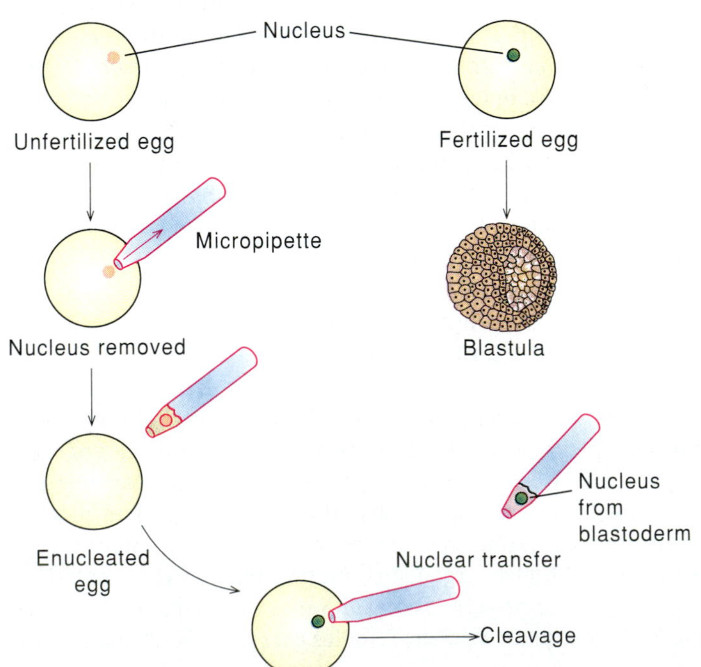

indicate that euchromatin is relatively active in RNA synthesis and heterochromatin is relatively inactive. Although the euchromatic state seems to be necessary for a high level of transcription, for most genes it is not sufficient. This has been demonstrated by autoradiography of cells grown briefly in the presence of ^{3}H-labeled RNA precursors. Those regions of the genome that are transcriptionally active pick up the label; the extent of labeling is proportional to the amount of RNA synthetic activity. Some euchromatic regions appear active by this test, whereas others do not.

Giant Chromosomes Permit Direct Visualization of Active Genes Polytene chromosomes are unusually large chromosomes found in the salivary glands of certain insects. Because of their size, they are convenient vehicles for studying the relationship between chromosome structure and activity. Polytene chromosomes are produced by the repeated replication of interphase chromosomes without separation, a process resulting in a large number of chromosomes that remain laterally aligned. For example, the DNA content in the giant salivary gland cells of *Drosophila melanogaster* may be as much as 1,000 times that of other cells in the fruit fly. Microscopically, the stained chromosomes appear as cross-banded extended bodies (fig. 31.13). There are about 5,000 bands in *Drosophila,* and it is tempting to associate single bands with single genes. However, the average DNA content found in each band is 30,000 base pairs, which is considerably more DNA than necessary to encode the average protein (about 1,000 base pairs).

Figure 31.13

A segment of giant chromosome from the midge larva of *Rhyncosciara* at different stages during development. The arrows and connecting lines indicate comparable bands. Changes in the extent of swelling of different regions reflect the activity of those regions in transcription. (Source: M.F. Breuer and C. Pavin, *Chromosoma* 7:275–280, 1946. Copyright © 1946 Springer-Verlag, Heidelberg, Germany.)

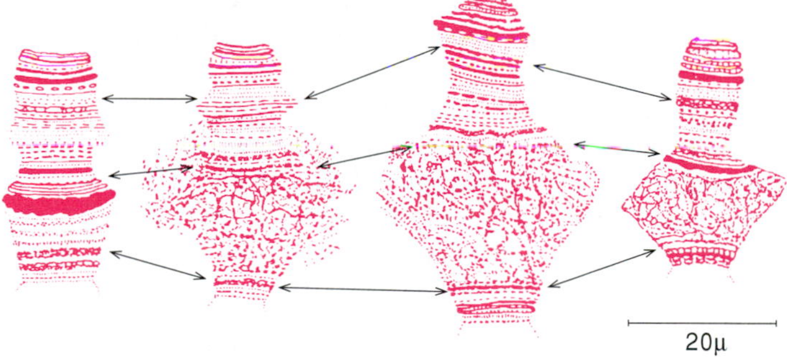

Transcription in polytene chromosomes usually is associated with local swellings of the bands, called puffs (see fig. 31.13). Sometimes puffing results in a broadening and lengthening, and sometimes the extended DNA projects laterally into loops that combine to form a large ringlike structure. As a rule, a puff originates from a single band, but it can result from swelling of one or more bands. The extent of incorporation of precursors into a puff is approximately proportional to the size of the puff. A detailed investigation of the puffing patterns as a function of the physiological state of the organism shows that different bands become activated, swollen, and transcriptionally active in a sequentially related, tissue-specific fashion.

The correlation of puffs with genetic functions has focused on two types of gene products, secretory proteins and proteins formed in response to heat shock. Certain puffs in *Drosophila* can be artificially induced by the insect hormone ecdysone, which is instrumental in regulating development. Some puffs are induced directly by exposure to the hormone, whereas other puffs are induced indirectly, depending for puffing on the gene products of the directly induced puffs. This dependence is shown by adding drugs that inhibit protein synthesis, with the result that secondary puffs do not form.

In Some Cases Entire Chromosomes Are Inactive In some cases entire chromosomes can be heterochromatized and stay that way through successive cell divisions. Male and female mammals are distinguished by the fact that females carry two X chromosomes, whereas males carry only one. Invariably, one of the X chromosomes in the somatic tissue of females is inactivated and condensed into a heterochromatic state known as a Barr body (fig. 31.14). The consequences of X-chromosome inactivation are readily observed in females who are heterozygous for an X-linked mutation.

For example, the enzyme glucose-6-phosphate dehydrogenase (G-6-PD) is encoded by an X-linked gene. A female that is heterozygous for this gene may carry two alleles that produce electrophoretically distinct forms of the same enzyme (A and B). When isolated cells from a skin biopsy are cloned,

Figure 31.14

Diagram of nuclei obtained from cells in the mucous membrane of the human mouth. (*a*) Nucleus from a female, showing one Barr body (arrow). (*b*) Nucleus from a male, with no Barr body. (*c*) Nucleus from an XXX female, showing two Barr bodies. (*d*) Nucleus from an XXXX female, showing three Barr bodies.

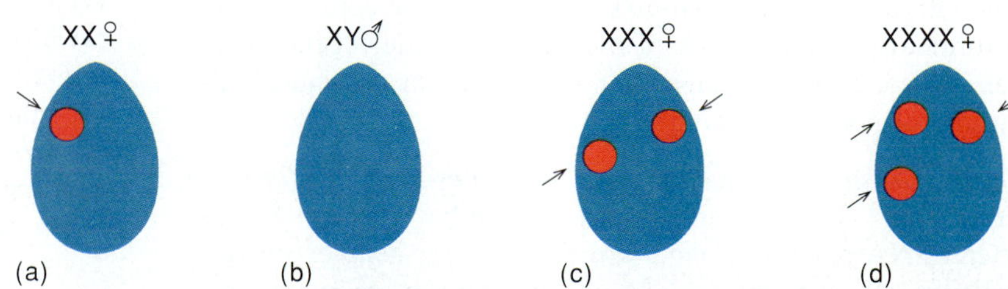

each clone contains either the A or the B form of the G-6-PD enzyme, but never both. If every skin cell of a single organism were to be analyzed for the enzyme, large homogeneous patches expressing one or the other of the G-6-PD alleles would be found. This pattern indicates that the decision as to which X chromosome of the female will be inactivated is made at the multicellular stage in the embryo. Once the decision is made, that X chromosome remains inactive through successive cell divisions. There is an equal chance that the X chromosome from either parent will be inactivated. In genetically abnormal cells that contain three or four X chromosomes, two or three Barr bodies, respectively, can be found (see fig. 31.14).

X-chromosome inactivation provides a simple means of maintaining equal amounts of active X-linked genes in both males and females. This form of so-called dosage compensation is not found in all species. For example, in *D. melanogaster,* neither of the X chromosomes in the female cells is condensed into a Barr body. In this species, dosage compensation, which regulates the activity of specific alleles, operates by a different mechanism so that many alleles in the female X are about 50% as active as the corresponding alleles in the male. However, not all alleles in the *Drosophila* X chromosome are regulated in this way, so that for certain genes twice as much gene product is produced in females as in males.

Chromosomal inactivation takes a different form in the ciliated protozoan *Tetrahymena.* In the vegetative phase each cell contains two nuclei, a micronucleus and a macronucleus. The micronucleus is transcriptionally inactive; it divides mitotically during normal cellular duplication and it undergoes meiosis during the sexual phase of the life cycle. The macronucleus contains much more DNA than the micronucleus, but only about 10% of the sequences present in the micronucleus are represented in the macronucleus. Clearly, the average sequence in the macronucleus is represented many times. The macronucleus divides in an imprecise fashion during normal cell divisions. It is completely lost during meiosis and therefore must be generated anew following each sexual phase.

All nuclear transcripts are made in the macronucleus. From its size and limited sequence complexity, it is clear that the macronucleus must contain many copies of selected regions of the genetic nucleus. How the number of copies made from different genes is regulated is unknown. Selective hybridization studies on rDNA show that the micronucleus contains about 20 copies of the rDNA genes, whereas the macronucleus contains about 200. These observations on *Tetrahymena* demonstrate that even in a unicellular eukaryote the percentage of the nuclear DNA that is transcriptionally active is quite small (about 10%). This intricate compartmentalization and amplification of transcriptionally active regions of the genome in the vegetative nucleus provides a mechanism for the organism to regulate the number of genes according to its needs.

Chromosomal Rearrangement Is Sometimes Required for Gene Expression There are infrequent cases in bacteria and in yeast where DNA rearrangements are essential for gene expression. For example, in yeast we saw that mating type requires DNA rearrangements, and the mechanisms are fairly well understood. We have also seen that in *Tetrahymena* wholesale rearrangements of the genome leading to a vegetative nucleus are essential for all forms of transcription.

In the cells of multicellular eukaryotes it is doctrine that the arrangement of genes in the chromosomes is not disturbed in going from the germ line to various differentiated cells. This simply means that we have not detected rearrangements for the most part. The possibility is still open that certain key rearrangements are involved in differentiation processes, and we must keep our eyes open for such possibilities. In one case, that of immunoglobulin genes, we know that intricate DNA rearrangements in the form of DNA splicing play a key role in determining which antibody genes are expressed. This topic is discussed in chapter 33.

Biochemical Differences between Active and Inactive Chromatin

Much of the preceding discussion on chromatin structure indicates that active chromatin exists in a more swollen state than inactive chromatin. Several biochemical changes accompany the transition from condensed to swollen chromatin. These changes include a redistribution of nucleosomes along the DNA duplex, chemical modification of histones, alteration in the pattern of nonhistone chromosomal protein binding, and chemical modification of the DNA.

DNA in Active Chromatin Is More Susceptible to DNase Degradation The greater accessibility of transcriptionally active chromatin has been elegantly demonstrated by Groudine and Weintraub for the hemoglobin and ovalbumin genes in the chicken. They isolated chromatin from chicken erythrocytes, in which the hemoglobin genes were recently very active, and the oviduct, in which the ovalbumin gene is very active. They then treated these two chromatins with DNase I, using an assay that measured gross DNA degradation as well as that of specific genes. In both cases the rate of DNA degradation from the active gene was much greater than from the average DNA. In erythrocyte chromatin the globin gene DNA was rapidly degraded, whereas in oviduct chromatin the ovalbumin gene DNA was rapidly degraded. Extensive regions surrounding the active genes were also quite susceptible to DNase I hydrolysis.

Histones Are Often Modified in Active Chromatin Histones undergo transient modifications that may have a general effect on transcription. The amino group in the side chain of lysine is subject to acetylation and methylation. Methylation also modifies certain arginine and histidine residues. Histones H3 and H4 are the main targets of these reactions, but H2A and H2B are also affected. Modification peaks during S phase, when the DNA is being duplicated. Certain serine groups in histone H1 are phosphorylated, with the maximum in phosphorylation occurring at a later time during mitosis. These covalent modifications all reduce the net positive charge of the histones, which could lower the affinity between the histones and the DNA.

Figure 31.15

Histone H2A can become linked to ubiquitin to form UH2A. The linkage involves the α-COOH on the C-terminal glycine of ubiquitin and the side-chain amino group on a lysine residue of the histone.

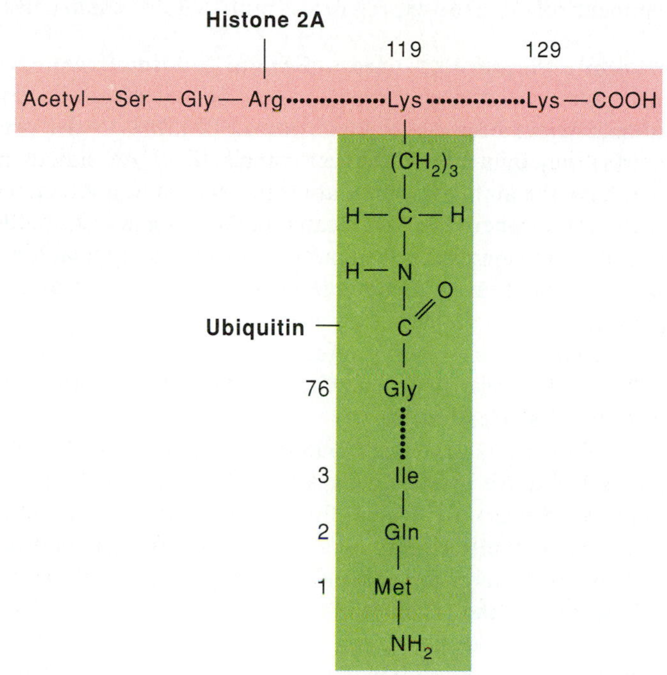

Figure 31.16

Hemimethylated DNA. The unmethylated C residue in the indicated sequence is destined to be methylated shortly after DNA replication.

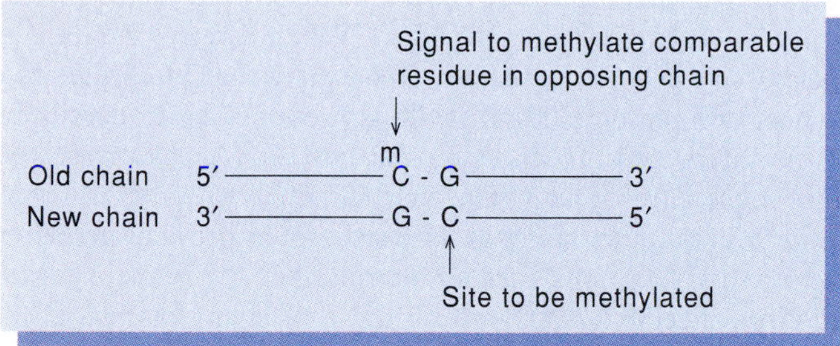

Some histone H2A is modified by combination with a 76-residue protein called ubiquitin. The complex forms between the C-terminus glycine residue of ubiquitin and the side-chain amine group of the lysine at position 119 of H2A (fig. 31.15). Between 5 and 15% of the H2A molecules exist in this type of complex. Although ubiquitin has a greater tendency to be associated with euchromatin, it is released from all chromatin at some time during mitosis. We have seen that ubiquitin is often associated with proteins prior to degradation (see chapter 28); however, there is no evidence that histones linked to ubiquitin are fated for degradation.

Active Chromatin Is Associated with High-Mobility Group (HMG) Proteins In addition to the histones and ubiquitin, chromatin is associated with a group of proteins known as the high-mobility group (HMG) proteins, so named because of their high electrophoretic mobility in low-pH polyacrylamide gels. The four major HMG proteins in calf thymus tissue—HMG1, 2, 14, and 17—have molecular weights of less than 30,000. The HMG proteins can be selectively removed from the chromatin with 0.35-M NaCl, evidence that they are bound less firmly than the histones. The major HMG proteins, HMG1 and HMG2, contain approximately 25% basic and acidic amino acid residues. Amino acid sequence studies show that the charge distribution within the HMG molecules is nonuniform, with major clusters of acidic amino acids. The HMG1 protein contains an extraordinary sequence of 41 continuous aspartic and glutamic acid residues.

The total nuclear concentration of HMGs is low relative to that of the histones (about 3%). Their distribution within the chromatin suggests that they are located in the transcriptionally active regions of the chromatin. Two of the nonhistone proteins, HMG14 and HMG17, are preferentially released from chromatin on partial digestion with DNase I. The notion that HMG proteins are associated with active chromatin regions is supported by reconstitution experiments. Selective removal of HMG from chicken erythrocyte chromatin by 0.35-M NaCl extraction eliminates the selective sensitivity of the globin gene to DNase I. The globin gene in such a preparation regains its DNase I sensitivity on reconstitution with HMG14 and HMG17. This result suggests that there are factors in the active regions of the genome that favor HMG protein binding, and that HMG protein binding enhances the activity of already active genes.

DNA Methylation Is Correlated with Inactive Chromatin Methylation of cytosine in CpG dinucleotides may play a regulatory role in the gene control of eukaryotes. In chapter 25 we noted that 5-methylcytosine (5-MC) is the main modified base in vertebrate DNA. Most of the 5-MC occurs in the dinucleotide CpG. Indeed, in mammals and birds approximately 50 to 70% of all such dinucleotide sequences are modified. In this connection it is noteworthy that CpG sequences occur much less frequently than we would expect on a statistical basis. The mechanism of methylation has been explored by DNA transfection of certain tissue-culture-grown cells. If the DNA used in transfection is methylated, the pattern of methylation is maintained through many cell duplications. Likewise, when unmethylated DNA is used in transfection, a nonmethylated pattern is maintained. These results indicate that under many circumstances methylation is passively maintained by a signal that recognizes hemimethylated DNA. If the C residue in a CpG dinucleotide is methylated immediately after semiconservative DNA replication, only the parental DNA chain is methylated; the passive methylation system signals methylation of the corresponding C residue in the new DNA chain (fig. 31.16).

The extent of methylation of a gene correlates with its ability to transcribe. A most elegant study of this correlation comes from the laboratory of Rudy Jaenisch, in work on the expression of Moloney murine leukemia virus (MuLV). MuLV is a tumor-producing RNA virus that belongs to the retrovirus family (see chapter 25, table 25.4). When a retrovirus infects a permissive cell, it produces a DNA copy that integrates into the host chromosome. Viral RNA is produced by transcription of this integrated DNA, using the host RNA polymerase II. Embryonal carcinoma (EC) cells infected with MuLV were found to contain as many as 100 integrated proviral genomes. Despite this, such cells were not permissive for virus replication, and virus-specific RNA synthesis did not occur. Analysis of the DNA in the infected EC cells showed that the proviral genomes had become highly methylated. Evidently, EC cells are different from many other cell lines in that they rapidly methylate unmethylated (or hypomethylated) DNA. Strong support for the idea that methylation is the cause of this inactivation came from experiments in which the EC-infected cells were pretreated with 5′-azacytidine, which inhibits DNA methylation; EC-infected cells treated with this inhibitor produced infectious viruses.

Given that DNA methylation usually reduces transcription, two important, closely related questions remain unanswered: How is methylation regulated *in vivo?* And how does methylation interfere with transcription? We know that methylation does not interfere with the elongation phase of RNA synthesis; therefore it seems likely that methylation blocks initiation. The binding of polymerase and other regulatory proteins at the initiation locus is sensitive to any modification of these nucleotides that may have taken place. The precise regulation and inhibition mechanisms, however, await further elucidation.

Thus far we have been dealing with aspects of chromosome structure that influence gene expression. You may be wondering how those differences in chromosome structure originate. We cannot give a complete answer, but it seems likely that *cis*-acting DNA sequences and *trans*-acting regulatory proteins are the primary factors that affect gene expression and chromosome structure in the interphase nucleus.

Influence of DNA-Regulatory Protein Interactions on Gene Expression

The primary sequences in the noncoding regions of eukaryotic DNA contain (1) signals for the binding of RNA polymerase, (2) gene regulatory proteins, (3) transcription start sites, and (4) potential coding regions for small peptides. In chapter 29 we saw that several proteins in addition to RNA polymerase II are required for the initiation of transcription at the adenovirus major late promoter; this promoter is believed to share many characteristics with other eukaryotic promoters. In addition to signals for regulatory protein binding in the immediate vicinity of the polymerase binding site, there are other signals, which bind at some distance from the promoter, that have a major influence on gene expression. In yeast, sequences of this type are termed UAS sequences (see our earlier discussion) since their action appears to be confined to regions located upstream from the promoters they influence. *Cis*-active sequences operating over great distances were first discovered in animal viruses. Such sequences are called enhancers because they enhance expression of their associated genes. Enhancer sequences are important and rather unusual transcription signals that play a prominent role in gene expression of multicellular eukaryotes.

Enhancer Sequences Are General Stimulators of Transcription Enhancer signals cannot be understood in conventional terms because of their location relative to the promoter(s) they influence. Enhancer signals, like UAS signals in yeast, have the ability to stimulate transcription at appreciable distances from where they are located in the genome (200–5,000 bp). Enhancer signals are even more versatile than typical UAS signals in that they are effective downstream as well as upstream from the promoters they stimulate. Moreover, some enhancer elements are equally effective in either orientation on the DNA. Thus inversion of an enhancer element frequently does not destroy its effectiveness.

Enhancers have been found in many systems; some enhancers are active in most cells, while others are quite tissue-specific. Enhancers are crucial for regulating the expression of many genes in multicellular eukaryotes. Possibly enhancers account for the difference between euchromatic and heterochromatic regions of the chromatin.

The first evidence for enhancers came from George Khoury's studies on simian virus 40 (SV40), which infects monkey cells. The SV40 genome is a circular duplex with about 5,200 base pairs. In SV40-infected cells, viral RNA synthesis is divided into early and late phases. Early transcription originates from a block of sites illustrated in figure 31.17. There is a TATA box, typical of many eukaryotic gene promoters (see chapter 28) about 27 bp upstream from the transcription start site. Immediately upstream of the TATA box the SV40 promoter contains a series of three repeated GC-rich regions that bind a regulatory protein known as SP1. SP1 protein is a widespread activator found in vertebrates. Further upstream from the transcription start site there is a tandemly repeated 72-bp sequence (-116 to -188 and -189 to -261 from the 5′ end of the messenger). Removal of one of these sequences has no effect on transcription, but removal of both of the 72-bp sequences drastically lowers early transcription.

Surprisingly, the precise position or orientation of the 72-bp segment is not critical for stimulating transcription. Thus the 72-bp segment remains effective after inversion or after translocation further upstream or downstream from the transcription start site. Foreign genes inserted into DNA containing the SV40 enhancer are frequently stimulated in the same way as the SV40 early region, a result demonstrating that this enhancer's effect is gene-specific.

Other animal viruses carry segments with a similar activating function, even though the sequences are quite different. Peter Gruss and his colleagues have shown that the SV40 72-bp repeat may be functionally replaced by a segment comparable in size from either end of the DNA cloned from Moloney murine sarcoma provirus (MSV). Enhancers have also been

Figure 31.17

Region in and around the early transcription start sites of the SV40 genome. The base-pair number on the circular genome is indicated. There are several transcription start sites. One cluster of start sites (early), used initially after infection, is located about 27 bp downstream from the TATA box. The other cluster of start sites (late early) is used at later times. This cluster of start sites is located upstream of the TATA box. One of the products of early transcription is the protein known as large T antigen. This protein binds in and around the core *ori*. T antigen inhibits early transcription. The upstream positive control *cis* elements include the tandem 72-bp enhancer sequences and three tandem 21-bp sites. The latter function in conjunction with host-encoded protein known as SP1.

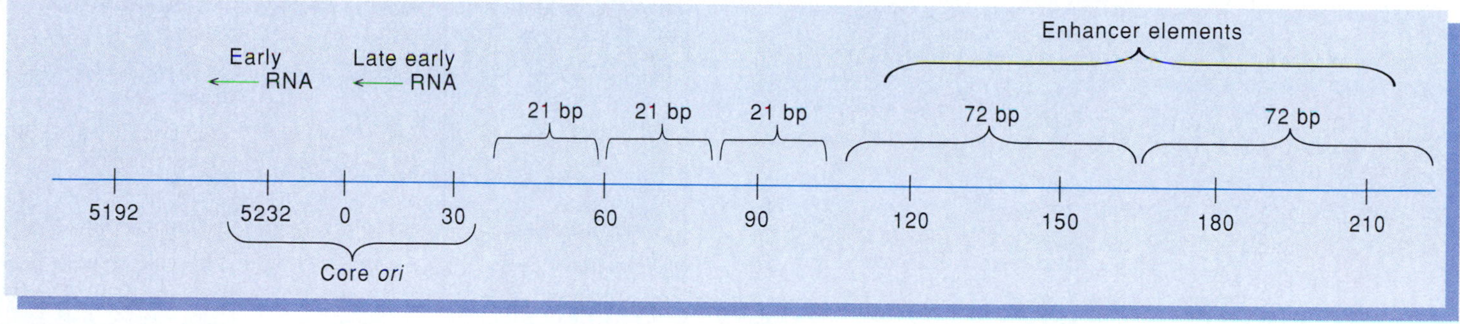

found in the host genome. For example, a tissue-specific enhancer has been found in the intron region of an immunoglobulin gene (see chapter 33); it is considered to be tissue-specific since it stimulates expression of nearby genes only in antibody-forming lymphocyte cells.

Further analysis of the 72-bp SV40 enhancer shows that it is composed of elements 15 to 20 bp in length that bind one or more protein factors specifically. These elements, called enhansons, are important when separated from one another. It is believed that different enhancers are composed of different combinations of enhansons. This system permits a great variety of responses in different cell types and for different genes in the same cell because many diverse combinations of enhansons can be generated from a limited number of enhansons. The general picture of the enhancer that is emerging is that of a complex multicomponent segment of DNA that can bind protein factors in a tissue-specific manner (fig. 31.18).

This type of modular construction of enhancer from enhansons may also be characteristic of promoters, which in addition to the polymerase binding site contain sequences where a number of protein factors that bind immediately next to or near to the polymerase. The basic distinction between enhancers and promoters is operational. An enhancer is able to stimulate transcription at a considerable distance; a promoter and its component modules are only able to stimulate transcription in the immediate vicinity.

Conventional regulatory proteins usually bind to specific sites on the DNA or to other specific DNA-binding proteins. It is not certain how the proteins that bind to enhancers influence transcription over long distances. However, several proposals have been made: (1) The enhancer element may function as a point of attachment to a structural component of the nucleus to stimulate transcription; (2) it may serve as an initial binding site for some factor required for transcription, which must subsequently move along the duplex to the initiation point for transcription; or (3) folding or looping of the chromosome may bring the *cis*-bound enhancer proteins into close contact with the gene it stimulates (fig. 31.19). Currently the third possibility is favored. But even if the third mechanism is the one most frequently used, the other two may be used in place of or in conjunction with the third mechanism for some genes.

Figure 31.18

Enhancers are frequently composed of two or more enhansons. Each enhanson contains two or more proteins binding side-by-side to specific sites on the DNA and with affinities for one another. An enhancer that contains two enhansons and two protein factors can be arranged in 16 ways. An enhancer composed of two enhansons with three possible protein factors can be arranged in 77 ways if all arrangements and combinations of protein factors are permitted. The power of combinations permits a great deal of variety in enhancer construction with a very limited number of protein factors.

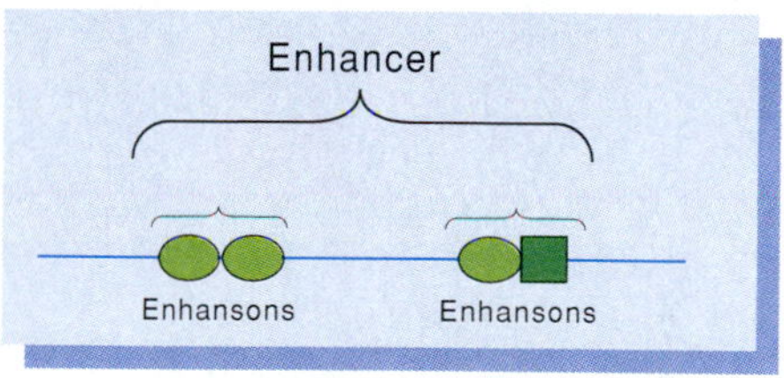

Many DNA Regulatory Proteins Contain Two Domains: One for Binding DNA and One for Binding to Another Protein

Many transcription activators and transcription repressors have two domains, one for binding to DNA and one for binding to a transcription-related protein. The two-domain structure was most effectively demonstrated by Ptashne's domain-swapping experiment (see fig. 31.5). Each domain comes in many varieties.

DNA-Binding Domains of Transcription Factors On the basis of current evidence, the helix-turn-helix is the most commonly used DNA-binding motif found in bacteria. This motif is also used extensively by eukaryotes (table 31.2). For example, it is found in a number of proteins that regulate transcription during

Figure 31.19

Possible mechanisms for enhancer action. (1) The enhancer draws the associated gene to the nuclear matrix, where it is more accessible to the transcription apparatus. (2) The enhancer is the initial binding site for an element that subsequently moves. (3) The enhancer folds or loops, depending on its polarity, to bind with other promoter elements.

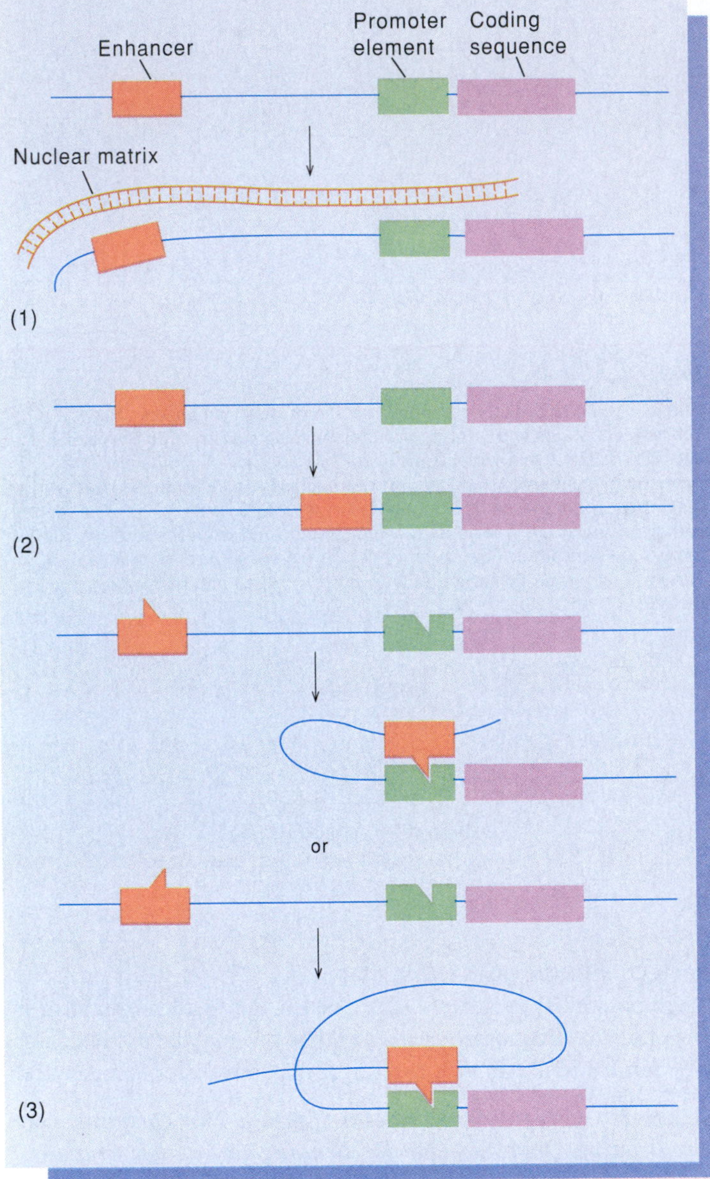

early development in *Drosophila*. These proteins carry a characteristic 180-base segment known as a homeobox (box 31A). Although most helix-turn-helix proteins in bacteria bind as dimers, the homeobox-containing proteins bind as monomers. Since dimers bind more strongly than monomers, other factors being equal, we must ask why monomers are used in some cases. Possibly it is because transcription activators that bind as monomers interact with one or more additional proteins in place of an identical monomer.

Table 31.2
DNA Regulatory Proteins Found in Eukaryotes

Regulatory Protein	Source	Function
Helix-turn-helix		
MATα1	Yeast	Activates α-specific genes
MATα2	Yeast	Inactivates a-specific genes
MATa1	Yeast	Combines with α2 to repress haploid-specific genes
Antennapedia	*Drosophila*	Homeotic gene
Ultrabithorax	*Drosophila*	Homeotic gene
Engrailed	*Drosophila*	Segment-polarity gene
Fushi tarazu	*Drosophila*	Pair-rule gene
Zinc finger[a]		
GAL4	Yeast	Galactose-dependent activator
TF IIIA	*Xenopus laevis*	Required for Pol III transcription of 5S RNA
Krüppel	*Drosophila*	Gap genes
Hunchback	*Drosophila*	Gap genes
Glucocorticoid steroid receptor	Vertebrates	Positive and negative acting; in many but not all cell types
Leucine zipper		
GCN4	Yeast	Activates genes for enzymes that make amino acids
C/EBP	Mammals	Activates genes in liver and other cells; limited cell distribution
c-Fos/c-Jun	Mammals	Growth regulation

[a]The category in which we should place a number of DNA regulatory proteins is uncertain because we do not know the precise three-dimensional structures of these proteins. There is a tendency to classify a protein in a specific category on a limited amount of composition and sequence data. For example, the GAL4 protein was once thought to be a zinc finger (see fig. 31.20), but current indications are that this protein involves two zincions and six cysteines in a more complex structure (see fig. 31.6).

In eukaryotes an even more popular DNA-binding motif is found in certain zinc metalloproteins. The role of zinc ion (Zn^{2+}) in this so-called zinc finger motif is to stabilize the tertiary structure of the recognition helix in the DNA-binding protein. This it does by forming a tetrahedral complex with histidine and cysteine residues. The most common complexes are formed with two histidines and two cysteines (the C_2-H_2 motif).

31A
BOX

Localization of eve *and* ftz *Transcripts in the Wild-Type Developing Embryo*

Localization of the *eve* and *ftz* transcripts in *Drosophila* embryos is determined by *in situ* labeling with a ^{32}P-labeled specific DNA probe. All embryos are oriented so that anterior is to the left and dorsal is up. In dark-field photomicrographs of tissue autoradiograms (see illustration), the labeling pattern is seen only near the dorsal and ventral edges of the embryos because the labeling is done on sections cut through the middle of the embryos. Keep in mind that cells are concentrated near the surface in the early embryo. *Eve* transcripts are located in adjacent anterior and posterior regions of every other segment, giving rise to seven bands in all (top). The same is true of *ftz* except that complementary regions are labeled (middle). This point is made most clear by the labeling pattern observed in a double-labeled section, where the embryo is simultaneously labeled with both probes (bottom). (Photomicrographs made by K. Harding and M. Levine.)

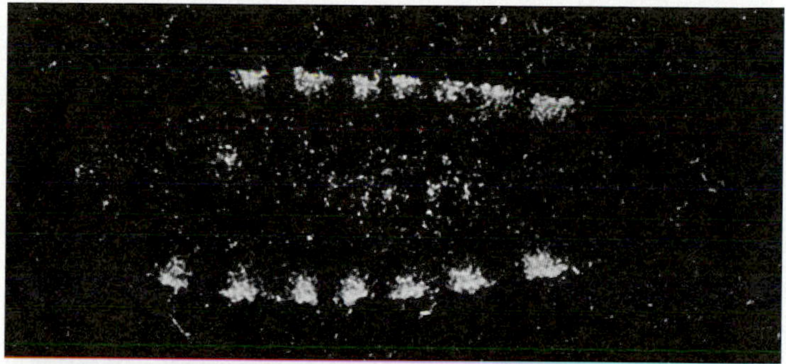

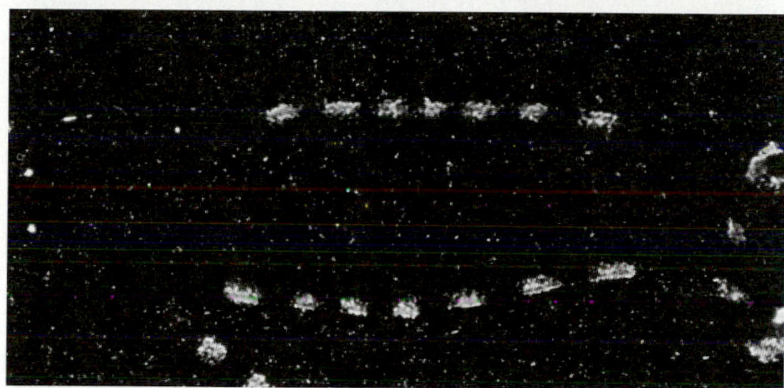

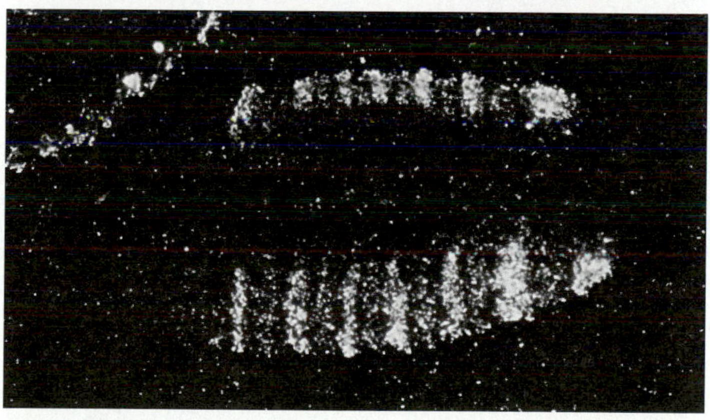

Complexes with four cysteines (the C_2-C_2 motif) are also well known. Zinc ion (Zn^{2+}) is a virtually unique divalent cation for making structural complexes with proteins for several reasons: (1) It has a strong affinity for polarizable nitrogens and sulfurs in amino acid side chains; (2) it is completely stable in the +2 valence state; and (3) in tetradentates, zinc ion (Zn^{2+}) always forms a complex with the precise geometry of a tetrahedron.

Proteins that use the C_2-H_2 and the C_2-C_2 motifs differ in several ways. The C_2-H_2 motif was first identified for the TFIIIA transcription factor that binds to the internal control region of the 5S rRNA gene in *Xenopus laevis* (see chapter 28). In this protein a sequence of 30 amino acids is repeated consecutively nine times. Two-dimensional NMR studies have shown that each 30-residue repeating unit folds to form an independent domain with a single zinc ion tetrahedrally coordinated between an antiparallel β sheet and a short thirteen-amino-acid α helix. Regulatory proteins with the C_2-H_2 motif have the same general tertiary structure (fig. 31.20*a*). In all cases the DNA-binding portion of the structure results from interaction

Figure 31.20

Schematic representation of three types of DNA-binding motifs found for regulatory proteins in eukaryotes. (*a*) The C_2-H_2 zinc finger found Xfin from *Xenopus laevis*. (Adapted from M. S. Lee et al., *Science* 245:645, 1989.) (*b*) The C_2-C_2 zinc finger found in the glucocorticoid receptor. (Adapted from T. Hard et al., *Science* 249:157, 1990.) (*c*) The leucine zipper motif. (Adapted from C. R. Vinson et al., *Science* 246:911, 1989.) The helical regions that make contact with the DNA are colored in red; the rest of the polypeptide chains are colored in green except for the zinc fingers, where cysteine sulfurs are colored in yellow and histidine nitrogens are colored in blue. The knobs on the leucine zipper represent leucine side chains.

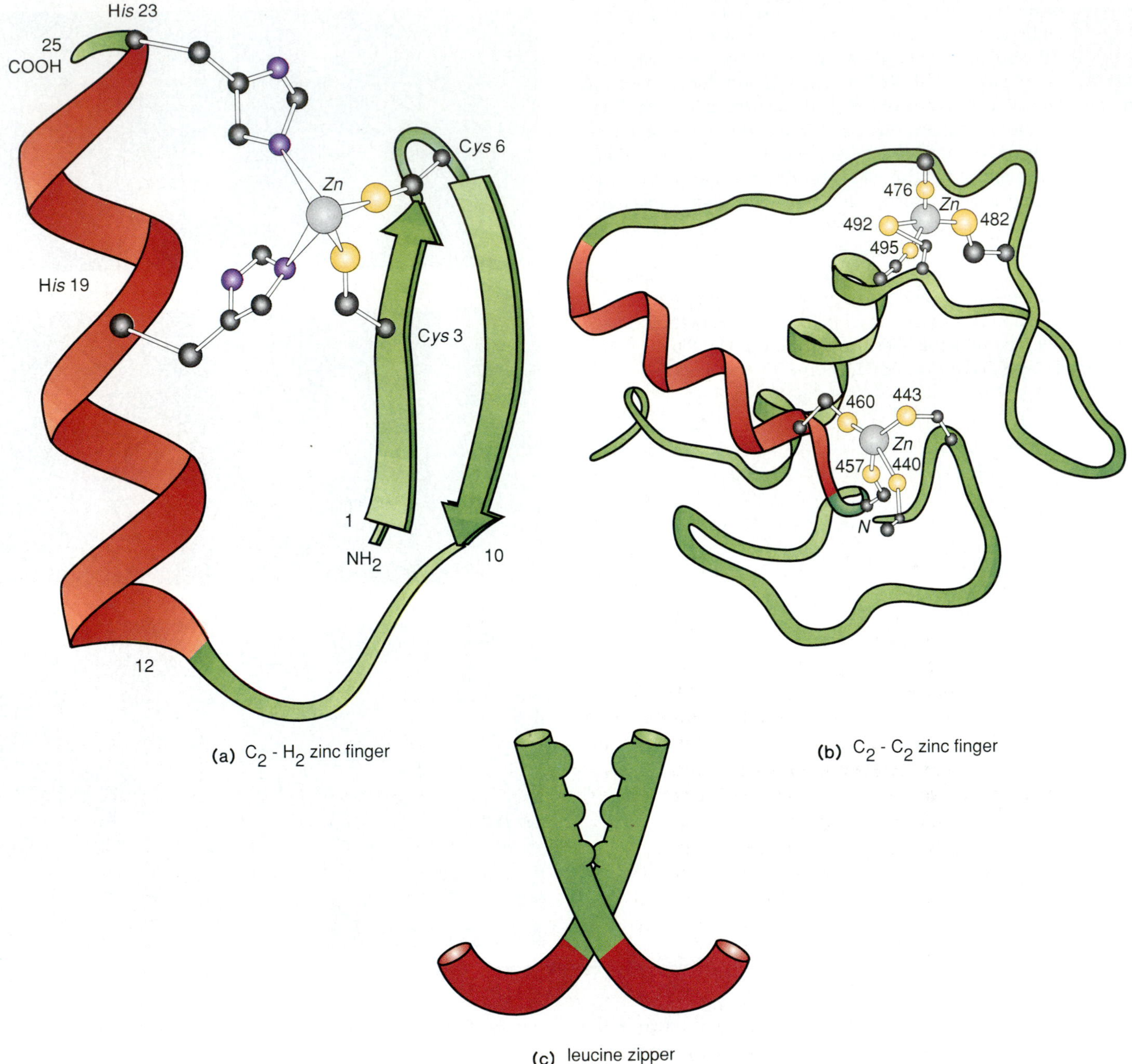

between the DNA base pairs in the large groove of the DNA and amino acid side chains on the exposed side of the recognition helix.

The C_2-C_2 motif is exemplified by the DNA-binding domains of steroid hormone receptors. There are several differences between proteins bearing this motif and those with the C_2-H_2 motif: (1) The antiparallel β sheet is used only in the C_2-H_2 motif; (2) in the C_2-C_2 motif the recognition helix extends beyond the zinc ligands, whereas in the C_2-H_2 motif it is located between the metal ligands; (3) the C_2-H_2 zinc fingers fold to form independent modules with no direct structural interaction between adjacent fingers, whereas in C_2-C_2 receptors two zinc-binding motifs are folded together to form a single structural domain; and (4) proteins containing the C_2-H_2 motif bind to DNA as monomers, whereas proteins containing the C_2-C_2 motif bind as dimers.

A third type of zinc binding motif has been found in a set of yeast activators, including GAL4, with two closely spaced zinc ions sharing six cysteines with their DNA-binding domains (see figure 31.6). It should be recalled that the yeast transcriptional activator GAL4 binds cooperatively to four related 17 base-pair sequences within an upstream activator sequence to activate the transcription of the GAL1 and GAL10 genes (see figure 31.4). The regulatory protein dimerizes through the interaction of two α-helical segments that form a short coiled-coil structure. The dimerization element makes contact with the narrow groove. The main recognition modules lie in the major groove separated by about one-and-one-half turns of the DNA helix and centered over the CCG triplets in the 17 base-pair consensus sequence (5′-CGGAGGACTGTCCTCCG-3′). It is conjectured that there may be another protein involved in the recognition complex.

Another type of DNA-binding domain was first described for the mammalian enhancer binding protein C-EBP. Proteins of this class show a primary sequence similarity consisting of a highly conserved stretch of about 30 amino acids with a substantial net basic charge, immediately followed by a region containing four leucine residues positioned at intervals of seven amino acids. The latter segment, named the leucine zipper by Steve McKnight, is required for dimerization and for DNA binding. Dimerization is favored by hydrophobic interactions between closely apposed leucine side chains (see fig. 31.20*c*). The adjacent helical regions, oriented in parallel, twist slightly around each other as in the classical coiled-coil conformation (see fig. 4.6). The basic regions adjacent to the leucine zipper interact in the large groove of the DNA; both proteins contribute their basic regions to form the DNA-binding domain. Dimerization is not confined to identical monomers. Many of the dimers that contain the leucine zipper motif are heterodimers. Thus the leucine zipper motif may be thought of as a general way of bringing two proteins together, but the effectiveness of this union will be determined by how and where they interact with specific segments on the DNA.

One of the best known heterodimers resulting from leucine zipper interaction is that formed between the Jun and Fos proteins. Both of these proteins are known to have the potential for activating a wide variety of host genes in mammals. The Jun protein can form homodimers or heterodimers, but the Fos protein can only form a heterodimer. Fos forms a heterodimer with Jun. Correlated with these dimer forming properties we find that Fos protein does not bind to the DNA except in the form of a heterodimer. By contrast the Jun protein can bind to DNA in the form of a homodimer or a heterodimer.

Thus far we have focused on the DNA-binding domains of regulatory proteins. The transcription activation domains of regulatory proteins are less well characterized. These domains may interact with the RNA polymerase directly or with other regulatory proteins. What makes an effective transcription activation domain? Different regulatory proteins appear to possess different types of groups for this purpose and some more than one type of grouping. Acidic, glutamine-rich, and proline-rich domains are implicated in the activation domains of a number of regulatory proteins.

In this section we have focused on some of the better known facts about regulatory proteins. This is an explosive field in which new discoveries are being made at a furious clip. It should be anticipated that we will know much more about this subject in the near future.

Posttranscriptional Modes of Regulating Gene Expression

Thus far we have concentrated on regulatory mechanisms for gene expression that occur at the transcriptional level. There are additional modes of regulation that are very important in eukaryotes that occur after transcription has taken place. These are divided into three categories: those that involve (1) RNA processing and those that involve (2) regulation of translation or (3) processing of the polypeptide chain.

Alternative Modes of Splicing Are Probably Very Widespread We discussed the basic mechanisms of RNA splicing in chapter 28. RNA splicing was first discovered in the transcription of adenovirus DNA, where different reading frames are connected to the same 5′ end. Thus we have known about the phenomenon of alternative splicing as long as we have known about splicing itself. Alternative splicing occurs for eukaryotic viruses such as SV40 and polyoma and it is also a common phenomenon in eukaryotic genes that contain multiple exons in their nascent transcripts. It is clearly a regulatory phenomenon in viruses, since we see a shift in the types of splicing as virus infection progresses. For eukaryotic genes it is also a regulatory phenomenon, since different modes of splicing are seen in different types of differentiated cells of the same multicellular organism.

The various patterns observed for alternative splicing are reviewed in fig. 31.21. Splicing usually involves segments of a single transcript, but on occasion we encounter transsplicing,

Figure 31.21

Patterns of alternative RNA splicing. Constitutive exons are shown in full color; alternative sequences (orange) and introns (solid lines) are spliced according to different pathways (dotted lines). Alternative promoters (TATA) and polyadenylation signals (AATAAA) are indicated. (Source: R. E. Breitbart et al., "Alternative splicing: A ubiquitous mechanism for the generation of multiple protein isoforms from single genes," in *Annual Review of Biochemistry* 56:467–495, 1987. Copyright © 1987 Annual Reviews Inc., Palo Alto, Calif.)

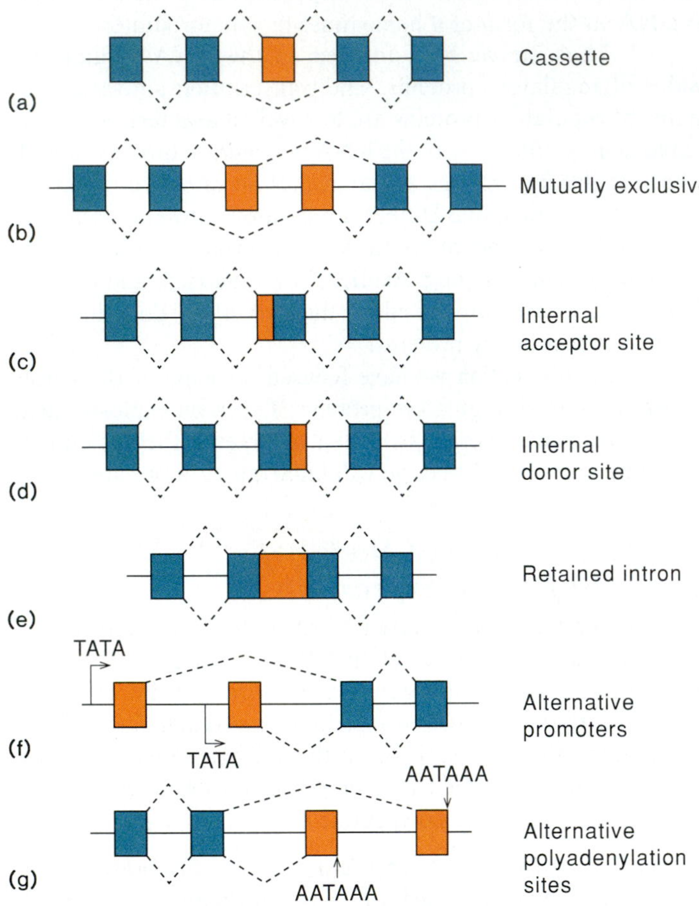

in which two transcripts participate in a common splicing operation. In the most common splicing situation, a promoter is present at one end of the transcript and the combination of exons that is used in the mature mRNA varies according to the pattern of splicing. In some cases one or more exons are excluded from the message by selective splicing. In other cases part of an intron is fused to one of the exons to make the final message. Splicing can also be influenced by the choice of the promoter or the choice of the polyadenylation site. In the latter two cases, parts of the upstream or downstream regions of the gene are excluded from the initial transcript and so the splicing of the transcript is limited to the RNA remaining in the nascent transcript.

The possibilities for alternative splicing are enormous and it is currently believed that this is a major mechanism for expanding the variety of proteins available to the organism in multicellular eukaryotes. One particularly elaborate example of this is seen in the gene for tropomyosin in vertebrates. Recall that tropomyosin is an important component of vertebrate striated muscle (see fig. 5.14). The polypeptide chain composition of the tropomyosin found in striated muscle involves seven splices in the nascent transcript from the gene (fig. 31.22). Variants of tropomyosin resulting from alternative splicing are found in other tissues of the same organism. It seems likely that the type of tropomyosin made in any given tissue is the type most suitable for the needs of that tissue. Mechanistically, it seems likely that some of the components involved in the splicing must also be tissue-specific.

Bingham and his co-workers have uncovered another role for splicing in *Drosophila.* In three different genes they have observed that most of the introns are removed from pre-mRNA molecules, leaving one or two introns intact. In this state the pre-mRNAs are nonfunctional. Only when the final introns are removed can they function as messages. Apparently the final splicing operations are regulated by the protein encoded by the messenger. If that protein is present above a certain concentration, then it prevents any further splicing of the pre-mRNAs. Bingham views this as a type of feedback inhibition that may be fairly common in *Drosophila* and perhaps other species.

Translation and Polypeptide Processing Also Affect Gene Expression After messenger formation there are still other ways in which the amount and types of finally synthesized protein can be affected. The initial polypeptide can be processed in various ways so that different polypeptides or proteins are expressed in specific tissues. We described such a possibility (see fig. 24.6) for the processing of the precursor polypeptide preproopiomelanocortin. This polypeptide is processed in different ways in the anterior and intermediate lobes of the pituitary gland to give rise to unique hormones in the two tissue types.

Because of the long lifetime of eukaryotic messengers it seems likely that translation-level controls play a major role in regulation of eukaryotic gene expression. Despite this belief, few mechanisms have yet been elucidated.

Inactivation of eukaryotic translation factors by covalent modification is one of the few mechanisms known to regulate the gross rate of translation. Specific protein kinases have been identified that phosphorylate and inactivate both eIF-2 and EF-2. The significance of the phosphorylation of EF-2 as a regulatory mechanism of the elongation rate is still not clear, but the phosphorylation of eIF-2 appears to be a general mechanism of controlling translation initiation in many cells.

The regulation of translation through the phosphorylation of eIF-2 is best understood as it operates in the rabbit reticulocyte. Two protein kinases specific for the *a* subunit of eIF-2 have been purified from reticulocytes. One of these kinases, termed the heme-regulated inhibitor repressor (HRI), serves to coordinate the rate of hemoglobin synthesis (more than 90% of the total protein synthesized in the reticulocyte is hemoglobin) with the availability of hemin (the precursor of the heme group in hemoglobin). Hemin binds to and inhibits the activity of HRI and this inhibition enhances the rate of globin synthesis (fig. 31.23).

Figure 31.22

Alternative modes of splicing of the tropomyosin gene transcript in different tissues. (Source: R. E. Breitbart et al., "Alternative splicing: A ubiquitous mechanism for the generation of multiple protein isoforms from single genes," in *Annual Review of Biochemistry* 56:467–495, 1987. Copyright © 1987 Annual Reviews Inc., Palo Alto, Calif.)

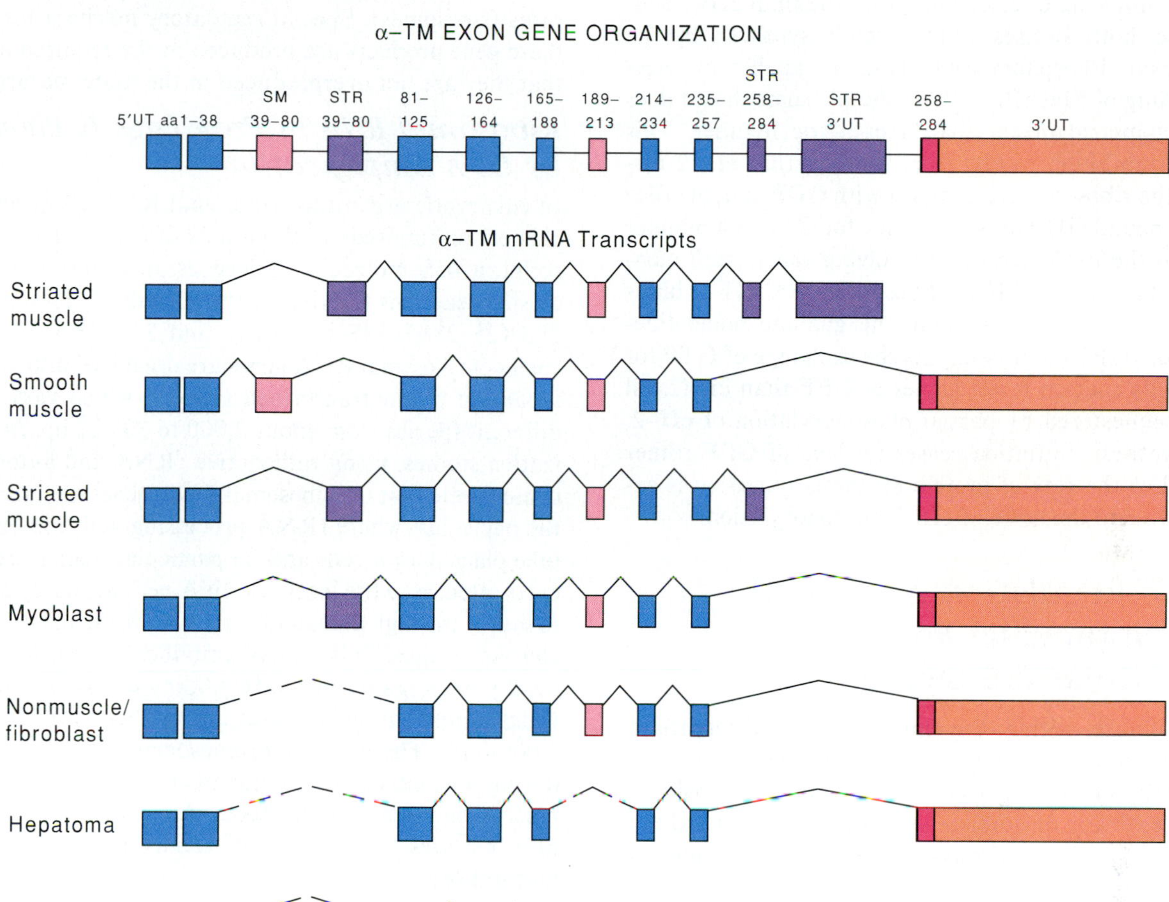

Figure 31.23

Regulation of protein synthesis in the rabbit reticulocyte. The vast majority of the protein synthesized in the rabbit reticulocyte is hemoglobin. The gross rate of protein synthesis in the reticulocyte is controlled indirectly by the concentration of heme. Heme inactivates a kinase that would otherwise inactivate the initiation complex involving eIF-2 and eIF-2B. The kinase phosphorylates the eIF-2 factor, making it impossible for the eIF-2–eIF-2B complex to exchange GDP for GTP.

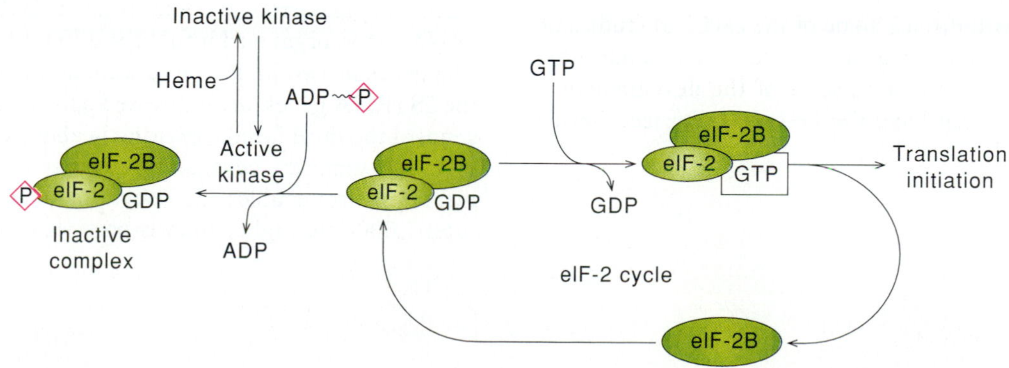

The second eIF-2-specific kinase appears to be present at low levels in most mammalian cells. This kinase is activated by double-stranded RNA and is therefore known as the double-stranded RNA-activated inhibitor (DAI). DAI may play a role in defending cells against invasion by viruses.

HRI and DAI are differnt proteins, but both phosphorylate the same amino acid residue in the *a* subunit of eIF-2 and, in consequence, both kinases inhibit protein synthesis by the same mechanism. Phosphorylated eIF-2 is capable of catalyzing the binding of Met-$tRNA_f^{Met}$ to the ribosome, but it does so in a stoichiometric rather than a catalytic manner. This mechanism of inhibition results from the fact that eIF-2 dissociates from the ribosome as a complex with GDP and, in order to recycle, the bound GDP must exchange for GTP in a manner very similar to the exchange cycle involving the *E. coli* elongation factors EF-Tu and EF-Ts. Phosphorylated eIF-2 binds to, but is unable to dissociate from, the guanine nucleotide-exchange factor (GEF) that catalyzes the exchange of GTP for GDP. Since cells contain fewer copies of GEF than eIF-2, all GEF can be sequestered by partial phosphorylation of eIF-2, and protein synthesis initiation ceases for lack of GEF rather than eIF-2. Thus the rate of protein synthesis initiation is exquisitely sensitive to the state of eIF-2 phosphorylation.

Patterns of Regulation Associated with Developmental Processes in Multicellular Eukaryotes

As cells proliferate within the embryo, they differentiate; changes are triggered by the genes that they express and the proteins that they synthesize. Cells that are initially capable of following any pathway of development become committed to a particular pathway. Most pathways have many branchpoints, so when partially committed cells reach a branchpoint they may differentiate further, down a more specialized pathway. This process of gradual commitment, repeating itself many times, gives rise to a complex pattern of cell lineages.

It seems likely that the pivotal events in the evolution of a differentiated cell reflect changes in gene expression that result from a complex hierarchy of controls. The key to understanding differentiation, therefore, is to identify the regulatory factors that are responsible for the controls involved in differentiation and to explain how they act.

We will focus on regulatory events that occur in development. First we will discuss some of the classical studies of gene expression in the embryogenesis of sea urchins and amphibians, and then we will turn to some of the developmental events in fruit flies, which have the best-characterized developmental system.

During Embryonic Development in the Amphibian, Specific Gene Products Are Required in Large Amounts

During early development, when cell divisions are occurring rapidly, there is an increased demand for certain products associated with the genome (histones) and the translation apparatus (ribosomes). Special regulatory mechanisms ensure that these gene products are produced in the required amounts and that they are not overproduced in the more mature organism.

Ribosomal RNA in Frog Eggs Is Elevated by DNA Amplification

In eukaryotic organisms, ribosomal RNA genes are present in clusters of hundreds to thousands of copies. The nascent transcript, a 45S molecule, undergoes an elaborate series of processing reactions to produce three kinds of ribosomal RNAs—28S (25–28S), 18S (17–18S), and 5.8S (5.5–5.8S). Multiple copies of ribosomal RNA genes are organized into tandem arrays separated by nontranscribed spacers, which vary in length in different species from about 2,000 to 30,000 bp. *In situ* hybridization studies, using radioactive rRNA and autoradiography, demonstrate that the ribosomal RNA genes are localized around the nucleolus, where rRNA processing and ribosome assembly take place. Germ cells and, in particular, immature eggs or oocytes often raise the level of rDNA per nucleus by amplification to satisfy the high demand for rRNA during the very rapid early cleavage stages. This type of amplification has been most thoroughly investigated in the frog *Xenopus laevis* (fig. 31.24), in which amplification increases the number of rRNA genes about 1,000-fold. These extrachromosomal genes are transcribed during oogenesis and subsequently discarded. The amplification mechanism is not understood. An entirely different mechanism elevates the level of 5S rRNA synthesis during oocyte maturation.

5S rRNA Synthesis in Frogs Requires a Regulatory Protein

Recall that eukaryotic ribosomes contain the small rRNA components (5S and 5.8S). While the 5.8S rRNA is contained in the 45S transcript with the other large rRNAs, the 5S exists independently in multiple gene copies that are organized into simple tandem multigene families. In amphibians the 5S rRNA genes are organized into different multigene families that are under developmental control. For instance, in *X. laevis* the normal haploid genome contains about 20,000 copies of the 5S rRNA genes, organized into three different multigenic families. The genes of two major families, comprising about 98% of all the 5S rRNA genes, are expressed only in growing oocytes. The genes of the third family, existing in about 400 copies, are active in most (somatic) cells and growing oocytes. This developmental control enables the oocyte to accumulate 5S rRNA at rates 1,000-fold higher than is possible in somatic cells.

Figure 31.24

Time course of rDNA amplification in the *Xenopus laevis* oocyte. The amount of rDNA per cell is plotted against time. Note that the ordinate is a logarithmic scale. The sharp drop in rDNA at fertilization is thought to be due to dilution of the amplified rDNA copies by cell division, rather than to their destruction. (Source: A. P. Bird, "Gene reiteration and gene amplification" in *Cell Biology*, vol. 3, *Gene Expression: The Production of RNAs*, edited by L. Goldstein and D. M. Prescott. Copyright © 1980 Academic Press, New York.)

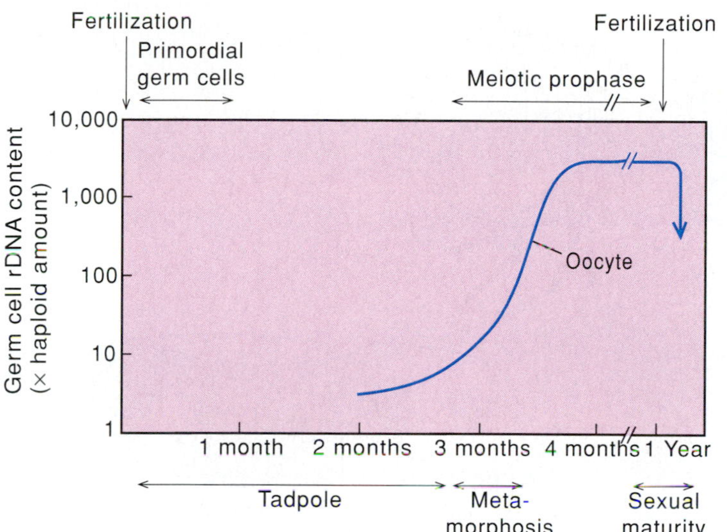

Figure 31.25

Molecular maps of histone gene repeat units of a sea urchin and a fruit fly. The histone gene organizational maps are calibrated in base pairs. The mRNA coding regions, including their leader and trailer sequences, are depicted by the colored boxes. The direction of transcription is indicated by the arrows.

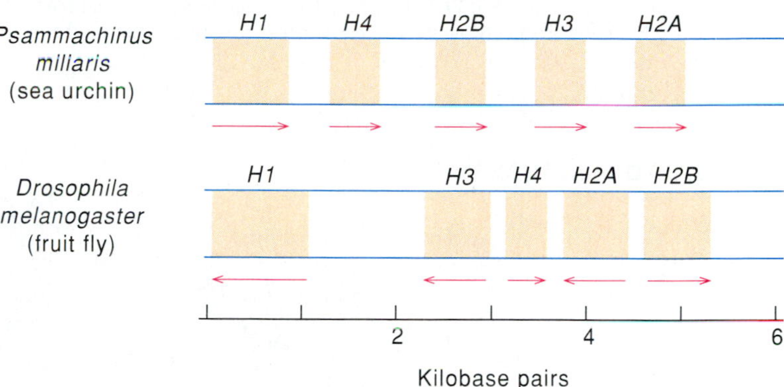

In chapter 28 we noted (see fig. 28.18*c*) that Pol III, which transcribes the 5S genes, requires the binding of three transcription factors to the 5S promoter. These factors are known as TFIIIA, TFIIIB, and TFIIIC. The abundance of TFIIIA appears to be much greater in the oocyte than in most somatic tissue. As a consequence, only in the oocyte is there enough TFIIIA to bind to all of the 5S promoters. Competition binding experiments in which both somatic and oocyte 5S gene DNAs are exposed to limited amounts of the TFIIIA protein show that the regulatory protein binds considerably more firmly to the somatic 5S genes. This fact probably explains why the somatic 5S genes are uniquely active in somatic tissues when the amounts of TFIIIA protein are much lower. Under a variety of conditions, large amounts of 5S rRNA are synthesized in the oocyte, but not in somatic cells; this difference is strongly related to the concentration of TFIIIA regulatory protein in the two situations.

Incidentally, it appears that TFIIIA protein has another function. It interacts with 5S rRNA to form a stable 7S nucleoprotein complex that accumulates in the cytoplasm of young oocytes. In growing oocytes, 5S rRNA is not incorporated immediately into ribosomes; rather, it is synthesized actively for several weeks before 18S and 28S rRNA synthesis begins. In addition to its storage function, it has been proposed that TFIIIA protein also serves a transfer function, facilitating the transport of 5S rRNA to the nucleolus, where ribosome assembly takes place.

Demands for Histones Are Met by Gene Amplification and Storage

Only a few gene families that encode mRNAs exist consistently in multiple copies. One such family includes the histone genes. Multiple copies of the histone genes are easy to rationalize in terms of the needs of most cells, which must make a large quantity of histones over short periods of time. Histones are usually but not always synthesized during S phase, when the DNA is being replicated in approximately equal-weight proportions to the histone. The five different types of histone genes invariably occur in clusters containing transcribed and nontranscribed regions (fig. 31.25). Each cluster is usually repeated many times in a tandem array.

Most of the detailed structural analysis of histone genes has been carried out in sea urchins. Cells of these organisms have several hundred copies of the genes for histones, whereas humans, mice, and *Drosophila* have only 10–50 copies. It seems likely that histone genes are reiterated in a very high frequency in the sea urchin because very large amounts of histones are needed during early embryogenesis, when cell cleavages occur once every 30 minutes. Indeed, those histone genes that are active during early embryogenesis constitute a distinct class. The set of histone genes that function late in embryogenesis of the sea urchin are present in only 5–12 copies per genome; unlike the clustered, tandemly arranged early histone genes, they are dispersed and irregularly arranged. Specific transcription factors are involved in class-switching between activating early and late histone genes.

In *X. laevis* it takes, on average, 14 min for the number of cells to increase from 50 to 5,000 during early embryogenesis. Since the frog has a larger genome than the sea urchin, its demand for histones should be even greater during early embryogenesis. Nevertheless, the reiteration of histone genes is less

Figure 31.26

Steps in the development of *D. melanogaster*. The fertilized egg contains a single zygotic nucleus (*a*). This divides (*b*) every ten minutes. After eight divisions, when there are 256 nuclei, nuclear migration toward the outer cortex structure begins (*c*). Eventually all the nuclei form a monolayer on the cortex surface. The first nuclei to become enclosed are the pole cells, which will become germ cells in the adult organism (*d*). The fully formed blastoderm contains about 6,000 cells, which form a monolayer around the cortex (*e*). Even before a cell membrane has formed around the nuclei, the embryo has become functionally divided into a segmented structure (*f*). During gastrulation (*g*), there is continued cell duplication, a folding of sheets of cells, and mass migration of segments. Eventually this structure hatches into the first larval stage (*h*). The remaining stages between the larva and the adult fly (*i*) are not illustrated. In (*f*) through (*i*), various segments are labeled. Three segments, Md, Mx, and Lb, fuse to make the head structure. Thoracic segments T1–T3 and abdominal segments A1–A8 retain their segmental appearance in the adult.

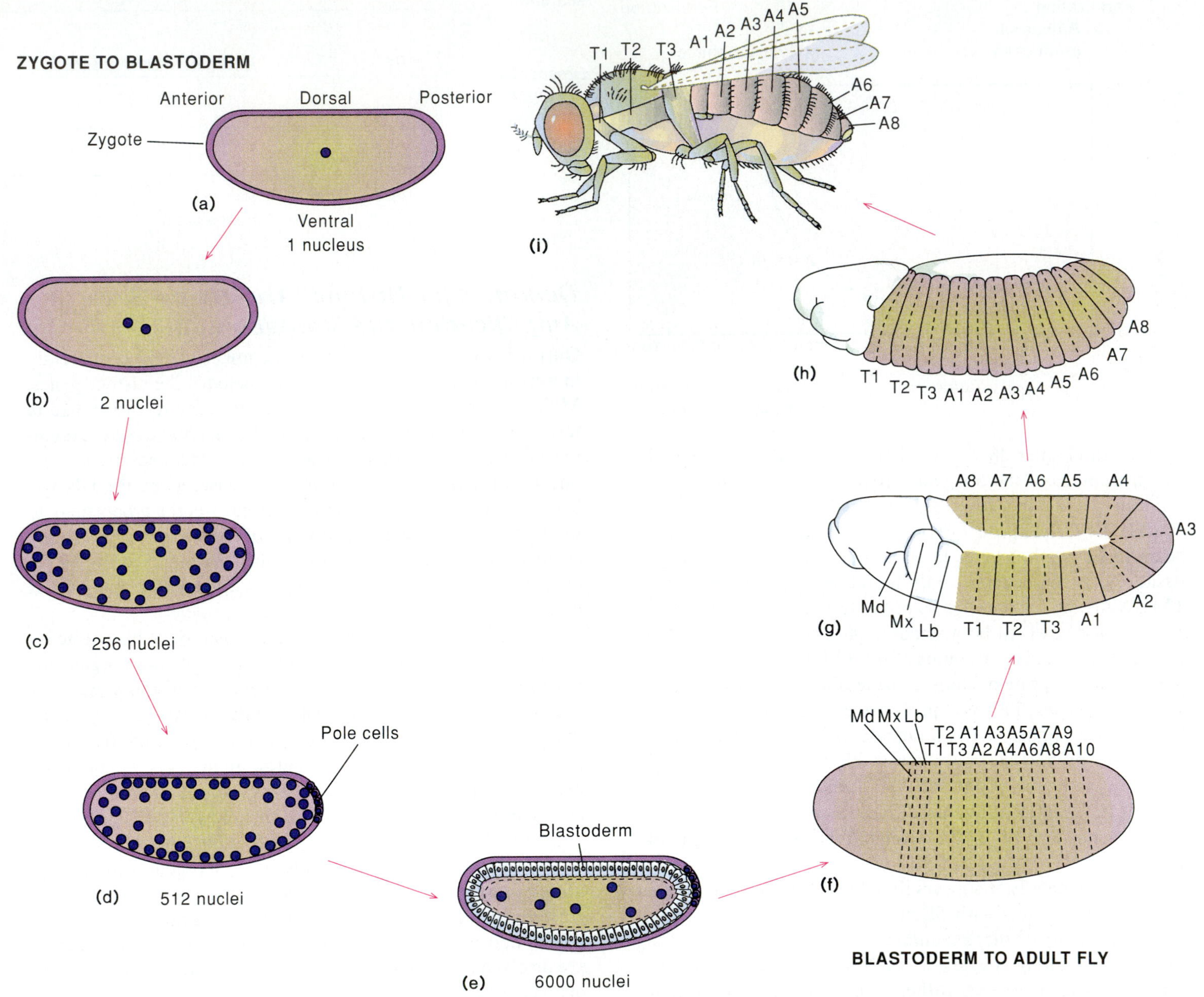

pronounced in the frog than in the sea urchin, because the histone requirements of the frog are met by a large accumulated store of histone proteins and histone mRNA in the egg.

Early Development in Drosophila *Leads to a Segmented Structure That Is Preserved to Adulthood*

Whereas sea urchins and frogs have proved valuable in the study of isolated embryonic events and the structure and composition of the embryo, they have not been very useful for understanding the overall pattern of early embryonic developments. For this purpose it has been necessary to turn to other organisms to identify the genes involved in regulating development. Many fundamental aspects of developmental biology have been studied using the techniques of genetics and molecular biology in *D. melanogaster.*

Some steps in the development of *Drosophila* are shown in figure 31.26. Beginning shortly after fertilization, the new diploid nucleus undergoes a rapid series of divisions with no segregation of nuclei into separate cells. After the eighth division, when there are 256 nuclei (see fig. 31.26*c*), the nuclei begin to migrate to the periphery of the egg cytoplasm. After another nuclear doubling, cell membranes form around a group of cells at the posterior end of the egg; these cells are progenitors of germ cells for the subsequent generation. The remaining nuclei continue to divide until there are about 6,000 nuclei at the periphery (see fig. 31.26*c*). At this point, about two hours after fertilization, membranes are formed, separating the nuclei into a monolayer of cells. The resulting structure, known as the blastoderm, is essentially a cell monolayer enclosing the yolk.

The blastoderm divides into fourteen different segments: Md, Mx, Lb, T1–T3, and A1–A8 (see fig. 31.26*f*). Three segments, Md, Mx, and Lb, become part of the head structure. The remaining segments become subdivided into two compartments, with anterior and posterior parts. During gastrulation there is a continued cell duplication, folding of sheets of cells, and mass migration of segments (see fig. 31.26*g*). The embryo eventually hatches into the first larval stage (see fig. 31.26*h*). Cells within the larva are of two types. About 80% of them are fully functional in the larva. The remaining 20% are embryonic precursors to adult tissues rather than larval tissues. These latter cells form packets within the larva, called imaginal disks, that are arrested in development until pupation (fig. 31.27), when a single hormone ecdysone triggers their differentiation into specific adult structures.

Early Development in Drosophila *Involves a Cascade of Regulatory Events*

There are two distinct phases in *Drosophila* embryogenesis; the first precedes cellularization of the blastoderm and is associated with a cascade of interacting regulators; the second occurs after cellularization and depends on intercellular signals that must be carried, at least in part, by membrane-bound proteins. We will focus on events occurring in the first phase because they are better understood.

Figure 31.27

Imaginal disks in a mature larva and the structures they lead to in the adult fly.

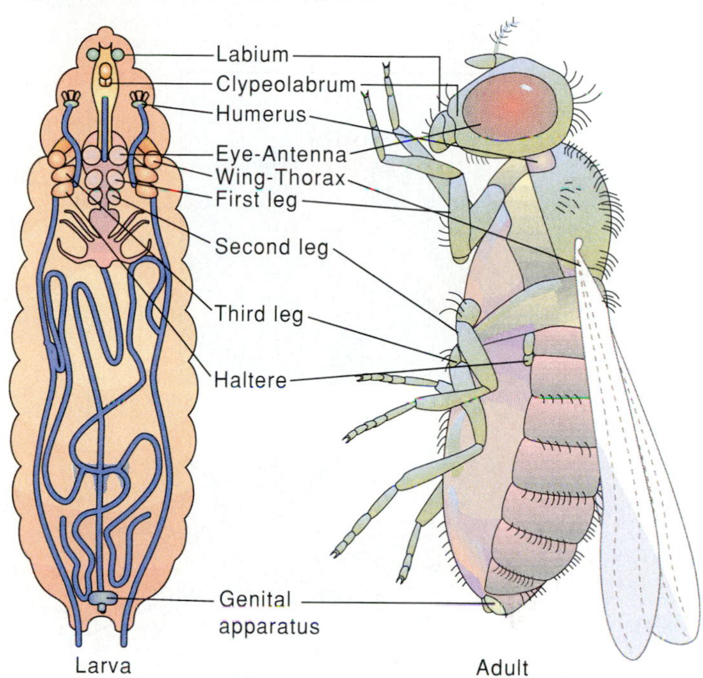

Recall that in the infectious cycle of the λ bacteriophage, each stage is characterized by the expression of specific gene products under the control of one or more regulatory proteins. A new phase in the development of the bacteriophage results from the synthesis of one or more new regulatory proteins. A hierarchy determines the order in which the different regulatory proteins make their appearance. Early development in *Drosophila* is very similar except for its greater complexity. In fact, some of the earliest regulatory proteins in the developing oocyte are supplied by surrounding cells. Another difference in *Drosophila* development is that the large size of the oocyte permits the establishment of gradients within a single cell. The regulatory protein encoded by a messenger RNA injected at the anterior end of the oocyte is likely to exist in highest concentration at the anterior end and at lowest concentration at the posterior end. Similarly, mRNAs injected into the oocyte at the posterior end establish regulatory-protein gradients in the opposite direction. These gradients are crucial for regulating the expression of genes involved in early development. Some gradients are also established in the perpendicular direction, on the dorsal-ventral axis, but we can overlook these for present purposes.

***Three Types of Regulatory Genes Are Involved in Early Segmentation Development in* Drosophila** The first insight into the genetic system directing *Drosophila* development came from the discovery, of bizarre mutations that affect the body plan (fig. 31.28). For example, the mutation *Antennapedia* results in a pair of extra legs sprouting from the head in place of antennae.

Figure 31.28

Abnormal phenotypes resulting from homeotic mutations. *Antennapedia* results in a pair of extra legs sprouting from the head in place of antennae. *Bithorax* results in an extra pair of wings appearing where normally halteres would appear.

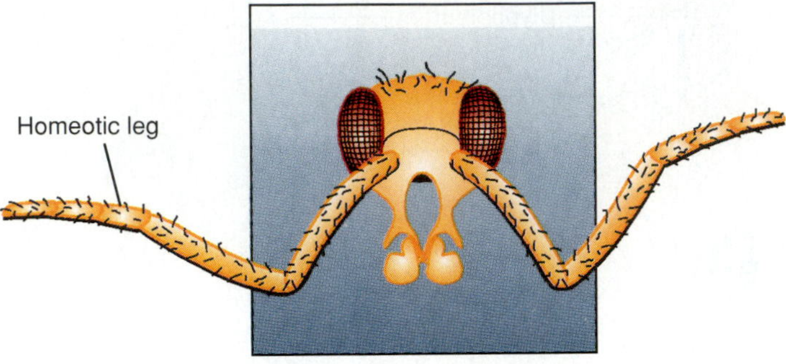

Table 31.3
Regulatory Genes Involved in Segmentation Development

Maternal-effect genes—establish gradients
Anterior
bicoid (*bcd*)
Posterior
oskar (*osk*)
Dorsal-ventral
dorsal (*dl*)
Segmentation genes
Gap genes—define four broad regions in the egg
hunchback (*hb*)
Krüppel (*Kr*)
Knirps (*kni*)
Pair-rule genes—define seven bands
runt (*runt*)
hairy (*h*)
fushi tarazu (*ftz*)
even skipped (*eve*)
paired (*prd*)
odd paired (*opa*)
Segment-polarity genes—define fourteen bands
engrailed (*en*)
wingless (*wg*)
gooseberry (*gb*)
Homeotic genes—specify structures associated with individual segments
BX-C locus
ANT-C locus

Another mutation, *Bithorax,* results in an extra pair of wings appearing where normally there should be much smaller appendages called halteres. These mutations occur in regulatory genes known as homeotic genes.

In addition to homeotic genes there are two other types of genes that influence the segmentation pattern; these are called maternal-effect genes and segmentation genes (table 31.3). Maternal-effect genes are so called because they affect the phenotype only according to the information present in the female parent. Maternal-effect mutations occur in genes responsible for establishing the anterior-posterior axes in the young embryo. Segmentation mutations affect the number and polarity of the body segments. Homeotic mutations are more specific than segmentation mutations; they result in changes in structures that are uniquely associated with individual segments or subsegments.

Analysis of the Genes That Control the Early Events of Drosophila *Embryogenesis*

Once we have found a mutation that appears to affect development, we must have a way to detect the presence or absence of the related gene product in the developing embryo. A principal tool for doing this is autoradiography using ^{32}P-labeled specific DNA probes. The embryo is labeled by *in situ* hybridization with the radioactive probe specific for the transcript of a gene (see box 31A). Both the location and the intensity of the labeling give an indication of the level of expression of the gene. In some cases the protein product relating to a specific transcript can also be detected by use of a specific antibody-labeling technique. In what follows we will trace the appearance of some of the major regulatory genes that are involved in early development.

Maternal-effect Gene Products for Oocytes Are Frequently Made in Helper Cells The first regulatory gene products to play a significant role in *Drosophila* development are active before fertilization. The bicoid (*bcd*) gene is a major determinant of the anterior-posterior pattern. Like many other maternally expressed genes, *bcd* is transcribed in the ovary in specialized cells that form a cluster around the future anterior pole of the developing oocyte. The bcd RNA synthesized in these so-called nurse cells passes into the oocyte through cytoplasmic canals and becomes localized at the anterior pole of the oocyte

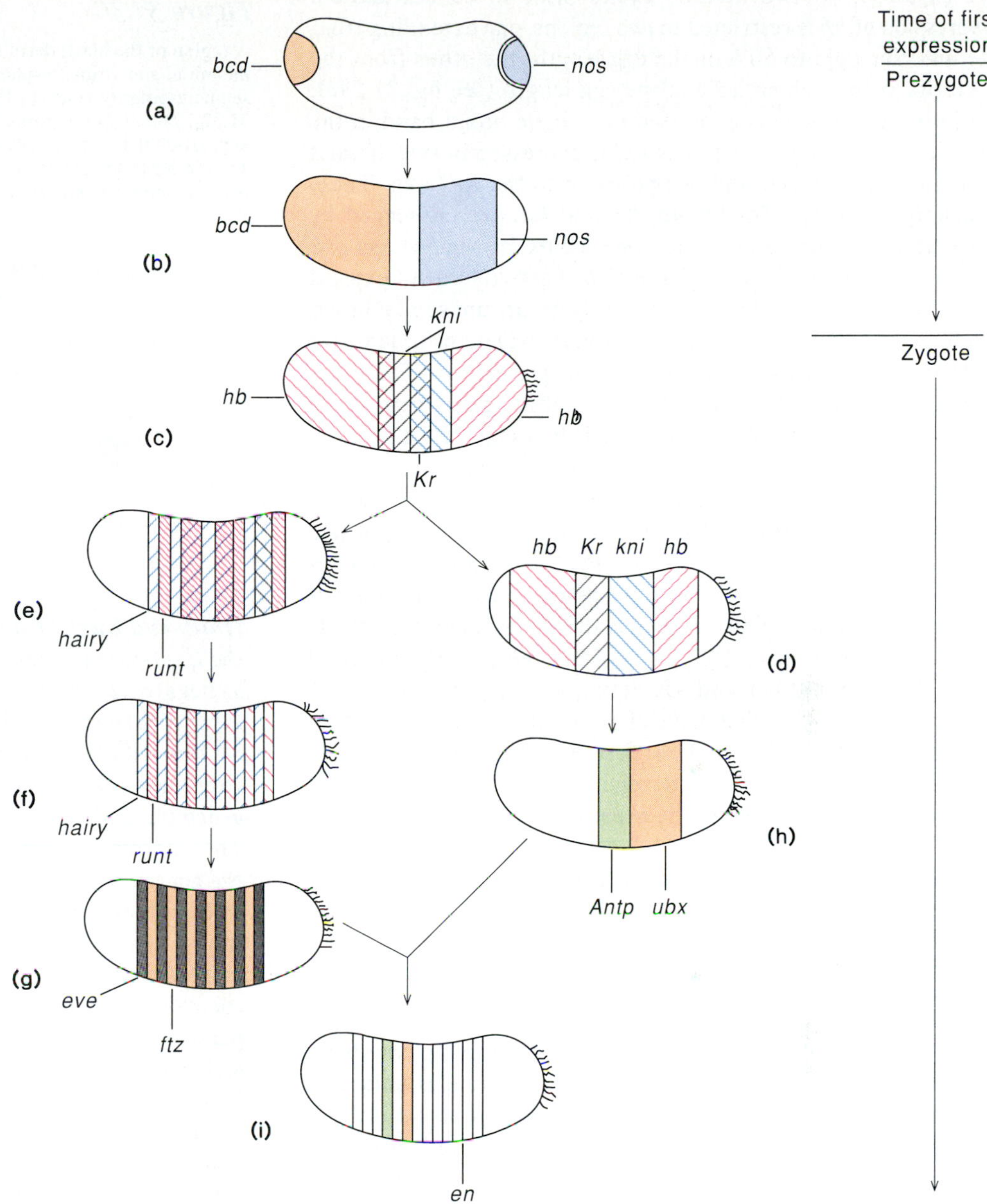

Figure 31.29

Key events in the expression of developmental regulatory genes in the blastoderm embryo. Events take place in the order shown over a period of a few hours. Measurements are based on the use of probes for detecting the transcripts of various regulatory genes. Description of genes is given in table 31.3. (Source: P. W. Ingham, "The molecular genetics of embryonic pattern formation in *Drosophila,*" in *Nature* 335:25–34, 1988. Copyright © 1988 Macmillan Magazines Ltd., London, England.)

(fig. 31.29). Similarly, the oskar (*osk*) group genes, such as *nos,* are involved in the synthesis and deposition of products that become localized at the posterior pole of the egg (see fig. 31.29*a*). Females lacking a functional copy of the *bcd* gene produce eggs that develop into embryos with no head or thorax. The same effect can be achieved by removal of material from the anterior pole of a normal egg. Embryos produced by female mutants for *nos* develop normal head and thoracic segments, but lack the entire abdomen.

The pattern-forming process continues upon fertilization; as the *bcd* RNA is translated, the protein diffuses from the anterior pole so that it becomes distributed over about half of the length of the egg (see fig. 31.29*b*). Simultaneously, information encoded and localized at the posterior pole during oogenesis by members of the *osk* group begins to move forward.

By the beginning of the precellular blastoderm stage there are gradients for at least two different maternally encoded products along the anterior-posterior axis of the embryo. This quantitative information is now transformed into qualitative differences in the form of region-specific gene expression, by a process requiring interaction between the maternally derived products and the zygotic genome that resulted from fertilization.

Gap Genes Are the First Segmentation Genes to Become Active The first segmentation genes to become active in the zygote are members of the gap class, so-called because their mutants lack major regions of the body, thereby creating a gap in the antero-posterior pattern. The three members of this class are hunchback (*hb*), Krüppel (*Kr*), and knirps (*kni*). These genes are expressed in two division cycles prior to cellularization. Expression of *hb* is restricted to two regions, one extending from the anterior pole to 50% of the egg length, the other from the posterior pole to about 25% of the egg length (see fig. 31.29*c*). Initially the *Kr* gene is expressed in a single broad band in the middle of the embryo, whereas *kni* is expressed in two distinct domains, one anterior and one posterior to the *Kr* band. These transcriptional domains for *hb, Kr,* and *kni* are influenced by the maternally derived information encoded by the *bcd* and *osk* group genes. Thus mutants that lack *bcd* activity do not express *hb,* whereas *Kr* is extended anteriorly in an unusually broad domain. The absence of the posterior maternal determinants results in the extension of the *Kr* domain posteriorly. These observations suggest that the *bcd* and *osk* group genes both act to repress transcription of *Kr* in the anterior and posterior regions, respectively, thereby restricting its region of expression to the central portion of the embryo. In contrast, *bcd* appears to act as a positive regulator of *hb,* and *osk* appears to be a positive regulator of *kni.* In normal embryos the transcriptional domains of *hb, Kr,* and *kni* narrow with time, giving rise to sharp boundaries of expression. This process is driven by negative effects between the gap genes (see fig. 31.29*d*), *hb* and *Kr* mutually repressing one another and *kni* acting as a negative regulator of *Kr.* Thus the establishment of stable domains of gap gene expression is a two-step process; first, a differential response to graded levels of maternal determinants, and second, a mutual repression effect, leading to the generation of stable boundaries between adjacent domains.

The gap gene products regulate the position-specific expression of other genes, those belonging to the pair-rule class of segmentation genes.

Periodic Gene Expression Is Initiated by Pair-Rule Genes A common feature of pair-rule genes is their transient expression in seven bands during the period of cellularization of the blastoderm. Despite this similarity, each pair-rule gene is unique in its pattern of expression. *Runt* and *hairy* are initially expressed throughout the embryo (see fig. 31.29*e*); a restricted pattern of expression of these two genes begins earlier than for other pair-rule genes.

By the beginning of the last interphase before cellularization, transcripts of both *hairy* and *runt* localize to two complementary series of seven bands that encircle the embryo (see fig. 31.29*f*). The generation of these banded patterns is a key event in the pattern-forming process. The sorts of interactions taking place probably involve the local enhancement of expression of one or another gene in response to local levels of one or more gap gene products. Such interactions result in a patchy pattern of *hairy* and *runt,* which is further refined and stabilized into a regular banded pattern by their mutual repression (see fig. 31.29*f*), in a manner analogous to the stabilization of the gap gene expression domains.

The complementary bands of *runt* and *hairy* are instrumental in the refinement of the spatial patterns of expression of other pair-rule genes.

Figure 31.30

A region of the blastoderm showing three of the fourteen parasegments and the initial and ultimate spheres of expression for certain pair-rule and segment-polarity genes (all fourteen parasegments are shown in figure 31.32). The dashed regions for *prd, opa, ftz,* and *eve* show a region of expression that occurs only at early stages, before the narrowing process. The circles represent cells in the early stage in the blastoderm, when the parasegments are only four cells in width.

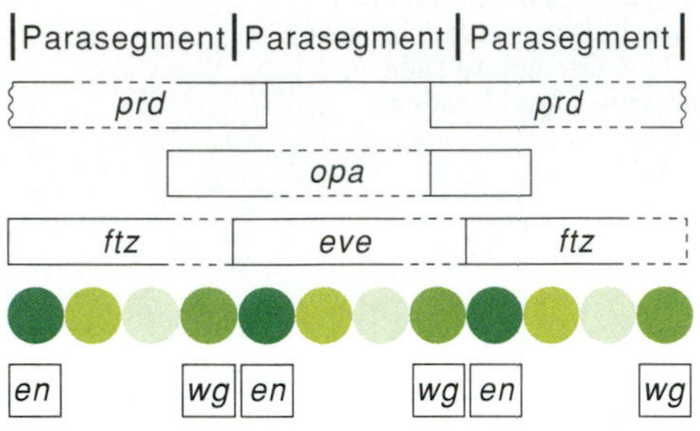

Hairy and Runt Proteins Function as Negative Regulators for the Expression of Other Pair-Rule Genes *Hairy* and *runt* act as negative regulators for the expression of the pair-rule genes *fushi tarazu* (*ftz*) and *even skipped* (*eve*) respectively (see fig. 31.29*g*). Both *ftz* and *eve* encode homeo-domain proteins, a strong indication that they act as transcriptional regulators of genes of the homeotic and segment-polarity class (box 31B). The interactions between *hairy* and *runt* and between *ftz* and *eve* generate complementary sets of seven bands expressing one or the other gene (the locations of *ftz* and *eve* expression are shown in figure 31.29). When they first become visible, each band is four nuclei wide. At this stage, each *ftz* and *eve* band coincides with the boundaries of a so-called parasegment. A parasegment is a unit that is out of register with a morphological segment but coincident with the domains of activity and function of the homeotic genes (fig. 31.30). At the same time as the *ftz* and *eve* patterns are resolved, the pattern of expression of a third homeobox-containing pair-rule gene called *paired* (*prd*), is also changing (see fig. 31.30 for the pattern of *prd* expression). Initially *prd* is expressed in seven broad bands also, but in this case the bands are each about six nuclei wide. This pattern is then transformed into one of fourteen bands by the elimination of expression from the central region of each of the original bands. As with *ftz* and *eve,* the generation of this pattern depends on the activity of *hairy* and *runt.* Thus, in a process resembling the establishment of the gap gene domains, information encoded by two components is transformed into a more complex form by the differential response of the various secondary pair-rule genes (*ftz, eve, prd*) to this information.

The Boundaries of the Parasegments Are Set by Wingless **(Wg)** ***and Engrailed*** **(en)** The combinations of pair-rule gene activities define the states of particular blastoderm cells. These cells subsequently serve as reference points for the elaboration of the pattern within individual parasegments during the period of

31B

BOX

The Homeobox Sequence

One common feature of all homeotic genes, and of some maternal-effect and segmentation genes as well, is that they possess a common 180-base segment in the 3′ exon. The base homology ranges between 60 and 80%, depending on the specific genes being compared, and the amino acid homology is even higher (up to 87%). The high basicity of the amino acid sequence in this region—called the homeo box—and other structural features suggest that this region encodes a DNA-binding protein that regulates gene expression by binding to specific sites on the DNA. Indeed, we see sufficient amino acid homology between the homeo box and the helix-turn-helix motif in many bacterial regulatory proteins to suggest that the gross structures of these very distantly related regulatory proteins are quite similar.

The structure of the homeodomain is built up from three helices connected by short loop regions. The three helices are formed by residues 10 to 21, 28 to 38, and 42–58. Helices 2 and 3 are part of the helix-turn-helix motif in which the structure of the region 30 to 50 is virtually identical to the helix-turn-helix motif found in bacteria.

One of the most exciting findings relating to the homeo box is that very similar sequences have been identified in many other animals, including frogs, mice, and even humans. This finding suggests that there are homeo box proteins in a wide range of organisms, possibly playing similar roles in regulating developmental processes.

Figure 1

The homeo box sequence found in three *Drosophila* regulatory proteins. The sequence of amino acids along the main, continuous line is that found in the *Antp* homeo box. At points where the sequence of ftz proteins is different, the differences are shown above the corresponding Antp proteins, and at points where the sequence of ubx proteins is different, they are shown below the corresponding Antp proteins. Certain proteins isolated from the human and mouse embryos have segments with similar sequences. Note the high basicity of the sequence, which should favor electrostatic binding to nucleic acid.

				5					10					15					20
			Thr																
Arg	Lys	Arg	Gly	Arg	Gln	Thr	Tyr	Thr	Arg	Tyr	Gln	Thr	Leu	Glu	Leu	Glu	Lys	Glu	Phe
Ser			Gly																
				25					30					35					40
					Ile							Asp			Asn			Ser	
His	Phe	Asn	Arg	Tyr	Leu	Thr	Arg	Arg	Arg	Arg	Ile	Glu	Ile	Ala	His	Ala	Leu	Cys	Leu
	Thr		His										Met		Tyr				
				45					50					55					60
Ser															Ser			Asp	Arg
Thr	Glu	Arg	Gln	Ile	Lys	Ile	Trp	Phe	Gln	Asn	Arg	Arg	Met	Lys	Trp	Lys	Lys	Glu	Asn
															Leu				Ile

limited cell proliferation that follows. In particular, the pair-rule genes have been shown to define the domains of expression of the segment polarity genes *en* and *wg*.

At the end of the cellularization process and the onset of gastrulation, transcripts of both the *en* and *wg* genes accumulate in fourteen narrow bands along the anterior-posterior axis of the embryo, patterns that represent the first evidence of its segmental organization. At this time each band of *en* expression is only one cell wide. Each of the *en* bands represents the anterior boundary of a parasegment. The establishment of alternate *en* bands requires the combined activities of different sets of pair-rule genes. In the case of the odd-numbered bands, the *en* expression cells are those that express both *eve* and *prd,* whereas the even-numbered bands require the expression of both *ftz* and another pair-rule gene, *odd-paired* (*opa*).

In contrast to this positive control of *en, wg* expression is repressed by the combination of eve and ftz proteins. At the end of blastoderm the domains of eve and ftz narrow, so that single-cell-wide bands come to separate them, and it is these cells that initiate expression of *wg*. The net result is the generation of two sets of cells that are adjacent to one another and mark the anterior and posterior limits of each of the parasegments (see fig. 31.30). These sets of cells, one expressing *en* and the other expressing *wg*, subsequently serve as reference points for the specification of position in each developing parasegment.

The positive and negative effects established between maternal-effect and pair-rule genes are summarized in figure 31.31. This diagram tells an incomplete story. The most serious gap in our knowledge is how the gradients of maternal-effect and gap genes help to deliver the signal for the establishment of the strict periodic patterns of expression seen for most of the segmentation genes. It seems unlikely that the signals in the form of gradients could ever institute such a pattern without other factors being involved. Possibly some of the signals for expression of pair-rule genes are established in the oocyte before fertilization.

Figure 31.31

Influence of developmental regulatory genes on one another. An arrow with a plus sign at the arrowhead indicates a positive effect on expression, whereas a minus sign at the arrowhead indicates a negative effect on expression. The gap genes are essential for the establishment of the hairy and runt banded patterns, but the precise way in which they do this is unclear.

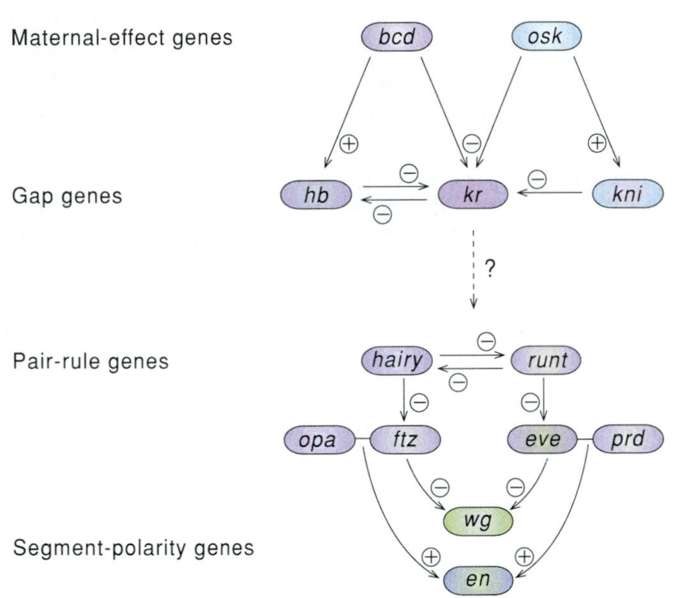

Homeotic Genes Specify Parasegment Character By the time cellularization of the blastoderm is complete, different cells express different combinations of pair-rule genes (fig. 31.32). At this stage, the embryo is subdivided into a series of repeating units. Despite their similarity, each parasegment is programmed to follow a unique pathway of differentiation as evidenced by subsequent events. Most probably the expression of the homeotic genes is influenced by the periodic information generated by the pair-rule and the segment-polarity genes, superimposed on the gradients of information deposited by the maternal genes and the gap genes. We do not know what information turns on specific homeotic genes in specific parasegments, but we do know a most exciting beginning of a story that should unfold in the near future.

Most of the homeotic genes are clustered in two giant loci, the *Antennapedia* (*ANT-C*) and *bithorax* (*BX-C*) complexes. The *ANT-C* locus occupies about 100 kb of the genome and the *BX-C* locus occupies about 300 kb of the genome. Both of these loci are situated on chromosome 3. The *BX-C* locus is subdivided into three sections, *Ubx, Abd-A,* and *Abd-B,* according to which body parts are affected by different mutations. The *Ubx* region affects thoracic segments T2 and T3 and the anterior portion of the A1 segment. The *Abd-A* region affects the posterior part of A1 and abdominal segments A2 to A4. Finally the *Abd-B* region affects the abdominal region from A5 to A8. The regions affected by different parts of the *ANT-C* locus include the head and the three thoracic segments (fig. 31.33). It is fascinating that the arrangement of the genes in

Figure 31.32

Pattern of expression of some pair-rule and segment-polarity genes. When the pattern of expression of maternal-effect genes and gap genes is superimposed on this regularly repeating pattern, it gives each region of the blastoderm a unique mixture of developmental regulatory genes.

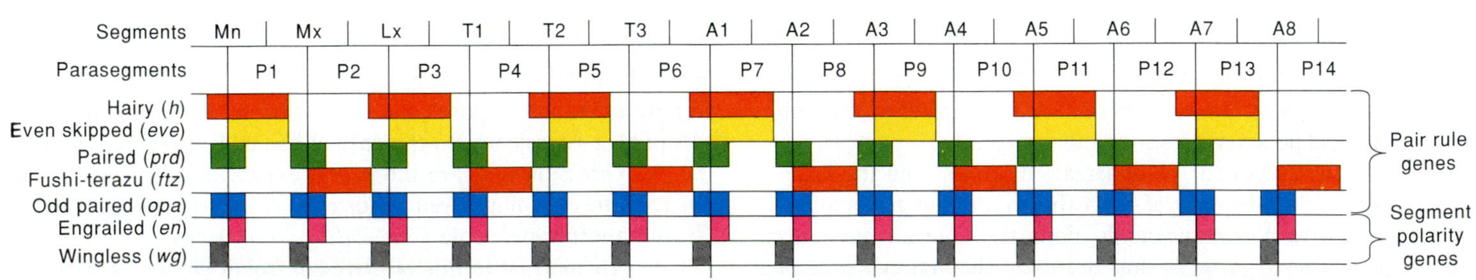

the two homeotic loci is colinear with the body plan; this colinearity may be related to the temporal mode of expression of the different genes in the two loci. Thus transcription of a region the size of *ANT-C* would take about two hours; the temporal expression of promoter proximal and promoter distal regions of the loci could be crucial in assuring the correct specification of body parts during development. Another aspect of gene arrangement and the body plan that is probably related to this factor is that a mutation at a given locus usually results in the conversion of the segment in question to the phenotype of the adjacent segment in the anterior direction. It is as though proceeding from the anterior to the posterior direction results in a stepwise increase in the amount of the locus that gets expressed.

The *ANT-C* and *BX-C* loci both contain very large introns and very few reading frames. These genes first become transcribed just prior to cellularization of the blastoderm, at about the same time as the pair-rule genes. Initially they are expressed at rather uniform levels in broad, overlapping domains. The boundaries of these domains are defined both by the maternal-effect organizing activities and by the gap genes. For example, transcripts from both of the known promoters of the *ANT-C* locus accumulate throughout the *Kr* domain. The longer transcript, P1, depends absolutely on *Kr* expression, whereas the shorter transcript, P2, appears to be defined by a combination of *bcd, osk,* and *hb* expression. Similarly, the initial *Ubx* domain is dependent on activation by the *osk* group activity and repression by *hb* activity. These broad regional differences are modulated by interactions with the early pair-rule pattern to generate a series of unique parasegmental states. Expression of *Scr, Antp,* and *Ubx* is specifically elevated in the cells of parasegments 2, 4, and 6 respectively, where the ftz protein is active. These enhanced levels of transcription are subsequently maintained throughout embryogenesis.

The activity of the homeotic gene domains is dependent in part on interactions between the genes themselves. In general, the most posteriorly expressed genes (such as *AbdA* and *AbdB* of the *BX-C*) repress the expression of the more anteriorly expressed genes (such as *Ubx* and *Antp*). In addition to generating differences between parasegments, homeotic genes are also expressed differentially between germ layers. The differential expression serves to modulate the basic pattern that underlies the organization of each segment.

Figure 31.33

Genetic maps of the antennapedia complex (*ANT-C*) and the bithorax (*BX-C*). The colored dashed lines specify the regions affected by mutations in a given locus. The bithorax complex is shown in greater detail. By lethal complementation assay, Sanchez-Herrero has shown that the *BX-C* is divided into three complementation groups: *Ubx, Abd-A,* and *Abd-B*. Within each of these complementation groups there are nonlethal point mutations (such as *abx, bx, bxd,* and *pbx* in *Ubx*), which cover specific phenotypes in the regions encompassed by the dashed lines. *Ubx* is required for the development of the posterior compartment of the mesothorax (T2p) through the anterior compartment of the first abdominal segment (A1a). The *Abd-A* function is required for morphogenesis of A1p through A4, and *Abd-B* is required for A5 through A8. At least three essential homeotic functions have been assigned to the antennapedia complex: *Dfd, Scr,* and *Antp*. These are shown on the genetic map below the schematic of the fly. The primary domains of *ANT-C* function are indicated. *Antp* function is required for proper segment morphogenesis of the thorax. Analysis of Scr^- and Dfd^- mutant embryos suggests that these genes are required for the differentiation of the prothorax and posterior head regions.

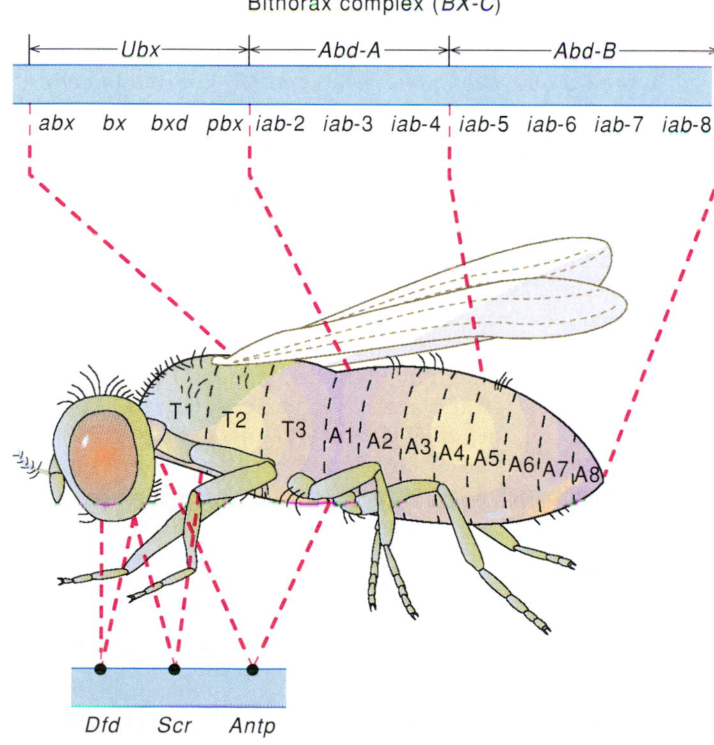

Cell–Cell Interaction Is Important in the Elaboration of the Developmental Pattern in Parasegments We have seen how the cells at the boundaries of each parasegment are specified, which is reflected by their expressing particular segment-polarity genes. Although it is possible that the intervening blastoderm cells likewise become determined in response to pair-rule gene signals, (genetic studies of specific mutants) suggest that they remain unspecified at this stage. Specification of cells within each parasegment would then occur as the cells divide in response to signals mediated by cell–cell interaction. For example, as cells divide they would assume states dependent on their apposition to *wg*-expressing (or *en*-expressing) cells. Further divisions would allow for the intercalation of additional cell states. Such a process requires the existence of signaling molecules and receptors capable of transducing and receiving information between cells. Although we do not know too much about the specific apparatus used for the transmission of such signals in *Drosophila,* it presumably employs the types of structures that are used in the transmission of hormonal signals (see chapter 24) as well as the type of signals that are involved in B-T cell interaction in antibody formation (see chapter 33).

Summary

Some mechanisms of gene expression are found in eukaryotes but are rarely, if ever, seen in prokaryotes. Still, *trans*-acting regulatory proteins that bind to *cis* effector sites on the genome are present in eukaryotic systems as well as in the *E. coli.* The best-understood unicellular eukaryote is the budding yeast *S. cerevisiae.* Gene regulation, particularly of development, can be quite complex in multicellular eukaryotes. Our discussion in this chapter has focused on the following points.

1. In eukaryotes, individual genes encode transcripts for a single polypeptide chain, such as the yeast genes for histidine biosynthesis and galactose catabolism. Although functionally related genes are clustered, each gene has its own promoter.
2. In yeast, genes of the *GAL* system are under the joint control of two *trans*-acting genes that encode regulatory proteins, *GAL4* and *GAL80.* GAL4 protein is an activator that binds at an upstream site and GAL80 is a repressor that inhibits GAL4 action by binding to it. The ability of the GAL80 protein to bind to GAL4 is lost in the presence of galactose, which binds to the GAL80 protein, causing a reversible allosteric change in its structure.
3. Many catabolic genes in yeast, *GAL* genes included, are subject to glucose effects, which selectively degrade enzymes and repress gene expression. The genes and gene products most affected by glucose fulfill comparable catabolic needs for the organism that can be more effectively met by the catabolism of glucose. Glucose-induced repression in yeast may be mediated by the phosphorylation of the relevant regulatory proteins.
4. Yeast has two haploid cells of opposite mating types and one nonmating diploid cell that results from the fusion of haploid cells of opposite mating type. The mating type is determined by the *MAT* locus. Information for the mating type is stored at other, silent loci and is expressed only if it is transposed to the *MAT* locus. Mating-type information is not expressed at the storage loci because of a complex repressor system of proteins interacting in *cis* fashion over a considerable distance at the control centers of the storage loci.
5. The greatest difference between the regulatory systems in yeast and *E. coli* is that yeast regulatory proteins can bind at a long distance from the RNA polymerase binding site and still be effective.
6. Complex multicellular eukaryotes differentiate irreversibly so that different cell types express a different profile of genes. Genes that are expressed are usually associated with swollen chromatin. Proteins found in active regions of the genome show characteristic modifications.
7. Enhancers are elements of the genome that generally stimulate transcription. They resemble yeast UAS sequences in yeast in that they can function over long distances. In fact, they can function over even greater distances than UAS sequences and they are effective in either orientation, either upstream or downstream from the promoter. A possible mechanism for enhancer and UAS function over long distances is that the chromosome folds to bring the proteins bound at the enhancer site in close proximity to other regulatory proteins or RNA polymerase bound at the promoter.
8. A typical enhancer is composed of a cluster (usually two or three) of *cis* sites called enhansons; each enhanson contains binding sites for a unique combination of regulatory proteins. Enhansons must be clustered to be effective.
9. Most regulatory proteins contain a domain for binding to a specific site on the DNA and another domain for binding to some other protein or proteins.
10. Special combinations of regulatory factors give rise to developmental patterns of cell differentiation. Some organisms must amplify genes to keep pace with the high demand for certain gene products, especially in early development. Histones, ribosomal proteins, and ribosomal RNA genes are amplified.
11. The existence of many kinds of regulatory mutants has helped to advance our understanding of early development in the fruit fly, *Drosophila melanogaster.* Regulatory gene products are proteins that activate or repress other genes. Early development in *Drosophila* is a sequence of events in which different regulatory proteins gradually come into play in cascade fashion, controlling a wide range of enzymes and structural proteins and also influencing each other. Up to the blastoderm stage the nuclei in a developing *Drosophila* embryo are not separated by cellular membranes. As a result, the regulatory proteins and other gene products may diffuse freely from their site of synthesis to other nuclei in the embryo. At the late blastoderm stage, the nuclei become cellularized. From this point on, the influence of regulatory proteins made in one cell must be exerted on another cell at the level of the cell membrane.

Selected Readings

Achneider, R. J., and T. Shenk, Impact of virus infection on host cell protein synthesis. *Ann. Rev. Biochem.* 56:317–332, 1987.

Atchison, M. L., Enhancers: mechanisms of action and cell specificity. *Ann. Rev. Cell. Biol.* 4:127–153, 1988.

Binétruy, B., T. Smeal, and M. Karin, Ha-Ras augments c-Jun activity and stimulates phosphorylation of its activation domain. *Nature* 351:122–127, 1991.

Bingham, P. M., T. Chou, I. Mims, and Z. Zachari, On/off regulation of gene expression at the level of splicing. *Trends Genet.* 4:134, 1988.

Brietbart, R. E., A. Andreadis, and B. Nadal-Ginard, Alternative splicing: a ubiquitous mechanism for the generation of multiple protein isoforms from single genes. *Ann. Rev. Biochem.* 56:467–495, 1987.

Brown, D. D., How a simple animal gene works. In *The Harvey Lectures,* Series 76, pp. 27–44. New York: Academic Press, 1982.

Chandler, V. L., B. A. Maler, and K. R. Yamamoto, DNA sequences bound specifically by glucocorticoid receptor *in vitro* render a heterologous promoter responsive *in vivo:* A steroid specific enhancer. *Cell* 33:489–499, 1983.

Chen, H-Z., T. Hoey, and G. Zubay, Purification and properties of the *Drosophila* zen protein. *Mol. Cell. Biochem.* 79:181–189, 1988. First evidence that a homeobox protein binds to DNA as a monomer.

Duncan, I., The bithorax complex. *Ann. Rev. Genet.* 21:285–320, 1987.

Evans, R. M., The steroid and thyroid hormone receptor superfamily. *Science* 240:889–895, 1988.

Forsburg, S. L., and L. Guarente, Communication between mitochondria and the nucleus in regulation of cytochrome genes in the yeast *Saccharomyces cerevisiae. Ann. Rev. Cell. Biol.* 5:153–180, 1989.

Gabrielsen, O. S., and A. Sentenac, RNA polymerase III (C) and its transcription factors. *Trends Biochem. Sci.* 16:412–416, 1991.

Gehring, U., Steroid hormone receptors: Biochemistry, genetics and molecular biology. *Trends Biochem. Sci.* 12:399–402, 1987.

Gehring, W. J., The molecular basis of development. *Sci. Am.* 253(4):152–162, 1985.

Gehring, W., Homeo boxes in the study of development. *Science* 236:1245–1252, 1987.

Giniger, E., and M. Ptashne, Cooperative binding of the yeast transcriptional activator *GAL4. Proc. Natl. Acad. Sci.* 85:382–386, 1988.

Gruenberg, D. A., S. Natesan, C. Alexandre, M. Z. Gilman. Human and *Drosophila* homeo domain proteins that enhance the DNA-binding activity of serum response factor. *Science* 257:1089–1095, 1992.

Guarente, L. P., Regulatory proteins in yeast. *Ann. Rev. Genet.* 21:425–452, 1987.

Guarente, L., UASs and enhancers: Common mechanism of transcriptional activation in yeast and mammals. *Cell* 52:303–305, 1988.

Gurdon, J., Egg cytoplasm and gene control in development. The Croonian Lecture, 1976. *Proc. R. Lond. B.* 198:211–247, 1977.

Hanes, S. D., and R. Brent, A genetic model for interaction of the homeodomain recognition helix with DNA. *Science* 251:426–430, 1991.

Harrison, S. C., A structural taxonomy of DNA-binding domains. *Nature* 353:715–719, 1991.

Herskowitz, I., Life cycle of the budding yeast *Saccharomyces cerevisiae. Microbiological Reviews* 53:536–553, 1988.

Hinnebusch, A. G., Involvement of an initiation factor and protein phosphorylation in translational control of GCN4 mRNA. *Trends Biochem. Sci.* 15:148–152, 1990.

Johnson, P. F., and S. L. McKnight, Eukaryotic transcriptional regulatory protein. *Ann. Rev. Biochem.* 58:799–839, 1989.

Karlsson, S., and A. W. Nienhius, Developmental regulation of human globin genes. *Ann. Rev. Biochem.* 54:1071–1108, 1985.

King, T., and R. Briggs, Serial transplantation of embryonic nuclei. *Cold Spring Harbor Symp. Quant. Biol.* 21:271–290, 1956.

Klevit, R. E., Recognition of DNA by Cys_2His_2 zinc fingers. *Science* 253:1367–1393, 1991.

Lai, E., and J. E. Darnell, Jr., Transcriptional control in hepatocytes: a window on development. *Trends Biochem. Sci.* 16:427–429, 1991.

Lamb, P., and S. L. McKnight, Diversity and specificity in transcriptional regulation: the benefits of heterotypic dimerization. *Trends Biochem. Sci.* 16:417–433, 1991.

Marmorstein, R., M. Carey, M. Ptashne, and S. C. Harrison, DNA recognition by GAL4: structure of a protein-DNA complex. *Nature* 356:408–414, 1992.

Marzluff, W. F., and N. B. Pandey, Multiple regulatory steps control histone messenger-RNA concentrations. *Trends Biochem. Sci.* 12:49–51, 1988.

McClintock, B., Controlling elements and the gene. *Cold Spring Harbor Symp. Quant. Biol.* 21:197–216, 1956.

McKinney, J. D., and N. Heintz, Transcriptional regulation in the eukaryotic cell cycle. *Trends Biochem. Sci.* 16:430–434, 1991.

Melton, D. A., Pattern formation during animal development. *Science* 252:234–241, 1991.

Miner, J. N., and K. R. Yamamoto, Regulatory crosstalk at composite response elements. *Trends Biochem. Sci.* 16:423–426, 1991.

Mitchell, P. J., and R. Tjian, Transcriptional regulation in mammalian cells by sequence-specific DNA binding proteins. *Science* 245:371–378, 1989.

Nevins, J. R., Transcriptional activation by viral regulatory proteins. *Trends Biochem. Sci.* 16:435–439, 1991.

Ptashne, M., How gene activators work. *Sci. Am.* 260(1):41–47, 1989.

Raghow, R., Regulation of messenger RNA turnover in eukaryotes. *Trends Biochem. Sci.* 12:3358–3360, 1987.

Roeder, R. G., The complexities of eukaryotic transcription initiation: regulation of preinitiation complex assembly. *Trends Biochem. Sci.* 16:402–407, 1991.

Schler, A. F., and W. J. Gehring, Direct-homeo domain-DNA interaction in the autoregulation of the *fushi tarazu* gene. *Nature* 356:804–806, 1992.

Schwabe, J. W. R., and D. Rhodes, Beyond zinc fingers: Steroid hormone receptors have a novel structural motif for DNA recognition. *Trends Biochem. Sci.* 16:291–296, 1991.

Singer, S. J., Intercellular communication and cell-cell adhesion. *Science* 255: 1671–1677, 1992. Considers aspects of regulation we didn't have time to cover.

Stark, G., M. Debatisse, E. Giulotto, and G. M. Wahl, Recent progress in understanding mechanisms of mammalian gene amplification. *Cell* 57:901–908, 1989.

Struhl, K., Molecular mechanisms of transcriptional regulation in yeast. *Ann. Rev. Biochem.* 58:1051–1077, 1989.

Struhl, K., Helix-turn-helix, zinc-finger and leucine-zipper motifs for eukaryotic transcriptional regulatory proteins. *Trends Biochem. Sci.* 14:137–140, 1989.

Weinberg, R. A., Finding the anti-oncogene. *Sci. Am.* 259(3):44–51, 1988.

Problems

1. On the basis of what you know about genes in prokaryotic and eukaryotic cells, define a gene. Make sure that your definition is brief and concise. Does your definition have any limitations or problems?
2. Why is attenuation control in eukaryotes unlikely?
3. A mutation in $GAL4^c$ of yeast leads to constitutive galactose fermentation in haploid cells. Describe the nature of this mutation at the molecular level. Would you expect expression of *GAL* genes to be responsive to glucose repression in this mutant?
4. Explain how starvation of yeast for any one of ten amino acids derepresses the synthesis of more than 30 enzymes in nine different amino acid biosynthetic pathways (general amino acid control).
5. How would you expect a deletion of *HMLE* to affect the expression of mating-type genes in yeast? Compare this effect with the deletion of the $\alpha 2$ gene from *MAT*α. Consider both homothallic and heterothallic backgrounds.

6. Expression of some genes in eukaryotic cells can be induced by treatment with 5-azacytidine. Explain how this happens.
7. Briggs and King were able to grow a differentiating embryo from an egg of *Rana pipiens* whose chromosomes were replaced with a single diploid nucleus from another embryo. What does this tell us about the state of the nucleus in the developing embryo?
8. Hemophilia is an X-chromosome-linked disorder. Bearing in mind the phenomenon of X-chromosome inactivation, suggest an explanation for the observation that females who are heterozygous for the defective gene are essentially asymptomatic.
9. High salt weakens the interaction of histones with DNA but has little affect on the binding of many regulatory proteins. Explain this observation in terms of how these molecules interact with DNA.
10. List some characteristics that distinguish active from inactive chromatin.
11. In the *Xenopus* oocyte a large number of ribosomes are made in a short time to handle the rapid demand for cell growth during cleavage stages. How is this large amount of rRNA made in such a short time?
12. When mammalian cells in culture are treated with the antifolate methotrexate (see chapter 20), cells can be selected that are resistant to high levels of this toxic compound. The enzyme dihydrofolate reductase becomes elevated about 1,000-fold in these resistant cells. What mechanism can account for such a large amount of this enzyme being made and how does this protect the cell from the toxic effects of methotrexate? How could you test for the molecular mechanism of this drug resistance?
13. Given that specific subsets of homeotic genes are required for the development of specific segments of the *Drosophila* embryo, suggest a possible mechanism whereby the necessary spatially restricted pattern of homeotic gene expression might be achieved. Incorporate the observed effect of homeotic mutations upon segment morphology into your model.
14. Explain how a DNA sequence (enhancer sequence) located 5,000 base pairs from a gene transcription start site can stimulate transcription even if its orientation is reversed.
15. Speculate on some possible models to explain tissue-specific alternative splicing such as that seen with variants of tropomyosin.
16. Discuss the types of structural motifs found in eukaryotic transcription factors.

Physiological Biochemistry

7

PART

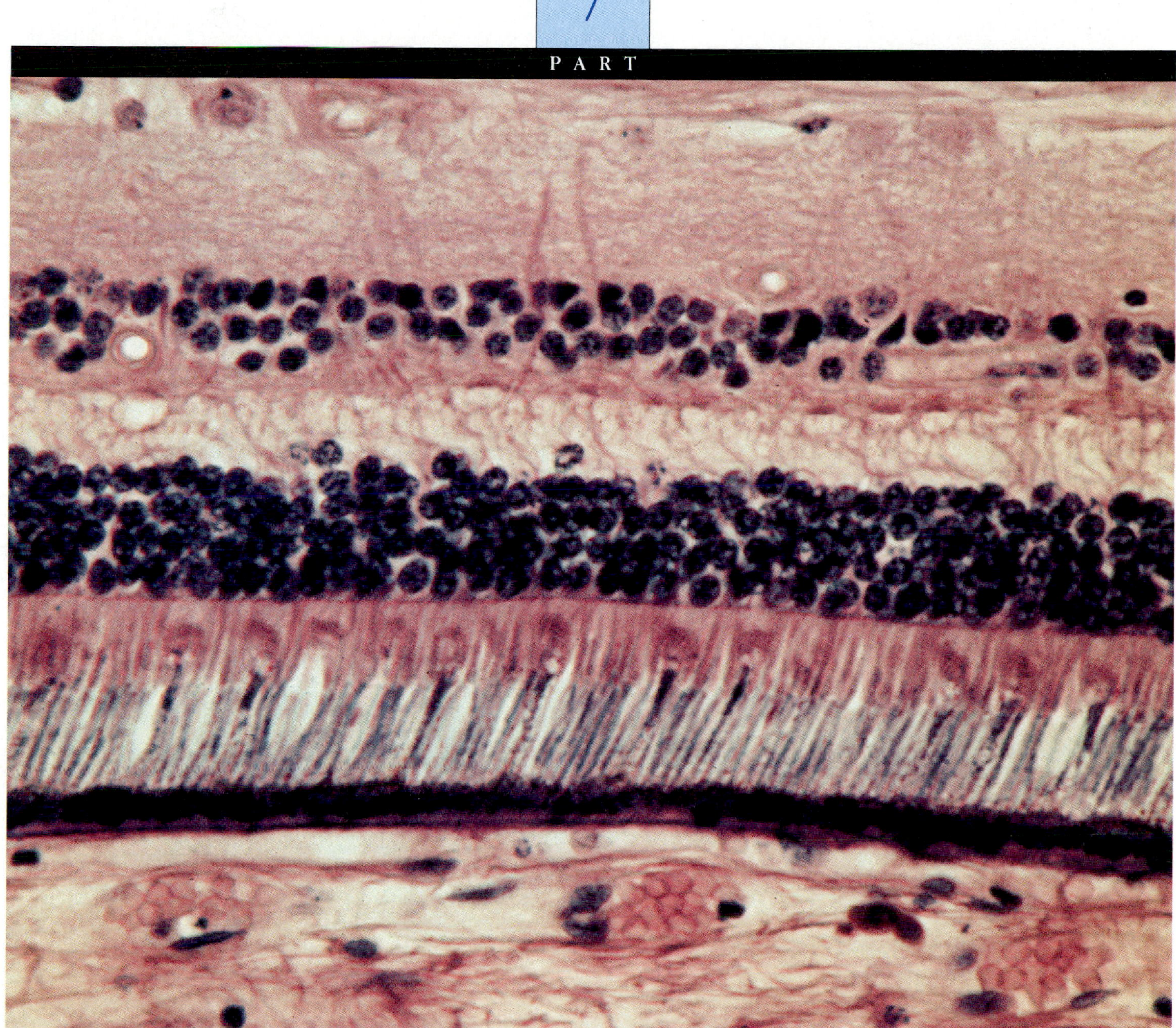

Light micrograph of a cross section of a human retina (100X). Ten distinct layers are visible. Starting from one end and going to the other we see pigment cells, rods and cones, outer limiting membrane, cell bodies of rods and cones, outer plaxiform layer, integrating bipolar cell layer, inner plaxiform layer, ganglion cell layer, optic nerve fibers, and finally inner limiting membrane. (© Ed Reschke/ Peter Arnold, Inc.)

The topics that we discuss in this part are, at first glance, quite diverse. However, they are all connected with the question of how cells interact in the whole organism.

Biochemical reactions can occur in which the interacting species are both soluble, in which one or both of the interacting species are membrane-bound, or in which substances or signals are transmitted across one or more membranes. In physiological biochemistry we are almost always concerned with reactions that involve both soluble and membrane phases.

Exchange of substances between the cytosol and the external environment commonly occurs in both prokaryotes and eukaryotes, while transmission of substances or signals between cells is most frequent in, but not exclusive to, differentiated multicellular eukaryotes. Although signal transmission sometimes requires direct cell-to-cell contact, it can also occur via secretion of a substance that is passed from one cell type to another, which may be located nearby or at a considerable distance. The diffusible substance can be a small molecule such as a steroid hormone, or a large molecule such as a protein growth factor. The small molecule is likely to exert its effect after passing through the cell membrane, whereas the large molecule usually exerts its effect by binding to the outer cell membrane, where it triggers a chain reaction starting on the inner membrane.

An understanding of the various types of cell–cell interactions is helping to throw light on such processes as membrane transport (chapter 32), hormone action (which we considered in chapter 24), antibody formation (chapter 33), carcinogenesis (chapter 34), neurotransmission (chapter 35), and vision (chapter 36).

32

CHAPTER

Mechanisms of Membrane Transport

In chapter 7 we considered the current state of knowledge about the structure of biological membranes. Two of the most important conclusions that we drew were that membranes are both dynamic and asymmetric structures. As we will see, these properties allow transmembrane processes to be carried out, often, but not always, in a unidirectional manner. The most prominent of these processes are the transport of nutrients and inorganic electrolytes into the cell and the export of toxic substances and waste materials out of the cytoplasm. In both instances, the problem is usually the same: How do small, hydrophilic metabolites traverse the hydrophobic lipid bilayer at rates sufficient to maintain the growth needs of the cell? Often this problem is solved by integral membrane proteins, which, by analogy with enzymes, specifically interact with their substrates, but which, unlike enzymes, have as their primary purpose the movement of a molecular species across the membrane rather than the catalysis of a chemical reaction.

The fairly recent development of general techniques for the purification of membrane proteins and of the powerful tools of molecular biology has greatly increased our understanding of the molecular mechanisms of transport proteins, or permeases, as they are often called. In this chapter, we will consider the theory, energetics, and mechanisms of transmembrane transport in cells.

A number of processes rely on transport, directly or indirectly, for their proper functioning. Examples of these seemingly diverse phenomena include nerve impulse propagation, cellular behavior and differentiation, and sensory reception and transduction. Indeed, it should become apparent that nearly all cellular and organismal functions are dependent on biological transport. As in chapter 7, examples from both prokaryotic and eukaryotic worlds will show us that many fundamental principles concerning the biochemistry of transport are probably valid throughout the living kingdom.

Figure 32.1

Different types of transport across membranes. The membrane—whether it is a plasma membrane or another type of membrane, such as those encompassing cellular organelles—imposes a barrier to the free movement of molecules. Five basically different types of processes are used to transport substances across membranes. The first, *simple diffusion,* does not require any special apparatus; it requires only that the substance being transported be soluble in the domain of the membrane bilayer. *Facilitated diffusion* is similar to free diffusion in not requiring energy. It differs from free diffusion in that a special pore or carrier (permease) is used to mediate the transport across the membrane. For transport processes that require energy, such as transport against a concentration gradient, energy is required. Such processes are referred to as active transport. In *primary active transport,* energy is directly expended in the process of transporting the substance across a membrane. This energy is donated in various ways in different active transport processes, for example, by ATP or PEP hydrolysis, by light energy, or by electron flow. In *secondary active transport,* a substance from one active transport process may be used to drive the transport of another substance. Finally, in *group translocation,* the transported substance is chemically modified by the membrane permease during the transport event. This type of transport generally requires energy as well, either directly or indirectly.

Although the focus in this figure is on transport of substances into the membrane-enclosed lumen, transport of substances in the reverse direction is frequently of great importance.

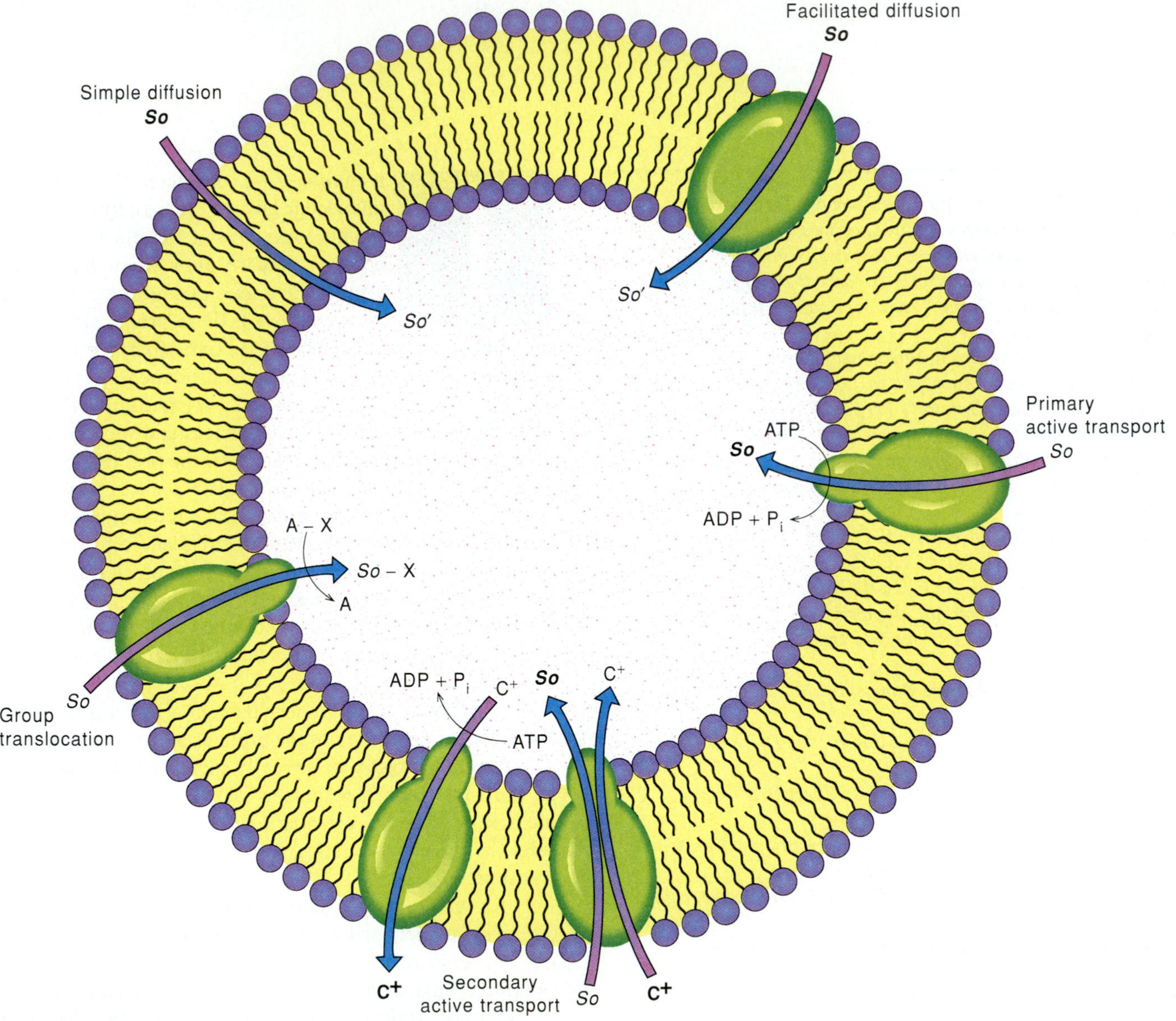

The Theory and Thermodynamics of Biological Transport

Before we consider specific examples of transport across biological membranes, we will discuss the thermodynamics of this process as well as specific ways in which such a transfer can occur. With this foundation, we can assign certain predicted characteristics to each kind of transport mechanism (fig. 32.1). And, on the basis of these predictions, we can design experiments that will give useful information regarding the events responsible for the transmembrane movement of a solute.

Most Solutes Are Transported by Specific Carriers

In principle, a solute could move across a barrier such as a membrane by either of two fundamentally different mechanisms: simple (or passive) diffusion or carrier-mediated transport. In the first mechanism, the molecule in question must be somewhat soluble in the hydrophobic phase of the bilayer, while in the second mechanism, the solute's journey across the membrane is facilitated by another molecule, in most cases a protein. For these reasons, simple diffusion is usually of importance in biological transport systems only for molecules having a largely hydrophobic character or in membrane systems possessing nonspecific aqueous pores, as we will see in a later section. An exception to this rule, however, is water, which traverses many biological membranes by simple diffusion. Most other hydrophilic molecules, which comprise the majority of species transported into and out of cells, are usually recognized by transport systems that include one or more protein carriers.

In order to illustrate the characteristics of simple diffusion, let us consider two dilute solutions of a solute *So*, one twice as concentrated as the other, which are separated by a barrier that is permeable to *So*, as in figure 32.2. As a consequence of random thermal motion, a certain proportion of the molecules originally present on one side of the barrier (one-quarter in figure 32.2) will move to the other side in a specified length of time t. This flux is bidirectional, so that after time t, the ratio of the concentration of *So* on the left side of the barrier to that on the right will have been reduced from 2:1 to 7:5. After many such time intervals, the ratio will approach 1:1, or equilibrium. The number of molecules of *So* that move from left to right in time interval t is therefore directly proportional to the initial concentration of *So* on the left side:

$$v \propto [So] \tag{1}$$

while a similar relationship holds true for movement in the opposite direction:

$$v' \propto [So'] \tag{2}$$

where v and v' refer to the rates of transfer from left to right and right to left, respectively, and $[So]$ and $[So']$ are the initial concentrations of the solute on the left and right sides of the barrier. The net rate of transfer V from the more concentrated to the more dilute solution is therefore the difference between these two component rates:

$$V = v - v' \quad \text{or} \quad V \propto [So] - [So'] \tag{3}$$

Hence, at equilibrium, $V = 0$.

If the barrier in figure 32.2 is, for example, a biological membrane of finite thickness l, we must also take this into account in the expression for overall transfer rate:

$$V \propto \frac{[So] - [So']}{l} \tag{4}$$

Finally, the rate of transfer of *So* will depend on such factors as temperature, viscosity of the solvent, and the solubilities of *So* in the solvent and in the membrane. In a given membrane system, these factors often can be held relatively constant, so that we can express relationship (4) as an equation:

$$V = \frac{D([So] - [So'])}{l} \tag{5}$$

where D is a constant, the diffusion coefficient, which has the dimensions of area per unit of time (e.g., cm^2/s). Equation (5), called Fick's law, shows that the net rate of diffusion is directly proportional to the concentration gradient of *So* across the membrane barrier.

It is apparent from equation (5) that in a system in which *So* traverses the membrane by simple diffusion, a plot of V versus $[So] - [So']$ should give a straight line that passes through the origin with a slope of D/l (fig. 32.3a). In practice, it is often possible to measure V as a function of the concentration gradient of *So* across a biological membrane by methods

Figure 32.2

Illustration of simple diffusion in a two-compartment system divided by a semipermeable membrane. After time t, one-fourth of the molecules present originally on one side of the membrane have diffused by random thermal motion to the other side. After many such intervals, equilibrium is approached. The molecules in this illustration are identical, but different colors are used to indicate their initial locations. In simple diffusion, the number of molecules diffusing from one side to the other in any given time interval is directly proportional to their concentration on the first side. (Source: K. D. Neame and T. G. Richards, *Elementary Kinetics of Membrane Carrier Transport*. Copyright © 1972 Blackwell Scientific Publications, Oxford, England.)

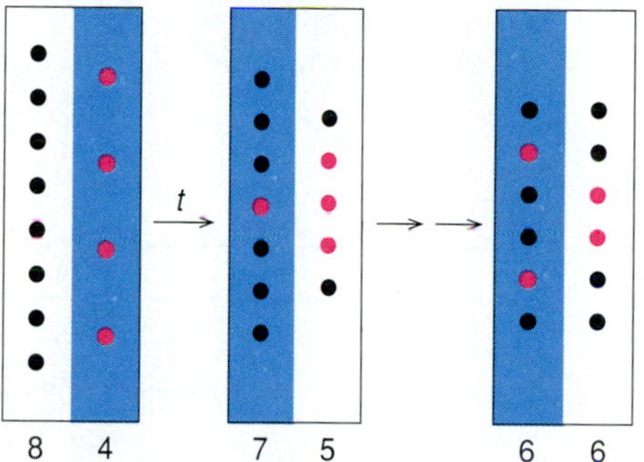

Figure 32.3

Initial rate kinetics of simple diffusion (*a*) and carrier-mediated transport (*b*). In (*b*), [*So*] and [*So′*] refer to the concentration of transported solute molecule on two sides of the membrane. Plots of velocity versus the difference in substrate concentrations on the two sides of the membrane give a straight line with a slope of *D/l* in (*a*) and a rectangular hyperbola approaching a maximum velocity (V_{max}) at high [*So*] − [*So′*] in (*b*).

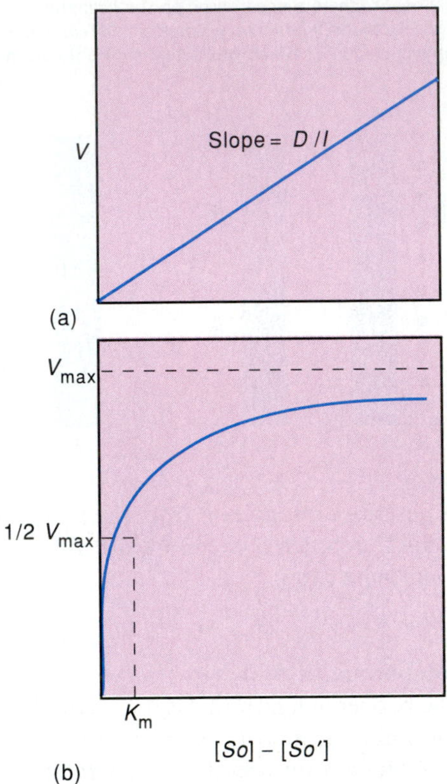

Figure 32.4

Double-reciprocal (Lineweaver-Burk) plots of carrier-mediated transport and simple diffusion. For diffusion, the line passes through the origin, while for carrier-mediated processes the *x* intercept is equal to $-1/K_m$ and the *y* intercept is equal to $1/V_{max}$, as shown in (*a*). In (*b*), the transport kinetics for a cell transporting a given solute by both simple diffusion and a carrier-mediated system are illustrated. The line shown is a combination of the processes plotted in (*a*) and passes through the origin. In the case of two carrier-mediated systems, a similar line would be obtained that would intersect the ordinate above the origin. (Source: K. D. Neame and T. G. Richards, *Elementary Kinetics of Membrane Carrier Transport*. Copyright © 1972 Blackwell Scientific Publications, Oxford, England.)

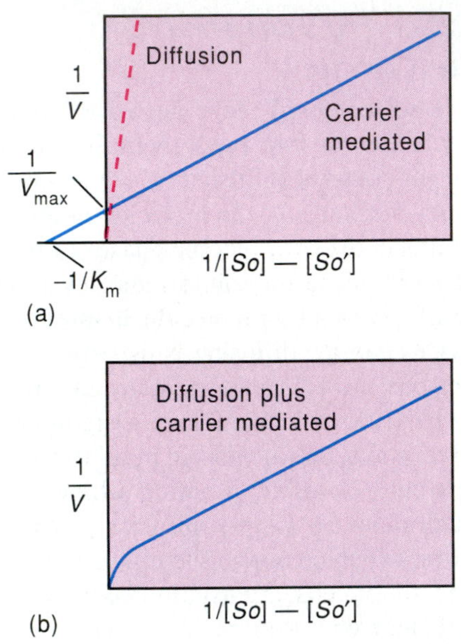

that we will consider later in this chapter. Provided that our data include a wide range of solute concentration gradients, a linear plot of *V* versus [*So* − *So′*] is evidence that *So* is taken up by simple diffusion.

The uptake kinetics shown in figure 32.3*a* for simple diffusion are, in fact, seen only in a minority of cases for transport of molecules across cytoplasmic membranes of cells. More often, the curve approximates that of a rectangular hyperbola, as illustrated in figure 32.3*b*. In this case, at high [*So*] − [*So′*] the transport rate approaches a limiting value that is termed V_{max} by analogy with steady-state enzyme kinetics (chapter 8). This saturation phenomenon implies that uptake of *So* is dependent on a discrete number of carriers in the membrane, all of which become filled at high concentration gradients of *So*. The situation is formally analogous to saturation of an enzyme by its substrate, and the rate equation for this process, carrier-mediated transport, is essentially the same as the Michaelis-Menten equation derived in chapter 8:

$$V = \frac{V_{max}[So]}{K_m + [So]} \qquad (6)$$

In the derivation of equation (6), [*So′*] is assumed to be negligible (just as is [*P*] in enzyme kinetics) and is taken to be the initial concentration of *So* inside the cell or vesicle being studied. Experimentally, this situation is usually easy to attain, so that [*So*] equals the concentration initially present on the outside, [*So′*] = 0, and *V* is an initial rate of transport of *So* into the system enclosed by the membrane. As in enzyme kinetics, K_m is the solute concentration at which $V = \frac{1}{2}V_{max}$ (fig. 32.3*b*). Note that the value of K_m, although often assumed to be a measure of the affinity of a carrier for its "substrate," is really a ratio of rate constants and is not necessarily equivalent to the dissociation constant of the solute from its carrier.

From what we have said, it is clear that in many cases, by studying the rate of transport, *V*, as a function of [*So*], we can distinguish simple diffusion from carrier-mediated transport. Often these data are plotted in a double-reciprocal fashion (Lineweaver-Burk plot, chapter 8). Plots of $1/V$ against $1/[So]$ yield straight lines in both instances, but the line for simple diffusion passes through the origin, while that for carrier-mediated transport intersects the ordinate at $1/V_{max}$ and the abscissa at $-1/K_m$ (fig. 32.4*a*). This difference, of course, reflects the fact that carrier systems are saturable (V_{max} is finite), while simple diffusion systems are not ($V_{max} = \infty$; $1/V_{max} = 0$).

Unfortunately for biochemists interested in studying a single biological transport system, double-reciprocal plots for the uptake of a specific solute into a cell, or into vesicles derived from cells, are sometimes nonlinear. Such behavior usually indicates that more than one process is involved in the transport of this substance into the cell. The different processes may involve both simple diffusion and carrier mediation, or both may be carrier-mediated (fig. 32.4*b*). In either case, however, it is first necessary to find experimental conditions under which only the system of interest is functional before we can draw conclusions concerning the transport mechanism. We may accomplish this by working at solute levels recognized only by one carrier, by studying mutants that lack one system, by using specific inhibitors, or even by purification and reconstitution of the carrier of interest (see the following discussion).

Finally, we can in many cases distinguish carrier-mediated transport from simple diffusion by its stereospecificity. For example, in a membrane system that is permeable to glucose only by simple diffusion, both D- and L-glucose will traverse the membrane at equal rates. In contrast, many biological membranes contain carriers specific for glucose that recognize only the D-stereoisomer. In these cells, the initial transport rate for D-glucose is much higher than that for the L isomer and shows saturation kinetics. In fact, the rate of transport of L-glucose is often used in such instances to correct the total rate observed for diffusion in order to obtain the "true" carrier-mediated rate for the D isomer. Stereospecificity in biological transport, as in enzyme catalyzed reactions, is a consequence of the highly selective configuration of the "active site" of the transport protein for its "substrate."

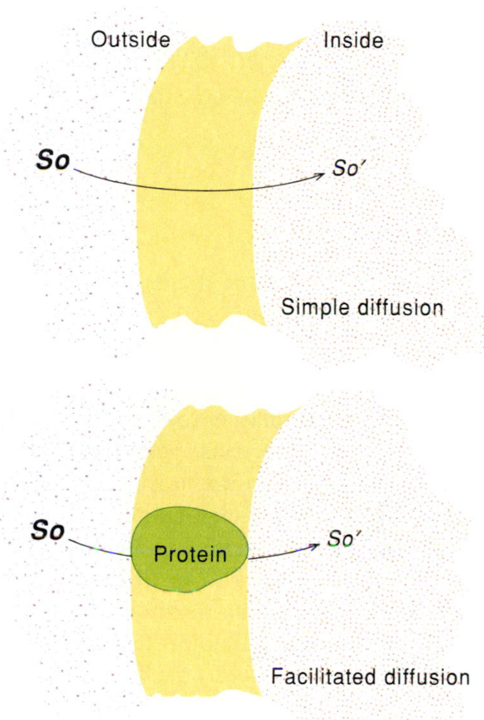

Figure 32.5

Schematic illustration of diffusion processes in biological transport. In both simple and facilitated diffusion, the concentrations of the solute on both sides of the membrane are the same at equilibrium, and no metabolic energy is expended.

Simple and Facilitated Diffusion Require No Energy

In the preceding section we saw that a solute will tend to diffuse "down" its concentration gradient (from a region of higher to a region of lower concentration) spontaneously, until equilibrium is reached. Such a process requires no outside input of energy into the system and thus occurs with a negative change in free energy ΔG. This change is a consequence of the fact that the equilibrium situation ($[So] = [So']$ in fig. 32.2) is more "random" or disordered than the starting situation (in which $[So] > [So']$) and thus has relatively more entropy S. Because $\Delta G = \Delta H - T\Delta S$, an increase in entropy results in a negative free energy change (chapter 2). The magnitude of ΔG can be calculated from the expression

$$\Delta G = 2.3RT \log \frac{[So']}{[So]} \tag{7}$$

for the situation illustrated in figure 32.2. Net transport from left to right will occur spontaneously in this system only if $[So']/[So] < 1$, while at equilibrium, $\Delta G = 0$ ($[So'] = [So]$).

In biological systems, both simple diffusion and facilitated diffusion are transport mechanisms that result in net transport of a solute only in the direction of negative ΔG, i.e., usually from higher to lower concentrations. In contrast to simple diffusion, facilitated diffusion is a carrier-mediated process and hence is saturable (fig. 32.5). Both it and simple diffusion generally result in equilibration of a solute across a membrane, unless the solubility properties of the solute are significantly different on both sides of the membrane or the solute has a net charge. If the solute is charged, both types of diffusion can lead to an equilibrium situation in which $[So] \neq [So']$ if there is an independently maintained electrical potential, $\Delta\psi$, across the membrane (also see chapter 15):

$$\Delta G = 2.3RT \log \frac{[So']}{[So]} + ZF\,\Delta\psi \tag{8}$$

In this equation, Z is the net charge on So, and F is the Faraday constant. Thus at equilibrium ($\Delta G = 0$), unequal concentrations of So can exist on either side of a membrane as long as the term $ZF\,\Delta\psi$ is not equal to zero (fig. 32.6). The relationship between free energy and membrane potential is very similar to that between free energy and electromotive potential discussed in chapter 15.

Transport against a Concentration Gradient Requires Energy

Both simple and facilitated diffusion are examples of energy-independent biological transport mechanisms. Although they occur in fundamentally different ways, neither requires the input of energy, nor can either lead to concentration of the solute on

Figure 32.6

At equilibrium, a *charged* solute, So^+, may be unequally distributed across a biological membrane if an independently maintained potential exists across it. This situation can occur even if So^+ is transported by simple or facilitated diffusion. Bold type indicates higher concentration.

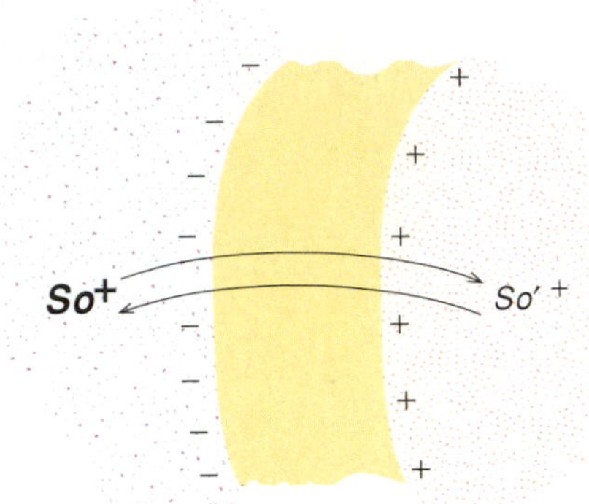

one side of the membrane relative to the other in the absence of a charged solute and an electrical potential across the membrane. Most cells, however, can maintain relatively high intracellular concentrations of some metabolites and ions relative to the extracellular environment. For example, many cells have an intracellular level of K^+ that is at least 30 times the concentration of this ion in the surrounding medium. It therefore follows that if carriers for accumulated molecules, such as K^+, exist in the cell membrane, they must be able to transport their substrates in the direction of a positive ΔG, i.e., up a concentration gradient of the solute. In order to accomplish this thermodynamically unfavorable movement, such processes must be coupled to events that have negative ΔG values, i.e., that can yield energy to "push" the solute "uphill." Concentrative accumulation of a solute inside a cell, organelle, or vesicle with the consequent expenditure of metabolic energy is termed active transport. In a related process, group translocation, energy for the transmembrane accumulation of a solute is provided by chemical modification of the molecule as it passes through the membrane.

Active Transport and Group Translocation Are Used to Concentrate Solutes

The use of metabolic energy to drive active transport implies that the transport system itself must somehow be able to harness the energy in a particular form and use it to do work, that is, to transport a solute against a concentration gradient. The exact molecular mechanisms by which these interconversions take place are not completely known for any system; however, a number of clues are emerging that we will consider later in this chapter. Because systems that carry out active transport are invariably carrier-mediated, it seems likely that in most cases integral membrane permeases themselves directly participate in this energy interconversion process. A major direction of research in membrane-transport biochemistry has therefore been to study the energetics of active transport and, whenever possible, to isolate and characterize the proteins that comprise a particular system.

Two basic types of active transport have been recognized. In primary active transport, energy is provided directly by the hydrolysis of ATP, by electrons flowing down an electron transport chain, or by light. Examples of these include the Na^+-K^+ ATPase of animal cells, which pumps K^+ into the cell and Na^+ out of the cell at the expense of ATP; the respiratory electron transport chain, which pumps H^+ out of the mitochondrion and many bacterial cells (chapter 15); and the light-driven H^+ pump of *Halobacterium,* bacteriorhodopsin. In secondary active transport, ion gradients across the membrane, themselves created by other active transport systems, are used to drive the concentrative uptake of other ions or metabolites. Because an ion gradient is a form of potential energy, a process involving its collapse releases this energy, which can be used to perform work. In mitochondria and bacteria, the potential energy stored in gradients of H^+ is used during respiratory metabolism to synthesize ATP (chapter 15). As we will see, gradients of H^+ and other ions also are an important source of energy for active transport in both prokaryotic and eukaryotic cells (also see chapters 15 and 16).

Related to primary active transport is the process of group translocation. While molecules transported by active systems are translocated across the membrane without chemical changes occurring, in group translocation the transported solute is chemically modified on its journey through the membrane. The best-documented example of group translocation occurs in many bacteria, which convert sugars taken up from the surrounding medium into sugar monophosphates simultaneously with their transport through the membrane. Energy is expended in the process, since the phosphoryl donor is the high-energy phosphate glycolytic intermediate, phosphoenolpyruvate. In this case the permease for the sugar is also an enzyme that catalyzes both the transport and the phosphorylation of its substrate. Figure 32.7 illustrates schematically the differences between primary and secondary active transport and group translocation, while table 32.1 lists representative well-characterized transport systems of both animal and bacterial cells.

Energy-Coupling Mechanisms in Biological Transport

It is the ultimate goal of the biochemist interested in transmembrane transport mechanisms to understand at the molecular level the physicochemical properties of a transport protein that allow it to carry out its function. In this section, we will briefly describe approaches commonly used to gain information about the structures of permease proteins and the relationship of these structures to the transmembrane transport processes catalyzed by them. We will also show how these systems can be used to investigate energy-coupling mechanisms at the molecular level. In doing so, we will place emphasis on some of the best-characterized biological transport systems. With this as a background, we will then venture one step further and attempt to explain how the structures and properties of integral membrane permeases permit them to carry out the myriad of transport processes so vital to the cell.

Figure 32.7

Schematic illustrations of primary active transport (*a*), secondary active transport (*b*), and group translocation (*c*). In (*a*), the energy source can be ATP, as shown, or electron transport or light can be used. In (*b*), active accumulation of *So* is energized by cotransport of an ion that flows down its concentration gradient. In (*c*), *So* is accumulated as a derivative, *So-X*, which is formed by a chemical reaction catalyzed by the carrier during transport though the membrane.

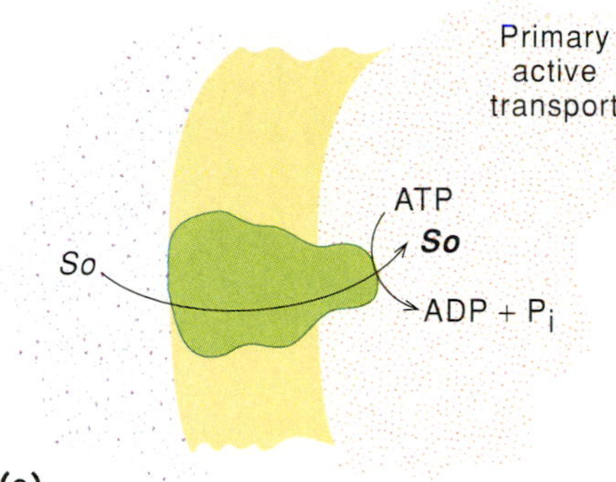

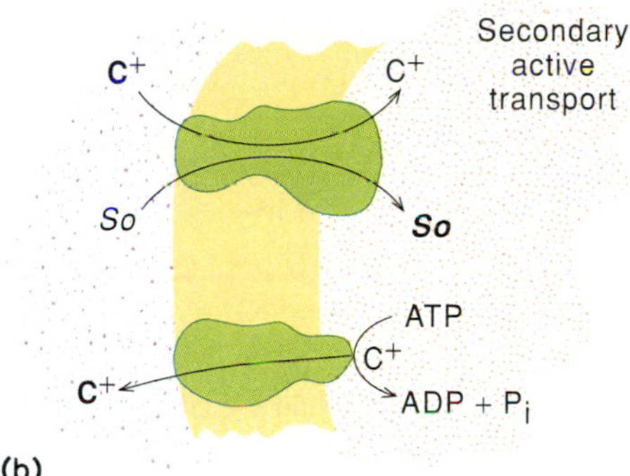

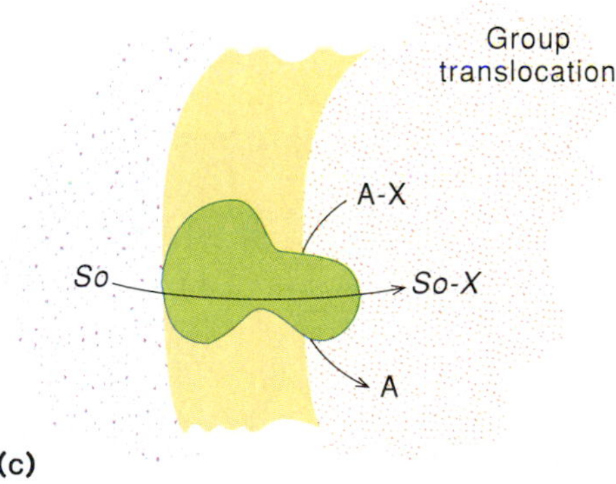

Table 32.1

Representative Well-Characterized Biological Solute Transport Systems

I. Nonstereospecific Channel Proteins
- A. Muscle acetylcholine receptor
- B. Nerve Na^+ channel
- C. Nerve K^+ channel
- D. Porins in the outer membranes of Gram negative bacteria
- E. Glycerol permease in *E. coli*
- F. Major intrinsic protein of mammalian lens

II. Single-species Facilitators (Uniporters or Antiporters)
- A. Glucose transporters in red cells, brain, liver, muscle, etc.
- B. Anion antiporters in red cells and various tissues
- C. Na^+/H^+ antiporters in bacteria and animal tissue cells
- D. ADP/ATP antiporters in mitochondria

III. Two-species Facilitators (Symporters)
- A. Lactose permease in *E. coli* (lactose:H^+)
- B. Melibiose permease in *Salmonella* (melibiose:Na^+)
- C. Xylose and arabinose permeases of *E. coli* (pentose:H^+)
- D. Galactose, glucose, and maltose permeases in yeast (sugar:H^+)
- E. Glucose transporters in certain epithelial cells (glucose:Na^+)
- F. α-Amino acid transporters in tissue cells (amino acid:Na^+)
- G. NaCl/KCl symporters of animal tissues

IV. ATP-driven Active Transporters
- A. Na^+, K^+, H^+, and Ca^{++} ATPases in eukaryotic cells and K^+ ATPases of bacteria (P-type)
- B. Vacuolar H^+ ATPases of eukaryotes; also found in bacteria (V-type)
- C. H^+ and Na^+ F_1F_0 ATPases in mitochondria and bacteria (F-type)
- D. ATP-driven, binding-protein-dependent permeases of bacteria (ABC-type)
- E. Multidrug resistance glycoproteins of animal cells (ABC-type)

V. Light and Electron Flow Driven H^+, Na^+, or Cl^- Transporters
- A. Bacteriorhodopsin of *Halobacterium* (H^+ translocating)
- B. Halorhodopsin of *Halobacterium* (Cl^- translocating)
- C. Cytochrome oxidases of mitochondria and bacteria (H^+ or Na^+ translocating)

VI. Group Translocation
- A. Mannitol, β-glucoside, and *N*-acetylglucosamine Enzyme II of the *E. coli* phosphotransferase system (one-component systems)
- B. Glucose, sucrose, and glucitol Enzyme II–III pairs of the *E. coli* phosphotransferase system (two-component systems)
- C. Mannose Enzyme II-P—II-M—III complex of the *E. coli* phosphotransferase system (three-component systems)

The Molecular Genetic Approach to Transport

Until recently, the amino acid sequences of very few permease proteins were known. With the advent of rapid gene cloning and DNA sequencing techniques (chapter 27), we now know the primary structures of literally dozens of biological solute transport proteins. These amino acid sequences first of all allow us to group these proteins into evolutionarily related families. We now know that the thousands of permeases probably fall into only about a dozen permease classes, and that representatives of these classes cross phylogenetic lines (see table 32.1). Second, the primary structures of these proteins permit the construction of models that allow us to predict the folding of these proteins in the membrane (see chapter 7). We can then test these predictions by a variety of methods. Third, comparison of the primary structures of related permeases allows us to predict which residues may be necessary for catalytic function. In general, those residues and regions of the protein that are most strongly conserved in a group of diverging proteins are believed to be important for tertiary structure and/or function. Finally, by employing the technique of site-directed mutagenesis, we can change residues that are predicted to be important for function to other aminoacyl residues and assess the effects of such changes on catalytic activity. The use of these new molecular genetic approaches, coupled to biochemical analyses, has greatly expanded our knowledge concerning transport protein mechanisms during the past decade.

Isotopes, Analogs, and Specially Prepared Vesicles Are Used to Study Transport

In order to measure the rate of transport of a solute into a cell, we must have some method for easily identifying it and distinguishing molecules that have indeed crossed the membrane. Most commonly, researchers use radioactively labeled solutes for this purpose, and time-dependent uptake of radioactivity into a cell is evidence that the compound is being transported through the cell membrane. With many animal cells grown in culture, it is often sufficient to separate intracellular label from that remaining in the medium by quickly washing the cells (which often adhere to the vessel in which they are grown) with buffer or medium that is free of the solute being tested. With bacteria and small, free-living eukaryotes such as yeast and protozoans, cells can be quickly separated from the medium by centrifugation or filtration.

One of the problems associated with measuring transport rates, especially in whole cells, is the fact that subsequent metabolism of the transport substrate may affect the rate observed. In order to uncouple transport from metabolism, nonmetabolizable analogs of some solutes have been used. Examples of compounds that have been employed to study the transport of glucose, a common carbon and energy source for many cells, are methyl-α-glucoside and 2-deoxyglucose (fig. 32.8). Both compounds cannot be metabolized beyond the hexose-phosphate stage by most cells. Another technique that has been used, especially in bacteria, is the isolation of mutants that are able to transport a given solute, but that lack one or more enzymes for the further metabolism of the compound. By these procedures, it is often possible to determine many characteristics of the transport system itself, without any complications owing to other reactions that take place after the solute enters the cell.

Finally, vesicles derived from the membrane system of interest are often used by biochemists interested in membrane transport phenomena. Cytoplasmic membrane vesicles can often be isolated from eukaryotic cells after mild homogenization or N_2 cavitation (chapter 7), which leave many of the intracellular organelles intact. In bacteria, the use of membrane vesicles was pioneered in the 1960s by H. R. Kaback, who developed a method, similar to that shown in figure 7.23, for preparing membrane vesicles from *E. coli.* Most of these vesicles (fig. 32.9) appear to be "right side out"; that is, the orientations of most integral membrane proteins are the same as those in the intact

Figure 32.8

Methyl-α-glucoside and 2-deoxyglucose are two analogs that are useful in the study of glucose transport in many cells. Usually they are not metabolized beyond the hexose-phosphate stage, which avoids the complications of subsequent reactions that might affect the measurement of transport rates.

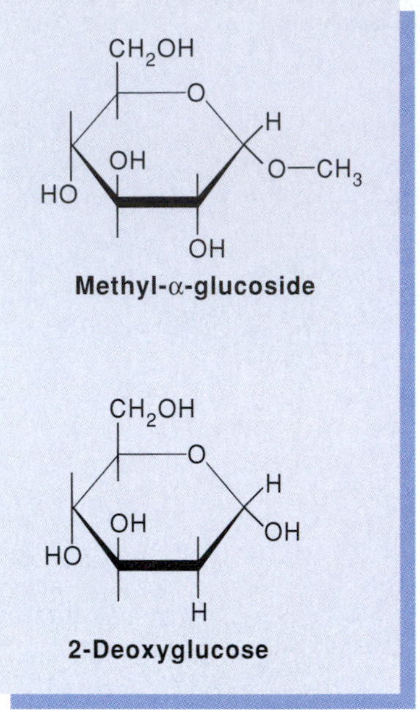

Figure 32.9

Electron micrograph of membrane vesicles obtained from *E. coli* spheroplasts subjected to osmotic lysis. These closed, spherical structures have been shown to transport a variety of compounds normally accumulated by whole cells as long as any required energy source is provided. They also have the advantage of lacking cytoplasmic components that could affect the transport rate (77,000×). (From H. R. Kaback, "Transport studies in bacterial membrane vesicles," *Science* 186:882, 1974. Copyright © 1974 by the AAAS.)

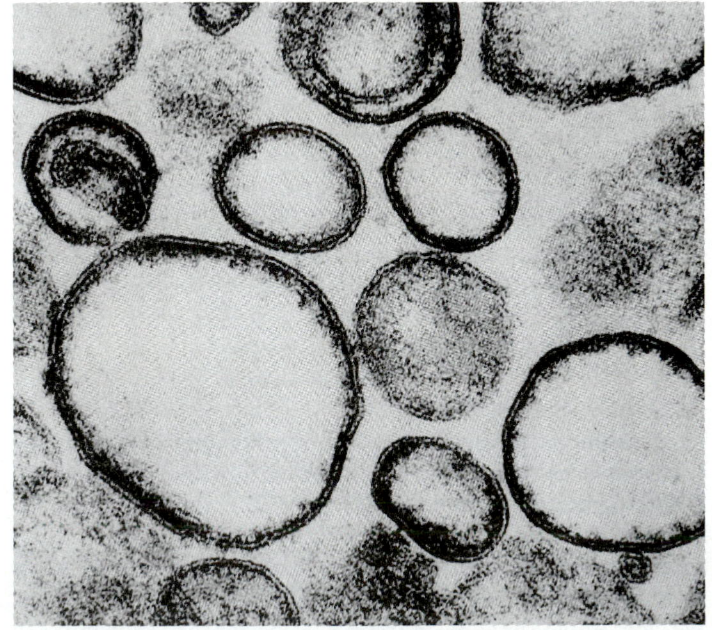

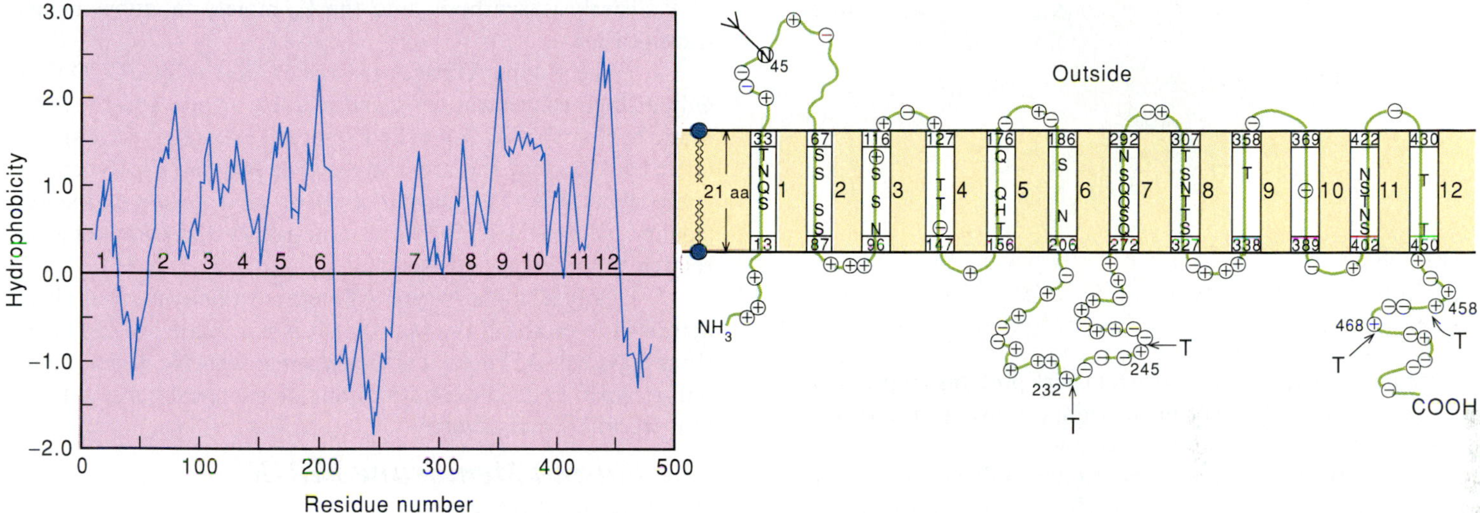

Figure 32.10

(*Left*) Hydropathy plot of the glucose transporter amino acid sequence. Hydropathy values are plotted with respect to position along the amino acid sequence. The numbers refer to putative membrane-spanning domains. (*Right*) Proposed model for the orientation of the glucose transporter in the membrane. The twelve putative membrane-spanning domains are numbered and shown as rectangles. The relative positions of acidic (Glu, Asp) and basic (Lys, Arg) amino acid residues are indicated by circled plus and minus signs, respectively. Uncharged polar residues within the membrane-spanning domains are indicated by their single-letter abbreviations: S, serine; T, threonine; H, histidine; N, asparagine; Q, glutamine. The predicted position of the N-linked oligosaccharide at Asn 45 is shown. The arrows point to the positions of known tryptic cleavage sites in the native, membrane-bound, erythrocyte glucose transporter. (Source: M. Mueckler et al., "Sequence and structure of a human glucose transporter," in *Science* 229:941–945, 1985. Copyright © 1985 American Association for the Advancement of Science, Washington, D.C.)

cell membrane. Kaback and others have shown that many substances transported by *E. coli* are also taken up by these vesicles as long as any required energy sources are provided. These vesicles have the advantage of lacking most cytoplasmic enzymes, so that the metabolism of the transport substrate is severely limited. Furthermore, they are especially suited for the study of energy-coupling mechanisms in active transport because the investigator can easily introduce various energy sources into the vesicle preparations and test their effects on the transport system of interest. Such experiments are sometimes difficult to interpret with whole cells, again because of the complications of multiple metabolic events occurring in the cytoplasm.

Examples of Facilitated Diffusion

Transport systems that appear to operate only by facilitated diffusion have been identified both in bacterial and animal cells. Glycerol, which can act as the sole carbon and energy source for many bacterial cells, has been shown to be transported in several bacteria by nonconcentrative systems. Studies using *E. coli* by E. Lin and collaborators have provided evidence for a "facilitator protein" that allows glycerol permeation in this organism. The facilitator increases the rate at which glycerol or any small, straight-chain, neutral molecule crosses the phospholipid bilayer but cannot accumulate these compounds within the cell.

The diameter of the pore formed by the glycerol facilitator is about 0.4 nm. Recently, the DNA sequence of the *glpF* gene encoding the glycerol facilitator was determined, and computer searches revealed that it is very similar in structure to two other potential pore-forming proteins, one from animal cells and one from plant cells. One of these proteins is called the major intrinsic protein (MIP) of the lens cells of the mammalian eye. It apparently facilitates the flow of electrolytes, small solutes, and fluid from the extracellular environment between two lens fiber cells into the cell cytoplasm. The second protein similar to the glycerol facilitator is called nodulin-26 and is found in the membranes of soybean nodules, which house bacteria and function in nitrogen fixation. It therefore appears that as we gain a clearer understanding of the mechanism of the *E. coli* glycerol facilitator, we may be better able to conceptualize the complex processes involved in cell–cell communication in higher organisms.

In animal cells, glucose, a common energy source, appears to be transported by a variety of mechanisms, depending on cell type. Both active (see the following discussion) and facilitated glucose transport systems have been characterized, sometimes both in the same type of cell. In human erythrocytes and hepatocytes, glucose is apparently transported only by facilitated diffusion. The carrier for this sugar has been isolated, and its gene has been cloned and sequenced. The purified glucose carrier has been reintroduced into phospholipid vesicles and was shown to rapidly equilibrate glucose across the bilayer, but no evidence was found consistent with any ability to carry out active transport. The primary amino acid sequence of the hepatocyte glucose facilitator, deduced from the nucleotide sequence of its gene, has allowed construction of models for how this glycoprotein may be folded in the membrane. One such model is shown in figure 32.10. This sequence is also very similar

Figure 32.11

Exchange diffusion of anions carried out by the human erythrocyte anion-channel (band 3) protein. This exchange is important in lung capillaries for the removal of CO_2 generated by cellular metabolism from the red cell cytoplasm.

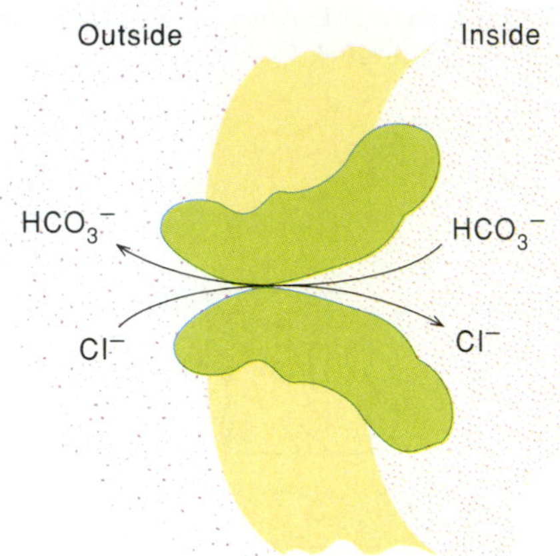

in certain aspects to those of transport proteins from bacteria and yeast that participate in secondary active transport of various carbohydrates, as we will later see.

A third example of facilitated diffusion in biological membranes is the process carried out by the anion-transport (band 3) protein of erythrocytes, whose structure we considered in chapter 7 (see fig. 7.33). The anion transporter allows CO_2 accumulated in the erythrocyte from respiring tissues to diffuse rapidly out of these cells in lung capillaries in the form of bicarbonate anion HCO_3^-. This process has been shown to be accompanied by an influx of Cl^- into the erythrocyte. Much work has established that the anion transporter catalyzes the one-for-one exchange of HCO_3^- for Cl^-. These two molecules always flow in a fashion determined by the sums of their concentration gradients, and thus this protein catalyzes a form of facilitated diffusion. Transport of HCO_3^- out of the cell is obligatorily coupled to the influx of a monovalent anion such as Cl^-, as shown in figure 32.11. This process, called exchange diffusion, is an example of a more general transport process termed antiport. Antiport involves the coupled transport of two different molecules that move in opposite directions across a membrane. "Antiporters" may carry out facilitated diffusion, as does the anion transport protein, or they can catalyze active transport, as we will discuss later.

Ion-Translocating ATPases Carry Out One Type of Primary Active Transport

Transmembrane, ATP-hydrolyzing enzymes use the energy in the terminal γ-phosphate anhydride bond of ATP to drive the active uptake or extrusion of an ion (or ions) into or out of a cell or organelle. Four types of these enzymes are now recognized: the P-type (plasma-membrane-localized), cation-translocating ATPases; the F-type (F_1/F_0), ion-translocating ATPases; the V-type (vacuolar), H^+-translocating ATPases; and the A-type, anion-translocating ATPases.

The P-type ATPases are sensitive to vanadate (a phosphate analog) and form an acyl phosphate intermediate during ATP hydrolysis. They are usually composed of monomeric or multimeric complexes in which a single polypeptide chain of about 100 kDa serves as the catalytic unit.

The F-type ATPases are inhibited by dicyclohexylcarbodiimide (DCCD; see chapter 15) and azide and do not form a phosphorylated enzyme derivative. They consist of multi-subunit F_1 and F_0 moieties in which the F_1 moiety functions in ATP hydrolysis/synthesis, and the F_0 moiety functions in ion translocation.

The V-type ATPases are also inhibited by DCCD (but only at high concentrations), and also by nitrate and the antibiotic bafilomycin A_1. Like the F-type ATPases they are usually large (>400 kDa) and are composed of two major polypeptide components of 70–89 kDa and of 60–65 kDa along with several smaller polypeptides. They are structurally and evolutionarily related to the F-type ATPases.

Only a single A-type ATPase is sufficiently well characterized to establish the fact that it is not related to the other three classes of ATPases; it is the oxyanion-specific ATPase that pumps toxic arsenic derivatives out of bacterial cells, a topic that we will discuss shortly.

The Plasma Membrane Na^+-K^+ ATPase of Eukaryotes

One of the first biological transport systems to be studied in detail was that responsible for maintaining the normally high concentrations of K^+ and low levels of Na^+ inside animal cells. In order for this to occur, both cations must be transported essentially unidirectionally against their concentration gradients. A breakthrough in explaining how this might be accomplished came in the 1950s, when J. C. Skou identified an enzyme from crab nerves that hydrolyzes ATP to ADP only in the presence of both Na^+ and K^+. This sodium-potassium ATPase has subsequently been shown to be a transmembrane protein that pumps Na^+ out of the cell and K^+ into the cell, accompanied by ATP hydrolysis. This coupling of chemical energy to concentrative transport is an example of primary active transport, as well as of an antiport system. The same protein catalyzes both ion translocations and the hydrolysis of ATP, as shown schematically in figure 32.12. Thus the energy released by ATP hydrolysis is "translated" by the Na^+-K^+ ATPase into potential energy in the form of Na^+ and K^+ electrochemical gradients across the plasma membrane. The Na^+-K^+ ATPase is an example of a P-type ATPase, of which we will discuss several other examples shortly.

An important tool in investigating the mechanism of the Na^+-K^+ ATPase has been the use of the specific inhibitor ouabain (fig. 32.13), a compound isolated from foxglove (*Digitalis*) and related plants. Ouabain binds to the outer surface of the enzyme and blocks both ion translocation and ATP hydrolysis by the Na^+-K^+ ATPase. Furthermore, it has been demonstrated that K^+ has a much higher affinity for the outside surface of the protein, while Na^+ binds much more tightly to the cytoplasmic domain. Finally, the site of ATP hydrolysis has

Figure 32.12

Primary active transport of Na^+ and K^+ by the sodium-potassium ATPase found in many animal cells. This enzyme maintains the normally high K^+ and low Na^+ concentrations within the cell at the expense of chemical energy stored in ATP. Each cycle of the enzyme results in the extrusion of three molecules of Na^+ out of the cell, the transport of two molecules of K^+ into the cell, and the hydrolysis of one molecule of ATP. The subunit structure and stoichiometry of this enzyme have been established. One αß heterodimer is the active catalytic unit. The α subunit (M_r = 112,200) is associated with all known catalytic properties of the enzyme complex, while the ß subunit (M_r = 57,000) is a glycoprotein of unknown function.

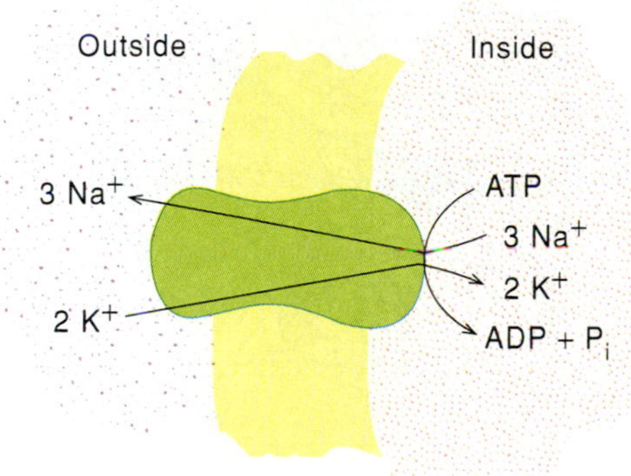

Figure 32.13

The structure of ouabain, a specific inhibitor of the Na^+-K^+ ATPase. Ouabain is a member of a class of compounds called *cardiotonic steroids. Digitalis,* the general name for a family of cardiotonic steroids that inhibit the Na^+-K^+ ATPase, indirectly causes an increase in intracellular Ca^{2+}, which is due to increased cytoplasmic Na^+ that arises because the cardiac sarcolemma contains a Na^+/Ca^{2+} antiport system. The increase in the intracellular Ca^{2+} concentration appears to directly stimulate contraction in cardiac muscle cells. This explains the use of digitalis in treating such conditions as congestive heart failure.

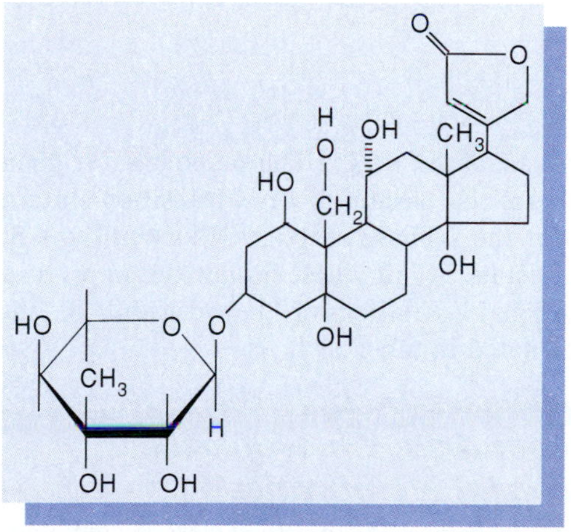

been localized to the inner surface of the cytoplasmic membrane. These observations support the schematic illustration shown in figure 32.12 and demonstrate that the Na^+-K^+ ATPase must be asymmetrically oriented in the plasma membrane.

The Na^+-K^+ ATPase exists in the membrane as a heterodimer consisting of 110-kDa α and 55-kDa β subunits. It has long been debated whether the functional unit is the heterotetramer ($\alpha_2\beta_2$) or the heterodimer ($\alpha\beta$). Recent evidence appears to favor the latter possibility. In the process of active Na^+ and K^+ transport, the cations become "trapped" or "occluded" in the structure of the protein. In this state, they exchange with the cations in the medium only very slowly. It has been found that the detergent-solubilized unit can occlude Rb^+ (a K^+ analog) or Na^+ with the same stoichiometry as the membrane-bound ATPase. These observations suggest that the cations are occluded in a cavity inside the $\alpha\beta$ unit, and that this unit therefore constitutes the minimal protein unit required for active Na^+ and K^+ transport.

Because the inhibitor ouabain binds on the outside of the Na^+-K^+ ATPase and yet inhibits ATPase activity, which resides on the cytoplasmic surface, information must be conducted through the protein by means of conformational changes in the enzyme. Studies with the purified Na^+-K^+ ATPase have shown further that the protein is covalently phosphorylated during its catalytic cycle, the site of phosphorylation being an aspartic acid residue. Phosphoryl Na^+-K^+ ATPase is formed in the presence of Na^+, ATP, and Mg^{2+}, while the dephosphorylation step is K^+-dependent. Ouabain binds tightly to the enzyme-phosphate intermediate and prevents dephosphorylation in the presence of K^+.

These observations and others have led to the formulation of the model depicted in figure 32.14 for the functioning of the Na^+-K^+ ATPase. Transport of Na^+ out of the cell and K^+ into the cytoplasm is thought to be mediated by conformational changes in the enzyme triggered by phosphorylation, dephosphorylation, and the binding of ligands at both surfaces. While figure 32.14 illustrates a reasonable mechanism for the Na^+-K^+ ATPase, it says little about how conformational changes in the protein can lead to unidirectional transport of ions through the membrane. This question has yet to be completely answered in any system, but, as we will see, a number of models have emerged for molecular mechanisms of such translocation events.

Other P-Type ATPases Exist for Ca^{2+} and H^+

Muscle cells contain a specialized endoplasmic reticular membrane system called the sarcoplasmic reticulum. This organelle acts as a repository for Ca^{2+} in resting muscle, and electrical stimulation of muscle cells causes rapid release of this ion from the sarcoplasmic reticulum into the sarcoplasm, where it triggers contraction of the myofibrils. Reuptake of Ca^{2+} into the reticular lumen is effected by a calcium ATPase, which catalyzes the following reaction:

$$2\ Ca^{2+}\ (out) + ATP\ (out) \xrightarrow{Mg^{2+}} 2\ Ca^{2+}\ (in) + ADP\ (out) + P_i\ (out)$$

In this reaction sequence, (out) refers to the outside of the sarcoplasmic reticulum (sarcoplasm), and (in) corresponds to the lumen of the system.

The Ca^{2+} ATPase has been extensively characterized, and like the Na^+-K^+ enzyme, it spans the membrane in an asymmetric fashion. In the presence of Ca^{2+} and ATP, formation of a covalent phosphoryl enzyme can also be demonstrated in this system. Thus many of the aspects of Ca^{2+} transport in the sarcoplasmic reticulum parallel those of Na^+ and K^+ translocation in the cytoplasmic membranes of animal cells. In fact,

Figure 32.14

Postulated mechanism of the functioning of the Na^+-K^+ ATPase. The yellow boundary separates events in which cation association or dissociation occurs on the inside and outside surfaces of the membrane, while the arrows relating E_1 and E_2 denote conformational changes in the enzyme, but are not meant to imply its physical translocation across the membrane. The enzyme is hypothesized to have at least two conformations, E_1 and E_2, both of which can exist in a phosphorylated state. E_1~P is a high-energy phosphoryl bond, in equilibrium with the ß-γ pyrophosphoryl bond of ATP. By contrast, E_2—P is of low energy, in equilibrium with inorganic phosphate. The changes in bond energies are presumed to be due to the free energies of Na^+, K^+, and Mg^{2+} binding. It should be noted that because ion binding may induce secondary conformational changes, both E_1 and E_2 may exist in both "empty" and "full" conformations. The "full" conformations, $E_2(K^+)$ and E_1~P(Na^+), contain occluded (2 K^+) and (3 Na^+) portions, respectively. E_1 and E_2—P have their transport sites facing in and out, respectively. Only E_1 possesses high affinity for Na^+, while only E_2 possesses high affinity for K^+. Phosphorylation of E_1 drives the cyclical process in a unidirectional fashion, causing Na^+ to be translocated from the inside to the outside. Conversion of E_1~P to E_2—P decreases the affinity of the enzyme for Na^+ and increases its affinity for K^+. Dephosphorylation of E_2 drives translocation of K^+ from the outside to the inside. Conversion of E_2 to E_1 decreases the affinity of the enzyme for K^+ and enhances its affinity for Na^+. Release of K^+ from E_2 is greatly accelerated by ATP binding. Translocation of Na^+ out of the cell and K^+ into the cell is the net consequence of this series of conformational changes in the protein. Ouabain inhibits by binding to E_2—P, preventing dephosphorylation in the presence of K^+.

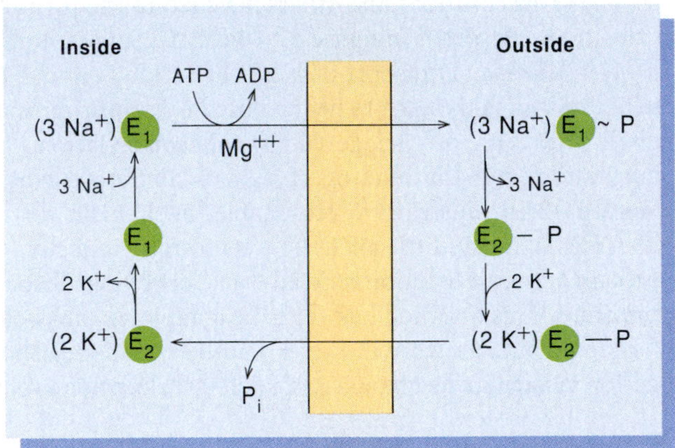

Figure 32.15

The F_1/F_0 ATP synthase or ATPase of bacterial, mitochondrial, and chloroplast membranes. (*a*) A proton gradient generated by electron transport can be used for ATP synthesis by means of chemiosmotic coupling. (*b*) Alternatively, in bacteria growing anaerobically, ATP hydrolysis is used to *generate* a proton gradient in the reverse reaction. This gradient can then be used to energize secondary active transport systems as well as for other processes in bacteria that use chemiosmotic energy, such as flagellar motility.

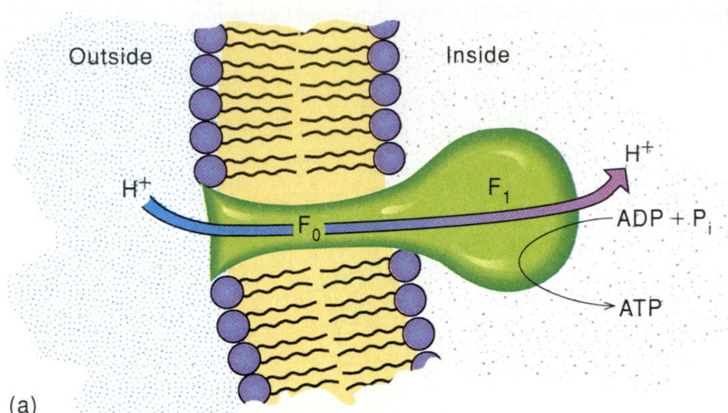

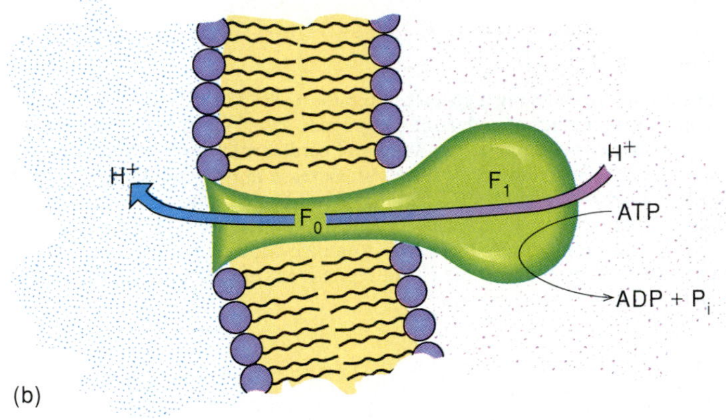

many animal cells also have a Ca^{2+} ATPase in their cytoplasmic membranes that is very similar in structure and mechanism to the sarcoplasmic reticular enzyme. The kinetic mechanism of the Ca^{2+} ATPase is thought to be analogous to that for the Na^+-K^+ enzyme shown in figure 32.14, with 2 Ca^{2+} replacing 3 Na^+, the polarity, however, being reversed. Again, at least two conformations of the protein are involved, and their interconversion is believed to be triggered by bound ligands and the state of phosphorylation of the enzyme. These structural and mechanistic similarities are expected since these proteins both belong to a single family of transport ATPases and presumably share a common evolutionary ancestor.

A third P-type ATPase is the H^+-translocating enzyme found in the plasma membrane of epithelial cells lining the stomach. This enzyme is responsible for acidification of the lumen and is structurally related to the Na^+-K^+ and Ca^{2+} ATPases. Each of these enzymes has a large catalytic polypeptide chain of about 100 kDa that is transiently phosphorylated during transport, and each is inhibited by the phosphate analog, pentavalent vanadate (VO_4^-). This anion has a trigonal bipyramidal structure and seems to be a transition-state analog of phosphate for the ATPase reactions, "locking" the enzyme in the transition state. Again, these similarities imply a reasonably close evolutionary relationship between all of the P-type ATPases, as noted in table 32.1.

F-Type ATPases Include the F_1/F_0 ATPases of Mitochondria, Chloroplasts, and Bacterial Membranes

In chapter 15 we discussed the second class of well-characterized ion-translocating ATPases (F-type). The F_1/F_0 ATPases (or ATP synthases) of mitochondria, chloroplasts, and bacteria appear to have a similar function in all three systems. Proton gradients formed as a result of electron transport are used to synthesize ATP from ADP and P_i as a consequence of H^+ flow down its electrochemical gradient through the F_1/F_0 ATPase (fig. 32.15*a*). It is also possible, however, for the F_1/F_0 ATPase

Table 32.2
Subunit Compositions of F_1 ATPases

Source	Molecular Weight ($\times 10^3$)	Subunit Sizes ($\times 10^3$)[a]				
		α	β	γ	δ	ϵ
E. coli	380	56	52	32	22	11
Thermophilic bacterium (PS3)	380	56	53	32	15	11
Beef heart mitochondrion	360	54	50	33	17	11
Chloroplast	325	59	56	37	17	13

Source: D. B. Wilson and J. B. Smith, "Bacterial transport proteins" in *Bacterial Transport,* edited by B. P. Rosen. Copyright © 1978 Dekker, New York, N.Y.
[a]The stoichiometry of subunits in the F_1 ATPase is $\alpha_3\beta_3\gamma\delta\epsilon$ for the bacterial enzymes.

to work in the reverse direction, i.e., to hydrolyze ATP and translocate protons against their electrochemical gradient (fig. 32.15*b*). This has been established in some bacteria under conditions where respiratory electron transport does not occur (e.g., anaerobiosis or the absence of a functional electron-transport chain in some strict anaerobes). Thus the F_1/F_o ATPase also can be classified as a protein complex carrying out primary active transport. Unlike the P-type ATPases, however, there is no evidence that the F_1/F_o ATPases are covalently phosphorylated during proton translocation.

We described the structure of the mitochondrial F_1/F_o ATPase in chapter 15; the chloroplast and bacterial enzymes have very similar structures. The F_1 portion, which possesses the ATPase activity, consists of five different subunits (named α through ϵ), each of which has a similar molecular weight in each of these three sources (table 32.2). The F_o portion, which spans the membrane, has been shown to be a proton channel. It consists of three different polypeptides (named *a, b,* and *c*) with molecular weights of 28,000, 19,000, and 8,500 respectively, in a subunit ratio of approximately 1:2:ca. 9. While subunit *b* is believed to function in binding F_o to F_1, the *a* and/or *c* subunits function as the proton channel. Removal of F_1 from membranes results in an increased permeability of the bilayer to protons, a fact suggesting that F_1 acts as a "cap" that couples proton translocation to ATP synthesis or hydrolysis. The F_o portion, which is rich in hydrophobic amino acids, has been shown to cause proton conductivity when incorporated into phospholipid vesicles.

Some elegant reconstitution studies by Y. Kagawa and co-workers first showed how the various subunits of the F_1/F_o ATPase may be arranged on the membrane. As a result of information gained by purifying each subunit of the F_1 portion from the thermophilic bacterium PS3 and recombining them in various ways, they proposed the structure depicted in figure 32.16*a*. In this model, the complex of α and β subunits ($\alpha_3\beta_3$), which possesses ATPase activity by itself, is bound to the F_o portion through subunits δ and ϵ. Subunit γ is believed to act as the "cap" or gate through which protons flow. Support for this model came from experiments showing that membranes containing F_o and the δ plus ϵ subunits of F_1 were permeable to protons, but addition of the γ subunit blocked proton conductance. Other workers have proposed models for the arrangement of the subunits of the F_o portion of the ATPase, one of which is shown in figure 32.16*b*. Because the subunit structures and functions of the F_1/F_o ATPases from a number of sources are very similar, it seems reasonable to expect that the functions and topographies of the individual subunits will be related as well.

Interestingly, in the strictly anaerobic bacterium *Propionigenium modestum* an F_1/F_o ATPase exists that normally couples Na^+ instead of H^+ translocation to ATP synthesis. The Na^+ translocation site is in the F_o portion of this protein, and a hybrid enzyme composed of F_o from *P. modestum* and F_1 from *E. coli* is a functional Na^+ pump. These results have important implications for the mechanism of ATP synthesis and hydrolysis by the F-type ATPases. All models in which protons play a role that can be performed only by protons must be incorrect. One such model that now appears unlikely is proton conduction through a "proton wire" consisting of a series of hydrogen-bonded residues within F_o (see the later discussion on bacteriorhodopsin). Rather, these observations are more in accord with a mechanism in which the binding of cations to sites on the external surface of F_o triggers a conformational change that exposes the cations to the inner membrane surface and simultaneously effects ATP synthesis by the F_1 moiety of the enzyme.

Vacuolar ATPases Pump Protons into Intracellular Vesicles and Vacuoles

The vacuolar ATPases (V-type), now recognized as a third distinct class of ATPases, are less well characterized than either the P-type or F-type enzymes. The V-type ATPases are present in the dynamic intracellular membrane networks of eukaryotic cells, which include the endoplasmic reticulum, Golgi, endosomes, coated vesicles, and lysosomes. They are also found in

Figure 32.16

Subunit structure of the F_1/F_0 ATPase. (*a*) A model for the arrangement of the subunits of the F_1 portion of the enzyme. The five subunits α, ß, γ, δ, and ε form a dissociable unit, which attaches to the integral membrane F_0 part of the complex. (Source: M. Yoshida et al., "Reconstitution of thermostable ATPase capable of energy coupling from its purified subunits" in *Proceedings. National Academy of Sciences USA* 74:936, 1977. Copyright © 1977 National Academy of Sciences, Washington, D.C.) (*b*) A schematic model of the F_0 part of the F_1F_0 ATPase of *E. coli,* shown both in side view (*top*) and top view (*bottom*). Subunits a, b, and c are indicated in their approximately correct stoichiometries. The essential carboxyl group (Asp 61, the DCCD binding site) is in contact with the lipid phase as shown. Note: Conformation is not indicated, as the "a" subunit probably traverses the membrane five times, while the "b" and "c" subunits pass through it one and two times, respectively. The F_0 part of the ATPase is more complex in eukaryotes and consists of four to six different subunits. (Reproduced from E. Schneider and K. Altendorf, "Bacterial adenosine 5′-triphosphate synthase (F_1F_0): Purification and reconstitution of F_0 complexes and biochemical and functional characterization of their subunits," in *Microbiology Reviews* 51:477–497, 1987.)

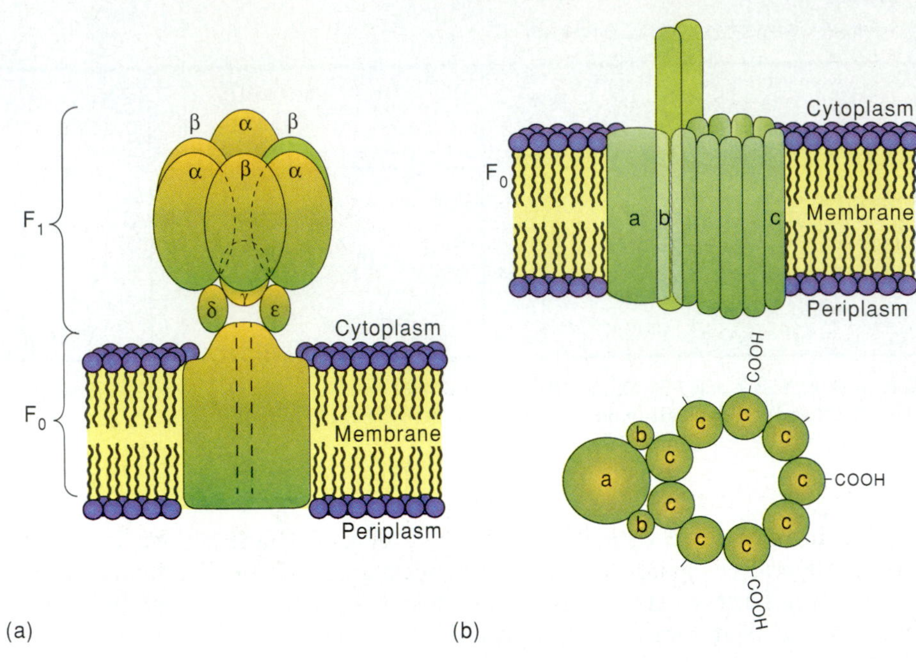

plant tonoplasts and the vacuoles of yeast and fungi. In all examples studied, these ATPases utilize cytoplasmic ATP to pump protons into the intracytoplasmic compartment. The vacuolar ATPases, of three or more distinct types of subunits, may be distantly related to the F_1/F_0 ATPases, as indicated by some of their biochemical properties. However, the vacuolar ATPases are not readily separable into F_1 and F_0 moieties, do not show a knoblike appearance by electron microscopy, and are solubilized only upon treatment with detergents. It therefore appears that this third class of ATPases and the F-type ATPases may have evolved from a common ancestor but diverged considerably. The recent demonstration of V-type ATPases in *Sulfolobus,* an archaebacterium, has led to the proposal that vacuolar ATPases of eukaryotes may have arisen by the internalization of plasma membrane V-type ATPases of an archaebacterial-like ancestral cell.

An Oxyanion-Specific ATPase Pumps Toxic Arsenic Derivatives Out of Bacterial Cells

The first member of a fourth class of ion-translocating ATPases has been discovered and characterized. This A-type ATPase is a plasmid-encoded enzyme complex that actively extrudes toxic oxyanions such as arsenite, antimonite, and arsenate from the cytoplasm of bacteria. Because the genes for the synthesis of this enzyme are encoded on a small, transferable plasmid, properties associated with the possession of this enzyme can be easily transferred from one population of bacteria to another. Since the plasmid confers on its host bacterium resistance to a variety of bacteriocidal agents, it serves as one kind of antibiotic resistance plasmid.

The arsenical resistance pump consists of three polypeptides, ArsA (63 kDa), ArsB (46 kDa), and ArsC (16 kDa). ArsB is an integral membrane protein that presumably serves as the oxyanion channel. ArsA is a peripheral membrane protein that is associated with ArsB and possesses the ATPase activity of the pump. The ArsC protein is also peripherally associated with the membrane, and it appears to function as a specificity protein. Surprisingly, sequence analyses suggest that the ArsA and ArsB proteins are related to enzymes involved in nitrogen fixation, which may have no transport function. It will be interesting to learn what functions any other proteins within this class of ATPases may possess.

Some Primary Active Transport Systems Are Driven by Electron Transport or Light

Membrane-bound electron-transport chains have as a primary function the extrusion of protons out of mitochondria and bacteria and into chloroplasts of plants and algae (chapter 16). These gradients can then be used for ATP synthesis as described earlier, or for secondary active transport processes linked to ion translocation (see the following discussion). The exact mechanisms of proton translocation during electron transport are still somewhat unclear, but both unidirectional proton channels and coenzyme Q are likely to be involved (chapter 16). In any case, energy derived from the oxidation-reduction reactions that take place within the membrane is more or less directly transformed into a chemical gradient of protons. For this reason, electron transport is generally thought of as a primary active transport event.

Light energy can likewise be used to drive active transport in a number of systems. For example, in photosynthesis, light energy absorbed by chlorophyll molecules is converted into a proton gradient at least partly by means of the electron-transport reactions that subsequently take place. The three-dimensional structure of the membrane-bound protein complex

Figure 32.17

Bacteriorhodopsin (BR). (*a*) Freeze-fracture electron micrograph of the membrane of *H. halobium*. Fine-grain regions are patches of purple membrane containing bacteriorhodopsin. (Magnification 48,000X.) (Courtesy of Walther Stoeckenius.) (*b*) Bacteriorhodopsin contains a covalently bound molecule of retinal, a derivative of vitamin A, attached by means of a Schiff base linkage to a lysine residue of the protein. Both the all-*trans* isomer and the 13-*cis* form can be isolated from the protein. Retinal is also the light-gathering chromophore of the vertebrate retina, where it exists as the 11-*cis* isomer attached by means of a Schiff base to the protein opsin (chapter 36). (*c*) The photochemical cycle of bacteriorhodopsin. Light absorption by the retinal moiety quickly leads to isomerization of the chromophore to the 13-*cis* isomer, followed by deprotonation of the Schiff base to the form absorbing maximally at 412 nm (BR_{412}). Two intermediate forms can be detected spectroscopically between the isomerization and deprotonation steps (BR_{625} and BR_{550}). Reprotonation and isomerization back to the all-*trans* form are slow steps involving two other intermediates (BR_{520} and BR_{640}). (Source: H. G. Khorana, "Bacteriorhodopsin, a membrane protein that uses light to translocate protons," in *Journal of Biological Chemistry* 263:7439–7442, 1988. Copyright © 1988 American Society for Biochemistry and Molecular Biology Inc., Bethesda, Md.)

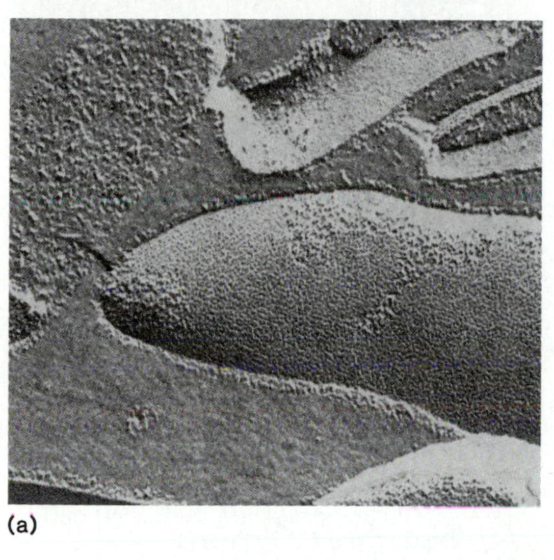

(a)

(b)

(c)

that initiates photosynthesis in the bacterium *Rhodopseudomonas viridis* was recently solved in Germany by J. Deisenhofer and H. Michel, a feat for which they were awarded the Nobel Prize. We described this structure in chapters 7 and 16, and it undoubtedly will be invaluable in determining the molecular mechanism involved in the transduction of light energy into a proton gradient in this photosynthetic system.

In halophilic bacteria, such as *Halobacterium halobium,* a remarkable light-driven proton pump, bacteriorhodopsin, has evolved in part to allow this organism to synthesize ATP under conditions of oxygen limitation. The structure of bacteriorhodopsin, which we considered in chapter 7, in some manner allows it to actively transport protons from inside the cell to the outside in response to light absorption. In 1971, Walther Stoeckenius and his co-workers showed that if O_2 was limited in cultures of *Halobacterium,* a purple membrane appeared on the surface of the cell in addition to its normally red, carotenoid-containing membrane (fig. 32.17*a*). The purple membrane is 75% protein by weight, and bacteriorhodopsin is the sole protein component. The purple color (λ_{max} = 570 nm) is due to the covalent association of one molecule of all-*trans* retinal, a derivative of vitamin A, to lysine 216 of bacteriorhodopsin by means of a Schiff base linkage (fig. 32.17*b*). It is the retinal moiety that absorbs photons and is responsible, along with aminoacyl residues of the protein, for the conversion of this form of energy into a proton gradient that can be used for ATP synthesis.

Although the photochemical reactions of bacteriorhodopsin are extremely complex, it is known that light absorption by retinal leads to deprotonation of the Schiff base to one of several intermediate forms, which absorbs light maximally at 412 nm. Associated with this process is the isomerization of the all-*trans* retinal (fig. 32.17*b*) to the 13-*cis* isomer. Reassociation of a proton with the retinal moiety then takes place more slowly, and the bacteriorhodopsin protein cycles back to the 570-nm (and all-*trans*) form through several other intermediate conformations that can be detected spectrophotometrically (fig. 32.17*c*). It is highly probable that this protonation-deprotonation

Figure 32.18

A helical-wheel model of bacteriorhodopsin, showing the relationships between the aspartic acid mutations that affect proton pumping and the amino acids that are believed to interact with the chromophore. The amino acids postulated to interact with retinal are shown in bold ovals. The aspartic residues in helices C and G whose substitution affects H^+ pumping are shown in red (D85, D96, and D212). (Source: H. G. Khorana, "Bacteriorhodopsin, a membrane protein that uses light to translocate proteins," in *Journal of Biological Chemistry* 263:7439–7442, 1988. Copyright © 1988 American Society for Biochemistry and Molecular Biology Inc., Bethesda, Md.)

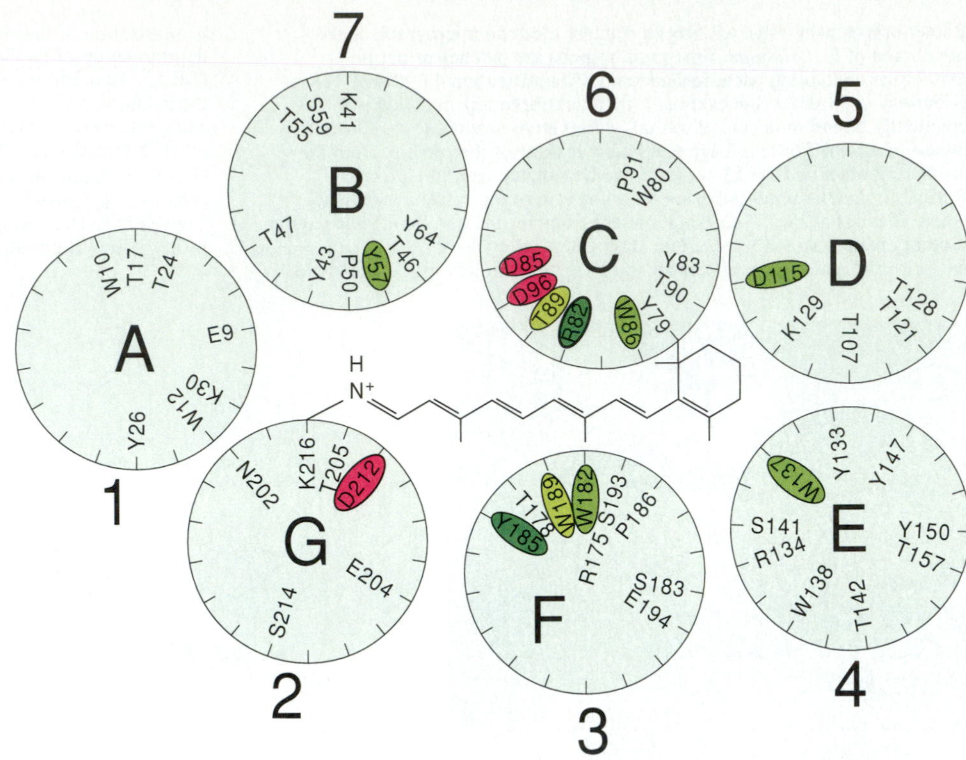

cycle involving multiple retinal and protein conformational states is the key to the mechanism of proton translocation and thus energy transformation carried out by bacteriorhodopsin.

Bacteriorhodopsin is one of several transport proteins that have been studied extensively in recent years by means of molecular genetic techniques. Through site-directed mutagenesis of specific amino acid residues, H. G. Khorana and coworkers have identified a number of residues that appear to be implicated in the proton-pumping mechanism. Several residues, which can reversibly accept and donate protons to form a proton path or "wire" through the interior of the molecule, have been tentatively identified by use of this technique, including Asp 85, Asp 96, Asp 212, and Tyr 185 (see fig. 7.35). The way in which these residues are situated with respect to the helical packing of bacteriorhodopsin and the retinal chromophore is shown in figure 32.18, which also shows residues that are believed to interact with the chromophore itself. The light-induced isomerization of retinal is therefore thought to initiate each cycle of proton conduction by allowing donation of the Schiff base proton to Asp 85 in the proton "wire," which ultimately leads to the release of a proton on the extracellular surface of the membrane. Reprotonation of the Schiff base is accomplished by Asp 96, which is protonated (directly or indirectly) via uptake of a proton from the cytoplasm. In this way, light energy can be used to drive the endergonic pumping of protons against their electrochemical gradient via a proton current involving amino acid residues of the protein, the Schiff base chromophore, and multiple conformational states of bacteriorhodopsin.

Some Bacterial Transport Systems Use Specific Binding Proteins

The last primary active transport mechanisms that we will consider are the so-called ATP-binding cassette (ABC-type) transport systems of the Gram negative bacteria. As we discussed in chapter 7, proteins that bind specific sugars, amino acids, and inorganic ions are found in the periplasmic spaces of Gram negative bacteria, and corresponding membrane-anchored proteins are found in Gram positive bacteria. Some of the compounds for which such binding proteins have been demonstrated are listed in table 32.3. Cold osmotic shock of Gram negative bacterial cells or the transformation of the cells to spheroplasts releases these proteins, and a concomitant decrease in the ability of these cells to transport binding-protein-linked substrates is observed. Thus these periplasmic proteins apparently are important either in concentrating their substrates at the cell surface or in interacting with the transmembrane permeases (or both). In all cases that have been studied, the binding proteins are not themselves responsible for the transmembrane transport of the molecules that they recognize. Rather, specific integral membrane proteins exist for this purpose (fig. 32.19).

The integral membrane proteins that, in part, comprise the binding-protein-dependent systems have been characterized in several instances. One such system is the maltose transport system in *E. coli*. It is known that four distinct proteins are required for the translocation of maltose across the cytoplasmic membrane. These are illustrated in figure 32.19. In addition to the maltose-binding protein in the periplasm (the E protein),

Table 32.3
Binding-Protein-Dependent or ATP-Binding Cassette (ABC-Type) Transport Systems

A. Binding-Protein-Dependent Uptake Systems in Enteric Bacteria	B. Related Binding-Protein-Independent Export Systems in Bacteria and Eukaryotes
I. **Oxyanions**	I. **Drugs**
A. Phosphate	A. Multidrug resistance (Mdr) in Animals
B. Sulfate	B. Drug resistance (Drr) in *Streptomyces pencetins*
II. **Amino Acids**	II. **Polysaccharides**
A. Glutamine	A. Polyribosylribitol phosphate (Bex) in *Hemophilus influenzae*
B. Lysine/arginine/ornithine	B. Polyneuraminic acid (Kps) in *E. coli*
C. Histidine	III. **Proteins and Peptides**
III. **Sugars**	A. α-Hemolysin (Hly) in *E. coli*
A. Arabinose	B. a-type Mating Pheromone (Ste6) in Yeast
B. Maltose	C. Antigenic peptides of the mammalian immune system (MHCII)
C. Galactose/glucose	IV. **Other Compounds**
D. Ribose	A. Chloride (the Cystic Fibrosis Transmembrane regulator (CFTR) in animals)
IV. **Other Compounds**	B. Opines in *Agrobacterium tumefaciens*
A. Oligopeptides	C. Arsenicals (arsenate, arsenite, antimonite) in bacteria
B. Dicarboxylic Acids	
C. Tricarboxylic Acids	
D. Vitamin B_{12}	
E. Iron and Iron Complexes	

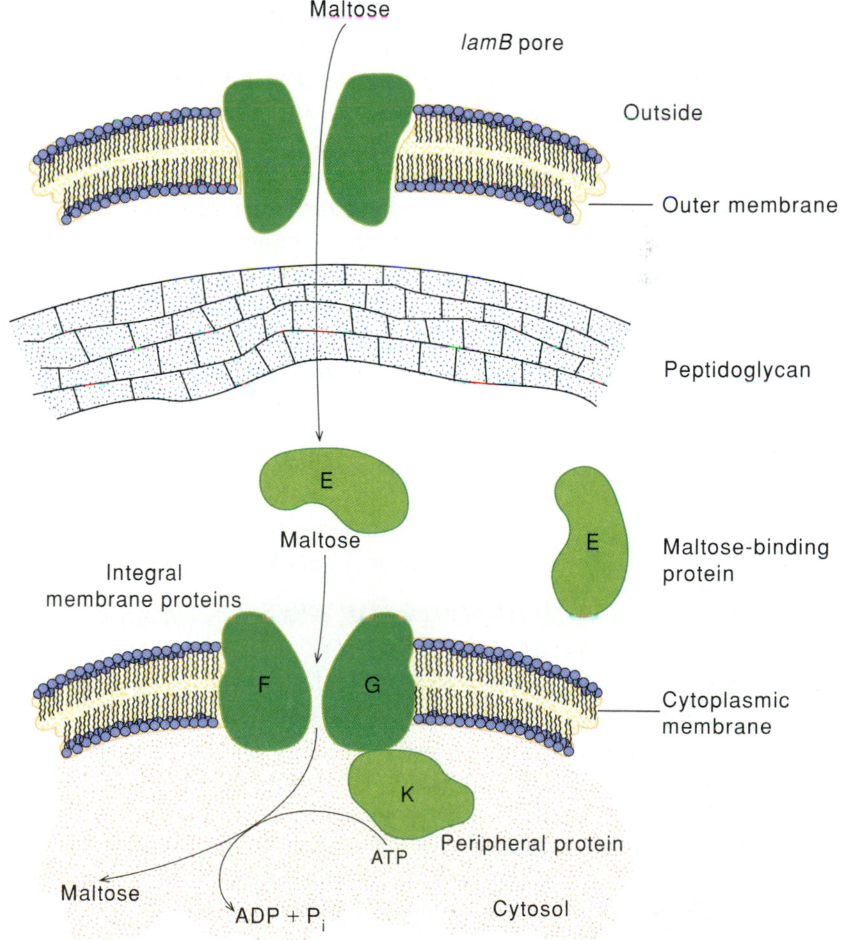

Figure 32.19

Schematic illustration of solute transport by the binding-protein systems of Gram negative bacteria using the maltose system as an example. Maltose passes through a pore in the outer membrane (the *lamB* gene product, which is also the receptor for bacteriophage λ), interacts with its periplasmic binding protein (E) and is then transported by inner membrane permeases (F and G) in a process dependent on ATP. The K protein is a peripheral membrane protein that is necessary for maltose transport, and is probably an ATPase (see text). The localizations and interactions of these proteins have been determined by biochemical and genetic analyses. The malE binding protein has been shown to interact both with the lamB protein pore in the outer membrane and with the maltose permease complex in the cytoplasmic membrane. It may act to facilitate and provide stereospecificity to the transport processes across both membranes.

two integral membrane proteins (F and G) span the membrane. While the E protein associates with the transmembrane transport components on the external surface, the K protein is localized to the inner membrane surface, probably owing to a specific association with the G protein (see fig. 32.19). The K protein is believed to function by energizing transport and also in the regulation of transport activity.

Energy for the active accumulation of binding-protein substrates is provided by ATP. Thus the peripheral membrane components of these transport systems that are present on the cytoplasmic side of the membrane (e.g., the MalK protein in figure 32.19) possess ATP-binding sites, and the energization of transport of both maltose and histidine has been shown to be fully provided by ATP in reconstituted systems *in vitro*. The mechanism whereby ATP hydrolysis allows the active accumulation of binding-protein substrates has yet to be determined, but again, protein conformational changes are undoubtedly involved.

A large number of homologous ATP-binding proteins are essential components of the periplasmic binding-protein-dependent permeases of bacteria. These structurally and functionally similar permeases are specific for such diverse substrates as peptides, amino acids, sugars, various intermediary metabolites, anions, and vitamins (as noted in table 32.3). Several of these systems have been characterized in some detail and, regardless of their specificities, they all have similar protein components, are similarly organized, and probably function by a common mechanism (see fig. 32.19).

Interestingly, ATP-binding proteins that are homologous to proteins comprising bacterial binding-protein systems have also been found in eukaryotic cells. The best characterized of these is the multidrug-resistance protein, or P-glycoprotein, of mammalian tumor cells. This protein functions to pump certain drugs out of tumor cells, and therefore the pump exhibits an opposite polarity for active transport as compared with the binding-protein systems. Other homologous proteins include the protein that is defective in patients suffering from cystic fibrosis (a genetic disease that is characterized by defective cellular Cl^- transport) and an export protein of yeast that mediates secretion of the mating pheromone, *a*-factor, from the cell. It is probable that all of these transport systems have a common evolutionary ancestor from which this class of transport proteins has diversified over many millions of years to accommodate many related functions involving primary active transport.

Secondary Active Transport Involves the Uptake of One Solute Coupled to the Transport of Another Solute

In chapter 15 we learned that a gradient of protons across a biological membrane is a form of potential energy that can be used to drive the synthesis of ATP from ADP and P_i. Expulsion of protons out of mitochondria or bacterial cells during electron transport results in both a difference in H^+ concentration (ΔpH) and in electrical charge ($\Delta\psi$; interior negative) across the membrane. Recall that both these components contribute to the so-called proton-motive force (pmf or Δp; see chapter 15), which can be expressed as

$$\Delta p = \Delta\psi - \frac{2.3RT}{F}\Delta\text{pH} \qquad (9)$$

The pmf, or Δp, is related to the free energy change experienced by protons, which tends to drive them across the cellular or organellar membrane. Since lipid bilayers are relatively impermeable to protons, such a flow must be mediated by membrane proteins. In the example of the F_1/F_o ATPase, the exergonic flow of H^+ through the enzyme, driven by Δp, is used to drive the endergonic synthesis of ATP. It should also be apparent that Δp could be used to drive active transport if the inward flow of H^+ were coupled to the uptake of a particular solute. Examples of such proton cotransport or symport mechanisms are found in mitochondria, bacteria, and probably all eukaryotic cells. Because the primary active transport of H^+ is in part responsible for Δp, active transport systems depending on Δp for energization are said to carry out secondary active transport by means of chemiosmotic coupling.

The best-characterized H^+ symport system is the lactose transport system of *E. coli*. This system has been extensively studied by P. Mitchell, T. H. Wilson, and H. R. Kaback, among many others. Evidence that the active accumulation of this sugar by the *lacY* gene product, the lactose permease, involves the cotransport of H^+ can be summarized as follows:

1. Agents such as 2,4-dinitrophenol (see fig. 15.22), which collapse proton gradients across membranes, inhibit lactose transport in whole cells and membrane vesicles.
2. Uptake of lactose, or the nonmetabolizable analog thiomethyl-β-D-galactoside (TMG) is stimulated by reducing the extracellular pH of energy-depleted cells.
3. A 1:1 ratio of entry of TMG and H^+ has been demonstrated in energy-depleted cells.
4. Lactose transport in *E. coli* membrane vesicles, which cannot synthesize ATP in the absence of a source of ADP, is greatly stimulated by compounds such as D-lactate, which are oxidized and donate electrons to the electron-transport chain in this system.
5. Mutations in the *lacY* gene have been isolated in which lactose uptake is not tightly coupled to H^+ entry. These mutant proteins are severely impaired in the active transport of lactose, but have an increased ability to facilitate its diffusion across the membrane compared to the normal permease.
6. The lactose permease has been purified and the transport function of the pure protein has been reconstituted in a vesicular phospholipid membrane. In this system, the active accumulation of lactose in response to an artificially imposed Δp has been demonstrated. In fact, the highly purified, reconstituted permease was found to exhibit essentially the same kinetic properties as the transport system in membrane vesicles from *E. coli*. Thus a single protein, the lactose permease, is both necessary and sufficient for catalysis of lactose:proton symport.

Figure 32.20

Transport events carried out by the bacterial lactose permeases. (*a*) In response to a proton gradient, lactose is actively accumulated in the cell. (*b*) In the presence of a lactose gradient, protons can be concentrated in the cytoplasm. (*c*) In the presence of the $\Delta\psi$, both lactose and protons can be actively accumulated. The physiological situation in respiring cells corresponds to (*a*), in which a $\Delta\psi$ is usually also present.

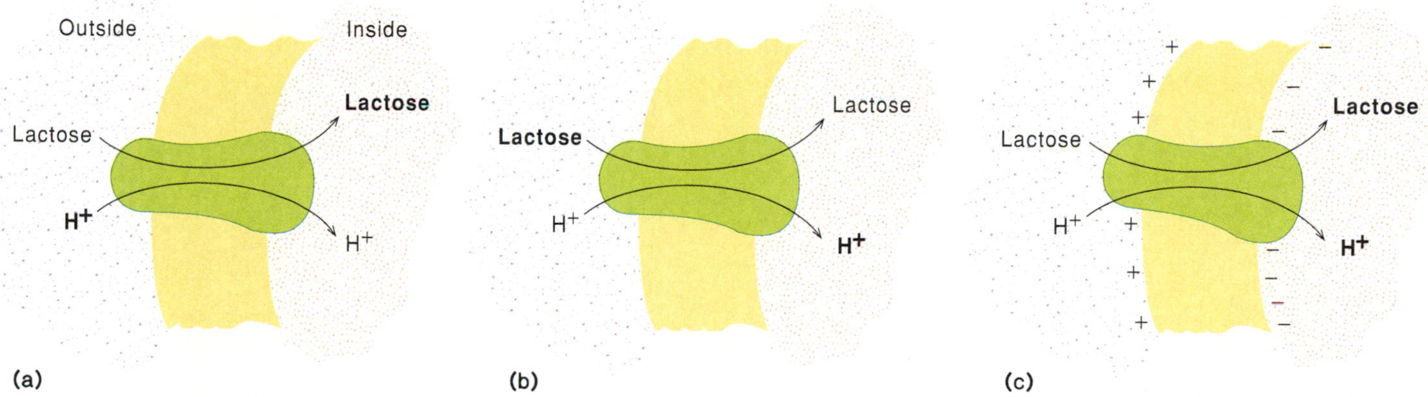

The preceding observations establish that the active accumulation of lactose in *E. coli* is obligatorily coupled to the simultaneous entry of H^+ flowing down its concentration and/or electrical gradients (fig. 32.20*a*). If this is true, then a gradient of lactose (or TMG) should produce a Δp in cells in which this value is initially zero. Indeed, it has been shown that addition of TMG to the outside of energy-depleted cells causes acidification of the cytoplasm as a result of H^+-TMG symport (fig. 32.20*b*).

The structure of the lactose permease is fairly well understood. The most significant advance leading to the elucidation of its structure resulted from the cloning and sequencing of the *lacY* gene, allowing deduction of the amino acid sequence of the lactose permease. Hydropathy analyses and other methods for studying membrane protein topography (see chapter 7) suggested that this protein consists of twelve extended hydrophobic segments, with a mean length of about 24 residues per segment. Both the C and N terminals appear to be localized to the cytoplasmic surface of the membrane. On the basis of circular dichroic measurements on the purified lactose permease protein, approximately 85% of the amino acid residues are probably arranged in helical secondary structure. Since about 70% of the 417 amino acid residues were found to be included within the twelve hydrophobic segments, it was predicted that the embedded segments are largely α-helical, with each helix extending the entire thickness of the membrane, as shown in figure 32.21.

As the result of both biochemical and molecular genetic approaches, the structure of the active site of the lactose permease is now coming to light. The galactoside-binding site resides within a segment of the protein that is embedded within the phospholipid bilayer. This conclusion was reached by the isolation and DNA sequence analysis of mutants of the *lacY* gene that encoded permease proteins with altered substrate specificities (e.g., that could recognize other disaccharides, such as maltose and sucrose, with a higher affinity than the wild-type protein). These studies revealed that Ala 177, Tyr 236, Thr 266, Ser 306, and Ala 389 of the native permease were likely to be at or near the substrate-binding site. Mutations in several other residues, including Arg 302, His 322, and Glu 325, have provided evidence that the side chains of these amino acids could possibly form a "charge-relay" system near the galactoside-binding site that could be responsible for the coupling of H^+ and lactose transport. This type of mechanism is reminiscent of the "proton wire" hypothesized for bacteriorhodopsin. It is reasonable to propose that the residues involved in sugar recognition/transport and in H^+ translocation might form a hydrophilic channel through the membrane. One such arrangement of six of the twelve transmembrane helices containing residues important for lactose permease function is shown in figure 32.22. Any such channel would have to be "gated" in the sense that lactose and H^+ translocation could normally occur only when both ligands are bound to the permease at the same time, presumably via a conformational change in the protein.

Transport systems such as the *E. coli* lactose permease, in which electrical charges are carried across the membrane without simultaneous compensation of the electrical potential so generated (for example, by the movement of other charged species across the membrane) are termed electrogenic. Electrogenic transport systems are always affected by the electrical potential $\Delta\psi$ across a biological membrane, even if the transported ion is not H^+. In the lactose transport system, a $\Delta\psi$ (interior negative) is sufficient to drive active accumulation of this sugar in the absence of a ΔpH because Δp has both $\Delta\psi$ and ΔpH components [equation (9)], as shown in fig. 31.19*c*. In animal cells and in some bacteria, such as *Halobacterium,* that live in high concentrations of Na^+, Na^+-solute symport systems are common means for the accumulation of salts, carbohydrates, and amino acids. Because these systems are also electrogenic, they are sensitive to $\Delta\psi$ and to $\Delta[Na^+]$, but not to ΔpH. For these systems

Figure 32.21

Secondary structure model of the lactose permease, based on the hydropathy profile of the protein. Hydrophobic segments are shown in boxes as transmembrane, α-helical domains connected by hydrophilic segments. The carboxyl terminus and hydrophilic segments 5 and 7 (with the amino terminus as hydrophilic segment 1) have been shown to be on the cytoplasmic surface of the membrane. Amino acids are represented by their single-letter codes. (Source: H. R. Kaback, "Use of site-directed mutagenesis to study the mechanism of a membrane transport protein," in *Biochemistry* 26:2071–2076, 1987. Copyright © 1987 American Chemical Society, Washington, D.C.)

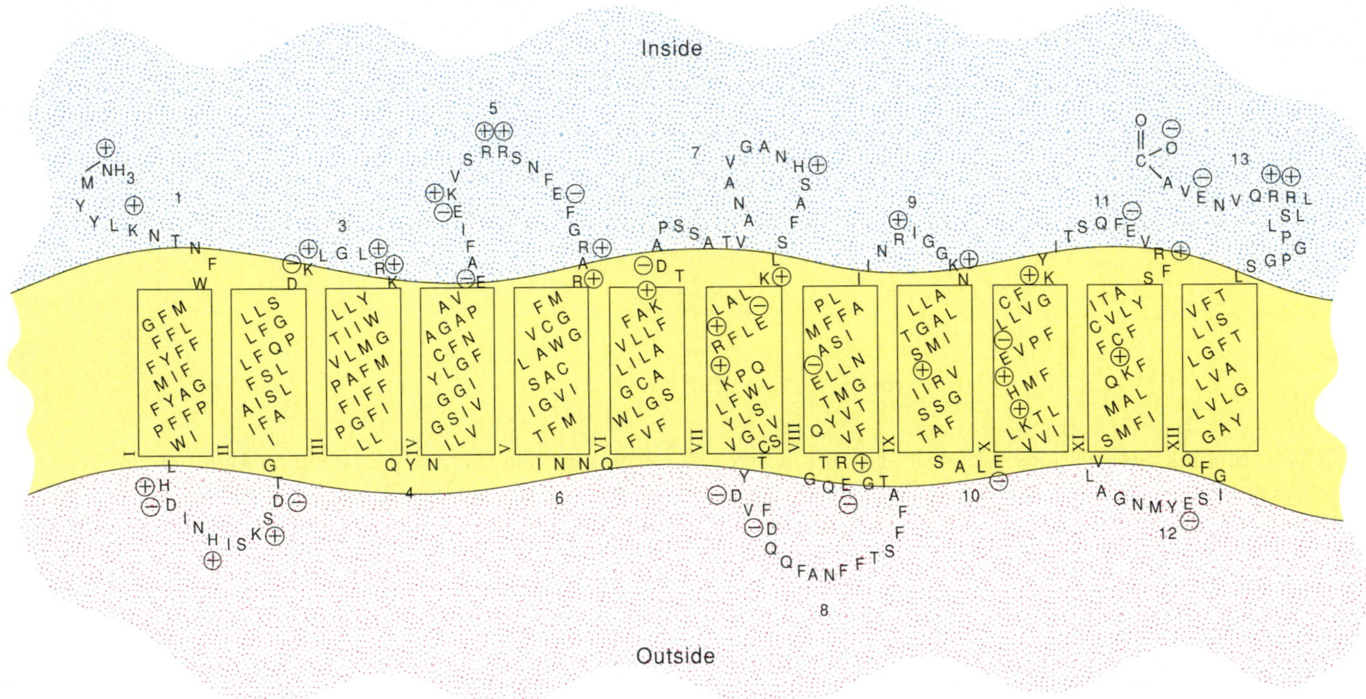

in animal cells, the Na^+-K^+ ATPase ensures that Na^+ flowing in through these Na^+-solute "symporters" is rapidly pumped back out to maintain the Na^+ gradient (fig. 32.23).

A list of some of the better-characterized secondary active transport systems is given in table 32.4. The amino acid sequences of many of the symporters listed are known. Some of these proteins show no homology to other known transport proteins or to each other (e.g., the *E. coli* lactose and melibiose permeases). On the other hand, the *E. coli* proton symporters for xylose and arabinose show some degree of homology to those for glucose, galactose, and maltose from yeast. Even more surprisingly, representatives of this latter group of permeases are homologous to parts of the glucose facilitators of human hepatocytes and rat brain cells. These observations suggest that the molecular mechanism of carrier-mediated facilitated diffusion may be similar to that of solute:cation symport except for the number of molecular species translocated in a single step (one versus two).

The Mitochondrial ATP/ADP Exchanger Is an Electrogenic Antiporter

Because ATP is synthesized within the mitochondria of eukaryotic cells, while metabolic processes that utilize ATP are largely confined to the cytoplasm or nucleus, a mechanism must exist for transport of this molecule across the inner mitochondrial membrane. The protein responsible for this function is an ATP/ADP exchanger that carries out exchange of intramitochondrial ATP for ADP formed from metabolic reactions in the cytoplasm. This antiport is electrogenic because, at pH 7, ATP molecules average about one more negative charge than do ADP molecules (fig. 32.24).

Despite the fact that the [ATP]/[ADP] ratio is usually higher in the cytoplasm than in the mitochondrion in actively respiring cells, the ATP/ADP exchanger still preferentially expels ATP from the organelle with the concomitant inward movement of ADP. This can be explained by the presence of a membrane potential $\Delta\psi$ (inside negative), resulting, in part, from the Δp formed across the mitochondrial inner membrane during respiration. This potential favors the outward transport of ATP and inward transport of ADP because this process results in the net translocation of one negative charge from inside to outside (see fig. 32.24).

The ATP/ADP exchanger, purified by M. Klingenberg and his associates, is a dimer of identical 30-kDa polypeptides. Each subunit is believed to span the membrane six times, and both the C and N terminals are localized to the cytoplasmic side of the membrane. Because the protein loses its binding affinity for nucleotides if it is dissociated into monomers, this dimer may be necessary for carrying out the exchange process. Biochem-

Figure 32.22

Hypothetical model for the tertiary structure of lactose permease. The model depicts twelve cylindrical transmembrane segments, in an α-helical conformation, connected by hydrophilic segments. Six of these helices are shown to form a hydrophilic channel through which protons and lactose might flow. Residues important for substrate binding and lactose: proton symport are shown projecting into the channel. Some of these include: Arg 302, His 322, and Glu 325, which have been proposed to be part of a charge-relay system for H^+ transport, and Ala 177, Tyr 236, Thr 266, Ser 306, and Ala 389, which are likely to be near the substrate-binding site. (Source: J. C. Collins et al., "Isolation and characterization of lactose permease mutants with an enhanced recognition of maltose and diminished recognition of cellobiose," in *Journal of Biological Chemistry* 264:14698–14703, 1989. Copyright © 1989 American Society for Biochemistry and Molecular Biology Inc., Bethesda, Md.)

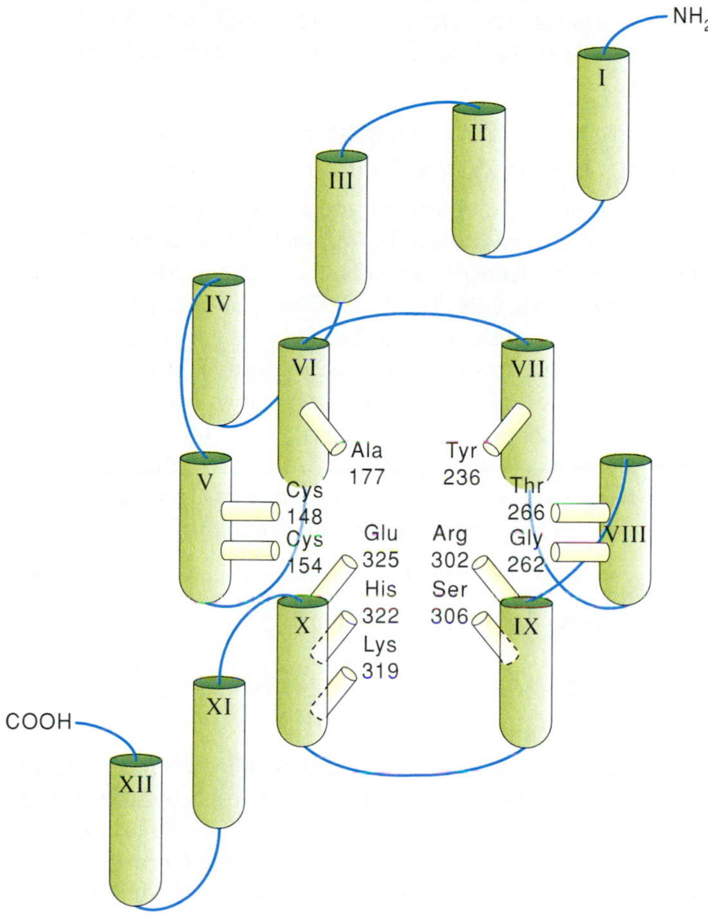

ical and kinetic studies on the ATP/ADP exchanger support a two-state "gated-pore" mechanism for this protein. The evidence for this mechanism can be summarized as follows:

1. There appears to be only one nucleotide-binding site per dimeric ATP/ADP exchanger.
2. The nucleotide-binding site has a higher affinity for ADP on the outer surface of the membrane than on the inner surface, while the opposite is true for ATP.
3. This site was shown never to be on both sides of the membrane simultaneously.
4. The respiratory poison atractylic acid binds to the form preferring ADP, while the antibiotic bongkrekic acid (see chapter 15) binds to the form that shows specificity for ATP.

Figure 32.23

Na^+ symport is a common means of actively transporting sugars and amino acids in animal cells. For example, glucose is actively accumulated by this mechanism in intestinal and renal epithelial cells. The Na^+-K^+ ATPase maintains the Na^+ gradient essential to these secondary active transport systems.

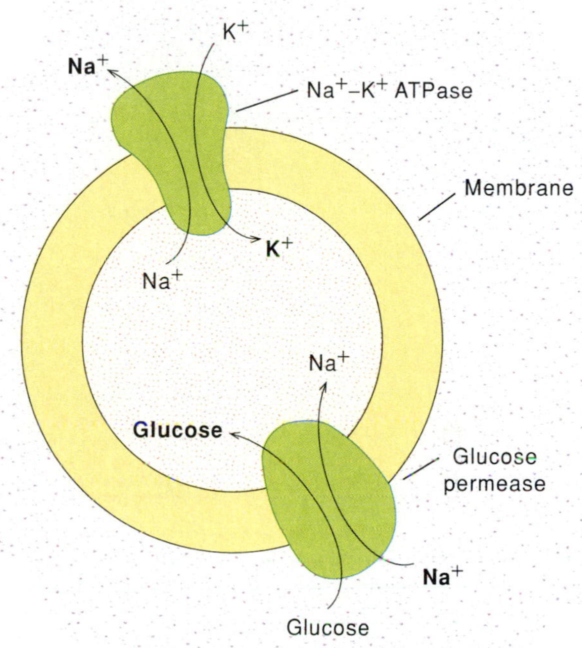

Table 32.4
Some Secondary Active Transport Systems

Substrate	Cotransported Ion	Organism/Tissue
Neutral amino acids	Na^+	Eukaryotic cells
Glucose	Na^+	Some animal cells (intestine, kidney, choroid plexus)
Lactose	H^+	*E. coli* and some other bacteria
Dicarboxylic acids	H^+	*E. coli*
Proline	Na^+	*E. coli*
Glutamate	Na^+	*E. coli*
Melibiose	Na^+ or H^+	*E. coli*
Xylose	H^+	*E. coli*
Arabinose	H^+	*E. coli*
Glucose	H^+	Yeast
Maltose	H^+	Yeast
Galactose	H^+	Yeast

Figure 32.24

The ATP/ADP exchanger of the mitochondrial inner membrane. This dimeric protein carries out the exchange of intramitochondrial ATP for ADP formed in the cytoplasm by metabolic reactions. A membrane potential (interior negative) favors this exchange because of its electrogenic nature at pH values near neutrality.

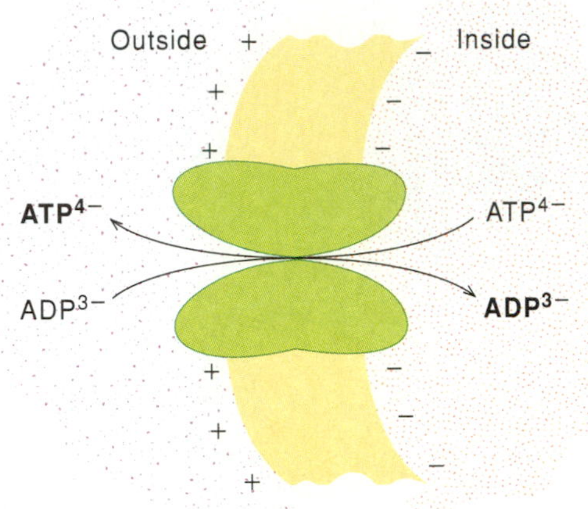

Figure 32.25

Molecular model for the functioning of the ATP/ADP exchanger of the mitochondrial inner membrane. The dimeric protein appears to have only one binding site for adenine nucleotides, perhaps involving amino acid residues from both subunits. When facing the outer surface of the membrane, this site has a high affinity for ADP (pentagon), and a low affinity for ATP (rectangle). The opposite specificities are observed when the site faces the lumen of the organelle. These states are interconvertible, even in the purified carrier, demonstrating that they arise from different conformations of the same protein. This model is analogous to that shown in figure 32.35 except that energy is apparently not required for the conformational interconversion. (Source: M. Klingenberg, "Membrane protein oligomeric structure and transport function," in *Nature* 290:449, 1981. Copyright © 1981 Macmillan Magazines Ltd., London, England.)

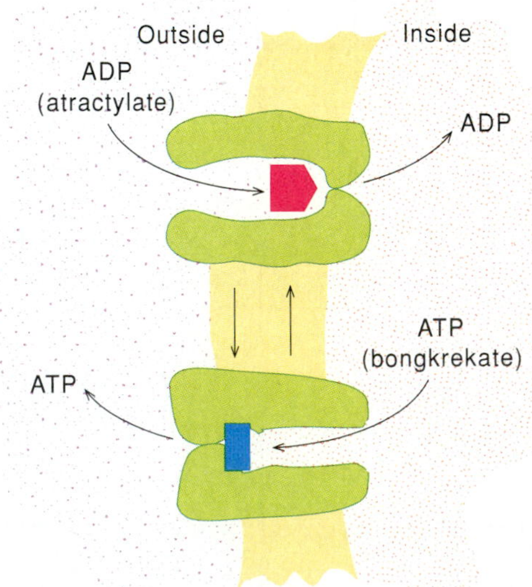

These results strongly support the transport model presented in figure 32.25. A single nucleotide-binding site per dimer is involved in transport by this protein. The configuration of this site, and thus its binding specificity, depends on the side of the membrane that it faces, and these two states are interconvertible through protein conformational changes that also result in translocation of any nucleotide bound to the protein. A conformational change can be triggered by binding of either ADP at the outside or ATP at the inside of the membrane surface. Once transported through the membrane, the nucleotide is now in a binding site that has little affinity for it, and it dissociates from the protein. The two forms of the adenine nucleotide exchanger appear to have quite different conformations indeed. Antibodies prepared against the atractylic acid-protein complex do not cross-react with the form that binds bongkrekic acid, and vice versa!

The Bacterial PEP:Sugar Phosphotransferase System Exemplifies Group Translocation

The final type of biological transport process that we will consider is group translocation. In this mechanism, chemical modification of the transported solute is tightly coupled to the translocation step (see fig. 32.7). The best-characterized example of group translocation is the phosphoenolpyruvate (PEP)-dependent sugar phosphotransferase system (PTS) in bacteria, which was discoverd by Saul Roseman and co-workers in 1964. The PTS both transports and phosphorylates sugars as they pass through the cytoplasmic membrane with PEP as the phosphoryl donor. It consists of several soluble and membrane-bound proteins in *E. coli* that catalyze the following set of reactions:

$$\text{PEP} + \text{enzyme I} \rightleftharpoons \text{enzyme I}{\sim}\text{P} + \text{pyruvate} \qquad \textbf{(10)}$$

$$\text{enzyme I}{\sim}\text{P} + \text{HPr} \rightleftharpoons \text{HPr}{\sim}\text{P} + \text{enzyme I} \qquad \textbf{(11)}$$

$$\text{HPr}{\sim}\text{P} + \text{sugar}^{a}_{(out)} \xrightarrow[\text{(Enzyme III}^{a})]{\text{Enzyme II}^{a}} \text{HPr} + \text{sugar}^{a}\text{-P}_{(in)} \qquad \textbf{(12)}$$

In this reaction scheme, enzyme I and HPr are soluble, cytoplasmic proteins that participate in phosphoryl transfer reactions common to all sugars transported by the PTS in *E. coli*. In contrast, enzyme II is an integral membrane protein that is usually specific for only one sugar (as indicated by the superscript) and acts as the permease as well as the phosphoryl transfer enzyme. Enzyme III, which is required for the transport of some but not all sugars, also is sugar-specific. The spatial arrangement of the reactions that occur during PTS sugar transport is depicted in figure 32.26. The PTS allows for unidirectional sugar transport because the sugar-phosphate product is very impermeable to the lipid bilayer and is "trapped" once it is transported into the cell. The cost to the cell is one ATP equivalent (in the form of PEP) for each sugar molecule accumulated by means of the PTS.

Figure 32.26

Schematic representation of the reactions carried out by the bacterial sugar phosphotransferase system (PTS). A phosphoryl moiety from PEP is sequentially transferred to enzyme I, HPr, and enzyme III^{glc} before its ultimate transfer to glucose by an integral membrane protein specific for this sugar (enzyme II^{glc}). Although the general PTS enzymes, enzyme I and HPr, are soluble proteins, there is some evidence that they are associated peripherally with the membrane as shown. In *E. coli* there are at least seven different enzymes II, each specific for the sugars recognized by the PTS in this organism: glucose, mannose, fructose, *N*-acetylglucosamine, mannitol, glucitol, and galactitol. HPr has been shown to be a monomer (M_r = 9,500), enzyme I is a dimer (subunit M_r = 70,000), and at least some of the enzymes II are dimers. The mannitol-specific enzyme II (or mannitol permease) has been shown to consist of a single kind of polypeptide chain (M_r = 68,000).

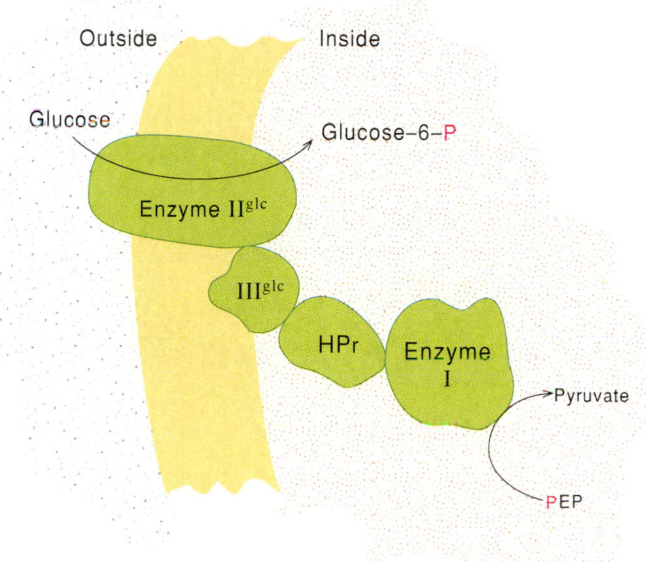

The phosphotransferase systems of *E. coli* and *Salmonella typhimurium* have been extensively studied. All the general proteins have been purified and shown to be transiently phosphorylated, as shown in reactions (10) through (12). The phosphate is attached to a single histidine residue in each of the soluble components (enzyme I, HPr, and enzyme III), as well as to one or two residues in each of the sugar-specific enzymes II, before being transferred to the incoming sugar. Such enzyme phosphorylation reactions may provide the driving force for transport by means of conformational changes in the enzymes II, as it apparently does for the ion-translocating ATPases.

One of the enzymes II, that specific for the sugar-alcohol mannitol, has been solubilized from the *E. coli* membrane, purified to homogeneity, and reconstituted in an artificial membrane. Additionally, the gene encoding the protein has been cloned and sequenced, so that like the lactose permease, the mannitol enzyme II is well understood structurally and functionally. The N-terminal half of the protein (M_r = 68,000) is strongly hydrophobic, traversing the membrane at least six times. In contrast, the C-terminal half of this protein exhibits the hydrophilic properties of a typical water-soluble protein and is localized to the cytoplasmic (inner) surface of the membrane. This particular protein, which functions without the aid of an enzyme III, is phosphorylated twice during catalytic turnover. A phosphoryl group is first transferred from phospho HPr to the N-3 position of a histidiyl residue in an enzyme III-like domain in the C-terminal part of the protein; next, the phosphoryl group is transferred to a cysteinyl residue also in the C-terminal half of the molecule; and finally, dephosphorylation of this residue is accompanied by the coupled phosphorylation and transport of the sugar across the membrane.

The DNA sequences of over a dozen genes encoding PTS permeases (enzymes II of enzyme II–III pairs) have been determined, and consequently the aminoacyl sequences of the encoded proteins are known. Computer analyses have revealed that most of these permeases are evolutionarily related, and show features that appear to be distinctive for the PTS permeases. For example, their N terminals can potentially form amphiphilic α helices with hydrophobic residues on one side of the helix and hydrophilic residues on the other side, a motif that may be important for insertion of these proteins into the membrane. In addition, site-specific mutagenesis studies and deletion analyses of some of these proteins have provided evidence for the involvement of specific residues in specific functions. Phosphoryl transfer between the individual PTS proteins/domains involves distinct catalytic residues, and as expected, the sugar-binding and translocation site is within the hydrophobic, transmembrane domains. These latter domains have been shown to exhibit similarities with some of the symporters and antiporters discussed previously in this chapter. Thus it is possible that the group-translocating permeases and the facilitators may share some common structural and mechanistic features.

Molecular Models of Transport Mechanisms

In the last section we examined the wide variety of mechanisms used by cells to accumulate and expel nutrients and ions. In order to fully understand transport at the molecular level, however, it is necessary to determine how the structures of transport permeases allow the often unidirectional translocation of hydrophilic molecules through the hydrophobic phospholipid bilayer. It is reasonable to assume that such proteins contain specific binding sites for their substrates by analogy with enzymes, and indeed this has been established for many transport permeases. Unlike most soluble enzymes, however, permeases also must be able to carry out a vectorial process: "picking up" a substrate on one side of the membrane and depositing it on the other. In no instance do we know the exact molecular details of how this process is accomplished. However, as more transport proteins are purified and their structures determined, clues are emerging that will undoubtedly allow this question to be answered for at least a few systems in the near future. In this section we will examine some of these clues and how they pertain to various models for the function of transport permeases.

Figure 32.27

In theory, a solute could be transported across a membrane by a mobile carrier that either rotates through or traverses the plane of the membrane (*a*), or by a more or less fixed channel or pore that may be controlled by a gate (*b*). Examples of both have been found among the ion-translocating antibiotics.

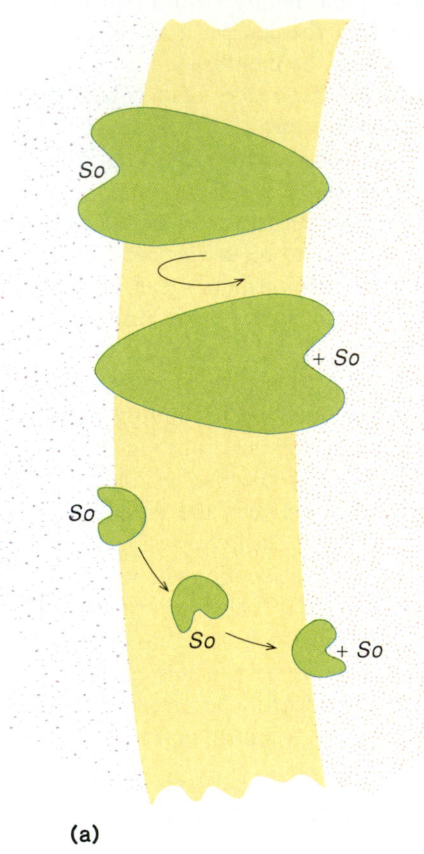

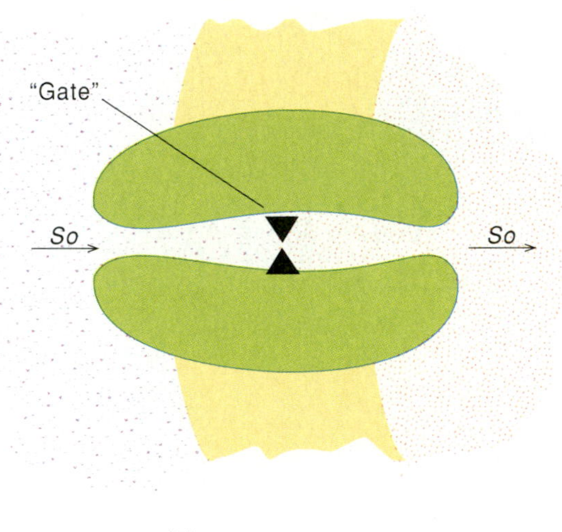

Mobile Carriers or Pores Could Explain Transport

At least two conceptual models could explain the transport of a substance through a biological membrane. In the mobile-carrier mechanism, the transporter or a solute-binding moiety of the permease is assumed to physically change its orientation in the membrane during translocation, either by shuttling back and forth across the bilayer or by rotation through the plane of the membrane (fig. 32.27*a*). On the other hand, the pore model assumes that the protein is more or less fixed in its intramembrane orientation and forms a hydrophilic pore that is stereospecific for its transport substrate. This pore is usually thought of as being gated in the sense that it opens only transiently in response to proper solute recognition (fig. 32.27*b*).

As model systems for studying biological transport, the ion-translocating antibiotics have provided evidence for both of the mechanisms illustrated in figure 32.27. Valinomycin, an antibiotic isolated from a species of the bacterium *Streptomyces,* is a cyclic peptide containing the sequence D-valine, L-lactate, L-valine, and D-hydroxyisovalerate repeated three times (see fig. 15.28). It is extremely selective for the binding of K^+ and makes both artificial bilayers and biological membranes permeable to this cation. This result can be explained by the fact that the valinomycin-K^+ complex effectively shields the hydrophilic groups of valinomycin on the interior of the molecule, while the hydrophobic side chains are exposed to solvent, as shown in figure 15.28. For this reason, the valinomycin-K^+ complex is soluble in lipid bilayers, while free K^+ is highly water-soluble. A second well-studied antibiotic is gramicidin A, also isolated from a bacterium (*Bacillus brevis*). It is a 15-amino-acid linear polypeptide that is a cation-specific ionophore with much less specificity than valinomycin for any particular cation (fig. 32.28).

The mechanisms of ion translocation carried out by both valinomycin and gramicidin A have been extensively studied. In one experiment, the conductance of an artificial phospholipid bilayer ($T_m = 41°$ C) to K^+ was measured as a function of temperature in the presence and absence of these antibiotics. The results showed that K^+ conductance was high in this model system throughout the temperature range studied with gramicidin A. In contrast, the K^+ permeability of the bilayer in the presence of valinomycin was low below 41° C, but sharply increased when the temperature was increased above this value.

Figure 32.28

The structure of gramicidin A, a linear antibiotic that facilitates the transmembrane transport of many cations.

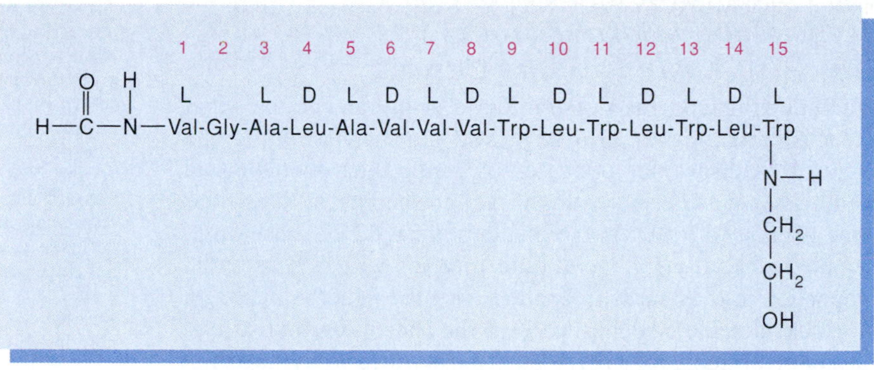

Figure 32.29

Mechanism of ion translocation by valinomycin. The carbonyl oxygens chelating K^+ are shown as "spokes" on the circular antibiotic molecule. In the aqueous phase, these oxygens are more freely available to the solvent; i.e., valinomycin undergoes a conformational change upon going from a lipid to an aqueous environment, as illustrated in the top of the figure. The lipid-soluble form is the one that complexes K^+ and traverses the membrane by diffusion, as shown in the lower part of the figure. (Source: Y. A. Ovchinnikov, "Physico-chemical basis of ion transport through biological membranes: Ionophores and ion channels," in *European Journal of Biochemistry* 94:321, 1979. Copyright © 1979 Springer-Verlag, Berlin, Germany.)

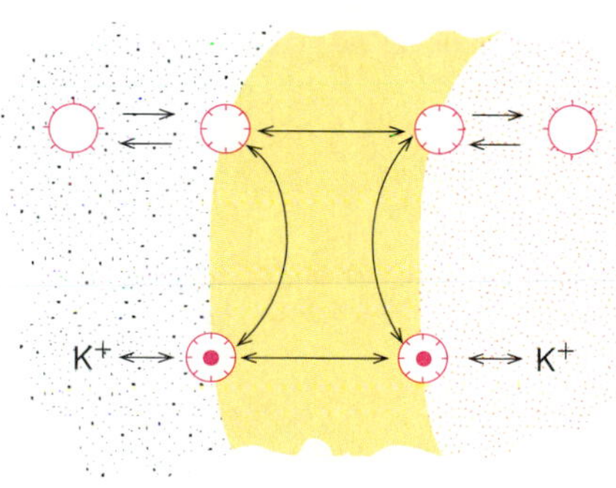

Hence ion translocation by valinomycin apparently depends on the physical state of the phospholipid bilayer, while the permeability induced by gramicidin A is insensitive to this parameter. This and many other experiments have led to the conclusion that valinomycin is a mobile carrier that can rapidly diffuse through the bilayer, while gramicidin A forms a static pore through the bilayer.

The mechanisms of ion translocation by valinomycin and gramicidin A as they are believed to occur are illustrated in figures 32.29 and 32.30. A single valinomycin-K^+ complex traverses the bilayer by simple diffusion. In contrast, physical measurements have shown that gramicidin A forms a transmembrane pore by means of head-to-head dimerization of helical monomers, each of which has a hydrophilic aqueous channel through the axis of the helix. The structure of each monomer is not an α helix because of the alternating D- and L-amino acid isomers in the chain. Rather, it is believed to be a so-called β helix as depicted in figure 32.30. Thus, at least in these model systems, both mobile-carrier and pore mechanisms of solute translocation can be readily demonstrated.

Figure 32.30

Model for the gramicidin A transmembrane channel. Gramicidin A molecules in a head-to-head dimer form a left-handed helix with a hydrogen-bonding pattern similar to that found in a parallel ß-pleated sheet. This hydrogen-bonding pattern is possible because of the alternating D and L configurations of the amino acid residues. The hollow, 4-Å core of the helix created by this structure is big enough to permit transport of alkali metal cations. (From Donald Voet and Judith G. Voet, *Biochemistry*. Copyright © 1991 John Wiley & Sons, Inc., New York. Reprinted by permission.)

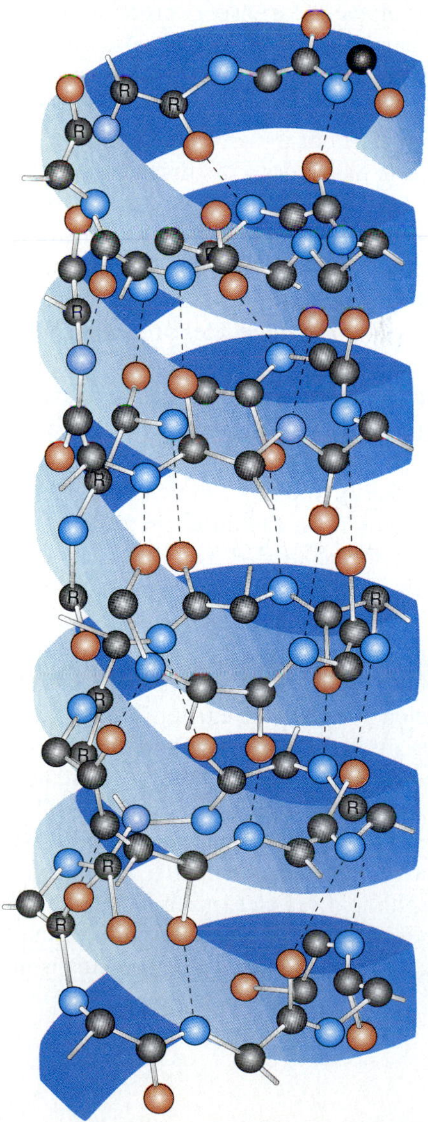

Most Biological Transport Systems Use Pores, Although Elements of the Mobile-Carrier Model May Also Apply in Some Cases

Studies on the topographies of integral membrane proteins that we considered in chapter 7 suggest a more or less static and asymmetric disposition of these polypeptides across biological membranes. Furthermore, any large movements, such as rotations through the membrane plane (see fig. 32.27*a*) are unfavorable, especially for membrane proteins with a substantial proportion of their mass in contact with the aqueous phase. In fact, considerable evidence leads to the conclusion that several classes of proteins that carry out transport do so by means of transmembrane pores or channels. These structures may or may not be static and/or gated. Nonspecific transport, where solute recognition is minimal, is presumably mediated by relatively static pore structures. By contrast, stereospecific solute permeation may involve conformational changes in the permease that transiently "open" the channel in response to binding of the solute recognized by the transport system, by energy-coupling mechanisms, or by both.

In most cases for which detailed structure-function information is available, the evidence is in favor of a pore-type mechanism for biological transport proteins, although elements of the mobile-carrier mechanism may also apply in some cases. In several instances, presumptive transmembrane pores formed by an integral membrane protein have been visualized by electron microscopy. The outer membrane of *E. coli* has been shown to be permeable to molecules with molecular weights below about 600. This permeability is conferred by a class of proteins called porins, which are integral constituents of the outer membrane of Gram negative bacteria, as well as of the outer membranes of mitochondria and chloroplasts.

In *E. coli,* porin molecules are arranged in a hexagonal lattice of trimers in the outer membrane, and optical filtration and computer processing of many negatively stained preparations viewed in the electron microscope give images that have been interpreted as representing transmembrane pores (fig. 32.31). Very recently, the three-dimensional structure of a similar porin molecule from the outer membrane of *Rhodobacter capsulatus* was solved at 3-Å resolution by x-ray crystallography. This structure gives definitive evidence for a pore through the center of each porin monomer as shown in fig. 32.32. Both the *E. coli* and *R. capsulatus* porins contain pores with a diameter of about 1 nm, which is consistent with the molecular weight exclusion limit given earlier. Unlike most transmembrane proteins studied to date, the membrane-spanning regions of porins consist almost entirely of β-type structures (β sheet and β barrel) as also shown in figure 32.32. That these molecules function as relatively nonspecific static pores has been shown by reconstitution experiments. When incorporated into artificial phospholipid vesicles, porin molecules allow rapid diffusion of most small hydrophilic molecules through the bilayer, and thus solute recognition by these proteins is minimal.

Figure 32.31

Arrangement of porin "pores" in the outer membrane of *E. coli*. Computer processing of many electron micrographs gives the image shown in (*a*), while a technique involving optical diffraction and filtering of micrographs gives the image shown in (*b*). A schematic interpretation of these images, shown in (*c*), suggests "triplet indentations" that are penetrated by the negative stain (dark circles) and which are the presumptive pores. A, B, and C are centers of local threefold symmetry and *L*, the lattice constant, is 7.7 nm. The center-to-center spacing *D* of the pores is 3 nm, and each is partially surrounded by a kidney-shaped region that excludes the negative stain and is probably part of a single molecule of porin. The view is perpendicular to the plane of the outer membrane. (From A. C. Steven, B. ten Heggeler, R. Müller, J. Kistler, and J. P. Rosenbusch, "Ultrastructure of a periodic protein layer in the outer membrane of *Escherichia coli*." Reproduced from the *Journal of Cell Biology,* 1977, 72:292, by copyright permission of the Rockefeller University Press.)

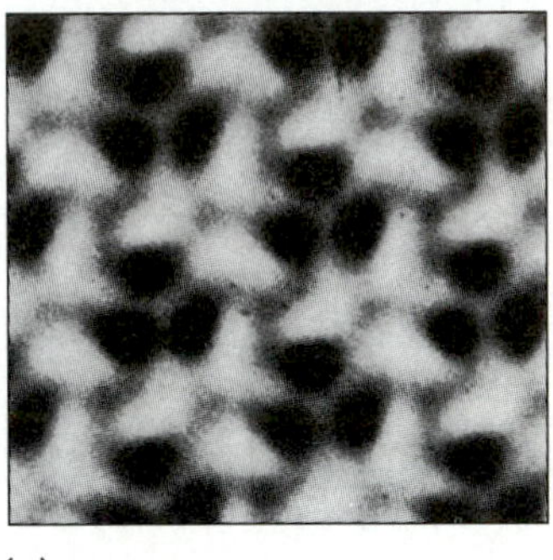

(a)

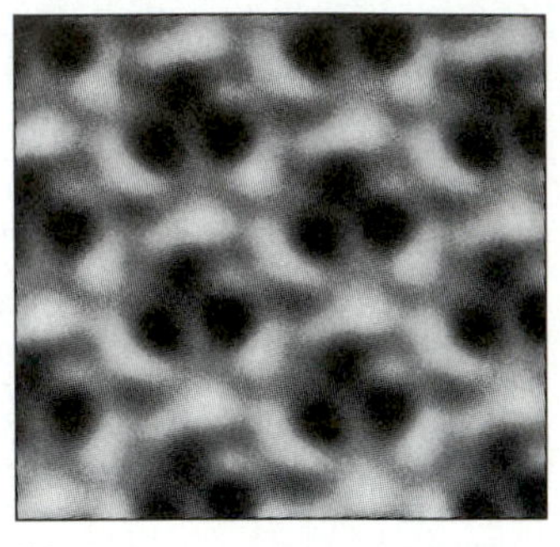

(b)

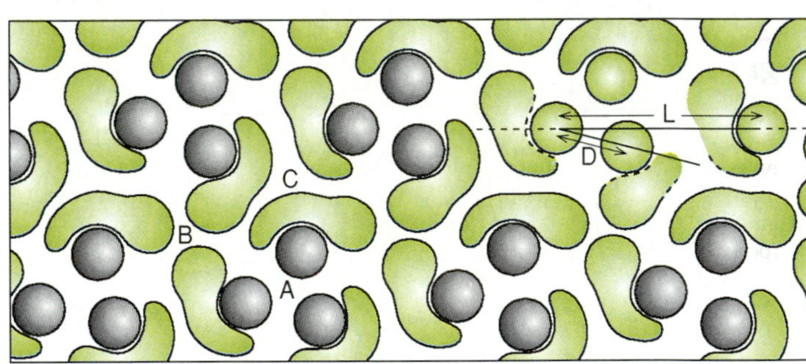

(c)

Nonspecific transmembrane pores also have been demonstrated in the plasma membranes of eukaryotic cells. Well-studied examples of these pores are the so-called gap junctions, which connect neighboring cells of similar types in tissues such as liver, intestine, kidney, brain, and cardiac muscle. In the region of the gap junction, the plasma membranes of two different cells are closely apposed and appear to have channels connecting them (fig. 32.33*a*). In the surface view of gap junctions, the region appears as a hexagonal lattice of protein hexamers, each of which seems to form a hydrophilic pore (fig. 32.33*b*). Ultrastructural studies have produced detailed models of the structure of the gap junction, one of which is shown in figure 32.34. The protein hexamers in each membrane are thought to span the bilayer, and apposition of two hexamers forms an aqueous channel between the cells.

Figure 32.32

Structure of the polypeptide backbone of the outer membrane porin of *Rhodobacter capsulatus*. (*a*) Side view of one subunit of the porin trimer. Note the extensive ß structures running perpendicular to (arrow) and at slight angles to the plane of the membrane. (*b*) Porin trimer viewed from the top (outside) of the membrane, showing how the ß structures form a type of ß barrel within which each pore resides. (From M. S. Weiss, T. Wacher, J. Weckesser, W. Welte, and G. E. Schultz, The three-dimensional structure of porin from *Rhodobacter capsulatus* at 3-Å resolution, *FEBS Lett.* 267:268–272, 1990.)

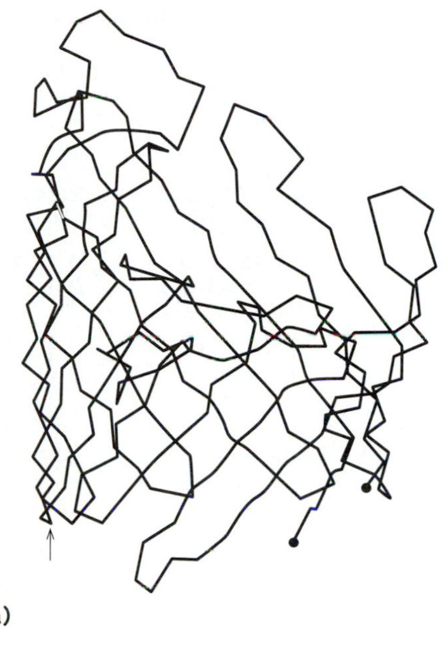

(a)

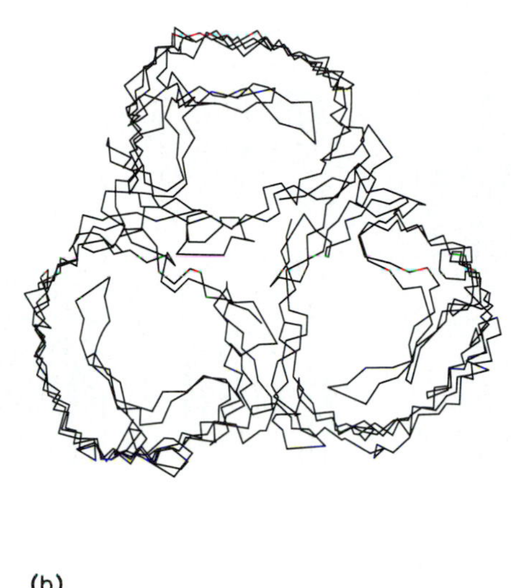

(b)

Figure 32.33

Isolated gap-junction sheets viewed in the electron microscope (*a*) parallel and (*b*) perpendicular to the planes of the two apposed membranes. The view in (*a*), part of which is outlined within a rectangle, arose from a curling up of a sheet on its edge and shows links (gaps or pores) penetrated by negative stain bridging the space between the two cell membranes. The perpendicular view in (*b*) shows annuli (rings), each composed of a protein hexamer surrounding a pore that has been filled with negative stain. The specimens were stained with uranyl acetate (300,000×). (From P. N. T. Unwin and G. Zampighi, "Structure of the junction between communicating cells." Reprinted by permission from *Nature* 283, 545; © 1980 Macmillan Magazines Limited.)

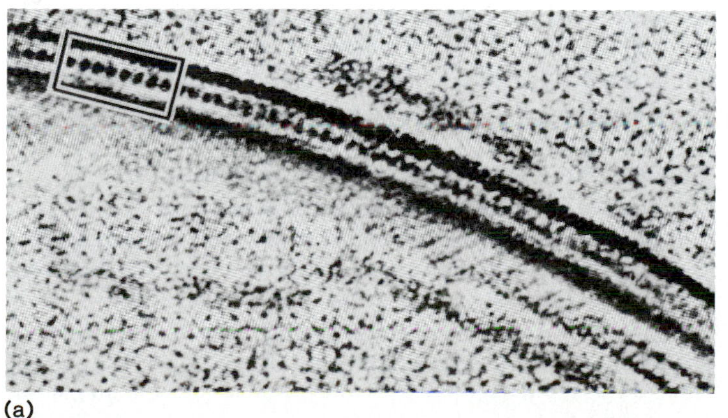

(a)

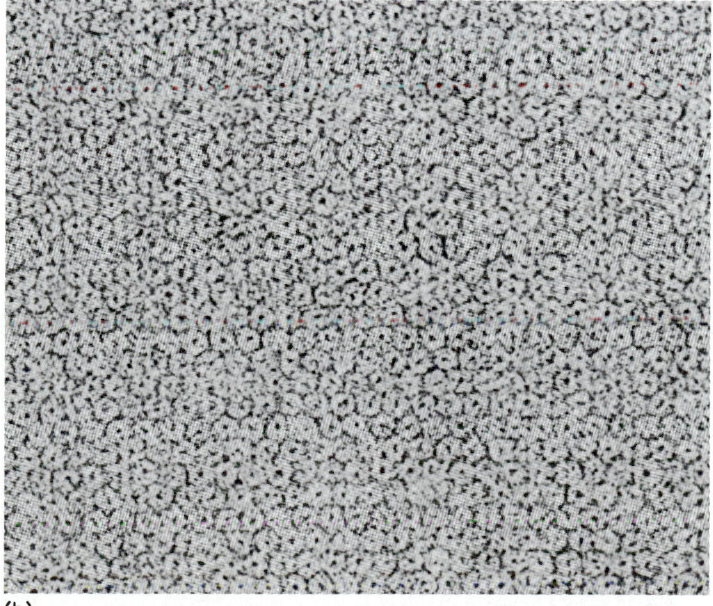

(b)

Figure 32.34

Molecular model of gap-junction pores as inferred from chemical, electron microscopic, and x-ray diffraction studies. A cross section, nearly perpendicular to the two membrane planes, is shown. The arrows, representing aqueous channels, are drawn through cross sections of two pores, formed by apposition of two hexamers (or *connexons*), each spanning its own lipid bilayer membrane. The end-on views of each connexon at the right and bottom of the illustration show that they protrude beyond the lipid bilayer on both sides of the plasma membrane. (Source: L. Makowski et al., "Gap junction structures. II. Analysis of the x-ray diffraction data," in *Journal of Cell Biology* 74:629, 1977. Copyright © 1977 Rockefeller University Press, New York, N.Y.)

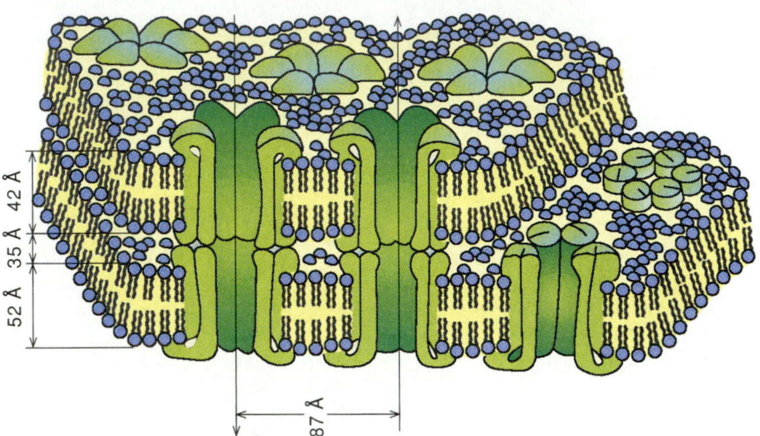

Gap junction pores have been shown to be permeable to hydrophilic molecules with masses as large as 1,000 to 2,000 daltons, depending on the cell type. They are believed to play roles in conducting electric impulses through particular cell types (nerve, muscle, and epithelia), as well as in providing avenues for the free flow of small metabolites between cells. The permeability of gap junction pores is regulated by the cytoplasmic concentration of Ca^{2+}; low concentrations ($<10^{-7}$ M) lead to open channels, while higher concentrations tend to close the channels in a graded manner. This sensitivity to $[Ca^{2+}]$ may play a role in the regulation of intracellular communication between cells connected by gap junctions.

Although porins and gap junctions are relatively nonspecific aqueous channels in biological membranes, it is reasonable to imagine that solute-specific pores can be formed by permease proteins as a result of their binding specificities. A number of workers have suggested similar models for how more specific permeases recognize and translocate their substrates. Aqueous channels between or within transmembrane subunits of transport proteins may respond to solute binding by undergoing conformational changes that alter the relative position of the solute-binding site with respect to the membrane and thereby, at least transiently, open the channels (fig. 32.35). This conformational change might be triggered solely by substrate binding, in which case facilitated diffusion would be the consequence. Alternatively, metabolic energy might expedite this process in one direction or the other, in which case active transport could result. Although conformational changes resulting in solute translocation are most easily envisioned between subunits of an oligomeric protein as depicted in figure 32.35, there is also evidence that at least some permeases may accomplish this process in a monomeric form.

Figure 32.35

Generalized molecular model for transport involving a conformational rearrangement of subunits of an oligomeric carrier protein. In active transport, energy (such as that from the hydrolysis of ATP) could be used to effect this conformational change as shown. Alternatively, binding of the transport substrate could be enough to trigger this rearrangement in systems carrying out facilitated diffusion. (Source: S. J. Singer, "The molecular organization of membranes," in *Annual Review of Biochemistry* 43:805, 1974. Copyright © 1974 Annual Reviews Inc., Palo Alto, Calif.)

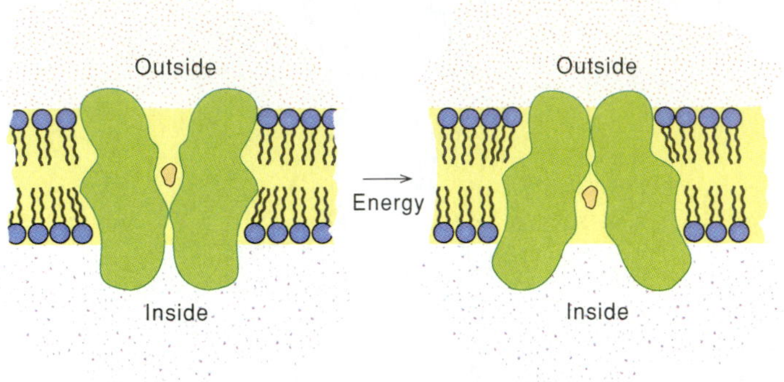

A Unifying Model for Transmembrane Transport

Although, as we have seen, there exists a wide variety of seemingly divergent transport mechanisms and permeases, there appear to be a few general principles which can be used to develop a unifying concept for transmembrane transport:

1. The translocation domain of a transport protein is a hydrophilic channel, or pore, at least in part formed by apposition of a number of transmembrane α helices (usually), or in the case of porins by transmembrane β structures.
2. This channel is usually gated for permeases recognizing specific substrates, such that the bound transport substrate is in equilibrium with the aqueous phase on only one side of the membrane at any given moment.
3. Translocation of the transport substrate involves a conformational change in the protein, exposing the binding site to the other membrane surface.
4. In facilitated diffusion, the conformational change is induced by substrate binding and serves only to equilibrate the concentrations of the transport substrate on both sides of the membrane. In this case, the binding affinity of the substrate to the permease would be approximately the same for binding sites exposed on either side of the membrane.

5. In active transport or group translocation into the cell, the binding affinity of the transport substrate is higher when the binding site is exposed extracellularly than when it is exposed intracellularly. This difference allows for binding of the substrate on the extracellular surface (and subsequently transport), even when there is a much higher concentration of the transport substrate inside the cell or organelle. Translocation of the "loaded" binding site to the inside surface therefore requires energy to disrupt the favorable interactions of substrate and permease as the binding site changes from a high-affinity to a low-affinity form. This energy may be provided by light (bacteriorhodopsin), ATP hydrolysis (ion-translocating ATPases; bacterial binding-protein transport systems), or cotransport of an ion down its electrochemical gradient (lactose permease).

These principles, and the unifying model derived from them, are depicted in figure 32.36. This model is consistent with most of the available evidence concerning the structures and mechanisms of the best-studied transport permeases and is also consistent with the evidence for evolutionary relatedness among even seemingly diverse transport systems. Although the detailed molecular mechanism will differ for each type of permease, the overall mechanisms are likely to be similar. For example, for bacteriorhodopsin the channel need only be very narrow (the proton "wire") and is probably gated by a rather subtle conformational change initiated by the light-induced isomerization of retinal. In this case, it would be predicted that proton pumping by bacteriorhodopsin would be rather insensitive to the physical state (gel or liquid crystalline) of the membrane, and this fact has been confirmed experimentally. On the other hand, conformational changes concomitant with translocation of larger solutes, such as nucleotides or sugars, would probably be more substantial to move the larger binding site within the channel to the other side of the membrane. This has been observed directly with the ATP/ADP exchanger as discussed earlier, and is consistent with the fact that the transport activity of the lactose permease is much more sensitive to temperature than is that of bacteriorhodopsin. Thus, although the pore model best describes most features of transport permeases, at least the substrate-binding sites of some of these proteins exhibit mobility through the permeability barrier of the membrane through protein conformational changes.

Figure 32.36

Models for facilitated diffusion (*a*) and active transport (*b*), incorporating many of the general features of most transport permeases. The substrate-binding site is comprised of transmembrane α helices (for clarity, only two are shown) with hydrophilic side chains projecting into the channel. These side chains interact with the substrate (filled triangle) when it binds to the outside of the membrane system (red lines). In facilitated diffusion, the affinity of the permease for the substrate is approximately the same on both sides of the membrane (*a*). In active transport, energy is used to drive the permease into a conformation in which the binding site "moves" across the membrane and simultaneously adopts a conformation with only a low affinity for the substrate (*b*). The situation depicted in (*b*) can lead to substrate accumulation.

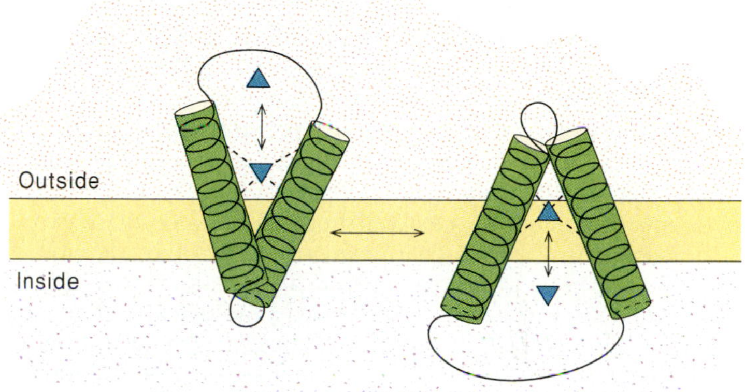

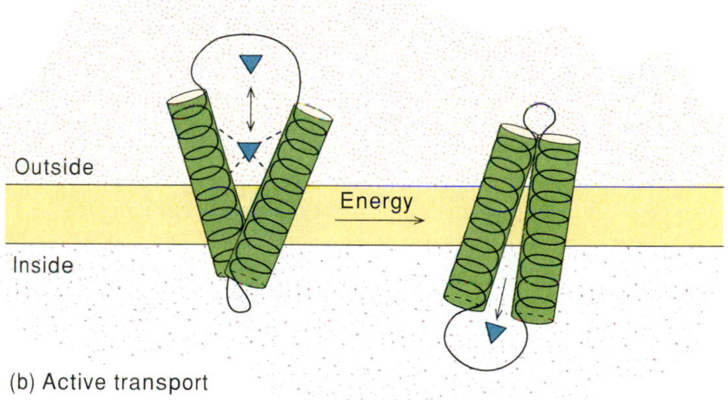

Perspectives for the Future

The advent of powerful techniques in molecular genetics has greatly increased our understanding of the probable structures of transport permeases during the past decade. These approaches, coupled with more classical biochemical and biophysical techniques, will undoubtedly continue to provide valuable insights into structure-function relationships in these proteins. Since methods have now been developed to crystallize membrane proteins, three-dimensional structures of transport permeases at atomic resolution will begin to emerge in the near future. The prospects are therefore bright that before the end of the century a combination of all of these techniques will allow us to determine, at a molecular level, the transport mechanisms of some of the better-characterized transport permeases described in this chapter. The recent recognition that most of these proteins are probably arranged similarly in membranes from a wide variety of sources will greatly simplify this task. The goal of understanding transmembrane processes vital to all cells and organisms, including nutrient transport, energy interconversion, photosynthesis, and intercellular communication, is clearly within reach.

Summary

In this chapter we have considered the theory, energetics, and mechanisms of transport in cells. The chief points that we discussed are as follows.

1. Cells and organelles must be capable of rapidly transporting hydrophilic solutes across the hydrophobic membrane barrier that bounds them. This process can occur by simple diffusion or can be mediated by a hydrophilic transmembrane pore or carrier. In simple diffusion, the solute must be appreciably soluble in the domain of the bilayer. The kinetics of its movement across the membrane obey Fick's law, and the net transport rate in either direction is directly proportional to the concentration gradient of the solute. In carrier-mediated transport, the transport rate is saturable with respect to solute concentration because only a finite number of permeases exist in the membrane. The kinetics of this process are described by the Michaelis-Menten equation when initial transport rates are measured. Thus kinetic measurements frequently distinguish between these two mechanisms, although in many cases cells may use more than a single system to transport a given solute.
2. Net transport of a molecule across a biological membrane in a given direction will occur spontaneously only if ΔG for the process is negative. For an uncharged molecule, this means that in the absence of energy input, the compound will reach equal concentrations on both sides of the membrane barrier at equilibrium. Simple and facilitated diffusion, therefore, can lead only to equilibration of an uncharged solute across the membrane. However, if the molecule being transported has a net charge, unequal concentrations of it can occur at equilibrium without the input of additional energy if there is an independently maintained electric potential across the membrane. Both simple and facilitated diffusion are therefore examples of energy-independent biological transport mechanisms.
3. Many cellular transport systems, however, can carry out the active accumulation of metabolites and ions. Because ΔG for solute accumulation is positive, the process must be coupled to an exergonic event in which ΔG is negative. Primary active transport systems may use the chemical energy of ATP, light energy, or electron flow to drive the uphill concentration of the solute. In secondary active transport, ion gradients, which themselves are maintained by primary active transport processes, are used to drive the active accumulation of a second solute. Finally, in group translocation, the transported substrate is chemically modified by the membrane permease during the transport event. This modification may result in the expenditure of metabolic energy, as in the case of many bacterial sugar transport systems, and may provide the driving force for solute accumulation.
4. Primary active transport systems that use ATP as a source of energy have been identified in both eukaryotic and prokaryotic cells. The Na^+-K^+ ATPase of animal cells is an asymmetrically oriented transmembrane protein that maintains high intracellular levels of K^+ by carrying out the active accumulation of this ion and the concomitant extrusion of Na^+. A second, well-studied ion-translocating ATPase, the Ca^{2+} ATPase of sarcoplasmic reticulum, catalyzes the active accumulation of Ca^{2+} into the lumen of this organelle by a similar mechanism. The F_1/F_o proton-translocating ATPase found in bacteria and eukaryotic organelles couples H^+ translocation with ATP synthesis or hydrolysis, thereby allowing the interconversion of chemiosmotic and chemical energy. At least four evolutionarily distinct classes of ion-translocating ATPases are now believed to exist. Other primary active transport systems include the ATP-binding cassette (ABC-type) solute transporters, the H^+-translocating carriers of electron-transport chains, and bacteriorhodopsin.
5. Na^+ and proton gradients across biological membranes are forms of potential energy that can be used to drive secondary active transport. In this type of transport, the flow of one of these cations down its electrochemical gradient can be coupled by cotransport, or symport, with the transport of another molecule against its concentration gradient. An example of this type of mechanism is the H^+-lactose symport system of *E. coli* and other bacteria.
6. The ATP/ADP exchanger found in the inner mitochondrial membrane represents a case in which membrane potential can drive the transport of a negatively charged molecule. The phosphoenolpyruvate (PEP)-dependent sugar phosphotransferase system (PTS) in bacteria is the best-studied group translocation system. From both sequence comparisons and structural analyses, it appears that many of these seemingly diverse transport permeases are structurally and mechanistically related.
7. A membrane permease could function mechanistically as a mobile carrier or a relatively static pore. Examples of both types of transport processes are found among the ion-translocating antibiotics. Examples of nonspecific pores in cells include the porins of Gram negative bacteria and gap junctions in animal cells. Specific transport permeases probably function as gated pores or channels in which substrate binding, and energy in active transport, serve to "move" the substrate-binding site from one side of the membrane to the other. These proteins therefore exhibit certain characteristics of both the pore and mobile-carrier models. From the information now available, it appears that a unifying model may explain transmembrane transport regardless of the many differences in detail of biological transport systems. The future appears bright for an understanding of biological transport mechanisms at the molecular level.

Selected Readings

Ames, G. F.-L., C. S. Mimura, S. R. Holbrook, and V. Shyamala, Traffic ATPases: A superfamily of transport proteins operating from *E. coli* to humans. *Adv. Enzymol.* 65:1–47, 1992. Excellent review of binding-protein-dependent transport systems in bacteria and related eukaryotic transport proteins that are also ATPases.

Boyer, P. D., A perspective of the binding change mechanism for ATP synthesis. *FASEB J.* 3:2164–2178, 1989. Reviews evidence for the mechanism of H^+-coupled ATP synthesis by the F_1/F_0 ATPase.

Bronner, F., and A. Kleinzeller (eds.), *Current Topics in Membranes and Transport.* New York: Academic Press. Continuing series reviewing some of the current problems in biological transport.

Dharmavaram, R. M., and J. Konisky, Characterization of a P-type ATPase of the archaebacterium *Methanococcus voltae. J. Biol. Chem.* 264:14085–14089, 1989. Recent discovery of a P-type ATPase in an archaebacterium, which may have significant evolutionary implications.

Friedlander, M., and M. Mueckler (eds.). Molecular biology of receptors and transporters, *Int. Rev. Cyto.* 137A, 1992. This recent volume contains 8 review articles on various types of transport systems in both eucaryotic and procaryotic cells.

Gennis, R. B., *Biomembranes: Molecular Structure and Function.* N.Y.: Springer-Verlag, 1989. Excellent book dealing with both membrane protein structure and function.

Higgins, C. F., M. P. Gallagher, M. L. Mimmack, and S. R. Pearce, A family of closely related ATP-binding subunits from prokaryotic and eukaryotic cells. *Bioessays* 8:111–118, 1988. An interesting summary of recent evidence for the structural and functional relatedness of seeming divergent proteins that bind ATP.

Inesi, G., D. Lewis, D. Nikic, A. Hussain, and M. Kirtley, Long-range intramolecular linked functions in the calcium transport ATPase. *Adv. Enzymol.* 65:182–215. Recent review of the structure and mechanism of the Ca^{2+}-ATPase, with an emphasis on how Ca^{2+} transport and ATP hydrolysis may be linked.

Kaback, H. R., E. Bibi, and P. D. Roepe, β-Galactoside transport in *E. coli:* a functional dissection of lac permease. *Trends. Biochem. Sci.* 15:309–314, 1990. Structure function relationships in the lactose permease.

Meadow, N. O., D. K. Fox, and S. Roseman, The bacterial phosphoenolpyruvate:glycose phosphotransferase system. *Ann. Rev. Biochem.* 59:497–542, 1990. Recent review of this carbohydrate transport system that is commonly found in anaerobic and facultatively anaerobic bacteria.

Saier, M. H., Jr., *Mechanisms and Regulation of Carbohydrate Transport in Bacteria.* New York: Academic Press, 1985. Reviews the various mechanisms of carbohydrate transport in bacteria, as well as transport regulation.

Silverman, M. Structure and function of hexose transporters. *Ann. Rev. Biochem.* 60:757–794, 1991. Excellent review of recent work on sugar transporters in eukaryotic cells.

Spudich, J. L., and R. A. Bogomolni, Sensory rhodopsins of halobacteria. *Ann. Rev. Biophys. Biophys. Chem.* 17:193–215, 1988. Review of structure and function of bacteriorhodopsin and other related proteins of halobacteria.

Problems

1. (a) Using Fick's law, show that the diffusion coefficient D has the dimensions of area per unit time.
   (b) The diameter of a pore channel is about 10^{-9} m and its length is 4×10^{-9} m. In planar membrane bilayers, glucose traverses this channel at the rate of about 50 molecules per channel per second at room temperature when the concentration of glucose is 3×10^{-6} M on one side of the membrane. Calculate the diffusion coefficient for glucose through the pore channels under these conditions.
2. The Nernst equation relates the electric potential $\Delta\psi$ resulting from an unequal distribution of a charged solute across a membrane permeable to that solute to the ratio between the concentration of solute on one side and on the other:

$$m\,\Delta\psi = \frac{-2.3RT}{F}\log\frac{[So]_1}{[So]_2}$$

where m is the charge on the solute, $2.3RT/F$ has a value of about 60 mV at 37° C, and $[So]_1$ and $[So]_2$ refer to the concentrations of solute on either side of the membrane. Consider a planar phospholipid bilayer separating two compartments of equal volume. Side 1 contains 50-mM KCl and 50-mM NaCl, while side 2 contains 100-mM KCl.
   (a) If the membrane is made permeable only to K^+, e.g., by addition of valinomycin, what will be the magnitude of $\Delta\psi$?
   (b) If the membrane is made permeable to H^+ and K^+, in which direction will H^+ initially flow?
   (c) If the membrane could be made selectively permeable to both K^+ and Cl^-, what would be the value of $\Delta\psi$ and the ion concentrations on both sides of the membrane at equilibrium? (Hint: Initially, K^+ would diffuse down its concentration gradient accompanied by an equivalent amount of Cl^-. Equilibrium would be established when the potentials due to K^+ and Cl^- were equal to each other and to the overall membrane potential.)
3. Membrane vesicles of *E. coli* that possess the lactose permease are preloaded with KCl and are suspended in an equal concentration of NaCl. It is observed that these vesicles actively, although transiently, accumulate lactose if valinomycin is added to the vesicle suspension. No such active uptake is observed if KCl replaces NaCl in the suspending medium. Explain these results in light of what you know about the mechanism of lactose transport and the properties of valinomycin.

4. Intracellular vacuoles in the yeast *Saccharomyces cerevisiae* are membrane-bounded organelles that are known to concentrate within them a variety of basic amino acids, including arginine (net charge = +1). Vesicles prepared from these vacuoles lack an electron-transport chain, and arginine uptake into them is dependent on extravesicular ATP. A membrane potential $\Delta\psi$ has no effect on ATP-dependent arginine uptake in the absence of a proton gradient, while proton ionophores and dicyclohexylcarbodiimide (a known inhibitor of the F_1/F_o ATPase) greatly inhibit accumulation of arginine by this system. Upon addition of ATP in the absence of arginine, the intravesicular pH of these vesicles drops. Describe a mechanism for the energization of arginine transport in this system, taking into account all these observations.
5. In *E. coli,* lactose is taken up by means of proton symport, maltose by means of a binding-protein-dependent (ABC-type) system, melibiose by means of Na^+ symport, and glucose by means of the phosphotransferase system (PTS). Although this bacterium normally does not transport sucrose, suppose that you have isolated a strain that does. How would you determine whether one of the four mechanisms just listed is responsible for sucrose transport in this mutant strain?
6. In some instances, the efflux of a radioactively labeled transport substrate out of preloaded cells or vesicles is transiently stimulated by addition of the same nonradioactive transport substrate to the outside. This phenomenon is known as trans-stimulation and occurs with transport systems that are reversible (i.e., can operate in either direction). Can you think of an explanation for trans-stimulation in view of what is known about the molecular mechanisms of transmembrane transport?
7. Outline a molecular mechanism by which, and the conditions under which, an H^+ symport system (such as the *E. coli* lactose permease system) might operate to actively accumulate a metabolite such as lactose.
8. Predict the effects of the following on the initial rate of glucose transport into vesicles derived from animal cells that accumulate this sugar by means of Na^+ symport. Assume that initially $\Delta\psi = 0$, $\Delta pH = 0$ (pH = 7), and the outside medium contains 0.2-M Na^+, while the vesicle interior contains an equivalent amount of K^+.
   (a) Valinomycin.
   (b) Gramicidin A.
   (c) Nigericin.
   (d) Preparing the membrane vesicles at pH 5 (in 0.2-M KCl), resuspending them at pH 7 (in 0.2-M NaCl), and adding 2,4-dinitrophenol.
9. You are growing some mammalian cells in culture and measure the uptake of D-glucose and L-glucose (see data below). What type of transport is observed with these sugars? (Hint: plot *V* versus [sugar] and 1/*V* versus 1/[sugar].) Explain the significance of these data.

	V (mM cm s^{-1}) × 10^7	
[Sugar] (mM)	**D-glucose**	**L-glucose**
0.100	166	4.8
0.167	252	8.0
0.333	408	16
1.000	717	50

10. If the cells in problem 9 are treated with $HgCl_2$ (mercury reacts with —SH groups in proteins), the rate of transport of D-glucose is the same as L-glucose. What is indicated about D-glucose transport?
11. The translocation of K^+ was studied using an artificial membrane system (this membrane system had a phase transition T_m of 41° C) with valinomycin and gramicidin. The results showed that K^+ translocation with gramicidin was high over a broad temperature range, while valinomycin translocated K^+ at a high rate only above 41° C. What models of translocation are suggested by these data for each of these polypeptide antibiotics?
12. While there are many different and diverse types of transport systems in cells, these systems seem to be evolutionarily related and can be accommodated within a unified model. Outline the overall mechanism that is similar in these transport systems.

33

CHAPTER

Immunobiology

When vertebrates are invaded by foreign agents, they can mobilize a versatile set of adaptive processes to form specifically reactive cells and proteins. These immune responses constitute the principal means of defense against pathogenic microorganisms and viruses, and probably also against host cells that undergo transformation into cancer cells.

Throughout most of this century, the subject of immunology has attracted some of the keenest minds in biology. As a result, the intricacies of this fascinating subject have been largely unraveled, and its understanding is providing a strong bridge between the fields of biochemistry and physiology. In this chapter we will give a brief introduction to some of the major findings in immunobiology and the experiments that have led to our current level of understanding.

Overview of the Immune System

The immune system was first studied in humans, but mice became a popular subject for immune system studies when researchers began to appreciate how close the mouse and human systems were in their organization and action. Currently both systems are under study in many different laboratories. Although we will focus here on mice, we will often refer to parallel observations on humans.

The immune system is an example of a developmental process that takes place in the mature organism. In the new cells that are constantly being generated, changes arise in the genes of the immune system. This variability results from DNA splicing and point mutations. It benefits the organism by providing the cells of the immune system with the widest possible range of specificities. As soon as the organism is invaded by foreign agents, usually viruses or bacteria, the immune system is activated. Those immune-system cells that carry the specific immune receptors for interacting with the foreign agent are stimulated to proliferate. In a matter of days, clones of immune-system cells with the appropriate specificity have been produced and the organism fends off the invader with the specific immunologic tools provided by those cells. The clones of cells tend to persist for some time, usually months to years, which accounts for the fact that the second time the same foreign invader makes its presence known, it is usually rejected promptly and without crisis.

There are two different classes of white cells or lymphocytes associated with the immune response; these are the B cells and the T cells. In mammals, B lymphocytes mature in the bone marrow and then migrate to secondary lymphoid organs (fig. 33.1). Upon exposure to foreign substances known as antigens, they proliferate and produce immunoglobulin proteins known as antibodies, which they secrete into the bloodstream.

Figure 33.1

B cells and T cells follow different pathways of development. In mammals, B lymphocytes mature in the bone marrow and then migrate to secondary lymphoid organs. Upon exposure to foreign substances known as antigens, they proliferate to produce immunoglobulins. T lymphocytes mature in the thymus gland. They also can be stimulated to proliferate by exposure to an appropriate antigen.

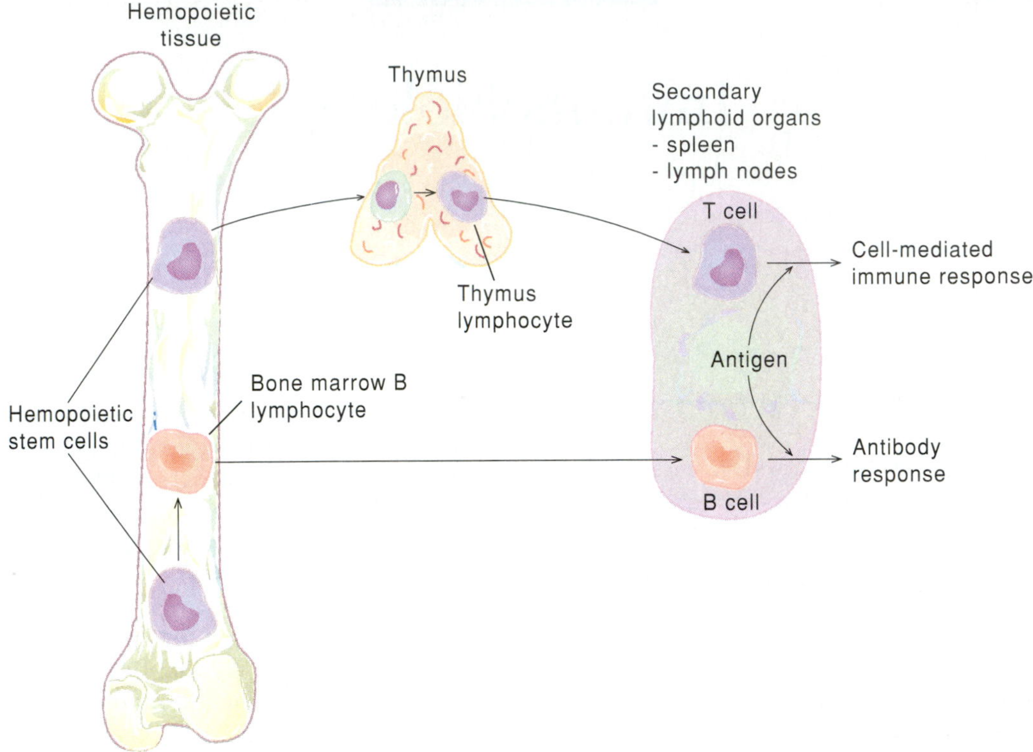

T lymphocytes, by contrast, mature in the thymus gland before migrating to the secondary lymphoid organs. They also can be stimulated to proliferate by exposure to an appropriate antigen, but their specific effector molecules remain firmly bound to the cellular membrane, as opposed to being secreted. Three main types of T cells have been recognized; T killer cells, which specifically destroy target cells; T helper cells, which promote the maturation of antigen-stimulated B and T cells; and T suppressor cells, which block the effects of T helper cells.

The Humoral Response: B Cells and T Cells Working Together

There are two basically different types of immune response. The first, called the humoral or antibody response, involves the concerted action of both B cells and T cells, and the active agents are the antibody or immunoglobulin proteins secreted by the B cells into the bloodstream. The combined B-T immune response is characteristic of most vertebrates. The second type of immune response, called the cell-mediated response, involves only T cells, and the active agent is the circulating T cell itself, which attacks the foreign agent. This type of immune response is limited to certain groups of vertebrates, including mammals. We will first discuss the B-T cell response mediated by antibodies.

Immunoglobulins Are Extremely Varied in Their Specificities

A detailed investigation of immunoglobulin structure provided the first leads as to how immunoglobulins of such a wide variety are synthesized. We have already provided a description of immunoglobulin structure (see fig. 5.25). For the purposes of this discussion, however, we need to describe certain features of the structure in greater detail. The most common type of immunoglobulin is the 7S molecule, known as immunoglobulin G (IgG). This immunoglobulin has a molecular weight of about 150,00 daltons. It is composed of four polypeptide chains, two heavy (H) chains with molecular weights of about 50,000 and two light (L) chains with molecular weights of about 25,000 (fig. 33.2).

To obtain a detailed understanding of immunoglobulin structure, it was necessary to isolate pure antibody proteins for sequencing. Fortunately, it was discovered that certain plasma cell tumors known as myelomas produce enormous amounts of pure immunoglobulins, which have the same gross structure as the mixed immunoglobulins isolated from serum. Each myeloma appears to originate from a single cell turned cancerous. As a result, each myeloma serves as the source of a pure antibody protein whose sequence and other properties can be determined. In some cases myelomas synthesize both heavy and

Figure 33.2

Structure of immunoglobulin G (IgG). The light (L) and heavy (H) chains have repeating domains, each with about 110 amino acid residues and an approximately 60-membered S—S bonded loop. Domains with variable sequences (V_L, V_H) are represented by blue bars; those with constant sequences (C_L, C_H) are represented by pink bars.

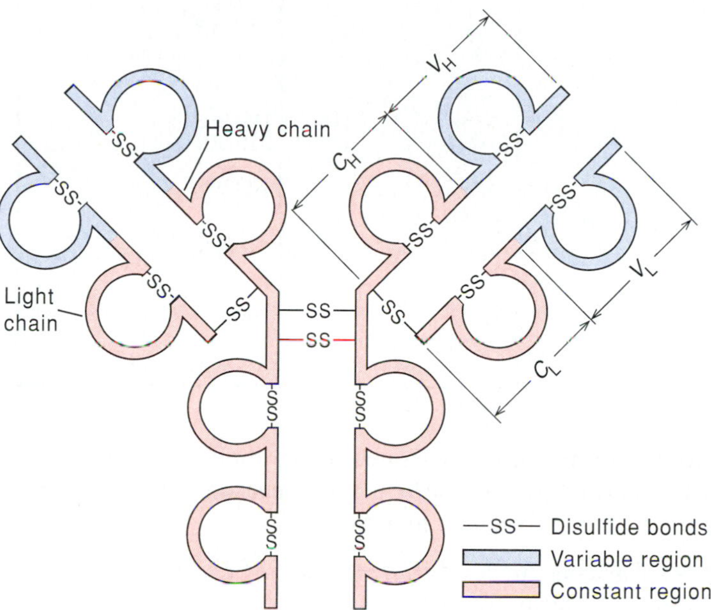

Figure 33.3

Approximate locations of the hypervariable regions in the heavy and light chains of IgG. Each hypervariable segment is believed to make up part of the site that binds to antigen, known as the complementarity-determining region (CDR). The CDR regions are located in the loop regions of the variable domains where they can make close contact with antigen (see fig. 5.25).

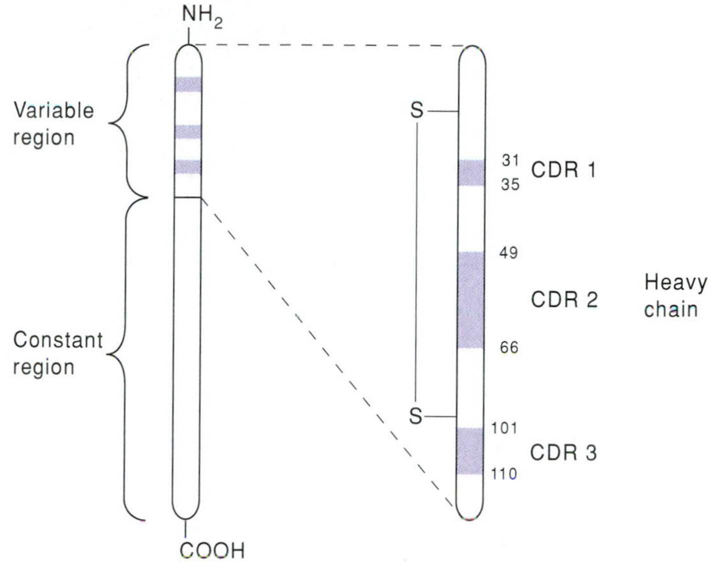

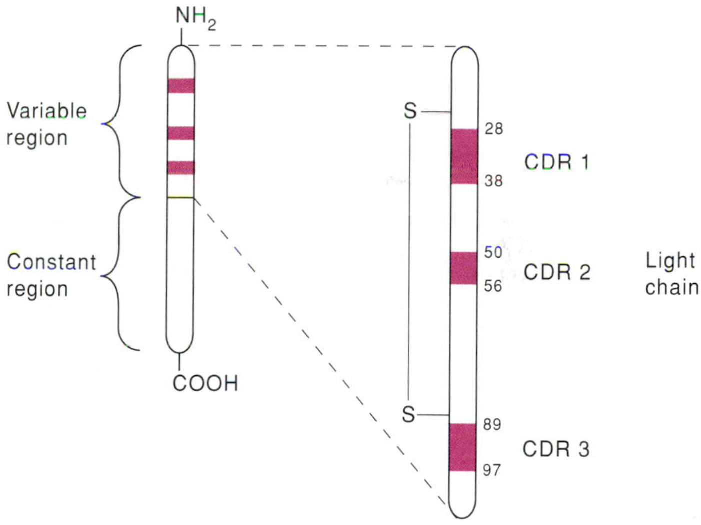

light chains, like normal antibody-forming cells, but in other cases they synthesize only one or the other type of chain. Comparison of the sequences from a number of different immunoglobins derived from myelomas has shown that both the heavy and the light chains for immunoglobulins of the same class are divided into regions of relatively constant sequence (the *C* segments) and regions of relatively variable sequence (the *V* segments). In the intact immunoglobulins the antigen-binding domain is composed exclusively of *V* segments originating from the H and L chains (see fig. 33.2), each tetramer containing two equivalent sites for the binding of a specific antigen.

As information on antibody sequences has accumulated, it has become increasingly clear that within the *V* regions of both the heavy and light chains there are three segments that account for most of the variability (fig. 33.3). These regions are called the hypervariable regions, and they are believed to be the parts of the antibody molecule in most direct contact with the antigen in the antigen-antibody complex.

Immunoglobulin G (IgG), the tetrameric species we have been discussing thus far, is not the only type of immunoglobulin found (table 33.1). Most of the other known antibodies—IgM, IgA, IgD, and IgE—also involve a closely related tetrameric structure, sometimes forming larger aggregates and always associated with different functions. For instance, IgM is a 19S antibody accounting for 5 to 10% of the serum Ig. It is an aggregate of five tetramers that is formed early during the immune reaction, soon to be diminished in quantity and overshadowed by large amounts of IgG. IgA occurs in various polymeric forms. It normally accounts for about 15% of the total immunoglobulin found in serum, and in addition it is the principal immunoglobulin in exocrine secretions. Each of the different classes of immunoglobulin possesses distinct heavy chains. Indeed, even within the human IgG class, antisera tests reveal four different types of IgG (IgG1 through IgG4), each with a distinctive H chain, comprising about 70, 19, 8, and 3%, respectively, of the total IgG proteins. The light chains, of which there are two types, are common to all classes of immunoglobulins.

Immunoglobulins that are produced in response to antigens are themselves highly immunogenic (that is, they stimulate antibody formation) when injected into genetically nonidentical organisms. The serological responses induced by

Table 33.1
Different Isotypes Found in Humans

Class	Heavy Chain	Subclasses	Light Chain	Molecular Formula	Molecular Weight (Daltons)
IgG	γ	$\gamma 1$, $\gamma 2$ $\gamma 3$, $\gamma 4$ $\alpha 1$, $\alpha 2$	κ or λ	$\gamma_2 \kappa_2$ $\gamma_2 \lambda_2$	150,000
IgA	α	None	κ or λ	$(\alpha_2 \kappa_2)^a_a$ $(\alpha_2 \lambda_2)^a_a$	160,000 320,000
IgM	μ	None	κ or λ	$(\mu_2 \kappa_2)_5$ $(\mu_2 \lambda_2)_5$	900,000
IgD	δ	None	κ or λ	$(\delta_2 \kappa_2)$ $(\delta_2 \lambda_2)$	185,000
IgE	ϵ	None	κ or λ	$(\epsilon_2 \kappa_2)$ $(\epsilon_2 \lambda_2)$	200,000

using immunoglobulins as antigens or immunogens have been useful in classifying them according to their antigenic determinants.

So-called isotypic determinants are shared by all immunoglobulin molecules of a given class. For example, the human IgG molecules are classified into four isotypes because they are recognized by heterologous antisera produced in one species against the immunoglobulins of another species. Within a given isotype the different immunoglobulins can usually be separated into sets, called allotypes, that are distinguished by minor antigenic differences. Allotypic differences usually reflect alternative amino acid substitutions within otherwise quite similar amino acid sequences of the *C* region of the antibody. Allotypes show typical Mendelian inheritance patterns. Antisera useful for discriminating between allotypes are usually made by injecting an individual who lacks a specific allotype with immunoglobulins from an individual who carries the allotype. Finally, idiotype refers to the specific antigenic determinant of the antibody. Some idiotypic determinants are limited to a single immunoglobulin; others are shared by a small number of immunoglobulins. Whereas isotypic and allotypic differences usually result from differences in the constant portion or *C* segments of the immunoglobulin polypeptide chain, idiotypic differences usually result from differences in the variable portion or *V* segment. As we know already, this is the region that contains the binding site for the foreign agent or antigen.

Antibody Diversity Is Augmented by Unique Genetic Mechanisms

Antibodies have been studied most extensively in the mouse, an ideal vertebrate for both genetic and biochemical manipulations. It is believed that mice can synthesize more than a million antibodies with different antigenic specificities. This enormous diversity of proteins is generated from a limited amount of genetic information with the help of two mechanisms: somatic recombination and somatic mutation.

DNA Splicing Brings Different Parts of the Antibody Gene Together The involvement of somatic recombination was first demonstrated by examining the structure of a specific antibody gene in embryonic and adult immunoglobulin-forming tissue. The adult tissues favored for many studies are myelomas. These tumorous tissues, which we mentioned earlier, produce a homogeneous population of polypeptide chains. They have served as a convenient source of pure immunoglobulin polypeptide chains as well as their mRNAs.

Recall that the generalized structure of the predominant serum antibody consists of two identical heavy chains and two identical light chains (see fig. 33.2). In the mouse there are two classes of light chains, which differ appreciably in the constant regions of the polypeptide chains. These are referred to as the kappa (κ) and the lambda (λ) light chains. Using highly inbred, genetically identical (isogenic) strains of mice, Tonegawa and his co-workers isolated the DNA from the embryo and from two different myeloma tumor cells, one that produces homogeneous λ light chains (strain H2020) and one that produces κ light chains (strain MOPC321). These DNAs were exhaustively digested with the *Eco*R1 restriction enzyme and electrophoresed on an agarose gel. The gels contained an enormous variety of restriction fragments representing total nuclear DNA, and consequently no discrete pattern of bands could be seen with a strain for nucleic acid. However, when the electrophoresed gels containing the DNA were denatured and hybridized with ^{32}P-labeled DNA containing the sequences found in the RNA for the λ chain, a specific pattern of bands showed up in autoradiographs (fig. 33.4). The R1 digest of DNA from

Figure 33.4

Analysis of DNA fragments containing λ_1 gene sequences from mouse embryo and myeloma cells. High-molecular-weight DNAs extracted from myeloma H2020, a λ-chain producer (A), from a 13-day-old BALB/c embryo (B), and from myeloma MOPC321, a κ-chain producer (C) were digested to completion with *Eco*R1, electrophoresed on agarose gel, and transferred to nitrocellulose membrane filters and hybridized with a nick-translated *Hha*I fragment of the plasmid B1 DNA. (Source: C. Brack et al., "A complete immunoglobulin gene is created by somatic recombination," in *Cell* 15:1–14, 1978. Copyright © 1978 Cell Press, Cambridge, Mass.)

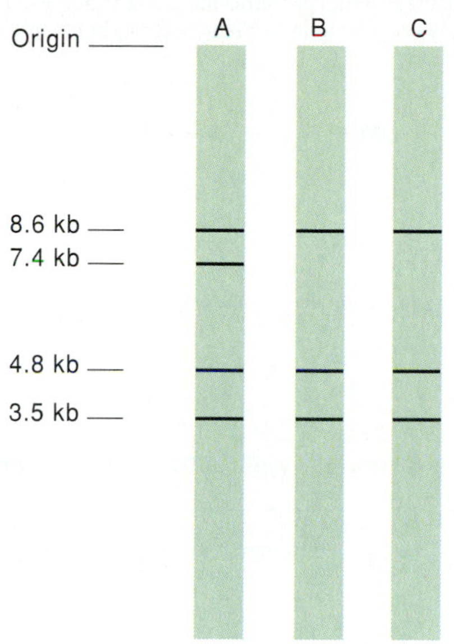

the λ-containing myeloma (H2020) showed four bands; the DNAs of the embryo and the κ-containing myeloma (MOPC321) showed three bands in common with the first DNA but were missing the fourth band.

These results strongly suggested that at some point during development from the embryonic state, a rearrangement had taken place in the H2020 myeloma cells on one of the homologous pairs of chromosomes carrying the λ-chain gene. Further analyses with more specific radioactive probes, containing sequences from either C_λ (the constant region of the λ chain) or V_λ (the variable region of the λ chain), were done. Only the 7.4-kb fragment originating from the λ myeloma cell hybridized to both probes. Thus the *Eco*R1 fragment must contain both V_λ and C_λ DNA. The 8.6-kb fragment hybridized to the C_λ but not the V_λ probe. Therefore this fragment must contain C_λ sequences but no V_λ sequences. Both the 3.5-kb fragment and the 4.8-kb fragment hybridized exclusively to the V_λ probe, so they must contain only V_λ sequences. Still further analyses indicated that the 7.4-kb fragment arose from a recombinational event between the 3.5-kb fragment and the 8.6-kb fragment (fig. 33.5). The 4.8-kb fragment originates from another V_λ sequence located in the same chromosome. The mouse has only two V_λ sequences in the embryo. As a rule, only one of the pairs of homologous chromosomes recombines to yield a productive antibody gene. This explains why the 3.5-kb and the 8.6-kb bands are still visible in the H2020 myeloma DNA preparation. The κ myeloma cell producer shows the same pattern as the embryonic cell. Had it been probed with a DNA carrying the V_κ antibody sequence, it would have shown differences from the embryonic pattern.

Figure 33.5

Arrangement of mouse λ_1 gene sequences in embryos and λ_1 chain-producing plasma cells. The vertical arrows point to *Eco*R1 restriction sites. The colored dashed diagonal lines point to hypothesized splice points that will explain the difference in structure of the region in the two cell types. The boxed regions represent coding regions, and I_1 and I_2 indicate first and second introns in the spliced gene.

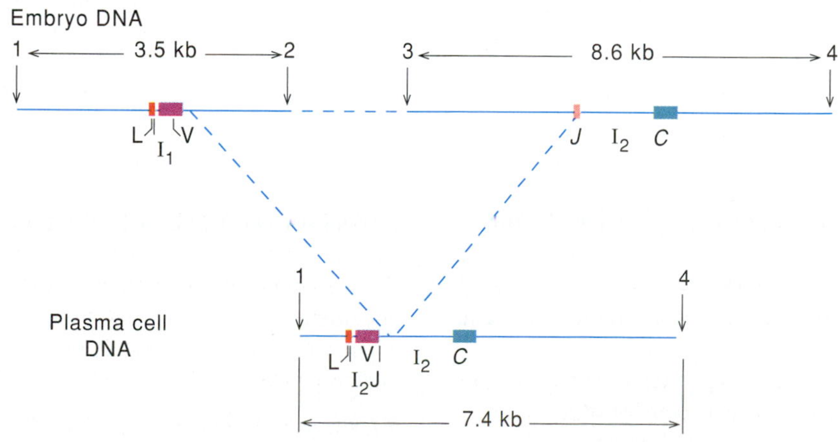

The success achieved by Tonegawa and others in this type of sequence analysis of the immunoglobulin genes led to a massive research effort and a general picture of antibody gene organization. All antibody polypeptide chains are derived from split genes. Antibody genes are encoded by three unlinked gene families located on different chromosomes. In the human, chromosomes 2, 22, and 14 encode those gene clusters associated with the kappa (κ) light chains, the lambda (λ) light chains, and the heavy chains, respectively.

The light chain is encoded by three distinct DNA elements—a variable (V_L), a joining (J_L), and a constant (C_L) element (fig. 33.6). During the differentiation of a B cell, a V_L element (encoding the first 95 amino-terminal residues) and a J_L segment (encoding about 13 amino acids) are joined by DNA splicing. A complete light-chain gene transcript consists of three exons and two introns, arranged in the following order; 5′L, I_1, *VJ*, I_2, *C*, 3′. Here L refers to sequences in the leader or signal peptide, which is removed in the mature immunoglobulin, and I_1 and I_2 refer to the first and second introns, respectively, which are removed by RNA splicing. The heavy chain is encoded by four distinct DNA elements—a V_H, a *D*, a J_H, and a C_H element. In heavy-chain variable region formation, a V_H element (encoding about 99 amino acids), a *D* element (encoding 1 to 15 amino acids), and a J_H element (encoding about 15 amino acids) are joined in two DNA-splicing steps. The heavy-chain gene family has several closely linked C_H genes, which determine the immunoglobulin class.

Class switching, involving an additional DNA-splicing operation, frequently occurs with heavy-chain genes. Thus an immature B cell, which initially expresses a μ chain, results in IgM antibodies. Subsequently the same cell can be induced to differentiate further so that the same V_H region of the genome becomes relocated next to a C_γ gene, causing the cell to produce an IgG antibody with the same V_H region associated with a different C_H region. In cases of class switching, the second DNA splicing involves removal of the intervening genes. Thus in the case of C_μ to C_γ switching, the region containing C_δ must be removed by the second splice reaction (fig. 33.7). In the case of C_μ to C_α class switching, all of the *C* genes between C_μ and C_α must be removed by the second splice reaction.

Class switching of the heavy-chain genes illustrates a temporally regulated process associated with the DNA rearrangements that can occur in the antibody-forming genes. Since any given cell normally synthesizes only one type of antibody, there must be a special type of regulatory mechanism that responds in a systematic way to antigenic stimulation. In B-cell development it is known that the V_H-D-J_H joining occurs first, leading to the formation of μ heavy chains. Subsequently V_L-J_L joining occurs, with the production of functional light chains.

As a rule, the DNA-splicing reactions leading to the joining process occur in only one of the alleles for each of these chains. The mechanism for this phenomenon, known by the name allelic exclusion, is not understood. Another type of exclusion process results in the production of only κ or λ light chains in a given B cell. This process is called isotypic exclusion. Indirect evidence that the κ gene family is expressed before the λ gene family comes from the examination of human B-cell lines that are active κ- or λ-chain producers. In all B cells producing λ chains, the C_κ genes are either deleted or rearranged. By contrast, in all cells producing κ chains, the C_λ genes are in the germ-line configuration. The obvious inference is that a B cell becomes a κ producer first and then becomes a λ producer. Direct proof of this order is still lacking. Other evidence exists for regulatory signals that may control isotypic exclusion. When a pre-B cell is fused with a myeloma cell that has been producing a functional heavy chain but no functional light chain, the hybrid cell is capable of expressing a new light chain as a result of a *V-J* joining of gene segments in the pre-B cell. If, however, the myeloma cell was already producing a functional light chain, no new light chains or DNA rearrangements result from the cell fusion process.

Figure 33.6

Organization and DNA splicing of the mouse immunoglobulin genes. The organization is depicted before and after somatic cell rearrangement. For each class only one example of a rearranged chromosome is shown. A light chain is encoded by three distinct DNA elements, a variable (*V*), a joining (*J*), and a constant (*C*) element. Boxes indicate exons; lines connecting boxes indicate introns. Somatic cell rearrangement involves the joining of a *V* and a *J* segment on the same chromosome. The heavy chain is encoded by four distinct types of DNA elements, a *V*, a *D*, a *J*, and a *C* element. In heavy-chain splicing a *V* element, a *D* element, and a *J* element are joined in two DNA splicing steps. The splice between *D* and *J* elements occurs first. Boxes that are filled are preserved after the DNA splicings. The splices shown provide only one example of many that might occur in this system.

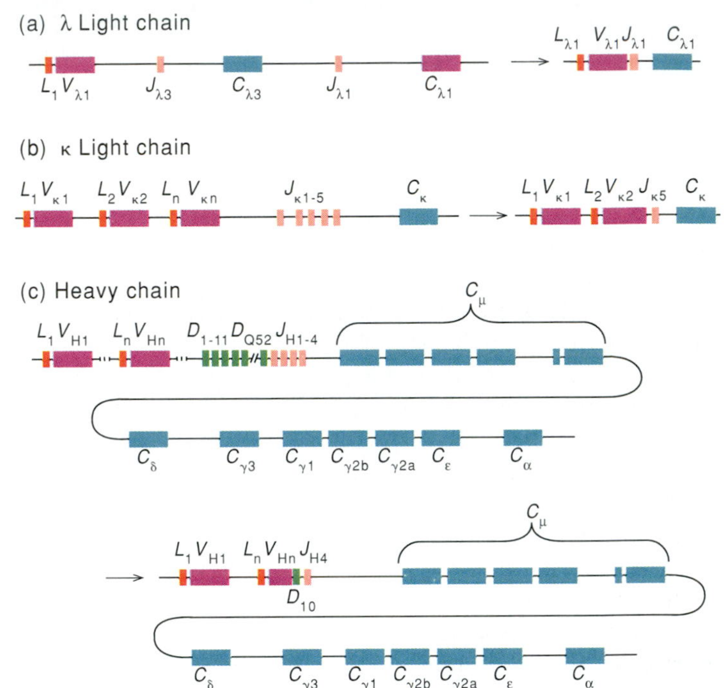

Alternative Pathways Exist for RNA Splicing of Heavy-chain mRNAs The heavy-chain transcript contains numerous exons and introns. In addition to the introns between the L_H and V_H exons and between the V_H-J_H and C_H exons, each C_H region

Figure 33.7

Class switching of heavy-chain genes involves additional DNA splicing. Switching from C_μ to $C_{\gamma l}$ is illustrated. The additional splicing involves removal of the continuous segment carrying the C_μ through the $C_{\gamma 3}$ genes.

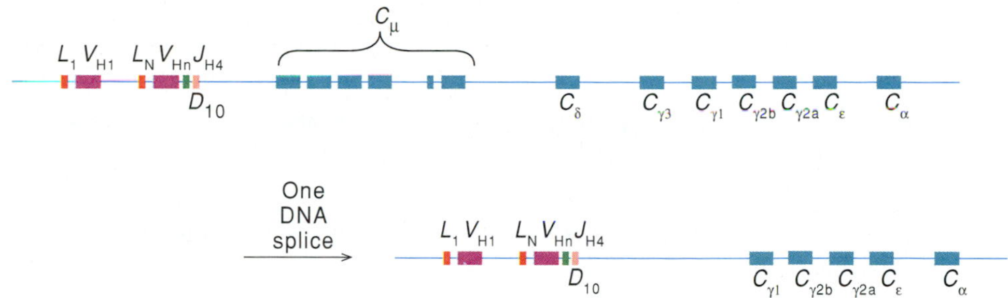

contains several introns. The hinge regions of γ chains (see fig. 5.25) and the intramembrane and cytoplasmic portions of all C_H chains are encoded by independent exons. Most of the RNA-splicing operations are presumed to occur by standard mechanisms similar to those described in chapter 28. In the case of the heavy-chain genes, it has been shown that alternative RNA-splicing patterns involving the C_H region can yield molecules that will function as membrane-bound or secreted forms of the antibody. Thus there are two membrane exons that have been detected in the 3′ flanking regions of both the C_μ and C_γ genes. Alternative pathways of RNA splicing give rise to two mRNAs, which encode separately the membrane-bound and the secreted forms of the heavy chains. It seems likely that the alternative splicing pattern is temporally regulated, since the membrane-bound forms are favored before and the secreted forms are favored after antigenic stimulation. Amino acid as well as nucleotide sequencing of the μ_M chain (μ_M stands for the membrane-bound form of the μ chain) has revealed a 26-residue hydrophobic segment at the COOH terminus, which anchors in the lipid bilayer of the membrane.

Somatic Mutation Contributes to Antibody Diversity Various estimates, some exceeding a million, have been put forward for the number of different antibodies an organism can make. A good deal of this diversity results from the different combinations of heavy and light chains that can be generated by making alternative splicings between the different gene segments that make up the light and heavy chains (table 33.2). Superimposed on the variation from this source, there is additional diversity provided by somatic mutation within the coding regions. Several subsets of κ and heavy chains and their germ-line V segments have been analyzed by cloning and sequencing. The results confirm that somatic mutations amplify the diversity encoded in the germ-line genome.

A $V\kappa$ probe prepared for the κ genes expressed in a particular myeloma, MOPC167, detected only one major band in the Southern gel blot analysis of total cellular DNA. This unusual situation permitted a relatively simple sequence comparison to be made between this segment, found in the germ-line, and its rearranged counterparts in two myeloma lines, MOPC167 and MOPC511. Four and five nucleotide differences were found, respectively, in the $V\kappa$ regions of these two lines. The changes were completely different in both lines. This investigation and other similarly directed investigations show that the regions in which mutations are introduced during somatic differentiation are highly restricted to the V_L-J_L or V_H-D-J_H segments. The density of mutations observed is about three times higher in the complementarity-determining regions (CDRs) of the variable segments than in the connecting framework regions (FRs) of the variable segments. (See figure 33.3 for the approximate location of the CDR regions.) The complementarity-determining regions are the regions in the immunoglobulin that play a critical role in the interaction with antigen. Mutations produced in these regions must occur either during or after DNA splicing, as they are not found in unrearranged (unspliced) DNA. Somatic mutation clearly increases the number of possible antibodies almost without limit. Most importantly, it leads to selectable changes in just those regions of the immunoglobulins that are responsible for antigen recognition. Several mechanisms for hypermutation have been proposed; none have been verified.

Table 33.2
Estimated Number of Germ-Line Gene Segments for Different Mouse Antibody Genes

	Light Chains		**Heavy Chains**
	κ	λ	H
C	1	4	8
V	90–300	2	100–200
J	5	4	4
D	—	—	12

From Geoffrey Zubay, *Genetics*. Copyright © 1987 Benjamin/Cummings Publishing Company Inc., Menlo Park, Calif. Reprinted by permission of the author.

Figure 33.8

DNA splicing of an antibody gene brings an enhancer element (E) close to the promoter region (P) of the gene. This step is depicted for a heavy-chain gene after the initial two splices (see fig. 33.6).

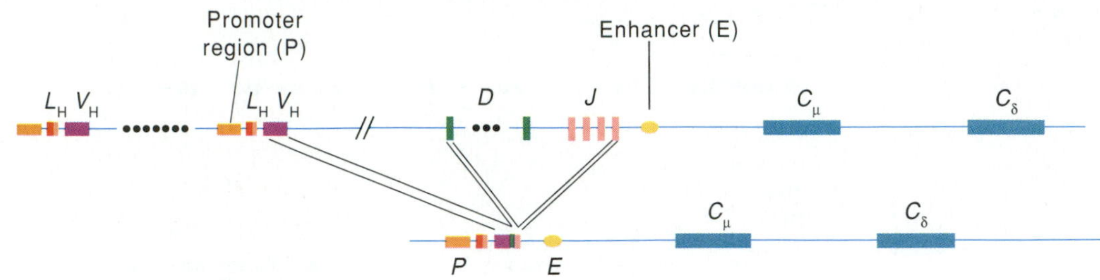

DNA Splicing Brings an Enhancer Element Close to the Promoter DNA sequences derived from the germ-line J_H-C_μ region are required for accurate and efficient transcription from a functionally rearranged V_H promoter (fig. 33.8). Similar to viral transcriptional enhancer elements (see chapters 28 and 34), these cellular sequences stimulate transcription from either the homologous V_H gene segment promoter or a heterologous SV40 promoter. They are active when placed on either the 5′ or the 3′ side of the rearranged V_H gene segment and they function when their orientation is reversed. However, unlike viral enhancers, the immunoglobulin gene enhancer appears to act in a tissue-specific manner, since it is active in mouse B cells but not in mouse fibroblasts.

The discovery of this enhancer suggests another benefit that results from splicing; it brings an enhancer in close proximity to the promoter just when that promoter activity becomes an important part of the cell's function.

Interaction of B Cells and T Cells Is Required for Antibody Formation

Thus far, our discussion of antibody formation has centered on those reactions that occur at the gene level and on the structure of the antibody. However, since the immune response *in vivo* involves reactions between whole cells, we must also consider what happens at the level of cell-to-cell interactions. Some of the pluripotent stem cells of the bone marrow replicate in an undifferentiated state, maintaining their pluripotency, while others replicate and differentiate at the same time. The latter cells can differentiate along various pathways (fig. 33.9). Some of those that become committed to the lymphocyte pathway differentiate further into B-cell and T-cell lineages. T cells mature in the thymus, whereas B cells mature in the marrow in most vertebrates. Subsequently both cell types relocate in secondary lymphatic organs, the spleen and the lymph nodes, and also more diffusely in all tissues. Here they remain dormant until they encounter specific antigens that trigger them to complete their differentiation into fully mature effector B cells and effector and regulator T cells. Each B cell contains immunoglobulins of a specific idiotype bound to its surface, usually of the IgM type and to a lesser extent of the IgD and IgG types.

Figure 33.9

Pluripotent stem cells in the marrow give rise to more stem cells and to various progenitors. The lymphocyte progenitors give rise to B-cell and T-cell lineages. The final differentiation step in both B-cell and T-cell development requires antigenic stimulation.

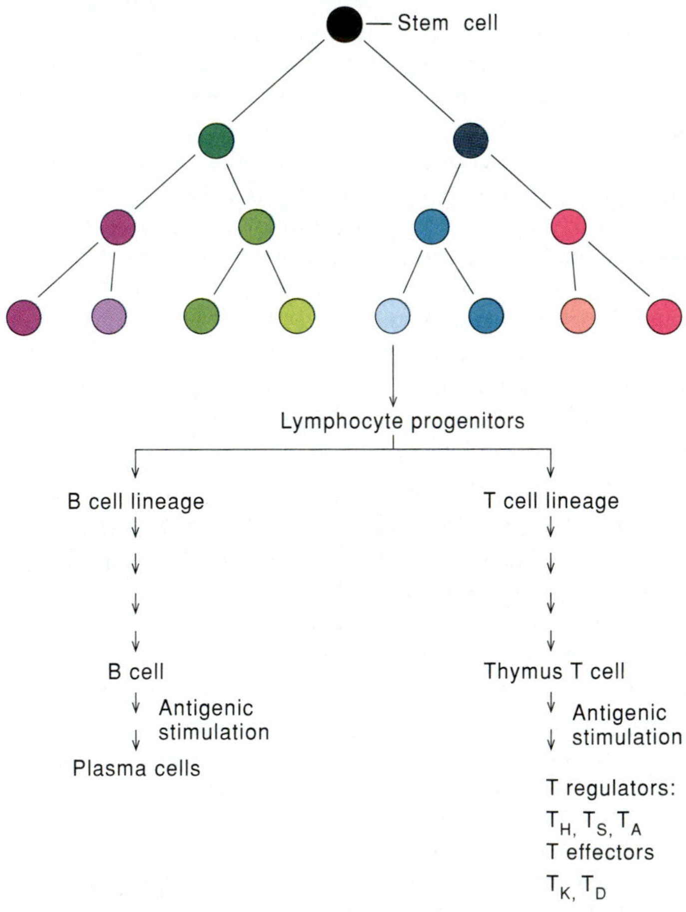

Antigens Stimulate the Formation of B-cell Clones The first time a foreign antigen is injected into the bloodstream, there is a long delay before specific antibody appears, and the response is fairly weak. The second time the same antigen is injected, the response is both more rapid and more intense (fig. 33.10). A

Figure 33.10

Kinetics of IgM and IgG appearance in the serum following a first and a second exposure to the same antigen. Upon first exposure there is a delay of several days before any antibody appears in the serum. IgM appears before IgG. After many days the serum level falls. If the system is exposed to a second equivalent dose of antigen, the appearance of IgM and IgG is much faster and the maximum level of IgG is much higher.

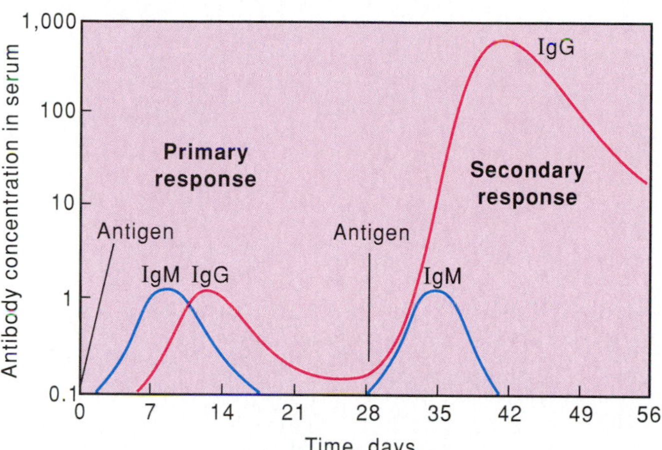

Figure 33.11

An immature B cell (*a*) and a fully developed antibody-producing B cell (*b*). (Electron micrographs courtesy of D. Zucker-Franklin, New York University Medical Center.)

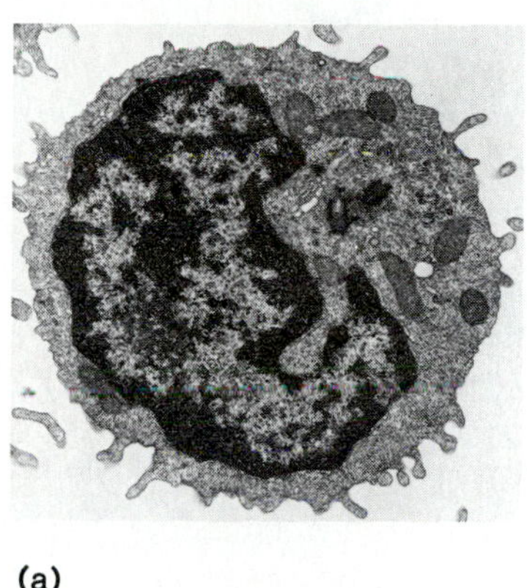

(a)

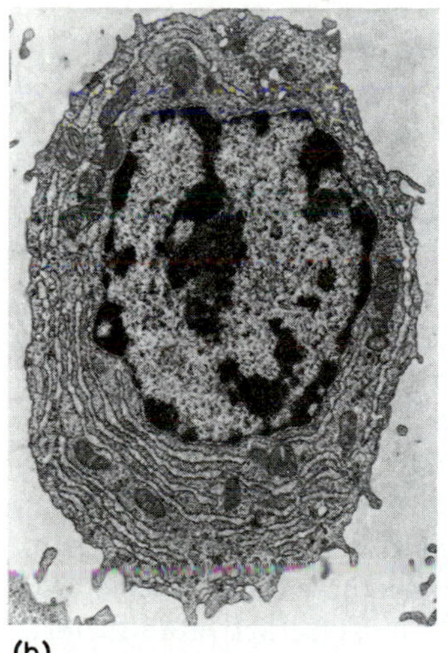

(b)

great deal of research effort has gone into seeking an explanation for this difference between primary and secondary responses. Specific labeling studies show that secondary lymphatic organs contain a very heterogeneous mixture of B cells, carrying membrane-bound antibodies of different idiotypes. When sufficient amounts of antigen bind to the surface antibody, the B cell becomes triggered to divide and make more cells of the same type. Some of these continue to divide, making clones of memory cells in the lymphatic organs, while others terminally differentiate into plasma cells richly laced with endoplasmic reticulum that is geared to the production of the antibody of the same idiotype as that which is membrane-bound (fig. 33.11). Whereas the idiotype and consequently the V_L and V_H parts of the antibody are the same as in the original, unstimulated B cell, the *C* regions change. The first secreted antibody is of the secretory IgM type. This is gradually replaced by the secretory IgG type, which dominates in the secondary response. As we have noted (see fig. 33.7), the replacement of IgM by IgG is indicative of an additional DNA splicing.

Helper T Cells Trigger B Cell Division and Differentiation

The importance of T cells in mediating the B-cell response to most antigens is demonstrated by the fact that thymectomized animals, which are depleted in T cells, usually show a greatly attenuated reaction to foreign antigen. A normal response can be restored by transplantation of thymus tissue from a genetically identical animal. Activated T helper cells interact with B cells whose antibodies can recognize the same antigen that stimulated the helper T cell. The complex between the T and B cell results in the clustering of a large number of transmembrane proteins causing a process known as CAP formation. CAP formation by some unknown means triggers the proliferation and differentiation of the B cells involved.

Figure 33.12

Stimulation of antibody-forming cells involves three types of cells. Typically antigen is phagocytosed by a macrophage. The phagocyte partially digests the antigen and "presents" the processed antigen on its outer plasma membrane. A specific helper T cell binds the antigen to a specific receptor bound to its outer plasma membrane. The T cell usually has many of the same types of receptors and binds many copies of the antigen in a similar way. This stimulates T-cell proliferation. These T cells interact with B cells that display similar processed antigen. Several B-cell receptors involved in a similar way become focused in a local region of the outer plasma membrane, a process known as CAP formation. CAP formation triggers the proliferation and differentiation of the B cells involved.

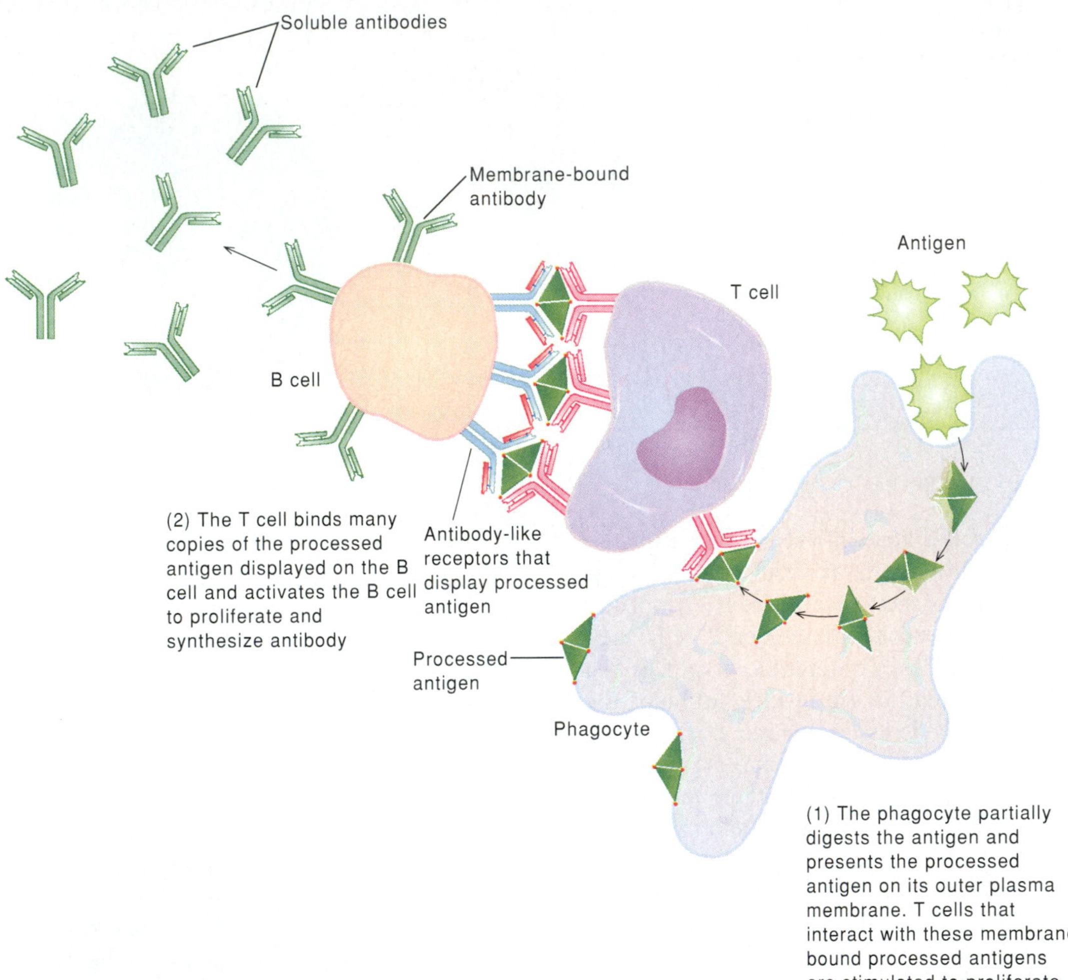

Soluble antibodies
Membrane-bound antibody
Antigen
T cell
B cell
(2) The T cell binds many copies of the processed antigen displayed on the B cell and activates the B cell to proliferate and synthesize antibody
Antibody-like receptors that display processed antigen
Processed antigen
Phagocyte
(1) The phagocyte partially digests the antigen and presents the processed antigen on its outer plasma membrane. T cells that interact with these membrane-bound processed antigens are stimulated to proliferate

Support for the role of T helper cells in focusing the antigen comes from the observation that T-independent stimulation of B cells is frequently elicited by polymeric antigens. Such polymeric antigens constitute another way of presenting a large amount of antigen at one location on the B-cell membrane and thereby focusing the B-cell receptors.

Before they can help other lymphocytes respond to antigen, helper T cells must be activated themselves. This activation occurs when a helper T cell recognizes a foreign antigen bound on the surface of a specialized antigen-presenting cell. The latter cells are found in most tissues; they are derived from bone marrow and comprise a heterogeneous group, including dendritic cells in lymphoid organs and certain types of macrophages. Together with B cells, which can present antigen to helper T cells, these are the main cell types that react with helper T cells. The sequence of events involving antigen-processing cell, T helper cell, and antibody-forming B cell is depicted in figure 33.12.

Note that the mechanism proposed for helper T-cell function requires that T cells have surface receptors that will recognize antigen-processing cells, antigen itself, and the appropriate B cell. We will shortly discuss the nature of T-cell receptors, as well as the surface structures they recognize, in a broader context.

The Complement System Facilitates Removal of Microorganisms and Antigen-Antibody Complexes

Many antigen-antibody complexes are eliminated by phagocytosis. Others are attacked by complement, a group of serum proteins that aids in the defense against microorganisms. Individuals with a deficiency in their complement system are subject to repeated bacterial infections, just as are individuals deficient in antibodies themselves.

Figure 33.13

The principal stages in complement activation. Complement activation occurs exclusively on the microbial cell membrane, where it is triggered by bound antibody or microbial envelope polysaccharides, both of which activate early complement components. There are two sets of early components belonging to two distinct pathways of complement activation. Activation of each complement system involves a cascade of proteolytic reactions. Each component of the complement system is a proenzyme that is activated by the preceding component of the chain by a limited proteolytic cleavage. The ultimate result of this chain reaction is the development of a complex that attacks the cell membrane.

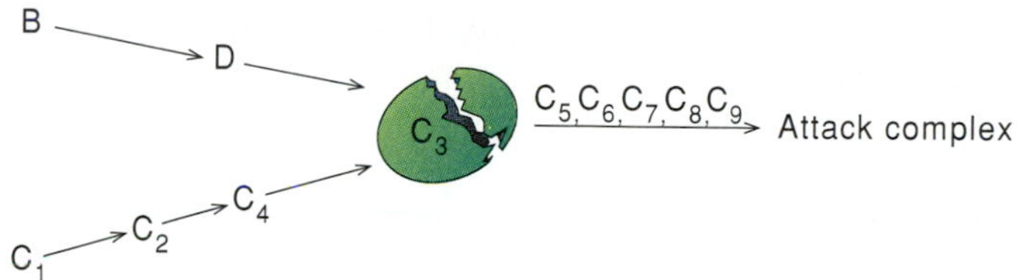

There are about twenty different proteins in the complement system—proteins C1–C9, factors B and D, and a series of regulatory proteins. All these proteins are made in the liver, and they circulate freely in the blood and extracellular fluid. Activation of the complement system involves a cascade of proteolytic reactions. In addition to forming membrane attack complexes, the proteolytic fragments released during the activation process promote dilation of blood vessels and the accumulation of phagocytes at the site of infection.

The activation of complement occurs exclusively on the microbial cell membrane, where it is triggered either by bound antibody or microbial envelope polysaccharides, both of which activate early complement components. There are two sets of early components belonging to two distinct pathways of complement activation: C1, C2, and C4 belong to the pathway that is triggered by antibody binding; factors B and D belong to the alternative pathway that is triggered by microbial polysaccharides (fig. 33.13). The early components of both pathways ultimately act on C3. The early components and C3 are proenzymes that are activated sequentially by limited proteolytic cleavage. As each proenzyme in the sequence is cleaved, it is activated to generate a serine protease, which cleaves the next proenzyme in the sequence. Many of these cleavages liberate small peptide fragments and expose a membrane-binding site on the larger fragment. The larger fragment binds tightly to the target cell membrane by its newly exposed membrane-binding site, and helps to carry out the next reaction in the sequence. In this way complement activation is confined largely to the cell surface, where it began.

The components of the complement attack complex have a very short half-life if released from the complex. This limitation confines their destructive action to the point of assembly, the surface membrane of the microorganism.

The Cell-Mediated Response: A Separate Response by T Cells

Most B cells are fixed in the secondary lymphatic organs, the spleen and the lymph nodes. They are stimulated by antigen only if the antigen is transported to the lymph nodes. In many cases, however, antigens are poorly transported. To rid the organism of antigens that are not readily transported, a second type of immune system exists that is independent of the B cells. This second type of immunity is not found in all vertebrates; it is restricted to mammals, birds, certain amphibians, and bony fishes. Long before the distinction between T and B cells was appreciated, it was known that certain types of immunity, which manifest a delayed-type hypersensitivity response, could be transferred from immunized animals to nonimmunized animals with leukocytes, but not with antibody-containing serum. Subsequently other forms of leukocyte-mediated immunity were discovered that involved increased phagocytic and bacteriocidal activity of macrophages, lysis of virus-infected cells and tumor cells, and the rejection of skin grafts from genetically dissimilar organisms.

There are two types of cell-mediated immunity, involving basically different types of T cells. In the delayed-type hypersensitivity response, the T cell, T_D, that reacts specifically with antigens secretes lymphokines (box 33A) that attract and activate macrophages or other leukocytes, thereby causing a slowly developing inflammatory response. A second type of T cell, known as the T killer cell, T_K, reacts specifically with antigen that is bound to target cells, causing their lysis.

Tolerance Prevents the Immune System from Attacking Self-Antigens

A thorough understanding of the immune system requires an explanation for how the immune system is able to discriminate between foreign antigens and its own antigens (self-antigens). The ability to recognize self-antigens and thus to avoid making an antagonistic response to them is called tolerance. Some serious diseases are believed to be due to a breakdown in the self-tolerance mechanism. Conversely, there are conditions under which a foreign antigen is able to establish itself so that it becomes tolerated, for either a short or a long time. Tolerance and rejection are so closely related that it seems likely that a full understanding of one will entail a full understanding of the other.

Several conditions favor the establishment of tolerance: (1) Tolerance is much easier to establish in the fetus or the newborn than in the adult. Thus if an organism is exposed as a fetus to a soluble foreign antigen or a foreign tissue graft, it may become indefinitely tolerant to the antigen. This situation is illustrated for two genetically nonidentical strains of mice in figure 33.14. Cells from the Y mouse are injected into a newborn X mouse. The same X mouse as an adult is the recipient of a skin graft (transplant) from a Y-type mouse. Normally such a graft would be rapidly rejected. However, because of the early exposure to the Y cells, the graft is tolerated indefinitely. This

33A
BOX

The Role of Lymphokines (Interleukins)

The interactions between immune and inflammatory cells are mediated in large part by certain proteins, called lymphokines or interleukins (IL), that are able to promote cell growth, differentiation, and functional activation. Interleukins resemble hormones, which also function as intercellular messengers. In contrast to hormones, however, they are secreted by isolated cells rather than discrete glands. Several interleukins have been described; each has unique biological activities as well as some that overlap with the others. For example, macrophages produce IL1 and IL6, whereas T cells produce IL2–IL6 and bone marrow stromal cells produce IL7. IL1 and IL6 not only play important roles in immune cell function, they also stimulate a spectrum of inflammatory cell types and induce fever. IL2 is a potent proliferative signal for T cells. IL1, IL3, IL4, and IL7 enhance the development of a variety of hematopoietic precursors. IL4–IL6 also serve to enhance B-cell proliferation and antibody production.

Figure 33.14

Establishment of tolerance to tissue grafts. If a newborn X mouse is injected with Y cells when it is young, it is tolerant to Y tissue transplants when it becomes an adult. If the X mouse is not exposed to Y cells at an early age, it will readily reject the tissue graft.

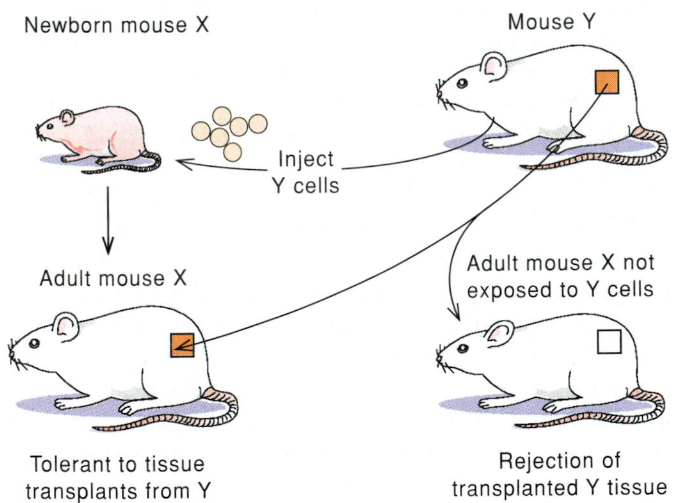

result indicates that the mouse that was exposed to the Y cells just after birth recognizes Y cells later in life as self-antigens. (2) Very high levels of antigen frequently favor the development of tolerance, whereas intermediate levels of antigen favor the development of an immune response. (3) Frequently the route of antigen administration is critical in the development of tolerance. An intravenously injected antigen is more likely to promote a tolerant condition than an antigen injected into the tissue.

Two cellular mechanisms may account for most forms of tolerance. In one of these the clones of T or B cells that could otherwise be activated by the antigen are destroyed or inactivated. In the other mechanism, the clones in question are still present but do not respond because they are blocked by T_S cells. In general, B cells are more difficult to make tolerant than T cells. This difference is evident from observations that longer and higher doses of antigen are required to establish tolerance to the B-dependent system. Moreover, the period of tolerance is generally shorter when B cells are involved.

Adult mice may be made tolerant to foreign antigens by destroying their immune system with whole-body irradiation and then supplying them with transplants of bone marrow and thymus from a foreign donor, thus refurbishing their system with competent B and T cells, respectively. With the new immune system the animal is now tolerant to any tissue from a mouse that is isogenic with the mouse used to supply the marrow and the thymus.

The irradiated host organism also supplies a useful means of testing the state of B and T cells from a nonirradiated isogenic mouse. B cells may be effectively transferred by bone marrow transplants, and T cells may be transferred by thymus tissue transplants. The donor cells may come from a normal, untreated isogenic strain. In this event the irradiated recipient will show normal immunologic behavior. Alternatively, the donor cells may come from a donor that has been made either sensitive or tolerant to a particular antigen. By doing varous combinations of experiments using different isogenic donors, it is possible to determine the relative importance of B-cell and T-cell involvement with different antigens.

T Cells Recognize a Combination of Self and Nonself

All T cells, whether they fight infection directly or regulate the activity of other effector cells, recognize antigen only in combination with a cell membrane surface. Moreover, the antigen must be associated with a cell from an organism that is genetically similar or identical to that of the T cell. This requirement was first demonstrated by an experiment in which T cells were removed from a mouse that had been immunized with a virus. Immunization meant that the extracted T cells had been activated to direct themselves against cells infected with this

Figure 33.15

Activated T cells will bind and kill only those cells that display a foreign antigen and a familiar cell-surface antigen.

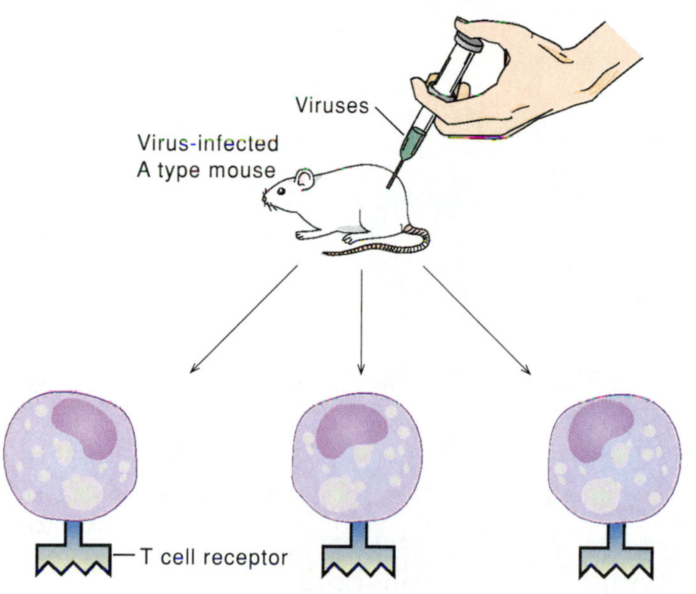

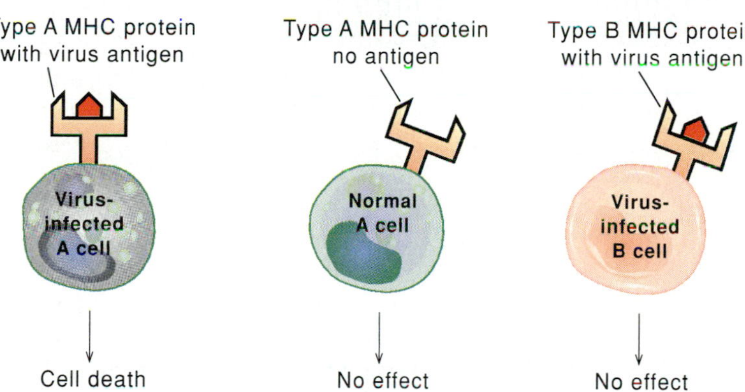

particular virus. Thus these T cells could kill fibroblasts infected with the virus in tissue culture. However, they could not kill fibroblasts infected with the virus if the fibroblasts came from a mouse of a different genetic background. Thus killer T cells do not recognize just any foreign antigens on any cell membrane surface; they recognize foreign antigens only in combination with determinants present on their host's own cells (fig. 33.15). These common determinants are a subset of the proteins encoded by the so-called major histocompatibility complex (MHC).

MHC Molecules Account for Graft Rejection

MHC molecules were recognized long before their major biological function was appreciated. They were initially defined as the main target antigens in transplantation reactions, and because of this they became known as major histocompatibility antigens. Experiments on mice demonstrated that graft rejection is an immune response to the surface antigens of the grafted cells. Subsequently it was shown that these reactions are mediated by T cells and that they are directed against a family of glycoproteins encoded by a group of genes called the major histocompatibility complex (MHC). MHC molecules are transmembrane proteins found on the cell membranes of all higher vertebrates.

A vertebrate does not normally need to be protected against invasion by foreign vertebrate cells, so the antagonistic reaction of its T cells to foreign MHC molecules was both an obstacle to transplant operations and an enigma to immunologists. The enigma was solved when it was discovered that MHC molecules contributed to the binding of T lymphocytes on those host cells that have foreign antigen on their surface, as in the case of the virus-infected cells we have just described (see fig. 33.15).

There Are Two Major Types of MHC Proteins, Class I and Class II

There are two major classes of MHC proteins which have similar types of structures (fig. 33.16). Class I molecules contain three external domains, each about 90 residues in length, a transmembrane region, and a cytoplasmic domain. The third external domain is noncovalently associated with a small polypeptide known as the β_2 microglobulin. Class II molecules are composed of two noncovalently associated polypeptide chains, α and β, whose overall structure resembles that of the class I complex.

There are several genes that encode each type of MHC protein, and there are several alleles that represent different MHC proteins for each of these genes. This gives rise to a tremendous variety of MHC proteins, with the consequence that it is extremely unlikely that any two members of the same species will possess the same assortment of MHC proteins. Thus two humans are not likely to carry the same MHC proteins on their cell membranes unless they are genetically identical twins. This is the reason why transplants between two individuals are almost always rejected—it is usually just a matter of time. Highly inbred strains of mice, which are genetically identical, will display the same MHC proteins and accordingly will take grafts from each other without any rejection.

The two classes of MHC proteins are displayed on different cell types. Class I MHC proteins are found on almost all nucleated cells including killer T cells. Class II MHC proteins are found mainly on cells involved in the immune response including antigen-presenting cells, B cells and T helper cells, but not T killer cells.

T-Cell Receptors (TCRs) Resemble Membrane-Bound Antibodies

The specific receptors found on the membranes of T cells resemble the highly specific antibodies found on the membranes of B cells. Specific T-cell receptors (TCRs) are composed of two disulfide-linked peptide chains (called α and β). Each of these chains share with antibodies the distinctive property of a variable amino-terminal region and a constant carboxyl-terminal region.

Figure 33.16

Structures of class I and class II molecules as they would appear in the lipid bilayers of the cell membrane. Models show striking similarities between the two types of molecules. S—S indicates disulfide bridges in the class I and class II molecules and the β_2 microglobulin.

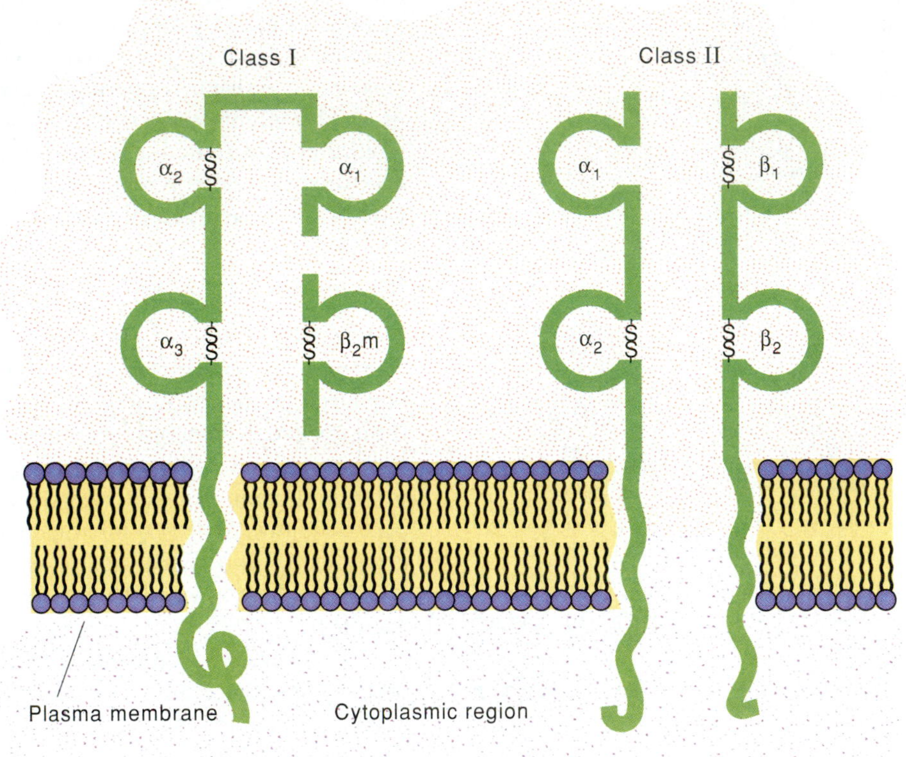

With one exception, all the mechanisms used by B cells to generate antibody diversity are also used by T cells to generate T-cell receptor diversity. The one mechanism that does not appear to operate in T-cell receptor diversification is somatic hypermutation. This is presumably because mutation would be likely to generate killer T cells that would wantonly attack self-molecules. This is much less of a problem for B cells, since most self-reactive B cells could not be activated without the aid of self-reactive helper T cells.

Additional Cell Adhesion Proteins Are Required to Mediate the Immune Response

In addition to membrane-bound antibodies, TCR proteins and MHC proteins, there are other transmembrane proteins required to obtain the specific cell–cell interactions necessary to mediate the immune response. These additional proteins are also important in a number of nonimmune reactions where cell–cell adhesion is important to the organism. We will only consider a limited number of these additional proteins because of the enormous complexity of this subject. Three transmembrane proteins that are directly involved in the T-cell reaction with other cells are members of a family of cell–cell adhesion proteins (the CD family of proteins). The CD3 protein is found on both killer and helper T cells; it forms a complex with the membrane-bound TCR proteins. The CD8 and CD4 proteins interact with this complex but unlike the CD3 proteins, CD8 is only found on killer T cells, while CD4 is only found on helper T cells. The presence of CD4 and CD8 proteins limits the range of binary complexes formed by T cells with other cells. This is because CD4 and CD8 have selective affinities for MHC class II and MHC class I proteins respectively. As a result killer T cells only form complexes with cells that display a processed foreign antigen in a complex with the MHC I protein; this includes most nucleated cells other than those that belong to the immune system. Similarly helper T cells are limited to reactions with cells, which display the appropriate processed antigen in a complex with the MHC II protein; this includes B cells or phagocytes (fig. 33.17).

Figure 33.17

Antigen-specific reactions involving T cells. (*a*) Killer T cells interact and destroy target cells that display complexes of processed antigen with an MHC I protein. Effective interaction requires that the TCR protein interact specifically with the processed antigen. All of the remaining reactions between the two cells are general reactions that occur between other killer T cells and other target cells. (*b*) Helper T cells interact specifically with antigen-presenting B cells, provided a specific complex is made between the TCR protein and the processed antigen complexed to the MHC II protein and the processed antigen complexed to the MHC II protein on the B cell. The unlabeled membrane proteins in this figure represent other nonspecific protein–protein interactions that strengthen the binary cell reaction. Processed antigen is indicated in red.

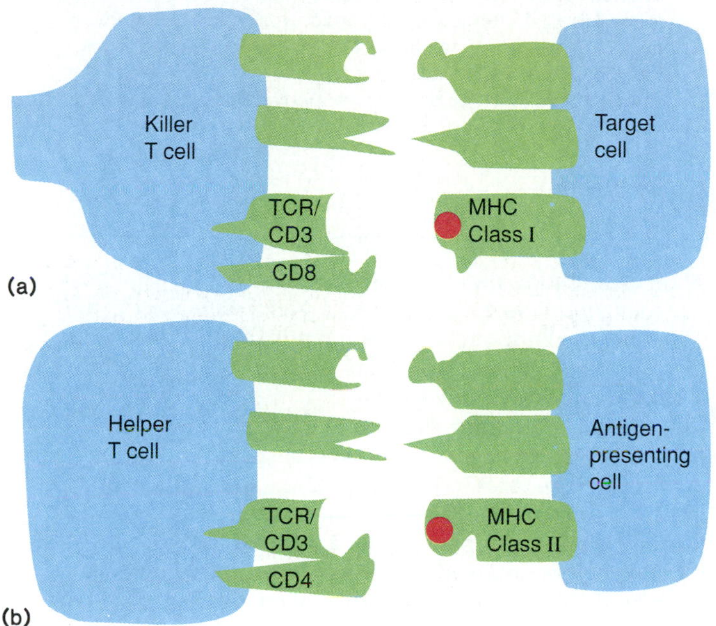

From this discussion we can see that a productive interaction between T cells and other cells requires a highly specific reaction between a TCR protein and a processed antigen that is supplemented by less specific interactions between other transmembrane proteins on the two cell types.

Left out of this discussion is an explanation for the source of the processed antigen on the antigen-presenting B cell. Immature B cells display a membane-bound antibody, which will interact with circulating complementary antigen. Once bound to the antibody the antigen is internalized, processed and ultimately displayed by the MHC II protein on the cell surface of the same B cell. It is the interaction between such a B cell and the appropriate helper T cell that triggers the B cell to divide and differentiate.

Immune Recognition Molecules Are Evolutionarily Related

There is a striking structural similarity between immune recognition proteins that almost certainly reflects their evolution from a common ancestor. All of them contain one or more immunoglobulinlike domains, each with about 90 amino acids stabilized by a conserved disulfide linkage. The simplest member of this family of genes is β_2 microglobulin. Possibly this or a similar protein was the ancestor for the remaining immune recognition proteins.

Summary

In this chapter we have considered the major findings that have led to our understanding of the immune system in vertebrates, expecially mice and humans. The following points are the highlights of our discussion.

1. There are two different classes of white cells (lymphocytes) associated with the immune response, the B cells and the T cells. B cells produce antibodies that are secreted and aggregate foreign substances known as antigens. T cells are of different types; some of them assist B cells to become antibody-forming cells, while others mount an attack on antigens of their own.
2. The most common type of antibody is a tetramer composed of two identical (H) chains and two identical light (L) chains. Each chain is divided into a relatively constant sequence (the *C* segments) and a relatively variable sequence (the *V* segments). Within the variable regions there are subsegments known as the hypervariable regions. These subsegments are the regions that specifically bind different antigens.
3. According to some estimates, each organism is capable of making more than a million antibodies from a limited amount of genetic information. This feat is accomplished with the help of somatic recombination and somatic mutation. Somatic recombination involves splicing at the DNA level. It brings specific *C* and *V* regions together to make a unit that can be transcribed. Somatic mutation is focused on those parts of the gene that encode the hypervariable regions.
4. T helper cells focus the antigen on an immature B cell. In response, some B cells turn into antibody-forming cells and some B cells turn into memory cells for production of antibodies at a later date. The first time an organism is exposed to a specific antigen it takes longer to mount an immunological response. That is because of the lack of memory cells for making a specific antibody.
5. Complement is a group of serum proteins that aids in the defense against microorganisms and removal of antibody-antigen complexes.
6. T cells give rise to an immune response of their own. There are two types of cell-mediated immunity, involving basically different types of T cells. In the delayed-type response, the T cell reacts specifically with antigens and secretes lymphokines. These are substances that attract macrophages or other leukocytes, thus producing a slowly developing inflammatory response. A second type of T cell reacts specifically with antigen bound to target cells and causes their lysis.
7. Tolerance prevents the immune system from attacking self-antigens. To understand tolerance we must appreciate how T cells work. T cells recognize a combination of self and nonself. The cell-surface antigens that are recognized by T cells are known as the major histocompatibility complex (MHC). If two organisms carry the same histocompatibility antigens, the tissues from the two organisms are completely compatible. The histocompatibility antigens resemble antibodies in structure.

Selected Readings

Ada, G. L., and G. Nossal, The clonal selection theory. *Sci. Am.* 257(2):62–69, 1987.

Honjo, T., and S. Habu, Origin of immune diversity: Genetic variation and selection. *Ann. Rev. Biochem.* 54:803–830, 1985.

Hood, L., M. Kronenberg, and T. Hunkapiller, T cell antigen receptors and the immunoglobulin super gene family. *Cell* 40:225–229, 1985.

Hunkapiller, T., and L. Hood, The growing immunoglobulin gene superfamily. *Nature* 323:15–17, 1986.

Kaappes, D., and J. L. Strominger, Human class II major histocompatibility genes and proteins. *Ann. Rev. Biochem.* 57: 991–1028, 1988.

Klein, J., *Immunology.* Blackwell Scientific Publications, 1990.

Leder, P., The genetics of antibody diversity. *Sci. Am.* 246(5):102–115, 1982.

Lieber, M. R., Site-specific recombination in the immune system. *FASEB J.* 5:2934–2944, 1991.

Milstein, C., Monoclonal antibodies. *Sci. Am.* 243(4):66–74, 1980. An exciting, tissue-culture technique for obtaining pure antibodies specific for an antigen of choice.

Mizel, S. B., The interleukins. *FASEB J.* 3:2379–2388, 1989.

Nowak, M. A., R. M. Anderson, A. R. McLean, T. F. W. Wolks, J. Goudsmit, and R. M. May, Antigenic diversity thresholds and the development of AIDS. *Science* 254:963–969, 1991.

Singer, S. J., Intercellular communication and cell–cell adhesion. *Science* 255:1671–1677, 1992. A comprehensive review of the major cell–cell interactions important in immune reactions.

Springer, T. A., Adhesion receptors of the immune system. *Nature* 346:425–434, 1990. An excellent review.

Tonegawa, S., Somatic generation of antibody diversity. *Nature* 302: 801–803, 1983. A classic paper.

Problems

1. Where do T cells and B cells mature in the body? Where do the mature T and B cells reside in the body?
2. Why does humoral immunity require both B and T cells?
3. What are the two types of cell-mediated immunity?
4. Explain the role of DNA recombination in generating antibody diversity. How does somatic mutation contribute to diversity?
5. How can DNA splicing help stimulate a higher level of transcription of an immunoglobulin gene?
6. What role does RNA splicing play in antibody production?
7. You have two strains of genetically nonidentical mice (strain A and strain B). How could an adult mouse from strain A be made to tolerate a transplant from strain B?
8. How could all the immune recognition proteins have evolved from a single common ancestor protein?
9. If an antibody and its antigen are mixed in a test tube at some optimum concentration, a precipitate is formed. If either the antibody or the antigen is in great excess, no precipitate is formed. Explain these results from what you know about antibody structure.
10. Recently antibodies with catalytic activity have been produced. How would you develop a catalytic antibody for a specific reaction? (Hint: How does an enzyme work?)
11. If you inject a pure protein into a rabbit and raise antibodies to a high titer, the antibodies produced will be a mixed population with various binding constants. Why would a single protein produce many different antibodies?
12. If you added a thiol compound such as 2-mercaptoethanol at a high concentration (1 mM) to an antibody and antigen mixture, no precipitation complex would form even if the concentration of antibody and antigen were optimal. Explain this observation. (Hint: How do thiol compounds work?)

34

CHAPTER

Carcinogenesis and Oncogenes

In the complex developmental process that leads to a mature human being from a fertilized egg, each cell takes up a role in a specialized organ or tissue. Each group of cells proliferates to a certain point and then stops. In the case of a solid organ such as the liver, we see cells of several different types that grow until the organ takes on a fixed size and shape and then cease to grow. Normally, further growth will occur only at the very slow rate required for replacement of aging cells. However, if part of the liver is amputated, a rapid regenerative process ensues so that the liver regrows to approximately the same size it had before the amputation. Other organs or tissues may show more or less growth than the liver, but as a rule, new cells will appear only as old or damaged cells are replaced. Furthermore, the cells of normal tissues do not as a rule mix.

Normally you do not find liver cells mixing with intestinal cells or any other types of cells, suggesting that cells have surface properties that favor interaction of like cells. This was demonstrated in the 1940s by experiments with embryonic tissues. If embryonic tissues are dissociated into single cells and allowed to reassociate in the presence of other cell types, they reassociate with cells of a similar type. It is now known that there are adhesion molecules located on cell surfaces that encourage such associations. This was alluded to in chapter 33 when we were describing the interaction of T cells with other cell types of the immune system. Cell adhesion molecules that favor the association of like cells are of at least two types: some require Ca^{2+} for adhesion, the cadherins, and others do not, the cell adhesion molecules (CAMs). When these molecules are not present, as in culture transformed cell lines, like cells adhere poorly to each other or to other cultured cells. It seems likely that, when tumor cells become invasive, the character of their cell adhesion molecules has been altered. We mention this in passing here as it is probably an important property of malignant cells, but we will have little more to say about it in this chapter as this aspect of malignancy is not well characterized.

Cancer cells are relatively easy to recognize when they occur in groups in the organism (fig. 34.1). A cluster of cancer cells is more rapidly dividing, and it usually clashes with the architecture of normal cells in its immediate surroundings. From cytological inspection alone, we might speculate that cancer cells are breaking the rules for normal cell growth and associations. It appears that the processes that regulate cells are disrupted; cell growth is uncontrolled, and patches of morphologically distinct cells show by their invasive character that they do not recognize normal territorial boundaries between cells of different types. Cancer cells are deadly because ultimately they invade and disrupt a vital area. In this chapter we present a brief overview of some aspects of carcinogenesis and oncogenes.

Cancers: Cells Out of Control

Cancer is not a stage in the orderly evolution of a complex organism; it is an organismic catastrophe. Perhaps it is only reasonable to expect that multicellular eukaryotes, with their many dividing cells, will occasionally produce a mutant cell that gets out of control and destroys the system. The notion that cancer cells are out of control was proposed more than half a century ago. In the intervening time we have learned a great deal about the biochemistry of growth regulation. This knowledge has helped us to understand the molecular basis of cancer in many cases. Conversely, the study of cancer has added to our knowledge of normal growth regulation.

Transformed Cells Are Closely Related to Cancer Cells

In the 1950s, systematic methods were developed for taking cells directly from the tissues of whole animals and inducing them to grow as single cells in liquid culture.

An outgrowth of this technology was the development of pure cell lines—cells that divide indefinitely. Primary cultures of dispersed single cells from tissue fragments usually die out after 50 or so duplications. However, during the growth of a primary culture, occasional altered cells appear that possess a somewhat different morphology; they frequently take over the culture, because they are capable of initiating new cultures with far fewer cells. A clone that is derived from one of these indefinitely dividing variants—which has in effect become immortal—is designated a cell line. The transition of primary culture cells to cells that can grow indefinitely is considered to be a step in the direction of becoming cancerous.

Typically, continuous cell lines will grow in culture only on a solid support or "anchor" (the surface of a Petri dish, for example) and in the presence of relatively high concentrations of nutrients. Even then, they will divide only as long as the culture is sparse. When the cell density increases beyond a critical point the growth rate decreases sharply; in fact, the cells of some continuous lines stop dividing altogether once they have formed a confluent monolayer.

At the density that normally halts growth, rare transformed cells continue to multiply and may reach cell densities up to twenty times higher than those of untransformed cells. By picking and subculturing such transformed cells, it is possible to establish clonal lines of transformants and to ask in what ways such cells differ from their untransformed progenitors.

The differences between transformed and untransformed cells can be classified in three ways: changes in cell growth, changes in cell-surface properties, and genetic alterations. These changes are too great in number and too diverse in quality to be caused directly by one or a few mutations. Clearly, then, most of the observed alterations must be secondary and tertiary effects that have occurred as a consequence of some primary event. It is the aim of much current work with tumor-causing agents, especially certain viruses, to discover the molecular nature of this primary event.

Figure 34.1

A benign glandular adenoma (*a*) and a malignant glandular adenocarcinoma (*b*). In both cases the tumor cells differ morphologically from the surrounding tissue and show a pattern of disorganized growth.

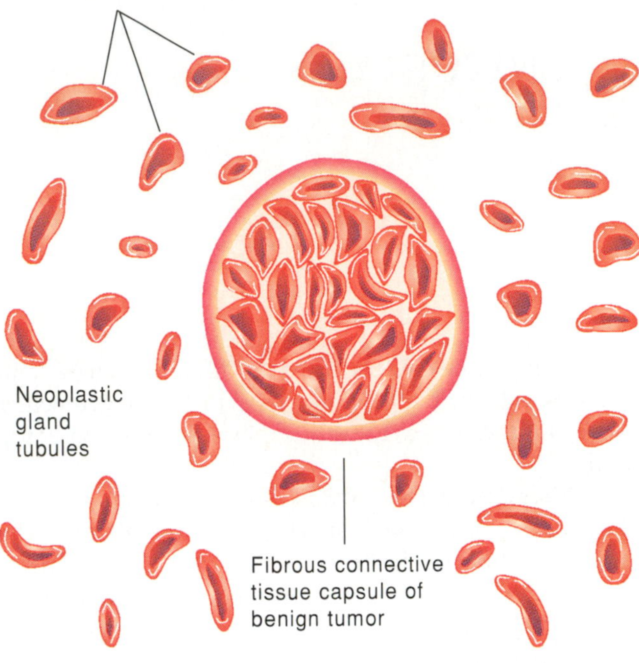

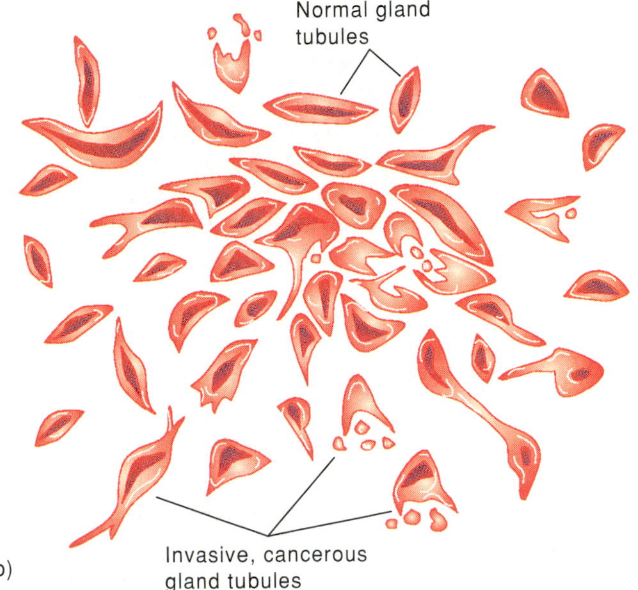

Figure 34.2

AIDS virion structure. Numbers associated with proteins are kilodalton masses (e.g., gp120 is a 120-kilodalton protein).

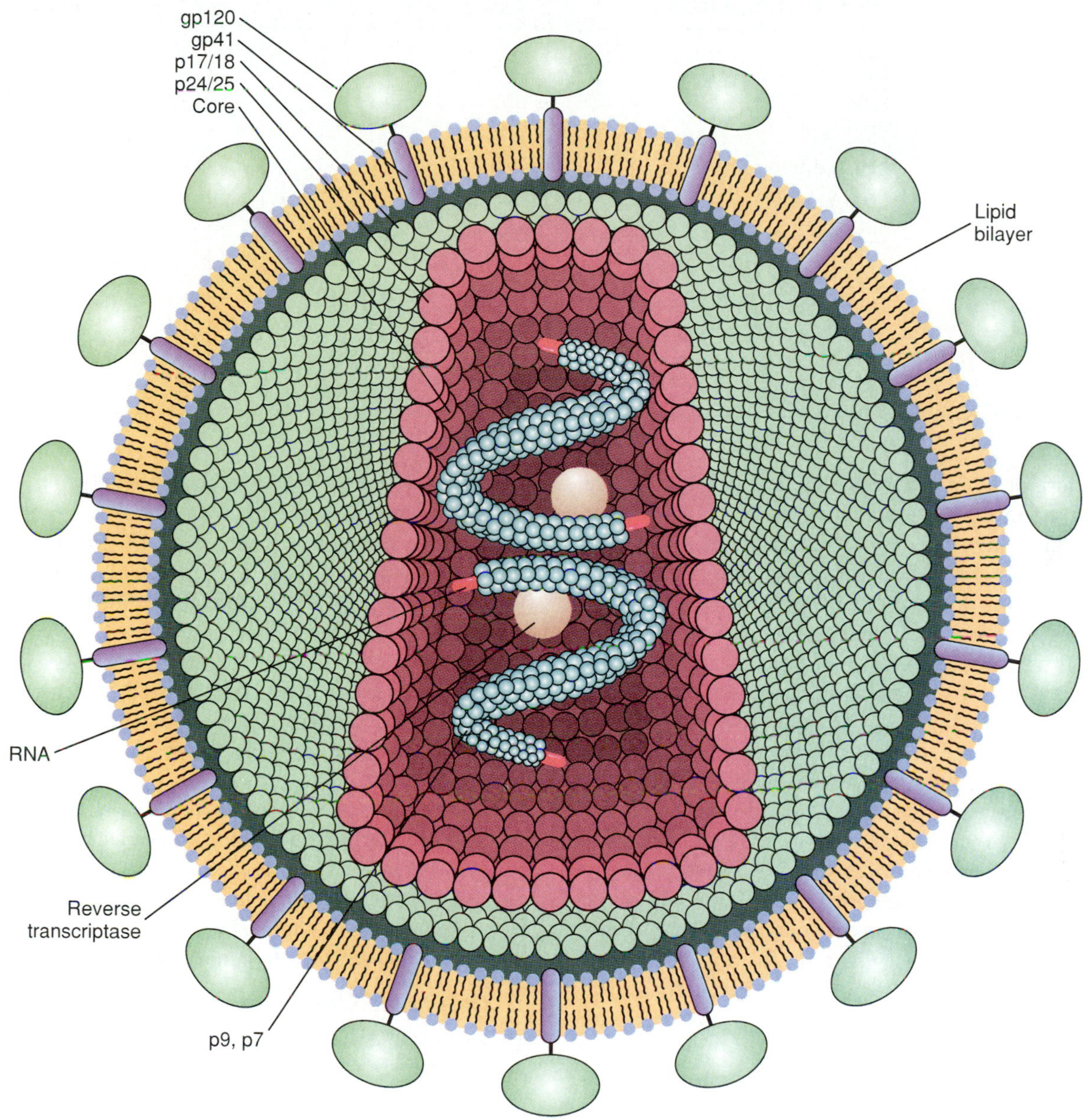

Transformed cells sometimes, but not always, give rise to tumors when injected into an isogenic animal. There may be numerous reasons for the failure to produce tumors on all occasions. One of the best-known reasons for lack of tumor production is immunological rejection. The transformed cell resulting from exposure to a tumor virus frequently displays viral antigens that are recognized as foreign to the animal host. This recognition excites a positive immune response that can lead to rejection of the transformed cells.

Just as transformed cells frequently do not develop into tumors when injected into whole animals, so is it that tumor cells do not always grow well when dispersed into tissue culture. Nevertheless, tumor cells usually adapt more readily to growth in tissue culture than normal cells taken from the whole animal.

Environmental Factors Influence the Incidence of Cancers

In seems very unlikely that we will find cures for many types of cancer in the foreseeable future. Therefore it would be a good idea to put considerable effort into cancer prevention, especially since there are reasons for believing that the vast majority of cancers are preventable. The frequency of different kinds of

Table 34.1
Variation in the Incidence of Cancers by Country

Type of Cancer	Country of High Incidence	Country of Low Incidence	Ratio (High/Low)
Skin	Australia	India	200
Esophagus	Iran	Nigeria	200
Lung	England	Nigeria	35
Prostate	United States	Japan	15
Liver	Mozambique	England	40
Breast	Canada	Israel	7
Rectum	Denmark	Nigeria	20
Colon	United States	Nigeria	10
Ovary	Denmark	Japan	6
Stomach	Japan	Uganda	25

cancer varies enormously in different human populations (table 34.1). For example, breast and colon cancer are much higher in the United States than in Japan, whereas stomach cancer is much higher in Japan. In these cases, diet is most likely the culprit. Liver cancer is highest in Third World countries. Once again, diet seems the most likely cause. Lung cancer is more prevalent in countries where smoking is popular, and indeed, it is well documented that heavy smokers are much more likely to get lung cancer than nonsmokers. Overall, it is clear that much of the variation in cancer incidence is environmental rather than genetic, since these differences tend to disappear from one generation to the next in migrant populations. For example, the high incidence of stomach cancer in Japan is not matched by a similar figure among Japanese-Americans in the United States.

There is a growing conviction that the carcinogenic agents in the environment are active as cancer-causing agents because they produce mutations. The hunt for carcinogens has been facilitated by Bruce Ames, who developed a test for carcinogens based on the mutagenic action of a compound on bacteria.

Ames tests have shown that many carcinogens originate from food and chemical pollutants. In addition to chemicals, ionizing radiations are carcinogens. For example, there are strong indications that the ionizing radiation we get from the sun is the major cause of skin cancer. Thus the incidence of skin cancer in the United States is much higher in the South than in the North. High dosages of radioactive materials or x-rays are also strongly correlated with the incidence of numerous forms of cancer.

In many cases, exposure to certain viruses is correlated with specific types of cancers. For example, liver cancer is 300 times more common in individuals who have previously been infected with hepatitis B virus. Although we can only speculate on the precise reasons for this, such numbers indicate that hepatitis resulting from hepatitus B virus infection is almost certainly the major cause of human liver cancer.

A cancer of the B lymphocytes known as Burkitt's lymphoma is strongly correlated with infection by the Epstein-Barr virus. Once again, we do not know the precise chain of events that lead to cancer. Hepatitis B and Epstein-Barr are DNA viruses. Certain RNA viruses have also been shown to be correlated with cancer. The best known is the human immunodeficiency virus (HIV-1, the AIDS virus; fig. 34.2). The AIDS virus is believed to cause cancer indirectly by incapacitating the helper T cells that are required for the humoral immune reaction (see chapter 33). Various types of cancer follow infection by the AIDS virus, a fact testifying to the key role that the immune system plays in protecting the body from cancer. The RNA virus known as human T-cell leukemia virus type I (HTLV-I) causes T-cell leukemia. It is believed that this virus carries a cancer-causing gene (an oncogene) whose continuous expression upsets the normal metabolism, causing infected cells to become transformed to cancer cells.

Cancerous Cells Are Genetically Abnormal

In the previous section we stated that there was a strong correlation between the mutagenic properties of a chemical and its potency as a carcinogen. This is consistent with other evidence that mutation often plays a causal role in carcinogenesis.

The most graphic illustrations of the correlation between genetic abnormality and cancer come from the frequent association of certain types of tumors with highly specific chromosomal translocations. Chromosomal translocations involve the movement of a segment of chromosome from one location to another. Frequently translocations occur between different chromosomes in a reciprocal fashion. This is the case for translocations between human chromosome 8 and the chromosomes that carry the antibody genes, 14, 2, and 22.

A reciprocal exchange of segments of chromosome from the ends of chromosomes 8 and 14 is frequently found in the genome of cancer cells of the Burkitt's lymphoma (fig. 34.3). Burkitt's lymphoma is a tumor of B lymphocytes; for unknown

Figure 34.3

Translocation associated with Burkitt's lymphoma. The protooncogene *c-myc,* on the end of chromosome 8q, is translocated to the end of chromosome 14q, adjacent to the immunoglobulin constant gene, *Ig-Cμ*.

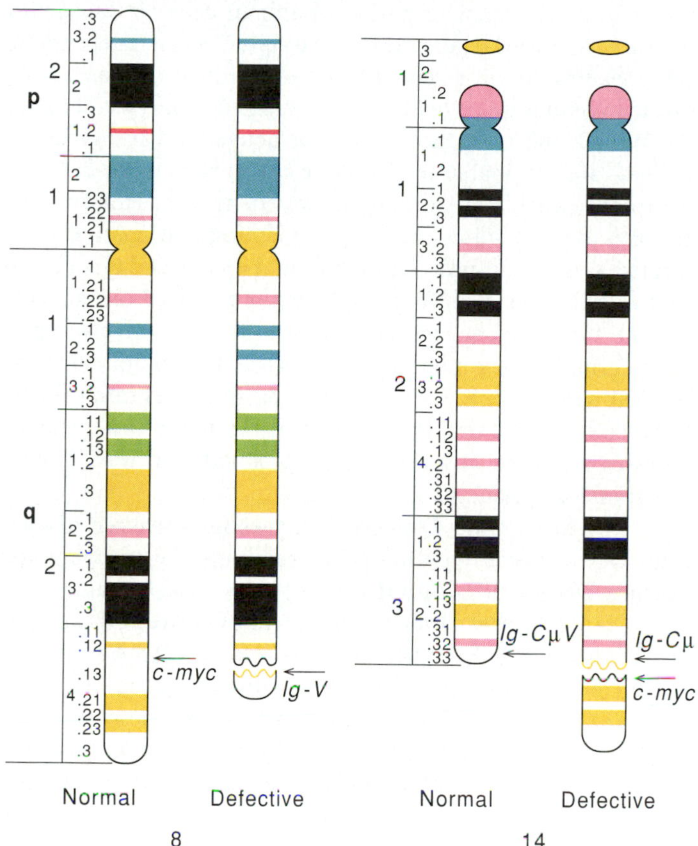

Figure 34.4

The Philadelphia chromosome arises by a reciprocal translocation between chromosomes 9 and 22. Note that the *abl* protooncogene is moved in the translocation process.

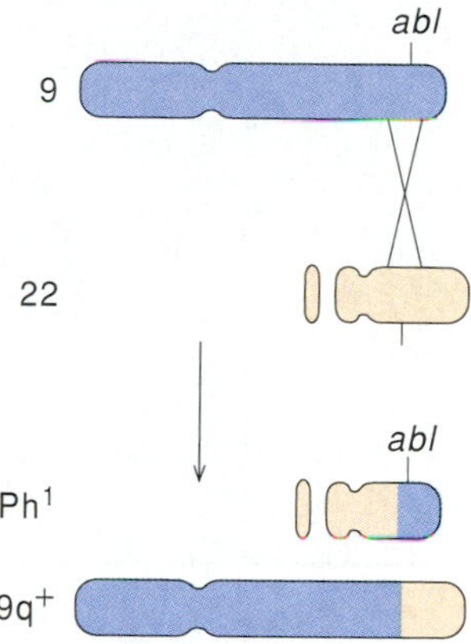

reasons, its occurrence is most common in Africa. The resulting tumor cells usually secrete antibodies. The fact that antibody genes, which are so active in B cells, map to the same chromosomal regions involved in the specific translocations in Burkitt's tumors led to the suggestion that a specific oncogene on chromosome 8 might be activated if it were placed near one of these antibody genes. This suggestion proved to be correct; the oncogene involved is *c-myc.* We will have more to say about *c-myc* later.

The first chromosome abnormality found to be associated with a cancer was the Philadelphia chromosome, which arises as the result of a reciprocal translocation between chromosomes 9 and 22 (fig. 34.4). The tumor associated with this translocation is chronic myelogenous leukemia and results from the activation of the *c-abl* oncogene, which is normally located on chromosome 9.

Whereas it is a common belief that tumor formation involves mutation, certain properties associated with the etiology of teratocarcinoma do not fit with this concept. Teratocarcinoma is a unique type of tumor in which the tumor is associated with a cluster of cells of different types including stem cells and differentiated cells of various types. Only the stem cells are oncogenic. These stem cells may be replicated *in vitro.* Upon subcutaneous injection these cells produce tumors, but upon injection into a blastocyst embryo these stem cells revert to normal cells, which can be identified in the adult animal by the genetic markers they carry. If these stem cells bear a mutation that accounts for their oncogenicity, then it is hard to see how this mutation could revert to wild type by placing it in the cytoplasmic environment of the embryo. Possibly the teratoma is an exceptional situation where the tumor is caused by abnormal development not necessitating a mutation.

Some Tumors Arise from Genetically Recessive Mutations

The oncogenes *c-myc* and *c-abl* are dominant to their wild-type alleles. Many tumors arise from oncogenes that show recessive behavior when compared with their wild-type alleles. Such is the case with the genetic locus implicated in the malady known as bilateral retinoblastoma, in which tumors develop in the retinas of both eyes. Human subjects with this condition usually inherit a genetically recessive mutation on chromosome 13. Sometimes the mutation is microscopically visible as a small deletion. In general, recessive mutations are not apparent unless they are present in the homozygous state. However, in the course of the very extensive cell duplication that ensues in retinal development, there is a high probability that at least one cell in each retina will mutate in the wild-type allele of the genetic locus in question. This change will result in a homozygous state and in the conversion of the normal cell to a tumorous cell that will proliferate at a rapid rate. As a result of this somatic mutation, both eyes usually develop tumors. It seems likely that genes that give rise to tumors when they are defective or lacking, as in retinoblastoma, are of the regulatory type. Some other tumors that result from recessive genetic lesions are listed in table 34.2.

Table 34.2
Recessive Genetic Lesions in Human Tumors

Tumor	Chromosomal Locus
Retinoblastoma and osteosarcoma	13 (q14)
Wilm's tumor (mephroblastoma)	11 (p13)
Embryonal tumor of the kidneys, muscle, liver, and adrenal glands	11p —
Bladder carcinoma	11p
Acoustic neuroma or meningioma	22

Many Tumors Arise by Mutational Events in Cellular Protooncogenes

A growing number of cellular genes have been implicated in a wide variety of cancers. These genes do not lead to cancers in their normal state, but only after a mutational event has somehow resulted in their altered expression. For that reason such genes are called protooncogenes in their normal state and oncogenes in their mutated, cancer-causing state. Most protooncogenes, including *c-abl* and *c-myc,* were not known until very similar genes were found as a result of studies of tumor-causing viruses.

Oncogenes Are Frequently Associated with Tumor-Causing Viruses

The first virus linked to tumor production was discovered in 1908 by Ellerman and Bang. Their finding was followed in 1911 by the better known discovery by Peyton Rous of a virus that produces sarcomas in chickens (now known as Rous sarcoma virus, or RSV). Later (1932) Shope showed that a papilloma virus produced cutaneous fibroma tumors in rabbits, and Lucke (1934) showed that adenoviruses produced renal adenocarcinoma in the frog. These and other discoveries made in the 40s and 50s indicated that exposure to certain viruses could result in tumor production. However, viral oncogenes had not yet been identified, and skeptics still classified cancer-causing viruses along with other carcinogens.

The development of tissue culture techniques in the 1950s permitted a systematic investigation of the causal link between tumor viruses and cellular transformation in culture. In many ways the properties of cells transformed in culture resembled those of tumor cells *in vivo,* and so the tissue culture system was considered to be a convenient way to explore the oncogenic properties of viruses. It was noted with some tumor viruses that the response of cells of different species was quite different. On cells that were normally permissive for virus replication, no tumors arose. This was hardly surprising, in view of the fact that the virus replication cycle, particularly for the DNA viruses, usually results in cell death. What was surprising was the finding that the same virus on another cell type, in which the virus could not replicate, would sometimes result in cell transformation. Observations of this sort led Renato Dulbecco to the hypothesis that there was a parallel between lysogeny of *E. coli* by λ and transformation of animal cells by oncogenic viruses. In λ, the viral genome can exist either as an independent entity, leading to virus replication and cell death, or as an integrated passive state, leading to a lysogenic cell (see chapter 30). With λ the two states—lytic or active, and lysogenic or passive—are attainable in the same cell type. Tumor-causing viruses are capable of existing in two states also. However, in this case the two states, actively replicating and passively integrated, are not normally attainable in the same cell type. Usually two cell types of different species are required. In one cell type the viral genome invariably exists in the actively replicating form, causing immediate cell death; in another cell type the viral genome can exist only in a passive integrated form. Dulbecco suggested that in the integrated state the viral genome expresses a limited number of gene products that lead to cellular transformation.

So far so good. Dulbecco's suggestion appeared to constitute a reasonable hypothesis for explaining the oncogenicity of certain DNA viruses. But it was also known that certain RNA viruses cause tumors. Indeed, one of the first known cancer-causing viruses, the Rous sarcoma virus, was an RNA virus. To explain the oncogenicity of RNA viruses, Dulbecco proposed that cancer-causing RNA viruses form a DNA copy of their genome that can be integrated into the host cellular genome. This bold proposal—that an RNA genome could be converted into a DNA copy—was confirmed by experiments in both David Baltimore's and Howard Temin's laboratories. Both groups found an enzyme, reverse transcriptase, uniquely associated with RNA tumor-causing viruses that could synthesize DNA from an RNA template.

The finding that many viruses cause cancer only after inserting their genome into the host genome has spurred new efforts in cancer research. More and more attention has been paid to viruses. With the help of genetic manipulations, the cancer-causing genes within the viruses have been identified and their gene products have been studied. The results of these studies are providing a broad platform of understanding about the kinds of biochemical processes that result in the conversion of normal cells to cancer cells.

The Role of DNA Viral Genes in Transformation Reflects Their Role in the Permissive Infectious Cycle

Usually tumor-causing DNA viruses possess oncogenes that bear no sequence relationships to any host genes. We can probably best understand the role of DNA viral oncogenes in cellular transformation by appreciating the role these genes play in the virus lytic cycle (table 34.3). First of all, the oncogenes we have found in the adenovirus, polyoma virus, and SV40 virus are all expressed early in virus infection. These genes also trigger a transient transformation process in most cells they infect, where

Table 34.3
Properties of DNA Virus Oncogenes

Viral Gene and/or Gene Product	Function in Lytic Infection	Function in Transformation	Other Properties
Adenovirus			
E1A 9 kDa, 26 kDA, 32 kDa	Required for expression of other early genes	Immortalizes cells; partial transformation	General transcription activator
E1B 9 kDa, 55 kDa	Required for lytic infection	Required for full transformation and tumorigenic cells Either 19 kDa or 55 kDa will suffice	Associates with p53[a] 55-kDa protein found primarily in nucleus; 19-kDa protein in endoplasmic reticulum and in nucleus
SV40			
Big T 90 kDa	Binds to specific sites; necessary, for DNA synthesis; inhibition of early RNA synthesis, and activation of late RNA synthesis	Immortalizes and transforms	Stimulates host DNA synthesis on microinjection; associates with p53[a]
Small t	None (?)	Required for full transformation on some cell lines	—
Polyoma			
Big T 100 kDa	Binds to DNA; necessary for DNA synthesis and regulation of early and late RNA synthesis	Immortalizes cells; only amino terminal 40% fragment is required	Associates with p53[a]; ATPase activity; gyrase activity
Middle T 56 kDa	None (?)	Transforms cells and makes them tumorigenic	Associates with pp60$^{c\text{-}src}$ [b]
Small t	None (?)	Transformed cells adhere less firmly when small t is present	Found in nucleus

[a]p53 is a cellular protein with unknown function. It is found in high concentrations in some tumors and it may be an oncogene.
[b]pp60$^{c\text{-}src}$ is a cellular protooncogene.

a permissive virus replication cycle is not possible. Thus it seems likely that these genes are primarily regulatory. In SV40 and polyoma the so-called big-T genes are clearly regulatory in productive infection, as they are involved in regulating both viral DNA and RNA synthesis. In adenovirus there is no comparable gene, but the early genes *E1A* and *E1B* may play a similar, albeit indirect role. *E1A* functions as a transcription activator.

Another lead as to how a DNA virus transforms cells is to look at the cellular proteins they influence. There is a striking parallel here between the functions of adenovirus *E1B,* SV40 big T, and polyoma big T in this regard. These three genes are required for immortalization or transformation, and they all form a tight association with the cellular protein p53, which is considered to be a good candidate for a protooncogene-related protein.* Finally, the polyoma middle-T protein, which is required for transformation by polyoma virus, forms a strong complex with the cellular protein encoded by the protooncogene *c-src.* It seems likely that these associations between viral oncogenes and host proteins result in changes in regulatory functions, and that if we knew what regulatory functions were being affected we would be on the threshold of understanding the biochemical mechanism(s) involved in transformation. We will have more to say about this after we have considered the retroviruses.

*In its normal protooncogenic form, p53 inhibits SV40 replication; in its mutated oncogenic state, it does not inhibit SV40 replication, nor does it bind to T antigen.

p53 Is the Most Common Gene Associated with Human Cancers

The p53 protein acts as a negative regulator of cell growth; it was first detected through its association with the SV40 big-T oncoprotein in virus-transformed cells. Viral oncoproteins like SV40 T antigen and adenovirus *E1B* sequester p53 protein in inactive complexes. Whereas this results in an increase in the steady-state concentration of p53, the p53 is no longer able to reach its normal site of action in the nucleus.

Many mutant *p53* alleles favor growth and transformation in cells that continue to carry intact, wild-type *p53* gene copies; accordingly, such mutant *p53* alleles act in a dominant fashion. The dominant behavior of mutant *p53* could be explained by the normal active form of the protein. In its active form p53 protein assembles into homotetramers and higher order homo-oligomeric structures. Defective subunits of such an oligomerizing protein (for example, mutant p53 molecules) may participate in forming a multi-subunit complex together with wild-type monomers and, in so doing, poison the function of the complex as a whole.

Figure 34.5

Hypothesis for how a retrovirus picks up a cellular protooncogene. The RNA-DNA cycle for retroviruses is shown in figure 26.28. The proviral DNA is believed to recombine the cellular DNA in such a way that it picks up a region of the cellular DNA containing a protooncogene.

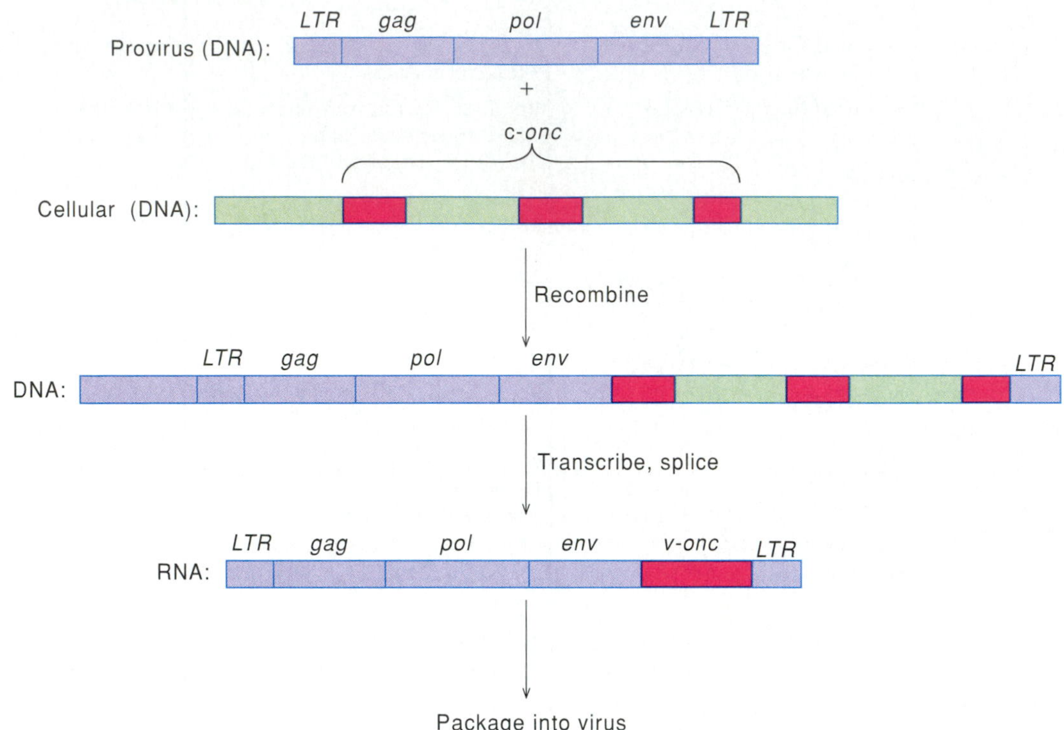

Normal p53 protein binds DNA in a sequence-specific manner and thus most likely regulates gene transcription. Cotransfection experiments show that wild-type p53 activates the expression of genes adjacent to a p53 DNA-binding site. Cells bearing oncogenic forms of p53 have lost this activity.

The best demonstration that the loss of active normal p53 explains the oncogenic behavior of mutant p53 comes from studies in which a null mutation was introduced into the gene by homologous recombination in murine embryonic stem cells. Mice homozygous for the null allele appear normal but are prone to the development of a variety of neoplasms by 6 months of age. These observations indicate that a normal *p53* gene is dispensable for embryonic development, but that its absence predisposes the animal to neoplastic disease. They also indicate that the oncogenicity of the mutant form of *p53* is most likely due to its negative effect on the normal protein.

Somatic mutations of *p53* have been implicated as causal events in the formation of a large and ever-increasing number of common tumors, including those involving the hematopoietic organs, bladder, liver, brain, breast, lung, and colon. In view of the fact that there seem to be so many other oncogenes, we must ask why *p53* is so often implicated in commonly occurring tumors. This appears to stem from its genetic and biochemical traits. Point mutations create carcinogenic *p53,* and such simple genetic changes occur readily. Unlike *ras,* for example, where point mutations productive for cancer are limited to specific changes in two or three codons, the cancer-favoring missense mutations of *p53* can occur in at least 30 distinct codons in its reading frame. In addition, these point mutations often create dominant alleles that produce shifts in cell phenotype even without a reduction to homozygosity.

Retroviral-Associated Oncogenes That Are Involved in Growth Regulation

The first intensively studied oncogene associated with a retrovirus was the RSV *src* gene. It is clear that the *src* gene found in RSV, *v-src,* is structurally quite similar to a normal host gene found in the chicken genome called *c-src*. Subsequent work has shown that at least 30 other animal retroviruses exist that have acquired a cellular oncogene during their evolution. This uncanny association of retroviruses with cancer-causing genes has made retroviruses useful devices for scanning the cellular genome for the presence of protooncogenes (fig. 34.5).

Table 34.4
Retroviral Oncogenes and Their Cellular Homologs

Oncogene	Viral Origin[a]	Viral Gene Product	Cellular Homolog	Activity	Subcellular Location
sis	Simian sarcoma virus	$p28^{sis}$	PDGF B-chain	PDGF agonist	Cytoplasm
src	Rous sarcoma virus	$pp60^{v\text{-}src}$	$pp60^{c\text{-}src}$	Tyrosine kinase	Plasma membrane
fps[b]	Fujinami sarcoma virus	$p140^{gag\text{-}fps}$			
abl	Abelson murine virus	$P120^{gag\text{-}abl}$	$p150^{c\text{-}abl}$	Tyrosine kinase	Plasma membrane
erbB	Avian erythroblastosis virus	$gp65^{erbB}$	Truncated EGF receptor	Tyrosine kinase	Plasma membrane
myc	Avian myelocytomatosis virus MC29	$P110^{gag\text{-}myc}$	$p58^{c\text{-}myc}$	Binds DNA	Nucleus
H-ras	Harvey murine sarcoma virus	$p21^{v\text{-}Hras}$	$p21^{c\text{-}H\text{-}ras}$	Threonine kinase binds GDP or GTP	Plasma membrane
K-ras	Kirsten murine sarcoma virus	$p21^{v\text{-}Kras}$	$p21^{v\text{-}Kras}$		
N-ras	See footnote *c*.		$p21^{c\text{-}N\text{-}ras}$		

[a]Only one example of a virus strain is given for each oncogene.
[b]*fps* and *fes* are homologous genes from chicken and cat, respectively.
[c]The transforming gene product, $p21^{N\text{-}ras}$, identified by transfection experiments.

A listing of some of the better-characterized retrovirus-associated cellular oncogenes appears in table 34.4. In this table the name of the oncogene is indicated in the leftmost column. Proceeding to the right in the table we have indicated: (1) the virus of origin, (2) the viral gene product, (3) the cellular homolog of the viral product; (4) the activity associated with the viral gene product, and (5) the subcellular location of the viral gene product, which in most cases is similar to the subcellular location of the cellular homolog.

The src *Gene Product*

The protein encoded by the *src* gene has a molecular weight of 60 kDa. Like many other oncogenic proteins, it is bound to the inner plasma membrane and it possesses a tyrosine phosphokinase activity; that is, it catalyzes the addition of a phosphate group to the tyrosine hydroxyl groups of various proteins. It is considered highly likely that the kinase activity of the src protein is associated with its transforming activity, since its loss in mutant *src* genes leads to loss of transforming activity. However, since many different proteins are phosphorylated by the src kinase, it is not clear which phosphorylations are crucial to the complex physiological response that triggers transformation.

The sis *Gene Product*

Many oncogenes are associated with tumors produced from a restricted class of differentiated cells. Such is the case with the *sis* oncogene, found in the simian sarcoma virus. This oncogene is believed to be closely related to the gene for platelet-derived growth factor, PDGF. PDGF is a small protein, synthesized in platelets, that stimulates the growth and division of target cells that carry a membrane-bound specific receptor for PDGF—in particular, the mesenchymal cells involved in wound healing. Tumors originate when the target cells carrying the PDGF receptor mutate to cells capable of synthesizing their own PDGF. This situation is believed to lead to uncontrolled growth because the factor continuously stimulates cell proliferation in the very cells in which it is synthesized (fig. 34.6).

The PDGF protein contains two polypeptide chains, A and B, with molecular weights of 28,000 and 32,000, respectively. The amino acid sequence of the PDGF protein is very similar to that of the predicted sequence of the transforming p23 sis protein of the SSV virus, indicating a common ancestral origin. Furthermore, it is believed that the SSV virus causes tumors by synthesizing large amounts of the viral p23 sis protein, which stimulates the unregulated proliferation of target cells that carry the PDGF receptor.

The erbB *Gene Product*

Epidermal growth factor (EGF) is a small mitogenic protein that stimulates the proliferation of cells carrying specific membrane-associated EGF receptors. The EGF receptor has a strong amino acid sequence homology with $gp65^{erbB}$, the transforming protein of avian erythroblastosis virus (AEV). The EGF receptor has tyrosine kinase activity, like the src protein. In addition, the EGF receptor contains an extracellular domain that binds the EGF growth factor.

The *v-erbB* oncogene acts to expand a pool of highly mitotic, undifferentiated erythroid precursor cells, but these are poorly tumorigenic, because they differentiate at high rates into post-mitotic, end-stage red cells. Another potential oncogene *v-erbA* on its own acts to block differentiation of these erythroid precursors, but creates no tumors because it is unable to provide the mitogenic impetus needed to expand the pool of stem cells.

Figure 34.6

Mechanism of mitogenesis in normal and transformed cells. (*a*) Schematic representation of the growth-factor-related mitogenic pathway in normal cells. Here, (1) represents the growth factor, (2) the growth factor receptor, and (3) the intracellular messenger system that transmits the mitogenic signal from the receptor to the nucleus. (*b*) Schematic representation of a possible perturbation of the growth-factor-related mitogenic pathway in transformed cells. Here, (1) represents endogenous production of growth factor that may stimulate the cell. The endogenously produced factor may be secreted and interact with growth factor receptors at the cell surface (as shown) or, alternatively, activate the receptor in an intracellular compartment.

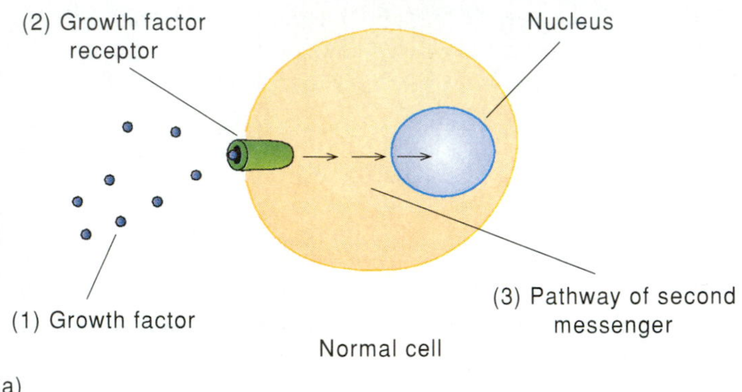

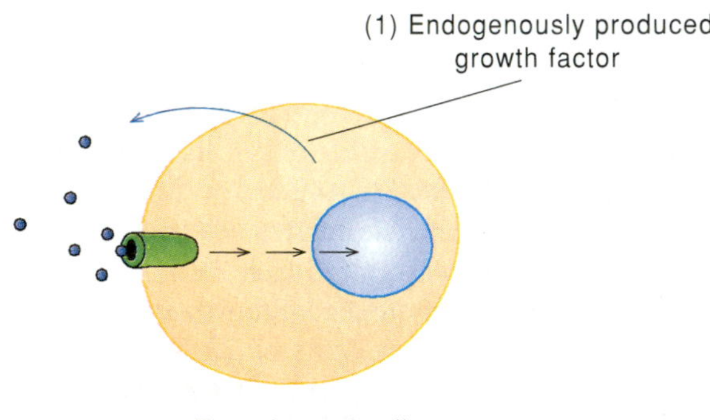

Figure 34.7

Interaction between the *ras* protooncogene and GDP. The protein is shown as a green ribbon, interacting with GDP (dark purple, yellow, and orange). The dark purple rectangle represents the guanosine; the dark yellow pentagon, the ribose; and the orange circles, the phosphates of GDP. The domains of the protein interacting with each of these parts of GDP are represented by sleeves on the protein ribbon, color-coded to match the corresponding part of GDP. The P domain of the protein contains glycine 12; this amino acid is in a critical position next to the phosphates of the nucleotide. The N and C termini of the protein are labeled.

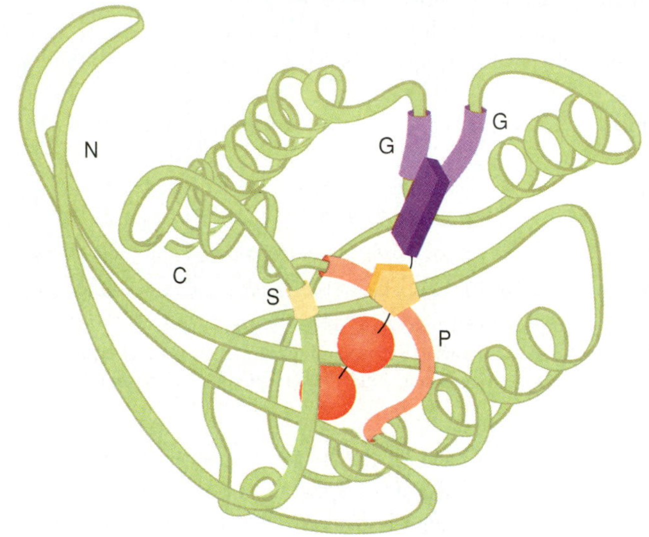

The two genes, *erbA* and *erbB,* carried into erythroid precursors by avian erythroblastosis virus, act in concert to create an aggressive erythroleukemia; *v-erbB* drives expansion of the pool of undifferentiated precursor cells, while *v-erbA* blocks their conversion by the differentiation pathway. The *v-erbA* allele that participates in formation of chicken erythroleukemias is a mutant version of a transcriptional regulatory protein, the chicken thyroid hormone (triiodothyronine) receptor. Function of the wild-type receptor protein is blocked in the presence of *v-erbA,* because the latter occupies critical DNA-binding sites in a way that precludes association by the wild-type receptor protein and inhibits transcription.

The ras *Gene Product*

One of the best understood protooncogenes is *ras*. The *ras* gene found in normal mammalian cells is closely related to the *ras* oncogenes of Harvey and Kirsten murine sarcoma viruses, *H-ras* and *K-ras,* respectively. *Ras* oncogenes have also been isolated with the help of DNA transfection techniques. Activated *ras* genes have been isolated from a wide variety of dissimilar neoplasms, including carcinomas, sarcomas, neuroblastomas, lymphomas, and leukemias. Some of these *ras* genes are associated with the murine sarcoma viruses (e.g., *N-ras* in table 34.4). All members of the *ras* gene family encode closely related proteins of approximately 21 kDa, which have been designated p21. Unlike the oncogenes associated with many retroviruses, the level of p21 expression is similar in normal cells and in many different human tumor-cell lines. Nucleotide sequence analysis of the *H-ras* transforming gene isolated from a human bladder carcinoma cell line has indicated that the transforming activity of this gene is the consequence of a point mutation altering amino acid 12 of wild-type p21 from glycine to valine (fig. 34.7).

Subcellular fractionation and immunofluorescence of normal cellular and viral ras proteins have indicated that p21 is localized on the inner face of the plasma membrane. Both the normal and the mutant ras proteins bind GTP specifically and strongly, but only the normal protein has GTPase activity.

We understand normal *ras* gene function better than that of most other protooncogenes because the yeast *Saccharomyces cerevisiae* carries *ras* genes, and many researchers have taken advantage of the powerful genetic tools available in yeast to study them. *S. cerevisiae* contains two closely related but distinct genes, *RAS1* and *RAS2*, that encode proteins that are highly homologous to the mammalian ras proteins. While neither *RAS1* or *RAS2* by itself is an essential gene, *RAS* function is required for the continued growth and viability of yeast cells. Thus *ras1*$^-$, *ras2*$^-$ double mutants are nonviable unless they carry a suppressor mutation *bcyL*. The *bcyL* mutation suppresses lethality in adenylate-cyclase-deficient yeast. We will see the significance of this fact presently. Yeast strains that are *ras2*$^-$ are noticeably depressed in cAMP levels. In *ras1*$^-$, *ras2*$^-$, *bcyL* triple mutants the levels of cAMP are virtually undetectable. But cells containing *RAS2 val19*, a *RAS2* allele with a missense mutation analogous to the one that activates the transforming potential of mammalian *ras* genes, have cAMP levels significantly higher than those observed in wild-type cells. Membranes from *ras1*$^-$, *ras2*$^-$, *bcyL* triple mutants lack the GTP-stimulated adenylate cyclase activity present in membranes from wild-type cells, and membranes from the *RAS2 val19* yeast strain have elevated levels of a GTP-independent adenylate cyclase activity. Mixing membranes from *ras1*$^-$, *ras2*$^-$ yeast with membranes from an adenylate-cyclase-deficient yeast leads to the reconstitution of a GTP-dependent adenylate cyclase. It appears, then, that *RAS* encodes a protein or proteins that somehow regulate activity of membrane-bound adenylate cyclase in a GTP-dependent manner. In the yeast mutant *RAS2 val19* and in the homologous mammalian mutant, the regulatory properties of this protein appear to be disrupted so that the adenylate cyclase is overactive and no longer requires GTP for activation. Further experiments have been done to show that the *ras* genes from yeast cells can function in mammalian cells and vice versa. Thus it has been shown that yeast carrying *ras1*$^-$, *ras2*$^-$ mutations remain viable if they carry the mammalian *H-ras* gene. Conversely, a mutant yeast *RAS* gene has been isolated that can bring about tumorigenic transformation in mammalian cells.

From what has been said, you might expect that ras proteins would be involved in activating adenylate cyclase in mammalian cells. However, there is no indication that this is the case. In mammals, other heterotrimeric GTP-binding proteins are involved in the adenylate cyclase pathway. Apparently, the *ras* system is a useful signaling module that has been adapted to various uses in different organisms; it could be likened to an electronic switching device, which could serve different functions depending upon the devices it is attached to. In this connection we note that a structural analysis indicates a similarity between p21^{ras} structure and the G-domain of the protein synthesis elongation factor Tu(EF-Tu) from *Escherichia coli*.

The myc *Gene Product*

The *c-myc* gene, identified originally as the cellular homolog of the transforming determinant carried by avian myelocytomatosis virus, is altered in association with a broad spectrum of neoplasms. Consistent with the observation that altered *c-myc* is associated with tumors of diverse origin, it has been observed that normal *c-myc* is expressed in a variety of tissues. Thus *c-myc* appears to encode a function associated with a ubiquitous metabolic pathway.

The *c-myc* gene encodes a 49-kDa protein that is highly concentrated in the nucleus of the cell. The concentration of c-myc protein normally varies appreciably with the metabolic state of the cell, increasing by more than an order of magnitude in cells just prior to chromosome duplication and cell division.

In most human tumors associated with the *c-myc* gene, the concentration of c-myc protein is greatly amplified. The Burkitt lymphoma, which we discussed earlier, is a notable exception. In this case the c-myc transcript is marginally, and in some cases not at all, increased by comparison with control lymphoblastoid cell lines. Recall that in the Burkitt lymphoma, reciprocal chromosomal translocations are found that involve a chromosome carrying *c-myc* protooncogene (chromosome 8) and one of the chromosome segments (usually chromosome 14) bearing immunoglobulin genes.

In at least one Burkitt lymphoma cell line, the structure of the amino-acid-coding portion of the translocated *c-myc* gene, and hence its predicted protein product, has not been altered, a fact suggesting that activation of *c-myc* must be mediated via a regulatory disturbance. Normally, c-myc protein synthesis is strongly regulated with respect to the cell cycle and tightly repressed in quiescent cells. The translocated *c-myc* gene appears to be somewhat deregulated with respect to its expression during various phases of the cell cycle.

Members of the *myc* gene family have been implicated in the control of normal cell proliferation as well as in neoplasia. A more direct role for *myc* genes in transformation is indicated by their ability to transform primary rat embryo fibroblasts in association with the *c-H-ras* oncogene.

An important clue to *c-myc* function was the discovery in the conserved carboxy-terminal region of three structural motifs, the leucine zipper (LZ), helix-loop-helix (HLH) and basic region (B). These motifs were originally defined in a number of other sequence-specific DNA-binding proteins but had not previously been found within a single protein.

The comparatively weak homo-oligomerization efficiency of the c-myc HLH/LZ suggested that a partner protein(s) might exist that hetero-oligomerizes with c-myc to form a specific DNA-binding complex. A protein termed max (for myc-associated "X" factor), was shown to interact with c-myc in a manner that required the integrity of the c-myc HLH/LZ motif. Max was also demonstrated to associate with N- and L-myc proteins, but not with the HLH/LZ proteins USF or AP-4, nor with several other HLH or leucine zipper proteins. In DNA-binding assays, c-myc-max specifically recognized a c-myc DNA-binding site (CACGTG) in a manner that required both an intact max basic region and the HLH/LZ motifs.

What is the molecular function of the c-myc-max complex *in vivo?* At present the characteristics of max do not suggest obvious new hints to myc function other than confirming a role for the B/HLH/LZ regions in DNA binding.

It has been reported that the myc amino-terminal and central regions can specifically interact with the retinoblastoma (RB) tumor suppressor protein *in vitro.* This finding prompts the suggestion that RB may directly regulate myc function, but it is not yet known whether the observed interaction reflects a physiologically relevant association *in vivo.*

The jun *and* fos *Gene Products*

The *jun* oncogene was discovered as a 0.93-kb insert in the genome of a replication-defective retrovirus, avian sarcoma virus 17 (ASV17), isolated from a chicken sarcoma. ASV17 causes fibrosarcomas in chickens and oncogenic transformation in cultured avian embryonic fibroblasts. The oncogenic potential of ASV17 is due to the presence of the *jun* gene, which is derived from the cell genome.

A great deal of excitement was generated by the discovery that the jun protein can substitute for the yeast transcription activator GCN4. Comparative analysis of the sequences of the two proteins showed that there was a strong homology over about one-third of their length in the DNA-binding parts. Subsequently it was discovered that the human transcription factor AP-1 also had very similar properties to both jun and GCN4. All three of these proteins bind to the same consensus sequence, TGACTCA.

The oncogene of the FBJ murine osteosarcoma virus (*fos*) also codes for a related nuclear protein that participates in transcriptional regulation. In human fibroblasts the fos protein is mostly associated with a cellular protein p39 with a molecular weight of 39 kDa. The fos-p39 complex binds specifically to DNA containing the AP-1 binding site. Since fos alone does not show specific DNA binding, it is believed that p39 is responsible for this affinity. Sequence studies showed that p39 and jun are identical proteins. Although jun, like GCN4, can form homodimers that bind to DNA, the heterodimers formed between fos and jun show a greater affinity for the AP-1 binding site. The heterodimers are also more effective in transcription activation; therefore the heterodimer is probably the functionally relevant state of the jun and fos proteins.

Structural analysis indicates that the fos and jun proteins belong to a class of DNA-binding proteins that share the conserved structural motif known as the leucine zipper (see fig. 31.20). Thus the dimerization of these two proteins is mediated by hydrophobic interaction between the leucine side chains of two leucine-zipper domains.

In its normal context as a cellular gene product, jun is not oncogenic. By contrast, as we have already mentioned, the viral *jun* gene of ASV17 is an effective carcinogen both in the animal and in cell culture. This difference appears to be due to the constitutively elevated expression of *jun* when the *jun* gene is in the virus.

The jun-fos protein complex interacts with regulatory regions of numerous genes. We have yet to find out which of these target genes are involved in aberrant cellular growth.

The Transition from Protooncogene to Oncogene

All of the oncogenes thus far discovered are associated with a cellular homolog that is required for normal growth. The transition of the protooncogene to an oncogene is accompanied by abnormal expression of the gene products. In many cases, e.g., *src, jun,* or *sis,* excessive amounts of the gene product are synthesized. In some other cases, e.g., *H-ras* or *K-ras,* normal amounts of the oncogenic product are synthesized but the protein encoded by the oncogene is altered so that it behaves differently. In still other cases, e.g., *myc,* the oncogene product is similar to the cellular homolog in amount as well as structure, but the time during the cell cycle when it is produced is altered.

Oncogene products assume specific locations within the cell. Usually they are associated with one of two locations: the plasma membrane or the nucleus. This specificity is consistent with the hypothesis that oncogene products are associated with normal cellular metabolism relating to regulation of cell proliferation; it seems likely that elements regulating cell proliferation would be found at the membrane and nucleus of the cell.

It is not surprising to find oncogenes with varying specificity for producing tumors. Some components of the cell proliferation regulatory apparatus are probably quite general. An oncogene like *myc* is probably associated with one of these and is therefore associated with tumors of widely varying origins. By contrast, an oncogene like *sis* is specifically associated with cells that are designed to be triggered into proliferation by PDGF. Hence tumors associated with this oncogene are limited to cell types possessing the PDGF membrane receptor.

Carcinogenesis Is a Multistep Process

A wide range of observations indicate that tumorigenesis is a multistep process involving several mutations, each of which results in discrete changes in the cellular metabolism. If this is the case, then we might expect that any particular oncogene would have the capacity for affecting only one step in the overall process. We have seen that a multistep biochemical pathway ordinarily requires a separate enzyme for each step. Enzymes that function in the different steps of a pathway are sometimes said to complement one another. The question is, in carcinogenesis do different oncogenes show a similar complementation? For example, can purely cellular oncogenes such as *N-ras* complement one or the other of the polyoma oncogenes? In fact, middle-T antigen and *N-ras* oncogene produced no new phenotypes when they were cotransfected into rat embryo fibroblasts (REFs), but large-T antigen and *ras* together achieved dramatic results, producing rapidly expanding foci. This study shows that the conversion of a normal cell to a tumor cell can be achieved by the complementary action of two distinct oncogenes; in this case, one is cellular and one is viral.

Similar studies have shown that the *H-ras* and the *myc* oncogenes can cooperate to produce dense foci of morphologically transformed cells from REF cultures. Thus two cellularly derived oncogenes have been shown to complement one another to produce a fully transformed phenotype. Experiments of this sort suggest that at a minimum, two different oncogene functions are required to convert a normal cell into a tumorigenic one, but it is too early to say that only two are required in general. Nevertheless, these results are very exciting because they are the beginning of genetic complementation assays that should help us to classify oncogenes into different complementary functions.

One further comment on the multistep nature of carcinogenesis: In our discussion here, we have focused on some of the early steps in the process that lead to uncontrolled growth, and have said nothing about those transitions that convert transformed cells into invasive cells. We have also concentrated on oncogenes related to cancer-causing viruses, because these are most likely to be the first understood in terms of their biochemical function.

Summary

From the time of their discovery, cancer cells have always appeared to be unruly. We have reached that stage in our understanding of cancer cells where we can point to specific aberrations in the genome. In this chapter we have taken the view that an understanding of cancer can be achieved by analyzing the regulatory pathways involved in growth control, because most cancers appear to originate from mutations in specific genes involved in growth regulation. The following points are the highlights of our discussion.

1. Many properties of transformed cells grown in tissue culture resemble cancer cells. The factors that lead to uncontrolled growth *in vivo* can be studied *in vitro* by the effects they have on tissue culture cells.
2. To judge by the frequency of occurrence of cancers in different countries, environmental factors have more influence on the incidence of cancers than genetic factors do. This conclusion is reinforced by studies of migrant populations.
3. Chromosomal translocations are frequently associated with specific types of cancer. This is direct evidence that genetic abnormalities can lead to cancer.
4. A number of tumors arise from recessive mutations in which the mutations appear to be in growth-control genes.
5. Growth-control genes that lead to cancers when they are altered in some way are referred to as protooncogenes. They become oncogenes, that is, cancer-causing genes, by mutation. Host protooncogenes are frequently very similar in structure to oncogenes carried by tumor-causing viruses.
6. Dulbecco proposed that cancer-causing viruses insert their oncogenes into the host genome. It appears that cancer-causing viruses are associated with a limited number of DNA viruses and RNA viruses known as retroviruses, which replicate through a DNA intermediate. It is this DNA intermediate that usually gets inserted into the host genome.
7. Abnormal expression accompanies the transition from protooncogene to oncogene. Three types of abnormalities occur: (1) excessive production of the gene product, (2) altered behavior of the gene product, such as a change in its regulatory properties, and (3) expression at a time during the cell cycle when the gene is not normally expressed.
8. A fully developed cancer appears to arise in steps, each step showing a breakdown in normal regulation.

Selected Readings

Aaronson, S. A., Growth factors and cancer. *Science* 254:1146–1152, 1991.

Donehower, L. A., M. Harvey, B. L. Slagle, M. J. McArthur, C. A. Montgomery, J. S. Butel, and A. Bradley, Mice deficient for p53 are developmentally normal but susceptible to spontaneous tumours. *Nature* 356:215–221, 1992. A remarkable new technique for obtaining null mutations demonstrates that embryogenesis is normal in the absence of p53. However, animals develop a variety of neoplasms in the first 6 months when p53 is lacking.

Downward, J., The ras superfamily of small GTP-binding proteins. *Trends Biochem. Sci.* 15:469–472, 1990.

Gallo, R. C., The AIDS virus. *Sci. Am.* 256(1):46–56, 1987.

Halauska, F. G., Y. Tsujimoto, and C. M. Croce, Oncogene activation by chromosome translocation in human malignancy. *Ann. Rev. Genet.* 21:321–345, 1987.

Hausen, H., Viruses in human cancers. *Science* 254:1167–1173, 1991.

Henderson, B. E., R. K. Ross, and M. C. Pike, Toward the primary prevention of cancer. *Science* 254:1131–1144, 1991.

Jacks, T., A. Fazeli, E. M. Schmitt, R. T. Bronson, M. A. Goodell, and R. A. Weinberg, Effects of an Rb mutation in the mouse. *Nature* 359:295–300, 1992.

Levine, A. J., The p53 protein and its interactions with the oncogene products of the small DNA tumor viruses. *Virology* 177:419–426, 1990.

Linzer, D. I. H., The marriage of oncogenes and anti-oncogenes. *Trends Genet.* 4:245–247, 1988.

Liotta, L. A., Cancer cell invasion and metastasis. *Sci. Am.* 266:54–63, 1992. A most important aspect of carcinogenesis that we have not dealt with in our short chapter.

Malkin, D., F. P. Li, L. C. Strong, J. F. Fraumeni, C. E. Nelson, D. H. Kim, J. Kassel, M. A. Gryka, F. Z. Bischoff, M. A. Tainsky, and S. H. Friend, Germ line p53 mutations in a familial syndrome of breast cancer, sarcomas, and other neoplasms. *Science* 250:1233–1238, 1990.

Milburn, M. V., L. Tong, A. M. deVos, A. Brunger, Z. Yamaizumi, S. Nishimura, and S. H. Kim, Molecular switch for signal transduction: structural differences between active and inactive forms of protooncogenic ras proteins. *Science* 247:939–945, 1990.

Solomon, E., J. Borrow, and A. D. Goddard, Chromosome aberrations and cancer. *Science* 254:1153–1160, 1991.

Varmus, H. E., Reverse transcription. *Sci. Am.* 257(3):56–64, 1987.

Weinberg, R. A., Finding the anti-oncogene. *Sci. Am.* 259(3):44–51, 1988.

Weinberg, R. A., The retinoblastoma gene and cell growth control. *Trends Biochem. Sci.* 15:199–202, 1990.

Weinberg, R. A., Tumor suppressor genes. *Science* 254:1138–1146, 1991. Provides an update on the mechanisms of action of a number of better known oncogenes.

Problems

1. What are some of the differences between a cancer cell and a normal cell?
2. Consider the following statements: "The vast majority of cancers are preventable" and "Cancer is a disease of aging, i.e., predominantly older people get cancer." Do you agree with either statement? Explain your answer.
3. The methionine analog L-ethionine (which has an ethyl instead of a methyl group attached to the sulfur) causes liver cancer in rats. This analog is a nonmutagenic carcinogen that inhibits cellular methylation. How might this analog cause cancer? (Hint: What role does DNA methylation play in gene expression?)
4. You have just gotten your first job in Houston, Texas, and are presented with the following observations:
   (a) West of Houston there is a very high incidence of a variety of skin cancers.
   (b) South of Houston the incidence of skin cancer is low.
   (c) The region west of Houston was settled by immigrants from Germany who are mostly farmers, while the region south of Houston was settled, in large part, by Hispanics, who also do mostly outdoor work.

   The Texas Health Department asks you to explain why there is such a high incidence of skin cancer west of Houston. (Give a biochemical explanation.)
5. Renato Dulbecco suggested that DNA viruses that cause cancer did so by a mechanism similar to lysogeny of *E. coli* by λ virus, that is, by integration of the viral genome into the host cell DNA. How can RNA viruses cause cancer?
6. Speculate on where and how retroviral oncogenes originated.
7. Why is an oncogene like *myc* found associated with many different kinds of cancer, while the oncogene *sis* is found only in cancers with the PDGF cell-surface receptor?
8. What are some of the ways a cellular protooncogene can be converted to an active oncogene?
9. How does overexpression of the *src* gene cause cancer?
10. The AIDS virus (HIV) can cause many types of cancers (Kaposi's sarcoma, B-cell lymphoma, cancers of the rectum and tongue, etc.), yet HIV does not have a viral oncogene. Explain how HIV can cause cancer.

35

CHAPTER

Neurotransmission

We considered the structures of biological membranes, as well as the mechanisms by which ions and metabolites are transported across them, in chapters 7 and 32. Clearly, however, cells must do much more than simply assimilate nutrients and expel waste materials. Unicellular, free-living organisms such as bacteria and protozoa must be able to navigate and successfully compete in environments that often contain many perils to their livelihood. Multicellular organisms have the additional problem of coordination and signaling among many diverse, differentiated cell types to ensure the efficient functioning of the organism as a whole. The necessary interactions of a cell with its environment and with other cells must obviously involve processes associated with the cytoplasmic membrane. In many cases, transmembrane transport is integrally involved in the signaling processes between cells and their environment.

In chapter 24 we dealt with one type of communication mechanism in animals. Hormones released in one part of the body interact with receptors, which are sometimes membrane-bound, to affect processes occurring in other cells that are often quite distant from the cells releasing the signal. This mechanism allows for efficient coordination of the wide variety of metabolic events continually taking place in higher organisms. Another, quite different phenomenon takes place in the cells of sensory organs—for example, in the retina cells of the eye of vertebrate animals. Light energy is converted by these cells into nerve impulses, which are translated by the brain into an image of our surroundings. This interconversion involves transmembrane movements of ions and will be considered in chapter 36.

In this chapter we will examine the mechanism by which nerve impulses are propagated, to illustrate how excitable tissues function and interact in higher organisms. As we will show, the fundamental mechanisms revealed by studies of membrane transport in both prokaryotic and eukaryotic cells (chapter 32) can account in large part for the seemingly more complex phenomena of nerve-impulse propagation and transmission.

Nerve-Impulse Propagation

As early as the late eighteenth century, experiments by Galvani and Volta suggested that the transmission of nerve impulses and muscular contraction involved electrical signals. In 1902, J. Bernstein first proposed that the unequal distribution of K^+ across the nerve-cell membrane and the selective permeability of this membrane for the potassium ion were responsible for a resting potential known to exist in nerve and muscle fibers. He further believed that excitation of a nerve cell involved a transient collapse in this selective permeability such that other ions were able to penetrate the nerve-cell membrane and abolish the resting potential. If these changes in ion permeability could move down the axon of a nerve cell and be transmitted to other cells, they could provide the basis for propagation of nerve impulses.

In the 1930s, the isolated squid giant nerve axon became available for experimentation, and its size made it especially amenable to electrophysiological measurements. Experiments pioneered by A. L. Hodgkin and A. F. Huxley soon established the essential ionic movements associated with impulse propagation, and the resultant local changes in membrane potential could be measured. It became clear that the transmembrane transport of ions was important in nerve cells for their signaling function and for the actual signal conductance itself. As we will see, two basic types of transmembrane transport proteins are intimately involved in these processes: ion channels, which participate in the propagation of the impulse down the axon, and neurotransmitter receptors (also ion channels), which are involved in the transmission of signals between nerve cells and their target cells.

An Unequal Distribution of Ionic Species Results in a Resting Transmembrane Potential

To understand how nerve impulses are generated along the axon of a nerve cell (fig. 35.1), we must first understand the basis for the electric potentials that exist across the neuronal membrane. An unequal distribution of ionic species across a biological membrane that is permeable to these molecules can result in a transmembrane electric potential, $\Delta\Psi$ (chapter 32). For a membrane system permeable to several ionic species, the numerical value of $\Delta\Psi$ can be approximated by the Goldman equation, derived by D. E. Goldman in 1943:

$\Delta\Psi$ ("in" relative to "out") =

$$\frac{2.3RT}{F} \log_{10} \left(\frac{\Sigma P_c[\mathrm{C}]_{out} + \Sigma P_a[\mathrm{A}]_{in}}{\Sigma P_c[\mathrm{C}]_{in} + \Sigma P_a[\mathrm{A}]_{out}} \right) \quad (1)$$

where C and A are univalent cations and anions, respectively, and P_c and P_a refer to their permeability coefficients* across the membrane of interest. Since multivalent ions are generally not quantitatively significant in contributing to $\Delta\Psi$ in resting neuronal membranes, we usually ignore them in calculating $\Delta\Psi$. If the membrane is selectively permeable to one ion only, for example C, equation (1) reduces to the familiar Nernst equation:

$$\Delta\Psi = E_c = \frac{2.3RT}{F} \log_{10} \frac{[\mathrm{C}]_{out}}{[\mathrm{C}]_{in}} \quad (2)$$

where E_c refers to the equilibrium electric potential of C.

It is the unequal distribution of protons and other cations that gives rise to transmembrane potentials that can drive ATP synthesis and secondary active transport in many cells (chapter 32). In nerve cells, a similar situation exists as summarized in table 35.1. Thus the external environment of the nerve cell contains a high concentration of sodium ions and a low concentration of K^+, while the reverse is true for the cytoplasm of the nerve cell, called the axoplasm. Furthermore, the extracellular concentration of Cl^- is 5 to 10 times that of the axoplasm.

*The permeability coefficient is equal to the diffusion coefficient D divided by the width of the membrane l.

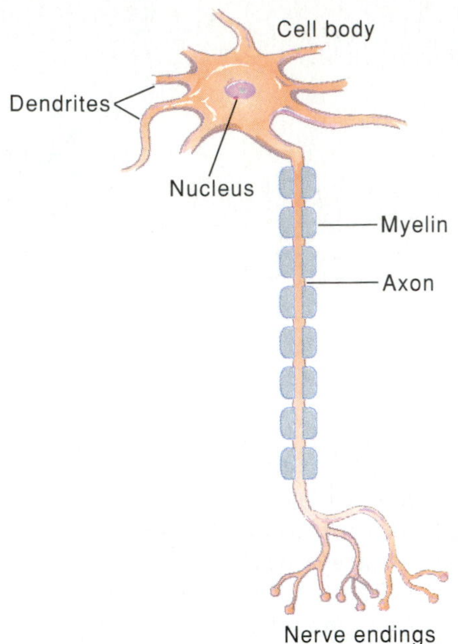

Figure 35.1

Schematic diagram of a typical motor neuron (a nerve cell conducting impulses to muscle cells).

A number of other impermeant anions—largely organic molecules, proteins, and nucleic acids—maintain an approximate charge neutrality in the axoplasm. Resting nerve cells are much more permeable to K^+ than to Na^+, and it is this selective permeability that allows an electric potential to develop across the membrane in the presence of a K^+ concentration gradient. It is, of course, the role of the Na^+-K^+ ATPase to maintain these unequal distributions of cations across the neuronal membrane, as we discussed in chapter 32.

To calculate the resting membrane potential $\Delta\Psi$ across the axonal membrane from equation (1), we can use the values in table 35.1 and the permeabilities of Na^+ and Cl^- relative to K^+ (0.04 and 0.45, respectively), assuming an intracellular Cl^- concentration of 50 mM:

$$\Delta\Psi = 60 \log \left(\frac{20 + 0.04(440) + 0.45(50)}{400 + 0.04(50) + 0.45(560)} \right) = -62 \text{ mV}$$

This value is close to the experimentally measured resting potential across a squid axonal membrane.

An Action Potential Is the Transient Change in Membrane Potential Occurring during Nerve Stimulation

The use of giant axons from squid nerve cells in electrophysiological experiments has greatly aided our understanding of the electrical events that take place during nerve stimulation. An experimental apparatus for measuring changes in the potential across the membrane of such an axon, which has a diameter of approximately 0.5 mm, is schematically illustrated in figure 35.2.

Table 35.1
Ionic Concentrations Inside (Axoplasm) and Outside (Blood) the Squid Giant Axon

Ion	Inside (mM)	Outside (mM)
Na^+	50	440
K^+	400	20
Cl^-	40–150	560

Source: S. W. Kuffler and J. G. Nicholls, *From Neuron to Brain.* © 1976 Sinauer Associates, Sunderland, Mass.

Figure 35.2

A device for eliciting and recording action potentials along the squid giant nerve axon. Brief closure of the switch connected to the stimulating electrode causes a current pulse into the axoplasm. If an impulse is generated, resultant potential changes can be detected by the recording electrode, which is connected to an oscilloscope or other recording device.

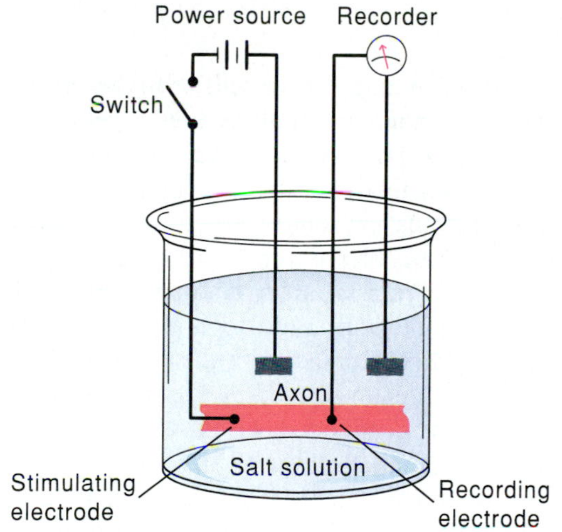

The apparatus consists of a pair of stimulating electrodes connected to a current source and a second pair of recording electrodes located slightly farther down the axonal segment. The latter electrodes are connected to a sensitive recording device, such as an oscilloscope, and can be used to measure time-dependent changes in the membrane potential.

Initially, the potential measured in the system shown in figure 35.2 is about −60 mV; i.e., the resting membrane potential (fig. 35.3). If a brief current is applied at the stimulating electrodes, a time-dependent change in the membrane potential may be recorded on the oscilloscope, as shown in figure 35.3*a*. This so-called action potential occurs only if the stimulus is sufficient to depolarize the membrane by about 20 mV (i.e., to about −40 mV). Weaker stimuli give small local potential changes, while current pulses greater than this threshold value give a curve similar in shape and height to that shown in figure 35.3,

Figure 35.3

A typical action potential V (blue) that might be recorded by the instrument in figure 35.2 if the stimulating current is sufficient to depolarize the membrane by at least 20 mV. The membrane potential returns to its resting value of about −60 mV within about 5 ms. Accompanying changes in relative Na^+ permeability, g_{Na^+} (red), and K^+ permeability, g_{K^+} (green), are also shown. Depolarization is correlated with an increase in g_{Na^+}, while repolarization is accompanied by a decrease in g_{Na^+} and a transient increase in g_{K^+}. Note that the membrane potential transiently becomes more negative than the resting value (hyperpolarizes) until g_{K^+} returns to its normal value. The time of appearance of the action potential after stimulation at $T = 0$ depends on the distance between the recording and stimulating electrodes. (Source: S. W. Kuffler and J. G. Nicholls, *From Neuron to Brain.* Copyright © 1976 Sinauer Associates, Sunderland, Mass.)

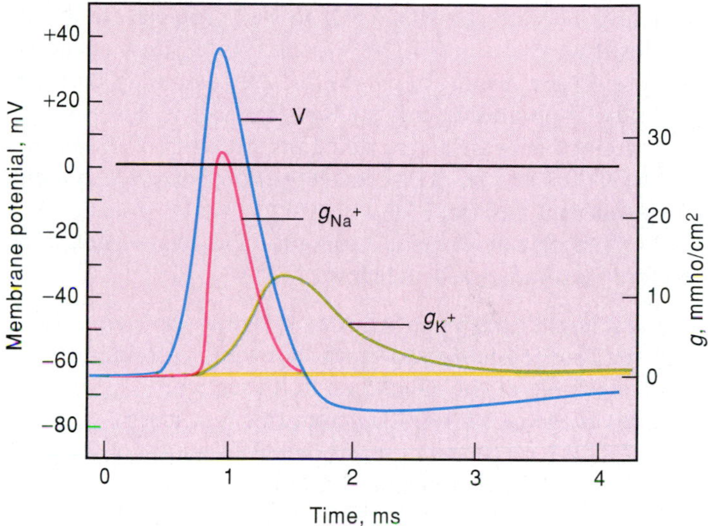

independent of the magnitude of the stimulus. During the development of the action potential, the value of $\Delta\Psi$ across the axonal membrane rises in about 1 ms to nearly +40 mV. This is followed by a somewhat slower return to the resting potential, during which time the membrane potential drops transiently below the resting value, to about −75 mV. This value, which is close to the one predicted by the Nernst equation if the membrane is permeable only to K^+ (the potassium equilibrium potential), marks the state referred to as hyperpolarization.

Classic experiments by A. L. Hodgkin and A. F. Huxley established that the changes in membrane potential occurring during nerve stimulation are due to transient changes in the permeability of the membrane to Na^+ and K^+ ions. As illustrated in figure 35.3, the rapid rise in $\Delta\Psi$ to a positive value is accompanied by a large increase in the relative permeability of Na^+, while the return to the resting potential is correlated with the inactivation of Na^+ permeability and a transient increase in K^+ permeability. An important conclusion from these observations is that the permeabilities of the axonal membrane to Na^+ and K^+ depend on the membrane potential. Thus depolarization above the threshold, leading to a more positive $\Delta\Psi$, first leads to an increased permeability of Na^+ followed by inactivation of this phenomenon and an increase in the membrane permeability of K^+. The latter events tend to hyperpolarize the membrane, increasing the negative $\Delta\Psi$ and decreasing the Na^+ permeability.

As we mentioned in the preceding section, the membrane potential depends on the relative permeabilities and concentration gradients of electrolytes across the membrane [equation (1)]. The resting potential is largely dependent on the K^+ gradient, since unstimulated nerve membranes have a high permeability only to this cation. At the height of depolarization, however, the membrane is much more permeable to Na^+ than to K^+. Because the Na^+ gradient is opposite to that of K^+, Na^+ influx rapidly causes the membrane potential to become positive, approaching but never reaching the value it would have if the membrane were permeable only to Na^+. You can see this if you substitute the values of $[Na^+]_{in}$ and $[Na^+]_{out}$ into equation (2). You will get a value of 57 mV for $\Delta\Psi$, assuming that the membrane is permeable only to Na^+. In reality, this value is never attained because an increased permeability of the membrane to K^+ follows the change in Na^+ permeability, which itself is only transient (see fig. 35.3).

The ionic movements leading to an action potential can therefore be summarized as follows:

1. Stimulation leads to an influx of Na^+ into the axoplasm, down its electrochemical gradient, owing to an increased permeability of the membrane to this cation.
2. The change in $\Delta\Psi$ resulting from this flow increases the membrane permeability of K^+, which flows out of the cell and down its electrochemical gradient; the result is the reestablishment of a negative $\Delta\Psi$, accompanied by an inactivation of the influx of Na^+.
3. The membrane potential eventually becomes sufficiently negative to return both K^+ and Na^+ permeabilities to their normal values, and $\Delta\Psi$ resumes its resting level.

The events shown in figure 35.3 record fluctuations in $\Delta\Psi$ and ionic currents at one point on the axonal membrane as a function of time during passage of an action potential through this point. The action potential, however, is conducted down the axon as a wave of depolarization-repolarization events through the following mechanism: Depolarization of a given area of the membrane causes current to flow in the axoplasm from the more positive (depolarized) region to neighboring regions. This flow, in turn, triggers an action potential across the neighboring section of the membrane, and so forth (fig. 35.4). Thus the action potential provides a mechanism whereby the transmitted signal is constantly amplified to maintain a constant amplitude. In a regular cable without amplification, the propagated pulse decreases with distance as a result of resistance and leakage. Without the action potential, a current pulse would therefore be reduced to an insignificant level after traveling a very short length along the axon. In nerve cells of invertebrate animals, as well as in many cells in vertebrates, nerve impulses are therefore conducted along axons and dendrites by these local currents and action potentials.

Figure 35.4

Nerve impulses in unmyelinated nerves are conducted through local current movements that propagate the action potential. A portion of the resting axonal membrane is shown at the top. Arrival of an action potential causes local depolarization of the membrane (*middle*), which is propagated from left to right by local currents shown by the arrows. (Source: A. L. Hodgkin, *Proceedings. Royal Society of London. Series B. Biological Sciences* 148:1, 1957. Copyright © 1957 Royal Society of London, London, England.)

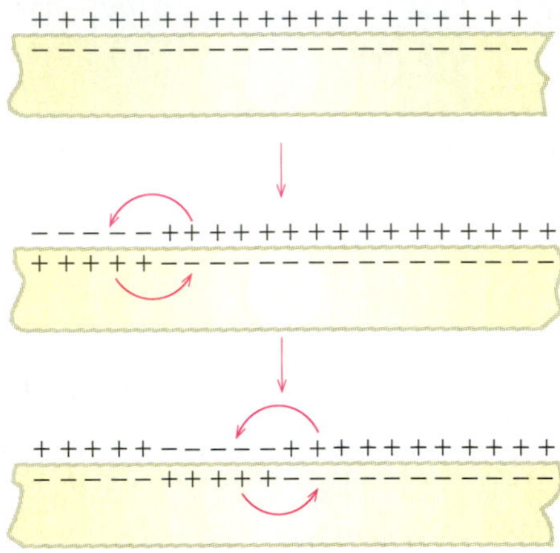

The axons of many nerve cells of higher animals, however, are also surrounded by a multilayered myelin sheath, each layer consisting of a typical lipid bilayer membrane (fig. 35.5). At intervals, the spacing of which depends on the fiber diameter, this myelin insulation is interrupted by the so-called nodes of Ranvier (fig. 35.6). Nerve impulses in these types of nerve fibers are conducted in a saltatory manner, that is, with action potentials "jumping" from node to node, where the axonal membranes are in direct contact with the extracellular fluid (see fig. 35.6). The insulation provided by the myelin sheath allows for efficient current conduction within the axoplasm by preventing signal loss, and as a result, this type of impulse propagation can be much more rapid than that observed in unmyelinated fibers of similar diameter. Indeed, impulse propagation velocities of over 100 m/s have been recorded in myelinated nerve fibers.

Gated Ion Channels

By a slight modification of the experimental setup shown in figure 35.2, it is possible to hold the membrane potential across an axonal membrane constant at any predetermined value. This is accomplished by connecting the recording electrodes to a device called a feedback amplifier. When any potential change is sensed by the electrodes, the feedback amplifier compensates for it by applying a current to keep the voltage constant. In such a voltage-clamped situation, ionic movements can be inferred from the amount of current necessary to hold $\Delta\Psi$ at its predetermined value. It was the use of a voltage clamp by Hodgkin and Huxley that allowed them to deduce the movements of Na^+ and K^+ that accompany the appearance of the action potential.

Figure 35.5

Cross section of a myelinated nerve axon from the superior cervical ganglion of the rabbit. Note the multilayered membrane of the myelin sheath that serves as an electric insulator. (Micrograph courtesy of Dr. T. Lentz.)

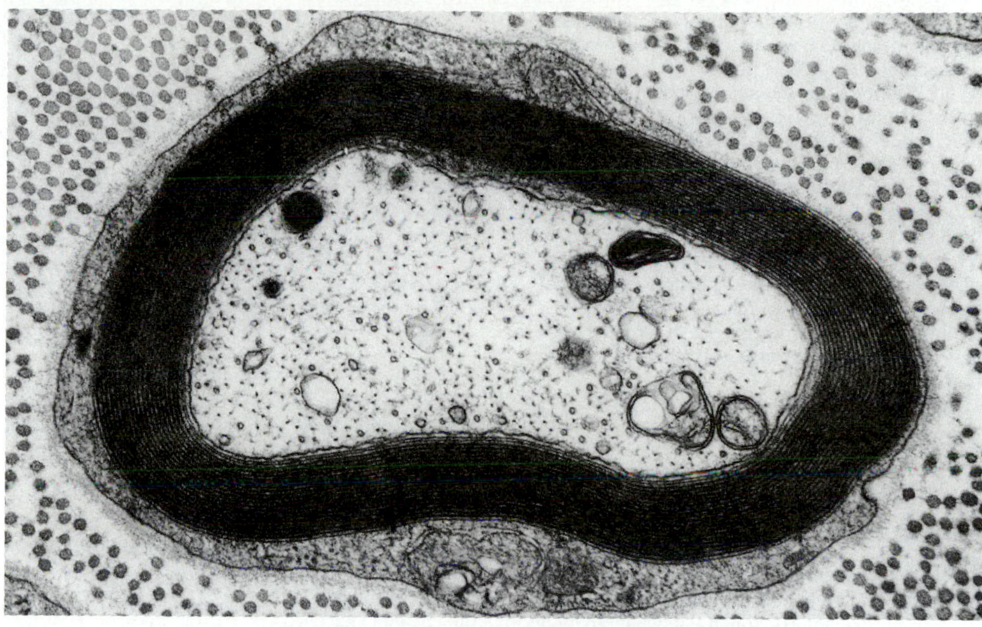

Figure 35.6

Schematic illustration of a longitudinal cross section of a myelinated axon. The sheath is interrupted at various intervals by the nodes of Ranvier, where the cytoplasmic membrane is exposed to the surrounding fluid. Nerve impulses are conducted in a saltatory manner in myelinated axons, as illustrated in this schematic diagram. Impulse propagation is from left to right and the arrows show the accompanying local current movements.

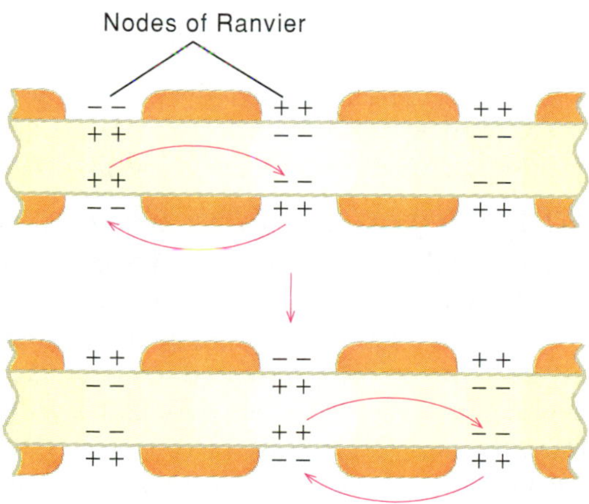

Because the changes in membrane permeability to Na^+ and K^+ were not superimposable in time, they tentatively concluded that two different "channels" were involved in the transmembrane movements of these ions.

Separate Channels for Na^+ and K^+ Have Been Found in Excitable Cell Membranes

Compounds that specifically block the conductance of the nerve membrane to either Na^+ or K^+ provided the first direct evidence for separate Na^+ and K^+ channels in the nerve-cell membrane. Tetrodotoxin and saxitoxin (fig. 35.7) are both potent nerve poisons that specifically inhibit the transmembrane movement of Na^+ by binding to the outside surface of nerve-cell membranes without affecting K^+ permeability, as measured in voltage-clamp experiments. Similarly, tetraethylammonium ions (fig. 35.8) have been shown to bind to the axoplasmic membrane surface and to specifically block the outward flow of K^+.

A large body of evidence has now accumulated that shows that K^+ and Na^+ movements through axonal membranes are mediated by separate channels that act as gated pores (chapter 32). During a typical action potential lasting about 5 ms, it has been calculated that at least 60 Na^+ ions pass through a single Na^+-specific channel (i.e., about 12,000 per second). Since the neuronal Na^+-K^+ ATPase normally pumps less than 200 Na^+ ions per second out of the cell, this pump cannot be directly involved in the development of the action potential. Instead, a process involving passive diffusion through a transmembrane pore best explains these rapid Na^+ fluxes. Furthermore, voltage-clamp studies by B. Hille established that molecules such as the K^+-monohydrate complex, with dimensions larger than about 0.4 nm^2 in cross section (the size of a monohydrated Na^+ ion), do not pass readily through the Na^+ channel. However, smaller molecules, such as monohydrated Li^+ ions, and other small cations such as hydroxylamine and hydrazine, do. This mechanism, that is, exclusion on the basis of size rather than chemical structure, is more characteristic of a pore than of a specific membrane permease (chapter 32).

Additional properties of the Na^+ channel have been inferred from many other experiments. For example, the pH dependence of Na^+ permeation, as well as other experimental observations, suggests that a negatively charged carboxylate

Figure 35.7

(*a*) The structure of tetrodotoxin, a compound that specifically blocks Na^+ channels in nerve cell membranes. This extremely toxic compound is found in the liver and ovaries of the Japanese puffer fish (*Spheroides rubripes*). (*b*) The structure of saxitoxin, a compound found in certain marine dinoflagellates ("plankton"), which are constituents of the so-called red tide. Mussels and clams that have fed upon these organisms are therefore extremely poisonous, and commercial shell fishing is banned in areas where these dinoflagellates appear. Saxitoxin is also a Na^+-channel blocking agent, and both compounds most probably interact with the channel through their positively charged guanidino groups.

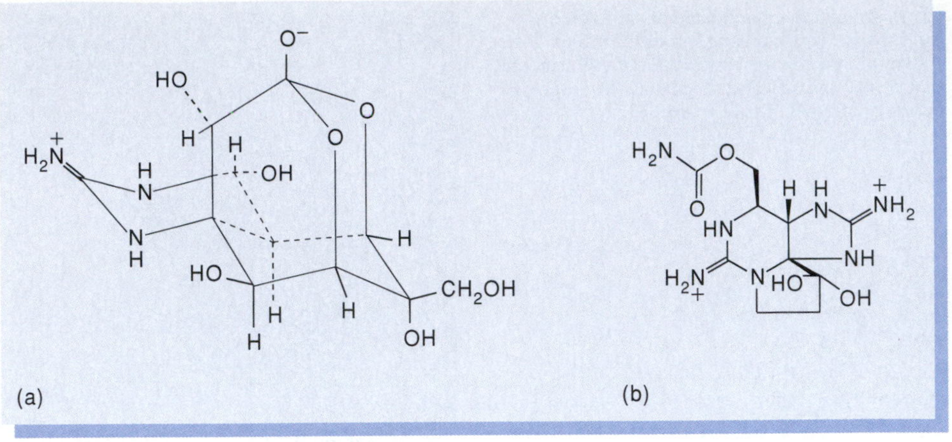

Figure 35.8

Tetraethylammonium ion, a compound that specifically inhibits K^+ channels in nerve cell membranes. The ethyl groups presumably sterically hinder passage of this cation through the channel, "plugging" the channel and therefore blocking the transport of K^+.

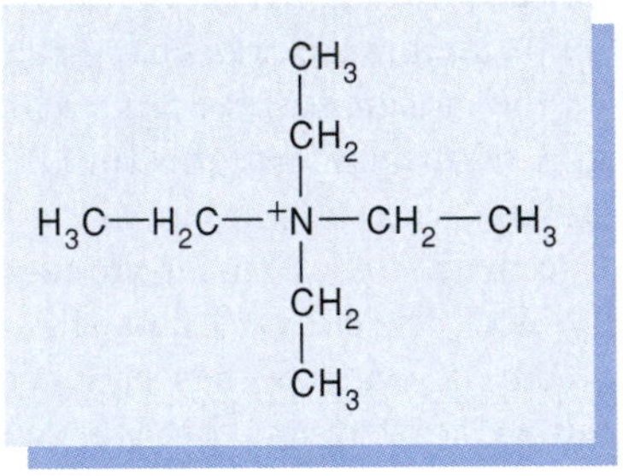

group ($-COO^-$) is present in the pore and is involved in interacting with the cationic molecules that can traverse the Na^+ channel. It is presumably this interaction that allows the positively charged guanidinium groups of tetrodotoxin and saxitoxin (see fig. 35.7) to bind to the channel, but the sizes of these poisons prevent their passage through the membrane, with the result that Na^+ conduction is blocked.

Estimates of the number of Na^+ channels in a variety of nerve types have been obtained by the use of radioactively labeled tetrodotoxin or saxitoxin molecules. From these studies, it is apparent that in unmyelinated nerve fibers, the density of Na^+ channels in the membrane is quite low. For example, the olfactory nerve of the garfish has only about 30 to 40 Na^+ channels per square micrometer of membrane, which corresponds to only about 0.2% of the total surface area of the phospholipid bilayer. In the squid giant axon, which has a fiber diameter some 2,500 times that of the garfish olfactory nerve, this value is increased only to several hundred Na^+ channels per square micrometer, or about 2% of the surface area. The situation is quite different, however, in myelinated nerves of vertebrate animals. In this case, Na^+ channels are found in significant numbers only at the nodes of Ranvier, as we might anticipate from the mechanism of impulse propagation in these types of nerve cells (see fig. 35.6). Here, the channel density is on the order of 10^4 channels per square micrometer, or about 60% of the total membrane surface area. Indeed, Na^+ fluxes at the nodes of Ranvier have been estimated to be 10 to 100 times larger than those in unmyelinated axons during propagation of the action potential.

In contrast to the situation with Na^+ channels, there are many fewer K^+ channels in the axonal membrane (about one-tenth the number in the squid giant axon). Nevertheless, it has been demonstrated that the K^+ pore is also quite specific, barring the passage of cations both smaller (Na^+) and larger (Cs^+; tetramethylammonium ions) than K^+. It seems likely, therefore, that both the diameter of the channel and specific interactions of channel components with the K^+-H_2O complex confer cation selectivity on this channel as well.

The Gating Properties of Ion Channels

How do channels selective for Na^+ and K^+ ions "sense" the membrane potential and react by opening or closing during various stages of action-potential propagation? Although this question is still far from being answered in molecular detail, a number of clues have emerged from work on the gating properties of these channels, their purification and reconstitution into artificial membrane systems, and the recent cloning and sequencing of genes encoding Na^+ and K^+ channels. The results of these investigations show that ion channels in nerve and muscle cell membranes can assume multiple conformations, one of which is more permeable to the ion in question than the others.

An important observation that bears on this question is the discovery of gating currents, first predicted by Hodgkin and Huxley. Under appropriate conditions, a small current opposite in direction to that carried by Na^+ during the opening of the Na^+ channels can be detected just before the development of the action potential. This current is short in duration (0.1 ms) and precedes the inward movement of Na^+ (fig. 35.9). One widely accepted proposal is that the Na^+ channel has a "built-in" voltage sensor that has a large dipole moment. Depolarization of the membrane, which elicits the action potential, would displace or rearrange the dipole in response to the change in the electric field, and this process would be detected as the gating current. The change in the orientation of the dipole could then be the trigger by which the Na^+ channel is opened, presumably by means of a conformational change in this protein. Indeed, a candidate for such a voltage sensor in both Na^+ and K^+ channels has recently been recognized from the primary structures of these proteins (see the following discussion).

Additional evidence for voltage-dependent conformational changes that allow the gating of the Na^+ channel has been obtained using nerve poisons from North African scorpions and the sea anemone. These toxins, which are basic polypeptides, bind to sites on the Na^+ channel that are distinct from those that bind the channel-blocking agents tetrodotoxin and saxitoxin. They appear to exert their physiological effects by slowing inactivation of the Na^+ channel after its initial opening during the action potential. Binding of scorpion toxin to Na^+ channels in nerve and muscle-cell membranes has been shown to be voltage-dependent, a fact suggesting that this toxin recognizes a conformation of the channel that also depends on the membrane potential (the "open" conformation). Toxin binding would thus reflect the conformational state of the channel, which, in turn, depends on $\Delta\Psi$. These and other experiments have led to the conclusion that both the Na^+ and K^+ channels can exist in at least three conformations: closed (resting $\Delta\Psi$), open, and inactive. The latter two states are responsible for the transient increase and then decrease in Na^+ and K^+ permeabilities, in sequence, observed during the action potential.

Further Evidence Concerning the Structure and Function of Ion Channels

One major approach to understanding structure-function relationships in the ion channels of nerve-cell membranes has been to purify these proteins and to study their properties in artificial membrane systems, as we described for transport permeases in chapter 32. The ability of Na^+ channels to tightly bind specific neurotoxins has been used as an assay in such purifications. Na^+ channels have been purified from the electric organ of the electric eel (*Electrophorus electricus*), which is rich in such channels, as well as from a number of higher vertebrate sources. A major component of all these preparations was a protein with a polypeptide-chain molecular weight of about 260,000, while variable amounts of a second polypeptide (M_r = ca. 37,000) were also found in some preparations. These two subunits have been referred to as the α and β subunits of the Na^+ channel, respectively.

Several highly purified preparations of the Na^+ channel have been reconstituted into proteoliposomes. In these experiments, reconstituted Na^+ channels were activated by channel activators and blocked by channel inhibitors such as saxitoxin and tetrodotoxin. Moreover, the cation specificity of reconstituted Na^+ channels was the same as that for the native channel:

Figure 35.9

Illustration of the gating current that precedes the inward-directed Na^+ current associated with depolarization during the action potential. The gating current is believed to be related to the voltage-dependent opening of the Na^+ channels and perhaps may reflect rearrangement of a dipolar "voltage sensor" associated with the channel gate. (Source: C. M. Armstrong and F. Bezanilla, "Charge movement associated with the opening and closing of the activation gates of the Na channels," in *Journal of General Physiology* 63:533, 1974. Copyright © 1974 Rockefeller University Press, New York, N.Y.)

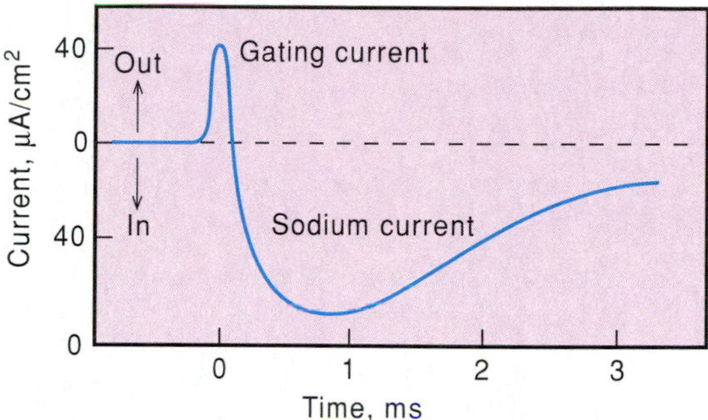

$Li^+ = Na^+ > K^+ > Rb^+ > Cs^+$. Further, in their purified form these channels retained a voltage dependency similar to that observed in the cells from which they were obtained. Since most of these properties could be attributed to the α subunit in many of these systems, it appears that this polypeptide alone is sufficient for most of the biological functions attributed to the Na^+ channel in nerve and muscle cells.

A major breakthrough in attempts to understand the structure and function of ion channels of excitable cells came with the cloning and sequencing of the cDNA that encodes the α subunit of the Na^+ channel from *Electrophorus electricus* by S. Numa and collaborators in 1984. The deduced amino acid sequence showed that this protein consists of 1,820 amino acid residues and has several remarkable features. First, the protein consists of four highly homologous domains, each of which may span the membrane six times, with both the N- and C-terminal regions on the cytoplasmic surface of the membrane (fig. 35.10). The membrane-spanning regions include both potential hydrophobic α helices (regions 1, 2, 5, and 6 in figure 35.10) and amphipathic α helices (regions 3 and 4). Strikingly, potential membrane-spanning region 4 (S4) of all four domains contains a stretch in which every third aminoacyl residue is basic (either lysine or arginine). A number of workers have proposed that this region may be the positively charged voltage sensor that moves outward in response to depolarization of the nerve-cell membrane, thus producing the gating current. In contrast, proposed membrane-spanning region 3 (S3) can form a highly amphipathic helix with a net negative charge. Synthetic peptides containing the amino acid sequence of S3 have been shown to form ion-selective channels in phospholipid bilayer membranes. Moreover, computer-assisted three-dimensional structural predictions have shown that four S3 peptides could form a tetrameric transmembrane pore with hydrophilic side chains lining this channel (fig. 35.11). The dimensions of the channel are strikingly similar to those deduced for the native Na^+ channel

Figure 35.10

A functional map of the Na^+ channel α subunit. The transmembrane folding model of the α subunit is depicted with experimentally demonstrated sites of cAMP-dependent phosphorylation (P), interaction of site-directed antibodies that define transmembrane orientation (≻), covalent attachment of α-scorpion toxins (ScTx), glycosylation (ΨΨ), and modulation of channel inactivation (h). (Source: W.A. Catterall, "Structure and function of voltage-sensitive ion channels," in *Science* 242:50, 1988.)

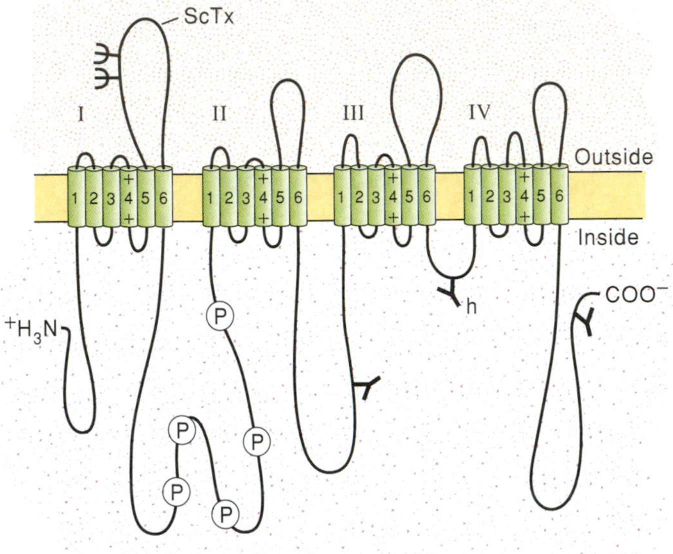

Figure 35.11

End view of energy-optimized parallel tetramer of the sodium channel S3 segment. Colors: light blue, α-carbon backbone; red, acidic residues; blue, basic residues; yellow, polar and neutral residues; and purple, lipophilic residues. This view is from the intracellular face of the membrane, at which the N terminus of each segment is predicted to be located.

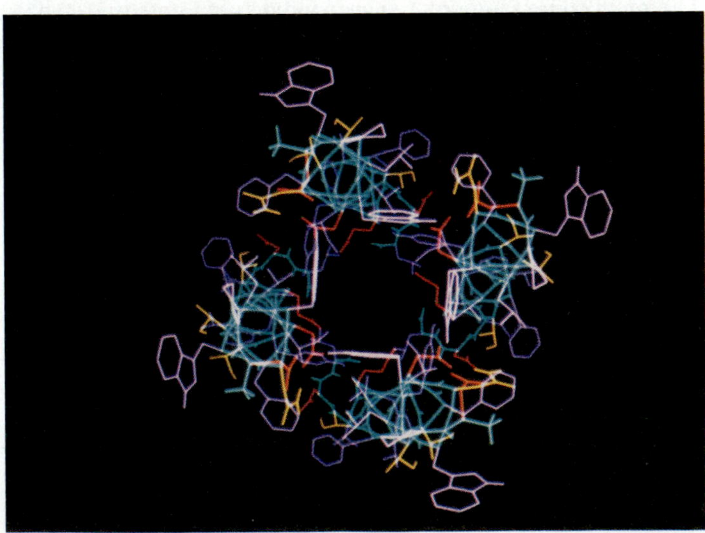

on the basis of its size selectivity. The negatively charged, acidic residues inside and at the entrances to the pore could account for its cation selectivity and also for the results of the pH-dependence studies of Na^+-channel function mentioned earlier. Additionally, the loop between S5 and S6 has recently been shown to have a role in channel function and may therefore comprise part of the pore. However, whether each of the four homologous domains of the Na^+ channel, or a complex including, for example, the S3 region and the S5/S6 loop from each of these domains, forms the functional channel remains to be determined. Finally, a region between homologous domains I and II that is exposed in the cytoplasm has been found to be phosphorylated in a cyclic-AMP-dependent manner (see fig. 35.10). Phosphorylation of Na^+ channels inhibits Na^+ influx, and therefore may have a regulatory role in excitable cells.

Genes encoding Na^+ channels from higher organisms have also been cloned and sequenced, and they have a structure that is highly homologous to the one we have just described. More recently, the gene corresponding to the *Shaker* locus of *Drosophila* has also been cloned and sequenced. True to its name, a mutation in this locus causes severe neurological malfunction in the affected fruit fly, which is due to a defect in one type of K^+ channel. The deduced amino acid sequence of this K^+ channel revealed a 616-residue polypeptide that showed significant primary and secondary structural homology to each of the four domains of the Na^+ channel. Furthermore, models of the intramembrane structure of this protein were very similar to that of a single Na^+ channel domain, complete with approximately six membrane-spanning regions, an S4-like voltage sensor domain, and an S3-like domain with an enrichment of negatively charged residues. These properties of the K^+ channel from *Drosophila,* and the fact that even unicellular eukaryotes, such as *Paramecium,* and bacteria, such as *E. coli,* have voltage-sensitive K^+-selective channels, have led to the suggestion that all such channels may have had a common evolutionary ancestor that was similar in structure to the present-day K^+ channels. Presumably, tandem duplication events involving such an ancestral gene could have given rise to the present-day Na^+ channel gene, consisting of four tandem segments that are highly homologous to each other and to the K^+-channel gene.

Very recently, the powerful tool of site-directed mutagenesis has been used to study structure-function relationships in ion channels. Such studies have been aided by the development of techniques to study specific mutant proteins expressed in a model system, oocytes from South American frogs of the genus *Xenopus*. These relatively large cells are easily manipulable, and mRNA transcribed from normal or mutant ion channel genes can be injected into the oocytes and expressed. Voltage-clamp studies of *Xenopus* oocytes, into which mRNAs encoding various ion channels have been injected, have shown that the channel proteins are expressed in a functional form in the oocyte membrane. Thus far, use of this technique has yielded its greatest benefits in the area of understanding the mechanism of ion channel inactivation.

Recent experiments by R. W. Aldrich and co-workers have provided strong evidence that ion channel inactivation may operate by a relatively simple "ball-and-chain" mechanism (fig. 35.12). These workers studied a number of site-directed mutants of the *Drosophila Shaker* gene for their inactivation properties after injecting the corresponding mRNA for the gene into *Xenopus* oocytes. Previous experiments had suggested that the extreme N terminus of the *Shaker* gene product (K^+ channel) may be involved in inactivation, so the focus was on mutations in this region. The results of the experiments can be summarized as follows:

1. Mutations within the first eleven amino acid residues of this protein that resulted in hydrophobic to polar substitutions gave a K^+ channel in which inactivation was slowed relative to the wild-type protein. Furthermore, mutations that neutralized or eliminated positively charged residues at positions 14, 16, 17, 18, and/or 19, immediately following the hydrophobic N terminus, also slowed the inactivation rate of the channel.
2. A region immediately C terminal to the positively charged cluster of amino acids (residues 23–37) was also studied. This region is rich in hydrophilic amino acids, and when deletion mutations were made in this region, inactivation was faster than in the wild-type protein. Conversely, insertion mutations in this region slowed inactivation of the channel.
3. Finally, a deletion mutant that lacked most of the first twenty aminoacyl residues failed to inactivate, as we would expect from the previous results. However, inactivation could be restored by adding back a synthetic peptide that had the same sequence as the amino-terminal twenty amino acid residues of the native *Shaker* channel. The degree of restoration of inactivation was directly proportional to the amount of peptide added.

The simplest interpretation of these results is that the first twenty amino acid residues of this K^+ channel comprise a "ball," with the positively charged residues on the surface and the hydrophobic residues forming the core. It is this structure that is responsible for K^+ channel inactivation by virtue of its binding to the inside surface of the pore, presumably at least partially by interacting with negatively charged residues near the pore opening. Mutations that disrupt this structure disrupt inactivation. The next twenty residues or so are hypothesized to be a "chain" that can "swing" the ball into or out of the pore opening. Shortening the chain by deletion mutation brings the ball closer to the opening and thus speeds up inactivation, while the reverse is true for insertion mutations. However, the free ball can inactivate a mutant protein lacking an attached ball, as long as it is provided in a high enough concentration. Perhaps the blocking effect of the triethylammonium cation, which we mentioned earlier, is simply due to its mimicking the positively charged region of the inactivation "ball." Figure 35.12 is thus a reasonable schematic model for how inactivation may work,

Figure 35.12

"Ball-and-chain" model for inactivation of the K^+ channel. (*a*) Open conformation, (*b*) inactive conformation. (Source: "Research News," in *Science* 250:507, 1990.)

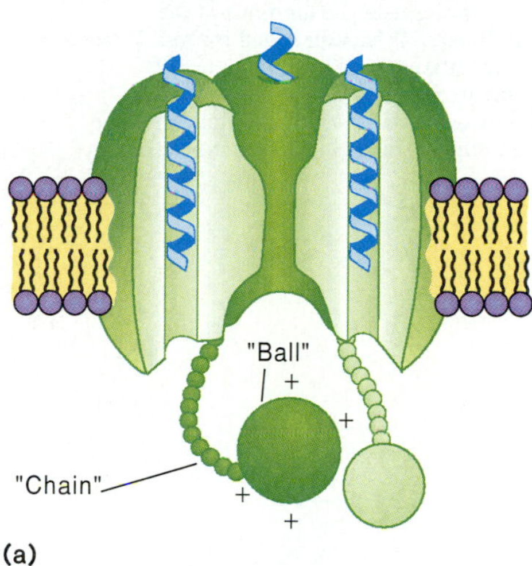

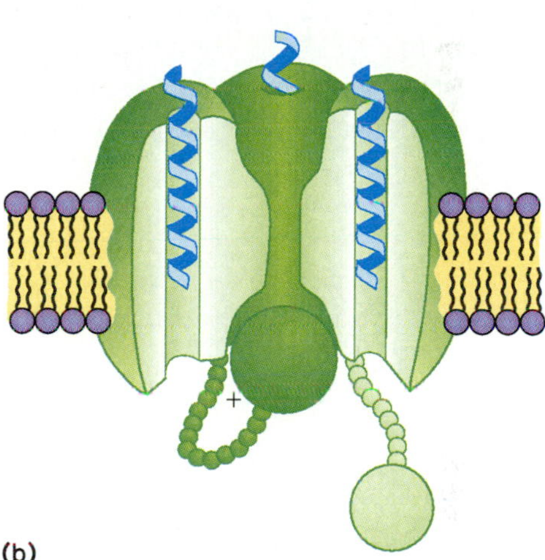

at least in the case of one ion channel. Application of similar molecular biological techniques to the cloned genes of other ion channels will no doubt cast further light on structure-function relationships in these ubiquitous proteins.

Synaptic Transmission: A Chemical Mechanism for Communication between Nerve Cell and Target Cell

Nerve impulses, propagated along the axon by mechanisms outlined in the preceding sections, must be transmitted between nerve cells or from nerve cells to muscle or glandular tissues in order for their effects (e.g., contraction or secretion) to take place. This type of intercellular communication can occur either by an electrical mechanism of coupling or by chemical transmission across specific connections, called synapses, between

Figure 35.13

Schematic diagram of a synaptic junction in which acetylcholine is the chemical transmitter. Arrival of an action potential at the terminus of the presynaptic cell (*top*) stimulates Ca^{2+} uptake, which triggers release of acetylcholine (ACh) from vesicles near the terminus of the presynaptic cell. Release is accomplished by vesicle fusion with the plasma membrane, and the interaction of acetylcholine with its receptors in the postsynaptic membrane triggers depolarization, thus propagating the action potential in the postsynaptic cell (*bottom*).

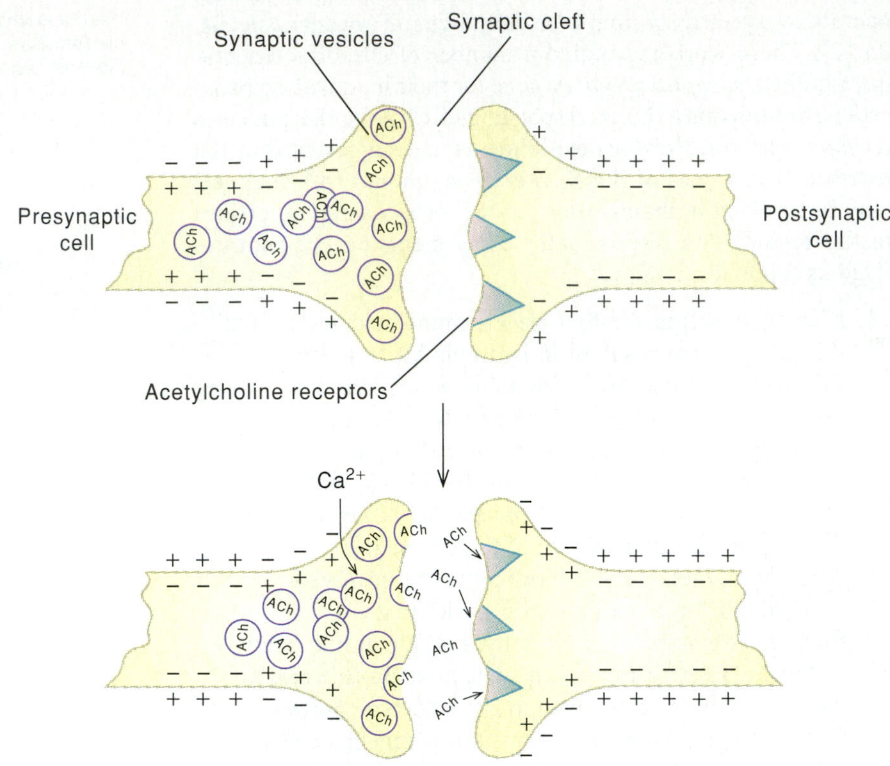

Figure 35.14

Acetylcholine is synthesized by cholineacetyltransferase, which esterifies choline with an acetyl group from acetyl-CoA.

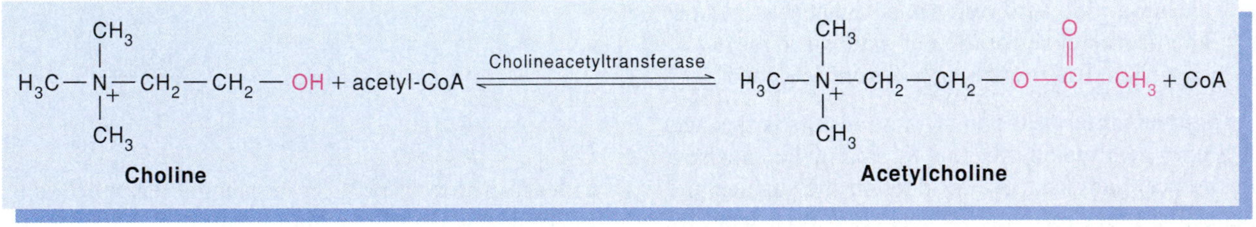

$$\underset{\textbf{Choline}}{H_3C-\overset{\overset{CH_3}{|}}{\underset{\underset{CH_3}{|}}{N^+}}-CH_2-CH_2-OH} + \text{acetyl-CoA} \xrightleftharpoons{\text{Cholineacetyltransferase}} \underset{\textbf{Acetylcholine}}{H_3C-\overset{\overset{CH_3}{|}}{\underset{\underset{CH_3}{|}}{N^+}}-CH_2-CH_2-O-\overset{\overset{O}{\|}}{C}-CH_3} + \text{CoA}$$

nerve cells and target cells. An example of electrical transmission was given in chapter 32. Direct cell-to-cell contact by means of gap junctions can allow action potentials to be transmitted between cells by ionic mechanisms. In another mechanism, specific chemicals called neurotransmitters are released at the presynaptic membrane of one cell in response to membrane depolarization. These chemicals then diffuse to receptors in the postsynaptic membrane of the recipient cell, where a new action potential may be generated.

Acetylcholine Is a Common Chemical Neurotransmitter

The process of synaptic transmission is depicted schematically in figure 35.13 for acetylcholine, the excitatory neurotransmitter at vertebrate neuromuscular junctions (motor end plates). Acetylcholine is synthesized in nerve cells by the enzyme cholineacetyltransferase (fig. 35.14) and is packaged in units of 10^3 to 10^4 molecules within synaptic vesicles, which are abundant near the cytoplasmic membrane of the presynaptic axon. The arrival of an action potential triggers a large increase in the permeability of the presynaptic membrane to Ca^{2+}, and this ion flows into the axoplasm down its chemical gradient. Fusion of the synaptic vesicles with the plasma membrane and the concomitant release of acetylcholine into the synaptic cleft are promoted by this increase in intracellular Ca^{2+}. In a typical neuromuscular synaptic junction, several hundred synaptic vesicles empty their contents into the synaptic cleft by this mechanism in response to a single action potential. The resultant large increase in the local concentration of acetylcholine is "sensed" by a protein, the acetylcholine receptor, located in the cytoplasmic membrane of the postsynaptic cell. Binding of the neurotransmitter to many receptor molecules triggers an action potential in the recipient cell. The acetylcholine is then rapidly degraded into acetate and choline by an enzyme in the synaptic cleft called acetylcholinesterase, and the resting potential of the postsynaptic membrane is soon restored.

How does the binding of a neurotransmitter to its receptor promote depolarization of the postsynaptic membrane? One of the first clues came from studies by B. Katz and collaborators, who showed that acetylcholine increases the ionic permeability of the postsynaptic membrane. Subsequent studies with radioactive tracers and by the voltage-clamp technique established that the membrane permeabilities to both Na^+ and K^+ were increased simultaneously by the action of acetylcholine. Because in the resting cell K^+ permeability is already quite high relative to Na^+ permeability, the major effect of acetylcholine binding and receptor channel opening is a substantial influx of Na^+ down its concentration gradient. The inward Na^+ current tends to collapse (or depolarize) the negative resting potential toward zero. This local perturbation in $\Delta\Psi$ is enough to initiate a new action potential in the recipient neuronal or muscular membrane as long as a sufficient number of receptor molecules bind the neurotransmitter. In fact, as we will see in a later section, the receptor itself contains the ion channel through which both Na^+ and K^+ can flow. The number of occupied receptors at a given moment thus dictates the magnitude of the inward flow of Na^+ and the resulting magnitude of the change in the membrane potential. Rapid hydrolysis of acetylcholine by acetylcholinesterase bound to the postsynaptic membrane then quickly reduces the number of transmitter-receptor complexes and repolarizes the membrane until a new action potential triggers the release of more neurotransmitter from the presynaptic membrane.

To elucidate the steps just outlined, researchers have made use of specific inhibitors of both acetylcholinesterase and the acetylcholine receptor. Compounds such as eserine (fig. 35.15*a*) block the hydrolytic enzyme and thus can be used to study the effects of acetylcholine in cholinergic systems (those using acetylcholine as a transmitter) under conditions where it cannot be hydrolyzed. Likewise, certain organophosphorus compounds efficiently inhibit acetylcholinesterase by forming stable covalent intermediates with an active-site serine in the enzyme. The widely used insecticides parathion and malathion are examples of this class of nerve poisons (fig. 35.15*b*). Neuromuscular junctions exposed to acetylcholinesterase inhibitors are paralyzed because the persistent presence of acetylcholine prevents repolarization of the postsynaptic membrane to restore its excitability. In fact, in such a situation the acetylcholine receptor eventually becomes desensitized, i.e., it remains closed to ion flow for long intervals even in the presence of the neurotransmitter.

Specific blocking agents of the acetylcholine receptor include *d*-tubocurarine, an active component of the neurotoxin curare (fig. 35.16), and the snake venom poisons α-bungarotoxin (from snakes of the genus *Bungarus*) and cobratoxin. The latter are small basic proteins with masses of around 7,000 daltons. All three of these substances interact noncovalently with the receptor and interfere with acetylcholine binding, thus blocking depolarization of the postsynaptic membrane. They are referred to as antagonists of the cholinergic systems. Another type of acetylcholine receptor inhibitor is exemplified by the divalent cation decamethonium (fig. 35.17), which "locks" the ion

Figure 35.15

Acetylcholinesterase inhibitors. (*a*) Eserine, or *physostigmine*, is an alkaloid that forms a relatively stable covalent carbamoyl intermediate with an active-site serine on the enzyme that is hydrolyzed only very slowly. (*b*) Parathion and malathion are organophosphorus compounds that also form stable covalent complexes with the active-site serine of acetylcholinesterase. They are widely used agriculturally as insecticides.

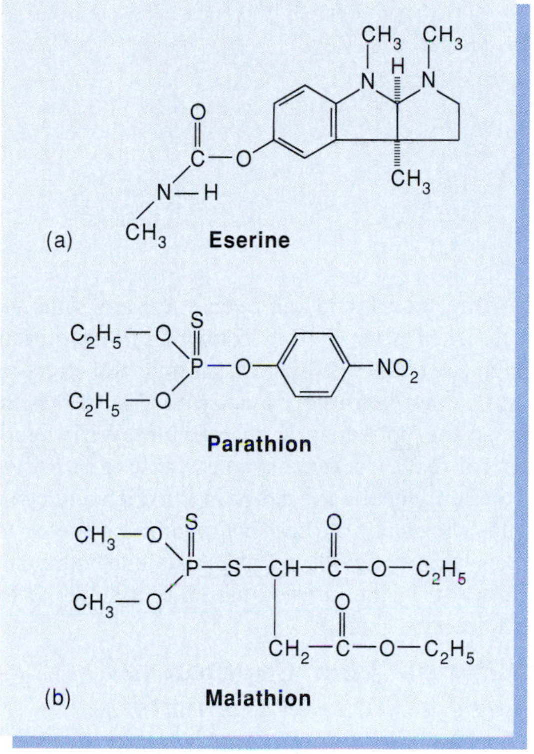

Figure 35.16

The structure of *d*(+)-tubocurarine, an active component of the neurotoxin curare. This compound binds to the acetylcholine-binding site on its receptor, preventing synaptic transmission and subsequent depolarization of the postsynaptic cell membrane.

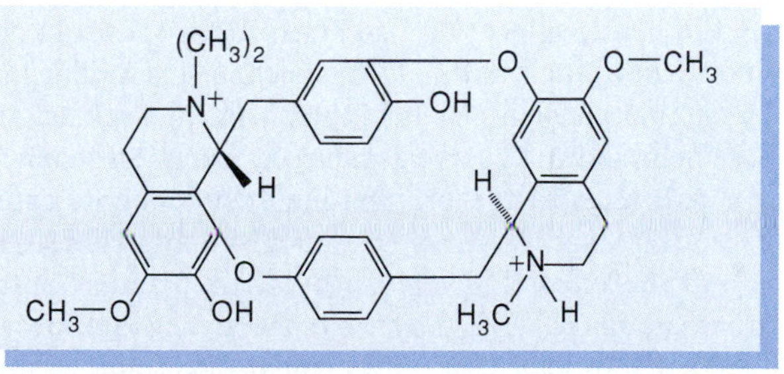

Figure 35.17

Decamethonium ion, an agonist of cholinergic systems. This compound binds to the acetylcholine receptor, but because it cannot be degraded, it causes persistent depolarization of the postsynaptic membrane.

$$CH_3-\overset{+}{N}(CH_3)_2-(CH_2)_{10}-\overset{+}{N}(CH_3)_2-CH_3$$

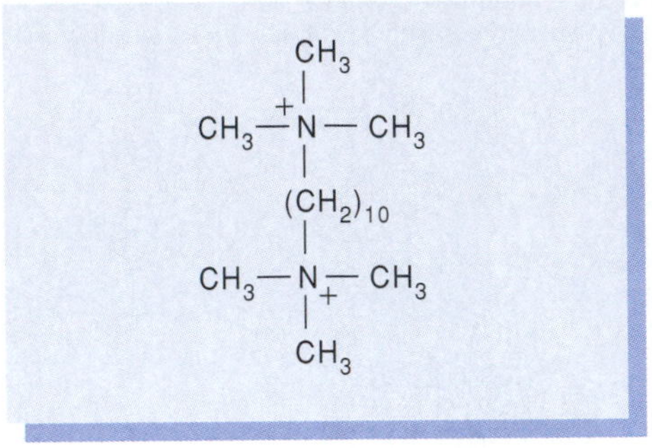

channel of the receptor in the open state and thus leads to a constant depolarization of the recipient cell membrane. Such compounds, referred to as agonists, mimic the effect of acetylcholine but cannot be rapidly inactivated, so they block resensitization of the postsynaptic membrane. With agonists and antagonists as tools, workers have been able to investigate properties of the acetylcholine receptor in the open and closed states and to study the effects of these compounds on the permeability of the postsynaptic membrane. Agonists and antagonists have also been useful in the purification of acetylcholine receptors, as we will describe shortly.

A Number of Other Compounds Also Serve as Neurotransmitters

The best-documented examples of chemical neuromessengers other than acetylcholine are the catecholamines, certain amino acids (and derivatives), and a variety of peptides. The catecholamines are all derived biosynthetically from L-tyrosine (fig. 35.18) and include the hormones norepinephrine and epinephrine (adrenaline). Because these compounds are also synthesized in the adrenal gland, neurons that use these substances as chemical transmitters are said to be adrenergic. Sympathetic nerve fibers that innervate smooth-muscle cells in internal organs such as the heart, spleen, and gut have been shown to release norepinephrine at their terminals by mechanisms similar to those used by cholinergic neurons. Norepinephrine and related amines also have been shown to serve as neurotransmitters in a number of nerve pathways in the brain. Like acetylcholine, catecholamines may be inactivated by chemical modifications. Inactivation may be effected by a methylation reaction or by an oxidation reaction catalyzed by the enzyme monoamine oxidase (see fig. 35.18). In some cases, they can also be resorbed through the presynaptic membrane after their release, providing an additional mechanism for removal from the synaptic cleft.

The hydroxylation of tyrosine, which is the first unique step in the biosynthesis of catecholamines, is catalyzed by tyrosine hydroxylase and yields the compound 3,4-dihydroxyphenylalanine (L-DOPA). The neurological disorder Parkinson's disease is associated with an underproduction in the human brain of the catecholamine transmitter dopamine, which is derived from L-DOPA by a decarboxylation reaction (see fig. 35.18). L-DOPA has therefore been found to be an effective drug in some instances in the treatment of Parkinson's disease. Interestingly, overproduction of dopamine in the brain also occasionally occurs and appears to be associated with psychological disorders such as schizophrenia. In these cases, dopamine-receptor blocking drugs, such as chlorpromazine (fig. 35.19), have been found to be useful therapeutic agents.

Amino acids that are believed to have roles as neurotransmitters include glutamic acid and glycine. The amino acid derivatives histamine (synthesized by the decarboxylation of histidine), 5-hydroxytryptamine (or serotonin, derived from tryptophan), and gamma-aminobutyric acid (or GABA, a decarboxylation product of glutamic acid) have all been shown to be transmitters in various systems as well. The reactions involved in the biosynthesis of these compounds are diagrammed in figure 35.20. GABA is used most often as an inhibitory transmitter. Interaction of this compound with its receptor on many postsynaptic membranes results in a large increase in the membrane permeability to Cl^- and/or K^+ ions. This change inhibits the postsynaptic cell, often by hyperpolarizing its membrane (recall, for example, that the equilibrium potential of K^+ is more negative than the resting potential). Most target tissues, in fact, are innervated by more than one type of nerve fiber, each using a different neurotransmitter. Thus different signals can be relayed to the recipient cells, some stimulatory and others inhibitory. Given the complexity of the nervous systems of higher animals, additional neurotransmitters are continually being discovered. Some compounds currently thought to be chemical transmitters are listed in table 35.2.

The Acetylcholine Receptor Is the Best-Understood Neurotransmitter Receptor

The structure and properties of the acetylcholine receptor are by far the best understood among all neurotransmitter receptors. One reason for this is their abundance in postsynaptic membranes found in the electric organs of the electric eel (*Electrophorus*) and the electric ray (*Torpedo*). These organs contain stacks of cells (electroplaxes), each cell of which receives nerve endings on one side but not on the other. Release of acetylcholine by the presynaptic nerve endings thus depolarizes the postsynaptic membrane by virtue of the binding of the neurotransmitter to the membrane-bound receptors, while the other face of the cell remains at its resting potential. In this way each cell, when stimulated, can attain a potential difference of over 100 mV between its two faces, and thousands of such stacked cells, present in the electric organ, can consequently emit an electric discharge of several hundred volts.

Acetylcholine-receptor-rich membranes can easily be isolated from the electric organs of *Electrophorus* and *Torpedo*. These membranes contain from 10 to 50 $\times$ 10^4 receptor sites per square micrometer, as measured by the binding of radioactively labeled antagonists such as α-bungarotoxin. Electron microscopy and x-ray diffraction have further revealed that such membrane fragments have a similar density of particles, each

Figure 35.18

Biosynthesis from L-tyrosine and inactivation by monoamine oxidase of catecholamine neurotransmitters. Norepinephrine, epinephrine, and dopamine are confirmed neurotransmitters in various systems, and L-DOPA is a probable neurochemical messenger (see table 35.2).

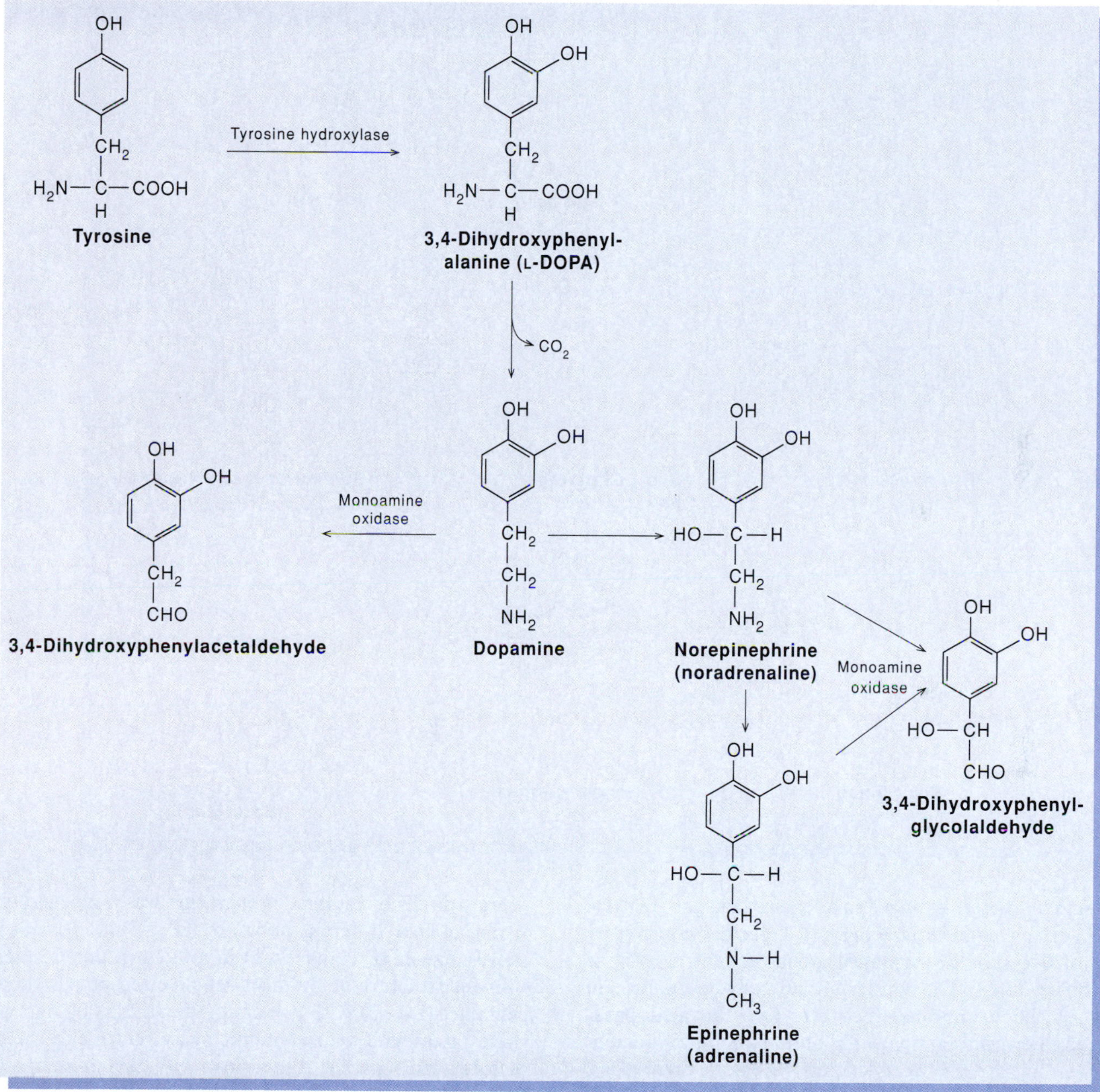

Figure 35.19

The structure of chlorpromazine, a drug that blocks dopamine receptors and has been used in the treatment of psychological disorders such as schizophrenia.

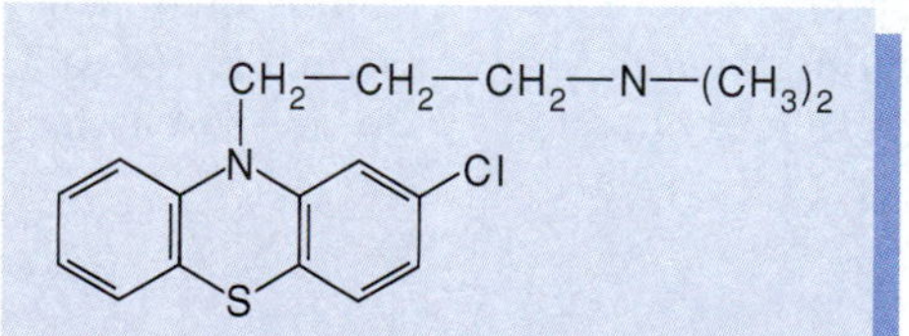

Figure 35.20

Biosynthesis of the neurotransmitters histamine, gamma-aminobutyrate (GABA), and serotonin (5-hydroxytryptamine).

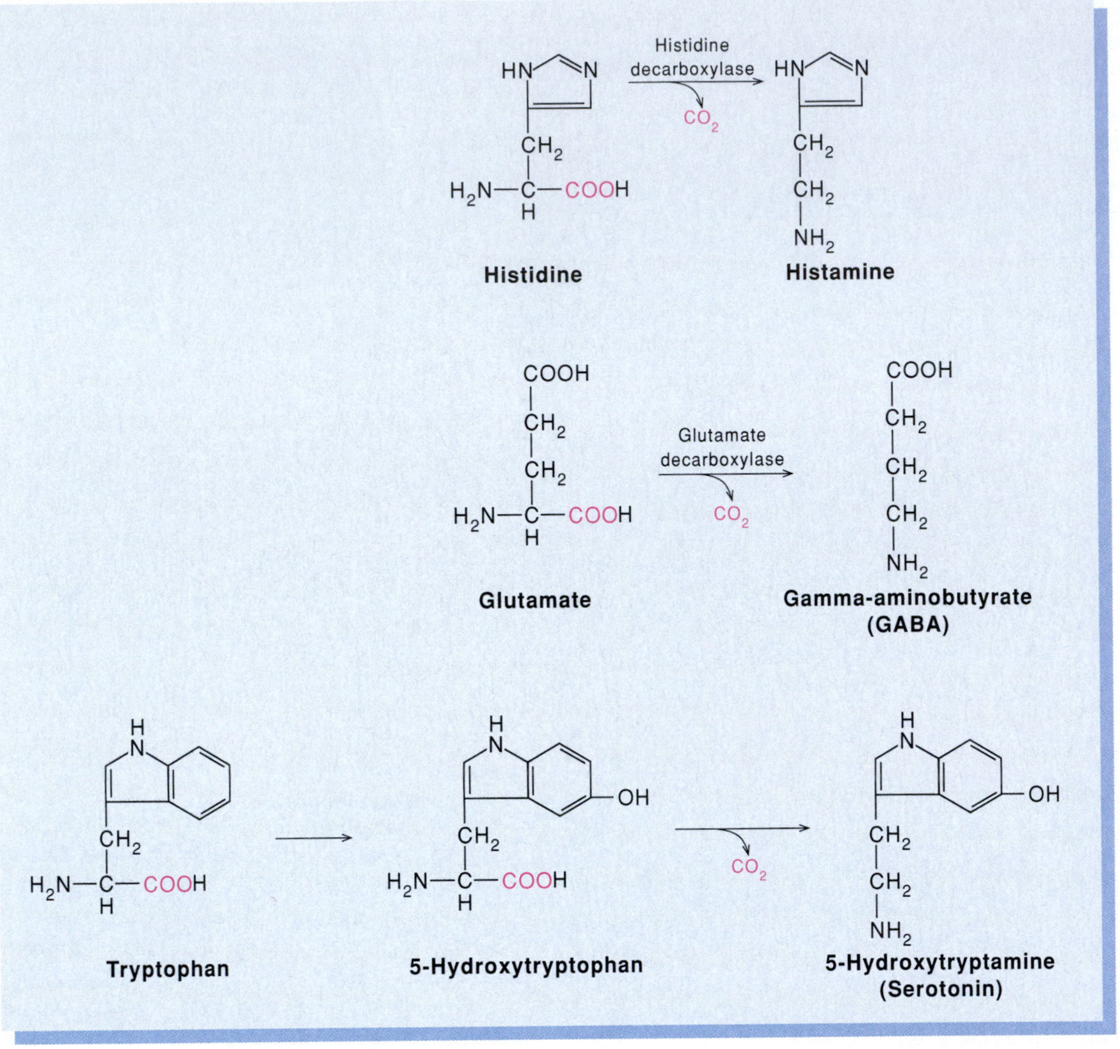

about 8 nm in diameter, arranged in a regular hexagonal lattice (fig. 35.21). Each particle is actually a "rosette," apparently consisting of five subunits arranged around a central axis, as shown in figure 35.21. The density of these oligomeric protein molecules on the membrane, their size (see the next paragraph), and their interaction with antibodies have demonstrated that they are the acetylcholine receptors themselves. Furthermore, closed membrane vesicles derived from such membrane fragments can be made permeable to Na^+ by adding agonists of the acetylcholine receptor, an effect that is blocked by antagonists such as the snake venom toxins.

The acetylcholine receptor was purified to homogeneity for the first time by solubilizing electroplax membranes of *Torpedo* with nonionic detergents and subjecting them to affinity chromatography on columns containing covalently bound cobratoxin. The resulting preparation was a glycoprotein consisting of four different subunits (M_r = 40, 50, 60, and 65 $\times 10^3$, named α, β, γ, and δ) in a molar ratio of 2:1:1:1. The simplest structure of the acetylcholine receptor is therefore a pentamer ($\alpha_2\beta\gamma\delta$) with a mass of about 255,000 daltons. This value agrees with hydrodynamic measurements on the purified receptor and with the dimensions and subunit composition of the particle seen by electron microscopy (see fig. 35.21). Interestingly, the four different subunits of this protein, all of which have been fully sequenced, exhibit extensive sequence homology, an indication that they all arose by duplication and divergence of a single ancestral gene. Computer analyses of these amino acid sequences suggest a common motif of secondary structure for each subunit inside the lipid bilayer (see the following discussion).

Table 35.2
Neurochemical Messengers

Compounds	Status[a]
Acetylcholine	C
Catecholamines	
Norepinephrine (noradrenaline)	C
Epinephrine (adrenaline)	C
L-DOPA	P
Dopamine	C
Octopamine	C
Amino acids (and derivatives)	
Glutamate	C
Aspartate	P
Glycine	C
Proline	Pos
Gamma-aminobutyrate (GABA)	C
Tyrosine	Pos
Taurine	P
Alanine	Pos
Cystathione	Pos
Histamine	C
Serotonin (5-hydroxytryptamine)	C
Peptides	
Substance P	P
Cholecystokinin	P
Neurotensin	P
Enkephalins	P
Somatostatin	P

[a]C = confirmed neurotransmitter; P = probable; Pos = possible.

Reactive affinity labels that covalently bind to the acetylcholine receptor have been prepared by chemically modifying known agonists and antagonists of cholinergic systems. In most cases examined, the α subunit was labeled by these treatments. These and other studies have shown that the α subunit contains the binding site for acetylcholine, and the residues most important for this interaction appear to be localized between aminoacyl residues 180 and 200 of this subunit. This region of the protein is believed to be part of an extracellular, hydrophilic domain of the receptor, as we might expect from its apparent function. The purified receptor from *Torpedo* has been functionally reconstituted into phospholipid vesicles. These proteoliposomes became permeable to Na^+ in the presence of an acetylcholine analog, while α-bungarotoxin blocked Na^+ permeability, as we would expect if the complex as purified comprised the authentic receptor molecule.

Figure 35.21

A view of the arrangement of acetylcholine receptors on electroplax membranes from *Torpedo californica*. (*a*) Membrane tubes formed spontaneously from membrane vesicles showing a crystalline surface lattice of acetylcholine receptors. Stained with uranyl acetate. (*b*) Computer filtration of micrographs similar to that in (*a*), showing the arrangement of protein density in a single receptor around the central depression (presumed to be the ion channel). (From J. Kistler, R. M. Stroud, M. W. Klymkowsky, R. A. Lalancette, and R. H. Fairclough. "Structure and function of an acetylcholine receptor." Reproduced from the *Biophysical Journal*, 1982, 37:371, by © permission of the *Biophysical* Society. Micrograph courtesy of Dr. R. Stroud.)

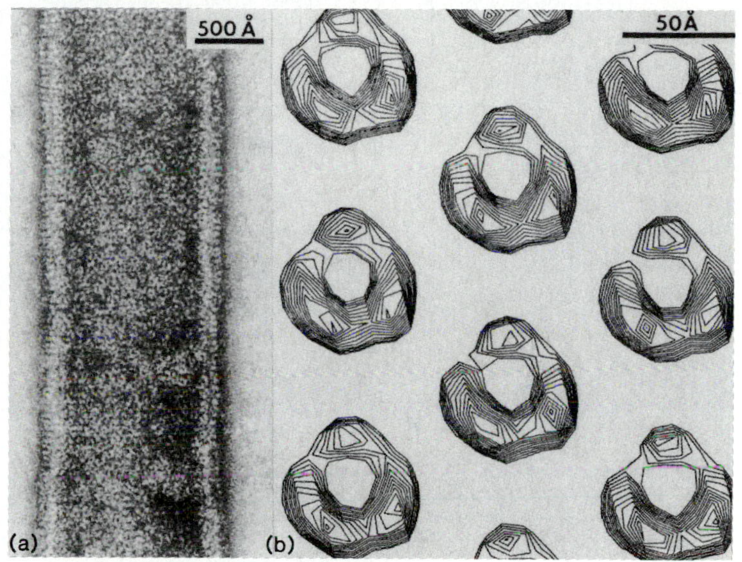

Because all five subunits of the acetylcholine receptor span the postsynaptic membrane, it was proposed that the channel corresponds to the central "hole" in the rosette seen by electron microscopy of electroplax membranes (see fig. 35.21). In fact, the five membrane-spanning subunits of the acetylcholine receptor have been recently resolved in three dimensions by electron microscopy of crystallized postsynaptic membranes of *Torpedo*. The complex consists of five rod-shaped structures, of about equal cross section, which lie largely perpendicular to the membrane. The five subunits are contained within a pentagonally symmetrical shell, about 140 Å long and 80 Å in diameter, which delineates a water-filled channel along the central axis of the cylinder. This cylinder extends about 40 Å into the cytoplasm and nearly 70 Å into the synaptic cleft. The channel opening to the synaptic cleft is wide, but narrows as it extends through the membrane into the interior of the cell (fig. 35.22). Treatment with an acetylcholine analog, carbamylcholine, produces a visible conformational change in this complex (see fig. 35.22). Thus, for the first time, we have a view of the changes in quaternary conformation that are probably the basis for opening and closing of the receptor ion channel.

The genes for all four of the acetylcholine receptor subunits have been cloned and sequenced. As we mentioned earlier, there is a great deal of amino acid sequence homology among

Figure 35.22

Conformational changes induced in the acetylcholine receptor upon binding of an agonist. *Top*: A view toward the face of the δ subunit before (N) and after (C) exposure to carbamylcholine. Note the change in tilt of the δ subunit (dotted line) and the outward movement of the γ subunit upon agonist binding. *Bottom*: Views of the unliganded and liganded receptor toward the face of the γ subunit.

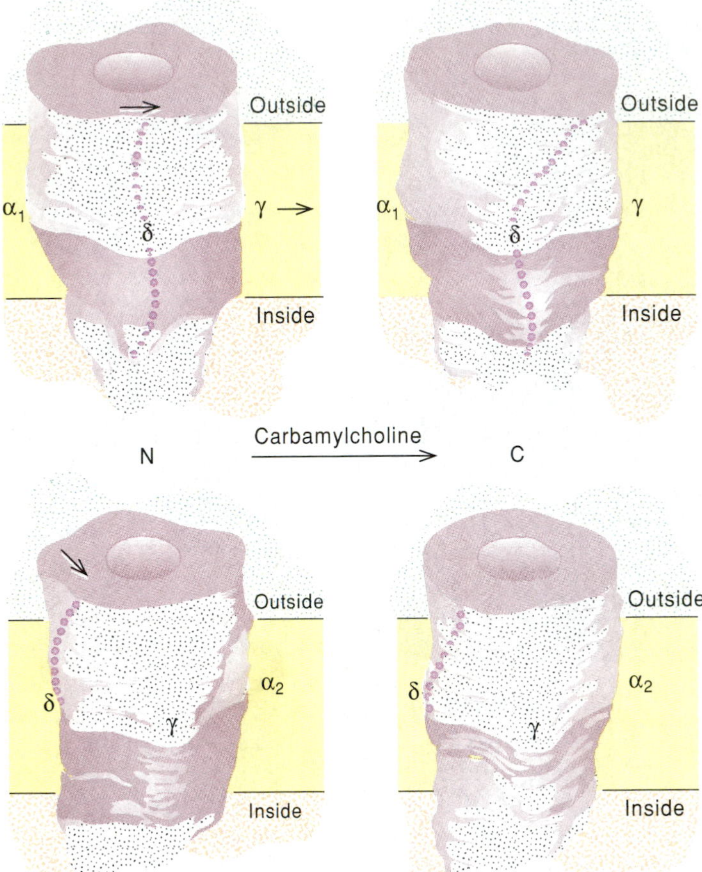

them. In fact, some workers have proposed that in all four subunits the polypeptides each span the membrane four times in a similar fashion as illustrated in figure 35.23. The potential membrane-spanning regions of each subunit (named M1 through M4 in figure 35.23) all contain polar residues, especially serine and threonine. With the *Xenopus* oocyte expression system described earlier for the Na^+ channel, it has been possible recently to use site-directed mutagenesis to explore regions important for channel function. The results have implicated membrane-spanning region M2, in particular, in channel formation and function. When serine and threonine residues within M2 were replaced with alanine residues, the ion conductance of the channel decreased, and when negatively charged or glutamine residues flanking the M2 regions of all four subunits were eliminated (see fig. 35.23), a similar effect was noted. Therefore it has been proposed that rings of negatively charged residues, contributed by all five M2 regions in the acetylcholine receptor complex, flank the cation-specific pore on both sides of the membrane, while hydrophilic residues within the pore provide the polar channel through which the ions can travel. In fact, a synthetic peptide corresponding in sequence to the M2 region of the δ subunit has been shown to form cation-selective channels in artificial phospholipid bilayers, and computer modeling studies have shown that a pentameric structure is the most likely arrangement of these helices in the membrane.

Thus considerable evidence points to a gated-pore-type mechanism for ion permeability conferred by the acetylcholine receptor, as is the case for the Na^+ and K^+ channels. At least two conformations of the protein, corresponding to open and closed states, have been recognized in the presence of agonists and antagonists, respectively. Furthermore, voltage-clamp studies of *Xenopus* oocytes, in which the acetylcholine receptor subunits have been expressed, have revealed current pulses having a square shape and a constant amplitude in the presence of cholinergic agonists. These properties are similar to those displayed by Na^+ and K^+ channels, as well as by bacterial porins and certain ionophore antibiotics as we discussed in chapter 32. It has been calculated that about 10^4 molecules of Na^+ flow through the receptor channel *in vivo* in the millisecond or so that it is open. A pore model allowing for more or less free diffusion of ions through the channel in its open state, rather than a slower, carrier-mediated mechanism, is therefore likely for the acetylcholine receptor.

Excitability Is Found in Many Different Cell Types

Although nerve cells are among the most extensively studied excitable cells, excitability has been demonstrated not only in other cell types in higher animals but also in organisms as simple as unicellular protozoans and in algae. In ciliated protozoa, for example, membrane depolarization events have been shown to regulate swimming behavior, while in certain algae localized transmembrane movements of ions appear to play a role in early development. Membrane excitability thus appears to be a common phenomenon in many different types of cells and to play a role in a variety of physiological processes. In the following chapter we will examine one such process, vision. In this case photons—impinging upon specialized nerve cells containing photoreceptors—initiate nerve impulses that are sent to the brain, which translates these signals into an image of our environment.

Figure 35.23

A model for the arrangement of the subunits of the acetylcholine receptor in the membrane. The amino acid sequences of each of the subunits (α, β, γ, δ, respectively, from left to right) at each position are shown for the proposed transmembrane (M1–M4) and flanking regions. Solid circles denote polar, uncharged residues; boxes indicate positively charged residues; and dotted boxes show negatively charged residues. The acetylcholine binding region in the α subunit is toward the N terminus of the transmembrane regions and is not shown.

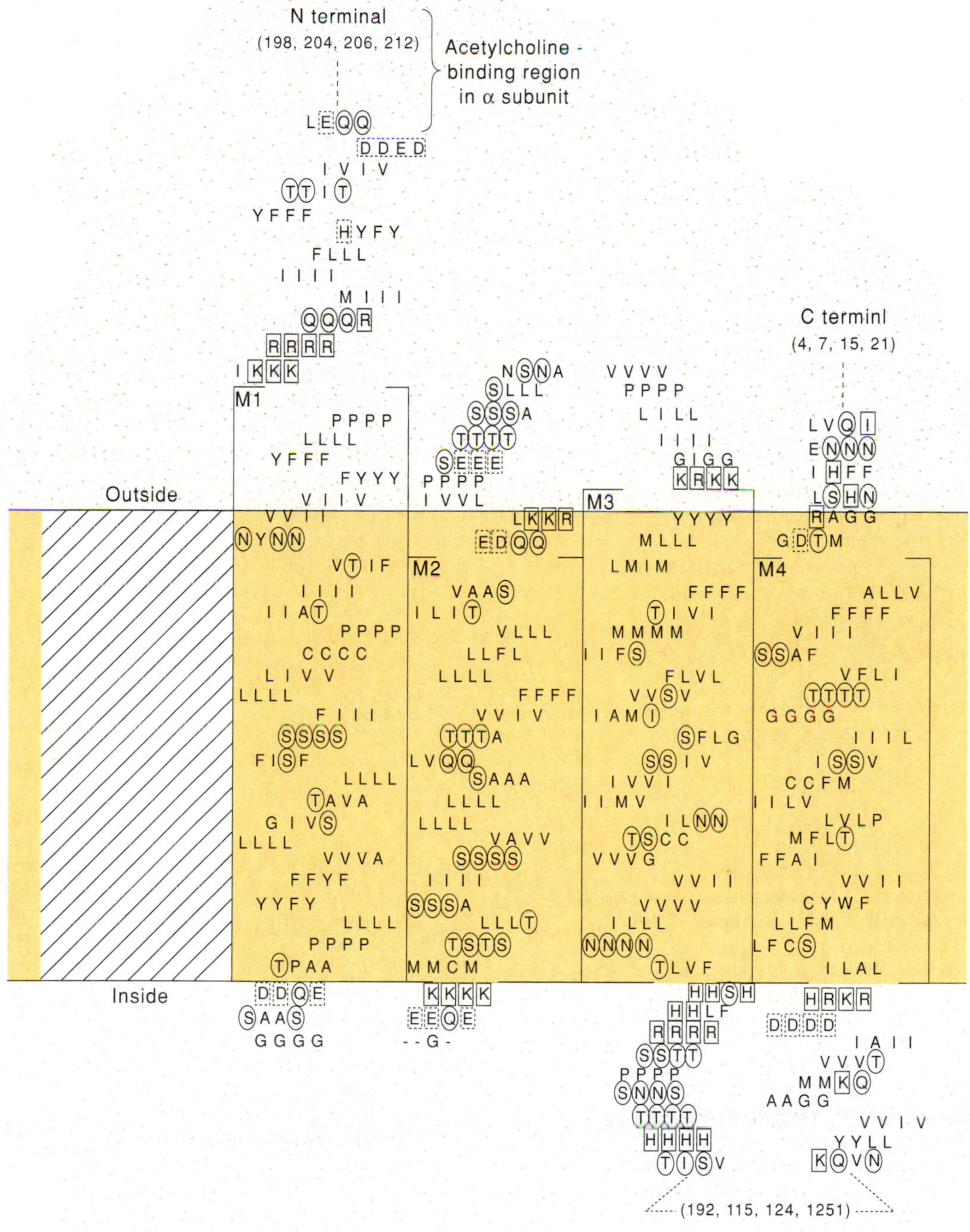

Summary

In this chapter we have examined the way in which nerve impulses are propagated along the neurons and are transmitted to receptor cells. The following points were central to our discussion.

1. Nerve cells are highly specialized for the reception of external signals and the transmission of those signals to other cells in the organism. Signal transmission within a neuron involves waves of membrane depolarization-repolarization events that travel the length of the axon to the nerve endings. The arrival of such a wave at a single region of the axon can be recorded as an action potential, which is generated by a transient influx of Na^+, followed by a transient efflux of K^+ and then a return of the membrane potential to its resting state.
2. Separate, voltage-gated channels for Na^+ and K^+ have been demonstrated in many types of excitable cells, and have been shown to be responsible for the ionic movements and changes in membrane potential associated with the action potential. Na^+ channels from a number of sources consist of an α subunit ($M_r = 260{,}000$) and variable amounts of a β subunit ($M_r = 37{,}000$). The best-characterized K^+ channel is that encoded by the *Drosophila Shaker* locus. It is a 616-residue polypeptide that has significant primary and secondary structural homology to each of the four domains of the Na^+ channel.
3. Transmission of the action potential from a neuron to another cell occurs most commonly by the release of chemical neurotransmitters at the synapse, triggered by the arrival of the action potential at the terminus of the axon. Neurotransmitter compounds include acetylcholine, the catecholamines such as epinephrine and norepinephrine, certain amino acids, and a variety of neuropeptides. Neurotransmitters bind to receptors in the postsynaptic membrane, initiating a response in the target cell that is, again, triggered by the influx of cations.
4. The best-studied neurotransmitter receptor is that for acetylcholine. It consists of four different polypeptide subunits (α, β, γ, and δ) in a molar ratio of 2:1:1:1 with a total mass of about 255,000 daltons. Biochemical and electron-microscopic techniques have established that all five subunits span the postsynaptic membrane. These subunits comprise a pentagonally symmetrical shell, which delineates a water-filled channel that is undoubtedly important in the translocation of cations initiated by ligand binding.
5. Recently, molecular biological techniques and modeling studies have shown how both the ion channel proteins and neurotransmitter receptors may be arranged in the membrane, and have identified regions in these proteins that are important for ion translocation and ligand binding. Site-directed mutagenesis studies of a *Drosophila* K^+ channel also provide evidence as to how the opening and closing of such channels may occur. Further application of these techniques will probably continue to increase our understanding of structure-function relationships in these proteins.

Selected Readings

Brisson, A., and P. N. T. Unwin, Quaternary structure of the acetylcholine receptor. *Nature* 315:457, 1985. This article describes the three-dimensional electron image analysis of tubular crystals of the acetylcholine receptor grown from native membrane vesicles.

Catterall, W., Structure and function of voltage-sensitive ion channels. *Science* 242:50, 1988. Excellent review of the subject through 1987.

Galzi, J-L., A. Devillers-Thiery, N. Hussy, S. Bertrand, J-P. Changeux, and D. Bertrand. Mutations in the channel domain of a neuronal nicotinic receptor convert ion selectivity from cationic to anionic. *Nature* 359:500–505, 1992. In this article, site-directed mutagenesis is used to probe amino acid residues in the acetylcholine receptor that may be important for ion selectivity.

Hille, B., *Ionic Channels in Excitable Membranes,* 2nd edition. Sunderland, MA:Sinauer, 1991. This book covers recent research on a variety of nerve and muscle ion channels. Theory of ion channel function and techniques of analysis are also reviewed.

Hoshi, T., W. N. Zagotta, and R. W. Aldrich, Biophysical and molecular mechanisms of *Shaker* potassium channel inactivation. *Science* 250:533, 1990. Evidence for the "ball-and-chain" mechanism of ion channel inactivation is presented in this article.

Katz, B., *Nerve, Muscle and Synapse.* New York: McGraw-Hill, 1966. Excellent overview of early and classical work on neurotransmission.

Kuffler, S. W., and J. G. Nicholls, *From Neuron to Brain: A Cellular Approach to the Function of the Nervous System.* Sunderland, Mass.: Sinauer Associates, 1984. This book covers aspects of neurotransmission and neurophysiology at the cellular level in an easily readable form.

Rehm, H., and B. L. Temple, Voltage-gated K^+ channels of the mammalian brain. *FASEB J.* 5:164, 1991. Excellent synopsis of recent work on structure-function relationships in the K^+ channel, including molecular genetic approaches.

Problems

1. (a) Calculate the membrane potential $\Delta\Psi$ across a resting nerve-cell membrane in the presence of tetrodotoxin. (Assume that the resting permeability to Na^+ is due to a small, steady-state level of "open" Na^+ channels.)
   (b) What would the value of $\Delta\Psi$ be if the resting axonal membrane were permeable only to Cl^-? (Assume an axoplasmic Cl^- concentration of 50 mM.)
2. Explain why a nerve-cell membrane exhibits an "all or none" response (i.e., action potential) independent of the magnitude of an electric or chemical stimulus (above a threshold value).
3. List the criteria for demonstrating that a particular compound acts as a neurotransmitter in a given system, assuming that the mechanism of synaptic transmission in most instances is analogous to that found in cholinergic systems.
4. In reconstituted transport systems, it is often important to demonstrate that the rate of transport is similar to that observed *in vitro,* that is, that the transport protein is fully functional in the reconstituted state. For systems in which *in vivo* fluxes are very rapid (e.g., cation flux through the acetylcholine receptor), it is often difficult to measure these rates directly in reconstituted vesicles. Thallous ion (Tl^+) is known to pass readily through the "open" state of the acetylcholine receptor. It also very efficiently quenches the fluorescence emission of the fluorophore 8-aminonaphthalene-1,3,6-trisulfonate (ANTS), which is relatively impermeable to phospholipid bilayers.
   (a) Using this information, outline a series of experiments to measure Tl^+ fluxes in reconstituted proteoliposomes containing purified acetylcholine receptors.
   (b) Actual Tl^+ fluxes into proteoliposomes containing an average of two acetylcholine receptor channels per vesicle in the presence of agonists have been measured to be 200 moles/(liter · s). If the average inner diameter of such vesicles is 400 Å, what is the number of Tl^+ ions transported per second by each activated acetylcholine receptor channel? How does this value compare with the rate of Na^+ flux measured *in vivo?*
5. Describe in your own words, and critique, the evidence that voltage-sensitive K^+ channels may be regulated by a "ball-and-chain" mechanism.
6. Patients with Parkinson's disease are treated with DOPA and inhibitors of dopa decarboxylase. This treatment will raise levels of dopamine in certain areas of the brain. However, dopamine given as a drug has no effect! How does this treatment of Parkinson's disease work? (Hint: Think about the blood-brain barrier and membrane permeability.)
7. Draw a chemical reaction mechanism for the reaction of the nerve poison parathion with acetylcholinesterase (see fig. 35.15*b*). How do these types of inhibitors prevent repolarization of the postsynaptic membrane?
8. What are the data which support a gated-pore-type model for ion permeability conferred by the acetylcholine receptor, as opposed to a carrier-mediated mechanism?
9. Earlier in this text, and in this chapter as well, we have cited evidence for protein structure-function relationships that has come by applying the powerful tools of molecular genetics. From what you have learned so far, what do you think are both the advantages and limitations of this kind of approach? (Use proteins described in this chapter as examples.)

Vision

The photochemistry of vision is different from that of photosynthesis. Perhaps this is not surprising. Animals that can see use light to obtain information; photosynthetic organisms use light as a source of energy. However, vision and photosynthesis both start with the excitation of an electron from one molecular orbital to another orbital of higher energy. The excited molecule must then undergo a transformation to a metastable product. Since chlorophyll can undergo such a transformation with a quantum yield near 1.0, it is not hard to imagine an eye that uses electron-transfer reactions like those that work so well in photosynthesis. In fact, there is a synthetic eye that works very much that way—a silicon-diode-array television camera tube. But the eyes of multicellular animals are different. Instead of chlorophyll, their light-sensitive cells contain a complex called rhodopsin, which consists of a protein, opsin, and a linear polyene, 11-*cis*-retinal. Instead of undergoing oxidation when it is excited with light, the retinal isomerizes.

In this chapter we will examine the biochemistry involved in this reaction to light. We will focus primarily on vertebrates, with only occasional references to invertebrate photochemistry, but we will conclude with a short section on the light-reactive substance found in one species of bacterium.

The Visual Pigments Found in Rod and Cone Cells

The eyes of vertebrates are marvelously complex organs, with many different types of specialized cells (fig. 36.1). Light rays entering the eye are refracted by the cornea, the clear tissue at the front of the eye. The light traverses an aqueous chamber and reaches the lens, which is densely packed with proteins called crystallins. Adjustments in the shape of the lens focus a sharp optical image onto the retina, a thin layer of tissue that lines the back of the eye. The retina is a neural tissue with several different layers of cells. Some of these cells, the rod and cone cells, contain the visual pigments. Other cells make synaptic connections to the rods or cones and to additional neural cells that carry impulses to the brain.

In both rods and cones, the light-sensitive molecules are collected in a layered system of membranes at one end of the cell (figs. 36.2 and 36.3). The membranes form by invaginations of the cytoplasmic (plasma) membrane near the middle of the cell. In cone cells, the membranes remain contiguous with the cytoplasmic membrane. In rods, they pinch off to form a stack of autonomous flattened vesicles, or disks, in the outer segment of the cell. The outer segment of each rod contains from 500 to 2,000 of these disks. This region of the cell is connected by a thin cilium to the inner segment, which is packed with mitochondria and ribosomes. The basal part of the cell ends in a synaptic junction with another neural cell called a bipolar cell. In a living rod cell, disks are constantly forming at the base of the outer segment and moving in a file to the tip of the outer segment, where they are sloughed off and phagocytosed by the underlying epithelial cells. It takes about ten days for a disk to make this journey. Exactly why the disks need to be replaced so rapidly is unclear.

Figure 36.1

Structure of the human eye.

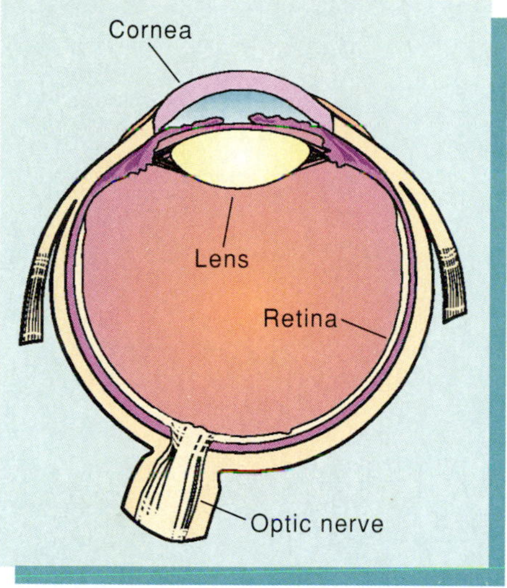

Figure 36.2

Thin-section electron micrograph (59,200×) of a portion of a rod cell in a rabbit retina. Part of the outer segment is shown at the top, and part of the inner segment at the bottom. (D = disks; M = mitochondrion; C = cilium.) (Courtesy of Dr. Ann H. Milam.)

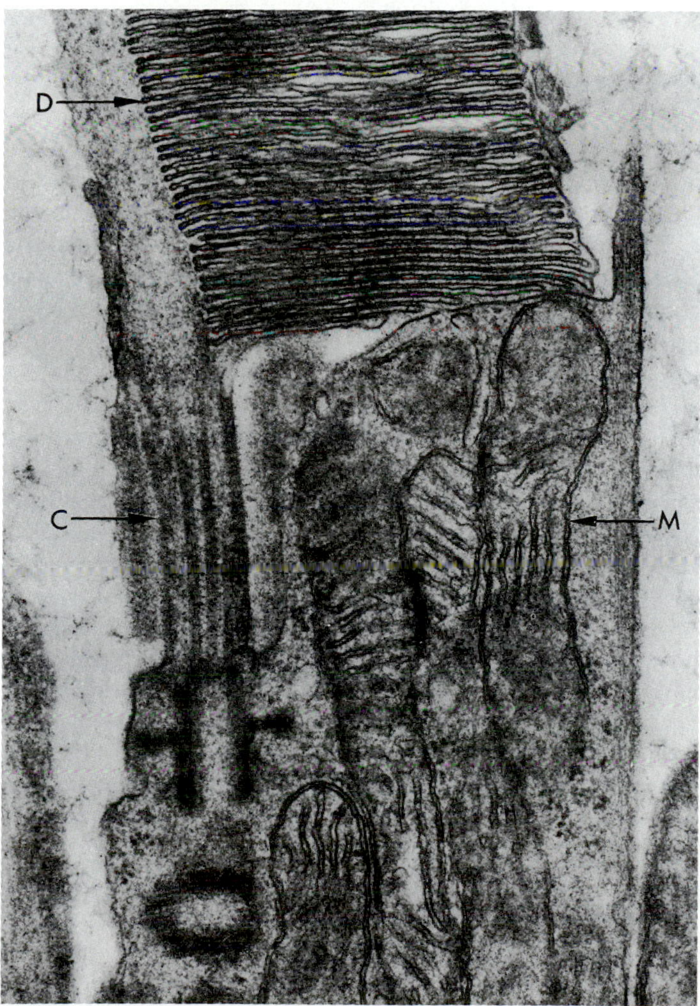

Figure 36.3

Schematic diagram of a rod cell. The orientation of the cell is the same as that in figure 36.2. In the retina, many rod cells are stacked side by side, with the outer segments all pointing out to the periphery of the eye. Light enters the cells end-on through the inner segment, after passing through several layers of other neural cells. In cone cells, the outer segments are shorter, and are conical rather than cylindrical in shape.

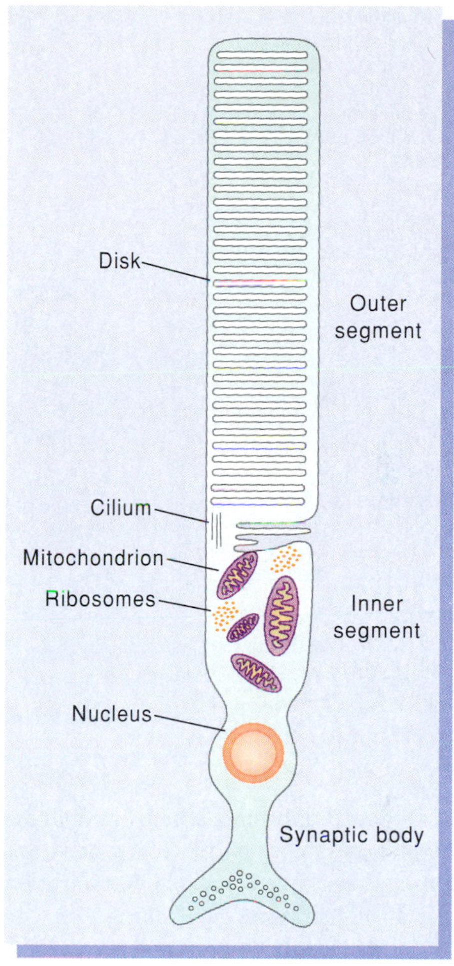

Rod cells are specially adapted for vision in dim light. Cones provide visual acuity in bright light, and also serve for perception of color. Animals such as owls, which have very high visual sensitivity in dim light but cannot distinguish colors, have only rod cells. Some animals, such as pigeons, have only cones and are skilled at distinguishing colors in bright light but inept at night vision. Primates have both rods and cones, with rods considerably outnumbering cones in all but the central portion of the retina.

*Rhodopsin Consists of 11-*cis*-Retinal Bound to a Protein, Opsin*

Figure 36.4 shows the structures of 11-*cis*-retinal and its more stable isomer all-*trans*-retinal. The retinals are related to the alcohol retinol, or vitamin A_1. These compounds cannot be synthesized *de novo* by mammals, but they can be formed from

Figure 36.4

Structures of retinals, retinols, and β-carotene. The structure of 11-*cis*-retinal (*top*) indicates the numbering system used for the carbons. For simplicity, the molecules are shown here as being planar, although the most stable conformer of 11-*cis*-retinal actually is twisted about the C-6—C-7 and C-12—C-13 single bonds to reduce steric crowding of the methyl groups. In rhodopsin, 11-*cis*-retinal is bound by a protonated Schiff's base linkage to a lysine of opsin.

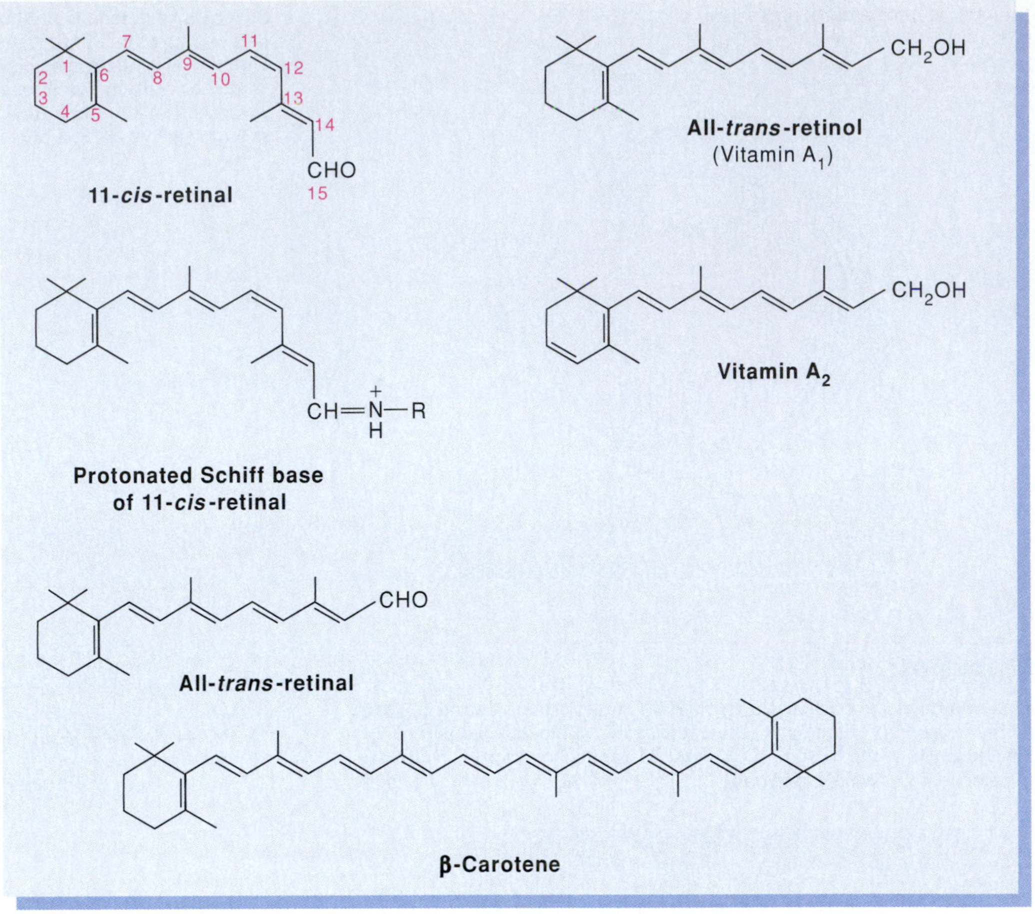

carotenoids, such as β-carotene, which are abundant in carrots and some other vegetables. A deficiency of vitamin A causes night blindness, along with a serious deterioration of the eyes and a number of other tissues.

There are animals with eyes in several different phyla, including Mollusca, Arthropoda, and Annelida, in addition to Chordata. The eyes of arthropods differ substantially from those of molluscs and chordates anatomically, and they apparently originated independently, after the phyla had separated in evolution. Some animals, such as sea turtles and amphibians in certain stages of development, use a retinal that has an additional double bond in the ring. The parent molecule in this case is vitamin A_2, or 3,4-dehydroretinal (see fig. 36.4). In all cases, however, the photochemically active protein complex is made from the 11-*cis* isomer of the aldehyde. 11-*cis*-Retinals must somehow be particularly fit for the task of responding to light. (Some unicellular organisms, such as *Euglena,* have light-sensitive organelles that contain a carotenoid instead of a retinal, but little is known of the biochemistry of these primitive receptors.)

Opsin is an intrinsic membrane protein with a molecular weight of about 38,000. It accounts for about 95% of the protein in the disk membranes. Labeling studies of isolated disks indicate that the carboxyl-terminal end of rhodopsin is exposed to the cytoplasmic space outside the disk, while the amino end is exposed to the space inside the disk. Seven hydrophobic α-helical stretches of the protein probably lace back and forth across the phospholipid bilayer. The retinal is bound to a lysyl-NH_2 group by a Schiff base, or aldimine linkage (see fig. 36.4), in a region that is buried in the bilayer. Rhodopsin also has several carbohydrate groups bound near the amino-terminal end.

In solution, 11-*cis*-retinal absorbs maximally near 380 nm, but in rhodopsin the peak is at 500 nm (fig. 36.5). The absorption spectrum of rhodopsin is essentially identical to the spectrum of the sensitivity of the rod cells (after correction for absorption in the cornea and lens). The perception of color by cones depends on the fact that there are three different types of cones with different absorption spectra. One of these absorbs blue light (440 nm) maximally, the second absorbs green (530 nm), and the third absorbs yellow (570 nm). In some species of birds, the third component absorbs maximally at 630 nm. However, all cones contain 11-*cis*-retinal bound to proteins that are similar to opsin. (Some fish use another strategy altogether for color vision. Their eyes have carotenoid-containing oil droplets of three colors, and these act as filters in front of the receptor cells.)

Figure 36.5

Absorption spectra of rhodopsin and of 11-*cis*-retinal in hexane solution. The absorption of rhodopsin in the 280-nm region is due mainly to the opsin.

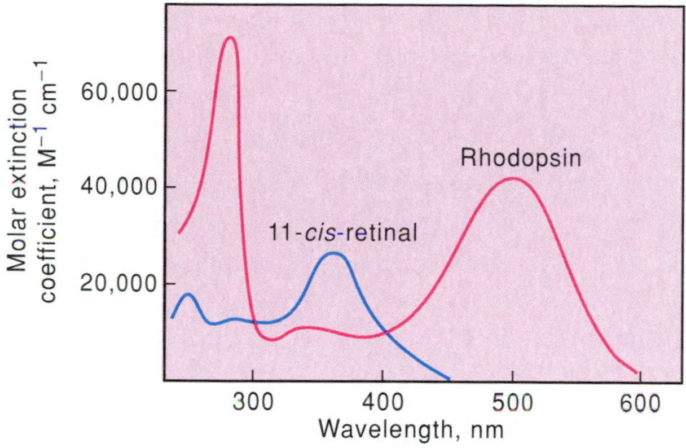

Light Isomerizes the Retinal of Rhodopsin to All-trans

The discovery that rhodopsin contains the 11-*cis* isomer of retinal came as a surprise. In the early 1950s, Ruth Hubbard, George Wald, and their co-workers found that rhodopsin decomposes into retinal and the apoprotein opsin when the retina is exposed to light. The retinal that was released was the all-*trans* isomer. When opsin was mixed with a crude preparation of retinal, rhodopsin was regenerated. Crystalline all-*trans*-retinal, however, did not bind to opsin. The regeneration achieved with the crude preparation proved to be due to a small amount of the 11-*cis* isomer in the mixture. Since rhodopsin binds 11-*cis*-retinal in the dark, but releases all-*trans*-retinal after illumination, Hubbard and Kropf concluded that the action of light in vision is to isomerize the chromophore about its C-11=C-12 double bond. This change in structure must somehow be translated into an electrophysiological signal that can be transmitted to the brain.

Transformations of Rhodopsin Can Be Detected by Changes in Its Absorption Spectrum

The release of all-*trans*-retinal from opsin takes several minutes and is too slow to be an obligatory step in visual perception. It appears to be a step in the regeneration of the active form of rhodopsin. To explore the changes in rhodopsin that precede the release of all-*trans*-retinal from the protein, Toru Yoshizawa and Wald measured the optical absorbance changes that occurred when they illuminated rhodopsin at low temperatures. Illumination at liquid N_2 temperature (77 K) caused the absorption band of the rhodopsin to shift from 500 to 543 nm. The product of this transformation is now called bathorhodopsin. Bathorhodopsin is stable indefinitely in the dark at 77 K, but if it is warmed above about 130 K it decays spontaneously to a species that absorbs maximally at 497 nm. This species is called lumirhodopsin. If the sample is warmed further, to about

Figure 36.6

Photochemical transformations of rhodopsin. The numbers in parentheses indicate the optical absorption maxima of the intermediates. The numbers in color on the right are approximate half-times for the conversions in rod outer segment membranes near 37° C. With isolated rhodopsin, some of the reactions are slower and are multiphasic. The steps following the formation of bathorhodopsin have progressively higher thermal activation energies. They can be blocked by lowering the temperature below the temperatures indicated on the left.

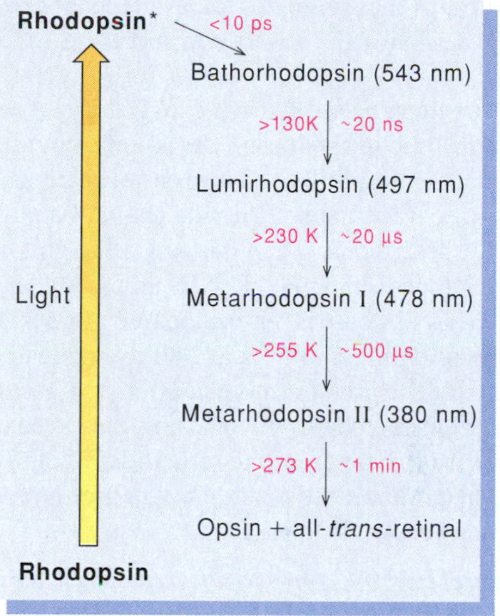

230 K, lumirhodopsin decays to metarhodopsin I, which absorbs at 478 nm. Above about 255 K, metarhodopsin I decays to metarhodopsin II, which absorbs at 380 nm. These transformations are outlined in figure 36.6.

Subsequent kinetic measurements by other investigators have shown that rhodopsin passes through the same series of states if it is excited with a short flash at physiological temperatures. The numbers on the right side of figure 36.6 give approximate half-times for the transformations at 37° C. Agreement between the kinetic studies and low-temperature experiments may seem routine, but it might not have turned out this way at all. The trapping of a metastable state at low temperature can depend on the thermodynamic properties of the state, as well as on its position in the kinetic sequence. For example, side products that normally are unimportant could accumulate when the normal pathway is blocked by lowering the temperature.

How can the absorption spectrum of rhodopsin go through such wild changes from 500 to 543 nm, and eventually to 380 nm? In thinking about this, we are reminded that rhodopsin's initial absorption spectrum is already shifted by 120 nm compared with the spectrum of free 11-*cis*-retinal (fig. 36.5). Further, the absorption maxima of the cone pigments vary from 450 to 630 nm, in spite of the fact that these pigments all contain 11-*cis*-retinal. The explanation for these spectral differences depends partly on the fact that the nitrogen atom of the retinylidine Schiff base linkage in rhodopsin is protonated and is therefore positively charged (fig. 36.7). The protonation state of the nitrogen has been demonstrated clearly by NMR and resonance Raman spectroscopy. When the retinal absorbs light and is raised to an excited state, the electron density on the nitrogen increases, and the positive charge moves to the opposite end of the molecule, as represented roughly in figure 36.7. Because of the redistribution of charge, the relative energies of the excited and ground states are extremely sensitive to the positions of other charged, dipolar, or polarizable groups nearby. An arrangement of charged groups that stabilizes the excited state relative to the ground state will shift the absorption spectrum to longer wavelengths.

There must be at least one charged group near the retinylidine Schiff base in rhodopsin, the anionic counterion that is needed to balance the positive charge on the nitrogen. The anion is probably the carboxylate of the side chain of a glutamate residue, Glu 113, because mutating this residue to glutamine causes a dramatic change in the pK_a of the Schiff base. In the mutant rhodopsin, the Schiff base is not protonated at physiological pH, and the absorption maximum is at 380 nm instead of 500 nm. The structure of the protein thus can affect the absorption spectrum of the pigment profoundly. Moving the carboxylate group of Glu 113 farther away from the retinylidine Schiff base would destabilize the ground state relative to the excited state, shifting the absorption spectrum to longer wavelengths.

Isomerization of the Retinal Causes Other Structural Changes in the Protein

Let us now consider bathorhodopsin, which appears to be the first metastable product of the photochemical reaction. If bathorhodopsin is excited with long-wavelength light at 77 K, it can be converted back to rhodopsin. Resonance Raman measurements support the view that the retinal in bathorhodopsin has isomerized to the all-*trans* form, but that it continues to be held as a protonated Schiff base. Additional evidence that bathorhodopsin contains all-*trans*-retinal has been obtained by studying a modified form of rhodopsin, isorhodopsin, which contains the 9-*cis* isomer of retinal. When isorhodopsin is illuminated, it gives rise to bathorhodopsin with 10^{-11} s, just as rhodopsin (11-*cis*) does. The only common product that could form directly from both the 9-*cis* and 11-*cis* isomers is the all-*trans* isomer.

Isomerization of the retinal Schiff base can occur when the molecule is excited with light, because the C-11=C-12 bond loses much of its double-bond character in the excited state. The valence bond diagrams of figure 36.7 illustrate this point qualitatively. In the ground state of rhodopsin, the potential energy barrier to rotation about the C-11=C-12 bond is probably on the order of 30 kcal/mole. This barrier essentially vanishes in the excited state. In fact, molecular orbital calculations suggest that the energy of the excited molecule is minimal when the C-11=C-12 bond is twisted by about 90° (fig. 36.8). The 11-*cis* and all-*trans* molecules thus have a common excited state. When the molecule decays from the excited state to a ground state, it can end up in either of these isomeric forms. It turns out that the excited state decays to bathorhodopsin (all-*trans*) about 67% of the time and to rhodopsin (11-*cis*) about 33% of the time. Isorhodopsin (9-*cis*) and other isomers are formed only in small amounts (about 1%). Similar ratios of products hold no matter whether we start by exciting rhodopsin or bathorhodopsin.

Figure 36.7

Excitation of the protonated Schiff's base causes a movement of positive charge from the nitrogen toward the ring. The C-11=C-12 bond loses much of its double-bond character. The valence bond diagrams shown here should not be taken too literally, but they give a good qualitative picture of the redistribution of electrons that occurs when the molecule is excited.

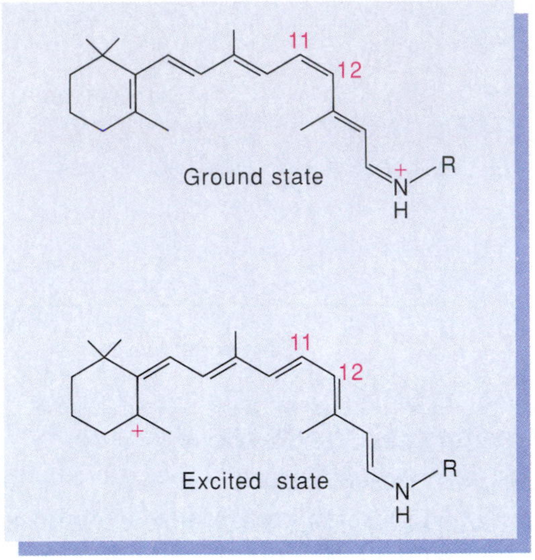

The high energy barrier to the isomerization of rhodopsin in the dark is physiologically important because it limits the "noise" in our perception of light. If the barrier were low enough to be overcome thermally, the discrimination of light from dark would be more difficult.

Because opsin binds 11-*cis*-retinal, but not all-*trans*-retinal, the all-*trans* molecule that is created in bathorhodopsin must find itself initially in a binding site that is tailored for the 11-*cis* structure. In solution, 11-*cis*-retinal is less stable than all-*trans*-retinal, because of steric repulsion between the methyl group on C-13 and the hydrogen atom on C-10 (see fig. 36.4). On the protein, there must be compensating interactions that decrease the free energy of the 11-*cis* complex relative to that of the all-*trans* complex. These interactions will be disrupted when the pigment is isomerized. The pronounced shift of the absorption spectrum to longer wavelengths in bathorhodopsin could be explained if the isomerization results in a movement of the Schiff base nitrogen, increasing the distance between the positively charged nitrogen and its counterion (fig. 36.9). Such a movement would destabilize the ground state of bathorhodopsin relative to the excited state, as we discussed earlier for rhodopsin. A repositioning of the Schiff base also is likely to lead to changes in the structure of the protein surrounding the binding site. If the positively charged nitrogen moves closer to a second anionic group, as illustrated in figure 36.9, a new hydrogen bond could form between this group and the Schiff base nitrogen, and we would expect the protein to respond by adjusting the positions of other nuclei nearby.

Figure 36.8

Potential energy surfaces of ground and excited states in rhodopsin, as functions of the angle of rotation of the bond between carbons 11 and 12. A rotational angle of 0° means that the retinyl group is 11-*cis* (rhodopsin); 180° means that it is all-*trans* (bathorhodopsin). In the ground state (lower curve), rhodopsin is stabilized with respect to bathorhodopsin, possibly because the positive charge of the protonated Schiff's base interacts more favorably with a negatively charged group of the protein (fig. 36.9). The energy of the excited state (upper curve) is less sensitive to the position of the Schiff's base than the energy of the ground state is, because the positive charge has moved to the opposite end of the retinal molecule (fig. 36.7). The energy of the excited state is minimal when the rotation angle is about 90°. Rhodopsin is raised from the ground state to the excited state by light (vertical arrow). It relaxes back to the ground state along a path like that indicated by the dashed arrow, ending up as bathorhodopsin.

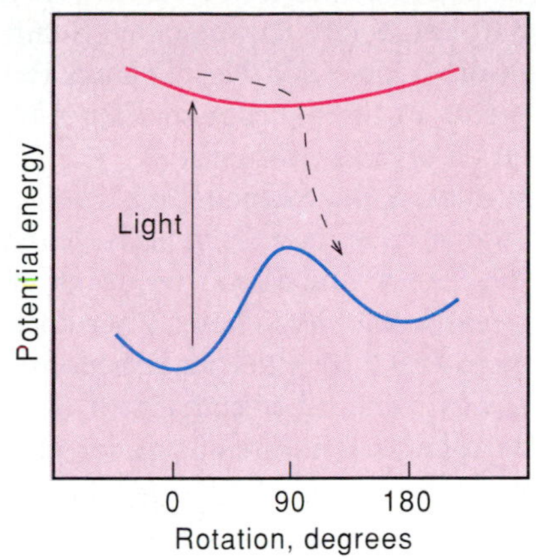

Figure 36.9

A model for the photochemical conversion of rhodopsin (*top*) to bathorhodopsin (*bottom*). In rhodopsin, the proton of the Schiff's base nitrogen is hydrogen bonded to a counterion, A_1^-. The retinal is twisted about the C-12—C-13 and C-6—C-7 single bonds. (As noted in figure 36.4, free retinal is twisted in this way, but the amount of twisting in rhodopsin is still uncertain.) When rhodopsin is converted to bathorhodopsin (*bottom*), the ring end of the retinal is presumed to be locked in position. Isomerization about the C-11═C-12 double bond flips the protonated nitrogen and its positive charge away from the counterion A_1^-. The separation of electrical charge could account for the shift of the absorption spectrum to longer wavelengths. The isomerization also can lead to proton movements. The proton of the Schiff's base could be passed to another basic group (A_2^-), and A_1^- might pick up a proton from another acidic group (A_3H). Other groups of the protein, including the lysine attached to the retinal, also must adjust to the new geometry of the retinyl Schiff's base. (For additional discussion, see B. Honig et al., An external point-charge model for wavelength regulation in visual pigments, *J. Am. Chem. Soc.* 101:7084, 1979, and Photoisomerization, energy storage, and charge separation: A model for light-energy transduction in visual pigments and bacteriorhodopsin, *Proc. Natl. Acad. Sci. USA* 76:2503, 1979.)

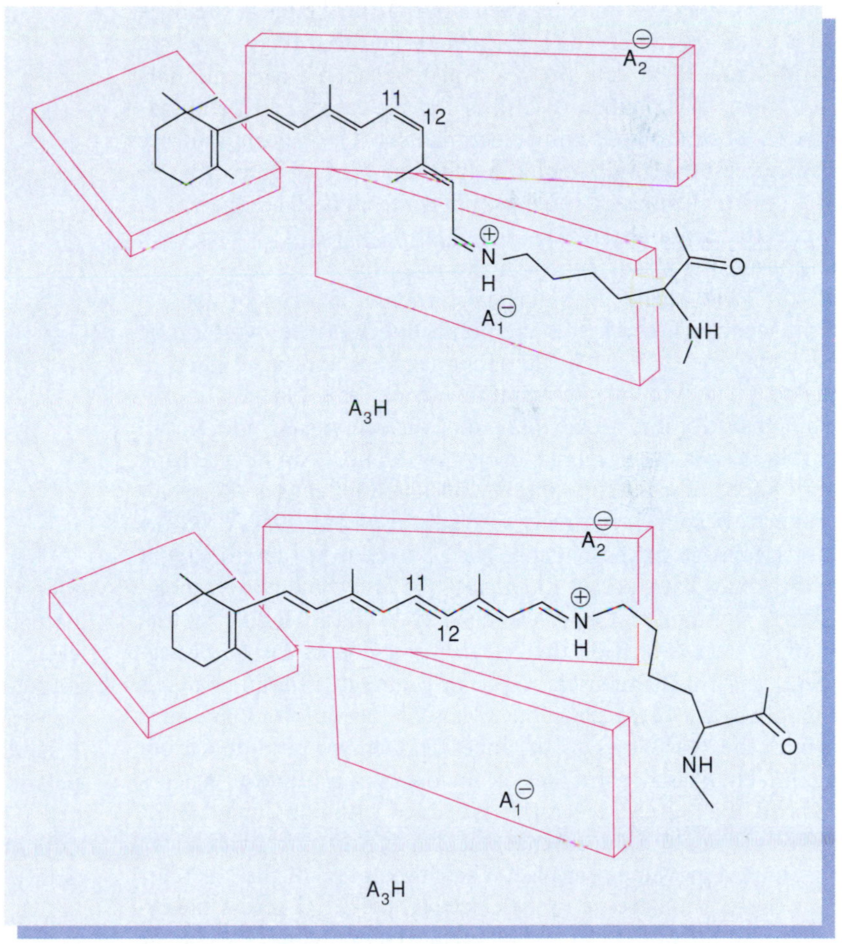

The reorganization of rhodopsin that is set in motion by isomerization of the retinylidine Schiff base continues as the system relaxes through the states lumirhodopsin, metarhodopsin I, and metarhodopsin II. Physical measurements indicate that isolated rhodopsin undergoes particularly substantial changes in protein conformation during the transition from metarhodopsin I to metarhodopsin II. The extent of these changes and the kinetics of formation of metarhodopsin II are sensitive to the type and number of phospholipids surrounding the protein. Conformation changes also undoubtedly occur in this step when the rhodopsin is in place in the disk membrane, but they are subtler and may be confined to small regions of the protein. The nitrogen of the retinyl Schiff base loses its proton as metarhodopsin II is formed, and a second, unidentified group takes up a proton from the solution. The loss of the positive charge on the nitrogen accounts for the shift of the absorption maximum to 380 nm, where unprotonated Schiff bases of all-*trans*-retinal absorb. The Schiff base linkage also becomes accessible to reagents in the aqueous solution during or shortly after the formation of metarhodopsin II.

The formation of metarhodopsin II is fast enough to be an obligatory step in visual transduction. It clearly is associated with changes in the interactions between rhodopsin and its surroundings. A reasonable hypothesis, therefore, is that the changes in protein structure allow metarhodopsin II to initiate an interaction with some other component of the disk membrane. We will explore the nature of this component in the following sections.

The Conductivity Change That Results from Absorption of a Photon

The human eye is amazingly sensitive. After being in the dark for a time, we can perceive continuous light that is so weak that an individual rod cell absorbs a photon, on the average, only once every 38 minutes. A flash of light is detectable if approximately six rods each absorb one photon. This means that the absorption of a single photon by any one of the approximately 3×10^7 molecules of rhodopsin in a rod must be sufficient to excite the cell and trigger a neuronal response. The combined responses of six rods can elicit the sensation of seeing.

The electrophysiological response of a rod or cone to light involves a change in the permeability of the cytoplasmic membrane to cations. In the inner segment and basal parts of the cell, the cytoplasmic membrane contains a Na^+-K^+ pump, which uses ATP to move Na^+ out of the cell and K^+ in. (Recall our discussion of ion pumps in chapter 32.) K^+ can diffuse back out of the cell relatively freely, and its efflux causes the membrane to become negatively charged by about 20 mV on the inside relative to the outside. Na^+, on the other hand, cannot readily pass back across the membrane into the inner segment. Electrophysiological measurements by William Hagins showed that Na^+ flows outside the cell to the region of the outer segment, and that it reenters the outer segment through channels that are selectively permeable to cations (fig. 36.10). Once back inside, the Na^+ flows to the inner segment to be pumped out again. The round trip takes on the order of a minute. When a rod cell of a vertebrate's retina is excited with light, some of the cation channels in the outer segment suddenly close, decreasing the inward movement of Na^+. The interruption of the Na^+ current causes an increase in the electrical potential across the cytoplasmic membrane. This hyperpolarization causes a change in the movement of an unidentified chemical transmitter to the bipolar cell that makes a synaptic junction with the rod. The bipolar cell then sends a signal to a ganglion cell, the third component in the hierarchy of retinal neurons.

The hyperpolarization of the rod and the response of the bipolar cell are graded effects. Absorption of a single photon causes the membrane potential to increase by about 1 mV, with the effect peaking about 1 s after the excitation. Up to about 100 photons per rod, the more light the cell absorbs, the larger

Figure 36.10

Na^+ is pumped out of the inner segment (solid arrows) and diffuses back into the cell through channels in the outer segment (dashed arrows). In rods of vertebrates, the Na^+ channels are held open in the dark, and they close in the light. In invertebrates, the channels open in the light.

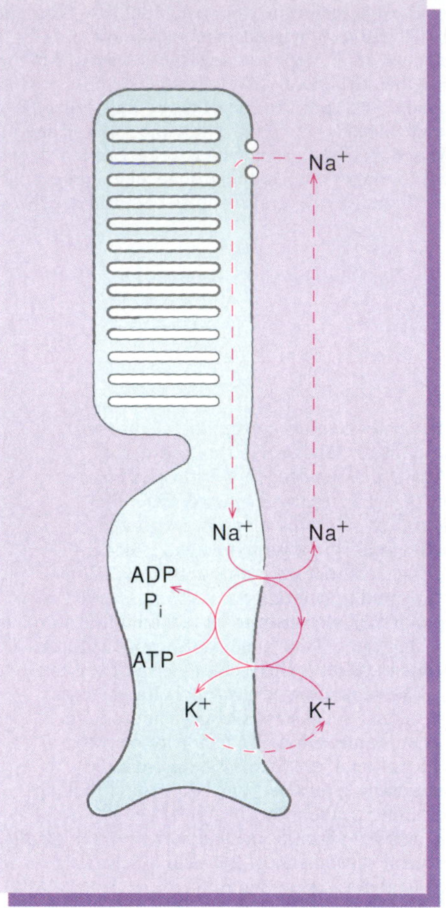

the amplitude of the hyperpolarization and the faster the hyperpolarization occurs. The reaction of the ganglion cell, however, is all-or-none. When the ganglion cell is triggered, it responds with an action potential that proceeds to the brain. Sensitivity to weak light is enhanced (with a sacrifice in spatial resolution) by having multiple rod cells connected to each bipolar cell and multiple bipolar cells connected to each ganglion cell. The output of cone cells is not summed in this way. This difference partly explains why the cones provide better visual acuity but lower sensitivity than the rods.

Analogous events occur in the eyes of invertebrates, except that light causes an increase in the cation permeability of the receptor cell rather than a decrease. In the rods of vertebrates, the absorption of a single photon decreases the current of Na^+ into the outer segment by about 3%. The inflow of approximately 10^6 Na^+ ions is transiently prevented. Exactly how many Na^+ channels have to close in order to achieve this effect is uncertain, but estimates have ranged from 25 to 1,000. Since only one molecule of rhodopsin is excited, the cell must have a mechanism for amplifying the effect of light by a substantial factor. Recall also that the cytoplasmic membrane of the rod

Figure 36.11

cGMP increases the electrical conductivity of the rod-cell cytoplasmic membrane. The conductivity was measured with a patch of membrane held at the tip of a micropipette. By varying the ions in the solution, it is possible to show that cGMP affects the movement of Na^+ and other cations across the membrane. Note the sigmoidal concentration dependence: the Hill coefficient is 1.6. (For more details, see E. E. Fesenko, S. S. Kolesnikov, and A. L. Lymbarsky, Induction by cyclic GMP of cationic conductance in plasma membrane of retinal rod outer segment, *Nature* 313:310, 1985.)

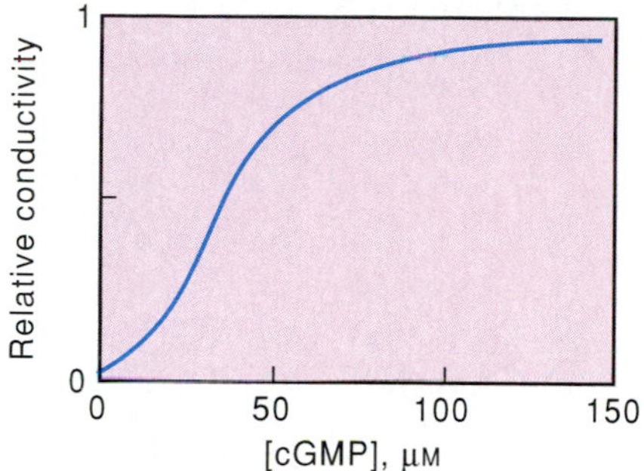

outer segment is not contiguous with the disk membranes that hold the rhodopsin (see fig. 36.3). This fact suggests that the excitation of rhodopsin causes a change in the concentration of a diffusible transmitter, which moves from the disk to the cytoplasmic membrane.

The Effect of Light Is Mediated by Guanine Nucleotides

There is strong evidence that the diffusible component that moves between the disk and the cytoplasmic membrane is 3′,5′-cyclic-GMP (cGMP). Electrophysiological measurements have shown that cGMP causes an increase in the permeability of the cytoplasmic membrane to Na^+ (fig. 36.11). This appears to be a direct effect of cGMP on the Na^+ channels, rather than an indirect effect mediated by a kinase, because it can be seen in the absence of ATP.

The disk membranes of the rod outer segment contain a phosphodiesterase, which hydrolyzes 3′,5′-cyclic-GMP (cGMP) to 5′-GMP. If the disk membranes are kept in the dark, the phosphodiesterase remains in a relatively inactive state. When the disks are illuminated and rhodopsin is converted to metarhodopsin II, the activity of the phosphodiesterase increases. For each molecule of rhodopsin that is excited, on the order of 500 molecules of the phosphodiesterase are activated. Illumination thus results in a decrease in the cGMP content of the cell. A drop in the cGMP concentration is in the right direction to cause a decrease in Na^+ permeability of the cell membrane (see fig. 36.11), and thus to cause a hyperpolarization of the membrane. In agreement with this scheme, injecting extra cGMP into rod outer segments temporarily inhibits the hyperpolarization caused by light.

Unlike rhodopsin, the phosphodiesterase is a peripheral membrane protein that can be readily solubilized. Its activation by light requires the presence of GTP and is associated with the binding of GTP to a second peripheral membrane protein called transducin. Transducin is a member of the family of G proteins that participate in the activation or inhibition of adenylate cyclase by hormones in other tissues (see chapter 24). Like other G proteins, transducin consists of three subunits, α, β, and γ (fig. 36.12). In the resting state, the α subunit contains a molecule of bound GDP. When rhodopsin is transformed to metarhodopsin II by light, it interacts with transducin, causing GTP to displace the bound GDP. Once GTP is attached, the α subunit separates from the β and γ subunits and binds to an inhibitory subunit of the phosphodiesterase. The removal of the inhibitory component activates the phosphodiesterase.

Each molecule of metarhodopsin II that is generated by light appears to be able to trigger about 500 molecules of transducin to bind GTP in place of GDP. This number agrees with the number of phosphodiesterase molecules that are activated, and accounts for much of the amplification in the response to light. Additional amplification results from the enzymatic action of each phosphodiesterase on many molecules of cGMP. These effects can be demonstrated with purified preparations of rhodopsin, transducin, and the phosphodiesterase. Purified, illuminated rhodopsin that has been incorporated into phospholipid vesicles is capable of activating transducin, and the complex of transducin's α subunit with GTP is capable of activating the phosphodiesterase in the absence of rhodopsin.

In order for the eye to respond rapidly to changing light intensities, the activation of the phosphodiesterase must be a transient response that quickly switches off again. The activation of transducin's α subunit is reversed by hydrolysis of the bound GTP to form bound GDP and free inorganic phosphate (see fig. 36.12). When this happens, the α subunit separates from the inhibitory polypeptide of the phosphodiesterase and recombines with the β and γ subunits of transducin. The inhibitory polypeptide then recombines with the phosphodiesterase, returning the phosphodiesterase to its resting, inactive state. We can prolong activation of the phosphodiesterase indefinitely if, instead of adding GTP, we add a nonhydrolyzable analog of GTP.

To switch off the response to light, two additional things must happen: Metarhodopsin II must decay to a form that is incapable of activating additional molecules of transducin, and the cGMP that was hydrolyzed must be resynthesized in order to reopen the cation channels in the plasma membrane. The inactivation of metarhodopsin appears to involve phosphorylation of the protein and binding of the phosphorylated rhodopsin to another protein called the 48K protein, or arrestin. The regeneration of the cGMP is carried out by a guanylate cyclase, which catalyzes the synthesis of cGMP from GTP. This enzyme is stimulated by still another regulatory protein, but is strongly inhibited by Ca^{2+}. Ca^{2+} enters the rod outer segment through the same cation channels that admit Na^+, and it is exported by a transport protein that exchanges Ca^{2+} for Na^+. The Ca^{2+} concentration in the cell thus decreases upon illumination, when the cation channels close. The drop in the Ca^{2+} concentration results in an activation of the guanylate cyclase. Changes in the intracellular Ca^{2+} concentration also appear to underlie the ability of the eye to adapt to large changes in illumination intensity.

Figure 36.12

The large arrows show the main steps in the activation of the cGMP phosphodiesterase by light. Light converts rhodopsin (R) to a metastable form (M_{II}, probably metarhodopsin II). M_{II} reacts catalytically with transducin, which contains GDP bound to one of its three subunits (α). The interaction with M_{II} changes the specificity of the nucleotide binding site so that GDP is released and GTP is bound. This causes the α subunit to dissociate from the other two subunits of transducin (ß and γ). In the dark, the phosphodiesterase (PDE) is inactive, because of the presence of an inhibitory polypeptide (I). The I polypeptide is removed by binding to the complex of the transducin α subunit and GTP. The active phosphodiesterase then can hydrolyze cGMP to 5′-GMP. M_{II} can go on to activate several hundred additional molecules of transducin, before slower enzymatic reactions (thin arrows) gradually convert it to an inactive form. The inactivation involves conversion of the rhodopsin to a phosphorylated form (R-P) and an interaction with another protein (48K). The GTP bound to transducin's α subunit eventually is hydrolyzed to GDP. When this happens, I dissociates and returns to the phosphodiesterase, the three subunits of transducin reassemble, and the system returns to its resting state. (For details, see M. Chabre, Trigger and amplification mechanisms in visual phototransduction, *Ann. Rev. Biophys. Biophys. Chem.* 14:331, 1985.)

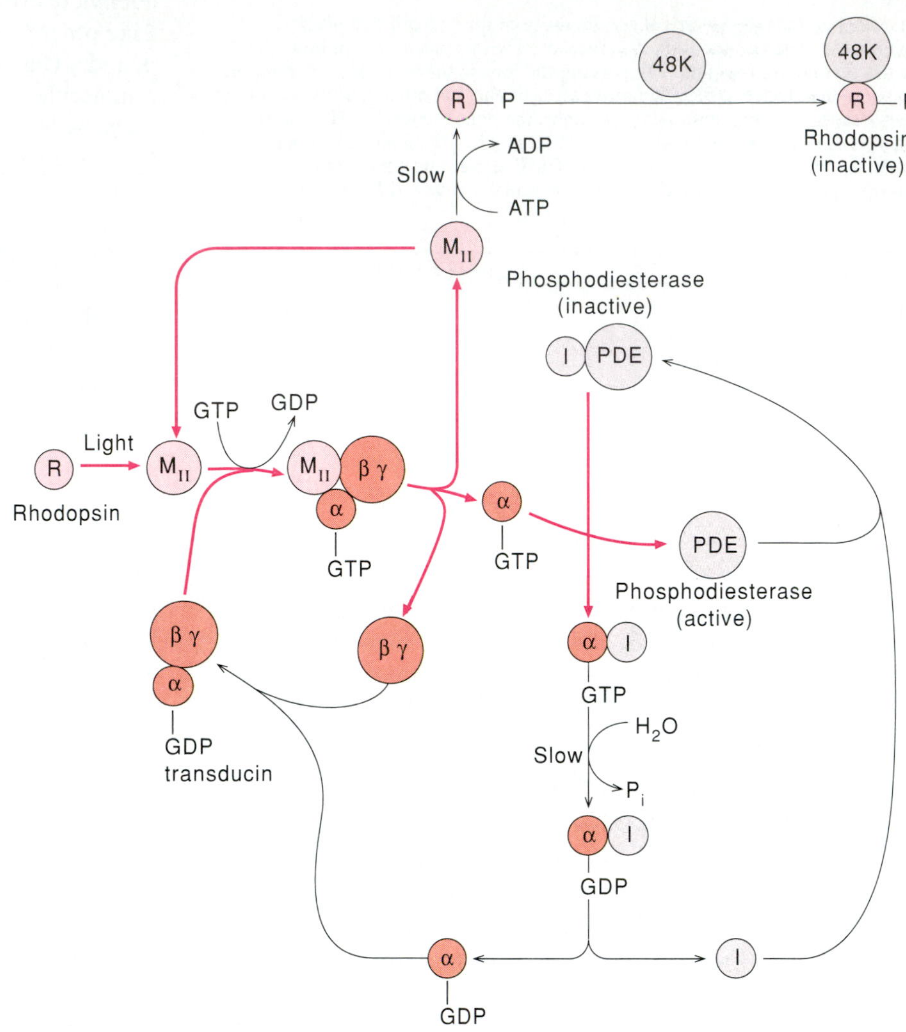

Regeneration of 11-*cis*-Retinal by Way of a Retinyl Ester

The transformations of rhodopsin begin with 11-*cis*-retinal bound to opsin and culminate in the release of all-*trans*-retinal from opsin. How is the 11-*cis*-isomer regenerated? The regeneration turns out to be more complex than we might have expected (fig. 36.13). The all-*trans*-retinal first is reduced to all-*trans*-retinol by NADH. The all-*trans*-retinol then leaves the retina and is taken up by the pigment epithelium, a separate tissue at the back of the eye, where it is esterified by an acyltransferase. This enzyme transfers a fatty acid from phosphatidylcholine to the retinol, forming an all-*trans*-retinyl ester. The retinyl ester then breaks down in an unusual, concerted reaction that couples isomerization of the retinol and the release of the free fatty acid. 11-*cis*-Retinal is regenerated by the oxidation of the 11-*cis*-retinol in the pigment epithelium. It then returns to the retina and is taken up again by opsin. The free energy that is needed to drive the conversion of retinol from all-*trans* to the less stable 11-*cis* isomer thus is provided by the breakdown of a phospholipid.

Rhodopsin Movement in the Disk Membrane

Each molecule of rhodopsin that is converted to metarhodopsin II is capable of causing some 500 molecules of transducin to bind GTP. Since this happens within about 0.5 s after the absorption of a photon, either rhodopsin or transducin, or both, must move about rapidly enough to encounter thousands of reaction partners per second. Rhodopsin is firmly embedded in the disk membrane and transducin is attached to the surface of the membrane, so they need to diffuse only in two dimensions in the plane of the membrane. But is it likely that they could move this rapidly? Actually, evidence that rhodopsin can diffuse rapidly within the membrane was obtained prior to the discovery of transducin and the phosphodiesterase.

The initial indications that rhodopsin moves around in the membrane came from studies of linear dichroism of rod outer segment disks. A material is said to exhibit linear dichroism if the strength of its optical absorption, when measured with polarized light, depends on the orientation of the polarizer. The polarization of a light beam is defined by the orientation of the

Figure 36.13

11-*cis*-retinal is regenerated from all-*trans*-retinal in a series of reactions that involve the formation of a retinyl ester as an intermediate. Most of these reactions occur in the pigment epithelium. (R_1 and R_2 = fatty acid side chains; R_3 = choline.)

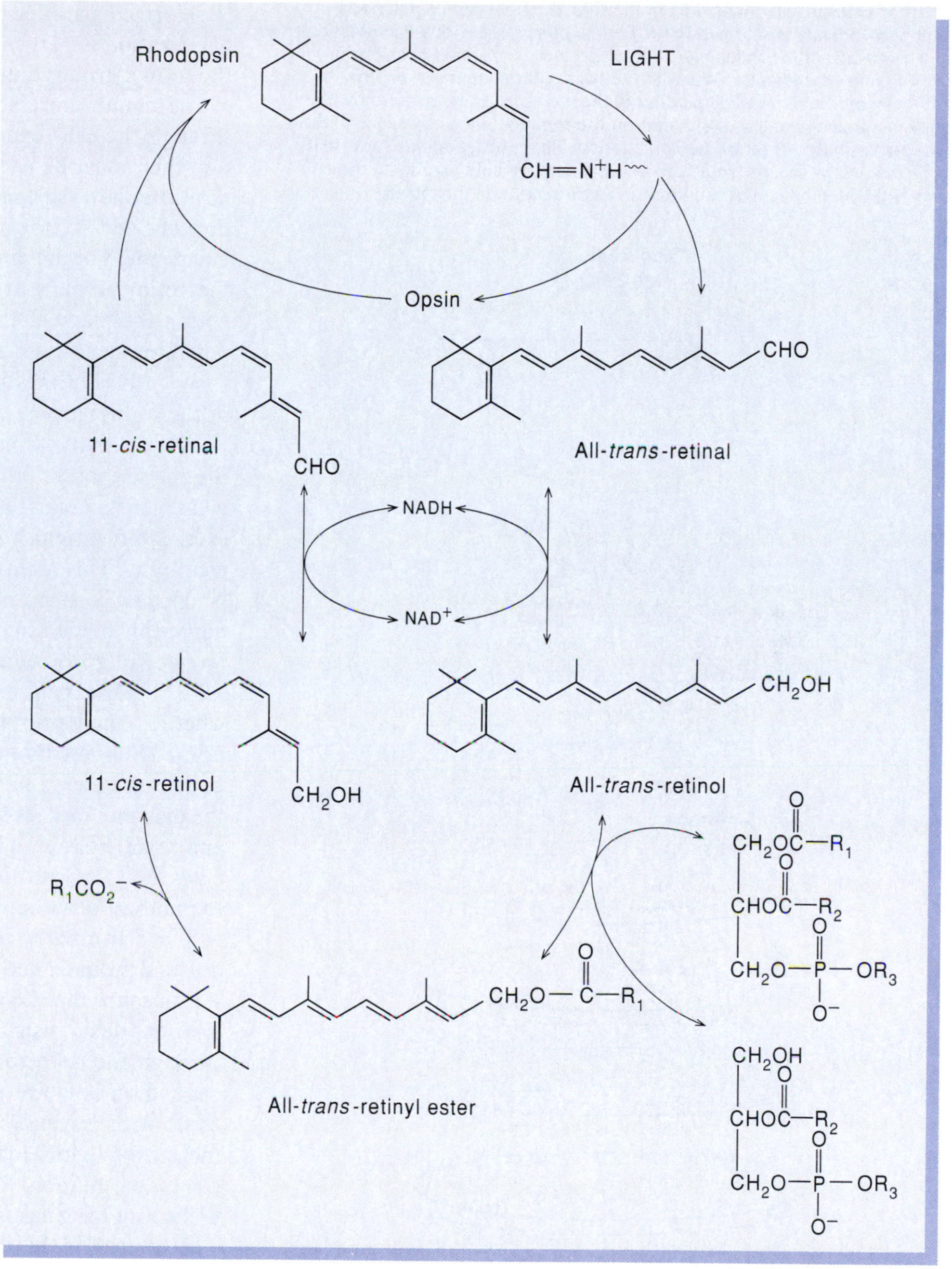

light's electric field, as indicated in figure 16.6. Molecules in solution usually do not exhibit linear dichroism because they are free to tumble about. At any given time, the solution contains molecules with all possible orientations relative to the polarizer, so the absorbance of the solution will not change if we rotate the polarizer. But if all the molecules are fixed in position with their molecular axes aligned, the system generally will exhibit linear dichroism. In the case of retinal or its Schiff base, light is absorbed best when the polarization makes the electric vector of the light parallel to the long axis of the molecule. You can get a feeling for this by examining the diagrams in figure 36.7. In order to convert the molecule from the ground state to the excited state, the electric field of the light must move electron density away from the ring end of the molecule and in the direction of the nitrogen.

The eyes of vertebrates are well suited for measurements of linear dichroism, because the rod cells are neatly aligned with their long axes perpendicular to the plane of the tissue. Suppose that we send a weak beam of polarized light through the rods from the side (fig. 36.14*a* and *b*). The polarization of the beam then could be made either parallel or perpendicular to the long axis of the cells. The planes of the disk membranes in the rod outer segments are aligned perpendicular to the cell axis, so the polarization of the light will be either

Figure 36.14

When rod cells are illuminated from the side, the light can be polarized either perpendicular (*a*) or parallel (*b*) to the planes of the disk membranes. The absorbance of the rhodopsin is much greater for parallel polarization than it is for perpendicular. When the cells are illuminated end-on, the light's electric field has to be parallel to the membranes, no matter how the light is polarized (*c* and *d*). With end-on illumination, the absorbance of the rhodopsin is independent of the polarization. These observations show that the 11-*cis*-retinyl groups (small arrows in disks) are held parallel to the plane of the membrane but can point in any direction in this plane.

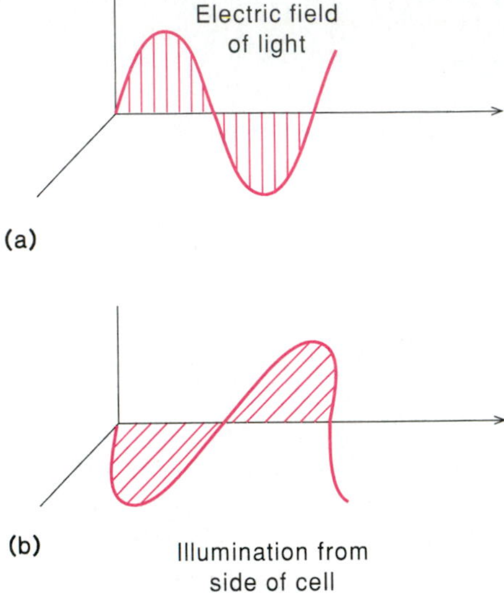

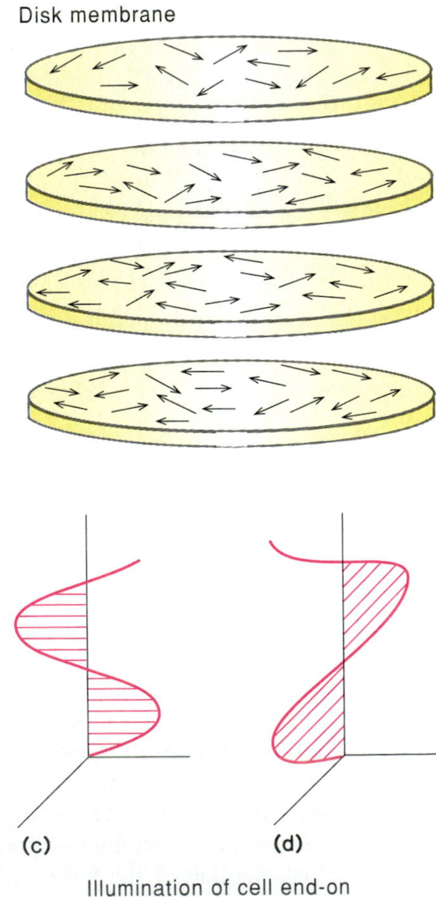

perpendicular or parallel to the membrane surface. It turns out that the rhodopsin absorbs much more strongly when the light is polarized parallel to the membranes than it does when the polarization is perpendicular. This means that the long axes of the retinyl groups must be held more or less parallel to the plane of the membrane. Now suppose that we send the light beam through the cells end-on (see fig. 36.14*c* and *d*). The polarization then must be parallel to the plane of the disk membranes, no matter how the polarizer is turned. With light coming through the cells end-on, the absorbance of the rhodopsin is found to be independent of the direction of the polarization. This means that the retinyl groups are not aligned in any particular direction within the plane of the membrane, but instead can take on all possible orientations in this plane.

It is interesting to note in passing that the orientation of the retinyl groups in the retinas of vertebrates optimizes our visual sensitivity for unpolarized light. Most of the light reaching the retinas passes through the rods end-on and is absorbed well, whatever its polarization (see fig. 36.14*c* and *d*). Insects, however, can distinguish between horizontally and vertically polarized light. This ability can be advantageous, because light that is reflected by smooth surfaces is partially polarized. Insects may make this distinction by having separate sets of cells with differently aligned retinyl groups.

How can measurements of linear dichroism reveal whether rhodopsin molecules move about within the membrane? Suppose that the retina is exposed to a flash of polarized light that enters the rods end-on. Some of the retinyl groups in the disk membranes will be oriented parallel to the polarization, and some will not. The light will *selectively* excite those molecules that are parallel to the polarization, because these have the highest absorbance. Most of the molecules that are excited will be converted to bathorhodopsin and the series of states that descend from it, and their absorption spectrum will change. If we measure the absorbance changes a short time after the excitation flash, using polarized measuring light that passes through the cells end-on, we see that the absorbance changes depend on how the polarization of the measuring light is oriented with respect to the excitation polarization. The polarized measuring light selectively measures molecules that have a particular orientation. Before the excitation flash, there is no linear dichroism for measurements made end-on; after the flash, there is. However, if the rhodopsin molecules can rotate in the plane of the membrane, the orientations of the molecules that were excited will in time become randomized, and the linear dichroism that is induced by the excitation will disappear. The kinetics of this disappearance will depend on how rapidly rhodopsin rotates.

Experiments of this sort were done by Richard Cone. Rhodopsin proved to rotate surprisingly rapidly. The time required for a rotation is about 2×10^{-5} s. This means that the environment of rhodopsin in the membrane must be highly fluid.

To determine whether rhodopsin can diffuse laterally in the membrane, Cone and Paul Leibman independently used a microspectrophotometric technique. They excited individual rod outer segments end-on near one edge of the disks through a microscope. Initially, the excitation caused absorbance changes only in the region that was illuminated, but within a few seconds after the excitation, the absorbance changes became uniformly distributed across the disk. This means that rhodopsin must be able to diffuse rapidly in the plane of the membrane. The diffusion coefficient is calculated to be about 5×10^{-9} cm^2/s. At this speed, rhodopsin molecules would collide with each other 10^5 to 10^6 times a second, and it would take only about a second for a single rhodopsin to collide with most of the molecules of transducin on the surface of the disk.

Bacteriorhodopsin: A Bacterial Pigment-Protein Complex That Resembles Rhodopsin

Halobacterium halobium is a red-colored, halophilic (salt-loving) bacterium that thrives in tidal salt flats. Its red color comes from two components of its cytoplasmic membrane: carotenoids and a purple pigment-protein complex that bears a striking resemblance to rhodopsin. The purple complex, bacteriorhodopsin, was first described by Walther Stoeckenius and his colleagues. Bacteriorhodopsin is not distributed randomly throughout the cell membrane, but rather is collected in patches that are held together by strong interactions among the bacteriorhodopsins. The patches, or purple membranes, can be isolated relatively simply, if the cells are disrupted by being placed in distilled water. Bacteriorhodopsin is the only protein in the purple membrane. It forms a highly ordered two-dimensional array that is almost crystalline in nature. Electron diffraction studies have shown that the individual proteins are folded into seven α-helical stretches extending from one side of the membrane to the other (see chapter 7).

Like rhodopsin, bacteriorhodopsin contains retinal bound to a lysine as a protonated Schiff base. The retinal isomerizes when it is excited with light. However, bacteriorhodopsin differs from rhodopsin in that the retinal starts out in the all-*trans* form rather than as the 11-*cis* isomer. The main product of the isomerization is the 13-*cis* isomer. Like rhodopsin, the excited bacteriorhodopsin progresses through a series of metastable states with different absorption spectra. The first of these states resembles bathorhodopsin in having an absorption spectrum that is shifted markedly to longer wavelengths. A later state resembles metarhodopsin II in that the retinyl Schiff base has lost its proton. A major difference, however, is that the transformations of bacteriorhodopsin are cyclic. Rather than dissociating from the protein, the retinal returns to the all-*trans* form, and the complex relaxes back to its original state and is ready to operate again.

The role of bacteriorhodopsin in *H. halobium* is to pump protons across the cytoplasmic membrane. We discussed some aspects of this pumping in chapter 32. As the excited bacteriorhodopsin undergoes its cyclic series of transformations, a proton is taken up from the solution inside the cell and released on the extracellular side of the membrane. This probably is the proton that is initially on the nitrogen atom of the Schiff base, because the Schiff base is deprotonated and then reprotonated in the course of the cycle. The isomerization of the retinal could move the proton with the nitrogen from a position where it equilibrates with the cytosol to a position where it equilibrates with the extracellular solution. Figure 36.9 will serve to illustrate this idea. The figure was drawn to show the isomerization of 11-*cis*-retinal to all-*trans*-retinal that occurs in rhodopsin, but the principles here are conceptually the same. The isomerization of the retinyl chain pushes the proton on the Schiff base nitrogen from the region of one functional group (A_1^- in the figure) to the region of another (A_2^-). From A_2H, a proton could be conducted to the solution on the extracellular side of the membrane. When the retinal then returns to its original state, the proton lost by the nitrogen could be replaced by one from a third group (A_3H) on the cytoplasmic side of the membrane.

As in other types of photosynthetic bacteria, the protons that are pumped across the membrane by bacteriorhodopsin in the light can reenter the cell by way of a proton-conducting ATP-synthase, generating ATP. *H. halobium* also contains a transport system that moves Na^+ ions out of the cell in exchange for protons coming in. Removal of Na^+ is a major chore for an organism that lives in water containing 4-M NaCl. The cytoplasmic membrane also has two other pigment-protein complexes that execute photochemical cycles similar to the transformations of bacteriorhodopsin. One of these, halorhodopsin, acts as a light-driven pump for Cl^-. The other, sensory rhodopsin, is involved in phototaxis, the tendency of the cells to swim in the direction of a source of light.

Summary

In this chapter we have described the biochemical mechanisms involved in vision. The main points of our discussion are as follows.

1. Light rays entering the eye of a vertebrate are refracted by the cornea and focused to form an image on the retina. The retina contains several layers of cells. Some of these cells, the rods and cones, contain the visual pigments that are responsible for the initial response to light; other cells make synaptic connections to rods or cones and to additional neural cells that carry impulses to the brain.
2. The light-sensitive protein complex in the rods, rhodopsin, consists of 11-*cis*-retinal bound as a Schiff base to a protein, opsin. Rhodopsin is an integral constituent of membranes that form a stack of disks at one end of the cell. Cone cells, which are responsible for the perception of color, contain similar complexes in infoldings of the plasma membrane.
3. When rhodopsin absorbs light, the retinal isomerizes to the all-*trans*-isomer. This change initiates a series of transformations of the pigment-protein complex that result in an interaction with another protein, transducin. The

interaction with rhodopsin causes one of transducin's subunits to take up a molecule of GTP in exchange for bound GDP, to dissociate from the other subunits, and to react with a phosphodiesterase that hydrolyzes cGMP. This reaction activates the phosphodiesterase. The resulting drop in cGMP concentration leads to a decrease in the Na^+ permeability of the cytoplasmic membrane. A hyperpolarization of the membrane then triggers the transmission of a signal at the rod's synaptic junction to the adjacent neural cell.

4. Bacteriorhodopsin, a bacterial pigment-protein complex, resembles rhodopsin (it contains all-*trans*-retinal), but serves a different function. Like rhodopsin, bacteriorhodopsin progresses through a series of metastable states when it is excited with light. In the course of these transformations, protons are taken up from the solution inside the cell and are released on the extracellular side of the membrane, generating an electrochemical potential gradient for protons across the membrane. Protons reenter the cell by way of an ATP-synthase, energizing the formation of ATP.

Selected Readings

Chabre, M., Trigger and amplification mechanisms in visual phototransduction. *Ann. Rev. Biophys. Biophys. Chem.* 14:331, 1985. A review covering the structure of rhodopsin and the roles of cGMP, GTP, the phosphodiesterase, and transducin in the response of the rod cell to light.

Cone, R., Rotational diffusion of rhodopsin in visual photoreceptor membrane. *Nature* 236:39, 1972. Excitation of the retina with polarized light creates a transient linear dichroism. The decay of the dichroism is explained by rotation of rhodopsin.

Fesenko, E. E., S. S. Kolesnikov, and A. L. Lymbarsky, Induction by cyclic GMP of cation conductance in plasma membrane of retinal rod outer segment. *Nature* 313:310, 1985. cGMP increases the Na^+ conductance of the cytoplasmic membrane.

Nathans, J., Rhodopsin: Structure, function, and genetics. *Biochem.* 31:4923–4932, 1992. A review of recent work on the structure and reactions of rhodopsin.

Stoeckenius, W., Purple membrane of halobacteria: A new light-energy converter. *Acc. Chem. Res.* 13:337, 1980. A review of the structure and function of bacteriorhodopsin and the purple membrane.

Wald, G., The molecular basis of visual excitation. *Nature* 219:800, 1968. Wald's Nobel Prize address, describing early work on the isomerization of retinal in rhodopsin.

Problems

1. What chemical characteristics would you look for in a compound that absorbs visible light?
2. The spectrophotometers in biochemistry laboratories are very sensitive because they use a photomultiplier tube as a detector and convert photons of light into an electrical signal. The signal subsequently is multiplied to produce a strong signal. How do our eyes multiply the signal derived from light to increase sensitivity?
3. Draw traces showing how the optical absorbance of a fresh suspension of rod outer segment disks might change as a function of time when the suspension is excited with a short flash of light at 37° C. Show the absorbance at (a) 545 nm and (b) 480 nm. Select the time scale for each trace judiciously, so that the traces illustrate the kinetics of the major absorbance changes that occur at the two wavelengths. (You may need to use two traces with different time scales at each wavelength to show both the initial absorbance change and its decay.)
4. If rhodopsin is illuminated at 500 nm at 77 K, the absorbance of the sample at 500 nm decreases. If the sample is then illuminated at 550 nm (still at 77 K), the absorbance at 500 nm increases again. Explain.
5. Why is the absorption spectrum of metarhodopsin II so different from that of metarhodopsin I?
6. Draw traces showing how the membrane potential of a rod changes with time when the cell is excited with (a) one photon, and (b) two photons. Assume that each photon converts one molecule of rhodopsin to bathorhodopsin and one to metarhodopsin II. (The probability of conversion actually is only about 0.67.)
7. Provide evidence and a model for the involvement of cGMP as a mediator in visual transduction.
8. A frog retina is excited with a flash of polarized light that passes through the rods end-on. The flash causes optical absorbance changes at 500 nm, which are measured with polarized light that also passes through the rods end-on.
   (a) Draw traces showing the kinetics of the absorbance changes measured with light polarized parallel to the excitation polarization and with light polarized perpendicular to the excitation polarization. The traces should cover the time period up to 100 μs after the excitation.
   (b) Similar experiments are done with a retina that has been treated with glutaraldehyde, which causes cross-linking of the rhodopsins in the membrane. Draw traces showing the kinetics of the absorbance changes expected in this case. Assume that the cross-linking immobilizes the rhodopsin molecules but does not otherwise affect their photochemical transformations.
9. In solution, 11-*cis*-retinal maximally absorbs light near 380 nm, but in rhodopsin the peak is at 500 nm. Explain the spectral shift in the retinal bound to the protein.
10. How do our eyes detect color? Why do we not see color in dim light?

A Student's Guide to Methods of Biochemical Analysis

Biochemistry is an experimental science, in which a great variety of techniques are used. We provide examples of the use of various methods throughout this text, usually in connection with the accounts of specific experiments. This approach has the advantage of allowing you to appreciate immediately the relevance of a particular technique. It does mean, however, that the examples are scattered throughout the text. Our aim here is to supply a guide to the locations of those examples within the framework of a unifying outline, so that you can see at a glance which methods of analysis are most useful for certain purposes and so that you can easily find the places in the text where those methods are described. In the outline, you are referred to the figures or boxes that contain such descriptions; you will sometimes find additional information in the adjoining text.

Our descriptions of techniques are brief. There are a number of excellent printed sources where you can find more complete descriptions, as well as comprehensive surveys of available techniques. For example, Academic Press has published more than two hundred books in their series *Methods in Enzymology,* and Oxford Press has published more than eighty books in their series *Biochemistry: A Practical Approach*. In addition, you will find descriptions of the procedures for manipulating DNA in the remarkable three-volume set *Molecular Cloning: A Laboratory Manual,* by J. Sambrook, E. F. Fritsch, and T. Maniatis, 2d ed. (Cold Spring Harbor, N.Y.: Cold Spring Harbor Laboratory, 1989).

I. Separation of major cellular components
  Comparison of differential centrifugation with isopycnic centrifigution (fig. 7.20).
  Separation of periplasmic proteins and inner and outer membranes of a Gram negative bacterium (fig. 7.23).

II. Subfractioning of major cellular components into their constituent molecules
  Methods of protein purification are described in the latter part of chapter 5, which covers differential precipitation with $(NH_4)_2SO_4$ (box 5B), column fraction procedures (box 5C), gel electrophoresis (box 5D), isoelectric focusing (box 5D), and differential centrifugation (box 5E). The following, related material appears in other chapters.
  Analysis of protein mixtures by polyacrylamide gel electrophoresis (fig. 7.31).
  Analysis of ribosomal proteins by two-dimensional gel electrophoresis (box 28A).
  Method for lipid fractionation (box 7A).
  Detergent solubilization of biological membranes (fig. 7.29).
  Separation of N-linked carbohydrates by lectin-affinity chromatography (fig. 6.26).

III. Structural characterization of individual cellular components
  A. **Amino acids and protein primary structure**
    Determination of p*K* for amino acids (fig. 3.2, 3.3, and 3.4).
    pH titration curve for a protein (fig. 3.6).
    Column method for analysis of amino acid mixtures (fig. 3.14).
    Determination of amino acid composition of a protein (figs. 3.15 and 3.16).
    Dansyl chloride method for N-terminal amino acid determination (box 3B).
    Determination of the amino acid sequence of the polypeptide chain(s) of a protein (figs. 3.17 through 3.22).

Separation of amino acid derivatives by thin-layer chromatography (fig. 3.23).
Measurement of ultraviolet absorption in solution (box 3A).
Determination of essential amino acids in rats and humans (table 18.2).

B. **Properties of intact proteins**
Most methods of determining protein conformation are described in chapter 4. These include predicting protein tertiary structure from the primary structure (fig. 4.37), X-ray diffraction methods for determining the structure of fibrous proteins (box 4A) and of globular proteins (box 4B), different ways of visualizing molecular structures of proteins (box 4C), and radiation techniques for examining protein structure (box 4D). The following, related material appears in other chapters.
Measuring the binding of inducer to repressor by equilibrium dialysis (box 30B).
Domain-swap experiment demonstrating bifunctional nature of a DNA-binding protein (fig. 31.5).
Electron microscopic examination of striated muscle indicating the structural changes that accompany muscular contraction (fig. 5.17).
Determination of the size of protein molecules in solution by sedimentation and diffusion (box 4E).

C. **Lipids and membranes**
Isolation and analysis of phospholipids (box 7A).
Hydrophobic interaction chromatography (fig. 7.30).
Hydropathy plot to determine segments of a polypeptide chain most likely to be membrane-bound (figs. 7.37 and 32.10).
Differential scanning calorimetry for estimating stability of a membrane structure (fig. 7.39).
Determination of the rate of diffusion of a molecule across a semipermeable membrane (figs. 32.2, 32.3 and 32.4).

D. **Carbohydrates**
Use of glycosidases for structural analysis of oligosaccharides (fig. 6.27).
Polarimetry (box 6A).
Nuclear magnetic resonance spectroscopy for structural analysis of oligosaccharides (box 6B).

E. **Nucleic acids**
Use of ultraviolet absorbance to study DNA denaturation and renaturation (fig. 25.22).
Use of reassociation kinetics to study sequence complexity of DNA (figs. 25.25 and 25.26).
Use of enzyme probes to investigate bacterial chromosome structure (fig. 25.28).
Assessment of supercoiling by ultracentrifugation (fig. 25.19) or electrophoresis (fig. 25.21).
X-ray diffraction analysis of DNA (box 25A).
Techniques using nucleic acid renaturation (box 25D).
Equilibrium density-gradient centrifugation for analysis of DNAs of different densities (box 25B).
Chromosome analysis by gel electrophoresis (box 25C).
Southern blot analysis to characterize DNA sequences in spliced immunoglobulin genes (fig. 33.4; also see fig. 27.27).
Hybridization followed by CsCl density-gradient sedimentation for structural analysis of phage transcripts (fig. 28.2).
Determination of DNA protein-binding sites by chemical protection (fig. 28.15).
Determination of intron size and location by selective annealing and electron microscopy (fig. 28.25).
Footprinting technique for determining location of site of protein binding to DNA (box 28B).
Techniques for manipulating DNA (see listing under part VI of this outline).

IV. **Characterization of individual reactions *in vitro***
Chapter 8 is concerned with kinetic methods for analyzing enzyme-catalyzed reactions. These include determination of the Michaelis constant, the turnover number, and the specificity constant, special kinetic considerations for reactions involving two substrates, and analysis of the effects of temperature, pH, and various types of inhibitors. Chapters 9 and 10 are concerned with methods for analyzing mechanisms of enzyme-catalyzed reactions. Chapter 10 deals exclusively with the unique properties of regulatory enzymes. The following, related material appears in other chapters.
Calculation of the equilibrium constant, K_{eq}, when the numbers of reactants and products are not equal (box 13A).
Calculation of the equilibrium constant for a reaction from the thermodynamic data (chapter 2).
Apparatus for measuring the difference between the standard redox potentials, E^0, of two redox couples (fig. 15.10).
Titration curves for redox couples (fig. 15.11).
Assay for the activity of fatty acid synthase (box 17A).
In vitro studies to determine properties of the recA enzyme (fig. 26.34).
Instrument for recording action potentials (figs. 35.2 and 35.3).
Isotope labeling technique for determining stereospecificity of an enzyme-catalyzed reaction (fig. 11.7 and associated text).

V. ***In vitro* or *in vivo* characterization of reactions in pathways**
Methods for pathway analysis are reviewed at the end of chapter 12. The following, related material appears in other chapters.
Difference spectra method for examining the redox states of respiratory carriers in intact mitochondria (fig. 15.13).

Inhibitors used to analyze electron transport in mitochondria (fig. 15.14).
Determination of the P/O ratio in respiring mitochondria (fig. 15.21).
Use of uncouplers to study the mechanism of ATP synthesis in respiring mitochondria (fig. 15.23).
Use of enzyme inhibitors to study ATP synthesis in mitochondria (fig. 15.24).
In vitro methods for studying the mechanism of ATP synthesis in respiring mitochondria (figs. 15.26, 15.27, 15.29, 15.32, and 15.34).
Analysis of O_2 production by illumination of chloroplasts (figs. 16.16, 16.20, and 16.23).
Spectral methods for analysis of photosystems I and II (fig. 16.21).
Manipulation of pH to study ATP synthesis of chloroplasts (fig. 16.26).
Spectral methods for the analysis of phytochrome behavior (fig. 16.23).
Use of equilibrium density-gradient centrifugation to determine overall mode of DNA replication (figs. 26.3 and 26.6).
Use of autoradiography to determine mode of DNA replication in eukaryotic chromosomes (figs. 26.4 and 26.6).
Use of pulse-chase labeling to follow precursor-product relationship in DNA synthesis (fig. 26.9).
Use of synthetic polynucleotides for determining genetic code assignments (fig. 29.7).
In vitro assay for translation release factors (fig. 29.20).
Device for recording action potentials (figs. 35.2 and 35.3).
Use of polarized light to determine orientation of optical pigments in rod cells (fig. 36.14).
Tissue transplantation technique used to establish the mechanism for immunologic tolerance (fig. 33.14).
Pathway analysis by genetic complementation (box 18A).
Use of isotopes as tracers to delineate the valine and isoleucine pathways (box 18C).
Use of isotopes and mutants to demonstrate that indole is not a true intermediate in the tryptophan pathway (box 18D).
Precursors of purine ring determined by isotope labeling (fig. 20.12).
Use of antibiotics to determine pathway for cell-wall biosynthesis in bacteria (figs. 21.16, 21.17, and 21.19).
Use of inhibitors to determine stages in N-linked carbohydrate synthesis (table 21.3).
Use of mutants to determine stages in N-linked and O-linked carbohydrate synthesis (table 21.4).
Assay for transit of glycoproteins between Golgi membranes (box 21D).
Use of substrates and inhibitors for demonstrating the circular nature of the Krebs cycle (fig. 14.3).
Demonstration of the prochiral nature of citrate in the Krebs cycle (fig. 14.9).

VI. Genetic methods

Chapter 27 is concerned with techniques of DNA manipulation. These include the Maxam-Gilbert and Sanger methods for sequencing DNAs and RNAs, the synthesis of oligonucleotides, the amplification of sequences by the polymerase chain reaction, the use of restriction enzymes to cut DNA into well-defined fragments, procedures for cloning DNA into *E. coli,* yeast, mammals, and plants, site-directed mutagenesis, and the use of recombinant DNA techniques for mapping the globin gene family and the cystic fibrosis gene. The following, related material appears in other chapters.
DNA transformation (fig. 25.2).
Demonstration that nucleic acids carry the genetic information in viruses (figs. 25.3 and 25.4).
Pathway analysis by genetic complementation (box 18A).
Use of mutants to determine stages in N-linked and O-linked carbohydrate synthesis (table 21.4).
Procedure for constructing a congenic strain (fig. 33.16).
Tissue transplantation technique to establish the mechanism for immunologic tolerance (fig. 33.14).
Genetic analysis of *lac* operon expression by mutant studies on haploid and diploid cells (table 30.1).
Mutant studies on expression of the tryptophan operon of *E. coli* (table 30.2).
Nuclear transplantation assay for totipotent nucleus (fig. 31.12).
Genetic analysis of the interaction between regulatory proteins involved in early *Drosophila* development (fig. 31.31).
Mutant analysis of genes responsible for characteristics of specific body segments (fig. 31.33).
Genetic analysis of genes regulating galactose metabolism in yeast (table 31.1).
Determination of the location of developmental gene expression in the *Drosophila* embryo by *in situ* labeling with specific probes followed by autoradiography (box 31A).
Analysis of key events in early *Drosophila* development through detection of specific transcripts at various times and locations (fig. 31.29).
Use of mutants to determine the significance of DNA polymerase I of *E. coli* (chapter 26).
Determination of the crucial amino acid residues in the hemoglobin molecule by observations on the invariant residues found in all hemoglobins (fig. 5.13) and the amino acid residues that lead to pathologic conditions (fig. 5.14).

Common Abbreviations in Biochemistry

A	adenine
ACh	acetylcholine
ACTH	adrenocorticotropic hormone
ADP	adenosine-5′-diphosphate
AIDS	acquired immune deficiency syndrome
Ala (A)	alanine
AMP	adenosine-5′-monophosphate
Asn (N)	asparagine
Asp (D)	aspartic acid
ATP	adenosine-5′-triphosphate
BChl	bacteriochlorophyll
bp	base pair
BPG	D-2,3-bisphosphoglycerate
C	cytosine
Cys (C)	cysteine
CAP	catabolite gene activator protein
CDP	cytidine-5′-diphosphate
CMP	cytidine-5′-monophosphate
CTP	cytidine-5′-triphosphate
CoA or CoASH	coenzyme A
CoQ	coenzyme Q (ubiquinone)
cAMP	adenosine 3′, 5′-cyclic monophosphate
cGMP	guanosine 3′, 5′-cyclic monophosphate
cyt	cytochrome
d	2′-deoxy-
DHAP	dihydroxyacetone phosphate
DHF	dihydrofolate
DHFR	dihydrofolate reductase
DMS	dimethyl sulfate
DNA	deoxyribonucleic acid
cDNA	complementary DNA
DNase	deoxyribonuclease
DNP	2,4-dinitrophenol
ER	endoplasmic reticulum
FAD	flavin adenine dinucleotide (oxidized form)
$FADH_2$	flavin adenine dinucleotide (reduced form)
FBP	fructose-1,6-bisphosphate
fMet	*N*-formylmethionine
FMN	flavin mononucleotide (oxidized form)
$FMNH_2$	flavin mononucleotide (reduced form)
F1P	fructose-1-phosphate
F6P	fructose-6-phosphate
G	guanine
G protein	guanine-nucleotide binding protein
GDP	guanosine-5′-diphosphate
Gly (G)	glycine
GMP	guanosine-5′-monophosphate
Gln (Q)	glutamine
Glu (E)	glutamic acid
GSH	glutathione
GSSG	glutathionine disulfide
GTP	guanosine-5′-triphosphate
Hb	hemoglobin
HDL	high-density lipoprotein
HETPP	hydroxyethylthiamine pyrophosphate
HGPRT	hypoxanthine-guanosine phosphoribosyl transferase
His (H)	histidine
HIV	human immunodeficiency virus
HMG-CoA	β-hydroxy-β-methylglutaryl-CoA
HPLC	high-performance liquid chromatography
IDL	intermediate-density lipoprotein
IF	initiation factor
IgG	immunoglobulin G
Ile (I)	isoleucine
IMP	inosine-5′-monophosphate
IP_1	inositol-1-phosphate
IP_3	inositol-1,4,5-triphosphate
IPTG	isopropylthiogalactoside
K_m	Michaelis constant
kb	kilobase pair
kDa	kilodaltons

LDL	low-density lipoprotein	RNA	ribonucleic acid
Leu (L)	leucine	hnRNA	heterogeneous nuclear RNA
Lys (K)	lysine	mRNA	messenger RNA
Man	mannose	rRNA	ribosomal RNA
MHC	major histocompatability complex	snRNA	small nuclear RNA
Met (M)	methionine	tRNA	transfer RNA
NAD^+	nicotinamide-adenine dinucleotide (oxidized form)	snRNP	small ribonucleoprotein
		RNase	ribonuclease
NADH	nicotinamide-adenine dinucleotide (reduced form)	Ru1,5P	ribulose-1,5-bisphosphate
		Ru5P	ribulose-5-phosphate
$NADP^+$	nicotinamide-adenine dinucleotide phosphate (oxidized form)	R5P	ribose-5′-phosphate
		RSV	Rous sarcoma virus
NADPH	nicotinamide-adenine dinucleotide phosphate (reduced form)	*s*	Svedberg constant
		SAM	*S*-adenosylmethionine
NDP	nucleoside-5′-diphosphate	SDS	sodium dodecyl sulfate
NAM	*N*-acetylmuramic acid	Ser (S)	serine
NMR	nuclear magnetic resonance	S7P	sedoheptulose-7-phosphate
NTP	nucleoside-5′-triphosphate	SRP	signal recognition particle
Phe (F)	phenylalanine	T	thymine
P_i	inorganic orthophosphate	THF	tetrahydrofolate
PEP	phosphoenolpyruvate	Thr (T)	threonine
PFK	phosphofructokinase	TLC	thin-layer chromatography
PG	prostaglandin	TMV	tobacco mosaic virus
2PG	2-phosphoglycerate	TPP	thiamine pyrophosphate
3PG	3-phosphoglycerate	Trp (W)	tryptophan
PIP_2	phosphatidylinositol-4,5-bisphosphate	TTP	thymidine-5′-triphosphate
PK	pyruvate kinase	Tyr (Y)	tyrosine
PLP	pyridoxal-5-phosphate	U	uracil
PP_i	inorganic pyrophosphate	UDP	uridine-5′-diphosphate
Pro (P)	proline	UDPG	UDP-glucose
PRPP	phosphoribosylpyrophosphate	UMP	uridine-5′-monophosphate
PS	photosystem	UQ	ubiquinone
Q	ubiquinone or plastoquinone	Val (V)	valine
QH_2	ubiquinol or plastoquinol	VLDL	very-low-density lipoprotein
RER	rough endoplasmic reticulum	XMP	xanthosine-5′-monophosphate
RF	release factor or replicative form	Xu5P	xylulose-5′-phosphate
RFLP	restriction-fragment length polymorphism		

Appendix A
Some Major Discoveries in Biochemistry

In this appendix we list, in chronological order, some of the most important discoveries made in biochemistry in the past two centuries. It is impossible, for reasons of space, to give credit to every worker who has made a significant contribution, but it is possible to identify certain events as milestones, and thus to show how progress in this field has accelerated with the passage of time.

1770–1774
Priestly showed that oxygen is produced by plants and consumed by animals.

1773
Rouelle isolated urea from urine.

1828
Wohler synthesized the first organic compound, urea, from inorganic components.

1838
Schleiden and Schwann proposed that all living things are composed of cells.

1854–1864
Pasteur proved that fermentation is caused by microorganisms.

1864
Hoppe-Seyler crystallized hemoglobin.

1866
Mendel demonstrated the segregation and independent assortment of alleles in pea plants.

1893
Ostwald showed that enzymes are catalysts.

1905
Knoop deduced the β oxidation mechanism for fatty acid degradation.

1907
Fletcher and Hopkins showed that lactic acid is formed quantitatively from glucose during anaerobic muscle contraction.

1910
Morgan discovered sex-limited inheritance in *Drosophila*.

1912
Warburg postulated a respiratory enzyme for the activation of oxygen.

1913
Michaelis and Menten developed a kinetic theory of enzyme action.

1922
McCollum showed that lack of vitamin D causes rickets.

1926
Sumner crystallized the first enzyme, urease.

1926
Jansen and Donath isolated vitamin B_1 (thiamine) from rice polishings.

1926–1930
Svedberg invented the ultracentrifuge and determined the molecular weights of macromolecules.

1928
Levene showed that nucleotides are the building blocks of nucleic acids.

1928
Szent-Gyorgyi isolated ascorbic acid (Vitamin C).

1928–1933
Warburg deduced the iron-prophyrin presence in the respiratory enzyme.

1931
Engelhardt discovered that phosphorylation is coupled to respiration.

1932
Warburg and Christian discovered the "yellow enzyme," a flavoprotein.

1933
Krebs and Henseleit discovered the urea cycle.

1933
Embden and Meyerhof demonstrated the intermediates in the glycolytic pathway.

1935
Schoenheimer and Rittenberg first used isotopes as tracers in the study of intermediary metabolism.

1935
Stanley first crystallized a virus, tobacco mosaic virus.

1937
Krebs discovered the citric acid cycle.

1937
Warburg showed how ATP formation is coupled to the dehydrogenation of glyceraldehyde-3-phosphate.

1938
Hill found that cell-free suspensions of chloroplasts yield oxygen when illuminated in the presence of an electron acceptor.

1939
C. Cori and G. Cori demonstrated the reversible action of glycogen phosphorylase.

1939
Lipmann postulated the central role of ATP in the energy-transfer cycle.

1939–1946
Szent-Gyorgyi discovered actin and actin-myosin complex.

1940
Beadle and Tatum deduced the one gene–one enzyme relationship.

1942
Bloch and Rittenberg discovered that acetate is the precursor of cholesterol. Subsequently Woodward and Bloch determined the complete pathway for cholesterol biosynthesis.

1943
Chance applied spectrophotometric methods to the study of enzyme–substrate interactions.

1943
Martin and Synge developed partition chromatography.

1944
Avery, MacLeod, and McCarty demonstrated that bacterial transformation is caused by DNA.

1947–1950
Lipmann and Kaplan isolated and characterized coenzyme A.

1948
Leloir discovered the role of uridine nucleotides in carbohydrate metabolism.

1948
Hogeboom, Schneider, and Palade refined the differential centrifugation method for fractionation of cell parts.

1948
Kennedy and Lehninger discovered that the tricarboxylic acid cycle, fatty acid oxidation, and oxidative phosphorylation all take place in mitochondria.

1950–1953
Chargaff discovered the base equivalences in DNA.

1951
Pauling and Corey proposed the α-helix structure for α-keratins.

1951
Lynen postulated the role of coenzyme A in fatty acid oxidation.

1952
Palade, Porter, and Sjostrand perfected thin sectioning and fixation methods for electron microscopy of intracellular structures.

1952–1954
Zamecnik and his colleagues developed the first cell-free systems for the study of protein synthesis.

1953
Sanger and Thompson determined the complete amino acid sequence of insulin.

1953
Horecker, Dickens, and Racker elucidated the 6-phosphogluconate pathway of glucose catabolism.

1953
Watson and Crick and Wilkins determined the double-helix structure of DNA.

1955
Ochoa and Grunberg-Manago discovered polynucleotide phosphorylase.

1956
Kornberg discovered the first DNA polymerase.

1956
Umbarger reported that the end product isoleucine inhibits the first enzyme in its biosynthesis from threonine.

1956
Ingram showed that normal and sickle-cell hemoglobin differ in a single amino acid residue.

1956
Anfinsen and White concluded that the three-dimensional conformation of proteins is specified by their amino acid sequence.

1957
Hoagland, Zamecnik, and Stephenson isolated tRNA and determined its function.

1957
Sutherland discovered cyclic AMP.

1958
Weiss, Hurwitz, and Stevens discovered DNA-directed RNA polymerase.

1958
Meselson and Stahl demonstrated that DNA is replicated by a semiconservative mechanism.

1960
Kendrew reported the x-ray analysis of the structure of myoglobin.

1961
Jacob and Monod proposed the operon hypothesis.

1961
Jacob, Monod, and Changeux proposed a theory of the function and action of allosteric enzymes.

1961
Mitchell postulated the chemiosmotic hypothesis for the mechanism of oxidative phosphorylation.

1961
Nirenberg and Matthaei reported that polyuridylic acid codes for polyphenylalanine.

1961
Marmur and Doty discovered DNA renaturation.

1962
Racker isolated F_1 ATPase from mitochondria and reconstituted oxidative phosphorylation in submitochondrial vesicles.

1966
Maizel introduced the use of sodium dodecylsulfate (SDS) for high-resolution electrophoresis of protein mixtures.

1966
Gilbert and Muller-Hill isolated the lac repressor.

1968
Meselson and Yuan discovered the first DNA restriction enzyme. Shortly thereafter Smith and Wilcox discovered the first restriction enzyme that cuts DNA at a specific sequence.

1969
Zubay and Lederman developed the first cell-free system for studying the regulation of gene expression.

1973
Cohen, Chang, Boyer, and Helling reported the first DNA cloning experiments.

1975
Sanger and Barrell developed rapid DNA-sequencing methods.

1977
Starlinger discovered the first DNA insertion element.

1977
Splicing of RNA simultaneously discovered in Broker's and Sharp's laboratories.

1978
Shortles and Nathans did the first experiments in directed mutagenesis.

1978
Tonegawa demonstrated DNA splicing for an immunoglobulin gene.

1981
Cech discovered RNA self-splicing.

1981
Steitz determined the structure of CAP protein.

1981–1982
Palmiter and Brinster produced transgenic mice.

1983
Mullis amplified DNA by the polymerase chain reaction (PCR) method.

1984
Schwartz and Cantor developed pulsed field gel electrophoresis for the separation of very large DNA molecules.

Appendix B

Answers to Selected Problems

Chapter 2

1. Reactions within the cell occur under a much more restricted range of temperature, pressure, pH, and reactant concentration than is possible *in vitro.*
3. Intensive thermodynamic parameters are independent of the amount of material in the state (e.g., temperature, density), whereas extensive parameters depend on the amount of material (energy, mass).
5. Free energy changes are state functions independent of pathway. A reaction mechanism defines a specific reaction path. You can predict free energy changes on the basis of a reaction mechanism but not vice versa.
7. The reaction will proceed toward oxaloacetate formation in the cell if the concentration of products is kept low. Oxidation of NADH by the mitochondrial electron-transport system and utilization of oxaloacetate in the formation of citrate shifts the malate–oxaloacetate reaction toward oxaloacetate production.
9. **(a)** For reaction (P1), $\Delta G°' = -2.4$ kcal/mole; **(b)** $\Delta G°'$ of ATP hydrolysis is -7.9 kcal/mole.
11. A minimum of 2.7 moles of protons is required per mole of ADP phosphorylated.
13. **(a)** If the ratio of NADH/NAD^+ is 1:10, the ratio of lipoamide/dihydrolipoamide is 0.98. **(b)** If the NADH/NAD^+ ratio is 10:1, the ratio is 9.8×10^{-3}.
15. K'_{eq} is 1.5×10^{-2} with pyrophosphatase absent; 4.8×10^3 with pyrophosphatase present.

Chapter 3

1. **(a)** $K_a = 1.11 \times 10^{-4}$; $pK_a = 3.95$; **(b)** pH = 3.47.
3. 1 mmole of NaOH
5. pH 2: Arg, His, Lys; pH 7: Arg, Asp, Glu, Lys; pH 12: Arg, Asp, Glu, Tyr, Cys
7. The two principal forms differ in degree of protonation of the R group (ϵ-amino group). Each will have a fully deprotonated carboxylate (net negative charge) and an unprotonated α-amino group (neutral). The equilibrium mixture will contain 0.028 mmoles protonated and 0.072 mmoles unprotonated ϵ-amino group.
9. Polyhistidine will bear little positive charge at pH 7.5 and even less at pH 10, hence the polymer will be insoluble. At pH 5.5, the fraction of imidazole groups protonated will be greater and the polymer more soluble.
11. At pH 6.1 the negatively charged glutamate will bind to the basic column, arginine will bind to the acidic column, and alanine will pass unretarded through both columns.
13. NH_2-Ala-Ala-Lys-Ala-Ala-Phe-Ala
15. NH_2-Ala-Ser-Lys-Phe-Gly-Lys-Tyr-Asp

Chapter 4

1. Consider the entropic effect of decreasing water organization by moving the hydrophobic residue side chains from an aqueous to a nonaqueous environment.
3. The α helix is a rodlike element that cannot easily change direction. Loops, β bends, and "random" structure break the helical structure and allow these directional changes.
5. An α helix broken at the Pro-Asn-Ala region with the hydrophobic residues on the exterior should insert into the membrane.
7. The metal at the active site must be aligned in reasonably precise geometry. Conservation of residues around the metal ligands preserves the geometry of metal binding.
9. The two proteins in question may share common epitopes and cross-react. Alternatively, the "pure" protein may have been contaminated with a trace of the other protein.
11. Dimer of 100,000-M_r units. Each 100,000-M_r unit is a dimer of a 25,000-M_r and a 75,000-M_r protein joined covalently by disulfide bonds.
13. Consider right-hand twist of extended α helices, right-hand crossovers, and right-hand twist of the β-sheet structure.
15. A_6, A_5B, A_4B_2, A_3B_3, A_2B_4, A_5B, B_6 held together by ionic or hydrophobic forces or by covalent (disulfide) bonds.

Chapter 5

1\.

Step	Specific Activity (U/mg)	Yield (%)	Purification (n-fold)
Cell extract	0.039	100	1
$((NH_4)_2SO_4)$ fractionation	0.091	85	2.3
Heat treatment	0.12	73	3.1
DEAE chromatography	4.3	62	110
CM-cellulose	29	50	740
Bio-Gel A	32	41	820

The DEAE chromatography (36-fold purification by this step) yields greatest purification. The enzyme was 0.12% of the protein in the initial extract.

3. The buffer pH must be between above pH 6 but below pH 9 to assure that the protein is negatively charged and the diethylaminoethyl groups are positively charged.
5. The volume following the heat treatment step is too large to allow efficient chromatography by gel exclusion (a column containing approximately 60 liters of resin would be required).
7. Contaminant B can be separated from both contaminant A and the ribonuclease by gel exclusion chromatography. Contaminant A and the ribonuclease can be separated by ion-exchange chromatography.
9. The characteristic absorbance features at 280 nm and 288 nm are consistent with the presence of tryptophan as a component of the amino acid composition of protein A, whereas protein B has little if any tryptophan.
11. The molecular weight is approximately 39,000.
13. Increasing the net positive charge is accomplished either by (**a**) substitution of a neutral amino acid for an acidic amino acid or (**b**) substitution of a basic amino acid for a neutral amino acid.
15. The quaternary structure is a tetramer of approximately 16,000-M_r subunits associated through ionic or electrostatic interactions.
17. [Hb] = 4.8 mM and is approximately equal to the concentration of GBP.

Chapter 6

1. D-mannose is the 2-epimer of D-glucose; D-galactose is the 4-epimer of D-glucose.
3. Carbohydrates as branched structures with multiple linkage sites conceivably could have attained the catalytic function now provided by proteins. However, there may be functional groups provided by amino acids (arginine, isoleucine, for example) that would be difficult to attain with carbohydrates.
5. Pure α or β anomers of the sugars are thermodynamically less stable than the mixture and mutarotate via opening and reclosing the sugar ring.
7. Cellulose is a linear polymer of $\beta(1,4)$-linked Glc residues. Polymer strands H bond with each other to form stable, insoluble fibrils suited for structural roles. Starch and glycogen are polymers of glucose in $\alpha(1,4)$ linkages with $\alpha(1,6)$ branches. The secondary structure is a hydrated helical coil with multiple nonreducing termini. These termini in glycogen provide the "rabid" release of stored glucose upon metabolic demand.
9. 2,3,4,6-tetra-*O*-methyl-D-mannose; 2,4,6-tri-*O*-methyl-D-glucose; 2,3,4-tri-*O*-methyl-D-galactose
11. D-Man $\alpha(1,6)$ / D-Man $\beta(1,4)$ > D-Gal (1,4) D-Glc
13. Carbohydrates are linked to the amide N of Asn in the sequence Asn-X-Ser(Thr). Other Asn residues not in that configuration will be exceedingly poor candidates for glycosylation.

Chapter 7

1. Plants provide a source of linolenic and linoleic acids, some of which is present in meat and milk of herbivores.
3. Triacylglycerols are neutral lipids, readily soluble in organic but not aqueous media. Phospholipids are amphipathic because the polar phosphoglyceryl portion is soluble in aqueous media, while the fatty acids esterified to the glyceryl portion are hydrophobic.
5. Review Haworth representations of Gal and Glc. The linkage to ceramide is a β-glycosidic linkage.
7. Triton X-100 is present above the critical micellar concentration and will probably not pass through the dialysis membrane. It is neutral and could be used in the ion-exchange chromatography. Sodium deoxycholate is present below CMC, should dialyze easily, and could be used for the gel exclusion experiment. The ionic nature of NaDOC would limit its use in ion exchange.
9. The hydrophobic amino acid side chains on the exterior of the integral membrane protein interact with the hydrophobic lipid of the membrane exterior and are stable in the nonaqueous environment. These residues pack in the interior, hydrophobic environment of globular proteins.
11. (**a**) Temperature above phase transition, greater lipid mobility. (**b**) Unsaturated fatty acid will decrease T_m. (**c**) Short-chain fatty acids will decrease T_m. (**d**) Integral proteins will increase and broaden temperature range of the phase change.
13. Glycosyl residues are hydrophilic and will not readily pass through the hydrophobic interior of a membrane.
15. Removal of the detergent has caused aggregation of the detergent-solubilized protein.

Chapter 8

1. Reaction order is the power to which a reactant concentration is raised in defining the rate equation. The example is first order in A and B, second order overall, second order in C.
3. K_m is $(k_2 + k_3)/k_1$, whereas K_s is k_2/k_1. K_m is equal to the [ES] dissociation constant K_s when the value of k_3 is much smaller than the value of k_2.
5. Nonlinearity develops because substrate concentration is less than saturating and velocity is proportional to substrate concentration. Substrate depletion causes decreased velocity; product concentration increases and reverse reaction becomes significant.
7. When [S] is 10^{-3} M, $v_0 = V_{max}$ (10^{-7} M min^{-1}).
When [S] is 10^{-5} M, $v_0 = 9.0 \times 10^{-8}$ M min^{-1} (90% V_{max}).
When [S] is 10^{-6} M, $v_0 = 5.0 \times 10^{-8}$ M min^{-1} (50% V_{max}).
When [S] is 2×10^{-2} M, $v_0 = V_{max}$, thus no increase in rate.

9. (a) $K_m = 10^{-4}$ M; (b) $V_{max} = 10^{-8}$ M s^{-1}; (c) $k_{cat} = 100$ s^{-1}; (d) $v_0 = 1.7 \times 10^{-9}$ M s^{-1}
11. Intersecting initial velocity plots are consistent with sequential (ordered or random) but not Ping-Pong mechanism.
13. $V_{max} = 3.3 \times 10^{-7}$ M s^{-1}; $K_m = 2.4 \times 10^{-4}$ M; $k_{cat} = 3.3 \times 10^4$ s^{-1}; specificity constant $= 1.4 \times 10^8$ M^{-1} s^{-1}
15. Irreversible inactivating reagents decrease the amount of active enzyme without altering the kinetic constants of the remaining active enzyme. Lineweaver-Burk analysis of the data reveals an apparent slope and *y*-intercept effect that may be confused with noncompetitive inhibition. Enzyme reversibly inhibited is reactivated upon dilution, while irreversibly inactivated enzyme is not.

Chapter 9

1. Each enzyme uses the "catalytic triad" (Asp-His-Ser) at the active site and an enzyme-bound intermediate is formed. The binding site for the side chain of the residue contributing the carboxyl to the bond that is cleaved differs in each case.
3. RNase activity depends on acid–base reaction catalyzed by two histidyl residues at the active site. Removal of the C′-2 OH proton is best accomplished by an unprotonated imidazole (pH > p*K*) and the addition of a proton to the C-5′ (—O$^-$) is accomplished by a protonated imidazole (pH < p*K*). The best compromise is pH at p*K* of imidazole (pH 6), where each of the histidines exists with the maximal fraction of each imidazole in the correct form. correct form.
5. Release of water during the conformational change initiated by binding the substrate provides a hydrophobic environment to stabilize the nonpolar side chain of the substrate.
7. Renaturation of denatured protein is dictated by the primary structure of the protein. The trypsin family of enzymes and carboxypeptidase A are synthesized as proenzymes that are proteolytically activated. The proteolyzed, active enzymes have primary structures different from the gene product and are not active upon renaturation. In addition, zinc is a cofactor required for carboxypeptidase A activity.
9. The structure of the transition-state analog is complementary to the structure of the active site. The analog thus binds tightly to the active site.
11. For the iron-dependent form, as an example,

$$Fe^{3+} + O_2^- \rightarrow Fe^{2+} + O_2$$

$$Fe^{2+} + O_2^- + 2\,H^+ \rightarrow Fe^{3+} + H_2O_2$$

13. (a) Substitution of Asp for Lys 86 markedly decreases activity and several explanations are possible: The lysine is a critical residue either at or near the active site or is essential for maintaining a catalytically competent conformation of the enzyme. (b) Lysines 21 and 101 are probably outside the catalytic site and may not be evolutionarily conserved. Their replacement with aspartate yielded no great change in enzymatic activity. (c) Lysine 86 is essential for enzymatic activity and would be conserved.
15. An intermediate covalently bound to the active site is an essential part of the mechanism in covalent catalysis. In noncovalent catalysis, the intermediates are bound through ionic, hydrophobic, or hydrogen bonds.

Chapter 10

1. Consider the reactions

$$\text{Protein(Ser)} + \text{ATP} \rightarrow \text{Protein(Ser-P}_i\text{)} + \text{ADP}$$

$$\text{Protein(Ser-P}_i\text{)} + \text{HOH} \rightarrow \text{Protein(Ser)} + \text{P}_i$$

$$\text{Sum: ATP} + \text{HOH} \rightarrow \text{ADP} + \text{P}_i$$

3. If G, H, and J are essential end products, each would be likely to inhibit its own production without inhibiting production of the others. The committed step (enzyme 1) may be cumulatively inhibited by each of the products. G and H might cumulatively inhibit enzyme 4. J could inhibit enzymes 7 and 8.
5. ATP binds at the active site and is a substrate, and also binds at an allosteric site, inhibiting the enzyme.
7. The substrate concentration required for half-maximal activity ($S_{0.5}$) of an allosterically regulated enzyme will depend on the cumulative effects of allosteric activators and/or inhibitors also present. Hence $S_{0.5}$ may be decreased with allosteric activators and may be increased with allosteric inhibitors.
9. Cooperativity in the tetrameric hemoglobin depends on the interaction of the two α and two β subunits. Separation of the tetramer into two α-β dimers would destroy the cooperativity exhibited by the tetramer. There should be more O_2 bound at lower O_2 tensions. 2,3-Bisphosphoglycerate should not stabilize the deoxyform of the dimer. The O_2-binding curve for the dimer should resemble that of myoglobin.
11. Multisubunit complexes respond cooperatively by transmitting the conformational change at one subunit to an adjacent subunit with subsequent changes in substrate binding. There may be local conformational changes in a subunit, as with LADH, that do not alter substrate binding to adjacent subunits.
13. The sulfhydryl groups on the activated enzymes must be reoxidized to the disulfide form. Whether the disulfide is formed by an enzymatically catalyzed process or occurs spontaneously via low-molecular-weight oxidants is not known. Irreversibly modified enzymes must in principle be degraded to allow recycling of the amino acids. Although the cell can reclaim the amino acids, there is still a large metabolic expense in *de novo* synthesis of the protein.
15. Free calmodulin would respond to Ca^{2+} signal by forming a Ca^{2+}-calmodulin complex, which then must collide with and bind to a target protein. This scenario requires the binding of three to four Ca^{2+} atoms per molecule of calmodulin, then a bimolecular reaction of Ca^{2+}-calmodulin with the target protein. Calmodulin complexed to the target protein is a single entity that must collide with Ca^{2+} atoms. The latter reaction should be kinetically less complex and would occur with a higher probability.

Chapter 11

1. The carboxyl group on the acyl substituent of biotin and lipoic acid forms an amide bond with the ϵ-amino groups of lysine in the cofactor-containing subunit. The cofactors are bound to flexible extended structures that allow the chemically reactive portion of the cofactor to move between several catalytic centers present on different subunits.

3. The two pyridine nucleotides (NADH, NADPH), although functionally similar, are specific coenzymes for dehydrogenases. The dehydrogenase-dependent specificity provides a mechanism frequently used to differentiate anabolic and catabolic processes. Pathways may be regulated by either NADH/NAD^+ or NADPH/$NADP^+$ ratios.

5. (a)

$$H-CH_2-\overset{\overset{\large O}{\|}}{C}-SCoA + CHO-COO^- \rightarrow \begin{array}{c} SCoA \\ | \\ C=O \\ | \\ CH_2 \\ | \\ (HO\,C\,H) \\ | \\ COO^- \end{array} \rightarrow CoASH + \text{L-malate}$$

(b) The thioester activates the carboxyl group of acetate and stabilizes the enolate formed upon enzyme-catalyzed abstraction of the α proton. Abstraction of an α proton from acetate, forming the dianion, is improbable. (c) Coenzyme A is a good leaving group and hydrolysis of the thioester is thermodynamically favorable. The products malate and CoASH are more stable when compared with the malylthioester.

7. Flavins are fully reduced upon accepting two electrons but may transfer electrons to one-electron acceptors (iron-sulfur proteins and hemoproteins). The free radical remaining after transfer of a single electron from FADH is resonance-stabilized by the isoalloxazine ring system.

9. (a) Reduced flavins in solution rapidly reduce O_2 to superoxide and H_2O_2, metabolites that are toxic to the cell. Enzyme-bound flavins are usually shielded from rapid oxidation by O_2. (b) Tightly bound NAD(P) is an advantage to enzymes catalyzing rapid $H:^-$ removal and readdition in a stereospecific fashion. Freely diffusing NADH is an advantage in transferring reducing equivalents among various enzyme-catalyzed reactions.

11. The products of the oxidation, NH_4^+ and H_2O_2, are toxic to most cells. In PLP-dependent transamination the amino group is retained in organic molecules of low toxicity.

13. (a) Flavin; (b) PLP; (c) biotin; (d) thiamine pyrophosphate.

15. NAD^+ is cleaved at the glycosidic bond between nicotinamide and the ribose ring. Adenosine diphosphate ribose is transferred to the acceptor and nicotinamide is released.

Chapter 12

1. Thermodynamics dictates the feasibility of a reaction and the direction in which a reaction proceeds to reach equilibrium, but does not specify the rate of the reaction or the reaction pathway. Kinetics describes that rate at which a reaction approaches equilibrium. The most thermodynamically favorable reaction is of no biological value if it is not adequately catalyzed.

3. Using different high-energy intermediates, or different portions of the same high energy intermediate, and different forms of reductant allows the cell ample sites to control catabolism and anabolism differentially.

5. The "committed step" in a reaction sequence steers the metabolite to a sequence of reactions whose intermediates have no other function in the cell. Control of the committed step prevents wasteful accumulation of these single-purpose intermediates and obviates the necessity of controlling each enzyme in a pathway.

Chapter 13

1. (a) Bacteria are adept at fermenting numerous sugars. (b) The NADH formed by glyceraldehyde-3-phosphate dehydrogenase is used to reduce pyruvate to lactate. There is no net oxidation, but the products are of lower free energy than substrates. Energy difference between products and substrates is available to drive the phosphorylation of ADP. (c) The culture could also be decarboxylating pyruvate to acetaldehyde and CO_2. The acetaldehyde is subsequently reduced to ethanol, using the NADH from the glyceraldehyde-3-phosphate dehydrogenase reaction. (d) Fluoride forms an inhibitor (fluorophosphate) of enolase. Inhibition of enolase would increase the amount of 2-phosphoglycerate, but the amount of PEP would decline as a result of pyruvate kinase activity.

3. (a) Hydrolysis rather than phosphorolysis of the thioester bound to glyceraldehyde-3-phosphate dehydrogenase releases 3-phosphoglycerate. (b) There would be no net yield of ATP if glycolysis began with glucose. (c) The obligate aerobe could oxidize the pyruvate, with the formation of sufficient ATP for growth.

5. Arsenate substitutes for phosphate at the glyceraldehyde-3-phosphate dehydrogenase reaction and forms a transient arsenocompound that spontaneously hydrolyzes, regenerating the arsenate. Arsenate acts catalytically rather than stoichiometrically in dispelling the energy of the thioester bond. Arsenate is ineffective at the pyruvate kinase step because the active phosphate was derived initially from ATP, not orthophosphate.

7. If the $\Delta G^{\circ\prime}$ for the hexokinase reaction is -5.0 kcal/mole, the K'_{eq} is 4,800. The glucose concentration in equilibrium with 1 mM glucose-6-phosphate must be at least 42 nM, a concentration easily attainable in the cell. However, the K_m of hexokinase for glucose is about 100 μM. The glucose concentration in actively metabolizing cells must be several orders of magnitude greater than 40 nM to have reasonable hexokinase activity.

9. (a) Alcohol dehydrogenase reduces acetaldehyde to ethanol to regenerate NAD^+ from the NADH produced in glycolysis. (b) Alcohol dehydrogenase oxidizes ethanol to acetaldehyde. Acetaldehyde is subsequently oxidized to acetyl CoA_2.

11. Consider enolase, aldolase, and phosphohexoisomerase.

13. (a) Review pyruvate carboxylase and PEP carboxykinase reactions. (b) $\Delta G^{\circ\prime}$ overall is app. -1.0 kcal/mole if the hydrolysis of PEP releases 14 kcal/mole and if hydrolysis of ATP (GTP) releases 7.5 kcal/mole. If the overall free

energy change for the coupled reactions is as calculated, the formation of PEP from pyruvate is metabolically feasible.

15. (a) PFK-2 will be active when the insulin/glucagon ratio is high (for example, when the animal is fed a carbohydrate-rich meal), or when adrenalin is released. FBPase-2 will be active when the insulin/glucagon ratio is low (blood glucose is low). (b) If fructose-2,6-bisphosphate were still present although not being actively synthesized, it would continue to stimulate PFK-1 and inhibit FBPase-1. (c) Activation of FBPase-2 causes the hydrolysis of fructose-2,6-bisphosphate, alleviates inhibition of FBPase-1, and prevents activation of PFK-1. (d) Insulin/glucagon ratio decreases. Thus FBPase-2, FBPase-1, pyruvate carboxylase, and PEPCK activities should increase. PFK-2 and PFK-1 activities will decline. (e) The administered insulin will restore the insulin/glucagon ratio. The cAMP level in the liver cell will decrease, as will the activities of FBPase-2, FBPase-1, and PEP carboxykinase.

Chapter 14

1. (a) False. Lipoamide transacetylase catalyzes reduction of the disulfide on lipoamide concomitantly with oxidation and transfer of the hydroxyethyl group from thiamine pyrophosphate. Dihydrolipoamide dehydrogenase catalyzes oxidation of dihydrolipoamide and reduction of NAD^+. (b) False. Hydrolysis of acetyl-CoA thioester should yield as much free energy as succinyl-CoA hydrolysis, a process coupled to ADP phosphorylation (succinate thiokinase). (c) False. The methyl group of acetyl-CoA could be derived from pyruvate, from β oxidation of long-chain fatty acids, or from amino acid metabolism. (d) False. If aconitase failed to discriminate between the ($-CH_2-COO^-$) groups, half the CO_2 would arise from oxaloacetate and half from the acetate carboxylate. The two CO_2 molecules released by oxidative decarboxylation of isocitrate and α-ketoglutarate are derived from the carboxyl groups of oxaloacetate with which the acetyl-CoA was condensed. (e) False. Malate can easily be dehydrated to fumarate by reversal of the fumarase reaction.
3. (a) Glyceraldehyde-3-phosphate, C-3; (b) acetyl-CoA, methyl of acetate; (c) citrate, C-2 (derived from methyl of acetyl-CoA); (d) α-ketoglutarate, C-4; (e) succinate, C-2, C-3; (f) malate, both C-2 and C-3.
5. (a) Hydroxypyruvate $\rightarrow$ B $\rightarrow$ D $\rightarrow$ C $\rightarrow$ A $\rightarrow$ pyruvate (b) Hydroxypyruvate + NADH + $H^+ \rightarrow$ pyruvate + NAD^+ + HOH (c) A = phosphoenolpyruvate; B = glycerate; C = 2-phosphoglycerate; D = 3-phosphoglycerate. A, C, and D are glycolytic intermediates. (d) ATP used to phosphorylate glycerate is regenerated from PEP.
7. Lipoic acid bound to the acyltransferase in the multienzyme complexes (a) provides a greater coenzyme concentration at the active site than may be seen if the cofactor must diffuse into and away from the complex; (b) decreases the diffusion path required for a product of one active site to move to the next catalytic center.
9. If the complexes were deficient in dihydrolipoamide dehydrogenase, dihydrolipoamide would be reoxidized to lipoamide more slowly than necessary to maintain activity of the complexes. ATP production would be lowered as a consequence of reduced TCA cycle activity.
11. In bacteria, acetate is used as a source of energy via the TCA cycle and oxidative phosphorylation and as a carbon source for synthesis of intermediates for cell growth and division. Competition between ICDH and isocitrate lyase for isocitrate is shifted to anabolism by phosphorylation of ICDH and shifted to catabolism upon dephosphorylating (reactivating) ICDH. The fraction of ICDH in the active form dictates the level of isocitrate available to isocitrate lyase. The isocitrate lyase is apparently not regulated at the metabolite level.
13. In plants, the TCA cycle and glyoxalate bypass are located in different subcellular organelles. The TCA cycle is located in mitochondria and the glyoxalate bypass is located in glyoxosomes. Communication between the compartments includes flow of succinate to the mitochondria for oxidation to malate.
15. In mammalian liver, acetyl-CoA provides reducing equivalents by its oxidation in the TCA cycle and subsequent oxidative phosphorylation to provide ATP. Acetyl-CoA is an obligatory activator of the pyruvate carboxylase, whose activity provides oxaloacetate for gluconeogenesis.

Chapter 15

1. Iron-sulfur protein: one electron, no H^+ transferred, no semiquinone; flavoprotein: isoalloxazine ring, two electrons, one H^+/electron, semiquinone formed; quinone: two electrons, one H^+/electron, semiquinone formed.
3. (a) $\Delta E^{\circ\prime} = +110$ mV, $\Delta G^{\circ\prime} = -2.5$ kcal/mole
(b) $\Delta E^{\circ\prime} = +580$ mV, $\Delta G^{\circ\prime} = -53$ kcal/mole O_2 reduced.
(c) $\Delta E^{\circ\prime} = +270$ mV, $\Delta G^{\circ\prime} = -12$ kcal/mole
(d) $\Delta E^{\circ\prime} = -430$ mV, $\Delta G^{\circ\prime} = +20$ kcal/mole
5. At +300 mV, the cytochrome *c* couple will be 91% oxidized and 9% reduced.
7. (a) Data are consistent with diffusion of cytochrome *c* between complex III and the cytochrome oxidase. Electrons are probably donated and accepted from the same area on the cytochrome *c* structure. (b) The lysines surrounding the heme crevice on cytochrome *c* should interact electrostatically with anionic groups (Glu, Asp) spatially located on complex III and cytochrome oxidase to facilitate interaction with the cytochrome.
9. (a) P/O = 2.3; (b) P/O = 0.2; (c) P/O = 2.7; (d) P/O = 0.5.
11. (a) Tightly coupled mitochondria exhibit an obligatory codependence of electron transfer and phosphorylation of ADP. In "perfectly" coupled mitochondria, no electron transport to O_2 should occur in the absence of ADP and phosphate. (b) Respiratory control is a function of coupled mitochondria. Oxidation of NADH and succinate is blocked when ATP levels are high; glycolysis and TCA cycle are inhibited. When ATP demand is high, ADP levels will stimulate the oxidation of succinate and NADH. Low ATP levels and high NAD^+/NADH ratio will stimulate glycolysis and the TCA cycle. (c) Brown-fat mitochondria contain the uncoupler thermogenin and the electron-transport system is partially uncoupled. The oxidation of fat leads to nonshivering thermogenesis, a process important to animals aroused from hibernation.

13. (a) Membrane potential is app. 180 mV, negative on matrix side of membrane. (b) An oxidative phosphorylation uncoupler allows H^+ equilibration across the membrane, collapsing both membrane potential and H^+ gradient.
15. (a) Transport of H^+ is in the same orientation in these SMPs as in the mitochondria, but the cytoplasmic aspect of the inner mitochondrial membrane is facing the interior of the vesicle. (b) Mechanical or chemical uncoupling allows free equilibration of H^+ across the SMP membrane. Succinate is oxidized without vectorial accumulation of H^+. (c) Atractyloside inhibits ADP transport into mitochondria. The SMP have the F_o-F_1 complex exposed to bulk solvent and do not have a transport barrier to ADP. The ADP/ATP transporter is not required, so atractyloside would have no effect in this system. SMP oxidize extravesicular NADH but intact mitochondria do not.

Chapter 16

1. (a) 33 kcal/Ein; (b) excited state reduction potential is -1.0 volt.
3. If oxidation of compound A requires operation of both photosystem I and photosystem II, oxidation of compound A will diminish if chloroplasts are illuminated with red light (>680 nm) and will be inhibited by DCMU. If compound A is oxidized by photosystem I, DCMU will not inhibit the oxidation, nor will the action spectrum of compound A oxidation exhibit this phenomenon of "red drop."
5. Light energy absorbed by antenna chlorophyll molecules is transferred to a photocenter trap (P_{680} or P_{700}) where an electron is transferred initially to pheophytin, then to bound plastoquinone. If the potential is decreased to 0 mV, the quinone pool will be reduced, electron transfer from pheophytin will diminish, and the steady-state level of activated chlorophyll will increase. The alternative energy dissipation pathway, fluorescence, increases.
7. The quantum yield of O_2 evolution from plants as a function of light absorbed parallels the absorption spectrum until the wavelength impinged is greater than approximately 700 nm. Although the chloroplasts absorb wavelengths greater than 700 nm, O_2 evolution declines because the oxygenic photosystem II is no longer activated at these wavelengths.
9. Photophosphorylation is the process of ATP generation driven by the proton gradient generated during the light-driven electron transfer (Z scheme). Noncyclic photophosphorylation requires net electron transfer from a donor (HOH) to an acceptor ($NADP^+$) with vectorial H^+-translocation. Cyclic photophosphorylation is driven by a proton gradient generated by electron transfer from photosystem I through the quinone pool and the electron-transfer system, back to photosystem I.
11. CO_2 and O_2 each bind to the catalytic site of the ribulose bisphosphate carboxylase/oxygenase (RubisCO) and react with the ribulose bisphosphate. The enzyme uses either CO_2 or O_2 as (alternative) substrate, and O_2 competes with CO_2 for the bound ribulose bisphosphate.
13. CO_2 is released by oxidative decarboxylation of malate in the bundle sheath cells of C_4 plants, causing the CO_2/HCO_3 concentration to exceed the CO_2/HCO_3^- concentration in the mesophyll cell. The larger concentration of CO_2 increases the fraction of ribulose bisphosphate that is carboxylated rather than oxygenated. Net photorespiration is lower in C_4 plants as compared with C_3 plants.
15. (a) $NADP^+$ is reduced to NADPH by the noncyclic electron flow in the chloroplast to $NADP^+$-ferredoxin oxidoreductase. (b) 6 moles of ATP are used to phosphorylate 6 moles of glycerate-3-phosphate to glycerate-1,3-bisphosphate. An additional 3 moles of ATP are consumed during phosphorylation of 3 moles of ribulose-5-phosphate to 3 moles of ribulose-1,5-bisphosphate. (c) The ATP consumed by the Calvin cycle remains unaltered, but 2 additional moles of ATP are used to regenerate PEP from pyruvate catalyzed by the pyruvate, phosphate dikinase. (d) NADPH is used in the mesophyll cell to reduce oxaloacetate to malate, but the NADPH is regenerated upon oxidative decarboxylation of malate to CO_2 and pyruvate.

Chapter 17

1. (a) Oxidation of 1 mole of glucose yields 32 moles of ATP. 350 kcal of energy are stored as ATP or 1.9 kcal/gm. (b) Oxidation of 1 mole of palmitate yields 106 moles of ATP. 1200 kcal of energy are stored as ATP, or 4.7 kcal/g. (c) Lipids are more highly reduced than are carbohydrates and supply more reducing equivalents to the electron-transport system than do carbohydrates. (d) Lipids have approximately 2.5 times greater energy storage per gram than do carbohydrates and are stored as compact, hydrophobic globules. Storage of an equivalent energy as carbohydrate would require at least 2.5 times the mass, not considering the water of hydration that would accompany the carbohydrate.
3. Carnitine acyltransferases (CAT) catalyze the reversible formation of fatty acyl carnitine esters from the acyl-CoA thioesters. The carnitine esters, but not the CoA derivatives, are transported across the inner mitochondrial membrane.
5. Propionyl-CoA, a product of β oxidation of odd-chain-length fatty acids and of pristanic acid, is metabolized to L-methylmalonyl-CoA in two steps. The L-methylmalonyl-CoA is substrate for the vitamin-B_{12}-dependent mutase. The product, succinyl-CoA, is metabolized in the TCA. A deficiency of vitamin B_{12} could limit the mutase reaction.
7. (a) Ketone body formation in liver supplies an easily transported, water-soluble, energy-rich metabolite that can be used in lieu of glucose in many nonhepatic tissues. (b) β-hydroxybutyrate supplies an additional hydride (2 reducing equivalents) compared with acetoacetate. (c) Consider the reactions catalyzed by β-hydroxybutyrate dehydrogenase, 3-ketoacyl-CoA transferase, and thiolase.
9. The NADPH-dependent dienoyl-CoA reductase reduces the Δ^2-*trans* $\Delta^{4,7}$-*cis* decatrienoyl-CoA produced after removal of the first 4 acetyl-CoA units from linolenyl-CoA in β oxidation. The product, Δ^3-*trans,* Δ^7-*cis* decadienoyl-CoA, is isomerized to the Δ^2-*trans,* Δ^7-*cis* decadienoyl-CoA by the enoyl-CoA isomerase.
11. (a) The ^{14}C-labeled methyl group of acetyl-CoA will be C-16 of palmitate. (b) Only one deuterium atom from each labeled malonyl-CoA will remain in the reduced lipid chain. Carbons 2, 4, 6, 8, 10, 12, and 14 of palmitic acid will each have one deuterium label. (c) The ^{14}C label will be lost by decarboxylation and no label will remain in the palmitate.

13. Glucose-6-phosphate dehydrogenase, 6-phosphogluconate dehydrogenase, and the $NADP^+$-malic enzyme are sources of the 14 moles of NADPH required for biosynthesis of palmitate.
15. The thioesterase activity of the fatty acid synthase prefers the palmitoyl acyl carrier protein thioester as substrate.

Chapter 18

1. Glutamine, the product of glutamine synthase, is the source of nitrogen required in the synthesis of a number of diverse, structurally unrelated compounds synthesized by different pathways. Total inhibition of the glutamine synthase by a single product would in turn inhibit the synthesis of all compounds requiring glutamine.
3. In the absence of serine hydroxymethyltransferase, the mutant would be unable to synthesize glycine. Glycine supplementation would be required. Metabolism of glycine contributes to the one-carbon pool, as does methionine. Addition of methionine should decrease the demand for glycine contribution to the one-carbon pool and stimulate growth of the mutant.
5. Pyruvate, from glycolysis of glucose, is carboxylated to oxaloacetate or oxidized to acetyl-CoA. These metabolites enter the Krebs cycle, are metabolized to α-ketoglutarate and oxaloacetate, then transaminated to aspartate or glutamate. Asn, Gln, and Pro are synthesized from Asp or Glu. The cycle replenishes intermediates via the anaplerotic reactions (e.g., carboxylation of pyruvate to form oxaloacetate).
7. Serine carbon: α carboxyl: C-3 or C-4 of glucose; α carbon: C-2 or C-5 of glucose; β carbon: C-1 or C-6 of glucose.
9. Glutamate is phosphorylated and reduced to L-glutamyl-γ-semialdehyde. The aldehyde condenses with the α-amino group on the same molecule to form Δ^1-pyrroline-5-carboxylate. Reduction of the latter compound yields proline. Transamination implies that the α-keto acceptor and the α-amino donor are separate molecules.
11. Tryptophan synthase catalyzes the cleavage of indole glycerol phosphate, with retention of indole as an enzyme-bound intermediate and release of glyceraldehyde-3-phosphate. The enzyme-bound indole replaces the hydroxyl group of serine to form tryptophan.
13. Hydroxyproline is formed by a posttranslational modification of proline residues in the protein. The ^{14}C-labeled hydroxyproline is not incorporated directly into the collagen because there is no genetic codon to specify the incorporation of hydroxyproline.

Chapter 19

1. Pyridoxal phosphate deficiency would curtail the transamination of α-amino group amino acids to recipient α-ketoacids. Disposition of the amino group via the urea cycle and subsequent catabolism of the carbon skeleton would be diminished.
3. Glutaminase catalyzes the hydrolytic cleavage of the amide group from glutamine, leaving glutamate. Glutamate dehydrogenase catalyzes the NAD^+-dependent oxidative deamination of glutamine to ammonium ion, α-ketoglutarate and NADH.
5. The α-ketoacid pyruvate, a product of glycolysis, is transaminated to form alanine and as such transports the α-amino groups, from amino acids catabolized in muscle, to the liver. In the liver, transamination of alanine with α-ketoglutarate yields pyruvate, a gluconeogenic substrate, and glutamate, an amino donor in urea biosynthesis.
7. Glutamate is oxidatively deaminated to α-ketoglutarate, a TCA cycle intermediate subsequently oxidized to oxaloacetate. Aspartate is transaminated directly to oxaloacetate. The TCA cycle regenerates oxaloacetate, preventing the complete oxidation of oxaloacetate directly. Oxaloacetate is reduced to L-malate, transported to the cytoplasm, and oxidatively decarboxylated by the $NADP^+$-dependent malic enzyme to pyruvate and NADPH. Pyruvate is completely oxidized in the mitochondria.
9. The amidino group of arginine is required for creatine biosynthesis. Creatine phosphate, a phosphoramide, is a "high-energy" phosphate reservoir for the muscle.
11. Each of these amino acids is catabolized to succinyl-CoA, a TCA cycle intermediate. Succinate is oxidized to L-malate, shuttled to the cytoplasm, and oxidized to oxaloacetate. Oxaloacetate is a gluconeogenic substrate.
13. Phenylalanine and tyrosine are catabolized to 4-hydroxyphenylpyruvate, whose further metabolism is blocked by the dioxygenase deficiency. A diet containing only maintenance levels of phenylalanine and deficient in tyrosine is recommended to prevent accumulation of the 4-hydroxyphenylpyruvate.
15. (a) L-glutathione is synthesized in successive steps catalyzed by γ-glutamyl cysteine synthase and glutathione synthase. Glutathione synthesis is directed by the substrate specificity of these enzymes. (b) Decreased glutathione synthesis would increase the probability of oxidative damage to the cell.

Chapter 20

3. Humans have two carbamoyl syntheses, one in the mitochondria (CPase I) and one in the cytosol (CPase II). The mitochondrial enzyme is used in urea and arginine synthesis, while the cytosolic enzyme (part of the multienzyme complex) is dedicated to the synthesis of pyrimidines and is the first enzyme in that pathway. Bacteria have only one CPase and the regulation is at the first committed enzyme in pyrimidine biosynthesis, ATCase—a form of regulation that makes biochemical sense.
5. Large amounts of dATP would accumulate and inhibit ribonucleotide reductase. The lymphocytes could not make the large amounts of dNTPs needed for DNA synthesis and cell division.
7. Because Lesch-Nyhan patients lack the enzyme hypoxanthine-guanine phosphoribosyltransferase, they accumulate high levels of PRPP, which stimulates purine biosynthesis to high levels, leading to large amounts of uric acid being made. The brain may not have high levels of *de novo* purine biosynthesis and probably relies on the salvage pathway enzymes for its purine nucleotides.

9. Antifolates indirectly inhibit the synthesis of thymidylate and prevent DNA synthesis; therefore cells that are growing rapidly may be killed by methotrexate. Normal cells that are actively dividing (in the cell cycle) and replicating their DNA are also sensitive to the toxic effects of antifolates. This is why we see harmful side effects of anticancer drugs.
11. GTP requires the hydrolysis of eight high-energy bonds, while CTP requires the hydrolysis of five high-energy bonds.
13. Allopurinol inhibits xanthine oxidase, preventing the oxidation of 6-mercaptopurine so that it is not as rapidly degraded.

Chapter 21

1. Lactose synthase is a galactosyltransferase using UDP-galactose and glucose to make the disaccharide lactose. The enzyme is composed of two peptides, the galactosyltransferase and α-lactalbumin. The galactosyltransferase is found in all tissues in the body, but the α-lactalbumin (which has no catalytic activity) is found only in milk. α-Lactalbumin modifies substrate specificity of the galactosyltransferase to use glucose so that lactose can be produced.
4. Sugars have large numbers of hydroxyl groups that form glycosidic bonds with the anomeric carbons of other sugars and can generate a large number of structures from a limited number of monosaccharides (over 80 different glycosidic linkages are known).
7. In mutants that lack Dol-P-Man synthase, the glycosylphosphatidylinositol (GPI) anchor is not made and the glycoproteins that would be on the plasma membrane are secreted outside the cell (not attached).
10. Patients with I-cell disease do not phosphorylate the mannose residues on the glycoproteins that are lysosome-bound. These "lysosomal hydrolases" are therefore secreted.

Chapter 22

1. Phospholipids are essential to living organisms, including humans. Any defects would be lethal in the early stages of development and would never be observed.
3. The lung surfactant dipalmitoylphosphatidylcholine is synthesized by removing the fatty acid in the C-2 position and then reacylating with palmitoyl-CoA. This process is called remodeling.
5. Some eukaryotic membranes contain significant amounts of ether-linked glycerophospholipids (for example, 50% of all phospholipids in heart tissue are ether-linked in the C-1 position). These phospholipids are called plasmalogens when they have a hydrocarbon chain linked to the SN-1 carbon by a vinyl ether linkage and are called alkylacylglycerophospholipids when linked by an ether linkage. Archaebacteria have ether-linked lipids in their cell membranes.
7. This person could still make PGE_2 because linoleic acid (which is considered an essential fatty acid in animals) can be desaturated and elongated to form arachidonic acid.
9. Arachidonic acid is stored in membranes as phospholipids with C_{20} polyunsaturated fatty acids in the SN-2 position. Phospholipase A_2 releases arachidonic acid, which is then used to synthesize prostaglandins, which induce inflammation.

Chapter 23

1. The major regulation occurs with HMG-CoA reductase. This enzyme is regulated at three levels: gene expression, rate of degradation of HMG-CoA reductase, and phosphorylation/dephosphorylation of HMG-CoA reductase. High cholesterol levels reduce HMG-CoA reductase mRNA, which reduces synthesis and increases degradation, and the enzyme already present is phosphorylated to reduce its activity. LDL receptors will be reduced in number.
3. Any of the enzymes in the synthesis of acetoacetate (acetoacetyl-CoA thiolase, HMG-CoA synthase, and HMG-CoA lyase) might be defective. Different isozymes are used for ketone bodies and cholesterol biosynthesis; therefore normal cholesterol biosynthesis would not be a clue for the deficient enzyme.
5. The patient could be treated with lovastatin and a bile-acid-binding resin, which together could lower serum cholesterol levels about 50%. Lovastatin inhibits HMG-CoA reductase, lowering cholesterol biosynthesis. Bile-acid-binding resins increase the excretion of bile acids, which stimulates the liver to convert more cholesterol to bile acids.
9. HDL and other lipoproteins would have a reduced level of cholesterol esters, but serum levels of free cholesterol would be elevated.

Chapter 24

2. Nothing will happen! Normally you would expect a person to go into insulin shock as a result of a drop in blood glucose levels, but a starving person's brain is using ketone bodies for energy and the blood glucose is already low.
4. Progestin and estrogen inhibit secretion of FSH and LH by the pituitary and thus prevent follicular growth and ovulation.
6. The patient most probably has a primary defect.
9. Vitamin D can be considered both a hormone and a vitamin. Its mode of action is like that of other steroid hormones and it is synthesized in the body. It can be given in the diet, for example, in supplemented milk, and would then be called a vitamin.

Chapter 25

2. Investigators used two different strains of TMV that had identifiable proteins. They made different reconstituted viruses and found that the progeny viral proteins were always determined by the type of RNA in the reconstituted virus.
4. At the pH extremes, some of the bases have either positive charges (low pH) or negative charges (high pH), which are disruptive of duplex structure.
6. 7-M urea, 90% formamide, pure water, and the T4 gene-32-encoded protein lower the T_m of the DNA. Increased salt raises the T_m of the DNA duplex. Urea and formamide disrupt hydrophobic forces that hold the duplex together (it was earlier thought that these compounds disrupted hydrogen bonding, but water is a better hydrogen-bonding reagent). Higher salt shields the anionic phosphate groups from each other, limiting the electrostatic repulsion of the DNA strands. The T4 protein binds to single-strand regions and pulls the equilibrium to the denatured state.

8. These proteins would bind to the Z DNA and pull the equilibrium to the Z structure. Therefore, even if the B form predominated in pure DNA at equilibrium, the binding of protein to the Z DNA would cause this region of the DNA to be in the Z configuration.
10. A is a DNA-RNA hybrid duplex.
12. T4 DNA will renature 25 times faster than *E. coli* DNA. Unsheared T4 DNA will anneal about 20 times faster than sheared DNA.

Chapter 26

2. The location of the origin is close to gene *c*, which has the highest number of gene copies. The pattern observed agrees with a bidirectional mode of replication.
4. Both ϕX174 and *E. coli* DNA replication are inhibited by rifampicin, which inhibits RNA polymerase. ϕX174 is inhibited during the replication of the RF since it is dependent on the synthesis of a viral protein A, which generates the 3′-OH by cleavage of the DNA. The synthesis of this protein is dependent on the transcription of gene *A* by RNA polymerase. The initiation of replication in *E. coli* is dependent on RNA polymerase to synthesize the RNA primer at *oriC*.
6. It would take about 33 min to replicate the chromosome, since you have a bidirectional mode of replication with two replication forks. If you had multifork replication (one round of replication starting before the other finished), then a 20-min division time would be possible.
9. This experiment by Cairns and DeLucia showed that DNA Pol I was not the primary DNA polymerase responsible for replicating DNA in *E. coli* and suggested that another enzyme was involved.
12. Reverse transcriptase is the unique viral protein that converts the viral RNA into DNA. A specific tRNA serves as a primer (all template-dependent DNA polymerases require a primer, even if they have a RNA template).
14. Cytosine spontaneously deaminates to uracil, which would then lead to a point mutation if not removed. The thymine base allows the organism to identify uracil produced by spontaneous deamination, which would not be possible if you had A-U base pairs in DNA.

Chapter 27

2. The indirect methods of sequencing would not detect many of the posttranscriptional modifications found in RNA, such as 2′-O-methylation, various base methylations, etc.
6. Digest the circular viral DNA with *Eco*R1 and ligate the structure with DNA ligase. This digestion would generate a hybrid virus with only one 72-bp sequence.
9. Human serum albumin is synthesized in the liver; therefore the levels of mRNA for serum albumin would be high. The cDNA made from liver would have many copies of albumin cDNA, so that the job of cloning this sequence would be easier. All human tissues should contain the albumin gene in one copy per haploid amount of DNA, so any tissue would serve to establish a genomic library.
11. If a fragment of DNA of interest has linked to it the thymidine kinase gene (*tk*), then if a cell has taken up this DNA it will also contain the *tk* gene. If the cell line used for cloning contains a mutant (tk^-) that does not allow it to grow in a selection media (HAT, or medium containing hypoxanthine, amethopterin, and thymidine), then only cells taking up the DNA with the *tk* marker will grow.

Chapter 28

1. The base composition of the newly synthesized RNA would be: C = 20%, G = 25%, U = 15%, and A = 40%.
3. mRNA has a short half-life (a couple of minutes), while tRNA and rRNA are very stable (half-life measured in hours) and accumulate to make up most of the RNA in the cell (95%).
5. The D and T loops in tRNA interact with each other to form the tertiary structure, leaving only the anticodon with a single-stranded loop able to be cleaved by RNase.
11. The 5′ terminus is capped (G^mpppX—), generating a guanosine nucleoside with 2′ and 3′ -OH groups and a 5′-5′ pyrophosphate linkage. The other end of the mRNA has poly(A) added, so the 3′ end is adenosine. Therefore most of the mRNA in the cell has a guanosine and an adenosine 2′ and 3′ -OH and no typical triphosphate ending (pppN—).
15. If an intron or part of an intron containing a stop signal for translation was not removed, this mRNA would be longer but would yield a shorter polypeptide. Alternative splicing would remove the intron or use an alternative splice site, generating a shorter mRNA and a longer polypeptide.

Chapter 29

2. The large size of tRNAs (larger than a three-nucleotide anticodon) results from the requirements of specific charging of the tRNA by its aminoacyl tRNA synthase. The enzyme must recognize its tRNA and attach the correct amino acid (a correspondence sometimes called the second genetic code). Also, the structure of the ribosome requires some unique spacing between the codon–anticodon interaction and the site of peptide bond formation.
4. If tRNAs were not charged correctly, the genetic code would have a high error rate. The correct charging of the tRNA with the correct amino acid is just as important as the codon–anticodon interaction that defines the genetic code.
12. PEST sequences are regions rich in proline, serine, and acetic amino acids that are common to short-lived eukaryotic proteins. Regulatory proteins generally have short half-lives, while structural proteins are more stable, as is required by their function in the cell.
16. The incorporation of the valine analog prevents the formation of the active tetramer structure of hemoglobin. This unassembled globin would be a preferred target for reaction with ubiquitin, which would lead to rapid degradation.

Chapter 30

2. **(a)** Both the wild type and z^- would express a low level of acetylase with no treatment. **(b)** When lactose is added, only the wild type would produce a lot of the acetylase. **(c)** IPTG would induce a high level of acetylase in both strains.
6. IPTG does not have to be metabolized by galactosidase to form an active inducer. It is stable and builds up to high levels in the cell.

8. At the enzyme level, the end product of the histidine pathway (histidine) inhibits the first enzyme in the pathway by simple linear feedback inhibition. Transcription is regulated by attenuation; no feedback repression is observed with this operon.
12. Bacterial genes are regulated in a reversible manner, depending on metabolic needs, while viral genes are turned on only once, in a predesignated sequence.
14. The expression of the *crp* gene is negatively autoregulated by cAMP-CAP. A divergent promoter is activated by cAMP-CAP, and the RNA produced binds to the 5′ end of the *crp* mRNA, producing a rho-independent terminator. The binding of this antisense RNA stops the transcription of the *crp* gene.

Chapter 31

2. In eukaryotes, transcription and processing of the mRNA occurs in the nucleus while translation occurs in the cytosol. Attenuation control is unlikely because you can't form a transcriptional-translational complex.
6. Hypomethylation of the 5′ flanking sequences is correlated with the expression of some genes in eukaryotes. 5-Azacytidine can be incorporated into DNA and generate hypomethylated regions in the DNA, leading to gene activation.
9. Histones bind to DNA by electrostatic interaction of basic amino acids with the negative charge on the phosphate in the deoxyribose phosphate backbone of the DNA duplex. While many regulatory proteins initially interact (nonspecific binding) with DNA by charge interactions, high-affinity binding (at specific binding sites) occurs by hydrophobic interactions, which are stabilized by high salt.
12. The gene for dihydrofolate reductase can be amplified to a high copy number, giving a large gene dose. The large amount of dihydrofolate reductase produced binds the methotrexate, leaving some active enzyme to meet the cell's need for tetrahydrofolate. A DNA probe for dihydrofolate reductase gene could be used to measure the amount of specific DNA amplified (the "dot blot" procedure with genomic DNA could be used to compare normal and resistant cells).
14. The enhancer sequence in the DNA can bind to tissue specific protein(s). The DNA can then loop to allow the enhancer proteins to come in close contact with the promoter of the gene being stimulated, even at great distances (5,000 bp).

Chapter 32

1. (b) $D = 5.7 \times 10^{-6}$ cm²/sec.
2. (a) $\Delta\Psi = +18$ mV; (b) from side 1 to side 2; (c) Side 1: 64.3 mM K^+, 50 mM Na^+, 144.3 mM Cl^-. Side 2: 85.7 mM K^+, 85.7 mM Cl^-. $\Delta\Psi = 7.5$ mV.
5. Sucrose uptake should be inhibited by a proton ionophore if uptake is by a proton symport. If a protein-binding system was operational, then membrane vesicles or cells subjected to osmotic shock would be defective in uptake. If a Na^+ symport was involved, then uptake would be dependent on extracellular Na^+. If a PTS was operational, then sucrose phosphorylation would be dependent on PEP and not ATP in a crude cell extract.
10. If Hg^{2+} inhibits the uptake of D-glucose, this suggests that the uptake is dependent on a protein permease (facilitated diffusion) that has —SH groups present.

Chapter 33

1. B cells mature in the bone marrow and then migrate to secondary lymphoid organs, while T cells mature in the thymus and then migrate to different areas of the body.
3. There is the delayed-type hypersensitivity response, in which T cells react with antigens and secrete lymphokines. A second type of T cell, known as a "killer cell," reacts with antigens bound to target cells and causes the death of the latter.
7. Irradiate the mouse from strain A with whole-body radiation to destroy the immune system and then transplant bone marrow and thymus from strain B. This new immune system is now tolerant to transplants from strain B mice.
10. Develop antibodies to the transition-state intermediate or a transition-state intermediate analog.
12. The thiol reagent would disrupt disulfide bonds and allow the peptides that form the antibody structure (four peptide chains) to dissociate.

Chapter 34

3. Expression of some genes is induced by hypomethylation of DNA on the 5′ side of the gene. Therefore some genes, not usually active in adult tissue, may be expressed. For example, some of the genes important in development can become oncogenes when expressed at the "wrong" time (oncofetal genes).
5. RNA viruses such as retroviruses can cause cancer by a reverse transcription mechanism, using reverse transcriptase to produce a DNA copy of their genome that can then integrate into the host-cell genome.
10. HIV can cause different cancers by destroying the immune system (T4 cells) and preventing immune surveillance.

Chapter 35

1. (a) $\Delta\Psi = -71$ mV; (b) $\Delta\Psi = -63$ mV.
3. Criteria are (i) excitation of proper postsynaptic cell, (ii) presence of the compound in presynaptic terminal, (iii) release upon stimulation, and (iv) receptors on postsynaptic membrane.
6. L-DOPA can pass the blood-brain barrier but dopamine cannot. Inhibitors of dopa carboxylase cannot pass the blood-brain barrier; therefore the DOPA is not converted into dopamine until it enters the brain.
8. The high rate of Na^+ flux can be accounted for only by a pore model, as opposed to the slower carrier-mediated mechanism.

Chapter 36

1. Compounds that absorb visible light have many alternating double bonds, i.e., conjugated double bonds.
4. Rhodopsin is converted to bathorhodopsin and then to rhodopsin.
5. The Schiff's base linkage between the retinal and opsin is protonated in metarhodopsin I but not in metarhodopsin II.
10. Cone cells contain 11-*cis*-retinal bound to proteins like opsin. There are three types of cones, which absorb blue, green, and yellow light, thereby allowing us to see color. Rod cells are connected to each bipolar cell and have multiple bipolar cells connected to each ganglion cell, thereby summing the signals from the rods. The cone cells are not connected in this manner, and therefore are not sensitive to dim light.

Glossary

A

A form. A duplex DNA structure with right-handed twisting in which the planes of the base pairs are tilted about 70° with respect to the helix axis.

Acetal. The product formed by the successive condensation of two alcohols with a single aldehyde. It contains two ether-linked oxygens attached to a central carbon atom.

Acetyl-CoA. Acetyl-coenzyme A, a high-energy ester of acetic acid that is important both in the tricarboxylic acid cycle and in fatty acid biosynthesis.

Actin. A protein found in combination with myosin in muscle and also found as filaments constituting an important part of the cytoskeleton in many eukaryotic cells.

Actinomycin D. An antibiotic that binds to DNA and inhibits RNA chain elongation.

Activated complex. The highest free energy state of a complex in going from reactants to products.

Active site. The region of an enzyme molecule that contains the substrate binding site and the catalytic site for converting the substrate(s) into product(s).

Active transport. The energy-dependent transport of a substance across a membrane.

Adenine. A purine base found in DNA or RNA.

Adenosine. A purine nucleoside found in DNA, RNA, and many cofactors.

Adenosine diphosphate (ADP). The nucleotide formed by adding a pyrophosphate group to the 5′-OH group of adenosine.

Adenosine triphosphate (ATP). The nucleotide formed by adding yet another phosphate group to the pyrophosphate group on ADP.

Adenylate cyclase. The enzyme that catalyzes the formation of cyclic 3′,5′-adenosine monophosphate (cAMP) from ATP.

Adipocyte. A specialized cell that functions as a storage depot for lipid.

Aerobe. An organism that utilizes oxygen for growth.

Affinity chromatography. A column chromatographic technique that employs attached functional groups that have a specific affinity for sites on particular proteins.

Alcohol. A molecule with a hydroxyl group attached to a carbon atom.

Aldehyde. A molecule containing a doubly bonded oxygen and a hydrogen attached to the same carbon atom.

Alleles. Alternative forms of a gene.

Allosteric enzyme. An enzyme whose active site can be altered by the binding of a small molecule at a nonoverlapping site.

Angstrom (Å). A unit of length equal to 10^{-8} cm.

Anomers. The sugar isomers that differ in configuration about the carbonyl carbon atom. This carbon atom is called the anomeric carbon atom of the sugar.

Antibiotic. A natural product that inhibits bacterial growth (is bacteriostatic) and sometimes results in bacterial death (is bacteriocidal).

Antibody. A specific protein that interacts with a foreign substance (antigen) in a specific way.

Anticodon. A sequence of three bases on the transfer RNA that pair with the bases in the corresponding codon on the messenger RNA.

Antigen. A foreign substance that triggers antibody formation and is bound by the corresponding antibody.

Antiparallel β-pleated sheet (β sheet). A hydrogen-bonded secondary structure formed between two or more extended polypeptide chains.

Apoactivator. A regulatory protein that stimulates transcription from one or more genes in the presence of a coactivator molecule.

Asexual reproduction. Growth and cell duplication that does not involve the union of nuclei from cells of opposite mating types.

Asymmetric carbon. A carbon that is covalently bonded to four different groups.

Attenuator. A provisional transcription stop signal.

Autoradiography. The technique of exposing film in the presence of disintegrating radioactive particles. Used to obtain information on the distribution of radioactivity in a gel or a thin cell section.

Autoregulation. The process in which a gene regulates its own expression.

Autotroph. An organism that can form its organic constituents from CO_2.

Auxin. A plant growth hormone usually concentrated in the apical bud.

Auxotroph. A mutant that cannot grow on the minimal medium on which a wild-type member of the same species can grow.

Avogadro's number. The number of molecules in a gram molecular weight of any compound (6.023×10^{23}).

B

B cell. One of the major types of cells in the immune system. B cells can differentiate to form memory cells or antibody-forming cells.

B form. The most common form of duplex DNA, containing a right-handed helix and about 10 (10.5 exactly) base pairs per turn of the helix axis.

β bend. A characteristic way of turning an extended polypeptide chain in a different direction, involving the minimum number of residues.

β oxidation. Oxidative degradation of fatty acids that occurs by the successive oxidation of the β-carbon atom.

β sheet. A sheetlike structure formed by the interaction between two or more extended polypeptide chains.

Base analog. A compound, usually a purine or a pyrimidine, that differs somewhat from a normal nucleic acid base.

Base stacking. The close packing of the planes of base pairs, commonly found in DNA and RNA structures.

Bidirectional replication. Replication in both directions away from the origin, as opposed to replication in one direction only (unidirectional replication).

Bilayer. A double layer of lipid molecules with the hydrophilic ends oriented outward, in contact with water, and the hydrophobic parts oriented inward toward each other.

Bile salts. Derivatives of cholesterol with detergent properties that aid in the solubilization of lipid molecules in the digestive tract.

Biochemical pathway. A series of enzyme-catalyzed reactions that results in the conversion of a precursor molecule into a product molecule.

Bioluminescence. The production of light by a biochemical system.

Blastoderm. The stage in embryogenesis when a unicellular layer at the surface surrounds the yolk mass.

Bond energy. The energy required to break a bond.

Branchpoint. An intermediate in a biochemical pathway that can follow more than one route in subsequent steps.

Buffer. A conjugate acid-base pair that is capable of resisting changes in pH when acid or base is added to the system. This tendency will be maximal when the conjugate forms are present in equal amounts.

C

cAMP. 3′,5′ cyclic adenosine monophosphate. The cAMP molecule plays a key role in metabolic regulation.

CAP. The catabolite gene activator protein, sometimes incorrectly referred to as the CRP protein. The latter term, in small letters (*crp*), should be used to refer to the gene but not to the protein.

Capping. Covalent modification involving the addition of a modified guanidine group in a 5′-5″ linkage. It occurs only in eukaryotes, primarily on mRNA molecules.

Carbohydrate. A polyhydroxy aldehyde or ketone.

Carboxylic acid. A molecule containing a carbon atom attached to a hydroxyl group and to an oxygen atom by a double bond.

Carcinogen. A chemical that can cause cancer.

Carotenoids. Lipid-soluble pigments that are made from isoprene units.

Catabolism. That part of metabolism that is concerned with degradation reactions.

Catabolite repression. The general repression of transcription of genes associated with catabolism that is seen in the presence of glucose.

Catalyst. A compound that lowers the activation energy of a reaction without itself being consumed.

Catalytic site. The site of the enzyme involved in the catalytic process.

Catenane. An interlocked pair of circular structures, such as covalently closed DNA molecules.

Catenation. The linking of molecules without any direct covalent bonding between them, as when two circular DNA molecules interlock like the links in a chain.

cDNA. Complementary DNA, made *in vitro* from the mRNA by the enzyme reverse transcriptase and deoxyribonucleotide triphosphates.

Cell commitment. That stage in a cell's life when it becomes committed to a certain line of development.

Cell cycle. All of those stages that a cell passes through from one cell generation to the next.

Cell line. An established clone originally derived from a whole organism through a long process of cultivation.

Cell lineage. The pedigree of cells resulting from binary fission.

Cell wall. A tough outer coating found in many plant, fungal, and bacterial cells that accounts for their ability to withstand mechanical stress or abrupt changes in osmotic pressure. Cell walls always contain a carbohydrate component and frequently also a peptide and a lipid component.

Chelate. A molecule that contains more than one binding site and frequently binds to another molecule through more than one binding site at the same time.

Chemiosmotic coupling. The coupling of ATP synthesis to an electrochemical potential gradient across a membrane.

Chimeric DNA. Recombinant DNA whose components originate from two or more different sources.

Chiral compound. A compound that can exist in two forms that are nonsuperimposable images of one another.

Chlorophyll. A green photosynthetic pigment that is made of a magnesium dihydroporphyrin complex.

Chloroplast. A chlorophyll-containing photosynthetic organelle, found in eukaryotic cells, that can harness light energy.

Chromatin. The nucleoprotein fibers of eukaryotic chromosomes.

Chromatography. A procedure for separating chemically similar molecules. Segregation is usually carried out on paper or in glass or metal columns with the help of different solvents. The paper or glass columns contain porous solids with functional groups that have limited affinities for the molecules being separated.

Chromosome. A threadlike structure, visible in the cell nucleus during metaphase, that carries the hereditary information.

Chromosome puff. A swollen region of a giant chromosome; the swelling reflects a high degree of transcription activity.

***Cis* dominance.** Property of a sequence or a gene that exerts a dominant effect on a gene to which it is linked.

Cistron. A genetic unit that encodes a single polypeptide chain.

Citric acid cycle. *See* tricarboxylic acid (TCA) cycle.

Clone. One of a group of genetically identical cells or organisms derived from a common ancestor.

Cloning vector. A self-replicating entity to which foreign DNA can be covalently attached for purposes of amplification in host cells.

Coactivator. A molecule that functions in conjunction with a protein apoactivator. For example, cAMP is a coactivator of the CAP protein.

Codon. In a messenger RNA molecule, a sequence of three bases that represents a particular amino acid.

Coenzyme. An organic molecule that associates with enzymes and affects their activity.

Cofactor. A small molecule required for enzyme activity. It could be organic in nature, like a coenzyme, or inorganic in nature, like a metallic cation.

Complementary base sequence. For a given sequence of nucleic acids, the nucleic acids that are related to them by the rules of base pairing.

Configuration. The spatial arrangement in which atoms are covalently linked in a molecule.

Conformation. The three-dimensional arrangement adopted by a molecule, usually a complex macromolecule. Molecules with the same configuration can have more than one conformation.

Consensus sequence. In nucleic acids, the "average" sequence that signals a certain type of action by a specific protein. The sequences actually observed usually vary around this average.

Constitutive enzymes. Enzymes synthesized in fixed amounts, regardless of growth conditions.

Cooperative binding. A situation in which the binding of one substituent to a macromolecule favors the binding of another. For example, DNA cooperatively binds histone molecules, and hemoglobin cooperatively binds oxygen molecules.

Coordinate induction. The simultaneous expression of two or more genes.

Cosmid. A DNA molecule with *cos* ends from λ bacteriophage that can be packaged *in vitro* into a virus for infection purposes.

Cot curve. A curve that indicates the rate of DNA-DNA annealing as a function of DNA concentration and time.

Cytidine. A pyrimidine nucleoside found in DNA and RNA.

Cytochromes. Heme-containing proteins that function as electron carriers in oxidative phosphorylation and photosynthesis.

Cytokinin. A plant hormone produced in root tissue.

Cytoplasm. The contents enclosed by the plasma membrane, excluding the nucleus.

Cytosine. A pyrimidine base found in DNA and RNA.

Cytoskeleton. The filamentous skeleton, formed in the cytoplasm, that is largely responsible for controlling cell shape.

Cytosol. The liquid portion of the cytoplasm, including the macromolecules but not including the larger structures, such as subcellular organelles or cytoskeleton.

D

D loop. An extended loop of single-stranded DNA displaced from a duplex structure by an oligonucleotide.

Dalton. A unit of mass equivalent to the mass of a hydrogen atom (1.66×10^{-24} g).

Dark reactions. Reactions that can occur in the dark, in a process that is usually associated with light, such as the dark reactions of photosynthesis.

***De novo* pathway.** A biochemical pathway that starts from elementary substrates and ends in the synthesis of a biochemical.

Deamination. The enzymatic removal of an amine group, as in the deamination of an amino acid to an α-keto acid.

Dehydrogenase. An enzyme that catalyzes the removal of a pair of electrons (and usually one or two protons) from a substrate molecule.

Denaturation. The disruption of the native folded structure of a nucleic acid or protein molecule; may be due to heat, chemical treatment, or change in pH.

Density-gradient centrifugation. The separation, by centrifugation, of molecules according to their density, in a gradient varying in solute concentration.

Dialysis. Removal of small molecules from a macromolecule preparation by allowing them to pass across a semipermeable membrane.

Diauxic growth. Biphasic growth on a mixture of two carbon sources in which one carbon source is used up before the other one is mobilized. For example, in the presence of glucose and lactose, *E. coli* will utilize the glucose before the lactose.

Difference spectra. Display comparing the absorption spectra of a molecule or an assembly of molecules in different states, for example, those of mitochondria under oxidizing or reducing conditions.

Differential centrifugation. Separation of molecules and/or organelles by sedimentation rate.

Differentiation. A change in the form and pattern of a cell and the genes it expresses as a result of growth and replication, usually during development of a multicellular organism. Also occurs in microorganisms (e.g., in sporulation).

Diploid cell. A cell that contains two chromosomes ($2N$) of each type.

Dipole. A separation of charge within a single molecule.

Directed mutagenesis. In a DNA sequence, an intentional alteration that can be genetically inherited.

Dissociation constant. An equilibrium constant for the dissociation of a molecule into two parts (e.g., dissociation of acetic acid into acetate anion and proton).

Disulfide bridge. A covalent linkage formed by oxidation between two SH groups either in the same polypeptide chain or in different polypeptide chains.

DNA. Deoxyribonucleic acid. A polydeoxyribonucleotide in which the sugar is deoxyribose; the main repository of genetic information in all cells and most viruses.

DNA cloning. The propagation of individual segments of DNA as clones.

DNA library. A mixture of clones, each containing a cloning vector and a segment of DNA from a source of interest.

DNA polymerase. An enzyme that catalyzes the formation of 3′-5′ phosphodiester bonds from deoxyribonucleotide triphosphates.

Domain. A segment of a folded protein structure showing conformational integrity. A domain could comprise the entire protein or just a fraction of the protein. Some proteins, such as antibodies, contain many structural domains.

Dominant. Describing an allele whose phenotype is expressed regardless of whether the organism is homozygous or heterozygous for that allele.

Double helix. A structure in which two helically twisted polynucleotide strands are held together by hydrogen bonding and base stacking.

Duplex. Synonymous with double helix.

Dyad symmetry. Property of a structure that can be rotated by 180° to produce the same structure.

E

Ecdysone. A hormone that stimulates the molting process in insects.

Edman degradation. A systematic method of sequencing proteins, proceeding by stepwise removal of single amino acids from the amino terminal of a polypeptide chain.

Eicosanoid. Any fatty acid with twenty carbons.

Electrophoresis. The movement of particles in an electrical field. A commonly used technique for analysis of mixtures of molecules in solution according to their electrophoretic mobilities.

Elongation factors. Protein factors uniquely required during the elongation phase of protein synthesis. Elongation factor G (EF-G) brings about the movement of the peptidyl-tRNA from the A site to the P site of the ribosome.

Eluate. The effluent from a chromatographic column.

Embryo. Plant or animal at an early stage of development.

Enantiomorphs. Isomers that are mirror images of one another.

Endergonic reaction. A reaction with a positive free energy change.

End-product (feedback) inhibition. The inhibition of the first enzyme in a pathway by the end product of that pathway.

Endocrine glands. Specialized tissues whose function is to synthesize and secrete hormones.

Endonuclease. An enzyme that breaks a phosphodiester linkage at some point within a polynucleotide chain.

Endopeptidase. An enzyme that breaks a polypeptide chain at an internal peptide linkage.

Endoplasmic reticulum. A system of double membranes in the cytoplasm that is involved in the synthesis of transported proteins. The rough endoplasmic reticulum has ribosomes associated with it. The smooth endoplasmic reticulum does not.

Energy charge. The fractional degree to which the AMP-ADP-ATP system is filled with high-energy phosphates (phosphoryl groups).

Enhancer. A DNA sequence that can stimulate transcription at an appreciable distance from the site where it is located. It acts in either orientation and either upstream or downstream from the promoter.

Entropy. The randomness of a system.

Enzyme. A protein that contains a catalytic site for a biochemical reaction.

Epimers. Two stereoisomers with more than one chiral center that differ in configuration at one of their chiral centers.

Equilibrium. In chemistry the point at which the concentrations of two compounds are such that the interconversion of one compound into the other compound does not result in any change in free energy.

Escherichia coli (E. coli). A Gram negative bacterium commonly found in the vertebrate intestine. It is the bacterium most frequently used in the study of biochemistry and genetics.

Established cell line. A group of cultured cells derived from a single origin and capable of stable growth for many generations.

Ether. A molecule containing two carbons linked by an oxygen atom.

Eukaryote. A cell or organism that has a membrane-bounded nucleus.

Excision repair. DNA repair in which a damaged region is replaced.

Excited state. An energy-rich state of an atom or a molecule, produced by the absorption of radiant energy.

Exergonic reaction. A chemical reaction that takes place with a negative change in free energy.

Exon. A segment within a gene that carries part of the coding information for a protein.

Exonuclease. An enzyme that breaks a phosphodiester linkage at one or the other end of a polynucleotide chain so as to release single or small nucleotide residues.

F

F factor. A large bacterial plasmid, known as the sex-factor plasmid because it permits mating between F^+ and F^- bacteria.

Facultative aerobe. An organism that can use molecular oxygen in its metabolism but that also can live anaerobically.

Fatty acid. A long-chain hydrocarbon containing a carboxyl group at one end. Saturated fatty acids have completely saturated hydrocarbon chains. Unsaturated fatty acids have one or more carbon–carbon double bonds in their hydrocarbon chains.

Feedback inhibition. *See* end-product inhibition.

Fermentation. The energy-generating breakdown of glucose or related molecules by a process that does not require molecular oxygen.

Fingerprinting. The characteristic two-dimensional paper chromatogram obtained from the partial hydrolysis of a protein or a nucleic acid.

Fluorescence. The emission of light by an excited molecule in the process of making the transition from the excited state to the ground state.

Frameshift mutations. Insertions or deletions of genetic material that lead to a shift in the translation of the reading frame. The mutation usually leads to nonfunctional proteins.

Free energy. That part of the energy of a system that is available to do useful work.

Furanose. A sugar that contains a five-membered ring as a result of intramolecular hemiacetal formation.

Futile cycle. *See* pseudocycle.

G

G_1 phase. That period of the cell cycle in which preparations are being made for chromosome duplication, which takes place in the S phase.

G_2 phase. That period of the cell cycle between S phase and mitosis (M phase).

Gametes. The ova and the sperm, haploid cells that unite during fertilization to generate a diploid zygote.

Gel exclusion chromatography. A technique that makes use of certain polymers that can form porous beads with varying pore sizes. In columns made from such beads, it is possible to separate molecules, which cannot penetrate beads of a given pore size, from small molecules that can.

Gene. A segment of the genome that codes for a functional product.

Gene amplification. The duplication of a particular gene within a chromosome two or more times.

Gene splicing. The cutting and rejoining of DNA sequences.

General recombination. Recombination that occurs between homologous chromosomes at homologous sites.

Generation time. The time it takes for a cell to double its mass under specified conditions.

Genetic map. The arrangement of genes or other identifiable sequences on a chromosome.

Genome. The total genetic content of a cell or a virus.

Genotype. The genetic characteristics of an organism (distinguished from its observable characteristics, or phenotype).

Globular protein. A folded protein that adopts an approximately globular shape.

Goldman equation. An equation expressing the quantitative relationship between the concentrations of charged species on either side of a membrane and the resting transmembrane potential.

Golgi apparatus. A complex series of double-membrane structures that interact with the endoplasmic reticulum and that serve as a transfer point for proteins destined for other organelles, the plasma membrane, or extracellular transport.

Gluconeogenesis. The production of sugars from nonsugar precursors such as lactate or amino acids. Applies more specifically to the production of free glucose by vertebrate livers.

Glycogen. A polymer of glucose residues in 1,4 linkage and 1,6 linkage at branchpoints.

Glycogenic. Describing amino acids whose metabolism may lead to gluconeogenesis.

Glycolipid. A lipid containing a carbohydrate group.

Glycolysis. The catabolic conversion of glucose to pyruvate with the production of ATP.

Glycoprotein. A protein linked to an oligosaccharide or a polysaccharide.

Glycosaminoglycans. Long, unbranched polysaccharide chains composed of repeating disaccharide subunits in which one of the two sugars is either *N*-acetylglucosamine or *N*-acetylgalactosamine.

Glycosidic bond. The bond between a sugar and an alcohol. Also the bond that links two sugars in disaccharides, oligosaccharides, and polysaccharides.

Glyoxylate cycle. A pathway that uses some of the enzymes of the TCA cycle and some enzymes whereby acetate can be converted into succinate and carbohydrates.

Glyoxysome. An organelle containing key enzymes of the glyoxylate cycle.

Gram molecular weight. For a given compound, the weight in grams that is numerically equal to its molecular weight.

Ground state. The lowest electronic energy state of an atom or a molecule.

Growth factor. A substance that must be present in the growth medium to permit cell proliferation.

Growth fork. The region on a DNA duplex molecule where synthesis is taking place. It resembles a fork in shape, since it consists of a region of duplex DNA connected to a region of unwound single strands.

Guanine. A purine base found in DNA or RNA.

Guanosine. A purine nucleoside found in DNA and RNA.

H

Hairpin loop. A single-stranded complementary region that folds back on itself and base-pairs into a double helix.

Half-life. The time required for the disappearance of one half of a substance.

Haploid cell. A cell containing only one chromosome of each type.

Heavy isotopes. Forms of atoms that contain greater numbers of neutrons (e.g., ^{15}N, ^{13}C).

Helix. A spiral structure with a repeating pattern.

Heme. An iron-porphyrin complex found in hemoglobin and cytochromes.

Hemiacetal. The product formed by the condensation of an aldehyde with an alcohol; it contains one oxygen linked to a central carbon in a hydroxyl fashion and one oxygen linked to the same central carbon by an ether linkage.

Henderson-Hasselbalch equation. An equation that relates the pK_a to the pH and the ratio of the proton acceptor (A^-) and the proton donor (HA) species of a conjugate acid-base pair.

Heterochromatin. Highly condensed regions of chromosomes that are not usually transcriptionally active.

Heteroduplex. An annealed duplex structure between two DNA strands that do not show perfect complementarity. Can arise by mutation, recombination, or the annealing of complementary single-stranded DNAs.

Heteropolymer. A polymer containing more than one type of monomeric unit.

Heterotroph. An organism that requires preformed organic compounds for growth.

Heterozygous. Describing an organism (a heterozygote) that carries two different alleles for a given gene.

Hexose. A sugar with a six-carbon backbone.

High-energy compound. A compound that undergoes hydrolysis with a high negative standard free energy change.

Histones. The family of basic proteins that is normally associated with DNA in most cells of eukaryotic organisms.

Holoenzyme. An intact enzyme containing all of its subunits with full enzymatic activity.

Homologous chromosomes. Chromosomes that carry the same pattern of genes, but not necessarily the same alleles.

Homopolymer. A polymer composed of only one type of monomeric building block.

Homozygous. Describing an organism (a homozygote) that carries two identical alleles for a given gene.

Hormone. A chemical substance made in one cell and secreted so as to influence the metabolic activity of a select group of cells located at other sites in the organism.

Hormone receptor. A protein that is located on the cell membrane or inside the responsive cell and that interacts specifically with the hormone.

Host cell. A cell used for growth and reproduction of a virus.

Hybrid (or chimeric) plasmid. A plasmid that contains DNA from two different organisms.

Hydrogen bond. A weak attractive force between one electronegative atom and a hydrogen atom that is covalently linked to a second electronegative atom.

Hydrolysis. The cleavage of a molecule by the addition of water.

Hydrophilic. Preferring to be in contact with water.

Hydrophobic. Preferring not to be in contact with water, as is the case with the hydrocarbon portion of a fatty acid or phospholipid chain.

Hydrophobic bonding. The association of nonpolar groups with each other in aqueous solution.

Hydroxyapatite. A calcium phosphate gel used, in the case of nucleic acids, to selectively absorb duplex DNA-RNA from a mixture of single-stranded and duplex nucleic acids.

I

Icosahedral symmetry. The symmetry displayed by a regular polyhedron that is composed of twenty equilateral triangular faces with twelve corners.

Imine. A molecule containing a nitrogen atom attached to a carbon atom by a double bond. The nitrogen is also covalently linked to a hydrogen.

Immunofluorescence. A cytological technique in which a specific fluorescent antibody is used to label an antigen. Frequently used to determine the location of an antigen in a tissue or a cell.

Immunoglobulin. A protein made in a B plasma cell and usually secreted; it interacts specifically with a foreign agent. Synonymous with antibody. It is composed of two heavy and two light chains linked by disulfide bonds. Immunoglobulins can be divided into five classes (IgG, IgM, IgA, IgD, and IgE) based on their heavy-chain component.

In vitro. Literally, "in glass," describing whatever happens in a test tube or other receptacle, as opposed to what happens in whole cells of the whole organism (*in vivo*).

Induced fit. A change in the shape of an enzyme that results from the binding of substrate.

Initiation factors. Those protein factors that are specifically required during the initiation phase of protein synthesis.

Intercalating agent. A chemical, usually containing aromatic rings, that can sandwich in between adjacent base pairs in a DNA duplex. The intercalation leads to an adjustment in the DNA secondary structure, as adjacent base pairs are usually close-packed.

Interferon. One of a family of proteins that are liberated by special host cells in the mammal in response to viral infection. The interferons attach to an infected cell, where they stimulate antiviral protein synthesis.

Intervening sequence. *See* intron.

Intron. A segment of the nascent transcript that is removed by splicing. Also refers to the corresponding region in the DNA. Synonymous with intervening sequence.

Inverted repeat. A chromosome segment that is identical to another segment on the same chromosome except that it is oriented in the opposite direction.

Ion-exchange resin. A polymeric resinous substance, usually in bead form, that contains fixed groups with positive or negative charge. A cation exchange resin has negatively charged groups and is therefore useful in exchanging the cationic groups in a test sample. The resin is usually used in the form of a column, as in other column chromatographic systems.

Isoelectric pH. The pH at which a protein has no net charge.

Isomerase. An enzyme that catalyzes an intramolecular rearrangement.

Isomerization. Rearrangement of atomic groups within the same molecule without any loss or gain of atoms.

Isozymes. Multiple forms of an enzyme that differ from one another in one or more of the properties.

K

K_m. *See* Michaelis constant.

Ketogenic. Describing amino acids that are metabolized to acetoacetate and acetate.

Ketone. A functional group of an organic compound in which a carbon atom is double-bonded to an oxygen. Neither of the other substituents attached to the carbon is a hydrogen. Otherwise the group would be called an aldehyde.

Ketone bodies. Refers to acetoacetate, acetone, and β-hydroxybutyrate made from acetyl-CoA in the liver and used for energy in nonhepatic tissue.

Ketosis. A condition in which the concentration of ketone bodies in the blood or urine is unusually high.

Kilobase. One thousand bases in a DNA molecule.

Kinase. An enzyme catalyzing phosphorylation of an acceptor molecule, usually with ATP serving as the phosphate (phosphoryl) donor.

Kinetochore. A structure that attaches laterally to the centromere of a chromosome; it is the site of chromosome tubule attachment.

Krebs cycle. *See* tricarboxylic acid (TCA) cycle.

L

Lampbrush chromosome. Giant diplotene chromosome found in the oocyte nucleus. The loops that are observed are the sites of extensive gene expression.

Law of mass action. The finding that the rate of a chemical reaction is a function of the product of the concentrations of the reacting species.

Leader region. The region of an mRNA between the 5′ end and the initiation codon for translation of the first polypeptide chain.

Lectins. Agglutinating proteins usually extracted from plants.

Ligase. An enzyme that catalyzes the joining of two molecules together. In DNA it joins 3′-OH to 5′ phosphates.

Linkers. Short oligonucleotides that can be ligated to larger DNA fragments, then cleaved to yield overlapping cohesive ends, suitable for ligation to other DNAs that contain comparable cohesive ends.

Linking number. The net number of times one polynucleotide chain crosses over another polynucleotide chain. By convention, right-handed crossovers are given a plus designation.

Lipid. A biological molecule that is soluble in organic solvents. Lipids include steroids, fatty acids, prostaglandins, terpenes, and waxes.

Lipid bilayer (*see* **Bilayer**). Model for the structure of the cell membrane based on the hydrophobic interaction between phospholipids.

Lipopolysaccharide. Usually refers to a unique glycolipid found in Gram negative bacteria.

Lyase. An enzyme that catalyzes the removal of a group to form a double bond, or the reverse reaction.

Lysogenic virus. A virus that can adopt an inactive (lysogenic) state, in which it maintains its genome within a cell instead of entering the lytic cycle. The circumstances that determine whether a lysogenic (temperate) virus will adopt an inactive state or an active lytic state are often subtle and depend on the physiological state of the infected cell.

Lysosome. An organelle that contains hydrolytic enzymes designed to break down proteins that are targeted to that organelle.

Lytic infection. A virus infection that leads to the lysis of the host cell, yielding progeny virus particles.

M

M phase. That period of the cell cycle when mitosis takes place.

Meiosis. Process in which diploid cells undergo division to form haploid sex cells.

Membrane transport. The facilitated transport of a molecule across a membrane.

Merodiploid. An organism that is diploid for some but not all of its genes.

Mesosome. An invagination of the bacterial cell membrane.

Messenger RNA (mRNA). The template RNA carrying the message for protein synthesis.

Metabolic turnover. A measure of the rate at which already existing molecules of the given species are replaced by newly synthesized molecules of the same type. Usually isotopic labeling is required to measure turnover.

Metabolism. The sum total of the enzyme-catalyzed reactions that occur in a living organism.

Metamorphosis. A change of form, especially the conversion of a larval form to an adult form.

Metaphase. That stage in mitosis or meiosis when all of the chromosomes are lined up on the equator (i.e., an imaginary line that bisects the cell).

Micelle. An aggregate of lipids in which the polar head groups face outward and the hydrophobic tails face inward; no solvent is trapped in the center.

Michaelis constant (K_m). The substrate concentration at which an enzyme-catalyzed reaction proceeds at one-half maximum velocity.

Michaelis-Menten equation (also known as the Henri-Michaelis-Menten equation). An equation relating the reaction velocity to the substrate concentration of an enzyme.

Microtubules. Thin tubules, made from globular proteins, that serve multiple purposes in eukaryotic cells.

Mismatch repair. The replacement of a base in a heteroduplex structure by one that forms a Watson-Crick base pair.

Missense mutation. A change in which a codon for one amino acid is replaced by a codon for another amino acid.

Mitochondrion. An organelle, found in eukaryotic cells, in which oxidative phosphorylation takes place. It contains its own genome and unique ribosomes to carry out protein synthesis of only a fraction of the proteins located in this organelle.

Mitosis. The process whereby replicated chromosomes segregate equally toward opposite poles prior to cell division.

Mobile genetic element. A segment of the genome that can move as a unit from one location on the genome to another, without any requirement for sequence homology.

Molecularity of a reaction. The number of molecules involved in a specific reaction step.

Monolayer. A single layer of oriented lipid molecules.

Mutagen. An agent that can bring about a heritable change (mutation) in an organism.

Mutagenesis. A process that leads to a change in the genetic material that is inherited in subsequent generations.

Mutant. An organism that carries an altered gene or change in its genome.

Mutarotation. The change in optical rotation of a sugar that is observed immediately after it is dissolved in aqueous solution, as the result of the slow approach of equilibrium of a pyranose or a furanose in its α and β forms.

Mutation. The genetically inheritable alteration of a gene or group of genes.

Myofibril. A unit of thick and thin filaments in a muscle fiber.

Myosin. The main protein of the thick filaments in a muscle myofibril. It is composed of two coiled subunits (M_r about 220,000) that can aggregate to form a thick filament, which is globular at each end.

N

Nascent RNA. The initial transcripts of RNA, before any modification or processing.

Negative control. Regulation of the activity by an inhibitory mechanism.

Nernst equation. An equation that relates the redox potential to the standard redox potential and the concentrations of the oxidized and reduced form of the couple.

Nitrogen cycle. The passage of nitrogen through various valence states, as the result of reactions carried out by a wide variety of different organisms.

Nitrogen fixation. Conversion of atmospheric nitrogen into a form that can be converted by biochemical reactions to an organic form. This reaction is carried out by a very limited number of microorganisms.

Nitrogenous base. An aromatic nitrogen-containing molecule with basic properties. Such bases include purines and pyrimidines.

Noncompetitive inhibitor. An inhibitor of enzyme activity whose effect is not reversed by increasing the concentration of substrate molecule.

Nonsense mutation. A change in the base sequence that converts a sense codon (one that specifies an amino acid) to one that specifies a stop (a nonsense codon). There are three nonsense codons.

Northern blotting. *See* Southern blotting.

Nuclease. An enzyme that cleaves phosphodiester bonds of nucleic acids.

Nucleic acids. Polymers of the ribonucleotides or deoxyribonucleotides.

Nucleohistone. A complex of DNA and histone.

Nucleolus. A spherical structure visible in the nucleus during interphase. The nucleolus is associated with a site on the chromosome that is involved in ribosomal RNA synthesis.

Nucleophilic group. An electron-rich group that tends to attack an electron-deficient nucleus.

Nucleosome. A complex of DNA and an octamer of histone proteins in which a small stretch of the duplex is wrapped around a molecular bead of histone.

Nucleotide. An organic molecule containing a purine or pyrimidine base, a five-carbon sugar (ribose or deoxyribose), and one or more phosphate groups.

Nucleus. In eukaryotic cells, the centrally located organelle that encloses most of the chromosomes. Minor amounts of chromosomal substance are found in some other organelles, most notably the mitochondria and the chloroplasts.

O

Okazaki fragment. A short segment of single-stranded DNA that is an intermediate in DNA synthesis. In bacteria, Okazaki fragments are 1,000–2,000 bases in length; in eukaryotes, 100–200 bases in length.

Oligonucleotide. A polynucleotide containing a small number of nucleotides. The linkages are the same as in a polynucleotide; the only distinguishing feature is the small size.

Oligosaccharide. A molecule containing a small number of sugar residues joined in a linear or a branched structure by glycosidic bonds.

Oncogene. A gene of cellular or viral origin that is responsible for rapid, unruly growth of animal cells.

Operon. A group of contiguous genes that are coordinately regulated by two *cis*-acting elements, a promoter and an operator. Found only in prokaryotic cells.

Optical activity. The property of a molecule that leads to rotation of the plane of polarization of plane-polarized light when the latter is transmitted through the substance. Chirality is a necessary and sufficient property for optical activity.

Organelle. A subcellular membrane-bounded body with a well-defined function.

Osmotic pressure. The pressure generated by the mass flow of water to that side of a membrane-bounded structure that contains the higher concentration of solute molecules. A stable osmotic pressure is seen in systems in which the membrane is not permeable to some of the solute molecules.

Oxidation. The loss of electrons from a compound.

Oxidative phosphorylation. The formation of ATP as the result of the transfer of electrons to oxygen.

Oxido-reductase. An enzyme that catalyzes oxidation-reduction reactions.

P

Palindrome. A sequence of bases that reads the same in both directions on opposite strands of the DNA duplex (e.g., GAATTC).

Pentose. A sugar with five carbon atoms.

Pentose phosphate pathway. The pathway involving the oxidation of glucose-6-phosphate to pentose phosphates and further reactions of pentose phosphates.

Peptide. An organic molecule in which a covalent amide bond is formed between the α-amino group of one amino acid and the α-carboxyl group of another amino acid, with the elimination of a water molecule.

Peptide mapping. Same as fingerprinting.

Peptidoglycan. The main component of the bacterial cell wall, consisting of a two-dimensional network of heteropolysaccharides running in one direction, cross-linked with polypeptides running in the perpendicular direction.

Periplasm. The space between the inner and outer membranes of a bacterium.

Permease. A protein that catalyzes the transport of a specific small molecule across a membrane.

Peroxisomes. Subcellular organelles that contain flavin-requiring oxidases and that regenerate oxidized flavin by reaction with oxygen.

Phenotype. The observable trait(s) that result from the genotype in cooperation with the environment.

Phenylketonuria. A human disease caused by a genetic deficiency in the enzyme that converts phenylalanine to tyrosine. The immediate cause of the disease is an excess of phenylalanine, which can be alleviated by a diet low in phenylalanine.

Pheromone. A hormonelike substance that acts as an attractant.

Phosphodiester. A molecule containing two alcohols esterified to a single molecule of phosphate. For example, the backbone of nucleic acids is connected by 5′-3′ phosphodiester linkages between the adjacent individual nucleotide residues.

Phosphogluconate pathway. Another name for the pentose phosphate pathway. This name derives from the fact that 6-phosphogluconate is an intermediate in the formation of pentoses from glucose.

Phospholipid. A lipid containing charged hydrophilic phosphate groups; a component of cell membranes.

Phosphorylation. The formation of a phosphate derivative of a biomolecule.

Photoreactivation. DNA repair in which the damaged region is repaired with the help of light and an enzyme. The lesion is repaired without excision from the DNA.

Photosynthesis. The biosynthesis that directly harnesses the chemical energy resulting from the absorption of light. Frequently used to refer to the formation of carbohydrates from CO_2 that occurs in the chloroplasts of plants or the plastids of photosynthetic microorganisms.

Pitch length (or pitch). The number of base pairs per turn of a duplex helix.

Plaque. A circular clearing on a lawn of bacterial or cultured cells, resulting from cell lysis and production of phage or animal virus progeny.

Plasma membrane. The membrane that surrounds the cytoplasm.

Plasmid. A circular DNA duplex that replicates autonomously in bacteria. Plasmids that integrate into the host genome are called episomes. Plasmids differ from viruses in that they never form infectious nucleoprotein particles.

Polar group. A hydrophilic (water-loving) group.

Polar mutation. A mutation in one gene that reduces the expression of a gene or genes distal to the promoter in the same operon.

Polarimeter. An instrument for determining the rotation of polarization of light as the light passes through a solution containing an optically active substance.

Polyamine. A hydrocarbon containing more than two amino groups.

Polycistronic messenger RNA. In prokaryotes, an RNA that contains two or more cistrons; note that only in prokaryotic mRNAs can more than one cistron be utilized by the translation system to generate individual proteins.

Polymerase. An enzyme that catalyzes the synthesis of a polymer from monomers.

Polynucleotide. A chain structure containing nucleotides linked together by phosphodiester (5′-3′) bonds. The polynucleotide chain has a directional sense with a 5′ and a 3′ end.

Polynucleotide phosphorylase. An enzyme that polymerizes ribonucleotide diphosphates. No template is required.

Polypeptide. A linear polymer of amino acids held together by peptide linkages. The polypeptide has a directional sense, with an amino and a carboxyl-terminal end.

Polyribosome (polysome). A complex of an mRNA and two or more ribosomes actively engaged in protein synthesis.

Polysaccharide. A linear or branched chain structure containing many sugar molecules linked by glycosidic bonds.

Porphyrin. A complex planar structure containing four substituted pyrroles covalently joined in a ring and frequently containing a central metal atom. For example, heme is a porphyrin with a central iron atom.

Positive control. A system that is turned on by the presence of a regulatory protein.

Posttranslational modification. The covalent bond changes that occur in a polypeptide chain after it leaves the ribosome and before it becomes a mature protein.

Primary structure. In a polymer, the sequence of monomers and the covalent bonds.

Primer. A structure that serves as a growing point for polymerization.

Primosome. A multiprotein complex that catalyzes synthesis of RNA primer at various points along the DNA template.

Prochiral molecule. A nonchiral molecule that may react with an enzyme so that two groups that have a mirror-image relationship to each other are treated differently.

Prokaryote. A unicellular organism that contains a single chromosome, no nucleus, no membrane-bound organelles, and has characteristic ribosomes and biochemistry.

Promoter. That region of the gene that signals RNA polymerase binding and the initiation of transcription.

Prophage. The silent phage genome. Some prophages integrate into the host genome; others replicate autonomously. The prophage state is maintained by a phage-encoded repressor.

Prophase. The stage in meiosis or mitosis when chromosomes condense and become visible as refractile bodies.

Proprotein. A protein that is made in an inactive form, so that it requires processing to become functional.

Prostaglandin. An oxygenated eicosanoid that has a hormonal function. Prostaglandins are unusual hormones in that they usually have effects only in that region of the organism where they are synthesized.

Prosthetic group. Synonymous with coenzyme except that a prosthetic group is usually more firmly attached to the enzyme it serves.

Protamines. Highly basic, arginine-rich proteins found complexed to DNA in the sperm of many invertebrates and fish.

Protein subunit. One of the components of a complex multicomponent protein.

Proteoglycan. A protein-linked heteropolysaccharide in which the heteropolysaccharide is usually the major component.

Protist. A relatively undifferentiated organism that can survive as a single cell.

Proton acceptor. A functional group capable of accepting a proton from a proton donor molecule.

Proton motive force (Δp). The thermodynamic driving force for proton translocation. Expressed quantitatively as $\Delta G_{H+}/F$ in units of volts.

Protooncogene. A cellular gene that can undergo modification to a cancer-causing gene (oncogene).

Pseudocycle. A sequence of reactions that can be arranged in a cycle but that usually do not function simultaneously in both directions. Also called a futile cycle, since the net result of simultaneous functioning in both directions would be the expenditure of energy without accomplishing any useful work.

Pulse-chase. An experiment in which a short labeling period is followed by the addition of an excess of the same, unlabeled compound to dilute out the labeled material.

Purine. A heterocyclic ring structure with varying functional groups. The purines adenine and guanine are found in both DNA and RNA.

Puromycin. An antibiotic that inhibits polypeptide synthesis by competing with aminoacyl-tRNA for the ribosomal binding site A.

Pyranose. A simple sugar containing the six-membered pyran ring.

Pyrimidine. A heterocyclic six-membered ring structure. Cytosine and uracil are the main pyrimidines found in RNA, and cytosine and thymine are the main pyrimidines found in DNA.

Pyrophosphate. A molecule formed by two phosphates in anhydride linkage.

Q

Quaternary structure. In a protein, the way in which the different folded subunits interact to form the multisubunit protein.

R

R group. The distinctive side chain of an amino acid.

R loop. A triple-stranded structure in which RNA displaces a DNA strand by DNA-RNA hybrid formation in a region of the DNA.

Rapid-start complex. The complex that RNA polymerase forms at the promoter site just before initiation.

Recombination. The transfer to offspring of genes not found together in either of the parents.

Redox couple. An electron donor and its corresponding oxidized form.

Redox potential (*E*). The relative tendency of a pair of molecules to release or accept an electron. The standard redox potential ($E°$) is the redox potential of a solution containing the oxidant and reductant of the couple at standard concentrations.

Regulatory enzyme. An enzyme in which the active site is subject to regulation by factors other than the enzyme substrate. The enzyme frequently contains a nonoverlapping site for binding the regulatory factor that affects the activity of the active site.

Regulatory gene. A gene whose principal product is a protein designed to regulate the synthesis of other genes.

Renaturation. The process of returning a denatured structure to its original native structure, as when two single strands of DNA are reunited to form a regular duplex, or an unfolded polypeptide chain is returned to its normal folded three-dimensional structure.

Repair synthesis. DNA synthesis following excision of damaged DNA.

Repetitive DNA. A DNA sequence that is present in many copies per genome.

Replica plating. A technique in which an impression of a culture is taken from a master plate and transferred to a fresh plate. The impression can be of bacterial clones or phage plaques.

Replication fork. The Y-shaped region of DNA at the site of DNA synthesis; also called a growth fork.

Replicon. A genetic element that behaves as an autonomous replicating unit. It can be a plasmid, phage, or bacterial chromosome.

Repressor. A regulatory protein that inhibits transcription from one or more genes. It can combine with an inducer (resulting in specific enzyme induction) or with an operator element (resulting in repression).

Resonance hybrid. A molecular structure that is a hybrid of two structures that differ in the locations of some of the electrons. For example, the benzene ring can be drawn in two ways, with double bonds in different positions. The actual structure of benzene is in between these two equivalent structures.

Restriction-modification system. A pair of enzymes found in most bacteria (but not eukaryotic cells). The restriction enzyme recognizes a certain sequence in duplex DNA and makes one cut in each unmodified DNA strand at or near the recognition sequence. The modification enzyme methylates (or modifies) the recognition sequence, thus protecting it from the action of the restriction enzyme.

Reverse transcriptase. An enzyme that synthesizes DNA from an RNA template, using deoxyribonucleotide triphosphates.

Rho factor. A protein involved in the termination of transcription of some messenger RNAs.

Ribose. The five-carbon sugar found in RNA.

Ribosomal RNA (rRNA). The RNA parts of the ribosome.

Ribosomes. Small cellular particles made up of ribosomal RNA and protein. They are the site, together with mRNA, of protein synthesis.

RNA (ribonucleic acid). A polynucleotide in which the sugar is ribose.

RNA polymerase. An enzyme that catalyzes the formation of RNA from ribonucleotide triphosphates, using DNA as a template.

RNA splicing. The excision of a segment of RNA, followed by a rejoining of the remaining fragments.

Rolling-circle replication. A mechanism for the replication of circular DNA. A nick in one strand allows the 3′ end to be extended, displacing the strand with the 5′ end, which is also replicated, to generate a double-stranded tail that can become larger than the unit size of the circular DNA.

S

S phase. The period during the cell cycle when the chromosome is replicated.

Salting in. The increase in solubility that is displayed by typical globular proteins upon the addition of small amounts of certain salts, such as ammonium sulfate.

Salting out. The decrease in protein solubility that occurs when salts such as ammonium sulfate are present at high concentrations.

Salvage pathway. A family of reactions that permits, for instance, nucleosides as well as purine and pyrimidine bases resulting from the partial breakdown of nucleic acids to be reutilized in nucleic acid synthesis.

Satellite DNA. A DNA fraction whose base composition differs from that of the main component of DNA, as revealed by the fact that it bands at a different density in a CsCl gradient. Usually repetitive DNA or organelle DNA.

Scissile. Capable of being cut smoothly or split easily.

Second messenger. A diffusible small molecule, such as cAMP, that is formed at the inner surface of the plasma membrane in response to a hormonal signal.

Secondary structure. In a protein or a nucleic acid, any repetitive folded pattern that results from the interaction of the corresponding polymeric chains.

Semiconservative replication. Duplication of DNA in which the daughter duplex carries one old strand and one new strand.

Sigma factor. A subunit of RNA polymerase that recognizes specific sites on DNA for initiation of RNA synthesis.

Single-copy DNA. A region of the genome whose sequence is present only once per haploid complement.

Somatic cell. Any cell of an organism that cannot contribute its genes to a subsequent generation.

SOS system. A set of DNA repair enzymes and regulatory proteins that regulate their synthesis so that maximum synthesis occurs when the DNA is damaged.

Southern blotting. A method for detecting a specific DNA restriction fragment, developed by Edward Southern. DNA from a gel electrophoresis pattern is blotted onto nitrocellulose paper; then the DNA is denatured and fixed on the paper. Subsequently the pattern of specific sequences in the Southern blot can be determined by hybridization to a suitable probe and autoradiography. A Northern blot is similar, except that RNA is blotted instead onto the nitrocellulose paper.

Splicing. *See* RNA splicing.

Sporulation. Formation from vegetative cells of metabolically inactive cells that can resist extreme environmental conditions.

Stacking energy. The energy of interaction that favors the face-to-face packing of purine and pyrimidine base pairs.

Steady state. In enzyme-kinetic analysis, the time interval when the rate of reaction is approximately constant with time. The term is also used to describe the state of a living cell where the concentrations of many molecules are approximately constant because of a balancing between their rate of synthesis and breakdown.

Stem cell. A cell from which other cells stem or arise by differentiation.

Stereoisomers. Isomers that are nonsuperimposable mirror images of each other.

Steroids. Compounds that are derivatives of a tetracyclic structure composed of a cyclopentane ring fused to a substituted phenanthrene nucleus.

Structural domain. An element of protein tertiary structure that recurs in many structures.

Structural gene. A gene encoding the amino acid sequence of a polypeptide chain.

Structural protein. A protein that serves a structural function.

Subunit. Individual polypeptide chains in a protein.

Supercoiled DNA. Supertwisted, covalently closed duplex DNA.

Suppressor gene. A gene that can reverse the phenotype of a mutation in another gene.

Suppressor mutation. A mutation that restores a function lost by an initial mutation and that is located at a site different from the initial mutation.

Svedberg unit (S). The unit used to express the sedimentation constant *s*: $1S = 10^{-13}$ s. The sedimentation constant *s* is proportional to the rate of sedimentation of a molecule in a given centrifugal field and is related to the size and shape of the molecule.

Synapse. The chemical connection for communication between two nerve cells or between a nerve cell and a target cell such as a muscle cell.

Synapsis. The pairing of homologous chromosomes, seen during the first meiotic prophase.

T

Tandem duplication. A duplication in which the repeated regions are immediately adjacent to one another.

TCA cycle. *See* tricarboxylic acid cycle.

Template. A polynucleotide chain that serves as a surface for the absorption of monomers of a growing polymer and thereby dictates the sequence of the monomers in the growing chain.

Termination factors. Proteins that are exclusively involved in the termination reactions of protein synthesis on the ribosome.

Terpenes. A diverse group of lipids made from isoprene precursors.

Tertiary structure. In a protein or nucleic acid, the final folded form of the polymer chain.

Tetramer. Structure resulting from the association of four subunits.

Thioester. An ester of a carboxylic acid with a thiol or mercaptan.

Thymidine. One of the four nucleosides found in DNA.

Thymine. A pyrimidine base found in DNA.

Topoisomerase. An enzyme that changes the extent of supercoiling of a DNA duplex.

Transamination. Enzymatic transfer of an amino group from an α-amino acid to an α-keto acid.

Transcription. RNA synthesis that occurs on a DNA template.

Transduction. Genetic exchange in bacteria that is mediated via phage.

Transfection. An artificial process of infecting cells with naked viral DNA.

Transfer RNA (tRNA). Any of a family of low-molecular-weight RNAs that transfer amino acids from the cytoplasm to the template for protein synthesis on the ribosome.

Transferase. An enzyme that catalyzes the transfer of a molecular group from one molecule to another.

Transformation. Genetic exchange in bacteria that is mediated via purified DNA. In somatic cell genetics the term is also used to indicate the conversion of a normal cell to one that grows like a cancer cell.

Transgenic. Describing an organism that contains transfected DNA in the germ line.

Transition state. The activated state in which a molecule is best suited to undergoing a chemical reaction.

Translation. The process of reading a messenger RNA sequence for the specified amino acid sequence it contains.

Transport protein. A protein whose primary function is to transport a substance from one part of the cell to another, from one cell to another, or from one tissue to another.

Tricarboxylic acid (TCA) cycle. The cyclical process whereby acetate is completely oxidized to CO_2 and water, and electrons are transferred to NAD^+ and FAD. The TCA cycle is localized to the mitochondria in eukaryotic cells and to the plasma membrane in prokaryotic cells. Also called the Krebs cycle.

Trypsin. A proteolytic enzyme that cleaves peptide chains next to the basic amino acids arginine and lysine.

Tryptic peptide mapping. The technique of generating a chromatographic profile characteristic of the fragments resulting from trypsin enzyme cleavage of the protein.

Tumorigenesis. The mechanism of tumor formation.

Turnover number. The maximum number of molecules of substrate that can be converted to product per active site per unit time.

U

Ultracentrifuge. A high-speed centrifuge that can attain speeds up to 60,000 rpm and centrifugal fields of 500,000 times gravity. Useful for characterizing and/or separating macromolecules.

Unidirectional replication. *See* bidirectional replication.

Unwinding proteins. Proteins that help to unwind double-stranded DNA during DNA replication.

Urea cycle. A metabolic pathway in the liver that leads to the synthesis of urea from amino groups and CO_2. The function of the pathway is to convert the ammonia resulting from catabolism to a nontoxic form, which is subsequently secreted.

UV irradiation. Electromagnetic radiation with a wavelength shorter than that of visible light (200–390 nm). Causes damage to DNA (mainly pyrimidine dimers).

V

van der Waals forces. Refers to two types of interactions, one attractive and one repulsive. The attractive forces are due to favorable interactions among the induced instantaneous dipole moments that arise from fluctuations in the electron charge densities of neighboring nonbonded atoms. Repulsive forces arise when noncovalently bonded atoms come too close together.

Viroids. Pathogenic agents, mostly of plants, that consist of short (usually circular) RNA molecules.

Virus. A nucleic-acid-protein complex that can infect and replicate inside a specific host cell to make more virus particles.

Vitamin. A trace organic substance required in the diet of some species. Many vitamins are precursors of coenzymes.

W

Watson-Crick base pairs. The type of hydrogen-bonded base pairs found in DNA, or comparable base pairs found in RNA. The base pairs are A-T, G-C, and A-U.

Wild-type gene. The form of a gene (allele) normally found in nature.

Wobble. A proposed explanation for base-pairing that is not of the Watson-Crick type and that often occurs between the 3′ base in the codon and the 5′ base in the anticodon.

X

X-ray crystallography. A technique for determining the structure of molecules from the x-ray diffraction patterns that are produced by crystalline arrays of the molecules.

Y

Ylid. A compound in which adjacent, covalently bonded atoms, both having an electronic octet, have opposite charges.

Z

Z form. A duplex DNA structure in which there is the usual type of hydrogen bonding between the base pairs but in which the helix formed by the two polynucleotide chains is left-handed rather than right-handed.

Zwitterion. A dipolar ion with spatially separated positive and negative charges. For example, most amino acids are zwitterions, having a positive charge on the α-amino group and a negative charge on the α-carboxyl group but no net charge on the overall molecule.

Zygote. A cell that results from the union of haploid male and female sex cells. Zygotes are diploid.

Zymogen. An inactive precursor of an enzyme. For example, trypsin exists in the inactive form trypsinogen before it is converted to its active form, trypsin.

Credits

Portions of this text have been adopted from Geoffrey Zubay, *Genetics*. To be published.

Line Art

Chapter 1

1.1: From Sylvia S. Mader, *Inquiry Into Life,* 6th ed. Copyright © 1991 Wm. C. Brown Communications, Inc., Dubuque, Iowa. All Rights Reserved. Reprinted by permission. **1.10, 1.12, 1.13, and 1.22:** Illustration copyright by Irving Geis. **1.15:** Illustration copyright by Irving Geis. Coordinates courtesy of D. C. Phillips, Oxford.

Chapter 3

3.1, 3.9, and 3.10: Illustration copyright by Irving Geis.

Chapter 4

4.1, 4.2, 4.4*a*, 4.5, 4.8, and 4.9*a*, *b*: Illustration copyright by Irving Geis. **4.11:** Line art courtesy of Karl A. Piez. **4.12–4.14, 4.18*c*, and 4.20:** Illustration copyright by Irving Geis. **4.21, 4.23, 4.24, and 4.28–4.34:** From Jane S. Richardson. **4.39*a* and 4.42:** From Lansing M. Prescott, et al., *Microbiology.* Copyright © 1990 Wm. C. Brown Communications, Inc. Dubuque, Iowa. All Rights Reserved. Reprinted by permission. **Box 4C: Figures 1–4:** Courtesy of Irving Geis.

Chapter 5

5.2, 5.7–5.11, 5.13 and 5.14: Illustration copyright by Irving Geis. **5.16:** From John W. Hole, Jr., *Human Anatomy and Physiology,* 5th ed. Copyright © 1990 Wm. C. Brown Communications, Inc., Dubuque, Iowa. All Rights Reserved. Reprinted by permission. **5.23 and 5.24:** From Jane S. Richardson. **5.25:** Illustration copyright by Irving Geis.

Chapter 6

6.16: Illustration copyright by Irving Geis. **6.21** (*top*): From B. Alberts, *Molecular Biology of the Cell.* Copyright © 1983 Garland Publishing Inc., New York, N.Y. Reprinted by permission.

Chapter 7

7.43*a*: Reprinted by permission from Dr. Vincent Marchesi. **7.43*b*:** From S. B. Shohet and S. E. Lux, *Hospital Practice 19* (Nov. 1984):90. Redrawn based on a drawing by Robert Margulies. Reprinted by permission.

Chapter 9

9.4, 9.5, 9.19, and 9.30: Illustration copyright by Irving Geis.

Chapter 15

15.27: Illustration copyright by Irving Geis.

Chapter 25

25.1 and 25.4: From Robert Tamarin, *Principles of Genetics,* 3d ed. Copyright © 1991 Wm. C. Brown Communications, Inc., Dubuque, Iowa. All Rights Reserved. Reprinted by permission. **25.9*a*, *b*:** Reprinted by permission from *Nature,* vol. 171, p. 737. Copyright © 1953 Macmillan Magazines Ltd. **25.13:** Illustration copyright by Irving Geis. **25.14:** From A. Wang et al., "Left-handed Double Helical DNA: Variations in the Backbone Conformation" in *Science,* 211. Copyright © 1981 by the American Association for the Advancement of Science. Reprinted by permission. **25.28:** From Geoffrey Zubay, *Genetics.* Copyright © 1987 Benjamin/Cummings Publishing Company Inc. Reprinted by permission of the author. **25.30*b*:** From Robert F. Weaver and Philip W. Hedrick, *Basic Genetics.* Copyright © 1991 Wm. C. Brown Communications, Inc., Dubuque, Iowa. All Rights Reserved. Reprinted by permission.

Chapter 26

26.12: From D. L. Ollis et al., "Structure of large fragment of *Escherichia coli* DNA polymerase I complexed with dTMP." Reprinted by permission from *Nature,* Vol. 313, pp. 762–766. Copyright © 1985 Macmillan Magazines Ltd. **26.13:** From C. M. Joyce and T. A. Steitz, "DNA polymerase I: From crystal structure to function via genetics" in *Trends in Biochemical Sciences,* 12:288, 1987. Copyright © 1987 Elsevier Trends Journals, Cambridge, England. Reprinted with permission. **26.27:** From J. Tooze, *The Molecular Biology of DNA Tumor Viruses.* Copyright © 1980 Cold Spring Harbor Press, New York. Reprinted by permission. **26.29:** From Robert F. Weaver and Philip W. Hedrick, *Basic Genetics.* Copyright © 1991 Wm. C. Brown Communications, Inc., Dubuque, Iowa. All Rights Reserved. Reprinted by permission. **26.33:** From Geoffrey Zubay, *Genetics.* Copyright © 1987 Benjamin/Cummings Publishing Co. Inc. Reprinted by permission of the author.

Chapter 27

27.11: From J. L. Marx, "Multiplying genes by leaps and bounds" in *Science,* 240:1408–1410, June 10, 1988. Copyright © 1988 by the American Association for the Advancement of Science. Reprinted by permission. **27.13:** From Robert F. Weaver and Philip W. Hedrick, *Basic Genetics.* Copyright © 1991 Wm. C. Brown Communications, Inc., Dubuque, Iowa. All Rights Reserved. Reprinted by permission. **27.34:** From J. R. Riordan et al., "Identification of the cystic fibrosis gene: Cloning and characterization of complementary DNA" in *Science,* 245:1066, Sept. 8, 1989. Copyright © 1989 by the American Association for the Advancement of Science. Reprinted by permission.

Chapter 28

28.2: From Geoffrey Zubay, *Genetics.* Copyright © 1987 Benjamin/Cummings Publishing Company Inc., Menlo Park, CA. Reprinted by permission of the author. **28.3*b*:** Illustration copyright by Irving Geis. **28.6:** From Robert F. Weaver and Philip W. Hedrick, *Genetics.* Copyright © 1989 Wm. C. Brown Communications, Inc., Dubuque, Iowa. All Rights Reserved. Reprinted by permission. **28.7:** Source: R. R. Gutell et al., "Comparative anatomy of 16S-like ribosomal RNA" in *Progress in Nucleic Acid Research and Molecular Biology,* 32:155–216, 1985. Copyright © 1985 Academic Press Inc., San Diego, CA. **Box 28A:** Courtesy of Bishwajit Nag and Robert R. Traut. **28.8 and 28.9:** From Robert F. Weaver and Philip W. Hedrick, *Genetics.* Copyright © 1989 Wm. C. Brown Communications, Inc., Dubuque, Iowa. All Rights Reserved. Reprinted by permission. **28.17*b*:** From Robert Tamarin, *Principles of Genetics,* 3d ed. Copyright © 1991 Wm. C. Brown Communications, Inc., Dubuque, Iowa. All Rights Reserved. Reprinted by permission.

Chapter 29

29.5*b*: From Quigley and Rich, *Science,* 185:436, 1974. Copyright © 1974 by the American Association for the Advancement of Science, Washington, DC. Reprinted by permission. **29.6:** From C. Bernabau and J. A. Lake, *Proceedings. National Academy of Sciences USA,* 79:3111–3115, 1982. Copyright © 1982 National Academy of Sciences, Washington, DC. Reprinted by permission of the author. **29.8:** From Robert F. Weaver and Philip W. Hedrick, *Genetics.* Copyright © 1989 Wm. C. Brown Communications, Inc., Dubuque, Iowa.

All Rights Reserved. Reprinted by permission. **29.11:** From L. H. Schulman and J. Abelson, "Recent excitement in understanding transfer RNA identity" in *Science,* 240:1591, June 17, 1988. Copyright © 1988 by the American Association for the Advancement of Science, Washington, DC. Reprinted by permission. **29.23:** From *Molecular Cell Biology,* 2nd edition. By James Darnell, Harvey Lodish, and David Baltimore. Copyright © 1990 by Scientific American Books. Reprinted with permission of W. H. Freeman and Company.

Chapter 31

31.16: From Geoffrey Zubay, *Genetics.* Copyright © 1987 Benjamin/Cummings Publishing Company Inc., Menlo Park, CA. Reprinted by permission of the author.

Chapter 32

32.16*b:* From E. Schneider and K. Altendorf, "Bacterial adenosine 5′-triphosphate synthase (F1FO): Purification and reconstitution of Fo complexes and biochemical and functional characterization of their subunits" in *Microbiological Reviews,* 51:477–497, 1987. Copyright © 1987 American Society for Microbiology, Washington, DC. Reprinted by permission. **32.31*c:*** From A. C. Steven et al., "Ultrastructure of a periodic protein layer in the outer membrane of *Escherichia coli."* Reproduced from the *Journal of Cell Biology,* 1977, vol. 72, p. 292 by copyright permission of the Rockefeller University Press. **32.32:** Source: M. S. Weiss et al., "The three-dimensional structure of porin from *Rhodobacter capsulatus* at 3 A resolution" in *FEBS Letters* 267:268–272, 1990. Copyright © 1990 Elsevier Science Publishers B.V., Amsterdam, Netherlands. **32.32:** Source: M. S. Weiss et al, "The three-dimensional structure of porin from *Rhodobacter capsulatus* at 3 A resolution" on *FEBS Letters* 267:268–272, 1990. © 1990 Elsevier Science Publishers B.V., Amsterdam, Netherlands.

Chapter 33

33.2, 33.6, 33.7, 33.10, and 33.16–33.19: From Geoffrey Zubay, *Genetics.* Copyright © 1987 Benjamin/Cummings Publishing Company Inc., Menlo Park, CA. Reprinted by permission of the author.

Chapter 34

34.1: From B. Alberts et al., *Molecular Biology of the Cell,* 2d ed. Copyright © 1989 Garland Publishing Inc., New York, NY. Reprinted by permission. **34.2:** © Scientific American, Inc., George V. Kelvin. **34.3:** From J. Yunis, "The chromosomal basis of human neoplasia" in *Science,* 221:227–236, Jan. 1, 1983. Copyright © 1983 by the American Association for the Advancement of Science. Reprinted by permission. **34.5:** From Robert F. Weaver and Philip W. Weaver, *Genetics.* Copyright © 1989 Wm. C. Brown Communications, Inc., Dubuque, Iowa. All Rights Reserved. Reprinted by permission. **34.6:** From Geoffrey Zubay, *Genetics.* Copyright © 1987 Benjamin/Cummings Publishing Company Inc., Menlo Park, CA. Reprinted by permission of the author. **34.7:** From L. Tong, et al., "Structure of the *ras* protein" in *Science,* 245:243, July 21, 1989. Copyright © 1989 by the American Association for the Advancement of Science. Reprinted by permission.

Chapter 35

35.10: From W. A. Catterall, "Structure and function of voltage-sensitive ion channels" in *Science* 242:50, Oct. 7, 1988. Copyright © 1988 American Association for the Advancement of Science. Reprinted by permission. **35.12:** Reprinted by permission from Suzanne Black. **35.22:** From N. Unwin et al., "Arrangement of the acetylcholine receptor subunits in the resting and desensitized states, determined by cryoelectron microscopy of crystallized *Torpedo* postsynaptic membranes." Reproduced from the *Journal of Cell Biology,* 1988, vol. 107, p. 1123 by copyright permission of the Rockefeller University Press. **35.23:** From M. Noda et al., "Structural homology of *Torpedo California* acetylcholine receptor subunits." Reprinted by permission from *Nature,* vol. 302, p. 528. Copyright © 1983 Macmillan Magazines Ltd.

Index

Page numbers followed by an "F" refer to illustrations; "T" indicates material in tables.

A

B

C

D

E

F

G

H

I

J

K

L

M

N

O

Q

R

S

T

U